FEATURES AND BENEFITS

GEOMETRY: Integration, Applications, Connections

1. **Integration** Integration helps students to view mathematics as a whole and not as compartmentalized areas of instruction. *Lessons* that connect the various areas of mathematics to geometry are included as they are appropriate (see Lesson 7-1). This integration also occurs *within* lessons. *Lesson Openers* often reflect integration (see page 397), as do *Examples* (see Example 2 on page 468) and *Exercises* (see Exercise 50 on page 685).

2. **Applications** Because the ability to grasp concepts and skills is greatly enhanced when they are tied to applications, many lessons open with a real-world application (see page 180). This is strengthened by including *Applications and Problem Solving* in every set of exercises (see page 318), as well as in *Examples* (see Example 2 on page 421).

3. **Connections** Connections to *interdisciplinary areas*, such as science, geography, history, music, and so on, enhance learning while increasing student interest. These connections appear as *Lesson Openers* (see page 163), *Examples* (see Example 3 on page 340), and as *Exercises* (see Exercise 56 on page 520).

4. **Proof** Throughout the text, a variety of types of proof are used.

Paragraph proof	See Lesson 1-5, page 39.
Two-column proof	See Lesson 2-4, page 94.
Flow proof	See Lesson 4-2, page 191.
Indirect proof	See Lesson 5-3, page 252.
Coordinate proof	See Lesson 12-4, page 666.

5. **Informal Geometry**. Assignment guides in the Teacher's Wraparound Edition provide a convenient option for teaching geometry without formal proof (see Lesson 3-4, page 150).

6. **Modeling** Students have the opportunity to bridge the gap between the concrete and the abstract through *hands-on experiences* in the *Modeling Mathematics* activities.

Modeling Mathematics Lessons	See Lesson 10-2A, page 522.
Modeling Mathematics Activities	See page 36.
Modeling Mathematics Exercises	See Exercise 5 on page 415.

7. **Technology** Although this program is *not* dependent upon graphing calculators, the graphing calculator is integrated throughout the program in various ways.

Graphing Technology Lessons	See Lesson 6-1A.
Graphing Technology Explorations	See page 491.
Graphing Technology Programs	See Exercise 32 on page 720.
Graphing Technology Exercises	See Exercise 43 on page 244.

Spreadsheets also play a role in the program, particularly with the integration of statistics (see the Exploration on page 630). *Technology Tips* are provided as appropriate.

8. **Problem Solving** Problem-solving strategies are *integrated* within lessons (see *make a model* in Example 1 on page 575). *Applications and Problem Solving* and *Critical Thinking* in *every* set of exercises illustrate the ongoing attention to problem solving (see Exercises 47-49 on page 129).

TEACHER'S WRAPAROUND EDITION

Glencoe

Geometry

Integration
Applications
Connections

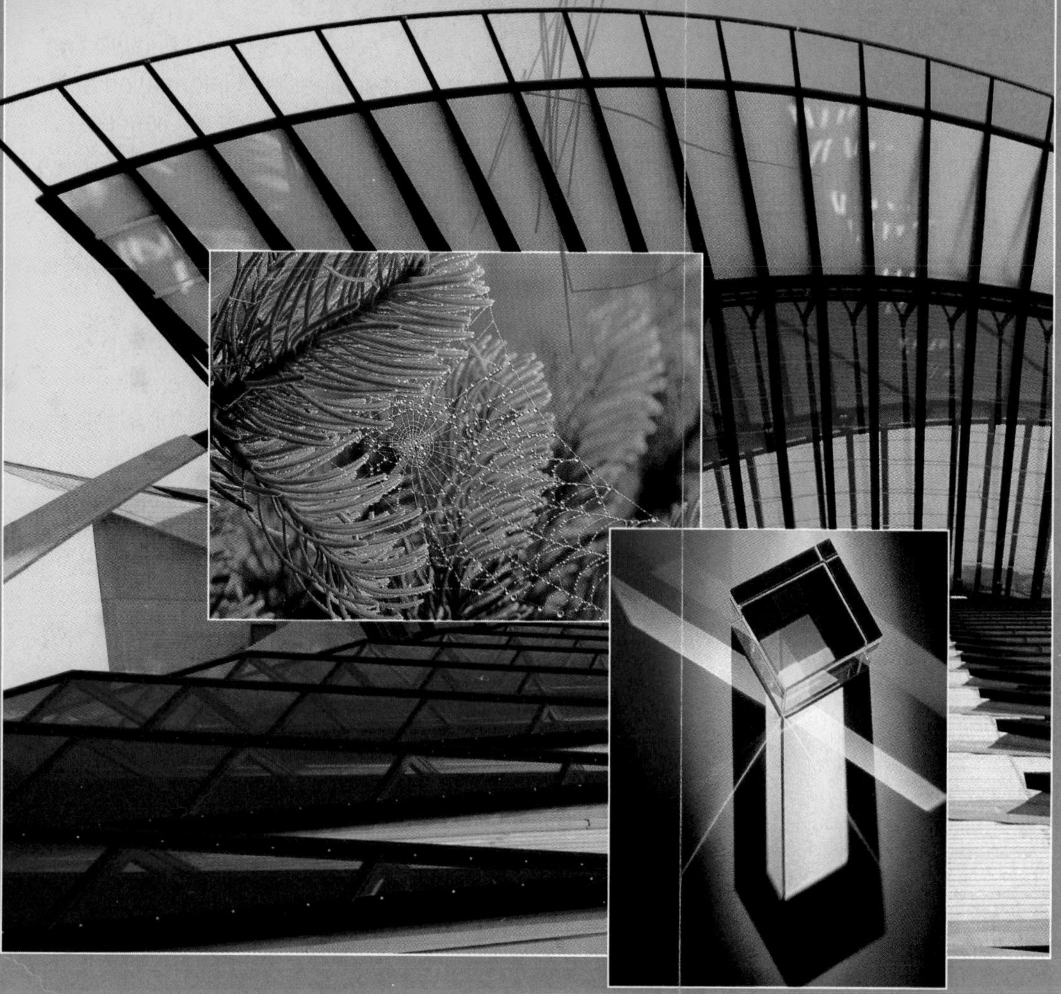

Glencoe
McGraw-Hill

New York, New York Columbus, Ohio Woodland Hills, California Peoria, Illinois

Glencoe/McGraw-Hill
A Division of The McGraw-Hill Companies

Send all inquiries to:
Glencoe/McGraw-Hill
936 Eastwind Drive
Westerville, OH 43081-3329

ISBN: 0-02-825275-6 (Student Edition)
0-02-825276-4 (Teacher's Wraparound Edition)

4 5 6 7 8 9 10 071/071 05 04 03 02 01 00 99 98

TEACHER'S WRAPAROUND EDITION

Table of Contents

Geometry

INTEGRATION ● APPLICATIONS ● CONNECTIONS

Glencoe Brings Geometry to Life!

INTEGRATION • APPLICATIONS • CONNECTIONS

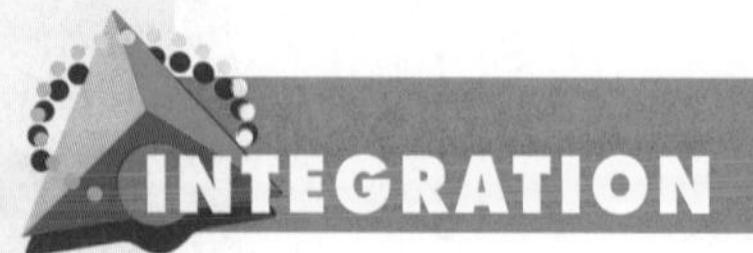

Students relate and apply geometric concepts to algebra, statistics, data analysis, probability, and discrete mathematics.

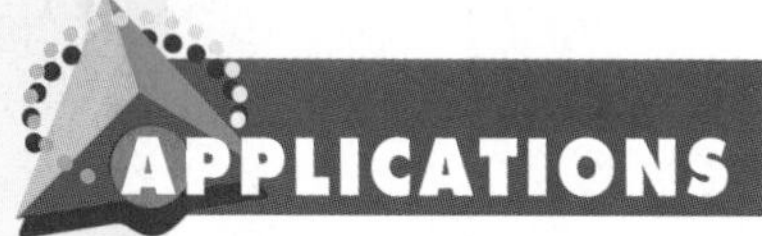

Realistic and relevant applications help answer the question "When am I ever going to use this stuff?" Sports, space, world cultures, and consumerism are just a few of the real-life problem settings that students explore.

Students connect mathematics to other topics they are studying, like biology, geography, art, history, and health, through problems that are rich in geometric content.

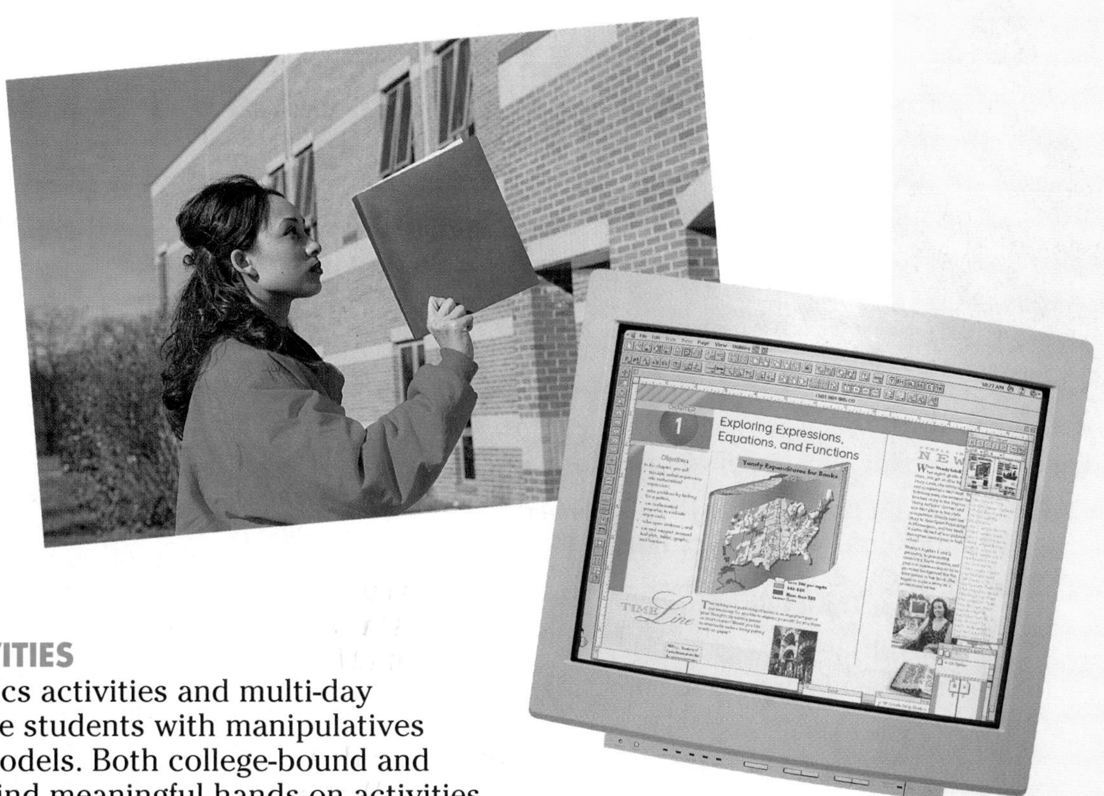

HANDS-ON ACTIVITIES

Modeling Mathematics activities and multi-day Investigations engage students with manipulatives and mathematical models. Both college-bound and tech-prep students find meaningful hands-on activities that relate to life and work.

TECHNOLOGY

Graphing calculators and CD-ROM multimedia technology provide tools for both problem solving and discovery. The variety of integrated technology supports diverse learning styles.

ASSESSMENT

A broad spectrum of evaluation and assessment tools, ranging from short-response questions to open-ended problems, lets all students demonstrate what they have learned.

ACCESSIBILITY

Practical strategies meet the needs of students with differing ability levels, rates of learning, and learning styles. Modeling, journal writing, cooperative learning projects, and technology tools help all students experience success.

NCTM STANDARDS

The sequence of topics, the content emphasis, integration with other mathematics areas, hands-on activities, and technology combine to form a program that reflects the NCTM Curriculum, Teaching, and Assessment Standards.

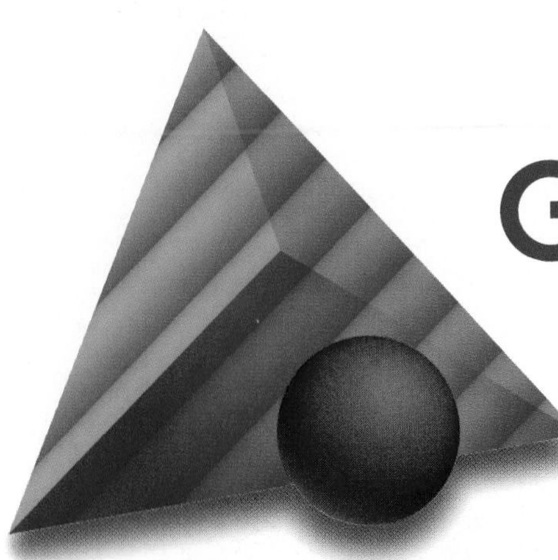

Glencoe's Dynamic Lesson Structure

Each lesson follows a straightforward format —

What You'll Learn/Why It's Important
Application/Connection/Integration
Examples
Check for Understanding
Exercises

— that focuses on concepts and applications.

WHAT YOU'LL LEARN/WHY IT'S IMPORTANT

Students quickly see what concepts are covered in the lesson and why it is beneficial to learn them.

APPLICATION/CONNECTION/INTEGRATION

A motivating lesson opener introduces the concept by posing a real-world application, an interdisciplinary connection, or a mathematical integration.

EXAMPLES

A set of completely worked-out examples with clear explanations supports the objectives and mirrors the exercises in the Guided Practice and Practice sections that follow.

MODELING MATHEMATICS/EXPLORATION

Activities with hands-on manipulatives and technology are fully integrated into appropriate lessons.

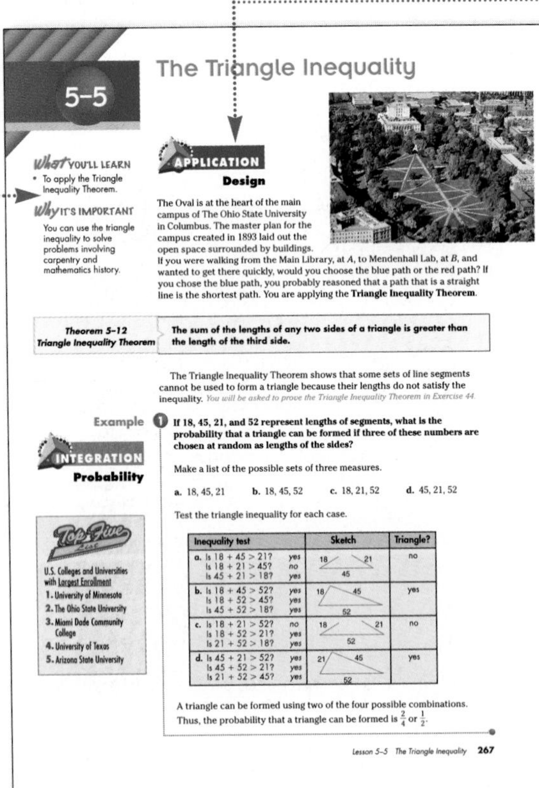

5-5 The Triangle Inequality

What YOU'LL LEARN
- To apply the Triangle Inequality Theorem.

Why IT'S IMPORTANT
You can use the triangle inequality to solve problems involving carpentry and mathematics history.

APPLICATION **Design**

The Oval is at the heart of the main campus of The Ohio State University in Columbus. The master plan for the campus created in 1893 laid out the open space surrounded by buildings. If you were walking from the Main Library, at A, to Mendenhall Lab, at B, and wanted to get there quickly, would you choose the blue path or the red path? If you chose the blue path, you probably reasoned that a path that is a straight line is the shortest path. You are applying the **Triangle Inequality Theorem**.

Theorem 5-12 Triangle Inequality Theorem	**The sum of the lengths of any two sides of a triangle is greater than the length of the third side.**

The Triangle Inequality Theorem shows that some sets of line segments cannot be used to form a triangle because their lengths do not satisfy the inequality. *You will be asked to prove the Triangle Inequality Theorem in Exercise 44.*

Example 1 **INTEGRATION Probability**

If 18, 45, 21, and 52 represent lengths of segments, what is the probability that a triangle can be formed if three of these numbers are chosen at random as lengths of the sides?

Make a list of the possible sets of three measures.

a. 18, 45, 21 **b.** 18, 45, 52 **c.** 18, 21, 52 **d.** 45, 21, 52

Test the triangle inequality for each case.

Inequality test		Sketch	Triangle?
a. Is 18 + 45 > 21? Is 18 + 21 > 45? Is 45 + 21 > 18?	yes no yes	18, 21, 45	no
b. Is 18 + 45 > 52? Is 18 + 52 > 45? Is 45 + 52 > 18?	yes yes yes	18, 45, 52	yes
c. Is 18 + 21 > 52? Is 18 + 52 > 21? Is 21 + 52 > 18?	no yes yes	18, 21, 52	no
d. Is 45 + 21 > 52? Is 45 + 52 > 21? Is 21 + 52 > 45?	yes yes yes	21, 45, 52	yes

A triangle can be formed using two of the four possible combinations. Thus, the probability that a triangle can be formed is $\frac{2}{4}$ or $\frac{1}{2}$.

Top Five List
U.S. Colleges and Universities with Largest Enrollment
1. University of Minnesota
2. The Ohio State University
3. Miami Dade Community College
4. University of Texas
5. Arizona State University

Lesson 5-5 The Triangle Inequality 267

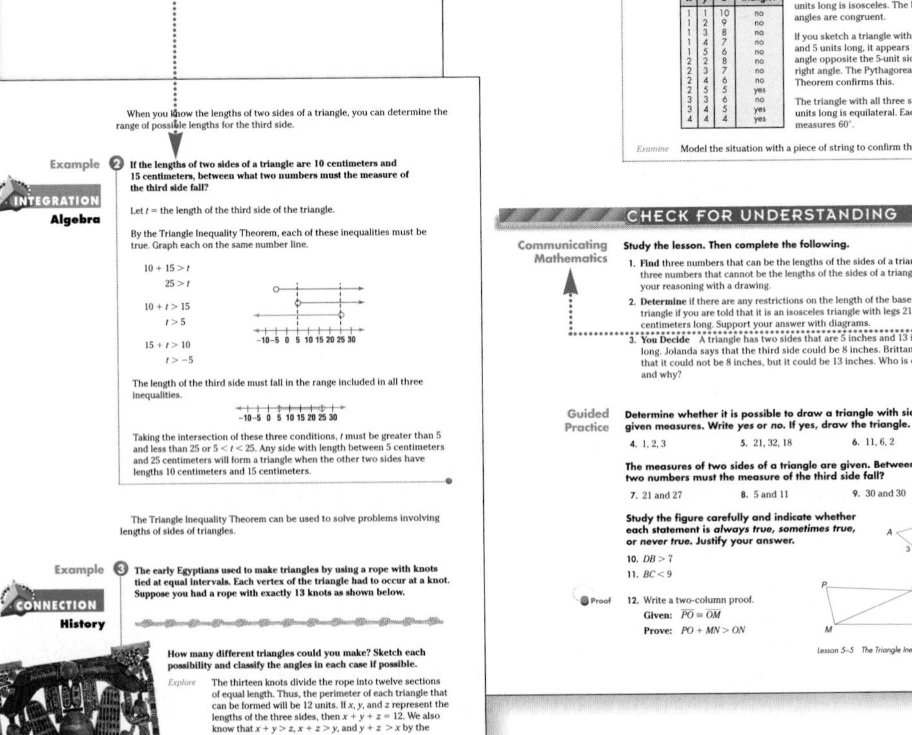

When you know the lengths of two sides of a triangle, you can determine the range of possible lengths for the third side.

Example 2 **INTEGRATION Algebra**

If the lengths of two sides of a triangle are 10 centimeters and 15 centimeters, between what two numbers must the measure of the third side fall?

Let t = the length of the third side of the triangle.

By the Triangle Inequality Theorem, each of these inequalities must be true. Graph each on the same number line.

$10 + 15 > t$
$25 > t$

$10 + t > 15$
$t > 5$

$15 + t > 10$
$t > -5$

The length of the third side must fall in the range included in all three inequalities.

Taking the intersection of these three conditions, t must be greater than 5 and less than 25 or $5 < t < 25$. Any side with length between 5 centimeters and 25 centimeters will form a triangle when the other two sides have lengths 10 centimeters and 15 centimeters.

The Triangle Inequality Theorem can be used to solve problems involving lengths of sides of triangles.

Example 3 **CONNECTION History**

The early Egyptians used to make triangles by using a rope with knots tied at equal intervals. Each vertex of the triangle had to occur at a knot. Suppose you had a rope with exactly 13 knots as shown below.

How many different triangles could you make? Sketch each possibility and classify the angles in each case if possible.

Explore The thirteen knots divide the rope into twelve sections of equal length. Thus, the perimeter of each triangle that can be formed will be 12 units. If x, y, and z represent the lengths of the three sides, then $x + y + z = 12$. We also know that $x + y > z$, $x + z > y$, and $y + z > x$ by the Triangle Inequality Theorem.

Plan Make a table of all values of x, y, and z that satisfy $x + y + z = 12$. Then check each possible combination with the Triangle Inequality Theorem.

268 Chapter 5 Applying Congruent Triangles

Solve There are three possible triangles. *How do we know that all the combinations have been checked?*

x	y	z	Triangle?
1	1	10	no
1	2	9	no
1	3	8	no
1	4	7	no
1	5	6	no
2	2	8	no
2	3	7	no
2	4	6	no
2	5	5	yes
3	3	6	no
3	4	5	yes
4	4	4	yes

The triangle with sides 2, 5, and 5 units long is isosceles. The base angles are congruent.

If you sketch a triangle with sides 3, 4, and 5 units long, it appears that the angle opposite the 5-unit side is a right angle. The Pythagorean Theorem confirms this.

The triangle with all three sides 4 units long is equilateral. Each angle measures 60°.

Examine Model the situation with a piece of string to confirm the solution.

CHECK FOR UNDERSTANDING

Communicating Mathematics **Study the lesson. Then complete the following.**

1. **Find** three numbers that can be the lengths of the sides of a triangle and three numbers that cannot be the lengths of the sides of a triangle. Justify your reasoning with a drawing.
2. **Determine** if there are any restrictions on the length of the base of a triangle if you are told that it is an isosceles triangle with legs 21 centimeters long. Support your answer with diagrams.
3. **You Decide** A triangle has two sides that are 5 inches and 13 inches long. Jolanda says that the third side could be 8 inches. Brittany says that it could not be 8 inches, but it could be 13 inches. Who is correct and why?

Guided Practice **Determine whether it is possible to draw a triangle with sides of the given measures. Write *yes* or *no*. If yes, draw the triangle.**

4. 1, 2, 3
5. 21, 32, 18
6. 11, 6, 2

The measures of two sides of a triangle are given. Between what two numbers must the measure of the third side fall?

7. 21 and 27
8. 5 and 11
9. 30 and 30

Study the figure carefully and indicate whether each statement is *always true, sometimes true,* or *never true.* Justify your answer.

10. $DB > 7$
11. $BC < 9$

Proof

12. Write a two-column proof.
Given: $\overline{PO} \cong \overline{OM}$
Prove: $PO + MN > ON$

Lesson 5-5 The Triangle Inequality 269

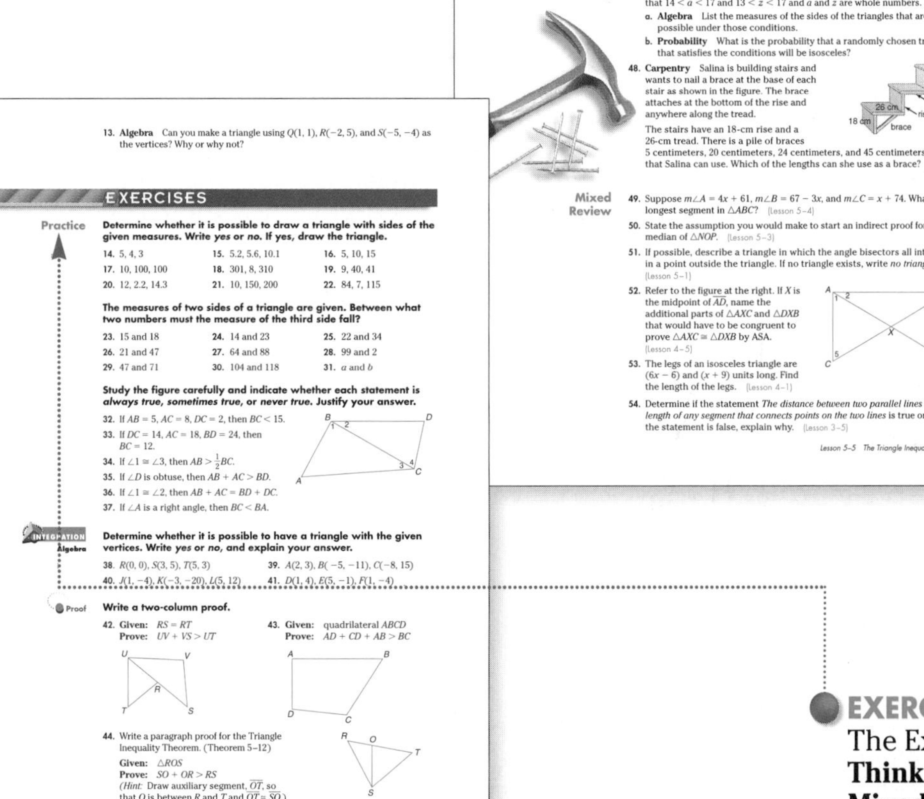

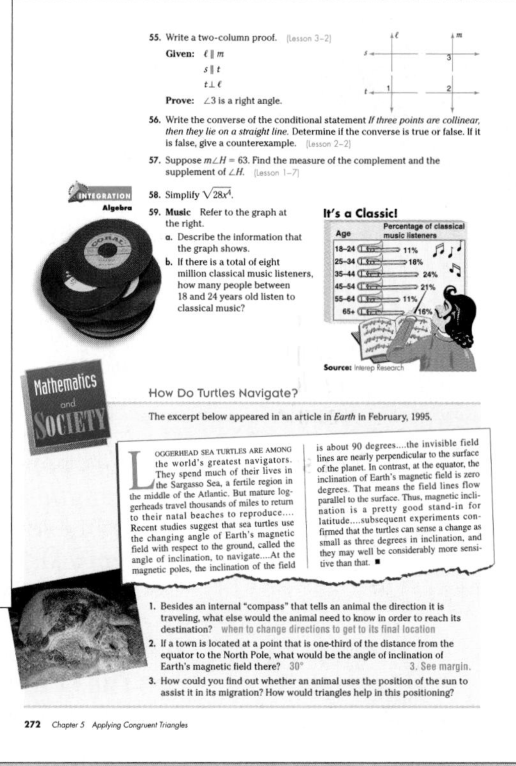

EXERCISES

The Exercises always include **Practice, Critical Thinking, Applications and Problem Solving,** and **Mixed Review.**

Practice

The Practice exercises are separated into A, B, and C sections, indicated only in the Teacher's Wraparound Edition. The A and B sections generally match the Guided Practice exercises in a 3:1 ratio. The C section offers challenging, thought-provoking questions.

Critical Thinking

Each lesson contains critical thinking exercises in which students explain, evaluate, and justify mathematical concepts and relationships.

Applications and Problem Solving

Students find numerous opportunities to apply concepts to both real-life and mathematical problem situations.

Mixed Review

This spiraled, cumulative review comprises about 15% of the total number of exercises in each lesson and includes two algebra concept problems.

CHECK FOR UNDERSTANDING

This section ensures that all students are engaged in the lesson and understand the concepts.

Communicating Mathematics

Students work in small groups or as a whole class to define, explain, describe, make lists, draw models, use symbols, or create graphs.

Modeling Mathematics and **Math Journal** activities, which further strengthen communication skills, appear frequently.

Guided Practice

Keyed to the lesson objectives, the Guided Practice exercises present a representative sample of the exercises in the Practice section.

Glencoe's Geometry Exemplifies the NCTM Standards

CONTENT EMPHASIS

Geometry: Integration, Applications, and Connections implements the shift from geometry as a course in proof to geometry as a representation of the world around us.

> *"In summary, synthetic geometry at the high school level should focus on more than deductive reasoning and proof. Equally important is the continued development of students' skills in visualization, pictorial representation, and the application of geometric ideas to describe and answer questions about natural, physical, and social phenomena."*
> ***NCTM Standards, page 160, 1989***

The study of geometry also encompasses its close relationship with algebra by using coordinate and algebraic means to verify the synthetic representations. In each chapter, students use algebraic tools to verify properties of figures presented on a coordinate plane.

> *"The interplay between geometry and algebra strengthens students' ability to formulate and analyze problems from situations both within and outside mathematics."* ***NCTM Standards, page 161, 1989***

Each lesson opener motivates students to master the content they need to solve application, connection, or integration problems presented in the lesson. Additional applications, connections, and integration in the exercises enable students to apply what they have learned.

Glencoe's ***Geometry*** incorporates graphing-calculator and computer software activities as well as hands-on manipulatives for discovery, problem solving, and modeling. The use of technology expands the possibilities of pencil-and-paper constructions with included analysis tools that lead students to discover the concepts presented in postulates and theorems. Hands-on activities make the visualization of abstract concepts real.

> *"In order to establish a discourse that is focused on exploring mathematical ideas, not just on reporting correct answers, the means of mathematical communication and approaches to mathematical reasoning must be broad and varied."* ***NCTM Professional Teaching Standards, page 52, 1991***

Applications, modeling activities, and open-ended projects encourage a diversity of approaches and engage today's students in Geometry.

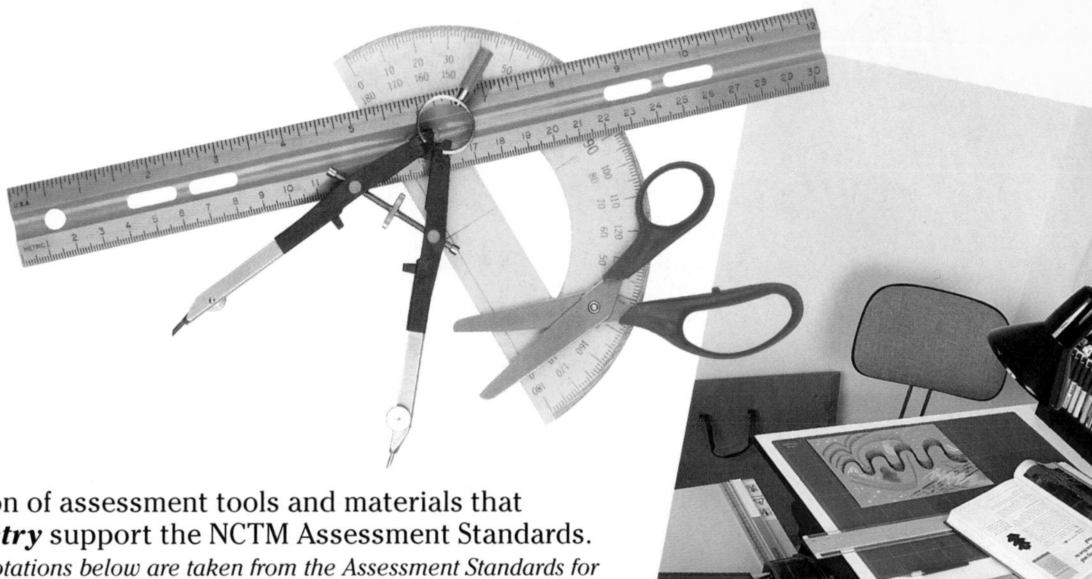

ASSESSMENT

The wide and varied collection of assessment tools and materials that accompany Glencoe's ***Geometry*** support the NCTM Assessment Standards. *[Note: The bulleted headings and quotations below are taken from the Assessment Standards for School Mathematics, NCTM, 1995, pages 11, 13, 15, 17, 19, and 21.]*

- **The Mathematics Standard** — "Assessment should reflect the mathematics that all students need to know and be able to do."
  Skills, procedural knowledge, factual knowledge, reasoning, applying mathematics, and problem solving are assessed through a broad range of short-response exercises and longer-term, open-ended activities.

- **The Learning Standard** — "Assessment should enhance mathematics learning."
  Assessment is an integral part of the Communicating Mathematics, Guided Practice, and Exercises sections of each of Glencoe's ***Geometry*** lessons. Students are able to evaluate their own learning by writing in their math journals and selecting portfolio items.

- **The Equity Standard** — "Assessment should promote equity."
  The wide variety of assessment tools enable all students to demonstrate what they know and can do. Performance Assessment tasks for each chapter include a scoring guide.

- **The Openness Standard** — "Assessment should be an open process."
  "What You'll Learn" tells students what they are expected to learn in each lesson. The examples, activities, and assessment match those objectives. Guided Practice prepares students to successfully complete the Exercises.

- **The Inferences Standard** — "Assessment should promote valid inferences about mathematics learning."
  The assortment of assessment tools provide many sources of evidence. Exercises, Mixed Review, and Chapter Tests are all correlated to the learning objectives and practice activities.

- **The Coherence Standard** — "Assessment should be a coherent process."
  The numerous, varied assessment tools are carefully designed to complement each other. They enable you to easily construct the assessment program that best fits your needs.

Learning Objectives & NCTM Standards Correlation

Lesson	Lesson Objectives	NCTM Standards
1-1	Graph ordered pairs on a coordinate plane. Identify collinear points.	1–5, 8, 10
1-2	Identify and model points, lines, and planes. Identify coplanar points and intersecting lines and planes. Solve problems by listing the possibilities.	1–5, 7, 8, 12
1-3	Solve problems by using formulas. Find maximum area of a rectangle for a given perimeter.	1–5, 6, 8, 12
1-4A	Use a TI-92 calculator to draw and measure segments.	1–5, 7, 8
1-4	Find the distance between two points on a number line and between two points in a coordinate plane. Use the Pythagorean Theorem to find the length of the hypotenuse.	1–5, 7, 8, 12
1-5	Find the midpoint of a segment. Complete proofs involving segment theorems.	1–5, 7, 8, 14
1-6	Identify angles and classify angles. Use the angle addition postulate to find the measures of angles. Identify and use congruent angles and the bisector of an angle.	1–5, 7, 8
1-7A	Use a TI-92 calculator to draw and measure angles.	1–5, 7, 8
1-7	Identify and use adjacent, vertical, complementary, supplementary, and linear pairs of angles, and perpendicular lines. Determine what information can and cannot be assumed from a diagram.	1–5, 7, 8, 14
2-1	Make conjectures based on inductive reasoning.	1–5, 8
2-2	Write a statement in *if-then* form. Write the converse, inverse, and contrapositive of an *if-then* statement. Identify and use basic postulates about points, lines, and planes.	1–5, 8
2-2B	Test the validity of conditional statements that involve mathematical sentences in one variable.	1–5
2-3	Use the Law of Detachment and the Law of Syllogism in deductive reasoning. Solve problems by looking for a pattern.	1–4
2-4	Use properties of equality in algebraic and geometric proofs.	1–5, 7, 8
2-5	Complete proofs involving segment theorems.	1–5, 7, 8
2-6	Complete proofs involving angle theorems.	1–5, 7, 8
3-1	Solve problems by drawing a diagram. Identify the relationships between two lines or two planes. Name angles formed by a pair of lines and a transversal.	1–4, 7
3-2A	Find the measures of the angles formed by two parallel lines and a transversal.	1–4, 7
3-2	Use the properties of parallel lines to determine angle measures.	1–4, 7
3-3	Find the slopes of lines. Use slope to identify parallel and perpendicular lines.	1–5, 7

Lesson	Lesson Objectives	NCTM Standards
3-4	Recognize angle conditions that produce parallel lines. Prove two lines are parallel based on given angle relationships.	1–5, 7
3-5	Recognize and use distance relationships among points, lines, and planes.	1–4, 7
3-5B	Find the distance between a point and a line.	1–4, 7
3-6	Identify points, lines, and planes in spherical geometry. Compare and contrast basic properties of plane and spherical geometry.	1–4, 7
4-1	Solve problems by drawing a diagram. Identify the relationships between two lines or two planes. Name angles formed by a pair of lines and a transversal.	1–4, 7
4-2A	Find the measures of the angles formed by two parallel lines and a transversal.	1–4, 7
4-2	Use the properties of parallel lines to determine angle measures.	1–4, 7
4-3	Find the slopes of lines. Use slope to identify parallel and perpendicular lines.	1–5, 7
4-4A	Recognize angle conditions that produce parallel lines. Prove two lines are parallel based on given angle relationships.	1–5, 7
4-4	Recognize and use distance relationships among points, lines, and planes.	1–4, 7
4-5	Find the distance between a point and a line.	1–4, 7
4-6	Identify points, lines, and planes in spherical geometry. Compare and contrast basic properties of plane and spherical geometry.	1–4, 7
5-1	Identify and use medians, altitudes, angle bisectors, and perpendicular bisectors in a triangle.	1–4, 7
5-2	Recognize and use tests for congruence of right triangles.	1–4, 7
5-3	Use indirect reasoning and indirect proof to reach a conclusion. Recognize and apply properties of inequalities to the measures of segments and angles. Solve problems by working backward.	1–5, 7
5-4	Recognize and apply relationships between sides and angles in a triangle.	1–5, 7
5-5A	Use a graphing calculator to investigate the relationship among the measures of the sides of a triangle.	1–5, 7
5-5	Apply the Triangle Inequality Theorem.	1–5, 7
5-6	Apply the SAS Inequality and the SSS Inequality.	1–5, 7
6-1A	Use a TI-92 calculator to draw parallelograms.	1–4, 7, 8
6-1	Recognize and apply the properties of a parallelogram. Find the probability of an event.	1–5, 7, 8
6-2	Recognize and apply the conditions that ensure a quadrilateral is a parallelogram. Identify and use subgoals in writing proofs.	1–5, 7, 8
6-3A	Use a TI-92 calculator to draw and explore characteristics of rectangles.	1–4, 7, 8
6-3	Recognize and apply the properties of rectangles.	1–5, 7, 8
6-4	Recognize and apply the properties of squares and rhombi.	1–5, 7, 8
6-4B	Construct quadrilaterals with exactly two distinct pairs of adjacent congruent sides.	1–4, 7
6-5	Recognize and apply the properties of trapezoids.	1–5, 7, 8
7-1	Recognize and use ratios and proportions. Apply the properties of proportions.	1–5
7-2	Identify similar figures. Solve problems involving similar figures.	1–5, 7
7-3	Identify similar triangles. Use similar triangles to solve problems.	1–5, 7
7-4	Use proportional parts of triangles to solve problems. Divide a segment into congruent parts.	1–5, 7

Lesson	Lesson Objectives	NCTM Standards
7-5	Recognize and use the proportional relationships of corresponding perimeters, altitudes, angle bisectors, and medians of similar triangles.	1–5, 7
7-6	Recognize and describe characteristics of fractals. Solve problems by solving a simpler problem.	1–5, 7
7-6B	Create fractal designs by using random numbers.	1–5, 7
8-1A	Use paper folding to develop the Pythagorean Theorem.	2, 4, 5, 7
8-1	Find the geometric mean between two numbers. Solve problems involving relationships between parts of a triangle and the altitude to its hypotenuse. Use the Pythagorean Theorem and its converse.	1–5, 7, 12
8-2	Use the properties of 45°-45°-90° and 30°-60°-90° triangles.	1–5, 7
8-3	Find trigonometric ratios using right triangles. Solve problems using trigonometric ratios.	1–5, 7, 9
8-4	Use trigonometry to solve problems involving angles of elevation or depression.	1–5, 7, 9
8-5	Use the Law of Sines to solve triangles.	1–5, 7, 9
8-6	Use the Law of Cosines to solve triangles. Choose the appropriate strategy for solving a problem.	1–5, 7, 9
9-1	Identify and use parts of circles. Solve problems involving the circumference of a circle.	1–4, 7
9-2	Recognize major arcs, minor arcs, semicircles, and central angles. Find measures of arcs and central angles. Solve problems by making circle graphs.	1–4, 7, 10
9-3	Recognize and use relationships among arcs, chords, and diameters.	1–4, 7
9-4	Recognize and find measures of inscribed angles. Apply properties of inscribed figures.	1–5, 7
9-5A	Use the TI-92 calculator to explore characteristics of tangents.	2, 3, 7
9-5	Recognize tangents and use properties of tangents.	1–4, 7
9-6	Find the measures of angles formed by intersecting secants and tangents in relation to intercepted arcs.	1–5, 7
9-7	Use properties of chords, secants, and tangents to solve segment measure problems.	1–5, 7
9-8	Write and use the equation of a circle in the coordinate plane.	1–5, 7, 8
10-1	Identify and name polygons. Find the sum of the measures of interior and exterior angles of convex polygons and measures of interior and exterior angles of regular polygons. Solve problems involving angle measures of polygons.	1–5, 7, 8
10-2A	Create tessellations using translations and rotations.	1–3, 7, 8
10-2	Identify regular and uniform (semi-regular) tessellations. Create tessellations with specific attributes. Solve problems by using guess and check.	1–5, 7, 8
10-3A	Compare the area of a rectangle with the area of a parallelogram having the same base length and the same parallel lines.	2–4, 7
10-3	Find areas of parallelograms.	1–5, 7, 8
10-4	Find areas of triangles, rhombi, and trapezoids.	1–5, 7, 8
10-5A	Create a regular *n*-gon.	2–4, 7
10-5	Find areas of regular polygons. Find areas of circles.	1–5, 7, 8
10-6	Use area to solve problems involving geometric probability.	1–5, 7, 8, 11
10-7	Recognize nodes and edges as used in graph theory. Determine if a network is traceable. Determine if a network is complete.	1–5, 7, 8

Lesson	Lesson Objectives	NCTM Standards
11-1A	Create cross sections and other slices of solids.	2, 3, 7
11-1	Use top, front, side, and corner views of three-dimensional solids to make models. Describe and draw cross sections and other slices of three-dimensional figures.	1–3, 7
11-1B	Construct a tetrahedron kite.	1–4, 7
11-2	Draw three-dimensional figures on isometric dot paper. Make two-dimensional nets for three-dimensional solids. Find surface areas.	1–5, 7
11-2B	Investigate Plateau's problem using soap film.	2, 3, 7
11-3	Find the lateral area and surface area of a right prism. Find the lateral area and surface area of a right cylinder.	1–5, 7
11-4	Find the lateral area and surface area of a regular pyramid. Find the lateral area and surface area of a right circular cone.	1–5, 7
11-5	Find the volume of a right prism. Find the volume of a right cylinder.	1–5, 7
11-6A	Compare the volumes of prisms and pyramids and the volumes of cylinders and cones.	2–5, 7
11-6	Find the volume of a pyramid. Find the volume of a circular cone.	1–5, 7
11-7	Recognize and define basic properties of spheres. Find the surface area and the volume of a sphere.	1–5, 7
11-8	Identify congruent or similar solids. State properties of congruent solids.	1–5, 7
12-1	Graph linear equations using the intercepts method and the slope-intercept method.	1–6
12-2A	Use a graphing calculator to determine an equation of the line that passes through plotted points.	2, 4, 5
12-2	Write an equation of a line given information about its graph. Solve problems by using equations.	1–6
12-3	Relate statistics and equations of lines to geometric concepts.	1–6, 10
12-4	Prove theorems using coordinate proofs.	1–5, 8
12-5A	Use a geometric model of motion.	2, 3
12-5	Find the magnitude and direction of a vector. Determine if two vectors are equal. Perform operations with vectors.	1–6, 8
12-6	Locate a point in space. Use the distance and midpoint formulas for points in space. Determine the center and radius of a sphere.	1–8
13-1	Locate, draw, and describe a locus in a plane or in space.	1–4, 7, 8
13-2	Find the locus of points that are solutions of a system of equations by graphing, by substitution, or by elimination.	1–5, 7, 8
13-3	Solve locus problems that satisfy more than one condition.	1–5, 7, 8
13-4	Recognize an isometry or congruence transformation. Solve problems by making a table.	1–5, 7, 8
13-5	Name, recognize, and draw reflection images, lines of symmetry, and points of symmetry.	1–5, 7, 8
13-6A	Use a geomirror to investigate translations.	1–4, 7, 8
13-6	Name and draw translation images of figures with respect to parallel lines.	1–5, 7, 8
13-6B	Use a TI-92 calculator to translate a triangle in the direction and magnitude of a given vector.	1–4, 7, 8
13-7	Name and draw rotation images of figures with respect to intersecting lines.	1–5, 7, 8
13-8	Use scale factors to determine if a dilation is an enlargement, a reduction, or a congruence transformation. Find the center and scale factor for a given dilation, and vice versa.	1–5, 7, 8

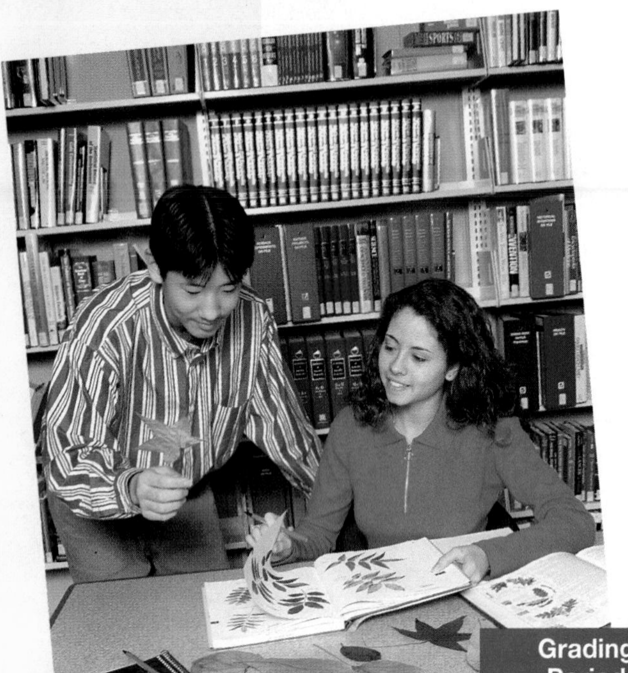

40- TO 50-MINUTE CLASS PERIODS

The chart below gives suggested pacing guides for two options, Standard and Honors, for four 9-week grading periods. The Standard option covers Chapters 1–12 while the Honors option covers Chapters 1–13. *A pacing guide that addresses both of these options is provided in each lesson of the Teacher's Wraparound Edition.*

The total days suggested for each option is about 165 days. This allows for teacher flexibility in planning due to school cancellation or shortened class periods.

Grading Period	Standard		Honors	
	Chapter	Days	Chapter	Days
1	1 2 3	13 12 14	1 2 3 4-1 to 4-3	13 11 13 5
2	4 5 6	14 14 13	4-4 to end 5 6 7-1 to 7-5	8 13 12 7
3	7 8 9	13 12 15	7-6 to end 8 9 10	5 11 13 14
4	10 11 12	15 18 12	11 12 13	16 11 13

For day-by-day lesson plans, please refer to Glencoe's Geometry Lesson Planning Guide.

BLOCK SCHEDULE, 90-MINUTE CLASS PERIODS

The chart below gives a suggested pacing guide for teaching Glencoe's ***Geometry*** using block scheduling in 1 semester (classes meet every day) or 1 year (classes meet about 85 days). The guide is based on teaching Chapters 1–11. *The pacing guide in each lesson of the Teacher's Wraparound Edition also addresses block scheduling.*

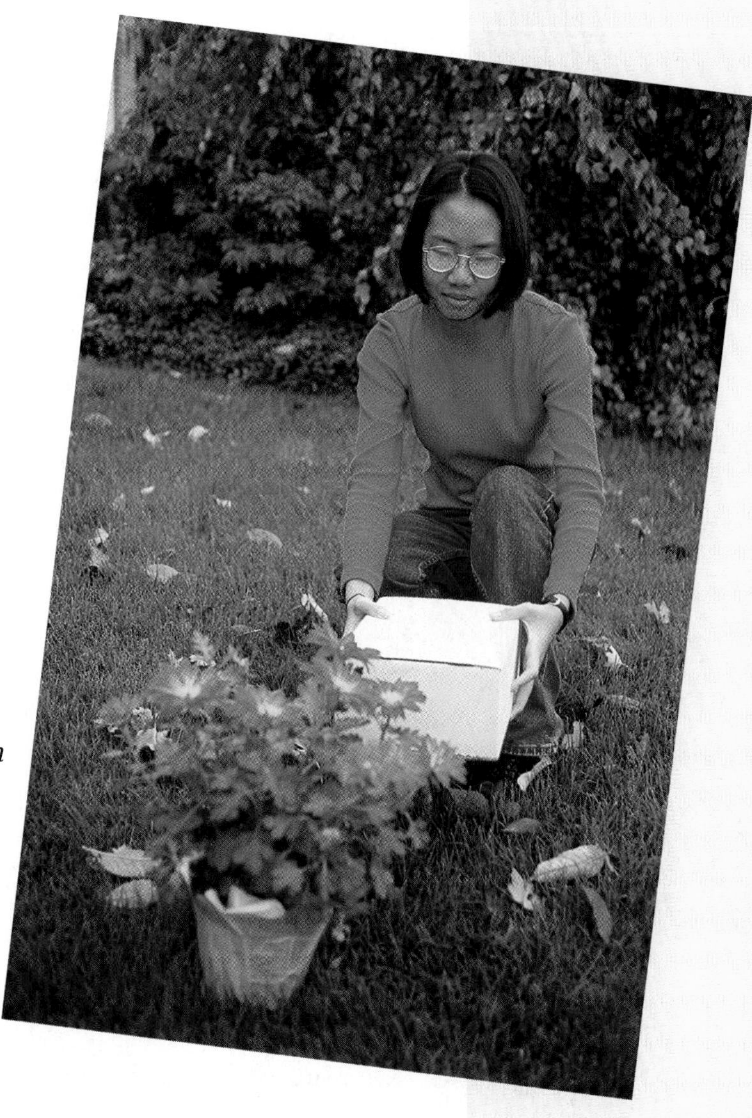

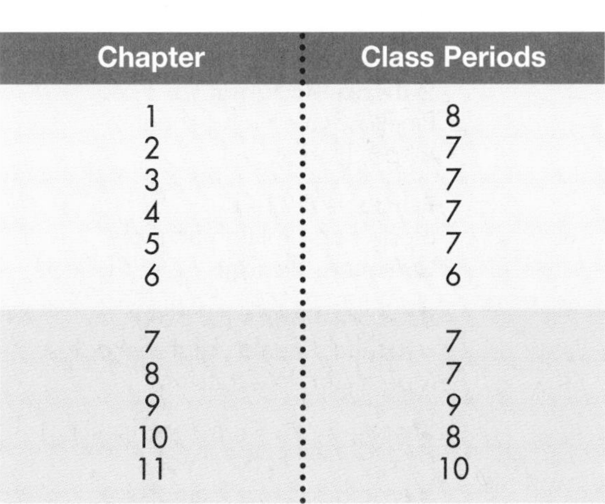

Chapter	Class Periods
1	8
2	7
3	7
4	7
5	7
6	6
7	7
8	7
9	9
10	8
11	10

For day-by-day lesson plans, please refer to Glencoe's Geometry Block Scheduling Booklet.

Classroom Vignettes

Glencoe's unique **Classroom Vignettes** are classroom-tested teaching suggestions made by experienced mathematics educators. Each vignette was written by a teacher who uses Glencoe's ***Geometry*** or by a Glencoe reviewer, consultant, or author. Glencoe thanks these outstanding mathematics educators for their unique contributions.

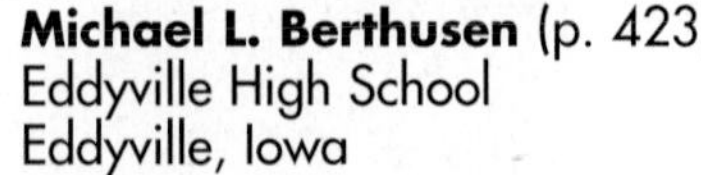

Michael L. Berthusen (p. 423)
Eddyville High School
Eddyville, Iowa

Cindy Boyd (pp. 21, 139)
Author
Glencoe *Geometry*

Gail Burrill (p. 190)
Author
Glencoe *Geometry*

Roberta L. Cook (p. 593)
Waterford Union High School
Waterford, Wisconsin

Jerry Cummins (pp. 87, 667)
Author
Glencoe *Geometry*

Karyn S. Cummins (p. 748)
Franklin Central High School
Indianapolis, Indiana

Diane DeRyckere (p. 371)
Bishop Foley High School
Madison Heights, Michigan

Loraine M. Duclos (p. 740)
Fairfield High School
Fairfield, Connecticut

Kathy A. Folske (p. 639)
Harrison High School
Farmington Hills, Michigan

Lucas Francois (p. 340)
Waterford Union High School
Waterford, Wisconsin

Karen L. George (p. 348)
Taunton High School
Taunton, Massachusetts

Ingrid Hart (p. 517)
E.O. Smith High School
Storrs, Connecticut

Mrs. Jean Herta (p. 398)
Powers Catholic High School
Flint, Michigan

Karen Jackson (p. 322)
Middletown High School
Middletown, Connecticut

Rick Kais (p. 93)
South Milwaukee High School
South Milwaukee, Wisconsin

Tim Kanold (p. 165)
Author
Glencoe *Geometry*

Nancy Keen (p. 242)
Martinsville High School
Martinsville, Indiana

Mrs. Janet Lovell (p. 379)
Osbourn High School
Manassas, Virginia

Carol Malloy (p. 467)
Author
Glencoe *Geometry*

George May (p. 553)
Salem High School
Salem, Virginia

Patricia McDaniel (p. 576)
Pinellas Park High School
Largo, Florida

Barbara O'Neil (p. 181)
North Forsyth High School
Cumming, Georgia

Rose M. Peters (p. 664)
Marshfield Senior High School
Marshfield, Wisconsin

Stacey S. M. Plummer (p. 448)
Merrimack High School
Merrimack, New Hampshire

Robert H. Schultz (p. 31)
Harrison High School
Farmington Hills, Michigan

Cheryl L. Walls (p. 260)
Milford High School
Milford, Delaware

Julianne Wanat (p. 315)
Milford High School
Milford, New Hampshire

Stan Williams (p. 526)
Rockmart High School
Rockmart, Georgia

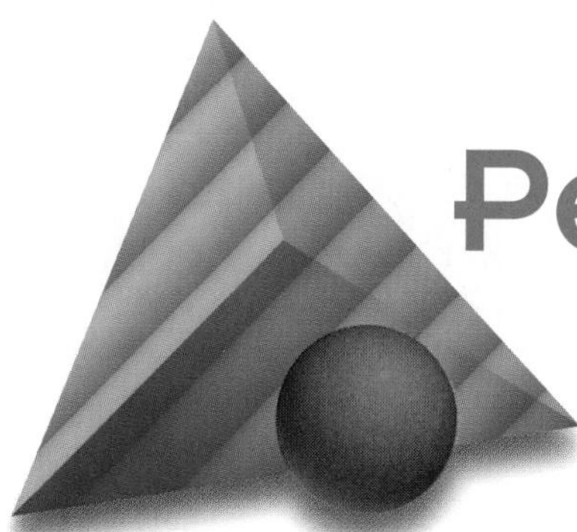

People in the News

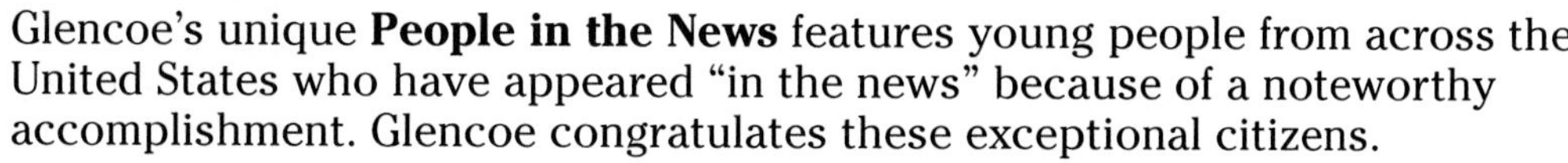

Glencoe's unique **People in the News** features young people from across the United States who have appeared "in the news" because of a noteworthy accomplishment. Glencoe congratulates these exceptional citizens.

Sabrina Alimahomed & Tara Church (p. 337)
California

Fred Bull (p. 445)
Alaska

James Fujita (p. 695)
California

Ebony Hood (p. 237)
District of Columbia

Brandi Hunt (p. 123)
Florida

Omar Lumkin (p. 5)
District of Columbia

Loui Maes (p. 179)
Colorado

Dara Marin (p. 289)
Colorado

Leander Morgan (p. 645)
New Mexico

Ryan Morgan (p. 513)
Maryland

Amit Paley (p. 69)
Massachusetts

J. R. Todd (p. 395)
Indiana

Adalee Velasquez (p. 573)
California

Internet Connections

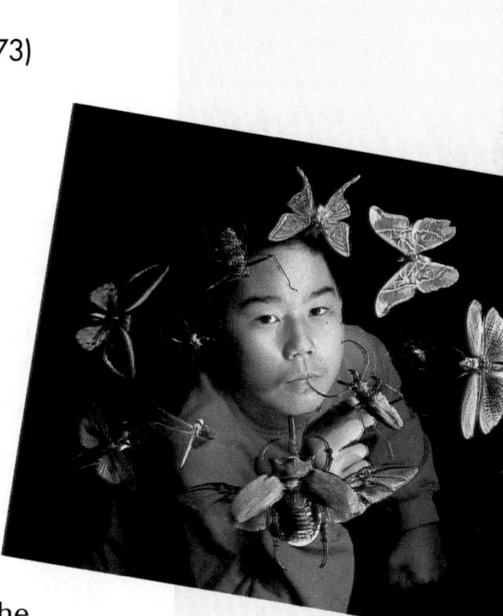

Glencoe's unique **Internet Connection** is a source for more information related to the **Chapter Project** and the interests of the person featured in **People in the News.**

The sites referenced in Glencoe's **Internet Connections** are not under the control of Glencoe. Therefore, Glencoe can make no representation concerning the content of these sites. Extreme care has been taken to list only reputable links by using educational and government sites whenever possible. Internet searches have been used that return only sites that contained no content apparently intended for mature audiences. The following list of mathematics Internet sites may prove helpful. Useful search tools include http://webcrawler.com/, http://www.yahoo.com/search.html, and http://www.microsoft.com/search

Internet Site	Comments
http://www.yahoo.com/Science/Mathematics	An ideal starting place that has links to other math sites.
http://www.tc.cornell.edu/Edu/MathSciGateway	Provides links to math and science resources; for grades 9–12.
http://forum.swarthmore.edu/dr.math/	Students in grades K–12 can ask Dr. Math their own questions.
http://www.enc.org/	This is the Eisenhower clearinghouse for K–12 math and science instructional materials.
http://forum.swarthmore.edu/mathmagic	K–12 telecommunications project in Texas that uses computers to increase problem-solving and communication skills.

Glencoe's Geometry Research Activities

Glencoe's ***Geometry: Integration, Applications, and Connections***, as well as the entire Glencoe Mathematics Series, is the product of ongoing, classroom-oriented research that involves students, teachers, curriculum supervisors, administrators, parents, and college-level mathematics educators.

The programs that make up the Glencoe Mathematics Series are currently being used by millions of students and tens of thousands of teachers. The key reason that Glencoe Mathematics programs are so successful in the classroom is the fact that each Glencoe author team is a mix of practicing classroom teachers, curriculum supervisors, and college-level mathematics educators. Glencoe's balanced author teams help ensure that Glencoe Mathematics programs are both practical and progressive.

Prior to publication of a Glencoe program, typical research activities include:

- a review of educational research and recommendations made by groups such as NCTM
- mail surveys of mathematics educators
- discussion groups involving mathematics teachers, department heads, and supervisors
- focus groups involving mathematics educators
- face-to-face interviews with mathematics educators
- telephone surveys of mathematics educators
- in-depth analyses of manuscript by a wide range of reviewers and consultants
- field tests in which students and teachers use pre-publication manuscript in the classroom

Feedback from teachers, curriculum supervisors, and even students who currently use Glencoe mathematics programs is also incorporated as Glencoe plans and publishes new and revised programs. For example, Classroom Vignettes, which are printed in the Teacher's Wraparound Editions, are one result of this feedback.

All of this research and information is used by Glencoe's authors and editors to publish the best instructional resources possible.

WHY IS GEOMETRY IMPORTANT?

Why do I need to study geometry? When am I ever going to have to use geometry in the real world?

Many people, not just geometry students, wonder why mathematics is important. ***Geometry*** is designed to answer those questions through **integration**, **applications**, and **connections**.

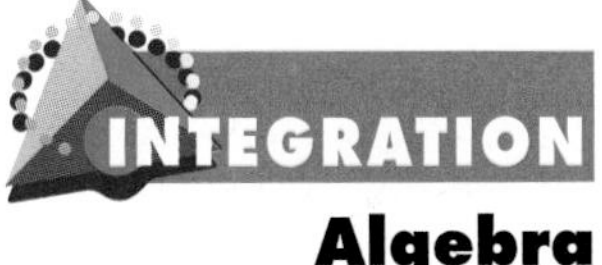

Algebra

Did you know that geometry and algebra are closely related? Topics from all branches of mathematics, like algebra and statistics, are integrated throughout the text.

You'll apply algebra to the midpoint formula. (Lesson 1–5, page 38)

Physical Therapy

Would you believe that medical personnel use angles and line segments in their work? Real-world uses of geometry are presented.

Physical therapists apply inequalities in two triangles to help people regain strength after injury by using special exercises. (Lesson 5–6, page 273)

Literature

What does literature have to do with mathematics? Mathematical topics are connected to other subjects you study.

Lewis Carroll, author of *Alice's Adventures in Wonderland,* frequently used the logic of mathematics in his writing. (Lesson 2–2, page 76)

Authors

Cindy J. Boyd teaches mathematics at Abilene High School in Abilene, Texas. She received her B.S.Ed. and M.Ed. degrees from Abilene Christian University. She has received numerous awards and is the first teacher to receive the Texas Presidential State Award for four consecutive years (1994–1997). Ms. Boyd was selected as the 1995–1996 Mathematics Teacher of the Year by Disney and McDonald's American Teacher Awards. Ms. Boyd is an active member of the National Council of Teachers of Mathematics, presenting *Skit-So-Phrenia!* at numerous conventions each year, and she serves on the regional Services Committee as the Southern-Region 2 representative. She is also a consultant on Glencoe's *Algebra 1* and *Algebra 2*.

"Glencoe Geometry is well suited for teachers and students because concepts are developed and built for deeper understanding. A strong problem-solving base can be developed and maintained by all students so they enter the workforce with the tools needed for whatever the twenty-first century holds."

Gail F. Burrill taught mathematics at Whitnall High School in Greenfield, Wisconsin. Ms. Burrill obtained her B.S. in Mathematics from Marquette University and her M.S. degree in Mathematics from Loyola University. Ms. Burrill received a Presidential Award for Excellence in Teaching Mathematics and Science in 1985. She is a past president of the Wisconsin Mathematics Council and received a Wisconsin Distinguished Mathematics Educator Award. Ms. Burrill is currently the president of the National Council of Teachers of Mathematics. She is a member of the National Board for Professional Teaching Standards and was Chair of the Mathematics Committee for Initial Teacher Certification for the Council of Chief State School Officers. She is a Fellow of the American Statistical Association, has spoken internationally, and has authored many articles on teaching math and statistics.

Jerry J. Cummins is a Staff Development Specialist for the Bureau of Education and Research. He is also a consultant on Systemic Leadership Initiatives in Mathematics and Science in the State of Illinois. He was the Chair of the Mathematics/Science Division at Lyons Township High School, in LaGrange, Illinois. Mr. Cummins obtained his B.S. degree in mathematics education and M.S. degree in educational administration and supervision from Southern Illinois University at Carbondale. He also holds an M.S. degree in mathematics education from the University of Oregon. Mr. Cummins has spoken at many local, state, and national mathematics conferences, and is a member of the Regional Services Committee of the National Council of Teachers of Mathematics. He is currently the First Vice-President of the National Council of Supervisors of Mathematics. Mr. Cummins received an Illinois State Presidential Award for Excellence in Teaching of Mathematics.

"Glencoe Geometry places the highest priority possible on student achievement. Motivation is one of the primary factors in learning. Students are able to see or experience reasons for studying each topic."

TIMOTHY D. KANOLD is the Director of Mathematics and a mathematics teacher at Adlai Stevenson High School in Lincolnshire, Illinois. Mr. Kanold obtained his B.S. degree in mathematics education and M.S. degree in mathematics from Illinois State University. He also holds a C.A.S. degree from the University of Illinois. In 1995, he received the Award of Excellence from the Illinois State Board of Education for outstanding contributions to Illinois education. A member of the National Council of Teachers of Mathematics, he served on NCTM's "Professional Standards for Teaching Mathematics" Commission, was a member of the Regional Services Committee, and served as a speaker for "New Dimensions in Leadership." He is a past president of the Council for Presidential Awardees of Mathematics. He is an author of numerous articles on effective classroom teaching practices including a chapter in NCTM's 1990 Yearbook. He is a nationally known, well-respected, motivational speaker.

"The text provides numerous formats and opportunities for students to clarify their thinking and formulate generalizations through guided discovery and investigations while maintaining a formal sense of rigor necessary in a contemporary geometry curriculum."

CAROL MALLOY is an assistant professor of mathematics education at the University of North Carolina at Chapel Hill. Dr. Malloy previously taught middle and high school mathematics. She received her B.S. in mathematics and education from West Chester State College, her M.S.T. in mathematics from Illinois Institute of Technology, and her Ph.D. in curriculum and instruction from the University of North Carolina at Chapel Hill. Dr. Malloy is the president of the Benjamin Banneker Association, Inc. Dr. Malloy is an active member of numerous local, state, and national professional organizations and boards, including the National Council of Teachers of Mathematics (NCTM) and Association for Supervision and Curriculum Development (ASCD). She frequently speaks at conferences and schools on the topics of mathematical problem solving, influence of culture on mathematics learning, and mathematics for all students.

"Geometry often has been an exciting course for many teachers and students. This textbook captures the excitement of learning and using geometry for students from all levels and experiences. Diverse approaches to learning and teaching from hands-on activities and technology provide a strong axiomatic structure in the development of deductive and inductive reasoning."

LEE E. YUNKER (1941–1994)

This edition of *Geometry: Integration, Applications, and Connections* is dedicated to the memory of Lee E. Yunker. For 31 years, he educated students and teachers alike. He was personally committed to life-long learning and strove to help all teachers best serve the needs of their students. His many contributions to mathematics education continue in this and the 15 other mathematics books he authored.

Consultants, Writers, and Reviewers

Consultants

Pamela Ann Chandler, Ed. D.
Mathematics Coordinator
Fort Bend ISD
Sugar Land, TX 77478

Eva Gates
Independent Mathematics Consultant
Pearland, Texas

Dr. Luis Ortiz-Franco
Consultant, Diversity
Associate Professor of Mathematics
Chapman University
Orange, California

Gilbert J. Cuevas
Professor of Mathematics Education
University of Miami
Coral Gables, Florida

Deborah Ann Haver, Ed. D.
Mathematics Consultant
Chesapeake, Virginia

Joan Estlow
Mathematics Teacher
Egg Harbor Township High School
Pleasantville, New Jersey

Melissa McClure
Mathematics Consultant
Teaching for Tomorrow
Fort Worth, Texas

Writers

David Foster
Writer, Investigations
Glencoe Author and Mathematics Consultant
Morgan Hill, California

Reviewers

Rhea Baldino
Mathematics Teacher
Charlottesville High School
Charlottesville, Virginia

Paul J. Bohney
Supervisor of Math/Science-Retired
Gary Community School Corporation
Gary, Indiana

Judith L. Conley
Geometry/Algebra II Teacher
Shades Valley High School
Birmingham, Alabama

Sandra Argüelles Daire
Mathematics Teacher
Miami Senior High School
Miami, Florida

Pamela M. Fey
Mathematics Teacher
Northside ISD
San Antonio, Texas

Linda M. Gordius
Mathematics Department Chair
Portland High School
Portland, Maine

Joy E. Hine
Mathematics Teacher
Carmel High School
Carmel, Indiana

Brenda Berg
Math Department Chair
Carbondale Community High School
Carbondale, Illinois

Marcia S. Chumas
Mathematics Teacher
East Mecklenburg High School
Charlotte, North Carolina

Amy L. Craig
Mathematics Teacher
Princeton High School
Cincinnati, Ohio

Walter S. Dewar
Mathematics Department Chair
Rowlett High School
Rowlett, Texas

John H. Geiger
Supervisor of Mathematics
Pasco County School District
Land O' Lakes, Florida

Barbara R. Gray
Mathematics Department Chair
West Forsyth High School
Clemmons, North Carolina

Craig Hochhaus
Mathematics Department Chair
Agoura High School
Agoura Hills, California

Mark Bleiler
Geometry Teacher
Breckenridge High School
Breckenridge, Michigan

Judy Cline
Math Teacher/Department Chair
North Mecklenburg High School
Huntersville, North Carolina

David Crowe
Mathematics Teacher
Thomas W. Harvey High School
Painesville, Ohio

Dollie Driver
Mathematics Teacher
Coronado High School
Lubbock, Texas

Norma L. Goldner
Mathematics Teacher
Benton Harbor High School
Benton Harbor, Michigan

Karen Sparks Gray
Teacher
Clay-Chalkville High School
Clay, Alabama

Nahid Mazdeh Huff
Mathematics Teacher
Colonel White High School
Dayton, Ohio

Judy Elza Johnson
Mathematics Teacher
North Laurel High School
London, Kentucky

Lynda B. (Lucy) Kay
Mathematics Department Chair
Martin Middle School
Raleigh, North Carolina

Nancy Keen
Geometry Teacher
Martinsville High School
Martinsville, Indiana

Jeff Killion
Math Challenge Tutor
Lower Merion School District
Ardmore, Pennsylvania

Nancy W. Kinard
Mathematics Teacher
Palm Beach Gardens High School
Palm Beach Gardens, Florida

Susan Rypkema Lambert
Mathematics Teacher
Plano Senior High School
Plano, Texas

Bruce Linn
Mathematics Teacher
Fort Zumwalt North High School
O'Fallon, Missouri

Kimberly K. Loomis
Mathematics Teacher
Southern Nevada Vocational-Technical Center
Las Vegas, Nevada

Gerald Martau
Deputy Superintendent-Retired
Lakewood City Schools
Lakewood, Ohio

David F. McReynolds
Mathematics Teacher
Boise City High School
Boise City, Oklahoma

Patrick Miller
Instructional Support Team
Fort Worth ISD
Fort Worth, Texas

Melleretha Moses-Johnson
Mathematics Coordinator
Saginaw Public Schools
Saginaw, Michigan

Kathleen Walker Murrell
Mathematics Teacher
J. Frank Dobie High School
Pasadena ISD
Houston, Texas

Maxine Nesbitt
Mathematics Teacher
Carmel High School
Carmel, Indiana

Roger O'Brien
Mathematics Supervisor
Polk County Schools
Bartow, Florida

Virginia A. Palmer
Mathematics Teacher
East Lyme High School
East Lyme, Connecticut

Joseph P. Pimental
Mathematics Teacher
Taunton High School
Taunton, Massachusetts

Brenda Price
Geometry Teacher
LaRue County High School
Hodgenville, Kentucky

Nancy Puhlmann
Mathematics Teacher
Logan High School
Logan, Utah

Scott Reed
Mathematics Teacher
Bogle Junior High School
Chandler, Arizona

Joseph "Joe" Ruffino
Mathematics Teacher
Covington High School
Covington, Louisiana

Linda C. Ruppenthal
Mathematics Department Chairperson
Hancock Middle-Senior High School
Hancock, Maryland

Lee Ann Russell
Mathematics Teacher
George Washington High School
Danville, Virginia

William Scott
Mathematics Supervisor (K–12)
Colorado Springs School District 11
Colorado Springs, Colorado

John Sico, Jr.
Vice Principal
Eastside High School
Paterson, New Jersey

Edith R. Simon
Teacher
McMain Magnet Secondary School
New Orleans, Louisiana

Mary Beth Smith
Mathematics Teacher
Dublin Coffman High School
Dublin, Ohio

Erik L. Thompson
Chemistry Teacher
Thomas Worthington High School
Worthington, Ohio

Margaretta L. Thornton
Mathematics Teacher
Sam Houston High School
Moss Bluff, Louisiana

John Tithof
Mathematics Teacher
Gaylord High School
Gaylord, Michigan

Peggy Turman
Mathematics Teacher
Cabell Midland High School
Ona, West Virginia

Connie G. Turner
Mathematics Teacher
East Rutherford High School
Forest City, North Carolina

Simonette Urbain
Mathematics Teacher
J. Sterling Morton High School, West Campus
Berwyn, Illinois

Susan K. White
Math Supervisor
West Chester Area School District
West Chester, Pennsylvania

Table of Contents

CHAPTER 2 Connecting Reasoning and Proof 68

CHAPTER 3 Using Perpendicular and Parallel Lines 122

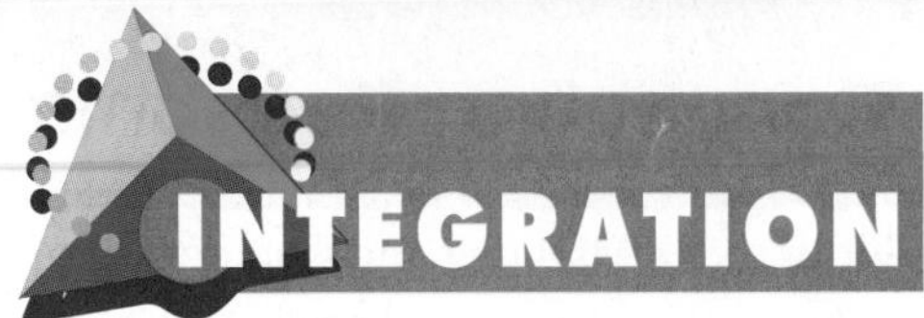

Content Integration

What does algebra have to do with geometry? Believe it or not, you can study most math topics from more than one point of view. Here are some examples.

▲ Probability You'll use probability to study properties of quadrilaterals (Lesson 6–1, Example 3, page 294) and the probability of a dart landing in a particular area of the target. (Lesson 10–6, page 554)

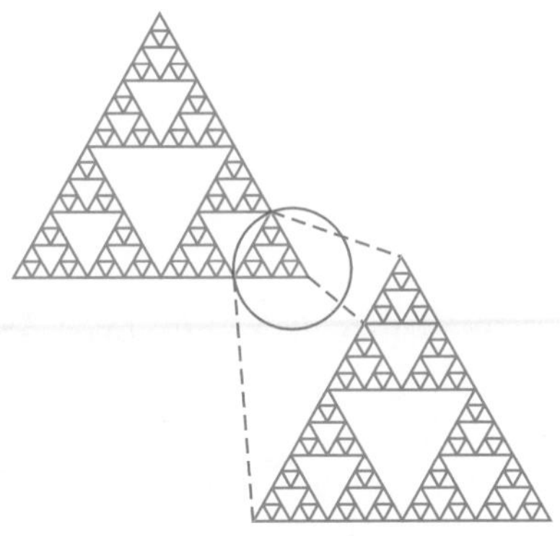

▲ Problem Solving You'll solve a simpler problem to find values in Pascal's triangle and look for a pattern in Sierpenski's triangle. (Lesson 7–6, pages 379 and 380)

LOOK BACK

You can refer to Lesson 1-4 for information on finding the distance between points.

Source: Lesson 3–5, page 156

Look Back refers you to skills and concepts that have been taught earlier in the book.

Discrete Mathematics ▶ You'll investigate sequences as you study geometric mean and Pythagorean triples. (Lesson 8–1, pages 397 and 401)

Statistics You'll learn ▶ how to display data on a circle graph. (Lesson 9–2, page 452)

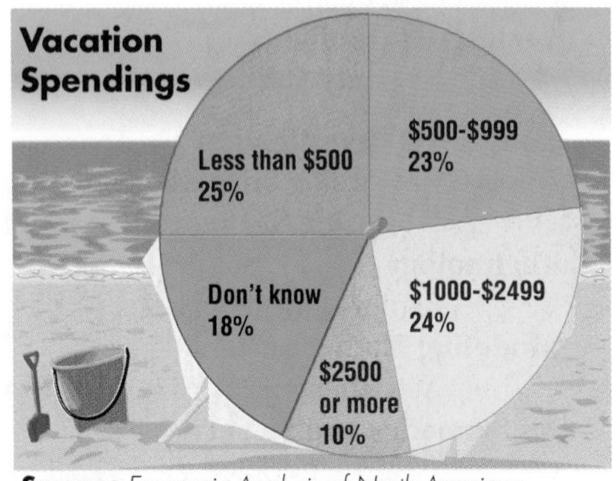

Source: *Economic Analysis of North American Ski Areas*

Algebra You'll use your algebra skills to prove that two triangles are congruent. (Lesson 4–4, page 206)

CHAPTER 4 Identifying Congruent Triangles 178

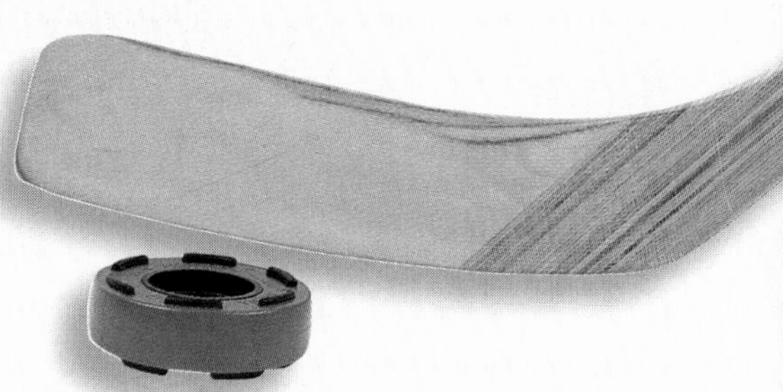

CHAPTER 5 Applying Congruent Triangles 236

CHAPTER 6 Exploring Quadrilaterals 288

CHAPTER 7 Connecting Proportion and Similarity 336

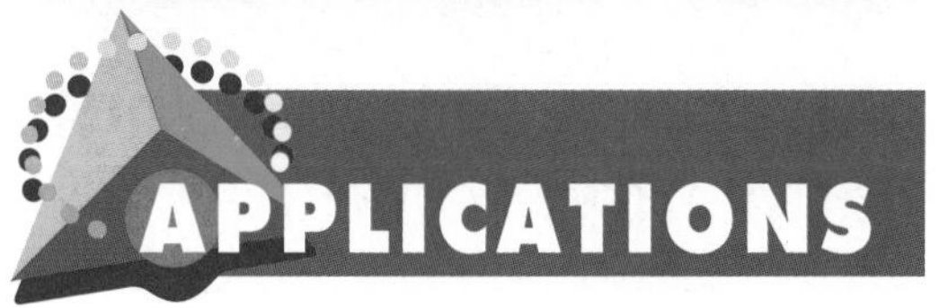

Real-Life Applications

Have you ever wondered if you'll ever actually use math? Every lesson in this book is designed to help show you where and when math is used in the real world. Since you'll explore many interesting topics, we believe you'll discover that math is relevant and exciting. Here are some examples.

Top Five List, FYI, and **Fabulous Firsts** contain interesting facts that enhance the applications.

Top Five List

Snowiest Cities in the U.S.	Mean Annual Snowfall (in.)
1. Blue Canyon, CA	240.8
2. Marquette, MI	128.6
3. Sault Ste. Marie, MI	116.7
4. Syracuse, NY	111.6
5. Caribou, ME	110.4

Source: Lesson 7–6, page 382

fabulous **FIRSTS**

Norma Merrick Sklarek (1928–)

Norma Merrick Sklarek was the first African-American woman to become a registered architect.

Source: Lesson 6–5, page 327

FYI

In 1996, the Olympic torch was carried 15,280 miles through 42 states before being carried to the opening ceremonies for the Centennial Summer Olympics in Atlanta, Georgia.

Source: Lesson 2–5, page 100

Construction You'll study an industrial technology application that involves congruent triangles formed by a power line and tower. (Lesson 5–2, page 246)

SHOE

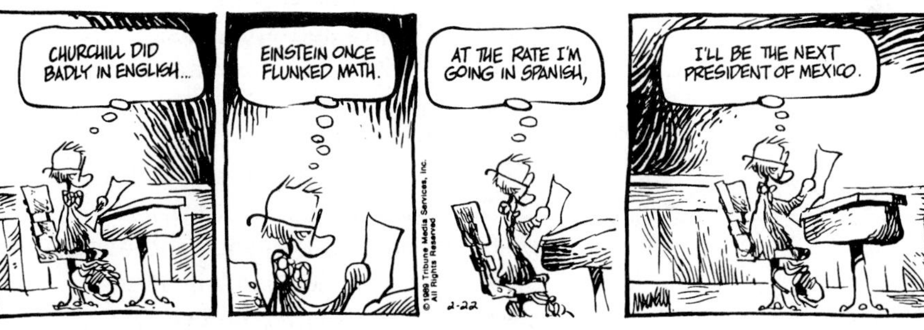

CHURCHILL DID BADLY IN ENGLISH...
EINSTEIN ONCE FLUNKED MATH.
AT THE RATE I'M GOING IN SPANISH,
I'LL BE THE NEXT PRESIDENT OF MEXICO.

▲ **Comics** You'll discuss a *Shoe*® cartoon and whether or not Skyler's conjecture is true. (Lesson 2–1, page 70)

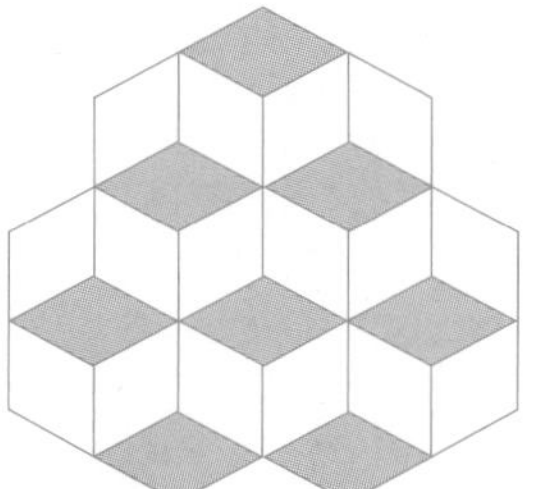

▲ **Optical Illusions** Rombi can be used to create an optical illusion. (Lesson 6–4, page 313)

Stealthy Geometry

What do radar waves and building design have to do with mathematics? Actual reprinted articles illustrate how mathematics is a part of our society. (Lesson 3–5, page 161)

CHAPTER 8 Applying Right Triangles and Trigonometry 394

CHAPTER 9 Analyzing Circles 444

CHAPTER 10 Exploring Polygons And Area 512

CHAPTER 11 Investigating Surface Area And Volume 572

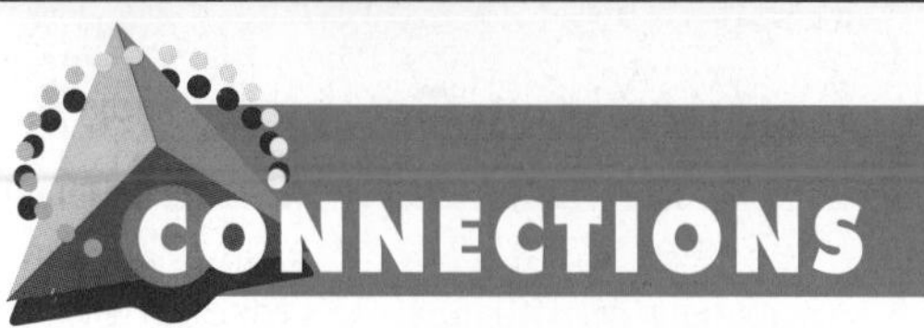

Interdisciplinary Connections

Did you realize that mathematics is used in botany? in history? in geography? Yes, it may be hard to believe, but mathematics is frequently connected to other subjects that you are studying.

Botany You'll study angles made by tree branches (Lesson 2–6, Exercise 14)

Geography You'll apply spherical geometry to find the flying distance between two cities. (Lesson 3–6, page 165)

Career Choices include information on interesting careers.

CAREER CHOICES

Architects have played an important role in history from the earliest cities to our modern skyscrapers. The word *architect* is from the Greek meaning "master builder." A licensed architect requires a bachelor's degree in architecture, participation in an internship of 3 years, and passing an exam.

For more information, contact:

Society of American Registered Architects
1245 S. Highland Ave.
Lombard, IL 60148

Source: Lesson 4–1, page 181

Evidence of Nine men's morris has been found in Ceylon, now Sri Lanka. It was also known to inhabitants of Troy and to the Vikings.

Source: Lesson 6–4, page 317

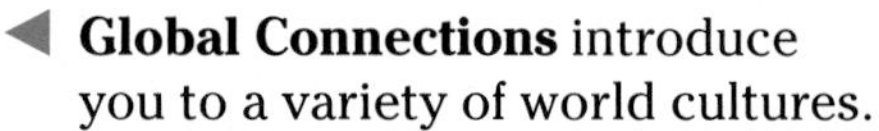

Global Connections introduce you to a variety of world cultures.

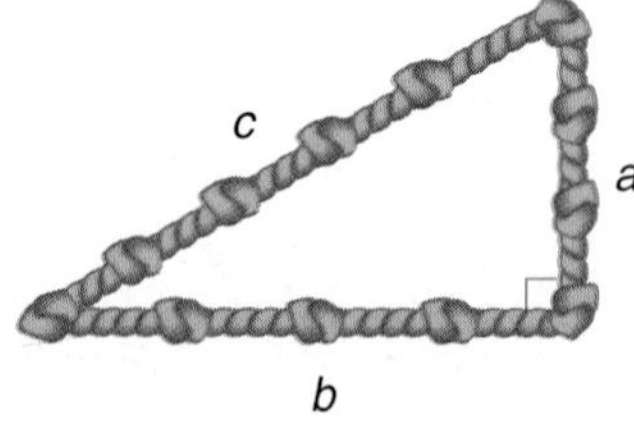

History You'll use Egyptian techniques to sketch and classify triangles. (Lesson 5–5, pages 268–269)

Math Journal exercises give you the opportunity to assess yourself and write about your understanding of key math concepts. (Lesson 2–3, Exercise 5)

5. Make up several logic puzzles. Have your friends try to solve them. Write the puzzles and solutions in your journal.

CHAPTER 12 Continuing Coordinate Geometry 644

CHAPTER 13 Investigating Loci and Coordinate Transformations 694

TECHNOLOGY

Do you know how to use computers and graphing calculators? If you do, you'll have a much better chance of being successful in today's high-tech society and workplace.

GRAPHING CALCULATORS

There are several ways in which graphing calculators are integrated.

- **Getting Acquainted with the Graphing Calculator** On pages 2–3, you'll get acquainted with the basic features and functions of a graphing calculator.
- **Graphing Calculator Explorations** You'll learn how to use a graphing calculator to explore slope-intercept form in Lesson 3–3.
- **Graphing Calculator Programs** In Lesson 2–2, Exercise 51 includes a graphing calculator program that can be used to find the volume of a cylinder.
- **Graphing Calculator Exercises** Many exercises are designed to be solved using a graphing calculator or a TI-92 calculator. For example, see Exercise 37 in Lesson 9–8.

- **Using Technology Lessons** On pages 26–27, you'll learn how to use a TI-92 calculator (or Cabri Geometry) to measure segments.

COMPUTER SOFTWARE

- **Spreadsheets** On page 21 of Lesson 1–3, a spreadsheet is used to calculate quantities determined by a formula.
- **Geometry Software** Cabri Geometry software on a computer or on a TI-92 calculator is used in Exercise 25 on page 526 to create a tessellation by reflecting a polygon.

Technology Tips, such as this one on page 516 of Lesson 10–1, are designed to help you make more efficient use of technology through practical hints and suggestions.

A spreadsheet can be used to calculate each sum. If cell B2 contains the number of sides, cell B3 can contain the formula, B2 − 2, and the formula for B4 is 180 * B3.

SYMBOLS

h	altitude*
$\angle$	angle
$\measuredangle s$	angles
a	apothem*
$\approx$	is approximately equal to, approximately
$\overset{\frown}{AB}$	minor arc with endpoints A and B
$\overset{\frown}{ACB}$	major arc with endpoints A and B
A	area of a polygon or circle* surface area of a sphere*
B	area of the base of a prism, cylinder, pyramid, or cone*
b	base of a triangle, parallelogram, or trapezoid*
$\odot P$	circle with center P
C	circumference*
$\cong$	is congruent to, congruent
$\leftrightarrow$	corresponds to
cos	cosine
$^\circ$	degree
d	diameter of a circle* distance*
AB	distance between points A and B*
$=$	equals, is equal to
$\neq$	is not equal to
$>$	is greater than
A'	the image of preimage A
$<$	is less than
L	lateral area*
$\overleftrightarrow{DE}$	line containing points D and E
$\rightarrow$	is mapped onto
$m\angle A$	degree measure of $\angle A$
$m\overset{\frown}{AB}$	degree measure of arc AB
$\sqrt{}$	nonnegative square root
(x, y)	ordered pair
(x, y, z)	ordered triple
$\parallel$	is parallel to, parallel
$\nparallel$	is not parallel to
$\square$	parallelogram
P	perimeter*
$\perp$	is perpendicular to, perpendicular
π	pi
n-gon	polygon with n sides
r	radius of a circle*
$\overrightarrow{PQ}$	ray with endpoint P passing through Q
$\overline{RS}$	segment with endpoints R and S
s	side of a regular polygon*
$\sim$	is similar to, similar
sin	sine
ℓ	line ℓ length of a rectangle slant height
m	slope
tan	tangent
T	total surface area
$\triangle$	triangle
$\overrightarrow{AB}$	vector from A to B
V	volume*

* indicates that this is the symbol for the measure of the item listed.

GETTING ACQUAINTED WITH THE GRAPHING CALCULATOR

What is it?
What does it do?
How is it going to help me learn math?

These are just a few of the questions many students ask themselves when they first see a graphing calculator. Some students may think, "Oh, no! Do we ***have*** to use one?", while others may think, "All right! We get to use these neat calculators!" There are as many thoughts and feelings about graphing calculators as there are students, but one thing is for sure: a graphing calculator can help you learn mathematics.

So what is a graphing calculator? Very simply, it is a calculator that draws graphs. This means that it will do all of the things that a "regular" scientific calculator will do, *plus* it will draw graphs of equations. In geometry, this capability is helpful as you analyze the graphs of equations.

But a graphing calculator can do more than just calculate and draw graphs. For example, you can program it or work with data to make statistical graphs and perform computations. If you need to generate random numbers, you can do that on the graphing calculator. You can even draw and manipulate geometric figures on some graphing calculators. It's really a very powerful tool—so powerful that it is often called a pocket computer.

As you may have noticed, graphing calculators have some keys that other calculators do not. The Texas Instruments TI-82, TI-83, or TI-92 calculators will be used throughout this text. Most of the keystrokes for the TI-82 and TI-83 are the same. The keys on the bottom half of the TI-82 and TI-83 are those found on scientific calculators. The keys located just below the screen are the graphing keys. You will also notice the up, down, left, and right arrow keys. These allow you to move the cursor around on the screen, to "trace" graphs that have been plotted, and to choose items from the menus. The other keys located on the top half of the calculator access the special features such as statistical computations and programming features.

A few of the keystrokes that can save you time when using the graphing calculator are listed below.

- The commands above the calculator keys are accessed with the [2nd] or [ALPHA] key. On a TI-82, the [2nd] key and the commands accessed by it are blue and the [ALPHA] key and its commands are gray. On a TI-83, the [2nd] key and its commands are yellow and the [ALPHA] and its commands are green.
- [2nd] [ENTRY] copies the previous calculation so you can edit and use it again.
- Pressing [ON] while the calculator is graphing stops the calculator from completing the graph.
- [2nd] [QUIT] will return you to the home (or text) screen.
- [2nd] [A-LOCK] locks the [ALPHA] key, which is like pressing "shift lock" or "caps locks" on a typewriter or computer. The result is that all letters will be typed and you do not have to repeatedly press the [ALPHA] key. (This is handy for programming.) Stop typing letters by pressing [ALPHA] again.
- [2nd] [OFF] turns the calculator off.

Some commonly used mathematical functions are shown in the table below. As with any scientific calculator, the graphing calculator observes the order of operations.

Mathematical Operation	Examples	Keys	Display
evaluate expressions	Find $2 + 5$.	2 [+] 5 [ENTER]	2+5 7
exponents	Find 3^5.	3 [^] 5 [ENTER]	3^5 243
multiplication	Evaluate $3(9.1 + 0.8)$.	3 [×] [(] 9.1 [+] .8 [)] [ENTER]	3(9.1+.8) 29.7
roots	Find $\sqrt{14}$.	[2nd] [√‾] 14 [ENTER]	√‾ 14 3.741657387
opposites	Enter -3.	[(-)] 3	-3

Graphing on the TI-82 or TI-83

Before graphing, we must instruct the calculator how to set up the axes in the coordinate plane. To do this, we define a **viewing window**. The viewing window for a graph is the portion of the coordinate grid that is displayed on the graphics screen of the calculator. The viewing window is written as [left, right] by [bottom, top] or [Xmin, Xmax] by [Ymin, Ymax]. A viewing window of $[-10, 10]$ by $[-10, 10]$ is called the **standard viewing window** and is a good viewing window to start with to graph an equation. The standard viewing window can be easily obtained by pressing [ZOOM] 6. Try this.

Move the arrow keys around and observe what happens. You are seeing a portion of the coordinate plane that includes the region from -10 to 10 on the x-axis and from -10 to 10 on the y-axis. Move the cursor, and you can see the coordinates of the point for the position of the cursor.

Any viewing window can be set manually by pressing the WINDOW key. The window screen will appear and display the current settings for your viewing window. Using the arrow and ENTER keys, move the cursor to edit the window settings. Xscl and Yscl refer to the x-scale and y-scale. This is the number of units between tick marks on the x- and y-axes. Xscl=1 means that there will be a tick mark for every unit of one along the x-axis.

Programming on the TI-82 or TI-83

The TI-82 and TI-83 have programming features that allow us to write and execute a series of commands to perform tasks that may be too complex or cumbersome to perform otherwise. Each program is given a name. Commands begin with a colon (:), which the calculator enters automatically, followed by an expression or an instruction. Most of the features of the calculator are accessible from program mode.

When you press PRGM, you see three menus: EXEC, EDIT, and NEW. EXEC allows you to execute a stored program, EDIT allows you to edit or change a program, and NEW allows you to create a program. The following tips will help you as you enter and run programs on the TI-82.

- To begin entering a new program, press PRGM ▶ ▶ ENTER.
- After a program is entered, press 2nd [QUIT] to exit the program mode and return to the home screen.
- To execute a program, press PRGM. Then use the down arrow key to locate the program name and press ENTER, or press the number or letter next to the program name.
- If you wish to edit a program, press PRGM ▶ and choose the program from the menu.
- To immediately re-execute a program after it is run, press ENTER when Done appears on the screen.
- To stop a program during execution, press ON.

Geometry on the TI-92

The TI-92 calculator includes a geometry program that allows you to perform operations on geometric figures. With TI-92 Geometry, you can draw, move, measure, graph, and alter figures. To access the TI-92 Geometry on your calculator, press APPS and choose 8: Geometry. If you are starting a new session, choose 3: New, press down on the cursor pad, enter a name for the construction (up to eight letters), and press ENTER.

The toolbar is organized into eight menus which are accessed by pressing the function keys F1 through F8. The list below describes the types of functions on the menu for each key.

Key	Menu Functions
F1	Freehand transformations
F2	Construction of points or linear objects
F3	Construction of curves and polygons
F4	Euclidean constructions and creating macros
F5	Transformational geometry constructions
F6	Measurements and calculations
F7	Labels and animation
F8	File operations and editing

All figures are drawn by choosing one or more points on the screen. To draw a figure, first choose the object you would like to draw from a menu, then create or select the points to define the object. For example to draw a triangle, press F3 3 to choose triangle. Then move the cursor to where you would like the first vertex and press ENTER. Move the cursor to locate the second vertex and press ENTER. Finally move the cursor to the third vertex and press ENTER to complete the triangle.

The procedures for some common operations you will use in the geometry application are listed below.

- To select an object that you have drawn, move the cursor toward the object until the object's name appears next to the cursor. Then press ENTER.
- If you would like to delete an object you have drawn, select the object and press ⬅ or F8 7.
- Move objects on the screen by choosing the pointer from the F1 menu. Then select the object, press and hold the grasping hand key, and use the arrow keys to move the object.

While a calculator cannot do everything, it can make some things easier. To prepare for whatever lies ahead, you should try to learn as much as you can. The future will definitely involve technology, and using a graphing calculator is a good start toward becoming familiar with technology. Who knows? Maybe one day you will be designing the next satellite, building the next skyscraper, or helping students learn mathematics with the aid of a graphing calculator!

CHAPTER 1

Discovering Points, Lines, Planes, and Angles

PREVIEWING THE CHAPTER

This chapter lays the foundation for the study of geometry by introducing key terms and concepts. By beginning with a review of the coordinate plane, the development uses this plane to make a logical transition from algebra to geometry. The undefined terms *point, line,* and *plane* are then used to define other geometric terms such as *angles, segments,* and *rays*. The chapter concludes by describing relationships among defined terms, extending the ideas they convey, and looking into how they relate to each other to extend geometric concepts.

Lesson (Pages)	Lesson Objectives	NCTM Standards	State/Local Objectives
1-1 (6–11)	Graph ordered pairs on a coordinate plane. Identify collinear points.	1–5, 8, 10	
1-2 (12–18)	Identify and model points, lines, and planes. Identify coplanar points and intersecting lines and planes. Solve problems by listing the possibilities.	1–5, 7, 8, 12	
1-3 (19–25)	Solve problems by using formulas. Find maximum area of a rectangle for a given perimeter.	1–5, 6, 8, 12	
1-4A (26–27)	Use a TI-92 calculator to draw and measure segments.	1–5, 7, 8	
1-4 (28–35)	Find the distance between two points on a number line and between two points in a coordinate plane. Use the Pythagorean Theorem to find the length of the hypotenuse.	1–5, 7, 8, 12	
1-5 (36–43)	Find the midpoint of a segment. Complete proofs involving segment theorems.	1–5, 7, 8, 14	
1-6 (44–51)	Identify angles and classify angles. Use the angle addition postulate to find the measures of angles. Identify and use congruent angles and the bisector of an angle.	1–5, 7, 8	
1-7A (52)	Use a TI-92 calculator to draw and measure angles.	1–5, 7, 8	
1-7 (53–60)	Identify and use adjacent, vertical, complementary, supplementary, and linear pairs of angles, and perpendicular lines. Determine what information can and cannot be assumed from a diagram.	1–5, 7, 8, 14	

ORGANIZING THE CHAPTER

A complete, 1-page lesson plan is provided for each lesson in the ***Lesson Planning Guide***. Answer keys for each lesson are available in the ***Answer Key Masters***.

You may want to refer to the **Course Planning Calendar** on page T12 for detailed information on pacing.
PACING: Standard—13 days; **Honors**—13 days; **Block**—8 days

LESSON PLANNING CHART

			Blackline Masters									
Lesson (Pages)	**Materials/ Manipulatives**	**Extra Practice (Student Edition)**	**Study Guide**	**Practice**	**Enrichment**	**Assessment & Evaluation**	**Modeling Mathematics**	**Multicultural Activity**	**Tech Prep Applications**	**Graphing Calc. & Computer**	**Real-World Applications**	**Teaching Transparencies**
1-1 (6–11)		p. 764	p. 1	p. 1	p. 1			p. 1		p. 1	1	1-1A 1-1B
1-2 (12–18)	colored paper scissors* tape	p. 764	p. 2	p. 2	p. 2	p. 16	p. 79		p. 1		2	1-2A 1-2B
1-3 (19–25)	spreadsheet software	p. 764	p. 3	p. 3	p. 3							1-3A 1-3B
1-4A (26–27)	TI-92 calculator									pp. 14–15		
1-4 (28–35)	compass* straightedge*	p. 765	p. 4	p. 4	p. 4	pp. 15, 16						1-4A 1-4B
1-5 (36–43)	patty paper ruler* compass*	p. 765	p. 5	p. 5	p. 5							1-5A 1-5B
1-6 (44–51)	compass* straightedge* patty paper	p. 765	p. 6	p. 6	p. 6	p. 17	pp. 15–18		p. 2			1-6A 1-6B
1-7A (52)	TI-92 calculator									p. 16		
1-7 (53–60)	compass* straightedge* patty paper	p. 766	p. 7	p. 7	p. 7	p. 17		p. 2				1-7A 1-7B
Study Guide/ Assessment (61–65)						pp. 1–14, 18–20						

*Included in Glencoe's High School Manipulative Kit and Overhead Manipulative Resources.

ORGANIZING THE CHAPTER

Other Chapter Resources

Student Edition
Chapter Opener, pp. 4–5
Mathematics and Society, p. 35

Teacher's Classroom Resources
Investigations and Projects Masters, pp. 25–28
Block Scheduling Booklet

Technology
Test and Review Software (IBM and Macintosh)
CD-ROM Multimedia Applications (Windows and Macintosh)
MindJogger Videoquizzes (VHS)

Professional Publications
Glencoe Mathematics Professional Series

Outside Resources

Books/Periodicals
Career Associates, *Career Choices - Mathematics*, Walker and Company.
National Council of Teachers of Mathematics, *Curriculum and Evaluation Standards for School Mathematics*, National Council of Teachers of Mathematics.
National Research Council, *Everybody Counts: A Report to the Nation on the Future of Mathematics Education*, National Academy Press.

Software
Cabri Géometre, Texas Instruments
Geometric Connectors: Coordinates, Sunburst

Videos/CD-ROMs
Computers: The Truth of the Matter, Disney Educational Productions, 500 S. Buena Vista St. Burbank, CA 91521
Computers: The Friendly Invasion, Disney Educational Productions, 500 S. Buena Vista St. Burbank, CA 91521

Assessment Resources

Student Edition
Math Journal, pp. 22, 58
Mixed Review, pp. 11, 18, 25, 35, 43, 51, 60
Self Test, p. 25
Chapter Highlights, p. 61
Chapter Study Guide and Assessment, pp. 62–64
Alternative Assessment, p. 65
Portfolio, p. 65
Chapter Test, p. 793

Teacher's Wraparound Edition
5-Minute Check, pp. 6, 12, 19, 28, 36, 44, 53
Check for Understanding, pp. 8, 15, 22, 32, 40, 49, 58
Closing Activity, pp. 11, 18, 24, 35, 43, 51, 60
Cooperative Learning, pp. 7, 20

Assessment and Evaluation Masters
Multiple-Choice Tests, Forms 1A (Honors), 1B (Average), 1C (Basic), pp. 1–6
Free-Response Tests, Forms 2A (Honors), 2B (Average), 2C (Basic), pp. 7–12
Calculator-Based Test, p. 13
Performance Assessment, p. 14
Mid-Chapter Test, p. 15
Quizzes A–D, pp. 16–17
Standardized Test Practice, p. 18
Cumulative Review, pp. 19–20

ENHANCING THE CHAPTER

Examples of some of the materials for enhancing Chapter 1 are shown below.

DIVERSITY

Multicultural Activity Masters, pp. 1, 2

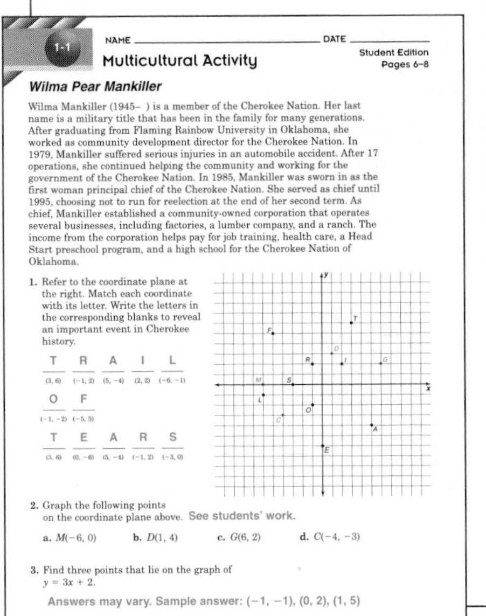

1-1 NAME ______ DATE ______

Multicultural Activity — Student Edition Pages 6–8

Wilma Pear Mankiller

Wilma Mankiller (1945–) is a member of the Cherokee Nation. Her last name is a military title that has been in the family for many generations. After graduating from Flaming Rainbow University in Oklahoma, she worked as community development director for the Cherokee Nation. In 1979, Mankiller suffered serious injuries in an automobile accident. After 17 operations, she continued helping the community and working for the government of the Cherokee Nation. In 1985, Mankiller was sworn in as the first woman principal chief of the Cherokee Nation. She served as chief until 1995, choosing not to run for reelection at the end of her second term. As chief, Mankiller established a community-owned corporation that operates several businesses, including factories, a lumber company, and a ranch. The income from the corporation helps pay for job training, health care, a Head Start preschool program, and a high school for the Cherokee Nation of Oklahoma.

1. Refer to the coordinate plane at the right. Match each coordinate with its letter. Write the letters in the corresponding blanks to reveal an important event in Cherokee history.

T	R	A	I	L
(3, 6)	(−1, 2)	(5, −4)	(2, 2)	(−6, −1)

O	F
(−1, −2)	(−5, 5)

T	E	A	R	S
(3, 6)	(0, −6)	(5, −4)	(−1, 2)	(−3, 0)

2. Graph the following points on the coordinate plane above. See students' work.

   a. $M(-6, 0)$ b. $D(1, 4)$ c. $G(6, 2)$ d. $C(-4, -3)$

3. Find three points that lie on the graph of $y = 3x + 2$.

   Answers may vary. Sample answer: (−1, −1), (0, 2), (1, 5)

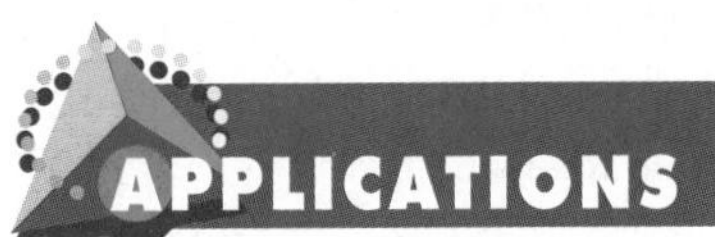

APPLICATIONS

Real-World Applications, 1, 2

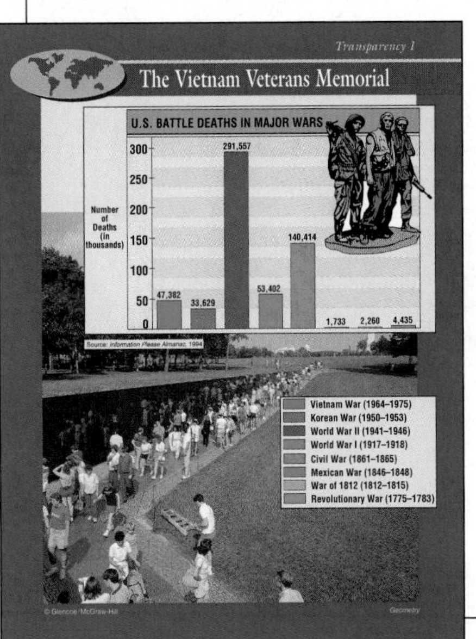

Transparency 1

The Vietnam Veterans Memorial

© Glencoe/McGraw-Hill Geometry

TECHNOLOGY

Graphing Calculator and Computer Masters, p. 1

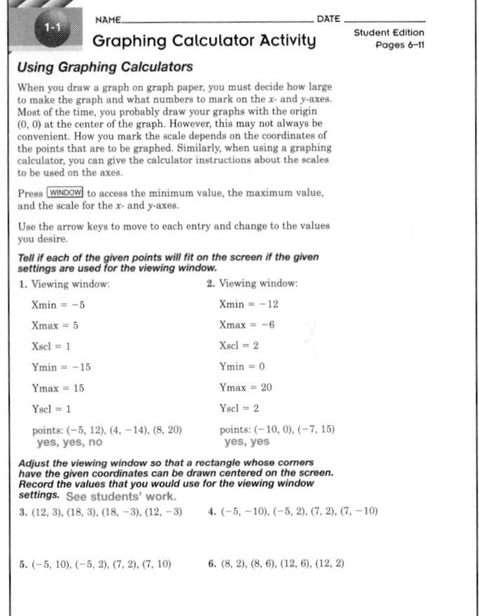

1-1 NAME ______ DATE ______

Graphing Calculator Activity — Student Edition Pages 6–11

Using Graphing Calculators

When you draw a graph on graph paper, you must decide how large to make the graph and what numbers to mark on the x- and y-axes. Most of the time, you probably draw your graphs with the origin (0, 0) at the center of the graph. However, this may not always be convenient. How you mark the scale depends on the coordinates of the points that are to be graphed. Similarly, when using a graphing calculator, you can give the calculator instructions about the scales to be used on the axes.

Press WINDOW to access the minimum value, the maximum value, and the scale for the x- and y-axes.

Use the arrow keys to move to each entry and change to the values you desire.

Tell if each of the given points will fit on the screen if the given settings are used for the viewing window.

1. Viewing window:

   Xmin = −5
   Xmax = 5
   Xscl = 1
   Ymin = −15
   Ymax = 15
   Yscl = 1

   points: (−5, 12), (4, −14), (8, 20)
   yes, yes, no

2. Viewing window:

   Xmin = −12
   Xmax = −6
   Xscl = 2
   Ymin = 0
   Ymax = 20
   Yscl = 2

   points: (−10, 0), (−7, 15)
   yes, yes

Adjust the viewing window so that a rectangle whose corners have the given coordinates can be drawn centered on the screen. Record the values that you would use for the viewing window settings. See students' work.

3. (12, 3), (18, 3), (18, −3), (12, −3)
4. (−5, −10), (−5, 2), (7, 2), (7, −10)
5. (−5, 10), (−5, 2), (7, 2), (7, 10)
6. (8, 2), (8, 6), (12, 6), (12, 2)

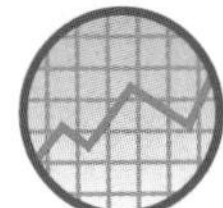

TECH PREP

Tech Prep Applications Masters, pp. 1, 2

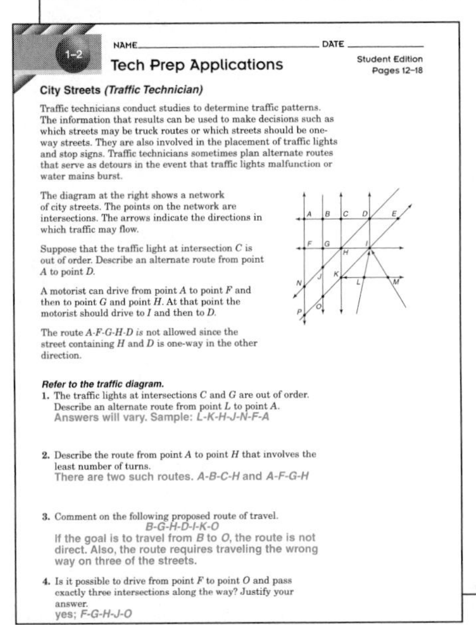

1-2 NAME ______ DATE ______

Tech Prep Applications — Student Edition Pages 12–18

City Streets *(Traffic Technician)*

Traffic technicians conduct studies to determine traffic patterns. The information that results can be used to make decisions such as which streets may be truck routes or which streets should be one-way streets. They are also involved in the placement of traffic lights and stop signs. Traffic technicians sometimes plan alternate routes that serve as detours in the event that traffic lights malfunction or water mains burst.

The diagram at the right shows a network of city streets. The points on the network are intersections. The arrows indicate the directions in which traffic may flow.

Suppose that the traffic light at intersection C is out of order. Describe an alternate route from point A to point D.

A motorist can drive from point A to point F and then to point G and point H. At that point the motorist should drive to I and then to D.

The route A-F-G-H-D is not allowed since the street containing H and D is one-way in the other direction.

Refer to the traffic diagram.

1. The traffic lights at intersections C and G are out of order. Describe an alternate route from point L to point A.
   Answers will vary. Sample: *L-K-H-J-N-F-A*
2. Describe the route from point A to point H that involves the least number of turns.
   There are two such routes. *A-B-C-H* and *A-F-G-H*
3. Comment on the following proposed route of travel.
   B-G-H-D-I-K-O
   If the goal is to travel from *B* to *O*, the route is not direct. Also, the route requires traveling the wrong way on three of the streets.
4. Is it possible to drive from point F to point O and pass exactly three intersections along the way? Justify your answer.
   yes; *F-G-H-J-O*

PROBLEM SOLVING

Problem-of-the-Week Cards, 1, 2, 3

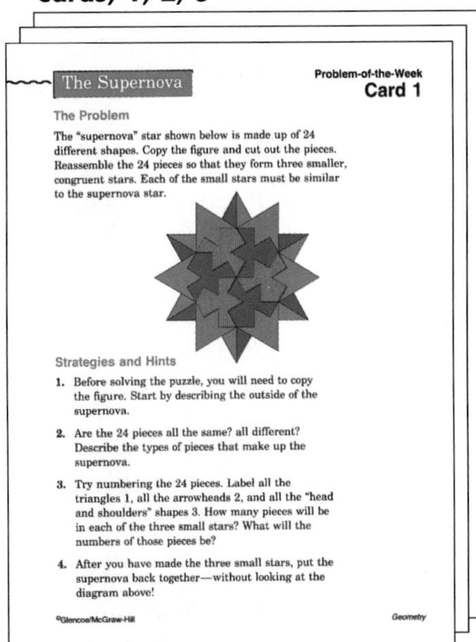

The Supernova — Problem-of-the-Week **Card 1**

The Problem

The "supernova" star shown below is made up of 24 different shapes. Copy the figure and cut out the pieces. Reassemble the 24 pieces so that they form three smaller, congruent stars. Each of the small stars must be similar to the supernova star.

Strategies and Hints

1. Before solving the puzzle, you will need to copy the figure. Start by describing the outside of the supernova.
2. Are the 24 pieces all the same? all different? Describe the types of pieces that make up the supernova.
3. Try numbering the 24 pieces. Label all the triangles 1, all the arrowheads 2, and all the "head and shoulders" shapes 3. How many pieces will be in each of the three small stars? What will the numbers of those pieces be?
4. After you have made the three small stars, put the supernova back together—without looking at the diagram above!

©Glencoe/McGraw-Hill Geometry

MAKING MATHEMATICS RELEVANT

This two-page introduction to the chapter provides students with an opportunity to explore an application of geometry that has historic roots as well as modern uses.

Background Information

Using Maps There are many types of maps. The purpose of the map usually dictates the type of features included on it. Modern wilderness backpackers and orienteers generally use *topographic maps*, which show natural features of the land such as elevation and relative steepness of the terrain, vegetation, and prominent natural or manmade features. *Political maps* show boundary lines, major cities, and major water masses. *Hydrographic charts* map large bodies of water, indicating tides, dangerous reefs, and navigational equipment.

CHAPTER 1

Discovering Points, Lines, Planes, and Angles

Objectives

In this chapter, you will:

- identify and model points, lines, and planes in space and on a coordinate plane,
- identify collinear points and coplanar points and lines,
- solve problems by listing the possibilities and by using formulas, finding maximum and minimum values for a given perimeter, and
- find the measures of segments and angles and the relationships that exist among them.

Geometry: Then and Now Maps have been developed for a wide variety of reasons. In the 19th century, seafaring peoples living in the Marshall Islands in the South Pacific developed maps made of stiff palm fibers for navigational purposes.

Today, advances in technology have enabled us to make highly accurate maps, many of which are computerized. Maps are used daily for personal and commercial navigation as well as for recreational purposes such as backpacking and orienteering. In what ways have you used maps?

TIME Line

876 B.C. The first use of a symbol for zero is made in India.

323 B.C. Euclid writes his work on geometry called *Elements*.

A.D. 1100 Chinese navigators use magnetic compasses to guide their ships.

1467 First ballad about Swiss hero William Tell is written.

TIME Line

Ask students to research and write a one-page report about the beginnings of geometry. Have students analyze what types of geometry have been written about in modern history.

interNET CONNECTION

Begin your research of orienteering, visit orienteering clubs, and read about upcoming events.

World Wide Web
http://www.cs.brown.edu/people/ssb/o2.html

PEOPLE IN THE NEWS

Orienteering is a competition in which people armed with maps and compasses visit a series of predetermined checkpoints in an unfamiliar area, usually in the wilderness, as quickly as possible. Boy Scouts like **Omar Lumkin** of Troop 316 in Washington, D.C., have learned to adapt orienteering to an urban setting. Rather than race against a clock, scouts compete for the highest score on a quiz about national landmarks such as sites on the National Mall, the Vietnam Veterans Memorials, and the Lincoln Memorial.

Chapter Project

Map reading is a vital skill that can help make a business trip or vacation a more enjoyable experience. You will work in cooperative groups investigating a map of your county or state.

- Locate your hometown on the map and approximate its coordinates. Choose another city on the map and approximate its coordinates. Devise a way to calculate the distance between these two cities by using their coordinates. Check your calculations by measuring. How accurate were your calculations?
- If you travel in a straight path from your town to the other city you chose, along what compass heading would you travel?
- Use the Internet or go to the library to research the sport of orienteering. How would a thorough knowledge of coordinate geometry and measures of segments and angles be useful to a competitor in this sport? Explain your reasoning and use your map to cite examples.
- Design an orienteering event for your hometown. Name specific checkpoints for participants to visit.

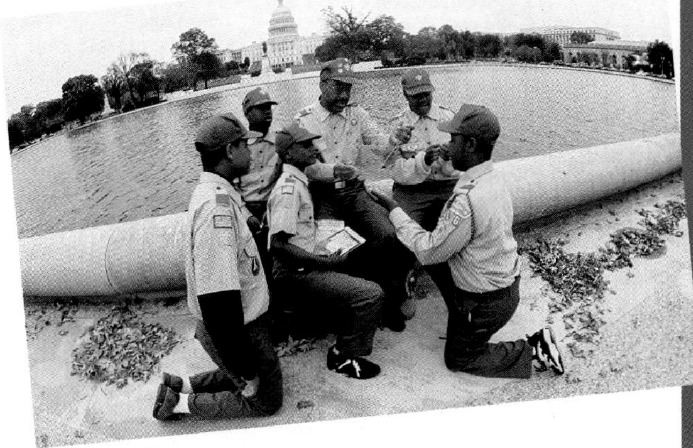

1848 Maria Mitchell becomes the first woman to join the American Academy of Science.

1944 The first issue of *Ebony* is published.

1540 1580 1620 1660 1700 1740 1780 1820 1860 1900 1940 1980 2020

1701 Yale University is founded.

1992 The Global Positioning System (GPS) is fully operational for use in Desert Storm.

PEOPLE IN THE NEWS

Ask students to compare orienteering with a scavenger hunt. Students can find out more information about orienteering by contacting the U.S. Orienteering Federation, P.O. Box 1444, Forest Park, Georgia 30051.

Chapter Project

Cooperative Learning Students might want to work in groups of three for this project. After students locate their hometown(s), each student should choose a different city for the distance calculation. Then students can discuss the methods they used to find the distance and check each other's work.

Investigations and Projects Masters, p. 25

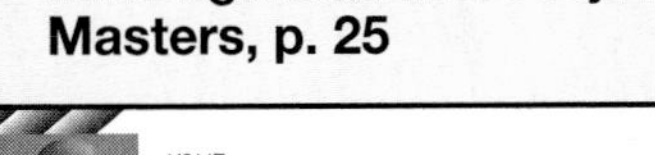

NAME________________ DATE________

Chapter 1 Project A

Student Edition Pages 4–65

Orienteering

1. Research the sport of orienteering. How are the race courses chosen? How would a participant in the race prepare for the event?
2. Work with your group to develop an orienteering course. Choose or draw a map of your city, neighborhood, favorite mall, or a national park as the location for your course. Be sure that your map includes a coordinate grid. Identify checkpoints by their coordinates.
3. Describe a route that a person might possibly take to complete your group's course. Include direction, distance, and coordinates used along the path taken.
4. Produce a brochure to advertise your group's race.

Alternative Chapter Projects

Two other chapter projects are included in the *Investigations and Projects Masters*. In Chapter 1 Project A, pp. 25–26, students extend the topic in the chapter opener. In Chapter 1 Project B, pp. 27–28, students explore fractal geometry.

1-1 LESSON NOTES

NCTM Standards: 1–5, 8, 10

Instructional Resources

- Study Guide Master 1-1
- Practice Master 1-1
- Enrichment Master 1-1
- Graphing Calculator and Computer Masters, p. 1
- Multicultural Activity Masters, p. 1
- Real-World Applications, 1

Transparency 1-1A contains the 5-Minute Check for this lesson; **Transparency 1-1B** contains a teaching aid for this lesson.

Recommended Pacing	
Standard Pacing	Day 1 of 13
Honors Pacing	Day 1 of 13
Block Scheduling*	Day 1 of 8

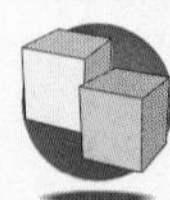

*For more information on pacing and possible lesson plans, refer to the *Block Scheduling Booklet*.

1 FOCUS

5-Minute Check

1. If $x = 18$ and $y = 6$, find the value of $x - 4y$. **−6**
2. What is the value of $x^2 + 3yz$ if $x = 3$, $y = 6$, and $z = 4$? **81**
3. Solve for y in $y = 8x - 5$ if
   a. $x = 2$ **11**
   b. $x = -1$ **−13**
4. Determine if $(-2, 5)$ is a solution of $3x + 4y = 14$. **yes**
5. The formula for acceleration of an object is $a = \frac{f - s}{t}$, where a is acceleration, f is the final velocity, s is the starting velocity, and t is the time. Find the starting velocity of a train that accelerated at a rate of 1.4 meters per second per second to a velocity of 6.8 meters in 4 seconds. **1.2 m/s**

1-1 Integration: Algebra The Coordinate Plane

What YOU'LL LEARN

- To graph ordered pairs on a coordinate plane, and
- to identify collinear points.

Why IT'S IMPORTANT

Coordinate systems are used to locate items in publishing, archeology, and aquatic explorations.

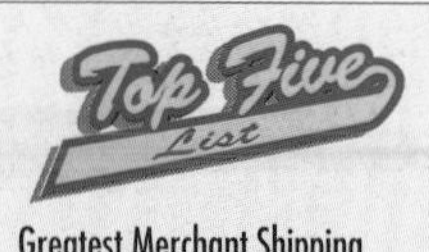

Greatest Merchant Shipping Losses in World War II

Country	Vessels
1. United Kingdom	4786
2. Japan	2346
3. Germany	1595
4. U.S.	578
5. Norway	427

Games

Have you ever played *Battleship?* The game was first introduced in 1967 by the Milton Bradley Company. In 1989, an electronic version was released that was powered by batteries. In *Battleship*, you try to guess the location of your opponents ships and sink them by calling out row and column coordinates. Then, by using logic and your opponent's responses, you guess more wisely with each turn, trying to sink all of your opponent's ships before he or she sinks yours.

The grid system used in playing *Battleship* is an example of plotting points in a coordinate system. In algebra, you used a **coordinate plane** in which a horizontal number line called the ***x*-axis** and a vertical number line called the ***y*-axis** intersect at their zero points. Their intersection is called the **origin**. The axes divide the coordinate plane into four **quadrants**, as shown in the figure at the right.

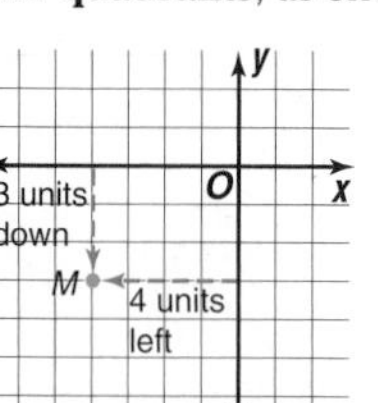

The notation M(−4, −3) indicates that point M is named by the ordered pair (−4, −3).

There are an infinite number of points on a coordinate plane. Each point can be named by an **ordered pair** of coordinates. The first coordinate of an ordered pair, usually called the ***x*-coordinate**, tells how far left or right of the *y*-axis the point is located. The second coordinate, the ***y*-coordinate**, tells how far up or down from the *x*-axis the point is. Point M is located in Quadrant III and has coordinates $(-4, -3)$. This means that the point is located 4 units left and 3 units down from the origin.

If the axes on a coordinate plane do not have numbers on them, you can assume that each square on the grid represents 1 unit.

Most of the U.S. losses were due to convoys traveling along the coast being backlighted by the coastal cities. German U-boats sat offshore and picked off all their targets. The U.S. mandated blackouts along the coast for the remainder of the war.

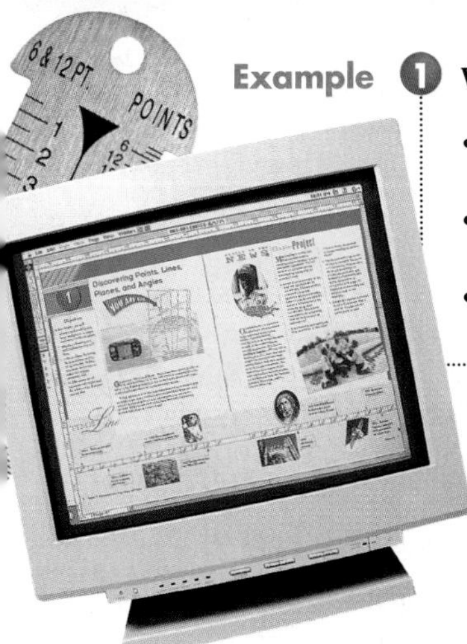

Example 1 **Write the coordinates for points P, Q, and R.**

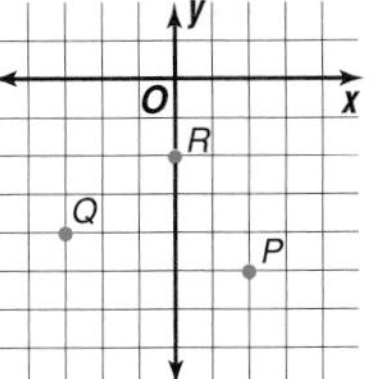

- Point P is 2 units right and 5 units down from the origin. Its coordinates are $(2, -5)$.
- Point Q is 3 units left and 4 units down from the origin. Its coordinates are $(-3, -4)$.
- Point R is at -2 on the y-axis. Its coordinates are $(0, -2)$.

There are many types of coordinate systems. In publishing, computers are often used to arrange type and pictures on the page. Photographs, artwork, and special text features are positioned by putting each item in a special box that can be moved around the page.

Example 2

In QuarkXPress™, a page layout software program, the position of the upper left corner of a box is expressed by X (how many picas right) and Y (how many picas down) it is from the upper left-hand corner of the page designated as 0. The drawing below shows a QuarkXPress layout screen. Identify which box is described by each location.

FYI

A *pica* is a unit of measure used in publishing that is equal to $\frac{1}{6}$ of an inch.

a. [X: 24, Y: 24]

This means the upper left corner of the box is located 24 picas right of 0 and 24 picas down. This is the location of box II.

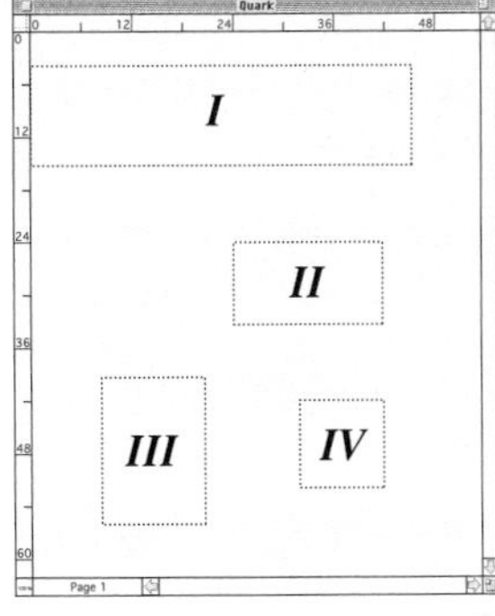

b. [X: 8, Y: 39]

The box that is located 8 picas right and 39 picas down is box III.

c. [X: 0, Y: 4]

The box located 0 units right and 4 units down is box I.

An algebraic relation can be represented in several ways.

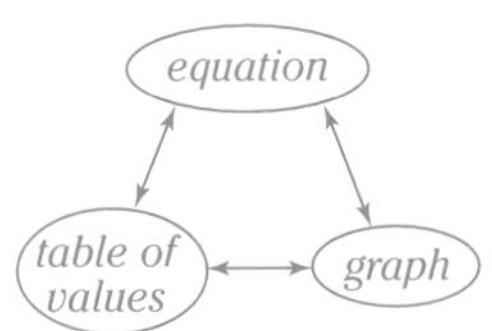

In algebra, you used a table to find values that satisfy an equation. Then you graphed those values and connected them with a line to draw the graph of the equation. For example, the table below shows five values that satisfy the equation $y = 2x - 1$. The coordinate plane shows the graphs of those points and the graph of the equation, which is a line.

$y = 2x - 1$		
x	y	(x, y)
−2	−5	$A(-2, -5)$
−1	−3	$B(-1, -3)$
0	−1	$C(0, -1)$
2	3	$D(2, 3)$
3	5	$E(3, 5)$

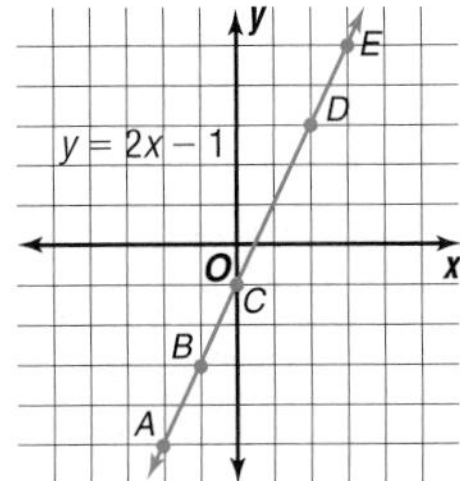

Motivating the Lesson

Hands-On Activity Give students copies of a map of a state or country. Have students describe the location of a city on the map without using the grid system. Then discuss how to use the grid to find a location. What advantages are there to having the grid on the map?

2 TEACH

In-Class Examples

For Example 1
Find the coordinates of points L, M, and N.

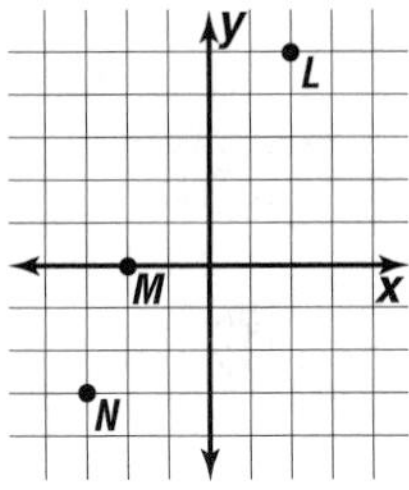

$L(2, 5)$, $M(-2, 0)$, $N(-3, -3)$

For Example 2
Using the figure in the *Student Edition*, which box has its upper right corner at each location?
a. X: 42, Y: 42 IV
b. X: 42, Y: 24 II

Teaching Tip Point out to students that the coordinate plane itself is often called a graph.

Teaching Tip When you are discussing the x- and y-axes, point out that they do not lie in any quadrant. Note that the numbers on the axes indicate how many units each block represents. Stress that in this book each block will represent one unit unless otherwise marked.

Publishers also use *points* as units of measure. There are 12 points in each pica, so a point is $\frac{1}{72}$ of an inch!

Cooperative Learning

Co-op Co-op Have groups of students create dot-to-dot drawings. Groups should first decide the subject of the drawings and then draw the pictures on coordinate grids. Next, they should make a list of the points to be connected to make the pictures. Have groups exchange lists and draw each other's pictures. For more information on the co-op co-op strategy, see *Cooperative Learning in the Mathematics Classroom*, one of the titles in the Glencoe Mathematics Professional Series, page 32.

Teaching Tip Point out that the axes are not always labeled x and y. For example, the horizontal axis could be labeled t for time and the vertical axis could be labeled d for distance.

In-Class Example

For Example 3
Find the coordinates of three points that lie on the graph of $6x - 2y = 12$. Graph the points and the line representing $6x - 2y = 12$. Then name the coordinates of a point not on the line. **Sample answer: (2, 0), (0, −6), (1, −3) are on the line, (−1, 0) is not.**

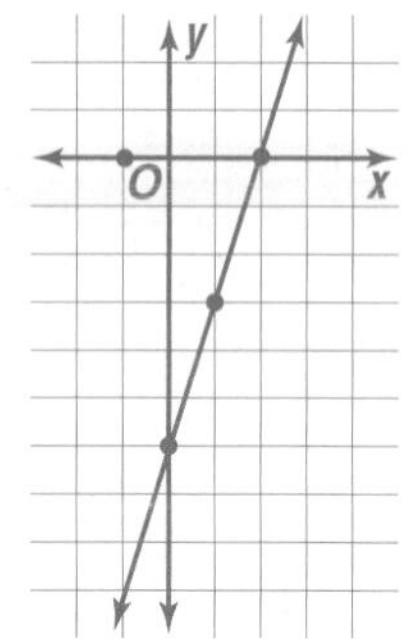

3 PRACTICE/APPLY

Check for Understanding

Exercises 1–10 are designed to help you assess your students' understanding through reading, writing, speaking, and modeling. You should work through Exercises 1–3 with your students and then monitor their work on Exercises 4–10.

Additional Answers

1. The first coordinate tells how far to go right (+) or left (−) of the origin. From that point, the second coordinate tells how far up (+) or down (−) to go before plotting the point.

3.

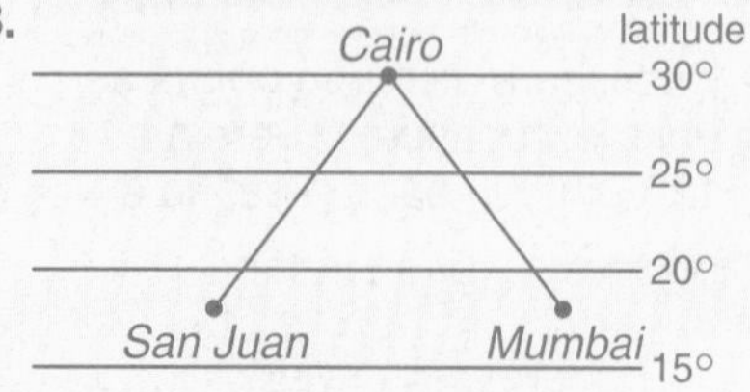

The notation ABCDE is sometimes used to show that points are collinear.

The coordinates of points A, B, C, D, and E satisfy the equation $y = 2x - 1$. The graph of $y = 2x - 1$ is a line, and points A, B, C, D, and E lie on that line. Thus, they are **collinear**. Is the point $M(4, 8)$ collinear with points A, B, C, D, and E? To determine if M lies on the same line with A, B, C, D, and E, see whether the coordinates of M satisfy the equation $y = 2x - 1$.

$$y = 2x - 1$$
$$8 \stackrel{?}{=} 2(4) - 1 \quad (x, y) = (4, 8)$$
$$8 \neq 7$$

If you graphed M(4, 8), you could also see that it does not lie on the line containing A, B, C, D, and E.

The coordinates of M do not satisfy the equation. Therefore, M does not lie on the line representing $y = 2x - 1$. Points A, B, C, D, E, and M are **noncollinear** points.

Example 3

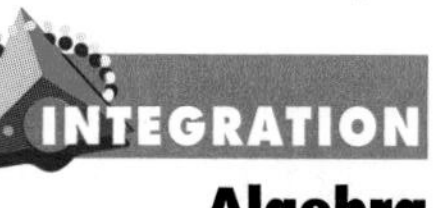

Algebra

a. Find the coordinates of three points that lie on the graph of $y = -3x + 3$.
b. Graph the points and draw the line representing $y = -3x + 3$.
c. Name the coordinates of one point not on the line.

a. Choose three values for x and make a table of values.

$y = -3x + 3$		
x	y	(x, y)
−1	6	(−1, 6)
0	3	(0, 3)
3	−6	(3, −6)

b. Graph the points from the table and draw the line representing the equation.

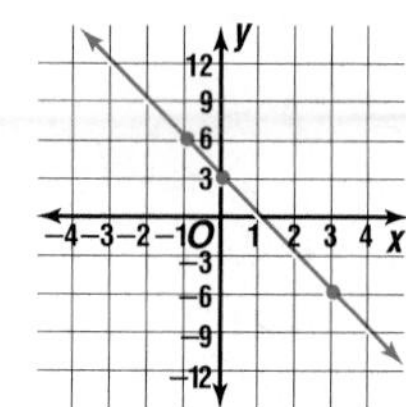

c. There are an infinite number of points in the coordinate plane that do not lie on the line representing $y = -3x + 3$. Visually, select any point that does not appear to lie on the line, for example, (1, 2). You can check your selection by testing its coordinates to make sure that they do not satisfy the equation.

$$y = -3x + 3$$
$$2 \stackrel{?}{=} -3(1) + 3$$
$$2 \neq 0$$

The point named by (1, 2) does not lie on the graph of $y = -3x + 3$.

CHECK FOR UNDERSTANDING

Communicating Mathematics

Study the lesson. Then complete the following. 1. See margin.

2. Sample answer: The vertical axis goes down, but is positive.

1. **Describe** how to graph a point if you know the coordinates of the point.
2. **Explain** some of the ways that the coordinate system in Example 2 differs from the coordinate plane used in mathematics.
3. **Draw** a diagram showing how you might determine whether the points on a map representing Cairo, Egypt; Mumbia (formerly Bombay), India; and San Juan, Puerto Rico, are noncollinear. See margin.

Reteaching

Using Models Draw a coordinate plane on the chalkboard. Blindfold a student and have him or her pick a spot on the coordinate plane (like Pin the Tail on the Donkey). Have students tell the coordinates of the point selected to the nearest whole number.

Guided Practice

Write the ordered pair for each point shown at the right.

4. T (−4, 4)
5. U (−4, −3)

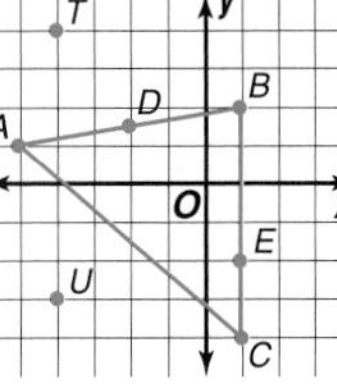

Graph each point on the same coordinate plane.

6. $R(3, -6)$ 6–7. See margin.
7. $S(0, 2)$

8. Use triangle ABC ($\triangle ABC$) shown in the coordinate plane above to answer each question.
   a. In which quadrant is point A located? II
   b. In which quadrant is point B located? I
   c. Which axis intersects the side connecting A and B? 8c. y-axis
   d. Which point, A, B, or D, has the greatest x-coordinate? B
   e. Which point, B, C, or E, has the greatest y-coordinate? B

9. Points $A(5, 7)$ and $B(-1, 1)$ lie on the graph of $y = x + 2$. Determine whether $Z(-3, -2)$ is collinear with A and B. noncollinear

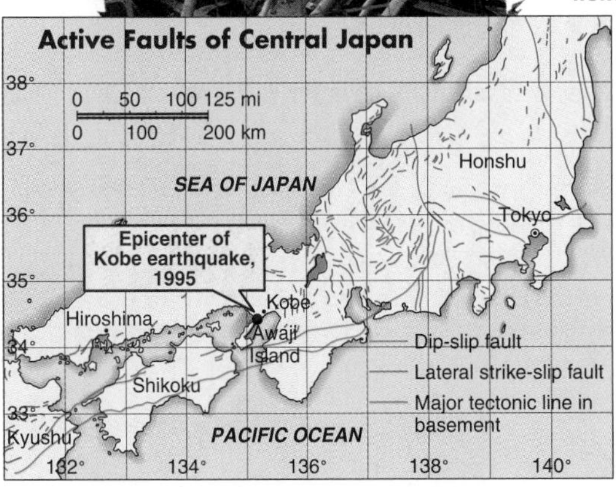

Active Faults of Central Japan

10. **Seismology** Mapmakers use letters and numbers to define sectors on a map. In January, 1995, a powerful earthquake struck the Japanese port of Kobe, killing more than 5100 people and destroying more than 45,000 buildings. The map at the left shows the active faults of central Japan.
   a. Use the longitude and latitude degree measures to write an ordered pair to approximate the location of Kobe earthquake on the map. (135°, 34.5°)
   b. Name the town located near the coordinates (139.5°, 35.5°). Tokyo

EXERCISES

Practice

Write the ordered pair for each point shown at the right.

11. M (4, 3)
12. A (0, 4)
13. T (−3, 2)
14. H 14. (−2, −3)
15. E (1, −4)
16. I (5, 0)

Graph each point on the same coordinate plane. 17–22. See margin.

17. $C(0, 5)$
18. $D(-3, -2)$
19. $E(4, -5)$
20. $F(-1, 4)$
21. $G(2.5, -3)$
22. $H(-4.5, 0)$

Points $M(1, -3)$ and $N(-2, -15)$ lie on the graph of $y = 4x - 7$. Determine whether each point is collinear with M and N.

23. $Q(0, -7)$ collinear
24. $R(0.5, -5)$ collinear
25. $S(-1, 11)$ 25. noncollinear

Assignment Guide

Core (with proof): 11–39 odd, 40–47
Core (informal): 11–39 odd, 40–47
Enriched: 12–34 even, 36–47

For **Extra Practice**, see p. 764.

The red A, B, and C flags, printed only in the Teacher's Wraparound Edition, indicate the level of difficulty of the exercises.

Additional Answers

6.–7.

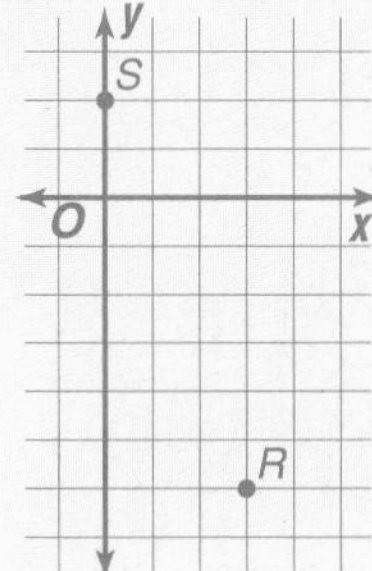

17.–22.

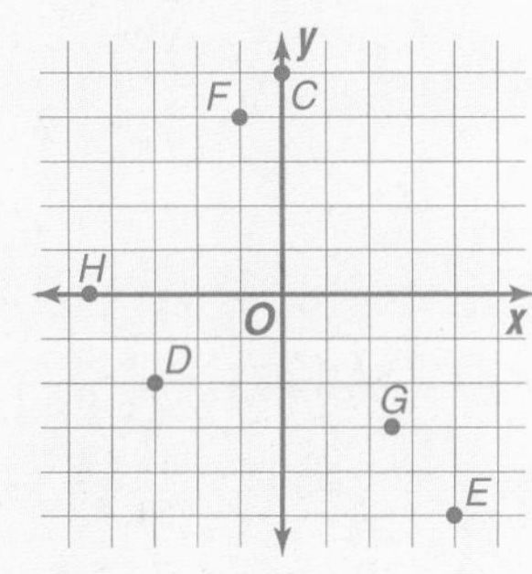

Study Guide Masters, p. 1

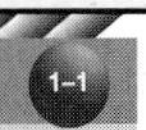

NAME ______ DATE ______

Study Guide

Student Edition Pages 6–11

Integration: Algebra
The Coordinate Plane

Every point in the coordinate plane can be denoted by an ordered pair consisting of two numbers. The first number is the **x-coordinate**, and the second number is the **y-coordinate**.

To determine the coordinates for a point, follow these steps.
1. Start at the origin and count the number of units to the right or left of the origin. The *positive direction* is to the right, and the *negative direction* is to the left.
2. Then count the number of units up or down. The positive direction is up, and the negative direction is down.
Note: If you do not move either right or left, the x-coordinate is 0. If you do not move up or down, the y-coordinate is 0.

Example: Write the ordered pair for each point shown at the right.
The ordered pair for R is (2, 4).
The ordered pair for S is (-3, 3).
The ordered pair for T is (-4, -2).
The ordered pair for U is (1, -4).
The ordered pair for W is (0, 2).
The ordered pair for X is (-2, 0).

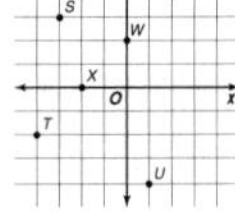

Write the ordered pair for each point shown at the right.

1. A (2, 3)
2. B (-3, 0)
3. C (-4, -3)
4. D (4, -4)
5. E (-4, 3)
6. F (2, -1)
7. G (0, -3)
8. H (0, 4)
9. I (-3, -4)

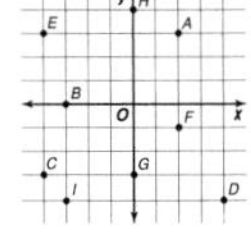

Graph each point on the coordinate plane.

10. $M(6, 4)$
11. $N(-5, 4)$
12. $P(-3, 5)$
13. $Q(6, 0)$
14. $J(0, -4)$
15. $K(7, -5)$
16. $Y(9, -3)$
17. $Z(-8, -5)$

Additional Answers

36. Plot two points in the plane and draw a line through them. Plot another point not on that line. Then draw all possible lines through any two of the points on the plane. Plot a fourth point not on any of these lines and then draw all possible lines through this point and any other plotted point on the plane. Finally plot a fifth point not on any of the lines. These five points must be noncollinear.

37. (see below)

38.

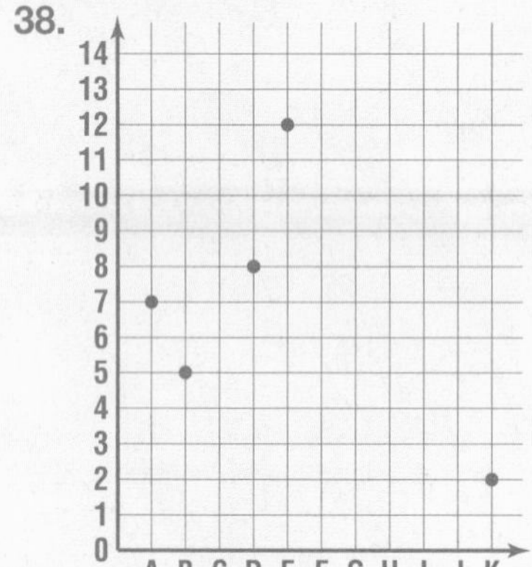

Practice Masters, p. 1

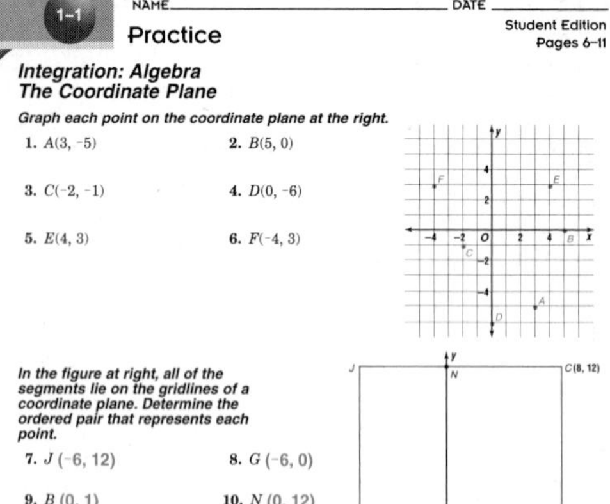

1-1 NAME______________ DATE ________

Practice

Student Edition Pages 6–11

Integration: Algebra
The Coordinate Plane

Graph each point on the coordinate plane at the right.

1. $A(3, -5)$ 2. $B(5, 0)$

3. $C(-2, -1)$ 4. $D(0, -6)$

5. $E(4, 3)$ 6. $F(-4, 3)$

In the figure at right, all of the segments lie on the gridlines of a coordinate plane. Determine the ordered pair that represents each point.

7. J (−6, 12)   8. G (−6, 0)

9. B (0, 1)   10. N (0, 12)

11. F (8, 1)   12. E (8, −2)

Points G (1, −1) and H (3, 3) lie on the graph of y = 2x − 3. Determine whether the each point is collinear with G and H.

13. $I(0, -3)$ yes   14. $J(2, 1)$ yes

15. $K(-3, -8)$ no   16. $L(5, 7)$ yes

17. $M(10, 16)$ no   18. $N(25, 48)$ no

19. $P(-10, -23)$ yes   20. $Q(-35, -67)$ no

In the figure at the right, the blue segments forming the rectangle lie on the gridlines of a coordinate plane. Determine the ordered pair that represents each point.

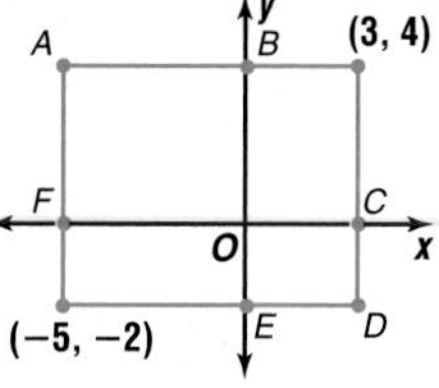

26. A (−5, 4)   27. B (0, 4)

28. C (3, 0)   29. D (3, −2)

30. E (0, −2)   31. F (−5, 0)

32–34. See Solutions Manual for graphs.

32. Sample points: (0, 5), (−5, 0), (−1, 4); not on line: (0, 0)

33. Sample points: (0, −9), (1, −3), (2, 3); not on line: (0, 0)

34. Sample points: (−3, −1), (−1, 2), (1, 5); not on line (0, 0)

35b. (2, 0), (6, 0)

Find the coordinates of three points that lie on the graph of each equation. Graph the points and the line representing the equation. Then name the coordinates of one point not on the line.

32. $y = x + 5$ 33. $y = 6x - 9$ 34. $3x = 2y - 7$

35. Refer to square $PQRS$ graphed at the right.

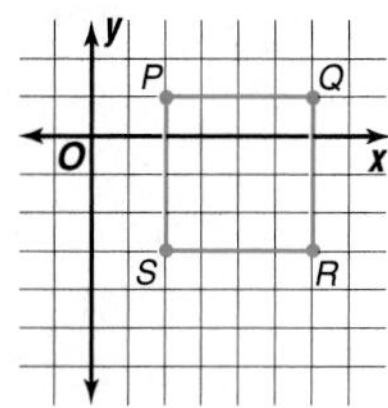

a. In how many quadrants does $PQRS$ lie? Name them. 2; I, IV
b. Name the coordinates of the points where segments PS and QR intersect the x-axis.
c. Name the coordinates of a point that would be collinear with S and R. Sample answer: (4, −3)
d. What is the y-coordinate of any point collinear with S and R? −3
e. What is the x-coordinate of any point collinear with P and S? 2

Critical Thinking

36. Explain how you can locate five points in a coordinate plane so that you are sure no three of these points will be collinear. Include a drawing with your explanation. See margin.

37. Find the coordinates of the intersection of the graphs of $y = 3x + 5$ and $y = -2x - 10$. Write a verbal argument to show that your answer is correct. See margin.

Applications and Problem Solving

38. **Aquatic Engineering** The Dutch Delta Plan, which controls the flow of the Atlantic Ocean on the southwest coast of the Netherlands, is one of the greatest achievements of engineering. The system was built using a grid system of nylon mattresses with graded gravel and rocks to support concrete piers and steel gates. The horizontal axis is labeled with letters and the vertical axis with numbers. Suppose the first five piers were placed at 5B, 2K, 8D, 7A, and 12E. Draw a graph showing the positions of the first five piers. See margin.

Additional Answer

37. (−3, −4); Substitute the −3 for x and −4 for y in each of the equations to make sure these coordinates satisfy both equations. Since they satisfy both equations, these are the coordinates of a point that lies on the graphs of both equations. In order for the point to lie on two distinct lines, it must be the point they have in common, or their intersection.

39. **Archaeology** Did you know that some archaeologists look beneath the surface of the sea for artifacts? Divers investigating the remains of the seaport city of Caesarea beneath the Mediterranean Sea, found statues, lamps, pieces of pottery, several ancient shipwrecks, and submerged portions of a port. The divers laid out ropes to form a grid system for mapping the ocean floor. As items were found, they were recorded on paper using a smaller grid. Suppose pieces of pottery were found at (4, 1), (2, 9), (8, 0), and (11, 6). Draw a graph showing the locations of the finds. See margin.

Mixed Review

Simplify.

40. $-2 + 3$ 1
41. $-4 - (-5)$ 1
42. $7(-2)$ -14
43. $-16 \div (-4)$ 4
44. $3(-5) + 2(7)$ -1
45. $-2(-7 + 4)$ 6

INTEGRATION **Algebra**

46. **Speed** The distance traveled by an automobile can be found by using the formula $d = rt$, where d is the distance, r is the average rate of speed, and t is the time traveled at that speed. Find the distance traveled if an automobile averages 63 mph for 3.5 hours. 220.5 miles

47. **Statistics** The graph below shows how much time parents say they spend per week helping their children with schoolwork.

Source: 20/20 Research for Shoney's Restaurants

a. Which response was most frequently given by parents? 5 or more hours

b. If 8000 parents were surveyed, how many said they do not help their children with their schoolwork? 1920 parents

c. Which category contains about one third of the parents polled in this survey? 1–4 hours

d. Keep track of how much time you get help with your schoolwork at home. In which category would your household fit? See students' work.

4 ASSESS

Closing Activity

Modeling Have each student use a coordinate plane to make the first letter of his or her name. Have them list the coordinates for each point in the order that the points should be connected to draw the letter.

Additional Answer

39.

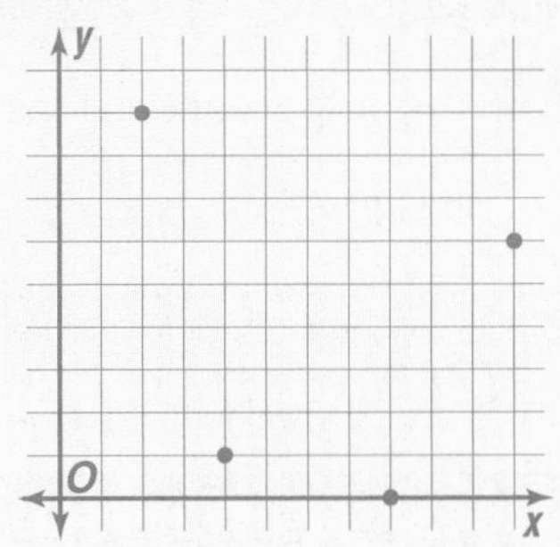

Extension

Problem Solving Make a table of values to determine five points that lie on the graph of $y = x^2 - 2x + 6$. Use the following values of x: 1, 2, 3, 4, and 5. Plot the five points. Do they appear to be collinear? Points (1, 5), (2, 6), (3, 9), (4, 14), (5, 21); no

Enrichment Masters, p. 1

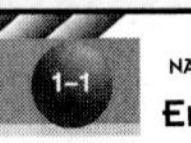

NAME__________ DATE ________

Enrichment

Student Edition Pages 6–11

Coordinate Pictures

By connecting the points for two ordered pairs. you can make a line segment. Line segments can be combined to make pictures.

Graph the two ordered pairs in each exercise, and connect them with a line segment. After you draw each segment, identify the picture.

1. (−2, 5.5) and (5, 5.5)
2. (−3, 4) and (4, 4)
3. (−2, 3) and (3, 3)
4. (−2, −0.5) and (3, −0.5)
5. (1, −1.5) and (3, −1.5)
6. (−3, −2) and (4, −2)
7. (−3, −3) and (4, −3)
8. (−4, −4) and (5.5, −4)
9. (−3, −4.5) and (4, −4.5)
10. (−3.5, −5) and (3.5, −5)
11. (−4, −5.5) and (3, −5.5)
12. (−4.5, −6) and (2.5, −6)
13. (−6.5, −6.5) and (3.5, −6.5)
14. (−6.5, −7) and (3.5, −7)
15. (−3, 4) and (−3, −3)
16. (−2, −0.5) and (−2, 3)
17. (3, −0.5) and (3, 3)
18. (4, −3) and (4, 4)
19. (5, −1.5) and (5, 5.5)
20. (3.5, −6.5) and (3.5, −7)
21. (5.5, −4) and (5.5, −4.5)
22. (−6.5, −6.5) and (−6.5, −7)
23. (4, −2) and (5, −0.5)
24. (4, −3) and (5, −1.5)
25. (−3, 4) and (−2, 5.5)
26. (4, 4) and (5, 5.5)
27. (−6.5, −6.5) and (−4, −4)
28. (5.5, −4) and (3.5, −6.5)
29. (5.5, −4.5) and (3.5, −7)
30. What does the picture show? A computer

1-2 LESSON NOTES

NCTM Standards: 1–5, 7, 8, 12

Instructional Resources

- Study Guide Master 1-2
- Practice Master 1-2
- Enrichment Master 1-2
- Assessment and Evaluation Masters, p. 16
- Modeling Mathematics Masters, p. 79
- Tech Prep Applications Masters, p. 1
- Real-World Applications, 2

Transparency 1-2A contains the 5-Minute Check for this lesson; **Transparency 1-2B** contains a teaching aid for this lesson.

Recommended Pacing	
Standard Pacing	Day 2 of 13
Honors Pacing	Day 2 of 13
Block Scheduling*	Day 2 of 8

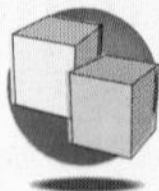

*For more information on pacing and possible lesson plans, refer to the *Block Scheduling Booklet*.

1 FOCUS

5-Minute Check
(over Lesson 1-1)

1. In a coordinate system, which quadrant is in the lower left-hand portion of the plane? **quadrant II**
2. Graph three points that lie on the graph of $y = 4x - 5$.

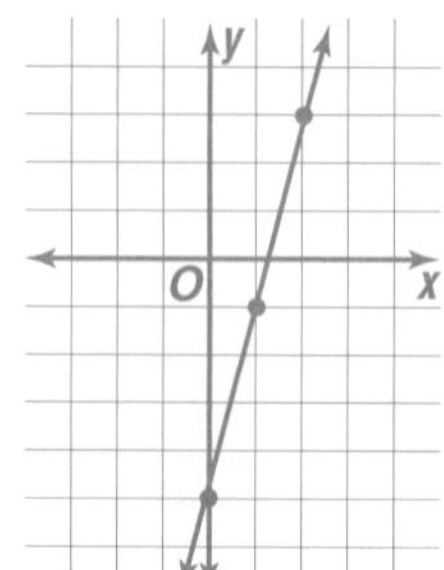

Sample answer: (0, −5), (1, −1), (2, 3)

3. Points $N(5, -2)$ and $M(2, -4)$ lie on the graph of $2x - 3y = 16$. Determine whether $P(8, 0)$ is collinear to N and M. **yes**

1-2 Points, Lines, and Planes

What YOU'LL LEARN
- To identify and model points, lines, and planes,
- to identify coplanar points and intersecting lines and planes, and
- to solve problems by listing the possibilities.

Why IT'S IMPORTANT

You can use points, lines, and planes to represent real-life objects.

APPLICATION
Art Design

Art can be functional as well as interesting from a geometric point of view. The chair shown at the right is a drawing of a design by Gerrit Rietveld in Holland in 1917. The artist used red and blue rectangles to represent **planes** extending into space. The black bars represented **lines** through space. Rietveld used yellow squares to represent **points** in space. Rietveld's chair demonstrates the effect of the cubism art movement.

FYI

Two leaders in the cubism art movement (1907–1914) were Spanish-born Pablo Picasso (1881–1973) and French artist Georges Braque (1882–1963).

You are already familiar with the terms *plane, line,* and *point* from your experiences with graphing in algebra. The terms have similar meaning in geometry.

Unlike the squares used to represent points in the chair design, in geometry, points do not have any actual size. They can, however, represent objects that do have size. A point is usually named by a capital letter, just like the graphs of ordered pairs in algebra. *All* geometric figures consist of points. **Space** is a boundless, three-dimensional set of all points.

In algebra, you found the equation for a line that passes through two points, and you know there are many points on that line. The same is true in geometry.

*A line **segment** is a section of a line. It has endpoints and does not extend indefinitely.*

A line has no thickness or width, although a picture of a line does. Arrows on each end of the line symbolize that the line extends indefinitely in both directions. A line is often named by a lowercase script letter. The line at the right is line n. If the names of two points on a line are known, then the line can be named by those points. In this example, points A and B lie on line n, so line n can be referred to in the following ways.

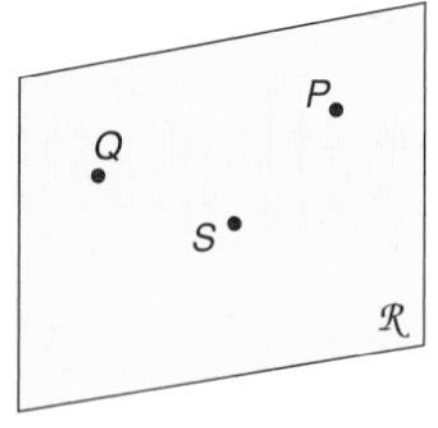

words	line *AB*	line *BA*
symbols	$\overleftrightarrow{AB}$	$\overleftrightarrow{BA}$

In algebra, you used a coordinate plane. This plane contained points and lines. In geometry, a plane is a flat surface that extends indefinitely in all directions. We use four-sided figures like the one at the right to model a plane. A plane can be named by a capital script letter or by three *noncollinear* points in the plane. The figure at the right can be named as plane $\mathcal{R}$ or plane *PQS*.

FYI

Cubism changed the way that artists represented three-dimensional objects. Several different sides of the subject may be represented at the same time. For example, a woman's head may be shown in profile and in a frontal view in the same painting.

There are often many names for the same line or plane, all of which are correct. One way to recognize all the names for a given figure is to **list the possibilities**.

Example  **List all of the possible names for each figure.**

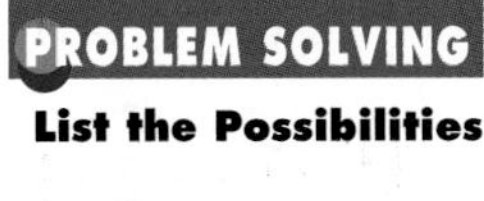

List the Possibilities

a. Line *RS*

Points *T* and *U* also lie on $\overleftrightarrow{RS}$. Choose two letters from the four named in the figure to name this line.

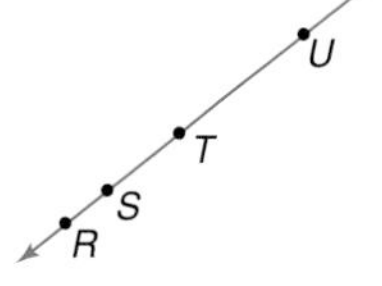

$\overleftrightarrow{RS}$	$\overleftrightarrow{SR}$	$\overleftrightarrow{RT}$	$\overleftrightarrow{TR}$	$\overleftrightarrow{RU}$	$\overleftrightarrow{UR}$
$\overleftrightarrow{ST}$	$\overleftrightarrow{TS}$	$\overleftrightarrow{SU}$	$\overleftrightarrow{US}$	$\overleftrightarrow{TU}$	$\overleftrightarrow{UT}$

b. Plane $\mathcal{M}$

Points *A, B,* and *C* lie on plane $\mathcal{M}$. Use different orders of these letters to name the plane.

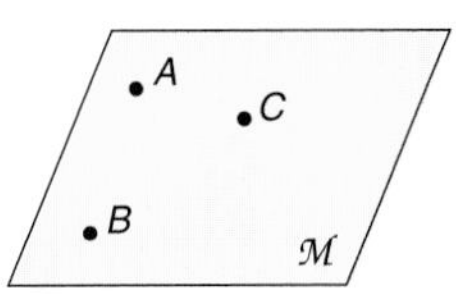

plane *ABC*	plane *ACB*	plane *BCA*
plane *BAC*	plane *CAB*	plane *CBA*

In geometry, the terms *point, line,* and *plane* are considered *undefined terms* because they have only been explained using examples and descriptions. Even though they are undefined, these terms can still be used to define other geometric terms and properties.

You may recall that *collinear* refers to points that lie on the same line. If points are **coplanar**, they lie on the same plane. All the points in a coordinate plane are coplanar. Sometimes it is difficult to identify coplanar points in space unless you understand what a drawing represents. In the figure in Example 2 shown below, dashed segments and lines are used to represent parts of the three-dimensional figure that are hidden from view. Often an artist will use different shades of color to denote different planes.

Example  **Refer to the figure at the right to answer each question.**

The figure shows a pyramid sitting on plane $\mathcal{N}$.

a. Are points *E, F,* and *C* collinear?

Since points *E, F,* and *C* lie on segment *EC* which is part of line *EC,* they are collinear.

b. Are points *A, C, D,* and *E* coplanar?

Points *A, C,* and *D* lie in plane $\mathcal{N}$, but point *E* does not lie in plane $\mathcal{N}$. Thus, the four points are not coplanar.

Points C, D, and E lie in plane CDE, but point A does not.

c. How many planes appear in this figure?

There are five planes: plane $\mathcal{N}$, plane *ABE,* plane *EBC,* plane *EDC,* and plane *ADE.* *Plane $\mathcal{N}$ can also be named by using any three of the four points named in the plane.*

Motivating the Lesson

Questioning Bring in pictures of art that contain clear examples of points, lines, and planes, such as pointillist paintings and geometric art. Have students point out the geometric figures.

2 TEACH

In-Class Examples

For Example 1

Jessie is choosing her lunch in the cafeteria. She may choose one vegetable from peas, carrots, or green beans; one fruit from watermelon, strawberries, or an apple; and the type of bread for her sandwich from white, wheat, or rye. How many different combinations of lunch items can Jessie choose? 27

For Example 2

Refer to the figure in the Student Edition.

a. Name a point that is not collinear to *F* and *C*. ***A, B,* or *D***

b. Identify a point that is not in plane $\mathcal{N}$. ***E* or *F***

c. What is the intersection of plane $\mathcal{ADE}$ and plane $\mathcal{N}$? **line *DA***

Teaching Tip Explain that a point has no dimension, a line has one dimension, and a plane has two dimensions.

Alternative Learning Styles

Auditory Have students brainstorm ideas for solving Example 1. This is a good lesson to demonstrate that sometimes bouncing ideas around a group can lead to a solution.

In-Class Example

For Example 3

Draw and label a figure showing lines *NP* and *QR* intersecting at point *S* for points *N*(3, −1), *P*(5, 2), *Q*(−3, 1), and *R*(0, −4).

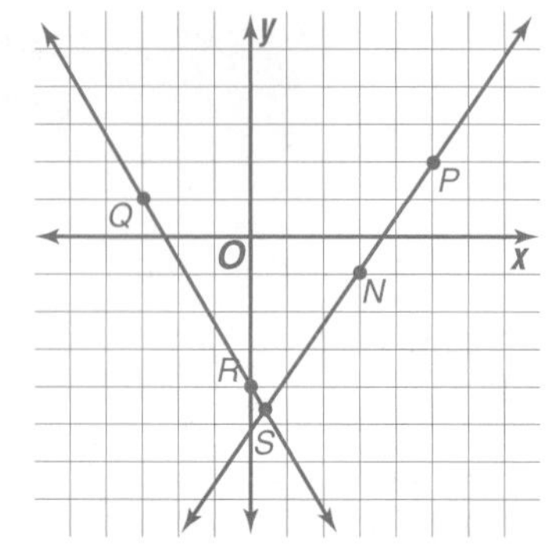

Teaching Tip Emphasize that two points determine a line, but a line contains an infinite number of points. Similarly, three noncollinear points determine a plane, but a plane contains an infinite number of points.

Teaching Tip Encourage students to think of everyday objects that can represent points, lines, planes, and intersecting lines and planes.

Note that information about measurement and equality cannot be determined by looking at a figure.

Figures play an important role in understanding geometric concepts. Drawing and labeling figures can help you model and visualize various geometric relationships. For example, the figures and descriptions below can help you visualize and write about some important relationships among points, lines, and planes.

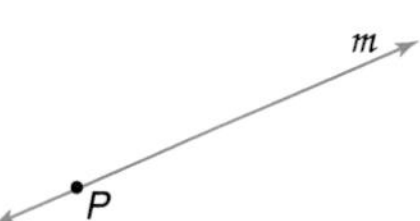

P is on *m*.

m contains *P*.

m passes through *P*.

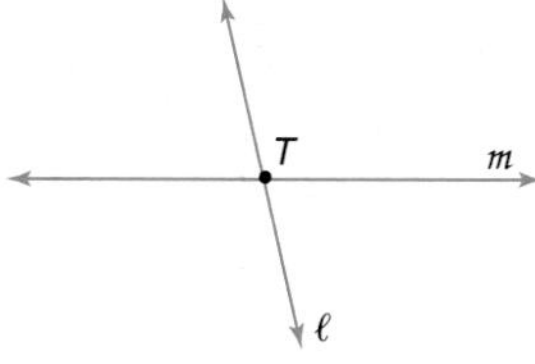

ℓ and *m* intersect in *T*.

ℓ and *m* both contain *T*.

T is the intersection of ℓ and *m*.

*The **intersection** of two figures is the set of points that are contained in both figures.*

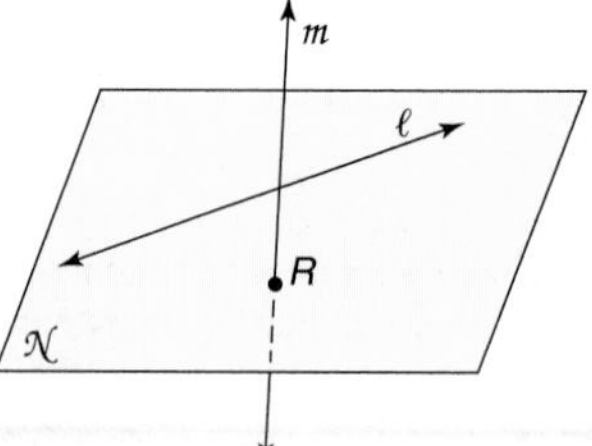

ℓ and *R* are in 𝒩.

𝒩 contains *R* and ℓ.

m intersects 𝒩 at *R*.

R is the intersection of *m* with 𝒩.

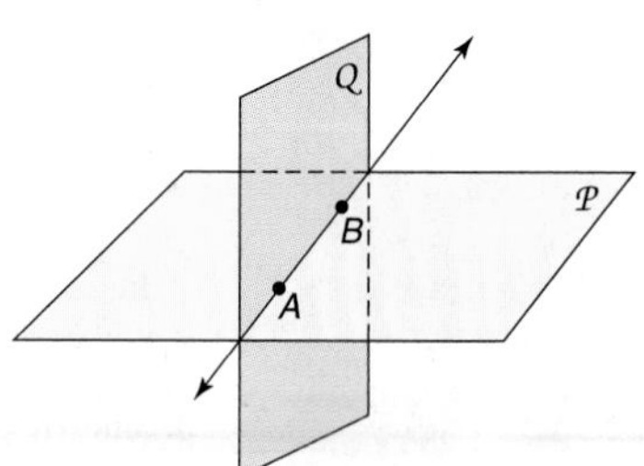

$\overleftrightarrow{AB}$ is in 𝒫 and is in 𝒬.

𝒫 and 𝒬 both contain $\overleftrightarrow{AB}$.

𝒫 and 𝒬 intersect in $\overleftrightarrow{AB}$.

$\overleftrightarrow{AB}$ is the intersection of 𝒫 and 𝒬.

The symbol ∩ is often used to denote intersection as in m ∩ 𝒩 = R.

Example 3 **Suppose four points on a coordinate plane are *A*(3, −4), *B*(−2, 3), *C*(4, 4), and *D*(0, −5). Draw and label a figure showing lines *AB* and *CD* intersecting at *F* and a point *G* that is coplanar with *A, B, C, D,* and *F*, but is not contained in either line.**

First graph the four given points on a coordinate plane.

Then draw line *AB* and line *CD*.

Label the point where the two lines intersect as *F*.

Place point *G* anywhere on the graph so that it does not lie on $\overleftrightarrow{AB}$ or $\overleftrightarrow{CD}$.

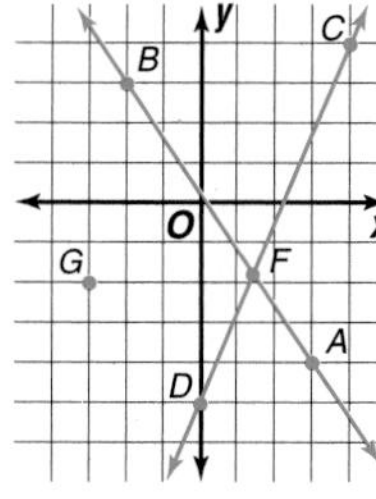

Sometimes it is helpful to make a model of a geometric situation in order to better visualize the information being presented. This is especially true of **three-dimensional figures**. You can use sheets of paper to model planes. The following activity shows you how to model two intersecting planes.

Alternative Teaching Strategies

Reading Geometry Have students look up the words *point, line,* and *plane* in the dictionary. Does the dictionary use examples and descriptions to define these words or does it define them in other ways? Discuss other words, such as *equals, number,* or *operation,* that cannot be defined easily.

Modeling Intersecting Planes

MODELING MATHEMATICS

Materials:  two sheets of different-colored paper  scissors  tape

Make a model of planes $\mathcal{M}$ and $\mathcal{N}$ that intersect in $\overleftrightarrow{AB}$. Point C lies in $\mathcal{M}$, but not in $\mathcal{N}$. Point D lies in $\mathcal{N}$, but not in $\mathcal{M}$. Point E lies in both $\mathcal{M}$ and $\mathcal{N}$.

- Label one sheet of paper as $\mathcal{M}$ and the other as $\mathcal{N}$. Hold the two sheets of paper together and cut a slit halfway through both sheets.

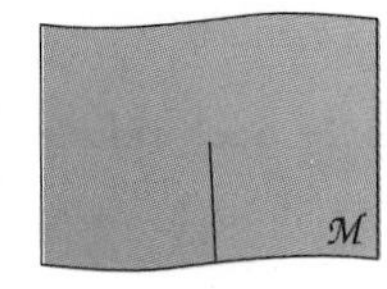
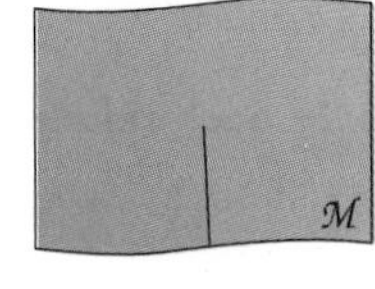

- Turn the papers so that the two slits meet and insert one sheet into the slit on the other sheet. Use tape to hold the two sheets together.

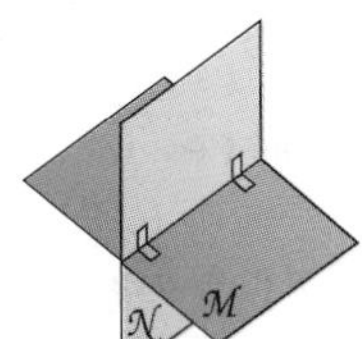

- The line where the two sheets meet is line AB. Draw the line and label points A and B.

a–d. For drawings, see margin.

Your Turn

a. Now draw point C so that it lies on $\mathcal{M}$ but not on $\mathcal{N}$. Can C lie on $\overleftrightarrow{AB}$? no

b. Draw point D so that it lies on $\mathcal{N}$ but not on $\mathcal{M}$. Can D lie on $\overleftrightarrow{AB}$? no

c. If point E lies in both $\mathcal{M}$ and $\mathcal{N}$, where would it lie? Draw point E. It lies on $\overleftrightarrow{AB}$.

d. Look at your model. Now draw a sketch of your model on your paper, labeling each point, line, and plane appropriately.

CHECK FOR UNDERSTANDING

Communicating Mathematics

Study the lesson. Then complete the following.

1. **Describe** how the walls in your classroom can represent planes, lines, and points. See margin.
2. **Name** three undefined terms listed in this lesson. point, line, plane
3. **Draw** plane Q with a line m intersecting Q at point E. See margin.
4. **You Decide** Malia told Lucia that she had figured out a pattern for computing the number of 3-letter names for any plane if she knew how many letters were named on the plane. For example, if there were four points, there were $4 \cdot 3 \cdot 2 \cdot 1$ names possible. Lucia said this pattern wouldn't work for all sets of letters. Who was right? Is there a pattern for n number of points? If so, describe it. Malia; yes; $n(n-1)(n-2)$
5. **List the possibilities** for naming a plane that contains points P, Q, and R. *PQR, PRQ, RPQ, RQP, QRP, QPR*
6. **MODELING MATHEMATICS** Fold a sheet of paper. Open the paper and fold it again in a different way. Open the paper and label the creases as lines m and n.
   a. Do the two creases intersect? If so, in how many points?
   b. Suppose the creases did not intersect on the paper. Would the lines they represent intersect?
   c. Do you think it is possible that the two lines represented by the creases would not intersect? Yes; parallel lines do not intersect.

6a. If they intersect, they intersect in one point.

6b. If the lines are not parallel, they would eventually intersect.

MODELING MATHEMATICS Students may wish to build a similar model to help them visualize the figure for Exercises 15–17.

Answer for Modeling Mathematics

Drawing for parts a–d.

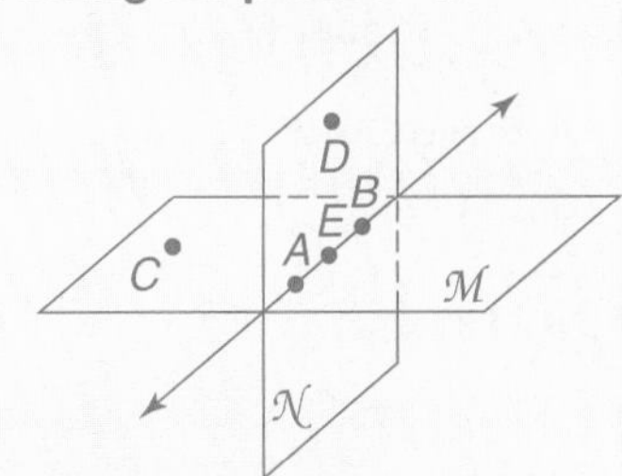

3 PRACTICE/APPLY

Check for Understanding

Exercises 1–18 are designed to help you assess your students' understanding through reading, writing, speaking, and modeling. You should work through Exercises 1–6 with your students and then monitor their work on Exercises 7–18.

Additional Answers

1. Each wall, floor, and ceiling represents a plane. Where the wall meets the ceiling, floor, or another wall represents a line. Where two walls and the floor or ceiling meet represents a point.

3.

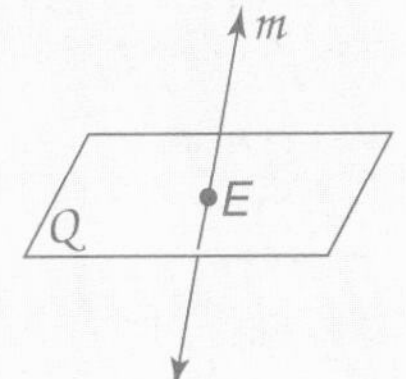

Reteaching

Using Models Use paper and a pen or pencil, and tape or glue to demonstrate what happens when a line and a plane intersect. Poke the pen or pencil through the paper to give students a visual demonstration of the concepts. Ask students to model other situations such as the intersection of two lines.

Assignment Guide

Core (with proof): 19–59 odd, 61–72
Core (informal): 19–59 odd, 61–72
Enriched: 20–54 even, 56–72

For **Extra Practice**, see p. 764.

The red A, B, and C flags, printed in the Teacher's Wraparound Edition, indicate the level of difficulty of the exercises.

Additional Answers

12.

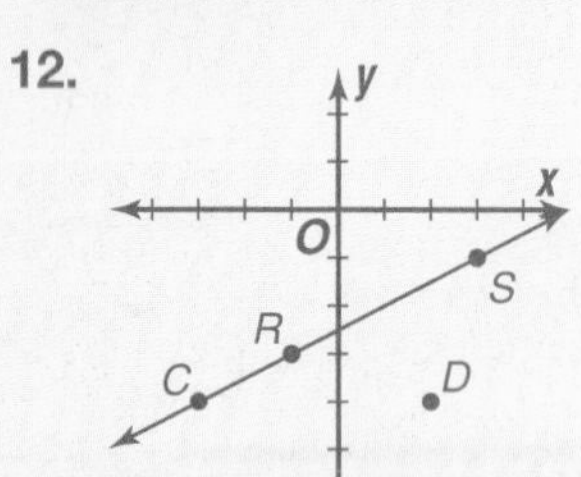

13.

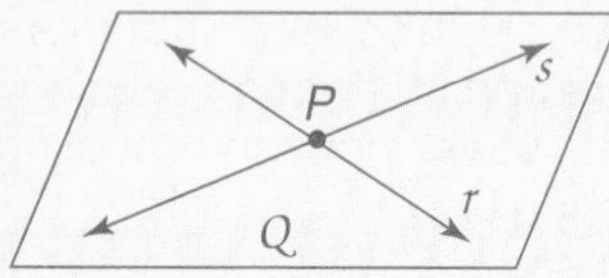

14.

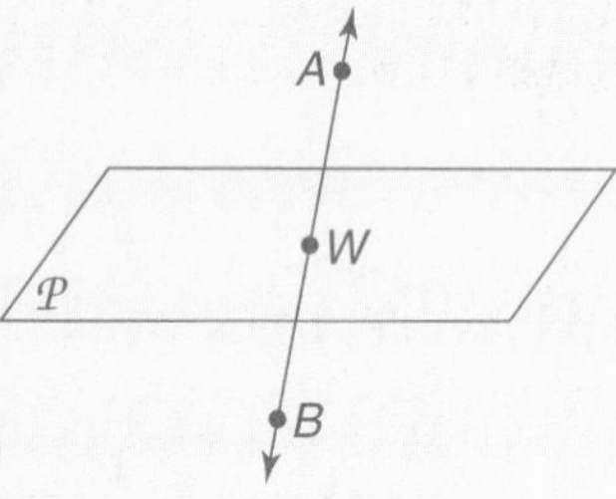

Study Guide Masters, p. 2

1-2 NAME ________ DATE ________

Study Guide — Student Edition Pages 12–18

Points, Lines, and Planes

Points, lines, and planes can be related in many different ways. Figures can be used to show these relationships. When two figures have one or more points in common, the figures are said to **intersect**. When points lie on the same line, the points are said to be **collinear**. When points lie in the same plane, the points are said to be **coplanar**.

Example: Draw and label a figure for each relationship.

Line k does not intersect plane $\mathcal{M}$.	Line t intersects plane $\mathcal{K}$ at point **S**.	Planes $\mathcal{R}$ and $\mathcal{S}$ intersect in line ***MN***.
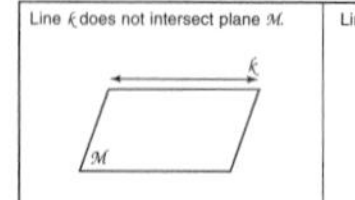	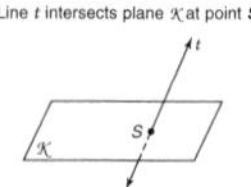	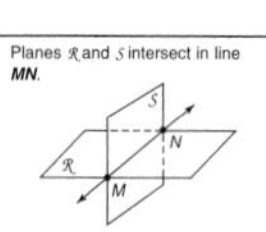

Draw and label a figure for each relationship.

1. Lines JK and EF are not in plane $\mathcal{M}$, but intersect plane $\mathcal{M}$ at X.

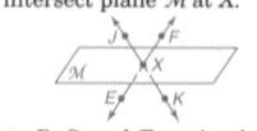

2. Lines m and n intersect at point Q.

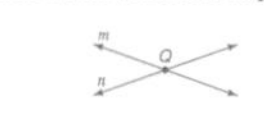

3. Points R, S, and T are in plane $\mathcal{M}$, but point W does not lie in plane $\mathcal{M}$.

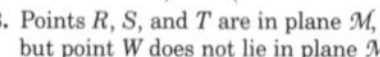

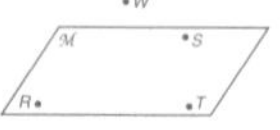

4. The intersection of planes $\mathcal{A}$, $\mathcal{B}$, and C is line EF.

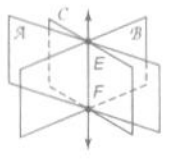

Refer to the figure at the right to answer each question.

5. Are points H, J, K, and L coplanar? yes
6. Name three lines that intersect at X. $\overleftrightarrow{WX}$, $\overleftrightarrow{KX}$, $\overleftrightarrow{XY}$
7. What points do plane $WXYZ$ and HW have in common? W
8. Are points W, X, and Y collinear? no
9. List the possibilities for naming a line contained in plane $WXKH$. $\overleftrightarrow{HK}$, $\overleftrightarrow{KH}$, $\overleftrightarrow{HW}$, $\overleftrightarrow{WH}$, $\overleftrightarrow{WX}$, $\overleftrightarrow{XW}$, $\overleftrightarrow{XK}$, $\overleftrightarrow{KX}$, $\overleftrightarrow{KW}$, $\overleftrightarrow{WK}$, $\overleftrightarrow{HX}$, $\overleftrightarrow{XH}$

Guided Practice

State whether each is best modeled by a *point, line,* or *plane.*

7. a star in the sky point
8. an ice skating rink plane
9. a telephone wire strung between two poles line

Refer to the figure at the right to name each of the following.

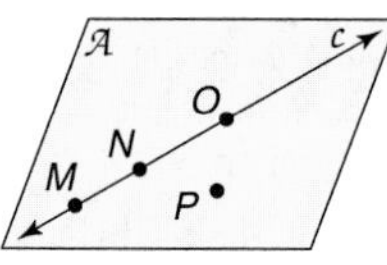

10. a line containing point N Sample answer: line c
11. a plane containing points P and M Sample answer: plane $\mathcal{A}$

Draw and label a figure for each relationship. 12–14. See margin.

12. A line passes through $C(-3, -4)$, $R(-1, -3)$, and $S(3, -1)$, but point D does not lie on $\overleftrightarrow{RS}$.
13. Plane $\mathcal{Q}$ contains lines r and s that intersect in P.
14. $\overleftrightarrow{AB}$ intersects plane $\mathcal{P}$ in W.

Refer to the figure at the right to answer each question.

15. Name the intersection of planes $\mathcal{A}$ and $\mathcal{B}$. $\overleftrightarrow{PR}$
16. Name another point that is collinear with points S and Q. T
17. Name a line that is coplanar with $\overleftrightarrow{VU}$ and point W. $\overleftrightarrow{PR}$
18. **List the Possibilities** The supermarket has a soft-drink machine that dispenses cans for 50¢ a can. The machine will only accept quarters, dimes, and nickels. The order in which the coins are placed in the machine does not matter. How many different combinations of coins must the machine be programmed to accept? 10: 2 quarters; 1 quarter, 2 dimes, 1 nickel; 1 quarter, 1 dime, 3 nickels; 1 quarter, 5 nickels; 5 dimes; 4 dimes, 2 nickels; 3 dimes, 4 nickels; 2 dimes, 6 nickels, 1 dime, 8 nickels: 10 nickels

EXERCISES

Practice

State whether each is best modeled by a *point, line,* or *plane.*

A

19. a taut piece of thread line
20. a knot in a piece of thread point
21. a piece of cloth plane
22. the corner of a room point
23. the rules on your notebook paper
24. your desktop plane
25. each color dot, or pixel, on a video-game screen point
26. a telecommunications beam to a satellite in space line
27. the crease in a folded sheet of wrapping paper line

23. lines

28–33. Sample names for figures are given.

Refer to the figure at the right to name each of the following.

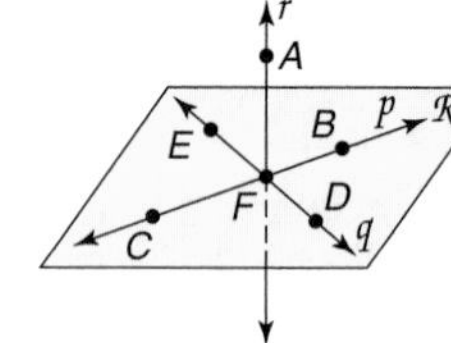

28. a line containing point A r
29. a line passing through B $\overleftrightarrow{BF}$
30. two points collinear with point D F, E
31. two points coplanar with point B F, C, E or D
32. a plane containing points B, C, and E BCE
33. a plane containing lines p and q $\mathcal{R}$

34–42. See Solutions Manual.

Draw and label a figure for each relationship.

34. Point S lies on $\overrightarrow{PR}$.
35. Points $A(2, 4)$, $B(2, -4)$, and C are collinear, but points F, A, B, and C are noncollinear.
36. $\overleftrightarrow{TU}$ lies in plane Q and contains point R.
37. $\overleftrightarrow{CD}$ and $\overleftrightarrow{RS}$ intersect at $P(3, 2)$ for $C(-1, 4)$ and $R(6, 4)$.
38. Line m contains A and B, but does not contain C.
39. Lines a, b, and c are coplanar, but do not intersect.
40. Planes $\mathcal{P}$ and $\mathcal{R}$ intersect in ℓ.
41. Point C and line m lie in Q. Line m intersects line n at T. Line m and C are coplanar but m, n, and C are not.
42. Lines a, b, and c are coplanar and meet at point Z.

Refer to the figure below to answer each question. The figure is a rectangular prism formed by six planes. Only a portion of each plane is shown. 43–51. Sample names for figures are given.

43. Name the six planes that form the rectangular prism. *AFH, BHD, CDE, EGF, AGC, FED*
44. Name two points that are coplanar with points B, H, and K. *A, F*
45. Name the lines that intersect at E. $\overleftrightarrow{GE}$, $\overleftrightarrow{DE}$, $\overleftrightarrow{FE}$
46. Which two planes intersect in $\overleftrightarrow{CD}$? *BHC, EDC*
47. Name two lines that lie in plane GEC. $\overleftrightarrow{GE}$, $\overleftrightarrow{GC}$
48. Name a plane and a line that intersect at A. *AGC*, $\overleftrightarrow{AF}$
49. Which point(s) do planes ABC, CDE, and AGE have in common? *G*
50. What do the dashed lines in the figure represent?
51. Are points B, H, E, and G coplanar? Explain your answer. Yes: you can draw $\overleftrightarrow{BG}$ and $\overleftrightarrow{HE}$ to form a plane that fits diagonally through the figure.

50. lines that cannot be seen from the perspective of the viewer

Draw and label a figure for each relationship. 52–55. See margin.

52. Planes $\mathcal{A}$ and $\mathcal{B}$ intersect, and planes $\mathcal{B}$ and $\mathcal{C}$ intersect, but planes $\mathcal{A}$ and $\mathcal{C}$ do not intersect.
53. Line t lies in planes $\mathcal{P}$, $\mathcal{Q}$, and $\mathcal{R}$.
54. Planes $\mathcal{P}$ and $\mathcal{Q}$ intersect each other. They both intersect plane $\mathcal{R}$.
55. The intersection of planes $\mathcal{A}$, $\mathcal{B}$, $\mathcal{C}$, and $\mathcal{D}$ is point E.

Critical Thinking

56. Think of your classroom as a model of six planes. The floor and ceiling are models of horizontal planes $\mathcal{A}$ and $\mathcal{B}$, and the four walls are models of vertical planes $\mathcal{C}$, $\mathcal{D}$, $\mathcal{E}$, and $\mathcal{F}$. Use your classroom as a model to explain your answers to the following questions.
    a. Can planes $\mathcal{A}$ and $\mathcal{B}$ intersect? No, they are parallel.
    b. Is it possible for any of the vertical planes to intersect?
    c. Are all lines in plane $\mathcal{C}$ vertical lines?
    d. What are the possiblities for the intersection of two of the vertical planes and plane $\mathcal{B}$? The intersection could be a point or two lines.
57. Four lines are coplanar. Draw figures to show the maximum number of intersection points and the minimum number of intersection points using the four lines. See margin.

56b. Yes, an example is where the front and side walls meet.
56c. No, there can be all types of lines in any plane.

Additional Answer

57. maximum: 6

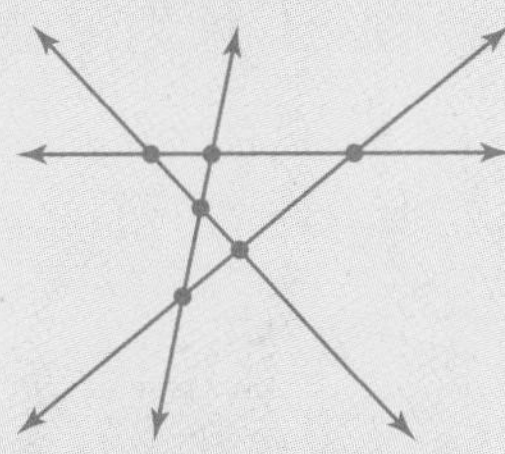

minimum: 0

Additional Answers

52.

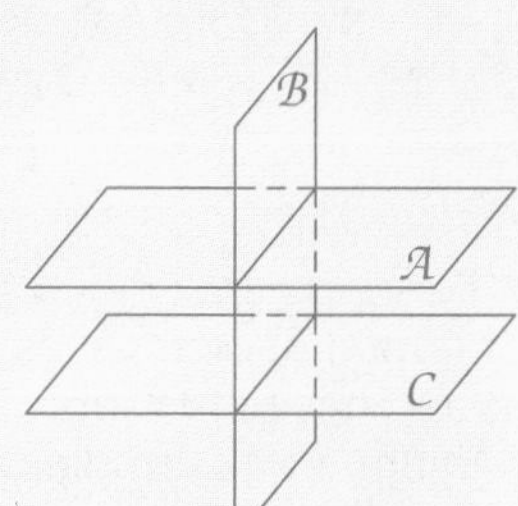

53.

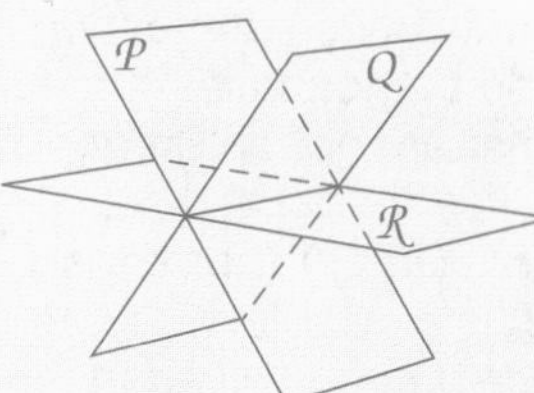

54.

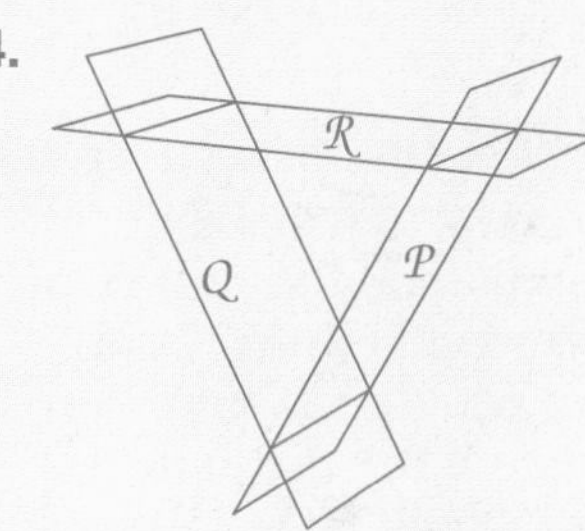

55.

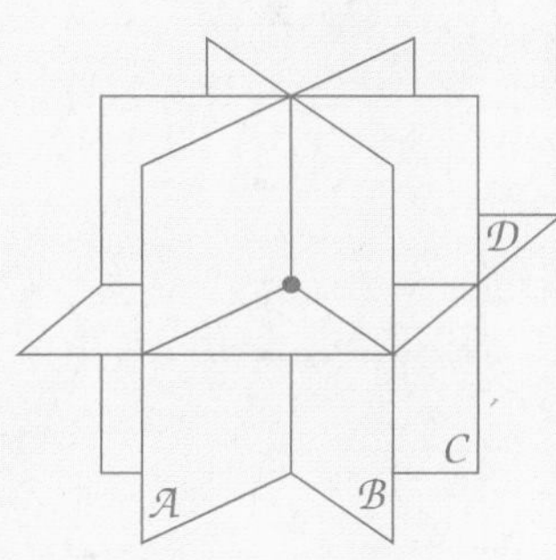

Practice Masters, p. 2

1-2 NAME ____ DATE ____ Student Edition Pages 12–18

Practice

Points, Lines, and Planes

Draw and label a figure for each relationship. Typical answers are given.

1. Lines ℓ, m and j intersect at P.

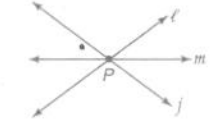

2. Plane $\mathcal{N}$ contains line ℓ.

3. Points A, B, C, and D are noncollinear.

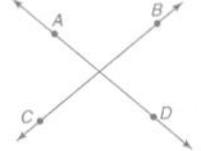

4. Points A, B, C, and D are noncoplanar.

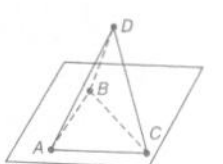

Refer to the figure at the right to answer each question.

5. Are points H, M, I, and J coplanar? yes
6. How many planes are shown? 7
7. Name the intersection of planes ABG and CHG. $\overline{BG}$
8. Name the intersection of plane ABC and $\overleftrightarrow{HL}$. C
9. Which segments are contained in all three of the planes GFH, CDI, and EDI? none
10. **List the Possibilities** Tim can choose from a tan shirt, a blue shirt, and a green shirt. He can choose from black slacks or blue jeans. He can choose from a windbreaker, a sweatshirt, or a jacket. How many different outfits can he wear if he will not wear the sweatshirt with the black slacks? 15

4 ASSESS

Closing Activity

Modeling Describe specific figures that contain points, lines, and planes and have students draw each figure. For example, you may have them draw a plane that contains two intersecting lines or a plane that contains three points that are not collinear.

Chapter 1 Quiz A (Lessons 1-1 and 1-2) is available in the *Assessment and Evaluation Masters*, p. 16.

Additional Answer

58. There are 16 possible outcomes on the first throw. Let D stand for any face down and 1, 2, 3, and 4 represent the number of dots on a face up disk. 1234, 234D, 134D, 124D, 123D, 12DD, 13DD, 14DD, 23DD, 24DD, 34DD, 1DDD, 2DDD, 3DDD, 4DDD, DDDD

Enrichment Masters, p. 2

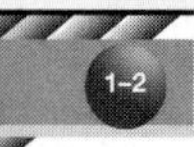

NAME__________ DATE ________

1–2 Enrichment

Student Edition Pages 12–18

Points and Lines on a Matrix

A **matrix** is a rectangular array of rows and columns. Points and lines on a matrix are not defined in the same way as in Euclidean geometry. A **point** on a matrix is a dot, which can be small or large. A **line** on a matrix is a path of dots that "line up." Between two points on a line there may or may not be other points. Three examples of lines are shown at the upper right. The broad line can be thought of as a single line or as two narrow lines side by side.

A dot-matrix printer for a computer uses dots to form characters. The dots are often called **pixels**. The matrix at the right shows how a dot-matrix printer might print the letter P.

Answers may vary. Sample answers are shown.
Draw points on each matrix to create the given figures.

1. Draw two intersecting lines that have four points in common.
2. Draw two lines that cross but have no common points.
3. Make the number 0 (zero) so that it extends to the top and bottom sides of the matrix.
4. Make the capital letter O so that it extends to each side of the matrix.

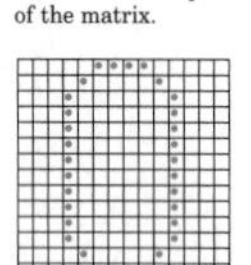

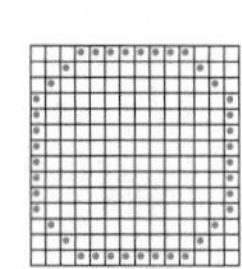

5. Using separate grid paper, make dot designs for several other letters. Which were the easiest and which were the most difficult? See students' work.

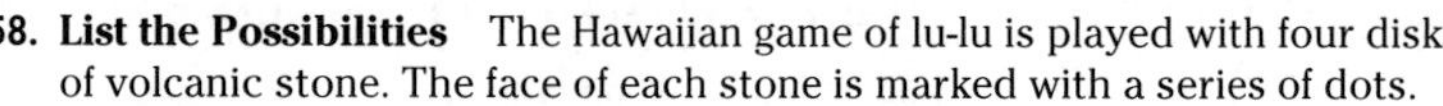

Applications and Problem Solving

58. **List the Possibilities** The Hawaiian game of lu-lu is played with four disks of volcanic stone. The face of each stone is marked with a series of dots.

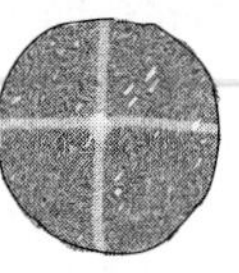
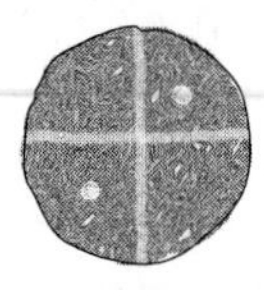
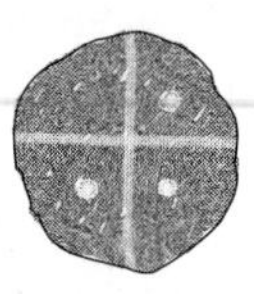
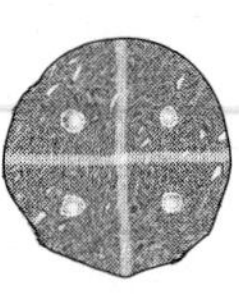

A player tosses the four disks and if they land all faceup, 10 points are scored, and the player tosses again. If any of the disks land facedown on the first toss, the player gets to toss those pieces again. The score is the total number of dots showing after the second toss. List the possible outcomes after the first toss. See margin.

59a–b. See students' work.

59. **Anatomy** Has anyone ever told you to stand up straight? If your posture is perfect, you should be able to draw a straight line from your ear to your ankle, running through your shoulder, hip, and knee.
   a. Study the posture of five of your friends or relatives. How many of them seem to have good posture according to the straight line rule?
   b. What percent of the people you observed have good posture?

60. **Art** In the 1950s, a movement called *op art* stressed pure abstract images rather than real-life images. Op art consists of carefully arranged colors and geometric patterns that create optical illusions, sometimes even movement, on a painted surface. The image at the right is an optical illusion. Use your knowledge of planes to describe what you see. Then look again to see if you can see the objects in a different perspective. Write about your observations. possible observations: a large cube with a small cube cut out; a small cube sitting in the corner of two walls and the floor

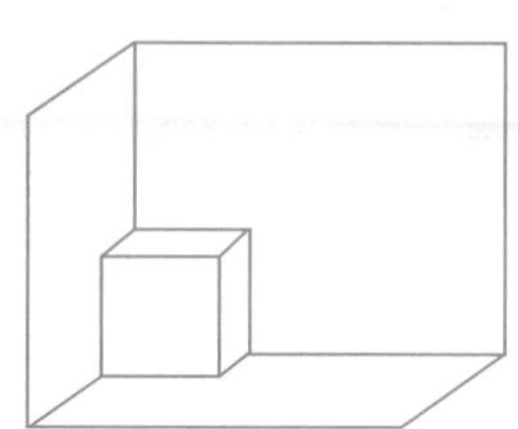

Mixed Review

61. Graph points $A(-2, 5)$, $B(3, -4)$, $C(0, -3)$, and $D(5, 5)$ on the same coordinate plane. (Lesson 1–1) See margin.

62. It is collinear because $-6 = 2(-3)$.

62. Points $F(2, 4)$ and $G(4, 8)$ lie on the graph of $y = 2x$. Determine whether $H(-3, -6)$ is collinear with F and G. (Lesson 1–1)

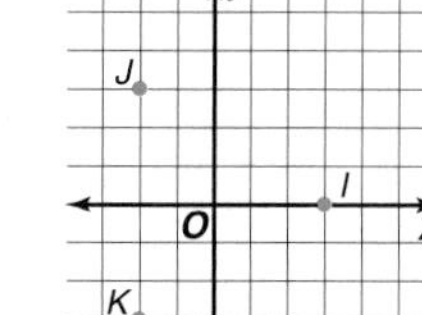

63. Write the ordered pair for each point shown on the coordinate plane at the right. (Lesson 1–1)
   a. I (3, 0) b. J (−2, 3) c. K (−2, −3)

Algebra

Solve each equation.

64. $x + 3 = 8$ 5
65. $4x = -44$ −11
66. $-y + 8 = -2$ 10
67. $3p + 4 = 2p$ −4
68. $-4c + 12 = 15$ $-\frac{3}{4}$
69. $3(x + 2) = -6$ −4

Evaluate each expression if $a = 3$, $b = -2$, and $c = 0$.

70. $ab + 6c$ −6
71. $a^2 + b$ 7
72. $bc + ac - c^2$ 0

Extension

Reasoning Ask one student to name a number. Then have the next student name a number that is less (or greater) than the previous number. Repeat this process until every student has had a turn. Ask students how long they could go on doing this. Explain that this demonstrates the concept of infinity.

Additional Answer

61.

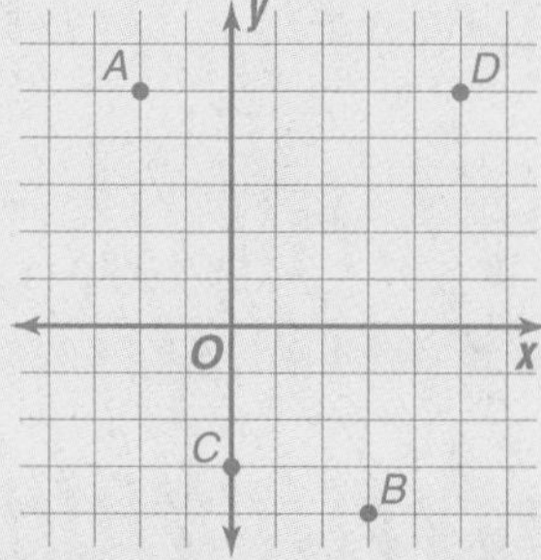

1-3 Integration: Algebra Using Formulas

What YOU'LL LEARN

- To solve problems by using formulas, and
- to find maximum area of a rectangle for a given perimeter.

Why IT'S IMPORTANT

You can use formulas to solve problems in science, social studies, and business as well as mathematics.

APPLICATION Photography

One of the easiest ways to "capture the moment" these days without investing in expensive equipment is with a single-use camera. A new type of single-use camera uses a panoramic lens. The diagram below shows what portion of a scene is captured by a 35-mm camera that has a panoramic mode, a standard single-use camera, and a panoramic single-use camera. How does the area of the view of a panoramic single-use camera compare with the area of the view of a 35-mm camera that has a panoramic mode? *This problem will be solved in Example 1.*

To be a good problem solver, you need to develop a plan for finding a solution. Four steps that can be used to solve any problem are listed below.

Problem-Solving Plan	1. **Explore the problem.** 2. **Plan the solution.** 3. **Solve the problem.** 4. **Examine the solution.**

To solve the problem presented above, we use the formula for the **area** of a rectangle since each photographic image is a rectangle.

Area of a Rectangle	**The formula for the area of a rectangle is $A = \ell w$, where A represents the area expressed in square units, ℓ represents the length, and w represents the width.**

Example 1

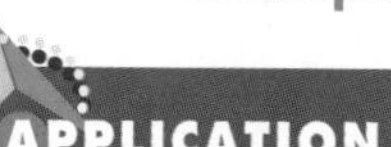

APPLICATION Photography

Refer to the application at the beginning of the lesson. Compare the area of the view of a panoramic single-use camera to the area of the view of a 35-mm camera that has a panoramic mode.

Explore The view of the panoramic single-use camera is much wider and taller than that of the 35-mm camera. To compare areas, we need to determine the dimensions of each view.

(continued on the next page)

FYI

The term 35-mm refers to the diagonal of the frame on the film. Some special panoramic lens cameras use various lengths of film for the image and are not classified as 35-mm cameras.

1-3 LESSON NOTES

NCTM Standards: 1–5, 6, 8, 12

Instructional Resources

- Study Guide Master 1-3
- Practice Master 1-3
- Enrichment Master 1-3

Transparency 1-3A contains the 5-Minute Check for this lesson; **Transparency 1-3B** contains a teaching aid for this lesson.

Recommended Pacing	
Standard Pacing	Day 3 of 13
Honors Pacing	Day 3 of 13
Block Scheduling*	Day 3 of 8

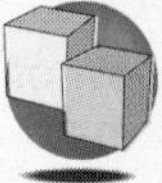

*For more information on pacing and possible lesson plans, refer to the *Block Scheduling Booklet*.

1 FOCUS

5-Minute Check (over Lesson 1-2)

1. Complete: ____ points determine a line. ____ points determine a plane. two, three
2. Give three names for a line that contains points *A* and *B*. Sample answer: line *AB*, $\overleftrightarrow{AB}$, and $\overleftrightarrow{BA}$

Refer to the figure below for Exercises 3 and 4.

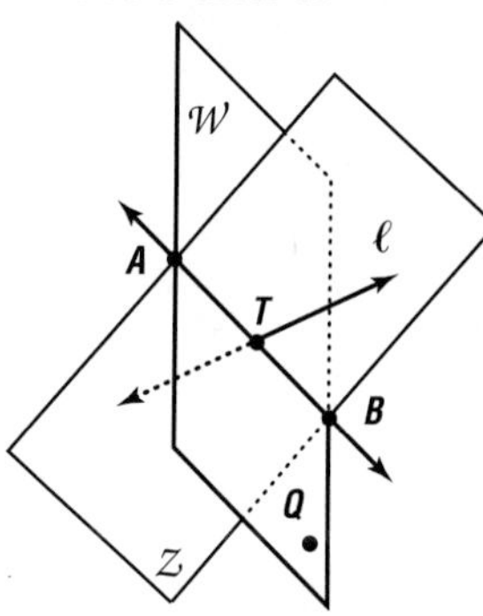

3. Name a line in plane $\mathcal{Z}$. line ℓ or line *AT*
4. How many lines are in plane $\mathcal{Z}$? infinitely many

Motivating the Lesson

Situational Problem Tell students that the sign at a local bank displays the temperature in Fahrenheit and Celsius. Ask if students think that the sign contains a Fahrenheit thermometer and a Celsius thermometer. Explain that the program running the sign probably contains a formula to convert a Fahrenheit temperature to Celsius or vice versa.

2 TEACH

In-Class Examples

For Example 1
Refer to the application at the beginning of the lesson. Compare the area of the view of a panoramic single-use camera. **The area of the view in a panoramic camera is about twice that of a standard single-use camera.**

For Example 2
Mr. and Mrs. Pinel are planning a rectangular vegetable garden. They want the garden to have an area of at least 15 square yards, but they have only 18 yards of wire fence to surround it. What are the possible dimensions of the garden if all the fence is used and the sides have whole-number lengths? **3 by 6 yards or 4 by 5 yards**

***fabulous* FIRSTS**

Moneta Sleet was born in 1926 in Owensboro, Kentucky. He graduated from Kentucky State College and New York University. He died of cancer in the fall of 1996 in New York City.

Moneta J. Sleet, Jr. (1926–1996)

The first African-American journalist to win a Pulitzer Prize for feature photography was Moneta Sleet Jr. in 1969. He won the award for his photograph of Coretta Scott King comforting her young daughter, Bernice, at the funeral of Martin Luther King, Jr.

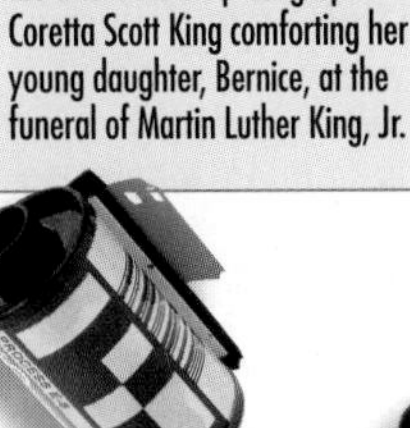

Plan Measure each rectangle using a millimeter ruler. Then use the measurements in the formula $A = \ell w$. Finally, compare the areas.

Solve

panoramic camera	*35-mm camera*
$A = \ell w$	$A = \ell w$
$= 80 \cdot 28$ or 2240	$= 38 \cdot 14$ or 532
The area is $2240\ \text{mm}^2$.	The area is $532\ \text{mm}^2$.

The view in the panoramic single-use camera has an area about 4 times that of the view in the 35-mm camera with panoramic mode.

Examine Suppose you draw a rectangle to model the view of the panoramic single-use camera. Then draw and cut out a rectangle to model the view of the 35-mm camera. See how many of the smaller rectangles will fit on the larger one.

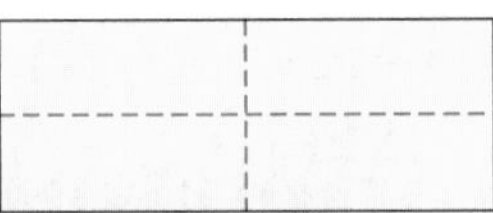

Four of the smaller rectangles are slightly larger than the area of the larger rectangle. Our estimate is correct.

There are many formulas that we will use in geometry. Another formula used with rectangles that you may remember is the **perimeter** formula. The perimeter is the distance around a figure. The formula for the perimeter P of a rectangle is $P = 2\ell + 2w$.

Perimeter of a Rectangle	**The formula for the perimeter of a rectangle is $P = 2\ell + 2w$, where P represents the perimeter, ℓ represents the length of the rectangle, and w represents the width of the rectangle.**

If you know the perimeter of a rectangle, you can find the maximum area that a rectangle with that perimeter can have.

Example 2

Pets

One type of pet enclosure is made up of 12 sections of fencing, connected by hinges. This enclosure folds for easy storage and is flexible for making different-sized areas in which your pet can play. Suppose the sections are positioned to make a rectangle. What are the dimensions of the rectangle that provide the maximum area for your pet?

Explore Since the perimeter will always be 12 units and fractions of a unit do not apply in this situation, we can draw all of the possibilities for rectangular shapes.

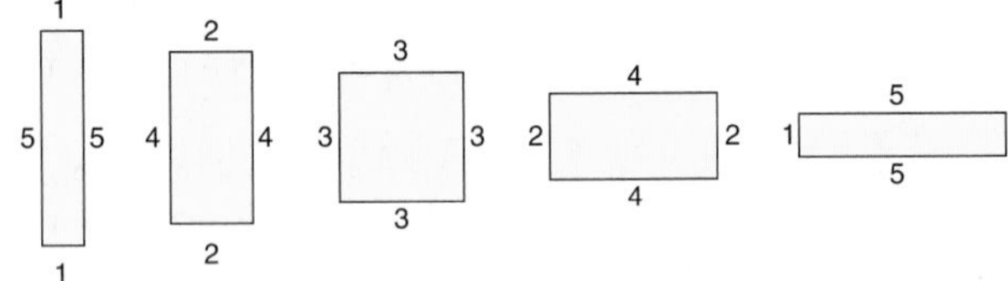

Cooperative Learning

Corners Separate students into groups of four. Assign each student a number from 1 to 4. Provide each group with the same four problems to solve using formulas. Student 1 solves equation 1, student 2 solves 2, and so on. After solving the equation, students go to the appropriate corner of the room to compare their answer with that of the other students with their number. Then students return to their groups to share the results. For more information on the corners strategy, see *Cooperative Learning in the Mathematics Classroom*, one of the titles in the Glencoe Mathematics Professional Series, pages 17–20.

Plan Use a table to find the area of each of the possible rectangles. Then select the one with the greatest area.

Solve Make a table that has columns for width w, length ℓ, and area A. It appears that the greatest area for the pet enclosure is 9 square units. This area occurs when the width is 3 units and the length is 3 units.

Why can't w be greater than or equal to 6?

w	ℓ	$A = \ell w$
1	5	5
2	4	8
3	3	9
4	2	8
5	1	5

The shape of the enclosure is a square.

Examine We can graph all ordered pairs (w, A) to look for a maximum.

w	A
1	5
2	8
3	9
4	8
5	5

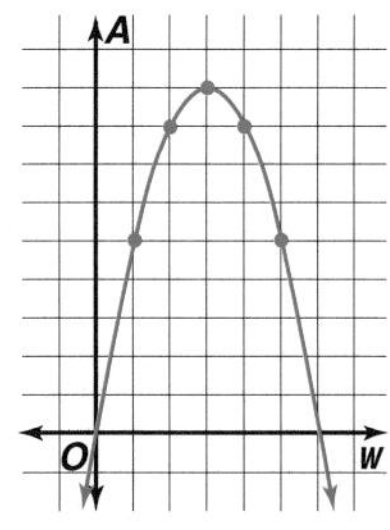

*The graph shows **all** the possible widths and areas, including fractional values.*

The graph is U-shaped and the greatest value of A appears to be when $w = 3$.

A spreadsheet can be used to quickly calculate quantities determined by a formula. This tool can help you look for patterns in the calculated data.

EXPLORATION — SPREADSHEETS

A spreadsheet is a table of cells that can contain text (labels) or numbers and formulas. Each cell is named by the column and row in which it is located. In the spreadsheet below, cells A1, B1, C1, and D1 contain labels.

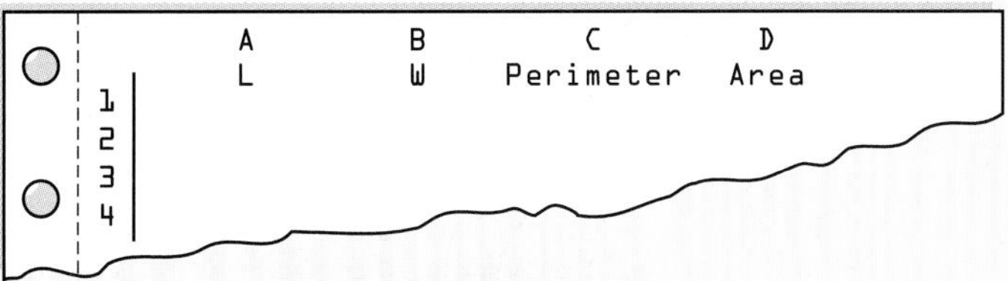

- In cell C2, enter the formula 2*A2 +2*B2, which represents the formula for the perimeter of a rectangle.
- Copy the formula to all of the cells of column C. When you do, the spreadsheet automatically uses the formula with the values in that row. For example, in cell D3, the formula becomes 2*A3 +2*B3 and in cell D4, it becomes 2*A4 +2*B4.

Your Turn

a. Enter a formula for the area of a rectangle in cell D2 and copy it to the other cells. **A2*B2**

b. Enter values in columns A and B and look for patterns in the values in columns C and D. **See students' work.**

EXPLORATION

Students may not be familiar with spreadsheets. Make sure they understand how each cell is named. Then explain that they can use the cell name instead of a variable in any formula to use a spreadsheet to help minimize calculation time.

Classroom Vignette

"You can investigate areas of rectangles using models. Have students form several rectangles of different dimensions by using square tiles, find the perimeter and area, then write a general formula. If commercial manipulatives are not available, small ceramic tiles from a builder's supply store are an inexpensive alternative."

Cindy Boyd
Author

In-Class Example

For Example 3
A car dealer advertised a loan for a $9000 used car with no money down and $179 per month for six years. What is the interest rate on the loan? 7.2%

Teaching Tip Remind students that perimeter is a linear measurement and area is measured in square units.

3 PRACTICE/APPLY

Check for Understanding

Exercises 1–14 are designed to help you assess your students' understanding through reading, writing, speaking, and modeling. You should work through Exercises 1–4 with your students and then monitor their work on Exercises 5–14.

Error Analysis
Some students will make mistakes when working with formulas because they try to take shortcuts or complete too many steps at one time. Encourage students to work in pairs and write all of the steps to solve each equation.

Additional Answers

1. P represents the perimeter of a rectangle, w its width, and ℓ its length.
2. In a square all sides have equal length, so ℓ and w can be replaced by s in each formula. $A = \ell w$ becomes $A = s \times s$ or s^2. $P = 2\ell + 2w$ becomes $P = 2s + 2s$ or $4s$.
3. Step 1: You *explore* what the problem is asking. Step 2: You *plan the solution* by determining what strategy you will use to solve the problem. Step 3: You *solve the problem* using the strategy you selected. Step 4: You *examine the solution* to determine if it makes sense in the original problem.

Many problems in mathematics use formulas. Sometimes it is necessary to solve a formula for a certain variable to find the answer to the problem. This requires some of the skills you learned in algebra.

Example 3

Money

A formula for computing simple interest is $I = prt$, where I is the interest earned, p is the principal (or the amount of money invested), r is the rate of interest (expressed as a decimal value), and t is the length of time in years it is invested. Soledad offered to lend her younger sister Lola $200 at a 3.5% interest rate. She told Lola that her loan would cost Lola $217.50 at the end of the loan period. How long is Soledad giving Lola to pay back the loan?

The interest on her loan is $217.50 – $200 or $17.50. There are two ways to determine the time of the loan.

Method 1

Solve the formula for t.

$I = prt$

$\frac{I}{pr} = t$ *Divide each side by pr.*

Now substitute 17.50 for I, 200 for p, and 0.035 for r. *3.5% = 0.035*

$t = \frac{I}{pr}$

$= \frac{17.50}{200 \cdot 0.035}$ or 2.5 *Use a calculator.*

Method 2

Substitute the known values into the formula and solve for t.

$I = prt$

$17.50 = 200 \cdot 0.035 \cdot t$

$17.50 = 7t$ *Simplify.*

$2.5 = t$ *Divide each side by 7.*

Regardless of the method used, Lola has 2.5 years to pay back the loan.

CHECK FOR UNDERSTANDING

Communicating Mathematics
1–3. See margin.
4. See students' work.

Study the lesson. Then complete the following.

1. **Describe** what each variable in the formula $P = 2\ell + 2w$ means.
2. **Demonstrate** how you could find the formulas for the perimeter and area of a square by using the formulas for a rectangle.
3. **Explain** the four-step plan for problem solving.

4. **Make a list** of other mathematical formulas you recall from previous mathematics or science courses. Explain what each variable represents.

Guided Practice

Find the perimeter and area of each rectangle. 6. 31 in., 59.5 in^2

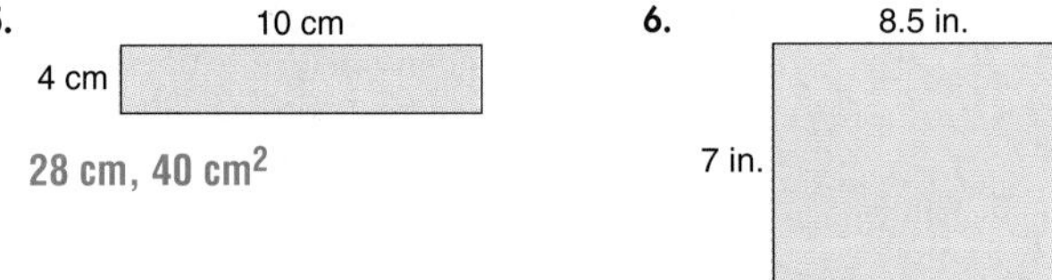

5. 28 cm, 40 cm^2

Find the missing measure in each formula.

7. $A = \ell w$; $\ell = 4$, $w = 7$, $A = ?$ 28
8. $P = 2\ell + 2w$; $\ell = 3$, $w = 5$, $P = ?$ 16
9. $I = prt$; $p = 350$, $r = 6\%$, $I = 42$, $t = ?$ 2

Reteaching

Using Models Have students use square tiles to form rectangles and visually determine the area and perimeter. Students can then use the formulas to find the area and perimeter and compare their findings. Students may wish to use the tiles as they complete the exercises.

10. $t = \frac{d}{r}$

11. $T = 5d + t$

10. **Distance** Solve the formula $d = rt$ for t. In the formula, d represents the distance, r the rate of speed, and t the time traveled.

11. **Meteorology** In the formula $\frac{T - t}{5} = d$, T represents the time when the flash of lightning occurs, t the time when the sound of the thunder begins, and d the distance the lightning is from you. Solve the formula for T.

Find the maximum area for the given perimeter of a rectangle. State the length and width of the rectangle. For Exercises 12–13, $\ell = w$.

12. 24 millimeters 36 mm^2; 6mm
13. 36 miles 81 mi^2; 9 mi

14. **Travel** Marguerite is going to Spain during spring break with the foreign language club. She receives an information sheet about the trip that says the average temperature during April is 18°C and the average rainfall is 8.6 centimeters. However, Marguerite is not familiar with metric units of measure.
   a. The formula for converting Celsius (C) temperatures to Fahrenheit (F) temperatures is $\frac{9}{5}C + 32 = F$. What is the average temperature in Spain in April in degrees Fahrenheit? about 64°
   b. The formula for estimating the number of inches in c centimeters is $\ell = \frac{c}{2.54}$. Find the average rainfall in inches. about 3.4 in.

EXERCISES

Practice **Find the perimeter and area of each rectangle.**

15. 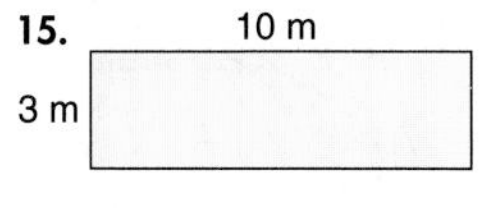

16. 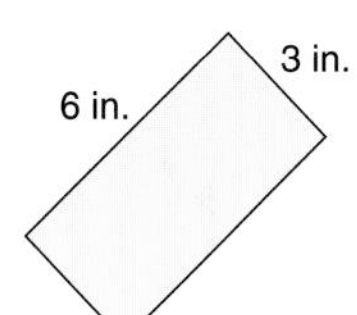

17. 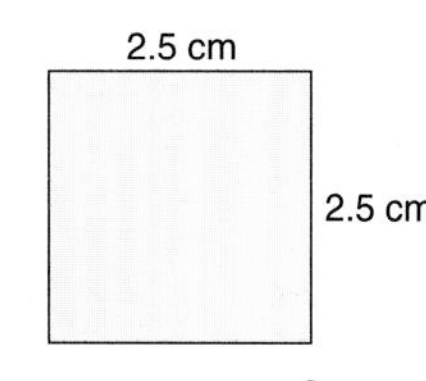

26m, 30m^2 | 18 in., 18 in^2 | 10 cm, 6.25 cm^2

18. 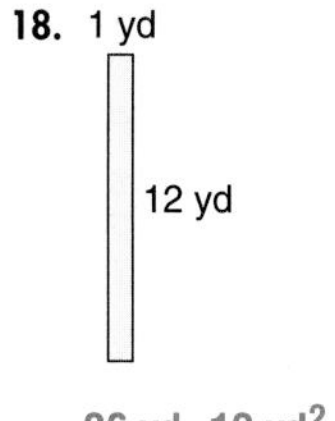

19. 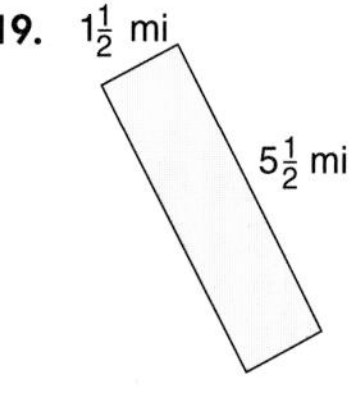

20. 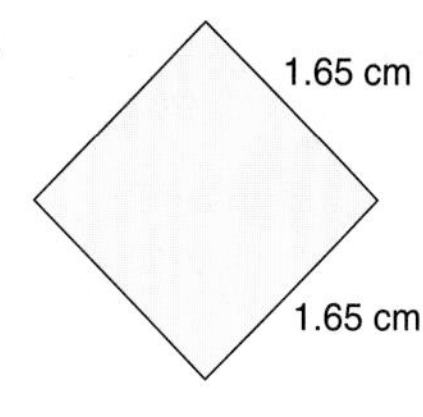

26 yd, 12 yd^2 | 14 mi, 8.25 mi^2 | 6.60 cm, 2.7225 cm^2

Find the missing measure in each formula if $P = 2\ell + 2w$ and $A = \ell w$.

21. $\ell = 7, w = 3, P = ?$ 20
22. $\ell = 4.5, w = 1.5, P = ?$ 12
23. $\ell = 8, w = 4, A = ?$ 32
24. $\ell = 2.2, w = 1.1, A = ?$ 2.42
25. $\ell = 6, A = 36, w = ?$ 6
26. $\ell = 12, A = 30, w = ?$ 2.5
27. $A = 34, w = 2, \ell = ?$ 17
28. $A = 3\frac{1}{2}, w = \frac{1}{2}, \ell = ?$ 7
29. $P = 84, \ell = 12, w = ?$ 30
30. $P = 13, w = 2.5, \ell = ?$ 4

Assignment Guide

Core (with proof): 15–39 odd, 40, 41, 43, 44–52
Core (informal): 15–39 odd, 40, 41, 43, 44–52
Enriched: 16–38 even, 40–52
All: Self Test, 1–10

For **Extra Practice**, see p. 764.

The red A, B, and C flags, printed only in the Teacher's Wraparound Edition, indicate the level of difficulty of the exercises.

Study Guide Masters, p. 3

NAME________ DATE________

Study Guide

Student Edition Pages 19–25

Integration: Algebra Using Formulas

The following four-step plan can be used to solve any problem.

Problem-Solving Plan	
1. *Explore* the problem.	Identify what you want to know.
2. *Plan* the solution.	Choose a strategy.
3. *Solve* the problem.	Use the strategy to solve the problem.
4. *Examine* the solution.	Check your answer.

When finding a solution, it may be necessary to use a formula. Two useful formulas are the area formula and perimeter formula for a rectangle.

Area of a Rectangle	The formula for the area of a rectangle is $A = \ell w$, where A represents the area expressed in square units, ℓ represents the length, and w represents the width.
Perimeter of a Rectangle	The formula for the perimeter of a rectangle is $P = 2\ell + 2w$, where P represents the perimeter, ℓ represents the length and w represents the width.

Examples

1 Find the perimeter and area of the rectangle at the right.

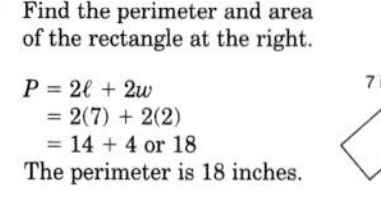

$P = 2\ell + 2w$
$= 2(7) + 2(2)$
$= 14 + 4$ or 18
The perimeter is 18 inches.

$A = \ell w$
$= 7 \cdot 2$ or 14
The area is 14 in^2.

2 Find the width of a rectangle whose area is 52 cm^2 and whose length is 13 cm.

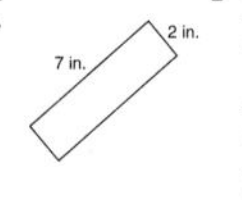

$A = \ell w$
$\frac{52}{13} = \frac{13w}{13}$
$4 = w$
The width is 4 cm.

Find the perimeter and area of each rectangle.

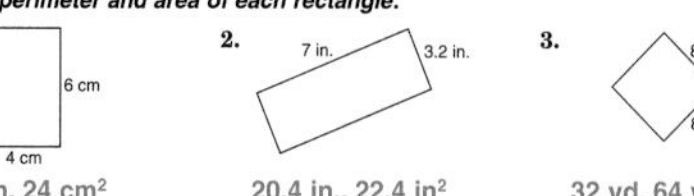

1. 20 cm, 24 cm^2
2. 20.4 in., 22.4 in^2
3. 32 yd, 64 yd^2

Find the missing measure in each formula.

4. $\ell = 3, w = 7, P = \underline{?}$ 20
5. $w = 5.2, \ell = 6.5, A = \underline{?}$ 33.8
6. $w = 4, A = 36, \ell = \underline{?}$ 9
7. $P = 65, \ell = 18, w = \underline{?}$ 14.5

4 ASSESS

Closing Activity

Speaking Ask students to name at least three advantages to using a formula to solve problems.

There is a technique of painting called pointillism that uses dots to form a picture. From a distance they blend together.

Additional Answer

40. $P = 12$ inches; Make a table of possible lengths and widths that yield an area of 36 square inches. The least perimeter is a square with sides 6 inches long.

Practice Masters, p. 3

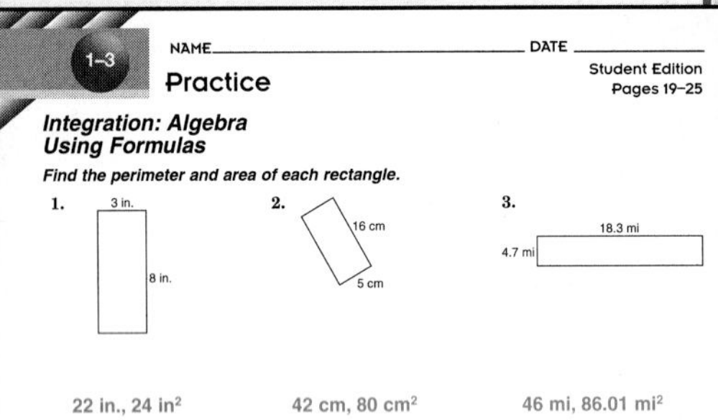

1-3 NAME ______ DATE ______
Practice — Student Edition Pages 19–25

Integration: Algebra Using Formulas

Find the perimeter and area of each rectangle.

1. 3 in. by 8 in. — 22 in., 24 in²
2. 16 cm by 5 cm — 42 cm, 80 cm²
3. 18.3 mi by 4.7 mi — 46 mi, 86.01 mi²

Find the missing measure in each formula.

4. $\ell = 4, w = 2, P = \underline{?}$ 12
5. $A = 32, w = 6.25, \ell = \underline{?}$ 5.12
6. $P = 27, \ell = 8, w = \underline{?}$ 5.5
7. $\ell = 3\frac{1}{2}, A = 5\frac{1}{4}, w = \underline{?}$ $1\frac{1}{2}$
8. $w = 7, \ell = 12, P = \underline{?}$ 38
9. $w = 15.5, P = 81, \ell = \underline{?}$ 25

Find the maximum area for the given perimeter of a rectangle. State the length and width of the rectangle.

10. 25 mm — 39 mm², 6.5 mm, 6 mm
11. 18 yards — 20 yd², 5 yd, 4 yd
12. 42 cm — 110 m², 11 m, 10 m

31. 49 in², 7 in.
32. 121 cm², 11 cm

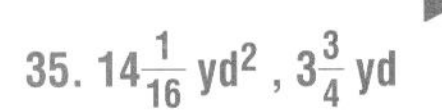

35. $14\frac{1}{16}$ yd², $3\frac{3}{4}$ yd

39. 17.5 units³

42. 5 · 8 · 1200 or 48,000 dots

Find the maximum area for the given perimeter of a rectangle. State the length and width of the rectangle. For Exercises 31–36, $\ell = w$.

31. 28 inches
32. 44 centimeters
33. 32 feet 64 ft², 8 ft
34. 26 meters 42.25 m², 6.5 m
35. 15 yards
36. 5 millimeters 1.5625 mm², 1.25 mm

Evaluate each formula for the values given.

37. **Area of a triangle:** $A = \frac{1}{2}bh$, if $b = 10$ and $h = 12$ 60 units²
38. **Changing °F to °C:** $C = \frac{5}{9}(F - 32)$ for 212°F 100°C
39. **Volume of a rectangular solid:** $V = \ell wh$, if $\ell = 3.5$, $w = 4$, and $h = 1.25$

Critical Thinking

40. Suppose the area of a rectangle is 36 square inches. Find the *minimum* perimeter for a rectangle with this area. Explain your procedure for finding the perimeter. See margin.

Applications and Problem Solving

41. **Ranching** A rancher is adding a corral to his barn so that the barn opens directly into the corral. He has 195 feet of fencing left over from another project. Find the greatest possible area for his corral using this length of fence. 4225 ft²

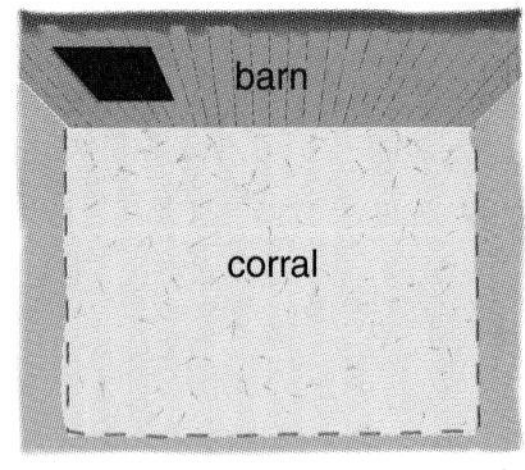

42. **Printing** Dots of four colors are used to break a photo down into its component parts for printing. Each photo is separated into four primary colors.

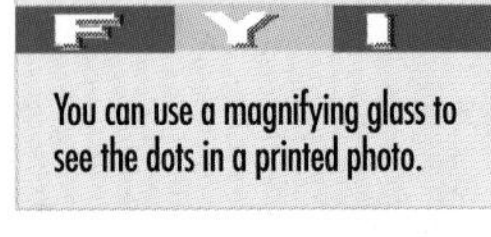

You can use a magnifying glass to see the dots in a printed photo.

black magenta cyan yellow

The photos in this book are printed at 1200 dpi (dots per square inch). What is the maximum number of different colored dots that could be used to print a 5-inch by 8-inch rectangular color photo?

43. **Energy Conservation** A house that has less exterior wall area loses less energy. You can compare the amount of exterior wall area on houses of the same height by comparing the possible perimeters of the house. A house with a greater perimeter will have a greater amount of exterior wall area.
   a. If the houses below have approximately the same amount of living area, which house will lose the least amount of energy through the exterior walls? B
   b. Which house actually has the greatest amount of living area? C

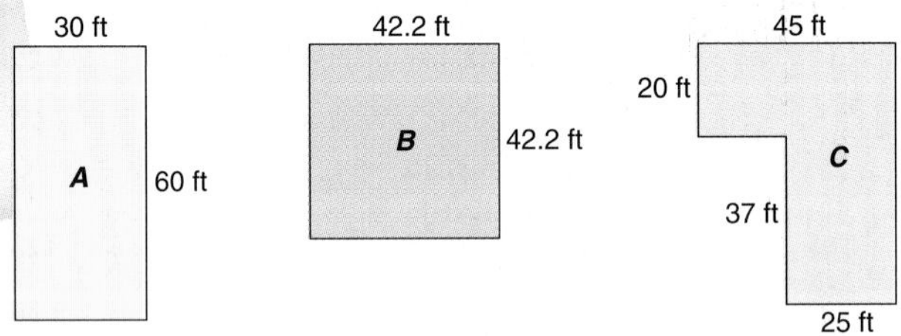

Extension

Problem Solving Challenge students to find all the possible pentominoes. A pentomino is made up of 5 squares so that every square has at least one edge in common with another of the 5 squares. Two pentominoes are shown at the right.

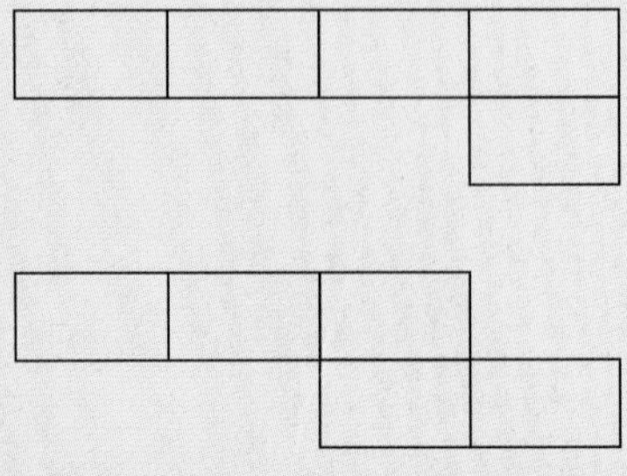

Mixed Review

44. Sample answer: W, X, U, or V
45. Sample answer: $\overleftrightarrow{ST}$, $\overleftrightarrow{SV}$, $\overleftrightarrow{SR}$

Refer to the figure at the right to answer each question. (Lesson 1–2)

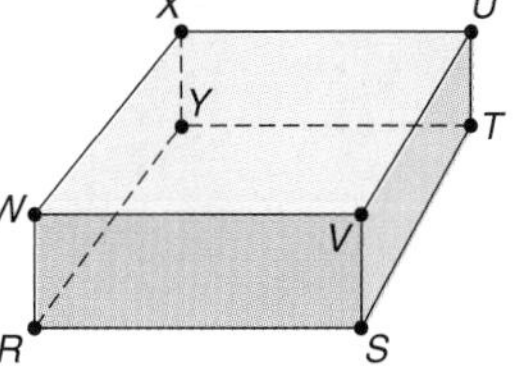

44. Name a point not coplanar with R, S, and T.

45. Name three lines that contain S.

46. Name the intersection of planes RWX and UTY. $\overleftrightarrow{XY}$

47. Name the ordered pair for each point graphed in the coordinate plane at the right. (Lesson 1–1)

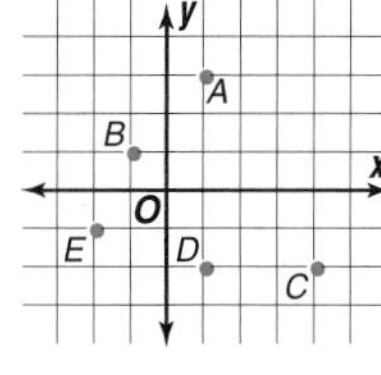

a. A (1, 3) b. B (−1, 1) c. C (4, −2)

48. Graph $A(-3, 4)$, $B(3, 4)$, $C(3, -4)$, and $D(-3, -4)$ on the same coordinate plane. (Lesson 1–1) See margin.

INTEGRATION
Algebra

Simplify each expression. 52. $4a^2b - b + a^2$

49. $4x(x + 3) + 3x^2 - 5x$ $7x^2 + 7x$ **50.** $3x + 4x - 9x + 2$ $-2x + 2$

51. $6c + 7d + 2c - 10d$ $8c - 3d$ **52.** $a^2b + 3a^2b - b + a^2$

SELF TEST

1. Name the ordered pair for each point graphed in the coordinate plane at the right. (Lesson 1–1)
   a. P (4, 0) b. Q (1, −3) c. R (−2, 1)
2. Graph each point on the same coordinate plane. (Lesson 1–1)
   a. $A(3, 3)$ b. $B(0, -5)$ c. $C(-3, -6)$ 2a–c. See margin.
3. List all the possible names for the figure below. (Lesson 1–2)

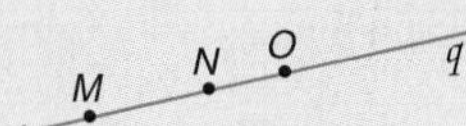

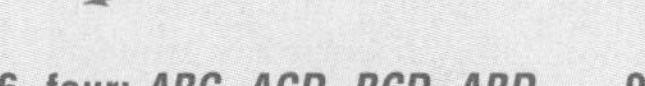

$\overleftrightarrow{MN}$, $\overleftrightarrow{NM}$, $\overleftrightarrow{OM}$, $\overleftrightarrow{MO}$, $\overleftrightarrow{ON}$, $\overleftrightarrow{NO}$, line q

6. four: ABC, ACD, BCD, ABD 9. 61 km

Refer to the figure at the right to answer each question. (Lesson 1–2)

4. Are points A, E, and D collinear? Explain. Yes, they all lie on $\overleftrightarrow{AD}$.
5. Are points A, B, C, and D coplanar? Explain. See margin.
6. How many planes appear in this figure? Name them.
7. Draw and label a figure showing lines a and b intersecting at T with line a in plane Q, but line b not in Q. (Lesson 1–2) See margin.

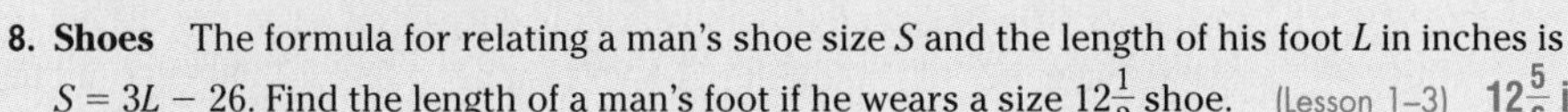

8. **Shoes** The formula for relating a man's shoe size S and the length of his foot L in inches is $S = 3L - 26$. Find the length of a man's foot if he wears a size $12\frac{1}{2}$ shoe. (Lesson 1–3) $12\frac{5}{6}$ in.
9. Find the perimeter of a rectangle that is 8 kilometers wide and 22.5 kilometers long. (Lesson 1–3)
10. What is the maximum area for a rectangle whose perimeter is 48 meters? (Lesson 1–3) 144 m^2

SELF TEST

The Self Test provides students with a brief review of the concepts and skills in Lessons 1-1 through 1-3. Lesson numbers are given at the end of problems or instruction lines so students may review concepts not yet mastered.

Additional Answer

48.

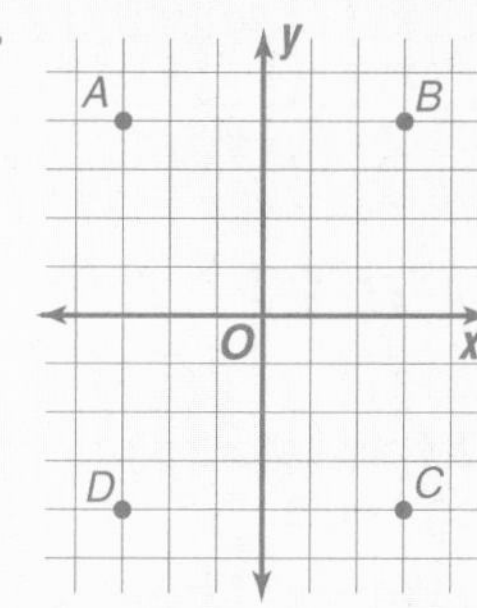

Answers for the Self Test

2a–2c.

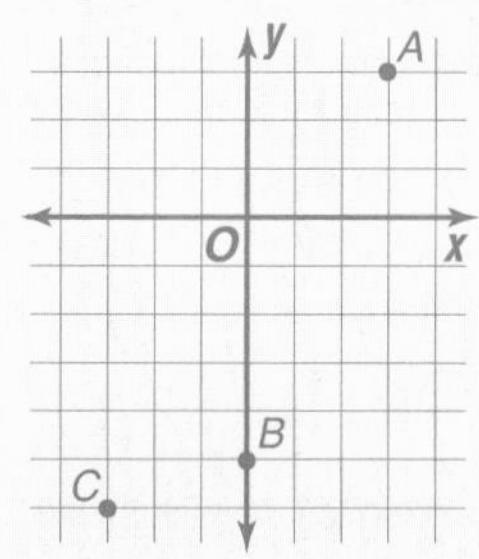

5. No; A, B, and C are coplanar, but D lies above plane $\mathcal{M}$.

7.

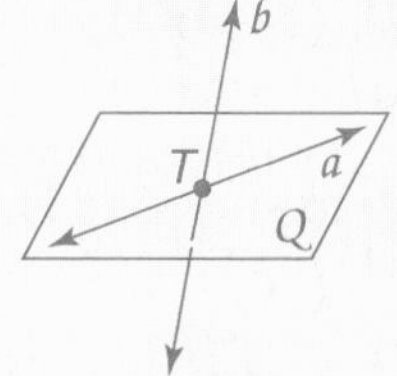

Enrichment Masters, p. 3

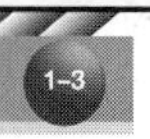

1-3 NAME______ DATE______

Enrichment

Student Edition Pages 19–25

Perimeter and Area of Irregular Shapes

Two formulas that are used frequently in mathematics are perimeter and area of a rectangle.

Perimeter: $P = 2\ell + 2w$,
Area: $A = \ell w$, where ℓ is the length and w is the width

However, many figures are combinations of two or more rectangles creating **irregular shapes**. To find the area of an irregular shape, it helps to separate the shape into rectangles, calculate the formula for each rectangle, then find the sum of the areas.

Example: Find the area of the figure at the right.

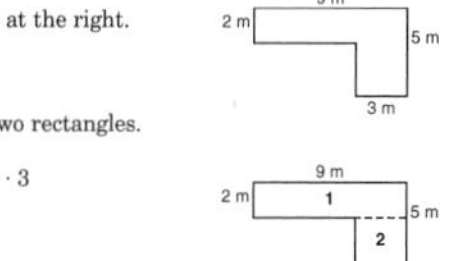

Separate the figure into two rectangles.

$A = \ell w$
$A_1 = 9 \cdot 2 = 18$ $A_2 = 3 \cdot 3 = 9$
$18 + 9 = 27$

The area of the irregular shape is 27 m^2.

Find the area of each irregular shape.

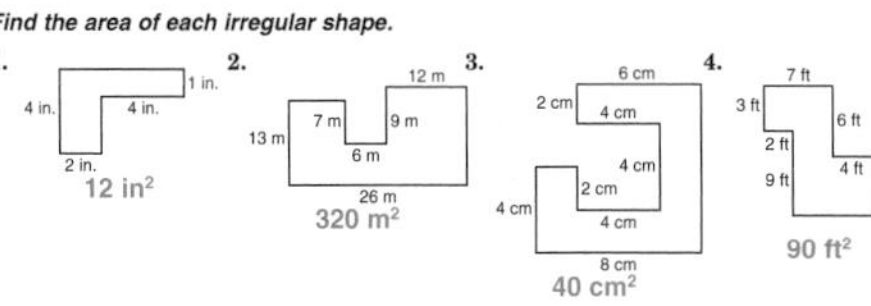

For questions 5–8, find the perimeter of the figures in Exercises 1–4.

5. 17 in. 6. 96 m 7. 44 cm 8. 48 ft

9. Describe the steps you used to find the perimeter in Exercise 1. See students' work.

1-4A LESSON NOTES

NCTM Standards: 1–5, 7, 8

Objective

Use a TI-92 calculator to draw and measure segments.

Recommended Time

25 minutes

Instructional Resources

Instructions for using the *Geometer's Sketchpad* for this activity are available in the *Graphing Calculator and Computer Masters*, pp. 14–15.

1 FOCUS

Motivating the Lesson

If this is the first time your students have used a TI-92 calculator, allow them some time to explore the calculators before beginning. Ask them to find any familiar terms in the F1 through F8 menus.

2 TEACH

Teaching Tip You may wish to have students work in pairs on this activity. One student can read the instructions while the other operates the calculator. Make sure every student has the opportunity to operate the calculator.

Encourage students to look at the commands in each menu. Stress that all of the TI-92 activities in this text can be done without specific keystroke sequences.

1–4A Using Technology Measuring Segments

A Preview of Lesson 1–4

This lesson can be done using Cabri software on a computer. Instead of keys, you use similar pulldown menus to perform the functions.

The TI-92 calculator is actually a small computer. It works very much like a TI-82 calculator in some respects, but has many more features, one of which is a keyboard like you find on a computer. The TI-92 contains certain features of *Cabri II* software. It allows you to construct geometric figures, measure them, and reposition them to study their characteristics. Many of the graphing calculator lessons in this book will use these geometry features of the TI-92.

To select an item on a menu, press down on the cursor pad until the selection is highlighted and then press ENTER. You can also select the item by entering the item's number on the menu.

To **access TI-92 Geometry**, press the applications button APPS, select 8:Geometry, and then 3:New. Press down on the cursor pad and type in any name for your work session by using the letter keys. Press ENTER twice. A screen like the one at the right appears. The cursor is a small + sign.

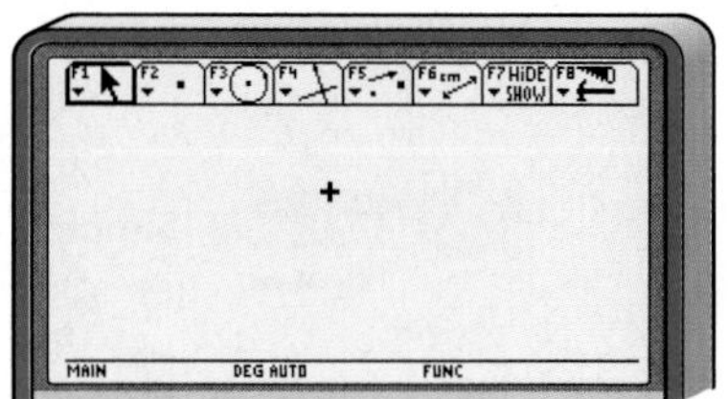

To **construct a segment**, press F2 (the drawing menu) and select 5:Segment. The cursor becomes a small pencil. Use the cursor pad to move the cursor to any location on the screen to start your drawing. Press ENTER to locate your first endpoint. If you wish to name your endpoint easily, do it now by pressing the shift key ⬆ while typing in your letter to name the point. *If you move the cursor before naming the point, the automatic label function will not work.*

Now move your cursor to wherever you wish your second endpoint to be. Press ENTER to locate the point. Name the other endpoint. The screen at the right shows a sample segment *AB*.

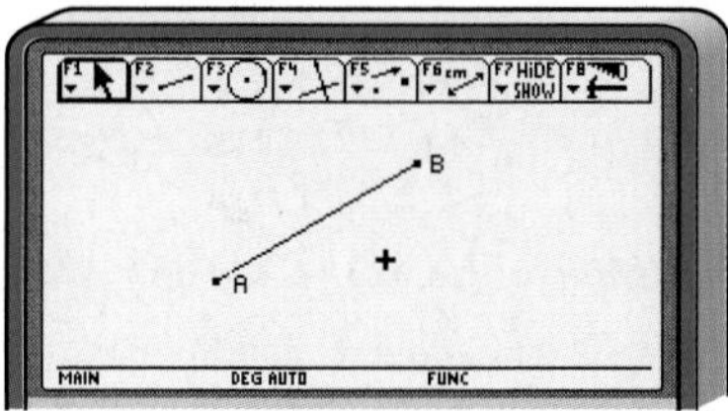

To select a different unit of measure, press F8 and select 9:Format. Select Length & Area on the menu. Press the cursor pad right. Select your preferred unit of measure.

TI-92 Geometry allows you to measure each segment you draw. Press F6 and select 1:Distance and Length. Move the cursor toward your segment until the message "LENGTH OF THIS SEGMENT" appears. Press ENTER and the measurement appears. You can **relocate the label** by moving the cursor toward the label until the message "THIS NUMBER" appears. Hold down the hand key and use the position pad to move the label to wherever you wish. Release the hand key to finalize its position.

Using Technology

This lesson offers an excellent opportunity for using technology in your geometry classroom. For more information on using technology, see *Graphing Calculators in the Mathematics Classroom*, one of the titles in the Glencoe Mathematics Professional Series.

All of the commands referenced are on the F2 *menu.*

You can select the items from the F2 menu to **draw additional points, segments, and lines** on the screen.

- Draw other segments by using the same method you used to draw $\overline{AB}$.
- Name a point of intersection by using 3:Intersection Point command. Once you select the command, move your cursor to where you think the intersection should be. When you are close enough to the point, the message "POINT AT THIS INTERSECTION" will appear. Press ENTER and name the point *E*.
- To name other points on a segment, select 2:Point on Object. Move the cursor to where you wish to place the point. The message "ON THIS SEGMENT" will appear. Press ENTER and name the point. Because you are in "draw a point" mode, you can put additional points on the segment without returning to the F2 menu. Just press ENTER at each location and name the point.
- To draw points not on a segment, select 1:Point, press ENTER, and name the point.
- To draw a line through two points, select 4:Line and move the cursor to one of the points through which the line will pass. When the message "THRU THIS POINT" appears, press ENTER and move the cursor to the second point. Press ENTER at the message and a line will appear.

You can **clear your screen** or **erase a figure** by using 7:Delete or 8:Clear All on the F8 menu.

EXERCISES

Analyze your drawing.

1. How many points could you put on a segment? any number
2. Which unit of measure seems more reasonable when drawing segments?
3. Try to measure the distance from *E* to *A*. What happens?
4. Move the cursor to the position of one of the endpoints of $\overline{AB}$. Hold down the hand key and use the cursor pad to move the endpoint to a new location. What happens to the measure of the segment? It changes with the length of the new segment.

2. centimeters or millimeters
3. The calculator shows the measurement of the entire segment containing $\overline{EA}$, not just $\overline{EA}$.

Construct each of the following. Sketch each drawing on your paper. 5–9. See margin.

5. Lines *PQ* and *RS* meet at *T*.
6. Segment *WT* intersects $\overline{TR}$.
7. Point *O* lies between *N* and *M* on $\overleftrightarrow{PQ}$.
8. Segments *AB* and *CD* do not intersect.
9. segment *XY* and its length
10. Experiment with the 4:Line command on the F2 menu when you do not draw two points. Write about your findings. See students' work.

If you have the TI GRAPH-LINK software and cable for the TI-92 available, you can print out your sketches from a computer.

3 PRACTICE/APPLY

Assignment Guide

Core (with proof): 1–10
Core (informal): 1–10
Enriched: 1–10

4 ASSESS

Observing students working with technology is an excellent method of assessment.

Additional Answers

5.

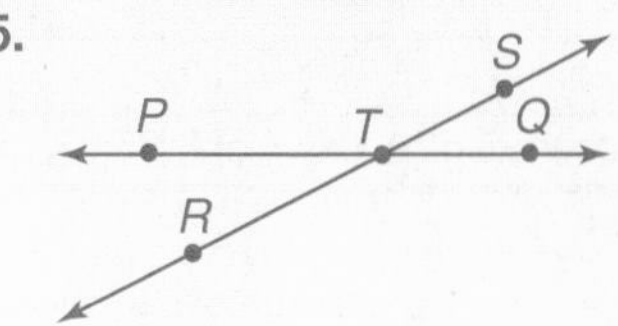

6.

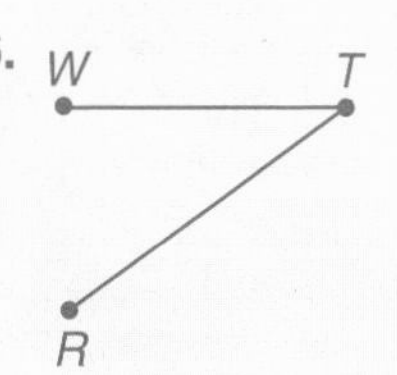

7.

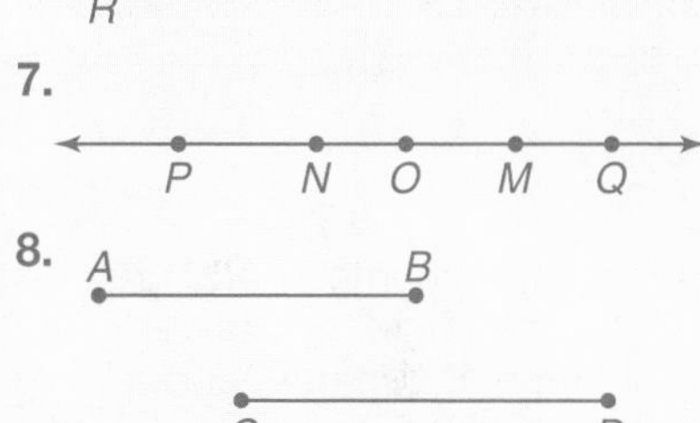

8.

9.

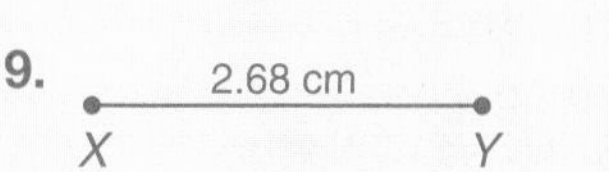

1-4 LESSON NOTES

NCTM Standards: 1–5, 7, 8, 12

Instructional Resources

- Study Guide Master 1-4
- Practice Master 1-4
- Enrichment Master 1-4
- Assessment and Evaluation Masters, pp. 15, 16

Transparency 1-4A contains the 5-Minute Check for this lesson; **Transparency 1-4B** contains a teaching aid for this lesson.

Recommended Pacing	
Standard Pacing	Day 5 of 13
Honors Pacing	Day 5 of 13
Block Scheduling*	Day 4 of 8

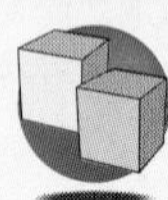

*For more information on pacing and possible lesson plans, refer to the *Block Scheduling Booklet*.

1 FOCUS

5-Minute Check
(over Lesson 1-3)

1. Find the perimeter and area of a rectangle that has a width of 6.2 meters and a length of 9.5 meters. **31.4 m; 58.9 m^2**
2. A rectangle has a perimeter of 56 inches. Find the maximum area of the rectangle. State the length and width of the rectangle. **196 in^2; 14 in. by 14 in.**
3. Use the formula $A = \frac{1}{2}Pa$ to find A if $P = 72$ and $a = 10.4$. **374.4**

1-4 Measuring Segments

What YOU'LL LEARN

- To find the distance between two points on a number line and between two points in a coordinate plane, and
- to use the Pythagorean Theorem to find the length of the hypotenuse of a right triangle.

Why IT'S IMPORTANT

Segment measures are used in discovering characteristics of many geometric shapes.

APPLICATION
Textiles

The woman in Guatemala shown in the photo at the right is weaving a colorful fabric by an age-old method. However, most fabric for your clothing was produced on an electronic high-speed loom.

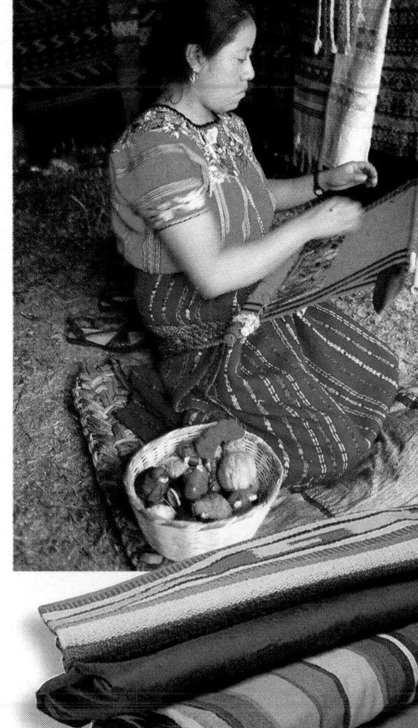

Fabric is measured in bolts. A bolt, which is 120 feet long, is only one of many units of linear measure used in the world. A piece of thread running the length of the fabric could be a model for a line segment. The ends of the thread could be thought of as the endpoints of the segment. In geometry, the length of a segment is the distance between its two endpoints. Segments can be defined by using the idea of *betweenness* of points.

THE FAR SIDE By GARY LARSON

"Well, lemme think. ... You've stumped me, son. Most folks only wanna know how to go the other way."

In the figure at the right, point B is **between** points A and C, while point D is *not* between A and C. For B to be between A and C, all three points must be collinear. Segment AB, written $\overline{AB}$, consist of points A and B and all points between A and B. The **measure** of $\overline{AB}$, written AB (no bar over the letters), is the distance between A and B. Thus, the measure of a segment is the same as the distance between its two endpoints.

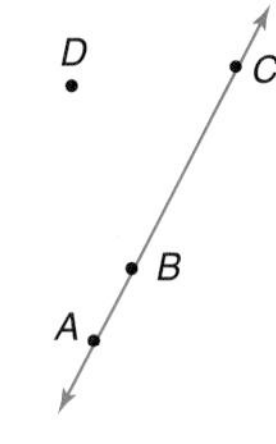

You have probably measured segments by using a centimeter- or inch-ruler. Centimeters and inches are just two examples of *units of measure.*

Whether you are trying to measure the distance from A to B or from B to A as in *The Far Side* cartoon at the left, the measure is the same. In fact, two points on any line can always be paired with real numbers so that one point is paired with zero and the other paired with a positive number. This correspondence suggests the **Ruler Postulate**.

*A **postulate** is a statement that is assumed to be true.*

Postulate 1–1 Ruler Postulate

The points on any line can be paired with real numbers so that, given any two points *P* and *Q* on the line, *P* corresponds to zero, and *Q* corresponds to a positive number.

One way to measure the distance between two points is to use a number line. The numbers on a ruler are a real-life example of a number line.

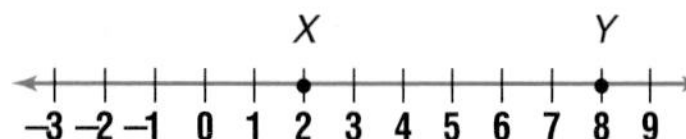

To find the measure of $\overline{XY}$, you first need to identify the coordinates of X and Y. The coordinate of X is 2, and the coordinate of Y is 8. One way to find the distance between the two points is to count the number of units between X and Y, which is 6. You could also use your algebra skills. Since measure is always a positive number, you can find the *absolute value* of the difference between the two coordinates. When you use absolute value, the order in which you subtract the coordinates does not matter.

distance from X to Y	distance from X to Y
$\|8 - 2\| = \|6\|$ or 6	$\|2 - 8\| = \|-6\|$ or 6

The measure of $\overline{XY}$ is 6, or $XY = 6$.

Example  **Find PQ, QR, and PR on the number line shown below.**

P Q R

–4 –3 –2 –1 0 1 2 3 4

$PQ = |-3.5 - (-2)|$
$= |-3.5 + 2|$
$= |-1.5|$ or 1.5

$QR = |-2 - 2.5|$
$= |-4.5|$ or 4.5

$PR = |-3.5 - 2.5|$
$= |-6|$ or 6

Examine the measures PQ, QR, and PR in Example 1. Notice that $1.5 + 4.5 = 6$. So $PQ + QR = PR$. This suggests the following postulate.

Postulate 1–2 Segment Addition Postulate

If *Q* is between *P* and *R*, then $PQ + QR = PR$.
If $PQ + QR = PR$, then *Q* is between *P* and *R*.

Example  **Find LM if L is between N and M, $NL = 6x - 5$, $LM = 2x + 3$, and $NM = 30$.**

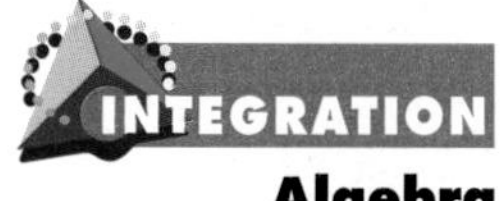

Algebra

A good way to begin solving this problem is to make a drawing of the given information. Draw L between N and M and label the measures of the segments.

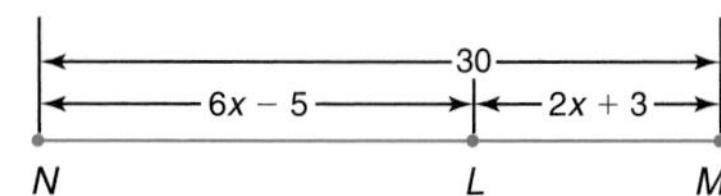

(continued on the next page)

Motivating the Lesson

Hands-On Activity Provide students with rulers and tell them to measure the length of their index fingers. They may soon begin asking where to begin measuring—at the knuckle or at the point where the index finger and middle finger split. Some may also ask how accurate you want the answer. Discuss the different aspects of measurement.

2 TEACH

Teaching Tip Before you discuss Example 1, point out that PQ represents a number and $\overline{PQ}$ represents a segment.

In-Class Examples

For Example 1
Find RS, ST, and RT on the number line shown below.

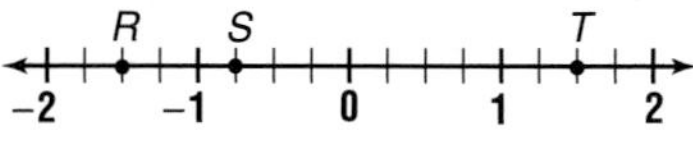

$RS = 0.75$, $ST = 2.25$, $RT = 3$

For Example 2
Find the measure of $\overline{MN}$ if N is between M and P, $MN = 3x + 2$, $NP = 18$, and $MP = 5x$.

Teaching Tip When discussing the distance between two points, make sure that students understand that any measure is always positive. Ask students if they could draw a segment with a measure of -5 inches, or if they could measure $-\frac{1}{2}$ cup of water and add it to a recipe.

FYI

The Rhind papyrus (1650 B.C.) and the Moscow papyrus (1850 B.C.) are the major sources that show the knowledge of mathematics attained by the ancient Egyptians.

In-Class Example

For Example 3
Find the distance between points $H(2, 3)$ and $K(-3, -1)$ by using the Pythagorean Theorem. $\sqrt{41}$ **or about 6.403**

FYI

In 2000 B.C., Egyptians used a loop of rope knotted at 12 equal intervals to determine the boundaries of their properties each year after the Nile flooded. They stretched the rope around three stakes so that the measures of the sides of the triangle were 3, 4, and 5. This right triangle was used to establish the corners of their fields.

Since L is between N and M, $NL + LM = NM$. Use this equation and the values we know to write a new equation.

$$NL + LM = NM$$
$$6x - 5 + 2x + 3 = 30 \quad \textit{NL = 6x − 5, LM = 2x + 3, NM = 30}$$
$$8x - 2 = 30 \quad \textit{Combine like terms.}$$
$$8x = 32 \quad \textit{Add 2 to each side.}$$
$$x = 4 \quad \textit{Divide each side by 8.}$$

Now use the value of x to find LM.

$$LM = 2x + 3$$
$$= 2(4) + 3 \text{ or } 11$$

You have probably encountered the **Pythagorean Theorem** in other math classes. The theorem is often expressed as $a^2 + b^2 = c^2$, where a and b are the measures of the shorter sides (legs) of the right triangle and c is the measure of the longest side (hypotenuse). *You will prove the Pythagorean Theorem in Lesson 8–1.*

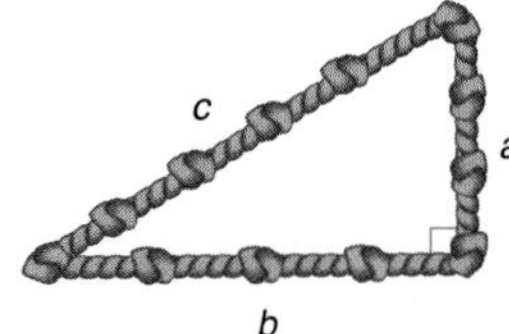

Pythagorean Theorem

In a right triangle, the sum of the squares of the measures of the legs equals the square of the measure of the hypotenuse.

If a and b represent the measures of the legs of a right triangle, and c represents the measure of the hypotenuse, then $a^2 + b^2 = c^2$.

Example 3 **Find the distance from $A(1, 2)$ to $B(6, 14)$ using the Pythagorean Theorem.**

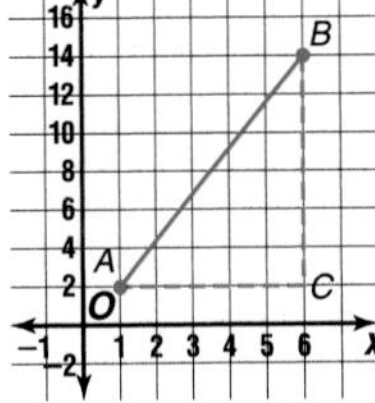

The grid lines on a coordinate plane meet at right angles, so we can draw segments along the grid lines to form a right triangle that has $\overline{AB}$ as its longest side. The coordinates of C are (6, 2).

Since the y-coordinates of A and C are the same, we can use the x-axis as the number line to find AC.

$$AC = |1 - 6| \text{ or } 5$$

Likewise, since the x-coordinates of B and C are the same, we can use the y-axis as the number line to measure the length of $\overline{BC}$.

$$BC = |14 - 2| \text{ or } 12$$

Now use the Pythagorean Theorem to find the length of $\overline{AB}$ if $AB = x$.

$$a^2 + b^2 = c^2$$
$$5^2 + 12^2 = x^2 \quad \textit{a = 5, b = 12, c = x}$$
$$25 + 144 = x^2$$
$$169 = x^2$$
$$\pm\sqrt{169} = x \quad \textit{Take the square root of each side.}$$
$$13 = x \quad \textit{Ignore the negative value of x.}$$

The distance from A to B is 13.

You will derive the distance formula in Chapter 8.

The process in Example 3 could become very tedious if you had to find a lot of distances by using the Pythagorean Theorem. A formula has been developed from the Pythagorean Theorem to find the distance between two points in a coordinate plane. This formula is called the **distance formula**.

Distance Formula	**The distance *d* between any two points with coordinates (x_1, y_1) and (x_2, y_2) is given by the formula** $d = \sqrt{(x_2 - x_1)^2 + (y_2 - y_1)^2}$.

The distance between two points is the same no matter which point is used as (x_1, y_1).

Example 4 **Find *PQ* for *P*(−3, −5) and *Q*(4, −6).**

Let (x_1, y_1) be (−3, −5) and (x_2, y_2) be (4, −6).

$$d = \sqrt{(x_2 - x_1)^2 + (y_2 - y_1)^2} \quad \textit{Distance formula}$$
$$PQ = \sqrt{[4 - (-3)]^2 + [-6 - (-5)]^2}$$
$$= \sqrt{7^2 + (-1)^2}$$
$$= \sqrt{49 + 1}$$
$$= \sqrt{50} \text{ or about } 7.07 \quad \textit{Use a calculator.}$$

When two segments have the same length, they are said to be **congruent** segments. For example, if $AB = CD$, then we write $\overline{AB} \cong \overline{CD}$, which is read "segment *AB* is congruent to segment *CD*."

We can use a compass and straightedge to construct a segment that is congruent to another segment.

CONSTRUCTION

Congruent Segments

A straightedge is any object that can be used to draw a straight line, such as a ruler or the edge of a protractor. A straightedge is not used to measure length.

Use a compass and a straightedge to construct a segment congruent to another segment.

1. Draw a segment *XY*.

2. Elsewhere on your paper, draw a line and a point on the line. Label the point *P*.

3. Place the compass at point *X* and adjust the compass setting so that the pencil is at point *Y*.

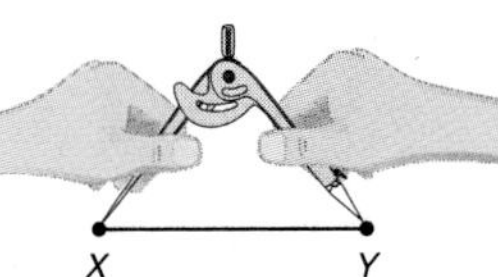

4. Using that setting, place the compass at point *P* and draw an arc that intersects the line. Label the point of intersection *Q*.

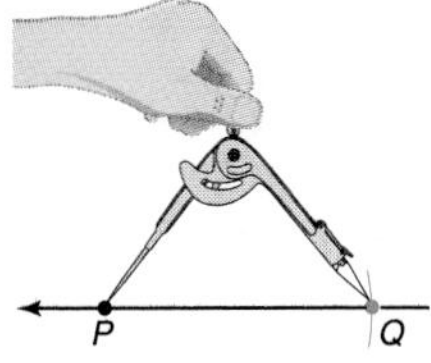

Conclusion: Since the compass setting used to construct $\overline{PQ}$ is the same as the distance from *X* to *Y*, $PQ = XY$. Thus, $\overline{PQ} \cong \overline{XY}$.

In-Class Example

For Example 4
Find *JK* for *J*(9, −5) and *K*(−6, 12). **$\sqrt{514}$ or about 22.6**

Teaching Tip If students are using the Triman safety compass from Glencoe's High School Manipulative Kit, they may need additional time to master using the compass before attempting a construction.

Classroom Vignette

"I teach all of the constructions as a unit. I end the unit with a final extra credit construction project in which the students make a creative construction design. These projects can make an attractive classroom display."

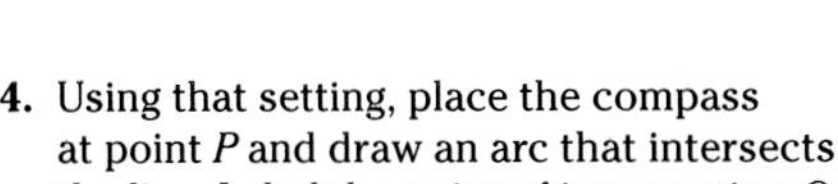

Robert H. Schultz
Harrison High School
Farmington Hills, Michigan

3 PRACTICE/APPLY

Check for Understanding

Exercises 1–15 are designed to help you assess your students' understanding through reading, writing, speaking, and modeling. You should work through Exercises 1–4 with your students and then monitor their work on Exercises 5–15.

Error Analysis

Students may have trouble determining which values to assign to x_1, x_2, y_1, and y_2. You may wish to have them write the formula using four different variables to illustrate that x_1 and x_2 are different. Also, students may have trouble subtracting negative numbers when finding a distance. Review how to subtract negative numbers, and demonstrate how to write subtraction involving negative numbers with parentheses.

Additional Answers

2. The measure of the longest side is *c*, and *a* and *b* are the measures of the other two sides. The designation of *a* and *b* to a particular side for the Pythagorean Theorem is not important.
4. *E* is between *D* and *F*.

When comparing measures of segments with no unit of measure given, assume that the units are the same.

Because the measures of segments are real numbers, they can be compared. To compare the measure of $\overline{AB}$ to the measures of $\overline{CD}$, $\overline{EF}$, and $\overline{GH}$ shown below, you can set your compass width to match the measure of $\overline{AB}$ and then compare this to the measure of each of the segments.

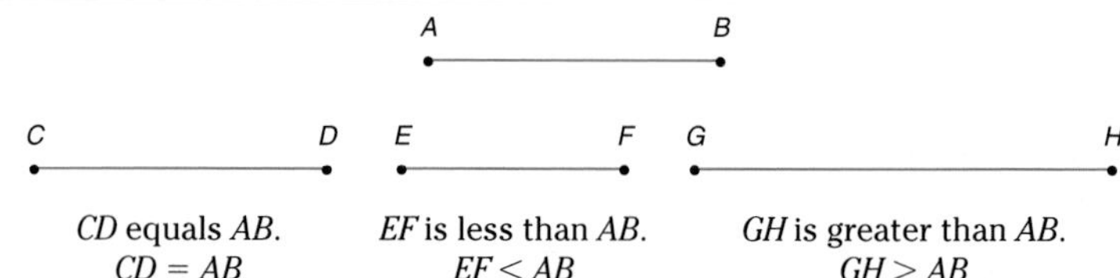

words	*CD* equals *AB*.	*EF* is less than *AB*.	*GH* is greater than *AB*.
symbols	$CD = AB$	$EF < AB$	$GH > AB$

CHECK FOR UNDERSTANDING

Communicating Mathematics

Study the lesson. Then complete the following.

1. **Write** an equation that relates the measures of $\overline{XY}$, $\overline{XZ}$, and $\overline{YZ}$, if point *Y* is between points *X* and *Z*. $XY + YZ = XZ$
2. **Describe** how you know which sides of a right triangle are represented by *a*, *b*, and *c* in the Pythagorean Theorem. See margin.
3. **Explain** the difference between $\overline{AB}$ and *AB*. *AB* is the measure of $\overline{AB}$.
4. If $DE + EF = DF$, explain what must be true of point *E*. See margin.

Guided Practice

Refer to the number line at the right to find each measure.

5. *AC* 8 units
6. *BD* 8 units
7. Find *HK* if *J* is between *H* and *K*, $HJ = 17$, and $JK = 6$. 23

Refer to the coordinate plane at the right to find each measure. Round your answers to the nearest hundredth.

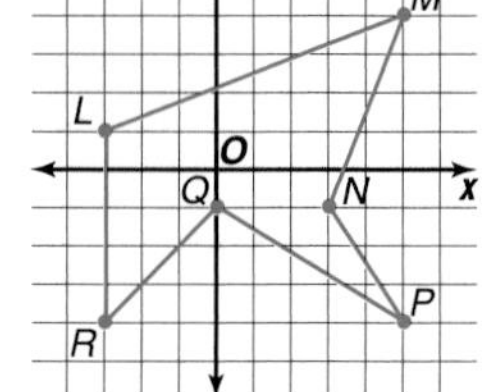

8. *MP* 8
9. *NP* 3.61
10. *QR* 4.24

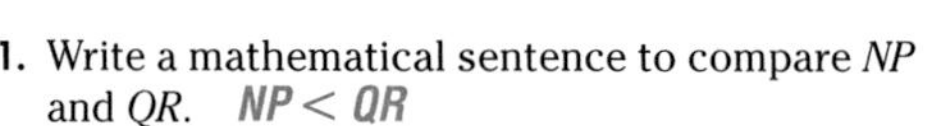

11. Write a mathematical sentence to compare *NP* and *QR*. $NP < QR$

12. Use the Pythagorean Theorem to find the missing side measure of $\triangle XYZ$. 13

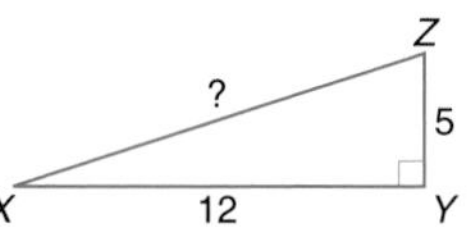

13. If *W* is between *R* and *S*, $RS = 7n + 8$, $RW = 4n - 3$, and $WS = 6n + 2$, find the value of *n* and *WS*. 3, 20

14. Use a compass to compare the length of the segments shown below. Then list the measures of the segments in order from least to greatest. 14. *PQ*, *TU*, *RS*

Pythagorus
(c. 580–500 B.C.)

Reteaching

Using Manipulatives Plot two points on a coordinate grid. Have students count the horizontal spaces between two points to find $|x_2 - x_1|$ and the vertical spaces to find $|y_2 - y_1|$. Then demonstrate that performing the subtractions using the coordinates gives the same distances. Complete the problem using the Pythagorean Theorem and distance formula and compare results.

15. **World Records** On September 8, 1989, the British Royal Marines stretched a rope from the top of Blackpool Tower (416 feet high) in Lancashire, Great Britain, to a fixed point on the ground 1128 feet from the base of the tower. Then Sgt. Alan Heward and Cpl. Mick Heap of the Royal Marines, John Herbert of Blackpool Tower, and TV show hosts Cheryl Baker and Roy Castle slid down the rope establishing the greatest distance recorded in a rope slide.
   a. Draw a right triangle to represent this event. See margin.
   b. How far did they slide? about 1202 feet

EXERCISES

Practice

Refer to the number line below to find each measure.

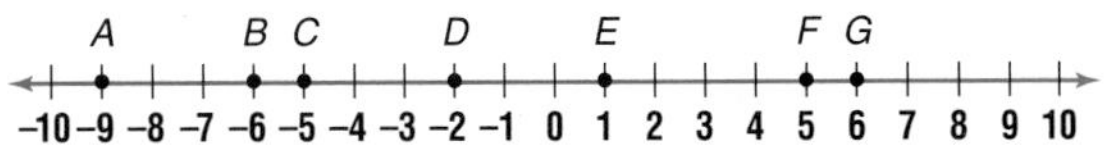

16. AE 10 17. BD 4 18. EC 6 19. EG 5 20. FC 10 21. CA 4

Given that *R* is between *S* and *T*, find each missing measure.

22. $RS = 6$, $TR = 4.5$, $TS =$ _?_ 10.5
23. $SR = 3\frac{2}{3}$, $RT = 1\frac{2}{3}$, $ST =$ _?_ $5\frac{1}{3}$
24. $ST = 15$, $SR = 6$, $RT =$ _?_ 9
25. $TS = 11.75$, $TR = 3.4$, $RS =$ _?_ 8.35

Refer to the coordinate plane at the right to find each measure. Round your answers to the nearest hundredth. 29. 1.41

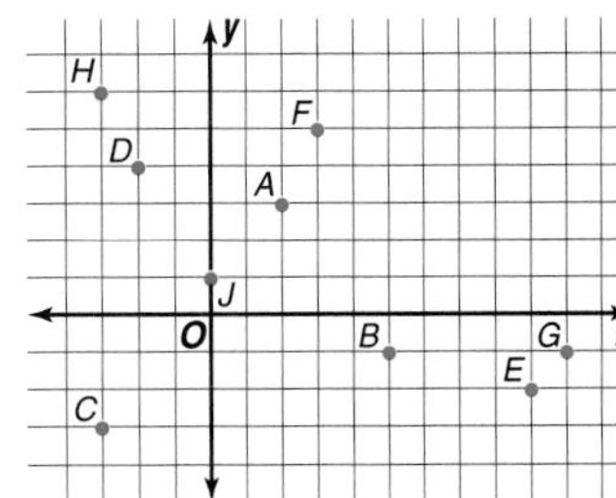

26. BG 27. HC 9 28. GH 14.76
29. EG 30. FJ 5 31. JC 5

Write a mathematical sentence to compare each pair of measures.

32. GB and GF $GB < GF$
33. FJ and JC $FJ = JC$
34. AC and AD $AC > AD$

Use the Pythagorean Theorem to find the missing length *x* in each right triangle.

35.
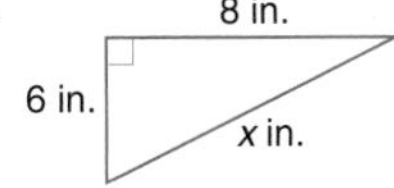

10

36.
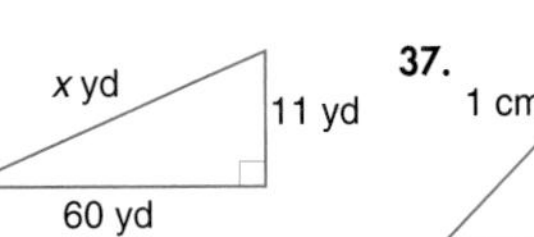

61

37. 1 cm, 1 cm, x cm
$\sqrt{2} \approx 1.414$

If *U* is between *T* and *B*, find the value of *x* and the measure of $\overline{TU}$.

38. $TU = 2x$, $UB = 3x + 1$, $TB = 21$ 4, 8
39. $TU = 4x - 1$, $UB = 2x - 1$, $TB = 5x$ 2, 7
40. $TU = 1 - x$, $UB = 4x + 17$, $TB = -3x$ −3, 4

Use $\overline{RS}$, $\overline{PQ}$, a compass, and a straightedge to construct $\overline{XY}$ for each set of measures.

P Q

R S

41–44. See margin.

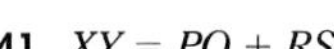

41. $XY = PQ + RS$ 42. $XY = 4(RS)$
43. $XY = PQ - RS$ 44. $XY = 3(PQ) - RS$

Assignment Guide

Core (with proof): 17–49 odd, 51–60
Core (informal): 17–49 odd, 51–60
Enriched: 16–46 even, 47–60

For **Extra Practice**, see p. 765.

The red A, B, and C flags, printed in the Teacher's Wraparound Edition, indicate the level of difficulty of the exercises.

Additional Answers

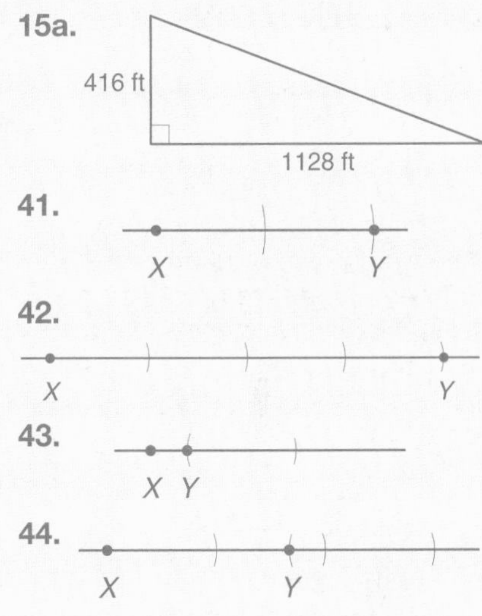

Study Guide Masters, p. 4

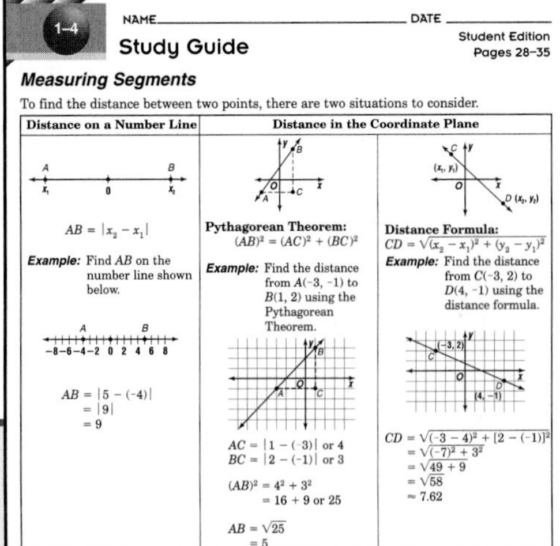

1-4 NAME ______ DATE ______

Study Guide — Student Edition Pages 28–35

Measuring Segments

To find the distance between two points, there are two situations to consider.

Distance on a Number Line	Distance in the Coordinate Plane	
$AB = \lvert x_2 - x_1 \rvert$	**Pythagorean Theorem:** $(AB)^2 = (AC)^2 + (BC)^2$	**Distance Formula:** $CD = \sqrt{(x_2 - x_1)^2 + (y_2 - y_1)^2}$
Example: Find AB on the number line shown below.	***Example:*** Find the distance from $A(-3, -1)$ to $B(1, 2)$ using the Pythagorean Theorem.	***Example:*** Find the distance from $C(-3, 2)$ to $D(4, -1)$ using the distance formula.
$AB = \lvert 5 - (-4) \rvert = \lvert 9 \rvert = 9$	$AC = \lvert 1 - (-3) \rvert$ or 4 $BC = \lvert 2 - (-1) \rvert$ or 3 $(AB)^2 = 4^2 + 3^2 = 16 + 9$ or 25 $AB = \sqrt{25} = 5$	$CD = \sqrt{(-3-4)^2 + [2-(-1)]^2} = \sqrt{(-7)^2 + 3^2} = \sqrt{49 + 9} = \sqrt{58} \approx 7.62$

Refer to the number line below to find each measure.

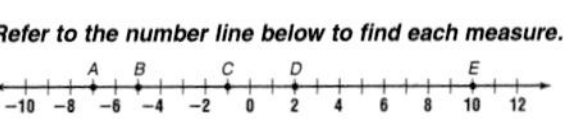

1. AC 6 2. BC 4 3. CD 3 4. AE 17
5. AB 2 6. DE 8 7. BE 15 8. CE 11

Refer to the coordinate plane at the right to find each measure. Round your measures to the nearest hundredth.

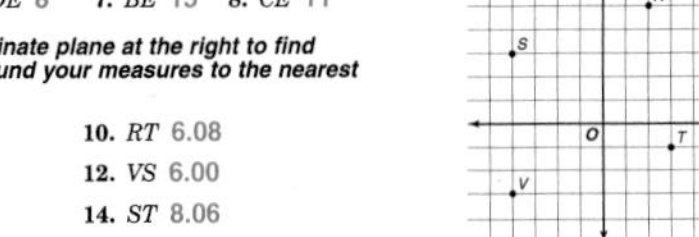

9. RS 6.32 10. RT 6.08
11. RV 10.00 12. VS 6.00
13. VT 7.28 14. ST 8.06

Using the Programming Exercises The program given in Exercise 46 is for use with a TI-82 or TI-83 graphing calculator. For other programmable calculators, have students consult their owner's manual for commands similar to those presented here.

Additional Answers

45a.

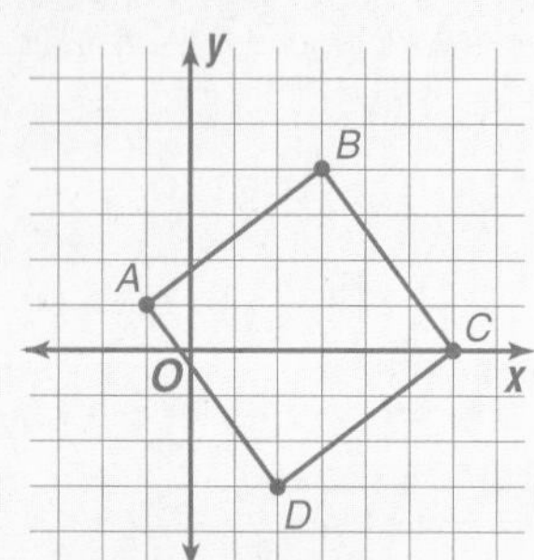

47b. The x-coordinate will be the same as the x-coordinate of F and the y-coordinate will be the same as the y-coordinate of E. Thus, $G(4, 4)$.

47c. $DG = GB$; use the distance formula to find DG and GB, which yields about 2.83.

49. The distance can be measured on the number line represented by the ruler. From 6 to 9 is 3 units.

Practice Masters, p. 4

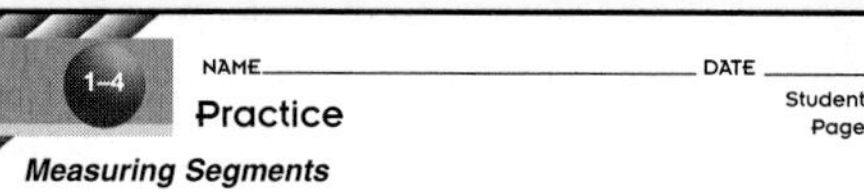

1-4 NAME______ DATE ______

Practice — Student Edition Pages 28–35

Measuring Segments

Given that B is between A and C, find each missing measure.

1. $AB = 5.3$, $BC = \underline{?}$, $AC = 6.7$ 1.4
2. $AB = 21$, $BC = 4.3$, $AC = \underline{?}$ 25.3
3. $AB = \underline{?}$, $BC = 18.9$, $AC = 23$ 4.1
4. $AB = 6\frac{3}{4}$, $BC = \underline{?}$, $AC = 10$ $3\frac{1}{4}$

If B is between A and C, find the value of x and the measure of BC.

5. $AB = 3x$, $BC = 5x$, $AC = 8$ 1, 5
6. $AB = 3(x + 7)$, $BC = 2(x - 3)$, $AC = 50$ 7, 8

Refer to the coordinate plane at the right to find each measure. Round your answers to the nearest hundredth.

7. AB 5.83
8. BD 9.43
9. AE 7.81
10. CE 10.44
11. AD 5
12. BE 9

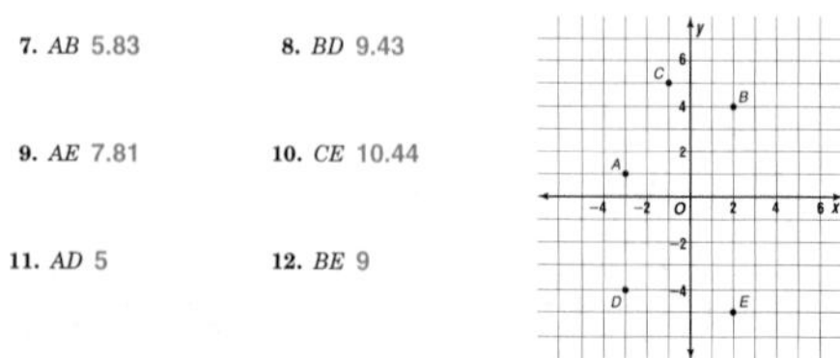

45. A rectangle has vertices $A(-1, 1)$, $B(3, 4)$, $C(6, 0)$, and $D(2, -3)$.
 a. Graph the rectangle. See margin.
 b. Find the area and perimeter of the rectangle.

45b. $A = 25$ units^2, $P = 20$ units

Programming

46. The TI-82/83 graphing calculator program at the right calculates the distance between two points given their coordinates as expressed in the distance formula.

```
PROGRAM:DISTANCE
: Disp "ENTER X1"
: Input A
: Disp "ENTER Y1"
: Input B
: Disp "Enter X2"
: Input C
: Disp "Enter Y2"
: Input D
: √((C − A)²+(D−B)²)→Y
: Disp "DISTANCE=",Y
: Stop
```

Use the program to find the distance between each pair of points.

 a. $(7, 11)$, $(-1, 5)$ 10
 b. $(-3, 5)$, $(12, -2)$ 16.55294536
 c. $(3, -2)$, $(0.67, -4)$ 3.070651397

Critical Thinking

47. In the figure, $BE = EC$ and $AF = FB$. $\overline{FG}$ and $\overline{GE}$ lie on grid lines of the coordinate plane. 47a. $E(6, 4)$, $F(4, 6)$
 a. Find the coordinates of E and F.
 b. How might you find the coordinates of G?
 c. How does DG relate to GB? Explain.

47b–47c. See margin.

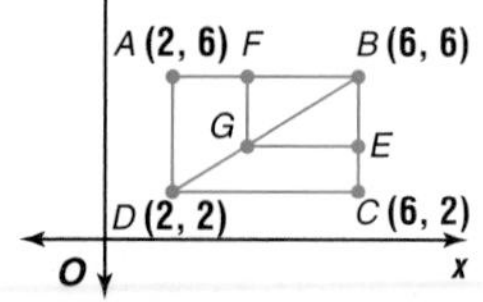

48. Draw a figure that satisfies all of the following conditions.
 - Points A, B, C, D, and E are collinear. See Solutions Manual.
 - Point A lies between points D and E.
 - Point C is next to point A, and $BD = DC$.

Applications and Problem Solving

49. See margin.

49. **Comics** Study the *Peanuts* comic strip shown below. Explain how Sally knew that the length of Snoopy's mouth was "lip to lip, three inches."

© 1980 United Feature Syndicate, Inc.

Pack	Durability	Comfort
Caribou Super Dealer Briefcase	5	2
Eastpak Day Pak' R	3	4
Eastpak X-Country Pak' R II	2	2
Esprit Big Bag	1	1
JanSport Collegian	3	3
JanSport Santa Fe	5	3
J.C. Penney Sport	2	3
Lands' End Bookpack	2	5
L.L. Bean Book Pack	3	4
Nike Elite Gear Bag	2	3

50. **Consumer Awareness** *Zillions* magazine rated backpacks based on durability and comfort. Packs were given a score of 1 to 5 stars for each category. One way to compare the rating to find the best backpack is to graph a point for each pack using the durability score as the x-coordinate and the comfort score as the y-coordinate. Suppose the point farthest from the origin represents the best-rated pack. Of the packs listed in the table at the left, which one was rated the best? See margin.

Extension

Reasoning Points D, E, and F are collinear. If $DE = x$, $EF = 4x - 6$, and $DF = 3x - 6$, which point is between the other two points? D is between E and F.

Additional Answer

50. JanSport Santa Fe

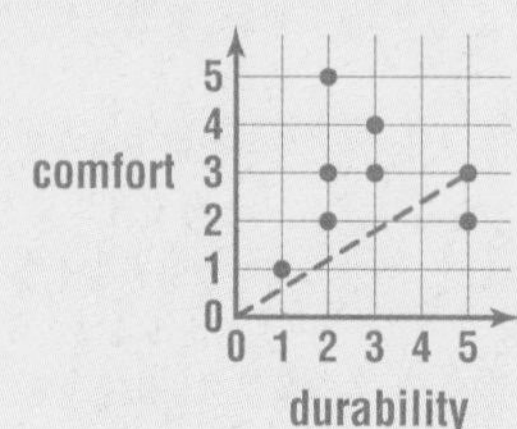

Mixed Review

51. Find the length of a rectangle whose area is 15 square centimeters and whose width is 2.5 centimeters. (Lesson 1–3) **6 cm**

52. Find the maximum area of a rectangle whose perimeter is 8 miles. (Lesson 1–3) **4 mi^2**

53. Draw a plane $\mathcal{B}$ containing line p. Points R, S, and T lie in $\mathcal{B}$, but only points R and S lie on p. (Lesson 1–2) **See margin.**

54. **List the Possibilities** Points A, B, C, D, and E lie on a circle. List all of the lines that contain exactly two of these five points. (Lesson 1–2)
$\overleftrightarrow{AE}$, $\overleftrightarrow{AD}$, $\overleftrightarrow{AC}$, $\overleftrightarrow{AB}$, $\overleftrightarrow{BC}$, $\overleftrightarrow{BD}$, $\overleftrightarrow{BE}$, $\overleftrightarrow{CD}$, $\overleftrightarrow{CE}$, $\overleftrightarrow{DE}$

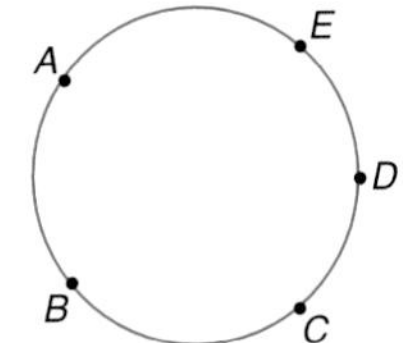

55. A five-sided figure is composed of segments that meet at $F(-2, 2)$, $G(0, 4)$, $H(3, 0)$, $I(2, -6)$, and $J(-3, -3)$. Graph these points and draw the figure. (Lesson 1–1) **See Solutions Manual.**

56. Name the quadrant in which each named point on the coordinate plane at the right lies. (Lesson 1–1)

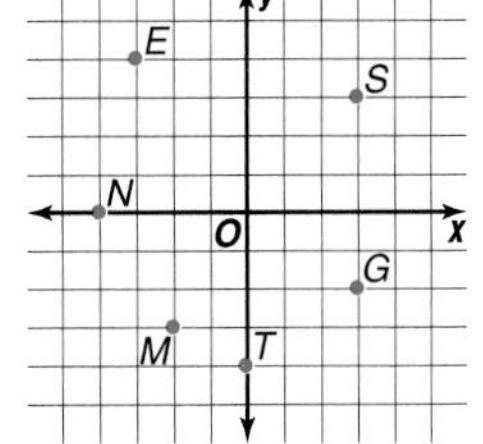

56. *S*: quadrant I, *E*: quadrant II, *M*: quadrant III, *G*: quadrant IV; points *N* and *T* lie on the axes and not in a quadrant.

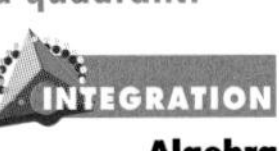

Algebra

Multiply.

57. $3x(x + 6)$ $3x^2 + 18x$

58. $(x + 2)(x + 2)$ $x^2 + 4x + 4$

59. $(x + 3)(x - 4)$ $x^2 - x - 12$

60. $(x - 4)(x + 4)$ $x^2 - 16$

2. The fine lines and printing would be very hard to duplicate. The result could be loss of detail, smudging, and smearing.

Mathematics and SOCIETY

Money Lines and Angles

The excerpt below appeared in an article in *Science News* on January 27, 1996.

IN 1994, THE SECRET SERVICE SEIZED $45.7 million in counterfeit bills before they entered U.S. circulation—though forgers sneaked $25.3 million in fakes into the economy.... The potential of desktop counterfeiting has guided the design, materials, and production decisions of the next generation of U.S. currency.... The new currency sports several security-enhancing features.... An enlarged, off-center portrait...increases recognition, reduces wear, and opens space for a watermark; extra detail stymies duplication.... Concentric, fine lines behind the portrait...frustrate replication.... A watermark...becomes visible only in transmitted light.... Numerals printed in color-shifting ink appear green when viewed straight on and black when seen from an angle.... The Secret Service will monitor the new features to see how well they deter counterfeiters and increase the capture of fakes. ■

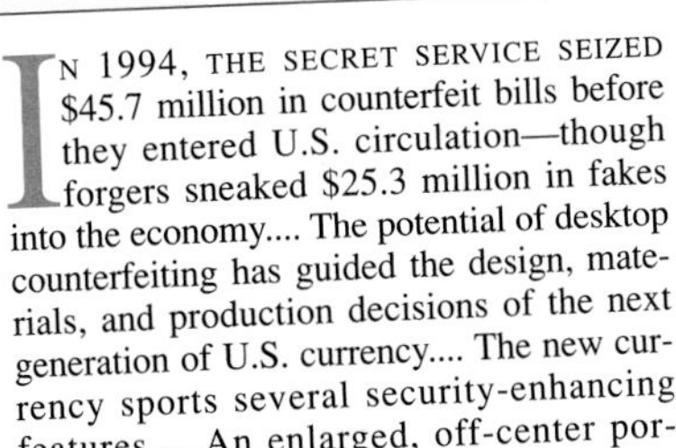

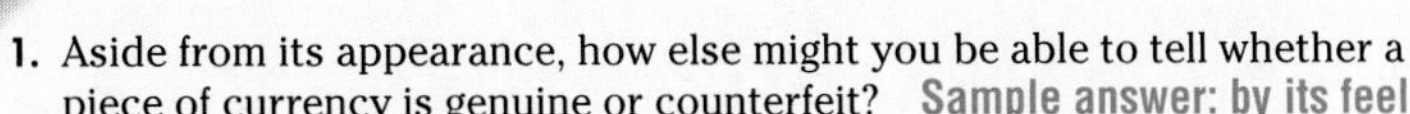

1. Aside from its appearance, how else might you be able to tell whether a piece of currency is genuine or counterfeit? **Sample answer: by its feel**
2. In addition to the fine, concentric lines, the new currency has tiny, microprinted words that cannot be read with the naked eye. What do you think would be the result if you tried to counterfeit these bills by photocopying them?
3. If you were designing currency, what other types of features could you use so it would be harder to counterfeit? **See margin.**

Additional Answer

53.

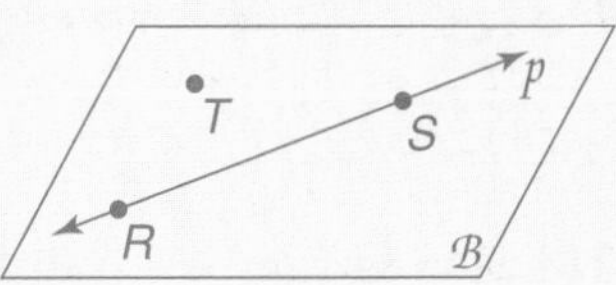

4 ASSESS

Closing Activity

Modeling Have students measure the length of an object such as a pen cap or an eraser. Have them draw a segment using a ruler and then construct another segment of the same length using a compass.

Chapter 1 Quiz B (Lessons 1-3 and 1-4) is available in the *Assessment and Evaluation Masters*, p. 16.

Mid-Chapter Test (Lessons 1-1 through 1-4) is available in the *Assessment and Evaluation Masters*, p. 15.

Enrichment Masters, p. 4

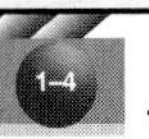

NAME_______ DATE_______

1-4 **Enrichment** Student Edition Pages 28–35

Lengths on a Grid

Evenly-spaced horizontal and vertical lines form a grid.

You can easily find segment lengths on a grid if the endpoints are grid-line intersections. For horizontal or vertical segments, simply count squares. For diagonal segments, use the Pythagorean Theorem (proven in Chapter 8). This theorem states that in any right triangle, if the length of the longest side (the side opposite the right angle) is c and the two shorter sides have lengths a and b, then $c^2 = a^2 + b^2$.

Example: Find the measure of $\overline{EF}$ on the grid at the right. Locate a right triangle with $\overline{EF}$ as its longest side.

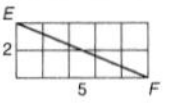

$EF = \sqrt{2^2 + 5^2} = \sqrt{29} \approx 5.4$ units

Find each measure to the nearest tenth of a unit.

1. $\overline{IJ}$ 3
2. $\overline{MN}$ 7
3. $\overline{RS}$ 4.2
4. $\overline{QS}$ 5.8
5. $\overline{IK}$ 7.6
6. $\overline{JK}$ 5
7. $\overline{LM}$ 4.1
8. $\overline{LN}$ 7.2

Use the grid above. Find the perimeter of each triangle to the nearest tenth of a unit.

9. $\triangle ABC$ 20.2
10. $\triangle QRS$ 18
11. $\triangle DEF$ 16.6
12. $\triangle LMN$ 18.3

Answers shown are found by rounding segment lengths before adding.

13. Of all the segments shown on the grid, which is longest? What is its length? $BC = 8.1$
14. On the grid, 1 unit = 0.5 cm. How can the answers above be used to find the measures in centimeters? divide by 2 or multiply by 0.5
15. Use your answer from exercise 8 to calculate the length of segment LN in centimeters. Check by measuring with a centimeter ruler. 3.6 cm
16. Use a centimeter ruler to find the perimeter of triangle IJK to the nearest tenth of a centimeter. 7.8 cm

Mathematics and SOCIETY

Historians believe that the development of paper money began in China in the 600s A.D. The Italian trader Marco Polo brought news of the use of paper money to Europe from China in the 1200s, but paper money was not accepted in European countries until the 1800s.

Answer for Mathematics and Society

3. Sample answers: use layers of material sandwiched together; blend particles of other materials into the paper; use special inks visible only in ultraviolet or infrared light; use holographic images; use encrypted codes to be read by a scanner.

1-5 LESSON NOTES

NCTM Standards: 1–5, 7, 8, 14

Instructional Resources

- Study Guide Master 1-5
- Practice Master 1-5
- Enrichment Master 1-5

Transparency 1-5A contains the 5-Minute Check for this lesson; **Transparency 1-5B** contains a teaching aid for this lesson.

Recommended Pacing	
Standard Pacing	Day 6 of 13
Honors Pacing	Day 6 of 13
Block Scheduling*	Day 5 of 8

*For more information on pacing and possible lesson plans, refer to the *Block Scheduling Booklet*.

1 FOCUS

5-Minute Check
(over Lesson 1-4)

1. Name two possible arrangements for G, H, and I on a segment if $GH + GI = HI$. **H, G, I or I, G, H**

Use the figure below to find each measure.

2. AC **4**
3. DE **9**
4. If M is between L and N, $LN = 3x - 1$, $LM = 4$, and $MN = x - 1$, find MN. **1**
5. What is the length of $\overline{ST}$ for $S(-1, -1)$ and $T(4, 6)$? **8.6**

1–5 Midpoints and Segment Congruence

What YOU'LL LEARN

- To find the midpoint of a segment, and
- to complete proofs involving segment theorems.

Why IT'S IMPORTANT

Finding the midpoint is often used to connect algebra to geometry.

In this lesson, you will study several methods for finding the midpoint of a line segment, including paper folding.

MODELING MATHEMATICS — **Locating the Midpoint of a Segment**

Materials:  patty paper  ruler

You can locate the midpoint of any segment by using paper folding.

- Draw points A and B anywhere on a sheet of patty paper.
- Connect the points to form segment AB.

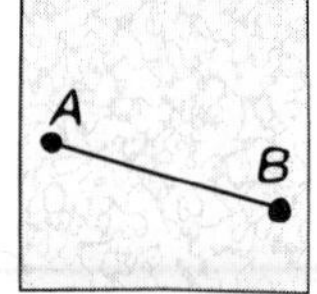

- Fold the paper so that the endpoints A and B lie on top of each other. Pinch the paper to make a crease on the segment.

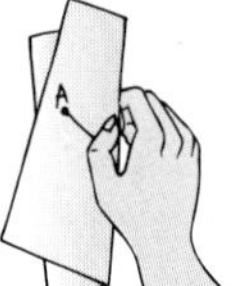

- Open the paper and label the point where the crease intersects $\overline{AB}$ as C. Point C is the midpoint of $\overline{AB}$.

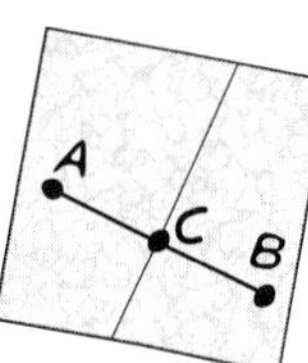

Your Turn

a. Use a ruler to measure $\overline{AC}$ and $\overline{CB}$. **AC should equal CB.**

b. Repeat the activity with two other segments. **See students' work.**

c. Write a sentence to summarize your observations.

c. Sample answer: The midpoint separates a segment into two segments of equal length.

In the Modeling Mathematics activity above, you discovered that the midpoint of a segment separates the segment into two segments that have equal measures.

Definition of Midpoint	The midpoint M of $\overline{PQ}$ is the point between P and Q such that $PM = MQ$.

Patty paper is the white paper squares used to separate hamburgers or cheese slices in the meat and dairy sections of your supermarket. It is available at restaurant supply warehouses.

You may remember from algebra that if a segment is graphed on a number line or coordinate plane, there is a formula for finding the coordinate(s) of the midpoint.

Midpoint Formulas	1. **On a number line, the coordinate of the midpoint of a segment whose endpoints have coordinates *a* and *b* is $\frac{a+b}{2}$.** 2. **In a coordinate plane, the coordinates of the midpoint of a segment whose endpoints have coordinates (x_1, y_1) and (x_2, y_2) are $\left(\frac{x_1+x_2}{2}, \frac{y_1+y_2}{2}\right)$.**

Example  **a. Use the number line to find the coordinate of the midpoint of $\overline{FG}$.**

The coordinate of *F* is −2, and the coordinate of *G* is 6.

Use these coordinates with the midpoint formula for a number line.

$$\frac{a+b}{2} = \frac{-2+6}{2}$$
$$= \frac{4}{2} \text{ or } 2$$

The coordinate of the midpoint of $\overline{FG}$ is 2.

b. Find the coordinates of *Q*, the midpoint of $\overline{RS}$, if the endpoints of $\overline{RS}$ are *R*(−3, −4) and *S*(5, 7).

Graph points *R* and *S* and connect them.

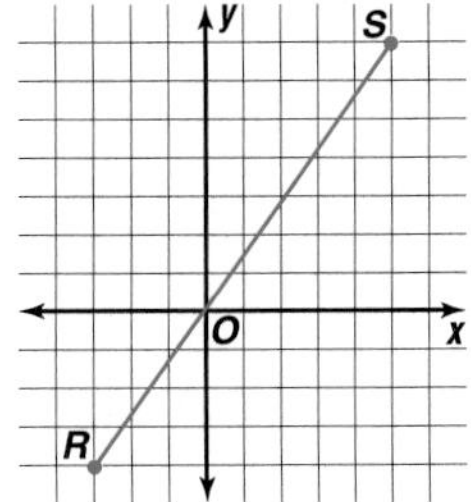

Use the midpoint formula to find the coordinates of *Q*.

$$\left(\frac{x_1+x_2}{2}, \frac{y_1+y_2}{2}\right) = \left(\frac{-3+5}{2}, \frac{-4+7}{2}\right) \quad (x_1, y_1) = (-3, -4) \quad (x_2, y_2) = (5, 7)$$
$$= \left(1, \frac{3}{2}\right)$$

The coordinates of midpoint *Q* are $\left(1, \frac{3}{2}\right)$.

You can use the midpoint formula to help you find one of the endpoints of a segment if you know the coordinates of one of its endpoints and its midpoint.

Motivating the Lesson

Hands-On Activity Have each student find the center of a piece of paper without measuring or marking with a pen or pencil. **Fold the piece of paper in half lengthwise and widthwise. The center is the intersection of the folds.**
Explain that, in mathematics, the midpoint of a segment is located by finding the halfway point of the segment.

2 TEACH

In-Class Example

For Example 1

a. If the coordinate of *H* is −5 and the coordinate of *J* is 4, what is the coordinate of the midpoint of $\overline{HJ}$? $-\frac{1}{2}$

b. Find the coordinates of the midpoint of $\overline{VW}$ for *V*(3, −6) and *W*(7, 2). **(5, −2)**

Teaching Tip It may help students to remember either midpoint formula if they remember that each coordinate is the average of the two values they have.

Alternative Learning Styles

Visual Bring a sewing pattern to class and show students the places on the pattern that are marked with arrows. Explain that when you sew the pieces together, you match these places to make sure the garment fits correctly. In geometry, hash marks are similar in that they mark segments of figures and objects that have the same measure.

In-Class Example

For Example 2

a. The midpoint of $\overline{RQ}$ is $P(4, -1)$. What are the coordinates of R if Q is at $(3, -2)$? **(5, 0)**

b. U is the midpoint of $\overline{XY}$. If $XY = 16x - 6$ and $UY = 4x + 9$, find the value of x and the measure of $\overline{XY}$. **3, 42**

Teaching Tip When discussing bisectors, draw a segment on the chalkboard or overhead projector and bisect it. Draw several bisectors to show that there can be an infinite number of bisectors and that each must pass through the midpoint.

Example 2

Algebra

a. Find the coordinates of point Q if $L(4, -6)$ is the midpoint of $\overline{NQ}$ and the coordinates of N are $(8, -9)$.

Graph L and N and connect them. Q will be located to the left of L. Use the midpoint formula. Let (x_1, y_1) be $(8, -9)$ and (x_2, y_2) be the coordinates of Q.

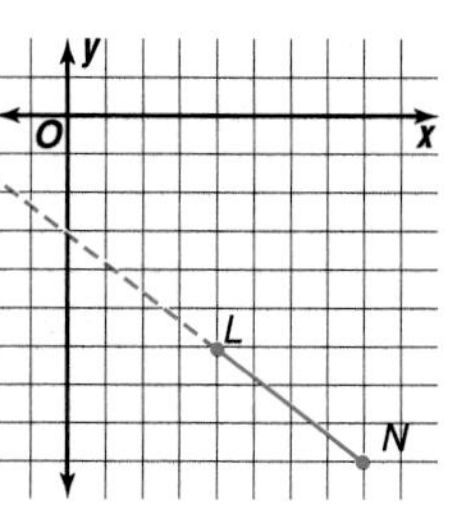

$$\left(\frac{x_1 + x_2}{2}, \frac{y_1 + y_2}{2}\right) = (4, -6)$$

$$\frac{x_1 + x_2}{2} = 4 \qquad \frac{y_1 + y_2}{2} = -6$$

$$\frac{8 + x_2}{2} = 4 \qquad \frac{-9 + y_2}{2} = -6$$

$$8 + x_2 = 8 \qquad -9 + y_2 = -12$$

$$x_2 = 0 \qquad y_2 = -3$$

The coordinates of Q are $(0, -3)$.

b. If Y is the midpoint of $\overline{XZ}$, $XY = 2a + 11$, and $YZ = 4a - 5$, find the value of a and the measure of $\overline{XZ}$.

Y is the midpoint of $\overline{XZ}$, so $XY = YZ$. Write an equation and solve for a.

$XY = YZ$

$2a + 11 = 4a - 5$

$11 = 2a - 5$ *Subtract 2a from each side.*

$16 = 2a$ *Add 5 to each side.*

$8 = a$ *Divide each side by 2.*

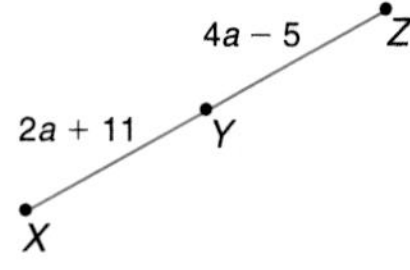

Now use the value of a to find XZ.

$XZ = XY + YZ$ *Segment Addition Postulate*

$= (2a + 11) + (4a - 5)$ *Substitution*

$= 6a + 6$

$= 6(8) + 6$ or 54

Any segment, line, or plane that intersects a segment at its midpoint is called a **segment bisector**. In the figure at the right, M is the midpoint of $\overline{PQ}$. Thus, point M, $\overline{TM}$, $\overleftrightarrow{RM}$, and plane $\mathcal{N}$ are all bisectors of $\overline{PQ}$ and are said to *bisect* $\overline{PQ}$.

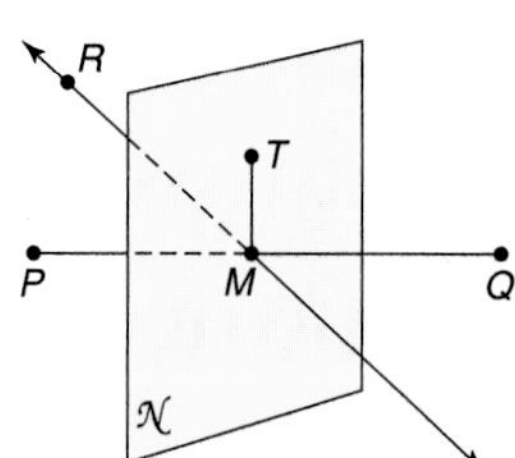

In the Modeling Mathematics activity at the beginning of this lesson, you used paper folding to find the midpoint of a segment. In Exercise 44, you will use a TI-92 calculator to construct a segment and find its midpoint. You can also use a compass and straightedge to find the midpoint of any segment, and thus, bisect the segment.

CONSTRUCTION

Bisecting a Segment

Use a compass and a straightedge to bisect a segment.

1. Draw a segment and name it $\overline{XY}$.
2. Place the compass at point X. Adjust the compass so that its width is greater than $\frac{1}{2}XY$.
3. Draw arcs above and below $\overline{XY}$.
4. Using the same compass setting, place the compass at point Y and draw arcs above and below $\overline{XY}$ so that they intersect the two arcs previously drawn. Label the points of the intersection of the arcs as P and Q.

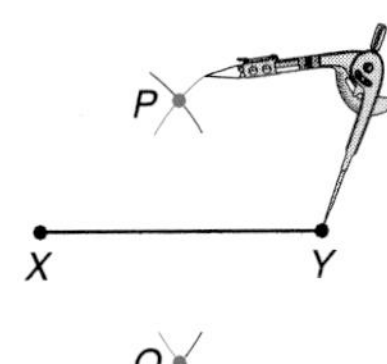

5. Use a straightedge to draw $\overleftrightarrow{PQ}$. Label the point where it intersects $\overline{XY}$ as M.

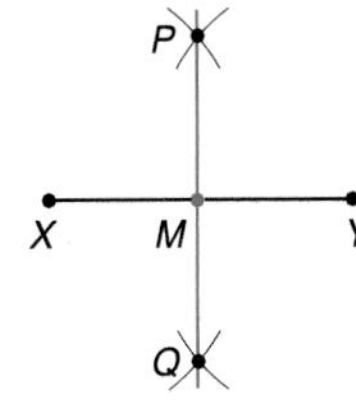

Conclusion: Point M is the midpoint of $\overline{XY}$, and $\overleftrightarrow{PQ}$ is a bisector of $\overline{XY}$. Also $XM = MY = \frac{1}{2}XY$.

In the study of geometry, definitions, postulates, and undefined terms are accepted as true without verification or proof. These three types of statements can be used to prove that other statements called **theorems** are true. The following theorem states that a congruence relationship exists between the segments formed by the midpoint of a segment.

Theorem 1-1 Midpoint Theorem	**If M is the midpoint of $\overline{AB}$, then $\overline{AM} \cong \overline{MB}$.**

You can use definitions, postulates, and, previously proven theorems to justify a statement.

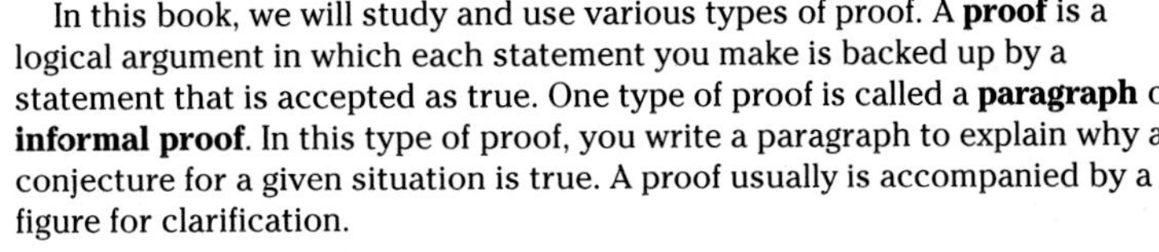

In this book, we will study and use various types of proof. A **proof** is a logical argument in which each statement you make is backed up by a statement that is accepted as true. One type of proof is called a **paragraph** or **informal proof**. In this type of proof, you write a paragraph to explain why a conjecture for a given situation is true. A proof usually is accompanied by a figure for clarification.

Proof of Theorem 1-1

Given that M is the midpoint of $\overline{AB}$, write a paragraph proof to show that $\overline{AM} \cong \overline{MB}$.

From the definition of the midpoint of a segment, we know that $AM = MB$. That means that $\overline{AM}$ and $\overline{MB}$ have the same measures. By the definition of congruence, if $\overline{AM}$ and $\overline{MB}$ have the same measure, they are congruent segments. Thus, $\overline{AM} \cong \overline{MB}$.

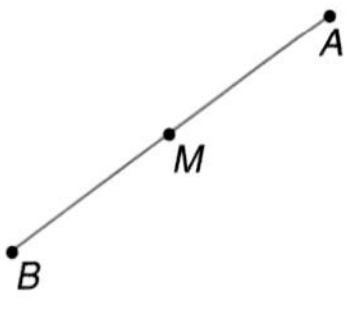

Teaching Tip As students become better at constructions, they may discover that parts of the construction may be omitted or modified. For example, in the construction of the bisector of a segment, they need not draw segment PQ; they could simply place a point on $\overline{XY}$ where their straightedge intersects it.

Proof Pointer

This is the first introduction of proof to the students. Other forms of proof, such as two-column proof and flow proof, will be introduced in later chapters.

Students are often intimidated by the word *proof*. Emphasize that a proof is just a verbal argument for saying something is true. You must have facts to back up what you say.

It may be helpful to have students work in pairs or cooperative groups in writing their first proofs.

In-Class Example

For Example 3

In the figure below, B is the midpoint of $\overline{AC}$, D is the midpoint of $\overline{AB}$, and E is the midpoint of $\overline{BC}$. Show that $AC = 4AD$.

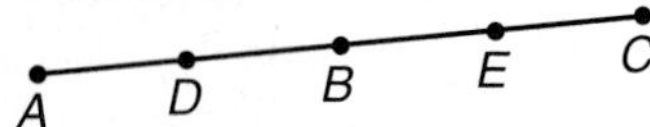

Since B is the midpoint of $\overline{AC}$, $AB = BC$. Suppose $AB = BC = x$. Then since D and E are the midpoints of $\overline{AB}$ and $\overline{BC}$, respectively, $AD = DE = \frac{1}{2}x$, and $BE = EC = \frac{1}{2}x$. Thus, $AD = DE = BE = EC$. The Segment Addition Postulate says that $AC = AD + DB + BE + EC$. Substitution gives us $AC = AD + AD + AD + AD$ or $AC = 4AD$.

3 PRACTICE/APPLY

Check for Understanding

Exercises 1–14 are designed to help you assess your students' understanding through reading, writing, speaking, and modeling. You should work through Exercises 1–5 with your students and then monitor their work on Exercises 6–14.

Additional Answer

1. You can (1) use geometric software to draw a segment and then locate its midpoint automatically; (2) draw a segment on paper and fold the paper so that the endpoints match so that the intersection of the fold and the segment is the midpoint; and (3) use compass and straightedge to construct the segment that bisects the segment. If the segment is graphed on a number line or coordinate plane, there are formulas you can use to find the coordinate(s) of the midpoint.

A paragraph proof can be used to show that other statements are also true.

Example  **In the figure at the right, T is the midpoint of $\overline{SE}$, and E is the midpoint of $\overline{TP}$. Show that $\overline{ST} \cong \overline{EP}$.**

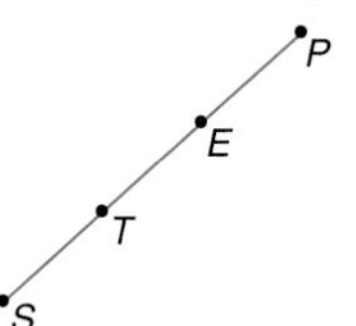

Since T is the midpoint of $\overline{SE}$, then $ST = TE$. Since E is the midpoint of $\overline{TP}$, then $TE = EP$. Substitute EP for TE in the statement $ST = TE$ to get $ST = EP$. Then, by the definition of congruence, $\overline{ST} \cong \overline{EP}$.

CHECK FOR UNDERSTANDING

Communicating Mathematics

Study the lesson. Then complete the following.

1. **Describe** three ways to locate the midpoint of a segment. See margin.
2. **You Decide** Jadine says that if $DE = EF$, then E is the midpoint of $\overline{DF}$. Misha says this is only true in a special case. Who is correct? Include drawings to demonstrate your argument. See margin.
3. **Write** a convincing argument to show that the number of bisectors a segment has is greater than the number of its midpoints. See margin.
4. **Explain** how you would find the value of x if R is the midpoint of $\overline{QS}$, $QR = 8 - x$, and $RS = 5x - 10$. See margin.

5. **Draw** a large rectangle on your paper. Determine how you could find the midpoint of each side with a minimum number of paper folds. Fold the rectangle matching the horizontal sides exactly and fold again, matching the vertical sides exactly.

Guided Practice

Use the number line below for Exercises 6–8.

6. Find the coordinate of the midpoint of $\overline{PQ}$. -0.5

Determine whether each statement is *true* or *false*. If false, state why.

7. Q is the midpoint of $\overline{SR}$ true
8. $PS = QR$ false; $PS = 2$, $QR = 3$, $2 \neq 3$

Use the coordinate plane at the right for Exercises 9–11. Find the coordinates of the midpoint of each segment.

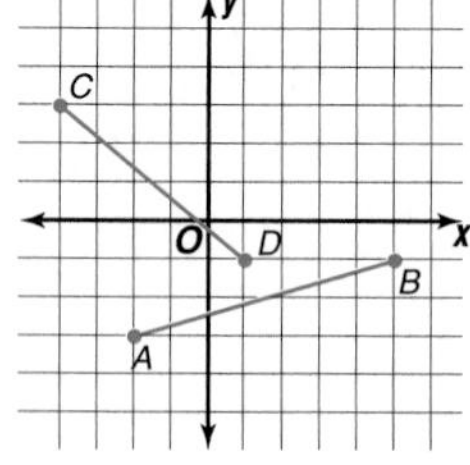

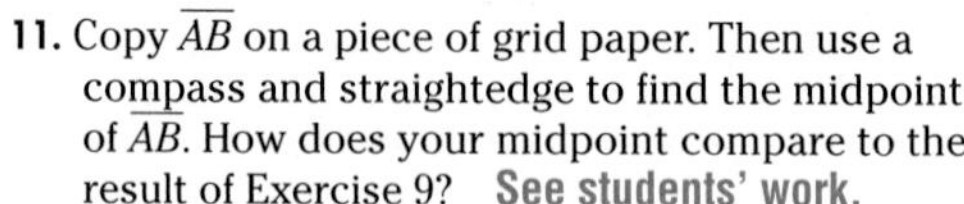

9. $\overline{AB}$ (1.5, −2)
10. $\overline{CD}$ (−1.5, 1)
11. Copy $\overline{AB}$ on a piece of grid paper. Then use a compass and straightedge to find the midpoint of $\overline{AB}$. How does your midpoint compare to the result of Exercise 9? See students' work.

Additional Answers

2. Jadine is correct only if D, E, and F are collinear. Angle DEF is a counterexample.

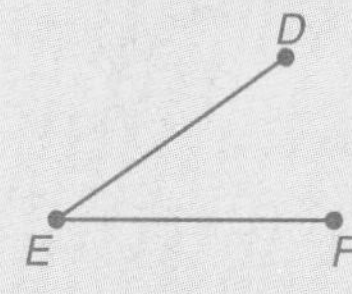

3. Every segment has only one midpoint. However, there are an infinite number of lines, segments, and planes that pass through the point that is the midpoint. So, each segment has an infinite number of bisectors.

Algebra

12. If $R(2, 5)$ is the midpoint of $\overline{ST}$ and the coordinates of T are $(-1, 8)$, find the coordinates of S. **(5, 2)**

13. In the figure at the right, $\overline{AC}$ and $\overline{BD}$ bisect each other at E. For each of the following, find the value of x and the measure of the indicated segment.
    a. $DE = 2x + 5$; $EB = 13 - 2x$; $\overline{DB}$ **2; 18**
    b. $AC = x + 3$, $EC = 3x - 1$; $\overline{EA}$ **1; 2**

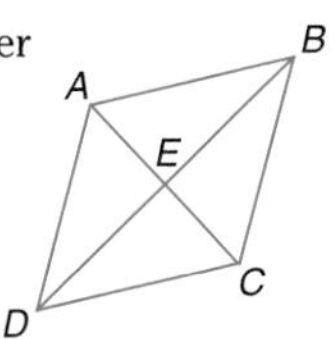

14. **Logical Reasoning** Refer to the figure for Exercise 13. Suppose that $\overline{DB}$ is extended to point F so that B is the midpoint of $\overline{DF}$. Write a paragraph proof to show that ED is one-fourth the length of $\overline{DF}$. **See margin.**

EXERCISES

Practice

Use the figure at the right for Exercises 15–23.

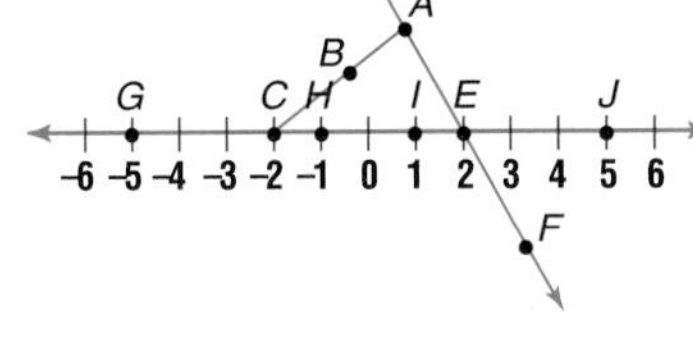

Find the coordinates of the midpoint of each segment.

15. $\overline{GI}$ **−2**
16. $\overline{HJ}$ **2**
17. $\overline{GC}$ **−3.5**

Determine whether each statement is *true* or *false*. If false, state why. **19. False; sample answer: E is the midpoint of $\overline{HJ}$.**

18. $\overline{GI} \cong \overline{JH}$ **true**
19. E is the midpoint of $\overline{CJ}$.
20. $\overline{AC}$ bisects $\overline{GE}$.
21. $\overline{EJ} \cong \overline{CI}$ **true**
22. $\overleftrightarrow{AE}$ bisects $\overline{HJ}$. **true**
23. $CH = \frac{1}{2}HI$ **true**

20. False; sample answer: $\overline{AC}$ bisects $\overline{GI}$.

W, R, and S are points on a number line, and W is the midpoint of $\overline{RS}$. For each pair of coordinates given, find the coordinate of the third point.

24. $R = 4, S = -6$ **$W = -1$**
25. $R = 12, W = -3$ **$S = -18$**
26. $W = -4, S = 2$ **$R = -10$**

Y is the midpoint of $\overline{XZ}$. For each pair of points given, find the coordinates of the third point.

27. $X(5, 5), Z(-1, 5)$
28. $X(-4, 3), Y(-1, 5)$
29. $Z(2, 8), Y(-2, 2)$
30. $Z(-3, 6), Y(0, 5.5)$
31. $X\left(\frac{2}{3}, -5\right), Y\left(\frac{5}{3}, 3\right)$
32. $X(2, -10), Y\left(\frac{1}{2}, -6\frac{1}{2}\right)$

27. $Y(2, 5)$
28. $Z(2, 7)$
29. $X(-6, -4)$
30. $X(3, 5)$
31. $Z\left(\frac{8}{3}, 11\right)$
32. $Z(-1, -3)$

In the figure below, $\overline{EC}$ bisects $\overline{AD}$ at C, and $\overline{EF}$ bisects $\overline{AC}$ at B. For each of the following, find the value of x and the measure of the indicated segment.

33. $AB = 3x + 6$, $BC = 2x + 14$; $\overline{AC}$ **8; 60**
34. $AC = 5x - 8$, $CD = 16 - 3x$; $\overline{AD}$ **3; 14**
35. $AD = 6x - 4$, $AC = 4x - 3$; $\overline{CD}$ **1; 1**
36. $AC = 3x - 1$, $BC = 12 - x$; $\overline{AB}$ **5; 7**
37. $AD = 5x + 2$, $BC = 7 - 2x$; $\overline{CD}$ **2; 6**
38. $AB = 4x + 17$, $CD = 25 + 5x$; $\overline{BC}$ **−3; 5**

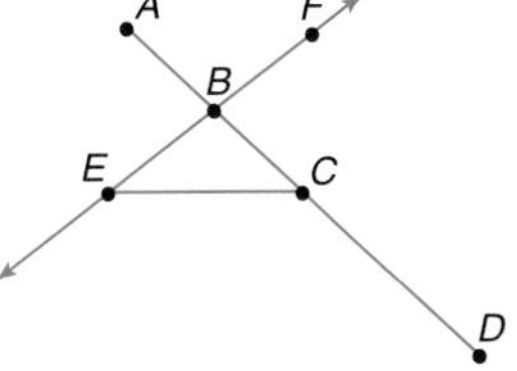

Assignment Guide

Core (with proof): 15–47 odd, 48–56
Core (informal): 15–41 odd, 45, 47, 48–56
Enriched: 16–44 even, 45–56

For **Extra Practice**, see p. 765.

The red A, B, and C flags, printed in the Teacher's Wraparound Edition, indicate the level of difficulty of the exercises.

Additional Answers

4. Since R is the midpoint of $\overline{QS}$, then $QR = RS$. Substitute $8 - x$ for QR and $5x - 10$ for RS and solve the equation. The solution is $x = 3$.
14. By the definition of midpoint, if B is the midpoint of $\overline{DF}$, then $DB = \frac{1}{2}DF$. By the same reason, since E is the midpoint of $\overline{DB}$, then $DE = \frac{1}{2}DB$. Using substitution, $DE = \frac{1}{2}\left(\frac{1}{2}DF\right)$ or $\frac{1}{4}DF$.

Reteaching

Using Manipulatives Have students graph a line segment on a coordinate plane. Then lay a piece of string along the segment to represent the length of the segment. Fold the string in half and mark the point of the fold. Next, lay the string along the line segment. The mark will indicate the midpoint of the segment. Use the midpoint formula to verify the result.

Study Guide Masters, p. 5

NAME__________ DATE ________

Study Guide

Student Edition Pages 36–43

Midpoints and Segment Congruence

There are two situations in which you may need to find the midpoint of a segment.

Midpoint on a Number Line	Midpoint in the Coordinate Plane
The coordinate of the midpoint of a segment whose endpoints have coordinates a and b is $\frac{a+b}{2}$.	The coordinates of the midpoint of a segment whose endpoints have coordinates (x_1, y_1) and (x_2, y_2) are $\left(\frac{x_1+x_2}{2}, \frac{y_1+y_2}{2}\right)$.
Example: R, S on number line −6 −4 −2 0 2 4 6 8 10. The coordinate of the midpoint of $\overline{RS}$ is $\frac{-3+9}{2}$ or 3.	***Example:*** 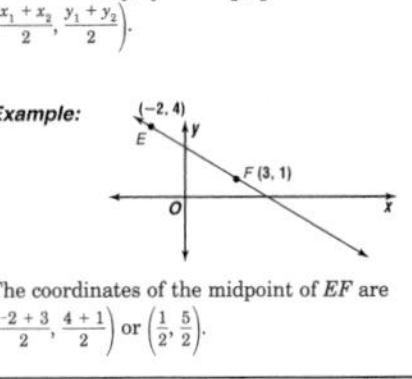 The coordinates of the midpoint of EF are $\left(\frac{-2+3}{2}, \frac{4+1}{2}\right)$ or $\left(\frac{1}{2}, \frac{5}{2}\right)$.

Use the number line below to find the coordinates of the midpoint of each segment.

A B C G D E F
−12 −10 −8 −6 −4 −2 0 2 4 6 8 10 12

1. $\overline{AB}$ $-\frac{19}{2}$
2. $\overline{BC}$ $-\frac{9}{2}$
3. $\overline{CE}$ 4
4. $\overline{DE}$ 8
5. $\overline{AE}$ −1
6. $\overline{FC}$ 5
7. $\overline{GE}$ $\frac{13}{2}$
8. $\overline{BF}$ $\frac{5}{2}$

Refer to the coordinate plane at the right to find the coordinates of the midpoint of each segment.

9. $\overline{JK}$ (−1, 4)
10. $\overline{KL}$ (−3, 3)
11. $\overline{LM}$ $\left(-3, -\frac{1}{2}\right)$
12. $\overline{MN}$ $\left(-\frac{1}{2}, -2\right)$
13. $\overline{NT}$ $\left(\frac{7}{2}, -\frac{5}{2}\right)$
14. $\overline{MT}$ $\left(0, -\frac{7}{2}\right)$

39. Draw a line segment about 5 inches long. Use a compass, a straightedge, and the segment bisector construction to find a segment that is $\frac{1}{8}$ the length of the original segment. **See margin.**

40. Point C lies on $\overline{AB}$ such that $AC = \frac{1}{4}AB$. If the endpoints of $\overline{AB}$ are $A(8, 12)$ and $B(-4, 0)$, find the coordinates of C. **(5, 9)**

41. Suppose $\overline{PQ}$ has endpoints $P(2, 3)$ and $Q(8, -9)$. Find the coordinates of R and S so that R lies between P and S and $\overline{PR} \cong \overline{RS} \cong \overline{SQ}$. **$R(4, -1)$, $S(6, -5)$**

42. In the figure at the right, E is the midpoint of $\overline{FG}$ and $\overline{HI}$, and $\overline{FG} \cong \overline{HI}$. Write a paragraph proof to show that $FE = HE$. **See margin.**

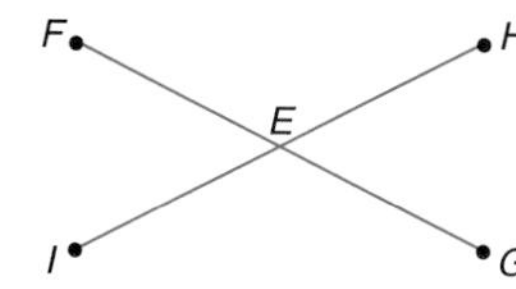

43. The notation $\overline{STEP}$ means that S, T, E, and P are collinear points in that order. Write a paragraph proof to show that $SP = ST + TE + EP$ for $\overline{STEP}$. Be sure to include a drawing with your proof. **See margin.**

Cabri Geometry

44. Use a TI-92 calculator to construct a segment AB. **44a–c. See students' work.**
    a. Locate midpoint M on $\overline{AB}$ by selecting 3:Midpoint from the F4 menu.
    b. Use the 1:Distance and Length command from the F6 menu to measure the distance from A to M and from M to B. What do you find?
    c. Use the hand key to move the endpoints of the segment to new locations. What do you notice about the measures of the two segments formed by the midpoint?

Critical Thinking

45. Draw a large triangle similar in shape to the one shown at the right. **45a–c. See students' work.**
    a. Find the midpoints of each side of the triangle. Label them D, E, and F.
    b. Connect the points to form triangle DEF.
    c. Measure the sides of triangles DEF and ABC.
    d. How do the perimeters of the two triangles compare?
    e. Make a conjecture about the areas of the two triangles. Explain how you could show that your conjecture is true. **See margin.**

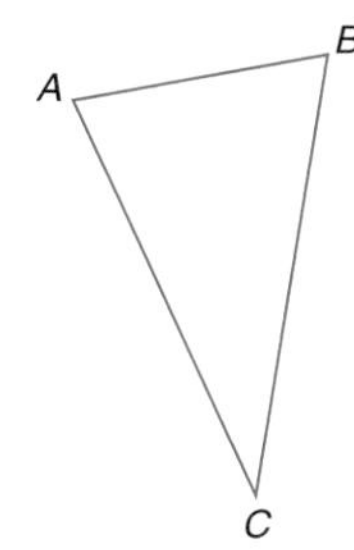

45d. The perimeter of the larger triangle is twice that of the smaller triangle.

Applications and Problem Solving

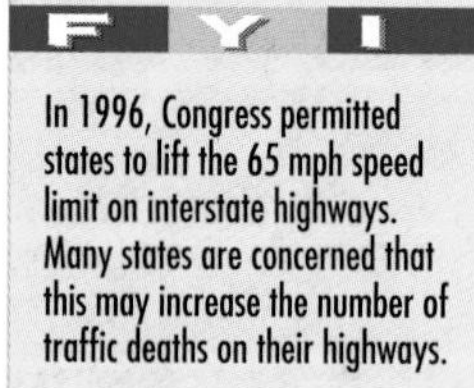
In 1996, Congress permitted states to lift the 65 mph speed limit on interstate highways. Many states are concerned that this may increase the number of traffic deaths on their highways.

46. **Music** A compact disc is actually a metallic disc covered with a plastic protective coating. The metal disc is pitted by an encoder that stores a small bit of information in each pit. A finished CD is 12 centimeters wide at its diameter. The hole in its center is 1.5 centimeters wide. Find the value of x in the picture. **5.25 cm**

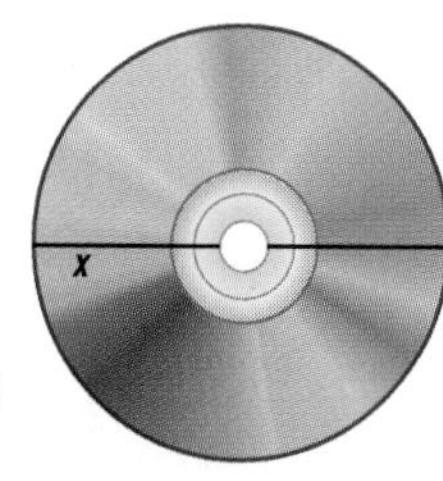

47. **Transportation** Interstate 70 passes through Kansas. Mile markers are used to name many of the exits. The exit for U.S. Route 283 North is Exit 128, and the exit to use U.S. Route 281 to Russell is Exit 184. The exit for Hays on U.S. Route 183 is 3 miles farther than halfway between Exits 128 and 184. What is the exit number for the Hays exit? **159**

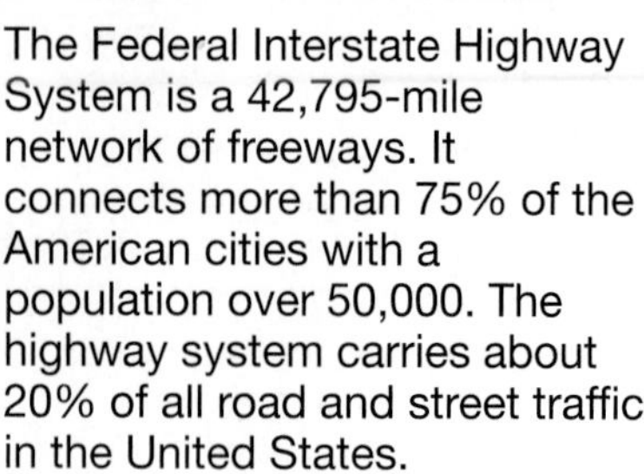
FYI

The Federal Interstate Highway System is a 42,795-mile network of freeways. It connects more than 75% of the American cities with a population over 50,000. The highway system carries about 20% of all road and street traffic in the United States.

Additional Answers

39. scale 2:5, $AB = 5$ in.,

$AC = \frac{1}{8}AB$

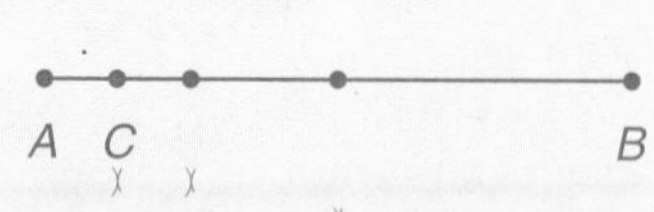

42. If E is the midpoint of $\overline{FG}$ and $\overline{HI}$, then $FE = \frac{1}{2}FG$ and $HE = \frac{1}{2}HI$. Since $\overline{FG} \cong \overline{HI}$, then $FG = HI$. If each side is multiplied by $\frac{1}{2}$, the result is $\frac{1}{2}FG = \frac{1}{2}HI$. Substitute values and the result is $FE = HE$.

Practice Masters, p. 5

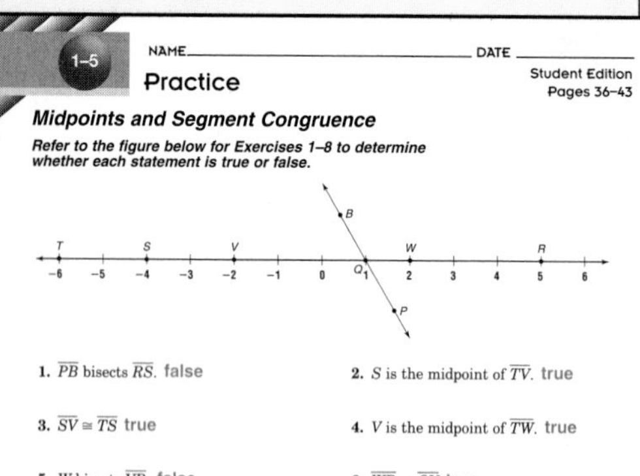
1-5 NAME________ DATE________

Practice

Student Edition Pages 36–43

Midpoints and Segment Congruence

Refer to the figure below for Exercises 1–8 to determine whether each statement is true or false.

1. $\overline{PB}$ bisects $\overline{RS}$. false
2. S is the midpoint of $\overline{TV}$. true
3. $\overline{SV} \cong \overline{TS}$ true
4. V is the midpoint of $\overline{TW}$. true
5. W bisects $\overline{VR}$. false
6. $\overline{WR} \cong \overline{QV}$ true
7. $\overline{SW}$ is longer than $\overline{VR}$. false
8. $VW \leq TV$ true

Given the coordinate of one endpoint of $\overline{AB}$ and its midpoint M, find the coordinates of the other endpoint.

9. $A(0, 9)$, $M(2, 5)$ $B(4, 1)$
10. $B(-5, 1)$, $M(1, -1)$ $A(7, -3)$
11. $A(-2, 3)$ $M(0.5, 0.5)$ $B(3, -2)$
12. $A(4, 2)$, $M(-2, 10)$ $B(-8, 18)$

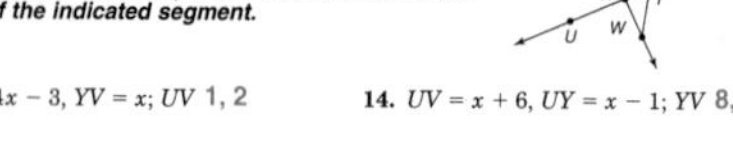
In the figure at the right, $\overline{WY}$ bisects $\overline{UV}$ at Y and $\overline{UY}$ bisects TW at X. For each situation, find the value of x and the measure of the indicated segment.

13. $UY = 4x - 3$, $YV = x$; UV 1, 2
14. $UV = x + 6$, $UY = x - 1$; YV 8, 7
15. $TX = 2x + 1$, $XW = x + 7$; TW 6, 26
16. $WX = x + 5$, $TW = 4x + 5$: TX 2.5, 7.5

Additional Answers

43.

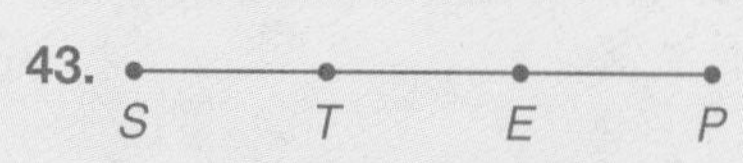

By the Segment Addition Postulate, we know that $SP = SE + EP$ and $SE = ST + TE$. By substituting for SE in the first equation, $SP = ST + TE + EP$.

45e. The area of the larger triangle is 4 times that of the smaller triangle. Sample answer: If you make four copies of the small triangle, you can fit them on the surface of the large triangle.

Mixed Review

48. Draw a segment with points A, B, C, and X that satisfy the following requirements. (Lesson 1–4) **See margin.**

- $AX = 10$
- $AB = 12$
- $XC = 6$

49. The coordinate grids below show three paths to get from point $A(-3, -1)$ to point $B(2, 2)$. (Lesson 1–4)

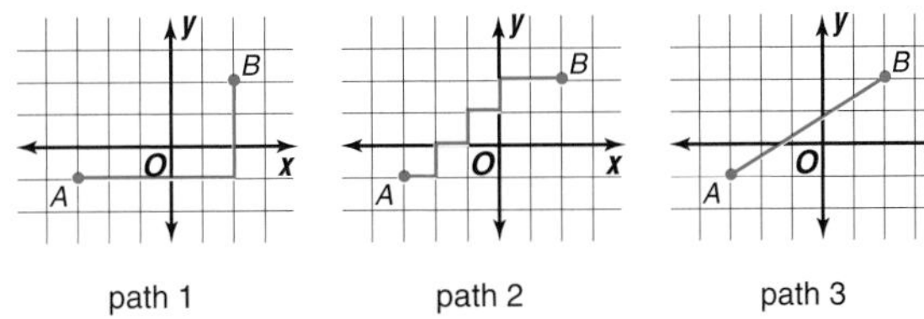

a. Find the distance along each path.

b. If the grid represents streets in a city, which path would a taxicab be least likely to take? Explain your reasoning.

49a. path 1: 8 units; path 2: 8 units; path 3: 5.83 units

49b. path 3, because you can't drive diagonally through city blocks

50. Carpentry The stringer is the diagonal board used in building a staircase. The riser is the height of each step, and the tread is the depth of each step. Study the diagram at the right and calculate the minimal length of board needed to make the stringer for this staircase. (Lesson 1–4) **47 inches**

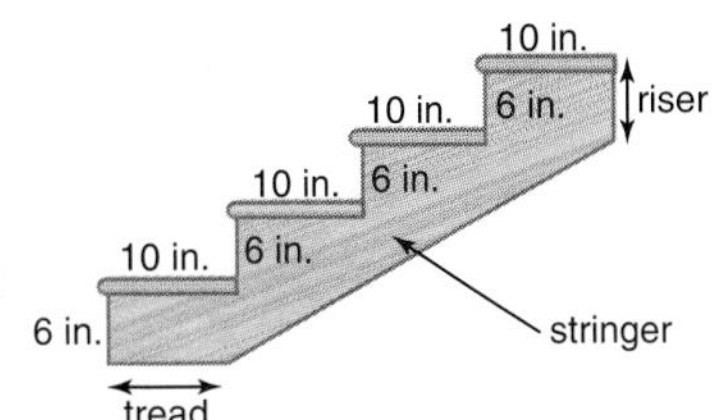

51. Photography The outer edge of a picture frame measures 17 inches by 14 inches. The sides of the frame are 2 inches wide. (Lesson 1–3)

a. Draw a picture to represent the frame, labeling all the information presented in the problem. **See margin.**

b. Find the area of a picture that will show in this frame. **130 in^2**

52. Line m contains points P, Q, R, and S. Write four other names for line m. (Lesson 1–2) **Sample answers: $\overleftrightarrow{PQ}$, $\overleftrightarrow{QR}$, $\overleftrightarrow{RS}$, $\overleftrightarrow{PS}$**

53. Draw and label figures to show the different ways in which planes $\mathcal{A}$, $\mathcal{B}$, and $\mathcal{C}$ can intersect. (Lesson 1–2) **See Solutions Manual.**

54. Graph points $A(3, -2)$, $B(-3, 5)$, and $C(-4, -6)$ on the same coordinate plane. (Lesson 1–1) **See margin.**

Algebra

55. Solve $3x + 4 = 7x - 12$. **$x = 4$**

56. Simplify $3(2x - 8) - 4(x + 2)$. **$2x - 32$**

4 ASSESS

Closing Activity

Writing Ask students to write the endpoints of three different line segments that have the midpoint of $(5, -2)$. **Sample answers: $(0, 0)$ and $(10, -4)$; $(4, -3)$ and $(6, -1)$; $(20, -12)$ and $(-10, 8)$**

Additional Answers

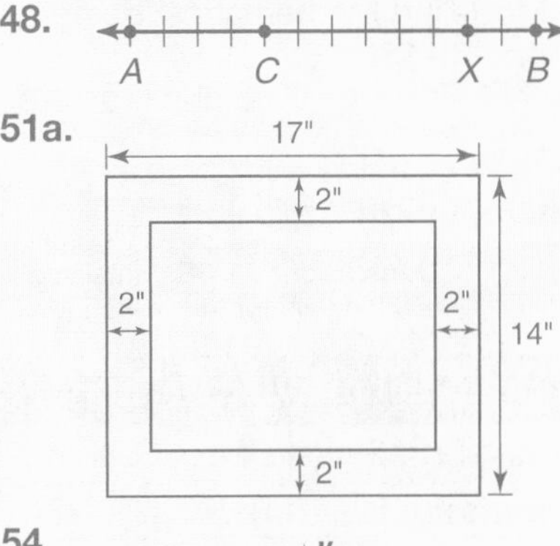

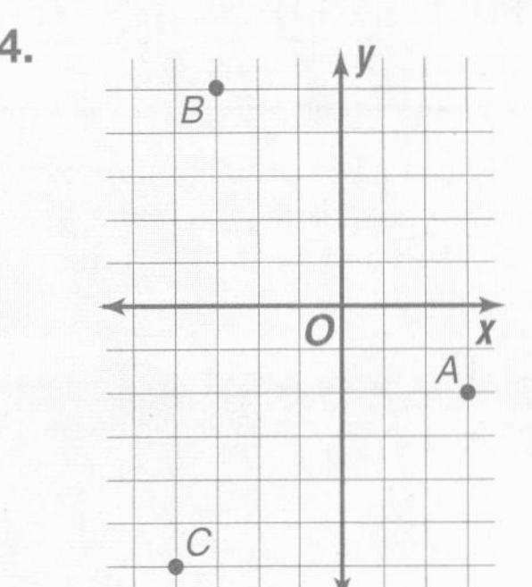

Enrichment Masters, p. 5

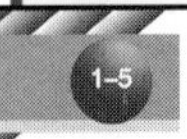

NAME ______ DATE ______

Enrichment

Student Edition Pages 36–43

Construction Problem

The diagram below shows segment AB adjacent to a closed region. The problem requires that you construct another segment XY to the right of the closed region such that points A, B, X, and Y are collinear. You are not allowed to touch or cross the closed region with your compass or straightedge.

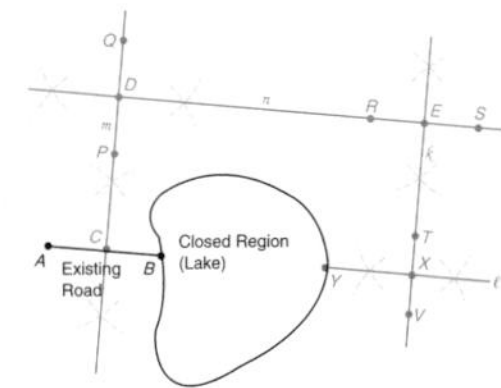

Follow these instructions to construct a segment XY so that it is collinear with segment AB.

1. Construct the perpendicular bisector of $\overline{AB}$. Label the midpoint as point C, and the line as m.
2. Mark two points P and Q on line m that lie well above the closed region. Construct the perpendicular bisector n of $\overline{PQ}$. Label the intersection of lines m and n as point D.
3. Mark points R and S on line n that lie well to the right of the closed region. Construct the perpendicular bisector k of $\overline{RS}$. Label the intersection of lines n and k as point E.
4. Mark point X on line k so that X is below line n and so that $\overline{EX}$ is congruent to $\overline{DC}$.
5. Mark points T and V on line k and on opposite sides of X, so that $\overline{XT}$ and $\overline{XV}$ are congruent. Construct the perpendicular bisector ℓ of $\overline{TV}$. Call the point where the line ℓ hits the boundary of the closed region point Y. $\overline{XY}$ corresponds to the new road.

Extension

Connections Have students write a graphing calculator program to find the midpoint of a line segment whose endpoints are entered into the calculator.

1-6 LESSON NOTES

NCTM Standards: 1–5, 7, 8

Instructional Resources

- Study Guide Master 1-6
- Practice Master 1-6
- Enrichment Master 1-6
- Assessment and Evaluation Masters, p. 17
- Modeling Mathematics Masters, pp. 15–18
- Tech Prep Applications Masters, p. 2

Transparency 1-6A contains the 5-Minute Check for this lesson; **Transparency 1-6B** contains a teaching aid for this lesson.

Recommended Pacing	
Standard Pacing	Days 7 & 8 of 13
Honors Pacing	Days 7 & 8 of 13
Block Scheduling*	Day 6 of 8

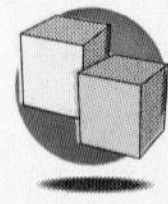

*For more information on pacing and possible lesson plans, refer to the *Block Scheduling Booklet*.

1 FOCUS

5-Minute Check
(over Lesson 1-5)

1. Name two segments whose midpoint is F in the figure below. **$\overline{EG}$, $\overline{DH}$**

   D E F G H

2. Find the coordinates of B if $M(-3, 2)$ is the midpoint of $\overline{AB}$ and A is at $(-1, -1)$. **$(-5, 5)$**
3. $\overline{RS}$ bisects $\overline{TU}$ at point Q. Find the value of x and the measure of $\overline{QU}$ if $TU = 7x - 5$ and $TQ = -5x + 6$. **1, 1**

1–6 Exploring Angles

What YOU'LL LEARN

- To identify angles and classify angles,
- to use the Angle Addition Postulate to find the measures of angles, and
- to identify and use congruent angles and the bisector of an angle.

Why IT'S IMPORTANT

Angles and their measures play a major part in many aspects of your study of geometry.

Art

The Japanese have long appreciated the beauty of nature through the art of *ikebana*. In ikebana, flowers and branches are arranged in a simple way that allows the beauty of each piece to be seen clearly. When creating an arrangement, the placement of each branch or flower is determined by forming an **angle** of a specific size.

A beam of light shooting into space is a model for a ray.

An angle is defined in terms of the two **rays** that form the angle. A ray extends indefinitely in *one* direction. A ray has exactly one endpoint and that point is always named first when naming the ray. Like segments, rays can be defined using betweenness of points.

Ray DE, written $\overrightarrow{DE}$, consists of the points on $\overline{DE}$ and all points F on $\overleftrightarrow{DF}$ such that E is between D and F.

*The figure formed by opposite rays is sometimes referred to as a **straight angle**. Unless otherwise specified, the term "angle" in this book means a nonstraight angle.*

If you choose any point on a line, that point determines exactly two rays, called **opposite rays.** The point is the common endpoint of the opposite rays. In the figure below, $\overrightarrow{GH}$ and $\overrightarrow{GI}$ are opposite rays, and G is the common endpoint. Opposite rays are collinear.

An *angle* is a figure formed by two *noncollinear* rays with a common endpoint. The two rays are called the **sides** of the angle. The common endpoint is called the **vertex**.

The sides of the angle at the right are $\overrightarrow{BA}$ and $\overrightarrow{BC}$, and the vertex is B. You could name this angle as $\angle B$, $\angle ABC$, $\angle CBA$, or $\angle 1$. When letters are used to name an angle, the letter that names the vertex is used either as the only letter or as the middle of three letters.

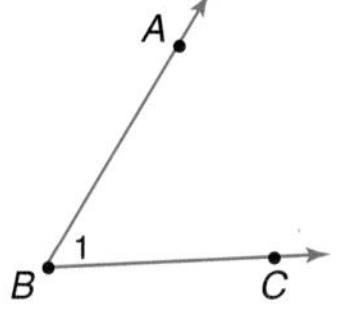

You should name an angle by a single letter only when there is no chance of confusion. For example, it is not obvious which angle shown at the right is $\angle P$ since there are three different angles that have P as a vertex. *Can you name the three angles?*

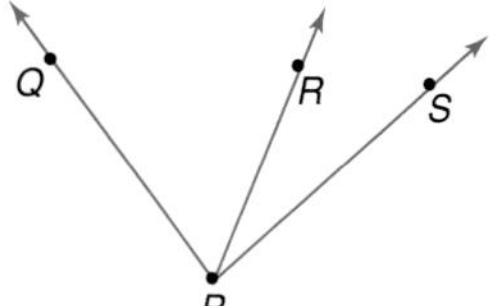

Example 1 **Refer to the figure at the right to answer each question.**

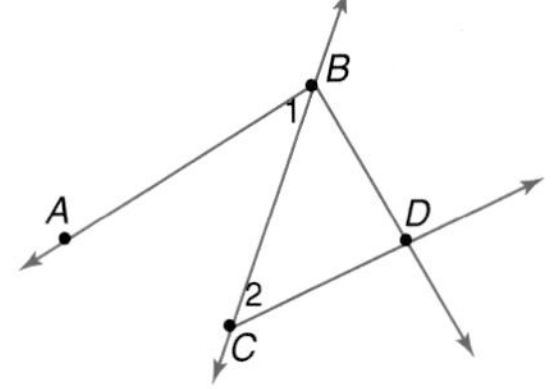

a. What other names could be used to identify $\angle BCD$?
$\angle 2$, $\angle C$, and $\angle DCB$ can be used to identify $\angle BCD$.

b. Name the vertex of $\angle 1$.
B is the vertex of $\angle 1$.

c. What are the sides of $\angle 2$?
$\overrightarrow{CB}$ and $\overrightarrow{CD}$ are the sides of $\angle 2$.

An angle separates a plane into three distinct parts, the **interior** of the angle, the **exterior** of the angle, and the angle itself. If a point does not lie on an angle and it lies on a segment whose endpoints are on each side of the angle, the point is in the interior of the angle. Neither of the endpoints of the segment can be the vertex of the angle.

In the figure at the right, point Z and all other points in the *blue* region are in the interior of $\angle X$. Any point that is not on the angle or in the interior of the angle is in the exterior of the angle. The *yellow* region is the exterior of $\angle X$.

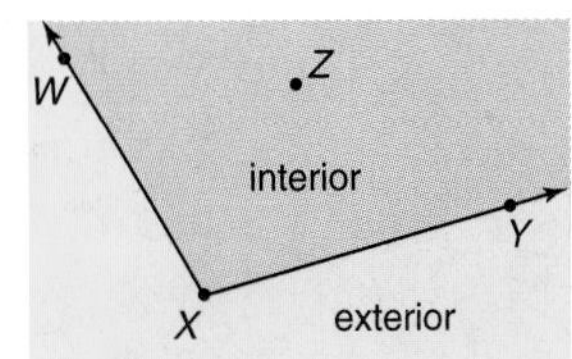

Angles can be measured in units called **degrees**. A *protractor* can be used to find the **measure** of an angle just as a ruler is used to find the measure of a segment. To find the measure of an angle, place the center point of the protractor over the vertex of the angle. Then align the mark labeled 0 on either side of the scale with one side of the angle.

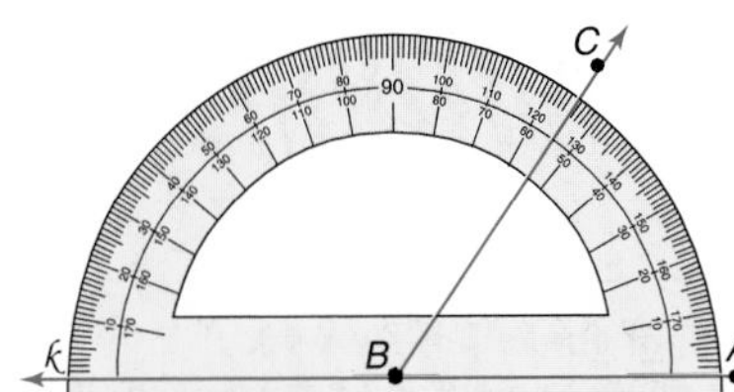

Using the inner scale of the protractor, you can see that $\angle B$ is a 55 degree (55°) angle. Thus, we say that the degree measure of $\angle ABC$ is 55 or $m\angle ABC = 55$.

The **Protractor Postulate** states that there can only be one 55° angle on either side of a given ray.

Motivating the Lesson

Questioning Ask students if they are familiar with the term *ray* and what it means. Draw a diagram of a ray and point out that it is called a ray because it has a starting point but no definite endpoint.

2 TEACH

Teaching Tip When discussing rays, point out that if you choose one point on a number line and cut off one end, the remainder is a ray.

In-Class Example

For Example 1
Refer to the figure below to answer each question.

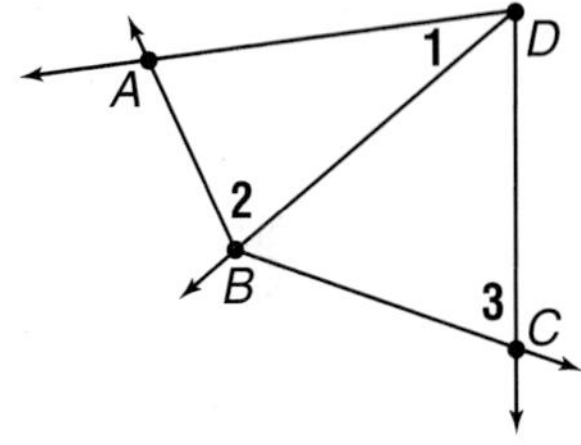

a. Give another name for $\angle ABD$. **$\angle 2$ or $\angle DBA$**
b. Name the vertex of $\angle 3$. **C**
c. What are the sides of $\angle 1$? **ray DA and ray DB**

Teaching Tip When you discuss the interior and the exterior of an angle, point out that just as the axes in a coordinate plane are not in any quadrant, the sides of the angle are not in the interior or the exterior of the angle.

Teaching Tip You can use a clear protractor with an overhead projector to demonstrate how to use a protractor. An overhead protractor is also provided in the Transparencies B binder and in the Overhead Manipulative Resources.

Teaching Tip When teaching students how to read a protractor, explain that you read the measure from the scale that shows 0 on the side to which the bottom ray points.

Alternative Learning Styles

Kinesthetic Have students use their arms to model rays. Have them lift one arm sideways. The shoulder represents the endpoint and the rest of the arm represents the extension of the ray. They can extend the ray by holding a ruler in the up-raised hand. Rays can be modeled by using one arm as one ray and the side of the body as the other ray.

In-Class Example

For Example 2
The third branch placed in an ikebana arrangement should form a 65° angle with the vertical on the opposite side from the other two branches. What is the measure of the angle formed by the first and third branches? 75°

It is generally believed that ikebana had its beginnings as a religious art with the offering of flowers to Buddha.

Postulate 1–3 Protractor Postulate

Given $\overrightarrow{AB}$ and a number r between 0 and 180, there is exactly one ray with endpoint A, extending on either side of $\overrightarrow{AB}$, such that the measure of the angle formed is r.

In Lesson 1–4, a measurement relationship between the lengths of segments, called the *Segment Addition Postulate*, was introduced. A similar relationship exists between the measure of angles.

Postulate 1–4 Angle Addition Postulate

If R is in the interior of $\angle PQS$, then $m\angle PQR + m\angle RQS = m\angle PQS$. If $m\angle PQR + m\angle RQS = m\angle PQS$, then R is in the interior of $\angle PQS$.

Example 2

Art

Refer to the application at the beginning of the lesson. The first branch placed in an ikebana arrangement is to be placed so that it forms a 10° angle with an imaginary vertical ray. The second branch is placed on the same side of the vertical as the first branch so that it forms an angle of 45° with the vertical. What is the measure of the angle formed by the first and second branches?

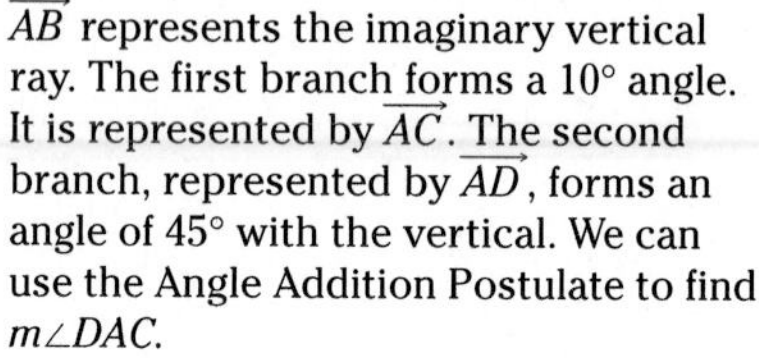

$\overrightarrow{AB}$ represents the imaginary vertical ray. The first branch forms a 10° angle. It is represented by $\overrightarrow{AC}$. The second branch, represented by $\overrightarrow{AD}$, forms an angle of 45° with the vertical. We can use the Angle Addition Postulate to find $m\angle DAC$.

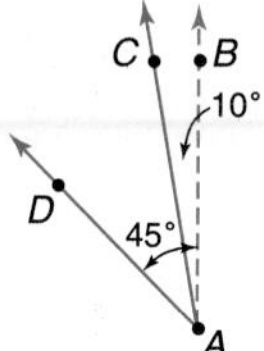

$m\angle DAB = m\angle DAC + \angle CAB$
$45 = m\angle DAC + 10$
$35 = m\angle DAC$

Therefore, the first and second branches form an angle of 35°.

FYI

Ikebana was developed in the Muromachi period, which occurred during the fourteenth to sixteenth centuries. Other arts and the Japanese tea ceremony also date from this time.

Angles can be classified by their measures.

The symbol ⌝ indicates a right angle.

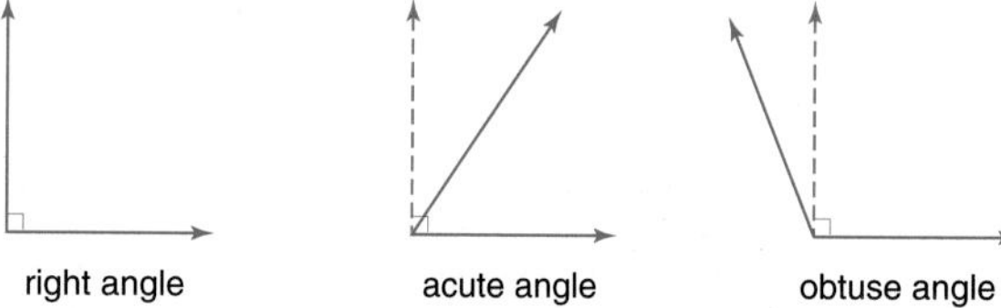

You can use a corner of a rectangular sheet of paper, which is a right angle, to help you determine if an angle is right, acute, or obtuse. The edge of the paper can be used to determine straight angles.

Definition of Right, Acute, and Obtuse Angles

$\angle A$ is a right angle if $m\angle A$ is 90.
$\angle A$ is an acute angle if $m\angle A$ is less than 90.
$\angle A$ is an obtuse angle if $m\angle A$ is greater than 90 and less than 180.

A straight angle has a measure of 180.

Tech Prep

Surveyor For students interested in surveying, you may wish to discuss how surveyors use transits to find the measures of angles between geographic features and to establish property lines. For more information on tech prep, see the *Teacher's Handbook*.

You know that congruent segments are segments that have the same length. Similarly, **congruent** angles are angles that have the same measure.

The definition of congruent angles tells us that statements such as $m\angle A = m\angle B$ and $\angle A \cong \angle B$ are equivalent. Therefore, we will use them interchangeably throughout this book.

Angle *NMP* and angle *RMQ* in the figure at the right each measure 70° Therefore, the two angles are congruent angles. You can write this as $\angle NMP \cong \angle RMQ$. Small arcs, like the ones in red shown at the right, are used to indicate congruent angles in a figure.

You can use a compass and straightedge to construct an angle that is congruent to a given angle without knowing the degree measure of the angle.

CONSTRUCTION

Congruent Angles

Construct an angle congruent to a given angle.

1. Draw an angle like $\angle P$ on your paper.

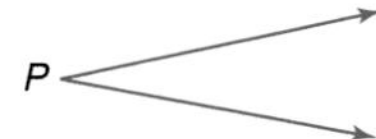

2. Use a straightedge to draw a ray on your paper. Label its endpoint *T*.

3. Place the tip of the compass at point *P* and draw a large arc that intersects both sides of $\angle P$. Label the points of intersection *Q* and *R*.

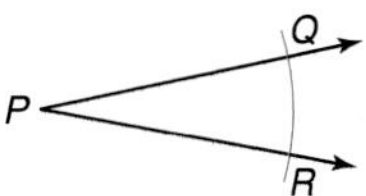

4. Using the same compass setting, put the compass at point *T* and draw a large arc that starts above the ray and intersects the ray. Label the point of intersection *S*.

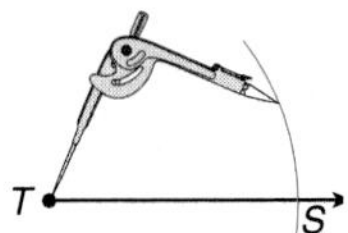

5. Place the point of your compass on *R* and adjust so that the pencil tip is on *Q*.

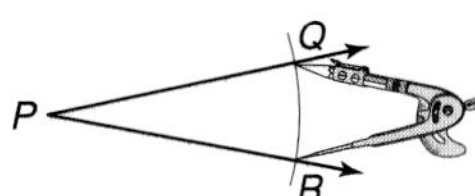

6. Without changing the setting, place the compass at point *S* and draw an arc to intersect the larger arc you drew in Step 4. Label the point of intersection *U*.

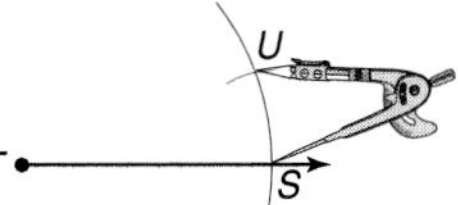

7. Use a straightedge to draw $\overrightarrow{TU}$.

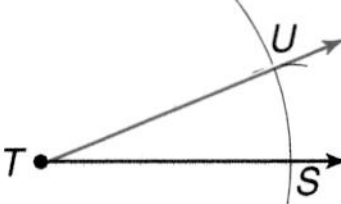

Conclusion: $m\angle QPR = m\angle UTS$ by construction. Thus, $\angle QPR \cong \angle UTS$ by definition of congruent angles.

Teaching Tip After students have constructed an angle congruent to a given acute angle, have them complete the construction for an obtuse and a right angle.

Teaching Tip After students have bisected a given acute angle, have them complete the construction for an obtuse and a right angle.

Teaching Tip Students can also use a geomirror to bisect angles. Have students place the reflector so that the reflection of one ray aligns with the other ray of the angle. The reflector is the bisector of the angle. Students can draw the bisector by drawing a line along the edge of the reflector.

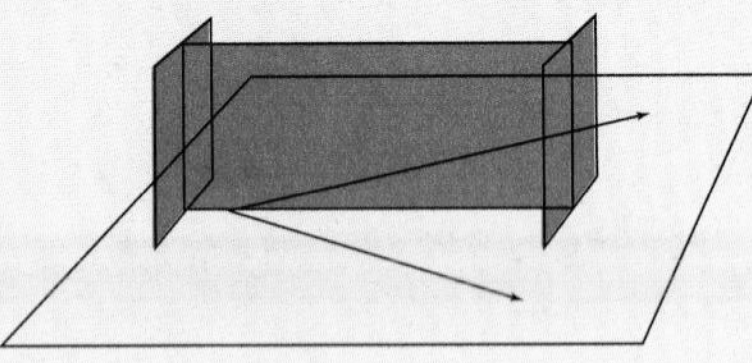

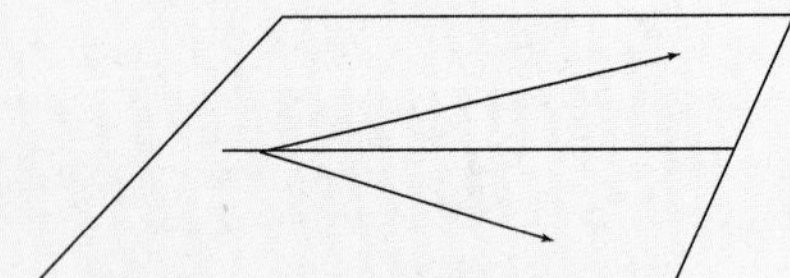

A segment that divides another segment into two congruent segments is called a *segment bisector*. A segment bisector intersects the midpoint of the segment that it bisects. Similarly, an **angle bisector** divides an angle into two congruent angles.

If $\overrightarrow{XZ}$ is the angle bisector of $\angle WXY$, point Z must be in the interior of $\angle WXY$ and $\angle WXZ \cong \angle ZXY$, as shown at the right.

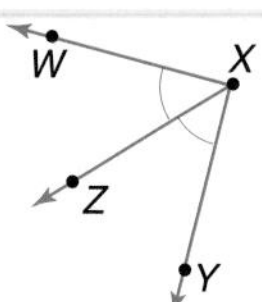

The angle bisector of a given angle can be constructed using the following procedure.

CONSTRUCTION

Angle Bisector

Construct the bisector of a given angle.

1. Draw an angle like $\angle A$ on your paper.

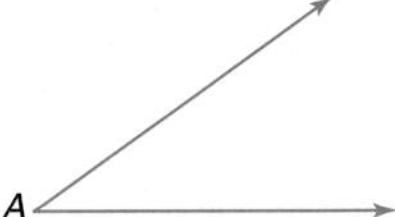

2. Put your compass at point A and draw a large arc that intersects both sides of $\angle A$. Label the points of intersection B and C.

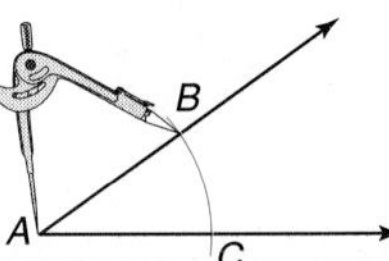

3. With the compass at point B, draw an arc in the interior of the angle.

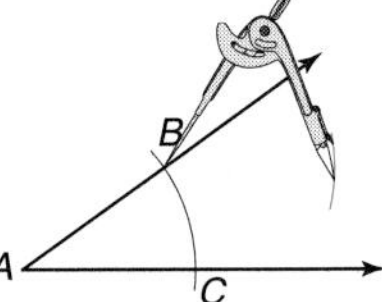

4. Keeping the same compass setting, place the compass at point C and draw an arc that intersects the arc drawn in Step 3. Label the point of intersection D.

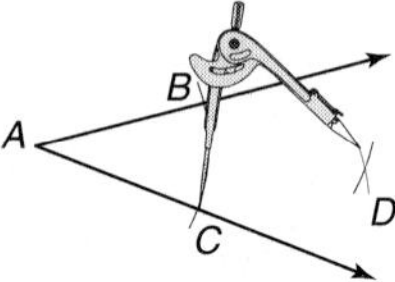

5. Draw $\overrightarrow{AD}$.

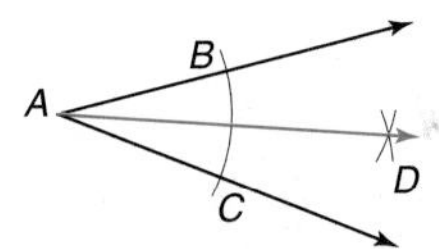

Conclusion: By construction, $\overrightarrow{AD}$ is the bisector of $\angle BAC$. $m\angle BAD = m\angle DAC$, and $\angle BAD \cong \angle DAC$.

CHECK FOR UNDERSTANDING

Communicating Mathematics

Study the lesson. Then complete the following. 1. two

1. **State** how many opposite rays are determined by a given point on a line.
2. **Write in your own words** why a straight angle does not fit our definition of an angle.
3. *True* or *false*: All right angles are congruent. Explain.
4. **Draw and label** a figure to show $\angle MNO$ with angle bisector $\overrightarrow{NQ}$ and $\angle MNQ$ with angle bisector $\overrightarrow{NR}$. Indicate any pairs of congruent angles. See margin.
5. **You Decide** Angeni says that for any three angles $\angle ABD$, $\angle DBC$, and $\angle ABC$ that share vertex B, it must be true that $m\angle ABD + m\angle DBC = m\angle ABC$. Reta says that the relationship is not necessarily true. Who is correct and why? Reta; for the relationship to be true, $\overrightarrow{BD}$ must be in the interior of $\angle ABC$.
6. Draw an angle on a piece of patty paper. Fold the paper through the vertex of the angle so that the sides of the angle match. Make a crease. Open the paper. What does the crease represent? Explain. The angle bisector; the angles formed are congruent.

2. A straight angle is formed by two collinear, not noncollinear, rays.

3. True, all right angles measure 90° and the definition of congruent angles is that they have the same measure.

Guided Practice

Refer to the figure below for Exercises 7–12.

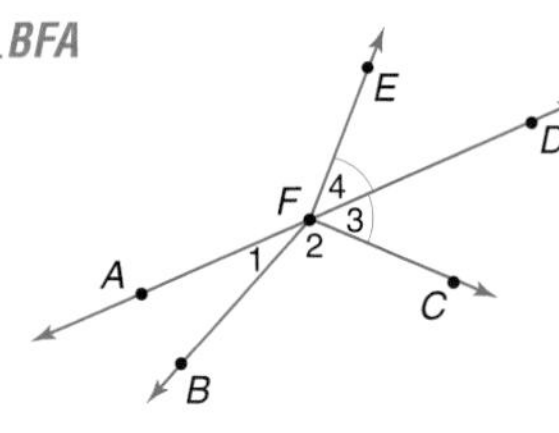

7. State two other names for $\angle 1$. $\angle AFB$, $\angle BFA$
8. Does $\angle EFC$ appear to be acute, right, obtuse, or straight? right
9. Name a ray that appears to bisect an angle and the angle that it appears to bisect. $\overrightarrow{FD}$; $\angle EFC$
10. Name all of the angles that have $\overrightarrow{FC}$ for a side. $\angle CFD$, $\angle CFE$, $\angle CFA$, $\angle CFB$
11. Complete: $m\angle DFB = m\angle 2 + \underline{\ ?\ }$ $m\angle 3$
12. Name a point in the exterior of $\angle AFB$. E, D, or C
13. Draw two angles that intersect in exactly two points. See margin.
14. If $\angle G$ is an acute angle and $m\angle G = 4x + 16$, write a compound inequality to describe all the possible values for x.
15. Suppose $\angle T \cong \angle S$, $m\angle T = 12n - 6$, and $m\angle S = 4n + 18$. Is $\angle T$ acute, right, obtuse, or straight? acute
16. **Aviation** Compass headings are used to indicate the directions of airplanes. A heading is stated as the measure of the angle formed by the flight path of the airplane and an imaginary path in the direction of due north. A pilot is on approach to land at a compass heading of 68° NW (northwest). The tower has instructed her to land on runway 9, which has a heading of 90° NW. How many degrees must the pilot turn her plane in order to land? 22°

14. $-4 < x < 18.5$

U.S. Navy Blue Angels

Reteaching

Using Mnemonics If students are having trouble identifying which angles are acute, right, obtuse, and straight, have them develop mnemonics to help them remember. For example, things that are cute are usually small (puppies, babies), and acute angles have smaller measures.

Additional Answer

13.

A D C B

3 PRACTICE/APPLY

Check for Understanding

Exercises 1–16 are designed to help you assess your students' understanding through reading, writing, speaking, and modeling. You should work through Exercises 1–6 with your students and then monitor their work on Exercises 7–16.

Error Analysis

Some students may be confused by the letters used to name angles when using the angle addition postulate. Encourage students to number the angles in a diagram first. Then the number can be used to write the relationship between the measures.

Additional Answer

4.

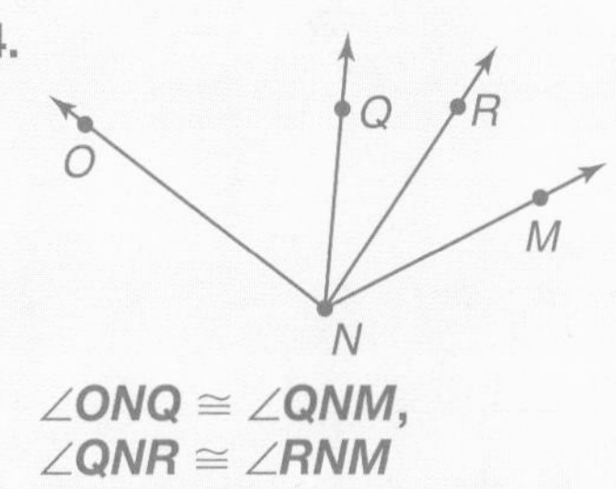

$\angle ONQ \cong \angle QNM$,
$\angle QNR \cong \angle RNM$

Study Guide Masters, p. 6

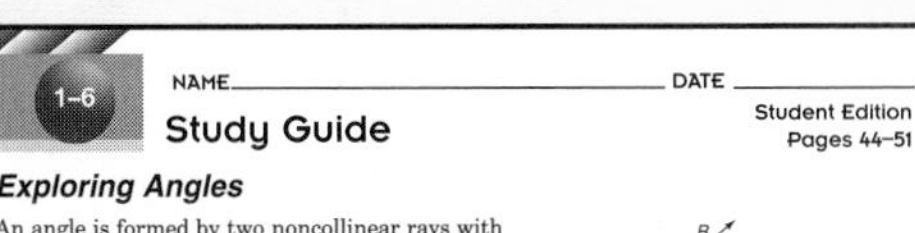

1-6 NAME ______ DATE ______

Study Guide — Student Edition Pages 44–51

Exploring Angles

An angle is formed by two noncollinear rays with a common endpoint. You could name the angle in the figure at the right as $\angle S$, $\angle RST$, $\angle TSR$, or $\angle 1$.

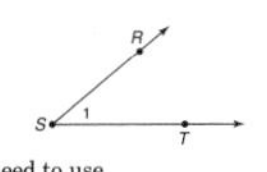

When two or more angles have a common vertex, you need to use either three letters or a number to name the angles. Make sure there is no doubt which angle your name describes.

A **right angle** is an angle whose measure is 90. Angles smaller than a right angle are **acute angles**. Angles larger than a right angle are **obtuse angles**. A **straight angle** has a measure of 180.

According to the Angle Addition Postulate, if D is in the interior of $\angle ABC$, then $m\angle ABD + m\angle DBC = m\angle ABC$.

Example: In the figure at the right, $m\angle ABC = 160$, $m\angle 1 = x + 14$, and $m\angle 2 = 3x - 10$. Find the value of x.

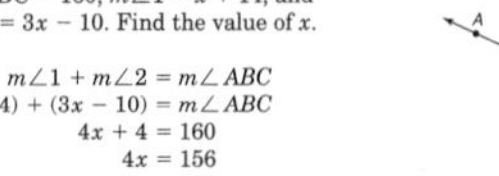

$$m\angle 1 + m\angle 2 = m\angle ABC$$
$$(x + 14) + (3x - 10) = m\angle ABC$$
$$4x + 4 = 160$$
$$4x = 156$$
$$x = 39$$

For Exercises 1–5, refer to the figure at the right.

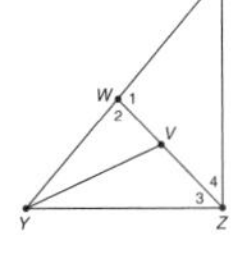

1. Do $\angle 3$ and $\angle Z$ name the same angle? Explain. Not necessarily; Z is the vertex of three angles, $\angle 3$, $\angle 4$, and $\angle YZX$.
2. List all the angles that have W as the vertex. $\angle YWX$, $\angle YWZ$, or $\angle 2$, $\angle XWZ$ or $\angle 1$
3. Name a straight angle. $\angle YWX$ or $\angle WVZ$
4. If $m\angle WYV = 4x - 2$, $m\angle VYZ = 2x - 5$, and $m\angle WYZ = 77$, find the measurements of $\angle WYV$ and $\angle VYZ$. 54, 23
5. Does $\angle YVW$ appear to be acute, obtuse, right, or straight? acute
6. In the figure at the right, if $\overrightarrow{QS}$ bisects $\angle RQP$, $m\angle RQS = 2x + 10$, and $m\angle SQP = 3x - 18$, find $m\angle SQR$. 66

R S T Q P

Assignment Guide

Core (with proof): 17–41 odd, 42–49
Core (informal): 17–41 odd, 42–49
Enriched: 18–38 even, 39–49

For **Extra Practice**, see p. 765.

The red A, B, and C flags, printed in the Teacher's Wraparound Edition, indicate the level of difficulty of the exercises.

Additional Answers

26. No, more than one angle has P as a vertex, so naming an angle as $\angle P$ would create confusion.
28. No, while both are rays that contain P and N, PN is a ray whose endpoint is P, and NP is a ray whose endpoint is N.

29. 30.

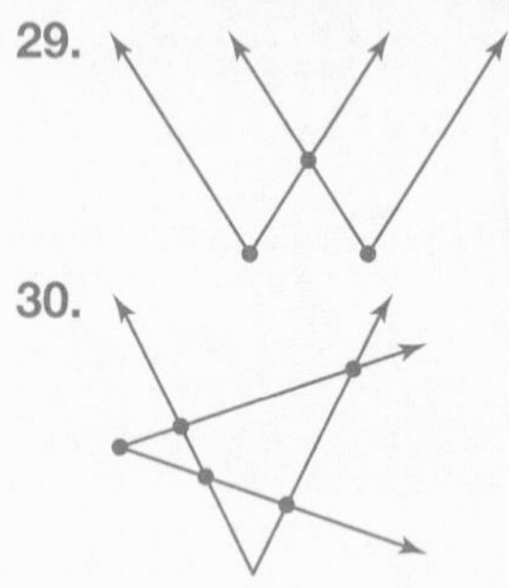

Practice Masters, p. 6

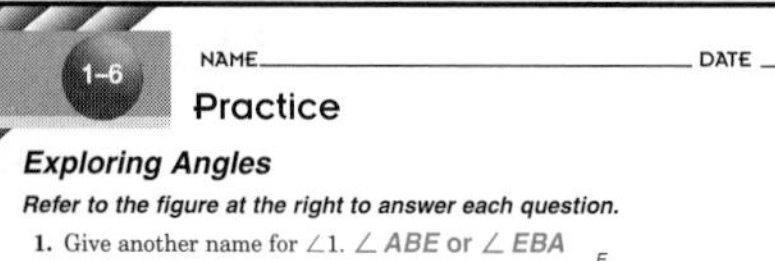

1-6 NAME________ DATE ______
Practice — Student Edition Pages 44–51

Exploring Angles

Refer to the figure at the right to answer each question.

1. Give another name for $\angle 1$. $\angle ABE$ or $\angle EBA$
2. Name the vertex of $\angle EBD$. B
3. Does $\angle ABE$ appear to be acute, obtuse, right, or straight? acute
4. If $m\angle DBE = 35$, what is the measure of $\angle EBF$? 125
5. Name a pair of opposite rays. $\overline{BA}$ and $\overline{BF}$
6. Name a point in the interior of $\angle EBF$. D
7. Name three angles with $\overline{BE}$ as a side? $\angle FBE$, $\angle DBE$, $\angle ABE$

In the figure, $\overrightarrow{XP}$ and $\overrightarrow{XT}$ are opposite rays and $\overrightarrow{XQ}$ bisects $\angle PXS$. For each situation, find the value of x and the measure of the indicated angle.

8. $m\angle SXT = 4x + 1$, $m\angle QXS = 2x - 2$, $m\angle QXT = 125$; $m\angle QXS$ 21, 40
9. $m\angle PXR = 3x$, $m\angle RXT = 5x + 20$, $m\angle RXT$ 20, 120
10. $m\angle RXQ = x + 15$, $m\angle RXS = 5x - 7$, $m\angle QXS = 3x + 5$; $m\angle RXS$ 27, 128
11. $m\angle RXQ = 2x + 7$, $m\angle RXP = 3x - 11$, $m\angle PXS = x + 37$; $m\angle QXS$ 5, 21
12. $m\angle TXS = x + 3$, $m\angle SXR = 2x + 9$, $m\angle RXP = 4x - 7$; $m\angle PXS$ 25, 152

EXERCISES

Practice

Refer to the figure below for Exercises 17–28.

17. Name two angles that have N as a vertex. $\angle ONM$, $\angle MNR$

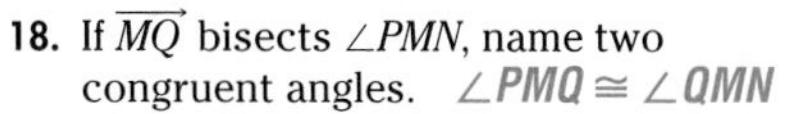

18. If $\overrightarrow{MQ}$ bisects $\angle PMN$, name two congruent angles. $\angle PMQ \cong \angle QMN$

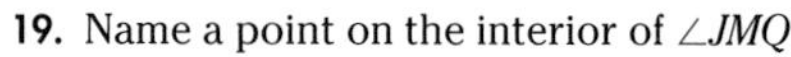

19. Name a point on the interior of $\angle JMQ$. 19. I or P

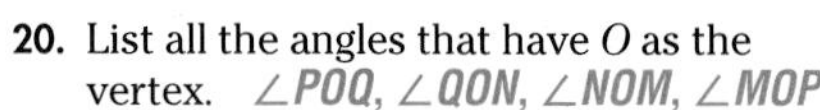

20. List all the angles that have O as the vertex. $\angle POQ$, $\angle QON$, $\angle NOM$, $\angle MOP$
21. Does $\angle QMJ$ appear to be acute, obtuse, right, or straight? obtuse
22. Name a pair of opposite rays. 22. Sample answer: $\overrightarrow{MJ}$ and $\overrightarrow{MN}$

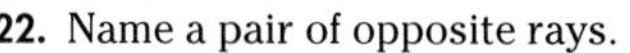

23. List all the angles that have $\overrightarrow{MN}$ as a side. $\angle NML$, $\angle NMK$, $\angle NMJ$, $\angle NMP$, $\angle NMO$
24. Name an angle that appears to be acute. Sample answer: $\angle LMN$
25. If $m\angle PMO = 55$ and $m\angle OMN = 65$, what is the measure of $\angle PMN$? 120
26. Is $\angle P$ a valid name for one of the angles? Explain. See margin.
27. Name a pair of angles that share exactly one point. $\angle JMQ$ and $\angle NMK$
28. Is $\overrightarrow{PN}$ the same as $\overrightarrow{NP}$? Explain. See margin.

Draw two angles that satisfy the following conditions.

29. The angles intersect in one point. See margin.
30. The angles intersect in four points. See margin.
31. The angles intersect in a segment. See margin.

32. Use a protractor to draw $\angle ABC$ so that it measures 80°.
    a. Use a compass and a straightedge to construct $\overrightarrow{BD}$, the bisector of the angle. See margin.
    b. Measure $\angle ABD$ and $\angle CBD$. What do you observe? $m\angle ABD = m\angle CBD = 40$

In the figure, $\overrightarrow{BA}$ and $\overrightarrow{BC}$ are opposite rays, and $\overrightarrow{BE}$ bisects $\angle ABD$.

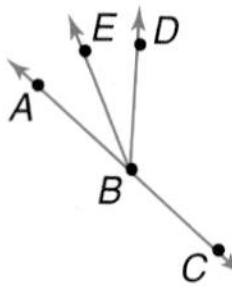

33. If $m\angle ABE = 6x + 2$ and $m\angle DBE = 8x - 14$, find $m\angle ABE$. 50
34. Given that $m\angle ABD = 2y$ and $m\angle DBC = 6y - 12$, find $m\angle DBC$. 132

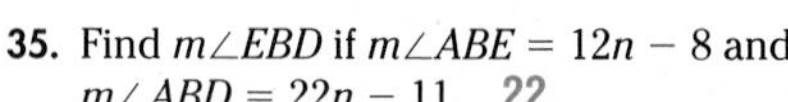

35. Find $m\angle EBD$ if $m\angle ABE = 12n - 8$ and $m\angle ABD = 22n - 11$. 22

36. If $m\angle ABE = 9x - 1$ and $m\angle DBC = 24x + 14$, find $m\angle EBD$. 35
37. Suppose $m\angle ABE = 15y$ and $m\angle DBC = 45y - 30$. Find the value of y. 2.8
38. If $\angle ABD$ is a right angle and $m\angle ABE = 13x - 7$, what is the value of x? 4

Critical Thinking

39. Each figure below shows noncollinear rays with a common endpoint.

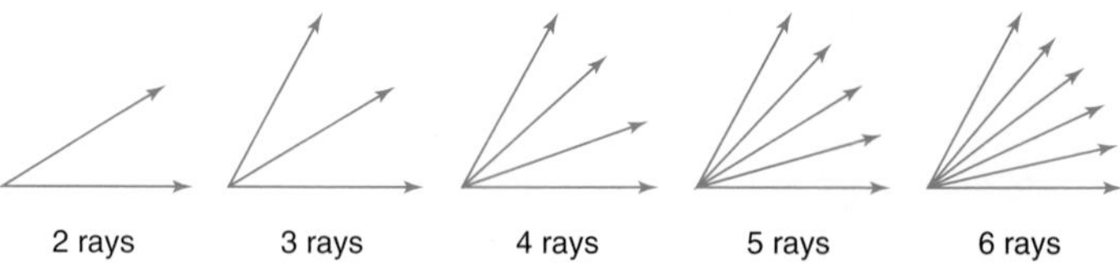

Additional Answers

31.

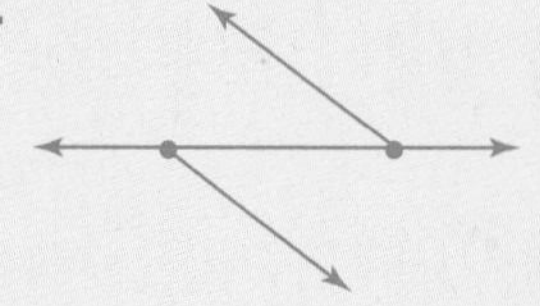

32a.

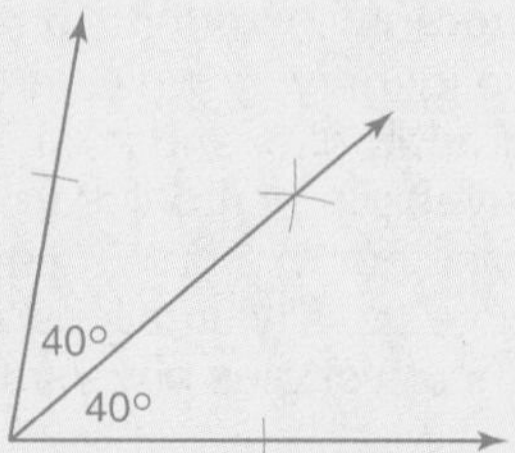

a. Count the number of angles in each figure. 1, 3, 6, 10, 15

b. Describe the pattern that exists between the number of rays and the number of angles in each figure. See Solutions Manual.

c. Predict the number of angles that are formed by seven rays and by 10 rays. 21, 45

d. Write a formula for the number of angles formed by n noncollinear rays with a common endpoint.
$a = \frac{n(n-1)}{2}$, for a = number of angles and n = number of rays

Applications and Problem Solving

40. Physics A ripple tank can be used to study the behavior of waves in two dimensions. As a wave strikes a barrier, it is reflected. The angle of incidence and the angle of reflection are congruent. In the diagram at the right, if $m\angle IBR = 78$, find the measure of the angle of reflection and $m\angle IBA$. 39; 51

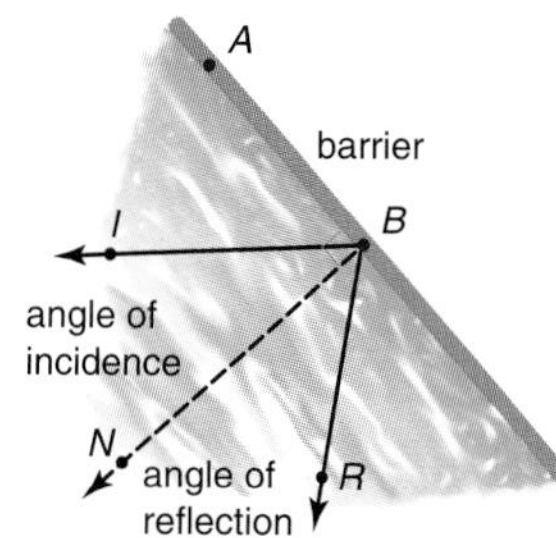

41. Entertainment John Trudeau used angles to help him design *The Flintstones* pinball machine. The angle that the machine tilts, or the pitch of the machine, determines the difficulty of the game. The angles at which the flipper will hit the ball are considered when the ramps, loops, and targets of the game are placed.

a. Mr. Trudeau recommends that *The Flintstones* machine be installed with a pitch of 6° to 7°. Is this an acute, right, straight, or obtuse angle? acute

b. Suppose a line is drawn down the center of a pinball machine. If the path of a ball hit by the flipper intersects that center line, an angle of between 5° and 20° is formed. If you drew an angle whose interior contained most of the paths of these balls, what would be the measure of that angle? 15

Mixed Review

42. Find B, if A is at $(5, -3)$ and $M(-4, 2)$ is the midpoint of $\overline{AB}$. (Lesson 1–5) $(-13, 7)$

43. G is between F and H. If $GH = 6$, $FG = 8x + 2$, and $FH = 16$, find FG. (Lesson 1–4) 10

44. What is the distance between $P(-3, 10)$ and $Q(12, -6)$? Round your answer to the nearest hundredth. (Lesson 1–4) 21.93

45. The perimeter of a rectangle is 27 millimeters. If the length is 8 millimeters, what is the width? (Lesson 1–3) 5.5 mm

46. Draw a figure for plane $\mathcal{C}$ containing line ℓ and point Y, not on ℓ. (Lesson 1–2) See margin.

47. Graph $R(-4, 7)$ in a coordinate plane. (Lesson 1–1) See margin.

Algebra

48. Find $\sqrt{49}$. 7

49. Solve $x^2 = 25$. ± 5

Extension

Reasoning Ask students why they think the protractor postulate needed only a number between 0 and 180. What does it mean when they hear "Keli did a 180 on her skateboard"? It may help them to look at a protractor and see that it forms a semicircle. What happens if they put two protractors together along the straight edge? How many degrees are in a circle? They form a circle; 360°.

Additional Answer

47.

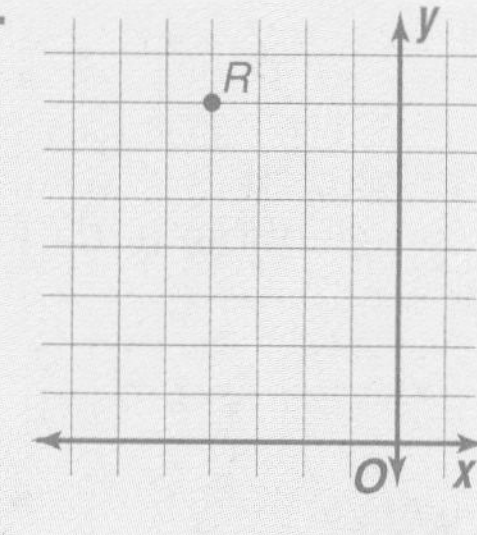

4 ASSESS

Closing Activity

Writing Have groups of students work together to write a set of instructions to complete one of the constructions or activities in the lesson. For example, one group may describe how to draw an angle with a specific measure, another may describe how to bisect an angle, and a third may describe how to determine if an angle is acute, right, obtuse, or straight.

Chapter 1 Quiz C (Lessons 1-5 and 1-6) is available in the *Assessment and Evaluation Masters*, p. 17.

Additional Answer

46.

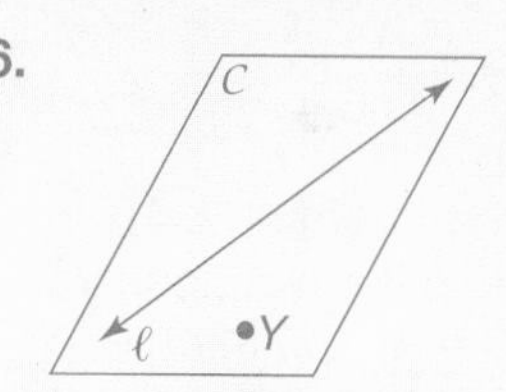

Enrichment Masters, p. 6

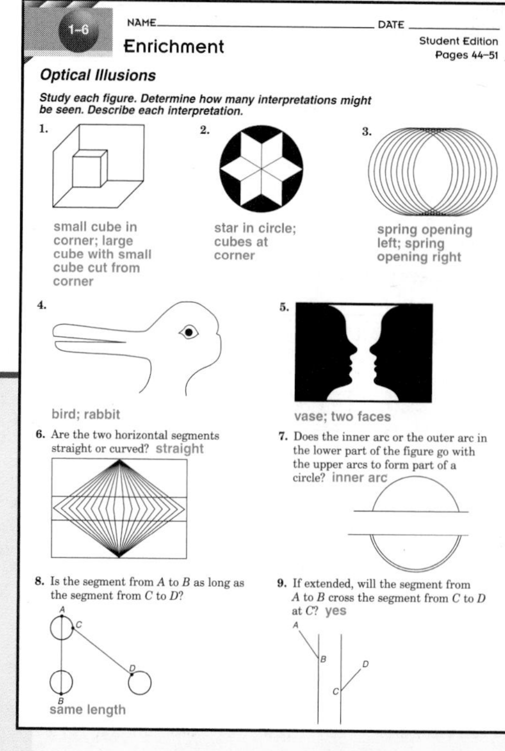
1-6 NAME_____ DATE_____

Enrichment

Student Edition Pages 44–51

Optical Illusions

Study each figure. Determine how many interpretations might be seen. Describe each interpretation.

1. small cube in corner; large cube with small cube cut from corner

2. star in circle; cubes at corner

3. spring opening left; spring opening right

4. bird; rabbit

5. vase; two faces

6. Are the two horizontal segments straight or curved? straight

7. Does the inner arc or the outer arc in the lower part of the figure go with the upper arcs to form part of a circle? inner arc

8. Is the segment from A to B as long as the segment from C to D? same length

9. If extended, will the segment from A to B cross the segment from C to D at C? yes

1-7A LESSON NOTES

NCTM Standards: 1–5, 7, 8

Objective

Use a TI-92 graphing calculator to measure angles and discover angle relationships when lines intersect.

Recommended Time

20 minutes

Instructional Resources

Instructions for using the *Geometer's Sketchpad* for this activity are available in the *Graphing Calculator and Computer Masters*, p. 16.

1 FOCUS

Motivating the Lesson

Have students tie two pieces of string together at the middle to form an X. Let this represent two intersecting lines. Then have them stretch out the strings on their desktops to show the angles formed by the strings. Ask them which angles appear to be equal in size.

2 TEACH

Teaching Tip You may wish to extend the activity by having students draw angles not formed by intersecting lines. Have them practice measuring the angles they draw.

3 PRACTICE/APPLY

Assignment Guide
Core (with proof): 1–6
Core (informal): 1–6
Enriched: 1–6

1–7A Using Technology
Angle Relationships

A Preview of Lesson 1–7

LOOK BACK

Refer to Lesson 1-4A to review how to start a new drawing screen and draw a segment.

In Lesson 1–4A, you learned to use the TI-92 calculator to draw and measure segments. You can also draw and measure angles on the calculator.

- Draw $\overleftrightarrow{AD}$.
- Draw $\overleftrightarrow{CE}$ so that it intersects $\overleftrightarrow{AD}$. Label the intersection point B.
- To find the measures of the angles created by the intersecting lines, press F6 and select 3:Angle. You will define the angle to be measured by selecting three points on the angle in the same order you would name the angle. Move the pointer to one of the points on the angle and press ENTER. Move to the next and press ENTER. Then move to the last and press ENTER. The measure of the angle to the nearest hundredth degree appears.
- Repeat this process to measure the other three angles.

This process can be used to measure any angle, regardless of how it was created.

The screen shows a sample result of this activity.

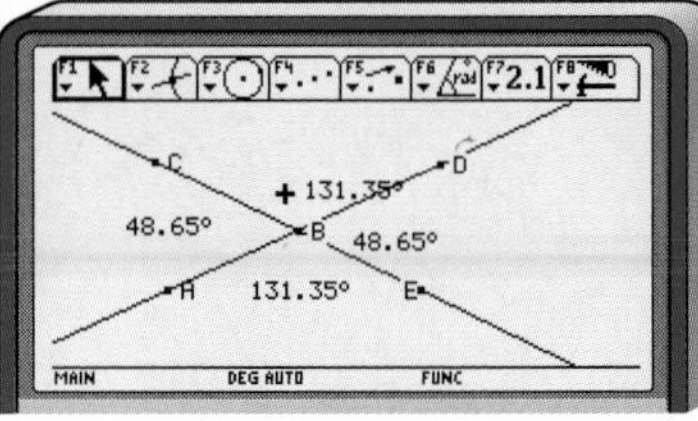

EXERCISES

Analyze your drawing. 1b. They are equal.

1. Angles that are formed by intersecting lines and share a common vertex are called *vertical angles.* 1a. $\angle CBA$ and $\angle DBE$; $\angle CBD$ and $\angle EBA$
   a. Name two pairs of vertical angles from your screen.
   b. Compare the measures of the vertical angles. What do you notice?
2. Angles that have a common side and vertex and whose noncommon sides are opposite rays are called a *linear pair.*
   a. Name the linear pairs of angles from your screen.
   b. Compare the measures of each linear pair. What do you notice?

2a. $\angle CBD$ and $\angle DBE$; $\angle DBE$ and $\angle EBA$; $\angle EBA$ and $\angle ABC$; $\angle ABC$ and $\angle CBD$

2b. The sum of the measures of the angles in each linear pair is 180.

Test your conclusion. 3–4. See students' work.

3. Move the cursor to one of the lines. Hold down the hand key and move the line. Record the angle measures after moving one of the lines.
4. Use the hand key to move the other line. Record the angle measures after this move.
5. Write a sentence to summarize the relationship you notice among the measures of each pair of vertical angles.
6. Write a sentence to summarize the relationship you notice among the measures of each linear pair of angles.

5. Regardless of the position of the lines, vertical angles have equal measures.

6. Regardless of the position of the lines, the sum of the measures of a linear pair of angles is 180.

4 ASSESS

Observing students working with technology is an excellent method of assessment.

Using Technology

This lesson offers an excellent opportunity for using technology in your geometry classroom. For more information on using technology, see *Graphing Calculators in the Mathematics Classroom*, one of the titles in the Glencoe Mathematics Professional Series.

1-7 Angle Relationships

What YOU'LL LEARN

- To identify and use adjacent, vertical, complementary, supplementary, and linear pairs of angles, and perpendicular lines, and
- to determine what information can and cannot be assumed from a diagram.

Why IT'S IMPORTANT

You can use the measures of angles to solve problems in geology, sports, and construction.

CONNECTION

Geology

Geophysicists at the U.S. Geological Survey are studying the way that the continents and seas have been formed. In order to describe a section of Earth's crust accurately, they measure the strike and dip of the area.

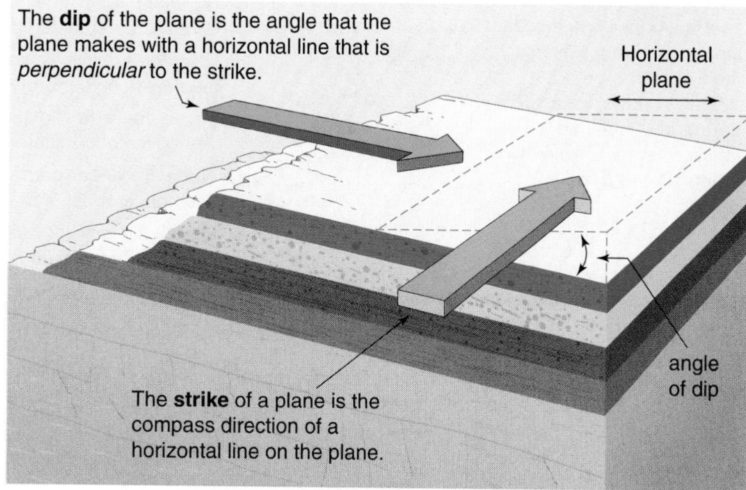

The **dip** of the plane is the angle that the plane makes with a horizontal line that is *perpendicular* to the strike.
Horizontal plane
angle of dip
The **strike** of a plane is the compass direction of a horizontal line on the plane.

CAREER CHOICES

A **geophysicist** combines geology and physics to study the physical properties and structure of Earth. They study electrical, thermal, gravitational, seismic, and magnetic phenomena.

For more information, contact:

American Geophysical Union
2000 Florida Avenue, NW
Washington, DC 20009

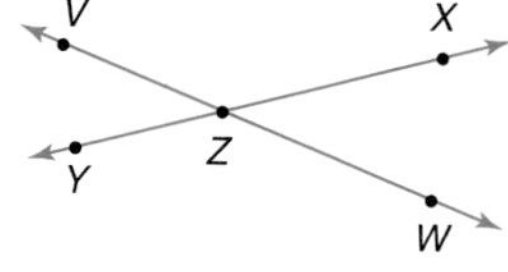

V
X
Z
Y
W

Perpendicular lines are special intersecting lines that form right angles. Not all lines that intersect are perpendicular lines. When two lines intersect, they form four angles. These angles are not necessarily right angles. In the figure at the left, $\overleftrightarrow{VW}$ intersects $\overleftrightarrow{XY}$ at Z forming $\angle VZX$, $\angle XZW$, $\angle WZY$, and $\angle YZV$. Certain pairs of angles formed by intersecting lines have special names that are used to describe the relationship between the angles. The chart below summarizes the names for special pairs of angles formed by these two lines.

Special Name	Definition	Examples
adjacent angles	angles in the same plane that have a common vertex and a common side, but no common interior points	$\angle VZX$ and $\angle XZW$ $\angle XZW$ and $\angle WZY$ $\angle WZY$ and $\angle YZV$ $\angle YZV$ and $\angle VZX$
vertical angles	two nonadjacent angles formed by two intersecting lines	$\angle VZY$ and $\angle XZW$ $\angle VZX$ and $\angle YZW$
linear pair	adjacent angles whose noncommon sides are opposite rays	$\angle VZX$ and $\angle XZW$ $\angle XZW$ and $\angle WZY$ $\angle WZY$ and $\angle YZV$ $\angle YZV$ and $\angle VZX$

CAREER CHOICES

Since the late 1960s, the study of geophysics has expanded to include all the planets. Scientific analyses comparing Earth's conditions with conditions on other planets have increased our understanding of the origin and evolution of Earth.

1-7 LESSON NOTES

NCTM Standards: 1–5, 7, 8, 14

Instructional Resources

- Study Guide Master 1-7
- Practice Master 1-7
- Enrichment Master 1-7
- Assessment and Evaluation Masters, p. 17
- Multicultural Masters, p. 2

Transparency 1-7A contains the 5-Minute Check for this lesson; **Transparency 1-7B** contains a teaching aid for this lesson.

Recommended Pacing	
Standard Pacing	Days 10 & 11 of 13
Honors Pacing	Days 10 & 11 of 13
Block Scheduling*	Day 7 of 8

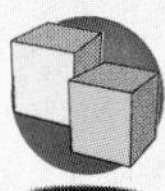

*For more information on pacing and possible lesson plans, refer to the *Block Scheduling Booklet*.

1 FOCUS

5-Minute Check
(over Lesson 1-6)

Refer to the figure below.

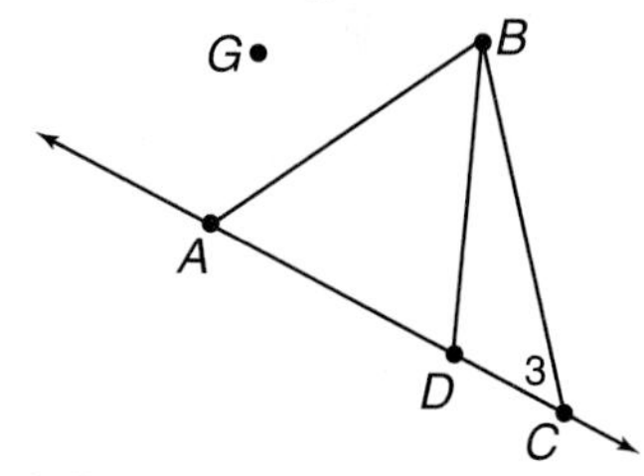

G
B
A
D
3
C

1. Which angles have D as their vertex? **$\angle ADB$, $\angle BDC$, $\angle ADC$**
2. Give two other names for $\angle ABD$. **$\angle B$, $\angle DBA$**
3. Does $\angle BDC$ appear to be acute, right, obtuse, or straight? **obtuse**
4. Find the value of x and $m\angle ABD$ if $m\angle ABC = 71$, $m\angle DBC = 2x - 3$, and $m\angle ABD = 3x + 4$. **14, 46**
5. Suppose $\overline{MN}$ bisects $\angle RMQ$, $m\angle RMN = x + 2$, and $m\angle NMQ = 2x - 3$. Find the value of x and $m\angle RMQ$. **5, 14**

Motivating the Lesson

Situational Problem Ask students to imagine they are constructing a kite. As they place the cross pieces to form the structure of the kite, one of the angles formed is a right angle. Do they need to check all four non-straight angles to make sure that they are all right? Explain. **No; if one of the angles is right, all four are right.**

2 TEACH

MODELING MATHEMATICS The method of folding paper to match vertices and rays can be used in any activity to determine if the angles appear to be congruent or of differing measures.

Answers for Modeling Mathematics

c. $\angle ACE$ and $\angle DCB$; $\angle DCA$ and $\angle BCE$; their measures are equal.

d. $\angle ACE$ and $\angle BCE$; $\angle BCE$ and $\angle DCB$; $\angle DCB$ and $\angle DCA$; $\angle DCA$ and $\angle ACE$; the sum of their measures is 180.

e. Regardless of the position of the lines, vertical angles are congruent and the sum of the measures of a linear pair is 180.

Teaching Tip When discussing vertical angles, point out that they have only one point in common, the vertex. Students may use this to help them remember what vertical angles are.

Teaching Tip Point out that *linear* means "dealing with lines" and *pair* means "two." Thus, a *linear pair* means "two angles that make a line."

In-Class Example

For Example 1

Refer to the figure in the Student Edition. Suppose $m\angle GIJ = 9x - 4$ and $m\angle JIH = 4x - 11$. Find the value of x and $m\angle KIH$. **15, 131**

The measures of vertical angles and linear pairs of angles have special relationships. The following activity will explore those relationships.

MODELING MATHEMATICS

Angle Relationships

Materials:  patty paper  protractor

You can create angles formed by intersecting lines by using paper folding.

- Fold a piece of patty paper so that it makes a crease across the paper. Open the paper and trace the crease with a pencil. Place two points, *A* and *B*, on the line.

- Fold the paper again so that the second crease intersects the first crease between points *A* and *B*. Trace this crease and label the intersection as *C*. Label two points, *D* and *E*, on the second line so that *C* is between *D* and *E*.

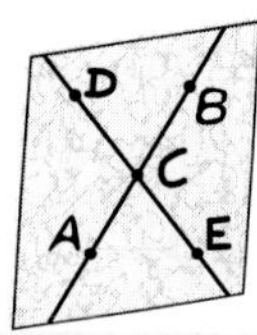

- Fold the paper again through point *C* so that $\angle ACD$ lies on top of $\angle ECB$. What do you notice?

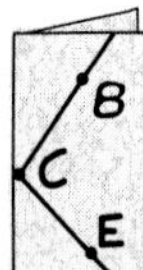

Your Turn **c–e. See margin.**

a. Fold the paper again through *C* so that $\angle ACE$ lies on $\angle DCB$. What do you notice? **$\angle ACE \cong \angle DCB$**

b. Use a protractor to measure each angle. Write the measures on your drawing. **See students' work.**

c. Name the vertical angles in this activity. What do you notice about their measures?

d. Name the linear pairs in this activity. What do you notice about their measures?

e. Repeat this activity with another piece of patty paper. What do you notice?

In the Modeling Mathematics activity, the following relationships among vertical angles and linear pairs are suggested. *You will be asked to prove these in Chapter 2.*

> **Vertical angles are congruent.**
> **The sum of the measures of the angles in a linear pair is 180.**

These relationships can help you solve algebraic problems related to geometry.

Example 

INTEGRATION Algebra

In the figure, $\overleftrightarrow{GH}$ and $\overleftrightarrow{JK}$ intersect at *I*. Find the value of x and the measure of $\angle JIH$.

$\angle GIJ$ and $\angle KIH$ are vertical angles, so $m\angle GIJ = m\angle KIH$.

$$m\angle GIJ = m\angle KIH$$
$$16x - 20 = 13x + 7$$
$$3x = 27$$
$$x = 9$$

$$m\angle GIJ = 16x - 20$$
$$= 16 \cdot 9 - 20 \text{ or } 124$$

G, J, $(16x - 20)°$, I, $(13x + 7)°$, K, H

Alternative Learning Styles

Kinesthetic Have students make two paper cones that are the same size. Then tape the tips of the cones together exactly opposite each other and flatten the cones. Trace along the edges of the cones. Using a protractor, measure each pair of vertical angles and discuss their relationship.

Since $\angle GIJ$ and $\angle JIH$ are a linear pair, $m\angle GIJ + m\angle JIH = 180$.

$m\angle GIJ + m\angle JIH = 180$
$124 + m\angle JIH = 180$
$m\angle JIH = 56$

The measure of $\angle JIH$ is 56.

Teaching Tip When discussing complementary angles, point out that obtuse angles do not have a complement because their measures are greater than 90.

In-Class Example

For Example 2
The measure of the complement of an angle is 3.5 times smaller than the measure of the supplement of the angle. Find the measure of the angle. 54

The sum of the measures of $\angle GIJ$ and $\angle JIH$ in Example 1 is 180. Two angles whose measures have a sum of 180 are called **supplementary angles**. When two angles are supplementary, each angle is said to be a *supplement* of the other angle. Similarly, if the sum of the measures of two angles is 90, the angles are called **complementary angles**. Whenever two angles are complementary, each angle is said to be the *complement* of the other angle.

Since we previously saw that the sum of the measures of a linear pair of angles is 180, we can now say that *any two angles that form a linear pair must be supplementary angles.*

Example 2

Algebra

The measure of the supplement of an angle is 60 less than three times the measure of the complement of the angle. Find the measure of the angle.

Explore Let x represent the measure of the angle. Then $180 - x$ is the measure of its supplement, and $90 - x$ is the measure of its complement.

Plan Write an equation.

The supplement *is* *60 less than three times the complement.*

$$180 - x = 3(90 - x) - 60$$

Solve
$180 - x = 3(90 - x) - 60$
$180 - x = 270 - 3x - 60$ *Use the distributive property.*
$180 - x = 210 - 3x$ *Combine like terms.*
$180 + 2x = 210$ *Add 3x to each side*
$2x = 30$ *Subtract 180 from each side.*
$x = 15$ *Divide each side by 2.*

The measure of the angle is 15.

Examine Is the measure of the supplement of a 15° angle 60 less than 3 times as large as the measure of its complement?

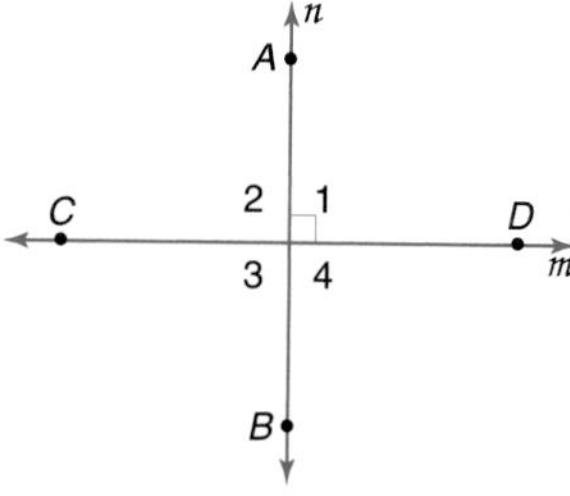

Perpendicular lines are two lines that intersect to form a right angle. In the figure at the left, lines m and n are perpendicular. To indicate this, we write $m \perp n$, which is read, "m is perpendicular to n." Line segments and rays can be perpendicular to lines or other line segments and rays if they intersect to form a right angle. For example, in the figure, $\overleftrightarrow{AB} \perp \overline{CD}$, $\overrightarrow{DC} \perp \overrightarrow{BA}$ and $\overline{AB} \perp \overline{DC}$.

Tech Prep

Construction Worker Have students find five examples of perpendicular line segments in your classroom. Point out that the construction workers who built the room had to know some geometry to be able to form the perpendiculars properly. Bring in a carpenter's T-square and show students how the tool is used. For more information on tech prep, see the *Teacher's Handbook*.

Teaching Tip When writing perpendicular statements, point out that order does not matter. For example, saying $m \perp n$ is the same as saying $n \perp m$.

Teaching Tip Students may experience some frustration when doing constructions because their end product may not be as accurate as they desire. Encourage students to use a sharp pencil and a lot of patience.

Let's examine the relationships among the measures of the angles formed by these two perpendicular lines.

- $\angle 1$ is a right angle. $\angle 1$ and $\angle 3$ are vertical angles. What can you conclude about $\angle 3$? *It is also a right angle.*
- Consider $\angle 2$ and $\angle 4$. $\angle 1$ forms a linear pair with $\angle 2$ and also with $\angle 4$. What can you conclude about $\angle 2$ and $\angle 4$? *They are both right angles.*

Thus, $\angle 1$, $\angle 2$, $\angle 3$, and $\angle 4$ are all right angles.

If you draw two different lines, ℓ and p, that are perpendicular, do you think the relationships between the four angles formed will be the same? Based on this example, we could make the following conclusion, which will be proved in Chapter 2.

Perpendicular lines intersect to form *four* right angles.

You can use a compass and a straightedge to construct a line perpendicular to a given line through a point on the line, *or* through a point *not* on the line.

CONSTRUCTION

Perpendicular Lines Through a Point on the Line

Construct a line perpendicular to line n and passing through point C on n.

1. Place the compass at point C. Using the same compass setting, draw arcs to the right and left of C, intersecting line n. Label the points of intersection A and B.

2. Open the compass to a setting greater than AC. Put the compass at point A and draw an arc above line n.

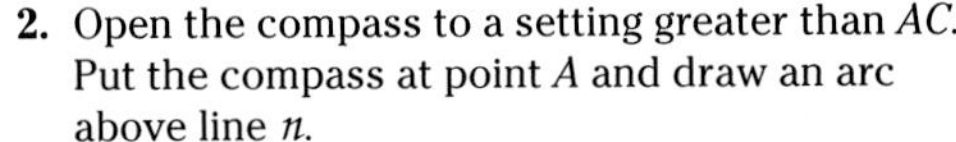

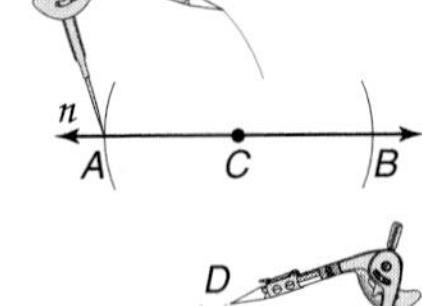

3. Using the same compass setting as in Step 2, place the compass at point B and draw an arc intersecting the arc previously drawn. Label the point of intersection D.

4. Use a straightedge to draw $\overleftrightarrow{CD}$.

Conclusion: By construction, $\overleftrightarrow{CD}$ is perpendicular to n at C.

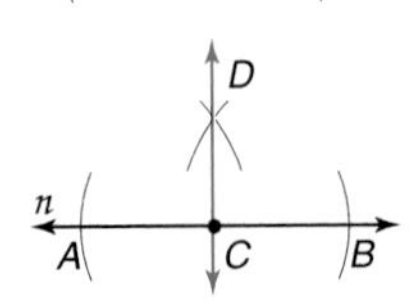

CONSTRUCTION

Perpendicular Lines Through a Point Not on the Line

Construct a line perpendicular to line m and passing through point Z *not* on m.

1. Place the compass at point Z. Draw an arc that intersects line m in two different places. Label the points of intersection X and Y.

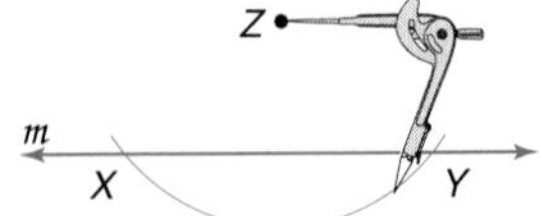

2. Open the compass to a setting greater than $\frac{1}{2}XY$. Put the compass at point X and draw an arc below line m.

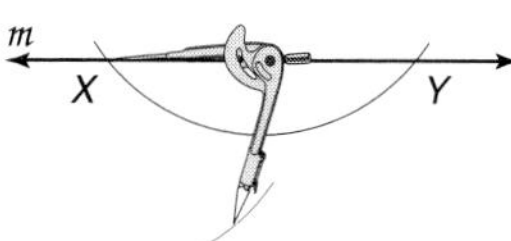

3. Using the same compass setting, place the compass at point Y and draw an arc intersecting the arc drawn in Step 2. Label the point of intersection A.

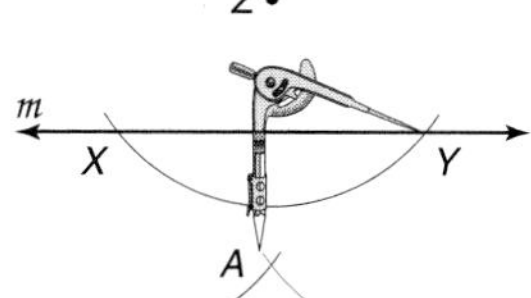

4. Use a straightedge to draw $\overleftrightarrow{ZA}$.

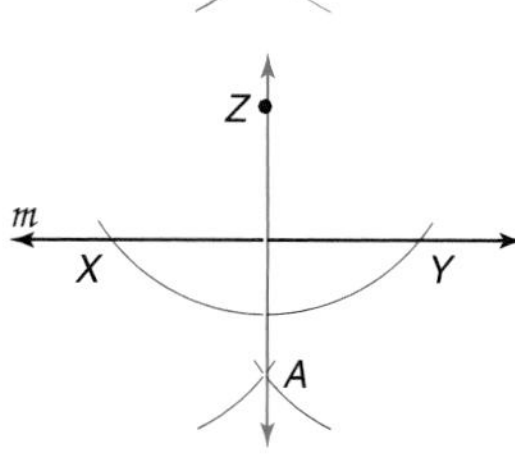

Conclusion: By construction, $\overleftrightarrow{ZA}$ is perpendicular to m.

The concept of perpendicularity can be extended to include planes. If a line is perpendicular to a plane, then the line must be perpendicular to every line in the plane that intersects it. Thus, in the figure at the right, $\overleftrightarrow{AB} \perp \mathcal{E}$ and $\overleftrightarrow{AB}$ must be perpendicular to every line in $\mathcal{E}$ that intersects it.

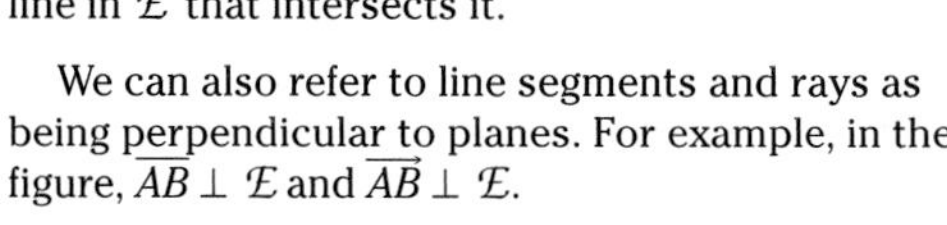

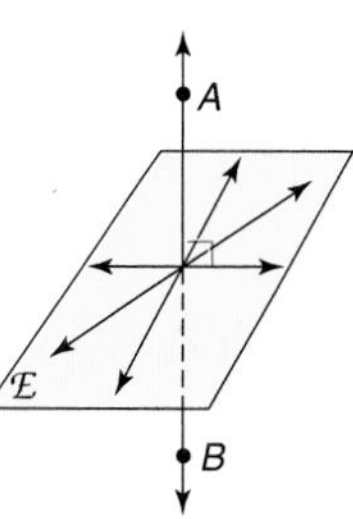

We can also refer to line segments and rays as being perpendicular to planes. For example, in the figure, $\overline{AB} \perp \mathcal{E}$ and $\overrightarrow{AB} \perp \mathcal{E}$.

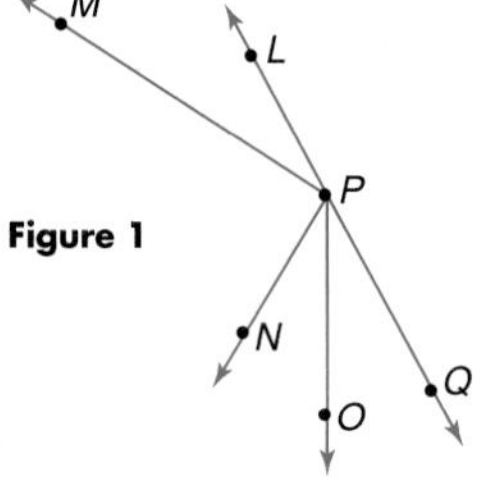

Figure 1

In this chapter, you have used figures to help describe or demonstrate different relationships among points, segments, lines, rays, and angles. Whenever you draw a figure, there are certain relationships that can be assumed from the figure and others that cannot be assumed. Consider the figure shown at the left.

Can Be Assumed from Figure 1	Cannot Be Assumed from Figure 1
All points shown are coplanar.	$\overrightarrow{PN} \perp \overrightarrow{PM}$
L, P and Q are collinear.	$\angle QPO \cong \angle LPM$
$\overrightarrow{PM}$, $\overrightarrow{PN}$, $\overrightarrow{PO}$, and $\overleftrightarrow{LQ}$ intersect at P.	$\overline{LP} \cong \overline{PQ}$
P is between L and Q.	$\overline{PQ} \cong \overline{PO}$
N is in the interior of $\angle MPO$.	$\angle QPO \cong \angle OPN$
$\angle LPQ$ is a straight angle.	$\angle OPN \cong \angle LPM$
$\angle LPM$ and $\angle MPN$ are adjacent angles.	$\overline{PO} \cong \overline{PN}$
$\angle LPN$ and $\angle NPQ$ are a linear pair.	$\overline{PN} \cong \overline{PL}$
$\angle QPO$ and $\angle OPL$ are supplementary.	

Teaching Tip When discussing the relationships between angles that can be assumed from Figure 1 on page 57, point out that other relationships exist among the angles that are not listed. You may want to have students name some of the unlisted relationships.

3 PRACTICE/APPLY

Check for Understanding

Exercises 1–14 are designed to help you assess your students' understanding through reading, writing, speaking, and modeling. You should work through Exercises 1–5 with your students and then monitor their work on Exercises 6–14.

Error Analysis

Some students may confuse supplementary and complementary angles. Encourage them to develop memory techniques to help remember the definitions. For example, 90 comes before 180 on the number line and *c* comes before *s* in the alphabet. So, associate *c* for complementary with 90 and *s* for supplementary with 180.

Additional Answer

1. Sample answer: Supplementary and complementary angles are both defined by the sum of the measures of the angles. The measures of supplementary angles have a sum of 180 and the measures of complementary angles have a sum of 90.

Study Guide Masters, p. 7

NAME______________________ DATE ____________

1-7 Study Guide Student Edition Pages 53–60

Angle Relationships

The following table identifies several different types of angles that occur in pairs.

Pairs of Angles		
Special Name	**Definition**	**Examples**
adjacent angles	angles in the same plane that have a common vertex and a common side, but no common interior points	∠EHF and ∠FHG are adjacent angles.
vertical angles	two nonadjacent angles formed by two intersecting lines (Vertical angles are congruent.)	∠1 and ∠3 are vertical angles. ∠2 and ∠4 are vertical angles. ∠1 ≅ ∠3, ∠2 ≅ ∠4
linear pair	adjacent angles whose noncommon sides are opposite rays	∠5 and ∠6 form a linear pair.
complementary angles	two angles whose measures have a sum of 90	30°, 60°
supplementary angles	two angles whose measures have a sum of 180	20°, 160°

Identify each pair of angles as adjacent, vertical, complementary, supplementary, and/or as a linear pair.

1. ∠1 and ∠2 adjacent
2. ∠1 and ∠4 vertical
3. ∠3 and ∠4 adjacent, complementary
4. ∠1 and ∠5 adjacent, supplementary, linear pair

Find the value of x.

5. 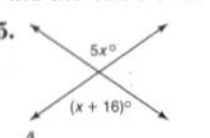

4

6. 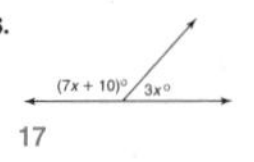

17

7. 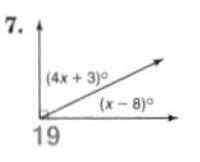

19

Compare Figure 2, shown at the right, with Figure 1 on the previous page. With these additional markings, some of the relationships that could *not* be assumed from Figure 1 *can* now be assumed from Figure 2.

$\overrightarrow{PN} \perp \overrightarrow{PM}$

$\angle QPO \cong \angle LPM$

$\overline{LP} \cong \overline{PQ}$

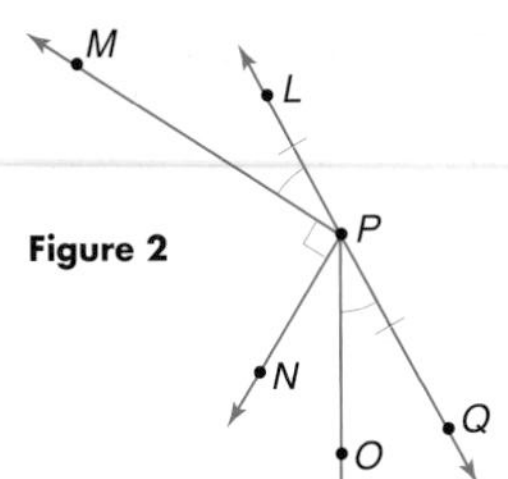

Figure 2

CHECK FOR UNDERSTANDING

Communicating Mathematics

Study the lesson. Then complete the following. 1. See margin.

1. **Compare and contrast** supplementary and complementary angles.
2. If two adjacent angles are supplementary, must they be a linear pair? Explain. See margin.
3. **Complete:** Line ℓ intersects plane C at point P and $\ell \perp C$. Thus, for any line n in C that passes through P, it must be true that __?__. $n \perp \ell$
4. Can you assume that K, L, and M are collinear from the figure at the right? Can you assume that $\overrightarrow{LN}$ bisects $\angle PLM$? that $m\angle KLP + m\angle PLN = m\angle KLN$? Explain your answers. See margin.

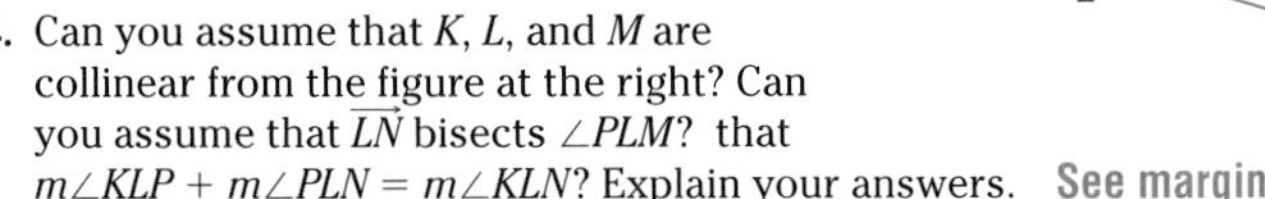

5. Write an explanation as to why you think a linear pair of angles is called "linear." See margin.

Guided Practice

Refer to the figure at the right to answer exercises 6–11.

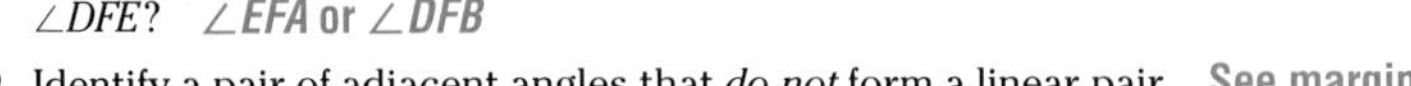

6. Identify a pair of obtuse vertical angles. ∠AFE, ∠BFD
7. Name a segment that is perpendicular to $\overrightarrow{FC}$. $\overleftrightarrow{AD}$
8. Which angle forms a linear pair with ∠DFE? ∠EFA or ∠DFB
9. Identify a pair of adjacent angles that *do not* form a linear pair. See margin.
10. Which angle is complementary to ∠CFB? ∠BFA
11. Can you assume that $\overrightarrow{FC}$ bisects $\overline{AD}$ from the figure? Explain. See margin.
12. $\angle N$ is a complement of $\angle M$, $m\angle N = 8x - 6$ and $m\angle M = 14x + 8$. Find the value of x and $m\angle M$. 4; 64
13. Find $m\angle S$ if $m\angle S$ is 20 more than four times its supplement. 148
14. **Sports** In the 1994 Winter Olympic Games, Espen Bredesen of Norway claimed the gold medal in the 90-meter ski jump. When a skier completes a jump, he or she tries to make the angle between his or her body and the front of his or her skis as small as possible. If Espen is aligned so that the front of his skis make a 15° angle with his body, what angle is formed by the tail of the skis and his body? 165° angle

Espen Bredesen

Reteaching

Using Cooperative Groups Give students various figures from other books. Have them work in small groups to name what they can assume from each figure. Have students who understand the material assist those who are having difficulty. Then have them summarize what types of information they can deduce from the figure itself.

Additional Answers

2. Yes; if the measures of the two angles have a sum of 180, their non-shared sides must be opposite rays, so they are a linear pair.
4. Yes, they lie on the same line; no, since no angle measures or markings are given; yes, because of the angle addition postulate.

EXERCISES

Practice

Refer to the figure at the right to answer Exercises 15–22.

15. Name a pair of vertical angles.
16. Can you assume that $\angle YUW$ and $\angle WUV$ are supplementary from the figure? Explain.
17. Which angle is complementary to $\angle YWT$? $\angle TWU$
18. Identify a pair of angles that are congruent.
19. Can you assume that $\overrightarrow{WT}$ is perpendicular to $\overline{YV}$? Explain. See margin.
20. Name a pair of angles that are noncongruent and supplementary.
21. Identify $\angle YWU$ and $\angle UWZ$ as congruent, adjacent, vertical, complementary, supplementary, and/or a linear pair.
22. Can you assume that $\overline{VZ} \cong \overline{WZ}$? Explain. Yes, slashes on the segments indicate that they are congruent.

15. $\angle YUW$ and $\angle XUV$ or $\angle YUX$ and $\angle VUW$

16. Yes, they form a linear pair.

18. $\angle YUW$ and $\angle XUV$ or $\angle YUX$ and $\angle VUW$

20. See margin.

21. congruent, adjacent supplementary, linear pair

Refer to the figure at the right to answer Exercises 23–26.

23. Can you determine that $\angle SRP$ and $\angle PRT$ are complementary from the figure? Explain.
24. Name two angles that are adjacent, but not complementary or supplementary.
25. Given that $\angle Q$ and $\angle QPS$ are supplementary, is $\angle Q$ obtuse, acute, or right? right
26. List four things that you *cannot* assume from the figure. Sample answer: $\overline{PT}$ bisects $\angle QPR$, $\overline{QT} \cong \overline{TR}$, $\angle SRQ$ is right, $\overline{PQ} \cong \overline{SR}$

23. No, there is no indication that $\angle SRT$ is a 90° angle.

24. $\angle SPR$ and $\angle RPT$, $\angle RPT$ and $\angle TPQ$, or $\angle SRP$ and $\angle PRT$

27. Fold a piece of patty paper to make a crease. Open the paper and label this as line a. Fold the paper again so that the two parts of line a lie on top of each other. Open the paper and label the second crease as line b.
    a. What types of lines are a and b? perpendicular
    b. Fold the paper to find the angle bisectors of two adjacent angles.
    c. What is the relationship of the lines containing the angle bisectors? Explain.
28. Find the measures of two complementary angles, $\angle A$ and $\angle B$, if $m\angle A = 7x + 4$ and $m\angle B = 4x + 9$. 53; 37
29. The measure of an angle is 44 more than the measure of its supplement. Find the measures of the angles. 112, 68
30. What are the measures of two complementary angles if the difference in the measures of the two angles is 12? 51, 39
31. Suppose $\angle T$ and $\angle U$ are complementary. Find $m\angle T$ and $m\angle U$ if $m\angle T = 16x - 9$ and $m\angle U = 4x + 3$. 67.8; 22.2
32. Find the measures of two supplementary angles, $\angle N$ and $\angle M$, if the measure of angle N is 5 less than 4 times the measure of angle M. 37, 143

33. $\angle R$ and $\angle S$ are complementary angles, and $\angle U$ and $\angle V$ are also complementary angles. If $m\angle R = y - 2$, $m\angle S = 2x + 3$, $m\angle U = 2x - y$, and $m\angle V = x - 1$, find the values of x, y, $m\angle R$, $m\angle S$, $m\angle U$, and $m\angle V$.

27b. See students' work.

27c. The angle bisectors are perpendicular, because the angles formed by the two lines measure 45° + 45° or 90°.

33. 36, 17, 15, 75, 55, 35

Assignment Guide

Core (with proof): 15–35 odd, 36, 37, 39–46
Core (informal): 15–33 odd, 36, 37, 39–46
Enriched: 16–34 even, 36–46

For **Extra Practice**, see p. 766.

The red A, B, and C flags, printed only in the Teacher's Wraparound Edition, indicate the level of difficulty of the exercises.

Additional Answers

5. The non-common sides of a linear pair of angles form a line.
9. $\angle AFB$ and $\angle BFC$, $\angle BFC$ and $\angle CFD$, or $\angle CFD$ and $\angle DFE$
11. No, because there are no markings or measures to indicate that the segments are congruent.
19. No, there are no measures or markings to indicate that right angles are present.
20. $\angle YWT$ and $\angle TWZ$, $\angle YUX$ and $\angle XUV$, $\angle XUV$ and $\angle VUW$, $\angle VUW$ and $\angle WUY$, or $\angle WUY$ and $\angle YUX$

Practice Masters, p. 7

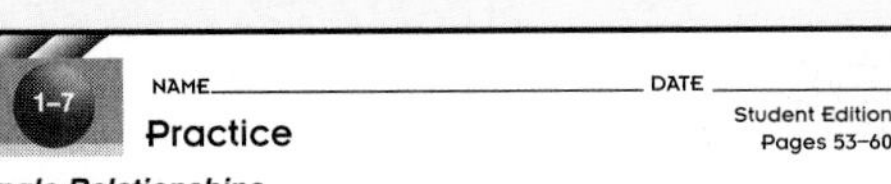

NAME ______ DATE ______

1-7 Practice — Student Edition Pages 53–60

Angle Relationships

Refer to the figure at the right to answer each question.

1. Name a pair of vertical angles. $\angle AED$ and $\angle BEC$
2. Can you assume that $\angle EFB$ is a right angle from the figure? no
3. Which angle is supplementary to $\angle FEB$? $\angle FED$
4. Can you assume $\overline{AE} \cong \overline{BE}$? yes
5. Can you assume that F bisects $\overline{AB}$ from the figure? no
6. Name an angle adjacent, but not supplementary, to $\angle DEA$. $\angle AEF$

Find the value of x and m∠ABC.

7. 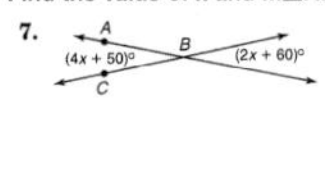

5, 70

8. 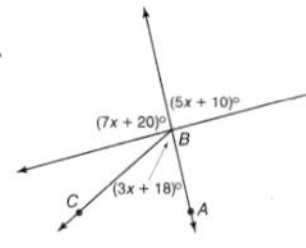

12.5, 55.5

For each figure, find the value of x. Then determine if $\overline{AB} \perp \overline{CD}$.

9. 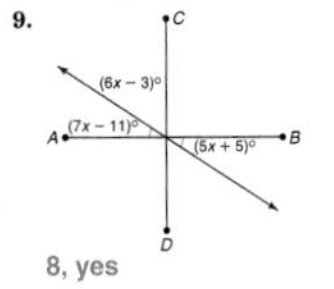

8, yes

10. 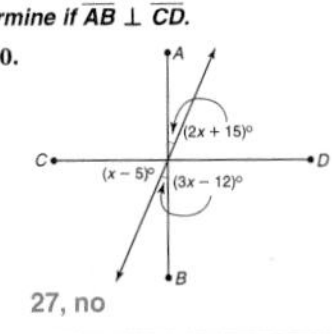

27, no

4 ASSESS

Closing Activity

Speaking Have students explain the steps in constructing a line perpendicular to line ℓ that passes through a point on line ℓ. Do this by having each student state the next step in the construction. Do the same for the construction of a line perpendicular to line ℓ and passing through a point not on ℓ.

Chapter 1 Quiz D (Lesson 1-7) is available in the *Assessment and Evaluation Masters*, p. 17.

Additional Answers

37. Sample answer: Complementary is defined as something that completes, and complementary angles complete a right angle. Supplementary is defined as something that completes or makes an addition, and supplementary angles add up to be a straight angle.

43.

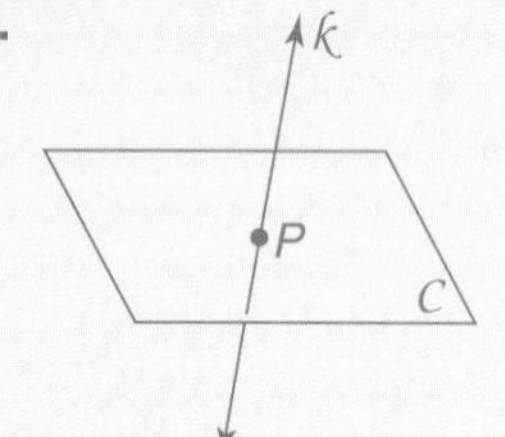

Enrichment Masters, p. 7

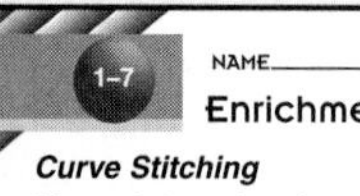

NAME____________ DATE ________

1-7 Enrichment

Student Edition Pages 53–60

Curve Stitching

The star design at the right was created by a method known as **curve stitching**. Although the design appears to contain curves, it is made up entirely of line segments.

To begin the star design, draw a 60° angle. Mark eight equally-spaced points on each ray, and number the points as shown below. Then connect pairs of points that have the same number.

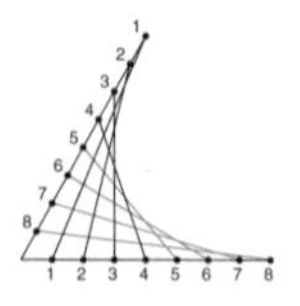

To make a complete star, make the same design in six 60° angles that have a common central vertex.

1. Complete the section of the star design above by connecting pairs of points that have the same number.

2. Complete the following design.

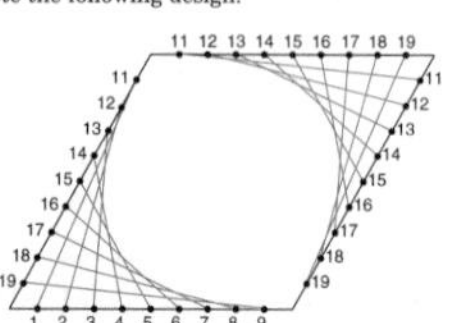

3. Create your own design. You may use several angles, and the angles may overlap. See students' work.

Proof

34. Suppose $\angle ABC$ and $\angle CBD$ form a linear pair, $\angle CBD$ and $\angle DBE$ form a linear pair, and $m\angle CBD = 30$. **34a–b. See Solutions Manual.**
   a. Draw a figure representing these angles.
   b. Write a paragraph proof showing that $\angle ABC \cong \angle DBE$.

35. $\angle PQR$ and $\angle RQS$ are complementary angles, and $m\angle PQR = 45$. Write a paragraph proof showing that $\angle PQR \cong \angle RQS$. **See Solutions Manual.**

Critical Thinking

36. Line ℓ in plane C contains point X. How many lines in the plane are perpendicular to line ℓ and pass through X? Justify your answer. **See Solutions Manual.**

Applications and Problem Solving

37. **Language** Look up the words *complementary* and *supplementary* in a dictionary. How do their everyday meanings relate to their mathematical meanings? **See margin.**

38. **Construction** A framer is installing a cathedral ceiling in a newly-built home. A protractor and a plumb bob are used to check the angle at the joint between the ceiling and wall. The wall is vertical, so the angle between the vertical plumb line and the ceiling is the same as the angle between the wall and the ceiling.

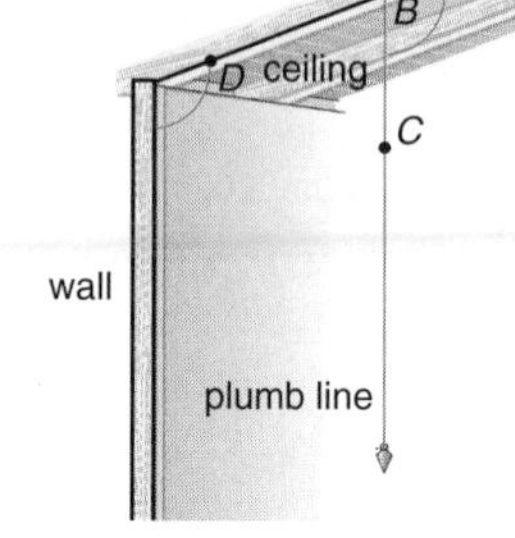

   a. How are $\angle ABC$ and $\angle CBD$ related?
   b. If $m\angle ABC = 110$, what is $m\angle CBD$? **70**
   38a. They form a linear pair.

Mixed Review

39. If $\angle Q \cong \angle R$, $m\angle Q = 8x - 17$, and $m\angle R = 7x - 3$, is $\angle Q$ acute, right, obtuse, or straight? (Lesson 1–6) **obtuse**

40. If $FG = 12x - 11$, $GH = 5x + 10$, and $FH = 36$, for what value of x are $\overline{FG}$ and $\overline{GH}$ congruent? (Lesson 1–5) **3**

41. **Safety** The National Safety Council recommends placing the base of a 15-foot ladder 5 feet from the wall. How high can the ladder safely reach? (Lesson 1–4) **about 14 ft 2 in.**

15 ft

5 ft

42. Find the width and the perimeter of a rectangle if its area is 45 square feet and its length is 18 feet. (Lesson 1–3) **2.5 ft; 41 ft**

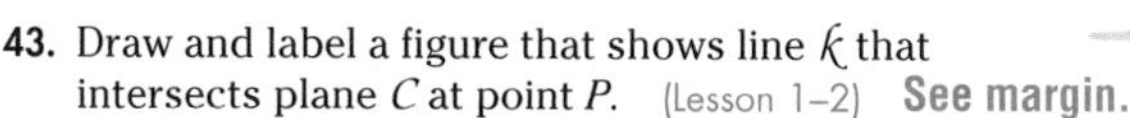

43. Draw and label a figure that shows line k that intersects plane C at point P. (Lesson 1–2) **See margin.**

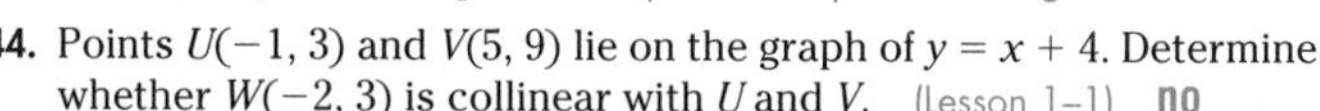

44. Points $U(-1, 3)$ and $V(5, 9)$ lie on the graph of $y = x + 4$. Determine whether $W(-2, 3)$ is collinear with U and V. (Lesson 1–1) **no**

INTEGRATION Algebra

45. Find the greatest common factor of $8ab^2$, $16a^2b$, and $12ab$. **$4ab$**

46. Factor $x^2 - 4x + 4$. **$(x - 2)^2$**

Extension

Reasoning Have students work in pairs to find the sum of the measures of the angles inside a square or rectangle. Have them draw a square or rectangle by constructing perpendicular lines. Then have them find the measures of the angles inside the figure and add them. Will this work for every square or rectangle? **yes** Have them explain why. **See students' work.**

CHAPTER 1 HIGHLIGHTS

VOCABULARY

After completing this chapter, you should be able to define each term, property, or phrase and give an example or two of each.

Geometry
- acute angle (p. 46)
- adjacent angles (p. 53)
- angle (p. 44)
- angle bisector (p. 48)
- area (p. 19)
- between (p. 28)
- collinear (p. 8)
- complementary angles (p. 55)
- congruent (pp. 31, 47)
- coplanar (p. 13)
- degree (p. 45)
- exterior of an angle (p. 45)
- informal proof (p. 39)
- interior of an angle (p. 45)
- intersection (p. 14)
- line (p. 12)
- linear pair (p. 53)
- measure (pp. 28, 45)
- midpoint (p. 36)
- noncollinear (p. 8)
- obtuse angle (p. 46)
- opposite rays (p. 44)
- paragraph proof (p. 39)
- perimeter (p. 20)
- perpendicular lines (p. 53)
- plane (p. 12)
- point (p. 12)
- postulate (p. 28)
- proof (p. 39)
- Protractor Postulate (p. 45)
- Pythagorean Theorem (p. 30)
- ray (p. 44)
- right angle (p. 46)
- Ruler Postulate (p. 28)
- segment (p. 12)
- segment bisector (p. 38)
- sides of an angle (p. 44)
- space (p. 12)
- straight angle (p. 44)
- supplementary angles (p. 55)
- theorem (p. 39)
- three-dimensional figure (p. 14)
- vertex of an angle (p. 44)
- vertical angles (p. 53)

Algebra
- coordinate plane (p. 6)
- distance formula (p. 31)
- midpoint formula (p. 37)
- ordered pair (p. 6)
- origin (p. 6)
- quadrants (p. 6)
- x-axis (p. 6)
- x-coordinate (p. 6)
- y-axis (p. 6)
- y-coordinate (p. 6)

Problem Solving
- list the possibilities (p. 13)
- problem-solving plan (p. 19)

UNDERSTANDING AND USING THE VOCABULARY

Choose the correct term to complete each sentence.

1. The x- and y-axes divide the coordinate plane into (*quadrants, perpendicular lines*).
2. An angle that measures 120° is an (*acute, obtuse*) angle.
3. Two angles whose measures have a sum of 180 are (*complementary, supplementary*).
4. Opposite rays form a (*plane, line*).
5. The intersection of two planes is a (*point, line*).
6. Adjacent angles share a (*ray, point*).
7. The intersection of the x- and y-axes is the (*origin, midpoint*).
8. A segment bisector divides a segment into (*two congruent segments, a linear pair*).
9. (*Adjacent, Vertical*) angles are congruent.
10. Statements that are assumed to be true are (*postulates, theorems*).

Using the CHAPTER HIGHLIGHTS

The Chapter Highlights begins with a listing of the new terms, properties, and phrases that were introduced in this chapter. Have students define each term and provide an example or two of it, if appropriate.

Assessment and Evaluation Masters, pp. 3–4

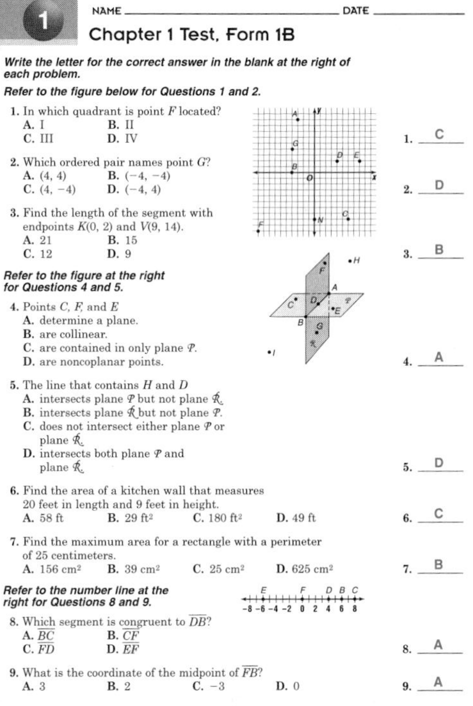

NAME ____ DATE ____

Chapter 1 Test, Form 1B

Write the letter for the correct answer in the blank at the right of each problem.

Refer to the figure below for Questions 1 and 2.

1. In which quadrant is point F located? A. I B. II C. III D. IV — 1. C
2. Which ordered pair names point G? A. (4, 4) B. (−4, −4) C. (4, −4) D. (−4, 4) — 2. D
3. Find the length of the segment with endpoints $K(0, 2)$ and $V(9, 14)$. A. 21 B. 15 C. 12 D. 9 — 3. B

Refer to the figure at the right for Questions 4 and 5.

4. Points C, F, and E A. determine a plane. B. are collinear. C. are contained in only plane $\mathcal{P}$. D. are noncoplanar points. — 4. A
5. The line that contains H and D A. intersects plane $\mathcal{P}$ but not plane $\mathcal{R}$. B. intersects plane $\mathcal{R}$ but not plane $\mathcal{P}$. C. does not intersect either plane $\mathcal{P}$ or plane $\mathcal{R}$. D. intersects both plane $\mathcal{P}$ and plane $\mathcal{R}$. — 5. D
6. Find the area of a kitchen wall that measures 20 feet in length and 9 feet in height. A. 58 ft B. 29 ft² C. 180 ft² D. 49 ft — 6. C
7. Find the maximum area for a rectangle with a perimeter of 25 centimeters. A. 156 cm² B. 39 cm² C. 25 cm² D. 625 cm² — 7. B

Refer to the number line at the right for Questions 8 and 9.

8. Which segment is congruent to $\overline{DB}$? A. $\overline{BC}$ B. $\overline{CF}$ C. $\overline{FD}$ D. $\overline{EF}$ — 8. A
9. What is the coordinate of the midpoint of $\overline{FB}$? A. 3 B. 2 C. −3 D. 0 — 9. A

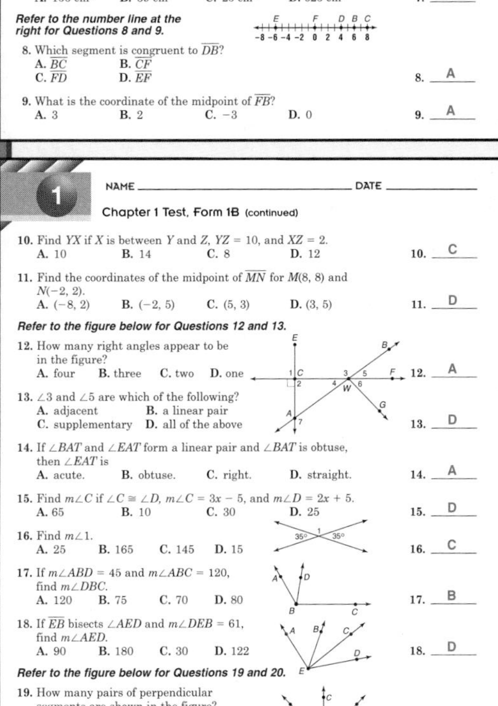

NAME ____ DATE ____

Chapter 1 Test, Form 1B (continued)

10. Find YX if X is between Y and Z, $YZ = 10$, and $XZ = 2$. A. 10 B. 14 C. 8 D. 12 — 10. C
11. Find the coordinates of the midpoint of $\overline{MN}$ for $M(8, 8)$ and $N(-2, 2)$. A. (−8, 2) B. (−2, 5) C. (5, 3) D. (3, 5) — 11. D

Refer to the figure below for Questions 12 and 13.

12. How many right angles appear to be in the figure? A. four B. three C. two D. one — 12. A
13. ∠3 and ∠5 are which of the following? A. adjacent B. a linear pair C. supplementary D. all of the above — 13. D
14. If $\angle BAT$ and $\angle EAT$ form a linear pair and $\angle BAT$ is obtuse, then $\angle EAT$ is A. acute. B. obtuse. C. right. D. straight. — 14. A
15. Find $m\angle C$ if $\angle C \cong \angle D$, $m\angle C = 3x - 5$, and $m\angle D = 2x + 5$. A. 65 B. 10 C. 30 D. 25 — 15. D
16. Find $m\angle 1$. A. 25 B. 165 C. 145 D. 15 — 16. C
17. If $m\angle ABD = 45$ and $m\angle ABC = 120$, find $m\angle DBC$. A. 120 B. 75 C. 70 D. 80 — 17. B
18. If $\overrightarrow{EB}$ bisects $\angle AED$ and $m\angle DEB = 61$, find $m\angle AED$. A. 90 B. 180 C. 30 D. 122 — 18. D

Refer to the figure below for Questions 19 and 20.

19. How many pairs of perpendicular segments are shown in the figure? A. 4 B. 3 C. 2 D. 1 — 19. A
20. Which angle is supplementary to $\angle FSC$? A. $\angle ASC$ B. $\angle CSD$ C. $\angle ASF$ D. $\angle DSF$ — 20. A

Bonus Suppose $\angle AXB$, $\angle BXC$, and $\angle CXA$ are congruent angles in the same plane with no common interior points. If the sum of their measures is 360 and $\overrightarrow{XY}$ is the bisector of $\angle AXB$, which of the following is true? A. $\angle AXY \cong \angle BXC$ B. $\angle BXC \cong \angle BXY$ C. $\overrightarrow{XC}$ and $\overrightarrow{XY}$ are opposite rays. D. $\angle AXC$ and $\angle YXB$ are vertical angles. — Bonus C

Instructional Resources

Three multiple-choice tests and three free-response tests are provided in the *Assessment and Evaluation Masters*. Forms 1A and 2A are for honors pacing, Forms 1B and 2B are for average pacing, and Forms 1C and 2C are for basic pacing. Chapter 1 Test, Form 1B, is shown at the right. Chapter 1 Test, Form 2B, is shown on the next page.

Postulates, Theorems, and Corollaries

A complete list of postulates, theorems, and corollaries begins on page 806.

Using the STUDY GUIDE AND ASSESSMENT

Skills and Concepts Encourage students to refer to the objectives and examples on the left as they complete the review exercises on the right.

Assessment and Evaluation Masters, pp. 9–10

NAME ______________________ DATE ____________

Chapter 1 Test, Form 2B

Graph each point in Questions 1 and 2 on the coordinate plane at the right.

1. $E(-3, -2)$
2. $A(3, 6)$

1–2. 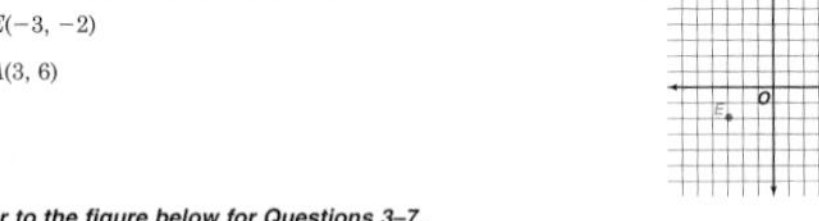

Refer to the figure below for Questions 3–7.

3. Name the intersection of $\overleftrightarrow{NO}$ and plane $\mathcal{P}$. — 3. $\overleftrightarrow{NO}$
4. If $\overleftrightarrow{SO}$ and $\overleftrightarrow{SN}$ were drawn, what would be their intersection? — 4. S
5. Name the intersection of line ℓ and $\overleftrightarrow{LM}$. — 5. none
6. Name four noncoplanar points. — 6. Sample answer: L, M, N, S
7. Write another name for plane $\mathcal{P}$. — 7. Sample answer: plane LMN
8. Find the perimeter and area of a rectangle 6 meters wide and 13 meters long. — 8. 38 m, 78 m^2

Refer to the number line below for Questions 9 and 10.

9. Find EF. — 9. $1\frac{1}{2}$
10. Find $AD - EF$. — 10. $1\frac{1}{2}$
11. Find the length of the segment with endpoints $K(-1, 1)$ and $H(3, 0)$. — 11. $\sqrt{17}$

Determine whether each statement in Questions 12–14 is true or false.

12. If C is between A and F, then $AC = CF - AF$. — 12. false
13. If D bisects $\overline{BC}$, then $DC = \frac{1}{2}CB$. — 13. true
14. If one segment bisects another, then the two segments have the same midpoint. — 14. false
15. Find the coordinates of the midpoint of segment MN given $M(10, 10)$ and $N(-2, 2)$. — 15. (4, 6)

NAME ______________________ DATE ____________

Chapter 1 Test, Form 2B (continued)

16. Find $m\angle 1$ if $m\angle 1 = 2x - 7$ and $m\angle 2 = 7x + 7$. — 16. 33
17. Find $m\angle PSR$ if $\overrightarrow{SQ}$ and $\overrightarrow{SR}$ bisect $\angle RST$ and $\angle PST$, respectively, $m\angle 1 = 5x - 11$, and $m\angle 2 = 3x + 5$. — 17. 58
18. You are enclosing a garden with prefabricated fencing you salvaged from a landfill. You found 16 pieces of fence that are each 5 feet long. What is the maximum rectangular area you can enclose with these pieces of fence? — 18. 400 ft^2

Refer to the figure below for Questions 19–21.

19. Name two right angles. — 19. $\angle STR$ and $\angle QRT$
20. Name two congruent angles that are not right angles. — 20. $\angle QRS$ and $\angle RST$
21. Name the obtuse angle. — 21. $\angle RQS$
22. The measure of an angle is two times the measure of its supplement. Find the measure of the angle. — 22. 120
23. Find the value of y in the figure at the right. — 23. 25

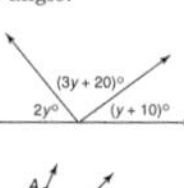

Refer to the figure below for Questions 24 and 25.

24. Find $m\angle DBC$ if $m\angle 1 = 20$, $m\angle 2 = 25$, and $m\angle ABC = 65$. — 24. 20
25. Find the value of x if $m\angle 1 = 20$, $m\angle 2 = 20 - x$, $m\angle 3 = 20 - 2x$, and $m\angle ABC = 20 + 17x$. — 25. 2

Bonus

A four-sided figure has vertices $Q(1, -2)$, $R(4, -1)$, $S(4, 2)$, and $T(1, 1)$. Find the midpoints of diagonals $\overline{QS}$ and $\overline{RT}$ and state whether the diagonals bisect each other. — Bonus $(2\frac{1}{2}, 0)$, $(2\frac{1}{2}, 0)$; yes

CHAPTER 1 STUDY GUIDE AND ASSESSMENT

SKILLS AND CONCEPTS

OBJECTIVES AND EXAMPLES

Upon completing this chapter, you should be able to:

- graph ordered pairs on a coordinate plane (Lesson 1–1)

Graph $P(2, -1)$, $Q(5, 0)$, and $R(-4, -3)$.

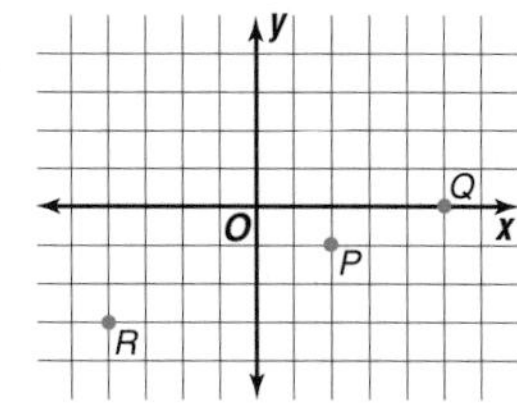

- identify collinear and coplanar points and intersecting lines and planes (Lesson 1–2)

Name the intersection of plane ACD and plane Q.

Line CD is the intersection of ACD and Q.

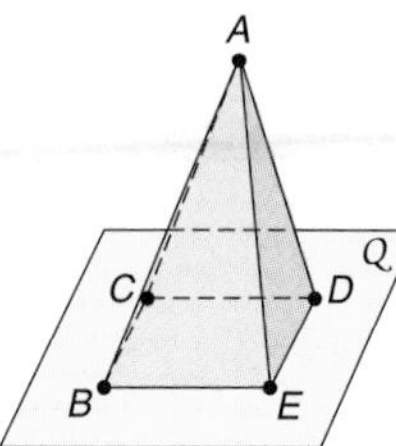

- solve problems using formulas (Lesson 1–3)

The formula $W = I \times V$ relates the electrical wattage (W), amperage (I), and voltage (V) of a circuit. If current flow through a lightbulb is 0.5 amps and it uses 150 volts, what is the wattage of the bulb?

The wattage of the bulb is 75 watts.

- find the distance between points on a number line (Lesson 1–4)

Find UW and VT on the number line shown below.

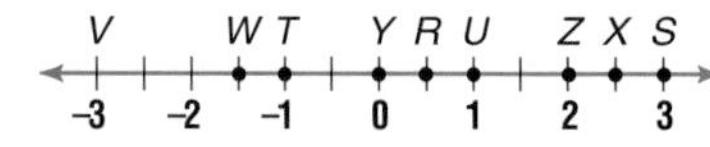

$$UW = \left|1 - \left(-1\frac{1}{2}\right)\right| \qquad VT = |-3 - (-1)|$$
$$= \left|1 + 1\frac{1}{2}\right| \qquad = |-3 + 1|$$
$$= \left|2\frac{1}{2}\right| \text{ or } 2\frac{1}{2} \qquad = |-2| \text{ or } 2$$

REVIEW EXERCISES

Use these exercises to review and prepare for the chapter test.

Graph each point on the same coordinate plane. **11–14. See margin.**

11. $S(1, 7)$
12. $T(-4, -1)$
13. $U(-8, -2)$
14. $V(0, 6)$
15. In which quadrant is the point $W(x, y)$ located if $x < 0$ and $y < 0$? **III**

Refer to the figure at the left to answer each question.

16. Are $\overline{BE}$ and $\overline{AD}$ coplanar? **no**
17. What points do planes ACB and ADE have in common? **point A**
18. Name two lines that intersect at point A. **Sample answer: $\overleftrightarrow{BA}$, $\overleftrightarrow{CA}$**

19. Use the formula $d = rt$ to find the rate r if the distance d is 135 miles and the time t is 3 hours. **45 mph**
20. Find the width of a rectangle if its perimeter is 45 inches and its length is 17 inches. **5.5 in.**
21. What is the greatest possible area of a rectangle if its perimeter is 30 centimeters? **21. 56.25 cm^2**

Refer to the number line at the left to answer each question.

22. Find the distance between X and T. **3.5**
23. What is the distance between Z and W? **3.5**
24. Which point(s) is 2 units from point U? **T, S**

GLENCOE Technology

Test and Review Software

You may use this software, a combination of an item generator and item bank, to create your own tests or worksheets. Types of items include free response, multiple choice, short answer, and open ended.

For IBM & Macintosh

Additional Answers

11–14.

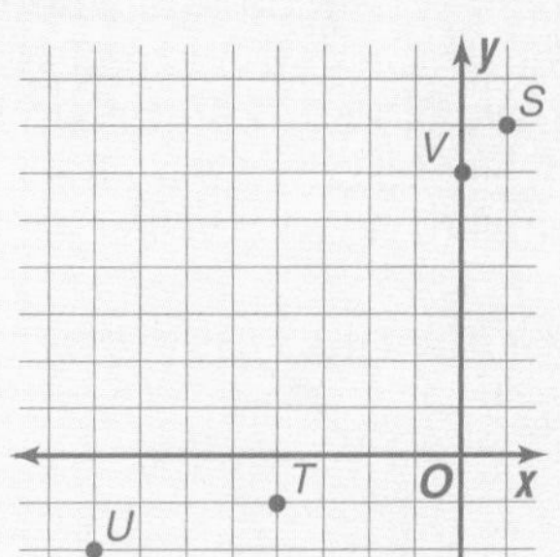

OBJECTIVES AND EXAMPLES

- find the distance between points on a coordinate plane (Lesson 1–4)

What is the distance between $F(-1, 4)$ and $G(6, -2)$?

$$d = \sqrt{(x_2 - x_1)^2 + (y_2 - y_1)^2}$$
$$= \sqrt{(6 - (-1))^2 + (-2 - 4)^2}$$
$$= \sqrt{7^2 + (-6)^2}$$
$$= \sqrt{49 + 36}$$
$$= \sqrt{85} \text{ or about } 9.22$$

REVIEW EXERCISES

Refer to the graph below to find each measure. If the measure is not a whole number, round the result to the nearest hundredth.

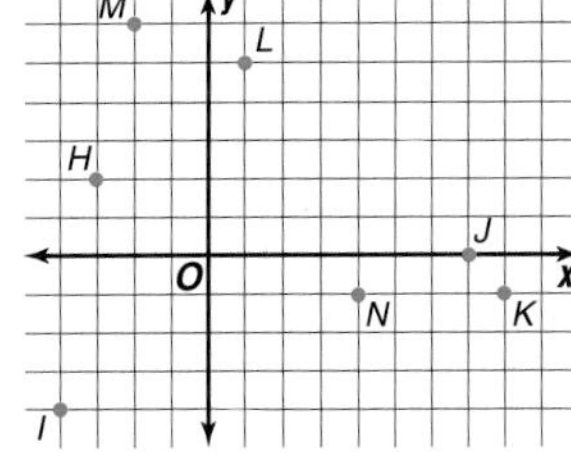

25. Find the distance between H and L. 5

26. What is the distance between J and H? 10.20

27. Find the distance between K and M. 12.21

28. What is the measure of segment IN? 8.54

- find the midpoint of a segment (Lesson 1–5)

For the graph above at the right, what is the midpoint of $\overline{HK}$?

Let $H(-3, 2)$ be (x_1, y_1).

Let $K(8, -1)$ be (x_2, y_2).

$$\left(\frac{x_1 + x_2}{2}, \frac{y_1 + y_2}{2}\right) = \left(\frac{-3 + 8}{2}, \frac{2 + (-1)}{2}\right)$$
$$= \left(\frac{5}{2}, \frac{1}{2}\right) \text{ or } \left(2\frac{1}{2}, \frac{1}{2}\right)$$

29. Refer to the graph above to find the midpoint of $\overline{NJ}$. (5.5, −0.5) **30. Sample answer: 1 and 5**

30. If the coordinate of point B on a number line is 3, name the coordinates of two points A and C such that B is the midpoint of $\overline{AC}$.

31. Algebra F is the midpoint of $\overline{EG}$. If $EF = 2x + 3$ and $EG = 6x - 3$, find FG. 12

- identify and use congruent segments (Lesson 1–5)

For the number line below, which segments are congruent to $\overline{AD}$?

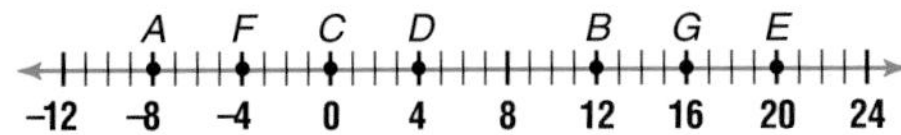

$AD = |-8 - 4|$

$= |-12|$ or 12

Since $BC = |12 - 0|$ or 12, $\overline{BC} \cong \overline{AD}$.

Since $DG = |4 - 16|$ or 12, $\overline{DG} \cong \overline{AD}$.

32. The coordinate of P on a number line is 6. If $QR = 18$, find the coordinate of S if $\overline{PS} \cong \overline{QR}$.

33. N, M, and O are collinear and $\overline{NO} \cong \overline{MO}$. Is O the midpoint of $\overline{NM}$? yes

34. Algebra $UX = 4x - 2$ and $WV = 2x + 8$. If $\overline{UX} \cong \overline{WV}$, what is the measure of $\overline{UX}$? 18

32. 24 or −12

Applications and Problem Solving Encourage students to work through the exercises in the Applications and Problem Solving section to strengthen their problem-solving skills.

Additional Answers

43.

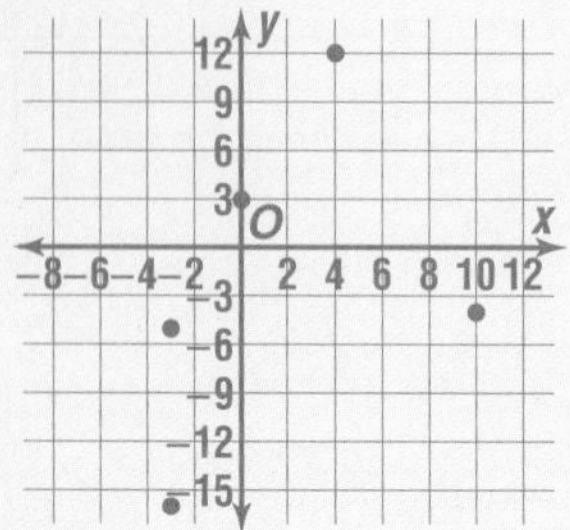

45a. As the loft of the club increases, the ball will travel higher in the air and for a shorter distance as long as the balls are struck with the same amount of force.

45b.

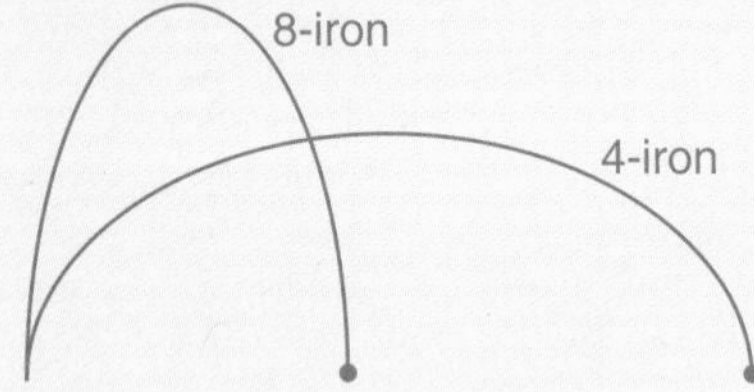

Answers to Alternative Assessment Cooperative Learning Project

See students' work for diagrams.

	(right ascension, declination)	
α	alpha	(2.09, 29.09)
δ	delta	(9.83, 30.86)
π	pi	(9.22, 33.72)
β	beta	(17.43, 35.62)
γ	gamma	(30.97, 42.33)
ν	nu	(12.45, 41.08)
υ	upsilon	(24.20, 41.41)

angular distances: $AD = 7.94$, $AP = 8.50$, $AB = 16.67$, $AG = 31.77$, $AN = 15.85$, $AU = 25.31$, $DP = 2.92$, $DB = 8.97$, $DG = 24.05$, $DN = 10.55$, $DU = 17.83$, $PB = 8.43$, $PG = 23.39$, $PN = 8.04$, $PU = 16.84$, $BG = 15.11$, $BN = 7.39$, $BU = 8.91$, $GN = 18.56$, $GU = 6.83$, $NU = 11.75$

Classification of angles: Sample answer for upsilon: obtuse: $\angle GUN$, $\angle GUB$; the 13 other angles are acute.

OBJECTIVES AND EXAMPLES

- use angle addition to find the measures of angles (Lesson 1–6)

If $m\angle AEG = 75$, $m\angle 1 = 25 - x$, and $m\angle 4 = 5x + 20$, find the value of x.

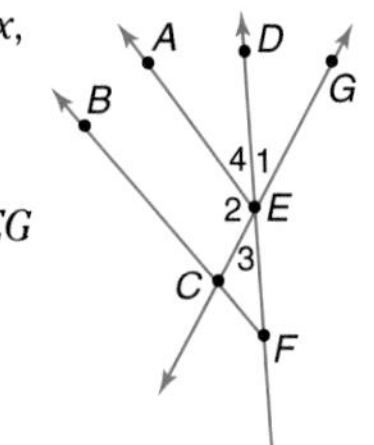

$$m\angle 1 + m\angle 4 = m\angle AEG$$
$$(25 - x) + (5x + 20) = 75$$
$$4x + 45 = 75$$
$$4x = 30$$
$$x = 7.5$$

REVIEW EXERCISES

Refer to the figure at the left for Exercises 35–37.

35. If $m\angle 3 = 32$, find $m\angle CED$. 148

36. If $m\angle 2 = 6x - 20$, $m\angle 4 = 3x + 18$, and $m\angle CED = 151$, find the value of x. 17

37. If $m\angle 1 = 49 - 2x$, $m\angle 4 = 4x + 12$, and $m\angle 2 = 15x$, find $m\angle 4$. 40

- identify and use adjacent angles, vertical angles, complementary angles, supplementary angles, linear pairs of angles, and perpendicular lines (Lesson 1–7)

Name a pair of adjacent angles and a pair of perpendicular lines.

$\angle NTP$ and $\angle PTQ$ are one pair of adjacent angles.

$\overrightarrow{TM}$ is perpendicular to $\overleftrightarrow{PS}$.

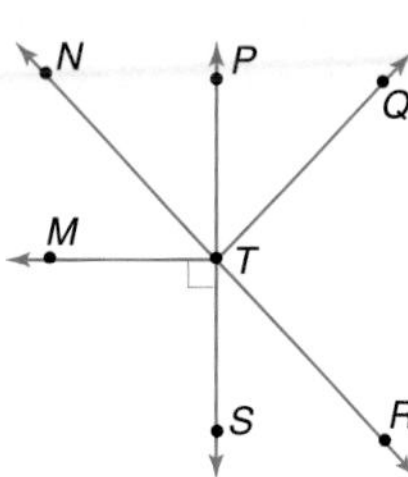

Refer to the figure at the left for Exercises 38–41. 39. $\angle PTN$ 40. They are congruent.

38. Name a pair of adjacent complementary angles. $\angle PTN$, $\angle NTM$

39. Which angle is a vertical angle to $\angle STR$?

40. If $\overline{QT}$ is perpendicular to $\overleftrightarrow{NR}$, what can you conclude about $\angle PTQ$ and $\angle NTM$?

41. **Algebra** The measure of $\angle PTR$ is twice the measure of $\angle NTP$. Find the measure of both angles. 120, 60

42. **Algebra** An angle measures 43° less than six times the measure of its complement. Find the measure of both angles. 71, 19

APPLICATIONS AND PROBLEM SOLVING

43. **Technology** Digital displays are made up of points turned on or off in a grid system. Draw a coordinate grid to represent points turned on at (0, 3), (4, 12), (−3, −16), (−3, −5), and (10, −4). (Lesson 1–1) See margin.

44. **Boat Safety** The Coast Guard recommends that the number of people that can safely occupy a boat equals the length of the boat multiplied by its width and then divided by 15. How many people can safely occupy a sailboat that is 25 feet long and 12 feet wide? (Lesson 1–3) 20 people

A practice test for Chapter 1 is provided on page 793.

45. **Sports** The face of a golf club is angled so that a ball will travel in different ways. The measure of the angle that the head of the club would form with a vertical ray is called the *loft*. Wood and iron clubs are numbered so that greater numbers indicate greater amounts of loft. For example, an 8-iron has more loft than a 4-iron. (Lesson 1–6) 45a–b. See margin.

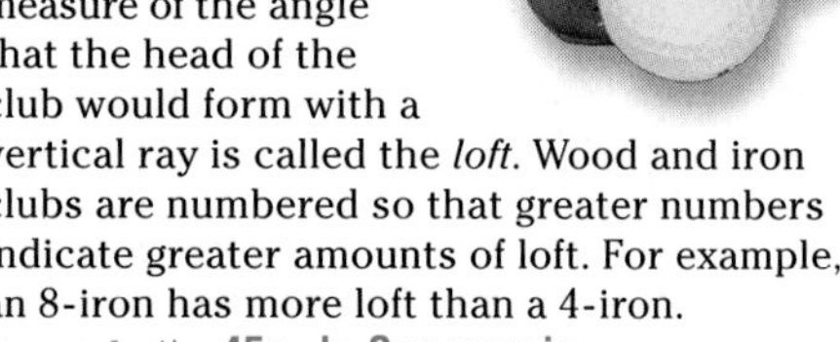

a. How do you think the loft of a club affects the path of a shot?

b. Draw a picture to show how the path of a shot hit with an 8-iron might differ from the path of a shot hit with a 4-iron.

Distances to each star (light years): alpha, 100; delta, 160, pi, 400; beta, 84; gamma, 93; nu, 440; upsilon, 53

There are 37 planes.

Answer for Thinking Critically

An ordered pair that has 0 as one of its coordinates is on an axis. The axes are not part of any quadrant; they define the quadrant boundaries

ALTERNATIVE ASSESSMENT

COOPERATIVE LEARNING PROJECT

Astronomy In this project, you will use one of the constellations in the night sky to apply the concepts you have learned in this chapter.

A star's position in the sky is given by two numbers, its *right ascension* and its *declination*. The right ascension is given in hours (h), minutes (m), and seconds (s). Each of the 24 hours is subdivided into 60 minutes, and each minute is subdivided into 60 seconds. Each hour is equivalent to 15°. For example, the star alpha Andromeda's right ascension is 0 h 8 m 23.2 s or about 2.09°.

A star's declination is given in degrees (°), minutes (′), and seconds (″). Each of the 360 degrees is subdivided into 60 minutes, and each minute is subdivided into 60 seconds. Alpha Andromeda's declination is +29° 5′ 26″ or about 29.09°.

Follow these steps to analyze the apparent geometry of the constellation Andromeda.

- Go to the library to find a book with detailed star charts. Good sources are *Starlist 2000* by Richard Dibon-Smith, *Cambridge Star Atlas 2000.0* by Wil Tirion, *The Star Guide* by Robin Kerrod, or a world atlas.
- Find the star chart for the constellation Andromeda, and locate the stars α (alpha), δ (delta), π (pi), β (beta), γ (gamma), ν (nu), and υ (upsilon). Draw a diagram of the seven stars with straight line segments connecting each pair of stars.
- Look up the right ascension and declination of each of the seven stars. Write this information as an ordered pair for each star, (right ascension, declination). Convert each measurement to the nearest hundredth degree.
- Use the Pythagorean Theorem to find the lengths of the line segments joining each pair of stars. This is known as the *angular distance* between stars.
- Use your eye to estimate whether each angle formed by each triple of stars is acute, right, or obtuse.

Follow these steps to analyze the real geometry of the constellation Andromeda.

- Look up the distances, in light years, to each of the seven stars.
- Since any three noncollinear points define a plane, any three stars define a plane in astronomical space. How many different planes are there in Andromeda? Use the strategy of listing all possibilities.

THINKING CRITICALLY

In which quadrant is the point (1, 0)? What about the points, (0, 1), (−1, 0), (0, −1), and (0, 0)? What can you conclude about any ordered pair that has 0 as one of its coordinates? Explain your answer. See margin.

PORTFOLIO

Find a photograph of a building with a large number of sides in different planes. Count the number of such planes. Keep the photograph in your portfolio.

SELF EVALUATION

Close your eyes and visualize a three-dimensional object. Imagine rotating the object slowly. Try to imagine the object from many perspectives.

Assess Yourself Did you find it easy or difficult to do this? Compare your attempt with other students. Discuss what difficulties you encountered.

Answer for Thinking Critically

An ordered pair that has 0 as one of its coordinates is on an axis. The axes are not part of any quadrant; they define the quadrant boundaries.

Assessment and Evaluation Masters, pp. 14, 25

NAME ____________ DATE ________

Chapter 1 Performance Assessment

Instructions: ***Demonstrate your knowledge by giving a clear, concise solution to each problem. Be sure to include all relevant drawings and to justify your answers. You may show your solution in more than one way or investigate beyond the requirements of the problem.***

1. Graph the points $A(3, 5)$, $B(-2, 1)$, $C(-5, 0)$, and $D(4, -4)$ on the coordinate plane at the right.
   a. Write the ordered pair that names point E.
   b. Of points A, B, C, D, and E, name three noncollinear points. Explain how you know that the points are noncollinear.
   c. List the line segments that could be drawn using the points A, B, C, D, or E as endpoints.
   d. Find the lengths of the segments between A and D and between B and C. Write a sentence comparing the lengths of the segments.
   e. Find the coordinates of the midpoint of $\overline{BD}$. Explain your method.
2. Terry is building a fence around his yard. The design of the back fence is shown below.

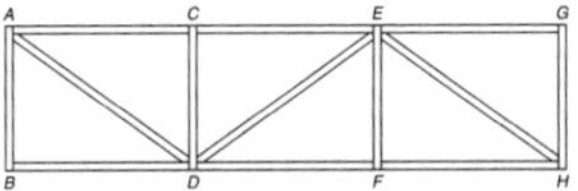

   a. Name two adjacent angles.
   b. Give two ways that $\angle BDA$ and $\angle ADF$ are related.
   c. Name two supplementary angles.
   d. Extend $\overrightarrow{DE}$ and $\overrightarrow{DF}$. Then name a point in the interior of $\angle EDF$.
   e. Name an acute angle, a right angle, an obtuse angle, and a straight angle.
   f. It appears that $\angle DEF \cong \angle HEF$. Can you assume that this is true? Explain.
3. Draw a line segment and construct the perpendicular bisector of the line segment.

Scoring Guide
Chapter 1
Performance Assessment

Level	Specific Criteria
3 Superior	• Shows thorough understanding of the concepts of *graphing ordered pairs; finding the distance between two points; finding the midpoint of a line segment; acute, right, obtuse, straight, adjacent, complementary, and supplementary angles;* and *interior of an angle.* • Computations are correct. • Written explanations are exemplary. • Graphs and diagram are accurate and appropriate. • Goes beyond requirements of some or all problems.
2 Satisfactory, with Minor Flaws	• Shows understanding of the concepts of *graphing ordered pairs; finding the distance between two points; finding the midpoint of a line segment; acute, right, obtuse, straight, adjacent, complementary, and supplementary angles;* and *interior of an angle.* • Computations are mostly correct. • Written explanations are effective. • Graphs and diagram are mostly accurate and appropriate. • Satisfies all requirements of some or all problems.
1 Nearly Satisfactory, with Serious Flaws	• Shows understanding of most of the concepts of *graphing ordered pairs; finding the distance between two points; finding the midpoint of a line segment; acute, right, obtuse, straight, adjacent, complementary, and supplementary angles;* and *interior of an angle.* • Computations are mostly correct. • Written explanations are satisfactory. • Graphs and diagram are mostly accurate and appropriate. • Satisfies most requirements of some or all problems.
0 Unsatisfactory	• Shows little or no understanding of the concepts of *graphing ordered pairs; finding the distance between two points; finding the midpoint of a line segment; acute, right, obtuse, straight, adjacent, complementary, and supplementary angles;* and *interior of an angle.* • Computations are incorrect. • Written explanations are not satisfactory. • Graphs and diagram are not accurate and appropriate. • Does not satisfy requirements of some or all problems.

Alternative Assessment

The Alternative Assessment section provides students with the opportunity to assess their own work by thinking critically, working with others, keeping a portfolio, and honestly evaluating their own progress. For more information on alternative forms of assessment, see *Alternative Assessment in the Mathematics Classroom,* one of the titles in the Glencoe Mathematics Professional Series.

Performance Assessment

Performance Assessment tasks for this chapter are included in the *Assessment and Evaluation Masters.* A scoring guide is also provided.

In•ves•ti•ga•tion

TEACHER NOTES

NCTM Standards: 1–4, 7, 8

This Investigation is designed to be completed over several days or weeks. It may be considered optional. You may want to assign the Investigation and the follow-up activities to be completed at the same time.

Objectives

Design a mural for a company lobby. Design the rest of the lobby to complement the mural.

Mathematical Overview

This Investigation will use the following mathematical skills and concepts from Chapters 2 and 3.

- using geometric patterns to create special visual effects
- using geometric shapes in art to direct a viewer's eye
- drawing three-dimensional objects using linear perspective

Recommended Time

Part	Pages	Time
Investigation	66–67	1 class period
Working on the Investigation	83, 114, 153	20 minutes each
Closing the Investigation	170	1 class period

Instructional Resources

Investigations and Projects Masters, pp. 1–4

A recording sheet, teacher notes, and scoring guide are provided for each Investigation in the *Investigations and Projects Masters*.

1 MOTIVATION

Bring in pictures of murals. Discuss styles and compositions of each. What do students like or dislike? What geometric shapes do they see? Where is their eye drawn when they first look at the mural? Why? Discuss the artist's possible reasoning for these various details.

LONG-TERM PROJECT

In•ves•ti•ga•tion

MATERIALS NEEDED

- markers
- sheet of paper 8½″x11″
- protractor
- ruler

Cygnus Software, Inc. is planning to renovate a floor of a high-rise office building to house their new corporate headquarters. The CEO (Chief Executive Officer) wants to design the office around a mural that will be placed on the wall of the lobby. Your company has been hired to create the mural and choose the furnishings, flooring, and wall coverings to complete the lobby.

FUNCTION AND FEELING

A good room design is determined by the way that the room will be used and the feeling that the designer wants to create. Discuss the image that you think Cygnus Software, Inc. would like to present to the people who visit their lobby. Use the following questions to guide your discussion.

- Should the company appear friendly, small, and warm?
- Would a professional, high-tech image be more appropriate?
- Will there be a reception desk or other furniture in the lobby?
- How much time will people spend in the lobby? Will people need to be comfortable waiting there for long periods of time?
- Will there be art other than the mural, such as smaller paintings or sculptures, in the lobby?

CHOOSING A STYLE

Art is a means of communicating. The way in which works of art communicate that message can be vastly different. Scholars classify art in

Cooperative Learning

This Investigation offers an excellent opportunity for using cooperative learning groups. For more information on cooperative learning strategies and group management, see *Cooperative Learning in the Mathematics Classroom*, one of the titles in the Glencoe Mathematics Professional Series.

different styles. Classical art, art deco, rococo, avant garde, Byzantine, Islamic, realism, and surrealism are a few examples of art styles. Classical art emphasizes order, balance and simplicity. Art deco became popular in the 1920s and 1930s and is characterized by geometric shapes. Repeated patterns of geometric shapes are common in Islamic art.

The style of art that you choose for your mural will dictate the look of the lobby. Research several different styles of art and architecture. You may wish to consult a reference book in a library or do some research on the Internet.

Study the characteristics of the style and the works of some prominent artists and architects who worked in that style. You may wish to contact an art instructor or an interior designer to help you in your research. Compare the artistic styles you are considering to the image that you believe the company would like to present. Which style communicates that message best?

Begin an Investigation folder to keep your ideas and materials organized throughout the artistic process.

You will continue working on this Investigation throughout Chapters 2 and 3.

Be sure to keep your research, discussion notes, and other materials in your Investigation folder.

Art for Art's Sake Investigation

Working on the Investigation
Lesson 2–2, p. 83

Working on the Investigation
Lesson 2–6, p. 114

Working on the Investigation
Lesson 3–4, p. 153

Closing the Investigation
End of Chapter 3, p. 170

2 SETUP

You may wish to have a student read the first four paragraphs of the Investigation to provide background information for the mural designing project. You may then wish to read the next three paragraphs that explain what the class's tasks will be. Discuss the activity with your students. Then separate the class into groups of five or more.

3 MANAGEMENT

Each group member should be responsible for a specific task.

Recorder Keeps track of information.

Designer Makes sure that the overall design correlates with the theme.

Illustrators Do the actual drawing.

Interior Designer Chooses the furniture to complement the mural.

Sample Answers

Answers will vary as they depend on mural themes chosen by each group.

Investigations and Projects Masters, p. 4

2, 3 NAME ______________ DATE ______________

Investigation, Chapters 2 and 3

Student Edition Pages 66–67, 83, 114, 153, 170

Art for Art's Sake

Work with your group to consider the following questions and to add others.

- What are the realistic choices for the style of the lobby design?
- Is there a lobby, mural, or business that you like and where you could get ideas?
- What color scheme will be used in the lobby?
- How much light is in the lobby? Is it sunlight or artificial light?
- Is there music playing in the lobby? If so, what kind?
- On which wall will the mural be painted: facing people as they walk in or beside people as they walk in?

Use the space below to list the characteristics of the artistic style that your group chose for the lobby design and mural.

Use this chart to develop a cost analysis for the lobby.

Materials and Tasks	Estimated Cost

Please keep this page and any other research in your Investigation Folder.

CHAPTER 2

Connecting Reasoning and Proof

PREVIEWING THE CHAPTER

This chapter lays the foundation for the concept of proof. The exploration of inductive and deductive reasoning strategies leads into a study of if-then statements and their logic. Conditional statements, their inverses, converses, and contrapositives prepare students for the more formal sections at the end of this chapter. Lessons 2-5 and 2-6 are devoted to developing proofs for line segments and angle theorems.

Lesson (Pages)	Lesson Objectives	NCTM Standards	State/Local Objectives
2-1 (70–75)	Make conjectures based on inductive reasoning.	1–5, 8	
2-2 (76–83)	Write a statement in if-then form. Write the converse, inverse, and contrapositive of an if-then statement. Identify and use basic postulates about points, lines, and planes.	1–5, 8	
2-2B (84)	Test the validity of conditional statements that involve mathematical sentences in one variable.	1–5	
2-3 (85–91)	Use the Law of Detachment and the Law of Syllogism in deductive reasoning. Solve problems by looking for a pattern.	1–4	
2-4 (92–99)	Use properties of equality in algebraic and geometric proofs.	1–5, 7, 8	
2-5 (100–106)	Complete proofs involving segment theorems.	1–5, 7, 8	
2-6 (107–114)	Complete proofs involving angle theorems.	1–5, 7, 8	

ORGANIZING THE CHAPTER

A complete, 1-page lesson plan is provided for each lesson in the ***Lesson Planning Guide***. Answer keys for each lesson are available in the ***Answer Key Masters***.

You may want to refer to the **Course Planning Calendar** on page T12 for detailed information on pacing.
PACING: Standard—12 days; **Honors**—11 days; **Block**—7 days

LESSON PLANNING CHART

Lesson (Pages)	Materials/ Manipulatives	Extra Practice (Student Edition)	Blackline Masters								Real-World Applications	Teaching Transparencies
			Study Guide	Practice	Enrichment	Assessment & Evaluation	Modeling Mathematics	Multicultural Activity	Tech Prep Applications	Graphing Calc. & Computer		
2-1 (70–75)	TI-92 calculator grid paper	p. 766	p. 8	p. 8	p. 8					p. 2		2-1A 2-1B
2-2 (76–83)	TI-82/83 graphing calculator	p. 766	p. 9	p. 9	p. 9	p. 44	pp. 19–22				3	2-2A 2-2B
2-2B (84)	TI-82/83 graphing calculator									pp. 17, 18		
2-3 (85–91)	patty paper straightedge*	p. 767	p. 10	p. 10	p. 10	pp. 43, 44		p. 3	p. 3			2-3A 2-3B
2-4 (92–99)		p. 767	p. 11	p. 11	p. 11		p. 80				4	2-4A 2-4B
2-5 (100–106)	ruler*	p. 767	p. 12	p. 12	p. 12	p. 45			p. 4			2-5A 2-5B
2-6 (107–114)	patty paper straightedge*	p. 768	p. 13	p. 13	p. 13	p. 45		p. 4				2-6A 2-6B
Study Guide/ Assessment (115–119)						pp. 29–42, 46–48						

*Included in Glencoe's High School Manipulative Kit and Overhead Manipulative Resources.

ORGANIZING THE CHAPTER

OTHER CHAPTER RESOURCES

Student Edition
Investigation, pp. 66–67
Chapter Opener, pp. 68–69
Mathematics and Society, p. 75
Working on the Investigation, pp. 83, 114

Teacher's Classroom Resources
Investigations and Projects Masters, pp. 29–32
Block Scheduling Booklet

Technology
Test and Review Software (IBM and Macintosh)
CD-ROM Multimedia Applications (Windows and Macintosh)
Mindjogger Videoquizzes (VHS)

Professional Publications
Glencoe Mathematics Professional Series

OUTSIDE RESOURCES

Books/Periodicals
"Technology and Reasoning in Algebra and Geometry," *Mathematics Teacher*, Feb., 1996
Geometry in the Middle Grades: Addenda Series, Grades 5–8, NCTM

Software
Exploring Geometry with the Geometer's Sketchpad, Key Curriculum Press
Geometric Probability, NCTM

Videos/CD-ROMs
Introduction to Logic, ETA, 620 Lakeview Parkway, Vernon Hills, IL 60061
Basic Math, The Teaching Company, 7405 Alban Station Court, Suite A107, Springfield, VA 22150

 ASSESSMENT RESOURCES

Student Edition
Math Journal, pp. 72, 80, 88
Mixed Review, pp. 74, 83, 90, 99, 106, 114
Self Test, p. 91
Chapter Highlights, p. 115
Chapter Study Guide and Assessment, pp. 116–118
Alternative Assessment, p. 119
Portfolio, p. 119
College Entrance Exam Practice, pp. 120–121
Chapter Test, p. 794

Teacher's Wraparound Edition
5-Minute Check, pp. 70, 76, 85, 92, 100, 107
Check for Understanding, pp. 72, 80, 88, 94, 103, 110
Closing Activity, pp. 75, 83, 91, 99, 106, 114
Cooperative Learning, pp. 79, 87

Assessment and Evaluation Masters
Multiple-Choice Tests, Forms 1A (Honors), 1B (Average), 1C (Basic), pp. 29–34
Free-Response Tests, Forms 2A (Honors), 2B (Average), 2C (Basic), pp. 35–40
Calculator-Based Test, p. 41
Performance Assessment, p. 42
Mid-Chapter Test, p. 43
Quizzes A–D, pp. 44–45
Standardized Test Practice, p. 46
Cumulative Review, pp. 47–48

ENHANCING THE CHAPTER

Examples of some of the materials for enhancing Chapter 2 are shown below.

DIVERSITY

Multicultural Activity Masters, pp. 3, 4

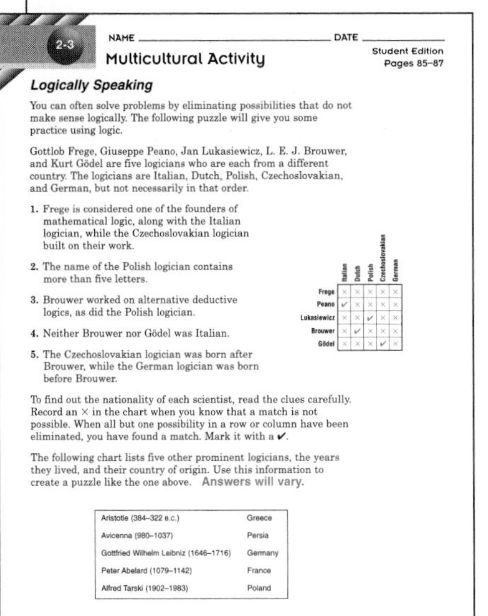

2-3 NAME ____________ DATE ________

Multicultural Activity

Student Edition Pages 85–87

Logically Speaking

You can often solve problems by eliminating possibilities that do not make sense logically. The following puzzle will give you some practice using logic.

Gottlob Frege, Giuseppe Peano, Jan Lukasiewicz, L. E. J. Brouwer, and Kurt Gödel are five logicians who are each from a different country. The logicians are Italian, Dutch, Polish, Czechoslovakian, and German, but not necessarily in that order.

1. Frege is considered one of the founders of mathematical logic, along with the Italian logician, while the Czechoslovakian logician built on their work.
2. The name of the Polish logician contains more than five letters.
3. Brouwer worked on alternative deductive logics, as did the Polish logician.
4. Neither Brouwer nor Gödel was Italian.
5. The Czechoslovakian logician was born after Brouwer, while the German logician was born before Brouwer.

	Italian	Dutch	Polish	Czechoslovakian	German
Frege	×	×	×	×	×
Peano	✓	×	×	×	×
Lukasiewicz	×	×	✓	×	×
Brouwer	×	✓	×	×	×
Gödel	×	×	×	✓	×

To find out the nationality of each scientist, read the clues carefully. Record an × in the chart when you know that a match is not possible. When all but one possibility in a row or column have been eliminated, you have found a match. Mark it with a ✓.

The following chart lists five other prominent logicians, the years they lived, and their country of origin. Use this information to create a puzzle like the one above. Answers will vary.

Aristotle (384–322 B.C.)	Greece
Avicenna (980–1037)	Persia
Gottfried Wilhelm Leibniz (1646–1716)	Germany
Peter Abelard (1079–1142)	France
Alfred Tarski (1902–1983)	Poland

APPLICATIONS

Real-World Applications, 3, 4

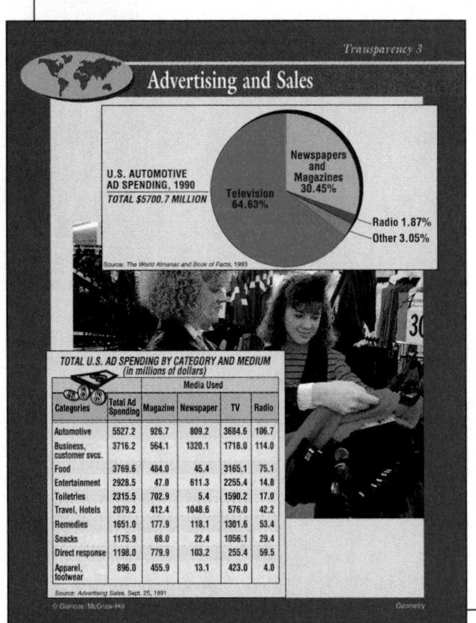

Transparency 3

Advertising and Sales

TOTAL U.S. AD SPENDING BY CATEGORY AND MEDIUM (in millions of dollars)

Categories	Total Ad Spending	Media Used: Magazine	Newspaper	TV	Radio
Automotive	5527.2	926.7	809.2	3684.6	106.7
Business, customer svcs.	3716.2	564.1	1320.1	1718.0	114.0
Food	3769.6	484.0	45.4	3165.1	75.1
Entertainment	2928.5	47.0	611.3	2255.4	14.8
Toiletries	2315.5	702.9	5.4	1590.2	17.0
Travel, Hotels	2079.2	412.4	1048.6	576.0	42.2
Remedies	1651.0	177.9	118.1	1301.6	53.4
Snacks	1175.9	68.0	22.4	1056.1	29.4
Direct response	1198.0	779.9	103.2	255.4	59.5
Apparel, footwear	896.0	455.9	13.1	423.0	4.0

Source: Advertising Sales, Sept. 25, 1991

© Glencoe/McGraw-Hill Geometry

TECHNOLOGY

Graphing Calculator and Computer Masters, p. 2

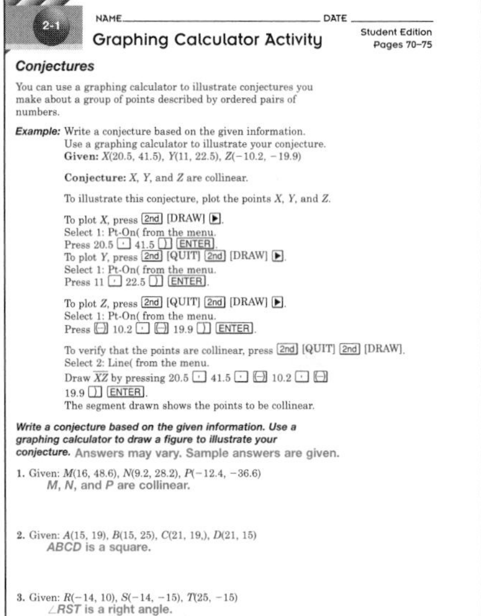

2-1 NAME ____________ DATE ________

Graphing Calculator Activity

Student Edition Pages 70–75

Conjectures

You can use a graphing calculator to illustrate conjectures you make about a group of points described by ordered pairs of numbers.

Example: Write a conjecture based on the given information. Use a graphing calculator to illustrate your conjecture.
Given: $X(20.5, 41.5)$, $Y(11, 22.5)$, $Z(-10.2, -19.9)$

Conjecture: X, Y, and Z are collinear.

To illustrate this conjecture, plot the points X, Y, and Z.

To plot X, press [2nd] [DRAW] [▶].
Select 1: Pt-On(from the menu.
Press 20.5 [,] 41.5 [)] [ENTER].
To plot Y, press [2nd] [QUIT] [2nd] [DRAW] [▶].
Select 1: Pt-On(from the menu.
Press 11 [,] 22.5 [)] [ENTER].

To plot Z, press [2nd] [QUIT] [2nd] [DRAW] [▶].
Select 1: Pt-On(from the menu.
Press [(-)] 10.2 [,] [(-)] 19.9 [)] [ENTER].

To verify that the points are collinear, press [2nd] [QUIT] [2nd] [DRAW].
Select 2: Line(from the menu.
Draw $\overline{XZ}$ by pressing 20.5 [,] 41.5 [,] [(-)] 10.2 [,] [(-)] 19.9 [)] [ENTER].
The segment drawn shows the points to be collinear.

Write a conjecture based on the given information. Use a graphing calculator to draw a figure to illustrate your conjecture. Answers may vary. Sample answers are given.

1. Given: $M(16, 48.6)$, $N(9.2, 28.2)$, $P(-12.4, -36.6)$
   M, N, and P are collinear.
2. Given: $A(15, 19)$, $B(15, 25)$, $C(21, 19,)$, $D(21, 15)$
   $ABCD$ is a square.
3. Given: $R(-14, 10)$, $S(-14, -15)$, $T(25, -15)$
   $\angle RST$ is a right angle.

TECH PREP

Tech Prep Applications Masters, pp. 3, 4

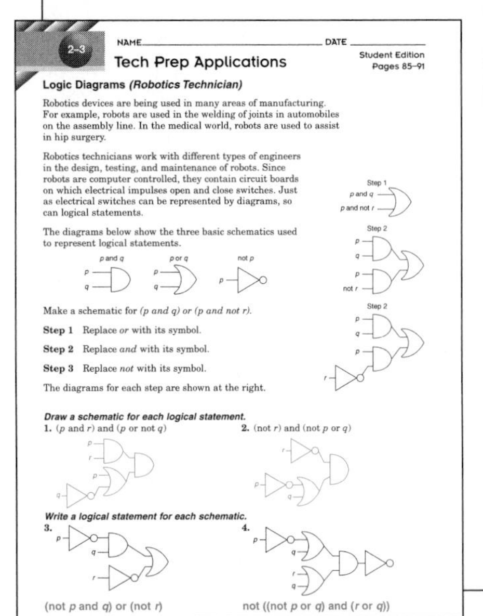

2-3 NAME ____________ DATE ________

Tech Prep Applications

Student Edition Pages 85–91

Logic Diagrams *(Robotics Technician)*

Robotics devices are being used in many areas of manufacturing. For example, robots are used in the welding of joints in automobiles on the assembly line. In the medical world, robots are used to assist in hip surgery.

Robotics technicians work with different types of engineers in the design, testing, and maintenance of robots. Since robots are computer controlled, they contain circuit boards on which electrical impulses open and close switches. Just as electrical switches can be represented by diagrams, so can logical statements.

The diagrams below show the three basic schematics used to represent logical statements.

Make a schematic for *(p and q) or (p and not r)*.

Step 1 Replace *or* with its symbol.

Step 2 Replace *and* with its symbol.

Step 3 Replace *not* with its symbol.

The diagrams for each step are shown at the right.

Draw a schematic for each logical statement.

1. (p and r) and (p or not q)
2. (not r) and (not p or q)

Write a logical statement for each schematic.

3. (not p and q) or (not r)
4. not ((not p or q) and (r or q))

PROBLEM SOLVING

Problem-of-the-Week Cards, 4, 5, 6

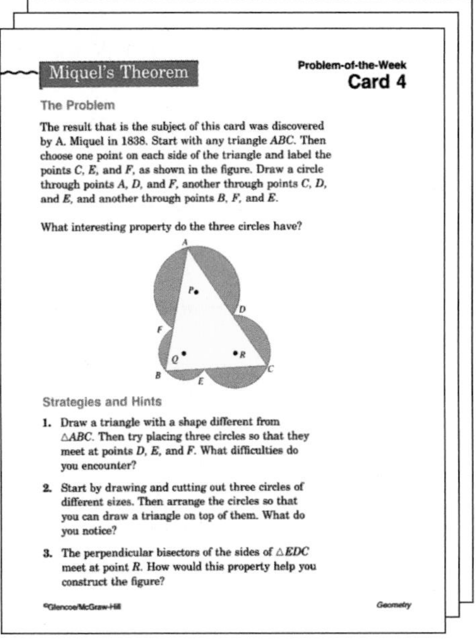

Miquel's Theorem

Problem-of-the-Week **Card 4**

The Problem

The result that is the subject of this card was discovered by A. Miquel in 1838. Start with any triangle ABC. Then choose one point on each side of the triangle and label the points C, E, and F, as shown in the figure. Draw a circle through points A, D, and F, another through points C, D, and E, and another through points B, F, and E.

What interesting property do the three circles have?

Strategies and Hints

1. Draw a triangle with a shape different from $\triangle ABC$. Then try placing three circles so that they meet at points D, E, and F. What difficulties do you encounter?
2. Start by drawing and cutting out three circles of different sizes. Then arrange the circles so that you can draw a triangle on top of them. What do you notice?
3. The perpendicular bisectors of the sides of $\triangle EDC$ meet at point R. How would this property help you construct the figure?

©Glencoe/McGraw-Hill Geometry

MAKING MATHEMATICS RELEVANT

This two-page introduction to the chapter provides students with an opportunity to explore an application of geometry that has historic roots as well as modern uses.

Background Information

In the Pursuit of Justice While the number of law school enrollments has remained steady over the past decade, the annual number of law school applications continues to increase. Competition for available seats in law schools is fierce. Students who are interested in a career in law should obtain a bachelor's degree before applying to law school. According to the *Occupational Outlook Handbook*, lawyers must be proficient in research, verbal communication, logical thinking, and data analysis. Undergraduate courses in economics, mathematics, public speaking, logic, government, philosophy, and history are recommended.

CHAPTER 2

Connecting Reasoning and Proof

Objectives

In this chapter, you will:

- make conjectures,
- use the laws of logic to make conclusions,
- solve problems by looking for a pattern,
- write algebraic proofs, and
- write proofs involving segment and angle theorems.

In the pursuit of
JUSTICE

Geometry: Then and Now A guild of lawyers and their apprentices first appeared in England in the 14th century. Since that time, the legal profession has spread around the world. According to the Occupational Outlook Handbook, lawyers and judges held approximately 716,000 jobs in the United States in 1992. Lawyers may act as legal advisers to or advocates for their clients. The details of the job depend on a lawyer's specialization. But no matter their role, all lawyers must interpret the law and apply it to their clients' situations. To interpret the law in specific situations, lawyers must be skillful in logical thinking and reasoning.

TIME Line

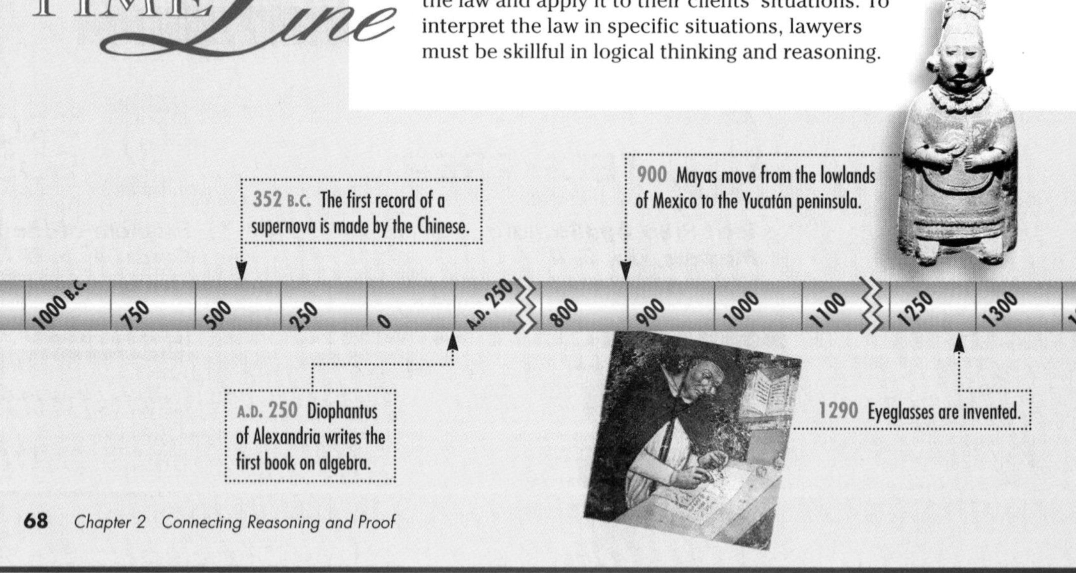

352 B.C. The first record of a supernova is made by the Chinese.
900 Mayas move from the lowlands of Mexico to the Yucatán peninsula.
1000 B.C.
750
500
250
0
A.D. 250
800
900
1000
1100
1250
1300
135
A.D. 250 Diophantus of Alexandria writes the first book on algebra.
1290 Eyeglasses are invented.

68 *Chapter 2 Connecting Reasoning and Proof*

TIME Line

Before the invention of eyeglasses in 1290, the science of optics had been developing for many centuries. The effect of lenses on light rays is describable by simple geometry.

interNET CONNECTION

Learn more about the Youth Voice Collaborative Campaign '96 Multimedia Project.

World Wide Web
http://alberti.mit.edu/projects/campaign96/about.html

PEOPLE IN THE NEWS

Fourteen-year-old **Amit Paley** of Boston, Massachusetts, is working to eliminate stereotyping of teens in the media. As a member of the *Youth Voice Collaborative* (YVC), he traveled with ten other teens to the Republican national convention in San Diego and the Democratic national convention in Chicago. Using full media credentials, these teens reported on the presidential candidates' speeches and investigated issues important to young people. Their articles and news segments were used by a major Boston newspaper and a cable news station. Amit thinks that by just being at the conventions, YVC was able to get rid of some stereotypes about young people.

Chapter Project

As you work through this chapter, you will learn how mathematicians organize their reasoning processes. Investigate the use of logical thinking and reasoning in the media. Begin by searching for examples of conditional statements in the articles and advertisements of popular magazines, newspapers, or television/radio news shows.

- If necessary, rewrite the examples you found in if-then form.
- Write the converse of each conditional statement. Are the converses true? Explain.
- Assume that each conditional is true. For each conditional, write (a) an example showing the correct use of the laws of logic to reach a valid conclusion, and (b) an example showing incorrect use of the laws of logic.
- Interview a lawyer in your hometown. How does he or she use reasoning to make logical arguments? Does he or she use inductive or deductive reasoning more often? Write a report explaining your findings.

1520 Chocolate is brought from Mexico to Spain.

1742 Charles Viner completes the 23-volume *Legal Encyclopaedia*.

1870 Esther Morris becomes the first female judge in the U.S.

1991 Whoopi Goldberg wins an Oscar for her performance in *Ghost*.

1400 1450 1500 1550 1600 1650 1700 1750 1800 1850 1900 1950 2000

PEOPLE IN THE NEWS

Encourage students to discuss the treatment of teenagers by the media in their hometowns. What are some positive ways that they could help to break negative stereotypes of teenagers?

Chapter Project

Cooperative Learning Students might want to work in groups of three or four for this project to expedite the required research. Ask students to share the examples of conditional statements that they found with the rest of the class. You may choose to invite a local attorney to speak to the entire class about using logical arguments in his or her work. If so, ask each group to think of several questions to ask the attorney.

Investigations and Projects Masters, p. 29

NAME ______ DATE ______

2 Chapter 2 Project A

Student Edition Pages 68–119

A Classroom Trial

1. Your class will conduct a mock trial. You may use one of the following scenarios or create one of your own.
   A. A teacher in your school has accused a student of cheating on an exam. The student denies the accusation, but the teacher says that he or she saw the student talking with another student and looking at the other student's paper.
   B. One student in your class accuses another student of stealing his or her calculator. The accused student denies the charge and insists that he or she bought the calculator at a store, but cannot produce a receipt.

   If you choose one of the events above, your class may need to add details (such as witnesses, background information, or extenuating circumstances) in order for the trial to take place.
2. Everyone in the class will play a part in the trial. Students should play the parts of defense attorney(s), prosecuting attorney(s), jurors, and witnesses (if any). The teacher will act as judge. Students should prepare individually for their role in the trial. Also, students should meet and prepare with other students who are involved with their part in the trial.
3. Conduct a complete trial, including having the jury panel reach a verdict.
4. After the trial, discuss as a class how logical reasoning was used in the case. What factors played the largest part in the trial? Do you think having different students play different roles would have changed the strategy or the outcome of the trial? Why or why not? What other factors do you think could have influenced the trial?

Alternative Chapter Projects

Two other chapter projects are included in the *Investigations and Projects Masters*. In Chapter 2 Project A, pp. 29–30, students extend the topic in the chapter opener. In Chapter 2 Project B, pp. 31–32, students research various inventions and the mathematics they utilize.

2-1 LESSON NOTES

NCTM Standards: 1–5, 8

Instructional Resources

- Study Guide Master 2-1
- Practice Master 2-1
- Enrichment Master 2-1
- Graphing Calculator and Computer Masters, p. 2

Transparency 2-1A contains the 5-Minute Check for this lesson; **Transparency 2-1B** contains a teaching aid for this lesson.

Recommended Pacing	
Standard Pacing	Day 1 of 12
Honors Pacing	Day 1 of 11
Block Scheduling*	Day 1 of 7

*For more information on pacing and possible lesson plans, refer to the *Block Scheduling Booklet*.

1 FOCUS

5-Minute Check
(over Chapter 1)

Graph each point on the same coordinate plane.

1. $A(4, -1)$
2. $B(0, 3)$
3. $C(-1, -2)$
4. $D(-2, 0)$

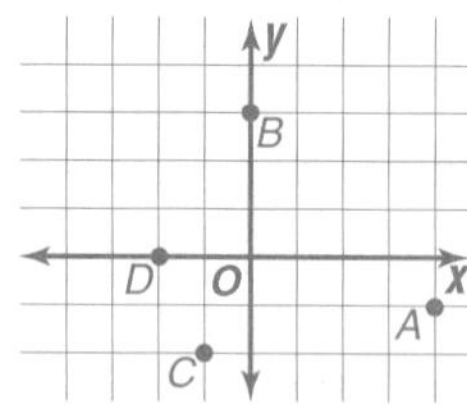

y
B
D
O
x
A
C

5. Find the length of a segment with endpoints $A(2, 5)$ and $B(-1, 1)$. **5**
6. If $\angle A$ is supplementary to $\angle B$ and the measure of $\angle A$ is twice the measure of $\angle B$, find the measure of both angles. **60, 120**

2-1 Inductive Reasoning and Conjecturing

What YOU'LL LEARN

- To make conjectures based on inductive reasoning.

Why IT'S IMPORTANT

Inductive reasoning and conjecturing help you arrive at valid conclusions.

APPLICATION

Comics

In the cartoon *Shoe* shown below, Skyler observed two specific situations in order to make the **conjecture**, "I'll be the next president of Mexico."

SHOE

CHURCHILL DID BADLY IN ENGLISH...
EINSTEIN ONCE FLUNKED MATH.

AT THE RATE I'M GOING IN SPANISH,

I'LL BE THE NEXT PRESIDENT OF MEXICO.

FYI

Albert Einstein did so poorly in school that his mother suggested he study music. He eventually became an accomplished violinist.

A conjecture is an educated guess. Looking at several specific situations to arrive at a conjecture is called **inductive reasoning**. For centuries, mathematicians have used inductive reasoning to develop the geometry that we study today.

Developing skills in games or sports usually involves some experimentation followed by conjectures about how to effectively participate in the activity. A game that involves striking a ball to make it go a specific direction is *carom billiards.* A carom billiard table is like a pool table without pockets. It is twice as long as it is wide, usually 10 feet by 5 feet.

Practice may improve your game. Knowledge of geometry could also be helpful in becoming a better player.

Example 1

APPLICATION

Billiards

This example does not take into account other factors that may affect the path of the ball, such as friction, spin, and the various materials of which the ball and table are made.

If you place the ball in one corner of a 10-foot by 5-foot carom billiard table and shoot it at a 45° angle with respect to the sides of the table, where will it go, assuming there are no other balls on the table? Use graph paper to draw the path of the ball. Make a conjecture about the path of the ball if it is shot at a 45° angle from any corner of the table.

Start

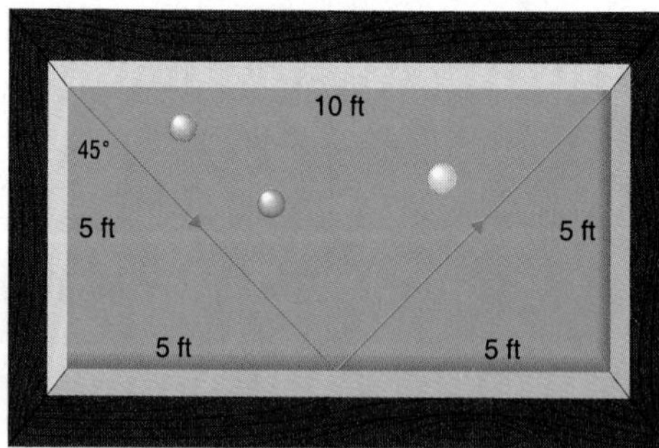

10 ft
45°
5 ft
5 ft
5 ft
5 ft

FYI

Albert's second cousin, Alfred Einstein, was a world authority on musicology. He received his doctorate in music from the University of Munich and wrote a wide variety of books on music.

Some conjectures are:

- The ball will strike the long side of the table at its midpoint.
- The ball will then bounce off the rail at the same angle.
- The ball will continue on a path and touch the opposite corner.

What happens to the path of the ball if we change the dimensions of the table?

In Chapter 1, you had some experience with basic geometric ideas. You can use some of these ideas to make conjectures in geometry.

Example  **For points A, B, and C, $AB = 10$, $BC = 8$, and $AC = 5$. Make a conjecture and draw a figure to illustrate your conjecture.**

Some conjectures can be made from a figure. However, you should not judge measurements from the appearance of a figure.

Given: Points A, B, and C, $AB = 10$, $BC = 8$, and $AC = 5$.

Conjecture: A, B, and C are noncollinear.

B, A, C; 10, 8, 5

Conjectures are made based on observations of a particular situation. You can use a TI-92 to make conjectures about the properties of a rectangle.

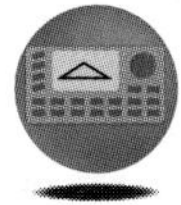

EXPLORATION CABRI GEOMETRY

Use 4:Polygon on the F3 menu to draw a rectangle named $ABCD$.

Press ENTER when you want to locate each corner of the rectangle.

Your conjecture is that you have drawn a rectangle with opposite sides that are equal in length and angles that are right angles. How could you test your conjecture?

Your Turn

a. Use the Distance & Length option on the F6 menu to find the measures of the sides of $ABCD$. Then measure the angle at each corner. Is your drawing a rectangle? Explain. **See students' work.**

b. Use the hand key to adjust the appearance of your figure so that it is a rectangle. Then draw $\overline{AC}$ and $\overline{BD}$. Make a conjecture about the relationship of the lengths $\overline{AC}$ and $\overline{BD}$. How could you test your conjecture? Is your conjecture true?

b. Sample answer: The lengths are equal. You could test your conjecture by using the Distance & Length option on the calculator to measure the lengths of $\overline{AC}$ and $\overline{BD}$. See students' work.

A conjecture based on several observations may be true or false.

Example  **Eric Pham was driving his friends to school when his car suddenly stopped two blocks away from school. Make a list of conjectures that Eric can make and investigate as to why his car stopped.**

APPLICATION
Transportation

Some conjectures are:

- The car ran out of gas.
- The battery cable lost its contact with the battery.

Can you make some more conjectures about what might have happened to Eric's car?

Motivating the Lesson

Questioning Tell students, "I have been to Pottersville three times in my life. Every time I went, it was raining. I think it must always rain in Pottersville." Ask students to discuss this type of reasoning.

2 TEACH

In-Class Examples

For Example 1
When a door is open, the angle the door makes with the door frame is complementary to the angle the door makes with the wall. Write a conjecture about the relationship of the measures of the two angles. **The measures add up to 90.**

For Example 2
For points A, B, C, and D, $AB = 5$, $BC = 10$, $CD = 8$, and $AD = 12$. Make a conjecture and draw a figure to illustrate your conjecture.
Given: Points A, B, C, and D, $AB = 5$, $BC = 10$, $CD = 8$, and $AD = 12$.
Conjecture: **The points form a four-sided figure.**

For Example 3
Rayna was preparing toast for breakfast. After a few minutes the bread popped up but was not toasted. Make a list of conjectures that Rayna can make as to why the bread was not toasted. **The toaster was set for very light toast. The heating element burned out.**

EXPLORATION

When using the TI-92, remind students that they must press ENTER to create a point. They can easily take this opportunity to conjecture about and measure the angles formed by $\overline{AC}$ and $\overline{BD}$ inside their rectangles.

Alternative Teaching Strategies

Reading Geometry Explain that the word *conjecture* can either be a noun or a verb. Typical synonyms for conjecture include *guess, proposition, speculation, conclusion, suspicion, opinion,* and *supposition*. Have students brainstorm examples of each term outside mathematics. Have students compare each example with the way mathematicians use *conjecture*.

In-Class Examples

For Example 4
Given that the points A, B, and C are collinear and B is between A and C, Juanita made a conjecture that B is the midpoint of $\overline{AC}$. Determine if her conjecture is *true* or *false*. Explain your answer. **False; although A, B, and C lie on the same segment, and B is between A and C, B could be closer to A than to C and, therefore, not be the midpoint.**

Teaching Tip After Example 4, ask students if one counterexample is enough to determine if a conjecture is false. Have them explain their answers.

3 PRACTICE/APPLY

Check for Understanding

Exercises 1–11 are designed to help you assess your students' understanding through reading, writing, speaking, and modeling. You should work through Exercises 1–5 with your students and then monitor their work on Exercises 6–11.

Additional Answers

1. **A conjecture is a guess, based on limited evidence, about what will *always* be true.**
2. **Sample answer: There is no way to draw a straight line through three points lying on the same circle.**
3. **Sample answer: Skyler looked at only two cases in history where great people did poorly academically. If he had looked at other cases, he could have made different observations.**
7. **If ℓ and m are perpendicular, then they form a right angle.**
10. **counterexample:**

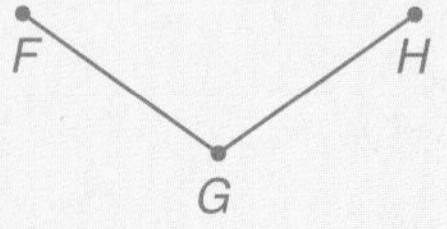

Like the character in the cartoon *Shoe*, we sometimes make a conjecture and later determine that the conjecture is false. It takes only one false example to show that a conjecture is not true. The false example is called a **counterexample**.

Example 4 **Given that points P, Q, and R are collinear, Joel made a conjecture that Q is between P and R. Determine if his conjecture is *true* or *false*. Explain your answer.**

Given: Points P, Q, and R are collinear.

Conjecture: Q is between P and R.

The figure at the right can can be used to disprove the conjecture. In this case, P, Q, and R are collinear and R is between Q and P. Since we can find a counterexample for the conjecture, the conjecture is false.

CHECK FOR UNDERSTANDING

Communicating Mathematics

Study the lesson. Then complete the following. 1–3. See margin.

1. **Explain** the meaning of *conjecture* in your own words.
2. **Describe** why three points on a circle could never be collinear.
3. **Explain** how Skyler's reasoning about Churchhill and Einstein in the application at the beginning of the lesson led him to a false conjecture. What could he have done differently to show that the conjecture was false?
4. **Explain** how you can prove that a conjecture is false. Give a counterexample.

5. **Assess Yourself** Describe a situation in which you had several experiences that led you to make a true conjecture. Then describe a situation where you had several experiences that led you to make a false conjecture. See students' work.

Guided Practice

6. Determine if the conjecture is *true* or *false* based on the given information. Explain your answer.
   Given: $\angle 1$ and $\angle 2$ are supplementary angles.
   $\angle 1$ and $\angle 3$ are supplementary angles.
   Conjecture: $\angle 2 \cong \angle 3$

6. True; if $m\angle 1 = 70$, then $m\angle 2 = 180 - 70$, or 110, and $m\angle 3 = 180 - 70$, or 110. Since $m\angle 2 = m\angle 3$, then $\angle 2 \cong \angle 3$.

Write a conjecture based on the given information. If appropriate, draw a figure to illustrate your conjecture.

7. **Given:** Lines ℓ and m are perpendicular. See margin.
8. **Given:** $A(-1, 0)$, $B(0, 2)$, $C(1, 4)$
9. Points H, I, and J are each located on different sides of a triangle. Make a conjecture about points H, I, and J. Points H, I, and J are noncollinear.
10. Determine if the conjecture is *true* or *false*. Explain your answer and give a counterexample if the conjecture is false.
    Given: $\overline{FG} \cong \overline{GH}$
    Conjecture: G is the midpoint of $\overline{FH}$.

8. Points A, B, and C are collinear; see Solutions Manual for figure.

10. False; see margin for counterexample.

Reteaching

Using Conjectures Have students write the letter of the true conjecture that can be made from the given information.

1. Given: $\angle A$ is complementary to $\angle B$.
   a. $m\angle A < m\angle B$
   b. $m\angle A + m\angle B = 180$
   c. $m\angle A + m\angle B = 90$ c
2. Given: $\overline{AB}$ is perpendicular to $\overline{BC}$ at B.
   a. A, B, and C are collinear.
   b. $m\angle ABC = 90$
   c. $AB < BC$ b

11. Sample answer: Earth is flat, Earth is the center of the universe.

11. **Astronomy** For thousands of years, astronomers believed that Earth, Mercury, Venus, Mars, Jupiter, and Saturn were the only planets in the Solar System. That conjecture was proved to be false when Uranus, Neptune, and Pluto were discovered during the last couple of centuries. Name some other famous conjectures that proved to be false.

EXERCISES

Practice

Determine if each conjecture is *true* or *false* based on the given information. Explain your answer.

12. **Given:** points D, E, F, and G
Conjecture: D, E, F, and G are noncollinear. See margin.

13. **Given:** collinear points X, Y, and Z; Z is between X and Y.
Conjecture: $XY + YZ = XZ$

13. False; $XZ + YZ = XY$ by the Segment Addition Post; see margin for figure.

14. **Given:** noncollinear points L, M, and N
Conjecture: $\overline{LM}$, $\overline{MN}$, and $\overline{LN}$ form a triangle.

14. True; any three noncollinear points can be the vertices of a triangle.

Write a conjecture based on the given information. If appropriate, draw a figure to illustrate your conjecture.

15. **Given:** Points A, B, and C are noncollinear. See margin.
16. **Given:** $AB = CD$ and $CD = EF$. $AB = EF$
17. **Given:** $\overleftrightarrow{XY}$ and $\overleftrightarrow{ZW}$ intersect at A. See margin.
18. **Given:** $\angle 1$ and $\angle 2$ are adjacent. See margin.
19. **Given:** $R(3, -4)$, $S(-2, -4)$, $T(0, -4)$ See margin.

INTEGRATION Algebra

20. **Given:** $x > y$, $y > 5$ $x > 5$

Make a conjecture about points *A*, *B*, *C*, and *D* based on the given information. 21. $m\angle ABD = m\angle CBD$

21. $\overrightarrow{DB}$ is an angle bisector of $\angle ABC$.

22. $ABCD$ is a square.

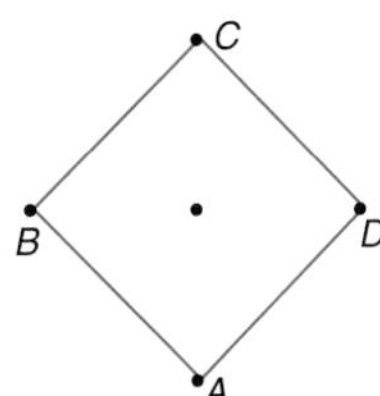

22. $\angle A$, $\angle B$, $\angle C$, and $\angle D$ are right angles.

Determine if each conjecture is *true* or *false*. Explain your answer and give a counterexample for any false conjecture.

23. **Given:** $WX = XY$
Conjecture: W, X, and Y are collinear. See Solutions Manual.

24. **Given:** $PQRS$ is a rectangle.
Conjecture: $PQ = RS$ and $QR = SP$ true

25. **Given:** $K(-1, 0)$, $L(1, 1)$, $M(5, 3)$ See Solutions Manual.
Conjecture: Points K, L, and M form a triangle.

26. **Given:** x is an integer.
Conjecture: $-x$ is negative.

26. False; counterexample: If $x = -2$, then $-x = -(-2) = 2$.

Assignment Guide

Core (with proof): 13–29 odd, 30, 31, 33–40
Core (informal): 13–29 odd, 30, 31, 33–40
Enriched: 12–28 even, 30–40

For **Extra Practice,** see p. 766.

The red A, B, and C flags, printed only in the Teacher's Wraparound Edition, indicate the level of difficulty of the exercises.

Additional Answers

12. false; counterexample:

D E F G

13. X Y Z

15. Points A, B, and C do not lie on a line.

A B C

17. X, Y, Z, and W are noncollinear.

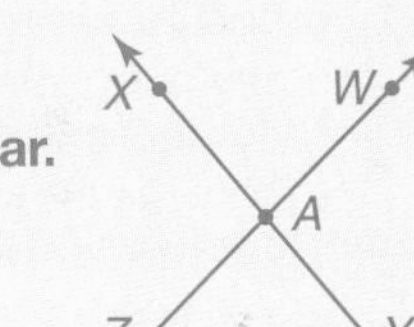

Study Guide Masters, p. 8

2–1 NAME______ DATE______

Study Guide

Student Edition Pages 70–75

Inductive Reasoning and Conjecturing

In daily life, you frequently look at several specific situations and reach a general conclusion based on these specific cases. For example, you might receive excellent service in a restaurant several times and conclude that the service is always good. Of course, you are not guaranteed that the service will be good when you return.

This type of reasoning, in which you look at several facts and then make an educated guess based on these facts, is called **inductive reasoning**. The educated guess is called a **conjecture**. Not all conjectures are true. When you find an example that shows the conjecture is false, this example is called a **counterexample**.

Example: Determine if the conjecture is *true* or *false* based on the given information. Explain your answer and give a counterexample if false.

Given: $\overline{AB} \cong \overline{BC}$
Conjecture: B is the midpoint of AC.

In the figure, $\overline{AB} \cong \overline{BC}$, but B is not the midpoint of $\overline{AC}$. So the conjecture is false.

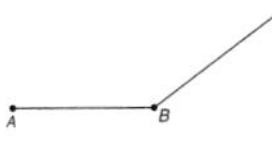

Determine if each conjecture is true or false based on the given information. Explain your answer and give a counterexample for any false conjecture.

1. **Given:** Collinear points D, E, and F.
Conjecture: $DE + EF = DF$. False; if F is between D and E (or if D is between E and F), then $DE + EF \neq DF$. D F E E D F

2. **Given:** $\angle A$ and $\angle B$ are supplementary.
Conjecture: $\angle A$ and $\angle B$ are adjacent angles. False; if $\angle A$ and $\angle B$ do not have the same vertex, they cannot be adjacent.

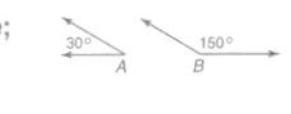

3. **Given:** $\angle D$ and $\angle F$ are supplementary.
$\angle E$ and $\angle F$ are supplementary.
Conjecture: $\angle D \cong \angle E$ True; if $m\angle F = 25$, them $m\angle D = 180 - 25$ or 155 and $m\angle E = 180 - 25$ or 155. Since $m\angle D = m\angle E$, then $\angle D \cong \angle E$.

4. **Given:** $\overline{AB}$ is perpendicular to $\overline{BC}$.
Conjecture: $\angle ABC$ is a right angle. True, if the sides of an angle are perpendicular, the angle is a right angle.

Additional Answers

18. $\angle 1$ and $\angle 2$ have a common side and a common vertex.

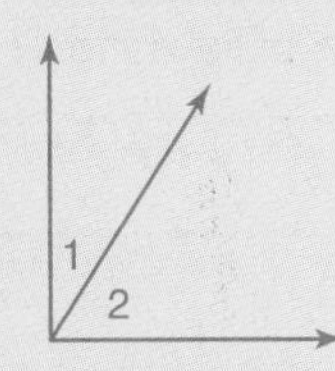

19. Points R, S, and T are collinear.

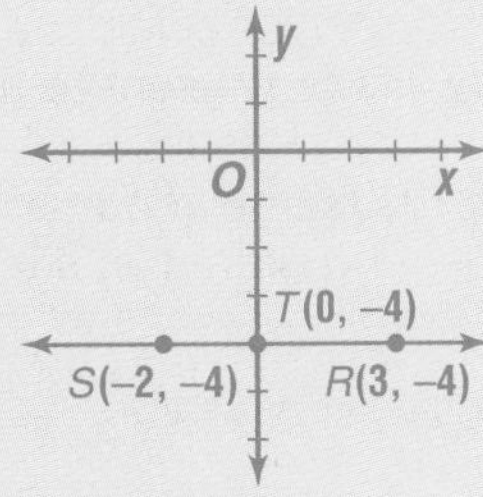

Additional Answers

27. *PQRST* is a pentagon.

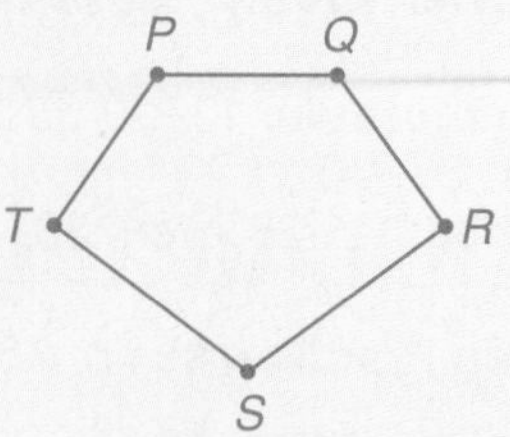

28. Segments $\overline{JK}$, $\overline{KL}$, $\overline{LM}$, $\overline{MN}$, and $\overline{NJ}$ form a quadrilateral.

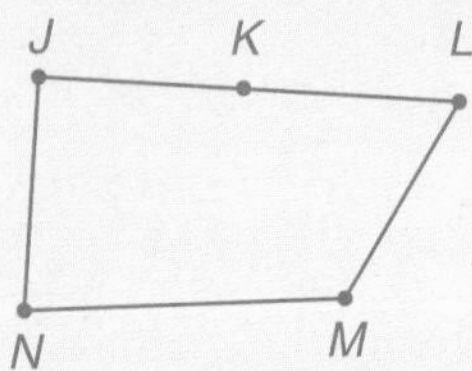

29b. Sample answer: Use the Angle option on the F6 menu to find the measures of $\angle PQS$ and $\angle QSR$.

32. Sample answer: The amount of aluminum cans being recycled is increasing; the amount of recycled cans is leveling off; something happened in 1982 that created an incentive for recycling aluminum or that made recycling of aluminum cans easier. To determine whether each conjecture is true or false, it might help to research the topic of aluminum can recycling.

Practice Masters, p. 8

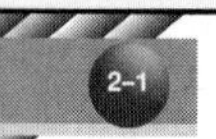

NAME__________ DATE ______

Practice

Student Edition Pages 70–75

Inductive Reasoning and Conjecturing

Determine if the conjecture is true or false based on the given information. Explain your answer and give a counterexample for any false conjecture.

1. Given: noncollinear points *A*, *B*, *C*, and *D*
Conjecture: *A*, *B*, *C*, and *D* are coplanar. false;

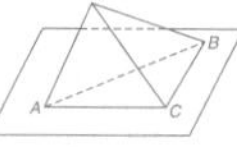

2. Given: *A*, *B*, *C*, and *D* are collinear points.
Conjecture: $AB + BC + CD = AD$
false; A C B D $AB + BC + CD > AD$

3. Given: $\angle A$ and $\angle B$ are right angles.
Conjecture: $\angle A \cong \angle B$ true; $m\angle A = 90$, $m\angle B = 90$

4. Given: Point *C* between *H* and *V*.
Conjecture: $\angle HCV$ is a straight angle. True; H C V
$\overrightarrow{CH}$ and $\overrightarrow{CV}$ are opposite rays.

Answers for Exercises 5–8 may vary. Sample answers are given.
Write a conjecture based on the given information. If appropriate, draw a figure to illustrate your conjecture.

5. $\overline{AB}$, $\overline{CD}$, and $\overline{EF}$ intersect at *X*. *A*, *B*, *C*, *D*, *E*, *F*, and *X* are noncollinear. (true)

6. $\angle MNO$ and $\angle PNO$ are adjacent angles. Point *O* is not between *M* and *P*. (true)

7. *A* (3, 1), *B* (3, -5), *C* (3, 7). Points *A*, *B*, and *C* lie on the same vertical line. (true)

8. *A*, *B*, *C*, and *D* are coplanar points. $\angle ABC$ and $\angle ABD$ are adjacent angles. (false)

Write a conjecture based on the given information. Draw a figure to illustrate your conjecture. Write a sentence or two to explain why you think your conjecture is true. 27–28. See margin.

27. points *P*, *Q*, *R*, *S*, and *T* with no three collinear

28. $\overline{JK}$, $\overline{KL}$, $\overline{LM}$, $\overline{MN}$, and $\overline{NJ}$ with only *J*, *K*, and *L* collinear

Cabri Geometry

29. Use a TI-92 to draw a square named *PQRS*. Draw $\overline{PR}$ and $\overline{QS}$.
   a. Make a conjecture about $\angle PQS$ and $\angle QSR$. They are congruent.
   b. How could you test your conjecture? See margin.
   c. Make another conjecture about the figure and test it. Write about your findings. See students' work.

Critical Thinking

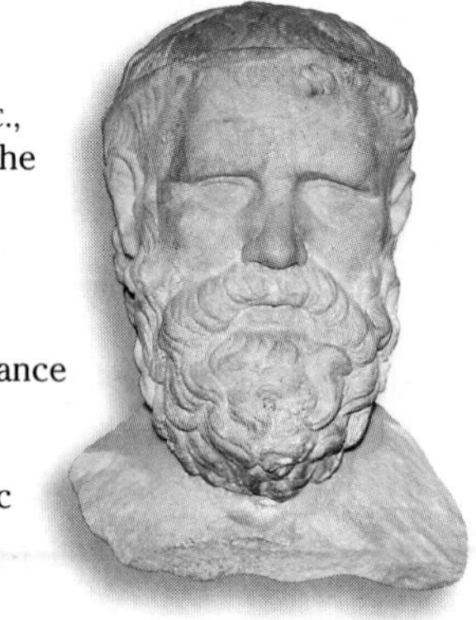

30. **Science** Eratosthenes, born around 274 B.C., made conjectures based on the position of the sun. At noon on June 21, the sun cast no shadows in the town of Syene. Eratosthenes concluded that the sun must be directly overhead. He used this conjecture and his knowledge of geometry to calculate the distance around Earth. Research various fields of science and give some examples where conjectures were helpful in proving scientific facts. See students' work.

31. The ball will strike two rails and then continue on a path to the corner on the opposite side of the table at the opposite end.

Applications and Problem Solving

31. **Billiards** Consider a carom billiard table with a length of 6 feet and a width of 3 feet. Suppose you start in the upper left-hand corner and shoot the ball at a 45° angle. Use graph paper to trace the path of the ball. Make a conjecture about shooting the ball from any corner of a table this size at a 45° angle.

No one knows for sure when or where billiard games originated. Some historians claim ancient Greece as the birthplace, while other scholars put the beginnings in 15th-century France.

32. **Recycling** The graph at the right shows the percent of aluminum cans in the U.S. recycled from 1974 to 1998. Make a list of conjectures that could be made from the data shown in the graph. What additional information might you need to know to determine the truth of your conjectures? See margin.

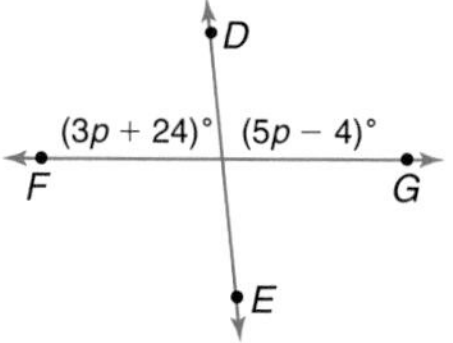

Mixed Review

33. Refer to the figure at the right. Find the value of *p* and determine if $\overline{DE} \perp \overline{FG}$. (Lesson 1–7) 20, no

34. Suppose $\angle MON$ is a right angle and *L* is in the interior of $\angle MON$. If $m\angle MOL$ is five times $m\angle LON$, find $m\angle LON$. (Lesson 1–6) 15

35. **Cartography** On a map of Ohio, Cincinnati is located at (3, 5), and Massillon, is located at (17, 19). If Columbus is halfway between the two cities, what ordered pair describes its position? (Lesson 1–5) (10, 12)

Extension

Communication Have students collect examples of true and false conjectures made in everyday life to present to the class as an oral report. Possible sources are advertising, newspaper or magazine articles, and detective stories.

There is a reference to billiards in Shakespeare's *Antony and Cleopatra* (1606–1607).

36. Use the distance formula to find the measure of $\overline{AB}$ with endpoints $A(5, -3)$ and $B(0, -5)$. (Lesson 1–4) $\sqrt{29}$

37. If the perimeter of the rectangle is 44 centimeters, find x and the dimensions of the rectangle. (Lesson 1–3) 3; 13 cm by 9 cm

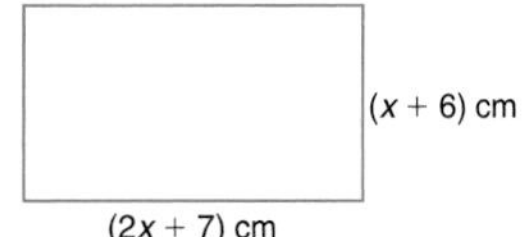

38. **List the Possibilities** Felicia and Marcus are making pizzas for the band's fall fund-raiser. A pizza can have thin or thick crust and either pepperoni, mushrooms, or green peppers. How many different pizzas can be made? (Lesson 1–2) 6 different pizzas

Algebra

39. Find $-\frac{1}{4} + \frac{2}{5}$. $\frac{3}{20}$

40. **Entertainment** Refer to the graph at the right.
   a. Describe the information that the graph shows.
   b. Explain what the ordered pair (7, 11.8) represents. 11.8% of all moviegoers attend movies in July.

40a. The graph shows the percentage of moviegoers who attend movies in different months of the year.

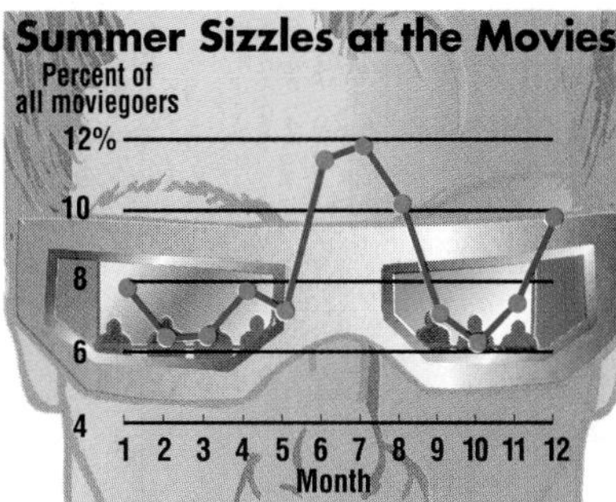

Source: *Frequent Moviegoer*

Finding a Better Fit

The excerpt below appeared in an article in *Science News* on May 18, 1996.

RESEARCHERS ARE DESIGNING SYSTEMS that can scan the surface of a person's body and produce an accurate three-dimensional image of it. These scanners offer the possibility of crafting quick, computerized representations that can be used to design a wide variety of items, from custom-fit clothing and shoes to automobile passenger compartments and crash helmets....To obtain a 3-D image for fitting clothes, a customer puts on a stretchy, tight-fitting garment and steps onto a platform surrounded by an array of lights and mirrors. The system passes harmless laser beams over the person's body from head to toe, while cameras record the patterns they make. The system can then put together a detailed surface map of the individual's shape in a process that takes only 17 seconds. ■

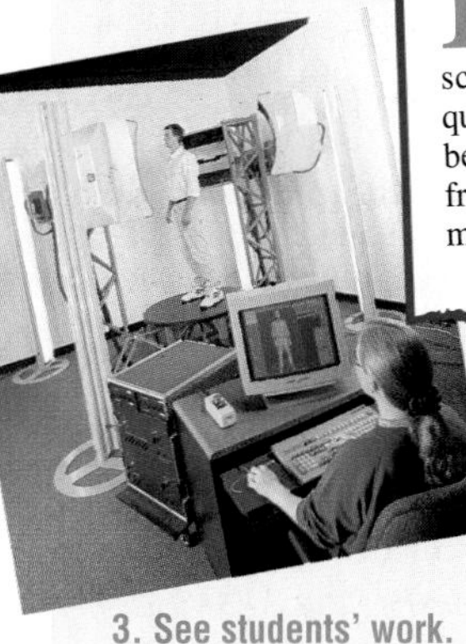

1. Make a conjecture about how the 3-D scanning process might be used to create an artificial limb for a person. See margin.
2. In the 3-D scanning process, why would you want to use cameras that view the person from several different angles and perspectives? See margin.
3. When you buy clothing, would you be willing to pay an extra charge for custom-fit clothing produced from 3-D scanner data? Why or why not?

3. See students' work.

4 ASSESS

Closing Activity

Speaking Ask volunteers to give you an example of a conjecture drawn from a given set of facts. The volunteers should decide to give a true conjecture or a false conjecture. The other students should respond by stating whether the conjectures are true or false and give reasons for their answers.

Mathematics and SOCIETY

Until recently, 3-D software has been expensive and available only for research. However, it is estimated that by 1998 commercial software will be available.

Enrichment Masters, p. 8

NAME ______ DATE ______

Enrichment — Student Edition Pages 70–75

Counterexamples

When you make a conclusion after examining several specific cases, you have used **inductive reasoning**. However, you must be cautious when using this form of reasoning. By finding only one **counterexample**, you disprove the conclusion.

Example: Is the statement $\frac{1}{x} \le 1$ true when you replace x with 1, 2, and 3? Is the statement true for all reals? If possible, find a counterexample.
$\frac{1}{1} = 1$, $\frac{1}{2} < 1$, and $\frac{1}{3} < 1$. But when $x = \frac{1}{2}$, then $\frac{1}{x} = 2$. This counterexample shows that the statement is not always true.

Answer each question.

1. The coldest day of the year in Chicago occurred in January for five straight years. Is it safe to conclude that the coldest day in Chicago is always in January? No
2. Suppose John misses the school bus four Tuesdays in a row. Can you safely conclude that John misses the school bus every Tuesday? No
3. Is the equation $\sqrt{k^2} = k$ true when you replace k with 1, 2, and 3? Is the equation true for all integers? If possible, find a counterexample. It is true for 1, 2, and 3. It is not true for negative integers. Sample: −2
4. Is the statement $2x = x + x$ true when you replace x with $\frac{1}{2}$, 4, and 0.7? Is the statement true for all real numbers? If possible, find a counterexample. It is true for all real numbers.
5. Suppose you draw four points A, B, C, and D and then draw $\overline{AB}$, $\overline{BC}$, $\overline{CD}$, and $\overline{DA}$. Does this procedure give a quadrilateral always or only sometimes? Explain your answers with figures. only sometimes
   Example: Counterexample:

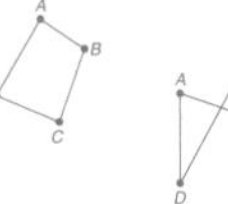

6. Suppose you draw a circle, mark three points on it, and connect them. Will the angles of the triangle be acute? Explain your answers with figures. no, only sometimes
   Example: Counterexample:

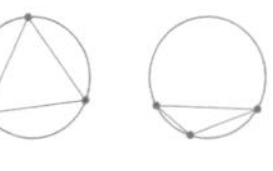

Answers for Mathematics and Society

1. If a person has lost an arm or leg, the system could scan the remaining limb and, using the computer, create a mirror image of it that could be used to make the replacement limb.
2. Using different angles and perspectives is more accurate than using only one or two views when you are dealing with three-dimensional objects. The locations of the numerous points on the body's surface can be precisely pinpointed by the process of triangulation.

2-2 LESSON NOTES

NCTM Standards: 1–5, 8

Instructional Resources

- Study Guide Master 2-2
- Practice Master 2-2
- Enrichment Master 2-2
- Assessment and Evaluation Masters, p. 44
- Modeling Mathematics Masters, pp. 19–22
- Real-World Applications, 3

Transparency 2-2A contains the 5-Minute Check for this lesson; **Transparency 2-2B** contains a teaching aid for this lesson.

Recommended Pacing	
Standard Pacing	Days 2 & 3 of 12
Honors Pacing	Day 2 of 11
Block Scheduling*	Day 2 of 7

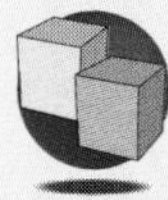

*For more information on pacing and possible lesson plans, refer to the *Block Scheduling Booklet*.

1 FOCUS

5-Minute Check
(over Lesson 2-1)

Determine if the conjecture is *true* or *false* based on the given information. Explain your answer and give a counterexample for any false conjecture.

1. **Given:** $\angle A$ and $\angle B$ are supplementary.
   Conjecture: $\angle A$ and $\angle B$ are not congruent. **False; each could measure 90°.**
2. **Given:** $m\angle A > m\angle B$; $m\angle B > m\angle C$
   Conjecture: $m\angle A > m\angle C$ **true**
3. **Given:** $\overline{AB}$, $\overline{BC}$, $\overline{AC}$
   Conjecture: A, B, and C are collinear. **False; the three segments may form a triangle.**
4. **Given:** $\angle A$ and $\angle B$ are vertical angles.
   Conjecture: $\angle A$ and $\angle B$ are congruent. **true**

2–2 If-Then Statements and Postulates

What YOU'LL LEARN

- To write a statement in if-then form,
- to write the converse, inverse, and contrapositive of an if-then statement, and
- to identify and use basic postulates about points, lines, and planes.

Why IT'S IMPORTANT

Understanding if-then statements helps determine the validily of conclusions.

CONNECTION Literature

Lewis Carroll, author of *Alice's Adventures in Wonderland* and *Through the Looking Glass,* was a mathematician as well as a writer. He was a master at creating puzzles and making connections between mathematics and literature. Following is an example of some of his statements.

Babies are illogical.
Nobody is despised who can manage a crocodile.
Illogical persons are despised.

The conclusion in the last statement may seem confusing. Rewriting each of statement in "if-then" form helps to make the progression of the logic easier to understand. *A conclusion will be written in Lesson 2–3, Exercise 37.*

CONNECTION Literature

Example 1 **Write these three Lewis Carroll statements in if-then form.**

Statement	If-Then Form
Babies are illogical.	If a person is a baby, then the person is not logical.
Nobody is despised who can manage a crocodile.	If a person can manage a crocodile, then that person is not despised.
Illogical persons are despised.	If a person is not logical, then the person is despised.

If-then statements can be used to clarify statements that may seem confusing as in Example 1 above.

If-then statements are called **conditional statements** or *conditionals*. The portion of the sentence following *if* is called the **hypothesis**, and the part immediately following *then* is called the **conclusion.** *$p \rightarrow q$ represents the conditional statement "if p, then q."*

Sometimes the word then is left out of a conditional statement

APPLICATION Broadcasting

Example 2 **The following statement is part of a message frequently played by radio stations across the nation. "If this had been an actual emergency, (then) the attention signal you just heard would have been followed by official information, news, or instruction." Identify the hypothesis and conclusion of the conditional.**

Hypothesis: this had been an actual emergency

Conclusion: the attention signal you just heard would have been followed by official information, news, or instruction

Note that "if" is not used when you write the hypothesis and "then" is not used when you write the conclusion.

Sometimes a conditional statement is written without using the "if" and "then" such as the Lewis Carroll statements in Example 1.

Sometimes you must add information to a statement when you write it in if-then form. For example, the statement *Perpendicular lines intersect* written in if-then form is *If two lines are perpendicular, then the lines intersect.* It was necessary to know that perpendicular lines come in pairs in order for the if-then statement to be clear.

Lewis Carroll

You can form another if-then statement by exchanging the hypothesis and conclusion of a conditional. This new statement is called the **converse** of the conditional. The converse of *If two lines are perpendicular, then they intersect* is *If two lines intersect, then they are perpendicular.*

The converse of $p \rightarrow q$ is $q \rightarrow p$.

It may be easier to write a conditional in if-then form first before writing the converse. The converse of a true statement is not necessarily true.

LOOK BACK

Refer to Lesson 1-7 to review adjacent and vertical angles.

Example 3 **Write the converse of the true conditional *Adjacent angles have a common side.* Determine if the converse is *true* or *false*. If false, give a counterexample.**

Explore The problem has three parts.

1. Write the converse.
2. Determine if it is true or false.
3. Give a counterexample if it is not true.

Plan First, write the conditional in if-then form. Then switch the hypothesis and conclusion to write the converse.

Solve
1. The statement in if-then form is *If two angles are adjacent, then they have a common side.*
2. **Converse:** If two angles have a common side, then they are adjacent.
3. The converse of the true conditional is false since two angles with a common side are not necessarily adjacent.

Examine Draw a figure to show that the converse is false.

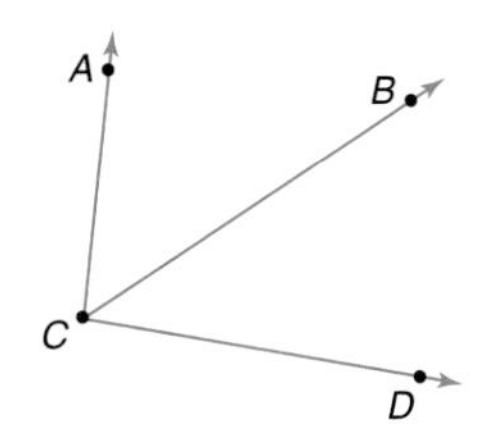

$\angle BCD$ and $\angle ACD$ have side $\overrightarrow{CD}$ in common. Yet, the two angles are not adjacent.

Motivating the Lesson

Situational Problem The coach has said, "If you miss three practices, you will be off the team," and you have just missed your third practice. What do you think will happen?

2 TEACH

In-Class Examples

For Example 1
Write the statement *An angle of 40° is acute* in if-then form. **If an angle measures 40°, then it is acute.**

For Example 2
Identify the hypothesis and conclusion of the conditional *If it is Tuesday, then Phil plays tennis.*
Hypothesis: it is Tuesday
Conclusion: Phil plays tennis

For Example 3
Write the converse of the true conditional *An angle that measures 120° is obtuse.* Determine if the converse is *true* or *false.* If false, give a counterexample. **Converse: If an angle is obtuse, then it measures 120°. This is a false statement. For example, an obtuse angle can measure 110°.**

Teaching Tip After reading the definition of *hypothesis*, ask students to recall another word related to this one. **hypothetical**

In-Class Examples

For Example 4
Write the inverse of the true conditional *A triangle has three sides*. Determine if the inverse is *true* or *false.* If false, give a counterexample. **If a polygon is not a triangle, then it does not have three sides, true.**

For Example 5
Write the contrapositive of the true conditional *If aliens have visited Earth, then there is life on other planets.* Determine if the contrapositive is *true* or *false.* **If there is not life on other planets, then aliens have not visited Earth, true.**

The denial of a statement is called a **negation**. For example, the negation of *An angle is obtuse* is *An angle is not obtuse*. If a statement is true, then its negation is false. If a statement is false, then its negation is true.

~p represents "not p" or the negation of p.

Given a conditional statement, its **inverse** can be formed by negating both the hypothesis and conclusion. The inverse of a true statement is not necessarily true. *The inverse of $p \to q$ is $\sim p \to \sim q$.*

Example 4 **Write the inverse of the true conditional *Vertical angles are congruent.* Determine if the inverse is *true* or *false.* If false, give a counterexample.**

First, write the conditional in if-then form.

The hypothesis is *two angles are vertical,* and the conclusion is *the angles are congruent.* So, the if-then form of the conditional is *If two angles are vertical, then they are congruent.*

Now, negate both the hypothesis and the conclusion to form the inverse of the conditional.

Inverse: If two angles are not vertical, then they are not congruent.

The inverse of the true conditional is false since two angles may be congruent without being vertical. A counterexample is shown below.

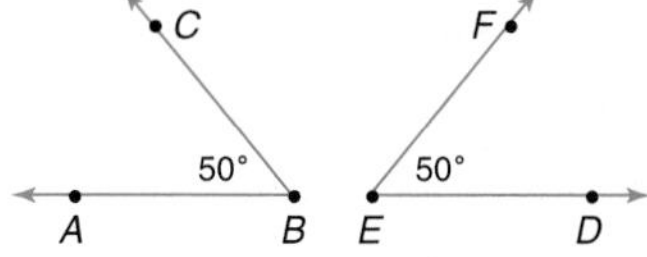

$\angle ABC \cong \angle DEF$, yet the two angles are not vertical.

Given a conditional statement, its **contrapositive** can be formed by negating the hypothesis and conclusion of the converse of the given conditional. When forming a contrapositive of a conditional it may be easier to write the converse first. *The contrapositive of $p \to q$ is $\sim q \to \sim p$.*

Example 5 **Write the contrapositive of the true conditional *If two angles are vertical, then they are congruent.* Determine if the contrapositive is *true* or *false.***

Statement: If two angles are vertical, then they are congruent.
Converse: If two angles are congruent, then they are vertical.
Contrapositive: If two angles are not congruent, then they are not vertical.

The contrapositive of the conditional is true.

The contrapositive of a true conditional is always true, and the contrapositive of a false conditional is always false.

Remember that postulates are principles that are accepted to be true without proof. The following postulates describe the ways that points, lines, and planes are related.

Postulate 2–1	**Through any two points, there is exactly one line.**

Through any 2 pts., there is 1 line.

Postulate 2–2	**Through any three points not on the same line, there is exactly one plane.**

Through any 3 noncollinear pts., there is 1 plane.

The relationships between points, lines, and planes can be used to solve problems.

Example 6

PROBLEM SOLVING

Draw a Diagram

Four people meet each other for the first time. Their first names are Kamaria (*K*), Juan (*J*), Colleen (*C*), and Mara (*M*). They each shake hands with each other once. How many handshakes will there be? Draw a diagram to illustrate the solution.

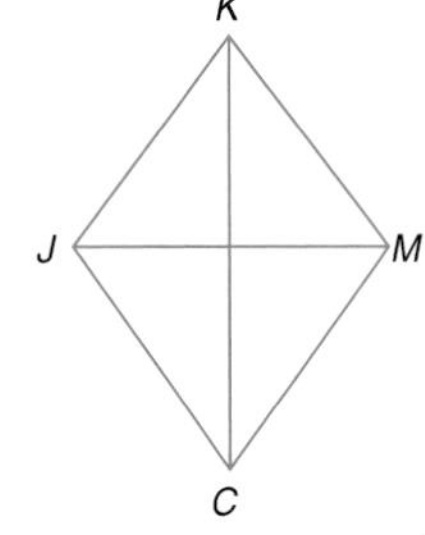

Let noncollinear points *K*, *J*, *C*, and *M* represent the people. For every two points there is exactly one line (or line segment). So for four points, there are six segments that can be drawn connecting the points.

In the figure, $\overline{KJ}$, $\overline{JC}$, $\overline{CM}$, $\overline{MK}$, $\overline{KC}$, and $\overline{JM}$ represent handshakes. Six handshakes will take place between the four people.

The next four postulates state more relationships among points, lines, and planes.

Postulate 2–3	**A line contains at least two points.**
Postulate 2–4	**A plane contains at least three points not on the same line.**
Postulate 2–5	**If two points lie in a plane, then the entire line containing those two points lies in that plane.**
Postulate 2–6	**If two planes intersect, then their intersection is a line.**

Binti and baby, Brookfield Zoo (IL)

Another type of diagram called a **Venn diagram** can be used to illustrate a conditional. A Venn diagram shows how different sets of data are related. The Venn diagram at the right shows that gorillas are a subset of the set of primates. Therefore, any animal that is a gorilla is also a primate. This diagram can also be an illustration of the conditional statement *If Binti is a gorilla, then she is a primate.*

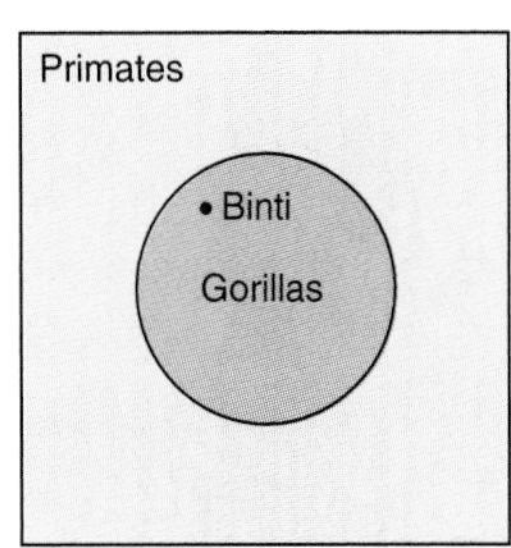

In-Class Example

For Example 6

Five people arrive at a meeting. Each one shakes hands one time with each of the others. How many different handshakes take place? Draw a diagram to illustrate the solution. 10

Teaching Tip When possible, have students act out problems. This is especially helpful in handshake questions.

Cooperative Learning

Brainstorming Have students work in small groups to invent five false conditional statements whose converses are true. You may wish to have the groups present their statements to the class to check if they are correct. For more information on the brainstorming strategy, see *Cooperative Learning in the Mathematics Classroom*, one of the titles in the Glencoe Mathematics Professional Series, page 30.

3 PRACTICE/APPLY

Check for Understanding

Exercises 1–18 are designed to help you assess your students' understanding through reading, writing, speaking, and modeling. You should work through Exercises 1–5 with your students and then monitor their work on Exercises 6–18.

Error Analysis

It is often difficult for students to convert a simple declarative sentence into if-then form. Tell students that the complete subject provides the information for the hypothesis while the complete predicate provides the information for the conclusion. Students can begin by writing "If (subject), then (predicate)," and then rewrite the sentence to make it read better. Example: Vertical angles have equal measure.

Write: If vertical angles, then equal measure.

Rewrite: If two angles are vertical angles, then they have equal measure.

Additional Answers

12. Converse: If you are a teenager, then you are 13 years old; false. Counterexample: If you are a teenager, you may be 15 years old. Inverse: If you are not 13 years old, then you are not a teenager; false. Counterexample: If you are not 13 years old, you may be a 15-year-old teenager. Contrapositive: If you are not a teenager, then you are not 13 years old; true.
13. Converse: If an angle is a right angle, then it measures 90°; true. Inverse: If an angle does not measure 90°, then it is not a right angle; true. Contrapositive: If an angle is not a right angle, then it does not measure 90°; true.

CHECK FOR UNDERSTANDING

Communicating Mathematics

1. If-then form makes statements clearer and easier to understand.
3. See Solutions Manual.

Study the lesson. Then complete the following.

1. **Explain** why writing statements in if-then form is helpful.
2. **Define** *conditional statement* in your own words and give an example of a conditional. See Solutions Manual.
3. **You Decide** Omar says that all squares are rectangles. Julia argues that all rectangles are squares. Minaku says that some rectangles are squares. Who is correct? Draw a diagram to help explain your answer.
4. **Discuss** similarities and differences between the inverse and the contrapositive of a conditional. See Solutions Manual.

5. Find some ads in magazines or newspapers that contain if-then statements. Tape the ads in your journal. Write the converse, inverse, and contrapositive of each statement in your journal. Do you think the messages these ads convey are valid or necessarily true? Explain. See students' work.

Guided Practice

6. Hypothesis: you can sell green toothpaste in this country Conclusion: you can sell opera
7. Hypothesis: three points lie on a line Conclusion: they are collinear
8. If a fish is a piranha, then it eats other fish.
9. If angles have the same measure, then they are congruent.

Identify the hypothesis and conclusion of each conditional statement.

6. "If you can sell green toothpaste in this country, you can sell opera." *(Sarah Caldwell, 1975)*
7. If three points lie on a line, then they are collinear.

Write each conditional statement in if-then form.

8. A piranha eats other fish.
9. Angles with the same measure are congruent.

Write the negation of each statement.

10. Three points are collinear Three points are noncollinear.
11. Four points are noncoplanar. Four points are coplanar.

Write the converse, inverse, and contrapositive of each conditional. Determine if the converse, inverse, and contrapositive are *true* or *false*. If false, give a counterexample. 12–13. See margin.

12. If you are 13 years old, then you are a teenager.
13. If an angle measures 90°, then it is a right angle.

In the figure at the right, *A*, *B*, and *C* are collinear. Points *A*, *B*, *C*, and *D* are in plane $\mathcal{N}$. Use the postulates you have learned to determine if each statement is *true* or *false*.

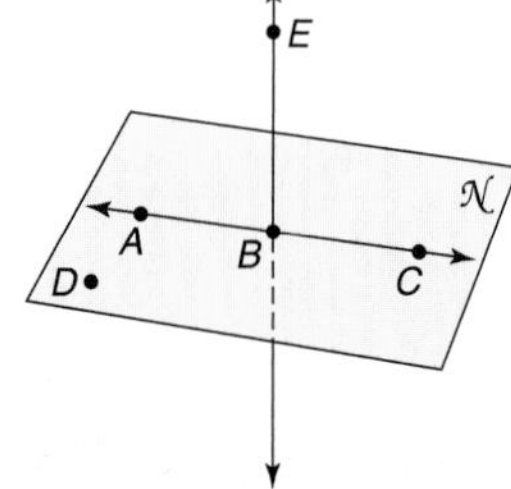

14. *A*, *B*, and *E* lie in plane $\mathcal{N}$. false
15. $\overleftrightarrow{BC}$ does not lie in plane $\mathcal{N}$. false
16. *A*, *B*, *C*, and *E* are coplanar. true
17. *A*, *B*, and *D* are collinear. false
18. **World Cultures** The Maoris were the first people to settle in New Zealand in the 900s. Write each statement in if-then form. Then write the converse, inverse, and contrapositive of each conditional. See Solutions Manual.
    a. The Maoris were outstanding sculptors.
    b. The Maori treaty house has pillars of carved wood.

Reteaching

Using Matching Ask students to match the hypotheses on the left with a conclusion at the right that makes a true conditional statement.

1. If two angles form a linear pair, c
2. If two angles are vertical angles, a
3. If two adjacent angles form a right angle, b

a. then the angles are congruent.
b. then the angles are complementary.
c. then the angles are supplementary.

EXERCISES

Practice

Identify the hypothesis and conclusion of each conditional statement.

A

19. "If a man hasn't discovered something that he will die for, he isn't fit to live." (*Martin Luther King, Jr., 1963*) 19–24. See margin.
20. "If you don't know where you are going, you will probably end up somewhere else." (*Laurence Peters, 1969*)
21. "If we would have new knowledge, we must get a whole world of new questions." (*Susanne K. Langer, 1957*)
22. If you are an NBA basketball player, then you are at least 5′ 2″ tall.
23. If $3x - 5 = -11$, then $x = -2$.
24. If you are an adult, then you are at least 21 years old.

Write each conditional in if-then form. 25–30. See margin.

B

25. "Happy people rarely correct their faults." (*La Rochefoucauld, 1678*)
26. "A champion is afraid of losing." (*Billie Jean King, 1970s*)
27. Adjacent angles have a common vertex.
28. Equiangular triangles are equilateral.
29. Angles whose measures are between 90 and 180 are obtuse angles.
30. Perpendicular lines form right angles.

Write the negation of each statement. 31. A book is not a mirror.

31. A book is a mirror.
32. Right angles are not acute angles.
33. Rectangles are not squares.
34. A cardinal is not a dog.
35. You live in Dallas. You do not live in Dallas.

32. Right angles are acute angles.

33. Rectangles are squares.

34. A cardinal is a dog.

Write the converse, inverse, and contrapositive of each conditional. Determine if the converse, inverse, and contrapositive are *true* or *false*. If false, give a counterexample. 36–41. See Solutions Manual.

36. All squares are quadrilaterals.
37. Three points not on the same line are noncollinear.
38. If a ray bisects an angle, then the two angles formed are congruent.
39. Acute angles have measures less than 90.
40. Vertical angles are congruent.
41. If you don't live in Chicago, then you don't live in Illinois.

In the figure at the right, *A*, *B*, and *C* are collinear. Points *A* and *X* lie in plane $\mathcal{M}$. Points *B* and *Z* lie in plane $\mathcal{N}$. Determine whether each statement is *true* or *false*.

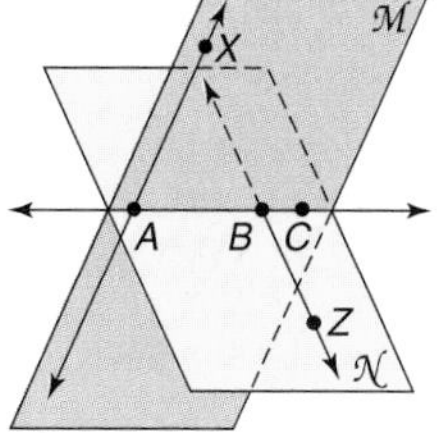

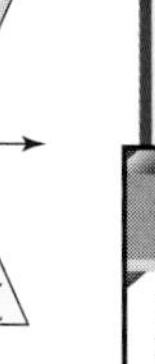

42. *B* lies in plane $\mathcal{M}$. true
43. *A*, *B*, and *C* lie in plane $\mathcal{M}$. true
44. *A*, *B*, *X*, and *Z* are coplanar. false
45. $\overleftrightarrow{BZ}$ lies in plane $\mathcal{N}$. true

46–48. See Solutions Manual for drawings.

State the number of lines that can be drawn containing each set of points taken two at a time. Draw a figure for each.

C

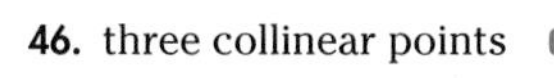

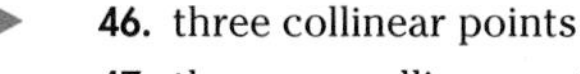

46. three collinear points one
47. three noncollinear points three
48. four points, no three of which are collinear six

Assignment Guide

Core (with proof): 19–55 odd, 57–67
Core (informal): 19–55 odd, 57–67
Enriched: 20–50 even, 52–67

For **Extra Practice,** see p. 766.

The red A, B, and C flags, printed only in the Teacher's Wraparound Edition, indicate the level of difficulty of the exercises.

Additional Answers

19. Hypothesis: a man hasn't discovered something that he will die for
    Conclusion: he isn't fit to live
20. Hypothesis: you don't know where you are going
    Conclusion: you will probably end up somewhere else
21. Hypothesis: we would have new knowledge
    Conclusion: we must get a whole world of new questions
22. Hypothesis: you are an NBA basketball player
    Conclusion: you are at least 5'2" tall
23. Hypothesis: $3x - 5 = -11$
    Conclusion: $x = -2$
24. Hypothesis: you are an adult
    Conclusion: you are at least 21 years old

Study Guide Masters, p. 9

2–2 NAME________ DATE ________

Study Guide

Student Edition Pages 76–83

If-Then Statements and Postulates

If-then statements are commonly used in everyday life. For example, an advertisement might say, "If you buy our product, then you will be happy." Notice that an if-then statement has two parts, a *hypothesis* (the part following "if") and a *conclusion* (the part following "then").

New statements can be formed from the original statement.

Statement	$p \to q$
Converse	$q \to p$
Inverse	$\sim p \to \sim q$
Contrapositive	$\sim q \to \sim p$

Example: Rewrite the following statement in if-then form. Then write the converse, inverse, and contrapositive.

All elephants are mammals.

If-then form:	If an animal is an elephant, then it is a mammal.
Converse:	If an animal is a mammal, then it is an elephant.
Inverse:	If an animal is not an elephant, then it is not a mammal.
Contrapositive:	If an animal is not a mammal, then it is not an elephant.

Identify the hypothesis and conclusion of each conditional statement.

1. If today is Monday, then tomorrow is Tuesday. H: today is Monday; C: tomorrow is Tuesday
2. If a truck weighs 2 tons, then it weighs 4000 pounds. H: a truck weighs 2 tons; C: it weighs 4000 pounds

Write each conditional statement in if-then form.

3. All chimpanzees love bananas. If an animal is a chimpanzee, then it loves bananas.
4. Collinear points lie on the same line. If points are collinear, then they lie on the same line.

Write the converse, inverse, and contrapositive of each conditional.

5. If an animal is a fish, then it can swim. Converse: If an animal can swim, then it is a fish. Inverse: If an animal is not a fish, then it cannot swim. Contrapositive: If an animal cannot swim, then it is not a fish.
6. All right angles are congruent. Converse: If angles are congruent, then they are right angles. Inverse: If angles are not right angles, then they are not congruent. Contrapositive: If angles are not congruent, then they are not right angles.

Additional Answers

25. If people are happy, then they rarely correct their faults.
26. If a person is a champion, then he is afraid of losing.
27. If angles are adjacent, then they have a common vertex.
28. If triangles are equiangular, then they are equilateral.
29. If angles have measures between 90 and 180, then they are obtuse.
30. If two lines are perpendicular, then they form right angles.

Using the Programming Exercises The program given in Exercise 51 is for use with a TI-82 or TI-83 graphing calculator. For other programmable calculators, have students consult their owner's manual for commands similar to those presented there.

Additional Answer

54.

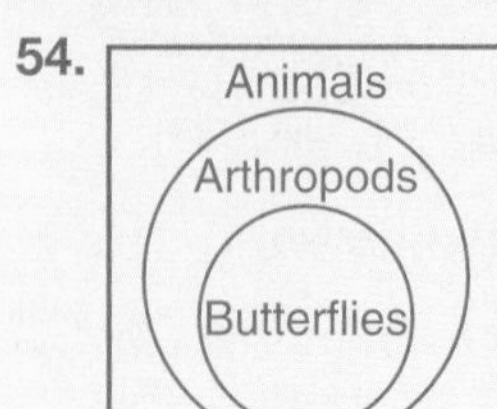

56.

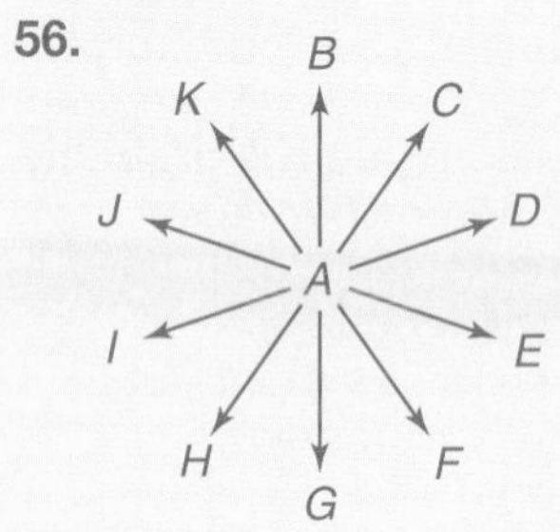

Practice Masters, p. 9

NAME ______ DATE ______

Practice

Student Edition Pages 76–83

If-Then Statements and Postulates

Identify the hypothesis and conclusion of each conditional statement.

1. If $3x - 1 = 7$, then $x = 2$. Hypothesis: $3x - 1 = 7$ Conclusion: $x = 2$
2. If Carl scores 85%, then he passes. Hypothesis: Carl scores 85% Conclusion: he passes

Write each conditional statement in if-then form.

3. All students like vacations. If a person is a student, then that person likes vacations.
4. The game will be played provided it doesn't rain. If it doesn't rain, then the game will be played.

Write the converse of each conditional. Determine if the converse is true or false. If it is false, give a counterexample.

5. If it rains, then it is cloudy. If it is cloudy, then it rains; false. It can be cloudy without raining.
6. If x is an even number, then x is divisible by 2. If x is divisible by 2, then x is an even number; true.

In the figure, P, Q, R, and S are in plane $\mathcal{N}$. Use the postulates you have learned to determine whether each statement is true or false.

7. R, S, and T are collinear. false
8. There is only one plane that contains all the points R, S, and Q. true
9. $\angle PQT$ lies in plane $\mathcal{N}$. false
10. $\angle SPR$ lies in plane $\mathcal{N}$. true
11. If X and Y are two points on line m, then $\overline{XY}$ intersects plane $\mathcal{N}$ at P. true
12. Point K is on plane $\mathcal{N}$. true
13. $\mathcal{N}$ contains $\overline{RS}$. true
14. T lies in plane $\mathcal{N}$. false
15. R, P, S, and T are coplanar. false
16. ℓ and m intersect. false

State the number of planes that can be drawn that contain the given set of points.

49. three noncollinear points one
50. a line and a point not on the line one

Programming

51. The TI-82/83 graphing calculator program at the right finds the volume of a cylinder, given the height and radius.

```
PROGRAM: CYLINDER
:Input "HEIGHT?",H
:Input "RADIUS?",R
:πR²H→V
:Disp "VOLUME IS",V
```

Suppose the height of a cylinder is 8 centimeters and the radius is 3 centimeters. Fill in the blanks by making a conjecture. Then use the program to test your conjectures.

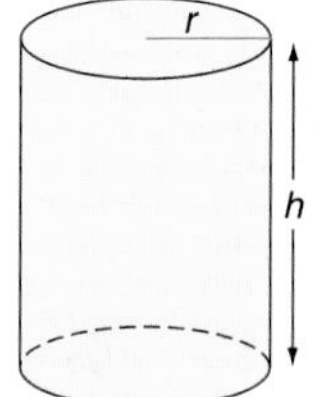

a. If the height of the cylinder is doubled and the radius remains the same, then the volume ______. doubles
b. If the radius of the cylinder is doubled and the height remains the same, then the volume ______.
c. If the radius and the height of the cylinder are both doubled, then the volume ______. is multiplied by 8

51b. quadruples

52. If two angles are adjacent, then they are not both acute. This is the original conditional.

53a. The Hatter is right; Alice exchanged the hypothesis and conclusion.

Critical Thinking

52. Consider the conditional *If two angles are adjacent, they are not both acute.* Write the converse of the contrapositive of the inverse of the conditional. Explain how the result is related to the original conditional.

53. **Literature** The following quote is from Lewis Carroll's *Alice's Adventures in Wonderland.*

> "Then you should say what you mean," the March Hare went on.
>
> "I do," Alice hastily replied; "at least—at least I mean what I say—that's the same thing, you know."
>
> "Not the same thing a bit!" said the Hatter. "Why, you might just as well say that 'I see what I eat' is the same thing as 'I eat what I see'!"

a. Who is right, Alice or the Hatter? Explain your reasoning.
b. How are the statements *say what you mean* and *mean what you say* related to each other? These statements are converses of each other.

Applications and Problem Solving

54. See margin.

55a. If you try Casa Fiesta, then you're looking for a fast, easy way to add some fun to your family's menu.

55b. They are a fast, easy way to add fun to your family's menu.

56. Sample answer: 10 collisions; see margin for diagram.

54. **Biology** Use a Venn diagram to illustrate the following conditional about the animal kingdom. *If an animal is a butterfly, then it is an arthropod.*

55. **Advertising** Advertising writers frequently use if-then statements to relay a message and promote their product. An ad for a type of Mexican food reads, *If you're looking for a fast, easy way to add some fun to your family's menu, try Casa Fiesta.*
a. Write the converse of the conditional.
b. What do you think the advertiser wants people to conclude about Casa Fiesta products?
c. Does the advertisement say that Casa Fiesta adds fun to your family's menu? No; the conclusion is implied.

56. **Draw a Diagram** Eleven students are riding bumper cars at a carnival. They are playing a game in which if one car bumps another car in the back, the car bumped in the back is out of the game. How many collisions will there be before there is a winner? Name the bumper cars with points A through K for your drawing.

Tech Prep

Biological Technician Many junior and community colleges offer two-year degrees for biological technicians. They work in laboratories doing such jobs as running lab tests. Median annual earnings are about $27,000. For more information on tech prep, see the *Teacher's Handbook*.

Mixed Review

57. Sample answer: $\angle C \cong \angle D$

57. Write a conjecture given that $\angle C$ and $\angle D$ are right angles. (Lesson 2–1)

58. Determine if $\overline{AB}$, $\overline{BC}$, and $\overline{AC}$ form a triangle, given that A, B, and C are collinear points. Explain and give a counterexample if they do not. (Lesson 2–1) False; collinear points cannot be the vertices of a triangle.

59. Write a conjecture if $WXYZ$ is a rectangle. (Lesson 2–1)

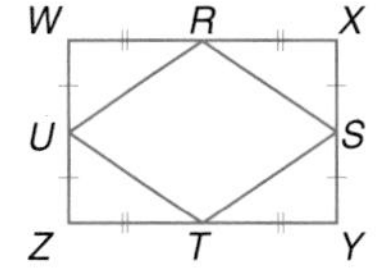

60. Angles AND and NOR are complementary. If $m\angle AND = 4m\angle NOR$, find the measures of the angles. (Lesson 1–7) $m\angle AND = 72$, $m\angle NOR = 18$

61. Is a 67° angle acute, obtuse, right, or straight? (Lesson 1–6) acute

62. Find the coordinates of the midpoint of $\overline{QP}$ with endpoints $Q(4, 8)$ and $P(-3, 0)$. (Lesson 1–5)

63. **Manufacturing** *Stars and Stripes Flag Company* manufactures small flags. The length of a flag must be 1.5 times the width, and the area of the flag must be between 60 square inches and 300 square inches. What are the possible whole-number dimensions for the flags? (Lesson 1–4)

64. Find the maximum area for a rectangle whose perimeter is 44 meters. (Lesson 1–3) 121 m^2

65. Points $R(-2, 3)$ and $S(4, 9)$ lie on the graph of $y = x + 5$. Determine whether point $T(-1, -4)$ is collinear to R and S. (Lesson 1–1) no

INTEGRATION **Algebra**

66. Write an algebraic expression for *15 increased by three times a number n.*

67. State whether $t^2 - 5 = 6$ is *true* or *false* for $t = 3$. false

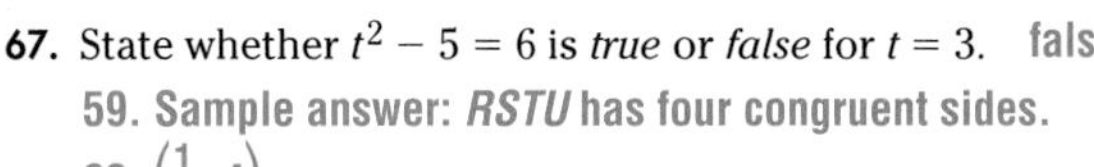

59. Sample answer: *RSTU* has four congruent sides.

62. $\left(\frac{1}{2}, 4\right)$

63. 12 in. by 8 in., 15 in. by 10 in., 18 in. by 12 in., 21 in. by 14 in.

66. $15 + 3n$

Refer to the Investigation on pages 66–67.

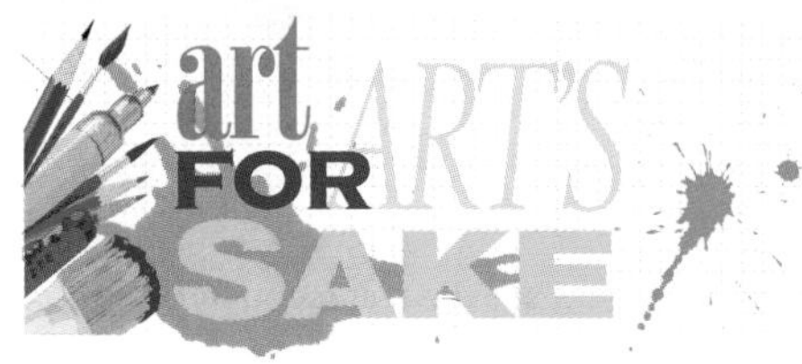

One popular style of geometric art is called op art, or optical art. Op art is abstract and uses straight lines or geometric patterns to create a special visual effect. Use the following steps to create an op art sample.

1. Draw three evenly-spaced points at the top, bottom, and sides of a piece of unlined paper. Choose a method for connecting some of the points. For example, connect the top corners of the page to each point on the bottom and opposite side.
2. Choose two markers and shade the regions bounded by the lines. No two regions that share a side should be the same color.
3. Create two other op art designs using different grid systems or coloring schemes.
4. Refer to your discussion notes about the mural project. Is op art a style that would be appropriate for the mural?

Add the results of your work to your Investigation Folder.

4 ASSESS

Closing Activity

Writing Have each student write a true if-then statement on a sheet of paper. Have students exchange papers, write the converse of the statement they are given, and tell if it is true or false. Select some examples for class discussion.

Chapter 2 Quiz A (Lessons 2-1 and 2-2) is available in the *Assessment and Evaluation Masters*, p. 44.

Enrichment Masters, p. 9

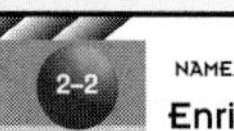

NAME ________ DATE ________

Enrichment

Student Edition Pages 76–83

Venn Diagrams

A type of drawing called a **Venn diagram** can be useful in explaining conditional statements. A Venn diagram uses circles to represent sets of objects.

Consider the statement "All rabbits have long ears." To make a Venn diagram for this statement, a large circle is drawn to represent all animals with long ears. Then a smaller circle is drawn inside the first to represent all rabbits. The Venn diagram shows that every rabbit is included in the group of long-eared animals.

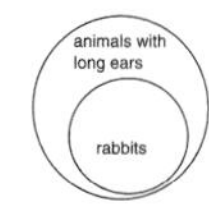

The set of rabbits is called a **subset** of the set of long-eared animals.

The Venn diagram can also explain how to write the statement, "All rabbits have long ears," in if-then form. Every rabbit is in the group of long-eared animals, so if an animal is a rabbit, then it has long ears.

For each statement, draw a Venn diagram. The write the sentence in if-then form.

1. Every dog has long hair.

If an animal is a dog, then it has long hair.

2. All rational numbers are real.

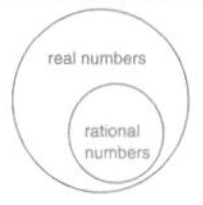

If a number is rational, then it is real.

3. People who live in Iowa like corn.

If a person lives in Iowa, then the person likes corn.

4. Staff members are allowed in the faculty lounge.

If a person is a staff member, then the person is allowed in the faculty cafeteria.

Extension

Reasoning Name two hypotheses that would let you conclude that two angles are congruent. Sample answers: if two angles are vertical angles; if two angles are formed by perpendicular lines; if each of two angles measures 30°

In·ves·ti·ga·tion

Working on the Investigation

The Investigation on pages 66–67 is designed to be a long-term project that is completed over several days or weeks. Encourage students to keep their materials in their Investigation Folder as they work on the Investigation.

2-2B LESSON NOTES

NCTM Standards: 1–5

Objective

Test the validity of conditional statements that involve mathematical sentences in one variable.

Recommended Time

20 minutes

Instructional Resources

Instructions for using the TI-81 and Casio calculators for this activity are also available in the *Graphing Calculator and Computer Masters*, pp. 17–18.

1 FOCUS

Motivating the Lesson

Have students discuss the following questions.

When is $2x - 2 < 0$ true? When is $2x > 2$ true?

Is the conditional *If* $2x - 2 < 0$, *then* $2x > 2$ true or false?

2 TEACH

Technology Tip
A function that returns only 0 or 1 is called a Boolean function, named after the mathematician George Boole.

Teaching Tip After the exercises, it might be helpful to set the y-range from 0 to 2. This displays the function more clearly.

3 PRACTICE/APPLY

Assignment Guide
Core (with proof): 1–5
Core (informal): 1–5
Enriched: 1–5

2–2B Using Technology
Testing Conditional Statements

An Extension of Lesson 2–2

A TI-82/83 graphing calculator can be helpful in testing the validity of some conditional statements that involve mathematical sentences in one variable.

Example **Given the conditional statement *If* $3x + 6 > 4x + 9$, *then* $x > -3$, use a graphing calculator to verify whether the conclusion is true.**

You can use the symbols on the TEST menu.

- Use the standard viewing window by pressing ZOOM 6.
- Clear the calculator's Y= list.
- Enter $3x + 6 > 4x + 9$ as Y1. To enter the $>$ symbol, press 2nd [TEST] 3.
- Press GRAPH. A screen like the one shown at the right should appear.
- Use the TRACE function to scan the Y values along the graph.

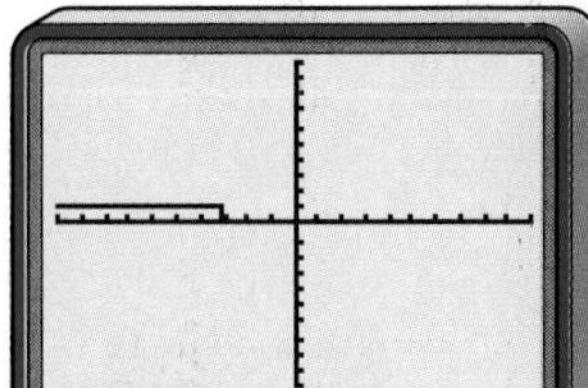

TECHNOLOGY Tip The TEST menu is the second function of MATH.

Notice that the value of Y is either 1 or 0. When the statement $3x + 6 > 4x + 9$ is true, the value of Y is 1. When it is false, the value of Y is 0. So, the hypothesis is true only when $x < -3$. Assuming the hypothesis $3x + 6 > 4x + 9$ is true, then the conclusion $x > -3$ is false. The correct conditional would be *If* $3x + 6 > 4x + 9$, *then* $x < -3$.

EXERCISES

Analyze the activity. 1. The result is the same, $x < -3$.

1. Solve the inequality in the hypothesis in the Example algebraically. How does it compare with the result from the calculator test?
2. Write a paragraph to explain how you could use the test menu to determine the validity of a solution of an equation. See students' work.

Use a graphing calculator to test the validity of the conclusion in each conditional statement. Rewrite the conditional if necessary to make a true statement.

3. If $x - 9 < 5x - 1$, then $x > -2$. true
4. If $4x + 19 \leq -8 + 7x$, then $x \leq 9$. False; if $4x + 19 \leq -8 + 7x$, then $x \geq 9$.
5. If $12 - 3x > 23 - 14x$, then $x < 1$. False; if $12 - 3x > 23 - 14x$, then $x > 1$.

4 ASSESS

Observing students working with technology is an excellent method of assessment.

Using Technology

This lesson offers an excellent opportunity for using technology in your geometry classroom. For more information on using technology, see *Graphing Calculators in the Mathematics Classroom*, one of the titles in the Glencoe Mathematics Professional Series.

2-3 Deductive Reasoning

What YOU'LL LEARN

- To use the Law of Detachment and the Law of Syllogism in deductive reasoning, and
- to solve problems by looking for a pattern.

Why IT'S IMPORTANT

You can use deductive reasoning to reach logical conclusions.

You can use paper folding to explore the measures of vertical angles.

MODELING MATHEMATICS — **Vertical Angles**

Materials: 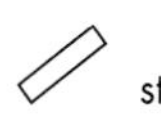 patty paper ⁄ straightedge

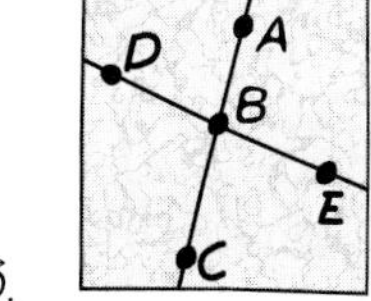

- Fold the paper and make a crease. Fold the paper again so that the second fold intersects the first.
- Open the paper and use a straightedge to draw the two lines formed by the creases. Label your paper similarly to the figure at the right.
- Fold the paper through B so that $\overrightarrow{BA}$ lies on top of $\overrightarrow{BD}$. What do you notice about $\angle DBC$ and $\angle ABE$? $\angle DBC \cong \angle ABE$
- Refold the paper through B so that $\overrightarrow{BA}$ lies on $\overrightarrow{BE}$. What do you notice about $\angle ABD$ and $\angle EBC$? $\angle ABD \cong \angle EBC$

Your Turn Sample answer: The angles opposite each other are congruent.

Make a conjecture about the opposite pairs of angles formed by any two intersecting lines.

In the Modeling Mathematics activity above, you discovered a pattern with vertical angles and made a conclusion based on the pattern. **Looking for a pattern** is a good way to help you make a conjecture.

Example 1 **Find the number of angles formed by 10 distinct rays with a common endpoint.**

PROBLEM SOLVING **Look for a Pattern**

Explore A figure drawn with 10 rays may look confusing. Instead, we can draw a series of figures with fewer rays.

no angle

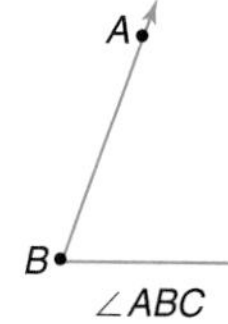

$\angle ABC$

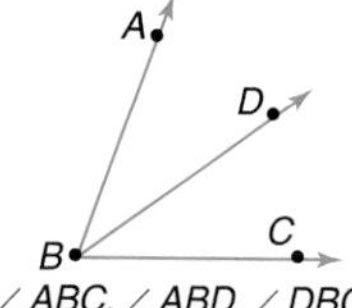

$\angle ABC$, $\angle ABD$, $\angle DBC$

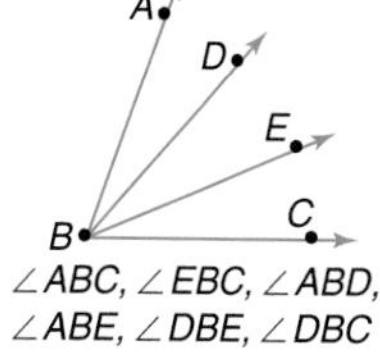

$\angle ABC$, $\angle EBC$, $\angle ABD$, $\angle ABE$, $\angle DBE$, $\angle DBC$

Plan Make a table. Look for a pattern in the number of angles formed.

Solve Record the number of angles in each figure.

number of rays	1	2	3	4	5	6
number of angles formed	0	1	3	6	10	?

+1 +2 +3 +4 +?

(continued on the next page)

2-3 LESSON NOTES

NCTM Standards: 1–4

Instructional Resources

- Study Guide Master 2-3
- Practice Master 2-3
- Enrichment Master 2-3
- Assessment and Evaluation Masters, pp. 43, 44
- Multicultural Activity Masters, p. 3
- Tech Prep Applications Masters, p. 3

Transparency 2-3A contains the 5-Minute Check for this lesson; **Transparency 2-3B** contains a teaching aid for this lesson.

Recommended Pacing	
Standard Pacing	Day 5 of 12
Honors Pacing	Day 4 of 11
Block Scheduling*	Day 3 of 7

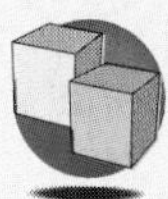

*For more information on pacing and possible lesson plans, refer to the *Block Scheduling Booklet*.

1 FOCUS

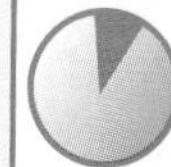

5-Minute Check (over Lesson 2-2)

Write each conditional statement in if-then form.

1. All triangles have three angles. If a figure is a triangle, then it has three angles.
2. Every Thursday Julia goes swimming. If it is Thursday, then Julia goes swimming.

Write the converse of each conditional. Determine if the converse is *true* or *false*. If it is false, give a counterexample.

3. If a figure is a rectangle, then it has four sides. If a figure has four sides, then it is a rectangle. False; a trapezoid has four sides and is not a rectangle.
4. If a number is divisible by 6, then it is divisible by 2. If a number is divisible by 2, then it is divisible by 6. False; 4 is divisible by 2 but not by 6.

MODELING MATHEMATICS This activity leads students to form conjectures about vertical and opposite angles. The activity can be revisited after instruction about transformations in order to reinforce the relationship among the angles.

Examine (continued)

After drawing a figure with 5 rays, you may conclude that a figure with 6 rays forms 10 + 5 or 15 angles, a figure with 7 rays forms 15 + 6 or 21 angles, and so on until a figure with 10 rays forms 36 + 9 or 45 angles.

Examine To check your conjecture that a figure with 10 rays forms 45 angles, you could draw a figure with 10 rays and count them all. Although this may be confusing and tedious, you would find that your conjecture is true. You would also discover that looking for a pattern is an easier way to solve the problem.

Looking for a pattern does not guarantee that a statement is always true. It is necessary to prove a statement in order to assume that it is absolutely true for all cases. Even if we listed 1000 examples in which vertical angles were congruent, it would not be enough to assume that it is true for all cases. We will prove the theorem in Lesson 2–6. In order to prove statements, we need some "tools" of the trade.

In the June 3, 1996 issue of *People* magazine, Della Reese made the following statement about Roma Downey of the TV drama *Touched by an Angel*.

If-Then Statement:
"If you can't look at Roma and see that she is sweet, caring, and tender, then I don't know what I can tell you."

True statement about the hypothesis:
A person can't look at Roma and see that she is sweet, caring, and tender.

True Conclusion: Della Reese doesn't know what she can tell that person.

This form of reasoning is used in proof and is called the **Law of Detachment**. The Law of Detachment offers us a way to draw conclusions from if-then statements. It states that whenever a conditional is true and its hypothesis is true, we can assume that its conclusion is true.

Law of Detachment	**If $p \rightarrow q$ is a true conditional and p is true, then q is true.**

Deductive reasoning uses a rule to make a conclusion. Inductive reasoning uses examples to make a conjecture or rule.

The Law of Detachment and other laws of logic can be used to provide a system for reaching logical conclusions. This system is called **deductive reasoning**. Deductive reasoning is one of the cornerstones of the study of geometry.

Example  ***If two numbers are odd, then their sum is even*** **is a true conditional, and 3 and 5 are odd numbers. Use the Law of Detachment to reach a logical conclusion.**

The hypothesis is *two numbers are odd*. 3 and 5 are indeed two odd numbers. Since the conditional is true and the given statement satisfies the hypothesis, the conclusion is true. So, the sum of 3 and 5 must be even.

Knowing that a conditional is true and that its conclusion is true does not allow us to say the hypothesis is true. Consider a counterexample from Example 2. The sum of 8 and 12 is 20, which is an even number, but 8 and 12 are not two odd numbers.

Motivating the Lesson

Situational Problem Present this example to the class: If Jamaal goes to the ball game, he will not go to the movies. Jamaal went to the ball game. What conclusion can you draw? **Jamaal did not go to the movies.**

Compare the above example with this: If Eiko goes to the ball game, she will not go to the movies. Eiko did not go to the movies. Can you conclude that Eiko went to the ball game? **No**

2 TEACH

In-Class Examples

For Example 1
Find the number of angles formed by 10 distinct lines with a common intersection point. **20**

For Example 2
If a vehicle is a car, then it has four wheels is a true conditional. A sedan is a car. Use the Law of Detachment to reach a valid conclusion. **A sedan has four wheels.**

Teaching Tip When reading the Law of Detachment, remind students that $p \rightarrow q$ can be read "p implies q" or "if p, then q."

Teaching Tip Some students may confuse the conclusion of a conditional with a logical conclusion reached using the Law of Detachment and the Law of Syllogism. Discuss these uses of the same word.

Alternative Learning Styles

Auditory Ask students to discuss a famous legal trial. Have them identify how reasoning is being used to evaluate evidence. See if they can find examples of the Law of Detachment and the Law of Syllogism in legal reasoning.

Example 3

APPLICATION Sports

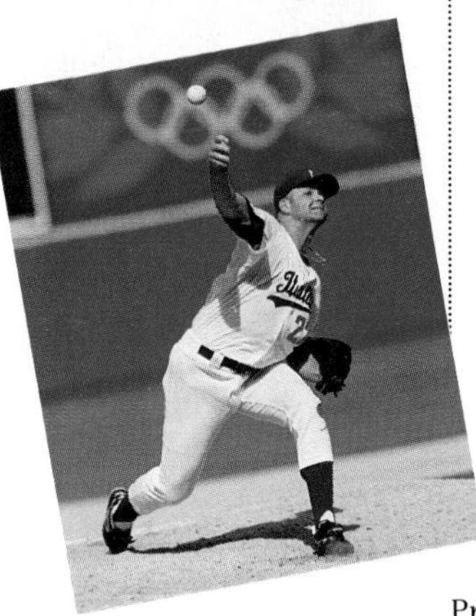

Determine if statement (3) follows from statements (1) and (2) by the Law of Detachment. *Assume that statements (1) and (2) are true.*

(1) If you played baseball in the '96 Summer Olympics, then you played baseball in Atlanta in the summer of '96.

(2) Greg Maddux (Atlanta Braves) played baseball in Atlanta in the summer of '96.

(3) Greg Maddux played baseball in the '96 Summer Olympics.

Hypothesis: you played baseball in the '96 Summer Olympics

Conclusion: you played baseball in Atlanta in the summer of '96

Given: Greg Maddux played baseball in Atlanta in the summer of '96.

A tendency may be to conclude that Greg Maddux played baseball in the '96 Summer Olympics. In fact, all of the Atlanta Braves players played baseball in Atlanta in the summer of '96, but none of them played in the Olympics. The given information satisfied the conclusion rather than the hypothesis. This led to the invalid conclusion even though the conditional was true and the given was true.

A second law of logic is the **Law of Syllogism**. It is similar to the Transitive Property of Equality that you should remember from your studies in algebra.

Law of Syllogism	**If $p \rightarrow q$ and $q \rightarrow r$ are true conditionals, then $p \rightarrow r$ is also true.**

Example 4

APPLICATION Ancient Artifacts

The Noks were one of Africa's earliest civilizations, dating back to 500 B.C. Determine if a valid conclusion can be reached from the two true statements *If it is a Nok sculpture, then it has hollowed-out eyes and mouth* and *If a sculpture has hollowed out eyes and mouth, then it has air vents that prevented cracking.*

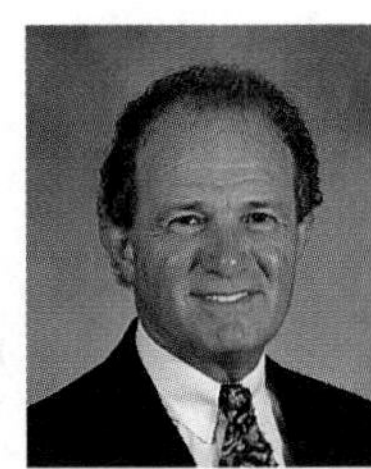

Let p, q, and r represent the parts of the statements.

p: it is a Nok sculpture

q: it has hollowed-out eyes and mouth

r: it has air vents that prevented cracking

Using these letters, the given statements can be represented as $p \rightarrow q$ and $q \rightarrow r$. Since the given statements are true, we can use the Law of Syllogism to conclude $p \rightarrow r$. That is, *If it is a Nok sculpture, then it has air vents that prevented cracking.*

CHECK FOR UNDERSTANDING

Communicating Mathematics

Study the lesson. Then complete the following.

1. **State** which property in algebra is similar to the Law of Syllogism. Explain.
2. "Those who choose Tint-and-Trim Hair Salon have impeccable taste; and you have impeccable taste" is an example of how an advertiser can misuse the Law of Detachment to make you come to an invalid conclusion.
   a. What conclusion do they want you to make?
   b. Write another example that illustrates incorrect logic. See margin.

1. Transitive Prop. (=)

2a. I should choose Tint-and-Trim Hair Salon.

In-Class Examples

For Example 3
Determine if a valid conclusion can be reached from the two true statements *If I watch TV, I will not do my homework* and *I did my homework* using the Law of Detachment. If a valid conclusion is possible, state it. **no conclusion**

For Example 4
Determine if a valid conclusion can be reached from the two true statements *If Elena takes the car to the store, she will stop at the post office* and *If Elena stops at the post office, she will buy stamps* using the Law of Syllogism. **If Elena takes the car to the store, she will buy stamps.**

Teaching Tip After reading the Law of Syllogism, remind students that in order to conclude $p \rightarrow r$, both $p \rightarrow q$ and $q \rightarrow r$ must be true.

Cooperative Learning

Trade-A-Problem Separate students into pairs. Have each student create if-then statements to illustrate the Law of Detachment and the Law of Syllogism and give them to the partner to apply the appropriate law. For more information on the trade-a-problem strategy, see *Cooperative Learning in the Mathematics Classroom*, one of the titles in the Glencoe Mathematics Professional Series, pages 25–26.

Classroom Vignette

"Students can find many applications and misapplications of deductive reasoning in the newspaper. Sometimes an illogical argument is what makes a comic funny. Advertisers often use statements that resemble logical arguments to convince consumers to buy. You may wish to create a bulletin board display of examples that students find."

Jerry Cummins
Author

3 PRACTICE/APPLY

Check for Understanding

Exercises 1–13 are designed to help you assess your students' understanding through reading, writing, speaking, and modeling. You should work through Exercises 1–5 with your students and then monitor their work on Exercises 6–13.

Error Analysis

Determining whether a valid conclusion follows can be difficult if students get caught up in the content of the statements. Encourage students to analyze each exercise by using letters to represent parts of statements. This should make the structure of the argument clearer.

Additional Answers

2b. Sample answer: Dogs have four legs. Kitty has four legs. Therefore Kitty is a dog.

4. Nina's reasoning is correct. Marlene's conclusion that she will necessarily make the team is invalid because the fact that she is practicing more on the weekends satisfies the conclusion of the conditional rather than the hypothesis.

3. See students' work.

3. **Write** your own example to illustrate the correct use of the Law of Detachment.

4. **You Decide** Marlene said, "The volleyball coach said that if I wanted to be on the varsity team, then I had to practice more on the weekends. Since I practice every weekend now, I'm going to make the team." Nina pointed out, "But you might not make the varsity team this season. You might have a better chance next year." Whose reasoning is correct? Explain. See margin.

5. Make up several logic puzzles. Have your friends try to solve them. Write the puzzles and solutions in your journal. See students' work.

Guided Practice

Determine if statement (3) follows from statements (1) and (2) by the Law of Detachment or the Law of Syllogism. If it does, state which law was used. If it does not, write *invalid.*

6. (1) If you plan to attend the University of Notre Dame, then you need to be in the top 10% of your class. yes; detachment
   (2) Rosita Nathan plans to attend Notre Dame.
   (3) Rosita Nathan needs to be in the top 10% of her class.
7. (1) If an angle is acute, then its measure is less than 90.
   (2) $m\angle A < 90$
   (3) $\angle A$ is acute. yes; syllogism
8. (1) Vertical angles are congruent.
   (2) $\angle 1 \cong \angle 2$
   (3) $\angle 1$ and $\angle 2$ are vertical. invalid

Determine if a valid conclusion can be reached from the two true statements using the Law of Detachment or the Law of Syllogism. If a valid conclusion is possible, state it and the law that is used. If a valid conclusion does not follow, write *no conclusion.*

9. Patricia Gorman should get 8 hours of sleep each day; detachment.
11. If the measure of an angle is less than 90, then it is not obtuse; syllogism.
12. There is exactly one line that contains points A and B; detachment.

9. (1) If you want good health, then you should get 8 hours of sleep each day.
   (2) Patricia Gorman wants good health.
10. (1) If $AB = BC$ and $BC = CD$, then $AB = CD$.
    (2) $AB = CD$ no conclusion
11. (1) If the measure of an angle is less than 90, then it is acute.
    (2) If an angle is acute, then it is not obtuse.
12. (1) If there are two points, then there is exactly one line that contains them.
    (2) There exist two points A and B.
13. **Music** The October/November, 1995 issue of *Zillions* magazine featured a music review of the Beatles album, *Live at the BBC.* It stated, "If you're a true Beatles fan, you'll love this, and if not, you'll become one." Determine if a possible conclusion can be reached, given each true statement. If a conclusion is possible, state it.
    a. You're not a true Beatles fan. You'll become a true Beatles fan.
    b. You'll love this album. no conclusion
    c. You'll become a true Beatles fan. no conclusion
    d. You're a true Beatles fan. You'll love this album.

Reteaching

Using Fill-in-the-Blanks Develop diagrams of the Law of Detachment and the Law of Syllogism. Write these on the chalkboard.

Law of Detachment

If ______, then ______.

______ is true.

Conclusion: ______.

Have students fill in the blanks with several different true conditionals.

Law of Syllogism

If ______, then ______.

If ______, then ....

Conclusion: If ______, then ....

Again, have students fill in the blanks with different statements to get a sense of the structure of this type of argument.

EXERCISES

Practice

Determine if statement (3) follows from statements (1) and (2) by the Law of Detachment or the Law of Syllogism. If it does, state which law was used. If it does not, write *invalid*.

A

14. (1) In-line skaters live dangerously.
(2) If you live dangerously, then you like to dance.
(3) If you are an in-line skater, then you like to dance. yes; syllogism

15. (1) If you drive safely, the life you save may be your own.
(2) Shani drives safely.
(3) The life she saves may be her own. yes; detachment

16. (1) If a figure is a rectangle, then its opposite sides are congruent.
(2) $\overline{AB} \cong \overline{DC}$ and $\overline{AD} \cong \overline{BC}$
(3) $ABCD$ is a rectangle. invalid

17. (1) Right angles are congruent.
(2) $\angle A \cong \angle B$
(3) $\angle A$ and $\angle B$ are right angles. invalid

18. (1) If an angle is obtuse, then it is not acute.
(2) $\angle 1$ is obtuse.
(3) $\angle 1$ is not acute. yes; detachment

19. (1) If you are a customer, then you are always right.
(2) If you are a teenager, then you are always right.
(3) If you are a teenager, then you are a customer. invalid

B

20. (1) Vertical angles are congruent.
(2) If two angles are congruent, then their measures are equal.
(3) If two angles are vertical, then their measures are equal. yes; syllogism

21. (1) If you do well, then you like criticism.
(2) If you resent criticism, then you do not do well.
(3) If you like criticism, then you do not do well. invalid

Determine if a valid conclusion can be reached from the two true statements using the Law of Detachment or the Law of Syllogism. If a valid conclusion is possible, state it and the law that is used. If a valid conclusion does not follow, write *no conclusion*.

22. (1) If two angles are vertical, then they do not form a linear pair.
(2) If two angles are vertical, then they are congruent. no conclusion

23. (1) If you eat to live, then you live to eat.
(2) Odina eats to live. Odina lives to eat; detachment.

24. (1) If a plane exists, then it contains at least three points not on the same line.
(2) Plane $\mathcal{N}$ contains points A, B, and C, which are not on the same line.

24. no conclusion

25. (1) If M is the midpoint of $\overline{AB}$, then $AM = MB$.
(2) If the measures of two segments are equal, then they are congruent.

25. If M is the midpoint of $\overline{AB}$, then $\overline{AM} \cong \overline{MB}$; syllogism.

26. (1) If an angle is obtuse, then its measure is greater than 90.
(2) $\angle 2$ is obtuse. $m\angle 2$ is greater than 90; detachment.

27. (1) If you spend money on it, then it's a business.
(2) If you spend money on it, then it's fun. no conclusion

28. (1) If two lines intersect to form a right angle, then they are perpendicular.
(2) ℓ and m are perpendicular. no conclusion

29. (1) If two planes intersect, then their intersection is a line.
(2) Planes $\mathcal{M}$ and $\mathcal{N}$ intersect.

29. Planes $\mathcal{M}$ and $\mathcal{N}$ intersect in a line; detachment.

Assignment Guide

Core (with proof): 15–39 odd, 41–49
Core (informal): 15–39 odd, 41–49
Enriched: 14–36 even, 37–49
All: Self Test, 1–10

For **Extra Practice,** see p. 767.

The red A, B, and C flags, printed only in the Teacher's Wraparound Edition, indicate the level of difficulty of the exercises.

Tech Prep

Broadcasting Technician For students interested in the field of music and broadcasting, you may wish to explain that broadcasting technicians handle the technical aspects of producing and transmitting musical recordings and/or radio and television programs. Courses at community colleges, as well as private broadcasting schools, provide the necessary training. For more information on tech prep, see the *Teacher's Handbook*.

Study Guide Masters, p. 10

2-3 NAME________ DATE________

Study Guide — Student Edition Pages 85–91

Deductive Reasoning

Two important laws used frequently in deductive reasoning are the **Law of Detachment** and the **Law of Syllogism**. In both cases you reach conclusions based on if-then statements.

Law of Detachment	Law of Syllogism
If $p \rightarrow q$ is a true conditional and p is true, then q is true.	If $p \rightarrow q$ and $q \rightarrow r$ are true conditionals, then $p \rightarrow r$ is also true.

Example: Determine if statement (3) follows from statements (1) and (2) by the Law of Detachment or the Law of Syllogism. If it does, state which law was used.
(1) If you break an item in a store, you must pay for it.
(2) Jill broke a vase in Potter's Gift Shop.
(3) Jill must pay for the vase.

Yes, statement (3) follows from statements (1) and (2) by the Law of Detachment.

Determine if a valid conclusion can be reached from the two true statements using the Law of Detachment or the Law of Syllogism. If a valid conclusion is possible, state it and the law that is used. If a valid conclusion does not follow, write no conclusion.

1. (1) If a number is a whole number, then it is an integer.
(2) If a number is an integer, then it is a rational number. If a number is a whole number, then it is a rational number; syllogism.

2. (1) If a dog eats Dogfood Delights, the dog is happy.
(2) Fido is a happy dog. no conclusion

3. (1) If people live in Manhattan, then they live in New York.
(2) If people live in New York, then they live in the United States. If people live in Manhattan, then they live in the United States; syllogism.

4. (1) Angles that are complementary have measures with a sum of 90.
(2) $\angle A$ and $\angle B$ are complementary. $m\angle A + m\angle B = 90$; detachment

5. (1) All fish can swim.
(2) Fonzo can swim. no conclusion

6. Look for a Pattern Find the next number in the list 83, 77, 71, 65, 59 and make a conjecture about the pattern. 53; each number is 6 less than the preceding one.

Additional Answers

34a. (2) I'm a careful bicycle rider.
(3) I wear a helmet.

34b. (2) If you wear a helmet, you will have fewer injuries.
(3) If you're a careful bicycle rider, then you will have fewer injuries.

35a. (2) I like pizza with everything.
(3) I'll like Jimmy's pizza.

35b. (2) If you like Jimmy's pizza, then you are a pizza connoisseur.
(3) If you like pizza with everything, then you are a pizza connoisseur.

36a. (2) $\angle A$ and $\angle B$ form a linear pair.
(3) $\angle A$ and $\angle B$ share a common ray.

36b. (2) If two angles share a common ray, then they are adjacent.
(3) If two angles form a linear pair, then they are adjacent.

Practice Masters, p. 10

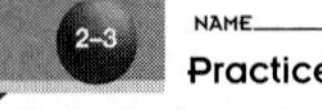

NAME__________ DATE ________

2-3 Practice — Student Edition Pages 85–91

Deductive Reasoning

Determine if a valid conclusion can be reached from the two true statements using the Law of Detachment or the Law of Syllogism. If a valid conclusion is possible, state it and the law that is used. If a valid conclusion does not follow, write no conclusion.

1. If Jim is a Texan, then he is an American.
Jim is a Texan. Jim is an American; detachment.
2. If Spot is a dog, then he has four legs.
Spot has four legs. no conclusion
3. If Rachel lives in Tampa, then Rachel lives in Florida.
If Rachel lives in Florida, then Rachel lives in the United States.
If Rachel lives in Tampa, then Rachel lives in the United States; syllogism.
4. If October 12 is a Monday, then October 13 is a Tuesday.
October 12 is a Monday. October 13 is a Tuesday; detachment.
5. If Henry studies his algebra, then he passes the test.
If Henry passes the test, then he will get a good grade. If Henry studies his algebra, then he will get a good grade; syllogism.

Determine if statement (3) follows from statements (1) and (2) by the Law of Detachment or the Law of Syllogism. If it does, state which law was used. If it does not, write invalid.

6. (1) If the measure of an angle is greater than 90, then it is obtuse.
(2) $m\angle T$ is greater than 90.
(3) $\angle T$ is obtuse. yes; detachment
7. (1) If Pedro is taking history, then he will study about World War II.
(2) Pedro will study about World War II.
(3) Pedro is taking history. invalid
8. (1) If Julie works after school, then she works in a department store.
(2) Julie works after school.
(3) Julie works in a department store. yes; detachment
9. (1) If William is reading, then he is reading a magazine.
(2) If William is reading a magazine, then he is reading a magazine about computers.
(3) If William is reading, then he is reading a magazine about computers. yes; syllogism
10. **Look for a Pattern** Tanya likes to burn candles. She has found that, once a candle has burned, she can melt 3 candle stubs, add a new wick, and have one more candle to burn. How many total candles can she burn from a box of 15 candles? 22

C

30. If two angles of a triangle are congruent, then the triangle is isosceles; syllogism.

31. Line t is perpendicular to line q; detachment.

30. (1) If two angles of a triangle are congruent, then the sides opposite these angels are congruent.
(2) If two sides of a triangle are congruent, then the triangle is isosceles.

31. (1) In a plane, if a line is perpendicular to one of two parallel lines, then it is perpendicular to the other.
(2) Line t is perpendicular to line p, which is parallel to line q.

32. (1) Cars are useful.
(2) Useful cars are practical. Cars are practical; syllogism.

33. (1) If two points lie in a plane, then the entire line containing those two points lie in that plane.
(2) Points X and Y lie in plane $\mathcal{P}$. $\overline{XY}$ lies in plane $\mathcal{P}$; detachment.

Using the given statement, create a second statement and a valid conclusion that illustrates the correct use of the Law of Detachment. Then write a statement and a conclusion that illustrate the correct use of the Law of Syllogism. 34–36. See margin.

34. If you're a careful bicycle rider, then you wear a helmet.
35. If you like pizza with everything, then you'll like Jimmy's Pizza.
36. If two angles form a linear pair, then they share a common ray.

Critical Thinking

37. The following if-then statements were used in the application at the beginning of Lesson 2–2. Assume all are true. Use deductive reasoning laws to write a true conclusion. You must use all three statements. Explain all steps used to arrive at your conclusion. (*Hint:* Remember if a conditional is true, then its contrapositive is true.) See Solutions Manual.
If a person is a baby, then the person is not logical.
If a person can manage a crocodile, then that person is not despised.
If a person is not logical, then the person is despised.

Applications and Problem Solving

38. (2) An unknown person attempted to give me an item to transport on my flight. (3) I did not accept it and notified airline personnel immediately.

40. If the statement *You can answer "yes" to these three questions* is true, then the conclusion, *You probably qualify for savings of up to 70% on your life insurance* is true.

38. **Airline Safety** The sign at the the right is posted in airports throughout the U.S. Provide information necessary to illustrate logical reasoning using the Law of Detachment with this if-then statement.

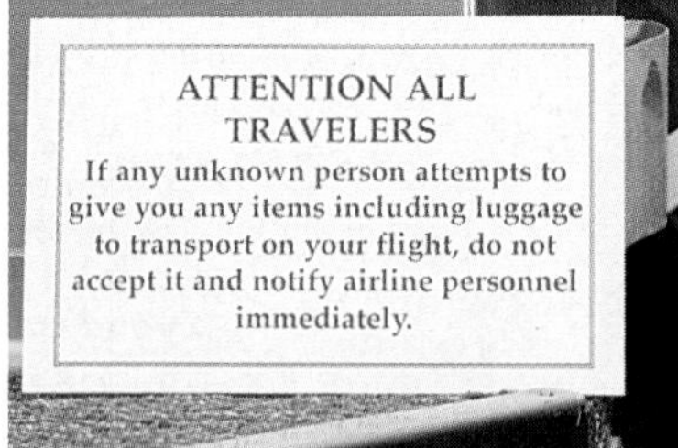

ATTENTION ALL TRAVELERS
If any unknown person attempts to give you any items including luggage to transport on your flight, do not accept it and notify airline personnel immediately.

39. **Look for a Pattern** What is the ones digit of 9^{46}? 1

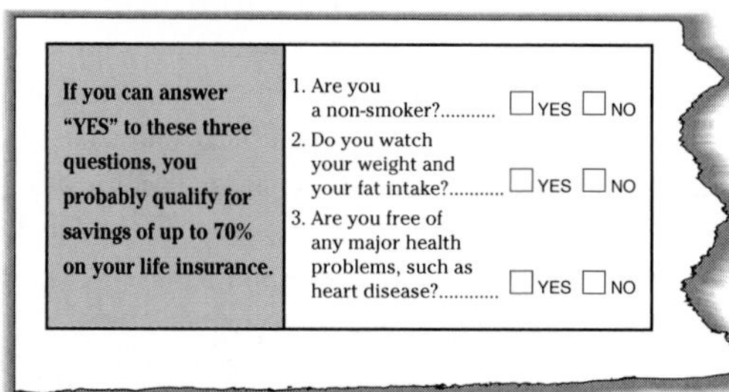

If you can answer "YES" to these three questions, you probably qualify for savings of up to 70% on your life insurance.
1. Are you a non-smoker?........... YES NO
2. Do you watch your weight and your fat intake?.......... YES NO
3. Are you free of any major health problems, such as heart disease?........... YES NO

40. **Advertising** The magazine ad at the left appeared in *Kiplinger's Personal Finance Magazine*. Describe how the Law of Detachment could be used to arrive at a logical conclusion, assuming the statements in the ad are true.

Mixed Review

41. Write the converse of the conditional *If two lines are perpendicular, then they intersect.* Determine if the converse is *true* or *false*. If false, give a counterexample. (Lesson 2–2) See Solutions Manual.

Extension

Problem Solving Arrange these statements so that, by the Law of Syllogism, you can conclude that Dexter has four legs.

A. All cats have four legs.
B. Dexter is in the basket.
C. If Alice has a pet, it is a cat.
D. If there is anything in the basket, it is Alice's pet.

D and B let you conclude that Dexter is Alice's pet. Then C and A let you conclude that Dexter has four legs.

42. Write the conditional *Two planes intersect in a line* in if-then form. (Lesson 2–2) If two planes intersect, then they intersect in a line.

43. Given $\overline{AB}$, $\overline{BC}$, and $\overline{CD}$, write a conjecture. Draw a figure to illustrate your conjecture. (Lesson 2–1) See Solutions Manual.

44. Find the value of z and $m\angle SUV$. (Lesson 1–7) 43, 140

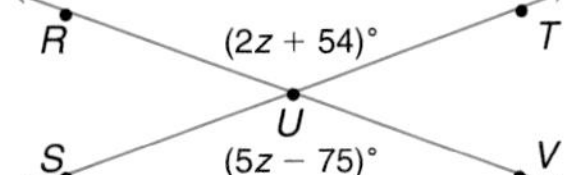

45. See students' work.

45. Use a protractor to draw a 65° angle. (Lesson 1–6)

46. Q is between R and S, $RQ = 6x - 1$, $QS = 2x + 4$, and $RS = 9x - 3$. Find RS. (Lesson 1–4) 51

47. Find the length of the segment with endpoints $B(4, -1)$ and $C(2, 5)$. (Lesson 1–4) $2\sqrt{10}$

Algebra

48. Find $3.8 + (-4.7)$. -0.9

49. **Sales** Carmen bought a pair of shoes that cost \$36.78 including tax. If she gave the clerk two twenty-dollar bills, how much should Carmen receive in change? \$3.22

SELF TEST

Determine if the conjecture is *true* or *false* based on the given information. Explain your answer. (Lesson 2–1) 1–2. See margin.

1. **Given:** points A, B, and C
   Conjecture: A, B, and C are collinear.

2. **Given:** $\angle A$ and $\angle C$ are complementary angles. $\angle B$ and $\angle C$ are complementary angles.
   Conjecture: $\angle A \cong \angle B$

3. **Botany** In the 1920s, some Japanese farmers observed that certain rice plants were growing taller and thinner than normal rice plants and then drooping over, making them impossible to harvest. (Lesson 2–1) a. Sample answer: insufficient light or water
   a. Make some conjectures about why the plants were drooping.
   b. The scientists who researched the problem discovered that fungus was growing on the drooping rice plants, while the healthy plants had no fungus. Does this observation prompt a new conjecture? Yes; the fungus is killing the plants.
   c. How might the scientists have tested the conjecture they made? See margin.

4. Identify the hypothesis and conclusion of the statement *If you're there before it's over, then you're on time.* (Lesson 2–2) Hypothesis: you're there before it's over, Conclusion: you're on time

5. Write the converse of the statement *If it is raining, then there are clouds.* (Lesson 2–2)

6. Write the inverse of the statement *If two lines are parallel, then they do not intersect.* (Lesson 2–2)

7. Write the contrapositive of the statement *If a figure is a square, then it has four sides.* (Lesson 2–2)
   5–7. See margin.

Determine if statement (3) follows from statements (1) and (2). If it does, state which law was used. If it does not, write *invalid*. (Lesson 2–3)

8. (1) If you are not satisfied with a tape, then you can return it within a week for a full refund.
   (2) Yong is not satisfied with a tape.
   (3) Yong can return the tape within a week for a full refund. yes; detachment

9. (1) If x is a real number, then x is an integer.
   (2) x is an integer.
   (3) x is a real number. invalid

10. (1) If fossil fuels are burned, then acid rain is produced.
    (2) If acid rain is produced, wildlife suffers.
    (3) If fossil fuels are burned, then wildlife suffers. yes; syllogism

SELF TEST

The Self Test provides students with a brief review of the concepts and skills in Lessons 2-1 through 2-3. Lesson numbers are given to the right of exercises or instruction lines so students can review concepts not yet mastered.

4 ASSESS

Closing Activity

Writing Ask students to compare and contrast the two laws of deductive reasoning that were presented in this lesson.

Chapter 2 Quiz B (Lesson 2-3) is available in the *Assessment and Evaluation Masters*, p. 44.

Mid-Chapter Test (Lessons 2-1 through 2-3) is available in the *Assessment and Evaluation Masters*, p. 43.

Answers for the Self Test

1. false; counterexample:

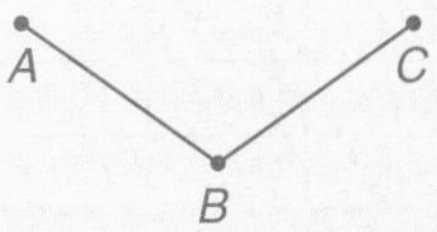

2. True; if $m\angle C = 35$, then $m\angle A = 90 - 35$ or 55, and $m\angle B = 90 - 35$ or 55. Since $m\angle A = m\angle B$, then $\angle A \cong \angle B$.

3c. Introduce the fungus to some healthy plants and see if they droop.

5. If there are clouds, then it is raining.
6. If two lines are not parallel, then they intersect.
7. If a figure does not have four sides, then it is not a square.

Enrichment Masters, p. 10

NAME ______ DATE ______

Enrichment

Student Edition Pages 85–91

Valid and Faulty Arguments

Consider the statements at the right. What conclusions can you make?

(1) Boots is a cat.
(2) Boots is purring.
(3) A cat purrs if it is happy.

From statements 1 and 3, it is correct to conclude that Boots purrs if it is happy. However, it is faulty to conclude from only statements 2 and 3 that Boots is happy. The if-then form of statement 3 is *If a cat is happy, then it purrs.*

Advertisers often use faulty logic in subtle ways to help sell their products. By studying the arguments, you can decide whether the argument is valid or faulty.

Decide if each argument is valid or faulty.

1. (1) If you buy Tuff Cote luggage, it will survive airline travel.
   (2) Justin buys Tuff Cote luggage.
   Conclusion: Justin's luggage will survive airline travel. valid

2. (1) If you buy Tuff Cote luggage, it will survive airline travel.
   (2) Justin's luggage survived airline travel.
   Conclusion: Justin has Tuff Cote luggage. faulty

3. (1) If you use Clear Line long distance service, you will have clear reception.
   (2) Anna has clear long distance reception.
   Conclusion: Anna uses Clear Line long distance service. faulty

4. (1) If you read the book *Beautiful Braids*, you will be able to make beautiful braids easily.
   (2) Nancy read the book *Beautiful Braids*.
   Conclusion: Nancy can make beautiful braids easily. valid

5. (1) If you buy a word processor, you will be able to write letters faster.
   (2) Tania bought a word processor.
   Conclusion: Tania will be able to write letters faster. valid

6. (1) Great swimmers wear AquaLine swimwear.
   (2) Gina wears AquaLine swimwear.
   Conclusion: Gina is a great swimmer. faulty

7. Write an example of faulty logic that you have seen in an advertisement. Answers will vary.

2-4 LESSON NOTES

NCTM Standards: 1–5, 7, 8

Instructional Resources

- Study Guide Master 2-4
- Practice Master 2-4
- Enrichment Master 2-4
- Modeling Mathematics Masters, p. 80
- Real-World Applications, 4

Transparency 2-4A contains the 5-Minute Check for this lesson; **Transparency 2-4B** contains a teaching aid for this lesson.

Recommended Pacing	
Standard Pacing	Day 6 of 12
Honors Pacing	Day 5 of 11
Block Scheduling*	Day 4 of 7

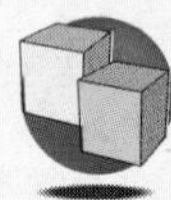

*For more information on pacing and possible lesson plans, refer to the *Block Scheduling Booklet*.

1 FOCUS

5-Minute Check
(over Lesson 2-3)

Draw a conclusion if possible.

1. All rectangles have congruent diagonals. *ABCD* is a rectangle. **ABCD has congruent diagonals.**
2. All squares have four congruent sides. *GHIJ* has four congruent sides. **no conclusion**
3. If ℓ is perpendicular to m, then $\angle ABC$ is a right angle. If $\angle ABC$ is a right angle, then $\angle BCD$ is complementary to $\angle DBA$. ℓ is perpendicular to m. **$\angle BCD$ is complementary to $\angle DBA$.**
4. If x is divisible by 4, it is an even number. If x is divisible by 3, it may be an even or odd number. x is an even number. **no conclusion**

2-4 Integration: Algebra
Using Proof in Algebra

What YOU'LL LEARN

- To use properties of equality in algebraic and geometric proofs.

Why IT'S IMPORTANT

You can use proof to solve problems in algebra.

APPLICATION

Exercise

A recent study by the University of Massachusetts reveals that in-line skating at moderate speeds burns as many Calories as running. The chart below shows the Calories burned per minute at various skating speeds for various body weights.

Calories Burned Per Minute

Weight (lb)	Skating Speed 8 mph	9 mph	10 mph	11 mph	12 mph
120	4.2	5.8	7.4	8.9	10.5
140	5.1	6.7	8.3	9.9	11.4
160	6.1	7.7	9.2	10.8	12.4
180	7.0	8.6	10.2	11.7	13.3
200	7.9	9.5	11.2	12.6	14.2

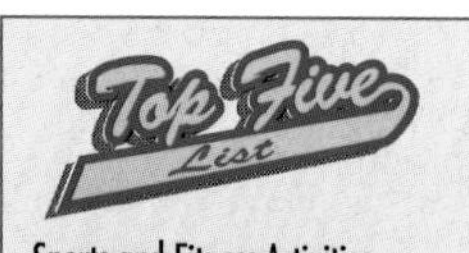

Sports and Fitness Activities with the Highest Percentage of Women Participants

1. Hard-court volleyball, 59%
2. Swimming, 57%
3. In-line skating, 45%
4. Cross-country skiing, 44%
5. Hiking/backpacking, 43%

You are familiar with many of the properties of algebra. These properties, along with various defined operations and sets of numbers, form a mathematical system. Working within the properties of the system allows you to perform algebraic operations.

For example, you can use the Multiplication Property of Equality to determine your skating speed in miles per hour. Use the formula $\frac{s}{d} = \frac{60}{t}$, where s represents the average speed (mph), d represents the distance (mi), 60 represents 60 minutes in an hour, and t represents time skating (min).

The Multiplication Property of Equality allows you to multiply each side of the equation by d to solve the equation in terms of s.

$$\frac{s}{d} \cdot d = \frac{60}{t} \cdot d$$

$$s = \frac{60}{t} \cdot d$$

So, if you skated 3 miles in 20 minutes, then the equation would become $s = \frac{60}{20} \cdot 3$ or $s = 9$ mph. You skated 9 mph for 20 minutes. Now use the table to determine the Calories burned. If you weighed 120 pounds, you would have burned $20 \cdot 5.8$ or 116 Calories.

Another example of a mathematical system can be found in geometry. Since geometry also deals with variables, numbers, and operations, many of the properties of algebra are also true in geometry. Some of the other important properties of algebra are listed in the following table.

Volleyball was invented in 1895 by William G. Morgan, a Y.M.C.A. director in Mass. He called it "mintonette" until someone suggested the name volleyball after seeing that the ball is volleyed, or hit without touching the ground.

Properties of Equality for Real Numbers	
Reflexive Property	For every number a, $a = a$.
Symmetric Property	For all numbers a and b, if $a = b$, then $b = a$.
Transitive Property	For all numbers a, b, and c, if $a = b$ and $b = c$, then $a = c$.
Addition and Subtraction Properties	For all numbers a, b, and c, if $a = b$, then $a + c = b + c$ and $a - c = b - c$.
Multiplication and Division Properties	For all numbers a, b, and c, if $a = b$, then $a \cdot c = b \cdot c$, and if $c \neq 0$, $\frac{a}{c} = \frac{b}{c}$.
Substitution Property	For all numbers a and b, if $a = b$, then a may be replaced by b in any equation or expression.
Distributive Property	For all numbers a, b, and c, $a(b + c) = ab + ac$.

Since segment measures and angle measures are real numbers, these properties from algebra can be used to discuss their relationships. Some examples of these applications are shown below.

Property	Segments	Angles
Reflexive	$PQ = PQ$	$m\angle 1 = m\angle 1$
Symmetric	If $AB = CD$, then $CD = AB$.	If $m\angle A = m\angle B$, then $m\angle B = m\angle A$.
Transitive	If $GH = JK$ and $JK = LM$, then $GH = LM$.	If $m\angle 1 = m\angle 2$ and $m\angle 2 = m\angle 3$, then $m\angle 1 = m\angle 3$.

Example 

Algebra

Name the property of equality that justifies each statement.

Statements	Reasons
a. If $AB + BC = DE + BC$, then $AB = DE$.	**a.** Subtraction Property (=)
b. $m\angle ABC = m\angle ABC$	**b.** Reflexive Property (=)
c. If $XY = PQ$ and $XY = RS$, then $PQ = RS$.	**c.** Substitution Property (=)
d. If $\frac{1}{3}x = 5$, then $x = 15$.	**d.** Multiplication Property (=)
e. If $2x = 9$, then $x = \frac{9}{2}$.	**e.** Division Property (=)

You can use these properties as reasons for the step-by-step solution of an equation.

Example 2

Algebra

Justify each step in solving $\frac{3x + 5}{2} = 7$.

Statements	Reasons
1. $\frac{3x + 5}{2} = 7$	**1.** Given
2. $2\left(\frac{3x + 5}{2}\right) = 2(7)$	**2.** Multiplication Property (=)
3. $3x + 5 = 14$	**3.** Distributive Property (=)
4. $3x = 9$	**4.** Subtraction Property (=)
5. $x = 3$	**5.** Division Property (=)

Classroom Vignette

"I allow students to work in groups as they learn about writing proofs. I give groups a completed proof with steps and reasons written in no particular order. The group must discuss the correct order of the proof and the appropriateness of the reasons given."

Rick Kais
South Milwaukee High School
South Milwaukee, Wisconsin

Motivating the Lesson

Questioning What can you conclude if you know that $\angle A$ measures 40° and $\angle B$ measures 40°? $\angle A \cong \angle B$

2 TEACH

In-Class Examples

For Example 1
Name the property of equality that justifies each statement.

a. If $3x = 120$, then $x = 40$. Division Prop. (=)
b. If $12 = AB$, then $AB = 12$. Symmetric Prop. (=)
c. If $AB = BC$, and $BC = CD$, then $AB = CD$. Transitive Prop. (=)
d. If $y = 75$ and $y = m\angle A$, then $m\angle A = 75$. Substitution Prop. (=)

For Example 2
Justify each step in solving $\frac{x}{3} + 4 = 1$.

Statements Reasons
1. $\frac{x}{3} + 4 = 1$ Given
2. $\frac{x}{3} = -3$ Subtraction Prop. (=)
3. $x = -9$ Multiplication Prop. (=)

Teaching Tip In Example 2, point out that statements and corresponding reasons are numbered for clarity. The first statement in a proof will always be some of the given information. Other pieces of given information may be introduced as needed. The last statement should be the conclusion you want to prove.

Proof Pointer

Most students subconsciously know why each step in their solution of an equation is accurate. A proof is a written record of their solution steps and those reasons why each step works.

In-Class Example

For Example 3
Justify the steps for the proof of the conditional *If ∠ABD and ∠DBC are complementary, then ∠ABC is a right angle.*

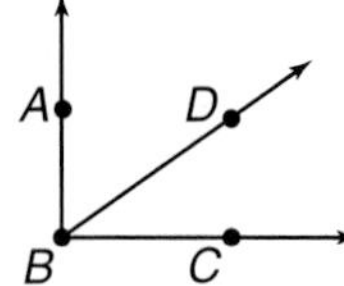

Given: ∠*ABD* and ∠*DBC* are complementary.
Prove: ∠*ABC* is a right angle.

Proof:
Statements Reasons
1. ∠*ABD* and ∠*DBC* are complementary. Given
2. $m\angle ABD + m\angle DBC = 90$ Definition of complementary angles
3. $m\angle ABD + m\angle DBC = m\angle ABC$ Angle Addition Postulate
4. $m\angle ABC = 90$ Substitution Prop. (=)
5. ∠*ABC* is a right angle. Definition of right angle

Teaching Tip After reading the properties of equality table, remind students that *a*, *b*, and *c* can be fractions or decimals for any of these properties.

3 PRACTICE/APPLY

Check for Understanding

Exercises 1–12 are designed to help you assess your students' understanding through reading, writing, speaking, and modeling. You should work through Exercises 1–5 with your students and then monitor their work on Exercises 6–12.

Error Analysis
Students may have difficulty differentiating between the Transitive Property and the Substitution Property. Tell them that the Transitive Property has to do with order: if $a = b$ and $b = c$, then $a = c$.

Example 2 is a proof of the conditional *If* $\frac{3x+5}{2} = 7$, *then* $x = 3$. The given information comes from the hypothesis of the conditional. It is the starting point of the proof. The conclusion, $x = 3$, is the end of the proof. Reasons (properties) listed for each step leading to the conclusion make this sequence a **proof**. This type of proof is called a **two-column proof**.

Proofs in geometry can be organized in the same manner. Algebra properties, definitions, postulates, and previously-proven theorems can be used for reasons. Proofs in geometry are usually written in two-column or in paragraph form. *Two-column proofs are sometimes called formal proofs.*

Example 3 **Justify the steps for the proof of the conditional *If PR = QS, then PQ = RS.*** *Remember that PR, QS, PQ, and RS represent real numbers.*

Given: $PR = QS$

Prove: $PQ = RS$

Proof:

P Q R S

Statements	Reasons
1. $PR = QS$	1. ?
2. $PQ + QR = PR$ $QR + RS = QS$	2. ?
3. $PQ + QR = QR + RS$	3. ? *Step 3 uses information from steps 1 and 2.*
4. $PQ = RS$	4. ?

Reason 1: Given (since it follows from the hypothesis)
Reason 2: Segment Addition Postulate
Reason 3: Substitution Property (=)
Reason 4: Subtraction Property (=)

CHECK FOR UNDERSTANDING

Communicating Mathematics

Study the lesson. Then complete the following.

1. **Describe** the parts of a two-column proof. *given and prove statements and two columns, one of statements, and one of reasons*
2. **State** the part of the conditional that is related to the *Given* statement of a proof. What part is related to the *Prove* statement? *hypothesis; conclusion*
3. **Write** a statement that illustrates the Substitution Property of Equality. *Sample answer: If AB = 5 and AB + CD = 8, then 5 + CD = 8.*
4. **List** the types of reasons that can be used to justify a statement in a geometric proof. *algebra properties, definitions, postulates, and previously-proven theorems*
5. **Choose** the number of the reason in the right column that best matches each statement in the left column.

Statements	Reasons
a. If $x - 7 = 12$, then $x = 19$. 2	(1) Distributive Property
b. If $MK = NJ$ and $BG = NJ$, then $MK = BG$. 4	(2) Addition Property (=)
c. If $m\angle 4 = m\angle 5$ and $m\angle 5 = m\angle 6$, then $m\angle 4 = m\angle 6$. 5	(3) Symmetric Property (=)
d. If $ST = UV$, then $UV = ST$. 3	(4) Substitution Property (=)
e. If $x = -3(2x - 4)$, then $x = -6x + 12$. 1	(5) Transitive Property (=)

Reteaching

Using Reasoning Review the general method for solving a simple linear equation. Write the justification for each step in solving $3x + 6 = 4$.

$3x = -2$ Addition and Subtraction Property (=)

$x = -\frac{2}{3}$ Multiplication and Division Property (=)

Guided Practice

Name the property of equality that justifies each statement.

6. If $2x = 3$, then $x = \frac{3}{2}$. Division Prop. (=)
7. If $XY - AB = WZ - AB$, then $XY = WZ$. Addition Prop. (=)

8. Substitution Prop. (=)

8. If $m\angle 1 + m\angle 2 = 90$ and $m\angle 2 = m\angle 3$, then $m\angle 1 + m\angle 3 = 90$.
9. For the proof below, the reasons in the right column are not in the proper order. Reorder the reasons to properly match the statements in the left column. 9c. 4 or 5 9d. 5 or 4

 Proof

Prove that if $m\angle AXC = m\angle DYF$ and $m\angle 1 = m\angle 3$, then $m\angle 2 = m\angle 4$.

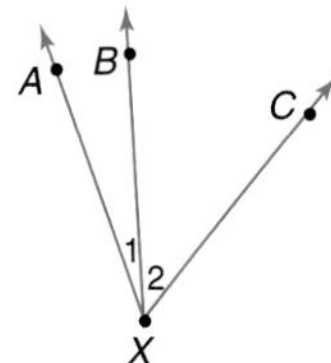

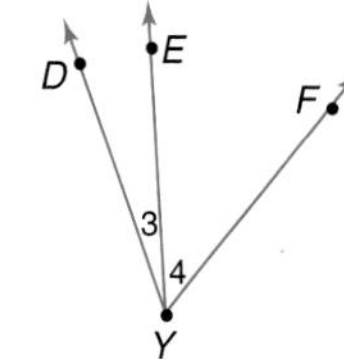

Given: $m\angle AXC = m\angle DYF$
$m\angle 1 = m\angle 3$

Prove: $m\angle 2 = m\angle 4$

Proof:

Statements	Reasons
a. $m\angle AXC = m\angle DYF$ $m\angle 1 = m\angle 3$ 1	(1) Given
b. $m\angle AXC = m\angle 1 + m\angle 2$ $m\angle DYF = m\angle 3 + m\angle 4$ 3	(2) Subtraction Property (=)
c. $m\angle 1 + m\angle 2 = m\angle 3 + m\angle 4$	(3) Angle Addition Postulate
d. $m\angle 3 + m\angle 2 = m\angle 3 + m\angle 4$	(4) Substitution Property (=)
e. $m\angle 2 = m\angle 4$ 2	(5) Substitution Property (=)

10. Copy the proof. Then name the property that justifies each statement.

Prove that if $-2x + \frac{3}{2} = 8$, then $x = -\frac{13}{4}$.

Given: $-2x + \frac{3}{2} = 8$

Prove: $x = -\frac{13}{4}$

Proof:

Statements	Reasons
a. $-2x + \frac{3}{2} = 8$	a. ? Given
b. $2\left(-2x + \frac{3}{2}\right) = 2(8)$	b. ? Multiplication Prop. (=)
c. $-4x + 3 = 16$	c. ? Distributive Prop.
d. $-4x = 13$	d. ? Subtraction Prop. (=)
e. $x = -\frac{13}{4}$	e. ? Division Prop. (=)

11. **Construction** When bridges, buildings, or roads are constructed, engineers must consider the expansion and contraction of materials during temperature changes. The coefficient of linear expansion k is different for every substance, and physicists find k by using the formula $k = \frac{\Delta\ell}{\ell(T - t)}$, where ℓ represents original length, $\Delta\ell$ represents change in length, T represents highest temperature, and t represents lowest temperature. Solve this formula for T and justify each step. See margin.

Additional Answer

11. Given: $k = \frac{\Delta\ell}{\ell(T - t)}$

Prove: $T = \frac{\Delta\ell}{k\ell} + t$

Proof:
Statements (Reasons)

1. $k = \frac{\Delta\ell}{\ell(T - t)}$ (Given)
2. $k(T - t) = \frac{\Delta\ell}{\ell}$ (Multiplication Prop. (=))
3. $T - t = \frac{\Delta\ell}{k\ell}$ (Division Prop. (=))
4. $T = \frac{\Delta\ell}{k\ell} + t$ (Addition Prop. (=))

Assignment Guide

Core (with proof): 13–33 odd, 34–41
Core (informal): 13–21 odd, 31, 33, 34–41
Enriched: 14–30 even, 31–41

For **Extra Practice,** see p. 767.

The red A, B, and C flags, printed only in the Teacher's Wraparound Edition, indicate the level of difficulty of the exercises.

12c. $m\angle 1 + m\angle 3 = m\angle 2 + m\angle 3$

12. Copy and complete the proof.

Prove that if $m\angle 1 = m\angle 2$, then $m\angle PXR = m\angle SXQ$.

Given: $m\angle 1 = m\angle 2$

Prove: $m\angle PXR = m\angle SXQ$

Proof:

Statements	Reasons
a. $m\angle 1 = m\angle 2$	a. ? Given
b. $m\angle 3 = m\angle 3$	b. ? Reflexive Property (=)
c. ?	c. Addition Property (=)
d. $m\angle PXR = m\angle 2 + m\angle 3$ $m\angle SXQ = m\angle 1 + m\angle 3$	d. ? Angle Addition Postulate
e. ? $m\angle PXR = m\angle SXQ$	e. Substitution Property (=)

EXERCISES

Practice

Name the property of equality that justifies each statement.

13. If $5 = 3x - 4$, then $3x - 4 = 5$. Symmetric Prop. (=)

14. If $3\left(x - \frac{5}{3}\right) = 1$, then $3x - 5 = 1$. Distributive Prop.

15. If $0.5AB = 0.5CD$, then $AB = CD$. Mult. Prop. (=) or Div. Prop. (=)

16. If $m\angle 1 = 90$ and $m\angle 2 = 90$, then $m\angle 1 = m\angle 2$. Substitution Prop. (=)

17. If $2m\angle ABC = 180$, then $m\angle ABC = 90$. Division Prop. (=)

18. For XY, $XY = XY$. Reflexive Prop. (=)

19. If $EF = GH$ and $GH = JK$, then $EF = JK$. Transitive Prop. (=)

20. If $m\angle 1 + 30 = 90$, then $m\angle 1 = 60$. Subtraction Prop. (=)

21. If $AB + IJ = MX + IJ$, then $AB = MX$. Subtraction Prop. (=)

For each proof, the reasons in the right column are not in the proper order. Reorder the reasons to properly match the statements in the left column.

22. Prove that if $2x - 7 = \frac{1}{3}x - 2$, then $x = 3$.

Given: $2x - 7 = \frac{1}{3}x - 2$

Prove: $x = 3$

Proof:

Statements	Reasons
a. $2x - 7 = \frac{1}{3}x - 2$ 1	(1) Given
b. $3(2x - 7) = 3\left(\frac{1}{3}x - 2\right)$ 4	(2) Distributive Property
c. $6x - 21 = x - 6$ 2	(3) Addition Property (=)
d. $5x - 21 = -6$ 6	(4) Multiplication Property (=)
e. $5x = 15$ 3	(5) Division Property (=)
f. $x = 3$ 5	(6) Subtraction Property (=)

23. Prove that if $\angle XWY \cong \angle XYW$, then $\angle AWX \cong \angle BYX$.

Given: $\angle XWY \cong \angle XYW$

Prove: $\angle AWX \cong \angle BYX$

Proof:

Statements	Reasons
a. $\angle XWY \cong \angle XYW$ 2	(1) Def. supplementary $\angle$s
b. $m\angle AWX + m\angle XWY = 180$ $m\angle BYX + m\angle XYW = 180$ 1	(2) Given
c. $m\angle AWX + m\angle XWY =$ $m\angle BYX + m\angle XYW$ 4 or 3	(3) Substitution Property (=)
d. $m\angle AWX + m\angle XWY =$ $m\angle BYX + m\angle XWY$ 3 or 4	(4) Substitution Property (=)
e. $m\angle AWX = m\angle BYX$ 5	(5) Subtraction Property (=)
f. $\angle AWX \cong \angle BYX$ 6	(6) Def. $\cong$ $\angle$s

Copy each proof. Then name the property that justifies each statement.

24. Prove that if $-\frac{1}{2}x = 9$, then $x = -18$.

Given: $-\frac{1}{2}x = 9$

Prove: $x = -18$

Proof:

Statements	Reasons
a. $-\frac{1}{2}x = 9$	a. ? Given
b. $-1x = 18$	b. ? Multiplication Prop. (=)
c. $x = -18$	c. ? Division Prop. (=) or Multiplication Prop. (=)

25. Prove that if $5 - \frac{2}{3}x = 1$, then $x = 6$.

Given: $5 - \frac{2}{3}x = 1$

Prove: $x = 6$

Proof:

Statements	Reasons
a. $5 - \frac{2}{3}x = 1$	a. ? Given
b. $3\left(5 - \frac{2}{3}x\right) = 3(1)$	b. ? Multiplication Prop. (=)
c. $15 - 2x = 3$	c. ? Distributive Prop.
d. $-2x = -12$	d. ? Subtraction Prop. (=)
e. $x = 6$	e. ? Division Prop. (=)

Study Guide Masters, p. 11

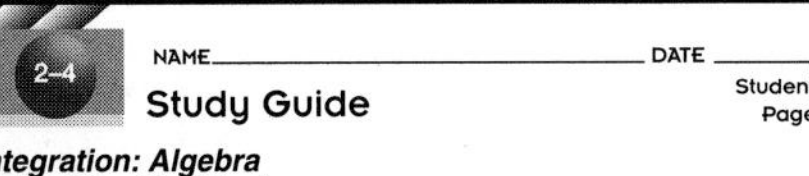

NAME ____________ DATE ________

Study Guide

Student Edition Pages 92–99

Integration: Algebra
Using Proof in Algebra

Many rules from algebra are used in geometry.

Properties of Equality for Real Numbers	
Reflexive Property	$a = a$
Symmetric Property	If $a = b$, then $b = a$.
Transitive Property	If $a = b$ and $b = c$, then $a = c$.
Addition Property	If $a = b$, then $a + c = b + c$.
Subtraction Property	If $a = b$, then $a - c = b - c$.
Multiplication Property	If $a = b$, then $a \cdot c = b \cdot c$.
Division Property	If $a = b$ and $c \neq 0$, then $\frac{a}{c} = \frac{b}{c}$.
Substitution Property	If $a = b$, then a may be replaced by b in any equation or expression.
Distributive Property	$a(b + c) = ab + ac$

Example: Prove that if $4x - 8 = -8$, then $x = 0$.

Given: $4x - 8 = -8$

Prove: $x = 0$

Proof:

Statements	Reasons
a. $4x - 8 = -8$	**a.** Given
b. $4x = 0$	**b.** Addition Property (=)
c. $x = 0$	**c.** Division Property (=)

Name the property that justifies each statement.

1. Prove that if $\frac{3}{5}x = -9$, then $x = -15$.

Given: $\frac{3}{5}x = -9$

Prove: $x = -15$

Proof:

Statements	Reasons
a. $\frac{3}{5}x = -9$	**a.** Given
b. $3x = -45$	**b.** Multiplication Property (=)
c. $x = -15$	**c.** Division Property (=)

2. Prove that if $3x - 2 = x - 8$, then $x = -3$.

Given: $3x - 2 = x - 8$

Prove: $x = -3$

Proof:

Statements	Reasons
a. $3x - 2 = x - 8$	**a.** Given
b. $2x - 2 = -8$	**b.** Subtraction Property (=)
c. $2x = -6$	**c.** Addition Property (=)
d. $x = -3$	**d.** Division Property (=)

26. Prove that if $AC = DF$ and $AB = DE$, then $BC = EF$.

Given: $AC = DF$ and $AB = DE$

Prove: $BC = EF$

Proof:

Statements	Reasons
a. $AC = DF$	a. ? Given
b. $AC = AB + BC$ $DF = DE + EF$	b. ? Segment Addition Postulate
c. $AB + BC = DE + EF$	c. ? Substitution Prop. (=)
d. $AB = DE$	d. ? Given
e. $BC = EF$	e. ? Subtraction Prop. (=)

Copy and complete the following proofs.

27. If $m\angle TUV = 90$, $m\angle XWV = 90$, and $m\angle 1 = m\angle 3$, then $m\angle 2 = m\angle 4$.

Given: $m\angle TUV = 90$, $m\angle XWV = 90$, and $m\angle 1 = m\angle 3$

Prove: $m\angle 2 = m\angle 4$

Proof:

Statements	Reasons
a. ?	a. Given
b. $m\angle TUV = m\angle XWV$	b. ? Substitution Prop. (=)
c. $m\angle TUV = m\angle 1 + m\angle 2$ $m\angle XWV = m\angle 3 + m\angle 4$	c. ? Angle Addition Postulate
d. ?	d. Substitution Property (=)
e. $m\angle 1 + m\angle 2 = m\angle 1 + m\angle 4$	e. ? Substitution Prop. (=)
f. ? $m\angle 2 = m\angle 4$	f. Subtraction Property (=)

27a. $m\angle TUV = 90$, $m\angle XWV = 90$, $m\angle 1 = m\angle 3$

27d. $m\angle 1 + m\angle 2 = m\angle 3 + m\angle 4$

28. If $4 - \frac{1}{2}x = \frac{7}{2} - x$, then $x = -1$.

Given: $4 - \frac{1}{2}x = \frac{7}{2} - x$

Prove: $x = -1$

Proof:

Statements	Reasons
a. $4 - \frac{1}{2}x = \frac{7}{2} - x$	a. ? Given
b. $2\left(4 - \frac{1}{2}x\right) = 2\left(\frac{7}{2} - x\right)$	b. ? Multiplication Prop. (=)
c. ? $8 - x = 7 - 2x$	c. Distributive Property
d. $1 - x = -2x$	d. ? Subtraction Prop. (=)
e. ? $1 = -x$	e. Addition Property (=)
f. $-1 = x$	f. ?
g. $x = -1$	g. ? Symmetric Prop. (=)

28f. Div. Prop. (=) or Mult. Prop. (=)

Practice Masters, p. 11

NAME ________ DATE ________

Practice — Student Edition Pages 92–99

Integration: Algebra
Using Proof in Algebra

Name the property of equality that justifies each statement.

1. If $m\angle A = m\angle B$, then $m\angle B = m\angle A$. symmetric
2. If $x + 3 = 17$, then $x = 14$. subtraction
3. $xy = xy$ reflexive
4. If $7x = 42$, then $x = 6$. division
5. If $XY - YZ = XM$, then $XM + YZ = XY$. addition
6. $2(x + 4) = 2x + 8$. distributive
7. If $m\angle A + m\angle B = 90$, and $m\angle A = 30$, then $30 + m\angle B = 90$. substitution
8. If $x = y + 3$ and $y + 3 = 10$, then $x = 10$. transitive

Complete each proof by naming the property that justifies each statement.

9. Prove that if $2(x - 3) = 8$, then $x = 7$.
Given: $2(x - 3) = 8$
Prove: $x = 7$
Proof:

Statements	Reasons
a. $2(x - 3) = 8$	a. Given
b. $2x - 6 = 8$	b. Distributive Property
c. $2x = 14$	c. Addition Property (=)
d. $x = 7$	d. Division Property (=)

10. Prove that if $3x - 4 = \frac{1}{2}x + 6$, then $x = 4$.
Given: $3x - 4 = \frac{1}{2}x + 6$
Prove: $x = 4$
Proof:

Statements	Reasons
a. $3x - 4 = \frac{1}{2}x + 6$	a. Given
b. $\frac{5}{2}x - 4 = 6$	b. Subtraction Property (=)
c. $\frac{5}{2}x = 10$	c. Addition Property (=)
d. $x = 4$	d. Multiplication Property (=)

Write a complete proof for each of the following

29. If $2x + 6 = 3 + \frac{5}{3}x$, then $x = -9$. See Solutions Manual.
30. If $AC = AB$, $AC = 4x + 1$, and $AB = 6x - 13$, then $x = 7$. See margin.

Critical Thinking

31. Discuss similarities and differences between the Transitive Property of Equality and the Transitive Property of Congruent Segments. Give an example of each property using segments and angles. See margin.

Applications and Problem Solving

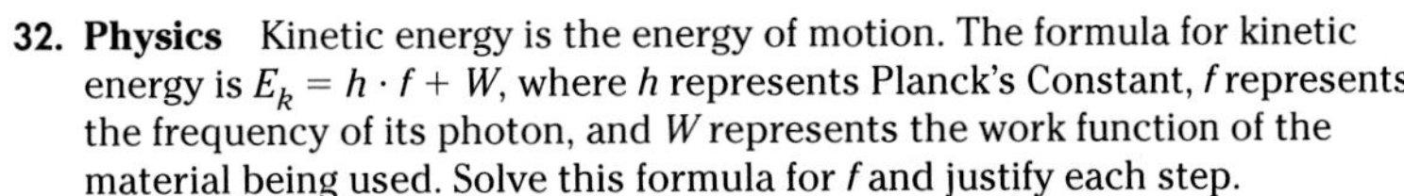

32. See Solutions Manual.

35. If $m\angle 1 \neq 27$, then $\angle 1$ is not acute; false, if $m\angle 1 = 32$, then $\angle 1$ is acute.

36. Sample answer: S, P, and T are collinear and X, P, and Y are collinear; see Solutions Manual for diagram.

32. **Physics** Kinetic energy is the energy of motion. The formula for kinetic energy is $E_k = h \cdot f + W$, where h represents Planck's Constant, f represents the frequency of its photon, and W represents the work function of the material being used. Solve this formula for f and justify each step.

33. **Photography** Film in a camera is fed through the camera by gears that peel the perforation in the film. The distance from the left edge of the film, A, to the right edge of the image, C, is the same as the distance from the left edge of the image, B, to the right edge of the film, D. Show that the two perforated strips are the same width. See Solutions Manual.

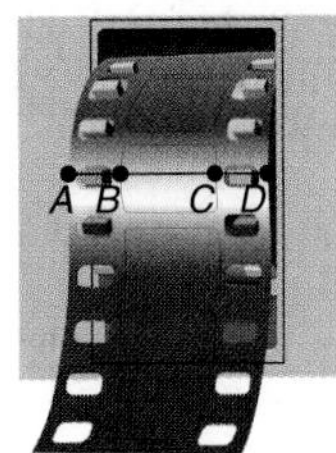

Mixed Review

34. **Ecology** Determine if a valid conclusion can be reached from the two true statements using the Law of Detachment or the Law of Syllogism. If a valid conclusion does not follow, write *no conclusion*. (Lesson 2–3)
    (1) Sponges belong to the phylum porifera.
    (2) Sponges are animals. no conclusion
35. Write the inverse of the statement *If $m\angle 1 = 27$, then $\angle 1$ is acute.* Determine if the inverse is true or false. If false, give a counterexample. (Lesson 2–2)
36. Write a conjecture if $\overleftrightarrow{XY}$ and $\overleftrightarrow{ST}$ intersect at P. Draw a diagram to illustrate your conjecture. (Lesson 2–1)
37. Find the measure of the complement and the supplement of an angle that measures 159°. (Lesson 1–7) Complement does not exist; 21.
38. If S is the midpoint of $\overline{RT}$, $RS = 3x + 5$, and $ST = 9x - 7$, find the value of x and the measure of $\overline{RT}$. (Lesson 1–5) 2; 22

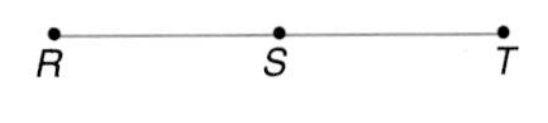

39. Draw and label a figure that shows lines ℓ and m intersecting at point P. (Lesson 1–2) See students' work.

INTEGRATION Algebra

40. Simplify $6 + 3(bc - 4a) + 2bc$. $6 + 5bc - 12a$
41. **Transportation** According to American Airlines, a layer of paint on a DC-10 would add 390 pounds to its weight. For every additional 100 pounds of weight, the fuel costs increase $300 each year. For this reason, the airline does not paint its DC-10 jets.
    a. Write a proportion to find out how much fuel costs for a DC-10 would increase each year if it were painted.
    b. Solve the proportion. $1170
    c. Suppose American Airlines flies an average of 23 DC-10s each year. How much can they save over a period of 10 years by not painting them?

41a. $\frac{100 \text{ lb}}{\$300} = \frac{390 \text{ lb}}{x}$

41c. $269,100

4 ASSESS

Closing Activity

Speaking Give students the statements "$\angle 1 \cong \angle 2$" and "$\angle 1$ and $\angle 2$ are a linear pair." Ask them to discuss what conclusion(s) they can draw and to describe how they would write a formal proof.

Additional Answers

30. Given: $AC = AB$, $AC = 4x + 1$, $AB = 6x - 13$
Prove: $x = 7$
Proof:
Statements (Reasons)
1. $AC = AB$, $AC = 4x + 1$, $AB = 6x - 13$ (Given)
2. $4x + 1 = 6x - 13$ (Substitution Prop. (=))
3. $1 = 2x - 13$ (Subtraction Prop. (=))
4. $14 = 2x$ (Addition Prop. (=))
5. $7 = x$ (Division Prop. (=))
6. $x = 7$ (Symmetric Prop. (=))

31. Sample answer: Both properties are transitive; equality relates numbers, congruence relates sets of points.

Enrichment Masters, p. 11

NAME______ DATE______

Enrichment — Student Edition Pages 92–99

Symmetric, Reflexive, and Transitive Properties

Equality has three important properties.

Symmetric $a = a$
Reflexive If $a = b$, then $b = a$.
Transitive If $a = b$ and $b = c$, then $a = c$.

Other relations have some of the same properties. Consider the relation "is next to" for objects labeled X, Y, and Z. Which of the properties listed above are true for this relation?

X is next to X. *False*
If X is next to Y, then Y is next to X. *True*
If X is next to Y and Y is next to Z, then X is next to Z. *False*

Only the symmetric property is true for the relation "is next to."

For each relation, state which properties (symmetric, reflexive, transitive) are true.

1. is the same size as — symmetric, reflexive, transitive
2. is a family descendant of — transitive
3. is in the same room as — symmetric, reflexive, transitive
4. is the identical twin of — symmetric
5. is warmer than — transitive
6. is on the same line as — symmetric, reflexive
7. is a sister of — none
8. is the same weight as — symmetric, reflexive transitive
9. Find two other examples of relations, and tell which properties are true for each relation. Answers will vary.

Extension

Connections In pipeline design the average velocity, u, is given by the equation $u = \frac{4q}{\pi D^2}$ where D is the pipe diameter (ft) and q is the volumetric flow rate $\left(\frac{ft^3}{s}\right)$. Solve the formula for q.

$q = \frac{\pi D^2 u}{4}$

2-5 LESSON NOTES

NCTM Standards: 1–5, 7, 8

Instructional Resources

- Study Guide Master 2-5
- Practice Master 2-5
- Enrichment Master 2-5
- Assessment and Evaluation Masters, p. 45
- Tech Prep Applications Masters, p. 4

Transparency 2-5A contains the 5-Minute Check for this lesson; **Transparency 2-5B** contains a teaching aid for this lesson.

Recommended Pacing	
Standard Pacing	Days 7 & 8 of 12
Honors Pacing	Days 6 & 7 of 11
Block Scheduling*	Day 5 of 7

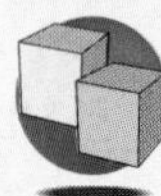

*For more information on pacing and possible lesson plans, refer to the *Block Scheduling Booklet*.

1 FOCUS

5-Minute Check
(over Lesson 2-4)

Name the property of equality that justifies each statement.

1. If $3x = 15$ and $5y = 15$, then $3x = 5y$. **Substitution Prop. (=)**
2. If $4z = -12$, then $z = -3$. **Division Prop. (=)**
3. If $AB = CD$, and $CD = EF$, then $AB = EF$. **Transitive Prop. (=)**
4. If $45 = m\angle A$, then $m\angle A = 45$. **Symmetric Prop. (=)**

FYI

During the original Olympics in ancient Greece, a sacred flame burned at the alter of the god Zeus. In the modern Olympics, the torch used to light the flame is first lit by the sun's rays in Olympia, Greece.

2-5 Verifying Segment Relationships

What YOU'LL LEARN

- To complete proofs involving segment theorems.

Why IT'S IMPORTANT

You can use segment relationships to solve problems in geography.

Geography

The U.S. mileage map below contains many line segments.

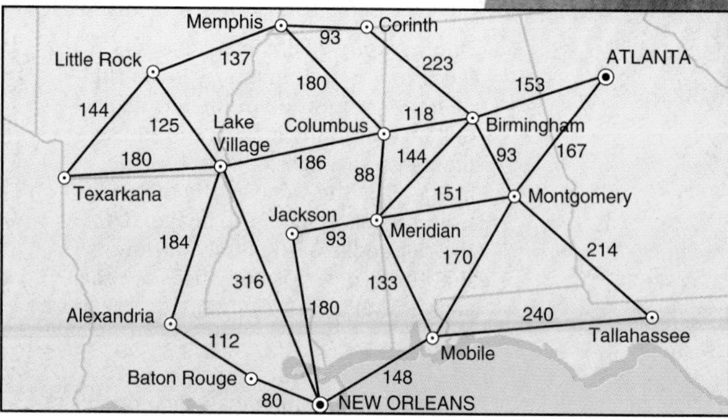

It is possible to use the mileage from city to city to illustrate some properties of equality and other facts about segments.

Geography

Example 1

a. Find two pairs of cities that illustrate the definition of congruent segments. Describe that relationship.

The distance from Jackson, Mississippi, to New Orleans, Louisiana, is 180 miles. The distance from Texarkana, Arkansas, to Lake Village, Arkansas, is 180 miles. Therefore, the segment connecting Jackson to New Orleans is congruent to the segment connecting Texarkana to Lake Village.

b. Find three pairs of cities that illustrate the Transitive Property of Equality. Describe that relationship.

City	City	Distance between Cities
Memphis, Tennessee	Corinth, Tennessee	93 miles
Birmingham, Alabama	Montgomery, Alabama	93 miles
Jackson, Mississippi	Meridian, Mississippi	93 miles

The Transitive Property of Equality can be illustrated by saying, *If the distance from Memphis to Corinth is the same as the distance from Birmingham to Montgomery, and the distance from Birmingham to Montgomery is the same as the distance from Jackson to Meridian, then the distance from Memphis to Corinth is the same as the distance from Jackson to Meridian.*

FYI

In 1996, the Olympic torch was carried 15,280 miles through 42 states before being carried to the opening ceremonies for the Centennial Summer Olympics in Atlanta, Georgia.

There are five essential parts in a good proof.

- State the theorem to be proved.
- List the given information.
- If possible, draw a diagram to illustrate the given information.
- State what is to be proved.
- Develop a system of deductive reasoning.

In order to use deductive reasoning to construct a valid proof, we must rely on statements that are accepted to be true. In geometry, those statements are definitions and postulates. We also depend on a list of undefined terms.

In Chapter 1, you learned that statements that are proved through deductive reasoning using definitions, postulates, and undefined terms are called *theorems.* Once a theorem is proved, it becomes another tool that we can use in the system. That is, proved theorems can be used in the proof of new theorems.

The first theorem we will prove is similar to properties of equality from algebra.

Theorem 2-1 — **Congruence of segments is reflexive, symmetric, and transitive.**

This theorem can be abbreviated as "Reflexive Prop. of ≅ Segments," "Symmetric Prop. of ≅ Segments," or "Transitive Prop. of ≅ Segments."

Theorem 2-1 can be written in symbols as follows.

Reflexive Property	$\overline{AB} \cong \overline{AB}$
Symmetric Property	If $\overline{AB} \cong \overline{CD}$, then $\overline{CD} \cong \overline{AB}$.
Transitive Property	If $\overline{AB} \cong \overline{CD}$ and $\overline{CD} \cong \overline{EF}$, then $\overline{AB} \cong \overline{EF}$.

The symmetric part of Theorem 2-1 is proved below. *You will be asked to prove the transitive and reflexive parts in Exercises 11 and 32, respectively.*

You can use the properties of algebra in geometric proofs. Notice that the symmetric property of equality is used in the proof of Theorem 2-1.

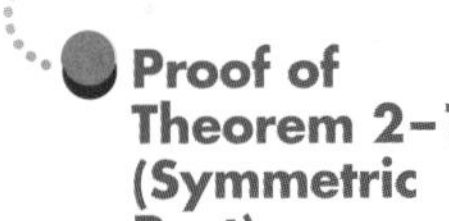

Proof of Theorem 2-1 (Symmetric Part)

Since measures of segments are real numbers, properties of algebra can be used to prove relationships with segment measures.

Given: $\overline{PQ} \cong \overline{RS}$

Prove: $\overline{RS} \cong \overline{PQ}$

P Q
R S

Proof:

Statements	Reasons
1. $\overline{PQ} \cong \overline{RS}$	1. Given
2. $PQ = RS$	2. Definition of congruent segments
3. $RS = PQ$	3. Symmetric Property (=)
4. $\overline{RS} \cong \overline{PQ}$	4. Definition of congruent segments

Motivating the Lesson

Hands-On Activity Have students place two bar magnets on a desk. Have them show that opposite poles attract and like poles repel. Ask them how they would prove this.

2 TEACH

In-Class Example

For Example 1
How would you illustrate the Symmetric Property using Example 1? **Any two cities can be used.**

Alternative Learning Styles

Visual Congruent figures have the same size and shape. Look around the classroom and identify any objects that you think are congruent. Especially look for items that could be represented by congruent segments. **Sample answer: the place where the wall and ceiling meet and the place where the same wall and floor meet**

In-Class Examples

For Example 2
Justify each step in the proof.

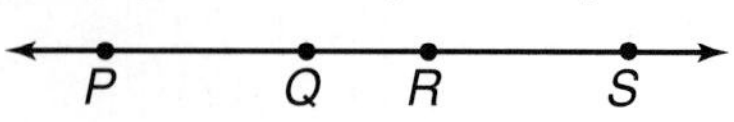

Given: Points P, Q, R, and S are collinear.
Prove: $PQ = PS - QS$

Proof:
Statements Reasons
1. Points P, Q, R, and S are collinear. Given
2. $PS = PQ + QS$ Segment Addition Postulate
3. $PS - QS = PQ$ Subtraction Prop. (=)
4. $PQ = PS - QS$ Symmetric Prop. (=)

For Example 3
Write a two-column proof.

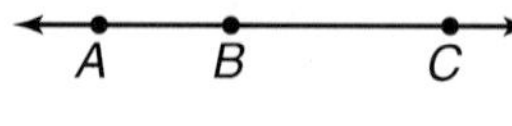

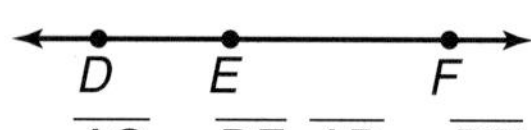

Given: $\overline{AC} \cong \overline{DF}$, $\overline{AB} \cong \overline{DE}$
Prove: $\overline{BC} \cong \overline{EF}$

Proof:
Statements (Reasons)
1. $\overline{AC} \cong \overline{DF}$, $\overline{AB} \cong \overline{DE}$ (Given)
2. $AC = DF$, $AB = DE$ (Def. congruent segments)
3. $AC - AB = DF - DE$ (Subtraction Prop. (=))
4. $AC = AB + BC$, $DF = DE + EF$ (Segment Addition Postulate)
5. $AC - AB = BC$, $DF - DE = EF$ (Subtraction Prop. (=))
6. $DF - DE = BC$ (Substitution Prop. (=))
7. $BC = EF$ (Substitution Prop. (=))
8. $\overline{BC} \cong \overline{EF}$ (Def. congruent segments)

Teaching Tip In Example 3, students may not recognize that, in statement 3, equal quantities are added to each side.

Example 

$\overline{ABCD}$ means A, B, C, and D are collinear.

Justify each step in the proof.

Given: $\overline{ABCD}$

Prove: $AD = AB + BC + CD$

Proof:

Statements	Reasons
1. $\overline{ABCD}$	1. ?
2. $AD = AB + BD$	2. ?
3. $BD = BC + CD$	3. ?
4. $AD = AB + BC + CD$	4. ?

Reason 1: Given
Reason 2: Segment Addition Postulate
Reason 3: Segment Addition Postulate
Reason 4: Substitution Property (=)

Proofs can be written in paragraph format or in two-column format.

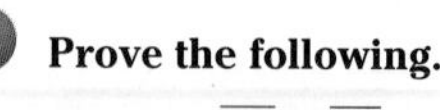

Prove the following.

Given: $\overline{AB} \cong \overline{XY}$
$\overline{BC} \cong \overline{YZ}$

Prove: $\overline{AC} \cong \overline{XZ}$

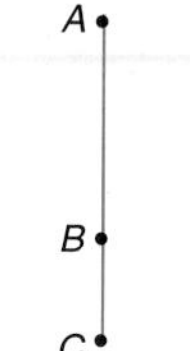

Method 1: Paragraph Proof

Since $\overline{AB} \cong \overline{XY}$ and $\overline{BC} \cong \overline{YZ}$, then by the definition of congruence, $AB = XY$ and $BC = YZ$. By the Addition Property of Equality, $AB + BC = XY + YZ$. From the Segment Addition Postulate, $AB + BC = AC$ and $XY + YZ = XZ$. Substituting AC for $AB + BC$ and substituting XZ for $XY + YZ$ in the statement $AB + BC = XY + YZ$, we get $AC = XZ$. Then, by the definition of congruence, $\overline{AC} \cong \overline{XZ}$.

Method 2: Two-Column Proof

Statements	Reasons
1. $\overline{AB} \cong \overline{XY}$, $\overline{BC} \cong \overline{YZ}$	1. Given
2. $AB = XY$, $BC = YZ$	2. Definition of congruent segments
3. $AB + BC = XY + YZ$	3. Addition Property (=)
4. $AB + BC = AC$ $XY + YZ = XZ$	4. Segment Addition Postulate
5. $AC = XZ$	5. Substitution Property (=) *twice*
6. $\overline{AC} \cong \overline{XZ}$	6. Definition of congruent segments

CHECK FOR UNDERSTANDING

Communicating Mathematics

Study the lesson. Then complete the following. 1–2. See margin.

1. **Explain** the five essential parts needed to construct a good proof.
2. **Describe** what a theorem is and explain why theorems are useful.

Proof Pointer

In Example 3, have students write each statement and its accompanying reason on a strip of paper. Have them shuffle the strips and then rewrite the proof by arranging the strips in a logical order. For higher level thinking skills, have students cut their strips in half so that the statement is on one part and the reasons are on the other. Then have them arrange all the parts logically to make a valid proof.

3. laws of logic and definitions, postulates, undefined terms and theorems

3. **Describe** in your own words what is meant by a system of deductive reasoning.
4. **Select** two cities from the U.S. mileage map. Describe the Reflexive Property of Equality using those two cities. See margin.

Guided Practice

Justify each statement with a property from algebra or a property of congruent segments.

5. If $\overline{KL} \cong \overline{MN}$, then $\overline{MN} \cong \overline{KL}$. Symmetric Prop. of ≅ Segments
6. If $AB = 10$ and $CD = 10$, then $AB = CD$. Substitution Prop. (=)
7. If $SY - 5 = RT - 5$, then $SY = RT$. Addition Prop. (=)

Write the given and prove statements you would use to prove each conjecture. Draw a figure if applicable.

8. If a figure is a square, then it has four right angles. See margin.
9. All whole numbers are integers.
10. Acute angles are angles with measures less than 90.

9. Given: x is a whole number. Prove: x is an integer.

10. Given: $\angle A$ is acute. Prove: $m\angle A < 90$

11. The reasons in the right column are not in the proper order. Reorder the reasons to properly match the statements in the left column for the proof of the transitive part of Theorem 2–1.

Given: $\overline{GH} \cong \overline{IJ}$
$\overline{IJ} \cong \overline{KL}$

Prove: $\overline{GH} \cong \overline{KL}$

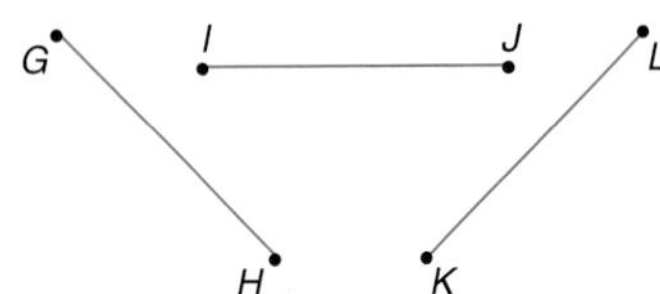

Proof:

Statements	Reasons
a. $\overline{GH} \cong \overline{IJ}$ $\overline{IJ} \cong \overline{KL}$ 3	(1) Definition of congruent segments
b. $GH = IJ$ $IJ = KL$ 1 or 4	(2) Transitive Property (=)
c. $GH = KL$ 2	(3) Given
d. $\overline{GH} \cong \overline{KL}$ 1 or 4	(4) Definition of congruent segments

12. Copy and complete the proof.

Given: $\overline{AB} \cong \overline{CD}$
$\overline{BD} \cong \overline{DE}$

Prove: $\overline{AD} \cong \overline{CE}$

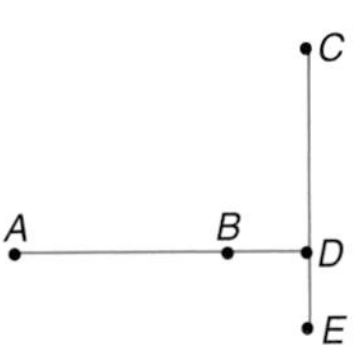

Proof:

Statements	Reasons
a. ? $\overline{AB} \cong \overline{CD}$? $\overline{BD} \cong \overline{DE}$	a. Given
b. $AB = CD$ $BD = DE$	b. ? Def. of ≅ Segments
c. $AD = AB + BD$ $CE = CD + DE$	c. ? Segment Addition Postulate
d. $AB + BD = CD + DE$	d. Addition Property (=)
e. ? $AD = CE$	e. Substitution Property (=)
f. $\overline{AD} \cong \overline{CE}$	f. ? Def. of ≅ Segments

3 PRACTICE/APPLY

Check for Understanding

Exercises 1–14 are designed to help you assess your students' understanding through reading, writing, speaking, and modeling. You should work through Exercises 1–4 with your students and then monitor their work on Exercises 5–14.

Additional Answers

1. Sample answer: State the theorem to be proved; list the given information; if possible, draw a diagram to illustrate the given; state what is to be proved; and develop a system of deductive reasoning.
2. Theorems are statements that are proved through deductive reasoning using definitions, postulates, and undefined terms and other proven theorems.
4. Sample answer: The distance from Atlanta to Montgomery AM is 167 miles; $\overline{AM} \cong \overline{AM}$.
8. Given: $ABCD$ is a square. Prove: Angles A, B, C, D are right angles.

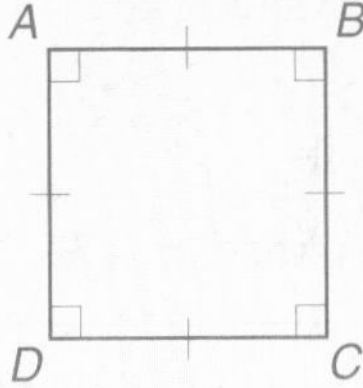

Reteaching

Using Cooperative Learning Have students work in small groups. Give each group three pieces of string or yarn, all the same length but different colors. Ask the groups to use the strings to demonstrate the reflexive, symmetric, and transitive properties of congruence of line segments.

Assignment Guide

Core (with proof): 15–37 odd, 38–46
Core (informal): 15–29 odd, 35, 37, 38–46
Enriched: 16–34 even, 35–46

For **Extra Practice,** see p. 767.

The red A, B, and C flags, printed only in the Teacher's Wraparound Edition, indicate the level of difficulty of the exercises.

Additional Answers

22.

24. Given: $\overline{AC}$ and $\overline{BD}$ intersect at E.
Prove: Four angles are formed; $\angle AEB$, $\angle AED$, $\angle BEC$, and $\angle DEC$ are formed.

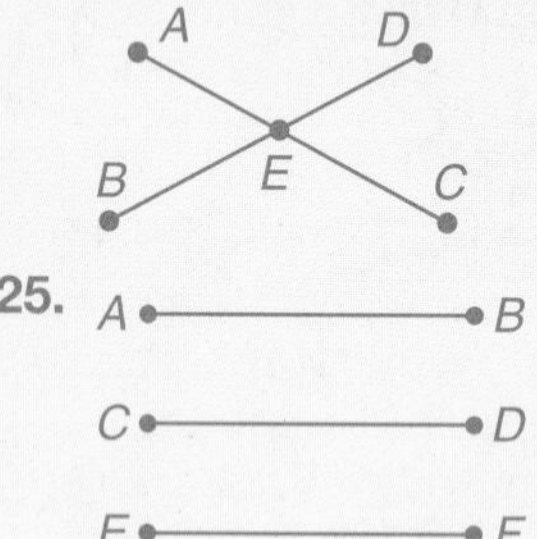

25. A———B

C———D

E———F

Study Guide Masters, p. 12

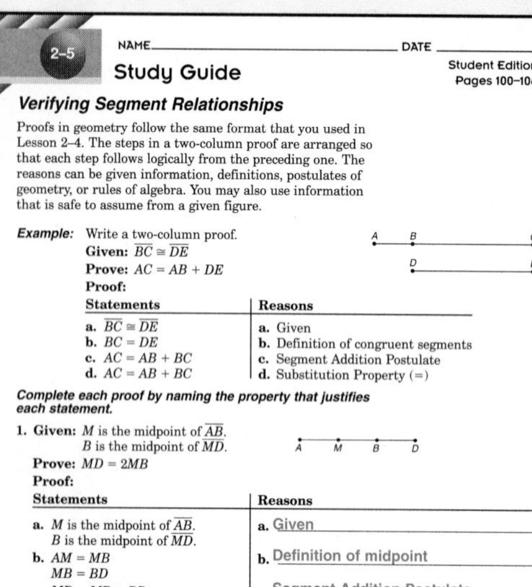

2-5 NAME ______ DATE ______

Study Guide Student Edition Pages 100–106

Verifying Segment Relationships

Proofs in geometry follow the same format that you used in Lesson 2-4. The steps in a two-column proof are arranged so that each step follows logically from the preceding one. The reasons can be given information, definitions, postulates of geometry, or rules of algebra. You may also use information that is safe to assume from a given figure.

Example: Write a two-column proof.
Given: $\overline{BC} \cong \overline{DE}$
Prove: $AC = AB + DE$
Proof:

Statements	Reasons
a. $\overline{BC} \cong \overline{DE}$	a. Given
b. $BC = DE$	b. Definition of congruent segments
c. $AC = AB + BC$	c. Segment Addition Postulate
d. $AC = AB + BC$	d. Substitution Property (=)

Complete each proof by naming the property that justifies each statement.

1. **Given:** M is the midpoint of $\overline{AB}$.
B is the midpoint of $\overline{MD}$.
Prove: $MD = 2MB$
Proof:

Statements	Reasons
a. M is the midpoint of $\overline{AB}$. B is the midpoint of $\overline{MD}$.	a. Given
b. $AM = MB$, $MB = BD$	b. Definition of midpoint
c. $MD = MB + BD$	c. Segment Addition Postulate
d. $MD = MB + MB$	d. Substitution Property (=)
e. $MD = 2MB$	e. Substitution Property (=)

2. **Given:** A, B, and C are collinear.
$AB = BD$
$BD = BC$
Prove: B is the midpoint of $\overline{AC}$.
Proof:

Statements	Reasons
a. A, B, and C are collinear. $AB = BD$, $BD = BC$	a. Given
b. $AB = BC$	b. Transitive Property (=)
c. B is the midpoint of $\overline{AC}$.	c. Definition of midpoint

13. Write a two-column proof. See Solutions Manual.
Given: $WX = XY$
Prove: $WY = 2XY$

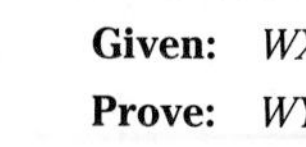

14a. Sample answer: The distance from Lake Village to Columbus is 186 miles (LC); from Memphis to Corinth is 93 miles (MC).

14. **Geography** Refer to the connection at the beginning of the lesson.
    a. Select two pairs of cities from a U.S. mileage map in which the distance between one pair is twice the distance between the other pair.
    b. Describe this relationship as an equality. $LC = 2MC$
    c. Divide each side of the equality by 2 and describe it again. $\frac{1}{2}LC = MC$

EXERCISES

Practice

Justify each statement with a property from algebra or a property of congruent segments.

15. $\overline{TJ} \cong \overline{TJ}$ Reflexive Prop. of $\cong$ Segments
16. If $2AB = 2WV$, then $AB = WV$. Division Prop. (=)
17. If $AB = XY$ and $DF = MN$, then $AB + DF = XY + MN$. Addition Prop. (=)
18. $\overline{PQ} \cong \overline{PQ}$ Reflexive Prop. of $\cong$ Segments
19. If $\overline{PO} \cong \overline{WE}$ and $\overline{WE} \cong \overline{QR}$, then $\overline{PO} \cong \overline{QR}$.
20. If $AB = AC + CB$, then $AB - AC = CB$. Subtraction Prop. (=)
21. $AB \cdot BC - PQ \cdot BC = (AB - PQ) \cdot BC$. Distributive Prop.

19. Transitive Prop. of $\cong$ Segments

Write the given and prove statements you would use to prove each conjecture. Draw a figure if applicable.

22. If two segments have equal measures, then they are congruent.
23. If a number greater than 2 is prime, then it is odd.
24. If two segments intersect at a point between the endpoints of the segments, then four angles are formed. See margin.
25. If two segments have measures equal to a third segment, then the two segments have equal measures.
26. If two segments are perpendicular, then they form a right angle. See margin.
27. If two points of a segment are on a line, then every point on that segment is on the line. See margin.
28. If a segment joins the midpoints of two sides of a triangle, then its measure is half the measure of the third side. See Solutions Manual.
29. All rational numbers are real numbers.
30. Rewrite the proof as a two-column proof. See Solutions Manual.
Given: $AC = BD$
Prove: $AB = CD$
Paragraph Proof:
We are given that $AC = BD$. From the Segment Addition Postulate, $AC = AB + BC$ and $BD = BC + CD$. Substituting $AB + BC$ for AC and $BC + CD$ for BD in the statement $AC = BD$, we get $AB + BC = BC + CD$. By the Subtraction Property of Equality, we can subtract BC from each side to get $AB = CD$.

22. Given: $AB = CD$, Prove: $\overline{AB} \cong \overline{CD}$; see margin for figure.

23. Given: $x > 2$ and x is prime. Prove: x is odd.

25. Given: $AB = CD$, $EF = CD$, Prove: $AB = EF$; see margin for figure.

29. Given: x is a rational number. Prove: x is a real number.

Additional Answers

26. Given: $\overline{AB} \perp \overline{CD}$
Prove: $\angle AED$ is a right angle.

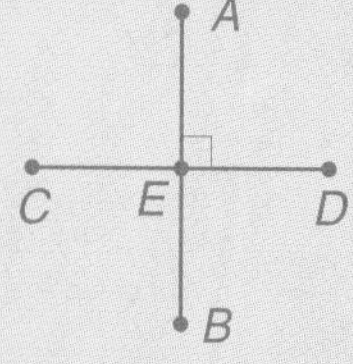

27. Given: A, B are on ℓ.
Prove: $\overline{AB}$ is on ℓ.

A B ℓ

31. Copy and complete the proof.

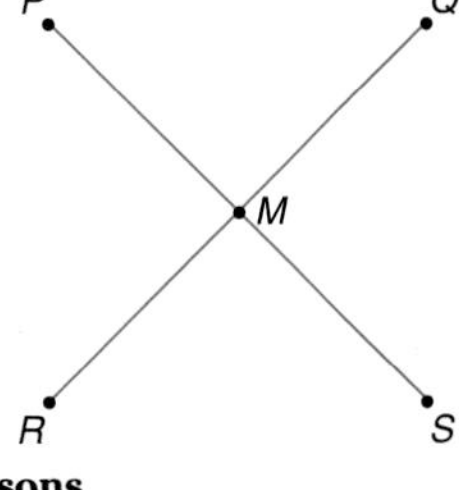

Given: $\overline{PS} \cong \overline{RQ}$
M is the midpoint of $\overline{PS}$.
M is the midpoint of $\overline{RQ}$.

Prove: $\overline{PM} \cong \overline{RM}$

Proof:

Statements	Reasons
a. $\overline{PS} \cong \overline{RQ}$; M is the midpoint of $\overline{PS}$. M is the midpoint of $\overline{RQ}$.	a. ? Given
b. $PS = RQ$	b. ? Def. ≅ Segments
c. ? $PM = MS$; ? $RM = MQ$	c. Definition of midpoint
d. $PS = PM + MS$; $RQ = RM + MQ$	d. ? Segment Addition Postulate
e. $PM + MS = RM + MQ$	e. ? Substitution Prop. (=)
f. $PM + PM = RM + RM$	f. ? Substitution Prop. (=)
g. $2PM = 2RM$	g. Substitution Property (=)
h. $PM = RM$	h. ? Division Prop. (=)
i. $\overline{PM} \cong \overline{RM}$	i. ? Def. ≅ Segments

32. Write a two-column proof of the reflexive part of Theorem 2–1. See margin.

Write a two-column proof. 33–34. See margin.

33. **Given:** $NL = NM$
$AL = BM$

Prove: $NA = NB$

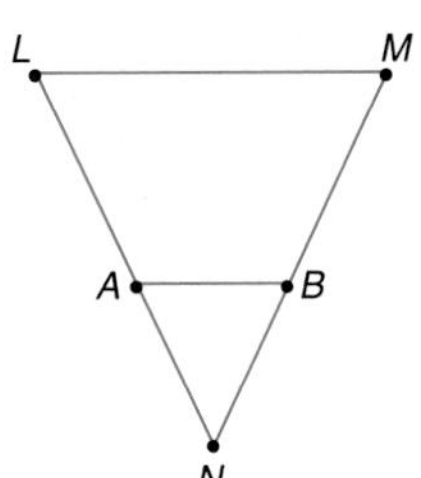

34. **Given:** $\overline{GR} \cong \overline{IL}$
$\overline{SR} \cong \overline{SL}$

Prove: $\overline{GS} \cong \overline{IS}$

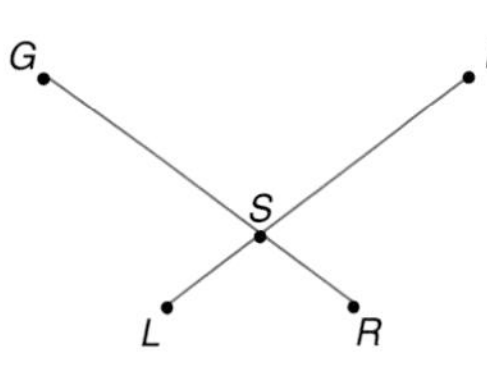

Critical Thinking

35. Sample answers: $\overline{LN} \cong \overline{QO}$ and $\overline{LM} \cong \overline{MN} \cong \overline{RS} \cong \overline{ST} \cong \overline{QP} \cong \overline{PO}$

35. Given that $\overline{LN} \cong \overline{RT}$, $\overline{RT} \cong \overline{QO}$, $\overline{LQ} \cong \overline{NO}$, $\overline{MP} \cong \overline{NO}$, M is the midpoint of $\overline{LN}$, S is the midpoint of $\overline{RT}$, and P is the midpoint of $\overline{QO}$, list three statements that you could prove using the postulates, theorems, and definitions that you have learned.

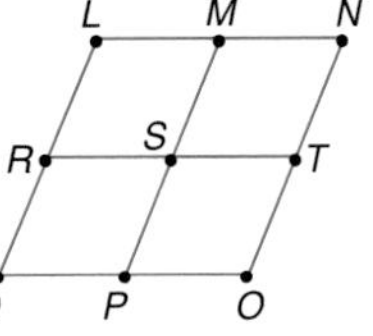

Additional Answers

32. Given: $\overline{AB}$
Prove: $\overline{AB} \cong \overline{AB}$

A B

Proof:
Statements (Reasons)
1. $\overline{AB}$ (Given)
2. $AB = AB$ (Reflexive Prop. (=))
3. $\overline{AB} \cong \overline{AB}$ (Def. ≅ segments)

33. Given: $NL = NM$
$AL = BM$
Prove: $NA = NB$
Proof:
Statements (Reasons)
1. $NL = NM$, $AL = BM$ (Given)
2. $NL = NA + AL$, $NM = NB + BM$ (Segment Addition Post.)
3. $NA + AL = NB + BM$ (Substitution Prop. (=))
4. $NA + BM = NB + BM$ (Substitution Prop. (=))
5. $NA = NB$ (Subtraction Prop. (=))

Practice Masters, p. 12

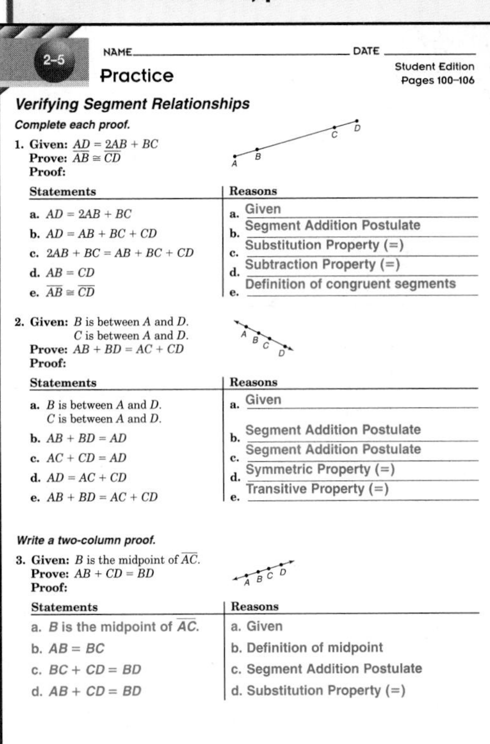

2–5 NAME ____ DATE ____ Student Edition Pages 100–106

Practice

Verifying Segment Relationships

Complete each proof.

1. Given: $AD = 2AB + BC$
Prove: $\overline{AB} \cong \overline{CD}$
Proof:

Statements	Reasons
a. $AD = 2AB + BC$	a. Given
b. $AD = AB + BC + CD$	b. Segment Addition Postulate
c. $2AB + BC = AB + BC + CD$	c. Substitution Property (=)
d. $AB = CD$	d. Subtraction Property (=)
e. $\overline{AB} \cong \overline{CD}$	e. Definition of congruent segments

2. Given: B is between A and D.
C is between A and D.
Prove: $AB + BD = AC + CD$
Proof:

Statements	Reasons
a. B is between A and D. C is between A and D.	a. Given
b. $AB + BD = AD$	b. Segment Addition Postulate
c. $AC + CD = AD$	c. Segment Addition Postulate
d. $AD = AC + CD$	d. Symmetric Property (=)
e. $AB + BD = AC + CD$	e. Transitive Property (=)

Write a two-column proof.

3. Given: B is the midpoint of $\overline{AC}$.
Prove: $AB + CD = BD$
Proof:

Statements	Reasons
a. B is the midpoint of $\overline{AC}$.	a. Given
b. $AB = BC$	b. Definition of midpoint
c. $BC + CD = BD$	c. Segment Addition Postulate
d. $AB + CD = BD$	d. Substitution Property (=)

Additional Answer

34. Given: $\overline{GR} \cong \overline{IL}$
$\overline{SR} \cong \overline{SL}$
Prove: $\overline{GS} \cong \overline{IS}$
Proof:
Statements (Reasons)
1. $\overline{GR} \cong \overline{IL}$, $\overline{SR} \cong \overline{SL}$ (Given)
2. $GR = IL$, $SR = SL$ (Def. ≅ segments)
3. $GR = GS + SR$, $IL = IS + SL$ (Segment Addition Post.)
4. $GS + SR = IS + SL$ (Substitution Prop. (=))
5. $GS + SL = IS + SL$ (Substitution Prop. (=))
6. $GS = IS$ (Subtraction Prop. (=))
7. $\overline{GS} \cong \overline{IS}$ (Def. ≅ segments)

4 ASSESS

Closing Activity

Modeling Use a Venn Diagram to prove that *Jack has four legs* given the following true statements:

All dogs have four legs.
Jack is a dog.

Show the relationship of this example to the Transitive Property.

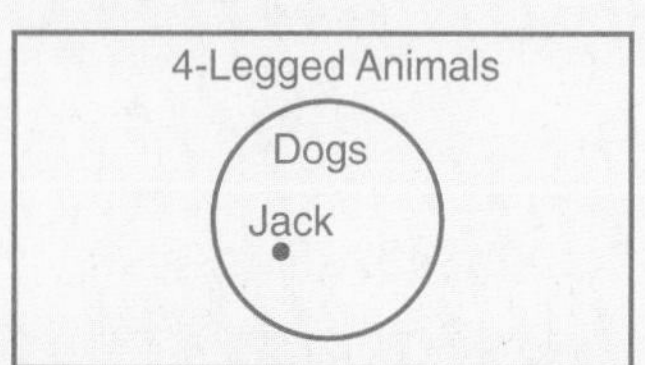

Chapter 2 Quiz C (Lessons 2-4 and 2-5) is available in the *Assessment and Evaluation Masters*, p. 45.

Additional Answer

44.

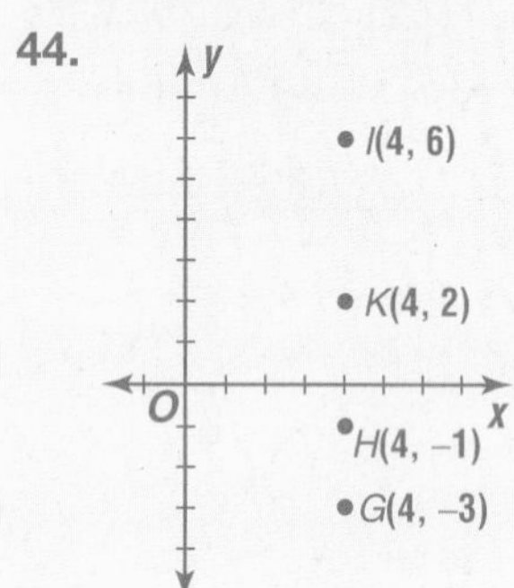

Enrichment Masters, p. 12

2-5 NAME______ DATE ______

Enrichment — Student Edition Pages 100–106

Geometry Crossword Puzzle

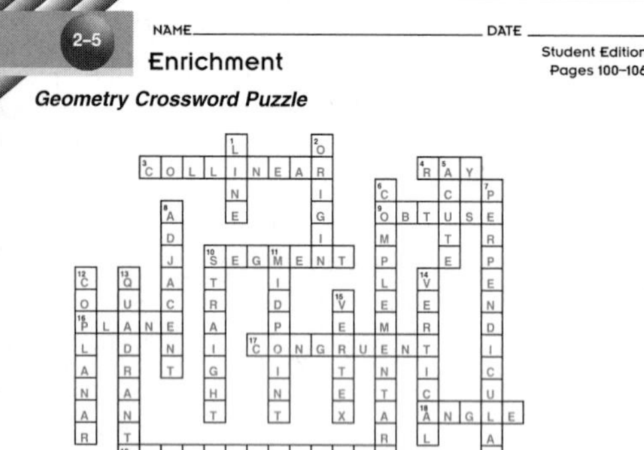

ACROSS

3. Points on the same line are ____.
4. A point on a line and all points of the line to one side of it.
9. An angle whose measure is greater than 90.
10. Two endpoints and all points between them.
16. A flat figure with no thickness that extends indefinitely in all directions.
17. Segments of equal length are ____ segments.
18. Two noncollinear rays with a common endpoint.
19. If $m\angle A + m\angle B = 180$, then $\angle A$ and $\angle B$ are ____ angles.

DOWN

1. The set of all points collinear to two points is a ____.
2. The point where the x- and y-axis meet.
5. An angle whose measure is less than 90.
6. If $m\angle A + m\angle D = 90$, then $\angle A$ and $\angle D$ are ____ angles.
7. Lines that meet at a 90° angle are ____.
8. Two angles with a common side but no common interior points are ____.
10. An "angle" formed by opposite rays is a ____ angle.
11. The middle point of a line segment.
12. Points that lie in the same plane are ____.
13. The four parts of a coordinate plane.
14. Two nonadjacent angles formed by two intersecting lines are ____ angles.
15. In angle ABC, point B is the ____.

Applications and Problem Solving

36. **Geography** On a U.S. mileage map, there is a scale that lists kilometers on the top and miles on the bottom. It appears that a segment can have two lengths, but the difference is the units.

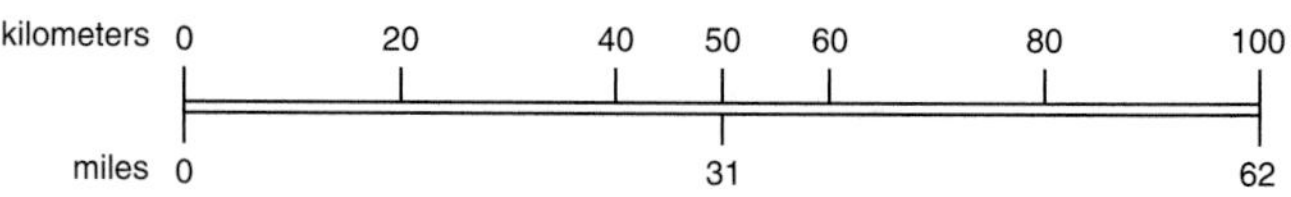

Suppose $\overline{AB}$ and $\overline{CD}$ are segments on the map with the scale shown above.

a. If $\overline{AB}$ is 100 kilometers long and $\overline{CD}$ is 62 miles long, is $\overline{AB} \cong \overline{CD}$? **yes**

b. It is common for charity fund-raising walks to be 5k or 10k (kilometers). How far is each distance in miles? **5k = 3.1 miles, 10k = 6.2 miles**

37. **Measure** Some rulers have centimeters on one edge and inches on the other edge. **a. about 15.2 cm**

a. About how long in centimeters is a segment that is 6 inches long?

b. Are the two segments congruent? Explain. **Yes; the scales are different.**

Mixed Review

38. Name the property of equality that justifies the statement *If 3(x – 1) = 10, then 3x – 3 = 10.* (Lesson 2–4) **Distributive Prop.**

39. Determine if a valid conclusion can be reached from the two true statements using the Law of Detachment or the Law of Syllogism. If a valid conclusion is possible, state it. (Lesson 2–3)

(1) Parallel lines do not intersect.

(2) If lines do not intersect, then they have no points in common.

39. Law of Syllogism; if lines are parallel, then they have no point in common.

40. **Advertising** A billboard reads *If you want a fabulous vacation, try Georgia.* (Lesson 2–2) **a. If you try Georgia, then you want a fabulous vacation.**

a. Write the converse of the conditional.

b. What do you think the advertiser wants you to conclude about vacations in Georgia? **They are fabulous.**

c. Does the advertisement say that vacations in Georgia are fabulous? **no**

41. $M(-3, 5)$ is the midpoint of $\overline{GH}$. If the coordinates of G are $(-7, 6)$, find the coordinates of H. (Lesson 1–5) **(1, 4)**

42. Use the Pythagorean Theorem to find the missing side measure in the triangle at the right. (Lesson 1–4) **65 cm**

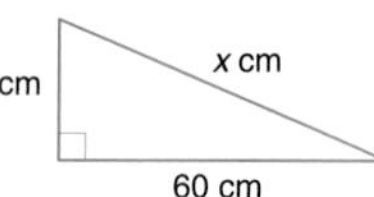

43. Find the missing measure in the formula $A = \ell w$ if $\ell = 17$ and $w = 8$. (Lesson 1–3) **136**

44. Graph points $G(4, -3)$, $H(4, -1)$, $I(4, 6)$, and $K(4, 2)$ on the same coordinate plane. What pattern do you notice? (Lesson 1–1) **See margin.**

Algebra

45. **Basketball** According to NBA standards, a basketball should bounce back $\frac{2}{3}$ of the distance from which it is dropped. How high should a basketball bounce when it is dropped from a height of $3\frac{3}{4}$ yards? **$2\frac{1}{2}$ yd**

46. Name the property illustrated by $(a + b)c = ac + bc$. **Distributive Prop.**

Extension

Reasoning Given points P, Q, R, S, and T are collinear, and $\overline{PQ} \cong \overline{ST}$, can you prove that $\overline{PR} \cong \overline{TR}$? Explain your answer.

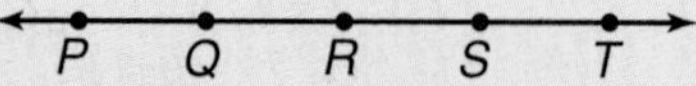

No; to prove $\overline{PR}$ and $\overline{TR}$ are congruent, you must know that $\overline{QR}$ and $\overline{RS}$ are congruent.

2-6 Verifying Angle Relationships

What YOU'LL LEARN
- To complete proofs involving angle theorems.

Why IT'S IMPORTANT

Understanding angle relationships is useful in solving problems involving art, nature, and architecture.

APPLICATION

Illusions

The following excerpt is from *Can You Believe Your Eyes*, by J. Richard Block and Harold Yuker.

> Illusions are misperceptions. Illusions sometimes involve angles. For example, the following two lines intersect to form two pairs of vertical angles. Hold the book parallel to the floor at eye level so the intersection point of the two lines is directly in front of you. A third line should appear. More angles are formed if you move the book so the third line goes through the intersection point. How many more angles do you see?

Two adjacent angles in the figure above form a linear pair. Theorem 2–2 states that if two angles form a linear pair, the angles are supplementary.

Theorem 2–2 Supplement Theorem	If two angles form a linear pair, then they are supplementary angles.

You will be asked to prove this theorem in Exercise 13.

Example  **If $\angle 1$ and $\angle 2$ form a linear pair and $m\angle 1 = 72$, find $m\angle 2$.**

$$m\angle 1 + m\angle 2 = 180$$
$$72 + m\angle 2 = 180$$
$$m\angle 2 = 108$$

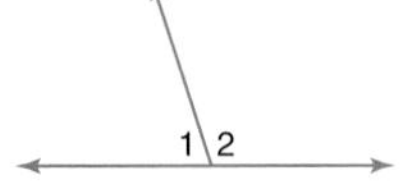

The properties of algebra that applied to the congruence of segments and the equality of their measures also hold true for the congruence of angles and the equality of their measures. The congruence relationships are stated in Theorem 2–3.

Theorem 2–3	Congruence of angles is reflexive, symmetric, and transitive.

This theorem can be abbreviated as "Reflexive Prop. of $\cong$ $\angle$s," "Symmetric Prop. of $\cong$ $\angle$s," or "Transitive Prop. of $\cong$ $\angle$s."

2-6 LESSON NOTES

NCTM Standards: 1–5, 7, 8

Instructional Resources

- Study Guide Master 2-6
- Practice Master 2-6
- Enrichment Master 2-6
- Assessment and Evaluation Masters, p. 45
- Multicultural Activity Masters, p. 4

Transparency 2-6A contains the 5-Minute Check for this lesson; **Transparency 2-6B** contains a teaching aid for this lesson.

Recommended Pacing	
Standard Pacing	Days 9 & 10 of 12
Honors Pacing	Days 8 & 9 of 11
Block Scheduling*	Day 6 of 7

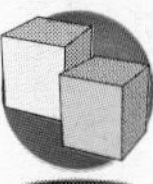

*For more information on pacing and possible lesson plans, refer to the *Block Scheduling Booklet*.

1 FOCUS

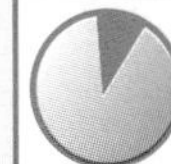

5-Minute Check
(over Lesson 2-5)

Write a reason for each statement.

1. If $\overline{AB} \cong \overline{CD}$, then $AB = CD$. **Def. congruent segments**
2. If $GH = JK$, then $GH + LM = JK + LM$. **Addition Prop. (=)**
3. R, D, and S are collinear. $RS = RD + DS$ **Segment Addition Postulate**
4. If $WX = YZ$, then $WX - UV = YZ - UV$. **Subtraction Prop. (=)**

Motivating the Lesson

Hands-On Activity Have students draw an angle using a protractor. Tell them to draw another angle of the same measure without using a protractor. How was the protractor helpful?

2 TEACH

In-Class Examples

For Example 1
The angle formed by a ladder and the ground measures 30°. Find the measure x of the larger angle formed by the ladder and the ground. **150**

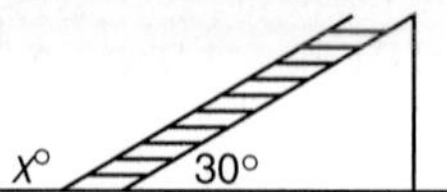

For Example 2
If $m\angle D = 30$ and $\angle D \cong \angle E$ and $\angle E \cong \angle F$, find $m\angle F$.
$m\angle F = 30$

The proof of the transitive portion of Theorem 2–3 is given below. *You will be asked to prove the reflexive and symmetric portions in Exercises 36 and 37, respectively.*

Proof of Theorem 2–3 (Transitive Part)

Given: $\angle 1 \cong \angle 2$
$\angle 2 \cong \angle 3$

Prove: $\angle 1 \cong \angle 3$

Proof:

Statements	Reasons
1. $\angle 1 \cong \angle 2$, $\angle 2 \cong \angle 3$	**1.** Given
2. $m\angle 1 = m\angle 2$, $m\angle 2 = m\angle 3$	**2.** Definition of congruent angles
3. $m\angle 1 = m\angle 3$	**3.** Transitive Property (=)
4. $\angle 1 \cong \angle 3$	**4.** Definition of congruent angles

Example 2

APPLICATION
Art

The gold disk shown at the right once belonged to the high rulers of the West African kingdom of Ghana, which was founded about A.D. 200. If $m\angle 1 = 40$ and $\angle 1 \cong \angle 2$ and $\angle 2 \cong \angle 3$, find $m\angle 3$.

Since congruence of angles is transitive, $\angle 1 \cong \angle 3$. Therefore, $m\angle 3 = 40$.

FYI
Early Africans had the technology for making steel tools as early as 100 B.C.

If two angles are supplementary to the same angle, what do you think is true about the angles? Draw several examples and make a conjecture.

Theorem 2–4	**Angles supplementary to the same angle or to congruent angles are congruent.**

This theorem can be abbreviated as "*s supp. to same ∠ or ≅ ∠s are ≅."*

Proof of Theorem 2–4

Given: $\angle 1$ and $\angle 3$ are supplementary.
$\angle 2$ and $\angle 3$ are supplementary.

Prove: $\angle 1 \cong \angle 2$

Proof:

Statements	Reasons
1. $\angle 1$ and $\angle 3$ are supplementary. $\angle 2$ and $\angle 3$ are supplementary.	**1.** Given
2. $m\angle 1 + m\angle 3 = 180$ $m\angle 2 + m\angle 3 = 180$	**2.** Definition of supplementary
3. $m\angle 1 + m\angle 3 = m\angle 2 + m\angle 3$	**3.** Substitution Property (=)
4. $m\angle 1 = m\angle 2$	**4.** Subtraction Property (=)
5. $\angle 1 \cong \angle 2$	**5.** Definition of congruent angles

Theorem 2–5 is a similar theorem for complementary angles. *You will be asked to prove Theorem 2–5 in Exercise 5.*

FYI

The main raw material for steel is iron ore, which was first reduced to a usable form about 3500 years ago in Armenia.

Theorem 2–5	Angles complementary to the same angle or to congruent angles are congruent.

This theorem can be abbreviated as "∡ compl. to same ∠ or ≅ ∡ are ≅."

You can create right angles and investigate congruent angles by paperfolding.

Right Angles

Materials:  patty paper  straightedge

- Fold the paper so that one corner is bent downward.
- Fold along the crease so that the top edge meets the side edge.
- Unfold the paper.

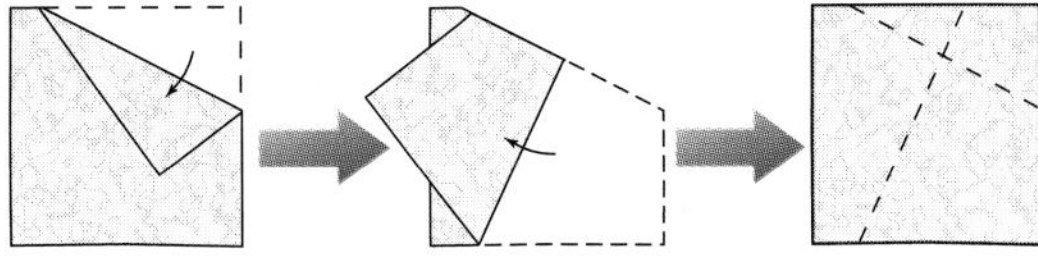

Your Turn

a. Measure each of the angles formed. **Each angle measures 90°.**

b. Repeat the activity three more times. **See students' work.**

c. Make a conjecture about what is true of all right angles. **Right angles all measure 90° and are therefore congruent.**

The following theorem supports the conjecture you formed in the Modeling Mathematics activity. *You will be asked to prove Theorem 2–6 in Exercise 38.*

Theorem 2–6	All right angles are congruent.

This theorem can be abbreviated as "All rt. ∡ are ≅."

In Lesson 2–3, you discovered that vertical angles are congruent. We stated that even 1000 cases would not represent a proof of the fact that vertical angles are congruent for every case. When we prove this as a theorem, it will be true for all cases. That is why theorems are so important.

Theorem 2–7	Vertical angles are congruent.

This theorem can be abbreviated as "Vert. ∡ are ≅."

Proof of Theorem 2–7

Given: $\angle 1$ and $\angle 3$ are vertical angles.

Prove: $\angle 1 \cong \angle 3$

Proof:

Statements	Reasons
1. $\angle 1$ and $\angle 3$ are vertical angles.	**1.** Given
2. $\angle 1$ and $\angle 2$ form a linear pair. $\angle 2$ and $\angle 3$ form a linear pair.	**2.** Definition of linear pair
3. $\angle 1$ and $\angle 2$ are supplementary. $\angle 2$ and $\angle 3$ are supplementary.	**3.** If 2 ∡ form a linear pair, then they are supp.
4. $\angle 1 \cong \angle 3$	**4.** ∡ supp. to same ∠ are ≅.

MODELING MATHEMATICS In this activity, students will use paper folding to create right angles. Encourage students to think of real-world situations in which this method could be useful.

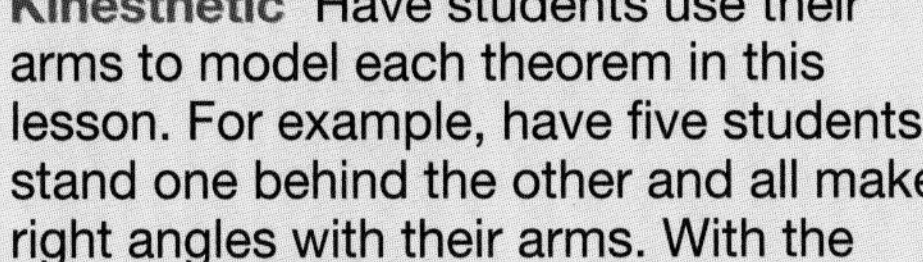

Alternative Learning Styles

Kinesthetic Have students use their arms to model each theorem in this lesson. For example, have five students stand one behind the other and all make right angles with their arms. With the help of the class, have each person change his or her position until all right angles match and are congruent.

In-Class Example

For Example 3
If $\angle A$ and $\angle C$ are vertical angles and $m\angle A = 3x - 2$ and $m\angle C = 2x + 4$, find $m\angle A$ and $m\angle C$.
$m\angle A = 16$, $m\angle C = 16$

3 PRACTICE/APPLY

Check for Understanding

Exercises 1–14 are designed to help you assess your students' understanding through reading, writing, speaking, and modeling. You should work through Exercises 1–10 with your students and then monitor their work on Exercises 11–14.

Additional Answers

3. If two angles are congruent, then their measures are equal. If two angles have equal measures, then they are congruent.
4. The next case you try may not be true. Sample answer: 11, 7, 5, and 3 are prime. Therefore, all prime numbers are odd. Since 2 is prime, it is a counterexample.
6. All angles of a square are congruent.

Example  **If $\angle 7$ and $\angle 8$ are vertical angles and $m\angle 7 = 3x + 6$ and $m\angle 8 = x + 26$, find $m\angle 7$ and $m\angle 8$.**

$m\angle 7 = m\angle 8$	*Vertical angles are congruent.*
$3x + 6 = x + 26$	*Substitution Property (=)*
$2x + 6 = 26$	*Substitution Property (=)*
$2x = 20$	*Substitution Property (=)*
$x = 10$	*Division Property (=)*

$m\angle 7 = 3x + 6$
$= 3(10) + 6$ or 36

$m\angle 8 = x + 26$
$= 10 + 26$ or 36

A theorem that follows from Theorem 2–7 is stated below.

Theorem 2–8  **Perpendicular lines intersect to form four right angles.**

This theorem can be abbreviated as "⊥ lines form 4 rt. ∠s."

CHECK FOR UNDERSTANDING

Communicating Mathematics

1. An illusion is a misperception.
2. All are transitive. The Transitive Property of Equality is with numbers; of congruent segments is with sets of points that form segments; of congruent angles is with sets of points that form angles.
4. See margin.

Study the lesson. Then complete the following.

1. **Explain** in your own words what an illusion is.
2. **Compare and contrast** the Transitive Property of Equality, the Transitive Property of Congruent Segments, and the Transitive Property of Congruent Angles.
3. The definition of congruent angles involves a statement and its converse. Write each one. See margin.
4. **Explain** why many true examples of a statement are not sufficient evidence to conclude that a statement is true for all cases. Give an illustration.
5. The reasons in the right column of the proof are not in the proper order. Reorder the reasons to properly match the statements in the left column.

Given: $\angle 1$ and $\angle 2$ are complementary.
$\angle 3$ and $\angle 2$ are complementary.

Prove: $\angle 1 \cong \angle 3$

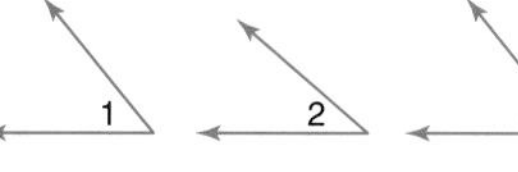

Proof:

Statements	Reasons
a. $\angle 1$ and $\angle 2$ are complementary. $\angle 3$ and $\angle 2$ are complementary. 2	(1) Subtraction Property (=)
b. $m\angle 1 + m\angle 2 = 90$ $m\angle 3 + m\angle 2 = 90$ 3	(2) Given
c. $m\angle 1 + m\angle 2 = m\angle 3 + m\angle 2$ 5	(3) Definition of complementary ∠s
d. $m\angle 1 = m\angle 3$ 1	(4) Definition of congruent ∠s
e. $\angle 1 \cong \angle 3$ 4	(5) Substitution Property (=)

MODELING MATHEMATICS

6. Fold a piece of patty paper so that all four corners lie on top of each other. Write a conjecture about the angles of a square. See margin.

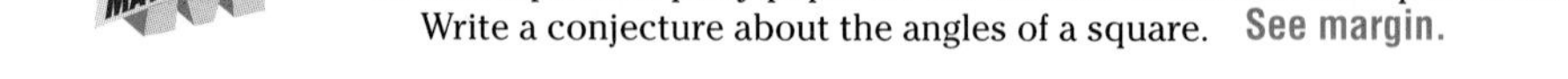

Reteaching

Using Diagrams Display the diagram at the right. Have students name as many pairs of congruent angles as they can.
$\angle AZB \cong \angle BZC \cong \angle CZD \cong \angle DZA$;
$\angle AZE \cong \angle FZC$; $\angle EZB \cong \angle FZD$;
$\angle AZF \cong \angle CZE$

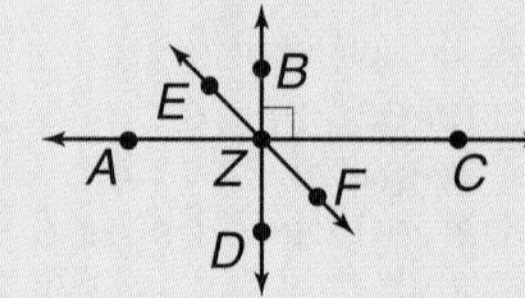

Complete each statement with *always, sometimes,* or *never.*

7. Two angles that are supplementary _?_ form a linear pair. sometimes
8. Two angles that form a linear pair are _?_ supplementary. always
9. Two angles that are congruent are _?_ right. sometimes
10. Two angles that are right are _?_ congruent. always

Guided Practice

11. $m\angle 1 = 121$, $m\angle 2 = 59$
12. $m\angle 3 = m\angle 4 = 57$

Find the measure of each numbered angle.

11. $m\angle 1 = 2x - 5$ and $m\angle 2 = x - 4$

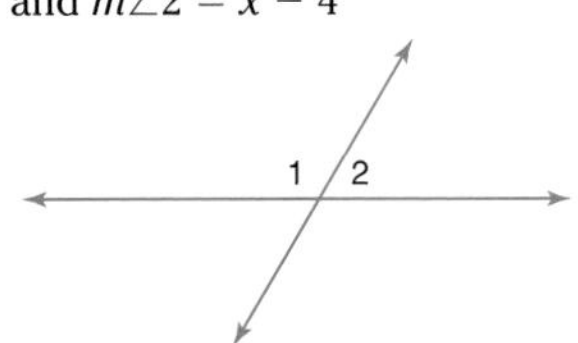

12. $m\angle 3 = 228 - 3x$ and $m\angle 4 = x$

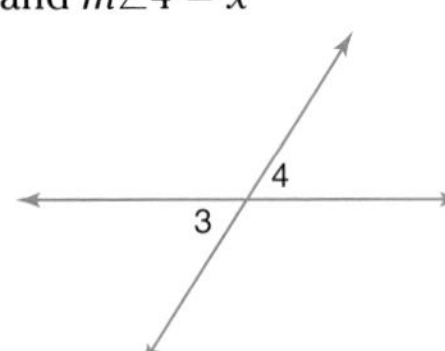

13. Copy and complete the proof of the Supplement Theorem (Theorem 2–2).

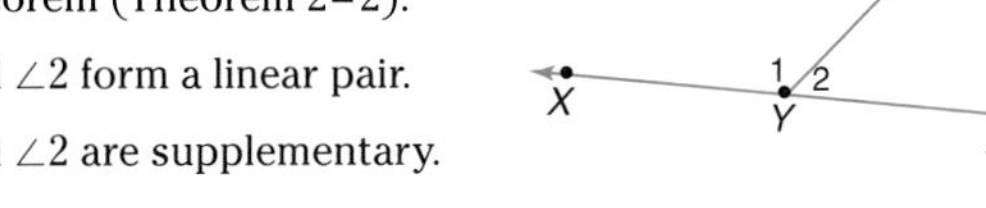

Given: $\angle 1$ and $\angle 2$ form a linear pair.

Prove: $\angle 1$ and $\angle 2$ are supplementary.

Proof:

Statements	Reasons
a. $\angle 1$ and $\angle 2$ form a linear pair.	a. Given
b. $\overrightarrow{YX}$ and $\overrightarrow{YZ}$ are opposite rays.	b. Definition of opposite rays
c. $m\angle XYZ = 180$	c. Definition of straight angle
d. _?_ $m\angle XYZ = m\angle 1 + m\angle 2$	d. Angle Addition Postulate
e. $m\angle 1 + m\angle 2 = 180$	e. _?_ Substitution Prop. (=)
f. $\angle 1$ and $\angle 2$ are supplementary.	f. _?_ Def. of supp. ∡s

CAREER CHOICES

A **botanist** studies plants, including their identification and classification, their geological record, and the structure and function of plant parts. The Ph.D. degree is generally required for biological scientists.

For more information, contact:

Business Office, Botanical Society of America
1725 Neil Ave.
Columbus, OH 43210

14. **Botany** In trees, features such as the angle that branches make with adjoining branches vary among species. In the drawing of the Mexican Mulberry tree branch at the right, $m\angle 1 = 55$. Find $m\angle 2$. 125

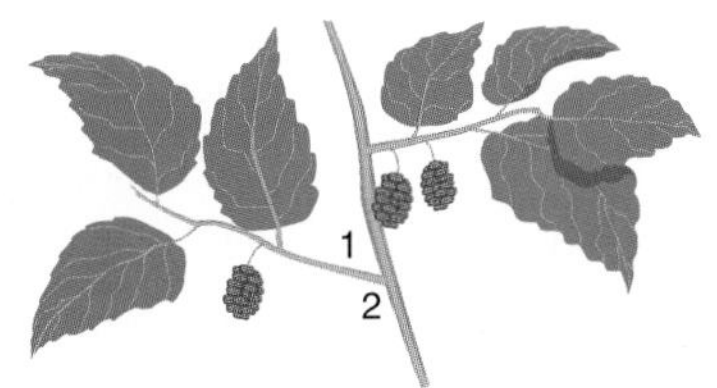

EXERCISES

Practice

Complete each statement in Exercises 15–26 with *always, sometimes,* or *never.* 15. sometimes 16. always

A

15. Two angles that are congruent are _?_ complementary to the same angle.
16. Two angles that are complementary to the same angle are _?_ congruent.
17. Two angles that are vertical are _?_ nonadjacent. always
18. Two angles that are nonadjacent are _?_ vertical. sometimes
19. Two angles that are complementary _?_ form a right angle. sometimes
20. Two angles that form a right angle are _?_ complementary. always
21. Two angles that form a linear pair are _?_ congruent. sometimes
22. Two angles that are supplementary are _?_ congruent. sometimes

Assignment Guide

Core (with proof): 15–39 odd, 40, 41, 43, 44–51
Core (informal): 15–33 odd, 40, 41, 43, 44–51
Enriched: 16–38 even, 40–51

For **Extra Practice,** see p. 768.

The red A, B, and C flags, printed only in the Teacher's Wraparound Edition, indicate the level of difficulty of the exercises.

CAREER CHOICES

Other areas that utilize botany are teaching, horticulture, agriculture, forestry, genetics, gardening, and wood production.

Additional Answers

34. Given: $m\angle ABC = m\angle DFE$
$m\angle 1 = m\angle 4$
Prove: $m\angle 2 = m\angle 3$
Proof:
Statements (Reasons)
1. $m\angle ABC = m\angle DFE$, $m\angle 1 = m\angle 4$ (Given)
2. $m\angle ABC = m\angle 1 + m\angle 2$, $m\angle DFE = m\angle 3 + m\angle 4$ (Angle Addition Post.)
3. $m\angle 1 + m\angle 2 = m\angle 3 + m\angle 4$ (Substitution Prop. (=))
4. $m\angle 4 + m\angle 2 = m\angle 3 + m\angle 4$ (Substitution Prop. (=))
5. $m\angle 2 = m\angle 3$ (Subtraction Prop. (=))

36. Given: $\angle A$ is an angle.
Prove: $\angle A \cong \angle A$

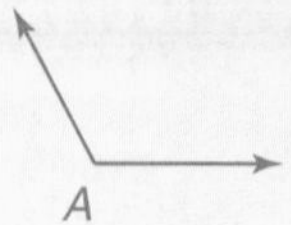

Proof:
Statements (Reasons)
1. $\angle A$ is an angle. (Given)
2. $m\angle A = m\angle A$ (Reflexive Prop. (=))
3. $\angle A \cong \angle A$ (Def. $\cong$ $\angle$s)

Study Guide Masters, p. 13

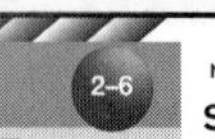

NAME______ DATE______

Study Guide Student Edition Pages 107–114

Verifying Angle Relationships

Many relationships involving angles can be proved by applying the rules of algebra, as well as the definitions and postulates of geometry.

Example: Given: $\angle EDG \cong \angle FDH$
Prove: $m\angle 1 = m\angle 3$
Proof:

Statements	Reasons
a. $\angle EDG \cong \angle FDH$	a. Given
b. $m\angle EDG = m\angle FDH$	b. Definition of congruent angles
c. $m\angle EDG = m\angle 1 + m\angle 2$, $m\angle FDH = m\angle 2 + m\angle 3$	c. Angle Addition Postulate
d. $m\angle 1 + m\angle 2 = m\angle 2 + m\angle 3$	d. Substitution Property (=)
e. $m\angle 1 = m\angle 3$	e. Subtraction Property (=)

Complete the following proofs.

1. Given: $\overline{AB} \perp \overline{BC}$
$m\angle 2 = m\angle 3$
Prove: $m\angle 1 + m\angle 3 = 90$
Proof:

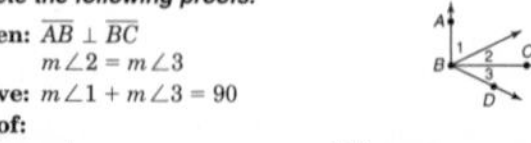
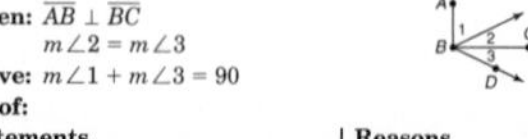

Statements	Reasons
a. $\overline{AB} \perp \overline{BC}$, $m\angle 2 = m\angle 3$	a. Given
b. ABC is a right angle.	b. Definition of perpendicular lines
c. $m\angle ABC = 90$	c. Definition of right angle
d. $m\angle ABC = m\angle 1 + m\angle 2$	d. Angle Addition Postulate
e. $m\angle 1 + m\angle 2 = 90$	e. Substitution Property (=)
f. $m\angle 1 + m\angle 3 = 90$	f. Substitution Property (=)

2. Given: $\angle 1$ and $\angle 2$ form a linear pair.
$m\angle 2 + m\angle 3 + m\angle 4 = 180$
Prove: $m\angle 1 = m\angle 3 + m\angle 4$
Proof:

Statements	Reasons
a. $\angle 1$ and $\angle 2$ form a linear pair. $m\angle 2 + m\angle 3 + m\angle 4 = 180$	a. Given
b. $\angle 1$ and $\angle 2$ are supplementary.	b. If 2 $\angle$s form a linear pair, they are supp.
c. $m\angle 1 + m\angle 2 = 180$	c. Definition of supplementary
d. $m\angle 1 + m\angle 2 = m\angle 2 + m\angle 3 + m\angle 4$	d. Substitution Property (=)
e. $m\angle 1 = m\angle 3 + m\angle 4$	e. Subtraction Property (=)

23. Two angles that are supplementary are __?__ complementary. never
24. Two right angles are __?__ supplementary. always
25. Vertical angles are __?__ complementary. sometimes
26. Angles with a common side and a common vertex __?__ form a linear pair. sometimes

Find the measure of each numbered angle.

27. $m\angle 5 = x$ and $m\angle 6 = 6x - 290$

27. $m\angle 5 = m\angle 6 = 58$

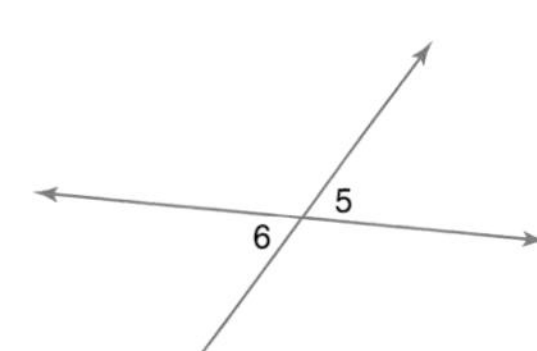
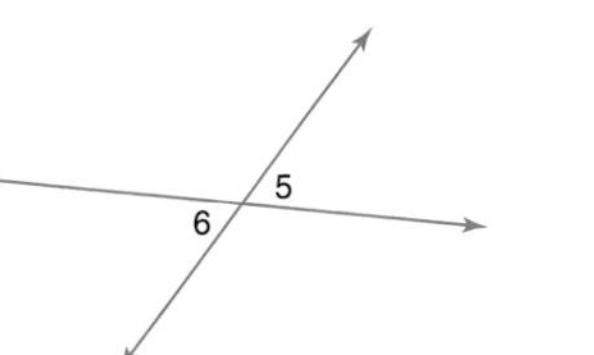

28. $m\angle 7 = 2x - 4$ and $m\angle 8 = 2x + 4$

28. $m\angle 7 = 86$, $m\angle 8 = 94$

29. $m\angle 1 = 4x$ and $m\angle 2 = 2x - 6$

29. $m\angle 1 = 124$, $m\angle 2 = 56$

30. $m\angle 3 = 2x + 7$ and $m\angle 4 = x + 30$

30. $m\angle 3 = m\angle 4 = 53$

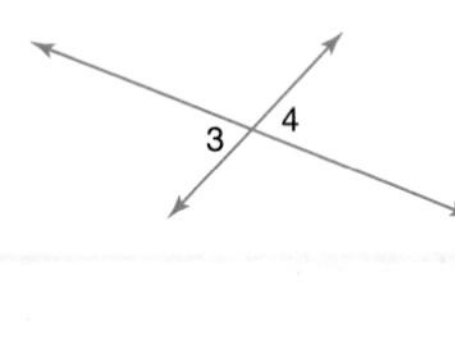

31. $\angle 5$ and $\angle A$ are complementary. $\angle 6$ and $\angle A$ are complementary. $m\angle 5 = 2x + 2$ and $m\angle 6 = x + 32$ $m\angle 5 = m\angle 6 = 62$

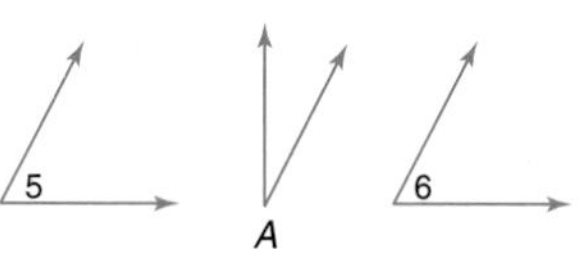

32. $m\angle 7 = x + 20$, $m\angle 8 = x + 40$, and $m\angle 9 = x + 30$

32. $m\angle 7 = m\angle 10 = 50$, $m\angle 8 = m\angle 11 = 70$, $m\angle 9 = m\angle 12 = 60$

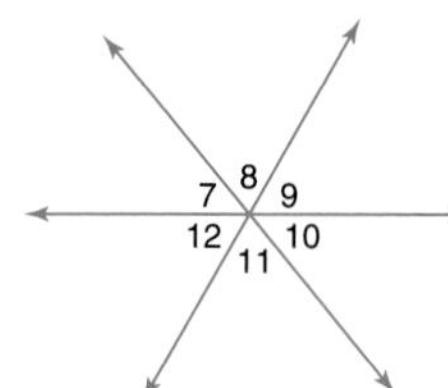

33. Refer to the figure at the right.
a. Use numbers to name four pairs of vertical angles.
b. Use letters to name eight different linear pairs of angles. $\angle AXB$ and $\angle BXE$, $\angle BXC$ and $\angle CXF$, $\angle CXD$ and $\angle DXH$, $\angle DXE$ and $\angle EXG$, $\angle EXF$ and $\angle FXA$, $\angle FXH$ and $\angle HXB$, $\angle HXG$ and $\angle GXC$, $\angle GXA$ and $\angle AXD$

33a. $\angle 1$ and $\angle 8$, $\angle 2$ and $\angle 6$, $\angle 3$ and $\angle 5$, $\angle 4$ and $\angle 7$

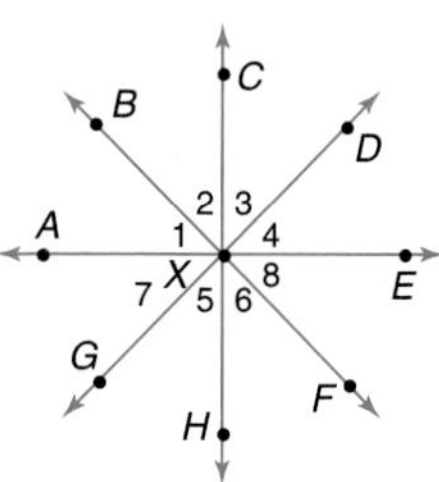

Write a two-column proof.

34. Given: $m\angle ABC = m\angle DFE$
$m\angle 1 = m\angle 4$
Prove: $m\angle 2 = m\angle 3$ See margin.

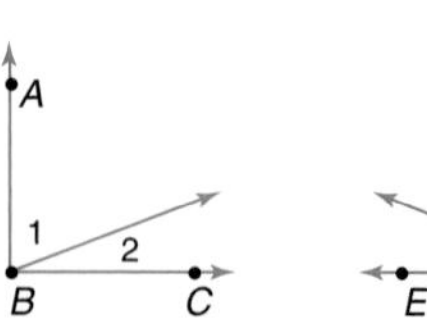

35. **Given:** $\angle ABD$ and $\angle CBD$ form a linear pair.
$\angle YXZ$ and $\angle WXZ$ form a linear pair.
$\angle ABD \cong \angle YXZ$

Prove: $\angle CBD \cong \angle WXZ$ See Solutions Manual.

For each theorem, name the given and prove statements and draw a figure. Then write a two-column proof. 36–38. See margin.

36. Congruence of angles is reflexive. (Theorem 2–3)
37. Congruence of angles is symmetric. (Theorem 2–3)
38. All right angles are congruent. (Theorem 2–6)
39. If one angle in a linear pair is a right angle, then the other angle is a right angle also. See Solutions Manual.

Critical Thinking

40. What conclusion can you draw about the sum of $m\angle 1$ and $m\angle 4$ if $m\angle 1 = m\angle 2$ and $m\angle 3 = m\angle 4$? Explain. See margin.

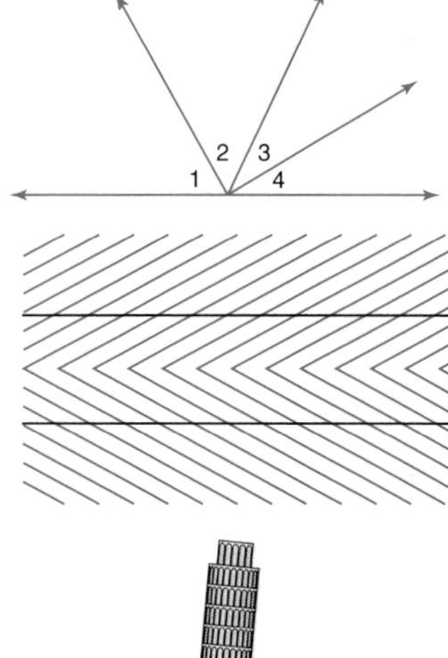

Applications and Problem Solving

41. **Illusions** At the right is an illusion created by lines and angles. If extended, will the long lines meet? Explain. Create your own illusion using the idea of angles. No, they are parallel; hold at an angle and look at it. See students' work for additional illusions.

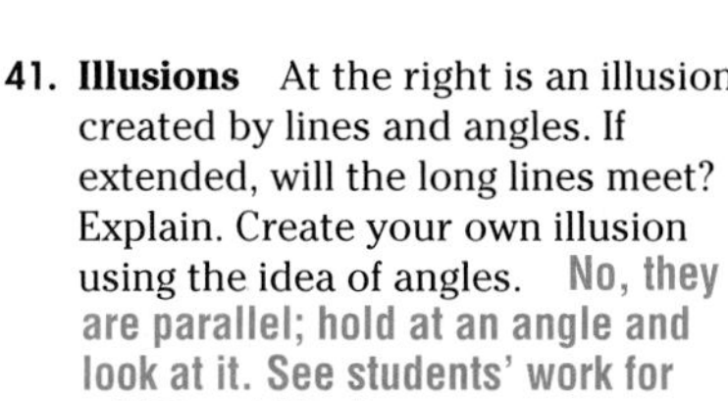

42. **Architecture** The Leaning Tower of Pisa in Italy makes an angle with the ground of about 84° on one side. If you look at the building as a ray and the ground as a line, then the angles that the tower forms with the ground form a linear pair. Find the measure x of the other angle that the tower makes with the ground. 96°

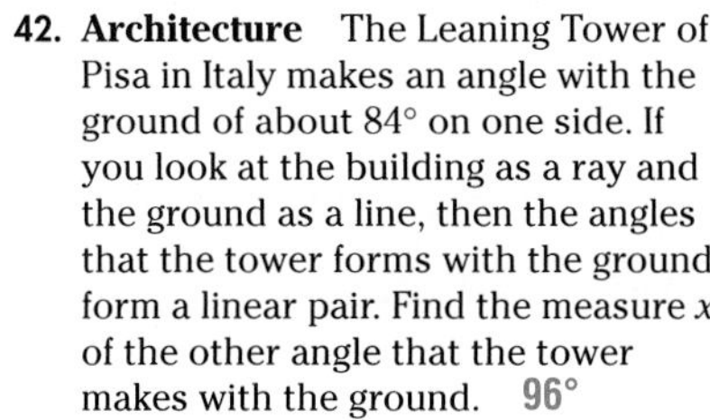

43. **Illusions** In 1879, psychologist Wilhelm Wundt of Germany used angular lines on the outside of two vertical lines to create the illusions at the right. a–c. See Solutions Manual.
    a. Describe what appears to be a relationship between the two vertical lines in each case.
    b. Tilt the page at an angle of about 45° so the vertical lines are pointing toward you. Redescribe what now appears to be a relationship between the two vertical lines.
    c. A close-up of the angular lines is shown at the right. If $\angle 4 \cong \angle 2$, prove that $\angle 3 \cong \angle 1$.

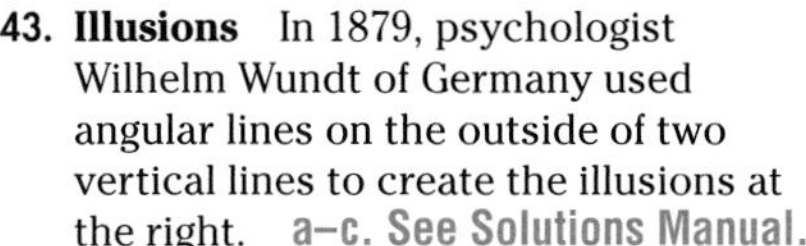

Wilhelm Wundt

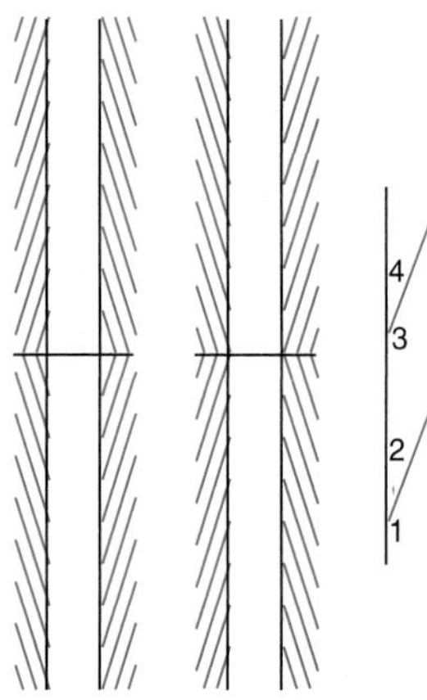

Additional Answers

37. Given: $\angle A \cong \angle B$
Prove: $\angle B \cong \angle A$

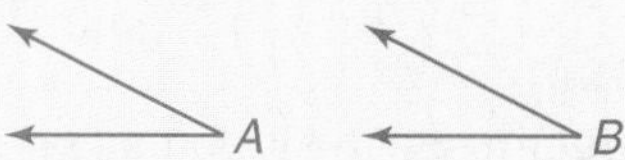

Proof:
Statements (Reasons)
1. $\angle A \cong \angle B$ (Given)
2. $m\angle A = m\angle B$ (Def. $\cong$ ∡s)
3. $m\angle B = m\angle A$ (Symmetric Prop. (=))
4. $\angle B \cong \angle A$ (Def. $\cong$ ∡s)

38. Given: $\angle X$ and $\angle Y$ are right angles.
Prove: $\angle X \cong \angle Y$

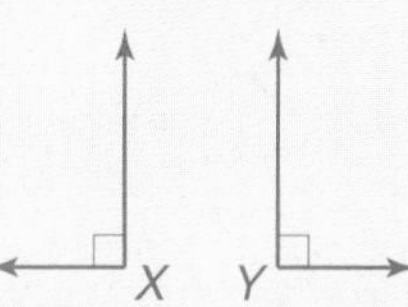

Proof:
Statements (Reasons)
1. $\angle X$ and $\angle Y$ are right angles. (Given)
2. $m\angle X = 90$, $m\angle Y = 90$ (Def. rt. ∡s)
3. $m\angle X = m\angle Y$ (Substitution Prop. (=))
4. $\angle X \cong \angle Y$ (Def. $\cong$ ∡s)

Practice Masters, p. 13

NAME________ DATE ______
2-6 Practice
Student Edition Pages 107–114

Verifying Angle Relationships

Complete each statement if $m\angle BGC = 43$ and $m\angle DGE = 56$.

1. $\angle FGA \cong$? $\angle DGC$
2. $\angle BGF$ and ? are supplementary. $\angle FGE$ or $\angle BGC$
3. $m\angle CGD =$? 81
4. $m\angle AGF =$? 81
5. $\angle EGC$ and ? are supplementary. $\angle BGC$ or $\angle FGE$
6. $m\angle AGB =$? 56
7. $m\angle AGC =$? 99

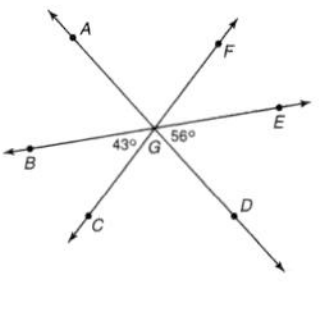

Write a two-column proof.

8. Given: $\angle AEC \cong \angle DEB$
Prove: $\angle AEB \cong \angle DEC$
Proof:

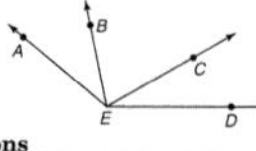

Statements	Reasons
a. $\angle AEC \cong \angle DEB$	a. Given
b. $m\angle AEC = m\angle DEB$	b. Definition of congruent angles
c. $m\angle AEC = m\angle AEB + m\angle BEC$, $m\angle DEB = m\angle DEC + m\angle BEC$	c. Angle Addition Postulate
d. $m\angle AEB + m\angle BEC = m\angle DEC + m\angle BEC$	d. Transitive Property (=)
e. $m\angle AEB = m\angle DEC$	e. Subtraction Property (=)
f. $\angle AEB \cong \angle DEC$	f. Definition of congruent angles

Extension

Connections Find $m\angle C$ if $\angle C$ is congruent to $\angle A$, $m\angle A = 3x$, $m\angle B = x + 20$ and $\angle A$ and $\angle B$ are supplementary angles. 120

Additional Answer

40. The sum of $m\angle 1$ and $m\angle 4$ is 90;
$$m\angle 1 + m\angle 2 + m\angle 3 + m\angle 4 = 180$$
$$m\angle 1 + m\angle 1 + m\angle 4 + m\angle 4 = 180$$
$$2m\angle 1 + 2m\angle 4 = 180$$
$$2(m\angle 1 + m\angle 4) = 180$$
$$m\angle 1 + m\angle 4 = 90$$

4 ASSESS

Closing Activity

Modeling Have students review each method of proving that vertical angles are congruent and that all right angles are congruent. Then have them use patty paper to physically prove each one.

Chapter 2 Quiz D (Lesson 2-6) is available in the *Assessment and Evaluation Masters*, p. 45.

Enrichment Masters, p. 13

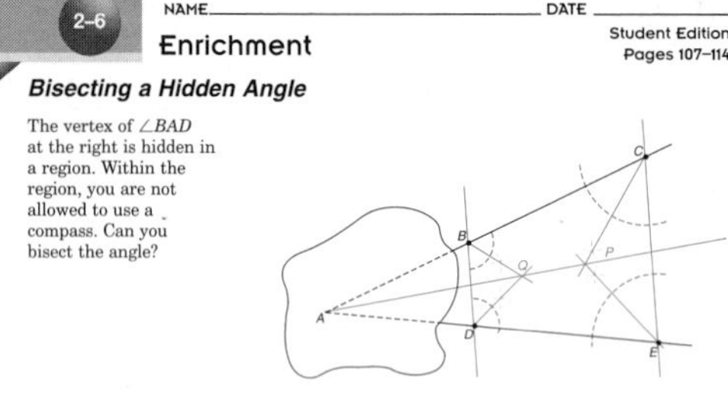

2-6 NAME________________ DATE ________

Enrichment

Student Edition Pages 107–114

Bisecting a Hidden Angle

The vertex of $\angle BAD$ at the right is hidden in a region. Within the region, you are not allowed to use a compass. Can you bisect the angle?

Follow these instructions to bisect $\angle BAD$.

1. Use a straightedge to draw lines CE and BD.
2. Construct the bisectors of $\angle DEC$ and $\angle BCE$.
3. Label the intersection of the two bisectors as point P.
4. Construct the bisectors of $\angle BDE$ and $\angle DBC$.
5. Label the intersection of the two previous bisectors as point Q.
6. Use a straightedge to draw line PQ, which bisects the hidden angle.
7. Another hidden angle is shown at right. Construct the bisector using the method above, or devise your own method. See students' work.

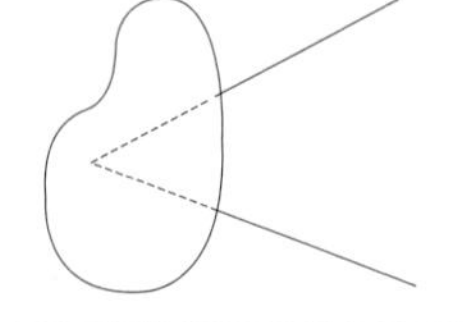

Mixed Review

44. Write the given and prove statements you would use to prove the statement *If an angle is a right angle, then its measure is 90.* (Lesson 2–5)
45. Name the property of equality that justifies the statement *If $y = 4x + 9$ and $x = 2$, then $y = 17$.* (Lesson 2–4) **Substitution Prop. (=)**
46. **Recycling** Write the following statement in if-then form. *A recycled aluminum can is remelted and back in the store within six weeks.* (Lesson 2–2)
47. If K is between J and L, $JK = 3x$, $KL = 2x - 1$, and $JL = 24$, find the value of x and the measure of $\overline{JK}$. (Lesson 1–4) **5, 15**
48. State whether a page of your geometry book is best modeled by a *point*, *line*, or *plane*. (Lesson 1–2) **plane**
49. **List the Possibilities** Emilio is planting a garden along the back of his house. The garden must have an area of at least 70 square feet. One side of the garden will be bounded by the house, but the other three sides will be bounded by 25 feet of fencing. What are the possible dimensions of Emilio's garden if he plans to use all the fencing and each side has a whole number length? (Lesson 1–2)
50. Evaluate $6(5) - 1 + 8 \cdot 4$. **61**
51. The graph at the right shows how many people watched prime-time TV on average each day of the week in 1996. If you were buying commercial time on one night of the week for your new product, which night would you choose? Why? **Thursday; the largest audience is available.**

Algebra

44. Given: an angle is a right angle

Prove: its measure is 90

46. If it is a recycled aluminum can, it is remelted and back in the store within six weeks.

49. 15 ft by 5 ft, 13 ft by 6 ft, 11 ft by 7 ft, 9 ft by 8 ft

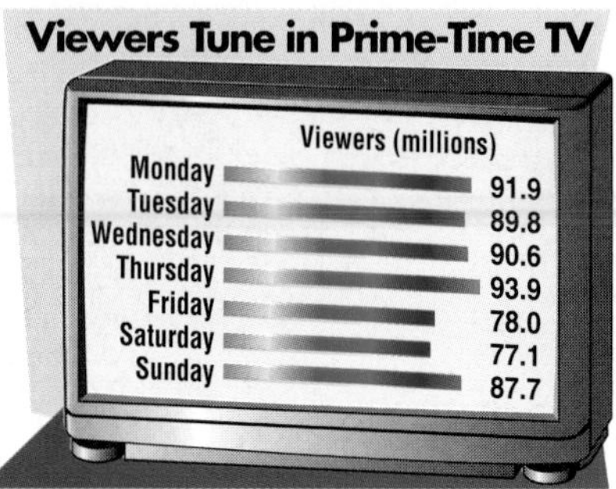

Source: Nielsen Media Research

WORKING ON THE In•ves•ti•ga•tion

Refer to the Investigation on pages 66–67.

Although you may not realize it, your eyes respond in different ways to the shapes within a painting. Artists make use of these patterns when they create their works. In general, diagonal lines give a painting energy while circles draw the eye toward the center.

1. Find several examples of paintings of different styles. You may wish to consult a reference book in a library or the Internet home page for a large museum.
2. Look at the paintings with your group members. Take turns looking at a painting and observing the ways that the observers' eyes move around the painting.
3. Discuss the theme of your mural. Do you want the mural to appear lively or comforting? Which geometric shape is best for arranging your mural to accomplish your goal?

Add the results of your work to your Investigation folder.

In•ves•ti•ga•tion

Working on the Investigation

The Investigation on pages 66–67 is designed to be a long-term project that is completed over several days or weeks. Encourage students to keep their materials in their Investigation Folder as they work on the Investigation.

HIGHLIGHTS

VOCABULARY

After completing this chapter, you should be able to define each term, property, or phrase and give an example or two of each.

Geometry

conclusion (p. 76)
conditional statement (p. 76)
conjecture (p. 70)
contrapositive (p. 78)
converse (p. 77)
counterexample (p. 72)
deductive reasoning (p. 86)
hypothesis (p. 76)
if-then statement (p. 76)
inductive reasoning (p. 70)
inverse (p. 78)
Law of Detachment (p. 86)
Law of Syllogism (p. 87)
negation (p. 78)
proof (p. 94)
two-column proof (p. 94)
Venn diagram (p. 79)

Algebra

Addition and Subtraction Properties (p. 93)
Distributive Property (p. 93)
Multiplication and Division Properties (p. 93)
Reflexive Property (p. 93)
Substitution Property (p. 93)
Symmetric Property (p. 93)
Transitive Property (p. 93)

Problem Solving

draw a diagram (p. 79)
look for a pattern (p. 85)

UNDERSTANDING AND USING THE VOCABULARY

Choose the letter of the term that best matches each statement or phrase. 1. k

1. the portion of a conditional statement immediately following *if*
2. using several examples to make a conjecture e
3. a statement formed by interchanging the hypothesis and conclusion b
4. When a conditional is true and its hypothesis is true, then its conclusion is true. g
5. a false example that shows a conjecture is not true c
6. a statement formed by negating the hypothesis and conclusion of the converse of the given conditional l
7. an educated guess about what will always be true a
8. the denial of a statement i
9. If $p \rightarrow q$ and $q \rightarrow r$ are true conditionals, then $p \rightarrow r$ is also true. h
10. a statement formed by negating both the hypothesis and conclusion f
11. a method of constructing a valid argument by listing reasons for each step j
12. using rules of logic to reach a conclusion d

a. conjecture
b. converse
c. counterexample
d. deductive reasoning
e. inductive reasoning
f. inverse
g. Law of Detachment
h. Law of Syllogism
i. negation
j. proof
k. hypothesis
l. contrapositive

Using the CHAPTER HIGHLIGHTS

The Chapter Highlights begins with a listing of the new terms, properties, and phrases that were introduced in this chapter. Have students define each term and provide an example or two of it, if appropriate.

Instructional Resources

Three multiple-choice tests and three free-response tests are provided in the *Assessment and Evaluation Masters*. Forms 1A and 2A are for honors pacing, Forms 1B and 2B are for average pacing, and Forms 1C and 2C are for basic pacing. Chapter 2 Test, Form 1B, is shown at the right. Chapter 2 Test, Form 2B, is shown on the next page.

Postulates, Theorems, and Corollaries

A complete list of postulates, theorems, and corollaries begins on page 806.

Assessment and Evaluation Masters, pp. 31–32

NAME ________ DATE ________

Chapter 2 Test, Form 1B

Write the letter for the correct answer in the blank at the right of each problem.

1. Which conjecture is always true based on the given information?
Given: $\overline{AC} \cong \overline{DF}$, $\overline{BC} \cong \overline{DE}$
A. $\overline{AB} \cong \overline{EF}$ B. $\overline{AB} \cong \overline{DE}$
C. $\overline{BC} \cong \overline{EF}$ D. $\overline{AC} \cong \overline{EF}$ 1. A
2. Identify a counterexample for the conjecture *All prime numbers are odd.*
A. 5 is a prime number. B. 2 is a prime number.
C. 89 is an odd number. D. none of these 2. B
3. Identify the converse of *If M is the midpoint of $\overline{PQ}$, then $MQ = \frac{1}{2}PQ$.*
A. If M is the midpoint of $\overline{PQ}$, then $PQ = 2MQ$.
B. If $MQ = \frac{1}{2}PQ$, then M is the midpoint of $\overline{PQ}$.
C. If $PQ = \frac{1}{2}MQ$, then M is the midpoint of $\overline{MQ}$.
D. If P is the midpoint of $\overline{MQ}$, then $PQ = \frac{1}{2}MQ$. 3. B
4. Identify the contrapositive of the statement *If $\angle R$ is acute, then $m\angle R$ is less than 90.*
A. If $\angle R$ is not acute, then $m\angle R$ is not less than 90.
B. If $m\angle R$ is less than 90, then $\angle R$ is acute.
C. If $m\angle R$ is not less than 90, then $\angle R$ is not acute.
D. none of these 4. C
5. Through any three noncollinear points, there is
A. exactly one plane.
B. exactly one line.
C. exactly two lines.
D. none of these 5. A
6. Identify the if-then form of the statement *All cats like tuna.*
A. If an animal likes tuna, then it is a cat.
B. If an animal is a cat, then it likes tuna.
C. If an animal is not a cat, then it does not like tuna.
D. none of these 6. B
7. Which statement represents the Law of Detachment?
A. If $p \rightarrow q$ is a true conditional and p is true, then q is true.
B. If $p \rightarrow g$ and $q \rightarrow r$ are true conditionals, then $p \rightarrow q$ is true.
C. If $a = b$ and $b = c$, then $a = c$.
D. none of these 7. A

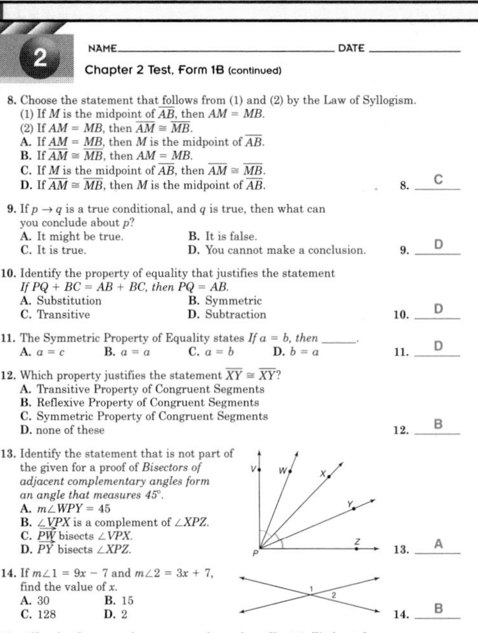

NAME ________ DATE ________

Chapter 2 Test, Form 1B (continued)

8. Choose the statement that follows from (1) and (2) by the Law of Syllogism.
(1) If M is the midpoint of $\overline{AB}$, then $AM = MB$.
(2) If $AM = MB$, then $\overline{AM} \cong \overline{MB}$.
A. If $AM = MB$, then M is the midpoint of $\overline{AB}$.
B. If $\overline{AM} \cong \overline{MB}$, then $AM = MB$.
C. If M is the midpoint of $\overline{AB}$, then $\overline{AM} \cong \overline{MB}$.
D. If $\overline{AM} \cong \overline{MB}$, then M is the midpoint of $\overline{AB}$. 8. C
9. If $p \rightarrow q$ is a true conditional, and q is true, then what can you conclude about p?
A. It might be true. B. It is false.
C. It is true. D. You cannot make a conclusion. 9. D
10. Identify the property of equality that justifies the statement *If $PQ + BC = AB + BC$, then $PQ = AB$.*
A. Substitution B. Symmetric
C. Transitive D. Subtraction 10. D
11. The Symmetric Property of Equality states *If $a = b$, then* ______.
A. $a = c$ B. $a = a$ C. $a = b$ D. $b = a$ 11. D
12. Which property justifies the statement $\overline{XY} \cong \overline{XY}$?
A. Transitive Property of Congruent Segments
B. Reflexive Property of Congruent Segments
C. Symmetric Property of Congruent Segments
D. none of these 12. B
13. Identify the statement that is not part of the given for a proof of *Bisectors of adjacent complementary angles form an angle that measures 45°.*
A. $m\angle WPY = 45$
B. $\angle VPX$ is a complement of $\angle XPZ$.
C. $\overrightarrow{PW}$ bisects $\angle VPX$.
D. $\overrightarrow{PY}$ bisects $\angle XPZ$. 13. A
14. If $m\angle 1 = 9x - 7$ and $m\angle 2 = 3x + 7$, find the value of x.
A. 30 B. 15
C. 128 D. 2 14. B
15. $\angle K$ and $\angle J$ are complementary angles and $m\angle K = 19$. Find $m\angle J$.
A. 71 B. 161 C. 90 D. 9 15. A

Bonus The contrapositive of a conditional $p \rightarrow q$ is *not q* $\rightarrow$ *not p*. The inverse of $p \rightarrow q$ is *not p* $\rightarrow$ *not q*. What is the inverse of the contrapositive of $p \rightarrow q$?
A. not $q \rightarrow$ not p B. $p \rightarrow q$
C. not $p \rightarrow$ not q D. $q \rightarrow p$ Bonus D

Using the STUDY GUIDE AND ASSESSMENT

Skills and Concepts Encourage students to refer to the objectives and examples on the left as they complete the review exercises on the right.

Assessment and Evaluation Masters, pp. 37–38

2 NAME ______ DATE ______

Chapter 2 Test, Form 2B

Determine if each conjecture is true or false based on the given information. Explain your answer and give a counterexample for any false conjecture.

1. **Given:** $\overrightarrow{PQ} \perp \overrightarrow{PR}$
   Conjecture: $\angle RPQ$ is a right angle. — 1. True; ⊥ lines form 4 rt. ∠s.

2. **Given:** Points J, K, and L are collinear.
   Conjecture: $JK + KL = JL$ — 2. False; J could be between K and L.

Write each conditional statement in if-then form. Identify the hypothesis and conclusion.

3. Perpendicular lines form four right angles. — 3. If two lines are perpendicular, then they form four right angles. hypothesis: two lines are perpendicular; conclusion: they form four right angles

4. Angles have measures greater than 0. — 4. If a figure is an angle, then it has a measure greater than 0. hypothesis: a figure is an angle; conclusion: it has a measure greater than zero

5. The product of two odd integers is odd. — 5. If two integers are odd, then their product is odd. hypothesis: two integers are odd conclusion: their product is odd

Write the converse of each conditional. Determine if the converse is true or false. If it is false, give a counterexample.

6. If $m\angle 1 = 42$ and $m\angle 2 = 138$, then $\angle 1$ and $\angle 2$ are supplementary. — 6. If $\angle 1$ and $\angle 2$ are supplementary, then $m\angle 1 = 42$ and $m\angle 2 = 138$; false; Let $m\angle 1 = 40$ and $m\angle 2 = 140$.

7. Congruent supplementary angles are right angles. — 7. If two angles are right angles, then they are congruent supplementary angles; true.

8. If c is negative and $a > b$, then $ac < bc$. — 8. If $ac < bc$, then c is negative and $a > b$; false. Let $a = 2$, $b = 3$, and $c = 1$.

In the figure at the right, C, E, and D are collinear and lie in plane $\mathcal{Q}$. Points A, E, and B are collinear and lie in plane $\mathcal{P}$. Determine whether each statement is true or false.

9. Plane $\mathcal{P}$ and plane $\mathcal{Q}$ intersect in $\overleftrightarrow{CD}$. — 9. true

10. A, E, and C lie in plane $\mathcal{Q}$. — 10. false

Determine if a valid conclusion can be reached from the two true statements. If it can, state it and the law that is used. If a valid conclusion does not follow, write no conclusion.

11. (1) If a mineral is gypsum, it has a hardness of 2.
    (2) If a mineral has a hardness of 2, then it can be scratched with a fingernail. — 11. If a mineral is gypsum, then it can be scratched with a fingernail; Law of Syllogism.

12. (1) Basalt is an igneous rock.
    (2) Granite is an igneous rock. — 12. no conclusion

13. (1) If you have a can opener, you can open a can.
    (2) Jon has a can opener. — 13. Jon can open a can; Law of Detachment.

2 NAME ______ DATE ______

Chapter 2 Test, Form 2B (continued)

14. If $p \to q$ is true and p is false, what must be true about q? — 14. It can be either true or false.

Name the property of equality that justifies each statement.

15. If $AC = AB + BC$ and $AC = BD$, then $BD = AB + BC$. — 15. Subst. Prop (=)

16. If $JL = JK + KL$ and $JK + KL = MN + NP$, then $JL = MN + NP$. — 16. Trans. Prop (=)

17. If $m\angle ABC + m\angle ABC = m\angle DEF + m\angle ABC$, then $m\angle ABC = m\angle DEF$. — 17. Subtr. Prop (=)

18. Find the measures of $\angle 1$ and $\angle 2$ if $m\angle 1 = 6x - 10$ and $m\angle 2 = 4x + 20$. — 18. 92, 88

Write a two-column proof.

19–20. Answers may vary. Sample answers follow.

19. **Given:** C is the midpoint of $\overline{AB}$. $\overline{DC} \cong \overline{CB}$
    Prove: $AC = DC$ — 19. a. C is the midpoint of $\overline{AB}$. $\overline{DC} \cong \overline{CB}$ (Given) b. $AC = CB$ (Midpoint Thm.) c. $DC = CB$ (Def. ≅ Seg.) d. $CB = DC$ (Symm. Prop. (=)) e. $AC = DC$ (Trans. Prop. (=))

20. **Given:** $\angle 3 \cong \angle 4$
    Prove: $\angle 1 \cong \angle 2$ — 20. a. $\angle 3 \cong \angle 4$ (Given) b. $\angle 1 \cong \angle 3$, $\angle 4 \cong \angle 2$ (Vert. ∠s are ≅.) c. $\angle 1 \cong \angle 4$ (Trans. Prop. ≅ ∠s) d. $\angle 1 \cong \angle 2$ (Trans. Prop. ≅ ∠s)

Bonus

The contrapositive of a conditional $p \to q$ is *not q* → *not p*. What is the converse of the contrapositive of $p \to q$? — Bonus: not q → not p

CHAPTER 2 STUDY GUIDE AND ASSESSMENT

SKILLS AND CONCEPTS

OBJECTIVES AND EXAMPLES

Upon completing this chapter, you should be able to:

- make conjectures based on inductive reasoning (Lesson 2–1)

To determine if a conjecture made from inductive reasoning is false, look at situations where the given information is true. Determine if there are situations where the given is true and the conjecture is false. A proof is required to determine if a conjecture is true.

REVIEW EXERCISES

Use these exercises to review and prepare for the chapter test. **13–14. See margin for explanations.**

Determine if each conjecture is *true* or *false* based on the given information. Explain your answer.

13. **Given:** X, Y, and Z are collinear and $XY = YZ$.
    Conjecture: Y is the midpoint of $\overline{XZ}$. **true**

14. **Given:** $\angle 1$ and $\angle 2$ are supplementary.
    Conjecture: $\angle 1 \cong \angle 2$ **false**

- write a conditional in if-then form (Lesson 2–2)

Write the statement *Adjacent angles have a common ray* in if-then form.

If angles are adjacent, then they have a common ray.

Write each conditional in if-then form.

15. Every cloud has a silver lining. **15–18. See margin.**
16. A rectangle has four right angles.
17. Obsidian is a glassy rock produced by a volcano.
18. The intersection of two planes is a line.

- write the converse, inverse, and contrapositive of an if-then statement (Lesson 2–2)

Given: If an angle measures 120°, then it is obtuse.

Converse: If an angle is obtuse, then it measures 120°.

Inverse: If an angle does not measure 120°, then it is not obtuse.

Contrapositive: If an angle is not obtuse, then it does not measure 120°.

Write the converse, inverse, and contrapositive of each statement. **19–22. See margin.**

19. If a rectangle has four congruent sides, then it is a square.
20. If three points are collinear, then they lie on a straight line.
21. If the month is January, then it has 31 days.
22. If an ordered pair for a point has 0 as its y-coordinate, then the point lies on the x-axis.

GLENCOE Technology

Test and Review Software

You may use this software, a combination of an item generator and item bank, to create your own tests or worksheets. Types of items include free response, multiple choice, short answer, and open ended.

For IBM & Macintosh

Additional Answers

13. definition of midpoint
14. If supplementary, $m\angle 1$ could be 50 and $m\angle 2$ could be 130, but they are not congruent.

OBJECTIVES AND EXAMPLES	REVIEW EXERCISES

- use the Law of Detachment and the Law of Syllogism in deductive reasoning (Lesson 2–3)

The Law of Detachment states that if $p \to q$ is a true conditional and p is true, then q is true.

The Law of Syllogism states that if $p \to q$ and $q \to r$ are true conditionals, then $p \to r$ is also true.

Determine if a valid conclusion can be reached from the two true statements using the Law of Detachment or the Law of Syllogism. If a valid conclusion is possible, state it and the law that is used. If a valid conclusion does not follow, write *no conclusion*. 23, 25. See margin.

23. (1) Angles that are supplementary have measures with a sum of 180.
    (2) $\angle A$ and $\angle B$ are supplementary.
24. (1) Well-known athletes appear on Wheaties™ boxes.
    (2) Michael Jordan appeared on a Wheaties™ box. no conclusion
25. (1) The sun is a star.
    (2) Stars are in constant motion.

- use properties of equality in algebraic and geometric proofs (Lesson 2–4)

Prove that if $ST = UV$, then $SU = TV$.

Given: $ST = UV$

Prove: $SU = TV$

Statements	Reasons
1. $ST = UV$	1. Given
2. $TU = TU$	2. Reflexive Property (=)
3. $ST + TU = TU + UV$	3. Addition Property (=)
4. $SU = ST + TU$ $TV = TU + UV$	4. Segment Addition Postulate
5. $SU = TV$	5. Substitution Property (=)

Name the property of equality that justifies each statement.

26. If $12x = 24$, then $x = 2$. Division Prop. (=)
27. For MN, $MN = MN$. Reflexive Prop. (=)
28. Copy the proof. Then name the property that justifies each statement.

Given: $MN = PN$
$NL = NO$

Prove: $ML = PO$

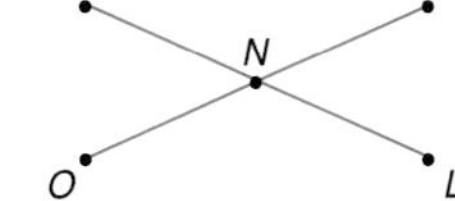

Proof:

Statements	Reasons
a. $MN = PN$, $NL = NO$	a. ? Given
b. $MN + NL = PN + NO$	b. ? Addition Prop. (=)
c. $ML = MN + NL$ $PO = PN + NO$	c. ? Segment Addition Postulate
d. $ML = PO$	d. ? Subst. Prop. (=)

- complete proofs involving segment theorems (Lesson 2–5)

Theorem 2–1 states that congruence of segments is reflexive, symmetric, and transitive.

29. Subtraction Prop. (=)
30. Symmetric Prop. of ≅ Segments

Justify each statement with a property from algebra or a property of congruence.

29. If $AB + BC = BC + CD$, then $AB = CD$.
30. If $\overline{XY} \cong \overline{OP}$, then $\overline{OP} \cong \overline{XY}$.
31. If $GH = 12$ and $GH + HI = GI$, then $12 + HI = GI$. Substitution Prop. (=)

Additional Answers

15. If something is a cloud, then it has a silver lining.
16. If a polygon is a rectangle, then it has four right angles.
17. If a rock is obsidian, then it is a glassy rock produced by a volcano.
18. If two planes intersect, then their intersection is a line.
19. Converse: If a rectangle is a square, then it has four congruent sides. Inverse: If a rectangle does not have four congruent sides, then it is not a square. Contrapositive: If a rectangle is not a square, then it does not have four congruent sides.
20. Converse: If three points lie on a straight line, then they are collinear. Inverse: If three points are noncollinear, then they do not lie on a straight line. Contrapositive: If three points do not lie on a straight line, then they are noncollinear.
21. Converse: If a month has 31 days, then it is January. Inverse: If the month is not January, then it does not have 31 days. Contrapositive: If a month does not have 31 days, then it is not January.
22. Converse: If a point lies on the x-axis, then an ordered pair for the point has 0 as its y-coordinate. Inverse: If an ordered pair for a point does not have 0 as its y-coordinate, then the point does not lie on the x-axis. Contrapositive: If a point does not lie on the x-axis, then an ordered pair for the point does not have 0 as its y-coordinate.
23. $\angle A$ and $\angle B$ have measures with a sum of 180; Law of Detachment.
25. The sun is in constant motion; Law of Syllogism.

Applications and Problem Solving Encourage students to work through the exercises in the Applications and Problem Solving section to strengthen their problem-solving skills.

Additional Answers

36. If you are a hardworking person, then you deserve a great vacation.
Hypothesis: you are a hardworking person
Conclusion: you deserve a great vacation
Converse: If you deserve a great vacation, then you are a hardworking person.
Inverse: If you are not a hardworking person, then you do not deserve a great vacation.
Contrapositive: If you do not deserve a great vacation, then you are not a hardworking person.

40. Statements (Reasons)
$t = 35d + 20$ (Given)
$t - 20 = 35d$ (Subtraction Prop. (=))
$\frac{t-20}{35} = d$ (Division Prop. (=))
$d = \frac{t-20}{35}$ (Symmetric Prop. (=))

OBJECTIVES AND EXAMPLES

- complete proofs involving angle theorems (Lesson 2–6)

Given: $\angle 1 \cong \angle 2$

Prove: $\angle 3 \cong \angle 4$

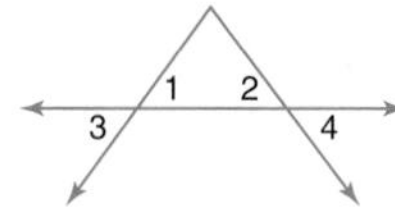

Proof:

Statements	Reasons
1. $\angle 1 \cong \angle 2$	1. Given
2. $\angle 3 \cong \angle 1$ $\angle 2 \cong \angle 4$	2. Vertical ∠s are ≅.
3. $\angle 3 \cong \angle 4$	3. Congruence of angles is transitive. *twice*

REVIEW EXERCISES

Complete each statement with *always*, *sometimes*, or *never*.

32. If two angles are right angles, they are __?__ adjacent. sometimes
33. If two angles are complementary, they are __?__ right angles. never
34. An angle is __?__ congruent to itself. always
35. Vertical angles are __?__ adjacent angles. never

APPLICATIONS AND PROBLEM SOLVING

36. **Advertising** Write the conditional *Hard-working people deserve a great vacation* in if-then form. Identify the hypothesis and the conclusion of the conditional. Then write the converse, inverse, and contrapositive. (Lesson 2–2) See margin.

37. **Look for a Pattern** Find the number of pairs of vertical angles determined by eight distinct lines passing through a point. (Lesson 2–3) 56 pairs

38. **Biology** If possible, write a valid conclusion that can be reached from the two true statements. State the law of logic that you used. (Lesson 2–3)
(1) A sponge is a sessile animal.
(2) A sessile animal is one that remains permanently attached to a surface for all of its adult life.
A sponge remains permanently attached to a surface for all of its adult life; syllogism.

39. **Algebra** Name the property of equality that justifies the statement *If $x + y = 3$ and $3 = w + v$, then $x + y = w + v$.* (Lesson 2–4) Substution Prop. (=) or Transitive Prop. (=)

40. **Geology** The underground temperature of rocks varies with their depth below Earth's surface. The deeper that a rock is, the hotter it is. The temperature t in degrees Celsius is estimated by the equation $t = 35d + 20$, where d is the depth in kilometers. Solve the formula for d and justify each step. (Lesson 2–4) See margin.

A practice test for Chapter 2 is provided on page 794.

ALTERNATIVE ASSESSMENT

COOPERATIVE LEARNING PROJECT

Anatomy The human brain consists of two hemispheres, the left and the right. Mental activities appear to be split between the two sides. The left hemisphere is responsible for the ways we think. These can be described as analytical, linear, sequential, verbal, and concrete. The right side is responsible for very different types of mental skills. These are often described as patterning, holistic, visual, spatial, intuitive, creative, symbolic, nonverbal, artistic, and spontaneous.

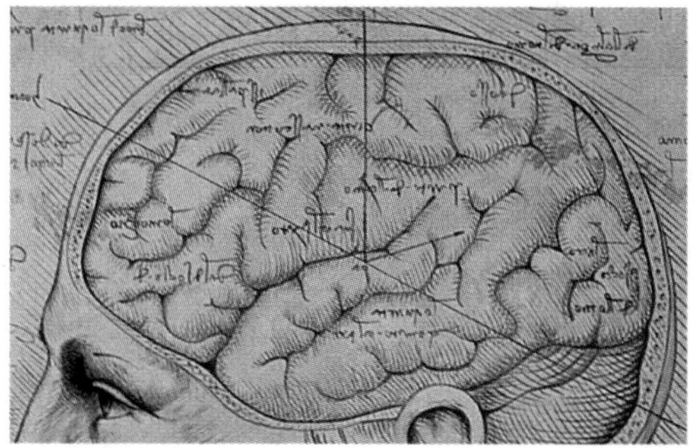

Follow these steps to complete your task.

Select three problems from this chapter.

- Choose one where you came up with the conjecture. Recall and write the steps you went through in your mind to arrive at your conjecture.
- Choose one where the conjecture was provided by the text. Recall and write the steps you went through in your mind to arrive at the proof when you were given the conjecture.
- Choose one where the entire proof was provided. Recall and write the steps you went through in your mind to understand the proof given in the text.

Complete the following table recording the steps you went through for each of the three problems.

Step	Left-Brain	Right-Brain
1		
2		
3		
4		

List each mental activity you performed in the order in which you performed it. Decide whether the activity is left-brain or right-brain. As a group, discuss which activities in mathematics are left-brain and which are right-brain.

THINKING CRITICALLY

Mathematical sentences use very few words. Many words can be replaced by symbols. *If A, then B* can be written as $A \Rightarrow B$. *A and B* is often replaced by $A \wedge B$, and *A or B* is often written as $A \vee B$.

- Do you think it is possible to replace all words in mathematics with symbols? no
- What would be some advantages and disadvantages of using all symbols?

PORTFOLIO

Select a proof from the chapter that you completed. Draw a diagram of the proof by putting each step of the proof in a circle. Then connect circles that are logically related. If, in the proof, the step in circle B used the step in circle A, draw a directed arrow from circle A to circle B. Try several ways to arrange the circles on the page. Select your favorite, and explain why it is your favorite. Include the graph in your portfolio.

SELF EVALUATION

Make a table with two columns. In one column, list the courses that you find easy and in the other column list those that are more challenging to you. Which courses use the right hemisphere of the brain, and which use the left hemisphere?

Assess Yourself Do you seem to be better at right-brain activities, left-brain activities, or equally good at both? Explain.

Alternative Assessment

The Alternative Assessment section provides students with the opportunity to assess their own work by thinking critically, working with others, keeping a portfolio, and honestly evaluating their own progress. For more information on alternative forms of assessment, see *Alternative Assessment in the Mathematics Classroom,* one of the titles in the Glencoe Mathematics Professional Series.

Performance Assessment

Performance Assessment tasks for this chapter are included in the *Assessment and Evaluation Masters.* A scoring guide is also provided.

Additional Answer for Thinking Critically

Sample answer: An advantage to using symbols would be less writing so statements would be more concise. A disadvantage would be if there were many symbols, then memorizing all of them would be like learning another language.

Assessment and Evaluation Masters, pp. 42, 53

NAME ____________ DATE ________

Chapter 2 Performance Assessment

Instructions: ***Demonstrate your knowledge by giving a clear, concise solution to each problem. Be sure to include all relevant drawings and justify your answers. You may show your solution in more than one way or investigate beyond the requirements of the problem.***

1. a. In your own words, explain the meaning of *conjecture*.

   b. Give an example of using inductive reasoning to make a correct conjecture.

   c. Give an example of using inductive reasoning to make an incorrect conjecture.

   d. Tell how you know the conjecture you wrote in part c is incorrect.

2. a. In your own words, explain the meaning of *deductive reasoning.*

   b. Give an example of the correct use of deductive reasoning to reach a conclusion. Tell why the reasoning is correct.

   c. Give an example of incorrect reasoning, using conditional statements.

3. State one property of equality you studied in this chapter. Show how this property can be used to make a conclusion about a geometric figure. Be sure to include a diagram.

4. $\overleftrightarrow{AD}$ and $\overleftrightarrow{BE}$ intersect at point C.

   a. Given that $\angle ACB \cong \angle BCD$, make a conjecture.

   b. Prove or disprove the conjecture you wrote in part a.

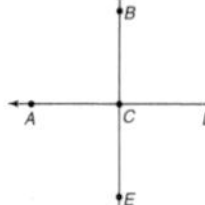

Scoring Guide
Chapter 2
Performance Assessment

Level	Specific Criteria
3 Superior	• Shows thorough understanding of the concepts of *conjecture, inductive and deductive reasoning, counterexample,* and *conditional statement.* • Uses appropriate strategies to construct proof. • Written explanations are exemplary. • Diagrams are accurate and appropriate. • Goes beyond requirements of some or all problems.
2 Satisfactory, with Minor Flaws	• Shows understanding of the concepts of *conjecture, inductive and deductive reasoning, counterexample,* and *conditional statement.* • Uses appropriate strategies to construct proof. • Written explanations are effective. • Diagrams are mostly accurate and appropriate. • Satisfies all requirements of some or all problems.
1 Nearly Satisfactory, with Serious Flaws	• Shows understanding of most of the concepts of *conjecture, inductive and deductive reasoning, counterexample,* and *conditional statement.* • May not use appropriate strategies to construct proof. • Written explanations are satisfactory. • Diagrams are mostly accurate and appropriate. • Satisfies most requirements of some or all problems.
0 Unsatisfactory	• Shows little or no understanding of the concepts of *conjecture, inductive and deductive reasoning, counterexample,* and *conditional statement.* • May not use appropriate strategies to construct proof. • Written explanations are not satisfactory. • Diagrams are not accurate or appropriate. • Does not satisfy requirements of some or all problems.

Using the COLLEGE ENTRANCE EXAM PRACTICE

These two pages review the skills and concepts presented in Chapters 1–2. This review is formatted to reflect new trends in college entrance testing.

Assessment and Evaluation Masters, pp. 47–48

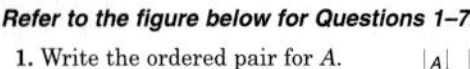

NAME ____________ DATE ________

2 Chapter 2 Cumulative Review

Refer to the figure below for Questions 1–7.

1. Write the ordered pair for A. 1. (−4, 4)
2. Write the y-coordinate of B. 2. −4
3. Find the length of $\overline{BD}$. 3. $8\sqrt{2}$
4. Name the point in Quadrant IV. 4. C
5. Write the x-coordinate of any point collinear with A and B. 5. −4
6. Find the coordinates of the midpoint of $\overline{BD}$. 6. (0, 0)
7. Which points are in the interior of $\angle C$? 7. A, O
8. Determine whether the statement *Four points are needed to determine a plane* is true or false. 8. false
9. If S is the midpoint of $\overline{RT}$, $RS = 7x - 13$, and $ST = 4x + 5$, find the measure of $\overline{RT}$. 9. 58
10. If $m\angle KSH = 4x - 10$, $m\angle LSH = 3x$, and $\overrightarrow{SH}$ bisects $\angle KSL$, find the measures of all the angles.

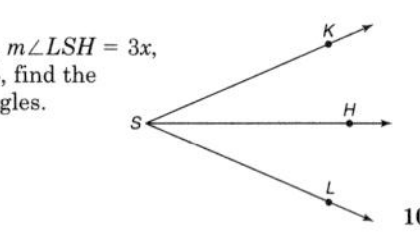

10. $m\angle KSH = 30$, $m\angle LSH = 30$, $m\angle KSL = 60$

11. If $\angle KPX$ and $\angle APX$ form a linear pair, then name a pair of opposite rays. 11. $\overrightarrow{PX}$, $\overrightarrow{PA}$
12. If $\angle M$ and $\angle N$ are complementary, $m\angle M = 4x - 3$, and $m\angle N = 2x + 9$, find $m\angle M$ and $m\angle N$. 12. 53, 37
13. The measure of an angle is 8 more than three times the measure of its supplement. Find the measures of both angles. 13. 43, 137

NAME ____________ DATE ________

2 Chapter 2 Cumulative Review (continued)

Tell whether each relationship can be assumed from the figure below. Write yes or no.

14. $\overline{BE} \perp \overline{BC}$ 14. no
15. E is the midpoint of $\overline{BD}$. 15. yes
16. $\angle ABE \cong \angle EDC$ 16. yes

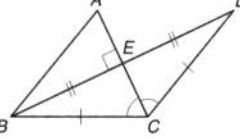

17. Which property of equality justifies the statement *If MN = PQ and RS = PQ, then MN = RS?* 17. Substitution Prop. (=)
18. Determine if a valid conclusion can be reached from the two true statements. If it can, state it and the law that is used. If a valid conclusion does not follow, write *no conclusion*.
(1) If $a = b$, then $b = a$.
(2) $a = b$ 18. $b = a$; Law of Detachment
19. Write the statement *Vertical angles are congruent* in if-then form. Then write the converse. 19. If two angles are vertical angles, then they are congruent. If two angles are congruent, then they are vertical angles.
20. Each of the following sentences was used in a correct proof of the statement *The angles formed by one pair of perpendicular lines are all congruent to those formed by any other pair of perpendicular lines.* Determine which sentences were in the *Statements* column and which were in the *Reasons* column.
A. All angles formed by ℓ and m are right angles.
B. Perpendicular lines intersect to form four right angles.
C. Each angle formed by ℓ and m is congruent to each angle formed by j and k.
D. All right angles are congruent. 20. statements: A, C; reasons: B, D

COLLEGE ENTRANCE EXAM PRACTICE

CHAPTERS 1–2

SECTION ONE: MULTIPLE CHOICE

There are ten multiple-choice questions in this section. After working each problem, write the letter of the correct answer on your paper.

1. If $5n + 5 = 10$, find the value of $11 - n$. **D**
   - A. -10
   - B. 0
   - C. 5
   - D. 10

2. Find the result when 14^{12} is divided by 14^{10}. **D**
   - A. 14^{22}
   - B. 14^{-2}
   - C. 14^{-22}
   - D. 14^{2}

3. Find the distance between X and Z. **A**

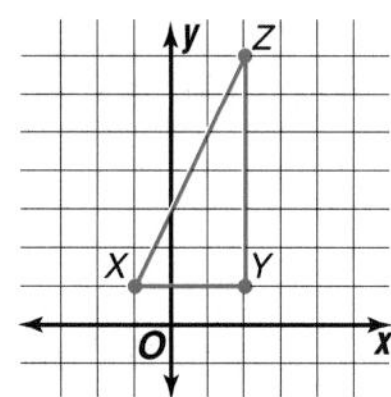

   - A. $3\sqrt{5}$ units
   - B. 12 units
   - C. $3\sqrt{3}$ units
   - D. 6 units

4. Find x if $0.08x = 32$. **B**
   - A. 800
   - B. 400
   - C. 40
   - D. 0.04

5. Find the perimeter of a square whose side is 8 meters long. **C**
   - A. 64 m
   - B. 64 m^2
   - C. 32 m
   - D. 32 m^2

6. If two planes intersect, their intersection can be **A**
   - I. a line.
   - II. three noncollinear points.
   - III. two intersecting lines.
   - A. I only
   - B. II only
   - C. III only
   - D. I and II only

7. If $\overrightarrow{BX}$ bisects $\angle ABC$, which of the following is true? **D**

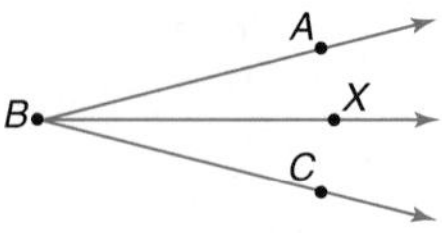

   - A. $m\angle ABX = m\angle XBC$
   - B. $m\angle ABX = \frac{1}{2}m\angle ABC$
   - C. $\frac{1}{2}m\angle ABC = m\angle XBC$
   - D. all of the above

8. Find x if $2x + 3 = \frac{1}{5}(10x + 15)$. **D**
   - A. 3
   - B. 2
   - C. no real number
   - D. any real number

9. 30% of 120 is one-third of what number? **C**
   - A. 10.8
   - B. 1200
   - C. 108
   - D. 12

10. Solve $y = mx + b$ for x. **B**
   - A. $x = \frac{y}{m} - b$
   - B. $x = \frac{y - b}{m}$
   - C. $x = \frac{b - y}{m}$
   - D. $x = \frac{y}{b} - m$

Standardized Test Practice Questions are also provided in the *Assessment and Evaluation Masters*, p. 46.

A more traditional cumulative review is provided in the *Assessment and Evaluation Masters*, pp. 47–48.

SECTION TWO: SHORT ANSWER

This section contains seven questions for which you will provide short answers. Write your answer on your paper.

11. In the graph below, the axes and the origin are not shown. If point M has coordinates $(-3, -4)$, what are the coordinates of point N? $N(-2, -8)$

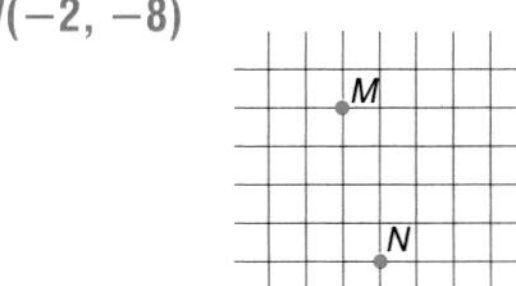

12. Find the expanded form of $(5x - 2)^2$. $25x^2 - 20x + 4$

13. There are 120 Calories in one cup of grapefruit juice. Write an equation to represent the number of Calories in four cups of grapefruit juice. $C = 120(4)$

14. A farmer fenced all but one side of a square field. If he has already used $3x$ meters of fencing, how many meters will he need for the last side? x

15. In the figure below, the hand pointing to 8 moves clockwise to the next numeral every hour. At the end of 24 hours, to which numeral will it point? 5

16. What is the greatest integer x such that $9x + 5 < 100$? 10

17. Westlake has 1600 public payphones. Three-fourths of the phones have dials. If one-third of the dial phones are replaced by push-button phones, how many dial phones remain? 800

SECTION THREE: COMPARISON

This section contains five comparison problems that involve comparing two quantities, one in column A and one in column B. In certain questions, information related to one or both quantities is centered above them. All variables used represent real numbers.

Compare quantities A and B below.

- Write A if quantity A is greater.
- Write B if quantity B is greater.
- Write C if the two quantities are equal.
- Write D if there is not enough information to determine the relationship.

	Column A		Column B	
18.		A B C D $AC = 60$, $CD = 12$, B is the midpoint of $\overline{AD}$.		A
	BC		CD	
19.	$\frac{1}{\frac{3}{2}}$		$\left(\frac{3}{2}\right)^2$	B
20.		$\frac{3}{4}x = -24$		A
	$\frac{1}{2}x$		$\frac{3}{2}x$	
21.		$a = -4$		C
	$a^2 + a$		12	
22.		$x°$ 60° 50°		A
	x		60	

TEST-TAKING TIP

Getting Ready

You may wish to give students the following suggestions.

Before you take any test, you should find out as much as you can about its form and content.

- Talk to several students you know who have taken the test to get their impressions of it.
- Ask teachers and counselors about the test. They can give you pointers about your strengths and weaknesses in areas covered by the test.
- Check out your school or public library. Books and computer programs are available to help you develop general test-taking skills and prepare for many of the common standardized tests.
- Take sample tests available from the testing agency itself.

CHAPTER 3

Using Perpendicular and Parallel Lines

PREVIEWING THE CHAPTER

This chapter examines the relationship between points, lines, angles, planes, and spheres. Parallel lines and their transversals are studied, as well as the angle measures formed when these lines intersect. In addition to angle measurements, slopes of parallel and perpendicular lines are found. Many postulates and theorems are introduced to relate lines and angles. The theorems are used to enhance the learning of geometric proofs. The chapter concludes with finding the distance between points and lines using the distance formula and with studying the properties of spherical geometry.

Lesson (Pages)	Lesson Objectives	NCTM Standards	State/Local Objectives
3-1 (124–129)	Solve problems by drawing a diagram. Identify the relationships between two lines or two planes. Name angles formed by a pair of lines and a transversal.	1–4, 7	
3-2A (130)	Find the measures of the angles formed by two parallel lines and a transversal.	1–4, 7	
3-2 (131–137)	Use the properties of parallel lines to determine angle measures.	1–4, 7	
3-3 (138–145)	Find the slopes of lines. Use slope to identify parallel and perpendicular lines.	1–5, 7	
3-4 (146–153)	Recognize angle conditions that produce parallel lines. Prove two lines are parallel based on given angle relationships.	1–5, 7	
3-5 (154–161)	Recognize and use distance relationships among points, lines, and planes.	1–4, 7	
3-5B (162)	Find the distance between a point and a line.	1–4, 7	
3-6 (163–169)	Identify points, lines, and planes in spherical geometry. Compare and contrast basic properties of plane and spherical geometry.	1–4, 7	

A complete, 1-page lesson plan is provided for each lesson in the ***Lesson Planning Guide***. Answer keys for each lesson are available in the ***Answer Key Masters***.

ORGANIZING THE CHAPTER

You may want to refer to the **Course Planning Calendar** on page T12 for detailed information on pacing.
PACING: Standard—14 days; **Honors**—13 days; **Block**—7 days

LESSON PLANNING CHART

			BLACKLINE MASTERS									
Lesson (Pages)	**Materials/ Manipulatives**	**Extra Practice (Student Edition)**	**Study Guide**	**Practice**	**Enrichment**	**Assessment & Evaluation**	**Modeling Mathematics**	**Multicultural Activity**	**Tech Prep Applications**	**Graphing Calc. & Computer**	**Real-World Applications**	**Teaching Transparencies**
3-1 (124–129)	colored pencils straightedge*	p. 768	p. 14	p. 14	p. 14		pp. 23–26	p. 5				3-1A 3-1B
3-2A (130)	TI-92 calculator									p. 19		
3-2 (131–137)	protractor* colored pencils straightedge*	p. 768	p. 15	p. 15	p. 15	p. 72			p. 5			3-2A 3-2B
3-3 (138–145)	TI-82/83 graphing calculator	p. 769	p. 16	p. 16	p. 16	pp. 71, 72			p. 6	p. 3	5	3-3A 3-3B
3-4 (146–153)	grid paper compass* straightedge* colored pencils	p. 769	p. 17	p. 17	p. 17		p. 81					3-4A 3-4B
3-5 (154–161)	cardboard grid paper scissors* straightedge* thumbtack	p. 769	p. 18	p. 18	p. 18	p. 73		p. 6			6	3-5A 3-5B
3-5B (162)	TI-82/83 graphing calculator									pp. 20, 21		
3-6 (163–169)	Styrofoam™ ball string ruler* markers	p. 770	p. 19	p. 19	p. 19	p. 73						3-6A 3-6B
Study Guide/ Assessment (171–175)						pp. 57–70, 74–76						

*Included in Glencoe's High School Manipulative Kit and Overhead Manipulative Resources.

ORGANIZING THE CHAPTER

Other Chapter Resources

Student Edition
Chapter Opener, pp. 122–123
Mathematics and Society, p. 161
Working on the Investigation, p. 153
Closing the Investigation, p. 170

Teacher's Classroom Resources
Investigations and Projects Masters, pp. 33–36
Block Scheduling Booklet

Technology
Test and Review Software (IBM and Macintosh)
CD-ROM Multimedia Applications (Windows and Macintosh)
Mindjogger Videoquizzes (VHS)

Professional Publications
Glencoe Mathematics Professional Series

Outside Resources

Books/Periodicals
Hands-On Geometry, ETA
Geometry Practice Exercises, NASCO

Software
Geometry Concepts, ETA
Geometry-Sensei Software, NASCO

Videos/CD-ROMs
Basic Geometry, Video Tutor, Inc. NASCO, 901 Janesville Ave., Fort Atkinson, WI 53538
An Introduction to the TI-92, TeleMath, P.O. Box 330777, Atlantic Beach, FL 32233

Assessment Resources

Student Edition
Math Journal, pp. 141, 149
Mixed Review, pp. 129, 137, 144, 153, 161, 169
Self Test, p. 145
Chapter Highlights, p. 171
Chapter Study Guide and Assessment, pp. 172–174
Alternative Assessment, p. 175
 Portfolio, p. 175
Chapter Test, p. 795

Teacher's Wraparound Edition
5-Minute Check, pp. 124, 131–132, 138, 146, 154, 163
Check for Understanding, pp. 127, 134, 141, 149, 157, 167
Closing Activity, pp. 129, 137, 145, 153, 161, 169
Cooperative Learning, pp. 125, 148

Assessment and Evaluation Masters
Multiple-Choice Tests, Forms 1A (Honors), 1B (Average), 1C (Basic), pp. 57–62
Free-Response Tests, Forms 2A (Honors), 2B (Average), 2C (Basic), pp. 63–68
Calculator-Based Test, p. 69
Performance Assessment, p. 70
Mid-Chapter Test, p. 71
Quizzes A–D, pp. 72–73
Standardized Test Practice, p. 74
Cumulative Review, pp. 75–76

ENHANCING THE CHAPTER

Examples of some of the materials for enhancing Chapter 3 are shown below.

DIVERSITY

Multicultural Activity Masters, pp. 5, 6

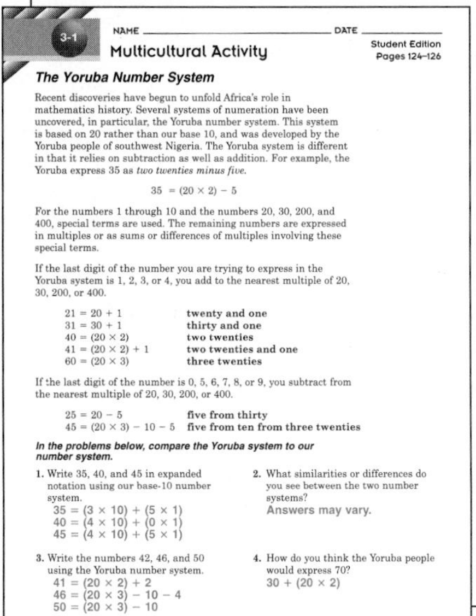

3-1 NAME ______________ DATE ______________

Multicultural Activity — Student Edition Pages 124–126

The Yoruba Number System

Recent discoveries have begun to unfold Africa's role in mathematics history. Several systems of numeration have been uncovered, in particular, the Yoruba number system. This system is based on 20 rather than our base 10, and was developed by the Yoruba people of southwest Nigeria. The Yoruba system is different in that it relies on subtraction as well as addition. For example, the Yoruba express 35 as *two twenties minus five.*

$$35 = (20 \times 2) - 5$$

For the numbers 1 through 10 and the numbers 20, 30, 200, and 400, special terms are used. The remaining numbers are expressed in multiples or as sums or differences of multiples involving these special terms.

If the last digit of the number you are trying to express in the Yoruba system is 1, 2, 3, or 4, you add to the nearest multiple of 20, 30, 200, or 400.

$21 = 20 + 1$	**twenty and one**
$31 = 30 + 1$	**thirty and one**
$40 = (20 \times 2)$	**two twenties**
$41 = (20 \times 2) + 1$	**two twenties and one**
$60 = (20 \times 3)$	**three twenties**

If the last digit of the number is 0, 5, 6, 7, 8, or 9, you subtract from the nearest multiple of 20, 30, 200, or 400.

$25 = 20 - 5$	**five from thirty**
$45 = (20 \times 3) - 10 - 5$	**five from ten from three twenties**

In the problems below, compare the Yoruba system to our number system.

1. Write 35, 40, and 45 in expanded notation using our base-10 number system.
   $35 = (3 \times 10) + (5 \times 1)$
   $40 = (4 \times 10) + (0 \times 1)$
   $45 = (4 \times 10) + (5 \times 1)$
2. What similarities or differences do you see between the two number systems?
   Answers may vary.
3. Write the numbers 42, 46, and 50 using the Yoruba number system.
   $41 = (20 \times 2) + 2$
   $46 = (20 \times 3) - 10 - 4$
   $50 = (20 \times 3) - 10$
4. How do you think the Yoruba people would express 70?
   $30 + (20 \times 2)$

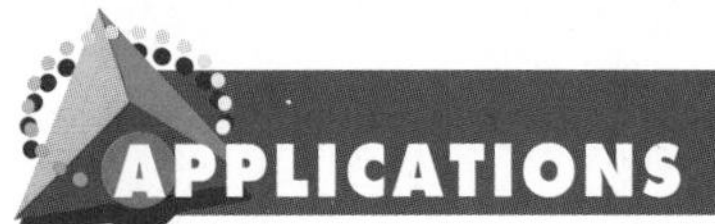

APPLICATIONS

Real-World Applications, 5, 6

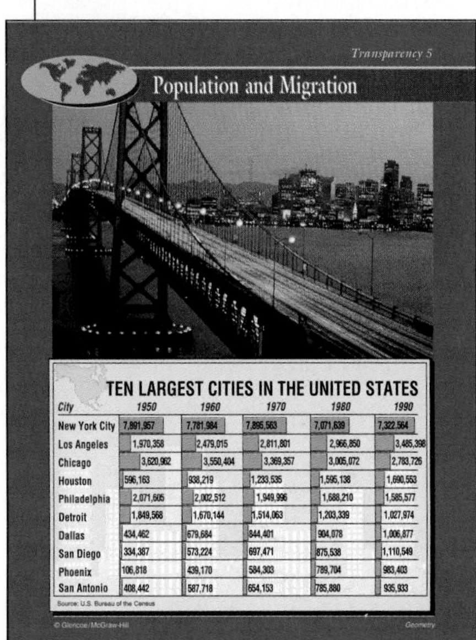

Transparency 5

Population and Migration

TEN LARGEST CITIES IN THE UNITED STATES

City	1950	1960	1970	1980	1990
New York City	7,891,957	7,781,984	7,895,563	7,071,639	7,322,564
Los Angeles	1,970,358	2,479,015	2,811,801	2,966,850	3,485,398
Chicago	3,620,962	3,550,404	3,369,357	3,005,072	2,783,726
Houston	596,163	938,219	1,233,535	1,595,138	1,690,553
Philadelphia	2,071,605	2,002,512	1,949,996	1,688,210	1,585,577
Detroit	1,849,568	1,670,144	1,514,063	1,203,339	1,027,974
Dallas	434,462	679,684	844,401	904,078	1,006,877
San Diego	334,387	573,224	697,471	875,538	1,110,549
Phoenix	106,818	439,170	584,303	789,704	983,403
San Antonio	408,442	587,718	654,153	785,880	935,933

Source: U.S. Bureau of the Census

 Geometry

TECHNOLOGY

Graphing Calculator and Computer Masters, p. 3

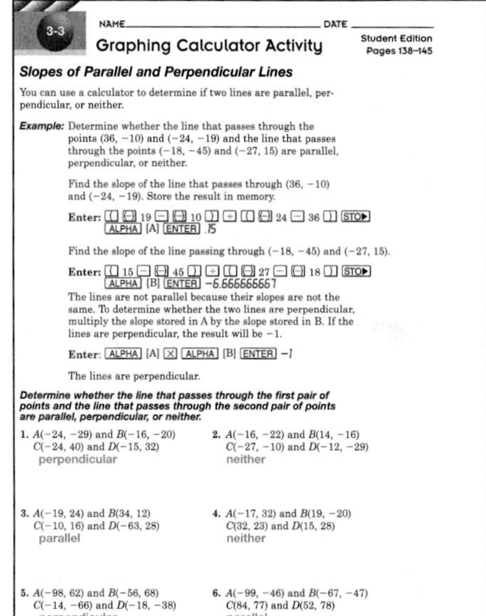

3-3 NAME ______________ DATE ______________

Graphing Calculator Activity — Student Edition Pages 138–145

Slopes of Parallel and Perpendicular Lines

You can use a calculator to determine if two lines are parallel, perpendicular, or neither.

Example: Determine whether the line that passes through the points (36, −10) and (−24, −19) and the line that passes through the points (−18, −45) and (−27, 15) are parallel, perpendicular, or neither.

Find the slope of the line that passes through (36, −10) and (−24, −19). Store the result in memory.

Enter: [(] [(-)] 19 [−] [(-)] 10 [)] [÷] [(] [(-)] 24 [−] 36 [)] [STO▶] [ALPHA] [A] [ENTER] .15

Find the slope of the line passing through (−18, −45) and (−27, 15).

Enter: [(] 15 [−] [(-)] 45 [)] [÷] [(] [(-)] 27 [−] [(-)] 18 [)] [STO▶] [ALPHA] [B] [ENTER] −6.666666667

The lines are not parallel because their slopes are not the same. To determine whether the two lines are perpendicular, multiply the slope stored in A by the slope stored in B. If the lines are perpendicular, the result will be −1.

Enter: [ALPHA] [A] [×] [ALPHA] [B] [ENTER] −1

The lines are perpendicular.

Determine whether the line that passes through the first pair of points and the line that passes through the second pair of points are parallel, perpendicular, or neither.

1. $A(-24, -29)$ and $B(-16, -20)$, $C(-24, 40)$ and $D(-15, 32)$ — perpendicular
2. $A(-16, -22)$ and $B(14, -16)$, $C(-27, -10)$ and $D(-12, -29)$ — neither
3. $A(-19, 24)$ and $B(34, 12)$, $C(-10, 16)$ and $D(-63, 28)$ — parallel
4. $A(-17, 32)$ and $B(19, -20)$, $C(32, 23)$ and $D(15, 28)$ — neither
5. $A(-98, 62)$ and $B(-56, 68)$, $C(-14, -66)$ and $D(-18, -38)$ — perpendicular
6. $A(-99, -46)$ and $B(-67, -47)$, $C(84, 77)$ and $D(52, 78)$ — parallel

TECH PREP

Tech Prep Applications Masters, pp. 5, 6

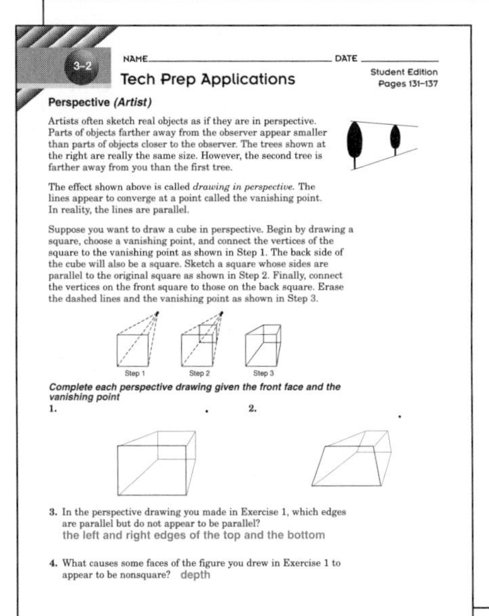

3–2 NAME ______________ DATE ______________

Tech Prep Applications — Student Edition Pages 131–137

Perspective *(Artist)*

Artists often sketch real objects as if they are in perspective. Parts of objects farther away from the observer appear smaller than parts of objects closer to the observer. The trees shown at the right are really the same size. However, the second tree is farther away from you than the first tree.

The effect shown above is called *drawing in perspective.* The lines appear to converge at a point called the vanishing point. In reality, the lines are parallel.

Suppose you want to draw a cube in perspective. Begin by drawing a square, choose a vanishing point, and connect the vertices of the square to the vanishing point as shown in Step 1. The back side of the cube will also be a square. Sketch a square whose sides are parallel to the original square as shown in Step 2. Finally, connect the vertices on the front square to those on the back square. Erase the dashed lines and the vanishing point as shown in Step 3.

Complete each perspective drawing given the front face and the vanishing point

1.
2.

3. In the perspective drawing you made in Exercise 1, which edges are parallel but do not appear to be parallel?
   the left and right edges of the top and the bottom
4. What causes some faces of the figure you drew in Exercise 1 to appear to be nonsquare? depth

PROBLEM SOLVING

Problem-of-the-Week Cards, 7, 8, 9

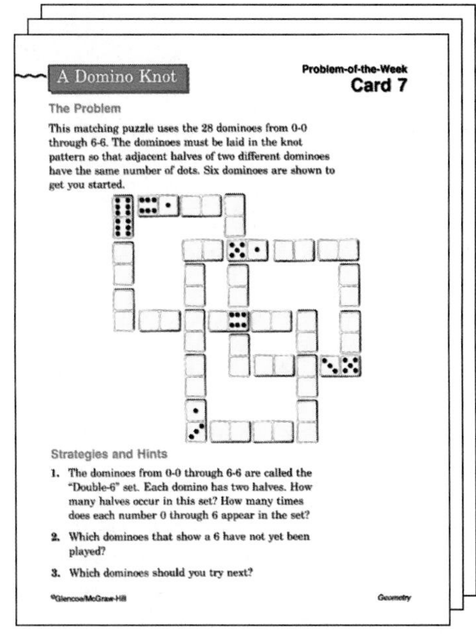

A Domino Knot — Problem-of-the-Week **Card 7**

The Problem

This matching puzzle uses the 28 dominoes from 0-0 through 6-6. The dominoes must be laid in the knot pattern so that adjacent halves of two different dominoes have the same number of dots. Six dominoes are shown to get you started.

Strategies and Hints

1. The dominoes from 0-0 through 6-6 are called the "Double-6" set. Each domino has two halves. How many halves occur in this set? How many times does each number 0 through 6 appear in the set?
2. Which dominoes that show a 6 have not yet been played?
3. Which dominoes should you try next?

 Geometry

MAKING MATHEMATICS RELEVANT

This two-page introduction to the chapter provides students with an opportunity to explore an application of geometry that has historic roots as well as modern uses.

Background Information

Skiing is on the Rise Snow skiing is usually classified into three types. Alpine skiing, also known as downhill skiing, usually involves skiing down steep slopes. Alpine ski racing events include downhill skiing, slalom racing, giant slalom, and super giant slalom. The second type of snow skiing is Nordic skiing, which involves cross-country skiing over relatively level surfaces and ski jumping competitions. The third type of snow skiing, freestyle, has become popular since the early 1980s. Freestyle skiing includes the disciplines of moguls and aerials.

Water skiing is an offshoot of snow skiing. Water skiers wearing special skis with fins are towed across the water by powerboats. It was invented in the United States in 1922 and was believed to be inspired by a popular amusement in which a horse towed a person wearing snow skis. Although water skiing has its own sanctioning body and competition events and rules, the basics of water skiing are similar to those of snow skiing. The key to traditional water skiing is still keeping the skis parallel.

CHAPTER 3

Using Perpendicular and Parallel Lines

Skiing is on the Rise

Objectives

In this chapter, you will:

- solve problems by drawing a diagram,
- use properties of parallel lines,
- use slope to identify parallel and perpendicular lines,
- prove lines parallel,
- apply distance relationships among points, lines, and planes, and
- analyze properties of spherical geometry.

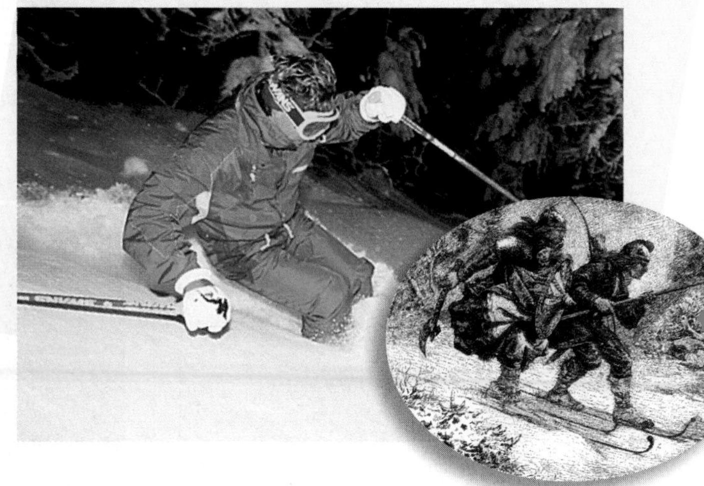

Geometry: Then and Now Do you like to ski? If so, you belong to a growing number of Americans who do. In ancient times, Scandinavians used skis as a way to travel over snow. The modern sport of skiing began in Norway in the nineteenth century and has been gaining popularity ever since. Today, nearly one fourth of U.S. skiers are between the ages of 7 and 17. One of the keys to the basic downhill movement, the schuss, is keeping the skis parallel.

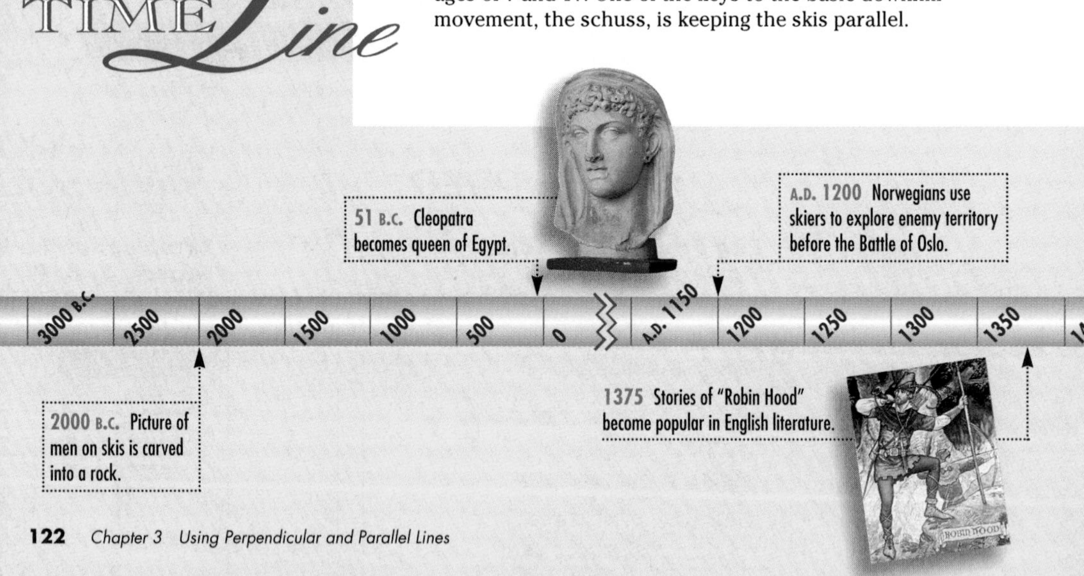
TIME Line
51 B.C. Cleopatra becomes queen of Egypt.
A.D. 1200 Norwegians use skiers to explore enemy territory before the Battle of Oslo.
3000 B.C.
2500
2000
1500
1000
500
0
A.D. 1150
1200
1250
1300
1350
2000 B.C. Picture of men on skis is carved into a rock.
1375 Stories of "Robin Hood" become popular in English literature.
ROBIN HOOD

Dürer was both a mathematician and an artist. Students might find it interesting to write a report on the relationship between geometry and art in Dürer's work.

interNET CONNECTION

Review water skiing techniques such as barefoot, arm chair, and kneeboarding at the "Water Skier's Web."

World Wide Web
http://waterski.net/

PEOPLE IN THE NEWS

At age 17, water-skier **Brandi Hunt** of Clermont, Florida, had already been ranked the number one junior water-skier in the world for two years in a row. Brandi has been waterskiing since she was seven years old and won her first national championship when she was nine. She hopes to become a professional water-skier. Brandi's philosophy about her success and advice to other athletes: "Compete against yourself, not other people. That way you know you'll be doing your own personal best."

Chapter Project

Skiing is not the only sport that involves the idea of parallels. Compile a list of sports that incorporate parallel lines in some form. Use the Internet to research these sports. Describe the aspect of the sport that requires parallels.

- For each sport on your list, discuss what happens if the lines you have described are not parallel.
- Choose two sports from your list. For each sport, carefully draw a diagram showing the use of parallel lines in the sport. If possible, draw your diagram to scale.
- Exchange your diagrams with other students. Verify that the lines drawn in the diagrams you receive are parallel.

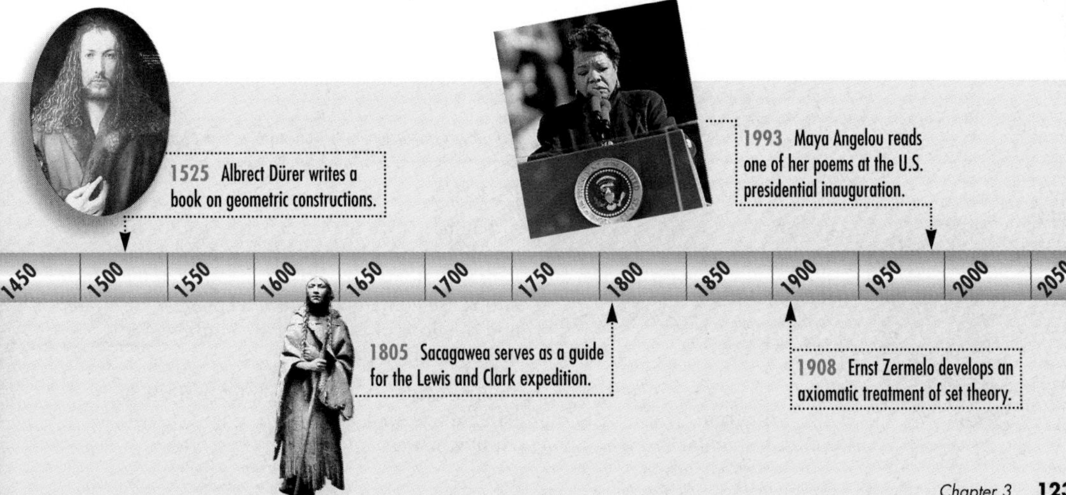

1525 Albrect Dürer writes a book on geometric constructions.
1993 Maya Angelou reads one of her poems at the U.S. presidential inauguration.
1450
1500
1550
1600
1650
1700
1750
1800
1850
1900
1950
2000
2050
1805 Sacagawea serves as a guide for the Lewis and Clark expedition.
1908 Ernst Zermelo develops an axiomatic treatment of set theory.

PEOPLE IN THE NEWS

Brandi trains year-round. During the school year, she wakes up at 6 A.M. for school. After school lets out at 3 P.M., she skis for two hours and then attends water-ski school from 6 P.M. to 8:30 P.M. After ski classes, she takes time for dinner and homework before retiring to bed at 10 P.M. Encourage a class discussion of time management for sports and hobbies during the school year.

Chapter Project

Cooperative Learning

Encourage students to work in small groups to brainstorm about aspects of parallel lines in various sports. Students might want to compare their lists with other groups before proceeding with the rest of the project.

Investigations and Projects Masters, p. 33

NAME____________ DATE_______

Student Edition Pages 122–175

Chapter 3 Project A

Skiing

1. Properties involving lines are evident in many aspects of the sport of skiing. For example, the equipment, the lifts, and the slopes themselves can be used to illustrate properties of lines. Make a list of how lines are involved in skiing.
2. Make a diagram of a skiing resort. Include at least two hills: one for beginners and one for advanced skiers. Explain why each hill is appropriate for each type of skier, using the idea of slope in your explanation.
3. Draw tow ropes or ski lifts for each hill in your ski resort. Explain the use of parallel lines and/or perpendicular lines in creating the ropes or lifts.
4. In addition to the skiing hills, draw a ski jump for the resort. Indicate the slope and length of the initial hill and the slope of the jump where the skiers take off. Explain the reasoning behind your ski jump design.

Alternative Chapter Projects

Two other chapter projects are included in the *Investigations and Projects Masters*. In Chapter 3 Project A, pp. 33–34, students extend the topic in the chapter opener. In Chapter 3 Project B, pp. 35–36, students investigate geometric patterns in art from other cultures.

3-1 LESSON NOTES

NCTM Standards: 1–4, 7

Instructional Resources

- Study Guide Master 3-1
- Practice Master 3-1
- Enrichment Master 3-1
- Modeling Mathematics Masters, pp. 23–26
- Multicultural Activity Masters, p. 5

Transparency 3-1A contains the 5-Minute Check for this lesson; **Transparency 3-1B** contains a teaching aid for this lesson.

Recommended Pacing	
Standard Pacing	Days 1 & 2 of 14
Honors Pacing	Day 1 of 13
Block Scheduling*	Day 1 of 7

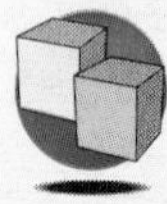

*For more information on pacing and possible lesson plans, refer to the *Block Scheduling Booklet*.

1 FOCUS

5-Minute Check
(over Chapter 2)

Refer to the figure below.

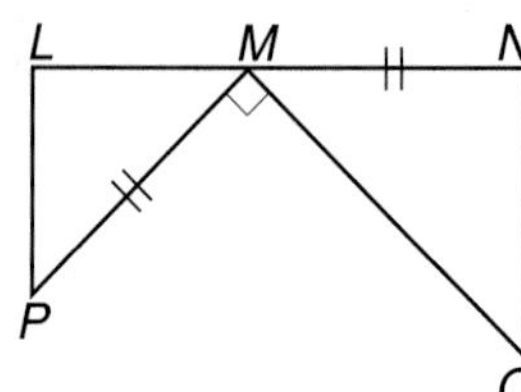

1. Are points L, M, and Q collinear? **no**
2. Find the measure of $\overline{MN}$ if $LM = 5x - 4$, $MN = 6x + 1$, and $LN = 30$. **$MN = 19$**
3. Can you determine if M is the midpoint of $\overline{LN}$? **no**
4. Name an acute angle, an obtuse angle, and a right angle in the figure. **Sample answer: $\angle QMN$ is acute, $\angle PMN$ is obtuse, and $\angle PMQ$ is right.**
5. Are $\angle LMP$ and $\angle NMQ$ complementary? **yes**

3-1 Parallel Lines and Transversals

What YOU'LL LEARN

- To solve problems by drawing a diagram,
- to identify the relationships between two lines or between two planes, and
- to name angles formed by a pair of lines and a transversal.

Why IT'S IMPORTANT

You can identify the relationship of lines and planes in architecture, agriculture, and air travel.

APPLICATION
Business

A corn harvest in Kansas and the multilevel shopping center in Hong Kong show examples of *parallel* planes, *parallel* lines, and *skew* lines.

Top Five List

Corn-Growing Countries	(million metric tons)
1. U.S.A.	240.8
2. China	95.3
3. Brazil	30.6
4. Mexico	15.0
5. France	14.6

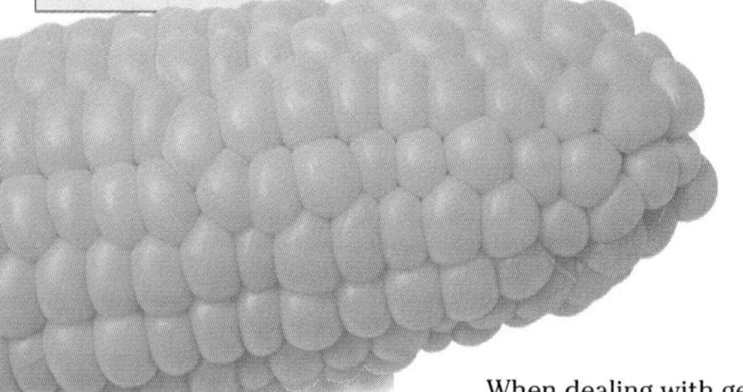

In the picture of the corn field, the rows are considered to be coplanar since they lie in the same plane. The rows represent lines in the plane that never meet. Two lines in a plane that never meet are called **parallel lines**.

In geometry, the symbol $\parallel$ means *is parallel to*. Red arrows are used in diagrams to indicate that lines are parallel. In the figure at the right, the arrows indicate that $\overleftrightarrow{BC}$ is parallel to $\overleftrightarrow{AD}$.

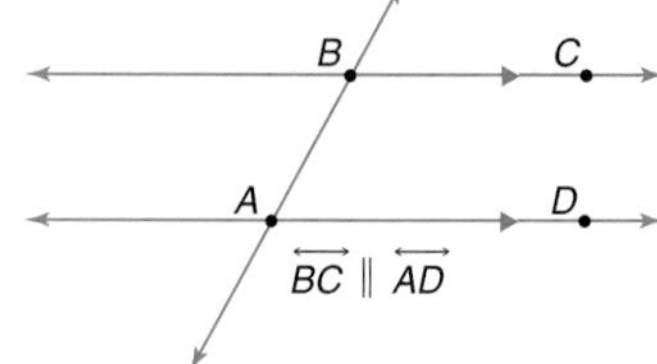

The term parallel and the notation $\parallel$ are used for lines, segments, rays, and planes. The symbol $\nparallel$ means "is not parallel to."

Similarly, two planes can intersect or be parallel. In the photograph of the shopping center in Hong Kong, the floors of each level of shops are contained in parallel planes. The walls and the floor of each level are contained in intersecting planes.

When dealing with geometric terms, it is often helpful to **draw a diagram**. The activity on the next page explains how to draw a three-dimensional figure involving parallel planes.

Corn can be grown between 42°S latitude and 58°N latitude.

MODELING MATHEMATICS — Drawing a Rectangular Prism

Materials:  plain paper  straightedge colored pencils

A rectangular prism resembles a shoe box or a cereal box. You can draw a diagram of a rectangular prism by drawing parallel planes.

- Draw two figures to represent the top and bottom of the prism. *These are similar to the figures used to represent parallel planes.*

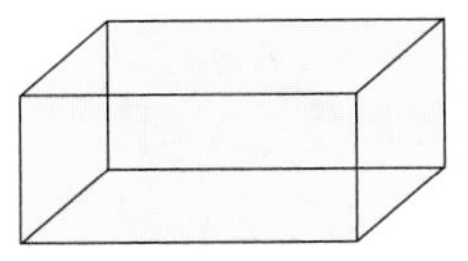

- Draw the vertical edges.

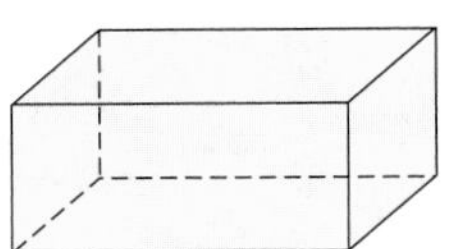

- Make the hidden edges of the prism dashed.
- Label the vertices of the prism.

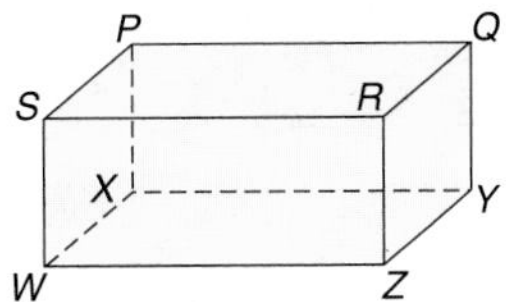

Your Turn

a. Name the parallel planes. **a–b. See margin.**

b. Name the planes that intersect plane PQR and name their intersection.

In the Modeling Mathematics activity, notice that $\overleftrightarrow{QR}$ and $\overleftrightarrow{WZ}$ do not intersect. Yet they are not parallel since they are not in the same plane. These lines are **skew lines**. The photograph at the right shows a cotton gin in Lubbock, Texas. Walkways $\overline{AB}$ and $\overline{CD}$ in the photograph are an example of skew lines.

Definition of Skew Lines	**Two lines are skew if they do not intersect and are not in the same plane.**

Example  Refer to the labeled planes and segments in the figure at the right.

a. Name all planes that are parallel to plane ABC.

plane TKG

b. Name all segments that intersect $\overline{AB}$.

$\overline{BC}$, $\overline{AC}$, $\overline{AD}$, $\overline{AT}$, and $\overline{BK}$

c. Name all segments that are parallel to $\overline{KG}$.

$\overline{BC}$, $\overline{AD}$, and $\overline{TH}$

d. Name all segments that are skew to $\overline{TK}$.

$\overline{CG}$, $\overline{DH}$, $\overline{AD}$, $\overline{AC}$, and $\overline{BC}$

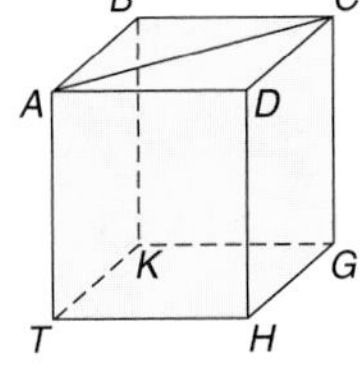

Motivating the Lesson

Hands-On Activity Bring in a rectangular box (a shoebox or a cereal box) and have students label all corners. Then have them list all planes and segments represented by the box.

2 TEACH

MODELING MATHEMATICS Suggest to students that there is another method of drawing a rectangular prism. They can begin by drawing two overlapping rectangles. This creates the front and back faces. Then they need to draw the horizontal edges.

In-Class Example

For Example 1
Refer to the figure in Example 1.

a. Name all planes parallel to the plane ADH. ***BCG***

b. Name all segments that intersect $\overline{AT}$. $\overline{AB}$, $\overline{AC}$, $\overline{AD}$

c. Name all segments that are parallel to $\overline{AT}$. $\overline{DH}$, $\overline{BK}$, $\overline{CG}$

d. Name all segments that are skew to $\overline{CG}$. $\overline{TK}$, $\overline{TH}$, $\overline{AD}$, $\overline{AB}$

Additional Answers for Modeling Mathematics

a. Sample answer: plane $RSW \parallel$ plane QPX; plane $RQY \parallel$ plane SPX; plane $PQR \parallel$ plane XYZ

b. Sample answer: plane RZY, $\overleftrightarrow{RQ}$; plane SWZ, $\overleftrightarrow{RS}$; plane PSW, $\overleftrightarrow{PS}$; plane PQY, $\overleftrightarrow{PQ}$

Cooperative Learning

Group Discussion Separate the class into small groups and have them use paper, cardboard, pencils, pipe cleaners, glue, tape, and other materials to represent planes or lines. Have them construct a figure that contains at least four of the geometric representations in this lesson. Have each group present its representations to the class and discuss the ideas. For more information on the group discussion strategy, see *Cooperative Learning in the Mathematics Classroom*, one of the titles in the Glencoe Mathematics Professional Series, page 31.

In-Class Examples

For Example 2

Identify each pair of angles as *alternate interior, alternate exterior, corresponding,* or *consecutive interior* angles.

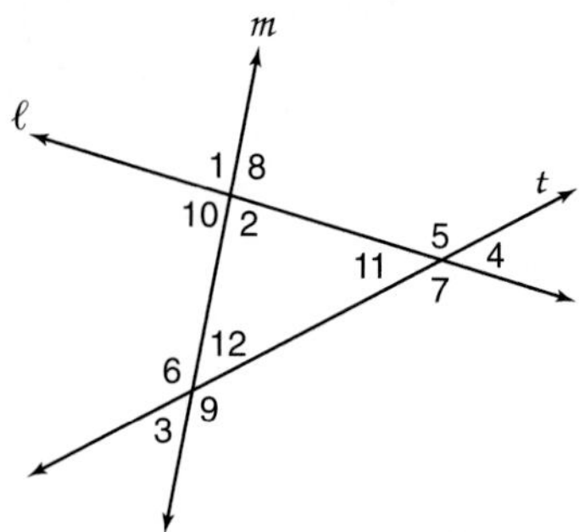

a. ∠6 and ∠10 consecutive interior
b. ∠9 and ∠11 alternate interior
c. ∠1 and ∠5 corresponding
d. ∠3 and ∠8 alternate exterior
e. ∠7 and ∠12 alternate exterior
f. ∠4 and ∠8 corresponding

For Example 3

Refer to the figure showing three lines and the angles formed by these lines.

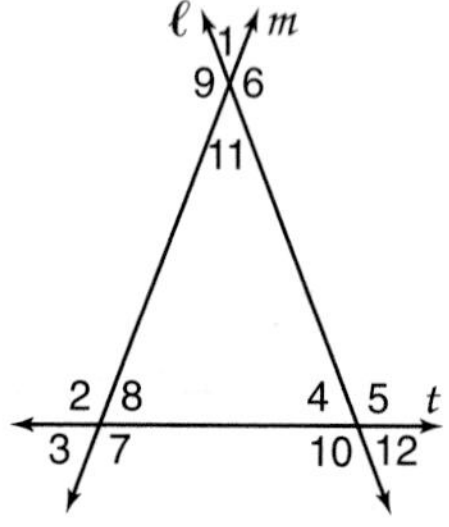

a. Identify the transversal to lines ℓ and m. t

b. Identify the special name given to each pair of angles.

∠7 and ∠12 corresponding

∠8 and ∠10 alternate interior

∠2 and ∠12 alternate exterior

Teaching Tip When defining *transversal*, note that the two lines that are intersected by the transversal do not need to be parallel.

A transversal can also be a ray or line segment.

In the photograph of the irrigated field, notice that line t intersects the rows of plants represented by lines ℓ and m. A line that intersects two or more lines in a plane at different points is called a **transversal**. The lines the transversal intersects need not be parallel.

When the transversal t intersects lines ℓ and m, eight angles are formed. These angles are given special names.

- **exterior angles** ∠1, ∠2, ∠7, ∠8
- **interior angles** ∠3, ∠4, ∠5, ∠6
- **consecutive interior angles** ∠3 and ∠5, ∠4 and ∠6
- **alternate exterior angles** ∠1 and ∠8, ∠2 and ∠7
- **alternate interior angles** ∠3 and ∠6, ∠4 and ∠5
- **corresponding angles** ∠1 and ∠5, ∠2 and ∠6, ∠3 and ∠7, ∠4 and ∠8

Consecutive interior angles are sometimes referred to as same-side interior angles.

Example 2

Refer to the figure below. Identify each pair of angles as *alternate interior, alternate exterior, corresponding,* or *consecutive interior* angles.

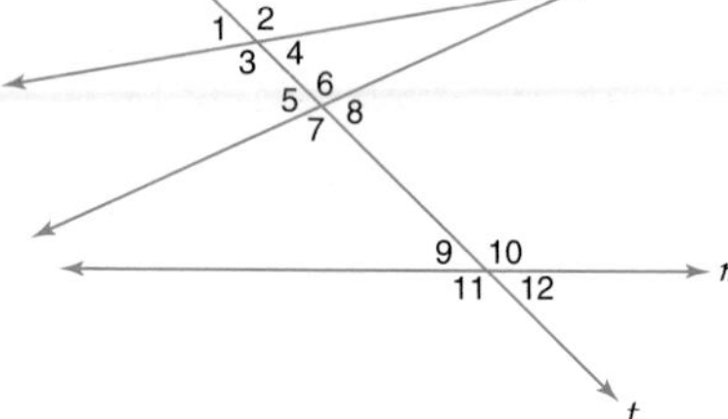

a. ∠1 and ∠8
alternate exterior

b. ∠7 and ∠10
alternate interior

c. ∠8 and ∠12
corresponding

d. ∠1 and ∠5
corresponding

e. ∠4 and ∠6
consecutive interior

f. ∠8 and ∠9
alternate interior

Example 3

Texas is the leading producer of livestock in the United States. The photograph shows a feed lot in Trent, Texas.

a. Identify all transversals to lines ℓ and m in the photo.

r, s, t, w

b. Identify the special name given to each pair of angles.

∠2 and ∠10	alternate interior angles
∠19 and ∠12	consecutive interior angles
∠15 and ∠17	corresponding angles

CHECK FOR UNDERSTANDING

Communicating Mathematics

Study the lesson. Then complete the following.

1. The word *parallel* comes from the Greek *parallelos* derived from *para*, beside, and *allelon*, of one another. Do you think these words describe parallel lines? Explain your reasoning.
2. **You Decide** Refer to the figure at the right. Jessica claims that ∠2 and ∠9 are corresponding angles. Enrique disagrees, stating that ∠2 and ∠6 are corresponding angles. Who is correct? Explain your reasoning.
3. **Write** a description of each figure using terms described in this lesson. a–c. See margin.

a. 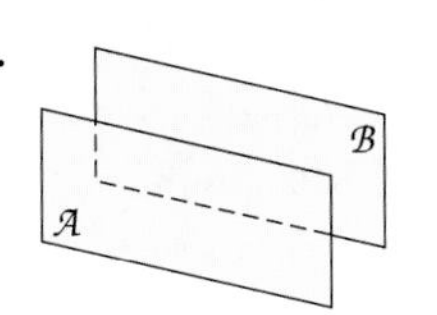

b. 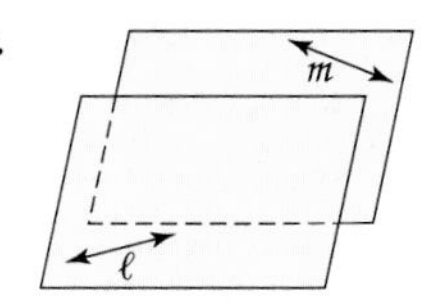

c. 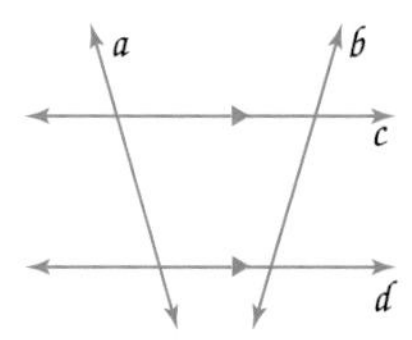

4. Refer to the drawing technique illustrated in the Modeling Mathematics activity on page 125. Draw a prism with triangular bases. See margin.

1. Sample answer: The definition describes parallel lines, because parallel lines are always beside each other and never intersect each other.

2. Both are correct depending on which line is the transversal.

Guided Practice

Describe each of the following as *intersecting, parallel,* or *skew.*

5. railroad crossing sign intersecting
6. opposite sides of a cereal box parallel

Determine whether each statement is *true* or *false*. Explain your reasoning.

7. A line that intersects two skew lines is a transversal.
8. If two lines do not intersect, they are parallel. False; the lines could be skew lines.

State the transversal that forms each pair of angles. Then identify the special name for the angle pair.

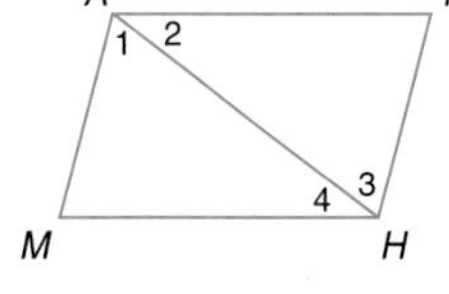

9. ∠1 and ∠3 $\overleftrightarrow{AH}$; alternate interior
10. ∠*ATH* and ∠*MAT* $\overleftrightarrow{AT}$; consecutive interior

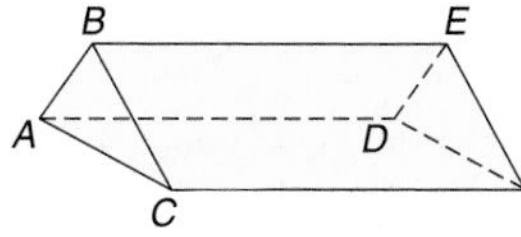

Name each of the following from the figure at the left.

11. all pairs of parallel planes
12. all pairs of skew lines

Draw a diagram to illustrate each of the following.

13. two parallel lines with two nonparallel transversals See margin.
14. two intersecting planes See margin.
15. **Air Traffic Control** For safety, airplanes heading eastbound are assigned an altitude level that is an odd number of thousands of feet and airplanes heading westbound are assigned an altitude level that is an even number of thousands of feet. If you are in an airplane flying northwest at 32,000 feet and your best friend is in an airplane flying east at 23,000 feet, describe the type of lines formed by the paths of the two planes. Explain your reasoning.

7. False; a transversal intersects 2 lines in a plane.

11. plane *ABC* and plane *DEF*

12. $\overleftrightarrow{AC}$ and $\overleftrightarrow{BE}$, $\overleftrightarrow{AC}$ and $\overleftrightarrow{DE}$, $\overleftrightarrow{AC}$ and $\overleftrightarrow{EF}$, $\overleftrightarrow{AB}$ and $\overleftrightarrow{CF}$, $\overleftrightarrow{AB}$ and $\overleftrightarrow{EF}$, $\overleftrightarrow{AB}$ and $\overleftrightarrow{DF}$, $\overleftrightarrow{BC}$ and $\overleftrightarrow{AD}$, $\overleftrightarrow{BC}$ and $\overleftrightarrow{DE}$, $\overleftrightarrow{BC}$ and $\overleftrightarrow{DF}$, $\overleftrightarrow{AD}$ and $\overleftrightarrow{EF}$, $\overleftrightarrow{CF}$ and $\overleftrightarrow{DE}$, $\overleftrightarrow{BE}$ and $\overleftrightarrow{DF}$

15. Skew lines; the planes are flying in different directions and at different altitudes.

3 PRACTICE/APPLY

Check for Understanding

Exercises 1–15 are designed to help you assess your students' understanding through reading, writing, speaking, and modeling. You should work through Exercises 1–4 with your students and then monitor their work on Exercises 5–15.

Additional Answers

3a. 2 parallel planes $\mathcal{A}$ and $\mathcal{B}$

3b. 2 skew lines ℓ and *m* in 2 parallel planes

3c. 2 parallel lines *c* and *d* and 2 non-parallel transversals *a* and *b*

4. 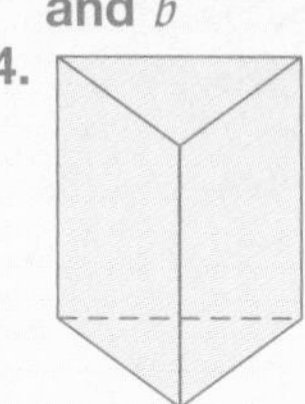

13. 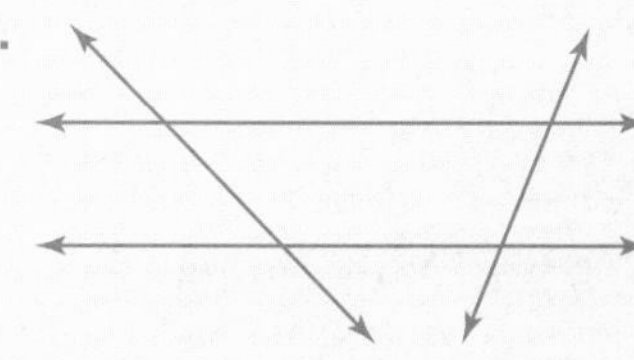

Study Guide Masters, p. 14

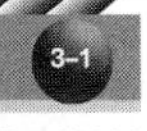

NAME ________ DATE ________

Study Guide

Student Edition Pages 124–129

Parallel Lines and Transversals

When planes do not intersect, they are said to be **parallel**. Also, when lines in the same plane do not intersect, they are parallel. But when lines are not in the same plane and do not intersect, they are **skew**. A line that intersects two or more lines in a plane at different points is called a **transversal**. Eight angles are formed by a transversal and two lines. These angles and pairs of them have special names.

Example: Planes *PQR* and *NOM* are parallel.
Segments *MO* and *RQ* are parallel.
Segments *MN* and *RQ* are skew.

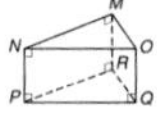

Example: Interior angles: ∠1, ∠2, ∠5, ∠6
Alternate interior angles: ∠1 and ∠6, ∠2 and ∠5
Consecutive interior angles: ∠1 and ∠5, ∠2 and ∠6
Exterior angles: ∠3, ∠4, ∠7, ∠8
Alternate exterior angles: ∠3 and ∠7, ∠4 and ∠8
Corresponding angles: ∠1 and ∠7, ∠2 and ∠8, ∠3 and ∠6, ∠4 and ∠5

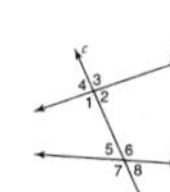

Refer to the figure in the first example.

1. Name two more pairs of parallel segments. $\overline{NM}$ and $\overline{PR}$, $\overline{NO}$ and $\overline{PQ}$, $\overline{NP}$ and $\overline{MR}$, $\overline{NP}$ and $\overline{OQ}$, $\overline{MR}$ and $\overline{OQ}$
2. Name two more segments skew to $\overline{NM}$. $\overline{PQ}$, $\overline{OQ}$
3. Name two transversals for parallel lines $\overleftrightarrow{NO}$ and $\overleftrightarrow{PQ}$. $\overleftrightarrow{NP}$, $\overleftrightarrow{OQ}$
4. Name a segment that is parallel to plane *MRQ*. $\overline{NP}$

Identify the special name for each pair of angles in the figure.

5. ∠2 and ∠6 consecutive interior angles
6. ∠4 and ∠8 alternate exterior angles
7. ∠4 and ∠5 corresponding angles
8. ∠2 and ∠5 alternate interior angles
9. Draw a diagram to illustrate two parallel planes with a line intersecting the planes.

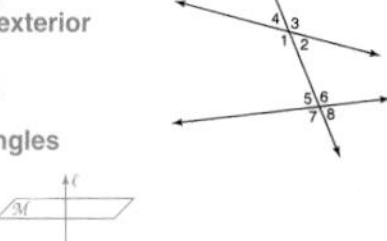

Reteaching

Using Diagrams Draw two lines cut by a transversal on the chalkboard or overhead. Have students name the relationships among the angles formed by the lines and the transversal.

Additional Answer

14. 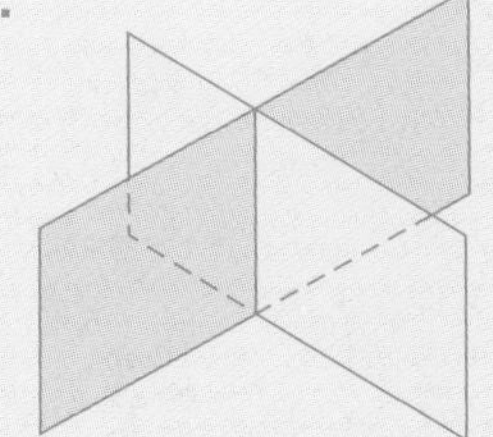

Assignment Guide

Core (with proof): 17–49 odd, 50–60
Core (informal): 17–49 odd, 52–60
Enriched: 16–44 even, 45–60

For **Extra Practice**, see p. 768.

The red A, B, and C flags, printed only in the Teacher's Wraparound Edition, indicate the level of difficulty of the exercises.

Additional Answers

22. m intersects 2 lines r and s in a plane.
23. The angles are not formed by 2 lines and a transversal.
24. The angles are exterior angles and on opposite sides of the transversal s.
25. The angles are in corresponding positions when the transversal m intersects r and s.
26. The angles are interior angles and on opposite sides of transversal ℓ.
27. The angles are formed by lines ℓ and m and transversal s.

Practice Masters, p. 14

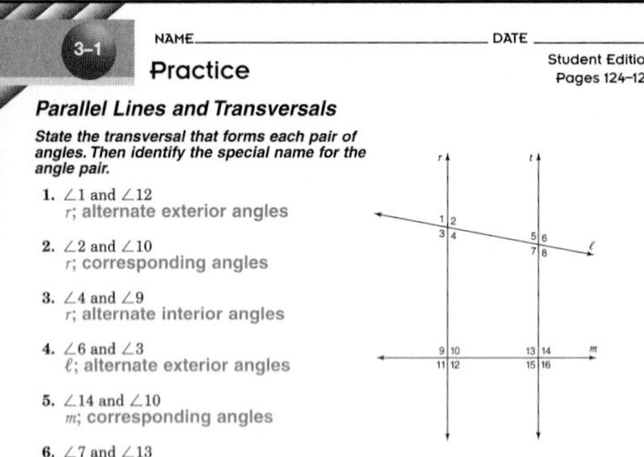

3-1 NAME ______________________ DATE __________

Practice

Student Edition Pages 124–129

Parallel Lines and Transversals

State the transversal that forms each pair of angles. Then identify the special name for the angle pair.

1. $\angle 1$ and $\angle 12$
   r; alternate exterior angles
2. $\angle 2$ and $\angle 10$
   r; corresponding angles
3. $\angle 4$ and $\angle 9$
   r; alternate interior angles
4. $\angle 6$ and $\angle 3$
   ℓ; alternate exterior angles
5. $\angle 14$ and $\angle 10$
   m; corresponding angles
6. $\angle 7$ and $\angle 13$
   t; consecutive interior angles

The three-dimensional figure shown at the right is called a right pentagonal prism.

7. Identify all segments joining points marked in plane JIH that appear to be skew to $\overline{EB}$.
   $\overline{FG}, \overline{FJ}, \overline{FH}, \overline{FI}, \overline{GH}, \overline{GI}, \overline{HJ}, \overline{IJ}$
8. Which segments seem parallel to $\overline{BG}$?
   $\overline{AF}, \overline{EJ}, \overline{DI}, \overline{CH}$
9. Which segments seem parallel to $\overline{GH}$? $\overline{BC}$
10. Identify all planes that appear parallel to plane FGH. plane ABC
11. **Draw a Diagram** At a town's bicentennial celebration, men dressed up as settlers and tipped their hats whenever they met another man. At a town meeting, ten men were present. How many times were pairs of hats tipped as two men met for the first time? 45

EXERCISES

Practice

Describe each of the following as *intersecting, parallel,* or *skew.*

A

16. yard lines on a football field
17. the sides of the Great Pyramid
18. ceiling and wall of a room
19. shelves of a bookcase parallel
20. a flag pole in a park and a road that runs along the edge of the park
21. the service lines on a tennis court parallel

16. parallel
17. intersecting
18. intersecting
20. skew

Determine whether each statement is *true* or *false*. Explain your reasoning. 22–27. See margin for reasons.

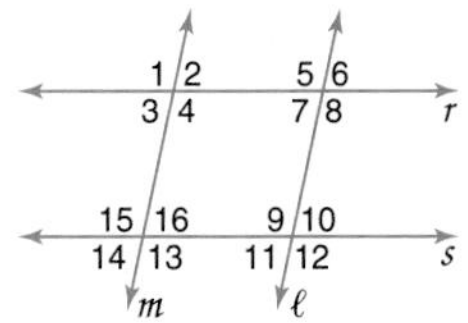

22. Line m is a transversal for lines r and s. true
23. $\angle 4$ and $\angle 9$ are consecutive interior angles. false
24. $\angle 14$ and $\angle 10$ are alternate exterior angles. true
25. $\angle 2$ and $\angle 16$ are corresponding angles. true
26. $\angle 7$ and $\angle 10$ are alternate interior angles. true
27. $\angle 13$ and $\angle 11$ are formed by lines ℓ and m and transversal r. false

State the transversal that forms each pair of angles. Then identify the special name for the angle pair.

B

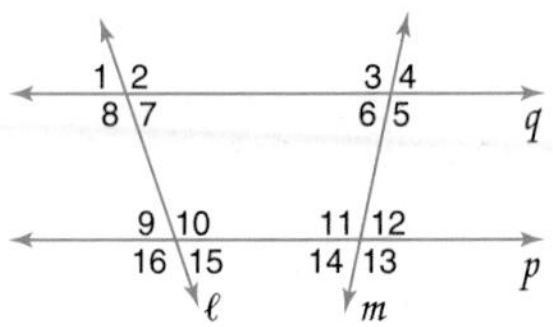

28. $\angle 6$ and $\angle 7$ q; consecutive interior
29. $\angle 16$ and $\angle 2$ ℓ; alternate exterior
30. $\angle 13$ and $\angle 5$ m; corresponding
31. $\angle 8$ and $\angle 10$ ℓ; alternate interior
32. $\angle 11$ and $\angle 15$ p; alternate interior
33. $\angle 4$ and $\angle 8$ q; alternate exterior

Name each of the following from the figure at the right.

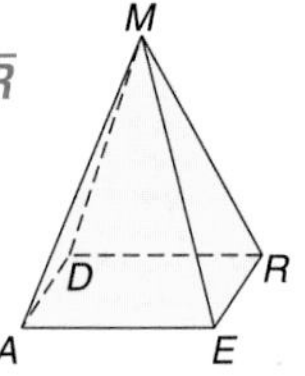

34. all pairs of intersecting planes See margin.
35. all pairs of parallel segments $\overline{AE}$ and $\overline{DR}$, $\overline{AD}$ and $\overline{ER}$
36. all pairs of skew segments See margin.
37. all pairs of parallel planes none
38. all points contained in four lines M
39. all planes intersecting with plane ADM

39. plane ADR, plane DRM, plane ERM, plane AEM

Draw a diagram to illustrate each of the following.

40. two parallel planes
41. two parallel planes containing two lines that are skew
42. three parallel planes with a line intersecting the planes
43. two parallel lines with a plane intersecting the lines

40–43. See Solutions Manual.

C

44. **Draw a Diagram** Use the technique in the Modeling Mathematics activity on page 125 to draw a prism with bases that are hexagons (6-sided figures).
    a. The figure at the right may show a different perspective than the prism you drew. Explain the difference between the "views" of the two prisms.
    b. Identify all segments skew to $\overline{TU}$.
    c. Identify all planes parallel to plane AFW. plane GDY

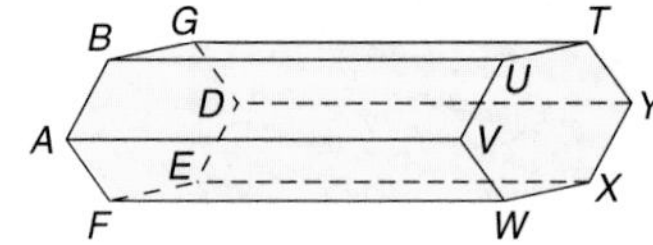

44a. Sample answer: One is standing up on one of its hexagon bases. The other is on its side.

44b. $\overline{DY}, \overline{EX}, \overline{FW}, \overline{AV}, \overline{AB}, \overline{GD}, \overline{DE}, \overline{AF}$

Additional Answers

34. plane ADM and plane DRM, plane ADM and plane AEM, plane ADM and plane ERM, plane ADM and plane ADR, plane DRM and plane ERM, plane DRM and plane AEM, plane DRM and plane ADR, plane ERM and plane AEM, plane ERM and plane ADR, plane AEM and plane ADR

36. $\overline{AD}$ and $\overline{EM}$, $\overline{AD}$ and $\overline{MR}$, $\overline{AE}$ and $\overline{DM}$, $\overline{AE}$ and $\overline{MR}$, $\overline{ER}$ and $\overline{AM}$, $\overline{ER}$ and $\overline{DM}$, $\overline{DR}$ and $\overline{AM}$, $\overline{DR}$ and $\overline{EM}$

Critical Thinking

45a. $m\angle 1 = m\angle 2 = 160$

45b. The measures of alternate interior angles are equal.

47. Sample answer: parallel circuits in electronics, parallel story lines in literature

45. Optical Illusion In the artist's drawing, lines ℓ and m are parallel, but appear to be bowed, due to the many transversals drawn through ℓ and m.

a. Use a protractor to measure $\angle 1$ and $\angle 2$.

b. Notice that $\angle 1$ and $\angle 2$ are a pair of alternate interior angles and make a conjecture about the relationship between the alternate interior angles if lines ℓ and m are parallel.

46. Suppose there is a line m and a point A not on the line. In space, how many lines can be drawn through A that do not intersect m? In space, how many lines can be drawn through A that are parallel to m? **infinite number; 1**

47. Music The word *parallel* is used in music to describe songs moving consistently by the same intervals such as harmony with parallel voices. Find at least two additional uses of the word *parallel* in other school subjects such as history, electronics, computer science, or English.

48. Graphics Locate an example of parallel lines with a transversal in a newspaper or magazine. Make a copy of the picture or diagram and label the lines and angles. **See students' work.**

49. Draw a Diagram Square dancing involves four couples. If each member of the square shakes hands with every other member of the square except his or her partner before the dance begins, what is the total number of handshakes? **See margin.**

Mixed Review

50. Write a two-column proof. (Lesson 2–5) **See margin.**

Given: $\overline{AB} \cong \overline{FE}$, $\overline{BC} \cong \overline{ED}$

Prove: $\overline{AC} \cong \overline{FD}$

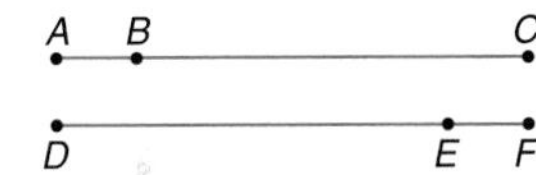

51. Write the given and prove statements that you would use to prove the statement *Opposite angles of a parallelogram are congruent.* Draw a figure. (Lesson 2–5) **See margin.**

52. Name the property of equality that justifies the statement *If $m\angle A = m\angle B$, then $m\angle B = m\angle A$.* (Lesson 2–4) **Symmetric Property (=)**

53. If something is a cloud, then it is composed of millions of water droplets.

53. Science Write *A cloud is composed of millions of water droplets* in if-then form. (Lesson 2–2)

54. The measure of an angle is $9x + 14$, and the measure of its supplement is $12x + 19$. Find the value of x. (Lesson 1–7) **7**

55. Aviation An airplane must fly from flight coordinates (7, 36) to (17, 0). What ordered pair describes the plane's position when it has completed half its trip? (Lesson 1–5) **(12, 18)**

56. T is between R and S. If $TS = 7$ and $RS = 20$, find RT. (Lesson 1–4) **13**

57. Explore the problem; plan the solution; solve the problem; and examine the solution.

57. Explain the four steps for solving any problem. (Lesson 1–3)

58. Graph points $A(-4, 7)$ and $B(3, -5)$ on the same coordinate plane. Name the quadrants in which each point lies. (Lesson 1–1) **See Solutions Manual.**

Algebra

59. Name the coefficient of p^2 in $\frac{2p^2}{5}$. $\frac{2}{5}$

60. State the additive inverse and absolute value of -16. **16; 16**

4 ASSESS

Closing Activity

Speaking Draw a picture of two lines cut by a transversal on the chalkboard or overhead and label the angles formed. Go around the room and have students name two angles and any relationships they have in the figure.

Additional Answers

49.

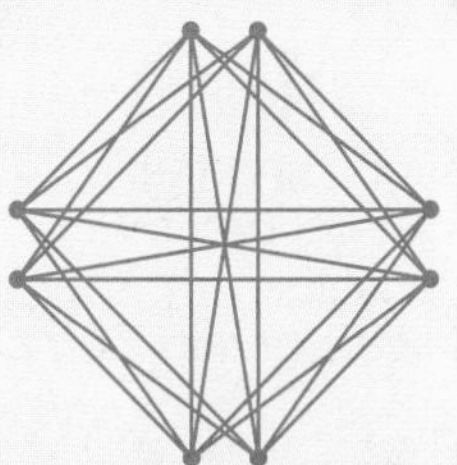

24 handshakes

50. Statements (Reasons)

1. $\overline{AB} \cong \overline{FE}$, $\overline{BC} \cong \overline{ED}$ (Given)
2. $AB = EF$, $BC = ED$ (Def. $\cong$ segments)
3. $AB + BC = FE + ED$ (Add. Prop. (=))
4. $AC = AB + BC$, $FD = FE + ED$ (Seg. Add. Post.)
5. $AC = FD$ (Subst. Prop. (=))
6. $\overline{AC} \cong \overline{FD}$ (Def. $\cong$ segments)

Enrichment Masters, p. 14

NAME________ DATE________

Enrichment

Student Edition Pages 124–129

Perspective Drawings

To draw three-dimensional objects, artists make **perspective drawings** such as the ones shown below. To indicate depth in a perspective drawing, some parallel lines are drawn as converging lines. The dotted lines in the figures below each extend to a **vanishing point**, or spot where parallel lines appear to meet.

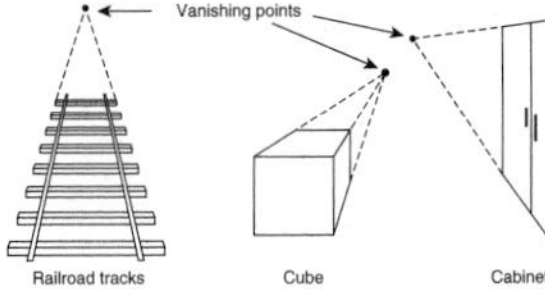

Draw lines to locate the vanishing point in each drawing of a box.

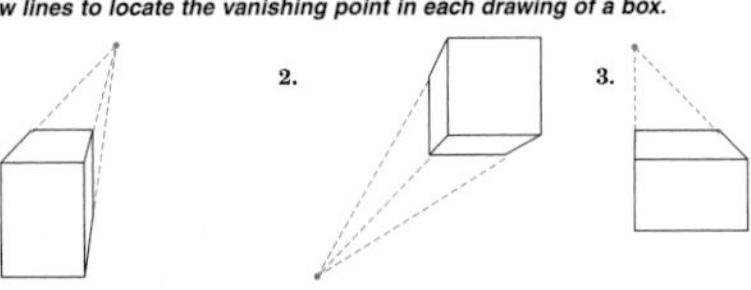

4. The fronts of two cubes are shown below. Using point P as the vanishing point for both cubes, complete the perspective drawings of the cubes.

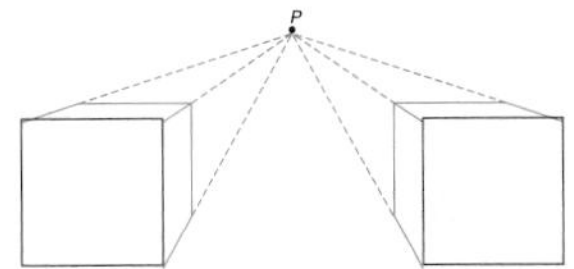

5. Find an example of a perspective drawing in a newspaper or magazine. Trace the drawing and locate a vanishing point. See students' work.

Extension

Reasoning Ask students if two lines can intersect at two points. Can they intersect at three points? Discuss the intersection of lines and point out that lines will intersect in either none, one, or an infinite number of points. Have students draw an example of each case.

Additional Answer

51. Given: $PARL$ is a parallelogram.
Prove: $\angle 1 \cong \angle 2$
$\angle 3 \cong \angle 4$

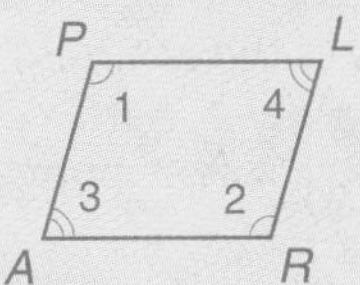

3-2A LESSON NOTES

NCTM Standards: 1–4, 7

Objective

Use a TI-92 to investigate the measures of the angles formed by two parallel lines and a transversal.

Recommended Time

20 minutes

Instructional Resources

Instructions for using the *Geometer's Sketchpad* for this activity are available in the *Graphing Calculator and Computer Masters*, p. 19.

1 FOCUS

Motivating the Lesson

Have half the class draw the lines on paper and half on the calculator. Which group was quicker? Which group was more accurate? Discuss advantages of using the calculator to draw parallel lines.

2 TEACH

Teaching Tip Have students try to prove their conjectures in Exercise 3.

3 PRACTICE/APPLY

Assignment Guide

Core (with proof): 1–5
Core (informal): 1–5
Enriched: 1–5

4 ASSESS

Observing students working with technology is an excellent method of assessment.

3–2A Using Technology
Angles and Parallel Lines

A Preview of Lesson 3–2

You can use a TI-92 calculator to investigate the measures of the angles formed by two parallel lines and a transversal.

- First clear your screen by using the 8:Clear All on the F8 menu.
- Draw any line on your screen. Label two points A and B on the line.
- You can draw a line parallel to $\overleftrightarrow{AB}$ by selecting 2:Parallel Line from the F4 menu. Move the cursor onto AB and the message "PARALLEL TO THIS LINE" appears. Press ENTER. Move the cursor to a new location and press ENTER. The parallel line will appear. Label two points C and D on the second line.
- Draw any line GH intersecting the two parallel lines and label the intersection points as E and F.

The screen at the right is a sample of two parallel lines cut by a transversal.

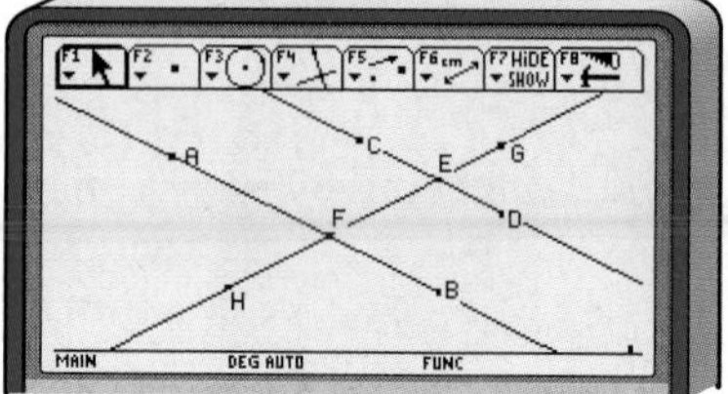

- Measure each of the angles by selecting 3:Angle on the F6 menu. Remember that you need to select three points on the angle to define it in order for the calculator to measure the angle.

EXERCISES

Analyze your drawing.

1. List pairs of angles by the special names you learned in Lesson 3–1. **1. See margin.**
2. Which pairs of angles listed in Exercise 1 are congruent? **2. $\angle GEC \cong \angle DEF \cong \angle EFA \cong \angle HFB$; $\angle GED \cong \angle CEF \cong \angle EFB \cong \angle HFA$**
3. Make a conjecture about the following pairs of angles formed by two parallel lines and a transversal.
   a. corresponding angles **Corresponding angles are congruent.**
   b. alternate interior angles **Alternate interior angles are congruent.**
   c. alternate exterior angles **Alternate exterior angles are congruent.**

Test your conjectures. **4. yes**

4. Rotate the transversal. Are the angles with equal measures in the same relative location as the angles with equal measures in the original drawing?
5. Rotate the transversal so that the measure of one of the angles is 90.
   a. What do you notice about the measures of the other angles? **5a. The measures of each of the eight angles is 90.**
   b. Make a conjecture about a transversal that is perpendicular to one of two parallel lines. **5b. A transversal that is perpendicular to one of two parallel lines is perpendicular to the other line.**

Using Technology

This lesson offers an excellent opportunity for using technology in your geometry classroom. For more information on using technology, see *Graphing Calculators in the Mathematics Classroom*, one of the titles in the Glencoe Mathematics Professional Series.

Additional Answer

1. consecutive interior angles: $\angle CEF$ and $\angle AFE$, $\angle DEF$ and $\angle BFE$
alternate exterior angles: $\angle CEG$ and $\angle BFH$, $\angle DEG$ and $\angle AFH$
alternate interior angles: $\angle CEF$ and $\angle BFE$, $\angle DEF$ and $\angle AFE$
corresponding angles: $\angle CEG$ and $\angle AFE$, $\angle DEG$ and $\angle BFE$, $\angle CEF$ and $\angle AFH$, $\angle DEF$ and $\angle BFH$

3-2 Angles and Parallel Lines

What YOU'LL LEARN

- To use the properties of parallel lines to determine angle measures.

Why IT'S IMPORTANT

You can apply properties of parallel lines in construction and interior decorating.

FYI

The New York Yankees have won 23 World Series—more than any other team.

APPLICATION Graphic Art

In the graph at the right, examine the method used by the artist to illustrate the number of teams making the playoffs. The parallel line segments illustrate the total number of teams, and a *transversal* is used to indicate the number of teams missing the playoffs.

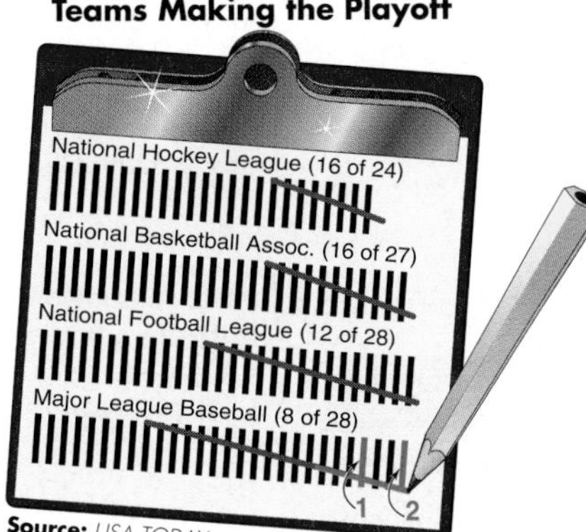

In the Major League Baseball display, ∠1 and ∠2 are a pair of *corresponding angles*. What is the relationship between a pair of corresponding angles formed by two parallel lines cut by a transversal?

MODELING MATHEMATICS Corresponding Angle Measures

Materials:  notebook paper  protractor 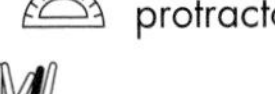 straightedge colored pencils

You can model parallel lines by using the lines printed on notebook paper.

- Use a pencil and straightedge to darken two lines on a piece of notebook paper. Use your straightedge to draw transversal t.

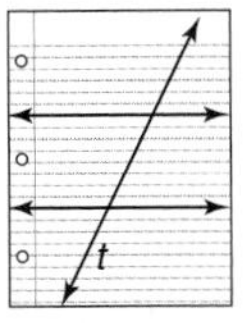

- Label each angle.

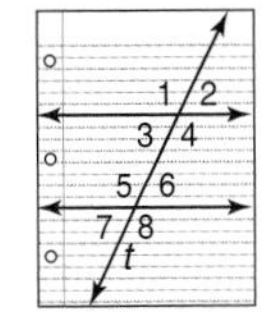

Your Turn

a. Use your protractor to measure each of the four pairs of corresponding angles. **See students' work.**

b. Make a conjecture about the corresponding angles formed by two parallel lines cut by a transversal. **The measures are equal.**

c. What appears to be true about alternate interior angles? consecutive interior angles? alternate exterior angles?

c. Measures are equal; sum of the measures is 180; measures are equal.

FYI

The New York Yankees became an American League franchise in 1901. Their 23rd World Series win was in 1996.

MODELING MATHEMATICS Remind students that their measurements are only approximations.

3-2 LESSON NOTES

NCTM Standards: 1–4, 7

Instructional Resources

- Study Guide Master 3-2
- Practice Master 3-2
- Enrichment Master 3-2
- Assessment and Evaluation Masters, p. 72
- Tech Prep Applications Masters, p. 5

Transparency 3-2A contains the 5-Minute Check for this lesson; **Transparency 3-2B** contains a teaching aid for this lesson.

Recommended Pacing	
Standard Pacing	Days 4 & 5 of 14
Honors Pacing	Days 3 & 4 of 13
Block Scheduling*	Day 2 of 7

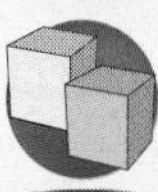

*For more information on pacing and possible lesson plans, refer to the *Block Scheduling Booklet*.

1 FOCUS

5-Minute Check *(over Lesson 3-1)*

Refer to the figure below to identify the special name for the given angle pair.

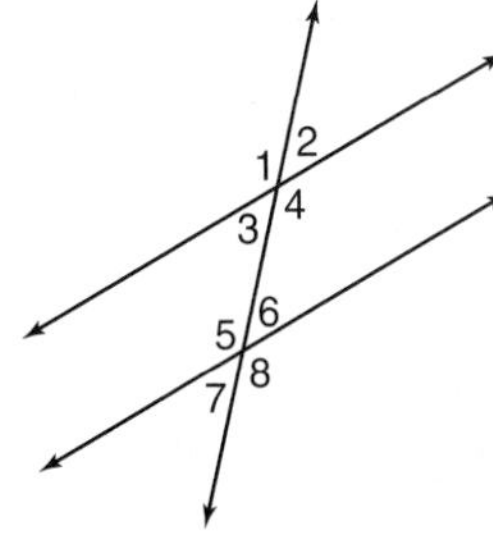

1. ∠1 and ∠8 **alternate exterior**
2. ∠4 and ∠5 **alternate interior**

(continued on the next page)

Draw a figure to illustrate each of the following.

3. two perpendicular planes

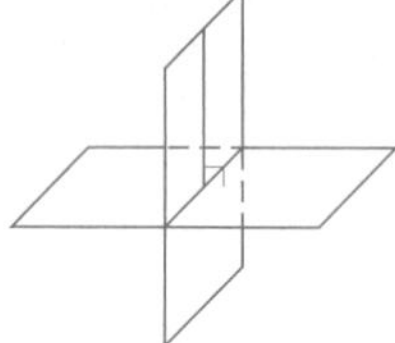

4. two parallel lines and a skew line

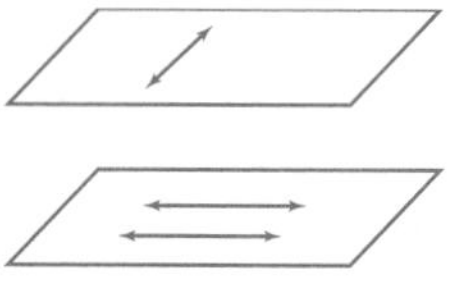

Motivating the Lesson

Hands-On Activity Have students draw two parallel lines and draw four or five transversals, each perpendicular to the parallel lines. They should then measure the perpendicular distance between the parallel lines.

2 TEACH

In-Class Example

For Example 1
Use the figure in Example 1 to find $m\angle 1$ if $m\angle 6 = 110$.
$m\angle 1 = 70$

Teaching Tip In Example 1 and in Theorem 3-3, you could first state that the vertical angles are congruent, use the corresponding angles theorem, and then use the transitive property.

Teaching Tip When discussing Theorem 3-3, it may help to rewrite the paragraph proof as a two-column proof.

The property relating corresponding angles shown in the Modeling Mathematics activity is accepted as a postulate.

Postulate 3-1 Corresponding Angles Postulate	**If two parallel lines are cut by a transversal, then each pair of corresponding angles is congruent.**

The postulate, combined with linear pair and vertical angle properties, helps to establish several angle relationships.

Example 1

The road displayed in the diagram divides a farm into two parts. The opposite edges of the field are parallel, and the road is a transversal. If $m\angle 6 = 115$, find $m\angle 3$.

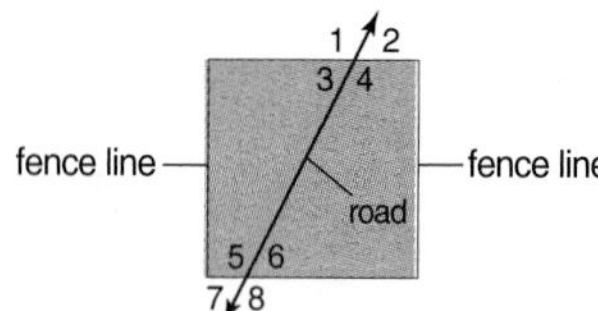

$m\angle 6 = m\angle 2$ *Corresponding Angle Postulate*

$m\angle 2 = m\angle 3$ *Vertical angles have equal measures.*

$m\angle 6 = m\angle 3$ *Transitive Property (=)*

$115 = m\angle 3$ *Substitution Property (=)*

Notice that in Example 1, ∠3 and ∠6 form a pair of alternate interior angles and were shown to have equal measures. This is an application of another of the special relationships between the angles formed by two parallel lines and a transversal. These relationships are summarized in Theorems 3–1, 3–2, and 3–3. *You will be asked to prove Theorems 3–1 and 3–2 in Exercises 44 and 45, respectively.*

Theorem 3-1 Alternate Interior Angles Theorem	**If two parallel lines are cut by a transversal, then each pair of alternate interior angles is congruent.**

Theorem 3-2 Consecutive Interior Angles Theorem	**If two parallel lines are cut by a transversal, then each pair of consecutive interior angles is supplementary.**

Theorem 3-3 Alternate Exterior Angles Theorem	**If two parallel lines are cut by a transversal, then each pair of alternate exterior angles is congruent.**

Alternative Learning Styles

Kinesthetic Have each student draw two parallel lines with a transversal. Have them label the angles from 1 to 8, using the same order as in the picture on page 133. Have them use a protractor to measure all eight angles. Record each student's measures on one big chart. Ask students to form conjectures about the angle relationships. **Corresponding, alternate interior, and alternate exterior angles are congruent. Consecutive interior angles are supplementary.**

Proof of Theorem 3-3

Given: $p \parallel q$

ℓ is a transversal of p and q.

Prove: $\angle 1 \cong \angle 8$, $\angle 2 \cong \angle 7$

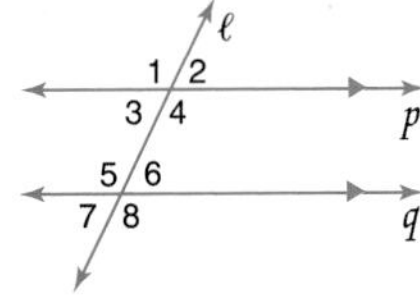

Paragraph Proof:

We are given that $p \parallel q$. If two parallel lines are cut by a transversal, corresponding angles are congruent. So, $\angle 1 \cong \angle 5$ and $\angle 2 \cong \angle 6$. $\angle 5 \cong \angle 8$ and $\angle 6 \cong \angle 7$ because vertical angles are congruent. Therefore, $\angle 1 \cong \angle 8$ and $\angle 2 \cong \angle 7$ since congruence of angles is transitive.

The special angle relationships between the angles formed by two parallel lines and a transversal can be used to solve for unknown values.

Example 2

Algebra

In the figure at the right, $\overleftrightarrow{MA} \parallel \overleftrightarrow{HT}$ and $\overleftrightarrow{NG} \parallel \overleftrightarrow{EL}$. Find the values of x, y, and z.

Find x. Since $\overleftrightarrow{MA} \parallel \overleftrightarrow{HT}$, $\angle MAT \cong \angle ATC$ by the Alternate Interior Angle Theorem.

$m\angle MAT = m\angle ATC$ *Def. ≅ ∠s*

$2x = 72$ *$m\angle MAT = 2x$, $m\angle ATC = 72$*

$x = 36$

Find z. Since $\overleftrightarrow{NG} \parallel \overleftrightarrow{EL}$, $\angle GMA$ and $\angle TAM$ are supplementary by the Consecutive Interior Angle Theorem.

$m\angle GMA + m\angle TAM = 180$ *Def. supplementary ∠s*

$4z + 2x = 180$ *$m\angle GMA = 4z$, $m\angle TAM = 2x$*

$4z + 2(36) = 180$ *Substitute 36 for x.*

$4z + 72 = 180$

$4z = 108$

$z = 27$

Find y. Since $\overleftrightarrow{NG} \parallel \overleftrightarrow{EL}$, $\angle GHO \cong \angle ATC$ by the Alternate Exterior Angle Theorem.

$m\angle GHO = m\angle ATC$ *Def. ≅ ∠s*

$5y + 2 = 72$ *$m\angle GHO = 5y + 2$, $m\angle ATC = 72$*

$5y = 70$

$y = 14$

Therefore, $x = 36$, $y = 14$, and $z = 27$.

There is a special relationship that occurs when one of two parallel lines is cut by a perpendicular line.

Theorem 3-4 Perpendicular Transversal Theorem

In a plane, if a line is perpendicular to one of two parallel lines, then it is perpendicular to the other.

You will be asked to prove this theorem in Exercise 43.

In-Class Example

For Example 2

In the figure, $\ell \parallel m$ and $c \parallel d$. Find the values of x, y, and z.

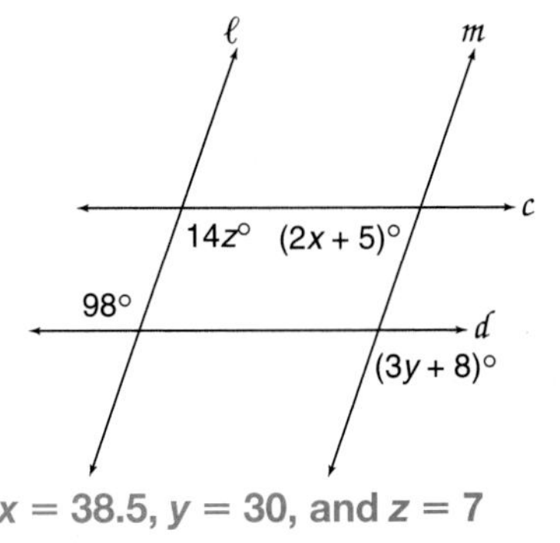

$x = 38.5$, $y = 30$, and $z = 7$

Proof Pointer

Your students may get confused as more and more postulates are introduced and explained. It may help to have them keep an ongoing list of all postulates and theorems given in the text, for an easy reference when writing proofs.

3 PRACTICE/APPLY

Check for Understanding

Exercises 1–16 are designed to help you assess your students' understanding through reading, writing, speaking, and modeling. You should work through Exercises 1–6 with your students and then monitor their work on Exercises 7–16.

Error Analysis

If students are having difficulty seeing the angle relationships in the figures, have them select an angle that is listed either in the given information or in the proof. Have them choose an angle relationship and see if it will help with the proof.

Additional Answers

2. Sample answers:
   Method 1: $m\angle 5 = m\angle 8$ because $\angle 5$ and $\angle 8$ are vertical angles. $m\angle 5 + m\angle 3 = 180$ because if 2 parallel lines are cut by a transversal, the consecutive interior angles are supplementary. Therefore, $130 + m\angle 3 = 180$ and $m\angle 3 = 50$.
   Method 2: $m\angle 1 = m\angle 8$ because if 2 parallel lines are cut by a transversal, then the alternate exterior angles are congruent. $\angle 1$ and $\angle 3$ are supplementary because $\angle 1$ and $\angle 3$ are a linear pair and if 2 angles form a linear pair, they are supplementary. So $m\angle 1 + m\angle 3 = 180$ by definition of supplementary. Therefore, $130 + m\angle 3 = 180$ and $m\angle 3 = 50$.
5. Since $\overline{SW} \parallel \overline{RK}$ and $\angle 4$ and $\angle 5$ are consecutive interior angles, $m\angle 4 + m\angle 5 = 180$. Substitute 110 for $m\angle 5$ and solve for $m\angle 4$. $m\angle 4 = 70$.

CHECK FOR UNDERSTANDING

Communicating Mathematics

1. If 2 parallel lines are cut by a transversal, consecutive interior angles are supplementary.

Study the lesson. Then complete the following.

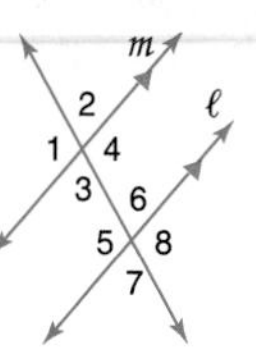

1. **Explain** why $\angle 3$ and $\angle 5$ must be supplementary.
2. **Describe** two different methods you could use to find $m\angle 3$ if $m\angle 8 = 130$. See margin.

Refer to the figure at the right for Exercises 3–5.

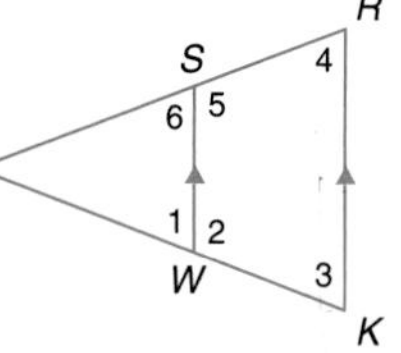

3. **Explain** what the arrowheads on the lines in the diagrams at the right indicate. The lines are parallel.
4. **Explain** why $\angle 1 \cong \angle 3$ and $\angle 6 \cong \angle 4$ if $\overline{SW} \parallel \overline{RK}$. Corresponding Angles Postulate
5. **Describe** how you could find $m\angle 4$ if $\overline{SW} \parallel \overline{RK}$ and $m\angle 5 = 110$. See margin.

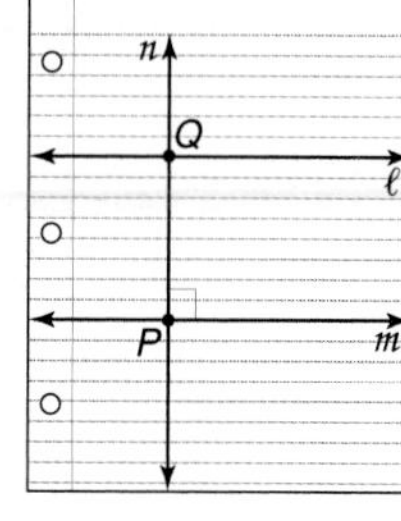

6. Draw and label parallel lines ℓ and m on notebook paper with point P on line m. Use a protractor to draw line n through P, perpendicular to line m. Label the point of intersection Q.
   a. Use your protractor to measure one of the four angles at Q. 90°
   b. Does the result verify the conclusion stated in the Perpendicular Transversal Theorem (Theorem 3–4)? Explain your reasoning. Yes; since the measure of the angle is 90, $n \perp \ell$.

Guided Practice

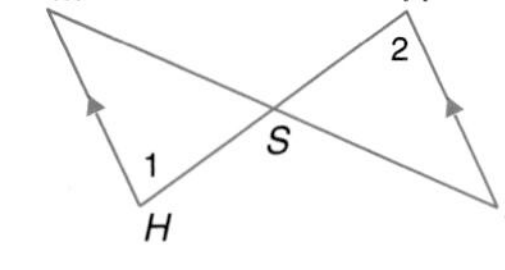

7. State the postulate or theorem that allows you to conclude that $\angle 1 \cong \angle 2$. Alternate Interior Angles Theorem

In the figure at the right, $p \parallel q$, $m\angle 1 = 107$, and $m\angle 11 = 48$. Find the measure of each angle.

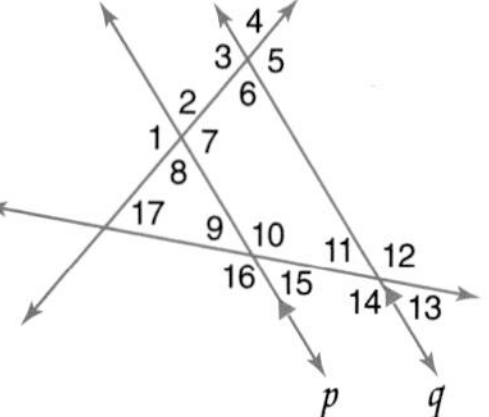

8. $\angle 3$ 107
9. $\angle 5$ 107
10. $\angle 13$ 48
11. $\angle 9$ 48
12. $\angle 15$ 48
13. $\angle 17$ 59

Find the values of x and y in each figure.

14.

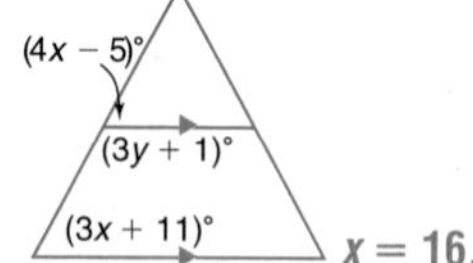

$x = 16$, $y = 40$

15.

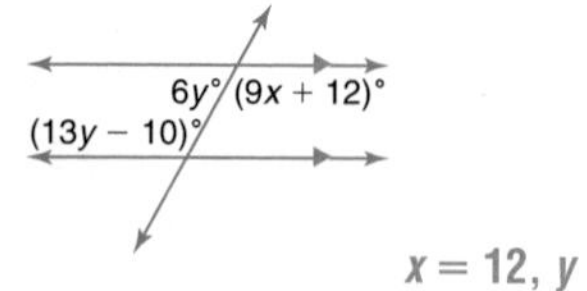

$x = 12$, $y = 10$

Reteaching

Using Modeling Use a geoboard and rubber bands to make two parallel lines cut by a transversal and number all of the angles formed. List all the corresponding angles that are congruent according to Postulate 3-1, the alternate interior angles that are congruent according to Theorem 3-1, the consecutive interior angles that are supplementary according to Theorem 3-2, and the alternate exterior angles that are congruent according to Theorem 3-3.

16. **Construction** A carpenter is building a flight of stairs. If the tops of the stairs are parallel to the floor and the stringer makes a 25° angle with the floor, find the measure of the angle formed by the tops of the steps and the stringer. 25

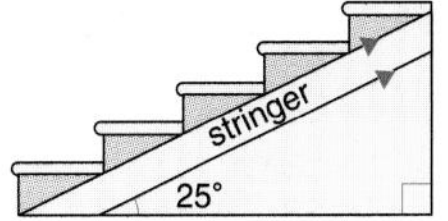

EXERCISES

Practice

In Exercises 17–19, state the postulate or theorem that allows you to conclude that $\angle 1 \cong \angle 2$. 19. Corresponding Angles Postulate

17. Alternate Interior Angles Theorem

18. Alternate Exterior Angles Theorem

17.
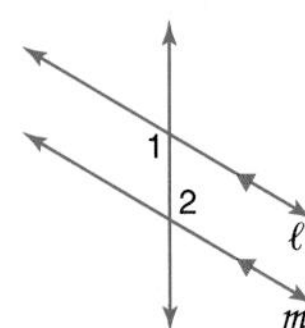

18.
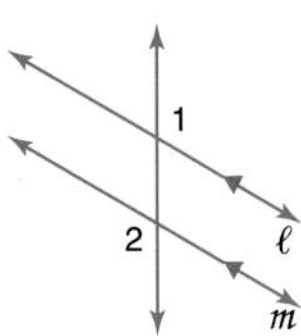

19.
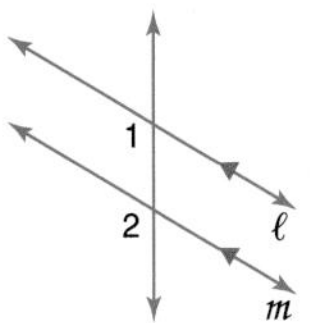

In the figure, $x \parallel y$, $\overline{ST} \parallel \overline{RQ}$, and $m\angle 1 = 131$. Find the measure of each angle.

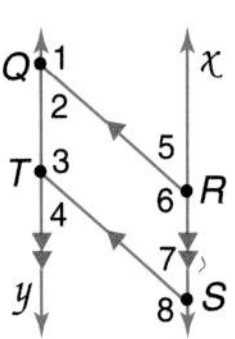

20. $\angle 6$ 131
21. $\angle 7$ 49
22. $\angle 4$ 49
23. $\angle 2$ 49
24. $\angle 5$ 49
25. $\angle 8$ 131

In the figure, $\overline{AB} \parallel \overline{EC}$, $m\angle 1 = 58$, $m\angle 2 = 47$, and $m\angle 3 = 26$. Find the measure of each angle.

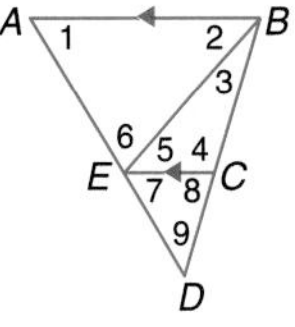

26. $\angle 7$ 58
27. $\angle 5$ 47
28. $\angle 6$ 75
29. $\angle 4$ 107
30. $\angle 8$ 73
31. $\angle 9$ 49

In the figure $\overline{BG} \parallel \overline{CE}$, $\overline{BE} \parallel \overline{CD}$, $\overrightarrow{BG}$ bisects $\angle EBA$, $m\angle 8 = 42$, and $m\angle 3 = 18$. Find the measure of each angle.

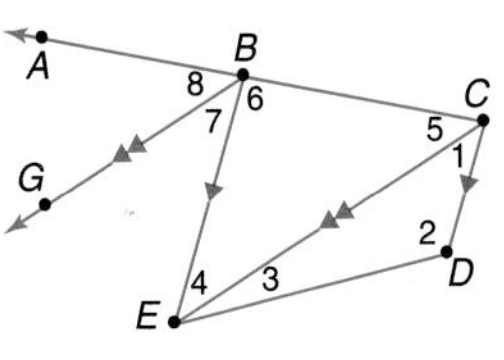

32. $\angle 7$ 42
33. $\angle 5$ 42
34. $\angle 1$ 42
35. $\angle 4$ 42
36. $\angle 6$ 96
37. $\angle 2$ 120

Find the values of x, y, and z in each figure.

38. $x = 34$, $y = 15$, $z = 142$

39. $x = 90$, $y = 15$, $z = 13.5$

40. $x = 14$, $y = 11$, $z = 73$

38.
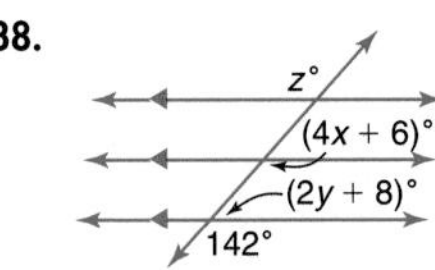

39.
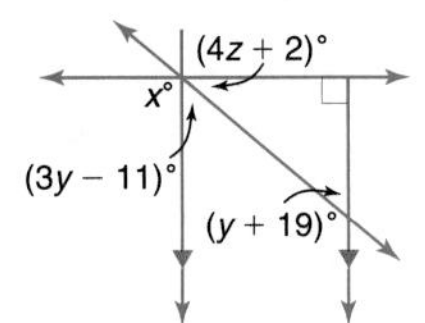

40.
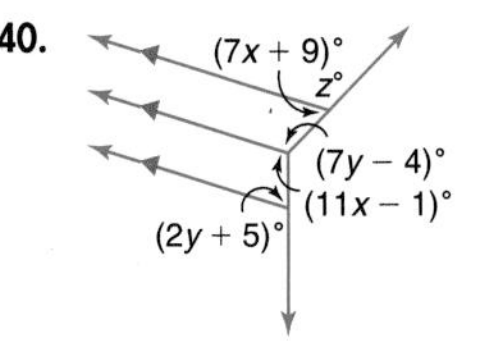

Assignment Guide

Core (with proof): 17–51 odd, 53–63
Core (informal): 17–41 odd, 49, 51, 53–55, 57–63
Enriched: 18–48 even, 49–63

For **Extra Practice**, see p. 768.

The red A, B, and C flags, printed only in the Teacher's Wraparound Edition, indicate the level of difficulty of the exercises.

Study Guide Masters, p. 15

3-2 NAME______ DATE______

Study Guide

Student Edition Pages 131–137

Angles and Parallel Lines

If two parallel lines are cut by a transversal, then the following pairs of angles are congruent.

corresponding angles | alternate interior angles | alternate exterior angles

If two parallel lines are cut by a transversal, then consecutive interior angles are supplementary.

Example: In the figure $m \parallel n$ and p is a transversal. If $m\angle 2 = 35$, find the measures of the remaining angles.

Since $m\angle 2 = 35$, $m\angle 8 = 35$ (corresponding angles).
Since $m\angle 2 = 35$, $m\angle 6 = 35$ (alternate interior angles).
Since $m\angle 8 = 35$, $m\angle 4 = 35$ (alternate exterior angles).

$m\angle 2 + m\angle 5 = 180$. Since consecutive interior angles are supplementary, $m\angle 5 = 145$, which implies that $m\angle 3$, $m\angle 7$, and $m\angle 1$ equal 145.

In the figure at the right $p \parallel q$, $m\angle 1 = 78$, and $m\angle 2 = 47$. Find the measure of each angle.

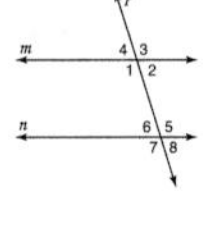

1. $\angle 3$ 102
2. $\angle 4$ 102
3. $\angle 5$ 78
4. $\angle 6$ 133
5. $\angle 7$ 47
6. $\angle 8$ 47
7. $\angle 9$ 133

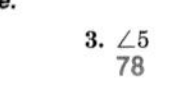
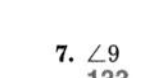

Find the values of x and y in each figure.

8.
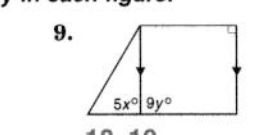

21, 60

9. 18, 10

10.
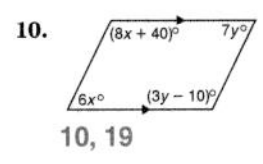

10, 19

Find the values of x, y and z in each figure.

11.
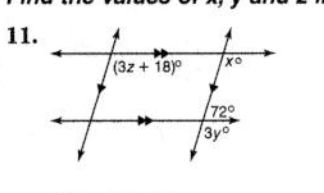

108, 36, 30

12.
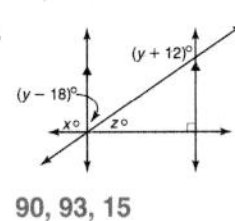

90, 93, 15

Additional Answers

41. Since ∠4 and ∠8 are alternate exterior angles formed when 2 parallel lines are cut by a transversal, $m\angle 4 = m\angle 8$ or $2x - 25 = x + 26$. Solving this equation, $x = 51$. Therefore, $m\angle 8 = 51 + 26$ or 77. Since ∠2 and ∠8 are vertical angles, $m\angle 2 = m\angle 8$ or $m\angle 2 = 77$.
42. Since ∠6 and ∠7 are consecutive interior angles formed when 2 parallel lines are cut by a transversal, $m\angle 6 + m\angle 7 = 180$ or $2x + 43 + 5x + 11 = 180$. Solving this equation, $x = 18$. Therefore, $m\angle 7 = 5(18) + 11$ or 101. Since ∠5 and ∠7 are corresponding angles formed when 2 parallel lines are cut by a transversal, $m\angle 5 = 101$.
47. Given: $\overline{MQ} \parallel \overline{NP}$
$\angle 4 \cong \angle 3$
Prove: $\angle 1 \cong \angle 5$
Proof:
Statements (Reasons)
1. $\overline{MQ} \parallel \overline{NP}$, $\angle 4 \cong \angle 3$ (Given)
2. $\angle 3 \cong \angle 5$ (Alt. Int. ⦞ Th.)
3. $\angle 4 \cong \angle 5$ (≅ of ⦞ is trans.)
4. $\angle 4 \cong \angle 1$ (Corr. Angles Post.)
5. $\angle 1 \cong \angle 5$ (≅ of ⦞ is trans.)

Practice Masters, p. 15

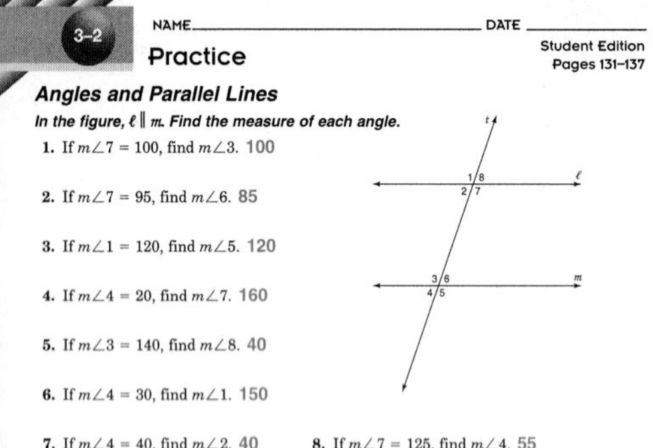
3–2 NAME__________ DATE ________

Practice — Student Edition Pages 131–137

Angles and Parallel Lines

In the figure, $\ell \parallel m$. Find the measure of each angle.

1. If $m\angle 7 = 100$, find $m\angle 3$. 100
2. If $m\angle 7 = 95$, find $m\angle 6$. 85
3. If $m\angle 1 = 120$, find $m\angle 5$. 120
4. If $m\angle 4 = 20$, find $m\angle 7$. 160
5. If $m\angle 3 = 140$, find $m\angle 8$. 40
6. If $m\angle 4 = 30$, find $m\angle 1$. 150
7. If $m\angle 4 = 40$, find $m\angle 2$. 40
8. If $m\angle 7 = 125$, find $m\angle 4$. 55
9. If $\ell \perp t$, find $m\angle 3$. 90
10. If $m\angle 1 + m\angle 3 = 230$, find $m\angle 6$. 65

In the figure, $s \parallel t$. Find the measure of each angle.

11. $m\angle 1$ 115
12. $m\angle 2$ 65
13. $m\angle 3$ 148
14. $m\angle 4$ 148
15. In the figure, $r \parallel s$. Find the value of x. 22

41–42. See margin for explanations.

Refer to the figure at the right for Exercises 41–42.

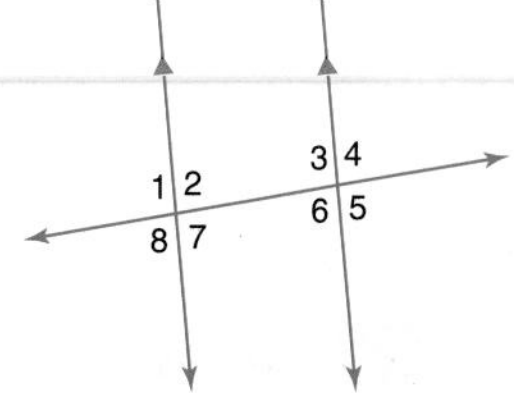

41. If $m\angle 4 = 2x - 25$ and $m\angle 8 = x + 26$, find $m\angle 2$. Explain your reasoning. 77

42. If $m\angle 6 = 2x + 43$ and $m\angle 7 = 5x + 11$, find $m\angle 5$. Explain your reasoning. 101

43. Copy and complete the proof of Theorem 3–4.

Given: $m \perp \ell$
$\ell \parallel p$

Prove: $m \perp p$

Proof:

Statements	Reasons
1. $m \perp \ell$, $\ell \parallel p$	1. ? Given
2. ∠1 is a right angle.	2. ? Definition of ⊥ lines
3. $m\angle 1 = 90$	3. ? Definition of right angle
4. $\angle 1 \cong \angle 2$	4. ?
5. $m\angle 1 = m\angle 2$	5. ?
6. $m\angle 2 = 90$	6. ? Substitution Property (=)
7. ∠2 is a right angle.	7. ? Definition of right angle
8. $m \perp p$	8. ? Definition of ⊥ lines

43. (Reason 4) Corresponding Angles Postulate
43. (Reason 5) Definition of congruent angles

44. Write a two-column proof of Theorem 3–1. See Solutions Manual.

45. Write a paragraph proof of Theorem 3–2. See Solutions Manual.

46. Find the measures x and y in the figure at the right, if $\overline{AF} \parallel \overline{BC}$ and $\overline{AE} \parallel \overline{CD}$. $x = 49$, $y = 76$

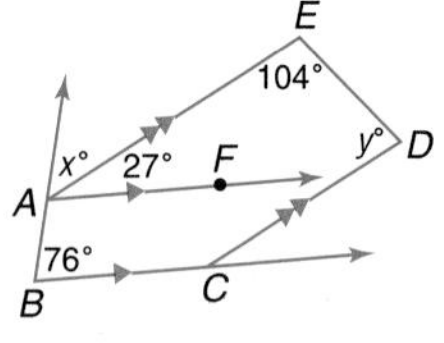

Proof

47. **Given:** $\overline{MQ} \parallel \overline{NP}$ See margin.
$\angle 4 \cong \angle 3$

Prove: $\angle 1 \cong \angle 5$

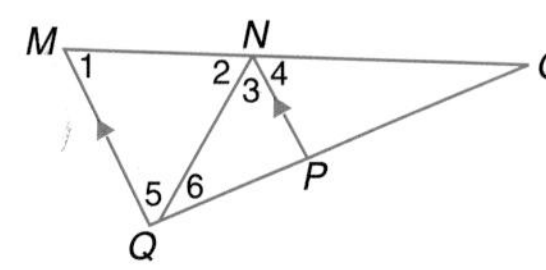

48. Find $m\angle TDK$ in the figure at the right. 67

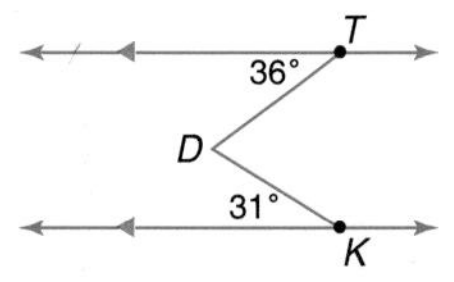

Critical Thinking

49. In the figure at the right, explain why you can conclude that $\angle 1 \cong \angle 4$, but you cannot state that ∠3 is necessarily congruent to ∠2. See margin.

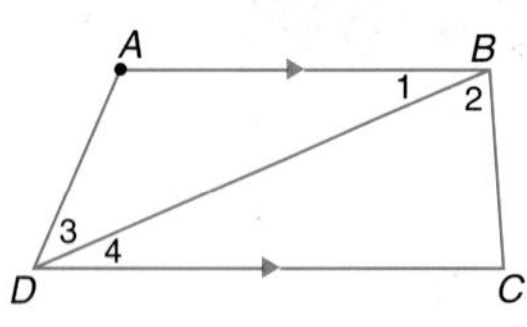

Additional Answers

49. Since $\overline{AB} \parallel \overline{DC}$, we know $\angle 1 \cong \angle 4$ by the Alternate Interior Angle Theorem. However, ∠3 and ∠2 are not alternate interior angles for a transversal and sides $\overline{AB}$ and $\overline{DC}$. They are alternate interior angles for a transversal and sides $\overline{AD}$ and $\overline{BC}$ which may or may not be parallel.

50. Since $\overline{MN} \parallel \overline{PO}$, we know ∠2 and ∠6 are supplementary by the Consecutive Interior Angles Theorem. However, ∠4 and ∠6 are not consecutive interior angles for a transversal and sides $\overline{MN}$ and $\overline{PO}$. They are consecutive interior angles for a transversal and sides $\overline{MP}$ and $\overline{NO}$ which may or may not be parallel.

51a. Exclude; the transversal was used to show the number of teams not making the playoffs.

50. In the figure at the right, explain why you can conclude that $\angle 2$ and $\angle 6$ are supplementary, but you cannot state that $\angle 4$ and $\angle 6$ are necessarily supplementary. See margin.

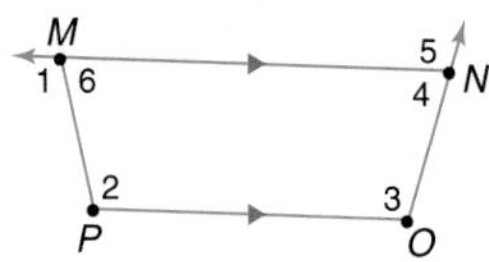

Applications and Problem Solving

51b. Sample answers: to make money, to create excitement during the season

52. The sides of the wallpaper are parallel, so if each new piece is parallel to the last one, they will all be vertical.

51. **Statistics** Refer to the application at the beginning of the lesson.
    a. Did the artist use the transversal segment to include or exclude the number of playoff teams? Explain your reasoning.
    b. Why do you think professional sports allow so many teams into the playoffs?

52. **Interior Decorating** Walls in houses are never perfectly vertical. To hang wallpaper, a true vertical line must be established so the pattern looks nice. The paperhanger uses a plumb line, which is a piece of string with a weight at the bottom, to make the vertical line for the first piece of wallpaper. How can she be sure that all of the pieces of wallpaper are vertical if she does not use the plumb line again?

Mixed Review

53. If $m \parallel n$ and $n \parallel p$, is $m \parallel p$? Explain. (Lesson 3–1) yes; Transitive Property

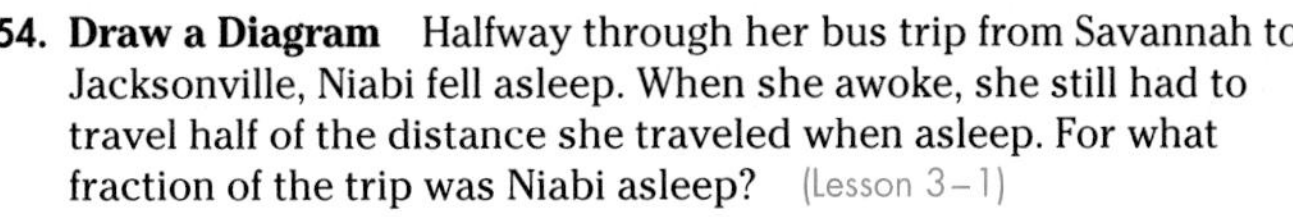

54. **Draw a Diagram** Halfway through her bus trip from Savannah to Jacksonville, Niabi fell asleep. When she awoke, she still had to travel half of the distance she traveled when asleep. For what fraction of the trip was Niabi asleep? (Lesson 3–1)

55. Find the value of x. (Lesson 2–6) 43

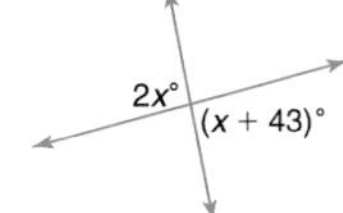

56. Write a complete proof of *If $5x - 7 = x + 1$, then $x = 2$.* (Lesson 2–4) See margin.

57. **Advertising** An ad for Wildflowers Gift Boutique says *When it has to be special, it has to be Wildflowers.* Catalina needs a special gift. (Lesson 2–3)
    a. Does it follow that she should go to Wildflowers? yes
    b. Which law was used to make this conclusion? Law of Detachment

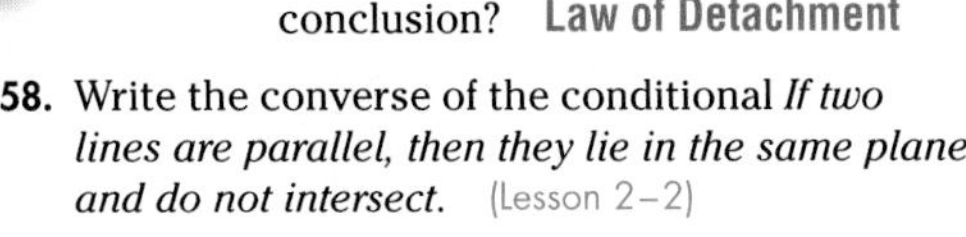

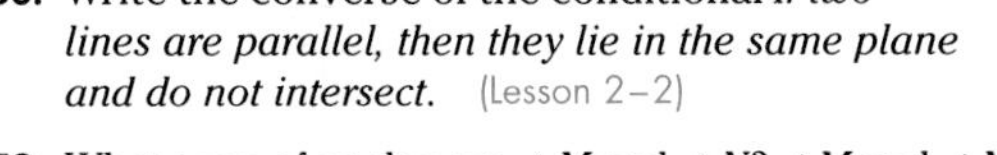

54. $\frac{1}{3}$; see Solutions Manual for diagram.

58. If two lines lie in the same plane and do not intersect, then they are parallel.

58. Write the converse of the conditional *If two lines are parallel, then they lie in the same plane and do not intersect.* (Lesson 2–2)

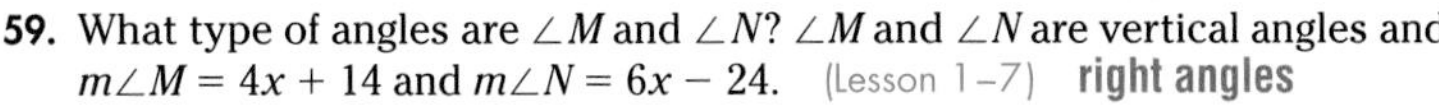

59. What type of angles are $\angle M$ and $\angle N$? $\angle M$ and $\angle N$ are vertical angles and $m\angle M = 4x + 14$ and $m\angle N = 6x - 24$. (Lesson 1–7) right angles

60. Point T is 6 units from point S on a number line. If the coordinate of point S is 3, what are the possible coordinates for point T? (Lesson 1–4) −3 or 9

61. Points W, X, Y, and Z are collinear. How many ways can you use two letters to name the line that contains these four points? (Lesson 1–2) 12

Algebra

62. Find $\sqrt{576}$. 24

63. Solve $\frac{r}{28} = \frac{5}{7}$. 20

4 ASSESS

Closing Activity

Writing Have each student draw two parallel lines cut by a transversal and number the angles in any fashion. Have students exchange papers and list five pairs of angles that are either congruent or supplementary, as well as the theorem or postulate that explains why.

Chapter 3 Quiz A (Lessons 3-1 and 3-2) is available in the *Assessment and Evaluation Masters*, p. 72.

Additional Answer

56. Given: $5x - 7 = x + 1$
Prove: $x = 2$
Proof:
Statements (Reasons)
1. $5x - 7 = x + 1$ (Given)
2. $4x - 7 = 1$ (Subt. Prop. (=))
3. $4x = 8$ (Add. Prop. (=))
4. $x = 2$ (Division Prop. (=))

Enrichment Masters, p. 15

3-2 NAME_______________ DATE_______
Enrichment
Student Edition Pages 131–137

More Optical Illusions

In drawings, diagonal lines may create the illusion of depth. For example, the figure at the right can be thought of as picturing a flat figure or a cube. The optical illusions on this page involve depth perception.

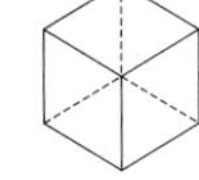

Answer each question.

1. How many cubes do you see in the drawing? 5 or 6

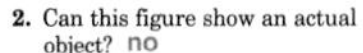

2. Can this figure show an actual object? no

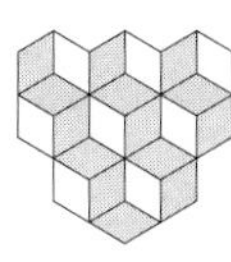

3.

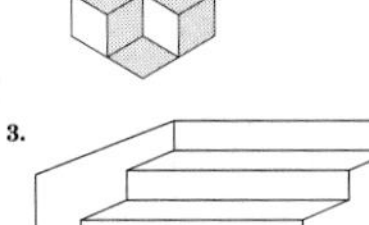

Does the drawing show a view from the top or the bottom of the stairs? Answers may vary.

4.

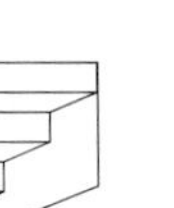

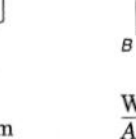

Which line segment is longer, $\overline{AB}$ or $\overline{CD}$? Measure to check your answer. AB

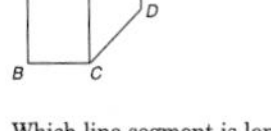

5. Which person in the drawing at the right appears to be tallest? Measure to check your answer. The person at the right appears tallest, but all are the same size.

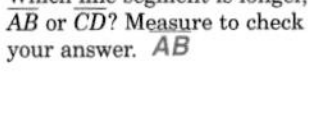

6. Draw two more objects the same size on the figure at the right. Does one appear larger than the other? Answers will vary.

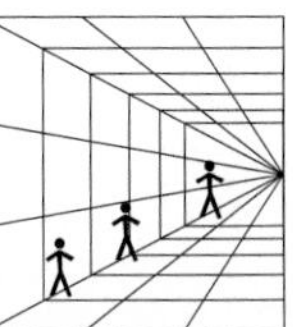

Extension

Connections Introduce students to non-Euclidean geometry. Discuss what "parallel" would mean in this case. You could demonstrate two non-Euclidean parallel lines by drawing a circle that contains two nonintersecting chords or by stretching two rubber bands around a ball, making sure they do not cross or touch.

Tech Prep

Graphic Designer In preparing a page of print advertising, the geometry of parallel lines and rectangles is used to design the layout of the page. For more information on tech prep, see the *Teacher's Handbook*.

3-3 LESSON NOTES

NCTM Standards: 1–4, 7

Instructional Resources

- Study Guide Master 3-3
- Practice Master 3-3
- Enrichment Master 3-3
- Assessment and Evaluation Masters, pp. 71, 72
- Graphing Calculator and Computer Masters, p. 3
- Real-World Applications, 5
- Tech Prep Applications Masters, p. 6

Transparency 3-3A contains the 5-Minute Check for this lesson; **Transparency 3-3B** contains a teaching aid for this lesson.

Recommended Pacing	
Standard Pacing	Day 6 of 14
Honors Pacing	Day 5 of 13
Block Scheduling*	Day 3 of 7

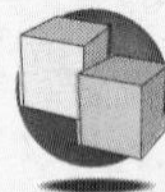

*For more information on pacing and possible lesson plans, refer to the *Block Scheduling Booklet*.

1 FOCUS

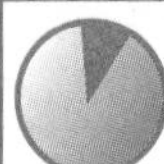

5-Minute Check
(over Lesson 3-2)

In the figure, $\ell \parallel m$, $a \parallel \overline{BD}$, and $m\angle 2 = 63$.

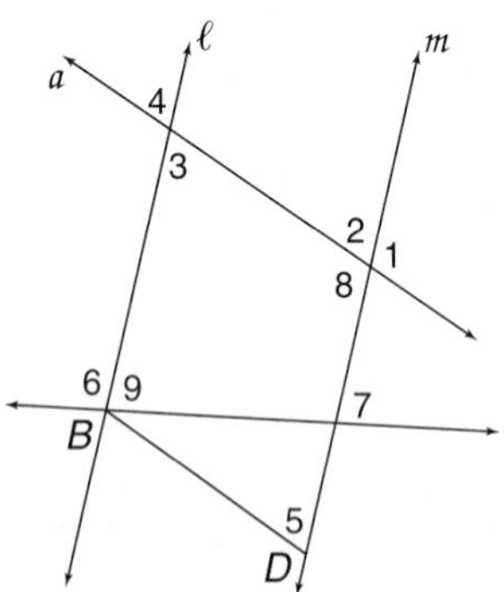

1. $\angle 3$ 63
2. $\angle 4$ 63
3. $\angle 8$ 117
4. $\angle 5$ 63

3-3 Integration: Algebra Slopes of Lines

What YOU'LL LEARN

- To find the slopes of lines, and
- to use slope to identify parallel and perpendicular lines.

Why IT'S IMPORTANT

The slopes of parallel and perpendicular lines will help you to investigate ways to prove lines are parallel and to find the distance between points and lines.

Skiing

Snowbird Ski Resort is located in the Wasatch-Cache National Forest outside Salt Lake City, Utah. The base of the resort is at an elevation of 7900 feet. The highest ski run starts on Hidden Peak at an elevation of 11,000 feet. A brochure for the Snowbird Resort claims that the vertical rise for the mountain at Snowbird is 11,000 − 7900 or 3100 feet. Skiers sometimes refer to this as the *slope* of the mountain.

In a coordinate plane, the **slope** of a line is the ratio of its vertical rise to its horizontal run.

$$\text{slope} = \frac{\text{vertical rise}}{\text{horizontal run}}$$

Definition of Slope

The slope m of a line containing two points with coordinate (x_1, y_1) and (x_2, y_2) is given by the formula

$$m = \frac{y_2 - y_1}{x_2 - x_1}, \text{ where } x_1 \neq x_2.$$

The slope of a vertical line, where $x_1 = x_2$, is undefined.

The slope of a line indicates whether the line rises to the right, falls to the right, or is horizontal.

Example 1 **Find the slope of each line.**

a.

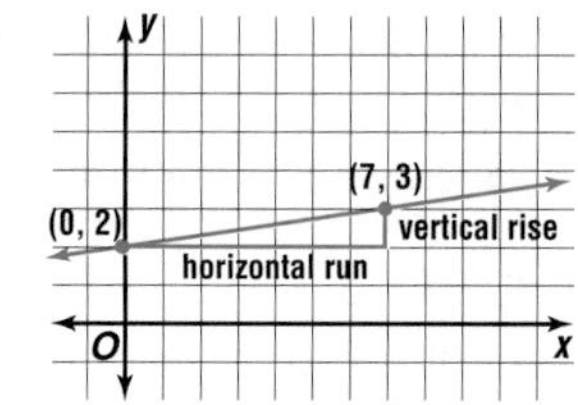

Let (0, 2) be (x_1, y_1) and (7, 3) be (x_2, y_2).

$$m = \frac{y_2 - y_1}{x_2 - x_1} = \frac{3 - 2}{7 - 0} \text{ or } \frac{1}{7}$$

Lines with positive slope rise as you move from left to right.

b.

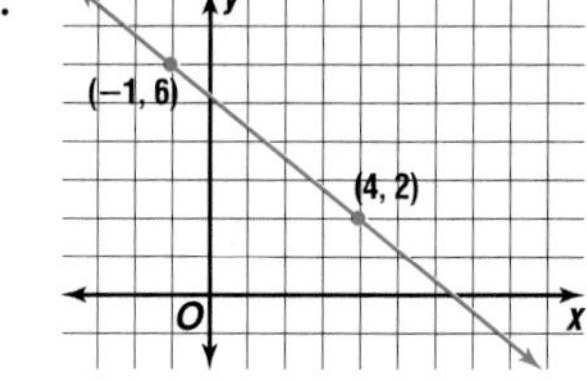

Let (−1, 6) be (x_1, y_1) and (4, 2) be (x_2, y_2).

$$m = \frac{y_2 - y_1}{x_2 - x_1} = \frac{2 - 6}{4 - (-1)} \text{ or } -\frac{4}{5}$$

Lines with negative slope fall as you move from left to right.

When finding slope, either point can be (x_1, y_1). Just be sure to subtract the coordinates in the same order.

c.

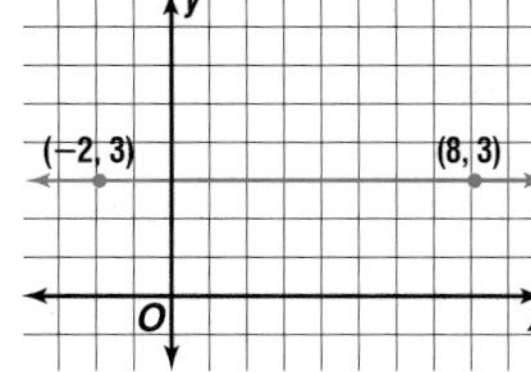

$$m = \frac{y_2 - y_1}{x_2 - x_1} = \frac{3 - 3}{8 - (-2)} = \frac{0}{10} \text{ or } 0$$

Lines with a slope of <u>0</u> *are horizontal.*

d.

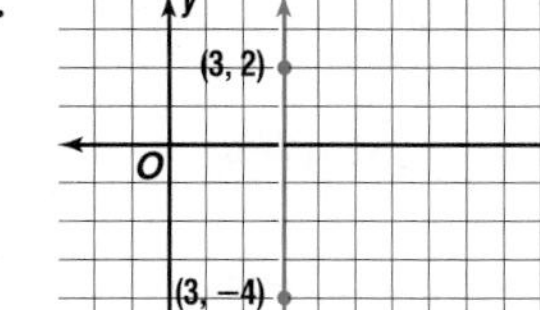

$$m = \frac{y_2 - y_1}{x_2 - x_1} = \frac{2 - (-4)}{3 - 3} = \frac{6}{0} \text{ or } \textit{undefined}$$

Lines with an <u>undefined</u> *slope are vertical.*

The graphs of lines r, s, and t are shown at the right. Lines r and s are parallel, and t is perpendicular to r and s. Let's investigate the slopes of these lines.

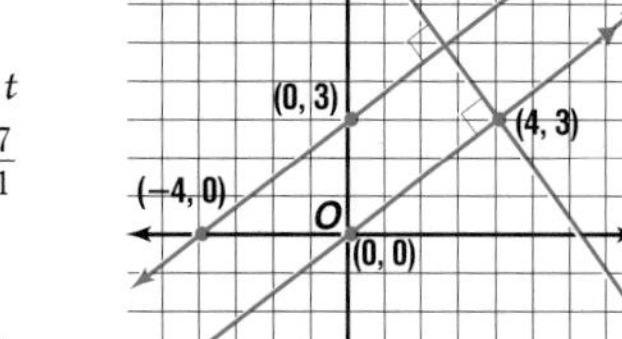

slope of r	slope of s	slope of t
$m = \frac{0 - 3}{-4 - 0}$	$m = \frac{0 - 3}{0 - 4}$	$m = \frac{3 - 7}{4 - 1}$
$= \frac{3}{4}$	$= \frac{3}{4}$	$= -\frac{4}{3}$

Lines r and s are parallel, and their slopes are the same. Line t is perpendicular to lines r and s, and its slope is the negative reciprocal of the slopes of r and s; that is, $-\frac{4}{3} \cdot \frac{3}{4} = -1$. These results suggest two important algebraic properties of parallel and perpendicular lines.

Postulate 3-2	**Two nonvertical lines have the same slope if and only if they are parallel.**
Postulate 3-3	**Two nonvertical lines are perpendicular if and only if the product of their slopes is −1.**

Note that Postulates 3–2 and 3–3 are written in **if and only if** form. If both a conditional and its converse are true, it can be written in *if and only if* form.

Example  **Given $A(-3, -2)$, $B(9, 1)$, $C(3, 6)$, and $D(5, -2)$, determine if $\overleftrightarrow{AB}$ is parallel or perpendicular to $\overleftrightarrow{CD}$.**

First find the slopes of $\overleftrightarrow{AB}$ and $\overleftrightarrow{CD}$.

slope of $\overleftrightarrow{AB} = \frac{1 - (-2)}{9 - (-3)} = \frac{3}{12} \text{ or } \frac{1}{4}$

slope of $\overleftrightarrow{CD} = \frac{-2 - 6}{5 - 3} = \frac{-8}{2} \text{ or } -4$

The product of the slopes for $\overleftrightarrow{AB}$ and $\overleftrightarrow{CD}$ is $\left(\frac{1}{4}\right)(-4)$ or -1. So, $\overleftrightarrow{AB} \perp \overleftrightarrow{CD}$.

Classroom Vignette

"Students often have trouble remembering whether a horizontal or a vertical line has no slope. You can use a ski slope to help them remember. A horizontal ski slope would be a big zero, but a skier on a vertical ski slope would say, 'Oh, no!'"

Cindy Boyd
Author

Motivating the Lesson

Situational Problem Ask students how they would solve this problem. A store needs to install a ramp so that people in wheelchairs can shop there. The ramp cannot have a slope of greater than $\frac{1}{12}$. Design a ramp that fits the specified condition.

2 TEACH

In-Class Examples

For Example 1
Find the slope of each line.

a. $-\frac{3}{4}$

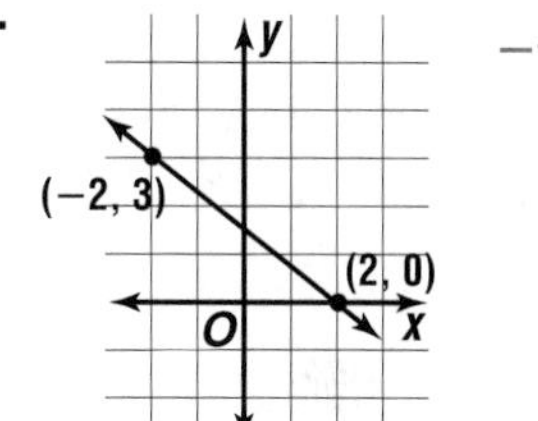

b. undefined

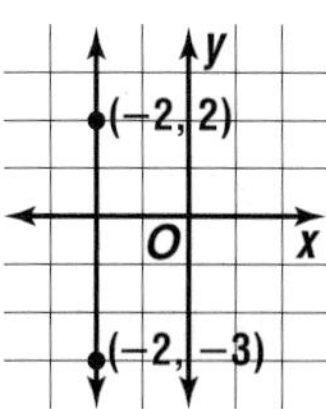

c. 0

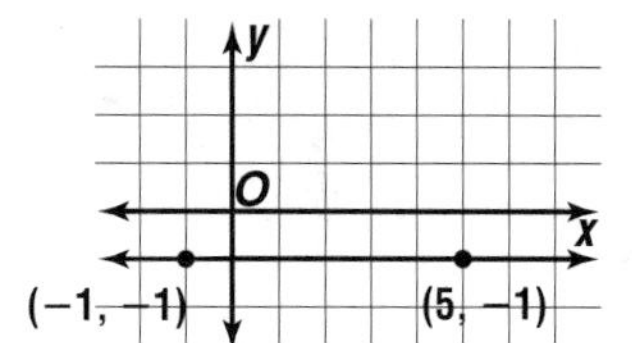

d. $\frac{5}{2}$

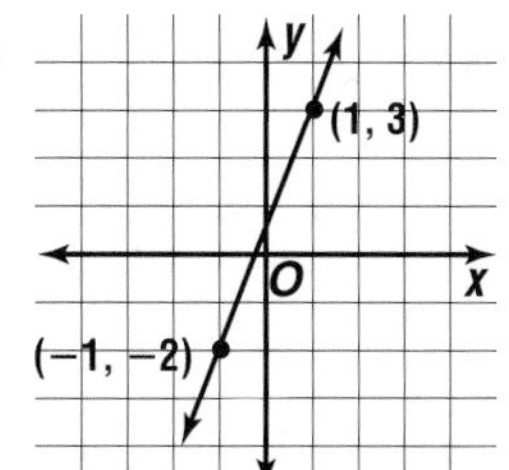

For Example 2
Given $P(-2, 2)$, $Q(2, 1)$, $R(1, -1)$, and $S(5, -2)$, determine if $\overline{PQ}$ is parallel to or perpendicular to $\overline{RS}$. **parallel**

Teaching Tip For clarification, compare "horizontal line" with the horizon.

In-Class Examples

For Example 3

Draw the segment with endpoints $A(0, 1)$ and $B(5, -2)$. Then draw a segment with endpoint $C(0, 4)$, parallel to $\overline{AB}$, and in the first quadrant only.

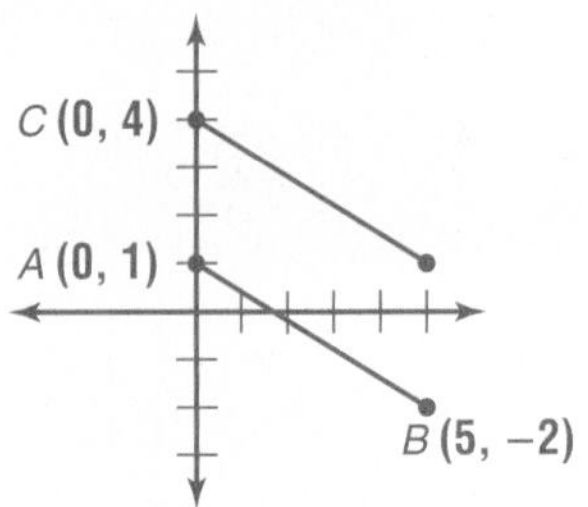

For Example 4

Find the value of x so the line that passes through $(x, 2)$ and $(3, 5)$ is perpendicular to the line that passes through $(0, 1)$ and $(2, 7)$. $x = 12$

Teaching Tip Have students list the steps they follow when answering Example 4. They can refer to this list for Exercises 42–44.

Answers for the Exploration

b. The points where the lines intersect the y-axis are 0, 3, and -2; the same as the values added to the x term.
c. The slope will be 1 and the y-intercept will be 2.
d. The slopes are the same and the y-intercepts are 0, 2, and -3 respectively.
e. The slope will be -1 and the y-intercepts will be -2.
f. The m value is the slope. If it is positive, the line rises from left to right. If it is negative, the line falls from left to right. The greater the absolute value of m, the steeper the line. The b value tells where the line crosses the y-axis.

Example 3

Transporation

You will learn to write equations of lines that are parallel or perpendicular to a given line in Lesson 12–2.

A maintenance crew is marking lines for parking spaces in a parking lot. Assume the parking lot is on an imaginary coordinate plane with each grid segment representing one yard. How can the maintainance crew draw $\overline{CD}$ parallel to $\overline{AB}$ for $C(0, 3)$, $A(0, 0)$, and $B(5, 2)$?

First, find the slope of $\overline{AB}$.

$$m = \frac{2-0}{5-0} \text{ or } \frac{2}{5}$$

Next, find point D so the slope of $\overline{CD}$ is the same as the slope of $\overline{AB}$ or $\frac{2}{5}$.

Since slope is $\frac{\text{vertical rise}}{\text{horizontal run}}$, start at point C.

Move vertically 2 yards, and then horizontally 5 yards.

Locate point D. Draw $\overline{CD}$.

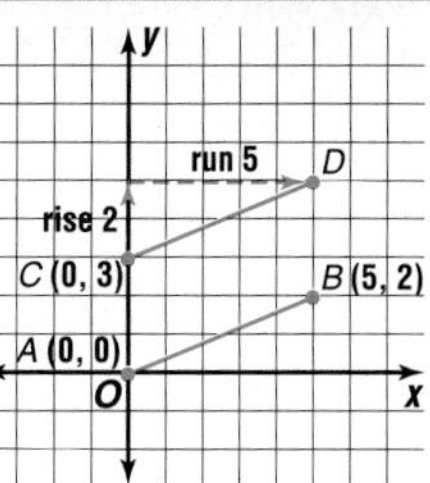

You can use what you know about slope to find missing values.

Example 4

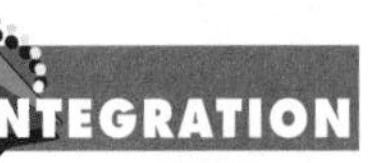

Algebra

Find the value of x so the line that passes through $(x, 5)$ and $(6, -1)$ is perpendicular to the line that passes through $(2, 3)$ and $(-3, -7)$.

First, find the slope of the line that passes through $(2, 3)$ and $(-3, -7)$.

$$m = \frac{y_2 - y_1}{x_2 - x_1} \quad \textit{Definition of slope}$$

$$= \frac{3-(-7)}{2-(-3)} \quad \textit{Let } (x_1, y_1) = (-3, -7) \textit{ and } (x_2, y_2) = (2, 3).$$

$$= \frac{10}{5} \text{ or } 2$$

The product of the slopes of the two perpendicular lines is -1. Since $2\left(-\frac{1}{2}\right) = -1$, the slope of the line through $(x, 5)$ and $(6, -1)$ is $-\frac{1}{2}$. Now use the formula for the slope of a line to find the value of x.

$$m = \frac{y_2 - y_1}{x_2 - x_1}$$

$$-\frac{1}{2} = \frac{5-(-1)}{x-6} \quad \textit{slope} = -\tfrac{1}{2}, (x_1, y_1) = (6, -1), \textit{ and } (x_2, y_2) = (x, 5)$$

$$\frac{1}{-2} = \frac{6}{x-6}$$

$$x - 6 = -12 \quad \textit{Cross multiply.}$$

$$x = -6$$

The line that passes through $(-6, 5)$ and $(6, -1)$ is perpendicular to the line that passes through $(2, 3)$ and $(-3, -7)$. To check your answer, graph the two lines.

Additional Answers

3. [graph of two perpendicular lines]

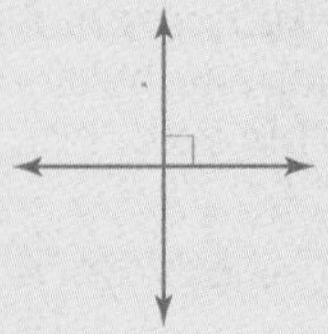

4. The resort starts at an elevation of 7900 feet and goes up to an elevation of 11,000 feet. Since 11,000 − 7900 = 3100, the vertical rise is 3100 feet.
5. Find the slope of each line. If the slope is the same, the lines are parallel. If the product of the slopes is −1, the lines are perpendicular. The lines are perpendicular.

An equation such as $y = x + 3$ is said to be in slope-intercept form. An equation can be graphed easily if it is in slope-intercept form.

EXPLORATION FAMILY OF LINES

Use a graphing calculator to graph $y = x$, $y = x + 3$, and $y = x - 2$. b–f. See margin.

Set the calculator on the standard viewing screen, [−10, 10] by [−10, 10]. Enter each equation in the Y= list and press GRAPH.

a. Compare the slopes of the lines. They are the same.

b. Compare the points where the lines intersect the y-axis.

c. Predict what the graph of $y = x + 2$ will look like. Then graph the equation to check your answer.

d. Graph $y = -x$, $y = -x + 2$, and $y = -x - 3$. Describe the similarities and differences among the graphs.

e. Predict what the graph of $y = -x - 2$ will look like. Check your answer.

f. The slope-intercept form of an equation of the line is $y = mx + b$, where m is the slope and b is the y-intercept. Write a paragraph explaining how the values of m and b affect the graph of the equation.

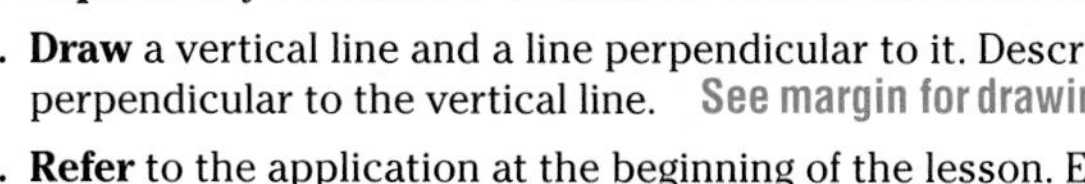

CHECK FOR UNDERSTANDING

Communicating Mathematics

1. A line whose slope is 0 is horizontal and a line whose slope is undefined is vertical.

2. The slope of a vertical line is undefined; yes.

Study the lesson. Then complete the following.

1. **Describe** a line whose slope is 0 and a line whose slope is undefined.
2. **Explain** why Postulate 3–2 excludes vertical lines. Are vertical lines parallel?
3. **Draw** a vertical line and a line perpendicular to it. Describe the line perpendicular to the vertical line. See margin for drawing; a horizontal line.
4. **Refer** to the application at the beginning of the lesson. Explain why there is a vertical rise of 3100 feet. See margin.
5. **Explain** how to determine if the lines in the graph at the right are parallel, perpendicular, or neither. Describe the relationship of these two lines. See margin.

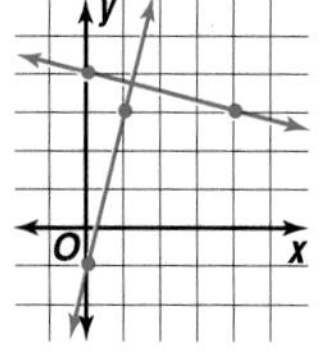

MATH JOURNAL

6a–c. See margin; see Solutions Manual for drawings.

6. **Compare and contrast** each set of lines. Determine which line is steeper and make a drawing to support your reasoning.
   a. a line with a slope of 2 and a line with a slope of $\frac{1}{3}$
   b. a line with a slope of $\frac{3}{4}$ and a line with a slope of $\frac{2}{3}$
   c. a line with a slope of −3 and a line with a slope of $\frac{1}{4}$

Guided Practice

Find the slope of the line passing through the given points. Then describe each line as you move from left to right as *rising, falling, horizontal,* or *vertical.* 8. undefined; vertical

7. $A(3, 2)$, $B(4, -3)$ −5; falling
8. $C(-3, 6)$, $D(-3, -2)$

Reteaching

Using Cooperative Groups List the coordinates of two points on the chalkboard. Have one half of the class use the first point as (x_1, y_1) and the second point as (x_2, y_2). The other half of the class should use the first point as (x_2, y_2) and the second point as (x_1, y_1). Instruct all students to find the slope of the line. Emphasize that it does not matter which point is selected as (x_1, y_1) or (x_2, y_2), but that both coordinates must be in the correct place in the formula for the slope.

EXPLORATION

In this Exploration, students use a graphing calculator to explore families of parallel lines. They will keep the slope constant while varying the y-intercept.

3 PRACTICE/APPLY

Check for Understanding

Exercises 1–16 are designed to help you assess your students' understanding through reading, writing, speaking, and modeling. You should work through Exercises 1–6 with your students and then monitor their work on Exercises 7–16.

Error Analysis

Students may have difficulty remembering that, when finding the slope of a line, the y-coordinates are in the numerator and the x-coordinates are in the denominator. If this is a problem, have them think of slope as $\frac{rise}{run}$. The y-axis rises (goes up and down), and the x-axis runs (goes left and right).

Additional Answers

6a. Both lines rise as you move from left to right. A line with a slope of 2 is steeper than a line with a slope of $\frac{1}{3}$.

6b. Both lines rise as you move from left to right. Since $\frac{3}{4} = \frac{9}{12}$ and $\frac{2}{3} = \frac{8}{12}$, a line with a slope of $\frac{3}{4}$ is steeper than a line with a slope of $\frac{2}{3}$.

6c. A line with a slope of −3 falls as you move from left to right, while a line with a slope of $\frac{1}{4}$ rises as you move from left to right. A line with a slope of −3 is steeper than a line with a slope of $\frac{1}{4}$.

Assignment Guide

Core (with proof): 17–47 odd, 48, 49, 51, 53–64
Core (informal): 17–47 odd, 48, 49, 51, 53, 54, 56–64
Enriched: 18–46 even, 48–64
All: Self Test, 1–10

For **Extra Practice**, see p. 769.

The red A, B, and C flags, printed only in the Teacher's Wraparound Edition, indicate the level of difficulty of the exercises.

Additional Answers

12.

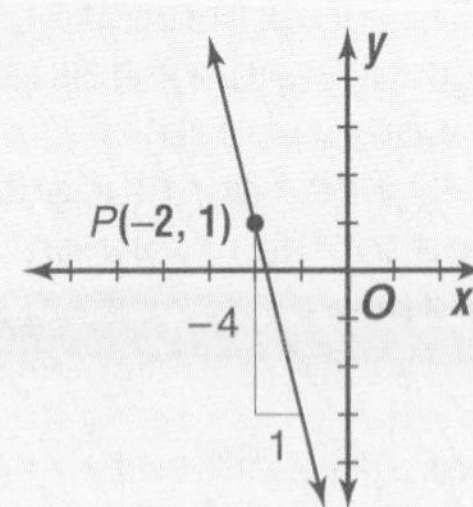

13.

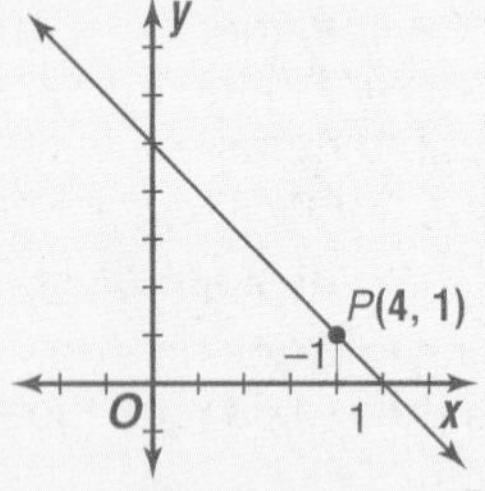

14. No; slope of $\overleftrightarrow{AB} = \frac{5}{6}$, slope of $\overleftrightarrow{CD} = -\frac{4}{3}$, and $\frac{5}{6}\left(-\frac{4}{3}\right) = -\frac{10}{9} \neq -1$.

32.

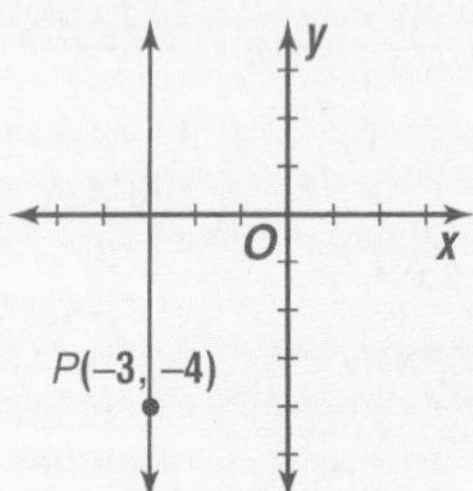

33.

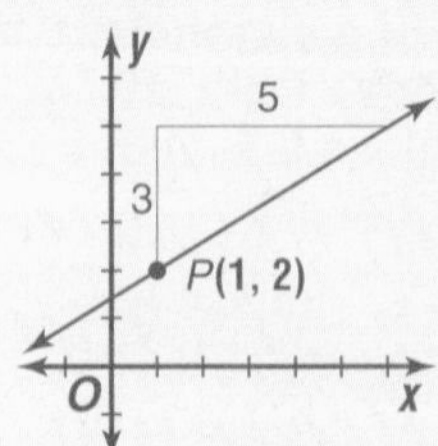

34.

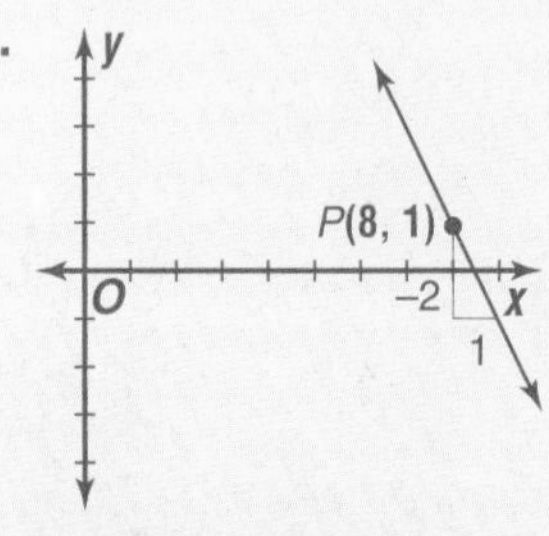

Determine the slope of each line named below.

9. q $\frac{3}{4}$
10. any line perpendicular to s $\frac{5}{2}$
11. any line parallel to r 0

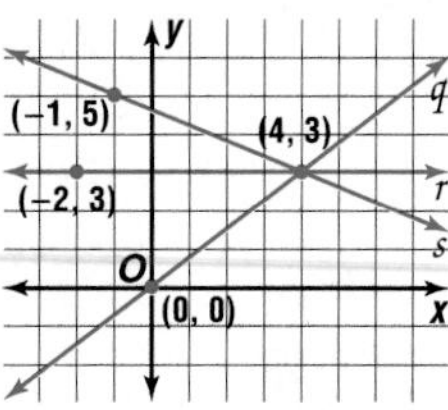

Graph the line that satisfies each description.

12. slope = −4, passes through $P(-2, 1)$ 12–13. See margin.
13. passes through $P(4, 1)$, perpendicular to $\overleftrightarrow{CD}$ with $C(0, 3)$ and $D(-3, 0)$
14. Given $A(16, 7)$, $B(4, -3)$, $C(1, 6)$, and $D(7, -2)$, determine if the intersection of $\overleftrightarrow{AB}$ and $\overleftrightarrow{CD}$ forms a right angle. Explain your reasoning. See margin.
15. Find the value of x so that the line passing through $(x, 6)$ and $(2, -3)$ is perpendicular to the line passing through $(1, 6)$ and $(7, 2)$. 8
16. **Aviation** An airplane passing over Richmond at an elevation of 33,000 feet begins its descent to land at Washington, D.C., 107 miles away. How many feet should the airplane descend per mile to land in Washington, which has an elevation of 25 feet? about 308.2 feet per mile

EXERCISES

Practice

Find the slope of the line passing through the given points. Then, describe each line as you move from left to right as *rising, falling, horizontal,* or *vertical.* 19. 0; horizontal

17. $A(0, 6)$, $B(4, 0)$
18. $C(-3, 8)$, $D(4, 2)$
19. $E(6, 3)$, $F(-6, 3)$
20. $G(8, 1)$, $H(8, -6)$
21. $I(-2, -3)$, $J(-6, -5)$
22. $K(-2, 5)$, $L(4, 9)$

17. $-\frac{3}{2}$; falling
18. $-\frac{6}{7}$; falling
20. undefined; vertical
21. $\frac{1}{2}$; rising
22. $\frac{2}{3}$; rising
23. $-\frac{5}{4}$
26. $\frac{1}{2}$
28. $\frac{1}{2}$
29. $\frac{4}{5}$

Determine the slope of each line named below.

23. $\overleftrightarrow{BD}$
24. $\overleftrightarrow{CD}$ 0
25. $\overleftrightarrow{AB}$ 1
26. $\overleftrightarrow{EO}$
27. any line parallel to $\overleftrightarrow{DE}$ undefined
28. any line parallel to $\overleftrightarrow{EO}$
29. any line perpendicular to $\overleftrightarrow{BD}$
30. any line perpendicular to $\overleftrightarrow{CD}$ undefined
31. any line perpendicular to $\overleftrightarrow{DE}$ 0

(0, 2) B
E (4, 2)
O
x
y
A(−4, −2)
C(−1, −3)
D (4, −3)

Graph the line that satisfies each description. 32–37. See margin.

32. undefined slope, passes through $P(-3, -4)$
33. slope = $\frac{3}{5}$, passes through $P(1, 2)$
34. slope = −2, passes through $P(8, 1)$
35. slope = 0, passes through $P(-7, 3)$
36. passes through $P(6, 4)$ and is perpendicular to $\overleftrightarrow{TK}$ with $T(0, 2)$ and $K(5, 0)$
37. passes through $P(-1, -3)$ and is parallel to $\overleftrightarrow{CR}$ with $C(-1, 7)$ and $R(5, 1)$

Additional Answers

35.

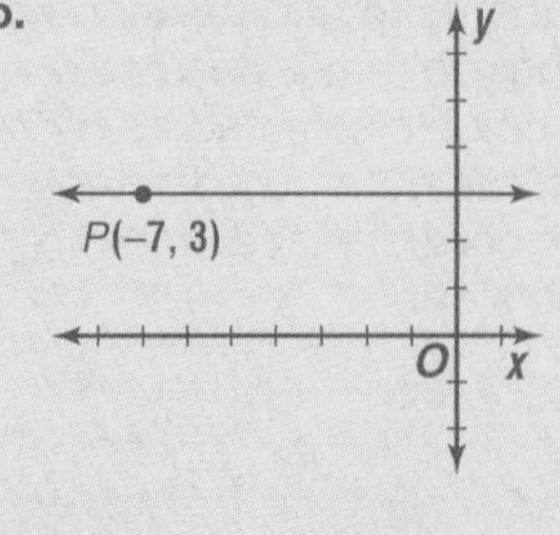

36.

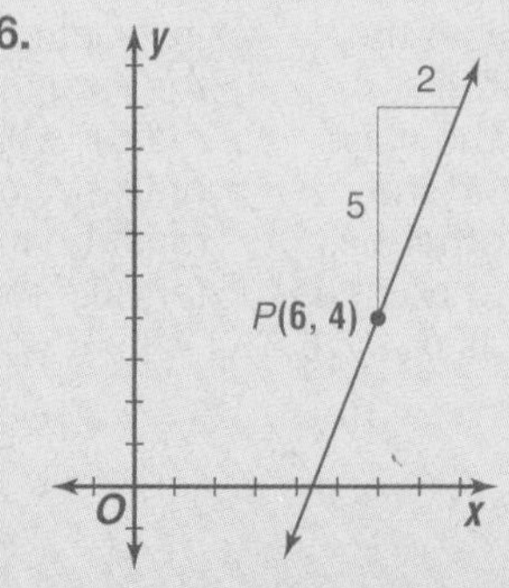

37.

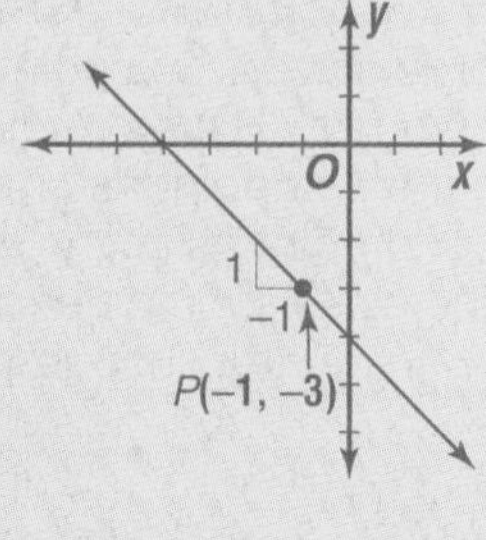

Given each set of points, determine if $\overleftrightarrow{MA} \parallel \overleftrightarrow{TH}$. Explain your reasoning.

38. Yes; slope of $\overleftrightarrow{MA} = \frac{7}{8}$, slope of $\overleftrightarrow{TH} = \frac{7}{8}$.

39. Yes; slope of $\overleftrightarrow{MA} = -\frac{3}{5}$, slope of $\overleftrightarrow{TH} = -\frac{3}{5}$.

40. Yes; slope of $\overleftrightarrow{PQ} = -\frac{1}{9}$, slope of $\overleftrightarrow{RS} = 9$, $-\frac{1}{9}(9) = -1$.

41. Yes; slope of $\overleftrightarrow{PQ} = \frac{1}{2}$, slope of $\overleftrightarrow{RS} = -2$, $\frac{1}{2}(-2) = -1$.

38. $M(-6, 1), A(2, 8), T(1, 1), H(9, 8)$
39. $M(-3, -4), A(7, -10), T(-1, 6), H(4, 3)$

Determine if the intersection of $\overleftrightarrow{PQ}$ and $\overleftrightarrow{RS}$ forms a right angle. Explain your reasoning.

40. $P(-9, 2), Q(0, 1), R(-1, 8), S(-2, -1)$
41. $P(3, 6), Q(-1, 4), R(4, 0), S(0, 8)$

Determine the value of x, so that a line through points with the given coordinates has the given slope. Draw a sketch of each situation. 42–43. See margin for sketch.

42. $(x, 2), (-4, -6)$, slope $= \frac{4}{5}$ 6
43. $(6, 2), (x, -1)$, slope $= -\frac{3}{7}$ 13
44. Find the value of x so that the line passing through the points at $(x, 2)$ and $(-4, 5)$ is perpendicular to the line that passes through the points at $(4, 8)$ and $(2, -1)$. 9.5

C

45. The vertices of figure $PQRS$ are $P(5, 2)$, $Q(1, 6)$, $R(-3, 2)$, and $S(1, -2)$.
   a. Show that the opposite sides are parallel.
   b. Show that the adjacent sides are perpendicular.
   c. Show that all sides are congruent. See margin.
   d. What type of figure is $PQRS$? square

45a. slope of $\overleftrightarrow{PQ} = -1$, slope of $\overleftrightarrow{RS} = -1$, slope of $\overleftrightarrow{QR} = 1$, slope of $\overleftrightarrow{PS} = 1$

45b. $-1(1) = -1$ and $1(-1) = -1$

46. A parallelogram is a four-sided figure whose opposite sides are parallel. Given $E(2, 3)$, $F(1, -6)$, and $G(-2, 5)$, find the coordinates of point H so that the points are the vertices of a parallelogram. (*Hint:* There is more than one location.) $(-1, 14), (-3, -4), (5, -8)$

Programming

47. The TI-82/83 graphing calculator program at the right finds the slope of a line containing two points at (x_1, y_1) and (x_2, y_2).
Use the program to find the slopes of each line.
   a. $\overleftrightarrow{AB}$ if $A(5, 3)$ and $B(-1, 8)$
   b. $\overleftrightarrow{GH}$ if $G(-3, -7)$ and $H(-5, -2)$
   c. $\overleftrightarrow{TS}$ if $T(3, -1)$ and $S(5, 4)$
   d. $\overleftrightarrow{VW}$ if $V(-5, 7)$ and $W(3, 1)$

```
PROGRAM: SLOPE
:Disp "ENTER THE","COORDINATES"
:Input "X1=", A
:Input "Y1=", B
:Input "X2=", C
:Input "Y2=", D
:(D-B)/(C-A)→M
:Disp "SLOPE=",M
:Stop
```

47a. $-\frac{5}{6}$
47b. $-\frac{5}{2}$
47c. $\frac{5}{2}$
47d. $-\frac{3}{4}$

Critical Thinking

48. A line contains the points at $(-3, 6)$ and $(1, 2)$. Using slope, write a convincing argument that the line intersects the x-axis at $(3, 0)$. Graph the points to verify your conclusion. See Solutions Manual.

Applications and Problem Solving

49. **Construction** According to the building code in Crystal Lake, Illinois, the slope of a stairway cannot be steeper than 0.88. The stairs in Li-Chih's home measure 11 inches deep and 7 inches high. Do the stairs in his home meet the code requirements? yes; $0.64 < 0.88$

Additional Answer

45c. $PQ = \sqrt{(5-1)^2 + (2-6)^2} = \sqrt{32}$
$QR = \sqrt{[1-(-3)]^2 + (6-2)^2} = \sqrt{32}$
$RS = \sqrt{(-3-1)^2 + [2-(-2)]^2} = \sqrt{32}$
$PS = \sqrt{(5-1)^2 + [2-(-2)]^2} = \sqrt{32}$

Using the Programming Exercises The program given in Exercise 47 is for use with a TI-82 or TI-83 graphing calculator. For other programmable calculators, have students consult their owner's manual for commands similar to those represented here.

Additional Answers

42.

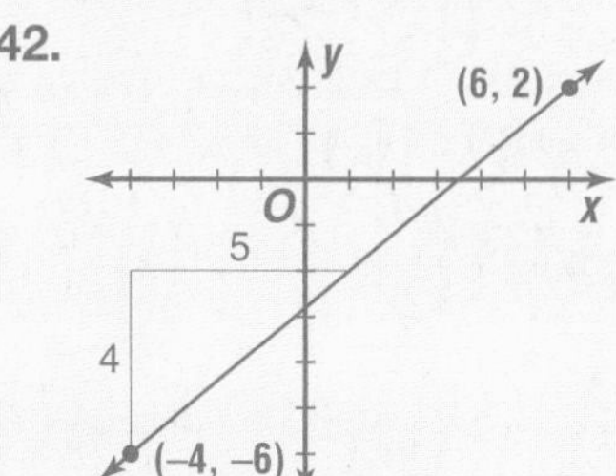

43.

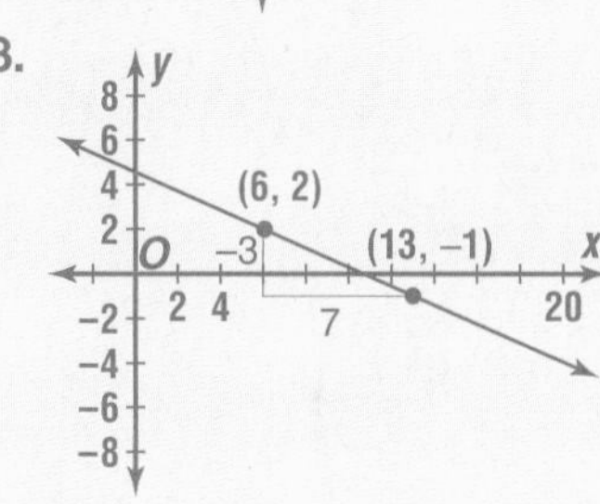

Study Guide Masters, p. 16

NAME ________ DATE ________

Study Guide

Student Edition Pages 138–145

Integration: Algebra
Slopes of Lines

To find the slope of a line containing two points with coordinates (x_1, y_1) and (x_2, y_2), use the following formula.

$$m = \frac{y_2 - y_1}{x_2 - x_1} \text{ where } x_1 \neq x_2$$

The slope of a vertical line, where $x_1 = x_2$, is undefined.

Two lines have the same slope if and only if they are parallel and nonvertical.

Two nonvertical lines are perpendicular if and only if the product of their slopes is -1.

Example: Find the slope of the line ℓ passing through $A(2, -5)$ and $B(-1, 3)$. State the slope of a line parallel to ℓ. Then state the slope of a line perpendicular to ℓ.

Let $(x_1, y_1) = (2, -5)$ and $(x_2, y_2) = (-1, 3)$.
Then $m = \frac{3-(-5)}{-1-2} = -\frac{8}{3}$.

Any line in the coordinate plane parallel to ℓ has slope $-\frac{8}{3}$.

Since $-\frac{8}{3} \cdot \frac{3}{8} = -1$, the slope of a line perpendicular to the line ℓ is $\frac{3}{8}$.

Find the slope of the line passing through the given points.

1. $C(-2, -4), D(8, 12)$ $\frac{8}{5}$
2. $J(-4, 6), K(3, -10)$ $-\frac{16}{7}$
3. $P(0, 12), R(12, 0)$ -1
4. $S(15, -15), T(-15, 0)$ $-\frac{1}{2}$
5. $F(21, 12), G(-6, -4)$ $\frac{16}{27}$
6. $L(7, 0), M(-17, 10)$ $-\frac{5}{12}$

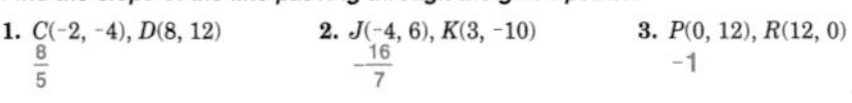

Find the slope of the line parallel to the line passing through each pair of points. Then state the slope of the line perpendicular to the line passing through each pair of points.

7. $I(9, -3), J(6, -10)$ $\frac{7}{3}, -\frac{3}{7}$
8. $G(-8, -12), H(4, -1)$ $\frac{11}{12}, -\frac{12}{11}$
9. $M(5, -2), T(9, -6)$ $-1, 1$

Additional Answers

52a. $\frac{34}{3} \approx 11.3$ million more packages shipped each year

52b. The rate of change equals the slope of a line that relates software sales over time.

55. Given: $\angle 1 \cong \angle 2$
Prove: $\angle 1 \cong \angle 3$
Proof:
Statements (Reasons)
1. $\angle 1 \cong \angle 2$ (Given)
2. $\angle 2 \cong \angle 3$ (Vertical ∡ are ≅.)
3. $\angle 1 \cong \angle 3$ (Congruence of angles is trans.)

57. $A = 2\pi r^2 + 2\pi rh$ (Given)
$A - 2\pi r^2 = 2\pi rh$ (Subt. Prop. (=))
$\frac{A - 2\pi r^2}{2\pi r} = h$ (Division Prop. (=))

Practice Masters, p. 16

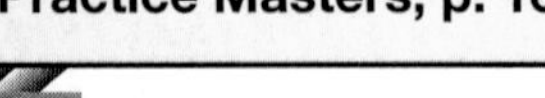

3-3 NAME______ DATE______
Practice
Student Edition Pages 138–145

Integration: Algebra
Slopes of Lines

Determine the slope of each line named below.

1. a $\frac{2}{3}$
2. b 3
3. c undefined
4. d −2
5. any line parallel to b 3
6. any line perpendicular to d $\frac{1}{2}$

Graph the line that satisfies each description.

7. slope = −3, passes through $P(1, -2)$
8. passes through $P(-1, 2)$ and is perpendicular to the line determined by $A(1, 4)$ and $B(2, 7)$

Determine the value of r so that a line through the points with the given coordinates has the given slope. Draw a sketch of each situation.

9. (5, 3), (r, 6); slope = 1 8
10. (2, r), (−2, 6); slope = $\frac{1}{2}$ 8

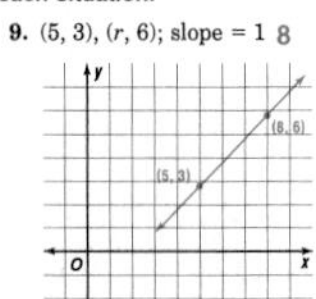

51a. $\frac{1}{18}$

50. **Construction** Measure the height and width of stairs in your school or home and determine the slope of those stairs. See students' work.

51. **Architecture** The *Americans with Disabilities Act* (ADA) of 1990 requires that ramps should have a least a 12-inch run for each rise of 1 inch, with a maximum rise of 30 inches. Your school needs to design a ramp 45 feet long.
   a. Determine the appropriate slope of the ramp in order to conform with the ADA.
   b. Find the maximum run in feet for each 1-inch rise. 1.5 feet

52. **Computer Software** Multimedia software on CD-ROM has been growing since 1992. Five million software packages were shipped in 1992, and 39 million packages were shipped in 1995. a–b. See margin.
   a. What is the average rate of change in software packages per year during this time period?
   b. How does the rate of change relate to the slope of a line?
   c. Predict how many CD-ROM software packages will be shipped in the year 2001. 107 million packages

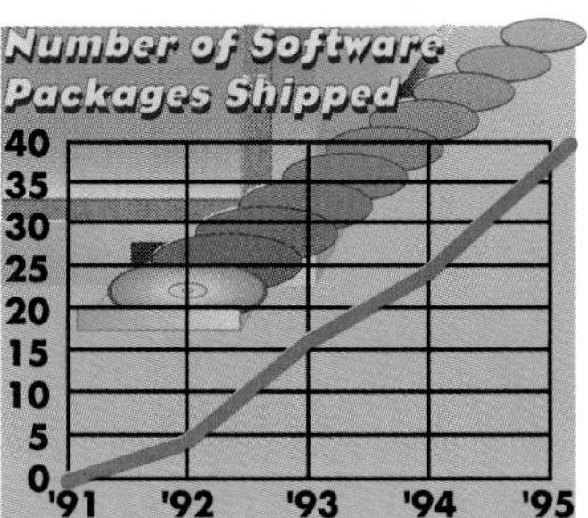

Source: *Information and Interactive Services Report, Dataquest*

Mixed Review

53. Find the values of x and y. (Lesson 3–2) $x = 16$, $y = 11$

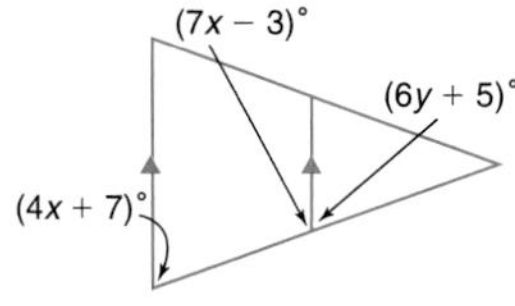

54c. intersecting

54. Identify each pair as *intersecting, parallel,* or *skew.* (Lesson 3–1)
   a. $\overline{BA}$ and $\overline{GH}$ parallel
   b. $\overline{EH}$ and $\overline{CD}$ skew
   c. plane EAB and plane GCB
   d. $\overline{HG}$ and plane EAB parallel
   e. plane HGC and $\overline{BC}$ intersecting

55. Write a two-column proof. (Lesson 2–6) See margin.
   Given: $\angle 1 \cong \angle 2$
   Prove: $\angle 1 \cong \angle 3$

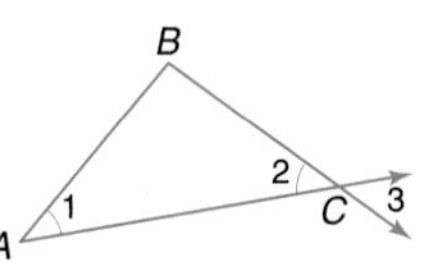

56. Justify each statement with a property from algebra or a property of congruent segments. (Lesson 2–5)
   a. If $JL + 5 = KM + 5$, then $JL = KM$. Subtraction Property (=)
   b. If $\overline{RS} \cong \overline{TU}$, then $\overline{TU} \cong \overline{RS}$. Congruence of segments is symmetric.
   c. If $\frac{2}{3}g = 12$, then $g = 18$. Multiplication Property (=)

57. The formula for finding the total surface area of a cylinder is $A = 2\pi r^2 + 2\pi rh$, where r is the radius of the base and h is the height. Solve the formula for h and justify each step. (Lesson 2–4) See margin.

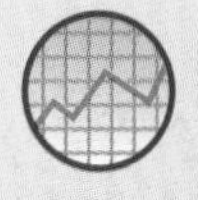

Tech Prep

Computer Artist Multimedia software makes extensive use of graphic design. Software such as Illustrator, Photoshop, and 3-D Studio are used in such multimedia systems. For more information on tech prep, see the *Teacher's Handbook.*

58. yes; Law of Detachment

59. Sample answer: $\left(\frac{1}{2}\right)^2 = \frac{1}{4}, \frac{1}{2} > \frac{1}{4}$

58. Recreation The sign in front of the Screaming Eagle Roller Coaster states *If you are over 48 inches tall, then you may ride the Screaming Eagle*. Rafael is 54 inches tall. Can he ride the Screaming Eagle? Which law of logic leads you to this conclusion? (Lesson 2–3)

59. Beng observed that 2^2 is greater than 2 and 5^2 is greater than 5, and made a conjecture that *The square of any real number is greater than the number.* Give a counterexample to this conjecture. (Lesson 2–1)

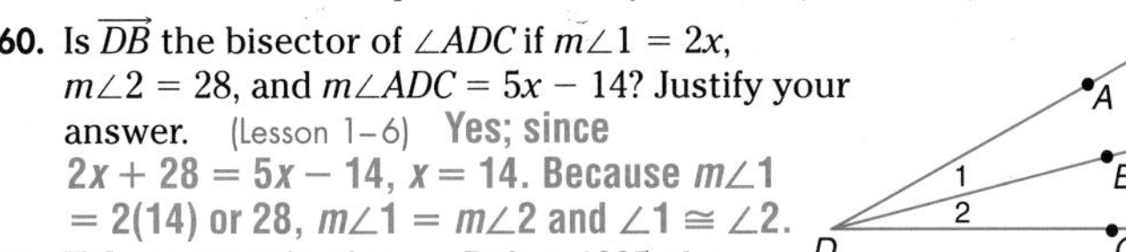

60. Is $\overrightarrow{DB}$ the bisector of $\angle ADC$ if $m\angle 1 = 2x$, $m\angle 2 = 28$, and $m\angle ADC = 5x - 14$? Justify your answer. (Lesson 1–6) **Yes; since $2x + 28 = 5x - 14$, $x = 14$. Because $m\angle 1 = 2(14)$ or 28, $m\angle 1 = m\angle 2$ and $\angle 1 \cong \angle 2$.**

61. Telecommunications. Before 1995, the telephone area codes in the United States and Canada were three-digit numbers in which the first digit was 2, 3, 4, 5, 6, 7, 8, or 9, the second digit was 0 or 1, and the third digit was any digit other than 0. Special numbers like 411 and 911 were excluded. How many different area codes started with the digit 6 prior to 1995? (Lesson 1–3) **18 area codes**

62. If $x < 0$ and $y < 0$, in which quadrant is the point $Q(x, y)$ located? (Lesson 1–1) **III**

INTEGRATION Algebra

63. Evaluate $4x^2 - 3x$ if $x = -2$. **22**

64. Solve $4n + 23 = 3 - 6n$. **−2**

SELF TEST

The three-dimensional figure shown at the right is called a *right-triangular prism*. (Lesson 3–1)

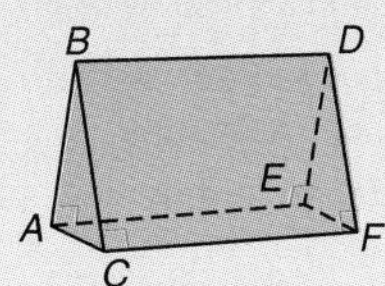

1. Name all segments parallel to $\overline{CF}$. **$\overline{AE}$, $\overline{BD}$**
2. Name all segments skew to $\overline{AC}$. **$\overline{DE}$, $\overline{BD}$, $\overline{DF}$**
3. Name all pairs of parallel planes. **plane *ABC* and plane *EDF***

Given $\ell \parallel m$, $m\angle 1 = 98$, and $m\angle 2 = 40$, find the measure of each angle. (Lesson 3–2)

4. $\angle 4$ **98**
5. $\angle 8$ **40**
6. $\angle 9$ **140**

Determine the slope of each line named below. (Lesson 3–3)

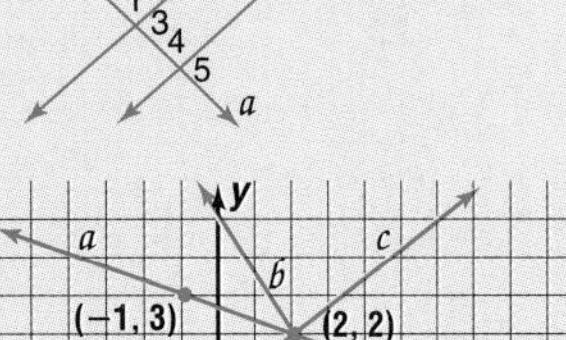

7. a $-\frac{1}{3}$
8. any line parallel to b $-\frac{3}{2}$
9. any line perpendicular to c $-\frac{4}{3}$

10. Road Construction Parallel pipes are laid on each side of Morris Road. A pipe under the road connects the two pipes. The pipe under the road makes a 115° angle with one of the pipes as shown in the diagram at the right. What angle does it make with the pipe on the other side of the road? (Lesson 3–2) **65°**

4 ASSESS

Closing Activity

Modeling Have each student state an example of a real object that has either a positive slope, a negative slope, a slope of 0, or an undefined slope. For example, a handrail on a staircase has a positive or negative slope depending on the viewpoint.

Chapter 3 Quiz B (Lesson 3-3) is available in the *Assessment and Evaluation Masters*, p. 72.

Mid-Chapter Test (Lessons 3-1 through 3-3) is available in the *Assessment and Evaluation Masters*, p. 71.

Extension

Reasoning Using the example in Motivating the Lesson, have students figure how long the ramp must be if it replaces steps that are two feet high. (The ramp cannot have a slope of greater than $\frac{1}{12}$.) **about 24.08 feet**

SELF TEST

The Self Test provides students with a brief review of the concepts and skills in Lessons 3-1 through 3-3. Lesson numbers are given to the right of exercises or instruction lines so students can review concepts not yet mastered.

Enrichment Masters, p. 16

NAME______________________ DATE__________

Student Edition Pages 138–145

Enrichment

The Möbius Strip

A Möbius strip is a special surface with only one side. It was discovered by August Ferdinand Möbius, a German astronomer and mathematician.

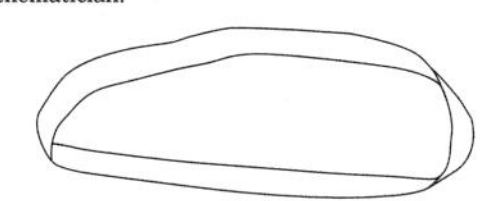

1. To make a Möbius strip, cut a strip of paper about 16 inches long and 1 inch wide. Mark the ends with the letters *A*, *B*, *C*, and *D* as shown below.

   Twist the paper once, connecting *A* to *D* and *B* to *C*. Tape the ends together on both sides. **See students' work.**
2. Use a crayon or pencil to shade one side of the paper. Shade around the strip until you get back to where you started. What happens? **The entire strip is shaded on both sides.**
3. What do you think will happen if you cut the Möbius strip down the middle? Try it. **Instead of two loops, you get one loop that is twice as long as the original one.**
4. Make another Möbius strip. Starting a third of the way in from one edge, cut around the strip, staying always the same distance in from the edge. What happens? **Two interlocking loops are formed.**
5. Start with another long strip of paper. Twist the paper twice and connect the ends. What happens when you cut down the center of this strip? **Two interlocking loops are formed.**
6. Start with another long strip of paper. Twist the paper three times and connect the ends. What happens when you cut down the center of this strip? **One two-sided loop with a knot is formed.**

3-4 LESSON NOTES

NCTM Standards: 1–5, 7

Instructional Resources

- Study Guide Master 3-4
- Practice Master 3-4
- Enrichment Master 3-4
- Modeling Mathematics Masters, p. 81

Transparency 3-4A contains the 5-Minute Check for this lesson; **Transparency 3-4B** contains a teaching aid for this lesson.

Recommended Pacing	
Standard Pacing	Days 7 & 8 of 14
Honors Pacing	Days 6 & 7 of 13
Block Scheduling*	Day 4 of 7

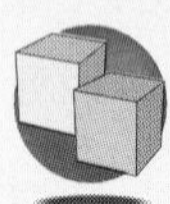

*For more information on pacing and possible lesson plans, refer to the *Block Scheduling Booklet*.

1 FOCUS

5-Minute Check
(over Lesson 3-3)

Refer to the figure below to find the slope of each line.

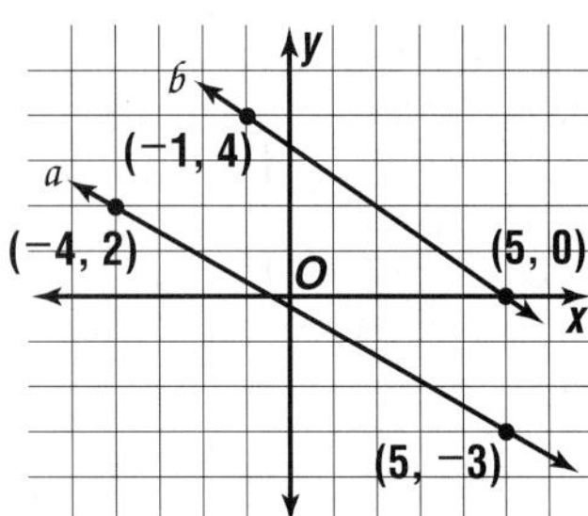

1. any line perpendicular to a $\frac{9}{5}$
2. any line parallel to b $-\frac{2}{3}$

Graph the line that satisfies each description.

3. slope = 4, passes through (1, 2)

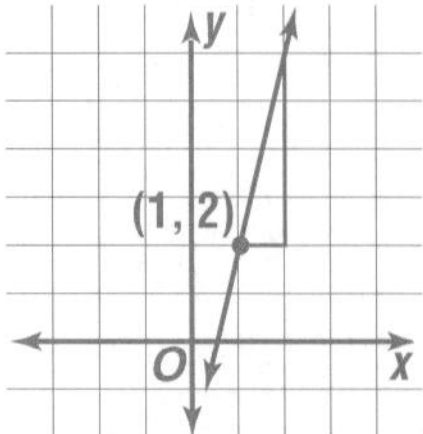

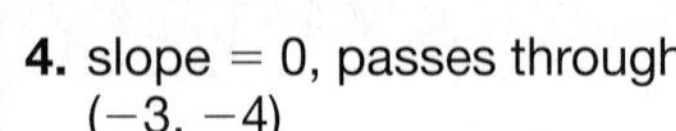

4. slope = 0, passes through (−3, −4)

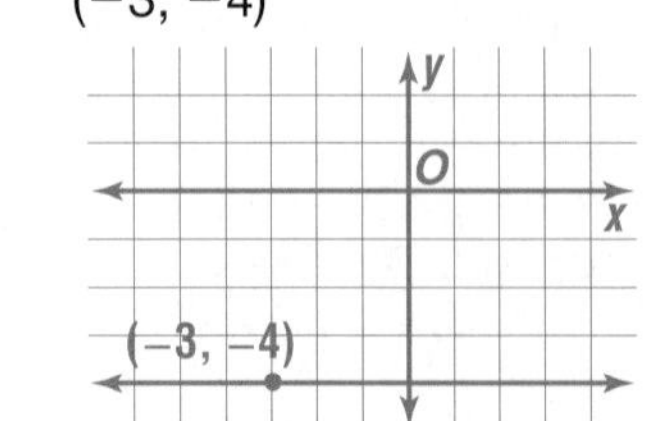

3-4 Proving Lines Parallel

What YOU'LL LEARN

- To recognize angle conditions that produce parallel lines, and
- to prove two lines are parallel based on given angle relationships.

Why IT'S IMPORTANT

You can use the various ways to prove lines are parallel in construction and physics.

APPLICATION
Farming

The orange trees on a farm in Polk County, Florida, appear to be planted in parallel lines. Using geometry, how could we prove the lines are parallel? The following Modeling Mathematics activity suggests one way to prove the lines are parallel.

MODELING MATHEMATICS — Construction of Parallel Lines

Materials:  graph paper 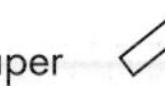 straightedge  compass  colored pencils

You can construct a line parallel to a given line through a point not on the line.

- Use a straightedge to draw line ℓ through $A(0, 0)$ and $R(2, 1)$. Locate $P(4, 5)$ not on ℓ.

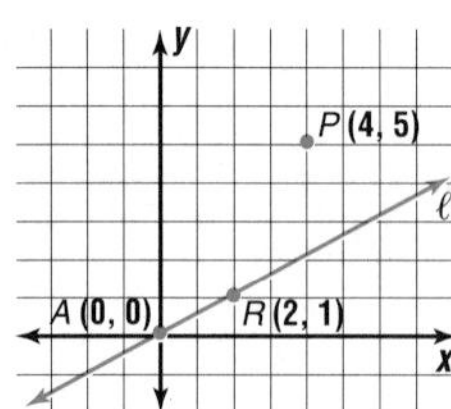

- Now draw a line through P that intersects line ℓ at R. Label $\angle 1$ as shown.

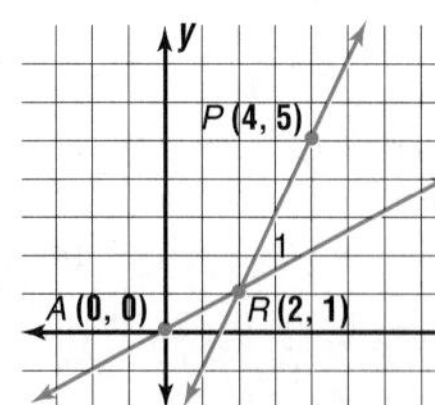

- Construct an angle congruent to $\angle 1$ using P as a vertex and $\overleftrightarrow{PR}$ as one side. Draw a line through P to form an angle congruent to $\angle 1$. Label the line m, the angle 2, and the point where m intersects the y-axis B.

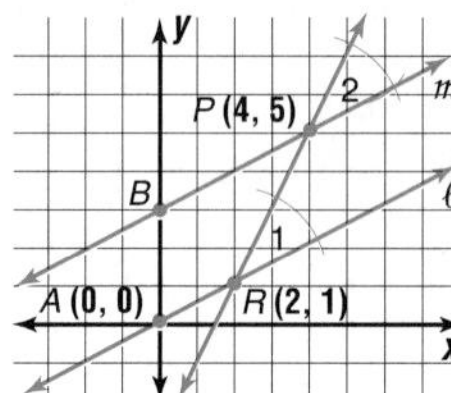

Your Turn

a. Identify the special angle pair name for $\angle 1$ and $\angle 2$. **corresponding**

b. Recall that $\angle 2$ was constructed congruent to $\angle 1$ and make a conjecture about ℓ and m.

c. Find the coordinates of point B where line m crosses the y-axis. Use points P and B to find the slope of line m.

d. Use points A and R to find the slope of ℓ. $\frac{1}{2}$

e. Do the results in parts c and d validate your conjecture about the relationship between lines ℓ and m? **Yes, the slopes are the same and $\ell \parallel m$.**

b. Lines ℓ and m appear to be $\parallel$. c. (0, 3); $\frac{1}{2}$

MODELING MATHEMATICS This activity draws together postulates that integrate geometry and algebra. Encourage students to identify the postulates used.

The Modeling Mathematics activity illustrates a postulate that helps to prove two lines are parallel. Notice that this postulate is the converse of Postulate 3–1.

Postulate 3–4	**If two lines in a plane are cut by a transversal so that corresponding angles are congruent, then the lines are parallel.**

This can be abbreviated as "If ⫲ and corr. $\angle$s are $\cong$, then the lines are $\parallel$."

The Modeling Mathematics activity establishes *at least* one line through P parallel to ℓ. In 1795, Scottish physicist and mathematician John Playfair provided the modern version of Euclid's famous **Parallel Postulate**, which states there is *exactly* one line parallel to a line through a given point not on the line.

Postulate 3–5 Parallel Postulate	**If there is a line and a point not on the line, then there exists exactly one line through the point that is parallel to the given line.**

There are sets of conditions other than Postulate 3–4 that prove that two lines are parallel. One of them is stated in Theorem 3–5.

Theorem 3–5	**If two lines in a plane are cut by a transversal so that a pair of alternate exterior angles are congruent, then the two lines are parallel.**

Proof of Theorem 3–5

Given: $\angle 1 \cong \angle 2$

Prove: $\ell \parallel m$

Proof:

Statements	Reasons
1. $\angle 1 \cong \angle 2$	**1.** Given
2. $\angle 2 \cong \angle 3$	**2.** Vertical angles are congruent.
3. $\angle 1 \cong \angle 3$	**3.** Congruence of angles is transitive.
4. $\ell \parallel m$	**4.** If ⫲ and corr. $\angle$s are $\cong$, then the lines are $\parallel$.

Theorems 3–6, 3–7, and 3–8 state three more ways to prove that two lines are parallel. *You will be asked to prove these theorems in Exercises 33, 13, and 34, respectively.*

Theorem 3–6	**If two lines in a plane are cut by a transversal so that a pair of consecutive interior angles is supplementary, then the lines are parallel.**
Theorem 3–7	**If two lines in a plane are cut by a transversal so that a pair of alternate interior angles is congruent, then the lines are parallel.**
Theorem 3–8	**In a plane, if two lines are perpendicular to the same line, then they are parallel.**

Motivating the Lesson

Questioning Draw two parallel lines on the chalkboard or overhead and ask students how they might prove the lines are parallel. Ask them how it might be done using angle relationships.

2 TEACH

Teaching Tip When discussing Postulate 3-5, point out that the word *exactly* means two things: there is not more than one line through the point, and there is *always* a line through the point.

Teaching Tip Have students discuss when it is best to use each of the four ways to prove two lines parallel.

Proof Pointer

Many students find it easier to understand an indirect proof of Theorem 3-5. Suppose $\ell \not\parallel m$. Then $\angle 1 \not\cong \angle 3$. Since $\angle 2 \cong \angle 3$, $\angle 1 \cong \angle 2$. This contradicts the assumption that $\angle 1 \not\cong \angle 2$. Therefore, $\ell \parallel m$.

Alternative Learning Styles

Auditory Discuss Postulate 3-4 with students and ask them to state its converse. Ask them to state the converses of each theorem in the lesson. Compare these with Postulate 3-5.

In-Class Examples

For Example 1

Find the value of x and $m\angle VST$ so that $\ell \parallel p$.

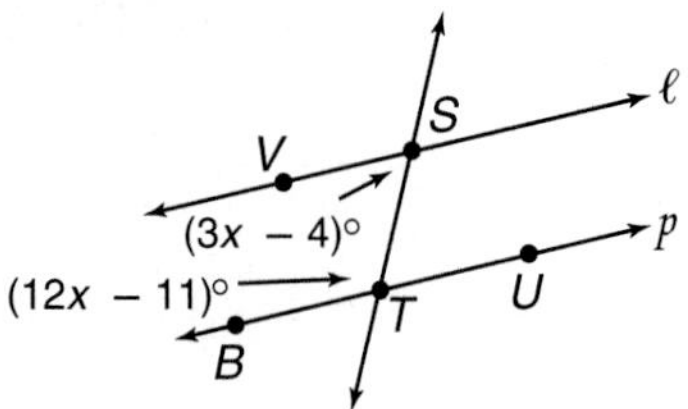

$x = 13$; $m\angle VST = 35$

For Example 2

If $\angle 2 \cong \angle 4$ and $\angle 3 \cong \angle 5$, which lines must be parallel? Explain.

$\overleftrightarrow{XW}$ is a transversal for $\overleftrightarrow{TX}$ and $\overleftrightarrow{WY}$. Since $\angle 2$ and $\angle 4$ are congruent alternate interior angles, $\overleftrightarrow{TX} \parallel \overleftrightarrow{WY}$. $\overleftrightarrow{XW}$ is a transversal for $\overleftrightarrow{TZ}$ and $\overleftrightarrow{SY}$. Since $\angle 3$ and $\angle 5$ are congruent alternate exterior angles, $\overleftrightarrow{TZ} \parallel \overleftrightarrow{SY}$.

For Example 3

Prove $\overleftrightarrow{XW} \parallel \overleftrightarrow{YZ}$ using two methods.

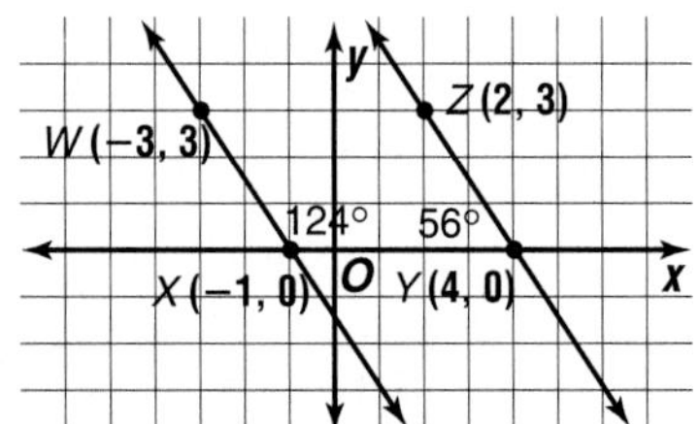

Method 1: Supplementary angles $\angle WXY$ and $\angle XYZ$ are consecutive interior angles, so $\overleftrightarrow{XW} \parallel \overleftrightarrow{YZ}$ by Theorem 3-6.
Method 2: slope $\overleftrightarrow{XW}$: $\frac{3-0}{-3-(-1)} = \frac{3}{-2}$ or $-\frac{3}{2}$, slope $\overleftrightarrow{YZ}$: $\frac{3-0}{2-4} = \frac{3}{-2}$ or $-\frac{3}{2}$. Since the slopes are the same, $\overleftrightarrow{XW} \parallel \overleftrightarrow{YZ}$.

Elements is a compilation of Euclid's predecessors' works. Euclid took their work and expanded upon it.

Algebra

Example 1 **Find the value of x and $m\angle ABC$ so that $p \parallel q$.**

Explore From the figure, you know that $m\angle ABC = 5x + 90$ and $m\angle BEF = 14x + 9$. You want to find the value of x and $m\angle ABC$ so that $p \parallel q$.

Plan If two lines are cut by a transversal so that corresponding angles are congruent, then the lines are parallel. So if $m\angle ABC = m\angle BEF$, then $p \parallel q$.

Solve

$m\angle ABC = m\angle BEF$

$5x + 90 = 14x + 9$ *Substitution*

$81 = 9x$ *Subtract 9 and 5x from each side.*

$9 = x$ *Divide each side by 9.*

Use the value of x to find $m\angle ABC$.

$m\angle ABC = 5x + 90$

$= 5(9) + 90$ or 135

Examine Since $m\angle BEF = 14x + 9$ and $x = 9$, $m\angle BEF = 14(9) + 9$ or 135. Therefore, $\angle ABC \cong \angle BEF$ and $p \parallel q$.

You can use relationships between angles to determine if lines are parallel.

Example 2 **If $\angle 1 \cong \angle 2$ and $\angle RAB \cong \angle CBS$, which lines must be parallel? Explain.**

$\overleftrightarrow{AE}$ is a transversal for $\overleftrightarrow{AB}$ and $\overleftrightarrow{DE}$. Since $\angle 1$ and $\angle 2$ are congruent alternate interior angles, $\overleftrightarrow{AB} \parallel \overleftrightarrow{DE}$.

$\overleftrightarrow{AS}$ is a transversal for $\overleftrightarrow{AR}$ and $\overleftrightarrow{BD}$. Since $\angle RAB$ and $\angle CBS$ are congruent corresponding angles, $\overleftrightarrow{AR} \parallel \overleftrightarrow{BD}$.

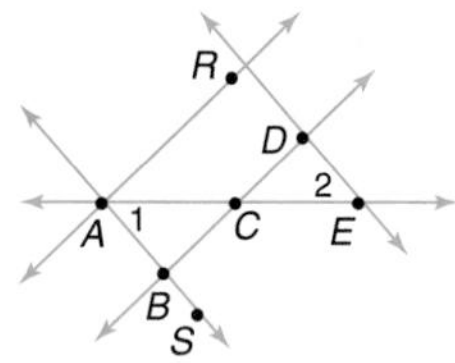

Some problems can be solved using either a geometric method or an algebraic method.

Example 3 **Prove $\overleftrightarrow{AB} \parallel \overleftrightarrow{CD}$ using two methods.**

Method 1: Using Geometric Theorems

$\angle BAD$ and $\angle CDA$ are congruent alternate interior angles, so $\overleftrightarrow{AB} \parallel \overleftrightarrow{CD}$ by Theorem 3–7.

Method 2: Using Slope

slope of $\overleftrightarrow{CD}$: $\frac{4-0}{4-3} = \frac{4}{1}$ or 4

slope of $\overleftrightarrow{AB}$: $\frac{0-(-4)}{-2-(-3)} = \frac{4}{1}$ or 4

Since the slopes are the same, $\overleftrightarrow{AB} \parallel \overleftrightarrow{CD}$.

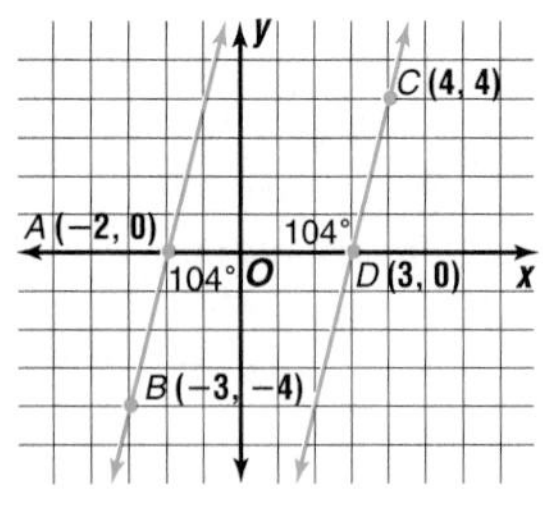

F Y I

Euclid's great book on geometry, *Elements,* has no reference to the word *geometry.*

Cooperative Learning

Co-op Co-op Separate the class into small groups and have each group attempt to construct a line parallel to a given line through a point not on the line. Have each group explain the steps it took to draw the line and why the two lines are parallel. For more information on the co-op co-op strategy, see *Cooperative Learning in the Mathematics Classroom*, one of the titles in the Glencoe Mathematics Professional Series, page 32.

CHECK FOR UNDERSTANDING

Communicating Mathematics

Study the lesson. Then complete the following. 1, 3, 4. See margin.

1. **Summarize** five different methods you can use to prove two lines are parallel.
2. **Explain** the meaning of *exactly* in the Parallel Postulate. Try to draw two lines parallel to a given line through a given point not on the line.

2. No more or no less than; such a drawing does not exist.

3. **You Decide** Using the figure at the right, Nancy claims $\overline{QU} \parallel \overline{AD}$, but Jasmine claims they definitely are not parallel. Who is right? Explain your answer.

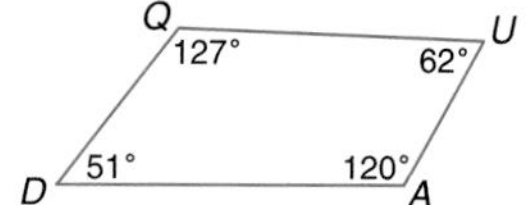

4. **Describe** two situations in your own life in which you encounter parallel lines. How could you guarantee the lines are parallel?

5. **Rearrange** the statements and their corresponding reasons in the following proof to write the seven steps of a correct proof.

Given: $r \parallel s$; $\angle 5 \cong \angle 6$

Prove: $\ell \parallel m$

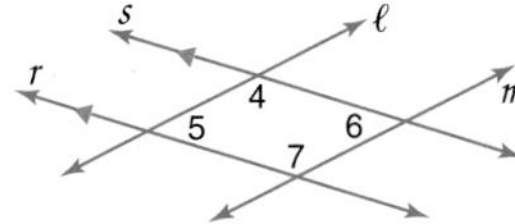

Proof:

Statements	**Reasons**
1. $m\angle 4 + m\angle 6 = 180$	a. Substitution Property (=)
2. $r \parallel s$, $\angle 5 \cong \angle 6$	b. Given
3. $m\angle 4 + m\angle 5 = 180$	c. Definition of supplementary
4. $\ell \parallel m$	d. If ⇻ and a pair of consecutive angles are supplementary, then the line are parallel.
5. $\angle 4$ and $\angle 6$ are supplementary.	e. Definition of supplementary angles
6. $\angle 4$ and $\angle 5$ are supplementary.	f. Conseutive Interior Angle Theorem.
7. $m\angle 5 = m\angle 6$	g. Definition of congruent angles

1. b; 2. f; 3. c; 4. g; 5. a; 6. e; 7. d

6. **Assess Yourself** Explain how you would construct parallel lines as described in Theorem 3–8. See margin.

Guided Practice

Given the following information, determine which lines, if any, are parallel. State the postulate or theorem that justifies your answer.

7. $a \parallel b$; If ⇻ and a pair of alt. int. ∡ are ≅, then the lines are ∥.

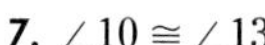

7. $\angle 10 \cong \angle 13$
8. $\angle 1 \cong \angle 6$ See margin.
9. $\angle 2 \cong \angle 13$ See margin.

Find the value of x so that $\ell \parallel m$.

10. 15

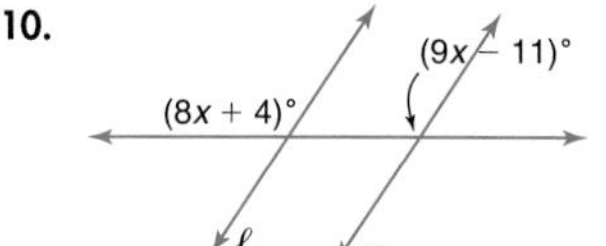

11. 7

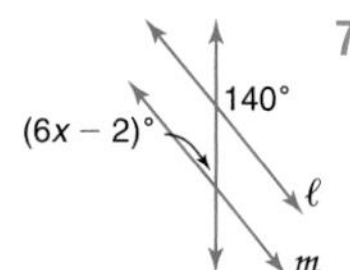

3 PRACTICE/APPLY

Check for Understanding

Exercises 1–14 are designed to help you assess your students' understanding through reading, writing, speaking, and modeling. You should work through Exercises 1–6 with your students and then monitor their work on Exercises 7–14.

Additional Answers

1. If two lines in a plane are cut by a transversal, the lines are ∥ if one of the following is true.
   (1) Corresponding angles are ≅.
   (2) Alternate exterior angles are ≅.
   (3) Consecutive interior angles are supplementary.
   (4) Alternate interior angles are ≅.
   (5) Both lines are ⊥ to the transversal.
3. Jasmine; since $127 + 51 = 178$ and $62 + 120 = 182$, the consecutive interior ∡ are not supplementary.
4. Sample answers: railroad tracks and yard lines on a football field; If a line is drawn across the tracks and a pair of alternate interior angles is ≅, the tracks are ∥. If the yard lines are ⊥ to one of the sidelines, the lines are ∥.
6. Construct a line ⊥ to a given line. Then construct another line ⊥ to the new line. This line and the given line are ∥.
8. $a \parallel m$; If ⇻ and a pair of alt. ext. ∡ is ≅, then the lines are ∥.
9. $\ell \parallel m$; If ⇻ and corr. ∡ are ≅, then the lines are ∥.

Reteaching

Using Theorems Have a student name an angle relationship, such as alternate interior angles or corresponding angles. Have students state the theorem from this lesson that contains that angle relationship and write it on the chalkboard or overhead. Diagram the theorem, labeling the relevant pair of angles and the parallel lines. You may also want to write the given information and what is to be proved.

Assignment Guide

Core (with proof): 15–41 odd, 43–52
Core (informal): 15–31 odd, 39, 41, 43–52
Enriched: 16–38 even, 39–52

For **Extra Practice,** see p. 769.

The red A, B, and C flags, printed only in the Teacher's Wraparound Edition, indicate the level of difficulty of the exercises.

Additional Answers

17. $\overleftrightarrow{JK} \parallel \overleftrightarrow{BL}$; If ⧣ and a pair of consecutive int. ∠s is supplementary, then the lines are ∥.
18. $\overleftrightarrow{EC} \parallel \overleftrightarrow{FH}$; In a plane, if 2 lines are ⊥ to the same line, then they are ∥.
19. $p \parallel q$; If ⧣ and a pair of alt. ext. ∠s is ≅, then the lines are ∥.
20. $\ell \parallel m$; If ⧣ and corr. ∠s are ≅, then the lines are ∥.
21. $p \parallel q$; If ⧣ and a pair of consecutive int. ∠s is supplementary, then the lines are ∥.
22. $\ell \parallel m$; If ⧣ and a pair of alt. int. ∠s is ≅, then the lines are ∥.
23. $\ell \parallel m$; If ⧣ and a pair of consecutive int. ∠s is supplementary, then the lines are ∥.

12. $q \parallel p$; In a plane if 2 lines are ⊥ to the same line, then they are ∥.

12. State which lines, if any, are parallel. State the postulate or theorem that justifies your answer.

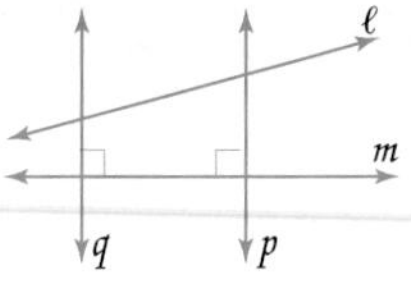

13. Copy and provide a reason for each step in the proof of Theorem 3–7.

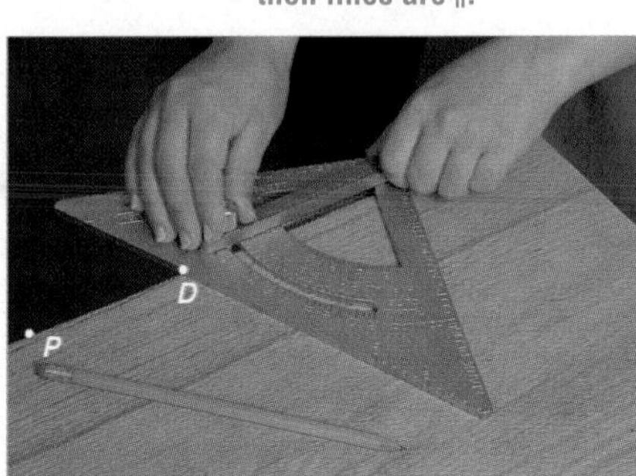

Given: $\angle 4 \cong \angle 6$

Prove: $\ell \parallel m$

Proof:

Statements	Reasons
1. $\angle 4 \cong \angle 6$	1. ? Given
2. $\angle 6 \cong \angle 7$	2. ? Vertical ∠s are ≅.
3. $\angle 4 \cong \angle 7$	3. ? Trans. Prop. ≅ ∠s
4. $\ell \parallel m$	4. ? If ⧣ and corr. ∠s are ≅, then lines are ∥.

14. Construction A carpenter uses a special instrument called an *engineer and carpenter square* to draw parallel line segments. Carmen wants to make two parallel cuts at an angle of 120° though points *D* and *P*. Explain why these lines will be parallel.
If ⧣ and corr. ∠s are ≅, then lines are ∥.

EXERCISES

Practice

Given the following information, determine which lines, if any, are parallel. State the postulate or theorem that justifies your answer. 17–23. See margin.

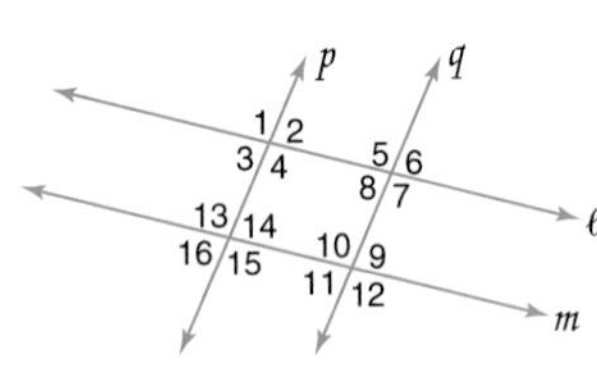

15. $\overleftrightarrow{EC} \parallel \overleftrightarrow{HF}$; If ⧣ and corr. ∠s are ≅, then the lines are ∥.

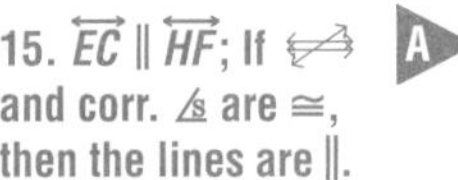

16. $\overleftrightarrow{EC} \parallel \overleftrightarrow{HF}$; If ⧣ and alt. int. ∠s are ≅, then the lines are ∥.

15. $\angle EAJ \cong \angle HGA$
16. $\angle BAD \cong \angle GDA$
17. $m\angle GAB + m\angle LBA = 180$
18. $\overleftrightarrow{EC} \perp \overleftrightarrow{BL}, \overleftrightarrow{FH} \perp \overleftrightarrow{BL}$
19. $\angle 1 \cong \angle 7$
20. $\angle 16 \cong \angle 3$
21. $m\angle 14 + m\angle 10 = 180$
22. $\angle 4 \cong \angle 13$
23. $m\angle 8 + m\angle 10 = 180$

24. Use slope to determine if $\ell \parallel m$. Verify your results by measuring $\angle 1$ and $\angle 2$. Which method do you prefer for showing whether ℓ is parallel to m?

Slope of $\ell = \frac{4}{3}$, slope of $m = \frac{4}{3}$; see students' work.

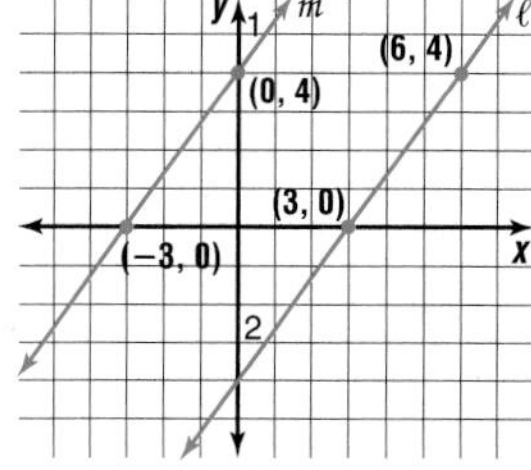

Find the value of x so that $\ell \parallel m$.

25.
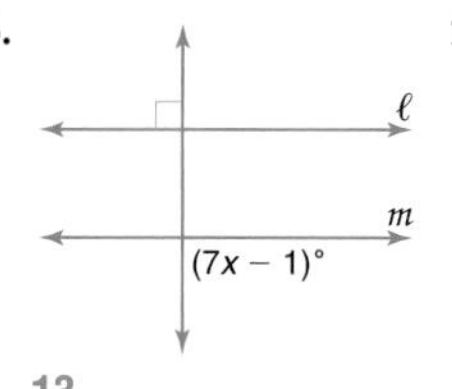

13

26.
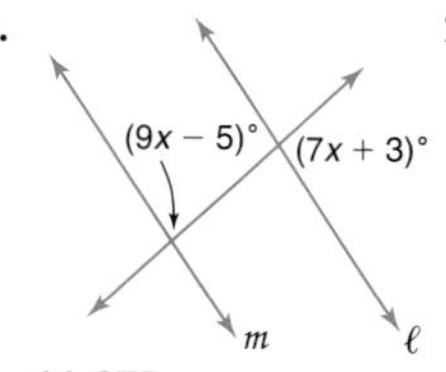

11.375

27.
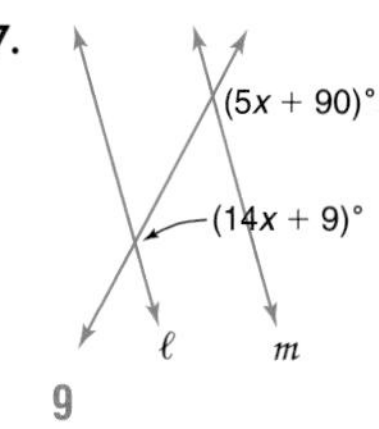

9

Find the values of x and y that make the blue lines parallel and the red lines parallel.

28.
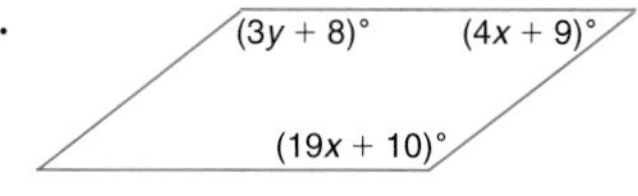

$x = 7$, $y = 45$

29.
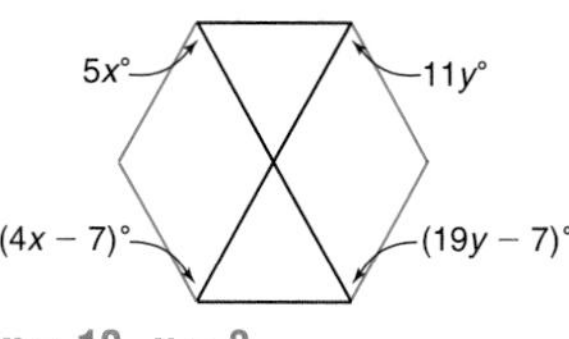

$x = 10$, $y = 3$

State which lines, if any, are parallel. State the postulate or theorem that justifies your answer.

Proof

30.
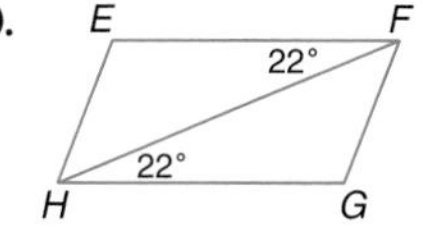

31.
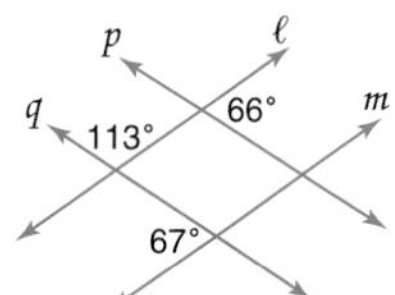

32.
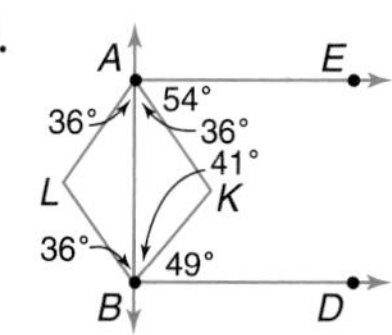

30. $\overline{EF} \parallel \overline{HG}$; If ⧣ and alt. int. ∡s are ≅, then the lines are ∥.

31. $\ell \parallel m$; If ⧣ and a pair of consecutive int. ∡s is supplementary, then the lines are ∥.

33. Copy and complete the proof of Theorem 3–6.

Given: $\angle 1$ and $\angle 2$ are supplementary.

Prove: $\ell \parallel m$

Proof:

Statements	Reasons
1. $\angle 1$ and $\angle 2$ are supplementary.	1. ? Given
2. $\angle 2$ and $\angle 3$ form a linear pair	2. ? Definition of linear pair
3. ? $\angle 2$ and $\angle 3$ are supplementary.	3. If 2 angles form a linear pair, they are supplementary.
4. $\angle 1 \cong \angle 3$	4. ?
5. $\ell \parallel m$	5. ?

33. (Reason 4) 2 ∡s supplementary to the same ∠ are ≅. (Reason 5) If ⧣ and corr. ∡s are ≅, then the lines are ∥.

34. Write a paragraph proof of Theorem 3–8. See margin.

Additional Answers

32. $\overrightarrow{AE} \parallel \overrightarrow{BD}$; In a plane if 2 lines are ⊥ to the same line, then they are ∥. $\overline{AK} \parallel \overline{LB}$; If ⧣ and a pair of alt. int. ∡s is ≅, then the lines are ∥.

34. Given: $\ell \perp t$
$m \perp t$
Prove: $\ell \parallel m$

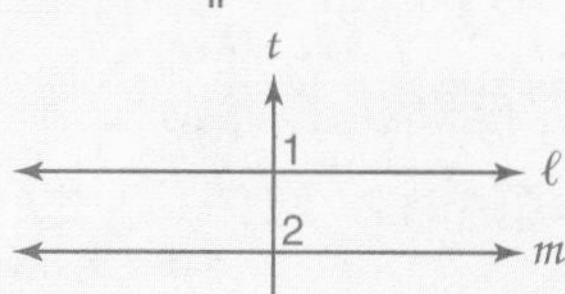

Proof:
We are given $\ell \perp t$ and $m \perp t$. Since ⊥ lines form right angles, $\angle 1$ and $\angle 2$ are right angles. $\angle 1 \cong \angle 2$ since right angles are ≅ to each other. If ⧣ and corr. ∡s are ≅, then the lines are ∥. Therefore, $\ell \parallel m$.

Study Guide Masters, p. 17

3-4 NAME ______ DATE ______

Study Guide

Student Edition Pages 146–153

Proving Lines Parallel

Suppose two lines in a plane are cut by a transversal. With enough information about the angles that are formed, you can decide whether the two lines are parallel.

IF	THEN
Corresponding angles are congruent, Alternate interior angles are congruent, Alternate exterior angles are congruent, Consecutive interior angles are supplementary, The lines are perpendicular to the same line,	the lines are parallel.

Example: If $\angle 1 = \angle 2$, which lines must be parallel? Explain.

$\overline{AC} \parallel \overline{BD}$ because a pair of corresponding angles are congruent.

Find the value of x so that a ∥ b.

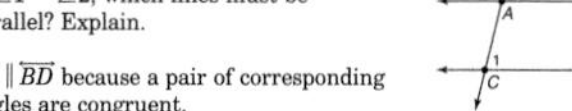

4. 17 5. 44 6. 15

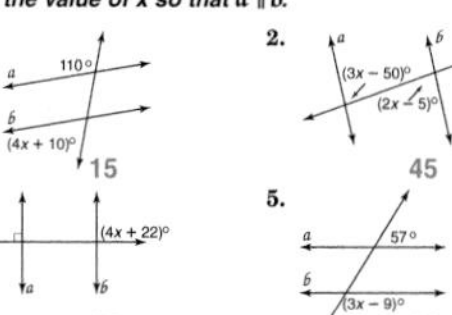

Given the following information, determine which lines, if any, are parallel. State the postulate or theorem that justifies your answer.

7. $\angle 1 \cong \angle 8$
$c \parallel d$, alternate exterior angles congruent

8. $\angle 4 \cong \angle 9$
$c \parallel f$, alternate interior angles congruent

9. $m\angle 7 + m\angle 13 = 180$
$e \parallel f$, consecutive interior angles supplementary

10. $\angle 9 \cong \angle 13$
$c \parallel d$, corresponding angles congruent

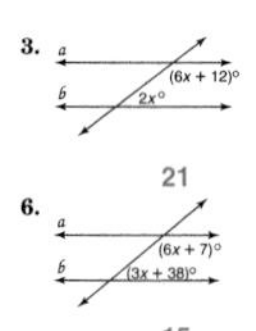

Additional Answers

35. Given: $\angle 2 \cong \angle 1$
$\angle 1 \cong \angle 3$
Prove: $\overline{ST} \parallel \overline{YZ}$
Proof:
Statements (Reasons)
1. $\angle 2 \cong \angle 1$, $\angle 1 \cong \angle 3$ (Given)
2. $\angle 2 \cong \angle 3$ (Congruence of ∠s is transitive.)
3. $\overline{ST} \parallel \overline{YZ}$ (If ∦ and a pair of alt. int. ∠s is ≅, then the lines are ∥.)

36. Given: $\overline{JO} \parallel \overline{KN}$
$\angle 1 \cong \angle 2$
$\angle 3 \cong \angle 4$
Prove: $\overline{KO} \parallel \overline{AN}$
Proof:
Statements (Reasons)
1. $\overline{JO} \parallel \overline{KN}$, $\angle 1 \cong \angle 2$, $\angle 3 \cong \angle 4$ (Given)
2. $\angle 1 \cong \angle 3$ (Corresponding Angles Postulate)
3. $\angle 3 \cong \angle 2$ (Congruence of ∠s is transitive.)
4. $\angle 2 \cong \angle 4$ (Congruence of ∠s is transitive.)
5. $\overline{KO} \parallel \overline{AN}$ (If ∦ and corr. ∠s are ≅, then the lines are ∥.)

Practice Masters, p. 17

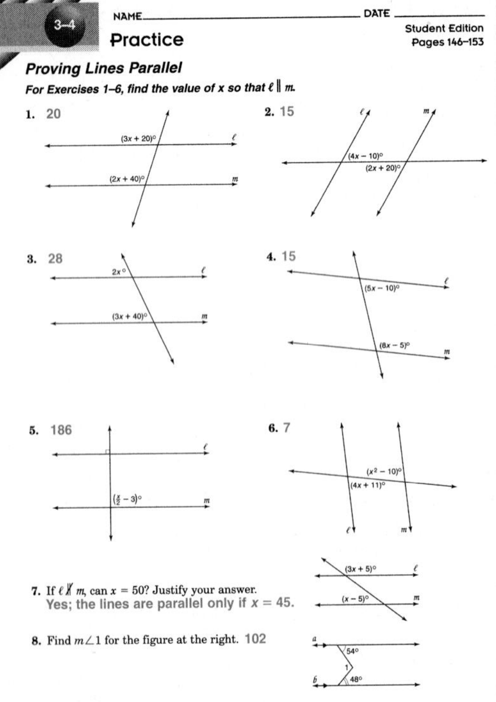

3-4 NAME ______ DATE ______

Practice — Student Edition Pages 146–153

Proving Lines Parallel

For Exercises 1–6, find the value of x so that ℓ ∥ m.

1. 20
2. 15
3. 28
4. 15
5. 186
6. 7
7. If ℓ ∦ m, can x = 50? Justify your answer. Yes; the lines are parallel only if x = 45.
8. Find m∠1 for the figure at the right. 102

Write a two-column proof. 35–37. See margin.

35. **Given:** $\angle 2 \cong \angle 1$
$\angle 1 \cong \angle 3$
Prove: $\overline{ST} \parallel \overline{YZ}$

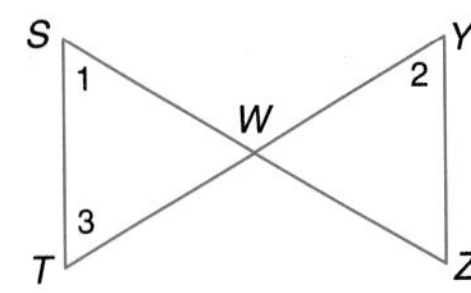

36. **Given:** $\overline{JO} \parallel \overline{KN}$
$\angle 1 \cong \angle 2$
$\angle 3 \cong \angle 4$
Prove: $\overline{KO} \parallel \overline{AN}$

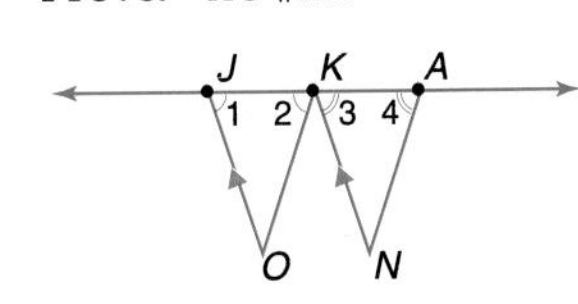

C

37. **Given:** $\overline{AU} \perp \overline{QU}$
$\angle 1 \cong \angle 2$
Prove: $\overline{DQ} \perp \overline{QU}$

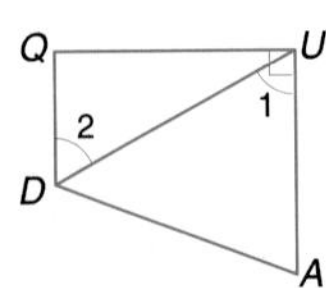

38. Use the figure below to determine the relationship between a and b so that $\ell \parallel m$. $ab = 20$

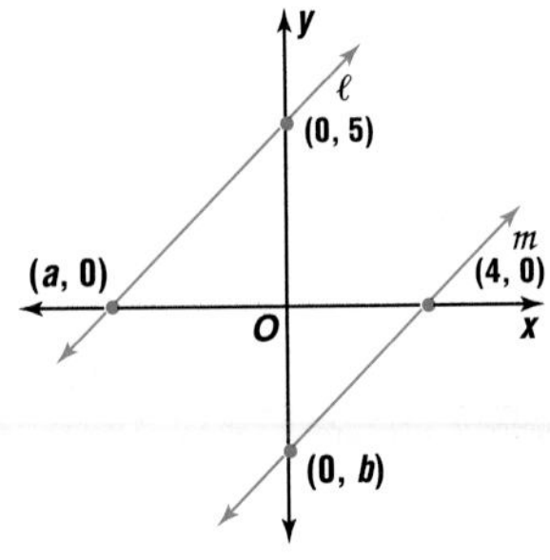

Critical Thinking

39. Suppose lines a, b, and c lie in the same plane and $a \parallel b$ and $a \parallel c$.
a. Draw a figure showing lines a, b, and c. a–b. See Solutions Manual.
b. Explain how you would prove that $b \parallel c$.

Applications and Problem Solving

40. **Construction** Carpenters use parallel lines in creating walls for construction projects. Copy the drawing at the right. Label your drawing and describe three different ways to guarantee that the wall studs are parallel. See margin.

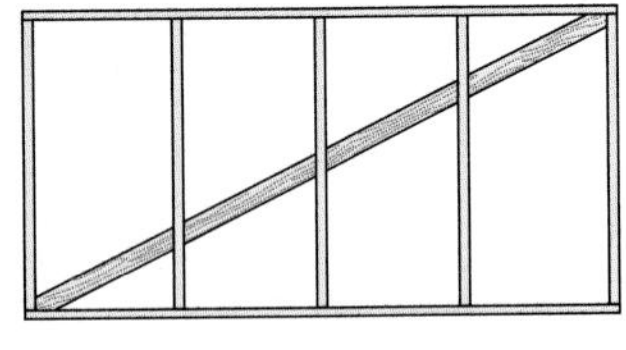

41. **Physics** The diagram represents a parabolic concave mirror as used in the Hubble Telescope. The Hubble Telescope gathers parallel light rays and directs them to a central focal point. Use a protractor to measure several of the angles shown in the diagram. Are the lines parallel? yes

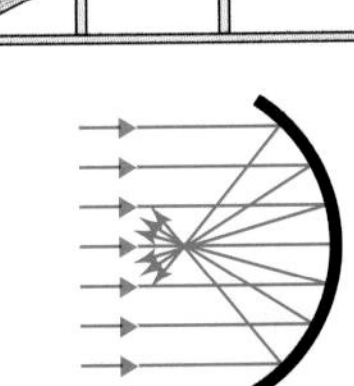

FYI

The Hubble Telescope orbits 380 miles above Earth allowing it to obtain images before they are distorted by Earth's atmosphere. It can observe objects that are only 2% as bright as those barely detected by telescopes on Earth.

42. **Architecture** Check with a classmate or teacher who participates in your school drafting classes. How do they draw parallel lines? If your school uses computer-aided design (CAD), determine the applications of parallel lines used in CAD. See students' work.

Mixed Review

43. Find the slope of the line that passes through points at $(-2, 3)$ and $(-7, -1)$. Then describe the line as you move from left to right as *rising*, *falling*, *horizontal*, or *vertical*. (Lesson 3–3)

43. $\frac{4}{5}$; rising

FYI

The Hubble Telescope is the only space-based astronomical observatory. It was launched in 1990.

Additional Answer

37. Given: $\overline{AU} \perp \overline{QU}$
$\angle 1 \cong \angle 2$
Prove: $\overline{DQ} \perp \overline{QU}$
Proof:
Statements (Reasons)
1. $\overline{AU} \perp \overline{QU}$, $\angle 1 \cong \angle 2$ (Given)
2. $\overline{AU} \parallel \overline{DQ}$ (If ∦ and a pair of alt. int. ∠s is ≅, then the lines are ∥.)
3. $\overline{DQ} \perp \overline{QU}$ (Perpendicular Transversal Theorem)

44. False; the alternate interior angles are congruent and the consecutive interior angles are supplementary.

44. True or false: *If two parallel lines are cut by a transversal, the alternate interior angles are supplementary.* If false, explain why. (Lesson 3–2)

45. Recreation Are the spokes on a wheel a model of parallel, intersecting, or skew lines? Explain. (Lesson 3–1) Intersecting; they all meet at the hub.

46. Find the measures of $\angle 1$ and $\angle 2$ if $m\angle 1 = 2x + 15$ and $m\angle 2 = 8x - 5$. (Lesson 2–6) 49; 131

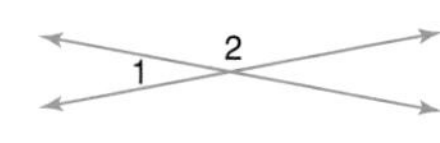

47. If possible, write a valid conclusion from the two true statements *If two lines are parallel, then they never meet* and *Lines p and m are parallel.* State the law of logic that you used. (Lesson 2–3)

48. Find the measure of the supplement of an angle that measures 159. (Lesson 1–7) 21

47. Lines p and m never meet; Law of Detachment.

49. Find the midpoint of the segment whose endpoints have coordinates (8, 11) and (−4, 7). (Lesson 1–5) (2, 9)

50. Education James forgot the combination to his gym locker. He remembers that the numbers are 18, 37, and 12, but doesn't remember the order. How many different combinations are possible? (Lesson 1–2) 6 combinations

Algebra

51. Factor $16x^2 - 22xy - 3y^2$. $(2x - 3y)(8x + y)$

52. Solve $f = \frac{W}{g} \cdot \frac{V^2}{R}$ for R. $R = \frac{WV^2}{fg}$

WORKING ON THE

In·ves·ti·ga·tion

Refer to the Investigation on pages 66–67.

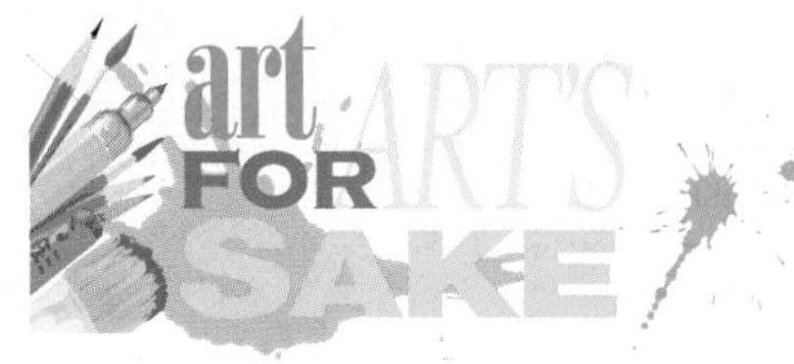

You may want to use a system for making the three-dimensional objects appear realistic in your mural by using linear perspective. Use the following steps to draw a room with a tile floor using linear perspective.

1. Draw a square with sides 15 centimeters long and separate the bottom edge into ten congruent segments.
2. Select a vanishing point V in the upper portion of the square, along the perpendicular bisector of the bottom edge of the square.
3. Mark points Z_1 and Z_2 ten centimeters to the left and right of V so that V, Z_1 and Z_2 are collinear and $\overline{Z_1Z_2}$ is parallel to the bottom edge of the square.
4. Draw lines connecting each point on the bottom edge with V. Lightly draw lines connecting the three rightmost points with Z_1 and lines connecting the three leftmost points with Z_2.
5. Start at the bottom of the square and draw ten horizontal lines between the leftmost and rightmost lines to V. The lines should pass through the points of intersection of the lines to V and the lines to Z_1 and Z_2.
6. Draw a line from the upper corners of the square to point V. Then draw two vertical lines from these lines to the leftmost and rightmost intersection points on the top horizontal line.
7. Finally, draw a horizontal line between the two verticals. Erase the lines to Z_1 and Z_2 and the lines within the small square that represents the back wall of your room to complete the picture.
8. You may wish to experiment with different methods of perspective, such as moving the vanishing point or changing the distance between V and Z_1 and Z_2.
9. Work with your group to begin creating your mural.

Add the results of your work to your Investigation folder.

4 ASSESS

Closing Activity

Modeling Have students construct a four-sided figure with each pair of opposite sides parallel. Ask them what the figure is called. parallelogram They will learn more about parallelograms in Chapter 6.

Additional Answer

40. Sample answer:

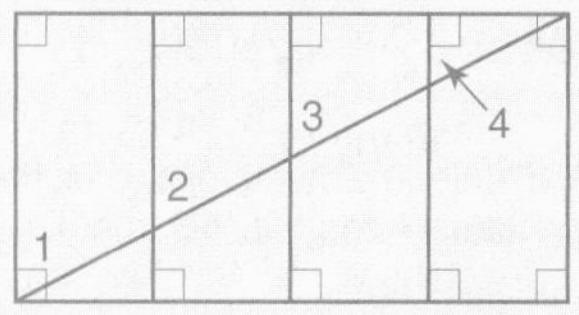

$m\angle 1 = m\angle 2 = m\angle 3 = m\angle 4$; all vertical lines should make right angles with the bottom board; all vertical lines should make right angles with the top board.

Enrichment Masters, p. 17

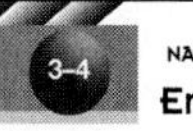

NAME__________ DATE ________

Enrichment

Student Edition Pages 146–153

Parallelism in Space

In space geometry, the concept of parallelism must be extended to include two planes and a line and a plane.

Definition: Two planes are parallel if and only if they do not intersect.

Definition: A line and a plane are parallel if and only if they do not intersect.

Thus, in space, two lines can be intersecting, parallel, or skew while two planes or a line and a plane can only be intersecting or parallel. In the figure at the right, $t \perp \mathcal{M}$, $t \perp \mathcal{P}$, $\mathcal{P} \parallel \mathcal{H}$, and ℓ and n are skew.

The following five statements are theorems about parallel planes.

Theorem: Two planes perpendicular to the same line are parallel.

Theorem: Two planes parallel to the same plane are parallel.

Theorem: A line perpendicular to one of two parallel planes is perpendicular to the other.

Theorem: A plane perpendicular to one of two parallel planes is perpendicular to the other.

Theorem: If two parallel planes each intersect a third plane, then the two lines of intersection are parallel.

Use the figure given above for Exercises 1 to 10. State yes or no to tell whether the statement is true.

1. $\mathcal{M} \parallel \mathcal{P}$ yes
2. $\ell \parallel n$ no
3. $m \parallel \mathcal{H}$ yes
4. $\ell \parallel \mathcal{P}$ yes
5. $\ell \perp t$ yes
6. $n \parallel \mathcal{H}$ yes
7. $\ell \perp \mathcal{P}$ no
8. $t \parallel \mathcal{H}$ no
9. $\mathcal{M} \perp t$ yes
10. $t \perp \mathcal{H}$ yes

For Exercises 11 to 14, make a small sketch at the bottom of the page to show that each statement is false.

11. If two lines are parallel to the same plane, then the lines are parallel.
12. If two planes are parallel, then any line in one plane is parallel to any line in the other plane.
13. If two lines are parallel, then any plane containing one of the lines is parallel to any plane containing the other line.
14. If two lines are parallel, then any plane containing one of the lines is parallel to the other line.

11. 12. 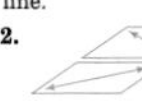 13. 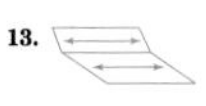 14. 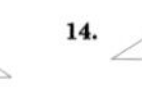

Extension

Communication Introduce students to the use of "if and only if," using theorems in Lessons 3-3 and 3-4. Point out that, in order to prove these statements, you must assume the first part and prove the second, and then assume the second part and prove the first.

In·ves·ti·ga·tion

Working on the Investigation

The Investigation on pages 66–67 is designed to be a long-term project that is completed over several days or weeks. Encourage students to keep their materials in their Investigation Folder as they work on the Investigation.

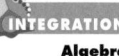

3-5 LESSON NOTES

NCTM Standards: 1–4, 7

Instructional Resources

- Study Guide Master 3-5
- Practice Master 3-5
- Enrichment Master 3-5
- Assessment and Evaluation Masters, p. 73
- Multicultural Activity Masters, p. 6
- Real-World Applications, 6

Transparency 3-5A contains the 5-Minute Check for this lesson; **Transparency 3-5B** contains a teaching aid for this lesson.

Recommended Pacing	
Standard Pacing	Day 9 of 14
Honors Pacing	Day 8 of 13
Block Scheduling*	Day 5 of 7

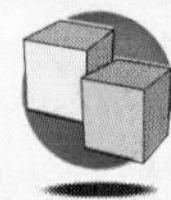

*For more information on pacing and possible lesson plans, refer to the *Block Scheduling Booklet*.

1 FOCUS

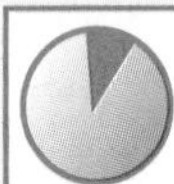

5-Minute Check
(over Lesson 3-4)

Given the following information, determine which segments, if any, are parallel. State the postulate or theorem that justifies your answer.

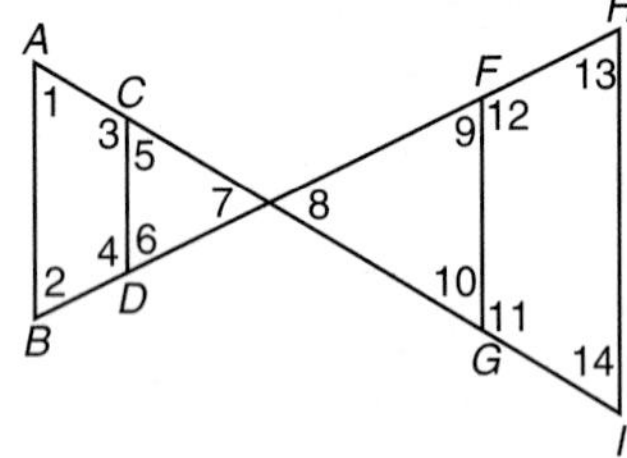

1. $\angle 1 \cong \angle 5$ **$\overline{AB} \parallel \overline{CD}$; if $\not\Leftarrow$ and corr. ⦣ are ≅, then the lines are ∥.**
2. $\angle 1 \cong \angle 6$ **none**
3. $m\angle 11 + m\angle 14 = 180$ **$\overline{FG} \parallel \overline{HI}$; if $\not\Leftarrow$ and consecutive int. ⦣ are supplementary, then the lines are ∥.**
4. Find the values of x and y so that $\overline{AB} \parallel \overline{CD}$ in the figure above if $m\angle 1 = 25$, $m\angle 5 = 7x - 10$, and $m\angle 3 = 2y + 3x$. **$x = 5, y = 70$**

3-5 Parallels and Distance

What **YOU'LL LEARN**

- To recognize and use distance relationships among points, lines, and planes.

Why **IT'S IMPORTANT**

You can find the correct distance between points and lines and between parallel lines and parallel planes.

The following Modeling Mathematics activity investigates the shortest distance between a line and a point not on the line.

MODELING MATHEMATICS — Distance Between a Line and a Point

Materials:  lined notebook paper  cardboard 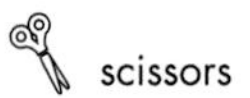 straightedge scissors  thumbtack

You can draw many line segments from a line to a point not on the line. But which one of the segments is the shortest one?

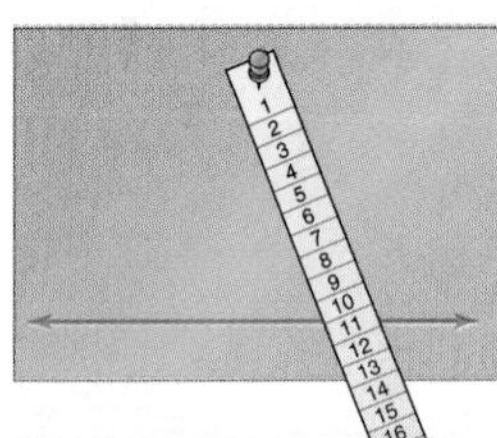

- Cut a vertical strip from a sheet of lined notebook paper, then number the lines from top to bottom.
- Draw a horizontal line on a piece of cardboard. Use a thumbtack to attach the numbered paper strip at a point above the horizontal line.
- Move the strip to the far right. Using the numbers on the strip as a scale, record the measure from the thumbtack to the horizontal line. Move the strip a little to the left and record the new distance. Continue moving the strip to the left and recording the measure until the strip is at the far left. You should record at least five measures.

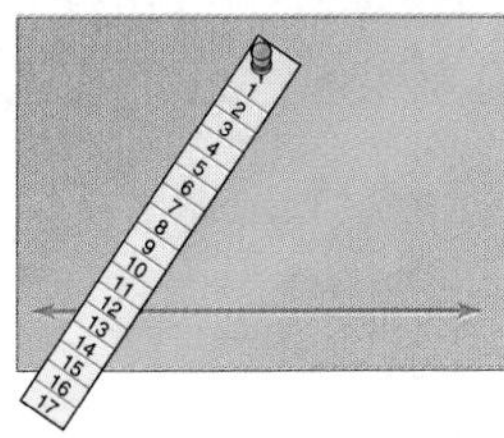

Your Turn

a. Move the strip to the position of your smallest measure. Can you locate any other position with a smaller measure? If so, move the strip to that position. **a. See students' work.**

b. The last position of the strip of paper in part a represents the shortest segment from the line to the point. What seems true about the strip of paper and the horizontal line? **They seem to be perpendicular.**

c. Make a conjecture about the shortest segment from a line to a point not on the line. **c. The shortest segment from a line to a point is perpendicular to the line.**

The shortest segment from a point to a line is the perpendicular segment from the point to the line.

Definition of the Distance Between a Point and a Line	**The distance from a line to a point not on the line is the length of the segment perpendicular to the line from the point.**

The measure of the distance between a line and a point on the line is zero.

MODELING MATHEMATICS In this activity, students use a try-all-possibilities approach to discover the shortest segment between a line and a point not on the line.

Example 1 **Draw the segment that represents the distance from R to $\overline{AB}$.**

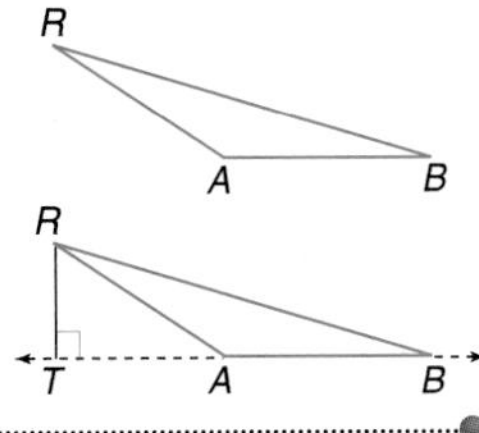

Since the distance from a line to a point not on the line is the length of the segment perpendicular to the line from the point, extend $\overline{AB}$ and draw $\overline{RT}$ so that $\overline{RT} \perp \overleftrightarrow{AB}$.

In Example 1, you can guarantee $\overline{RT} \perp \overleftrightarrow{AB}$ when you draw the segment by using the construction of a line perpendicular to a line through a point not on the line.

Example 2 **Construct a line perpendicular to line r through $A(-4, 0)$, not on r. Then find the distance from A to r.**

Copy line r and point A. Place the compass point at point A. Make the setting wide enough so that when an arc is drawn, it intersects r in two places. Label these points of intersection B and C.

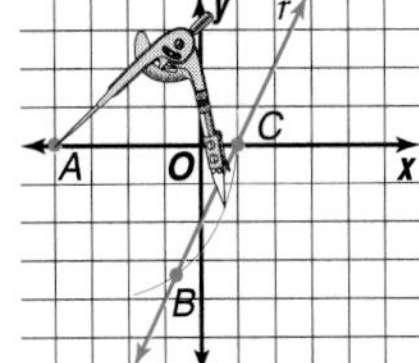

Put the compass at point B and draw an arc below line r. (*Hint:* Any compass setting greater than $\frac{1}{2}$ BC will work.)

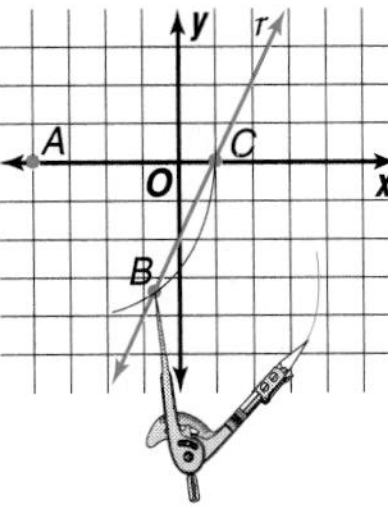

Using the same compass setting, put the compass at point C and draw an arc to intersect the one drawn in the previous step. Label the point of intersection D.

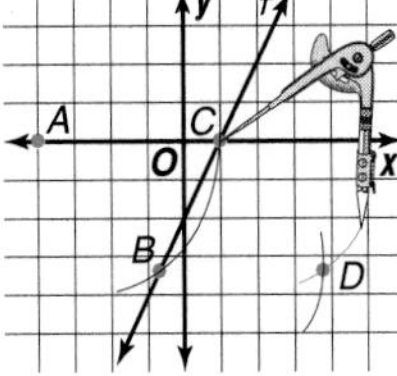

Draw $\overleftrightarrow{AD}$. $\overleftrightarrow{AD} \perp r$. Label point E at the intersection of $\overleftrightarrow{AD}$ and r. *Use the slopes of $\overleftrightarrow{AD}$ and r to verify that the lines are perpendicular.*

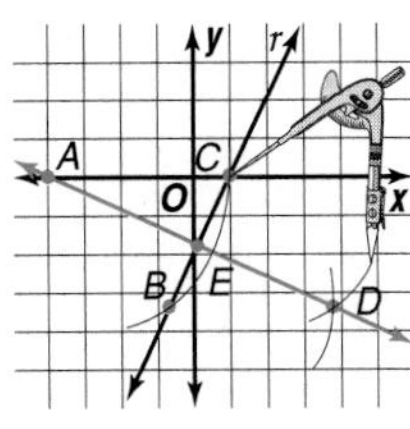

(continued on the next page)

Motivating the Lesson

Questioning Have each student draw a triangle and label the three vertices. Ask them how far each vertex is from the opposite side. Have them draw the segments they measured and describe how they determined which segments were the best ones to measure.

2 TEACH

In-Class Examples

For Example 1

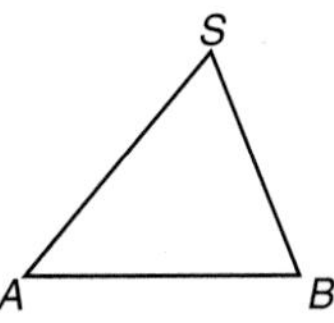

Draw the segment that represents the distance from S to $\overline{AB}$.

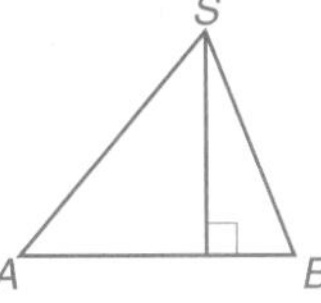

For Example 2

Construct a line perpendicular to line w through $H(2, 1)$ not on w. Then find the distance from H to w. **about 2.2 units**

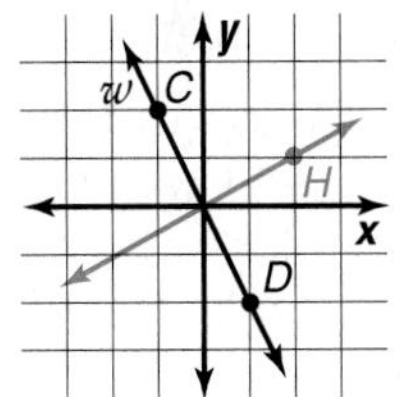

In-Class Example

For Example 3

Find the distance between the parallel lines whose equations are $y = x + 6$ and $y = x - 10$. $\sqrt{128}$ **or about 11.31 units**

Teaching Tip When finding the distance between two parallel lines, remind students that distance is always positive.

LOOK BACK

You can refer to Lesson 1-4 for information on finding the distance between points.

The line segment constructed from point $A(-4, 0)$ perpendicular to the line r intersects line r at $E(0, -2)$. We can use the distance formula to find the distance between point A and line r.

$$d = \sqrt{(x_2 - x_1)^2 + (y_2 - y_1)^2}$$
$$= \sqrt{(-4 - 0)^2 + (0 - (-2))^2}$$
$$= \sqrt{20}$$

The distance between A and r is $\sqrt{20}$ or about 4.47 units.

Remember that the distance from a point to a line is the perpendicular distance from the point to the line.

In this chapter, we have learned how to prove two lines parallel using slopes in the coordinate plane or by using properties of angle relationships formed by a transversal to the parallel lines. Yet, how could you show two lines parallel if there were no coordinate plane and no transversal to the lines?

According to the definition, two parallel lines do not intersect. An alternate definition states that two lines in a plane are parallel if they are everywhere **equidistant**. Equidistant means that the distance between two lines measured along a perpendicular line to the line is always the same. To measure the distance between two parallel lines, we can measure the distance between one of the lines and any point on the other line.

Definition of the Distance Between Parallel Lines	**The distance between two parallel lines is the distance between one of the lines and any point on the other line.**

Example 3

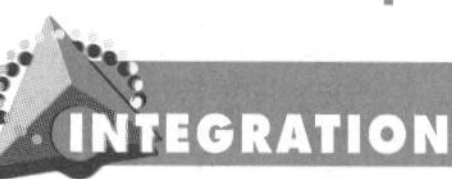

Algebra

Find the distance between the parallel lines ℓ and m whose equations are $y = 2x + 2$ and $y = 2x - 3$, respectively.

Explore You know the equations of two parallel lines and wish to find the distance between the lines.

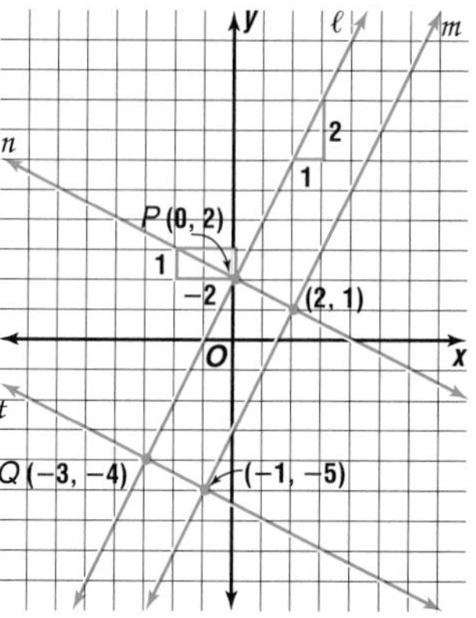

Plan Graph the lines on a coordinate plane. Pick a point such as $P(0, 2)$ on line ℓ. Draw a line perpendicular to ℓ through P and use the distance formula to find the distance between ℓ and P. *Recall that the product of the slopes of perpendicular lines is −1.*

Solve Since line ℓ has a vertical change of 2 units for a horizontal change of 1 unit, its slope is 2. The slope of a line perpendicular to ℓ is $-\frac{1}{2}$. From point P, go left 2 units and up 1 unit to $(-2, 3)$. Draw line n through $(0, 2)$ and $(-2, 3)$. Line n is perpendicular to lines ℓ and m and appears to intersect line m at $(2, 1)$. Find the distance between $(0, 2)$ and $(2, 1)$.

Alternative Teaching Strategies

Reading Geometry Have students think of words that begin with *equ* and look them up in the dictionary. Most of the definitions will contain words like "equal," "same," "balanced," etc. Have students conjecture what *equidistant* means and look it up. Explain how it relates to parallel lines.

$$d = \sqrt{(x_2 - x_1)^2 + (y_2 - y_1)^2}$$
$$= \sqrt{(0 - 2)^2 + (2 - 1)^2}$$
$$= \sqrt{5}$$

The distance between the lines is $\sqrt{5}$ or about 2.24 units.

Examine Pick another point $Q(-3, -4)$ on line ℓ. Since $\ell \parallel m$, the distance from Q to m should be the same, $\sqrt{5}$. Draw a line perpendicular to line m through Q. This line intersects m at $(-1, -5)$. Use the distance formula to find the distance between $Q(-3, -4)$ and $(-1, -5)$.

$$d = \sqrt{(x_2 - x_1)^2 + (y_2 - y_1)^2}$$
$$= \sqrt{(-3 - (-1))^2 + (-4 - (-5))^2}$$
$$= \sqrt{5}$$

In both cases, the distance between the lines is $\sqrt{5}$, so the answer checks.

For figures not on a coordinate plane, you can actually measure the distance between lines at several different points to determine if they are parallel.

Example 4

CONNECTION

Art

Use a ruler to determine whether the lines shown in black by the artist are parallel.

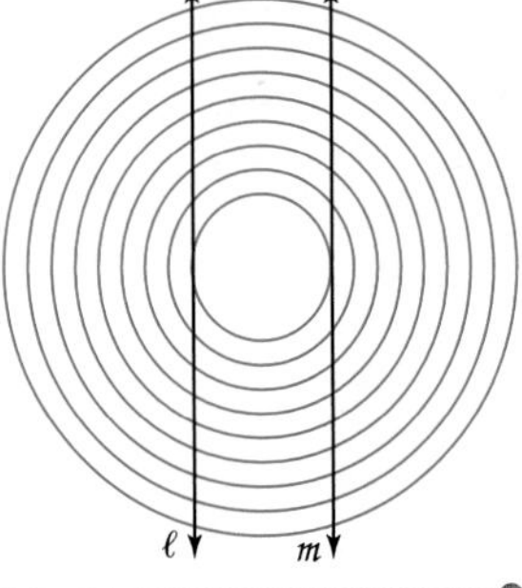

Choose several points on either line, such as line ℓ, and measure the perpendicular distance from ℓ to m with a ruler. The measurements will verify that ℓ and m are everywhere equidistant.

CHECK FOR UNDERSTANDING

Communicating Mathematics

Study the lesson. Then complete the following. 1–3. See margin.

1. **Compare** the following definition for *parallel* from the *American Heritage Dictionary* and the alternate definition given in this lesson. How could you verify the dancers are in "parallel rows"?

   parallel being equal distance apart everywhere: *dancers in two parallel rows*

2. **Examine** the linear equations used for lines ℓ and m in Example 3. Describe the coefficient of the x term in each equation. What information does the coefficient provide regarding the slope of the lines?
3. **Make up a problem** involving an everyday situation where you need to find the distance between a point to a line or the distance between two lines. For example, find the shortest path from the garage of a house to the street to minimize the length of a driveway and material used in its construction.

In-Class Example

For Example 4

Find two lines that appear to be parallel. Use a ruler to determine whether the lines are parallel. **Answers will vary.**

3 PRACTICE/APPLY

Check for Understanding

Exercises 1–13 are designed to help you assess your students' understanding through reading, writing, speaking, and modeling. You should work through Exercises 1–5 with your students and then monitor their work on Exercises 6–13.

Additional Answers

1. **Both definitions define parallel as being equidistant. You could measure the perpendicular distance between the rows of dancers to see if they are everywhere equidistant.**
2. **In both equations, the coefficient of the x term is 2. The coefficient is the slope of the lines.**
3. **Sample answer: You are camping and need to find the shortest path to the river to get water.**

Reteaching

Using Models Tape a piece of string to the chalkboard. Draw a line and have a student find the shortest length of string needed to reach the line. Have the student trace over the string to draw a segment from the point to the line. What do the students notice about the line and the segment? Have other students do the same with different points and lines to test the conjecture. Also, have a student measure the distance between two parallel lines at several different locations. What does the student notice? Have several students repeat the activity to test the conjecture.

Assignment Guide

Core (with proof): 15–37 odd, 38, 39, 41–49
Core (informal): 15–37 odd, 38, 39, 41–49
Enriched: 14–36 even, 38–49

For **Extra Practice,** see p. 769.

The red A, B, and C flags, printed only in the Teacher's Wraparound Edition, indicate the level of difficulty of the exercises.

Additional Answers

4. You can show that 2 lines in a plane are parallel by comparing the slopes, by using a transversal and angle relationships, and by measuring to see if the lines are everywhere equidistant. To compare slopes, you need to know the vertical and horizontal change. To use transversals and angle relationships, you need a transversal. You can always measure the distance between the lines.
5. Pick any point on one line. Construct a perpendicular segment from the point to the other line.
9. 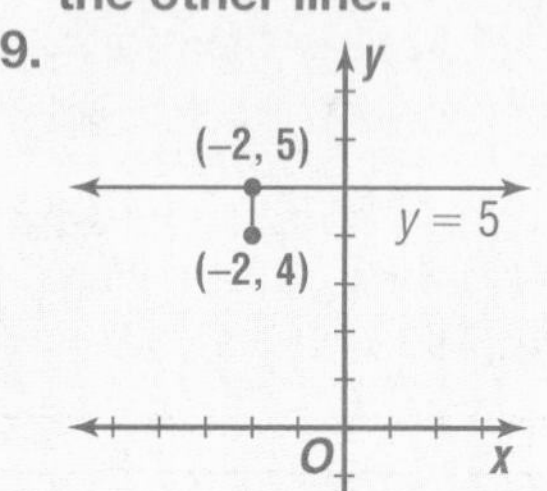

10. 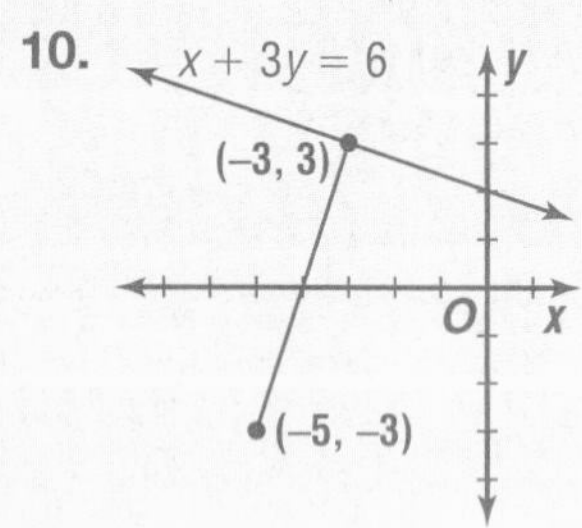

4. **Describe** three distinct methods you can use to show two lines in a plane are parallel. Compare and contrast the methods. See margin.

MODELING MATHEMATICS

5. **Explain** how to construct a segment that represents the distance between two parallel lines, not on a coordinate plane. See margin.

Guided Practice

Copy each figure and draw the segment that represents the distance indicated.

6. Q to $\overleftrightarrow{AD}$

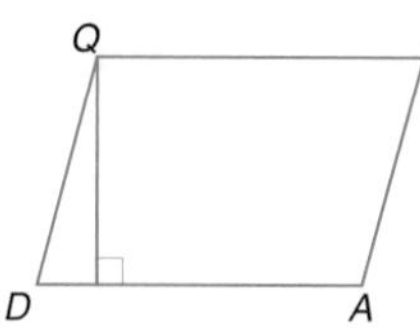

7. s to t

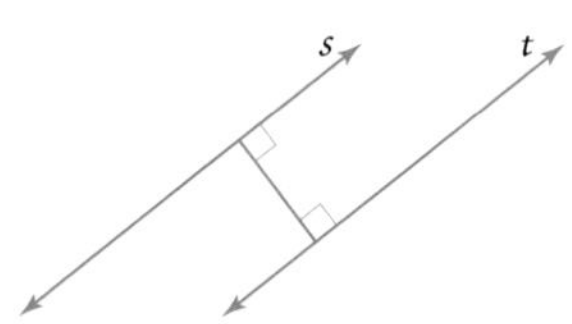

8. **Art** Use a ruler to determine whether the black lines are parallel. Write *yes* or *no*. Explain your answer. Yes; the lines are everywhere equidistant.

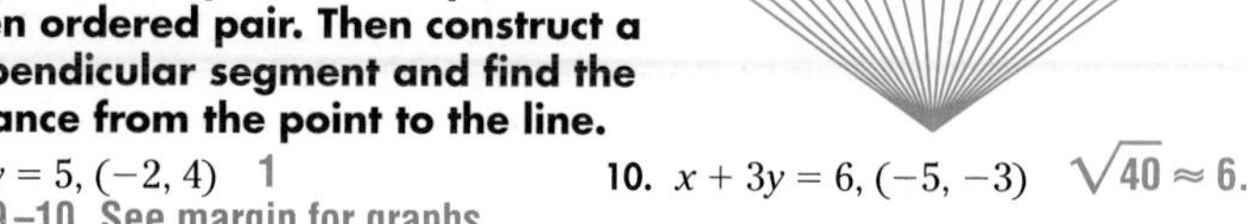

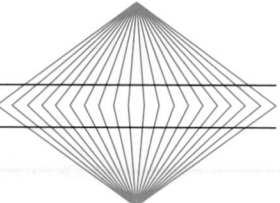

Graph each equation and plot the given ordered pair. Then construct a perpendicular segment and find the distance from the point to the line.

9. $y = 5, (-2, 4)$ 1
10. $x + 3y = 6, (-5, -3)$ $\sqrt{40} \approx 6.32$

9–10. See margin for graphs.

In the figure at the right, $\overline{PS} \perp \overline{SQ}$, $\overline{PQ} \perp \overline{QR}$, and $\overline{QR} \perp \overline{SR}$. Name the segment whose length represents the distance between the following points and lines.

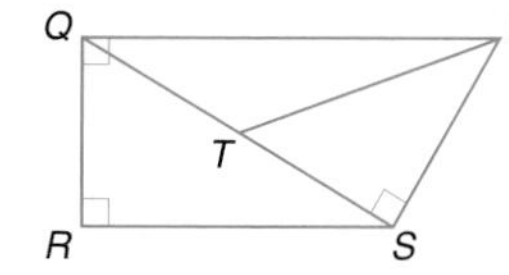

11. P to $\overleftrightarrow{SQ}$ $\overline{PS}$
12. S to $\overleftrightarrow{QR}$ $\overline{RS}$

13. **Interior Design** Terrell is installing a curtain rod on the wall above the window. In order to ensure the rod is parallel to the ceiling, he measures and marks 9 inches below the ceiling in several places. If he installs the rod at these markings centered over the window, how does he know the curtain rod will be parallel to the ceiling? It is everywhere equidistant.

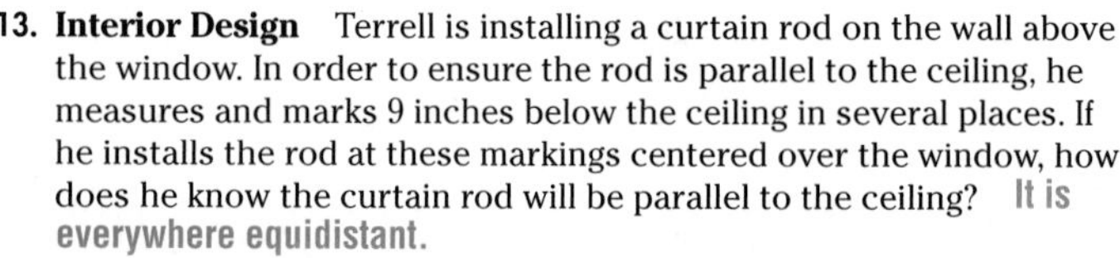

EXERCISES

Practice

Copy each figure and draw the segment that represents the distance indicated.

14. G to $\overleftrightarrow{EA}$

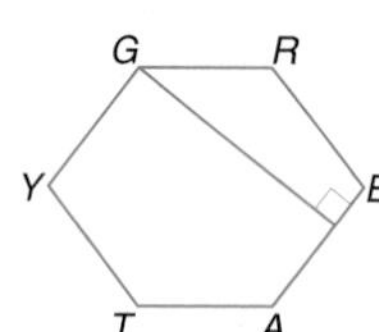

15. R to $\overleftrightarrow{LM}$

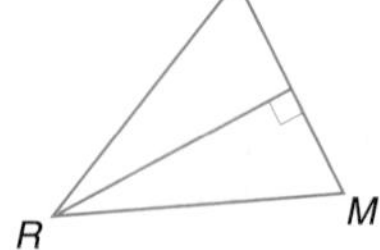

16. A to $\overleftrightarrow{HT}$

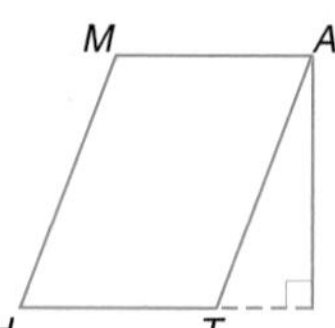

17. T to $\overleftrightarrow{AP}$

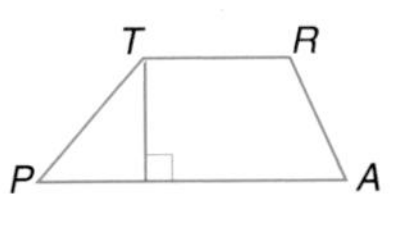

18. S to $\overleftrightarrow{PT}$

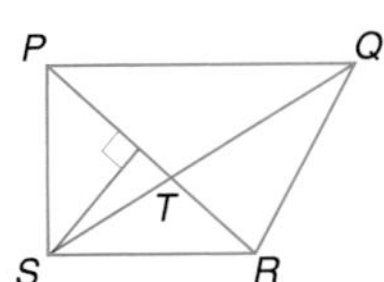

19. G to $\overleftrightarrow{AD}$

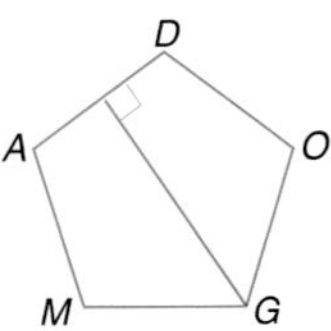

Use a ruler to determine whether the black lines are parallel. Explain your reasoning. 20–21. Yes; the lines are everywhere equidistant.

20.

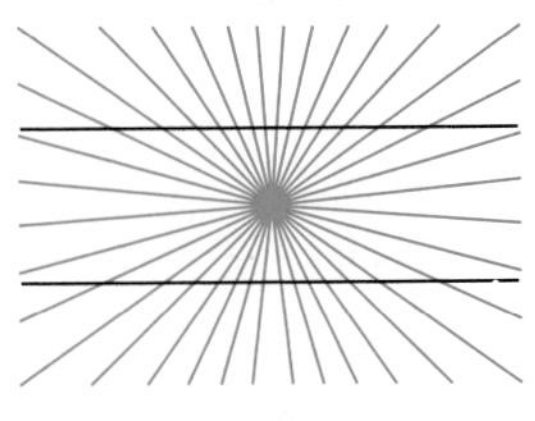

21.

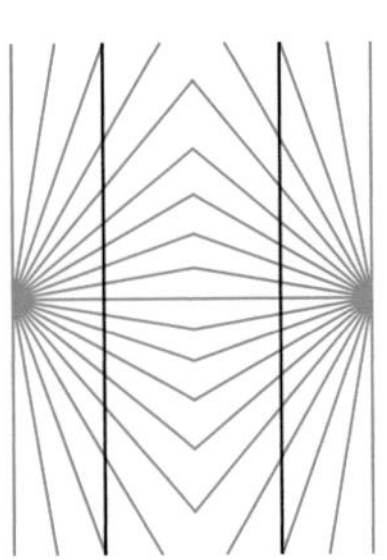

Using graph paper, graph each equation and plot the given ordered pair. Then construct a perpendicular segment and find the distance from the point to the line. 22–27. See margin for graphs.

24. $\sqrt{17} \approx 4.12$

B

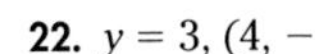

22. $y = 3$, $(4, -1)$ 4 **23.** $x = -2$, $(3, 2)$ 5 **24.** $y = 4x$, $(5, 3)$

25. $2x - y = 3$, $(2, 6)$ $\sqrt{5} \approx 2.24$ **26.** $3x + 4y = 1$, $(2, 5)$ 5 **27.** $2x - 3y = -9$, $(2, 0)$ $\sqrt{13} \approx 3.61$

In the figure at the right, $\overline{MT} \perp \overline{HM}$, $\overline{AT} \perp \overline{HT}$, and $\overline{AT} \perp \overline{AM}$. Name the segment whose length represents the distance between the following points and lines.

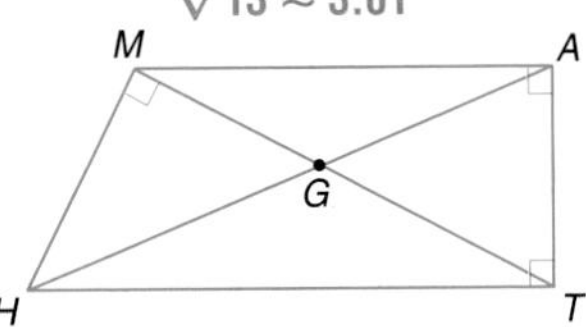

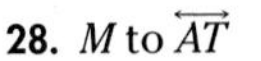

28. M to $\overleftrightarrow{AT}$ $\overline{AM}$ **29.** T to $\overleftrightarrow{HM}$ $\overline{MT}$

30. H to $\overleftrightarrow{AT}$ $\overline{HT}$ **31.** G to $\overleftrightarrow{HM}$ $\overline{GM}$

32. T to $\overleftrightarrow{AM}$ $\overline{AT}$ **33.** H to $\overleftrightarrow{MT}$ $\overline{HM}$

Proof

34. Provide the reason for each statement.

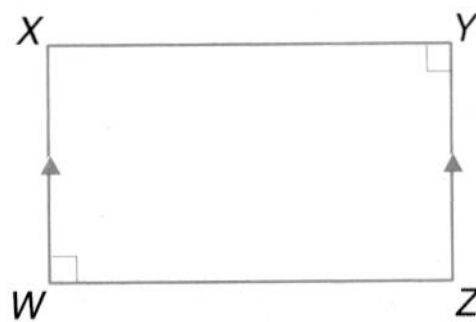

Given: $\overline{XW} \parallel \overline{YZ}$
$\overline{XW} \perp \overline{WZ}$
$\overline{XY} \perp \overline{YZ}$

Prove: $\overline{XY} \parallel \overline{WZ}$

Proof:

Statements	Reasons
1. $\overline{XW} \parallel \overline{YZ}$, $\overline{XW} \perp \overline{WZ}$, $\overline{XY} \perp \overline{YZ}$	1. ? Given
2. $\overline{WZ} \perp \overline{YZ}$	2. ? $\perp$ Transversal Theorem
3. $\overline{XY} \parallel \overline{WZ}$	3. ?

34. (Reason 3) In a plane, 2 lines $\perp$ to the same line are $\parallel$.

Additional Answers

22.

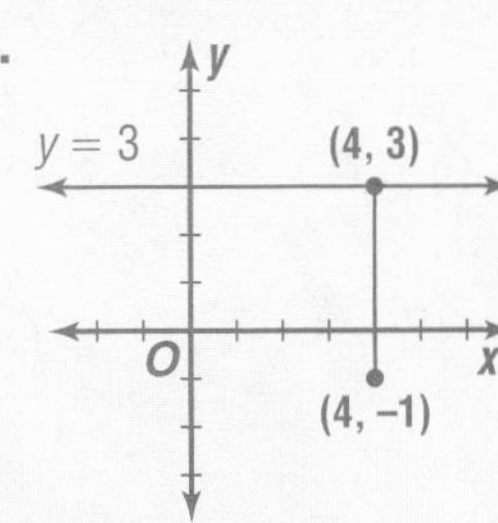

23.

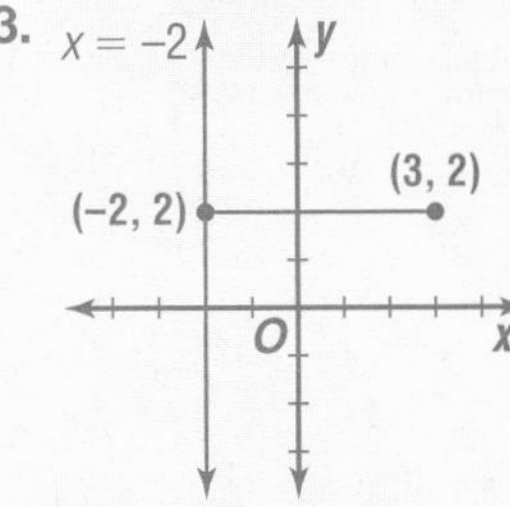

24.

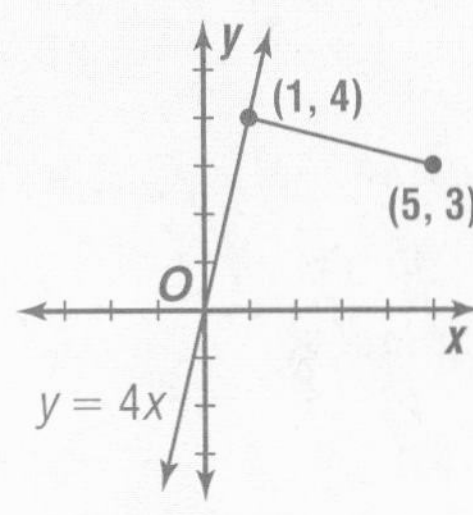

Study Guide Masters, p. 18

NAME______________________ DATE ________

Study Guide

Student Edition Pages 154–161

Parallels and Distance

The shortest segment from a point to a line is the perpendicular segment from the point to the line.

Distance Between a Point and a Line	The distance from a line to a point not on the line is the length of the segment perpendicular to the line from the point.
Distance Between Parallel Lines	The distance between two parallel lines is the distance between one of the lines and any point on the other line.

Example 1: Draw the segment that represents the distance indicated.

E to $\overleftrightarrow{AF}$

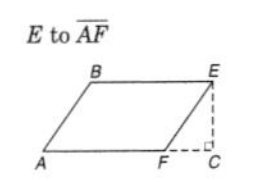

EC represents the distance from E to AF.

Example 2: Use a ruler to determine whether the lines are parallel.

The lines are not everywhere equidistant, therefore they are not parallel.

Draw the segment that represents the distance indicated.

1. K to $\overleftrightarrow{HJ}$ **2.** A to $\overleftrightarrow{BC}$ **3.** T to $\overleftrightarrow{VW}$ **4.** R to $\overleftrightarrow{NP}$

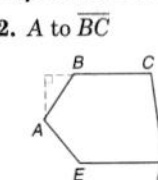

Use a ruler to determine whether the lines are parallel. Explain your reasoning.

5.

Yes, the lines are 1 cm apart.

6.

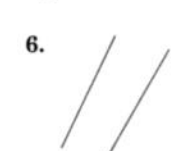

No, the lines are not everywhere equidistant.

7. Use a ruler to draw a line parallel to the given line through the given point.

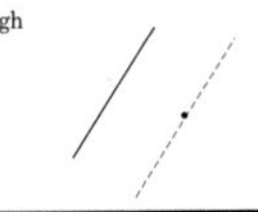

Additional Answers

25.

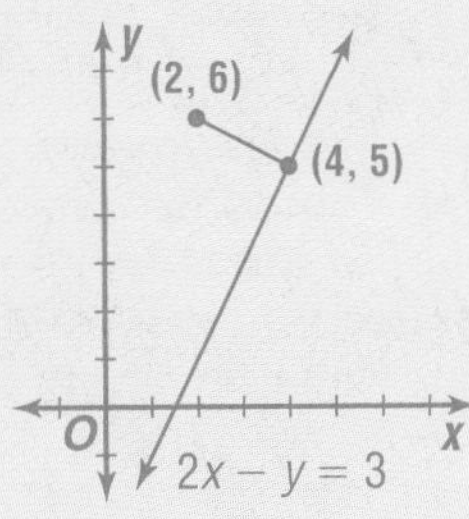

26.

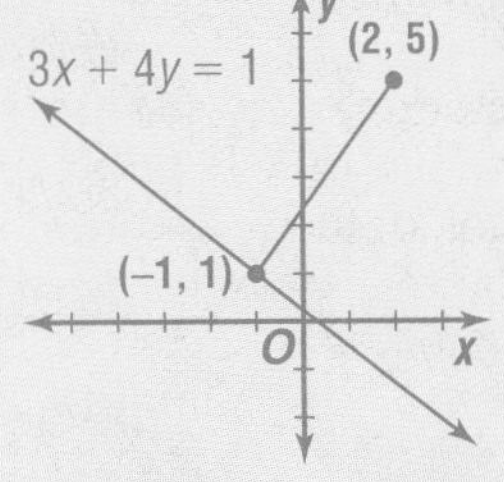

27.

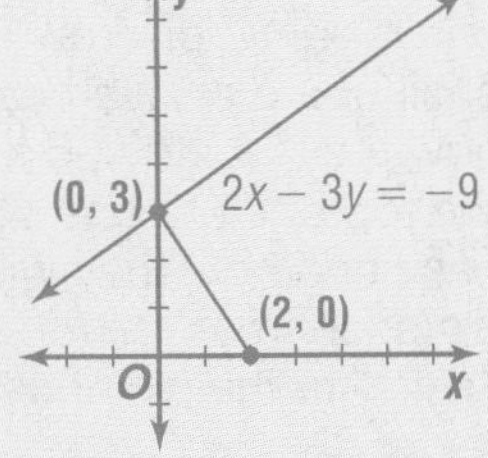

Additional Answer

38b. Sample Answer: Pass a pencil through parallel sides of a cereal box.

Practice Masters, p. 18

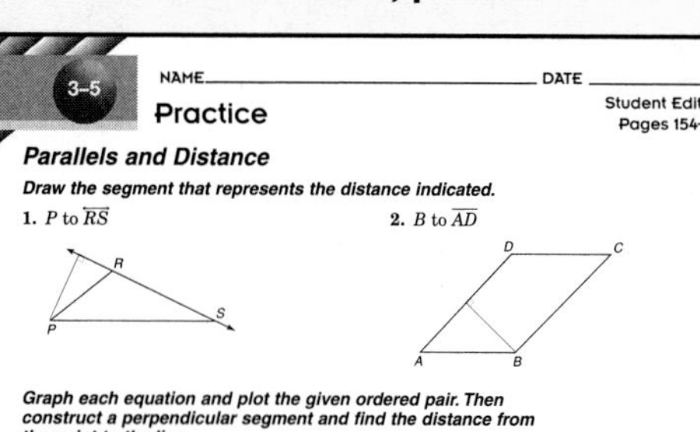
3–5 NAME____________ DATE ____________

Practice

Student Edition Pages 154–161

Parallels and Distance

Draw the segment that represents the distance indicated.

1. P to $\overline{RS}$
2. B to $\overline{AD}$

Graph each equation and plot the given ordered pair. Then construct a perpendicular segment and find the distance from the point to the line.

3. $y = x + 2$, $(2, -2)$ $3\sqrt{2}$
4. $x + y = 2$, $(3, 3)$ $2\sqrt{2}$

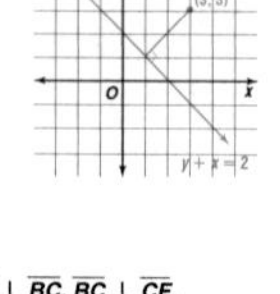

In the figure below, $\overline{BH} \perp \overline{AE}$, $\overline{CF} \perp \overline{AE}$, $\overline{BH} \perp \overline{BC}$, $\overline{BC} \perp \overline{CF}$, and $\overline{GD} \perp \overline{CE}$. Name the segment whose length represents the distance between the following points and lines.

5. B to $\overline{AE}$ $\overline{BH}$ or $\overline{CF}$
6. G to $\overline{CE}$ $\overline{GD}$
7. C to $\overline{BH}$ $\overline{CB}$ or $\overline{FH}$
8. F to $\overline{BC}$ $\overline{FC}$ or $\overline{BH}$

35. Find the distance between the two parallel lines shown at the right.
$\sqrt{24.5} \approx 4.95$

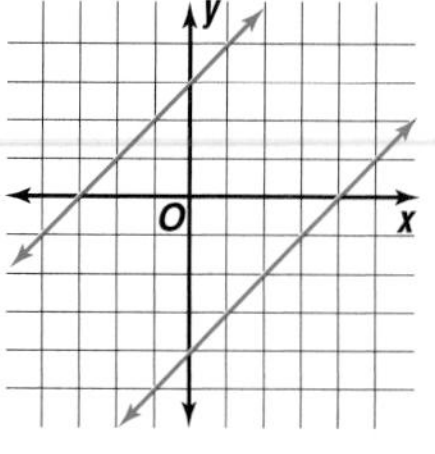

36. In the figure at the right, plane ABC is parallel to plane EFG. Explain how you could find the distance between two parallel planes. Find the length of a perpendicular segment from plane ABC to plane EFG.

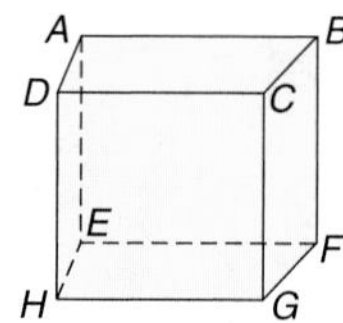

37. $\dfrac{|3 \cdot 2 + 4 \cdot 5 - 1|}{\sqrt{3^2 + 4^2}} = 5$
Yes; both equal 5.

37. The distance from a point $P(x_1, y_1)$ to a line ℓ with equation $Ax + By + C = 0$ can be found by the formula $d = \dfrac{|Ax_1 + By_1 + C|}{\sqrt{A^2 + B^2}}$. Use the formula to find the distance between the line $3x + 4y = 1$ and the point $P(2, 5)$. Check your answer against your solution to Exercise 26. Did the formula verify your answer? Explain.

Critical Thinking

38a. Yes, $\ell \perp Q$. If a line is perpendicular to one of 2 parallel planes, it is perpendicular to the other.

38. In the figure at the right, line ℓ is perpendicular to plane $\mathcal{P}$. Planes $\mathcal{P}$ and Q are parallel.
  a. Is line ℓ perpendicular to plane Q? Explain your reasoning.
  b. Create a three-dimensional model that verifies your conjecture. See margin.

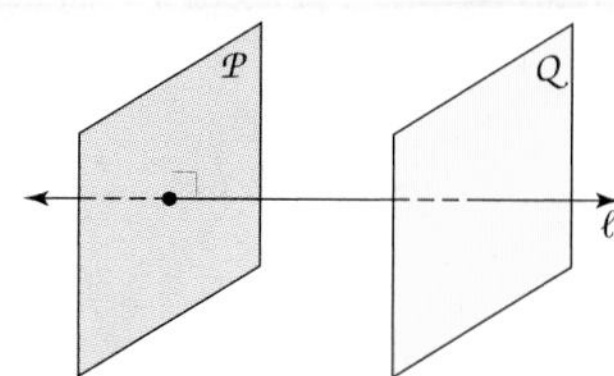

Applications and Problem Solving

39. Graphic Arts A graphic artist used the bar graph to illustrate statistics that compare the number of words used. Use a ruler to measure the distances between the bars.

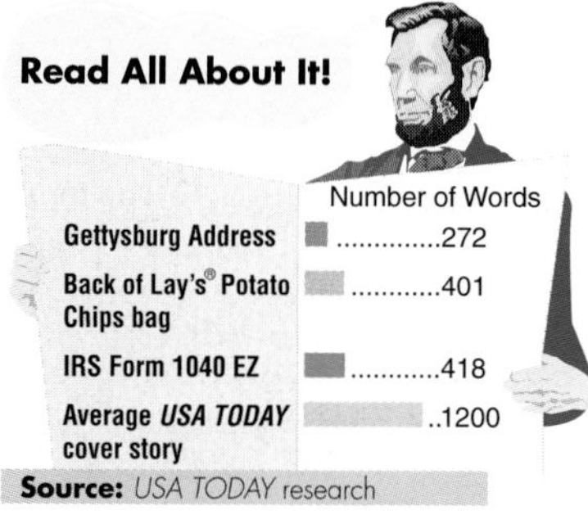

  a. Are the bars parallel? yes
  b. Are the bars the same distance apart? no
  c. Do the statistical results surprise you? Explain. See students' work.

40. Construction Dominique wants to install vertical boards to strengthen the handrail structure on her deck. How can she guarantee the vertical boards will be parallel? Attach the boards at equal distances.

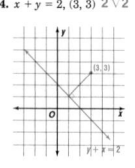

Extension

Problem Solving Bring in an example of a pyramid or draw one on the chalkboard or overhead. Ask students how they would find the distance from the top point to one of the edges of the base and how they would find the distance from the top point to the base itself. Are the distances the same?

To find the distance from the top point to an edge, draw a perpendicular segment from the point to the edge. To find the distance from the top point to the base itself, draw a segment perpendicular to the base. The distances will not be the same.

Mixed Review

Refer to the figure at the right for Exercises 41–43.

41. If $m\angle 6 = 50$ and $m\angle 7 = 120$, what can you say about lines p and q? Explain. (Lesson 3–4) $p \nparallel q$; cons. int. ∠s are not supp.

42. If p and q are parallel, what can you say about their slopes? (Lesson 3–3) The slopes are the same.

43. If p and q are parallel, what can you say about $m\angle 2$ and $m\angle 4$? Explain. (Lesson 3–2) $m\angle 2 = m\angle 4$; They are corresponding angles.

44. Sports During the winter, Alma is a ski instructor. When helping a novice skier, should she describe the correct position for holding the skis as parallel, intersecting, or skew lines? (Lesson 3–1) parallel lines

45. Two angles are congruent and supplementary. What else do you know about these angles? (Lesson 2–6) They are right angles.

46. Recreation While putting together a puzzle, Shaniqua noticed that one piece did not seem to fit. She made a conjecture that she must be missing some pieces. Do you think this is a valid conjecture? Explain. (Lesson 2–1)

46. Sample answer: No; the piece could be from another puzzle.

47. Modeling Draw and label a figure to show plane $\mathcal{R}$ contains point A and line ℓ. (Lesson 1–2) See margin.

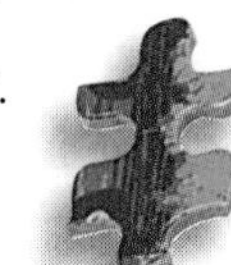

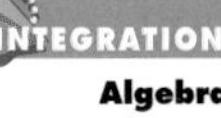

INTEGRATION **Algebra**

48. Find $68 \div (-17)$. -4

49. Solve $15 - 2.7a = 22.29 - 3.6a$. 8.1

1. Radar waves can be reflected or absorbed by an object.
3. This refers to the image of the building as it appears on a radar screen.

Mathematics and SOCIETY

Stealthy Geometry

The excerpt below appeared in an article in *Popular Mechanics* in September, 1995.

A BRAND-NEW BUILDING AT LONDON'S Heathrow Airport sports unique architecture that renders it virtually invisible to radar. Airport authorities were worried that traffic-control radar would ricochet off the new British Airways nerve center and interfere with runway operation. To squelch the problem, architect Norman Grimshaw incorporated distinctive geometry. The building's glass face looms outward as it rises, reaching an angle of 21° at the top. This curvature, combined with a series of external baffles and ribs, scatters radio energy down into an adjoining parking lot, which is paved with blocks rather than radar-reflecting asphalt. These features have shaved the building's radar signature by 99%. ■

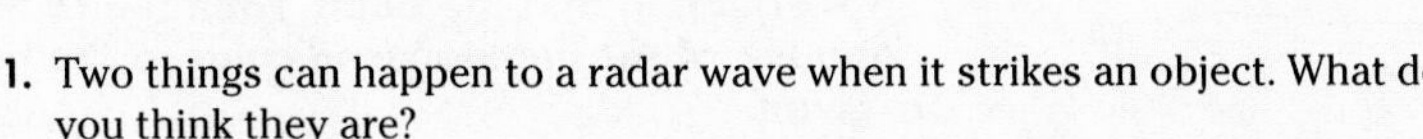

1. Two things can happen to a radar wave when it strikes an object. What do you think they are?
2. What did designers need to know about the reflection of the radar waves off the new building in order to properly design it? See margin.
3. The article refers to the building's *radar signature.* What do you think is meant by this term?

4 ASSESS

Closing Activity

Writing Have students find a model of a line and a point not on that line in your classroom. Have them measure the distance between the two. Next have them locate an example of two parallel lines in the classroom and measure the distance between them. Have students write descriptions of the measuring process used in each situation.

Chapter 3 Quiz C (Lessons 3-4 and 3-5) is available in the *Assessment and Evaluation Masters*, p. 73.

Additional Answer

47.

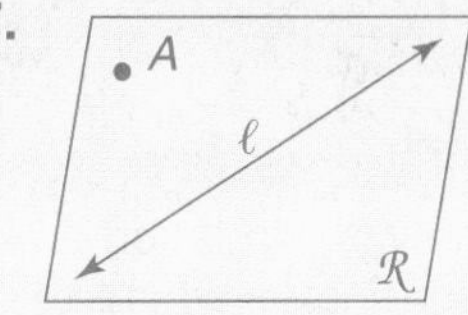

Enrichment Masters, p. 18

NAME______________ DATE ______

3-5 Enrichment

Student Edition Pages 154–161

Toothpick Puzzles

For these puzzles, you will need toothpicks or short straws of equal length.

1. Make four triangles with six toothpicks. Make a pyramid with a triangular base.
2. Make six triangles and a hexagon with six toothpicks.
3. Use 15 toothpicks to make this arrangement. Then take away three toothpicks to leave three squares.

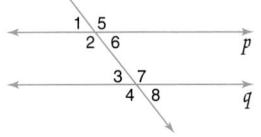

4. Use 16 toothpicks to make this arrangement. Then move two toothpicks to make four squares.
5. Use 24 toothpicks to make this arrangement. Then take away six toothpicks to leave three squares.

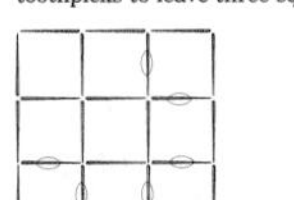

6. Use four toothpicks and a marble (or other small object) to make this fork arrangement. Move two toothpicks to make another fork, but with the marble no longer between the prongs of the fork.

7. Use nine toothpicks to make this arrangement. Then move two toothpicks to make a figure that creates the illusion of a cube viewed from one corner.

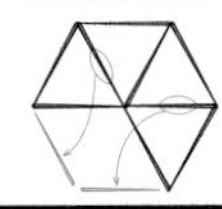

8. Make your own toothpick puzzle. See students' work.

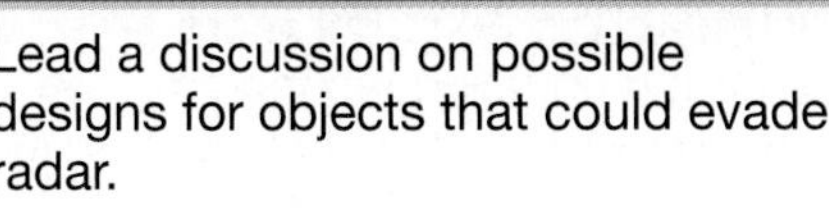

Lead a discussion on possible designs for objects that could evade radar.

Answer for Mathematics and Society

2. angles at which the radar waves would reflect off the building (this would indicate the path of the reflected waves)

3-5B LESSON NOTES

NCTM Standards: 1–4, 7

Objective

Use a graphing calculator to find the distance between a point and a line.

Recommended Time

20 minutes

Instructional Resources

Instructions for using the TI-81 and Casio calculators for this activity are available in the *Graphing Calculator and Computer Masters*, pp. 20, 21.

1 FOCUS

Motivating the Lesson

Graph the line with equation $y = 2x$ on coordinates where the units on the x-axis and y-axis are the same. Then graph it on coordinates where the units on one axis are twice the length of the units on the other axis.

2 TEACH

Teaching Tip Point out that the two lines on the graphing calculator do not look perpendicular even though they actually are, because you are looking at something in a rectangular box that is supposed to be in a square box. Have students look at the tick marks on the x-axis and y-axis. Even though they are both the same scale, the tick marks are not the same distance apart on each axis.

3 PRACTICE/APPLY

Assignment Guide
Core (with proof): 1–4
Core (informal): 1–4
Enriched: 1–4

3–5B Using Technology
Distance Between a Point and a Line

An Extension of Lesson 3–5

You can use a graphing calculator to find the distance between a point and a line.

Example

Find the distance between the line whose equation is $y = 2x - 5$ and the point at $(7, -1)$. The equation of the line perpendicular to the first line through $(7, -1)$ is $y = -\frac{1}{2}x + \frac{5}{2}$.

The standard viewing window can be selected in the ZOOM menu.

Clear the graphics screen and select the standard viewing window. Graph the lines by entering the equations in the Y= list.

Enter:

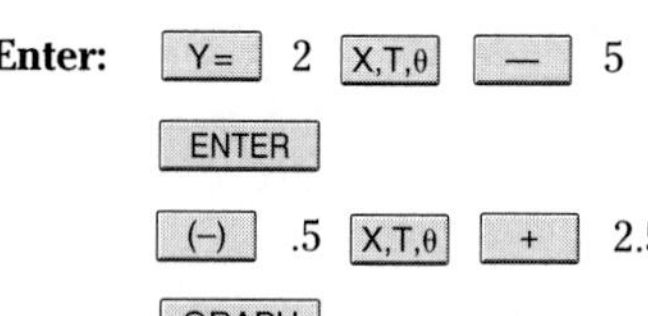

[Y=] 2 [X,T,θ] [−] 5 [ENTER]

[(−)] .5 [X,T,θ] [+] 2.5 [GRAPH]

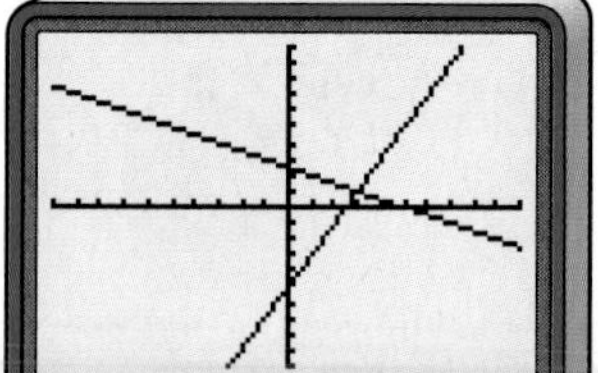

You can select the 5:Intersect command in the CALC menu to find the coordinates of the intersection of two graphed lines.

Notice the messages that apppear on the screen before each [ENTER]. By pressing [ENTER], you answer yes to the prompt.

Enter: [2nd] [CALC] 5 [ENTER] [ENTER] [ENTER].

The coordinates at the bottom of the screen are X=3 and Y=1, or (3, 1).

Use your calculator to find the distance between the intersection point and the given point.

$$d = \sqrt{(x_2 - x_1)^2 + (y_2 - y_1)^2} \text{ or } \sqrt{(7 - 3)^2 + (-1 - 1)^2}$$

Enter: [2nd] [$\sqrt{\ }$] [(] [(] 7 [−] 3 [)] [x^2] [+] [(] [(−)] 1 [−] 1 [)] [x^2] [)] [ENTER]

4.472135955

The distance between the point and the line is about 4.47.

EXERCISES

Use a graphing calculator to find the distance between the graph of each equation and the point, rounded to the nearest hundredth. The equation of the line perpendicular to the first line through the point is given.

1. $y = x + 7$, $A(2, 7)$ **1.41**
   perpendicular: $y = -x + 9$
2. $y = \frac{2}{5}x + 3$, $B(-8, -6)$ **5.39**
   perpendicular: $y = -\frac{5}{2}x - 26$
3. $y = 3x - 13$, $C(1, 0)$ **3.16**
   perpendicular: $y = -\frac{1}{3}x + \frac{1}{3}$
4. $y = \frac{3}{2}x$, $D(-4, 7)$ **7.21**
   perpendicular: $y = -\frac{2}{3}x + \frac{13}{3}$

4 ASSESS

Observing students working with technology is an excellent method of assessment.

Using Technology

This lesson offers an excellent opportunity for using technology in your geometry classroom. For more information on using technology, see *Graphing Calculators in the Mathematics Classroom*, one of the titles in the Glencoe Mathematics Professional Series.

3–6 Integration: Non-Euclidean Geometry
Spherical Geometry

What YOU'LL LEARN
- To identify points, lines, and planes in spherical geometry, and
- to compare and contrast basic properties of plane and spherical geometry.

Why IT'S IMPORTANT

You can understand the relationship of locations on the surface of Earth.

CONNECTION Geography

This world map uses a system of coordinates to locate places on the surface of Earth.

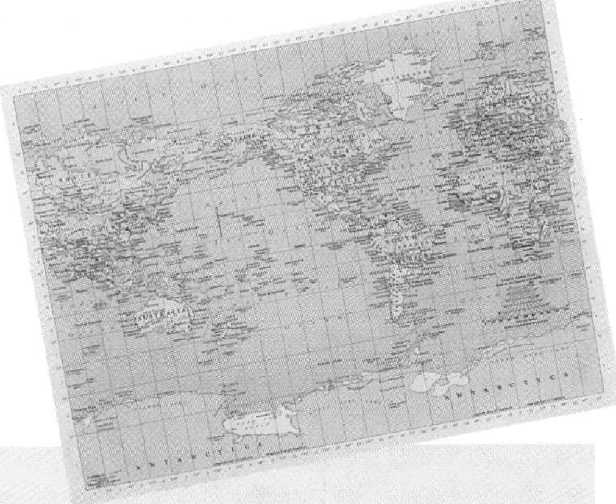

Midway between the North and South poles, a great circle called the *equator* divides Earth into the Northern and Southern hemispheres.

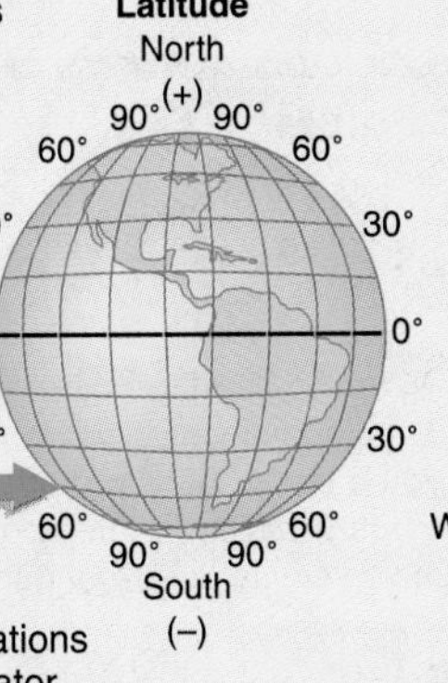

Latitude provides the locations north or south of the equator. Latitude is measured by angles ranging from 0° at the equator to 90° at the poles.

Longitude provides the locations east or west of the prime meridian. Longitude is measured in angles ranging from 0° at the *prime meridian* to 180° at the international date line.

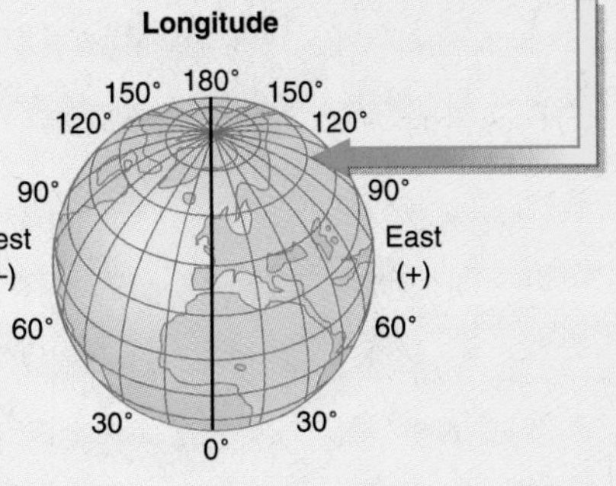

Example 1 **CONNECTION Geography**

Find a major city in the United States whose location is near the point with coordinates of 90° W and 35° N.

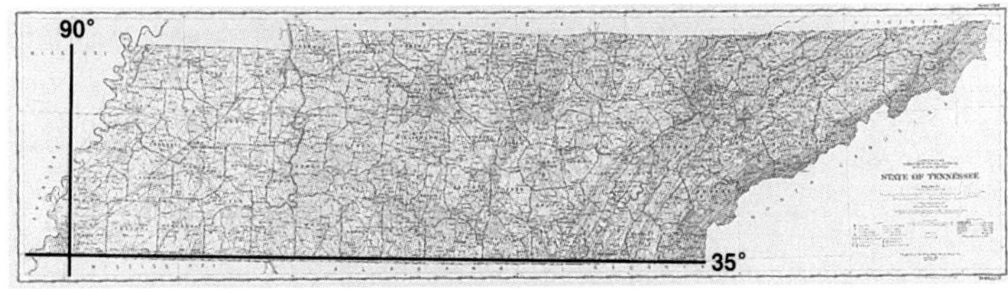

Memphis, Tennessee, is located near the intersection of 90° W and 35° N.

3-6 LESSON NOTES

NCTM Standards: 1–4, 7

Instructional Resources

- Study Guide Master 3-6
- Practice Master 3-6
- Enrichment Master 3-6
- Assessment and Evaluation Masters, p. 73

Transparency 3-6A contains the 5-Minute Check for this lesson; **Transparency 3-6B** contains a teaching aid for this lesson.

Recommended Pacing	
Standard Pacing	Days 11 & 12 of 14
Honors Pacing	Days 10 & 11 of 13
Block Scheduling*	Day 6 of 7

*For more information on pacing and possible lesson plans, refer to the *Block Scheduling Booklet*.

1 FOCUS

5-Minute Check
(over Lesson 3-5)

In the figure below, name the segment whose length represents the distance between the following points and lines.

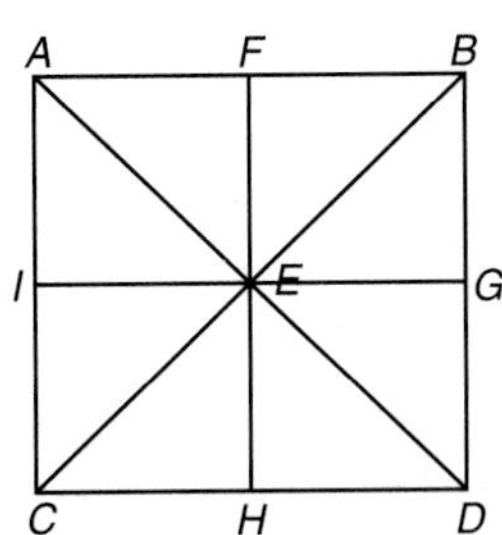

1. from C to $\overline{AD}$ $\overline{CE}$
2. from E to $\overline{BD}$ $\overline{EG}$
3. from F to $\overline{CD}$ $\overline{FH}$
4. from $\overline{AB}$ to $\overline{CD}$ $\overline{AC}$, $\overline{FH}$, or $\overline{BD}$

Motivating the Lesson

Situational Problem Using a globe of the world, point out to students that there are many points on the surface of the globe, but no lines. Ask students to decide what is the closest thing to a line. lines of latitude and longitude

2 TEACH

In-Class Example

For Example 1
Find the city in the United States whose location has coordinates 84°W and 36°N. **Knoxville, Tennessee**

Teaching Tip Compare finding a location to graphing a point. Longitude is the *x*-coordinate and latitude is the *y*-coordinate.

MODELING MATHEMATICS Point out to students that the inside of the sphere is "off limits" because we cannot dig through Earth to get from one point on the surface to another.

FYI

Many 19th-century maps had different coordinate grids until 1884, when an International Prime Meridian passing through Greenwich Observatory, near London, was officially designated.

The equator is called a *great circle* of Earth because it divides Earth into two equal halves. Similarly in spherical geometry, a line is defined as a great circle of a sphere. A line is a circle that divides the sphere into two equal *half-spheres*. A plane is the sphere itself.

So far in our studies of **plane Euclidean geometry**, we studied a system of points, lines, and planes. In **spherical geometry**, we study a system of points, great circles (lines), and spheres (planes). Spherical geometry is one type of **non-Euclidean geometry**.

Plane Euclidean Geometry	Spherical Geometry
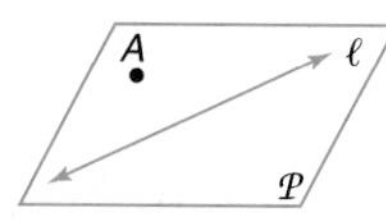	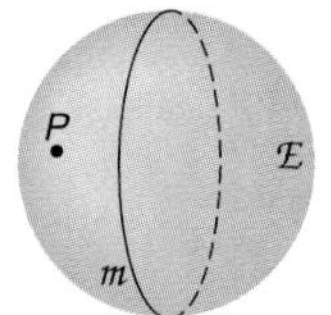
Plane 𝒫 contains line ℓ and point A not on ℓ.	Sphere ℰ contains great circle m and point P not on m. m is a line on sphere ℰ.

MODELING MATHEMATICS — Shortest Path between Two Points

Materials: Styrofoam™ ball, ruler, string, markers

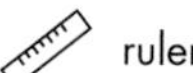

- Locate and label any two points *A* and *B* on your sphere.

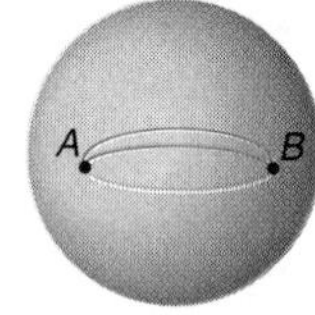

- How many curved paths are there from *A* to *B*?
- Use string to show several curved paths from *A* to *B*. Measure the length of each string to the nearest millimeter. Make a conjecture about the shortest distance between two points in spherical geometry.
- Wrap your string from *A* through *B* around the sphere until you return to point *A*. What appears to be true about the shortest path from point *A* to point *B*? What is true about the relationship between the sphere and the string?

Your Turn

a. Choose two new points *C* and *D* so that they appear to be directly opposite each other on your sphere. These are called *polar points* of the sphere. Draw a figure showing the sphere and points *C* and *D*. **See Solutions Manual.**

b. In how many different ways can you measure the shortest path from *C* to *D*? **Infinitely many great circles pass through the polar points.**

The polar points discussed in the Modeling Mathematics activity are points of intersection of a line through the center of a sphere with the sphere.

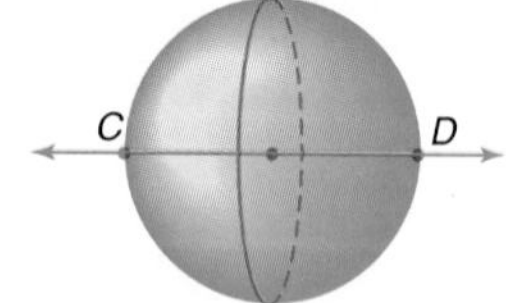

FYI

The Royal Greenwich Observatory was established in 1675. By the 1940s, it became unusable.

Alternative Learning Styles

Visual Have two students hold up a world map and a globe. Compare the latitude and longitude lines on each. The map is a representation of the world in plane Euclidean geometry. The globe is a realistic view in non-Euclidean spherical geometry. Discuss the advantages of each. Have students try to wrap the map around the globe. Discuss the disadvantages of each representation.

The table below compares and contrasts lines in the system of plane Euclidean geometry and lines (great circles) in spherical geometry.

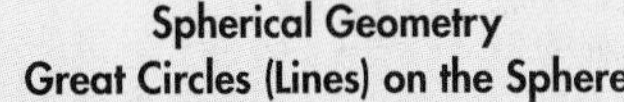

Plane Euclidean Geometry Lines on the Plane	Spherical Geometry Great Circles (Lines) on the Sphere
1. A line segment is the shortest path between two points.	**1.** An arc of a great circle is the shortest path between two points.
2. There is a unique straight line passing through any two points.	**2.** There is a unique great circle passing through any pair of nonpolar points.
3. A straight line is infinite.	**3.** A great circle is finite and returns to its original starting point.
4. If three points are collinear, exactly one is between the other two. 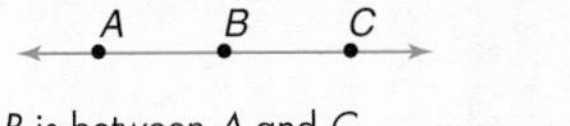 *B* is between *A* and *C*.	**4.** If three points are collinear, any one of the three points is between the other two. *A* is between *B* and *C*. *B* is between *A* and *C*. *C* is between *A* and *B*. 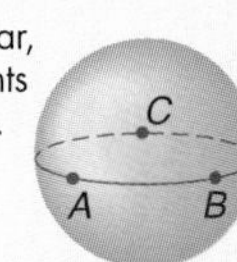

The Segment Addition Postulate holds true in spherical geometry as well as plane Euclidean geometry.

Example 2 CONNECTION **Geography**

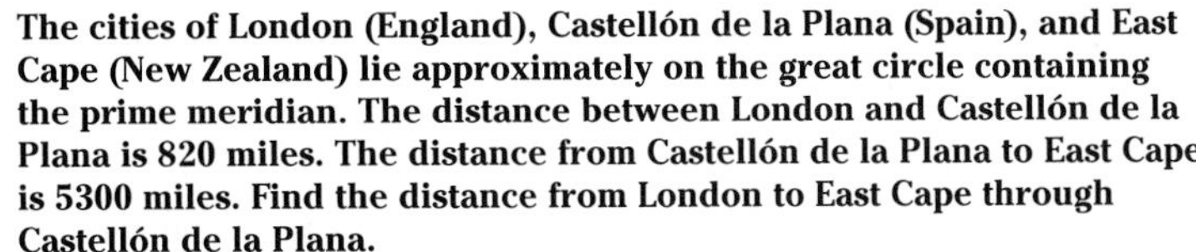

The cities of London (England), Castellón de la Plana (Spain), and East Cape (New Zealand) lie approximately on the great circle containing the prime meridian. The distance between London and Castellón de la Plana is 820 miles. The distance from Castellón de la Plana to East Cape is 5300 miles. Find the distance from London to East Cape through Castellón de la Plana.

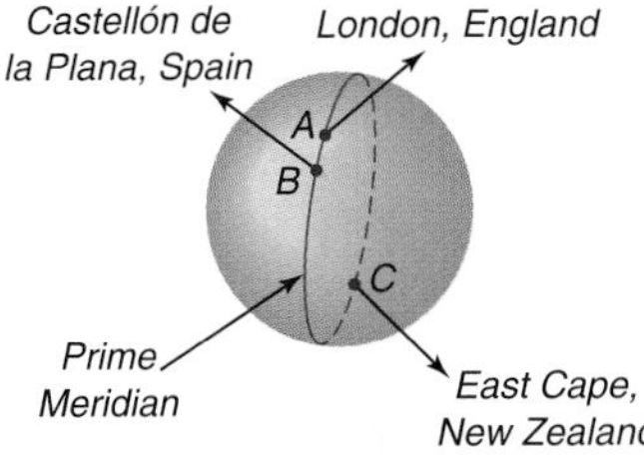

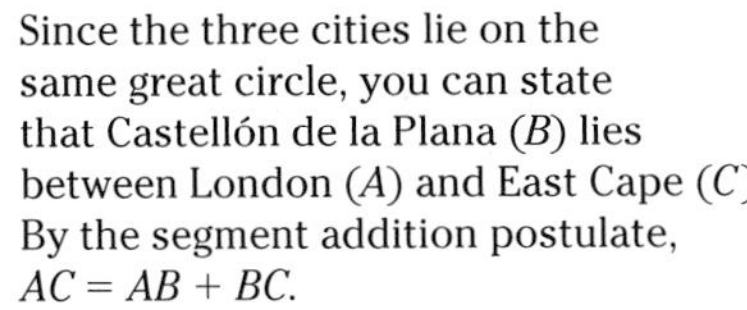

Since the three cities lie on the same great circle, you can state that Castellón de la Plana (B) lies between London (A) and East Cape (C). By the segment addition postulate, $AC = AB + BC$.

$$\begin{aligned} AC &= AB + BC \\ &= 820 + 5300 \\ &= 6120 \end{aligned}$$

Therefore, the distance from London to East Cape through Castellón de la Plana is approximately 6120 miles.

In-Class Example

For Example 2

Pickle Lake in Canada, New Orleans in the United States, and Sisal in Mexico lie approximately on the great circle containing the 90°W longitude. The distance between Pickle Lake and New Orleans is about 1500 miles. The distance between New Orleans and Sisal is about 600 miles. Find the distance between Pickle Lake and Sisal. **about 2100 miles**

Teaching Tip Remind students that the distances they are adding are not linear; they are spherical.

Classroom Vignette

"When working with spherical geometry, I draw *great circles* and mark *poles* on an old softball. This helps students to visualize concepts taught in the lesson. It greatly helps their understanding of this non-Euclidean geometry in which the Parallel Postulate fails to exist."

Tim Kanold
Author

Teaching Tip Encourage students to use models to answer Example 3.

In-Class Examples

For Example 3

For each property listed from plane Euclidean geometry, write a corresponding statement for non-Euclidean spherical geometry.

a. If two lines are perpendicular to a given line, the two lines are parallel. **There exist no parallel lines.**

b. One and only one line can be drawn through two points. **At least one great circle can be drawn through two points.**

c. If two lines intersect and form a right angle, they are perpendicular. **If two great circles intersect to form a right angle, they are perpendicular.**

For Example 4

In the figure, lines a, b, and c intersect at point P.

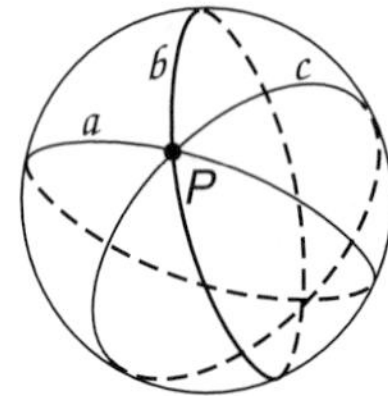

a. In plane Euclidean geometry, will any line a, b, or c intersect one of the other lines at a point other than P? **no**

b. Is this true in spherical non-Euclidean geometry? Why or why not? **In spherical geometry, the lines (great circles) will intersect at another point besides P.**

FYI

Spherical Geometry, also known as *Riemannian geometry*, was discovered by German mathematician G.F.B. Reimann about 1860.

In spherical geometry, Euclid's first four postulates (Postulates 2–1 through 2–4) and their related theorems hold true. However, theorems that depend on the parallel postulate may not be true.

The definition of parallel lines states that they lie in the same plane and never intersect. In spherical geometry, the sphere is the plane, and a line must be a great circle. Every great circle containing A intersects ℓ. Thus, there exists no line through point A that is parallel to ℓ.

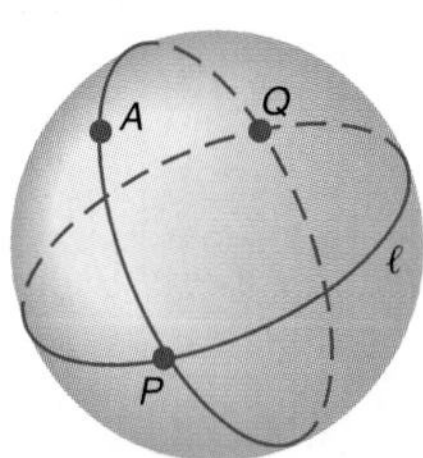

Every great circle of a sphere intersects all other great circles of a sphere in exactly two points. In the figure at the left, one possible line through point A intersects line ℓ at P and Q.

If two great circles divide a sphere into four congruent regions, the lines are perpendicular to each other.

Example 3 For each property listed from plane Euclidean geometry, write a corresponding statement for non-Euclidean spherical geometry.

a. Perpendicular lines intersect at one point.

Perpendicular great circles intersect at two points.

b. Perpendicular lines form four right angles.

Perpendicular great circles form eight right angles.

c. Perpendicular lines divide a plane into four infinite regions.

Perpendicular great circles divide the sphere into four finite regions.

Example  **In the figure at the right, both lines ℓ and m are perpendicular to line s.**

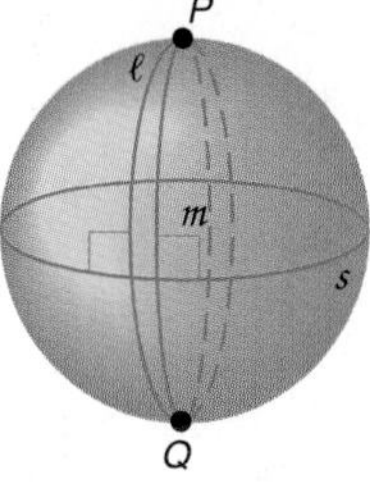

a. In plane Euclidean geometry, what relationship must exist between lines ℓ and m if they are both perpendicular to line s?

In plane Euclidean geometry, Theorem 3–8 states that two lines perpendicular to the same line are parallel.

b. Is this true in spherical non-Euclidean geometry? Why or why not?

In spherical geometry, this cannot be true since parallel lines (great circles) fail to exist. In this case, although $\ell \perp s$ and $m \perp s$, ℓ and m do intersect at points P and Q.

Reimann developed the theory of differential geometry, which was later used by Einstein in his Theory of Relativity.

CHECK FOR UNDERSTANDING

Communicating Mathematics

1. Two polar points have many great circles passing through them.

3. Since 24,900 − 6120 = 18,780, it is shorter to go through Castellón de la Plana.

Study the lesson. Then complete the following.

1. In spherical geometry, there is a unique great circle passing through any pair of nonpolar points. Explain why the points must be nonpolar.
2. **Compare** lines in plane Euclidean geometry with lines in spherical geometry. b. plane Euclidean
   a. Which lines divide the plane into two finite congruent regions? spherical
   b. Which lines have one point of intersection instead of two?
3. **Refer** to Example 2 on page 165. If the approximate distance around Earth is 24,900 miles, explain whether or not it would be shorter to go from London to East Cape without going through Castellón de la Plana.

4a. Circle r does not divide the sphere into 2 equal halves.

4. a. **Draw** a great circle on a ball. Also, draw point P not on ℓ. Draw a circle r on the sphere that passes through P and appears to be parallel to ℓ. Why is the circle r not a great circle of the sphere?
   b. **Draw** a line (great circle) m perpendicular to ℓ passing through P. Can any lines other than m be drawn through point P and be perpendicular to ℓ? Explain. No; see margin for explanation.

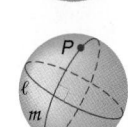

Guided Practice

6. False; in spherical geometry, perpendicular lines form 8 right angles.

CAREER CHOICES

Cartographers prepare maps using information provided by aerial photographs, geodetic surveys, and satellite data. A bachelor's degree in engineering or physical science is necessary for the field, with mathematics, geography, and computer science training very helpful.

For more information, contact:

American Congress on Surveying and Mapping
5410 Grosvenor Lane
Bethesda, MD 20814-2122

If spherical points are restricted to be nonpolar points, determine if each statement from Euclidean geometry is also *true* in spherical geometry. If *false*, explain your reasoning.

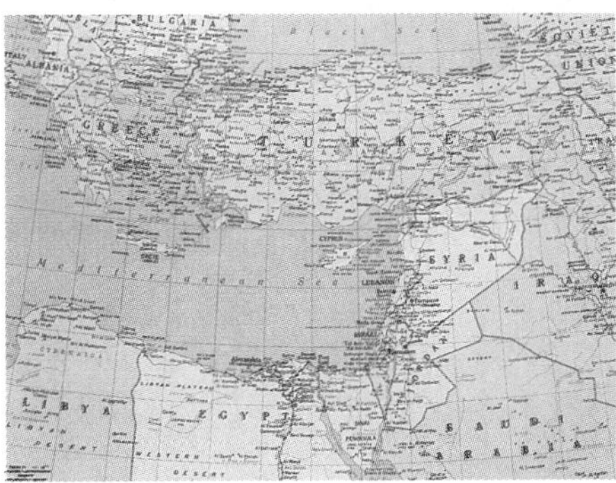

5. Any two distinct points determine exactly one line. true
6. Two perpendicular lines create four right angles.
7. Use a map of southwestern Asia to name the latitude and longitude of Istanbul, Turkey. 41° N, 29° E
8. Use a map to name the city and country located near 25° N, 55° E. Dubai, Saudi Arabia

For each property from plane Euclidean geometry, write a corresponding statement for non-Euclidean spherical geometry.

10. A curved path on the great circle passing through 2 points is the shortest distance between the 2 points.

9. The straight line is infinite. The great circle is finite.
10. A straight line segment is the shortest path between two points.
11. **Geography** Suppose the town where you live is a pole of Earth.
    a. Use a globe to state the latitude and longitude of your home town.
    b. State the latitude and longitude of the location of your opposite pole. What city is near it? 11a–b. See students' work.

3 PRACTICE/APPLY

Check for Understanding

Exercises 1–11 are designed to help you assess your students' understanding through reading, writing, speaking, and modeling. You should work through Exercises 1–4 with your students and then monitor their work on Exercises 5–11.

CAREER CHOICES

Evidence shows that cartography began in the Stone Age. Early cartographers used bone, wood, bark, mammoth tusk, and walls for mapmaking. The world's oldest urban plan was found on a wall in Turkey and dated to around 6200 B.C.

Additional Answer

4b. Unless ℓ is the equator and P is a pole, only one great circle can be drawn through P and perpendicular to ℓ.

Reteaching

Using Tables Construct a table listing properties in Euclidean geometry in one column and a corresponding property in spherical geometry in another column. Through class discussion, have students complete the table.

Study Guide Masters, p. 19

NAME ______________ DATE ________

Study Guide — Student Edition Pages 163–169

Integration: Non-Euclidean Geometry Spherical Geometry

Spherical geometry is one type of **non-Euclidean geometry**. A line is defined as a great circle of a sphere that divides the sphere into two equal half-spheres. A plane is the sphere itself.

Plane Euclidean Geometry Lines on the Plane	Spherical Geometry Great Circles (Lines) on the Sphere
1. A line segment is the shortest path between two points.	1. An arc of a great circle is the shortest path between two points.
2. There is a unique straight line passing through any two points.	2. There is a unique great circle passing through any pair of nonpolar points.
3. A straight line is infinite.	3. A great circle is finite and returns to its original starting point.
4. If three points are collinear, exactly one is between the other two. A B C — B is between A and C.	4. If three points are collinear, any one of the three points is between the other two. A is between B and C. B is between A and C. C is between A and B.

Latitude and **longitude**, measured in degrees, are used to locate places on a world map. Latitude provides the locations north or south of the equator. Longitude provides the locations east or west of the prime meridian (0°).

Example: Find a city located near the point with coordinates 29°N and 95°W.

The city near these coordinates is Houston, Texas.

Longitude West: 125° 115° 105° 95° 85° 75° 65°; 50°N 45°N 40°N 35°N 30°N 25°N; Seattle, Boston, Houston, Miami

Decide which statements from Euclidean geometry are true in spherical geometry. If false, explain your reasoning.

1. Given a point Q and a line r, where Q is not on r, exactly one line perpendicular to r passing through Q can be drawn. True
2. Two lines equidistant from each other are parallel. False; there are no parallel lines in spherical geometry.

Use a globe or world map to name the latitude and longitude of each city.

3. Havana, Cuba 23°N, 82°W
4. Beira, Mozambique 18°S, 35°E
5. Kabul, Afghanistan 35°N, 69°E

Use a globe or world map to name the city located near each set of coordinates.

6. 39°N, 73°W New York City, USA
7. 59°N, 18°E Stockholm, Sweden
8. 42°S, 146°E Hobart, Tasmania

Assignment Guide

Core (with proof): 13–37 odd, 38–46
Core (informal): 13–37 odd, 38–41, 43–46
Enriched: 12–34 even, 35–46

For **Extra Practice**, see p. 770.

The red A, B, and C flags, printed only in the Teacher's Wraparound Edition, indicate the level of difficulty of the exercises.

Additional Answers

12. False; in spherical geometry, if 3 points are collinear, any point can be between the other 2 points.
25. There exists no parallel lines.
26. Two distinct great circles intersect in exactly 2 points.
27. A pair of perpendicular great circles divides the sphere into 4 finite congruent regions.
28. A pair of perpendicular great circles intersects twice and creates 8 right angles.
29. There exists no parallel lines.
30. There are 2 distances between 2 points.

EXERCISES

Practice

16. False; in spherical geometry, there are no parallel lines.

A

If spherical points are restricted to be nonpolar points, decide which statements from Euclidean geometry are *true* in spherical geometry. If *false*, explain your reasoning. 12. See margin.

12. If three points are collinear, exactly one point is between the other two.
13. A line has infinite length. False; in spherical geometry, a line has finite length.
14. Given a line ℓ and point P not on ℓ, there exists exactly one line parallel to ℓ passing through P. False; in spherical geometry, there are no parallel lines.
15. Given a line ℓ and a point P not on ℓ, there exists exactly one line perpendicular to ℓ passing through P. true
16. Two lines perpendicular to the same line are parallel to each other.
17. Two intersecting lines divide the plane into four regions. true

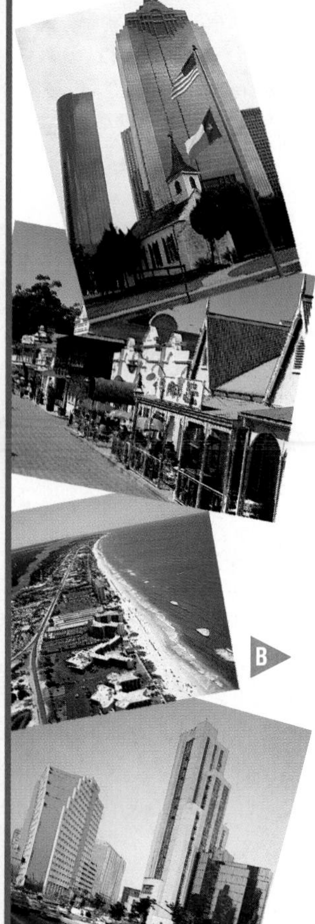

Use a globe or world map to name the latitude and longitude of each city. 20. 27° S, 28° E

18. Houston, Texas 29° N, 95° W
19. Panama City, Florida 30° N, 85° W
20. Johannesburg, South Africa
21. Mexico City, Mexico 19° N, 99° W

Use a globe or world map to name the city located near each set of coordinates.

22. 30° N, 90° W New Orleans, Louisiana
23. 40° N, 120° W Reno, Nevada
24. 32° S, 116° E Perth, Australia

B

For each property listed from plane Euclidean geometry, write a corresponding statement for non-Euclidean spherical geometry.

25. Two distinct lines with no point of intersection are parallel.
26. Two distinct intersecting lines intersect in exactly one point.
27. A pair of perpendicular straight lines divides the plane into four infinite regions.
28. A pair of perpendicular straight lines intersects once and creates four right angles.
29. Parallel lines have infinitely many common perpendicular lines.
30. There is only one distance that can be measured between two points.
31. Draw two great circles, ℓ and m, on a sphere as shown. Draw as many great circles as possible that are perpendicular to both ℓ and m. How many perpendiculars were you able to draw? 1

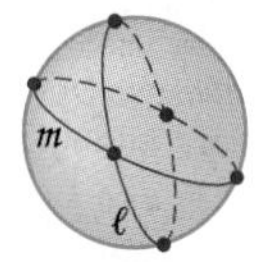

25–30. See margin.
32b. one-fourth the length of the great circle
33a. They are each perpendicular to each other.
33b. They lie at the intersection of the other 2 circles.

C

32. Compare each of the following lengths to the length of a great circle of a sphere. a. half the length of the great circle
    a. distance between any pair of pole points
    b. distance between any pole point and its equator
33. Draw two perpendicular great circles and construct the common perpendicular. See margin for drawing.
    a. What relationship exists between the great circles?
    b. Describe the location of polar points for each great circle.

Additional Answer

33.

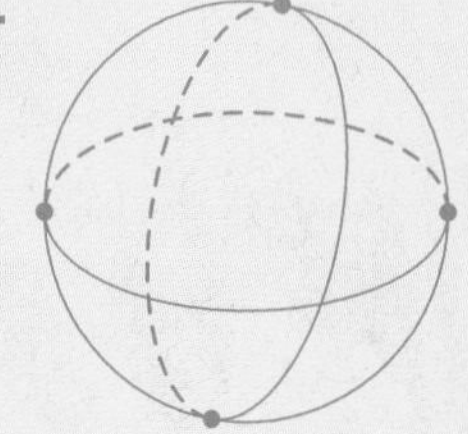

Practice Masters, p. 19

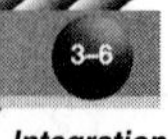

NAME________________ DATE ________

Practice

Student Edition Pages 163–169

Integration: Non-Euclidean Geometry Spherical Geometry

Use a globe or world map to name the latitude and longitude of each city.

1. Bogota, Colombia 5°N, 75°W
2. Cairo, Egypt 30°N, 32°E
3. Melbourne, Australia 37°S, 145°E
4. Fairbanks, Alaska 63°N, 148°W
5. Nairobi, Kenya 1°S, 37°E
6. London, England 50°N, 0°
7. Cape Town, South Africa 33°S, 18°E
8. Julianehab, Greenland 61°N, 46°W
9. La Paz, Bolivia 16°S, 68°W

Use a globe or world map to name the city located near each set of coordinates.

10. 34°S, 59°W Buenos Aires, Argentina
11. 30°N, 90°W New Orleans, Louisiana
12. 48°N, 123°W Vancouver, Canada
13. 7°S, 107°E Jakarta, Indonesia
14. 9°N, 38°E Adis Abeba, Ethiopia
15. 51°N, 14°E Berlin, Germany
16. 23°S, 46°W São Paulo, Brazil
17. 18°N, 99°W Mexico City, Mexico
18. 35°S, 173°E Auckland, New Zealand

For each property listed from plane Euclidean geometry, write a corresponding statement for non-Euclidean spherical geometry.

19. Two lines intersecting to form four right angles are perpendicular. Two great circles intersecting to form eight right angles are perpendicular.
20. Through any two points in a plane there is a unique and infinite straight line. Through any two points on a sphere there is a unique and finite great circle.
21. An infinite number of lines can be drawn through a point in a plane. An infinite number of great circles can be drawn through a point on a sphere.

34. Consider another version of the Parallel Postulate: *Given a line and a point not on the line, there are many lines that pass through the point and that are parallel to the given line.* a. See margin.
   a. Draw a model that would show how this might be true in space.
   b. Do some research to find the name of this type of non-Euclidean geometry. hyperbolic geometry

Critical Thinking

35. a. Explain why $\triangle ABC$ could not exist in plane Euclidean geometry. See margin.
   b. Could $\triangle ABC$ exist in non-Euclidean spherical geometry? Explain your reasoning. Include a drawing. See margin.

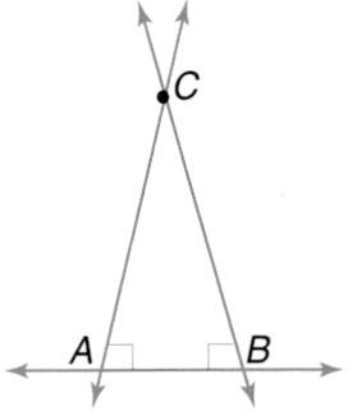

Applications and Problem Solving

36. **Climate** Locate Philadelphia, Pennsylvania, in an atlas.
   a. Determine its approximate latitude above the equator. 40° N
   b. Use a globe to locate two other cities in the Northern Hemisphere that lie on the same latitude. Sample answers: Madrid, Spain, and Beijing, China
   c. Is there a connection between latitude and climate? Explain your reasoning. Yes; distance from the Equator affects the closeness of the sun.

37. **Space Travel** According to *Einstein's Spherical Universe*, geometric properties of space are similar to those on the surface of a sphere. Based on this model, what conclusion follows about the path of a spaceship along a straight line? Explain your reasoning. See margin.

Albert Einstein

39. The 2 lines must be perpendicular to the transversal.

Mixed Review

38. Find the distance between point $P(6, -2)$ and the graph of line ℓ whose equation is $y = 7$. (Lesson 3–5) 9

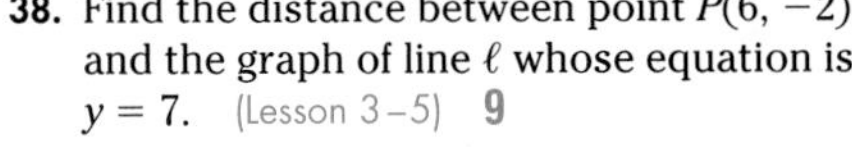

39. Two lines are cut by a transversal and the consecutive interior angles are congruent. What must be true for the lines to be parallel? (Lesson 3–4)

40. Find the slope of the line that passes through (3, 0) and (8, −2). (Lesson 3–3) $-\frac{2}{5}$

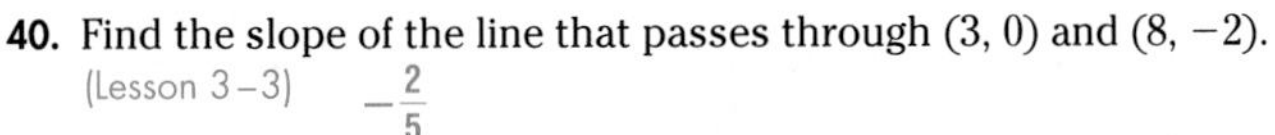

41. **Draw a Diagram** Hector and Shantel are building steps to their new shed. It takes one concrete block for one step, three blocks for two steps, and six blocks for three steps. How many concrete blocks will it take to build six steps? (Lesson 3–1) 21

42. Write a two-column proof. (Lesson 2–6)
   Given: $RS = ST$
   Prove: $RT = 2ST$ See margin.

   R S T

43. Write a conditional statement about angle measures. Identify the hypothesis and conclusion. (Lesson 2–2) See Solutions Manual.

45. Commutative Prop. (+)

44. Find AB, given $A(-1, 4)$ and $B(6, -20)$. (Lesson 1–4) 25

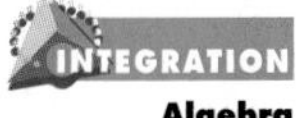

INTEGRATION **Algebra**

45. Name the property illustrated by 15 + 41 = 41 + 15.

46. Solve $-8 - m = 13$. −21

4 ASSESS

Closing Activity

Speaking Have students explain why there are no parallel lines in spherical geometry.

Chapter 3 Quiz D (Lesson 3-6) is available in the *Assessment and Evaluation Masters*, p. 73.

Additional Answers

34a.

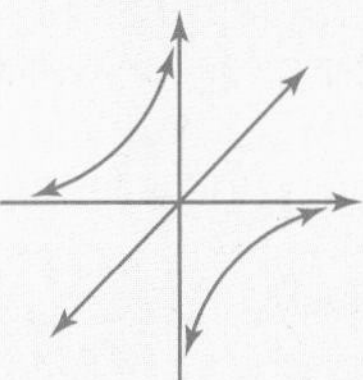

35a. In a plane, if 2 lines are perpendicular to the same line, then they are parallel.

35b. Yes; 2 intersecting great circles can both be ⊥ to another great circle.

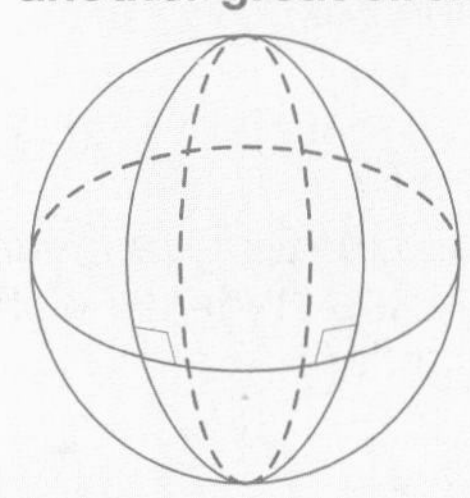

Enrichment Masters, p. 19

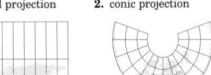

3-6 NAME____________ DATE________

Enrichment

Student Edition Pages 163–169

Map Projections

A **map projection** is a medium for transferring the surface of Earth to a flat surface map. Projection maps often distort the shapes of countries and bodies of water. You cannot measure distance accurately on such maps. Countries of equal area often do not appear to be the same size. The lines of latitude and longitude become the guidelines for your orientation in reading the map.

Three of the basic types of map projections are shown here. Each type of projection can be visualized by imagining a piece of film in a certain shape being exposed to light coming from within the globe.

cylindrical projection

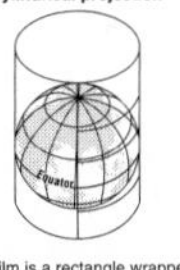

The film is a rectangle wrapped around the globe. It touches the globe at the equator.

conic projection

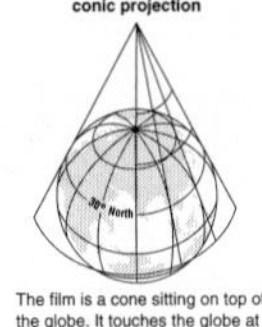

The film is a cone sitting on top of the globe. It touches the globe at one of the latitude lines.

azimuthal projection

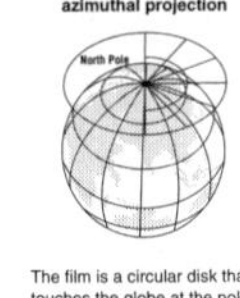

The film is a circular disk that touches the globe at the pole.

Sketch an example of what the film would look like for each type of map projection.

1. cylindrical projection
2. conic projection
3. azimuthal projection

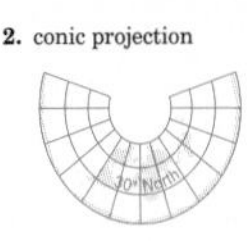

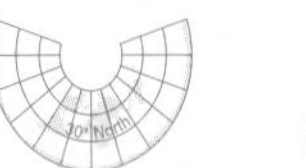

Study each type of projection. At what point(s) would the distortion to the actual shapes of the countries be the least?

4. cylindrical projection near the equator
5. conic projection at the latitude line where it touches the globe
6. azimuthal near the pole

Extension

Connections Have students research the connections between non-Euclidean geometry and modern cosmology. They might want to look at Morris Kline's *Mathematics in Western Culture*.

Additional Answers

37. Sample answer: The spaceship will eventually return to its starting point.

42. Given: $RS = ST$
   Prove: $RT = 2ST$
   Proof:
   Statements (Reasons)
   1. $RS = ST$ (Given)
   2. $RT = RS + ST$ (Seg. Add. Post.)
   3. $RT = 2ST$ (Subst. Prop. (=))

Closing the Investigation

This activity provides students an opportunity to bring their work on the Investigation to a close. For each Investigation, students should present their findings to the class. Here are some ways students can display their work.

- Conduct and report on an interview or survey.
- Write a letter, proposal, or report.
- Write an article for the school or local paper.
- Make a display, including graphs and/or charts.
- Plan an activity.

Assessment

To assess students' understanding of the concepts and topics explored in the Investigation and its follow-up activities, you may wish to examine students' Investigation Folders.

The scoring guide provided in the *Investigations and Projects Masters*, p. 3, provides a means for you to score students' work on the Investigation.

Investigations and Projects Masters, p. 3

Scoring Guide
Chapters 2 and 3
Investigation

Level	Specific Criteria
3 Superior	• Shows thorough understanding of the concepts of *artistic style, artistic design, geometric art, optical art, geometric patterns in art, linear perspective, mural,* and *cost analysis.* • Uses appropriate strategies to solve problems. • Computations are correct. • Written explanations are exemplary. • Designs, drawings, and furnishings are appropriate and sensible. • Goes beyond requirements of all or some problems.
2 Satisfactory, with Minor Flaws	• Shows understanding of the concepts of *artistic style, artistic design, geometric art, optical art, geometric patterns in art, linear perspective, mural,* and *cost analysis.* • Uses appropriate strategies to solve problems. • Computations are mostly correct. • Written explanations are effective. • Designs, drawings, and furnishings are appropriate and sensible. • Satisfies all requirements of problems.
1 Nearly Satisfactory, with Obvious Flaws	• Shows understanding of most of the concepts of *artistic style, artistic design, geometric art, optical art, geometric patterns in art, linear perspective, mural,* and *cost analysis.* • May not use appropriate strategies to solve problems. • Computations are mostly correct. • Written explanations are satisfactory. • Designs, drawings, and furnishings are appropriate and sensible. • Satisfies most requirements of problems.
0 Unsatisfactory	• Shows little or no understanding of the concepts of *artistic style, artistic design, geometric art, optical art, geometric patterns in art, linear perspective, mural,* and *cost analysis.* • Does not use appropriate strategies to solve problems. • Computations are incorrect. • Written explanations are not satisfactory. • Designs, drawings, and furnishings are not appropriate or sensible. • Does not satisfy requirements of problems.

Refer to the Investigation on pages 66–67.

Commercial artists create artwork for specific uses for their clients. Many art pieces are used in advertising, as logos, or in decorating. Creating art for a client is different than working for yourself, since the client's tastes, goals, and budget must be considered.

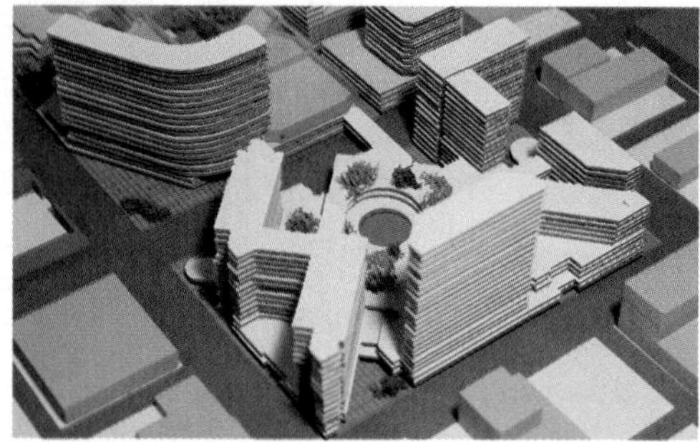

Analyze

You have chosen a style for the lobby and created your mural around a chosen theme. It is now time to analyze your work and finalize the project.

PORTFOLIO ASSESSMENT

You may want to keep your work on this Investigation in your portfolio.

1. Look over your mural and the style you have chosen for the lobby furnishings. Do they harmonize well? Do they communicate the client's image?
2. Refer to your notes about the uses for the lobby. Compile a list of suggested furniture, flooring, wallcovering, and any additional artwork for the lobby.
3. Acquire samples of wallcoverings and photographs of furniture and flooring materials for the lobby. Place these and the mural design in a portfolio for the client.
4. You now need to do a cost analysis and determine what to charge the client. First determine the total number of hours your team spent on the project, including the time you would spend installing the mural and furniture in the lobby. Consult a reference book or a working artist to determine a reasonable hourly rate. Add the cost of materials and furnishings to create a final bill.

Present

It is time to unveil the lobby. Choose one or more members of your team to present the lobby to the CEO and the board members of Cygnus Software, Inc.

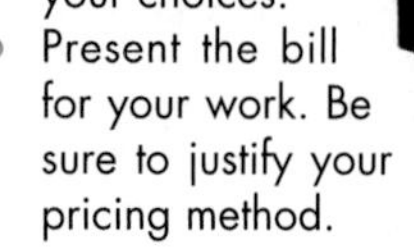

5. Give a short speech describing the process of the design, the goals, the image that the lobby is to present, and the reasons for your choices.
6. Present the bill for your work. Be sure to justify your pricing method.

CHAPTER 3

HIGHLIGHTS

VOCABULARY

After completing this chapter, you should be able to define each term, property, or phrase and give an example or two of each.

Geometry

- alternate exterior angles (p. 126)
- alternate interior angles (p. 126)
- consecutive interior angles (p. 126)
- corresponding angles (p. 126)
- distance between a point and a line (p. 154)
- distance between parallel lines (p. 156)
- equidistant (p. 156)
- exterior angle (p. 126)
- if and only if (p. 139)
- interior angle (p. 126)
- non-Euclidean Geometry (p. 164)
- parallel lines (p. 124)
- Parallel Postulate (p. 147)
- plane Euclidean geometry (p. 164)
- skew lines (p. 125)
- spherical geometry (p. 164)
- transversal (p. 126)

Algebra

- slope (p. 138)

Problem Solving

- draw a diagram (p. 124)

UNDERSTANDING AND USING THE VOCABULARY

Refer to the figure below and choose the letter of the term that best matches each statement or phase.

1. $\angle 3$, $\angle 4$, $\angle 5$, and $\angle 6$ f
2. line c l
3. $\angle 3$ and $\angle 7$ d
4. $\angle 2$ and $\angle 8$ a
5. lines a and b g
6. $\angle 1$, $\angle 2$, $\angle 7$, and $\angle 8$ e
7. $\angle 5$ and $\angle 4$ c
8. There is exactly one line parallel to line a through point M. h
9. $\angle 5$ and $\angle 3$ b
10. line c and a line perpendicular to the plane of this page passing through point M i
11. the ratio of the vertical rise to the horizontal run of line b j

a. alternate exterior angles
b. alternate interior angles
c. consecutive interior angles
d. corresponding angles
e. exterior angles
f. interior angles
g. parallel lines
h. parallel postulate
i. skew lines
j. slope
k. spherical geometry
l. transversal

Instructional Resources

Three multiple-choice tests and three free-response tests are provided in the *Assessment and Evaluation Masters*. Forms 1A and 2A are for honors pacing, Forms 1B and 2B are for average pacing, and Forms 1C and 2C are for basic pacing. Chapter 3 Test, Form 1B, is shown at the right. Chapter 3 Test, Form 2B, is shown on the next page.

Postulates, Theorems, and Corollaries

A complete list of postulates, theorems, and corollaries begins on page 806.

Using the CHAPTER HIGHLIGHTS

The Chapter Highlights begins with a listing of the new terms, properties, and phrases that were introduced in this chapter. Have students define each term and provide an example or two of it, if appropriate.

Assessment and Evaluation Masters, pp. 59–60

NAME ______ DATE ______

Chapter 3 Test, Form 1B

Write the letter for the correct answer in the blank at the right of each problem.

1. **Draw a Diagram** Kathy wants to put 16 rectangular pictures of the same size on the bulletin board. If she wants them all to be seen and wants to have a tack in each corner of each picture, how many tacks must she have? (A tack can be in the corner of more than one picture.)
   A. 20 B. 24 C. 25 D. 28 — 1. C
2. What symbol is used to indicate two angles are congruent?
   A. = B. $\parallel$ C. $\perp$ D. $\cong$ — 2. D

Refer to the figure below for Questions 3–6.

3. Identify the special angle pair name for $\angle 1$ and $\angle 3$.
   A. consecutive interior
   B. corresponding
   C. vertical
   D. alternate interior — 3. B
4. Given $\ell \parallel m$ and $m\angle 4 = 63$, find $m\angle 6$.
   A. 27 B. 63
   C. 105 D. 117 — 4. D
5. Given $\ell \parallel m$, $m\angle 11 = 2x - 3$, and $m\angle 9 = 71$, find the value of x.
   A. 28 B. 37
   C. 71 D. 98 — 5. B
6. Given $m\angle 10 = 3x - 2$ and $m\angle 12 = 70$, find the value of x so that $\ell \parallel m$.
   A. 72 B. 70 C. 68 D. 24 — 6. D
7. Suppose that h and k are intersected by transversal t. If one of the angles formed by h and t is congruent to every angle formed by k and t, which of the following must be true?
   A. $h \parallel t$ B. $t \perp k$ C. $k \perp h$ D. $k \parallel t$ — 7. B
8. Which figure shows what you assume is "given" in a proof of the theorem *If two lines in a plane are cut by a transversal so that alternate interior angles are congruent, then the lines are parallel*? (It should show *only* the "given.")
   A. B. C. D.

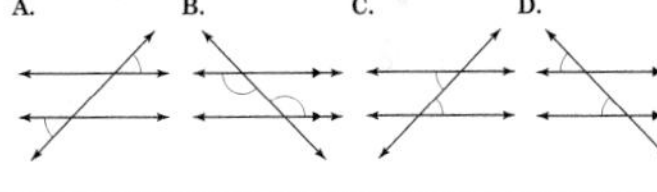

8. C

NAME ______ DATE ______

Chapter 3 Test, Form 1B (continued)

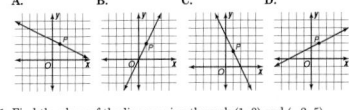

9. The slope of line k is undefined. Which of the following is a true statement about line k?
   A. k is horizontal. B. k falls to the right.
   C. k falls to the left. D. k is vertical. — 9. D
10. Which figure shows the line with a slope of 2 passing through $P(1, 2)$?
    A. B. C. D. — 10. B
11. Find the slope of the line passing through (1, 3) and (−2, 5).
    A. $\frac{2}{3}$ B. $-\frac{2}{3}$ C. $\frac{3}{2}$ D. $-\frac{3}{2}$ — 11. B
12. Find the slope of any line parallel to the line passing through (1, 5) and (−1, 0).
    A. $\frac{5}{2}$ B. $-\frac{5}{2}$ C. $\frac{2}{5}$ D. $-\frac{2}{5}$ — 12. A
13. What is the distance from the point at (3, −2) to the line $y = -5$?
    A. 3 B. 7 C. −3 D. 1 — 13. C
14. Which one of the following statements from Euclidean geometry is also true in spherical geometry?
    A. A segment is the shortest path between two points.
    B. Perpendicular lines intersect at one point.
    C. Perpendicular lines form four right angles.
    D. A line has infinite length. — 14. A
15. What figures of Euclidean geometry are comparable to the spheres of spherical geometry?
    A. points B. lines
    C. planes D. segments — 15. C

Bonus

What is the slope of any line parallel to any line perpendicular to the line $x = 3$?
A. 0 B. 3 C. −3 D. undefined — Bonus A

Using the STUDY GUIDE AND ASSESSMENT

Skills and Concepts Encourage students to refer to the objectives and examples on the left as they complete the review exercises on the right.

Assessment and Evaluation Masters, pp. 65–66

NAME ________________ DATE ________

Chapter 3 Test, Form 2B

1. Consider the three-dimensional figure shown at the right. How many pairs of skew segments are shown? 1. 8

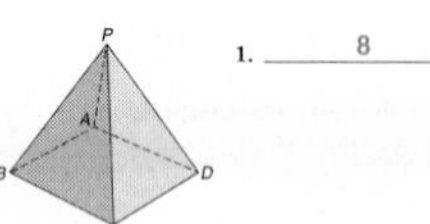

2. **Draw a Diagram** The Lee family put a fence around their garden, which was in the shape of a square. There was a post at each corner. If there were 7 posts on each side of the square, how many posts were there altogether? 2. 24

State the transversal that forms each pair of angles in the figure below. Then identify the special angle pair name.

3. ∠5 and ∠7 3. ℓ; consecutive interior
4. ∠1 and ∠8 4. $\overleftrightarrow{AC}$; corresponding
5. ∠4 and ∠10 5. ℓ; alternate exterior
6. ∠1 and ∠3 6. $\overleftrightarrow{AB}$; alternate interior

In the figure below, ℓ ∥ m, $\overleftrightarrow{AB} \parallel \overleftrightarrow{CE}$, $\overrightarrow{CD}$ bisects ∠ACE, and m∠1 = 130. Find the measure of each angle.

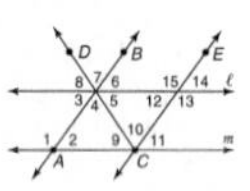

7. ∠3 7. 50
8. ∠11 8. 50
9. ∠13 9. 130
10. ∠9 10. 65

Find the value of x so that p ∥ q.

11. 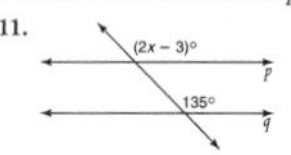

12. (2x + 7)°, 103°

11. 69
12. 35

NAME ________________ DATE ________

Chapter 3 Test, Form 2B (continued)

13. Tell whether a line whose slope is undefined is horizontal, vertical, or neither. 13. vertical

Find the slope of the line passing through the given points.

14. $A(-3, 0)$, $B(2, -1)$ 14. $-\frac{1}{5}$
15. $C(5, -1)$, $D(3, 0)$ 15. $-\frac{1}{2}$
16. $E(8, -6)$, $F(9, 2)$ 16. 8

Find the slope of a line parallel to the line passing through the given points.

17. $G(5, -3)$, $H(8, 2)$ 17. $\frac{5}{3}$
18. $J(-4, 0)$, $K(1, -5)$ 18. −1
19. $M(0, -6)$, $N(-2, -3)$ 19. $-\frac{3}{2}$

Find the slope of a line perpendicular to the line passing through the given points.

20. $P(-1, 4)$, $Q(8, 1)$ 20. 3
21. $R(3, -2)$, $S(5, 1)$ 21. $-\frac{2}{3}$
22. $T(4, -3)$, $U(1, -1)$ 22. $\frac{3}{2}$

23. Draw and label a figure that shows three lines, $\overleftrightarrow{AB}$, $\overleftrightarrow{CD}$, and $\overleftrightarrow{KL}$ such that each is a transversal of the other two. 23.

For each property listed from plane Euclidean geometry, write a corresponding statement for non-Euclidean spherical geometry.

24. There is a unique straight line passing through any two points. 24. There is a unique great circle passing through any pair of non-polar points.
25. Perpendicular lines intersect at one point. 25. Perpendicular great circles intersect at two points.

Bonus

Suppose ℓ, m, and n are three different lines in space. Suppose that ℓ ∥ n, ℓ and m intersect at P, and m does not intersect n. Is m parallel to n, skew to n, or neither? Explain. Bonus *Skew:* By the Parallel Postulate m and n cannot be in the same plane, or we would have two lines through P parallel to n.

SKILLS AND CONCEPTS

OBJECTIVES AND EXAMPLES	REVIEW EXERCISES

Upon completing this chapter, you should be able to:

- describe the relationships between two lines, between two planes, and between angles formed by two lines and a transversal (Lesson 3–1)

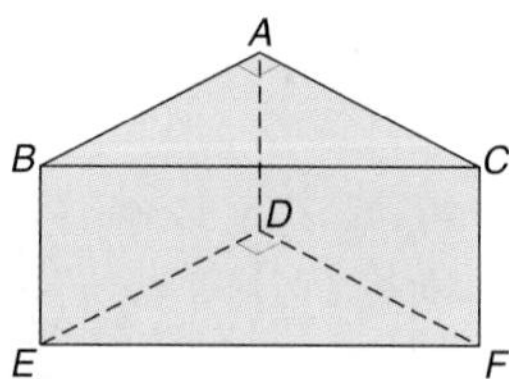

Plane *ABC* and plane *DEF* are parallel.
Segments *BC* and *EF* are parallel.
Segments *AB* and *DF* are skew.

Use these exercises to review and prepare for the chapter test.

Refer to the figure at the right to find an example of each of the following.

12. two parallel lines
13. two skew lines *t* and *m*
14. two parallel lines and a transversal
15. two intersecting lines
16. two noncoplanar lines *t* and *m*

12. *t* and *n*
14. *t* and *n* with transversal ℓ
15. Sample answer: ℓ and *n*

- use the properties of parallel lines to determine angle measures (Lesson 3–2)

List the conclusions that can be drawn if ℓ ∥ *m*.

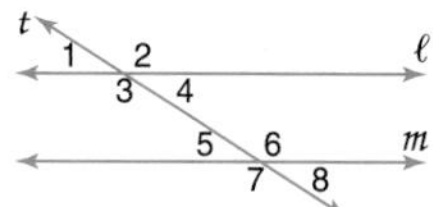

corresponding angles ∠1 ≅ ∠5, ∠2 ≅ ∠6, ∠3 ≅ ∠7, and ∠4 ≅ ∠8

alternate interior angles ∠3 ≅ ∠6, and ∠4 ≅ ∠5

consecutive interior angles ∠3 and ∠5, and ∠4 and ∠6 are supplementary.

alternate exterior angles ∠1 ≅ ∠8, and ∠2 ≅ ∠7

Refer to the figure below for Exercises 17–20.

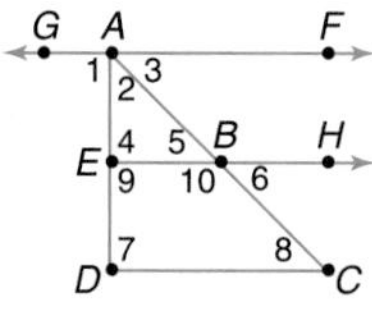

17. If $\overline{GF} \parallel \overline{EH}$, which angles are congruent?
18. If $\overline{GF} \parallel \overline{DC}$, which angles are congruent?
19. If $\overline{GF} \parallel \overline{DC}$, which angles are supplementary to ∠*CAG*? ∠3, ∠8
20. If $\overline{EH} \parallel \overline{DC}$, which angles are supplementary to ∠*BCD*? ∠10, ∠*ABH*

17. ∠3 ≅ ∠5, ∠3 ≅ ∠6, ∠1 ≅ ∠4, ∠5 ≅ ∠6, ∠*GAB* ≅ ∠10, ∠*GAB* ≅ ∠*ABH*, ∠*FAE* ≅ ∠9
18. ∠3 ≅ ∠8, ∠1 ≅ ∠7, ∠5 ≅ ∠6

- find the slopes of lines (Lesson 3–3)

Find the slope of the line passing through (5, −2) and (−1, −5). Describe the line.

$$m = \frac{y_2 - y_1}{x_2 - x_1} = \frac{-2 - (-5)}{5 - (-1)} \text{ or } \frac{1}{2}$$

Since the slope is positive, the line rises from left to right.

21. 6; rising 22. undefined; vertical

Find the slope of the line passing through points having the given coordinates. Then describe the line as you move from left to right as *rising, falling, horizontal, or vertical.*

21. (0, 4), (−1, −2)
22. (2, 0), (2, −6)
23. (11, 2), (5, 4) $-\frac{1}{3}$; falling
24. (−1, −5), (3, −7) $-\frac{1}{2}$; falling

GLENCOE Technology

Test and Review Software

You may use this software, a combination of an item generator and item bank, to create your own tests or worksheets. Types of items include free response, multiple choice, short answer, and open ended.

For IBM & Macintosh

OBJECTIVES AND EXAMPLES

- Use slope to identify parallel and perpendicular lines (Lesson 3–3)

Find the slopes of the lines parallel to and perpendicular to a line passing through (4, −2) and (5, 3).

The slope of the line passing through (4, −2) and (5, 3) is $\frac{-2-3}{4-5}$ or 5.

A parallel line has slope 5, and a perpendicular line has slope $-\frac{1}{5}$.

REVIEW EXERCISES

Find the slopes of the lines parallel and perpendicular to the line through points having the given coordinates.

25. (−3, 7), (4, −2) $-\frac{9}{7}; \frac{7}{9}$
26. (0, 6), (3, 6) 0; undefined
27. (7, −2), (1, −3) $\frac{1}{6}; -6$
28. (9, −2), (−1, 4) $-\frac{3}{5}; \frac{5}{3}$

- prove two lines are parallel based on given angle relationships (Lesson 3–4)

Ways to Show Two Lines Parallel

- Corresponding angles are congruent.
- Alternate exterior angles are congruent.
- Consecutive interior angles are supplementary.
- Alternate interior angles are congruent.
- Two lines are perpendicular to a third line.

29. $\overline{AD} \parallel \overline{EF}$; $\angle 1 \cong \angle 4$ since vertical angles are congruent. So, $\angle 4$ and $\angle 2$ are supplementary. The lines are parallel because consecutive interior angles are supplementary.

Refer to the figure below for Exercises 29–31.

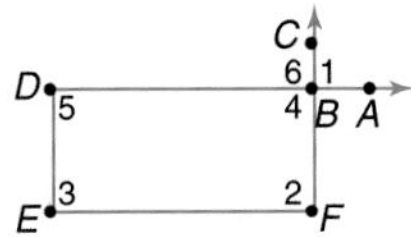

29. Given $\angle 1$ and $\angle 2$ are supplementary, which lines are parallel and why?

30. Given $\angle 5 \cong \angle 6$, which lines are parallel and why? $\overline{DE} \parallel \overline{CF}$; alt. int. ⦣s are ≅.

31. Given $\angle 6 \cong \angle 2$, which lines are parallel and why? $\overline{AD} \parallel \overline{EF}$; corr. ⦣s are ≅.

- recognize and use distance relationships among points, lines, and planes (Lesson 3–5)

The distance from B to $\overline{AD}$ is BD.

The distance from $\overline{BA}$ to $\overline{CD}$ is BC.

In the figure below, $\overline{PS} \perp \overline{SQ}$, $\overline{PQ} \perp \overline{QR}$, $\overline{PM} \perp \overline{RM}$, and $\overline{QR} \perp \overline{SR}$. Name the segment whose length represents the distance between the following points and lines.

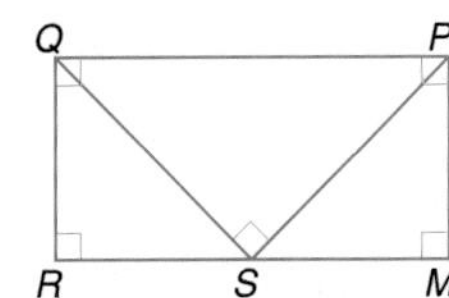

32. from P to $\overleftrightarrow{SQ}$ $\overline{PS}$
33. from R to $\overleftrightarrow{PQ}$ $\overline{RQ}$
34. from $\overleftrightarrow{RM}$ to $\overleftrightarrow{QP}$ $\overline{RQ}$ or $\overline{MP}$
35. from $\overleftrightarrow{RQ}$ to $\overleftrightarrow{MP}$ $\overline{PQ}$ or $\overline{MR}$

Applications and Problem Solving Encourage students to work through the exercises in the Applications and Problem Solving section to strengthen their problem-solving skills.

Additional Answers

41.

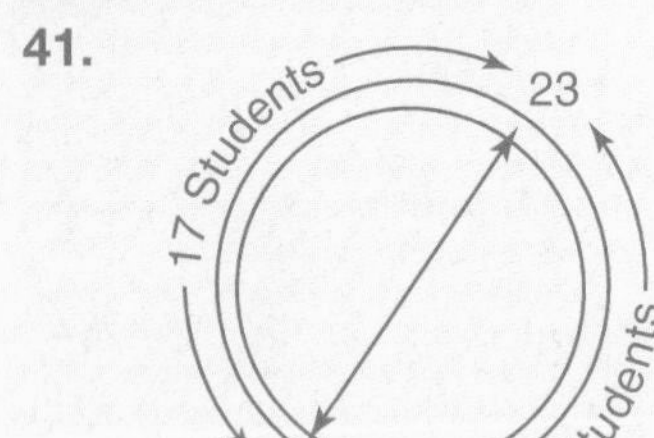

36 students

42a.

2 ft

24 ft

OBJECTIVES AND EXAMPLES

- compare and contrast basic properties of plane and spherical geometry (Lesson 3–6)

Plane Euclidean geometry studies a system of points, lines, and planes.

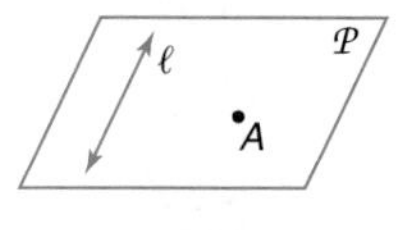

Spherical geometry studies a system of points, great circles, and spheres.

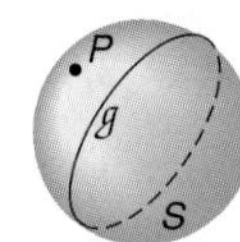

REVIEW EXERCISES

For each property listed from plane Euclidean geometry, write a corresponding statement for spherical geometry.

36. If three points are collinear, exactly one point is between the other two.
37. The intersection of two lines creates four angles.
38. If two lines are parallel to a given line, the lines are parallel to each other.
39. The shortest path between two points is a straight line segment.
40. There is exactly one line passing through two points.

36. If three points are collinear, any of the points can be between the other 2 points.
37. The intersection of 2 great circles creates 8 angles. 38. There exists no parallel lines.
39. The shortest path between 2 points is an arc on the great circle passing through the points.
40. There is at least one great circle passing through 2 points.
41. 36 students; see margin for diagram.

APPLICATIONS AND PROBLEM SOLVING

41. **Draw a Diagram** The students in a gym class are standing in a circle. When they count off, the students with numbers 5 and 23 are standing exactly opposite one another. Assuming the students are evenly spaced around the circle, how many students are in the class? (Lesson 3–1)

42. **Construction** A ramp was installed to give handicapped people access to the new public library. The top of the ramp is two feet higher than the bottom. The lower end of the ramp is 24 feet from the door of the library. (Lesson 3–3)
    a. Draw a diagram of the ramp. See margin.
    b. Find the slope of the ramp. $\frac{1}{12}$

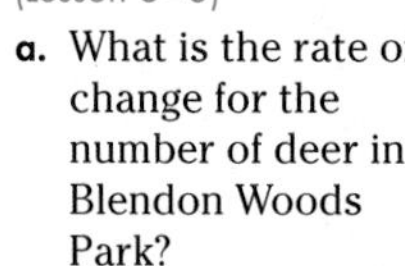

Sculpture of a book located in front of Whetstone Library in Columbus, Ohio

A practice test for Chapter 3 is available on page 795.

43. **Travel** At 10:00 A.M., Luisa had completed 195 miles of her cross-country trip. By 2:00 P.M., she had traveled a total of 455 miles. Use slope to determine Luisa's average rate of travel. (Lesson 3–3) 65 mph

44. **Nature Studies** Hong is a park ranger and she estimates that there are 6000 deer in the Blendon Woods Park. She also estimates that one year ago there were 6100 deer in the park. (Lesson 3–3)

a. What is the rate of change for the number of deer in Blendon Woods Park?
b. At the same rate, how many deer will there be in the park in ten years? 5000 deer
c. Without using a graph, determine if the graph of this rate would rise or fall. Explain. Decline; since the slope is negative the line will fall as you move fromleft to right.

44a. −100 deer per year

ALTERNATIVE ASSESSMENT

COOPERATIVE LEARNING PROJECT

Construction In this project, you will explore parallel lines in the real world. Read the project completely before beginning the first task.

Follow these steps to accomplish your task.

- Look around you to locate five examples of parallel lines.
- Measure the shortest distance between the parallel segments at their endpoints and at their midpoints.
- Record your measurements in a table.
- Are the lines parallel, as defined in this chapter?
- Discuss the following questions with your group. Why may the lines not be exactly parallel? Is it due to poor workmanship? Could the builder have made the lines parallel?

Follow these steps to accomplish your next task.

- Select three pairs of parallel lines you measured in the activity. Stand between the lines near one of the endpoints and photograph them so that all of both lines are included in the image. Measure the distance from your standpoint to the nearest endpoint. (Instead of taking a picture, you could sketch the lines from this perspective.)
- On the photograph or drawing, measure the shortest distance between the lines at each endpoint and at the midpoint.
- Record the measurements in a table.

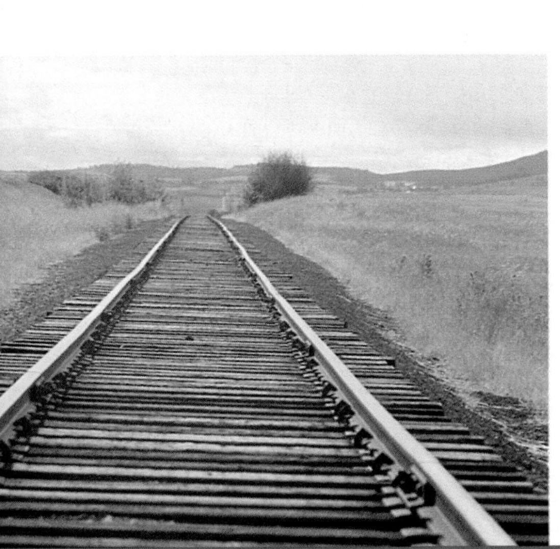

- Discuss the following questions with your group. Do you think you could find a mathematical function to model the decreasing distance between the lines as they are farther away? Do you think you could find a mathematical function relating your distance from the nearest endpoint to the mathematical function mentioned in the previous question? Could you use this relationship to determine whether two converging lines in a photograph or drawing are parallel or not?

THINKING CRITICALLY

- If $\mathcal{A}$ and $\mathcal{B}$ are parallel planes, ℓ is a line on $\mathcal{A}$, and m is a line on $\mathcal{B}$, is it true that ℓ and m are parallel? no
- If ℓ is a line parallel to the plane $\mathcal{A}$, and $\mathcal{B}$ is a plane containing the line ℓ, is it true that $\mathcal{A}$ and $\mathcal{B}$ are parallel? no

PORTFOLIO

Use word processing software to write a one-page report on the history of projective geometry and its relationship to perspective drawing. The Internet may help you to gather information. Try to answer these questions.

- Who developed the ideas first, mathematicians or artists?
- How did artists handle the problem of perspective before this development?

Place your report in your portfolio.

SELF EVALUATION

Some people are concrete thinkers while others think abstractly. Concrete thinkers use visual or physical models to think about abstract ideas.

Assess Yourself Do you prefer to draw a picture to solve a geometry problem, or do you prefer to think without paper and reason abstractly?

Alternative Assessment

The Alternative Assessment section provides students with the opportunity to assess their own work by thinking critically, working with others, keeping a portfolio, and honestly evaluating their own progress. For more information on alternative forms of assessment, see *Alternative Assessment in the Mathematics Classroom,* one of the titles in the Glencoe Mathematics Professional Series.

Performance Assessment

Performance Assessment tasks for this chapter are included in the *Assessment and Evaluation Masters.* A scoring guide is also provided.

Assessment and Evaluation Masters, pp. 70, 81

NAME ____________ DATE ______

Chapter 3 Performance Assessment

Instructions: ***Demonstrate your knowledge by giving a clear, concise solution to each problem. Be sure to include all relevant drawings and to justify your answers. You may show your solutions in more than one way or investigate beyond the requirements of the problems.***

1. Horatio is building a shed. The framing diagram for the shed is shown below.

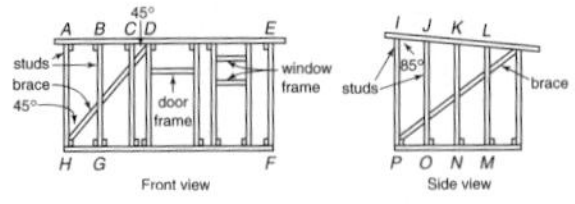

a. Horatio wants the roof line of the front to be parallel to the floor. Describe at least two ways that he can make sure they are parallel.

b. Is the roof line of the side parallel to the floor? Explain.

c. What is the measure of $\angle KJO$? How do you know?

d. What is the measure of $\angle KLM$? How do you know?

e. What is the measure of $\angle HDE$? How do you know?

f. Are the studs all parallel? Why or why not?

g. Prove that the sum of the measures of the angles of quadrilateral $IJOP$ is 360.

2. a. Draw line ℓ through $A(-1, 2)$ and $B(3, 5)$ on the coordinate plane at the right. Find the slope of the line.

b. Draw line m through $C(-2, 3)$ parallel to line ℓ. Show your work and explain each step.

c. Draw line n through $C(-2, 3)$ perpendicular to line ℓ. Show your work and explain each step.

d. Draw line p with a slope of $\frac{2}{3}$ on the coordinate plane at the right. Explain how you determined that the line has a slope of $\frac{2}{3}$.

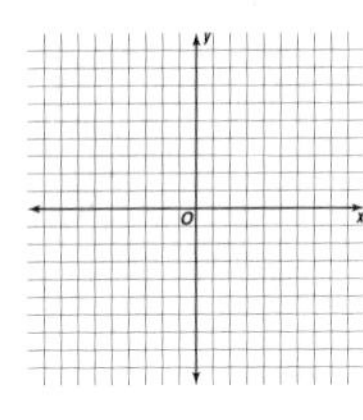

Scoring Guide
Chapter 3
Performance Assessment

Level	Specific Criteria
3 Superior	• Shows thorough understanding of the properties of parallel and perpendicular lines. • Uses appropriate strategies to construct proof. • Computations are correct. • Written explanations are exemplary. • Graphs are accurate and appropriate. • Goes beyond requirements of some or all problems.
2 Satisfactory, with Minor Flaws	• Shows understanding of the properties of parallel and perpendicular lines. • Uses appropriate strategies to construct proof. • Computations are mostly correct. • Written explanations are effective. • Graphs are mostly accurate and appropriate. • Satisfies all requirements of some or all problems.
1 Nearly Satisfactory, with Serious Flaws	• Shows understanding of most of the properties of parallel and perpendicular lines. • May not use appropriate strategies to construct proof. • Computations are mostly correct. • Written explanations are satisfactory. • Graphs are mostly accurate and appropriate. • Satisfies most requirements of some or all problems.
0 Unsatisfactory	• Shows little or no understanding of the properties of parallel and perpendicular lines. • May not use appropriate strategies to construct proof. • Computations are incorrect. • Written explanations are not satisfactory. • Graphs are not accurate or appropriate. • Does not satisfy requirements of some or all problems.

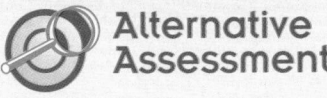

In•ves•ti•ga•tion

TEACHER NOTES

NCTM Standards: 1–5, 7, 8

This Investigation is designed to be completed over several days or weeks. It may be considered optional. You may want to assign the Investigation and the follow-up activities to be completed at the same time.

Objective

Design a poster display, slide show, or video to demonstrate a naturally occurring pattern.

Mathematical Overview

This Investigation will use the following mathematical skills and concepts from Chapters 4 and 5.

- identifying types of triangles
- identifying mathematical patterns
- using proportional reasoning
- applying mathematical patterns to natural objects

Recommended Time		
Part	**Pages**	**Time**
Investigation	176–177	1 class period
Working on the Investigation	187, 228, 265, 279	20 minutes each
Closing the Investigation	280	1 class period

Instructional Resources

Investigations and Projects Masters, pp. 5–8

A recording sheet, teacher notes, and scoring guide are provided for each Investigation in the *Investigations and Projects Masters*.

1 MOTIVATION

Have students look around the room and try to find any man-made patterns. Discuss different types of patterns and why they are useful. Discuss whether patterns occur in nature.

LONG-TERM PROJECT

In•ves•ti•ga•tion

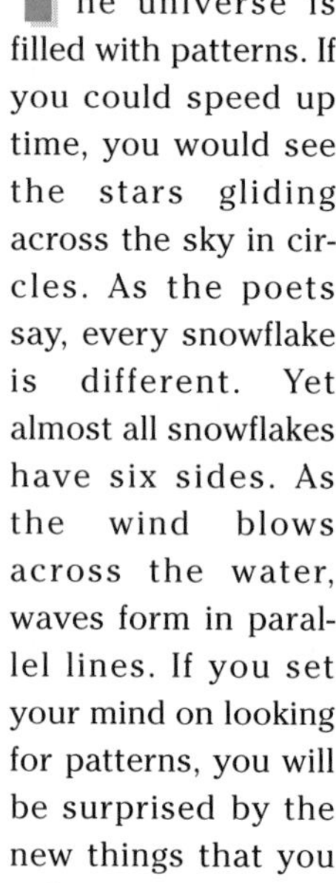

MATERIALS NEEDED

- camera
- slide film
- posterboard
- markers
- scissors
- glue
- video camera
- videotape
- soft pencil
- clear tape

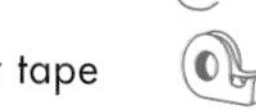

The universe is filled with patterns. If you could speed up time, you would see the stars gliding across the sky in circles. As the poets say, every snowflake is different. Yet almost all snowflakes have six sides. As the wind blows across the water, waves form in parallel lines. If you set your mind on looking for patterns, you will be surprised by the new things that you will see.

In this Investigation, you will work with your group to investigate a pattern found in nature. You will describe where the pattern occurs and any mathematical formulas that can be applied to the pattern. You will complete the project by presenting your findings in a poster display, slide show, or video.

PATTERNS TO FOLLOW

Begin your project by choosing a pattern to investigate. Use reference books from the library or do some research on the Internet to read about the patterns that have been observed. Some areas you may decide to study in this Investigation are described below.

Symmetry Symmetric objects show a one-to-one correspondence among their parts. For example, things like butterflies that could be folded so that the halves match are symmetric. Another type of symmetry involves rotating an object so that its parts correspond. An apple blossom has this type of rotational symmetry.

Packing Natural objects are almost always formed so that space is used efficiently. For example, the cells of a honeycomb are arranged to use the available space most efficiently. How do you think three-dimensional space is used most efficiently?

Cooperative Learning

This Investigation offers an excellent opportunity for using cooperative learning groups. For more information on cooperative learning strategies and group management, see *Cooperative Learning in the Mathematics Classroom*, one of the titles in the Glencoe Mathematics Professional Series.

Cracking Have you ever noticed the way that dry soil or concrete cracks? The same patterns are repeated many times. What are the measures of the angles formed when two, three, or more cracks meet?

Spirals The horns of a ram, a nautilus shell, hurricanes, and galaxies all form spirals. You can see a spiral by filling a sink with water, then releasing it and observing the path the water takes as it drains.

Motion When you pluck a guitar string, the string moves back and forth in a regular pattern called an *oscillation.* You can also see an oscillating pattern in the way that an animal's limbs move when it walks. For example, when a horse trots its front left and back right legs hit the ground together. On the next step, the front right and back left legs hit the ground together.

Fibonacci Numbers and the Golden Section The sequence 1, 1, 2, 3, 5, 8, 13, 21, ... is called the *Fibonacci sequence* in honor of its discoverer Leonardo Fibonacci. The numbers in this sequence, called *Fibonacci numbers*, are found in many natural objects such as the rows of seeds in a sunflower and the numbers of leaves on plant stems. The ratios between successive Fibonacci numbers approach the *Golden Section.* This special irrational number is also associated with many natural objects.

Spheres and Near-Spheres A beachball is a good model for a sphere. As raindrops fall, they form spheres. Bubbles are also spherical. The 1996 Nobel Prize for Chemistry was awarded to Harold Kroto, Robert Curl, and Richard Smalley for their work on an unusual carbon molecule that is nearly spherical. The molecule is roughly shaped like a soccer ball and is called *buckminsterfullerene.*

There are many other patterns displayed in nature. Also, more than one of the patterns can often be found in any one object. After you have chosen the pattern you will study, begin looking for information in books or on the Internet. Gather photographs for a poster display or slide presentation. If the equipment is available, you may wish to make a video presentation of your findings.

You will continue working on this Investigation throughout Chapters 4 and 5.

Be sure to keep your research, designs, and other materials in your Investigation Folder.

It's Only Natural Investigation

2 SETUP

Have a student read the first two paragraphs and the paragraphs about patterns in the Investigation to provide background information for the patterns research project. Read the third paragraph and the last paragraph that explains the class's tasks. Discuss the activity with students. Separate the class into groups of four or five.

3 MANAGEMENT

Each group member should be responsible for a specific task.

Recorder Keeps track of the pattern chosen and related information.

Researcher Keeps track of what sources have been used.

Photographer Takes pictures of patterns.

Layout Editor Designs the poster display.

Sample Answers

Answers will vary as they depend on the type of pattern chosen by each group.

Investigations and Projects Masters, p. 8

NAME ____________ DATE ________

Investigation, Chapters 4 and 5

Student Edition Pages 176–177, 187, 228, 265, 279, 280

It's Only Natural

Overview

Use the space below for the thumbprints from the members in your group.

Use this chart to record the natural objects that contain the pattern your group selected.

Object	Where Found	Notes/Comments

Work with your group to consider these questions.

- Were examples of your pattern easier or more difficult to find than you anticipated?
- Was there any place in particular that contained many examples of your selected pattern?
- What percentage of examples had more than one pattern?

Please keep this page and any other research in your Investigation Folder.

CHAPTER

4

Identifying Congruent Triangles

PREVIEWING THE CHAPTER

This chapter provides an intensive study of triangles. The chapter begins by classifying triangles according to angles and sides. Interior and exterior angles and their measures are explored. This leads to the coverage of triangle congruence including SSS, SAS, and ASA postulates as well as the AAS Theorem. The chapter concludes with a study of isosceles and equilateral triangles. Compass and straightedge constructions permeate the chapter.

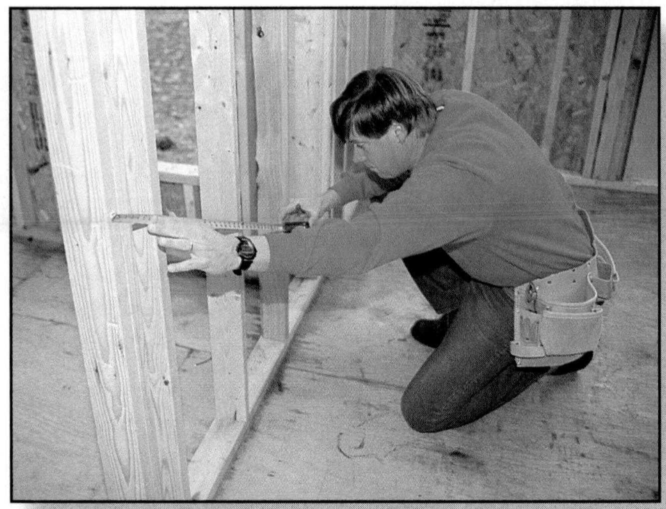

Lesson (Pages)	Lesson Objectives	NCTM Standards	State/Local Objectives
4-1 (180–187)	Identify the parts of triangles and classify triangles by their parts.	1–4, 7, 8	
4-2A (188)	Investigate the relationship of the angles of a triangle.	1–4, 7	
4-2 (189–195)	Apply the Angle Sum Theorem. Apply the Exterior Angle Theorem.	1–4, 7	
4-3 (196–203)	Name and label corresponding parts of congruent triangles.	1–4, 7, 8	
4-4A (204–205)	Draw congruent triangles.	2, 4, 7	
4-4 (206–213)	Use SSS, SAS, and ASA postulates to test for triangle congruence.	1–5, 7, 8	
4-5 (214–221)	Use the AAS Theorem to test triangle congruence. Solve problems by eliminating the possibilities.	1–4, 7, 8	
4-6 (222–228)	Use properties of isosceles and equilateral triangles.	1–5, 7, 8	

ORGANIZING THE CHAPTER

A complete, 1-page lesson plan is provided for each lesson in the ***Lesson Planning Guide***. Answer keys for each lesson are available in the ***Answer Key Masters***.

You may want to refer to the **Course Planning Calendar** on page T12 for detailed information on pacing.
PACING: Standard—14 days; **Honors—**13 days; **Block—**7 days

LESSON PLANNING CHART

			Blackline Masters									
Lesson (Pages)	**Materials/ Manipulatives**	**Extra Practice (Student Edition)**	**Study Guide**	**Practice**	**Enrichment**	**Assessment & Evaluation**	**Modeling Mathematics**	**Multicultural Activity**	**Tech Prep Applications**	**Graphing Calc. & Computer**	**Real-World Applications**	**Teaching Transparencies**
4-1 (180–187)	patty paper TI-82/83 graphing calculator protractor*	p. 770	p. 20	p. 20	p. 20							4-1A 4-1B
4-2A (188)	straightedge* scissors*						p. 92					
4-2 (189–195)		p. 770	p. 21	p. 21	p. 21	p. 100						4-2A 4-2B
4-3 (196–203)	grid paper scissors*	p. 771	p. 22	p. 22	p. 22	pp. 99, 100	pp. 27–30		p. 7	p. 4	7	4-3A 4-3B
4-4A (204–205)	compass* straightedge* protractor* scissors*						p. 93					
4-4 (206–213)	ruler*	p. 771	p. 23	p. 23	p. 23			p. 7				4-4A 4-4B
4-5 (214–221)	patty paper ruler* protractor* compass* straws push pins tape	p. 771	p. 24	p. 24	p. 24	p. 101	p. 82				8	4-5A 4-5B
4-6 (222–228)	patty paper compass* straightedge* TI-82/83 graphing calculator	p. 772	p. 25	p. 25	p. 25	p. 101		p. 8	p. 8			4-6A 4-6B
Study Guide/ Assessment (229–233)	construction paper transparent tape					pp. 85–98, 102–104						

*Included in Glencoe's High School Manipulative Kit and Overhead Manipulative Resources.

ORGANIZING THE CHAPTER

Other Chapter Resources

Student Edition
Investigation, pp. 176–177
Chapter Opener, pp. 178–179
Mathematics and Society, p. 213
Working on the Investigation, pp. 187, 228

Teacher's Classroom Resources
Investigations and Projects Masters, pp. 37–40
Block Scheduling Booklet

Technology
Test and Review Software (IBM and Macintosh)
CD-ROM Multimedia Applications (Windows and Macintosh)
Mindjogger Videoquizzes (VHS)

Professional Publications
Glencoe Mathematics Professional Series

Outside Resources

Books/Periodicals
Notes on a Triangle, Dale Seymour Publications
Quilt Design Masters, Dale Seymour Publications

Software
Tangrams Puzzler, ETA

Videos/CD-ROMs
Introduction to Geometric Terms, Angles, and Triangles, NASCO, 901 Janesville Ave., Fort Atkinson, WI 53538

Assessment Resources

Student Edition
Math Journal, pp. 199, 208
Mixed Review, pp. 187, 195, 202, 212, 221, 228
Self Test, p. 203
Chapter Highlights, p. 229
Chapter Study Guide and Assessment, pp. 230–232
Alternative Assessment, p. 233
Portfolio, p. 233
College Entrance Exam Practice, pp. 234–235
Chapter Test, p. 796

Teacher's Wraparound Edition
5-Minute Check, pp. 180, 189, 196, 206, 214, 222
Check for Understanding, pp. 183, 192, 199, 208, 216, 224
Closing Activity, pp. 187, 195, 203, 213, 221, 228
Cooperative Learning, pp. 209, 223

Assessment and Evaluation Masters
Multiple-Choice Tests, Forms 1A (Honors), 1B (Average), 1C (Basic), pp. 85–90
Free-Response Tests, Forms 2A (Honors), 2B (Average), 2C (Basic), pp. 91–96
Calculator-Based Test, p. 97
Performance Assessment, p. 98
Mid-Chapter Test, p. 99
Quizzes A–D, pp. 100–101
Standardized Test Practice, p. 102
Cumulative Review, pp. 103–104

ENHANCING THE CHAPTER

Examples of some of the materials for enhancing Chapter 4 are shown below.

Multicultural Activity Masters, pp. 7, 8

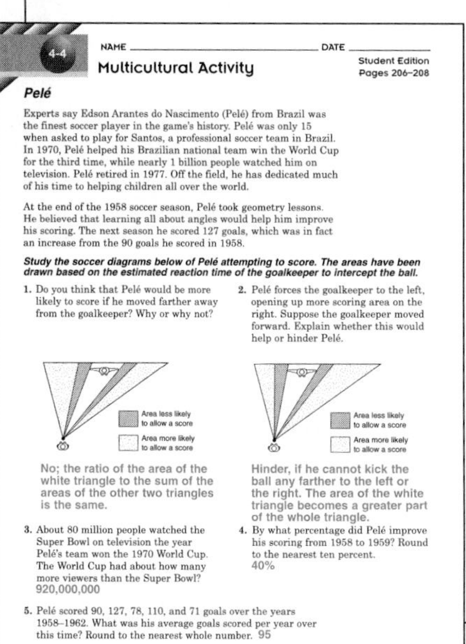

4-4 NAME ______ DATE ______

Multicultural Activity — Student Edition Pages 206–208

Pelé

Experts say Edson Arantes do Nascimento (Pelé) from Brazil was the finest soccer player in the game's history. Pelé was only 15 when asked to play for Santos, a professional soccer team in Brazil. In 1970, Pelé helped his Brazilian national team win the World Cup for the third time, while nearly 1 billion people watched him on television. Pelé retired in 1977. Off the field, he has dedicated much of his time to helping children all over the world.

At the end of the 1958 soccer season, Pelé took geometry lessons. He believed that learning all about angles would help him improve his scoring. The next season he scored 127 goals, which was in fact an increase from the 90 goals he scored in 1958.

Study the soccer diagrams below of Pelé attempting to score. The areas have been drawn based on the estimated reaction time of the goalkeeper to intercept the ball.

1. Do you think that Pelé would be more likely to score if he moved farther away from the goalkeeper? Why or why not?
   No; the ratio of the area of the white triangle to the sum of the areas of the other two triangles is the same.
2. Pelé forces the goalkeeper to the left, opening up more scoring area on the right. Suppose the goalkeeper moved forward. Explain whether this would help or hinder Pelé.
   Hinder, if he cannot kick the ball any farther to the left or the right. The area of the white triangle becomes a greater part of the whole triangle.

Area less likely to allow a score — Area more likely to allow a score

3. About 80 million people watched the Super Bowl on television the year Pelé's team won the 1970 World Cup. The World Cup had about how many more viewers than the Super Bowl? 920,000,000
4. By what percentage did Pelé improve his scoring from 1958 to 1959? Round to the nearest ten percent. 40%
5. Pelé scored 90, 127, 78, 110, and 71 goals over the years 1958–1962. What was his average goals scored per year over this time? Round to the nearest whole number. 95

Real-World Applications, 7, 8

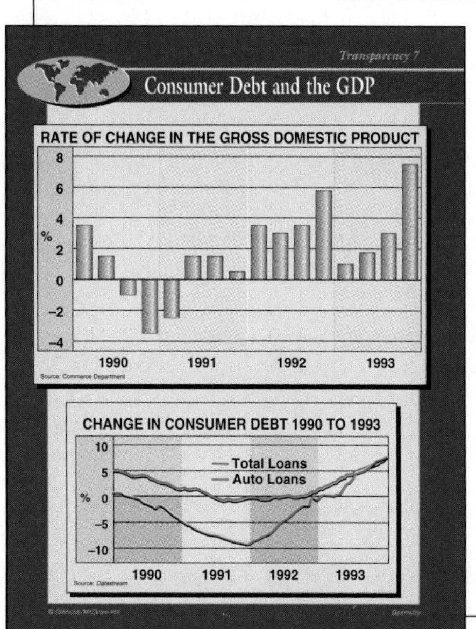

Graphing Calculator and Computer Masters, p. 4

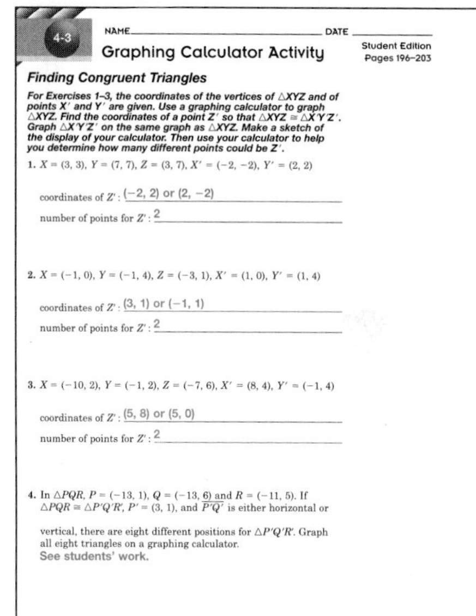

4-3 NAME ______ DATE ______

Graphing Calculator Activity — Student Edition Pages 196–203

Finding Congruent Triangles

For Exercises 1–3, the coordinates of the vertices of $\triangle XYZ$ and of points X' and Y' are given. Use a graphing calculator to graph $\triangle XYZ$. Find the coordinates of a point Z' so that $\triangle XYZ \cong \triangle X'Y'Z'$. Graph $\triangle X'Y'Z'$ on the same graph as $\triangle XYZ$. Make a sketch of the display of your calculator. Then use your calculator to help you determine how many different points could be Z'.

1. $X = (3, 3)$, $Y = (7, 7)$, $Z = (3, 7)$, $X' = (-2, -2)$, $Y' = (2, 2)$
   coordinates of Z': $(-2, 2)$ or $(2, -2)$
   number of points for Z': 2
2. $X = (-1, 0)$, $Y = (-1, 4)$, $Z = (-3, 1)$, $X' = (1, 0)$, $Y' = (1, 4)$
   coordinates of Z': $(3, 1)$ or $(-1, 1)$
   number of points for Z': 2
3. $X = (-10, 2)$, $Y = (-1, 2)$, $Z = (-7, 6)$, $X' = (8, 4)$, $Y' = (-1, 4)$
   coordinates of Z': $(5, 8)$ or $(5, 0)$
   number of points for Z': 2
4. In $\triangle PQR$, $P = (-13, 1)$, $Q = (-13, 6)$ and $R = (-11, 5)$. If $\triangle PQR \cong \triangle P'Q'R'$, $P' = (3, 1)$, and $\overline{P'Q'}$ is either horizontal or vertical, there are eight different positions for $\triangle P'Q'R'$. Graph all eight triangles on a graphing calculator.
   See students' work.

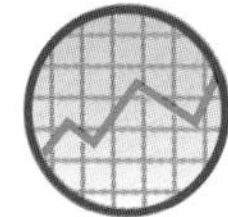

Tech Prep Applications Masters, pp. 7, 8

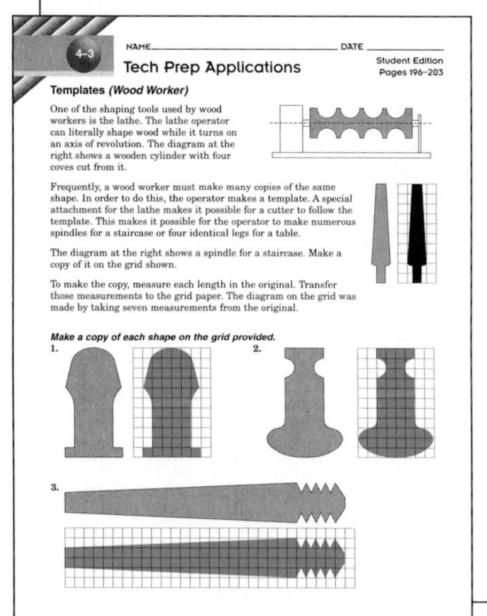

4-3 NAME ______ DATE ______

Tech Prep Applications — Student Edition Pages 196–203

Templates *(Wood Worker)*

One of the shaping tools used by wood workers is the lathe. The lathe operator can literally shape wood while it turns on an axis of revolution. The diagram at the right shows a wooden cylinder with four coves cut from it.

Frequently, a wood worker must make many copies of the same shape. In order to do this, the operator makes a template. A special attachment for the lathe makes it possible for a cutter to follow the template. This makes it possible for the operator to make numerous spindles for a staircase or four identical legs for a table.

The diagram at the right shows a spindle for a staircase. Make a copy of it on the grid shown.

To make the copy, measure each length in the original. Transfer those measurements to the grid paper. The diagram on the grid was made by taking seven measurements from the original.

Make a copy of each shape on the grid provided.

1. 2. 3.

Problem-of-the-Week Cards, 10, 11

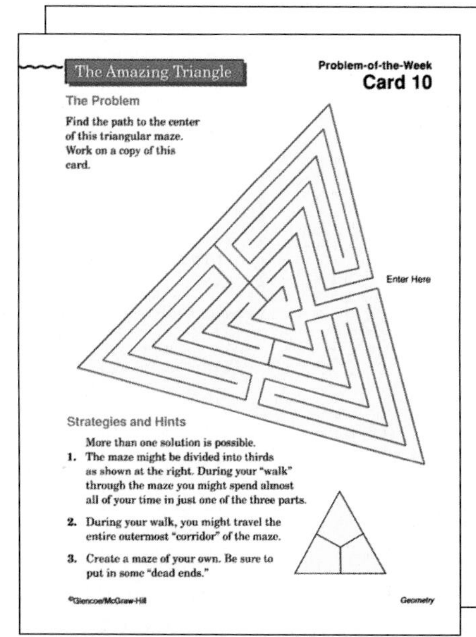

The Amazing Triangle — Problem-of-the-Week **Card 10**

The Problem

Find the path to the center of this triangular maze. Work on a copy of this card.

Enter Here

Strategies and Hints

More than one solution is possible.

1. The maze might be divided into thirds as shown at the right. During your "walk" through the maze you might spend almost all of your time in just one of the three parts.
2. During your walk, you might travel the entire outermost "corridor" of the maze.
3. Create a maze of your own. Be sure to put in some "dead ends."

©Glencoe/McGraw-Hill — Geometry

MAKING MATHEMATICS RELEVANT

This two-page introduction to the chapter provides students with an opportunity to explore an application of geometry that has historic roots as well as modern uses.

Background Information

Rock Art Native American petroglyphs can be found across the North American continent, but seem to be concentrated most heavily in the western and southwestern portions of the United States. Petroglyph National Monument, near Albuquerque, New Mexico, was created in 1990 to protect one of the largest prehistoric rock carving collections in the world. This federally protected area contains 15,000–17,000 rock carvings, the oldest of which are geometric patterns carved over 3000 years ago.

CHAPTER 4

Identifying Congruent Triangles

Objectives:

In this chapter, you will:

- classify triangles by their parts,
- apply the Angle Sum Theorem and the Exterior Angle Theorem,
- use CPCTC, SSS, SAS, ASA, and AAS to test triangle congruence,
- solve problems by eliminating the possibilities, and
- use properties of isosceles and equilateral triangles.

ROCK ART

Geometry: Then and Now The triangle is the first geometric shape you will study. The use of this shape has a long history. The triangle played a practical role in the lives of ancient Egyptians and Chinese as an aid to surveying land. The shape of a triangle also played an important role in ancient art forms. Native Americans often used inverted triangles to represent the torso of human beings in paintings or carvings. Many Native American rock carvings called *petroglyphs*, like this one near Moab, Utah, show these stylized figures. Today, triangles are frequently used in architecture. What examples of triangles can you see in your community?

TIME Line

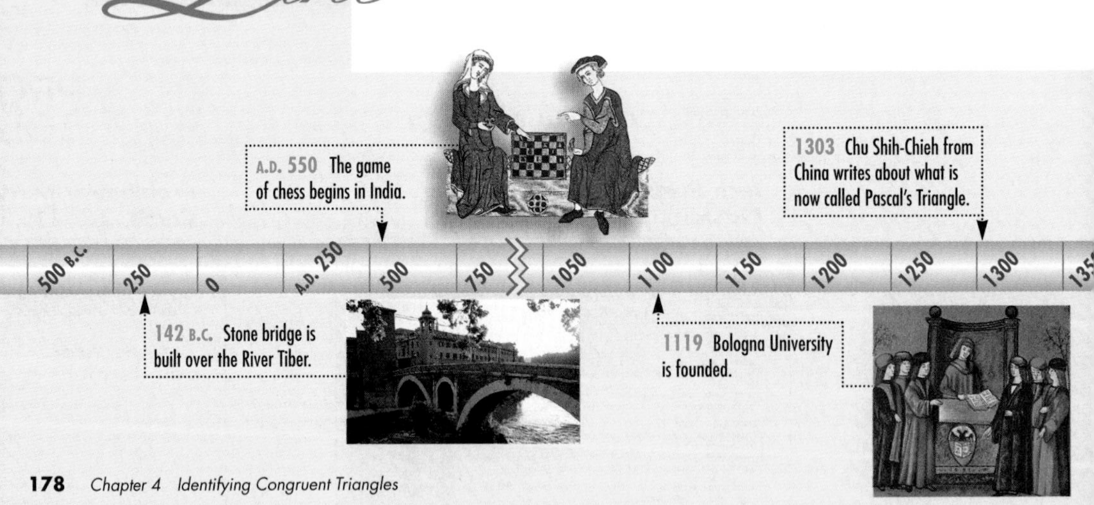

A.D. 550 The game of chess begins in India.
1303 Chu Shih-Chieh from China writes about what is now called Pascal's Triangle.
500 B.C.
250
0
A.D. 250
500
750
1050
1100
1150
1200
1250
1300
142 B.C. Stone bridge is built over the River Tiber.
1119 Bologna University is founded.

TIME Line

Mathematics has often been compared to chess. Each legal move is like one step in reasoning. Students might find it interesting to investigate the similarities between theorems and postulates in geometry and rules and reasoning in chess.

internet CONNECTION

Explore petroglyphs and rock art from sites around the world at the "Rock-Links" web page.

World Wide Web
http://www.geocities.com/Tokyo/2384/links.html

PEOPLE IN THE

NEWS

Not only does **Loui Maes** enjoy learning about his Native American culture, he uses it as an inspiration for his art. The 16-year-old artist from Denver, Colorado, chisels soft pieces of sandstone into designs resembling the ancient petroglyphs found throughout the southwestern United States. Loui uses his own imagination and searches reference materials for new ideas for his art. So far, his interest in petroglyph pieces has paid off. His petroglyph pieces have been sold in home-decorating stores and art galleries in Denver.

Chapter Project

Many works of art are based on geometric shapes. Work in small groups to investigate the use of triangles in art.

- Use the Internet or library to investigate other cultures that use triangular shapes for human forms in their art. In what art forms did these stylized forms appear?
- Research the use of geometric shapes, especially triangles, in more recent works of art. Which shapes are typical? How are the shapes used? What do these shapes help to accomplish?

- Find several examples of different types of works of art that incorporate triangles or triangular shapes. Discuss how the triangles are used in each example.
- On a piece of paper, trace the triangular shape found in each work of art. Classify each triangle by its angles and by its sides. Share your examples with the rest of the class.

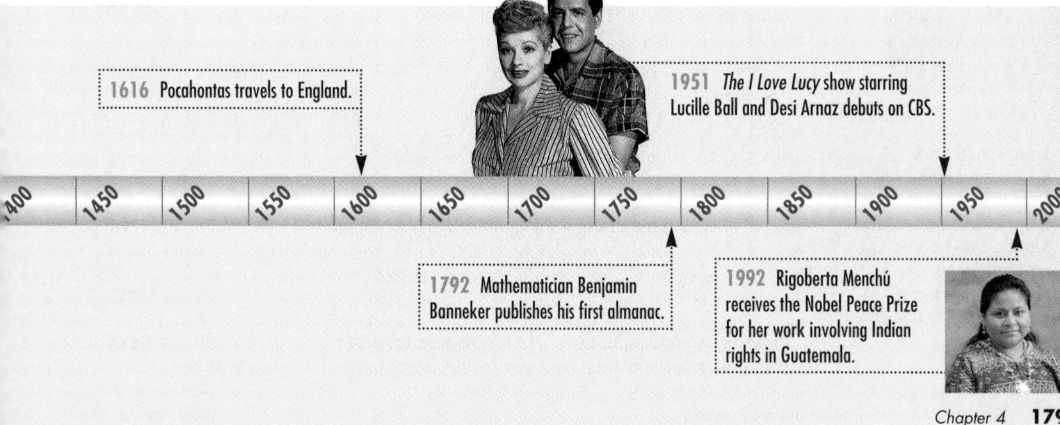
1616 Pocahontas travels to England.
1951 The I Love Lucy show starring Lucille Ball and Desi Arnaz debuts on CBS.
1400
1450
1500
1550
1600
1650
1700
1750
1800
1850
1900
1950
2000
1792 Mathematician Benjamin Banneker publishes his first almanac.
1992 Rigoberta Menchú receives the Nobel Peace Prize for her work involving Indian rights in Guatemala.

PEOPLE IN THE

NEWS

Loui uses a chisel and hammer or mallet to create his petroglyphs. Most ancient petroglyphs were made on rock surfaces, such as sandstone, granite, or volcanic basalt, by a method called pecking. Pecking was done either by striking a softer rock surface with a sharp harder stone or by hitting a chisel stone with a second stone used as a hammer. The pecked design normally started as a group of dots. To fill in portions of the design, ancient artists scraped the rock surface with a harder stone.

Chapter Project

Cooperative Learning Students might want to work in groups of three or six for this project. In setting up teams, it would be helpful to include at least one student who has a background or interest in art. Students might want to divide up the research so that one student or pair researches ancient art, one researches the use of triangles in more recent art, and the other finds several examples of art for analysis by the group.

Investigations and Projects Masters, p. 37

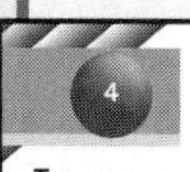

NAME____________ DATE ______

Chapter 4 Project A

Student Edition Pages 178–233

Tangrams

The seven shapes below are pieces from a Chinese puzzle commonly referred to as a *tangram*. These seven pieces can be arranged to form many different shapes and geometric designs.

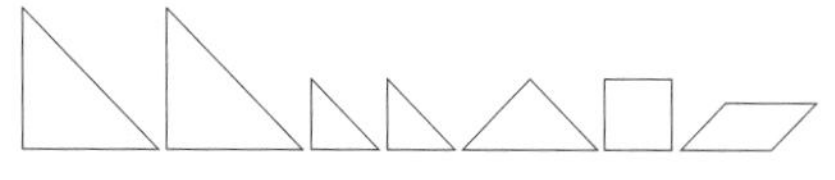

For example, you can put the pieces together to make a square, as shown below.

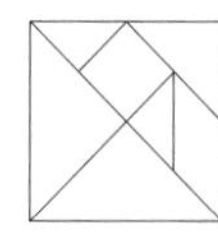

1. Classify each of the five triangle pieces above. Describe each triangle completely. Which tangram triangles are congruent?
2. Copy the seven shapes on thick paper or cardboard and cut them out. Rearrange the tangram shapes to form the shape below. Make a sketch of how you arranged the pieces.

3. Create at least three of your own geometric designs, using all seven pieces in each design. Try to create geometric shapes or shapes of figures or objects.
4. Draw only the outline of each design, then exchange your designs with a classmate. Try to recreate your classmate's designs with the seven tangram pieces while he or she recreates your designs. Tell which design was easiest for you to solve and explain why.

Alternative Chapter Projects

Two other chapter projects are included in the *Investigations and Projects Masters*. In Chapter 4 Project A, pp. 37–38, students extend the topic in the chapter opener. In Chapter 4 Project B, pp. 39–40, students research geometric patterns in fabric or clothing from around the world.

4-1 LESSON NOTES

NCTM Standards: 1–4, 7, 8

Instructional Resources

- Study Guide Master 4-1
- Practice Master 4-1
- Enrichment Master 4-1

Transparency 4-1A contains the 5-Minute Check for this lesson; **Transparency 4-1B** contains a teaching aid for this lesson.

Recommended Pacing	
Standard Pacing	Day 1 of 14
Honors Pacing	Day 1 of 13
Block Scheduling*	Day 1 of 7

*For more information on pacing and possible lesson plans, refer to the *Block Scheduling Booklet*.

1 FOCUS

5-Minute Check
(over Chapter 3)

Refer to the figure below.

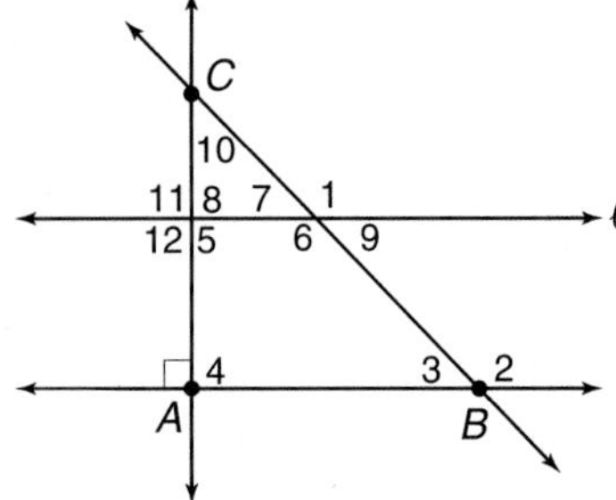

1. Name a pair of consecutive interior angles. **Sample answer: ∠6 and ∠3**
2. If $\ell \parallel \overleftrightarrow{AB}$, name a pair of congruent angles and state why they are congruent. **Sample answer: ∠7 and ∠3 because they are corresponding angles**
3. If $\ell \parallel \overleftrightarrow{AB}$, name a pair of supplementary angles. **Sample answer: ∠3 and ∠6**
4. If $\overleftrightarrow{AB}$ represents the x-axis and $\overleftrightarrow{AC}$ represents the y-axis, is the slope of $\overline{CB}$ positive, negative, zero, or undefined? **negative**

4-1 Classifying Triangles

What YOU'LL LEARN

- To identify the parts of triangles and classify triangles by their parts.

Why IT'S IMPORTANT

You can classify triangles found in architecture and crafts.

APPLICATION

Architecture

EPCOT Center, near Orlando, Florida, includes one of the world's most recognized structures. It looks like a giant golf ball and is the first completely spherical *geodesic dome* ever built. It is also the largest, standing 180 feet or 18 stories tall. The framework of the geodesic dome consists of 1450 steel beams covered with waterproof neoprene sheeting. The external cladding of nearly 1000 *triangular* aluminum panels is bolted to the steel framework. The basic structure is drawn below.

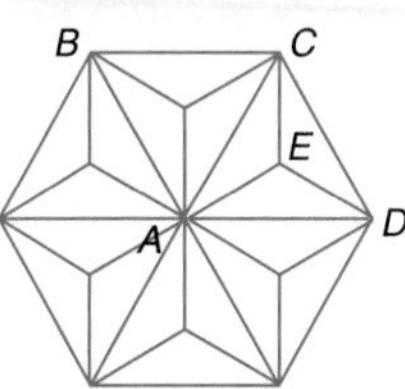

There are two types of triangles in a geodesic dome. A **triangle** is a three-sided polygon. A **polygon** is a closed figure in a plane that is made up of segments, called **sides**, that intersect only at their endpoints, called **vertices**.

The vertices of the triangle can be named in any order.

Triangle *CDE*, written $\triangle CDE$, has the following parts.

sides: $\overline{CD}$, $\overline{DE}$, $\overline{CE}$

vertices: C, D, E

angles: $\angle CDE$ or $\angle D$, $\angle CED$ or $\angle E$, $\angle DCE$ or $\angle C$

The side *opposite* $\angle C$ is $\overline{DE}$. The angle *opposite* $\overline{CE}$ is $\angle D$. $\angle E$ is opposite $\overline{CD}$.

FYI

Inventor and designer R. Buckminster (Bucky) Fuller (1895–1983) used geometry to create the geodesic dome. Recently, a new molecule, C_{60}, was discovered and named *buckminsterfullerene*, or *buckyball* for short, after the famous designer.

One way of classifying triangles is by their angles. All triangles have at least two acute angles, but the third angle, which is used to classify the triangle, can be acute, right, or obtuse. In the basic structure of the geodesic dome, $\triangle CED$ is an obtuse triangle because $\angle CED$ is obtuse.

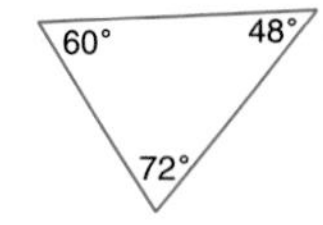

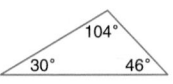

In an **acute triangle**, all the angles are acute.

104° 30° 46°

In an **obtuse triangle**, one angle is obtuse.

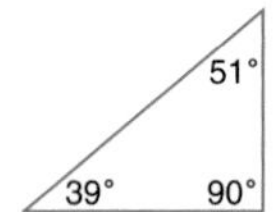

In a **right triangle**, one angle is right.

5. If $\ell \parallel \overleftrightarrow{AB}$, is $\overleftrightarrow{AC} \perp \ell$? Explain. **Yes; $\overleftrightarrow{AC} \perp \overleftrightarrow{AB}$, so if a line is ⊥ to one of 2 ∥ lines, then it is ⊥ to the other.**

FYI

Buckminster Fuller holds more than 2000 patents. He has also written more than 25 books. Best known is the book *Operating Manual for Spaceship Earth*, published in 1969.

An **equiangular triangle** is an acute triangle in which all angles are congruent. In the figure showing the basic structure of the geodesic dome, $\triangle ABC$ is an equiangular triangle.

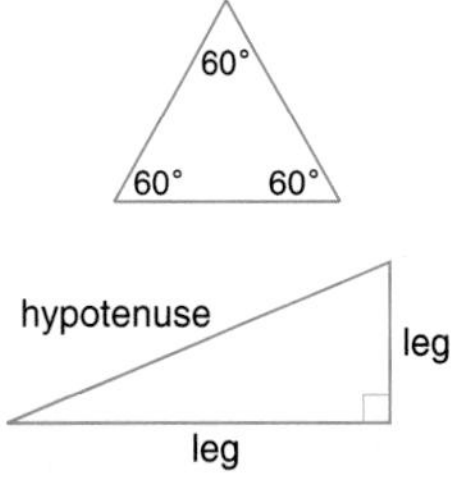

In Chapter 1, you learned that the sides of a right triangle have special names. The side opposite the right angle is called the *hypotenuse*. The two sides that form the right angle are called the *legs*.

hypotenuse
leg
leg

Example 1 APPLICATION **Architecture**

The Alcoa Office Building shown at the right is located in San Francisco, California. Triangular bracings help to secure the building in the event of high winds or an earthquake. Classify $\triangle ABC$, $\triangle BCD$, and $\triangle BCE$ as acute, obtuse, right, or equiangular.

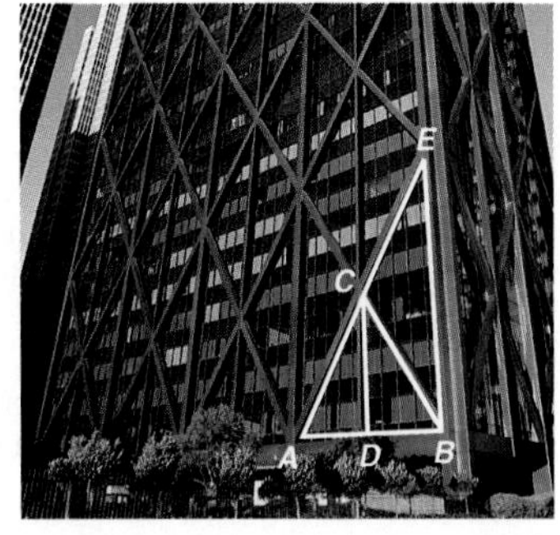

$\triangle ABC$ appears to be an equiangular triangle.

$\triangle BCD$ appears to be a right triangle.

$\triangle BCE$ appears to be an obtuse triangle.

CAREER CHOICES

Architects have played an important role in history from the earliest cities to our modern skyscrapers. The word *architect* is from the Greek meaning "master builder." A licensed architect requires a bachelor's degree in architecture, participation in an internship of 3 years, and passing an exam.

For more information, contact:

Society of American Registered Architects
1245 S. Highland Ave.
Lombard, IL 60148

Triangles can also be classified according to the number of congruent sides they have. An equal number of slashes on sides of a triangle indicate that those sides are congruent.

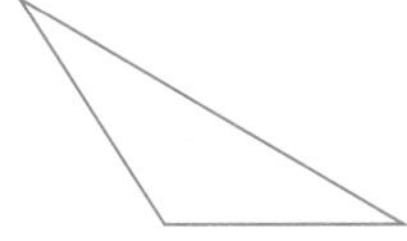

No two sides of a **scalene triangle** are congruent.

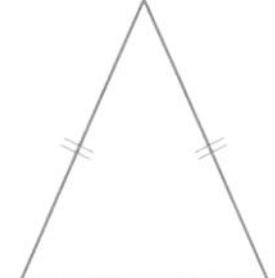

At least two sides of an **isosceles triangle** are congruent.

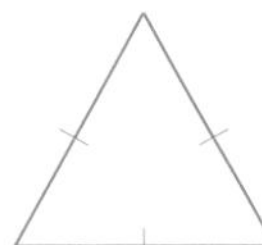

All the sides of an **equilateral triangle** are congruent.

MODELING MATHEMATICS — **Making Triangles**

Materials:  patty paper

Make an equilateral triangle.

- Take three pieces of patty paper and align them as indicated below. Make a dot at *A* with your pencil.

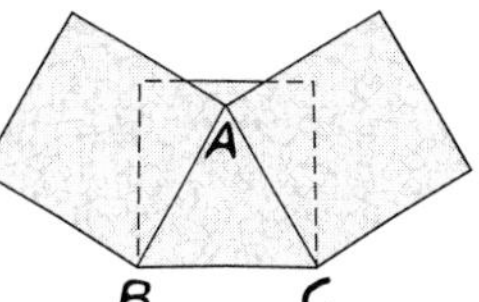

- Make a fold from *C* through *A* and from *B* through *A*. $\triangle ABC$ is equilateral.

Your Turn c. All edges of paper are equal length.

a. Make a triangle that is isosceles but not equilateral. a–b. See students' work.

b. Make a triangle that is scalene.

c. How do you know that $\triangle ABC$ is equilateral?

Classroom Vignette

"One of our favorite class activities is to have students construct equilateral triangles from dowel rods to make tetrahedron kites. The activity takes two 50-minute class periods and students really enjoy it."

Barbara O'Neil
North Forsyth High School
Cumming, Georgia

Motivating the Lesson

Hands-On Activity Ask students to draw a triangle on a sheet of paper and then display what they have drawn. Ask if all types of triangles have been included. After some discussion, have students draw a sketch of any type of triangle that was not included.

2 TEACH

In-Class Examples

For Example 1

The Camel-Back Truss is a style of bridge construction. Classify $\triangle FGH$, $\triangle ADC$, and $\triangle EFG$ as acute, obtuse, right, or equiangular.

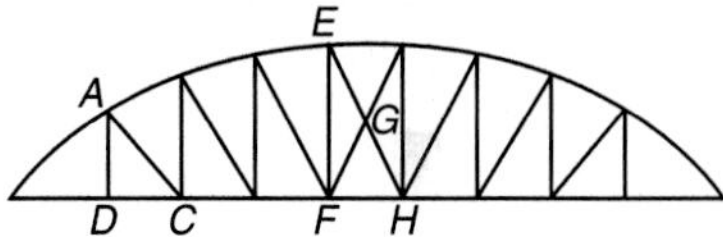

$\triangle FGH$ appears to be an acute and an equiangular triangle. $\triangle ADC$ appears to be a right triangle. $\triangle EFG$ appears to be an obtuse triangle.

Teaching Tip After reading the classifications of triangles, ask students if an equilateral triangle may also be called isosceles. yes

CAREER CHOICES

Architects must take the first year of the "engineering track" math. But, more importantly, they must have a good spatial sense. Many of their drafting and structure classes require a thorough knowledge of geometry and its applications.

MODELING MATHEMATICS Encourage students to explore other possibilities. Have them cut the two side papers to different heights and see what kind of triangle can be formed.

In-Class Examples

For Example 2
Triangle PQR is an equilateral triangle. One side measures $2x + 5$ and another side measures $x + 35$. Find the length of each side. **65**

For Example 3
Given $\triangle STU$ with vertices $S(2, 3)$, $T(4, 3)$, and $U(3, -2)$, use the distance formula to prove $\triangle STU$ is isosceles.
$SU = \sqrt{26}$, $TU = \sqrt{26}$; $\triangle STU$ is isosceles since at least two sides have the same length.

Teaching Tip In Example 3, point out that, when using the distance formula, it is important to keep track of which x-coordinate goes with which y-coordinate.

Like the right triangle, the parts of an isosceles triangle have special names.

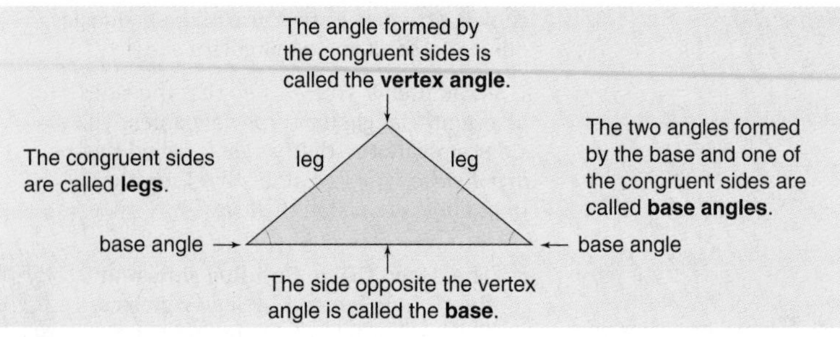

Example 2

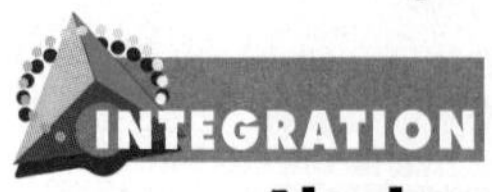

INTEGRATION Algebra

Triangle RST is an isosceles triangle. $\angle R$ is the vertex angle, $RS = x + 7$, $ST = x - 1$, and $RT = 3x - 5$. Find x, RS, ST, and RT.

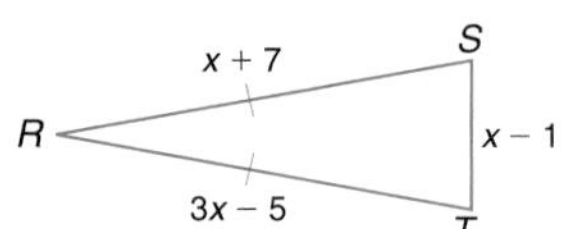

Since $\angle R$ is the vertex angle, the side opposite $\angle R$, $\overline{ST}$, is the base of the triangle. The congruent legs are $\overline{RS}$ and $\overline{RT}$. So, $RS = RT$.

$$
\begin{aligned}
RS &= RT \\
x + 7 &= 3x - 5 && \textit{Substitution Property (=)} \\
12 &= 2x && \textit{Addition Property (=)} \\
6 &= x && \textit{Division Property (=)}
\end{aligned}
$$

If $x = 6$, then $RS = 6 + 7$ or 13. Since $RS = RT$, $RT = 13$. Since $ST = x - 1$, $ST = 6 - 1$ or 5. The legs of the isosceles triangle are each 13 units long, and the base is 5 units long.

Triangles can be graphed on the coordinate plane.

Example 3

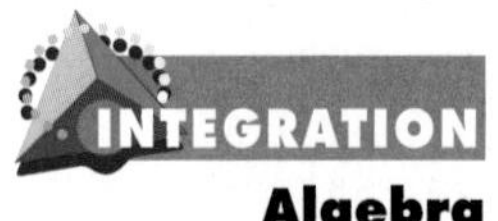

INTEGRATION Algebra

Given $\triangle DAR$ with vertices $D(2, 6)$, $A(4, -5)$, and $R(-3, 0)$, use the distance formula to show that $\triangle DAR$ is scalene.

According to the distance formula, the distance between the points at (x_1, y_1) and (x_2, y_2) is $\sqrt{(x_2 - x_1)^2 + (y_2 - y_1)^2}$.

$$
\begin{aligned}
DR &= \sqrt{(-3 - 2)^2 + (0 - 6)^2} \\
&= \sqrt{25 + 36} \text{ or } \sqrt{61} \\
AD &= \sqrt{(2 - 4)^2 + (6 - (-5))^2} \\
&= \sqrt{4 + 121} \text{ or } \sqrt{125} \\
RA &= \sqrt{(4 - (-3))^2 + (-5 - 0)^2} \\
&= \sqrt{49 + 25} \text{ or } \sqrt{74}
\end{aligned}
$$

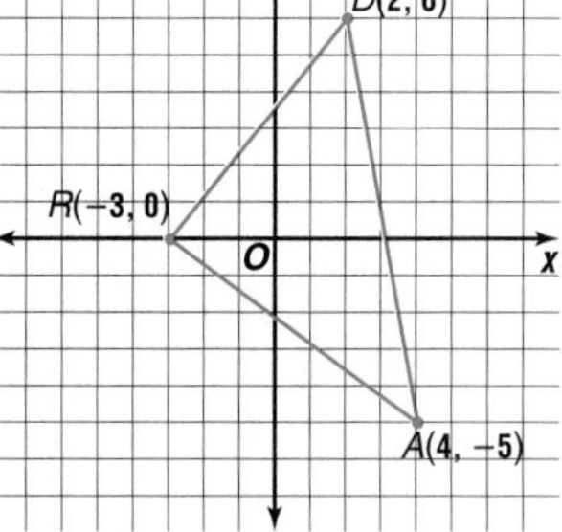

Since no two sides have the same length, the triangle is scalene.

LOOK BACK
You can refer to Lesson 1-4 for information on finding the distance between points.

You can draw a triangle on a graphing calculator by plotting its vertices and connecting them.

Alternative Learning Styles

Kinesthetic Provide students with strips of heavy paper and paper fasteners. Ask them to use these to investigate the following:

1. Can a triangle include a right angle and an obtuse angle? **no**
2. Can a triangle include two obtuse angles? **no**
3. If a triangle includes a right angle, what must be true of the other angles? **both acute**

EXPLORATION — GRAPHING CALCULATOR

Use a graphing calculator to graph △ABC whose vertices are A(0, 2), B(6, 2), and C(5, 4).

Set your calculator to the standard viewing screen and make sure the Y= list is cleared. Use the 2:Line command on the Draw menu to draw the segment connecting two of the points. Enter the points using Line (x_1, y_1, x_2, y_2) followed by ENTER. Select the Line command again and use the arrow keys to move the cursor to one endpoint of the segment. Press ENTER. Then move the cursor to the third vertex, watching the coordinates given at the bottom of the screen to approximate the position of the point. Press ENTER twice. Finally, move the cursor to the other endpoint of the segment, and press ENTER to complete the triangle.

Your Turn

a. Clear the screen. Select the standard viewing window and graph △*ABC* with vertices *A*(−2, −2), *B*(4, −2), and *C*(1, 3.2). Classify the triangle by its appearance. isosceles

b. Regraph △*ABC* using the square viewing window. What type of triangle does it appear to be? equilateral

c. Why is it unwise to use the appearance of a triangle on a graphing calculator screen to classify a triangle?

c. Sample answer: The triangle takes on different appearances depending on the type of scale used for the axes.

CHECK FOR UNDERSTANDING

Communicating Mathematics

Study the lesson. Then complete the following.

1. **Name** any terms at the right that describe each triangle whose angle measures are given.
   a. 30, 20, 130 obtuse
   b. 45, 45, 90 right
   c. 60, 60, 60 acute, equiangular
   d. 85, 45, 50 acute

   acute
   equiangular
   obtuse
   right

2. **Describe** three types of triangles that are classified by their sides. 2. See margin.
3. **You Decide** Kanisha says the figure at the right is an isosceles triangle, but Juana says it is a right triangle. Who is correct? Explain.

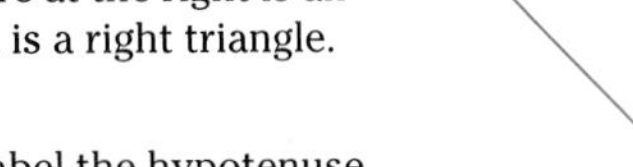

3. Both are correct. If the triangle is classified according to its sides, it is isosceles. If it is classified according to its angles, it is right.

4. **Draw** a scalene right triangle and label the hypotenuse and legs. See margin.

MODELING MATHEMATICS

5. Use three pieces of patty paper to create an equilateral triangle. Use your protractor to measure each of the angles. Compare your results with a partner. What can you conclude? See margin.

Guided Practice

Refer to the figure at the right for Exercises 6–8.

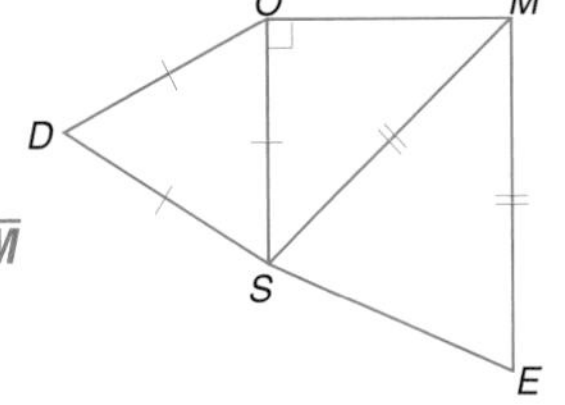

6. Identify an equilateral triangle. △*OSD*
7. Name the hypotenuse of the right triangle. $\overline{SM}$
8. Name the legs of the isosceles triangle that is not equilateral. $\overline{SM}$, $\overline{ME}$

EXPLORATION

Students might try varying the *y*-coordinate of the point *C*(5, 4) one unit at a time. They can visually identify the type of triangle for each value. They can try to form a conjecture based on any observed pattern.

3 PRACTICE/APPLY

Check for Understanding

Exercises 1–15 are designed to help you assess your students' understanding through reading, writing, speaking, and modeling. You should work through Exercises 1–5 with your students and then monitor their work on Exercises 6–15.

Additional Answers

2. A scalene triangle has no two sides congruent. An isosceles triangle has at least two sides congruent. An equilateral triangle has all three sides congruent.
4.

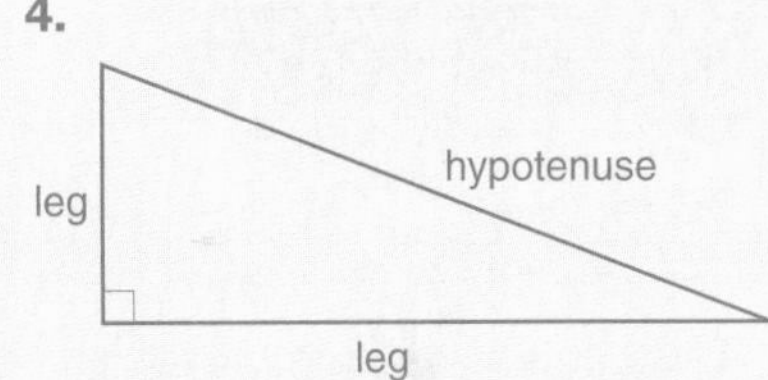

5. See students' work. The measure of each angle is 60. An equilateral triangle is also an equiangular triangle.

Reteaching

Using Cooperative Learning Ask students to work in a cooperative group to make a chart of the triangle classifications. The chart should include the name of each type of triangle, a brief description, and a drawing.

Assignment Guide

Core (with proof): 17–49 odd, 51–62
Core (informal): 17–41 odd, 45, 47, 49, 51–62
Enriched: 16–46 even, 47–62

For **Extra Practice**, see p. 770.

The red A, B, and C flags, printed only in the Teacher's Wraparound Edition, indicate the level of difficulty of the exercises.

Additional Answers

9. Sample answer:

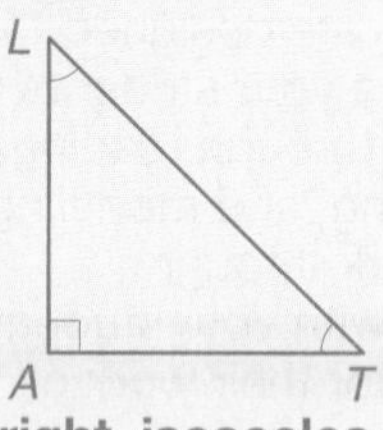

right, isosceles

10. Sample answer:

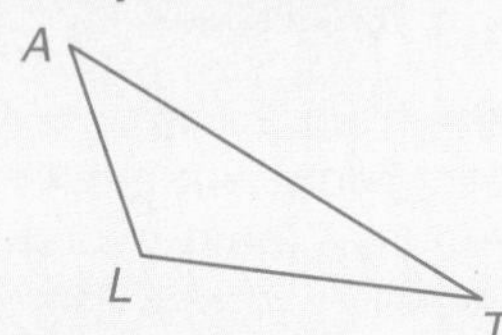

scalene

12. $AB = \sqrt{(3-0)^2 + (0-\sqrt{27})^2} = \sqrt{9+27} = \sqrt{36} = 6$

$BC = \sqrt{(-3-3)^2 + (0-0)^2} = \sqrt{36+0} = \sqrt{36} = 6$

$AC = \sqrt{(-3-0)^2 + (0-\sqrt{27})^2} = \sqrt{9+27} = \sqrt{36} = 6$

equilateral

15b. Sample answer:

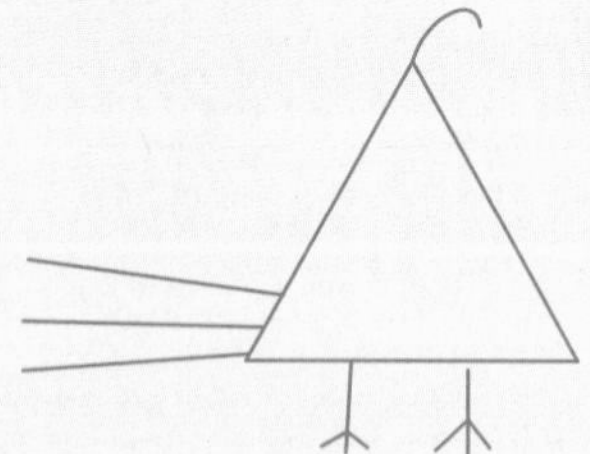

15c. Sample answer:

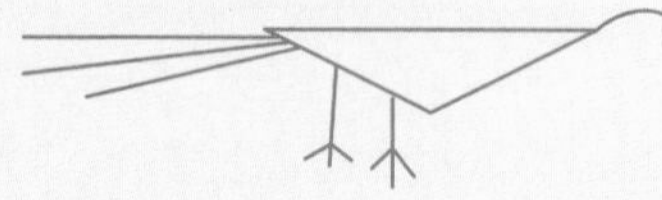

Use a protractor and ruler to draw $\triangle ALT$ using the given conditions. If possible, classify each triangle by the measures of its angles and sides.

9. $m\angle A = 90$ and $m\angle L = m\angle T$
10. $AL < LT < AT$ 9–10. See margin.
11. $\triangle EQU$ is an equilateral triangle. Find x and the measure of each side if $EQ = 4x - 3$ and $QU = 3x + 4$. $x = 7$; $EQ = 25$, $QU = 25$, $EU = 25$
12. Given $\triangle ABC$ with vertices $A(0, \sqrt{27})$, $B(3, 0)$, and $C(-3, 0)$, use the distance formula to classify the triangle by the lengths of its sides. See margin.

Copy each sentence. Fill in the blank with *sometimes, always,* or *never.*

13. Isosceles triangles are _?_ equilateral. sometimes
14. Right triangles are _?_ obtuse. never
15. **Crafts** The Zuni Indians are a branch of the Pueblo Indians who live in the western part of New Mexico. They are known for their beautiful pottery, basketry, jewelry, and weavings. The rain-bird is frequently found on the water jars they make. Two examples of rain-birds are shown at the right.

   a. Describe the triangles used to draw these two rain-birds.

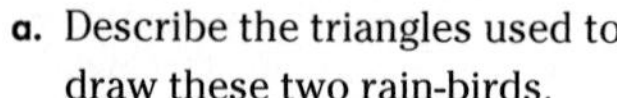

   b. Draw a rain-bird using an equilateral triangle. See margin.

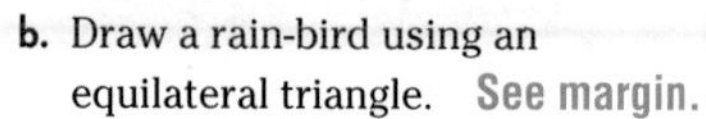

   c. Draw a rain-bird using an obtuse, isosceles triangle. See margin.

15a. right, scalene; obtuse, scalene

EXERCISES

Practice

In the figure at the right, $\triangle BLM$ is isosceles with base $\overline{ML}$. Refer to the figure for Exercises 16–23.

16. Identify an acute triangle. $\triangle BLM$
17. Name the hypotenuse. $\overline{LM}$
18. Name the vertex angle. $\angle B$
19. Name the side opposite $\angle C$. $\overline{LM}$
20. Name the angle opposite $\overline{MB}$. $\angle BLM$
21. Name the base angles. $\angle BLM$, $\angle BML$
22. Name the vertices of the right triangle. L, C, M
23. Name the legs of the isosceles triangle. $\overline{BL}$, $\overline{BM}$

Use a protractor and ruler to draw $\triangle BQS$ using the given conditions. If possible, classify each triangle by the measures of its angles and sides.

24. $m\angle B < 90$ and $\overline{BQ}$ is the hypotenuse. 24–29. See Solutions Manual.
25. $SB = SQ$ and $m\angle S = 90$.
26. $m\angle S > 90$ and $SQ < BQ$.
27. $SB > SQ > QB$
28. $SB = SQ = QB$
29. $\angle S$ is obtuse and $\angle BQS$ is isosceles.

Alternative Teaching Strategies

Reading Geometry Have students study the origins of the prefix *tri*. They can also make a list of other words beginning with *tri*.

30. $x = 3$; $BC = 10$, $CD = 10$, $BD = 5$

30. $\triangle BCD$ is isosceles with $\angle C$ as the vertex angle. Find x and the measure of each side if $BC = 2x + 4$, $BD = x + 2$, and $CD = 10$. (*Hint:* Draw a diagram.)

31. $\triangle HKT$ is equilateral. Find x and the measure of each side if $HK = x + 7$ and $HT = 4x - 8$. $x = 5$; $HK = 12$, $HT = 12$, $KT = 12$

32. $\triangle ABC$ is isosceles with $\angle A$ as the vertex angle. AC is five less than two times a number. AB is three more than the number. BC is one less than the number. Find the measure of each side. $AC = 11$, $AB = 11$, $BC = 7$

Use the distance formula to classify each triangle by the measures of its sides. 33–35. See margin.

33. $\triangle PQR$ with vertices $P(0, 6)$, $Q(3, 6)$, and $R(3, 0)$
34. $\triangle SUV$ with vertices $S(-3, -1)$, $U(2, 1)$, and $V(2, -3)$
35. $\triangle KLM$ with vertices $K(4, 0)$, $L(-2, 0)$, and $M(1, 5)$

Copy each sentence. Fill in the blank with *sometimes, always,* or *never.*

36. Equilateral triangles are __?__ isosceles. always
37. Scalene triangles are __?__ isosceles. never
38. Right triangles are __?__ acute. never
39. Acute triangles are __?__ equilateral. sometimes
40. Obtuse triangles are __?__ scalene. sometimes
41. Equiangular triangles are __?__ acute. always
42. $\triangle RST$ is equilateral, and V lies on $\overline{RS}$ so that $\overline{TV} \perp \overline{RS}$. Classify $\triangle TVS$ by the measures of its angles and its sides. right, scalene

43. Copy and complete the proof.

Given: $\angle RGH$ is a right angle.
$\overline{TS} \parallel \overline{HG}$

Prove: $\triangle RST$ is a right triangle.

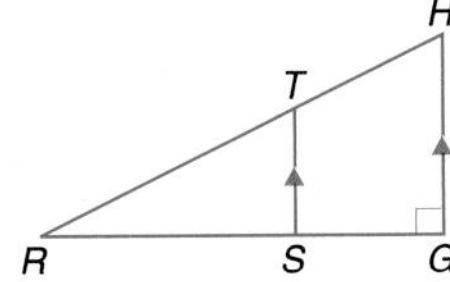

Proof:

Statements	Reasons
1. $\angle RGH$ is a right angle. $\overline{TS} \parallel \overline{HG}$	1. Given
2. $\overline{RG} \perp \overline{GH}$	2. __?__ Definition of perpendicular
3. $\overline{RS} \perp \overline{ST}$	3. __?__ Perpendicular Transversal Th.
4. __?__ $\angle RST$ is a rt. $\angle$.	4. Definition of perpendicular lines
5. __?__ $\triangle RST$ is a rt. $\triangle$.	5. Definition of right triangle

44. Write a two-column proof.

Given: $m\angle LGM = 25$

Prove: $\triangle LGS$ is an obtuse triangle. See Solutions Manual.

L
25°
M G S

45. $\triangle DEF$ is isosceles with a perimeter between 23 and 32 units. Which angle is the vertex angle? Explain your answer. See margin.

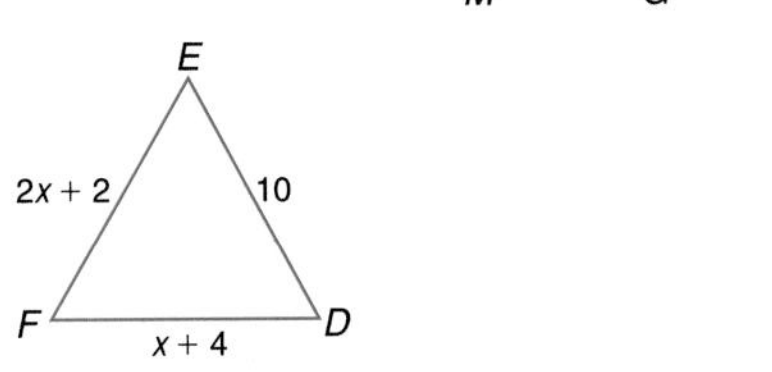

Additional Answer

45. $\angle E$; since $23 < 2x + 2 + 10 + x + 4 < 32$, $\frac{7}{3} < x < \frac{16}{3}$. If $2x + 2 = 10$, $x = 4$. If $2x + 2 = x + 4$, $x = 2$. If $x + 4 = 10$, $x = 6$. The only value of x that satisfies the conditions is $x = 4$, so $\overline{EF}$ and $\overline{ED}$ are the legs of the isosceles triangle and $\angle E$ is the vertex.

Additional Answers

33. $PQ = \sqrt{(3-0)^2 + (6-6)^2} = \sqrt{9+0} = \sqrt{9} = 3$

$QR = \sqrt{(3-3)^2 + (0-6)^2} = \sqrt{0+36} = \sqrt{36} = 6$

$PR = \sqrt{(3-0)^2 + (0-6)^2} = \sqrt{9+36} = \sqrt{45} = 3\sqrt{5}$

scalene

34. $SU = \sqrt{[2-(-3)]^2 + [1-(-1)]^2} = \sqrt{25+4} = \sqrt{29}$

$UV = \sqrt{(2-2)^2 + (-3-1)^2} = \sqrt{0+16} = \sqrt{16} = 4$

$SV = \sqrt{[2-(-3)]^2 + [-3-(-1)]^2} = \sqrt{25+4} = \sqrt{29}$

isosceles

35. $KL = \sqrt{(-2-4)^2 + (0-0)^2} = \sqrt{36+0} = \sqrt{36} = 6$

$LM = \sqrt{[1-(-2)]^2 + (5-0)^2} = \sqrt{9+25} = \sqrt{34}$

$KM = \sqrt{(1-4)^2 + (5-0)^2} = \sqrt{9+25} = \sqrt{34}$

isosceles

Study Guide Masters, p. 20

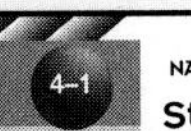

NAME________________ DATE________

Study Guide

Student Edition Pages 180–187

Classifying Triangles

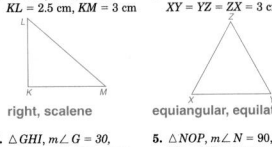

Triangles are classified in two different ways, either by their angles or by their sides.

Classification of Triangles			
Angles		Sides	
acute	three acute angles	scalene	no two sides congruent
obtuse	one obtuse angle	isosceles	at least two sides congrent
right	one right angle	equilateral	three sides congruent
equiangular	three congruent angles		

Examples: Classify each triangle by its angles and by its sides.

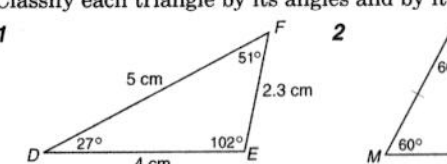

$\triangle DEF$ is obtuse and scalene. $\triangle MNO$ is equiangular and equilaterial.

Use a protractor and ruler to draw triangles using the given conditions. If possible, classify each triangle by the measures of its angles and sides.

1. $\triangle KLM$, $m\angle K = 90$, $KL = 2.5$ cm, $KM = 3$ cm

L K M

right, scalene

2. $\triangle XYZ$, $m\angle X = 60$, $XY = YZ = ZX = 3$ cm

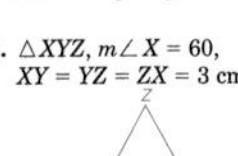

equiangular, equilateral

3. $\triangle DEF$, $m\angle D = 150$, $DE = DF = 1$ inch

F D E

obtuse, isoceles

4. $\triangle GHI$, $m\angle G = 30$, $m\angle H = 45$, $GH = 4$ cm

5. $\triangle NOP$, $m\angle N = 90$, $NO = NP = 2.5$ cm

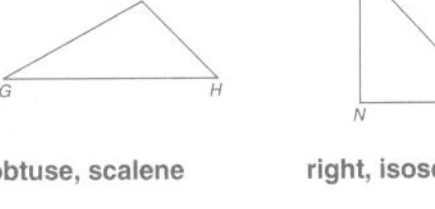

obtuse, scalene

right, isosceles

6. $\triangle QRS$, $m\angle Q = 100$, $QS = 1$ inch, $QR = 1\frac{1}{2}$ inches

obtuse, scalene

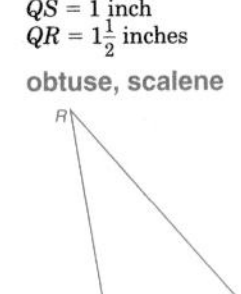

Additional Answers

46. Equiangular; since $m \parallel n$ and $\angle 1$ and $\angle 5$ are consecutive interior angles, $\angle 1$ and $\angle 5$ are supplementary and $m\angle 1 + m\angle 5 = 180$. Therefore $m\angle 5 = 60$. Since $m\angle 6 + m\angle 5 + m\angle 4 = 180$ and $m\angle 4 = m\angle 6$, $m\angle 4 = 60$, and $m\angle 6 = 60$. Since $\angle 6$ and $\angle 3$ are corresponding angles, $\angle 3 \cong \angle 6$ and $m\angle 3 = 60$. Since $\angle 1$ and $\angle 2$ are linear pairs, $m\angle 1 + m\angle 2 = 180$ and $m\angle 2 = 60$. The triangle is equiangular.

47. See students' work. If a triangle is an acute triangle, the square of the length of the longest side is less than the sum of the squares of the lengths of the other two sides. If a triangle is an obtuse triangle, the square of the length of the longest side is greater than the sum of the squares of the lengths of the other two sides.

50a. (triangular array of 10 dots: rows of 1, 2, 3, 4)

Practice Masters, p. 20

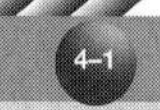

NAME ______ DATE ______

Practice — Student Edition Pages 180–187

Classifying Triangles

For Exercises 1–7, refer to the figure at the right. Triangle ABC is isosceles with AB > AC and AB > BC. Also, $\overleftrightarrow{XY} \parallel \overline{AB}$. Name each of the following.

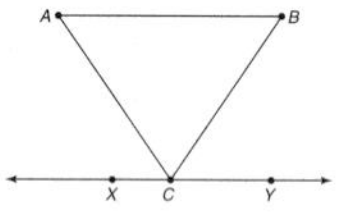

1. sides of the triangle $\overline{AB}, \overline{BC}, \overline{AC}$
2. angles of the triangle $\angle A, \angle B, \angle ACB$
3. vertex angle $\angle ACB$
4. base angles $\angle A, \angle B$
5. side opposite $\angle BCA$ $\overline{AB}$
6. congruent sides $\overline{AC}, \overline{BC}$
7. angle opposite $\overline{AC}$ $\angle B$

Use a protractor and ruler to draw triangles using the given conditions. Classify each triangle by the measures of its angles and sides.

8. $\triangle BHE$, $BE = 1$ inch, $m\angle E = 60$, $HE = \frac{1}{2}$ inch right, scalene

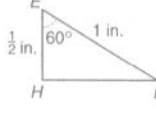

9. $\triangle QTR$, $m\angle T = 60$, $QT = TR = 4$ cm equiangular, equilateral

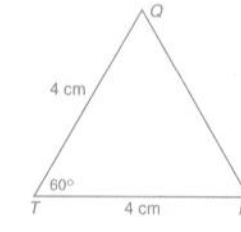

10. Find the measures of the legs of isosceles triangle ABC if $AB = 2x + 4$, $BC = 3x - 1$, $AC = x + 1$, and the perimeter of $\triangle ABC$ is 34 units. 14 units

46. In the figure at the right, $m \parallel n$, $\angle 4 \cong \angle 6$, and $m\angle 1 = 120$. Classify $\triangle RST$ by the measure of its angles. Justify your answer. **See margin.**

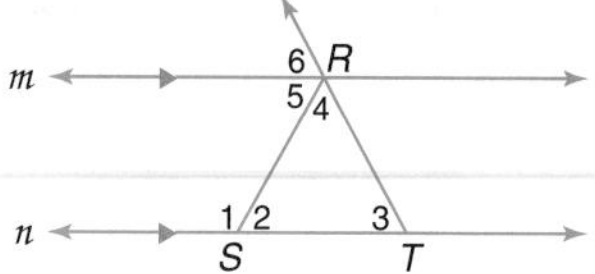

Critical Thinking

47. The Pythagorean Theorem states that in a right triangle, the square of the measure of the hypotenuse is equal to the sum of the squares of the measures of the legs. Use patty paper to create several obtuse and acute triangles. Measure the sides of each. Square the measure of each side. Compare the square of the measure of the longest side to the sum of the squares of the measures of the other two sides. Make a conjecture about the relationship of the square of the measure of the longest side compared to the sum of the squares of the measures of the two shorter sides. **See margin.**

Applications and Problem Solving

48. **Architecture** Consider a figure of the basic structure of the geodesic dome. How many equilateral triangles are in the figure. How many obtuse triangles are in the figure? **6; 18**

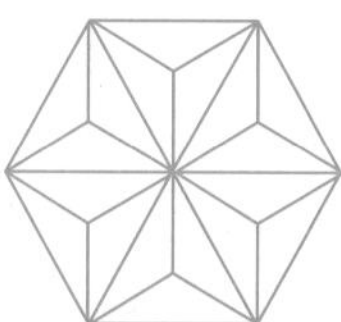

49. **Architecture** At the right is a diagram of an A-frame house.
    a. Identify the triangles that appear to be equilateral.
    b. Identify the triangles that appear to be isosceles, but not equilateral. **none**
    c. Identify the triangles that appear to be right triangles. ***△BED, △CFG, △BJH, △CKM, △DIH, △GLM, △BPN, △CQS, △DON, △GRS***

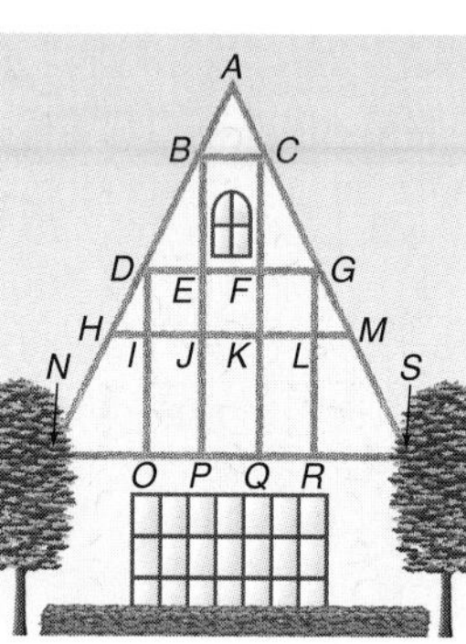

49a. $\triangle ABC$, $\triangle ADG$, $\triangle AHM$, $\triangle ANS$

50. **Look for a Pattern** Consider the pattern formed by the dots.

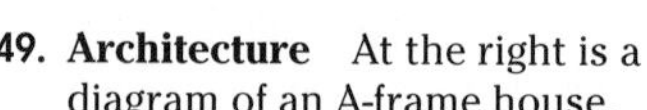

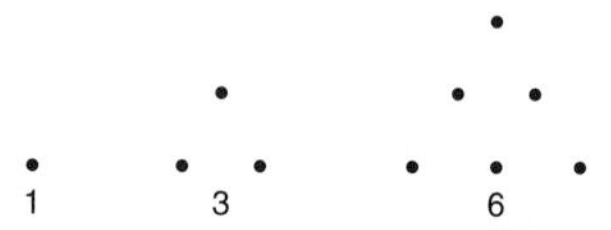

The numbers used to describe each array of dots are called *triangular numbers*. The third triangular number is 6.

a. Draw the array for the fourth triangular number. **See margin.**
b. How many dots will be in the array for the eighth triangular number? **36**

Mixed Review

51. In spherical geometry, if points M and N are polar points, how many great circles can be drawn through M and N? (Lesson 3–6) **infinite number**

52. Copy the figure and draw a line segment that represents the shortest distance from C to $\overleftrightarrow{ED}$. (Lesson 3–5)

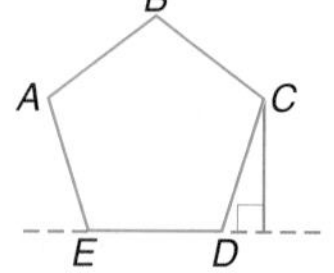

Extension

Reasoning Triangles can be classified by their angle measures or by the measures of their sides. Therefore, any triangle has two labels. Which combinations of labels are possible and which are impossible?

Possible: right scalene, right isosceles, acute scalene, acute isosceles, acute equilateral, obtuse scalene, obtuse isosceles
Impossible: right equilateral, obtuse equilateral

53. See margin.

53. Name five ways to prove that two lines are parallel. (Lesson 3–4)

54. **Algebra** Determine the value of r so that a line through the points at $(r, 2)$ and $(4, -6)$ has a slope of $-\frac{8}{3}$. (Lesson 3–3) 1

55. Sample answer: the line formed by the intersection of the floor and a wall of a room and the line formed by two walls on the opposite side of the room

55. Give a real-world example of two skew lines. (Lesson 3–1)

56. Name the congruence property that justifies the statement $\angle A \cong \angle A$. (Lesson 2–6) Congruence of angles is reflexive.

57. **Algebra** Which property of equality is similar to the Law of Syllogism? Explain. (Lesson 2–3) See margin.

58. hypothesis: you want a pizza; conclusion: go to Pizza Haven

58. State the hypothesis and conclusion of the statement *If you want pizza, go to Pizza Haven.* (Lesson 2–2)

59. The measure of an angle is one-third the measure of its supplement. Find the measure of the angle. (Lesson 1–7) 45°

60. Find the value of a so that the distance between $M(5, 6)$ and $N(a, 10)$ is 5 units. (Lesson 1–4) 2 or 8

Algebra

61. Solve $-6 = 5u + 9$. −3

62. If y varies inversely as x and $y = 27$ when $x = 40$, find x when $y = 10$. 108

Refer to the Investigation on pages 176–177.

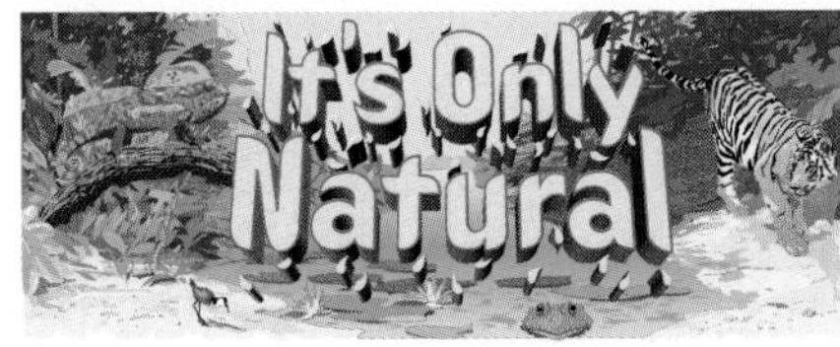

Many people are afraid of spiders. But of the more than 30,000 different species of spiders, only a few are actually dangerous to humans. Spiders feed on insects. And while all spiders spin silk, not all of them make webs to trap their food. All web-spinning spiders that are of the same species spin the same type of web.

1. The triangle spider spins its web between two twigs. Identify the type of triangle formed by the web shown at the right.
2. It is possible for a triangle spider to spin a web that is a different type of triangle than the one shown? If so, when would this occur?
3. Investigate the different types of webs spun by different spiders. Do any of the web types fit a pattern described on pages 176–177 of the Investigation?
4. If a type of spider web fits the pattern you have chosen to research, add a photograph of the web to your presentation. Also include an explanation of the pattern in the web.

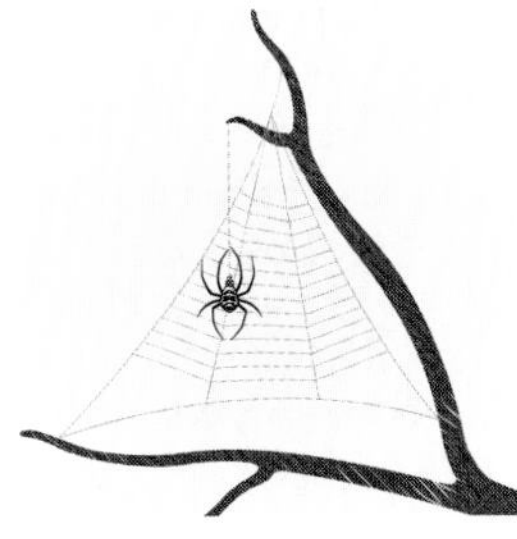

Add the results of your work to your Investigation Folder.

4 ASSESS

Closing Activity

Modeling Use straws or strips of paper to model each type of triangle described in this lesson.

Additional Answers

53. (1) If two lines in a plane are cut by a transversal so that corr. ∡s are ≅, then the lines are ∥.
(2) If two lines in a plane are cut by a transversal so that a pair of alt. int. ∡s is ≅, then the lines are ∥.
(3) If two lines in a plane are cut by a transversal so that a pair of consecutive int. ∡s is supplementary, then the lines are ∥.
(4) If two lines in a plane are cut by a transversal so that a pair of alt. ext. ∡s is ≅, then the lines are ∥.
(5) In a plane, if two lines are perp. to the same line, then they are ∥.

57. Transitive Property of Equality; Sample answer: The Law of Syllogism says if $p \to q$ and $q \to r$, then $p \to r$. The Transitive Property of Equality says if $p = q$ and $q = r$, then $p = r$.

Enrichment Masters, p. 20

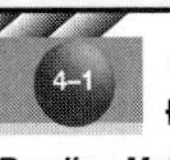

NAME ______ DATE ______

Enrichment

Student Edition Pages 180–187

Reading Mathematics

When you read geometry, you may need to draw a diagram to make the text easier to understand.

Example: Consider three points, A, B, and C on a coordinate grid. The y-coordinates of A and B are the same. The x-coordinate of B is greater than the x-coordinate of A. Both coordinates of C are greater than the corresponding coordinates of B. Is triangle ABC acute, right, or obtuse?

To answer this question, first draw a sample triangle that fits the description.

Side AB must be a horizontal segment because the y-coordinates are the same. Point C must be located to the right and up from point B.

From the diagram you can see that triangle ABC must be obtuse.

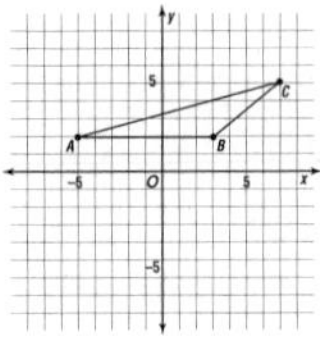

Answer each question. Draw a simple triangle on the grid above to help you.

1. Consider three points, R, S, and T on a coordinate grid. The x-coordinates of R and S are the same. The y-coordinate of T is between the y-coordinates of R and S. The x-coordinate of T is less than the x-coordinate of R. Is angle R of triangle RST acute, right, or obtuse? acute
2. Consider three noncollinear points, J, K, and L on a coordinate grid. The y-coordinates of J and K are the same. The x-coordinates of K and L are the same. Is triangle JKL acute, right, or obtuse? right
3. Consider three noncollinear points, D, E, and F on a coordinate grid. The x-coordinates of D and E are opposites. The y-coordinates of D and E are the same. The x-coordinate of F is 0. What kind of triangle must $\triangle DEF$ be: scalene, isosceles, or equilateral? isosceles
4. Consider three points, G, H, and I on a coordinate grid. Points G and H are on the positive y-axis, and the y-coordinate of G is twice the y-coordinate of H. Point I is on the positive x-axis, and the x-coordinate of I is greater than the y-coordinate of G. Is triangle GHI scalene, isosceles, or equilateral? scalene

In•ves•ti•ga•tion

Working on the Investigation

The Investigation on pages 176–177 is designed to be a long-term project that is completed over several days or weeks. Encourage students to keep their materials in their Investigation Folder as they work on the Investigation.

Sample Answers for Working on the Investigation

1. acute isosceles
2. Yes; a right or obtuse triangle could occur when the branches of the tree form a larger angle.
3. An orb web resembles a spiral.
4. See students' work.

4-2A LESSON NOTES

NCTM Standards: 1–4, 7

Objective

Investigate the relationship of the angles of a triangle.

Recommended Time

Demonstration and discussion: 15 minutes; Exercises: 30 minutes

Instructional Resources

For each student or group of students
Modeling Mathematics Masters
• p. 92 (worksheet)
For teacher demonstration
Algebra and Geometry Overhead Manipulative Resources

1 FOCUS

Motivating the Lesson

Have each student draw a triangle, measure its angles, and add the measures. Does everyone get the same sum? What does this say about the angles of a triangle? Sum of the measures is 180.

2 TEACH

Teaching Tip Point out to students that they do not need to cut the angles, but can tear them.

3 PRACTICE/APPLY

Assignment Guide
Core (with proof): 1–4 **Core (informal):** 1–4 **Enriched:** 1–4

4 ASSESS

Observing students working in cooperative groups is an excellent method of assessment.

4–2A Angles of Triangles

Materials: 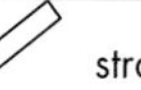 straightedge  scissors

There are special relationships among the angles of a triangle.

A Preview of Lesson 4–2

Activity 1 Find the relationships among the measures of the angles of a triangle.

- Use a straightedge to draw an acute triangle.
- Cut out your acute triangle and label the angles 1, 2, and 3.
- Tear off each angle and arrange the angles to form three adjacent angles as shown.

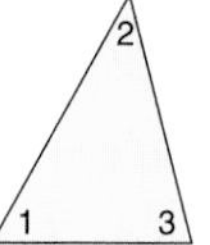

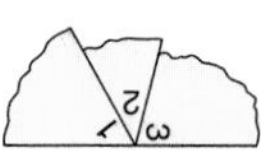

What seems to be true about the sum of the measures of the angles of a triangle?

In the figure at the right, ∠4 is called an **exterior angle** of the triangle, and ∠2 and ∠3 are called its **remote interior angles**.

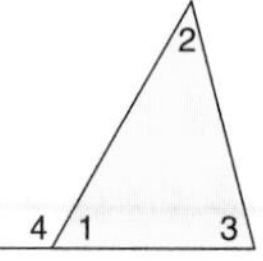

Activity 2 Find the relationship among m∠4, m∠2, and m∠3.

- Arrange the angles from Activity 1 back into a triangle.
- Use a straightedge to draw an exterior angle of your triangle at the vertex of ∠1.
- Place ∠2 and ∠3, the remote interior angles, over the exterior angle you drew. What do you observe?
- Draw exterior angles for angles 2 and 3.

Make a conjecture about the relationship of the remote interior angles of a triangle and their exterior angles.

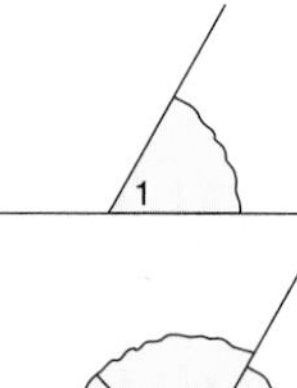

Model 1–2. See students' work.

1. Draw an obtuse triangle. Cut out the triangle and tear off each angle.
   a. Arrange the angles to form three adjacent angles.
   b. Draw three exterior angles of the triangle. Compare each exterior angle with its remote interior angles.
2. Draw a right triangle. Cut out the triangle and tear off each angle.
   a. Arrange the angles to form three adjacent angles.
   b. Draw three exterior angles of the triangle. Compare each exterior angle with its remote interior angles.

3. The sum of the measures of the angles of a triangle is 180.

Write **Refer to your observations in Activities 1 and 2.**

3. Write a statement about the sum of the measures of the angles of a triangle.
4. Write a statement about the measure of an exterior angle and the sum of the measures of the two remote interior angles. See margin.

Additional Answer

4. The measure of an exterior angle is equal to the sum of the measures of the two remote interior angles.

Using Cooperative Learning

This lesson offers an excellent opportunity for using cooperative learning groups. For more information on cooperative learning strategies and group management, see *Cooperative Learning in the Mathematics Classroom*, one of the titles in the Glencoe Mathematics Professional Series.

4-2 Measuring Angles in Triangles

What YOU'LL LEARN
- To apply the Angle Sum Theorem, and
- to apply the Exterior Angle Theorem.

Why IT'S IMPORTANT

You can determine the measures of angles in triangles involved in countruction and designs.

APPLICATION

Astronomy

The Summer Triangle can be viewed during the month of June. It consists of three bright stars from different constellations. They are Vega in the constellation Lyra, Altair in the constellation Aquile, and Deneb in the constellation Cygnus. If the measure of the angle at Vega is 74 and the measure of the angle at Altair is 41, what is the measure of the angle at Deneb? To answer this question, you need to use the Angle Sum Theorem. *This problem will be solved in Example 1.*

Theorem 4–1 Angle Sum Theorem	The sum of the measures of the angles of a triangle is 180.

In order to prove the Angle Sum Theorem, we will need to draw an **auxiliary line**. An auxiliary line is a line or line segment added to a diagram to help in a proof. These are shown as dashed lines in the diagram. Be sure it is possible to draw auxiliary lines that you use.

Proof of Theorem 4–1

Given: $\triangle PQR$

Prove: $m\angle R + m\angle 2 + m\angle Q = 180$

Proof:

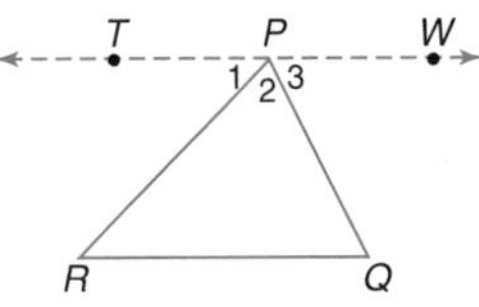

Statements	Reasons
1. $\triangle PQR$	1. Given
2. Draw $\overleftrightarrow{TW}$ through P parallel to $\overleftrightarrow{RQ}$.	2. Parallel Postulate
3. $\angle 1$ and $\angle RPW$ form a linear pair.	3. Definition of a linear pair
4. $\angle 1$ and $\angle RPW$ are supplementary.	4. If two angles form a linear pair, they are supplementary.
5. $m\angle 1 + m\angle RPW = 180$	5. Definition of supplementary angles
6. $m\angle RPW = m\angle 2 + m\angle 3$	6. Angle Addition Postulate
7. $m\angle 1 + m\angle 2 + m\angle 3 = 180$	7. Substitution Property (=)
8. $\angle 1 \cong \angle R$, $\angle 3 \cong \angle Q$	8. Alternate Interior Angles Theorem
9. $m\angle 1 = m\angle R$, $m\angle 3 = m\angle Q$	9. Definition of congruent angles
10. $m\angle R + m\angle 2 + m\angle Q = 180$	10. Substitution Property (=)

fabulous FIRSTS

Carolyn Shoemaker (1929–)

Carolyn Shoemaker is the first person to discover more than 30 comets. So far, she has found 32 comets, all of which bear her name. Her most famous comet, Shoemaker-Levy 9, was discovered with her husband Eugene Shoemaker and her coworker, David Levy, in 1993. This comet looked like a string of pearls when it collided with Jupiter in 1994.

fabulous FIRSTS

A comet is a ball of dust and ice traveling through space. Most comets are 10 kilometers in diameter. Comet Hale-Bopp, which passed near Earth in April 1997, was estimated at 30–40 kilometers in diameter.

4-2 LESSON NOTES

NCTM Standards: 1–4, 7

Instructional Resources

- Study Guide Master 4-2
- Practice Master 4-2
- Enrichment Master 4-2
- Assessment and Evaluation Masters, p. 100

Transparency 4-2A contains the 5-Minute Check for this lesson; **Transparency 4-2B** contains a teaching aid for this lesson.

Recommended Pacing	
Standard Pacing	Day 3 of 14
Honors Pacing	Day 3 of 13
Block Scheduling*	Day 2 of 7

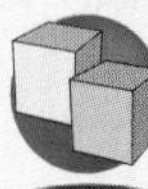

*For more information on pacing and possible lesson plans, refer to the *Block Scheduling Booklet*.

1 FOCUS

5-Minute Check
(over Lesson 4-1)

Determine whether each statement is *true* or *false*. If false, explain why.

1. A right triangle has one angle that measures 90°. **true**
2. A scalene triangle has no two sides with the same measure. **true**
3. An obtuse triangle has three angles measuring less than 90°. **False; one angle must measure more than 90°.**
4. In an isosceles triangle, all three sides have the same measure. **False; at least two sides have the same measure.**

Solve.

5. The legs of an isosceles triangle measure $2x + 5$ and $3x - 1$. Find the value of x. **6**
6. An equilateral triangle has a perimeter of 54 centimeters. One side measures $5x - 2$. Find the value of x. **4**

Motivating the Lesson

Situational Problem Use a map of the stars and identify several triangles connecting the stars. Use a protractor to measure the angles.

2 TEACH

In-Class Example

For Example 1

A surveyor has drawn a triangle on a map. One angle measures 42° and another measures 53°. Find the measure of the third angle. 85

Teaching Tip After reading Theorem 4-3, point out that each triangle has six possible exterior angles. The theorem applies to any of them.

If you know the measures of two angles of a triangle, you can find the measure of the third angle.

Example 1

Astronomy

Refer to the application at the beginning of the lesson. Find the measure of the angle at Deneb.

Explore Let V, A, and D be the vertices of the triangle. We are given that $m\angle V = 74$ and $m\angle A = 41$. We are asked to find the $m\angle D$.

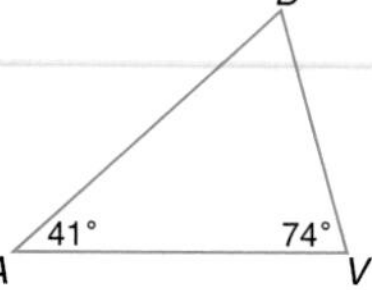

Plan Use the Angle Sum Theorem to write an equation. Then solve the equation for the measure of the missing angle.

Solve

$m\angle V + m\angle A + m\angle D = 180$ *Angle Sum Theorem*

$74 + 41 + m\angle D = 180$ *Substitution Property (=)*

$m\angle D = 65$ *Subtraction Property (=)*

The measure of the angle at Deneb is 65.

Examine Find the sum of the measures of the three angles of the triangle.

$$74 + 41 + 65 = 180$$

Since the sum is 180, the answer is correct.

The Angle Sum Theorem leads to a useful theorem about the angles in two triangles. *You will be asked to prove this theorem in Exercise 36.*

Theorem 4–2 Third Angle Theorem	**If two angles of one triangle are congruent to two angles of a second triangle, then the third angles of the triangles are congruent.**

In the figure at the right, $\angle CBD$ is an **exterior angle** of $\triangle ABC$. An exterior angle is formed by one side of a triangle and the extension of another side. The interior angles of the triangle not adjacent to a given exterior angle are called **remote interior angles** of the exterior angle. In the figure, $\angle CBD$ is an exterior angle with $\angle A$ and $\angle C$ as its remote interior angles. Theorem 4–3 relates the measure of the exterior angle of a triangle to the sum of the measures of the two remote interior angles.

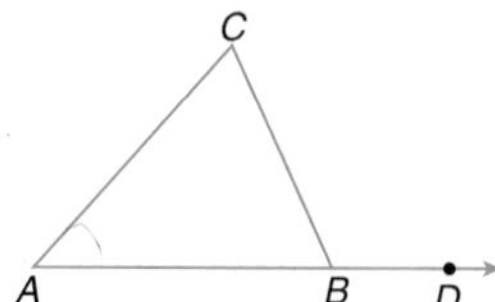

Theorem 4–3 Exterior Angle Theorem	**The measure of an exterior angle of a triangle is equal to the sum of the measures of the two remote interior angles.**

Classroom Vignette

"I like to use statistics to lead students to the Angle Sum Theorem. Have each student draw a triangle and measure its angles. Then record the sums of the measures in a stem-and-leaf plot. Ask students what they observe about the sums. Then discuss the Angle Sum Theorem."

Gail F. Burrill

Gail Burrill
Author

To prove this theorem, we will use a **flow proof**. A flow proof organizes a series of statements in logical order, starting with the given statements. Each statement along with its reason is written in a box, and arrows are used to show how each statement leads to another.

Proof of Theorem 4–3

A flow proof can be drawn in a horizontal or a vertical direction.

Given: $\triangle ABC$

Prove: $m\angle CBD = m\angle A + m\angle C$

Flow Proof:

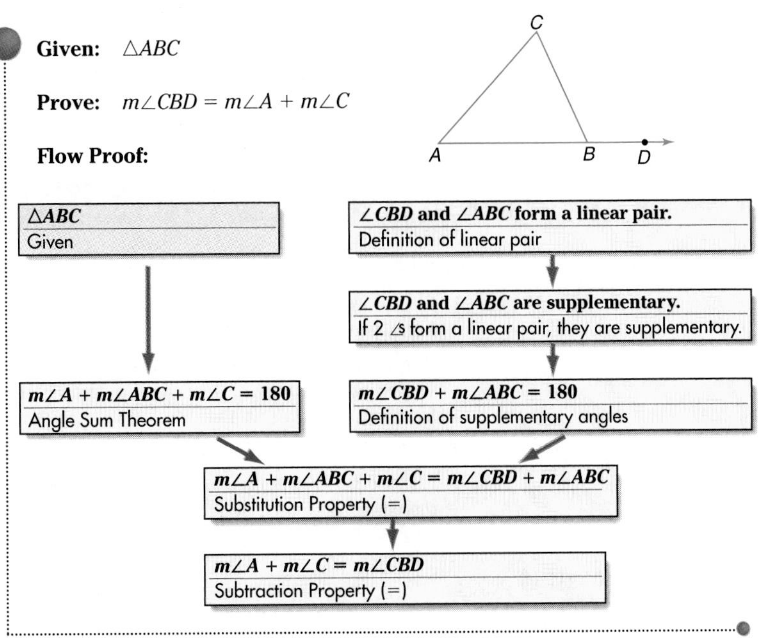

C
A
B
D
$\triangle ABC$
Given
$\angle CBD$ and $\angle ABC$ form a linear pair.
Definition of linear pair
$\angle CBD$ and $\angle ABC$ are supplementary.
If 2 ∠s form a linear pair, they are supplementary.
$m\angle A + m\angle ABC + m\angle C = 180$
Angle Sum Theorem
$m\angle CBD + m\angle ABC = 180$
Definition of supplementary angles
$m\angle A + m\angle ABC + m\angle C = m\angle CBD + m\angle ABC$
Substitution Property (=)
$m\angle A + m\angle C = m\angle CBD$
Subtraction Property (=)

Sometimes the measures of angles can be determined by using a combination of theorems.

Example 2 **Find the measure of each numbered angle in the figure at the right.**

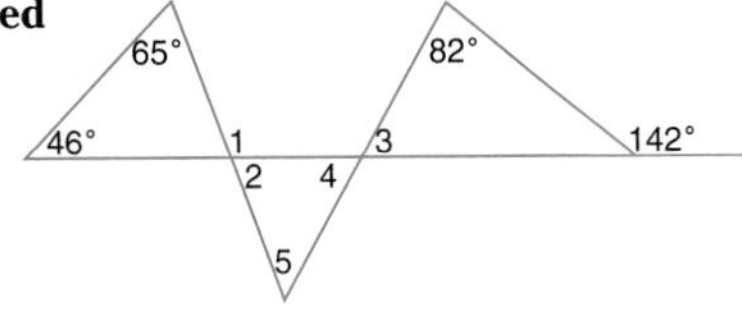

$m\angle 1 = 46 + 65$ *Exterior Angle Theorem*
$= 111$

$m\angle 1 + m\angle 2 = 180$ *If 2 ∠s form a linear pair, they are supplementary.*
$111 + m\angle 2 = 180$ *Substitution Property (=)*
$m\angle 2 = 69$ *Subtraction Property (=)*

$m\angle 3 + 82 = 142$ *Exterior Angle Theorem*
$m\angle 3 = 60$ *Subtraction Property (=)*

Since $\angle 3$ and $\angle 4$ are vertical angles, $m\angle 4 = 60$.

$m\angle 2 + m\angle 4 + m\angle 5 = 180$ *Angle Sum Theorem*
$69 + 60 + m\angle 5 = 180$ *Substitution Property (=)*
$m\angle 5 = 51$ *Subtraction Property (=)*

Therefore, $m\angle 1 = 111$, $m\angle 2 = 69$, $m\angle 3 = 60$, $m\angle 4 = 60$, and $m\angle 5 = 51$.

In-Class Example

For Example 2
Find the measure of each numbered angle in the figure if $\overleftrightarrow{AB} \parallel \overleftrightarrow{CD}$.

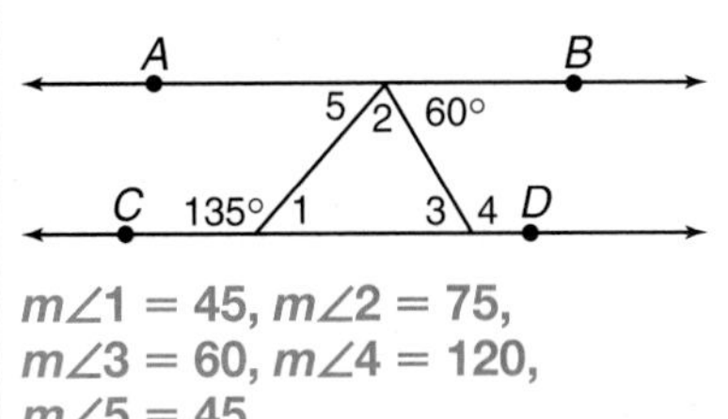

$m\angle 1 = 45$, $m\angle 2 = 75$, $m\angle 3 = 60$, $m\angle 4 = 120$, $m\angle 5 = 45$

Teaching Tip After reading Corollary 4-1, ask a volunteer to outline a strategy for proving the statement.

3 PRACTICE/APPLY

Check for Understanding

Exercises 1–13 are designed to help you assess your students' understanding through reading, writing, speaking, and modeling. You should work through Exercises 1–5 with your students and then monitor their work on Exercises 6–13.

Additional Answers

2. The exterior angles at a vertex are the two angles that are not the angle of the triangle or the vertical angle to that angle.

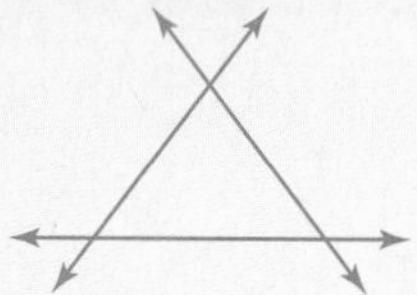

4. Sample answer: 40 and 50; the acute angles of a right triangle are complementary.

A statement that can be easily proved using a theorem is often called a **corollary** of that theorem. A corollary, just like a theorem, can be used as a reason in a proof. *You will be asked to prove Corollaries 4–1 and 4–2 in Exercises 37 and 38, respectively.*

Corollary 4–1	**The acute angles of a right triangle are complementary.**
Corollary 4–2	**There can be at most one right or obtuse angle in a triangle.**

CHECK FOR UNDERSTANDING

Communicating Mathematics

1. Subtract the sum of the measures of the two angles from 180.
3. The sum of the measures of two obtuse angles is greater than 180. This contradicts the Angle Sum Theorem.

Study the lesson. Then complete the following.

1. **Explain** in your own words how to find the third angle of a triangle if you know the measure of the other two angles.
2. **Draw** a triangle and extend all of its sides. There are four angles at each vertex of the triangle. Describe how you can distinguish exterior angles from other angles at a vertex. See margin.
3. **State** why it is impossible to have two obtuse angles in a triangle.
4. **Choose** two angle measures other than the right angle that could be measures of angles of a right triangle. Explain why you know these two angles along with the right angle are angles of a right triangle. See margin.

5. See students' work.
5a. The sum of the measures of the acute angles of a right triangle is 90.

5. Fold a rectangular piece of paper along a diagonal from A to C. Then cut along the fold to form a right triangle ABC. Write the name of each angle on the inside of the triangle. Tear off angles A and C. Place the torn angles at $\angle B$.
   a. What seems to be true about the sum of the measures of the acute angles of a right triangle?
   b. Do your results seem to confirm or deny Corollary 4–1? confirm

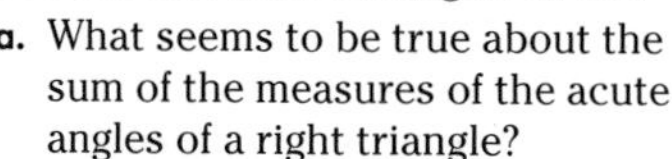

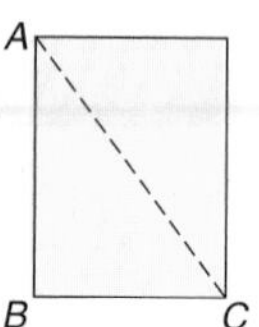

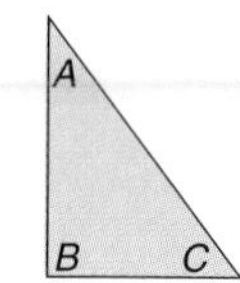

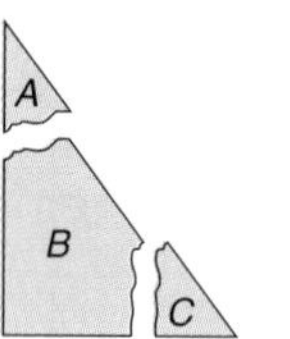

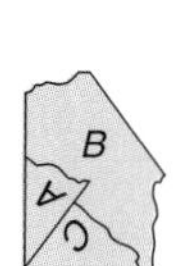

Guided Practice

Find the value of x.

6. 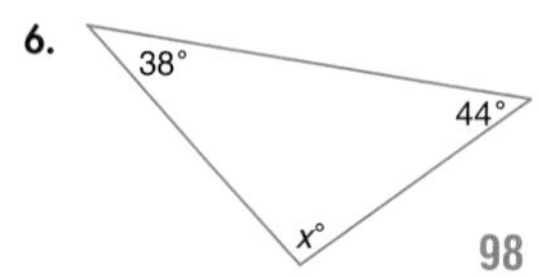

98

7. 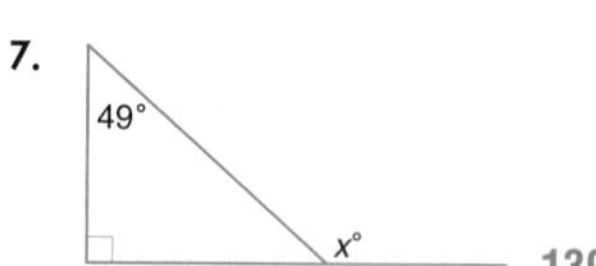

139

If $\overline{KH}$ is parallel to $\overline{JI}$, find the measure of each angle in the figure at the right.

8. $\angle 1$ 66
9. $\angle 2$ 54
10. $\angle 3$ 90

Reteaching

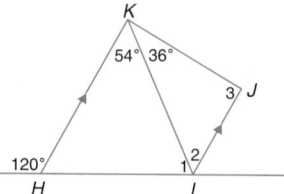

Using Diagrams Have students answer the following questions about this diagram.

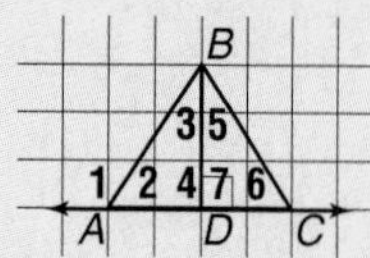

1. Name an exterior angle. Sample answer: $\angle 1$
2. Name a pair of complementary angles. $\angle 2$ and $\angle 3$ or $\angle 5$ and $\angle 6$
3. $m\angle 1$ is equal to the sum of the measures of which two angles? Sample answer: $\angle 3$ and $\angle 4$
4. If $m\angle 3 = m\angle 5$, does $m\angle 2 = m\angle 6$? Why or why not? Yes; each triangle has a right $\angle$, and since two $\angle$s of one triangle are congruent to two $\angle$s of another, the third $\angle$s are $\cong$.

11. Refer to the figure at the right.
   a. Find x. 45
   b. Find $m\angle A$. 90
   c. Classify the triangle by the measure of its angles. right triangle

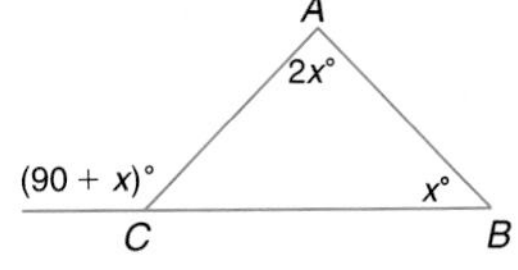

12. Yes; the two right angles would be at the same vertex.

12. Is it possible to have two right angles as exterior angles of a triangle? Explain your reasoning.

13. **Astronomy** The Big Dipper is a part of the larger constellation Ursa Major, which means *great bear*. Three of the brighter stars in the constellation form a triangle *SRA*. If $m\angle S = 109$ and $m\angle R = 41$, find $m\angle A$. 30

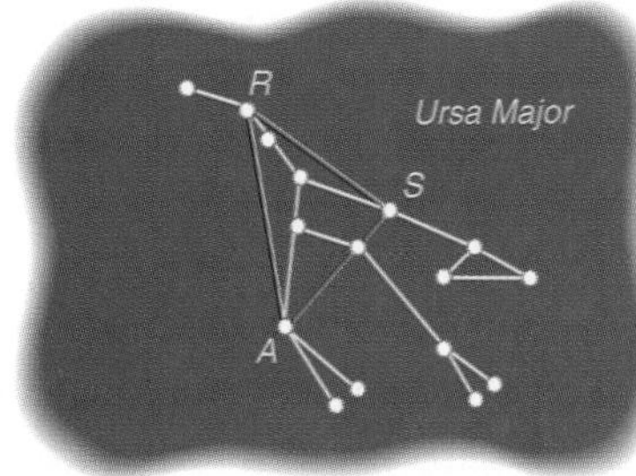

EXERCISES

Practice

Find the value of x.

14. 93

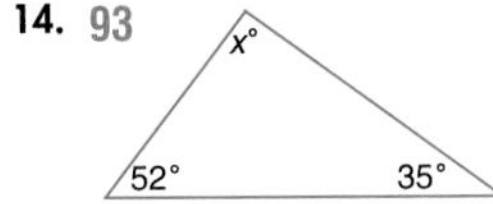

15. 91

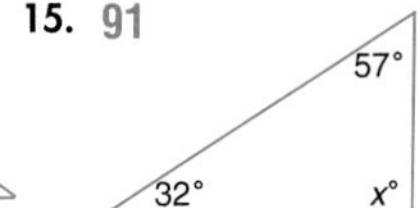

16. 115

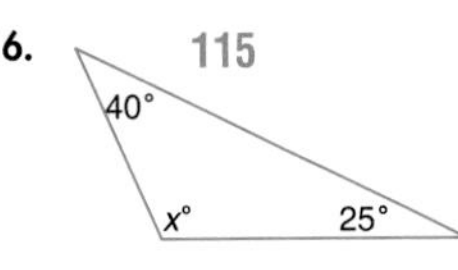

17. 115

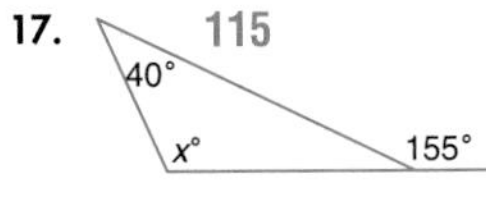

18. 58

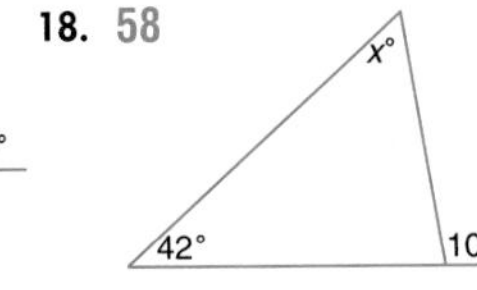

19. 75

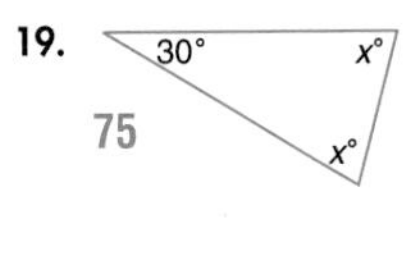

If $\overline{AB}$ is perpendicular to $\overline{BC}$, find the measure of each angle in the figure below.

20. $\angle 1$ 76
21. $\angle 2$ 68
22. $\angle 3$ 76
23. $\angle 4$ 40
24. $\angle 5$ 64
25. $\angle 6$ 26
26. $\angle 7$ 140
27. $\angle 8$ 14

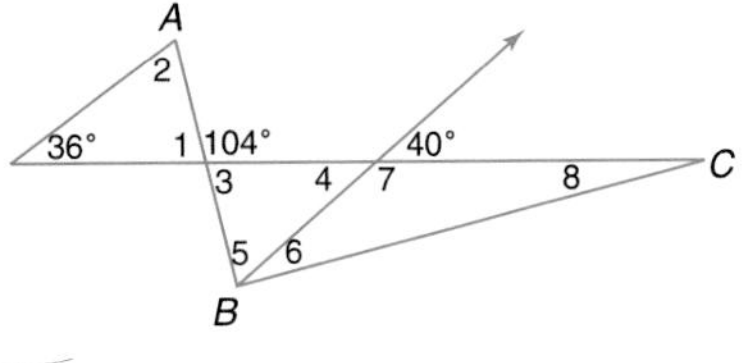

For each triangle, find $m\angle A$.

28.

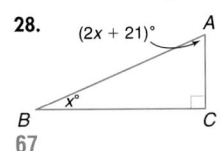

67

29.

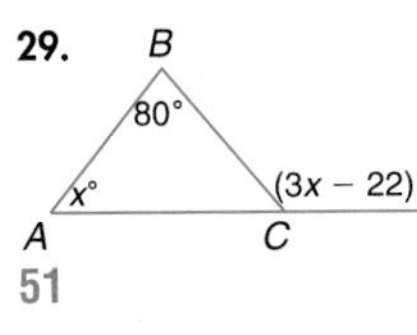

51

30.

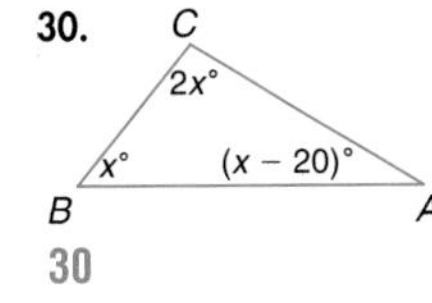

30

31. Scalene; none of the angles are congruent so none of the sides are congruent.

31. Refer to $\triangle ABC$ in Exercise 29. Is this triangle scalene, equilateral, isosceles, or right? Explain.

32. In $\triangle ABC$, $m\angle A$ is 16 more than $m\angle B$, and $m\angle C$ is 29 more than $m\angle B$.
   a. Write an equation relating the measures. $(x + 16) + x + (x + 29) = 180$
   b. Find the measure of each angle. $m\angle A = 61$, $m\angle B = 45$, $m\angle C = 74$

Assignment Guide

Core (with proof): 15–43 odd, 44–53
Core (informal): 15–31 odd, 39, 41, 43–53
Enriched: 14–40 even, 41–53

For **Extra Practice**, see p. 770.

The red A, B, and C flags, printed only in the Teacher's Wraparound Edition, indicate the level of difficulty of the exercises.

Study Guide Masters, p. 21

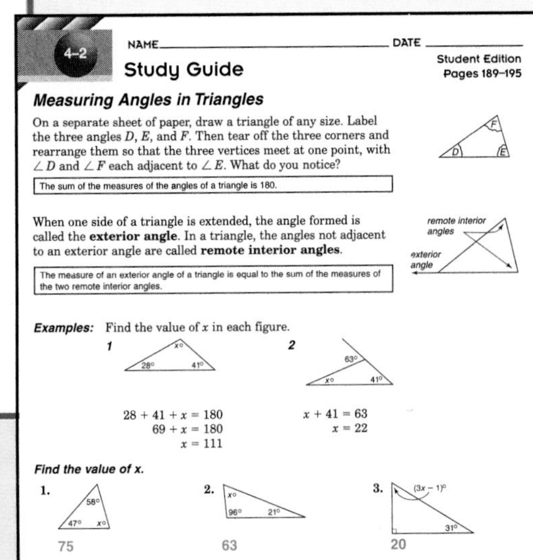

4-2 NAME ______ DATE ______

Study Guide

Student Edition Pages 189–195

Measuring Angles in Triangles

On a separate sheet of paper, draw a triangle of any size. Label the three angles D, E, and F. Then tear off the three corners and rearrange them so that the three vertices meet at one point, with $\angle D$ and $\angle F$ each adjacent to $\angle E$. What do you notice?

The sum of the measures of the angles of a triangle is 180.

When one side of a triangle is extended, the angle formed is called the **exterior angle**. In a triangle, the angles not adjacent to an exterior angle are called **remote interior angles**.

The measure of an exterior angle of a triangle is equal to the sum of the measures of the two remote interior angles.

remote interior angles
exterior angle

Examples: Find the value of x in each figure.

1. $28 + 41 + x = 180$; $69 + x = 180$; $x = 111$
2. $x + 41 = 63$; $x = 22$

Find the value of x.

1. 75
2. 63
3. 20
4. 55
5. 23
6. 17

Find the measure of each angle.

7. $\angle 1$ 112
8. $\angle 2$ 68
9. $\angle 3$ 90
10. $\angle 4$ 90
11. $\angle 5$ 22
12. $\angle 6$ 158

Additional Answer

35. Given: $\triangle RED$ is equiangular.
Prove: $m\angle R = m\angle E = m\angle D = 60$

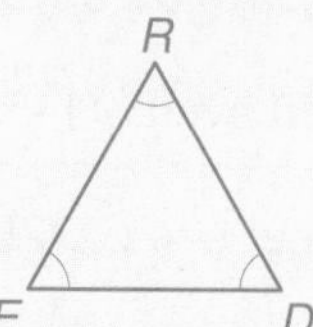

Proof:
Statements (Reasons)
1. $\triangle RED$ is equiangular. (Given)
2. $\angle R \cong \angle E \cong \angle D$ (Def. equiangular $\triangle$)
3. $m\angle R = m\angle E = m\angle D$ (Def. $\cong$ $\angle$s)
4. $m\angle R + m\angle E + m\angle D = 180$ (Angle Sum Th.)
5. $m\angle R + m\angle R + m\angle R = 180$ (Subst. Prop. (=))
6. $3m\angle R = 180$ (Subst. Prop. (=))
7. $m\angle R = 60$ (Div. Prop. (=))
8. $m\angle R = m\angle E = m\angle D = 60$ (Subst. Prop. (=))

Practice Masters, p. 21

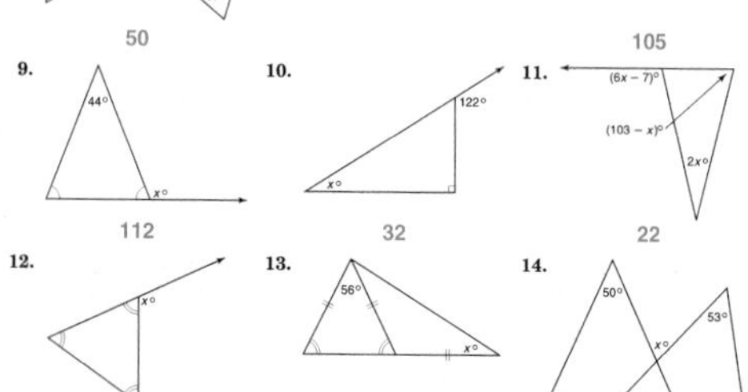
4-2 NAME ______ DATE ______
Practice — Student Edition Pages 189–195
Measuring Angles in Triangles
Find the value of x.

1. 85
2. 30
3. 29
4. 147
5. 60
6. 27
7. 50
8. 105
9. 112
10. 32
11. 22
12. 120
13. 31
14. 65

Proof

33. The statements and reasons used in a flow proof to prove $\angle V \cong \angle W$ are listed below. Put the boxes in the correct order and add the arrows. See Solutions Manual.

Given: $\angle RUW \cong \angle VSR$

Prove: $\angle W \cong \angle V$

Proof:

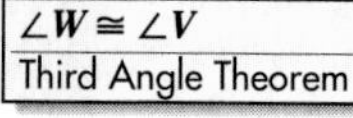
$\angle W \cong \angle V$ — Third Angle Theorem

$\angle RUW \cong \angle VSR$ — Given

$\angle URW \cong \angle VRS$ — Congruence of angles is reflexive.

34. 135; The third angle in the triangle measures 65°. The largest exterior angle of the triangle measures $65 + 70 = 135$.

34. The measures of two angles of a triangle are 45 and 70. What is the largest exterior angle of the triangle? Explain your reasoning.

Proof

35. Write a two-column proof to show that if a triangle is equiangular, the measure of each angle is 60. See margin.

36–38. See Solutions Manual.

36. Write a two-column proof for the Third Angle Theorem. (Theorem 4–2).

37. Write a flow proof to show that the acute angles of a right triangle are complementary. (Corollary 4–1).

38. Write a paragraph proof to show there can be at most one right or obtuse angle in a triangle. (Corollary 4–2).

39. What is the sum of the interior angles of the quadrilateral at the right? (*Hint*: Think in terms of triangles.) 360

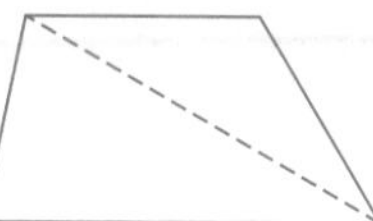

Cabri Geometry

40. Use a TI-92 calculator to draw a triangle and a line through a vertex of the triangle that is parallel to the side opposite that vertex.
- **a.** Calculate all the angle measures in the figure. See students' work.
- **b.** Drag a vertex around on the screen. Describe how the measure changes as you move the location of the vertex. See students' work.

Critical Thinking

41. In spherical geometry, triangles can be drawn on a sphere. Sides of a triangle can consist of part of the equator and any great circle that passes through the two poles. In spherical geometry, any great circle passing through the two poles is perpendicular to the equator. Three such triangles are drawn below. a–e. See margin.

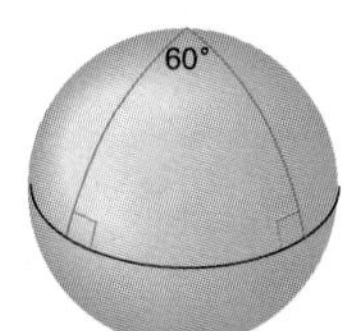

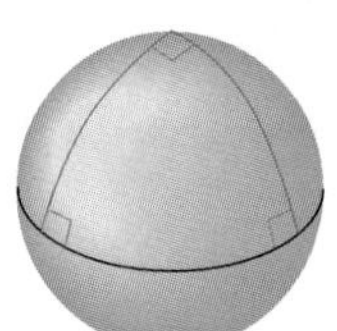
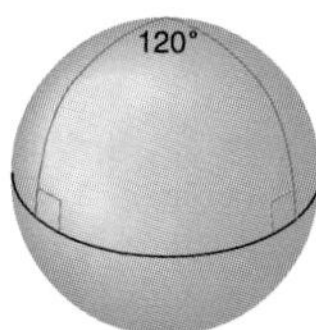

LOOK BACK
You can refer to Lesson 3-6 for information on spherical geometry.

- **a.** Is it possible for the sum of the measures of the angles of a triangle in this system to be 180? Explain your reasoning.
- **b.** What is the possible range of the sum the measures of angles of a triangle in this system? Explain your reasoning.
- **c.** How would you define an acute triangle in this system?
- **d.** How would you define an obtuse triangle in this system?
- **e.** How would you define a right triangle in this system?

Additional Answers

41a. No; two of the $\angle$s are rt. $\angle$s, so the sum of their measures is 180.

41b. The sum of the measures is > 180 and ≤ 360. The sum of the measures of two $\angle$s is 180. The measure of the third $\angle$ is > 0 and ≤ 180.

41c. If the measure of the $\angle$ at the pole is < 90, the $\triangle$ is an acute $\triangle$.

41d. If the measure of the $\angle$ at the pole is > 90, the $\triangle$ is an obtuse $\triangle$.

41e. If the measure of the $\angle$ at the pole is 90, the $\triangle$ is a right $\triangle$.

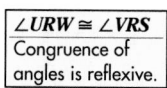

Applications and Problem Solving

42. **Astronomy** Leo is a constellation that represents a lion. Three of the brighter stars in the constellation form a triangle *LEO*. If the angles have measures indicated in the figure at the right, find $m\angle L$. **66**

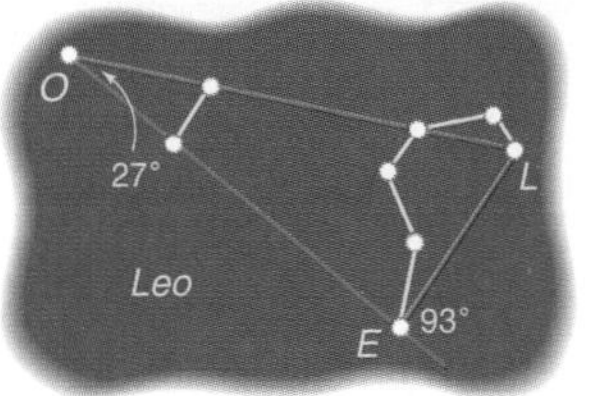

43. **Construction** The roof support at the right is shaped like a triangle. Two angles each have a measure of 25. Find $m\angle R$. **130**

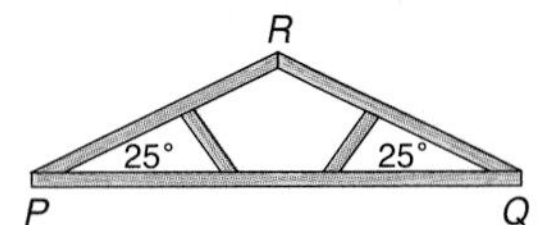

Mixed Review

44. The measures of the angles of a triangle are 65, 95, and 20. Classify the triangle by its angles. (Lesson 4–1) **obtuse**

45. Determine if $\overleftrightarrow{RS} \parallel \overleftrightarrow{LM}$ given that $m\angle 1 = 42$ and $m\angle 5 = 48$. Justify your answer. (Lesson 3–4) **No; If they are parallel, $m\angle 1 = m\angle 5$.**

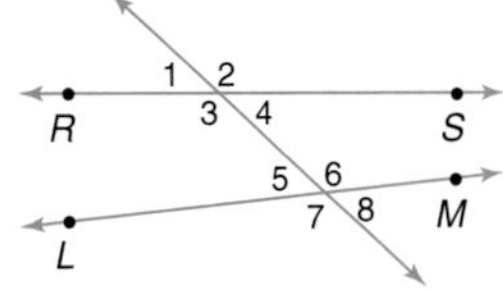

46. 2800 people per year

46. **Demographics** The population of Greensboro, North Carolina, was 156,000 in 1980 and 184,000 in 1990. What was the average annual rate of change for the population of Greensboro from 1980 to 1990? (Lesson 3–3)

47. **Draw a Diagram** The Cedarville Club is holding its annual women's tennis tournament. If 8 women are registered for the tournament, and each woman plays every other woman exactly once, how many games will be played? (Lesson 3–1) **28 games**

48. Write the given and prove statements that you would use to prove the theorem *The acute angles of a right triangle are complementary.* Then draw a figure. (Lesson 2–5) **See margin.**

49a–b. Derrick will receive an A by the Law of Detachment.

49. a. Determine if a valid conclusion can be made from these two statements.
    (1) If a student scores a 92 or above on the geometry test, he or she will receive an A.
    (2) Derrick scored 94 on the geometry test.

    b. State the law of logic you used. (Lesson 2–3)

50. Copy segment LN. Then use a compass and straightedge to bisect the segment. (Lesson 1–5) **See margin.**

L ———— N

51. In which quadrant is $P(-4, -6)$ located? (Lesson 1–1) **III**

Algebra

52. Solve $\frac{6}{t} = \frac{2}{7}$. **21**

53. **Mechanics** A nationwide poll of 500 women was taken to determine how many women know how to take care of their cars. The poll results are shown at the right.

    a. How many of the women surveyed knew how to charge a car battery?

    b. How many of the women surveyed knew how to check the oil?

53a. 260 women

53b. 395 women

Women Care for Their Cars

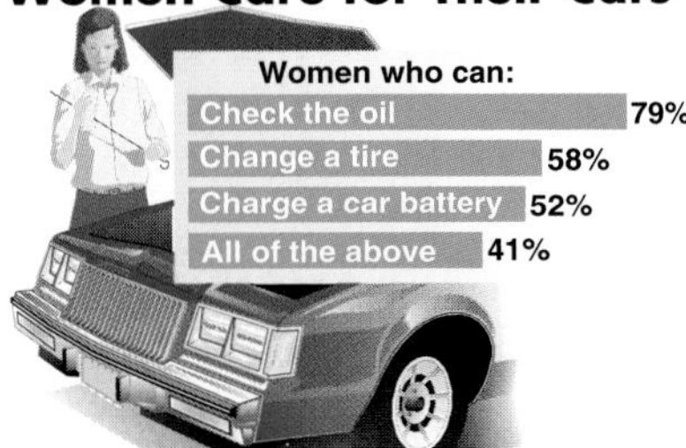

Source: *EDK Forecast*

4 ASSESS

Closing Activity

Writing Have students write a summary, including diagrams, of the relationships among angle measures in triangles that were presented in this lesson.

Chapter 4 Quiz A (Lessons 4-1 and 4-2) is available in the *Assessment and Evaluation Masters*, p. 100.

Additional Answers

48. Given: $\angle B$ is a right angle.
Prove: $\angle A$ and $\angle C$ are complementary.

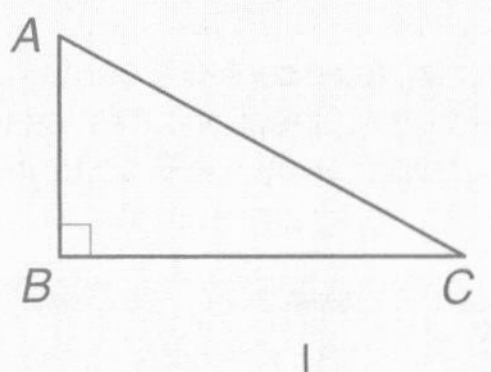

50.

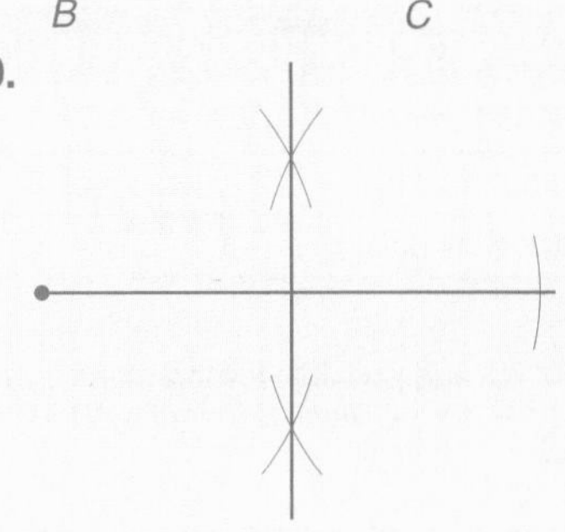

Extension

Reasoning 1. If the sum of the measures of the two remote interior angles of a triangle is less than 90, how would you classify the triangle?
obtuse

2. If the sum of the measures of the two remote interior angles of a triangle is more than 90, is the triangle acute?
not necessarily

Enrichment Masters, p. 21

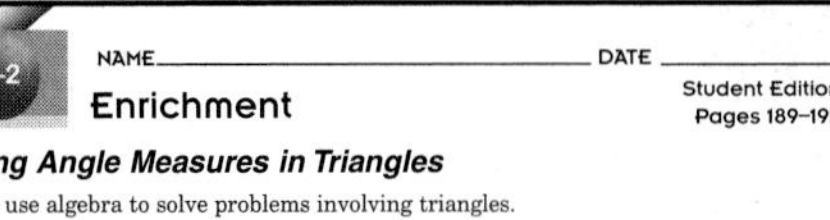

4-2 NAME______ DATE______
Enrichment
Student Edition Pages 189–195

Finding Angle Measures in Triangles

You can use algebra to solve problems involving triangles.

Example: In triangle ABC, $m\angle A$, is twice $m\angle B$, and $m\angle C$ is 8 more than $m\angle B$. What is the measure of each angle?

Write and solve an equation. Let $x = m\angle B$.

$$m\angle A + m\angle B + m\angle C = 180$$
$$2x + x + (x + 8) = 180$$
$$4x + 8 = 180$$
$$4x = 172$$
$$x = 43$$

So, $m\angle A = 2(43)$ or 86, $m\angle B = 43$, and $m\angle C = 43 + 8$ or 51.

Solve each problem.

1. In triangle DEF, $m\angle E$ is three times $m\angle D$, and $m\angle F$ is 9 less than $m\angle E$. What is the measure of each angle? $m\angle D = 27$, $m\angle E = 81$, $m\angle F = 72$

2. In triangle RST, $m\angle T$ is 5 more than $m\angle R$, and $m\angle S$ is 10 less than $m\angle T$. What is the measure of each angle? $m\angle R = 60$, $m\angle S = 55$, $m\angle T = 65$

3. In triangle JKL, $m\angle K$ is four times $m\angle J$, and $m\angle L$ is five times $m\angle J$. What is the measure of each angle? $m\angle J = 18$, $m\angle K = 72$, $m\angle L = 90$

4. In triangle XYZ, $m\angle Z$ is 2 more than twice $m\angle X$, and $m\angle Y$ is 7 less than twice $m\angle X$. What is the measure of each angle? $m\angle X = 37$, $m\angle Y = 67$, $m\angle Z = 76$

5. In triangle GHI, $m\angle H$ is 20 more than $m\angle G$, and $m\angle G$ is 8 more than $m\angle I$. What is the measure of each angle? $m\angle G = 56$, $m\angle H = 76$, $m\angle I = 48$

6. In triangle MNO, $m\angle M$ is equal to $m\angle N$, and $m\angle O$ is 5 more than three times $m\angle N$. What is the measure of each angle? $m\angle M = m\angle N = 35$, $m\angle O = 110$

7. In triangle STU, $m\angle U$ is half $m\angle T$, and $m\angle S$ is 30 more than $m\angle T$. What is the measure of each angle? $m\angle S = 90$, $m\angle T = 60$, $m\angle U = 30$

8. In triangle PQR, $m\angle P$ is equal to $m\angle Q$, and $m\angle R$ is 24 less than $m\angle P$. What is the measure of each angle? $m\angle P = m\angle Q = 68$, $m\angle R = 44$

9. Write your own problems about measures of triangles. See students' work.

4-3 LESSON NOTES

NCTM Standards: 1–4, 7, 8

Instructional Resources

- Study Guide Master 4-3
- Practice Master 4-3
- Enrichment Master 4-3
- Assessment and Evaluation Masters, pp. 99, 100
- Graphing Calculator and Computer Masters, p. 4
- Modeling Mathematics Masters, pp. 27–30
- Real-World Applications, 7
- Tech Prep Applications Masters, p. 7

Transparency 4-3A contains the 5-Minute Check for this lesson; **Transparency 4-3B** contains a teaching aid for this lesson.

Recommended Pacing	
Standard Pacing	Days 4 & 5 of 14
Honors Pacing	Days 4 & 5 of 13
Block Scheduling*	Day 3 of 7

*For more information on pacing and possible lesson plans, refer to the *Block Scheduling Booklet*.

1 FOCUS

5-Minute Check
(over Lesson 4-2)

Refer to the figure below.

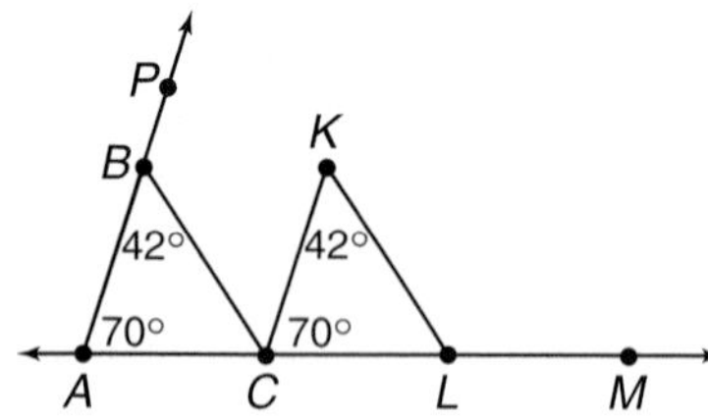

1. Find $m\angle ACB$. 68
2. Find $m\angle CLK$. 68
3. Find $m\angle CBP$. 138
4. Which two angles are the remote interior angles for $\angle MLK$? $\angle LCK$ and $\angle LKC$
5. Find $m\angle BCK$. 42
6. Find $m\angle MLK$. 112

4-3 Exploring Congruent Triangles

What YOU'LL LEARN

- To name and label corresponding parts of congruent triangles.

Why IT'S IMPORTANT

You can identify congruent triangles and their corresponding parts involved in crafts, arts, and construction.

Crafts

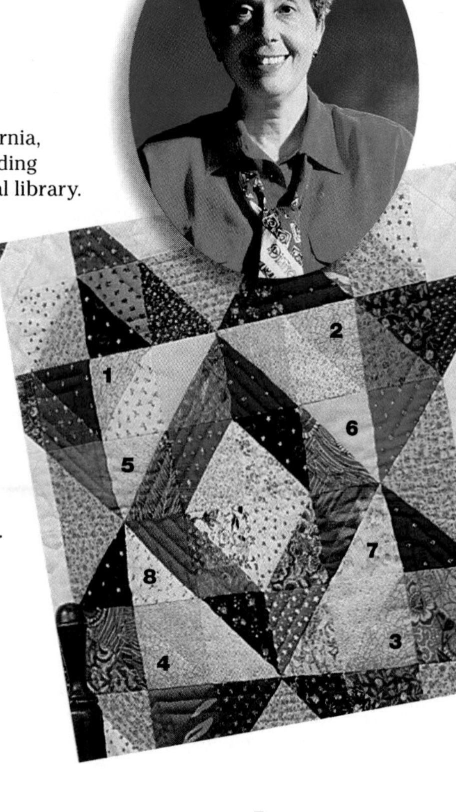

Diane Leighton of Yuba City, California, learned the craft of quilting by reading books on the subject from her local library. She teaches quilting while maintaining her career as a registered nurse. One of Ms. Leighton's quilts, which includes several isosceles right triangles, is pictured.

The numbered triangles are all the same size and same shape. Triangles that are the same size and same shape are **congruent triangles**. Each triangle has six parts, three angles and three sides. If the corresponding six parts of one triangle are congruent to the six parts of another triangle, then the triangles are congruent.

FYI

To celebrate the 1989 centennial of North Dakota, 7000 citizens made the largest quilt in the world. It is 85 feet by 134 feet.

In the figure below, $\triangle DEF$ is congruent to $\triangle ABC$. If you *slide* $\triangle DEF$ up and to the right, $\triangle DEF$ is still congruent to $\triangle ABC$.

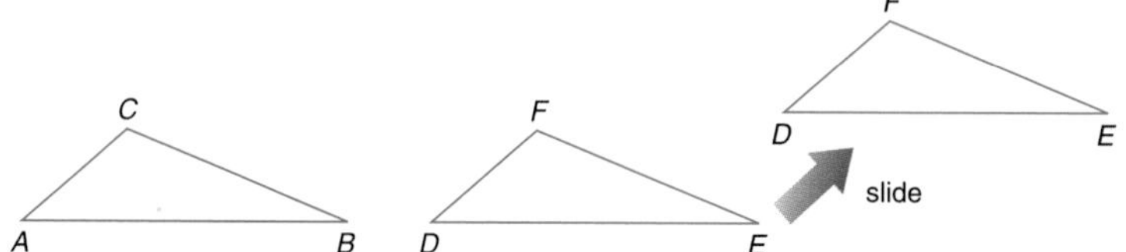

Also, if you *rotate* $\triangle DEF$, $\triangle DEF$ remains congruent to $\triangle ABC$. If you *flip* $\triangle DEF$, $\triangle DEF$ remains congruent.

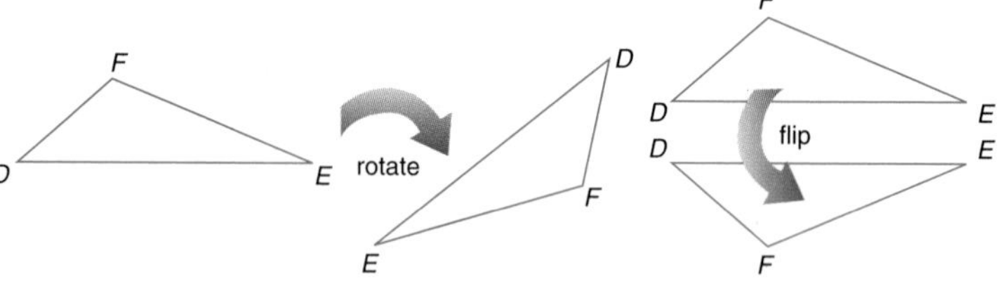

FYI

In 1996, the AIDS quilt was spread on the Mall in Washington, D.C. It covered an area equal to 29 football fields. If you walked all the pathways between the sections, you would have walked 21 miles.

If you slide, rotate, or flip a figure, congruence will not change. These three transformations are called **congruence transformations**.

Example 1

Crafts

A design for a quilt has been drawn on graph paper. Identify the triangles that appear to be congruent.

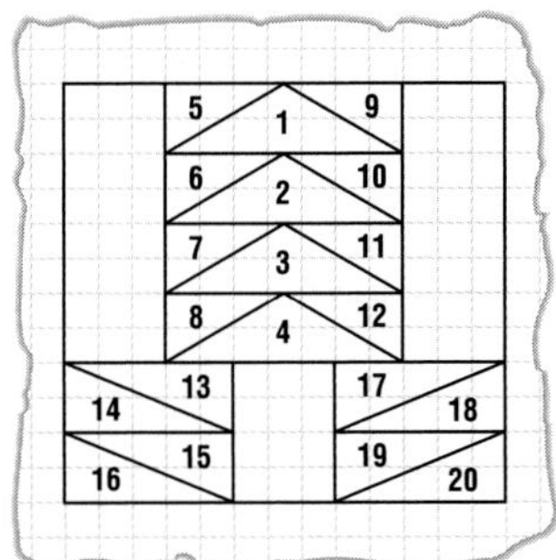

Triangles 1 through 4 appear to be congruent.

Triangles 5 through 12 appear to be congruent.

Triangles 13 through 20 appear to be congruent.

The symbol ↔ means "corresponds to."

If $\triangle LMN$ is congruent to $\triangle PQR$ ($\triangle LMN \cong \triangle PQR$), the vertices of the two triangles correspond in the same order as the letters naming the triangles.

$L \leftrightarrow P$ $\qquad$ $M \leftrightarrow Q$ $\qquad$ $N \leftrightarrow R$

This correspondence of vertices can be used to name the corresponding congruent sides and angles of the two triangles.

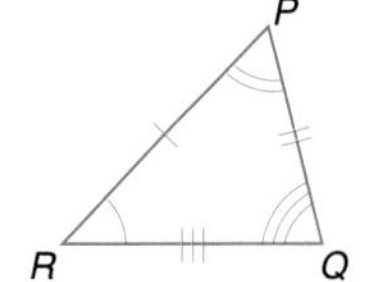

$\angle L \cong \angle P$ $\qquad$ $\angle M \cong \angle Q$ $\qquad$ $\angle N \cong \angle R$

$\overline{LM} \cong \overline{PQ}$ $\qquad$ $\overline{MN} \cong \overline{QR}$ $\qquad$ $\overline{LN} \cong \overline{PR}$

The corresponding sides and angles can be determined from any congruence statement by following the order of the letters. For example, $\triangle ABC \cong \triangle FGH$ indicates the following congruences.

$\angle A \cong \angle F$ $\qquad$ $\angle B \cong \angle G$ $\qquad$ $\angle C \cong \angle H$

$\overline{AB} \cong \overline{FG}$ $\qquad$ $\overline{BC} \cong \overline{GH}$ $\qquad$ $\overline{AC} \cong \overline{FH}$

It is important that you list the letters of the vertices in the correct order whenever you write a congruence statement.

Definition of Congruent Triangles (CPCTC)	Two triangles are congruent if and only if their corresponding parts are congruent.

The abbreviation CPCTC means Corresponding Parts of Congruent Triangles are Congruent. "If and only if" is used to show that the conditional and its converse are both true.

Motivating the Lesson

Questioning Ask students to think of as many different ways as they can of producing two triangles that are the same size and shape. **Sample answers: Use tracing paper, use ruler and protractor, use dot paper.**

2 TEACH

In-Class Example

For Example 1

Here is another quilt design. Identify the triangles that appear to be congruent.

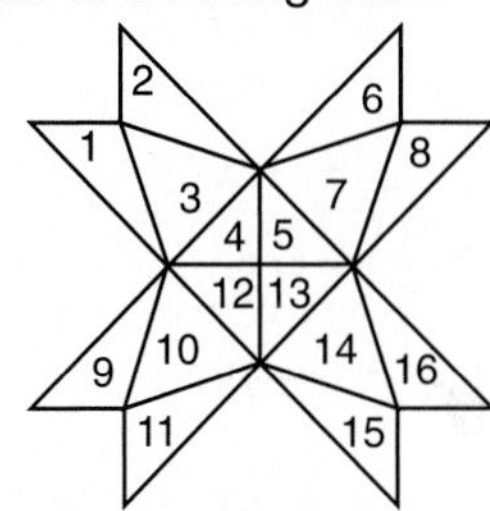

Triangles 1, 2, 6, 8, 9, 11, 15, and 16 appear to be congruent. Triangles 3, 7, 10, and 14 appear to be congruent. Triangles 4, 5, 12, and 13 appear to be congruent.

Teaching Tip Have students compare triangles by looking at the portions of grid squares that are in each triangle. If two triangles have the same portions of squares in the same places, they are probably congruent.

In-Class Example

For Example 2
If $\triangle ABC \cong \triangle DEF$, name the corresponding congruent angles and sides.

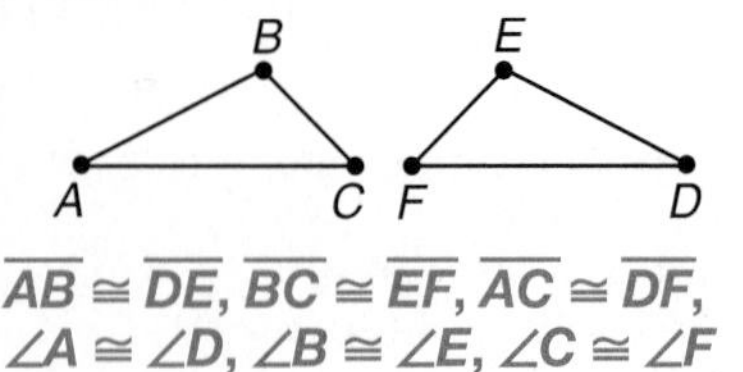

$\overline{AB} \cong \overline{DE}$, $\overline{BC} \cong \overline{EF}$, $\overline{AC} \cong \overline{DF}$, $\angle A \cong \angle D$, $\angle B \cong \angle E$, $\angle C \cong \angle F$

Teaching Tip Remind students to never call an angle by one letter if it can be confused. Always refer to it by its three letter name. For example, $\angle RTS$ or $\angle RTU$, not $\angle T$.

Example 2 APPLICATION Construction

The bridge in the picture below uses a simple triangular truss design. The vertices of two triangles are labeled and $\triangle TRU \cong \triangle TSU$. Name the corresponding congruent angles and sides.

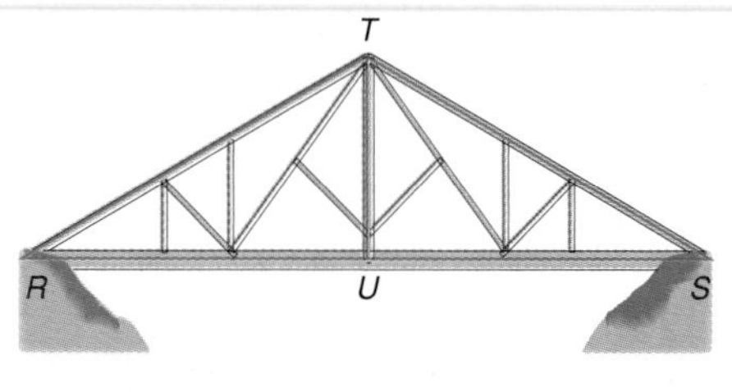

$\angle RTU \cong \angle STU$ $\quad \angle R \cong \angle S$ $\quad \angle TUR \cong \angle TUS$

$\overline{TR} \cong \overline{TS}$ $\quad \overline{RU} \cong \overline{SU}$ $\quad \overline{TU} \cong \overline{TU}$

Congruence of triangles, like congruence of segments and angles, is reflexive, symmetric, and transitive. This is stated in Theorem 4–4. The proof of the transitive part of the theorem is given. *You will be asked to prove the reflexive and symmetric parts of the theorem in Exercises 9 and 22, respectively.*

Theorem 4–4	**Congruence of triangles is reflexive, symmetric, and transitive.**

Proof of Theorem 4–4 (Transitive Part)

Given: $\triangle ABC \cong \triangle DEF$
$\triangle DEF \cong \triangle GHI$

Prove: $\triangle ABC \cong \triangle GHI$

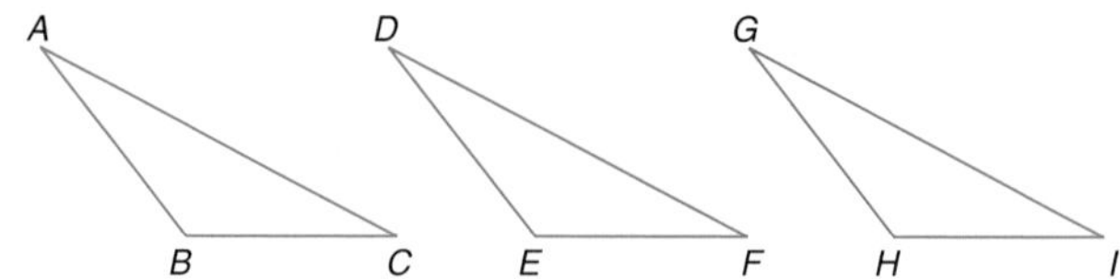

Proof:

Statements	**Reasons**
1. $\triangle ABC \cong \triangle DEF$	1. Given
2. $\angle A \cong \angle D$, $\angle B \cong \angle E$, $\angle C \cong \angle F$ $\overline{AB} \cong \overline{DE}$, $\overline{BC} \cong \overline{EF}$, $\overline{AC} \cong \overline{DF}$	2. CPCTC
3. $\triangle DEF \cong \triangle GHI$	3. Given
4. $\angle D \cong \angle G$, $\angle E \cong \angle H$, $\angle F \cong \angle I$ $\overline{DE} \cong \overline{GH}$, $\overline{EF} \cong \overline{HI}$, $\overline{DF} \cong \overline{GI}$	4. CPCTC
5. $\angle A \cong \angle G$, $\angle B \cong \angle H$, $\angle C \cong \angle I$	5. Congruence of angles is transitive.
6. $\overline{AB} \cong \overline{GH}$, $\overline{BC} \cong \overline{HI}$, $\overline{AC} \cong \overline{GI}$	6. Congruence of segments is transitive.
7. $\triangle ABC \cong \triangle GHI$	7. Definition of congruent triangles

CHECK FOR UNDERSTANDING

Communicating Mathematics

Study the lesson. Then complete the following.

1. **Draw** two triangles that are the same size and shape. What is the name given to these two triangles? congruent
2. **Explain** in your own words the meaning of *congruence of triangles is reflexive*. A triangle is congruent to itself.
3. **Name** the corresponding angles if $\triangle BCD \cong \triangle EFG$.
4. **Describe** how a congruence transformation would affect the perimeter of a triangle. Explain your reasoning. See margin.

5. **Look** around your school or community. Describe where congruent triangles are used as part of the architecture. Identify each set of congruent triangles as acute, obtuse, right, equiangular, scalene, isosceles, and/or equilateral. See students' work.

3. $\angle B \leftrightarrow \angle E$, $\angle C \leftrightarrow \angle F$, $\angle D \leftrightarrow \angle G$

Guided Practice

6. In the quilt design at the right, indicate which triangles appear to be congruent. 1, 8, 3, 4; 2, 5, 6, 7
7. If $\triangle ABC \cong \triangle XYZ$, name all the corresponding angles and corresponding sides. Use $\leftrightarrow$ to indicate each correspondence. Draw a figure showing the two triangles, and mark the corresponding parts.
8. Refer to the figure at the right. Complete the congruence statement.

   $\triangle LMN \cong \triangle$? ZYX

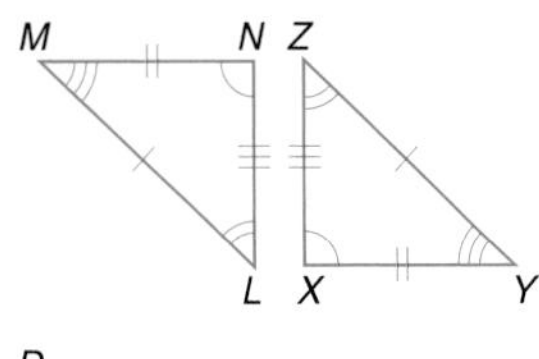

7. $\angle A \leftrightarrow \angle X$, $\angle B \leftrightarrow \angle Y$, $\angle C \leftrightarrow \angle Z$; $\overline{AB} \leftrightarrow \overline{XY}$, $\overline{AC} \leftrightarrow \overline{XZ}$, $\overline{BC} \leftrightarrow \overline{YZ}$; see margin for drawing.

9. Copy and complete the proof that congruence of triangles is reflexive.

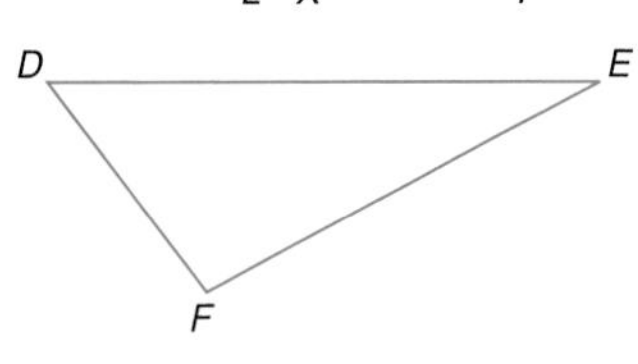

Given: $\triangle DEF$

Prove: $\triangle DEF \cong \triangle DEF$

Proof:

Statements	Reasons
1. $\triangle DEF$	1. ? Given
2. $\overline{DE} \cong \overline{DE}$, $\overline{EF} \cong \overline{EF}$, $\overline{FD} \cong \overline{FD}$	2. ?
3. $\angle D \cong \angle D$, $\angle E \cong \angle E$, $\angle F \cong \angle F$	3. ?
4. $\triangle DEF \cong \triangle DEF$	4. ?

10. Write a flow proof that congruence of triangles is reflexive.
11. Given $\triangle CAT \cong \triangle DOG$, $CA = 14$, $AT = 18$, $TC = 21$, and $DG = 2x + 7$.
    a. Draw and label a figure showing the congruent triangles. See margin.
    b. Find the value of x. 7
12. Draw two noncongruent triangles that have the same perimeter.

9. (reason 2) Congruence of segments is reflexive. (reason 3) Congruence of angles is reflexive. (reason 4) Definition of congruent triangles

10. See Solutions Manual.

12. See margin.

Reteaching

Using Diagrams Have students name the corresponding sides and angles in the figure below.

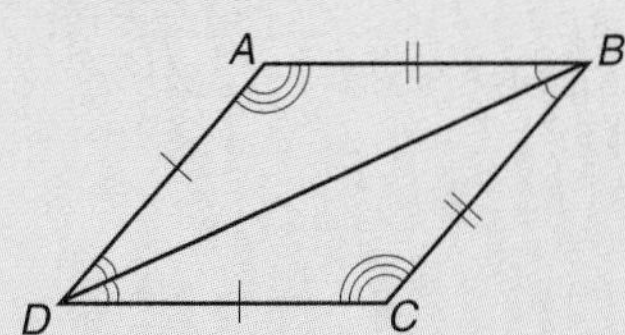

$\overline{AB} \leftrightarrow \overline{CB}$; $\overline{BD} \leftrightarrow \overline{BD}$; $\overline{DA} \leftrightarrow \overline{DC}$; $\angle ABD \leftrightarrow \angle CBD$; $\angle BDA \leftrightarrow \angle BDC$; $\angle DAB \leftrightarrow \angle DCB$

3 PRACTICE/APPLY

Check for Understanding

Exercises 1–14 are designed to help you assess your students' understanding through reading, writing, speaking, and modeling. You should work through Exercises 1–5 with your students and then monitor their work on Exercises 6–14.

Error Analysis

Students may confuse the corresponding parts of congruent triangles when the triangles are not drawn in the same position. These students may benefit from resketching the triangles on paper so that they have the same orientation.

Additional Answers

4. The perimeter would remain the same. A congruence transformation preserves size and shape.

7.

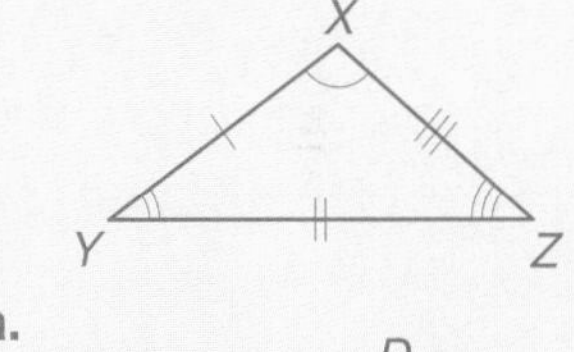

11a.

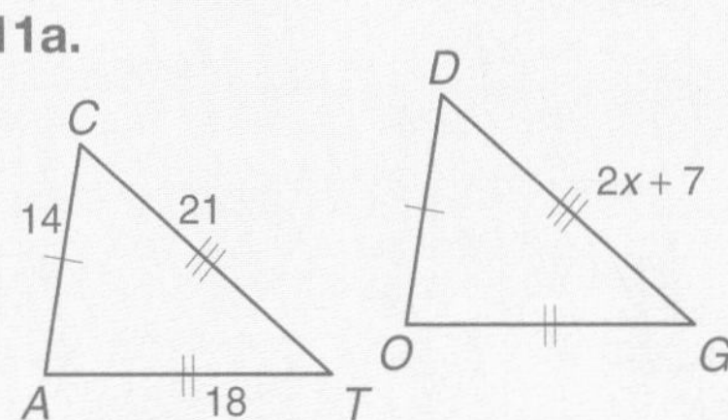

12. Sample answer:

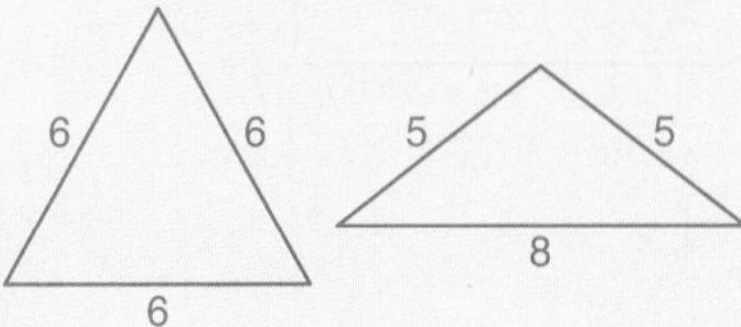

Robert Mangold was born in North Tonawanda, New York, in 1937. He studied at the Cleveland Institute of Art and at Yale University, where he received his B.A. and M.F.A degrees.

Assignment Guide

Core (with proof): 15–41 odd, 42–51
Core (informal): 15–21 odd, 25–35 odd, 39, 41–51
Enriched: 16–38 even, 39–51
All: Self Test, 1–10

For **Extra Practice**, see p. 771.

The red A, B, and C flags, printed only in the Teacher's Wraparound Edition, indicate the level of difficulty of the exercises.

Additional Answers

14. Sample answer: $\triangle CAE \cong \triangle EBG$, $\triangle CAD \cong \triangle EAD \cong \triangle EBF \cong \triangle GBF$

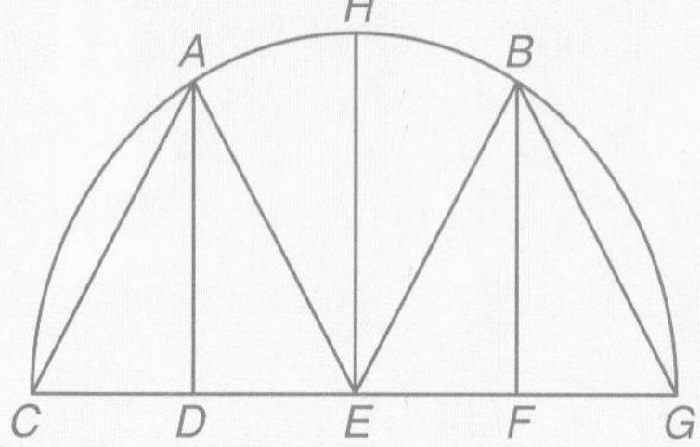

16. $\overline{BI} \leftrightarrow \overline{DE}$, $\overline{BG} \leftrightarrow \overline{DN}$, $\overline{IG} \leftrightarrow \overline{EN}$, $\angle B \leftrightarrow \angle D$, $\angle I \leftrightarrow \angle E$, $\angle G \leftrightarrow \angle N$

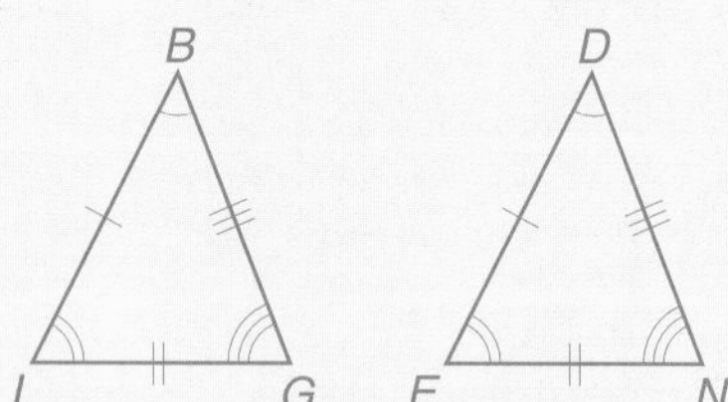

17. $\overline{PQ} \leftrightarrow \overline{RS}$, $\overline{PR} \leftrightarrow \overline{RT}$, $\overline{QR} \leftrightarrow \overline{ST}$, $\angle P \leftrightarrow \angle TRS$, $\angle Q \leftrightarrow \angle S$, $\angle PRQ \leftrightarrow \angle T$

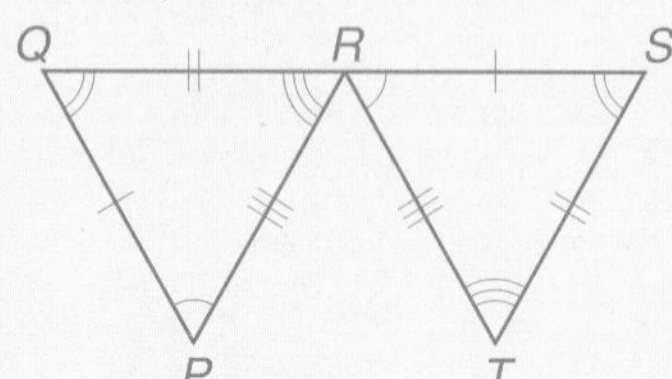

13. Write either a two-column or a flow proof for the following. See Solutions Manual.

Given: $\triangle MXR$ is a right isosceles triangle with $\angle X$ the vertex angle.
$\overline{XY} \perp \overline{MR}$
Y is the midpoint of $\overline{MR}$.
$\angle M \cong \angle R$
$\overline{YX}$ bisects $\angle MXR$.

Prove: $\triangle MXY \cong \triangle RXY$

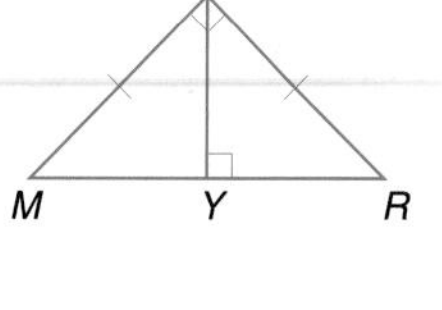

FYI

Robert Mangold (1937–) became interested in geometric art while he was working as a guard at the Museum of Modern Art in New York City.

14. **Art** Draw a sketch of the painting by Robert Mangold shown at the right. Then label the intersections of lines and name the triangles that appear to be congruent. See margin.

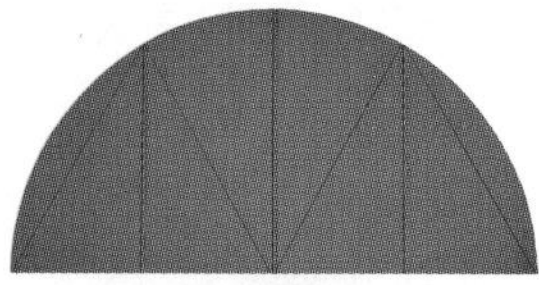

Half–W Series

EXERCISES

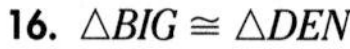

A Practice

15. In the quilt design at the right, indicate which triangles appear to be congruent. $\triangle HGA$, $\triangle KLA$, $\triangle JBA$, $\triangle IEA$, $\triangle HEA$, $\triangle KGA$, $\triangle JLA$, $\triangle IBA$; $\triangle HDE$, $\triangle KFG$, $\triangle JML$, $\triangle ICB$; $\triangle EDA$, $\triangle GFA$, $\triangle LMA$, $\triangle BCA$

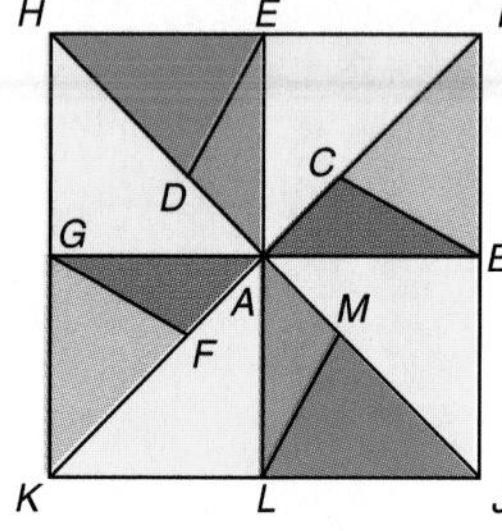

For each pair of congruent triangles, name all the corresponding sides and angles. Use ↔ to indicate each correspondence. Draw a figure for each pair of triangles, and mark the corresponding parts. 16–18. See margin.

16. $\triangle BIG \cong \triangle DEN$
17. $\triangle PQR \cong \triangle RST$
18. $\triangle EGO \cong \triangle PGO$

Complete each congruence statement.

19. $\triangle YZW \cong \triangle$ _?_ WXY
20. $\triangle MQN \cong \triangle$ _?_ OPN
21. $\triangle ARZ \cong \triangle$ _?_ ERG

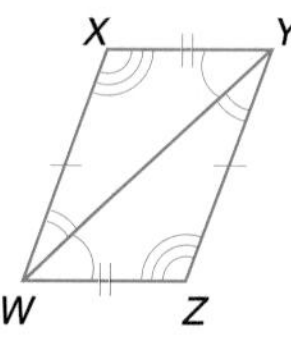

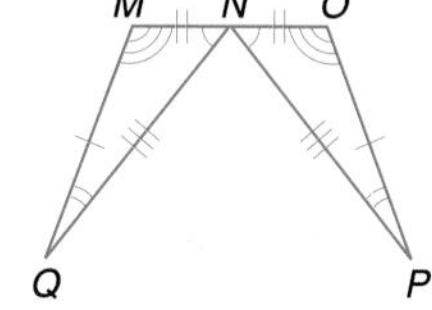

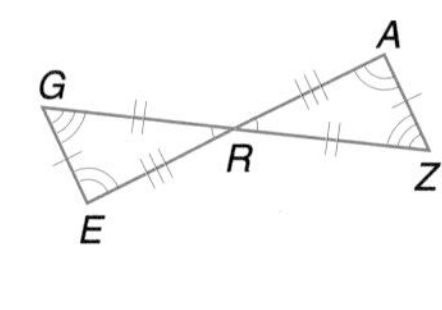

B Proof

22. Copy the flow proof of *Congruence of triangles is symmetric*. Provide the reasons for each statement. See margin.

Given: $\triangle RST \cong \triangle XYZ$

Prove: $\triangle XYZ \cong \triangle RST$

Proof:

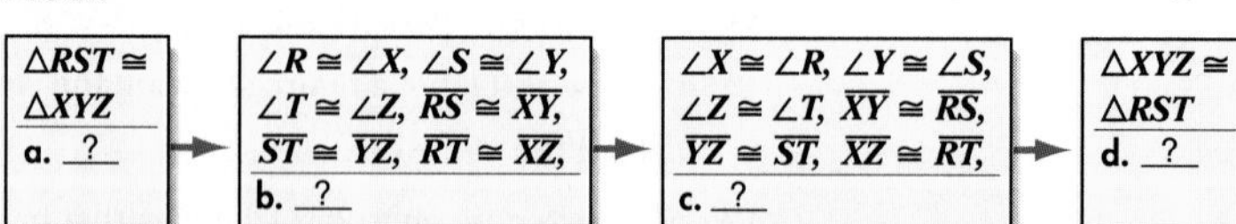

Additional Answers

18. $\overline{EG} \leftrightarrow \overline{PG}$, $\overline{EO} \leftrightarrow \overline{PO}$, $\overline{GO} \leftrightarrow \overline{GO}$, $\angle E \leftrightarrow \angle P$, $\angle EGO \leftrightarrow \angle PGO$, $\angle POG \leftrightarrow \angle EOG$

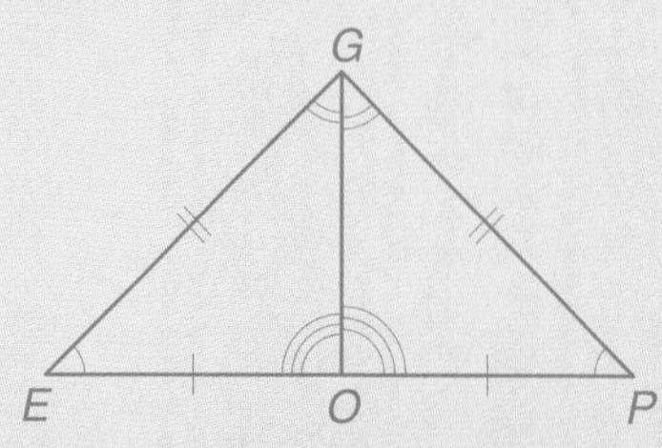

22. a. Given
b. CPCTC
c. Congruence of △s is symmetric. Congruence of segments is symmetric.
d. Definition of congruent triangles

23. Copy the flow proof and provide the reasons for each statement.

Given: $\overline{AB} \cong \overline{CD}$, $\overline{AD} \cong \overline{CB}$, $\overline{AD} \perp \overline{DC}$, $\overline{AB} \perp \overline{BC}$, $\overline{AD} \parallel \overline{BC}$, $\overline{AB} \parallel \overline{CD}$

Prove: $\triangle ACD \cong \triangle CAB$

Proof:

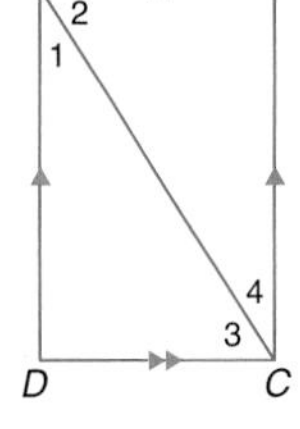

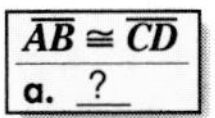

$\overline{AB} \cong \overline{CD}$
a. ?

$\overline{AD} \cong \overline{CB}$
b. ?

$\overline{AC} \cong \overline{CA}$
c. ?

$\overline{AD} \perp \overline{DC}$
d. ? → $\angle D$ is a rt. $\angle$.
e. ?

$\overline{AB} \perp \overline{BC}$
f. ? → $\angle B$ is a rt. $\angle$.
g. ?

$\angle D \cong \angle B$
h. ?

$\overline{AD} \parallel \overline{BC}$
i. ? → $\angle 1 \cong \angle 4$
j. ?

$\overline{AB} \parallel \overline{CD}$
k. ? → $\angle 2 \cong \angle 3$
l. ?

$\triangle ACD \cong \triangle CAB$
m. ?

23a. Given
23b. Given
23c. Congruence of segments is reflexive.
23d. Given
23e. Def. ⊥ lines
23f. Given
23g. Def. ⊥ lines
23h. All right ∠s are ≅.
23i. Given
23j. Alternate Interior Angle Theorem
23k. Given
23l. Alternate Interior Angle Theorem
23m. Def. ≅ △

24. Given $\triangle ABC \cong \triangle DEF$, $BC = 12$, $AB = 10$, $AC = 6$, and $EF = 2x - 4$.

a. Draw and label a figure to show the congruent triangles.

b. Find the value of x. 8

24a., 25a. See Solutions Manual.

25. Given $\triangle JKL \cong \triangle ABC$, $m\angle J = 36$, $m\angle B = 64$, and $m\angle C = 3x + 52$.

a. Draw and label a figure to show the congruent triangles.

b. Find the value of x.

25b. $\frac{28}{3}$

26. If $\triangle PQR \cong \triangle CDE$, PQ is 10 less than 3 times a number, PR is 2 less than twice the number, CE is 5 more than the number, CD is 4 more than the number, find PQ and CE. $PQ = 11$, $CE = 12$

Draw two triangles that have the following properties.

27–29. See margin.

27. equal areas and are congruent

28. equal perimeters and are congruent

29. equal areas but are not congruent

If $\triangle DNB \cong \triangle ANC$, determine if each statement is *true or not necessarily true.* Explain your reasoning. 30–35. See margin.

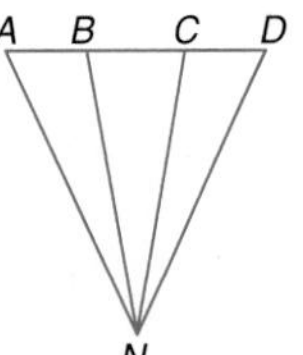

30. $\overline{DB} \cong \overline{AC}$

31. $\overline{DC} \cong \overline{AB}$

32. $\overline{DN} \cong \overline{CN}$

33. $\overline{CN} \cong \overline{BN}$

34. $\angle NAB \cong \angle NDC$

35. $\angle ANB \cong \angle DNC$

Additional Answers

27. Sample answer:

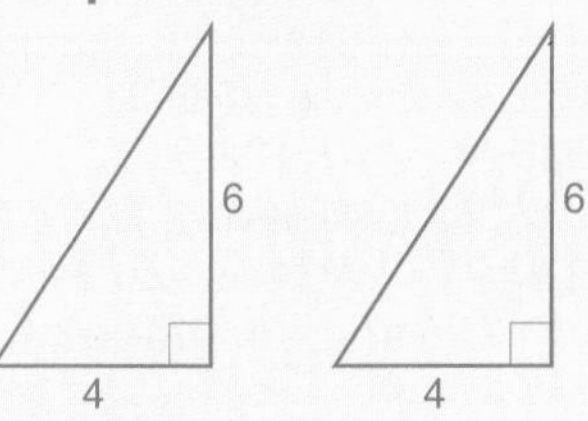

28. Sample answer:

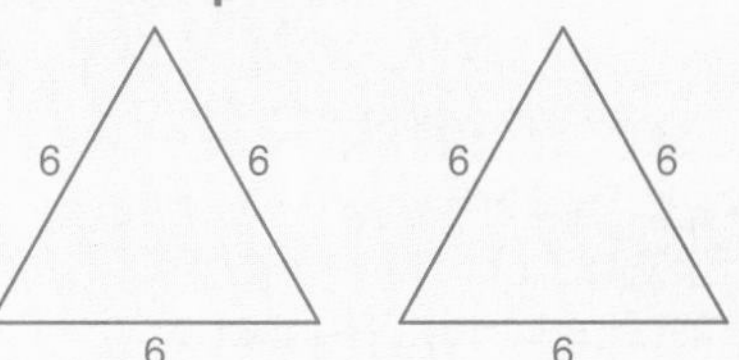

29. Sample answer:

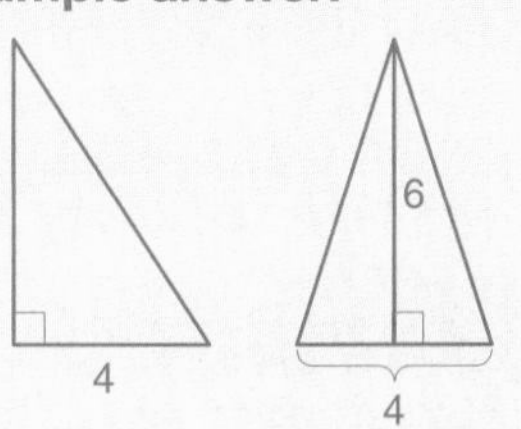

Study Guide Masters, p. 22

Exploring Congruent Triangles

When two figures have exactly the same shape and size, they are said to be congruent. For two congruent triangles there are three pairs of corresponding sides and three pairs of corresponding angles. To write a correspondence statement about congruent triangles, you should name corresponding angles in the same order. Remember that congruent parts are marked by identical markings.

Example: Write a correspondence statement for the triangles in the diagram.

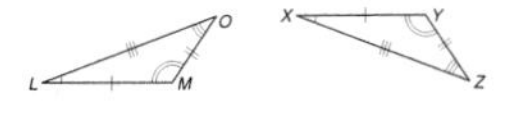

$\triangle LMO \cong \triangle XYZ$

Complete each correspondence statement.

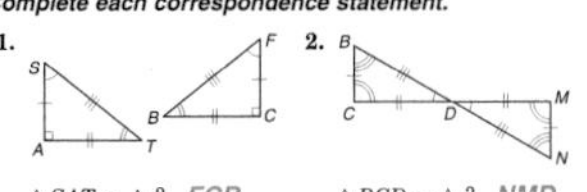

1. $\triangle SAT \cong \triangle$? FCB
2. $\triangle BCD \cong \triangle$? NMD
3. $\triangle GHK \cong \triangle$? GTK

Write a congruence statement for each pair of congruent triangles.

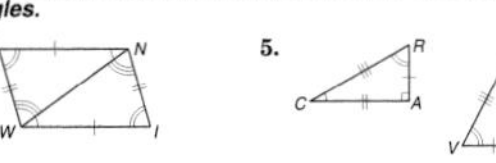

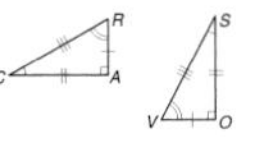

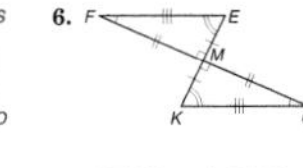

4. $\triangle TWN \cong \triangle INW$
5. $\triangle CAR \cong \triangle SOV$
6. $\triangle FEM \cong \triangle GKM$

Draw triangles △EDG and △QRS. Label the corresponding parts if △EDG ≅ △QRS. Then complete each statement.

7. $\angle E \cong$? $\angle Q$
8. $\overline{DG} \cong$? $\overline{RS}$
9. $\angle EDG \cong$? $\angle QRS$
10. $\overline{GE} \cong$? $\overline{SQ}$
11. $\overline{ED} \cong$? $\overline{QR}$
12. $\angle EGD \cong$? $\angle QSR$

Drawings may vary.

Additional Answers

30. true; CPCTC

31. true; Since $AD = AB + BD$ and $AD = AC + CD$, $AB + BD = AC + CD$. By CPCTC $\overline{BD} \cong \overline{AC}$, so $BD = AC$. By Subtraction Property (=), $AB = CD$ and $\overline{AB} \cong \overline{CD}$ or $\overline{DC} \cong \overline{AB}$.

32. Not necessarily true; they are not corresponding parts.

33. true; CPCTC

34. true; CPCTC

35. true; Since $m\angle AND = m\angle ANB + m\angle BND$ and $m\angle AND = m\angle ANC + m\angle CND$, $m\angle ANB + m\angle BND = m\angle ANC + m\angle CND$. By CPCTC $\angle BND \cong \angle ANC$, so $m\angle BND = m\angle ANC$. By Subtraction Property (=), $m\angle ANB = m\angle CND$ and $\angle ANB \cong \angle CND$.

Additional Answers

36. No; the sides of $\triangle MNO$ are not necessarily congruent to the sides of $\triangle PQR$.

38. Given: $\triangle ANG \cong \triangle NGA$
$\triangle NGA \cong \triangle GAN$
Prove: $\triangle AGN$ is equilateral and equiangular.

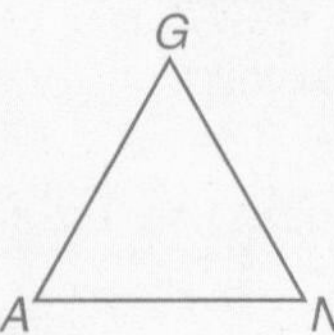

Proof:
Statements (Reasons)
1. $\triangle ANG \cong \triangle NGA$ (Given)
2. $\overline{AN} \cong \overline{NG}$, $\angle A \cong \angle N$ (CPCTC)
3. $\triangle NGA \cong \triangle GAN$ (Given)
4. $\overline{NG} \cong \overline{GA}$, $\angle N \cong \angle G$ (CPCTC)
5. $\overline{AN} \cong \overline{NG} \cong \overline{GA}$ (Congruence of segments is transitive.)
6. $\triangle AGN$ is equilateral. (Def. equilateral $\triangle$)
7. $\angle A \cong \angle N \cong \angle G$ ($\cong$ of $\angle$s is trans.)
8. $\triangle AGN$ is equiangular. (Def. equiangular $\triangle$)

Practice Masters, p. 22

NAME ______________________ DATE ____________

Practice

Student Edition Pages 196–203

Exploring Congruent Triangles

Label the corresponding parts if $\triangle RST \cong \triangle ABC$. Use the figures to complete each statement.

1. $\angle C \cong$? $\angle T$
2. $\angle R \cong$? $\angle A$
3. $\overline{AC} \cong$? $\overline{RT}$
4. $\overline{ST} \cong$? $\overline{BC}$
5. $\overline{RS} \cong$? $\overline{AB}$
6. $\angle B \cong$? $\angle S$

Write a congruence statement for the congruent triangles in each diagram. Sample triangle names are given.

7.

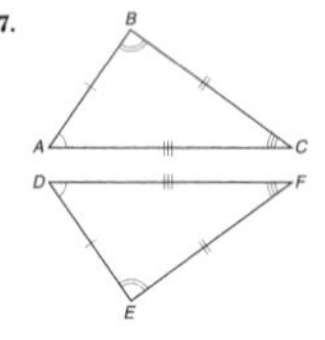

$\triangle ABC \cong \triangle DEF$

8.

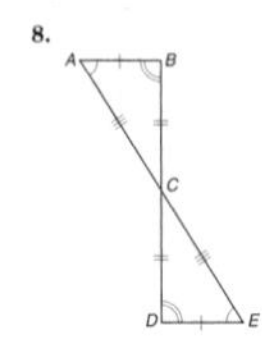

$\triangle ACB \cong \triangle ECD$

9. Given $\triangle ABC \cong \triangle DEF$, $AB = 15$, $BC = 20$, $AC = 25$, and $FE = 3x - 7$, find x. 9

10. Given $\triangle ABC \cong \triangle DEF$, $DE = 10$, $EF = 13$, $DF = 16$, and $AC = 4x - 8$, find x. 6

36. $\triangle MNO$ is both equilateral and equiangular. $\triangle PQR$ has three congruent sides, $m\angle P = 60$, and $m\angle Q = 60$. Is $\triangle MNO \cong \triangle PQR$? Explain your reasoning. See margin.

Proof

37. If $\triangle SIO \cong \triangle OIS$, prove $\triangle SIO$ is isosceles by using a flow proof. See Solutions Manual.

38. If $\triangle ANG \cong \triangle NGA$ and $\triangle NGA \cong \triangle GAN$, prove $\triangle AGN$ is both equilateral and equilangular by using a two-column proof. See margin.

Critical Thinking

39. On graph paper, draw six separate congruent right scalene triangles. Cut out the triangles. Arrange the triangles so that congruent sides fit together. Try several different arrangements.
   a. How many different shapes with four sides can you make? 4
   b. How many different shapes with three sides can you make? 0

Applications and Problem Solving

40. **Crafts** Use graph paper to design a quilt using congruent triangles. Classify the triangles used by name and identify those that can be proved congruent. See students' work.

41. $\triangle RQU \cong \triangle RSU$; $\triangle QUP \cong \triangle SUT$; $\triangle RUP \cong \triangle RUT$

41. **Construction** The figure below is a drawing of a truss roof. Using the given information below, name all triangles that are congruent.

Given: $\triangle RPT$ is isosceles with vertex $\angle R$.
U is the midpoint of $\overline{PT}$.
Q is the midpoint of $\overline{PR}$.
S is the midpoint of $\overline{RT}$.
$\overline{RU} \perp \overline{PT}$
$\angle PQU \cong \angle TSU$
$\overline{RU}$ bisects $\angle PRT$.
$\angle RPT \cong \angle RTP$

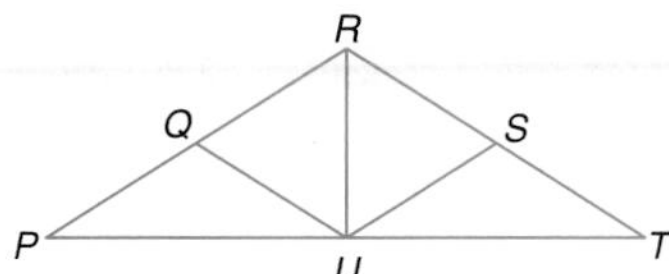

Mixed Review

42. **Construction** A brace shown at the right is used to keep a shelf perpendicular to the wall. If $m\angle CBA = 35$, find $m\angle BCD$. (Lesson 4–2) 125

43. Find the perimeter of equilateral $\triangle JLK$ if $JK = x + 3$ and $KJ = 2x - 5$. (Lesson 4–1) 33

44. Copy the figure and draw the segment that represents the distance from R to $\overline{PQ}$. (Lesson 3–5)

45. List the conclusions that can be drawn from the figure at the right. (Lesson 3–2) See margin.

46. Draw a figure to illustrate two intersecting lines with a plane parallel to both lines. (Lesson 3–1) See margin.

Additional Answers

45. $\angle 1 \cong \angle 7 \cong \angle 9$, $\angle 10 \cong \angle 8$, $\angle 3 \cong \angle 5 \cong \angle 6$, $\angle 2 \cong \angle 4$; and the following pairs of angles are supp.: $\angle 1$ and $\angle 10$, $\angle 7$ and $\angle 10$, $\angle 7$ and $\angle 8$, $\angle 8$ and $\angle 9$, $\angle 9$ and $\angle 10$, $\angle 1$ and $\angle 8$, $\angle 2$ and $\angle 6$, $\angle 2$ and $\angle 5$, $\angle 5$ and $\angle 4$, $\angle 4$ and $\angle 3$, $\angle 3$ and $\angle 2$, $\angle 6$ and $\angle 4$.

46.

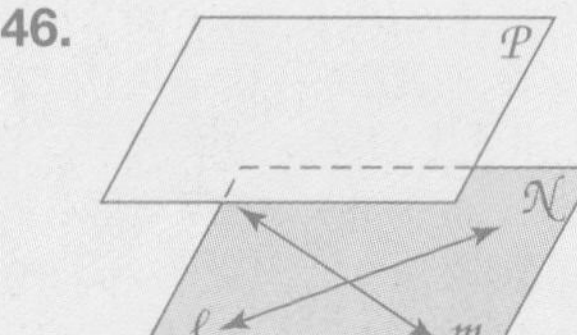

48. false; Sample counterexample: $a = -2, -2 < -\frac{1}{2}$

47. Algebra Name the algebraic property that justifies the statement *If $a = x$ and $a = y$, then $x = y$.* (Lesson 2–4) Substitution Property (=)

48. Determine if the following conjecture is *true* or *false*. Explain your answer and give a counterexample if false. (Lesson 2–1)

Given: a is a real number.

Conjecture: $a > \frac{1}{a}$

Algebra

49. $\overrightarrow{QT}$ and $\overrightarrow{QS}$ are opposite rays. Describe $\angle TQS$. (Lesson 1–6) a straight angle

50. Government The chart at the right shows the number of countries in the world listed in various sources.

a. Write a conclusion from the data given in the table. Justify your statement.

b. Would the average of these numbers have any meaning? Explain.

50a. Sample answer: More countries belong to the U.N. than the World Bank.

50b. No; since the data is not connected, the average of the numbers would have no meaning.

51. Solve $-10 < 3x - 1 \leq 5$. $\{x \mid -3 < x \leq 2\}$

The Number of Countries

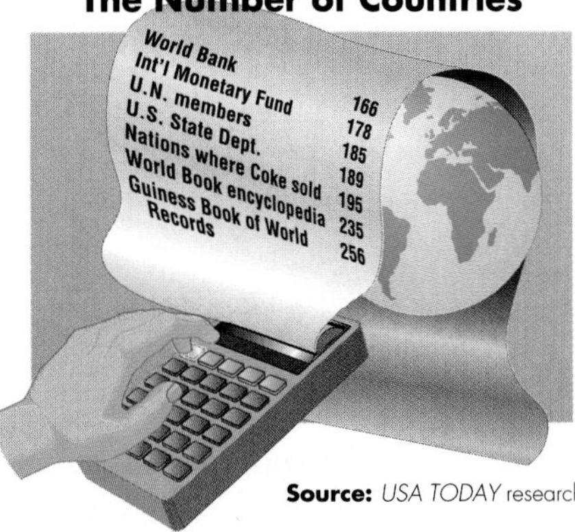

World Bank 166
Int'l Monetary Fund 178
U.N. members 185
U.S. State Dept. 189
Nations where Coke sold 195
World Book encyclopedia 235
Guiness Book of World Records 256

Source: *USA TODAY* research

SELF TEST

In figure *ACDE*, $\angle E$ and $\angle ADC$ are right angles and the congruent parts are indicated. (Lesson 4–1)

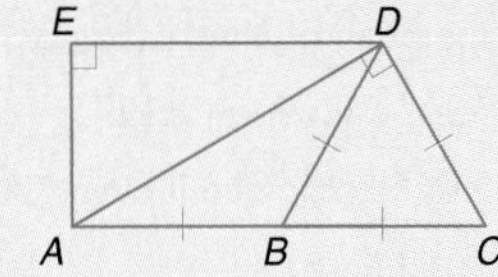

1. Identify any isosceles triangles. $\triangle DBC, \triangle ABD$
2. Identify any obtuse triangles. $\triangle ABD$
3. Identify any segments that are hypotenuses. $\overline{AD}, \overline{AC}$
4. Which segments are opposite $\angle C$? $\overline{BD}, \overline{AD}$

Find the value of *x*. (Lesson 4–2)

5. 107

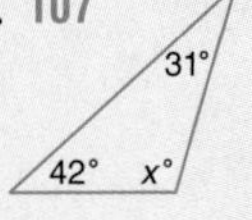

6. 51

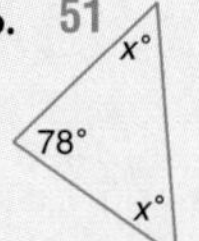

7. 125

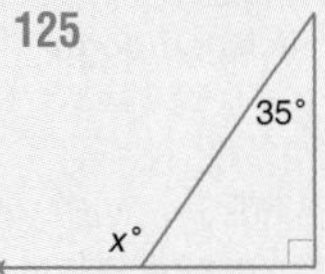

Complete each congruence statement. (Lesson 4–3)

8. $\triangle CDB \cong$ ___?___ $\triangle EAD$

9. $\triangle PTL \cong$ ___?___ $\triangle YZX$

B, A, C, D, E

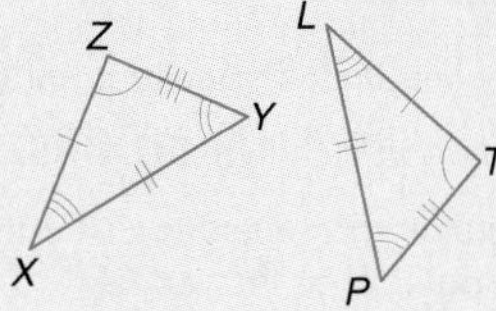

10. Given $\triangle BCD \cong \triangle RST$, $m\angle C = 57$, $m\angle R = 64$, and $m\angle D = 5x + 4$, find the value of x. (Lesson 4–3) 11

4 ASSESS

Closing Activity

Modeling Ask students to work in pairs; each student should have a ruler and protractor. One partner draws a triangle without the other seeing it. The other partner, using a ruler and protractor, draws a congruent triangle by following the verbal directions of the first partner. When the second triangle is complete, students compare the triangles.

Chapter 4 Quiz B (Lesson 4-3) is available in the *Assessment and Evaluation Masters*, p. 100.

Mid-Chapter Test (Lessons 4-1 through 4-3) is available in the *Assessment and Evaluation Masters*, p. 99.

Extension

Problem Solving Given $\triangle ABC \cong \triangle DEF$, $AB = 2x$, $AC = y$, $DE = y$, and $DF = 6 - x$. Find the values of x and y. $x = 2, y = 4$

SELF TEST

The Self Test provides students with a brief review of the concepts and skills in Lessons 4-1 through 4-3. Lesson numbers are given to the right of exercises or instruction lines so students can review concepts not yet mastered.

Enrichment Masters, p. 22

4-3 NAME ________ DATE ________

Enrichment — Student Edition Pages 196–203

Congruent Parts of Regular Polygonal Regions

Congruent figures are figures that have exactly the same size and shape. There are many ways to divide regular polygonal regions into congruent parts. Three ways to divide an equilateral triangular region are shown. You can verify that the parts are congruent by tracing one part, then rotating, sliding, or reflecting that part on top of the other parts.

1. Divide each square into four congruent parts. Use three different ways. Answers will vary. Sample answers are shown.

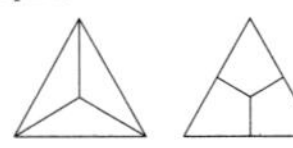

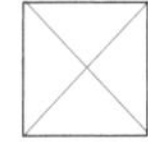
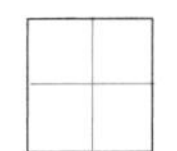
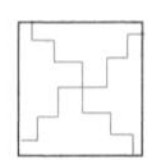

2. Divide each pentagon into five congruent parts. Use three different ways. Answers will vary. Sample answers are shown.

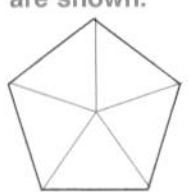
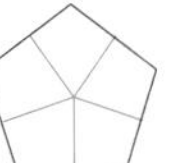
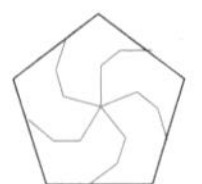

3. Divide each hexagon into six congruent parts. Use three different ways. Answers will vary. Sample answers are shown.

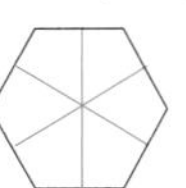

4. What hints might you give another student who is trying to divide figures like those into congruent parts? See students' work.

4-4A LESSON NOTES

NCTM Standards: 2, 4, 7

Objective
Draw congruent triangles.

Recommended Time
Demonstration and discussion: 30 minutes; Exercises: 30 minutes

Instructional Resources
For each student or group of students
Modeling Mathematics Masters
• p. 93 (worksheet)
For teacher demonstration
Algebra and Geometry Overhead Manipulative Resources

1 FOCUS

Motivating the Lesson
Ask students to list all the ways they could reproduce a triangle congruent to a given triangle. What materials do they need for each?

2 TEACH

Teaching Tip Point out to students that each step in an activity can be seen as a step in a proof.

4–4A Constructing Congruent Triangles

Materials: 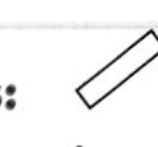 straightedge  protractor  compass scissors

A Preview of Lesson 4–4

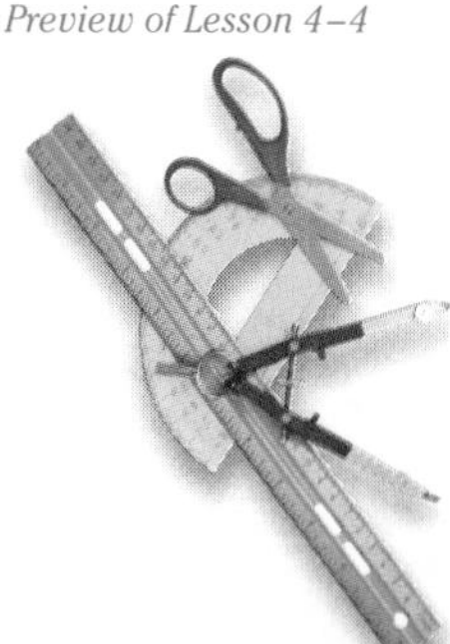

How can you draw congruent triangles? To investigate this question, use your protractor to draw a triangle with angle measures of 20, 40, and 120. Label your triangle ABC as shown at the right. How does your triangle compare to triangles drawn by other students?

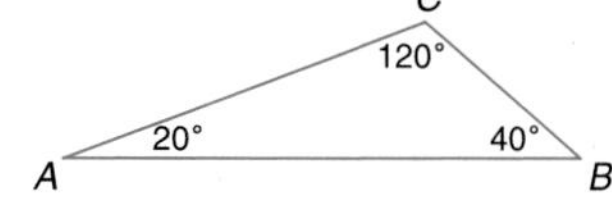

Use your $\triangle ABC$ to complete the activities.

Activity 1 **Construct a triangle in which the three sides are congruent to the three sides of $\triangle ABC$ and compare it to $\triangle ABC$.**

- Use a straightedge to draw any line ℓ and select a point D.
- Use a compass to construct $\overline{DE}$ on ℓ such that $\overline{DE} \cong \overline{AB}$.
- Using D as the center, draw an arc with radius equal to AC.
- Using E as the center, draw an arc with radius equal to BC.
- Let F be the point of intersection of the two arcs.
- Draw $\overline{DF}$ and $\overline{EF}$ to form $\triangle DEF$.
- Cut out $\triangle DEF$ and place it over $\triangle ABC$. How does $\triangle DEF$ compare to $\triangle ABC$? $\triangle DEF \cong \triangle ABC$

In a triangle, the angle formed by two given sides is called the **included angle** of the sides. In $\triangle ABC$, $\angle A$ is the included angle of sides $\overline{AB}$ and $\overline{AC}$.
$\angle B$ is also the included angle of sides $\overline{AB}$ and $\overline{CB}$, and $\angle C$ is the included angle of sides $\overline{AC}$ and $\overline{BC}$.

Activity 2 **Construct a triangle in which two sides are congruent to two sides of $\triangle ABC$ and the included angle is congruent to the included angle in $\triangle ABC$. Compare the triangle to $\triangle ABC$.**

- Use a straightedge to draw any line m and select a point G.
- Use a compass to construct $\overline{GH}$ on m such that $\overline{GH} \cong \overline{AB}$.
- Use a compass to construct an angle congruent to $\angle A$ using $\overline{GH}$ as a side of the angle and point G as the vertex.
- Use a compass to construct $\overline{GI}$ on the new side of the angle such that $\overline{GI} \cong \overline{AC}$.

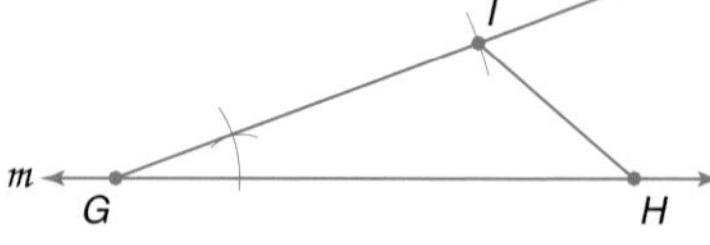

Using Cooperative Learning
This lesson offers an excellent opportunity for using cooperative learning groups. For more information on cooperative learning strategies and group management, see *Cooperative Learning in the Mathematics Classroom*, one of the titles in the Glencoe Mathematics Professional Series.

- Draw $\overline{HI}$ to complete $\triangle GHI$.
- Cut out $\triangle GHI$ and place it over $\triangle ABC$. How does $\triangle GHI$ compare to $\triangle ABC$? **$\triangle GHI \cong \triangle ABC$**

The side of a triangle that forms a side of two of its angles is called the **included side** of the angles. In $\triangle ABC$, $\overline{AB}$ is the included side of $\angle A$ and $\angle B$.
$\overline{AC}$ is the included side of $\angle A$ and $\angle C$.
$\overline{BC}$ is the included side of $\angle B$ and $\angle C$.

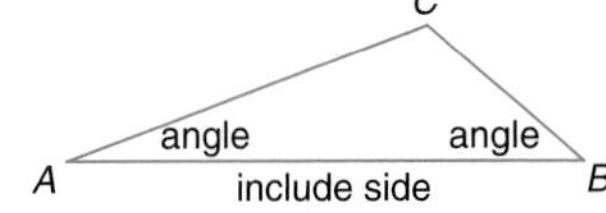

Activity 3 **Construct a triangle in which two angles are congruent to two angles of $\triangle ABC$ and the included side is congruent to the included side of $\triangle ABC$.**

- Use a straightedge to draw any line m and select a point J.
- Use a compass to draw $\overline{JK}$ such that $\overline{JK} \cong \overline{AB}$.
- Use a compass to construct an angle congruent to $\angle A$ using $\overrightarrow{JK}$ as a side of the angle and point J as the vertex.
- Use a compass to construct an angle congruent to $\angle B$ using $\overrightarrow{KJ}$ as a side of the angle and point K as the vertex.
- Label the point where the new sides of the angles meet L.
- Cut out $\triangle JKL$ and place it over $\triangle ABC$. How does $\triangle JKL$ compare to $\triangle ABC$? **$\triangle JKL \cong \triangle ABC$**

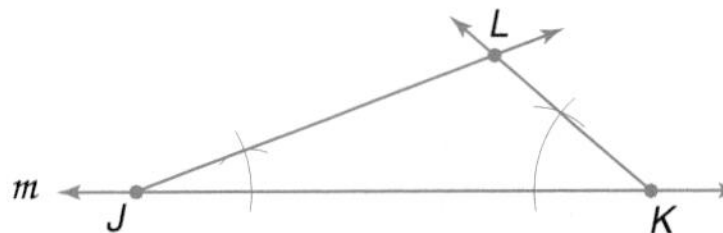

Model **Use $\triangle ABC$ from the beginning of the lesson.**

1. Construct a triangle in which two sides are congruent to $\overline{AC}$ and $\overline{BC}$ and the included angle is congruent to $\angle C$. How does this triangle compare to $\triangle ABC$? **See students' work. It is congruent to $\triangle ABC$.**
2. Construct a triangle in which two sides are congruent to $\overline{AB}$ and $\overline{BC}$ and the included angle is congruent to $\angle B$. How does this triangle compare to $\triangle ABC$? **See students' work. It is congruent to $\triangle ABC$.**
3. Construct a triangle in which two angles are congruent to $\angle A$ and $\angle C$ and the included side is congruent to $\overline{AC}$. How does this triangle compare to $\triangle ABC$? **See students' work. It is congruent to $\triangle ABC$.**
4. Construct a triangle in which two angles are congruent to $\angle B$ and $\angle C$ and the included side is congruent to $\overline{BC}$. How does this triangle compare to $\triangle ABC$? **See students' work. It is congruent to $\triangle ABC$.**

Write

5. Write a conjecture about two triangles in which the sides of one triangle are congruent to the sides of the other triangle. **If the sides of one triangle are congruent to the sides of another triangle, the triangles are congruent.**
6. Write a conjecture about two triangles in which two sides and the included angle of one triangle are congruent to two sides and the included angle of the other triangle. **See margin.**
7. Write a conjecture about two triangles in which two angles and the included side of one triangle are congruent to two angles and the included side of the other triangle. **See margin.**
8. What can you say about two triangles in which the angles of one triangle are congruent to the angles of the other triangle? **See margin.**

3 PRACTICE/APPLY

Assignment Guide

Core (with proof): 1–8
Core (informal): 1–8
Enriched: 1–8

Additional Answers

6. If two sides and the included angle of one triangle are congruent to two sides and the included angle of another triangle, the triangles are congruent.
7. If two angles and the included side of one triangle are congruent to two angles and the included side of another triangle, the triangles are congruent.
8. If the angles of one triangle are congruent to the angles of another triangle, the triangles have the same shape, but not necessarily the same size.

4 ASSESS

Observing students working in cooperative groups is an excellent method of assessment. You may wish to ask a student from each group to explain how to model and solve a problem. Also watch for and acknowledge students who are helping others understand the concept being taught.

4-4 LESSON NOTES

NCTM Standards: 1–5, 7, 8

Instructional Resources

- Study Guide Master 4-4
- Practice Master 4-4
- Enrichment Master 4-4
- Multicultural Activity Masters, p. 7

Transparency 4-4A contains the 5-Minute Check for this lesson; **Transparency 4-4B** contains a teaching aid for this lesson.

Recommended Pacing	
Standard Pacing	Days 7 & 8 of 14
Honors Pacing	Days 7 & 8 of 13
Block Scheduling*	Day 4 of 7

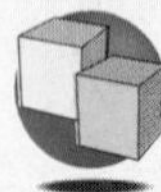

*For more information on pacing and possible lesson plans, refer to the *Block Scheduling Booklet*.

1 FOCUS

5-Minute Check
(over Lesson 4-3)

Refer to the figures below. $\triangle LMN \cong \triangle QRS$. Find each measure.

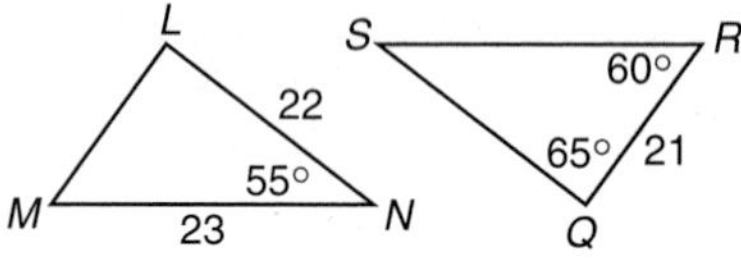

1. $m\angle L$ 65 **2.** $m\angle S$ 55
3. LM 21 **4.** QS 22
5. RS 23 **6.** $m\angle M$ 60

4-4 Proving Triangles Congruent

What YOU'LL LEARN

- To use SSS, SAS, and ASA Postulates to test for triangle congruence.

Why IT'S IMPORTANT

You can use postulates to determine if two triangles are congruent.

APPLICATION

Construction

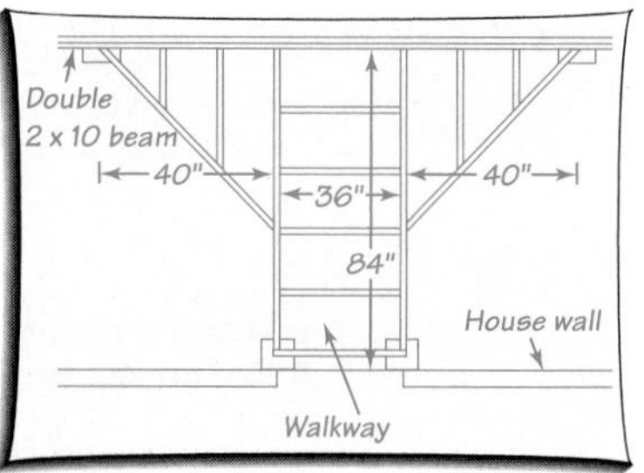

Is it always necessary to show that all of the corresponding parts of two triangles are congruent to be sure that the two triangles are congruent? The design of a deck at the right offers a carpenter many opportunities to use congruent triangles.

One method of guaranteeing a square corner in construction is called the 3-4-5 method. The 3-4-5 method lets the carpenter check a right angle using a tape measure instead of a framing square. The carpenter marks a point 3 feet from a corner along one side. Then he marks another point 4 feet from the corner along the other side. If these points are 5 feet apart, the corner forms a right angle. The 3-4-5 method is most useful when a carpenter is doing a large-scale layout that is beyond the accuracy of a framing square.

The carpenter measures several 3-4-5 triangles in the deck. All of these triangles are congruent. Only the measures of the sides were necessary to have congruent triangles.

Postulate 4–1 SSS Postulate (Side-Side-Side)	**If the sides of one triangle are congruent to the sides of a second triangle, then the triangles are congruent.**

The SSS Postulate can be used to prove that two triangles are congruent.

Example

INTEGRATION

Algebra

Given $\triangle STU$ with vertices $S(0, 5)$, $T(0, 0)$, and $U(-2, 0)$ and $\triangle XYZ$ with vertices $X(4, 8)$, $Y(4, 3)$, and $Z(6, 3)$, determine if $\triangle STU \cong \triangle XYZ$.

Use the distance formula to show that the corresponding sides are congruent.

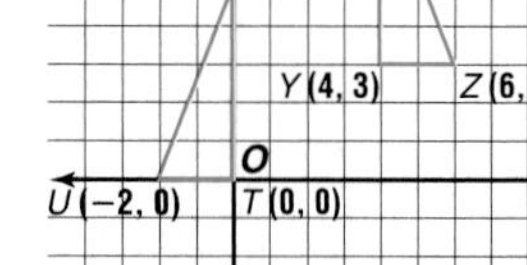

$ST = \sqrt{(0-0)^2 + (0-5)^2} = \sqrt{25}$ or 5
$TU = \sqrt{(-2-0)^2 + (0-0)^2} = \sqrt{4}$ or 2
$SU = \sqrt{(-2-0)^2 + (0-5)^2} = \sqrt{29}$

Notice that $\triangle XYZ$ is a flip and slide of $\triangle STU$.

$XY = \sqrt{(4-4)^2 + (3-8)^2} = \sqrt{25}$ or 5
$YZ = \sqrt{(6-4)^2 + (3-3)^2} = \sqrt{4}$ or 2
$XZ = \sqrt{(6-4)^2 + (3-8)^2} = \sqrt{29}$

$ST = XY$, $TU = YZ$, and $SU = XZ$, or by definition of congruent segments, all corresponding segments are congruent. Therefore, $\triangle STU \cong \triangle XYZ$ by SSS.

Will any other combinations of corresponding congruent sides and angles determine a unique triangle? Suppose you were given the measures of two sides and the angle that they form, which is called the **included angle**. According to the SAS Postulate, only one triangle can be formed with these given conditions.

Postulate 4–2 ***SAS Postulate*** ***(Side-Angle-Side)***	**If two sides and the included angle of one triangle are congruent to two sides and the included angle of another triangle, then the triangles are congruent.**

The following proof uses the SAS Postulate.

Example 2  Proof

Write a proof for the following.

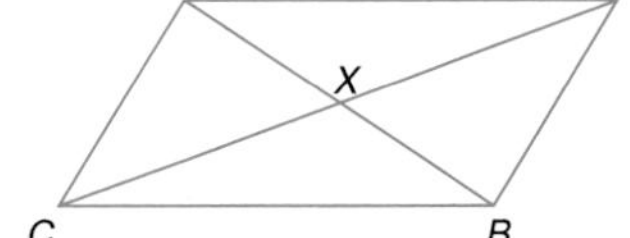

Given: X is the midpoint of $\overline{BD}$.
X is the midpoint of $\overline{AC}$.

Prove: $\triangle DXC \cong \triangle BXA$

Method 1: Two-Column Proof

Statements	**Reasons**
1. X is the midpoint of $\overline{BD}$.	**1.** Given
2. $\overline{DX} \cong \overline{BX}$	**2.** Definition of midpoint
3. X is the midpoint of $\overline{AC}$.	**3.** Given
4. $\overline{CX} \cong \overline{AX}$	**4.** Definition of midpoint
5. $\angle DXC \cong \angle BXA$	**5.** Vertical angles are congruent.
6. $\triangle DXC \cong \triangle BXA$	**6.** SAS

Method 2: Flow Proof

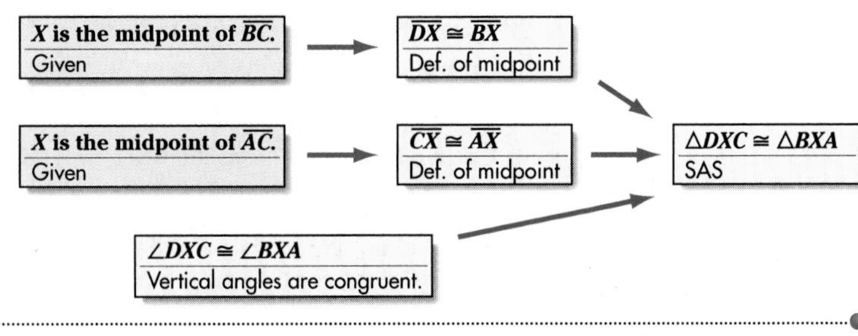

Suppose you were given the measures of two angles and the side of the triangle that forms a side of the two angles called the **included side**. According to the ASA Postulate, only one triangle can be formed with these given conditions.

Postulate 4–3 ***ASA Postulate*** ***(Angle-Side-Angle)***	**If two angles and the included side of one triangle are congruent to two angles and the included side of another triangle, the triangles are congruent.**

Motivating the Lesson

Hands-On Activity Give each student three straws measuring 4 cm, 5 cm, and 6 cm. Ask students to form as many different triangles from the three straws as they can. Have students compare their triangles with one another's. Ask them what conjecture they can make based on this activity. **If three sides of one triangle are congruent to three sides of another triangle, the triangles are congruent.**

2 TEACH

Teaching Tip Before students read through Example 1, you may want to write the distance formula on the chalkboard, explaining what each variable represents.

In-Class Examples

For Example 1

Given $\triangle PQR$ with vertices $P(3, 4)$, $Q(2, 2)$, and $R(7, 2)$ and $\triangle STU$ with vertices $S(6, -3)$, $T(4, -2)$, and $U(4, -7)$, show that $\triangle PQR \cong \triangle STU$. **The distance formula shows that $PQ = \sqrt{5}$, $ST = \sqrt{5}$; $PR = 2\sqrt{5}$, $SU = 2\sqrt{5}$; $QR = 5$, $TU = 5$. So $\triangle PQR \cong \triangle STU$ by SSS.**

For Example 2

Write a two-column proof.

Given: $\angle 1$ and $\angle 2$ are right angles, $\overline{ST} \cong \overline{TP}$.

Prove: $\triangle STR \cong \triangle PTR$

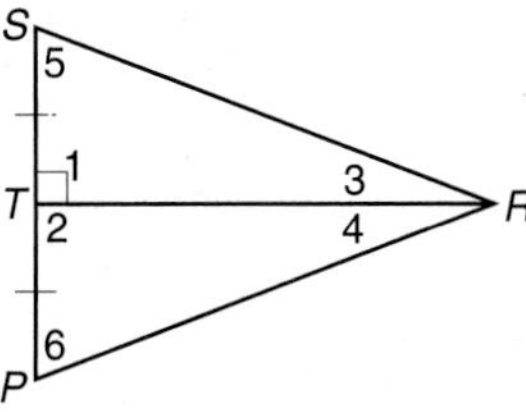

Statements (Reasons)

1. **$\angle 1$ and $\angle 2$ are right angles, $\overline{ST} \cong \overline{TP}$. (Given)**
2. **$\angle 1 \cong \angle 2$ (All rt. ∠s are ≅.)**
3. **$\overline{TR} \cong \overline{TR}$ (≅ of segments is reflexive.)**
4. **$\triangle STR \cong \triangle PTR$ (SAS)**

Alternative Learning Styles

Auditory Ask students to describe the difference between proving two triangles congruent by SAS and proving them congruent by ASA. **SAS requires 2 ≅ sides and a ≅ included angle. ASA requires 2 ≅ ∠s and a ≅ included side.**

Teaching Tip To be sure that students understand what is meant by the included angle, sketch a few triangles on the chalkboard and ask volunteers to mark two sides and their included angles.

In-Class Example

For Example 3

Write a two-column proof.

Given: $\overline{BE}$ bisects $\overline{AD}$, $\angle A \cong \angle D$.

Prove: $\overline{AB} \cong \overline{DC}$

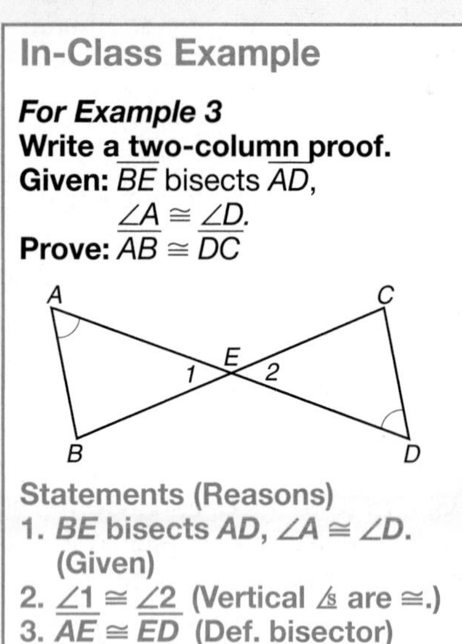

Statements (Reasons)

1. **$\overline{BE}$ bisects $\overline{AD}$, $\angle A \cong \angle D$. (Given)**
2. **$\angle 1 \cong \angle 2$ (Vertical ⦣s are ≅.)**
3. **$\overline{AE} \cong \overline{ED}$ (Def. bisector)**
4. **$\triangle AEB \cong \triangle DEC$ (ASA)**
5. **$\overline{AB} \cong \overline{DC}$ (CPCTC)**

3 PRACTICE/APPLY

Check for Understanding

Exercises 1–13 are designed to help you assess your students' understanding through reading, writing, speaking, and modeling. You should work through Exercises 1–5 with your students and then monitor their work on Exercises 6–13.

Error Analysis

Some students will think that two triangles can be proved congruent if any three parts of one are congruent to any three parts of another. Exercises 6–8 and 14–22 give practice in discerning the necessary conditions for two triangles to be proved congruent.

Some proofs ask you to show that a pair of corresponding parts of two triangles are congruent. Often you can do this by first proving that the two triangles are congruent. Then the definition of congruent triangles (CPCTC) can be used to show that the corresponding parts are congruent.

Example 3 Proof

Write a two-column proof.

Given: $\overline{VR} \perp \overline{RS}$
$\overline{UT} \perp \overline{SU}$
$\overline{RS} \cong \overline{US}$

Prove: $\overline{VR} \cong \overline{TU}$

Proof:

Statements	Reasons
1. $\overline{VR} \perp \overline{RS}$, $\overline{UT} \perp \overline{SU}$	1. Given
2. $\angle R$ is a right angle. $\angle U$ is a right angle.	2. Perpendicular lines form four right angles.
3. $\angle R \cong \angle U$	3. All right angles are congruent.
4. $\overline{RS} \cong \overline{US}$	4. Given
5. $\angle RSV \cong \angle UST$	5. Vertical angles are congruent.
6. $\triangle VRS \cong \triangle TUS$	6. ASA
7. $\overline{VR} \cong \overline{TU}$	7. CPCTC

CHECK FOR UNDERSTANDING

Communicating Mathematics

Study the lesson. Then complete the following.

1. **Draw** a triangle that has sides with lengths 3 inches, 4 inches, and 5 inches. Identify the triangle according to its angles. See students' work; right triangle.
2. **Explain** how CPCTC is useful when proving that parts of two triangles are congruent.

2. You can first prove the triangles congruent, then use the definition of congruent triangles to prove the corresponding parts are congruent.

3. **You Decide** Patty observes that the two triangles at the right have two pairs of congruent sides and a pair of congruent angles. She says the triangles are congruent by the SAS Postulate. Mirna says the triangles are not necessarily congruent. Who is correct? Explain your reasoning. See margin.

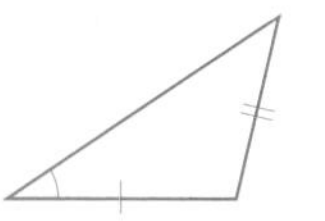
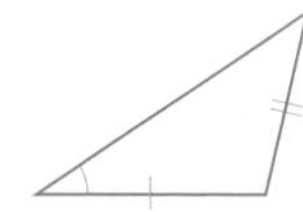

4. Refer to Example 3. Describe $\triangle TUS$ as a congruence transformation of $\triangle VRS$. $\triangle TUS$ is a rotation of $\triangle VRS$.

5. **Assess Yourself** Compare and contrast the SSS Postulate, the SAS Postulate, and the ASA Postulate. Give an example of each. See margin.

Additional Answers

3. Mirna is correct. The SAS postulate requires 2 sides and an included angle.
5. All are used to prove triangles congruent and require 3 parts congruent to 3 parts. One requires 3 sides, one 2 sides and an angle, and one 2 angles and a side. For sample examples, see Solutions Manual.

Reteaching

Using Reasoning Have students make up three examples, proving two triangles congruent by SSS, SAS, and ASA. Have them use a different sheet of paper for each example. Encourage them to make careful drawings for each. Ask them to keep these sheets for future reference.

Guided Practice

Determine which postulate can be used to prove the triangles are congruent. If it is not possible to prove that they are congruent, write *not possible*.

6.  ASA

7. 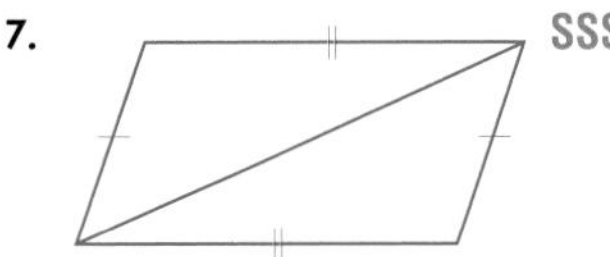 SSS

8. If the vertices of $\triangle ACM$ are $A(0, 4)$, $C(0, 0)$, and $M(3, 0)$ and the vertices of $\triangle SCR$ are $S(-4, 0)$, $C(0, 0)$, and $R(0, -4)$, determine if $\triangle ACM \cong \triangle SCR$. Justify your answer. See margin.

Proof

9. Supply the reason for each step in the two-column proof.

Given: $\overline{AD}$ bisects $\overline{BE}$.
$\overline{AB} \parallel \overline{DE}$

Prove: $\triangle ABC \cong \triangle DEC$

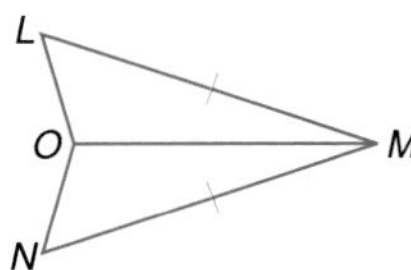

Proof:

Statements	Reasons
1. $\overline{AD}$ bisects $\overline{BE}$.	1. ? Given
2. $\overline{BC} \cong \overline{EC}$	2. ? Def. seg. bis.
3. $\overline{AB} \parallel \overline{DE}$	3. ? Given
4. $\angle B \cong \angle E$	4. ? Alt. Int. ∠ Th.
5. $\angle BCA \cong \angle ECD$	5. ? Vertical angles are ≅.
6. $\triangle ABC \cong \triangle DEC$	6. ? ASA

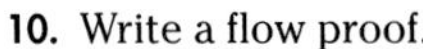

10. Write a flow proof.

Given: $\overline{OM}$ bisects $\angle LMN$.
$\overline{LM} \cong \overline{NM}$

Prove: $\triangle MOL \cong \triangle MON$

See Solutions Manual.

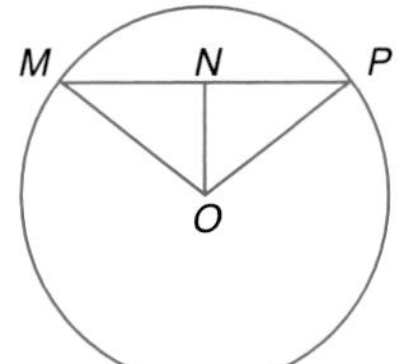

Use the figure at the right for Exercises 11–12. 11–12. See margin.

11. **Given:** $\overline{MO} \cong \overline{PO}$
$\overline{NO}$ bisects $\overline{MP}$.

Prove: $\triangle MNO \cong \triangle PNO$

12. **Given:** $\overline{NO}$ bisects $\angle POM$.
$\overline{NO} \perp \overline{MP}$

Prove: $\triangle MNO \cong \triangle PNO$

13. **Construction** According to the SSS Postulate, only one triangle can be formed with three fixed segments lengths. For this reason, a triangle is rigid and is frequently used to reinforce structures.

a. Which of the following frameworks is rigid? the two middle frameworks

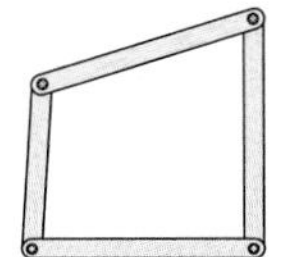
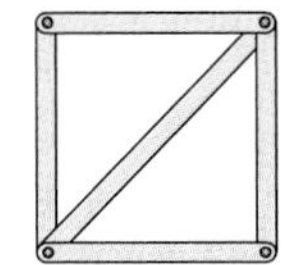
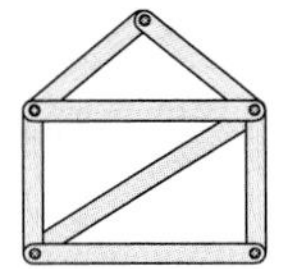
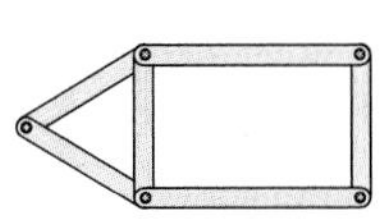

13b. See students' work.

b. Find three examples in which triangles are used to reinforce a structure.

Cooperative Learning

Think-Pair-Share Separate the class into groups of four or five. With ruler and protractor, have each person in the group produce a triangle based on SAS using 5 cm, 30°, and 6 cm. Have group members compare triangles. Repeat this activity for ASA using 35°, 4 cm, 65°; SSA using 5 cm, 6 cm, 45°. Ask which resulted in congruent triangles. SAS and ASA For more information on the think-pair-share strategy, see *Cooperative Learning in the Mathematics Classroom*, one of the titles in the Glencoe Mathematics Professional Series, pages 24–25.

Proof Pointer

Remind students that when they have difficulty with a flow proof, they can write a 2-column proof first and then use those steps to organize their flow proof.

Additional Answers

8. $AC = \sqrt{(0-0)^2 + (0-4)^2} = \sqrt{16} = 4$
$CM = \sqrt{(3-0)^2 + (0-0)^2} = \sqrt{9} = 3$
$AM = \sqrt{(3-0)^2 + (0-4)^2} = \sqrt{25} = 5$
$SC = \sqrt{[0-(-4)]^2 + (0-0)^2} = \sqrt{16} = 4$
$CR = \sqrt{(0-0)^2 + (-4-0)^2} = \sqrt{16} = 4$
$SR = \sqrt{[0-(-4)]^2 + (-4-0)^2} = \sqrt{32} = 4\sqrt{2}$
No; $\overline{AC} \cong \overline{SC}$, but the other sides are not congruent.

11. Given: $\overline{MO} \cong \overline{PO}$
$\overline{NO}$ bisects $\overline{MP}$.
Prove: $\triangle MNO \cong \triangle PNO$
Proof:
Statements (Reasons)
1. $\overline{MO} \cong \overline{PO}$, $\overline{NO}$ bisects $\overline{MP}$. (Given)
2. $\overline{MN} \cong \overline{PN}$ (Def. segment bisector)
3. $\overline{NO} \cong \overline{NO}$ (≅ of segments is reflexive.)
4. $\triangle MNO \cong \triangle PNO$ (SSS)

12. Given: $\overline{NO}$ bisects $\angle POM$.
$\overline{NO} \perp \overline{MP}$
Prove: $\triangle MNO \cong \triangle PNO$
Proof:
Statements (Reasons)
1. $\overline{NO}$ bisects $\angle POM$. (Given)
2. $\angle MON \cong \angle PON$ (Def. ∠ bisector)
3. $\overline{NO} \perp \overline{MP}$ (Given)
4. $\angle MNO$ and $\angle PNO$ are rt. ∡. (⊥ lines form 4 rt. ∡.)
5. $\angle MNO \cong \angle PNO$ (All rt. ∡ are ≅.)
6. $\overline{NO} \cong \overline{NO}$ (≅ of segments is reflexive.)
7. $\triangle MNO \cong \triangle PNO$ (ASA)

Assignment Guide

Core (with proof): 15–37 odd, 38–46
Core (informal): 15–23 odd, 35, 37–46
Enriched: 14–34 even, 35–46

For **Extra Practice**, see p. 771.

The red A, B, and C flags, printed only in the Teacher's Wraparound Edition, indicate the level of difficulty of the exercises.

Additional Answers

20. $RT = \sqrt{(5-2)^2 + (2-5)^2} = \sqrt{18} = 3\sqrt{2}$
$TY = \sqrt{(1-5)^2 + (1-2)^2} = \sqrt{17}$
$RY = \sqrt{(1-2)^2 + (1-5)^2} = \sqrt{17}$
$MG = \sqrt{[-7-(-4)]^2 + (1-4)^2} = \sqrt{18} = 3\sqrt{2}$
$GE = \sqrt{[-3-(-7)]^2 + (0-1)^2} = \sqrt{17}$
$ME = \sqrt{[-3-(-4)]^2 + (0-4)^2} = \sqrt{17}$
Yes; $\overline{RT} \cong \overline{MG}$, $\overline{TY} \cong \overline{GE}$, and $\overline{RY} \cong \overline{ME}$, so $\triangle RTY \cong \triangle MGE$ by SSS.

21. $PQ = \sqrt{[0-(-1)]^2 + [6-(-1)]^2} = \sqrt{50} = 5\sqrt{2}$
$QR = \sqrt{(2-0)^2 + (3-6)^2} = \sqrt{13}$
$PR = \sqrt{[2-(-1)]^2 + [3-(-1)]^2} = \sqrt{25} = 5$
$XY = \sqrt{(5-3)^2 + (3-1)^2} = \sqrt{8} = 2\sqrt{2}$
$YZ = \sqrt{(8-5)^2 + (1-3)^2} = \sqrt{13}$
$XZ = \sqrt{(8-3)^2 + (1-1)^2} = \sqrt{25} = 5$
No; $\overline{QR} \cong \overline{YZ}$ and $\overline{PR} \cong \overline{XZ}$, but $\overline{PQ}$ is not congruent to $\overline{XY}$.

EXERCISES

Practice

Determine which postulate can be used to prove the triangles are congruent. If it is not possible to prove that they are congruent, write *not possible*.

14. 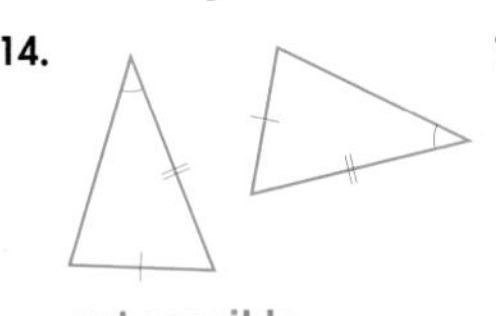 not possible

15. SAS

16. 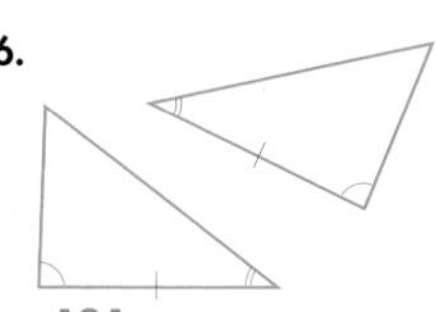 ASA

17. 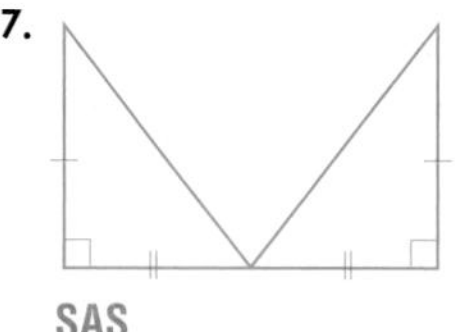 SAS

18.  not possible

19. 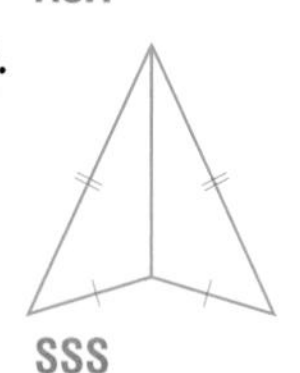 SSS

Determine if each pair of triangles is congruent. Justify your answer.

20. The vertices of $\triangle RTY$ are $R(2, 5)$, $T(5, 2)$, and $Y(1, 1)$, and the vertices of $\triangle MGE$ are $M(-4, 4)$, $G(-7, 1)$, and $E(-3, 0)$. See margin.

21. The vertices of $\triangle PQR$ are $P(-1, -1)$, $Q(0, 6)$, and $R(2, 3)$, and the vertices of $\triangle XYZ$ are $X(3, 1)$, $Y(5, 3)$, and $Z(8, 1)$. See margin.

22. The vertices of $\triangle TSR$ are $T(-1, -1)$, $S(-2, -2)$, and $R(-5, -1)$, and the vertices of $\triangle HND$ are $H(2, -1)$, $N(3, -2)$, and $D(2, -5)$. See margin.

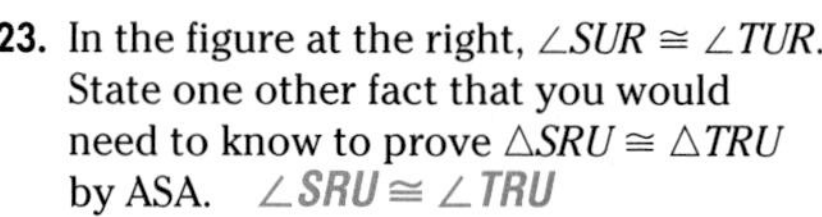

23. In the figure at the right, $\angle SUR \cong \angle TUR$. State one other fact that you would need to know to prove $\triangle SRU \cong \triangle TRU$ by ASA. $\angle SRU \cong \angle TRU$

Proof

24. Justify each step in the flow proof.

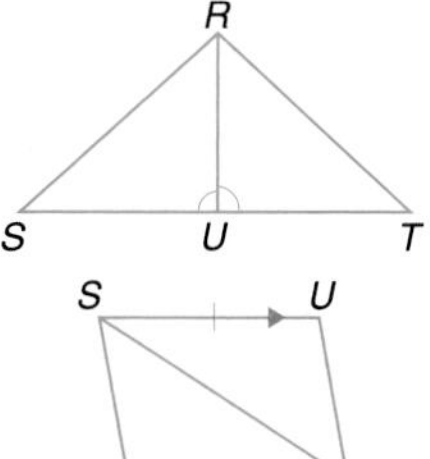

Given: $\overline{SU} \parallel \overline{AC}$
$\overline{SU} \cong \overline{AC}$

Prove: $\triangle SUA \cong \triangle ACS$

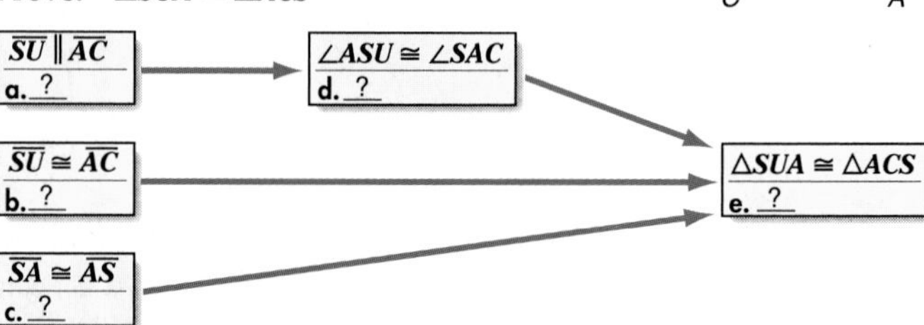

24a. Given
24b. Given
24c. Reflexive Prop. ≅ Segments
24d. Alt. Int. ∠ Th.
24e. SAS

Write a proof. 25–26. See Solutions Manual.

25. **Given:** $\angle J \cong \angle L$
B is the midpoint of $\overline{JL}$.
Prove: $\triangle JHB \cong \triangle LCB$

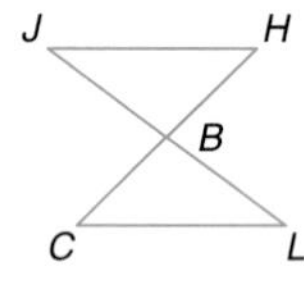

26. **Given:** $\overline{JL}$ bisects $\overline{HC}$.
$\overline{HC}$ bisects $\overline{JL}$.
Prove: $\triangle JHB \cong \triangle LCB$

Additional Answer

22. $TS = \sqrt{[-2-(-1)]^2 + [-2-(-1)]^2} = \sqrt{2}$
$SR = \sqrt{[-5-(-2)]^2 + [-1-(-2)]^2} = \sqrt{10}$
$TR = \sqrt{[-5-(-1)]^2 + [-1-(-1)]^2} = \sqrt{16} = 4$
$HN = \sqrt{(3-2)^2 + [-2-(-1)]^2} = \sqrt{2}$
$ND = \sqrt{(2-3)^2 + [-5-(-2)]^2} = \sqrt{10}$
$HD = \sqrt{(2-2)^2 + [-5-(-1)]^2} = \sqrt{16} = 4$
Yes; $\overline{TS} \cong \overline{HN}$, $\overline{SR} \cong \overline{ND}$, and $\overline{TR} \cong \overline{HD}$, so $\triangle TSR \cong \triangle HND$ by SSS.

For over 1000 years, the ancient Japanese art of *origami* has used small pieces of square or rectangular paper folded into artistic reproductions of birds, boats, hats, frogs or a thousand other designs. When step-by-step diagrams are shown to illustrate the progression of folds necessary to complete a figure, it is easy to see the myriad types of triangles used in origami to achieve the desired effect.

Write a two-column proof. 27–28. See margin.

27. Given: $\triangle MGR$ is an isosceles triangle with vertex $\angle MGR$.
K is the midpoint of $\overline{MR}$.

Prove: $\triangle MGK \cong \triangle RGK$

28. Given: $\overline{GK} \perp \overline{MR}$
$\overline{GK}$ bisects $\overline{MR}$.

Prove: $\triangle MGK \cong \triangle RGK$

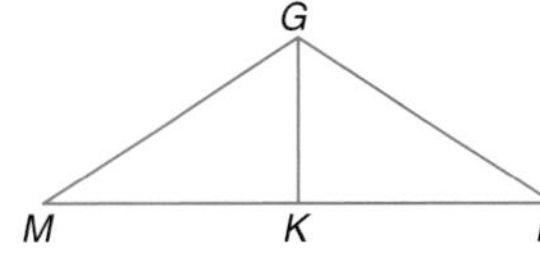

Use quadrilateral *CDRL* for Exercises 29–30. 29–30. See Solutions Manual.

29. Given: $\overline{RL} \parallel \overline{DC}$
$\overline{LC} \parallel \overline{RD}$

Prove: $\angle R \cong \angle C$

30. Given: $\angle 4 \cong \angle 2$
$\overline{DR} \cong \overline{LC}$

Prove: $\overline{RL} \cong \overline{CD}$

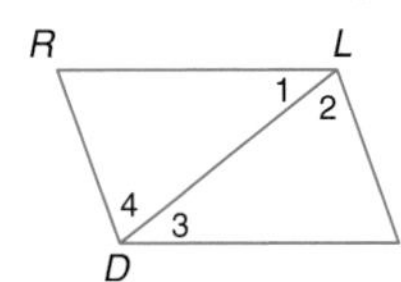

Use $\triangle LRW$ for Exercises 31–34. 31–34. See Solutions Manual.

31. Given: $\angle 1 \cong \angle 2$
$\angle 3 \cong \angle 4$
$\overline{LA} \cong \overline{RU}$

Prove: $\triangle WLU \cong \triangle WRA$

32. Given: $\angle LWU \cong \angle RWA$
$\angle 3 \cong \angle 4$
$\overline{LW} \cong \overline{RW}$

Prove: $\triangle LWA \cong \triangle RWU$

33. Given: $\angle 5 \cong \angle 7$
$\angle 3 \cong \angle 4$
$\overline{LW} \cong \overline{RW}$

Prove: $\overline{LU} \cong \overline{RA}$

34. Given: $\overline{LU} \cong \overline{RA}$
$\angle 3 \cong \angle 4$
$\overline{LW} \cong \overline{RW}$

Prove: $\overline{AW} \cong \overline{UW}$

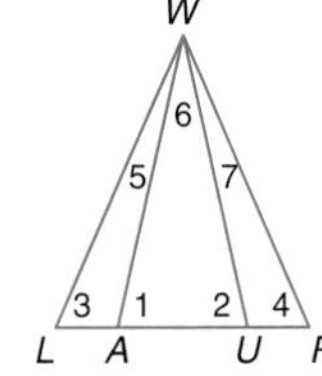

Critical Thinking

35. $\angle A$ is not the included angle.

35. Explain why the following proof is not valid.

Given: $\overline{TI} \cong \overline{TN}$

Prove: $\triangle ATI \cong \triangle ATN$

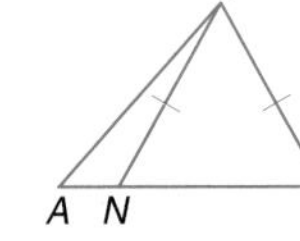

Proof:

Statements	Reasons
1. $\overline{TI} \cong \overline{TN}$	1. Given
2. $\angle A \cong \angle A$	2. Congruence of angles is reflexive.
3. $\overline{AT} \cong \overline{AT}$	3. Congruence of segments is reflexive.
4. $\triangle ATI \cong \triangle ATN$	4. SAS

GLOBAL CONNECTIONS

There are two categories of origami. One is used in ceremonial etiquette and the other to represent real objects. There are many books about origami, and the paper is available in many stores.

Additional Answer

27. Given: $\triangle MGR$ is an isosceles triangle with vertex $\angle MGR$.
K is the midpoint of $\overline{MR}$.
Prove: $\triangle MGK \cong \triangle RGK$
Proof:
Statements (Reasons)
1. $\triangle MGR$ is an isosceles triangle with vertex $\angle MGR$. (Given)
2. $\overline{GM} \cong \overline{GR}$ (Def. isosceles $\triangle$)
3. K is the midpoint of $\overline{MR}$. (Given)
4. $\overline{MK} \cong \overline{RK}$ (Mdpt. Th.)
5. $\overline{GK} \cong \overline{GK}$ ($\cong$ of segments is reflexive.)
6. $\triangle MGK \cong \triangle RGK$ (SSS)

Study Guide Masters, p. 23

NAME______________ DATE________

4-4 **Study Guide** Student Edition Pages 206–213

Proving Triangles Congruent

You can show two triangles are congruent with the following:

SSS Postulate (Side–Side–Side)	Three sides of one triangle are congruent to the sides of a second triangle.
SAS Postulate (Side–Angle–Side)	Two sides and the included angle of one triangle are congruent to two sides and an included angle of another triangle.
ASA Postulate (Angle–Side–Angle)	Two angles and the included side of one triangle are congruent to two angles and the included side of another triangle.

Examples: Determine whether each pair of triangles are congruent. If they are congruent, indicate the postulate that can be used to prove their congruence.

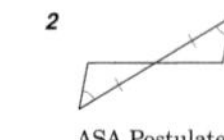

1 SAS Postulate 2 ASA Postulate 3 not congruent

Determine which postulate can be used to prove the triangles are congruent. If it is not possible to prove that they are congruent, write not possible.

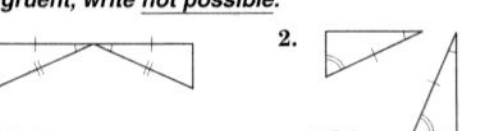

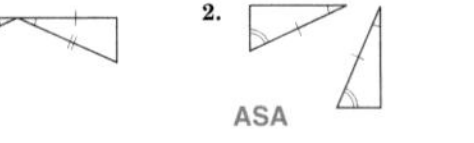

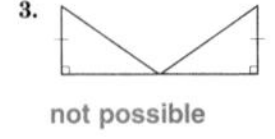

1. SAS 2. ASA 3. not possible

Mark all congruent parts in each figure, complete the prove statement, and identify the postulate that proves their congruence.

4.

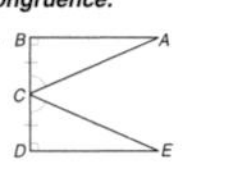

Given: $\angle BCA \cong \angle DCE$
$\angle B$ and $\angle D$ are right angles.
$\overline{BC} \cong \overline{CD}$
Prove: $\triangle CAB \cong \triangle CED$
ASA

5. X Y Z P Q

Given: $\overline{XY} \cong \overline{YZ}$
$\overline{PY} \cong \overline{QY}$
$\overline{XP} \cong \overline{ZQ}$
Prove: $\triangle XYP \cong \triangle ZYQ$
SSS

Additional Answer

28. Given: $\overline{GK} \perp \overline{MR}$
$\overline{GK}$ bisects $\overline{MR}$.
Prove: $\triangle MGK \cong \triangle RGK$
Proof:
Statements (Reasons)
1. $\overline{GK} \perp \overline{MR}$ (Given)
2. $\angle GKM$ and $\angle GKR$ are rt. $\angle$s. ($\perp$ lines form 4 rt. $\angle$s.)
3. $\angle GKM \cong \angle GKR$ (All rt. $\angle$s are $\cong$.)
4. $\overline{GK}$ bisects $\overline{MR}$. (Given)
5. $\overline{MK} \cong \overline{RK}$ (Def. segment bisector)
6. $\overline{GK} \cong \overline{GK}$ ($\cong$ of segments is reflexive.)
7. $\triangle MGK \cong \triangle RGK$ (SAS)

Additional Answers

37. Since Jamal is perpendicular to the ground, two right triangles are formed and the two right angles are congruent. The angles of sight are the same and his height is the same, so the triangles are congruent by ASA. By CPCTC, the distances are the same and the method is valid.

38a.

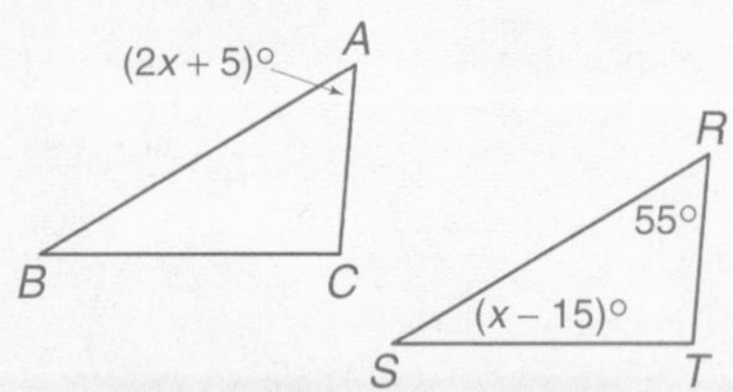

42. Converse: If an angle is acute, then its measure is 60; false; sample counterexample: An acute angle could measure 20. Inverse: If the measure of an angle is not 60, then the angle is not acute; false; sample counterexample: An angle that measures 20 is acute. Contrapositive: If an angle is not acute, then its measure is not 60; true.

Practice Masters, p. 23

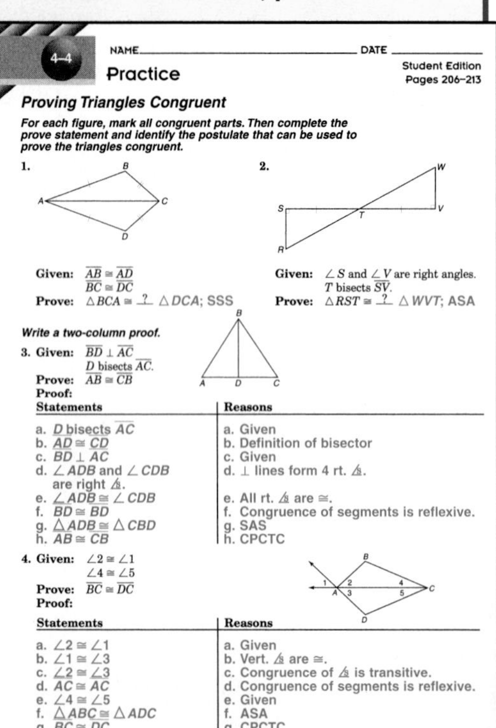

4-4 NAME__________ DATE ______

Practice — Student Edition Pages 206–213

Proving Triangles Congruent

For each figure, mark all congruent parts. Then complete the prove statement and identify the postulate that can be used to prove the triangles congruent.

1. **Given:** $\overline{AB} \cong \overline{AD}$, $\overline{BC} \cong \overline{DC}$
Prove: $\triangle BCA \cong$ _?_ $\triangle DCA$; SSS

2. **Given:** $\angle S$ and $\angle V$ are right angles. T bisects $\overline{SV}$.
Prove: $\triangle RST \cong$ _?_ $\triangle WVT$; ASA

Write a two-column proof.

3. Given: $\overline{BD} \perp \overline{AC}$, D bisects $\overline{AC}$.
Prove: $\overline{AB} \cong \overline{CB}$
Proof:

Statements	Reasons
a. D bisects $\overline{AC}$	a. Given
b. $\overline{AD} \cong \overline{CD}$	b. Definition of bisector
c. $\overline{BD} \perp \overline{AC}$	c. Given
d. $\angle ADB$ and $\angle CDB$ are right ∠s.	d. ⊥ lines form 4 rt. ∠s.
e. $\angle ADB \cong \angle CDB$	e. All rt. ∠s are ≅.
f. $\overline{BD} \cong \overline{BD}$	f. Congruence of segments is reflexive.
g. $\triangle ADB \cong \triangle CBD$	g. SAS
h. $\overline{AB} \cong \overline{CB}$	h. CPCTC

4. Given: $\angle 2 \cong \angle 1$, $\angle 4 \cong \angle 5$
Prove: $\overline{BC} \cong \overline{DC}$
Proof:

Statements	Reasons
a. $\angle 2 \cong \angle 1$	a. Given
b. $\angle 1 \cong \angle 3$	b. Vert. ∠s are ≅.
c. $\angle 2 \cong \angle 3$	c. Congruence of ∠s is transitive.
d. $\overline{AC} \cong \overline{AC}$	d. Congruence of segments is reflexive.
e. $\angle 4 \cong \angle 5$	e. Given
f. $\triangle ABC \cong \triangle ADC$	f. ASA
g. $\overline{BC} \cong \overline{DC}$	g. CPCTC

Applications and Problem Solving

36. Recreation Tapatan is a game played in the Philippines on a square board as shown at the right. The players take turns placing each of their three pieces on a different point of intersection. After all the pieces have been played, the players take turns moving a piece along a line to another intersection. A piece cannot jump over another piece. A player who gets all his or her pieces in a straight line wins the game. Point E bisects all four line segments that pass through it. **a. $\triangle GHE \cong \triangle CBE$ by SAS**

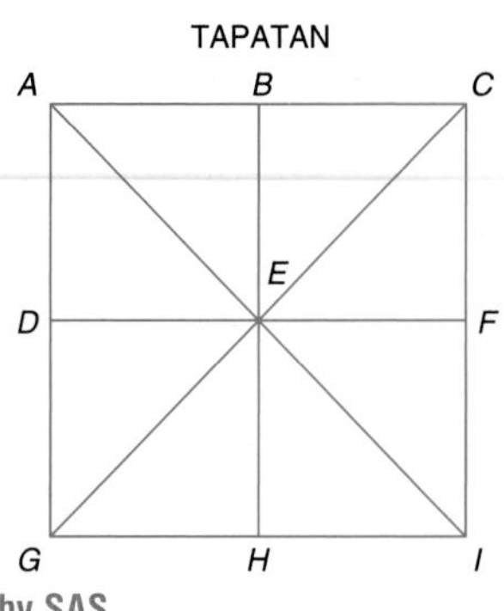

a. What can you say about $\triangle GHE$ and $\triangle CBE$? Explain your reasoning.
b. What can you say about $\triangle AEG$ and $\triangle IEG$? Explain your reasoning.
c. What can you say about $\triangle ACI$ and $\triangle CAG$? Explain your reasoning. **$\triangle ACI \cong \triangle CAG$ by SAS**

36b. $\triangle AEG \cong \triangle IEG$ by SSS

37. Estimation Jamal observes a duckling on the other side of the river. He wishes to estimate his distance from the duckling. Jamal adjusts the visor of his cap so that it is in line with his line of sight to the duckling. He keeps his neck stiff and turns his body. Using the visor of his cap, he uses the same line of sight to observe another point on the ground. He then paces out the distance to the new point. Jamal concludes that his distance to the duckling was the same as the distance he just paced out. Do you agree with Jamal's method of estimation? Explain your reasoning. **See margin.**

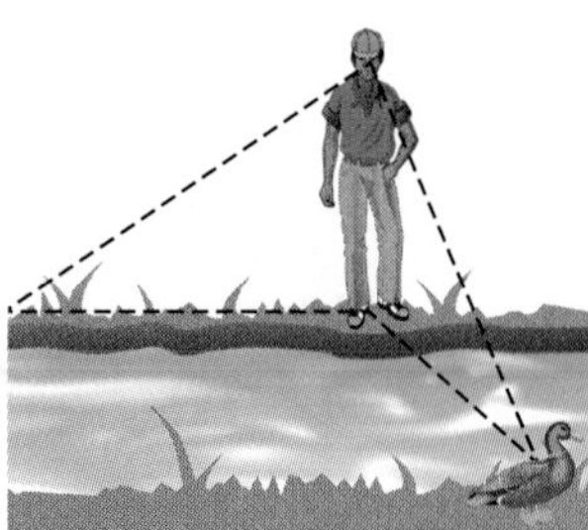

Mixed Review

38. Given $\triangle ABC \cong \triangle RST$, $m\angle A = 2x + 5$, $m\angle S = x - 15$, and $m\angle R = 55$. (Lesson 4–3) **a. See Solutions Manual.**
a. Draw and label a figure to show the congruent triangles.
b. Find the value of x. **25**
c. Find the measure of each of the angles in the triangles. **$m\angle A = 55$, $m\angle B = 10$, $m\angle C = 115$, $m\angle R = 55$, $m\angle S = 10$, $m\angle T = 115$**

39. Refer to the figure at the right. Find the value of x. (Lesson 4–2) **30**

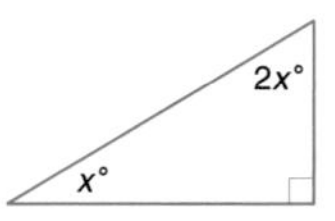

40. Triangle ROM is an isosceles triangle with the congruent sides as marked. Name each of the following. (Lesson 4–1)
a. sides **$\overline{RO}$, $\overline{RM}$, $\overline{OM}$**
b. angles **$\angle R$, $\angle O$, $\angle M$**
c. vertex angle **$\angle R$**
d. base angles **$\angle O$, $\angle M$**
e. side opposite $\angle R$ **$\overline{OM}$**
f. angle opposite $\overline{OR}$ **$\angle M$**

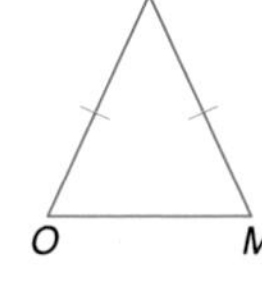

Tech Prep

Travel Agent Students interested in travel may wish to learn about a career as a licensed travel agent. Travel agents book travel agendas for their customers. This requires knowledge of geography and computers. Community colleges offer 2-year programs or travel schools offer 16–18 week concentrated courses. For more information on tech prep, see the *Teacher's Handbook*.

41. $\frac{1}{10}$

41. **Algebra** State the slope of the line parallel to the line passing through (3, 9) and (−7, 8). (Lesson 3–3)

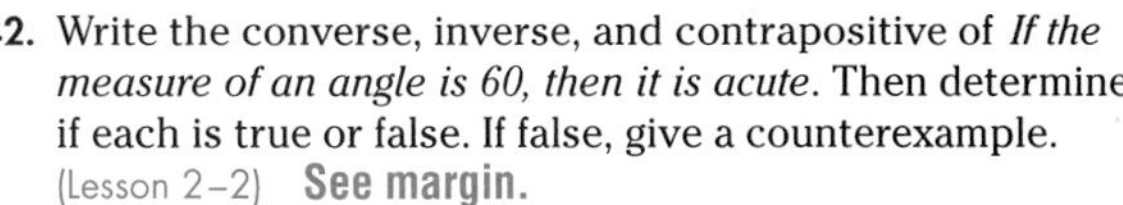

42. Write the converse, inverse, and contrapositive of *If the measure of an angle is 60, then it is acute.* Then determine if each is true or false. If false, give a counterexample. (Lesson 2–2) See margin.

43. **Botany** Delaney noticed that several potted plants on her porch were dying. All of these plants were supposed to receive partial sun. Make a conjecture about why the plants were dying and describe a method for testing this conjecture. (Lesson 2–1)

44. Refer to the figure at the right. (Lesson 1–2)

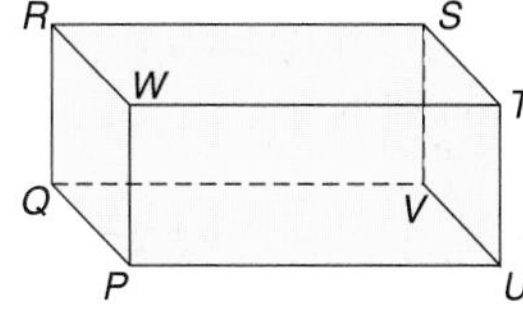

a. Are *R*, *W*, *P*, and *U* coplanar? no

b. What is the intersection of *RW* and *PW*? *W*

c. What is the intersection of plane *QPV* and plane *STV*? $\overleftrightarrow{VU}$

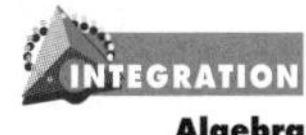

Algebra

43. Sample answer: The plants need more sunlight to live. Place them somewhere with more sun and see if they survive.

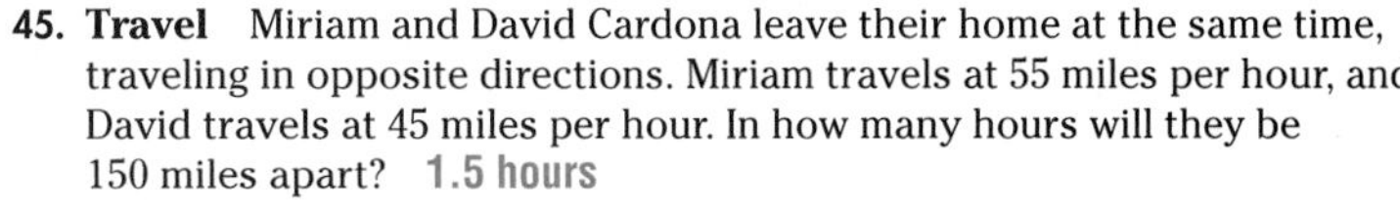

45. **Travel** Miriam and David Cardona leave their home at the same time, traveling in opposite directions. Miriam travels at 55 miles per hour, and David travels at 45 miles per hour. In how many hours will they be 150 miles apart? 1.5 hours

46. Use substitution to solve the system of equations. If the system does not have exactly one solution, state whether it has *no* solution, or *infinitely many* solutions. $c = 4d + 2$, infinitely many solutions

$c = 4d + 2$
$3c - 12d = 6$

Mathematics and SOCIETY

Light Shows in the Sky

The excerpt below appeared in an article in *Earth* in October, 1996.

THE INTERPLAY OF SUNLIGHT AND ICE crystals in the sky can produce spectacular displays of halos and other phenomena. Many people have seen a ring of light around the Sun or Moon—a halo that forms when light passes through ice crystals in the air. But on occasion, halos, arcs and pillars of light can fill the sky, creating bright geometrical patterns....What are these beautiful apparitions? How do they form?....The location of the Sun and the geometry of the ice crystals determine which halos, arcs or pillars form in the sky ....Scientists now know that the ice crystals that play a crucial role in halo formation are not intricate patterns like snowflakes, but rather simple six-sided crystals shaped like plates and columns....there are dozens of possible light paths through the crystals, and these explain the enormous variety of displays. ■

1. same shapes; reflect light in same direction

1. As the light passes through a prism and is reflected, its path outlines a triangle. What characteristics might two crystals have so they produce congruent triangular light paths?
2. Would you expect to see more of these light displays where you live or near the South Pole? Why? See margin.
3. Which would be more likely to produce one of these displays, ice crystals tumbling at random or crystals falling so that one of their sides remains horizontal as they fall? See margin.

4 ASSESS

Closing Activity

Speaking Write SSS, SAS, and ASA on three index cards. Have volunteers come to the front of the room one at a time, choose a card, and tell the class an example of the postulate. The speaker may use the chalkboard or overhead projector to draw appropriate diagrams.

Answers for Mathematics and Society

2. Near the South Pole. These displays are caused by light that is refracted and reflected from ice crystals, and these crystals would be much more plentiful near the South Pole.
3. The crystal with one side that remains horizontal. The stable orientation of crystals like these permits patterns to form, but a randomly tumbling crystal would cause light to be scattered in all directions, rather than being focused or concentrated in a specific area.

Enrichment Masters, p. 23

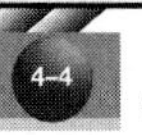

NAME ____________ DATE ______

Enrichment

Student Edition Pages 206–213

Congruent Triangles in the Coordinate Plane

If you know the coordinates of the vertices of two triangles in the coordinate plane, you can often decide whether the two triangles are congruent. There may be more than one way to do this.

1. Consider $\triangle ABD$ and $\triangle CDB$ whose vertices have coordinates $A(0, 0)$, $B(2, 5)$, and $D(7, 0)$. Briefly describe how you can use what you know about congruent triangles and the coordinate plane to show that $\triangle ABD \cong \triangle CDB$. You may wish to make a sketch to help get you started. Sample answer: Show that the slopes of $\overline{AB}$ and $\overline{CD}$ are equal and that the slopes of $\overline{AD}$ and $\overline{BC}$ are equal. Conclude that $\overline{AB} \parallel \overline{CD}$ and $\overline{BD} \parallel \overline{AD}$. Use the angle relationships for parallel lines and a transversal and the fact that $\overline{BD}$ is a common side for the triangles to conclude that $\triangle ABD \cong \triangle CDB$ by ASA.

2. Consider $\triangle PQR$ and $\triangle KLM$ whose vertices are the following points.

$P(1, 2)$	$Q(3, 6)$	$R(6, 5)$
$K(-2, 1)$	$L(-6, 3)$	$M(-5, 6)$

Briefly describe how you can show that $\triangle PQR \cong \triangle KLM$. Use the distance formula to find the lengths of the sides of both triangles. Conclude that $\triangle PQR \cong \triangle KLM$ by SSS.

3. If you know the coordinates of all the vertices of two triangles, is it *always* possible to tell whether the triangles are congruent? Explain. Yes; you can use the distance formula and SSS.

Extension

Problem Solving Can two triangles be proven congruent by SSA (two sides and an angle not included between those sides)? Give reasons or a counterexample for your answer. No; students should draw two triangles that are not congruent even though they each satisfy the SSA condition.

Mathematics and SOCIETY

A rainbow is produced when light is bent as it passes from air into another medium. Students who are interested in this topic should talk to their physics or earth science teacher for more information.

4-5 LESSON NOTES

NCTM Standards: 1–4, 7, 8

Instructional Resources

- Study Guide Master 4-5
- Practice Master 4-5
- Enrichment Master 4-5
- Assessment and Evaluation Masters, p. 101
- Modeling Mathematics Masters, p. 82
- Real-World Applications, 8

Transparency 4-5A contains the 5-Minute Check for this lesson; **Transparency 4-5B** contains a teaching aid for this lesson.

Recommended Pacing	
Standard Pacing	Days 9 & 10 of 14
Honors Pacing	Day 9 of 13
Block Scheduling*	Day 5 of 7

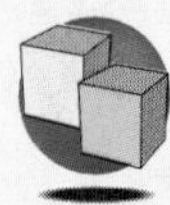

*For more information on pacing and possible lesson plans, refer to the *Block Scheduling Booklet*.

1 FOCUS

5-Minute Check
(over Lesson 4-4)

Give reasons for each step in this proof.

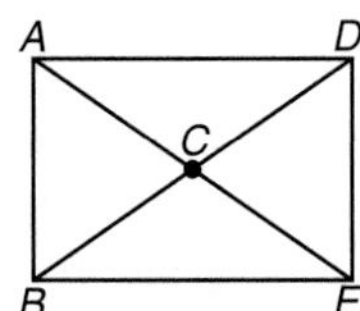

Given: C is the midpoint of $\overline{AE}$. C is the midpoint of $\overline{DB}$.
Prove: $\overline{AB} \cong \overline{ED}$

Statements (Reasons)

1. C is the midpoint of $\overline{AE}$. C is the midpoint of $\overline{DB}$. (Given)
2. $\overline{AC} \cong \overline{CE}$ (Mdpt. Theorem)
3. $\overline{DC} \cong \overline{CB}$ (Mdpt. Theorem)
4. $\angle ACB \cong \angle DCE$ (Vert. ∠s are ≅.)
5. $\triangle ACB \cong \triangle ECD$ (SAS)
6. $\overline{AB} \cong \overline{ED}$ (CPCTC)

4-5 More Congruent Triangles

What YOU'LL LEARN

- To use the AAS Theorem to test triangle congruence, and
- to solve problems by eliminating the possibilities.

Why IT'S IMPORTANT

You can determine if two triangles are congruent by eliminating the possibilities.

You have studied different methods for proving that two triangles are congruent. Is proving that two angles and a nonincluded side of two triangles are congruent, or AAS, sufficient for proving the triangles congruent?

MODELING MATHEMATICS — Angle-Angle-Side

Materials:  patty paper  straightedge

You can draw a triangle given two angles and a nonincluded side.

- Draw a triangle on a piece of patty paper. Label the vertices A, B, and C.

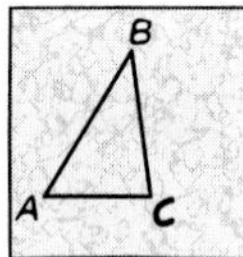

- Copy $\overline{AB}$, $\angle B$, and $\angle C$ on another piece of patty paper.
- Cut out the side and two angles.

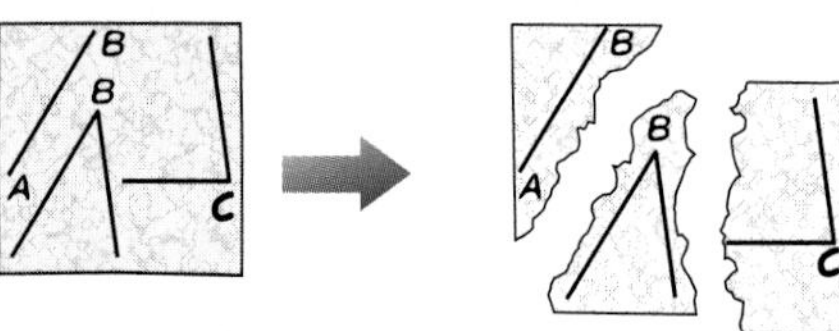

- Place them together to form a triangle in which the side is not the included side of the angles.
- Place another piece of patty paper over the triangle and trace it.
- Place this patty paper over $\triangle ABC$. How does the new triangle compare to $\triangle ABC$? It is congruent to $\triangle ABC$.

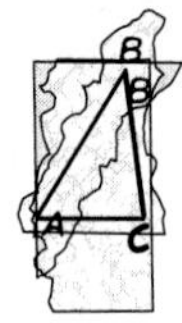

Your Turn a. See students' work. It is congruent to $\triangle ABC$.

a. Copy $\overline{BC}$, $\angle A$, and $\angle B$. Use these to form a triangle in which the side is not the included side of the angles. How does this triangle compare to $\triangle ABC$?

b. Write a conjecture about two triangles in which two angles and the nonincluded side of one triangle are congruent to two angles and the nonincluded side of the other triangle.

b. If 2 angles and a nonincluded side of a triangle are congruent to the corresponding angles and side of another triangle, the triangles are congruent.

This Modeling Mathematics activity suggests Theorem 4–5.

Theorem 4–5 AAS (Angle-Angle-Side)	**If two angles and a nonincluded side of one triangle are congruent to the corresponding two angles and side of a second triangle, the two triangles are congruent.**

MODELING MATHEMATICS Draw line segment PQ. At point P, create an acute angle. Locate point R so that $\triangle QRP$ is a right triangle with hypotenuse $\overline{PQ}$. How many locations are possible for point R? 1 How does the triangle change as you change the location of R? $\angle QRP$ changes size.

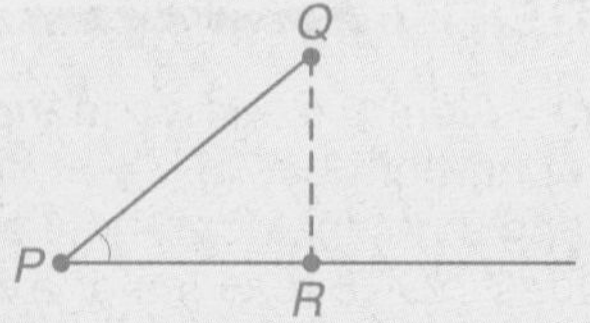

Proof of Theorem 4–5

Given: $\angle N \cong \angle D$
$\angle G \cong \angle I$
$\overline{AN} \cong \overline{SD}$

Prove: $\triangle ANG \cong \triangle SDI$

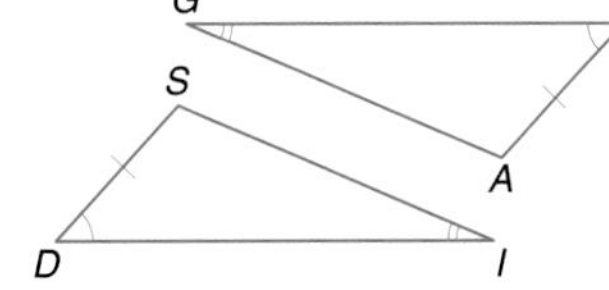

Proof:

Statements	Reasons
1. $\angle N \cong \angle D$, $\angle G \cong \angle I$, $\overline{AN} \cong \overline{SD}$	1. Given
2. $\angle A \cong \angle S$	2. Third Angle Theorem
3. $\triangle ANG \cong \triangle SDI$	3. ASA

Sometimes the triangles we would like to prove congruent are overlapping. Then it is helpful to slide the two triangles separately or to use different colors to distinguish each triangle.

Example 1

Proof

Write a paragraph proof.

Given: $\angle PSU \cong \angle PTR$
$\overline{SU} \cong \overline{TR}$

Prove: $\overline{SP} \cong \overline{TP}$

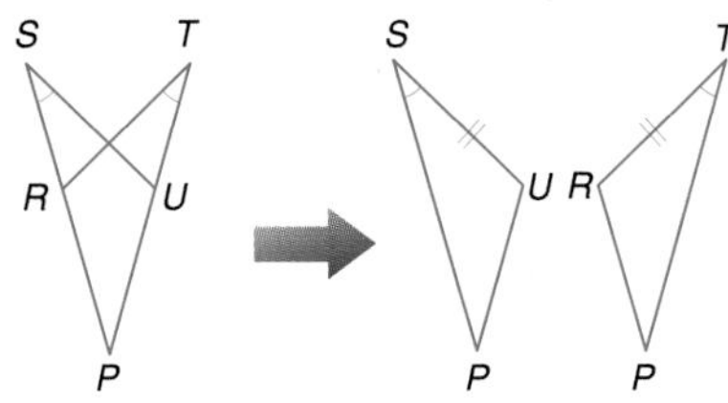

Plan for Proof:

Slide $\triangle SUP$ and $\triangle TRP$ to show two separate triangles. Prove that these two triangles are congruent, and then use CPCTC to prove that $\overline{SP} \cong \overline{TP}$.

Paragraph Proof:

We are given that $\angle PSU \cong \angle PTR$ and $\overline{SU} \cong \overline{TR}$. By the Reflexive Property of Congruent Angles, $\angle P \cong \angle P$. Then $\triangle SUP \cong \triangle TRP$ by AAS and $\overline{SP} \cong \overline{TP}$ by CPCTC.

One important problem-solving strategy is to **eliminate the possibilities**.

Example 2

PROBLEM SOLVING

Eliminate the Possibilities

Some of the measurements of $\triangle ABC$ and $\triangle DEF$ are given at the right. Can you determine if the two triangles are congruent from this information?

C, 2.5 cm, 30°, A, 4 cm, B

F, 2.5 cm, 30°, D, 4 cm, E

Explore We are given three measurements of each triangle. We are asked if we can determine that the triangles are congruent from the three given measurements.

Plan Since the $m\angle A = 30$ and $m\angle D = 30$, by definition $\angle A \cong \angle D$. Since $AB = DE$ and $BC = EF$, by definition $\overline{AB} \cong \overline{DE}$ and $\overline{BC} \cong \overline{EF}$. We know five methods to prove that two triangles are congruent: the definition of congruent triangles, SSS, SAS, ASA, and AAS. We can check each of these possibilities.

(continued on the next page)

Motivating the Lesson

Situational Problem Ask students to name as many real-life examples of uses of triangles as they can. **Sample answers: bridges, roofs, signs, clothing design**

2 TEACH

Teaching Tip For Example 1, draw the figure on the chalkboard and use colored chalk to outline the two triangles being considered.

In-Class Examples

For Example 1
Write a two-column proof.
Given: $\overline{AD} \cong \overline{AE}$, $\angle ACD \cong \angle ABE$
Prove: $\triangle ABC$ is isosceles.

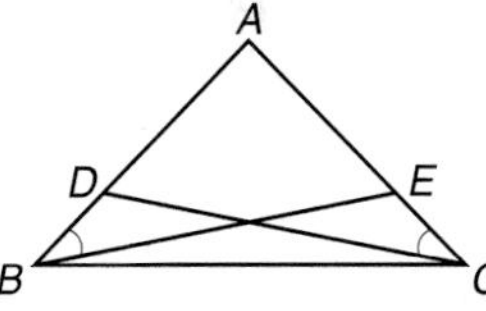

Proof:
Statements (Reasons)
1. $\overline{AD} \cong \overline{AE}$, $\angle ACD \cong \angle ABE$ (Given)
2. $\angle A \cong \angle A$ (Congruence of angles is reflexive.)
3. $\triangle ACD \cong \triangle ABE$ (AAS)
4. $\overline{AC} \cong \overline{AB}$ (CPCTC)
5. $\triangle ABC$ is isosceles. (Def. isosceles triangle)

For Example 2
Some of the measurements of $\triangle ABC$ and $\triangle DEF$ are given below. Can you determine if the triangles are congruent?

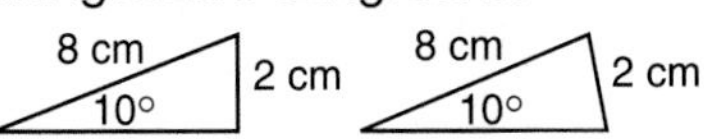

None of the five methods for proving congruence of triangles—CPCTC, SSS, SAS, ASA, or AAS—are applicable. So, you cannot determine if the two are congruent.

Alternative Learning Styles

Visual Ina found a triangular piece of scrap wood that would make a good bookend. She wants to cut another just like it. What other measurement does she need in order to make a congruent bookend? **either AC, $m\angle B$, or $m\angle C$** Have students make a model of the given triangle to help them find the answer. Without the missing information, are all student triangles the same? **No** Compare the different examples.

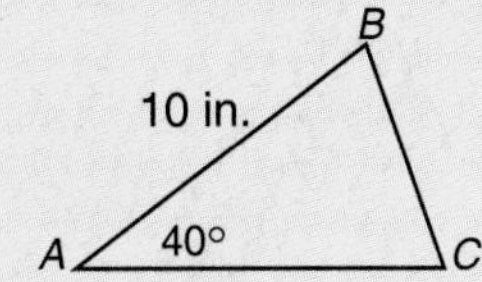

3 PRACTICE/APPLY

Check for Understanding

Exercises 1–14 are designed to help you assess your students' understanding through reading, writing, speaking, and modeling. You should work through Exercises 1–5 with your students and then monitor their work on Exercises 6–14.

Error Analysis

If diagrams are complex, students often have difficulty visualizing the pertinent information. When students are trying to write proofs that involve overlapping triangles to be proved congruent, have them redraw the two triangles separately and mark the congruent parts on the separated triangles. This will enable them to see the corresponding parts more clearly, to determine the reason for congruency, and to determine which parts are congruent.

Additional Answers

1. In AAS the side is not included, but in ASA the side is included.
2. Both refer to three parts of a triangle. AAS requires 2 angles and a nonincluded side. SSA requires 2 sides and a nonincluded angle. AAS is a triangle congruence theorem, but SSA is not.
3. It is often helpful to use a congruence transformation to draw the two triangles separately when writing a proof involving overlapping triangles.

Solve

Method	Qualifications	Does It Apply?
Definition of Congruent Triangles	All six parts of one triangle must be congruent with all six parts of the other triangle.	No, we only know about three parts of each triangle.
SSS	The three sides of one triangle must be congruent to the three sides of the other triangle.	No, we only know about two of the sides of the triangles.
SAS	Two sides and the included angle of one triangle must be congruent to two sides and the included angle of the other triangle.	No, we know that two sides are congruent, but we do not know about the included angle.
ASA	Two angles and the included side of one triangle must be congruent to two angles and the included side of the other triangle.	No, we only know about one angle of each triangle.
AAS	Two angles and a nonincluded side of one triangle must be congruent to the corresponding two angles and side of the other triangle.	No, we only know about one angle of each triangle.

The triangles cannot be proven to be congruent by just the information we are given and the methods we know at this time.

Examine Use a ruler, protractor, and a compass to try to draw a triangle as described in this example. First draw a segment 4 centimeters long. At one end of the segment, draw an angle that is 30°. Open a compass so that it has a radius of 2.5 centimeters. Place the point of the compass on the other end of the 4-centimeter segment and draw an arc to find the other vertex of the triangle. Notice that there are two different vertices that could satisfy the conditions. With the given information, the triangles are not necessarily congruent.

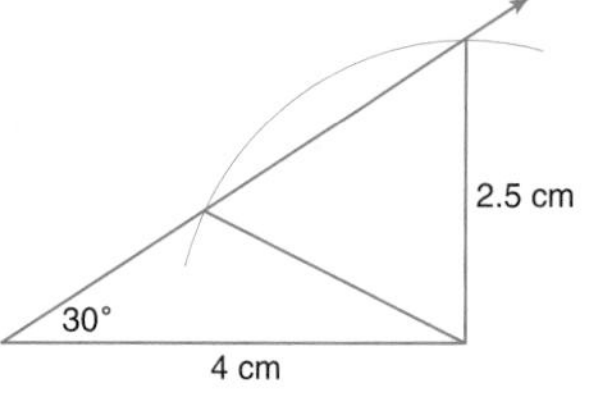

Example 2 shows that more than one triangle can be drawn with two given sides and a nonincluded angle. By this counterexample, we know that side-side-angle (SSA) cannot be used as a proof of congruent triangles.

CHECK FOR UNDERSTANDING

Communicating Mathematics

Study the lesson. Then complete the following. 1–3. See margin.

1. **Explain** why AAS is different than ASA.
2. **Compare and contrast** AAS and SSA.
3. **Describe** how a congruence transformation might help you to write a proof involving overlapping triangles.

Reteaching

Using Writing In Lesson 4-4, students were directed to make summary sheets for proving congruence by SSS, SAS, and ASA. Have students add a summary sheet for AAS that includes careful drawings.

4. **List** the methods used to show that two triangles are congruent. For each method, draw a diagram showing the congruent parts needed to show the congruence. See margin.

5a–b. See students' work.

5d. No; more than one triangle can be formed with 2 sides and a nonincluded angle congruent.

5. **Draw** a horizontal line using a straightedge.

   a. Connect two straws of unequal length with a pin, then position the straws so that the two straws form a triangle with the segment. Anchor one of the straws with tape, and label the vertices of the triangle *A*, *B*, and *C*.

   b. Now swing the straw representing $\overline{BC}$ so that the end of the straw touches the segment again. Label this point *D*.

   c. Is $\triangle ABC$ congruent to $\triangle ABD$? no

   d. Is SSA a valid test for determining triangle congruence? Explain your reasoning.

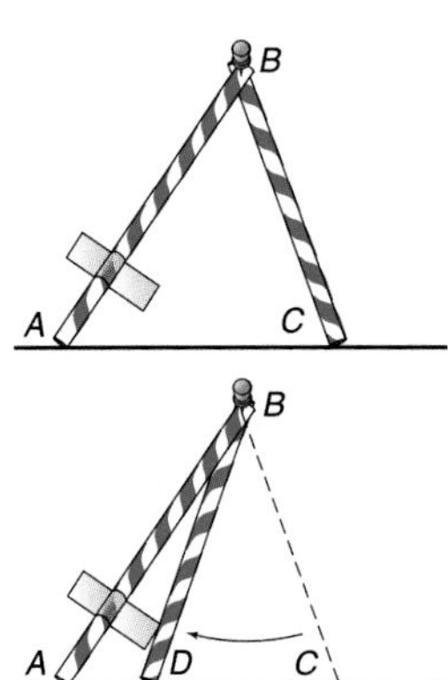

Guided Practice

Refer to the figure at the right for Exercises 6–9.

6. Name the included side for $\angle 1$ and $\angle 5$. $\overline{DC}$

7. Name a pair of angles in which $\overline{DE}$ is not included. Sample answer: $\angle 1$ and $\angle 5$

8. If $\angle 6 \cong \angle 10$, and $\overline{DC} \cong \overline{VC}$, then $\triangle DCA \cong \triangle$ _?_ by _?_. *VCE*; ASA

9. Given $\angle 7 \cong \angle 11$, $\overline{AD} \cong \overline{EV}$, and $\overline{DC} \cong \overline{VC}$, can $\triangle ADC$ and $\triangle EVC$ be congruent? Explain. Yes; see margin for explanation.

10. The statements and reasons used in a flow proof to prove $\triangle CTJ \cong \triangle HTN$ are listed below. Rearrange the boxes and add arrows to complete the flow proof.

    Given: $\overline{JC} \parallel \overline{NH}$ See Solutions Manual.
    $\overline{CH}$ bisects $\overline{JN}$.

    Prove: $\triangle CTJ \cong \triangle HTN$

    Flow Proof:

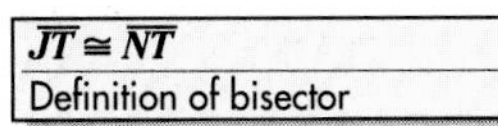

$\overline{JT} \cong \overline{NT}$
Definition of bisector

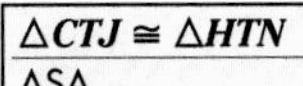

$\triangle CTJ \cong \triangle HTN$
ASA

$\overline{JC} \parallel \overline{NH}$
Given

$\angle CJT \cong \angle HNT$
Alternate Interior Angle Theorem

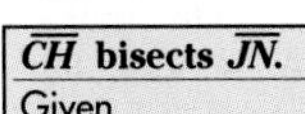

$\overline{CH}$ bisects $\overline{JN}$.
Given

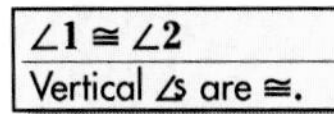

$\angle 1 \cong \angle 2$
Vertical ∠s are ≅.

11. Supply the reason for each step in the two-column proof.

    Given: $\angle E \cong \angle C$
    $\overline{AE} \cong \overline{DC}$

    Prove: $\overline{EB} \cong \overline{CB}$

    Proof:

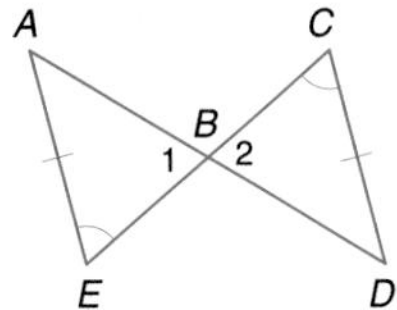

Statements	**Reasons**
1. $\angle E \cong \angle C$ $\overline{AE} \cong \overline{DC}$	1. _?_ Given
2. $\angle 1 \cong \angle 2$	2. _?_ Vertical ∠s are ≅.
3. $\triangle EBA \cong \triangle CBD$	3. _?_ AAS
4. $\overline{EB} \cong \overline{CB}$	4. _?_ CPCTC

Additional Answers

4. CPCTC

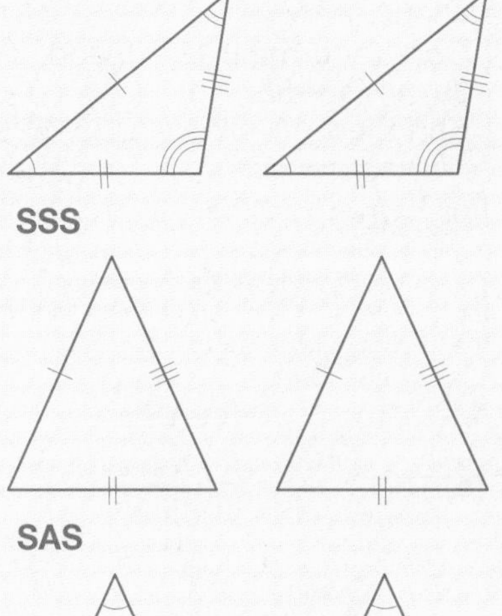

SAS

ASA

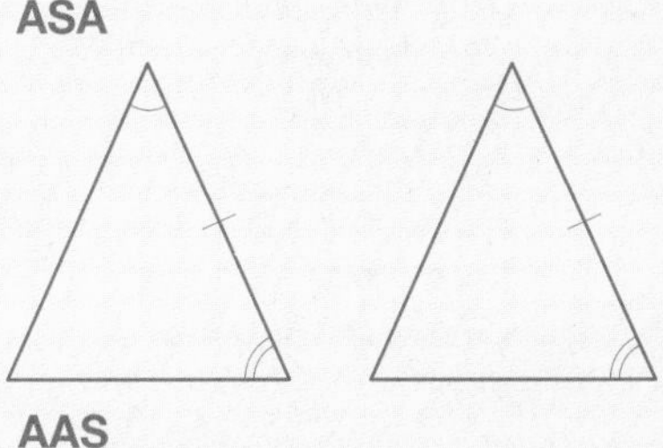

AAS

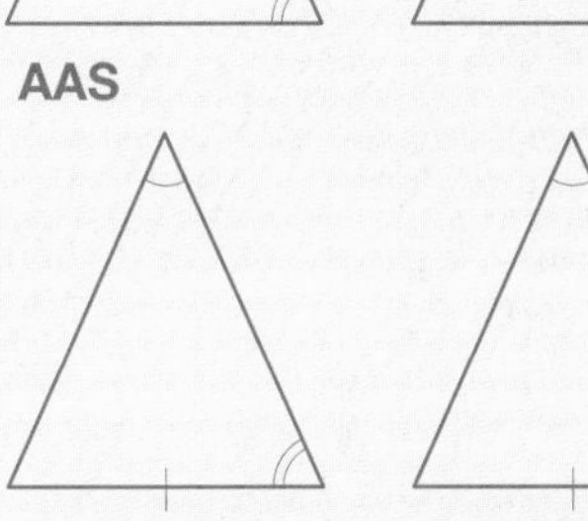

9. Yes; Since $\angle 7 \cong \angle 11$, $\overline{AD} \parallel \overline{EV}$. Then $\angle 6 \cong \angle 10$ by the Alt. Interior Angle Theorem. Then you have two sides and the included angle congruent.

Assignment Guide

Core (with proof): 15–41 odd, 43–52
Core (informal): 15–25 odd, 39, 41, 43–52
Enriched: 16–38 even, 39–52

For **Extra Practice**, see p. 771.

The red A, B, and C flags, printed only in the Teacher's Wraparound Edition, indicate the level of difficulty of the exercises.

Additional Answers

12. Proof:
Statements (Reasons)
1. X is the midpoint of $\overline{RD}$. (Given)
2. $\overline{RX} \cong \overline{XD}$ (Mdpt. Th.)
3. $\overline{RM} \parallel \overline{DN}$ (Given)
4. $\angle 1 \cong \angle 2$, $\angle 3 \cong \angle 4$ (Alt. Int. ∠ Th.)
5. $\triangle RXM \cong \triangle DXN$ (AAS)
6. $\overline{MX} \cong \overline{NX}$ (CPCTC)

13. Proof:
Statements (Reasons)
1. $\overline{NM}$ bisects $\overline{RD}$. (Given)
2. $\overline{RX} \cong \overline{XD}$ (Def. segment bisector)
3. $\angle 7 \cong \angle 8$ (Given)
4. $\angle 6 \cong \angle 5$ (Vert. ∠s are ≅.)
5. $\triangle DXM \cong \triangle RXN$ (AAS)
6. $\overline{MD} \cong \overline{NR}$ (CPCTC)

29. Proof:
Since $\angle EBC \cong \angle ECB$, by def. of ≅ ∠s $m\angle EBC = m\angle ECB$. Since $\angle ABE$ and $\angle EBC$ are linear pairs, the ∠s are supp. and $m\angle ABE + m\angle EBC = 180$. Likewise, $m\angle DCE + m\angle ECB = 180$. By subst., $m\angle ABE + m\angle EBC = m\angle DCE + m\angle ECB$ and $m\angle ABE + m\angle ECB = m\angle DCE + m\angle ECB$. Using the Subtr. Prop. (=), $m\angle ABE = m\angle DCE$. By the def. of ≅ ∠s, $\angle ABE \cong \angle DCE$. Since we are given that $\angle A \cong \angle D$ and $\overline{AE} \cong \overline{DE}$, so $\triangle ABE \cong \triangle DCE$ by AAS.

30. Proof:
Statements (Reasons)
1. $\triangle TEN$ is an isosceles △ with base $\overline{TN}$. (Given)
2. $\overline{TE} \cong \overline{NE}$ (Def. isosceles △)
3. $\angle 2 \cong \angle 3$, $\angle T \cong \angle N$ (Given)
4. $\triangle TEA \cong \triangle NEC$ (AAS)

Refer to the figure at the right for Exercises 12–13. 12–13. See margin.

12. **Given:** X is the midpoint of $\overline{RD}$.
$\overline{RM} \parallel \overline{DN}$
Prove: $\overline{MX} \cong \overline{NX}$

13. **Given:** $\overline{NM}$ bisects $\overline{RD}$.
$\angle 7 \cong \angle 8$
Prove: $\overline{MD} \cong \overline{NR}$

14. $\triangle RTE$ is a right triangle with right angle T. $\triangle ANG$ is a right triangle with right angle N. Suppose $m\angle A$ is 26, $m\angle R$ is 14 less than three times $m\angle E$, and $\overline{RE} \cong \overline{AG}$.
   a. Find $m\angle R$ and $m\angle E$. $m\angle R = 64$, $m\angle E = 26$
   b. Are the triangles congruent? Explain your reasoning. Yes; they are congruent by AAS.

EXERCISES

Practice

Refer to the figure at the right for Exercises 15–26.

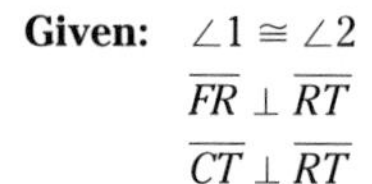

15. Name the included side for $\angle 1$ and $\angle 4$. — 15. $\overline{FD}$
16. Name the included side for $\angle 7$ and $\angle 8$. — 16. $\overline{DS}$
17. Name a nonincluded side for $\angle 5$ and $\angle 6$. — 17. $\overline{DR}$ or $\overline{RS}$
18. Name a nonincluded side for $\angle 9$ and $\angle 10$. — 18. $\overline{DT}$ or $\overline{CT}$
19. $\overline{CT}$ is included between what two angles? — 19. $\angle 10$ and $\angle 11$
20. $\overline{SR}$ is included between what two angles? $\angle 3$ and $\angle 6$
21. In $\triangle FDR$, name a pair of angles so that $\overline{FR}$ is not included. — 21. $\angle 1$ and $\angle 4$ or $\angle 2$ and $\angle 4$
22. In $\triangle SDT$, name a pair of angles so that $\overline{ST}$ is not included. — 22. $\angle 7$ and $\angle 12$ or $\angle 7$ and $\angle 8$
23. If $\angle 1 \cong \angle 6$, $\angle 4 \cong \angle 3$, and $\overline{FR} \cong \overline{DS}$, then $\triangle FDR \cong \triangle$ _?_ by _?_. *SRD*; AAS
24. If $\angle 5 \cong \angle 7$, $\angle 6 \cong \angle 8$, and $\overline{DS} \cong \overline{DS}$, then $\triangle TDS \cong \triangle$ _?_ by _?_. *RDS*; ASA
25. If $\angle 4 \cong \angle 9$, what sides would need to be congruent to show $\triangle FDR \cong \triangle CDT$? — 25. $\overline{FD} \cong \overline{CD}$ and $\overline{DR} \cong \overline{DT}$
26. If $\overline{RS} \cong \overline{TS}$ and $\overline{DR} \cong \overline{DT}$, name a pair of angles that would create an SSA relationship. $\angle 5 \cong \angle 7$ or $\angle 6 \cong \angle 8$

Proof

27. Justify each step.

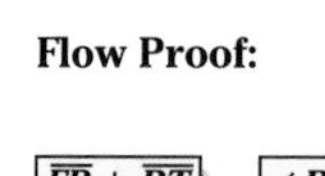

Given: $\angle 1 \cong \angle 2$
$\overline{FR} \perp \overline{RT}$
$\overline{CT} \perp \overline{RT}$
$\overline{FR} \cong \overline{CT}$
Prove: $\overline{RO} \cong \overline{TO}$

Flow Proof:

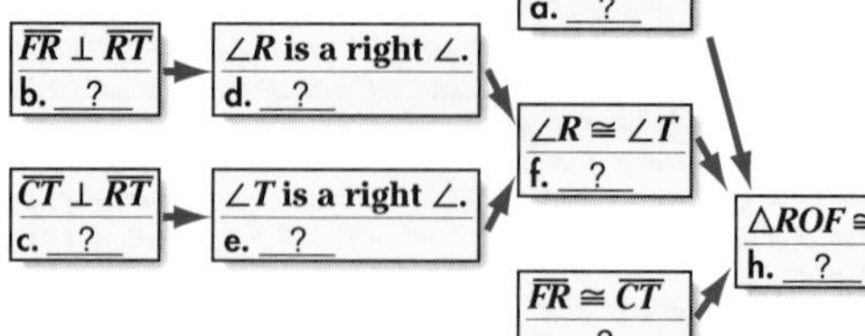

27a. Given
27b. Given
27c. Given
27d. ⊥ lines form 4 rt. ∠s.
27e. ⊥ lines form 4 rt. ∠s.
27f. All rt. ∠s are ≅.
27g. Given
27h. AAS
27i. CPCTC

28. Supply the reason for each step in the two-column proof.

Given: $\overline{VZ} \parallel \overline{WX}$
$\overline{VY} \cong \overline{WY}$
$\angle V \cong \angle W$
$\angle X$ is a right angle.

Prove: $\overline{VZ} \cong \overline{WX}$

Proof:

Statements	Reasons
1. $\angle X$ is a right angle.	1. ? Given
2. $\overline{WX} \perp \overline{ZX}$	2. ?
3. $\overline{VZ} \parallel \overline{WX}$	3. ? Given
4. $\overline{VZ} \perp \overline{ZX}$	4. ?
5. $\angle Z$ is a right angle.	5. ?
6. $\angle X \cong \angle Z$	6. ? All right $\angle$s are $\cong$.
7. $\overline{VY} \cong \overline{WY}$	7. ? Given
8. $\angle V \cong \angle W$	8. ? Given
9. $\triangle ZYV \cong \triangle XYW$	9. ? AAS
10. $\overline{VZ} \cong \overline{WX}$	10. ? CPCTC

28. (reason 2) Definition of perpendicular lines
(reason 4) If a line is $\perp$ to one of 2 $\parallel$ lines, then it is $\perp$ to the other.
(reason 5) Definition of perpendicular lines

29. Write a paragraph proof for the following. See margin.

Given: $\angle A \cong \angle D$
$\angle EBC \cong \angle ECB$
$\overline{AE} \cong \overline{DE}$

Prove: $\triangle ABE \cong \triangle DCE$

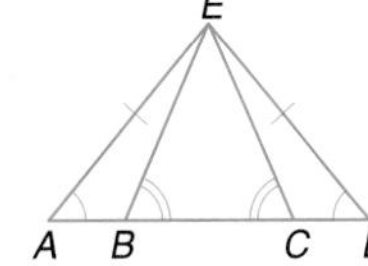

Refer to $\triangle TEN$ for Exercises 30–31. 30–31. See margin.

30. **Given:** $\triangle TEN$ is an isosceles triangle with base $\overline{TN}$.
$\angle 2 \cong \angle 3$
$\angle T \cong \angle N$

Prove: $\triangle TEA \cong \triangle NEC$

31. **Given:** $\angle 3 \cong \angle 2$
$\angle T \cong \angle N$
$\overline{TC} \cong \overline{NA}$

Prove: $\triangle TEA \cong \triangle NEC$

Refer to quadrilateral *FLMP* for Exercises 32–33. 32–33. See margin.

32. **Given:** $\angle F \cong \angle M$
$\angle 1 \cong \angle 2$

Prove: $\overline{FP} \cong \overline{ML}$

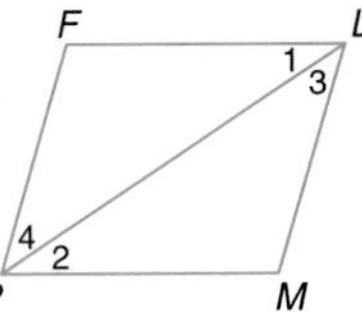

33. **Given:** $\overline{FP} \parallel \overline{ML}$
$\overline{FL} \parallel \overline{MP}$

Prove: $\overline{PM} \cong \overline{LF}$

Additional Answers

31. Proof:
Statements (Reasons)
1. $\angle 3 \cong \angle 2$, $\angle T \cong \angle N$, $\overline{TC} \cong \overline{NA}$ (Given)
2. $TC = NA$ (Def. $\cong$ segments)
3. $TA = TC + CA$, $NC = NA + AC$ (Segment Add. Post.)
4. $TA = NA + AC$ (Subst. Prop. (=))
5. $TA = NC$ (Subst. Prop. (=))
6. $\overline{TA} \cong \overline{NC}$ (Def. $\cong$ segments)
7. $\triangle TEA \cong \triangle NEC$ (ASA)

32. Proof:
Statements (Reasons)
1. $\angle F \cong \angle M$, $\angle 1 \cong \angle 2$ (Given)
2. $\overline{LP} \cong \overline{PL}$ ($\cong$ of segments is reflexive.)
3. $\triangle FLP \cong \triangle MPL$ (AAS)
4. $\overline{FP} \cong \overline{ML}$ (CPCTC)

33. Proof:
Statements (Reasons)
1. $\overline{FP} \parallel \overline{ML}$ (Given)
2. $\angle 3 \cong \angle 4$ (Alt. Int. $\angle$ Th.)
3. $\overline{FL} \parallel \overline{MP}$ (Given)
4. $\angle 1 \cong \angle 2$ (Alt. Int. $\angle$ Th.)
5. $\overline{PL} \cong \overline{LP}$ ($\cong$ of segments is reflexive.)
6. $\triangle FLP \cong \triangle MPL$ (ASA)
7. $\overline{PM} \cong \overline{LF}$ (CPCTC)

Study Guide Masters, p. 24

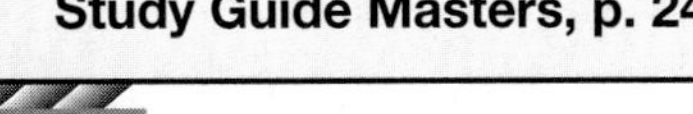

4-5 NAME ________ DATE ________

Study Guide

Student Edition Pages 214–221

More Congruent Triangles

In the previous lesson, you learned three postulates for showing that two triangles are congruent: Side–Side–Side (SSS), Side–Angle–Side (SAS), and Angle–Side–Angle (ASA).

Another test for triangle congruence is the Angle–Angle–Side theorem (AAS).

If two angles and a non-included side of one triangle are congruent to the corresponding two angles and a side of a second triangle, the two triangles are congruent.

Example: In $\triangle ABC$ and $\triangle DBC$, $\overline{AC} \cong \overline{DC}$, and $\angle ACB \cong \angle DCB$. Indicate the additional pair of corresponding parts that would have to be proved congruent in order to use AAS to prove $\triangle ACB \cong \triangle DCB$.

You would need to prove $\angle ABC \cong \angle DBC$ in order to prove that $\triangle ACB \cong \triangle DCB$.

Draw and label triangles ABC and DEF. Indicate the additional pairs of corresponding parts that would have to be proved congruent in order to use the given postulate or theorem to prove the triangles congruent. Drawings will vary.

1. $\angle B \cong \angle E$ and $\overline{BC} \cong \overline{EF}$ by ASA
$\angle C \cong \angle F$

2. $\overline{AC} \cong \overline{DF}$ and $\overline{CB} \cong \overline{FE}$ by SSS
$\overline{AB} \cong \overline{DE}$

Eliminate the possibilities. Determine which postulates show that the triangles are congruent.

3. AAS

4.

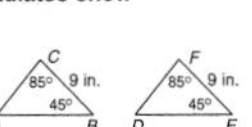

ASA

Write a paragraph proof.

5. **Given:** $\overline{HK}$ bisects $\angle GKN$.
$\angle G \cong \angle N$
Prove: $\overline{GK} \cong \overline{NK}$

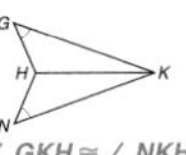

We are given that $\overline{HK}$ bisects $\angle GKN$. So $\angle GKH \cong \angle NKH$. We also are given that $\angle G \cong N$. $\overline{HK} \cong \overline{HK}$ since congruence of segments is reflexive. Therefore, $\triangle GKH \cong \triangle NKH$ by AAS. So, $\overline{GK} \cong \overline{NK}$ by the definition of congruent triangles (CPCTC).

Refer to pentagon *GRSTV* for Exercises 34–35.

C

34. Given: $\angle 1 \cong \angle 2$ See margin.
$\angle 3 \cong \angle 4$
$\angle 1 \cong \angle 4$
$\overline{GV} \cong \overline{TV}$

Prove: $\triangle RVS$ is an isosceles triangle.

35. Given: $\triangle GVR$ is an isosceles triangle with base $\overline{GR}$.
$\triangle TVS$ is an isosceles triangle with base $\overline{TS}$.
$\overline{GV} \cong \overline{TV}$
$\angle 5 \cong \angle 6$

Prove: $\overline{GR} \cong \overline{TS}$ See margin.

Refer to quadrilateral *PLXT* for Exercises 36–37. 36–37. See Solutions Manual.

36. Given: $\angle 1 \cong \angle 2$
$\angle 3 \cong \angle 4$

Prove: $\overline{PT} \cong \overline{LX}$

37. Given: $\overline{PX} \cong \overline{LT}$
$\triangle PRL$ is an isosceles triangle with base $\overline{PL}$.

Prove: $\triangle TRX$ is an isosceles triangle.

Cabri Geometry

38. Draw an obtuse, scalene triangle. Label it $\triangle ABC$ with $\angle A$ as the obtuse angle and $AC < AB$. a–c. See students' work.

a. Find the measures of the sides and angles of the triangle.

b. Draw a circle with center at A and radius equal to AC. The circle intersects $\overline{BC}$ at two points. Label the new point D.

c. Find BD, AD, $m\angle BDA$, and $m\angle BAD$.

d. Consider $\triangle ABC$ and $\triangle ABD$. How many sides of $\triangle ABD$ have the same length as a side of $\triangle ABC$? How many angles of $\triangle ABD$ are the same measure as an angle of $\triangle ABC$? 2; 1

e. Is $\triangle ABD$ congruent to $\triangle ABC$? What does this say about SSA as a proof of congruent triangles? No; SSA is not a proof for congruent triangles.

Critical Thinking

39. Can two triangles be proved congruent by AAA (Angle-Angle-Angle)? Justify your answer completely. See margin.

Applications and Problem Solving

40. History It is said that Thales determined the distance from the shore to enemy Greek ships during an early war by sighting the angle to the ship from a point P on the shore, walking a distance to point Q, and then sighting the angle to the ship from that point. He then reproduced the angles on the other side of line PQ and continued these lines until they intersected.

a. How did he determine the distance to the ship in this way? He created congruent triangles.

b. Why does this method work? See margin.

Additional Answer

34. Given: $\angle 1 \cong \angle 2$
$\angle 3 \cong \angle 4$
$\angle 1 \cong \angle 4$
$\overline{GV} \cong \overline{TV}$

Prove: $\triangle RVS$ is an isosceles triangle.

Proof:

Statements (Reasons)

1. $\angle 1 \cong \angle 2$, $\angle 3 \cong \angle 4$, $\angle 1 \cong \angle 4$ (Given)
2. $\angle 2 \cong \angle 1$ ($\cong$ of ∠s is symmetric.)
3. $\angle 2 \cong \angle 4$ ($\cong$ of ∠s is transitive.)
4. $\angle 4 \cong \angle 3$ ($\cong$ of ∠s is symmetric.)
5. $\angle 2 \cong \angle 3$ ($\cong$ of ∠s is transitive.)
6. $\overline{GV} \cong \overline{TV}$ (Given)
7. $\triangle GRV \cong \triangle TSV$ (AAS)
8. $\overline{RV} \cong \overline{SV}$ (CPCTC)
9. $\triangle RVS$ is an isosceles triangle. (Def. isosceles $\triangle$)

Practice Masters, p. 24

NAME__________ DATE________

Practice

Student Edition Pages 214–221

More Congruent Triangles

Draw and label triangles MNO and XYZ. Indicate the additional pairs of corresponding parts that would have to be proved congruent in order to use the given postulate or theorem to prove the triangles congruent.

1. $\angle N \cong \angle Y$ and $\overline{NO} \cong \overline{YZ}$ by ASA

$\angle O \cong \angle Z$

2. $\angle O \cong \angle Z$ and $\angle M \cong \angle X$ by AAS

$\angle \overline{NO} \cong \overline{YZ}$ or $\overline{MN} \cong \overline{XY}$

3. $\angle O \cong \angle Z$ and $\overline{MO} \cong \overline{XZ}$ by AAS

$\angle N \cong \angle Y$

4. $\angle N \cong \angle Y$ and $\angle M \cong \angle X$ by ASA

$\overline{MN} \cong \overline{XY}$

5. The statements in the following proof are *not* in logical order. Rearrange them in a correct sequence and give the reasons.

Given: $\angle A \cong \angle D$
$\overline{AB} \cong \overline{DE}$

Prove: $\overline{CA} \cong \overline{CD}$

Proof:

One order of steps: b, c, e, a, d

Statements	Reasons
a. $\triangle CAB \cong \triangle CDE$	a. AAS
b. $\angle A \cong \angle D$	b. Given
c. $\angle BCA \cong \angle ECD$	c. Vertical ∠s are ≅.
d. $\overline{CA} \cong \overline{CD}$	d. CPCTC
e. $\overline{BA} \cong \overline{ED}$	e. Given

6. Eliminate the Possibilities Barky, Spot, and Tiger are dogs. One is a black labrador retriever, one is a multi-colored collie, and one is a spotted dalmatian. One is owned by a doctor, one is owned by a lawyer, and one is owned by an insurance salesperson. The lawyer's dog does not have spots. The doctor's dog has more than one color. The insurance salesperson's dog is solid in color. Who owns which dog?
doctor—dalmatian; lawyer—collie; insurance salesperson—labrador retriever

Additional Answer

35. Proof:

Statements (Reasons)

1. $\triangle GVR$ is isosceles with base $\overline{GR}$. (Given)
2. $\overline{RV} \cong \overline{GV}$ (Def. isosceles $\triangle$)
3. $\triangle TVS$ is isosceles with base $\overline{TS}$. (Given)
4. $\overline{TV} \cong \overline{SV}$ (Def. isosceles $\triangle$)
5. $\overline{GV} \cong \overline{TV}$ (Given)
6. $\overline{RV} \cong \overline{TV}$ ($\cong$ of segments is transitive.)
7. $\overline{RV} \cong \overline{SV}$ ($\cong$ of segments is transitive.)
8. $\angle 5 \cong \angle 6$ (Given)
9. $\triangle GRV \cong \triangle TSV$ (SAS)
10. $\overline{GR} \cong \overline{TS}$ (CPCTC)

41. The guy wires are all the same length because the triangles are congruent by AAS.

41. **Broadcasting** In order to stabilize a television antenna so that it remains perpendicular to the ground, three guy wires are attached as shown at the right. If the guy wires are attached to the antenna at the same height from the ground and the measures of $\angle A$, $\angle B$, and $\angle C$ are the same, what can you say about the length of the guy wires? Explain your reasoning.

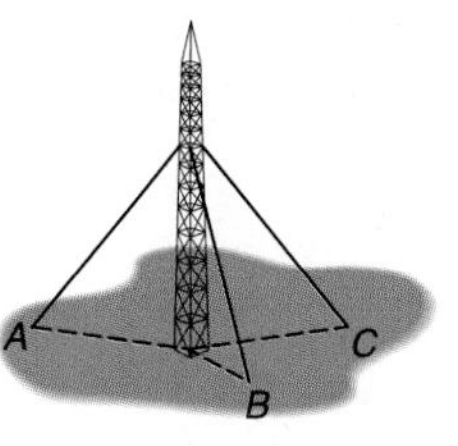

42. Sample A, halite; Sample B, feldspar; Sample C, biotite; Sample D, hematite; Sample E, jasper

42. **Eliminate the Possibilities** On a recent geology test, Raul was given five different mineral samples to identify using the portion of a mineral characteristics chart shown at the right. Raul made the following observations about the samples.

- Sample C is softer than glass.
- Samples D and E are red and Sample C is brown.
- Samples B and E are harder than glass.

Identify each of the samples.

Mineral	Color	Hardness
Biotite	brown or black	softer than glass
Halite	white	softer than glass
Hematite	red	softer than glass
Feldspar	white, pink, or green	harder than glass
Jasper	red	harder than glass

Mixed Review

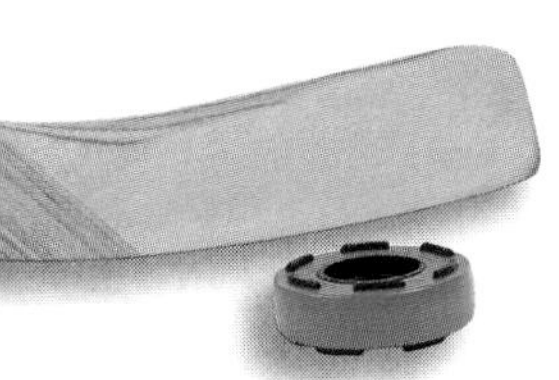

43. **Hockey** The width of the opening of a hockey net is 2 meters. A player shoots from a point 9 meters from point A and 8 meters from point B. Another player shoots from a point 8 meters from point A and 9 meters from point B. Which player has a greater angle of possible shots and therefore the better chance of making a score? Explain your reasoning. (Lesson 4–4)

43. Neither player has a greater angle. The triangles formed are congruent by SSS and therefore the angles are congruent.

44. If $\triangle HRT \cong \triangle MNP$, complete each statement. (Lesson 4–3)

a. $\angle R \cong$ ___?___ $\angle N$ b. $\overline{HT} \cong$ ___?___ $\overline{MP}$ c. $\angle P \cong$ ___?___ $\angle T$

45. Decide whether it is possible to have a triangle with no acute angles. (Lesson 4–2) **no**

46. Draw an acute, scalene triangle. (Lesson 4–1) **See margin.**

47. **Algebra** The slope of line n is 9. What is the slope of any line perpendicular to n? (Lesson 3–3)

47. $-\frac{1}{9}$

48. If $\angle A$ is complementary to $\angle B$, $\angle C$ is complementary to $\angle B$, and $m\angle A = 34$, find $m\angle B$ and $m\angle C$. (Lesson 2–6) **56; 34**

49. $\overrightarrow{QT}$ and $\overrightarrow{QS}$ are opposite rays. Describe $\angle TQS$. (Lesson 1–7)

49. a straight angle

50. The length of a side of a square is s centimeters. Find the possible whole number values of s if the perimeter of the square must be less than 50 centimeters and the area of the square must be greater than 50 square centimeters. (Lesson 1–3) **8 cm, 9 cm, 10 cm, 11 cm, 12 cm**

INTEGRATION Algebra

51. State whether $18 = rs$ represents an inverse variation or a direct variation. Then find the constant of variation. **inverse; 18**

52. Which ordered pairs are solutions for the equation $4x - y = 7$? **c and d**

a. (2, 6) b. (3, 0) c. (2, 1) d. (0, −7)

Extension

Communication Is this statement true or false? Explain your answer. "To prove that two triangles are congruent, you can use any two pairs of congruent, corresponding sides and a pair of corresponding, congruent angles, or you can use any two pairs of corresponding, congruent angles and a pair of corresponding, congruent sides."

The first part of the statement is not true. If two pairs of corresponding, congruent sides are given, the *included* corresponding angles must be congruent in order to prove the triangles congruent (SAS). The second part is true, since you can prove triangles congruent by ASA and by AAS.

4 ASSESS

Closing Activity

Writing Have students write a paragraph describing how AAS can be used to prove that two triangles are congruent.

Chapter 4 Quiz C (Lessons 4-4 and 4-5) is available in the *Assessment and Evaluation Masters*, p. 101.

Additional Answers

39. No; two equiangular triangles are an example of AAA, but the sides are not necessarily congruent.

40b. The distance from the shore to where the lines intersect is the distance from the shore to the ship. The triangle formed by the ship and P and Q is congruent to the triangle formed by the intersection point and P and Q by ASA.

46. Sample answer:

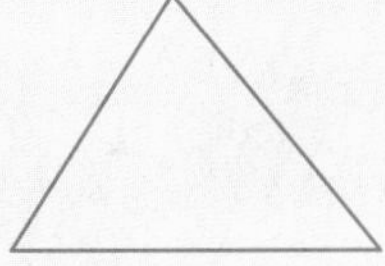

Enrichment Masters, p. 24

4-5 NAME________ DATE________

Enrichment — Student Edition Pages 214–221

How Many Triangles?

Each puzzle below contains many triangles. Count them carefully. Some triangles overlap other triangles.

How many triangles in each figure?

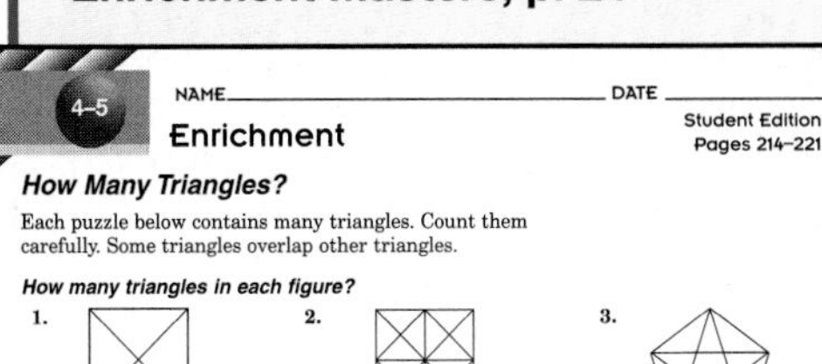

4.

How many triangles can you form by joining points on each circle? List the vertices of each triangle.

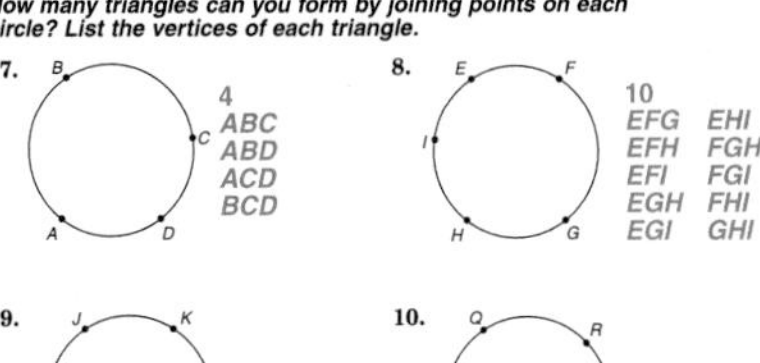

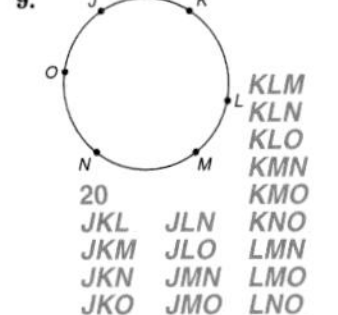

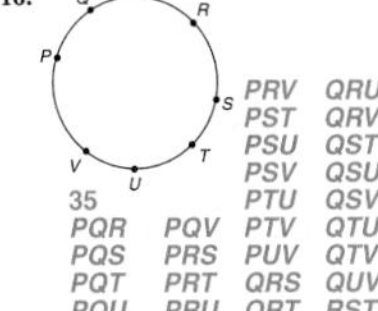

4-6 LESSON NOTES

NCTM Standards: 1–5, 7, 8

Instructional Resources

- Study Guide Master 4-6
- Practice Master 4-6
- Enrichment Master 4-6
- Assessment and Evaluation Masters, p. 101
- Multicultural Activity Masters, p. 8
- Tech Prep Applications Masters, p. 8

Transparency 4-6A contains the 5-Minute Check for this lesson; **Transparency 4-6B** contains a teaching aid for this lesson.

Recommended Pacing	
Standard Pacing	Days 11 & 12 of 14
Honors Pacing	Days 10 & 11 of 13
Block Scheduling*	Day 6 of 7

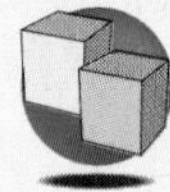

*For more information on pacing and possible lesson plans, refer to the *Block Scheduling Booklet*.

1 FOCUS

5-Minute Check
(over Lesson 4-5)

Write a congruence statement for each pair of triangles. Which postulate or theorem proves their congruence?

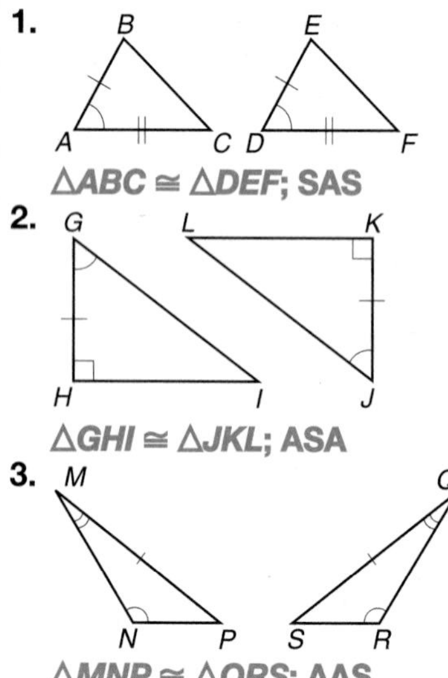

1. $\triangle ABC \cong \triangle DEF$; SAS
2. $\triangle GHI \cong \triangle JKL$; ASA
3. $\triangle MNP \cong \triangle QRS$; AAS

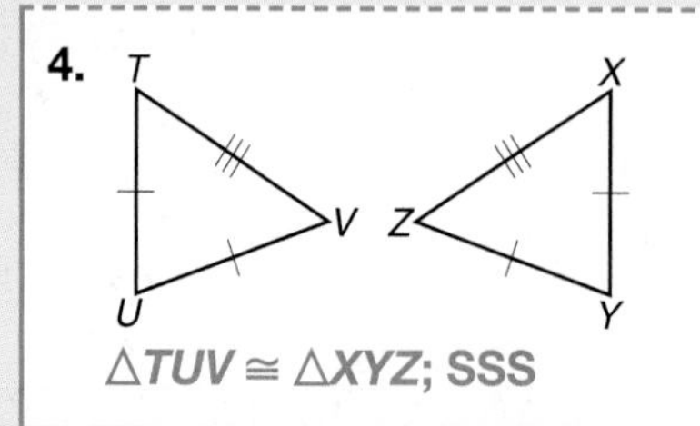

4. $\triangle TUV \cong \triangle XYZ$; SSS

MODELING MATHEMATICS Encourage students to notice the correspondence between the steps in this activity and the proof of Theorem 4-6.

4–6 Analyzing Isosceles Triangles

What **YOU'LL LEARN**

- To use properties of isosceles and equilateral triangles.

Why **IT'S IMPORTANT**

You can use the properties of isosceles and equilateral triangles involved in designs, carpentry, and navigation.

APPLICATION

Art and Design

Many teens like to collect pennants displaying the names of their school or favorite sports teams. Pennants are designed by artists and usually are in the shape of isosceles triangles.

MODELING MATHEMATICS

Isosceles Triangles

Materials:  patty paper  straight edge  compass

Make an isosceles triangle and compare its base angles.

- Draw an acute angle with vertex C on a piece of patty paper.

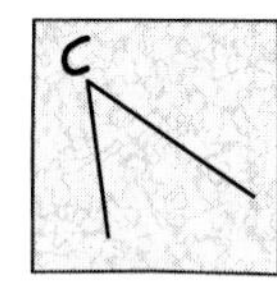

- Mark equal lengths on the sides of $\angle C$. Name the points A and B. Draw $\overline{AB}$. $\triangle ABC$ is isosceles with base $\overline{AB}$.

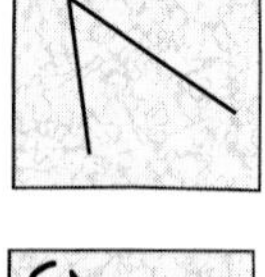

- Fold the patty paper through C so that A coincides with point B. What do you notice about $\angle A$ and $\angle B$? $\angle A \cong \angle B$

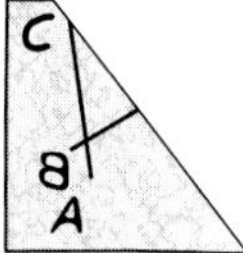

Your Turn

a. Draw an obtuse, isosceles triangle. Compare the base angles. They are congruent.

b. Draw a right, isosceles triangle. Compare the base angles. They are congruent.

c. Use three pieces of patty paper to form an equilateral triangle. Compare the three angles of the triangle. They are congruent.

The Modeling Mathematics activity suggests Theorem 4–6.

Theorem 4–6 Isosceles Triangle Theorem	**If two sides of a triangle are congruent, then the angles opposite those sides are congruent.**

Proof of Theorem 4–6

Given: $\triangle PQR$, $\overline{PQ} \cong \overline{RQ}$

Prove: $\angle P \cong \angle R$

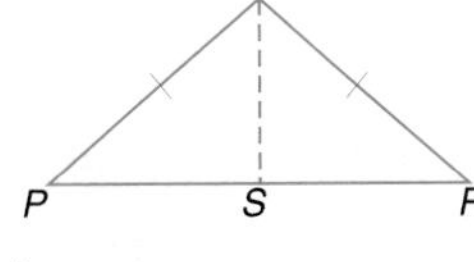

Proof:

Statements	Reasons
1. Let S be the midpoint of $\overline{PR}$.	**1.** Every segment has exactly one midpoint.
2. Draw auxiliary segment $\overline{QS}$.	**2.** Through any two points there is one line.
3. $\overline{PS} \cong \overline{RS}$	**3.** Definition of midpoint
4. $\overline{QS} \cong \overline{QS}$	**4.** Congruence of segments is reflexive.
5. $\overline{PQ} \cong \overline{RQ}$	**5.** Given
6. $\triangle PQS \cong \triangle RQS$	**6.** SSS
7. $\angle P \cong \angle R$	**7.** CPCTC

You can use Theorem 4–6 and algebra to find missing measures of an isosceles triangle.

Example 1 **In isosceles $\triangle ISO$ with base $\overline{SO}$, $m\angle S = 5x - 18$ and $m\angle O = 2x + 21$. Find the measure of each angle of the triangle.**

INTEGRATION Algebra

$m\angle S = m\angle O$ *Isosceles Triangle Theorem*

$5x - 18 = 2x + 21$

$3x = 39$

$x = 13$

$m\angle S = 5x - 18 = 5(13) - 18$ or 47

$m\angle O = 2x + 21 = 2(13) + 21$ or 47

$m\angle S + m\angle O + m\angle I = 180$

$47 + 47 + m\angle I = 180$

$m\angle I = 86$

The measures of $\angle S$, $\angle O$, and $\angle I$ are 47, 47, and 86, respectively.

The converse of the Isosceles Triangle Theorem is also true.

Theorem 4–7 **If two angles of a triangle are congruent, then the sides opposite those angles are congruent.**

Example 2 **Write a plan for a two-column proof for Theorem 4–7.**

Draw an auxiliary ray that is the bisector of $\angle ABC$ and let D be the point where the bisector intersects $\overline{AC}$. Show that $\triangle ABD \cong \triangle CBD$ by AAS using the given information and the auxiliary ray. Then, $\overline{AB} \cong \overline{CB}$ by CPCTC.

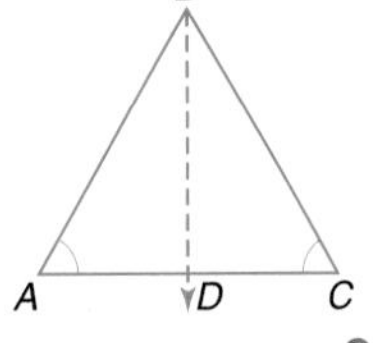

You will be asked to prove this theorem in Exercise 13.

Motivating the Lesson

Hands-On Activity Have students fold a sheet of paper in half and use it to draw and cut out an isosceles triangle. Ask them to make a conjecture about angles 1, 2, 3, and 4. $\angle 1 \cong \angle 2$ and $\angle 3 \cong \angle 4$

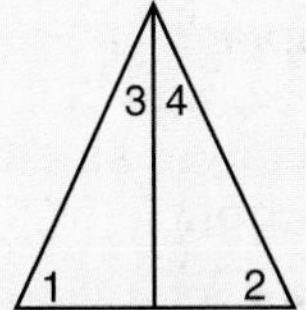

2 TEACH

In-Class Examples

For Example 1
In isosceles $\triangle DEF$, $\angle D$ is the vertex angle. If $m\angle E = 2x + 40$ and $m\angle F = 3x + 22$, find the measure of each angle of the triangle. $m\angle D = 28$, $m\angle E = 76$, $m\angle F = 76$

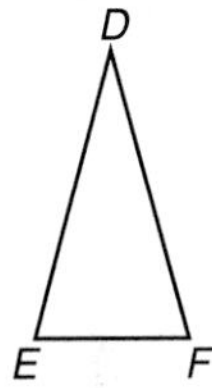

For Example 2
Write a plan for a proof of Corollary 4-4.
Given: $\triangle XYZ$ is equilateral.
Prove: The measure of each angle of $\triangle XYZ$ is 60.

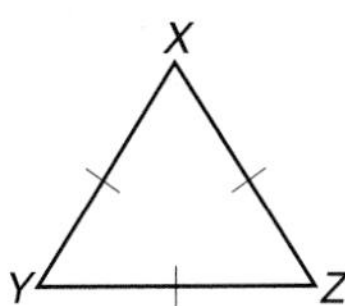

Since $\triangle XYZ$ is equiangular and the sum of the measures of the angles is 180, use substitution to show that the measure of each angle is 60.

Cooperative Learning

Brainstorming Have students work in groups to decide if the following statement is true or false. Be sure they give reasons for their answers. "Base angles of an isosceles triangle are always acute." True; if the base angles were right or obtuse, the sum of the measures of the angles would be greater than 180, which is impossible in a triangle. For more information on the brainstorming strategy, see *Cooperative Learning in the Mathematics Classroom*, one of the titles in the Glencoe Mathematics Professional Series, page 30.

The Isosceles Triangle Theorem leads us to some interesting corollaries. *You will be asked to prove Corollaries 4–3 and 4–4 in Exercises 42 and 43, respectively.*

Corollary 4–3	**A triangle is equilateral if and only if it is equiangular.**
Corollary 4–4	**Each angle of an equilateral triangle measures 60°.**

Recall that corollaries can be used as reasons in proofs just like theorems.

Example  **Proof**

Write a paragraph proof.

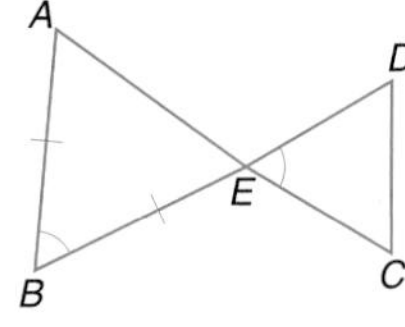

Given: $\overline{AB} \cong \overline{EB}$
$\angle DEC \cong \angle B$

Prove: $\triangle ABE$ is equilateral.

Proof:

Since $\overline{AB} \cong \overline{EB}$, $\angle A \cong \angle AEB$ by the Isosceles Triangle Theorem. Since vertical angles are congruent, $\angle AEB \cong \angle DEC$. But $\angle DEC \cong \angle B$, so $\angle AEB \cong \angle B$ because congruence of angles is transitive. Also, $\angle A \cong \angle AEB \cong \angle B$ for the same reason. By definition of equiangular triangles, $\triangle AEB$ is equiangular. However, a triangle that is equiangular is also equilateral, so $\triangle AEB$ is an equilateral triangle.

CHECK FOR UNDERSTANDING

Communicating Mathematics

Study the lesson. Then complete the following. 2. $\overline{ST} \cong \overline{RT}$, $\angle R \cong \angle S$

1. **Combine** Theorems 4–6 and 4–7 to form one *if and only if* statement.
2. **Name** the congruent sides and angles of isosceles $\triangle RST$ with base $\overline{RS}$.
3. $\triangle ABC$ with right angle C is flipped over $\overleftrightarrow{BC}$ to form $\triangle A'BC$ as shown at the right.
   a. Since a flip is a congruence transformation, what is true about $\overline{AB}$ and $\overline{A'B}$? $\overline{AB} \cong \overline{A'B}$
   b. What is true about $\angle A$ and $\angle A'$? $\angle A \cong \angle A'$
   c. Identify $\triangle ABA'$. isosceles

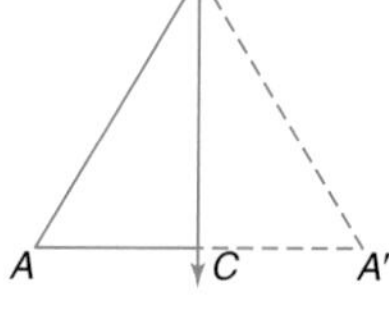

1. Two sides of a triangle are congruent if and only if the angles opposite those sides are congruent.

4. Use a compass and a straightedge to construct an equilateral triangle, an isosceles triangle that is not equilateral, and a scalene triangle. Cut out each triangle.
   a. How many ways can an equilateral triangle be folded to form two congruent triangles? 3 ways
   b. How many ways can an isosceles triangle that is not equilateral be folded to form two congruent triangles? 1 way
   c. How many ways can a scalene triangle be folded to form two congruent triangles? none

Guided Practice

Refer to the figure at the right for Exercises 5–7.

5. If $\overline{NL} \cong \overline{SL}$, name two congruent angles. $\angle 5 \cong \angle 11$
6. If $\angle 1 \cong \angle 4$, name two congruent segments.
7. If $\angle 9 \cong \angle 10$, name two congruent segments.
8. Draw $\triangle ABC$ with point D on $\overline{BC}$ such that $\overline{AD} \perp \overline{BC}$, but $\overline{BD}$ and $\overline{DC}$ are not congruent. See margin.

6. $\overline{ML} \cong \overline{MN}$
7. $\overline{MT} \cong \overline{MR}$

Teaching Tip Point out that Corollary 4-3 is an "if and only if" statement. Ask students to state this corollary another way. A triangle is equiangular if and only if it is equilateral.

In-Class Example

For Example 3
Rewrite Example 3 in the Student Edition as a two-column proof.

Given: $\overline{AB} \cong \overline{EB}$
$\angle DEC \cong \angle B$
Prove: $\triangle ABE$ is equilateral.
Proof:
Statements (Reasons)
1. $\overline{AB} \cong \overline{EB}$, $\angle DEC \cong \angle B$ (Given)
2. $\angle A \cong \angle AEB$ (Isosceles Triangle Th.)
3. $\angle AEB \cong \angle DEC$ (Vert. ∠s are ≅.)
4. $\angle A \cong \angle AEB \cong \angle B$ (≅ of ∠s is trans.)
5. $\triangle AEB$ is equiangular. (Def. equiangular △)
6. $\triangle AEB$ is equilateral. (If a △ is equiangular, it is equilateral.)

3 PRACTICE/APPLY

Check for Understanding

Exercises 1–14 are designed to help you assess your students' understanding through reading, writing, speaking, and modeling. You should work through Exercises 1–4 with your students and then monitor their work on Exercises 5–14.

Error Analysis
Students often fail to recognize isosceles triangles when the triangle is oriented so that it is resting on one of its congruent sides. Display isosceles triangles in different orientations and have students identify the congruent sides.

Additional Answer

8. Sample answer:

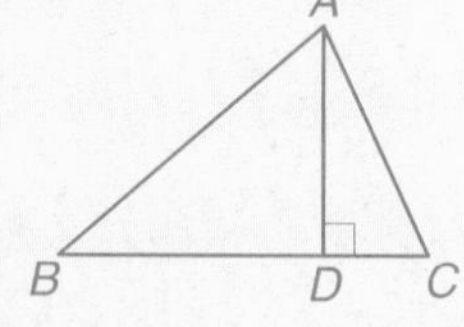

Reteaching

Using Diagrams Find the missing measures in each triangle.

1. $AC = 12$
$m\angle A = 36$
$m\angle B = 72$

2. 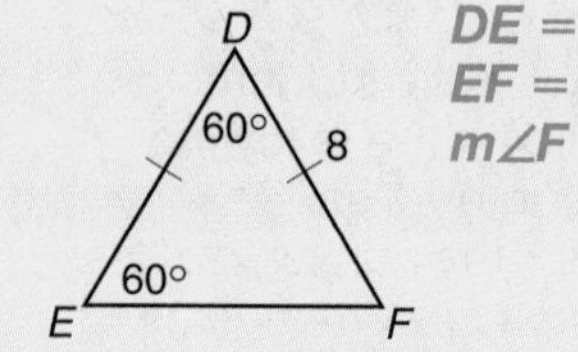

$DE = 8$
$EF = 8$
$m\angle F = 60$

Find the value of x.

9. 56

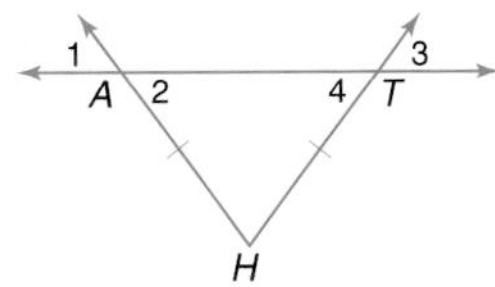

10. 7.5

11. $\triangle ABC$ has vertices $A(2, 5)$, $B(5, 2)$, and $C(2, -1)$. a. See margin
   a. Use the distance formula to show that $\triangle ABC$ is isosceles.
   b. Name the pair of congruent angles. $\angle A \cong \angle C$

Proof

12. **Given:** $\overline{HT} \cong \overline{HA}$

Prove: $\angle 1 \cong \angle 3$
See margin.

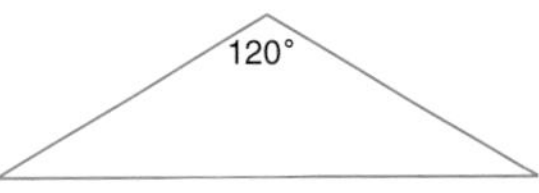

13. Write a two-column proof for Theorem 4–7. If two angles of a triangle are congruent, then the sides opposite those angles are congruent. See margin.

14. **Construction** A frame for a roof is in the form of an isosceles triangle. If the measure of the vertex angle is 120, find the measures of the other two angles. 30, 30

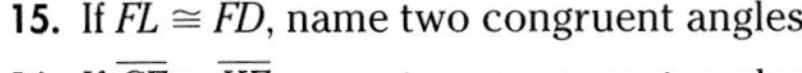

EXERCISES

Practice

15. $\angle 10 \cong \angle 6$
16. $\angle 3 \cong \angle 2$
17. $\angle 11 \cong \angle 9$
18. $\angle OEF \cong \angle EOF$
19. $\angle OEF \cong \angle OFE$
20. $\overline{LF} \cong \overline{LD}$
21. $\overline{EF} \cong \overline{EO}$
22. $\overline{DF} \cong \overline{DL}$

Refer to the figure at the right for Exercises 15–23.

15. If $\overline{FL} \cong \overline{FD}$, name two congruent angles.
16. If $\overline{CF} \cong \overline{KF}$, name two congruent angles.
17. If $\overline{FA} \cong \overline{FB}$, name two congruent angles.
18. If $\overline{OF} \cong \overline{EF}$, name two congruent angles.
19. If $\overline{FO} \cong \overline{OE}$, name two congruent angles.
20. If $\angle 10 \cong \angle DFL$, name two congruent segments.
21. If $\angle EOF \cong \angle EFO$, name two congruent segments.
22. If $\angle 6 \cong \angle LFD$, name two congruent segments.
23. If $\angle 12 \cong \angle 8$, name two congruent segments. $\overline{FB} \cong \overline{FA}$

For Exercise 24–26, draw $\triangle MTN$ with point X on $\overline{TN}$ such that the following conditions are satisfied. 24–25. See margin.

24. $\overline{XM}$ bisects $\angle TMN$, but $\overline{MT}$ is not congruent to $\overline{MN}$.
25. $\overline{TX} \cong \overline{XN}$, but $\overline{XM}$ does not bisect $\angle NMT$.
26. $\overline{TX} \cong \overline{XN}$, but $\overline{MX}$ is not perpendicular to $\overline{TN}$. See Solutions Manual.

Find the value of x.

27. 60

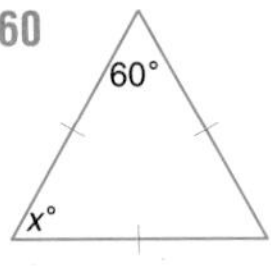

28. 15

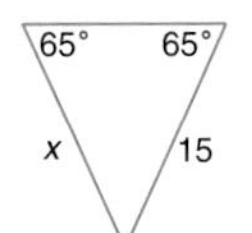

29. 45

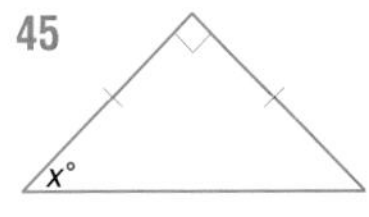

Additional Answers

24. Sample answer:

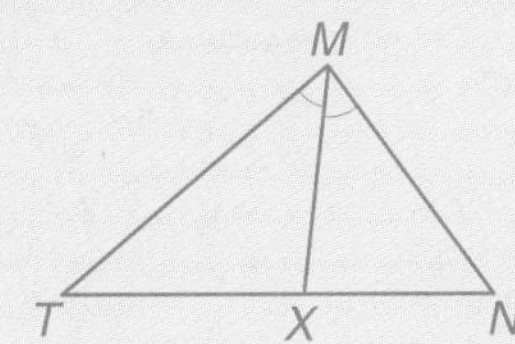

25. Sample answer:

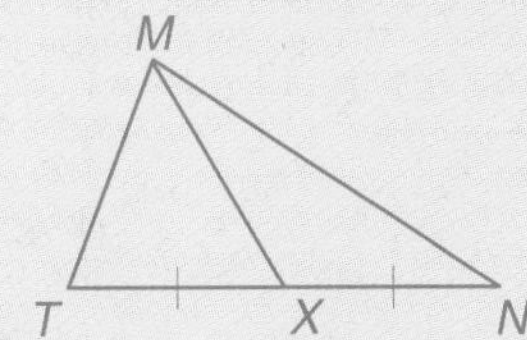

Assignment Guide

Core (with proof): 15–47 odd, 49–59
Core (informal): 15–35 odd, 45, 47, 49–54, 56–59
Enriched: 16–44 even, 45–59

For **Extra Practice**, see p. 772.

The red A, B, and C flags, printed only in the Teacher's Wraparound Edition, indicate the level of difficulty of the exercises.

Additional Answers

11a. $AB = \sqrt{(5-2)^2 + (2-5)^2} = \sqrt{18} = 3\sqrt{2}$
$BC = \sqrt{(2-5)^2 + (-1-2)^2} = \sqrt{18} = 3\sqrt{2}$
$AC = \sqrt{(2-2)^2 + (-1-5)^2} = \sqrt{36} = 6$
$\overline{AB} \cong \overline{BC}$

12. Given: $\overline{HT} \cong \overline{HA}$
Prove: $\angle 1 \cong \angle 3$
Proof:
Statements (Reasons)
1. $\overline{HT} \cong \overline{HA}$ (Given)
2. $\angle 2 \cong \angle 4$ (Isosceles $\triangle$ Th.)
3. $\angle 1 \cong \angle 2$ (Vert. $\angle$s are $\cong$.)
4. $\angle 1 \cong \angle 4$ ($\cong$ of $\angle$s is trans.)
5. $\angle 4 \cong \angle 3$ (Vert. $\angle$s are $\cong$.)
6. $\angle 1 \cong \angle 3$ ($\cong$ of $\angle$s is trans.)

13. Given: $\triangle ABC$
$\angle A \cong \angle C$
Prove: $\overline{AB} \cong \overline{CB}$

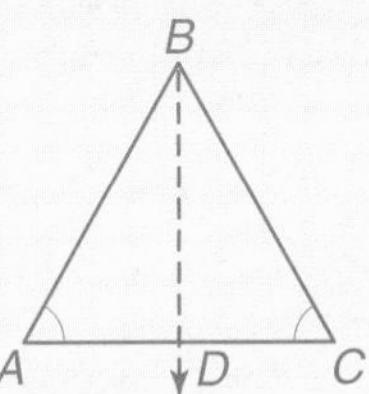

Proof:
Statements (Reasons)
1. Let $\overrightarrow{BD}$ bisect $\angle ABC$. (Protractor Post.)
2. $\angle ABD \cong \angle CBD$ (Def. $\angle$ bisector)
3. $\angle A \cong \angle C$ (Given)
4. $\overline{BD} \cong \overline{BD}$ ($\cong$ of segments is reflexive.)
5. $\triangle ABD \cong \triangle CBD$ (AAS)
6. $\overline{AB} \cong \overline{CB}$ (CPCTC)

Additional Answers

36. Given: $\angle 3 \cong \angle 4$
Prove: $\overline{MA} \cong \overline{MC}$
Proof:
Statements (Reasons)
1. $\angle 3 \cong \angle 4$ (Given)
2. $\angle 2$ and $\angle 4$, $\angle 1$ and $\angle 3$ form a linear pair. (Def. linear pair)
3. $\angle 2$ and $\angle 4$ are supp., $\angle 1$ and $\angle 3$ are supp. (If 2 ∠s form a linear pair, they are supp.)
4. $\angle 2 \cong \angle 1$ (∠s that are supp. to ≅ ∠s are ≅.)
5. $\overline{MA} \cong \overline{MC}$ (If 2 ∠s of a △ are ≅, then the sides opp. those ∠s are ≅.)

37. Given: $\angle 5 \cong \angle 6$
$\overline{FR} \cong \overline{GS}$
Prove: $\angle 4 \cong \angle 3$
Proof:
Statements (Reasons)
1. $\angle 5 \cong \angle 6$, $\overline{FR} \cong \overline{GS}$ (Given)
2. $\angle 1 \cong \angle 2$ (Vert. ∠s are ≅.)
3. $\triangle FXR \cong \triangle GXS$ (AAS)
4. $\overline{FX} \cong \overline{GX}$ (CPCTC)
5. $\angle 4 \cong \angle 3$ (Isosceles △ Th.)

Find the value of x.

30. 18

31. 18

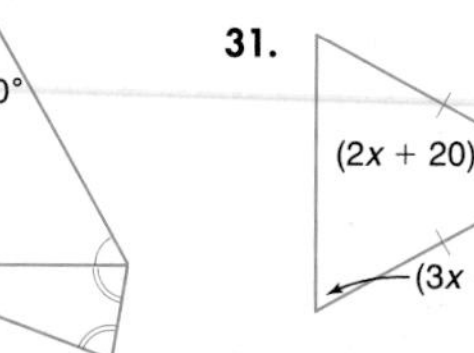

32. 30

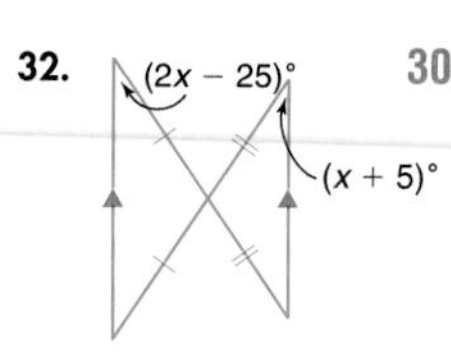

33. If $\triangle DEF$ has vertices $D(-4, -3)$, $E(-2, -1)$, and $F(0, -3)$, use the distance formula and slopes of the congruent sides to show that $\triangle DEF$ is a right isosceles triangle. See Solutions Manual.

34. In $\triangle ABD$, $\overline{AB} \cong \overline{BD}$, $m\angle A$ is 12 less than 3 times a number, and $m\angle D$ is 13 more than twice the same number. Find $m\angle B$. 54

35. Find each angle measure if $m\angle 1 = 30$.
$m\angle 2 = 120$, $m\angle 3 = 30$, $m\angle 4 = 60$, $m\angle 5 = 60$, $m\angle 6 = 60$, $m\angle 7 = 120$, $m\angle 8 = 90$

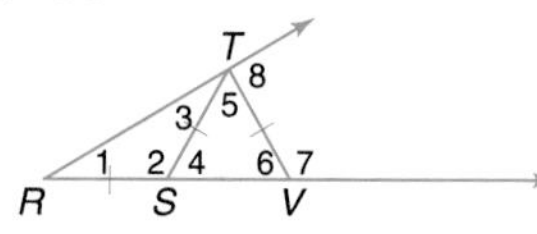

Proof

Write a proof for each of the following. 36–38. See margin.

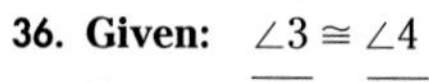

36. Given: $\angle 3 \cong \angle 4$
Prove: $\overline{MA} \cong \overline{MC}$

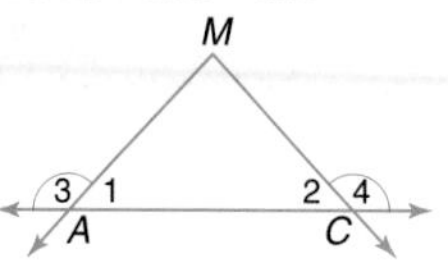

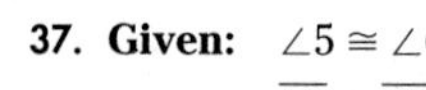

37. Given: $\angle 5 \cong \angle 6$
$\overline{FR} \cong \overline{GS}$
Prove: $\angle 4 \cong \angle 3$

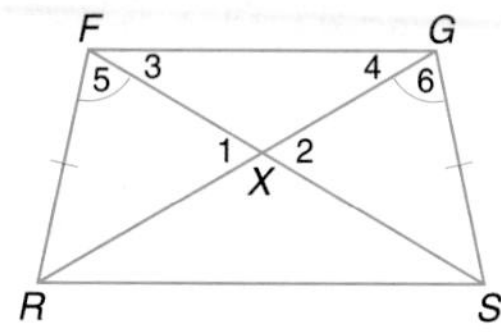

38. Given: $\angle 1 \cong \angle 4$
$\overline{NA} \cong \overline{TC}$
Prove: $\angle 3 \cong \angle 2$

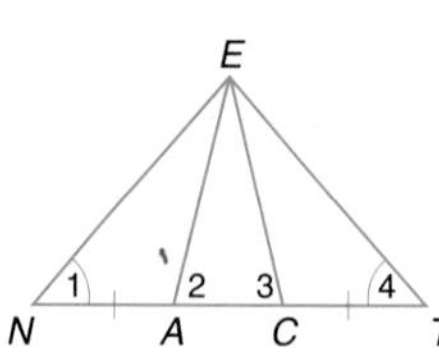

39. Given: $\triangle CAN$ is an isosceles triangle with vertex $\angle N$.
$\overline{CA} \parallel \overline{BE}$
Prove: $\triangle NEB$ is an isosceles triangle.

39–41. See Solutions Manual.

C

40. Given: $\overline{PH}$ bisects $\angle YHX$.
$\overline{HP} \perp \overline{YX}$
Prove: $\triangle YHX$ is an isosceles triangle.

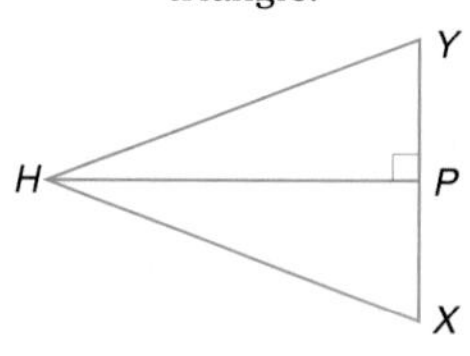

41. Given: $\triangle IOE$ is an isosceles triangle with base $\overline{OE}$.
$\overline{AO}$ bisects $\angle IOE$.
$\overline{AE}$ bisects $\angle IEO$.
Prove: $\triangle AEO$ is an isosceles triangle.

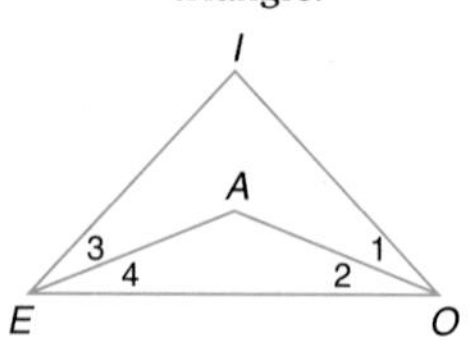

Study Guide Masters, p. 25

NAME______________ DATE________

Study Guide — Student Edition Pages 222–228

Analyzing Isosceles Triangles

Remember that two sides of an isosceles triangle are congruent. Two important theorems about isosceles triangles are as follows.

If two sides of a triangle are congruent, then the angles opposite those sides are congruent.
If two angles of a triangle are congruent, then the sides opposite those angles are congruent.

Example: Find the value of x.

Since $\overline{AB} \cong \overline{BC}$, the angles opposite $\overline{AB}$ and $\overline{BC}$ are congruent. So $m\angle A = m\angle C$.

Therefore, $3x - 10 = 2x + 6$
$x = 16$

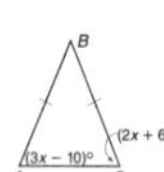

Find the value of x.

1. 20
2. 8
3. 76
4. 15

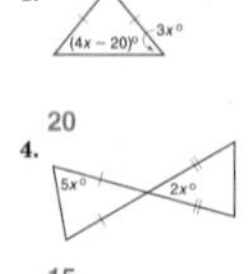

5. 16

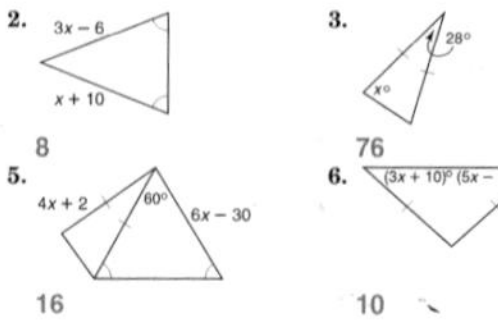

6. 10

Write a two-column proof.

7. **Given:** $\angle 1 \cong \angle 4$
Prove: $\overline{DE} \cong \overline{FE}$

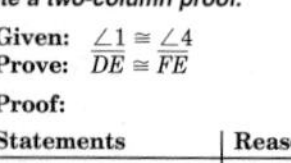

Proof:

Statements	Reasons
a. $\angle 1 \cong \angle 4$	a. Given
b. $\angle 2 \cong \angle 4$	b. Vertical angles are congruent.
c. $\angle 1 \cong \angle 2$	c. Congruence of angles is transitive.
d. $\overline{DE} \cong \overline{FE}$	d. If two angles are congruent, then the sides opposite those angles are congruent.

Additional Answer

38. Given: $\angle 1 \cong \angle 4$
$\overline{NA} \cong \overline{TC}$
Prove: $\angle 3 \cong \angle 2$
Proof:
Statements (Reasons)
1. $\angle 1 \cong \angle 4$, $\overline{NA} \cong \overline{TC}$ (Given)
2. $\overline{NE} \cong \overline{TE}$ (If 2 ∠s of a △ are ≅, then the sides opp. those ∠s are ≅.)
3. $\triangle NEA \cong \triangle TEC$ (SAS)
4. $\overline{EA} \cong \overline{EC}$ (CPCTC)
5. $\angle 3 \cong \angle 2$ (Isosceles △ Th.)

42. A triangle is equilateral if and only if it is equiangular. (Corollary 4–3) (*Hint*: Since the statement contains if and only if, the proof must show an equilateral triangle is equiangular and that an equiangular triangle is equilateral.) **See margin.**

43. Each angle of an equilateral triangle measures 60°. (Corollary 4–4) **See Solutions Manual.**

Programming

44. Use the TI-82/83 program to find the measures of the base angles of an isosceles triangle that has a vertex angle of the given measure.
   a. 26 **77, 77** b. 120 **30, 30**
   c. 78 **51, 51** d. 101 **39.5, 39.5**

```
PROGRAM:ISOTRI
: Input "VERTEX ANGLE =", A
: (180−A)/2→B
: Disp "BASE ANGLE =", B
: Stop
```

Critical Thinking

45. Draw an isosceles triangle ABC with vertex angle at A. Find the midpoints of each side. Label the midpoint of $\overline{AB}$ point D, the midpoint of $\overline{BC}$ point E, and the midpoint of $\overline{AC}$ point F. Draw $\triangle DEF$. **a–b. See Solutions Manual.**
   a. Name a pair of congruent triangles. Explain your reasoning.
   b. Name three isosceles triangles. Explain your reasoning.

Applications and Problem Solving

46. **Design** Tiled patterns formed by repeating figures to fill a plane without gaps or overlaps are called *tessellations*. Tessellations are found in woven rugs, pottery, and flooring. An example of a tessellation using octagons and squares is given at the right.
   a. Draw a tessellation using only isosceles triangles. **a–c. See Solutions Manual.**
   b. Draw a tessellation using squares and equilateral triangles.
   c. Draw a tessellation using other types of figures.

47. **Carpentry** Before the invention of the bubble level, carpenters used a device called a *plumb level* to verify that a surface was level. This level consisted of a frame in the shape of an isosceles triangle with the midpoint of the base (point M in the figure at the right) marked. A plumb line was suspended from the vertex angle. To use this instrument, the carpenter would hold it upright with the base resting on the surface to be leveled. If the surface were level, over what point on the base do you think the plumb line would hang? Explain your reasoning. **See Solutions Manual.**

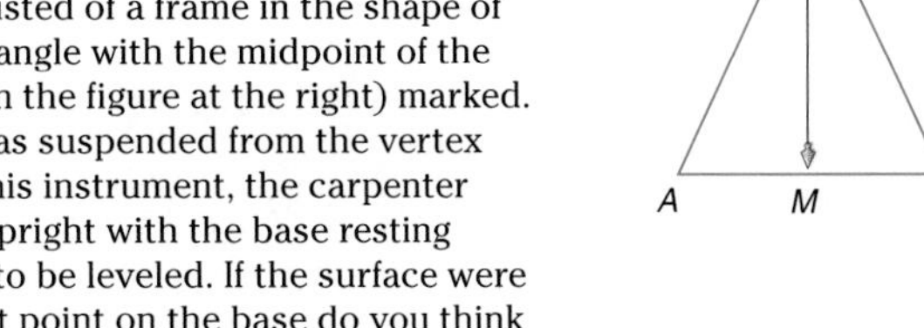

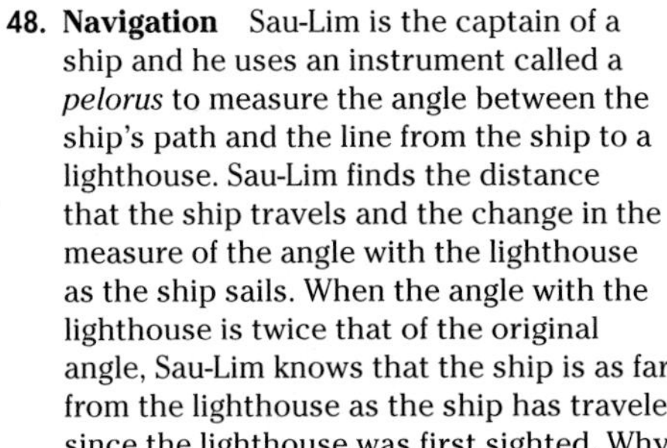

48. **Navigation** Sau-Lim is the captain of a ship and he uses an instrument called a *pelorus* to measure the angle between the ship's path and the line from the ship to a lighthouse. Sau-Lim finds the distance that the ship travels and the change in the measure of the angle with the lighthouse as the ship sails. When the angle with the lighthouse is twice that of the original angle, Sau-Lim knows that the ship is as far from the lighthouse as the ship has traveled since the lighthouse was first sighted. Why? **See Solutions Manual.**

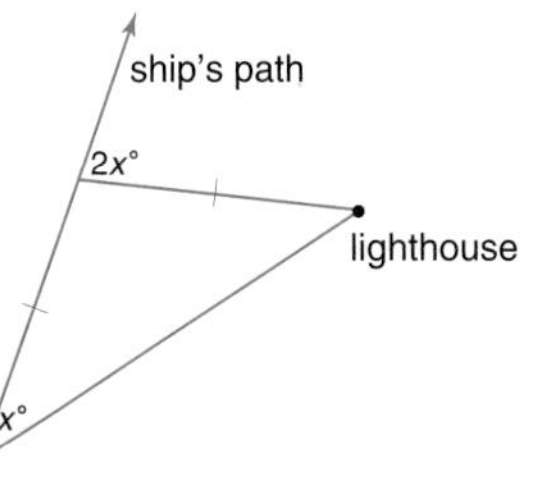

Tech Prep

Carpenter Carpenters need to have a thorough understanding of supplementary and complementary angles. They miter corners by cutting complementary angles. A corner cabinet would probably be an isosceles triangle. For more information on tech prep, see the *Teacher's Handbook*.

Using the Programming Exercises

The program given in Exercise 44 is for use with a TI-82 or TI-83 graphing calculator. For other programmable calculators, have students consult their owner's manual for commands similar to those presented here.

Additional Answer

42. Given: $\triangle ABC$ is an equilateral triangle.
Prove: $\triangle ABC$ is an equiangular triangle.

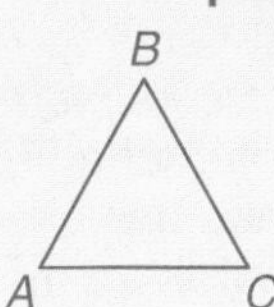

Proof:
Statements (Reasons)
1. $\triangle ABC$ is an equilateral triangle. (Given)
2. $\overline{AB} \cong \overline{AC} \cong \overline{BC}$ (Def. equilateral $\triangle$)
3. $\angle A \cong \angle B \cong \angle C$ (Isosceles $\triangle$ Th.)
4. $\triangle ABC$ is an equiangular triangle. (Def. equiangular)

Given: $\triangle ABC$ is an equiangular triangle.
Prove: $\triangle ABC$ is an equilateral triangle.
Proof:
Statements (Reasons)
1. $\triangle ABC$ is an equiangular triangle. (Given)
2. $\angle A \cong \angle B \cong \angle C$ (Def. equiangular)
3. $\overline{AB} \cong \overline{AC} \cong \overline{BC}$ (If 2 ∠s of a $\triangle$ are ≅, then the sides opp. those ∠s are ≅.)
4. $\triangle ABC$ is an equilateral triangle. (Def. equilateral)

Practice Masters, p. 25

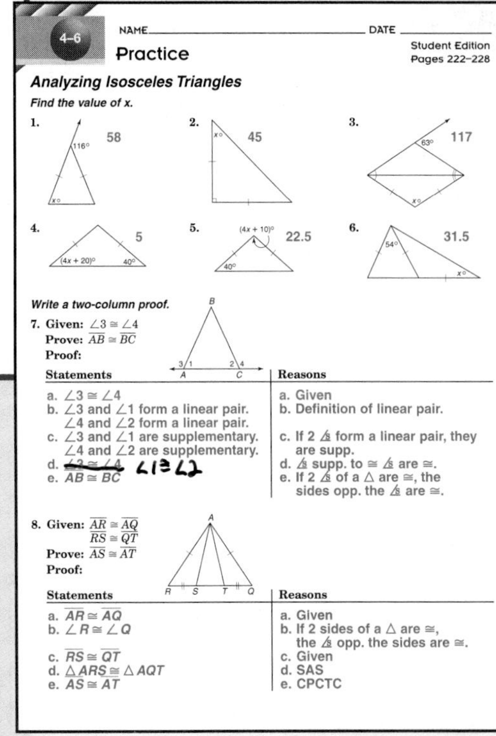

NAME ____________ DATE ________

4-6 Practice — Student Edition Pages 222–228

Analyzing Isosceles Triangles

Find the value of x.

1. 116°, x° — 58
2. x° — 45
3. 63°, x° — 117
4. (4x + 20)°, 40° — 5
5. (4x + 10)°, 40° — 22.5
6. 54°, x° — 31.5

Write a two-column proof.

7. Given: $\angle 3 \cong \angle 4$
Prove: $\overline{AB} \cong \overline{BC}$
Proof:

Statements	Reasons
a. $\angle 3 \cong \angle 4$	a. Given
b. ∠3 and ∠1 form a linear pair. ∠4 and ∠2 form a linear pair.	b. Definition of linear pair.
c. ∠3 and ∠1 are supplementary. ∠4 and ∠2 are supplementary.	c. If 2 ∠s form a linear pair, they are supp.
d. ~~$\angle 3 \cong \angle 4$~~ $\angle 1 \cong \angle 2$	d. ∠s supp. to ≅ ∠s are ≅.
e. $\overline{AB} \cong \overline{BC}$	e. If 2 ∠s of a △ are ≅, the sides opp. the ∠s are ≅.

8. Given: $\overline{AR} \cong \overline{AQ}$, $\overline{RS} \cong \overline{QT}$
Prove: $\overline{AS} \cong \overline{AT}$
Proof:

Statements	Reasons
a. $\overline{AR} \cong \overline{AQ}$	a. Given
b. $\angle R \cong \angle Q$	b. If 2 sides of a △ are ≅, the ∠s opp. the sides are ≅.
c. $\overline{RS} \cong \overline{QT}$	c. Given
d. $\triangle ARS \cong \triangle AQT$	d. SAS
e. $\overline{AS} \cong \overline{AT}$	e. CPCTC

4 ASSESS

Closing Activity

Modeling Have students use ruler, protractor, scissors, and heavy paper to construct an isosceles triangle and an equilateral triangle. Ask them which properties they used to construct their models.

Chapter 6 Quiz D (Lesson 4-6) is available in the *Assessment and Evaluation Masters*, p. 101.

Additional Answer

49. SSA means 2 sides and a nonincluded ∠ are ≅ to the corr. sides and ∠. It cannot be used as a proof for ≅ triangles. SAS means 2 sides and the included ∠ are ≅ to 2 sides and the included ∠. It can be used as a proof for ≅ triangles.

Sample Answers for Working on the Investigation

6. Some thumbprints exhibit a spiral pattern.
7. Whorls, loops, and arches are used in fingerprinting techniques.

Enrichment Masters, p. 25

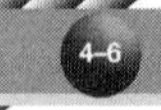

NAME______________________ DATE ________

4-6 Enrichment

Student Edition Pages 222–228

Triangle Challenges

Some problems include diagrams. If you are not sure how to solve the problem, begin by using the given information. Find the measures of as many angles as you can, writing each measure on the diagram. This may give you more clues to the solution.

1. Given: $BE = BF$ and $ABFG$ and $BCDE$ each have opposite sides parallel and congruent. Find $m\angle ABC$.

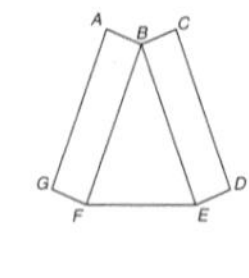

148

2. Given: $AC = AD$, and $\overline{AB} \perp \overline{BD}$, $m\angle DAC = 44$ and $\overline{CE}$ bisects $\angle ACD$. Find $m\angle DEC$.

78

3. Given: $m\angle UZY = 90$, $\triangle YZU \cong \triangle VWX$, $UVXY$ is a square (all sides congruent, all angles right angles). Find $m\angle WZY$.

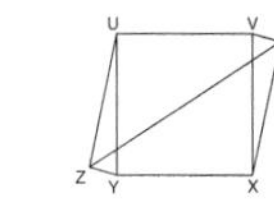

45

4. Given: $m\angle N = 120$, $\overline{JN} \cong \overline{MN}$, $\triangle JNM \cong \triangle KLM$. Find $m\angle JKM$.

15

Mixed Review

49. Explain the difference between SSA and SAS. (Lesson 4–5) See margin.
50. **Eliminate the Possibilities** Four friends have birthdays in January, February, August, and September. Amy was not born in the winter. Kiana celebrates her birthday during the summer vacation. Timothy's birthday is the month after Pablo's. Give the month of each person's birthday. (Lesson 4–5) Amy, Sept.; Kiana, Aug.; Timothy, Feb.; Pablo, Jan.
51. Suppose $\triangle CDE \cong \triangle PQR$. List all the pairs of corresponding parts. (Lesson 4–3) See Solutions Manual.
52. The measures of two interior angles of a triangle are 54 and 79. What is the measure of the exterior angle opposite these angles? (Lesson 4–2) 133
53. **Algebra** The legs of an isosceles triangle are $(6x - 6)$ units and $(x + 9)$ units long. Find the length of the legs. (Lesson 4–1) 12 units
54. Find the slope of the line passing through $A(7, -3)$ and $B(6, -1)$. (Lesson 3–3) -2
55. Write a two-column proof. (Lesson 2–5) See Solutions Manual.

Given: $\overline{AB} \cong \overline{EF}$
$\overline{EF} \cong \overline{JK}$
$\overline{BC} \cong \overline{HJ}$

Prove: $\overline{AC} \cong \overline{HK}$

56. Write an if-then statement for *Every whole number is an integer.* (Lesson 2–2) If a number is a whole number, then it is an integer.
57. Find the length and midpoint of the segment with endpoints $C(7, 4)$ and $D(-3, 8)$. (Lesson 1–4) $2\sqrt{29} \cong 10.77$; (2, 6)

58. D = {7, 4, −19, 5}; R = {−2, 0, −1, 5}

INTEGRATION Algebra

58. State the domain and range of the relation {(7, −2), (4, 0), (−19, −1), (5, 5)}.
59. Find the x- and y-intercepts of the graph of $9x - 2y = 18$. 2; −9

WORKING ON THE In·ves·ti·ga·tion

Refer to the Investigation on pages 176–177.

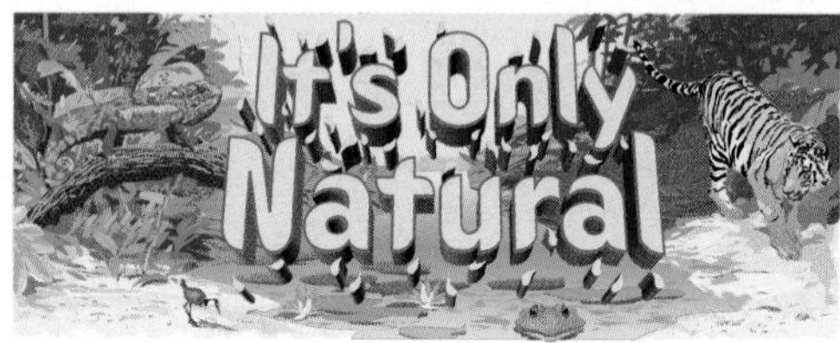

Because each person has unique fingerprints, they are often used to identify people. Even though each fingerprint is different, there are patterns that can be observed in fingerprints.

1 Rub a pencil with soft lead on a piece of paper.
2 Roll your thumb through the pencil mark a few times to coat the thumb with dust.
3 Press the thumb on a piece of clear tape to create a thumbprint.
4 Place the tape on a recording sheet and note your name.
5 Create a thumbprint for each person in your group.
6 Look for similarities and differences among the thumbprints. Are any of the natural patterns described on pages 176–177 displayed in any of the thumbprints? 6–7. See margin.
7 Research fingerprinting techniques. What patterns are used?

Add the results of your work to your Investigation Folder.

Extension

Communication Discuss the statement, "You need to know the measure of only one angle in an isosceles triangle in order to find the measure of all three." To what extent is this statement true? Given one measure, you also need to know whether that angle is a base angle or the vertex angle.

In·ves·ti·ga·tion

Working on the Investigation

The Investigation on pages 176–177 is designed to be a long-term project that is completed over several days or weeks. Encourage students to keep their materials in their Investigation Folder as they work on the Investigation.

HIGHLIGHTS

VOCABULARY

After completing this chapter, you should be able to define each term, property, or phrase and give an example or two of each.

Geometry

acute triangle (p. 180)
auxiliary line (p. 189)
base (p. 182)
base angles (p. 182)
congruence transformation (p. 197)
congruent triangles (p. 196)
corollary (p. 192)
equiangular triangle (p. 181)
equilateral triangle (p. 181)
exterior angle (pp. 188, 190)
flow proof (p. 191)
included angle (pp. 204, 207)
included side (pp. 205, 207)
isosceles triangle (p. 181)
legs (p.182)
obtuse triangle (p. 180)
polygon (p. 180)
remote interior angles (pp. 188, 190)
right triangle (p. 180)
scalene triangle (p. 181)
sides (p. 180)
triangle (p. 180)
vertex angle (p. 182)
vertices (p. 180)

Problem Solving

eliminate the possibilities (p. 215)

UNDERSTANDING AND USING THE VOCABULARY

State whether each sentence is *true* or *false*. If false, replace the underlined word(s) to make a true statement. 9. false; flow proof 11. true

1. A <u>triangle</u> is a polygon with three sides. true
2. The side opposite the right angle of a triangle is called the <u>leg</u>. false; hypotenuse
3. An isosceles triangle has two <u>vertex angles</u>. false; base angles
4. A triangle in which no two sides are congruent is called a <u>scalene</u> triangle. true
5. AAS refers to two angles and their <u>included side</u>. false; nonincluded
6. The endpoints of the sides of a polygon are called <u>vertices</u>. true
7. SSS, SAS, ASA, and AAS are ways to prove that two triangles are <u>congruent</u>. true
8. A triangle with <u>two</u> acute angles is called acute. false; three
9. A <u>paragraph proof</u> uses boxes to show the steps of a proof and arrows to show the order of the steps.
10. An <u>equiangular triangle</u> is defined as a triangle with three congruent sides. false; equilateral triangle
11. The sum of the measures of two remote interior angles is equal to the measure of the <u>exterior angle</u>.
12. An equilateral triangle is also an <u>isosceles triangle</u>. true
13. In an isosceles triangle, the side opposite the vertex angle is the <u>hypotenuse</u>. false; base
14. An <u>auxiliary line</u> is a statement that can be easily proved using a theorem. false; corollary
15. A triangle with angle measures 95, 35, and 50 is an <u>obtuse triangle</u>. true

Instructional Resources

Three multiple-choice tests and three free-response tests are provided in the *Assessment and Evaluation Masters*. Forms 1A and 2A are for honors pacing, Forms 1B and 2B are for average pacing, and Forms 1C and 2C are for basic pacing. Chapter 4 Test, Form 1B, is shown at the right. Chapter 4 Test, Form 2B, is shown on the next page.

Postulates, Theorems, and Corollaries

A complete list of postulates, theorems, and corollaries begins on page 806.

Using the CHAPTER HIGHLIGHTS

The Chapter Highlights begins with a listing of the new terms, properties, and phrases that were introduced in this chapter. Have students define each term and provide an example or two of it, if appropriate.

Assessment and Evaluation Masters, pp. 87–88

4 NAME________ DATE ________

Chapter 4 Test, Form 1B

Write the letter for the correct answer in the blank at the right of each problem.

1. If $\triangle XYZ$ is isosceles and $XY = YZ$, which statement must be true?
A. $XY = XZ$ B. $\angle Z \cong \angle X$ C. $XZ \neq YZ$ D. none of these — 1. B
2. $\triangle ABC$ is an isosceles triangle with vertex angle B, $AB = 5x - 28$, $AC = x + 5$, and $BC = 2x + 11$. Find the length of the base.
A. 13 B. 18 C. 37 D. 39 — 2. B
3. In $\triangle PQR$, what is the side opposite $\angle R$?
A. $\overline{PQ}$ B. $\overline{PR}$ C. $\overline{QR}$ D. $\angle P$ — 3. A
4. In $\triangle LMN$, if $m\angle L > 90$ and $m\angle M = m\angle N$, then $\triangle LMN$ is
A. acute and scalene. B. right and scalene. C. right and isosceles. D. obtuse and isosceles. — 4. D
5. The measures of two angles of a triangle are 52 and 89. What is the measure of the third angle?
A. 29 B. 39 C. 128 D. 141 — 5. B
6. For $\triangle MNO$ and $\triangle RST$, if $\angle N \cong \angle S$ and $\angle R \cong \angle M$, then
A. $\triangle MNO \cong \triangle RST$. B. $\overline{MN} \cong \overline{RS}$. C. $\overline{ON} \cong \overline{TR}$. D. $\angle T \cong \angle O$. — 6. D
7. If $\triangle PQR \cong \triangle DEF$, then $\overline{RP}$ is congruent to
A. $\overline{PQ}$. B. $\overline{DE}$. C. $\overline{FD}$. D. $\overline{EF}$. — 7. C
8. Given $\triangle ABC \cong \triangle XYZ$, $AB = 38$, $YZ = 28$, and $XY = 5x + 8$. Find the value of x.
A. 30 B. 20 C. 6 D. 4 — 8. C
9. Given $\triangle RST \cong \triangle HIJ$, $m\angle R = 97$, $m\angle J = 37$, and $m\angle S = 4x + 14$, find the value of x.
A. 8 B. 32 C. 46 D. 134 — 9. A
10. If $\triangle ABC \cong \triangle DEF \cong \triangle GHI$, what must be true about AC and GI?
A. $AC > GI$ B. $AC < GI$ C. $AC = GI$ D. no conclusion possible — 10. C
11. Which of the following is not a postulate used to prove the congruence of triangles?
A. AAS B. SSS C. AAA D. ASA — 11. C
12. In $\triangle QXR$, what angle is opposite $\overline{XQ}$?
A. $\angle X$ B. $\angle R$ C. $\overline{XR}$ D. $\angle Q$ — 12. B
13. Find the value of x.
A. $16\frac{4}{7}$ B. $18\frac{3}{7}$ C. $25\frac{3}{7}$ D. $36\frac{5}{7}$ — 13. B
(7x − 9)° 45° 75°

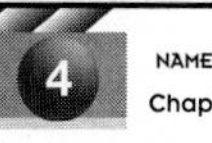

NAME________ DATE ________

Chapter 4 Test, Form 1B (continued)

14. Name one additional pair of corresponding parts that need to be congruent in order to prove that $\triangle FIG \cong \triangle TOM$ by SAS.
A. $\overline{FG} \cong \overline{MT}$ B. $\angle F \cong \angle T$ C. $\overline{IG} \cong \overline{OM}$ D. $\overline{FI} \cong \overline{OT}$ — 14. C
F G M T I O
15. Determine which triangles in the figure are congruent if $\angle BAE \cong \angle BCD$ and $\overline{BD} \cong \overline{BE}$.
A. $\triangle ABE \cong \triangle CBD$
B. $\triangle ADF \cong \triangle CEF$
C. $\triangle ADC \cong \triangle CEA$
D. $\triangle ABC \cong \triangle AFC$ — 15. A
B D E F A C
16. **Eliminate the Possibilities** Marsha, Brad, and Joann each play a different sport. One plays water polo, one plays ice hockey, and one plays basketball. Marsha cannot swim or ice skate. Joann does not play ice hockey. What sport does Brad play?
A. water polo B. ice hockey C. basketball D. none of these — 16. B
17. In isosceles $\triangle ABC$, $\angle C$ is the vertex angle. If $m\angle A = 3x - 6$ and $m\angle C = 2x$, find the measure of $\angle B$.
A. 24 B. 46 C. 48 D. 66 — 17. D
18. Classify the triangle DEF with vertices $D(1, 5)$, $E(-3, 1)$, and $F(-1, -3)$.
A. right B. isosceles C. scalene D. equilateral — 18. C
19. The measure of a base angle of an isosceles triangle is 52. What is the measure of the vertex angle?
A. 52 B. 76 C. 104 D. 128 — 19. B
20. What conclusion can be drawn from the given information?
A. $\angle BAC = \angle BCA$
B. $\angle DAF = \angle ECF$
C. $\angle ADC = \angle CEA$
D. $\angle FAB = \angle FCB$ — 20. C
B D E F A C

Bonus

Which of the following might be the coordinates of the vertices of $\triangle PQR$ if $\triangle PQR$ is a right isosceles triangle?
A. $P(3, 4)$, $Q(7, 6)$, $R(5, 1)$ B. $P(7, 6)$, $Q(11, 6)$, $R(10, 10)$ C. $P(4, 4)$, $Q(10, 7)$, $R(7, -2)$ D. $P(5, 5)$, $Q(15, 5)$, $R(10, 15)$ — Bonus C

Using the STUDY GUIDE AND ASSESSMENT

Skills and Concepts Encourage students to refer to the objectives and examples on the left as they complete the review exercises on the right.

Assessment and Evaluation Masters, pp. 93–94

NAME__________ DATE __________

Chapter 4 Test, Form 2B

Determine whether each statement is true or false. If false, draw a counterexample.

1. A scalene triangle is never isosceles. — 1. true
2. An isosceles triangle can be equilateral. — 2. true
3. A scalene triangle is never right. — 3. False; see students' work.
4. All equilateral triangles are acute. — 4. true

Use the given conditions to classify each triangle by its angles and by its sides.

5. $\triangle HEY$, $m\angle E > 90$, $HE \neq EY$, $EY \neq YH$, $HE \neq YH$ — 5. obtuse, scalene
6. $\triangle TIE$, $m\angle T > 90$, $m\angle I = m\angle E$ — 6. obtuse, isosceles

Find the value of x.

7. 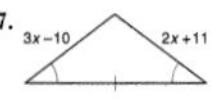

8. 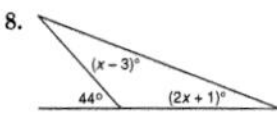

7. 21
8. 14

Write a congruence statement for the triangles in each diagram.

9. 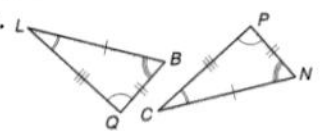

10. 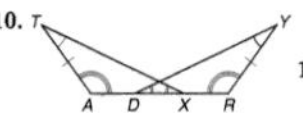
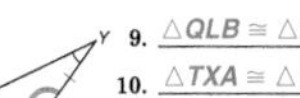

9. $\triangle QLB \cong \triangle PCN$
10. $\triangle TXA \cong \triangle YDR$

Determine which postulate or theorem can be used to prove the triangles congruent. If it is not possible to prove them congruent, write not possible.

11. $\triangle PQR \cong \triangle PTS$

12. 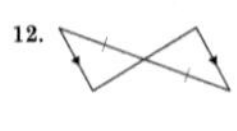

11. SSS
12. ASA

NAME__________ DATE __________

Chapter 4 Test, Form 2B (continued)

The following proof counts for Questions 13–18. Write the missing statement or reason in each blank.

Given: $\overline{AE} \cong \overline{DE}$
$\angle EBC \cong \angle ECB$
Prove: $\overline{AB} \cong \overline{DC}$

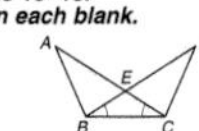

Statements	Reasons
a. $\overline{AE} \cong \overline{DE}$	a. Given
b. $\angle AEB \cong \angle DEC$	b. (Question 13)
c. $\angle EBC \cong \angle ECB$	c. Given
d. (Question 14)	d. (Question 15)
e. (Question 16)	e. (Question 17)
f. $\overline{AB} \cong \overline{DC}$	f. (Question 18)

13. Vert. ∠s are ≅.
14. $\overline{BE} \cong \overline{CE}$
15. Isos. △ Th.
16. $\triangle ABE \cong \triangle DCE$
17. SAS
18. CPCTC

19. **Eliminate the Possibilities** Alex, John, and Heather each live in different cities. One lives in Cleveland, one lives in San Diego, and one lives in Salt Lake City. Alex does not live in Utah. Heather does not live west of the Mississippi River. Who lives in San Diego? — 19. Alex

20. ***Write a two-column proof on a separate sheet of paper.***
Given: $\angle BAC \cong \angle BCA$
$\overline{BD} \cong \overline{BE}$
Prove: $\angle BDC \cong \angle BEA$

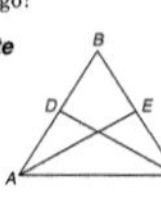

20. a. $\angle BAC \cong \angle BCA$ (Given)
b. $\overline{AB} \cong \overline{BC}$ (Isos. △ Th.)
c. $\angle B \cong \angle B$ (Refl. Prop. of ≅ angles)
d. $\overline{BD} \cong \overline{BE}$ (Given)
e. $\triangle ABE \cong \triangle CBD$ (SAS)
f. $BDC \cong \angle BEA$ (CPCTC)

Bonus

If $\triangle MNP \cong \triangle PNM$, classify $\triangle MNP$ by its sides. — Bonus sosceles

CHAPTER 4 STUDY GUIDE AND ASSESSMENT

SKILLS AND CONCEPTS

OBJECTIVES AND EXAMPLES

Upon completing this chapter, you should be able to:

- identify parts of triangles and classify triangles by their parts (Lesson 4–1)

Types of Triangles

Classification by Angles	
acute	three acute angles
obtuse	one obtuse angle
right	one right angle
equiangular	three congruent angles
Classification by Sides	
scalene	no two sides congruent
isosceles	at least two sides congruent
equilateral	three sides congruent

REVIEW EXERCISES

Use these exercises to review and prepare for the chapter test.

In the figure, $\angle STU$ and $\angle SVU$ are right angles, and the congruent parts are indicated.

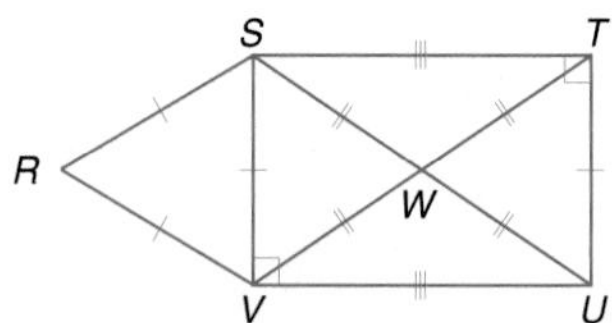

16. Identify the equilateral triangle(s). $\triangle RSV$
17. Identify the right triangle(s). $\triangle SUV$, $\triangle STU$
18. Identify the acute triangle(s).
19. Identify the obtuse triangle(s). $\triangle SWT$, $\triangle VWU$
20. Identify the triangles that are isosceles but not equilateral. $\triangle SWV$, $\triangle SWT$, $\triangle TWU$, $\triangle UWV$
21. Name the side opposite $\angle R$. $\overline{SV}$
22. Name the angle opposite $\overline{RV}$. $\angle RSV$

18. $\triangle RSV$, $\triangle SWV$, $\triangle TWU$

- apply the Angle Sum Theorem and Exterior Angle Theorem (Lesson 4–2)

Find the values of x and y.

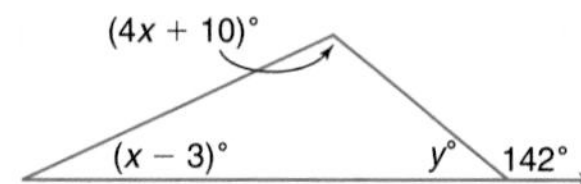

By the Exterior Angle Theorem,

$$(x - 3) + (4x + 10) = 142$$
$$5x = 135$$
$$x = 27$$

By the Angle Sum Theorem,

$$(x - 3) + (4x + 10) + y = 180$$
$$(27 - 3) + [4(27) + 10] + y = 180$$
$$142 + y = 180$$
$$y = 38$$

In the figure, $\overline{RK} \parallel \overline{SL}$ and $\overline{RS} \perp \overline{SL}$. Find the measure of each angle.

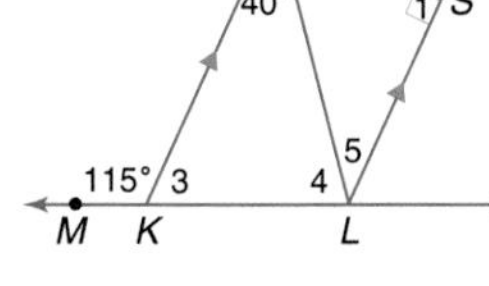

23. $m\angle 1$ 90
24. $m\angle 2$ 50
25. $m\angle 3$ 65
26. $m\angle 4$ 75
27. $m\angle 5$ 40

Refer to the figure at the right to find each value.

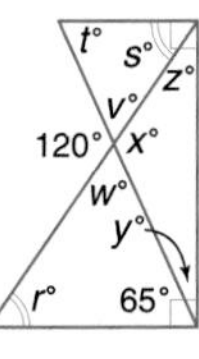

28. x 120
29. r 55
30. w 60
31. y 25
32. z 35
33. s 55
34. v 60
35. t 65

GLENCOE Technology

Test and Review Software

You may use this software, a combination of an item generator and item bank, to create your own tests or worksheets. Types of items include free response, multiple choice, short answer, and open ended.

For IBM & Macintosh

OBJECTIVES AND EXAMPLES	REVIEW EXERCISES

- name and label corresponding parts of congruent triangles (Lesson 4–3)

Definition of Congruent Triangles

Two triangles are congruent if and only if their corresponding parts are congruent.

Draw $\triangle MNO$ and $\triangle RST$. Label the corresponding parts if $\triangle MNO \cong \triangle RST$. Use the figures to complete each statement.

36. $\angle T \cong$ __?__ $\angle O$
37. $\overline{MO} \cong$ __?__ $\overline{RT}$
38. $\overline{SR} \cong$ __?__ $\overline{NM}$
39. $\angle TSR \cong$ __?__ $\angle ONM$
40. $\overline{TR} \cong$ __?__ $\overline{OM}$
41. $\angle NOM \cong$ __?__ $\angle STR$

- use SSS, SAS, and ASA Postulates to test for triangle congruence (Lesson 4–4)

Given: $\overline{LN}$ and $\overline{OP}$ bisect each other at M.

Prove: $\angle O \cong \angle P$

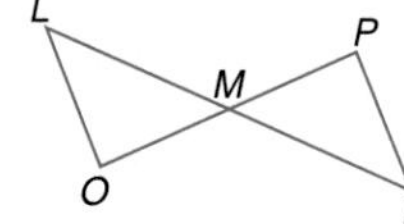

Proof:

Statements	Reasons
1. $\overline{LN}$ and $\overline{OP}$ bisect each other at M.	1. Given
2. $\overline{LM} \cong \overline{MN}$, $\overline{PM} \cong \overline{MO}$	2. Definition of bisector
3. $\angle LMO \cong \angle NMP$	3. Vertical ∠s are ≅.
4. $\triangle LMO \cong \triangle NMP$	4. SAS
5. $\angle O \cong \angle P$	5. CPCTC

Write a flow proof. 42. See Solutions Manual.

42. **Given:** $\overline{EF} \cong \overline{GH}$, $\overline{EH} \cong \overline{GF}$

Prove: $\triangle EFH \cong \triangle GHF$

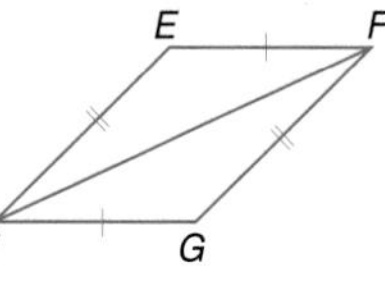

Write a two-column proof.

43. **Given:** $\overline{AM} \parallel \overline{CR}$ See margin.
B is the midpoint of $\overline{AR}$.

Prove: $\overline{AM} \cong \overline{RC}$

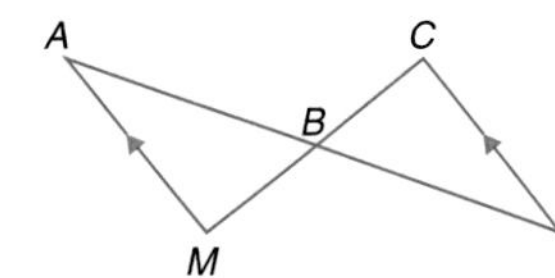

- use AAS Theorem to test for triangle congruence (Lesson 4–5)

AAS

If two angles and a nonincluded side of one triangle are congruent to the corresponding two angles and side of a second triangle, the two triangles are congruent.

Write a two-column proof.

44. **Given:** $\overline{BC} \cong \overline{DC}$ See margin.
$\angle A \cong \angle E$
$\angle 1 \cong \angle 2$

Prove: $\overline{AB} \cong \overline{ED}$

45. **Given:** $\overline{AC} \cong \overline{EC}$ See margin.
$\angle 1 \cong \angle 2$
$\overline{BC} \cong \overline{DC}$

Prove: $\angle B \cong \angle D$

Additional Answers

43. Given: $\overline{AM} \parallel \overline{CR}$
B is the midpoint of $\overline{AR}$.
Prove: $\overline{AM} \cong \overline{RC}$
Proof:
Statements (Reasons)
1. $\overline{AM} \parallel \overline{CR}$ (Given)
2. $\angle A \cong \angle R$ (Alt. Int. ∠ Th.)
3. B is the midpoint of $\overline{AR}$. (Given)
4. $\overline{AB} \cong \overline{RB}$ (Mdpt. Th.)
5. $\angle ABM \cong \angle RBC$ (Vertical ∠s are ≅.)
6. $\triangle ABM \cong \triangle RBC$ (ASA)
7. $\overline{AM} \cong \overline{RC}$ (CPCTC)

44. Given: $\overline{BC} \cong \overline{DC}$
$\angle A \cong \angle E$
$\angle 1 \cong \angle 2$
Prove: $\overline{AB} \cong \overline{ED}$
Proof:
Statements (Reasons)
1. $\overline{BC} \cong \overline{DC}$, $\angle A \cong \angle E$, $\angle 1 \cong \angle 2$ (Given)
2. $\triangle BAC \cong \triangle DEC$ (AAS)
3. $\overline{AB} \cong \overline{ED}$ (CPCTC)

45. Given: $\overline{AC} \cong \overline{EC}$
$\angle 1 \cong \angle 2$
$\overline{BC} \cong \overline{DC}$
Prove: $\angle B \cong \angle D$
Proof:
Statements (Reasons)
1. $\overline{AC} \cong \overline{EC}$, $\angle 1 \cong \angle 2$, $\overline{BC} \cong \overline{DC}$ (Given)
2. $\triangle ABC \cong \triangle EDC$ (SAS)
3. $\angle B \cong \angle D$ (CPCTC)

Applications and Problem Solving Encourage students to work through the exercises in the Applications and Problem Solving section to strengthen their problem-solving skills.

Additional Answer

48. Given: $\angle 2 \cong \angle 1$
$\angle 4 \cong \angle 3$
Prove: $\overline{AM} \cong \overline{AO}$
Proof:
Statements (Reasons)
1. $\angle 2 \cong \angle 1$ (Given)
2. $\overline{AM} \cong \overline{AE}$ (If 2 ∡ of a △ are ≅, then the sides opp. those ∡ are ≅.)
3. $\angle 4 \cong \angle 3$ (Given)
4. $\overline{AE} \cong \overline{AO}$ (If 2 ∡ of a △ are ≅, then the sides opp. those ∡ are ≅.)
5. $\overline{AM} \cong \overline{AO}$ (≅ of segments is trans.)

OBJECTIVES AND EXAMPLES

- use the properties of isosceles and equilateral triangles (Lesson 4–6)

In isosceles triangle XYZ, $\angle Z$ is the vertex angle. If $m\angle X = 4a + 5$ and $m\angle Y = 2a + 27$, find the measure of each angle of the triangle.

Since $\angle XYZ$ is isosceles and $\angle Z$ is the vertex angle, $m\angle X = m\angle Y$.

$$4a + 5 = 2a + 27 \qquad m\angle X = 4a + 5$$
$$2a = 22 \qquad = 4(11) + 5$$
$$a = 11 \qquad = 49$$

$$m\angle X + m\angle Y + m\angle Z = 180$$
$$49 + 49 + m\angle Z = 180$$
$$m\angle Z = 82$$

The measures of the angles are 49, 49, and 82.

REVIEW EXERCISES

Find the value of x.

46. 6

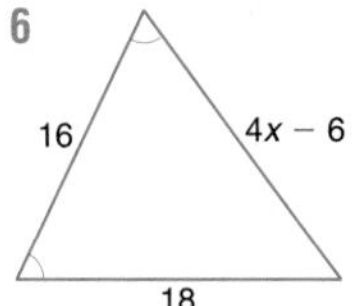

47. 32

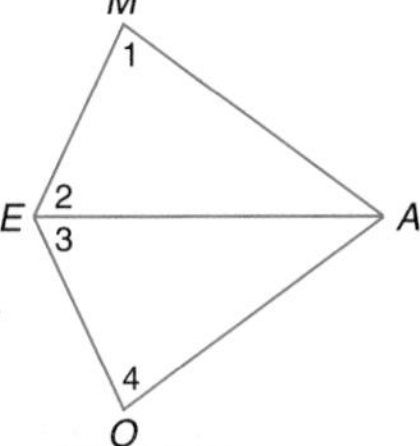

Write a proof.

48. **Given:** $\angle 2 \cong \angle 1$
$\angle 4 \cong \angle 3$
Prove: $\overline{AM} \cong \overline{AO}$
See margin.

M 1 E 2 3 A 4 O

50. Sample 1, mercury; Sample 2, gallium; Sample 3, lithium; Sample 4, calcium

APPLICATIONS AND PROBLEM SOLVING

49. **Sports** The sail for a sailboat resembles a right triangle. If the angle at the top of the sail measures 54°, what is the measure of the acute angle at the bottom? (Lesson 4–2) 36

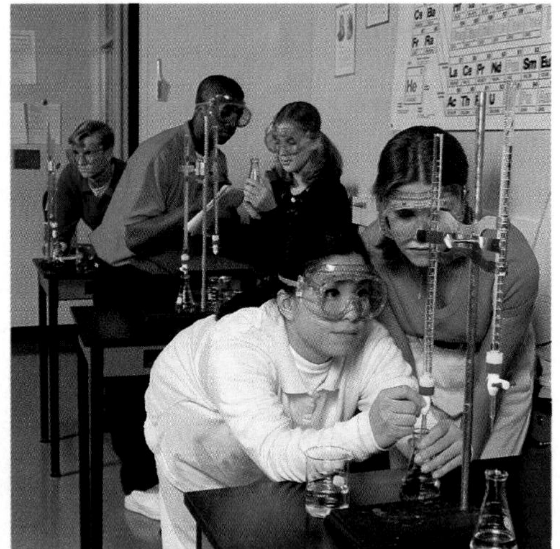

50. **Eliminate the Possibilities** In chemistry lab, Ebony was given four different metal samples to identify. They were gallium, a metal that melts when held in the hand; mercury, which is liquid at room temperature; lithium, which will float in water; and calcium, which bubbles slowly when placed in water. When Ebony held Sample 1, it was a liquid. Sample 2 was a solid and sank to the bottom when dropped in a beaker of water. Bubbling action occurred when Ebony dropped Sample 4 in the water. Identify each sample. (Lesson 4–5)

51. **Hang Gliding** The sail of a certain type of hang glider consists of two congruent isosceles triangles joined along a keel so that a 90° angle is formed at the nose of the sail. In order to construct such a sail, what must be the measure of $\angle BCD$? (Lesson 4–6) 135

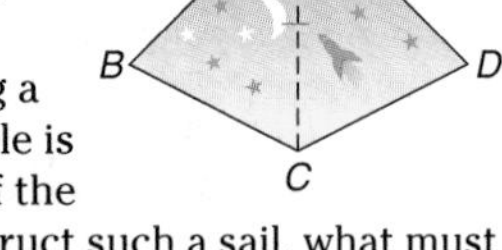

A practice test for Chapter 4 is available on page 796.

ALTERNATIVE ASSESSMENT

COOPERATIVE LEARNING PROJECT

Three-Dimensional Design In this project, you will apply your knowledge of congruent triangles to construct a variety of geometric sculptures.

Follow these steps to explore geometric sculptures.

- You will need several pieces of heavy construction paper and a roll of transparent tape.
- Each group member will cut out ten congruent equilateral triangles. (Sides measuring 4 inches are a convenient size.)
- See how many different solids you can construct using various numbers of triangles. Be creative!
- For each solid you construct, record the number of triangles you used, the number of edges, and the number of sides.
- Discuss the following questions with your group. Are there any numerical relationships between the number of triangles, edges, and sides?

Follow these steps to explore more geometric sculptures.

- Cut out ten right triangles, with sides 3 inches, 4 inches, and 5 inches.
- Construct a variety of solids out of these triangles. Be creative!
- For each solid you construct, record the number of triangles, the number of edges, and the number of sides.
- Discuss the following questions with your group. Are there any numerical relationships between the number of triangles, edges, and sides? Was it harder or easier to construct solids from the non-equilateral triangles than equilateral triangles? Why?

THINKING CRITICALLY

Imagine intersecting a sphere with three planes, each of which passes through the center of the sphere. Note that each forms a circle on the surface of the sphere. Adjust the planes so that the three circles intersect to form a three-sided figure on the surface of the sphere. Is it possible to create a two-sided figure on the surface using two circles? yes

PORTFOLIO

Travel around your school or neighborhood to find examples of triangles. Make a particular effort to find objects that contain congruent triangles. Compile photographs or drawings. Organize the images by triangle type; obtuse, right, or acute. Place your collection in your portfolio.

SELF EVALUATION

So far in this book, you have used the problem-solving strategies—list the possibilities, look for a pattern, draw a diagram, and eliminate the possibilities. Which strategy did you find most helpful? Why?

Assess Yourself You have learned to prove a statement using paragraph proof, two-column proof, and flow proof. Do you understand each type of proof? Can you write any type of proof without help, or is it easier for you to write one type of proof and then convert that proof to another type? Which type of proof do you like best?

Assessment and Evaluation Masters, pp. 98, 109

NAME________________ DATE ________

Chapter 4 Performance Assessment

Instructions: Demonstrate your knowledge by giving a clear, concise solution to each problem. Be sure to include all relevant drawings and to justify your answers. You may show your solution in more than one way or investigate beyond the requirements of the problem.

1. Rosa is constructing a stained glass window. The design she is creating is shown below.

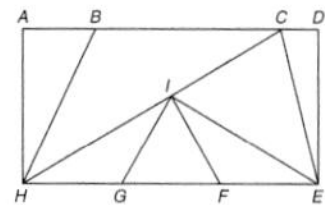

   a. In your own words, define a scalene triangle. Name a triangle in the window above that appears to be scalene.

   b. Describe an equilateral triangle. Name a triangle in the window that appears to be equilateral.

   c. Write the definition of an obtuse triangle. Name a triangle in the window that appears to be obtuse.

   d. Can a triangle have both an obtuse angle and a right angle? Explain.

   e. Give at least three postulates or theorems that give the conditions sufficient for two triangles to be congruent.

   f. Name two triangles in the window that appear to be congruent.

2. Design a stained glass window. The window should contain two obtuse triangles congruent by SAS, a scalene triangle, an isosceles triangle, and two triangles congruent by AAS. Label the triangles, mark the congruent sides and angles. Tell which triangles are congruent.

Scoring Guide
Chapter 4
Performance Assessment

Level	Specific Criteria
3 Superior	• Shows thorough understanding of the concepts of *acute, right, obtuse, scalene, isosceles, equilateral,* and *congruent triangles.* • Uses appropriate strategies to prove triangles congruent. • Written explanations are exemplary. • Diagram is accurate and appropriate. • Goes beyond requirements of some or all problems.
2 Satisfactory, with Minor Flaws	• Shows understanding of the concepts of *acute, right, obtuse, scalene, isosceles, equilateral,* and *congruent triangles.* • Uses appropriate strategies to prove triangles congruent. • Written explanations are effective. • Diagram is mostly accurate and appropriate. • Satisfies all requirements of some or all problems.
1 Nearly Satisfactory, with Serious Flaws	• Shows understanding of most of the concepts of *acute, right, obtuse, scalene, isosceles, equilateral,* and *congruent triangles.* • May not use appropriate strategies to prove triangles • congruent. • Written explanations are satisfactory. • Diagram is mostly accurate and appropriate. • Satisfies most requirements of some or all problems.
0 Unsatisfactory	• Shows little or no understanding of the concepts of *acute, right, obtuse, scalene, isosceles, equilateral,* and *congruent triangles.* • May not use appropriate strategies to prove triangles congruent. • Written explanations are not satisfactory. • Diagram is not accurate or appropriate. • Does not satisfy requirements of some or all problems.

Alternative Assessment

The Alternative Assessment section provides students with the opportunity to assess their own work by thinking critically, working with others, keeping a portfolio, and honestly evaluating their own progress. For more information on alternative forms of assessment, see *Alternative Assessment in the Mathematics Classroom,* one of the titles in the Glencoe Mathematics Professional Series.

Performance Assessment

Performance Assessment tasks for this chapter are included in the *Assessment and Evaluation Masters.* A scoring guide is also provided.

Using the COLLEGE ENTRANCE EXAM PRACTICE

These two pages review the skills and concepts presented in Chapters 1–4. This review is formatted to reflect new trends in college entrance testing.

Assessment and Evaluation Masters, pp. 103–104

NAME________________ DATE ________

Cumulative Review (Chapters 1–4)

For Questions 1–6, refer to the figure at the right.

1. Which of the numbered angles appear to be obtuse? 1. $\angle 7$ and $\angle 9$
2. Are $\angle 2$ and $\angle 6$ vertical angles? 2. no
3. Are $\angle 7$ and $\angle 8$ supplementary angles? 3. yes
4. If $\angle 6 \cong \angle 8$, which lines are parallel and why? 4. $r \parallel s$, alt. int $\angle$s
5. If $r \parallel s$ and $m\angle 5 = 27$, find $m\angle 3$. 5. 63
6. If $r \parallel s$ and $m\angle 5 = 15$, find $m\angle 7$. 6. 105

For Questions 7–10, refer to the figure at the right.

7. Classify $\triangle ABC$. 7. obtuse isosceles
8. Find $m\angle ACD$. 8. 76
9. Find $m\angle ACB$. 9. 104
10. Find BC. 10. 11
11. Can an isosceles triangle be an acute triangle? 11. yes
12. Name the congruence property that justifies the statement "If $\angle A \cong \angle B$, then $\angle B \cong \angle A$." 12. symmetric
13. The measure of an angle is 18 less than twice its complement. Find the measure of the angle. 13. 54

Refer to the figure to complete each statement. Justify your answers.

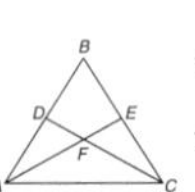

14. If $\overline{DC} \cong \overline{AE}$ and $\angle ACD \cong \angle EAC$, then $\triangle$ ____ $\cong \triangle$ ____. 14. $\triangle ACD \cong \triangle ACE$ by SAS
15. If $\overline{BD} \cong \overline{BE}$ and $\angle BDC \cong \angle BEA$, then $\triangle$ ____ $\cong \triangle$ ____. 15. $\triangle BDC \cong \triangle BEA$ by ASA
16. If $\angle FAC \cong \angle FCA$ and $\overline{DF} \cong \overline{EF}$, then $\triangle$ ____ $\cong \triangle$ ____. 16. $\triangle ADF \cong \triangle ECF$ by SAS

NAME________________ DATE ________

Cumulative Review (Chapters 1–4 continued)

17. John, Joe, and Jim each play an instrument. The instruments are the drums, the sax, and the trombone. Joe and the drummer are in the same Geometry class. The sax player and the drummer go to Jim's house to practice. Who plays the drum? 17. John
18. The faces of a cube are numbered with consecutive numbers. Three of the numbers are shown. What are the possible sums of the numbers on all faces of the cube? 18. 123; 129

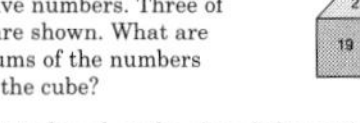

19. Write a conjecture based on the given information. Draw a figure to illustrate your conjecture. Given: M is the midpoint of $\overline{AB}$. 19. Sample answer: $AM + MB = AB$; see students' work.

Write the converse of each conditional. Determine if the converse is true or false.

20. If an angle measures 128°, then it is obtuse. 20. If an angle is obtuse, then it measures 128°
21. If two lines have the same slope, then they are parallel and nonvertical. 21. If 2 lines are $\parallel$ and non vertical, they have the same slope; true.

For Questions 22–25, refer to the figure at the right.

22. Classify $\triangle ABC$. 22. right scalene
23. What is the slope of any line parallel to $\overline{BC}$? 23. -1
24. B is the midpoint of $\overline{AD}$. What are the coordinates of D? 24. $(-6, -3)$
25. What is the length of any segment congruent to $\overline{AC}$? 25. $\sqrt{82} \approx 9.06$

COLLEGE ENTRANCE EXAM PRACTICE

CHAPTERS 1–4

SECTION ONE: MULTIPLE CHOICE

There are eight multiple-choice questions in this section. After working each problem, write the letter of the correct answer on your paper.

1. What is the slope of a line that passes through the points at $(3, -2)$ and $(-6, 4)$? B

   A. $-\frac{3}{2}$ C. $\frac{3}{2}$
   B. $-\frac{2}{3}$ D. $\frac{2}{3}$

2. If $0 < x < \frac{1}{2}$, which of the following could not be x? C

   A. $\frac{1}{4}$ C. $\frac{5}{8}$
   B. $\frac{3}{10}$ D. $\frac{7}{16}$

3. In the figure below, if the three equilateral triangles have a common vertex, then find the value of $a + b + c$. C

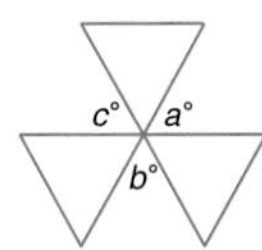

   A. 120 C. 180
   B. 90 D. 360

4. In the figure below, if $\ell_1 \parallel \ell_2$, $\ell_2 \parallel \ell_3$, and $\ell_1 \perp \ell_4$, which of the following statements must be true? C

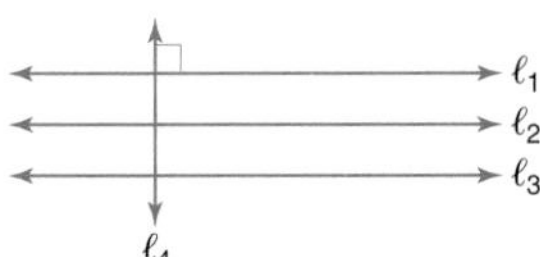

   I. $\ell_1 \parallel \ell_3$
   II. $\ell_2 \perp \ell_4$
   III. $\ell_3 \perp \ell_4$

   A. I only C. I, II, and III
   B. II and III only D. none

5. 0.08 is the ratio of 8 to which number? B

   A. 1000 C. 10
   B. 100 D. $\frac{1}{100}$

6. If $\frac{6}{n}$ is an odd integer, which of the following could be a value of n? A

   A. $\frac{6}{5}$ C. $\frac{3}{4}$
   B. $\frac{1}{3}$ D. $\frac{4}{3}$

7. In the figure below, find x. D

   A. 60
   B. 120
   C. 45
   D. 30

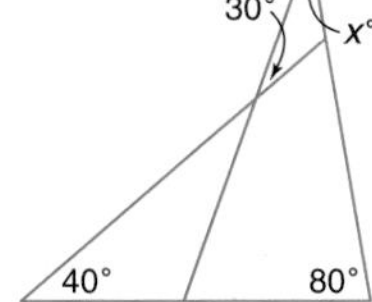

8. In the figure below, which of the following must be equal to 180? B

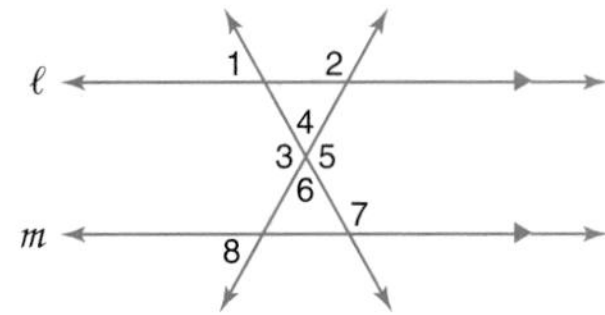

   I. $m\angle 3 + m\angle 5$
   II. $m\angle 4 + m\angle 6$
   III. $m\angle 1 + m\angle 7$
   IV. $m\angle 2 + m\angle 8$
   V. $m\angle 7 + m\angle 8$

   A. I and II only C. V only
   B. III and IV only D. I, II, III, and IV only

Standardized Test Practice Questions are also provided in the *Assessment and Evaluation Masters*, p. 102.

A more traditional cumulative review is provided in the *Assessment and Evaluation Masters*, pp. 103–104.

SECTION TWO: SHORT ANSWER

This section contains seven questions for which you will provide short answers. Write your answer on your paper.

9. What is the slope of a line parallel to $5x - y = 10$? **5**

10. The product of 4, 5, and 6 is equal to twice the sum of 10 and what number? **50**

11. If $\frac{x}{2} + \frac{x}{3} + \frac{x}{6} = kx$, then find k if $x \neq 0$. **1**

12. Find $10x - 5$ if $10x - 4 = 8$. **7**

13. In the figure below, if $\triangle ABC$ is equilateral, what is the ratio of AC to CD? **1**

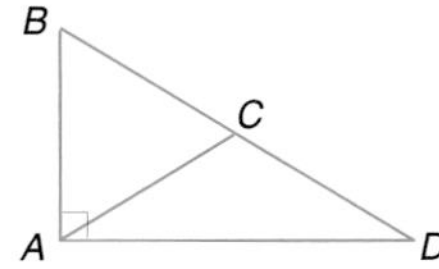

14. If $x = 1$, $y = -1$, and $z = -2$, then find $\frac{x^2y}{(x-z)^2}$. $-\frac{1}{9}$

15. In the figure below, lines r and s are parallel. List three angles all having the same measure. $\angle 1$, $\angle 2$, $\angle 4$

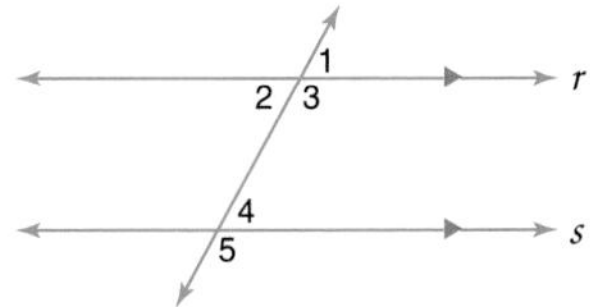

SECTION THREE: COMPARISON

This section contains five comparison problems that involve comparing two quantities, one in column A and one in column B. In certain questions, information related to one or both quantities is centered above them. All variables used represent real numbers.

Compare quantities A and B below.

- Write A if quantity A is greater.
- Write B if quantity B is greater.
- Write C if the two quantities are equal.
- Write D if there is not enough information to determine the relationship.

Column A	Column B
A jacket that was priced at \$36.50 is sold at a 30% discount.	
16. the price of the coat after the discount	\$25.50 **A**
$x \neq 0$	**B**
17. $3x^2$	$(3x)^2$
$\frac{3}{5} = \frac{x}{20}$, $\frac{4}{8} = \frac{y}{24}$	**C**
18. x	y
19. $2(9 + 6) - 4(10 \div 2)$	$2 \times 9 + 6 - 4 \times 10 \div 2$ **19. A**
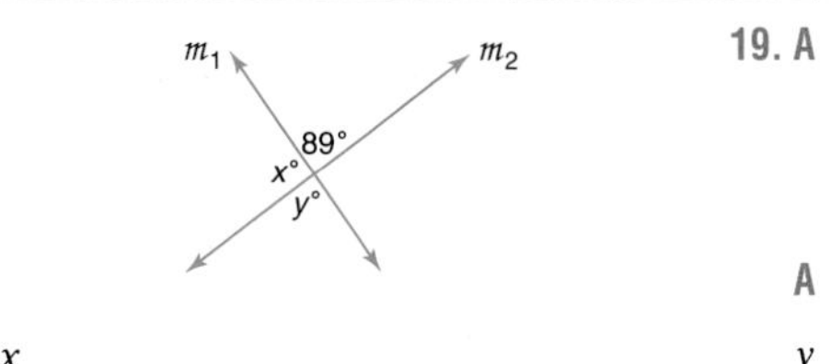	**A**
20. x	y

TEST-TAKING TIP

Managing Time

College entrance exams are timed tests, so good use of time is essential. Remind students to pace themselves as they work and to try not to spend too much time on any one problem. Develop a game plan before taking the test. Three possible plans are

- Answer all questions in order, taking as much time as needed for each one.
- Answer all of the easy questions first. Mark more difficult problems with a star and then go back to finish them after all of the easy ones are finished. This plan may allow you to answer many of the questions in a short period of time and spend more time on the difficult ones.
- Answer all of the difficult questions first, then go back and do the easy ones. This approach may allow you to complete the difficult questions when your mind is fresh and leave the easy problems for the end of the test time when you may be tired or rushed.

Whatever plan is used to attack the test, remind students to organize their work and write legibly as they solve problems. This will allow time to check work easily if time remains after finishing all of the problems.

CHAPTER 5

Applying Congruent Triangles

PREVIEWING THE CHAPTER

This chapter applies congruence to different types and parts of triangles. Students identify and use medians, altitudes, angle bisectors, and perpendicular bisectors. Tests for congruence of right triangles are used. Indirect reasoning and indirect proofs are used to reach conclusions and solve problems. Properties of inequalities are applied to the measures of segments and angles. Relationships between sides and angles in a triangle are modeled and investigated using a graphing calculator. The Triangle Inequality Theorem, SAS Inequality, and SSS Inequality are applied to different triangles.

Lesson (Pages)	Lesson Objectives	NCTM Standards	State/Local Objectives
5-1 (238–244)	Identify and use medians, altitudes, angle bisectors, and perpendicular bisectors in a triangle.	1–4, 7	
5-2 (245–251)	Recognize and use tests for congruence of right triangles.	1–4, 7	
5-3 (252–258)	Use indirect reasoning and indirect proof to reach a conclusion. Recognize and apply properties of inequalities to the measures of segments and angles. Solve problems by working backward.	1–5, 7	
5-4 (259–265)	Recognize and apply relationships between sides and angles in a triangle.	1–5, 7	
5-5A (266)	Use a graphing calculator to investigate the relationship among the measures of the sides of a triangle.	1–5, 7	
5-5 (267–272)	Apply the Triangle Inequality Theorem.	1–5, 7	
5-6 (273–279)	Apply the SAS Inequality and the SSS Inequality.	1–5, 7	

A complete, 1-page lesson plan is provided for each lesson in the ***Lesson Planning Guide***. Answer keys for each lesson are available in the ***Answer Key Masters***.

ORGANIZING THE CHAPTER

You may want to refer to the **Course Planning Calendar** on page T12 for detailed information on pacing.
PACING: Standard—14 days; **Honors**—13 days; **Block**—7 days

LESSON PLANNING CHART

			BLACKLINE MASTERS									
Lesson (Pages)	**Materials/ Manipulatives**	**Extra Practice (Student Edition)**	**Study Guide**	**Practice**	**Enrichment**	**Assessment & Evaluation**	**Modeling Mathematics**	**Multicultural Activity**	**Tech Prep Applications**	**Graphing Calc. & Computer**	**Real-World Applications**	**Teaching Transparencies**
5-1 (238–244)	patty paper ruler* TI-92 calculator	p. 772	p. 26	p. 26	p. 26		pp. 31–34 83		p. 9			5-1A 5-1B
5-2 (245–251)	TI-92 calculator	p. 772	p. 27	p. 27	p. 27	p. 128		p. 9		p. 5	9	5-2A 5-2B
5-3 (252–258)		p. 773	p. 28	p. 28	p. 28	pp. 127, 128						5-3A 5-3B
5-4 (259–265)	ruler* protractor*	p. 773	p. 29	p. 29	p. 29			p. 10				5-4A 5-4B
5-5A (266)	TI-82/83 graphing calculator									pp. 22, 23		
5-5 (267–272)	TI-82/83 graphing calculator	p. 773	p. 30	p. 30	p. 30	p. 129			p. 10			5-5A 5-5B
5-6 (273–279)	rubber band centimeter ruler* ball-bearing compass protractor*	p. 774	p. 31	p. 31	p. 31	p. 129					10	5-6A 5-6B
Study Guide/ Assessment (281–285)						pp. 113–126, 130–132						

*Included in Glencoe's High School Manipulative Kit and Overhead Manipulative Resources.

ORGANIZING THE CHAPTER

Other Chapter Resources

Student Edition
Chapter Opener, pp. 236–237
Mathematics and Society, p. 272
Working on the Investigation, pp. 265, 279
Closing the Investigation, p. 280

Teacher's Classroom Resources
Investigations and Projects Masters, pp. 41–44
Block Scheduling Booklet

Technology
Test and Review Software (IBM and Macintosh)
CD-ROM Multimedia Applications (Windows and Macintosh)
Mindjogger Videoquizzes (VHS)

Professional Publications
Glencoe Mathematics Professional Series

Outside Resources

Books/Periodicals
Problem-Solving Experiences in Geometry, ETA
Geometry from Multiple Perspectives, NCTM
Notes on a Triangle, Dale Seymour Publications
Visualized Geometry: A Van Hiele Level Approach, NASCO

Software
Geometry Connections: Inequalities, William K. Bradford Publishing Co.
Geometry Concepts, ETA

Videos/CD-ROMs
The Alhambra Past and Present: A Geometer's Odyssey, Dale Seymour Publications, P.O. Box 10888, Palo Alto, CA 94303
Introduction to Geometric Terms, Angles, and Triangles, NASCO, 901 Janesville Ave., Fort Atkinson, WI 53538

Assessment Resources

Student Edition
Math Journal, pp. 248, 255
Mixed Review, pp. 244, 251, 258, 265, 271–272, 279
Self Test, p. 258
Chapter Highlights, p. 281
Chapter Study Guide and Assessment, pp. 282–284
Alternative Assessment, p. 285
Portfolio, p. 285
Chapter Test, p. 797

Teacher's Wraparound Edition
5-Minute Check, pp. 238, 245, 252, 259, 267, 273
Check for Understanding, pp. 241, 248, 255, 262, 269, 276
Closing Activity, pp. 244, 251, 258, 265, 272, 279
Cooperative Learning, pp. 247, 254

Assessment and Evaluation Masters
Multiple-Choice Tests, Forms 1A (Honors), 1B (Average), 1C (Basic), pp. 113–118
Free-Response Tests, Forms 2A (Honors), 2B (Average), 2C (Basic), pp. 119–124
Calculator-Based Test, p. 125
Performance Assessment, p. 126
Mid-Chapter Test, p. 127
Quizzes A–D, pp. 128–129
Standardized Test Practice, p. 130
Cumulative Review, pp. 131–132

ENHANCING THE CHAPTER

Examples of some of the materials for enhancing Chapter 5 are shown below.

DIVERSITY

Multicultural Activity Masters, pp. 9, 10

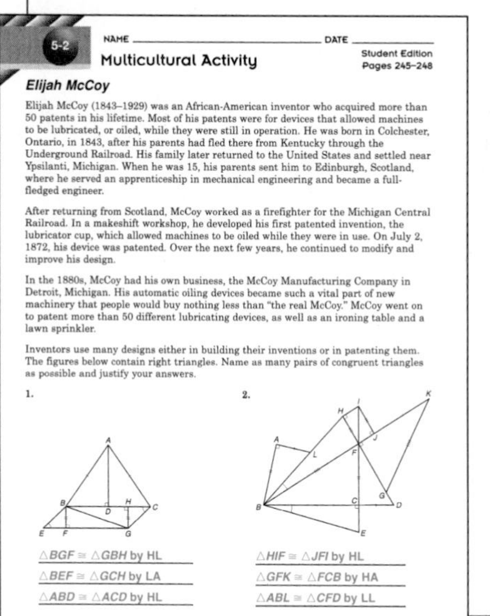

5-2 NAME ______ DATE ______

Multicultural Activity — Student Edition Pages 245–248

Elijah McCoy

Elijah McCoy (1843–1929) was an African-American inventor who acquired more than 50 patents in his lifetime. Most of his patents were for devices that allowed machines to be lubricated, or oiled, while they were still in operation. He was born in Colchester, Ontario, in 1843, after his parents had fled there from Kentucky through the Underground Railroad. His family later returned to the United States and settled near Ypsilanti, Michigan. When he was 15, his parents sent him to Edinburgh, Scotland, where he served an apprenticeship in mechanical engineering and became a full-fledged engineer.

After returning from Scotland, McCoy worked as a firefighter for the Michigan Central Railroad. In a makeshift workshop, he developed his first patented invention, the lubricator cup, which allowed machines to be oiled while they were in use. On July 2, 1872, his device was patented. Over the next few years, he continued to modify and improve his design.

In the 1880s, McCoy had his own business, the McCoy Manufacturing Company in Detroit, Michigan. His automatic oiling devices became such a vital part of new machinery that people would buy nothing less than "the real McCoy." McCoy went on to patent more than 50 different lubricating devices, as well as an ironing table and a lawn sprinkler.

Inventors use many designs either in building their inventions or in patenting them. The figures below contain right triangles. Name as many pairs of congruent triangles as possible and justify your answers.

1. $\triangle BGF \cong \triangle GBH$ by HL
   $\triangle BEF \cong \triangle GCH$ by LA
   $\triangle ABD \cong \triangle ACD$ by HL

2. $\triangle HIF \cong \triangle JFI$ by HL
   $\triangle GFK \cong \triangle FCB$ by HA
   $\triangle ABL \cong \triangle CFD$ by LL

APPLICATIONS

Real-World Applications, 9, 10

Transparency 9

Ships and Oceans

OCEAN AREAS AND AVERAGE DEPTH

	Area (square miles)	Average Depth (feet)
Pacific Ocean	64,000,000	13,215
Atlantic Ocean	31,815,000	12,880
Indian Ocean	25,300,000	13,002
Arctic Ocean	5,440,200	3,953

Source: *Information Please Almanac*, 1994

ROMAN GRAIN SHIP ABOUT A.D. 200

© Glencoe/McGraw-Hill — Geometry

TECHNOLOGY

Graphing Calculator and Computer Masters, p. 5

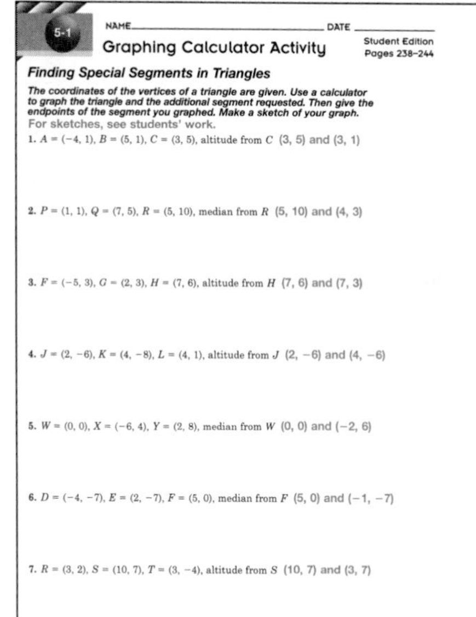

5-1 NAME ______ DATE ______

Graphing Calculator Activity — Student Edition Pages 238–244

Finding Special Segments in Triangles

The coordinates of the vertices of a triangle are given. Use a calculator to graph the triangle and the additional segment requested. Then give the endpoints of the segment you graphed. Make a sketch of your graph.

For sketches, see students' work.

1. $A = (-4, 1)$, $B = (5, 1)$, $C = (3, 5)$, altitude from C (3, 5) and (3, 1)
2. $P = (1, 1)$, $Q = (7, 5)$, $R = (5, 10)$, median from R (5, 10) and (4, 3)
3. $F = (-5, 3)$, $G = (2, 3)$, $H = (7, 6)$, altitude from H (7, 6) and (7, 3)
4. $J = (2, -6)$, $K = (4, -8)$, $L = (4, 1)$, altitude from J (2, −6) and (4, −6)
5. $W = (0, 0)$, $X = (-6, 4)$, $Y = (2, 8)$, median from W (0, 0) and (−2, 6)
6. $D = (-4, -7)$, $E = (2, -7)$, $F = (5, 0)$, median from F (5, 0) and (−1, −7)
7. $R = (3, 2)$, $S = (10, 7)$, $T = (3, -4)$, altitude from S (10, 7) and (3, 7)

TECH PREP

Tech Prep Applications Masters, pp. 9, 10

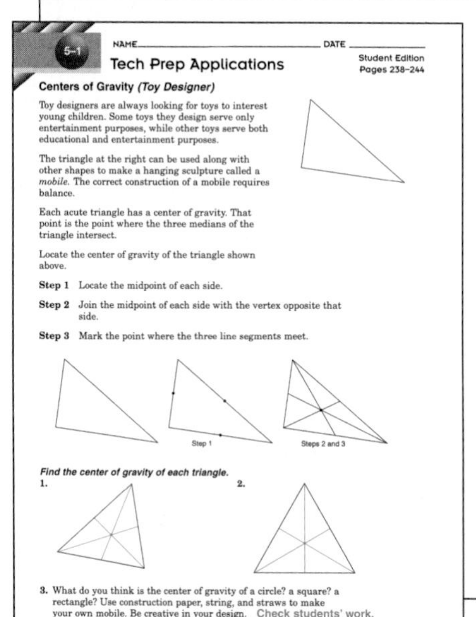

5-1 NAME ______ DATE ______

Tech Prep Applications — Student Edition Pages 238–244

Centers of Gravity *(Toy Designer)*

Toy designers are always looking for toys to interest young children. Some toys they design serve only entertainment purposes, while other toys serve both educational and entertainment purposes.

The triangle at the right can be used along with other shapes to make a hanging sculpture called a *mobile*. The correct construction of a mobile requires balance.

Each acute triangle has a center of gravity. That point is the point where the three medians of the triangle intersect.

Locate the center of gravity of the triangle shown above.

Step 1 Locate the midpoint of each side.

Step 2 Join the midpoint of each side with the vertex opposite that side.

Step 3 Mark the point where the three line segments meet.

Step 1 — Steps 2 and 3

Find the center of gravity of each triangle.

1.

2.

3. What do you think is the center of gravity of a circle? a square? a rectangle? Use construction paper, string, and straws to make your own mobile. Be creative in your design. Check students' work.

PROBLEM SOLVING

Problem-of-the-Week Cards, 12, 13, 14

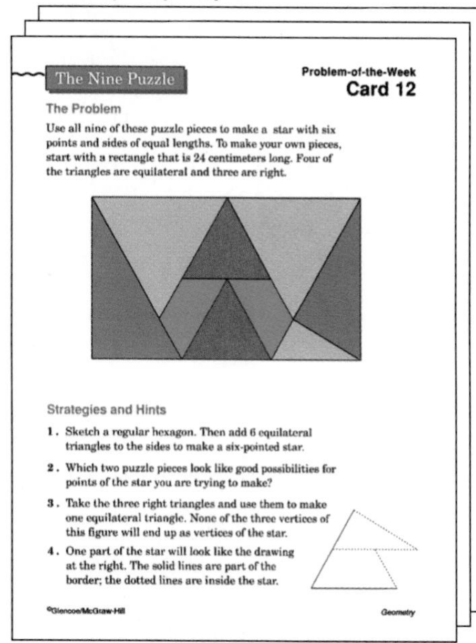

The Nine Puzzle — Problem-of-the-Week **Card 12**

The Problem

Use all nine of these puzzle pieces to make a star with six points and sides of equal lengths. To make your own pieces, start with a rectangle that is 24 centimeters long. Four of the triangles are equilateral and three are right.

Strategies and Hints

1. Sketch a regular hexagon. Then add 6 equilateral triangles to the sides to make a six-pointed star.
2. Which two puzzle pieces look like good possibilities for points of the star you are trying to make?
3. Take the three right triangles and use them to make one equilateral triangle. None of the three vertices of this figure will end up as vertices of the star.
4. One part of the star will look like the drawing at the right. The solid lines are part of the border; the dotted lines are inside the star.

©Glencoe/McGraw-Hill — *Geometry*

MAKING MATHEMATICS RELEVANT

This two-page introduction to the chapter provides students with an opportunity to explore an application of geometry that has historic roots as well as modern uses.

Background Information

It's an Art "Regular division of the plane" is what M.C. Escher called his repeating designs that left no empty spaces. He described his techniques for regular division of the plane in mathematical terms in his book *Regelmatige vlakverdeling* published in 1958 by De Roos Foundation in Utrecht, Netherlands. His *Regular Division of the Plane IV* and its diagram shown in the Chapter Project appeared in this book. Escher's regular division of the plane designs have become popular themes for scarves, neckties, wrapping paper, and even computer screen backgrounds.

CHAPTER 5

Applying Congruent Triangles

Objectives

In this chapter, you will:

- identify and use the special segments in triangles,
- prove right triangles congruent, and
- recognize and apply relationships between the sides and angles in a triangle.

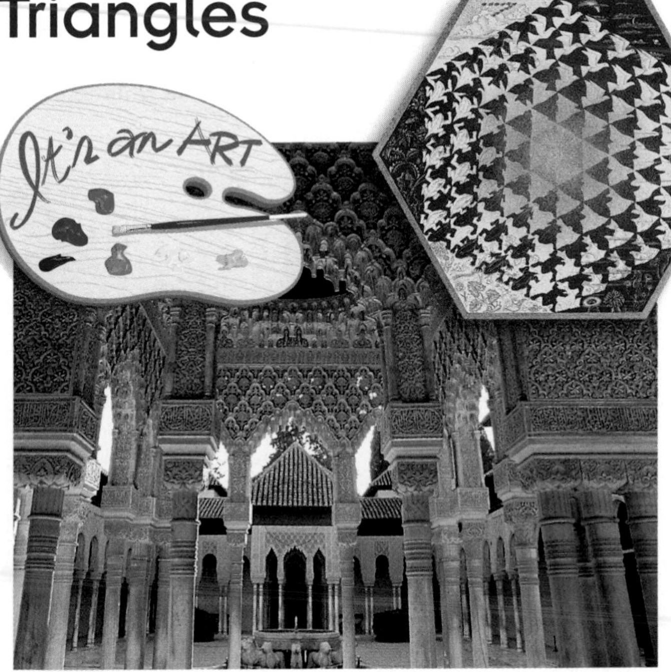

Geometry: Then and Now The Alhambra, a palace built between 1248 and 1254 in Granada, Spain, contains some of the finest examples of Islamic art. The patterns of tiles in The Alhambra fascinated Dutch artist M.C. Escher. Escher built upon the ideas he saw in the tile patterns to create more than 150 designs of repeating patterns that, if continued indefinitely, would fill a plane. In addition to being beautiful pieces of art, some of Escher's patterns have been used to produce unique fabric patterns for clothing and accessories. Do you see how congruent triangles are used in the Escher print shown above?

TIME*Line*

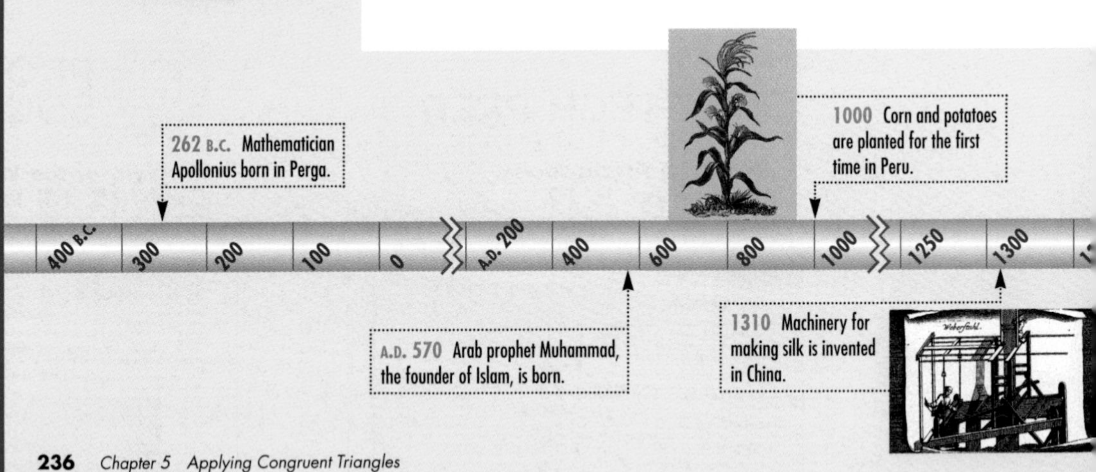

262 B.C. Mathematician Apollonius born in Perga.
1000 Corn and potatoes are planted for the first time in Peru.
400 B.C.
300
200
100
0
A.D. 200
400
600
800
1000
1250
1300
A.D. 570 Arab prophet Muhammad, the founder of Islam, is born.
1310 Machinery for making silk is invented in China.

TIME*Line*

Much Islamic art uses geometric design. In art, architecture, and garden design, repeated geometric patterns are often used. Students might find it interesting to find photographs of such works.

interNET CONNECTION

View more than 50 of M.C. Escher's pictures and download full-size images of those you find most interesting.

World Wide Web
http://magic.caltech.edu/ESCHER/

PEOPLE IN THE NEWS

Currently, over 13 million Americans run home-based businesses full time, and an additional 14 million do so part time. According to these figures, **Ebony Hood** of Washington, D.C., fits right in. This 18-year-old entrepreneur started her own business selling stylish patterned scarves and fashion pins. She sells her products in friends' homes and at school. A special business course at school taught her what she needed to know to launch her business and keep records. Ebony enjoys being her own boss and earning extra money that she plans to use for college.

Chapter Project

Study Escher's *Regular Division of the Plane VI* shown below. He used the diagram at the right to plan his work. Compare the diagram to the completed design.

- Make a list of triangles in the diagram that appear to be congruent. How could you prove that they are congruent?
- Choose a pair of triangles from your list. Write a formal proof exercise involving the congruence of the pair for a partner to solve. Determine what information should be given and use a ruler or protractor to verify the information that you state as given. Exchange proofs with your partner and solve.

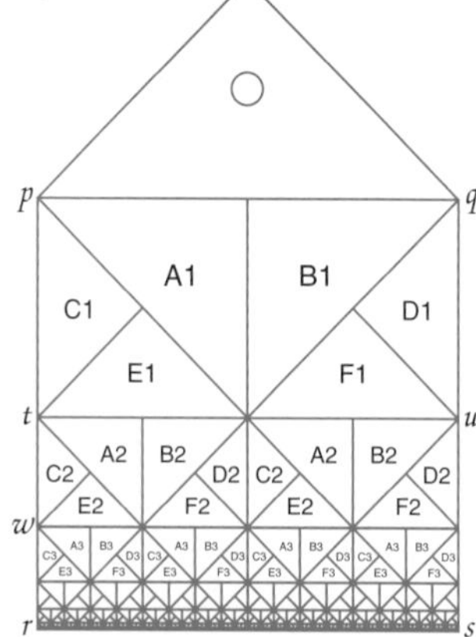

- Use software or paper and pencil to create a simple design based on congruent triangles that could be repeated to fill a plane.

1400 1450 1500 1550 1600 1650 1700 1750 1800 1850 1900 1950 2000

1492 The Alhambra, the last stronghold of the Moors in Spain, is captured by Ferdinand and Isabella.

1621 Massasoit, chief of the Wampanoags, enters into a treaty with the Pilgrims.

1972 M.C. Escher dies in Baarn, Holland.

1995 The computer animated movie *Toy Story* is released.

PEOPLE IN THE NEWS

Ask students if any of them have ever started their own businesses. Encourage a class discussion of how different types of math are used in running a business.

Chapter Project

Cooperative Learning Have students work in pairs for this project. Encourage students to recognize that the triangles in the figure are right triangles. Ask them to verify the right angles and then use this information when writing their proof exercises. Also consider asking students to write proof exercises that involve inequalities.

Investigations and Projects Masters, p. 41

NAME______________________ DATE __________

Chapter 5 Project A

Student Edition Pages 236–285

The Works of M. C. Escher

1. Work in groups of three. Study the works of M. C. Escher and find out how he created some of his drawings or patterns. You may want to use the library, interactive encyclopedias, or the Internet as a resource.
2. Within your group, select one of Escher's works and analyze it. What geometric shapes are used? Can the work be divided into a diagram that explains how the work was created? What interesting features does the picture contain?
3. Present your picture and analysis to the class. Invite other classmates' comments or observations about the picture you chose.
4. Put your ideas in a written report about the Escher work you chose. Include a reproduction of the work.

Alternative Chapter Projects

Two other chapter projects are included in the *Investigations and Projects Masters*. In Chapter 5 Project A, pp. 41–42, students extend the topic in the chapter opener. In Chapter 5 Project B, pp. 43–44, students investigate how triangles are used in the construction of houses.

5-1 LESSON NOTES

NCTM Standards: 1–4, 7

Instructional Resources

- Study Guide Master 5-1
- Practice Master 5-1
- Enrichment Master 5-1
- Modeling Mathematics Masters, pp. 31–34, 83
- Tech Prep Applications Masters, p. 9

Transparency 5-1A contains the 5-Minute Check for this lesson; **Transparency 5-1B** contains a teaching aid for this lesson.

Recommended Pacing	
Standard Pacing	Days 1 & 2 of 14
Honors Pacing	Day 1 of 13
Block Scheduling*	Day 1 of 7

*For more information on pacing and possible lesson plans, refer to the *Block Scheduling Booklet.*

1 FOCUS

5-Minute Check
(over Chapter 4)

1. In $\triangle ABC$, $m\angle A = 2x + 5$, $m\angle B = 3x - 15$, and $m\angle C = 5x - 10$. Find the value of x and the measure of each angle. What kind of triangle is $\triangle ABC$? **20; 45, 45, 90; right isosceles**
2. Given that $\triangle DEF \cong \triangle GHI$, complete each statement.
   - **a.** $\angle D \cong$ ___ $\angle G$
   - **b.** $\angle E \cong$ ___ $\angle H$
   - **c.** $\angle F \cong$ ___ $\angle I$
   - **d.** $\overline{GH} \cong$ ___ $\overline{DE}$
   - **e.** $\overline{HI} \cong$ ___ $\overline{EF}$
   - **f.** $\overline{GI} \cong$ ___ $\overline{DF}$

State the postulate or theorem you could use to prove $\triangle JKL$ and $\triangle MNO$ congruent given the following.

3. $\overline{JK} \cong \overline{MN}$, $\angle K \cong \angle N$, and $\angle L \cong \angle O$ **AAS**
4. $\overline{JK} \cong \overline{MN}$, $\overline{KL} \cong \overline{NO}$, and $\overline{JL} \cong \overline{MO}$ **SSS**

5-1 Special Segments in Triangles

What YOU'LL LEARN

- To identify and use medians, altitudes, angle bisectors, and perpendicular bisectors in a triangle.

Why IT'S IMPORTANT

You can use the special segments in triangles to solve problems involving engineering, sports, and physics.

Triangles have four types of special segments. You can use paper folding to model some of these segments.

MODELING MATHEMATICS

Special Segments of a Triangle

Materials: patty paper 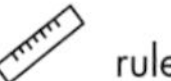 ruler

- Draw a triangle like $\triangle ABC$ on a piece of patty paper. Fold the paper so that point A falls on point B and make a crease through $\overline{AB}$. Unfold the paper and label the point where the crease intersects $\overline{AB}$ as M and the point where it intersects $\overline{AC}$ as P. $\overleftrightarrow{MP}$ is a **perpendicular bisector**.

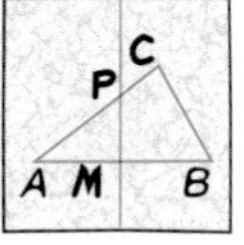

- Fold the paper and make a crease from point C to M. $\overline{CM}$ is a **median**.

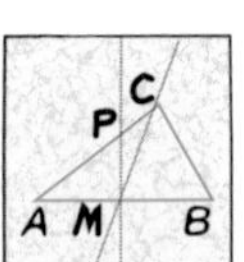

- Next, fold the paper so that B lies on $\overline{AB}$ and the fold passes through point C. Make a crease. Label the point where the crease intersects $\overline{AB}$ as T. $\overline{CT}$ is an **altitude**.

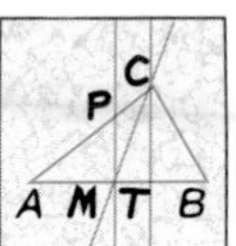

Your Turn

Write a list of observations about each of the three special segments you have folded.

Sample answer: Both the perpendicular bisector and the altitude are perpendicular to the side of the triangle. They are parallel to each other. The perpendicular bisector and the median intersect at the midpoint of the side.

The first special segment you found in the Modeling Mathematics activity was the *perpendicular bisector* of the side of a triangle. A line or line segment that passes through the midpoint of a side of a triangle *and* is perpendicular to that side is the perpendicular bisector of the side of the triangle. The red lines in the figure at the right are the perpendicular bisectors of the sides of $\triangle GHI$. Since there are three sides in every triangle, every triangle has three perpendicular bisectors.

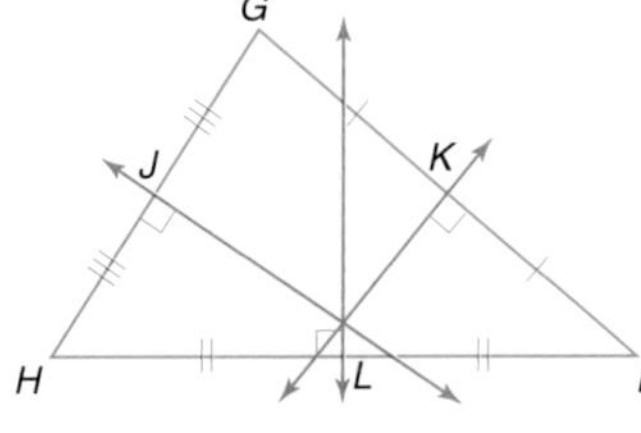

Perpendicular bisectors of segments have some special properties. These properties are described in Theorems 5–1 and 5–2. *You will be asked to prove these theorems in Exercises 35 and 36, respectively.*

Theorem 5–1	**Any point on the perpendicular bisector of a segment is equidistant from the endpoints of the segment.**
Theorem 5–2	**Any point equidistant from the endpoints of a segment lies on the perpendicular bisector of the segment.**

MODELING MATHEMATICS This activity introduces students to special segments in triangles. Challenge students to try to create the same folds in a different order.

You also folded a median and an altitude of $\triangle ABC$. A *median* is a segment that connects a vertex of a triangle to the midpoint of the side opposite that vertex. For $\triangle JKL$, $\overline{JM}$, $\overline{KN}$, and $\overline{LD}$ are the medians. Every triangle has three medians.

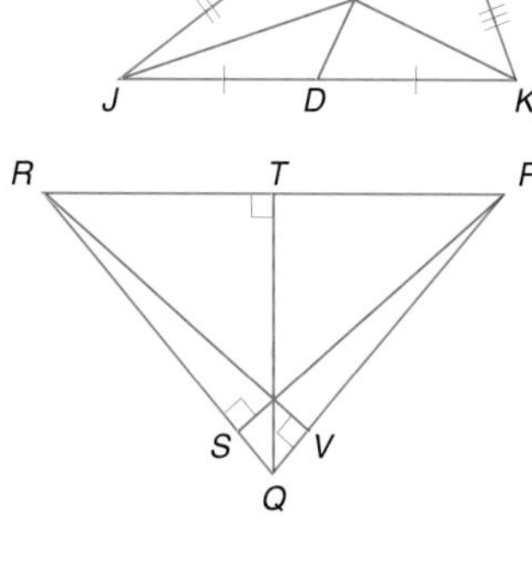

An *altitude* has one endpoint at a vertex of a triangle and the other on the line that contains the side opposite that vertex so that the segment is perpendicular to this line. For acute triangle PQR, $\overline{PS}$, $\overline{QT}$, and $\overline{RV}$ are the altitudes. Every triangle has three altitudes.

The diagrams below show the altitudes for a right triangle and an obtuse triangle. The legs of a right triangle are two of the altitudes. For an obtuse triangle, two of the altitudes lie outside the triangle.

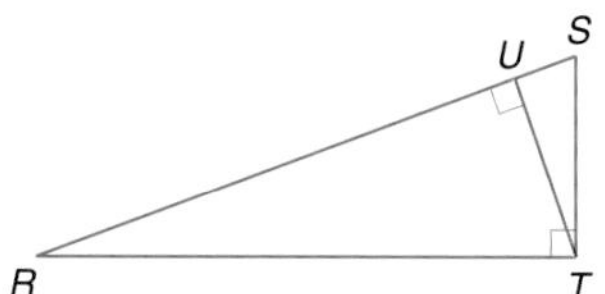

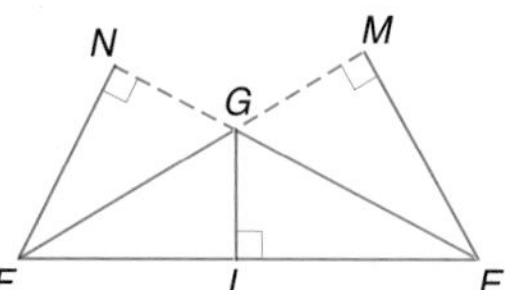

In right triangle TRS, $\overline{ST}$ is the altitude from S to $\overline{RT}$, $\overline{RT}$ is the altitude from R to $\overline{TS}$, and $\overline{TU}$ is the altitude from T to $\overline{RS}$.

In obtuse triangle EFG, $\overline{FN}$, $\overline{GL}$, and $\overline{EM}$ are the altitudes. Notice that $\overline{FN}$ and $\overline{EM}$ are outside $\triangle EFG$.

Algebra

Example 1 **$\triangle ABC$ has vertices $A(-3, 10)$, $B(9, 2)$, and $C(9, 15)$.**

a. Determine the coordinates of point P on $\overline{AB}$ so that $\overline{CP}$ is a median of $\triangle ABC$.

b. Determine if $\overline{CP}$ is an altitude of $\triangle ABC$.

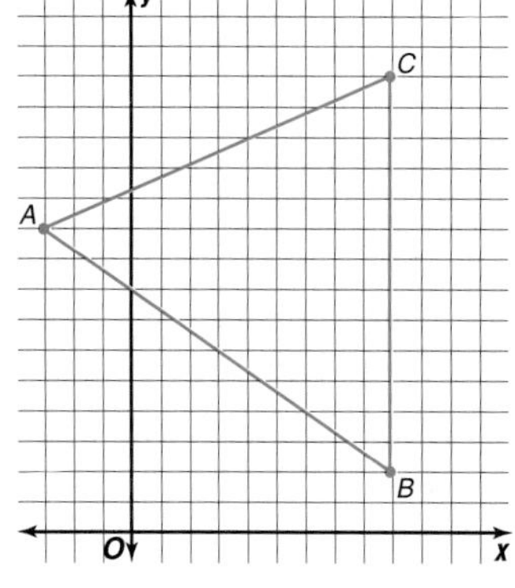

LOOK BACK

You can review finding the midpoint in a coordinate plane in Lesson 1-5.

a. By the definition of median, $\overline{CP}$ is a median of $\triangle ABC$ if P is the midpoint of $\overline{AB}$.

$$P\left(\frac{x_1 + x_2}{2}, \frac{y_1 + y_2}{2}\right) = \left(\frac{-3 + 9}{2}, \frac{10 + 2}{2}\right)$$ *Substitute $A(-3, 10)$ for (x_1, y_1) and $B(9, 2)$ for (x_2, y_2).*

$$= (3, 6)$$

The coordinates of P are $(3, 6)$.

(continued on the next page)

Motivating the Lesson

Questioning Ask students what they think of when you say *median.* Some might say the middle of the freeway or the middle score on a test for one class. Explain what a median of a triangle is. Ask them what they think of when you say the word *altitude.* Students may say it is how high something is. Explain what an altitude of a triangle is. Do the same with *angle bisector* and *perpendicular bisector.*

2 TEACH

In-Class Example

For Example 1

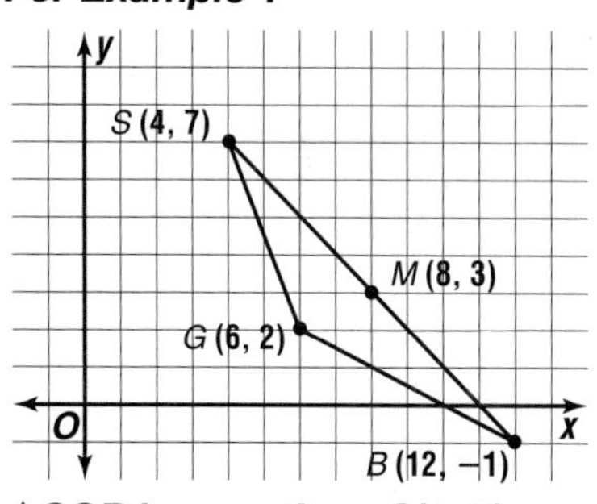

$\triangle SGB$ has vertices $S(4, 7)$, $G(6, 2)$, and $B(12, -1)$.

a. Determine the coordinates of point J on $\overline{GB}$ so that $\overline{SJ}$ is a median of $\triangle SGB$. $\left(9, \frac{1}{2}\right)$

b. Point M has coordinates $(8, 3)$. Is $\overline{GM}$ an altitude of $\triangle SGB$? **no**

Alternative Teaching Strategies

Reading Geometry Point out to students that the word "tri-angle" means "three-angles." Encourage students to think up other words that begin with the prefix "tri."

In-Class Example

For Example 2
Prove that if a triangle is equilateral, then an angle bisector is also a median.

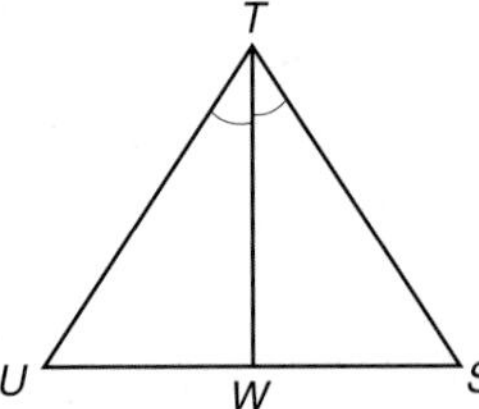

Given: $\triangle STU$ is equilateral. $\overline{TW}$ is an angle bisector of $\triangle STU$.
Prove: $\overline{TW}$ is a median of $\triangle STU$.
Proof:
Statements (Reasons)
1. $\triangle STU$ is equilateral. $\overline{TW}$ is an angle bisector of $\triangle STU$. (Given)
2. $\angle UTW \cong \angle STW$ (Def. angle bisector)
3. $\overline{TW} \cong \overline{TW}$ (Cong. of segments is reflexive.)
4. $\overline{TU} \cong \overline{TS}$ (Def. equilateral triangle)
5. $\triangle TUW \cong \triangle TSW$ (SAS)
6. $\overline{UW} \cong \overline{SW}$ (CPCTC)
7. W is the midpoint of $\overline{US}$. (Def. midpoint)
8. $\overline{TW}$ is a median of $\triangle STU$. (Def. median)

Teaching Tip Remind students to include every step when writing a proof.

Proof Pointer

It may be difficult for some students to begin a proof for a statement that is general, as in Example 2. You may want to give examples of other statements to prove that do not involve specifically named segments and have students draw the figure from which their proof would be based.

b. For $\overline{CP}$ to be an altitude of $\triangle ABC$, $\overline{CP}$ must be perpendicular to $\overline{AB}$. This means that the product of the slopes of $\overline{CP}$ and $\overline{AB}$ must equal -1.

$$\text{slope of } \overline{CP} = \frac{15-6}{9-3} = \frac{9}{6} \text{ or } \frac{3}{2}$$

$$\text{slope of } \overline{AB} = \frac{10-2}{-3-9} = \frac{8}{-12} \text{ or } -\frac{2}{3}$$

$$\text{product of slopes} = \frac{3}{2} \cdot -\frac{2}{3} \text{ or } -1$$

Since the product of these slopes is -1, $\overline{CP}$ is perpendicular to $\overline{AB}$. Thus, $\overline{CP}$ is an altitude of $\triangle ABC$.

The fourth special segment of a triangle is an **angle bisector**. In Lesson 1–6, you learned that an angle bisector is a ray that divides an angle into two congruent angles. An angle bisector of a triangle is a segment that bisects an angle of the triangle and has one endpoint at a vertex of the triangle and the other endpoint at another point on the triangle. For $\triangle LMN$, $\overline{LA}$, $\overline{MB}$, and $\overline{NC}$ are the angle bisectors. Since there are three angles in every triangle, every triangle has three angle bisectors.

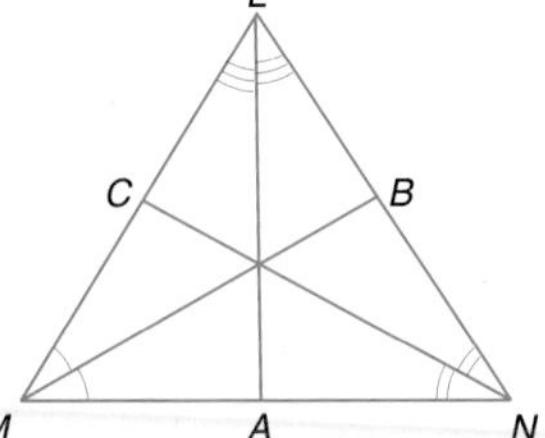

Theorems 5–3 and 5–4 involve special properties related to angle bisectors.

Theorem 5–3	**Any point on the bisector of an angle is equidistant from the sides of the angle.**
Theorem 5–4	**Any point on or in the interior of an angle and equidistant from the sides of an angle lies on the bisector of the angle.**

You will be asked to prove Theorem 5–3 in Exercise 37.

The special segments of different types of triangles have special properties.

Example 2 **Prove that if a triangle is isosceles, then the bisector of the vertex angle of the triangle is also a median.**

Proof

Begin by drawing and labeling a diagram of the situation.

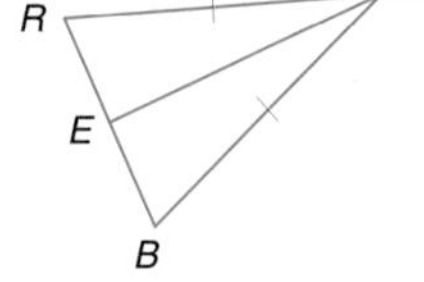

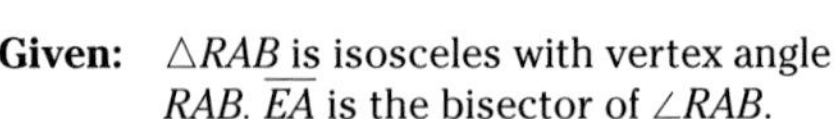

Given: $\triangle RAB$ is isosceles with vertex angle RAB. $\overline{EA}$ is the bisector of $\angle RAB$.

Prove: $\overline{EA}$ is a median.

Alternative Learning Styles

Auditory Without showing an example, tell students to draw an acute triangle with a perpendicular line from each vertex to the opposite side. Check their drawings for correctness. Explain that they just found the altitudes of an acute triangle. Practice with the altitudes of a right triangle and of an obtuse triangle.

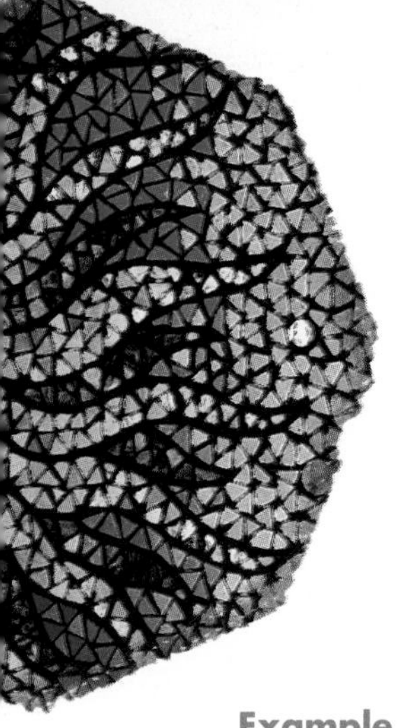

Proof:

Statements	Reasons
1. $\triangle RAB$ is isosceles. $\overline{EA}$ is the bisector of $\angle RAB$.	1. Given
2. $\overline{RA} \cong \overline{BA}$	2. Def. isosceles $\triangle$
3. $\angle RAE \cong \angle BAE$	3. Def. $\angle$ bisector
4. $\overline{AE} \cong \overline{AE}$	4. Congruence of segments is reflexive.
5. $\triangle AER \cong \triangle AEB$	5. SAS
6. $\overline{ER} \cong \overline{EB}$	6. CPCTC
7. $\overline{AE}$ is a median.	7. Def. median

The special segments of a triangle are often used to make patterns involving triangles more interesting. Triangles are also used in designing bridges and buildings because they are the most rigid shape.

Example 3

APPLICATION
Civil Engineering

Study the picture of the Waddell "A" Truss Bridge carefully. Write at least one conclusion that you can make from each statement.

a. $\overline{LO}$ is an altitude of $\triangle JKL$.

By the definition of an altitude, $\overline{LO} \perp \overline{JK}$. $\angle LOJ \cong \angle LOK$ because perpendicular lines form right angles and all right angles are congruent.

b. $\overline{JM} \cong \overline{ML}$

M is the midpoint of $\overline{JL}$. Thus, $\overline{OM}$ is a median of $\triangle JOL$.

c. $\overline{NQ}$ is an angle bisector and an altitude of $\triangle ONK$.

Since $\overline{NQ}$ is an angle bisector, $\angle ONQ \cong \angle KNQ$. $\overline{NQ} \perp \overline{OK}$ because $\overline{NQ}$ is an altitude. $\angle OQN \cong \angle KQN$ because perpendicular lines form right angles and all right angles are congruent. Congruence of segments is reflexive, so $\overline{NQ} \cong \overline{NQ}$. Therefore, by ASA, $\triangle ONQ \cong \triangle KNQ$. By CPCTC, $\overline{ON} \cong \overline{KN}$. Thus, $\triangle ONK$ is isosceles.

FYI

The Waddell "A" Truss type of bridge was patented in 1894. J.A.L. Waddell taught engineering in Tokyo, Japan in the 1880s.

CHECK FOR UNDERSTANDING

Communicating Mathematics

Study the lesson. Then complete the following.

1. **Compare and contrast** a perpendicular bisector and an altitude of a triangle. See margin.
2. **Determine** which of the four special segments might occur outside of a triangle. Draw an example of each.
3. **Find a counterexample** to the statement *An altitude and an angle bisector of a triangle are never the same segment.* See margin.
4. Draw $\triangle EFG$ with altitude $\overline{FH}$. What type of triangle is $\triangle EFH$? right

5. Draw a right triangle ABC on patty paper with the right angle at B. Fold the paper to create the median, altitude, and the angle bisector from vertex B, and the perpendicular bisector of $\overline{AC}$. Are any of the segments parallel? Justify your answer by making one or more additional folds in the paper.

2. Altitudes may occur outside of a triangle; see students' drawings.
5. The altitude and the perpendicular bisector are parallel. See students' justifications.

In-Class Example

For Example 3
Which statement appears to be true of the figure in Example 3?

a. $\overline{ON}$ is a median of $\triangle LOK$.
b. $\overline{MP}$ is a median of $\triangle JLO$.
c. $\overline{MO}$ is a perpendicular bisector of $\triangle JLO$. a

3 PRACTICE/APPLY

Check for Understanding

Exercises 1–17 are designed to help you assess your students' understanding through reading, writing, speaking, and modeling. You should work through Exercises 1–5 with your students and then monitor their work on Exercises 6–17.

Error Analysis

Students may confuse the meanings of the terms in this lesson. Have them think of similar words to help them remember each term. They may also use mnemonic devices to remember what the terms mean (such as **m**edian goes to the **m**idpoint or **m**iddle).

Additional Answers

1. Both a perpendicular bisector and an altitude are perpendicular to a side of the triangle. However, a perpendicular bisector passes through the midpoint of the side to which it is perpendicular. A perpendicular bisector does not necessarily pass through a vertex of the triangle. An altitude passes through the vertex of the triangle opposite the side to which the altitude is parallel. An altitude does not necessarily pass through the midpoint of the side to which it is perpendicular. One of the perpendicular bisectors in an isosceles triangle is also an altitude.
3. The altitude and the angle bisector to the vertex of the vertex angle of an isosceles triangle are the same segment.

Reteaching

Using Diagrams Have students draw an example of each special segment after you have explained it to them. Have them compare and contrast each segment. For example, an altitude and a perpendicular bisector are both perpendicular to the opposite segments; or, a perpendicular bisector must have an endpoint on the segment, while the altitude may not.

FYI

A truss is a planar structure composed of a combination of triangular structures that form a rigid framework.

Assignment Guide

Core (with proof): 19–43 odd, 44, 45, 47–56
Core (informal): 19–33 odd, 43, 44, 45, 47–50, 52–56
Enriched: 18–42 even, 44–56

For **Extra Practice**, see p. 772.

The red A, B, and C flags, printed only in the Teacher's Wraparound Edition, indicate the level of difficulty of the exercises.

Additional Answer

17. The flag is located at the intersection of the angle bisector between the West and Shore roads and the perpendicular bisector of the segment joining the lookout tower to the entrance.

Guided Practice

6–9. See Solutions Manual.

Draw and label a figure to illustrate each situation.

6. $\overline{AB}$ is a median of $\triangle BOC$, and A is between O and C.
7. $\overline{RA}$ is an altitude and a median in $\triangle RST$. A is between S and T.
8. $\overline{TU}$ is an altitude in $\triangle TUL$.
9. $\overline{QL}$ is an angle bisector in $\triangle PQR$, and L is between P and R.

Refer to $\triangle RES$. Write at least one conclusion you can make from each statement. Sample answers given for Exercises 10–14.

11. N is the midpoint of $\overline{SE}$; $\overline{RN}$ is a median of $\triangle RES$.
12. $\overline{SM} \perp \overline{RE}$; $\overline{RM} \cong \overline{ME}$; $\overline{SM}$ is the perpendicular bisector of $\overline{RE}$.

10. $\overline{SM}$ is an altitude to $\overline{RE}$. $\overline{SM} \perp \overline{RE}$
11. $\overline{SN} \cong \overline{NE}$
12. M is equidistant from R and E, and $\angle RMS$ is a right angle.
13. $\angle ERN \cong \angle SRN$ $\overline{NR}$ bisects $\angle SRE$.
14. $\overline{EL} \perp \overline{SR}$ $\overline{EL}$ is an altitude of $\triangle RES$.

INTEGRATION
Algebra

15b. $\frac{7}{5}$

15d. No; $AT \approx 17.2$ and $AB \approx 21.5$ by using the distance formula.

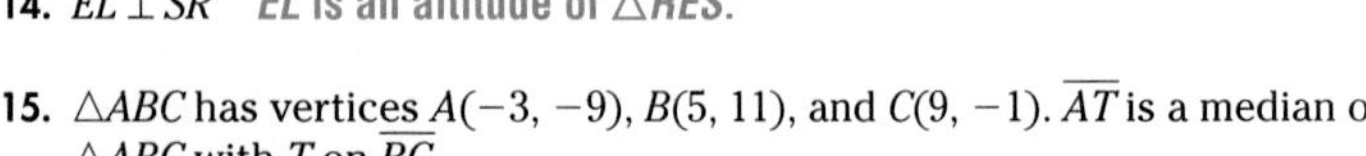

15. $\triangle ABC$ has vertices $A(-3, -9)$, $B(5, 11)$, and $C(9, -1)$. $\overline{AT}$ is a median of $\triangle ABC$ with T on $\overline{BC}$.
    a. What are the coordinates of T? (7, 5)
    b. Find the slope of $\overline{AT}$.
    c. Is $\overline{AT}$ an altitude of $\triangle ABC$? Explain. No; because $\frac{7}{5} \cdot -3 \neq -1$.
    d. Determine if $\overline{AT}$ is longer than $\overline{AB}$. Explain your findings.

Proof

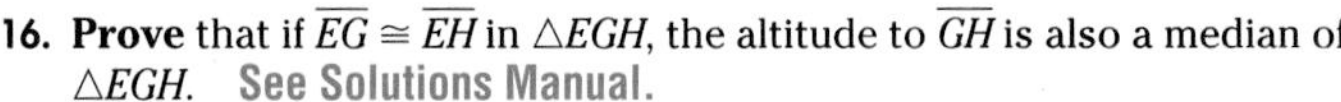

16. **Prove** that if $\overline{EG} \cong \overline{EH}$ in $\triangle EGH$, the altitude to $\overline{GH}$ is also a median of $\triangle EGH$. See Solutions Manual.

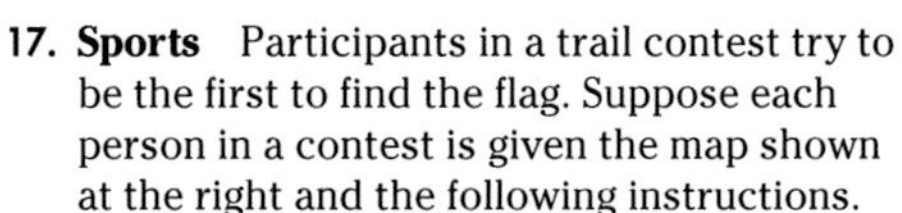

17. **Sports** Participants in a trail contest try to be the first to find the flag. Suppose each person in a contest is given the map shown at the right and the following instructions.
    - The flag is as far from the Lookout Tower as it is from the park entrance.
    - If you walk from West Road to the flag or from Shore Road to the flag, you would walk the same distance.

    Describe the position of the flag. See margin.

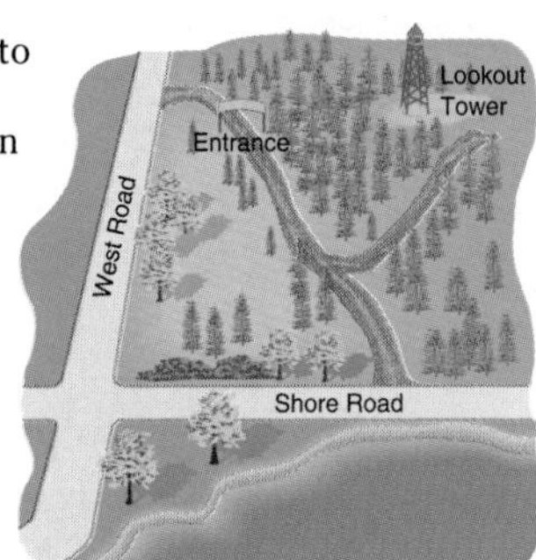

EXERCISES

Practice

A

18–23. See Solutions Manual.

Draw and label a figure to illustrate each situation.

18. $\overline{RS}$ is an angle bisector and an altitude of $\triangle RQP$.
19. $\overline{AB}$ and $\overline{CD}$ are angle bisectors of $\triangle ACT$ and intersect at X.
20. $\overline{LT}$ and $\overline{OS}$ are altitudes of $\triangle LOC$ and intersect at M outside of $\triangle LOC$.
21. $\triangle EFG$ is a right triangle with right angle F. $\overline{FT}$ is both an altitude and a median of $\triangle EFG$.
22. $\overline{PM}$ is a median and an angle bisector of $\triangle PQR$, and $\overline{RP}$ is an altitude.
23. $\triangle JKL$ is a right triangle with right angle at L. $\overline{LM}$ is a median of $\triangle JKL$, and $\overline{MN}$ is the perpendicular bisector of $\overline{JK}$.

Study Guide Masters, p. 26

5-1 NAME______ DATE______
Study Guide
Student Edition Pages 238–244

Special Segments in Triangles

Four special types of segments are associated with triangles.

- A **median** is a segment that connects a vertex of a triangle to the midpoint of the opposite side.
- An **altitude** is a segment that has one endpoint at a vertex of a triangle and the other endpoint on the line containing the opposite side so that the altitude is perpendicular to that line.
- An **angle bisector** of a triangle is a segment that bisects an angle of the triangle and has one endpoint at the vertex of that angle and the other endpoint on the side opposite that vertex.
- A **perpendicular bisector** is a segment or line that passes through the midpoint of a side and is perpendicular to that side.

Examples:

1 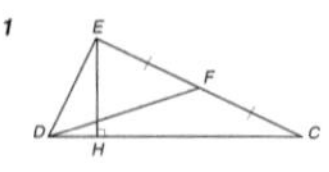

2 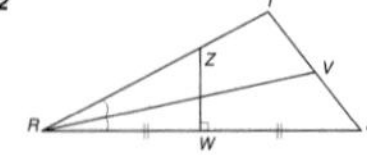

$\overline{DF}$ is a median of $\triangle DEC$. $\overline{EH}$ is an altitude of $\triangle DEC$.

$\overline{RV}$ is an angle bisector of $\triangle RST$. $\overline{WZ}$ is a perpendicular bisector of side $\overline{RS}$.

Draw and label a figure to illustrate each situation. Sample answers are given.

1. $\overline{OQ}$ is a median and an altitude of $\triangle POM$.

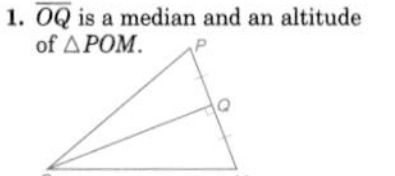

2. $\overline{KT}$ is an altitude of $\triangle KLM$, and L is between T and M.

3. $\overline{HS}$ is an angle bisector of $\triangle GHI$, and S is between G and I.

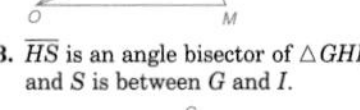

4. $\triangle NRW$ is a right triangle with right angle at N. $\overline{NX}$ is a median of $\triangle NRW$. $\overline{YX}$ is a perpendicular bisector of $\overline{WR}$.

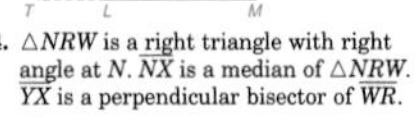

5. $\triangle TRE$ has vertices $T(3, 6)$, $R(-3, 10)$, and $E(-9, 4)$. Find the coordinates of point M if $\overline{TM}$ is a median of $\triangle TRE$. (−6, 7)

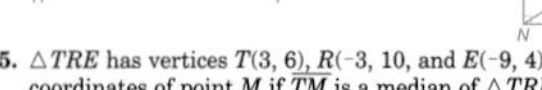

Classroom Vignette

"I have students construct the points of concurrency of the special segments in acute, equilateral, obtuse, and right triangles. We name the points and examine the properties. This activity allows us to discover new concepts and review constructions."

Nancy Keen
Martinsville High School
Martinsville, Indiana

Refer to $\triangle ABC$. Write at least one conclusion you can make from each statement.

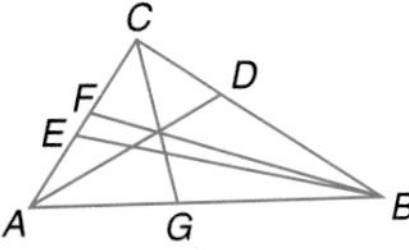

24. $\overline{AD}$ is an altitude. $\overline{AD} \perp \overline{BC}$
25. $\overline{BE}$ is an altitude and a median.
26. $\overline{AF} \cong \overline{FC}$ $\overline{BF}$ is a median.
27. $\overline{AD}$ is an angle bisector. $\angle CAD \cong \angle DAB$

25. $\overline{AE} \cong \overline{EC}$; $\overline{BE} \perp \overline{AC}$; $\overline{AB} \cong \overline{BC}$

If possible, describe a triangle for which each statement is true. If no triangle exists, write *no such triangle*.

28. The three angle bisectors of a triangle intersect at a point inside the triangle. any triangle
29. The three altitudes of a triangle intersect on the triangle.
30. The three angle bisectors of a triangle will not intersect. no such triangle
31. The perpendicular bisectors intersect outside of a triangle. an obtuse triangle
32. $\overline{RT}$ is a median in $\triangle RLB$ with points $R(3, 8)$, $T(12, 3)$, and $B(9, 12)$.
    a. What are the coordinates of L? $(15, -6)$
    b. Is $\overline{RT}$ an altitude of $\triangle RLB$?
    c. The graph of point S is at $(4, 13)$. $\overline{SC}$ intersects $\overline{RB}$ at C. If C is at $(6, 10)$, is $\overline{SC}$ a perpendicular bisector of $\overline{RB}$?

29. In a right triangle the altitudes intersect at the vertex of the right angle.

INTEGRATION **Algebra**

32b. No; $-\frac{5}{9} \cdot -3 \neq -1$.

32c. Yes; $\frac{2}{3} \cdot -\frac{3}{2} = -1$, and C is the midpoint of $\overline{RB}$.

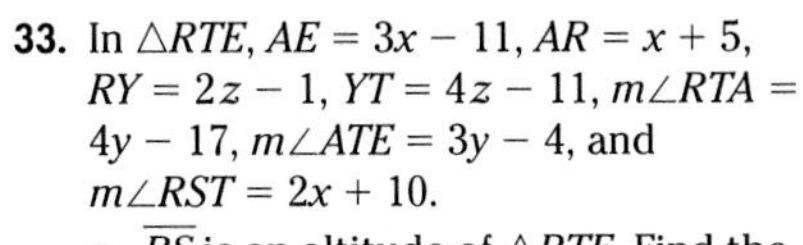
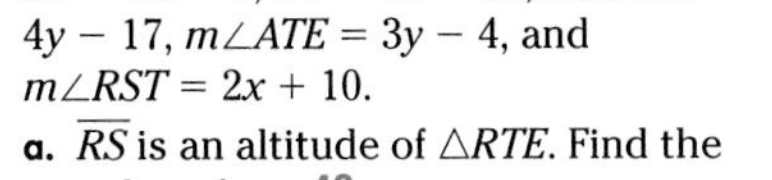

33. In $\triangle RTE$, $AE = 3x - 11$, $AR = x + 5$, $RY = 2z - 1$, $YT = 4z - 11$, $m\angle RTA = 4y - 17$, $m\angle ATE = 3y - 4$, and $m\angle RST = 2x + 10$.
    a. $\overline{RS}$ is an altitude of $\triangle RTE$. Find the value of x. 40
    b. If $\overline{TA}$ is an angle bisector of $\angle RTE$, find $m\angle RTA$. 35
    c. $\overline{EY}$ is a median of $\triangle RTE$. Find RT. 18

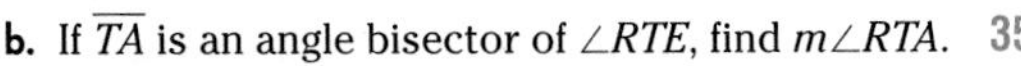

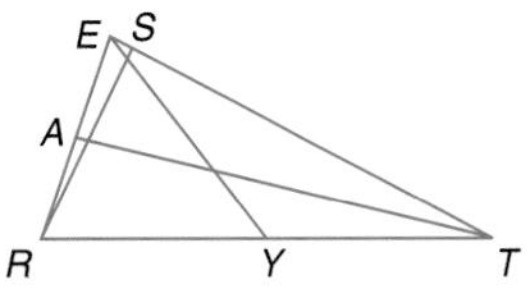

Draw and label a figure for each statement. List the information that is given and the statement to be proved in terms of your figure. Then write a proof of the statement. 36–41. See Solutions Manual.

C Proof

34–35. See margin.

34. If an angle bisector of a triangle is also an altitude, then the triangle is isosceles.
35. A point on the perpendicular bisector of a segment is equidistant from the endpoints of the segment. (Theorem 5–1)
36. A point equidistant from the endpoints of a segment lies on the perpendicular bisector of the segment. (Theorem 5–2)
37. A point on the bisector of an angle is equidistant from the sides of the angle. (Theorem 5–3)
38. The medians drawn to the congruent sides of an isosceles triangle are congruent.
39. The median to the base of an isosceles triangle bisects the vertex angle.
40. Corresponding angle bisectors of congruent triangles are congruent.
41. Corresponding medians of two congruent triangles are congruent.

Algebra

42. $\triangle ABC$ has vertices $A(-2, 3)$, $B(10, 13)$, and $C(4, 17)$. Draw $\triangle ABC$.
    a. Draw the medians of $\triangle ABC$. Label the point of intersection of the medians D. What are the coordinates of D? $(4, 11)$
    b. Show that the distance from A to D is one-third of the distance from the midpoint of $\overline{AB}$ to C.

42b. The distance from C to the midpoint of $\overline{AB}$ is 9. The distance from $(4, 11)$ to $(4, 8)$ is 3.

Additional Answers

34. Given: $\overline{CD}$ is an angle bisector. $\overline{CD}$ is an altitude.
Prove: $\triangle ABC$ is isosceles.

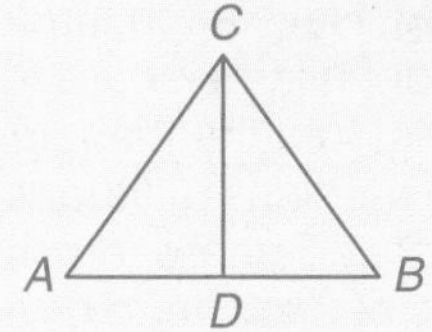

Proof:
Statements (Reasons)
1. $\overline{CD}$ is an angle bisector. $\overline{CD}$ is an altitude. (Given)
2. $\angle ACD \cong \angle BCD$ (Def. bisector)
3. $\overline{CD} \perp \overline{AB}$ (Def. altitude)
4. $\angle CDA$ and $\angle CDB$ are rt. ∆s. ($\perp$ lines form 4 rt. ∆s.)
5. $\angle CDA \cong \angle CDB$ (All rt. ∆s are $\cong$.)
6. $\overline{CD} \cong \overline{CD}$ (Cong. of segments is reflexive.)
7. $\triangle ACD \cong \triangle BCD$ (ASA)
8. $\overline{AC} \cong \overline{BC}$ (CPCTC)
9. $\triangle ACB$ is isosceles. (Def. isosceles $\triangle$)

35.

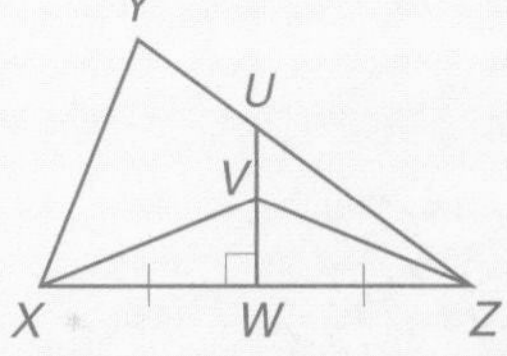

(continued in the bottom margin)

Practice Masters, p. 26

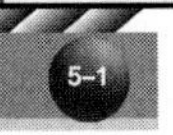

NAME________ DATE ________

Practice

Student Edition Pages 238–244

Special Segments in Triangles

1. Find AB if $\overline{BD}$ is a median of $\triangle ABC$. 13

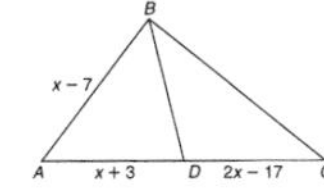

2. Find BC if $\overline{AD}$ is an altitude of $\triangle ABC$. 96

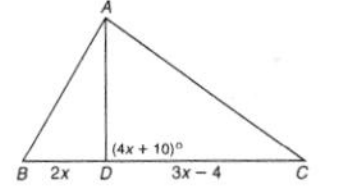

3. Find $m\angle ABC$ if $\overline{BD}$ is an angle bisector of $\triangle ABC$. 30

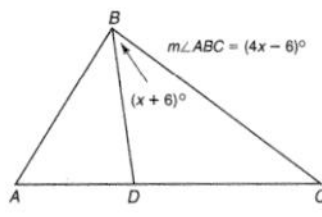

In Exercises 4–6, A(2, 5), B(12, −1), and C(−6, 8) are the vertices of $\triangle ABC$.

4. What are the coordinates of K if $\overline{CK}$ is a median of $\triangle ABC$? $(7, 2)$

5. What is the slope of the perpendicular bisector of $\overline{AB}$? What is the slope of $\overline{CL}$ if $\overline{CL}$ is the altitude from point C?
$\frac{5}{3}$, $\frac{5}{3}$

6. Point N on $\overline{BC}$ has coordinates $\left(\frac{8}{5}, \frac{21}{5}\right)$. Is $\overline{NA}$ an altitude of $\triangle ABC$? Explain your answer.
Yes; N is on the line that contains $\overline{BC}$, and the product of the slope of $\overline{BC}$, $-\frac{1}{2}$, and the slope of $\overline{AN}$, 2, is -1.

Additional Answer *(continued)*

35. Given: $\overline{UW}$ is the perpendicular bisector of $\overline{XZ}$
Prove: For any point V on $\overline{UW}$, $VX = VZ$.

Statements (Reasons)
1. $\overline{UW}$ is the $\perp$ bisector of $\overline{XZ}$. (Given)
2. W is the midpoint of $\overline{XZ}$. (Def. $\perp$ bisector)
3. $\overline{XW} \cong \overline{WZ}$ (Def. midpoint)
4. $\overline{UW} \perp \overline{XZ}$ (Def. $\perp$ bisector)
5. $\angle XWV$, $\angle ZWV$ are right ∆s. ($\perp$ lines form 4 rt. ∆s.)
6. $\angle XWV \cong \angle ZWV$ (All rt. ∆s are $\cong$.)
7. $\overline{VW} \cong \overline{VW}$ (Cong. of segments is reflexive.)
8. $\triangle XWV \cong \triangle ZWV$ (SAS)
9. $\overline{VX} \cong \overline{VZ}$ (CPCTC)
10. $VX = VZ$ (Def. $\cong$ segments)

4 ASSESS

Closing Activity

Modeling Have each student draw and cut out a large acute triangle. Have them fold the triangle to find all of the medians, altitudes, angle bisectors, and perpendicular bisectors of each vertex.

Additional Answers

43. Acute triangle: all segments meet inside the triangle; right triangle: altitudes meet at the vertex of right angle, other segments meet inside the triangle; obtuse triangle: altitudes meet outside the triangle, rest meet inside
44. The altitude will be the same for both triangles and the bases will be congruent, so the areas will be equal.
51. Statements (Reasons)
    1. $\overline{PQ}$ bisects $\overline{AB}$ at M. (Given)
    2. M is the midpoint of $\overline{AB}$. (Def. bisector)
    3. $AM = MB$ (Def. midpoint)
    4. $\overline{AM} \cong \overline{MB}$ (Def. $\cong$ seg.)

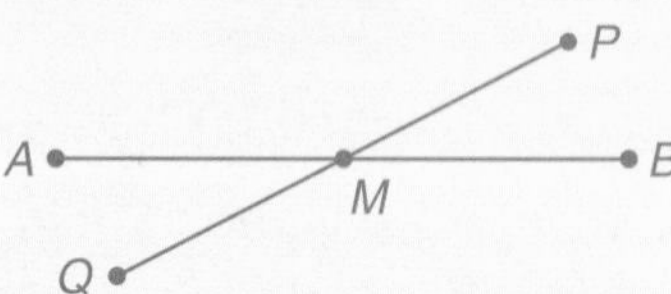

Enrichment Masters, p. 26

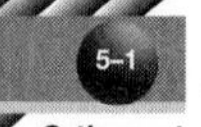

NAME ______________ DATE ______

Enrichment

Student Edition Pages 238–244

Orthocenter

The lines containing the altitudes of a triangle are concurrent, that is, they all pass through a common point. The point of concurrency is called the **orthocenter** of the triangle. In the figure at the right, Q is the orthocenter of $\triangle ABC$. Note that for altitude $\overline{AY}$, $m\angle BAY = 90 - m\angle ABC$ and $m\angle CAY = 90 - m\angle ACB$.

For any given triangle, the orthocenter can be located by using a protractor to draw the altitudes of the triangle.

Example: Find the orthocenter of $\triangle RST$ if $m\angle S = 36$, $m\angle T = 68$, and $m\angle R = 76$. Label the orthocenter Q.

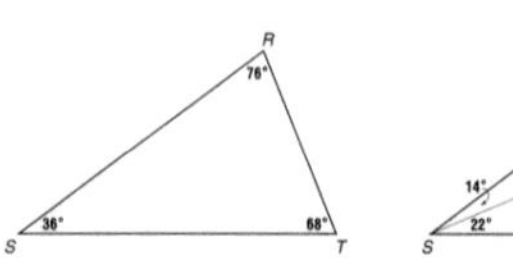

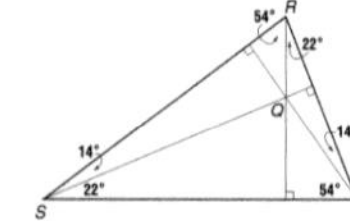

For each triangle, use a protractor to determine the measure of each angle. Then find the orthocenter by using the relationship described above. Label the orthocenter Q.

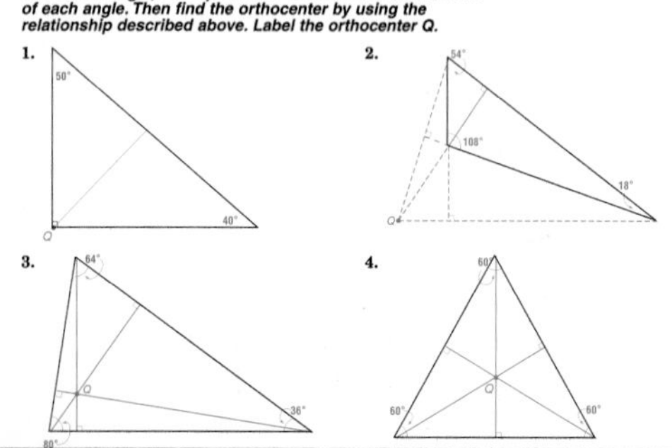

Cabri Geometry

43. Use a TI-92 calculator to draw different kinds of triangles and the special segments in each triangle. Determine where the special types of segments for each kind of triangle meet. **See students' work; see margin.**

Critical Thinking

44. Draw any $\triangle ABC$ with median $\overline{AD}$ and altitude $\overline{AE}$. Recall that the area of a triangle is one-half the product of the measures of the base and the altitude. What conclusion can you make about the relationship between the areas of $\triangle ABD$ and $\triangle ACD$? **See margin.**

Applications and Problem Solving

45. **Statistics** $\triangle TSR$ has vertices $T(-6, 12)$, $S(2, 4)$, and $R(16, 8)$. Graph $\triangle TSR$.
    a. Find $\bar{x}$, the average of the x-coordinates for T, S, and R. **4**
    b. Find $\bar{y}$, the average of the y-coordinates for T, S, and R. **8**
    c. Plot $M(\bar{x}, \bar{y})$ on the graph of $\triangle TSR$. **See Solutions Manual.**
    d. Draw the medians of $\triangle TSR$. What do you observe?
46. **Physics** Physicists often make calculations based on the center of gravity of an object. Follow the steps and answer the questions below to investigate the center of gravity of a triangle.
    a. Draw a large acute triangle that is not isosceles. **See students' work.**
    b. Construct the three medians of the triangle. What do you notice?
    c. Construct the three angle bisectors of the triangle. What do you notice?
    d. Construct the three altitudes of the triangle. What do you notice?
    e. Construct the three perpendicular bisectors of the sides of the triangle. What do you notice?
    f. Cut out the triangles you made for parts a–e. Place the point where the segments intersect on the flat end of a pencil for each of the triangles. What do you observe?
    g. What changes would occur in the construction in parts b–e if the triangle were right or obtuse instead of acute?

45d. The medians intersect at M.

46b–e. They intersect in one point.

46f. The triangle balances on the intersection of the medians.

46g. The perpendicular bisectors and the altitudes would intersect at a point on or outside of the triangle.

48. Raul: pet store; Molesha: fast food; Jennifer: day camp

49. Yes, both have slope $-\frac{2}{3}$.

Mixed Review

47. Graph $\triangle BAY$ with vertices $B(-1, 1)$, $A(-5, 4)$, and $Y(3, 4)$. Prove that $\triangle BAY$ is isosceles. (Lesson 4–6) **$BA = 5$, $AY = 8$, $BY = 5$**
48. **Eliminate the Possibilities** Raul, Molesha, and Jennifer just began after-school jobs. One works at a pet store, one at a day camp, and one at a fast-food restaurant. Raul takes his sister to the day camp on his way to work. Jennifer is allergic to pet hair. Molesha receives free meals from her job. Who works at which job? (Lesson 4–5)
49. Without graphing, determine if $\overleftrightarrow{PQ} \parallel \overleftrightarrow{RS}$ given points $P(0, 0)$, $Q(3, -2)$, $R(1, -7)$, and $S(-5, -3)$. Explain. (Lesson 3–4)
50. Draw a figure to illustrate two parallel lines with a plane perpendicular to both lines. (Lesson 3–1) **See Solutions Manual.**
51. Write a two-column proof for *If $\overline{PQ}$ bisects $\overline{AB}$ at point M, then $\overline{AM} \cong \overline{MB}$.* (Lesson 2–5) **See margin.**
52. **Algebra** Write a conjecture based on $12x - 42 = 54$. (Lesson 2–1) **$x = 8$**
53. If S is between R and T, $RS = 13$, $ST = 2x + 7$, and $RT = 3x + 8$, find the value of x and RT. (Lesson 1–4) **12, 44**
54. Draw and label a figure to show points A, B, C, and D so that no more than two points are collinear. (Lesson 1–2) **See Solutions Manual.**

Algebra

55. **Oceanography** A scuba diver was exploring mineral deposits at a depth of 75 meters. She went up 30 meters. At what depth was she then?
56. Simplify $12r - (-10r)$. **$22r$**

55. 45 meters

Extension

Reasoning Do an indirect proof to show that every triangle cannot have more than three medians. Explain to students that an indirect proof assumes the conclusion to be false so that when a contradiction occurs, it proves the conclusion must be true.
Assume a triangle can have more than three medians. Then one vertex must contain at least two medians since there are only three vertices in a triangle. By definition, both medians must be connected to the opposite side of the vertex. However, each segment has only one midpoint, so each vertex can be an endpoint of only one median.

5-2 Right Triangles

What YOU'LL LEARN
- To recognize and use tests for congruence of right triangles.

Why IT'S IMPORTANT
Right triangles are used in trigonometry, a branch of mathematics you will study later in this course.

Windsurfing

Harnessing the wind may seem like it would take a great deal of strength. But Rhonda Smith, one of the pioneers in women's windsurfing, says that it's more a matter of proper technique than strength. She should know—Rhonda has won five world windsurfing championships! She now runs a windsurfing school in Hood River, Oregon.

A light-wind sail for a sailboard is a right triangle. Suppose right triangles *DEF* and *RST* model the sails for two of Rhonda's boards. What would be required to prove the triangles congruent by SAS? The conditions necessary for this congruence are stated in Theorem 5–5.

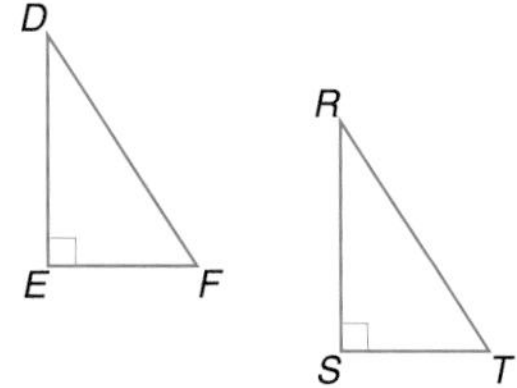

Theorem 5–5 LL	**If the legs of one right triangle are congruent to the corresponding legs of another right triangle, then the triangles are congruent.**

Proof of Theorem 5–5

Given: $\triangle DEF$ and $\triangle RST$ are right triangles.
$\angle E$ and $\angle S$ are right angles.
$\overline{EF} \cong \overline{ST}$
$\overline{ED} \cong \overline{SR}$

Prove: $\triangle DEF \cong \triangle RST$

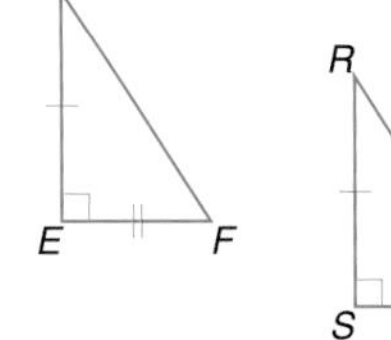

Paragraph Proof:

We are given that $\overline{EF} \cong \overline{ST}$, $\overline{ED} \cong \overline{SR}$, and $\angle E$ and $\angle S$ are right angles. Since all right angles are congruent, $\angle E \cong \angle S$. Therefore, by SAS, $\triangle DEF \cong \triangle RST$.

FYI

The longest sailboard in the world is 165 feet long. It was constructed in Fredrikstad, Norway, and first sailed June 28, 1986.

Suppose the hypotenuse and an acute angle of one sail are congruent to the hypotenuse and corresponding acute angle of the other sail. Is that enough information to prove that the sails are congruent?

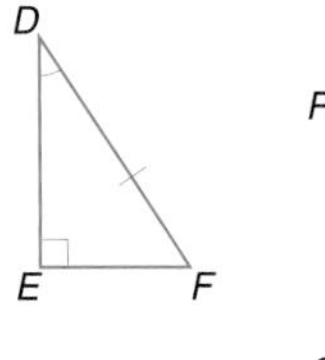

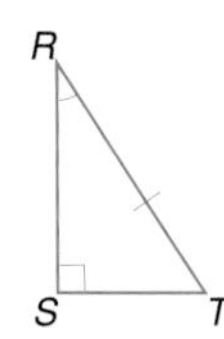

The right angles in each triangle are congruent. Knowing that another angle and the hypotenuse are congruent indicates that two angles and a non-included side are congruent. Thus, the triangles are congruent by AAS. The two sails would be the same size. This leads us to Theorem 5–6. *You will be asked to prove Theorem 5–6 in Exercise 28.*

Sailboarding is now an Olympic event. In the U.S. the Windsurfer is a registered trademark.

4. The altitude of a triangle contains the midpoint of the opposite side. **sometimes**
5. The medians of a triangle contain the midpoints of the opposite sides. **always**

5-2 LESSON NOTES

NCTM Standards: 1–4, 7

Instructional Resources

- Study Guide Master 5-2
- Practice Master 5-2
- Enrichment Master 5-2
- Assessment and Evaluation Masters, p. 128
- Graphing Calculator and Computer Masters, p. 5
- Multicultural Activity Masters, p. 9
- Real-World Applications, 9

Transparency 5-2A contains the 5-Minute Check for this lesson; **Transparency 5-2B** contains a teaching aid for this lesson.

Recommended Pacing	
Standard Pacing	Day 3 of 14
Honors Pacing	Day 2 of 13
Block Scheduling*	Day 2 of 7

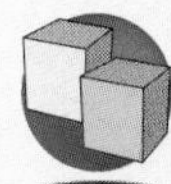

*For more information on pacing and possible lesson plans, refer to the *Block Scheduling Booklet.*

1 FOCUS

5-Minute Check
(over Lesson 5-1)

Draw and label a figure to illustrate each situation.

1. $\overline{SL}$ is an altitude of $\triangle RST$ and a perpendicular bisector of side $\overline{RT}$.
2. $\triangle HIJ$ is an isosceles triangle with vertex angle at J. $\overline{JN}$ is an altitude of $\triangle HIJ$.

1. 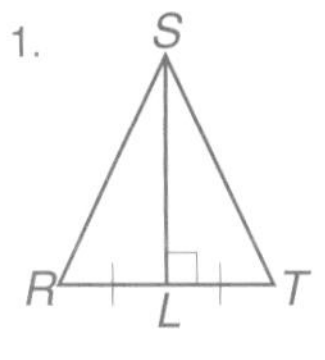

2. 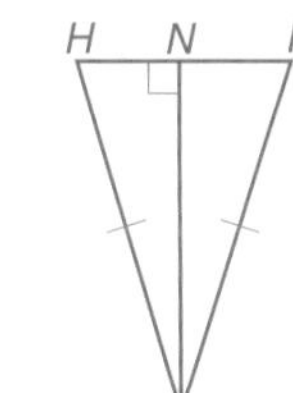

State whether each sentence is *always*, *sometimes*, or *never* true.

3. The perpendicular bisectors of a triangle intersect at more than one point. **never**

Motivating the Lesson

Situational Problem Have each student name a place or thing that has a right angle or a right triangle in it. After all students have given an example, talk about how prevalent right angles and right triangles are in our surroundings.

2 TEACH

In-Class Example

For Example 1

The metal bars on the side of a fold-up stroller are put together so that the middle bar is a perpendicular bisector of the handlebar. Prove that the two triangles formed are congruent.

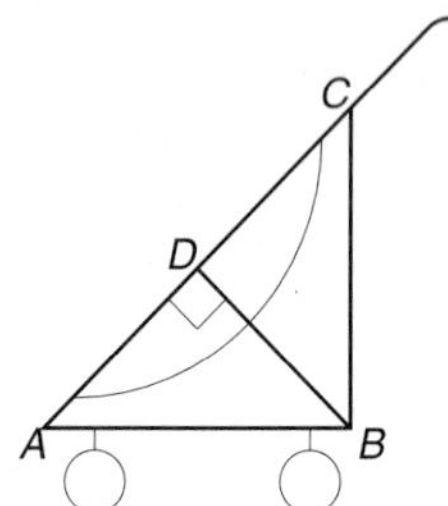

Given: $\overline{BD}$ is a perpendicular bisector of $\triangle ABC$.

Prove: $\triangle ABD \cong \triangle CBD$

Proof:

Statements (Reasons)

1. **$\overline{BD}$ is a perpendicular bisector of $\triangle ABC$. (Given)**
2. **$\overline{BD} \perp \overline{AC}$ (Def. perpendicular bisector)**
3. **$\angle BDA$ and $\angle BDC$ are right angles. ($\perp$ lines form four rt. $\angle$s.)**
4. **$\triangle ABD$ and $\triangle CBD$ are right triangles. (Def. rt. triangle)**
5. **$\overline{BD} \cong \overline{BD}$ (Cong. of segments is reflexive.)**
6. **$\overline{DA} \cong \overline{DC}$ (Def. perpendicular bisector)**
7. **$\triangle ABD \cong \triangle CBD$ (LL)**

Theorem 5–6 *HA*	**If the hypotenuse and an acute angle of one right triangle are congruent to the hypotenuse and corresponding acute angle of another right triangle, then the two triangles are congruent.**

You can use the HA theorem to complete proofs involving right triangles.

Example 1

Part of the brace used to support electrical power lines is shaped like an isosceles triangle with an altitude from the vertex to the base. Prove that the two right triangles formed by the altitude are congruent.

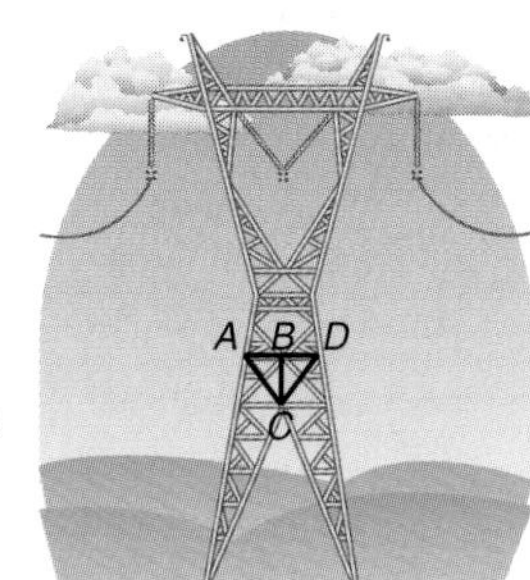

Given: $\overline{CB}$ is an altitude of $\triangle ACD$.
$\triangle ACD$ is an isosceles triangle with legs $\overline{CA}$ and $\overline{CD}$.

Prove: $\triangle ABC \cong \triangle DBC$

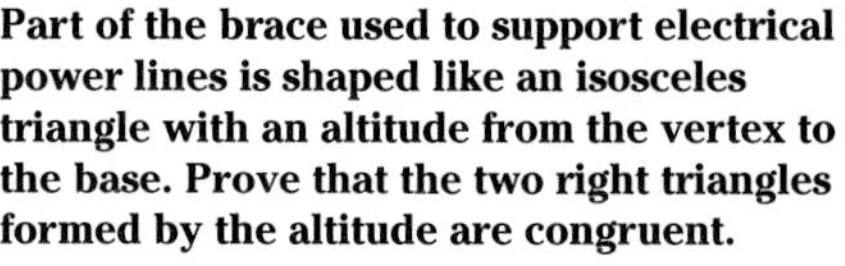

Two-column Proof:

Statements	Reasons
1. $\triangle ACD$ is an isosceles triangle.	**1.** Given
2. $\overline{CA} \cong \overline{CD}$	**2.** Def. isos. $\triangle$
3. $\angle BAC \cong \angle BDC$	**3.** Isosceles Triangle Theorem
4. $\overline{CB}$ is an altitude of $\triangle ACD$.	**4.** Given
5. $\overline{BC} \perp \overline{AD}$	**5.** Def. altitude
6. $\angle ABC$ and $\angle DBC$ are right angles.	**6.** $\perp$ lines form 4 rt. $\angle$s.
7. $\triangle ABC$ and $\triangle DBC$ are right triangles.	**7.** Def. rt. $\triangle$
8. $\triangle ABC \cong \triangle DBC$	**8.** HA

We have already determined that right triangles are congruent if corresponding legs are congruent (LL) or if the hypotenuses and corresponding acute angles are congruent (HA). As a third possibility, suppose a leg and an acute angle of one sail are congruent to the corresponding leg and acute angle of the second sail. Will the two sails be congruent under these conditions? There are two different cases to consider.

Case 1

The leg is included between the acute angle and the right angle.

Case 2

The leg is not included between the acute angle and the right angle.

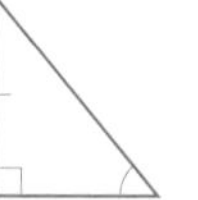

The triangles are congruent in both cases. For Case 1, the triangles are congruent by ASA. For Case 2, the triangles are congruent by AAS. This observation suggests Theorem 5–7. *You will be asked to prove Theorem 5–7 in Exercise 29.*

Theorem 5–7
LA

If one leg and an acute angle of one right triangle are congruent to the corresponding leg and acute angle of another right triangle, then the triangles are congruent.

Example 2 **INTEGRATION Algebra**

Find the values of x and y so that $\triangle MNO$ is congruent to $\triangle RST$.

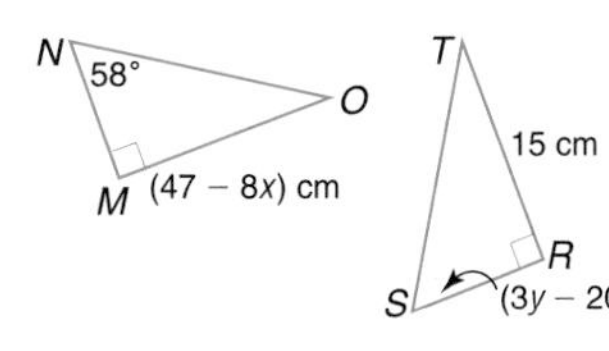

Assume $\triangle MNO \cong \triangle RST$. Then $\angle N \cong \angle S$ and $\overline{MO} \cong \overline{RT}$.

$m\angle N = m\angle S$ | $MO = RT$
$58 = 3y - 20$ | $47 - 8x = 15$
$78 = 3y$ | $-8x = -32$
$26 = y$ | $x = 4$

By LA, $\triangle MNO \cong \triangle RST$ for $x = 4$ and $y = 26$.

You can use a TI-92 graphing calculator to investigate the following question.

EXPLORATION CABRI GEOMETRY

If the hypotenuse and one leg of one right triangle are congruent to the hypotenuse and corresponding leg of a second right triangle, are the triangles congruent?

- Use a TI-92 calculator to draw a right triangle.
- Draw two perpendicular rays in another place on the screen.
- Measure one of the legs of the right triangle. Then, use 9:Measurement Transfer on F4 to transfer the measurement to one of the perpendicular rays to create a congruent leg.
- Measure the hypotenuse of the right triangle. Then transfer the measurement to the endpoint of the segment created above that is not the vertex of the right angle. Position the segment being drawn so that the vertex lies on the perpendicular ray in order to form a right triangle.
- Measure the angles and sides of both triangles.

Your Turn a. All corresponding measurements are equal.

a. How do the measurements of the two right triangles compare?

b. Write a conjecture about two right triangles that have one pair of corresponding legs congruent and congruent hypotenuses.
They are congruent.

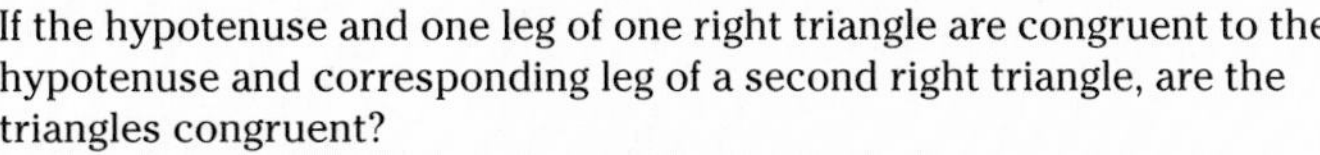

The results of the Exploration lead us to Postulate 5–1.

Postulate 5–1
HL

If the hypotenuse and a leg of one right triangle are congruent to the hypotenuse and corresponding leg of another right triangle, then the triangles are congruent.

Since this is a postulate, it need not be proved.

In-Class Example

For Example 2
Find the values of x and y so that $\triangle JKL$ is congruent to $\triangle MLK$.

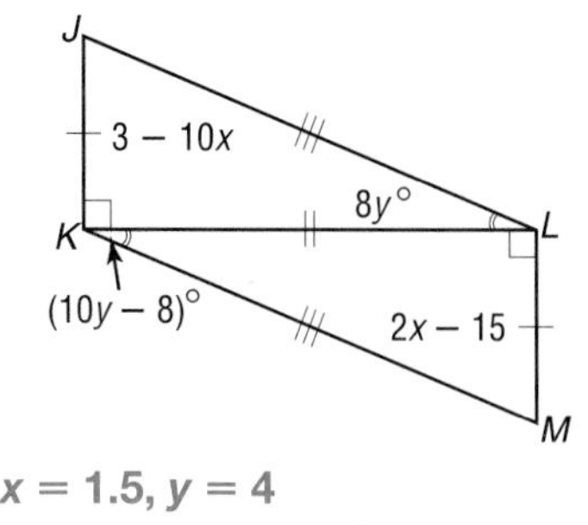

$x = 1.5$, $y = 4$

Teaching Tip When discussing the theorems in this lesson, point out that they are an extension of previous theorems or postulates, such as AAS or SAS. This is because all right triangles have one angle that is congruent, namely the right angle. Thus, you only need two other parts to prove two right angles congruent.

Teaching Tip If students are having difficulty visualizing the concepts in this lesson, have them make models of the problems.

EXPLORATION

In this activity, students construct a copy of part of a right triangle in order to verify congruence. The procedure imitates a compass-and-straightedge construction.

Cooperative Learning

Group Discussion Have students list all of the parts of a right triangle. Have them list all of the possible combinations of two parts. For example, they could combine a leg and an acute angle, a leg and the right angle, and so on. Have them find what combinations will prove two right triangles congruent by trying to find counterexamples or by proving the statements true. For more information on the group discussion strategy, see *Cooperative Learning in the Mathematics Classroom*, one of the titles in the Glencoe Mathematics Professional Series, page 31.

In-Class Example

For Example 3
Write the proof in Example 3 as a two-column proof.

Proof:
Statements (Reasons)
1. $\overline{PR} \perp \overline{QR}$, $\overline{PS} \perp \overline{QS}$ (Given)
2. $\angle QSP$ is a rt. $\angle$, $\angle QRP$ is a rt. $\angle$. ($\perp$ lines form rt. $\angle$s.)
3. $\triangle QSP$ is a rt. $\triangle$, $\triangle QRP$ is a rt. $\triangle$. (Def. rt. $\triangle$)
4. $\overline{QR} \cong \overline{QS}$ (Given)
5. $\overline{PQ} \cong \overline{PQ}$ (Cong. of segments is reflexive.)
6. $\triangle QSP \cong \triangle QRP$ (HL)
7. $\angle SQP \cong \angle RQP$ (CPCTC)
8. $\overline{QP}$ bisects $\angle SQR$. (Def. $\angle$ bisector)

Proof Pointer

A proof with more givens often has more columns when written as a flow proof.

3 PRACTICE/APPLY

Check for Understanding

Exercises 1–13 are designed to help you assess your students' understanding through reading, writing, speaking, and modeling. You should work through Exercises 1–5 with your students and then monitor their work on Exercises 6–13.

Error Analysis
If students are having difficulty understanding why only four combinations of segments and angles can be used to prove right triangles congruent, give a counterexample to show why each of the other combinations will not work.

Example 3 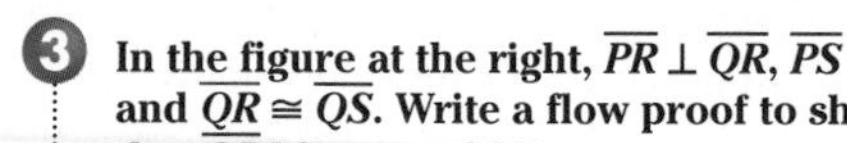 **In the figure at the right, $\overline{PR} \perp \overline{QR}$, $\overline{PS} \perp \overline{QS}$, and $\overline{QR} \cong \overline{QS}$. Write a flow proof to show that $\overline{QP}$ bisects $\angle SQR$.**

Proof

Given: $\overline{PR} \perp \overline{QR}$, $\overline{PS} \perp \overline{QS}$, $\overline{QR} \cong \overline{QS}$

Prove: $\overline{QP}$ bisects $\angle SQR$

Flow Proof:

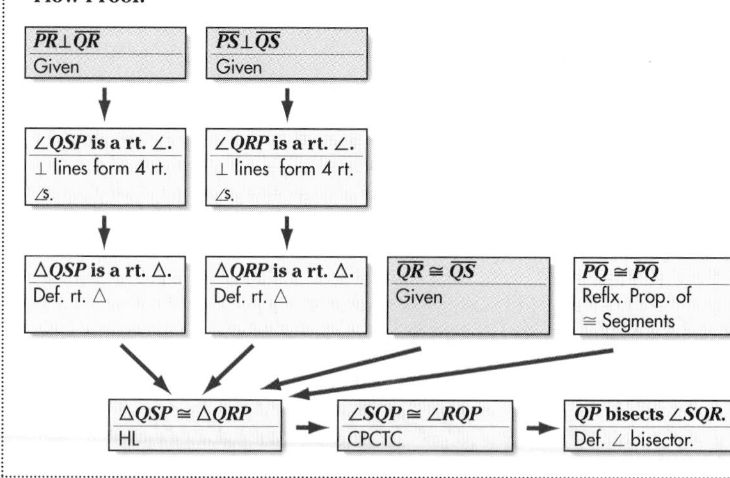

CHECK FOR UNDERSTANDING

Communicating Mathematics

Study the lesson. Then complete the following. 2–4. See margin.

1. **Explain** why there are only two requirements to prove two right triangles congruent while tests for other triangles have three.
   1. Right triangles always have one part, a right angle, in common.
2. **Compare and contrast** the tests for triangle congruence (SAS, AAS, ASA) and the tests for congruence of right triangles (LL, HA, and LA).
3. **Show** why two cases must be considered when proving LA.
4. **You Decide** Carlos said that you could justify the Leg-Leg test for congruent right triangles by SSS. Jeannie disagrees. Who is correct and why?

5. Write a short paragraph describing each of the tests for congruence of right triangles presented in this lesson. Use word processing software if it is available. See students' work.

Guided Practice

State the additional information needed to prove each pair of triangles congruent by the given theorem or postulate.

6. HL
7. LL $\overline{ST} \cong \overline{TU}$
8. LA

6. $\angle B$ and $\angle D$ are right angles.
8. $\overline{LN} \cong \overline{QR}$ or $\overline{NM} \cong \overline{RP}$

Reteaching

Using Diagrams Have students draw two triangles that are congruent by SAS. Then have them draw two right triangles that are congruent by SAS. If you take out the A for angle, you are left with SS. In a right triangle, this is the same as LL, and since all right angles are congruent, LL for right triangles is actually the same as SAS for any triangle. Do the same for the other theorems in this lesson.

INTEGRATION Algebra

For each figure, find the value of x so that each pair of triangles is congruent by the indicated theorem or postulate.

9. HA 6

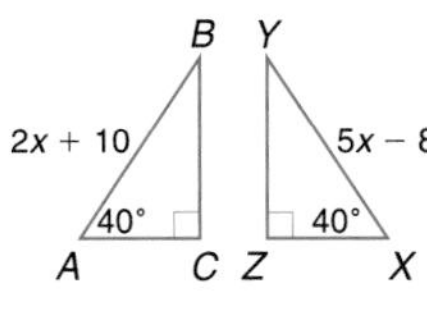

10. LA 6

11. HL 9

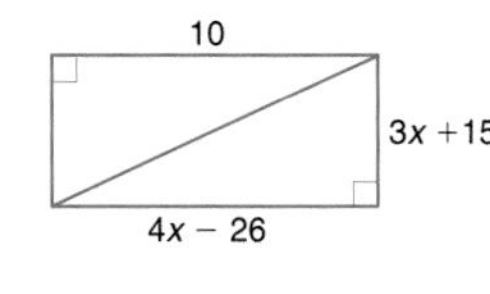

Proof

12. Use the figure at the right to write a two-column proof. See margin.

Given: $\angle RLT$ and $\angle SLT$ are right angles.
$\angle LTR \cong \angle LTS$

Prove: $\overline{LR} \cong \overline{LS}$

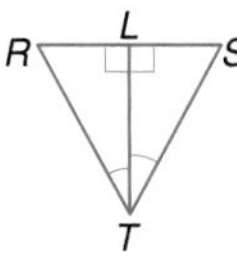

13. **Art** *The Sculptured Space* is a sculpture built by a team of sculptors headed by Helen Escobedo. The sculpture is made up of triangular pieces of concrete surrounding a circle of petrified lava. One leg of each right triangle is 5 meters long, and the other leg is 7.31 meters long. Can you conclude that all the triangles are congruent? Explain. yes, LL

The Sculptured Space

EXERCISES

Practice

State the additional information needed to prove each pair of triangles congruent by the given theorem or postulate.

A

14. LL $\overline{UX} \cong \overline{XW}$

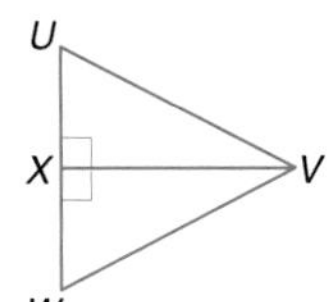

15. HA

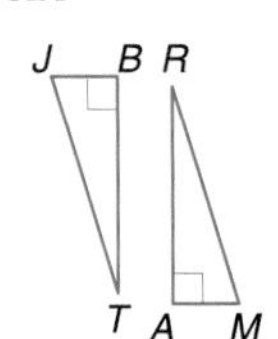

16. HL

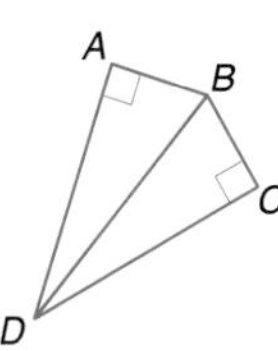

15. $\overline{JT} \cong \overline{MR}$ and $\angle J \cong \angle M$ or $\overline{JT} \cong \overline{MR}$ and $\angle T \cong \angle R$

16. $\overline{AD} \cong \overline{CD}$ or $\overline{AB} \cong \overline{CB}$

17. LL

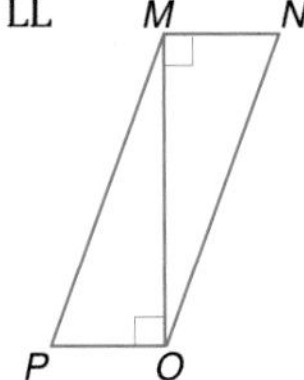

18. HL

19. LA

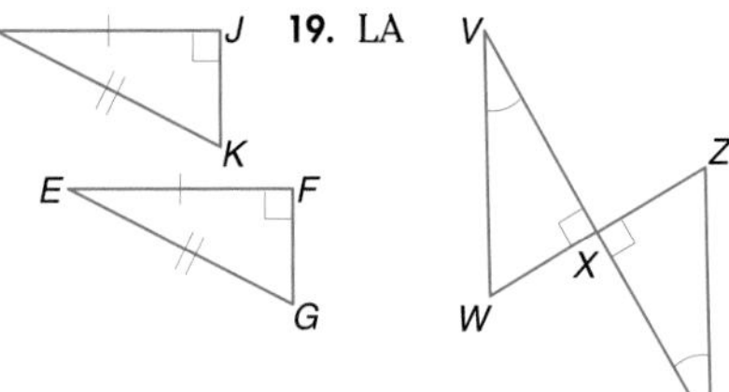

17. $\overline{MN} \cong \overline{OP}$

18. no extra information needed

19. $\overline{VX} \cong \overline{XY}$ or $\overline{WX} \cong \overline{XZ}$

INTEGRATION Algebra

B

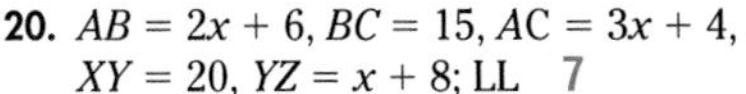

Find the value of x so that $\triangle ABC \cong \triangle XYZ$ by the indicated theorem or postulate.

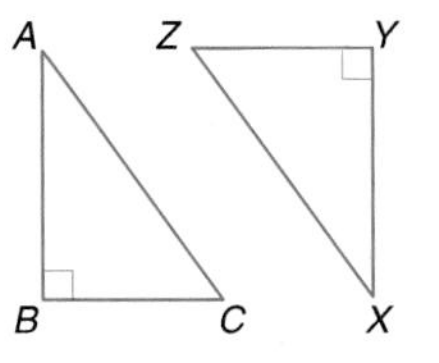

20. $AB = 2x + 6$, $BC = 15$, $AC = 3x + 4$, $XY = 20$, $YZ = x + 8$; LL 7

21. $m\angle Z = 55°$, $BC = 15x + 2$, $m\angle C = 55°$, $AB = 24$, $YZ = 4x + 13$; LA 1

22. $YX = 21$, $m\angle X = 9x + 9$, $AB = 21$, $m\angle A = 11x - 3$; LA 6

23. $AC = 28$, $AB = 7x + 4$, $ZX = 9x + 1$, $YX = 5(x + 2)$; HL 3

Assignment Guide

Core (with proof): 15–33 odd, 34, 35, 37–45
Core (informal): 15–23 odd, 33, 34, 35, 37–45
Enriched: 14–32 even, 34–45

For **Extra Practice**, see p. 772.

The red A, B, and C flags, printed only in the Teacher's Wraparound Edition, indicate the level of difficulty of the exercises.

Additional Answer

12. Given: $\angle RLT$ and $\angle SLT$ are right angles.
$\angle LTR \cong \angle LTS$
Prove: $\overline{LR} \cong \overline{LS}$
Proof:
Statements (Reasons)
1. $\angle RLT$ and $\angle SLT$ are right angles. $\angle LTR \cong \angle LTS$ (Given)
2. $\triangle RLT$ and $\triangle SLT$ are right triangles. (Def. rt. $\triangle$)
3. $\overline{LT} \cong \overline{LT}$ (Congruence of segments is reflexive.)
4. $\triangle RLT \cong \triangle SLT$ (LA)
5. $\overline{LR} \cong \overline{LS}$ (CPCTC)

Study Guide Masters, p. 27

5–2 NAME_____ DATE _____

Study Guide

Student Edition Pages 245–251

Right Triangles

Two right triangles are congruent if one of the following conditions exist.

Theorem 5-5 LL	If the legs of one right triangle are congruent to the corresponding legs of another right triangle, then the triangles are congruent.
Theorem 5-6 HA	If the hypotenuse and an acute angle of one right triangle are congruent to the hypotenuse and corresponding acute angle of another right triangle, then the two triangles are congruent.
Theorem 5-7 LA	If one leg and an acute angle of one right triangle are congruent to the corresponding leg and acute angle of another right triangle, then the triangles are congruent.
Postulate 5-1 HL	If the hypotenuse and a leg of one right triangle are congruent to the hypotenuse and corresponding leg of another right triangle, then the triangles are congruent.

State the additional information needed to prove each pair of triangles congruent by the given theorem or postulate.

1. HL

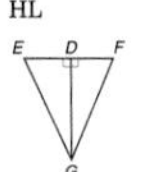

$\overline{EG} \cong \overline{FG}$

2. HA

$\overline{AB} \cong \overline{DB}$

3. LL

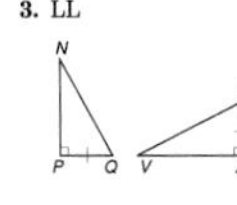

$\overline{NP} \cong \overline{VX}$

4. LA

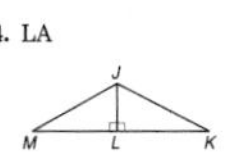

$\angle M \cong \angle K$ or $\angle MJL \cong \angle KJL$

5. HA

no extra information needed

6. LA

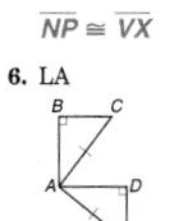

$\overline{BC} \cong \overline{DF}$ or $\overline{AB} \cong \overline{AD}$, and $\angle BAC \cong \angle DAF$ or $\angle C \cong \angle F$

Additional Answers

2. LL is the same as SAS since the congruent right angles are between the congruent sides, the legs. LA is the same as ASA or AAS since the leg may be included between the right angle and the acute angle or it may be not included. HA is the same as AAS.
3. The leg may be included between the right angle and the acute angle or it may be not included.
4. Jeannie is correct. LL is the same as SAS because the hypotenuse is not given as congruent, but the right angle is.

Additional Answers

24. Statements (Reasons)
    1. $\overline{NO} \perp \overline{MN}$, $\overline{NO} \perp \overline{OP}$, $\angle M \cong \angle P$ (Given)
    2. $\angle MNO$ and $\angle NOP$ are rt. $\angle$s. ($\perp$ lines form 4 rt. $\angle$s.)
    3. $\triangle MNO$ and $\triangle NOP$ are rt. $\triangle$s. (Def. rt. $\triangle$)
    4. $\overline{NO} \cong \overline{NO}$ (Cong. of segments is reflexive.)
    5. $\triangle MNO \cong \triangle NOP$ (LA)
    6. $\overline{MO} \cong \overline{NP}$ (CPCTC)
25. Statements (Reasons)
    1. $\angle M$ and $\angle P$ are right angles. $\overline{MN} \parallel \overline{OP}$ (Given)
    2. $\triangle MNO$ and $\triangle PON$ are rt. $\triangle$s. (Def. rt. $\triangle$)
    3. $\angle MNO \cong \angle PON$ (Alt. Int. $\angle$s Th.)
    4. $\overline{NO} \cong \overline{NO}$ (Cong. of segments is reflexive.)
    5. $\triangle MNO \cong \triangle PON$ (HA)
    6. $\overline{MN} \cong \overline{OP}$ (CPCTC)
26. Statements (Reasons)
    1. $\overline{AP} \perp$ plane $\mathcal{M}$, $\overline{AB} \cong \overline{AC}$ (Given)
    2. $\angle APB$ and $\angle APC$ are rt. $\angle$s. ($\perp$ lines form 4 rt. $\angle$s.)
    3. $\triangle APB$ and $\triangle APC$ are rt. $\triangle$s. (Def. rt. $\triangle$)
    4. $\overline{AP} \cong \overline{AP}$ (Cong. of segments is reflexive.)
    5. $\triangle APC \cong \triangle APB$ (HL)
    6. $\overline{PB} \cong \overline{PC}$ (CPCTC)
    7. $\triangle BPC$ is isosceles. (Def. isosceles $\triangle$)

Practice Masters, p. 27

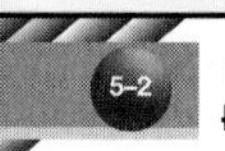

5-2 NAME________ DATE________

Practice Student Edition Pages 245–251

Right Triangles

For each figure, find the values of x and y so that $\triangle DEF \cong \triangle PQR$ by the indicated theorem or postulate.

1. HA 2, 30

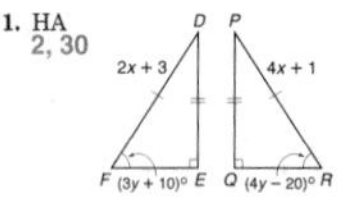

2. LL 26, 16

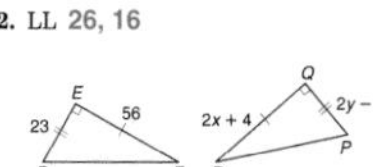

3. LA 5, 15

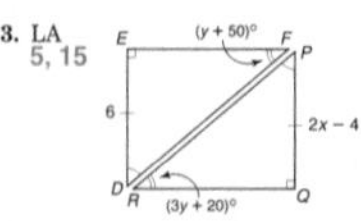

4. HL 19, 23

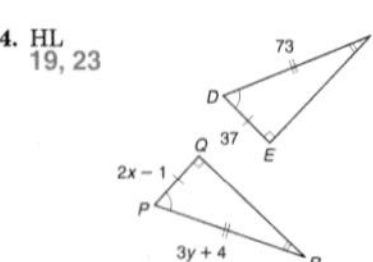

5. Write a two-column proof.
Given: $\overline{AB}$ bisects $\angle DAC$. $\angle C$ and $\angle D$ are right angles.
Prove: $\overline{BC} \cong \overline{BD}$
Proof:

Statements	Reasons
a. $\overline{AB}$ bisects $\angle DAC$.	a. Given
b. $\angle 1 \cong \angle 2$	b. Definition of bisector
c. $\angle C$ and $\angle D$ are right angles.	c. Given
d. $\triangle ACB$ and $\triangle ADB$ are right triangles.	d. Definition of right triangle
e. $\overline{AB} \cong \overline{AB}$	e. Congruence of segments is reflexive.
f. $\triangle ACB \cong \triangle ADB$	f. HA
g. $\overline{BC} \cong \overline{BD}$	g. CPCTC

Use the figure at the right to write a two-column proof.

Proof 24–27. See margin.

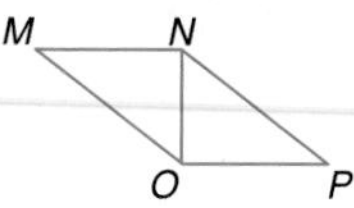

24. **Given:** $\overline{NO} \perp \overline{MN}$
$\overline{NO} \perp \overline{OP}$
$\angle M \cong \angle P$
Prove: $\overline{MO} \cong \overline{NP}$

25. **Given:** $\angle M$ and $\angle P$ are right angles.
$\overline{MN} \parallel \overline{OP}$
Prove: $\overline{MN} \cong \overline{OP}$

Use the figure at the right to write a two-column proof.

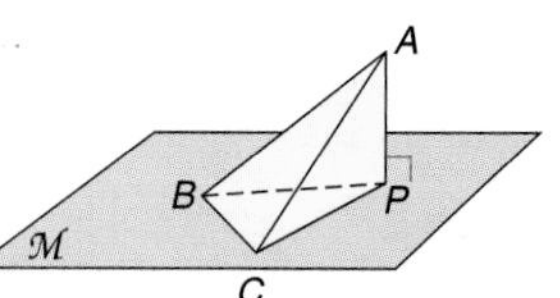

26. **Given:** $\overline{AP} \perp$ plane $\mathcal{M}$
$\overline{AB} \cong \overline{AC}$
Prove: $\triangle BPC$ is isosceles.

27. **Given:** $\angle PBC \cong \angle PCB$
$\overline{AP} \perp$ plane $\mathcal{M}$
Prove: $\angle ABC \cong \angle ACB$

Draw and label a figure for each statement. List the information that is given and the statement to be proved in terms of your figure. Then write a paragraph proof. 28–32. See Solutions Manual.

28. If the hypotenuse and an acute angle of one right triangle are congruent to the hypotenuse and corresponding acute angle of another right triangle, then the two triangles are congruent. (Theorem 5–6)

29. If one leg and an acute angle of one right triangle are congruent to the corresponding leg and acute angle of another right triangle, then the triangles are congruent. (Theorem 5–7)

30. Corresponding altitudes of congruent triangles are congruent.

C 31. The two segments that have the midpoint of each leg of an isosceles triangle as one endpoint and are perpendicular to the base of the triangle are congruent.

32. Obtuse triangle RST with obtuse angle at S shares side $\overline{ST}$ with acute triangle UTS. $\overline{RS}$ is congruent to $\overline{UT}$, and $\angle RST$ is supplementary to $\angle UTS$. Prove that the altitude from R to $\overline{ST}$ in $\triangle RST$ is congruent to the altitude from U to $\overline{ST}$ in $\triangle UST$.

Cabri Geometry

33. Use a TI-92 to explore two right triangles that have congruent corresponding angles. What do you observe? The corresponding sides are proportional, but not necessarily congruent.

Critical Thinking

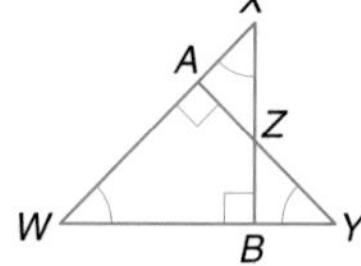

34. In the figure at the right, $m\angle W = m\angle X = m\angle Y = 45$, $\overline{XB} \perp \overline{WY}$, $\overline{YA} \perp \overline{WX}$. If $WZ = 10$, find XY. 10

Additional Answer

27. Statements (Reasons)
    1. $\angle PBC \cong \angle PCB$, $\overline{AP} \perp$ plane $\mathcal{M}$ (Given)
    2. $\angle APB$ and $\angle APC$ are rt. $\angle$s. ($\perp$ lines form 4 rt. $\angle$s.)
    3. $\triangle APB$ and $\triangle APC$ are rt. $\triangle$s. (Def. rt. $\triangle$)
    4. $\overline{BP} \cong \overline{CP}$ (Isosceles Triangle Th.)
    5. $\overline{AP} \cong \overline{AP}$ (Cong. of segments is reflexive.)
    6. $\triangle APC \cong \triangle APB$ (LL)
    7. $\overline{AC} \cong \overline{AB}$ (CPCTC)
    8. $\angle ABC \cong \angle ACB$ (Isosceles Triangle Th.)

Applications and Problem Solving

35. Transportation Many tractor-trailer trucks have airfoils on the top of the cab to make the truck more fuel efficient. An airfoil is made in a wedge shape from sheets of metal. One side of an airfoil is shown at the right. List the possible different sets of information that could be used to produce a congruent piece for the other side of the airfoil. See margin.

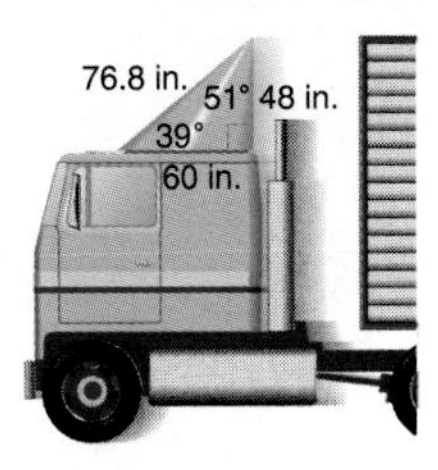

36. All the triangles are right, and they are all congruent by LL.

36. Construction A wooden bridge is being constructed over a stream in a park. The braces for the support posts came from the lumberyard already cut. The carpenter measures from the top of the support post to a point on the post to find where to attach the brace to the post. Explain why only one measurement must be made to ensure that all of the braces will be in the same relative position.

Mixed Review

37. Draw and label a figure to illustrate that $\overline{PT}$ and $\overline{RS}$ are medians of $\triangle PQR$ and intersect at V. (Lesson 5–1) See margin.

38. Given $\triangle ADN \cong \triangle DEO$, $AN = 15$, $ND = 19$, $AD = 27$, and $EO = 4x - 1$, find the value of x. (Lesson 4–3) 5

39. Draw and label $\triangle LNA$ with $m\angle L > 90$ and $NL = AL$. Then classify the triangle by its angles and sides. (Lesson 4–1) See margin.

40. Alt. int. ∠s are ≅; corr. ∠s are ≅; alt. ext. ∠s are ≅; cons. ext. ∠s are ≅; cons. int. ∠s are supplementary; two lines in a plane are parallel to a third line.

40. Name five ways to prove that two lines are parallel. (Lesson 3–4)

41. Find the slope of the line that passes through points at $(6, -11)$ and $(4, 9)$. (Lesson 3–3) -10

42. Name the properties of equality that justify the statement *If $2x + 6 = 8$, then $x + 3 = 4$.* (Lesson 2–4) Distributive Property and Division Property (=)

43. Explain how to form the converse, inverse, and contrapositive of a conditional statement. (Lesson 2–2) See Solutions Manual.

Algebra

44. Write $\frac{3}{20}$ as a percent. 15%

45. {(1938, 50,000), (1939, 40,000), (1938, 5200), (1938, 4200), (1939, 4020)}; D = {1938, 1939}; R = {50,000, 40,000, 5200, 4200, 4020}

45. **Collectibles** The table below shows the most valuable editions of Superman comic books in near-mint condition. Express the relation of the year and value shown in the table as a set of ordered pairs. Then state the domain and range of the relation.

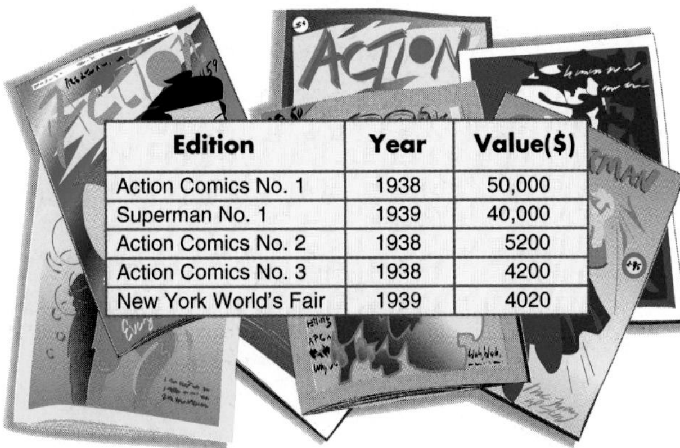

Edition	Year	Value($)
Action Comics No. 1	1938	50,000
Superman No. 1	1939	40,000
Action Comics No. 2	1938	5200
Action Comics No. 3	1938	4200
New York World's Fair	1939	4020

4 ASSESS

Closing Activity

Writing Have students write the four pairs of segments and angles of right triangles that can be used to prove two right triangles congruent. Have them draw a picture and label the congruent segments or angles that would be used in the proof.

Chapter 5 Quiz A (Lessons 5-1 and 5-2) is available in the *Assessment and Evaluation Masters*, p. 128.

Additional Answers

35. legs: 48 in. and 60 in.; hypotenuse and an angle: 76.8 in. and 51° or 76.8 in. and 39°; leg and an angle: 60 in. and 51°, or 60 in. and 39°, 48 in. and 51°, or 48 in. and 39°; hypotenuse and a leg: 76.8 in. and 60 in. or 76.8 in. and 48 in.

37.

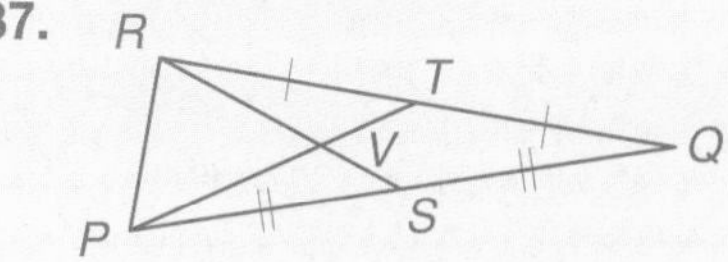

Enrichment Masters, p. 27

NAME ______ DATE ______

5-2 Enrichment

Student Edition Pages 245–251

Tangrams

The tangram puzzle is composed of seven pieces that form a square, as shown at the right. This puzzle has been a popular amusement for Chinese students for hundreds and perhaps thousands of years.

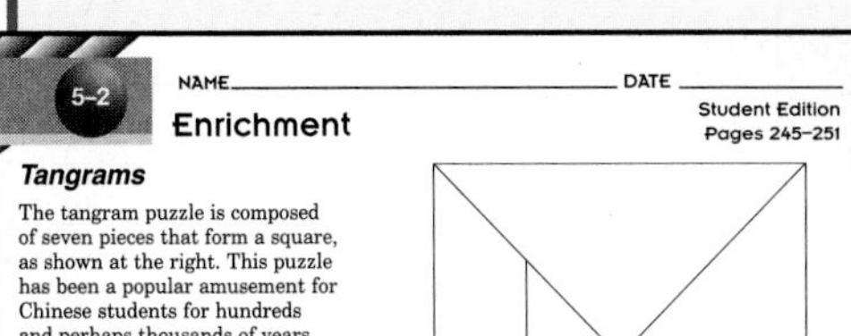

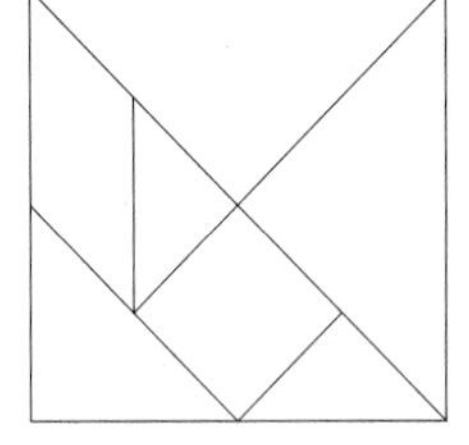

Make a careful tracing of the figure above. Cut out the pieces and rearrange them to form each figure below. Record each answer by drawing lines within each figure.

1. 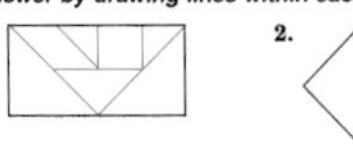 2. 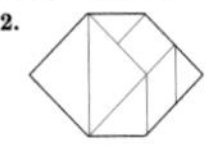 3. 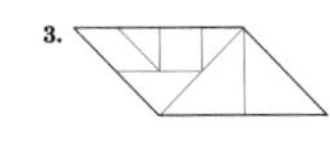

4.  5. 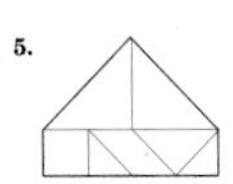 6. 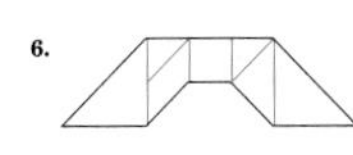

7. Create a different figure using the seven tangram pieces. Trace the outline. Then challenge another student to solve the puzzle. See students' work.

Extension

Reasoning If a and b represent the lengths of the two legs of a right triangle and c represents the length of the hypotenuse, explain why $a + b > c$ and why $c > a$.

Additional Answer

39. obtuse, isosceles

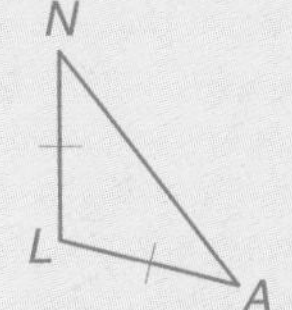

5-3 LESSON NOTES

NCTM Standards: 1–5, 7

Instructional Resources

- Study Guide Master 5-3
- Practice Master 5-3
- Enrichment Master 5-3
- Assessment and Evaluation Masters, pp. 127, 128

Transparency 5-3A contains the 5-Minute Check for this lesson; **Transparency 5-3B** contains a teaching aid for this lesson.

Recommended Pacing	
Standard Pacing	Days 4 & 5 of 14
Honors Pacing	Days 3 & 4 of 13
Block Scheduling*	Day 3 of 7

*For more information on pacing and possible lesson plans, refer to the *Block Scheduling Booklet.*

1 FOCUS

5-Minute Check
(over Lesson 5-2)

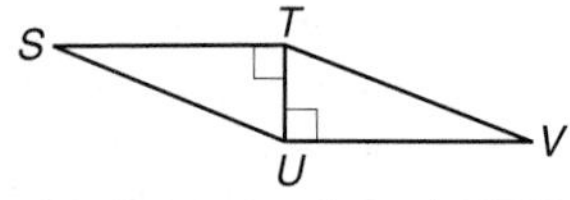

Determine whether $\triangle STU \cong \triangle VUT$, using the given information. Justify.

1. $\angle S \cong \angle V$ yes, by LA
2. $\overline{SU} \cong \overline{VT}$ yes, by HL
3. $\angle STU$ and $\angle VUT$ are right angles. no

Write a two-column proof.

4. **Given:** $\angle STU$ and $\angle VUT$ are right angles; $\overline{SU} \cong \overline{VT}$.
   Prove: $\angle S \cong \angle V$
   Proof:
   Statements (Reasons)
   1. $\angle STU$ and $\angle VUT$ are right angles; $\overline{SU} \cong \overline{VT}$. (Given)
   2. $\triangle STU$ and $\triangle VUT$ are right triangles. (Def. rt. $\triangle$)
   3. $\overline{TU} \cong \overline{UT}$ (Cong. of segments is reflexive.)
   4. $\triangle STU \cong \triangle VUT$ (HL)
   5. $\angle S \cong \angle V$ (CPCTC)

5-3 Indirect Proof and Inequalities

What YOU'LL LEARN

- To use indirect reasoning and indirect proof to reach a conclusion,
- to recognize and apply properties of inequalities to the measures of segments and angles, and
- to solve problems by working backward.

Why IT'S IMPORTANT

Indirect reasoning is often used in the legal system and in advertising.

CONNECTION
Health

Alzheimer's disease is the fourth leading cause of death in the United States. Unfortunately, doctors have a difficult time diagnosing the disease. When a patient exhibits symptoms that may be Alzheimer's, doctors test for other diseases with similar symptoms. If the tests for the other diseases are negative, the doctor concludes that the patient has Alzheimer's disease. This is an example of **indirect reasoning**.

You have used direct reasoning in the proofs you have encountered up to this point. When using direct reasoning, you start with a true hypothesis and prove that the conclusion is true. With indirect reasoning, you assume that the conclusion is false and then show that this assumption leads to a contradiction of the hypothesis or some other accepted fact, like a postulate, theorem, or corollary. Then, since your assumption has been proved false, the conclusion must be true.

The following steps summarize the process of indirect reasoning. Follow these steps when doing an **indirect proof**.

Steps for Writing an Indirect Proof	**1. Assume that the conclusion is false.** **2. Show that the assumption leads to a contradiction of the hypothesis or some other fact, such as a postulate, theorem, or corollary.** **3. Point out that the assumption must be false and, therefore, the conclusion must be true.**

Example 1 **State the assumption you would make to start an indirect proof of each statement. Do *not* write the proofs.**

a. The number 117 is divisible by 13.

b. Akita is the best candidate in the election.

c. $\overline{AB}$ is a median of $\triangle ACD$.

d. $m\angle C < m\angle D$

LOOK BACK

You may wish to review direct reasoning and the Law of Detachment in Lesson 2-3.

The first step in writing an indirect proof is to assume that the conclusion is false.

a. The number 117 is not divisible by 13.

b. Akita is not the best candidate in the election.

c. $\overline{AB}$ is not a median of $\triangle ACD$.

d. $m\angle C \geq m\angle D$ *Recall that if $m\angle C$ is not less than $m\angle D$, then it could be greater than or equal to $m\angle D$.*

You have learned about a relationship between exterior angles of a triangle and their remote interior angles. The following theorem is a result of this relationship. This theorem can be proved using an indirect proof.

Theorem 5–8 Exterior Angle Inequality Theorem	**If an angle is an exterior angle of a triangle, then its measure is greater than the measure of either of its corresponding remote interior angles.**

The following statement can be used to define the inequality relationship between two numbers. This definition is often used in proofs.

Definition of Inequality	**For any real numbers a and b, $a > b$ if and only if there is a positive number c such that $a = b + c$.**

Proof of Theorem 5–8

Given: $\angle 1$ is an exterior angle of $\triangle MNP$.

Prove: $m\angle 1 > m\angle 4$

$m\angle 1 > m\angle 3$

Indirect Proof:

Step 1: Make the assumption that $m\angle 1 \ngtr m\angle 3$ and $m\angle 1 \ngtr m\angle 4$. Thus, $m\angle 1 \leq m\angle 3$ and $m\angle 1 \leq m\angle 4$.

Step 2: We will only show that the assumption $m\angle 1 \leq m\angle 3$ leads to a contradiction, since the argument for $m\angle 1 \leq m\angle 4$ uses the same reasoning.

$m\angle 1 \leq m\angle 3$, means that either $m\angle 1 = m\angle 3$ or $m\angle 1 < m\angle 3$. So, we need to consider both cases.

Case 1: $m\angle 1 = m\angle 3$

Since $m\angle 3 + m\angle 4 = m\angle 1$ by the Exterior Angle Theorem, we have $m\angle 3 + m\angle 4 = m\angle 3$ by substitution. Then $m\angle 4 = 0$, which contradicts the fact that the measure of an angle is greater than 0.

Case 2: $m\angle 1 < m\angle 3$

By the Exterior Angle Theorem, $m\angle 3 + m\angle 4 = m\angle 1$. Since angle measures are positive, the definition of inequality implies $m\angle 1 > m\angle 3$ and $m\angle 1 > m\angle 4$. This contradicts the assumption that $m\angle 1 \leq m\angle 3$ and $m\angle 1 \leq m\angle 4$.

Step 3: In both cases, the assumption leads to the contradiction of a known fact. Therefore, the assumption that $m\angle 1 \leq m\angle 3$ must be false, which means that $m\angle 1 > m\angle 3$ must be true. Likewise, $m\angle 1 > m\angle 4$.

Motivating the Lesson

Questioning Ask students which of the following was a German mathematician.

a. Stephen Hawking **American scientist**
b. Georgia O'Keefe **American artist**
c. Alfred Adler **correct answer**
d. John Lennon **British musician**

Ask them what they think the answer is and why. Explain that they have just exhibited indirect reasoning.

2 TEACH

In-Class Example

For Example 1
State the assumption you would make to start an indirect proof of each statement. Do not write the proofs.

a. $\overline{AB}$ bisects $\angle A$. **$\overline{AB}$ does not bisect $\angle A$.**
b. $\triangle XTZ$ is isosceles. **$\triangle XTZ$ is not isosceles.**
c. $m\angle 1 < m\angle 2$ **$m\angle 1 \geq m\angle 2$**

Teaching Tip Relate this lesson to Lesson 5-2. In a direct proof, you solve the problem by working from the beginning, or forward. In an indirect proof you solve the problem by working backward, or by assuming the conclusion is false.

Teaching Tip Theorem 5-8 can also be proven directly. You may want to have students write the direct proof as a project.

Proof Pointer

Show students that an indirect proof can also be written as a flow proof. Write the proof of Theorem 5-8 as a flow proof.

In-Class Example

For Example 2
Mrs. Miller spent \$72.16 at the grocery store. While there, she used a gift certificate worth \$10 in free groceries that she won in a game that the store was running. Sales tax of 5.25% was then added, and Mrs. Miller turned in \$6.35 worth of coupons, which the store doubled. How much did Mrs. Miller's groceries total before any deductions? **\$90.63**

Teaching Tip When discussing the chart, remind students that if $c < 0$, it is a negative number, and if $c > 0$, it is a positive number. Review rules for operations using positive and negative numbers.

Teaching Tip You can build the Properties of Inequality chart with students. Have them pick numbers for a, b, and c and then devise the statements in the chart. Have them use several examples to get an idea of what is happening. Ask students to write the statements in mathematical terms.

Some of the properties of inequalities that you encountered in algebra were used in the proof of Theorem 5–8. For example, in Step 1 of the proof, it was stated that if $m\angle 1 \not> m\angle 3$, then $m\angle 1 < m\angle 3$ or $m\angle 1 = m\angle 3$. This statement is an application of the Comparison or Trichotomy Property which states that for any two numbers a and b, either $a > b$, $a < b$, or $a = b$. The following chart gives a list of properties of inequalities you studied in algebra.

Properties of Inequality for Real Numbers	
For all numbers *a, b,* and *c,*	
Comparison Property	$a < b$, $a = b$, or $a > b$.
Transitive Property	**1.** If $a < b$ and $b < c$, then $a < c$. **2.** If $a > b$ and $b > c$, then $a > c$.
Addition and Subtraction Properties	**1.** If $a > b$, then $a + c > b + c$ and $a - c > b - c$. **2.** If $a < b$, then $a + c < b + c$ and $a - c < b - c$.
Multiplication and Division Properties	**1.** If $c > 0$ and $a < b$, then $ac < bc$ and $\frac{a}{c} < \frac{b}{c}$. **2.** If $c > 0$ and $a > b$, then $ac > bc$ and $\frac{a}{c} > \frac{b}{c}$. **3.** If $c < 0$ and $a < b$, then $ac > bc$ and $\frac{a}{c} > \frac{b}{c}$. **4.** If $c < 0$ and $a > b$, then $ac < bc$ and $\frac{a}{c} < \frac{b}{c}$.

Indirect proofs use a problem-solving strategy called **working backward**. After assuming that the conclusion is false, you work backward from the assumption to show that, for the given information, the assumption is false. The strategy of working backward can be used in many other problem-solving situations.

Example 2

Banking

Hama's used car loan costs \$207.98 per month. This includes 9% interest paid to the bank for borrowing the money. When Hama bought the car, she was able to negotiate a 10% discount and made a down payment of \$2000. If the loan is for 48 months, what was the asking price of the car?

Explore We know the monthly payment, the interest rate, the down payment, the discount, and the length of the loan. We are looking for the asking price of the car.

Plan Work backward through each step to find the asking price.

Solve ***Interest*** The monthly payment is \$207.98 per month including 9% interest. Let p = the price per month before interest is added.

$$\underbrace{p}_{\text{price without interest}} + \underbrace{0.09p}_{\text{9\% interest}} = \underbrace{207.98}_{\text{price with interest}}$$
$$1.09p = 207.98$$
$$p \approx 190.81$$

The monthly payment without the interest payment is \$190.81.

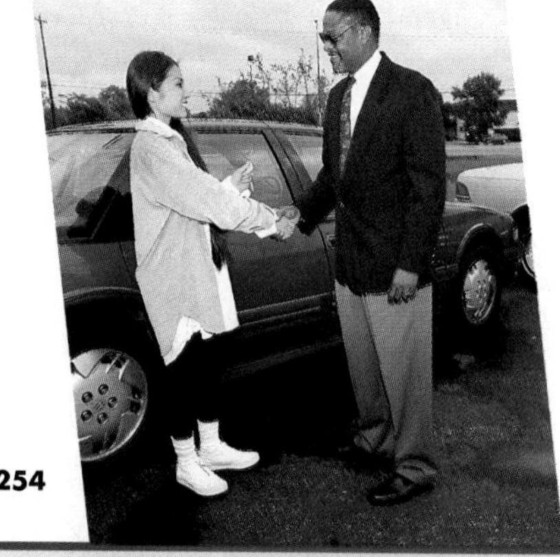

254

Cooperative Learning

Brainstorming Brainstorm ideas with students where working backward is necessary to solve a problem. You can give them a list of places where they might need to work backward (in school, at home, playing sports, and so on). For more information on the brainstorming strategy, see *Cooperative Learning in the Mathematics Classroom*, one of the titles in the Glencoe Mathematics Professional Series, page 30.

Loan amount Let l = the loan amount. The loan amount is the total of the money paid, less the interest.

$$\underbrace{\text{number of payments}}_{48} \times \underbrace{\text{payment amount}}_{190.81} = \underbrace{\text{loan amount}}_{l}$$
$$9158.88 = l$$

The loan amount is \$9158.88.

Down payment The down payment was subtracted from the sale price s to determine the loan amount. So to find the sale price, we should add the down payment to the loan amount.

$$\underbrace{\text{loan amount}}_{9158.88} + \underbrace{\text{down payment}}_{2000} = \underbrace{\text{sale price}}_{s}$$
$$11{,}158.88 = s$$

The sale price was \$11,158.88.

Discount Hama negotiated a 10% discount. Let a represent the asking price.

$$\underbrace{\text{asking price}}_{a} - \underbrace{\text{discount}}_{0.10a} = \underbrace{\text{sale price}}_{11{,}158.88}$$
$$0.90a = 11{,}158.88$$
$$a = \$12{,}398.76$$

The asking price of the car was \$12,399.

Examine Check the solution by starting with the asking price and working forward to find the monthly payment with interest.

CHECK FOR UNDERSTANDING

Communicating Mathematics

Study the lesson. Then complete the following. 3–4. See students' work.

1. **Draw** $\triangle ABC$ with an exterior angle at A named $\angle 1$. Which angles have measures less than the measure of $\angle 1$? $\angle B$ and $\angle C$
2. **Analyze** the argument that given $x > y$ and $z > w$, then $xz > yw$.
3. **Compare and contrast** the properties for equality of real numbers that were reviewed in Lesson 2–4 to the properties of inequality in this lesson.
4. **Assess Yourself** Describe a situation in your life when you have used indirect reasoning or the work-backward problem-solving strategy.

2. The statement is not always true. $10 > -1$ and $-5 > -10$, but -50 is not greater than 10.

MATH JOURNAL

Guided Practice

State the assumption you would make to start an indirect proof of each statement.

5. Lines ℓ and m intersect at point X. Lines ℓ and m do not intersect at point X.
6. If the alternate interior angles formed by two lines and a transversal are congruent, the lines are parallel.
7. Sabrina ate the leftover pizza. Sabrina did not eat the leftover pizza.

6. If the alt. int. $\angle$s formed by two lines and a transversal are $\cong$, the lines are not parallel.

3 PRACTICE/APPLY

Check for Understanding

Exercises 1–16 are designed to help you assess your students' understanding through reading, writing, speaking, and modeling. You should work through Exercises 1–4 with your students and then monitor their work on Exercises 5–16.

Error Analysis

Students may have difficulty understanding how an indirect proof actually proves the conclusion true. Point out that *all* possibilities are looked at and *all* of the incorrect conclusions must be ruled out before you can say that the original conclusion is true.

Reteaching

Using Writing Work through the steps for writing an indirect proof using one or more examples. Make up real-world examples so that students can better comprehend the logic involved. For example, prove that England has been ruled by both kings and queens. Indirect proof: England has been ruled only by kings or only by queens. England has been ruled by Queen Elizabeth I and King Henry VIII. Therefore, England has been ruled by kings and by queens. Explain again that *all* possibilities must be looked at to prove that only one answer is possible by the process of elimination.

Assignment Guide

Core (with proof): 17–41 odd, 42, 43, 45–55
Core (informal): 17–33 odd, 42, 43, 45, 47–55
Enriched: 18–40 even, 42–55
All: Self Test, 1–5

For **Extra Practice**, see p. 773.

The red A, B, and C flags, printed only in the Teacher's Wraparound Edition, indicate the level of difficulty of the exercises.

Additional Answer

14. Statements (Reasons)
1. B, C, and D, lie on $\overleftrightarrow{AE}$. $\overline{BD} \cong \overline{AC}$ (Given)
2. $BD = AC$ (Def. ≅ segments)
3. $BE = BD + DE$ (Segment Add. Post.)
4. $BD < BE$ (Def. inequality)
5. $AC < BE$ (Subst. Prop. Inequality)

Use the figure at the right for Exercises 8–10.

8. List all the angles whose measures are less than $m\angle 1$. $\angle 2, \angle 3, \angle 5, \angle 6, \angle 7$
9. List all the angles whose measures are greater than $m\angle 2$. $\angle 1, \angle 4, \angle 8$
10. Which is greater, $m\angle 7$ or $m\angle 8$? Justify your answer.

10. $\angle 8$; by the Transitive Property (=); $m\angle 8 > m\angle 4$; $m\angle 4 > m\angle 7$ by the Exterior Angle Inequality Theorem in both cases.

13. Transitive Property (≠)

Name the property that justifies each statement.

11. If $-3x < 24$, then $x > -8$. Division Property (≠)
12. If $AB < CD$, then $AB + RS < CD + RS$. Addition Property (≠)
13. If $m\angle R > m\angle S$ and $m\angle S > m\angle T$ and $m\angle T > m\angle U$, then $m\angle R > m\angle U$.
14. Write a two-column proof.

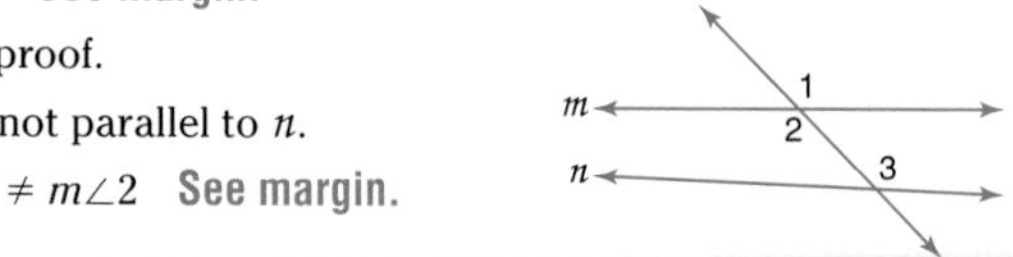

Given: B, C, and D lie on $\overleftrightarrow{AE}$. $\overline{BD} \cong \overline{AC}$

Prove: $AC < BE$ See margin.

15. Write an indirect proof.

Given: m is not parallel to n.

Prove: $m\angle 3 \neq m\angle 2$ See margin.

16. **Sales** Mrs. Pinel is offering her students the opportunity to buy discounted graphing calculators. She was able to get the calculators on sale for 20% off at the store, and then a 10% school discount. The cost for each calculator is \$97.05 including the 7% sales tax. What was the original price of each calculator? \$125.97

EXERCISES

Practice

State the assumption you would make to start an indirect proof of each statement.

17. $\overline{AB} \cong \overline{CD}$ $\overline{AB}$ is not congruent to $\overline{CD}$.
18. An altitude of an isosceles triangle is also a median.
19. The disk is defective. The disk is not defective.
20. The suspect is guilty. The suspect is not guilty.
21. If two altitudes of a triangle are congruent, then the triangle is isosceles.
22. If two lines are cut by a transversal and corresponding angles are congruent, then the lines are parallel.

18. The altitude of an angle in an isosceles triangle is not a median of the triangle.
21. If two altitudes of a triangle are congruent, then the triangle is not isosceles.
22. If two lines are cut by a transversal and corresponding angles are congruent, then the lines are not parallel.

Use the figure at the right for Exercises 23–28.

B

23. Which is greater, $m\angle 7$ or $m\angle 5$? $\angle 7$
24. Which is greater, $m\angle 4$ or $m\angle 1$? $m\angle 1$
25. Name an angle whose measure is greater than $m\angle 6$. Sample answer: $\angle 1$
26. List all the angles whose measures are greater than $m\angle 2$. $\angle 1, \angle 7\ \angle 8$
27. List all the angles whose measures are less than $m\angle 7$. $\angle 2, \angle 3, \angle 4, \angle 5, \angle 6$
28. If $m\angle 2 > m\angle 6$, which is greater, $m\angle 8$ or $m\angle 1$? $m\angle 8$

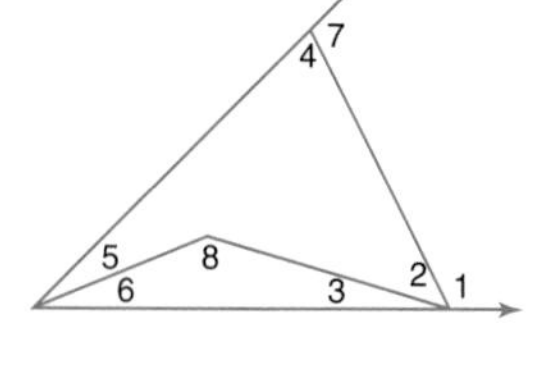

Study Guide Masters, p. 28

NAME ______ DATE ______

Study Guide — Student Edition Pages 252–258

Indirect Proof and Inequalities

A type of proof called **indirect proof** is sometimes used in geometry. In an indirect proof you assume that the conclusion is false and work backward to show that this assumption leads to a contradiction of the original hypothesis or some other known fact, such as a postulate, theorem, or corollary.

The following theorem can be proved by an indirect proof. (See page 253 in your book.)

Exterior Angle Inequality Theorem	If an angle is an exterior angle of a triangle, then its measure is greater than the measure either of its corresponding remote interior angles.

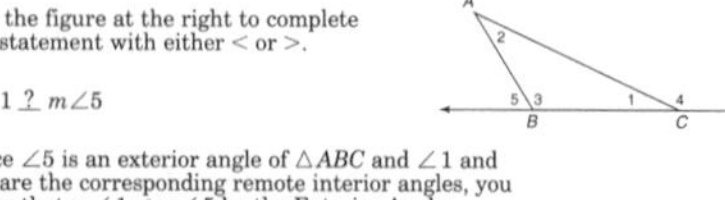

Example: Use the figure at the right to complete the statement with either < or >.

$m\angle 1 \ ? \ m\angle 5$

Since $\angle 5$ is an exterior angle of $\triangle ABC$ and $\angle 1$ and $\angle 2$ are the corresponding remote interior angles, you know that $m\angle 1 < m\angle 5$ by the Exterior Angle Inequality Theorem.

Use the figure at the right to complete each statement with < or >.

1. $m\angle 1$ > $m\angle 6$
2. $m\angle 2$ < $m\angle 1$
3. $m\angle 6$ > $m\angle 3$
4. $m\angle 4$ < $m\angle 6$
5. Use the problem-solving strategy of working backward to complete the indirect proof in paragraph form.
Given: $m\angle 1 \neq m\angle 2$
Prove: BD is not an altitude of $\triangle ABC$.

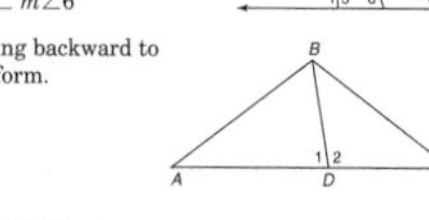

Proof:

a. Assume that $\overline{BD}$ is an altitude of $\triangle ABC$.
b. Then $\overline{BD} \perp \overline{AC}$ by definition of altitude.
c. Since perpendicular segments form four right angles, $\angle 1$ and $\angle 2$ are right angles.
d. Since all right angles are congruent, $\angle 1 \cong \angle 2$.
e. Since $\angle 1 \cong \angle 2$, $m\angle 1 = m\angle 2$.
f. But it is given that $m\angle 1 \neq m\angle 2$.
g. So our assumption is incorrect. Therefore, $\overline{BD}$ is not an altitude of $\triangle ABC$.

Additional Answers

15. Step 1: Assume $m\angle 3 = m\angle 2$.
Step 2: $\angle 3$ and $\angle 2$ are alt. int. ∠s. If ∦, alt. int. ∠s are ≅, then the lines are ∥. This means that $m \parallel n$. However, that contradicts the given statement.
Step 3: Therefore, since the assumption leads to a contradiction, the assumption must be false. Thus, $m\angle 3 \neq m\angle 2$.

35. Since if an ∠ is an ext. ∠ of a △ its measure is > the measure of either corr. int. ∠, we can say that $m\angle 7 > m\angle 8$ and $m\angle 8 > m\angle 6$. Thus, by the Trans. Prop. of Inequality, $m\angle 7 > m\angle 6$.

Name the property that justifies each statement.

29. If $5x < 25$, then $x < 5$. Division Property ($\neq$)

30. Transitive Property ($\neq$)

30. Given $m\angle A < m\angle B$ and $m\angle C < m\angle A$, then $m\angle C < m\angle B$.

31. If $RT \neq AB$, then $RT > AB$ or $RT < AB$. Comparison Property ($\neq$)

32. Given $x + 5 > 9$, then $x > 4$. Subtraction Property ($\neq$)

33. If $m\angle X < m\angle Y$, then $\frac{m\angle X}{2} < \frac{m\angle Y}{2}$. Division Property ($\neq$)

34. If $m\angle X \not> m\angle Y$, then $m\angle X < m\angle Y$ or $m\angle X = m\angle Y$. Comparison Prop. ($\neq$)

Use the figure at the right to write a paragraph proof.

Proof

35–36. See margin.

35. **Given:** $\triangle XYV$, $\triangle XZU$
Prove: $m\angle 7 > m\angle 6$

36. **Given:** $\triangle XYV$, $\triangle XZU$
$\angle 4 \cong \angle 1$
Prove: $m\angle 5 > m\angle 2$

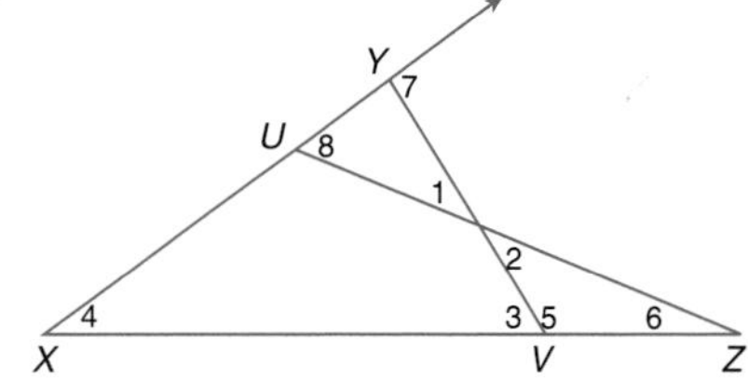

Use the figure at the right to write a two-column proof.

37. See margin.

37. **Given:** $\overline{FD} \perp \overline{AB}$
$\overline{GE} \perp \overline{AB}$
Prove: $m\angle 5 > m\angle 2$

38. See Solutions Manual.

38. **Given:** $\overline{AC} \cong \overline{BC}$
$\overline{FD} \perp \overline{AB}$
$\overline{GE} \perp \overline{AB}$
Prove: $m\angle 7 > m\angle 4$

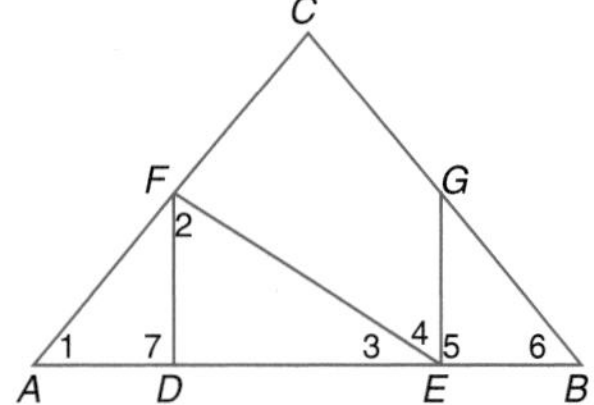

Write an indirect proof. 39–41. See Solutions Manual.

39. **Given:** $\overline{MO} \cong \overline{ON}$
$\overline{MP}$ is not congruent to $\overline{NP}$.
Prove: $\angle MOP$ is not congruent to $\angle NOP$.

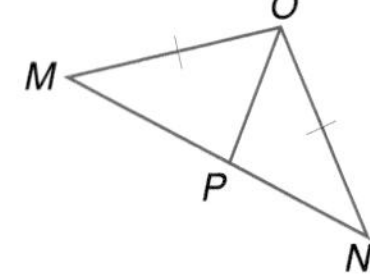

C

40. A triangle can have no more than one acute exterior angle.

41. If no two angles in a triangle are congruent, then no two sides are congruent.

Critical Thinking

42. Mr. Mendez was checking on the date he had attended a four-day conference a year ago. The page in his record book was torn and all that remained of the date for the meeting was *ber 31*. What was the month of the first day of the conference? October

Applications and Problem Solving

43. **Work Backward** A guard's prisoner is given the choice of opening one of two doors. Each door leads either to freedom or to the dungeon. A sign on the door on the right reads "This door leads to freedom and the other door leads to the dungeon." The door on the left has a sign that reads "One of these doors leads to freedom and the other leads to the dungeon." A guard tells the prisoner that one of the signs is true and the other is false. Which door should the prisoner choose? Why? See margin.

Additional Answers

36. We are given that $\angle 4 \cong \angle 1$. The def. of $\cong$ ∡s allows us to say that $m\angle 4 = m\angle 1$. Since if an $\angle$ is an ext. $\angle$ of a $\triangle$ its measure > the measure of either corr. int. $\angle$, we can say that $m\angle 5 > m\angle 4$. By the Subst. Prop. of Inequaltity, $m\angle 5 > m\angle 1$. $\angle 1 \cong \angle 2$ because they are vert. ∡s. Thus by the def. of $\cong$ ∡s, $m\angle 1 \cong m\angle 2$. Then by the Subst. Prop. of Inequality, $m\angle 5 > m\angle 2$.

37. Statements (Reasons)
   1. $\overline{FD} \perp \overline{AB}$, $\overline{GE} \perp \overline{AB}$ (Given)
   2. $\angle 7$ and $\angle 5$ are rt. ∡s. ($\perp$ lines form 4 rt. ∡s.)
   3. $m\angle 7 > m\angle 2$ (Ext. $\angle$ Inequality Th.)
   4. $\angle 7 \cong \angle 5$ (All rt. ∡s are $\cong$.)
   5. $m\angle 7 = m\angle 5$ (Def. $\cong$ ∡s)
   6. $m\angle 5 > m\angle 2$ (Subst. Prop. of Inequality)

Practice Masters, p. 28

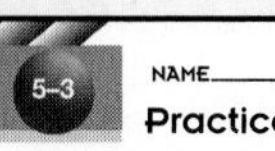

NAME_______________ DATE_______

5-3 Practice

Student Edition Pages 252–258

Indirect Proof and Inequalities

Use the figure at the right to complete each statement with either < or >.

1. $\angle 4$ _?_ $\angle 8$ >
2. $\angle 13$ _?_ $\angle 11$ >
3. $\angle 17$ _?_ $\angle 8$ <
4. If $m\angle 15 = m\angle 7$ then $m\angle 11$ _?_ $m\angle 7$. <

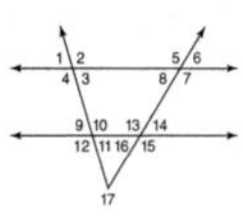

Write an indirect proof.

5. **Given:** $m\angle 1 \neq m\angle 2$
Prove: $\triangle ABC$ is not an isosceles triangle with vertex B.
Proof: Assume $\triangle ABC$ is isosceles, with vertex B. Then $AB = BC$ by the definition of isosceles triangle. This implies that $m\angle 1 = m\angle 2$, since angles opposite congruent sides of a triangle are congruent. This contradicts the given information. Therefore, $\triangle ABC$ cannot be an isosceles triangle with vertex B.

6. **Given:** $m\angle 1 + m\angle 2 \neq 180$
Prove: $a \nparallel b$
Proof: Assume $a \parallel b$. Then $m\angle 1 = m\angle 3$, since they are corresponding angles formed by parallel lines and a transversal. $\angle 2$ and $\angle 3$ are a linear pair, so $m\angle 2 + m\angle 3 = 180$. By substitution, $m\angle 2 + m\angle 1 = 180$. This contradicts given information. Therefore, $a \nparallel b$.

7. **Work Backward** Joe spent half of the money in his wallet on a table for his computer printer. He then spent \$8.12 for a printer ribbon. Then he spent half of what he had left on supplies for his office. He then had \$14.50 remaining. How much money did Joe have to start with? \$74.24

Extension

Problem Solving Have students act as owners of their own computer stores. You can give them a wholesale price and have them calculate what the list price should be. If they want to advertise a discount, will they still make a profit? Do they want to sell a lot of computers at a lower price, or do they want a higher profit?

Additional Answer

43. The door on the left. If the sign on the door on the right were true, then both signs would be true. But one sign is false, so the sign on the door on the right must be false.

4 ASSESS

Closing Activity

Writing Have students write three mathematical statements. Have them exchange papers and write the assumptions they would make to start an indirect proof.

Chapter 5 Quiz B (Lesson 5-3) is available in the *Assessment and Evaluation Masters*, p. 128.

Mid-Chapter Test (Lessons 5-1 through 5-3) is available in the *Assessment and Evaluation Masters*, p. 127.

Additional Answer

44. By assuming it was each of them, the police would find a contradiction from their past that would eliminate all but the actual murderer.

fabulous FIRSTS

George Ruffin was also elected to the Massachusetts state legislature.

Enrichment Masters, p. 28

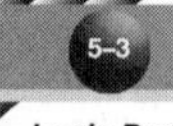

NAME__________ DATE ________

5-3 Enrichment — Student Edition Pages 252–258

Logic Problems

The following problems can be solved by eliminating possibilities. It may be helpful to use charts such as the one shown in the first problem. Mark an X in the chart to eliminate a possible answer.

Solve each problem.

1. Nancy, Olivia, Mario, and Kenji each have one piece of fruit in their school lunch. They have a peach, an orange, a banana, and an apple. Mario does not have a peach or a banana. Olivia and Mario just came from class with the student who has an apple. Kenji and Nancy are sitting next to the student who has a banana. Nancy does not have a peach. Which student has each piece of fruit?

	Nancy	Olivia	Mario	Kenji
Peach	X	X	X	
Orange	X	X		X
Banana	X		X	X
Apple		X	X	X

Nancy-apple, Olivia-banana, Mario-orange, Kenji-peach

2. Victor, Leon, Kasha, and Sheri each play one instrument. They play the viola, clarinet, trumpet, and flute. Sheri does not play the flute. Kasha lives near the student who plays flute and the one who plays trumpet. Leon does not play a brass or wind instrument. Which student plays each instrument? Victor-flute, Leon-viola, Kasha-clarinet, Sheri-trumpet

3. Mr. Guthrie, Mrs. Hakoi, Mr. Mirza, and Mrs. Riva have jobs of doctor, accountant, teacher, and office manager. Mr. Mirza lives near the doctor and the teacher. Mrs. Riva is not the doctor or the office manager. Mrs. Hakoi is not the accountant or the office manager. Mr. Guthrie went to lunch with the doctor. Mrs. Riva's son is a high school student and is only seven years younger than his algebra teacher. Which person has each occupation? Mr. Guthrie-teacher, Mrs. Hakoi-doctor, Mr. Mirza-office manager, Mrs. Riva-accountant

4. Yvette, Lana, Boris, and Scott each have a dog. The breeds are collie, beagle, poodle, and terrier. Yvette and Boris walked to the library with the student who has a collie. Boris does not have a poodle or terrier. Scott does not have a collie. Yvette is in math class with the student who has a terrier. Which student has each breed of dog? Yvette, poodle; Lana, collie; Boris, beagle; Scott, terrier

fabulous FIRSTS

George Lewis Ruffin (1834–1886)

George Lewis Ruffin became the first African-American to graduate from Harvard University Law School in 1869. Later, he became judge of the District Court in Charlestown, Massachusetts.

44. **Literature** In Agatha Christie's famous mystery *And Then There Were None*, all ten people on an island were murdered. Because of a storm, no one else had been able to land on or leave the island during the time the murders took place. The murderer thought the police might be able to solve the crime because they knew that the murderer had no background in murder and that, at a certain period of time, there were only four people present. Of those four people, only one person had no background in murder. Explain how the police could use indirect reasoning to determine the identity of the murderer. See margin.

45. **Law** The defense attorney said to the jury, "My client is not guilty. According to the police, the crime occurred on July 14 at 6:30 P.M. in Boston. I can prove that at that time my client was attending a business meeting in New York City. A verdict of not guilty is the only possible verdict." Is this an example of indirect reasoning? Explain why or why not. See Solutions Manual.

Mixed Review

46. Prove that if two altitudes of a triangle are congruent, then the triangle is isosceles. (Lesson 5–2) See Solutions Manual.

47. Draw and label $\triangle ABC$ with $\overline{AD}$, which is an altitude and a median. (Lesson 5–1) See Solutions Manual.

48. Draw $\triangle JKL$ with point M on $\overline{KL}$ such that $\overline{JM}$ bisects $\angle KJL$, but $\overline{KM}$ and $\overline{ML}$ are not congruent. (Lesson 4–6) $\angle 1 \cong \angle 2$, $\overline{KM} \ncong \overline{LM}$

49. If $\triangle NQD$ has vertices $N(2, -1)$, $Q(-4, -1)$, and $D(-1, -3)$, describe $\triangle NQD$ in terms of its angles and sides. (Lesson 4–1) obtuse, isosceles

50. Determine the value of r so that a line through the points at $(5, r)$ and $(2, 3)$ has a slope of 2. (Lesson 3–3) 9

51. Are the pieces of glass in a double-paned window a model of parallel or intersecting planes? (Lesson 3–1) parallel

52. Write the statement *A median of a triangle bisects one side of the triangle* in if-then form. (Lesson 2–2)

52. If a segment is a median of a triangle, then it bisects one side of the triangle.

53. **Draw a Diagram** Draw a tree diagram to determine the number of lunch specials at Lucky's Restaurant if each special comes with chicken, turkey, or fish; salad or soup; a baked potato or fries; and pie, frozen yogurt, or fruit. (Lesson 1–3) 36 lunch specials; see Solutions Manual for diagram.

INTEGRATION Algebra

54. Find the next three terms in the sequence $-2x, 6x, 14x, \ldots$. $22x, 30x, 38x$

55. Evaluate $2ab - \frac{c}{d}$ if $a = -9.32$, $b = -0.04$, $c = 0.932$, and $d = 2.5$. 0.3728

SELF TEST

Refer to the figure at the right for Exercises 1–4. 5. See Solutions Manual.

1. What is the value of x if $\overline{AD}$ is a median of $\triangle ABC$? (Lesson 5–1) 5

2. Find the value of y if $\overline{AD}$ is an altitude of $\triangle ABC$. (Lesson 5–1) 48

3. Suppose $\overline{AD}$ is the perpendicular bisector of $\overline{BC}$. Is $\triangle ABD \cong \triangle ACD$? Justify your answer. (Lesson 5–2) yes; by LL or SAS

4. Given that $\overline{AD}$ is an altitude of $\triangle ABC$, what additional information do you need to prove that $\triangle ABD \cong \triangle ACD$ by HL? (Lesson 5–2) $\overline{AB} \cong \overline{AC}$

5. Write an indirect proof for *If $\overleftrightarrow{AB}$ and $\overleftrightarrow{PQ}$ are skew lines, then $\overleftrightarrow{AP}$ and $\overleftrightarrow{BQ}$ are skew lines.* (Lesson 5–3)

SELF TEST

The Self Test provides students with a brief review of the concepts and skills in Lessons 5-1 through 5-3. Lesson numbers are given to the right of exercises or instruction lines so students can review concepts not yet mastered.

Tech Prep

Legal Assistant Programs for legal secretary and legal assistant are available at many two-year colleges and vocational schools. Legal reasoning is very similar to reasoning in geometry proofs. For more information on tech prep, see the *Teacher's Handbook*.

5-4 Inequalities for Sides and Angles of a Triangle

What YOU'LL LEARN

- To recognize and apply relationships between sides and angles in a triangle.

Why IT'S IMPORTANT

You can use the inequalities in triangles to solve problems involving transportation and sports.

In Chapter 4, you learned that in a triangle if two sides are congruent, then the angles opposite those sides are congruent. In the following activity, you will investigate the relationship between the measures of two angles of a triangle if the sides opposite those angles are not congruent.

MODELING MATHEMATICS

Inequalities for Sides and Angles of Triangles

Materials: ruler protractor

- Draw $\triangle ABC$.
- Measure each side of the triangle and record the measures in a table like the one at the right.
- Measure each angle of the triangle. Record each measure in the table.

Side	Measure
$\overline{AB}$ $\overline{BC}$ $\overline{AC}$	
Angle	**Measure**
$\angle A$ $\angle B$ $\angle C$	

Your Turn

a. Describe the measure of the angle opposite the longest side of the triangle in terms of the other angles. **It is the greatest measure.**

b. Describe the measure of the angle opposite the shortest side of the triangle in terms of the other angles. **It is the least measure.**

c. Repeat the activity for another triangle. **See students' work.**

d. Make a conjecture about the relationship between the measures of sides and angles of a triangle.

d. Sample answer: The measures of the angles opposite the sides are in the same order as the lengths of the respective sides.

The Modeling Mathematics activity suggests the following theorem.

Theorem 5-9	**If one side of a triangle is longer than another side, then the angle opposite the longer side has a greater measure than the angle opposite the shorter side.**

This theorem can be abbreviated as "If one side of a $\triangle$ is longer than another, the $\angle$ opp. the longer side > the $\angle$ opp. the shorter side."

The converse of the theorem is also true.

Theorem 5-10	**If one angle of a triangle has a greater measure than another angle, then the side opposite the greater angle is longer than the side opposite the lesser angle.**

This theorem can be abbreviated as "If an $\angle$ of a $\triangle$ > another, the side opp. the greater $\angle$ is longer than the side opp. the lesser $\angle$."

A proof of Theorem 5-9 is shown on the next page. *You will be asked to write a proof of Theorem 5-10 in Exercise 34.*

MODELING MATHEMATICS This modeling activity leads students to conjectures that are formulated in Theorems 5-9 and 5-10.

5-4 LESSON NOTES

NCTM Standards: 1–5, 7

Instructional Resources

- Study Guide Master 5-4
- Practice Master 5-4
- Enrichment Master 5-4
- Multicultural Activity Masters, p. 10

Transparency 5-4A contains the 5-Minute Check for this lesson; **Transparency 5-4B** contains a teaching aid for this lesson.

Recommended Pacing	
Standard Pacing	Days 6 & 7 of 14
Honors Pacing	Days 5 & 6 of 13
Block Scheduling*	Day 4 of 7

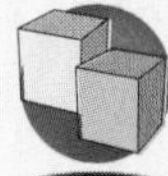

*For more information on pacing and possible lesson plans, refer to the *Block Scheduling Booklet.*

1 FOCUS

5-Minute Check
(over Lesson 5-3)

1. State the first step in an indirect proof. **Assume the conclusion is false.**
2. Decrease a number by 3, divide the result by 10, then multiply by 4, and add 21. If the final result is 33, with what number did you start? **33**
3. What is the least number of colors needed to paint the regions in the picture? No two regions that share a border can be painted the same color. **3**

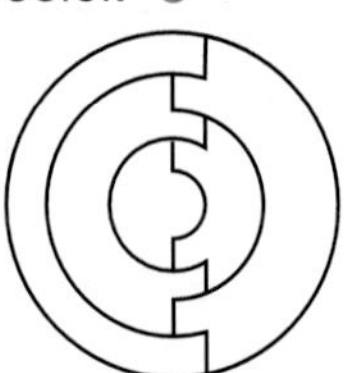

Fill in the blank to make a true statement.

4. If $a < b$ and $b < c$, then _____. $a < c$
5. If $c < 0$ and $a < b$, then _____. $ac > bc$ and $\frac{a}{c} > \frac{b}{c}$

Motivating the Lesson

Hands-On Activity Have students cut an isosceles triangle from a piece of paper. Have them measure the sides and angles.

2 TEACH

In-Class Example

For Example 1
Refer to the figure in Example 1 in the student edition.
Given: $m\angle A > m\angle D$
Prove: $BD > AB$
Proof:
Statements (Reasons)
1. $m\angle A + m\angle E + m\angle EBA = 180$; $m\angle C + m\angle D + m\angle CBD = 180$ (Angle Sum Theorem)
2. $m\angle EBA = m\angle CBD$ (Vert. ∠s are ≅.)
3. $m\angle A + m\angle E = 180 - m\angle EBA$; $m\angle C + m\angle D = 180 - m\angle CBD$ (Subtraction Prop. (=))
4. $m\angle A + m\angle E = m\angle C + m\angle D$ (Substitution Prop. (=))
5. $m\angle A > m\angle D$ (Given)
6. $m\angle C > m\angle E$ (Subtraction Prop. (=))
7. $BD > AB$ (If an ∠ of a △ > another, the side opp. the greater ∠ is longer than the side opp. the lesser ∠.)

Teaching Tip Have students rewrite the paragraph proof of Theorem 5-9 as a flow proof.

Teaching Tip Theorems 5-9 and 5-10 can be summarized by saying that the shorter segment is opposite the smaller angle and the longer segment is opposite the larger angle.

Compare the proofs of Theorems 5-9 and 5-10. Sometimes the proof of the converse of a theorem can be written by just writing the proof of the original theorem backward. However, point out to students that this is not always true.

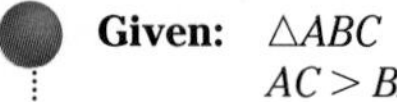

Given: $\triangle ABC$
$AC > BA$

Prove: $m\angle C < m\angle ABC$

Paragraph Proof:

Draw auxiliary segment, $\overline{BF}$, shown in red, so that F is between A and C and $\overline{AF} \cong \overline{AB}$.

If two sides of a triangle are congruent, the angles opposite those sides are congruent. Therefore, $\angle 1 \cong \angle 2$.

Notice that $\angle 1$ is an exterior angle of $\triangle BFC$. Since the measure of an exterior angle is greater than the measure of either of its corresponding remote interior angles, $m\angle C < m\angle 1$.

The Angle Addition Postulate allows us to say that $m\angle 2 + m\angle 3 = m\angle ABC$. By the definition of inequality, $m\angle 2 < m\angle ABC$. Thus, $m\angle 1 < m\angle ABC$ by the Substitution Property of Equality.

We can apply the Transitive Property of Inequality to the inequalities $m\angle C < m\angle 1$ and $m\angle 1 < m\angle ABC$ to conclude that $m\angle C < m\angle ABC$.

Theorems 5–9 and 5–10 can be used to prove statements about angle or side relationships in other figures.

Example 1  Proof

Write a two-column proof.

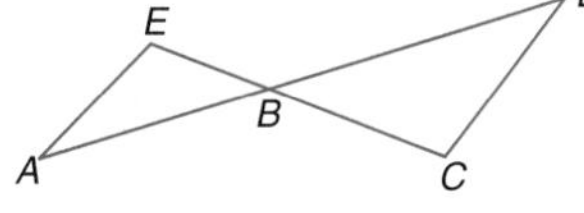

Given: $m\angle E > m\angle A$
$m\angle C > m\angle D$

Prove: $AD > EC$

Proof:

Statements	Reasons
1. $m\angle E > m\angle A$ $m\angle C > m\angle D$	1. Given
2. $AB > EB$ $BD > BC$	2. If one ∠ of a △ is greater than another, then the side opp. the greater ∠ is longer than the side opp. the lesser ∠.
3. $AB + BD > EB + BC$	3. Addition Property (≠)
4. $AB + BD = AD$ $EB + BC = EC$	4. Segment Addition Postulate
5. $AD > EC$	5. Substitution Property (=)

Example 2

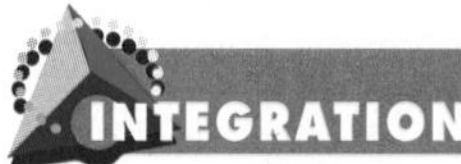
Algebra

Draw $\triangle RST$ with vertices $R(-2, 4)$, $S(-5, -8)$, and $T(6, 10)$. List the angles in order from the greatest measure to least measure.

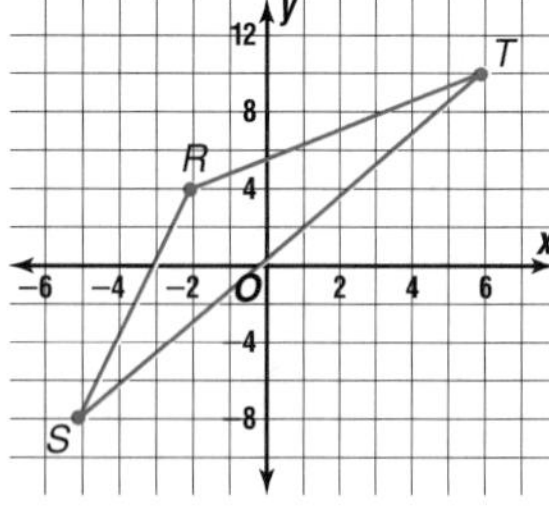

To compare the measures of the angles in $\triangle RST$, you can determine the lengths of the sides and then apply Theorem 5–9. Use the distance formula to find the length of each side of $\triangle RST$.

Classroom Vignette

"I begin this lesson by giving each student a strip of paper about 12 inches long which they cut into three pieces of any size. Then I ask them to make a triangle using the three strips for sides. Because some of the combinations form triangles and others don't, this leads into the discussion of the Triangle Inequality Theorem."

Cheryl L. Walls

Cheryl L. Walls
Milford High School
Milford, Delaware

$$RS = \sqrt{[-2-(-5)]^2 + [4-(-8)]^2}$$
$$= \sqrt{(3)^2 + (12)^2}$$
$$= \sqrt{153} \approx 12.37$$

$$ST = \sqrt{(-5-6)^2 + (-8-10)^2}$$
$$= \sqrt{(-11)^2 + (-18)^2}$$
$$= \sqrt{445} \approx 21.10$$

$$RT = \sqrt{(-2-6)^2 + (4-10)^2}$$
$$= \sqrt{(-8)^2 + (-6)^2}$$
$$= \sqrt{100} = 10$$

Because $ST > RS > RT$, by Theorem 5–9, $m\angle R > m\angle T > m\angle S$. So, the angles in order from largest to smallest are $\angle R$, $\angle T$, $\angle S$.

You can review the distance formula in Lesson 1-4.

In the figure at the right, T is a point not on line m, and R is the point on m such that $\overline{TR} \perp m$. In Lesson 3–7, we defined the distance from T to m as TR. It was stated, without proof, that $\overline{TR}$ was the shortest segment from T to m. Assuming this is true, if L is any point on m other than R, $\overline{TL}$ would have to be longer than $\overline{TR}$. *Why?*

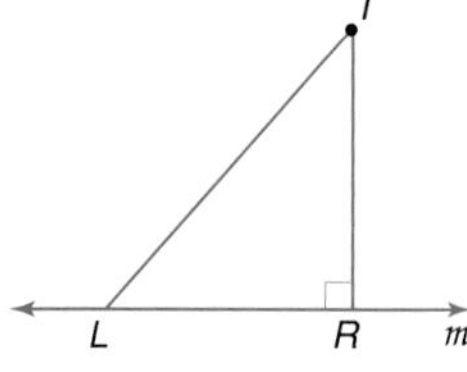

We will restate our assumption about the shortest segment from a point to a line as a theorem and show that it is true as a direct result of Theorem 5–10.

Theorem 5–11 **The perpendicular segment from a point to a line is the shortest segment from the point to the line.**

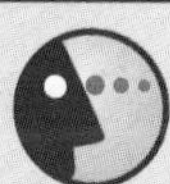

Proof of Theorem 5–11

Given: $\overline{TR} \perp m$
$\overline{TL}$ is any segment from T to m that is different from $\overline{TR}$.

Prove: $TL > TR$

T, L, R, m, 1, 2, 3

Proof:

Statements	Reasons
1. $\overline{TR} \perp m$	1. Given
2. $\angle 2$ and $\angle 3$ are right angles.	2. $\perp$ lines form four rt. $\angle$s.
3. $\angle 2 \cong \angle 3$	3. All rt. $\angle$s are $\cong$.
4. $m\angle 2 = m\angle 3$	4. Def. $\cong$ $\angle$s
5. $m\angle 3 > m\angle 1$	5. Exterior Angle Inequality Theorem
6. $m\angle 2 > m\angle 1$	6. Substitution Prop. (=)
7. $TL > TR$	7. If an $\angle$ of a $\triangle$ > another, then the side opp. the greater $\angle$ is longer than the side opp. the lesser $\angle$.

In-Class Example

For Example 2
Draw $\triangle JKL$ with vertices $J(-4, 2)$, $K(4, 3)$, $L(1, -3)$. List the angles in order from least measure to greatest measure.
$\angle L$, $\angle K$, $\angle J$

Teaching Tip When discussing Theorem 5-11 and Corollary 5-1, you may want to review Lesson 3-5 and the distance between two points and a line.

Alternative Learning Styles

Visual Instruct students to draw a triangle and have them use a protractor to measure each angle and a ruler to measure each side. Have them list the angles and the sides in order from greatest measure to least measure. What do they notice about the order? **The order of the angles corresponds to the order of the opposite sides.** Does everyone's triangle fit the pattern? **yes**

3 PRACTICE/APPLY

Check for Understanding

Exercises 1–14 are designed to help you assess your students' understanding through reading, writing, speaking, and modeling. You should work through Exercises 1–5 with your students and then monitor their work on Exercises 6–14.

Additional Answers

1. See students' drawings. The 40° angle is opposite the 5 cm side, the 60° angle is opposite the 6.7 cm side, and the 80° angle is opposite the 7.7 cm side.
2. No; the difference between the *y*-intercepts will not be the distance along a perpendicular between the two lines unless the lines are horizontal. If the lines are not horizontal, this method finds a distance longer than the distance between the lines.
3. In an obtuse triangle, there is one obtuse angle and it must be the largest angle of the triangle. The longest side must be opposite the largest angle.

4a. $\angle Z \cong \angle Y$
4b. $XZ > XY$
4c. $m\angle Z < m\angle Y$

13. Given: $\angle B$ is a right angle. $\overline{BD}$ is an altitude.
Prove: $BD < AB$ and $BD < BC$

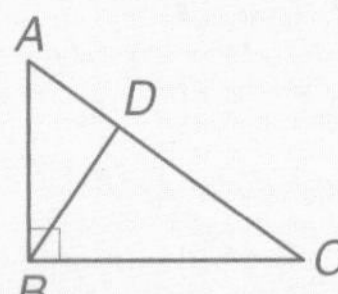

Proof:
We are given that $\overline{BD}$ is an altitude. By the def. of altitude, $\overline{BD} \perp \overline{AC}$. Since $\perp$ lines form 4 rt. angles, $\angle ADB$ and $\angle CDB$ are rt. $\angle$s. $\triangle ADB$ and $\triangle CDB$ are rt. $\triangle$s by the def. of rt. $\triangle$s. $\angle BAD$ and $\angle DBA$, and $\angle BCD$ and $\angle DBC$, are complementary since acute $\angle$s of rt. $\triangle$s are complementary. By the def. of complementary $\angle$s, $m\angle BAD + m\angle DBA = 90$ and $m\angle BCD + m\angle DBC = 90$. The def. of inequality allows that

Corollary 5–1 follows directly from Theorem 5–11. *You will be asked to prove the corollary in Exercise 35.*

Corollary 5–1	**The perpendicular segment from a point to a plane is the shortest segment from the point to the plane.**

CHECK FOR UNDERSTANDING

Communicating Mathematics

Study the lesson. Then complete the following. 1–4. See margin.

1. **Draw and label** a triangle that has sides measuring about 5 centimeters, 6.7 centimeters and 7.7 centimeters. Its angle measures are about 40°, 60°, and 80°. Label the angles. Explain how you knew where to place the labels.
2. **Analyze** finding the distance between two lines in a coordinate plane. Could you use the difference between the *y*-intercepts as the distance? Explain.
3. **Write a convincing argument** for the statement *In an obtuse triangle, the longest side is opposite the obtuse angle.*
4. **State** the conclusions you can draw about $\triangle XYZ$ for each relationship.
   a. $XY = XZ$
   b. $m\angle Y > m\angle Z$
   c. $XY < XZ$

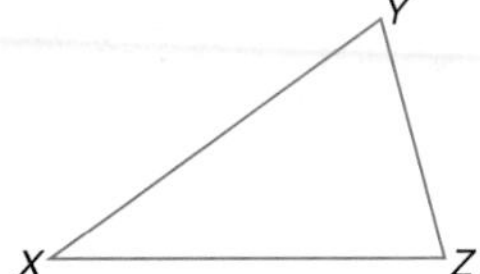

5. Draw obtuse $\triangle ABC$. Construct the distance from A to the line containing $\overline{BC}$ and the distance from B to the line containing $\overline{AC}$. Explain what each distance represents. The distance is an altitude of $\triangle ABC$.

Guided Practice

Refer to the figure at the right for Exercises 6–8.

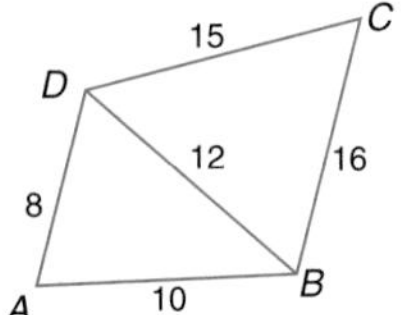

6. Which is greater, $m\angle CBD$ or $m\angle CDB$? 6. $m\angle CDB$
7. Is $m\angle ADB > m\angle DBA$? yes
8. Which is greater, $m\angle CDA$ or $m\angle CBA$? $m\angle CDA$

Given the angles in the figure at the right, answer each question.

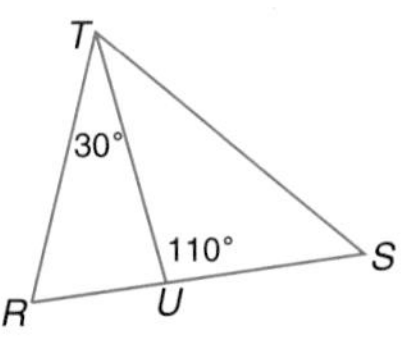

9. Which side of $\triangle RTU$ is the longest? $\overline{TU}$
10. Name the side of $\triangle UST$ that is the longest. $\overline{TS}$
11. If $\overline{TU}$ is an angle bisector, which side of $\triangle RST$ is the longest? $\overline{TS}$
12. Find the value of x and list the sides of $\triangle ABC$ in order from shortest to longest if $m\angle A = 3x + 20$, $m\angle B = 2x + 37$, and $m\angle C = 4x + 15$. $x = 12$; $m\angle A = 56$, $m\angle B = 61$, $m\angle C = 63$; so $BC < AC < AB$

Proof

13. Write a paragraph proof for the statement *The altitude to the hypotenuse of a right triangle is shorter than either of the legs.* See margin.

13. *(continued)*
$m\angle BAD < 90$ and $m\angle BCD < 90$. $m\angle ADB = 90$ and $m\angle CDB = 90$ by the def. of a rt. $\angle$. By subst. $m\angle BAD < m\angle ADB$ and $m\angle BCD < m\angle CDB$. Therefore, in $\triangle ABD$, $BD < AB$; and in $\triangle BCD$, $BD < BC$.

Reteaching

Using Manipulatives Students may have difficulty understanding that the shortest segment from a point to a line is the perpendicular segment. Have them stretch a rubber band to reach different points on a line. Where is the rubber band least taut?

14. The prediction will be the least accurate for plane C, which goes about 413 mph and costs $1800 per hour to operate. The line predicts about $2250/h.

14. Transportation The line graphed below represents average fuel cost as a function of the speed of a plane. Points A, B, and C represent a DC-10, a DC-9, and a Boeing 737, respectively. For which plane is the prediction the worst? Justify your claim.

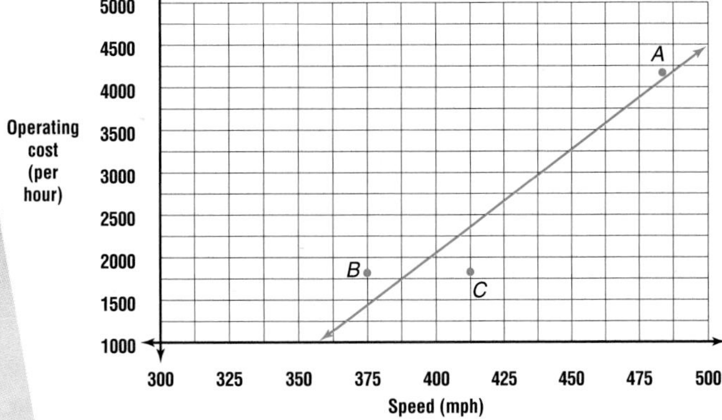

Assignment Guide

Core (with proof): 15–35 odd, 36, 37, 39–47
Core (informal): 15–27 odd, 36, 37, 39–47
Enriched: 16–34 even, 36–47

For **Extra Practice**, see p. 773.

The red A, B, and C flags, printed only in the Teacher's Wraparound Edition, indicate the level of difficulty of the exercises.

Additional Answer

23. In $\triangle ECD$, $\overline{EC}$ is longest. In $\triangle BEC$, $\overline{BE}$ is longest. In $\triangle ABE$, $\overline{BE}$ is longest. Thus, $\overline{BE}$ is the longest segment.

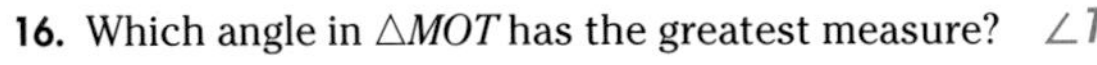

EXERCISES

Practice

A

Refer to the figure at the right for Exercises 15–17.

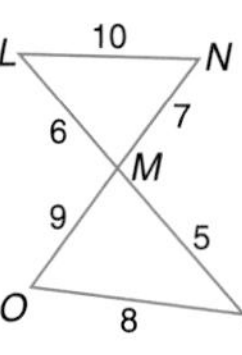

15. Name the angle with the least measure in $\angle LMN$. $\angle N$
16. Which angle in $\triangle MOT$ has the greatest measure? $\angle T$
17. Name the greatest of the six angles in the two triangles, LMN and MOT. $\angle T$

Refer to the figure at the right for Exercises 18–20.

18. Name the angles with the least and the greatest measure in $\triangle XYZ$. $\angle Z$, $\angle XYZ$
19. What is the angle with the least measure in $\triangle WXY$? $\angle WYX$
20. Suppose $\overline{XY}$ bisects $\angle WYZ$. List the angles of $\triangle XYZ$ and $\triangle WXY$ in order from the least to the greatest measure.

20. $\angle Z$, $\angle YXZ$, $\angle XYZ$, $\angle WYX$, $\angle WXY$, $\angle W$ or $\angle Z$, $\angle YXZ$, $\angle WYX$, $\angle XYZ$, $\angle WXY$, $\angle W$

Refer to the figure at the right for Exercises 21–25.

B

21. What is the longest segment in $\triangle CED$? $\overline{CE}$
22. Find the longest segment in $\triangle ABE$. $\overline{BE}$
23. Find the longest segment in the figure. Justify your choice. $\overline{BE}$; See margin.
24. What is the shortest segment in $BCDE$? $\overline{ED}$
25. Is the figure drawn to scale? Explain. No; the segments do not have the correct relative lengths.

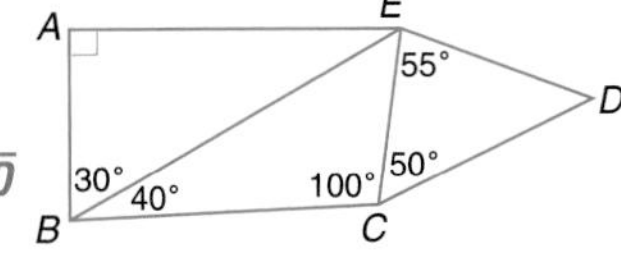

INTEGRATION **Algebra**

Find the value of x and list the sides of $\triangle ABC$ in order from shortest to longest if the angles have the indicated measures.

26. $m\angle A = 9x + 29$, $m\angle B = 93 - 5x$, $m\angle C = 10x + 2$
27. $m\angle A = 9x - 4$, $m\angle B = 4x - 16$, $m\angle C = 68 - 2x$
28. $m\angle A = 12x - 9$, $m\angle B = 62 - 3x$, $m\angle C = 16x + 2$

26. $x = 4$; $m\angle A = 65$, $m\angle B = 73$, $m\angle C = 42$, so $AB < BC < AC$.
27. $x = 12$; $m\angle A = 104$, $m\angle B = 32$, $m\angle C = 44$; so $AC < AB < BC$.
28. $x = 5$; $m\angle A = 51$, $m\angle B = 47$, $m\angle C = 82$; so $AC < BC < AB$.

Study Guide Masters, p. 29

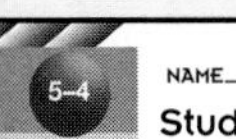

5-4 NAME ______ DATE ______
Study Guide
Student Edition Pages 259–265

Inequalities for Sides and Angles of a Triangle

Two theorems are very useful for determining relationships between sides and angles of triangles.

- If one side of a triangle is longer than another side, then the angle opposite the longer side is greater than the angle opposite the shorter side.
- If one angle of a triangle is greater than another angle, then the side opposite the greater angle is longer than the side opposite the lesser angle.

Examples: 1 List the angles in order from least to greatest measure. 2 List the sides in order from shortest to longest.

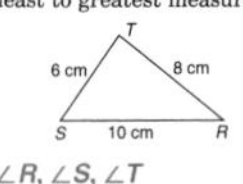

$\angle R$, $\angle S$, $\angle T$

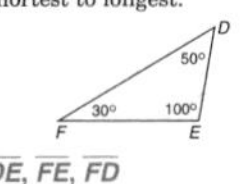

$\overline{DE}$, $\overline{FE}$, $\overline{FD}$

For each triangle, list the angles in order from least to greatest measure.

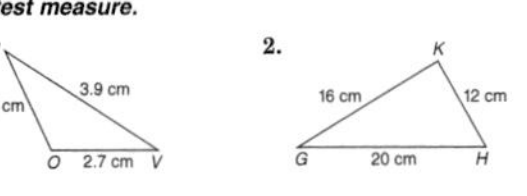

1. $\angle V$, $\angle M$, $\angle O$
2. $\angle G$, $\angle H$, $\angle K$

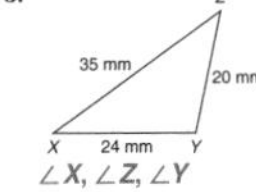

3. $\angle X$, $\angle Z$, $\angle Y$

For each triangle, list the sides in order from shortest to longest.

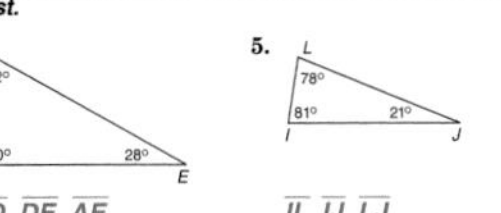

4. $\overline{AD}$, $\overline{DE}$, $\overline{AE}$
5. $\overline{IL}$, $\overline{IJ}$, $\overline{LJ}$

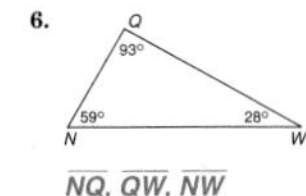

6. $\overline{NQ}$, $\overline{QW}$, $\overline{NW}$

List the sides of $\triangle ABC$ in order from longest to shortest if the angles of $\triangle ABC$ have the indicated measures.

7. $m\angle A = 5x + 2$, $m\angle B = 6x - 10$, $m\angle C = x + 20$ $\overline{AC}$, $\overline{BC}$, $\overline{AB}$
8. $m\angle A = 10x$, $m\angle B = 5x - 17$, $m\angle C = 7x - 1$ $\overline{BC}$, $\overline{AB}$, $\overline{AC}$

Additional Answer

35. Given: $\overline{PQ} \perp$ plane $\mathcal{M}$
Prove: $\overline{PQ}$ is the shortest segment from P to plane $\mathcal{M}$.

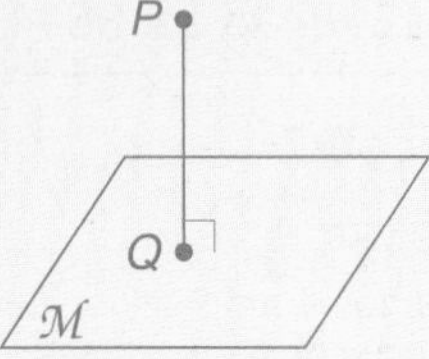

Proof:
By definition, $\overline{PQ}$ is perpendicular to plane $\mathcal{M}$ if it is perpendicular to every line in $\mathcal{M}$ that intersects it. But since the perpendicular segment from a point to a line is the shortest segment from the point to the line, that perpendicular segment is the shortest segment from the point fo each of these lines. Therefore, $\overline{PQ}$ is the shortest segment from P to $\mathcal{M}$.

The International Olympic Committee was formed on June 23, 1894. It governs the Olympic games and recognizes over 160 national Olympic committees.

Practice Masters, p. 29

5–4 NAME_______________ DATE_______

Practice

Student Edition Pages 259–265

Inequalities for Sides and Angles of a Triangle

Refer to the figure on the right for Exercises 1–4.

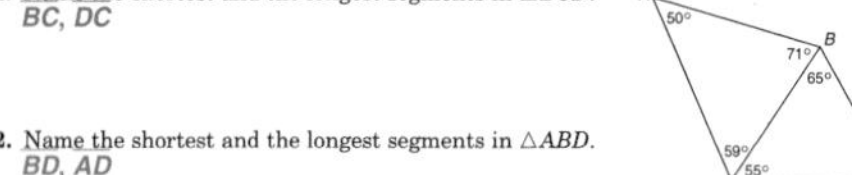

1. Name the shortest and the longest segments in $\triangle BCD$. BC, DC
2. Name the shortest and the longest segments in $\triangle ABD$. BD, AD
3. Find the shortest segment in the figure. $\overline{BC}$
4. How many of the segments in the figure are longer than $\overline{BD}$? 3
5. List the angles in order from least to greatest. $\angle E$, $\angle G$, $\angle F$

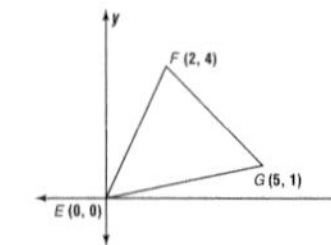

6. List the sides of $\triangle MNO$ in order from longest to shortest if $m\angle M = 4x + 20$, $m\angle N = 2x + 10$, and $m\angle O = 3x - 20$. NO, MO, MN
7. List the angles of $\triangle KLM$ in order from least to greatest if $KL = x - 4$, $LM = x + 4$, $KM = 2x - 1$, and the perimeter of $\triangle KLM$ is 27. $\angle M$, $\angle K$, $\angle L$

Proof

Write a two-column proof. 29–32. See Solutions Manual.

29. **Given:** $\overline{RQ}$ bisects $\angle SRT$.
Prove: $m\angle SQR > m\angle SRQ$

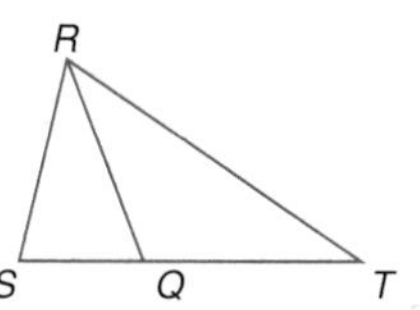

30. **Given:** $\overline{AB} \perp$ plane BCD
Prove: $AD > BD$

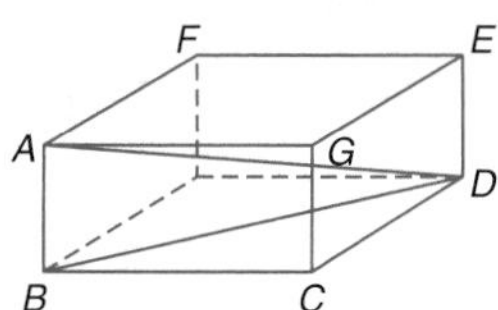

31. **Given:** D is on $\overline{AC}$.
$\angle BDC$ is acute.
Prove: $BA > BD$

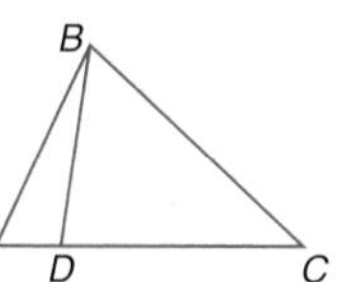

32. **Given:** $\overline{ON} \cong \overline{OM}$
$\overline{OM} \cong \overline{LM}$
Prove: $m\angle MON > m\angle OMN$

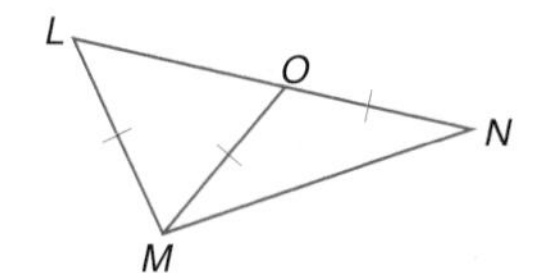

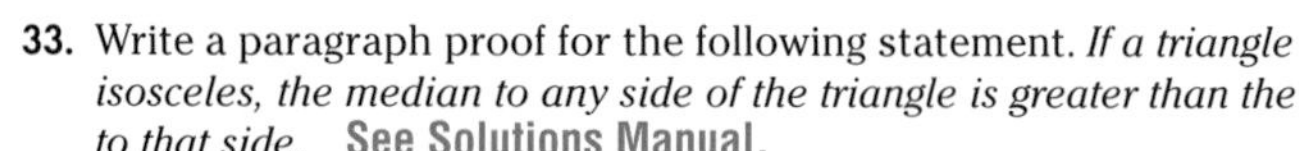

33. Write a paragraph proof for the following statement. *If a triangle is not isosceles, the median to any side of the triangle is greater than the altitude to that side.* See Solutions Manual.
34. Write an indirect proof for Theorem 5–10: *If one angle of a triangle has a greater measure than another angle, then the side opposite the greater angle is longer than the side opposite the lesser angle.* See Solutions Manual.
35. Write a proof for Corollary 5–1: *The perpendicular segment from a point to a plane is the shortest segment from the point to the plane.* See margin.

Critical Thinking

36. Find the coordinates of point D on line AB such that $\overline{CD}$ is the shortest distance from C to $\overleftrightarrow{AB}$ for $A(9, 9)$, $B(-1, -11)$, and $C(-2, 2)$. (4, −1)

Applications and Problem Solving

37b. The equation is the best predictor for Henry Rono's record in 1978. It is the worst predictor for Ron Clarke's record in 1965.

The International Amateur Athletic Federation (IAAF) governs the world records for 65 track and field events. The IAAF has 180 member countries.

37. **Sports** Analysts have found that the world record for the 10,000-meter run has been decreasing steadily since 1940. The trend has been described by the equation $y = 30.18 - 0.07x$, where x is the number of years since 1940 and y is the record time in minutes. a. See margin.
    a. The table below shows several world record times for the 10,000-meter run. Graph the equation including a point to represent each record.
    b. For whose record is the equation the best and worst predictor? Explain.

RECORD HOLDER, COUNTRY	YEAR	TIME
EMIL ZATOPEK, CZECHOSLOVAKIA	1949	29:21.6
VLADIMIR KUTS, SOVIET UNION	1956	28:30.4
RON CLARKE, AUSTRALIA	1965	27:39.4
HENRY RONO, KENYA	1978	27:22.3
ARTURO BARRIOS, MEXICO	1989	27:08.28
HAILE GEBRSELASSIE, ETHIOPIA	1995	26:43.53

38. **Recreation** Mr. Ramirez is assembling an intricate framework that is part of a jungle gym for his daughter. The instructions say to attach bars to a given bar whose ends are marked as A and B so that $\angle A$ is a 30° angle and $\angle B$ is a 60° angle. The bars available are two different lengths. Which bars should he attach at end A and end B? Attach the longer bar at end A.

Additional Answer

37a.

Tech Prep

Athletic Trainer Fitness instructor and sports medicine training programs are available at many two-year colleges. Analysis of fitness and health utilizes scientific and mathematical methods. For more information on tech prep, see the *Teacher's Handbook*.

Mixed Review

39. **Work Backward** Songan, Lucy, and Tamoko are playing a card game. They have a rule that when a player loses a hand, he or she must subtract enough points from his or her score to double each of the other players' scores. First Songan loses a hand, then Tamoko, then Lucy. If each player now has 8 points, who lost the most points? (Lesson 5–3) **Songan**

40. Suppose $\triangle ABC$ and $\triangle XYZ$ are right triangles with right angles at $\angle C$ and $\angle Z$. Determine whether $\triangle ABC \cong \triangle XYZ$ if it is given that $\angle A \cong \angle X$ and $\angle B \cong \angle Y$. Explain. (Lesson 5–2) **No; the sides may not be congruent.**

41. Draw an acute $\triangle DEF$ and construct the angle bisectors of the triangle. (Lesson 5–1) **See students' work.**

42. Is the conclusion $\triangle STL \cong \triangle SNL$ valid based on the information given in the figure at the right? Justify your answer. (Lesson 4–5)

42. Yes; the triangles are congruent by SAS.

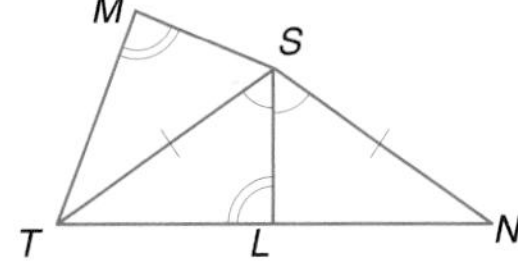

43. Determine if the statement *A right triangle is equilateral* is *true* or *false*. If it is false, draw a counterexample. (Lesson 4–1)

43. False; the hypotenuse must be longer than the other two sides.

44. If possible, write a conclusion. State the law of logic used. (Lesson 2–3)
   (1) $\overline{AX}$ is an altitude of $\triangle ABC$ if $\overline{AX} \perp \overline{BC}$ and X is on $\overline{BC}$.
   (2) $\overline{AX} \perp \overline{BC}$ and X is on $\overline{BC}$.

44. $\overline{AX}$ is an altitude of $\triangle ABC$; Law of Detachment.

45. Make a table of values to determine the coordinates of three points that lie on the graph of $y = 2x - 6$. Use the headings x, $2x - 6$, and (x, y). Then use these three points to draw the line in a coordinate plane. (Lesson 1–1)

45. See Solutions Manual.

INTEGRATION **Algebra**

46. Solve $\frac{1}{3} + \frac{5}{x} = \frac{1}{x}$. **−12**

47. **Nutrition** One kiwi fruit has 230% of the recommended daily allowance of vitamin C. Write this percent as a decimal. **2.3**

WORKING ON THE In•ves•ti•ga•tion

Refer to the Investigation on pages 176–177.

If you read *Alice's Adventures in Wonderland,* you may remember that Alice grows very big and very small. Suppose it was possible to increase the size of a living thing in this way. You would find that the following relationship holds true.

$$\overset{\textit{Volume}}{\frac{\text{space enclosed by new organism}}{\text{space enclosed by old organism}}} > \overset{\textit{Surface Area}}{\frac{\text{amount of skin on new organism}}{\text{amount of skin on old organism}}}$$

1 Some small animals like frogs can respire through their skin, but a larger animal like a human could not support its volume with its surface area. Research human respiration.

2 Most mammals stay cool by releasing heat through their skin and by perspiring. What features of an elephant compensate for its small surface area compared to its volume?

3 Cacti are the trees and shrubs of the hot, dry desert. The surface area of a cactus is minimized to reduce water loss, yet it functions like a tree's branches and leaves. How does the cactus achieve this?

Add the results of your work to your Investigation Folder.

4 ASSESS

Closing Activity

Modeling Have students use toothpicks to make a triangle and have them list the sides in order from shortest to longest and the angles from least to greatest.

Enrichment Masters, p. 29

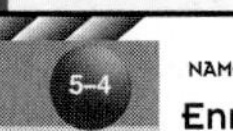

5–4 NAME____________ DATE ________

Enrichment Student Edition Pages 259–265

Letter Puzzles

An **alphametic** is a computation puzzle using letters instead of digits. Each letter represents one of the digits 0–9, and two different letters cannot represent the same digit. Some alphametic puzzles have more than one answer.

Example: Solve the alphametic puzzle at the right.

$$\begin{array}{r} \text{FOUR} \\ +\ \ \text{ONE} \\ \hline \text{FIVE} \end{array}$$

Since R + E = E, the value of R must be 0. Notice that the thousands digit must be the same in the first addend and the sum. Since the value of I is 9 or less, O must be 4 or less. Use trial and error to find values that work.

F = 8, O = 3, U = 1, R = 0
N = 4, E = 7, I = 6, and V = 5.

$$\begin{array}{r} 8310 \\ +\ \ 347 \\ \hline 8657 \end{array}$$

Can you find other solutions to this puzzle?

Find a value for each letter in each alphametic.

1. $\begin{array}{r} \text{HALF} \\ +\ \ \text{HALF} \\ \hline \text{WHOLE} \end{array}$ $\begin{array}{r} 9703 \\ +\ \ 9703 \\ \hline 19406 \end{array}$

A sample answer is shown.
H = 9 A = 7 L = 0
F = 3 W = 1 O = 4
E = 6

2. $\begin{array}{r} \text{TWO} \\ +\ \ \text{TWO} \\ \hline \text{FOUR} \end{array}$ $\begin{array}{r} 734 \\ +\ \ 734 \\ \hline 1468 \end{array}$

A sample answer is shown.
T = 7 W = 3 O = 4
F = 1 U = 6 R = 8

3. $\begin{array}{r} \text{THREE} \\ \text{THREE} \\ +\ \ \text{ONE} \\ \hline \text{SEVEN} \end{array}$ $\begin{array}{r} 43277 \\ 43277 \\ +\ \ 517 \\ \hline 87071 \end{array}$

T = 4 H = 3 R = 2
E = 7 O = 5 N = 1
S = 8 V = 0

4. $\begin{array}{r} \text{SEND} \\ +\ \ \text{MORE} \\ \hline \text{MONEY} \end{array}$ $\begin{array}{r} 9567 \\ +\ \ 1085 \\ \hline 10652 \end{array}$

S = 9 E = 5 N = 6
D = 7 M = 1 O = 0
R = 8 Y = 2

5. Do research to find more alphametic puzzles, or create your own puzzles. Challenge another student to solve them. See students' work.

Extension

Connections Ask students *In an isosceles triangle, how can you tell from angle measures if the third side is longer or shorter than the two equal sides?* **If the base angle measures greater than 60°, then the third side will be shorter; if less than 60°, then the third side will be longer.**

In•ves•ti•ga•tion

Working on the Investigation

The Investigation on pages 176–177 is designed to be a long-term project that is completed over several days or weeks. Encourage students to keep their materials in their Investigation Folder as they work on the Investigation.

5-5A LESSON NOTES

NCTM Standards: 1–5, 7

Objective

Use a graphing calculator to investigate the relationship among the measures of the sides of a triangle.

Recommended Time

20 minutes

Instructional Resources

Instructions for using the TI-81 and Casio calculators for this activity are available in the *Graphing Calculator and Computer Masters*, pp. 22, 23.

1 FOCUS

Motivating the Lesson

Encourage students to pick three numbers between 0 and 10. Then have them try to draw a triangle with sides of these lengths.

2 TEACH

Teaching Tip Work through each line of the program with students and explain the logic. Some of the ideas here are very important in computer programming and can give students a good understanding of what the computer is doing. After you have explained the program itself, have students check it by picking three numbers and "pretending to be the computer." Have them follow each step as the computer would and do the necessary computations. This can be used as a problem-solving strategy when programming.

3 PRACTICE/APPLY

Assignment Guide
Core (with proof): 1–4 **Core (informal):** 1–4 **Enriched:** 1–4

5–5A Using Technology
The Sides of a Triangle

A Preview of Lesson 5–5

There is a special relationship among the measures of the sides of a triangle. The TI-82/83 graphing calculator program below will allow you to investigate the relationship. The program will generate three random numbers between 0 and 100 and determine if segments with these measures can form a triangle.

```
PROGRAM: TRIANGLE
:int (100rand)→A
:int (100rand)→B
:int (100rand)→C
:Disp "SIDES = ", A, B, C
:If A+B≤C or A+C≤B or B+C≤A
:Then
:Disp "NO TRIANGLE"
:Else
:Disp "TRIANGLE EXISTS"
:Stop
```

Run the program at least ten times. Record the sets of side measures that do or do not form a triangle in a table like the one below. Several sample sets are shown below.

No Triangle	Triangle Exists
63, 8, 97	19, 26, 27
51, 40, 73	28, 79, 95
4, 33, 99	
54, 85, 97	

EXERCISES

Work in groups of three or four.

1. Compare the numbers in each column of your table with the numbers that the other members of your group recorded. **See students' work.**
2. Suppose the first number in each set is S1, the second number is S2, and the third number is S3. Find S1 + S2, S2 + S3, and S1 + S3 for the first three sets of numbers in each column. **See students' work.**
3. Using the sums from Exercise 2, compare S1 + S2 to S3, S2 + S3 to S1, and S1 + S3 to S2. What do you observe? Compare your results to those of other members of your group.
4. Write a conjecture about the measures of the sides of a triangle based on your observations.

3. In the column of sets that do not make a triangle, either $S1 + S2 \leq S3$, $S2 + S3 \leq S1$, or $S1 + S3 \leq S2$. In the column of sets that do make a triangle, $S1 + S2 > S3$, $S2 + S3 > S1$, and $S1 + S3 > S2$.

4. Sample answer: In a triangle, the sum of the measures of any two sides is greater than the measure of the third side.

4 ASSESS

Observing students working with technology is an excellent method of assessment.

Using Technology

This lesson offers an excellent opportunity for using technology in your geometry classroom. For more information on using technology, see *Graphing Calculators in the Mathematics Classroom*, one of the titles in the Glencoe Mathematics Professional Series.

5-5 The Triangle Inequality

- To apply the Triangle Inequality Theorem.

Why IT'S IMPORTANT

You can use the triangle inequality to solve problems involving carpentry and mathematics history.

Design

The Oval is at the heart of the main campus of The Ohio State University in Columbus. The master plan for the campus created in 1893 laid out the open space surrounded by buildings. If you were walking from the Main Library, at A, to Mendenhall Lab, at B, and wanted to get there quickly, would you choose the blue path or the red path? If you chose the blue path, you probably reasoned that a path that is a straight line is the shortest path. You are applying the **Triangle Inequality Theorem**.

Theorem 5-12 ***Triangle Inequality Theorem***	**The sum of the lengths of any two sides of a triangle is greater than the length of the third side.**

The Triangle Inequality Theorem shows that some sets of line segments cannot be used to form a triangle because their lengths do not satisfy the inequality. *You will be asked to prove the Triangle Inequality Theorem in Exercise 44.*

Example 1

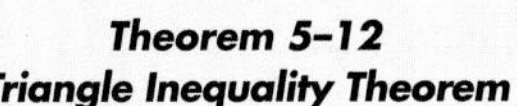

Probability

If 18, 45, 21, and 52 represent lengths of segments, what is the probability that a triangle can be formed if three of these numbers are chosen at random as lengths of the sides?

Make a list of the possible sets of three measures.

a. 18, 45, 21 **b.** 18, 45, 52 **c.** 18, 21, 52 **d.** 45, 21, 52

Test the triangle inequality for each case.

Inequality test		Sketch	Triangle?
a. Is 18 + 45 > 21? Is 18 + 21 > 45? Is 45 + 21 > 18?	*yes* *no* *yes*	18, 21, 45	no
b. Is 18 + 45 > 52? Is 18 + 52 > 45? Is 45 + 52 > 18?	*yes* *yes* *yes*	18, 45, 52	yes
c. Is 18 + 21 > 52? Is 18 + 52 > 21? Is 21 + 52 > 18?	*no* *yes* *yes*	18, 21, 52	no
d. Is 45 + 21 > 52? Is 45 + 52 > 21? Is 21 + 52 > 45?	*yes* *yes* *yes*	21, 45, 52	yes

A triangle can be formed using two of the four possible combinations. Thus, the probability that a triangle can be formed is $\frac{2}{4}$ or $\frac{1}{2}$.

U.S. Colleges and Universities with Largest Enrollment

1. University of Minnesota
2. The Ohio State University
3. Miami Dade Community College
4. University of Texas
5. Arizona State University

The University of Minnesota has four separate campuses—Crockston, Duluth, Mervis, and Twin-Cities. It was founded in 1851 at the Twin-Cities campus.

5-5 LESSON NOTES

NCTM Standards: 1–5, 7

Instructional Resources

- Study Guide Master 5-5
- Practice Master 5-5
- Enrichment Master 5-5
- Assessment and Evaluation Masters, p. 129
- Tech Prep Applications Masters, p. 10

Transparency 5-5A contains the 5-Minute Check for this lesson; **Transparency 5-5B** contains a teaching aid for this lesson.

Recommended Pacing	
Standard Pacing	Days 9 & 10 of 14
Honors Pacing	Days 8 & 9 of 13
Block Scheduling*	Day 5 of 7

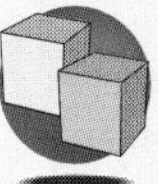

*For more information on pacing and possible lesson plans, refer to the *Block Scheduling Booklet.*

1 FOCUS

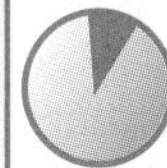

5-Minute Check
(over Lesson 5-4)

For Exercises 1–3, refer to the figure below.

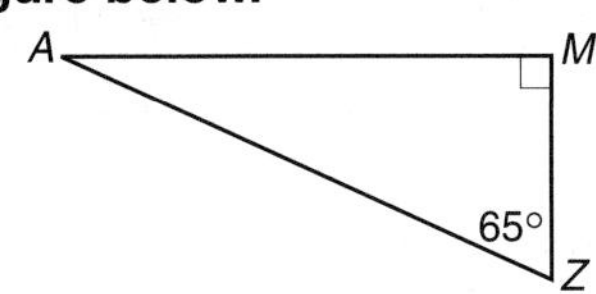

1. If $m\angle Z$ is 65, list the angles in order from least measure to greatest measure. $\angle A$, $\angle Z$, $\angle M$
2. List the sides in order from shortest to longest. $\overline{MZ}$, $\overline{MA}$, $\overline{AZ}$
3. Is $\overline{MZ}$ the shortest segment to MA from point Z? Justify your answer. **Yes, because $\overline{MZ} \perp \overline{MA}$ and by Corollary 5-1, $\overline{MZ}$ must be the shortest segment.**
4. If $m\angle D = 9x + 14$, $m\angle G = 4x - 2$, and $m\angle P = 3x - 8$, list the sides of $\triangle DGP$ in order from longest to shortest. $\overline{GP}$, $\overline{DP}$, $\overline{DG}$

Motivating the Lesson

Hands-On Activity Give students rubber bands and have them make a triangle with their fingers. Have them stretch and distort the rubber band to make different triangles. Ask them what they can say about the sides of the triangles.

2 TEACH

Teaching Tip Encourage students to model problems with pieces of paper.

In-Class Examples

For Example 1
If Mrs. Bailey gave Jane four pieces of copper tubing measuring 6 m, 7 m, 9 m, and 16 m, how many different triangles could Jane make? 1

For Example 2
The lengths of two sides of a triangle are 8 and 13. What are the possible lengths of the third side? **The third side must be greater than 5 and less than 21.**

For Example 3
Given a rope 10 units in length, how many triangles can be made? The lengths of the sides must be in whole units. 2

Teaching Tip In Example 3, ask students how many triangles are possible if the sides don't have to be in whole units. **infinite number**

When you know the lengths of two sides of a triangle, you can determine the range of possible lengths for the third side.

Example 2

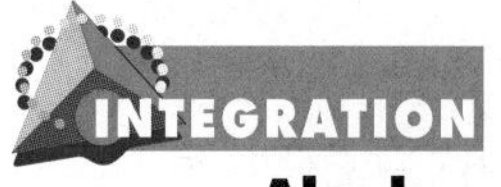

Algebra

If the lengths of two sides of a triangle are 10 centimeters and 15 centimeters, between what two numbers must the measure of the third side fall?

Let t = the length of the third side of the triangle.

By the Triangle Inequality Theorem, each of these inequalities must be true. Graph each on the same number line.

$10 + 15 > t$

$25 > t$

$10 + t > 15$

$t > 5$

$15 + t > 10$

$t > -5$

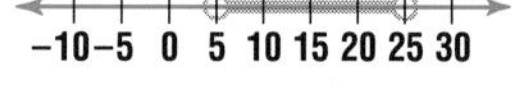

The length of the third side must fall in the range included in all three inequalities.

−10 −5 0 5 10 15 20 25 30

Taking the intersection of these three conditions, t must be greater than 5 and less than 25 or $5 < t < 25$. Any side with length between 5 centimeters and 25 centimeters will form a triangle when the other two sides have lengths 10 centimeters and 15 centimeters.

The Triangle Inequality Theorem can be used to solve problems involving lengths of sides of triangles.

Example 3

History

The early Egyptians used to make triangles by using a rope with knots tied at equal intervals. Each vertex of the triangle had to occur at a knot. Suppose you had a rope with exactly 13 knots as shown below.

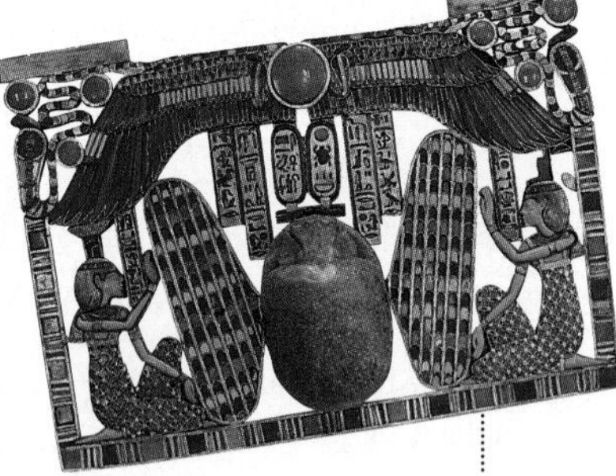

How many different triangles could you make? Sketch each possibility and classify the angles in each case if possible.

Explore The thirteen knots divide the rope into twelve sections of equal length. Thus, the perimeter of each triangle that can be formed will be 12 units. If x, y, and z represent the lengths of the three sides, then $x + y + z = 12$. We also know that $x + y > z$, $x + z > y$, and $y + z > x$ by the Triangle Inequality Theorem.

Plan Make a table of all values of x, y, and z that satisfy $x + y + z = 12$. Then check each possible combination with the Triangle Inequality Theorem.

Solve There are three possible triangles. *How do we know that all the combinations have been checked?*

x	y	z	Triangle?
1	1	10	no
1	2	9	no
1	3	8	no
1	4	7	no
1	5	6	no
2	2	8	no
2	3	7	no
2	4	6	no
2	5	5	yes
3	3	6	no
3	4	5	yes
4	4	4	yes

The triangle with sides 2, 5, and 5 units long is isosceles. The base angles are congruent.

If you sketch a triangle with sides 3, 4, and 5 units long, it appears that the angle opposite the 5-unit side is a right angle. The Pythagorean Theorem confirms this.

The triangle with all three sides 4 units long is equilateral. Each angle measures 60°.

Examine Model the situation with a piece of string to confirm the solution.

CHECK FOR UNDERSTANDING

Communicating Mathematics

Study the lesson. Then complete the following.

1. **Find** three numbers that can be the lengths of the sides of a triangle and three numbers that cannot be the lengths of the sides of a triangle. Justify your reasoning with a drawing. See students' work.
2. **Determine** if there are any restrictions on the length of the base of a triangle if you are told that it is an isosceles triangle with legs 21 centimeters long. Support your answer with diagrams.
3. **You Decide** A triangle has two sides that are 5 inches and 13 inches long. Jolanda says that the third side could be 8 inches. Brittany says that it could not be 8 inches, but it could be 13 inches. Who is correct and why?
   Brittany is correct. $5 + 8$ is not greater than 13, but $5 + 13 > 13$.

2. The base must be less than 42 cm long.

Guided Practice

Determine whether it is possible to draw a triangle with sides of the given measures. Write *yes* or *no*. If yes, draw the triangle.

5. Yes; see margin.

4. 1, 2, 3 no
5. 21, 32, 18
6. 11, 6, 2 no

The measures of two sides of a triangle are given. Between what two numbers must the measure of the third side fall?

7. 21 and 27 6 and 48
8. 5 and 11 6 and 16
9. 30 and 30 0 and 60

Study the figure carefully and indicate whether each statement is *always true, sometimes true,* or *never true*. Justify your answer.

10. $DB > 7$ never true; $AD + AB > DB$
11. $BC < 9$ always true; $DB < 7$ and $BC < 2 + DB$

Proof

12. Write a two-column proof.
    Given: $\overline{PO} \cong \overline{OM}$
    Prove: $PO + MN > ON$ See margin.

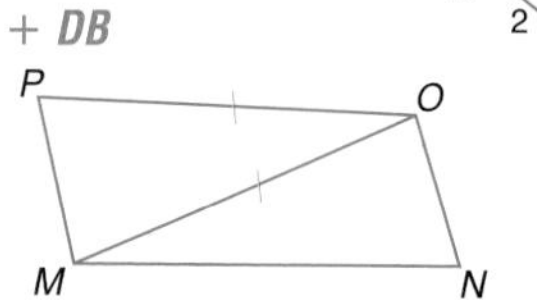

3 PRACTICE/APPLY

Check for Understanding

Exercises 1–13 are designed to help you assess your students' understanding through reading, writing, speaking, and modeling. You should work through Exercises 1–3 with your students and then monitor their work on Exercises 4–13.

Error Analysis
Students may have difficulty understanding why three inequalities need to be checked to see if three segments can form a triangle. Use several examples with different lengths for sides and have students use straws of proportional length to test the measures.

Additional Answers

5.

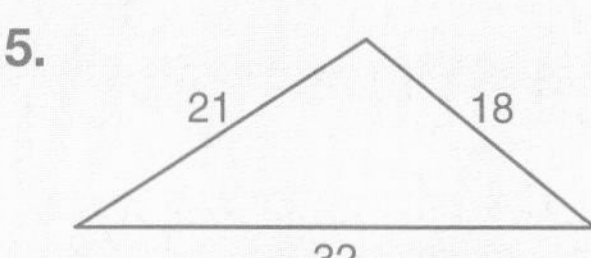

12. Given: $\overline{PO} \cong \overline{OM}$
    Prove: $PO + MN > ON$
    Proof:
    Statements (Reasons)
    1. $\overline{PO} \cong \overline{OM}$ (Given)
    2. $PO = OM$ (Def. ≅ segments)
    3. $OM + MN > ON$ (Triangle Inequality)
    4. $PO + MN > ON$ (Substitution Prop. (=))

Reteaching

Using Exploration Ask students for three numbers. Have them test these numbers to see if they can be the lengths of the sides of a triangle. Repeat this process until students begin to grasp the idea of the Triangle Inequality Theorem.

Assignment Guide

Core (with proof): 15–45 odd, 46, 47, 49–59
Core (informal): 15–41 odd, 45–47, 49–54, 56–59
Enriched: 14–44 even, 46–59

For **Extra Practice**, see p. 773.

The red A, B, and C flags, printed only in the Teacher's Wraparound Edition, indicate the level of difficulty of the exercises.

Additional Answers

32. $BC < AB + AC$
33. $BD + DC < BC$ and $BC + DC > BC$, so $10 < BC < 38$. 12 is one of the acceptable values for BC.
34. $AB = AC$ and $BC < AB + AC$, $BC < 2AB$, $\frac{1}{2}BC > AB$
35. $AB + AC > BC$ and $BC + DC > BD$, so $AB + AC > BD - DC$, then $AB + AC > BD$
36. This is true if $\triangle ABC \cong \triangle DBC$.
37. $\overline{BA}$ is the perpendicular, so it is the shortest distance from B to $\overline{AC}$.

Study Guide Masters, p. 30

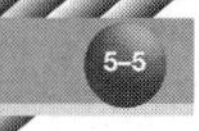

NAME__________ DATE ________

Study Guide — Student Edition Pages 267–272

The Triangle Inequality

If you take three straws that are 8 inches, 4 inches, and 3 inches in length, can you use these three straws to form a triangle? Without actually trying it, you might think it is possible to form a triangle with the straws. If you try it, however, you will notice that the two smaller straws are too short. This example illustrates the following theorem.

Triangle Inequality Theorem	The sum of the lengths of any two sides of a triangle is greater than the length of the third side.

Example: If the lengths of two sides of a triangle are 7 centimeters and 11 centimeters, between what two numbers must the measure of the third side fall?

Let x = the length of the third side.

By the Triangle Inequality Theorem, each of these inequalities must be true.

$x + 7 > 11$ $x + 11 > 7$ $11 + 7 > x$
$x > 4$ $x > -4$ $18 > x$

Therefore, x must be between 4 centimeters and 18 centimeters.

Determine whether it is possible to draw a triangle with sides of the given measures. Write yes or no.

1. 15, 12, 9 yes
2. 23, 16, 7 no
3. 20, 10, 9 no
4. 8.5, 6.5, 13.5 yes
5. 47, 28, 70 yes
6. 28, 41, 13 no

The measures of two sides of a triangle are given. Between what two numbers must the measure of the third side fall?

7. 9 and 15 6 and 24
8. 11 and 20 9 and 31
9. 23 and 14 9 and 37
10. Suppose you have three different positive numbers arranged in order from greatest to least. Which sum is it most crucial to test to see if the numbers could be the lengths of the sides of a triangle? the sum of the two smaller numbers

13. **Algebra** Can you make a triangle using $Q(1, 1)$, $R(-2, 5)$, and $S(-5, -4)$ as the vertices? Why or why not? Yes. The measures of the sides are 5, $\sqrt{61}$, and $\sqrt{90}$ and the Triangle Inequality Theorem applies.

EXERCISES

Practice

14, 15, 17, 19. See Solutions Manual.

Determine whether it is possible to draw a triangle with sides of the given measures. Write *yes* or *no*. If yes, draw the triangle.

14. 5, 4, 3 yes
15. 5.2, 5.6, 10.1 yes
16. 5, 10, 15 no
17. 10, 100, 100 yes
18. 301, 8, 310 no
19. 9, 40, 41 yes
20. 12, 2.2, 14.3 no
21. 10, 150, 200 no
22. 84, 7, 115 no

The measures of two sides of a triangle are given. Between what two numbers must the measure of the third side fall?

24. 9 and 37
27. 24 and 152
30. 14 and 222
31. $|a - b|$ and $a + b$

23. 15 and 18 3 and 33
24. 14 and 23
25. 22 and 34 12 and 56
26. 21 and 47 26 and 68
27. 64 and 88
28. 99 and 2 97 and 101
29. 47 and 71 24 and 118
30. 104 and 118
31. a and b

Study the figure carefully and indicate whether each statement is *always true, sometimes true,* or *never true.* Justify your answer.

32–37. See margin for justifications.
32. always true
34. always true
35. always true

32. If $AB = 5$, $AC = 8$, $DC = 2$, then $BC < 15$.
33. If $DC = 14$, $AC = 18$, $BD = 24$, then $BC = 12$. sometimes true
34. If $\angle 1 \cong \angle 3$, then $AB > \frac{1}{2}BC$.
35. If $\angle D$ is obtuse, then $AB + AC > BD$.
36. If $\angle 1 \cong \angle 2$, then $AB + AC = BD + DC$. sometimes true
37. If $\angle A$ is a right angle, then $BC < BA$. never true

Algebra

Determine whether it is possible to have a triangle with the given vertices. Write *yes* or *no*, and explain your answer.

38–41. See margin for explanations.

38. $R(0, 0)$, $S(3, 5)$, $T(5, 3)$ yes
39. $A(2, 3)$, $B(-5, -11)$, $C(-8, 15)$ yes
40. $J(1, -4)$, $K(-3, -20)$, $L(5, 12)$ no
41. $D(1, 4)$, $E(5, -1)$, $F(1, -4)$ yes

Proof

Write a two-column proof. 42–43. See Solutions Manual.

42. **Given:** $RS = RT$
Prove: $UV + VS > UT$

43. **Given:** quadrilateral $ABCD$
Prove: $AD + CD + AB > BC$

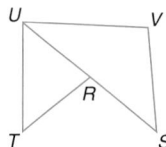

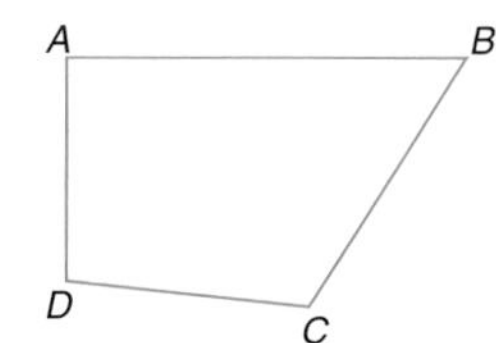

44. See margin.

44. Write a paragraph proof for the Triangle Inequality Theorem. (Theorem 5–12)
Given: $\triangle ROS$
Prove: $SO + OR > RS$
(Hint: Draw auxiliary segment, $\overline{OT}$, so that O is between R and T and $\overline{OT} \cong \overline{SO}$.)

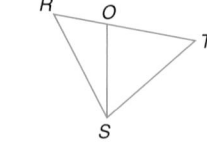

Additional Answers

38. $RS = \sqrt{34} \approx 5.8$, $RT = \sqrt{34} \approx 5.8$, $ST = \sqrt{8} \approx 2.8$
These measures satisfy the Triangle Inequality Theorem.
39. $AB = \sqrt{245} \approx 15.7$, $AC = \sqrt{244} \approx 15.6$, $BC = \sqrt{685} \approx 26.2$
These measures satisfy the Triangle Inequality Theorem.
40. $JK = \sqrt{272} \approx 16.5$, $JL = \sqrt{272} \approx 16.5$, $KL = \sqrt{1088} \approx 33.0$
These measures do not satisfy the Triangle Inequality Theorem.
41. $DE = \sqrt{41} \approx 6.4$, $DF = 8$, $EF = 5$
These measures satisfy the Triangle Inequality Theorem.

Programming

45. The TI-82/83 graphing calculator program at the right tests to see if three measures can form the sides of a triangle.

```
PROGRAM: TESTER
:Disp "ENTER MEASURES"
:Input "SIDE 1:", A
:Input "SIDE 2:", B
:Input "SIDE 3:", C
:If A+B>C and A+C>B and
 B+C>A
:Then
:Disp "TRIANGLE EXISTS"
:Else
:Disp "NO TRIANGLE"
:Stop
```

Use the program to test each set of measures.

a. 5, 7, 12 no triangle
b. 4.5, 8.85, 6.25 triangle exists
c. 9.87, 12.32, 32.90 no triangle
d. 112, 89, 75 triangle exists
e. 256, 219, 311 triangle exists

Critical Thinking

46. Is it true that the difference between any two sides of a triangle is less than the third side? Explain your reasoning. It is true; $a + b > c$, $a > c - b$ or $c - b < a$.

Applications and Problem Solving

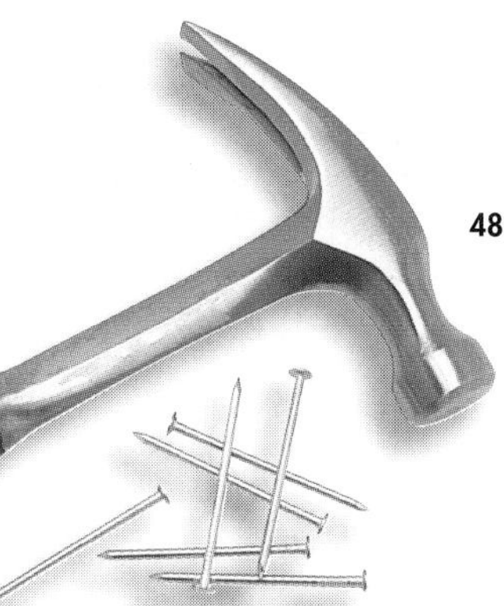

47. One side of a triangle is 2 centimeters long. Let a represent the measure of the second side and z represent the measure of the third side. Suppose that $14 < a < 17$ and $13 < z < 17$ and a and z are whole numbers.
   a. **Algebra** List the measures of the sides of the triangles that are possible under those conditions. See Solutions Manual.
   b. **Probability** What is the probability that a randomly chosen triangle that satisfies the conditions will be isosceles? $\frac{2}{5}$

48. **Carpentry** Salina is building stairs and wants to nail a brace at the base of each stair as shown in the figure. The brace attaches at the bottom of the rise and anywhere along the tread.

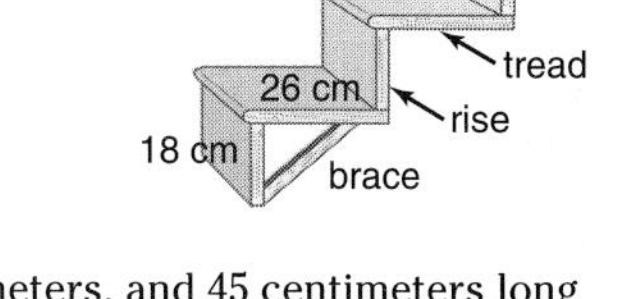

The stairs have an 18-cm rise and a 26-cm tread. There is a pile of braces 5 centimeters, 20 centimeters, 24 centimeters, and 45 centimeters long that Salina can use. Which of the lengths can she use as a brace? She can use the 20 and the 24 cm braces.

Mixed Review

49. Suppose $m\angle A = 4x + 61$, $m\angle B = 67 - 3x$, and $m\angle C = x + 74$. What is the longest segment in $\triangle ABC$? (Lesson 5–4) $\overline{AC}$

50. State the assumption you would make to start an indirect proof for $\overline{NM}$ is a median of $\triangle NOP$. (Lesson 5–3) $\overline{NM}$ is not a median of $\triangle NOP$.

51. If possible, describe a triangle in which the angle bisectors all intersect in a point outside the triangle. If no triangle exists, write *no triangle*. (Lesson 5–1) no triangle

52. Refer to the figure at the right. If X is the midpoint of $\overline{AD}$, name the additional parts of $\triangle AXC$ and $\triangle DXB$ that would have to be congruent to prove $\triangle AXC \cong \triangle DXB$ by ASA. (Lesson 4–5) $\angle 1 \cong \angle 6$

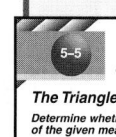

53. The legs of an isosceles triangle are $(6x - 6)$ and $(x + 9)$ units long. Find the length of the legs. (Lesson 4–1) 12 units

54. Determine if the statement *The distance between two parallel lines is the length of any segment that connects points on the two lines* is true or false. If the statement is false, explain why. (Lesson 3–5)

54. False; the segment must be perpendicular to the lines to represent the distance.

Extension

Connections Tell students that the Triangle Inequality Theorem can be rewritten as "The difference between any sides of a triangle is less than the length of the third side." Why is this true?
(side 1) + (side 2) > side 3
(side 2) + (side 3) > side 1
(side 1) + (side 3) > side 2
If you subtract (side 1) from each side of the first inequality, (side 2) from each side of the second inequality, and (side 3) from each side of the third inequality, you get
side 2 > (side 3) − (side 1)
side 3 > (side 1) − (side 2)
side 1 > (side 2) − (side 3)

Using the Programming Exercises The program given in Exercise 45 is for use with a TI-82 or TI-83 graphing calculator. For other programmable calculators, have students consult their owner's manual for commands similar to those presented here.

Additional Answer

44. The Ruler Post. allows us to construct $\overline{OT}$ so that O is between R and T and $\overline{OS} \cong \overline{OT}$. $OS = OT$ by def. of $\cong$ segments. If any two sides of a triangle are $\cong$, then the $\angle$s opp. the sides are $\cong$, so $\angle 2 \cong \angle 3$. $m\angle 2 = m\angle 3$ by the def. of $\cong$ $\angle$s. The Angle Add. Post. allows us to say that $m\angle 1 + m\angle 2 = m\angle RST$. By substitution, $m\angle 1 + m\angle 3 = m\angle RST$. By the def. of inequality, $m\angle 3 < m\angle RST$. If the measure of an $\angle$ of a $\triangle$ is greater than the measure of the other $\angle$, then the side opp. the larger $\angle$ is longer than the side opp. the smaller angle, so $RS < RT$. $RT = RO + OT$ by the Segment Add. Post. By substitution, $RS < RO + OT$. Then by substitution, $RS < RO + OS$.

Practice Masters, p. 30

NAME________ DATE ________

Practice

Student Edition Pages 267–272

The Triangle Inequality

Determine whether it is possible to draw a triangle with sides of the given measure. Write yes or no.

1. 3, 3, 3 yes
2. 2, 3, 4 yes
3. 1, 2, 3 no
4. 8.9, 9.3, 18.3 no
5. 16.5, 20.5, 38.5 no
6. 19, 19, 0.5 yes

Determine whether it is possible to have a triangle with the given vertices. Write yes or no, and explain your answer.

7. $A(-2, -2)$, $B(-1, 1)$, $C(1, 4)$ Yes; $AB = \sqrt{10}$, $BC = \sqrt{13}$, $AC = 3\sqrt{5}$ and all triangle inequalities are satisfied.
8. $A(-4, 2)$, $B(-2, 1)$, $C(2, -1)$ no; $AB = \sqrt{5}$, $BC = 2\sqrt{5}$, $AC = 3\sqrt{5}$ $AB + BC = AC$
9. $A(2, 5)$, $B(-3, 5)$, $C(6, -1)$ yes; $AB = 5$, $BC = 3\sqrt{13}$, $AC = 2\sqrt{13}$ and all triangle inequalities are satisfied.
10. $A(3, -6)$, $B(1, 2)$, $C(-2, 10)$ yes; $AB = 2\sqrt{17}$, $BC = \sqrt{73}$, $AC = \sqrt{281}$ and all triangle inequalities are satisfied.

Two sides of a triangle are 21 and 24 inches long. Determine whether each measurement can be the length of the third side.

11. 3 inches no
12. 40 inches yes
13. 56 inches no

If the sides of a triangle have the following lengths, find all possible values for x.

14. $AB = 2x + 5$, $BC = 3x - 2$, $AC = 4x - 8$ $\{x \mid x > 3\}$
15. $PQ = 3x$, $QR = 4x - 7$, $PR = 2x + 9$ $\left\{x \mid x > \frac{16}{5}\right\}$

4 ASSESS

Closing Activity

Speaking Separate students into small groups. Have members of each group discuss and define the Triangle Inequality Theorem in their own words. Have each group present its definition to the class.

Chapter 5 Quiz C (Lessons 5-4 and 5-5) is available in the *Assessment and Evaluation Masters*, p. 129.

Additional Answer

55. Statements (Reasons)
1. $\ell \parallel m$, $s \parallel t$, $t \perp \ell$ (Given)
2. $\angle 1$ is a right angle. ($\perp$ lines form four rt. $\angle$s.)
3. $m\angle 1 = 90$ (Def. rt. $\angle$s)
4. $\angle 1 \cong \angle 2$ (Corr. $\angle$s Post.)
5. $m\angle 2 = 90$ (Subst. Prop. (=))
6. $\angle 2$ and $\angle 3$ are supp. (Consecutive Int. $\angle$ Th.)
7. $m\angle 2 + m\angle 3 = 180$ (Def. supp.)
8. $90 + m\angle 3 = 180$ (Subst. Prop. (=))
9. $m\angle 3 = 90$ (Subtr. Prop. (=))
10. $\angle 3$ is a rt. $\angle$. (Def. rt. $\angle$)

Enrichment Masters, p. 30

5-5 NAME______ DATE______

Enrichment Student Edition Pages 267–272

Triangles in Construction

A **rigid** figure will hold its shape. Triangles are rigid but many other shapes are not. The diagrams below show figures made from cardboard strips and paper fasteners. Suppose the figure in the middle is pushed at the location of the arrow. The figure will collapse as shown at the right.

This triangle is rigid. This quadrilateral is not rigid.

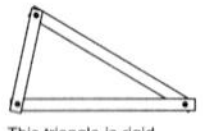

To make the quadrilateral rigid, a diagonal strip can be added. This forms two rigid triangles. When buildings are constructed, diagonal supports are used to make the frame stronger.

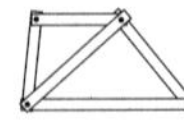

State whether each figure is rigid or not rigid. If a figure is not rigid, show where strips can be added to make the figure rigid.

1. not rigid
2. rigid
3. not rigid

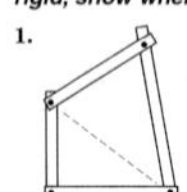

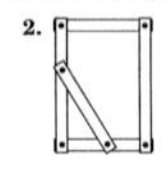

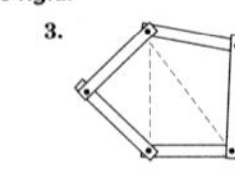

4. rigid
5. not rigid
6. not rigid

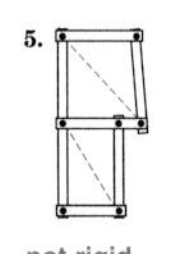

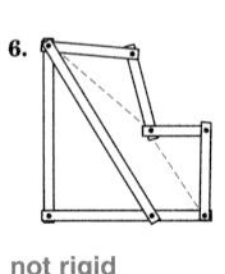

7. A geodesic dome is formed from triangles. Do research on the construction of domes. See students' work.

272 Chapter 5

55. Write a two-column proof. (Lesson 3–2)

Given: $\ell \parallel m$
$s \parallel t$
$t \perp \ell$

Prove: $\angle 3$ is a right angle. See margin.

56. Write the converse of the conditional statement *If three points are collinear, then they lie on a straight line.* Determine if the converse is true or false. If it is false, give a counterexample. (Lesson 2–2)

56. If three points lie on a straight line, then they are collinear; true.

57. Suppose $m\angle H = 63$. Find the measure of the complement and the supplement of $\angle H$. (Lesson 1–7) 27; 117

Algebra

58. Simplify $\sqrt{28x^4}$. $2x^2\sqrt{7}$

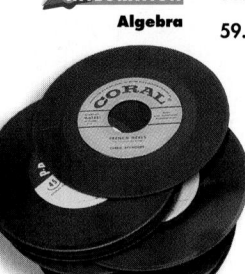

59. **Music** Refer to the graph at the right.
a. Describe the information that the graph shows.
b. If there is a total of eight million classical music listeners, how many people between 18 and 24 years old listen to classical music? 880,000 people

59a. the percentage of classical music listeners who are in different age groups

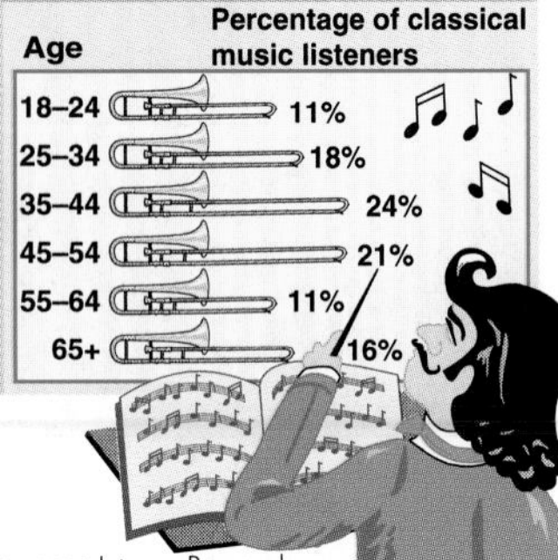

Mathematics and Society

How Do Turtles Navigate?

The excerpt below appeared in an article in *Earth* in February, 1995.

> Loggerhead sea turtles are among the world's greatest navigators. They spend much of their lives in the Sargasso Sea, a fertile region in the middle of the Atlantic. But mature loggerheads travel thousands of miles to return to their natal beaches to reproduce.... Recent studies suggest that sea turtles use the changing angle of Earth's magnetic field with respect to the ground, called the angle of inclination, to navigate....At the magnetic poles, the inclination of the field is about 90 degrees....the invisible field lines are nearly perpendicular to the surface of the planet. In contrast, at the equator, the inclination of Earth's magnetic field is zero degrees. That means the field lines flow parallel to the surface. Thus, magnetic inclination is a pretty good stand-in for latitude....subsequent experiments confirmed that the turtles can sense a change as small as three degrees in inclination, and they may well be considerably more sensitive than that. ■

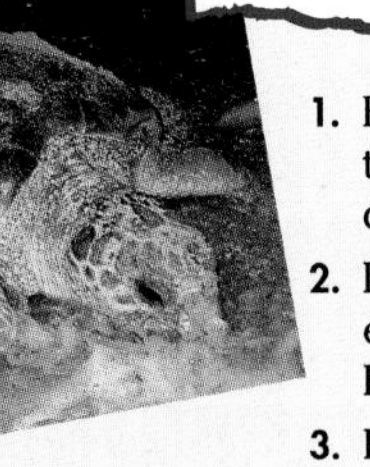

1. Besides an internal "compass" that tells an animal the direction it is traveling, what else would the animal need to know in order to reach its destination? when to change directions to get to its final location
2. If a town is located at a point that is one-third of the distance from the equator to the North Pole, what would be the angle of inclination of Earth's magnetic field there? 30°
3. How could you find out whether an animal uses the position of the sun to assist it in its migration? How would triangles help in this positioning? 3. See margin.

Mathematics and Society

Discuss other animals that migrate, such as monarch butterflies, salmon, and Canada geese. What obstacles do these animals need to overcome? What helps them in their migration?

Answer for Mathematics and Society

3. Conduct an experiment where there is no sun to see if the animal can still find its way. Also, you could create an artificial sun out of a light source and observe how the animal responds to it.

5-6 Inequalities Involving Two Triangles

What YOU'LL LEARN

- To apply the SAS Inequality and the SSS Inequality.

Why IT'S IMPORTANT

You can use the inequalities in triangles to solve problems involving physical therapy and biology.

APPLICATION Physical Therapy

Physical therapists help people use exercise to regain their strength after an injury or surgery. One exercise a physical therapist may recommend as a part of a rehabilitation program following arm surgery is called a *curl*. To perform a curl, hold a weight in your hand with your arm at your side and your palm facing upward. Raise your forearm by bending at the elbow as far as possible and slowly return to the starting position.

Observe the positions of your arm as you complete a curl. Notice that the measure of the angle between the upper and lower parts of your arm decreases as you raise your hand. What happens to the distance between your hand and your shoulder? A model can help clarify the relationships.

CAREER CHOICES

If you like to help people, **physical therapy** may be for you!

A career in physical therapy requires a bachelor's degree. Studies include anatomy, physics, and mathematics.

For more information, contact:

American Physical Therapy Assn.
1111 N. Fairfax St.
Alexandria, VA 22314

MODELING MATHEMATICS — Inequalities in Two Triangles

Materials:  rubber band  ball-bearing compass  centimeter ruler  protractor

- Place a rubber band over the tips of a ball-bearing compass as shown. Imagine that the sides of the compass and the rubber band form a triangle, which we will call $\triangle PRT$.
- Measure $\angle R$ and $\overline{PT}$.
- Spread the compass arms farther apart to form a different triangle, $\triangle P'R'T'$.
- Measure $\angle R'$ and $\overline{P'T'}$.

Your Turn

a. Compare $m\angle R$ to $m\angle R'$ and PT to $P'T'$.

b. Change the position of the compass arms again. Make a conjecture about how $m\angle R$ and PT will change. Check your conjecture.

c. Suppose $\overline{AB} \cong \overline{DE}$ and $\overline{BC} \cong \overline{EF}$ in $\triangle ABC$ and $\triangle DEF$. If $m\angle B > m\angle E$, how are AC and DF related? **$AC > DF$**

a. $m\angle R < m\angle R'$; $PT < P'T'$

b. If $m\angle R$ becomes greater than $m\angle R'$, then $PT > P'T'$. Otherwise the relationship stays the same.

The Modeling Mathematics activity suggests the following two theorems.

Theorem 5-13 SAS Inequality (Hinge Theorem)	**If two sides of one triangle are congruent to two sides of another triangle, and the included angle in one triangle is greater than the included angle in the other, then the third side of the first triangle is longer than the third side in the second triangle.**

CAREER CHOICES

In 1994, the median annual salary for physical therapists was $37,596. The top 10% earned over $61,776.

MODELING MATHEMATICS Ask students what happens when angle R is 180° and when it is greater than 180°.

5-6 LESSON NOTES

NCTM Standards: 1–5, 7

Instructional Resources

- Study Guide Master 5-6
- Practice Master 5-6
- Enrichment Master 5-6
- Assessment and Evaluation Masters, p. 129
- Real-World Applications, 10

Transparency 5-6A contains the 5-Minute Check for this lesson; **Transparency 5-6B** contains a teaching aid for this lesson.

Recommended Pacing	
Standard Pacing	Days 11 & 12 of 14
Honors Pacing	Days 10 & 11 of 13
Block Scheduling*	Day 6 of 7

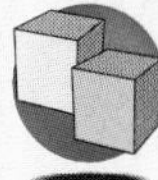

*For more information on pacing and possible lesson plans, refer to the *Block Scheduling Booklet.*

1 FOCUS

5-Minute Check *(over Lesson 5-5)*

Determine whether it is possible to draw a triangle with sides of the given measures.

1. 4, 5, 6 yes
2. 6.1, 8.3, 14.4 no
3. 99, 210, 310 no

Two sides of a triangle are 5 cm and 9 cm in length. Determine whether each measurement can be the length of the third side.

4. 3 cm no
5. 13 cm yes

Motivating the Lesson

Questioning Review the Triangle Inequality Theorem from the previous lesson and ask students to relate it to two triangles. What would the triangles need to have in common to demonstrate the theorem? What would need to be different?

2 TEACH

In-Class Example

For Example 1

Refer to the figure below to relate how each pair of angle measures is related.

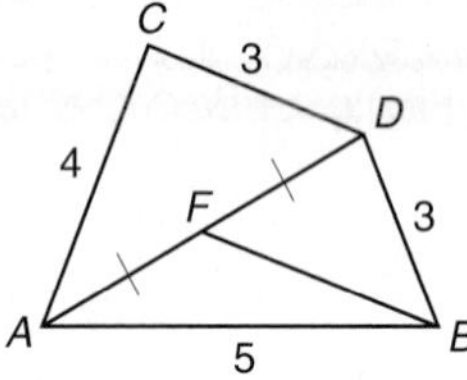

a. $m\angle ADC$ and $m\angle ADB$
$m\angle ADC < m\angle ADB$

b. $m\angle AFB$ and $m\angle BFD$
$m\angle AFB > m\angle BFD$

Teaching Tip It may help students if you point out that in Theorems 5-13 and 5-14 the longer side is in the same triangle as the greater angle.

Teaching Tip It may be unclear which angle is being discussed in the SSS Theorem. Stress that it will be the angle between the two congruent sides, or the angle opposite the longest side. Students can also look at it as the angle that would be necessary to use the SAS Theorem.

Theorem 5–14 SSS Inequality	**If two sides of one triangle are congruent to two sides of another triangle and the third side in one triangle is longer than the third side in the other, then the angle between the pair of congruent sides in the first triangle is greater than the corresponding angle in the second triangle.**

An indirect proof for the SSS Inequality is given below. *You will be asked to write a proof for the SAS Inequality in Exercise 34.*

Proof of Theorem 5–14

Given: $\overline{RS} \cong \overline{PW}$
$\overline{ST} \cong \overline{WV}$
$RT > PV$

Prove: $m\angle S > m\angle W$

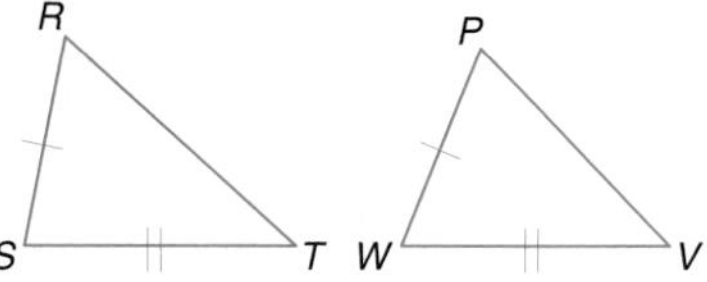

Indirect Proof:

Step 1: Assume $m\angle S \leq m\angle W$.

Step 2: If $m\angle S \leq m\angle W$, then either $m\angle S = m\angle W$ or $m\angle S < m\angle W$.

Case 1: If $m\angle S = m\angle W$, then $\triangle RST \cong \triangle PWV$ by SAS, and $\overline{RT} \cong \overline{PV}$ by CPCTC. Thus, $RT = PV$.

Case 2: If $m\angle S < m\angle W$, then $RT < PV$ by the SAS Inequality.

Step 3: In both cases, our assumptions led to a contradiction of the given information that $RT > PV$. Therefore, the assumption must be false, and the conclusion, $m\angle S > m\angle W$, must be true.

The following examples illustrate geometric and algebraic applications of the SAS Inequality and the SSS Inequality.

Example  **In the figure at the right, $EF = 2.9$ and $AD = 2.8$. Describe how each pair of angle or segment measures is related.**

a. $m\angle AED$, $m\angle DEF$

b. DE, EC

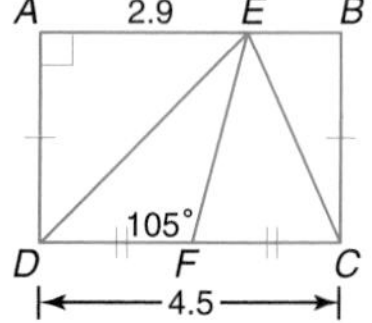

a. In $\triangle AED$ and $\triangle DEF$, $\overline{ED} \cong \overline{ED}$, since congruence of segments is reflexive. $EF = 2.9$ and $AE = 2.9$, so $\overline{EF} \cong \overline{AE}$. $DF = 4.5 \div 2$ or 2.25 and $AD = 2.8$. Then the SSS Inequality allows us to conclude that $m\angle AED > m\angle DEF$.

b. Compare $\triangle DEF$ and $\triangle CEF$. $\overline{EF} \cong \overline{EF}$ since congruence of segments is reflexive. The diagram indicates that $\overline{DF} \cong \overline{FC}$. $\angle DFE$ and $\angle EFC$ form a linear pair, so they are supplementary. Then $m\angle DFE + m\angle EFC = 180$. Thus, $105 + m\angle EFC = 180$ or $m\angle EFC = 75$. Therefore, $m\angle DFE > m\angle EFC$. Hence by the SAS Inequality, $DE > EC$.

Example 2   Proof 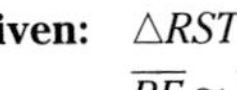

Write a two-column proof.

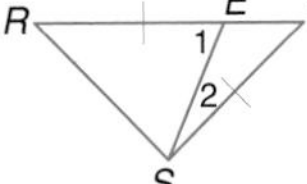

Given: $\triangle RST$

$\overline{RE} \cong \overline{ST}$

Prove: $RS > TE$

Proof:

Statements	Reasons
1. $\overline{RE} \cong \overline{ST}$	1. Given
2. $\overline{ES} \cong \overline{ES}$	2. Congruence of segments is reflexive.
3. $\angle 1$ is an exterior angle of $\triangle EST$.	3. Def. exterior angle
4. $m\angle 1 > m\angle 2$	4. If an $\angle$ is an ext. angle of a $\triangle$, then its measure > the measure of either of its corr. remote int. $\angle$s.
5. $RS > TE$	5. SAS Inequality

You can use algebra to relate the measures of the angles and sides of two triangles.

Example 3

Algebra

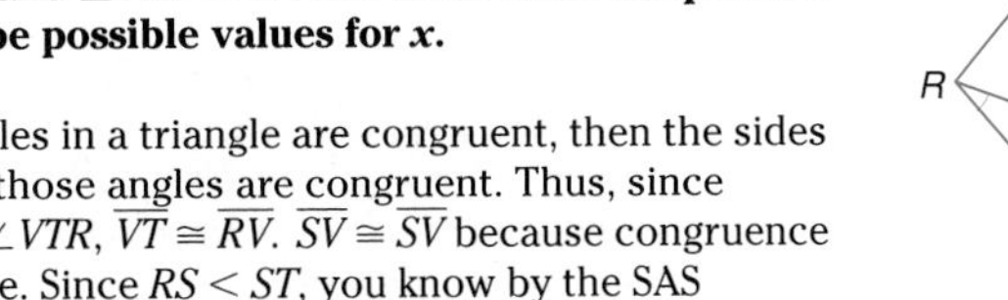

In the figure, $m\angle SVR = 5x + 15$, $m\angle TVS = 10x - 20$, $RS < ST$, and $\angle VRT \cong \angle VTR$. Write three inequalities to describe possible values for x.

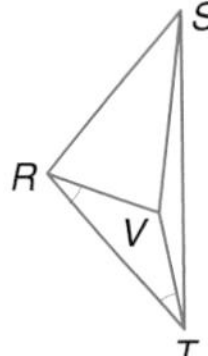

If two angles in a triangle are congruent, then the sides opposite those angles are congruent. Thus, since $\angle VRT \cong \angle VTR$, $\overline{VT} \cong \overline{RV}$. $\overline{SV} \cong \overline{SV}$ because congruence is reflexive. Since $RS < ST$, you know by the SAS Inequality that $m\angle SVR < m\angle TVS$.

Since both $\angle SVR$ and $\angle TVS$ are angles in a triangle, you know that their measures are less than 180. Thus, you can write the following three inequalities.

$m\angle SVR < 180$	$m\angle TVS < 180$	$m\angle SVR < m\angle TVS$
$5x + 15 < 180$	$10x - 20 < 180$	$5x + 15 < 10x - 20$
$5x < 165$	$10x < 200$	$35 < 5x$
$x < 35$	$x < 20$	$7 < x$

The range of possible values for x is $7 < x < 20$.

CHECK FOR UNDERSTANDING

Communicating Mathematics

Study the lesson. Then complete the following.

1. **Draw** an isosceles triangle with legs 5 centimeters long and a base 8 centimeters long. Then draw another isosceles triangle with legs 5 centimeters long and a base 2 centimeters long.
   a. Measure the vertex angle of each triangle. Which triangle has the larger vertex angle? **the triangle with base of 8 cm**
   b. What theorem does this demonstrate? **SSS Inequality**

In-Class Examples

For Example 2
Write a two-column proof.

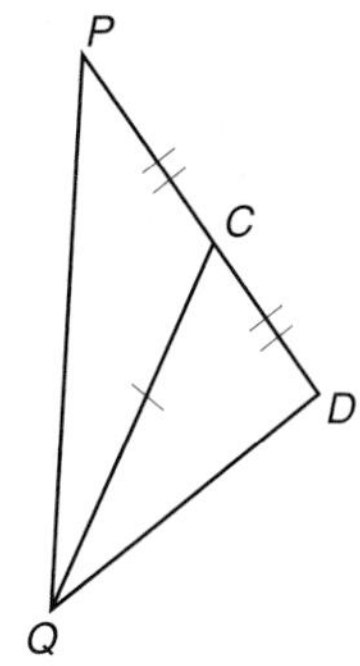

Given: $\overline{QC}$ bisects $\overline{PD}$.
$m\angle PCQ > m\angle DCQ$
Prove: $m\angle D > m\angle P$
Proof:
Statements (Reasons)
1. $m\angle PCQ > m\angle DCQ$ (Given)
2. $\overline{CQ} \cong \overline{CQ}$ (Cong. of segments is reflexive.)
3. $\overline{QC}$ bisects $\overline{PD}$. (Given)
4. $\overline{PC} \cong \overline{CD}$ (Def. bisector)
5. $PQ > QD$ (SAS Inequality)
6. $m\angle D > m\angle P$ (If one side of a $\triangle$ is larger than another side, then the $\angle$ opp. the larger side is greater than the $\angle$ opp. the shorter side.)

For Example 3

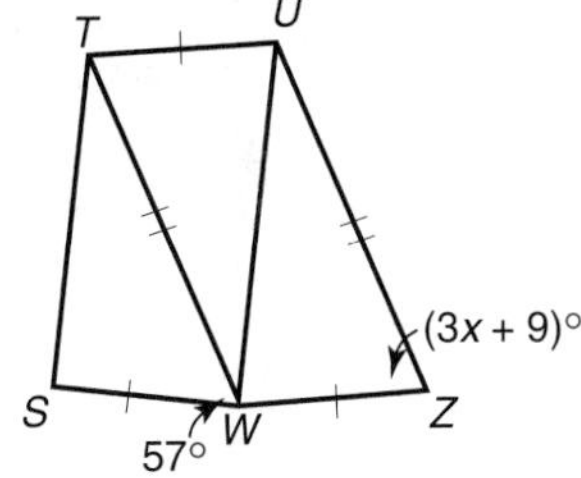

In the figure, $\overline{SW} \cong \overline{WZ} \cong \overline{TU}$, $\overline{TW} \cong \overline{UZ}$, $m\angle UTW > m\angle SWT$, and $m\angle SWT = 57$. Write two inequalities to describe the possible values for x.
$x > 16$, $x < 57$

Alternative Learning Styles

Kinesthetic Have students use their hands, their elbows, and their noses as vertices of a triangle. Tell them to move their hands toward their nose and then away from their nose while keeping their elbow in the same place. What happens to the angle formed by their elbow as their hand moves away from their nose? What happens to the angle as their hand moves toward their nose?

3 PRACTICE/APPLY

Check for Understanding

Exercises 1–13 are designed to help you assess your students' understanding through reading, writing, speaking, and modeling. You should work through Exercises 1–4 with your students and then monitor their work on Exercises 5–13.

Additional Answers

2. The SAS Inequality Theorem compares the third side of a triangle for which two sides are congruent and the included angle is different. The SAS Postulate states that two triangles that have two sides and an included angle congruent are congruent.

12. Given: $\overline{OT} \cong \overline{TV}$
T is the midpoint of $\overline{SW}$.
$m\angle STO > m\angle WTV$
$\overline{RO} \cong \overline{RV}$
Prove: $RS > RW$
Proof:
Statements (Reasons)
1. $\overline{OT} \cong \overline{TV}$
T is the midpoint of $\overline{SW}$.
$m\angle STO > m\angle WTV$
$\overline{RO} \cong \overline{RV}$ (Given)
2. $ST = TW$ (Def. midpoint)
3. $OS > VW$ (SAS Inequality)
4. $RO = RV$ (Def. $\cong$ segments)
5. $OS + RO > RV + VW$ (Addition Prop. (=))
6. $RS = OS + RO$
$RW = RV + VW$ (Segment Addition Post.)
7. $RS > RW$ (Substitution Prop. (=))

2. **Compare and contrast** the SAS Inequality Theorem to the SAS Postulate you studied in Chapter 4. See margin.
3. **Explain** why you think the SAS Inequality Theorem is subtitled the Hinge Theorem.

3. Sample answer: As the angle between the edges of a hinge gets larger, the distance between the ends of the sides gets larger.

4a. As the string gets shorter, the angle gets smaller.

4. Tie a loop at one end of a piece of string. Then run the other end of the string through two straws and back through the loop to form a triangle as shown in the figure. Measure the sides of the triangle formed and the angle between the straws. Next, gently pull on the long end of the string to form a different triangle. Measure the sides of the new triangle and the angle between the straws.

a. Describe what happens to the angle between the straws as the side determined by the string changes.
b. What theorem does this activity demonstrate? SSS Inequality

Guided Practice

Refer to the figure at the right to write an inequality relating the given pair of angle or segment measures.

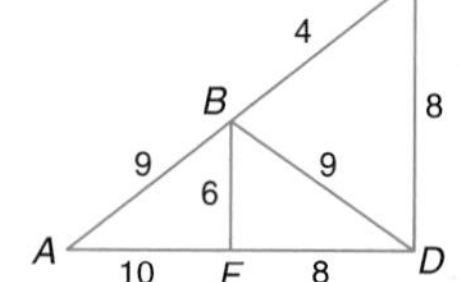

5. $m\angle DBF$, $m\angle BFA$ $m\angle DBF < m\angle BFA$
6. AB, FD $AB > FD$
7. $m\angle FDB$, $m\angle BDC$ $m\angle FDB > m\angle BDC$

In the figure, $\overline{SO}$ is a median in $\triangle SLN$, $\overline{OS} \cong \overline{NP}$, $m\angle 1 = 3x - 50$, and $m\angle 2 = x + 30$. Determine if each statement is *always true, sometimes true,* or *never true.*

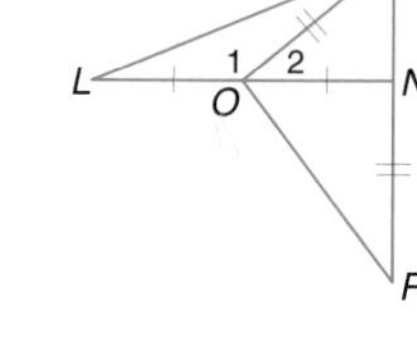

8. $LS > SN$ always
9. $SN < OP$ sometimes
10. $x = 45$ never

11. Write an inequality or pair of inequalities to describe the possible values of x.
$2.8 < x$ and $x < 12$

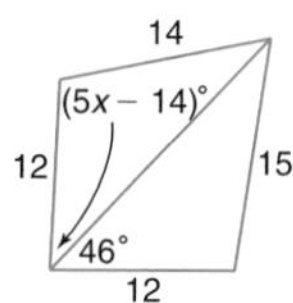

Proof

12. Write a two-column proof. See margin.
Given: $\overline{OT} \cong \overline{TV}$
T is the midpoint of $\overline{SW}$.
$m\angle STO > m\angle WTV$
$\overline{RO} \cong \overline{RV}$
Prove: $RS > RW$

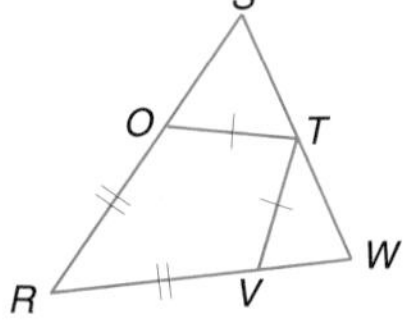

13. **Physical Therapy** Range of motion is the amount that a limb can be moved from the straight position. To determine the range of motion of a patient's arm, find the distance between the palm and the shoulder when the elbow is bent as far as possible. For one patient, the left palm was 8 inches from the shoulder, and the right palm was 3 inches from the shoulder. Which arm has the greater range of motion? right arm

Reteaching

Using Predictions Give each student two straws of different lengths. Have students use the straws as two sides of a triangle. As they vary the size of the angle included between the two straws, have them record the measure of the angle and the measure of the distance between the other ends of the straws (the length of the third side of the triangle). Have them predict a relationship between the angle and the third side.

EXERCISES

Practice **Refer to each figure to write an inequality relating the given pair of angle or segment measures.**

14. $m\angle 1, m\angle 2$ $m\angle 1 > m\angle 2$

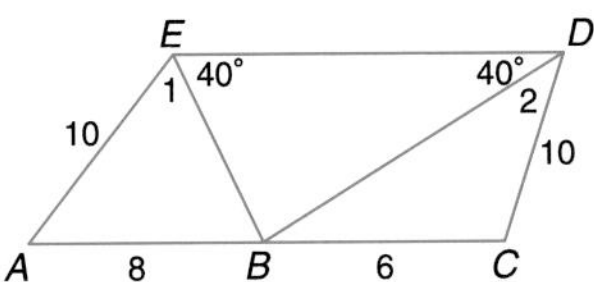

15. TM, RS $TM < RS$

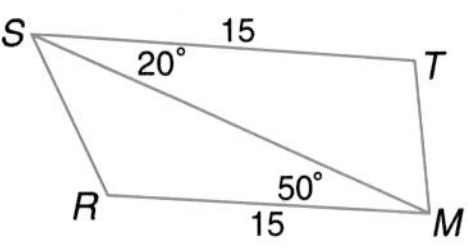

16. $m\angle ACB > m\angle ACD$

16. $m\angle ACB, m\angle ACD$

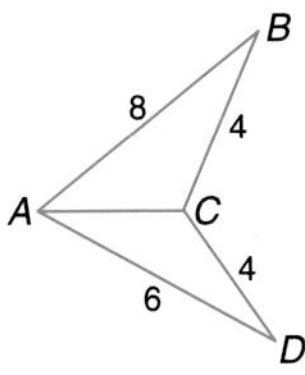

17. FH, GE $FH > GE$

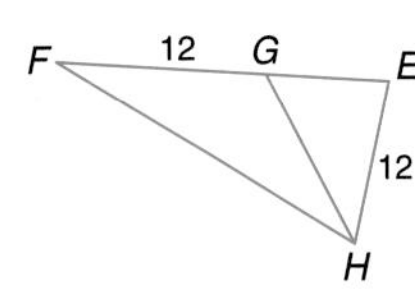

19. $m\angle AOD > m\angle AOB$

18. OC, AO $OC < AO$

19. $m\angle AOD, m\angle AOB$

20. DC, AD $DC < AD$

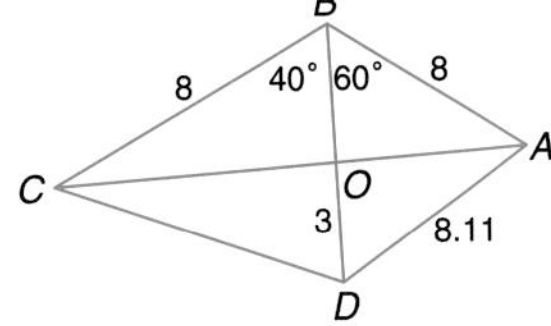

In $\triangle ABC$, $\overline{CM}$ is a median. Determine if each statement is *always true, sometimes true,* or *never true.*

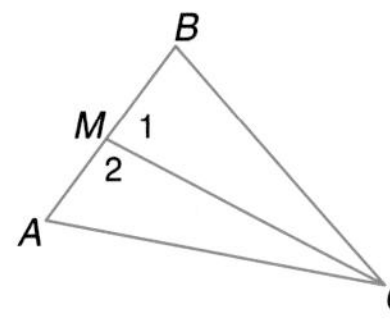

21. If $m\angle 2 < m\angle 1$, then $BC > AC$. always
22. If $m\angle B > m\angle A$, then $\angle 1$ is obtuse. sometimes
23. If $\angle 2$ is acute, then $m\angle A > m\angle B$. always
24. If $m\angle B < m\angle A$, then $m\angle 1 < m\angle 2$. never
25. If $m\angle 2 = m\angle 1$, then $BC > AC$. never
26. If $m\angle B = 90$, then $AC < BC$. never

Algebra

Write an inequality or pair of inequalities to describe the possible values of *x*.

27. $x < 14$
28. $2\frac{1}{3} < x$
29. $x > 4$

27.

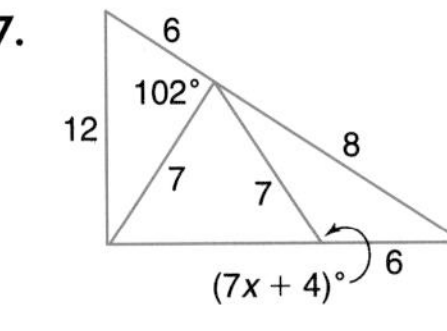

28.

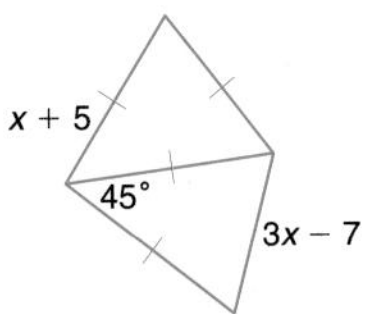

29.

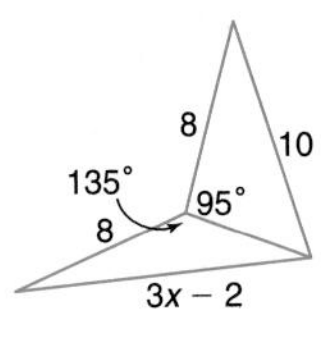

Assignment Guide

Core (with proof): 15–37 odd, 38–46
Core (informal): 15–29 odd, 35, 37–46
Enriched: 14–34 even, 35–46

For **Extra Practice,** see p. 774.

The red A, B, and C flags, printed only in the Teacher's Wraparound Edition, indicate the level of difficulty of the exercises.

Study Guide Masters, p. 31

5-6 NAME______ DATE______
Study Guide
Student Edition Pages 273–279

Inequalities Involving Two Triangles

The following two theorems are useful in determining relationships between sides and angles in triangles.

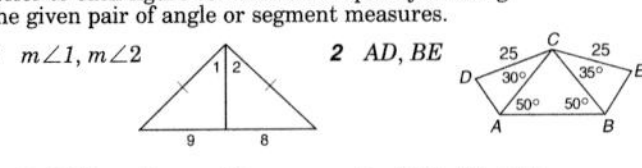

SAS Inequality (Hinge Theorem)	If two sides of one triangle are congruent to two sides of another triangle, and the included angle in one triangle is greater than the included angle in the other, then the third side of the first triangle is longer than the third side in the second triangle.
SSS Inequality	If two sides of one triangle are congruent to two sides of another triangle and the third side in one triangle is longer than the third side in the other, then the angle between the pair of congruent sides in the first triangle is greater than the corresponding angle in the second triangle.

Examples: Refer to each figure to write an inequality relating the given pair of angle or segment measures.

1 $m\angle 1, m\angle 2$ 2 AD, BE

By SSS, $m\angle 1 > m\angle 2$. By SAS, $AD < BE$.

Refer to each figure to write an inequality relating the given pair of angle or segment measures.

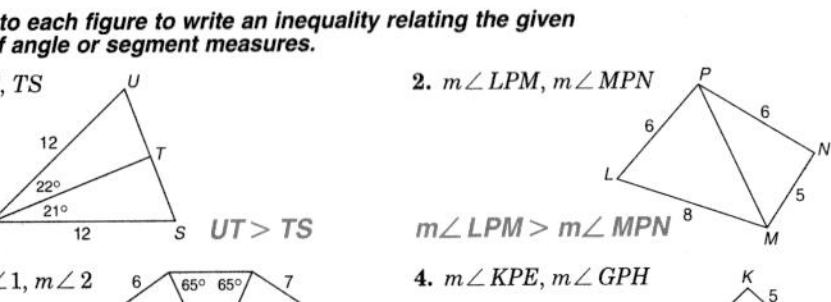

1. UT, TS $UT > TS$
2. $m\angle LPM, m\angle MPN$ $m\angle LPM > m\angle MPN$
3. $m\angle 1, m\angle 2$ $m\angle 1 < m\angle 2$
4. $m\angle KPE, m\angle GPH$ $m\angle KPE < m\angle GPH$

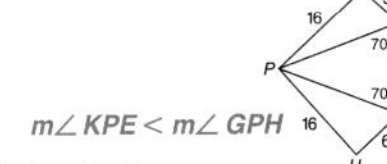

Write an inequality or pair of inequalities to describe the possible values of x.

5. $x > 7$

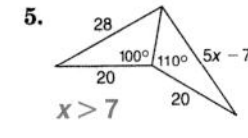

6. $6 < x$ and $x < 20$

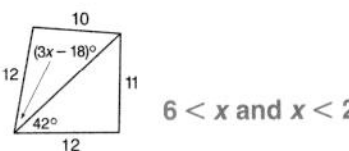

Additional Answers

30. Statements (Reasons)
 1. $\overline{DB}$ is a median of $\triangle ABC$. $m\angle 1 > m\angle 2$ (Given)
 2. D is the midpoint of $\overline{AC}$. (Def. median)
 3. $\overline{AD} \cong \overline{DC}$ (Midpoint Theorem)
 4. $\overline{DB} \cong \overline{DB}$ (Cong. of segments is reflexive.)
 5. $AB > BC$ (SAS Inequality)
 6. $m\angle C > m\angle A$ (If one side of a $\triangle$ is longer than another, the $\angle$ opp. the longer side > the $\angle$ opp. the shorter side.)

31. Statements (Reasons)
 1. $\overline{MN} \cong \overline{QR}$, $\overline{MN} \parallel \overline{QR}$, $m\angle MPQ > m\angle QPR$ (Given)
 2. $\angle MNQ \cong \angle NQR$ (Alt. Int. $\angle$s Th.)
 3. $\angle MPN \cong \angle QPR$ (Vertical $\angle$s are $\cong$.)
 4. $\triangle MPN \cong \triangle RPQ$ (AAS)
 5. $\overline{MP} \cong \overline{RP}$ (CPCTC)
 6. $\overline{PQ} \cong \overline{PQ}$ (Congruence of segments is reflexive.)
 7. $MQ > QR$ (SAS Inequality)

Practice Masters, p. 31

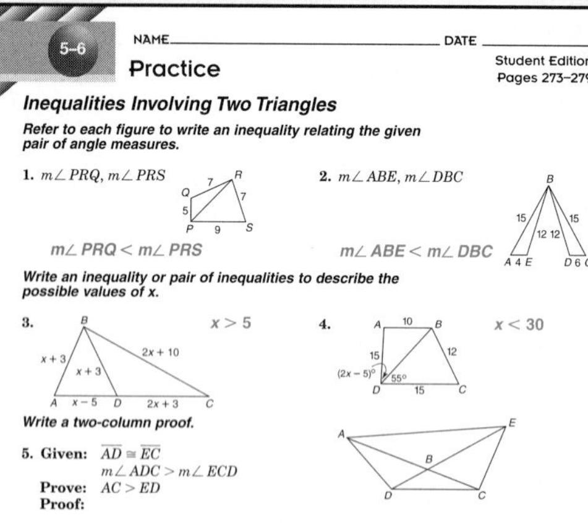
5–6 NAME______________________ DATE __________

Practice Student Edition Pages 273–279

Inequalities Involving Two Triangles

Refer to each figure to write an inequality relating the given pair of angle measures.

1. $m\angle PRQ$, $m\angle PRS$ — $m\angle PRQ < m\angle PRS$

2. $m\angle ABE$, $m\angle DBC$ — $m\angle ABE < m\angle DBC$

Write an inequality or pair of inequalities to describe the possible values of x.

3. $x > 5$

4. $x < 30$

Write a two-column proof.

5. Given: $\overline{AD} \cong \overline{EC}$, $m\angle ADC > m\angle ECD$
Prove: $AC > ED$
Proof:

Statements	Reasons
a. $AD \cong EC$	a. Given
b. $DC \cong DC$	b. Congruence of segments is reflexive.
c. $m\angle ADC > m\angle ECD$	c. Given
d. $AC > ED$	d. SAS inequality

6. Given: D is the midpoint of $\overline{AC}$. $BC > AB$
Prove: $m\angle 1 > m\angle 2$
Proof:

Statements	Reasons
a. D is the midpoint of $\overline{AC}$.	a. Given
b. $\overline{AD} \cong \overline{DC}$	b. Definition of midpoint
c. $\overline{BD} \cong \overline{BD}$	c. Congruence of segments is reflexive.
d. $\overline{BC} > \overline{AB}$	d. Given
e. $m\angle 1 > m\angle 2$	e. SSS Inequality

Proof

Write a two-column proof. 30–31. See margin.

30. Given: $\overline{DB}$ is a median of $\triangle ABC$.
$m\angle 1 > m\angle 2$
Prove: $m\angle C > m\angle A$

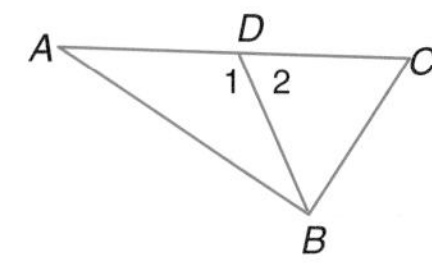

31. Given: $\overline{MN} \cong \overline{QR}$
$\overline{MN} \parallel \overline{QR}$
$m\angle MPQ > m\angle QPR$
Prove: $MQ > QR$

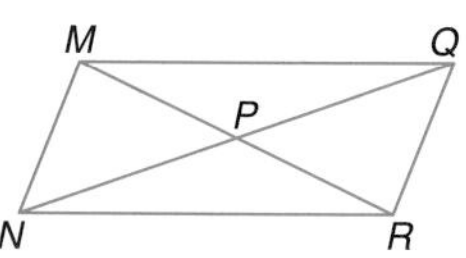

Write a paragraph proof. 32. See margin. 33. See Solutions Manual.

C

32. Given: $\overline{RM} \cong \overline{ST}$
$\overline{RM} \parallel \overline{ST}$
$ST > RS$
Prove: $m\angle ROM > m\angle MOT$

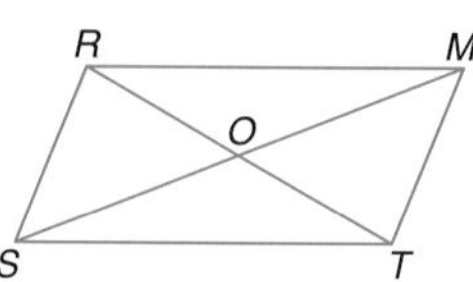

33. Given: $\triangle ABC$ is equilateral.
$\triangle ABD \cong \triangle CBD$
$BD > AD$
Prove: $m\angle BCD > m\angle DAC$

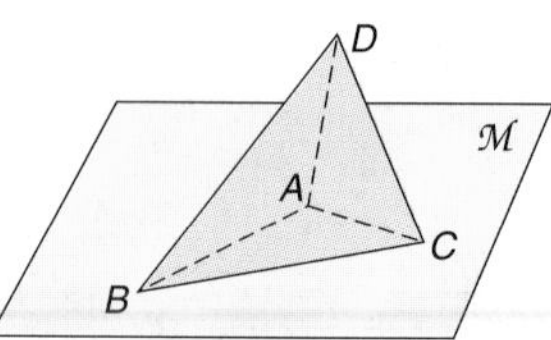

34. See Solutions Manual.

34. Write an indirect proof for the SAS Inequality. (Theorem 5–13)
Given: $\triangle ABC$ and $\triangle DEF$
$\overline{AC} \cong \overline{DF}$
$\overline{BC} \cong \overline{EF}$
$m\angle F > m\angle C$
Prove: $DE > AB$

Critical Thinking

35. Suppose that plane $\mathcal{F}$ bisects $\overline{AC}$ at B and for a point D in $\mathcal{F}$, $DC > DA$. What can you conclude about the relationship between $\overline{AC}$ and plane $\mathcal{F}$? $\overline{AC}$ is not perpendicular to plane F.

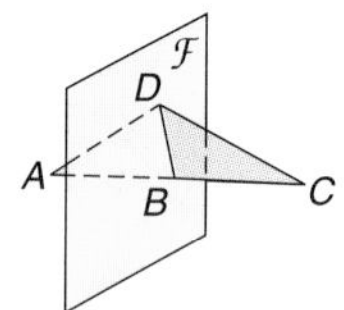

Applications and Problem Solving

36b. As the length of the stride increases, the angle formed at the hip increases.

36. Biology The fastest human can run about 23 mph for a very short distance, but many animals can run much faster. The formula $v = \frac{0.78s^{1.67}}{h^{1.17}}$, where v is the speed of the animal in meters per second, s is the length of the animal's stride in meters, and h is the height of the animal's hip in meters, can be used to estimate the speed (velocity) of an animal.

a. Use a calculator to find the velocities of two animals that each have a hip height of 1.08 meters and that have strides of 2.26 meters and 2.40 meters. 2.78 m/s, 3.07 m/s

b. Draw a mathematical model of the triangles formed by the two animals in part a if one point of the triangle represents the position of each animal's hip and the other two points represent the beginning and end of each animal's stride. Then discuss how this model is related to either the SAS Inequality or the SSS Inequality.

Additional Answer

32. We are given that $\overline{RM} \cong \overline{ST}$, $\overline{RM} \parallel \overline{ST}$, and $ST > RS$. If two parallel lines are cut by a transversal, then alternate interior angles are congruent. Thus, $\angle MRT \cong \angle RTS$ and $\angle RMS \cong \angle MST$. Therefore, $\triangle ROM \cong \triangle TOS$ by ASA. Then $\overline{RO} \cong \overline{OT}$ by CPCTC. $\overline{SO} \cong \overline{SO}$ because congruence of segments is reflexive. Therefore, $m\angle SOT > m\angle ROS$ by the SSS Inequality. Vertical angles are congruent, so $\angle SOT \cong \angle ROM$ and $\angle ROS \cong \angle MOT$. By the definition of congruent angles, $m\angle SOT = m\angle ROM$ and $m\angle ROS = m\angle MOT$. Substitution allows us to say that $m\angle ROM > m\angle MOT$.

37. **Physics** The discovery of the lever allowed ancient people to accomplish great tasks like building Stonehenge and the Egyptian pyramids. A lever multiplies the force applied to an object. One example of a lever is the nutcracker at the right. Use the SAS or SSS Inequality to explain how to operate the nutcracker. See margin.

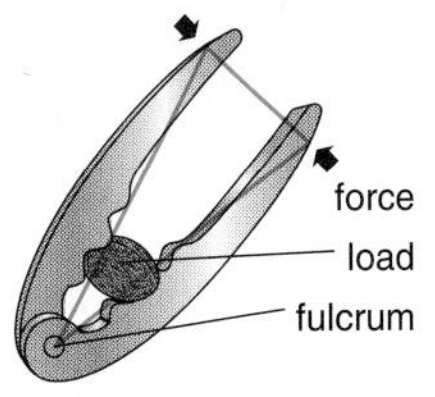

Mixed Review

38. Do the points $A(3, 1)$, $B(9, 9)$, and $C(4, 7)$ form a triangle? Explain. (Lesson 5–5)

39. **Algebra** Find the value of x and list the sides of $\triangle FGH$ in order from shortest to longest if $m\angle F = 6x + 25$, $m\angle G = 14x - 18$, and $m\angle H = 65 - 2x$. (Lesson 5–4) 6; $\overline{FG}$, $\overline{GH}$, $\overline{FH}$

40. **Sailing** A tripping line is often attached to an anchor to allow it to be dislodged from the bottom easily. What ensures that the boat will always be the same distance from the tripping line marker if both the tripping line and the anchor line are taut? (Lesson 5–2)

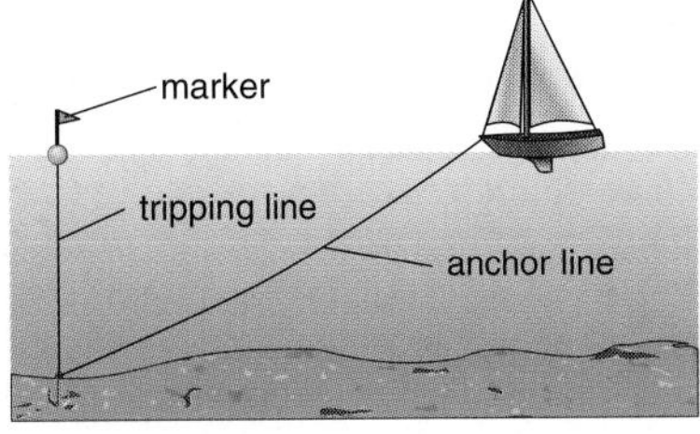

41. **Algebra** $\triangle JKL$ has vertices $J(6, 3)$, $K(-2, 7)$, and $L(9, 0)$. If $\overline{JT}$ is a median of $\triangle JKL$, what are the coordinates of T? (Lesson 5–1)

42. Refer to the figure at the right. If $m\angle NBC = 34$, find $m\angle ANB$. (Lesson 4–2) 107

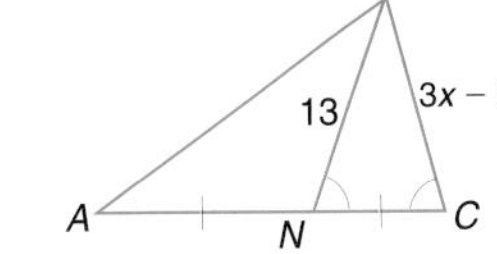

43. **Algebra** If $\triangle NQD$ has vertices $N(2, -1)$, $Q(-4, -1)$, and $D(-1, 3)$, describe $\triangle NQD$ in terms of its angles and sides. (Lesson 4–1)

44. If possible, write a conclusion. State the law of logic that you use. (Lesson 2–3)
(1) If $2x + 4 = 4x + 8$, then x can be any real number.
(2) $2x + 4 = 4x + 8$

Algebra

45. Use the distributive property to rewrite $4(8y - 5)$ without parentheses.

46. Solve $|w - 4| = 6$. $\{-2, 10\}$

45. $32y - 20$

38. Yes; the measures of $\overline{AB}$, $\overline{BC}$, and $\overline{AC}$ satisfy the Triangle Inequality.

40. The anchor line is the hypotenuse and the tripping line is the leg of a right triangle. Any triangle formed will be congruent to any other triangle formed by HL.

41. $\left(3\frac{1}{2}, 3\frac{1}{2}\right)$

43. acute, isosceles

44. x can be any real number; Law of Detachment.

WORKING ON THE In·ves·ti·ga·tion

Refer to the Investigation on pages 176–177.

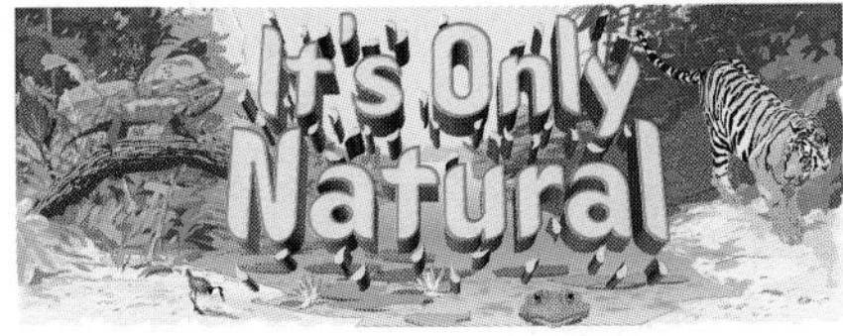

People have long been fascinated with the shape of snowflakes. The oldest recorded statement on snowflakes is found in Chinese writings from the second century B.C. Use reference material or an Internet search to research snowflakes.

1. What is the general structure of a snowflake?
2. How do chemists explain this structure?
3. How is the shape of a snowflake affected by the weather conditions?
4. Compare snowflake structure to the natural patterns described on pages 176–177. Are any of the patterns displayed in snowflakes?

Add the results of your work to your Investigation Folder.

4 ASSESS

Closing Activity

Speaking Have three to five of your students say the SAS Inequality Theorem and the SSS Inequality Theorem in their own words.

Chapter 5 Quiz D (Lesson 5-6) is available in the *Assessment and Evaluation Masters*, p. 129.

Additional Answer

37. The nutcracker is an example of an application of the SAS Inequality. As force is applied to the arms of the lever, the distance between the ends of the arms of the lever is decreased. According to the SSS Inequality, this makes the angle between the arms of the lever get smaller. This is the same angle as the one in the triangle formed by the arms of the lever and the segment between the points where the nut meets the arms of the lever. As this angle gets smaller, the segment between the points where the nut meets the arms of the lever gets shorter, thereby crushing the nut.

Enrichment Masters, p. 31

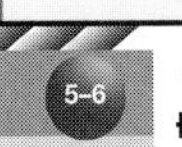

NAME__________ DATE __________

Enrichment

Student Edition Pages 273–279

Inscribed and Circumscribed Circles

The three angle bisectors of a triangle intersect in a single point called the **incenter**. This point is the center of a circle that just touches the three sides of the triangle. Except for the three points where the circle touches the sides, the circle is inside the triangle. The circle is said to be inscribed in the triangle.

With a compass and a straightedge, construct the inscribed circle for $\triangle PQR$ by following the steps below.

1. **Step 1** Construct the bisectors of $\angle P$ and $\angle Q$. Label the point where the bisectors meet A.
Step 2 Construct a perpendicular segment from A to $\overline{RQ}$. Use the letter B to label the point where the perpendicular segment intersects $\overline{RQ}$.
Step 3 Use a compass to draw the circle with center at A and radius $\overline{AB}$.

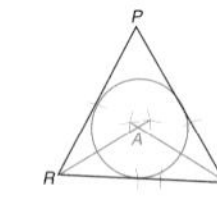

Construct the inscribed circle in each triangle.

2.

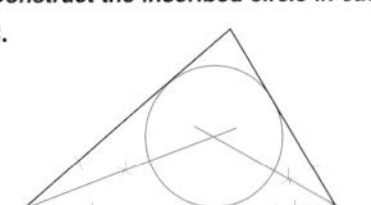

3.

The three perpendicular bisectors of the sides of a triangle also meet in a single point. This point is the center of the circumscribed circle, which passes through each vertex of the triangle. Except for the three points where the circle touches the triangle, the circle is outside the triangle.

Follow the steps below to construct the circumscribed circle for $\triangle FGH$.

4. **Step 1** Construct the perpendicular bisectors of $\overline{FG}$ and $\overline{FH}$. Use the letter A to label the point where the perpendicular bisectors meet.
Step 2 Draw the circle that has center A and radius $\overline{AF}$.

G
A
F
H

Construct the circumscribed circle for each triangle.

5.

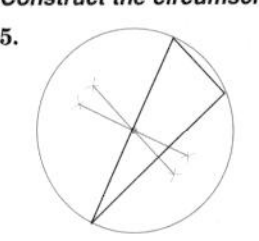

6.

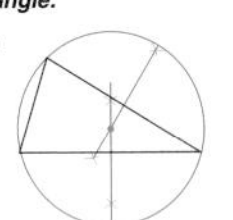

Extension

Communication The SAS Inequality Theorem is also called the Hinge Theorem. Bring in a hinge or use the hinge on a door in your classroom and have your students explain why the inequality would be called the Hinge Theorem.

In·ves·ti·ga·tion

Working on the Investigation

The Investigation on pages 176–177 is designed to be a long-term project that is completed over several days or weeks. Encourage students to keep their materials in their Investigation Folder as they work on the Investigation.

Closing the Investigation

This activity provides students an opportunity to bring their work on the Investigation to a close. For each Investigation, students should present their findings to the class. Here are some ways students can display their work.

- Conduct and report on an interview or survey.
- Write a letter, proposal, or report.
- Write an article for the school or local paper.
- Make a display, including graphs and/or charts.
- Plan an activity.

Assessment

To assess students' understanding of the concepts and topics explored in the Investigation and its follow-up activities, you may wish to examine students' Investigation Folders.

The scoring guide provided in the *Investigations and Projects Masters,* p. 7, provides a means for you to score students' work on the Investigation.

Investigations and Projects Masters, p. 7

Scoring Guide
Chapters 4 and 5
Investigation

Level	Specific Criteria
3 Superior	• Shows thorough understanding of the concepts of *patterns in nature, types of triangles, volume, surface area,* and *chemical structure.* • Uses appropriate strategies to solve problems. • Computations are correct. • Written explanations are exemplary. • Drawing, descriptions, information, and presentation are appropriate and sensible. • Goes beyond requirements of all or some problems.
2 Satisfactory, with Minor Flaws	• Shows understanding of the concepts of *patterns in nature, types of triangles, volume, surface area,* and *chemical structure.* • Uses appropriate strategies to solve problems. • Computations are mostly correct. • Written explanations are effective. • Drawing, descriptions, information, and presentation are appropriate and sensible. • Satisfies all requirements of problems.
1 Nearly Satisfactory, with Obvious Flaws	• Shows understanding of most of the concepts of *patterns in nature, types of triangles, volume, surface area,* and *chemical structure.* • May not use appropriate strategies to solve problems. • Computations are mostly correct. • Written explanations are satisfactory. • Drawing, descriptions, information, and presentation are appropriate and sensible. • Satisfies most requirements of problems.
0 Unsatisfactory	• Shows little or no understanding of the concepts of *patterns in nature, types of triangles, volume, surface area,* and *chemical structure.* • Does not use appropriate strategies to solve problems. • Computations are incorrect. • Written explanations are not satisfactory. • Drawing, descriptions, information, and presentation are not appropriate or sensible. • Does not satisfy requirements of problems.

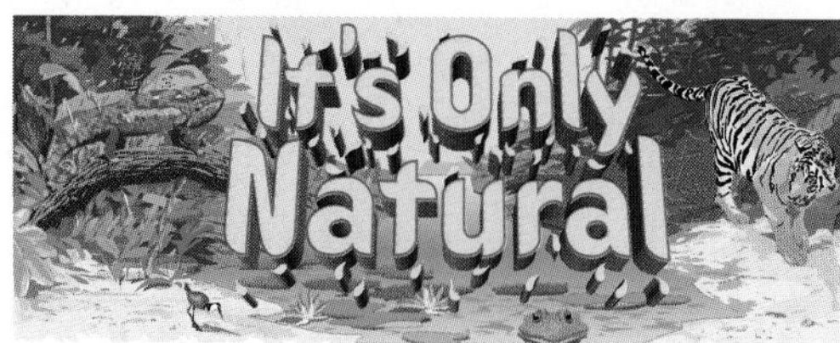

Refer to the Investigation on pages 166–167.

Once you have recognized a pattern, it is easy to find any irregularities. Scientists have spent centuries studying patterns and then finding irregularities to make discoveries about nature. For example, the planets were discovered because they broke the circular pattern of the movements of the stars. Because of this break from the pattern, the Greeks chose their name, *planetes,* meaning wanderer. Some of the patterns of nature are easy to observe. But we are just learning to recognize some of the patterns that at first appeared to be only random events.

Analyze

You have looked for and researched a pattern found in nature. It is now time to compile your findings and prepare your presentation.

PORTFOLIO ASSESSMENT

You may want to keep your work on this Investigation in your portfolio.

1 Look over the information you have gathered. If you have not already decided, determine whether you present your findings in a poster display, a slide presentation, or a video format.

2 Write a general description of the pattern you observed. Place this description at the beginning of the script for your presentation.

3 Make a list of the places that you observed your pattern. Decide upon the order for each example in your presentation.

4 Write the script to accompany your presentation. Be sure to make the script entertaining as well as informative. If possible, include information on the scientists who have done research in the field.

5 Complete the preparation for the presentation by putting together your poster, compiling your slides, or filming your video.

6 Write a brief summary of your findings to accompany the presentation.

Present

Present your findings to your class.

7 Introduce your presentation by telling how you conducted your research.

8 Explain your poster presentation, show your slides, or play your video.

9 Allow the class members to ask questions or explain how your observations apply to the pattern that they investigated.

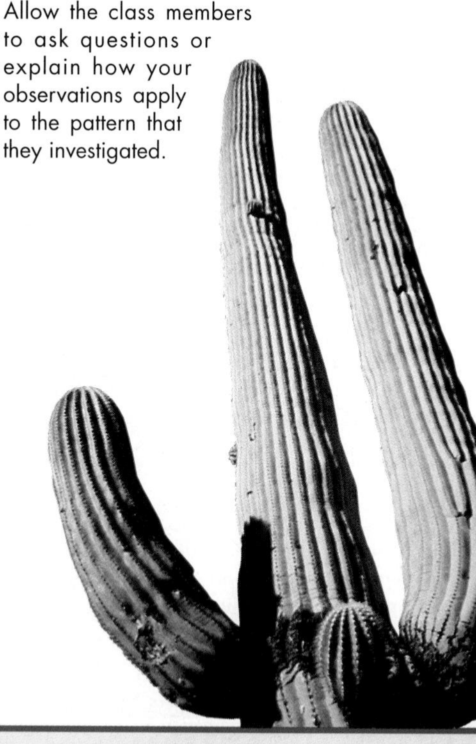

CHAPTER 5 HIGHLIGHTS

VOCABULARY

After completing this chapter, you should be able to define each term, property, or phrase and give an example or two of each.

Geometry

altitude of a triangle (p. 238)
angle bisector of a triangle (p. 240)
Exterior Angle Inequality Theorem (p. 253)
Hinge Theorem (p. 273)
Hypotenuse-Angle Theorem (p. 246)
Hypotenuse-Leg Postulate (p. 247)
indirect reasoning (p. 252)
indirect proof (p. 252)
Leg-Angle Theorem (p. 247)
Leg-Leg Theorem (p. 245)
median of a triangle (p. 238)
perpendicular bisector of a triangle (p. 238)
SAS Inequality Theorem (p. 273)
SSS Inequality Theorem (p. 274)
Triangle Inequality Theorem (p. 267)

Algebra

Addition Property of Inequality (p. 254)
Comparison Property of Inequality (p. 254)
Division Property of Inequality (p. 254)
inequality (p. 254)
Multiplication Property of Inequality (p. 254)
Subtraction Property of Inequality (p. 254)
Transitive Property of Inequality (p. 254)

Problem Solving

work backward (p. 254)

UNDERSTANDING AND USING THE VOCABULARY

Choose the term from the list that best completes each statement or phrase.

1. The _?_ verifies that if $4x > 28$, then $x > 7$. d
2. In $\triangle ABC$, D is the midpoint of $\overline{BC}$. $\overline{AD}$ is a _?_ of $\triangle ABC$. e
3. The _?_(s) and the _?_(s) of a triangle are perpendicular to the sides of the triangle. b and l
4. In right triangles XYZ and MNO, $\angle X$ and $\angle M$ are the right angles. If $\overline{XY} \cong \overline{MN}$ and $\angle Z \cong \angle O$, the _?_ justifies that the triangles are congruent. f
5. According to the _?_, in $\triangle TUV$, $TV + TU > UV$. h
6. Eliminating three of four possible answers for a multiple-choice question and then choosing the fourth choice is an example of _?_. j
7. The statement *If* $m\angle R > m\angle T$ *and* $m\angle T > m\angle V$, *then* $m\angle R > m\angle V$ is justified by the _?_. m
8. The _?_(s) of a triangle can intersect outside of the triangle. b
9. The _?_ can be proved using the Angle-Angle-Side Theorem. k
10. In $\triangle MNP$ and $\triangle IJK$, $\overline{MN} \cong \overline{IJ}$, $\overline{PN} \cong \overline{KJ}$, and $MP > IK$. Then according to the _?_, $m\angle N > m\angle J$. g

a. Multiplication Property of Inequality
b. altitude
c. Comparison Property of Inequality
d. Division Property of Inequality
e. median
f. Leg-Angle Theorem
g. SSS Inequality
h. Triangle Inequality Theorem
i. work backward
j. indirect reasoning
k. Hypotenuse-Angle Theorem
l. perpendicular bisector
m. Transitive Property of Inequality

Instructional Resources

Three multiple-choice tests and three free-response tests are provided in the *Assessment and Evaluation Masters*. Forms 1A and 2A are for honors pacing, Forms 1B and 2B are for average pacing, and Forms 1C and 2C are for basic pacing. Chapter 5 Test, Form 1B is shown at the right. Chapter 5 Test, Form 2B is shown on the next page.

Postulates, Theorems, and Corollaries

A complete list of postulates, theorems, and corollaries begins on page 806.

Using the CHAPTER HIGHLIGHTS

The Chapter Highlights begins with a listing of the new terms, properties, and phrases that were introduced in this chapter. Have students define each term and provide an example or two of it, if appropriate.

Assessment and Evaluation Masters, pp. 115–116

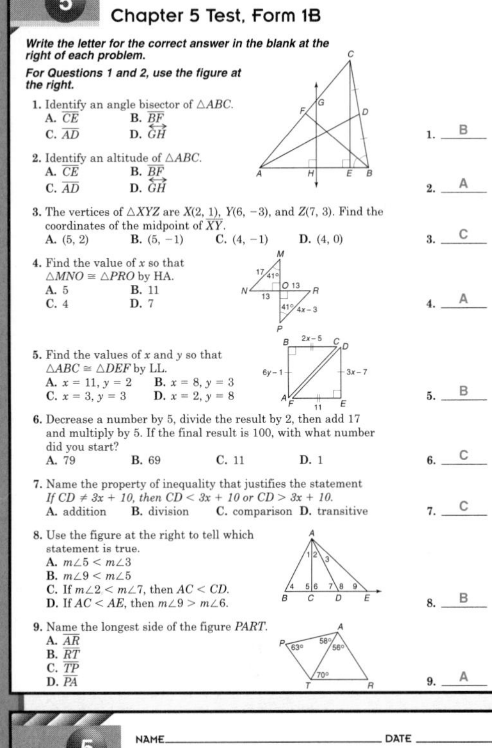

5 NAME________________ DATE ________

Chapter 5 Test, Form 1B

Write the letter for the correct answer in the blank at the right of each problem.

For Questions 1 and 2, use the figure at the right.

1. Identify an angle bisector of $\triangle ABC$.
A. $\overline{CE}$ B. $\overline{BF}$
C. $\overline{AD}$ D. $\overleftrightarrow{GH}$ 1. B

2. Identify an altitude of $\triangle ABC$.
A. $\overline{CE}$ B. $\overline{BF}$
C. $\overline{AD}$ D. $\overleftrightarrow{GH}$ 2. A

3. The vertices of $\triangle XYZ$ are $X(2, 1)$, $Y(6, -3)$, and $Z(7, 3)$. Find the coordinates of the midpoint of $\overline{XY}$.
A. (5, 2) B. (5, −1) C. (4, −1) D. (4, 0) 3. C

4. Find the value of x so that $\triangle MNO \cong \triangle PRO$ by HA.
A. 5 B. 11
C. 4 D. 7 4. A

5. Find the values of x and y so that $\triangle ABC \cong \triangle DEF$ by LL.
A. $x = 11, y = 2$ B. $x = 8, y = 3$
C. $x = 3, y = 3$ D. $x = 2, y = 8$ 5. B

6. Decrease a number by 5, divide the result by 2, then add 17 and multiply by 5. If the final result is 100, with what number did you start?
A. 79 B. 69 C. 11 D. 1 6. C

7. Name the property of inequality that justifies the statement *If* $CD \neq 3x + 10$, *then* $CD < 3x + 10$ *or* $CD > 3x + 10$.
A. addition B. division C. comparison D. transitive 7. C

8. Use the figure at the right to tell which statement is true.
A. $m\angle 5 < m\angle 3$
B. $m\angle 9 < m\angle 5$
C. If $m\angle 2 < m\angle 7$, then $AC < CD$.
D. If $AC < AE$, then $m\angle 9 > m\angle 6$. 8. B

9. Name the longest side of the figure $PART$.
A. $\overline{AR}$
B. $\overline{RT}$
C. $\overline{TP}$
D. $\overline{PA}$ 9. A

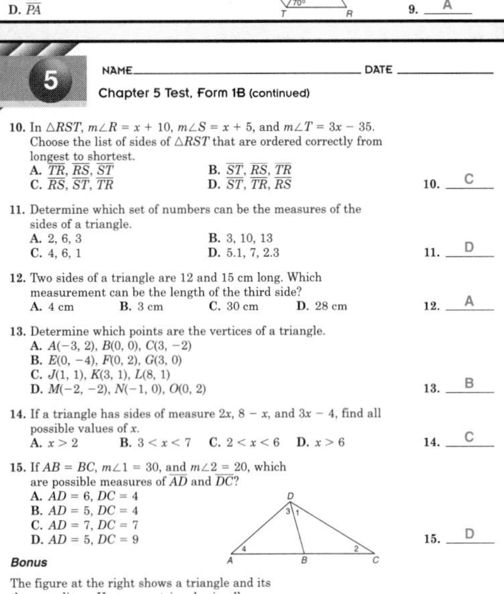

5 NAME________________ DATE ________

Chapter 5 Test, Form 1B (continued)

10. In $\triangle RST$, $m\angle R = x + 10$, $m\angle S = x + 5$, and $m\angle T = 3x - 35$. Choose the list of sides of $\triangle RST$ that are ordered correctly from longest to shortest.
A. $\overline{TR}, \overline{RS}, \overline{ST}$ B. $\overline{ST}, \overline{RS}, \overline{TR}$
C. $\overline{RS}, \overline{ST}, \overline{TR}$ D. $\overline{ST}, \overline{TR}, \overline{RS}$ 10. C

11. Determine which set of numbers can be the measures of the sides of a triangle.
A. 2, 6, 3 B. 3, 10, 13
C. 4, 6, 1 D. 5.1, 7, 2.3 11. D

12. Two sides of a triangle are 12 and 15 cm long. Which measurement can be the length of the third side?
A. 4 cm B. 3 cm C. 30 cm D. 28 cm 12. A

13. Determine which points are the vertices of a triangle.
A. $A(-3, 2)$, $B(0, 0)$, $C(3, -2)$
B. $E(0, -4)$, $F(0, 2)$, $G(3, 0)$
C. $J(1, 1)$, $K(3, 1)$, $L(8, 1)$
D. $M(-2, -2)$, $N(-1, 0)$, $O(0, 2)$ 13. B

14. If a triangle has sides of measure $2x$, $8 - x$, and $3x - 4$, find all possible values of x.
A. $x > 2$ B. $3 < x < 7$ C. $2 < x < 6$ D. $x > 6$ 14. C

15. If $AB = BC$, $m\angle 1 = 30$, and $m\angle 2 = 20$, which are possible measures of $\overline{AD}$ and $\overline{DC}$?
A. $AD = 6, DC = 4$
B. $AD = 5, DC = 4$
C. $AD = 7, DC = 7$
D. $AD = 5, DC = 9$ 15. D

Bonus

The figure at the right shows a triangle and its three medians. How many triangles in all are in the figure?

A. 7 B. 8
C. 13 D. 16 Bonus D

Using the STUDY GUIDE AND ASSESSMENT

Skills and Concepts Encourage students to refer to the objectives and examples on the left as they complete the review exercises on the right.

Assessment and Evaluation Masters, pp. 121–122

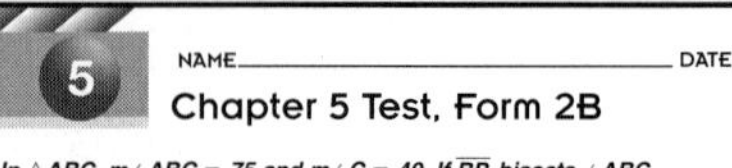

5 NAME__________ DATE ________

Chapter 5 Test, Form 2B

In $\triangle ABC$, $m\angle ABC = 75$ and $m\angle C = 40$. If $\overline{BP}$ bisects $\angle ABC$ and $\overline{BQ}$ is an altitude, find the measure of each angle.

1. $\angle PBC$ — 1. 37.5
2. $\angle CAB$ — 2. 65
3. $\angle BQA$ — 3. 90
4. If $\overline{AR}$ is a median, $BR = 2y + 8$, and $RC = 7y - 7$, find BC. — 4. 28

For Questions 5 and 6, find the values of x and y so that the triangles are congruent by the indicated theorem or postulate.

5. $\triangle PQR \cong \triangle VST$ by HA
6. $\triangle PQR \cong \triangle VST$ by LL

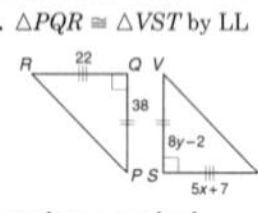

5. 6, 10
6. 3, 5

7. George used his monthly allowance to buy a notebook for \$2.50, a T-shirt for \$12.00, a movie ticket for \$3.75, and batteries for \$4.25. If George had \$3.50 left, how much was his allowance? — 7. \$26.00

Name the property of inequality that justifies each statement.

8. If $m\angle Y > 37$, then $m\angle Y \neq 37$ and $m\angle Y \not< 37$. — 8. comparison
9. If $x < 15$, then $5x < 75$. — 9. multiplication
10. If $PQ - BC > RS - BC$, then $PQ > RS$. — 10. addition
11. If $m\angle 1 > m\angle 2$ and $m\angle 3 > m\angle 1$, then $m\angle 3 > m\angle 2$. — 11. transitive

For Questions 12 and 13, refer to the figure at the right.

12. List the sides of $\triangle ABD$ in order from shortest to longest. — 12. $\overline{AD}, \overline{AB}, \overline{DB}$
13. Name the longest side of the figure $ABCD$. — 13. $\overline{DC}$
14. Determine whether it is possible to draw a triangle with sides 14 cm, 16 cm, and 15 cm. Write *yes* or *no*. — 14. yes
15. Is it possible for the points $T(3, 1)$, $U(0, -3)$, and $V(0, 2)$ to be the vertices of a triangle? — 15. yes

5 NAME__________ DATE ________

Chapter 5 Test, Form 2B (continued)

The following proof counts for Questions 16–19. Write the missing statement or reason in each blank.

Given: $\overline{AD} \cong \overline{BD}$
Prove: $DC > AD$

Statements	Reasons
a. $\overline{AD} \cong \overline{BD}$	a. Given
b. $\angle 1 \cong \angle 2$	b. (Question 16)
c. $m\angle 1 = m\angle 2$	c. Def. $\cong$ $\angle$s
d. (Question 17)	d. Exterior Angle Inequality Theorem
e. $m\angle 1 > m\angle 3$	e. (Question 18)
f. $DC > AD$	f. (Question 19)

16. Isos. △ Thm.
17. $m\angle 2 > m\angle 3$
18. Subst. Prop. (=)
19. If one ∠ of a △ is greater than another ∠, then the side opp. the greater ∠ is longer than the side opp. the lesser ∠.

20. Write a two-column proof on a separate sheet of paper.
Given: $\overline{PQ} \cong \overline{QR}$
$m\angle P > m\angle R$
Prove: $m\angle 2 > m\angle 1$

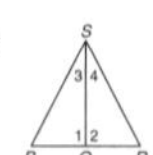

20. Sample answer:
a. $\overline{PQ} \cong \overline{QR}$ (Given)
b. $\overline{SQ} \cong \overline{SQ}$ (Congruence of segments is refl.)
c. $m\angle P > m\angle R$ (Given)
d. $SR > SP$ (If one ∠ of a △ is greater than another ∠, then the side opp. the greater ∠ is longer than the side opp. the lesser ∠.)
e. $m\angle 2 > m\angle 1$ (SSS Inequality)

Bonus

State whether the sentence is *always, sometimes,* or *never* true: "If $\triangle ABC$ is a triangle with $AB < AC < BC$, then $\angle C$ is obtuse." — Bonus never true

CHAPTER 5 STUDY GUIDE AND ASSESSMENT

SKILLS AND CONCEPTS

OBJECTIVES AND EXAMPLES

Upon completing this chapter, you should be able to:

- identify and use medians, altitudes, angle bisectors, and perpendicular bisectors of a triangle (Lesson 5–1)

If $\overline{NP}$ is an altitude of $\triangle NOQ$, then $\overline{NP} \perp \overline{OQ}$.

If $\overline{MS}$ is a median of $\triangle MNQ$, then $\overline{SN} \cong \overline{SQ}$.

If $\overline{SR}$ is an angle bisector of $\triangle MSQ$, then $\angle 1 \cong \angle 2$.

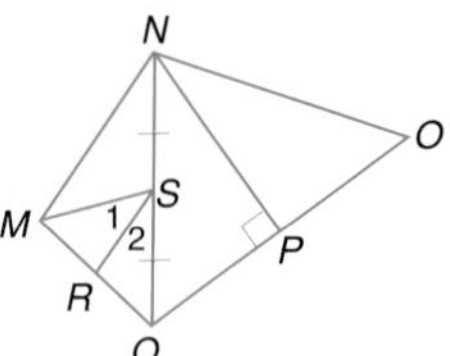

REVIEW EXERCISES

Use these exercises to review and prepare for the chapter test.

Refer to the figure at the left for Exercises 11–13. 12. 7, 58 13. No; $m\angle MSQ \neq 90$.

11. $\overline{NP}$ is an altitude and an angle bisector of $\triangle NOQ$. If $m\angle QNP = 33$, find the measures of the three angles of $\triangle NOQ$.
12. Find the value of x and $m\angle 2$ if $\overline{MS}$ is an altitude of $\triangle MNQ$, $m\angle 1 = 3x + 11$, and $m\angle 2 = 7x + 9$.
13. If $\overline{MS}$ is a median of $\triangle MNQ$, $QS = 3x - 14$, $SN = 2x - 1$, and $m\angle MSQ = 7x + 1$, is $\overline{MS}$ also an altitude of $\triangle MNQ$? Explain.

11. $m\angle NQO = m\angle NOQ = 57$, $m\angle ONQ = 66$

- recognize and use tests for congruence of right triangles (Lesson 5–2)

Find the values of x and y so that $\triangle ABC \cong \triangle DEF$.

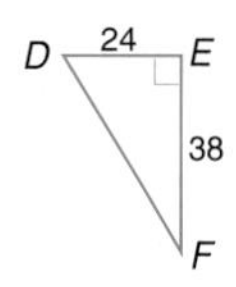

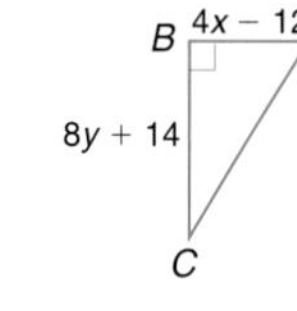

$$\begin{aligned} AB &= DE \\ 4x - 12 &= 24 \\ 4x &= 36 \\ x &= 9 \end{aligned} \qquad \begin{aligned} BC &= EF \\ 8y + 14 &= 38 \\ 8y &= 24 \\ y &= 3 \end{aligned}$$

By LL, $\triangle ABC \cong \triangle DEF$ for $x = 9$ and $y = 3$.

Find the values of *x* and *y* so that $\triangle ABC \cong \triangle XYZ$ by the indicated theorem or postulate.

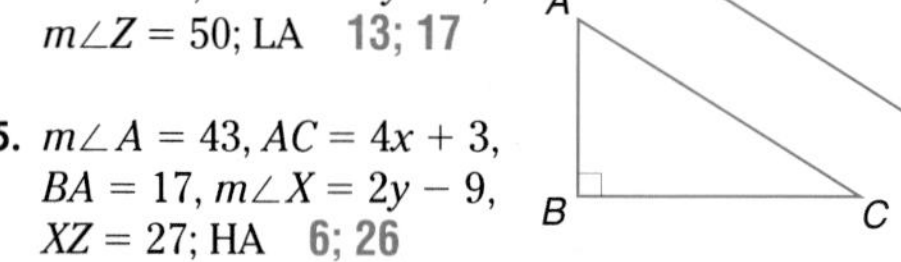

14. $AB = 40$, $XY = 3x + 1$, $AC = 36$, $m\angle C = 3y - 1$, $m\angle Z = 50$; LA 13; 17
15. $m\angle A = 43$, $AC = 4x + 3$, $BA = 17$, $m\angle X = 2y - 9$, $XZ = 27$; HA 6; 26
16. $XY = 42$, $m\angle Z = 68$, $AB = 5x - 8$, $m\angle A = 14y - 6$; LA 10; 2
17. $AC = 12y + 1$, $AB = x + 3$, $XY = 4x - 3$, $YZ = 12$, $XZ = 10y + 3$; HL 2; 1

- use indirect reasoning and indirect proof to reach a conclusion (Lesson 5–3)

Steps for Writing an Indirect Proof

1. Assume that the conclusion is false.
2. Show that the assumption leads to a contradiction of the hypothesis or some other fact, such as a postulate, theorem, or corollary.
3. Conclude that the assumption must be false, and therefore the conclusion must be true.

Write an indirect proof. 18–19. See margin.

18. **Given:** $\overline{MJ}$ does not bisect $\angle NML$. $\overline{MN} \cong \overline{ML}$
Prove: $\overline{MJ}$ is not a median of $\triangle NML$.

19. **Given:** $\triangle MJN \cong \triangle MJL$ $\overline{MJ}$ does not bisect $\angle NML$.
Prove: $\overline{MJ}$ is an altitude of $\triangle NML$.

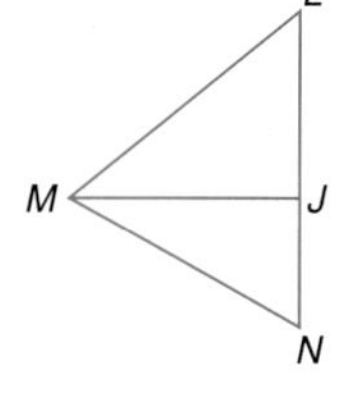

GLENCOE Technology

Test and Review Software

You may use this software, a combination of an item generator and item bank, to create your own tests or worksheets. Types of items include free response, multiple choice, short answer, and open ended.

For IBM & Macintosh

OBJECTIVES AND EXAMPLES

- recognize and apply the properties of inequalities to the measures of segments and angles (Lesson 5–3)

Find the value of x if $AC = AB$, $AX = 16$, and $AC = 3x - 2$.

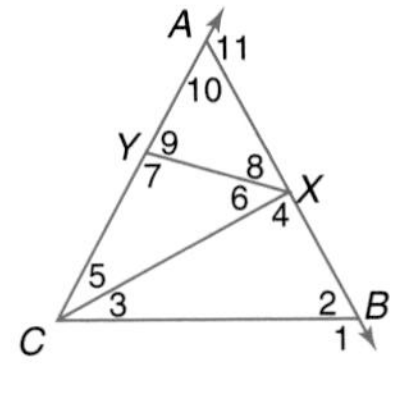

$AC = AB$

$AC = AX + XB$

$AC > AX$

$3x - 2 > 16$

$3x > 18$

$x > 6$

REVIEW EXERCISES

Refer to the figure at the left for Exercises 20–24.

20. Which is greater, $m\angle 3$ or $m\angle ACB$? $m\angle ACB$
21. Name an angle whose measure is greater than $m\angle 10$. Sample Answer: $\angle 7$
22. Which is greater, $m\angle 6$ or $m\angle 11$? $m\angle 11$
23. List all the angles whose measures are greater than $m\angle 5$. $\angle YXB$, $\angle ACB$, $\angle 9$, $\angle 4$, $\angle 11$
24. If $m\angle 7 < m\angle 4$, which is greater, $m\angle 1$ or $m\angle 8$? $m\angle 1$

- recognize and apply relationships between sides and angles in a triangle (Lesson 5–4)

What is the longest segment in figure *PQRS*? *The figure is not drawn to scale.*

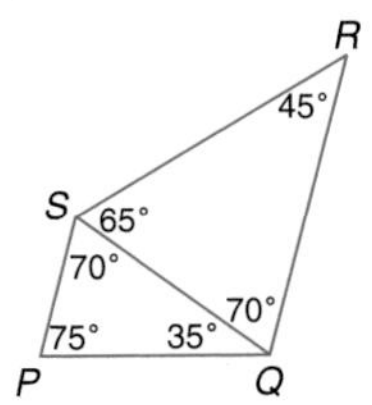

The longest side of $\triangle PQS$ is the side opposite the 75° angle, $\overline{SQ}$. This side is also the shortest side of $\triangle QRS$ since it is opposite the 45° angle. Thus, the longest side is the side opposite the 70° angle in $\triangle QRS$, $\overline{RS}$.

25. Name the shortest segment in the figure at the left. $\overline{SP}$
26. List the sides of $\triangle ABC$ in order from shortest to longest. $\overline{AC}$, $\overline{BC}$, $\overline{AB}$

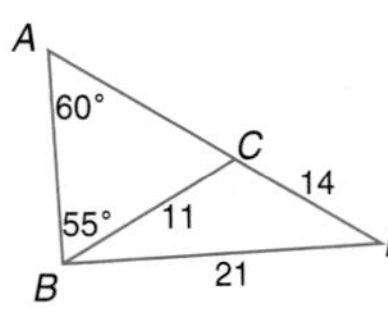

27. Write a two-column proof.

Given: $FG < FH$

Prove: $m\angle 1 > m\angle 2$

See margin.

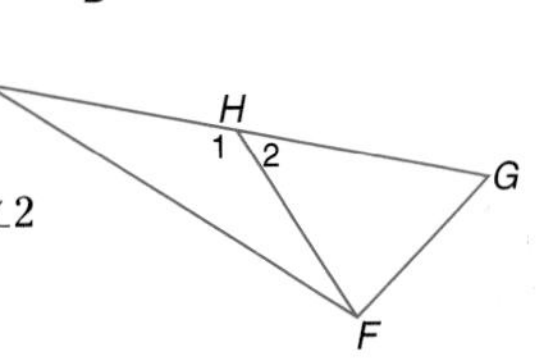

- apply the Triangle Inequality Theorem (Lesson 5–5)

In $\triangle TRI$, $TR = 27$ and $RI = 22$. Between what two numbers is TI?

By the Triangle Inequality Theorem, the following inequalities must be true.

$TI + 27 > 22$	$27 + 22 > TI$	$TI + 22 > 27$
$TI > -5$	$49 > TI$	$TI > 5$

Therefore, TI must be between 5 and 49.

The measures of two sides of a triangle are given. Between what two numbers must the measure of the third side fall?

28. 5 and 11 6, 16
29. 24 and 7 17, 31

Is it possible to have a triangle with the given vertices? Write *yes* or *no*, then explain your answer. 30–31. See margin for explanations.

30. $A(-4, 13)$, $B(5, -2)$, $C(1, 1)$ yes
31. $D(-5, 2)$, $E(1, -7)$, $F(-3, -1)$ no
32. How many different scalene triangles are possible if the measures of the three sides must be selected from the measures 2, 3, 4, and 5? 3

Additional Answers

18. We assume that $\overline{MJ}$ is a median of $\triangle LMN$. By the definition of a median, J is the midpoint of $\overline{LN}$, and hence by the definition of midpoint, $\overline{LJ} \cong \overline{JN}$. We are given that $\overline{ML} \cong \overline{MN}$ and we know that $\overline{MJ} \cong \overline{MJ}$ since congruence of segments is reflexive. Therefore, $\triangle MJL \cong \triangle MJN$ by SSS and $\angle LMJ \cong \angle NMJ$ by CPCTC. By the definition of angle bisector, $\overline{MJ}$ bisects $\triangle LMN$. But this contradicts are given statement. So, our assumption must be false, and $\overline{MJ}$ is not a median of $\triangle LMN$.
19. We assume that $\overline{MJ}$ is not an altitude of $\triangle LMN$. This means that $\overline{MJ}$ is not perpendicular to $\overline{LN}$ by the definition of altitude. So, by the definition of perpendicular, this means $\angle MJN$ is not a right angle. Therefore, $\angle MJN$ is either acute or obtuse. Since $\angle MJL$ and $\angle MJN$ form a linear pair, these two angles are suppementary. Hence, if $\angle MJL$ is acute, $\angle MJN$ is obtuse (or vice versa), by the definition of supplementary. In either case, $\triangle MJL \cong \triangle MJN$. This is a contradiction, and hence our assumption must be false. Therefore $\overline{MJ}$ is an altitude of $\triangle LMN$.
27. Statements (Reasons)
    1. $FG < FH$ (Given)
    2. $m\angle FGH > m\angle 2$ (If one side of a $\triangle$ is longer than another, the $\angle$ opp. the longer side > the $\angle$ opp. the shorter side.)
    3. $m\angle 1 > m\angle FGH$ (Exterior Angle Inequality Th.)
    4. $m\angle 1 > m\angle 2$ (Trans. Prop. of Inequality)
30. The distances between the points satisfy the triangle inequality.
31. The distances between the points do not satisfy the triangle inequality.

Applications and Problem Solving Encourage students to work through the exercises in the Applications and Problem Solving section to strengthen their problem-solving skills.

Additional Answers

37. Given: $AD = BC$
Prove: $AC > DB$
Proof:
Statements (Reasons)
1. $AD = BC$ (Given)
2. $\overline{AD} \cong \overline{BC}$ (Def. $\cong$ segments)
3. $\overline{BA} \cong \overline{BA}$ (Congruence of segments is reflexive.)
4. $m\angle CBA > m\angle DAB$ (Ext. Angle Inequality Th.)
5. $AC > DB$ (SAS Inequality Th.)

38. The brace represents the hypotenuse of a right triangle as shown in the diagram below, assuming that the deck will be attached at a right angle to the wall. If each brace is attached at the same distance from the wall, this distance represents a leg of the triangle. HL says that the four triangles will be congruent and hence all will be attached at the same distance on the wall since that distance represents the other leg of the triangle.

OBJECTIVES AND EXAMPLES

- apply the SAS Inequality and the SSS Inequality (Lesson 5–6)

Refer to the figure below to write an inequality relating AB and AC.

$m\angle AXC = 75 + 18$
$= 93$
$m\angle AXB = 180 - 93$
$= 87$

Since $m\angle AXC > m\angle AXB$, by the SAS Inequality, $AB < AC$.

REVIEW EXERCISES

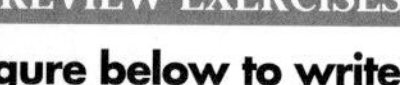

Refer to the figure below to write an inequality relating each pair of angles.

33. $m\angle ALK$, $m\angle ALN$ $\quad$ 33. $m\angle ALK < m\angle ALN$
34. $m\angle ALK$, $m\angle NLO$ $\quad$ 34. $m\angle ALK < m\angle NLO$
35. $m\angle OLK$, $m\angle NLO$ $\quad$ 35. $m\angle OLK > m\angle NLO$
36. $m\angle KLO$, $m\angle ALN$ $\quad$ $m\angle KLO = m\angle ALN$

37. Write a two-column proof. See margin.
Given: $AD = BC$
Prove: $AC > DB$

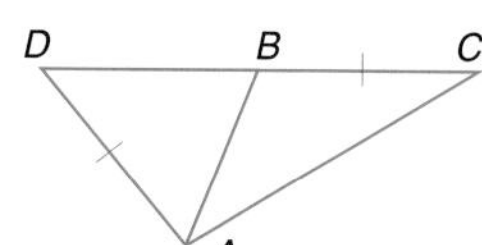

APPLICATIONS AND PROBLEM SOLVING

38. **Construction** Four wood braces of equal length are to be used to support a deck near where the deck meets the wall of a house. One end of each brace will be attached to the bottom of the deck at the same distance from the wall. Explain why the other end of each brace will be attached to the wall at the same distance from the bottom of the deck. (Lesson 5–2) See margin.

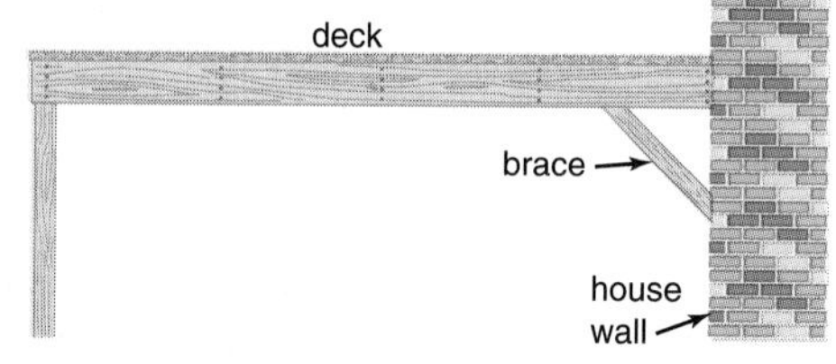

39. **Work Backward** On a game show, all the contestants begin with the same number of points. They are awarded points for questions answered correctly and points are deducted for questions answered incorrectly. Derice answered six 200-point questions correctly. Then he answered a 400-point question and a 500-point question incorrectly. In the final-round question, Derice tripled his score and won with a score of 1500 points. How many points did each player have at the beginning of the game? (Lesson 5–3) 200 points

40. **Kitchen Design** When a designer lays out a kitchen, he or she analyzes the work triangle formed by the sink, the stove, and the refrigerator. In the most efficient kitchen designs, the perimeter of the work triangle is less than 26 feet and more than 12 feet and none of the sides of the work triangle are less than 4 feet or more than 9 feet long. Mrs. Alomar would like to move the sink when the kitchen is remodeled. What are the possible distances from the stove to the sink that would meet the recommendations? (Lesson 5–5) $4 < y < 9$

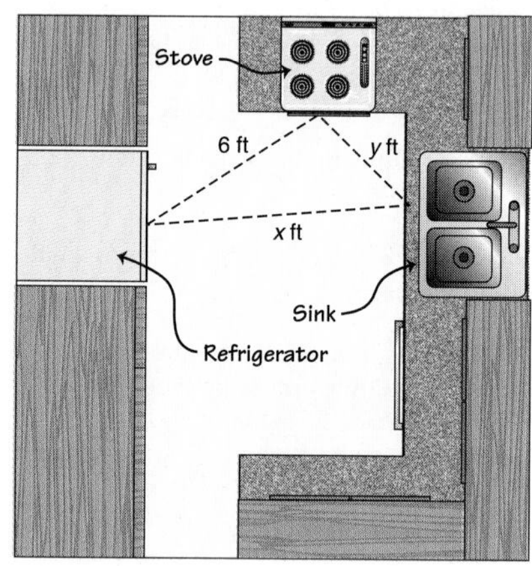

A practice test for Chapter 5 is provided on page 797.

ALTERNATIVE ASSESSMENT

COOPERATIVE LEARNING PROJECT

Projective Geometry In this project, you will apply your knowledge of congruent triangles to projections of triangles.

Follow these steps to explore projections of triangles.

- You will need a projector or similar light source, a flat wall, and construction paper.
- Cut a variety of triangles of different shapes from the construction paper.
- Position the light source so the beam is parallel to the floor.
- Hold two triangles of similar shape in front of the light source, perpendicular to the light source and parallel to the wall. Vary the distance of the triangles from the light source. See if you can find a distance at which the projected images are congruent.

Follow these steps to explore more projections of triangles.

- Now hold two cutouts in front of the projector. Keep one triangle perpendicular to the light source. Try to tilt the other triangle so that its projection is congruent to the projection of the first triangle.
- If you are unable to do this with the first triangle, try the other cutouts.
- If none of them work, note what would have to be different to make it work.
- Try to cut out a triangle that will work.

As a group, discuss what worked and what did not. Try to figure out why it worked and arrive at some general rules to follow. Can you state these rules in terms of numerical relationships between lengths of sides?

Write a report and present your findings to the class. Make your report as attractive as possible, using word-processing and drawing software if available.

THINKING CRITICALLY

Suppose that you proved a theorem from three postulates. Suppose also that you proved the negation of the theorem from the same three postulates. What would you conclude? See margin.

PORTFOLIO

Objects of the same height will project shadows of the same length at a given time of day. Have classmates of the same height stand outside in the sunlight. Make measurements of each person's height and the length of each shadow. Draw and label a triangle for each object and its shadow like the one shown below.

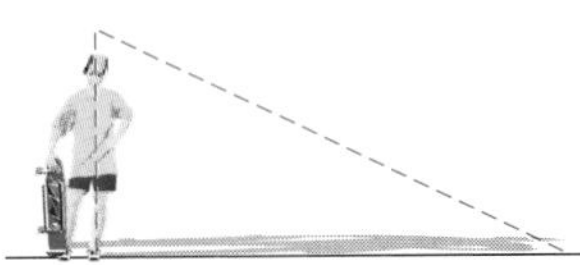

Make measurements at three different times of the day. Compile the data in a table. Determine which triangles are congruent. Keep the results in your portfolio.

SELF EVALUATION

Many theorems in geometry are stated without reference to numerical values, such as measures of angles or lengths. These are referred to as *qualitative*. One such theorem is the Hypotenuse-Leg Theorem.

Other theorems are stated using numerical values. These are referred to as *quantitative*. The SAS Inequality Theorem is a quantitative theorem.

Assess Yourself Do you find it easier to think qualitatively or quantitatively? Think about how your answer applies to other courses you study.

Answer for Thinking Critically

If you can prove a theorem and its negation from three postulates, then the postulates must contradict each other. Then one or more of the postulates should not be assumed to be true.

Assessment and Evaluation Masters, pp. 126, 137

5 NAME________________ DATE ________

Chapter 5 Performance Assessment

Instructions: ***Demonstrate your knowledge by giving a clear, concise solution to each problem. Be sure to include all relevant drawings and to justify your answers. You may show your solution in more than one way or investigate beyond the requirements of the problem.***

1. Sandra is giving Timothy information about roof trusses and braces made up of right triangles. In drafting, a T-square and triangle are used to draw right angles. Timothy wonders what information he needs to ensure that the right triangle he draws with a compass, ruler, and protractor is unique.

   a. If Sandra says that the measure of the horizontal leg is 3 and the measure of the hypotenuse is 6, does the information determine a unique right triangle? Explain. If it does, describe how Timothy can construct the triangle.

   b. Timothy knows that the measures of the acute angles of a right triangle are 47° and 43°. Does this information determine a unique triangle? Tell how you know. If it does, describe how to construct the triangle.

   c. Is a unique right triangle determined when the length of the vertical leg is 10 cm and the angle opposite the leg measures 35°? Justify your answer. If it does, describe how to construct the triangle.

   d. If Timothy is told that the two legs of a right triangle measure $4\frac{1}{2}$ in. and $2\frac{3}{4}$ in., is a unique triangle determined? Explain. If it does, tell how to construct the triangle.

   e. Sandra says that the hypotenuse of a right triangle measures 8.4 cm and the top acute angle measures 72°. Is a unique triangle determined? Tell how you know. If it is, tell how Timothy can construct the triangle.

2. a. A sculptor is using a triangular base for one of her works. To maximize stability, she wishes to attach the sculpture to the base at a point P equidistant from the sides. Draw a triangle and show how she can find P. Explain each step.

   b. As part of the sculpture, an acute triangle will be hung from a point Q equidistant from its vertices. Draw a triangle and show how Q can be located. Explain each step.

Scoring Guide
Chapter 5
Performance Assessment

Level	Specific Criteria
3 Superior	• Shows thorough understanding of the concepts of *acute angle, right triangle, legs and hypotenuse of a right triangle, angle bisector,* and *perpendicular bisector of a line segment.* • Uses appropriate strategies to justify conclusions. • Written explanations are exemplary. • Diagrams are accurate and appropriate. • Goes beyond requirements of some or all problems.
2 Satisfactory, with Minor Flaws	• Shows understanding of the concepts of *acute angle, right triangle, legs and hypotenuse of a right triangle, angle bisector,* and *perpendicular bisector of a line segment.* • Uses appropriate strategies to justify conclusions. • Written explanations are effective. • Diagrams are mostly accurate and appropriate. • Satisfies all requirements of some or all problems.
1 Nearly Satisfactory, with Serious Flaws	• Shows understanding of most of the concepts of *acute angle, right triangle, legs and hypotenuse of a right triangle, angle bisector,* and *perpendicular bisector of a line segment.* • May not use appropriate strategies to justify conclusions. • Written explanations are satisfactory. • Diagrams are mostly accurate and appropriate. • Satisfies most requirements of some or all problems.
0 Unsatisfactory	• Shows little or no understanding of the concepts of *acute angle, right triangle, legs and hypotenuse of a right triangle, angle bisector,* and *perpendicular bisector of a line segment.* • May not use appropriate strategies to justify conclusions. • Written explanations are not satisfactory. • Diagrams are not accurate or appropriate. • Does not satisfy requirements of some or all problems.

Alternative Assessment

The Alternative Assessment section provides students with the opportunity to assess their own work by thinking critically, working with others, keeping a portfolio, and honestly evaluating their own progress. For more information on alternative forms of assessment, see *Alternative Assessment in the Mathematics Classroom,* one of the titles in the Glencoe Mathematics Professional Series.

Performance Assessment

Performance Assessment tasks for this chapter are included in the *Assessment and Evaluation Masters.* A scoring guide is also provided.

In•ves•ti•ga•tion

TEACHER NOTES

NCTM Standards: 1–4, 7, 8

This Investigation is designed to be completed over several days or weeks. It may be considered optional. You may want to assign the Investigation and the follow-up activities to be completed at the same time.

Objective

Design a proposal and budget for the use of six acres of parkland.

Mathematical Overview

This Investigation will use the following mathematical skills and concepts from Chapters 6 and 7.

- finding area of various polygons
- formulating a budget
- drawing a master plan
- building a three-dimensional model from a master plan

Recommended Time		
Part	Pages	Time
Investigation	286–287	1 class period
Working on the Investigation	328, 345, 377	20 minutes each
Closing the Investigation	386	1 class period

Instructional Resources

Investigations and Projects Masters, pp. 9–12

A recording sheet, teacher notes, and scoring guide are provided for each Investigation in the *Investigations and Projects Masters*.

1 MOTIVATION

Ask students what outdoor activities they enjoy. Do nearby parks meet their needs? Why or why not? How would they change the parks if given a chance?

LONG-TERM PROJECT

In•ves•ti•ga•tion

MATERIALS NEEDED

calculator 

posterboard 

construction paper 

glue 

grid paper 

markers 

modeling clay 

paint

ruler

scissors 

masking tape

According to historians, the ancient Sumerians of Mesopotamia created the first parks around 2300 B.C. These early parks were private areas set aside by the wealthy for hunting or gardening. The first public parks were probably constructed in Greece. In 1634, the first public park in colonial North America, Boston Common, was established. This park is still a popular recreational area. The National Park Service was founded in 1916 to administer the growing park system in the United States. The system oversees 366 sites, including historic sites, battlefields, lakeshores, memorials, and recreation areas. State and local governments also administer parkland in their area.

Mason County requires that a developer provide one acre of park land for every 100 acres of land that is used for residential or commercial building. The Heflin Development Company is planning a 600-acre residential and business complex on the county's north side, so six acres will be used as park land. The land that has been chosen was previously used as farmland, so it is flat with few trees. The company will allow the designer to choose the shape of the six-acre lot.

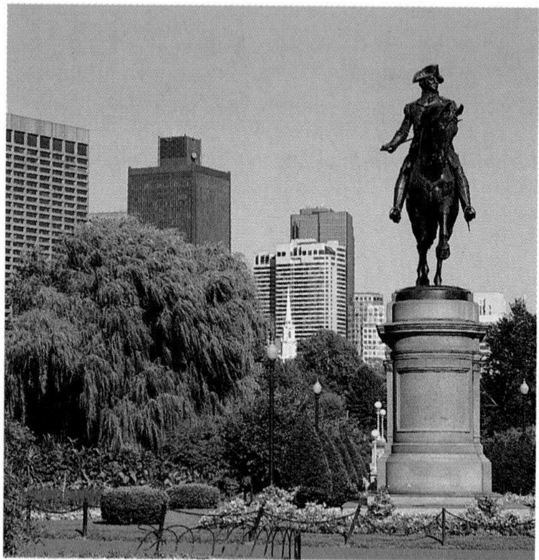

Cooperative Learning

This Investigation offers an excellent opportunity for using cooperative learning groups. For more information on cooperative learning strategies and group management, see *Cooperative Learning in the Mathematics Classroom*, one of the titles in the Glencoe Mathematics Professional Series.

BACKGROUND INFORMATION

The county commissioners have conducted a survey of 1000 residents on possible park features. The results of the survey are given in the chart.

Park Feature	Percent of Respondents Interested
Basketball courts	37%
Bike/jogging path	81%
Community gardens	31%
Fishing pond	6%
Frisbee golf course	4%
Nature trails	74%
Open space	22%
Picnic area	68%
Playground equipment	78%
Racquetball courts	8%
Rose garden	12%
Soccer/football fields	66%
Softball fields	42%
Swimming pool	51%
Tennis courts	48%

Your team of designers has been asked to submit a proposal on how to use the parkland. Your design should incorporate as many of the requested features as possible and meet the desires of as many residents as possible. The budget for the project is limited to $2 million. This budget will pay for design and completion of all elements of the park. If your company's design is chosen, you will supervise the construction of the park.

RESEARCHING THE SITE

Begin by researching the project. Some questions to consider are:

- How much does each park feature cost to construct?
- How much space is available?
- What are the dimensions of each feature?
- Can any park features be combined, such as using the same area for softball and soccer fields during different seasons?
- Will any of the park features require special lighting or safety equipment?
- What features are required that are not on the list, such as restroom facilities and parking space?
- How much maintenance will be required for the park after construction is completed?
- What will be the staff requirements for the park? Will you need rangers, lifeguards, or maintenance personnel? Can any of the positions be filled by volunteers?
- What is a reasonable fee for you to charge for your design and supervisory work?

You will continue working on this Investigation throughout Chapters 6 and 7.

Be sure to keep your research, designs, and other materials in your Investigation Folder.

This Land Is Your Land Investigation

Working on the Investigation
Lesson 6–5, p. 328

Working on the Investigation
Lesson 7–1, p. 345

Working on the Investigation
Lesson 7–5, p. 377

Closing the Investigation
End of Chapter 7, p. 386

2 SETUP

Have one student read aloud the first three paragraphs and park feature box of the Investigation to provide background information for the park designing project. Read the remaining paragraphs that explain the class's research tasks. Discuss the activity with students. Separate the class into groups of four of five.

3 MANAGEMENT

Each group member should be responsible for a specific task.

Recorder Keeps track of all information.

Architect Oversees the master plan design.

Researcher Coordinates the group's research efforts.

Builder Oversees the building of the three-dimensional model.

Sample Answers

Answers will vary as they depend on group designs.

Investigations and Projects Masters, p. 12

6, 7 NAME ______________________ DATE __________

Investigation, Chapters 6 and 7

Student Edition Pages 286–287, 328, 345, 377, 386

This Land Is Your Land

Work with your group to consider these and other possible questions.

- Are all features of the park designed to be accessible to handicapped people?
- What are the building or fire codes for each feature?
- Where will telephones be located?
- Must drinking fountains or a water source be provided?

Use these charts to record the calculated costs.

Required Features

Feature	Dimensions	Area Required	Estimated Cost

Desirable Features

Feature	Dimensions	Area Required	Estimated Cost

Use the space below to list other important considerations for design of the park.

____

____

____

____

Please keep this page and any other research in your Investigation Folder.

CHAPTER 6

Exploring Quadrilaterals

PREVIEWING THE CHAPTER

In this chapter, students are introduced to parallelograms, rhombi, rectangles, squares, and trapezoids. Students use technology to investigate these quadrilaterals. Problems are solved by identifying subgoals. Students use the properties of parallelograms, rhombi, rectangles, squares, and trapezoids to solve problems. Students also apply these quadrilaterals to real examples such as kites.

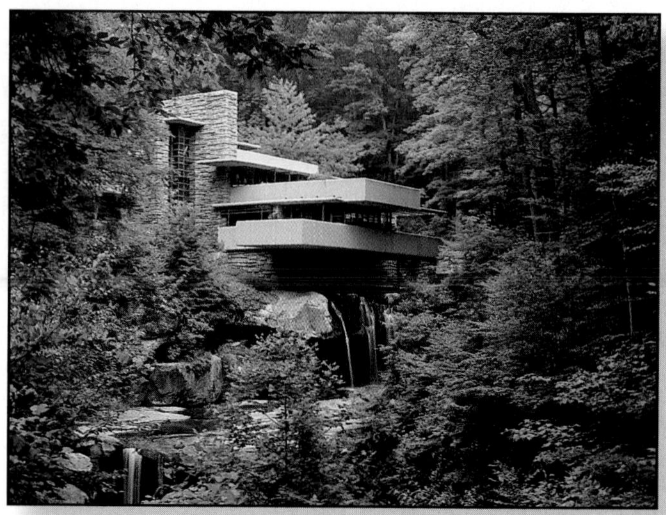

Lesson (Pages)	Lesson Objectives	NCTM Standards	State/Local Objectives
6-1A (290)	Use a TI-92 calculator to draw parallelograms.	1–4, 7, 8	
6-1 (291–297)	Recognize and apply the properties of a parallelogram. Find the probability of an event.	1–5, 7, 8	
6-2 (298–304)	Recognize and apply the conditions that ensure a quadrilateral is a parallelogram. Identify and use subgoals in writing proofs.	1–5, 7, 8	
6-3A (305)	Use a TI-92 calculator to draw and explore characteristics of rectangles.	1–4, 7, 8	
6-3 (306–312)	Recognize and apply the properties of rectangles.	1–5, 7, 8	
6-4 (313–319)	Recognize and apply the properties of squares and rhombi.	1–5, 7, 8	
6-4B (320)	Construct quadrilaterals with exactly two distinct pairs of adjacent congruent sides.	1–4, 7	
6-5 (321–328)	Recognize and apply the properties of trapezoids.	1–5, 7, 8	

ORGANIZING THE CHAPTER

A complete, 1-page lesson plan is provided for each lesson in the ***Lesson Planning Guide***. Answer keys for each lesson are available in the ***Answer Key Masters***.

You may want to refer to the **Course Planning Calendar** on page T12 for detailed information on pacing.
PACING: Standard—13 days; **Honors**—12 days; **Block**—6 days

LESSON PLANNING CHART

			Blackline Masters									
Lesson (Pages)	**Materials/ Manipulatives**	**Extra Practice (Student Edition)**	**Study Guide**	**Practice**	**Enrichment**	**Assessment & Evaluation**	**Modeling Mathematics**	**Multicultural Activity**	**Tech Prep Applications**	**Graphing Calc. & Computer**	**Real-World Applications**	**Teaching Transparencies**
6-1A (290)	TI-92 calculator									p. 24		
6-1 (291–297)	patty paper straightedge*	p. 774	p. 32	p. 32	p. 32		pp. 35–38					6-1A 6-1B
6-2 (298–304)	straws scissors* pipe cleaners ruler* compass* straightedge*	p. 774	p. 33	p. 33	p. 33	p. 156		p. 11	p. 11		11	6-2A 6-2B
6-3A (305)	TI-92 calculator									p.25		
6-3 (306–312)	spreadsheet software compass* straightedge*	p. 775	p. 34	p. 34	p. 34	pp. 155,156						6-3A 6-3B
6-4 (313–319)	compass* straightedge* ruler*	p. 775	p. 35	p. 35	p. 35	p. 157	p. 84				12	6-4A 6-4B
6-4B (320)	compass* ruler* protractor*						p.94					
6-5 (321–328)	compass* straightedge* calculator	p. 775	p. 36	p. 36	p. 36	p. 157		p. 12	p. 12	p. 6		6-5A 6-5B
Study Guide/ Assessment (329–335)						pp. 141–154, 158–160						

*Included in Glencoe's High School Manipulative Kit and Overhead Manipulative Resources.

ORGANIZING THE CHAPTER

OTHER CHAPTER RESOURCES

Student Edition
Investigation, pp. 286–287
Chapter Opener, pp. 288–289
Mathematics and Society, p. 304
Working on the Investigation, p. 328

Teacher's Classroom Resources
Investigations and Projects Masters, pp. 45–48
Block Scheduling Booklet

Technology
Test and Review Software (IBM and Macintosh)
CD-ROM Multimedia Applications (Windows and Macintosh)
Mindjogger Videoquizzes (VHS)

Professional Publications
Glencoe Mathematics Professional Series

OUTSIDE RESOURCES

Books/Periodicals
Visualized Geometry, ETA
Geometry from Multiple Perspectives, NCTM

**Software**
Geometry Inventor, Sunburst
Shape Up, Sunburst

Videos/CD-ROMs
The Geometry of Parallel Lines, Geometric Figures, the Parallelogram, and Circles, NASCO, 901 Janesville Ave., Fort Atkinson, WI 53538
The Graphing Calculator: Building New Models, Teaching Mathematics with Calculators: A National Workshop, Education Division, 2809 Ross Avenue, Dallas, TX 75201

ASSESSMENT RESOURCES

Student Edition
Math Journal, pp. 301, 309, 316
Mixed Review, pp. 297, 303, 312, 319, 327
Self Test, p. 312
Chapter Highlights, p. 329
Chapter Study Guide and Assessment, pp. 330–332
Alternative Assessment, p. 333
Portfolio, p. 333
College Entrance Exam Practice, pp. 334–335
Chapter Test, p. 798

Teacher's Wraparound Edition
5-Minute Check, pp. 291–292, 298, 306, 313, 321
Check for Understanding, pp. 295, 301, 309, 316, 324
Closing Activity, pp. 297, 304, 312, 319, 328
Cooperative Learning, pp. 292, 314

Assessment and Evaluation Masters
Multiple-Choice Tests, Forms 1A (Honors), 1B (Average), 1C (Basic), pp. 141–146
Free-Response Tests, Forms 2A (Honors), 2B (Average), 2C (Basic), pp. 147–152
Calculator-Based Test, p. 153
Performance Assessment, p. 154
Mid-Chapter Test, p. 155
Quizzes A–D, pp. 156–157
Standardized Test Practice, p. 158
Cumulative Review, pp. 159–160

ENHANCING THE CHAPTER

Examples of some of the materials for enhancing Chapter 6 are shown below.

DIVERSITY

Multicultural Activity Masters, pp. 11, 12

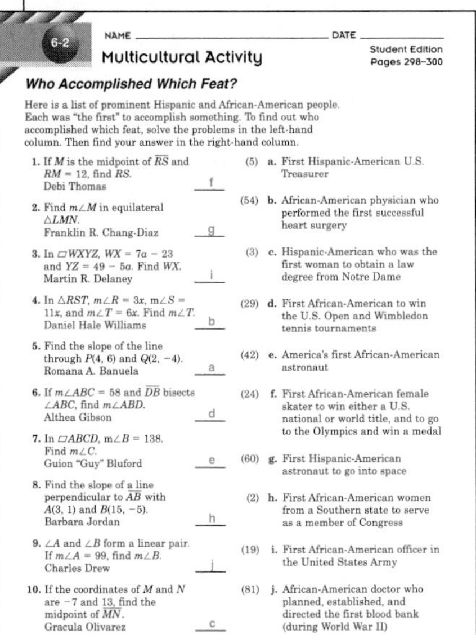

6-2 NAME ______________ DATE ______

Multicultural Activity — Student Edition Pages 298–300

Who Accomplished Which Feat?

Here is a list of prominent Hispanic and African-American people. Each was "the first" to accomplish something. To find out who accomplished which feat, solve the problems in the left-hand column. Then find your answer in the right-hand column.

1. If M is the midpoint of $\overline{RS}$ and $RM = 12$, find RS. Debi Thomas ___f___
2. Find $m\angle M$ in equilateral $\triangle LMN$. Franklin R. Chang-Diaz ___g___
3. In $\square WXYZ$, $WX = 7a - 23$ and $YZ = 49 - 5a$. Find WX. Martin R. Delaney ___i___
4. In $\triangle RST$, $m\angle R = 3x$, $m\angle S = 11x$, and $m\angle T = 6x$. Find $m\angle T$. Daniel Hale Williams ___b___
5. Find the slope of the line through $P(4, 6)$ and $Q(2, -4)$. Romana A. Banuela ___a___
6. If $m\angle ABC = 58$ and $\overline{DB}$ bisects $\angle ABC$, find $m\angle ABD$. Althea Gibson ___d___
7. In $\square ABCD$, $m\angle B = 138$. Find $m\angle C$. Guion "Guy" Bluford ___e___
8. Find the slope of a line perpendicular to $\overline{AB}$ with $A(3, 1)$ and $B(15, -5)$. Barbara Jordan ___h___
9. $\angle A$ and $\angle B$ form a linear pair. If $m\angle A = 99$, find $m\angle B$. Charles Drew ___j___
10. If the coordinates of M and N are -7 and 13, find the midpoint of $\overline{MN}$. Gracula Olivarez ___c___

(5) **a.** First Hispanic-American U.S. Treasurer

(54) **b.** African-American physician who performed the first successful heart surgery

(3) **c.** Hispanic-American who was the first woman to obtain a law degree from Notre Dame

(29) **d.** First African-American to win the U.S. Open and Wimbledon tennis tournaments

(42) **e.** America's first African-American astronaut

(24) **f.** First African-American female skater to win either a U.S. national or world title, and to go to the Olympics and win a medal

(60) **g.** First Hispanic-American astronaut to go into space

(2) **h.** First African-American women from a Southern state to serve as a member of Congress

(19) **i.** First African-American officer in the United States Army

(81) **j.** African-American doctor who planned, established, and directed the first blood bank (during World War II)

APPLICATIONS

Real-World Applications, 11, 12

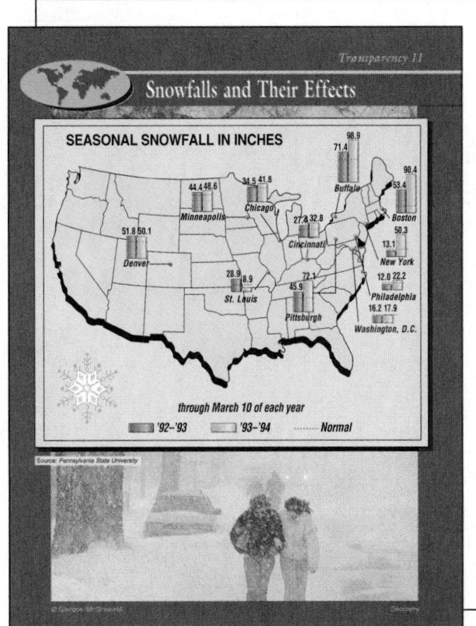

TECHNOLOGY

Graphing Calculator and Computer Masters, p. 6

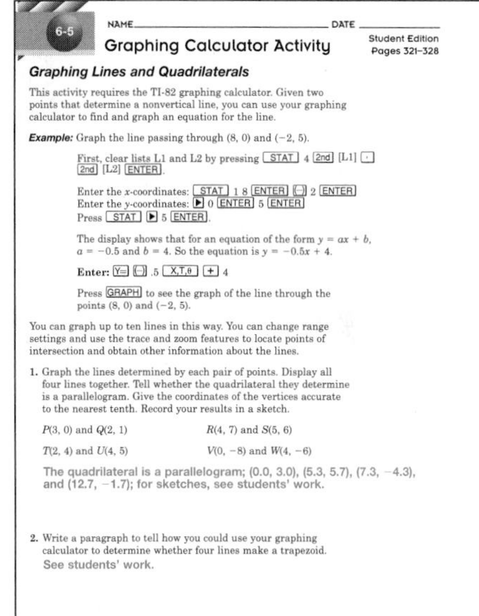

6-5 NAME ______________ DATE ______

Graphing Calculator Activity — Student Edition Pages 321–328

Graphing Lines and Quadrilaterals

This activity requires the TI-82 graphing calculator. Given two points that determine a nonvertical line, you can use your graphing calculator to find and graph an equation for the line.

Example: Graph the line passing through (8, 0) and (−2, 5).

First, clear lists L1 and L2 by pressing [STAT] 4 [2nd] [L1] [,] [2nd] [L2] [ENTER].

Enter the x-coordinates: [STAT] 1 8 [ENTER] [(−)] 2 [ENTER]
Enter the y-coordinates: [▶] 0 [ENTER] 5 [ENTER]
Press [STAT] [▶] 5 [ENTER].

The display shows that for an equation of the form $y = ax + b$, $a = -0.5$ and $b = 4$. So the equation is $y = -0.5x + 4$.

Enter: [Y=] [(−)] .5 [X,T,θ] [+] 4

Press [GRAPH] to see the graph of the line through the points (8, 0) and (−2, 5).

You can graph up to ten lines in this way. You can change range settings and use the trace and zoom features to locate points of intersection and obtain other information about the lines.

1. Graph the lines determined by each pair of points. Display all four lines together. Tell whether the quadrilateral they determine is a parallelogram. Give the coordinates of the vertices accurate to the nearest tenth. Record your results in a sketch.

   $P(3, 0)$ and $Q(2, 1)$ $R(4, 7)$ and $S(5, 6)$

   $T(2, 4)$ and $U(4, 5)$ $V(0, -8)$ and $W(4, -6)$

   The quadrilateral is a parallelogram; (0.0, 3.0), (5.3, 5.7), (7.3, −4.3), and (12.7, −1.7); for sketches, see students' work.

2. Write a paragraph to tell how you could use your graphing calculator to determine whether four lines make a trapezoid. See students' work.

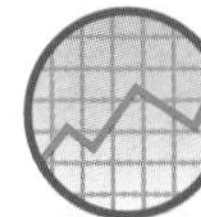

TECH PREP

Tech Prep Applications Masters, pp. 11, 12

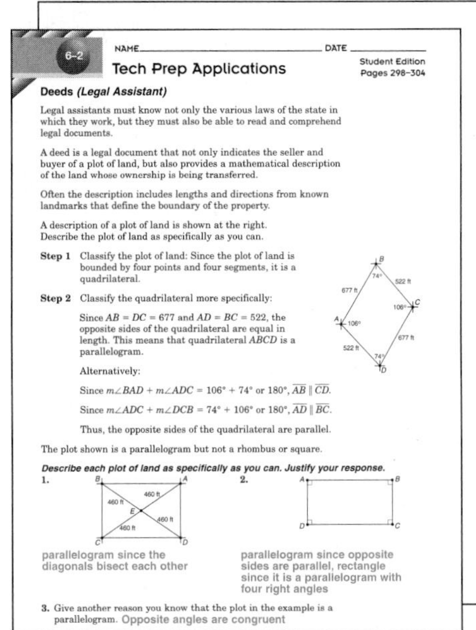

6–2 NAME ______________ DATE ______

Tech Prep Applications — Student Edition Pages 298–304

Deeds *(Legal Assistant)*

Legal assistants must know not only the various laws of the state in which they work, but they must also be able to read and comprehend legal documents.

A deed is a legal document that not only indicates the seller and buyer of a plot of land, but also provides a mathematical description of the land whose ownership is being transferred.

Often the description includes lengths and directions from known landmarks that define the boundary of the property.

A description of a plot of land is shown at the right. Describe the plot of land as specifically as you can.

Step 1 Classify the plot of land: Since the plot of land is bounded by four points and four segments, it is a quadrilateral.

Step 2 Classify the quadrilateral more specifically:

Since $AB = DC = 677$ and $AD = BC = 522$, the opposite sides of the quadrilateral are equal in length. This means that quadrilateral $ABCD$ is a parallelogram.

Alternatively:

Since $m\angle BAD + m\angle ADC = 106° + 74°$ or $180°$, $\overline{AB} \parallel \overline{CD}$.

Since $m\angle ADC + m\angle DCB = 74° + 106°$ or $180°$, $\overline{AD} \parallel \overline{BC}$.

Thus, the opposite sides of the quadrilateral are parallel.

The plot shown is a parallelogram but not a rhombus or square.

Describe each plot of land as specifically as you can. Justify your response.

1. parallelogram since the diagonals bisect each other
2. parallelogram since opposite sides are parallel, rectangle since it is a parallelogram with four right angles
3. Give another reason you know that the plot in the example is a parallelogram. Opposite angles are congruent

PROBLEM SOLVING

Problem-of-the-Week Cards, 15, 16, 17

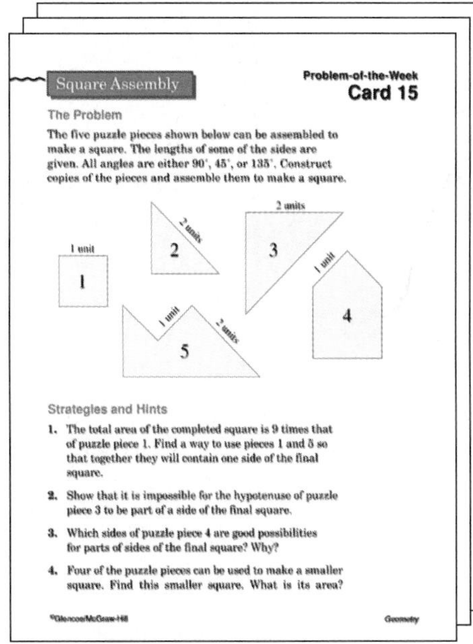

Square Assembly — Problem-of-the-Week **Card 15**

The Problem

The five puzzle pieces shown below can be assembled to make a square. The lengths of some of the sides are given. All angles are either 90°, 45°, or 135°. Construct copies of the pieces and assemble them to make a square.

Strategies and Hints

1. The total area of the completed square is 9 times that of puzzle piece 1. Find a way to use pieces 1 and 5 so that together they will contain one side of the final square.
2. Show that it is impossible for the hypotenuse of puzzle piece 3 to be part of a side of the final square.
3. Which sides of puzzle piece 4 are good possibilities for parts of sides of the final square? Why?
4. Four of the puzzle pieces can be used to make a smaller square. Find this smaller square. What is its area?

 Geometry

MAKING MATHEMATICS RELEVANT

This two-page introduction to the chapter provides students with an opportunity to explore an application of geometry that has historic roots as well as modern uses.

Background Information

A Stitch in Time Quilting refers to the stitching together of two layers of cloth with a layer of soft, fluffy material, such as batting, between the layers. Patchwork quilt tops are often pieced together from quilt blocks of a similar pattern. For instance, the design at the right is a patchwork quilt block called *Bachelor's Puzzle* whose pattern was first published in the *Kansas City Star* in 1931. Ask students to identify and classify the various geometric shapes used in this quilt block.

CHAPTER 6

Exploring Quadrilaterals

Objectives

In this chapter, you will:

- recognize and define parallelograms, rhombi, rectangles, squares, and trapezoids,
- solve problems by identifying subgoals, and
- use the properties of parallelograms, rhombi, rectangles, squares, and trapezoids to solve problems.

A Stitch in Time

Geometry: Then and Now Quiltmaking has become a popular art form. However, it began for practical purposes. Chinese, Russian, and Central American people first used quilted clothing for warmth and protection. The idea of quilting was adopted from the Europeans after the Crusades of the Middle Ages and was later carried to the American colonies. New fabric was scarce in the colonies, so quilting flourished. Quiltmakers stitched together small bits of fabric, often scraps or parts of worn-out clothing, to make patterned quilt tops. Today, quilters use new fabric, special quilting templates, and sewing machines to construct their creations.

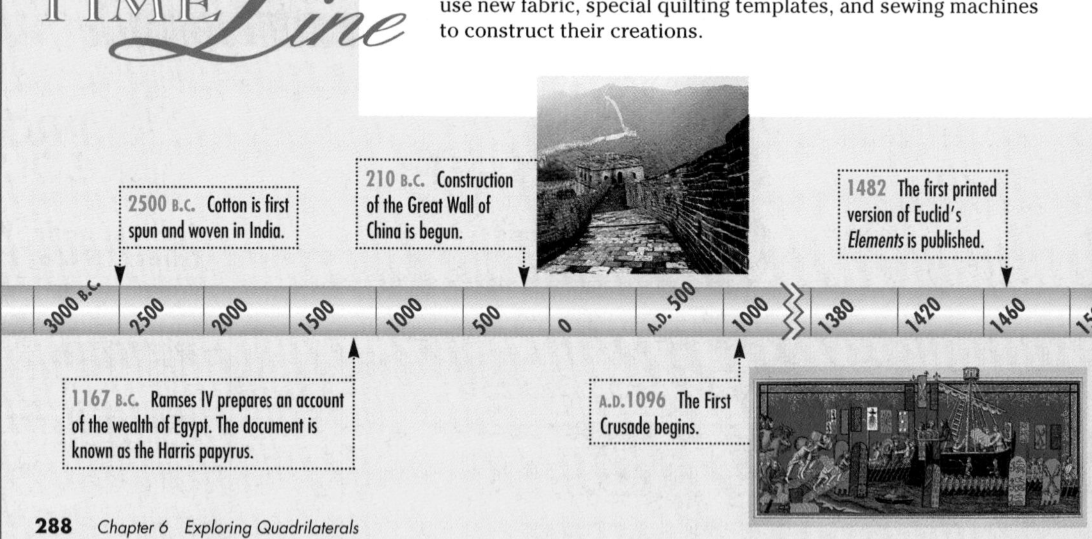
TIME Line
2500 B.C. Cotton is first spun and woven in India.
210 B.C. Construction of the Great Wall of China is begun.
1482 The first printed version of Euclid's *Elements* is published.
3000 B.C.
2500
2000
1500
1000
500
0
A.D. 500
1000
1380
1420
1460
1167 B.C. Ramses IV prepares an account of the wealth of Egypt. The document is known as the Harris papyrus.
A.D. 1096 The First Crusade begins.

TIME Line

Modern sewing machines create a variety of stitches. Most have a simple geometric pattern. Students might want to investigate the types of stitches for a particular sewing machine. Have them describe each using geometric language.

interNET CONNECTION

Visit the Smithsonian Institute to learn the history of quilting and view quilts from the 18th to 20th centuries.

World Wide Web
http://www.si.edu/organiza/museums/nmah/docs/quilts/quilt.html

PEOPLE IN THE NEWS

After participating in a class project to make a quilt for a needy family, **Dara Marin** of Englewood, Colorado, decided that one quilt wasn't enough. Dara went on to launch a quilting crusade benefiting the homeless of Denver. She publicized her cause with a television interview and flyers asking for donations of materials. Local stores contributed fabric remnants to make quilt tops and damaged sheets for quilt backings. With the help of quilting groups, family members, friends, and classmates, Dara has made 13 full-sized quilts for the homeless. Her goal is to make at least 20 quilts per year.

Chapter Project

Work in cooperative groups to design a patchwork quilt top.

- Research patchwork quilt block designs. Choose a traditional design or create your own quilt block design based on quadrilaterals.
- Make a full-sized pattern for a 12-inch quilt block of your design.
- Make a drawing to represent the entire patchwork quilt top as it would look when completed.
- Estimate how much of each type of fabric you would need to complete the design.
- Describe the shapes used in your design in geometric terms.

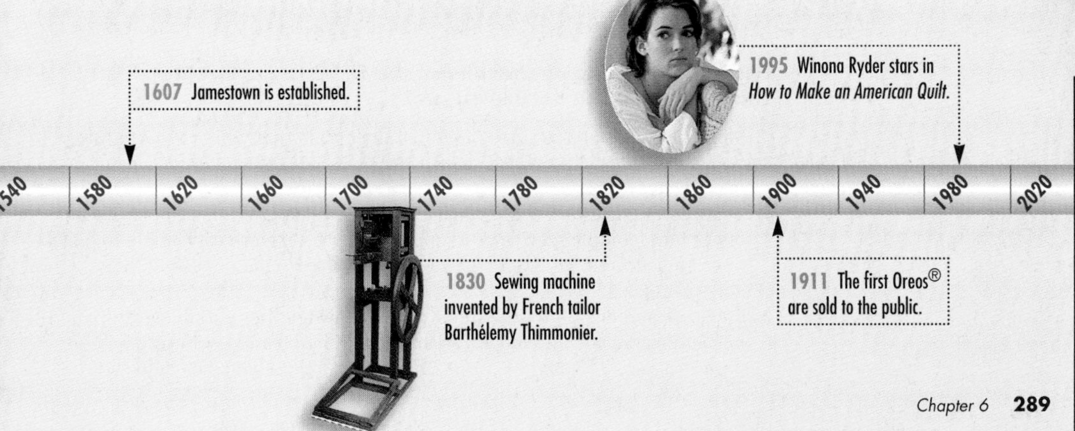

1607 Jamestown is established.
1995 Winona Ryder stars in *How to Make an American Quilt.*
1540
1580
1620
1660
1700
1740
1780
1820
1860
1900
1940
1980
2020
1830 Sewing machine invented by French tailor Barthélemy Thimmonier.
1911 The first Oreos® are sold to the public.

PEOPLE IN THE NEWS

The idea of using quilts in conjunction with charity work is not new. Around the turn of the century autograph quilts were made as fundraisers for charity. All kinds of people, including local and/or national celebrities, were asked to sign a quilt block and donate a small sum of money, such as 10¢. Each signature on these blocks was then embroidered over and sewn into a patchwork quilt top. The resulting quilt was usually raffled and the proceeds of the raffle went to charity.

Chapter Project

Cooperative Learning If possible when setting up cooperative teams, include a student who has experience with making clothing from a pattern. Encourage students to use the construction techniques for special quadrilaterals outlined in this chapter to make their quilt block patterns. Suggest that certain group members work together to construct a preliminary pattern and others test each quadrilateral in the pattern before finalizing the pattern.

Investigations and Projects Masters, p. 45

6

NAME ____________ DATE ________

Chapter 6 Project A

Student Edition Pages 288–333

Quiltmakers

1. Many people around the world make and design quilts. Find a person who makes quilts in your community and who is willing to be interviewed. You may be able to find quiltmakers by investigating quilting or sewing clubs, fabric stores, or friends who like to sew.
2. Interview the quiltmaker. Record a brief biography as well as how and when he or she began making quilts. Here are some other questions you may want to ask.
   - Who taught you how to make quilts?
   - Do you use patterns? What patterns are your favorites?
   - Why do you enjoy quiltmaking?
   - How many quilts do you think you have made?
   - How long does it usually take to make one quilt?
   - What is the process involved in making a quilt?
3. Look at some of the patterns or books that your quiltmaker has. How many patterns can you find that involve quadrilaterals?
4. Write a newspaper story about the quiltmaker you interviewed. Many papers offer profiles of local artisans; look for these in your local paper and use them to help organize your story. Illustrate your story with a quilting pattern that your subject uses.

Alternative Chapter Projects

Two other chapter projects are included in the *Investigations and Projects Masters*. In Chapter 6 Project A, pp. 45–46, students extend the topic in the chapter opener. In Chapter 6 Project B, pp. 47–48, students discover what quadrilaterals are present in various sports.

6-1A LESSON NOTES

NCTM Standards: 1–4, 7, 8

Objective

Use a TI-92 calculator to draw parallelograms.

Recommended Time

20 minutes

Instructional Resources

Instructions for using the *Geometer's Sketchpad* for this activity are available in the *Graphing Calculator and Computer Masters*, p. 24.

1 FOCUS

Motivating the Lesson

Have students create a square with four strips of construction paper or cardboard as the sides. Connect each corner with a paper fastener. Point out how the structure is not rigid.

2 TEACH

Teaching Tip Have students compare each conjecture for Exercise 2 with what they know about squares and rectangles.

3 PRACTICE/APPLY

Assignment Guide
Core (with proof): 1–3
Core (informal): 1–3
Enriched: 1–3

4 ASSESS

Observing students working with technology is an excellent method of assessment.

6–1A Using Technology Exploring Parallelograms

A Preview of Lesson 6–1

A **parallelogram** is a four-sided figure with both pairs of opposite sides parallel. You can draw a parallelogram with the geometry program of a TI-92 by drawing two pairs of intersecting parallel lines.

Step 1 Draw one line anywhere on the screen. Press F2 and choose Line from the menu. Use the arrow key to position the cursor anywhere on the screen and press ENTER. Then move to another place on the screen and press ENTER to position the line.

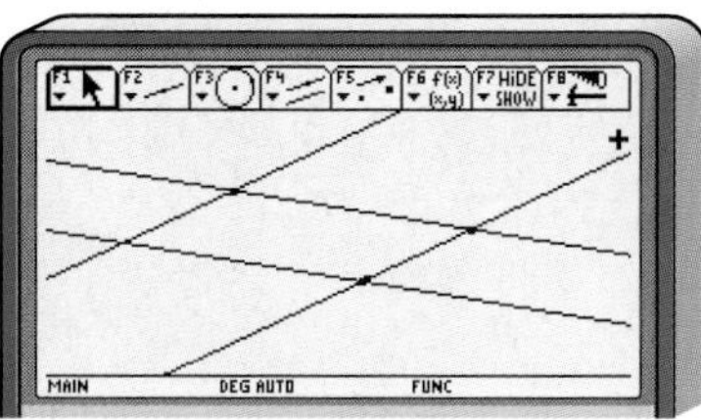

Step 2 Next, use the procedure in Step 1 to draw any line that intersects the first line.

Step 3 To draw a line parallel to the first line, press F4 and choose Parallel Line from the menu. Move the cursor to a point on the first line so that the prompt *parallel to this line* appears. Press ENTER. Then move the cursor to a point through which you would like the line to pass and press ENTER.

Step 4 Use the procedure in step 3 to draw a line parallel to the second line that you drew.

The figure formed by the segments whose endpoints are the points of intersection of the parallel lines is a parallelogram.

EXERCISES

Use the measuring capabilities of the TI-92 to explore the characteristics of a parallelogram.

1. Measure the opposite sides of the parallelogram that you drew. What do you observe? **Sample answer: The opposite sides are congruent.**
2. Make conjectures about the angles in the parallelogram that appear to be congruent, complementary, or supplementary. Measure each angle to check your conjectures. **Sample answer: The opposite angles are congruent. Pairs of consecutive angles are supplementary.**
3. Clear the screen and follow steps 1–4 to draw another parallelogram. Then measure each side and angle. Do the same relationships hold true for this parallelogram? **Sample answer: Yes, the same relationships hold true.**

Using Technology

This lesson offers an excellent opportunity for using technology in your geometry classroom. For more information on using technology, see *Graphing Calculators in the Mathematics Classroom*, one of the titles in the Glencoe Mathematics Professional Series.

6-1 Parallelograms

What YOU'LL LEARN

- To recognize and apply the properties of a parallelogram, and
- to find the probability of an event.

Why IT'S IMPORTANT

You can use the properties of parallelograms to solve problems involving transportation and interior design.

APPLICATION

Interior Design

Dalia Berlin, ASID, is the owner of the award-winning interior design company Berlin Designs of Hollywood, Florida. Ms. Berlin has designed commercial buildings as well as homes in South America and the United States. When Ms. Berlin designs a room, she tries to create an attractive, comfortable, and functional area. She carefully selects each item to suit the room's purpose and overall mood. Important elements she considers are the style of the room, the form of the furniture, color, light, scale, pattern, and texture.

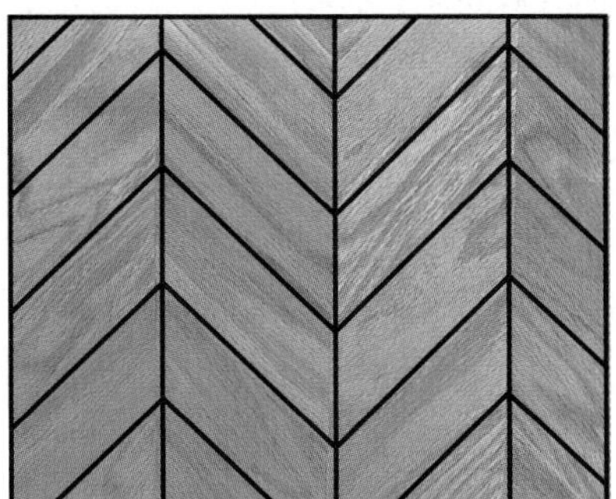

Ceramic tile and wood flooring are often designers' favorite choices for homes. Many styles and colors are available and the floor can be designed in many ways. The wood flooring at the left is laid out in a herringbone pattern.

The herringbone pattern is created using a pattern of **quadrilaterals**. Quadrilaterals are four-sided polygons. Below are some examples and non-examples of quadrilaterals.

CAREER CHOICES

Interior designers need creativity to assemble beautiful and functional rooms. They also need math skills to determine measurements and costs.

For more information, contact:

American Society of Interior Designers (ASID)
608 Massachusetts Ave., NE
Washington, DC 20002

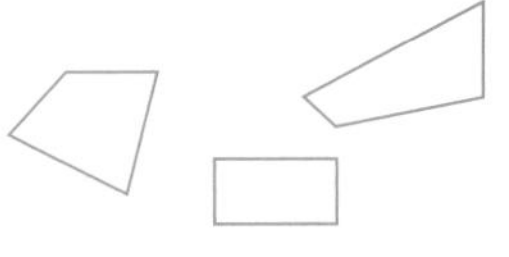

Examples

Non-examples

The special quadrilaterals in the floor shown above are called **parallelograms**. A parallelogram is a quadrilateral with *both pairs of opposite sides parallel.*

The parallelogram at the right has vertices M, N, P, and Q. A symbol for parallelogram $MNPQ$ is $\square MNPQ$. $\overline{MN}$ and $\overline{PQ}$, and $\overline{MQ}$ and $\overline{NP}$ are opposite sides of $\square MNPQ$. The opposite angles are $\angle M$ and $\angle P$, and $\angle N$ and $\angle Q$.

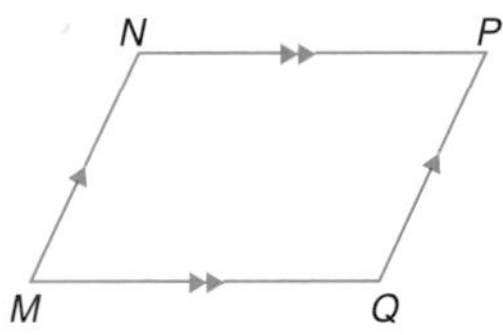

What conjectures can you make about the sides and angles of a parallelogram?

CAREER CHOICES

Many of the same skills are required by landscape designers. A knowledge of both geometry and biology is needed.

6-1 LESSON NOTES

NCTM Standards: 1–5, 7, 8

Instructional Resources

- Study Guide Master 6-1
- Practice Master 6-1
- Enrichment Master 6-1
- Modeling Mathematics Masters, pp. 35–38

Transparency 6-1A contains the 5-Minute Check for this lesson; **Transparency 6-1B** contains a teaching aid for this lesson.

Recommended Pacing	
Standard Pacing	Days 1 & 2 of 13
Honors Pacing	Days 1 & 2 of 12
Block Scheduling*	Day 1 of 6

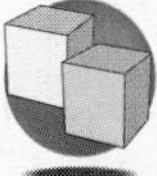

*For more information on pacing and possible lesson plans, refer to the *Block Scheduling Booklet*.

1 FOCUS

5-Minute Check
(over Chapter 5)

Refer to the figure below for Exercises 1–3.

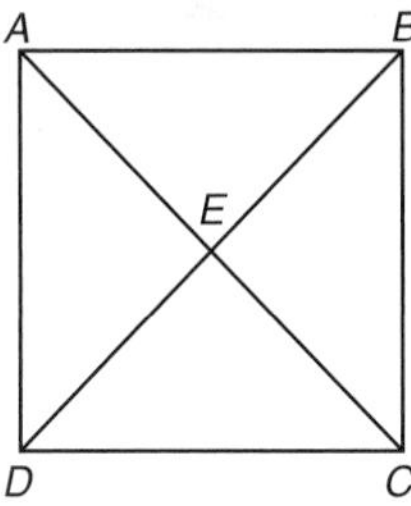

Given: $\overline{BD}$ is a perpendicular bisector of $\overline{AC}$.

1. Can you say that $\triangle AED \cong \triangle CED$? **yes, by LL**
2. Can you say that $\triangle AED \cong \triangle CEB$? **no**
3. If you want to prove that $m\angle BCD = m\angle CBA$ by an indirect proof, what is your assumption? **Assume that $m\angle BCD < m\angle CBA$ or $m\angle BCD > m\angle CBA$.**

(continued on next page)

Refer to the figure below for Exercises 4–5.

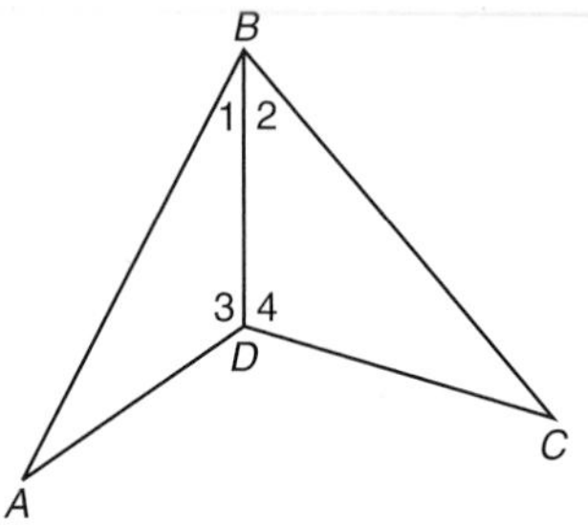

4. If $m\angle 4 > m\angle 2$, what can you say about their opposite segments? **$BC > DC$; if one $\angle$ of a $\triangle$ has a greater measure than another $\angle$, then the side opp. the greater $\angle$ is longer than the side opp. the lesser $\angle$.**
5. If $m\angle 2 > m\angle 1$ and $\overline{AB} \cong \overline{BC}$, what can you say about the segments in the triangles? **$DC > AD$ by the SAS Inequality Theorem**

Motivating the Lesson

Situational Problem Have each student make a simplified line drawing of a picture of a bridge. Use pictures of local bridges and famous bridges. Have the class share drawings and discuss the different types of polygons shown and their properties.

2 TEACH

Students can compare their results by performing the same operations on a rectangle.

Teaching Tip A good way to investigate parallelograms is by having students cut out examples and use paperfolding to demonstrate the different properties mentioned in Theorems 6-1, 6-2, 6-3, and 6-4.

Parallelogram Properties

Materials:  patty paper  straightedge

- Use a straightedge to draw two sets of intersecting parallel lines on your patty paper like the ones shown below. Label the parallelogram *ABCD*.

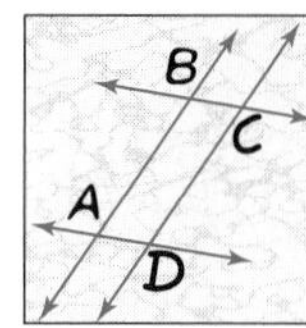

- Place a second patty paper over the first and trace *ABCD*. Label the second parallelogram *PQRS*.

Your Turn **a. They are congruent.**

a. Move the second patty paper over the first and compare side $\overline{RS}$ to side $\overline{AB}$. How do the lengths of the opposite sides compare?
b. Rotate and move the second patty paper over the first to compare opposite angles *R* and *A* or angles *S* and *B*. How do the opposite angles compare? **They are congruent.**
c. Rotate the second patty paper around the first to compare $\angle S$ and $\angle A$. What is their relationship? What conclusion can be drawn? **They form a linear pair. The consecutive angles are supplementary.**

The Modeling Mathematics activity provides insight into three important properties about parallelograms.

Theorem 6–1	**Opposite sides of a parallelogram are congruent.**
Theorem 6–2	**Opposite angles of a parallelogram are congruent.**
Theorem 6–3	**Consecutive angles in a parallelogram are supplementary.**

You will be asked to prove Theorems 6–1 and 6–2 in Exercises 38 and 39, respectively.

Proof of Theorem 1–1

Write a paragraph proof of Theorem 6–3.

Given: $\square GEOM$

Prove: $\angle G$ and $\angle E$ are supplementary.
$\angle E$ and $\angle O$ are supplementary.
$\angle O$ and $\angle M$ are supplementary.
$\angle M$ and $\angle G$ are supplementary.

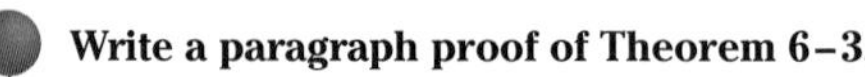

Paragraph Proof:

By the definition of a parallelogram, $\overline{GE} \parallel \overline{MO}$ and $\overline{GM} \parallel \overline{EO}$. For parallels $\overline{GE}$ and $\overline{MO}$, $\overline{GM}$ and $\overline{EO}$ are transversals, and for parallels $\overline{GM}$ and $\overline{EO}$, $\overline{GE}$ and $\overline{MO}$ are transversals. Thus, the consecutive interior angles on the same side of a transversal are supplementary. Therefore, $\angle G$ and $\angle E$, $\angle E$ and $\angle O$, $\angle O$ and $\angle M$, and $\angle M$ and $\angle G$ are supplementary.

Cooperative Learning

Co-op Co-op Have groups of three or four take a sheet of rectangular dot paper and divide it into several 4 by 4 arrays. Ask them to draw all possible noncongruent parallelograms that are not rectangles or rhombi. Have the groups compare their answers and the methods they used to organize their counts. For more information on the co-op co-op strategy, see *Cooperative Learning in the Mathematics Classroom*, one of the titles in the Glencoe Mathematics Professional Series, page 32.

Polygons with more than three sides have diagonals. The **diagonals** of a figure are the segments that connect any two vertices of the polygon.

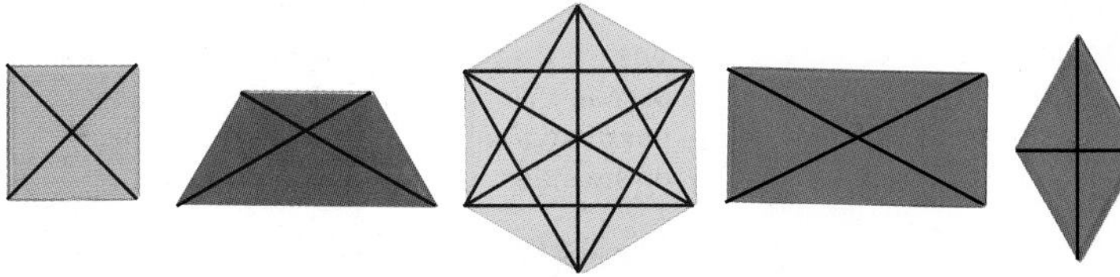

In parallelogram *KLMN* at the right, the dashed line segments, $\overline{KM}$ and $\overline{LN}$, are diagonals. There is a special relationship between the diagonals of a parallelogram. This relationship is stated in Theorem 6–4. *You will be asked to prove this theorem in Exercise 40.*

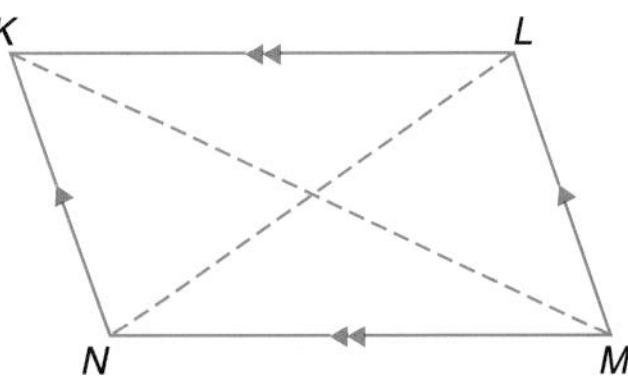

Theorem 6-4	**The diagonals of a parallelogram bisect each other.**

Example 1 **WXYZ is a parallelogram, $m\angle ZWX = b$, and $m\angle WXY = d$. Find the values of a, b, c, and d.**

Algebra

Since the diagonals of a parallelogram bisect each other, $a = 15$.

Opposites angles of a parallelogram are congruent. So $m\angle ZWX = m\angle XYZ$.

$m\angle ZWX = m\angle XYZ$

$b = 31 + 18 \text{ or } 49$

Opposite sides of a parallelogram are congruent. Therefore, $\overline{WX} \cong \overline{ZY}$.

$WX = ZY$

$2c = 22$

$c = 11$

Since consecutive angles of a parallelogram are supplementary, $\angle WXY$ and $\angle XYZ$ are supplementary.

$m\angle WXY + m\angle XYZ = 180$

$d + (31 + 18) = 180$ *$m\angle WXY = d$*

$d + 49 = 180$

$d = 131$

You can also use slope to verify properties of a parallelogram.

For Example 1

MATH is a parallelogram. Find the values of *w*, *x*, *y*, and *z*.

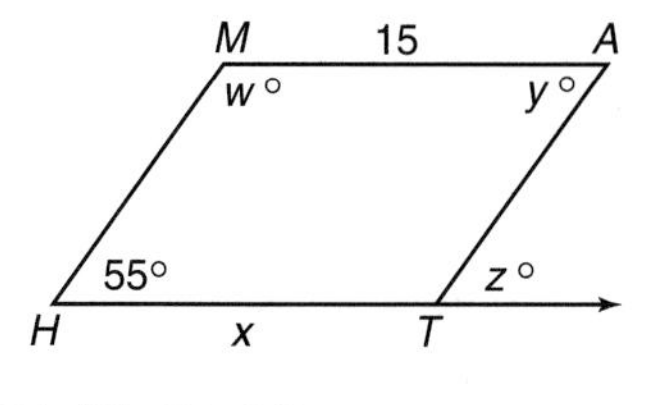

125, 15, 55, 55

In-Class Examples

For Example 2

Given $\angle 1 \cong \angle 3$ and $\angle 2 \cong \angle 4$, determine if quadrilateral $ABCD$ is a parallelogram.

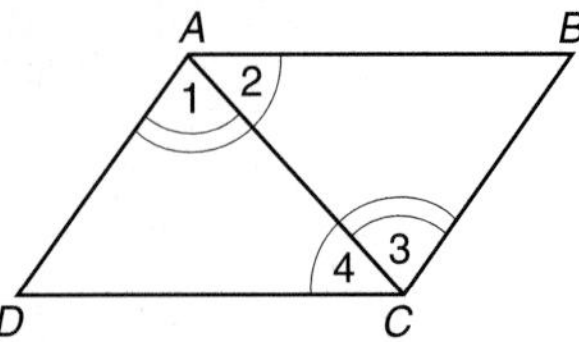

By Postulate 3-2, $\overline{AB} \parallel \overline{DC}$ and $\overline{BC} \parallel \overline{AD}$. So, by definition of a parallelogram, $ABCD$ is a parallelogram.

For Example 3

For ▱$ABCD$ in Example 3, what is the probability that two randomly chosen angles are congruent? List the possible pairs of angles and determine which are congruent.

$\angle A$, $\angle B$	not congruent
$\angle A$, $\angle C$	congruent
$\angle A$, $\angle D$	not congruent
$\angle B$, $\angle C$	not congruent
$\angle B$, $\angle D$	congruent
$\angle C$, $\angle D$	not congruent

The probability of choosing two angles at random that are congruent is $\frac{2}{6}$ or $\frac{1}{3}$.

Teaching Tip Remind students that the probability of something *not* occurring and the probability of it occurring add up to 1.

Additional Answers

1. Quadrilateral

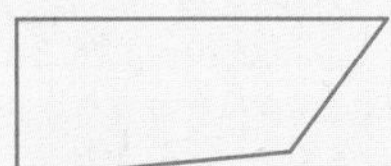

This is a quadrilateral because it is in a plane, its sides intersect exactly two other sides, one at each endpoint, no two sides with a common endpoint are collinear, and it has four sides.

Not a quadrilateral

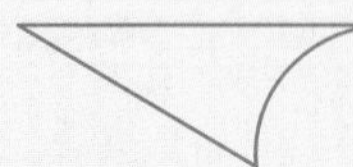

2. A general quadrilateral is a polygon with four sides. A parallelogram is a special quadrilateral with opposite sides parallel.

Example 2

Algebra

The coordinates of the vertices of *RSTV* are *R*(1, 1), *S*(3, 6), *T*(8, 8), and *V*(6, 3). Determine if *RSTV* is a parallelogram.

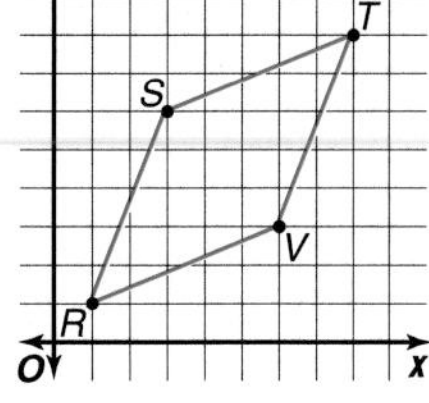

The opposite sides of a parallelogram are parallel. We can determine if $RSTV$ is a parallelogram by comparing the slopes of the sides.

slope of $\overline{RS} = \frac{6-1}{3-1}$ or $\frac{5}{2}$

slope of $\overline{TV} = \frac{3-8}{6-8}$ or $\frac{5}{2}$

slope of $\overline{ST} = \frac{8-6}{8-3}$ or $\frac{2}{5}$ slope of $\overline{RV} = \frac{3-1}{6-1}$ or $\frac{2}{5}$

Since the opposite sides have the same slope, $\overline{RS} \parallel \overline{TV}$ and $\overline{ST} \parallel \overline{RV}$. Therefore, $RSTV$ is a parallelogram.

The **probability** of an event is the ratio of the number of favorable outcomes to the total number of possible outcomes. For example, the probability of rolling an even number on a die is $\frac{3}{6}$ or $\frac{1}{2}$ because there are 3 even numbers out of 6 possible numbers to roll.

Example 3

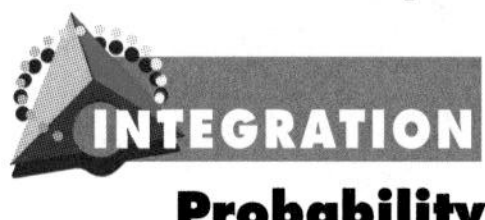

Probability

Two sides of ▱*ABCD* are chosen at random. What is the probability that the two sides are *not* congruent?

List the possible pairs of sides and determine which are not congruent.

$\overline{AB}$, $\overline{BC}$	not congruent
$\overline{AB}$, $\overline{CD}$	congruent
$\overline{AB}$, $\overline{AD}$	not congruent
$\overline{BC}$, $\overline{CD}$	not congruent
$\overline{BC}$, $\overline{AD}$	congruent
$\overline{CD}$, $\overline{AD}$	not congruent

Four of 6 possibilities are not congruent. So the probability of choosing two sides at random that are not congruent is $\frac{4}{6}$ or $\frac{2}{3}$.

CHECK FOR UNDERSTANDING

Communicating Mathematics

Study the lesson. Then complete the following. 1–4. See margin.

1. **Draw** a quadrilateral and explain why it is a quadrilateral. Then draw a non-example of a quadrilateral.
2. **Compare and contrast** a general quadrilateral and a parallelogram.
3. **Verify** that the diagonals of $RSTV$ in Example 2 bisect each other.
4. **Summarize** five properties that hold true for any parallelogram.

Additional Answers

3. $\overline{RT}$ and $\overline{SV}$ intersect at $M(4.5, 4.5)$. $RM = \sqrt{(1-4.5)^2 + (1-4.5)^2}$ or $\sqrt{24.5}$; $TM = \sqrt{(8-4.5)^2 + (8-4.5)^2}$ or $\sqrt{24.5}$; $SM = \sqrt{(3-4.5)^2 + (6-4.5)^2}$ or $\sqrt{4.5}$; $VM = \sqrt{(6-4.5)^2 + (3-4.5)^2}$ or $\sqrt{4.5}$. Thus since $RM = TM$ and $SM = VM$, the diagonals bisect each other.

4. The opposite sides are parallel, the opposite sides are congruent, the opposite angles are congruent, the consecutive angles are supplementary, and the diagonals bisect each other.

5a. Two pairs of triangles are congruent. $\triangle QRP \cong \triangle STP$ and $\triangle RSP \cong \triangle TQP$.

5. Draw a parallelogram $QRST$ on a piece of paper. Draw diagonal $\overline{QS}$ with a red marker and diagonal $\overline{TR}$ with a blue marker. Label the intersection of $\overline{QS}$ and $\overline{TR}$, P. Cut along each side and diagonal to form four triangles.
   a. Move and compare the four triangles to determine whether any of the triangles are congruent.
   b. How do the lengths of all the blue segments compare? How do the lengths of all the red segments compare? See margin.
   c. What relationship appears to exist for the two diagonals of a parallelogram? The diagonals bisect each other.

Guided Practice

Complete each statement about ▱CDFG. Then name the theorem or definition that justifies your answer.

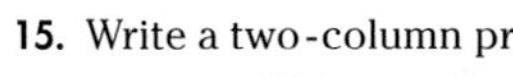

6. $\overline{CH} \cong$ _?_ $\overline{HF}$
7. $\overline{GF} \parallel$ _?_ $\overline{DC}$
8. $\angle DCG \cong$ _?_ $\angle DFG$
9. $\overline{DC} \cong$ _?_ $\overline{GF}$
10. $\angle DCG$ is supplementary to _?_. $\angle CDF$ and $\angle CGF$
11. $\triangle HGC \cong$ _?_ $\triangle HDF$
12. In parallelogram $ABCD$, $AB = 2x + 5$, $m\angle BAC = 2y$, $m\angle B = 120$, $m\angle CAD = 21$, and $CD = 21$. Find the values of x and y. $x = 8$, $y = 19.5$
13. Quadrilateral $WXYZ$ is a parallelogram with $m\angle W = 47$. Find the measure of angles X, Y, and Z. $m\angle Y = 47$, $m\angle X = 133 = m\angle Z$
14. $PQRT$ has vertices $P(-4, 7)$, $Q(3, 0)$, $R(2, -5)$, and $T(-5, 2)$. Determine if $PQRT$ is a parallelogram. See margin.

6. Diagonals of ▱ bisect each other.
7. Opp. sides of ▱ are ∥.
8. Opp. ∡ of ▱ are ≅.
9. Opp. sides of ▱ are ≅.
10. Consecutive ∡ of ▱ are supplementary.
11. SSS: diagonals of a ▱ bisect each other ($\overline{HC} \cong \overline{HF}$ and $\overline{HG} \cong \overline{HD}$); opp. sides of ▱ are ≅ ($\overline{GC} \cong \overline{DF}$).

Proof

15. Write a two-column proof.
   Given: $PRSV$ is a parallelogram. $\overline{PT} \perp \overline{SV}$, $\overline{QS} \perp \overline{PR}$
   Prove: $\triangle PTV \cong \triangle SQR$

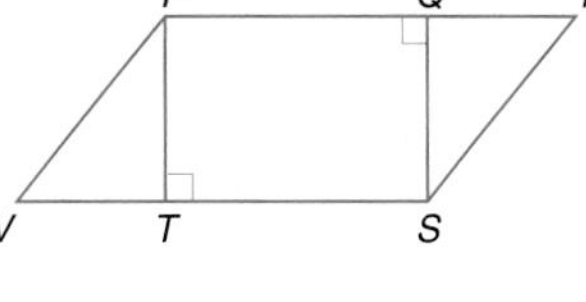

15. See Solutions Manual.

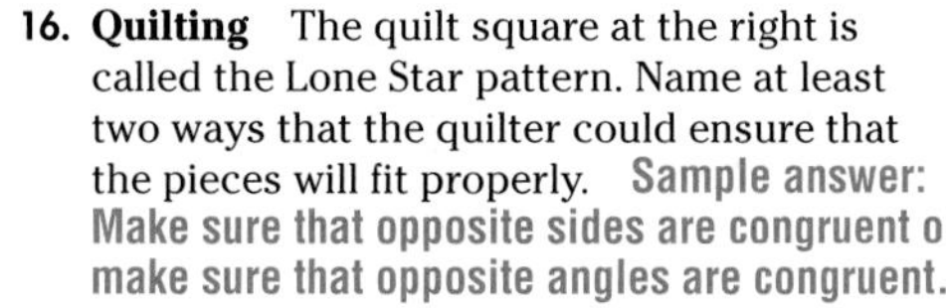

16. **Quilting** The quilt square at the right is called the Lone Star pattern. Name at least two ways that the quilter could ensure that the pieces will fit properly. Sample answer: Make sure that opposite sides are congruent or make sure that opposite angles are congruent.

EXERCISES

Practice

Complete each statement about ▱MARK. Then name the theorem or definition that justifies your answer. 17–26. See Solutions Manual for justifications.

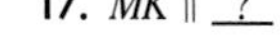

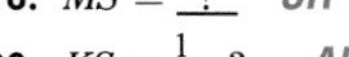

A
17. $\overline{MK} \parallel$ _?_ $\overline{AR}$
18. $\overline{MS} \cong$ _?_ $\overline{SR}$
19. $\angle MKR \cong$ _?_
20. $KS = \frac{1}{2}$ _?_ AK

B
21. $\triangle MAR \cong$ _?_
22. $\overline{AS} \cong$ _?_ $\overline{SK}$
23. $\overline{AM} \parallel$ _?_ $\overline{RK}$
24. $\angle ARK$ and _?_ are supplementary. $\angle MAR$ or $\angle RKM$
25. $\triangle SAM \cong$ _?_
26. $\angle RKA \cong$ _?_ $\angle KAM$

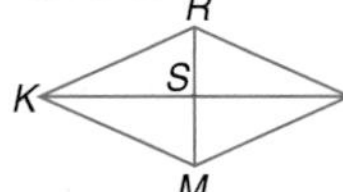

19. $\angle MAR$
21. $\triangle RKM$
25. $\triangle SKR$

Reteaching

Using Modeling Have students cut out two pairs of congruent strips of cardboard, one pair longer than the other. Have them use paper fasteners to attach the strips to form a parallelogram. The angles of the parallelogram are adjustable. Have students use a protractor to adjust the sides so that the measure of one angle is 75 and then measure the other three angles and note their relationships to the 75° angle. They can reset the angle to 45° and repeat the activity. You may also have students check the diagonals with a ruler to demonstrate that the diagonals bisect each other.

3 PRACTICE/APPLY

Check for Understanding

Exercises 1–16 are designed to help you assess your students' understanding through reading, writing, speaking, and modeling. You should work through Exercises 1–5 with your students and then monitor their work on Exercises 6–16.

Assignment Guide

Core (with proof): 17–41 odd, 42, 43, 45, 47–56
Core (informal): 17–35 odd, 41, 42, 43, 45, 47–56
Enriched: 18–40 even, 42–56

For **Extra Practice,** see p. 774.

The red A, B, and C flags, printed only in the Teacher's Wraparound Edition, indicate the level of difficulty of the exercises.

Additional Answers

5b. All red segments are congruent and all blue segments are congruent.

14. Slope of $\overline{PT} = 5$, slope of $\overline{QR} = 5$, slope of $\overline{QP} = -1$, slope of $\overline{TR} = -1$. Thus, the opposite sides are parallel and $PQRT$ is a parallelogram.

Study Guide Masters, p. 32

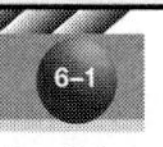

6–1 NAME________ DATE______

Study Guide

Student Edition Pages 291–297

Parallelograms

Any four-sided polygon is called a **quadrilateral**. A segment joining any two nonconsecutive vertices in a quadrilateral is called a **diagonal**. A special kind of quadrilateral in which both pairs of opposite sides are parallel is called a **parallelogram**.

The following theorems all concern parallelograms.

- Opposite sides of a parallelogram are congruent.
- Opposite angles of a parallelogram are congruent.
- Consecutive angles in a parallelogram are supplementary.
- The diagonals of a parallelogram bisect each other.

Example: If the quadrilateral in the figure is a parallelogram, find the values of x, y, and z.

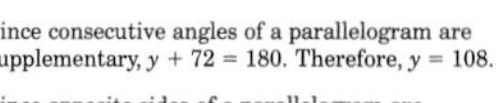

Since opposite angles of a parallelogram are congruent, $x = 72$.

Since consecutive angles of a parallelogram are supplementary, $y + 72 = 180$. Therefore, $y = 108$.

Since opposite sides of a parallelogram are congruent, $z = 8$.

If each quadrilateral is a parallelogram, find the values of x, y, and z.

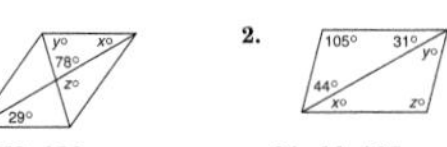

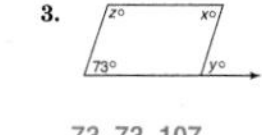

1. 29, 73, 102
2. 31, 44, 105
3. 73, 73, 107
4. In parallelogram $ABCD$, $m\angle A = 3x$ and $m\angle B = 4x + 40$. Find the measure of angles A, B, C, and D. $m\angle A = 60$, $m\angle B = 120$, $m\angle C = 60$, $m\angle D = 120$
5. In parallelogram $RSTV$, diagonals $\overline{RT}$ and $\overline{VS}$ intersect at Q. If $RQ = 5x + 1$ and $QT = 3x + 15$, find QT. $QT = 36$

Explain why it is impossible for each figure to be a parallelogram.

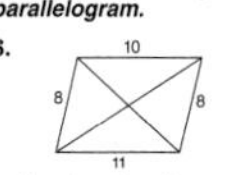

6. Each pair of opposite sides should be congruent.
7. The opposite angles should be congruent.

Additional Answers

36. Given: *SRWV* and *TVXY* are parallelograms.
Prove: $\angle Y \cong \angle R$
Proof:
Statements (Reasons)
1. *SRWV* and *TVXY* are parallelograms. (Given)
2. $\angle R \cong \angle V$, $\angle V \cong \angle Y$ (Opp. ∠s of a ▱ are ≅.)
3. $\angle R \cong \angle Y$ (Congruence of ∠s is transitive.)

37. Given: ▱*TEAM*
$MS = FS$
Prove: $\angle F \cong \angle TEA$
Proof:
Statements (Reasons)
1. ▱*TEAM*, $MS = FS$ (Given)
2. $\angle F \cong \angle M$ (Isosceles △ Th.)
3. $\angle M \cong \angle TEA$ (Opp. ∠s of a ▱ are ≅.)
4. $\angle F \cong \angle TEA$ (Congruence of ∠s is transitive.)

Practice Masters, p. 32

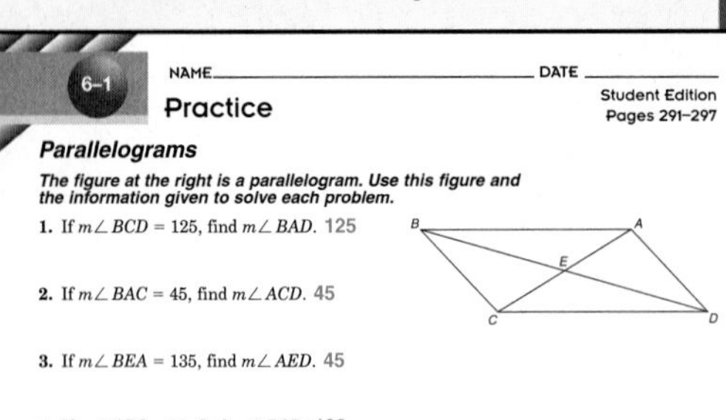

6-1 NAME______ DATE______
Practice Student Edition Pages 291–297

Parallelograms

The figure at the right is a parallelogram. Use this figure and the information given to solve each problem.

1. If $m\angle BCD = 125$, find $m\angle BAD$. 125
2. If $m\angle BAC = 45$, find $m\angle ACD$. 45
3. If $m\angle BEA = 135$, find $m\angle AED$. 45
4. If $m\angle ABC = 50$, find $m\angle BCD$. 130
5. If $AB = 5x - 3$ and $CD = 2x + 9$, find AB. 17
6. If $m\angle DAB = 2x - 10$ and $m\angle ADC = 3x$, find $m\angle DAB$. 66
7. If $m\angle BAD = 3x - 12$ and $m\angle BDC = x + 40$, find $m\angle BAD$. 66
8. Write a two-column proof.
Given: $ABCD$ is a parallelogram, $\overline{BE} \cong \overline{AD}$
Prove: $\angle 1 \cong \angle C$

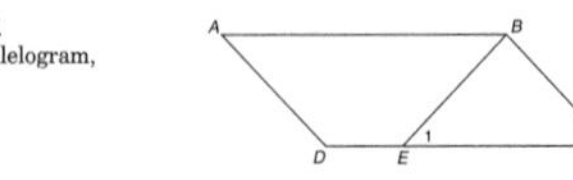

Proof:

Statements	Reasons
a. *ABCD* is a parallelogram.	a. Given
b. $\overline{AD} \cong \overline{BC}$	b. Opposite sides of a parallelogram are congruent.
c. $\overline{BE} \cong \overline{AD}$	c. Given
d. $\overline{BC} \cong \overline{BE}$	d. Congruence of segments is transitive.
e. $\angle 1 \cong \angle C$	e. If two sides of a triangle are congruent, then the angles opposite those sides are congruent.

27. $x = 118$; $y = 62$; $z = 118$
28. $x = 70$; $y = 42$; $z = 20$
29. $x = 53$; $y = 15$; $z = 53$

30. $CM = ME = \sqrt{8}$; $DM = FM = \sqrt{20}$
31. The diagonals are not congruent.
32. Slope of $\overline{CD} = 2$; slope of $\overline{CF} = 0$; thus, the consecutive sides are not perpendicular.

For each parallelogram, find the values of *x*, *y*, and *z*.

27. 28. 29.

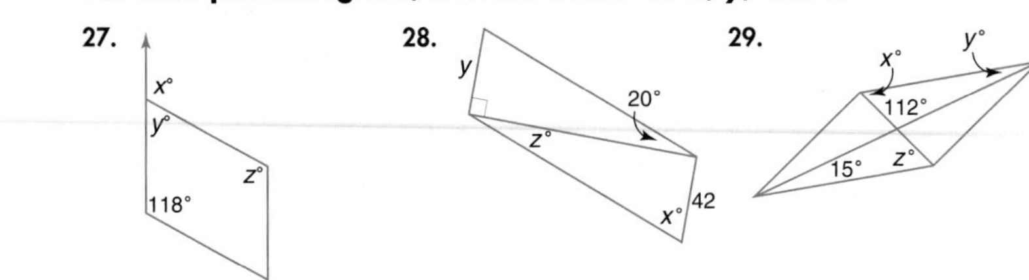

Refer to ▱*CDEF* at the right for Exercises 30–32.

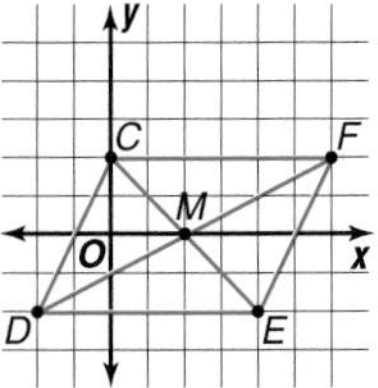

30. Use the distance formula to verify that the diagonals bisect each other.
31. Determine if the diagonals of this parallelogram are congruent.
32. Find the slopes of $\overline{CD}$ and $\overline{CF}$. Are the consecutive sides of *CDEF* perpendicular? Explain.

C
33. Find all the possible coordinates for the fourth vertex of a parallelogram with vertices $T(4, -1)$, $D(-4, 1)$, and $K(0, 8)$. (0, −8), (8, 6), or (−8, 10)
34. *NCTM* is a parallelogram with diagonals $\overline{NT}$ and $\overline{MC}$ that intersect at point *Q*. If $NQ = 3a + 18$, $NT = 12a$, $QC = a + 2b$, and $QM = 3b + 1$, find *a*, *b*, and *CM*. $a = 6$, $b = 5$, $CM = 32$
35. If *ABCD* and *PQRS* are parallelograms, find $m\angle APS$. 19

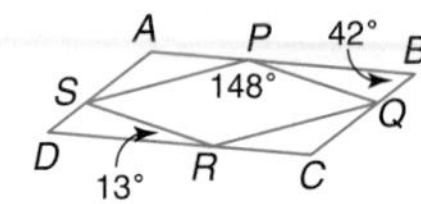

Write a two-column proof. 36–37. See margin.

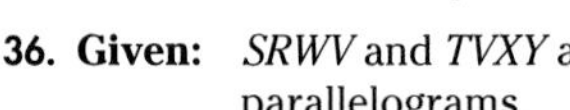

Proof

36. **Given:** *SRWV* and *TVXY* are parallelograms.
Prove: $\angle Y \cong \angle R$

37. **Given:** ▱*TEAM*
$MS = FS$
Prove: $\angle F \cong \angle TEA$

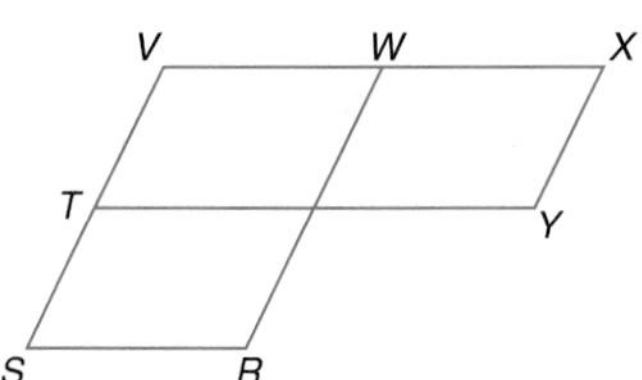

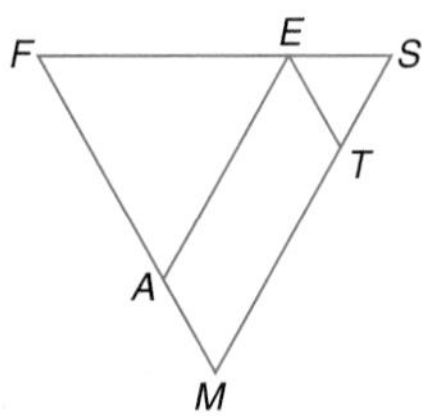

38. Write a paragraph proof of Theorem 6–1. See margin.
39. Write a two-column proof of Theorem 6–2. See Solutions Manual.
40. Write a paragraph proof of Theorem 6–4. See Solutions Manual.
41. Draw a parallelogram *MNPQ* with diagonal $\overline{MP}$. Make a conjecture about the sum of the measures of the angles of a parallelogram. Do you think your conjecture holds true for any quadrilateral? Explain. Sum = 360; yes, two triangles are always formed by any quadrilateral and a diagonal.

Critical Thinking

42. Consider parallelogram *RSTV*.
a. As the measure of angle *R* decreases, what must happen to the measure of angle *V*? $m\angle V$ must increase. b. See margin.
b. What is the maximum measure for angle *V*? Explain your reasoning.

Additional Answer

38. Given: ▱*PQRS*
Prove: $\overline{PQ} \cong \overline{RS}$
$\overline{QR} \cong \overline{SP}$

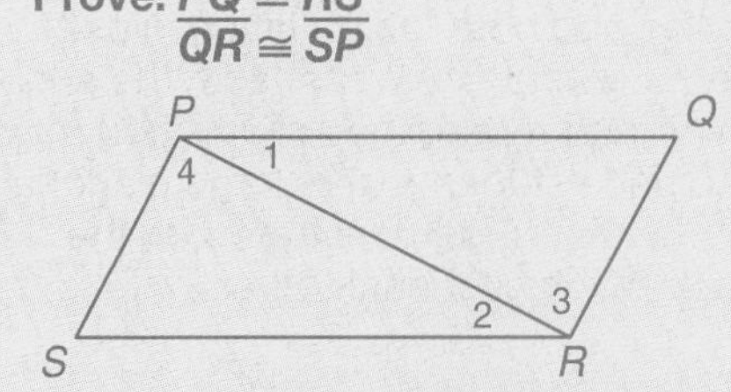

Proof:
Draw an auxiliary segment, $\overline{PR}$, and label the angles 1, 2, 3, and 4 as shown. Since the opposite sides of a parallelogram are parallel, $\angle 1 \cong \angle 2$, and $\angle 3 \cong \angle 4$ because they are alternate interior angles. Since congruence of segments is reflexive, $\overline{PR} \cong \overline{PR}$. So $\triangle QPR \cong \triangle SRP$ by ASA. $\overline{PQ} \cong \overline{RS}$ and $\overline{QR} \cong \overline{SP}$ by CPCTC.

Applications and Problem Solving

43. **Language** In a commercial for CompuServe Computer Discount House, the announcer says that he thought a parallelogram was a "telegram for gymnasts at the Olympics." Look up the suffix *gram* in a dictionary. Then make a conjecture about why a parallelogram is named as it is.

44. **Interior Design** Create a unique colorful pattern using parallelograms that could be used to arrange tiles in your kitchen or to create a quilt top.

45. **Transportation** Two tugboats are pulling a ship into harbor. The force exerted by each boat can be represented by a *vector*. A vector is a ray whose length is proportional to a force and whose direction indicates the direction of the force. In the diagram at the right, two forces, A and B, are acting on an object. The net force, or resultant vector R is found by drawing a parallelogram. The resultant is represented by the diagonal that begins at the endpoint of the vectors. Copy vectors X and Y that represent the forces of the tugboats and find the resultant. See margin.

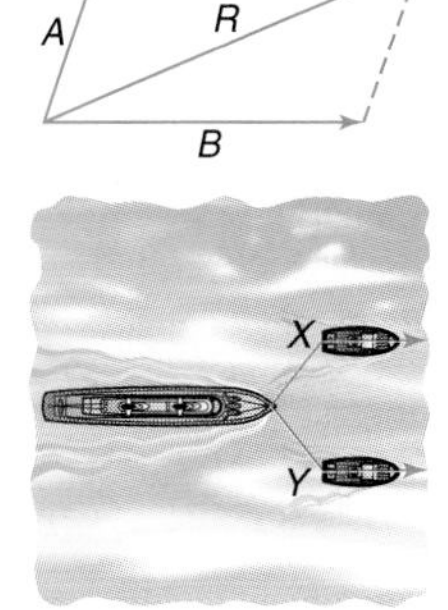

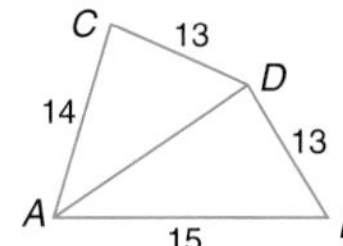

46. **Probability** Suppose $MNOP$ is a parallelogram, but not a rectangle. If you chose two interior angles at random, what is the probability that they would be congruent? $\frac{1}{3}$

43–44. See students' work.

Mixed Review

47. Write an inequality relating the measures of $\angle ADC$ and $\angle ADB$. (Lesson 5–6) $m\angle ADC < m\angle ADB$

C 13 D 14 13 A 15 B

48. Determine whether it is possible to draw a triangle with sides of 6 inches, 9 inches, and 14 inches. Explain. (Lesson 5–5)

48. yes; satisfies triangle inequality

49. Given $\triangle KLM$, with $K(-3, 7)$, $L(2, 5)$, and $M(0, -2)$, list the angles in order from least to greatest measure. (Lesson 5–4) $\angle M$, $\angle K$, $\angle L$

50. If in $\triangle FGH$ and $\triangle JKL$, $\angle H \cong \angle L$ and $GH \cong KL$, what else must be congruent in order to prove that $\triangle FGH \cong \triangle JKL$ by AAS? (Lesson 4–5) $\angle F \cong \angle J$

51. If $m\angle F = 120$ in $\triangle FGH$, describe $\triangle FGH$ as *acute*, *obtuse*, or *right*. Could $\triangle FGH$ be isosceles? Could $\triangle FGH$ be equiangular? Explain your answer. (Lesson 4–1)

51. Obtuse; yes; $m\angle G$ and $m\angle H$ could be 30 each; no, $m\angle F \neq 60$.

52. Are the edges of a ruler a model of lines that are *intersecting*, *parallel*, or *skew*? (Lesson 3–1) parallel

53. Write the conditional *A parallelogram is a quadrilateral with opposite sides parallel* in if-then form. (Lesson 2–2)

53. If a quadrilateral has opposite sides parallel, then it is a parallelogram.

54. Refer to the number line to find the coordinate of the midpoint of $\overline{PR}$. (Lesson 1–5) $\frac{1}{2}$

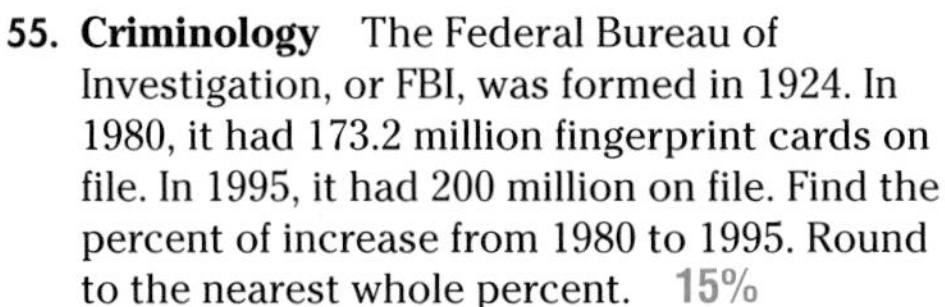

Algebra

55. **Criminology** The Federal Bureau of Investigation, or FBI, was formed in 1924. In 1980, it had 173.2 million fingerprint cards on file. In 1995, it had 200 million on file. Find the percent of increase from 1980 to 1995. Round to the nearest whole percent. 15%

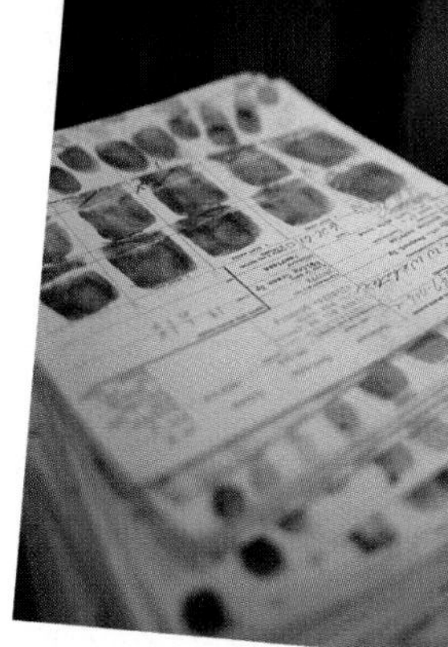

56. Solve $\frac{d-4}{3} = 5$. 19

4 ASSESS

Closing Activity

Modeling Ask students to draw and label a parallelogram and write which angles and segments are congruent, which segments are parallel, and which angles are supplementary.

Additional Answers

42b. $m\angle V < 180$; Because consecutive interior angles are supplementary, each of these angles must measure less than 180.

45. Sample answer:

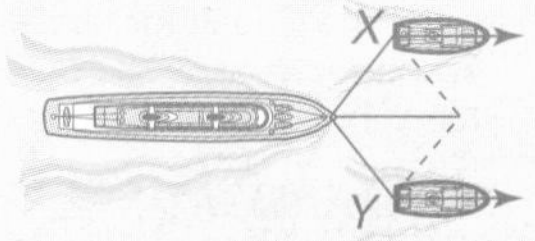

Enrichment Masters, p. 32

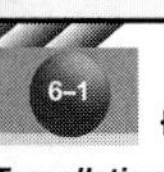

NAME______ DATE______

Enrichment

Student Edition Pages 291–297

Tessellations

A **tessellation** is a tiling pattern made of polygons. The pattern can be extended so that the polygonal tiles cover the plane completely with no gaps. A checkerboard and a honeycomb pattern are examples of tessellations. Sometimes the same polygon can make more than one tessellation pattern. Both patterns below can be formed from an isosceles triangle.

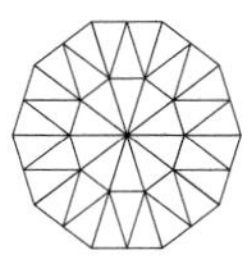

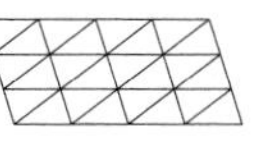

Draw a tessellation using each polygon.

1. 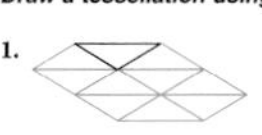
2. 
3. 

Draw two different tessellations using each polygon.

4. 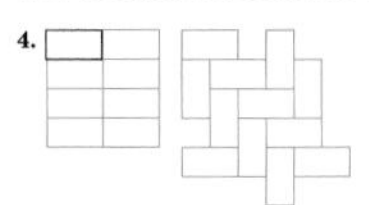
5. 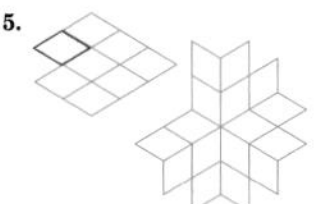

Extension

Connections An interesting numerical pattern that describes some phenomena in nature is the Fibonacci sequence: 1, 1, 2, 3, 5, 8, 13, 21, 34, . . . The center of a sunflower is arranged so that the number of spirals in one direction and the number of spirals in the other direction are two consecutive Fibonacci numbers. Name the next two numbers in the sequence. 55 and 89

6-2 LESSON NOTES

NCTM Standards: 1–5, 7, 8

Instructional Resources

- Study Guide Master 6-2
- Practice Master 6-2
- Enrichment Master 6-2
- Assessment and Evaluation Masters, p. 156
- Multicultural Activity Masters, p. 11
- Real-World Applications, 11
- Tech Prep Applications Masters, p. 11

Transparency 6-2A contains the 5-Minute Check for this lesson; **Transparency 6-2B** contains a teaching aid for this lesson.

Recommended Pacing	
Standard Pacing	Days 3 & 4 of 13
Honors Pacing	Days 3 & 4 of 12
Block Scheduling*	Day 2 of 6

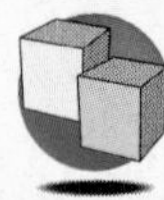

*For more information on pacing and possible lesson plans, refer to the *Block Scheduling Booklet*.

1 FOCUS

5-Minute Check
(over Lesson 6-1)

Find the missing measures in parallelogram *WXYZ*.

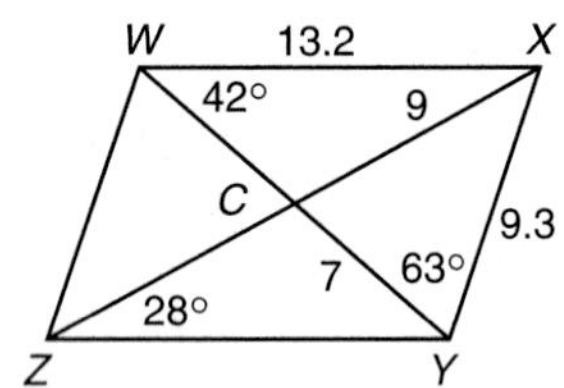

1. $m\angle XWZ$ 105
2. $m\angle WXY$ 75
3. $m\angle WZC$ 47
4. $m\angle WCX$ 110
5. YZ 13.2
6. XZ 18
7. WY 14
8. WZ 9.3

6-2 Tests for Parallelograms

What YOU'LL LEARN

- To recognize and apply the conditions that ensure a quadrilateral is a parallelogram, and
- to identify and use subgoals in writing proofs.

Why IT'S IMPORTANT

You can use parallelograms to solve problems involving engineering and the arts.

You know that the definition of a parallelogram allows you to prove that a quadrilateral is a parallelogram by proving that opposite sides are parallel. However, there are other tests that can be used.

MODELING MATHEMATICS — Testing for a Parallelogram

Materials:  straws, scissors 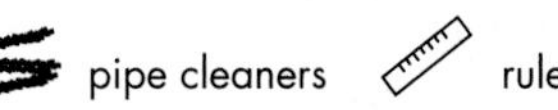 pipe cleaners, ruler

- Cut two straws to one length and two straws to a different length.
- Insert each pipe cleaner in one end of each size of straw. Then form a quadrilateral like the one shown at the right.

Your Turn

a. Shift the position of the sides to form quadrilaterals of different shapes. Measure the distance between opposite sides of the quadrilateral in at least three places. Repeat for several figures. What seems to be true about the opposite sides? **They are parallel.**

b. What type of quadrilaterals are you forming? **parallelograms**

c. How do the measures of pairs of opposite sides compare?

d. What conditions does this activity suggest are sufficient for showing a quadrilateral to be a parallelogram? **Opposite sides are congruent.**

c. They are congruent.

The Modeling Mathematics activity leads to Theorem 6–5. Theorems 6–6 and 6–7 are other tests for parallelograms. *You will be asked to prove Theorems 6–6 and 6–7 in Exercises 34 and 35, respectively.*

Theorem 6–5	**If both pairs of opposite sides of a quadrilateral are congruent, then the quadrilateral is a parallelogram.**
Theorem 6–6	**If both pairs of opposite angles of a quadrilateral are congruent, then the quadrilateral is a parallelogram.**
Theorem 6–7	**If the diagonals of a quadrilateral bisect each other, then the quadrilateral is a parallelogram.**

Sometimes when you solve a problem, it is helpful to identify the smaller steps needed to solve the larger problem. **Identifying subgoals** can help you to organize proofs.

Example  **Write a two-column proof for Theorem 6–5.**

Explore We know that both pairs of opposite sides of a quadrilateral are congruent. We need to prove that the quadrilateral must be a parallelogram.

Have students trace each figure they make to keep track of what they have tried. Discuss different shapes that can't be formed with this setup.

Plan Use quadrilateral *GEOM*.

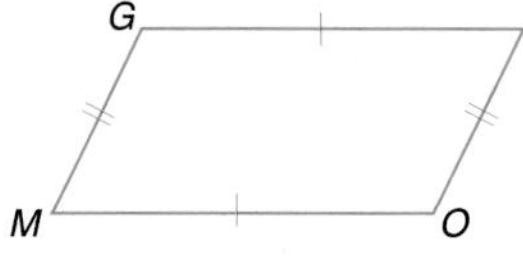

Given: $\overline{GM} \cong \overline{EO}$

$\overline{GE} \cong \overline{MO}$

Prove: *GEOM* is a parallelogram.

Let's set some subgoals before we start the proof. Reasoning backward, we can show that *GEOM* is a parallelogram if we can show that $\overline{GM} \parallel \overline{EO}$ and $\overline{GE} \parallel \overline{MO}$. To show opposite sides parallel, we need a transversal or diagonal that could create alternate interior angles congruent. This can be proved by showing the two triangles formed by the diagonal are congruent and using CPCTC.

Our subgoals are:

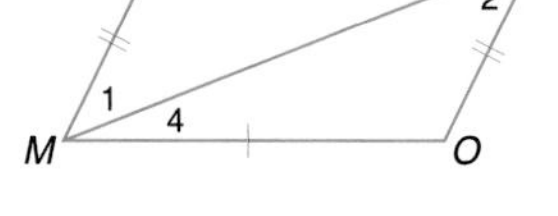

1. Draw diagonal $\overline{EM}$ and prove $\triangle GEM \cong \triangle OME$ by SSS.
2. Use CPCTC to show $\angle 1 \cong \angle 2$ and $\angle 3 \cong \angle 4$.
3. Use alternate interior angles to show that $\overline{GM} \parallel \overline{EO}$ and $\overline{GE} \parallel \overline{MO}$.
4. Show that *GEOM* is a parallelogram by the definition of parallelogram.

Proof:

Solve

Statements	Reasons
1. $\overline{GM} \cong \overline{EO}$ $\overline{GE} \cong \overline{MO}$	1. Given
2. Draw $\overline{EM}$.	2. Through any 2 pts. there is 1 line.
3. $\overline{EM} \cong \overline{EM}$	3. Congruence of segments is reflexive.
4. $\triangle GEM \cong \triangle OME$	4. SSS
5. $\angle 1 \cong \angle 2$ $\angle 3 \cong \angle 4$	5. CPCTC
6. $\overline{GM} \parallel \overline{EO}$ $\overline{GE} \parallel \overline{MO}$	6. If ⟷ and alt. int. ∠s are ≅, then the lines are ∥.
7. *GEOM* is a parallelogram.	7. Def. parallelogram

Examine The proof uses a general quadrilateral and shows that if opposite sides are congruent, then the quadrilateral is a parallelogram.

Theorem 6–8 states another test for parallelograms. *You will be asked to prove this theorem in Exercise 36.*

Theorem 6–8 **If one pair of opposite sides of a quadrilateral is both parallel and congruent, then the quadrilateral is a parallelogram.**

We can use the distance formula and the slope formula to determine if a quadrilateral in the coordinate plane is a parallelogram.

Motivating the Lesson

Hands-On Activity Ask students to draw $\overline{AB}$ and $\overline{CD}$ so they both share a midpoint, *M*. Have them use a ruler to draw *ACBD*. Ask what conjecture they might make about *ACBD*. **Sample answer: If you begin with two diagonals that bisect each other, you get a parallelogram.**

2 TEACH

In-Class Example

For Example 1

The coordinates of the vertices of quadrilateral *ABCD* are *A*(−1, 3), *B*(2, 1), *C*(9, 2), and *D*(6, 4). Determine if quadrilateral *ABCD* is a parallelogram. **Find *BC* and *AD* to determine if this pair of opposite sides are congruent.** $BC = \sqrt{(2-9)^2 + (1-2)^2}$ **or** $\sqrt{50}$. $AD = \sqrt{(-1-6)^2 + (3-4)^2}$ **or** $\sqrt{50}$ **Since** $BC = AD$, $\overline{BC} \cong \overline{AD}$. **Find the slopes of** $\overline{BC}$ **and** $\overline{AD}$ **to determine if these opposite sides are parallel.**

slope of $\overline{BC} = \frac{1-2}{2-9}$ **or** $\frac{1}{7}$

slope of $\overline{AD} = \frac{3-4}{-1-6}$ **or** $\frac{1}{7}$

Since the slopes of $\overline{BC}$ **and** $\overline{AD}$ **are equal, the sides are parallel. Since one pair of sides are both congruent and parallel, *ABCD* is a parallelogram.**

Teaching Tip Encourage students to draw pictures to keep track of information.

Alternative Learning Styles

Visual Have students make a chart summarizing the ways in which a quadrilateral can be proven to be a parallelogram. Display the chart where students can see it as a reference.

In-Class Example

For Example 2
The coordinates of the vertices of quadrilateral *EFGH* are $E(6, 5)$, $F(6, 11)$, $G(14, 18)$, and $H(14, 12)$. Determine if quadrilateral *EFGH* is a parallelogram.
Find the lengths of $\overline{FG}$ and $\overline{EH}$.

$FG = \sqrt{(14-6)^2 + (18-11)^2}$
$= \sqrt{113}$
$EH = \sqrt{(14-6)^2 + (12-5)^2}$
$= \sqrt{113}$

Since $FG = EH$, $\overline{FG} \cong \overline{EH}$.
Find the slopes of $\overline{FG}$ and $\overline{EH}$.

slope of $\overline{FG} = \frac{18-11}{14-6} = \frac{7}{8}$
slope of $\overline{EH} = \frac{12-5}{14-6} = \frac{7}{8}$

Since two opposite sides are congruent and parallel, *EFGH* is a parallelogram.

Teaching Tip After discussing Example 2, ask students how they might have used Theorem 6-5 to show that quadrilateral *ABCD* is a parallelogram. Use the distance formula on all four sides and show that both pairs of opposite sides are congruent.

Additional Answers

1. One pair of congruent angles does not guarantee that the opposite sides will be parallel.

2a. The quadrilateral is a parallelogram because the opposite pairs of angles are congruent.
2b. No; none of the tests for parallelograms is fulfilled.
5. Theorems 6-1, 6-2, and 6-4 and Theorems 6-5, 6-6, and 6-7 are converses of each other.
12. No; the slope of $\overline{GH} = -\frac{2}{3}$ and the slope of $\overline{JK} = \frac{4}{7}$. If *GHJK* were a parallelogram, these segments would have the same slope.

Example 2

Algebra

The coordinates of the vertices of quadrilateral *PQRS* are $P(-5, 3)$, $Q(-1, 5)$, $R(6, 1)$, and $S(2, -1)$. Determine if quadrilateral *PQRS* is a parallelogram.

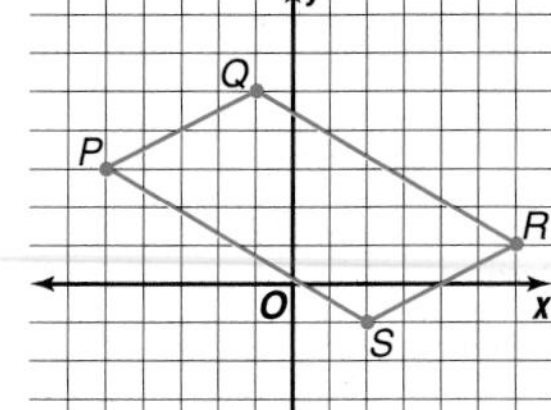

First, find *QR* and *PS* to determine if these opposite sides are congruent.

$QR = \sqrt{(-1-6)^2 + (5-1)^2}$ or $\sqrt{65}$
$PS = \sqrt{(-5-2)^2 + (3-(-1))^2}$ or $\sqrt{65}$
Since $QR = PS$, $\overline{QR} \cong \overline{PS}$.

Next, find the slopes of $\overline{QR}$ and $\overline{PS}$ to determine if $\overline{QR} \parallel \overline{PS}$.

Slope of $\overline{QR} = \frac{5-1}{-1-6}$ or $-\frac{4}{7}$

Slope of $\overline{PS} = \frac{3-(-1)}{-5-2}$ or $-\frac{4}{7}$

Since the slopes are the same, $\overline{QR} \parallel \overline{PS}$.

One pair of opposite sides of *PQRS* are both congruent and parallel, so quadrilateral *PQRS* is a parallelogram.

Here is a summary of the tests to show that a quadrilateral is a parallelogram.

A quadrilateral is a parallelogram if any one of the following is true.
1. Both pairs of opposite sides are parallel. (Definition)
2. Both pairs of opposite sides are congruent. (Theorem 6–5)
3. Both pairs of opposite angles are congruent. (Theorem 6–6)
4. Diagonals bisect each other. (Theorem 6–7)
5. A pair of opposite sides is both parallel and congruent. (Theorem 6–8)

CHECK FOR UNDERSTANDING

Communicating Mathematics

Study the lesson. Then complete the following. 1–2. See margin.

1. **Draw a diagram** that illustrates why only one pair of opposite angles congruent is not enough to prove a quadrilateral is a parallelogram.
2. **Explain** why each quadrilateral is or is not a parallelogram.

a.

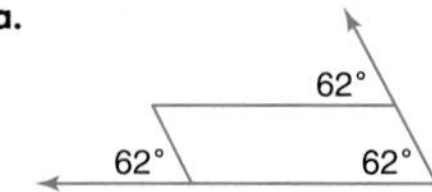

b.

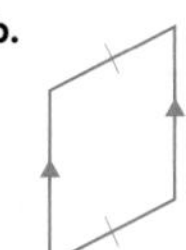

3. **You Decide** Nida says "If a quadrilateral is *not* a parallelogram, then neither pair of opposite sides are parallel." Celina disagrees. Who is correct and why? Celina; one pair of sides could be parallel.
4. **Explain** how slope can be used to identify parallelograms in the coordinate plane.

4. Since parallel lines have same slope, find the slopes of each pair of sides and see if they are the same.

5. **Compare and contrast** Theorems 6–1, 6–2, and 6–4 with Theorems 6–5, 6–6, and 6–7. See margin.

6. Write a memory tool you can use to remember the tests for a parallelogram. See students' work.

Guided Practice

7. Yes; the triangles are congruent by SAS, so both pairs of opposite sides are congruent.
8. No; none of the tests for parallelograms is fulfilled.

Determine if each quadrilateral is a parallelogram. Justify your answer.

7. 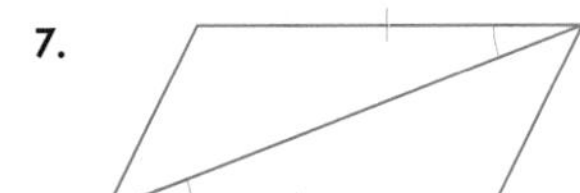

8. 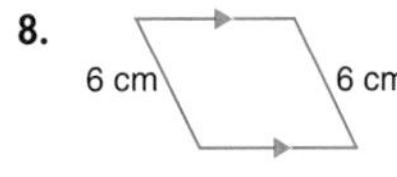

Find the values of x and y that ensure each quadrilateral is a parallelogram.

9. 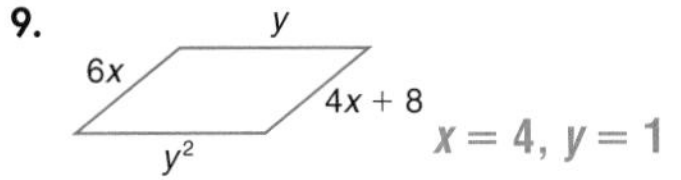

$x = 4$, $y = 1$

10. 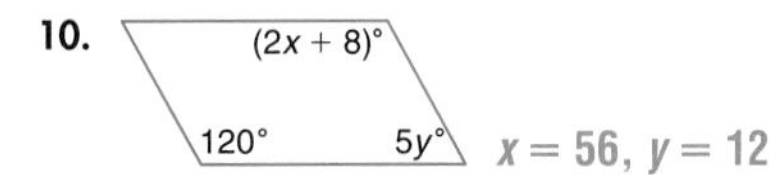

$x = 56$, $y = 12$

11. False; the diagonals being congruent does not ensure that opposite sides will be congruent or parallel.

11. Determine if the conditional *If the diagonals of a quadrilateral are congruent, then the quadrilateral is a parallelogram* is true or false. Explain your reasoning.

12. Quadrilateral *GHJK* has vertices $G(-2, 8)$, $H(4, 4)$, $J(6, -3)$, and $K(-1, -7)$. Determine whether *GHJK* is a parallelogram by Theorem 6–8. Justify your answer. See margin.

Proof

13. Complete a two-column proof. See margin.

Given: $\overline{HD} \cong \overline{DN}$
$\angle DHM \cong \angle DNT$

Prove: Quadrilateral *MNTH* is a parallelogram.

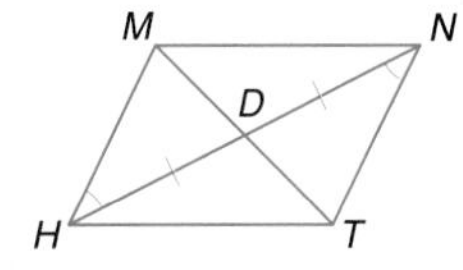

14. **Engineering** Deshon uses an expandable gate to keep his new puppy in the kitchen. As the gate expands or collapses, the shapes that form the gate (such as *ABCD*) always remain parallelograms. Explain why this is true.

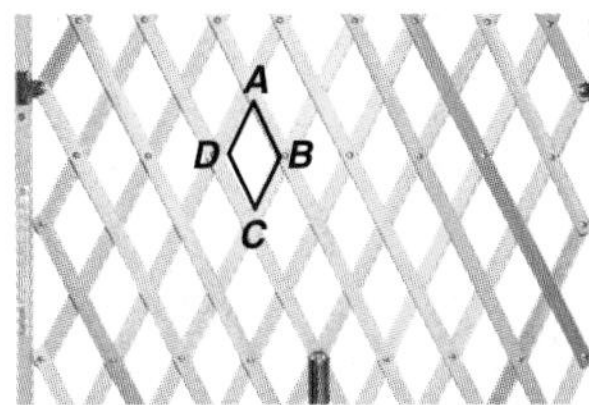

EXERCISES

Practice

14. The lengths of the sides of each part of the gate are nonchanging. Thus both pairs of opposite sides remain congruent and the quadrilateral is always a parallelogram.

A

B

Determine if each quadrilateral is a parallelogram. Justify your answer. 15–20. See Solutions Manual for justifications.

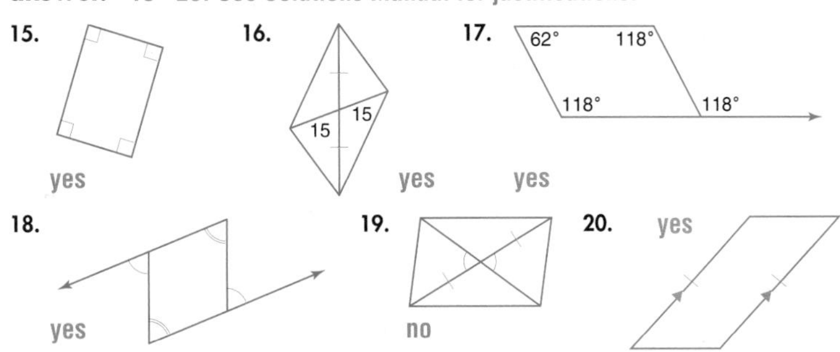

Reteaching

Using Modeling Each theorem in the lesson can be modeled using straws and tape. Have students model the conditions of each theorem. (To be sure that pairs of straws are parallel, students may use grid or dot paper.) In each case, have them determine that the resulting figure is a parallelogram.

3 PRACTICE/APPLY

Check for Understanding

Exercises 1–14 are designed to help you assess your students' understanding through reading, writing, speaking, and modeling. You should work through Exercises 1–6 with your students and then monitor their work on Exercises 7–14.

Error Analysis

Without sufficient information, a student might decide that a quadrilateral is a parallelogram. For example, a student might decide a figure is a parallelogram based on one pair of congruent opposite angles. When such an error occurs, it may help to have the student draw a counterexample, such as a trapezoid.

Assignment Guide

Core (with proof): 15–37 odd, 38, 39, 41, 42–50
Core (informal): 15–33 odd, 38, 39, 41, 42–50
Enriched: 16–36 even, 38–50

For **Extra Practice**, see p. 774.

The red A, B, and C flags, printed only in the Teacher's Wraparound Edition, indicate the level of difficulty of the exercises.

Additional Answer

13. Given: $\overline{HD} \cong \overline{DN}$
$\angle DHM \cong \angle DNT$
Prove: Quadrilateral *MNTH* is a parallelogram.
Proof:
Statements (Reasons)
1. $\overline{HD} \cong \overline{DN}$, $\angle DHM \cong \angle DNT$ (Given)
2. $\angle MDH \cong \angle TDN$ (Vert. ∠s are ≅.)
3. $\triangle MDH \cong \triangle TDN$ (ASA)
4. $\overline{MH} \cong \overline{TN}$ (CPCTC)
5. $\overline{MH} \parallel \overline{TN}$ (If ⫲ and alt. int. ∠s are ≅ then the lines are ∥.)
6. *MNTH* is a parallelogram. (If one pair of opp. sides of a quad. are both ∥ and ≅, then the quad. is a parallelogram.)

Additional Answers

27. false; counterexample:

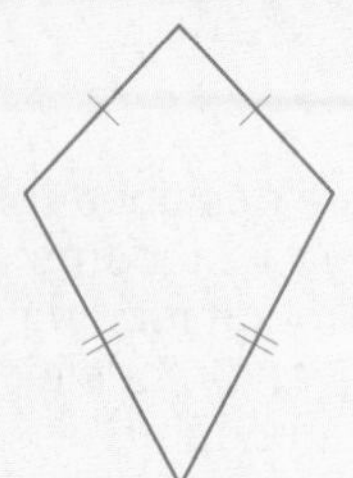

29. false; counterexample:

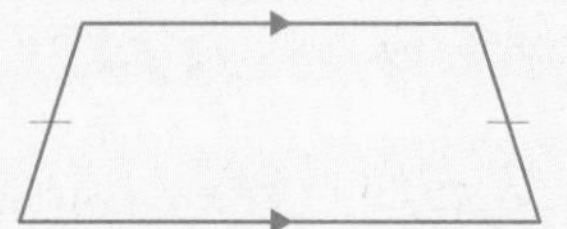

30. not a parallelogram; $AB = \sqrt{52}$, $CD = \sqrt{34}$, $BC = \sqrt{26}$, $DA = \sqrt{68}$. Since opposite sides are not congruent, $ABCD$ is not a parallelogram.

31. a parallelogram; slope of $\overline{FG} = -\frac{1}{4}$, slope of $\overline{HJ} = -\frac{1}{4}$, $FG = \sqrt{17}$, $HJ = \sqrt{17}$. Since one pair of opposite sides is both parallel and congruent, $FGHJ$ is a parallelogram.

32. a parallelogram; midpoint of $\overline{KM} = (2.5, 2.5)$ and midpoint of $\overline{LN} = (2.5, 2.5)$. Since the diagonals bisect each other, $KLMN$ is a parallelogram.

Study Guide Masters, p. 33

6-2 NAME________ DATE______

Study Guide Student Edition Pages 298–304

Tests for Parallelograms

You can show that a quadrilateral is a parallelogram if you can show that one of the following is true.

1. Both pairs of opposite sides are parallel.
2. Both pairs of opposite sides are congruent.
3. Diagonals bisect each other.
4. Both pairs of opposite angles are congruent.
5. A pair of opposite sides are both parallel and congruent.

Example: Find the values of x and y that ensure the quadrilateral is a parallelogram.

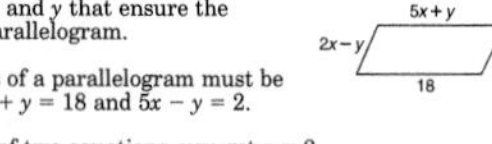

Since opposite sides of a parallelogram must be congruent, then $5x + y = 18$ and $5x - y = 2$.

Solving the system of two equations, you get $x = 2$ and $y = 8$.

Determine if each quadrilateral is a parallelogram. Justify your answer.

1. 
2. 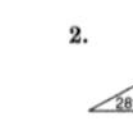 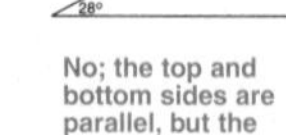
3. 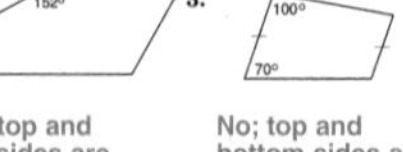

1. Yes; both pairs of opposite sides are congruent.
2. No; the top and bottom sides are parallel, but the other pair may not be.
3. No; top and bottom sides are not parallel.

Find the values of x and y that ensure each quadrilateral is a parallelogram.

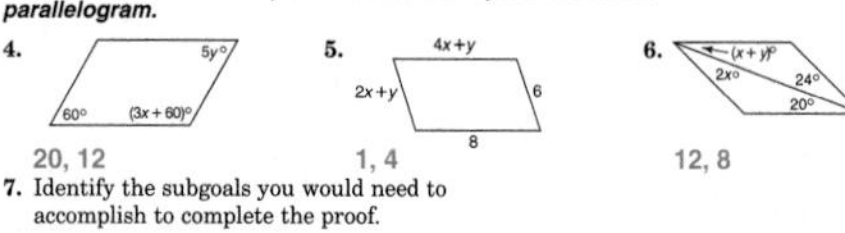

4. 20, 12
5. 1, 4
6. 12, 8

7. Identify the subgoals you would need to accomplish to complete the proof.

Given: $\overline{AB} \perp$ plane BCD.
$\overline{DB} \cong \overline{CB}$

Prove: $\angle DAB \cong \angle CAB$

Prove $\triangle ADB \cong \triangle ACB$, then show $\angle DAB \cong \angle CAB$.

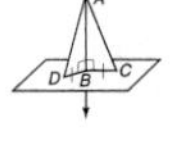

Find the values of x and y that ensure each quadrilateral is a parallelogram.

21. $x = 5$, $y = 16$
22. $x = 19$, $y = 8$
23. $x = 11$, $y = 12$

21. 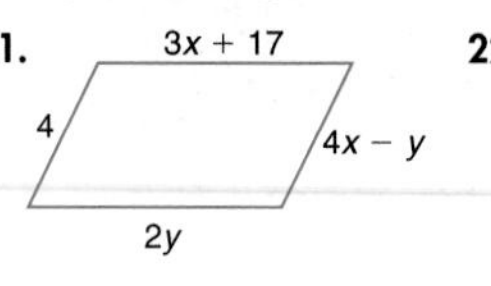

22. 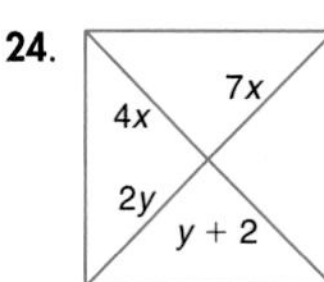

23. 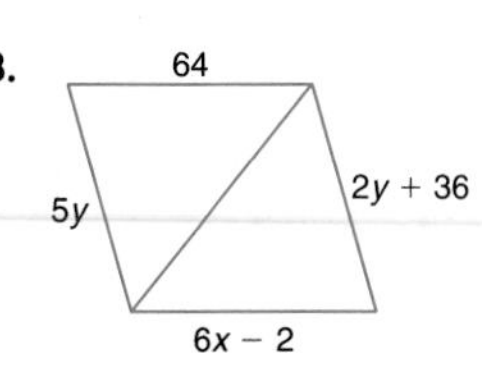

24. $x = 4$, $y = 14$
25. $x = 64$, $y = 23.5$
26. $x = 3$ or -2, $y = 7$ or -7

24. 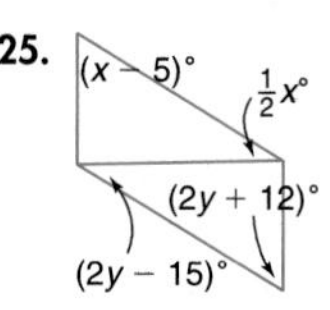

25. 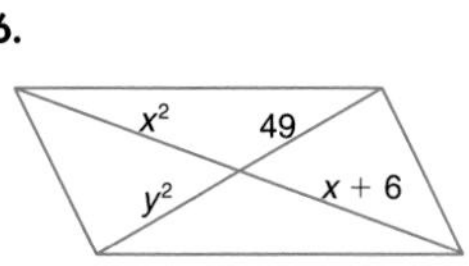

26.

Determine if each conditional is true or false. Explain your reasoning.

27. If two pairs of consecutive sides of a quadrilateral are congruent, then the quadrilateral must be a parallelogram. See margin.

28. If all four sides of a quadrilateral are congruent, then the quadrilateral is a parallelogram.

28. True; if all four sides are congruent, then both pairs of opposite sides are congruent and the quadrilateral is a parallelogram.

29. If a quadrilateral has one pair of congruent sides and one pair of parallel sides, then the quadrilateral must be a parallelogram. See margin.

Determine whether the quadrilateral with the given vertices is a parallelogram by the indicated theorem. 30–32. See margin.

30. $A(5, 6)$, $B(9, 0)$, $C(8, -5)$, $D(3, -2)$; Theorem 6–5

31. $F(-7, 3)$, $G(-3, 2)$, $H(0, -4)$, $J(-4, -3)$; Theorem 6–8

32. $K(-1, 9)$, $L(3, 8)$, $M(6, 2)$, $N(2, 3)$; Theorem 6–7

33. Using a compass and a straightedge, construct a parallelogram *QRST* with one angle congruent to $\angle P$ and sides congruent to $\overline{AB}$ and $\overline{CD}$. See margin.

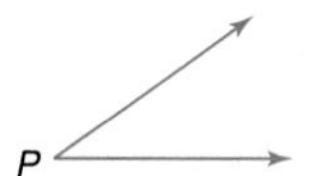

34. Write a paragraph proof of Theorem 6–6. 34–37. See Solutions Manual.

35. Write a two-column proof of Theorem 6–7.

36. Prove Theorem 6–8.

37. A *regular hexagon* is a six-sided polygon with all sides and all angles congruent. If *ABCDEF* is a regular hexagon, identify the subgoals you would use to prove *FDCA* is a parallelogram. Then write a proof.

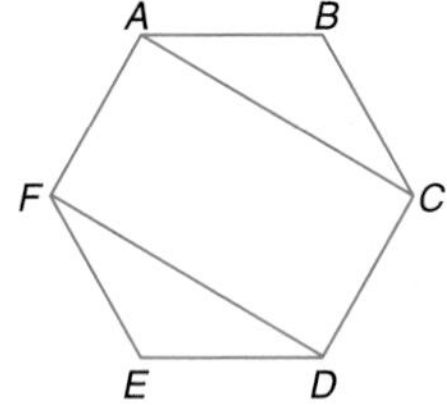

Additional Answer

33.

Critical Thinking

38. Ellen claims she has invented a new geometry theorem: *A diagonal of a parallelogram bisects its angles.* She gives the following proof.

Given: $\square MATH$ with diagonal $\overline{MT}$

Prove: $\overline{MT}$ bisects $\angle AMH$ and $\angle ATH$

Proof: Since $MATH$ is a parallelogram, $\overline{MH} \cong \overline{AT}$ and $\overline{MA} \cong \overline{HT}$. Since $\overline{MT} \cong \overline{MT}$, $\triangle MHT \cong \triangle MAT$ by SSS. Therefore, $\angle 1 \cong \angle 2$ and $\angle 3 \cong \angle 4$.

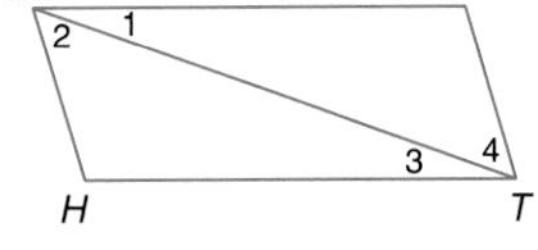

a. Do you think Ellen's new theorem is true? no

b. Is her proof correct? Explain your reasoning. No; the conclusion should be $\angle 1 \cong \angle 3$ and $\angle 2 \cong \angle 4$.

Applications and Problem Solving

39. **Construction** Wood lattice panels are made in a configuration of parallelograms.

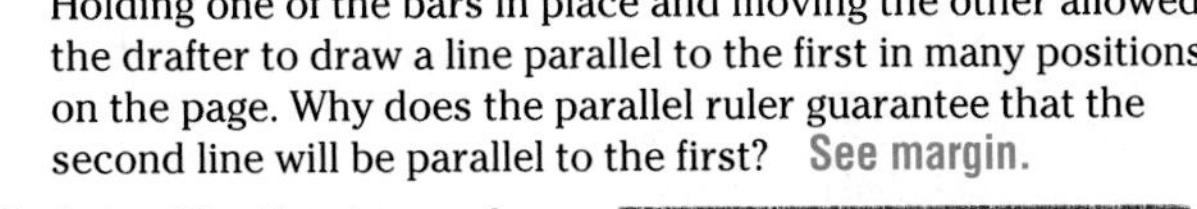

a. Explain how the person who manufactured the panels could verify that the overlapped boards form parallelograms.

b. Look up the word *lattice* in the dictionary. Is the word *crisscross* used in the definition? Why do you think a crisscross pattern results in a series of parallelograms? See margin.

39a. Sample answer: Measure each pair of opposite segments.

FYI

The Navajo reservation in Arizona, New Mexico, and Utah is the largest American Indian reservation, with a population exceeding 140,000 people.

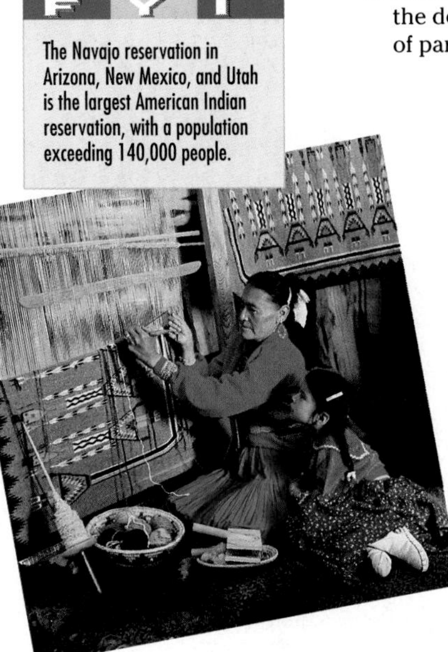

40. **Drafting** Before computer drawing programs became available, blueprints for buildings or mechanical parts were drawn by hand. One of the tools drafters used, a parallel ruler, is shown at the right. Holding one of the bars in place and moving the other allowed the drafter to draw a line parallel to the first in many positions on the page. Why does the parallel ruler guarantee that the second line will be parallel to the first? See margin.

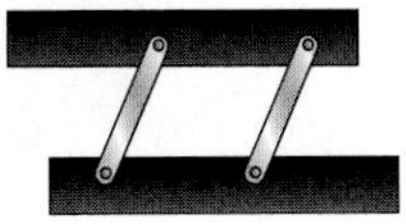

41. **Arts** The Navajo people are well known for their skill in weaving. The design at the right, known as the Eye-Dazzler, became popular with Navajo weavers in the 1880s. How many parallelograms are in the pattern? 47

Mixed Review

42. In $\square ABCD$, $AB = 2x + 5$, $CD = y + 1$, $AD = y + 5$, and $BC = 3x - 4$. Find the measure of each side. (Lesson 6–1)

43. If the sides of a triangle have measures of $3x + 2$, $8x + 10$, and $5x + 8$, find all possible values of x. (Lesson 5–5) no solution

44. **Work Backward** Juana purchased a jacket for \$190.80. She received a 20% discount off of the original price and was charged a 6% sales tax. What was the original price of the jacket? (Lesson 5–3) \$225.00

45. If $\triangle DOG \cong \triangle CAT$, what segment in $\triangle CAT$ is congruent to $\overline{GO}$ in $\triangle DOG$? (Lesson 4–3) $\overline{TA}$

46. Find the slope of the line that passes through $A(-4, 8)$ and $B(3, 0)$. (Lesson 3–3)

42. $AB = CD = 31$; $AD = BC = 35$

46. $-\frac{8}{7}$

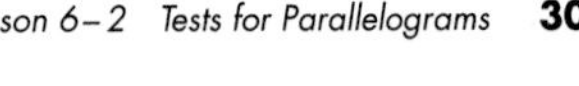

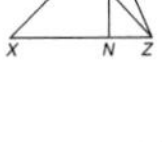

Tech Prep

Draftsperson A draftsperson requires knowledge of geometry to prepare and read blueprints. Two-year degrees in Civil Engineering Technology can lead to opportunities in this field. For more information on tech prep, see the *Teacher's Handbook*.

FYI

The Navajo also occupy the largest territory in area, consisting of 25,000 square miles.

Additional Answers

39b. Sample answer: Yes; the pattern yields quadrilaterals with guaranteed parallel opposite sides.

40. Both pairs of opposite sides are congruent, so the sides and arms of the ruler always form a parallelogram.

Practice Masters, p. 33

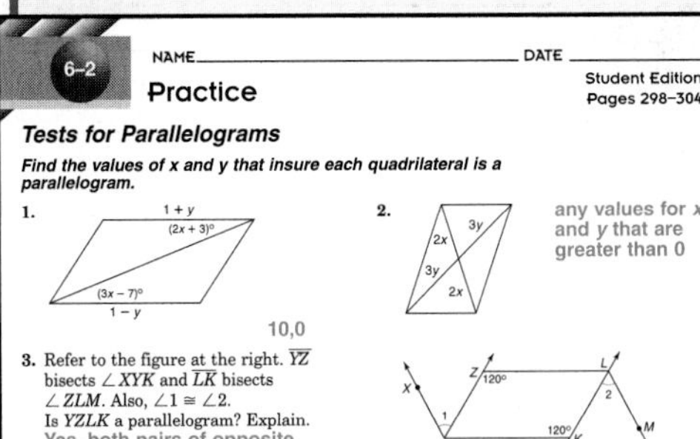

6–2 NAME________ DATE ________

Practice — Student Edition Pages 298–304

Tests for Parallelograms

Find the values of x and y that insure each quadrilateral is a parallelogram.

1. (figure: $1 + y$, $(2x + 3)°$, $(3x - 7)°$, $1 - y$) 10,0

2. (figure: $2x$, $3y$, $3y$, $2x$) any values for x and y that are greater than 0

3. Refer to the figure at the right. $\overline{YZ}$ bisects $\angle XYK$ and $\overline{LK}$ bisects $\angle ZLM$. Also, $\angle 1 \cong \angle 2$. Is $YZLK$ a parallelogram? Explain. Yes, both pairs of opposite angles are congruent.

4. Refer to the figure at the right. $\overline{AC} \cong \overline{WV}$ and $\overline{BD} \cong \overline{WV}$. Also, $\overline{AB} \cong \overline{XY}$ and $\overline{CD} \cong \overline{XY}$. Is $ABCD$ a parallelogram? Explain. Yes, both pairs of opposite sides are congruent.

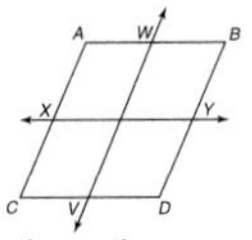

Determine whether quadrilateral ABCD with the given vertices is a parallelogram. Explain.

5. $A(2, 5)$, $B(5, 9)$, $C(3, -1)$, $D(6, 3)$ Yes, opposite sides have equal lengths and are therefore congruent.

6. $A(-1, 6)$, $B(2, -3)$, $C(5, 9)$, $D(2, 7)$ No, slope $\overline{AD} = \frac{1}{3}$ and slope $\overline{BC} = 4$, so these two opposite sides are not parallel.

7. **Identify Subgoals** Identify the subgoals you would need to accomplish to complete the proof.

Given: $\overline{YN} \perp \overline{XZ}$, $\overline{ZM} \perp \overline{XY}$, $\overline{XZ} \cong \overline{XY}$, $\overline{XM} \cong \overline{XN}$

Subgoals: $\triangle XZM \cong \triangle XYN$, $\angle XMZ \cong \angle XNY$

4 ASSESS

Closing Activity

Speaking Draw a parallelogram on the chalkboard and label its vertices. Ask a volunteer to invent the minimum amount of data about the measures of the figure that would prove that it is a parallelogram. Ask another student to do the same but with different data measures. Repeat until all five methods have been presented.

Chapter 6 Quiz A (Lessons 6-1 and 6-2) is available in the *Assessment and Evaluation Masters*, p. 156.

Mathematics and SOCIETY

Skyscrapers have to be built to move with the wind. Computers and wind tunnels test scale models to determine stress points and potential problems.

Enrichment Masters, p. 33

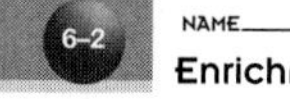

NAME________________ DATE ________

Enrichment Student Edition Pages 298–304

Polygon Puzzles

1. How many nonrectangular parallelograms can you draw that have vertices on the dots below? (Hint: Label the dots and make a list.)

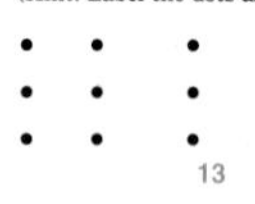

13

2. How many ways can the large rectangle be divided into five congruent rectangles? One way is shown.

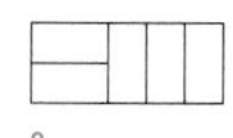

9

3. What polygons can be formed by drawing a single straight line across a square? For example, the line shown forms two isosceles right triangles.

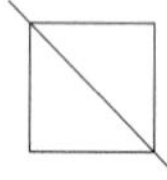

pentagon, scalene triangle, rectangle, trapezoid, isosceles right triangle

4. How many rectangles can be formed by joining points on the circle? The points are equally spaced around the circle. For example, one rectangle is *ABFG*.

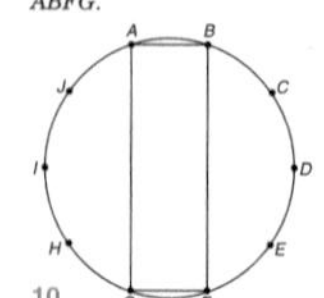

10

5. Suppose *ABDE* is a parallelogram and that in the figure, $m\angle BFC$ is 4 more than twice $m\angle EFC$. Find the measure of each angle of each triangle in the figure.

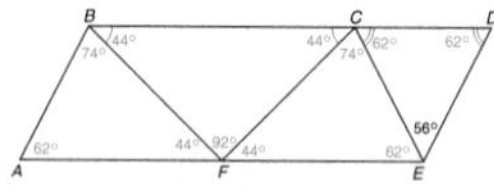

47. Write a conjecture based on the statement *Both pairs of opposite sides of a quadrilateral are congruent.* Draw a figure to illustrate your conjecture. (Lesson 2–1) **The quadrilateral is a parallelogram; see students' work.**

48. If $\angle B \cong \angle L$, $m\angle B = 7x + 29$, and $m\angle L = 9x - 1$, is $\angle B$ acute, right, or obtuse? (Lesson 1–6) **obtuse**

INTEGRATION **Algebra**

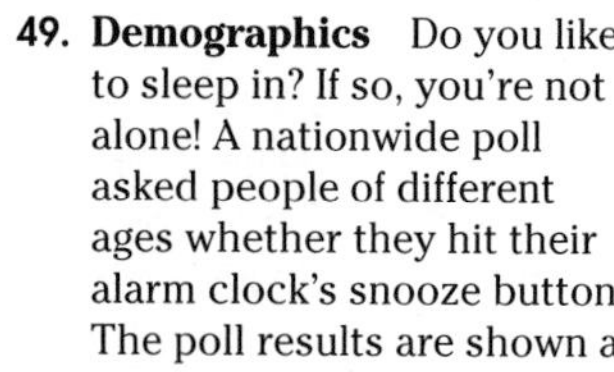

49. **Demographics** Do you like to sleep in? If so, you're not alone! A nationwide poll asked people of different ages whether they hit their alarm clock's snooze button. The poll results are shown at the right.

Source: Opinion Research for Select Comfort

a. What percent of the people aged 35–44 hit the snooze button? **36%**

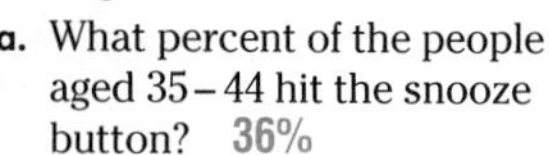

b. The U.S. Census Bureau projects that there are 25,465,000 people 18 to 24 years old in the United States. How many of those people do you predict hit the snooze button each morning? **13,241,800**

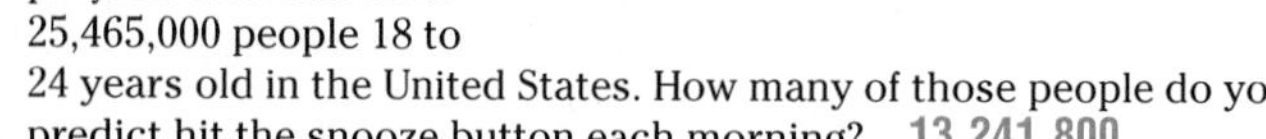

50. **Business** How much coffee that costs \$6 a pound should be mixed with 10 pounds of coffee that costs \$7.25 a pound to obtain a mixture that costs \$7 a pound? **2.5 lbs**

3. **The concrete building; it would not flex and stretch as much as a steel one.**

Skyscraper Geometry

The excerpt below appeared in an article in *American Scientist* in July–August, 1996.

TODAY, MOST OF THE TALLEST BUILDINGS in the world are being proposed for locations such as Tokyo, Taiwan, Hong Kong and mainland China. And they are not only being proposed: they are being built, with the tallest building in the world recently being topped out at 1,482 feet in Kuala Lumpur, Malaysia....the pair of buildings known as the Petronas Twin Towers...have risen to become the world's tallest buildings....The tapering at the top of the building demanded some especially tricky structural engineering, and its geometry necessitated the installation of a wide variety of different-size glass panels....The Twin Towers required about a million and a half square feet of stainless steel cladding and glass, in the form of 32,000 windows, to form a so-called curtain wall. ■

1. The *footprint* or base of each of the Twin Towers has the shape of an eight-pointed star with intermediate arcs joining the points. Is this shape a type of quadrilateral? Explain. **No; it's not a four-sided polygon.**
2. If a building's foundation is not secure, the building can settle and begin to tilt, so its walls are no longer vertical. What could happen if the angle of tilt continued to increase each year? **The building could eventually collapse.**
3. If a steel building and a concrete building were built in the shape of identical parallelograms, which would retain its shape better when exposed to high winds? Why?

Extension

Reasoning Given quadrilateral *XYZW* with vertices $X(-2, 3)$, $Y(-2, -3)$, $Z(3, 2)$, and $W(3, 8)$. Write a plan using the fact that if the diagonals of a quadrilateral bisect each other, then the quadrilateral is a parallelogram, for a proof that *XYZW* is a parallelogram.

The diagonals are $\overline{XZ}$ and $\overline{YW}$. If you can show that they have the same midpoint, you can show that they bisect each other. Therefore, Theorem 6-7 applies, and the figure is a parallelogram.

6–3A Using Technology Exploring Rectangles

A Preview of Lesson 6–3

A quadrilateral with four right angles is a **rectangle**. A TI-92 is a useful tool for exploring some of the characteristics of a rectangle. Use the following steps to draw a rectangle.

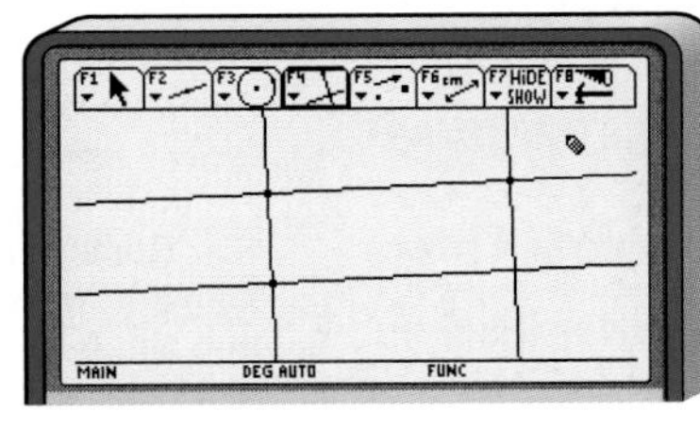

Step 1 Press F2 and choose Line from the menu to draw a line anywhere on the screen. Use the arrow key to position the cursor anywhere on the screen and press ENTER. Move to another place on the screen and press ENTER to position the line.

Step 2 Next, draw a line perpendicular to the first line. Press F4 and choose Perpendicular Line from the menu. Position the cursor somewhere on the first line so that the prompt *perpendicular to this line* appears. Press ENTER. Now move the cursor to another point you would like the line to pass through and press ENTER.

Step 3 Repeat the procedure in step 2 to draw a line perpendicular to the second line. Then draw a line perpendicular to the third line.

A rectangle is formed by the segments whose endpoints are the points of intersection of the lines.

EXERCISES

Use the measuring capabilities of the TI-92 to explore the characteristics of a rectangle.

1. What appears to be true about the opposite sides of the rectangle that you drew? Make a conjecture and then measure each side to check your conjecture. The opposite sides of a rectangle are congruent.
2. Draw the diagonals of the rectangle by first pressing F2 and choosing Segment from the menu. Then position the cursor over one of the vertices of the rectangle. When the prompt *point at this intersection* appears, press ENTER. Move the cursor to the opposite vertex. Press ENTER when the prompt *point at this intersection* appears. Repeat to draw the other diagonal. a. The diagonals of a rectangle are congruent.
   a. Measure each diagonal. What do you observe?
   b. What is true about the triangles formed by the sides of the rectangle and a diagonal? Justify your conclusion.
3. Clear the screen and follow steps 1–3 to draw another rectangle. Do the relationships you found for the first rectangle you drew hold true for this rectangle also?

2b. The triangles are congruent by SSS.

3. Yes; the opposite sides are congruent and the diagonals are congruent.

6-3A LESSON NOTES

NCTM Standards: 1–4, 7, 8

Objective

Use a TI-92 calculator to draw and explore characteristics of rectangles.

Recommended Time

20 minutes

Instructional Resources

Instructions for using the *Geometer's Sketchpad* for this activity are available in the *Graphing Calculator and Computer Masters*, p. 25. Answers may vary from those given.

1 FOCUS

Motivating the Lesson

Have students list ten objects in or near the classroom that are rectangles.

2 TEACH

Teaching Tip Remind students that when they use the Perpendicular Line option from the menu, they are forming a right angle.

Using Technology

This lesson offers an excellent opportunity for using technology in your geometry classroom. For more information on using technology, see *Graphing Calculators in the Mathematics Classroom*, one of the titles in the Glencoe Mathematics Professional Series.

3 PRACTICE/APPLY

Assignment Guide

Core (with proof): 1–3
Core (informal): 1–3
Enriched: 1–3

4 ASSESS

Observing students working with technology is an excellent method of assessment.

6-3 LESSON NOTES

NCTM Standards: 1–5, 7, 8

Instructional Resources

- Study Guide Master 6-3
- Practice Master 6-3
- Enrichment Master 6-3
- Assessment and Evaluation Masters, pp. 155, 156

Transparency 6-3A contains the 5-Minute Check for this lesson; **Transparency 6-3B** contains a teaching aid for this lesson.

Recommended Pacing	
Standard Pacing	Day 6 of 13
Honors Pacing	Day 6 of 12
Block Scheduling*	Day 3 of 6

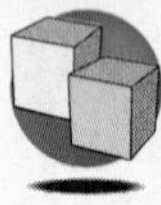

*For more information on pacing and possible lesson plans, refer to the *Block Scheduling Booklet*.

1 FOCUS

5-Minute Check
(over Lesson 6-2)

Write a reason for each statement in the proof.
Given: $\triangle PRS \cong \triangle RPQ$
Prove: $PQRS$ is a parallelogram.

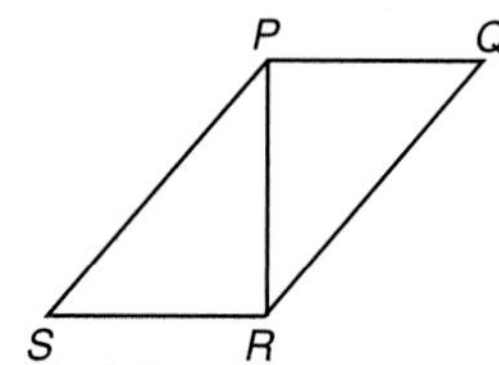

Statements (Reasons)

1. $\triangle PRS \cong \triangle RPQ$ (Given)
2. $\angle SRP \cong \angle QPR$ (CPCTC)
3. $\overline{PQ} \parallel \overline{RS}$ (If two lines are cut by a transversal and alt. int. $\angle$s are $\cong$, then the lines are $\parallel$.)
4. $\overline{PQ} \cong \overline{RS}$ (CPCTC)
5. $PQRS$ is a parallelogram. (If a pair of opp. sides of a quad. are $\cong$ and $\parallel$, it is a $\square$.)

6-3 Rectangles

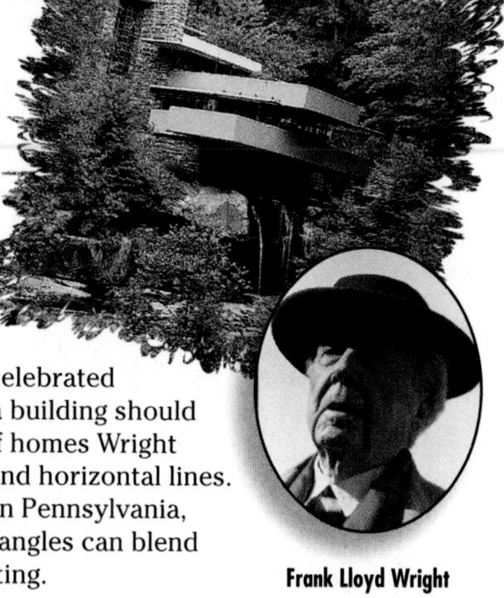

What YOU'LL LEARN
- To recognize and apply the properties of rectangles.

Why IT'S IMPORTANT

You can use properties of rectangles to solve problems involving architecture and sports.

APPLICATION
Architecture

Frank Lloyd Wright is one of the most celebrated American architects. He believed that a building should "grow" from its site. The prairie style of homes Wright created emphasizes natural materials and horizontal lines. His "Falling Water" design at Bear Run in Pennsylvania, shows how a structure made up of rectangles can blend into and become a part of a natural setting.

Frank Lloyd Wright

A **rectangle** is a quadrilateral with four right angles. It follows that since both pairs of opposite angles are congruent, a rectangle is a special type of parallelogram. Thus, a rectangle has all the properties of a parallelogram.

However, the diagonals of a rectangle have an additional special relationship. You can use a spreadsheet to explore this relationship.

EXPLORATION — SPREADSHEETS

Form a rectangle $MNOP$ on a coordinate grid by choosing any point $M(a, b)$ for the first vertex. The other three vertices are $N(a, 0)$, $O(0, 0)$, and $P(0, b)$.

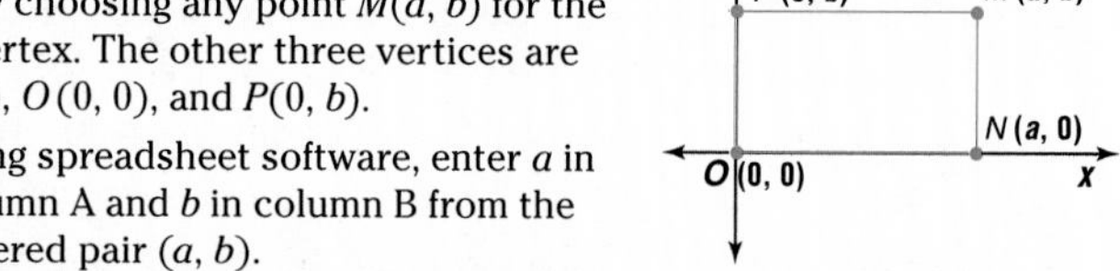

- Using spreadsheet software, enter a in column A and b in column B from the ordered pair (a, b).
- Enter the formula SQR((A1 − 0)^2 + (B1 − 0)^2) in cell C1. This formula will find the distance between M and O. Copy the formula into the other cells in column C.
- Write a similar formula to find the distance between N and P. Enter the formula in the cells in column D.

Your Turn b. They are congruent

a. Use the spreadsheet to compare the lengths of the diagonals for at least seven different rectangles. See students' work.

b. What appears to be true about the diagonals of a rectangle?

This exploration leads us to the following theorem. *You will be asked to prove this Theorem in Exercise 14.*

Theorem 6-9	**If a parallelogram is a rectangle, then its diagonals are congruent.**

You can construct a rectangle using right angles and then verify that the diagonals are congruent.

EXPLORATION

Compare this result with what students know about the diagonals of a parallelogram. The diagonals of a rectangle are congruent, but the diagonals of a parallelogram are not congruent.

CONSTRUCTION

Rectangle

Construct a rectangle with a length of 6 centimeters and a width of 3 centimeters.

1. Use a straightedge to draw line ℓ. Label a point J on ℓ. With your compass set at 6 centimeters, place the point at J and locate point K on ℓ so that $JK = 6$. Now construct lines perpendicular to ℓ through J and through K. Label them m and n.

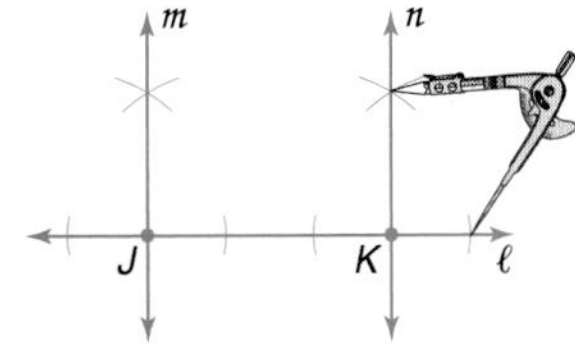

2. Set your compass at 3 centimeters. Place the compass point at J and mark off a segment on m. Using the same compass setting, place the compass at K and mark a segment on n. Label these points O and P.

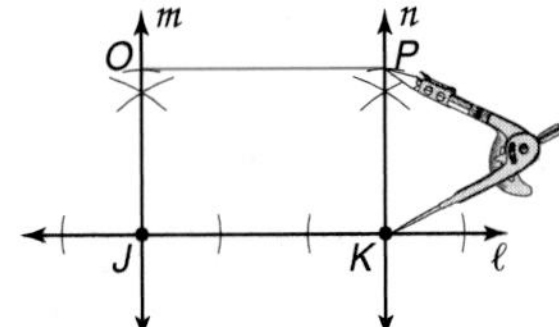

3. Draw $\overline{OP}$.

Quadrilateral $JKPO$ is a parallelogram, and all angles are right angles. Therefore, quadrilateral $JKPO$ is a rectangle with a length of 6 centimeters and a width of 3 centimeters.

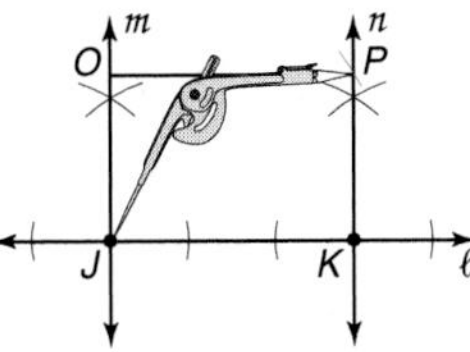

4. Locate the compass setting that represents JP and compare to the setting for OK. The measures should be the same.

Another method of constructing a rectangle often used in the building industry is to use the converse of Theorem 6–9. If the diagonals of a parallelogram are congruent, the parallelogram is a rectangle.

Example 1

Construction

The Mazdrons are building an addition to their house. Mr. Mazdron is cutting an opening for a new window. If he has measured to see that the opposite sides are congruent and that the diagonals are congruent, can Mr. Mazdron be sure that the window opening is a rectangle?

Explore Draw a diagram and label the vertices of the window opening A, B, C, and D. According to Mr. Mazdron's measurements, $\overline{AB} \cong \overline{CD}$, $\overline{BC} \cong \overline{AD}$, and $\overline{AC} \cong \overline{BD}$.

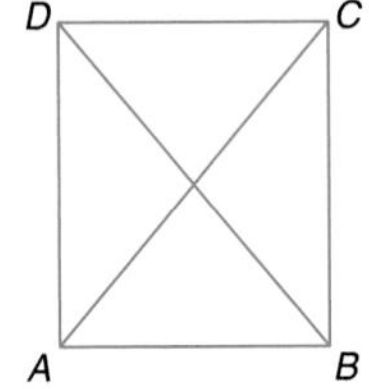

Plan We can show that $ABCD$ is a rectangle by first proving that it is a parallelogram and then proving that it has four right angles.

Solve **Given:** $\overline{AB} \cong \overline{CD}$

$\overline{BC} \cong \overline{AD}$

$\overline{AC} \cong \overline{BD}$

Prove: $ABCD$ is a rectangle.

(continued on the next page)

Motivating the Lesson

Questioning Ask students how they would make sure the shelves in a bookcase are rectangles.

2 TEACH

Teaching Tip For the construction, demonstrate each step by using the chalkboard or overhead projector.

In-Class Example

For Example 1

Write a paragraph proof.

Given: $\square RECT$

$\overline{RE} \perp \overline{RT}$

Prove: $\square RECT$ is a rectangle.

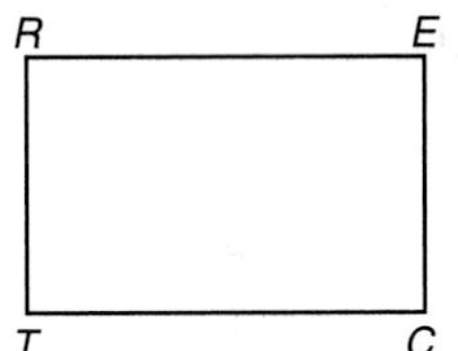

Since $\overline{RE} \perp \overline{RT}$, $\angle ERT$ is a right angle. Since $RECT$ is a parallelogram, opposite angles are congruent and consecutive angles are supplementary. Therefore, $\angle ECT$, $\angle RTC$, and $\angle REC$ are all right angles. Since all four angles are right, $\square RECT$ is a rectangle.

Alternative Learning Styles

Auditory Ask students to discuss whether it can be proven that a rectangle is a parallelogram, given only that a rectangle is a quadrilateral with four right angles. **Yes; since both pairs of opposite angles are congruent, the figure is a parallelogram.**

In-Class Example

For Example 2
If quadrilateral *JKLM* is a rectangle, $LP = 3x + 7$, and $MK = 26$, find the value of x.

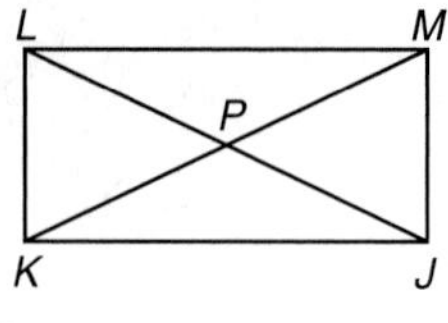

2

Paragraph proof:

Since congruence of segments is reflexive, $\overline{AB} \cong \overline{AB}$. So, $\triangle ADB \cong \triangle BCA$ by SSS. Thus, $\angle DAB \cong \angle CBA$ by CPCTC. *ABCD* is a parallelogram since the opposite sides are congruent. Therefore, since consecutive angles of a parallelogram are supplementary, $\angle DAB$ and $\angle CBA$ are supplementary. So, $m\angle DAB + m\angle CBA = 180$ and $m\angle DAB = m\angle CBA$ by the definitions of supplementary and congruent angles.

By the Substitution Property of Equality, $2m\angle DAB = 180$, so $m\angle DAB = 90$. Therefore, $\angle DAB$ and $\angle CBA$ are right angles. Since the opposite angles of a parallelogram are congruent, $\angle DCB$ and $\angle CDA$ are right angles. By definition, *ABCD* is a rectangle.

As a result, Mr. Mazdron can be sure that the window opening is a rectangle if the opposite sides are congruent and the diagonals are congruent.

Examine Try to draw a quadrilateral that is not a rectangle, but has opposite sides congruent and diagonals congruent.

Example 1 proves the converse of Theorem 6–9, which is Theorem 6–10.

Theorem 6–10	**If the diagonals of a parallelogram are congruent, then the parallelogram is a rectangle.**

Recall that perpendicular lines have slopes whose product is -1. You can use this fact to verify that a quadrilateral in the coordinate plane is a rectangle.

Example 2

Algebra

Determine whether parallelogram *ABCD* is a rectangle, given $A(-6, 9)$, $B(5, 10)$, $C(6, -1)$, and $D(-5, -2)$.

Method 1: Using slopes

slope of $\overline{AB} = \frac{10 - 9}{5 - (-6)}$ or $\frac{1}{11}$

slope of $\overline{CD} = \frac{-1 - (-2)}{6 - (-5)}$ or $\frac{1}{11}$

slope of $\overline{AD} = \frac{9 - (-2)}{-6 - (-5)}$ or -11

slope of $\overline{BC} = \frac{10 - (-1)}{5 - 6}$ or -11

Thus, $\overline{AB} \parallel \overline{CD}$ and $\overline{AD} \parallel \overline{BC}$. In addition, the product of the slopes of consecutive sides is -1. Thus, $\overline{AB} \perp \overline{BC}$, $\overline{BC} \perp \overline{CD}$, $\overline{CD} \perp \overline{AD}$, and $\overline{AD} \perp \overline{AB}$, creating four right angles. Therefore, quadrilateral *ABCD* is a rectangle.

Method 2: Using diagonals

$AC = \sqrt{(-6 - 6)^2 + (9 - (-1))^2}$
$= \sqrt{144 + 100}$
$= \sqrt{244}$

$\sqrt{244} \approx 15.62$

$BD = \sqrt{(5 - (-5))^2 + (10 - (-2))^2}$
$= \sqrt{100 + 144}$
$= \sqrt{244}$

Since *ABCD* is a parallelogram and the diagonals are congruent, *ABCD* is a rectangle.

Here is a summary of the properties of a rectangle.

If a quadrilateral is a rectangle, then the following properties hold true.
1. Opposite sides are congruent and parallel. 2. Opposite angles are congruent. 3. Consecutive angles are supplementary. 4. Diagonals are congruent and bisect each other. 5. All four angles are right angles.

CHECK FOR UNDERSTANDING

Communicating Mathematics

1. All rectangles are parallelograms, but not all parallelograms are rectangles.

Study the lesson. Then complete the following.

1. **Explain** why a rectangle is a special type of parallelogram.
2. **Draw** an example of a quadrilateral with congruent diagonals that is *not* a rectangle. See margin.
3. **You Decide** Kalere claims that if the diagonals of a quadrilateral are congruent and bisect each other, then the quadrilateral is a rectangle. Amy says the quadrilateral could be a non-rectangular parallelogram. Who is correct and why? See margin.

4. **Assess Yourself** Look for examples of rectangles in the objects around you. Could the objects you see be another shape and still be effective? For example, would a circular window be as useful as a rectangular one? Explain your reasoning. See students' work.

Guided Practice

Quadrilateral *MNOP* is a rectangle. Find the value of *x*.

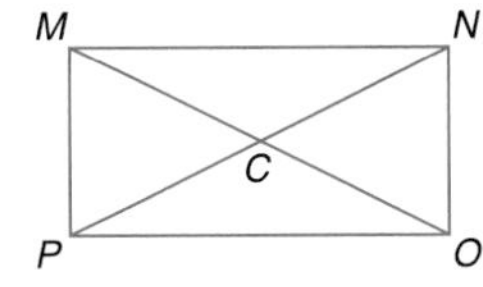

5. $MO = 2x - 8$; $NP = 23$ $x = 15.5$
6. $CN = x^2 + 1$; $CO = 3x + 11$ $x = 5$ or -2
7. $MO = 4x - 13$; $PC = x + 7$ $x = 13.5$
8. *True* or *false:* If a quadrilateral has opposite sides that are congruent, then it is a rectangle. Justify your answer.
9. *ABCD* has vertices at $A(-3, 1)$, $B(4, 8)$, $C(7, 5)$ and $D(0, -2)$. Use slopes to determine if *ABCD* is a rectangle. See margin.

8. False; if a quadrilateral has opposite sides that are congruent, then it is a parallelogram, but it may not be a rectangle.

Use rectangle *KLMN* and the given information to solve each problem.

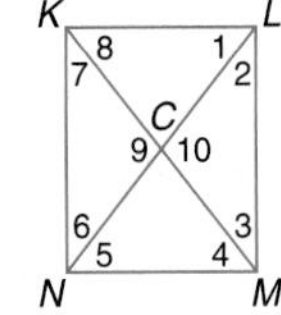

10. $m\angle 1 = 70$. Find $m\angle 2$, $m\angle 5$ and $m\angle 6$. 20; 70; 20
11. $m\angle 9 = 128$. Find $m\angle 6$, $m\angle 7$ and $m\angle 8$. 26; 26; 64
12. $m\angle 5 = 36$. Find $m\angle 2$ and $m\angle 3$. 54; 54
13. **Sports** Jonesboro, Georgia, was the site for the beach volleyball competition of the 1996 Summer Olympics. A beach volleyball court is a rectangle 60 feet long and 30 feet wide. When making the court, a contractor placed stakes and strings to mark the boundaries with the corners at *J*, *K*, *L*, and *M*. To make sure that quadrilateral *JKLM* was a rectangle, the contractor measured $\overline{MK}$ and $\overline{JL}$. If $MK < JL$, describe how she should have moved stakes *F* and *G* to make *JKLM* a rectangle. Explain your answer. Left; see margin for explanation.

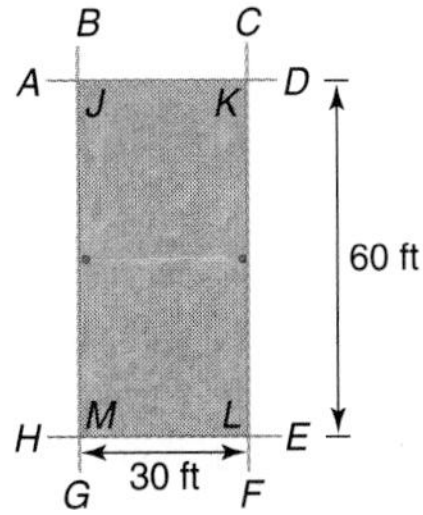

FYI

Beach volleyball made its first appearance as an Olympic sport in the 1996 Atlanta Games.

Reteaching

Using Reasoning Draw rectangle *ABCD* on the chalkboard and draw diagonals $\overline{AC}$ and $\overline{BD}$. Label $AB = 14$ and $BC = 9$. Label the point of intersection of the diagonals *E*, and label $AE = 8.3$. Have students find all the other measures of the rectangle. $BE = DE = CE = 8.3$, $CD = 14$, $AD = 9$

Teaching Tip After reading the list of properties of a rectangle, ask students to identify which are properties of all parallelograms and which are true only for rectangles.

3 PRACTICE/APPLY

Check for Understanding

Exercises 1–14 are designed to help you assess your students' understanding through reading, writing, speaking, and modeling. You should work through Exercises 1–4 with your students and then monitor their work on Exercises 5–14.

The net in volleyball is a rectangle with a width of 3 feet 3 inches.

Additional Answers

2.

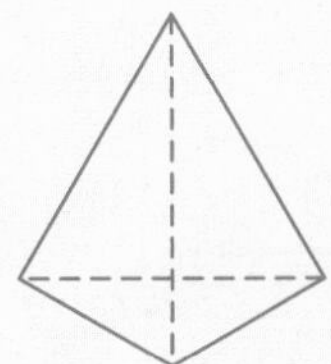

3. Sample answer: Kalere is correct. If the diagonals bisect each other, then the quadrilateral is a parallelogram. If the diagonals of a parallelogram are congruent, then it is a rectangle.
9. *ABCD* is a rectangle, slope of $\overline{AB} = 1$; slope of $\overline{BC} = -1$; slope of $\overline{CD} = 1$; slope of $\overline{DA} = -1$
13. When stakes *F* and *G* are moved left an appropriate distance, $\overline{MK}$ lengthens and its midpoint will coincide with the midpoint of $\overline{JL}$, which becomes shorter after the move.

Assignment Guide

Core (with proof): 15–41 odd, 42, 43, 45–54
Core (informal): 15–37 odd, 42, 43, 45–54
Enriched: 16–40 even, 42–54
All: Self Test, 1–10

For **Extra Practice**, see p. 775.

The red A, B, and C flags, printed only in the Teacher's Wraparound Edition, indicate the level of difficulty of the exercises.

Additional Answers

21. false; a counterexample:

22. True; consider parallelogram *ABCD*. Since opposite angles of a parallelogram are ≅, if ∠*A* of ▱*ABCD* is right, then ∠*C* is right. Since consecutive angles of a parallelogram are supplementary, then if ∠*A* is right, then ∠*B* and ∠*D* are right. Thus, if one angle of a parallelogram is right, then all angles are right. Therefore, ▱*ABCD* is a rectangle.

Study Guide Masters, p. 34

6-3 NAME________________ DATE ________

Study Guide — Student Edition Pages 306–312

Rectangles

A **rectangle** is a quadrilateral with four right angles. Since both pairs of opposite angles are congruent, a rectangle is a parallelogram and has all the properties of a parallelogram. The following list summarizes the properties of a rectangle.

- Opposite sides are congruent.
- Opposite angles are congruent.
- Consecutive angles are supplementary.
- Diagonals bisect each other.
- All four angles are right angles.
- Diagonals are congruent.

Example: Quadrilateral *EFGH* is a rectangle. If $EM = 5x + 1$ and $HF = 42$, find the value of x.

Since the diagonals of a rectangle bisect each other and are congruent, you know that $5x + 1 = \frac{1}{2}(42)$.

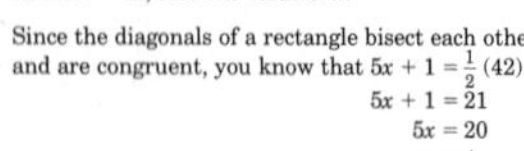

$$5x + 1 = 21$$
$$5x = 20$$
$$x = 4$$

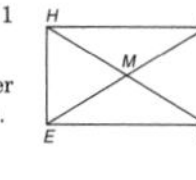

Quadrilateral RSTV is a rectangle. Find the values of x and y.

1. $VW = 2x + y$, $WS = 36$, $RS = x - y$, $VT = 9$ 15, 6
2. $VR = y$, $TS = x + 11$, $VT = y - 3x$, $RS = x + 2$ 3, 14

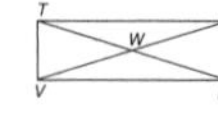

Quadrilateral ABCD is a rectangle. Find the value of x.

3. $m\angle DAC = 4x + 8$, $m\angle CAB = 5x - 8$ 10
4. $AC = x^2$, $DB = 6x - 8$ 2 or 4

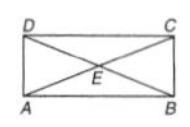

Determine whether ABCD is a rectangle. Justify your answer.

5. $A(10, 4)$, $B(10, 8)$, $C(-4, 8)$, $D(-4, 4)$ Yes; opposite sides are parallel and all angles are right angles.
6. $A(3, 7)$, $B(10, 7)$, $C(11, 12)$, $D(4, 12)$ no; not all right angles

Proof

14. Complete the proof for Theorem 6-9.

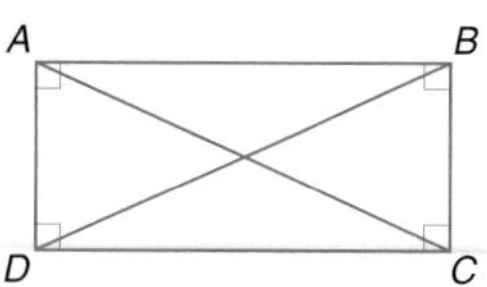

Given: *ABCD* is a rectangle with diagonals $\overline{AC}$ and $\overline{BD}$.

Prove: $\overline{AC} \cong \overline{BD}$

Proof:

Statements	Reasons
1. *ABCD* is a rectangle with diagonals $\overline{AC}$ and $\overline{BD}$.	1. Given
2. $\overline{DC} \cong \overline{DC}$	2. ? Congruence of segments is reflexive.
3. ? $\overline{AD} \cong \overline{BC}$	3. Opp. sides of a ▱ are ≅.
4. ∠*ADC* and ∠*BCD* are rt. ∠s.	4. Def. rectangle
5. ∠*ADC* ≅ ∠*BCD*	5. ? All rt. ∠s are ≅.
6. ? △*ADC* ≅ △*BCD*	6. SAS
7. $\overline{AC} \cong \overline{BD}$	7. ? CPCTC

EXERCISES

Practice

Quadrilateral *QUAD* is a rectangle. Find the value of x.

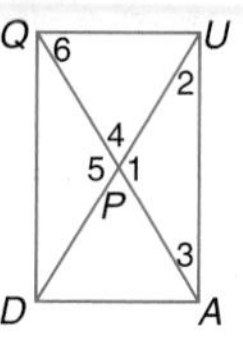

15. $DU = 26$, $QP = 2x + 7$ 3
16. $m\angle 2 = 52$, $m\angle 3 = 16x - 12$ 4
17. $m\angle 4 = 6x - 16$, $m\angle 2 = 2x + 4$ 12
18. $DP = 4x + 1$, $QP = x + 13$ 4
19. $m\angle 3 = 70 - 4x$, $m\angle 6 = 18x - 8$ 2

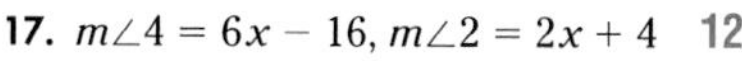

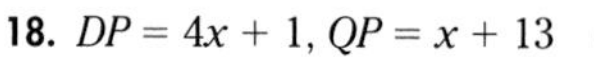

Determine whether each statement is *true* or *false*. Explain.

20. If a quadrilateral is a rectangle, then it is a parallelogram.
21. If two sides of a quadrilateral are perpendicular, then it is a rectangle.
22. If a parallelogram has a right angle, then it is a rectangle.
23. If all four angles of a quadrilateral are congruent, then it is a rectangle.

20. True; all rectangles are parallelograms.
21–23. See margin.

Use rectangle *QRST*, parallelogram *QZRC*, and the given information to solve each problem.

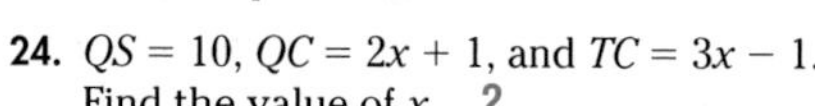

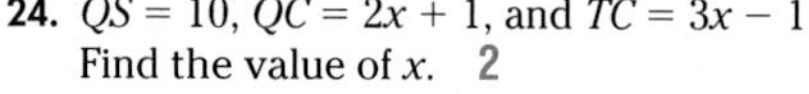

24. $QS = 10$, $QC = 2x + 1$, and $TC = 3x - 1$. Find the value of x. 2
25. $m\angle TQC = 70$. Find $m\angle QZR$. 140
26. $m\angle RCS = 35$. Find $m\angle RTS$. 17.5
27. $RT = 3x^2$ and $QC = 5x + 4$. What is the value of x? $-\frac{2}{3}$ or 4
28. $RZ = 6x$, $ZQ = 3x + 2y$, and $CS = 14 - x$. Find the values of x and y.
29. $m\angle QRT = m\angle TRS$. Find $m\angle TCQ$. 90

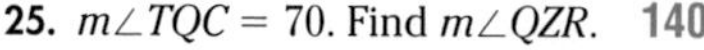

28. $x = 2$, $y = 3$

Additional Answer

23. True; the sum of the measures of the angles of a quadrilateral is 360. If all four angles are congruent, then the measure of each angle is $\frac{360}{4}$ or 90. Thus, all four angles are right and the definition of a rectangle is a quadrilateral with four right angles.

Use rectangle *STUV* and the given information to find each measure.

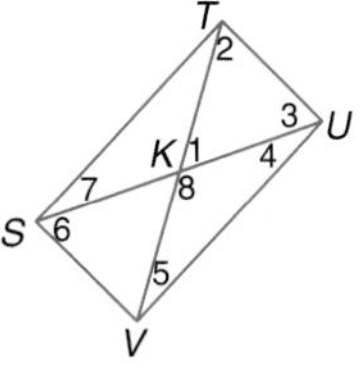

30. If $m\angle 1 = 30$, find $m\angle 2$. **75**
31. If $m\angle 6 = 57$, what is $m\angle 4$? **33**
32. If $m\angle 8 = 133$, find $m\angle 2$. **66.5**
33. If $m\angle 5 = 16$, what is $m\angle 3$? **74**

36. Yes; opposite sides are parallel and consecutive sides are perpendicular.

37b. midpoint of $\overline{AC} = \left(\frac{1}{2}, -1\frac{1}{2}\right)$; midpoint of $\overline{BD} = \left(3\frac{1}{2}, 1\frac{1}{2}\right)$

37c. *ABCD* is not a rectangle because it is not a parallelogram.

Determine if *PQRS* is a rectangle. Justify your answer.

34. $P(9, -1), Q(9, 5), R(-6, 5), S(-6, 1)$ **No; $\overline{PS}$ and $\overline{QR}$ are not parallel.**
35. $P(6, 2), Q(8, -1), R(10, 6), S(12, 3)$ **no; not a quadrilateral**
36. $P(-4, -3), Q(-5, 8), R(6, 9), S(7, -2)$
37. Graph *ABCD* for $A(2, 4), B(-2, 0), C(-1, -7)$, and $D(9, 3)$.
    a. Use the distance formula to find *AC* and *BD*. **$AC = \sqrt{130}$; $BD = \sqrt{130}$**
    b. Use the midpoint formula to locate the midpoints of $\overline{AC}$ and $\overline{BD}$.
    c. Explain why you cannot conclude that *ABCD* is a rectangle.
38. Graph *WXYZ* with vertices $W(-7, -3), X(0, 4), Y(3, 1)$, and $Z(-4, -7)$.
    a. Describe two different methods for determining if *WXYZ* is a rectangle.
    b. Is *WXYZ* a rectangle? Justify your reasoning. **38a–b. See margin.**

Proof C

Write a two-column proof. **39–41. See Solutions Manual.**

39. **Given:** *ACDE* is a rectangle. *ABCE* is a parallelogram.
    Prove: $\triangle ABD$ is isosceles.

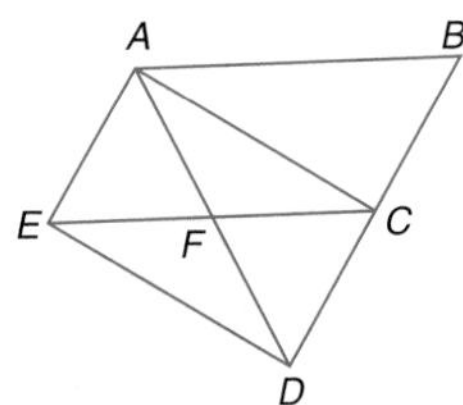

40. **Given:** *HJLM* is a rectangle. $\overline{KJ} \cong \overline{NM}$
    Prove: $\overline{HK} \cong \overline{LN}$

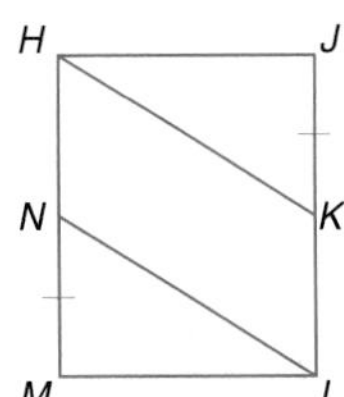

41. **Given:** *PQMO* and *RQMN* are rectangles. $\angle SVT \cong \angle UTV$. $\overline{SU}$ and $\overline{TV}$ intersect at *W*.
    Prove: *STUV* is a parallelogram.

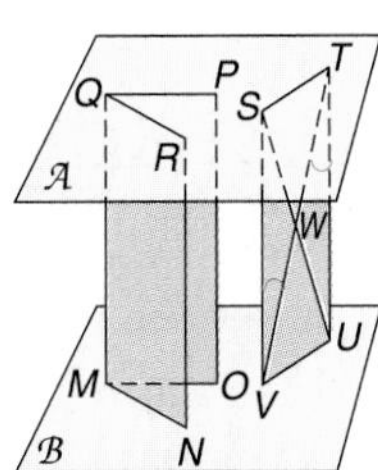

Critical Thinking

42. **Spherical Geometry** The figure shows a *Saccheri quadrilateral* on a sphere. Notice it has four sides with $\overline{CT} \perp \overline{TR}$ and $\overline{AR} \perp \overline{TR}$ and $\overline{CT} \cong \overline{AR}$. *Recall that all line segments are parts of the great circles.* **a. obtuse**

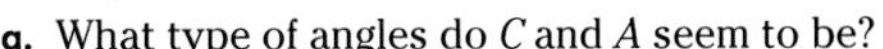

    a. What type of angles do *C* and *A* seem to be?
    b. Is $\overline{CT}$ parallel to $\overline{AR}$? Explain. **See margin.**
    c. How does *AC* compare to *TR*? **$AC < TR$**
    d. Can a rectangle exist in spherical geometry? Explain. **See margin.**

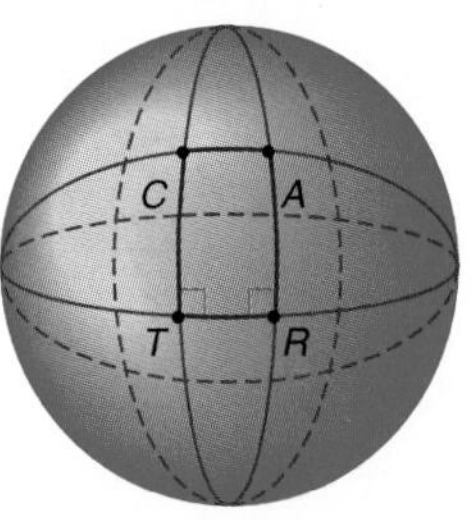

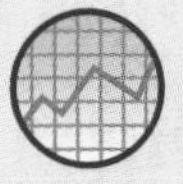
LOOK BACK
You can review spherical geometry in Lesson 3-6.

Additional Answers

38. [graph of WXYZ]

38a. Sample answer: Test the slopes of opposite pairs of sides to see that they are parallel and test a pair of consecutive sides to see that they are perpendicular; test the diagonals to see that they are congruent and bisect each other.
38b. No; opposite sides are not parallel.
42b. No; there are no parallel lines in spherical geometry.
42d. No; you cannot have parallel sides.

Practice Masters, p. 34

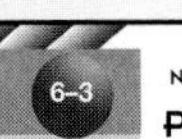
6-3

NAME______________ DATE ________

Practice Student Edition Pages 306–312

Rectangles

Use rectangle ABCD and the given information to solve each problem.

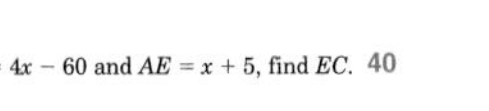

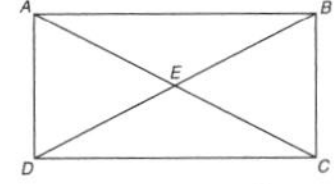

1. If $AC = 4x - 60$ and $BD = 30 - x$, find *BD*. 12
2. If $AC = 4x - 60$ and $AE = x + 5$, find *EC*. 40
3. If $m\angle BAC = 4x + 5$ and $m\angle CAD = 5x - 14$, find $m\angle CAD$. 41
4. If $AE = 2x + 3$ and $BE = 12 - x$, find *BD*. 18
5. If $m\angle BAC = 3x + 5$ and $m\angle ACD = 40 - 2x$. Find $m\angle AED$. 52

Determine whether PQRS is a rectangle. Justify your answer.

6. $P(2, 3), Q(5, 9), R(11, 6), S(8, 0)$ yes; opposite sides parallel and all right angles
7. $P(-1, 4), Q(3, 6), R(9, -3), S(5, -5)$ no; not all right angles
8. $P(1, 3), Q(4, 7), R(6, 2), S(2, 4)$ no; opposite sides not parallel
9. $P(-1, -3), Q(-4, 6), R(8, 10), S(11, 1)$ yes; opposite sides parallel and all right angles
10. $P(-1, -2), Q(5, 2), R(13, -10), S(7, -14)$ yes; opposite sides parallel and all right angles

Tech Prep

Aviation Technician Navigation requires knowledge of spherical geometry and spherical trigonometry. Aviation technician training is available through many two-year colleges. For more information on tech prep, see the *Teacher's Handbook*.

Extension

Problem Solving Given rectangle *LMNP*, $LI = 3x - 2$, $MI = 2x + 3$. Find *LN*. **26** Explain the method for solving the problem.

4 ASSESS

Closing Activity

Modeling Show the class two Popsicle sticks or pens of the same length. Mark the midpoint of each and cross them at their midpoints. Stretch a rubber band around the four endpoints. Ask students to explain why the rubber band forms a rectangle.

Chapter 6 Quiz B (Lesson 6-3) is available in the *Assessment and Evaluation Masters*, p. 156.

Mid-Chapter Test (Lessons 6-1 through 6-3) is available in the *Assessment and Evaluation Masters*, p. 155.

Additional Answers

43. Sample answer: A rectangular package fits better in a box or display unit than a circular package would.

44a. Sample answers: In architecture, the golden rectangle is used in many structures such as the Parthenon. In marketing, many products such as credit cards are approximately golden rectangles.

Enrichment Masters, p. 34

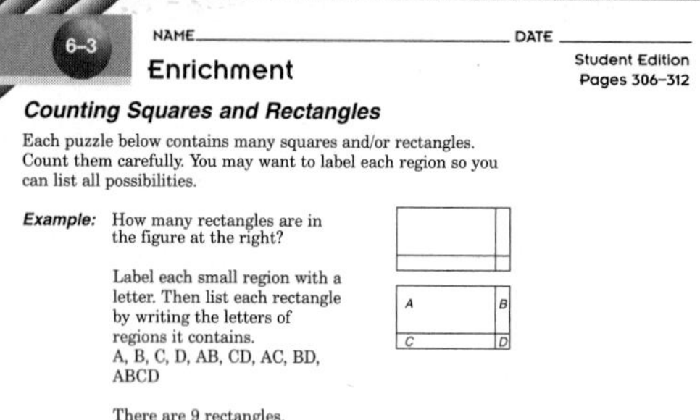

6-3 NAME ____________ DATE ________

Enrichment — Student Edition Pages 306–312

Counting Squares and Rectangles

Each puzzle below contains many squares and/or rectangles. Count them carefully. You may want to label each region so you can list all possibilities.

Example: How many rectangles are in the figure at the right?

Label each small region with a letter. Then list each rectangle by writing the letters of regions it contains.
A, B, C, D, AB, CD, AC, BD, ABCD

There are 9 rectangles.

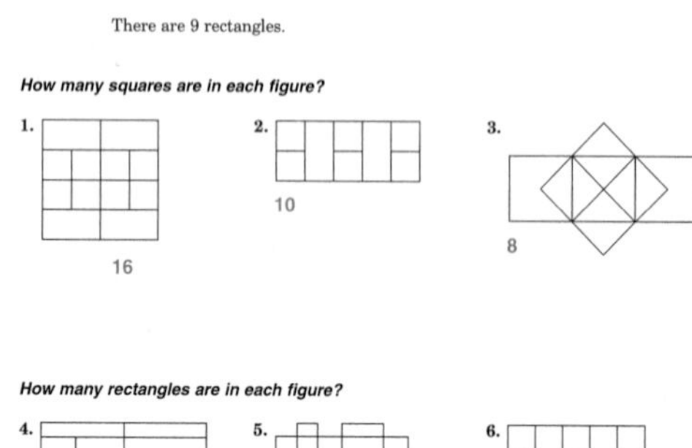

How many squares are in each figure?

1. 16
2. 10
3. 8

How many rectangles are in each figure?

4. 19
5. 24
6. 21

Applications and Problem Solving

43. See margin.

43. **Music** Compact discs (CDs) are circular in shape, but packaged in rectangular cases. Why do you think rectangular packaging is used?

44. **Art** *Golden rectangles* have been used by famous artists such as George Seurat, Leonardo DiVinci, and Salvador Dali. In a golden rectangle, the ratio of the length to the width is about 1.618, or the *golden ratio*.
    a. Use your library or search the Internet for information on two other examples of golden rectangles. **See margin.**
    b. Find the length of the diagonal of a golden rectangle whose sides measure 1 unit and 1.618 units. **The diagonal is about 1.90 units long.**

Mixed Review

45. Quadrilateral $ABCD$ has vertices $A(0, -2)$, $B(-4, 6)$, $C(5, 6)$, and $D(9, 4)$. Is $ABCD$ a parallelogram? (Lesson 6–2) **no**

46. In parallelogram $ABCD$, $AB = 4x + 9$, $m\angle BAC = 5y + 1$, $m\angle D = 75$, $m\angle ACD = 56$, and $CD = 45$. Find the values of x and y. (Lesson 6–1) **46. $x = 9$, $y = 11$**

47. Name the postulates and theorems that can be used to prove two right triangles congruent. (Lesson 5–2) **HA, LL, LA, and HL**

48. **Algebra** A median of $\triangle XYZ$ separates side $\overline{XZ}$ into $\overline{XQ}$ and $\overline{QZ}$. If $XQ = r - 3$ and $XZ = 7r - 15$, what is the value of r? (Lesson 5–1) **48. $\frac{9}{5}$**

49. In $\triangle STV$ and $\triangle WXY$, $\angle S \cong \angle W$ and $\overline{ST} \cong \overline{WX}$. What additional information is needed to prove $\triangle STV \cong \triangle WXY$ by ASA? (Lesson 4–5) **$\angle T \cong \angle X$**

50. The angles in a triangle have measures $7x - 8$, $3x + 3$, and $18x - 11$. Is the triangle *acute, obtuse,* or *right*? Explain. (Lesson 4–1) **50. obtuse; one obtuse angle**

51. What algebraic property allows us to say that if $m\angle 1 = m\angle 2$ and $m\angle 2 = m\angle 3$, then $m\angle 1 = m\angle 3$? (Lesson 2–4) **Transitive Property (=)**

52. Find the measure of the supplement of $\angle Q$ if $m\angle Q = 167$. (Lesson 1–7) **13**

INTEGRATION Algebra

53. Solve $-6 = 5u + 9$. **-3**

54. If x varies directly as y and $x = 13$ when $y = 1.3$, find x when $y = 8$. **80**

SELF TEST

Complete each statement about parallelogram *LMNP*. Then name the theorem or definition that justifies your answer. (Lesson 6–1)

1. $\overline{LM} \parallel$? **$\overline{PN}$**
2. $\angle LMN \cong$? **$\angle LPN$**
3. $\overline{MN} \cong$? **$\overline{LP}$**
4. $\triangle LMQ \cong$? **$\triangle NPQ$**

1–4. See margin for theorem or definition.

Determine if the quadrilateral with the given vertices is a parallelogram. (Lesson 6–2)

5. $A(9, 0)$, $B(2, -5)$, $C(6, -5)$, $D(13, 0)$ **yes**
6. $E(-1, -1)$, $F(1, 1)$, $G(6, -6)$, $H(-6, 0)$ **no**
7. $H(5, 0)$, $I(0, -5)$, $J(-5, 0)$, $K(0, 5)$ **yes**
8. $L(-2, -1)$, $M(2, 5)$, $N(-10, 13)$, $P(-14, 7)$ **yes**

Write a proof. (Lesson 6–3) **9–10. See Solutions Manual.**

9. **Given:** $\square WXZY$
   $\angle 1$ and $\angle 2$ are complementary.
   Prove: $WXZY$ is a rectangle.

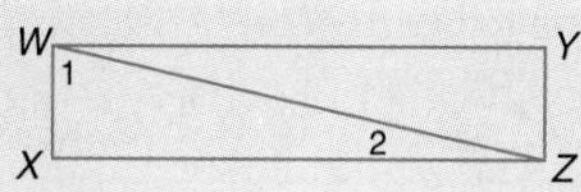

10. **Given:** $\square KLMN$
    Prove: $PQRS$ is a rectangle.

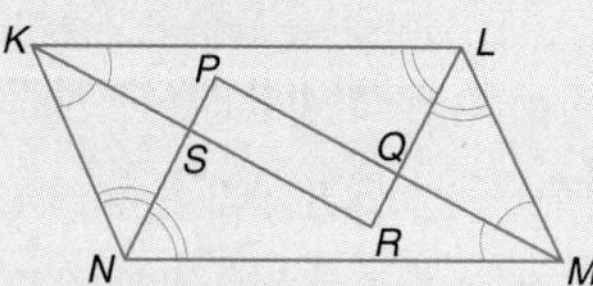

SELF TEST

The Self Test provides students with a brief review of the concepts and skills in Lessons 6-1 through 6-3. Lesson numbers are given to the right of exercises or instruction lines so students can review concepts not yet mastered.

Answers for the Self Test

1. Opposite sides of a parallelogram are parallel by definition.
2. Opposite angles of a parallelogram are congruent.
3. Opposite sides of a parallelogram are congruent.
4. SSS with diagonals of a parallelogram bisect each other, and opposite sides of a parallelogram are congruent.

6–4 Squares and Rhombi

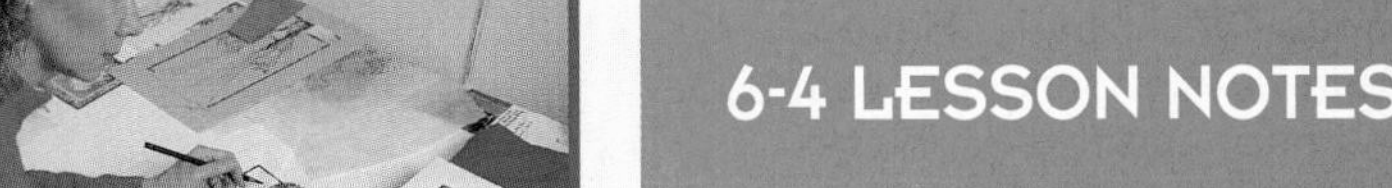

What **YOU'LL LEARN**

- To recognize and apply the properties of squares and rhombi.

Why **IT'S IMPORTANT**

You can use properties of squares and rhombi to solve problems involving art and construction.

APPLICATION
Optical Illusions

Have you ever looked down a long road and noticed that it seems to grow narrower in the distance? Or have you painted a room a lighter color and found that the room seems bigger than it did before? Then you have experienced an optical illusion. Artists, interior decorators, and clothing designers take advantage of optical illusions to make things appear different than they are.

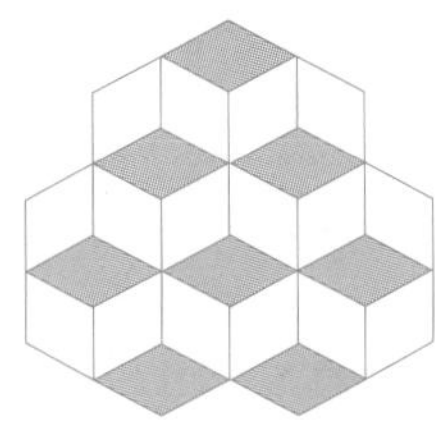

Study the figure at the right. You may be able to see six cubes or seven cubes depending on which quadrilaterals are considered to be the tops of the cubes. The optical illusion is created using special quadrilaterals called **rhombi** (pronounced rom-bye). A **rhombus** is a quadrilateral with four congruent sides. Since opposite sides of a rhombus are congruent, a rhombus is a parallelogram.

You will be asked to prove Theorems 6–11 and 6–12 in Exercises 19 and 51, respectively.

In addition to all of the properties of a parallelogram, a rhombus has two special relationships that are described in the following theorems. The proof of Theorem 6–13 is given.

Theorem 6–11	**The diagonals of a rhombus are perpendicular.**
Theorem 6–12	**If the diagonals of a parallelogram are perpendicular, then the parallelogram is a rhombus.**
Theorem 6–13	**Each diagonal of a rhombus bisects a pair of opposite angles.**

Proof of Theorem 6–13

Given: *ABCD* is a rhombus.

Prove: Each diagonal bisects a pair of opposite angles.

Proof:

Statements	Reasons
1. *ABCD* is a rhombus.	**1.** Given
2. *ABCD* is a parallelogram.	**2.** Def. rhombus
3. $\angle ABC \cong \angle ADC$, $\angle BAD \cong \angle BCD$	**3.** Opp. ∡ of a ▱ are ≅.
4. $\overline{AB} \cong \overline{BC} \cong \overline{CD} \cong \overline{DA}$	**4.** Def. rhombus
5. $\triangle ABC \cong \triangle ADC$	**5.** SAS
6. $\angle 5 \cong \angle 6$, $\angle 7 \cong \angle 8$	**6.** CPCTC
7. $\triangle BAD \cong \triangle BCD$	**7.** SAS
8. $\angle 1 \cong \angle 2$, $\angle 3 \cong \angle 4$	**8.** CPCTC
9. Each diagonal bisects a pair of opposite angles.	**9.** Def. ∠ bisector

6-4 LESSON NOTES

NCTM Standards: 1–5, 7, 8

Instructional Resources

- Study Guide Master 6-4
- Practice Master 6-4
- Enrichment Master 6-4
- Assessment and Evaluation Masters, p. 157
- Modeling Mathematics Masters, p. 84
- Real-World Applications, 12

Transparency 6-4A contains the 5-Minute Check for this lesson; **Transparency 6-4B** contains a teaching aid for this lesson.

Recommended Pacing	
Standard Pacing	Days 7 & 8 of 13
Honors Pacing	Days 7 & 8 of 12
Block Scheduling*	Day 4 of 6

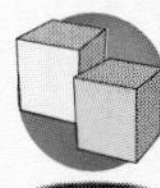

*For more information on pacing and possible lesson plans, refer to the *Block Scheduling Booklet*.

1 FOCUS

5-Minute Check
(over Lesson 6-3)

Quadrilateral *SURE* is a rectangle. Find the values of *v*, *w*, *x*, *y*, and *z*.

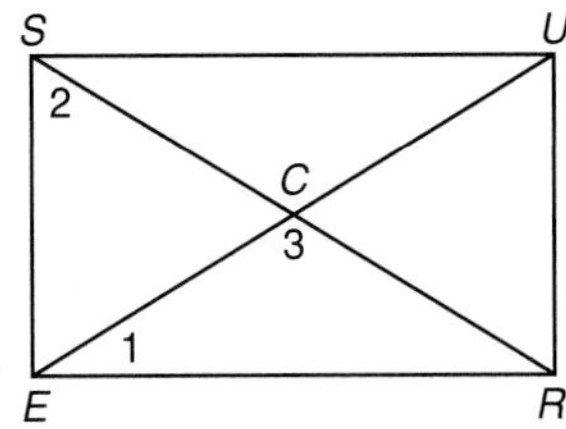

1. $SR = 4v + 2$, $EU = 6v - 8$ 5
2. $EC = 2w + 3$, $CU = 3w - 1$ 4
3. $m\angle 1 = x$, $m\angle 2 = 2x$ 30
4. $UR = 6y - 7$, $SE = 4y - 1$ 3
5. $m\angle 1 = 2z$, $m\angle 3 = 8z$ 15

Motivating the Lesson

Hands-On Activity Have students form a square with paper fasteners and four strips of cardboard as the sides. Have them alter the angles so that they are not right angles. Ask them to describe the new figure. **It is a parallelogram with four equal sides and is called a rhombus.**

2 TEACH

Teaching Tip Before reading Example 1, ask students to recall which properties of diagonals are common to all parallelograms, which are common to rectangles, and which are common to rhombi.

In-Class Example

For Example 1
Use rhombus *DLMP* with *DM* = 26 to determine whether each statement is *true* or *false*. Justify your answers.

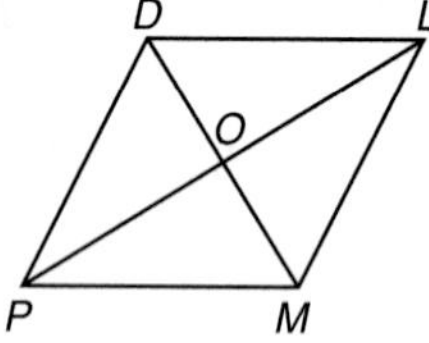

a. $OM = 13$ **True; since a rhombus is a parallelogram, its diagonals bisect each other. Thus, if $DM = 26$, $OM = \frac{1}{2}(26)$ or 13.**

b. $\overline{MD} \cong \overline{PL}$ **False; the diagonals of a rhombus are not congruent unless the rhombus is a square.**

c. $m\angle DLO = m\angle LDO$ **False; these angles of the rhombus are complementary.**

Teaching Tip Have a student do the construction at the chalkboard or at the overhead projector so that the class can see each step.

You can use the properties of a rhombus to solve problems.

Example 1 **Use rhombus *BCDE* and the given information to find each missing value.**

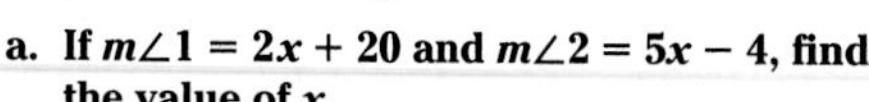

a. If $m\angle 1 = 2x + 20$ and $m\angle 2 = 5x - 4$, find the value of x.

The diagonals of a rhombus bisect a pair of opposite angles. So, $\overline{CE}$ bisects $\angle BCD$ and $m\angle 1 = m\angle 2$.

$$m\angle 1 = m\angle 2$$
$$2x + 20 = 5x - 4$$
$$24 = 3x$$
$$8 = x$$

b. If $BD = 15$, find BF.

Since a rhombus is a parallelogram, its diagonals bisect each other. Thus, if $BD = 15$, $BF = \frac{1}{2}(15)$ or 7.5.

c. If $m\angle 3 = y^2 + 26$, find y.

The diagonals of a rhombus are perpendicular, so $\angle 3$ is a right angle.

$$m\angle 3 = 90$$
$$y^2 + 26 = 90$$
$$y^2 = 64$$
$$y = 8 \text{ or } -8$$

It is possible to construct a rhombus using a compass and a straightedge.

CONSTRUCTION

Rhombus

Construct a rhombus.

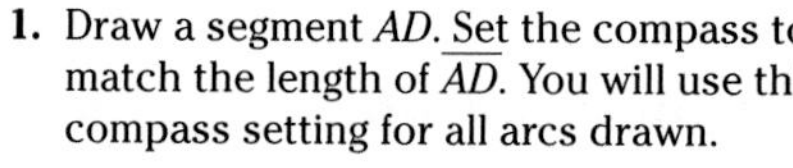

1. Draw a segment AD. Set the compass to match the length of $\overline{AD}$. You will use this compass setting for all arcs drawn.

A D

2. Place the compass at point A and draw an arc above $\overline{AD}$. Choose any point on that arc and label it B. Place the compass at point B and draw an arc to the right of B.

3. Then place the compass at point D and draw an arc to intersect the arc drawn from point B. Label the point of intersection C.

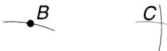

4. Use a straightedge to draw $\overline{AB}$, $\overline{BC}$, and $\overline{CD}$.

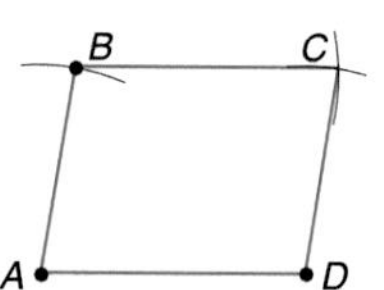

Conclusion: Since all of the sides are congruent, quadrilateral $ABCD$ is a rhombus.

Cooperative Learning

Group Discussion Separate the class into groups of four. Have each group make up four statements like the following that use the words *all* or *some*: "Some rhombi are squares." "All rectangles are parallelograms." Have the groups exchange statements and discuss them. For more information on the group discussion strategy, see *Cooperative Learning in the Mathematics Classroom*, one of the titles in the Glencoe Mathematics Professional Series, page 31.

If a quadrilateral is both a rhombus and a rectangle, it is a **square**. A square is a quadrilateral with four right angles and four congruent sides. You can construct a square in the same way that you constructed a rhombus, but a square must have a right angle.

You will be asked to construct a square in Exercise 48.

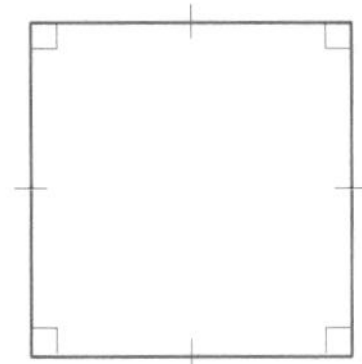

Example 

INTEGRATION **Algebra**

Determine whether parallelogram *WXYZ* is a rhombus, a rectangle, or a square for *W*(1, 10), *X*(−4, 0), *Y*(7, 2), and *Z*(12, 12).

We can determine if *WXYZ* is a rhombus, a rectangle, or a square by examining its diagonals. The diagonals of a rhombus are perpendicular. So if $\overline{WY} \perp \overline{XZ}$, then *WXYZ* is a rhombus. If the diagonals are congruent, then *WXYZ* is a rectangle. If it is both a rhombus and a rectangle, *WXYZ* is a square.

slope of $\overline{WY} = \frac{10-2}{1-7} = -\frac{8}{6}$ or $-\frac{4}{3}$ slope of $\overline{XZ} = \frac{0-12}{-4-12} = \frac{12}{16}$ or $\frac{3}{4}$

The product of the slopes of the diagonals is $\left(-\frac{4}{3}\right)\left(\frac{3}{4}\right)$ or −1. Therefore, the diagonals are perpendicular, and *WXYZ* is a rhombus.

$WY = \sqrt{(1-7)^2 + (10-2)^2}$ $XZ = \sqrt{(-4-12)^2 + (0-12)^2}$

$= \sqrt{100}$ or 10 $= \sqrt{400}$ or 20

Since diagonals of *WXYZ* are not congruent, it is not a rectangle. Since a square is a special rectangle, *WXYZ* is not a square.

WXYZ is a rhombus.

Geometric art is one of the ways that geometry is part of everyday life.

Example 3

Art

The artwork at the right is Dorthea Rockburne's *Golden Section Painting*. The diagram below shows the shapes of the pieces of fabric that make up the artwork. Use a ruler or protractor to determine which type of quadrilateral is represented by each figure.

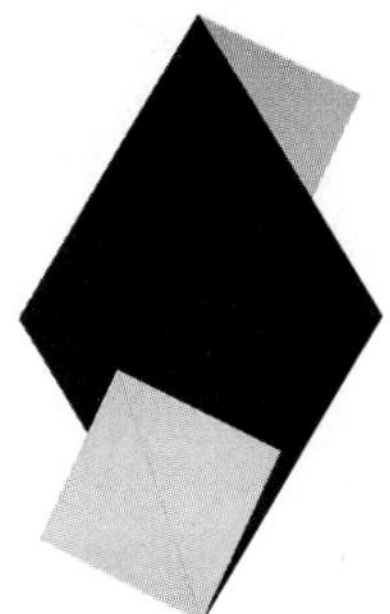

a. *ABCD*

ABCD is a square because the diagonals are both congruent and perpendicular.

b. *AEFJ*

All four sides of *AEFJ* are congruent. But the consecutive sides are not perpendicular, so *AEFJ* is a rhombus.

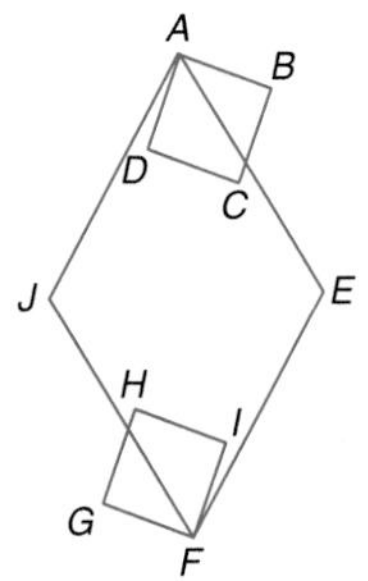

In-Class Examples

For Example 2

Determine whether the parallelogram *ABCD* is a rhombus, a rectangle, or a square for *A*(−4, 3), *B*(−2, 3), *C*(−4, 1), and *D*(−2, 1). **square**

For Example 3

Use a ruler or protractor to determine which type of quadrilateral is represented by *PQRS* for *P*(0, 0), *Q*(2, 3), *R*(6, 3), and *S*(4, 0). **parallelogram**

Teaching Tip Encourage students to draw pictures and mark congruent sides and angles.

Classroom Vignette

"I like to discuss area and perimeter as we study each type of figure. For example, I have students make a table of possible lengths and widths if the perimeter of a rectangle is 24 centimeters. Then they find the area of each possible rectangle and graph the area as a function of the width. From the graph, they can determine that a square has the greatest possible area for a rectangle with a given perimeter."

Julianne Wanat

Julianne Wanat
Milford High School
Milford, New Hampshire

3 PRACTICE/APPLY

Check for Understanding

Exercises 1–20 are designed to help you assess your students' understanding through reading, writing, speaking, and modeling. You should work through Exercises 1–5 with your students and then monitor their work on Exercises 6–20.

Error Analysis

Students may confuse the properties of parallelograms, rectangles, rhombi, and squares. It may help if they make a chart to refer to as they work.

Additional Answers

1. In a rhombus: the sides are all congruent, the diagonals are perpendicular, the opposite angles are congruent, the diagonals bisect a pair of opposite angles, the diagonals bisect each other, and consecutive angles are supplementary. In a square: the sides are all congruent, the diagonals are perpendicular, the opposite angles are congruent, the diagonals bisect a pair of opposite angles, the diagonals bisect each other, consecutive angles are supplementary, the diagonals are congruent, and all four angles are right. A square has all of the characteristics of a rhombus because a square is a rhombus.
2. Both definitions are correct. A square has all four sides congruent and all four angles right. A rhombus has all four sides congruent, so if it is also equiangular, its angles will be all right and it is a square. A rectangle has all four right angles. If its sides are all congruent, then it will be a square.
3. Sample answer, not drawn to scale:

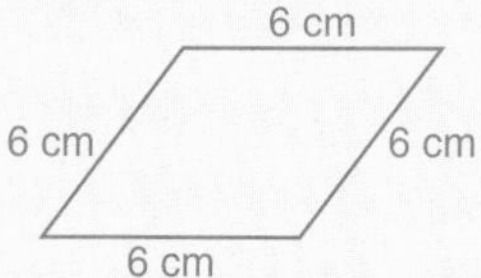

CHECK FOR UNDERSTANDING

Communicating Mathematics

Study the lesson. Then complete the following. 1–3. See margin.

1. **Compare and contrast** the properties of a rhombus and a square. How are they different? How are they the same?
2. One geometry book defines a square as "*an equiangular rhombus.*" Another book defines a square as "*an equilateral rectangle.*" Which definition is correct and why?

3. **Construct** a rhombus with sides that are 6 centimeters long using a compass and straightedge.
4. Copy and complete the summary of the diagonal properties that exist for each of the four types of quadrilaterals in the table below. Write *yes* or *no* in each cell of the table.

Property	Parallelogram	Rectangle	Rhombus	Square
The diagonals bisect each other.	yes	yes	yes	yes
The diagonals are congruent.	no	yes	no	yes
Each diagonal bisects a pair of opposite angles.	no	no	yes	yes
The diagonals are perpendicular.	no	no	yes	yes

5. Locate two examples of a rhombus or a square in newspaper illustrations. Explain how you know that the shape is a rhombus or a square.
See students' work.

Guided Practice

Use rhombus *PLAN* for Exercises 6–9. Justify your answers.

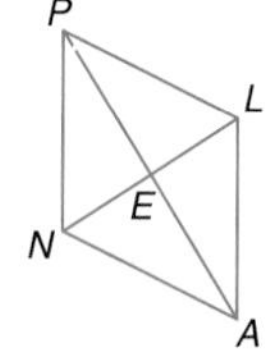

6. What type of triangle is $\triangle PLA$?
7. What type of triangle is $\triangle PEN$?
8. Is $\triangle PEN \cong \triangle AEL$?
9. Is it true that $\overline{PA} \cong \overline{NL}$? Explain.

6. Isosceles; all four sides of $PLAN$ are congruent, so $\overline{PL} \cong \overline{LA}$.
7. Right; the diagonals of a rhombus are perpendicular, so $\angle PEN$ is right.
8. Yes; the diagonals of a parallelogram bisect each other so $\overline{PE} \cong \overline{EA}$ and $\overline{EN} \cong \overline{EL}$. $\angle PEN \cong \angle AEL$ since they are vertical angles. Thus, $\triangle PEN \cong \triangle AEL$ by SAS.
9. No; the diagonals of a rhombus are not congruent unless the rhombus is a square.

Use rhombus *RSTV* and the given information to find each value.

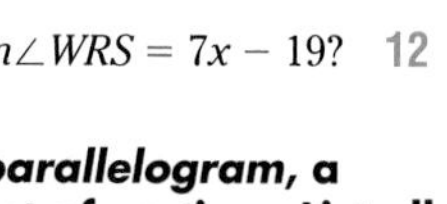

10. If $m\angle RST = 67$, find $m\angle RSW$. 33.5
11. Find $m\angle SVT$ if $m\angle STV = 135$. 22.5
12. If $m\angle SWT = 2x + 8$, find the value of x. 41
13. What is the value of x if $m\angle WRV = 5x + 5$ and $m\angle WRS = 7x - 19$? 12

Determine whether quadrilateral *PARK* is a *parallelogram*, a *rectangle*, a *rhombus*, or a *square* for each set of vertices. List all that apply. 14. parallelogram, rectangle, rhombus, square

14. $P(-1, 0), A(1, -1), R(2, 1), K(0, 2)$
15. $P(-1, 4), A(-1, 10), R(14, 10), K(14, 4)$ parallelogram, rectangle
16. $P(2, 11), A(-3, 1), R(8, 3), K(13, 3)$ none

Reteaching

Using Diagrams Have students draw diagrams to show diagonals of a parallelogram, a rectangle, a rhombus, and a square. Have them mark and discuss congruent sides and angles in each figure.

Alternative Learning Styles

Kinesthetic Have students work in groups to design and build a simple kite. Encourage them to create kites in the shapes of each quadrilateral discussed. Discuss which kites will fly and which won't.

Name all the quadrilaterals—*parallelogram, rectangle, rhombus,* or *square*—that have each property.

17. All angles are congruent. rectangle, square

18. All sides are congruent. rhombus, square

Proof

19. Write a paragraph proof of Theorem 6–11.

Given: *ABCD* is a rhombus.

Prove: $\overline{AC} \perp \overline{BD}$

See Solutions Manual.

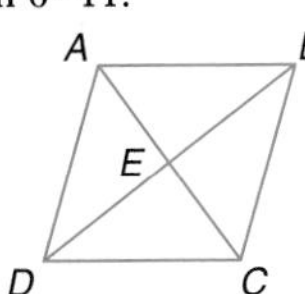

GLOBAL CONNECTIONS

Evidence of Nine men's morris has been found in Ceylon, now Sri Lanka. It was also known to inhabitants of Troy and to the Vikings.

20. **Games** *Nine men's morris* is a game for two players that dates from about 1400 B.C. The game board was one of seven found cut into the roof tiles of the temple in Karnak, Egypt. The game board is made from three squares that share the same center as shown at the right. Use a compass and a straightedge to construct a Nine men's morris board. See students' work.

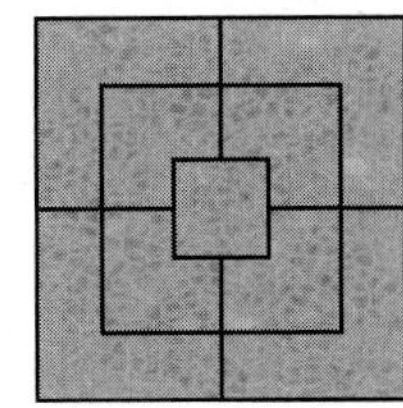

EXERCISES

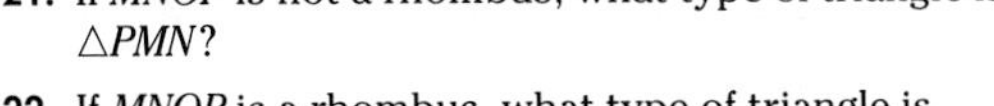

Practice

Use parallelogram *MNOP* for Exercises 21–27. Justify your answers.

21. If *MNOP* is not a rhombus, what type of triangle is $\triangle PMN$?

22. If *MNOP* is a rhombus, what type of triangle is $\triangle PQM$? Right; the diagonals are perpendicular.

23. Is $\triangle PQM \cong \triangle NQM$ if *MNOP* is a square?

24. Is it true that $\overline{PQ} \cong \overline{NQ}$ if *MNOP* is a square?

25. If *MNOP* is a rhombus, must it also be a square?

26. If $\angle NQO$ is right, is *MNOP* a rhombus?

27. If $\triangle PON$ is isosceles, must *MNOP* be a square?

M N Q P O

21. Scalene; the consecutive sides of *MNOP* are not congruent. (A)

23. Yes; the sides of *MNOP* are congruent and the diagonals are perpendicular, so the triangles are congruent by SAS. (B)

24. Yes; the diagonals of a parallelogram bisect each other.

25. No; all squares are rhombi, but not all rhombi are squares.

26. Yes; if the diagonals of a parallelogram are perpendicular, then the parallelogram is a rhombus.

27. No; *MNOP* could be a rhombus if $\overline{PO} \cong \overline{ON}$, but the conditions are not sufficient for it to be a square.

Use rhombus *ABCD* and the given information to find each value.

28. If $m\angle BAF = 28$, find $m\angle ACD$. 28

29. Find the value of x if $m\angle AFB = 16x + 6$. 5.25

30. If $m\angle ACD = 34$, find $m\angle ABC$. 112

31. Find the value of x if $m\angle BFC = 120 - 4x$. 7.5

32. What is the value of x if $m\angle BAC = 4x + 6$ and $m\angle ACD = 12x - 18$? 3

33. If $m\angle DCB = x^2 - 6$ and $m\angle DAC = 5x + 9$, find the value of x. 12

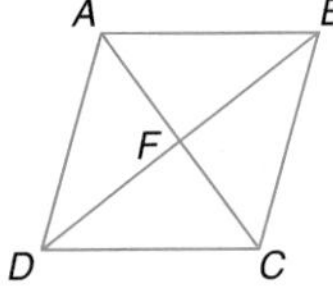

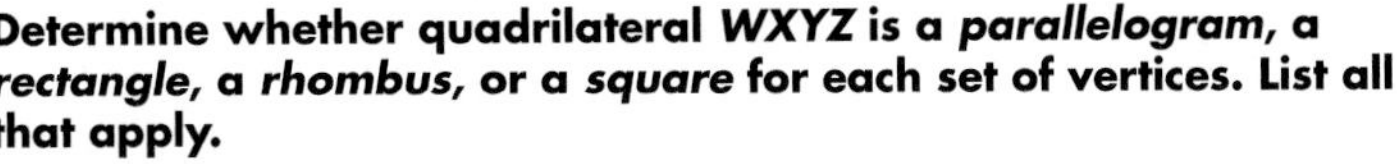

Determine whether quadrilateral *WXYZ* is a *parallelogram*, a *rectangle*, a *rhombus*, or a *square* for each set of vertices. List all that apply.

34. $W(5, 6), X(7, 5), Y(9, 9), Z(7, 10)$ parallelogram, rectangle

35. $W(-3, -3), X(1, -6), Y(5, -3), Z(1, 0)$ parallelogram, rhombus

36. $W(-6, 11), X(-11, -7), Y(-7, -2), Z(-2, -6)$ none

37. $W(10, 6), X(6, 10), Y(10, 14), Z(14, 10)$

37. parallelogram, rectangle, rhombus, square

Assignment Guide

Core (with proof): 21–51 odd, 52, 53, 55, 56–67
Core (informal): 21–47 odd, 52, 53, 55, 57–67
Enriched: 22–50 even, 52–67

For **Extra Practice**, see p. 775.

The red A, B, and C flags, printed only in the Teacher's Wraparound Edition, indicate the level of difficulty of the exercises.

Study Guide Masters, p. 35

NAME______ DATE______ Student Edition Pages 313–319

Study Guide

Squares and Rhombi

A **rhombus** is a quadrilateral with four congruent sides. A **square** is a quadrilateral with four right angles and four congruent sides.

The diagonals of a rhombus have two special relationships.

- The diagonals of a rhombus are perpendicular.
- Each diagonal of a rhombus bisects a pair of opposite angles.

Example: *ABCD* is a rhombus. If $m\angle ADB = 27$, find $m\angle ADC$.

Since each diagonal of a rhombus bisects a pair of opposite angles, $m\angle ADC = 2(m\angle ADB)$. So $m\angle ADC = 2(27)$ or 54.

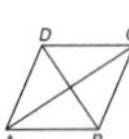

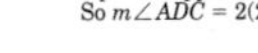

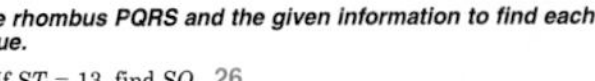

Use rhombus PQRS and the given information to find each value.

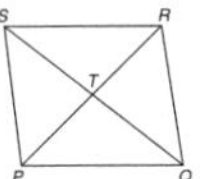

1. If $ST = 13$, find SQ. 26
2. If $m\angle PRS = 17$, find $m\angle QRS$. 34
3. Find $m\angle STR$. 90
4. If $SP = 4x - 3$ and $PQ = 18 + x$, find the value of x. 7

Determine whether each quadrilateral with the given vertices is a parallelogram, a rectangle, a rhombus, or a square. List all that apply.

5. $M(1, 5), N(6, 5), O(6, 10), P(1, 10)$ parallelogram, rectangle, rhombus, square

6. $W(-4, -2), X(5, -2), Y(8, 4), Z(-1, 4)$ parallelogram

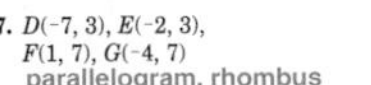

7. $D(-7, 3), E(-2, 3), F(1, 7), G(-4, 7)$ parallelogram, rhombus

8. $R(0, 0), E(10, 0), S(10, 5), T(0, 5)$ parallelogram, rectangle

Sri Lanka is an island located in the Indian Ocean off the southern tip of India.

Additional Answers

48. Sample answer, not drawn to scale:

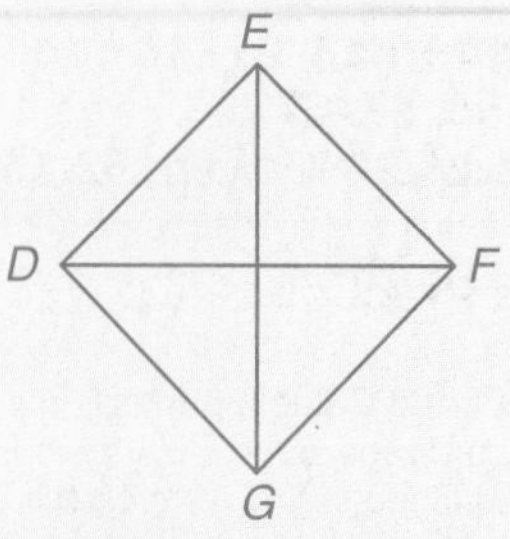

49. Statements (Reasons)
    1. *JKLM* is a square. (Given)
    2. *JKLM* is a rectangle. (Def. square)
    3. $\overline{JL} \cong \overline{KM}$ (If a ▱ is a rect., then its diagonals are ≅.)

50. Statements (Reasons)
    1. Rhombus *MTRN* (Given)
    2. $\overline{MR} \perp \overline{TN}$ (The diagonals of a rhom. are ⊥.)
    3. ∠*MST* is a right angle. (Def. ⊥)
    4. $m\angle MST = 90$ (Def. rt. ∠)
    5. $m\angle MST + m\angle 1 + m\angle 2 = 180$ (Angle Sum Th.)
    6. $90 + m\angle 1 + m\angle 2 = 180$ (Substitution Prop. (=))
    7. $m\angle 1 + m\angle 2 = 90$ (Subtraction Prop. (=))
    8. ∠1 and ∠2 are complementary. (Def. comp. ∠s)

39. parallelogram, rectangle, rhombus, square

42. True; the set of squares is a subset of the set of rhombi.

43. False; some rhombi are not squares.

44. False; some rectangles are not squares.

45. True; the set of squares is a subset of the set of rectangles.

46. True; the set of rhombi is a subset of the set of parallelograms.

Name all the quadrilaterals—*parallelogram, rectangle, rhombus,* or *square*—that have each property.

38. Diagonals are congruent. rectangle, square
39. One pair of opposite sides are congruent and parallel.
40. All sides congruent *and* all angles are congruent. square
41. The diagonals are perpendicular. rhombus, square

Use the Venn diagram at the right to determine whether each statement is *true* or *false*. Explain your reasoning.

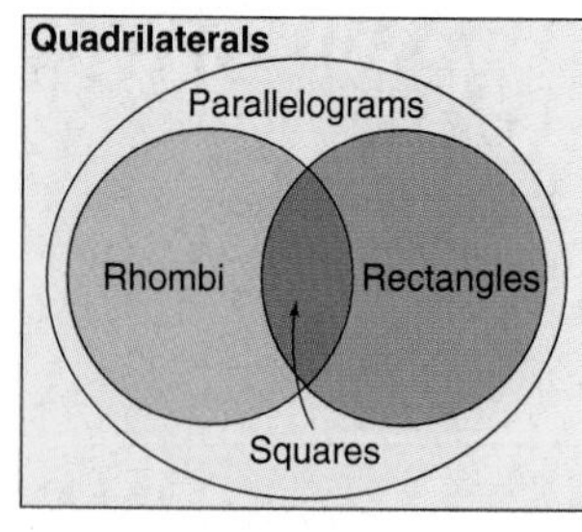

42. Every square is a rhombus.
43. Every rhombus is a square.
44. Every rectangle is a square.
45. Every square is a rectangle.
46. All rhombi are parallelograms.
47. Every parallelogram is a rectangle. False; some parallelograms are not rectangles.

48. Construct square *DEFG* with diagonal congruent to $\overline{AB}$. See margin.

Proof C

Write a two-column proof. 49–50. See margin.

49. **Given:** *JKLM* is a square.
    Prove: $\overline{JL} \cong \overline{KM}$

K L N J M

50. **Given:** Rhombus *MTRN*
    Prove: ∠1 and ∠2 are complementary.

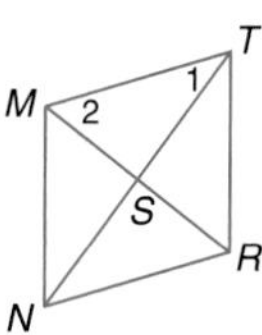

51. Write a paragraph proof of Theorem 6–12. See Solutions Manual.

Critical Thinking

52. Use a ruler to draw a quadrilateral with perpendicular diagonals that is not a rhombus. See Solutions Manual.

Applications and Problem Solving

53. **Flags** Study the flags shown below. Use a ruler and protractor to determine if any of the flags contain parallelograms, rectangles, rhombi, or squares.

Denmark

St. Vincent and The Grenadines

Trinidad and Tobago

54. **Construction** The opening for the reinforced doors of a storage shed is shaped like a square.
    a. Identify the quadrilaterals that make up the doors.
    b. Explain why you think the doors are shaped as they are.

53. See Solutions Manual.

54a. Each door is a rectangle made up of two congruent squares and one of their diagonals.

54b. Sample answer: Triangles are rigid and strong for construction, so using a square and a diagonal will make a sturdy door.

Practice Masters, p. 35

NAME ______ DATE ______

6-4 Practice — Student Edition Pages 313–319

Squares and Rhombi

Use square ABCD and the given information to find each value.

1. If $m\angle AEB = 3x$, find x. 30
2. If $m\angle BAC = 9x$, find x. 5
3. If $AB = 2x + 4$ and $CD = 3x - 5$, find BC. 22
4. If $m\angle DAC = y$ and $m\angle BAC = 3x$, find x. 15
5. If $AB = x^2 - 15$ and $BC = 2x$, find x. 5

Use rhombus ABCD and the given information to find each measure.

6. $m\angle BCE$ 31
7. $m\angle BEC$ 90
8. AC 24 cm
9. $m\angle ABD$ 59
10. AD 14 cm

Determine whether EFGH is a parallelogram, a rectangle, a rhombus, or a square for each set of vertices. List all that apply.

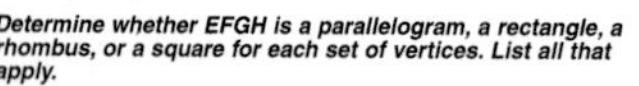

11. $E(0, -3), F(-3, 0), G(0, 3), H(3, 0)$ parallelogram, rectangle, rhombus, square
12. $E(2, 1), F(3, 4), G(7, 2), H(6, -1)$ parallelogram

Alternative Teaching Strategies

Reading Geometry Have students use a dictionary to look up words that begin with *quad*. What do these words have in common? They all refer to four of something. What would a quadricycle be? a four-wheeled vehicle powered by a pedal Have students make up new words using *quad*.

55. **Sports** In the game of racquetball, the server stands between the service line and the short line as he or she delivers the serve. The receiver stands behind the short line. Study the diagram of a racquetball court at the right. If the court is a rectangle, identify each quadrilateral and explain how you determine the shape. See Solutions Manual.

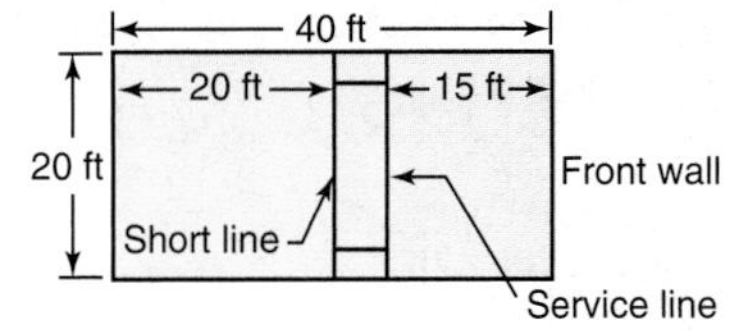

Mixed Review

56. See Solutions Manual.

57. The legs are made so that they will bisect each other, so the quadrilateral formed by the ends of the legs is always a parallelogram. Therefore, the table top is parallel to the floor.

59. Yes; if V is (12, 3), $TU = UV$.

62. No; slope of $\overleftrightarrow{AB} = 3$ and slope of $\overleftrightarrow{CD} = 3$, but $\overleftrightarrow{AB}$ and $\overleftrightarrow{CD}$ are the same line since A, B, C, and D are collinear.

64. See margin.

56. Write a two-column proof. (Lesson 6–3)

Given: $WXYZ$

$\angle 1$ and $\angle 2$ are complementary.

Prove: $WXYZ$ is a rectangle.

57. **Music** Why will the keyboard stand shown at the right always remain parallel to the floor? (Lesson 6–2)

58. *True* or *false:* The altitude of a triangle can sometimes be located outside the triangle. (Lesson 5–1) true

59. $\triangle TUV$ is an isosceles triangle with two vertices $T(7, 10)$ and $U(2, 3)$. If $\angle T$ is the vertex angle, could V be at (12, 3)? Explain. (Lesson 4–6)

60. If $\triangle HJK \cong \triangle MNO$, name a segment congruent to $\overline{KH}$. (Lesson 4–3) $\overline{OM}$

61. *True* or *false*: An obtuse triangle can be isosceles. (Lesson 4–1) true

62. Determine if $\overleftrightarrow{AB} \parallel \overleftrightarrow{CD}$ for $A(-4, -6)$, $B(-1, 3)$, $C(-2, 0)$, and $D(1, 9)$. Justify your answer. (Lesson 3–3)

63. *True* or *false*: If two lines do not intersect, then the lines must be parallel. (Lesson 3–1) false

64. **Civics** Identify the hypothesis and the conclusion of the conditional *If a U.S. citizen is over 18 years old, then he or she may vote.* (Lesson 2–2)

65. Suppose B is between A and C, $AB = 9$, $BC = 4x - 7$, and $AC = 18$. Find the value of x and the measure of $\overline{BC}$. (Lesson 1–4) 4; 9

Algebra

66. domain: $\{-9, 5, 7, 8\}$; range: $\{-2, -1, 6\}$

66. State the domain and range of the relation $\{(8, -2), (7, 6), (-9, -1), (5, 6)\}$.

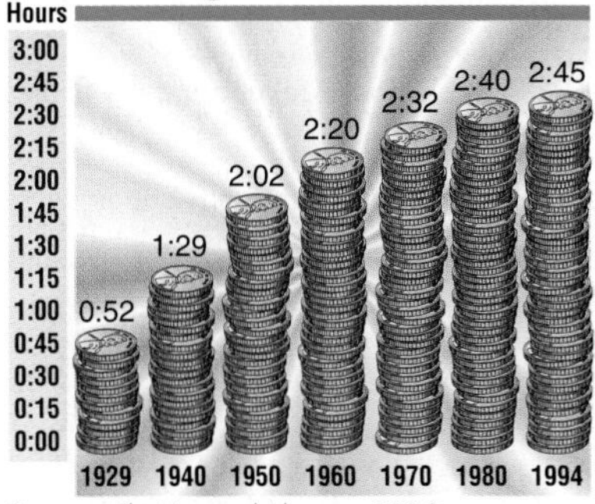

Source: *The Universal Almanac*, 1995

67. **Money** The graph at the left shows the amount of time in an eight-hour workday that it takes to earn enough money to pay a day's worth of taxes. Find the percent of change from 1929 to 1994. Round to the nearest whole percent. 217%

4 ASSESS

Closing Activity

Writing Ask students to copy rhombus *RHOM* and its diagonals.

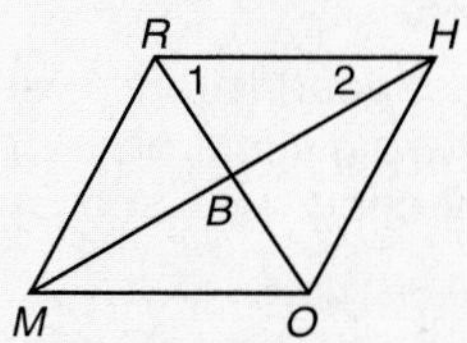

Ask them to write how they would find the measure of $\angle 1$ if they knew the measure of $\angle 2$. Sample answer: Since $\triangle RHB$ is a right triangle, the acute angles are complementary. Therefore, $90 - m\angle 2 = m\angle 1$.

Chapter 6 Quiz C (Lesson 6-4) is available in the *Assessment and Evaluation Masters*, p. 157.

Additional Answer

64. hypothesis: a U.S. citizen is over 18 years old; conclusion: he or she may vote

Enrichment Masters, p. 35

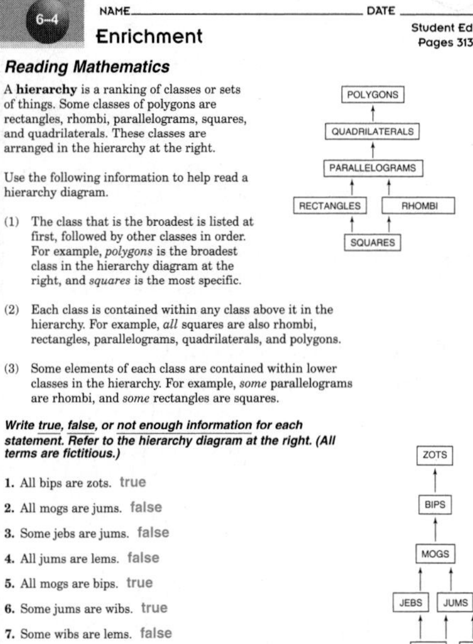
6-4 NAME______________ DATE ________

Enrichment

Student Edition Pages 313–319

Reading Mathematics

A **hierarchy** is a ranking of classes or sets of things. Some classes of polygons are rectangles, rhombi, parallelograms, squares, and quadrilaterals. These classes are arranged in the hierarchy at the right.

POLYGONS → QUADRILATERALS → PARALLELOGRAMS → RECTANGLES, RHOMBI → SQUARES

Use the following information to help read a hierarchy diagram.

(1) The class that is the broadest is listed at first, followed by other classes in order. For example, *polygons* is the broadest class in the hierarchy diagram at the right, and *squares* is the most specific.

(2) Each class is contained within any class above it in the hierarchy. For example, *all* squares are also rhombi, rectangles, parallelograms, quadrilaterals, and polygons.

(3) Some elements of each class are contained within lower classes in the hierarchy. For example, *some* parallelograms are rhombi, and *some* rectangles are squares.

Write true, false, or not enough information for each statement. Refer to the hierarchy diagram at the right. (All terms are fictitious.)

ZOTS, BIPS, MOGS, JEBS, JUMS, WIBS, LEMS

1. All bips are zots. true
2. All mogs are jums. false
3. Some jebs are jums. false
4. All jums are lems. false
5. All mogs are bips. true
6. Some jums are wibs. true
7. Some wibs are lems. false
8. Draw a hierarchy diagram to show these classes: equilateral triangles, polygons, isosceles triangles, triangles. See students' diagrams. Diagrams should indicate equilateral triangles → isosceles triangles → triangles → polygons.

Extension

Communication A device called a pantograph is used to copy, enlarge, or reduce images. Thomas Jefferson used one at Monticello to make copies of his letters. Have students investigate the pantograph and discuss what mathematical properties are at work in the device. The pantograph creates similar figures by using congruent angles formed in parallelograms.

6-4B LESSON NOTES

NCTM Standards: 1–4, 7

Objective

Construct quadrilaterals with exactly two distinct pairs of adjacent congruent sides.

Recommended Time

Demonstration and discussion: 15 minutes; Exercises: 30 minutes

Instructional Resources

For each student or group of students
Modeling Mathematics Masters
- p. 94 (worksheet)

For teacher demonstration
Algebra and Geometry Overhead Manipulative Resources

1 FOCUS

Motivating the Lesson

Have students bring pictures of various styles and shapes of kites to class. Identify the polygons in each kite.

2 TEACH

Teaching Tip Have students place a kite on a coordinate plane and determine the coordinates for the points *A*, *B*, *C*, and *D*.

3 PRACTICE/APPLY

Assignment Guide

Core (with proof): 1–5
Core (informal): 1–5
Enriched: 1–5

4 ASSESS

Observing students working in cooperative groups is an excellent method of assessment.

MODELING MATHEMATICS

6–4B Kites

Materials:  compass  ruler 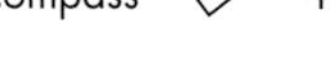 protractor

An Extension of Lesson 6–4

A **kite** is a quadrilateral with exactly two distinct pairs of adjacent congruent sides. You can draw a kite by drawing two noncongruent, nonoverlapping isosceles triangles that share the same base.

Activity **Draw a kite *ABCD*.**

Step 1 Draw a segment *BD* on a piece of paper.

Step 2 Place the tip of your compass at point *B* and draw an arc above $\overline{BD}$. Then without changing the compass setting, move the compass to point *D* and draw an arc to intersect the first one. Label the point of intersection *A*.

Step 3 Change the compass setting. Place the tip of the compass at *B* and draw an arc below $\overline{BD}$. Then, using the same compass setting, draw an arc from point *D* to intersect the first one. Label the point of intersection *C*.

Step 4 Draw *ABCD*.

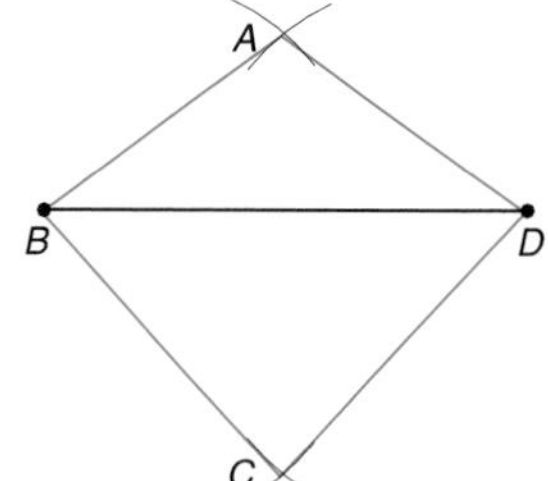

Model

1. Draw $\overline{AC}$ in kite *ABCD*. Use a protractor to measure the angles formed by the intersection of $\overline{AC}$ and $\overline{BD}$. **1, 3, 4. See students' work.**
2. Measure the interior angles of kite *ABCD*. Are any congruent? $\angle ABC \cong \angle ADC$
3. Label the intersection of $\overline{AC}$ and $\overline{BD}$ as point *E*. Find the lengths of $\overline{AE}$, $\overline{BE}$, $\overline{CE}$, and $\overline{DE}$. How are they related?

Draw

4. Draw another kite *JKLM*. Then repeat Exercises 1–3 using kite *JKLM*.

Write

5. Write as many conjectures about kites as you can. Justify your conjectures. **See margin.**

Using Cooperative Learning

This lesson offers an excellent opportunity for using cooperative learning groups. For more information on cooperative learning strategies and group management, see *Cooperative Learning in the Mathematics Classroom*, one of the titles in the Glencoe Mathematics Professional Series.

Additional Answer

5. Sample answer: For kite *ABCD*, $\angle B \cong \angle D$; $\overline{AC} \perp \overline{BD}$; $\overline{AC}$ bisects $\overline{BD}$, $\overline{AB} \cong \overline{AD}$; $\overline{BC} \cong \overline{CD}$; $\triangle ABC \cong \triangle ADC$. See students' justifications.

6-5 Trapezoids

What YOU'LL LEARN

- To recognize and apply the properties of trapezoids.

Why IT'S IMPORTANT

You can use properties of trapezoids to solve problems involving sailing and engineering.

FYI

Qin Shi Huangdi, who began the Great Wall, was the first emperor of China. Qin reigned from 221 to 210 B.C.

APPLICATION

Civil Engineering

The greatest accomplishment in the field of civil engineering stretches 2150 miles across China. Millions of soldiers and workers spent more than 1800 years creating the Great Wall of China. All of the materials for the wall were carried by human power. The wall is the only human-made object that could be seen by the astronauts walking on the moon.

Terra Cotta Army in tomb of Qin Shi Huangdi

Because it was built over a long period of time, different sections of the Great Wall are made of different materials. Most of the wall that remains is masonry built during the Ming Dynasty. The diagram below shows a cross section of a masonry portion of the wall.

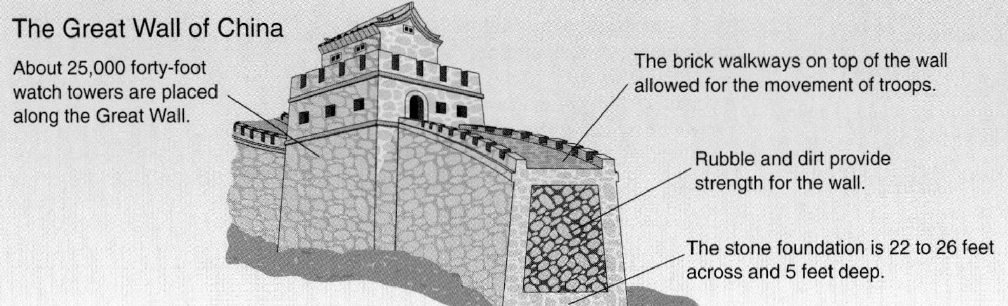

The Great Wall of China
About 25,000 forty-foot watch towers are placed along the Great Wall.
The brick walkways on top of the wall allowed for the movement of troops.
Rubble and dirt provide strength for the wall.
The stone foundation is 22 to 26 feet across and 5 feet deep.

The cross section of the wall is shaped like a **trapezoid**. A trapezoid is a quadrilateral with exactly one pair of parallel sides. The parallel sides are called **bases**. The nonparallel sides are called **legs**. In trapezoid *TRAP* at the right, $\angle T$ and $\angle R$ are one pair of **base angles**. $\angle P$ and $\angle A$ are the other pair.

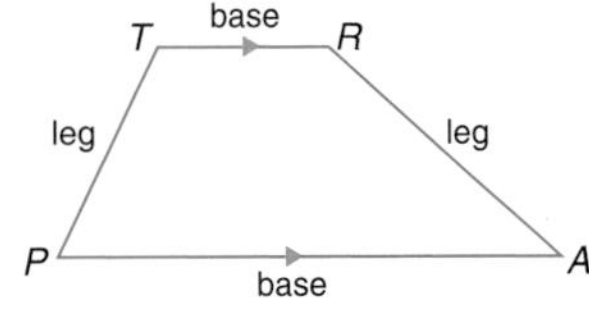

T
base
R
leg
leg
P
base
A

If the legs of a trapezoid are congruent, then the trapezoid is an **isosceles trapezoid**. Isosceles trapezoids have two special relationships that hold true. The proof of Theorem 6–14 is given. *You will be asked to prove Theorem 6–15 in Exercise 38.*

Theorem 6–14	**Both pairs of base angles of an isosceles trapezoid are congruent.**
Theorem 6–15	**The diagonals of an isosceles trapezoid are congruent.**

FYI

In 221 B.C., the western state of Qin united all of China. The king of Qin was Qin Shi Huangdi who became the first emperor and began the first Imperial Dynasty. A unified system of weights and measures was developed during this time.

6-5 LESSON NOTES

NCTM Standards: 1–5, 7, 8

Instructional Resources

- Study Guide Master 6-5
- Practice Master 6-5
- Enrichment Master 6-5
- Assessment and Evaluation Masters, p. 157
- Graphing Calculator and Computer Masters, p. 6
- Multicultural Activity Masters, p. 12
- Tech Prep Applications Masters, p. 12

Transparency 6-5A contains the 5-Minute Check for this lesson; **Transparency 6-5B** contains a teaching aid for this lesson.

Recommended Pacing	
Standard Pacing	Days 10 & 11 of 13
Honors Pacing	Day 10 of 12
Block Scheduling*	Day 5 of 6

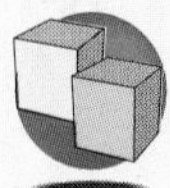

*For more information on pacing and possible lesson plans, refer to the *Block Scheduling Booklet*.

1 FOCUS

5-Minute Check
(over Lesson 6-4)

Using P for parallelogram, R for rectangle, S for square, and Rh for rhombus, write the letters of all the quadrilaterals that have these properties.

1. perpendicular diagonals S, Rh
2. diagonals that bisect each other P, R, S, Rh
3. congruent diagonals R, S
4. two pairs of congruent opposite angles P, R, S, Rh
5. supplementary consecutive angles P, R, S, Rh
6. diagonals that bisect opposite angles S, Rh
7. four right angles R, S
8. two pairs of opposite parallel sides P, R, S, Rh

Motivating the Lesson

Questioning Ask each student to draw four quadrilaterals that are not parallelograms. Have students compare their examples and determine if any of them have special names. **Sample answers: trapezoid, kite**

2 TEACH

In-Class Example

For Example 1

Find the length of each diagonal to verify that Theorem 6-14 is true for isosceles trapezoid *MATH* with vertices *M*(0, 0), *A*(0, 3), *T*(4, 4), and *H*(4, −1). **Graph the trapezoid to find the measures of the diagonals.**

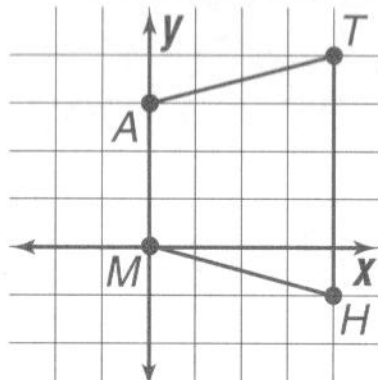

***MT* = $4\sqrt{2}$, *AH* = $4\sqrt{2}$. Therefore, *MT* = *AH* and Theorem 6-14 is true for trapezoid *MATH*.**

Teaching Tip Point out that Theorems 6-14 and 6-15 apply only to isosceles trapezoids while Theorem 6-16 applies to all trapezoids.

Proof of Theorem 6-14

Given: *ABCD* is an isosceles trapezoid.
$\overline{BC} \parallel \overline{AD}$
$\overline{AB} \cong \overline{CD}$

Prove: $\angle A \cong \angle D$
$\angle ABC \cong \angle DCB$

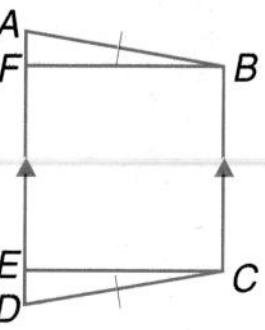

Paragraph Proof:

Draw auxiliary segments so that $\overline{BF} \perp \overline{AD}$ and $\overline{CE} \perp \overline{AD}$. Since $\overline{BC} \parallel \overline{AD}$ and parallel lines are everywhere equidistant, $\overline{BF} \cong \overline{CE}$. Perpendicular lines form right angles, so $\angle BFA$ and $\angle CED$ are right angles. $\triangle BFA$ and $\triangle CED$ are right triangles by definition. Therefore, $\triangle BFA \cong \triangle CED$ by HL. $\angle A \cong \angle D$ by CPCTC.

Since $\angle CBF$ and $\angle BCE$ are right angles and all right angles are congruent, $\angle CBF \cong \angle BCE$. $\angle ABF \cong \angle DCE$ by CPCTC. So, $\angle ABC \cong \angle DCB$ by the Angle Addition Postulate.

You can use algebra to investigate trapezoids on the coordinate plane.

Example 1

Algebra

Verify that isosceles trapezoid *TRAP* with vertices *T*(−1, 1), *R*(−5, −3), *A*(−4, −10), and *P*(6, 0) has congruent diagonals and congruent base angles.

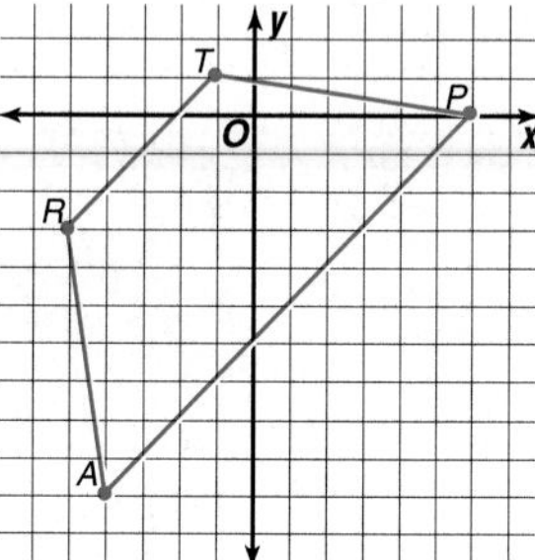

First, show that the diagonals are congruent by using the distance formula to find the lengths of $\overline{RP}$ and $\overline{AT}$.

$$RP = \sqrt{(-5-6)^2 + (-3-0)^2} = \sqrt{(-11)^2 + (-3)^2} \text{ or } \sqrt{130}$$

$$AT = \sqrt{(-4-(-1))^2 + (-10-1)^2} = \sqrt{(-3)^2 + (-11)^2} \text{ or } \sqrt{130}$$

Since *RP* = *AT*, the diagonals are congruent.

Next, use triangle congruence to show that the base angles are congruent.

We know that *TRAP* is an isosceles trapezoid, so by the definition of isosceles trapezoid, $\overline{RA} \cong \overline{TP}$. Congruence of segments is reflexive, so $\overline{AP} \cong \overline{PA}$. We have shown that $\overline{RP} \cong \overline{AT}$ using the distance formula. Therefore, $\triangle RAP \cong \triangle TPA$ by SSS. Then we can conclude that $\angle RAP \cong \angle TPA$ by CPCTC.

You can use a similar method with $\triangle TRA$ and $\triangle RTP$ to prove that $\angle R \cong \angle T$.

The **median** of a trapezoid is the segment that joins the midpoints of its legs.

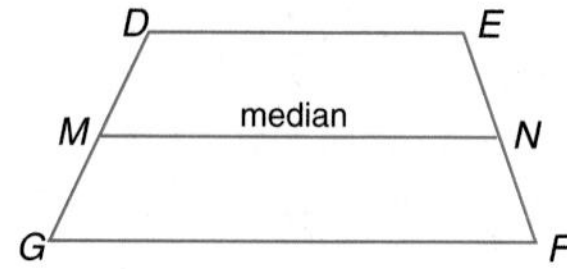

Classroom Vignette

"As an enrichment for this chapter, I discuss tessellations. Students then create their own tessellation drawings, which we display in the classroom."

Karen Jackson

Karen Jackson
Middletown High School
Middletown, Connecticut

You can construct the median of a trapezoid using a compass and a straightedge.

CONSTRUCTION

Median of a Trapezoid

Construct the median of trapezoid *ZOID*.

Step 1 Draw a trapezoid of any shape or size. Label the vertices *ZOID* with legs $\overline{ZD}$ and $\overline{OI}$.

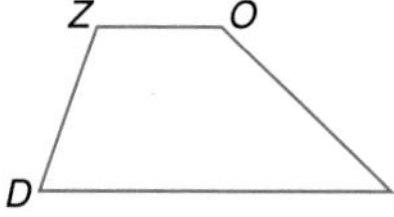

Step 2 Bisect legs $\overline{ZD}$ and $\overline{OI}$. Label the midpoints *M* and *N* as shown.

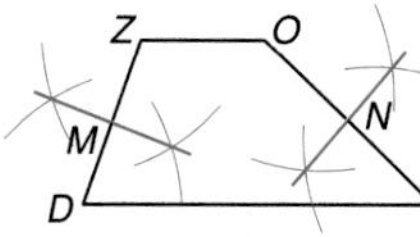

Step 3 Draw line segment *MN*.

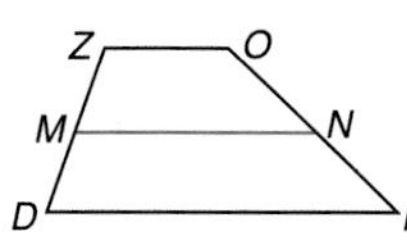

The median has a special relationship to the bases. Use a ruler to measure $\overline{ZO}$, $\overline{ID}$, and $\overline{MN}$ in the figure you constructed. What do you observe about the lengths? This leads to Theorem 6–16.

Theorem 6–16	**The median of a trapezoid is parallel to the bases, and its measure is one-half the sum of the measures of the bases.**

The length of the median can be used to find the length of the bases.

Example 2

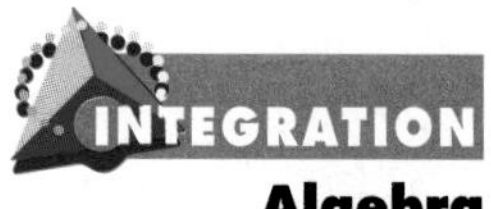

Given trapezoid *EZOI* with median $\overline{AB}$, find the value of *x*.

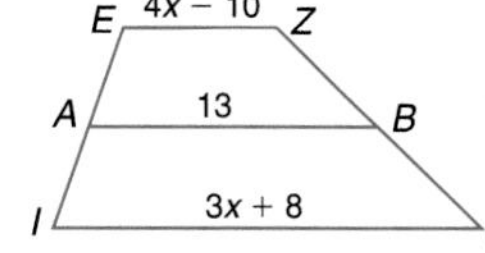

By Theorem 6–16, *AB* equals half the sum of *EZ* and *OI*. We can find the value of *x* by writing and solving an equation.

$AB = \frac{1}{2}(EZ + OI)$

$13 = \frac{1}{2}(4x - 10 + 3x + 8)$ *Substitution Property (=)*

$13 = \frac{1}{2}(7x - 2)$

$26 = 7x - 2$ *Multiplication Property (=)*

$28 = 7x$

$4 = x$

The value of *x* is 4. *Check by substituting 4 for x in the expressions for EZ and OI.*

Teaching Tip After reading Theorem 6-16, have students draw a few examples of trapezoids on grid or dot paper in such a way that it is easy to count to find the lengths of the bases and the median.

In-Class Example

For Example 2

Given trapezoid *GHIJ* with median $\overline{KL}$, find the value of *x*.

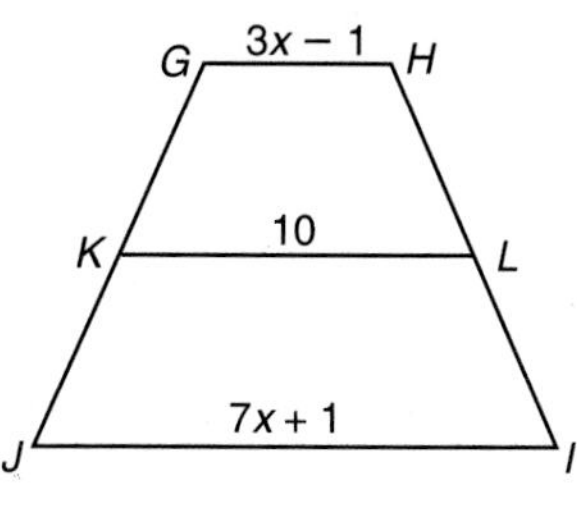

$KL = \frac{1}{2}(GH + JI)$

$10 = \frac{1}{2}(3x - 1 + 7x + 1)$

$10 = \frac{1}{2}(10x)$

$10 = 5x$

$2 = x$

Teaching Tip After Example 2, have students use $x = 4$ to find the measure of each base of *EZOI*.

3 PRACTICE/APPLY

Check for Understanding

Exercises 1–15 are designed to help you assess your students' understanding through reading, writing, speaking, and modeling. You should work through Exercises 1–4 with your students and then monitor their work on Exercises 5–15.

Error Analysis

Some students may subtract the measures of the bases instead of finding their sum when finding the median. Remind them that the measure of the median is the average of the measures of the bases.

Additional Answers

1.

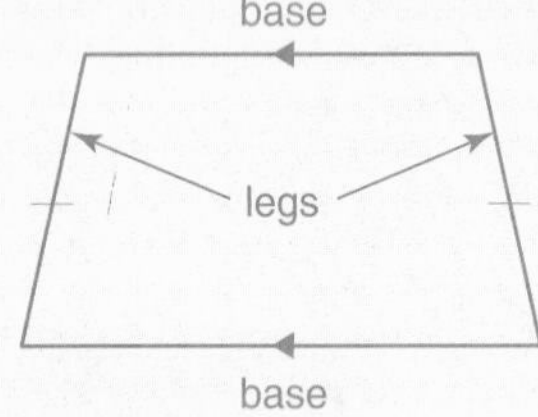

2. midpoint of $\overline{RA}$ = $\left(\frac{-5+(-4)}{2}, \frac{-3+(-10)}{2}\right)$, or $M\left(-4\frac{1}{2}, -6\frac{1}{2}\right)$;
midpoint of $\overline{PT}$ = $\left(\frac{6+(-1)}{2}, \frac{0+1}{2}\right)$ or $N\left(2\frac{1}{2}, \frac{1}{2}\right)$;
slope of $\overline{MN}$ = $\left(\frac{-6\frac{1}{2}-\frac{1}{2}}{-4\frac{1}{2}-2\frac{1}{2}}\right)$ or 1; slope of $\overline{TR} = \frac{1-(-3)}{-1-(-5)}$ or 1; slope of $\overline{AP} = \frac{-10-0}{-4-6}$ or 1; since the slopes are all the same, $\overline{MN} \parallel \overline{TR} \parallel \overline{AP}$.

CHECK FOR UNDERSTANDING

Communicating Mathematics

Read and study the lesson to answer each question.

1–2. See margin.

1. **Draw** an isosceles trapezoid and label the legs and the bases.
2. Refer to Example 1 on page 322. Find the midpoints of $\overline{RA}$ and $\overline{PT}$ and label them M and N, respectively. Use slopes to show that $\overline{MN}$ is parallel to bases $\overline{TR}$ and $\overline{AP}$.
3. Decide whether each statement is *true* or *false*. Explain your reasoning.
   a. A trapezoid is a parallelogram. 3a. False; a trapezoid has exactly one pair of parallel sides, and a parallelogram has two pairs of parallel sides.
   b. The length of the median of a trapezoid is one-half the sum of the lengths of the bases. True; this is a statement of Theorem 6–16.
   c. The bases of any trapezoid are parallel. True; this is true by definition.
   d. The legs of a trapezoid are always congruent. False; the legs of a trapezoid are congruent only if the trapezoid is an isosceles trapezoid.

4. Darken two lines on a piece of lined notebook paper to show two parallel lines. Place the point of a compass on a point R on one of the lines, and draw an arc that intersects the second line. Label the point where the arc intersects the line as point S and draw $\overline{RS}$. Place the point of the compass at another point U on the first line, and use the same compass setting to mark a point T. Draw $\overline{TU}$. Make sure that $\overline{SR} \nparallel \overline{TU}$.

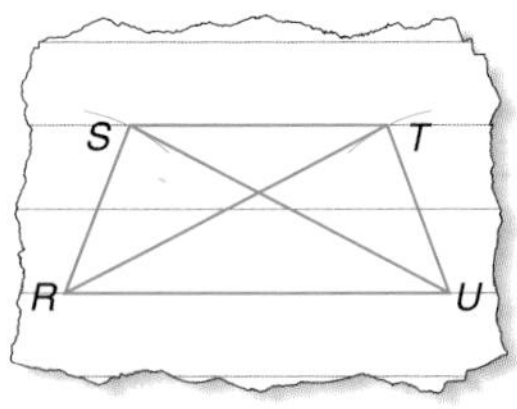

   a. Measure $\angle R$, $\angle S$, $\angle T$, and $\angle U$ of trapezoid $RSTU$. What do you observe about the base angles? Base angles are congruent.
   b. Draw $\overline{RT}$ and $\overline{SU}$. Then use a centimeter ruler to find RT and SU to the nearest millimeter. The diagonals are congruent.
   c. What type of trapezoid is $RSTU$? an isosceles trapezoid

Guided Practice

***ABCD* is an isosceles trapezoid. Decide whether each statement is *true* or *false*. Explain your reasoning.**

5. $AC = BD$
6. $\overline{AD} \cong \overline{CB}$
7. $\overline{CA}$ and $\overline{BD}$ bisect each other.

5. True; the diagonals of an isosceles trapezoid are congruent.
6. True; the legs of an isosceles trapezoid are congruent.
7. False; if the diagonals bisected each other, *ABCD* would be a parallelogram not a trapezoid.

***QUAD* is an isosceles trapezoid with bases $\overline{QD}$ and $\overline{AU}$, and median $\overline{EF}$. Use the given information to solve each problem.**

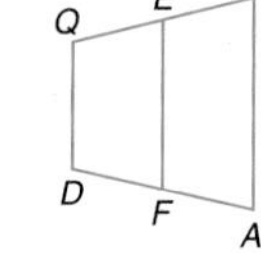

8. If $QU = x^2$ and $AD = 16$, find the value of x. 4 or −4
9. If $QE = 17$, find AF. 17
10. If $m\angle QUA = 62$, find $m\angle ADQ$. 118

Consider *MNPQ* with vertices at *M*(0, 7), *N*(3, 2), *P*(7, 4) and *Q*(5, 9).

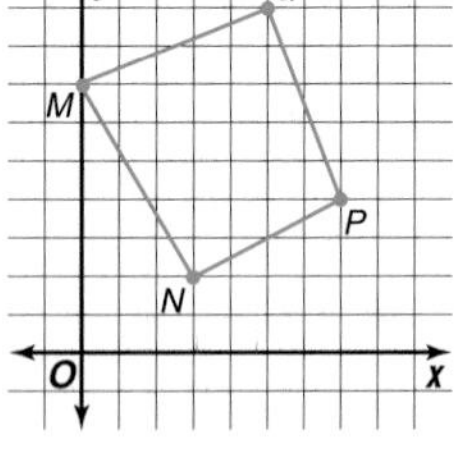

11. Find the coordinates of points R and S if they are the midpoints of $\overline{NP}$ and $\overline{MQ}$. 11. (5, 3), $(2\frac{1}{2}, 8)$
12. Find the length of $\overline{RS}$. $\sqrt{31.25} \approx 5.59$
13. Determine whether $RS = \frac{1}{2}(NP + MQ)$. Explain the result. 13. $RS \neq \frac{1}{2}(NP + MQ)$, because *MNPQ* is not a trapezoid.

Reteaching

Using Diagrams Draw this trapezoid on the chalkboard or overhead projector. Find the missing measures with students.

1. $m\angle IRD$ 75
2. YR 3
3. DR 13
4. AC 10

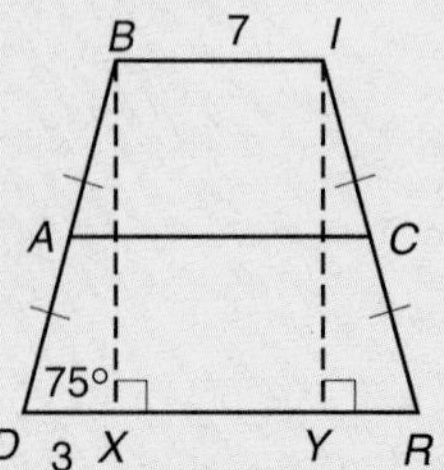

Proof

14. Write a flow proof. See Solutions Manual.

Given: $KLMN$ is an isosceles trapezoid with bases $\overline{KL}$ and $\overline{MN}$.

Prove: $\angle 1 \cong \angle 2$

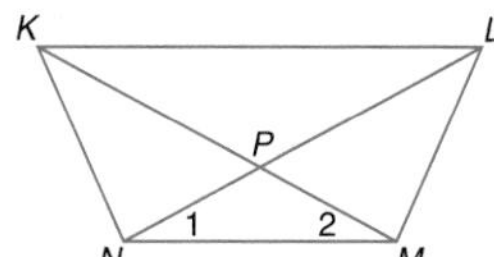

15. **Gardening** A cold frame is placed over plants on the ground to protect them during cold weather. The top of the frame is glass or plastic to allow the sunlight in and to hold in the heat. The cover slants downward to the south to let in as much sunlight as possible. Identify the shape of each quadrilateral used to construct the cold frame shown at the right. Sides are trapezoids; front, back & tops are rectangles.

EXERCISES

Practice

RSTV is an isosceles trapezoid. Decide whether each statement is *true* or *false*. Explain your reasoning.

16. $\overline{TR} \perp \overline{SV}$

16. False; the diagonals of a rhombus are perpendicular, but the diagonals of a trapezoid are not.

17. $\angle TRV \cong \angle VST$ See margin.
18. $\angle RVT \cong \angle STV$ See margin.
19. $\angle SRV$ and $\angle TVR$ are supplementary. See margin.

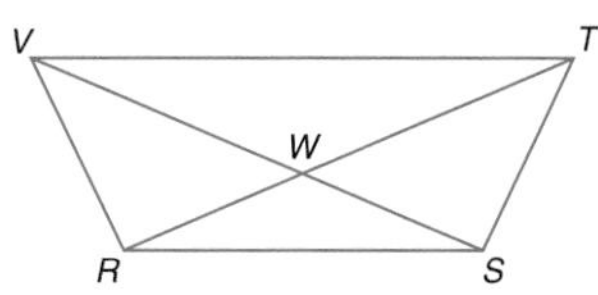

WXYZ is an isosceles trapezoid with bases $\overline{WZ}$ and $\overline{XY}$ and median $\overline{MN}$. Use the given information to solve each problem.

B

20. Find MN if $WZ = 11$ and $XY = 3$. 7
21. Find $m\angle XMN$ if $m\angle WZN = 78$. 78
22. If $MN = 10$ and $WZ = 14$, find XY. 6
23. What is the value of x if $m\angle MWZ = 15x - 5$ and $m\angle WZN = 90 - 4x$? 5
24. If $XY = 21.7$ and $ZW = 93.6$, find MN. 57.65
25. If $m\angle XWZ = 2x - 7$ and $m\angle XYZ = 117$, find the value of x. 35
26. If $MN = 60$, $XY = 4x - 1$, and $WZ = 6x + 11$, find the value of x. 11
27. If $MN = 10x + 3$, $WZ = 11$, and $XY = 8x + 19$, find the value of x. 2
28. If $MN = 2x + 1$, $XY = 8$, and $WZ = 3x - 3$, find the value of x. 3

For Exercises 29–32, consider trapezoid RSTV with bases $\overline{RS}$ and $\overline{TV}$.

29. Show $\overline{RS} \parallel \overline{TV}$.
30. Is $RSTV$ an isosceles trapezoid? Why or why not? See margin.
31. Find the coordinates of the endpoints of median $\overline{AB}$ for $RSTV$. Show that $\overline{AB} \parallel \overline{RS}$.
32. Find AB, RS, and TV. Verify that $AB = \frac{1}{2}(RS + TV)$. See margin.

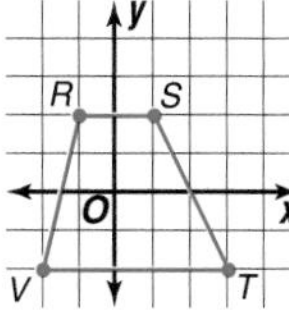

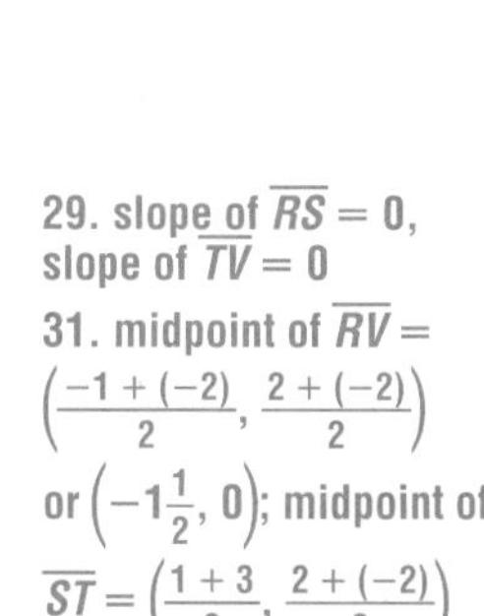

29. slope of $\overline{RS} = 0$, slope of $\overline{TV} = 0$

31. midpoint of $\overline{RV} = \left(\frac{-1 + (-2)}{2}, \frac{2 + (-2)}{2}\right)$ or $\left(-1\frac{1}{2}, 0\right)$; midpoint of $\overline{ST} = \left(\frac{1+3}{2}, \frac{2 + (-2)}{2}\right)$ or $(2, 0)$; slope of $\overline{AB} = 0$; slope of $\overline{RS} = 0$

Assignment Guide

Core (with proof): 17–45 odd, 47–56
Core (informal): 17–37 odd, 41, 43, 45, 47–56
Enriched: 16–40 even, 41–56

For **Extra Practice**, see p. 775.

The red A, B, and C flags, printed only in the Teacher's Wraparound Edition, indicate the level of difficulty of the exercises.

Additional Answers

17. True; $\overline{TV} \cong \overline{TV}$ because congruence of segments is reflexive, $\angle TVR \cong \angle VTS$ because base angles of an isosceles trapezoid are congruent, and $\overline{TR} \cong \overline{VS}$ because diagonals of an isosceles trapezoid are congruent; thus $\triangle TRV \cong \triangle VST$ by SAS and $\angle TRV \cong \angle VST$ by CPCTC.

18. True; base angles of an isosceles trapezoid are congruent.

19. True; $\overline{TV} \parallel \overline{SR}$ and consecutive interior angles are supplementary.

30. $RSTV$ is not an isosceles trapezoid because $RV = \sqrt{[-1 - (-2)]^2 + [2 - (-2)]^2}$ or $\sqrt{17}$, $ST = \sqrt{(1 - 3)^2 + [2 - (-2)]^2}$ or $\sqrt{20}$, and $RV \neq ST$.

32. $AB = \sqrt{\left(-1\frac{1}{2} - 2\right)^2 + (0 - 0)^2}$ or $3\frac{1}{2}$; $RS = \sqrt{(-1 - 1)^2 + (2 - 2)^2}$ or 2; $TV = \sqrt{[3 - (-2)]^2 + [-2 - (-2)]^2}$ or 5; $\frac{5 + 2}{2} = 3\frac{1}{2}$

Using the Programming Exercises The program given in Exercise 41 is for use with a TI-82 or TI-83 graphing calculator. For other programmable calculators, have students consult their owner's manual for commands similar to those presented here.

Additional Answer

39. Given: *DEFH* is a trapezoid with bases $\overline{DE}$ and $\overline{FH}$. $\overline{DE} \cong \overline{FG}$

Prove: *DEFG* is a parallelogram.

Proof:

Statements (Reasons)

1. *DEFH* is a trapezoid with bases $\overline{DE}$ and $\overline{FH}$. $\overline{DE} \cong \overline{FG}$ (Given)
2. $\overline{DE} \parallel \overline{FH}$. (Def. base of a trapezoid)
3. *DEFG* is a parallelogram. (If a pair of opp. sides of a quad. are ≅ and ∥, it is a ▱.)

Study Guide Masters, p. 36

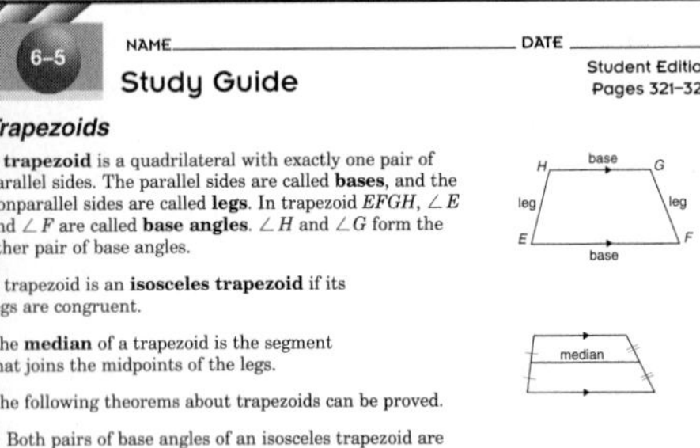

6–5 NAME ______ DATE ______

Study Guide — Student Edition Pages 321–328

Trapezoids

A **trapezoid** is a quadrilateral with exactly one pair of parallel sides. The parallel sides are called **bases**, and the nonparallel sides are called **legs**. In trapezoid *EFGH*, ∠*E* and ∠*F* are called **base angles**. ∠*H* and ∠*G* form the other pair of base angles.

A trapezoid is an **isosceles trapezoid** if its legs are congruent.

The **median** of a trapezoid is the segment that joins the midpoints of the legs.

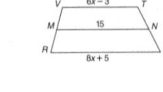

The following theorems about trapezoids can be proved.

- Both pairs of base angles of an isosceles trapezoid are congruent.
- The diagonals of an isosceles trapezoid are congruent.
- The median of a trapezoid is parallel to the bases, and its measure is one-half the sum of the measures of the bases.

Example: Given trapezoid *RSTV* with median $\overline{MN}$, find the value of *x*.

$$MN = \frac{1}{2}(VT + RS)$$
$$15 = \frac{1}{2}(6x - 3 + 8x + 5)$$
$$15 = \frac{1}{2}(14x + 2)$$
$$15 = 7x + 1$$
$$14 = 7x$$
$$2 = x$$

***HJKL* is an isosceles trapezoid with bases $\overline{HJ}$ and $\overline{LK}$, and median $\overline{RS}$. Use the given information to solve each problem.**

1. If $LK = 30$ and $HJ = 42$, find *RS*. 36
2. If $RS = 17$ and $HJ = 14$, find *LK*. 20
3. If $RS = x + 5$ and $HJ + LK = 4x + 6$, find *RS*. 7
4. If $m\angle LRS = 66$, find $m\angle KSR$. 66
5. Find the length of the median of a trapezoid with vertices at *C*(3, 1), *D*(10, 1), *E*(7, 9), and *F*(5, 9). $\frac{9}{2}$

Determine whether each figure below is a *trapezoid*, a *parallelogram*, a *rectangle*, or a *quadrilateral*. Choose the most specific term.

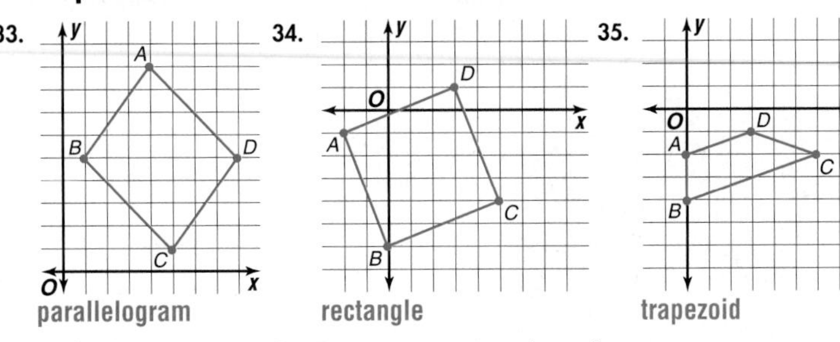

33. parallelogram 34. rectangle 35. trapezoid

36. In the figure at the right, *P* is a point not in plane $\mathcal{A}$, and △*XYZ* is an equilateral triangle in plane $\mathcal{A}$. Plane $\mathcal{B}$ is parallel to plane $\mathcal{A}$ and intersects $\overline{PX}$, $\overline{PY}$, and $\overline{PZ}$ in points *J*, *I*, and *K*, respectively.
   a. How many trapezoids are formed? Name them. **3; *JKZX*, *IJXY*, *IKZY***
   b. If the trapezoids are isosceles, are they congruent? Justify your answer.
   c. Where could *P* be if none of the trapezoids are congruent? **Answers may vary.**

36b. Yes: the bases and legs must be congruent if they are isosceles.

37. In isosceles trapezoid *WXYZ*, $\overline{WZ} \cong \overline{YX}$. If $m\angle W = x$, what is $m\angle Z$? **180 − *x***

C Proof

38. Write a paragraph proof for Theorem 6–15. **See Solutions Manual.**

Write a two-column proof. **39–40. See margin.**

39. **Given:** *DEFH* is a trapezoid with bases $\overline{DE}$ and $\overline{FH}$. $\overline{DE} \cong \overline{FG}$

    Prove: *DEFG* is a parallelogram.

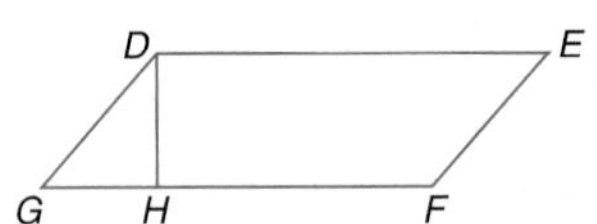

40. **Given:** *JKLM* is an isosceles trapezoid.

    Prove: △*MNL* is isosceles.

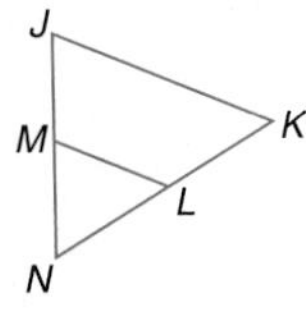

Programming

41. The TI-82/83 graphing calculator program at the right will find the measure of the median of a trapezoid *ABCD* with legs $\overline{AB}$ and $\overline{CD}$.

```
PROGRAM: MEDIAN
: For (A,1,4)
: Disp "ENTER", "COORDINATES:"
: Input "X",X
: X→L1(A)
: Input "Y",Y
: Y→L2(A)
: End
: (L1(1)+L1(2))/2→L1(5)
: (L2(1)+L2(2))/2→L2(5)
: (L1(3)+L1(4))/2→L1(6)
: (L2(3)+L2(4))/2→L2(6)
: √((L1(5)−L1(6))²+(L2(5)−
  L2(6))²→M
: Disp "MEDIAN MEASURE=",M
: Stop
```

Use the program to find the measure of the median of the trapezoid with the given vertices. Round your answers to the nearest hundredth.

   a. *A*(7, 5), *B*(7, 6), *C*(3, 5), *D*(−1, 3) **6.18**
   b. *A*(2, 10), *B*(−4, 8), *C*(−1, 7), *D*(8, 8) **4.74**
   c. *A*(−5, 7), *B*(0, 12), *C*(18, −6), *D*(2, −22) **26.6**

Additional Answer

40. Given: *JKLM* is an isosceles trapezoid.

Prove: △*MNL* is isosceles.

Proof:

Statements (Reasons)

1. *JKLM* is an isosceles trapezoid. (Given)
2. ∠*J* ≅ ∠*K* (Base ∠s of an isosceles trap. are ≅.)
3. $\overline{KN} \cong \overline{JN}$ (If 2 ∠s of a △ are ≅, the sides opp. the ∠s are ≅.)
4. △*MNL* is isosceles. (Def. isosceles △)

Critical Thinking

42. a. Draw a trapezoid with two right angles. Label the bases and the legs.
    b. Try to draw an isosceles trapezoid with two right angles. Is it possible? Explain why or why not. a–b. See margin.

43. The median of a trapezoid is 20 millimeters long. a–c. See margin.
    a. List two possible integral values for the measures of the bases.
    b. How many integral value combinations are possible for the bases? Explain.
    c. How many values are possible for the measures of the bases if the restriction of integral values were removed? Explain.

Applications and Problem Solving

44. **Sailing** Many large sailboats have a *keel* to keep the boat stable in high winds. A keel is shaped like a trapezoid with its top and bottom parallel. The distance across the top of the keel is called the *root chord*, the distance across the bottom of the keel is the *tip chord*, and the distance from the midpoint of the front of the keel to the midpoint of the back of the keel is the *mid-chord*. The chart above shows the root chord and the tip chord for selected types of sailboats. Find the length of the mid-chord for each boat type. See margin.

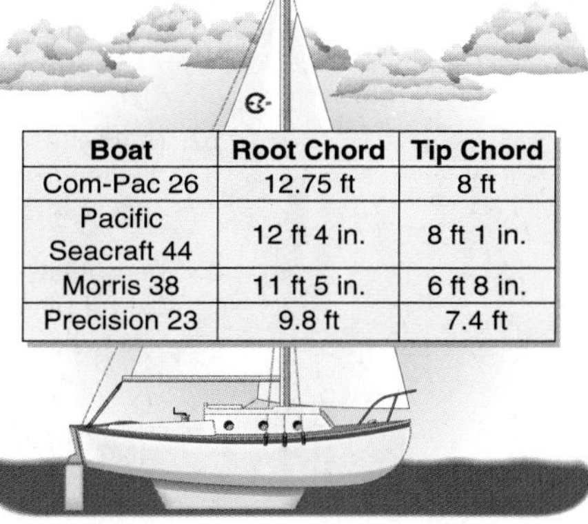

Boat	Root Chord	Tip Chord
Com-Pac 26	12.75 ft	8 ft
Pacific Seacraft 44	12 ft 4 in.	8 ft 1 in.
Morris 38	11 ft 5 in.	6 ft 8 in.
Precision 23	9.8 ft	7.4 ft

45. Top is a square; sides are isosceles trapezoids.

fabulous **FIRSTS**

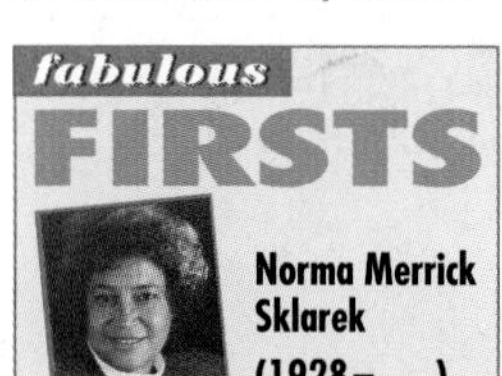

Norma Merrick Sklarek (1928–)

Norma Merrick Sklarek was the first African-American woman to become a registered architect.

45. **Architecture** Tray ceilings make a room appear to be larger than it actually is. The trays are made using wood or drywall panels. What type of quadrilaterals are used to make the tray ceiling shown at the right?

46. **Make a Table** Copy the table below. Write *yes* or *no* in each table cell according to whether each figure has the given characteristic.

	Parallelogram	Rectangle	Square	Isosceles Trapezoid
diagonals congruent	no	yes	yes	yes
two pairs of opposite sides congruent	yes	yes	yes	no
one pair of opposite sides congruent	yes	yes	yes	yes
diagonals perpendicular	no	no	yes	no
one pair of opposite sides parallel and congruent	yes	yes	yes	no

Additional Answers

42a.

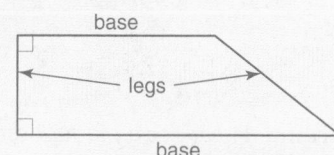

42b. It is not possible. Since pairs of base angles of an isosceles trapezoid are congruent, if two angles are right, all four angles will be right. Then the quadrilateral would be a rectangle, not a trapezoid.

43a. Sample answer: 10 and 30

43b. A measure cannot be negative and the average of the measures of the bases is 20. So the possible measures of the bases are 1 and 39, 2 and 38, . . . and 19 and 21. The measures of the bases cannot be the same or the figure would be a parallelogram. Thus, there are 19 possible combinations.

Practice Masters, p. 36

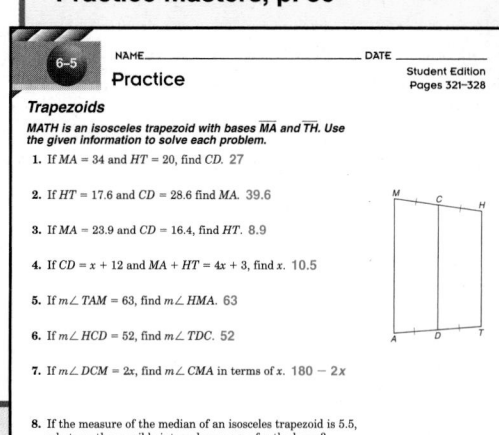

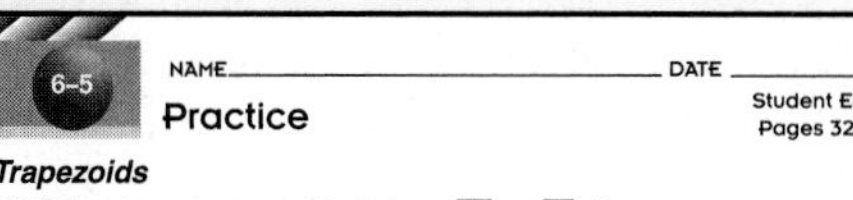

6–5 NAME______ DATE ______

Practice

Student Edition Pages 321–328

Trapezoids

MATH is an isosceles trapezoid with bases $\overline{MA}$ and $\overline{TH}$. Use the given information to solve each problem.

1. If $MA = 34$ and $HT = 20$, find CD. 27
2. If $HT = 17.6$ and $CD = 28.6$ find MA. 39.6
3. If $MA = 23.9$ and $CD = 16.4$, find HT. 8.9
4. If $CD = x + 12$ and $MA + HT = 4x + 3$, find x. 10.5
5. If $m\angle TAM = 63$, find $m\angle HMA$. 63
6. If $m\angle HCD = 52$, find $m\angle TDC$. 52
7. If $m\angle DCM = 2x$, find $m\angle CMA$ in terms of x. $180 - 2x$

8. If the measure of the median of an isosceles trapezoid is 5.5, what are the possible integral measures for the bases? 1, 10; 2, 9; 3, 8; 4, 7; 5, 6

9. $\overline{VW}$ is the median of a trapezoid that has bases $\overline{MN}$ and $\overline{PO}$, with V on $\overline{OM}$ and W on $\overline{PN}$. If the vertices of the trapezoid are $M(2, 6)$, $N(4, 6)$ $P(10, 0)$, and $O(0, 0)$, find the coordinates of V and W. $V(1, 3)$, $W(7, 3)$

10. $\overline{VW}$ is the median of a trapezoid that has bases $\overline{MN}$ and $\overline{PO}$, with V on $\overline{PM}$ and W on $\overline{ON}$. If four of the points are $M(5, 10)$, $N(9, 10)$, $V(3, 7)$, and $W(11, 7)$, find the coordinates of P and O. $P(1, 4)$, $O(13, 4)$

fabulous **FIRSTS**

There are over eight hundred African American architects registered in the United States.

Additional Answers

43c. There are an infinite number of combinations of possible measures if the integer restriction is removed because there is an infinite number of positive values with an average of 20.

44. Com-Pac 26: 10.375 ft; Pacific Seacraft 44: 10 ft 2.5 in.; Morris 38: 9 ft 0.5 in.; Precision 23: 8.6 ft

4 ASSESS

Closing Activity

Writing Have each student create a problem about finding missing measures in an isosceles trapezoid and in a trapezoid that is not isosceles. Have students exchange their problems for solution.

Chapter 6 Quiz D (Lesson 6-5) is available in the *Assessment and Evaluation Masters*, p. 157.

Additional Answers

48. Since all the angles and the opposite sides are congruent, the bricks are interchangeable and can be installed in rows easily.

52. $\overline{AC} \cong \overline{ST}$ or $\overline{BC} \cong \overline{RT}$

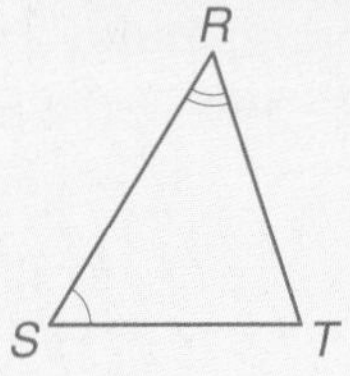

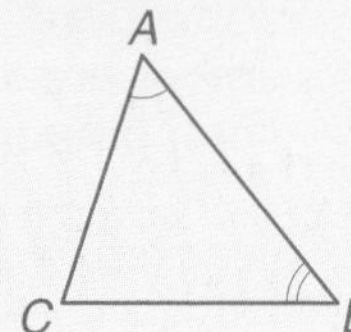

Enrichment Masters, p. 36

NAME________________ DATE ________

Enrichment — Student Edition Pages 321–328

Kites

A kite is a special type of quadrilateral that is neither a parallelogram nor a trapezoid. A quadrilateral is a kite if and only if it has exactly two distinct pairs of congruent consecutive sides and its diagonals intersect. In the figure, quadrilateral *ABCD* is a kite.

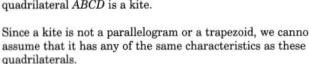

Since a kite is not a parallelogram or a trapezoid, we cannot assume that it has any of the same characteristics as these quadrilaterals.

Determine if each statement is true for all kites. Write yes or no.

1. Both pairs of opposite sides are parallel. no
2. Both pairs of sides are congruent. no
3. The diagonals bisect each other. no
4. Each diagonal bisects a pair of opposite angles. no
5. The diagonals are perpendicular. yes
6. Both pairs of opposite angles are congruent. no
7. One pair of opposite angles are congruent. yes
8. In kite *ABCD*, $\overline{AB} \cong \overline{AD}$ and $\overline{BC} \cong \overline{DC}$. If $AB = 3x - 4$, $BC = x + 8$, $CD = y + 7$, and $DA = 2y + 5$, find the perimeter of the kite. 96 units

Prove each theorem about kites. Write your proofs on a separate sheet.

9. ***Theorem:*** If a quadrilateral is a kite, then one of its diagonals bisects a pair of opposite angles.

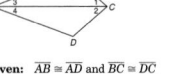

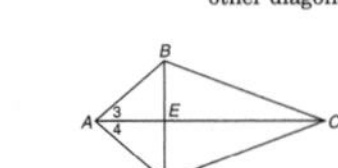

Given: $\overline{AB} \cong \overline{AD}$ and $\overline{BC} \cong \overline{DC}$
Prove: $\angle 1 \cong \angle 2$ and $\angle 3 \cong \angle 4$
See students' work.

10. ***Theorem:*** If a quadrilateral is a kite, then one of its diagonals bisects the other diagonal.

Given: $\overline{AB} \cong \overline{AD}$ and $\overline{BC} \cong \overline{DC}$
Prove: $\overline{AC}$ bisects $\overline{BD}$.
See students' work.

Mixed Review

47. In rhombus *BCDE*, $m\angle B = 68$. Find $m\angle E$. (Lesson 6–4) **112**

48. **Building Materials** Explain why the rectangle is used as the shape of bricks in the construction of walls. (Lesson 6–3) **See margin.**

49. Draw several quadrilaterals to determine if the statement *If the midpoints of the sides of any quadrilateral are connected in clockwise order, they form a parallelogram* is true or false. (Lesson 6–2) **true**

50. Write the assumption you would make to start an indirect proof of the statement *If two lines intersect, then no more than one plane contains them.* (Lesson 5–3) **Two lines intersect and more than one plane contains them.**

51. Determine if the figure contains a pair of congruent triangles. Justify your answer. (Lesson 5–2) **yes; LA or AAS**

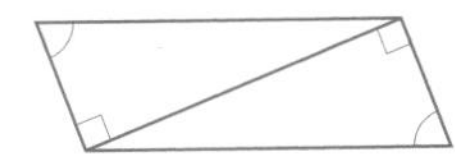

52\. See margin.

52. Draw and label $\triangle RST$ and $\triangle ABC$ with $\angle S \cong \angle A$ and $\angle R \cong \angle B$. Indicate the additional pair of corresponding parts that would have to be proved congruent in order to prove the triangles congruent by AAS. (Lesson 4–5)

53. Determine if statement (3) follows from statements (1) and (2) by the Law of Detachment or the Law of Syllogism. If it does, state which law was used. (Lesson 2–3) **yes; detachment**

(1) If a quadrilateral has exactly two opposite sides parallel, then it is a trapezoid.

(2) *ABCD* has exactly two opposite sides parallel.

(3) *ABCD* is a trapezoid.

54. $\angle N$ and $\angle M$ are supplementary. If $\angle N$ is acute, is $\angle M$ acute, right, obtuse, or straight? (Lesson 1–7) **obtuse**

55\. 123,454,321

INTEGRATION **Algebra**

55. **Patterns** Use your calculator to find the values of 11^2, 111^2, and 1111^2. Without calculating, use a pattern to determine the value of $11{,}111^2$.

56. Solve $\frac{12}{k} = \frac{2}{9}$. **54**

WORKING ON THE In·ves·ti·ga·tion

Refer to the Investigation on pages 286–287.

1 Complete your list of the dimensions of each possible feature of the park.

2 Determine which features you would like to incorporate. You may wish to make two lists: one of features that you feel are necessary and one of features that would be nice if space and money are available. Note the budget for each feature on the lists.

3 Find the area required for each feature you are considering.

Add the results of your work to your Investigation Folder.

Extension

Problem Solving Three vertices of a trapezoid *TWIN* are $T(-2, 3)$, $W(2, 3)$, and $I(6, -1)$. Find the coordinates of *N* if the coordinates of the endpoints of the median are $(-2, 1)$ and $(4, 1)$. $N(-2, -1)$

In·ves·ti·ga·tion

Working on the Investigation

The Investigation on pages 286–287 is designed to be a long-term project that is completed over several days or weeks. Encourage students to keep their materials in their Investigation Folder as they work on the Investigation.

CHAPTER 6 HIGHLIGHTS

VOCABULARY

After completing this chapter, you should be able to define each term, property, or phrase and give an example or two of each.

Geometry
- base (p. 321)
- base angle (p. 321)
- diagonal (p. 293)
- isosceles trapezoid (p. 321)
- kite (p. 320)
- leg (p. 321)
- median (p. 322)
- parallelogram (pp. 290, 291)
- quadrilateral (p. 291)
- rectangle (pp. 305, 306)
- rhombus (p. 313)
- square (p. 315)
- trapezoid (p. 321)

Algebra
- probability (p. 294)

Problem Solving
- identify subgoals (p. 298)

7. False; a square has all of the characteristics of a parallelogram, a rectangle, and a rhombus, but not the characteristics of a trapezoid.

12. False; the diagonals of an isosceles trapezoid are congruent.

UNDERSTANDING AND USING THE VOCABULARY

Determine whether each statement is *true* or *false*. If the statement is false, rewrite it to make it true.

1. Every quadrilateral is a parallelogram. False; every parallelogram is a quadrilateral.
2. If quadrilateral $ABCD$ is a parallelogram, then $\overline{AB} \parallel \overline{CD}$. true
3. If both pairs of opposite angles in a quadrilateral are congruent, then the quadrilateral is a parallelogram. true
4. If $NMOP$ is a rectangle, then it is a parallelogram. true
5. You can prove that a quadrilateral is a rectangle by proving that the diagonals are congruent. False; you can prove that a parallelogram is a rectangle by proving that the diagonals are congruent.
6. If a quadrilateral is a rhombus or a square, then the diagonals are perpendicular. true
7. A square has all of the characteristics of a parallelogram, a rectangle, a rhombus, and a trapezoid.
8. If a quadrilateral has four right angles, then it must be a rectangle. true
9. The legs of an isosceles parallelogram are congruent. False; the legs of an isosceles trapezoid are congruent.
10. The median of a trapezoid is parallel to the bases of the trapezoid and its measure is half the sum of the measures of the bases. true
11. If $QUAD$ is a square, then it is also a parallelogram, a rectangle, a rhombus, a quadrilateral, and a trapezoid. False; if $QUAD$ is a square, then it is also a parallelogram, a rectangle, a rhombus, and a quadrilateral. But $QUAD$ is not a trapezoid.
12. The diagonals of a trapezoid are congruent.

Instructional Resources

Three multiple-choice tests and three free-response tests are provided in the *Assessment and Evaluation Masters*. Forms 1A and 2A are for honors pacing, Forms 1B and 2B are for average pacing, and Forms 1C and 2C are for basic pacing. Chapter 6 Test, Form 1B, is shown at the right. Chapter 6 Test, Form 2B, is shown on the next page.

Postulates, Theorems, and Corollaries

A complete list of postulates, theorems, and corollaries begins on page 806.

Using the CHAPTER HIGHLIGHTS

The Chapter Highlights begins with a listing of the new terms, properties, and phrases that were introduced in this chapter. Have students define each term and provide an example or two of it, if appropriate.

Assessment and Evaluation Masters, pp. 143–144

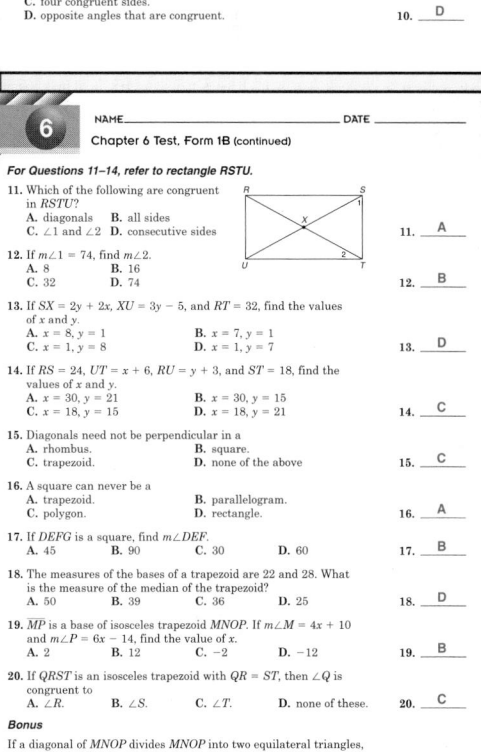

6 NAME__________ DATE __________

Chapter 6 Test, Form 1B

Write the letter for the correct answer in the blank at the right of each question.

1. If $ABCD$ is a parallelogram, $m\angle D = x$, and $m\angle A = 3x + 4$, find the value of x.
   A. 46 B. 45 C. 44 D. 43 — 1. C
2. $XYZW$ is a parallelogram with diagonals $\overline{XZ}$ and $\overline{YW}$ that intersect at point A. If $YA = 2t$, $WA = 3t - 4$, and $XZ = 5t$, find XA.
   A. 20 B. 10 C. 9 D. 4 — 2. B
3. Opposite angles are *not* always congruent in a
   A. trapezoid. B. rhombus. C. rectangle. D. parallelogram. — 3. A
4. Consecutive angles are *not* always supplementary in a
   A. rectangle. B. trapezoid. C. parallelogram. D. rhombus. — 4. B
5. One way to prove that a quadrilateral is a parallelogram is to show that
   A. it has one pair of parallel sides.
   B. the diagonals are congruent.
   C. it has one pair of congruent sides.
   D. both pairs of opposite angles are congruent. — 5. D

For Questions 6–9, refer to parallelogram ACFG.

6. If $m\angle GAC = 125$, find $m\angle ACF$.
   A. 125 B. 55 C. 35 D. 30 — 6. B
7. If $m\angle CFG = 2w + 30$ and $m\angle ACF = w + 15$, find $m\angle AGF$.
   A. 45 B. 60 C. 75 D. 120 — 7. B
8. If $AC = 2x + 2$ and $GF = 4x - 14$, find the value of x.
   A. 8 B. 6 C. $\frac{8}{3}$ D. 2 — 8. A
9. If $GX = 2y + 2$ and $XC = 3y - 3$, find the value of y.
   A. 1 B. −1 C. 5 D. −5 — 9. C
10. All parallelograms have
    A. opposite angles that are supplementary.
    B. diagonals that are congruent.
    C. four congruent sides.
    D. opposite angles that are congruent. — 10. D

6 NAME__________ DATE __________

Chapter 6 Test, Form 1B (continued)

For Questions 11–14, refer to rectangle RSTU.

11. Which of the following are congruent in $RSTU$?
    A. diagonals B. all sides C. $\angle 1$ and $\angle 2$ D. consecutive sides — 11. A
12. If $m\angle 1 = 74$, find $m\angle 2$.
    A. 8 B. 16 C. 32 D. 74 — 12. B
13. If $SX = 2y + 2x$, $XU = 3y - 5$, and $RT = 32$, find the values of x and y.
    A. $x = 8, y = 1$ B. $x = 7, y = 1$ C. $x = 1, y = 8$ D. $x = 1, y = 7$ — 13. D
14. If $RS = 24$, $UT = x + 6$, $RU = y + 3$, and $ST = 18$, find the values of x and y.
    A. $x = 30, y = 21$ B. $x = 30, y = 15$ C. $x = 18, y = 15$ D. $x = 18, y = 21$ — 14. C
15. Diagonals need not be perpendicular in a
    A. rhombus. B. square. C. trapezoid. D. none of the above — 15. C
16. A square can never be a
    A. trapezoid. B. parallelogram. C. polygon. D. rectangle. — 16. A
17. If $DEFG$ is a square, find $m\angle DEF$.
    A. 45 B. 90 C. 30 D. 60 — 17. B
18. The measures of the bases of a trapezoid are 22 and 28. What is the measure of the median of the trapezoid?
    A. 50 B. 39 C. 36 D. 25 — 18. D
19. $\overline{MP}$ is a base of isosceles trapezoid $MNOP$. If $m\angle M = 4x + 10$ and $m\angle P = 6x - 14$, find the value of x.
    A. 2 B. 12 C. −2 D. −12 — 19. B
20. If $QRST$ is an isosceles trapezoid with $QR = ST$, then $\angle Q$ is congruent to
    A. $\angle R$. B. $\angle S$. C. $\angle T$. D. none of these. — 20. C

Bonus

If a diagonal of $MNOP$ divides $MNOP$ into two equilateral triangles, what kind of figure must $MNOP$ be?
A. isosceles trapezoid B. rhombus C. square D. none of these — Bonus B

Using the STUDY GUIDE AND ASSESSMENT

Skills and Concepts Encourage students to refer to the objectives and examples on the left as they complete the review exercises on the right.

Assessment and Evaluation Masters, pp. 149–150

NAME________________ DATE ________

Chapter 6 Test, Form 2B

For Questions 1–8, determine whether the statement is always, sometimes, or never true.

1. Diagonals of a trapezoid are congruent. 1. sometimes
2. Opposite sides of a rectangle are congruent. 2. always
3. Base angles of an isosceles trapezoid are congruent. 3. always
4. Opposite angles of a rhombus are supplementary and congruent. 4. sometimes
5. Diagonals of a square are congruent. 5. always
6. A square is not a rhombus. 6. never
7. All angles of a parallelogram are congruent. 7. sometimes
8. The diagonals of a parallelogram are perpendicular. 8. sometimes

For Questions 9 and 10, refer to the figure at the right.

120° 60° 120°

9. State the theorem that would prove the quadrilateral is a parallelogram. 9. If both pairs of opp. ∠s of a quadrilateral are ≅, then the quadrilateral is a parallelogram.
10. Write a subgoal you would use to prove the quadrilateral is a parallelogram. 10. Sample answer: Use corr. ∠s to show the lines are ∥.

For Questions 11–13, refer to parallelogram ABCD.

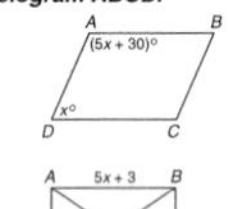

11. $\overline{AD} \parallel$ _?_ 11. $\overline{BC}$
12. $\overline{CD} \cong$ _?_ 12. $\overline{AB}$
13. Find $m\angle C$. 13. 155
14. Find the values of x and y in rectangle $ABCD$. 14. $x = 3, y = 2$

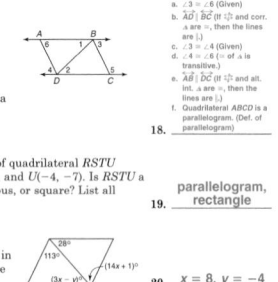

NAME________________ DATE ________

Chapter 6 Test, Form 2B (continued)

For Questions 15–17, refer to trapezoid CAMR and its median $\overline{OL}$.

15. If $RM = 4$ and $CA = 8$, find OL. 15. 6
16. If $CA = 12$ and $OL = 11$, find RM. 16. 10
17. If $OR = 5g$ and $CO = g + 12$, find CR. 17. 30
18. Write a two-column proof on a separate sheet of paper.

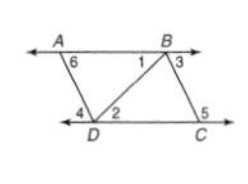

Given: $\angle 3 \cong \angle 6$
$\angle 3 \cong \angle 4$
Prove: Quadrilateral $ABCD$ is a parallelogram.

18. Sample answer:
a. $\angle 3 \cong \angle 6$ (Given)
b. $\overleftrightarrow{AD} \parallel \overleftrightarrow{BC}$ (If ∥s and corr. ∠s are ≅, then the lines are ∥.)
c. $\angle 3 \cong \angle 4$ (Given)
d. $\angle 4 \cong \angle 6$ (≅ of ∠s is transitive.)
e. $\overleftrightarrow{AB} \parallel \overleftrightarrow{DC}$ (If ∥s and alt. int. ∠s are ≅, then the lines are ∥.)
f. Quadrilateral $ABCD$ is a parallelogram. (Def. of parallelogram)

19. The coordinates of the vertices of quadrilateral $RSTU$ are $R(-5, -2)$, $S(5, 0)$, $T(6, -5)$, and $U(-4, -7)$. Is $RSTU$ a parallelogram, rectangle, rhombus, or square? List all that apply. 19. parallelogram, rectangle
20. What values must x and y have in order for the quadrilateral at the right to be a parallelogram? 20. $x = 8, y = -4$

28° 113° $(14x + 1)°$ $(3x - y)°$

Bonus

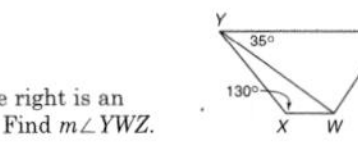

Figure $XYZW$ at the right is an isosceles trapezoid. Find $m\angle YWZ$. Bonus 95

CHAPTER 6 STUDY GUIDE AND ASSESSMENT

SKILLS AND CONCEPTS

OBJECTIVES AND EXAMPLES	REVIEW EXERCISES

Upon completing this chapter, you should be able to:

- recognize and apply the properties of a parallelogram (Lesson 6–1)

If $BCDE$ is a parallelogram, then you can make the following statements.

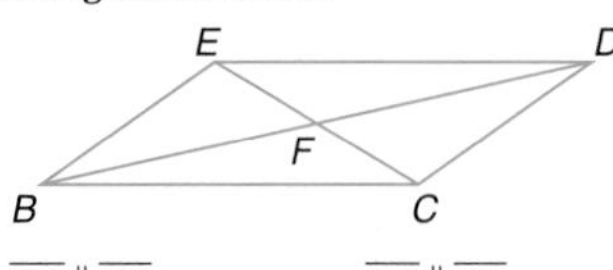

$\overline{BC} \parallel \overline{ED}$ $\quad$ $\overline{BE} \parallel \overline{CD}$

$\overline{BC} \cong \overline{ED}$ $\quad$ $\overline{BE} \cong \overline{CD}$

$\angle EBC \cong \angle CDE$ $\quad$ $\angle BCD \cong \angle DEB$

$\angle EBC$ is supplementary to $\angle BED$.

$\angle EDC$ is supplementary to $\angle DCB$.

$\angle BED$ is supplementary to $\angle EDC$.

$\angle EBC$ is supplementary to $\angle BCD$.

$\overline{BD}$ and $\overline{EC}$ bisect each other.

Use these exercises to review and prepare for the chapter test.

Complete each statement about ▱WXYZ. Then name the theorem or definition that justifies your answer. 13–20. See margin for theorem or definition.

13. $\angle WZY \cong$ _?_ $\angle WXY$
14. $\overline{WX} \cong$ _?_ $\overline{ZY}$
15. $\overline{XE} \cong$ _?_ $\overline{EZ}$
16. $\overline{XY} \cong$ _?_ $\overline{WZ}$
17. $\triangle XYZ \cong$ _?_ $\triangle ZWX$
18. $\angle 1 \cong$ _?_ $\angle 2$
19. $\overline{YE} \cong$ _?_ $\overline{EW}$
20. $\angle WZY$ and _?_ are supplementary. $\angle ZYX$ or $\angle ZWX$

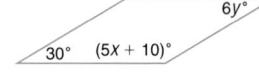

- recognize and apply the conditions that ensure a quadrilateral is a parallelogram (Lesson 6–2)

If $OQ = 20$, $MQ = 4y$, $PQ = x^2$, and $NQ = 36$, find the values of x and y in order for $MNOP$ to be a parallelogram.

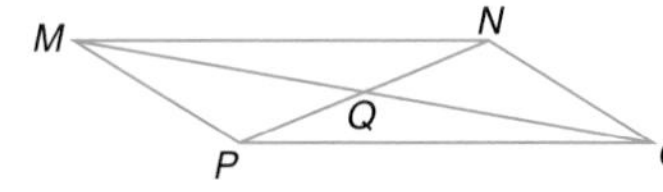

If the diagonals of a quadrilateral bisect each other, then it is a parallelogram. So, $OQ = MQ$ and $PQ = NQ$.

$OQ = MQ$ $\quad$ $PQ = NQ$

$20 = 4y$ $\quad$ $x^2 = 36$

$5 = y$ $\quad$ $x = 6$ or -6

To ensure that $MNOP$ is a parallelogram, $x = -6$ or 6 and $y = 5$.

Find the values of x and y that ensure each quadrilateral is a parallelogram.

21. $x = 28$, $y = 5$

22. $x = 8$ or -2, $y = 5$ or -5

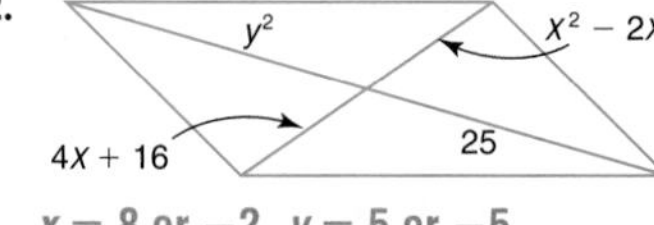

Write a two-column proof.

23. **Given:** ▱$ABCD$
$\overline{AE} \cong \overline{CF}$
Prove: Quadrilateral $EBFD$ is a parallelogram.
See margin.

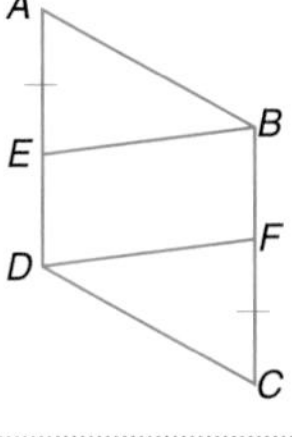

GLENCOE Technology

Test and Review Software

You may use this software, a combination of an item generator and item bank, to create your own tests or worksheets. Types of items include free response, multiple choice, short answer, and open ended.

For IBM & Macintosh

OBJECTIVES AND EXAMPLES

- recognize and apply the properties of rectangles (Lesson 6–3)

If a quadrilateral is a rectangle, then the following properties hold true.

1. Opposite sides are congruent and parallel.
2. Opposite angles are congruent.
3. Consecutive angles are supplementary.
4. Diagonals are congruent and bisect each other.
5. All four angles are right angles.

REVIEW EXERCISES

Quadrilateral *EFGH* is a rectangle. Find the value of *x*.

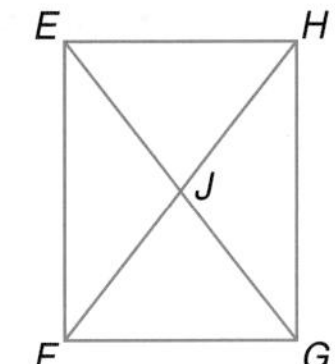

24. $m\angle HEG = 12x + 1$
$m\angle GEF = 6x - 1$ 5

25. $HF = 5x - 4$
$EG = 6x - 10$ 6

26. $JF = 8x + 4$
$EG = 24x - 8$ 2

27. $EF = x^2$
$HG = 3x - 2$ 1 or 2

28. $m\angle FGH = 10x^2$
$m\angle GHE = 8x^2 + 18$ 3 or −3

- recognize and apply the properties of squares and rhombi (Lesson 6–4)

Determine whether quadrilateral *FGHI* with $F(5, 0)$, $G(0, 0)$, $H(2, 3)$, and $I(7, 3)$ is a *parallelogram*, a *rectangle*, a *rhombus*, or a *square*. List all that apply.

$FG = \sqrt{(5-0)^2 + (0-0)^2}$ or 5

$GH = \sqrt{(0-2)^2 + (0-3)^2}$ or $\sqrt{13}$

$HI = \sqrt{(2-7)^2 + (3-3)^2}$ or 5

$IF = \sqrt{(7-5)^2 + (3-0)^2}$ or $\sqrt{13}$

The opposite sides are congruent, so *FGHI* is a parallelogram. All sides are not congruent, so it is not a rhombus or a square. Slopes can determine if *FGHI* is a rectangle.

slope of $\overline{FG} = \frac{0-0}{5-0}$ or 0

slope of $\overline{GH} = \frac{0-3}{0-2}$ or $\frac{3}{2}$

Consecutive sides are not perpendicular, so *FGHI* is a parallelogram, but not a rectangle.

38. $x = 4$, $y = 6$

Determine whether quadrilateral *KLMN* is a *parallelogram, a rectangle, a rhombus,* or a *square* for each set of vertices. List all that apply. 29–31. See margin.

29. $K(4, 8)$, $L(0, 9)$, $M(-2, 1)$, $N(2, 0)$

30. $K(12, 0)$, $L(6, -6)$, $M(0, 0)$, $N(6, 6)$

31. $K(5, 4)$, $L(3, -6)$, $M(0, -10)$, $N(2, 0)$

32. $K(1, 5)$, $L(8, 6)$, $M(15, 5)$, $N(8, 4)$
rhombus, parallelogram

Use rhombus *PQRS* and the given information to find each value.

33. If $PT = 14$, find PR. 28

34. If $m\angle PQT = 34$, find $m\angle PQR$. 68

35. What is $m\angle STP$? 90

36. If $RQ = 4x - 1$ and $PQ = 20 + x$, find the value of x. 7

37. What is $m\angle PQT$ if $m\angle QPT = 52$? 38

38. Find the values of x and y if $PT = 4x - 8$, $QT = 6y - 9$, $TR = 16 - 2x$, and $TS = 3y + 9$.

Additional Answers

13. Opposite ∠s in a ▱ are ≅.
14. Opp. sides of a ▱ are ≅.
15. Diagonals of a ▱ bisect each other.
16. Opposite sides of a ▱ are ≅.
17. SAS or SSS
18. Alt. Int. ∠ Th.
19. Diagonals of a ▱ bisect each other.
20. Cons. ∠s in a ▱ are supp.
23. Given: ▱*ABCD*
$\overline{AE} \cong \overline{CF}$
Prove: Quadrilateral *EBFD* is a parallelogram.
Proof:
Statements (Reasons)
    1. ▱*ABCD*, $\overline{AE} \cong \overline{CF}$ (Given)
    2. $\overline{BA} \cong \overline{DC}$ (Opp. sides of a ▱ are ≅.)
    3. $\angle A \cong \angle C$ (Opp. ∠s of a ▱ are ≅.)
    4. $\triangle BAE \cong \triangle DCF$ (SAS)
    5. $\overline{BE} \cong \overline{DF}$, $\angle BEA \cong \angle DFC$ (CPCTC)
    6. $\overline{BC} \parallel \overline{AD}$ (Def. ▱)
    7. $\angle DFC \cong \angle FDE$ (Alt. Int. ∠s Th.)
    8. $\angle BEA \cong \angle FDE$ (Cong. of ∠s is trans.)
    9. $\overline{BE} \parallel \overline{DF}$ (If ∦ and corr. ∠s are ≅, the lines are ∥.)
    10. Quadrilateral *EBFD* is a parallelogram. (If a pair of opp. sides of a quad are ≅ and ∥, it is a ▱.)
29. rectangle, parallelogram
30. square, rhombus, rectangle, parallelogram
31. parallelogram

Applications and Problem Solving Encourage students to work through the exercises in the Applications and Problem Solving section to strengthen their problem-solving skills.

Additional Answer

45. the kite flaps and the back are all trapezoids

OBJECTIVES AND EXAMPLES

- recognize and apply the properties of trapezoids (Lesson 6–5)

If a quadrilateral is a trapezoid, then the following properties hold true.
1. The bases are parallel. **2.** The median is parallel to the bases and its measure is half of the sum of the measures of the bases.

If a quadrilateral is an isosceles trapezoid, then the following additional properties hold true.
1. Both pairs of base angles are congruent. **2.** The diagonals are congruent.

REVIEW EXERCISES

STUVis a trapezoid with bases $\overline{ST}$ and $\overline{UV}$. Use the figure and the given information to solve each problem.

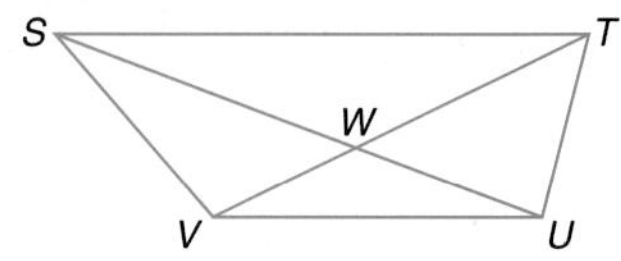

39. If $m\angle STV = 45$, find $m\angle UVT$. 45

40. If $m\angle TSW = 35$, find $m\angle SUV$. 35

41. If $m\angle SUV = 47$, find $m\angle TSU$. 47

42. What is the measure of the median of *STUV* if $ST = 23$ and $UV = 19$? 21

43. If $m\angle WUV = 23$ and $m\angle TWS = 127$, find $m\angle WVU$. 30

APPLICATIONS AND PROBLEM SOLVING

44. Architecture Oakland-Alameda County Coliseum uses rings of parallel and intersecting concrete columns around the outside of the coliseum to support the roof so that no supports are needed inside. What shape are the quadrilaterals formed by the support columns? (Lesson 6–4) rhombi

45. Kites Make a kite like those made by children on the island of Molokái, Hawaii. See margin.

a. Fold a sheet of notebook paper in half lengthwise.
b. Fold again along a diagonal such as $\overline{AB}$.
c. Tape a stick along $\overline{CD}$.
d. Turn the kite over and fold the flap back and forth until it stands up straight.
e. Complete the kite with a tail at *B* and a string to hold about 2 inches from *A* in the flap.

A D C B

Identify each quadrilateral formed in construction of the kite. (Lessons 6–1 to 6–5)

46. Art The painting *Either...Or*, by Josef Albers is shown at the right. Trace the shapes in the painting and label each vertex. Identify each quadrilateral in the painting. (Lessons 6–1 to 6–5) See margin.

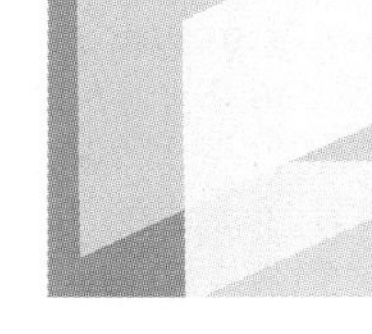

47. Civil Engineering The world's longest single-span suspension bridge is the Humber Bridge in Humberside County, England. The hollow boxes that form the bridge's deck are shaped like trapezoids. Which surfaces of the bridge are parallel? (Lesson 6–5) the deck top and the road surface

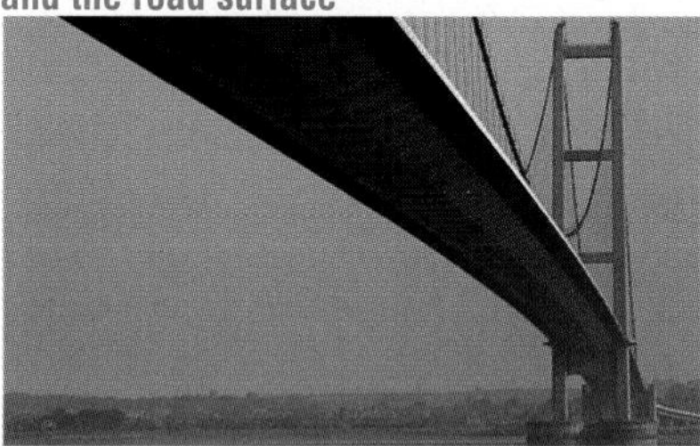

A practice test for Chapter 6 is provided on page 798.

Additional Answer

46.

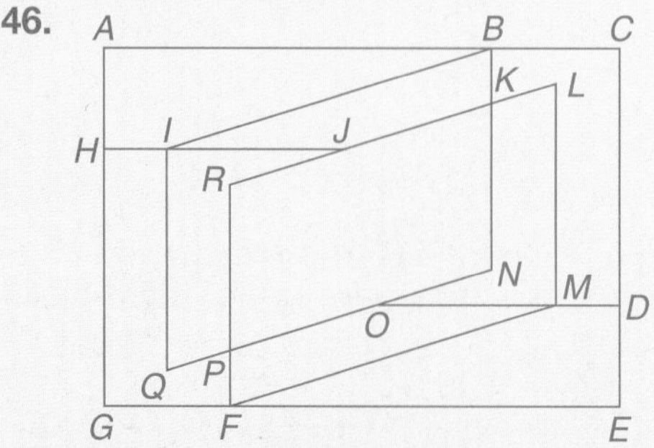

ACEG–rectangle;
IBNQ–parallelogram;
RLMF–parallelogram;
RKNP–parallelogram;
IBKJ–trapezoid;
OPFM–trapezoid;
ABIH–trapezoid;
MDEF–trapezoid

ALTERNATIVE ASSESSMENT

COOPERATIVE LEARNING PROJECT

Architecture In this chapter, you learned about a variety of quadrilaterals. Quadrilaterals can be classified into those with right angles and those without right angles. Most buildings have floors, walls, and roofs that are quadrilaterals having four right angles. Can you imagine a building made entirely of quadrilaterals with no right angles?

In this project, you will build a scale model, or make a detailed drawing, of a building constructed almost entirely out of quadrilaterals without right angles.

Follow these steps to complete your building design.

- In the first design, floors and ceilings may be rectangular.
- Design a living room with one wall that is all glass, oriented so it can take in the noonday sun.
- Design a family room with one wall that inclines out over a patio or deck. This provides shelter from the rain for your outdoor enjoyment.
- Design three second floor bedrooms so that each one has two all glass walls, some looking up toward the sky and some looking down for a scenic view of the yard.
- Remember, none of the walls in your house are rectangular.
- As a group, discuss the advantages of your design.

Follow these steps to complete your next building design.

- In this design, there are no rectangles—no rectangular floors, no rectangular ceilings, or roofs. Even more, not even the steps may use any rectangles.
- As a group, discuss what you will design to walk from one floor to the next.
- Be creative, but make sure everything functions correctly.
- As a group, discuss the advantages and disadvantages of your design.

THINKING CRITICALLY

Find the measure of $\overline{MD}$ in rectangle *ABCD*. (*Hint:* Draw an auxiliary segment.) **about 9.2 units**

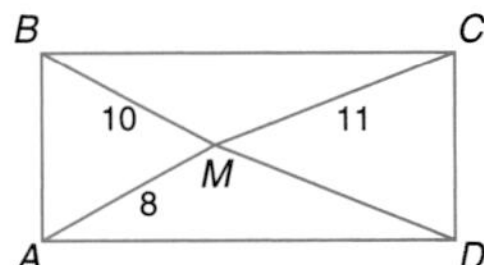

PORTFOLIO

Look through magazines that contain photographs of buildings. Find examples of quadrilaterals in the architecture. Collect these images and organize them by type of quadrilateral. Keep this collection in your portfolio.

SELF EVALUATION

In this chapter, you learned the strategy, identify subgoals.

Assess Yourself What would you like to accomplish in the next twenty years? Write these down as your long-term goals. Write down what you need to accomplish to reach these long-term goals. These are your subgoals. Identify the value of each of the courses you are taking to reaching your subgoals.

Alternative Assessment

The Alternative Assessment section provides students with the opportunity to assess their own work by thinking critically, working with others, keeping a portfolio, and honestly evaluating their own progress. For more information on alternative forms of assessment, see *Alternative Assessment in the Mathematics Classroom,* one of the titles in the Glencoe Mathematics Professional Series.

Performance Assessment

Performance Assessment tasks for this chapter are included in the *Assessment and Evaluation Masters.* A scoring guide is also provided.

Assessment and Evaluation Masters, pp. 154, 165

6 NAME__________ DATE ________

Chapter 6 Performance Assessment

Instructions: ***Demonstrate your knowledge by giving a clear, concise solution to each problem. Be sure to include all relevant drawings and to justify your answers. You may show your solution in more than one way or investigate beyond the requirements of the problem.***

1. Two quadrilaterals are congruent when their corresponding sides and angles are congruent. Tell whether each of the following are sufficient conditions for two quadrilaterals to be congruent. If they are, construct a proof. If not, give a counterexample.
   a. Parallelograms *ABCD* and *EFGH* have all corresponding sides congruent.
   b. Parallelograms *ABCD* and *EFGH* have all corresponding angles congruent.
   c. In parallelograms *ABCD* and *EFGH*, $\overline{AB} \cong \overline{EF}$, $\overline{BC} \cong \overline{FG}$, and $\angle B \cong \angle F$.
   d. In parallelograms *ABCD* and *EFGH*, $\overline{AB} \cong \overline{EF}$, $\overline{BC} \cong \overline{FG}$, and $\overline{BD} \cong \overline{FH}$.
   e. In rectangles *ABCD* and *EFGH*, $\overline{AB} \cong \overline{EF}$.
   f. In trapezoids *ABCD* and *EFGH*, corresponding bases $\overline{AD}$ and $\overline{EH}$, and $\overline{BC}$ and $\overline{FG}$ are congruent.
2. Miguel is making a pen and ink drawing for art class. He decided to start his drawing with a square, then connect the midpoints of the adjacent sides of the square to form a smaller figure inside. He found the midpoints of the sides of the smaller figure he constructed and repeated this procedure several times to make a drawing like the one shown at the right.

   a. Miguel thinks that all of the inside figures are squares. Prove or disprove this conjecture.
   b. In a second drawing, Miguel used a compass to mark a point a given distance from each vertex of a square and then connected the points on adjacent sides. What type of figure was formed inside the square in this drawing? Justify your answer.

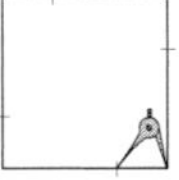

Scoring Guide
Chapter 6
Performance Assessment

Level	Specific Criteria
3 Superior	• Shows thorough understanding of the concepts of *parallelogram, rhombus, rectangle, square,* and *trapezoid.* • Uses appropriate strategies to construct proofs or find counterexamples. • Written explanations are exemplary. • Diagrams are accurate and appropriate. • Goes beyond requirements of some or all problems.
2 Satisfactory, with Minor Flaws	• Shows understanding of the concepts of *parallelogram, rhombus, rectangle, square,* and *trapezoid.* • Uses appropriate strategies to construct proofs or find counterexamples. • Written explanations are effective. • Diagrams are mostly accurate and appropriate. • Satisfies all requirements of some or all problems.
1 Nearly Satisfactory, with Serious Flaws	• Shows understanding of most of the concepts of *parallelogram, rhombus, rectangle, square,* and *trapezoid.* • May not use appropriate strategies to construct proofs or find counterexamples. • Written explanations are satisfactory. • Diagrams are mostly accurate and appropriate. • Satisfies most requirements of some or all problems.
0 Unsatisfactory	• Shows little or no understanding of the concepts of *parallelogram, rhombus, rectangle, square,* and *trapezoid.* • May not use appropriate strategies to construct proofs or find counterexamples. • Written explanations are not satisfactory. • Diagrams are not accurate or appropriate. • Does not satisfy requirements of some or all problems.

Using the COLLEGE ENTRANCE EXAM PRACTICE

These two pages review the skills and concepts presented in Chapters 1–6. This review is formatted to reflect new trends in college entrance testing.

Assessment and Evaluation Masters, pp. 159–160

6 NAME__________ DATE ______

Chapter 6 Cumulative Review

1. If N is the midpoint of $\overline{PQ}$, $NQ = 2x + 2$, and $PQ = 3x + 9$, find the value of x. 1. 5

For Questions 2–5, refer to the figure at the right.

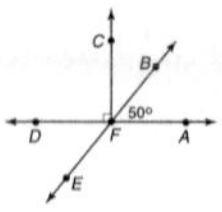

2. Name a pair of adjacent acute angles. 2. ∠CFB and ∠BFA
3. Name a pair of congruent supplementary angles. 3. ∠DFC and ∠CFA
4. Complete the statement: $\angle BFA$ ____ $\angle DFE$ 4. ≅
5. Complete the statement: $m\angle EFA =$ ____ 5. 130
6. How many lines can be drawn containing three collinear points? 6. one

For Questions 7 and 8, name the property of equality that justifies the statement.

7. If $y = 4x + 1$ and $x = 4$, then $y = 4(4) + 1$. 7. substitution
8. If $AC = BD$, then $BD = AC$. 8. symmetric
9. Are a table top and the floor a model of parallel planes? 9. yes

For Questions 10–12, refer to the figure at the right. If $\overleftrightarrow{AB} \parallel \overleftrightarrow{CD}$, find the measure of each angle.

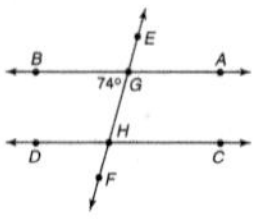

10. $\angle DHF$ 10. 74
11. $\angle CHG$ 11. 74
12. $\angle AGH$ 12. 106
13. Determine the value of r so that a line through the points with the coordinates $(r, 4)$ and $(-2, 3)$ has slope $\frac{1}{4}$. 13. 2

6 NAME__________ DATE ______

Chapter 6 Cumulative Review (continued)

For Questions 14–17, determine whether the statement is true or false.

14. A right triangle can be isosceles. 14. true
15. All scalene triangles are acute. 15. false
16. Every triangle has three medians. 16. true
17. It is possible to draw a triangle with sides of measures 4.1, 3.5, and 2.6. 17. true

For Questions 18–20, assume that $\triangle ABC \cong \triangle DEF$.

18. $\angle A \cong$ ____ 18. $\angle D$
19. $\angle BCA \cong$ ____ 19. $\angle EFD$
20. $\overline{CA} \cong$ ____ 20. $\overline{FD}$
21. Which postulate can be used to prove that the triangles at the right are congruent? 21. SSS

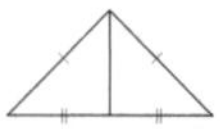

For Questions 22–23, refer to isosceles trapezoid WXYZ.

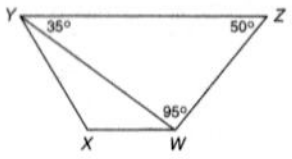

22. Name the least and greatest angles in $\triangle YXW$. 22. $\angle XYW$, $\angle YXW$
23. Name the shortest and longest sides in $\triangle YZW$. 23. $\overline{WZ}$, $\overline{YZ}$

For Questions 24–25, refer to parallelogram MNOP.

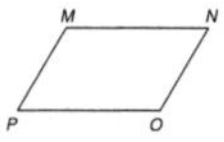

24. $\angle PMN \cong$ ____ 24. $\angle NOP$
25. $\overline{PM} \cong$ ____ 25. $\overline{ON}$

COLLEGE ENTRANCE EXAM PRACTICE

CHAPTERS 1–6

There are eight multiple-choice questions in this section. After working each problem, write the letter of the correct answer on your paper.

1. If a right triangle has two sides of measures 1 and $\sqrt{2}$, which of the following could be the measure of the third side? D

I. 1

II. $\sqrt{2}$

III. $\sqrt{3}$

A. I only
B. II only
C. III only
D. I and III only

2. Find 5% of 6%. B

A. 0.11%
B. 0.3%
C. 3%
D. 11%

3. Which of the following is the graph of the solution set of $|x - 3| < 2$? A

A. 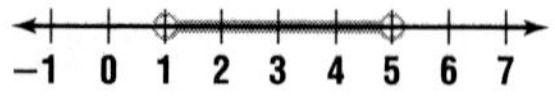

B. 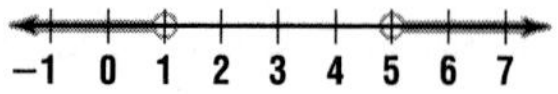

C. 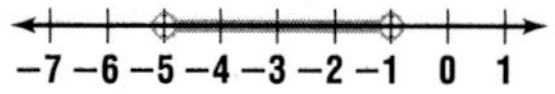

D. 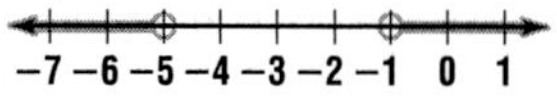

4. What is the value of 3^{-2}? C

A. 9
B. -6
C. $\frac{1}{9}$
D. $-\frac{1}{9}$

5. In the diagram below, which type of segment is $\overline{AD}$? C

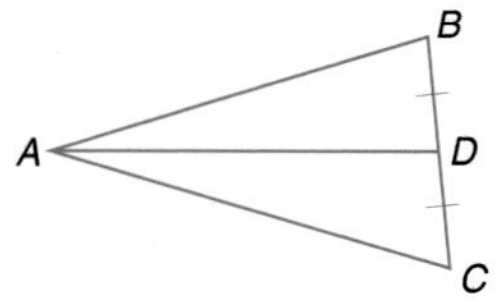

A. angle bisector
B. perpendicular bisector
C. median
D. altitude

6. Which of the following is a false statement? B

A. All squares are rectangles.
B. All rhombi are squares.
C. All rectangles are quadrilaterals.
D. All trapezoids are polygons.

7. In right triangle QRS, $m\angle Q = 30$, $m\angle R = 60$, and $QS = 6$. What is the measure of $\overline{QR}$? A

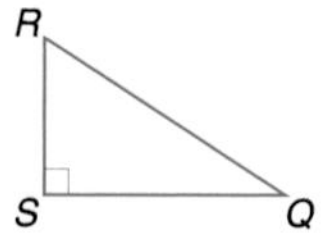

A. $4\sqrt{3}$
B. $6\sqrt{3}$
C. $3\sqrt{3}$
D. $3\sqrt{2}$

8. What is the simplified form of the expression $\frac{a^{-3}bc^2}{a^{-4}b^2c^{-3}}$? B

A. $\frac{c^5}{a^7b}$
B. $\frac{ac^5}{b}$
C. $\frac{a^7c^5}{b}$
D. $\frac{a^7}{bc^5}$

Standardized Test Practice Questions are also provided in the *Assessment and Evaluation Masters*, p. 158.

A more traditional cumulative review is provided in the *Assessment and Evaluation Masters*, pp. 159–160.

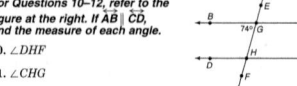

SECTION TWO: SHORT ANSWER

This section contains seven questions for which you will provide short answers. Write your answer on your paper.

9. What is the lowest common denominator for the fractions $\frac{5}{4x^2y}$, $\frac{7}{6x^2y}$, and $\frac{-4}{15xy}$? $60x^2y$

10. Find the sum of the perimeters of a square with sides of length x and an equilateral triangle with sides of length y. $4x + 3y$

11. If y varies inversely as x and y is 10 when x is $\frac{1}{2}$, find y when x is $\frac{2}{3}$. $\frac{15}{2}$

12. In the triangle below, what is the measure of $\angle BAC$? 60

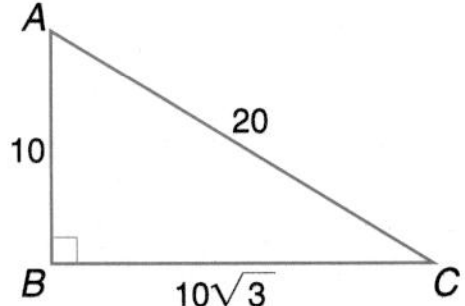

13. Find the solution set for $3 - (x - 5) = 2x - 3(4 - x)$. $\left\{\frac{10}{3}\right\}$

14. For all $x \neq 0$, find $\frac{x^6 + x^6 + x^6}{x^2}$. $3x^4$

15. In the figure below, $\angle ABC$ is a straight angle, and $\overline{DB}$ is perpendicular to $\overline{BE}$. If $\angle ABD$ measures x degrees, write an expression to represent the measure of $\angle CBE$. $90 - x$

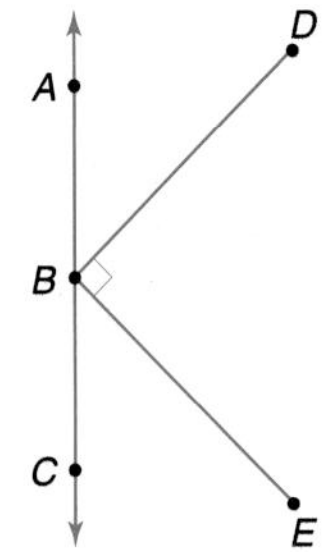

SECTION THREE: COMPARISON

This section contains five comparison problems that involve comparing two quantities, one in column A and one in column B. In certain questions, information related to one or both quantities is centered above them. All variables used represent real numbers.

Compare quantities A and B below.

- Write A if quantity A is greater.
- Write B if quantity B is greater.
- Write C if the two quantities are equal.
- Write D if there is not enough information to determine the relationship.

16. B 17. B 18. A 19. D 20. C

	Column A	Column B
	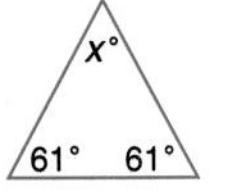	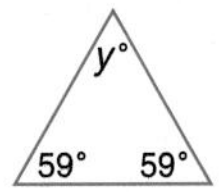
16.	x	y
17.	$\frac{1}{\frac{2}{3}}$	$\frac{2}{\frac{1}{3}}$
	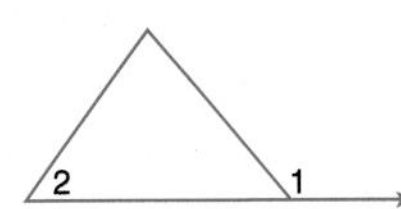	
18.	$m\angle 1$	$m\angle 2$
	In $\triangle ABC$, $m\angle A > m\angle B$ and $m\angle A > m\angle C$.	
19.	AC	AB
	$3x - 4y = -2$ $4x + 2y = 12$	
20.	x	y

TEST-TAKING TIP

Quantitative Comparisons

Even though students don't know if they are dealing with an equality or an inequality in a quantitative comparison, they can use the rules for inequalities to simplify the comparison. In algebra, students learned that they can add or subtract a quantity from both sides of an inequality without changing the direction of the inequality sign. They can also multiply or divide by positive values and maintain the direction of the inequality. Study the example below.

Column A	Column B
x is a positive integer, and $0 < k < 1$	
x	$\frac{x}{k}$

Since k is a positive number, you can multiply the value in each column by k.

kx	x

Then you can divide each value by x.

k	1

Since the instructions say that $0 < k < 1$, the quantity in Column B must be greater.

Remember that multiplying or dividing by a negative number will reverse an inequality sign. *Never assume* that variables represent positive numbers if that is not stated in the instructions.

CHAPTER

7 Connecting Proportion and Similarity

PREVIEWING THE CHAPTER

In this chapter, students use algebra skills to solve proportions. Then they identify similar figures and solve problems using proportions. They apply this practice to similar triangles. The proportional parts of similar triangles are used to solve problems and to divide segments into congruent parts. The proportional relationships between perimeters, altitudes, angle bisectors, and medians of similar triangles are investigated. Similar figure relationships are extended to fractals, and the chapter wraps up with a game to explore patterns that are randomly generated.

Lesson (Pages)	Lesson Objectives	NCTM Standards	State/Local Objectives
7-1 (338–345)	Recognize and use ratios and proportions. Apply the properties of proportions.	1–5	
7-2 (346–353)	Identify similar figures. Solve problems involving similar figures.	1–5, 7	
7-3 (354–361)	Identify similar triangles. Use similar triangles to solve problems.	1–5, 7	
7-4 (362–369)	Use proportional parts of triangles to solve problems. Divide a segment into congruent parts.	1–5, 7	
7-5 (370–377)	Recognize and use the proportional relationships of corresponding perimeters, altitudes, angle bisectors, and medians of similar triangles.	1–5, 7	
7-6 (378–383)	Recognize and describe characteristics of fractals. Solve problems by solving a simpler problem.	1–5, 7	
7-6B (384–385)	Create fractal designs by using random numbers.	1–5, 7	

ORGANIZING THE CHAPTER

A complete, 1-page lesson plan is provided for each lesson in the ***Lesson Planning Guide***. Answer keys for each lesson are available in the ***Answer Key Masters***.

You may want to refer to the **Course Planning Calendar** on page T12 for detailed information on pacing.
PACING: Standard—13 days; **Honors**—12 days; **Block**—7 days

LESSON PLANNING CHART

Lesson (Pages)	Materials/ Manipulatives	Extra Practice (Student Edition)	Blackline Masters								Real-World Applications	Teaching Transparencies
			Study Guide	Practice	Enrichment	Assessment & Evaluation	Modeling Mathematics	Multicultural Activity	Tech Prep Applications	Graphing Calc. & Computer		
7-1 (338–345)	spreadsheet software	p. 776	p. 37	p. 37	p. 37			p. 13	p. 13	p. 7	13	7-1A 7-1B
7-2 (346–353)	ruler* blank transparency overhead projector grid paper compass*	p. 776	p. 38	p. 38	p. 38	p. 184						7-2A 7-2B
7-3 (354–361)	ruler* protractor*	p. 776	p. 39	p. 39	p. 39	pp. 183, 184	pp. 39–42, 85					7-3A 7-3B
7-4 (362–369)	compass* ruler*	p. 777	p. 40	p. 40	p. 40							7-4A 7-4B
7-5 (370–377)		p. 777	p. 41	p. 41	p. 41	p. 185					14	7-5A 7-5B
7-6 (378–383)	isometric dot paper ruler* grid paper	p. 777	p. 42	p. 42	p. 42	p. 185		p. 14	p. 14			7-6A 7-6B
7-6B (384–385)	grid paper calculator die or spinner overhead transparency						p. 95					
Study Guide/ Assessment (387–391)						pp. 169–182, 186–188						

*Included in Glencoe's High School Manipulative Kit and Overhead Manipulative Resources.

ORGANIZING THE CHAPTER

Other Chapter Resources

Student Edition
Chapter Opener, pp. 336–337
Mathematics and Society, p. 369
Working on the Investigation, pp. 345, 377
Closing the Investigation, p. 386

Teacher's Classroom Resources
Investigations and Projects Masters, pp. 49–52
Block Scheduling Booklet

Technology
Test and Review Software (IBM and Macintosh)
CD-ROM Multimedia Applications (Windows and Macintosh)
Mindjogger Videoquizzes (VHS)

Professional Publications
Glencoe Mathematics Professional Series

Outside Resources

Books/Periodicals
Fractals in the Classroom, NCTM
Geometry and Fractals with Tangrams, ETA

Software
Koyn Fractal Studio, William K. Bradford Publishing Company

Videos/CD-ROMs
Similarity, NCTM, 1906 Association Drive, Reston, VA 20191
Chaos, Fractals, and Dynamics: Computer Experiences in Mathematics, Dale Seymour Publications, P.O. Box 10888, Palo Alto, CA 94303
What Is Mathematical Modeling?, 2809 Ross Avenue, Dallas, TX 75201

Assessment Resources

Student Edition
Math Journal, pp. 341, 366
Mixed Review, pp. 345, 353, 360, 369, 377, 383
Self Test, p. 361
Chapter Highlights, p. 387
Chapter Study Guide and Assessment, pp. 388–390
Alternative Assessment, p. 391
Portfolio, p. 391
Chapter Test, p. 799

Teacher's Wraparound Edition
5-Minute Check, pp. 338, 346, 354, 362, 370, 378
Check for Understanding, pp. 341, 349, 357, 366, 372, 381
Closing Activity, pp. 345, 353, 361, 369, 377, 383
Cooperative Learning, pp. 339, 356

Assessment and Evaluation Masters
Multiple-Choice Tests, Forms 1A (Honors), 1B (Average), 1C (Basic), pp. 169–174
Free-Response Tests, Forms 2A (Honors), 2B (Average), 2C (Basic), pp. 175–180
Calculator-Based Test, p. 181
Performance Assessment, p. 182
Mid-Chapter Test, p. 183
Quizzes A–D, pp. 184–185
Standardized Test Practice, p. 186
Cumulative Review, pp. 187–188

ENHANCING THE CHAPTER

Examples of some of the materials for enhancing Chapter 7 are shown below.

Multicultural Activity Masters, pp. 13, 14

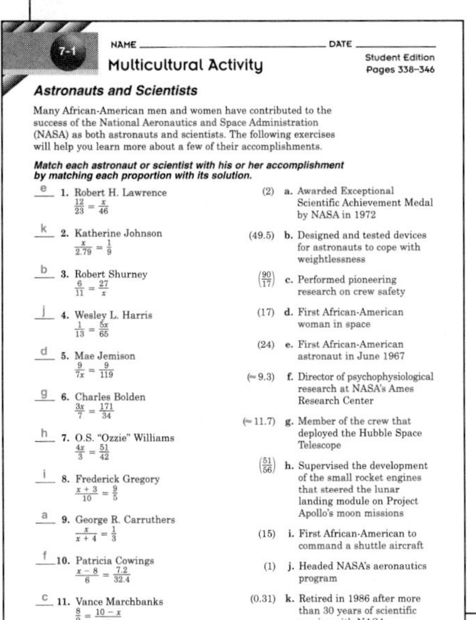

NAME ______ DATE ______

7-1 Multicultural Activity — Student Edition Pages 338–346

Astronauts and Scientists

Many African-American men and women have contributed to the success of the National Aeronautics and Space Administration (NASA) as both astronauts and scientists. The following exercises will help you learn more about a few of their accomplishments.

Match each astronaut or scientist with his or her accomplishment by matching each proportion with its solution.

e 1. Robert H. Lawrence $\frac{12}{23} = \frac{x}{46}$

k 2. Katherine Johnson $\frac{x}{2.79} = \frac{1}{9}$

b 3. Robert Shurney $\frac{6}{11} = \frac{27}{x}$

j 4. Wesley L. Harris $\frac{1}{13} = \frac{5x}{65}$

d 5. Mae Jemison $\frac{9}{7x} = \frac{9}{119}$

g 6. Charles Bolden $\frac{3x}{7} = \frac{171}{34}$

h 7. O.S. "Ozzie" Williams $\frac{4x}{3} = \frac{51}{42}$

i 8. Frederick Gregory $\frac{x+3}{10} = \frac{9}{5}$

a 9. George R. Carruthers $\frac{x}{x+4} = \frac{1}{3}$

f 10. Patricia Cowings $\frac{x-8}{6} = \frac{7.2}{32.4}$

c 11. Vance Marchbanks $\frac{8}{9} = \frac{10-x}{x}$

(2) a. Awarded Exceptional Scientific Achievement Medal by NASA in 1972

(49.5) b. Designed and tested devices for astronauts to cope with weightlessness

$\left(\frac{90}{17}\right)$ c. Performed pioneering research on crew safety

(17) d. First African-American woman in space

(24) e. First African-American astronaut in June 1967

($\approx$ 9.3) f. Director of psychophysiological research at NASA's Ames Research Center

($\approx$ 11.7) g. Member of the crew that deployed the Hubble Space Telescope

$\left(\frac{51}{56}\right)$ h. Supervised the development of the small rocket engines that steered the lunar landing module on Project Apollo's moon missions

(15) i. First African-American to command a shuttle aircraft

(1) j. Headed NASA's aeronautics program

(0.31) k. Retired in 1986 after more than 30 years of scientific service with NASA

Real-World Applications, 13, 14

TWO DAYS OF CURRENCY TRADING

EXCHANGE RATES
Wednesday, April 13, 1994

Country	U.S. $ equiv. Wed.	U.S. $ equiv. Tues.	Currency Per U.S. $ Wed.	Currency Per U.S. $ Tues.
Argentina (Peso)	1.01	1.01	0.99	0.99
Bahrain (Dinar)	2.6525	2.6525	0.3770	0.3770
Britain (Pound)	1.4775	1.4755	0.6768	0.6777
Ireland (Punt)	1.4334	1.4291	0.6976	0.6997
Jordan (Dinar)	1.4474	1.4474	0.6909	0.6909
Kuwait (Dinar)	3.3572	3.3572	0.2979	0.2979
Malta (Lira)	2.5674	2.5674	0.3895	0.3895
Peru (New Sol)	0.4737	0.4737	2.11	2.11
Poland (Zloty)	0.00004444	0.00004463	22504.00	22408.00

Source: *The New York Times*

© Glencoe/McGraw-Hill — Geometry

TECHNOLOGY

Graphing Calculator and Computer Masters, p. 7

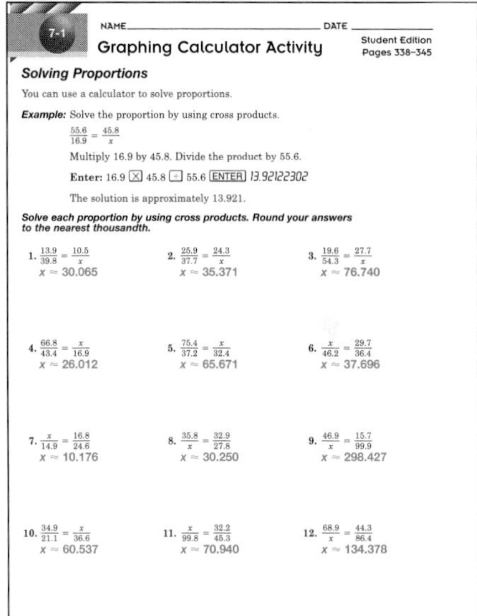

NAME ______ DATE ______

7-1 Graphing Calculator Activity — Student Edition Pages 338–345

Solving Proportions

You can use a calculator to solve proportions.

Example: Solve the proportion by using cross products.

$\frac{55.6}{16.9} = \frac{45.8}{x}$

Multiply 16.9 by 45.8. Divide the product by 55.6.

Enter: 16.9 [×] 45.8 [÷] 55.6 [ENTER] 13.92122302

The solution is approximately 13.921.

Solve each proportion by using cross products. Round your answers to the nearest thousandth.

1. $\frac{13.9}{39.8} = \frac{10.5}{x}$ $x \approx 30.065$
2. $\frac{25.9}{37.7} = \frac{24.3}{x}$ $x \approx 35.371$
3. $\frac{19.6}{54.3} = \frac{27.7}{x}$ $x \approx 76.740$
4. $\frac{66.8}{43.4} = \frac{x}{16.9}$ $x \approx 26.012$
5. $\frac{75.4}{37.2} = \frac{x}{32.4}$ $x \approx 65.671$
6. $\frac{x}{46.2} = \frac{29.7}{36.4}$ $x \approx 37.696$
7. $\frac{x}{14.9} = \frac{16.8}{24.6}$ $x \approx 10.176$
8. $\frac{35.8}{x} = \frac{32.9}{27.8}$ $x \approx 30.250$
9. $\frac{46.9}{x} = \frac{15.7}{99.9}$ $x \approx 298.427$
10. $\frac{34.9}{21.1} = \frac{x}{36.6}$ $x \approx 60.537$
11. $\frac{x}{99.8} = \frac{32.2}{45.3}$ $x \approx 70.940$
12. $\frac{68.9}{x} = \frac{44.3}{86.4}$ $x \approx 134.378$

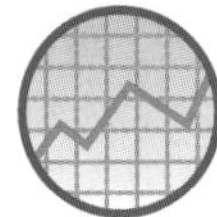

TECH PREP

Tech Prep Applications Masters, pp. 13, 14

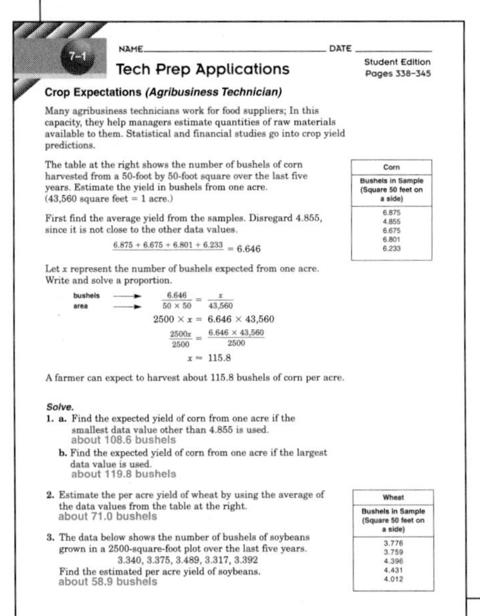

NAME ______ DATE ______

7-1 Tech Prep Applications — Student Edition Pages 338–345

Crop Expectations *(Agribusiness Technician)*

Many agribusiness technicians work for food suppliers; In this capacity, they help managers estimate quantities of raw materials available to them. Statistical and financial studies go into crop yield predictions.

The table at the right shows the number of bushels of corn harvested from a 50-foot by 50-foot square over the last five years. Estimate the yield in bushels from one acre. (43,560 square feet = 1 acre.)

Corn
Bushels in Sample (Square 50 feet on a side)
6.875
4.855
6.675
6.801
6.233

First find the average yield from the samples. Disregard 4.855, since it is not close to the other data values.

$$\frac{6.875 + 6.675 + 6.801 + 6.233}{} = 6.646$$

Let x represent the number of bushels expected from one acre. Write and solve a proportion.

bushels → / area →

$$\frac{6.646}{50 \times 50} = \frac{x}{43{,}560}$$
$$2500 \times x = 6.646 \times 43{,}560$$
$$\frac{2500x}{2500} = \frac{6.646 \times 43{,}560}{2500}$$
$$x \approx 115.8$$

A farmer can expect to harvest about 115.8 bushels of corn per acre.

Solve.

1. a. Find the expected yield of corn from one acre if the smallest data value other than 4.855 is used.
   about 108.6 bushels

   b. Find the expected yield of corn from one acre if the largest data value is used.
   about 119.8 bushels

2. Estimate the per acre yield of wheat by using the average of the data values from the table at the right.
   about 71.0 bushels

3. The data below shows the number of bushels of soybeans grown in a 2500-square-foot plot over the last five years.
   3.340, 3.375, 3.489, 3.317, 3.392
   Find the estimated per acre yield of soybeans.
   about 58.9 bushels

Wheat
Bushels in Sample (Square 50 feet on a side)
3.776
3.759
4.396
4.431
4.012

Problem-of-the-Week Cards, 18, 19, 20

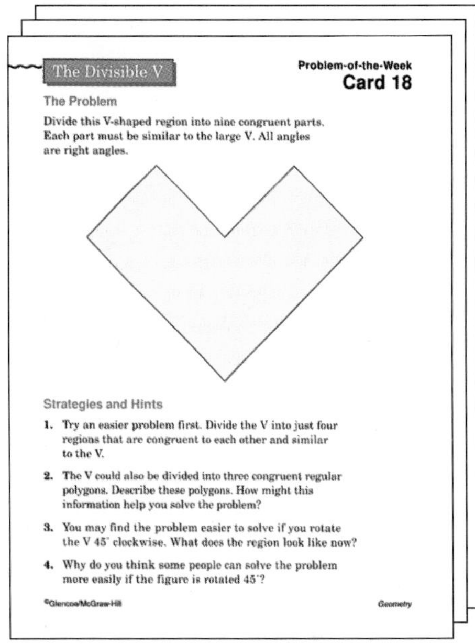

The Divisible V — Problem-of-the-Week Card 18

The Problem

Divide this V-shaped region into nine congruent parts. Each part must be similar to the large V. All angles are right angles.

Strategies and Hints

1. Try an easier problem first. Divide the V into just four regions that are congruent to each other and similar to the V.
2. The V could also be divided into three congruent regular polygons. Describe these polygons. How might this information help you solve the problem?
3. You may find the problem easier to solve if you rotate the V 45° clockwise. What does the region look like now?
4. Why do you think some people can solve the problem more easily if the figure is rotated 45°?

©Glencoe/McGraw-Hill — Geometry

MAKING MATHEMATICS RELEVANT

This two-page introduction to the chapter provides students with an opportunity to explore an application of geometry that has historic roots as well as modern uses.

Background Information

Identifying Trees Trees are popularly classified as either evergreen (leaves are kept year round) or deciduous (leaves are shed before winter). There are 60,000 to 70,000 different species of trees distributed throughout the world. Roughly 850 species are native to the United States, and another 60 species originally from Europe or Asia have been naturalized in the United States. In addition, over 200 foreign species are commonly grown in the United States. A tree identification specialist is called a dendrologist.

CHAPTER 7

Connecting Proportion and Similarity

Objectives

In this chapter, you will:

- recognize and use ratios and proportions,
- identify similar figures and use the properties of similar figures to solve problems,
- use proportional parts of triangles to solve problems,
- solve problems by first solving a simpler problem, and
- recognize and describe characteristics of fractals.

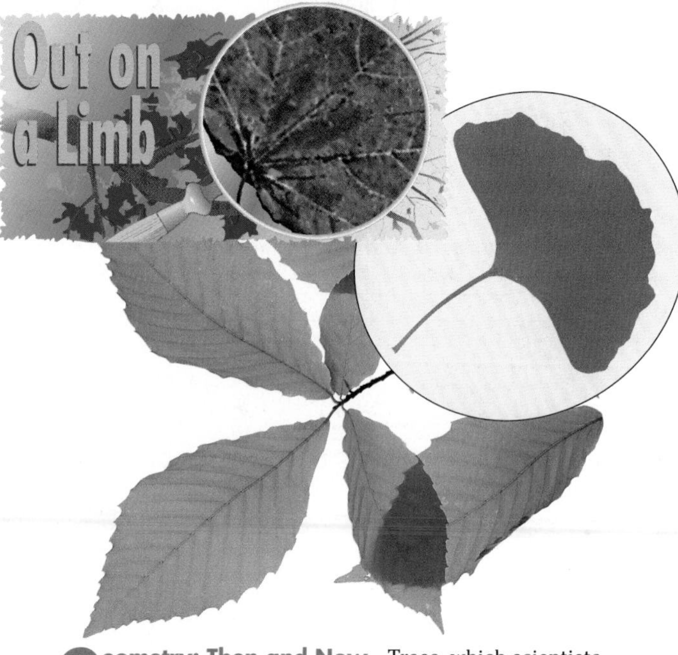

Geometry: Then and Now Trees, which scientists estimate developed about 200 million years ago, are much older than the grasses. Today, there are more than 500 species of trees. Have you ever noticed that leaves of the same tree often have the same shape (or nearly the same shape) no matter what the size of the leaf? This phenomenon is very useful in identifying the species of a tree. A tree can be classified based on the shape of any of its leaves because leaves are similar figures; that is, they have the same shape, but not necessarily the same size.

TIME*Line*

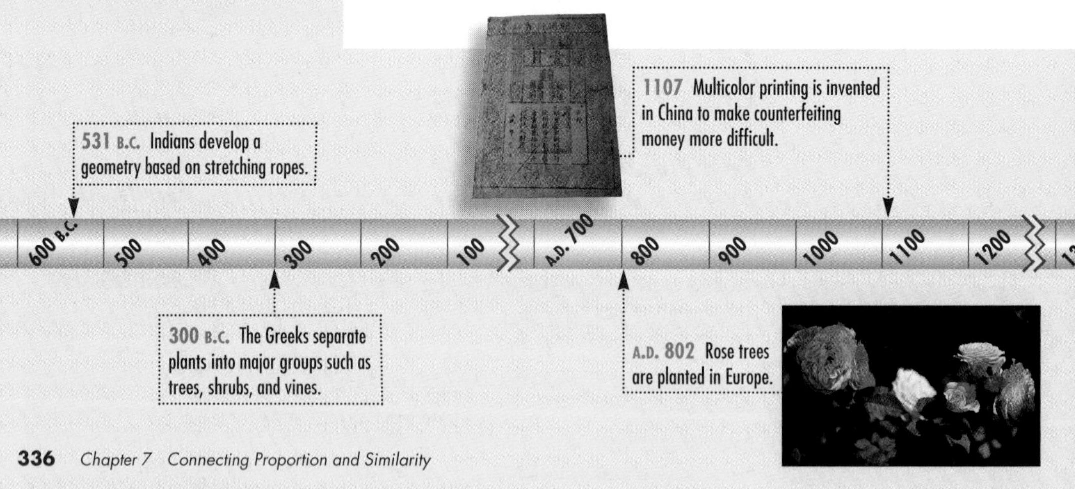

531 B.C. Indians develop a geometry based on stretching ropes.
1107 Multicolor printing is invented in China to make counterfeiting money more difficult.
600 B.C.
500
400
300
200
100
A.D. 700
800
900
1000
1100
1200
1350
300 B.C. The Greeks separate plants into major groups such as trees, shrubs, and vines.
A.D. 802 Rose trees are planted in Europe.

Students may find it interesting to research the development of descriptive geometry. While it began with Gaspard Monge in 1763, it has continued to develop to this day. It is now a part of CAD systems, Computer-Aided Design.

interNET CONNECTION

Browse the air, water, flora and fauna, fire, and earth areas at the Envirolink library for environmental information.

World Wide Web
http://www.envirolink.org/EnviroLink_Library/

PEOPLE IN THE NEWS

As 8-year-old members of a Brownie Troop, **Sabrina Alimahomed** and **Tara Church** of El Segundo, California, planted a sycamore tree to make up for using paper plates on a camping trip. Planting that tree was the birth of a youth-run, nonprofit organization called *Tree Musketeers*. Under their guidance, the *Tree Musketeers* aimed to plant a pollution barrier of trees around El Segundo to protect it from the effects of a nearby refinery, sewage treatment plant, and the Los Angeles International Airport. Since 1987, the *Tree Musketeers* have planted over 700 trees in El Segundo and have motivated others to plant thousands more all over Southern California.

Chapter Project

In this project, you will use materials in your library or on the Internet to help you identify trees in your community.

- Collect two leaves from each of five different trees on your school's campus or near your home.
- Use the materials that you found to identify each tree based on its leaves. Make a table summarizing the basic characteristics of the trees you investigated.
- For each pair of leaves from the same tree, find the ratios of lengths of corresponding parts of the leaves. What can you conclude?

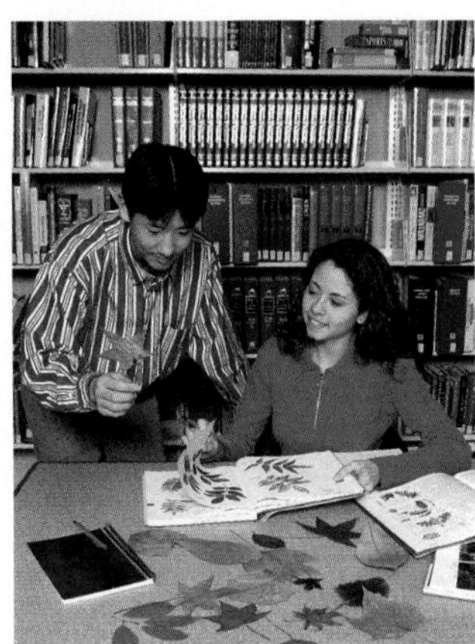

- Did you have any pairs of leaves for which the ratios of lengths of parts were not all the same? How different were these proportions? Explain why these differences may have occurred.

1565 Flemish painter Pieter Brueghel uses similar figures in his painting *The Harvesters.*

1994 Disney's *The Lion King* uses similar cartoon figures to create herds of jungle animals.

1400 1450 1500 1550 1600 1650 1700 1750 1800 1850 1900 1950 2000

1763 Gaspard Monge develops descriptive geometry, which remained a French military secret for 32 years.

1923 Garrett Morgan patents a three-way automatic traffic signal.

Alternative Chapter Projects

Two other chapter projects are included in the *Investigations and Projects Masters*. In Chapter 7 Project A, pp. 49–50, students extend the topic in the chapter opener. In Chapter 7 Project B, pp. 51–52, students learn how to make perspective drawings.

PEOPLE IN THE NEWS

Tara Church says, "Don't ever let anyone tell you that you can't change the world." For information about the *Tree Musketeers*, including how to start a local chapter, contact *Tree Musketeers* at 136 Main Street, Department P, El Segundo, CA 90245. The *Grassroots Youth Magazine* can be contacted on the Internet at http://www.envirolink.org/products/grassroots

Chapter Project

Cooperative Learning
Encourage students to mount or photocopy their pairs of leaves and label them generically ("Tree A," "Tree B," and so on) for reference before identifying them. Suggested characteristics to summarize for each tree: location, species, average species height, average species trunk diameter, and whether the tree generally bears flowers, fruit, or cones.

Investigations and Projects Masters, p. 49

NAME ______________________ DATE ____________

7 **Chapter 7 Project A** — Student Edition Pages 336–391

Similar Trees

While there are many ways to classify trees, one way is to classify them into two main groups: *conifers,* or cone-bearing trees and *deciduous,* or flowering trees. Within these two groups, there are classifications of different types. Below is a list of some of the different types of trees in the two groups.

Conifers	Deciduous	
Eucalyptus	Cherry	Oak
Pine	Magnolia	Beech
Fir	Elm	Poplar
Spruce	Maple	Birch
	Palm	

1. Select two of the classifications of trees, one from each list above, or select two other types of trees not listed but classified into the two different groups. For each type of tree, research why the trees are classified into that particular group. Include such characteristics as leaves, buds, fruit, and bark.
2. Discuss how size and shape are involved in the characteristics of each tree. How are the trees within each group similar? How are they different? Which trees keep similar shape as they grow older?
3. Find one of these trees in your neighborhood and ask its owner if you could have four leaves from the tree. Compare the leaves and determine if each leaf is similar to the others.
4. Make a drawing of each leaf and mark points on the tips of the leaves. For each leaf, connect the points to form a polygon. Determine if the polygons are similar or close to being similar, and what scale factor relates the leaves.
5. Record your findings in a list or report. Include your drawings and pictures in the report.

7-1 LESSON NOTES

NCTM Standards: 1–5

Instructional Resources

- Study Guide Master 7-1
- Practice Master 7-1
- Enrichment Master 7-1
- Graphing Calculator and Computer Masters, p. 7
- Multicultural Activity Masters, p. 13
- Real-World Applications, 13
- Tech Prep Applications Masters, p. 13

Transparency 7-1A contains the 5-Minute Check for this lesson; **Transparency 7-1B** contains a teaching aid for this lesson.

Recommended Pacing	
Standard Pacing	Day 1 of 13
Honors Pacing	Day 1 of 12
Block Scheduling*	Day 1 of 7

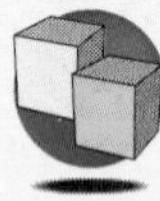

*For more information on pacing and possible lesson plans, refer to the *Block Scheduling Booklet.*

1 FOCUS

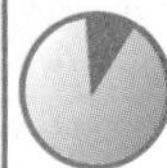

5-Minute Check
(over Chapter 6)

Complete.

1. In a parallelogram, opposite sides and angles are _____. congruent
2. If one pair of opposite sides of a quadrilateral are congruent and parallel, then the quadrilateral is a _____. parallelogram
3. In rectangle $ABCD$, $AC = 4x + 6$, and $BD = 9x - 14$. Find the value of x. 4
4. In rhombus $DEFG$, the diagonals intersect at X. If $DX = 7$, find DF. 14
5. In trapezoid $ABCD$, $\overline{AB} \parallel \overline{CD}$. X is the midpoint of $\overline{DA}$, and Y is the midpoint of $\overline{BC}$. If $AB = 6$ and $CD = 10$, find XY. 8

7-1 Integration: Algebra Using Proportions

What YOU'LL LEARN

- To recognize and use ratios and proportions, and
- to apply the properties of proportions.

Why IT'S IMPORTANT

You can use proportions to solve problems involving movie props, literature, and meteorology.

APPLICATION
Art

In 1871, American painter James Whistler painted a portrait of his mother called *Arrangement in Black and Gray*. Suppose Whistler drew a preliminary sketch in which his mother's height was 10 inches and the width of the widest part of her gown was $8\frac{1}{3}$ inches. In the actual painting, his mother's height is 50 inches. How wide is the widest part of her gown in the actual painting?

You will be asked to solve this problem in Exercise 16.

You can use **ratios** to solve problems like the one in the application above. A ratio is a comparison of two quantities. The ratio of a to b can be expressed as $\frac{a}{b}$, where b is not zero. The ratio can also be written as a:b.

Example 1

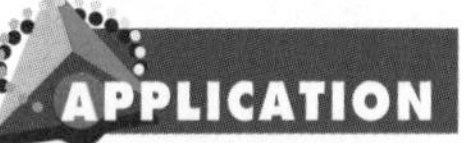

Entertainment

Special effects in movies are often created using miniature models. In the movie *Jurassic Park*, a model jeep 22 inches long was created to look like a real $14\frac{2}{3}$ foot jeep. What is the ratio of the length of the model compared to the length of the real jeep?

$$\frac{\text{length of model}}{\text{length of real jeep}} = \frac{22 \text{ inches}}{176 \text{ inches}} \qquad 14\tfrac{2}{3} \text{ feet} = 176 \text{ inches}$$

$$= \frac{1}{8}$$

FYI

Some authorities attribute the word *jeep* to the letters G.P., which represent general purpose vehicle. The jeep, introduced in 1940 by the U.S. Army, was 11 feet long and about 5 feet wide.

The ratio comparing the two lengths is $\frac{1}{8}$ or 1:8. The length of the model is $\frac{1}{8}$ the length of a real jeep.

A special ratio often found in nature and used by artists and architects is the *golden ratio*. It is approximately 1:1.618. The *Fibonacci sequence* is related to the golden ratio. You can use a spreadsheet to explore this relation.

In the military, the jeep was used as a command car; as a reconnaissance car; and for light weapons, ammunition, and personnel. It also was amphibious because it had a waterproof hull and propeller.

EXPLORATION SPREADSHEETS

Recall that in a spreadsheet each cell is named by the column and row in which it is located (A1 means column A, row 1). Suppose we set up a spreadsheet to evaluate the formula for the Fibonacci sequence and find the ratio between the pairs of consecutive terms of the sequence. Enter the labels in rows 1 and 2 as shown below. Enter 1 in each cell in row 3.

- In cell A4, enter A3 + 1. Copy this formula to cells A5 through A22.
- In cell B5, enter B3 + B4. Copy this formula to cells B6 through B22.
- In cell C3, enter B4/B3. Copy this formula to cells C4 through C22.

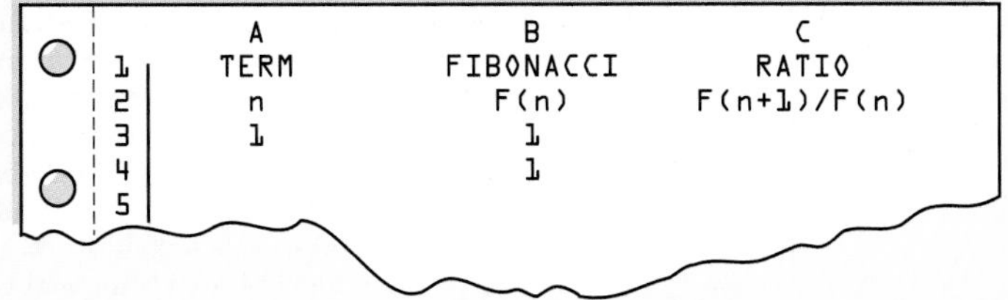

	A	B	C
1	TERM	FIBONACCI	RATIO
2	n	F(n)	F(n+1)/F(n)
3	1	1	
4		1	
5			

Your Turn

a. Write the first twenty terms of the Fibonacci sequence.

b. Describe the relationship between the terms of the Fibonacci sequence.

c. Find the ratio between each pair of consecutive terms of the sequence. (Round to seven decimal places.) What do you notice about these ratios?

a. 1, 1, 2, 3, 5, 8, 13, 21, 34, 55, 89, 144, 233, 377, 610, 987, 1597, 2584, 4181, 6765

b. A term is the sum of the two previous terms.

c. 1, 2, 1.5, 1.6666667, 1.6, 1.625, 1.6153846, 1.6190476, 1.6176471, 1.6181818, 1.6179775, 1.6180556, 1.6180258, 1.6180371, 1.6180328, 1.6180344, 1.6180338, 1.6180341, 1.6180340; they are increasingly closer approximations of the golden ratio.

Ratios, like fractions can be simplified. According to Nielsen Media Research, in 1995 there were about 224 televisions for every 100 homes. One radio station reported this as 336 televisions for every 150 homes. How do the two ratios compare? When they are simplified, both ratios are equivalent to $\frac{2.24}{1}$.

$$\frac{224}{100} = \frac{2.24}{1} \quad (\div 100) \qquad \frac{336}{150} = \frac{2.24}{1} \quad (\div 150)$$

An equation stating that two ratios are equal is a **proportion**. So, $\frac{224}{100} = \frac{336}{150}$ is a proportion. Every proportion has two **cross products**. In the proportion $\frac{224}{100} = \frac{336}{150}$, the cross products are 224 times 150 and 100 times 336. The **extremes** of the cross product are 224 and 150. The **means** are 100 and 336. The cross products of a proportion are equal.

$$\frac{224}{100} = \frac{336}{150}$$
$$224(150) = 100(336)$$
extremes — means
$$33{,}600 = 33{,}600$$

Consider the general case.

$$\frac{a}{b} = \frac{c}{d}$$
$$(bd)\frac{a}{b} = (bd)\frac{c}{d} \quad \textit{Multiply each side by bd.}$$
$$da = bc \quad \textit{Simplify.}$$
$$ad = bc \quad \textit{Commutative Property } (\times)$$

Equality of Cross Products	**For any numbers a and c and any nonzero numbers b and d, $\frac{a}{b} = \frac{c}{d}$ if and only if $ad = bc$.**

The product of the means equals the product of the extremes.

Motivating the Lesson

Questioning Ask students what the phrase *The average number of children in a family is 2.4* means to them. How can a family have 2.4 children? Explain that this is an average number and that a ratio was used to calculate the number.

2 TEACH

In-Class Example

For Example 1

Students in a geometry class measured the distance from the base of their wrists to the top of their middle fingers and found the average size to be about 15.9 centimeters. They also measured their height and found an average height of 167 centimeters. What is a good estimate for the ratio of height to length of hand? $\frac{167}{15.9} \approx \frac{10.5}{1}$

Teaching Tip When expressing a ratio $\frac{a}{b}$, b cannot be zero because you cannot divide by zero (or the denominator of a fraction cannot be zero).

Teaching Tip When considering the general case of $\frac{a}{b} = \frac{c}{d}$, your students may wonder why each side of the equation was multiplied by bd. Explain that the purpose was to eliminate the fractions.

EXPLORATION

Point out to students that the value of the terms increases very quickly after a few terms.

Cooperative Learning

Co-op Co-op Have groups of students research the nutritional information on several packages of food. The groups should set up ratios and proportions. Relate the ratios to one serving and also to the entire package. Groups should exchange data and results and verify each other's results. For more information on the co-op co-op strategy, see *Cooperative Learning in the Mathematics Classroom*, one of the titles in the Glencoe Mathematics Professional Series, page 32.

Teaching Tip Encourage students to include units when setting up a proportion to make it easier to be consistent.

In-Class Examples

For Example 2

Solve $\frac{x+3}{8} = \frac{4}{3}$ by using cross products. $x = \frac{23}{3}$

For Example 3

If one U.S. dollar is worth 102.70 Spanish pesetas, and a shirt costs $19.95 in U.S. currency, how much would it cost in Spanish pesetas? **2048.87 pesetas**

For Example 4

The ratio of the measures of two consecutive angles between the parallel sides of a trapezoid is 5:7. The ratio of the other two consecutive angle measures between the parallel sides is 1:3. Find the measures of the angles in the trapezoid. **75, 105, 45, and 135**

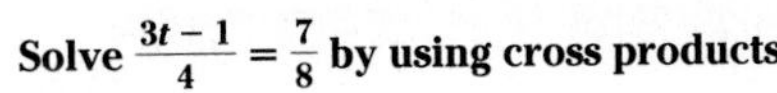

Example 2 **Solve $\frac{3t-1}{4} = \frac{7}{8}$ by using cross products.**

$$\frac{3t-1}{4} = \frac{7}{8}$$

$(3t-1)(8) = (4)\,7$ *Find the cross products*

$24t - 8 = 28$ *Distributive Property*

$24t = 36$ *Add 8 to each side.*

$t = \frac{36}{24}$ or $\frac{3}{2}$ *Divide each side by 24.*

Example 3 **In Chapter 2, you learned that *Alice's Adventures in Wonderland* by Lewis Carroll contains deductive reasoning. It also contains examples of proportions. When Alice ate or drank something, her size changed. ". . . she came suddenly to an open place with a little house in it four feet high . . . . she began nibbling at the right hand bit again, and did not venture to go near the house till she had brought herself down to nine inches high." If Alice was normally about 50 inches tall, how tall would the house have been in Alice's normal world?**

You can write a proportion to show the relationship between the measures in the normal world and those in the shrunken world.

$\frac{48}{9} = \frac{x}{50}$ *4 feet = 48 inches*

$(48)(50) = 9x$ *Find the cross products.*

$2400 = 9x$

$266\frac{2}{3} = x$ *Divide each side by 9.*

Note that the numerators are measures associated with the house and the denominators are measures associated with Alice.

The house would have been $266\frac{2}{3}$ inches or about 22 feet high in a normal world.

Ratios can also be used to compare three or more numbers. The expression $a{:}b{:}c$ means that the ratio of the first two numbers is $a{:}b$, the ratio of the last two numbers is $b{:}c$, and the ratio of the first and last numbers is $a{:}c$.

Example 4 **In a triangle, the ratio of the measures of three sides is 8:7:5 and its perimeter is 240 centimeters. Find the measure of each side of the triangle.**

Let $8x$, $7x$, and $5x$ represent the measures of the sides of the triangle. The perimeter of a triangle is the sum of the measures of its sides. Write an equation to represent the perimeter.

$8x + 7x + 5x = 240$

$20x = 240$

$x = 12$ *Divide each side by 20.*

Since $x = 12$, the measures of the sides of the triangle are 8(12) or 96 centimeters, 7(12) or 84 centimeters, and 5(12) or 60 centimeters.

Classroom Vignette

"I give each student a secret ballot to vote for his or her favorite musical group. Then I collect the ballots and display the ratio for each group. For example, if $\frac{3}{25}$ of the class voted for Garth Brooks, then I have the students use a proportion like $\frac{3}{25} = \frac{x}{500}$ to predict the number of fans he has in the entire school."

Lucas Francois
Waterford Union High School
Waterford, Wisconsin

CHECK FOR UNDERSTANDING

Communicating Mathematics

1a. Men earn \$1.31 for every dollar women earn.

1b. The proportion is $\frac{522}{399} = \frac{1.31}{1}$. It is a proportion because the ratios are equal.

Study the lesson. Then complete the following.

1. In 1994, the median weekly earnings for men was \$522 and for women was \$399. The ratio of the weekly earnings of men to women was $\frac{522}{399} \approx \frac{1.31}{1}$.
   a. **Describe** what the ratio $\frac{1.31}{1}$ represents.
   b. **Identify** the proportion and explain why it is a proportion.
2. **Explain** how you would solve the proportion $\frac{22}{30} = \frac{14}{x}$. See margin.
3. **Write** possible proportions if the means of a proportion are 8 and 10, and the extremes are 5 and 16. See margin.
4. **Determine** which proportions below are equivalent. Explain your reasoning.
   a. $\frac{7}{8} = \frac{x}{y}$ b. $\frac{y}{x} = \frac{8}{7}$ c. $\frac{y}{7} = \frac{x}{8}$ d. $\frac{7}{x} = \frac{8}{y}$
   a, b, and d are equivalent because they have the same cross products.

5. **Assess Yourself** Skim *Alice's Adventures in Wonderland* to find other references to ratio. Use the information in the story to make up several problems about ratio that you think Alice might have used to help her understand her surroundings.
   See students' work.

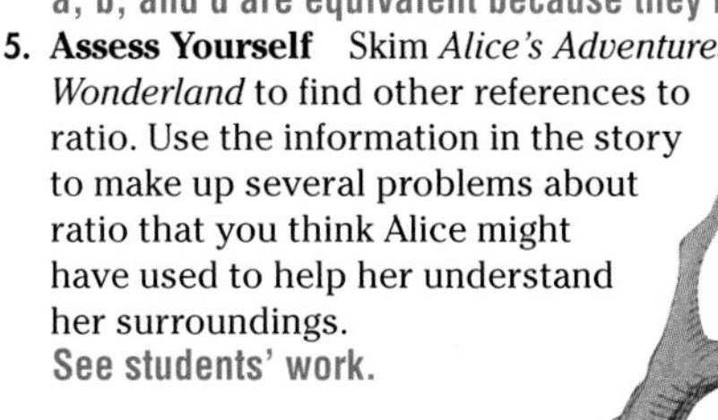

Guided Practice

Express each ratio as a fraction in simplest form.

6. 2 inches on a map represent 150 miles. Find a ratio involving 1 inch. $\frac{1}{75}$
7. 340 South African rands is equivalent to 18 American dollars. Find the ratio of rands to dollars. $\frac{170}{9}$

Solve each proportion by using cross products.

8. $\frac{x}{5} = \frac{11}{35}$ $\frac{11}{7} \approx 1.57$
9. $\frac{13}{49} = \frac{26}{7x}$ 14
10. $\frac{x-2}{x} = \frac{3}{8}$ $\frac{16}{5} = 3.2$

Use the number line at the right to determine if the given ratios are equal.

11. $\frac{CD}{CE} = \frac{EF}{EG}$ yes
12. $\frac{DE}{CD} = \frac{GI}{EG}$ yes
13. $\frac{CE}{FG} = \frac{EH}{GH}$ no
14. In $\triangle ALD$, $DE = ET = TL$. b. $\frac{1}{1}$
    a. Find the ratio of DE to EL. $\frac{1}{2}$
    b. If $\triangle ALD$ is an isosceles triangle with $\overline{AD} \cong \overline{AL}$, find the ratio of $m\angle ADL$ to $m\angle ALD$.

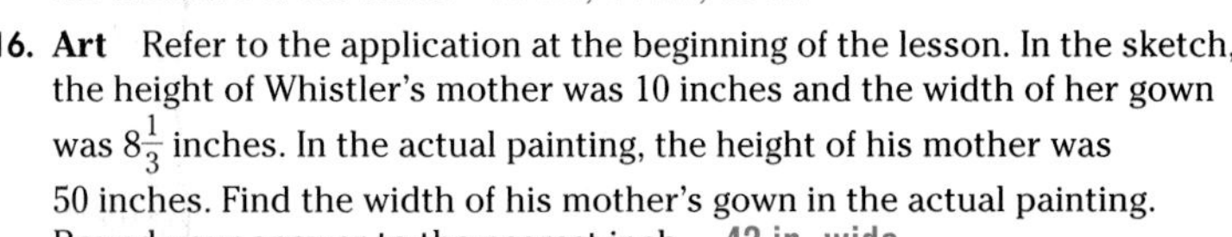
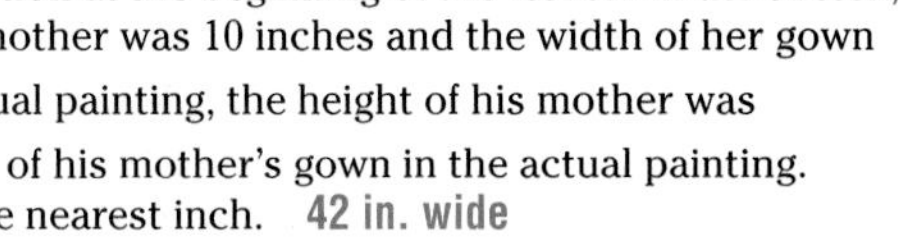

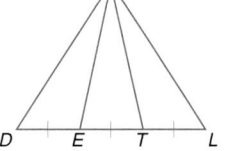

15. The perimeter of a triangle is 72 inches, and the ratio of the measures of the sides is 3:4:5. Find the measure of the sides. 18 in., 24 in., 30 in.
16. **Art** Refer to the application at the beginning of the lesson. In the sketch, the height of Whistler's mother was 10 inches and the width of her gown was $8\frac{1}{3}$ inches. In the actual painting, the height of his mother was 50 inches. Find the width of his mother's gown in the actual painting. Round your answer to the nearest inch. 42 in. wide

James Whistler

3 PRACTICE/APPLY

Check for Understanding

Exercises 1–16 are designed to help you assess your students' understanding through reading, writing, speaking, and modeling. You should work through Exercises 1–5 with your students and then monitor their work on Exercises 6–16.

Error Analysis

Remind students that they can find cross products only when there is an equals sign between the ratios.

Additional Answers

2. Find the product of the means, (30)(14), and of the extremes, 22x, and set them equal. Divide each side by 22.
3. $\frac{16}{10} = \frac{8}{5}$, $\frac{16}{8} = \frac{10}{5}$, $\frac{5}{10} = \frac{8}{16}$, $\frac{5}{8} = \frac{10}{16}$

Reteaching

Using Research Have students think of a situation that involves ratios or proportions like batting averages from the sports page or scales on a map. Then have them set up a ratio and a proportion for some of the examples.

Assignment Guide

Core (with proof): 17–51 odd, 53–63
Core (informal): 17–51 odd, 53–63
Enriched: 18–44 even, 45–63

For **Extra Practice**, see p. 776.

The red A, B, and C flags, printed only in the Teacher's Wraparound Edition, indicate the level of difficulty of the exercises.

Additional Answers

39.
$$\frac{a+b}{b} = \frac{c+d}{d}$$
$$(a+b)d = (c+d)b$$
$$ad + bd = cb + db$$
$$ad + bd = cb + bd$$
$$ad + bd - bd = cb + bd - bd$$
$$ad = cb$$
$$\frac{a}{b} = \frac{c}{d}$$

40.
$$\frac{a-b}{b} = \frac{c-d}{d}$$
$$(a-b)d = (c-d)b$$
$$ad - bd = cb - db$$
$$ad - bd = cb - bd$$
$$ad - bd + bd = cb - bd + bd$$
$$ad = cb$$
$$\frac{a}{b} = \frac{c}{d}$$

EXERCISES

Practice

Find each ratio and express it as a fraction in simplest form.

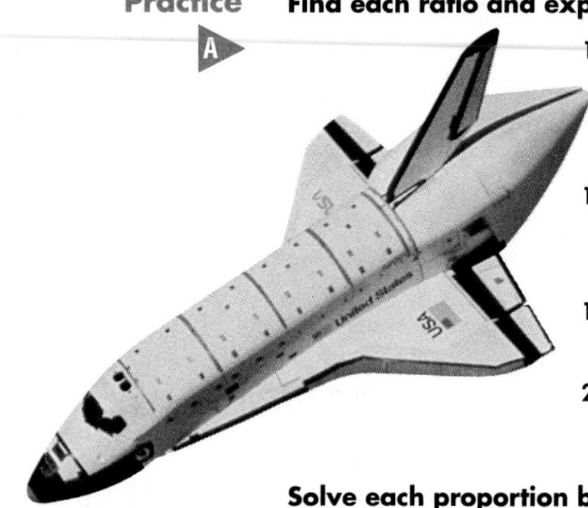

17. By 2010, it may be possible for people to travel in space, at a cost of about $150,000 for a 150-pound person. Find the ratio of cost to the number of pounds. $\frac{1000}{1}$
18. By 2000, for every 2 new U.S. male workers, there will be 3 new U.S. female workers. Find the ratio of male workers to female workers. $\frac{2}{3}$
19. A designated hitter made 8 hits in 10 games. Find the ratio of hits to games. $\frac{4}{5}$
20. There are 76 boys and 89 girls in the sophomore class. Find the ratio of boys to girls. $\frac{76}{89}$

Solve each proportion by using cross products.

22. $-\frac{1}{7} \approx -0.14$

21. $\frac{a}{5.18} = \frac{1}{4}$ 1.295
22. $\frac{5}{n+3} = \frac{7}{4}$
23. $\frac{7}{11} = \frac{11}{x}$ $\frac{121}{7} \approx 17.29$
24. $\frac{3x}{23} = \frac{48}{92}$ 4
25. $\frac{a+1}{a-1} = \frac{5}{6}$ −11
26. $\frac{2}{3x+1} = \frac{1}{x}$ −1
27. A cable that is 42 feet long is divided into lengths in the ratio of 3:4. What are the two lengths into which the cable is divided? 18 ft, 24 ft
28. In $\triangle ACE$, $\frac{AB}{BC} = \frac{CD}{DE} = \frac{FE}{AF}$. If $AB = 4$, $DE = 9$, $FE = 6$, and $BC = 12$, find the lengths of each side of $\triangle ACE$. $AE = 24$, $AC = 16$, $CE = 12$

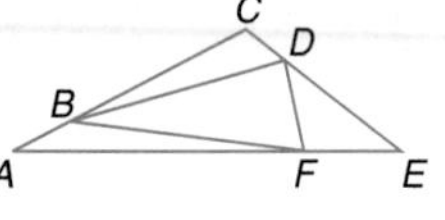

Corresponding sides of polygon *ABCD* are proportional to the sides of polygon *EFGH*.

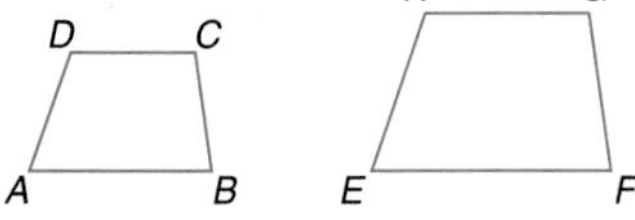

29. If $AB = 14$, $BC = 2.6$, and $EF = 21$, find FG. 3.9
30. If $GH = 40$, $FG = 32$, and $DC = 25$, find CB. 20
31. If $AD = \frac{2}{3}$, $BC = \frac{3}{4}$, and $FG = \frac{1}{2}$, find EH. $\frac{4}{9}$
32. In the figure at the right, $\frac{AB}{BC} = \frac{AD}{DE}$. Use proportions to complete the table.

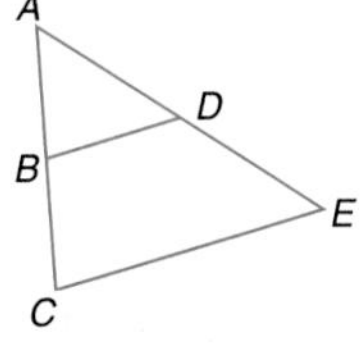

AB	BC	AD	DE	AC	AE
5	8	6.25	10	13	16.25
5.5	14.5	8.8	23.2	20	32
9.6	6.4	12	8	16	20
16.5	16.5	17	17	33	34

33. 40 in., 60 in., 90 in.

33. The ratio of the measures of three sides of a triangle is 4:6:9, and its perimeter is 190 inches. Find the measure of each side of the triangle.

34. In a triangle, the ratio of the measures of three angles is 2:5:8. Find the measure of each angle in the triangle. 24, 60, 96

Proportions can be used to find the solutions to percent problems. For example, to find 23% of 15, you can solve the proportion $\frac{23}{100} = \frac{x}{15}$. Use a proportion to solve each problem.

35. Find 32% of 156. 49.92

36. Find 175% of 42. 73.5

37. What percent of 56 is 14? 25%

38. 32 is 60% of what? $53\frac{1}{3}$

Use cross products to show each of the following is true. Assume that *b* and *d* are not zero. 39–40. See margin.

39. $\frac{a}{b} = \frac{c}{d}$ if $\frac{a+b}{b} = \frac{c+d}{d}$.

40. $\frac{a}{b} = \frac{c}{d}$ if $\frac{a-b}{b} = \frac{c-d}{d}$.

An expression of the form *a* = *b* = *c* means *a* = *b*, *b* = *c*, and *a* = *c*. Solve for *x* and *y*.

41. $\frac{x}{20} = \frac{6}{5} = \frac{y}{10}$ $x = 24, y = 12$

42. $\frac{4}{3} = \frac{x+1}{x+3} = \frac{y+1}{y}$ $x = -9, y = 3$

43. This is true because one of the angles measures 90° and the other two measure 45°. The ratio of 45:90 is 1:2.

43. The ratio of the measures of two angles in an isosceles right triangle is 1:2. Explain why this is true.

44. Sandra reduced a rectangle that is 21.3 centimeters by 27.5 centimeters so that it would fit in a 10-centimeter by 10-centimeter area.

a. Use a proportion to find the maximum dimensions of the reduced rectangle. approximately 7.75 cm by 10 cm

b. What is the percent of reduction of the length? approximately 64%

Critical Thinking

45. Gas Mileage At each gas stop on a recent trip, Keshia recorded the distance she traveled since the last stop and the amount of gas she purchased for her midsize car.

Stops	Distance (mi)	Gas (gal)
1	351	13
2	275	11
3	362.5	12.5
4	372	12
5	260	12.8
6	294.4	10

a. Find the ratios of the distances she drove to the gas she used for each stop.

b. Can you use the results of part a to find the average number of miles per gallon for her car? Why or why not? See margin.

a. The ratios (miles per gallon) are 27, 25, 29, 31, 20.3125, 29.44.

Applications and Problem Solving

46. Meteorology Some people are concerned about the weather when they choose a place to live. The table at the right contains the average number of days per year for precipitation (either rain or snow) and the average number of days that are clear.

City	Days with Precipitation	Clear Days
Albuquerque, NM	59	172
Boston, MA	128	99
Buffalo, NY	168	55
Fresno, CA	44	200
Lubbock, TX	60	164
Montgomery, AL	109	107

Source: *Places Rated Almanac*

a. Find the ratio of the number of days with precipitation to the number of clear days for each city. Express each ratio as a decimal rounded to the nearest hundredth. 0.34, 1.29, 3.05, 0.22, 0.37, 1.02

b. Choose one of the ratios and explain what it means. See margin.

c. Use the information in the table to rank the cities according to precipitation. See margin.

CAREER CHOICES

A **meteorologist** studies the atmosphere, including its physical characteristics, and its effects on the environment. A bachelor's degree in meteorology is required, but advanced degrees are often advantageous.

For more information, contact:

American Meteorological Society
45 Beacon Street
Boston, MA 02108-3693

CAREER CHOICES

Meteorologists gather their information from computer models, radars, weather balloons, and satellites. Dynamic meteorology deals with motion in the atmosphere and airflow. Synaptic meteorology deals with the weather.

Additional Answers

45b. You cannot find the average number of miles per gallon by adding the number of miles per gallon at each stop and dividing by 6 because Keshia drove a different distance between each stop. To find the average, you must divide the total distance by the total amount of gas for all six stops. The average is about 26.86 miles per gallon.

46b. Sample answer: 3.05 means that there are usually 3 days with precipitation for every clear day in Buffalo.

Additional Answer

46c. Sample answer: ranking the cities from greatest precipitation to least precipitation: Buffalo, Boston, Montgomery, Lubbok, Albuquerque, Fresno

Study Guide Masters, p. 37

NAME________________ DATE ________

Study Guide — Student Edition Pages 338–345

Integration: Algebra Using Proportions

A **ratio** is a comparison of two quantities. The ratio of a to b can be expressed as $\frac{a}{b}$, where b is not 0. The ratio can also be written a:b.

An equation stating that two ratios are equal is a **proportion**. Therefore, $\frac{a}{b} = \frac{c}{d}$ is a proportion for any numbers a and c and any nonzero numbers b and d. In any true proportion, the cross products are equal. So, $\frac{a}{b} = \frac{c}{d}$ if and only if $ad = bc$.

Example: Solve $\frac{11}{16} = \frac{44}{x}$ by using cross products.

$$\frac{11}{16} = \frac{44}{x}$$
$$11x = 16 \cdot 44$$
$$11x = 704$$
$$x = 64$$

For Exercises 1–4, use the table to find the ratios. Express each ratio as a decimal rounded to three places.

Teams	Wins	Losses
Hawks	16	13
Tigers	15	14
Mustangs	12	16

1. games won to games lost for Hawks 1.231

2. games won by the Hawks to games won by Tigers 1.067

3. games won to games played for Tigers 0.517

4. games won to games played for Mustangs 0.429

Solve each proportion by using cross products.

5. $\frac{9}{28} = \frac{x}{84}$ 27

6. $\frac{3}{18} = \frac{4x}{7}$ $\frac{7}{24}$

7. $\frac{x+5}{7} = \frac{x+3}{5}$ 2

Use a proportion to solve each problem.

8. If two cassettes cost $14.50, how much will 15 cassettes cost? $108.75

9. If a 6-foot post casts a shadow that is 8 feet long, how tall is an antenna that casts a 60-foot shadow at the same time? 45 ft

Additional Answers

52b. An increase of more than 20 blue-collar jobs is predicted for a loss of every white-collar job in Pittsburgh.

52c. New York City will lose jobs. A loss of 6 blue-collar jobs is predicted for every 100 lost white-collar jobs.

Practice Masters, p. 37

NAME______________________ DATE __________

Practice

Student Edition Pages 338–345

Integration: Algebra Using Proportions

Solve each proportion using cross products.

1. $\frac{3}{5} = \frac{x}{15}$ 9
2. $\frac{20 - x}{x} = \frac{6}{4}$ 8
3. $\frac{x + 1}{5} = \frac{x - 1}{2}$ $\frac{7}{3}$
4. $\frac{x}{x - 3} = \frac{x + 4}{x}$ 12
5. $\frac{x + 1}{6} = \frac{x - 1}{x}$ 2, 3
6. $\frac{1}{x} = \frac{6}{x + 9}$ $\frac{9}{5}$
7. $\frac{x}{x + 8} = \frac{2}{3}$ 16
8. $\frac{4}{12} = \frac{x + 2}{2x + 5}$ −1

In the figure at the right, $\frac{AC}{CD} = \frac{CE}{CB}$. Use proportions to complete the table.

	AC	BC	AB	CE	ED	DC
9.	10	4	6	8	3	5
10.	12	7.5	4.5	10	1	9

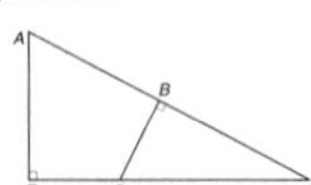

Use a proportion to solve each problem.

11. The ratio of seniors to juniors in the Math Club is 2:3. If there are 21 juniors, how many seniors are in the club? 14 seniors

12. A 15-foot building casts a 9-foot shadow. How tall is a building that casts a 30-foot shadow at the same time? 50 ft

13. A photo that is 3 inches wide and 5 inches high was enlarged so that it is 12 inches wide. How high is the enlargement? 20 in.

14. Philip has been eating 2 hamburgers every 5 days. At that rate, how many hamburgers will he eat in 30 days? 12 hamburgers

47b. You could find the ratio of people per movie screen to see which is smallest; Kansas City is the lowest at 10,862 people per screen, followed by Ann Arbor at 11,049 people per screen.

49a. 6,658,400 lb

49b. 16.4 lb

51a. See Solutions Manual.

51b. Sample answer: No, it costs the most to make a penny since the ratio is the greatest at 0.8 and it costs the least to make a dollar bill since the ratio is the smallest at 0.03.

52a. An increase of almost 2 blue-collar jobs is predicted for every increase in a white-collar job in Seattle.

47. **Business** Suppose you were the field agent for a company that operates movie theaters around the United States and you were thinking about building a theater in one of the six metropolitan areas listed below. The number of movie screens and the size of the population for each area is given in the table.

Metropolitan Area	Population	Number of Movie Screens
Ann Arbor, MI	497,197	45
Kansas City, MO	1,629,241	150
New Haven, CT	538,643	28
Santa Fe, NM	129,841	10
Provo, UT	299,084	22

Source: *Places Rated Almanac, 1993*

   a. Find the ratio of people per screen for each metropolitan area. Round your answer to next whole number. **11,049; 10,862; 19,238; 12,985; 13,595**

   b. How could you use this information to help you decide where to build a theater?

48. **Cartography** The scale on a map indicates that 1.5 centimeters represents 200 miles. If the distance on the map between Norfolk, Virginia, and Atlanta, Georgia, measures 2.4 centimeters, how many miles apart are the cities? **320 miles**

49. **Food** There were approximately 255,082,000 people in the United States in 1992. According to figures from the United States Census, they consumed about 4,183,344,800 pounds of ice cream that year.

   a. If there were 406,000 people in the city of Des Moines, Iowa, about how much ice cream might they have been expected to consume?

   b. What was the approximate consumption of ice cream per person?

50. **Geography** The area of Texas is 262,017 square miles. The area of Alaska is 570,017 square miles. Find the ratio of Texas area to Alaska area. **0.46**

51. **Money** The graph at the right shows how much it costs to make different types of currency.

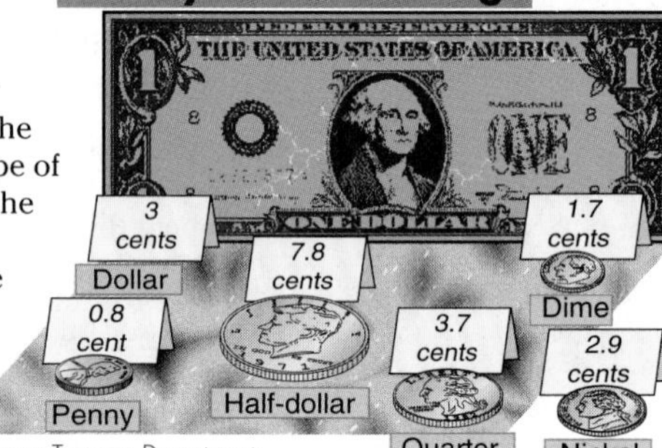

Source: Treasury Department

   a. Make a table showing the cost of making each type of currency in cents and the value of each type of currency. Then find the ratio of the cost to the value for each type of currency.

   b. Are the ratios equal? Explain.

   c. How much would it cost to make a $5 bill if the cost were proportional to that of making a nickel? to that of making a quarter? **$2.90, $0.74**

52. **Economics** White-collar jobs are usually salaried employees whose duties do not require them to wear work clothes or protective clothing, whereas blue-collar jobs are usually hourly wage workers who do require special clothing. As the economy changes, the number of jobs in these areas shifts. The ratio of projected blue-collar jobs to white-collar jobs for the year 1998 is given below. Explain what the ratio indicates for each city.

   a. Seattle, WA: 1.75:1

   b. Pittsburgh, PA: 20.45:−1 **b–c. See margin.**

   c. New York, NY: −6:−100

Extension

Connections Show that if $\frac{a}{b} = \frac{c}{d} = \frac{e}{f}$, then $\frac{a}{b} = \frac{c}{d} = \frac{e}{f} = \frac{a + c + e}{b + d + f}$.

Set each ratio equal to x. So, $a = bx$, $c = dx$, and $e = fx$. Then, $a + c + e = (b + d + f)x$. So, $\frac{a + c + e}{b + d + f} = \frac{a}{b} = \frac{c}{d} = \frac{e}{f}$.

53. trapezoid

Mixed Review

53. Name a quadrilateral that has exactly one pair of parallel sides. (Lesson 6–5)

54. If *JKLM* is a parallelogram and $m\angle J = 72$, find the measures of $\angle K$, $\angle L$, and $\angle M$. (Lesson 6–1) $m\angle K = 108$, $m\angle L = 72$, $m\angle M = 108$

55. It is less than 13 and greater than 3.

55. If two sides of a triangle have measures of 5 and 8, what must be true about the measure of the third side? (Lesson 5–5)

56. For $\triangle FGH$, list the angles in order of measure from greatest to least. (Lesson 5–4) $\angle F$, $\angle H$, $\angle G$

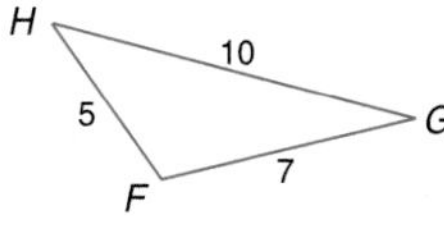

57. In a right triangle, one angle measures 38°. Find the measures of the other two angles. (Lesson 4–2) **90, 52**

60. If a mineral sample is quartz, then it can scratch glass; Law of Syllogism.

58. Determine whether the statement *A right triangle is never scalene* is true or false. Explain. (Lesson 4–1) **false**

59. Find the values of *a* and *b* in the figure at the right. (Lesson 3–2) **75, 105**

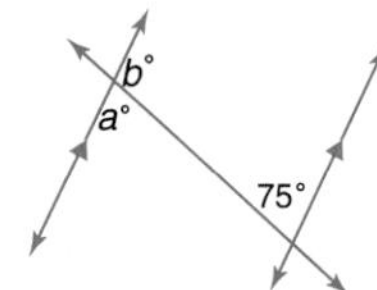

60. **Geology** Determine if a valid conclusion can be reached from the two true statements using the Law of Detachment or the Law of Syllogism. If a valid conclusion is possible, state it and the law of logic that is used. (Lesson 2–3)
   (1) If a mineral sample is a sample of quartz, then the sample has a hardness factor of 7.
   (2) If a mineral sample has a hardness factor greater than 5, then it can scratch glass.

61. Given that $AP = PB$, is the conjecture that *P* is the midpoint of $\overline{AB}$ *true* or *false*? Explain. (Lesson 2–1) **False; *A*, *B*, and *P* are not necessarily collinear.**

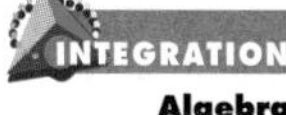

Algebra

62. Solve $2n + 6 < n - 1$. $\{n \mid n < -7\}$

63. Solve $x^2 + 6x - 16 = 0$. $-8, 2$

Refer to the Investigation on pages 286–287.

1 Use grid paper to draw a plan for the park. Begin by drawing the boundary lines of the park.

2 Refer to your list of features. Determine the best location for each feature and add it to the plan.

3 As you add features to the plan, compile a list of costs. Make sure that your project will be in line with the budget. Don't forget to include the costs of roadways in your budget.

Add the results of your work to your Investigation Folder.

4 ASSESS

Closing Activity

Writing Have each student measure the distance around his or her clenched fist. How does it compare with the length of his or her foot? Write a proportion. Compare students' proportions. Is there a trend?

In·ves·ti·ga·tion

Working on the Investigation

The Investigation on pages 286–287 is designed to be a long-term project that is completed over several days or weeks. Encourage students to keep their materials in their Investigation Folder as they work on the Investigation.

Enrichment Masters, p. 37

NAME ____________ DATE ________

Enrichment Student Edition Pages 338–345

Using a Grid to Enlarge a Drawing

Here is method of enlarging a drawing or picture.

1. Lay a grid pattern of small squares over the picture.
2. On separate paper, draw a larger grid with the same number of squares. (If you want to double the dimensions of the picture, the sides of the squares of the larger grid should be twice as long as the sides of the original.)
3. Draw the contents of each square of the original picture on the corresponding square of the larger grid.

Enlarge each drawing.

1.

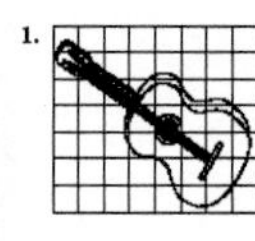

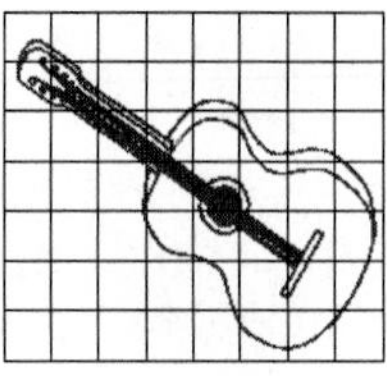

2.

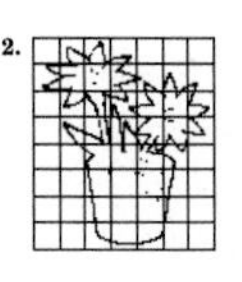

3. ... ge the picture by reproducing each square on a larger grid.
See students' work.

7-2 LESSON NOTES

NCTM Standards: 1–5, 7

Instructional Resources

- Study Guide Master 7-2
- Practice Master 7-2
- Enrichment Master 7-2
- Assessment and Evaluation Masters, p. 184

Transparency 7-2A contains the 5-Minute Check for this lesson; **Transparency 7-2B** contains a teaching aid for this lesson.

Recommended Pacing	
Standard Pacing	Day 2 of 13
Honors Pacing	Day 2 of 12
Block Scheduling*	Day 2 of 7

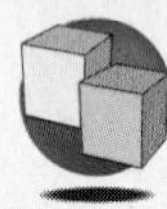

*For more information on pacing and possible lesson plans, refer to the *Block Scheduling Booklet*.

1 FOCUS

5-Minute Check
(over Lesson 7-1)

Determine if each pair of ratios forms a proportion.

1. $\frac{2}{3} \stackrel{?}{=} \frac{16}{24}$ yes

2. $\frac{9}{11} \stackrel{?}{=} \frac{81}{88}$ no

Solve each proportion by using cross products.

3. $\frac{10}{11} = \frac{11}{x}$ 12.1

4. $\frac{2}{5} = \frac{9x}{13}$ 0.58

5. $\frac{x-1}{9} = \frac{3x+2}{21}$ −6.5

Motivating the Lesson

Hands-On Activity Draw several pairs of similar polygons on the chalkboard or overhead along with several pairs of polygons that are not similar. Draw pairs of polygons that have a different number of sides. Ask students what they notice about the pairs.

7-2 Exploring Similar Polygons

What YOU'LL LEARN

- To identify similar figures, and
- to solve problems involving similar figures.

Why IT'S IMPORTANT

You can use similar polygons to solve problems involving cartography, gardening, and construction work.

Art

In 1928, Charles Demuth used properties of similar polygons to create proportional numeral 5s in his painting *I Saw the Figure 5 in Gold* shown at the right. The numerals are the same shape, but different sizes.

When figures have the same shape but are different sizes, they are called **similar figures**.

Definition of Similar Polygons	Two polygons are similar if and only if their corresponding angles are congruent and the measures of their corresponding sides are proportional.

The definition states that in order for two figures to be similar, the corresponding angles of the two figures must be congruent, and the measures of the corresponding sides must be proportional. The quadrilaterals below are similar.

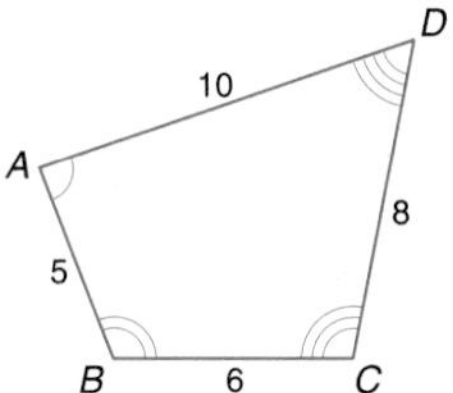

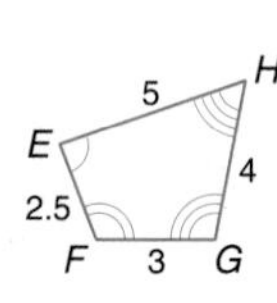

The symbol ~ means *is similar to*. We write quadrilateral *ABCD* ~ quadrilateral *EFGH*. Just as in congruence, the order of the letters indicates the vertices that correspond. We can make the following statements about quadrilaterals *ABCD* and *EFGH*.

$$\angle A \cong \angle E \qquad \angle B \cong \angle F \qquad \angle C \cong \angle G \qquad \angle D \cong \angle H$$

$$\frac{AB}{EF} = \frac{BC}{FG} = \frac{CD}{GH} = \frac{DA}{HE} = \frac{2}{1}$$

The ratio of the lengths of two corresponding sides of two similar polygons is called the **scale factor**. The scale factor of quadrilateral *ABCD* to quadrilateral *EFGH* is 2. The scale factor of quadrilateral *EFGH* to quadrilateral *ABCD* is $\frac{1}{2}$.

You can use an overhead projector and transparencies to investigate characteristics of similar polygons.

MODELING MATHEMATICS

Similar Polygons

Materials:  ruler  compass  blank transparency  overhead projector

b. The measures of the corresponding sides of the original triangle and its projection are proportional.

You can use an overhead projector to create similar polygons.

- Use a ruler and compass to draw a triangle with sides measuring 4 centimeters, 7 centimeters, and 9 centimeters on a blank transparency.
- Project the triangle on a screen by using an overhead projector.
- Measure the sides and the angles of the triangle projected on the screen.

Your Turn

a. What is the ratio between the 4-centimeter side and its projection? between the 7-centimeter side and its projection? between the 9-centimeter side and its projection? **See students' work.**

b. What conclusion can you make?

c. What else do you have to check to verify that the original triangle you placed on the screen and its projection are similar?

c. You have to check that the corresponding angles are congruent.

The properties of proportions can be used to solve problems involving similar polygons.

Example  **Polygon *RSTUV* is similar to polygon *ABCDE*.**

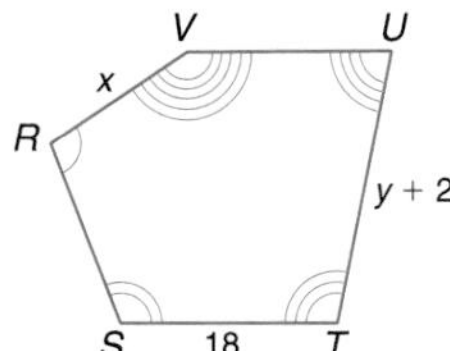

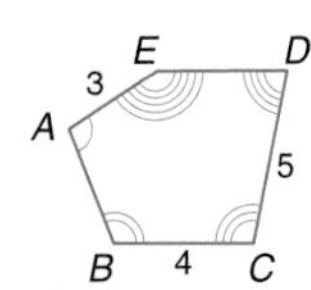

a. Find the scale factor of polygon *RSTUV* to polygon *ABCDE*.

The scale factor is the ratio of the lengths of two corresponding sides.

$$\text{scale factor} = \frac{ST}{BC} = \frac{18}{4} \text{ or } \frac{9}{2}$$

b. Find the values of *x* and *y*.

Because the polygons are similar, the corresponding sides are proportional. Thus, we can write proportions to find the values of *x* and *y*.

Write a proportion that involves numbers and the variable *x*.

$\frac{ST}{BC} = \frac{VR}{EA}$

$\frac{18}{4} = \frac{x}{3}$ *ST = 18, BC = 4, VR = x, EA = 3*

$54 = 4x$ *Cross products*

$13.5 = x$

Write a proportion that involves numbers and the variable *y*.

$\frac{ST}{BC} = \frac{UT}{DC}$

$\frac{18}{4} = \frac{y + 2}{5}$ *ST = 18, BC = 4, UT = y + 2, DC = 5*

$90 = 4y + 8$ *Cross products*

$82 = 4y$

$20.5 = y$

(continued on the next page)

2 TEACH

 Point out to students that the transparency must be perpendicular to the direction of the light source. Ask them how the image would change if the transparency were not perpendicular. You may need to remind students how to construct a triangle for three given measures for its sides.

In-Class Example

For Example 1

Trapezoid *PQRS* is similar to trapezoid *UTWV*. Find the value of *x*. **15**

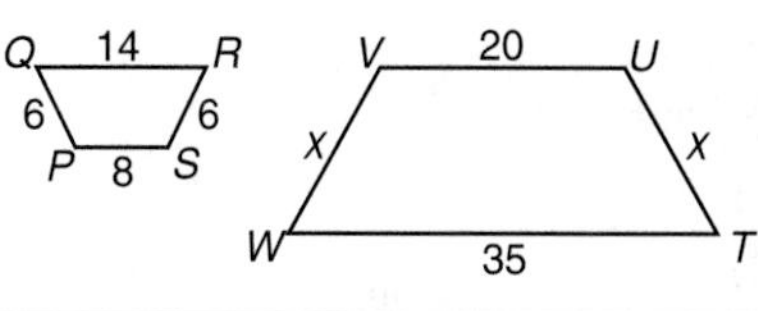

Teaching Tip When discussing scale factor, point out that there is more than one scale factor for two similar polygons. For example, the scale factor of parallelogram *ABCD* to parallelogram *PQRS* is 2, and the scale factor of parallelogram *PQRS* to parallelogram *ABCD* is $\frac{1}{2}$.

Teaching Tip After defining similar polygons, you may want to ask students if all polygons of different types are similar. For example, are all rectangles similar? **no** Are all regular hexagons similar? **yes**

Alternative Teaching Strategies

Reading Geometry Discuss the phrase "if and only if" with students. Ask them what it means in the definition of similar polygons. Remind them that in a definition or theorem, the phrase "if and only if" means that the hypothesis leads to the conclusion *and* the conclusion leads to the hypothesis.

In-Class Examples

For Example 2
How long would it take to drive the north to south dimension of Delaware? It measures 3.3 cm on the map and you would average 50 miles per hour.
1.65 hours or 1 hour 39 minutes

For Example 3
A square has vertices $D(0, 0)$, $E(5, 0)$, $F(0, 5)$, and $G(5, 5)$. If the coordinates of each vertex are multiplied by 3, will the new figure be similar to the original?
yes

Check: $\frac{VR}{EA} = \frac{x}{3}$ $\qquad$ $\frac{TU}{CD} = \frac{y+2}{5}$

$= \frac{13.5}{3}$ $\qquad$ $= \frac{20.5+2}{5}$

$= 4.5 \text{ or } \frac{9}{2}$ $\qquad$ $= 4.5 \text{ or } \frac{9}{2}$

Since both ratios equal the value of the scale factor, the answers are correct.

Because longitude and latitude lines are often used to determine boundaries, many states are shaped like polygons.

Example 2 APPLICATION Cartography

Delaware is a very small state, shaped similarly to a five-sided polygon. The scale on a map of Delaware is approximately 2 centimeters = 50 miles. The distance on the map across Delaware from east to west is 1.4 centimeters. How long would it take to drive across Delaware if you drove at approximately 50 miles per hour?

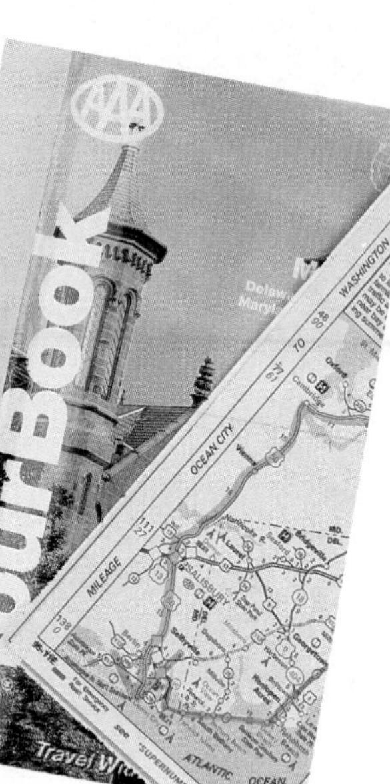

Create a proportion relating the measurements to the scale to find the distance in miles.

$$\begin{aligned} \text{centimeters} \rightarrow \frac{2.0}{50} &= \frac{1.4}{x} \leftarrow \text{centimeters} \\ \text{miles} \rightarrow \quad & \quad \leftarrow \text{miles} \\ 2x &= 70 \quad \textit{Cross products} \\ x &= 35 \quad \textit{Divide each side by 2.} \end{aligned}$$

The distance across Delaware is approximately 35 miles. Use the formula $d = rt$ to find the time.

$$\begin{aligned} d &= rt \\ 35 &= 50t \quad d = 35 \text{ miles}, r = 50 \text{ mph} \\ \frac{35}{50} &= t \\ 0.7 &= t \end{aligned}$$

It would take 0.7 of an hour or about 42 minutes to drive across the state of Delaware at 50 miles per hour.

Some transformations will produce similar figures. A **dilation** is a transformation that reduces or enlarges a figure.

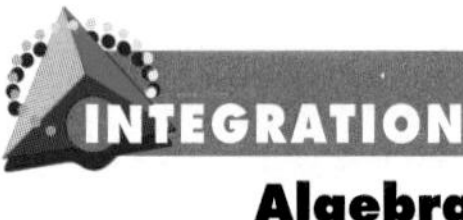

Example 3 INTEGRATION Algebra

Triangle ABC has vertices $A(0, 0)$, $B(8, 0)$, and $C(2, 7)$. If the coordinates of each vertex are multipled by 2, will the new figure be similar to the original?

Explore Find the coordinates of the new triangle. Graph each triangle.

Coordinates	New Coordinates
$A(0, 0)$	$(2(0), 2(0))$ or $(0, 0)$
$B(8, 0)$	$(2(8), 2(0))$ or $(16, 0)$
$C(2, 7)$	$(2(2), 2(7))$ or $(4, 14)$

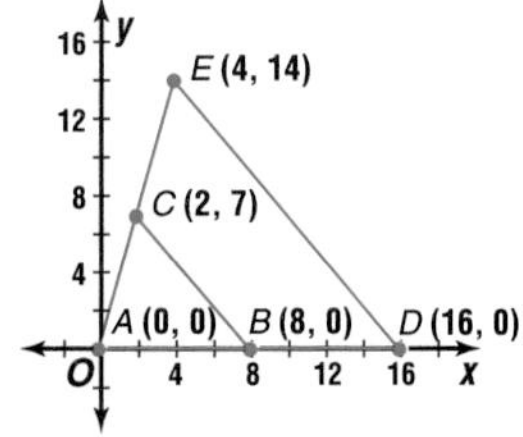

$\triangle ADE$ will be similar to $\triangle ABC$ if the corresponding angles are congruent and the measures of the corresponding sides are proportional.

Classroom Vignette

"As a long-term project, I have students create a geoboard. They use a square board at least 8 inches wide and paint it or cover it with cloth. Then they hammer nails or small brads into the board and create a symmetric design using yarn, string, or elastic bands. When the projects are completed, I allow the classes to vote for their three favorites. The winners will receive prizes."

Karen L. George
Taunton High School
Taunton, Massachusetts

Plan To determine if the angles are congruent, we can investigate the slopes to see if $\overline{CB} \parallel \overline{ED}$. Then we can use the properties of parallel lines to determine whether the sides are proportional and the distance formula to find the lengths of each side.

Solve The formula for slope is $\frac{y_2 - y_1}{x_2 - x_1}$.

$$\text{slope of } \overline{CB} = \frac{7-0}{2-8} = -\frac{7}{6}$$

$$\text{slope of } \overline{ED} = \frac{14-0}{4-16} = \frac{14}{-12} \text{ or } -\frac{7}{6}$$

Because the slopes are equal and the lines do not contain a common point, $\overline{CB} \parallel \overline{ED}$. If two parallel lines are cut by a transversal, the corresponding angles are congruent, which makes $\angle ABC \cong \angle ADE$ and $\angle ACB \cong \angle AED$. $\angle A \cong \angle A$ because congruence of angles is reflexive. Thus, the corresponding angles in $\triangle ABC$ are congruent to the corresponding angles in $\triangle ADE$.

Now use the distance formula, $d = \sqrt{(x_2 - x_1)^2 + (y_2 - y_1)^2}$, to find the lengths of the sides of each triangle.

In $\triangle ABC$	*In* $\triangle ADE$
$AB = \sqrt{(8-0)^2 + (0-0)^2} = \sqrt{64} \text{ or } 8$	$AD = \sqrt{(16-0)^2 + (0-0)^2} = \sqrt{256} \text{ or } 16$
$BC = \sqrt{(8-2)^2 + (0-7)^2} = \sqrt{85}$	$DE = \sqrt{(16-4)^2 + (0-14)^2} = \sqrt{340} \text{ or } 2\sqrt{85}$
$AC = \sqrt{(2-0)^2 + (7-0)^2} = \sqrt{53}$	$AE = \sqrt{(4-0)^2 + (14-0)^2} = \sqrt{212} \text{ or } 2\sqrt{53}$

$$\frac{AB}{AD} = \frac{8}{16} = \frac{1}{2} \qquad \frac{BC}{DE} = \frac{\sqrt{85}}{2\sqrt{85}} = \frac{1}{2} \qquad \frac{AC}{AE} = \frac{\sqrt{53}}{2\sqrt{53}} = \frac{1}{2}$$

Therefore, $\frac{AB}{AD} = \frac{BC}{DE} = \frac{AC}{AE}$.

Examine Because the measures of the corresponding sides are proportional and the corresponding angles are congruent, $\triangle ABC$ is similar to $\triangle ADE$. If the coordinates of each vertex are multiplied by 2, the new figure is similar to the original figure.

CHECK FOR UNDERSTANDING

Communicating Mathematics

1a. Yes; all the corresponding angles are congruent and the corresponding sides are in the ratio of 1:1.

Study the lesson. Then complete the following. 1b. See margin.

1. **a. Explain** whether two figures that are congruent are also similar.
   b. Explain whether two figures that are similar are also congruent.
2. **You Decide** Yvonne calculated the scale factor between two similar polygons to be $\frac{5}{3}$. Trina found the scale factor to be $\frac{3}{5}$ for the same polygon. Explain how this could happen and who is right. See margin.

3 PRACTICE/APPLY

Check for Understanding

Exercises 1–12 are designed to help you assess your students' understanding through reading, writing, speaking, and modeling. You should work through Exercises 1–5 with your students and then monitor their work on Exercises 6–12.

Additional Answers

1b. No; their corresponding sides may not be congruent.

2. They both could be right; it depends on whether you are going from a smaller polygon to a larger one or the other way around.

Reteaching

Using Hands-On Activity Have each student use a straightedge to draw and label a polygon of any type and size. Have them measure each side and list the measures. Have them multiply each measure by $\frac{1}{2}$ and use these calculated values to construct a similar polygon with a scale factor of $\frac{1}{2}$.

Additional Answer

3. See students' work for drawings. The drawings will be similar since the measures of the corresponding sides will be proportional and the corresponding angles will be congruent.

3. See margin.

4. The corresponding angles must be congruent: $\angle A \cong \angle M$; $\angle B \cong \angle N$; $\angle C \cong \angle O$; $\angle D \cong \angle P$.

3. **Draw** the state of Delaware using a scale of 2 centimeters = 30 miles. Is your drawing similar to the one in Example 2? Explain how you know.

4. Every segment in quadrilateral *MNOP* is twice as long as the corresponding segment in quadrilateral *ABCD*. List additional information you would need to determine whether the figures are similar.

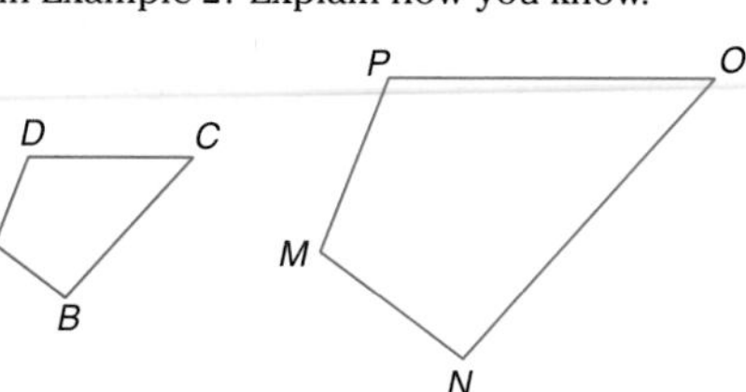

5b. $\frac{2}{15}$

5. A parallelogram with sides 4 centimeters and 8 centimeters long is projected onto a screen. The ratio between a 4-centimeter side and its projection is 2:17.
   a. What is the measurement of the longer side of the parallelogram as it is projected on the screen? 68 cm
   b. If you move the projector, the measurement of the shorter side of the parallelogram becomes 30 centimeters. What is the new scale factor?

Guided Practice

6. Yes; angles are congruent and sides are proportional.

6. Determine whether the figures below are similar. Justify your answer.

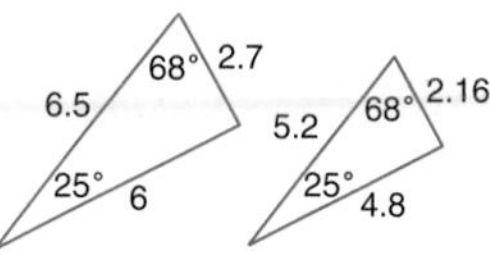

7. The polygons below are similar. Find the values of x and y. 12, 12

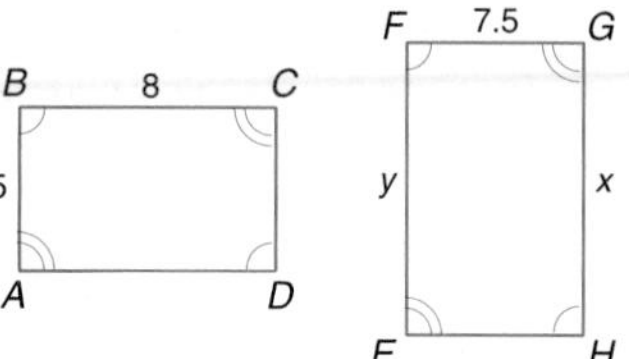

For each statement, write *A* if the statement is *always* true, *S* if the statement is *sometimes* true, and *N* if the statement is *never* true. Draw figures to support your answer. 8–9 See students' drawings.

8. Two right triangles are similar. S
9. Two congruent quadrilaterals are similar. A
10. Quadrilateral *RSTV* is similar to quadrilateral *LMNO*. The sides of *RSTV* are 6, 10, 12, and 14 inches long. The shortest side of *LMNO* is 9 inches long.
    a. What is the scale factor of *RSTV* to *LMNO*? $\frac{2}{3}$
    b. Find the two longest sides of *LMNO*. 21 in. and 18 in.
    c. Find the perimeter of *LMNO*. 63 in.
    d. What is the ratio of the perimeters of *RSTV* and *LMNO*? $\frac{2}{3}$

11. Yes; the corresponding angles are congruent and the corresponding sides are proportional with a scale factor of $\frac{1}{3}$.

INTEGRATION
Algebra

11. $\triangle ABC$ has vertices $A(0, 0)$, $B(3, 0)$ and $C(0, 4)$. If the coordinates of each vertex are multiplied by 3, will the new triangle be similar to $\triangle ABC$? Why or why not?

12. **Photography** A picture is enlarged by a scale factor of $\frac{5}{4}$ and then enlarged again by the same factor.
    a. If the original picture was 2.5 inches by 4 inches, how large was it after both enlargements? 3.9 in. by 6.3 in.
    b. Write an equation describing the enlargement process. $E = \frac{5}{4}\left(\frac{5}{4}x\right)$
    c. By what scale factor was the original picture enlarged? $\frac{25}{16}$

EXERCISES

Practice

13. No; the side on the second triangle measuring 18.4 should be 21.6.

14. Yes; the corresponding angles are congruent and $\frac{2.8}{8.4} = \frac{3.1}{9.3} = \frac{4}{12}$ or $\frac{1}{3}$.

Determine whether each pair of figures is similar. Justify your answer.

13. 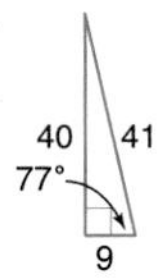 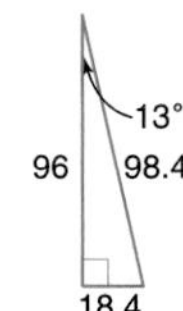

14. 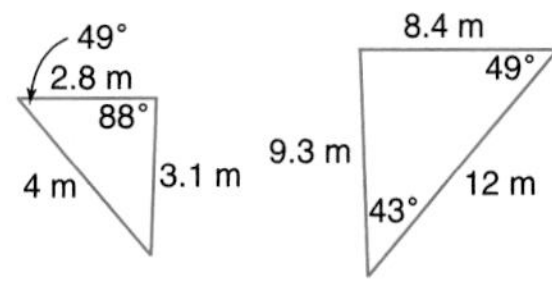

Each pair of polygons is similar. Find the values of x and y.

15. 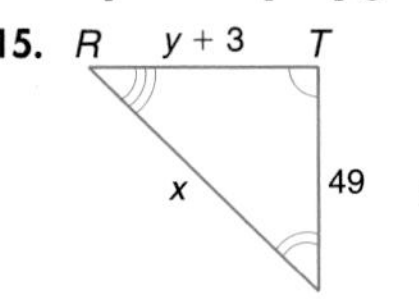 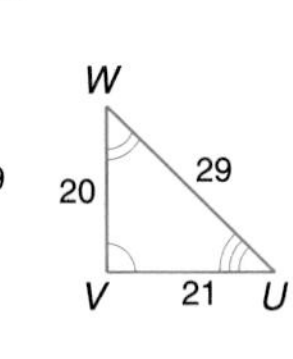

71.05, 48.45

16. 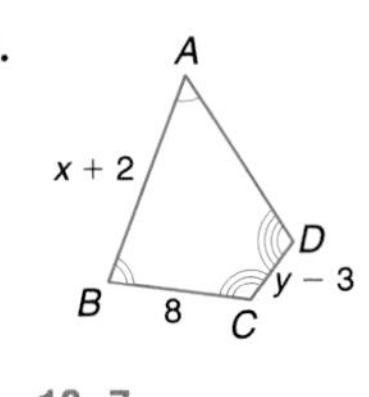 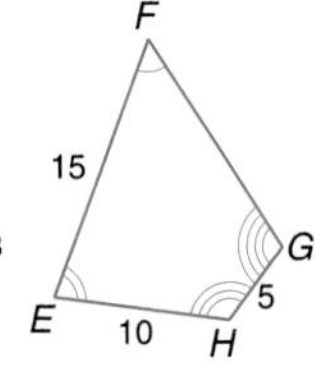

10, 7

17. 18. 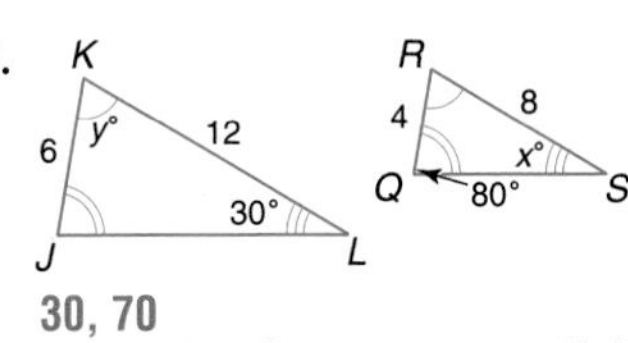

91, 30 — 30, 70

For each statement, write A if the statement is *always* true, S if the statement is *sometimes* true, and N if the statement is *never* true. Draw figures to support your answer.

19. Two rectangles are similar. S
20. Two squares are similar. A
21. A triangle is similar to a quadrilateral. N
22. Two isosceles triangles are similar. S
23. Two rhombi are similar. S
24. Two obtuse triangles are similar. S
25. Two equilateral triangles are similar. A

Triangle *RST* is similar to triangle *EGF*.

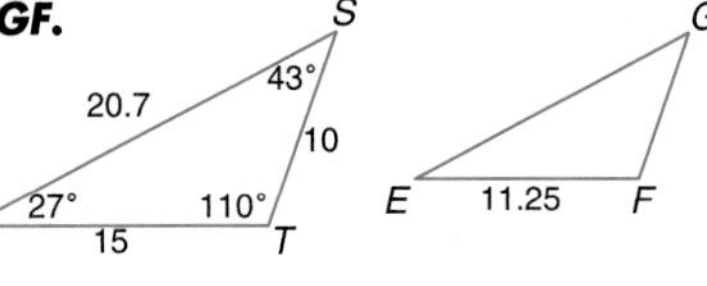

26. Find the length of the shortest side of $\triangle EFG$. 7.5

27. What is the ratio of RS to EG? $\frac{4}{3}$

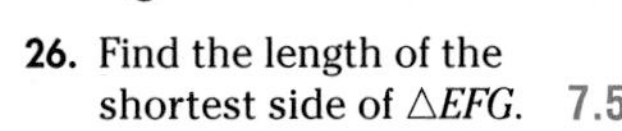

Given trapezoid *ABCD* ~ trapezoid *AEFG*, $m\angle AGF = 108$, $GF = 14$, $AD = 12$, $DG = 4.5$, $EF = 8$, and $AB = 26$, find each of the following.

28. scale factor of *ABCD* to *AEFG* $\frac{8}{5}$

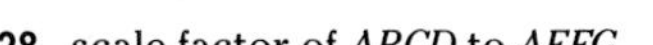

29. a. AG 7.5 b. DC 22.4
c. $m\angle ADC$ 108 d. BC 12.8

30. a. perimeter of *ABCD* 73.2
b. perimeter of *AEFG* 45.75
c. ratio of the perimeters of *ABCD* and *AEFG*

30c. $\frac{8}{5}$

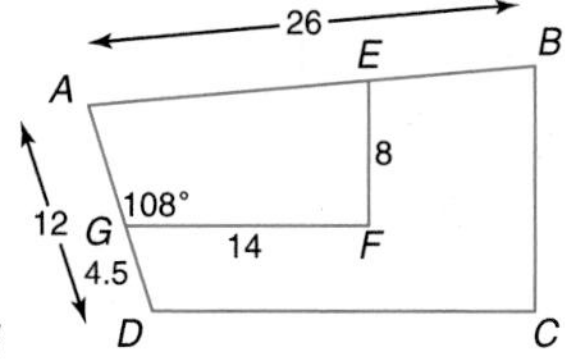

Assignment Guide

Core (with proof): 13–41 odd, 43–52
Core (informal): 13–41 odd, 43–52
Enriched: 14–38 even, 39–52

For **Extra Practice**, see p. 776.

The red A, B, and C flags, printed only in the Teacher's Wraparound Edition, indicate the level of difficulty of the exercises.

Teaching Tip For Exercises 13–18, you may suggest that students redraw figures so that the corresponding parts are oriented in the same direction.

Study Guide Masters, p. 38

7-2 NAME ________ DATE ________
Study Guide
Student Edition Pages 346–353

Exploring Similar Polygons

Two polygons are **similar** if and only if their corresponding angles are congruent and the measures of their corresponding sides are proportional.

The symbol ~ means *is similar to.*

The ratio of the lengths of two corresponding sides of two similar polygons is called the **scale factor**.

Example: Find x if $\triangle RST \sim \triangle XYZ$.

The corresponding sides are proportional, so we can write a proportion to find the value of x.

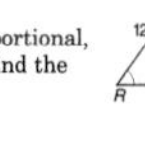 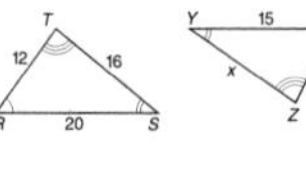

$\frac{16}{x} = \frac{20}{15}$
$20x = 240$
$x = 12$

If quadrilateral ABCD is similar to quadrilateral EFGH, find each of the following.

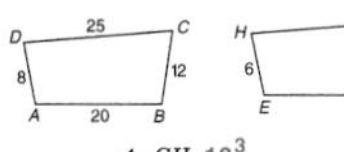

1. scale factor of *ABCD* to *EFGH* $\frac{4}{3}$
2. *EF* 15
3. *FG* 9
4. *GH* $18\frac{3}{4}$
5. perimeter of *ABCD* 65
6. perimeter of *EFGH* $48\frac{3}{4}$
7. ratio of perimeter of *ABCD* to perimeter of *EFGH* $\frac{4}{3}$

Each pair of polygons is similar. Find the values of x and y.

8. 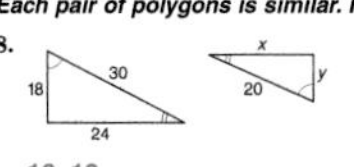
16, 12

9. 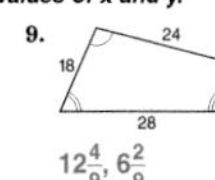 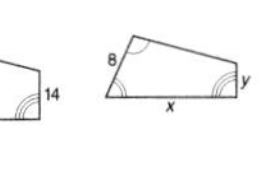
$12\frac{4}{9}$, $6\frac{2}{9}$

Additional Answers

33. Students should draw a rectangle $5\frac{1}{4}$ in. by $3\frac{1}{8}$ in.
34. Students should draw a rectangle 91 mm by 46 mm.
35. Students should draw a rectangle $4\frac{1}{2}$ in. by $9\frac{3}{4}$ in.

36g. Yes; all of the sides are changed by a scale factor of 2, so they remain in the same proportion and none of the angles are changed.

39. In triangles *ABC* and *DEF*, $\overline{AC}$ and $\overline{DF}$ both have slope $\frac{3}{2}$ and $\overline{BC}$ and $\overline{EF}$ both have slope $-\frac{3}{2}$. Thus, $\overline{AC} \parallel \overline{DF}$ and $\overline{BC} \parallel \overline{EF}$. The horizontal axis is a transversal for both pairs of parallel lines. Thus, corresponding angles *A* and *D* are ≅. In the same way, corresponding angles *B* and *E* are ≅. $\angle C \cong \angle F$ because if 2 ∠s in a △ are ≅ to two ∠s in another △, the third ∠s are also ≅. Using the distance formula, $\frac{AC}{DF} = \frac{AB}{DE} = \frac{BC}{EF} = \frac{3}{2}$. Corresponding sides have the same ratio, and the angles are congruent, so the figures are similar.

Practice Masters, p. 38

7-2 NAME ______ DATE ______

Practice — Student Edition Pages 346–353

Exploring Similar Polygons

In the figure at the right, △ABC is similar to △DEF.

1. Write three equal ratios to show corresponding sides are proportional. $\frac{AB}{DE} = \frac{AC}{DF} = \frac{BC}{EF}$
2. Find the value of *x*. 6
3. Find the value of *y*. 12
4. Find the ratio $\frac{m\angle A}{m\angle D}$. $\frac{1}{1}$

In the figure at the right, quadrilateral ABCD is similar to quadrilateral EFGH.

5. Write four equal ratios to show corresponding sides are proportional. $\frac{AB}{EF} = \frac{BC}{FG} = \frac{CD}{GH} = \frac{DA}{HE}$
6. Find *AB*. $2\frac{2}{3}$
7. Find *HG*. $3\frac{3}{4}$
8. Find *FG*. 3
9. The sum of the measures of ∠*A* and ∠*C* equals the sum of the measures of which two angles of quadrilateral *EFGH*? ∠*E* and ∠*G*

32a. Yes; the new triangle is congruent to the original, but shifted to the right two units and up two units.

32b. Yes; it is reflected over the line $y = 4$ and each side is three times larger.

Algebra

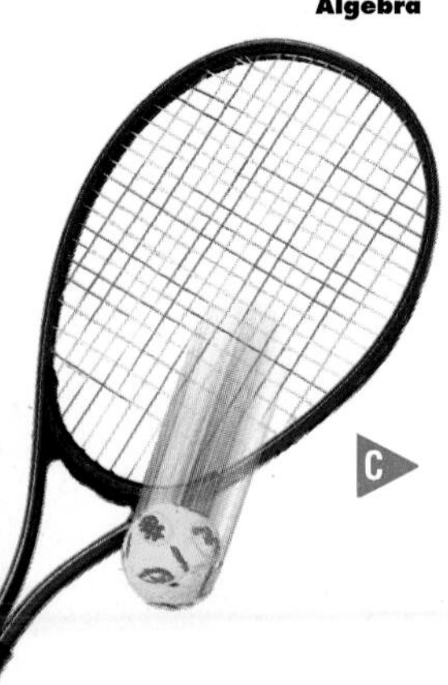

31. Determine which of the following right triangles are similar. Justify your answer. $\triangle ABC \sim \triangle IHG \sim \triangle JLK$ and $\triangle NMO \sim \triangle RPS$

32. A triangle has vertices $E(-2, 4)$, $F(6, 4)$, and $H(3, 9)$.
    a. If the coordinates of each vertex are increased by 2, describe the new figure. Is it similar to $\triangle EFH$?
    b. If the coordinates of each vertex are multiplied by -3, describe the new figure. Is it similar to $\triangle EFH$?

33–35. See margin.

Make a scale drawing of each using the given scale.

33. A basketball court is 84 feet by 50 feet. scale: $\frac{1}{4}$ in. = 4 ft
34. A soccer field is 91 meters by 46 meters. scale: 1 mm = 1 m
35. A tennis court is 36 feet by 78 feet. scale: $\frac{1}{8}$ in. = 1 ft
36. The figure at the right is a block building made out of cubes (each block has the same length, width, and height). The edge of each cube is 1 unit long.

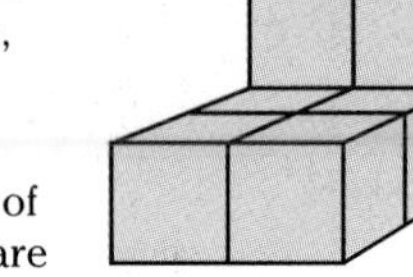

    a. Find the total area of the outside surfaces of the building, including the bottom, in square units. 28 square units
    b. Find the perimeter of the base of the building in units. 10 units
    c. Find the volume of the building in cubes. 8 cubic units
    d. The builder decides to increase the length, width, and height by 2 times the original measures. What is the ratio of the surface area of the new building to the surface area of the smaller building? 4:1
    e. What is the ratio of the perimeter of the base of the new building to the base of the smaller building? 2:1
    f. What is the ratio of the volume of the new building to the volume of the smaller building? 8:1
    g. Is the new building similar to the first one? Explain why or why not. See margin.

37. $L(16, 8)$ and $O(8, 8)$ or $L(16, -8)$ and $O(8, -8)$

37–38. See Solutions Manual for drawings.

Plot the given points on graph paper. Draw *ABCD* and $\overline{MN}$. Find the coordinates for vertices *L* and *O* such that *ABCD* is similar to *NLOM*.

37. $A(2, 0)$, $B(4, 4)$, $C(0, 4)$, $D(-2, 0)$; $M(4, 0)$, $N(12, 0)$
38. $A(0, 0)$, $B(5, 6)$, $C(0, 8)$, $D(-4, 4)$; $M(-2, 14)$, $N(10, 2)$
    $L(25, 20)$ and $O(10, 26)$ or $L(-8, -7)$ and $O(-14, 2)$

Critical Thinking

39. Use what you know about slope and distance to show that the two triangles are similar. Triangle *ABC* has vertices $A(0, 0)$, $B(12, 0)$, and $C(6, 9)$. Triangle *DEF* has vertices $D(18, 0)$, $E(26, 0)$, and $F(22, 6)$. See margin.

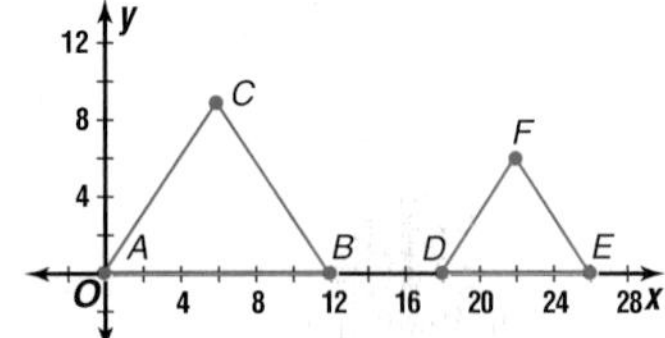

40a. four times as many straws or 64 straws because the sides are in ratio of 4:1

40. Two rectangular solids are similar with a ratio of 4:1 between the corresponding sides.

a. Suppose you use 16 straws to build the smaller rectangular solid. How many straws would you need to build the larger solid? Explain.

b. If you were to paint each solid, would the larger solid take four times as much paint? Explain. **No; it would take 16 times as much paint.**

c. Would the larger solid hold exactly four times as much water as the smaller solid? Explain. **No; it would hold 64 times as much.**

Applications and Problem Solving

41. Construction A floor plan is given for the first floor of a new house. One inch represents 18 feet. Use the information in the plan to find the dimensions.

a. living room

b. deck **22.5 ft by 6.75 ft**

41a. 13.5 ft by 11.25 ft

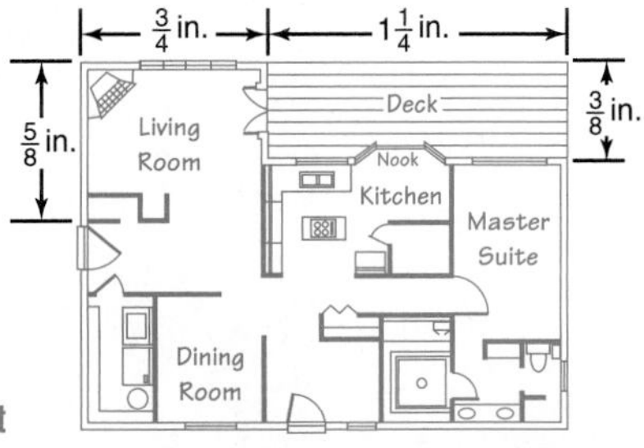

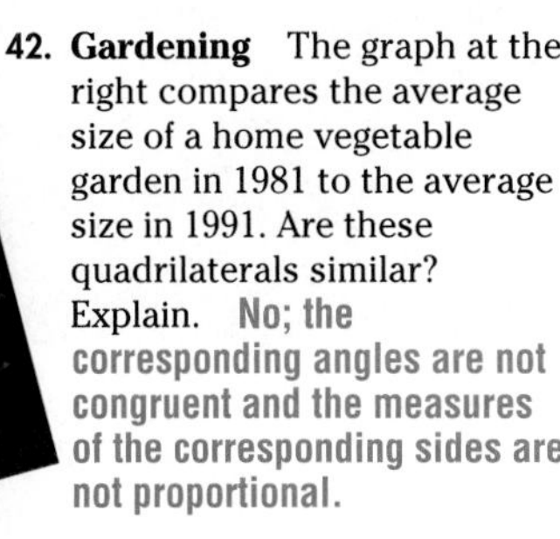

42. Gardening The graph at the right compares the average size of a home vegetable garden in 1981 to the average size in 1991. Are these quadrilaterals similar? Explain. **No; the corresponding angles are not congruent and the measures of the corresponding sides are not proportional.**

Source: National Garden Association/Gallup Inc.

Mixed Review

43. Solve $\frac{m+3}{12} = \frac{5}{4}$ by using cross products. (Lesson 7–1) **12**

44. Classify the statement *The diagonals of a parallelogram are congruent* as always, sometimes, or never true. (Lesson 6–3) **sometimes**

45. Is quadrilateral *JKLM* a parallelogram? Justify your answer. (Lesson 6–2)

45. No; one pair of sides parallel and the other pair of sides congruent is not a test for a parallelogram.

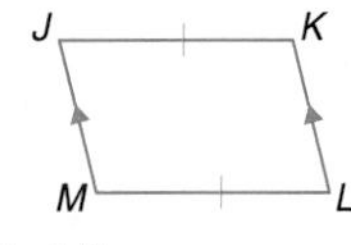

46. In $\triangle PQR$, $PQ > PR > QR$. List the angles in $\triangle PQR$ in order of measure from greatest to least. (Lesson 5–5) **$\angle R$, $\angle Q$, $\angle P$**

47. Work Backward A number is divided by 2, decreased by 17, multiplied by 3, and subtracted from 400. If the result is 187, what was the original number? (Lesson 5–3) **176**

48. True or false: *AAS is a triangle congruence test.* (Lesson 4–5) **true**

49. Suppose two parallel lines are cut by a transversal and $\angle 1$ and $\angle 2$ are alternate exterior angles. If $m\angle 2 = 84$, find $m\angle 1$. (Lesson 3–2) **84**

50. Write the negation of the statement *Springer spaniels are not dogs.* (Lesson 2–2) **Springer spaniels are dogs.**

Algebra

51. Write $y + 6 = -\frac{5}{8}(x - 1)$ in standard form, $Ax + By = C$, where A, B, and C are integers. **$5x + 8y = -43$**

52. If $f(x) = 5x + 2$, find $f(-1)$. **−3**

4 ASSESS

Closing Activity

Modeling Give students the approximate dimensions of their school building and ask them to determine a scale factor that would allow them to sketch a floor plan on an 8.5" × 11" sheet of paper. Have students build a scale model out of cardboard.

Chapter 7 Quiz A (Lessons 7-1 and 7-2) is available in the *Assessment and Evaluation Masters*, p. 184.

Extension

Connections If quadrilateral *NTSU* is similar to quadrilateral *JXBF* and the scale factor of quadrilateral *NTSU* to quadrilateral *JXBF* is 1, what is the scale factor of quadrilateral *JXBF* to quadrilateral *NTSU*? What can you say about the two quadrilaterals? **1; they are congruent.**

Enrichment Masters, p. 38

NAME ____________ DATE ________

Enrichment — Student Edition Pages 346–353

Scale Drawings

Scale drawings are helpful for planning furniture arrangements. The shapes in the drawing below are similar to the bases of the actual objects, but smaller in size. In the drawing, the scale is $\frac{1}{4}$ inch = 1 foot, and the side of each small square is $\frac{1}{4}$ inch.

Example: What are the dimensions of the bed in inches?

$3\frac{1}{2} \times 12 = 42$

$6\frac{1}{2} \times 12 = 78$

The bed is about 42 inches wide and 78 inches long.

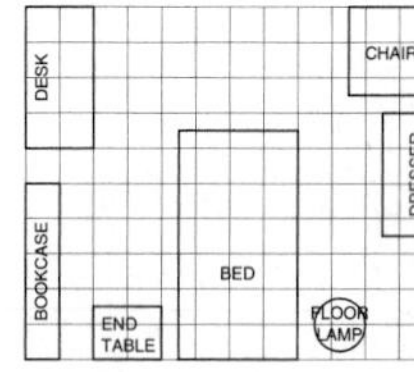

Use the scale drawing above to find the approximate size of each object.

1. size of room in feet 12 ft by 10 ft
2. length and width of bookcase in feet 5 ft by 1 ft
3. length and width of dresser in inches 42 in. by 18 in.
4. length and width of end table in inches 24 in. by 18 in.
5. length and width of chair in inches 30 in. by 30 in.
6. diameter of the floor lamp in inches 18 in.

Show each object on the grid at the right. Use the scale $\frac{1}{4}$ in. = 1 ft.

7. sofa, 72 in. by 36 in.
8. coffee table, 24 in. by 48 in.
9. chair, 24 in. by 30 in.
10. cabinet, 36 in. by 18 in.
11. television, 24 in. by 18 in.
12. On separate paper, make a scale drawing of a bedroom or living room of your home. Show the furniture.

7–12. See students' work.

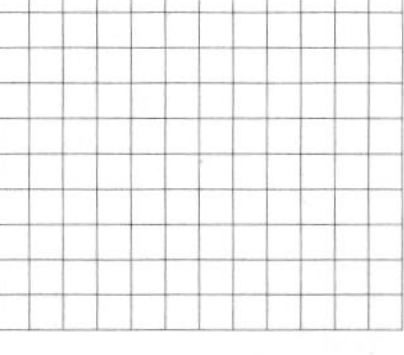

7-3 LESSON NOTES

NCTM Standards: 1–5, 7

Instructional Resources

- Study Guide Master 7-3
- Practice Master 7-3
- Enrichment Master 7-3
- Assessment and Evaluation Masters, pp. 183, 184
- Modeling Mathematics Masters, pp. 39–42, 85

Transparency 7-3A contains the 5-Minute Check for this lesson; **Transparency 7-3B** contains a teaching aid for this lesson.

Recommended Pacing	
Standard Pacing	Days 3 & 4 of 13
Honors Pacing	Days 3 & 4 of 12
Block Scheduling*	Day 3 of 7

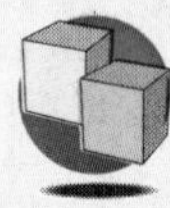

*For more information on pacing and possible lesson plans, refer to the *Block Scheduling Booklet.*

1 FOCUS

5-Minute Check
(over Lesson 7-2)

Refer to the figure below. Pentagon *ABCDH* ~ pentagon *DEFGH*.

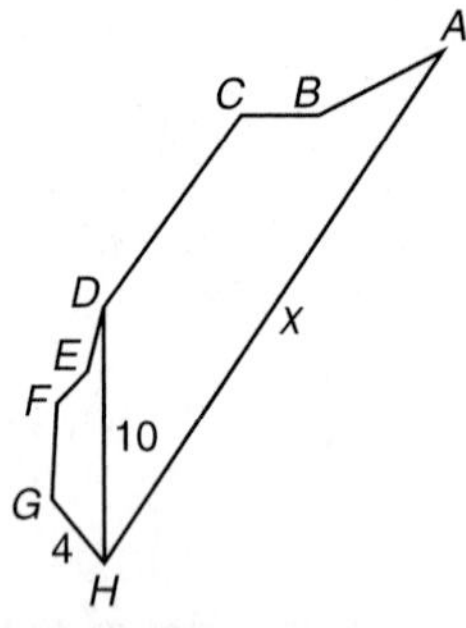

1. Find the scale factor of pentagon *ABCDH* to pentagon *DEFGH*. $\frac{4}{10}$ or $\frac{2}{5}$
2. Find the value of *x*. 25
3. If $m\angle EDH = 12$, $m\angle DCB = 96$, and $BC = 4$, find $m\angle GFE$, FE, and $m\angle BAH$. 96, 1.6, 12

7-3 Identifying Similar Triangles

What YOU'LL LEARN

- To identify similar triangles, and
- to use similar triangles to solve problems.

Why IT'S IMPORTANT

Identifying similar triangles is the first step in solving problems that apply similar triangles.

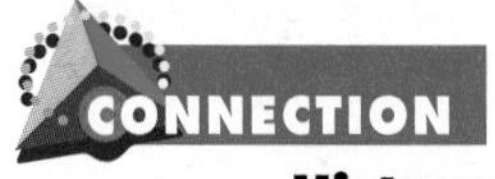

History

The Greek mathematician Thales was the first to measure the height of a pyramid by using geometry. He showed that the ratio of a pyramid to a staff was equal to the ratio of one shadow to the other.

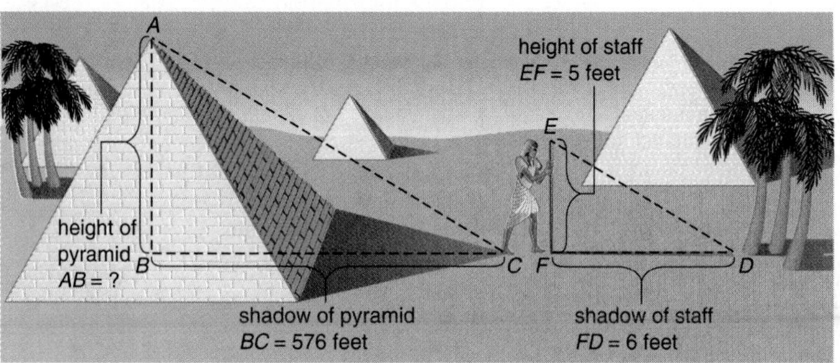

Using the information above, can you find the height of the pyramid?
You will be asked to solve this problem in Exercise 12.

Similar triangles can help you solve problems like the one above. How do you know whether two triangles are similar? In Chapter 4, you learned several tests to determine whether two triangles are congruent. There are also tests to determine whether two triangles are similar.

Similar Triangles

Materials: ruler  protractor

One way to determine whether triangles are similar is to see if the measures of corresponding sides are proportional.

- Draw $\triangle DEF$ with $m\angle D = 35$, $m\angle F = 80$, and $DF = 4$ centimeters.
- Draw $\triangle RST$ with $m\angle T = 35$, $m\angle S = 80$, and $ST = 7$ centimeters.
- Measure $\overline{EF}$, $\overline{ED}$, $\overline{RS}$, and $\overline{RT}$.
- Calculate the ratios $\frac{FD}{ST}$, $\frac{EF}{RS}$, and $\frac{ED}{RT}$.
- Record your findings in a chart like the one below.

	EF	ED	RS	RT	$\frac{FD}{ST}$	$\frac{EF}{RS}$	$\frac{ED}{RT}$
$\triangle DEF$, $\triangle RST$							

Your Turn

a. What can you conclude about all the ratios? They equal about 0.57.

b. Draw two more triangles with the same angle measures and different side measures. Complete the chart for these two triangles. b–d. See students' work.

c. Repeat the process with two more triangles.

d. Do all of the triangles appear to be similar?

Compare this modeling activity with the modeling activity in Lesson 7-2. In the first activity students use proportion to *verify* similarity. In the second activity students use proportion to *create* similarity.

The Modeling Mathematics activity leads to the following postulate.

Postulate 7-1 AA Similarity (Angle-Angle)

If two angles of one triangle are congruent to two angles of another triangle, then the triangles are similar.

You can use AA Similarity to find missing measurements in triangles.

Example 1

INTEGRATION Algebra

Given $\overline{AB} \parallel \overline{CD}$, $AB = 4$, $AE = 3x + 4$, $CD = 8$, and $ED = x + 12$, find AE and DE.

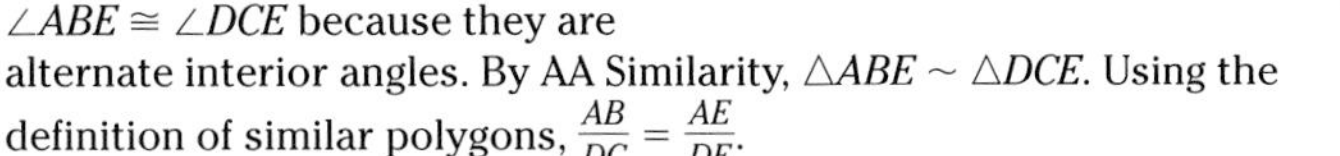

Since $\overline{AB} \parallel \overline{CD}$, $\angle BAE \cong \angle CDE$ and $\angle ABE \cong \angle DCE$ because they are alternate interior angles. By AA Similarity, $\triangle ABE \sim \triangle DCE$. Using the definition of similar polygons, $\frac{AB}{DC} = \frac{AE}{DE}$.

$$\frac{AB}{DC} = \frac{AE}{DE}$$

$$\frac{4}{8} = \frac{3x+4}{x+12} \quad \text{Substitution Property (=)}$$

$$4(x+12) = 8(3x+4) \quad \text{Cross products}$$

$$4x + 48 = 24x + 32 \quad \text{Distributive Property (=)}$$

$$-20x = -16 \quad \text{Subtract 24x and 48 from each side.}$$

$$x = \frac{16}{20} \text{ or } \frac{4}{5} \quad \text{Divide each side by } -20.$$

Now find AE and ED.

$AE = 3x + 4 = 3\left(\frac{4}{5}\right) + 4$ or $6\frac{2}{5}$

$ED = x + 12 = \frac{4}{5} + 12$ or $12\frac{4}{5}$

FYI

Thales, sometimes called the first geometrician, is credited with predicting the eclipse of 585 B.C.

It is also possible to prove triangles similar by testing the measures of corresponding sides for proportionality.

Theorem 7-1 SSS Similarity (Side-Side-Side)

If the measures of the corresponding sides of two triangles are proportional, then the triangles are similar.

Proof of Theorem 7-1

Given: $\frac{PQ}{AB} = \frac{QR}{BC} = \frac{RP}{CA}$

Prove: $\triangle BAC \sim \triangle QPR$

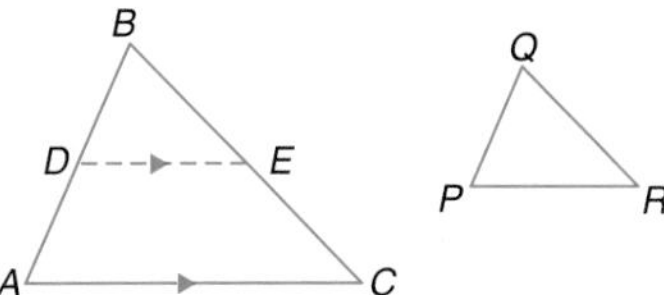

Locate D on $\overline{AB}$ so that $\overline{DB} \cong \overline{PQ}$ and draw $\overline{DE}$ so that $\overline{DE} \parallel \overline{AC}$.

Paragraph Proof:

Since $\overline{DB} \cong \overline{PQ}$, the given proportion will become $\frac{DB}{AB} = \frac{QR}{BC} = \frac{RP}{CA}$. Since $\overline{DE} \parallel \overline{AC}$, $\angle BDE \cong \angle A$ and $\angle BED \cong \angle C$. By AA Similarity, $\triangle BDE \sim \triangle BAC$. By the definition of similar polygons, $\frac{DB}{AB} = \frac{BE}{BC} = \frac{ED}{CA}$. Using the two proportions and substitution, $\frac{QR}{BC} = \frac{BE}{BC}$ and $\frac{RP}{CA} = \frac{ED}{CA}$. This means that $QR = BE$ and $RP = ED$ or $\overline{QR} \cong \overline{BE}$ and $\overline{RP} \cong \overline{ED}$. With these congruences and $\overline{DB} \cong \overline{PQ}$, $\triangle BDE \cong \triangle QPR$ by SSS Postulate. By CPCTC, $\angle B \cong \angle Q$ and $\angle BDE \cong \angle P$. But $\angle BDE \cong \angle A$, so $\angle A \cong \angle P$. By AA Similarity, $\triangle BAC \sim \triangle QPR$.

Alternative Learning Styles

Visual Have students draw a triangle and a line parallel to one side of the triangle that intersects the other two sides. Have them make conjectures about what is necessary to prove the two triangles similar. Have them draw other triangles to test their conjectures.

FYI

Thales was the first metaphysician, speculating that everything is formed out of water and consists of water.

Motivating the Lesson

Situational Problem Show students examples of nested objects, such as different sized frames. Use 4" × 5", 8" × 10", and 16" × 20" frames. Ask students if they can identify a pattern.

2 TEACH

In-Class Example

For Example 1

In the figure below, $\overline{AB} \parallel \overline{DE}$, $DA = 2$, $CA = 8$, and $CE = 3$.

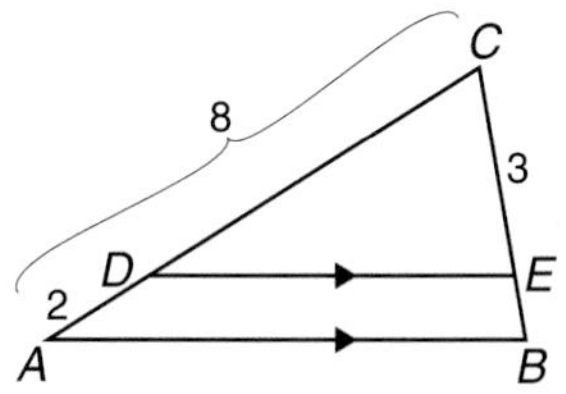

Show that $CB = 4$.

Since $\overline{AB} \parallel \overline{DE}$, $\angle CDE \cong \angle CAB$ and $\angle CED \cong \angle CBA$. Thus, $\triangle CDE \sim \triangle CAB$ by AA Similarity, and $\frac{CD}{CA} = \frac{CE}{CB}$. By the Segment Addition Postulate, $CA = CD + DA$ and $CB = CE + EB$. By substitution, the proportion becomes

$$\frac{CD}{CD + DA} = \frac{CE}{CE + EB}$$

$$\frac{6}{8} = \frac{3}{3 + EB}$$

$$6(3 + EB) = 3 \cdot 8$$

$$18 + 6(EB) = 24$$

$$6(EB) = 6$$

$$EB = 1$$

$$CB = CE + EB$$

$$CB = 3 + 1$$

$$CB = 4$$

Teaching Tip When beginning this lesson, write the definition of similar polygons on the chalkboard or overhead. Ask students if all of those conditions need to be met to prove two triangles similar.

Teaching Tip When stating Theorem 7-1, emphasize that the conditions must be true for all three pairs of sides.

In-Class Examples

For Example 2
$\triangle STU$ has point X in its interior. If $\angle TXS$, $\angle SXU$, and $\angle TXU$ each measure 120°, $SX = 16$, $XU = 4$, and $XT = 8$, determine which triangles in the figure are similar.

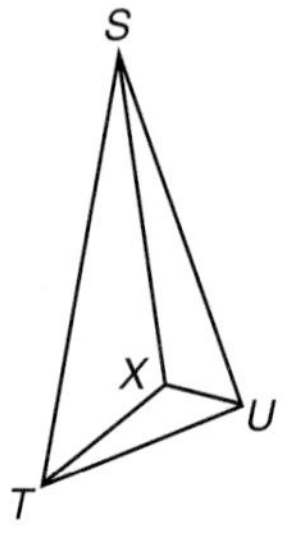

$\triangle SXT \sim \triangle TXU$ by SAS Similarity.

For Example 3
Using the same method as in Example 3, how tall is a building if the scout has to move to 0.4 meters from the mirror to see the top? **150.9 meters**

The giant sequoia is a member of the redwood family. There were once forty species of redwood in the northern hemisphere. There are now only three.

The next theorem describes another test for similarity of triangles. *You will be asked to prove this theorem in Exercise 33.*

Theorem 7–2 SAS Similarity (Side-Angle-Side)	**If the measures of two sides of a triangle are proportional to the measures of two corresponding sides of another triangle and the included angles are congruent, then the triangles are similar.**

You can use SAS Similarity to determine whether two triangles are similar.

Example  **In the figure below, $\overline{FG} \cong \overline{EG}$, $BE = 15$, $CF = 20$, $AE = 9$, $DF = 12$. Determine which triangles in the figure are similar.**

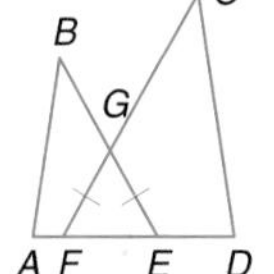

$\overline{FG} \cong \overline{EG}$ implies $\angle GFE \cong \angle GEF$ because if two sides of a triangle are congruent, the angles opposite those sides are congruent. If the corresponding sides that include the angle are proportional, then the triangles are similar.

$\frac{AE}{DF} = \frac{9}{12}$ or $\frac{3}{4}$ and $\frac{BE}{CF} = \frac{15}{20}$ or $\frac{3}{4}$.

By the Transitive Property of Equality, $\frac{AE}{DF} = \frac{BE}{CF}$.

Thus, we have the measures of two sides of one triangle proportional to the measures of two sides of another and the included angles congruent. So by SAS Similarity, $\triangle ABE$ is similar to $\triangle DCF$.

Similar triangles can also be used to solve problems involving distance.

Example 3

The giant sequoia trees in California reach heights of more than 100 meters.

To estimate the height of a tree, a Girl Scout sights the top of the tree in a mirror that is 34.5 meters from the tree. The mirror is on the ground and faces upward. The scout is 0.75 meters from the mirror, and the distance from her eyes to the ground is about 1.75 meters. How tall is the tree?

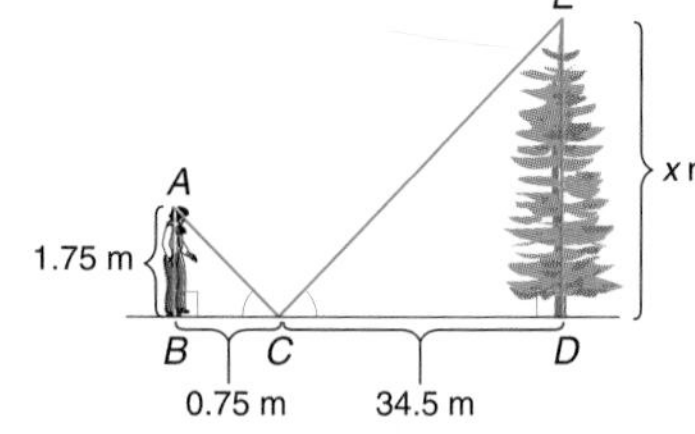

Let x represent the height of the tree. The angle between the ground and the line of sight is congruent to the angle between the ground and the line from the mirror to the tree. If we can show $\triangle ABC$ and $\triangle EDC$ are similar, we can set up a proportion of corresponding sides and solve for x.

Assume $\angle ABC$ and $\angle EDC$ are right angles and, thus, are congruent. Since $\angle ACB \cong \angle ECD$, $\triangle ABC \sim \triangle EDC$ by AA Similarity. Therefore, $\frac{AB}{ED} = \frac{BC}{DC}$.

$\frac{1.75}{x} = \frac{0.75}{34.5}$ *Substitution Property (=)*

$60.375 = 0.75x$ *Cross products*

$80.5 = x$ *Divide each side by 0.75.*

The tree is 80.5 meters high.

Cooperative Learning

Brainstorming Have groups of students brainstorm to discover real-world applications where similar triangles are used. They should determine how they would guarantee that the triangles are similar. For more information on the brainstorming strategy, see *Cooperative Learning in the Mathematics Classroom*, one of the titles in the Glencoe Mathematics Professional Series, page 30.

Like congruence of triangles, similarity of triangles is reflexive, symmetric, and transitive.

reflexive $\triangle ABC \sim \triangle ABC$

symmetric If $\triangle ABC \sim \triangle DEF$, then $\triangle DEF \sim \triangle ABC$.

transitive If $\triangle ABC \sim \triangle DEF$ and $\triangle DEF \sim \triangle GHI$, then $\triangle ABC \sim \triangle GHI$.

Theorem 7-3	**Similarity of triangles is reflexive, symmetric, and transitive.**

You will be asked to prove this theorem in Exercise 35.

CHECK FOR UNDERSTANDING

Communicating Mathematics

Study the lesson. Then complete the following.

1. **Compare** the tests to prove triangles similar with the tests to prove triangles congruent. How are they alike and how are they different? See margin.

2. If you know two triangles are right triangles, how many of the acute angles must be congruent before you know the two triangles are similar? Explain your answer. Only one; you already know the right angles are congruent.

3. Use an example to explain what it means to say that similarity of triangles is reflexive, symmetric, and transitive. See margin.

4. They both are. In both cases, the cross products will produce $rm = ks$.

4. **You Decide** Lacretia and Song were working with the similar triangles shown at the right. Lacretia used the proportion $\frac{r}{s} = \frac{k}{m}$ for her calculations, and Song used $\frac{r}{k} = \frac{s}{m}$ for hers. Who is right, and why?

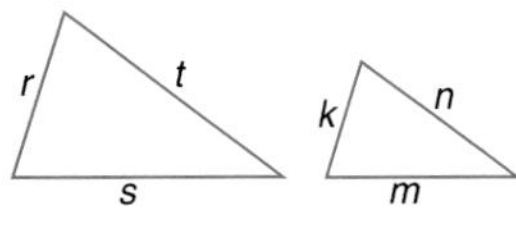

5. Draw $\triangle JKL$ with $m\angle J = 60$, $JK = 3$ centimeters, and $JL = 6$ centimeters. Draw $\triangle QRS$ with $m\angle Q = 60$, $QR = 4.5$ centimeters, and $QS = 9$ centimeters. Measure $\overline{KL}$ and $\overline{RS}$ and calculate the ratios of the corresponding sides. What can you conclude about the ratios? See students' drawings. Both ratios equal $\frac{2}{3}$.

Guided Practice

Determine whether each pair of triangles is similar. If similarity exists, write a mathematical sentence relating the two triangles. Give a reason for your answer.

6. yes; $\triangle ABC \sim \triangle EDF$; AA Similarity

6. 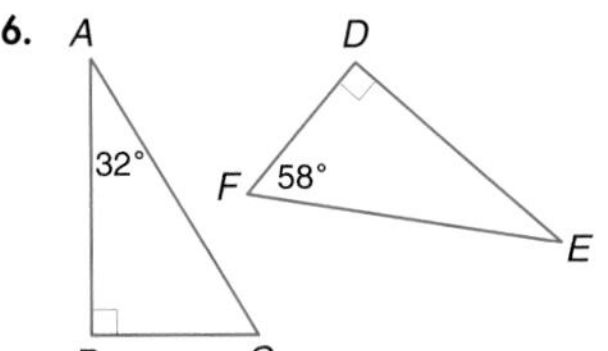

7. G, H, J, K, L, I — 25°, 25° — no; not enough information

8. Identify the following statement as true or false. If false, state why. *If the measures of the sides of one triangle are three times the measures of the sides of a second triangle, the two triangles are similar.* true

3 PRACTICE/APPLY

Check for Understanding

Exercises 1–12 are designed to help you assess your students' understanding through reading, writing, speaking, and modeling. You should work through Exercises 1–5 with your students and then monitor their work on Exercises 6–12.

Error Analysis

Students may have difficulty seeing which triangles are similar in complicated figures. Have them mark all angles that are congruent and then check the sides for proportionality if necessary.

Additional Answers

1. Sample answer: Two triangles are congruent and similar by SSS Similarity and SAS Similarity. To be congruent, the parts must be congruent; to be similar, the sides must be proportional and the angles congruent. Two triangles can also be shown to be congruent by ASA Similarity and AAS Similarity. Two triangles can be shown to be similar by AA Similarity.
3. Sample answer:

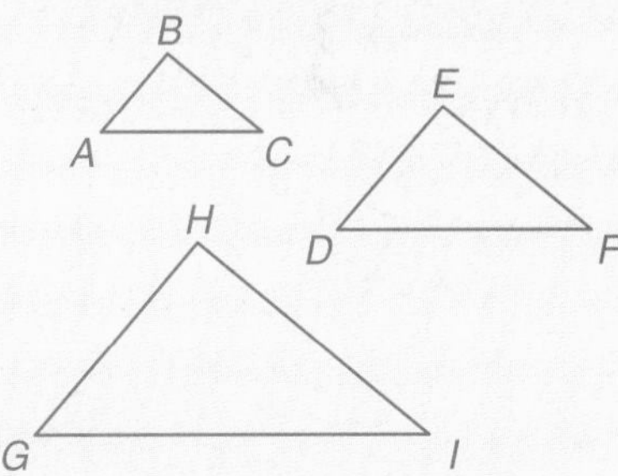

Given $\triangle ABC$, then $\triangle ABC \sim \triangle ABC$. (reflexive)
Given $\triangle ABC \sim \triangle DEF$, then $\triangle DEF \sim \triangle ABC$. (symmetric)
Given $\triangle ABC \sim \triangle DEF$ and $\triangle DEF \sim \triangle GHI$, then $\triangle ABC \sim \triangle GHI$. (transitive)

Reteaching

Using Hands-On Activity Have each student use a straightedge to draw a triangle. Have them measure the sides of the triangle and draw a similar triangle with a scale factor of $\frac{1}{2}$. Ask students how they know the triangles are similar. because the sides are proportional Explain that this is a demonstration of SSS Similarity.

Assignment Guide

Core (with proof): 13–35 odd, 36, 37, 39–47
Core (informal): 13–27 odd, 36, 37, 39–47
Enriched: 14–34 even, 36–47
All: Self Test, 1–5

For **Extra Practice**, see p. 776.

The red A, B, and C flags, printed only in the Teacher's Wraparound Edition, indicate the level of difficulty of the exercises.

Additional Answer

28a.

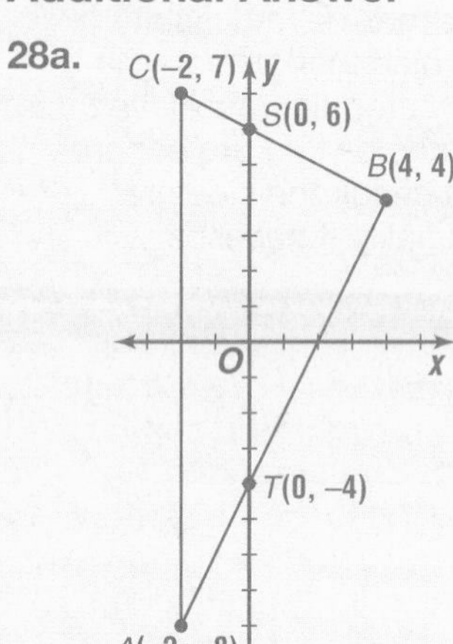

$AB = \sqrt{6^2 + 12^2} = \sqrt{180}$ or $6\sqrt{5}$

$BC = \sqrt{6^2 + 3^2} = \sqrt{45}$ or $3\sqrt{5}$

$CA = |7 - (-8)| = 15$

$ST = |6 - (-4)| = 10$

$TB = \sqrt{8^2 + 4^2} = \sqrt{80}$ or $4\sqrt{5}$

$BS = \sqrt{2^2 + 4^2} = \sqrt{20}$ or $2\sqrt{5}$

$\frac{CA}{ST} = \frac{15}{10}$ or $\frac{3}{2}$, $\frac{AB}{TB} = \frac{6\sqrt{5}}{4\sqrt{5}}$ or $\frac{3}{2}$, $\frac{BC}{BS} = \frac{3\sqrt{5}}{2\sqrt{5}}$ or $\frac{3}{2}$

Since $\frac{CA}{ST} = \frac{AB}{TB} = \frac{BC}{BS}$, $\triangle ABC \sim \triangle TBS$ by SSS Similarity.

9. Identify the similar triangles in the figure at the right. Explain your answer. $\triangle AEC \sim \triangle BDC$; AA Similarity

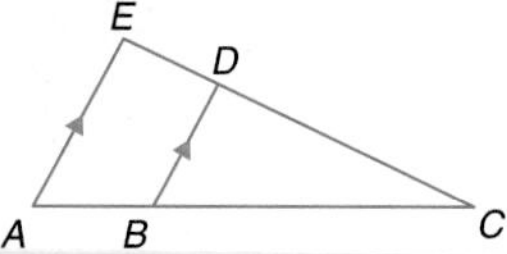

Explain why each pair of triangles is similar and use the given information to find each measure.

10. If 2 ∥ lines are cut by a transversal, corres. ∡s are ≅; $\angle S \cong \angle VTU$; $\angle U \cong \angle U$, so by AA Similarity, $\triangle RSU \sim \triangle VTU$; $x = 4$, $y = 5$.

10.

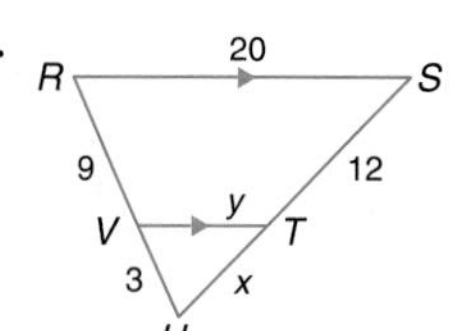

11.

SAS Similarity; $x = 4.5$

12. **History** Refer to the connection at the beginning of the lesson. If $BC = 576$ feet, $FD = 6$ feet and $EF = 5$ feet, use similar triangles to find the height of the pyramid. 480 feet

EXERCISES

Practice

Determine whether each pair of triangles is similar. If similarity exists, write a mathematical sentence relating the two triangles. Give a reason for your answer.

13. yes; $\triangle MNO \sim \triangle PQR$; SSS Similarity

14. yes; $\triangle STU \sim \triangle XVW$; SAS Similarity

15. No; SSA is not valid.

16. No; sides are not proportional.

13.

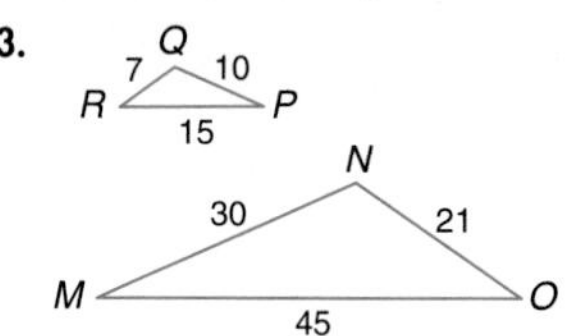

14.

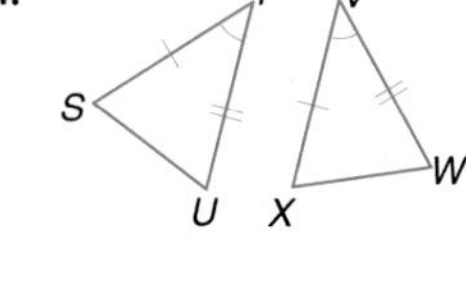

15.

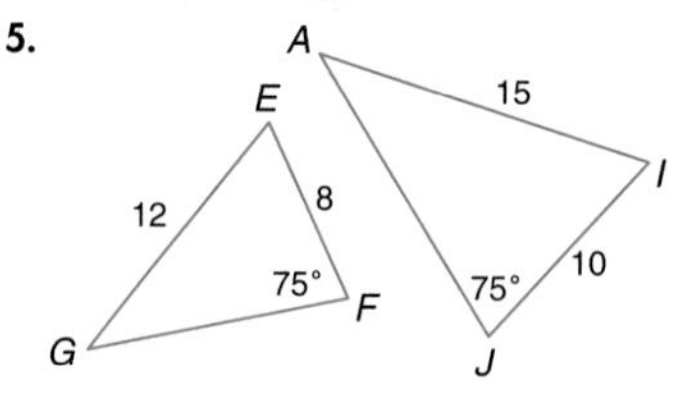

16.

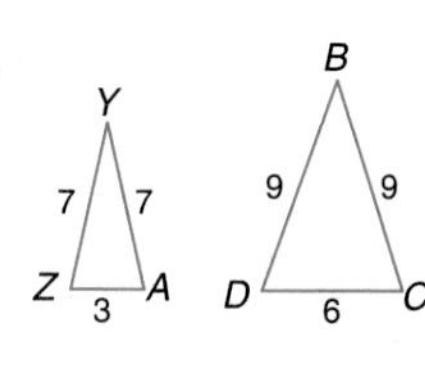

17.

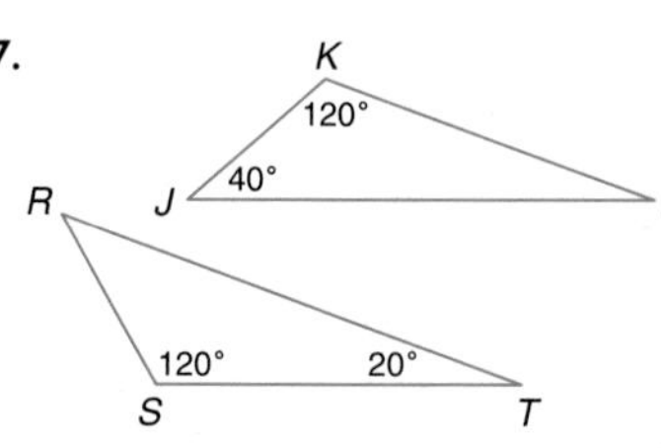

yes; $\triangle RST \sim \triangle JKL$; AA Similarity

18.

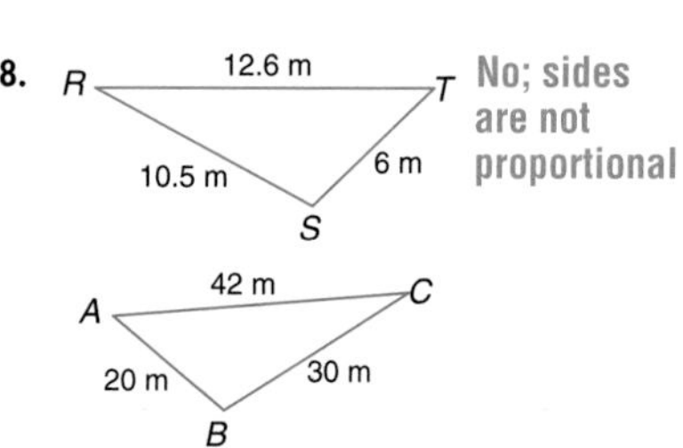

No; sides are not proportional

Identify each statement as *true* or *false*. If false, state why.

19. False; this is not true for equilateral or isosceles triangles.

19. For every pair of similar triangles, there is only one correspondence of vertices that will give you correct angle correspondence and segment proportions.

21. If the measures of the sides of a triangle are different lengths x, y, and z and the measures of the sides of a second triangle are $x + 1$, $y + 1$ and $z + 1$, the two triangles are similar. **False; the proportions are not the same.**

Identify the similar triangles in each figure. Explain your answer.

B

22. $\triangle EAB \sim \triangle EFC \sim \triangle AFD$ (AA Similarity)

23. $\triangle QRS \sim \triangle QTR$ (AA Similarity); $\triangle QRS \sim \triangle RTS$ (AA Similarity); $\triangle QTR \sim \triangle RTS$ (Transitive Prop. of $\sim \triangle$s)

24. $EC = 5$, $GC = 4$, $EF = 6$

25. $AE = 5$, $AC = 15$, $CF = 13\frac{1}{3}$, $CB = 20$, $CD = 12$, $EF = 16\frac{2}{3}$

22.

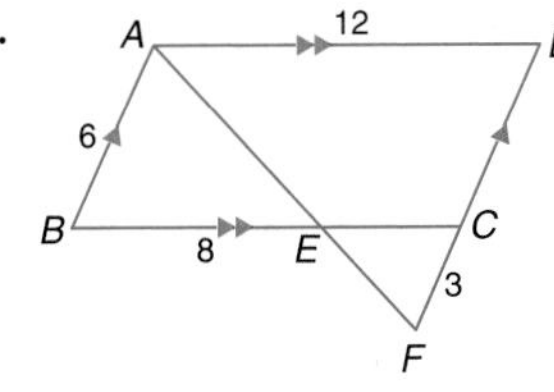

23.

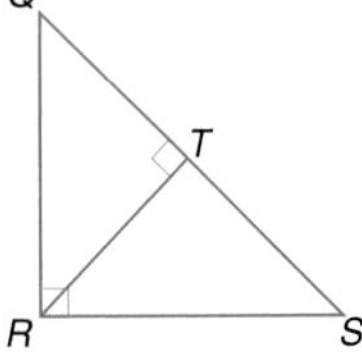

Use the given information to find each measure.

24. If $\overline{ED} \parallel \overline{AB}$, $AB = 10$, $BC = 6$, $AC = 8$, $CD = 5$, and $GE = 3$, find EC, GC, and EF.

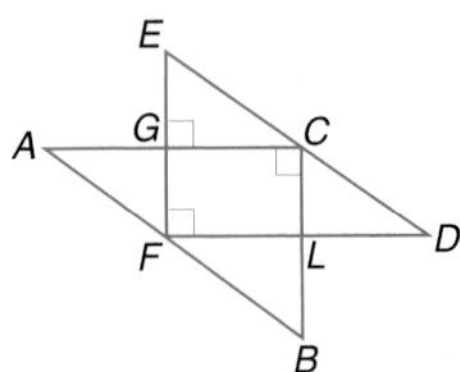

25. If $\overline{EF} \parallel \overline{AB}$, $AD = 9$, $DB = 16$, $EC = 2(AE)$, find AE, AC, CF, CB, CD, and EF.

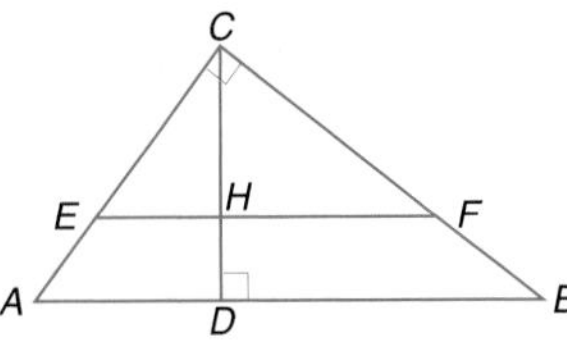

C

26. $m\angle CFD = 50$, $m\angle GKF = 70$, $m\angle GFK = 50$, $m\angle G = 60$, $m\angle GKJ = 110$

27. $m\angle CDE = 43$, $m\angle A = 43$, $m\angle ABD = 47$, $m\angle DBE = 43$, $m\angle BDE = 47$

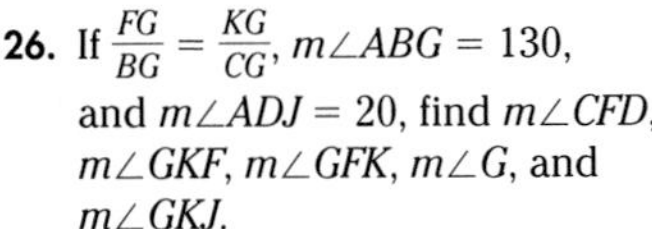

26. If $\frac{FG}{BG} = \frac{KG}{CG}$, $m\angle ABG = 130$, and $m\angle ADJ = 20$, find $m\angle CFD$, $m\angle GKF$, $m\angle GFK$, $m\angle G$, and $m\angle GKJ$.

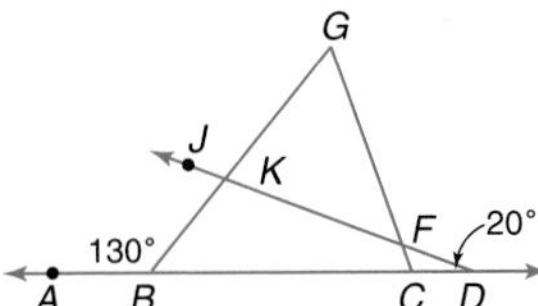

27. If $\angle ABC$ is a right angle, $\overline{BD} \perp \overline{AC}$, $\overline{DE} \perp \overline{BC}$, and $m\angle ACB = 47$, find $m\angle CDE$, $m\angle A$, $m\angle ABD$, and $m\angle BDE$.

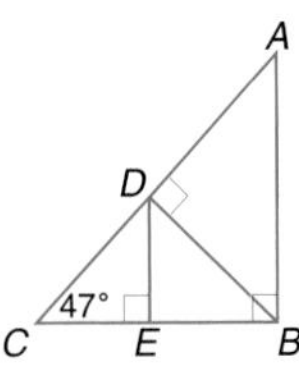

Algebra

28. Graph $\triangle ABC$ and $\triangle TBS$ with vertices $A(-2, -8)$, $B(4, 4)$, $C(-2, 7)$, $T(0, -4)$, and $S(0, 6)$.

a. Show that $\triangle ABC \sim \triangle TBS$. **See margin.**

b. Find the ratio of the perimeters of the two triangles. $\frac{3}{2}$

Write a two-column proof. **29–30. See margin.**

29. **Given:** $\overline{LP} \parallel \overline{MN}$

Prove: $\frac{LJ}{JN} = \frac{PJ}{JM}$

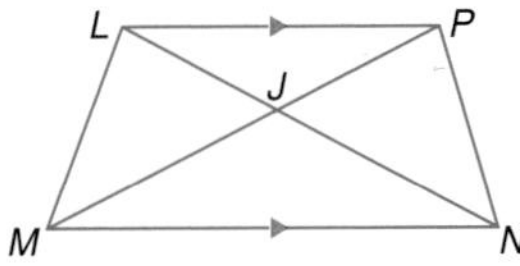

30. **Given:** $\overline{EB} \perp \overline{AC}$, $\overline{BH} \perp \overline{AE}$, and $\overline{CJ} \perp \overline{AE}$

Prove: $\triangle ABH \sim \triangle DCB$

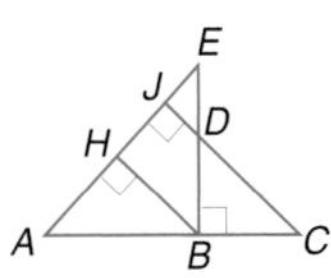

Additional Answers

29. Proof:

Statements (Reasons)

1. $\overline{LP} \parallel \overline{MN}$ (Given)
2. $\angle PLN \cong \angle LNM$, $\angle LPM \cong \angle PMN$ (Alt. Int. $\angle$ Th.)
3. $\triangle LPJ \sim \triangle NMJ$ (AA Similarity)
4. $\frac{LJ}{JN} = \frac{PJ}{JM}$ (Corr. sides of $\sim$ $\triangle$s are proportional.)

30. Proof:

Statements (Reasons)

1. $\overline{EB} \perp \overline{AC}$, $\overline{BH} \perp \overline{AE}$, $\overline{CJ} \perp \overline{AE}$ (Given)
2. $\angle AHB$, $\angle AJC$, $\angle EBC$ are rt. $\angle$s. ($\perp$ lines form 4 rt. $\angle$s.)
3. $\angle AHB \cong \angle AJC$, $\angle AJC \cong \angle EBC$ (All rt. $\angle$s are $\cong$.)
4. $\angle A \cong \angle A$, $\angle C \cong \angle C$ (Congruence of $\angle$s is reflexive.)
5. $\triangle AHB \sim \triangle AJC$, $\triangle CBD \sim \triangle CJA$ (AA Similarity)
6. $\triangle ABH \sim \triangle DCB$ (Transitive Prop. (=))

Study Guide Masters, p. 39

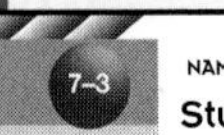

NAME______________ DATE__________

Study Guide Student Edition Pages 354–361

Identifying Similar Triangles

There are three ways to determine whether two triangles are similar.

- Show that two angles of one triangle are congruent to two angles of the other triangle. (AA Similarity)
- Show that the measures of the corresponding sides of the triangles are proportional. (SSS Similarity)
- Show that the measure of two sides of a triangle are proportional to the measures of the corresponding sides of the other triangle and that the included angles are congruent. (SAS Similarity)

Example: Determine whether the triangles are similar. Explain your answer.

Since $\frac{15}{12} = \frac{25}{20} = \frac{20}{16}$, the triangles are similar by SSS Similarity.

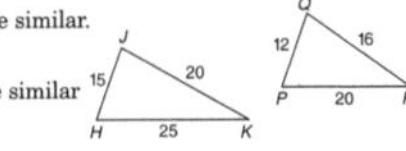

Determine whether each pair of triangles is similar. Give a reason for your answer.

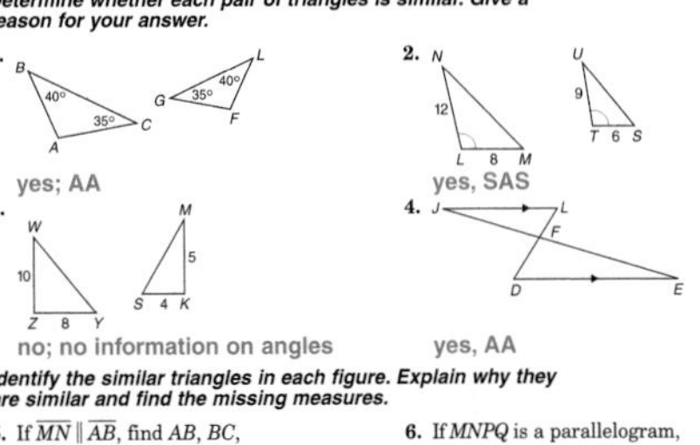

1. yes; AA
2. yes, SAS
3. no; no information on angles
4. yes, AA

Identify the similar triangles in each figure. Explain why they are similar and find the missing measures.

5. If $\overline{MN} \parallel \overline{AB}$, find AB, BC, and BN.

$\triangle CMN \sim \triangle CAB$; AA; $18\frac{2}{3}$, $9\frac{1}{3}$, $5\frac{1}{3}$

6. If $MNPQ$ is a parallelogram, find RN, RP, and SP.

$\triangle QSM \sim \triangle PSR \sim \triangle NMR$; AA; $33\frac{1}{3}$, $13\frac{1}{3}$, 20

Additional Answers

31. Since $\triangle JFM \sim \triangle EFB$ and $\triangle LFM \sim \triangle GFB$, then by the definition of similar triangles, $\frac{JF}{EF} = \frac{MF}{BF}$ and $\frac{MF}{BF} = \frac{LF}{GF}$. By the Trans. Prop. (=), $\frac{JF}{EF} = \frac{LF}{GF}$. Then, by SAS Similarity, $\triangle JFL \sim \triangle EFG$.

32. Since $\overline{JM} \parallel \overline{EB}$ and $\overline{LM} \parallel \overline{GB}$, then $\angle MJF \cong \angle BEF$ and $\angle FML \cong \angle FBG$ because if 2 lines are ∥, corr. ∡ are ≅. $\angle EFB \cong \angle EFB$ and $\angle BFG \cong \angle BFG$ since congruence of segments is reflexive. Then $\triangle EFB \sim \triangle JFM$ and $\triangle FBG \sim \triangle FML$ by AA Similarity. Then $\frac{JF}{EF} = \frac{MF}{BF}$ and $\frac{MF}{BF} = \frac{LF}{GF}$ by the def. of ~ △s. $\frac{JF}{EF} = \frac{LF}{GF}$ by the Trans. Prop. (=) and $\triangle EFG \cong \triangle EFG$ since cong. of ∡ is reflexive. Thus, $\triangle JFL \sim \triangle EFG$ by SAS Similarity and $\angle FJL \cong \angle FEG$ by def. of ~ △s. $\overline{JL} \parallel \overline{EG}$ because if ≠ so the corr. ∡ are ≅, then the lines are ∥.

Practice Masters, p. 39

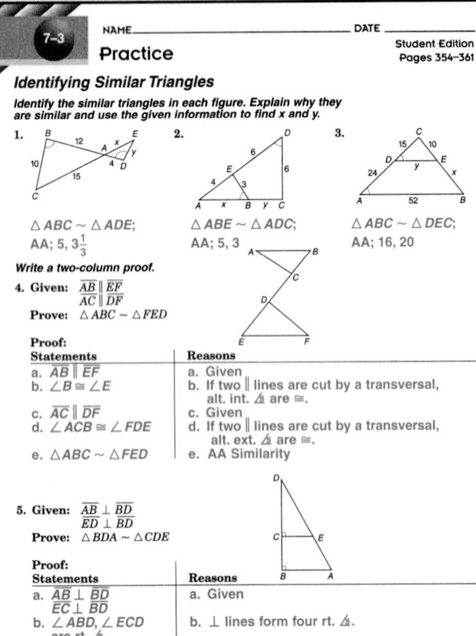

7-3 NAME________ DATE________

Practice — Student Edition Pages 354–361

Identifying Similar Triangles

Identify the similar triangles in each figure. Explain why they are similar and use the given information to find x and y.

1. △ABC ~ △ADE; AA; 5, $3\frac{1}{3}$

2. △ABE ~ △ADC; AA; 5, 3

3. △ABC ~ △DEC; AA; 16, 20

Write a two-column proof.

4. Given: $\overline{AB} \parallel \overline{EF}$, $\overline{AC} \parallel \overline{DF}$
Prove: △ABC ~ △FED

Proof:

Statements	Reasons
a. $\overline{AB} \parallel \overline{EF}$	a. Given
b. ∠B ≅ ∠E	b. If two ∥ lines are cut by a transversal, alt. int. ∡ are ≅.
c. $\overline{AC} \parallel \overline{DF}$	c. Given
d. ∠ACB ≅ ∠FDE	d. If two ∥ lines are cut by a transversal, alt. ext. ∡ are ≅.
e. △ABC ~ △FED	e. AA Similarity

5. Given: $\overline{AB} \perp \overline{BD}$, $\overline{ED} \perp \overline{BD}$
Prove: △BDA ~ △CDE

Proof:

Statements	Reasons
a. $\overline{AB} \perp \overline{BD}$, $\overline{EC} \perp \overline{BD}$	a. Given
b. ∠ABD, ∠ECD are rt. ∡.	b. ⊥ lines form four rt. ∡.
c. ∠ABD ≅ ∠ECD	c. All rt. ∡ are ≅.
d. ∠D ≅ ∠D	d. Congruence of ∡ is reflexive.
e. △BDA ~ △CDE	e. AA Similarity

Use the figure below to write a paragraph proof.

31. **Given:** △JFM ~ △EFB, △LFM ~ △GFB — See margin.
Prove: △JFL ~ △EFG

32. **Given:** $\overline{JM} \parallel \overline{EB}$, $\overline{LM} \parallel \overline{GB}$
Prove: $\overline{JL} \parallel \overline{EG}$ — See margin.

33–35. See Solutions Manual.

33. Prove that if the measures of two sides of a triangle are proportional to the measures of two corresponding sides of another triangle and the included angles are congruent, the triangles are similar. (Theorem 7–2)

34. Prove that if the measures of the legs of two right triangles are proportional, the triangles are similar.

35. Prove that similarity of triangles is reflexive, symmetric, and transitive. (Theorem 7–3)

Critical Thinking

36. Is it possible that △ABC is not similar to △RST and that △RST is not similar to △EFG, but that △ABC is similar to △EFG? Explain. Yes; the Transitive Property does not hold true for negative statements.

Applications and Problem Solving

37. **Surveying** Mr. Cardona uses a carpenter's square, an instrument used to draw right angles, to find the distance across a stream. He puts the square on top of a pole that is high enough to sight along $\overline{OL}$ to point P across the river. Then he sights along $\overline{ON}$ to point M. If MK is 2.5 feet and OK = 5.5 feet, find the distance KP across the stream. 12.1 feet

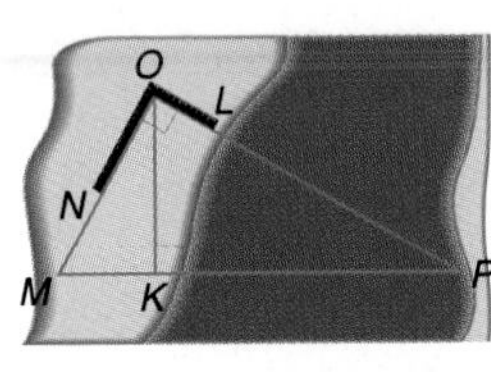

38. **Forestry** A hypsometer as shown below can be used to estimate the height of a tree. Benito looks through the straw to the top of the tree and obtains the readings given below. If Benito's eye was 165 centimeters from the ground and he was 14 meters from the tree, find the height of the tree. (*Hint:* 1 m = 100 cm) 10.05 meters

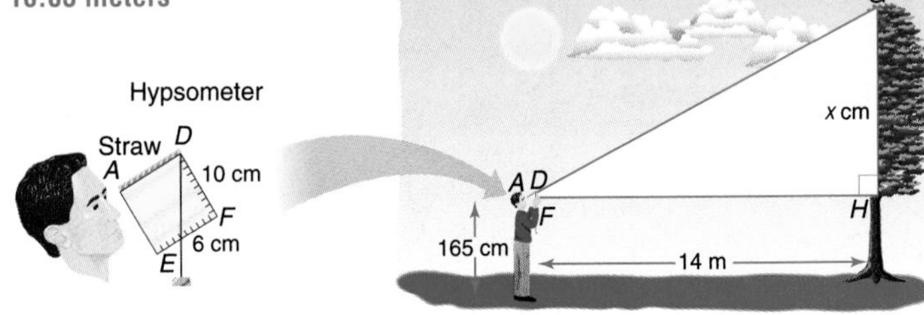

Mixed Review

39. Determine whether the pair of figures at the right is similar. Justify your answer. (Lesson 7–2) Yes; corresponding ∡ are ≅ and $\frac{2.0}{3.0} = \frac{1.6}{2.4} = \frac{1.8}{2.7}$.

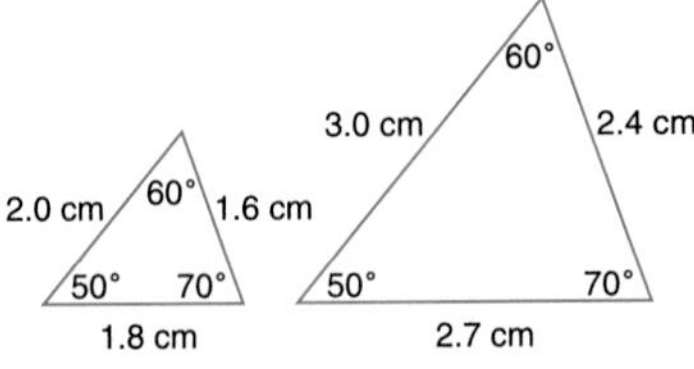

Tech Prep

Surveyor Surveying is crucial to navigation, architecture, and engineering. One method used is triangulation. Triangulation measures an area by breaking it into triangular regions. Then the surveyor measures the angles in each. For more information on tech prep, see the *Teacher's Handbook*.

40. Suppose the measures of corresponding sides of the polygons at the right are proportional. If $EF = 8$, $EH = 6$, and $AB = 15$, find AD. (Lesson 7–1) **11.25**

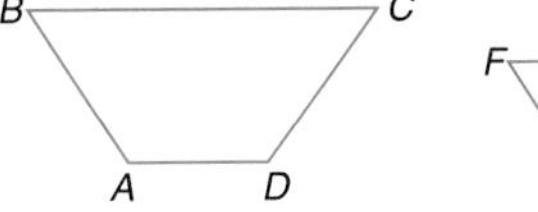

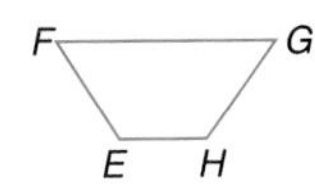

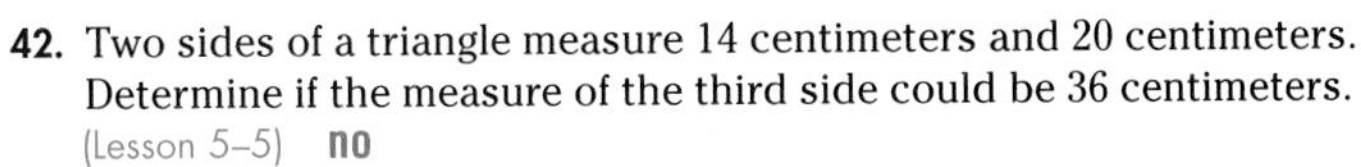
41. Determine whether the statement *Consecutive sides of a rhombus are congruent* is true or false. (Lesson 6–4) **true**

42. Two sides of a triangle measure 14 centimeters and 20 centimeters. Determine if the measure of the third side could be 36 centimeters. (Lesson 5–5) **no**

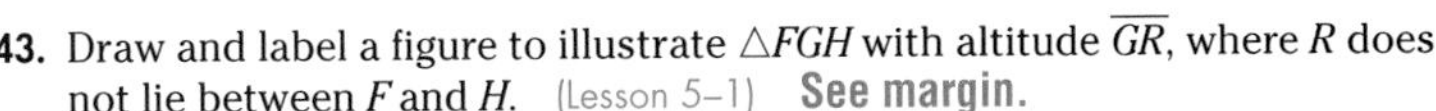
43. Draw and label a figure to illustrate $\triangle FGH$ with altitude $\overline{GR}$, where R does not lie between F and H. (Lesson 5–1) **See margin.**

44. Find the slope of a line that is perpendicular to the line passing through points $G(1, 5)$ and $H(-2, 4)$. (Lesson 3–3) **−3**

45. What property justifies the statement $\overline{LM} \cong \overline{LM}$? (Lesson 2–5) **Reflexive Property ≅ Segments**

INTEGRATION Algebra

46. **Production** In one day, 37,000,000 Tootsie Rolls® are produced by the company's four plants. Express this number in scientific notation.

46. 3.7×10^7

47. Solve $3t - 2 \geq t + 8$. $\{t \mid t \geq 5\}$

SELF TEST

1. Solve $\frac{1}{3} = \frac{t}{8-t}$ by using cross products. (Lesson 7–1) **2**

2. Determine whether the pair of figures at the right is similar. Justify your answer. (Lesson 7–2) **No; corresponding ∠s are not congruent.**

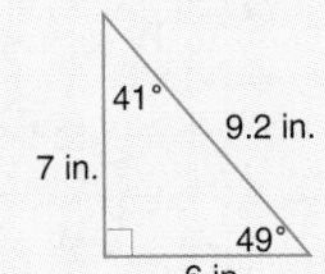

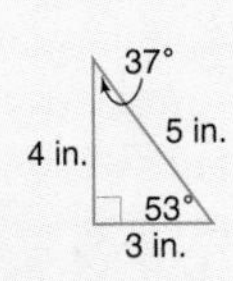

3. **Consumerism** The average American drinks about 4 soft drinks every 3 days. About how many soft drinks will the average person drink in a year? (Lesson 7–1) **about 487 soft drinks**

Determine whether each pair of triangles is similar using the given information. Explain your answer. (Lesson 7–3)

4. **yes; AA Similarity**

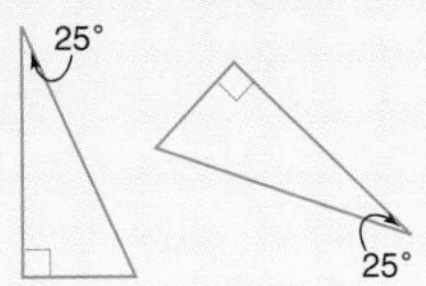

5. **yes; SAS Similarity**

4 ASSESS

Closing Activity

Speaking Choose two or three students to lead a discussion about the ways to test for similarity of triangles. What are the requirements for each? Are there any short cuts?

Chapter 7 Quiz B (Lesson 7-3) is available in the *Assessment and Evaluation Masters*, p. 184.

Mid-Chapter Test (Lessons 7-1 through 7-3) is available in the *Assessment and Evaluation Masters*, p. 183.

Additional Answer

43.

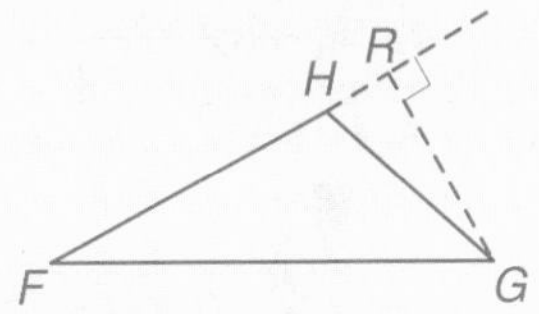

Enrichment Masters, p. 39

7-3 NAME ____ DATE ____ Student Edition Pages 354–361

Enrichment

Ratio Puzzles with Triangles

If you know the perimeter of a triangle and the ratios of the sides, you can find the lengths of the sides.

Example: The perimeter of a triangle is 84 units. The sides have lengths r, s, and t. The ratio of s to r is 5:3, and the ratio of t to r is 2:1. Find the length of each side.

Since both ratios contain r, rewrite one or both ratios to make r the same. You can write the ratio of t to r as 6:3. Now you can write a three-part ratio.

$r:s:t = 3:5:6$

There is a number x such that $r = 3x$, $s = 5x$, and $t = 6x$. Since you know the perimeter, 84, you can use algebra to find the lengths of the sides

$r + s + t = 84$
$3x + 5x + 6x = 84$
$14x = 84$
$x = 6$
$13x = 18, 5x = 30, 6x = 36$

So $r = 18$, $s = 30$, and $t = 36$.

Find the lengths of the sides of each triangle.

1. The perimeter of a triangle is 75 units. The sides have lengths a, b, and c. The ratio of b to a is 3:5, and the ratio of c to a is 7:5. Find the length of each side. $a = 25$, $b = 15$, $c = 35$

2. The perimeter of a triangle is 88 units. The sides have lengths d, e, and f. The ratio of e to d is 3:1, and the ratio of f to e is 10:9. Find the length of each side. $d = 12$, $e = 36$, $f = 40$

3. The perimeter of a triangle is 91 units. The sides have lengths p, q, and r. The ratio of p to r is 3:1, and the ratio of q to r is 5:2. Find the length of each side. $p = 42$, $q = 35$, $r = 14$

4. The perimeter of a triangle is 68 units. The sides have lengths g, h, and j. The ratio of j to g is 2:1, and the ratio of h to g is 5:4. Find the length of each side. $g = 16$, $h = 20$, $j = 32$

5. Write a problem similar to those above involving ratios in triangles. See students' work.

Extension

Problem Solving Recall that similarity of triangles is reflexive, symmetric, and transitive. Such a relation is called an equivalence relation. What are other equivalence relations? Have students find out what is required to qualify as an equivalence relation.

SELF TEST

The Self Test provides students with a brief review of the concepts and skills in Lessons 7-1 through 7-3. Lesson numbers are given to the right of exercises or instruction lines so students can review concepts not yet mastered.

7-4 LESSON NOTES

NCTM Standards: 1–5, 7

Instructional Resources

- Study Guide Master 7-4
- Practice Master 7-4
- Enrichment Master 7-4

Transparency 7-4A contains the 5-Minute Check for this lesson; **Transparency 7-4B** contains a teaching aid for this lesson.

Recommended Pacing	
Standard Pacing	Days 5 & 6 of 13
Honors Pacing	Day 5 of 12
Block Scheduling*	Day 4 of 7

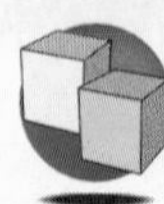

*For more information on pacing and possible lesson plans, refer to the *Block Scheduling Booklet.*

1 FOCUS

5-Minute Check
(over Lesson 7-3)

1. Are the triangles below similar? Explain why or why not. **yes; SSS Similarity**

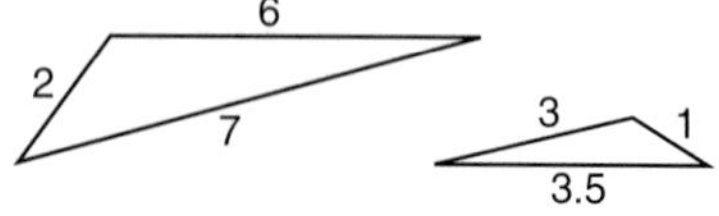

In the figure below, $\overline{QN} \parallel \overline{RP}$.

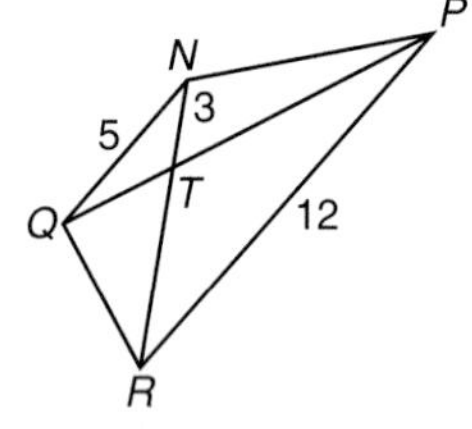

2. Identify the similar triangles. Explain why the triangles are similar. **$\triangle QNT \sim \triangle PRT$; AA Similarity**
3. Find TR. **7.2**

7–4 Parallel Lines and Proportional Parts

What YOU'LL LEARN

- To use proportional parts of triangles to solve problems, and
- to divide a segment into congruent parts.

Why IT'S IMPORTANT

You can use proportions to find the lengths of segments determined by parallel lines.

CONNECTION

Music

A harp is a triangular-shaped instrument whose strings can be thought of as parallel lines and whose body and neck as transversals.

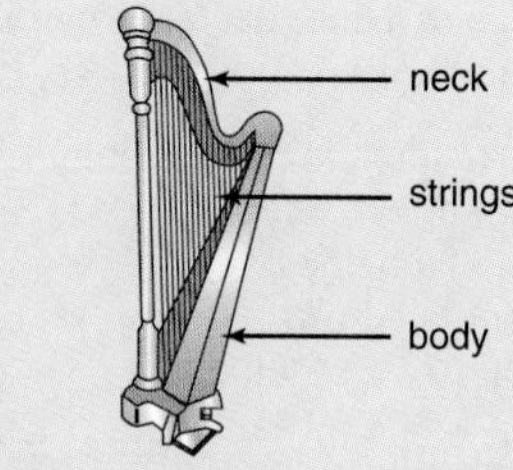

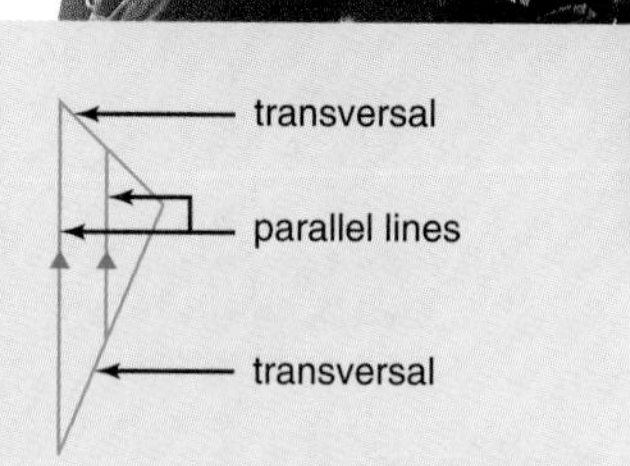

Each string divides the body and the neck proportionally. This is stated in the following theorem.

Theorem 7–4 Triangle Proportionality	**If a line is parallel to one side of a triangle and intersects the other two sides in two distinct points, then it separates these sides into segments of proportional lengths.**

Proof of Theorem 7–4

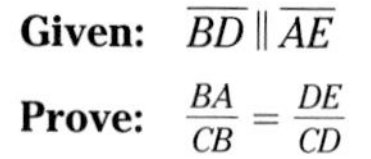

Given: $\overline{BD} \parallel \overline{AE}$

Prove: $\frac{BA}{CB} = \frac{DE}{CD}$

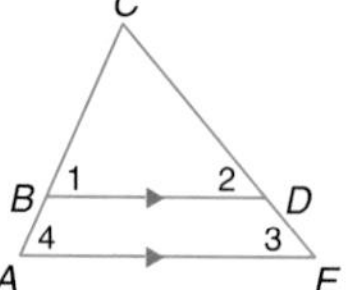

Paragraph Proof:

Since $\overline{BD} \parallel \overline{AE}$, $\angle 4 \cong \angle 1$ and $\angle 3 \cong \angle 2$. Then by AA Similarity, $\triangle ACE \sim \triangle BCD$. From the definition of similar polygons, $\frac{CA}{CB} = \frac{CE}{CD}$. Since B is between A and C, $CA = BA + CB$, and since D is between C and E, $CE = DE + CD$. Substituting for CA and CE in the ratio, we get the following proportion.

$$\frac{BA + CB}{CB} = \frac{DE + CD}{CD}$$

$$\frac{BA}{CB} + \frac{CB}{CB} = \frac{DE}{CD} + \frac{CD}{CD}$$

$$\frac{BA}{CB} + 1 = \frac{DE}{CD} + 1 \quad \textit{Substitution Property}(=)$$

$$\frac{BA}{CB} = \frac{DE}{CD} \quad \textit{Subtraction Property}(=)$$

Likewise, proportional parts of a triangle can be used to prove the converse of Theorem 7–4. *You will be asked to prove this theorem in Exercise 32.*

GLOBAL CONNECTIONS

Harps were used in Mesopotamia as early as 3000 B.C.

GLOBAL CONNECTIONS

The modern orchestral harp stands 170 centimeters tall and has a range of more than $6\frac{1}{2}$ octaves.

Theorem 7-5	**If a line intersects two sides of a triangle and separates the sides into corresponding segments of proportional lengths, then the line is parallel to the third side.**

Example 1

In $\triangle EFG$, $EG = 15$, $EH = 5$, and LG is twice FL. Determine whether $\overline{HL} \parallel \overline{EF}$.

From the Segment Addition Postulate, $EG = EH + HG$.
Substitute the known measures.

$$EG = EH + HG$$
$$15 = 5 + HG$$
$$10 = HG$$

Let $x = FL$. Then $2x = LG$. To show $\overline{HL} \parallel \overline{EF}$, we must show that $\frac{EH}{HG} = \frac{FL}{LG}$. Using the information above, $\frac{EH}{HG} = \frac{5}{10}$ or $\frac{1}{2}$ and $\frac{FL}{LG} = \frac{x}{2x}$ or $\frac{1}{2}$. Since the sides have proportional lengths, $\overline{HL} \parallel \overline{EF}$.

The proof of the following theorem is based on Theorems 7–4 and 7–5. *You will be asked to prove this theorem in Exercise 33.*

Theorem 7-6	**A segment whose endpoints are the midpoints of two sides of a triangle is parallel to the third side of the triangle, and its length is one-half the length of the third side.**

In $\triangle RST$, suppose M is the midpoint of $\overline{RS}$ and L is the midpoint of $\overline{RT}$.
By Theorem 7–6, $\overline{ML} \parallel \overline{ST}$ and $ML = \frac{1}{2}ST$.
This can also be expressed as $2ML = ST$.

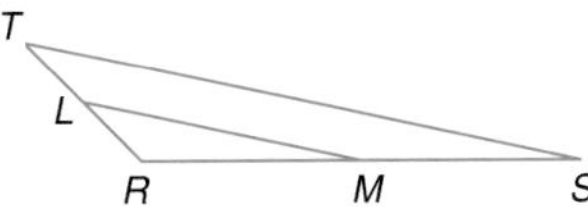

Example 2

INTEGRATION Algebra

LOOK BACK Refer to Lesson 1-5 to review the midpoint formula.

Triangle ABC has vertices $A(0, 2)$, $B(12, 0)$, and $C(2, 10)$.

a. Find the coordinates of D, the midpoint of $\overline{AB}$, and E, the midpoint of $\overline{CB}$.
b. Show that $\overline{DE}$ is parallel to $\overline{AC}$.
c. Show that $2DE = AC$.

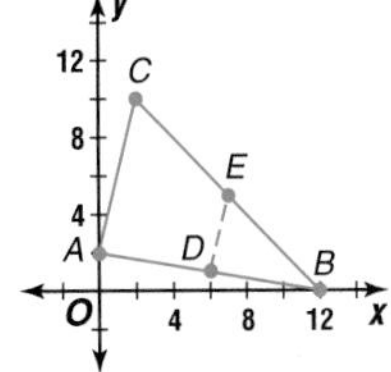

a. Use the midpoint formula to find the midpoints of $\overline{AB}$ and $\overline{CB}$.

$$D\left(\frac{x_1 + x_2}{2}, \frac{y_1 + y_2}{2}\right) = D\left(\frac{0 + 12}{2}, \frac{2 + 0}{2}\right) \text{ or } D(6, 1)$$

$$E\left(\frac{x_1 + x_2}{2}, \frac{y_1 + y_2}{2}\right) = E\left(\frac{12 + 2}{2}, \frac{0 + 10}{2}\right) \text{ or } E(7, 5)$$

Thus, the midpoint of $\overline{AB}$ has coordinates (6, 1), and the midpoint of $\overline{CB}$ has coordinates (7, 5).

(continued on the next page)

Motivating the Lesson

Situational Problem Ask students how many of them have ever gone camping and had to pitch a tent. Discuss how the poles supporting the tent were arranged. **Sample answer: Two poles and the ground formed a triangle.** Note that the use of triangles makes a structure sturdy.

2 TEACH

In-Class Examples

For Example 1
In the figure below, $CA = 10$, $CE = 2$, $DA = 6$, and $BA = 12$. Determine if $\overline{ED} \parallel \overline{CB}$.

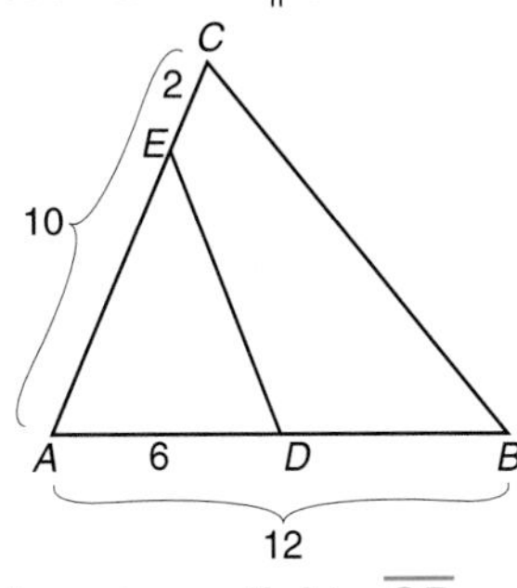

$\overline{ED}$ is not parallel to $\overline{CB}$.

For Example 2
Triangle DEF has vertices $D(1, 2)$, $E(7, 4)$, and $F(3, 6)$.

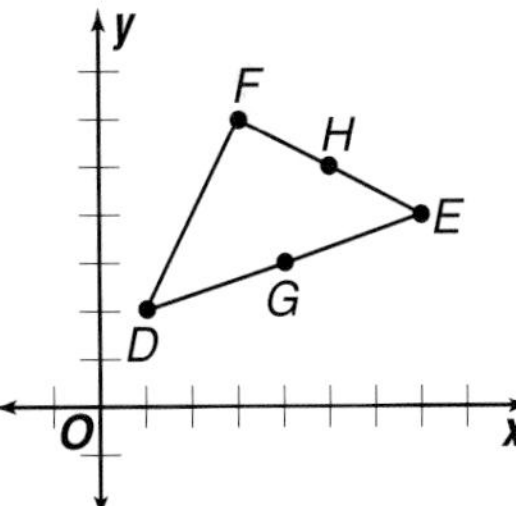

a. Find the coordinates of G, the midpoint of $\overline{DE}$, and H, the midpoint of $\overline{FE}$. **$G(4, 3)$, $H(5, 5)$**
b. What two lines are parallel? **$\overline{DF} \parallel \overline{GH}$**

Alternative Learning Styles

Kinesthetic Have students make a sector compass using two rulers that each have a hole punched in the middle. Put the rulers on top of each other and use a rubber band to tie them together. Have students draw a segment that can be measured with this sector compass and construct a segment that is some multiple of the length of the original segment. Discuss the fact that each longer segment is similar to the original. Have them use the sector compass to make similar triangles.

In-Class Example

For Example 3

You want to plant a row of pine trees along a slope with the spacing as shown. Find the spacing needed for planting the trees if the length of the slope is 80 feet. All measurements are in feet.

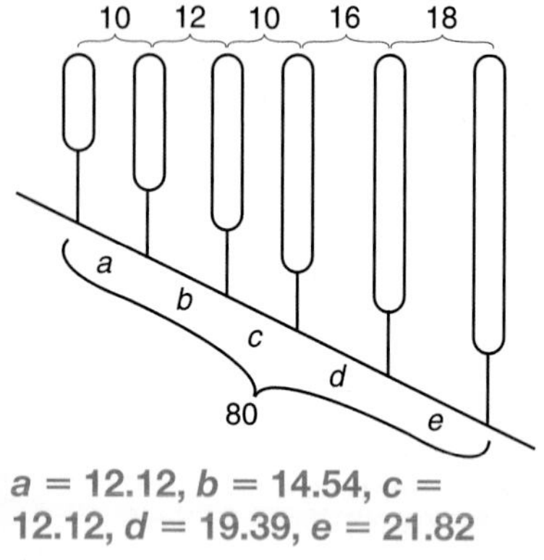

a = 12.12, *b* = 14.54, *c* = 12.12, *d* = 19.39, *e* = 21.82

Teaching Tip Draw an example to illustrate Corollary 7-2. Make sure that students understand that each transversal contains one set of congruent segments and that all segments on the transversal are congruent.

b. To show that $\overline{AC} \parallel \overline{DE}$, you can show that the slopes of $\overline{AC}$ and $\overline{DE}$ are equal.

slope of $\overline{AC} = \frac{2-10}{0-2}$ or 4 $\qquad$ slope of $\overline{DE} = \frac{1-5}{6-7}$ or 4

Because the slopes of $\overline{AC}$ and $\overline{DE}$ are equal, $\overline{AC} \parallel \overline{DE}$.

c. To show that $2DE = AC$, we must find the length of $\overline{DE}$ and $\overline{AC}$. Use the distance formula.

$$AC = \sqrt{(0-2)^2 + (2-10)^2} = \sqrt{4+64} = \sqrt{68} \text{ or } 2\sqrt{17}$$

$$DE = \sqrt{(6-7)^2 + (1-5)^2} = \sqrt{1+16} \text{ or } \sqrt{17}$$

Therefore, $2DE = AC$.

We have seen that parallel lines cut the sides of a triangle into proportional parts. Three or more parallel lines separate transversals into proportional parts as stated in the next two corollaries.

Corollary 7-1 **If three or more parallel lines intersect two transversals, then they cut off the transversals proportionally.**

Corollary 7-2 **If three or more parallel lines cut off congruent segments on one transversal, then they cut off congruent segments on every transversal.**

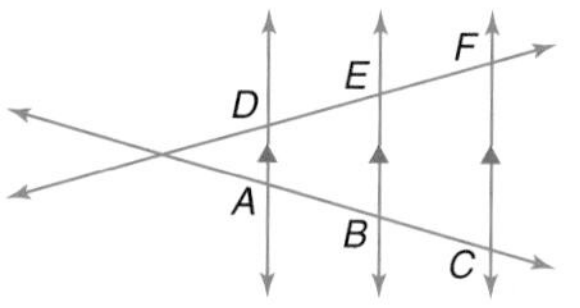

In the figure at the left, $\overleftrightarrow{AD} \parallel \overleftrightarrow{EB} \parallel \overleftrightarrow{FC}$. The transversals $\overleftrightarrow{AC}$ and $\overleftrightarrow{DF}$ have been separated into proportional segments. Sample proportions are listed below.

$$\frac{AB}{BC} = \frac{DE}{EF}, \frac{AC}{DF} = \frac{BC}{EF}, \frac{AC}{BC} = \frac{DF}{EF}$$

The theorems and corollaries about proportionality of triangles can be used to solve practical problems.

Example 3

Real Estate

In Lake Creek, the lots on which houses are to be built are laid out as shown. What is the lake frontage for each of the five lots if the total frontage is 135.6 meters?

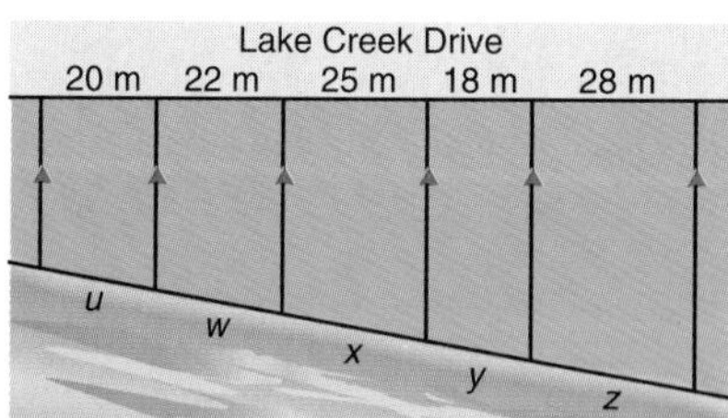

Explore By Corollary 7–1, if three or more parallel lines intersect two transversals, they cut off the transversals proportionally. Thus, $\frac{u}{20} = \frac{w}{22} = \frac{x}{25} = \frac{y}{18} = \frac{z}{28}$.

Plan Use these proportions to find a value for each frontage in terms of one variable, for example, u.

$$\frac{u}{20} = \frac{w}{22} \qquad \frac{u}{20} = \frac{x}{25} \qquad \frac{u}{20} = \frac{y}{18} \qquad \frac{u}{20} = \frac{z}{28}$$

$$w = \frac{22}{20}u \qquad x = \frac{25}{20}u \qquad y = \frac{18}{20}u \qquad z = \frac{28}{20}u$$

Solve

$u + w + x + y + z = 135.6$ *Length of the frontage is 135.6 m.*

$u + \frac{22}{20}u + \frac{25}{20}u + \frac{18}{20}u + \frac{28}{20}u = 135.6$ *Substitution Property (=)*

$20u + 22u + 25u + 18u + 28u = 135.6(20)$ *Multiply each side by 20.*

$113u = 2712$ *Simplify.*

$u = 24$ *Divide each side by 113.*

Therefore, $w = \frac{22}{20}u \qquad x = \frac{25}{20}u \qquad y = \frac{18}{20}u \qquad z = \frac{28}{20}u$

$$w = \frac{22}{20}(24) \qquad x = \frac{25}{20}(24) \qquad y = \frac{18}{20}(24) \qquad z = \frac{28}{20}(24)$$

$$w = 26.4 \qquad x = 30 \qquad y = 21.6 \qquad z = 33.6$$

The individual lake frontage for each lot is 24 meters, 26.4 meters, 30 meters, 21.6 meters, and 33.6 meters.

Examine By substitution, $\frac{u}{20} = \frac{w}{22} = \frac{x}{25} = \frac{y}{18} = \frac{z}{28} = 1.2$. Since the proportions are equal, the values are correct.

It is possible to separate a segment into two congruent parts by constructing the perpendicular bisector of a segment. However, a segment cannot be separated into three congruent parts by constructing perpendicular bisectors. To do this, you must use parallel lines and the similarity relationships.

Trisecting a Segment

Separate a segment into three congruent parts.

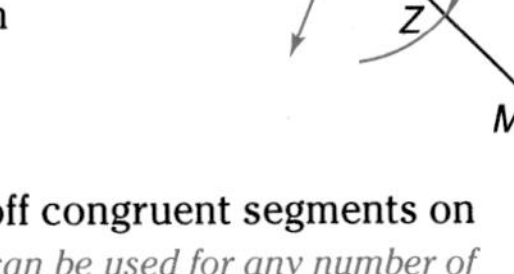

1. Copy $\overline{AB}$ and draw $\overline{AM}$.
2. With a compass point at A, mark off an arc that intersects $\overline{AM}$ at X. Then construct $\overline{XY}$ and $\overline{YZ}$ so that $\overline{AX} \cong \overline{XY} \cong \overline{YZ}$.
3. Draw $\overline{ZB}$. Then construct lines through Y and X that are parallel to $\overline{ZB}$. Call P and Q the intersection points on $\overline{AB}$.

Conclusion: Because parallel lines cut off congruent segments on transversals, $\overline{AP} \cong \overline{PQ} \cong \overline{QB}$. *This process can be used for any number of congruent parts.*

Teaching Tip For step 3 of the construction, emphasize that the far endpoints must be connected first.

3 PRACTICE/APPLY

Check for Understanding

Exercises 1–12 are designed to help you assess your students' understanding through reading, writing, speaking, and modeling. You should work through Exercises 1–5 with your students and then monitor their work on Exercises 6–12.

Error Analysis

Students often confuse the proportional parts in figures like the one below.

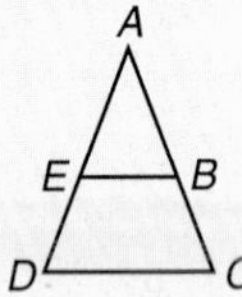

They might think $\frac{AE}{ED} = \frac{AB}{BC} = \frac{EB}{DC}$. Ask them to identify each segment in terms of a triangle. If the descriptions of compared parts don't match, the proportion is incorrect.

Additional Answers

1.

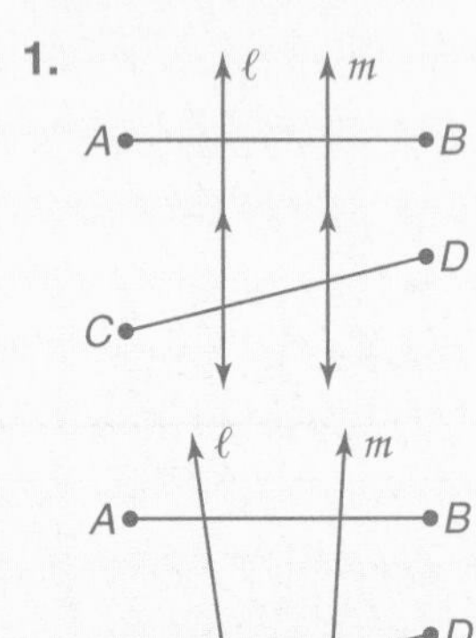

2. Sample answer: Given three or more parallel lines intersecting two transversals, Corollary 7-1 states that the parts of the transversals are proportional. Corollary 7-2 states that if the parts of one transversal are congruent, then the parts of every transversal are congruent.

3.

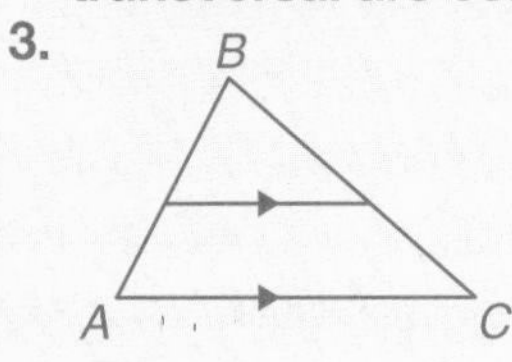

CHECK FOR UNDERSTANDING

Communicating Mathematics

Study the lesson. Then complete the following.

1. **Make a drawing** to show two segments intersected by two lines so that the parts are proportional. Then draw a counterexample. See margin.
2. **Explain** the difference between Corollary 7–1 and Corollary 7–2. *2. See margin.*
3. **Draw** a segment from one side of a triangle to another so that it is parallel to the third side. When will the segment be half the length of the third side of the triangle? Explain why this is the case. *3. When the segment is drawn through the midpoints of the two sides; see margin for sample drawing.*
4. **Describe** how to separate a segment into four congruent parts. How should this process change to separate a segment into five congruent parts? See margin.

5. **List** at least four methods to prove two lines parallel, including one from this lesson. See margin.

Guided Practice

Refer to $\triangle RST$ for Exercises 6 and 7. $\overline{LW} \parallel \overline{TS}$.

6. Complete each statement.
   a. $\frac{RW}{WS} = \frac{RL}{?}$ LT
   b. $\frac{RW}{RS} = \frac{?}{RT}$ RL
7. Determine whether each statement is *true* or *false*. If false, explain why.
   a. If $TL = LR$, then $SW = \frac{1}{2}SR$. true
   b. If $TR = 8$, $LR = 3$, and $RW = 6$, then $WS = 16$. false; $RS = 16$

8. Find the values of x and y. 2, 12

Refer to the figure at the right for Exercises 9–11. Determine whether each conclusion is valid under the given conditions. If the conclusion is valid, give a reason.

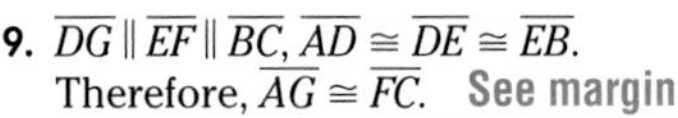

9. $\overline{DG} \parallel \overline{EF} \parallel \overline{BC}$, $\overline{AD} \cong \overline{DE} \cong \overline{EB}$. Therefore, $\overline{AG} \cong \overline{FC}$. See margin.
10. $\overline{DG} \parallel \overline{BC}$. Therefore, $\frac{AD}{DE} = \frac{AG}{GF}$. no; must have $\overline{DG} \parallel \overline{EF}$
11. $\overline{DG} \parallel \overline{EF} \parallel \overline{BC}$; $DE = x$, $EB = 20$, $GF = x - 5$, $FC = 15$. Therefore, $x = 20$. *11. See margin.*
12. **Construction** Hai was building a large open stairway and used wood strips as a decoration along the inside wall. If the strips were spaced along the bottom as shown in the diagram at the right, at what distance should he attach the top of the strips if the strips are to be parallel? $x = 3\frac{1}{3}$ ft, $y = 2\frac{2}{3}$ ft, $z = 2$ ft

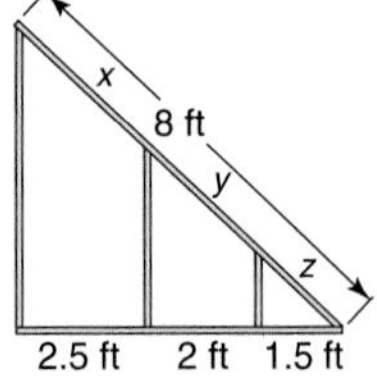

Reteaching

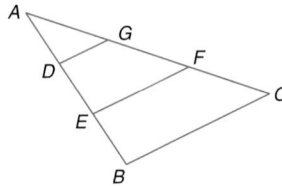

Using Vocabulary Review some of the terminology contained in the theorems of this lesson, such as parallel, midpoint, and transversal. Go through the theorems word by word (or phrase by phrase) and draw a picture illustrating the meaning of the word (or phrase). The final picture should be an illustration of the theorem.

EXERCISES

Practice

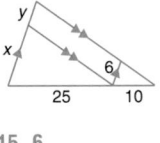

In the figure at the right, $\overleftrightarrow{AB} \parallel \overleftrightarrow{CD} \parallel \overleftrightarrow{EF}$. Complete each statement.

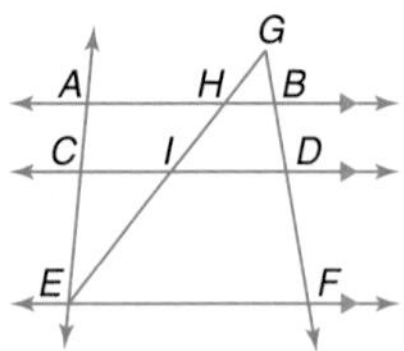

13. $\frac{AC}{BD} = \frac{CE}{?}$ DF

14. $\frac{CE}{IE} = \frac{?}{HI}$ AC

15. $\frac{GH}{GE} = \frac{?}{GF}$ GB

16. $\frac{CE}{?} = \frac{AC}{HI}$ IE

In $\triangle ABC$, $\overline{ED} \parallel \overline{AC}$. Determine whether each statement is *true* or *false*. If false, explain why.

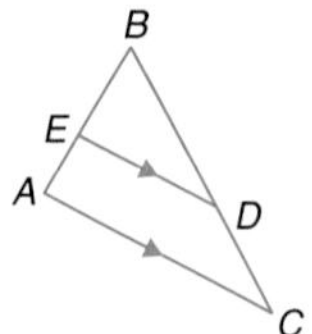

17. If $BE = 4$, $EA = 3$, and $BD = 5$, then $DC = 3\frac{3}{4}$. true

18. If $BE = \frac{3}{5}AB$, then $DC = \frac{2}{5}BC$. true

19. If $AE = \frac{3}{5}AB$, then $\frac{BD}{DC} = \frac{3}{2}$. false; $\frac{BD}{DC} = \frac{2}{3}$

INTEGRATION **Algebra**

Find the values of *x* and *y*.

20. 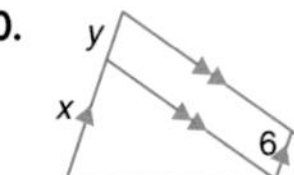

15, 6

21. 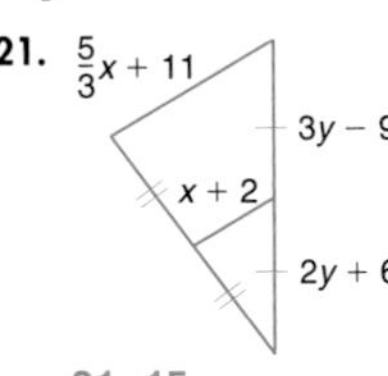

21, 15

22. 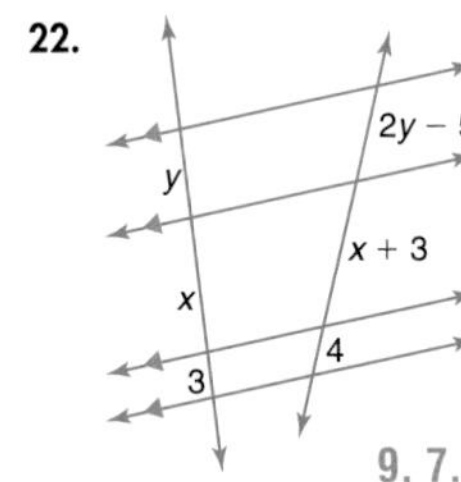

9, 7.5

In $\triangle ACE$, find *x* so that $\overline{DB} \parallel \overline{AE}$.

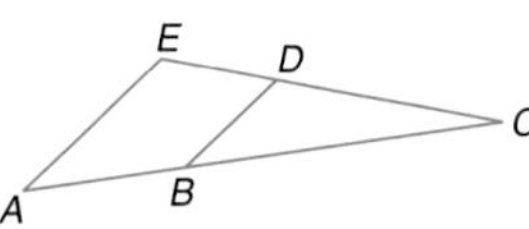

23. $ED = 8$, $DC = 20$, $BC = 25$, $AB = x$ 10

24. $BC = 12$, $AB = 6$, $ED = 8$, $DC = x - 4$ 20

25. $ED = x - 5$, $DC = 15$, $CB = 18$, $AB = x - 4$ 10

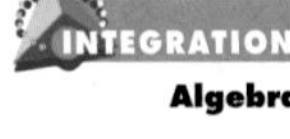

26. Write a paragraph proof to show that $\overline{SM}$ is parallel to $\overline{TW}$ and that $\overline{SM}$ divides the sides $\overline{RT}$ and $\overline{RW}$ in a ratio of 1 to 2. See margin.

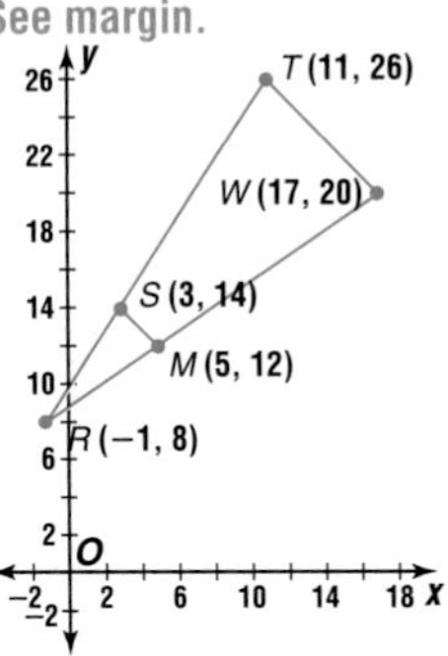

27. If $\overline{BC} \parallel \overline{DE}$ and $2(DE) = BC$, find the coordinates of *B* and *C*. *B*(5, −2); *C*(11, 2)

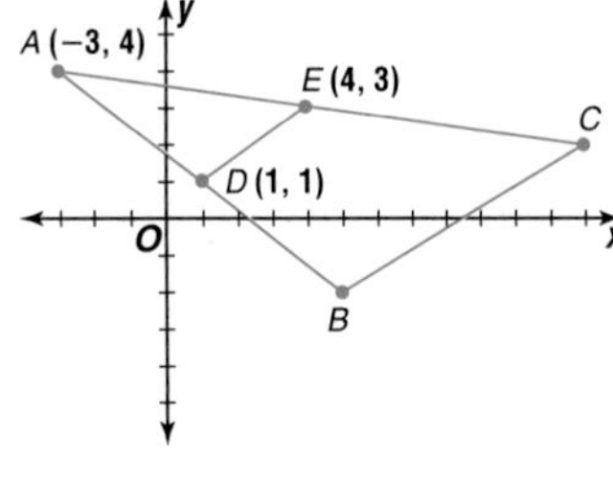

A, B, and C are the midpoints of sides $\overline{DF}$, $\overline{DE}$, and $\overline{FE}$, respectively.

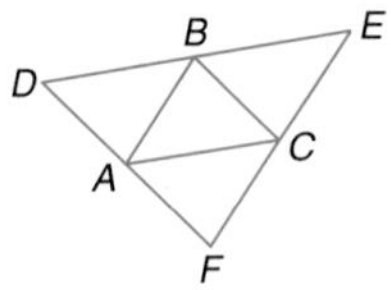

28. If $BC = 11$, $AC = 13$, and $AB = 15$, find the perimeter of $\triangle DEF$. 78

29. If $DE = 18$, $DA = 10$, and $FC = 7$, find *AB*, *BC*, and *AC*.

29. $AB = 7$, $BC = 10$, $AC = 9$

Additional Answers

9. Yes; if 3 or more ∥ lines cut off ≅ segments on one transversal, then they cut off ≅ segments on every transversal.
11. Yes; if 3 or more ∥ lines intersect 2 transversals, then they cut off the transversals proportionately.

26. The slope of $\overline{SM}$ = slope of $\overline{TW} = -1$. $RS = RM = 2\sqrt{13}$; $ST = MW = 4\sqrt{13}$, so $\frac{RS}{ST} = \frac{RM}{MW} = \frac{2\sqrt{13}}{4\sqrt{13}} = \frac{1}{2}$.

Assignment Guide

Core (with proof): 13–39 odd, 40–48
Core (informal): 13–31 odd, 35, 37, 39, 40–48
Enriched: 14–34 even, 36–48

For **Extra Practice**, see p. 777.

The red A, B, and C flags, printed only in the Teacher's Wraparound Edition, indicate the level of difficulty of the exercises.

Additional Answers

4. Draw a ray extending from one of the endpoints of the segment. Mark off an arc on the ray. Then mark off three more arcs of equal lengths. Construct four lines connecting these points. To separate a segment into five congruent parts, mark off five arcs of equal lengths.
5. Sample answer: alternate interior angles; corresponding angles; lines perpendicular to same line; if a line intersects two sides of a triangle and separates sides into corresponding segments of proportional lengths, then it is parallel to the third side of the triangle.

Study Guide Masters, p. 40

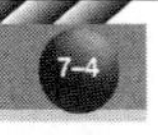

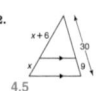

NAME ____________ DATE ________

Study Guide

Student Edition Pages 362–369

Parallel Lines and Proportional Parts

The following theorems involve proportional parts of triangles.

- If a line is parallel to one side of a triangle and intersects the other two sides, then it separates these sides into segments of proportional lengths.
- If a line intersects two sides of a triangle and separates the sides into corresponding segments of proportional lengths, then the line is parallel to the third side.
- A segment whose endpoints are the midpoints of two sides of a triangle is parallel to the third side of the triangle and its length is one-half the length of the third side.

Example: In $\triangle ABC$, $\overline{EF} \parallel \overline{CB}$, find the value of x.

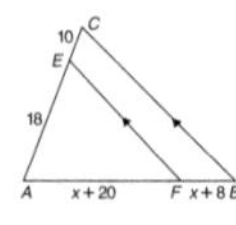

$\overline{EF} \parallel \overline{CB}$ implies that $\frac{AF}{FB} = \frac{AE}{EC}$.

Rewrite the proportion and solve.

$$\frac{x+20}{x+8} = \frac{18}{10}$$
$$10x + 200 = 18x + 144$$
$$56 = 8x$$
$$7 = x$$

Find the value of x.

1. 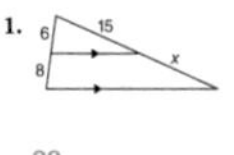

20

2. 4.5

3. 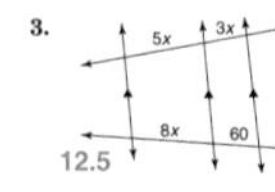

12.5

In $\triangle ABC$, find x so that $\overline{DE} \parallel \overline{CB}$.

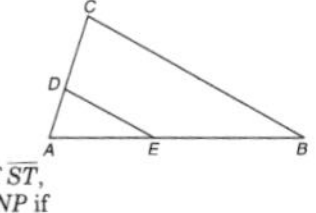

4. $DC = 18$, $AD = 6$, $AE = 12$, $EB = x - 3$ 39

5. $AC = 30$, $AD = 10$, $AE = 22$, $EB = x + 4$ 40

6. In $\triangle RST$, *M* is the midpoint of $\overline{RS}$, *N* is the midpoint of $\overline{ST}$, and *P* is the midpoint of $\overline{RT}$. Find the perimeter of $\triangle MNP$ if $RS = 28$, $ST = 34$, and $RT = 30$. 46

Additional Answers

30.

31.

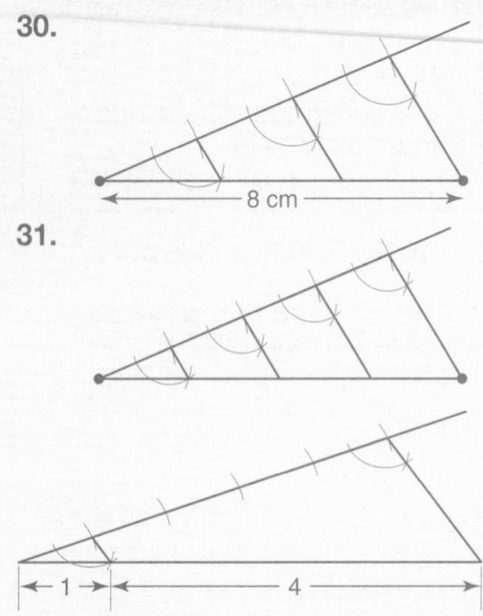

38a. The bar connects the midpoints of each leg of the letter and is parallel to the base. Therefore, the length of the bar is one-half the length of the base because if a segment has endpoints at the midpoints of two sides of a triangle, it is parallel to the third side of the triangle and its length is one-half the length of the third side.

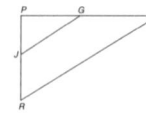

30. Draw a segment 8 centimeters long. By construction, separate the segment into three congruent segments. See margin.

31. Draw a segment and by construction, separate it into four congruent parts. Draw a segment and separate it into two segments whose ratios are 1 to 4. See margin.

Proof

Write a two-column proof for Exercises 32 and 33.

32–33. See Solutions Manual.

32. If a line intersects two sides of a triangle and separates the sides into corresponding segments of proportional lengths, then the line is parallel to the third side. (Theorem 7–5)

33. A segment whose endpoints are the midpoints of two sides of a triangle is parallel to the third side of the triangle and its length is one-half the length of the third side. (Theorem 7–6)

34. Given $A(2, 12)$ and $B(5, 0)$, find P such that P separates $\overline{AB}$ into two parts with a ratio of 2 to 1. $(3, 8)$ or $(4, 4)$

35. In $\triangle LMN$, $\overline{PR}$ divides sides $\overline{NL}$ and $\overline{MN}$ proportionally. If the points are $N(8, 20)$, $P(11, 16)$, and $R(3, 8)$ and $\frac{LP}{PN} = \frac{2}{1}$, find the coordinates of L and M.

35. $L(17, 8)$; $M(-7, -16)$

Critical Thinking

36. Draw any quadrilateral $ABCD$ and connect the midpoints E, F, G, H of the sides in order.
   a. Determine what kind of figure $EFGH$ will be. Use the information from this lesson to prove your claim. Rhombus; see students' work.
   b. Will the same reasoning work with five-sided polygons? Explain why or why not.

36b. Sample answer: no, there will be an odd number of sides so you cannot get all sets of sides to be parallel.

37. Draw segment $\overline{AB}$ so it is 7 centimeters long. If $r = 2$ centimeters, $s = 3$ centimeters, and $t = 4$ centimeters, divide $\overline{AB}$ into segments x, y, and z such that $\frac{x}{r} = \frac{y}{s} = \frac{z}{t}$. $x = \frac{14}{9}$ cm, $y = \frac{7}{3}$ cm, $z = \frac{35}{9}$ cm

Applications and Problem Solving

38. **History** In the fifteenth century, mathematicians and artists tried to construct the perfect letter by using certain proportions and geometric instructions. Damiano da Moile used a square as a frame to design the letter "A" at the right. The thickness of the major stroke of the letter was to be $\frac{1}{12}$ of the height of the letter (including descenders). a. See margin.

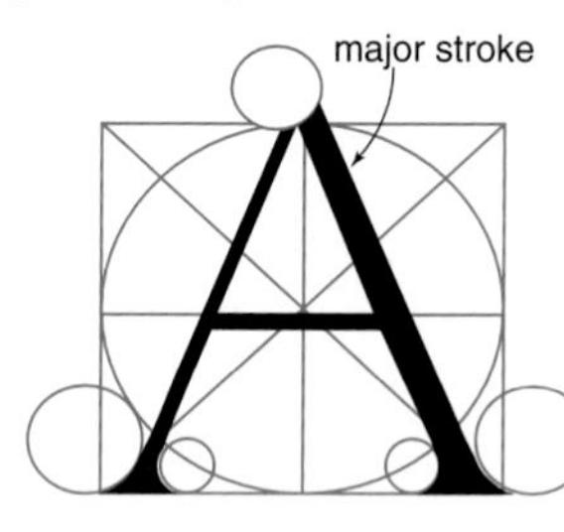

   a. Explain how you can tell that the bar through the middle of the A is half the length between the outside bottom corners of the sides of the letter.
   b. If the letter were 3 centimeters tall, how wide would the major stroke of the A be? 0.25 cm

39. $\triangle ABC \sim \triangle ADE$ by SAS Similarity, so $\frac{AD}{AB} = \frac{DE}{BC}$. $AD = 60$ and $AB = 100$, so $\frac{60}{100} = \frac{DE}{BC}$. $\frac{3}{5} = \frac{DE}{BC}$ and $\frac{3}{5}BC = DE$.

39. **Navigation** The sector compass was an instrument used in the seventeenth and eighteenth centuries to help explorers find their way on land and sea. A sector compass consisted of two arms, fastened at one end by a pivot joint. A scale was marked on each arm as shown in the diagram. To draw a segment three fifths of the length of a given segment, the 100-marks are placed at the endpoints of the given line. A segment drawn between the two 60-marks is three fifths of the length of the given segment. Explain why.

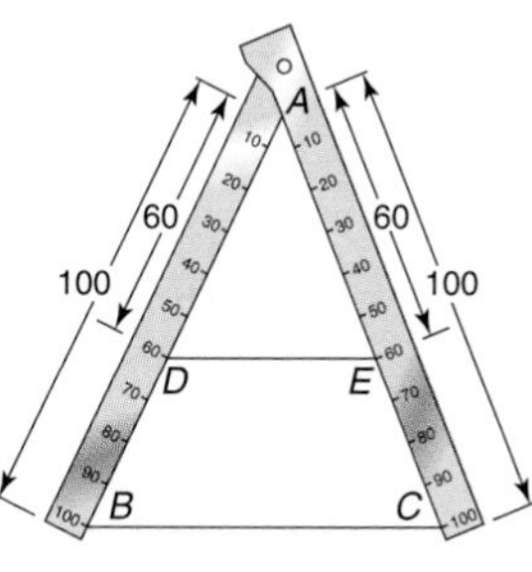

Practice Masters, p. 40

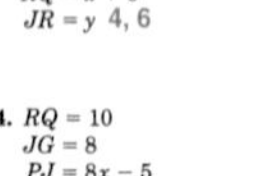

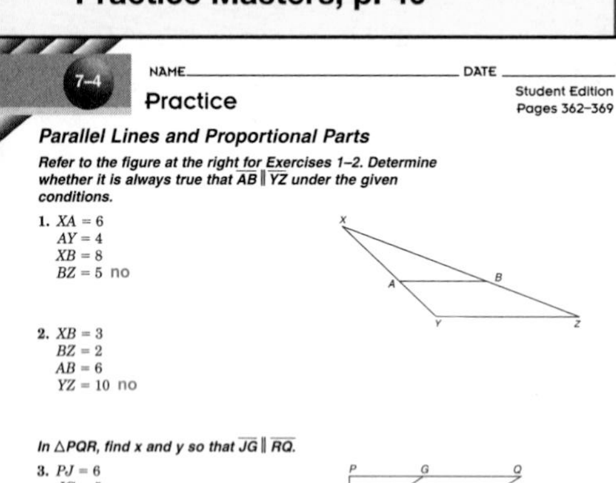

NAME__________ DATE________

7-4 Practice

Student Edition Pages 362–369

Parallel Lines and Proportional Parts

Refer to the figure at the right for Exercises 1–2. Determine whether it is always true that $\overline{AB} \parallel \overline{YZ}$ under the given conditions.

1. $XA = 6$, $AY = 4$, $XB = 8$, $BZ = 5$ no

2. $XB = 3$, $BZ = 2$, $AB = 6$, $YZ = 10$ no

In $\triangle PQR$, find x and y so that $\overline{JG} \parallel \overline{RQ}$.

3. $PJ = 6$, $JG = 5$, $PG = 4$, $GQ = x$, $RQ = x + 6$, $JR = y$ 4, 6

4. $RQ = 10$, $JG = 8$, $PJ = 8x - 5$, $JR = x$, $PG = 3y + 2$, $QG = y$ $\frac{5}{4}$, 2

5. In the figure at the right, $\overline{YA} \parallel \overline{OE} \parallel \overline{BR}$. Find the values of x and y if $YO = 4$, $ER = 16$, and $AR = 24$. 8, 8

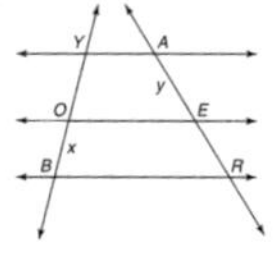

Extension

Connections Using $\triangle SRT$ below Theorem 7-6, draw the angle bisector of $\angle R$, the altitude of $\triangle SRT$ from point R, and the median of $\triangle SRT$ from point R. What relationship does $\overline{ML}$ have to these segments?

$\overline{ML}$ bisects all three segments.

Mixed Review

40. Determine if the triangles at the right are similar using the given information. Explain. (Lesson 7–3) **yes; SSS Similarity**

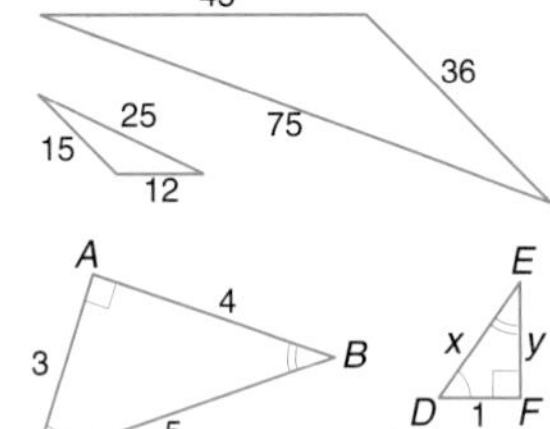

41. Two similar polygons are shown at the right. Find the values of x and y. (Lesson 7–2) $1\frac{2}{3}, 1\frac{1}{3}$

42. Yes, if it is a square.

42. Explain whether a rectangle can be a rhombus. (Lesson 6–4)

43. Determine if the statement *A median of a triangle always lies in the interior of the triangle* is true or false. (Lesson 5–1) **true**

44. If $\triangle MNO \cong \triangle FGH$ and $\triangle FGH \cong \triangle ABC$, what conclusion can you make? (Lesson 4–3) $\triangle MNO \cong \triangle ABC$

45. Write a conjecture based on the information that $\angle RST$ and $\angle DSE$ are vertical angles and $\angle DSE \cong \angle ABC$. (Lesson 2–1) $\angle RST \cong \angle ABC$

46. Find the coordinates of point B if $M(6, -1)$ is the midpoint of $\overline{AB}$ and the coordinates of A are $(-2, 3)$. (Lesson 1–5) **(14, −5)**

Algebra

47. Factor $k^2 + 3k - 18$. $(k - 3)(k + 6)$

48. **Statistics** Many drinks advertised and labeled as juice drinks contain as little as 5% fruit juice. Below is the number of grams of sugar contained in a single-serving size of 10 different fruit juices.

33, 23, 7, 29, 23, 27, 40, 29, 34, 22

Find the mean, median, and mode of the data. **26.7; 28; bimodal: 23, 29**

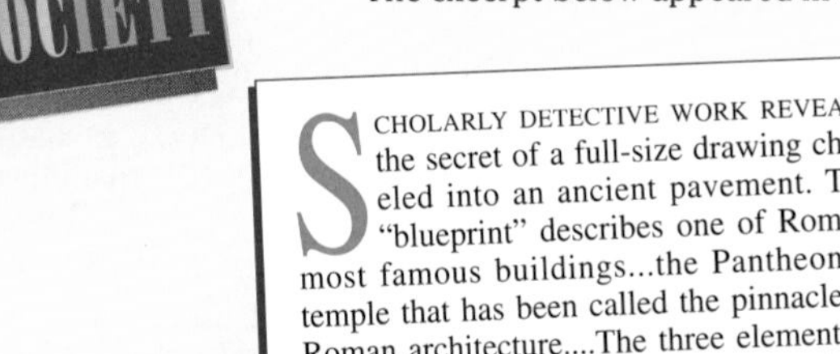

Interpreting an Ancient Blueprint

The excerpt below appeared in an article in *Scientific American* in June, 1995.

SCHOLARLY DETECTIVE WORK REVEALS the secret of a full-size drawing chiseled into an ancient pavement. The "blueprint" describes one of Rome's most famous buildings...the Pantheon, a temple that has been called the pinnacle of Roman architecture....The three elementary dimensions of the Pantheon's facade—diameter of the columns... "clear distance," or space between the columns...and height of the columns—are in the ratio of 1 to 2 to 9.5. This formula is among those described by Hermogenes, one of the most celebrated architects of the Hellenistic age, as belonging to an ideal facade....For the first time in the exploration of Roman architecture, a numerical recipe for beauty, as known from a textual source, can be tied to an existing monument. ■

1. The chiseled stone blueprint for the Pantheon was drawn at full scale; its dimensions were the same as those of the actual temple. What advantages and disadvantages are there in using this approach? **1–3. See margin.**
2. The full-size stone blueprint was identified as the Pantheon by comparing its angles and dimensions. How could you find the size of a large building without drawings or measuring directly?
3. The granite and marble blocks used to build the temple weighed many tons. Why do you think the courtyard, which contained the blueprint and was the place where the stones were measured and cut, was located close to the Tiber River?

4 ASSESS

Closing Activity

Speaking On the chalkboard or overhead, draw a picture similar to the figure in Example 1. Go around the room and have each student name parts of the two triangles that are either congruent or proportional to each other. Have them justify their statements.

Answers for Mathematics and Society

1. This approach avoids the possibility of making mathematical errors that could result from using a reduced scale. Also, pieces of stone could be measured directly against the blueprint lines as they were cut, to ensure proper size and fit.
2. You could use a smaller "reference" object whose size you know. For example, you could photograph a familiar person or object that is next to the building and use the photograph to determine the height and width of the building. Or you could measure shadows cast by the building and by a reference object.

Enrichment Masters, p. 40

NAME ____________ DATE ________

Enrichment

Student Edition Pages 362–369

Proportions for Similar Triangles

Recall that if a line crosses two sides of a triangle and is parallel to the third side, then the line separates the two sides that it crosses into segments of proportional lengths.

You can write many proportions by identifying similar triangles in the following diagram. In the diagram, $\overleftrightarrow{AM} \parallel \overleftrightarrow{BN}$, $\overleftrightarrow{AE} \parallel \overleftrightarrow{FL} \parallel \overleftrightarrow{MR}$, and $\overleftrightarrow{DQ} \parallel \overleftrightarrow{ER}$.

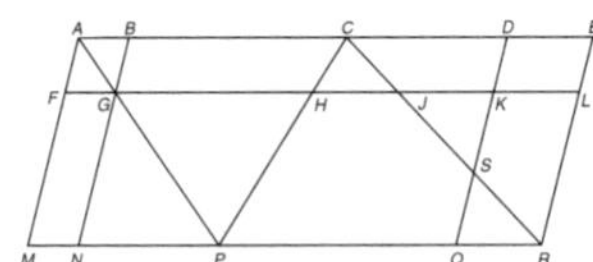

Answer each question. Use the diagram above.

1. Name a triangle similar to $\triangle GNP$. $\triangle AMP$ or $\triangle GBA$ or $\triangle AFG$
2. Name a triangle similar to $\triangle CJH$. $\triangle CRP$
3. Name two triangles similar to $\triangle JKS$. $\triangle JLR$, $\triangle CDS$ (others are possible)
4. Name a triangle similar to $\triangle ACP$. $\triangle GHP$

Complete each proportion.

5. $\frac{AG}{AP} = \frac{AF}{?}$ AM
6. $\frac{CP}{CH} = \frac{CR}{?}$ CJ
7. $\frac{JS}{JR} = \frac{?}{JL}$ JK
8. $\frac{PH}{PC} = \frac{PG}{?}$ PA
9. $\frac{ER}{LR} = \frac{?}{JR}$ CR
10. $\frac{MN}{MP} = \frac{?}{AP}$ GA

Solve.

11. If $CJ = 16$, $JR = 48$, and $LR = 30$, find EL. 10
12. If $DK = 5$, $KS = 7$, and $CJ = 8$, find JS. 11.2
13. If $MN = 12$, $NP = 32$, and $AP = 48$, find AG. Round to the nearest tenth. 13.1
14. If $CH = 18$, $HP = 82$, and $CR = 130$, find CJ. 23.4
15. Write three more problems that can be solved using the diagram above. See students' work.

Mathematics and SOCIETY

Modern reconstructive surgeons also use mathematical proportions in their work. Ideally, the distance from the tip of the chin to the tip of the nose, the distance from the tip of the nose to the eyebrows, and the distance from the eyebrows to the hairline should be equal.

Answer for Mathematics and Society

3. The rough blocks could be floated down the river from a quarry site to the blueprint area where the measuring and cutting took place. This would have been much easier than trying to transport the stones over land. It also avoided potential problems caused by rough terrain and bad weather.

7-5 LESSON NOTES

NCTM Standards: 1–5, 7

Instructional Resources

- Study Guide Master 7-5
- Practice Master 7-5
- Enrichment Master 7-5
- Assessment and Evaluation Masters, p. 185
- Real-World Applications, 14

Transparency 7-5A contains the 5-Minute Check for this lesson; **Transparency 7-5B** contains a teaching aid for this lesson.

Recommended Pacing	
Standard Pacing	Days 7 & 8 of 13
Honors Pacing	Days 6 & 7 of 12
Block Scheduling*	Day 5 of 7

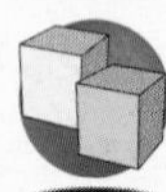

*For more information on pacing and possible lesson plans, refer to the *Block Scheduling Booklet.*

1 FOCUS

5-Minute Check
(over Lesson 7-4)

In the figure below, $\overleftrightarrow{TM} \parallel \overleftrightarrow{XN}$.

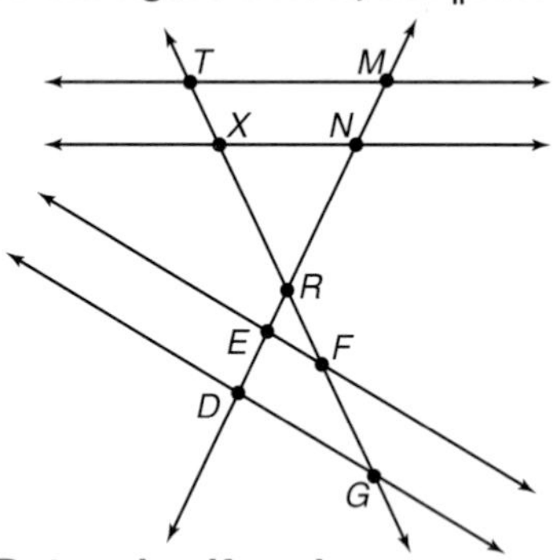

Determine if each statement is *true* or *false*.

1. $\frac{TX}{TR} = \frac{MN}{MR}$ true
2. $\frac{TR}{TX} = \frac{MR}{NR}$ false

Determine whether $\overleftrightarrow{EF} \parallel \overleftrightarrow{DG}$ under the given condition.

3. $\frac{EF}{RE} = \frac{DG}{RF}$ no
4. $\frac{RE}{EF} = \frac{RD}{DG}$ no
5. If $RN = 5$, $MN = 3$, and $TX = 6$, find RX. $\frac{5}{3} = \frac{RX}{6}$; $RX = 10$

7-5 Parts of Similar Triangles

What YOU'LL LEARN

- To recognize and use the proportional relationships of corresponding perimeters, altitudes, angle bisectors, and medians of similar triangles.

Why IT'S IMPORTANT

You can use proportional measures to solve problems involving photography, design, and art.

Games

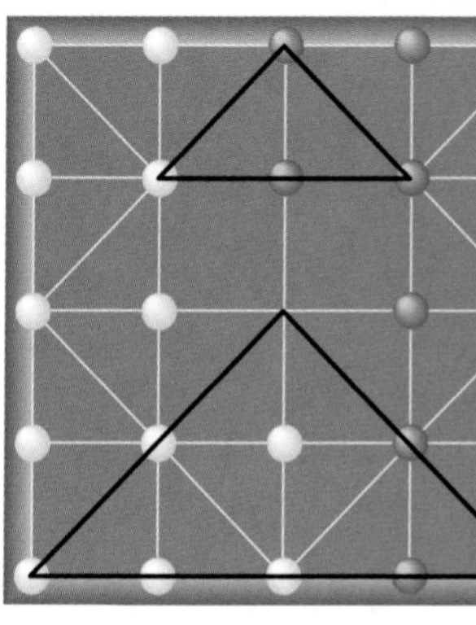

Alquerque is a Spanish game similar to *Checkers.* Two triangular paths are shown on the game board at the right. Notice that the paths are similar in shape and each side of the larger triangle is twice as long as each corresponding side of the smaller triangle. Will a game piece that moves along the path of the larger triangle move twice as far as a game piece that moves along the smaller triangle?

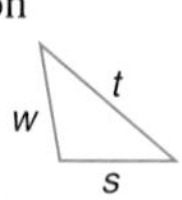

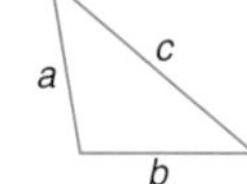

When two triangles are similar, there is a common ratio r such that $\frac{w}{a} = \frac{s}{b} = \frac{t}{c} = r$.

So, $w = ar$, $s = br$, $t = cr$.

The perimeter of the smaller triangle P_1 is $w + s + t$, and the perimeter of the larger triangle P_2 is $a + b + c$.

$$\frac{P_1}{P_2} = \frac{w+s+t}{a+b+c}$$

$$= \frac{ar+br+cr}{a+b+c} \quad \textit{Substitution Property (=)}$$

$$= \frac{r(a+b+c)}{a+b+c} \quad \textit{Distributive Property}$$

$$= r$$

So, the ratio of the perimeters of the two triangles is the same as the ratio between the corresponding sides. Thus, a game piece that moves along the path of the larger triangle will move twice as far as a game piece that moves along the shorter path.

Theorem 7–7 Proportional Perimeters	**If two triangles are similar, then the perimeters are proportional to the measures of corresponding sides.**

Example  **If $\triangle LMN \sim \triangle QRS$, $QR = 40$, $RS = 41$, $SQ = 9$, and $LM = 9$, find the perimeter of $\triangle LMN$.**

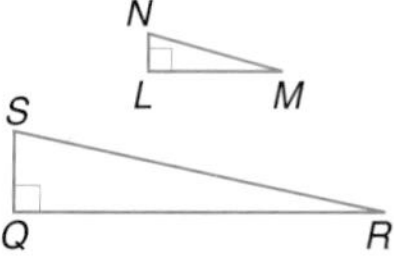

Let x represent the perimeter of $\triangle LMN$. The perimeter of $\triangle QRS = 40 + 41 + 9$ or 90 units.

$\frac{LM}{QR} = \frac{x}{\text{perimeter of } \triangle QRS}$ *Proportional perimeters*

$\frac{9}{40} = \frac{x}{90}$ *Substitution Property (=)*

$x = 20.25$

The perimeter of $\triangle LMN$ is 20.25 units.

When two triangles are similar, the measures of their corresponding sides are proportional. What about the measures of their corresponding altitudes, medians, and angle bisectors? The following theorem states one relationship. *You will be asked to prove this theorem in Exercise 36.*

Theorem 7–8 **If two triangles are similar, then the measures of the corresponding altitudes are proportional to the measures of the corresponding sides.**

Example 2 **In the figure at the right, $\triangle ABC \sim \triangle DEF$. If $\overline{BG}$ is an altitude of $\triangle ABC$, and $\overline{EH}$ is an altitude of $\triangle DEF$, then complete the following.**

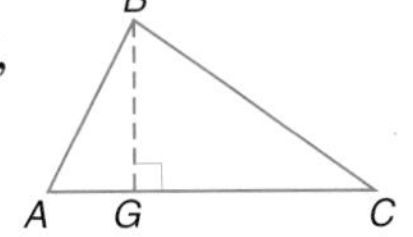

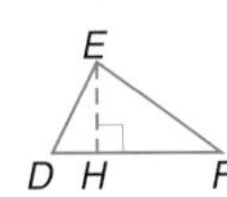

a. $\frac{BG}{EH} = \frac{?}{DE}$ The corresponding measure is *AB*.

b. $\frac{BG}{EH} = \frac{BC}{?}$ The corresponding measure is *EF*.

There is a similar theorem involving angle bisectors.

Theorem 7–9 **If two triangles are similar, then the measures of the corresponding angle bisectors of the triangles are proportional to the measures of the corresponding sides.**

Proof of Theorem 7–9

Given: $\triangle RTS \sim \triangle EGF$

$\overline{TA}$ is an angle bisector of $\angle RTS$.

$\overline{GB}$ is an angle bisector of $\angle EGF$.

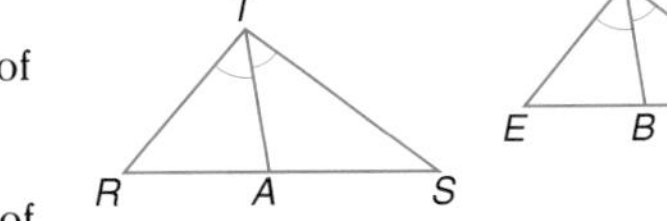

Prove: $\frac{TA}{GB} = \frac{RT}{EG}$

Paragraph Proof:

Since $\triangle RTS \sim \triangle EGF$, $\angle R$ and $\angle E$ are corresponding angles, and $\angle RTS$ and $\angle EGF$ are corresponding angles. Thus, $\angle R \cong \angle E$ and $\angle RTS \cong \angle EGF$. If $\overline{TA}$ and $\overline{GB}$ bisect $\angle RTS$ and $\angle EGF$, respectively, then $\angle RTA \cong \angle EGB$. $\triangle RTA \sim \triangle EGB$ by AA Similarity. Thus, $\frac{TA}{GB} = \frac{RT}{EG}$.

The following theorem states the relationship between the medians of two similar triangles. *You will be asked to prove this theorem in Exercise 39.*

Theorem 7–10 **If two triangles are similar, then the measures of the corresponding medians are proportional to the measures of the corresponding sides.**

The theorems about the relationships of special segments in similar triangles can be used to solve practical problems.

Classroom Vignette

"I have students use *The Geometric Supposer* to discover that corresponding altitudes and medians in similar triangles have the same ratio as the sides. Measures of angles, segments, perimeters, and areas of polygons can be computed by the program. To encourage student involvement and collaboration, students worked in groups to complete a worksheet. Students can use the software to test their conjectures."

Diane DeRyckere
Bishop Foley High School
Madison Heights, Michigan

Motivating the Lesson

Questioning On the chalkboard or overhead, draw two similar triangles whose sides have a common ratio of 2. Have students make conjectures about the relationship of the perimeters of the two triangles. **They are proportional to the lengths of the corresponding sides.**

2 TEACH

In-Class Examples

For Example 1

In the figure below, $\triangle ABD \sim \triangle ADC$. If $AD = 16$, $AC = 31$, and $DC = 23$, find the perimeter of $\triangle ABD$. **36.1**

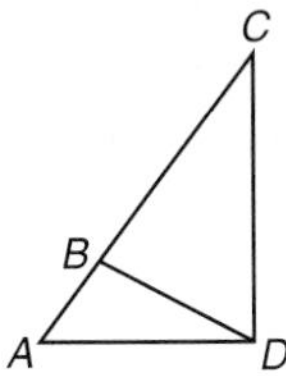

For Example 2

In the figures below, $\triangle PQR \sim \triangle TUV$. If $\overline{QO}$ is an altitude of $\triangle PQR$, and $\overline{US}$ is an altitude of $\triangle TUV$, then complete the following.

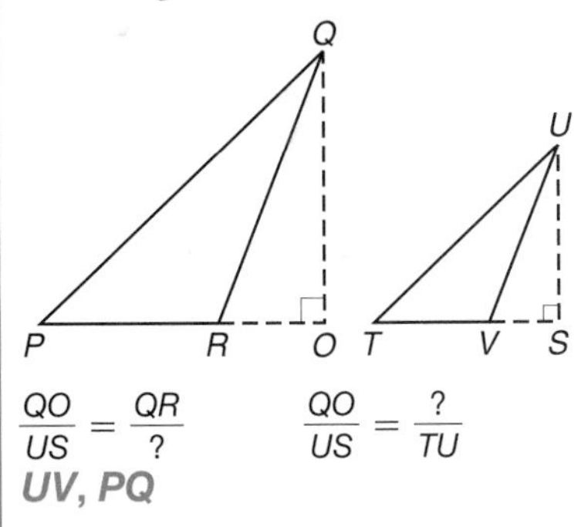

$\frac{QO}{US} = \frac{QR}{?}$ $\frac{QO}{US} = \frac{?}{TU}$

UV, PQ

Teaching Tip Theorem 7-9 also states that $\frac{TA}{GB} = \frac{TS}{GF}$ and that $\frac{TA}{GB} = \frac{RS}{EF}$. In addition, you can choose any altitude; the ratios will stay the same.

Teaching Tip Theorems 7-8, 7-9, and 7-10 are true for any of the three pairs of corresponding sides.

In-Class Example

For Example 3
Using the diagram for Example 3, suppose the film is 1.8 centimeters from the lens and a person being photographed is 360 centimeters from the camera. Find the height of the image on the film if the height of the person is 260 centimeters. **1.3 cm**

3 PRACTICE/APPLY

Check for Understanding

Exercises 1–11 are designed to help you assess your students' understanding through reading, writing, speaking, and modeling. You should work through Exercises 1–4 with your students and then monitor their work on Exercises 5–11.

Error Analysis
Students may have difficulty setting up ratios with the correct corresponding sides. Have them mark which sides correspond in the figure or write down which sides correspond before writing the ratios. It may also be helpful to redraw triangles separately when they overlap in the original figure.

Example 3 APPLICATION **Photography**

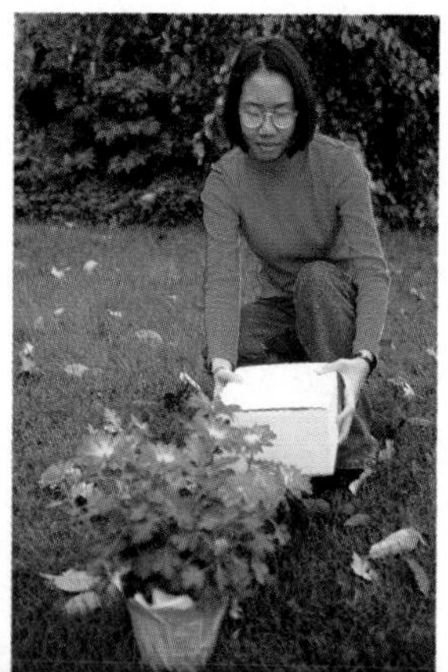

In photography class, Sua made a camera from a box by poking a small hole P in the center of one side of the box. Suppose Sua photographed a flowering bush that was 24 inches tall. How tall will the image of the bush be on the film, if the film is 5 inches from the lens and the camera is 40 inches from the bush?

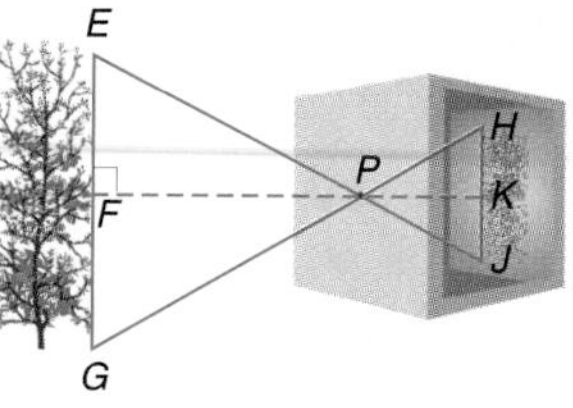

The dashed lines are the altitudes of $\triangle EGP$ and $\triangle JHP$.

Let b represent the height of the bush on the film. Assume that $\overline{EG} \parallel \overline{HJ}$. $\angle PEF \cong \angle PJH$ and $\angle PGE \cong \angle PHJ$ since they are alternate interior angles. Thus, $\triangle EPG \sim \triangle JPH$ by AA Similarity.

The measures of the corresponding altitudes of similar triangles are proportional to the measures of corresponding sides.

$$\frac{PF}{PK} = \frac{EG}{JH}$$

$$\frac{40}{5} = \frac{24}{b} \quad \text{Substitution Property (=)}$$

$$40b = 120 \quad \text{Cross products}$$

$$b = 3$$

The image of the bush on the film will be 3 inches tall.

There is also a relationship between an angle bisector and the side of the triangle opposite the angle.

Theorem 7–11 ***Angle Bisector Theorem***	**An angle bisector in a triangle separates the opposite side into segments that have the same ratio as the other two sides.**

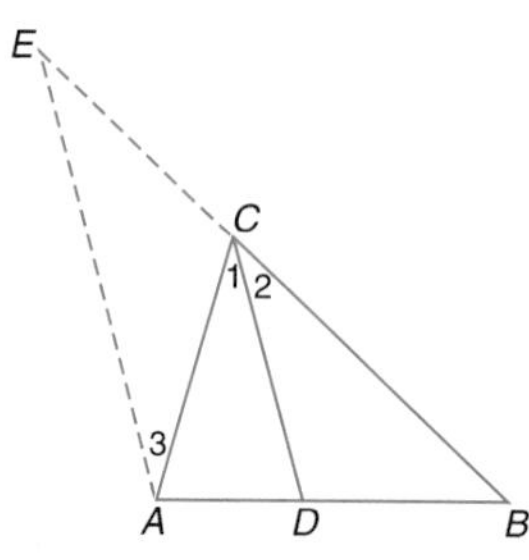

$\overline{CD}$ is the bisector of $\angle ACB$ in $\triangle ABC$. According to Theorem 7–11, $\frac{AD}{DB} = \frac{AC}{BC}$. To prove this, construct a line through point A parallel to $\overline{DC}$ and meeting $\overleftrightarrow{BC}$ at E. You can prove that $\frac{AD}{DB} = \frac{EC}{BC}$ and $CA = CE$. *You will be asked to prove this theorem in Exercise 40.*

1. 3 to 1; by Proportional Perimeter Theorem

CHECK FOR UNDERSTANDING

Communicating Mathematics

Study the lesson. Then complete the following.

1. **Write** the ratio of the perimeters of two similar triangles if the ratio of their corresponding sides is 3 to 1. Explain how you know.
2. **Explain** what must be true about $\triangle EFG$ and $\triangle RST$ before you can conclude that $\frac{FH}{SW} = \frac{EF}{RS}$.

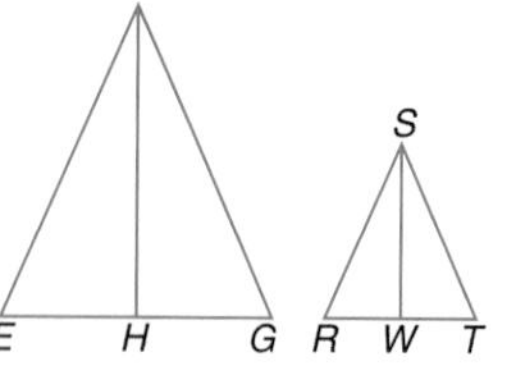

2. $\triangle EFG \sim \triangle RST$ and one of the following: $\overline{FH}$ and $\overline{SW}$ are altitudes, angle bisectors, or medians.

Reteaching

Using Models Use the zoom feature on the copy machine to make similar triangles. Separate the class into groups and give each group a set of triangles. Assign each group a theorem from this lesson. Have each group make a detailed example for the theorem and present its example to the class.

3. Sample answer: Two of the three sets of altitudes are also sides of the triangles.

3. **Describe** how Theorem 7–8 applies to the three sets of altitudes in two similar right triangles.

4. **You Decide** Ayashe was given the figure at the right. She concluded that $\frac{AB}{BC} = \frac{DC}{AD}$ by the Angle Bisector Theorem. Was she correct? Why or why not? No; the correct ratio is $\frac{AD}{DC} = \frac{AB}{BC}$.

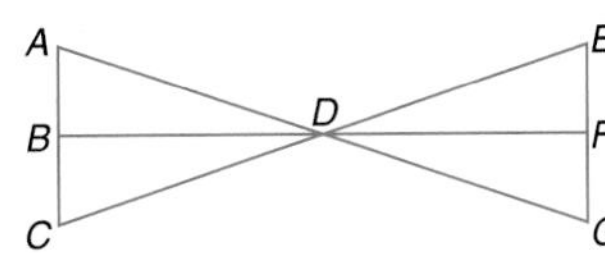

Guided Practice

In the figure at the right, $\triangle ADC \sim \triangle GDE$. Complete each proportion. Give a reason for your answer.

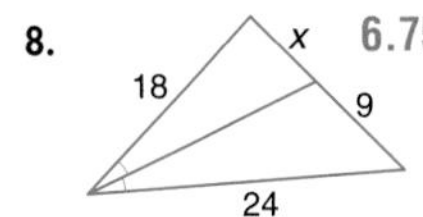

5. If BD bisects $\angle ADC$, $\frac{AD}{DC} = \frac{?}{BC}$. *AB*; Angle Bisector Theorem

6. *DF*; the medians of two similar triangles are proportional to the corresponding sides.

6. If B and F are midpoints, $\frac{DC}{DE} = \frac{BD}{?}$.

Find the value of x.

7. 15

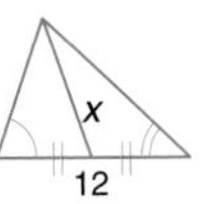

8. 6.75

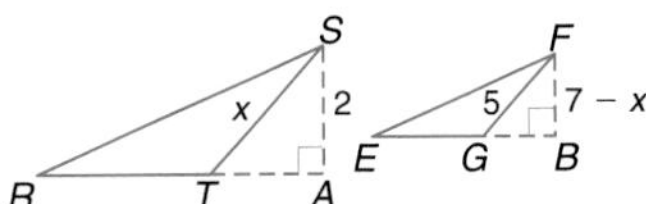

9. In the figure at the right, $\triangle RST \sim \triangle EFG$, and $\overline{SA}$ and $\overline{FB}$ are altitudes. Find FB. 5 or 2

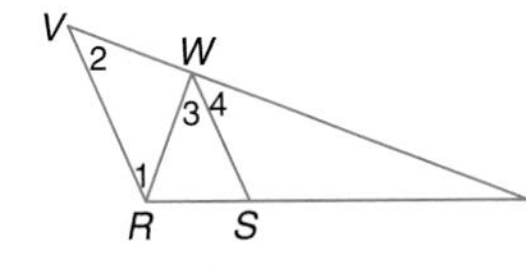

10. Write a two-column proof.

Given: $\overline{VR} \parallel \overline{WS}$

$\overline{WS}$ bisects $\angle RWT$.

Prove: $\frac{RW}{WT} = \frac{RS}{ST}$ See margin.

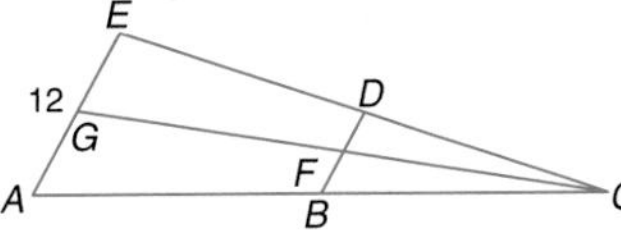

11. **Eliminate the Possibilities** The perimeter of one triangle is 24 centimeters, and the perimeter of a second triangle is 36 centimeters. Find measures for the sides of the two triangles so the triangles will be similar. Sample answer: 6, 8, 10 and 9, 12, 15

EXERCISES

Practice

Refer to $\triangle AEC$ for Exercises 12–17. $\overline{EA} \parallel \overline{DB}$, $EA = 12$, and D and B are midpoints of $\overline{EC}$ and $\overline{AC}$, respectively.

E
12
D
G
F
A
B
C

12. If $\angle DCF \cong \angle BCF$, $DF = 2$, and $BC = 8$, $DC = \underline{?}$. 4

13. If $\overline{GC} \perp \overline{EA}$ and $FC = 4.3$, $GC = \underline{?}$. 8.6

14. If F and G are midpoints and $GC = 14$, $FC = \underline{?}$. 7

Additional Answer

10. Given: $\overline{VR} \parallel \overline{WS}$, $\overline{WS}$ bisects $\angle RWT$.

Prove: $\frac{RW}{WT} = \frac{RS}{ST}$

Proof:

Statements (Reasons)

1. $\overline{VR} \parallel \overline{WS}$, $\overline{WS}$ bisects $\angle RWT$. (Given)
2. $\frac{VW}{WT} = \frac{RS}{ST}$ ($\triangle$ proportionality)
3. $\angle 3 \cong \angle 4$ (Def. $\angle$ bisector)
4. $\angle 3 \cong \angle 1$ (Alt. Int. $\angle$ Theorem)
5. $\angle 4 \cong \angle 1$ (Trans. Prop. (=))
6. $\angle 4 \cong \angle 2$ (Corr. $\angle$s Post.)
7. $\angle 1 \cong \angle 2$ (Trans. Prop. (=))
8. $\overline{VW} \cong \overline{RW}$ (If 2 $\angle$s of a $\triangle$ are $\cong$, the sides opp. those $\angle$s are $\cong$.)
9. $VW = RW$ (Def. $\cong$ segments)
10. $\frac{RW}{WT} = \frac{RS}{ST}$ (Subst. Prop. (=))

Alternative Learning Styles

Auditory Have students work in small groups to draw two similar triangles. They should measure the sides to find the perimeters and examine the ratios of perimeters and sides. As a group, they should discuss what patterns they observe and try to formulate conjectures.

Assignment Guide

Core (with proof): 13–41 odd, 42, 43, 45, 47–57
Core (informal): 13–33 odd, 41, 42, 43, 45, 47–57
Enriched: 12–40 even, 42–57

For **Extra Practice**, see p. 777.

The red A, B, and C flags, printed only in the Teacher's Wraparound Edition, indicate the level of difficulty of the exercises.

Additional Answer

34. Given: $\overline{RU}$ bisects $\angle SRT$. $\overline{VU} \parallel \overline{RT}$

Prove: $\frac{SV}{VR} = \frac{SR}{RT}$

Proof:

Statements (Reasons)

1. $\overline{RU}$ bisects $\angle SRT$, $\overline{VU} \parallel \overline{RT}$ (Given)
2. $\angle S \cong \angle S$ (Reflexive Prop. of $\cong$ ∠s)
3. $\angle SUV \cong \angle STR$ (Corr. ∠s Postulate)
4. $\triangle SUV \sim \triangle STR$ (AA Similarity)
5. $\frac{SV}{VU} = \frac{SR}{RT}$ (Def. $\sim$ △s)
6. $\angle URT \cong \angle VUR$ (Alt. Int. ∠ Theorem)
7. $\angle VRU \cong \angle URT$ (Def. ∠ bisector)
8. $\angle VUR \cong \angle VRU$ (Trans. Prop. (=))
9. $\overline{VU} \cong \overline{VR}$ (If 2 ∠s of a △ are $\cong$, the sides opp. these ∠s are $\cong$.)
10. $\frac{SV}{VR} = \frac{SR}{RT}$ (Substitution Prop. (=))

15. If $FG = 18$ and $ED = 16$, $FC = \underline{\ ?\ }$. 18

16. If $\overline{CF} \perp \overline{BD}$ and $\overline{CG} \perp \overline{EA}$, $\frac{CF}{CG} = \frac{?}{CE}$. CD

17. If $\overline{CF}$ bisects $\angle BCD$ and $\overline{CG}$ bisects $\angle ACE$, $\frac{CG}{CF} = \frac{EA}{?}$. DB

Find the value of x.

18. 15.5

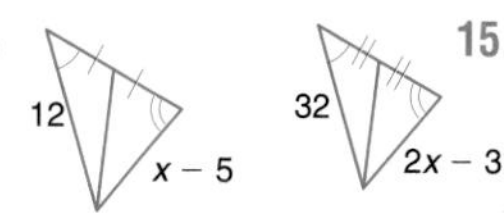

19. $11\frac{1}{5}$

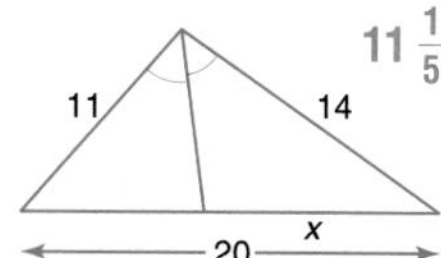

20. $8\frac{2}{11}$

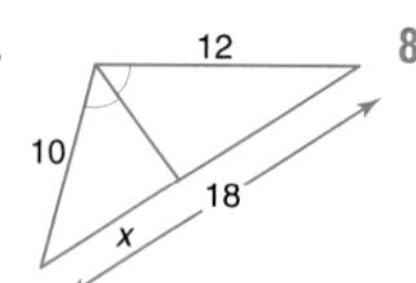

21. 36.68

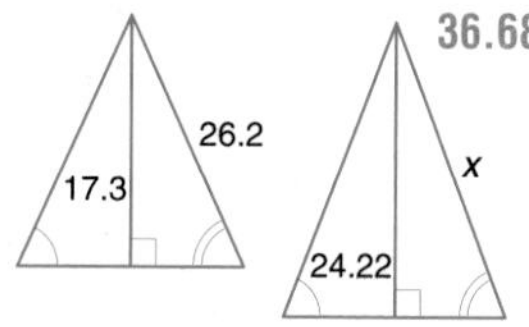

22. 6

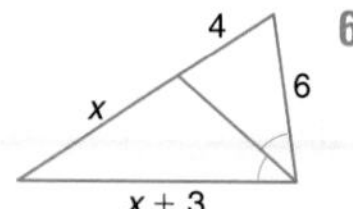

23. 6

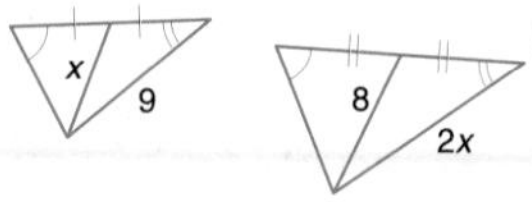

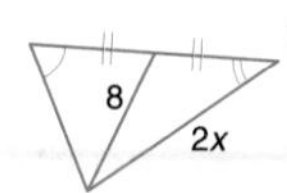

Refer to the figure at the right for Exercises 24–26. △*ABS* ~ △*RTS*.

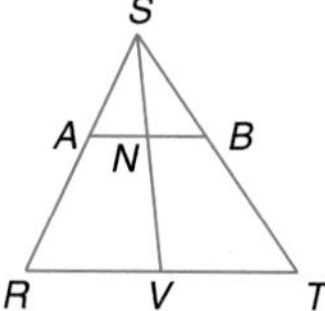

24. $\overline{SV}$ bisects $\angle ASB$. If $AS = 7$ and $AR = 11$, and $SV = 9$, find SN. 3.5

25. N is the midpoint of $\overline{AB}$ and V is the midpoint of $\overline{RT}$. If $AB = 16$, $SN = 13$, and $RT = 36$, find SV. 29.25

26. $142\frac{2}{9}$

26. If $AB = 22$, $SB = 24$, $AS = 18$, and $RS = 40$, find the perimeter of $\triangle RTS$.

27. In the figure at the right, $\triangle RST \sim \triangle UVW$, $\overline{TA}$ and $\overline{WB}$ are medians. If $TA = 8$, $RA = 3$, $WB = 3x - 6$, and $UB = x + 2$, find UB. 36

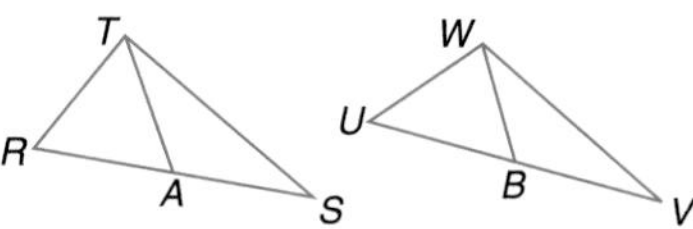

28. The two triangles in the figure at the right are similar. If the perimeter of $\triangle ABC$ is 28, find the value of x. 15.25

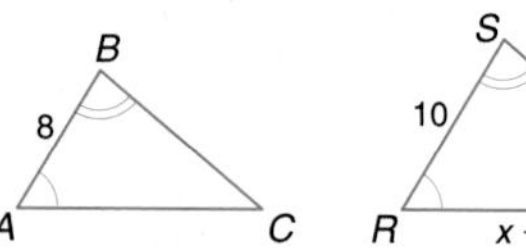

29. In the figure at the right, $\overline{BF}$ bisects $\angle ABC$ and $\overline{AC} \parallel \overline{ED}$. If $BA = 6$, $BC = 7.5$, $AC = 9$, and $DE = 9$, find CF and BD. 5, 13.5

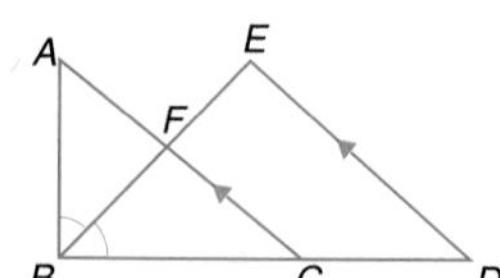

30. In the figure at the right, $\overline{BC} \parallel \overline{EF}$ and $\triangle ABC$ has vertices $A(0, 0)$, $B(6, 0)$ and $C(0, 8)$. Find the perimeter of $\triangle DEF$. Explain your reasoning. 36; $\triangle ABC \sim \triangle DEF$ by AA Similarity with a ratio of 1:2; perimeter of $\triangle DEF$ = 2(perimeter of $\triangle ABC$).

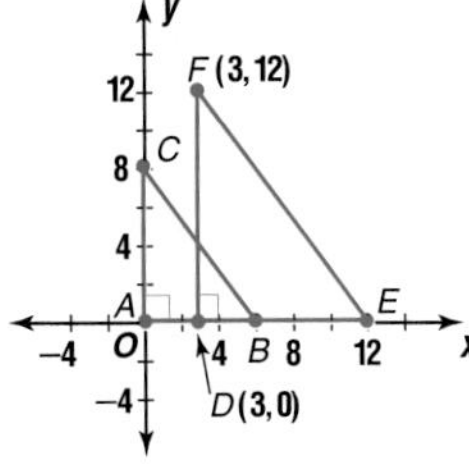

31. The measures of the sides of a triangle are 20, 24, and 30. Find the measures of segments formed where the bisector of the smallest angle meets the opposite side. $8\frac{8}{9}$, $11\frac{1}{9}$

32. If the bisector of an angle of a triangle bisects the opposite side, what is the ratio of the measures of the other two sides of the triangle? 1

33. If $\triangle ABC \sim \triangle RST$, $AB = 10.2$, $RS = 12.24$, and the perimeter of $\triangle RST$ is 32, find the perimeter of $\triangle ABC$. $26\frac{2}{3}$

Write a two-column proof. 34–35. See Margin.

34. **Given:** $\overline{RU}$ bisects $\angle SRT$.
$\overline{VU} \parallel \overline{RT}$

Prove: $\frac{SV}{VR} = \frac{SR}{RT}$

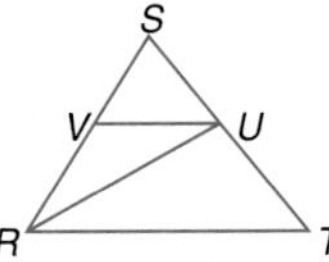

35. **Given:** $\overline{JF}$ bisects $\angle EFG$.
$\overline{EH} \parallel \overline{FG}$, $\overline{EF} \parallel \overline{HG}$

Prove: $\frac{EK}{KF} = \frac{GJ}{JF}$

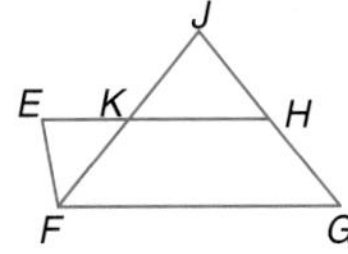

36–38. See Solutions Manual.

36. **Given:** $\triangle ABC \sim \triangle PQR$
$\overline{BD}$ is an altitude of $\triangle ABC$.
$\overline{QS}$ is an altitude of $\triangle PQR$.

Prove: $\frac{BD}{QS} = \frac{BA}{QP}$

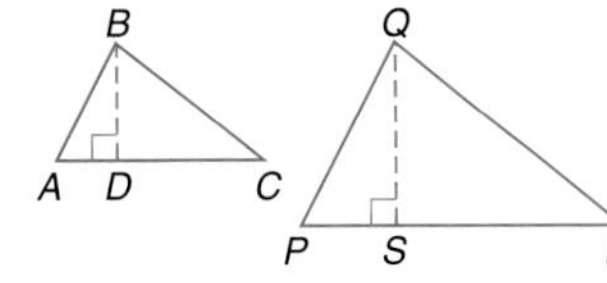

37. Write a flow proof.

Given: $\angle C \cong \angle BDA$

Prove: $\frac{AC}{DA} = \frac{AD}{BA}$

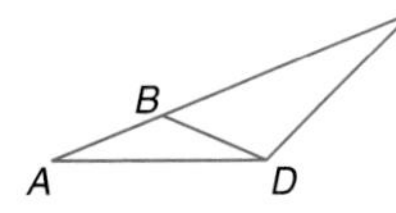

38. Prove using any method.

Given: $\triangle RST \sim \triangle ABC$
W and D are midpoints of $\overline{TS}$ and $\overline{CB}$, respectively.

Prove: $\triangle RWS \sim \triangle ADB$

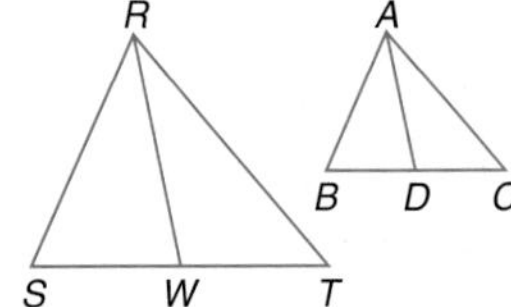

Write a paragraph proof. 39–40. See Solutions Manual.

39. If two triangles are similar, then the measures of the corresponding medians are proportional to the measures of the corresponding sides. (Theorem 7–10)

40. An angle bisector in a triangle separates the opposite side into segments that have the same ratio as the other two sides. (Theorem 7–11)

Additional Answer

35. Given: $\overline{JF}$ bisects $\angle EFG$.
$\overline{EH} \parallel \overline{FG}$
$\overline{EF} \parallel \overline{HG}$

Prove: $\frac{EK}{KF} = \frac{GJ}{JF}$

Proof:

Statements (Reasons)

1. $\overline{JF}$ bisects $\angle EFG$, $\overline{EH} \parallel \overline{FG}$, $\overline{EF} \parallel \overline{HG}$ (Given)
2. $\angle EFK \cong \angle KFG$ (Def. $\angle$ bisector)
3. $\angle KFG \cong \angle JKH$ (Corr. $\angle$ Post.)
4. $\angle JKH \cong \angle EKF$ (Vert. $\angle$s are $\cong$.)
5. $\angle EFK \cong \angle EKF$ (Transitive Prop. (=))
6. $\angle EFK \cong \angle FJH$ (Alt. Int. $\angle$ Theorem)
7. $\angle EKF \cong \angle FJH$ (Transitive Prop. (=))
8. $\triangle EKF \sim \triangle GJF$ (AA Similarity)
9. $\frac{EK}{KF} = \frac{GJ}{JF}$ (Def. $\sim \triangle$s)

Study Guide Masters, p. 41

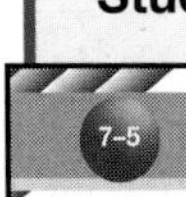

NAME ____________ DATE ____

Study Guide

Student Edition Pages 370–377

Parts of Similar Triangles

Each mathematical word in the following list has a different meaning in everyday usage than when it is used in a mathematical context. When you read mathematics, you must be sure that you are using each word in the correct context.

altitude	right	mean	scale
extreme	product	segment	ruler
plane	obtuse	acute	
meter	yard	median	

Each definition below describes an everyday usage of one of the words in the list above. For each definition, write the correct word from the list above in the blanks at the right.

1. an instrument for measuring meter
2. a tool for smoothing a wood surface plane
3. the opposite of left right
4. differing widely from the ordinary extreme
5. keen in perception acute
6. dull obtuse
7. contemptible mean
8. fragment segment
9. the dividing strip down the middle of a highway median
10. height above sea level altitude
11. a person in charge of a country ruler
12. the grounds of a building yard
13. an item that is manufactured product
14. a small plate forming part of the external covering of a fish scale
15. Choose five words from the list above. Compare their mathematical definition and their everyday usage. Describe how the definitions are similar and how they are different. See students' work.

Using the Programming Exercises The program given in Exercise 41 is for use with a TI-82 or TI-83 graphing calculator. For other programmable calculators, have students consult their owner's manual for commands similar to those presented here.

Additional Answer

41. Sample answer: If the ratio of the measures of the sides of two similar triangles is $a:b$, then the ratio of their areas is $a^2:b^2$.

Programming

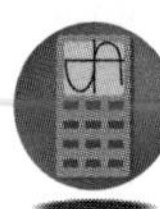

41. The graphing calculator program at the right finds the area of a triangle.

```
PROGRAM: AREA
:Input "SIDE 1 =", A
:Input "SIDE 2 =", B
:Input "SIDE 3 =", C
:If A+B≤C
:Goto 1
:If A+C≤B
:Goto 1
:If B+C≤A
:Goto 1
:(A+B+C)/2→S
:√(S(S−A)(S−B)(S−C))→K
:Disp"AREA=",K
:End
:Lbl 1
:Disp "NO TRIANGLE"
:Stop
```

Use the program to find the areas of similar triangles whose sides have the given measures. Then find the ratio of the measures of the sides and the ratio of the areas of the triangles. Write a conjecture about the relationship between the areas of two similar triangles.

a. 3, 4, 5 and 6, 8, 10 6; 24; 1:2; 1:4

b. 8, 11, 17 and 24, 33, 51

c. 5.1, 6.3, 8.1 and 30.6, 37.8, 48.6 16.1; 578.3; 1:6; 1:36

See margin for conjecture.

41b. 35.5; 319.5; 1:3; 1:9

42a. $\frac{s}{a} = \frac{h}{b}$ and $\frac{h}{b} = \frac{w}{c}$

Critical Thinking

42b. proportional to corresponding length, width, or height relationship

43. Yes; the enlarged picture will take approximately 109.2 cm of piping.

42. Consider the two rectangular prisms shown at the right.

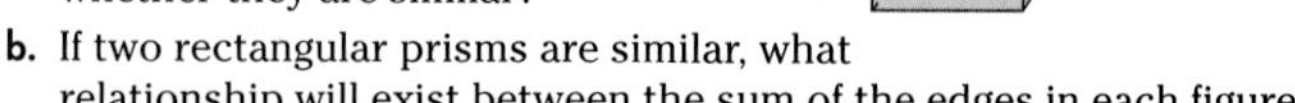

a. What ratios are necessary to determine whether they are similar?

b. If two rectangular prisms are similar, what relationship will exist between the sum of the edges in each figure?

c. If the first prism has dimensions that are three times as large as the second, will the volumes also have a ratio of 1 to 3? Explain why or why not. No; the ratio of volumes is 1 to 27.

Applications and Problem Solving

43. **Design** Julian had a picture 18 centimeters by 24 centimeters that he wanted enlarged by 30% and then have the inside of the frame edged with navy blue piping. The store only had 110 centimeters of navy blue piping in stock. Will this be enough piping to fit on the inside edge of the frame? Explain.

44. 330 cm

44. **Photography** A camera is 10 centimeters long and can have an image no more than 5 centimeters high. Estrella is 165 centimeters tall. How far from the camera should Estrella stand in order to have a full-length picture?

45. **Physical Fitness** Two triangular jogging paths are laid out in a park as shown. The dimensions of the inner path are 300 meters, 350 meters, and 550 meters. The shortest side of the outer path is 600 meters. Will a jogger on the inner path run half as far as the one on the outer path? Explain. yes; proportional perimeters

46. **Art** The design at the right is an abstract pentagonal design that illustrates a sequence of ratios in the relationships among the segments. Copy the figure and measure the segments. Use your measurements to investigate the similar figures in the design. Find any angle bisectors, altitudes or medians and determine if they are illustrations of the theorems in this lesson. See students' work.

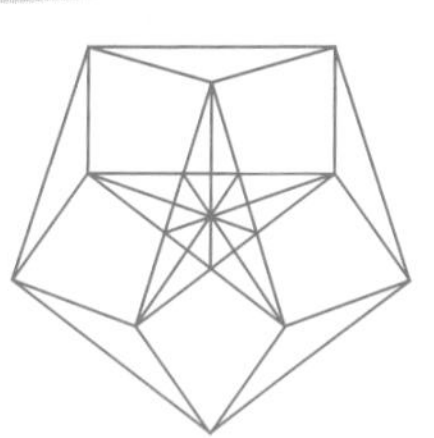

Practice Masters, p. 41

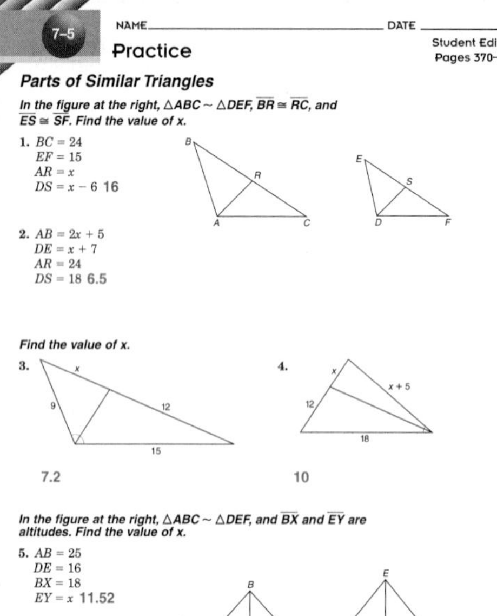

7-5 NAME________ DATE ________

Practice

Student Edition Pages 370–377

Parts of Similar Triangles

In the figure at the right, $\triangle ABC \sim \triangle DEF$, $\overline{BR} \cong \overline{RC}$, and $\overline{ES} \cong \overline{SF}$. Find the value of x.

1. $BC = 24$, $EF = 15$, $AR = x$, $DS = x - 6$ 16

2. $AB = 2x + 5$, $DE = x + 7$, $AR = 24$, $DS = 18$ 6.5

Find the value of x.

3. 7.2

4. 10

In the figure at the right, $\triangle ABC \sim \triangle DEF$, and $\overline{BX}$ and $\overline{EY}$ are altitudes. Find the value of x.

5. $AB = 25$, $DE = 16$, $BX = 18$, $EY = x$ 11.52

6. $AB = 30$, $DE = 25$, $BX = 2x + 5$, $EY = x + 10$ 8.75

Extension

Reasoning Determine if the areas of two similar triangles are proportional in any way. If so, find a proportion.

Let A_1 = area of smaller triangle, A_2 = area of larger triangle, r = common ratio, and $A_1 = \frac{1}{2}bh$. By Theorem 7-8, the measures of the corresponding altitudes are proportional to the measures of the corresponding bases. The base of the larger triangle is rb, and the altitude of the triangle is rh. So, $A_2 = \frac{1}{2}(rb)(rh)$ or $A_2 = \frac{1}{2}bh(r^2)$. Substituting, you obtain the equation $A_2 = A_1(r^2)$, or $\frac{A_2}{A_1} = r^2$.

Mixed Review

47. Find the value of x. (Lesson 7–4) **8.4**

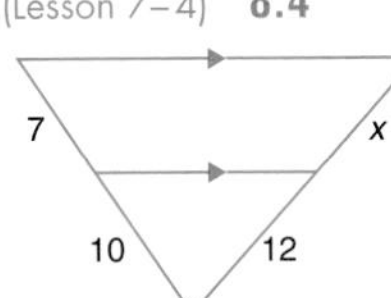

48. Determine whether the pair of triangles below is similar. Explain your answer. (Lesson 7–3) **yes; SSS Similarity**

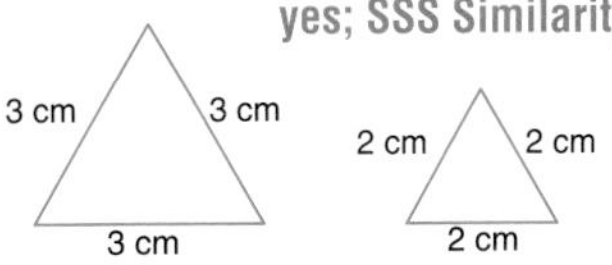

49. Sports The ratio of seniors to juniors on a powder puff football team is 2:3. If there are 21 juniors, how many seniors are on the team? (Lesson 7–1)

50. Determine whether the statement *The diagonals of an isosceles trapezoid are congruent* is true or false. (Lesson 6–5) **true**

51. Determine whether quadrilateral $ABCD$ must be a parallelogram. Justify your answer. (Lesson 6–2)

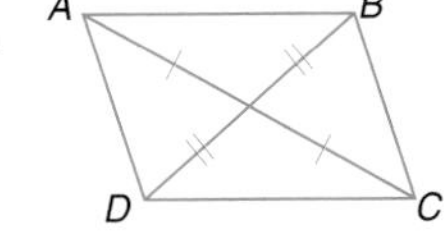

52. Draw and label $\triangle QRS$ with altitudes $\overline{RA}$ and $\overline{QB}$. (Lesson 5–1) **See margin.**

53. If $\triangle ABC$ has sides with measures 4, 8, and 10 and $\triangle RST$ has sides with measures 2, 4, and 5, is $\triangle ABC \cong \triangle RST$? Explain. (Lesson 4–4)

54. Two lines have slopes -4 and $\frac{1}{2}$. State whether or not the lines are perpendicular. Explain. (Lesson 3–3)

55. Write the statement *An 89 is an above average score on the geometry test* in if-then form. (Lesson 2–2)

Algebra

56. Solve $|3 - t| \leq 1$. $\{t \mid 2 \leq t \leq 4\}$

57. Entertainment In 1996, the group *Hootie and the Blowfish* discovered that scalpers bought the first ten rows of seats for two of their New York concerts. The tickets originally priced at \$25 were being sold for as much as \$150. At what percent of the original price were the tickets sold? **600%**

49. 14 seniors

51. Yes; diagonals bisect each other.

53. No; the measures of corresponding sides are not equal.

54. No; the product of the slopes is not −1.

55. If a geometry test score is 89, then it is above average.

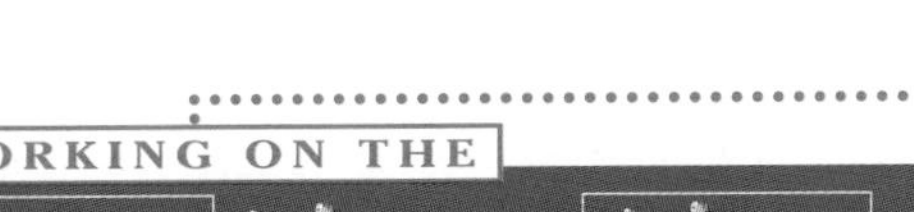

Refer to the Investigation on pages 286–287.

1 Review your plan. Make sure that all of the features you wanted to include are placed on the drawing.

2 Check to make sure that your proposal meets the budget requirements. If you are over budget, brainstorm methods for cutting costs. You may wish to cut less popular features, use volunteer labor for construction of less complicated features, or consider building the park in phases when more money is available.

3 Build a three-dimensional model of the park. Draw the boundary lines of the park on a piece of posterboard. Then use modeling clay, construction paper, paint, markers, and whatever else you need to create a three-dimensional model of your design.

Add the results of your work to your Investigation Folder.

4 ASSESS

Closing Activity

Writing Draw a triangle on the chalkboard or overhead. Have each student copy the figure and use it to demonstrate one of the theorems in this lesson. Have them write the given information and what is to be proved, based upon the figure.

Chapter 7 Quiz C (Lessons 7-4 and 7-5) is available in the *Assessment and Evaluation Masters*, p. 185.

Additional Answer

52.

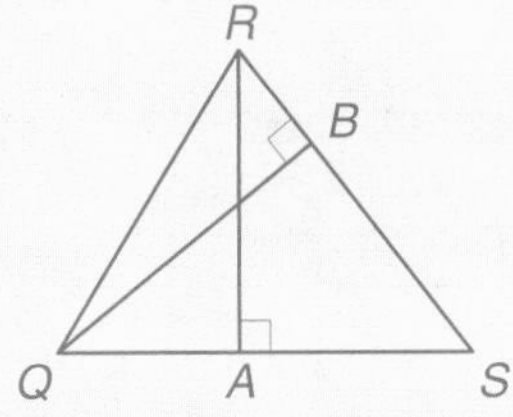

Enrichment Masters, p. 41

NAME ______ DATE ______ Student Edition Pages 370–377

Enrichment

Constructing Similar Polygons

Here are four steps for constructing a polygon that is similar to and with sides twice as long as those of an existing polygon.

Step 1 Choose any point either inside or outside the polygon and label it O.

Step 2 Draw rays from O through each vertex of the polygon.

Step 3 For vertex V, set the compass to length OV. Then locate a new point V' on ray OV such that $VV' = OV$. Thus, $OV' = 2(OV)$.

Step 4 Repeat Step 3 for each vertex. Connect points V', W', X' and Y' to form the new polygon.

Two constructions of polygons similar to and with sides twice those of $VWXY$ are shown below. Notice that the placement of point O does not affect the size or shape of $V'W'X'Y'$, only its location.

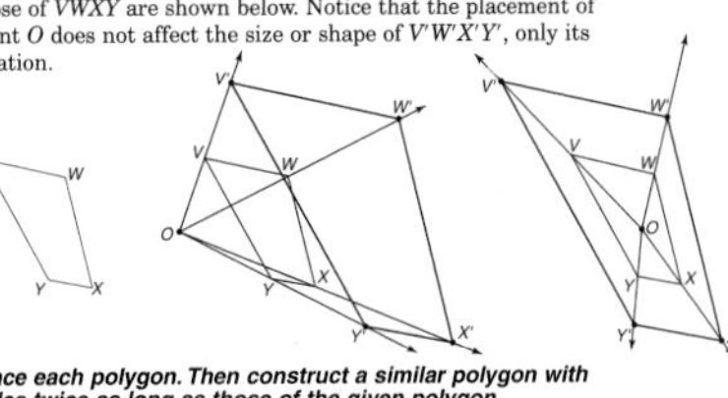

Trace each polygon. Then construct a similar polygon with sides twice as long as those of the given polygon.

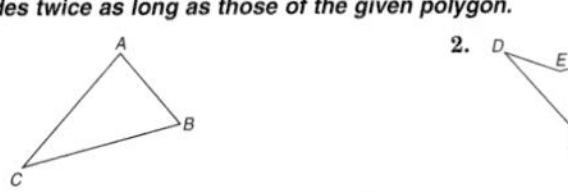

1–2. See students' constructions.

3. Explain how to construct a similar polygon with sides three times the length of those of polygon $HIJKL$. Then do the construction. See students' work.

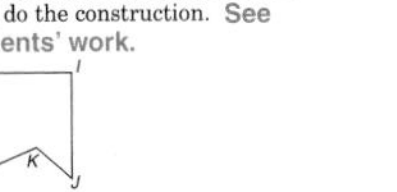

4. Explain how to construct a similar polygon $1\frac{1}{2}$ times the length of those of polygon $MNPQRS$. Then do the construction. See students' work.

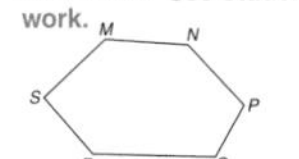

Tech Prep

Physical Therapist Assistant Physical therapists and their assistants measure the range of motion angles for joints on the body to determine results of injury or illness and to track a patient's recovery. For more information on tech prep, see the *Teacher's Handbook*.

In•ves•ti•ga•tion

Working on the Investigation

The Investigation on pages 286–287 is designed to be a long-term project that is completed over several days or weeks. Encourage students to keep their materials in their Investigation Folder as they work on the Investigation.

7-6 LESSON NOTES

NCTM Standards: 1–5, 7

Instructional Resources

- Study Guide Master 7-6
- Practice Master 7-6
- Enrichment Master 7-6
- Assessment and Evaluation Masters, p. 185
- Multicultural Activity Masters, p. 14
- Tech Prep Applications Masters, p. 14

Transparency 7-6A contains the 5-Minute Check for this lesson; **Transparency 7-6B** contains a teaching aid for this lesson.

Recommended Pacing	
Standard Pacing	Days 9 & 10 of 13
Honors Pacing	Days 8 & 9 of 12
Block Scheduling*	Day 6 of 7

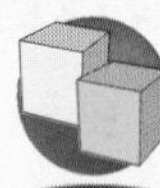

*For more information on pacing and possible lesson plans, refer to the *Block Scheduling Booklet.*

1 FOCUS

5-Minute Check
(over Lesson 7-5)

In the figure below, $\triangle AFD \sim \triangle BEC$.

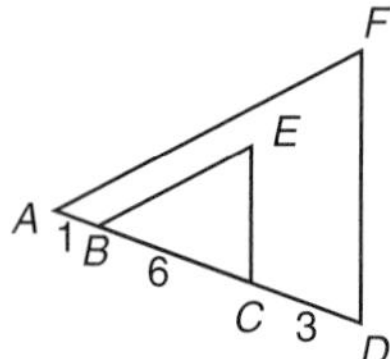

1. If $AB = 1$, $BC = 6$, $CD = 3$, and $AF = 14$, find BE. **8.4**
2. If $\overline{EL}$ is an altitude of $\triangle BEC$ and $\overline{FM}$ is an altitude of $\triangle AFD$, is it true that $\frac{EL}{BC} = \frac{FM}{AD}$? Why or why not? **Yes; if 2 $\triangle$s are similar, then the measures of the corr. altitudes are proportional to the measures of the corr. sides.**
3. If $\overline{EL}$ bisects $\angle E$ and $\overline{FM}$ bisects $\angle F$, is it true that $\frac{EL}{FM} = \frac{BD}{AC}$? Why or why not? **No; $\overline{BD}$ and $\overline{AC}$ are not corr. sides.**
4. If $\overline{EL}$ is a median of $\triangle BEC$ and $\overline{FM}$ is a median of $\triangle AFD$, is it true that $\frac{EL}{FM} = \frac{BL}{AM}$? **yes**

7-6 Fractals and Self-Similarity

What YOU'LL LEARN

- To recognize and describe characteristics of fractals, and
- to solve problems by solving a simpler problem.

Why IT'S IMPORTANT

You can observe patterns in nature that are examples of fractal geometry.

One interesting characteristic that can be found in mathematics is that orderly patterns and structure sometimes result from processes that seem to be random and chaotic.

The following activity shows a result from a process called **iteration**. Iteration is a process of repeating the same procedure over and over again.

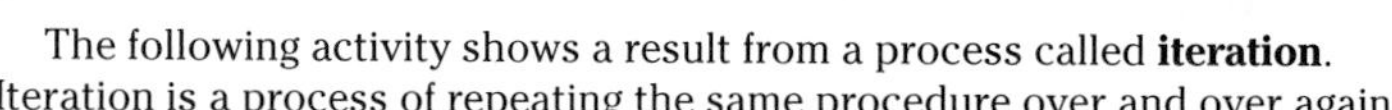

MODELING MATHEMATICS — Sierpinski Triangle

Materials:  isometric dot paper  ruler

- Draw an equilateral triangle on isometric dot paper where each side is 16 units long (stage 0).
- Connect the midpoints of each side to form another triangle. Shade the center triangle (stage 1).
- Repeat the process using the three nonshaded triangles. Connect the midpoints of each side to form other triangles (stage 2).

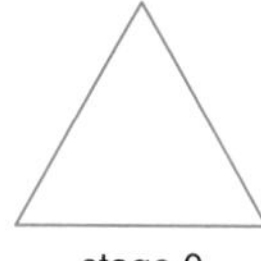

stage 0

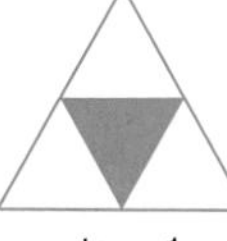

stage 1

stage 2

Your Turn **b. stage 0: 48 units, stage 1: 24 units, stage 2: 12 units, stage 3: 6 units, stage 4: 3 units**

a. Continue the process through stage 4. How many nonshaded triangles do you have at stage 4? **81**

b. What is the perimeter of a nonshaded triangle in each of the stages?

c. If you continue the process indefinitely, describe what will happen to the perimeter of each nonshaded triangle. **The perimeter will approach 0.**

The triangle that will result if you continue to repeat the process in the Modeling Mathematics activity indefinitely is called a **Sierpinski triangle**, named after Polish mathematician Waclaw Sierpinski. A Sierpinski triangle is an example of a mathematical object called a **fractal**. A fractal is a geometric figure that is created using iteration. It is infinite in structure.

Fractal geometry is the geometry of things in nature that are irregular in shape such as clouds, coastlines, or the growth of a tree. Look carefully at the tree in the top picture at the left and observe how the branches have developed. The picture below it shows veins and arteries in a human kidney.

One characteristic of fractals is that they have **self-similar** shapes. That is, the smaller and smaller details of a shape have the same geometrical characteristics as the original, larger form.

MODELING MATHEMATICS Have students duplicate the iterative process with other shapes on dot paper. Is there a trend to follow?

Example  **Prove that a triangle formed in stage 2 of a Sierpinski triangle is similar to the triangle formed in stage 0.**

The argument will be the same for any triangle in stage 2, so we will use only $\triangle CGJ$ from stage 2.

Given: $\triangle ABC$ is equilateral.
D, E, and F are midpoints of sides $\overline{AB}$, $\overline{BC}$, and $\overline{CA}$, respectively.
G, J, and H are midpoints of sides $\overline{FC}$, $\overline{CE}$, and $\overline{FE}$, respectively.

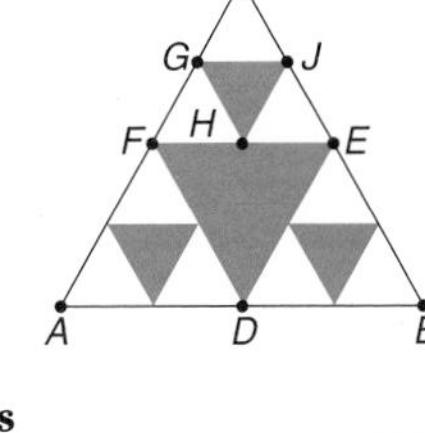

Prove: $\triangle ABC \sim \triangle GJC$

Proof:

Statements	Reasons
1. $\triangle ABC$ is equilateral; D, E, F are midpoints of $\overline{AB}$, $\overline{BC}$, $\overline{CA}$; G, J, and H are midpoints of $\overline{FC}$, $\overline{CE}$, $\overline{FE}$	1. Given
2. $\overline{FE} \parallel \overline{AB}$	2. A segment whose endpoints are the midpoints of 2 sides of a $\triangle$ is $\parallel$ to the third side.
3. $\angle CFH \cong \angle CAB$; $\angle CEF \cong \angle CBA$	3. Corresponding ∠s Postulate
4. $\triangle FEC \sim \triangle ABC$	4. AA Similarity
5. $\overline{GJ} \parallel \overline{FE}$	5. A segment whose endpoints are the midpoints of 2 sides of a $\triangle$ is $\parallel$ to the third side.
6. $\angle CGJ \cong \angle CFE$; $\angle CJG \cong \angle CEF$	6. Corresponding ∠s Postulate
7. $\triangle GJC \sim \triangle FEC$	7. AA Similarity
8. $\triangle ABC \sim \triangle GJC$	8. Transitive Property ($\sim$)

Thus, using the same reasoning, every triangle in stage 2 is similar to the original triangle in stage 0.

A figure is **strictly self-similar** if any of its parts, no matter where they are located or what size is selected, contain the same figure as the whole. The Sierpinski triangle is strictly self-similar.

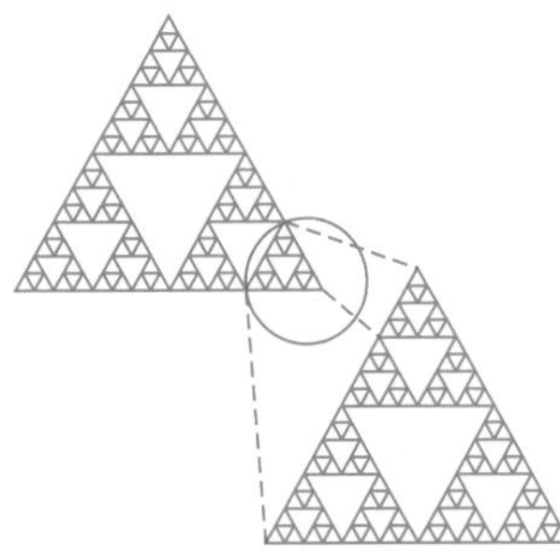

Blaise Pascal

Other repeated patterns are not self-similar. Pascal's triangle is a famous array of numbers that is generated according to a certain pattern. The first number in the initial row is 1. The number in each successive row is the sum of the two numbers above it. How could you find the sum of values in the tenth row?

When you are working on a complicated or unfamiliar problem, it is often helpful to first **solve a simpler problem** that is similar to the original. Then, use the strategy that worked on that simpler problem to solve the original problem. You can use this strategy to find the sum of the values in the tenth row of Pascal's triangle.

Motivating the Lesson

Hands-On Activity Have students collect objects from nature. They should look for objects that cannot be described in terms of regular polygons. Discuss the variety of shapes represented.

2 TEACH

In-Class Example

For Example 1
Is the fractal below self-similar?
yes

Teaching Tip Point out to students that the fractal structures generated in mathematics are idealized structures. Objects in the natural world that seem to have a fractal structure are not nearly as regular or symmetric.

Classroom Vignette

"I give each student a piece of isometric dot paper with the starting points for Sierpinski's Triangle highlighted. Students then draw the triangle and color attractively. The completed triangles make a nice bulletin board display. Sierpinski's Carpet and the Koch Snowflake are also nice projects for tying the material in the chapter to self-similarity."

Mrs. Janet Lovell
Mrs. Janet Lovell
Osbourn High School
Manassas, Virginia

In-Class Examples

For Example 2

a. Find the formula for the number of L's at each stage of the process pictured here.

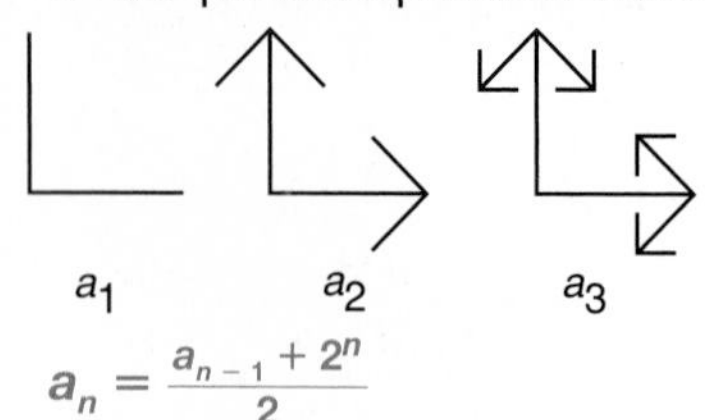

a_1 a_2 a_3

$a_n = \frac{a_{n-1} + 2^n}{2}$

b. How many L's are at stage 5? 63

For Example 3

Describe the pattern formed by $a_n = 2n - 1$.

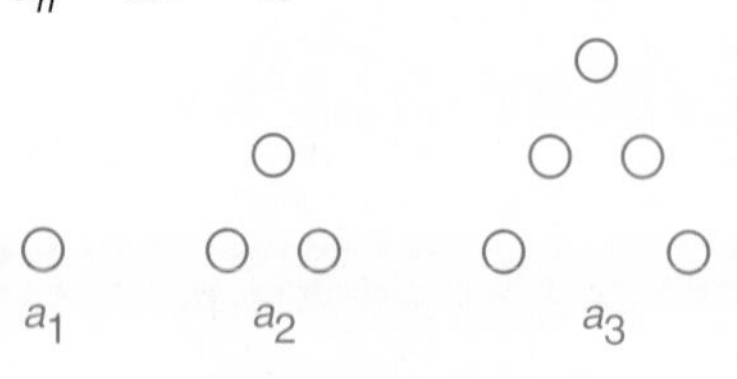

a_1 a_2 a_3

Teaching Tip Writing formulas from a pattern may be unfamiliar to some students. You may want to develop the formulas for odd $(2n + 1)$ or even $(2n)$ numbers to get them thinking in these terms.

Example 

PROBLEM SOLVING

Solve a Simpler Problem

a. Find a formula in terms of the row number for the sum of the values in any row in the triangle.

b. What is the sum of the values in the tenth row of Pascal's triangle?

a. To find the sum of the values in the tenth row, we can investigate a simpler problem. What is the sum of values in the first four rows of the triangle?

Pascal's Triangle

Row		Sum
1	1	$1 = 2^0$
2	1 1	$2 = 2^1$
3	1 2 1	$4 = 2^2$
4	1 3 3 1	$8 = 2^3$
5	1 4 6 4 1	$16 = 2^4$

It appears that the sum of any row is a power of 2. The formula is 2 to a power that is one less than the row number: $S_n = 2^{n-1}$.

b. The sum of the values in the tenth row will be 2^{10-1} or 512.

There is a relationship between Pascal's triangle and the Sierpinski triangle.

Example 

Replace each of the even numbers in Pascal's triangle with a 0 and each of the odd numbers with a 1. Then color each 1 and leave the 0s uncolored. Describe the picture you will have if you generate eight rows of Pascal's triangle and color the cells according to the rule.

First, generate eight rows of the triangle and replace each number as even or odd.

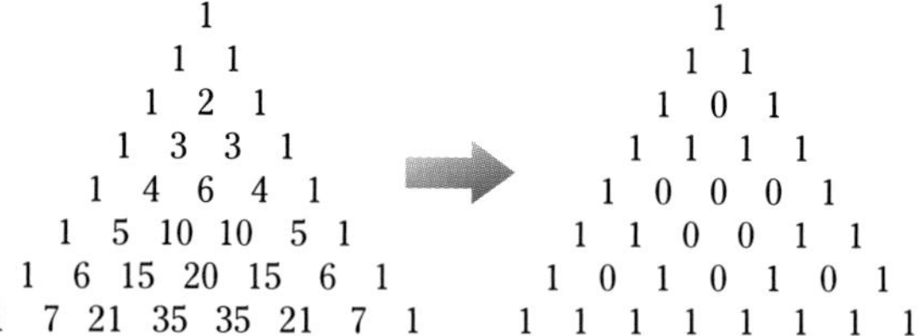

Now color each 1. The result looks like a stage 3 Sierpinski Triangle.

You can generate many other fractal images using an iterative process. For example, trisect a segment of a given length. Replace the middle segment with two segments each the same length as the segment removed, as shown in the diagram at the right. Repeat the process on each of the four segments in stage 2, then continue to repeat the process. The fractal image generated by this process is called a *Koch curve*.

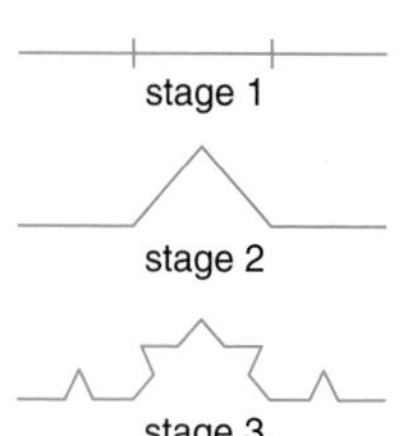

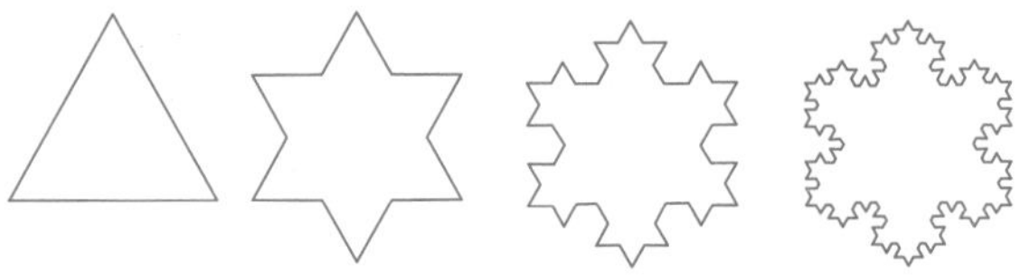

If you generate a Koch curve on each side of an equilateral triangle, you will produce an image called *Koch's snowflake.*

CHECK FOR UNDERSTANDING

Communicating Mathematics

Study the lesson. Then complete the following.

1. **Describe** a specific iterative process. Then draw examples of four stages of that iteration. See students' work.
2. **Describe** two ways you can find the sum of a row in Pascal's triangle.
3. **State** the number of segments that are in each of the first three stages of Koch's snowflake. What pattern can you use to predict the number of segments in stage 4?

2. See margin.

3. Stage 1: 3, stage 2: 12, stage 3: 48; multiply the previous number by 4, so there would be 192 segments in stage 4.

Use grid paper to draw a square that is 27 units long on each side. Connect the trisection points on each side to make nine squares. Then shade the middle square. This is stage 1. Repeat this process on the eight outside squares. The middle square remains a "hole."

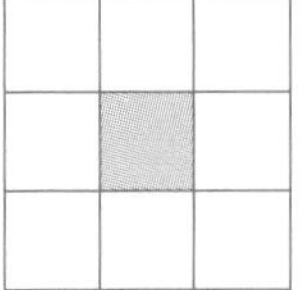

4. Draw stage 2 and stage 3. How many holes are there in each stage?
5. If you continue the process indefinitely, will the figure you obtain be strictly self-similar? Explain.

4. 9 holes, 73 holes; see Solutions Manual for drawings.

5. Yes; any part contains the same figure as the whole, 9 squares with the middle one shaded.

Guided Practice

Count the number of dots in each arrangement. These numbers are called *triangular numbers.* The second triangular number is 3 because there are three dots in the array.

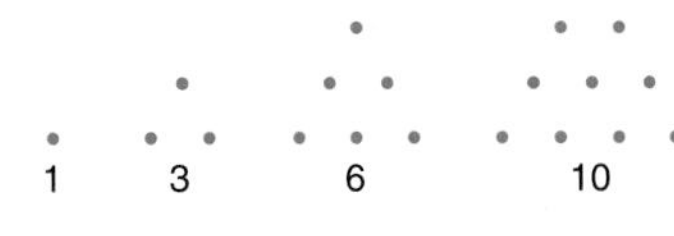

6. How many dots will be in the seventh triangular number? 28
7. Generate Pascal's triangle and look at the third diagonal from either side. Describe the pattern you see. How does it relate to the triangular numbers?
8. **Science** Find at least two examples in nature excluding those mentioned in the lesson in which a figure appears to be self-similar.

7. 1, 3, 6, 10, 15, ...; each difference is 1 more than the preceding difference. The triangular numbers are the numbers in the diagonal.

8. Sample answer: broccoli, leaf veins

EXERCISES

Practice

Draw an equilateral triangle, trisect the three sides, and connect the points. Shade the three inside triangles as shown at the right.

9. Prove that one of the nonshaded triangles is similar to the original triangle. See Solutions Manual.
10. Using the figure from Exercise 9, trisect the three sides in each of the nonshaded triangles and connect the points as in the original figure. Repeat the process once more on this figure.
   a. Is the new figure strictly self-similar? Explain.
   b. How many nonshaded triangles are in stages 1 and 2? stage 1: 6, stage 2: 36

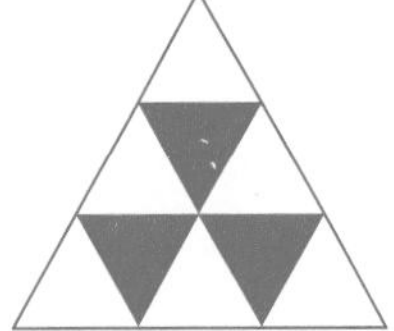

10a. Yes; any part will be exactly the same as the whole.

3 PRACTICE/APPLY

Check for Understanding

Exercises 1–8 are designed to help you assess your students' understanding through reading, writing, speaking, and modeling. You should work through Exercises 1–5 with your students and then monitor their work on Exercises 6–8.

Additional Answer

2. You can use the formula $S_n = 2^{n-1}$ from the lesson or you can generate the sum of the next row by multiplying the previous sum by 2.

Study Guide Masters, p. 42

NAME______________________ DATE__________

Study Guide

Student Edition Pages 378–383

Fractals and Self-Similarity

Fractal geometry is the geometry of things in nature that are irregular in shape. A **fractal** is a geometric shape created using a process called **iteration**. Iteration is a process of repeating the same procedure over and over again. **Self-similarity** is a characteristic of fractals. The smaller and smaller details of a shape have the same geometrical characteristics as the original, larger form.

Example:

Stage 1

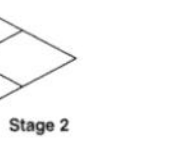

Stage 2

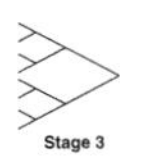

Stage 3

1. Follow the iteration process to produce a figure.
   - Draw a square.
   - Attach the hypotenuse of an isosceles right triangle to one side of the square. The hypotenuse should be the same length as the side of the square.
   - Attach a square to each leg of the triangle. The sides of the squares should be the same length as the legs of the triangle.

   This is Stage 1.

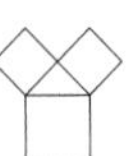

2. Describe the next step in the iteration process. Attach an isosceles right triangle to each of the two smaller squares.
3. Draw Stage 3 of the figure in Exercise 1.

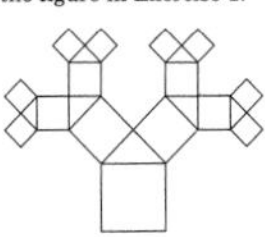

4. Is the figure produced in Exercise 3 self-similar? Yes

Reteaching

Using Diagrams Draw a square. Explore several ways of successively subdividing the square. Each leads to a different fractal.

Assignment Guide

Core (with proof): 9–21 odd, 23–29
Core (informal): 11–21 odd, 23–29
Enriched: 10–18 even, 19–29

For **Extra Practice**, see p. 777.

The red A, B, and C flags, printed only in the Teacher's Wraparound Edition, indicate the level of difficulty of the exercises.

Additional Answers

13a.

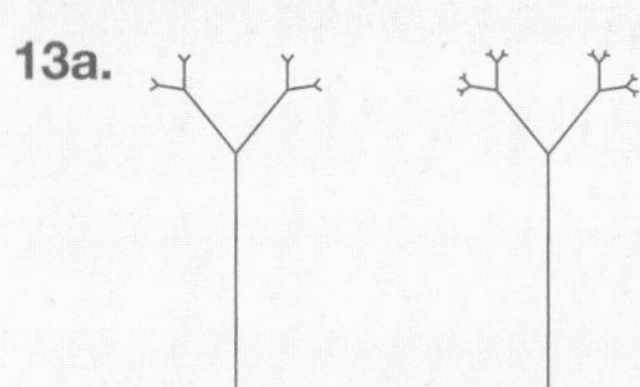

stage 1: 2, stage 2: 6, stage 3: 14, stage 4: 30

13b. Sample answer: At stage n, the number of branches would be $S_n = 2(1 - 2^n)$.

16b. stage 1: 1 unit, stage 2: $\frac{1}{3}$ unit, stage 3: $\frac{1}{9}$ unit, stage 4: $\frac{1}{27}$ unit; as the stages increase, the length of the segments will approach zero.

Practice Masters, p. 42

7-6 NAME________ DATE________

Practice Student Edition Pages 378–383

Fractals and Self-Similarity

Use the drawings below of three stages of a variation on the Koch curve to complete Exercises 1 and 2.

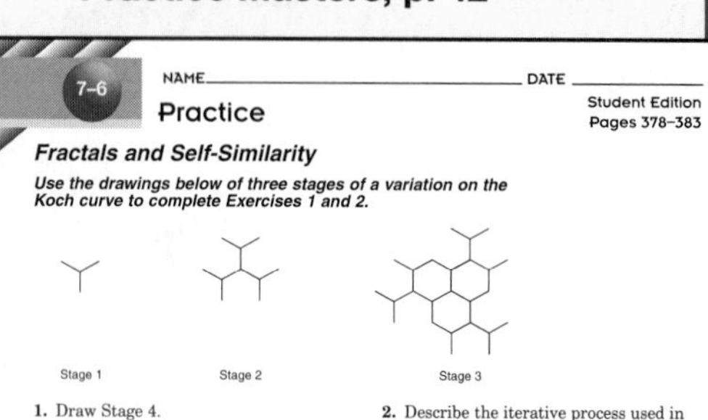

1. Draw Stage 4.

2. Describe the iterative process used in this variation. Two similar branches are drawn at the end of one segment of the previous branch.

3. Draw an equilateral triangle. Divide each side into fourths, and connect the points to form three lines parallel to each side of the triangle.

4. Find the number of similar triangles within the figure in Exercise 3. 27

5. Create a figure and draw stages 1–3 of iteration. See student's work.

6. Describe the iterative process used in Exercise 5. See student's work.

7. **Solve a Simpler Problem** How many diagonals can be drawn for a polygon with 15 sides? 90

---

11. Find the value of each expression. Then, use that value as the next x in the expression. Repeat the process until you can make some observations. Describe what happens in each of the iterations.
    a. $\sqrt{x}$, where x initially equals 12. converges to 1
    b. $\frac{1}{x}$, where x initially equals 5. alternates between 0.2 and 5.0
    c. $x^{\frac{1}{3}}$, where x initially equals 0.3 converges to 1

12. Use a calculator to find the square root of 2. Then take the square root of the result. Continue to repeat the process.
    a. What was your result after you repeated the process many times? 1
    b. Was the process you used an iterative process? Explain.

12b. Yes; the procedure of taking the square root is repeated over and over.

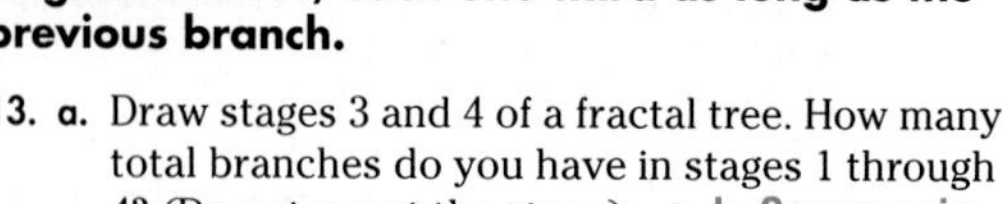

A "fractal tree" can be drawn by making two new branches from the endpoint of each original branch, each one-third as long as the previous branch.

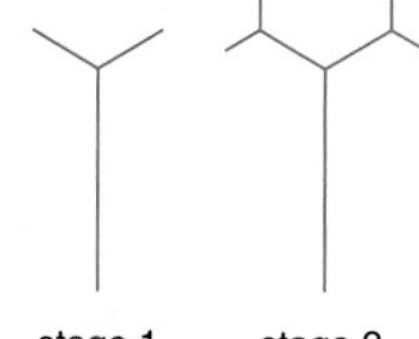

B

13. a. Draw stages 3 and 4 of a fractal tree. How many total branches do you have in stages 1 through 4? (Do not count the stem.) a–b. See margin.
    b. Look for a pattern that could be used to predict the number of branches at each stage.

14. Is a fractal tree strictly self-similar? Explain.

15. Generate Pascal's triangle. Divide every entry by 3. If the remainder is 1 or 2, shade the number cell black. If the remainder is 0, leave the cell unshaded. Describe the pattern that emerges.

14. No; the base of the tree or a segment of a branch without an end would not contain a replica of the entire tree.

15. It is similar to stage 1 of the Sierpinski triangle variation from Exercise 10.

C

16. Refer to the Koch curve on page 380. b. See margin.
    a. What is a formula for the number of segments in terms of the stage number? Use your formula to predict the number of segments in stage 8 of a Koch curve. $S_n = 4^{n-1}$; 16,384
    b. If the length of the original segment is 1 unit, how long will the segments be in each of the first four stages? What will happen to the length of each segment as the number of stages continues to increase?

Top Five List

Snowiest Cities in the U.S.	Mean Annual Snowfall (in.)
1. Blue Canyon, CA	240.8
2. Marquette, MI	128.6
3. Sault Ste. Marie, MI	116.7
4. Syracuse, NY	111.6
5. Caribou, ME	110.4

17. Refer to the Koch snowflake on page 381. At stage 1, the length of each side is 1 unit. a–b. See margin.
    a. What is the perimeter at each of the first four stages of a Koch snowflake?
    b. What is a formula for the perimeter in terms of the stage number? Describe the perimeter as the number of stages continues to increase.

18. Consider a Koch snowflake in stage 1. Write a paragraph proof to show that the triangles generated on the sides are similar to the original triangle. See margin.

Critical Thinking

19. The fractal in the pictures at the right is the space-filling *Hilbert curve*.
    a. Study the pictures carefully. Then define the iterative process used to generate the curve. See margin.
    b. Why do you think it is called "space filling"? See students' work.

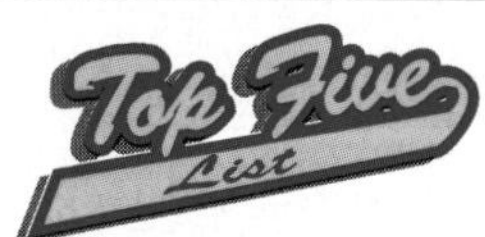

The most snowfall in one day was 76 inches recorded in Silver Lake, Colorado, in April, 1921. The most snowfall in one year was 1224 inches recorded on Mt. Rainier, Washington, from Feb. 1971–Feb. 1972.

Applications and Problem Solving

20. Figure 1 is real trees, Figure 2 is a fractal mountainscape, Figure 3 is a fractal representation of Yellowstone lake, and Figure 4 is a real creek; all the images possess self-similarity.

Fractal Web

20. Nature Some of these pictures are of real objects, and others are fractal images of objects. Compare the pictures and identify those you think are of real objects. Describe the characteristics of fractals that are shown in the images.

21. Solve a Simpler Problem Look at the diagonals in Pascal's triangle.

a. Find the sum of the first 25 numbers in the outside diagonal. 25

b. Find the sum of the first 50 numbers in the second diagonal. 1275

22. Art Describe how artist Edward Berko used iteration and self-similarity in his painting *Fractal Web*, shown at the left. See students' work.

Mixed Review

23. In the figure, $\triangle RST \sim \triangle WVU$. If $UV = 500$, $VW = 400$, $UW = 300$, and $ST = 1000$, find the perimeter of $\triangle RST$. (Lesson 7–5) 2400 units

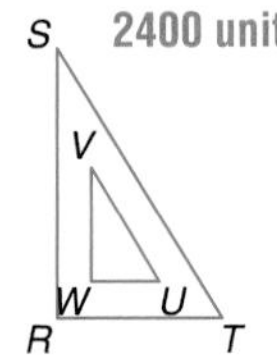

24. Find the values of x and y in the figure below. (Lesson 7–4) 24, 30

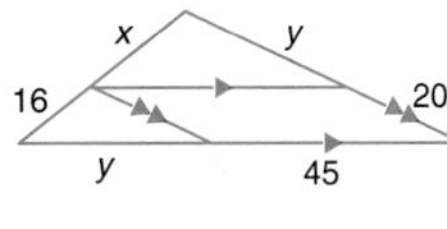

25. $-\frac{2}{7}$

25. Graph $\triangle RST$ with vertices $R(-1, -4)$, $S(1, 3)$, and $T(-6, 2)$. Find the slope of the perpendicular bisector of $\overline{RS}$. (Lesson 5–1)

26. State whether 35, 45, and 55 are possible angle measures of a triangle. Explain. (Lesson 4–2) No; the sum of the angle measures does not equal 180.

27. Advertising Determine if a valid conclusion can be reached from the two true statements found in an advertisement using the Law of Detachment or the Law of Syllogism. If a conclusion is possible, state it and the law that is used. If a conclusion does not follow, write *no conclusion*. (Lesson 2–3)

(1) If you are an avid sailor, then you need a Sail Sun Boat.

(2) Tamika has a Sail Sun Boat. no conclusion

29. $\{(-1, -1), (0, 2), (2, 8), (4, 14)\}$

Algebra

28. Biology Hair grows at a rate of 0.00000001 miles per hour. Express this number in scientific notation. 1.0×10^{-8}

29. Find the solution set for $y = 3x + 2$ if the domain is $\{-1, 0, 2, 4\}$.

4 ASSESS

Closing Activity

Modeling Cut lengths of wood or cardboard into increasingly smaller pieces: one 8" piece, two 4" pieces, four 2" pieces, and so on. Build a fractal pattern by gluing or taping the pieces together.

Chapter 7 Quiz D (Lesson 7-6) is available in the *Assessment and Evaluation Masters*, p. 185.

Additional Answers

17a. stage 1: 3 units

stage 2: $3 \cdot \frac{4}{3}$ or 4 units

stage 3: $3 \cdot \frac{4}{3} \cdot \frac{4}{3} = 3\left(\frac{4}{3}\right)^2$ or $5\frac{1}{3}$ units

stage 4: $3 \cdot \frac{4}{3} \cdot \frac{4}{3} \cdot \frac{4}{3} = 3\left(\frac{4}{3}\right)^3$ or $7\frac{1}{9}$ units

17b. $P = 3\left(\frac{4}{3}\right)^{n-1}$; as the stages increase, the perimeter increases and will approach infinity.

18. The original triangle and the new triangles are equilateral and thus, all of the angles are equal to 60. By AA Similarity, the triangles are similar.

Enrichment Masters, p. 42

NAME ______ DATE ______

Enrichment

Student Edition Pages 378–383

Golden Rectangles

Use a straightedge, compass, and the instructions below to construct a golden rectangle.

Check students' construction marks. The final figure should look like the one below.

1. Construct square *ABCD* with sides of 2 cm.
2. Construct the midpoint of $\overline{AB}$. Call the midpoint *M*.
3. Draw $\overline{AB}$. Set your compass at an opening equal to $\overline{MC}$. Use *M* as the center to draw an arc that intersects $\overline{AB}$. Call the point of intersection *P*.
4. Construct a line through *P* that is perpendicular to $\overline{AB}$.
5. Draw $\overline{DC}$ so that it intersects the perpendicular line in step 4. Call the intersection point *Q*. *APQD* is a **golden rectangle** because the ratio of its length to its width is 1.618. Check this conclusion by finding the value of $\frac{QP}{AP}$. Rectangles whose sides have this ratio are, it is said, the most pleasing to the human eye.

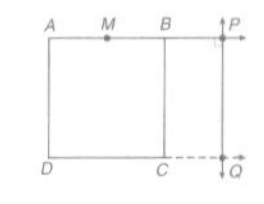

A figure consisting of similar golden rectangles is shown below. Use a compass and the instructions below to draw quarter-circle arcs that form a spiral like that found in the shell of a chambered nautilus.

6. Using *A* as a center, draw an arc that passes through *B* and *C*.
7. Using *D* as a center, draw an arc that passes through *C* and *E*.
8. Using *F* as a center, draw an arc that passes through *E* and *G*.
9. Using *H* as a center, draw an arc that passes through *G* and *J*.
10. Using *K* as a center, draw an arc that passes through *J* and *L*.
11. Using *M* as a center, draw an arc that passes through *L* and *N*.

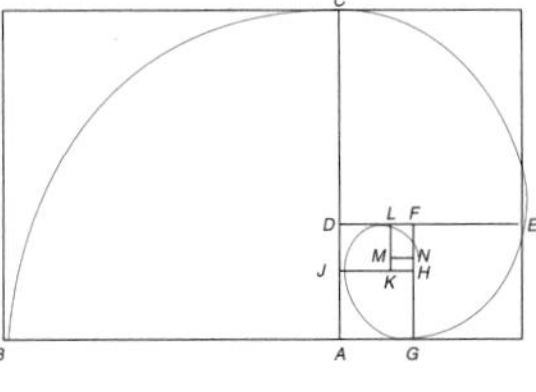

Extension

Communication Have students research the applications of fractals to the study of nature. Mandelbrot's books are an excellent source. Have students present their reports to the class and discuss the topics.

Additional Answer

19a. Trisect each of the three edges; replace the middle section on the center edge with three segments of length equal to the length removed; replace the first section on each of the outside edges with three segments of length equal to that removed. Repeat.

7-6B LESSON NOTES

NCTM Standards: 1–5, 7

Objective

Create fractal designs by using random numbers.

Recommended Time

Demonstration and discussion: 30 minutes; Exercises: 30 minutes

Instructional Resources

For each student or group of students

Modeling Mathematics Masters
- p. 1 (grid paper)
- p. 95 (worksheet)

For teacher demonstration

Algebra and Geometry Overhead Manipulative Resources

1 FOCUS

Motivating the Lesson

Flip a coin twenty or more times. Record the number of heads and the number of tails. Note that a pattern begins to emerge—about one-half heads and one-half tails.

2 TEACH

Teaching Tip Point out to students that while the die tossing is random, the choice of only three clearly defined moves is not random.

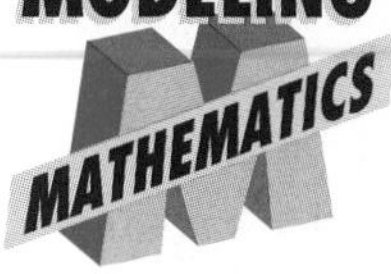

An Extension of Lesson 7–6

7–6B Chaos Game

Materials: Grid paper  calculator  dice  spinner

Fractal designs show that patterns can result from purely random events. The *Chaos game* shows how fractal designs can be created by using random numbers.

Activity 1 **On a coordinate plane, graph the three vertices of an equilateral triangle, *A*(0, 0), *B*(1, 0), and *C*(0.5, 0.9). Let each square represent 0.1 unit.**

- Graph any point *P* inside of the triangle.
- Generate a random number 1, 2, or 3 by using a calculator, tossing a die, or using a spinner that has three congruent parts.
- For a 1, move halfway to *A* and mark the midpoint; for a 2, move halfway to *B* and mark the midpoint, and for a 3, move halfway to *C* and mark the midpoint.

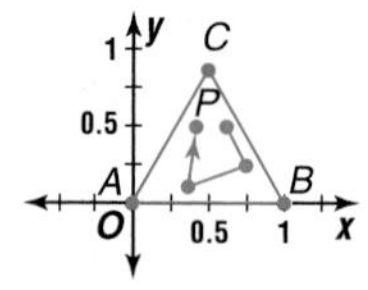

- After you have moved, connect the point to the newly located midpoint. Use that midpoint as your starting point and repeat the process.
- Play the game again with a triangle made on overhead transparency using the same scale as everyone else in class. This time mark only the successive midpoints.
- Play the game to obtain about 50 midpoints. Combine your set of midpoints with others in class by placing your transparencies in a stack. What conclusion can you make?

In Activity 1, you should have discovered that when all of the transparencies are stacked on the overhead projector, a pattern will appear that will be similar to a Sierpinski triangle. You can also use a Sierpinski triangle to determine the probablility of finding certain midpoints.

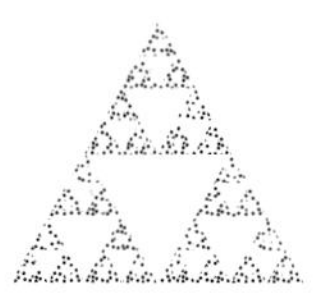

Activity 2 **The triangles in a Sierpinski triangle can be labeled by using *T* for top triangle, *R* for right triangle, and *L* for left triangle as shown in the figure. What is the probability that you will mark a midpoint in triangle *T* after the first random number has been generated? after the second random number has been generated?**

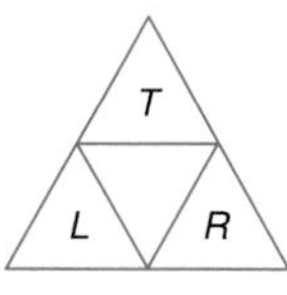

- Make a tree diagram to list the three possible outcomes and probabilities for the first outcome and follow each by the three possible outcomes and probabilities for the second outcome.

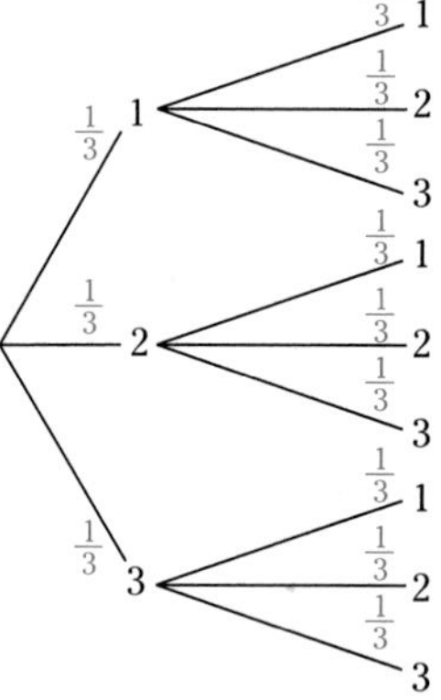

Cooperative Learning

This lesson offers an excellent opportunity for using cooperative learning groups. For more information on cooperative learning strategies and group management, see *Cooperative Learning in the Mathematics Classroom*, one of the titles in the Glencoe Mathematics Professional Series.

- Look at the pattern in the diagram on the previous page. No matter what initial point you begin with, you have a one-third chance to randomly generate each of the numbers 1, 2, or 3. Suppose 1, 2, and 3 are replaced by *T*, *L*, and *R*, respectively, as shown at the right. If 1 is the number associated with triangle *T*, then you will have a $\frac{1}{3}$ probability of marking a midpoint in triangle *T* after the first random number has been generated. You will have $\frac{1}{3} \cdot \frac{1}{3}$ probability of generating two 1s in a row or *TT*, so you will have a $\frac{1}{9}$ probability of marking a midpoint in triangle *T* after two moves.

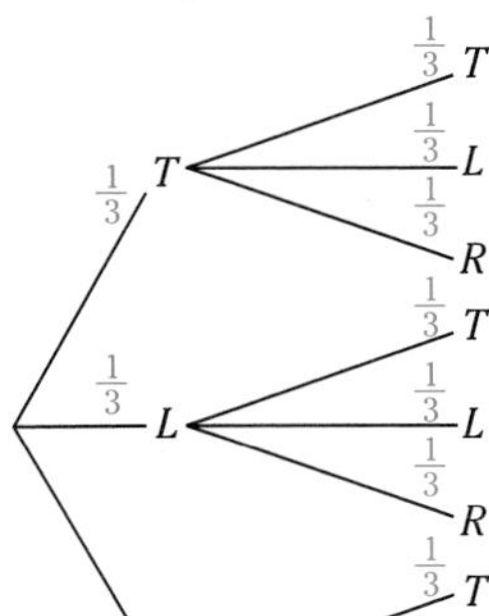

- There are three ways you can end up in triangle *T*. They are *LT*, *RT*, or *TT*. The probability of ending up in *LT* is $\frac{1}{9}$, in *RT*, $\frac{1}{9}$, and in *TT*, $\frac{1}{9}$. Therefore, the probability of ending up in triangle *T* is $\frac{1}{9} + \frac{1}{9} + \frac{1}{9} = \frac{3}{9}$ or $\frac{1}{3}$.

2a. The midpoint of $\overline{PA}$ is (0.15, 0.25) in $\triangle L$; the midpoint of $\overline{PB}$ is (0.65, 0.25) in $\triangle R$; the midpoint of $\overline{PC}$ is (0.4, 0.7) in $\triangle T$.
2b. The first midpoint toward *A* is (0.15, 0.25) in $\triangle L$; the second midpoint toward *A* is (0.075, 0.125) in $\triangle L$; the third midpoint toward *B* is (0.5375, 0.0625) in $\triangle R$.
4. See students' work.

Model

1. Suppose you are playing the Chaos game. Given triangle *ABC* with vertices *A*(0, 0), *B*(1, 0), and *C*(0.5, 0.9), show that when you begin with each of the following points, the midpoint you generate in stage 1 will never end up in the center triangle.
   a. *P*(0.2, 0.4) **See margin.**
   b. *P*(0.5, 0.2) **See margin.**
   c. What is the probability you will end up in triangle *T* after three moves? $\frac{1}{27}$
   d. What is the probability you will end up in triangle *L* after two moves? $\frac{1}{9}$
   e. What is the probability you will mark midpoints in triangle *T*, then in triangle *L*? $\frac{1}{9}$

2. Given $\triangle ABC$ from Exercise 1, begin the game from the point *P*(0.3, 0.5).
   a. Show that in the first move you do not mark a midpoint in the center region.
   b. If you generated the random numbers 1, 1, 2, state the coordinates of each midpoint that you would mark. Indicate whether the midpoint is in $\triangle L$, $\triangle R$, $\triangle T$, or the center triangle.

3. If you are playing the Chaos game, state the probability that you will end up in each location.
   a. *TT* in stage 2 $\frac{1}{9}$
   b. *T* in stage 3 $\frac{1}{3}$
   c. *TTT* in stage 3 $\frac{1}{27}$

Write

4. Prove that in playing the Chaos game, if you begin from a point on the perimeter of the original triangle, the midpoint you mark will be on the perimeter of the interior triangle.

3 PRACTICE/APPLY

Assignment Guide

Core (with proof): 1–4
Core (informal): 1–4
Enriched: 1–4

4 ASSESS

Observing students working in cooperative groups is an excellent method of assessment. You may wish to ask a student from each group to explain how to model and solve a problem. Also watch for and acknowledge students who are helping others understand the concept being taught.

Additional Answers

1a. The midpoint of $\overline{PA}$ is (0.1, 0.2) in $\triangle L$; the midpoint of $\overline{PB}$ is (0.6, 0.2) in $\triangle R$; the midpoint of $\overline{PC}$ is (0.35, 0.65) in $\triangle T$.

1b. The midpoint of $\overline{PA}$ is (0.25, 0.1) in $\triangle L$; the midpoint of $\overline{PB}$ is (0.75, 0.1) in $\triangle R$; the midpoint of $\overline{PC}$ is (0.5, 0.55) in $\triangle T$.

Closing the Investigation

This activity provides students an opportunity to bring their work on the Investigation to a close. For each Investigation, students should present their findings to the class. Here are some ways students can display their work.

- Conduct and report on an interview or survey.
- Write a letter, proposal, or report.
- Write an article for the school or local paper.
- Make a display, including graphs and/or charts.
- Plan an activity.

Assessment

To assess students' understanding of the concepts and topics explored in the Investigation and its follow-up activities, you may wish to examine students' Investigation Folders.

The scoring guide provided in the *Investigations and Projects Masters,* p. 11, provides a means for you to score students' work on the Investigation.

Investigations and Projects Masters, p. 11

Scoring Guide
Chapters 6 and 7
Investigation

Level	Specific Criteria
3 Superior	• Shows thorough understanding of the concepts of *percentage, dimensions, area, budget, blueprint, three-dimensional model, bid,* and *finances.* • Uses appropriate strategies to solve problems. • Computations are correct. • Written explanations are exemplary. • Blueprint, three-dimensional model, bid, and presentation are appropriate and sensible. • Goes beyond requirements of all or some problems.
2 Satisfactory, with Minor Flaws	• Shows understanding of the concepts of *percentage, dimensions, area, budget, blueprint, three-dimensional model, bid,* and *finances.* • Uses appropriate strategies to solve problems. • Computations are mostly correct. • Written explanations are effective. • Blueprint, three-dimensional model, bid, and presentation are appropriate and sensible. • Satisfies all requirements of problems.
1 Nearly Satisfactory, with Obvious Flaws	• Shows understanding of most of the concepts of *percentage, dimensions, area, budget, blueprint, three-dimensional model, bid,* and *finances.* • May not use appropriate strategies to solve problems. • Computations are mostly correct. • Written explanations are satisfactory. • Blueprint, three-dimensional model, bid, and presentation are appropriate and sensible. • Satisfies most requirements of problems.
0 Unsatisfactory	• Shows little or no understanding of the concepts of *percentage, dimensions, area, budget, blueprint, three-dimensional model, bid,* and *finances.* • Does not use appropriate strategies to solve problems. • Computations are incorrect. • Written explanations are not satisfactory. • Blueprint, three-dimensional model, bid, and presentation are not appropriate or sensible. • Does not satisfy requirements of problems.

Refer to the Investigation on pages 276–277.

The competition for a contract to do work for a government agency is regulated by federal, state, and local laws. Each company must submit a bid with a detailed plan for how the work will be completed. The bids and plans are then considered, and the contract is awarded.

Analyze

You have made a master plan and three-dimensional model of your design. A budget has also been completed. Now analyze your design and verify your conclusions.

PORTFOLIO ASSESSMENT

You may want to keep your work on this Investigation in your portfolio.

1 Look over your plan and verify the dimensions and placement of each park feature.

2 Compare your plan to your three-dimensional model. Make sure that all features are accurately represented on both the plan and the model.

3 Organize your financial bid for the park design. The bid should include the cost of supplies, labor, building materials, and a margin of profit for your design company.

Present

Select ten members of your class to represent the county commissioners. The rest of your class represents the members of the general public who have come to speak out on the issues related to the park. Present your proposed park design.

4 Begin your presentation by explaining which park features you chose to incorporate. Explain how you chose the features that you did.

5 Explain how you developed your design. Note any difficulties that you encountered or compromises that had to be made because of space or budget requirements.

6 Discuss the project budget. Explain any overruns and options for reducing the cost of the park.

7 State the advantages of your park design. Then summarize your presentation with a few statements about why your company would be the best one for the job.

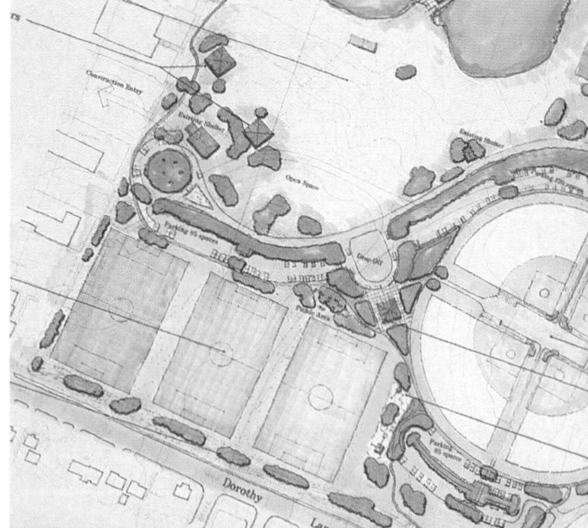

CHAPTER 7 HIGHLIGHTS

VOCABULARY

After completing this chapter, you should be able to define each term, property, or phrase and give an example or two of each.

Geometry	*Algebra*
dilation (p. 348)	cross products (p. 339)
fractal (p. 378)	extremes (p. 339)
iteration (p. 378)	means (p. 339)
scale factor (p. 346)	proportion (p. 339)
self-similar (p. 378)	ratio (p. 338)
Sierpinski triangle (p. 378)	
similar figures (p. 346)	
similar polygons (p. 346)	***Problem Solving***
strictly self-similar (p. 379)	solve a simpler problem (p. 379)

UNDERSTANDING AND USING THE VOCABULARY

State whether each sentence is *true* or *false*. If false, replace the underlined word or expression to make a true sentence. 1. false; proportional 3. false; two, two

1. If the measures of the corresponding sides of two triangles are <u>equal</u>, then the triangles are similar.
2. The product of the means <u>equals</u> the product of the extremes. true
3. If <u>one</u> angle of a triangle is congruent to <u>one</u> angle of another triangle, then the triangles are similar.
4. A <u>proportion</u> is a comparison of two quantities. false; ratio
5. If two triangles are similar, then the measures of the corresponding altitudes are proportional to the measures of the corresponding <u>sides</u>. true
6. In the proportion $\frac{15}{9} = \frac{5}{3}$, 15 and 3 are called the <u>means</u>. false; extremes
7. Two polygons are similar if and only if their corresponding sides are <u>proportional</u>. true
8. If a line is parallel to one side of a triangle and intersects the other two sides in two distinct points, then it separates these sides into segments of <u>equal</u> lengths. false; proportional
9. <u>Three</u> or more parallel lines separate transversals into proportional parts. true
10. An equation stating that two ratios are equal is a <u>proportion</u>. true
11. For any numbers a and c and any nonzero numbers b and d, $\frac{a}{b} = \frac{c}{d}$ if and only if <u>$ab = cd$</u>. false; $ad = bc$
12. A fractal is a geometric figure that has a <u>self-similar</u> shape. true

Using the CHAPTER HIGHLIGHTS

The Chapter Highlights begins with a listing of the new terms, properties, and phrases that were introduced in this chapter. Have students define each term and provide an example or two of it, if appropriate.

Assessment and Evaluation Masters, pp. 171–172

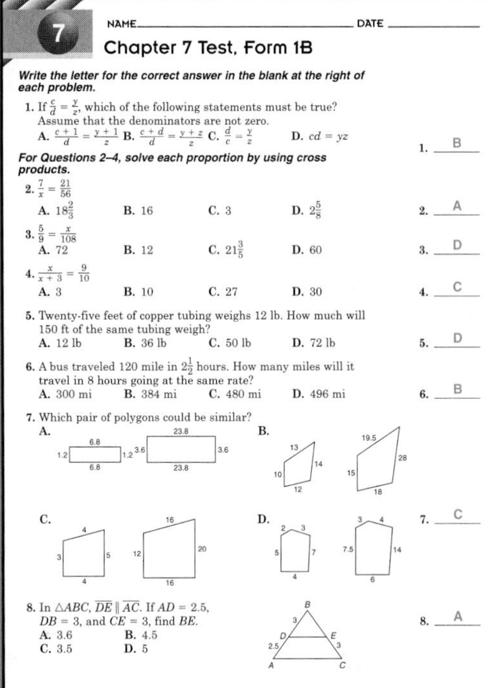

NAME ____ DATE ____

Chapter 7 Test, Form 1B

Write the letter for the correct answer in the blank at the right of each problem.

1. If $\frac{c}{d} = \frac{y}{z}$, which of the following statements must be true? Assume that the denominators are not zero.
A. $\frac{c+1}{d} = \frac{y+1}{z}$ B. $\frac{c+d}{d} = \frac{y+z}{z}$ C. $\frac{d}{c} = \frac{y}{z}$ D. $cd = yz$ — 1. B

For Questions 2–4, solve each proportion by using cross products.

2. $\frac{7}{x} = \frac{21}{56}$
A. $18\frac{2}{3}$ B. 16 C. 3 D. $2\frac{5}{8}$ — 2. A
3. $\frac{5}{9} = \frac{x}{108}$
A. 72 B. 12 C. $21\frac{3}{5}$ D. 60 — 3. D
4. $\frac{x}{x+3} = \frac{9}{10}$
A. 3 B. 10 C. 27 D. 30 — 4. C
5. Twenty-five feet of copper tubing weighs 12 lb. How much will 150 ft of the same tubing weigh?
A. 12 lb B. 36 lb C. 50 lb D. 72 lb — 5. D
6. A bus traveled 120 mile in $2\frac{1}{2}$ hours. How many miles will it travel in 8 hours going at the same rate?
A. 300 mi B. 384 mi C. 480 mi D. 496 mi — 6. B
7. Which pair of polygons could be similar?
A. B. C. D. — 7. C
8. In $\triangle ABC$, $\overline{DE} \parallel \overline{AC}$. If $AD = 2.5$, $DB = 3$, and $CE = 3$, find BE.
A. 3.6 B. 4.5 C. 3.5 D. 5 — 8. A

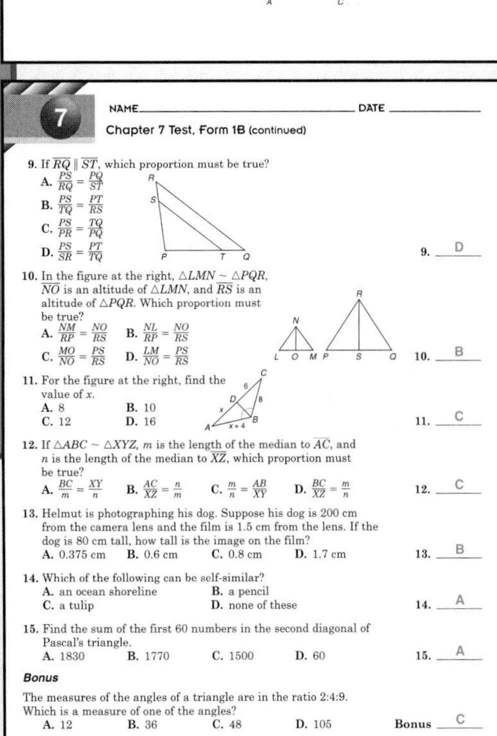

NAME ____ DATE ____

Chapter 7 Test, Form 1B (continued)

9. If $\overline{RQ} \parallel \overline{ST}$, which proportion must be true?
A. $\frac{PS}{RQ} = \frac{PQ}{ST}$ B. $\frac{PS}{TQ} = \frac{PT}{RS}$ C. $\frac{PS}{PR} = \frac{TQ}{PQ}$ D. $\frac{PS}{SR} = \frac{PT}{TQ}$ — 9. D
10. In the figure at the right, $\triangle LMN \sim \triangle PQR$, $\overline{NO}$ is an altitude of $\triangle LMN$, and $\overline{RS}$ is an altitude of $\triangle PQR$. Which proportion must be true?
A. $\frac{NM}{RP} = \frac{NO}{RS}$ B. $\frac{NL}{RP} = \frac{NO}{RS}$ C. $\frac{MO}{NO} = \frac{PS}{RS}$ D. $\frac{LM}{NO} = \frac{PS}{RS}$ — 10. B
11. For the figure at the right, find the value of x.
A. 8 B. 10 C. 12 D. 16 — 11. C
12. If $\triangle ABC \sim \triangle XYZ$, m is the length of the median to $\overline{AC}$, and n is the length of the median to $\overline{XZ}$, which proportion must be true?
A. $\frac{BC}{m} = \frac{XY}{n}$ B. $\frac{AC}{XZ} = \frac{n}{m}$ C. $\frac{m}{n} = \frac{AB}{XY}$ D. $\frac{BC}{XZ} = \frac{m}{n}$ — 12. C
13. Helmut is photographing his dog. Suppose his dog is 200 cm from the camera lens and the film is 1.5 cm from the lens. If the dog is 80 cm tall, how tall is the image on the film?
A. 0.375 cm B. 0.6 cm C. 0.8 cm D. 1.7 cm — 13. B
14. Which of the following can be self-similar?
A. an ocean shoreline B. a pencil C. a tulip D. none of these — 14. A
15. Find the sum of the first 60 numbers in the second diagonal of Pascal's triangle.
A. 1830 B. 1770 C. 1500 D. 60 — 15. A

Bonus

The measures of the angles of a triangle are in the ratio 2:4:9. Which is a measure of one of the angles?
A. 12 B. 36 C. 48 D. 105 — Bonus C

Instructional Resources

Three multiple-choice tests and three free-response tests are provided in the *Assessment and Evaluation Masters*. Forms 1A and 2A are for honors pacing, Forms 1B and 2B are for average pacing, and Forms 1C and 2C are for basic pacing. Chapter 7 Test, Form 1B, is shown at the right. Chapter 7 Test, Form 2B, is shown on the next page.

Postulates, Theorems, and Corollaries

A complete list of postulates, theorems, and corollaries begins on page 806.

Using the STUDY GUIDE AND ASSESSMENT

Skills and Concepts Encourage students to refer to the objectives and examples on the left as they complete the review exercises on the right.

Assessment and Evaluation Masters, pp. 177–178

NAME______________________ DATE ________

Chapter 7 Test, Form 2B

For Questions 1–4, solve each proportion by using cross products.

1. $\frac{14}{17} = \frac{70}{x}$ 1. 85
2. $\frac{21}{x} = \frac{168}{256}$ 2. 32
3. $\frac{3x}{5} = \frac{72}{12}$ 3. 10
4. $\frac{3}{2} = \frac{x+1}{x}$ 4. 2
5. A pitcher allowed 20 earned runs in 30 innings. At this rate, how many runs would the pitcher allow in 9 innings? 5. 6 runs
6. A map is scaled so that 1.5 cm represents 60 km. Two cities are 6 cm apart on the map. How far apart are the cities? 6. 240 km
7. The legs of a right triangle are 8.2 in. and 10 in. long. The shorter leg of a second triangle, similar to the first, is 15.4 in. long. Find the ratio of the measures of the hypotenuses. 7. 8.2 : 15.4
8. Identify the following statement as *true* or *false*: All isosceles right triangles are similar. 8. true

For Questions 9 and 10, determine whether each pair of triangles is similar. If they are similar, give a reason.

9. 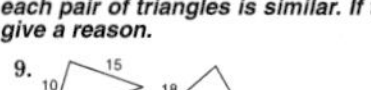

9. yes; SSS Similarity

10. 

10. no

11. In the figure at the right, find the value of x.
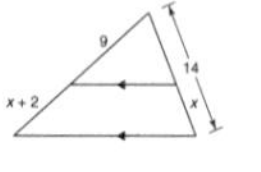

11. 7

NAME______________________ DATE ________

Chapter 7 Test, Form 2B (continued)

12. In the figure at the right, $\triangle ABC \sim \triangle DEF$, $\overline{BX}$ is a median of $\triangle ABC$, and $\overline{EY}$ is a median of $\triangle DEF$. If $BX = 6$, $AX = 4$, $DF = x + 1$, and $EY = 2x - 1$, find DY. 12. 1.2
13. What is the maximum number of diagonals that can be drawn in a polygon with 90 sides? 13. 3915 diagonals
14. The figure at the right shows an iteration of angle bisectors with the measure of the initial angle equal to 180°. If the pattern continues indefinitely, would the figure be self-similar? Explain. 14. No; every portion of the figure will not resemble the entire figure.

The following proof counts for Questions 15–20. Write the missing statement or reason in each blank.

Given: Quadrilateral $GHIJ$ with $\angle JIG \cong \angle IGH$

Prove: $GK \cdot JK = IK \cdot HK$

Proof:

Statements	Reasons
a. $\angle JIG \cong \angle IGH$	a. Given
b. (Question 15)	b. If 2 lines are cut by a transversal and alt. int. $\angle$s are $\cong$, then the lines are $\parallel$.
c. $\angle IJH \cong \angle JHG$	c. (Question 16)
d. $\triangle JIK \sim \triangle HGK$	d. (Question 17)
e. $\frac{JK}{HK} = \frac{IK}{GK}$	e. (Question 18)
f. (Question 19)	f. (Question 20)

15. $\overline{JI} \parallel \overline{GH}$
16. Alt. Int. $\angle$ Thm.
17. AA Similarity
18. Def. $\sim \triangle$s
19. $GK \cdot JK = IK \cdot HK$
20. Equality of cross products

Bonus

Quadrilateral $RSTU \sim$ quadrilateral $JKLM$. Complete the following proportion to make it true.

$\frac{RS + JK}{JK} = \frac{TU + ?}{LM}$ Bonus LM

SKILLS AND CONCEPTS

OBJECTIVES AND EXAMPLES	REVIEW EXERCISES
Upon completing this chapter, you should be able to:	Use these exercises to review and prepare for the chapter test.

- recognize and use ratios and proportions (Lesson 7–1)

To solve the equation $\frac{x}{6} = \frac{18}{4}$, find the cross products.

$$\frac{x}{6} = \frac{18}{4}$$
$$4x = 6 \cdot 18$$
$$4x = 108$$
$$x = 27$$

Find each ratio and express it as a fraction in simplest form. 13. $\frac{1}{2}$ 14. $\frac{3}{11}$

13. Loretta's wrist and neck measure 15 centimeters and 30 centimeters, respectively. Find the ratio of her wrist measure to her neck measure.
14. If there are 15 rear sprocket teeth and 55 front sprocket teeth on a bicycle, find the gear ratio, which is the ratio of the number of rear sprocket teeth to the number of front sprocket teeth.

Solve by using cross products.

15. $\frac{x}{8} = \frac{6}{15}$ 3.2
16. $\frac{7}{x} = \frac{2}{3}$ 10.5
17. $\frac{1}{a} = \frac{5}{a+5}$ 1.25
18. $\frac{k+3}{4} = \frac{5k-2}{9}$ $\frac{35}{11}$

- apply and use the properties of proportions (Lesson 7–1)

A telephone pole casts a 33-foot shadow. Nearby a girl 4 feet tall casts a 6-foot shadow. To find the height of the telephone pole, write a proportion.

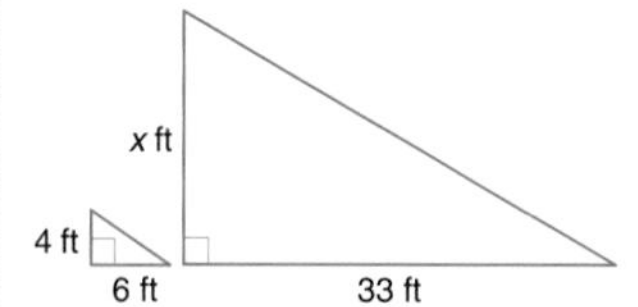

$$\frac{x}{33} = \frac{4}{6}$$
$$6x = 4 \cdot 33$$
$$6x = 132$$
$$x = 22$$

The pole is 22 feet tall.

Corresponding sides of $\triangle ABC$ are proportional to the sides of $\triangle JKL$.

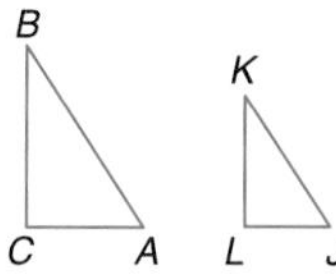

19. If $JK = 7$, $JL = 6$, and $AC = 8$, find AB. $9\frac{1}{3}$
20. If $KL = 7$, $JL = 6$, and $AC = 14$, find BC. $16\frac{1}{3}$

- solve problems involving similar figures (Lesson 7–2)

If $\square WXYZ \sim \square QRST$, find the value of x.

$$\frac{x}{18} = \frac{6}{12}$$
$$x \cdot 12 = 18 \cdot 6$$
$$12x = 108$$
$$x = 9$$

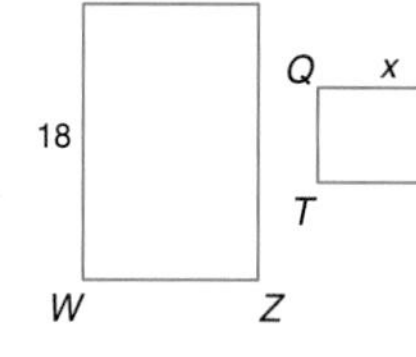

Determine whether each statement is *true* or *false*.

21. All similar triangles are congruent. false
22. All congruent triangles are similar. true

In the figure below, $\triangle LMN \sim \triangle TUV$.

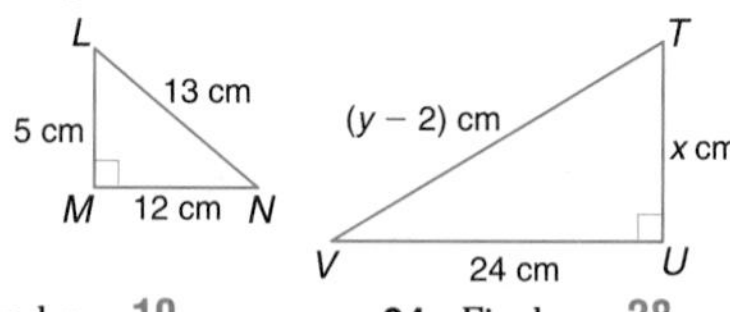

23. Find x. 10
24. Find y. 28

GLENCOE Technology

Test and Review Software

You may use this software, a combination of an item generator and item bank, to create your own tests or worksheets. Types of items include free response, multiple choice, short answer, and open ended.

For IBM & Macintosh

OBJECTIVES AND EXAMPLES

- identify similar triangles (Lesson 7–3)

There are three ways to prove that two triangles are similar.

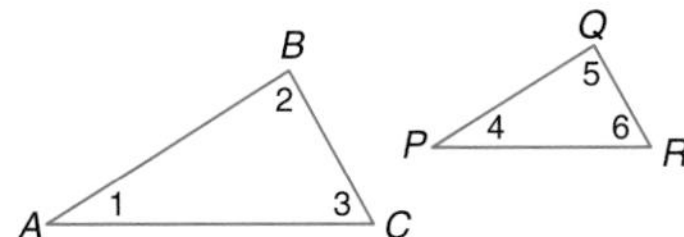

AA Similarity $\angle 1 \cong \angle 4$, $\angle 3 \cong \angle 6$

SSS Similarity $\frac{AB}{PQ} = \frac{BC}{QR} = \frac{CA}{RP}$

SAS Similarity $\frac{AB}{PQ} = \frac{AC}{\text{PR}}$, $\angle 1 \cong \angle 4$

REVIEW EXERCISES

Determine whether each pair of triangles is similar. If similarity exists, write a mathematical sentence relating the two triangles. Give a reason for your answer.

25.

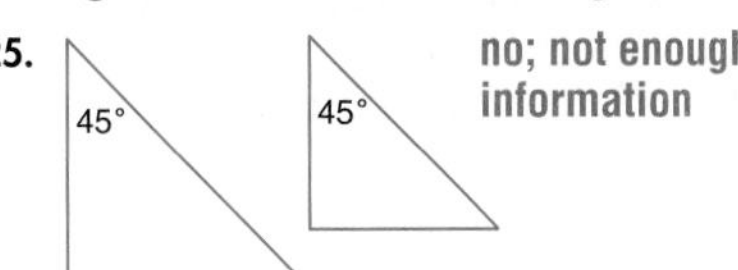

no; not enough information

26.

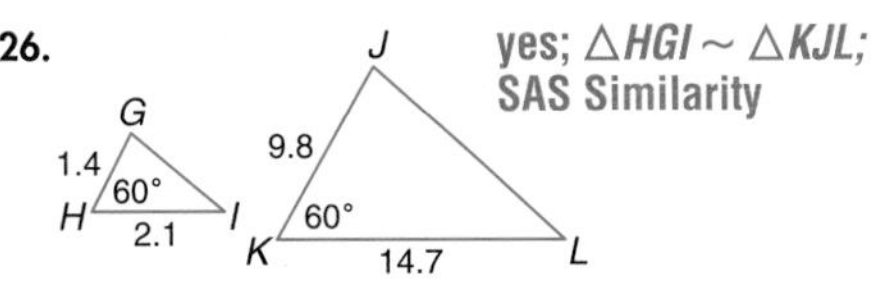

yes; $\triangle HGI \sim \triangle KJL$; SAS Similarity

OBJECTIVES AND EXAMPLES

- use proportional parts of triangles to solve problems (Lesson 7–4)

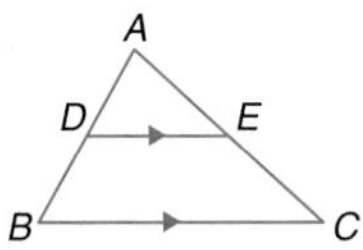

If $\overline{DE} \parallel \overline{BC}$, then $\frac{AD}{DB} = \frac{AE}{EC}$.

REVIEW EXERCISES

Use the figure and the given information to find the value of x.

27. $\overline{PQ} \parallel \overline{DF}$
$EQ = 3$
$DP = 12$
$QF = 8$
$PE = x + 2$ 2.5

28. $\overline{SV} \parallel \overline{PR}$
$TS = 5 + x$
$TV = 8 + x$
$VP = 4$
$SR = 3$ 4

OBJECTIVES AND EXAMPLES

- recognize and use the proportional relationships of corresponding perimeters, altitudes, angles bisectors, and medians of similar triangles (Lesson 7–5)

$\triangle ABC \sim \triangle PQR$

Perimeter $\frac{AB + BC + CA}{PQ + QR + RP} = \frac{CA}{RP}$

Altitude $\frac{BD}{QS} = \frac{BA}{QP}$

Angle bisector $\frac{AE}{PT} = \frac{CA}{RP}$

Median $\frac{CF}{RU} = \frac{CA}{RP}$

REVIEW EXERCISES

29. If $\triangle ABC \sim \triangle XYZ$, find YN. $9\frac{1}{3}$

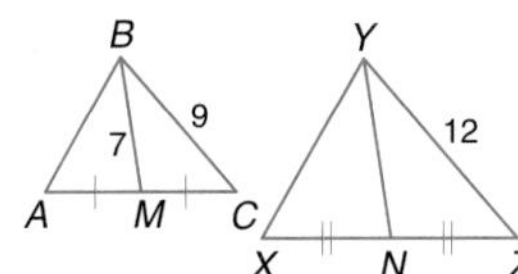

30. If $\triangle STV \sim \triangle PQM$, find the perimeter of $\triangle PQM$. 27.5

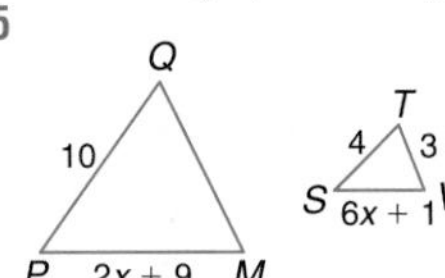

Applications and Problem Solving Encourage students to work through the exercises in the Applications and Problem Solving section to strengthen their problem-solving skills.

Additional Answer

31.

Stage 3

OBJECTIVES AND EXAMPLES

- recognize and describe characteristics of fractals (Lesson 7–6)

The fractal below is formed by trisecting the sides of a square, which forms nine smaller squares, and then removing the middle square. Each stage contains a similar, smaller pattern of the previous stage.

Stage 1

Stage 2

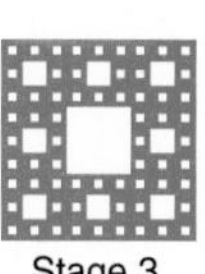
Stage 3

REVIEW EXERCISES

31. Draw stage 3 of the fractal shown below. See margin.

Stage 1 Stage 2

32. Explain how self-similarity is evident in a head of cauliflower. Sample answer: the smaller pieces are duplicates of the whole head.

APPLICATIONS AND PROBLEM SOLVING

33. Probability If the probability of rolling a sum of 7 with 2 dice is $\frac{1}{6}$, how many sums of 7 would you expect to get if you rolled the dice 174 times? (Lesson 7–1) 29

34. Travel A map is scaled so that 1 centimeter represents 15 kilometers. How far apart are two towns if they are 7.9 centimeters apart on the map? (Lesson 7–1) 118.5 km

35. Hobbies A twin-jet airplane especially suited for medium-range flights has a length of 78 meters and a wingspan of 90 meters. If a scale model is made with a wingspan of 36 centimeters, find its length. (Lesson 7–2) 31.2 cm

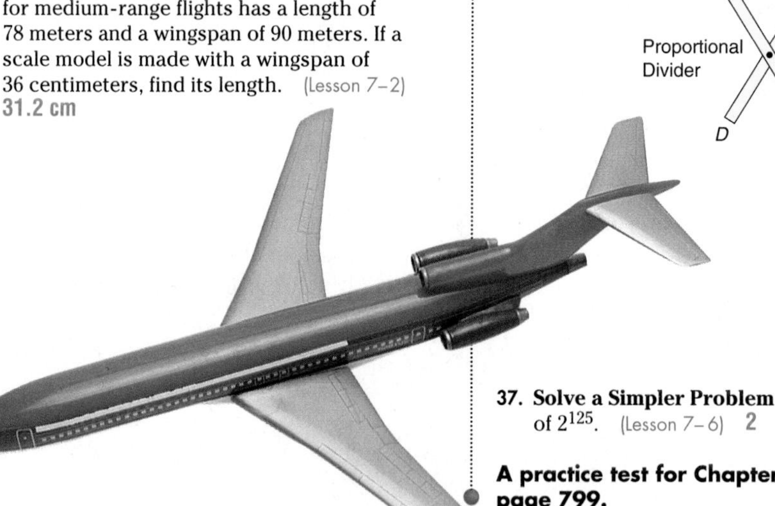

36. Drafting A proportional divider is a drafting instrument that is used to enlarge or decrease a drawing. A screw at T keeps $\overline{AT} \cong \overline{TC}$ and $\overline{DT} \cong \overline{TB}$. $\overline{AB}$ and $\overline{CD}$ can be set so they are divided proportionally at T in any desired ratio. How would you adjust the divider in order to enlarge a design in the ratio 7 to 4? (Lesson 7–3) $AT = TC = 4$ and $DT = TB = 7$

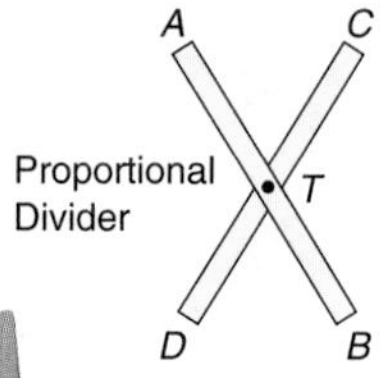

Proportional Divider

37. Solve a Simpler Problem Find the units digit of 2^{125}. (Lesson 7–6) 2

A practice test for Chapter 7 is provided on page 799.

ALTERNATIVE ASSESSMENT

COOPERATIVE LEARNING PROJECT

Fractals In this chapter, you learned about fractals. In this project, you will explore fractal patterns in the natural world. You will explore the natural world and discover as many patterns as you can that can be modeled with fractals.

Your goal will be to build a complete collection of natural fractals. If you find more than one of a certain type, share it with others in your group. It may help another student complete their collection. You are encouraged to search wherever you can. Photocopies of images contained in books of nature, photography, biology, astronomy, geology, your own photographs or drawings, or even books on fractal geometry can be used.

Follow these steps to complete your collection of natural fractals.

- Collect objects from the natural world: leaves, branches, rocks, and so on. When a real object is not available (cloud, river), find a photograph of it, or photograph or draw it. Find only objects that cannot be represented as conventional geometric objects—quadrilaterals, circles, ellipses, triangles, and so on.
- Refer to a book by Mandelbrot or others to find fractal patterns that look like the objects you found. When available, record the names, or equations, for these patterns.
- Organize all the objects, or images of objects, into the categories defined in the previous step.
- For each category, write one paragraph describing how the pattern is generated. Analyze the basic geometric figure used and the rule for repeating the pattern. Explain whether the repeated geometric figures are similar or congruent.

THINKING CRITICALLY

You are preparing a four-bean salad for lunch. The recipe you followed last time called for 2 cups of beans, and the salad filled a medium-sized bowl. This time you want to make enough salad to fill your largest bowl. The two bowls are shaped similarly, but the diameter of the larger bowl is 30% larger than the diameter of the medium-sized bowl.

The volume of a sphere can be found using the formula $V = \frac{4}{3}\pi r^3$, where r is the radius. The volume of each bowl is half of a sphere or $\frac{2}{3}\pi r^3$ cubic units. How many cups of beans do you need so that the recipe fills the larger bowl? **4.394 cups**

PORTFOLIO

Find a photograph from a book or magazine showing the horizon. Make a photocopy of it. Notice that there is an imaginary point where parallel lines appear to meet. This is a *vanishing point*. There is also an imaginary horizontal line where the sky seems to meet Earth. This is called the *horizon line*. On the photocopy, draw the vanishing point and the horizon line.

Now try to find a photo of the same image with a different vanishing point. Photocopy it, and draw the vanishing point and horizon line on the photocopy. Keep both images in your portfolio.

SELF EVALUATION

In this chapter, you learned the strategy *solving a simpler problem*. This approach can also be applied in everyday life.

Assess Yourself What have you done when you've faced a problem you did not know how to solve? Did you compare it to a simpler problem and try to apply those ideas? What other methods have you tried when faced with a problem you could not think of a way to solve?

Assessment and Evaluation Masters, pp. 182, 193

NAME ________________ DATE ________

Chapter 7 Performance Assessment

Instructions: *Demonstrate your knowledge by giving a clear, concise solution to each problem. Be sure to include all relevant drawings and to justify your answers. You may show your solution in more than one way or investigate beyond the requirements of the problem.*

1. For each of the following, identify the pair of triangles that are similar and justify your answer.

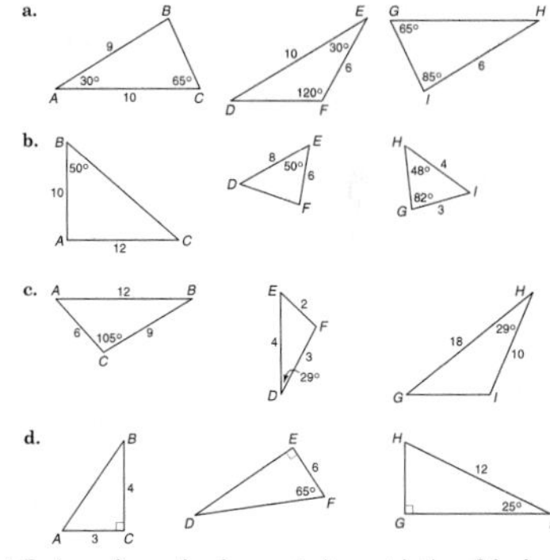

2. Engineers frequently solve problems graphically by drawing similar figures to scale. The truss below is drawn to the scale 1 in. = 5 ft. Find the lengths of the sides of the truss.

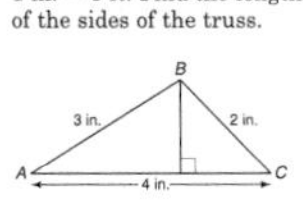

3. At a certain time of the day, Lisa's shadow is 8 ft long. Lisa is 5 ft 6 in. tall, and the oak tree in her yard is 44 ft tall. If Lisa stands in the shadow of the tree so that the end of her shadow coincides with the end of the tree's shadow, how far from the tree will Lisa be standing?

Scoring Guide
Chapter 7
Performance Assessment

Level	Specific Criteria
3 Superior	• Shows thorough understanding of the concepts of *ratio, proportion,* and *similar triangles.* • Uses appropriate strategies to solve problems. • Computations are correct. • Written explanations are exemplary. • Goes beyond requirements of some or all problems.
2 Satisfactory, with Minor Flaws	• Shows understanding of the concepts of *ratio, proportion,* and *similar triangles.* • Uses appropriate strategies to solve problems. • Computations are mostly correct. • Written explanations are effective. • Satisfies all requirements of some or all problems.
1 Nearly Satisfactory, with Serious Flaws	• Shows understanding of most of the concepts of *ratio, proportion,* and *similar triangles.* • May not use appropriate strategies to solve problems. • Computations are mostly correct. • Written explanations are satisfactory. • Satisfies most requirements of some or all problems.
0 Unsatisfactory	• Shows little or no understanding of the concepts of *ratio, proportion,* and *similar triangles.* • May not use appropriate strategies to solve problems. • Computations are incorrect. • Written explanations are not satisfactory. • Does not satisfy requirements of some or all problems.

Alternative Assessment

The Alternative Assessment section provides students with the opportunity to assess their own work by thinking critically, working with others, keeping a portfolio, and honestly evaluating their own progress. For more information on alternative forms of assessment, see *Alternative Assessment in the Mathematics Classroom,* one of the titles in the Glencoe Mathematics Professional Series.

Performance Assessment

Performance Assessment tasks for this chapter are included in the *Assessment and Evaluation Masters.* A scoring guide is also provided.

In•ves•ti•ga•tion

TEACHER NOTES

NCTM Standards: 1–5, 7, 8

This Investigation is designed to be completed over several days or weeks. It may be considered optional. You may want to assign the Investigation and the follow-up activities to be completed at the same time.

Objectives

Design a permanent dwelling place for Mars. Design a public relations program.

Mathematical Overview

This Investigation will use the following mathematical skills and concepts from Chapters 8 and 9.

- finding the area of a polygon
- estimating
- using proportional reasoning
- solving equations
- approximating percentages

Recommended Time		
Part	**Pages**	**Time**
Investigation	392–393	1 class period
Working on the Investigation	411, 436, 465, 497	20 minutes each
Closing the Investigation	504	1 class period

Instructional Resources

Investigations and Projects Masters, pp. 13–16

A recording sheet, teacher notes, and scoring guide are provided for each Investigation in the *Investigations and Projects Masters*.

1 MOTIVATION

Bring in examples of imagined space stations. Discuss with students how life would be different if we could travel between planets as we can between states.

LONG-TERM PROJECT

In•ves•ti•ga•tion

MATERIALS NEEDED

- calculator
- protractor
- ruler
- grid paper
- markers
- modeling clay
- paint
- scissors
- tape

The year is 2125, and your team has been chosen for a mission to Mars. Your mission will be to establish a permanent dwelling place on Mars so that people may live there for an indefinite amount of time. After your mission is completed, scientists will travel to Mars to live, and tourists may visit for weeks or months at a time. You will also design a public relations program to gain backing from Congress and the American public for the mission.

One of the things that the scientists who will follow you to Mars intend to study is the evidence of life on Mars that was discovered in the 1990s. That discovery came from a study of a potato-sized meteorite that was chipped from the surface of Mars by the impact of an asteroid. That meteorite, called ALH84001, contained what is suspected to be microscopic fossils of primitive, bacteria-like organisms. According to their findings, NASA scientists stated that life may have existed on Mars more than 3.6 billion years ago.

NASA has given you the information shown on the next page that has been gathered about Mars to help you in your mission.

Yours will not be the first encounter that the space program has had with Mars. The first spacecraft to approach Mars from Earth was the unoccupied *U.S. Mariner 4.* It traveled within 6118 miles of the planet in 1965. The *Mariner 9* orbited Mars at a distance of about 1000 miles and photographed the entire surface. The *U.S. Viking 1* and *Viking 2* probes landed on Mars in 1976. Both probes transmitted photographs and analyses of the atmosphere and soil back to Earth.

Cooperative Learning

This Investigation offers an excellent opportunity for using cooperative learning groups. For more information on cooperative learning strategies and group management, see *Cooperative Learning in the Mathematics Classroom*, one of the titles in the Glencoe Mathematics Professional Series.

MARS FACTS

- The fourth planet from the Sun, Mars is next beyond Earth.
- The mean distance from Mars to the Sun is 141,600,000 miles, compared with about 93,000,000 miles for Earth.
- At its closest, Mars is 34,600,000 miles from Earth.
- The diameter of Mars is 4223 miles, a little over half the size of Earth.
- Mars orbits the Sun in an elliptical, or oval-shaped, path. The distance from Mars to the Sun varies between about 128,400,000 to 154,800,000 miles.
- Mars takes about 687 Earth-days to make one orbit of the Sun.
- It takes 24 hours, 37 minutes for Mars to rotate once. Earth takes 23 hours, 56 minutes.
- The temperature on the surface of Mars varies from $-225°$ to $63°F$.
- The atmosphere of Mars consists of carbon dioxide, nitrogen, argon, oxygen, carbon monoxide, neon, krypton, xenon, and water vapor.
- Mars has two moons, Phobos and Deimos. Phobos is about 5800 miles from the center of Mars, and it travels around Mars once every $7\frac{1}{2}$ hours. Deimos is about 14,600 miles from Mars and completes an orbit every 30 hours.
- The atmospheric pressure on Mars is about 0.1 pound per square inch. That is less than one-hundredth the pressure felt on Earth.

RESEARCH

Brainstorm a list of concerns that need to be addressed by the team. Begin your discussion with the questions below.

- Will it be possible for people to breathe on Mars, or will something have to be done to accommodate respiration?
- What is the best place to locate the space station on Mars?
- How can energy be produced on Mars?
- Where is the best place on Earth to conduct tests of equipment that will be sent to Mars?
- What factors of daily life will be different on Mars? For example, how will the difference in the length of a day and the length of a year affect life?
- How correct are the public perceptions of life on Mars?

You may wish to visit the library or do some research on the NASA Internet home page regarding photographs of Mars and other information that has been gathered about the planet.

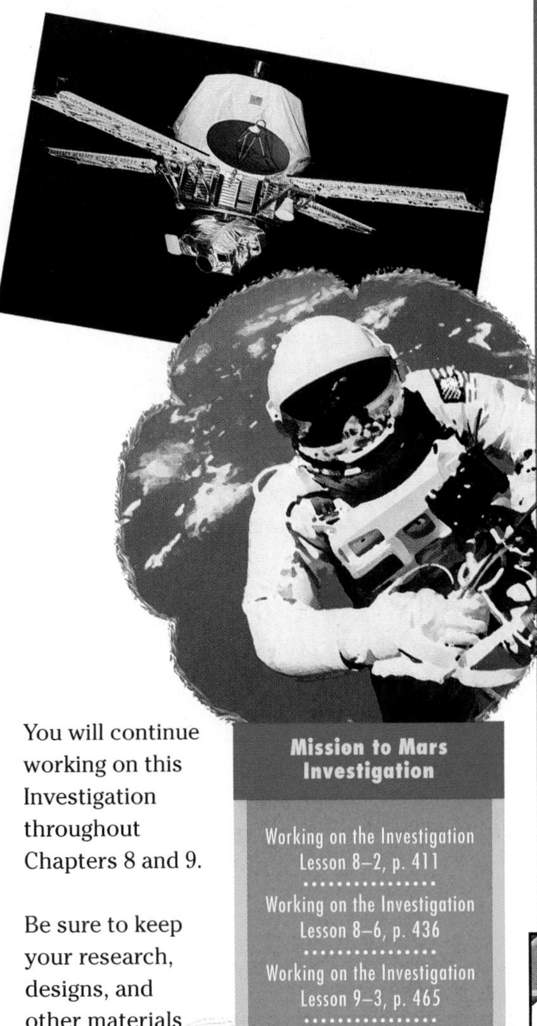

You will continue working on this Investigation throughout Chapters 8 and 9.

Be sure to keep your research, designs, and other materials in your Investigation Folder.

Mission to Mars Investigation

Working on the Investigation Lesson 8–2, p. 411

Working on the Investigation Lesson 8–6, p. 436

Working on the Investigation Lesson 9–3, p. 465

Working on the Investigation Lesson 9–7, p. 497

Closing the Investigation End of Chapter 9, p. 504

2 SETUP

Have one student read aloud the first four paragraphs of the Investigation and another student read the Mars Facts box to provide background information for the Mars dwelling design project. Read the remaining paragraphs that explain the class's research tasks. Discuss the activity with students. Separate the class into groups of four or five.

3 MANAGEMENT

Each group member should be responsible for a specific task.

Recorder Keeps track of all information.

Editor Edits information for organization and readability.

Researcher Coordinates the group's researching efforts.

Bookkeeper Keeps track of bibliography.

Sample Answers

Answers will vary depending on group designs.

Investigations and Projects Masters, p. 16

8, 9 NAME ________________ DATE ________

Investigation, Chapters 8 and 9

Student Edition Pages 392–393, 411, 436, 465, 497, 504

Mission to Mars

Use the space below to record your research questions.

Use the space below to record your ideas for your Martian calendar.

Use the space below to show where the outpost stations will be

Mars

Please keep this page and any other research in your Investigation Folder.

CHAPTER

8 Applying Right Triangles and Trigonometry

PREVIEWING THE CHAPTER

In this chapter, students investigate and use the Pythagorean Theorem. They use paper to model the Pythagorean Theorem. Students find geometric means. They use geometric means to solve triangles using the altitude to the hypotenuse. They use the properties of special right triangles. Students extend their skills to trigonometry and use ratios to solve for specific sides or angles. Trigonometry is applied to the angles of elevation and depression through word problems. Finally, students generalize their skills and work with the laws of sines and cosines. They also work on problem-solving strategies.

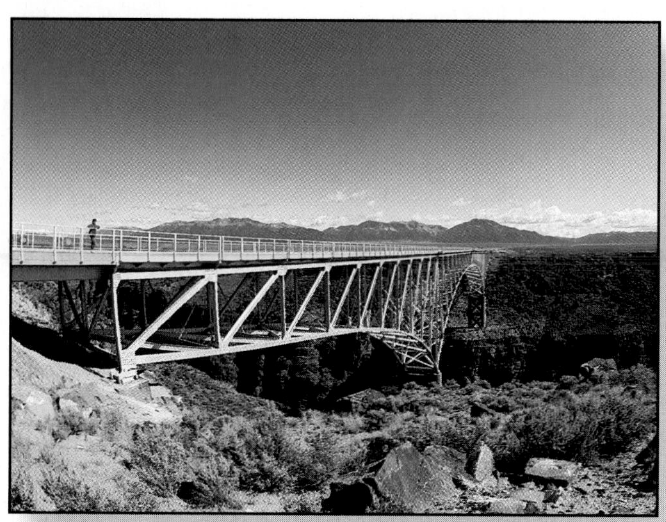

Lesson (Pages)	Lesson Objectives	NCTM Standards	State/Local Objectives
8-1A (396)	Use paper folding to develop the Pythagorean Theorem.	2, 4, 5, 7	
8-1 (397–404)	Find the geometric mean between two numbers. Solve problems involving relationships between parts of a triangle and the altitude to its hypotenuse. Use the Pythagorean Theorem and its converse.	1–5, 7, 12	
8-2 (405–411)	Use the properties of 45°-45°-90° and 30°-60°-90° triangles.	1–5, 7	
8-3 (412–419)	Find trigonometric ratios using right triangles. Solve problems using trigonometric ratios.	1–5, 7, 9	
8-4 (420–425)	Use trigonometry to solve problems involving angles of elevation or depression.	1–5, 7, 9	
8-5 (426–430)	Use the Law of Sines to solve triangles.	1–5, 7, 9	
8-6 (431–436)	Use the Law of Cosines to solve triangles. Choose the appropriate strategy for solving a problem.	1–5, 7, 9	

ORGANIZING THE CHAPTER

A complete, 1-page lesson plan is provided for each lesson in the ***Lesson Planning Guide***. Answer keys for each lesson are available in the ***Answer Key Masters***.

You may want to refer to the **Course Planning Calendar** on page T12 for detailed information on pacing.
PACING: Standard—12 days; **Honors**—11 days; **Block**—7 days

LESSON PLANNING CHART

Lesson (Pages)	Materials/ Manipulatives	Extra Practice (Student Edition)	Blackline Masters								Real-World Applications	Teaching Transparencies
			Study Guide	Practice	Enrichment	Assessment & Evaluation	Modeling Mathematics	Multicultural Activity	Tech Prep Applications	Graphing Calc. & Computer		
8-1A (396)	patty paper ruler* colored pencils						p. 96					
8-1 (397–404)	TI-92 calculator TI-82/83 graphing calculator geoboard or dot paper*	p. 778	p. 43	p. 43	p. 43		p. 86	p. 15	p. 15		15	8-1A 8-1B
8-2 (405–411)	grid paper	p. 778	p. 44	p. 44	p. 44	p. 212						8-2A 8-2B
8-3 (412–419)	scissors* ruler* scientific calculator	p. 778	p. 45	p. 45	p. 45	pp. 211, 212	pp. 43–46	p. 16				8-3A 8-3B
8-4 (420–425)	scientific calculator	p. 779	p. 46	p. 46	p. 46				p. 16		16	8-4A 8-4B
8-5 (426–430)	scientific calculator	p. 779	p. 47	p. 47	p. 47	p. 213				p. 8		8-5A 8-5B
8-6 (431–436)	scientific calculator	p. 779	p. 48	p. 48	p. 48	p. 213					17	8-6A 8-6B
Study Guide/ Assessment (437–441)						pp. 197–210, 214–216						

*Included in Glencoe's High School Manipulative Kit and Overhead Manipulative Resources.

ORGANIZING THE CHAPTER

Other Chapter Resources

Student Edition
Investigation, pp. 392–393
Chapter Opener, pp. 394–395
Mathematics and Society, p. 404
Working on the Investigation, pp. 411, 436

Teacher's Classroom Resources
Investigations and Projects Masters, pp. 53–56
Block Scheduling Booklet

Technology
Test and Review Software (IBM and Macintosh)
CD-ROM Multimedia Applications (Windows and Macintosh)
Mindjogger Videoquizzes (VHS)

Professional Publications
Glencoe Mathematics Professional Series

Outside Resources

Books/Periodicals
The Pythagorean Theorem, Dale Seymour Publications
Maneuvers with Triangles, Dale Seymour Publications
Building Success in Math, Dale Seymour Publications

Software
Meridian Math Coach: Trigonometry, William K. Bradford Publishing Company

Videos/CD-ROMs
The Theorem of Pythagoras, NCTM, 1906 Association Drive, Reston, VA 20191
Sines and Cosines, Part 2, NCTM, 1906 Association Drive, Reston, VA 20191

Assessment Resources

Student Edition
Math Journal, pp. 422, 433
Mixed Review, pp. 403–404, 411, 418–419, 425, 430, 436
Self Test, p. 419
Chapter Highlights, p. 437
Chapter Study Guide and Assessment, pp. 438–440
Alternative Assessment, p. 441
Portfolio, p. 441
College Entrance Exam Practice, pp. 442–443
Chapter Test, p. 800

Teacher's Wraparound Edition
5-Minute Check, pp. 397, 405, 412, 420, 426, 431
Check for Understanding, pp. 400, 408, 415, 422, 428, 433
Closing Activity, pp. 404, 411, 419, 425, 430, 436
Cooperative Learning, pp. 427, 434

Assessment and Evaluation Masters
Multiple-Choice Tests, Forms 1A (Honors), 1B (Average), 1C (Basic), pp. 197–202
Free-Response Tests, Forms 2A (Honors), 2B (Average), 2C (Basic), pp. 203–208
Calculator-Based Test, p. 209
Performance Assessment, p. 210
Mid-Chapter Test, p. 211
Quizzes A–D, pp. 212–213
Standardized Test Practice, p. 214
Cumulative Review, pp. 215–216

ENHANCING THE CHAPTER

Examples of some of the materials for enhancing Chapter 8 are shown below.

Multicultural Activity Masters, pp. 15, 16

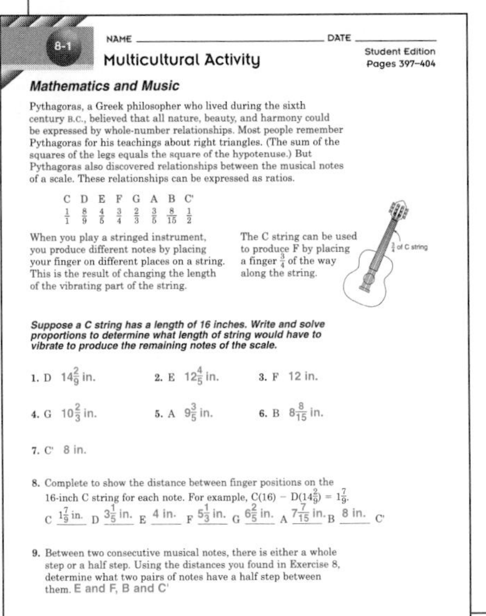

8-1 NAME ______ DATE ______

Multicultural Activity — Student Edition Pages 397–404

Mathematics and Music

Pythagoras, a Greek philosopher who lived during the sixth century B.C., believed that all nature, beauty, and harmony could be expressed by whole-number relationships. Most people remember Pythagoras for his teachings about right triangles. (The sum of the squares of the legs equals the square of the hypotenuse.) But Pythagoras also discovered relationships between the musical notes of a scale. These relationships can be expressed as ratios.

C	D	E	F	G	A	B	C'
$\frac{1}{1}$	$\frac{8}{9}$	$\frac{4}{5}$	$\frac{3}{4}$	$\frac{2}{3}$	$\frac{3}{5}$	$\frac{8}{15}$	$\frac{1}{2}$

When you play a stringed instrument, you produce different notes by placing your finger on different places on a string. This is the result of changing the length of the vibrating part of the string.

The C string can be used to produce F by placing a finger $\frac{3}{4}$ of the way along the string.

$\frac{3}{4}$ of C string

Suppose a C string has a length of 16 inches. Write and solve proportions to determine what length of string would have to vibrate to produce the remaining notes of the scale.

1. D $14\frac{2}{9}$ in.
2. E $12\frac{4}{5}$ in.
3. F 12 in.
4. G $10\frac{2}{3}$ in.
5. A $9\frac{3}{5}$ in.
6. B $8\frac{8}{15}$ in.
7. C' 8 in.
8. Complete to show the distance between finger positions on the 16-inch C string for each note. For example, $C(16) - D(14\frac{2}{9}) = 1\frac{7}{9}$.
   C $1\frac{7}{9}$ in. D $3\frac{1}{5}$ in. E 4 in. F $5\frac{1}{3}$ in. G $6\frac{2}{5}$ in. A $7\frac{7}{15}$ in. B 8 in. C'
9. Between two consecutive musical notes, there is either a whole step or a half step. Using the distances you found in Exercise 8, determine what two pairs of notes have a half step between them. E and F, B and C'

Real-World Applications, 15, 16, 17

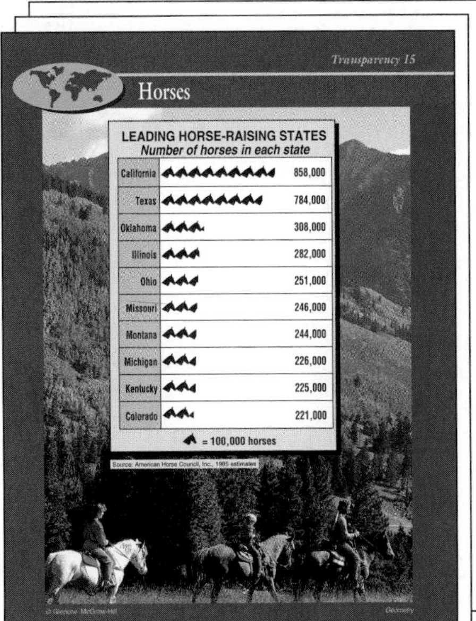

Transparency 15

Horses

LEADING HORSE-RAISING STATES

Number of horses in each state

California	858,000
Texas	784,000
Oklahoma	308,000
Illinois	282,000
Ohio	251,000
Missouri	246,000
Montana	244,000
Michigan	226,000
Kentucky	225,000
Colorado	221,000

= 100,000 horses

Source: American Horse Council, Inc., 1985 estimates

© Glencoe/McGraw-Hill — Geometry

TECHNOLOGY

Graphing Calculator and Computer Masters, p. 8

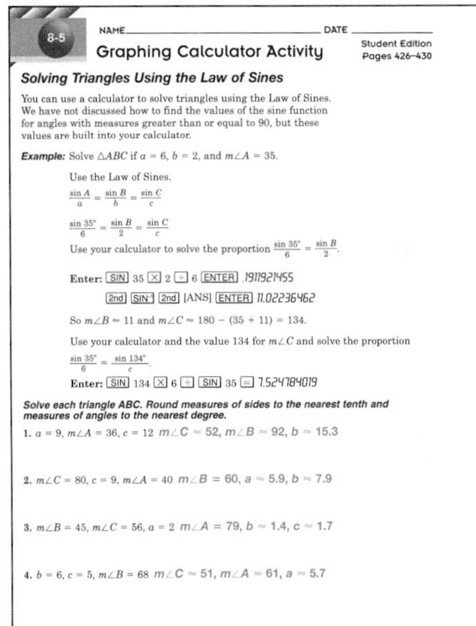

8-5 NAME ______ DATE ______

Graphing Calculator Activity — Student Edition Pages 426–430

Solving Triangles Using the Law of Sines

You can use a calculator to solve triangles using the Law of Sines. We have not discussed how to find the values of the sine function for angles with measures greater than or equal to 90, but these values are built into your calculator.

Example: Solve $\triangle ABC$ if $a = 6$, $b = 2$, and $m\angle A = 35$.

Use the Law of Sines.

$\frac{\sin A}{a} = \frac{\sin B}{b} = \frac{\sin C}{c}$

$\frac{\sin 35°}{6} = \frac{\sin B}{2} = \frac{\sin C}{c}$

Use your calculator to solve the proportion $\frac{\sin 35°}{6} = \frac{\sin B}{2}$.

Enter: [SIN] 35 [×] 2 [÷] 6 [ENTER] .1911921455

[2nd] [SIN⁻¹] [2nd] [ANS] [ENTER] 11.02236462

So $m\angle B \approx 11$ and $m\angle C \approx 180 - (35 + 11) = 134$.

Use your calculator and the value 134 for $m\angle C$ and solve the proportion $\frac{\sin 35°}{6} = \frac{\sin 134°}{c}$.

Enter: [SIN] 134 [×] 6 [÷] [SIN] 35 [=] 7.524784019

Solve each triangle ABC. Round measures of sides to the nearest tenth and measures of angles to the nearest degree.

1. $a = 9$, $m\angle A = 36$, $c = 12$ $m\angle C \approx 52$, $m\angle B \approx 92$, $b \approx 15.3$
2. $m\angle C = 80$, $c = 9$, $m\angle A = 40$ $m\angle B = 60$, $a \approx 5.9$, $b \approx 7.9$
3. $m\angle B = 45$, $m\angle C = 56$, $a = 2$ $m\angle A = 79$, $b \approx 1.4$, $c \approx 1.7$
4. $b = 6$, $c = 5$, $m\angle B = 68$ $m\angle C \approx 51$, $m\angle A \approx 61$, $a \approx 5.7$

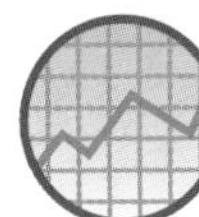

TECH PREP

Tech Prep Applications Masters, pp. 15, 16

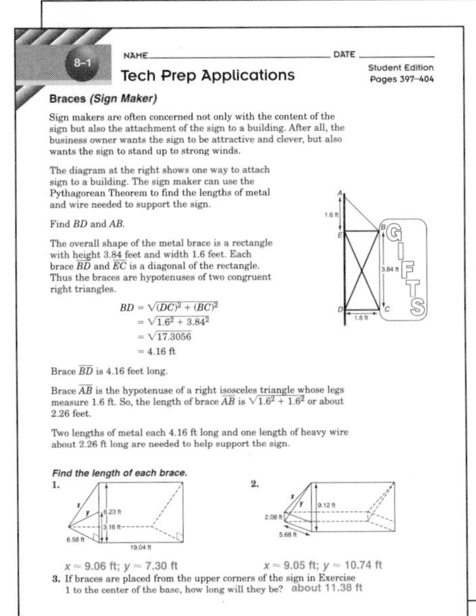

8-1 NAME ______ DATE ______

Tech Prep Applications — Student Edition Pages 397–404

Braces *(Sign Maker)*

Sign makers are often concerned not only with the content of the sign but also the attachment of the sign to a building. After all, the business owner wants the sign to be attractive and clever, but also wants the sign to stand up to strong winds.

The diagram at the right shows one way to attach sign to a building. The sign maker can use the Pythagorean Theorem to find the lengths of metal and wire needed to support the sign.

Find BD and AB.

The overall shape of the metal brace is a rectangle with height 3.84 feet and width 1.6 feet. Each brace $\overline{BD}$ and $\overline{EC}$ is a diagonal of the rectangle. Thus the braces are hypotenuses of two congruent right triangles.

$$BD = \sqrt{(DC)^2 + (BC)^2} = \sqrt{1.6^2 + 3.84^2} = \sqrt{17.3056} = 4.16 \text{ ft}$$

Brace $\overline{BD}$ is 4.16 feet long.

Brace $\overline{AB}$ is the hypotenuse of a right isosceles triangle whose legs measure 1.6 ft. So, the length of brace $\overline{AB}$ is $\sqrt{1.6^2 + 1.6^2}$ or about 2.26 feet.

Two lengths of metal each 4.16 ft long and one length of heavy wire about 2.26 ft long are needed to help support the sign.

Find the length of each brace.

1. $x \approx 9.06$ ft; $y \approx 7.30$ ft
2. $x \approx 9.05$ ft; $y \approx 10.74$ ft
3. If braces are placed from the upper corners of the sign in Exercise 1 to the center of the base, how long will they be? about 11.38 ft

Problem-of-the-Week Cards, 21, 22, 23

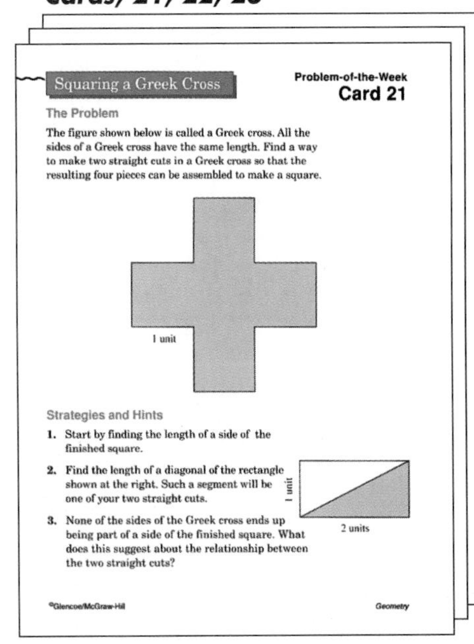

Squaring a Greek Cross — Problem-of-the-Week Card 21

The Problem

The figure shown below is called a Greek cross. All the sides of a Greek cross have the same length. Find a way to make two straight cuts in a Greek cross so that the resulting four pieces can be assembled to make a square.

1 unit

Strategies and Hints

1. Start by finding the length of a side of the finished square.
2. Find the length of a diagonal of the rectangle shown at the right. Such a segment will be one of your two straight cuts.
3. None of the sides of the Greek cross ends up being part of a side of the finished square. What does this suggest about the relationship between the two straight cuts?

1 unit — 2 units

©Glencoe/McGraw-Hill — Geometry

MAKING MATHEMATICS RELEVANT

This two-page introduction to the chapter provides students with an opportunity to explore an application of geometry that has historic roots as well as modern uses.

Background Information

Fast Cars The National Hot Rod Association (NHRA) is the sanctioning body for drag racing events in the United States. The drag racer shown at the right is in the top fuel dragster class. These cars, fueled with nitromethane, are designed to accelerate very quickly from a standing start to achieve final speeds over 300 miles per hour on a straight quarter-mile-long track called a drag strip. At the end of the race, drag racers must release a parachute to slow down the car. Ask students to calculate how long a drag race on a quarter-mile strip lasts, assuming that the *average* speed for the race is 187.5 miles per hour.

CHAPTER 8

Applying Right Triangles and Trigonometry

Objectives

In this chapter, you will:

- use the geometric mean to solve problems,
- use the Pythagorean Theorem and its converse,
- use the properties of 45°-45°-90° and 30°-60°-90° triangles,
- use trigonometry to solve triangles, and
- choose the appropriate strategy for solving a problem.

Geometry: Then and Now You already know that, of all polygons, the triangle is the most rigid. This is why the triangle is used so often in the designs of bridges, buildings, and even cars. Notice the triangular design of the Benz carriage, the forerunner of the Mercedes-Benz designed in the 1880s. Today, the use of triangles in car designs can be seen most often in the frame, or chassis, of a race car. The rigidity of triangles gives a race-car chassis strength. To make the strongest and lightest chassis possible, race-car designers must have a thorough understanding of trigonometry.

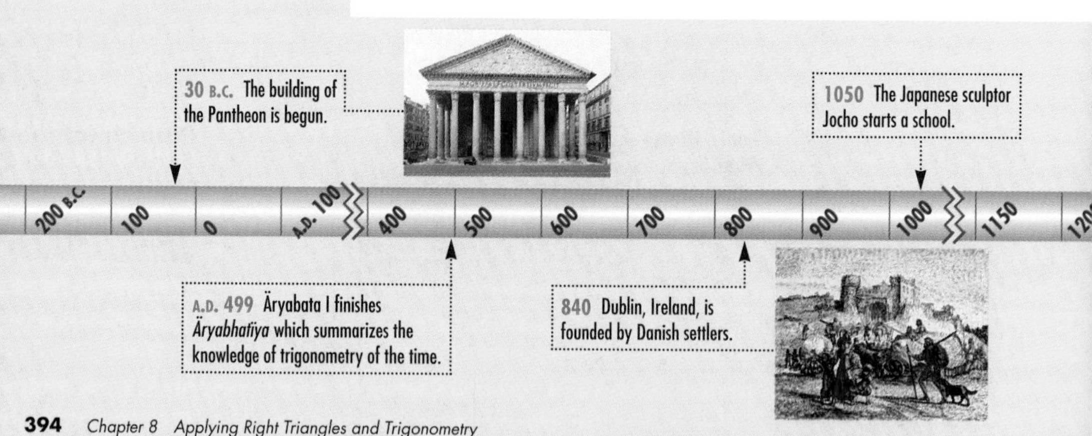

30 B.C. The building of the Pantheon is begun.
1050 The Japanese sculptor Jocho starts a school.
200 B.C.
100
0
A.D. 100
400
500
600
700
800
900
1000
1150
A.D. 499 Āryabata I finishes *Āryabhatīya* which summarizes the knowledge of trigonometry of the time.
840 Dublin, Ireland, is founded by Danish settlers.

Students might find it interesting to research the history of geometry and trigonometry around the world. Contributions were made by many cultures.

interNET CONNECTION

Minimize your pitstop time with information on the NHRA, NASCAR, and other racing associations.

World Wide Web
http://www.goracing.com/index.html

PEOPLE IN THE

NEWS

Eight- to seventeen-year-old girls and boys who dream of driving race cars can live their dreams in the National Hot Rod Association (NHRA) Junior Drag Racing League. These young drivers, like fourteen-year-old **J.R. Todd** of Lawrenceburg, Indiana, compete at league drag meets around the country. Their half-scale cars are otherwise just like the top fuel dragsters that professionals drive. In junior drag racing, pairs of drivers race their cars down an eighth-mile strip at speeds up to 60 miles per hour. Each driver must complete a drag race within a certain range of time estimated for his or her car. Going faster than or slower than the estimated time can lose the race. "What do I like? . . . It's the people, the excitement," says J.R.

Chapter Project

Suppose you are designing and building the chassis of a drag racer. Refer to the side view of the chassis in the diagram below. The lengths are given in inches.

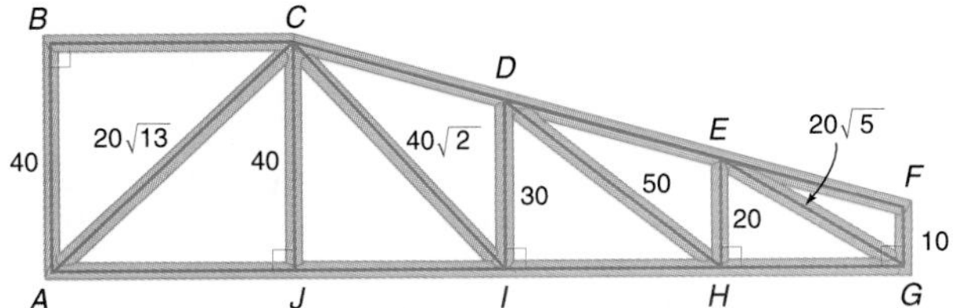

- Suppose you plan to build this chassis from steel tubing. What length of tubing is needed for the bottom edge from *A* to *G* of the chassis assembly? How much tubing is needed for the top from *B* to *F*?
- If steel tubing costs $7.50 per foot, how much will the tubing to build this chassis assembly cost?
- To build the chassis, pieces of steel tubing must be welded together at the joints. Before welding you will need to check that the pieces of tubing are in the correct positions. Find the measures of the angles formed by the two pieces of tubing for each angle: $\angle CAJ$, $\angle CIJ$, $\angle DHI$, and $\angle EGH$.
- Use triangles to design your own chassis.

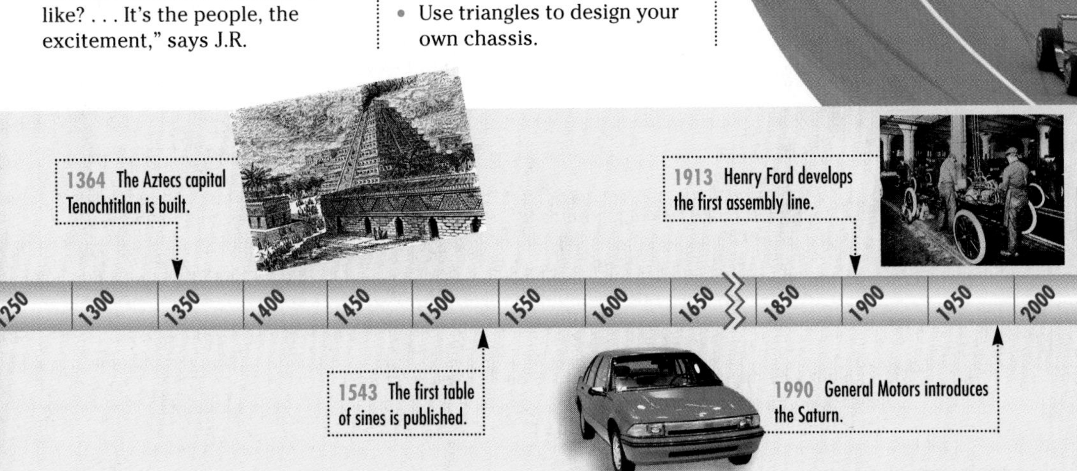

PEOPLE IN THE

NEWS

In junior drag racing the primary concern is safety. At each league drag meet, young drivers are instructed on safety issues and racing procedures. Each driver wears a helmet, fire- and abrasion-resistant suit and gloves, and a neck collar. Dragsters are equipped with roll cage, deflector plate, and driver restraint systems. Car equipment, safety gear, and track conditions are all closely regulated by league and track officials. For more information on the NHRA Jr. Drag Racing League, write P.O. Box 5555, Glendora, CA 91740-0950 or call 818-914-4761.

Chapter Project

Cooperative Learning Students should work together to develop a plan for solving these problems and coming to a consensus. Encourage students to research prices of steel tubing at local suppliers to estimate the cost of building the assembly. Groups might want to separate the tasks into geometry and accounting tasks.

Investigations and Projects Masters, p. 53

NAME______________________ DATE ____________

Chapter 8 Project A

Student Edition
Pages 394–441

Start Your Engines

1. Make a rough sketch of the side view of a car of your choice. It may be a car from the past, a car of today, or a car that might be made in the future.
2. Car chassis are sometimes made of steel tubing, which is both strong and lightweight. Use your sketch to draw what you think would be the chassis of steel supports. The foundation of the chassis should be composed of right triangles.
3. Decide on the length of tubing that is needed for the bottom edge and the height of the chassis, then calculate all of the other lengths needed for the steel tubing of the chassis.
4. Make a scale model of the side view of your chassis on a piece of paper or cardboard, using toothpicks, pipe cleaners, or other materials. Include the length of each piece of tubing with the model. You may also want to draw the car exterior over the chassis to illustrate the finished product.

Alternative Chapter Projects

Two other chapter projects are included in the *Investigations and Projects Masters*. In Chapter 8 Project A, pp. 53–54, students extend the topic in the chapter opener. In Chapter 8 Project B, pp. 55–56, students explore how geometry is visible in sculpture.

8-1A LESSON NOTES

NCTM Standards: 2, 4, 5, 7

Objective

Use paper folding to develop the Pythagorean Theorem.

Recommended Time

Demonstration and discussion: 15 minutes; Exercises: 30 minutes

Instructional Resources

For each student or group of students
Student Manipulative Kit
• ruler
Modeling Mathematics Masters
• p. 96 (worksheet)
For teacher demonstration
Algebra and Geometry Overhead Manipulative Resources

1 FOCUS

Motivating the Lesson

Draw several different right triangles. Use a ruler to measure all three sides. Ask students if they see a pattern. Depending on accuracy, they may see $a^2 + b^2 = c^2$.

2 TEACH

Teaching Tip Measured lengths must be very accurate and angles must be 90° in order for results to be accurate. If students are having a problem visualizing Step 6, have them cut each square apart and take out the corresponding regions.

3 PRACTICE/APPLY

Assignment Guide
Core (with proof): 1–3
Core (informal): 1–3
Enriched: 1–3

MODELING MATHEMATICS

A Preview of Lesson 8–1

8–1A The Pythagorean Theorem

Materials:  patty paper  ruler  colored pencils

In Chapter 1, you learned that the Pythagorean Theorem relates the measures of the legs and the hypotenuse of a right triangle. Throughout the world, you can find evidence of ancient cultures that used the Pythagorean Theorem before it was officially named in 1909.

You too can discover this relationship among the measures of the sides of a right triangle by using patty paper and algebra.

Activity **Use paper folding to develop the Pythagorean Theorem.**

Step 1 On a piece of patty paper, make a mark along one side so that the two resulting segments are not congruent. Call one length a and the other length b.

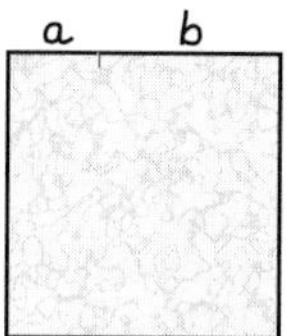

Step 2 Copy these measures on the other sides in the order shown at the right. Fold the paper to divide the square into four sections. Label the area of each section.

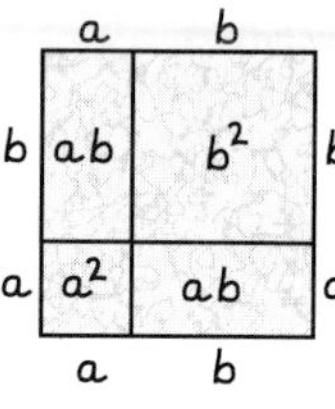

Step 3 On another sheet of patty paper, mark the same lengths a and b on the sides in the different pattern shown at the right.

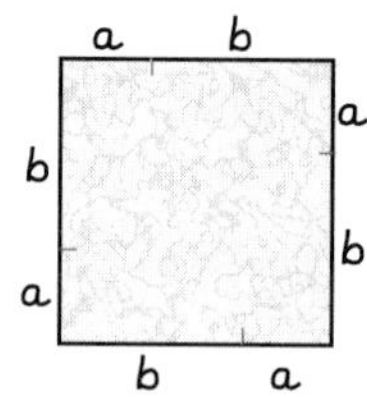

Step 4 Use your straightedge and pencil to connect the marks as shown at the right. Let c represent the length of each hypotenuse.

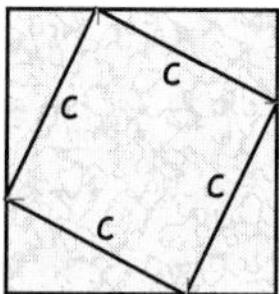

Step 5 Label the area of each section, which is $\frac{1}{2}ab$ for each triangle and c^2 for the square.

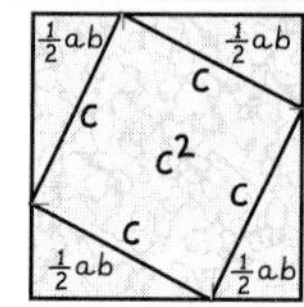

Step 6 Place the squares side by side and color the corresponding regions that have the same area. For example, $ab = \frac{1}{2}ab + \frac{1}{2}ab$.

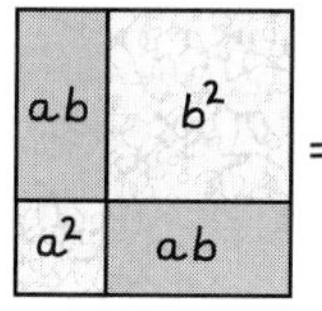

=
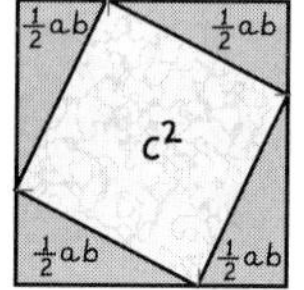

The parts that are not shaded tell us $a^2 + b^2 = c^2$.

Model

1. Use your ruler to find actual measures for a, b, and c. Do these measures confirm that $a^2 + b^2 = c^2$? **yes**
2. Repeat the activity with different a and b values. What do you notice? $a^2 + b^2 = c^2$

Write

3. **Explain** why the drawing at the right is an illustration of the Pythagorean Theorem. **The sum of the areas of the two smaller squares is equal to the area of the largest square.**

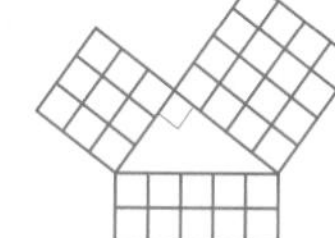

4 ASSESS

Observing students working in cooperative groups is an excellent method of assessment.

Using Cooperative Learning

This lesson offers an excellent opportunity for using cooperative learning groups. For more information on cooperative learning strategies and group management, see *Cooperative Learning in the Mathematics Classroom*, one of the titles in the Glencoe Mathematics Professional Series.

8-1 Geometric Mean and the Pythagorean Theorem

* To find the geometric mean between two numbers,
* to solve problems involving relationships between parts of a triangle and the altitude to its hypotenuse, and
* to use the Pythagorean Theorem and its converse.

You can use relationships to find missing measures of parts of right triangles.

Discrete Mathematics

A geometric sequence is one in which each consecutive number is found by multiplying the previous number by a given factor. For example, suppose the first number of a sequence is 5 and the factor is 3. You can find the next four numbers in the sequence as shown below.

$$5 \xrightarrow{\times 3} 15 \xrightarrow{\times 3} 45 \xrightarrow{\times 3} 135 \xrightarrow{\times 3} 405$$

Consecutive numbers of the sequence form proportions.

$$\frac{5}{15} = \frac{15}{45} \qquad \frac{15}{45} = \frac{45}{135} \qquad \frac{45}{135} = \frac{135}{405}$$

Note that the denominator of one fraction is the numerator of the next. The **geometric mean** between two positive numbers a and b is the positive number x where $\frac{a}{x} = \frac{x}{b}$. By cross multiplying, we see that $x^2 = ab$ or $x = \sqrt{ab}$. Note that in the proportion, x and x represent the *means* and a and b represent the *extremes*.

Example  **Find the geometric mean between 2 and 10.**

Let x represent the geometric mean.

$\frac{2}{x} = \frac{x}{10}$ *Definition of geometric mean*

$x^2 = 20$ *Cross multiply.*

$x = \sqrt{20}$ or about 4.47 *Take the square root of each side.*

The geometric mean between 2 and 10 is about 4.47.

Consider right triangle ABC with altitude $\overline{CD}$ drawn from the right angle C to the hypotenuse $\overline{AB}$. A special relationship exists for the three right triangles ABC, ACD, and CBD.

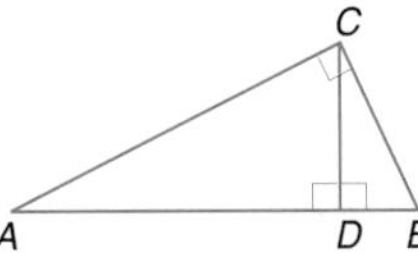

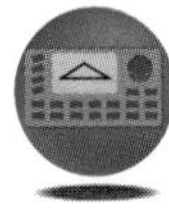

EXPLORATION CABRI GEOMETRY

Use a TI-92 calculator to draw right triangle ABC shown above. Explore the relationships among the three right triangles by finding the measures of $\angle A$, $\angle ACB$, $\angle B$, $\angle ADC$, $\angle ACD$, $\angle BDC$, and $\angle BCD$.

a. What is the relationship between the measures of $\angle A$ and $\angle BCD$? What is the relationship between the measures of $\angle B$ and $\angle ACD$?

b. Drag point C to another position. Describe the relationship between the measures of $\angle A$ and $\angle BCD$ and between the measures of $\angle B$ and $\angle ACD$.

c. Make a conjecture about $\triangle ABC$, $\triangle ACD$, and $\triangle CBD$.

a. They are the same; they are the same.

b. They are the same; they are the same.

c. $\triangle ABC \sim \triangle ACD \sim \triangle CBD$

8-1 LESSON NOTES

NCTM Standards: 1–5, 7, 12

Instructional Resources

* Study Guide Master 8-1
* Practice Master 8-1
* Enrichment Master 8-1
* Modeling Mathematics Masters, p. 86
* Multicultural Activity Masters, p. 15
* Real-World Applications, 15
* Tech Prep Applications Masters, p. 15

Transparency 8-1A contains the 5-Minute Check for this lesson; **Transparency 8-1B** contains a teaching aid for this lesson.

Recommended Pacing	
Standard Pacing	Days 2 & 3 of 12
Honors Pacing	Day 2 of 11
Block Scheduling*	Day 1 of 7

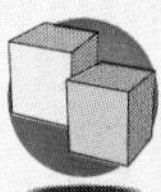

*For more information on pacing and possible lesson plans, refer to the *Block Scheduling Booklet*.

1 FOCUS

5-Minute Check *(over Chapter 7)*

1. Solve $\frac{10}{t} = \frac{6}{t-1}$ using cross products. 2.5
2. If two similar parallelograms have a scale factor of 4.5 and one angle of the larger parallelogram measures 65°, what is the measure of the corresponding angle in the smaller parallelogram? 65

Refer to the figure below.

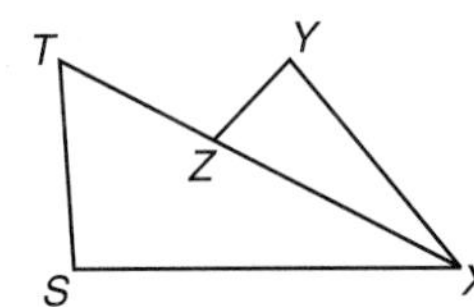

3. If $\overline{TX}$ bisects $\angle SXY$ and $\angle S \cong \angle Y$, can you say the two triangles are similar? Why or why not? yes; AA Similarity
4. If $\triangle STX \sim \triangle YZX$, $TZ = 3$, $ZX = 7$, $TS = 6$, and $SX = 14$, what is the perimeter of $\triangle YZX$? 21 units

EXPLORATION

This Exploration leads students to visualize Theorem 8-1.

Motivating the Lesson

Hands-On Activity Have each student draw a right triangle on grid paper. Have them draw a square on each side so that one side of the square is a side of the triangle. Have them cut out the three squares, cut the two smaller squares into pieces, and arrange them so that they cover the larger square.

2 TEACH

In-Class Examples

For Example 1
Find the geometric mean between 5 and 25. 11.18

For Example 2
To find the height of the tree in her backyard, Wendie held a book near her eye so that the top and bottom of the tree were in line with the edges of the cover. If Wendie's eye is 5 feet off the ground and she is standing approximately 14 feet from the tree, how tall is the tree? Assume that the tree is perpendicular to the ground and that the edges of the cover of the book are at right angles. 44.2 feet

Teaching Tip Have students use the method in Example 2 to find the height of various things around the school, if possible.

Using Technology
You can have students use a TI-92 or the *Geometer's Sketchpad* to explore the concepts presented in Theorems 8-1 and 8-2.

The results of the Exploration suggest that the following angle relationships are true for right $\triangle ABC$ with right $\angle C$ and altitude $\overline{CD}$.

$\triangle ABC$		$\triangle ACD$		$\triangle CBD$
$\angle A$	$\cong$	$\angle A$	$\cong$	$\angle BCD$
$\angle B$	$\cong$	$\angle ACD$	$\cong$	$\angle B$
$\angle ACB$	$\cong$	$\angle ADC$	$\cong$	$\angle CDB$

This means that all three triangles are similar by the Angle-Angle Postulate.

$$\triangle ABC \sim \triangle ACD \sim \triangle CBD$$

This is more generally stated in Theorem 8–1. *You will be asked to prove this theorem in Exercise 36.*

Theorem 8–1	**If the altitude is drawn from the vertex of the right angle of a right triangle to its hypotenuse, then the two triangles formed are similar to the given triangle and to each other.**

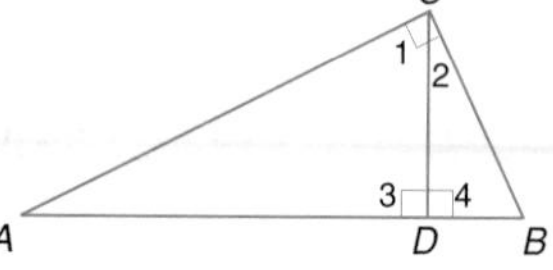

In the figure at the right, $\triangle ADC \sim \triangle CDB$. So $\frac{AD}{CD} = \frac{CD}{BD}$ because corresponding sides of similar triangles are proportional. Thus, CD is the geometric mean between AD and BD. This is stated in Theorem 8–2. *You will be asked to prove this theorem in Exercise 37.*

Theorem 8–2	**The measures of the altitude drawn from the vertex of the right angle of a right triangle to its hypotenuse is the geometric mean between the measures of the two segments of the hypotenuse.**

The geometric mean can be used to estimate hard-to-measure distances.

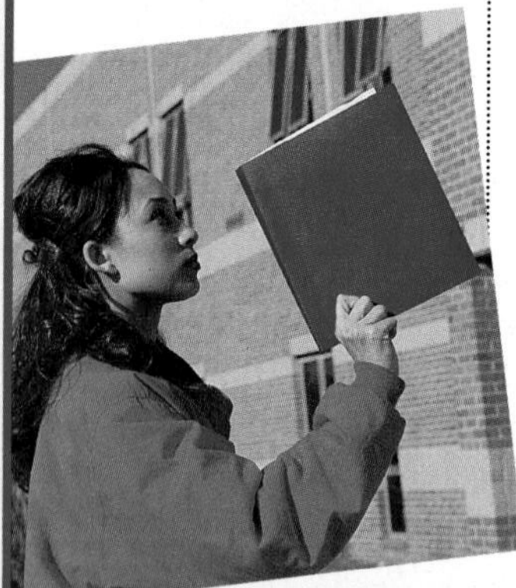

2 **To find the height of her school building, Mieko held a book near her eye so that the top and bottom of the building were in line with the edges of the cover. If Mieko's eye level is 5 feet off the ground and she is standing about 10 feet from the building, how tall is the building? *Assume the building is perpendicular to the ground and the edges of the cover of the book form right angles.***

Draw a diagram of the situation. $\overline{CD}$ is the altitude drawn from the right angle of $\triangle ABC$.

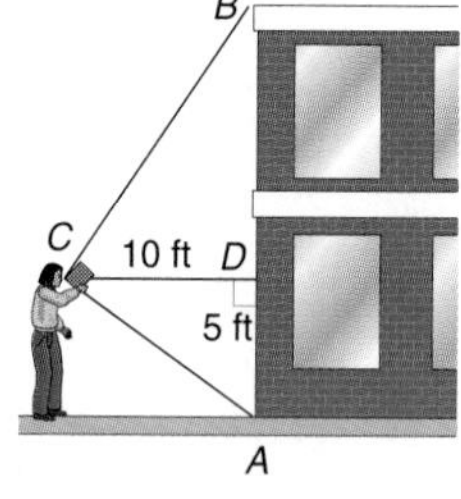

$\frac{AD}{CD} = \frac{CD}{BD}$ *Theorem 8–2*

$\frac{5}{10} = \frac{10}{BD}$ *AD = 5, CD = 10*

$5BD = 100$ *Cross multiply.*

$BD = 20$ *Division Property (=)*

Mieko estimates that the building is 20 + 5 or 25 feet tall.

Classroom Vignette

"At the beginning of this lesson, I have students draw three large right triangles in their notebooks. As we discuss the relationships, they highlight the sides of the triangles in two different colors to show which length is the geometric mean between the other two. The triangles can serve as a quick reference as students study or complete their homework."

J Herta

Mrs. Jean Herta
Powers Catholic High School
Flint, Michigan

The altitude to the hypotenuse of a right triangle determines another relationship between segments.

Theorem 8–3	**If the altitude is drawn to the hypotenuse of a right triangle, then the measure of a leg of the triangle is the geometric mean between the measures of the hypotenuse and the segment of the hypotenuse adjacent to that leg.**

You will be asked to prove this theorem in Exercise 38.

Example 3 **Find a and b in $\triangle TGR$.**

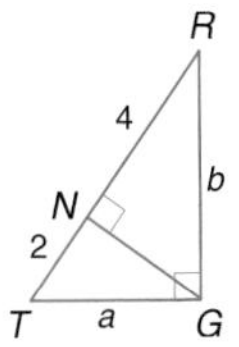

According to Theorem 8–3, we can write the following proportions.

LOOK BACK
You can refer to Lesson 1-4 for information on the Pythagorean Theorem.

$\frac{TN}{TG} = \frac{TG}{TR}$	$\frac{NR}{RG} = \frac{RG}{RT}$	
$\frac{2}{a} = \frac{a}{6}$	$\frac{4}{b} = \frac{b}{6}$	*$TG = a$, $TN = 2$, $NR = 4$, $TR = 6$, $RG = b$*
$a^2 = 12$	$b^2 = 24$	*Cross multiply.*
$a = \sqrt{12}$	$b = \sqrt{24}$	*Take the square root of each side.*
$a \approx 3.46$	$b \approx 4.90$	

The geometric-mean relationships can be used to prove one of the most important theorems in mathematics, the **Pythagorean Theorem**. Its name comes from Pythagoras, a Greek mathematician from the sixth century B.C. who is said to have been the first to write a proof of the theorem.

Theorem 8–4 Pythagorean Theorem	**In a right triangle, the sum of the squares of the measures of the legs equals the square of the measure of the hypotenuse.**

If c is the measure of the hypotenuse and a and b are the measures of the legs, then $a^2 + b^2 = c^2$.

Proof of Theorem 8–4

Given: right $\triangle ABC$ with right angle at C

Prove: $a^2 + b^2 = c^2$

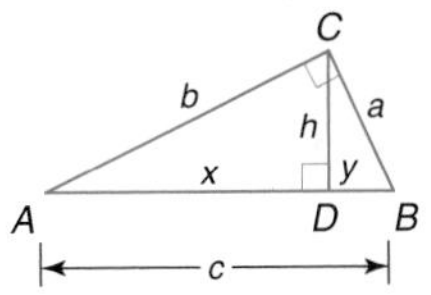

Proof:

Draw the altitude from C to $\overline{AB}$. Let $AB = c$, $AC = b$, $BC = a$, $AD = x$, $DB = y$, and $CD = h$. Two geometric means now exist.

$\frac{c}{a} = \frac{a}{y}$ and $\frac{c}{b} = \frac{b}{x}$

$a^2 = cy$ and $b^2 = cx$ *Cross multiply.*

Add the equations.

$a^2 + b^2 = cy + cx$

$a^2 + b^2 = c(y + x)$ *Factor.*

$a^2 + b^2 = c^2$ *Since $c = y + x$, substitute c for $(y + x)$.*

In-Class Example

For Example 3
Find AB and BC.

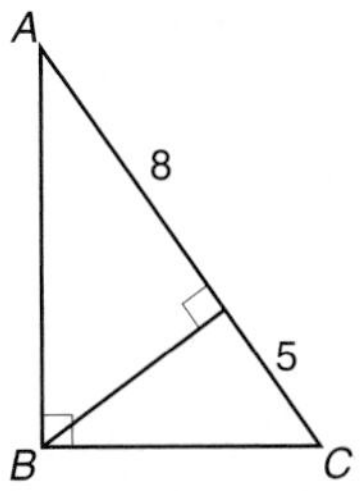

$AB \approx 10.2$, $BC \approx 8.1$

Proof Pointer

In the proof for Theorem 8-4, point out that this is an example of using an auxiliary line to help with a proof.

Alternative Teaching Strategies

Reading Geometry The Pythagorean Theorem was known in India and China before Pythagoras proved the theorem. Have students research the work of Pythagoras and his followers. Be careful: there was a sculptor named Pythagoras who was born in the same town a hundred years before the mathematician.

In-Class Example

For Example 4
Determine whether a triangle with side measures of 40, 96, and 104 forms a right triangle. That is, do 40, 96, and 104 form a Pythagorean triple? yes

3 PRACTICE/APPLY

Check for Understanding

Exercises 1–14 are designed to help you assess your students' understanding through reading, writing, speaking, and modeling. You should work through Exercises 1–6 with your students and then monitor their work on Exercises 7–14.

Error Analysis
Some students may not apply the Pythagorean Theorem and its converse correctly because they do not begin by determining which side is the hypotenuse. Remind students that the longest side in a right triangle is the hypotenuse and that its square must equal the sum of the squares of the other two sides.

Additional Answers

1. If x is the geometric mean, $\frac{5}{x} = \frac{x}{7}$. To find the mean, solve the equation for x.
2.

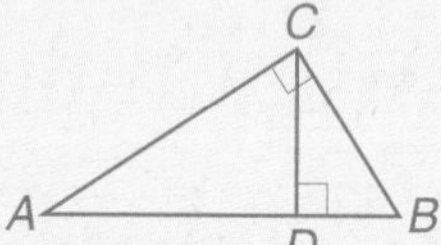

$\angle A \cong \angle A$ and $\angle CDA \cong \angle ACB$, so $\triangle ACD \sim \triangle ABC$ by AA. $\angle B \cong \angle B$ and $\angle ACB \cong \angle CDB$, so $\triangle ABC \sim \triangle CBD$ by AA. $\triangle ACD \sim \triangle ABC \sim \triangle CBD$ by the Transitive Property.

In Chapter 1, you used the Pythagorean Theorem to find the measures of the sides of a right triangle. The converse of the Pythagorean Theorem is useful to determine if three measures of the sides of a triangle are those of a right triangle. *You will be asked to prove this theorem in Exercise 39.*

Theorem 8–5 Converse of the Pythagorean Theorem	**If the sum of the squares of the measures of two sides of a triangle equals the square of the measure of the longest side, then the triangle is a right triangle.**

A **Pythagorean triple** is a group of three whole numbers that satisfies the equation $a^2 + b^2 = c^2$, where c is the greatest number. One common Pythagorean triple is 3, 4, and 5. If the measures of a right triangle are whole numbers, the measures of the sides are a Pythagorean triple.

Example 4 **Determine if a triangle with side measures of 74, 91, and 52 forms a right triangle. That is, do 74, 91, and 52 form a Pythagorean triple?**

The measure of the longest side is 91. Use the converse of the Pythagorean Theorem.

$$a^2 + b^2 = c^2$$
$$74^2 + 52^2 \stackrel{?}{=} 91^2$$
$$5476 + 2704 \stackrel{?}{=} 8281$$
$$8180 \neq 8281$$

Since $8180 \neq 8281$, the triangle is not a right triangle, and the three numbers do not form a Pythagorean triple.

CHECK FOR UNDERSTANDING

Communicating Mathematics

Study the lesson. Then complete the following. 1. See margin.

1. **Describe** how you would find the geometric mean between 5 and 7.
2. **Draw and label** a right triangle with an altitude from the right angle. Explain why the original triangle and the two triangles created by the altitude are similar. See margin.
3. **Explain** why a Pythagorean triple can represent the measure of the sides of a right triangle.
4. **Draw and label** a right triangle with an altitude from the right angle. Use your drawing to explain what is meant by "the hypotenuse and the segment of the hypotenuse adjacent to a leg" in Theorem 8–3. See margin.
5. **You Decide** Molly draws an equilateral $\triangle RST$ with altitude $\overline{SU}$ as shown at the right. She says that SU is the geometric mean between RU and UT. Sierra disagrees. Who is correct? Explain. See margin.

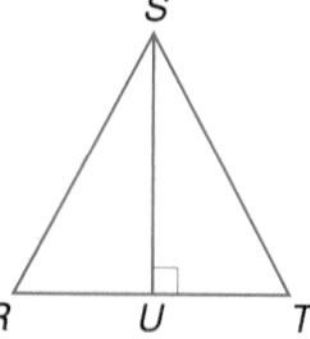

3. Since the numbers in a Pythagorean triple satisfy the equation $a^2 + b^2 = c^2$, they represent the sides of a right triangle by the converse of the Pythagorean Theorem.

Reteaching

Using Cooperative Groups Separate students into groups of five with each student responsible for one of these questions about $\triangle LMN$. Ask each student to present the answer and rationale to the group and to round answers to the nearest tenth.

1. If $\ell = 8$ and $n = 12$, find m. 14.4
2. Find ℓ if $n = 3$ and $m = 9$. 8.5
3. Find n if $\ell = 4$ and $m = 6$. 4.5
4. If $m = 50$ and $n = 40$, find ℓ. 30
5. If $\ell = 18$ and $n = 24$, can $m = 42$? no

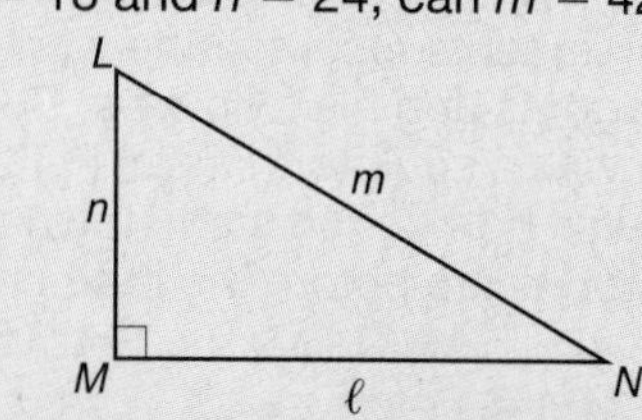

6b–c. See margin.

6. On a geoboard or dot paper, form a right triangle with legs one unit long. Build squares on the two legs and the hypotenuse as shown on the right.
   a. What is the area of each square? 1; 1; 2
   b. What is the relationship between the area of the squares built on the legs and the area of the square built on the hypotenuse?
   c. Form another right triangle and build squares on the two legs and the hypotenuse. How are the areas of the squares related?

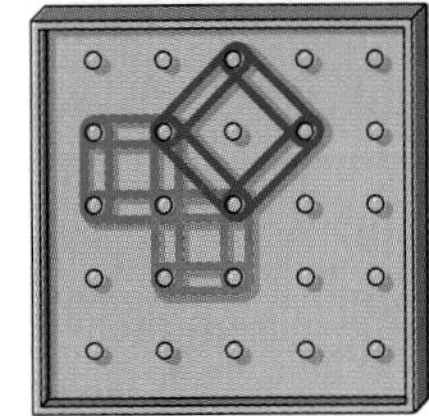

Guided Practice

7. Find the geometric mean between 4 and 25. 10
8. Refer to right triangle PTG. a. $\triangle PTG \sim \triangle PGA \sim \triangle GTA$
   a. Name three similar triangles.
   b. Name two angles congruent to $\angle PGT$. $\angle PAG$, $\angle TAG$

Find the values of *x* and *y*. 9. $\sqrt{10} \approx 3.16$; $\sqrt{14} \approx 3.74$

10. $2\sqrt{13} \approx 7.21$; 13
11. $\sqrt{51} \approx 7.14$

9. 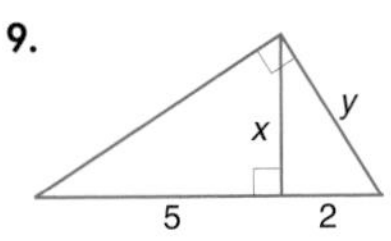

10. 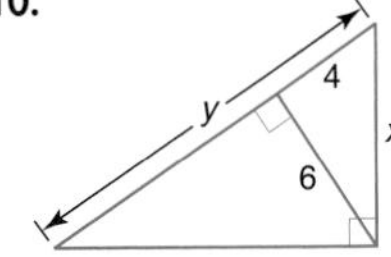

11. 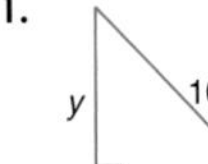

12. yes

12. Determine if 4, 7.5, and 8.5 are measures of the sides of a right triangle.
13. **Look for a Pattern** Complete the chart of Pythagorean triples.
   a. A *primitive* Pythagorean triple is set of three numbers with no common factors except 1. Does the chart contain any primitive Pythagorean triples? yes; 3, 4, 5
   b. Describe the pattern that relates these sets of Pythagorean triples.
   c. Why do you think these Pythagorean triples are called a *family*?
   d. Are the triangles described by a family of Pythagorean triples similar? Explain.

a	*b*	*c*
3	4	5
6	8	10
9	12	15
12	16	20

13b. Each triple is a multiple of the triple 3, 4, and 5.
13c. Sample answer: The triples are all multiples of the triple 3, 4, and 5.
13d. Yes; the measures of the sides are always multiples of 3, 4, and 5.

14. **Architecture** The distance between the base of the Leaning Tower of Pisa and the top of the tower is 180 feet. The tower is leaning 16 feet off the perpendicular. Find the distance of the top of the tower from the ground. about 179.29 feet

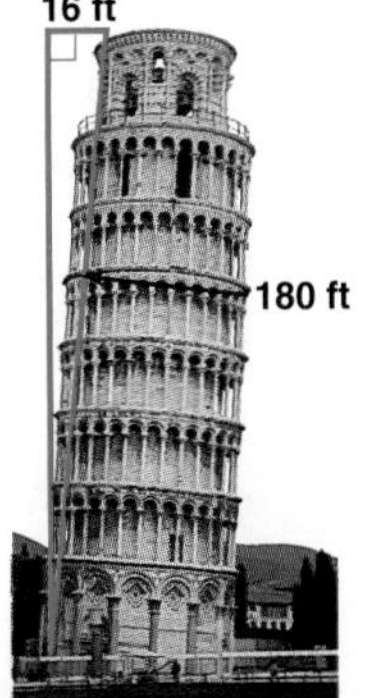

FYI

The construction of the bell tower, which is now known as the Leaning Tower of Pisa, began in 1173. It is actually one of three buildings that form the Cathedral of Pisa.

EXERCISES

Practice 

Find the geometric mean between each pair of numbers.

15. 4 and 9 6
16. 4 and $\frac{1}{9}$ $\frac{2}{3}$
17. $\frac{2}{3}$ and $\frac{1}{3}$ $\frac{\sqrt{2}}{3} \approx 0.47$

FYI

The tower leans an additional 0.043 inches per year. So far, efforts to reinforce the tower have not been effective. Eventually, it will fall.

Additional Answers

4. For leg $\overline{AC}$, $\overline{AB}$ is the hypotenuse and $\overline{AD}$ is the segment adjacent to the leg. For leg $\overline{BC}$, $\overline{AB}$ is the hypotenuse and $\overline{BD}$ is the segment adjacent to the leg.
5. Sierra; SU is the geometric mean between RS and TS only if the triangle is a right triangle and $\overline{SU}$ is the altitude from the right angle.

6b. The sum of the areas of the squares built on the legs equals the area of the square built on the hypotenuse.

6c. See students' work; the sum of the areas of the squares built on the legs equals the area of the square built on the hypotenuse.

Assignment Guide

Core (with proof): 15–41 odd, 42, 43, 45–56
Core (informal): 15–35 odd, 41–43, 45–50, 52–56
Enriched: 16–40 even, 42–56

For **Extra Practice**, see p. 778.

The red A, B, and C flags, printed only in the Teacher's Wraparound Edition, indicate the level of difficulty of the exercises.

Additional Answer

40. $(CB)^2 + (AC)^2 = (AB)^2$
(Pythagorean Theorem)
$|x_2 - x_1|^2 + |y_2 - y_1|^2 = d^2$
$(CB = x_2 - x_1, AC = y_2 - y_1, AB = d)$
$(x_2 - x_1)^2 + (y_2 - y_1)^2 = d^2$
$(|x_2 - x_1|^2 = (x_2 - x_1)^2, |y_2 - y_1|^2 = (y_2 - y_1)^2)$
$\sqrt{(x_2 - x_1)^2 + (y_2 - y_1)^2} = d$
(Take the square root of each side.)

Study Guide Masters, p. 43

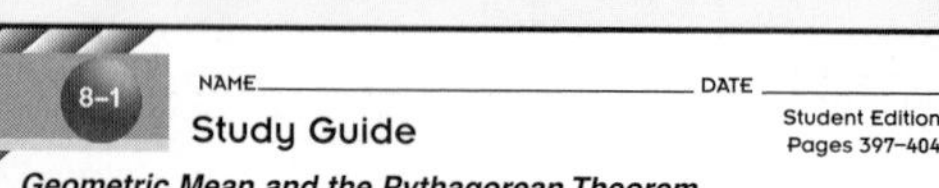

8–1 NAME______________ DATE________

Study Guide

Student Edition Pages 397–404

Geometric Mean and the Pythagorean Theorem

The geometric mean between two positive numbers a and b is the positive number x where $\frac{a}{x} = \frac{x}{b}$.

If $\triangle ABC$ is a right triangle with altitude $\overline{BD}$, then the following relationships hold true.

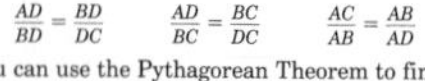

$\frac{AD}{BD} = \frac{BD}{DC}$ $\quad \frac{AD}{BC} = \frac{BC}{DC}$ $\quad \frac{AC}{AB} = \frac{AB}{AD}$

You can use the Pythagorean Theorem to find missing measures for right triangles.

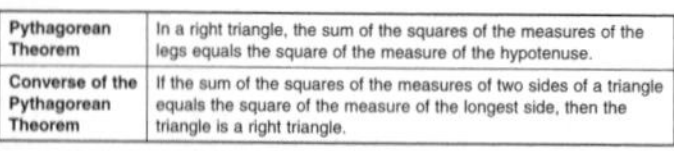

Pythagorean Theorem	In a right triangle, the sum of the squares of the measures of the legs equals the square of the measure of the hypotenuse.
Converse of the Pythagorean Theorem	If the sum of the squares of the measures of two sides of a triangle equals the square of the measure of the longest side, then the triangle is a right triangle.

Use $\triangle ABC$ above for the following examples.

Examples: *1* Find a.
You can use geometric mean relationships.
$\frac{AD}{BC} = \frac{BC}{DC}$
$\frac{20}{a} = \frac{a}{50}$
$a^2 = 1000$
$a = \sqrt{1000}$
$a \approx 31.6$

2 Find b.
You can use the Pythagorean Theorem.
$b^2 + a^2 = 70^2$
$b^2 + 1000 = 4900$
$b^2 = 3900$
$b = \sqrt{3900}$
$b \approx 62.4$

Find the geometric mean between each pair of numbers.

1. 3 and 10 $\sqrt{30} \approx 5.5$ **2.** 10 and 20 $\sqrt{200} \approx 14.1$ **3.** 10 and 40 20

Find the values of x and y. Round to the nearest tenth.

4.
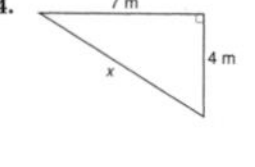
8.1 m

5.
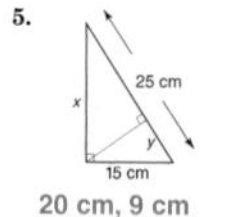
20 cm, 9 cm

6.
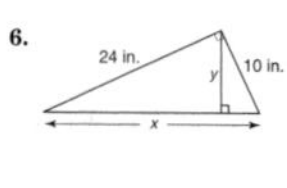
26 in., 9.2 in.

Determine if the given measures are measures of the sides of a right triangle.

7. 18, 24, 30 yes **8.** 20, 30, 40 no **9.** 4.5, 6, 7.5 yes

---

Refer to right triangle *ASH* for Exercises 18–20.

18. Name the three similar triangles. $\triangle ASH \sim \triangle ALS \sim \triangle SLH$

19. Name two segments such that the given segment is the geometric mean.
a. $\overline{AS}$ $\overline{AH}, \overline{SH}$ **b.** $\overline{AH}$ $\overline{AS}, \overline{SH}$ **c.** $\overline{LH}$ c. $\overline{SH}, \overline{LS}$

20. Name at least one angle congruent to each angle.
a. $\angle SLH$ $\angle ASH, \angle ALS$ **b.** $\angle A$ $\angle LSH$ **c.** $\angle 2$ $\angle H$

Find the values of *x* and *y*. 21. $2\sqrt{6} \approx 4.90$; $\sqrt{33} \approx 5.74$

21.
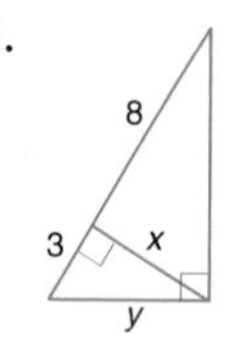

22.
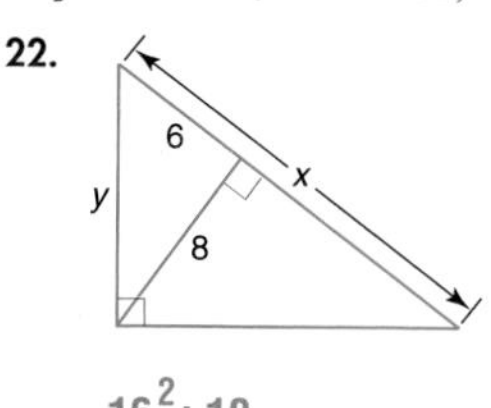
$16\frac{2}{3}$; 10

23.
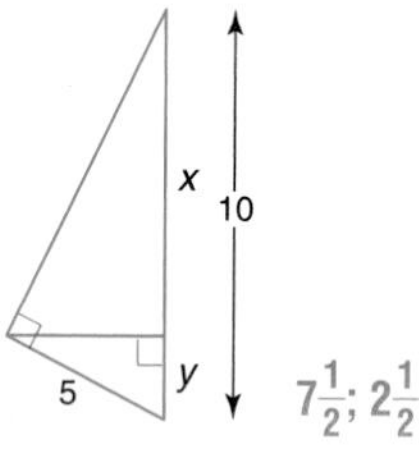
$7\frac{1}{2}$; $2\frac{1}{2}$

24.
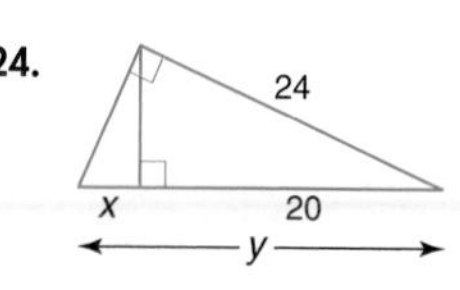
8.8; 28.8

25.
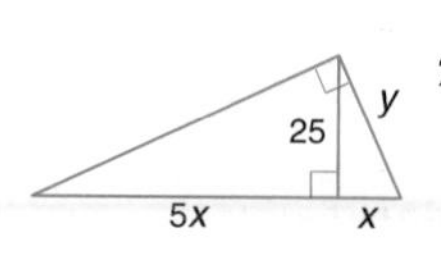
25. $5\sqrt{5} \approx 11.18$; $5\sqrt{30} \approx 27.39$

26.
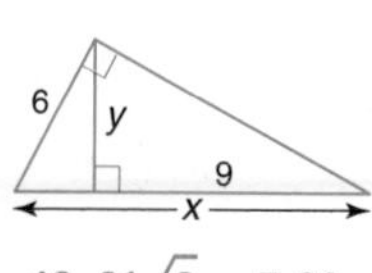
12; $3\sqrt{3} \approx 5.20$

27.
4.5

28.
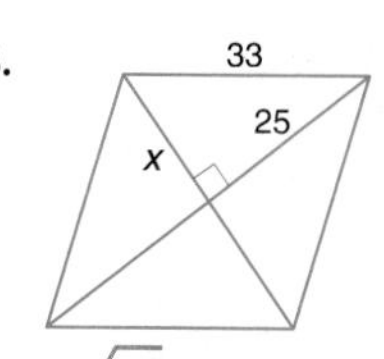
$4\sqrt{29} \approx 21.54$

29.
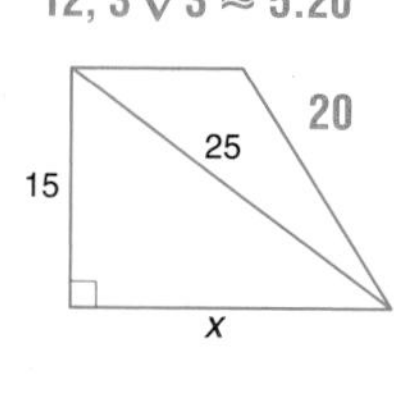
20

Determine if the given measures are measures of the sides of a right triangle.

30. 5, 12, 13 yes **31.** 20, 21, 28 no **32.** 37, 12, 34 no

Refer to Exercise 13. For each Pythagorean triple, find two triples in the same family. 35. Sample answers: 18, 80, 82; 27, 120, 123

33. 8, 15, 17 **34.** 7, 24, 25 **35.** 9, 40, 41

33. Sample answers: 16, 30, 34; 24, 45, 51
34. Sample answers: 14, 48, 50; 21, 72, 75

C Proof

36. Write a two-column proof of Theorem 8–1. 36–39. See Solutions Manual.

37. Write a paragraph proof of Theorem 8–2.

38. Write a two-column proof of Theorem 8–3.

39. Write a paragraph proof of the converse of the Pythagorean Theorem (Theorem 8–5).

40. Use the Pythagorean Theorem and the figure at the right to show $d = \sqrt{(x_2 - x_1)^2 + (y_2 - y_1)^2}$. That is, show the distance formula is true. See margin.

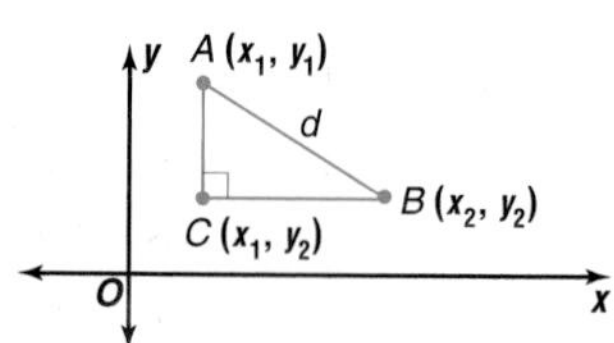

Programming

41a. 3, 4, 5; 6, 8, 10; 12, 16, 20; 24, 32, 40; 27, 36, 45

Euclid

41. The program at the right uses a procedure for finding *Pythagorean triples* that was developed by Euclid around 320 B.C.

```
PROGRAM: PYTHTRIP
: For (X,2,6)
: For (Y,1,5)
: If X>Y
: Then
: int(X²–Y²+0.5)→A
: 2XY→B
: int(X²+Y²+0.5)→C
: If A>B
: Then
: Disp B,A,C
: Else
: Disp A,B,C
: End
: End
: Pause
: Disp " "
: End
: End
: Stop
```

Run the program to generate a list of Pythagorean triples.

a. List all the members of the 3-4-5 family that are generated by the program.

b. Ricardo made the conjecture that if three whole numbers are a Pythagorean triple, then their product is divisible by 60. Does Ricardo's conjecture hold true for each triple that is produced by the program? **yes**

Critical Thinking

42. Draw an acute triangle and an obtuse triangle. In the acute triangle, draw an altitude to the longest side. In the obtuse triangle, draw an altitude from the obtuse angle. In either case, are the triangles formed by the altitude similar to the original? Explain. **See margin.**

Applications and Problem Solving

43. **Construction** In the United States, most building codes limit the steepness of the slope of a roof to a 3-4-5 right triangle. A model of this type of roof is shown at the right. A builder wants to put a support brace from point C perpendicular to $\overline{AP}$. Find the length of such a brace. **2.4 yd**

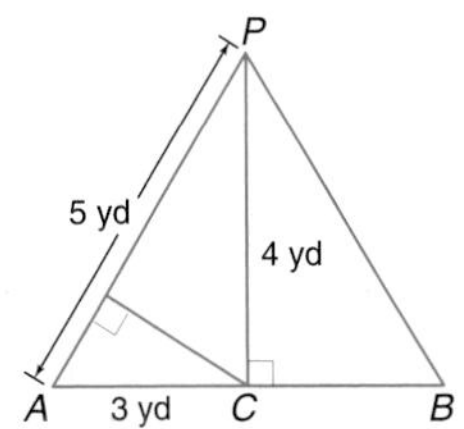

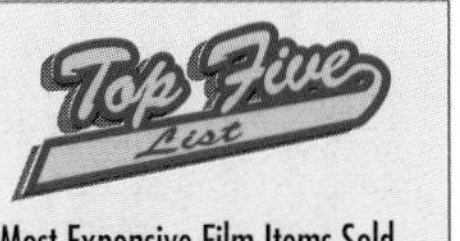

Most Expensive Film Items Sold at Auction

1. Aston Martin DB5 from *Goldfinger*, $275,000
2. Scripts for *Citizen Kane* and *The American*, $231,000
3. Ruby slippers from *The Wizard of Oz*, $165,000
4. Piano from *Casablanca*, $154,000
5. Charlie Chaplin's cane and hat, $135,300

44. **Sailing** The mast of a sailboat is supported by wires called shrouds. In the diagram at the right, the shrouds are shown in red. What is the total length of wire needed to form these shrouds? **about 72.83 ft**

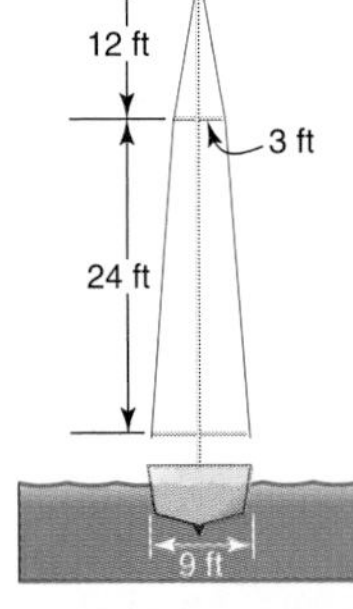

45. **Motion Pictures** In the movie *The Wizard of Oz*, the Scarecrow is looking for a brain. When the Wizard presents him with a Doctor of Thinkology degree, the Scarecrow immediately announces "The sum of the square roots of any two sides of an isosceles triangle is equal to the square root of the remaining side." Do you agree with the "Scarecrow Theorem"? Explain. **See margin.**

Mixed Review

46. **Solve a Simpler Problem** Donte must finish John Steinbeck's *The Pearl* and write a report by Friday. He is starting to read page 71, and the book ends on page 197. How many pages does Donte have yet to read? (Lesson 7–6) **127 pages**

Using the Programming Exercises The program given in Exercise 41 is for use with a TI-82 or TI-83 graphing calculator. For other programmable calculators, have students consult their owner's manual for commands similar to those represented here.

The newest Aston Martin, the DB7, sells for $120,000.

Additional Answer

42.

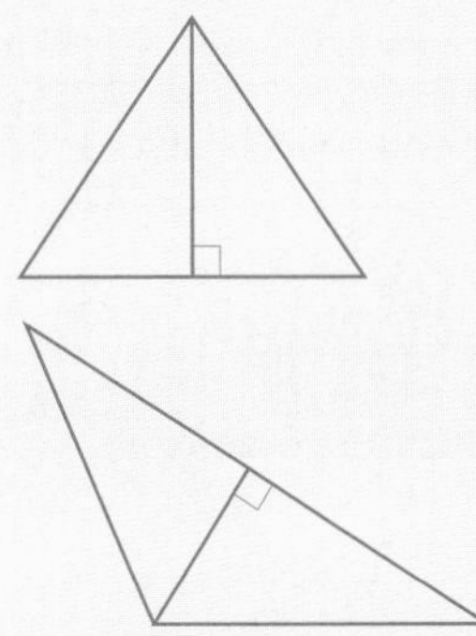

No; the smaller triangles formed when the altitudes are drawn are right triangles but the original triangles are not. Therefore, they cannot be similar.

Practice Masters, p. 43

8–1 NAME______ DATE ______

Practice — Student Edition Pages 397–404

Geometric Mean and the Pythagorean Theorem

Find the geometric mean between each pair of numbers.

1. 5 and 10 $5\sqrt{2} \approx 7.1$
2. 3 and 27 9
3. 6 and $\frac{1}{2}$ $\sqrt{3} \approx 1.7$
4. 16 and $\frac{1}{9}$ $\frac{4}{3}$

Find the values of x and y.

5. 3.2, 8.2
6. $4\sqrt{6} \approx 9.8$, $4\sqrt{2} \approx 5.7$
7. $4, 3\sqrt{5} \approx 6.7$
8. $3\sqrt{2} \approx 4.2$, $3\sqrt{6} \approx 7.3$
9. 7.5
10. 2.8
11. 34
12. 25

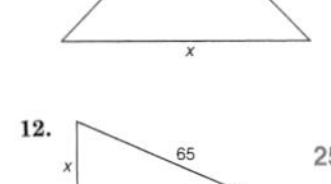

Determine if the given measures are measures of the sides of a right triangle.

13. 14, 48, 50 yes
14. 50, 75, 85 no
15. 15, 36, 39 yes
16. 45, 60, 80 no

Tech Prep

Mechanic A variety of two-year programs serve the automotive industry. Automotive Body Repairer, Automotive Mechanic, and Automotive Engineering Technologist all require a good mathematics background. For more information on tech prep, see the *Teacher's Handbook*.

Additional Answer

45. No; sample counterexample: isosceles triangle with sides measuring 4, 4, and 7.

$$\sqrt{4} + \sqrt{4} \stackrel{?}{=} \sqrt{7}$$

$$2 + 2 \stackrel{?}{=} 2.65$$

$$4 \neq 2.65$$

4 ASSESS

Closing Activity

Writing Ask each student to write one problem that can be solved using the Pythagorean Theorem and one problem that can be solved using its converse. They should include a solution for each problem.

Mathematics and SOCIETY

Encourage students to speculate about the following question. How could the senses of touch and sound be added to the visor so they are also 3-D?

Answer for Mathematics and Society

3. See students' work. Responses could include many types of teaching/training environments, architecture, engineering, and airline flight training.

Enrichment Masters, p. 43

8-1 NAME__________ DATE________

Enrichment — Student Edition Pages 397–404

Converse of a Right Triangle Theorem

You have learned that the measure of the altitude from the vertex of the right angle of a right triangle to its hypotenuse is the geometric mean between the measures of the two segments of the hypotenuse. Is the converse of this theorem true? In order to find out, it will help to rewrite the original theorem in if-then form as follows:

If $\triangle ABQ$ is a right triangle with right angle at Q, then QP is the geometric mean between AP and PB, where P is between A and B and $\overline{QP}$ is perpendicular to $\overline{AB}$.

1. Write the converse of the if-then form of the theorem. If QP is the geometric mean between AP and PB, where P is between A and B and $\overline{QP} \perp \overline{AB}$, then $\triangle ABQ$ is a right triangle with right angle at Q.

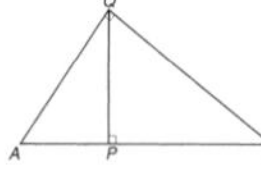

2. Is the converse of the original theorem true? Refer to the figure at the right to explain your answer. Yes; $(PQ)^2 = (AP)(PB)$ implies that $\frac{PQ}{AP} = \frac{PB}{PQ}$. Since both $\angle APQ$ and $\angle QPB$ are right angles, they are congruent. Therefore $\triangle APQ \sim \triangle QPB$ by SAS similarity. So $\angle A \cong \angle PQB$ and $\angle AQP \cong \angle B$. But the acute angles of $\triangle AQP$ are complementary and $m\angle AQB = m\angle AQP + m\angle PQB$. Hence $m\angle AQB = 90$ and $\triangle AQB$ is a right triangle with right angle at Q.

You may find it interesting to examine the other theorems in Lesson 8-1 of your book to see whether their converses are true or false. You will need to restate the theorems carefully in order to write their converses.

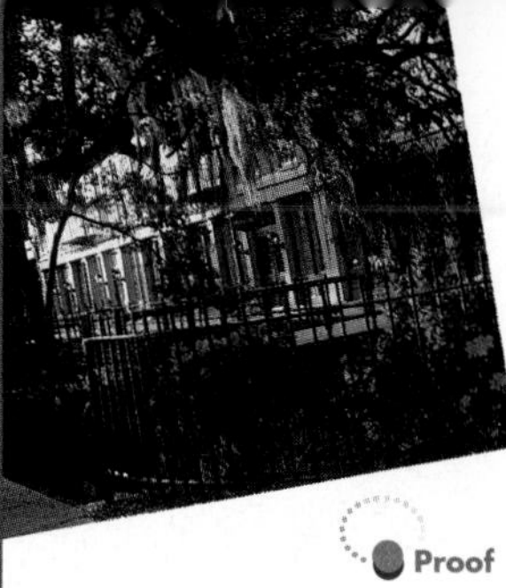

47. **Travel** A Georgia map is drawn so that 1 centimeter represents 20 miles. If Savannah and Atlanta are 12.7 centimeters apart on the map, how far apart are the cities? (Lesson 7–1) **254 mi**
48. The bases of an isosceles trapezoid measure 10 inches and 22 inches. Find the length of the median of the trapezoid. (Lesson 6–5) **16 in.**
49. In square $LMNP$, $LN = 3x - 2$ and $MP = 2x + 3$. Find LN. (Lesson 6–4) **13**
50. Determine whether the statement *The diagonals of a rectangle bisect the opposite angles* is true or false. Justify your answer. (Lesson 6–3) **False; the diagonals of a rhombus bisect opposite angles.**

Proof

51. Write a paragraph proof. (Lesson 5–6) **See Solutions Manual.**

Given: $\overline{GH} \cong \overline{HJ}$
$GL < JL$

Prove: $m\angle 1 < m\angle 2$

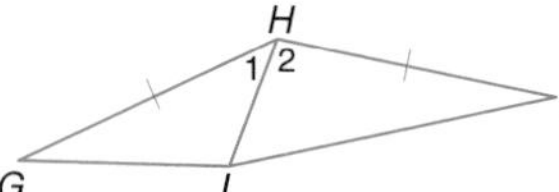

52. $\left\{x \mid \frac{8}{11} < x < 2\right\}$

52. If the sides of a triangle have lengths of $(2x + 7)$ meters, $(5x - 3)$ meters, and $(8x + 2)$ meters, find all possible values for x. (Lesson 5–5)
53. The measure of the angles in $\triangle XYZ$ are in the ratio 4:8:12. What are the measures of the angles? (Lesson 4–2) **30, 60, 90**
54. Two vertical angles have measures of $7x + 12$ and $4x + 42$. Find x. (Lesson 2–6) **10**

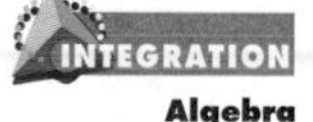

Algebra

55. Graph $y = 3x + 1$ by making a table of ordered pairs. **See Solutions Manual.**
56. Find $(6t + 7a) - (3t + 12a)$. **$3t - 5a$**

Mathematics and SOCIETY

2. The viewer can see a more close-up image; the viewer can constantly be changing her or his position to get the best view or the view of most interest.

Virtual Reality Uses 3D Geometry

The excerpt below appeared in *Popular Mechanics* in July, 1995.

STEP INTO THE "STAR TREK" HOLODECK and you can saunter about a 3-dimensional world, furnished and peopled with lifelike holograms....researchers today are racing toward a similar idea. At the University of North Carolina, computer scientists are developing 3D virtual environments stocked not with computer-wrought scenery, but with real-life images from a remote location. The key: A concept the researchers call a "sea of cameras." At the remote site, video cameras stud the walls and ceiling. Each has a view from a slightly different perspective....At the viewing site, everyone wears virtual-reality visor displays that are fitted with head-tracking sensors. As the participants walk around or turn their heads, the sensors maintain a fix on their positions and viewing directions....Using this technique, researchers have built a 3D image of an operating room. The goal: to let a large audience watch world-class surgeons at work without cramping their style. ■

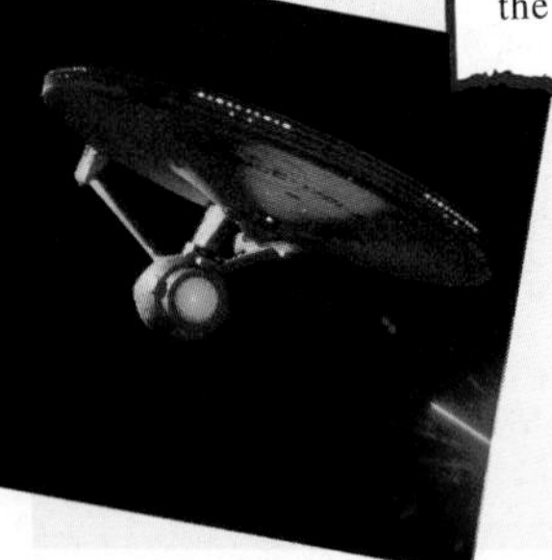

1. The computer needs to accurately track the location of each person's virtual-reality visor. How many dimensions or coordinates are required to determine the location of an object in a room? **3**
2. What advantages would this virtual-reality viewing have over watching the operation in person or on a videotape?
3. Can you think of other situations or applications where this type of virtual-reality viewing could be used? **See margin.**

Extension

Communication In ancient Egypt, builders would use a rope that had knots tied at 3 units, 7 units, and 12 units from one end. Write a paragraph explaining how you think they used such a rope to form right angles.

If stakes are placed in the ground at the end of the rope, at the 3-unit mark, and at the 7-unit mark, and if the 12-unit mark is brought back to the end of the rope, a 3-4-5 triangle has been formed.

8-2 Special Right Triangles

What YOU'LL LEARN

- To use the properties of 45°-45°-90° and 30°-60°-90° triangles.

Why IT'S IMPORTANT

You can use properties of special right triangles to find missing measures involved in sports, city planning, and landscaping.

Baseball

Years ago, baseball teams often played on asymmetric fields. However, today many teams play in new stadiums with symmetric fields such as the one at Dodger Stadium. A diagram of Dodger Stadium in Los Angeles is shown below.

Suppose a center fielder is standing on the point where a line passing through home plate and second base intersects with a line passing through the foul-ball poles. Is the center fielder closer to dead center field or second base? *This problem will be solved in Example 1.*

DODGER STADIUM (Los Angeles)

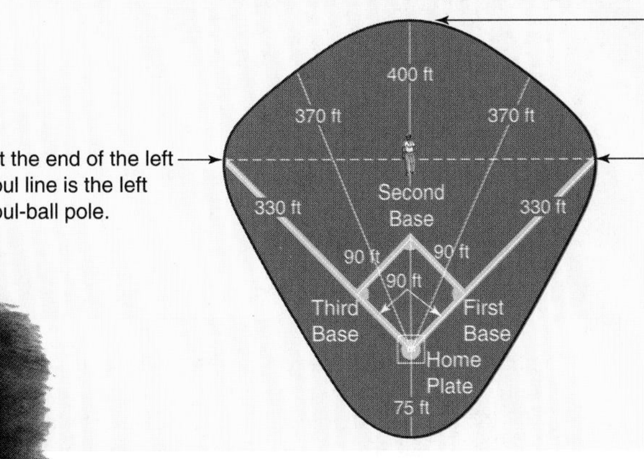

400 ft
370 ft
370 ft
330 ft
330 ft
Second Base
90 ft
90 ft
90 ft
Third Base
First Base
Home Plate
75 ft
Dead center field is the end of the outfield on a straight line drawn through home plate and second base.
At the end of the left foul line is the left foul-ball pole.
At the end of the right foul line is the right foul-ball pole.

In order to solve this problem, we must be able to solve a 45°-45°-90° triangle. The Pythagorean Theorem allows us to discover the special relationship that exists among the sides of a 45°-45°-90° triangle.

Draw a diagonal of a square. The two triangles formed are isosceles right triangles. Let x represent the measure of each side and d represent the measure of the hypotenuse.

$d^2 = x^2 + x^2$ *Pythagorean Theorem*

$d^2 = 2x^2$

$d = \sqrt{2x^2}$ *Take the square root of each side.*

$d = x\sqrt{2}$ *Ignore the negative root, since d is a measure.*

So the length of the hypotenuse of any 45°-45°-90° triangle is $\sqrt{2}$ times the length of a leg.

Theorem 8-6	**In a 45°-45°-90° triangle, the hypotenuse is $\sqrt{2}$ times as long as a leg.**

8-2 LESSON NOTES

NCTM Standards: 1–5, 7

Instructional Resources

- Study Guide Master 8-2
- Practice Master 8-2
- Enrichment Master 8-2
- Assessment and Evaluation Masters, p. 212

Transparency 8-2A contains the 5-Minute Check for this lesson; **Transparency 8-2B** contains a teaching aid for this lesson.

Recommended Pacing	
Standard Pacing	Days 4 & 5 of 12
Honors Pacing	Days 3 & 4 of 11
Block Scheduling*	Day 2 of 7

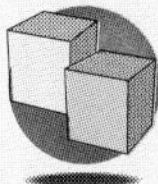

*For more information on pacing and possible lesson plans, refer to the *Block Scheduling Booklet*.

1 FOCUS

5-Minute Check
(over Lesson 8-1)

Refer to the figure. Round answers to the nearest tenth.

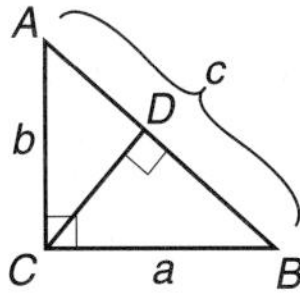

A
c
D
b
C
a
B

1. If $a = 8$ and $b = 6$, find c. 10
2. Find b if $a = 6$ and $c = 8$. 5.3
3. If the altitude $\overline{CD}$ is drawn in $\triangle ABC$ so that $AD = 3$ and $BD = 4$, find a and b. $a \approx 5.29$, $b \approx 4.58$

Determine if the given measures are measures for the sides of a right triangle.

4. 12.6, 16.8, 21 yes
5. 4, 7.5, 8.5 yes
6. 6, 10, 16 no

Motivating the Lesson

Situational Problem Ask students to describe the shape of a baseball diamond. Have them help you draw the shape and label the angle and distance measures on the chalkboard or overhead. Some students may be surprised to find out that the diamond is actually a square.

2 TEACH

In-Class Example

For Example 1

Find the value of *x*.

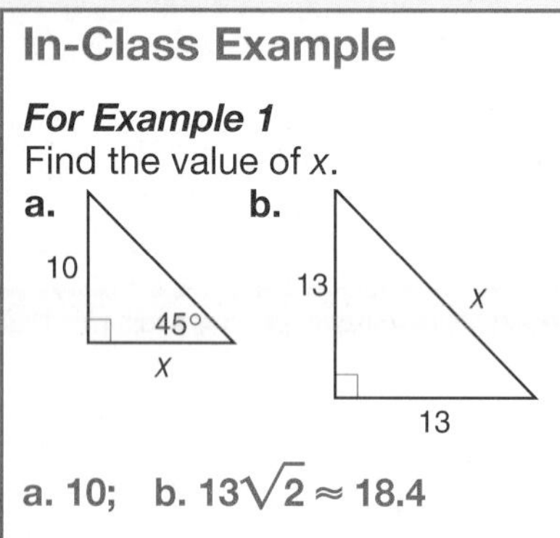

a. 10; **b.** $13\sqrt{2} \approx 18.4$

Teaching Tip After Example 1, point out that $\sqrt{2} \approx 1.414$. To check answers by estimating, think "a little less than $1\frac{1}{2}$ times."

Example 1

Baseball

Refer to the application at the beginning of the lesson. Is the center fielder closer to dead center field or second base?

First, find the distance from home plate to second base. By Theorem 8–6, this distance d is $90\sqrt{2}$ or about 127.28 feet.

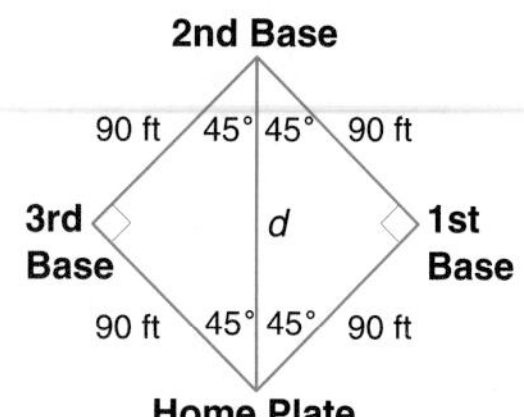

Now, consider the whole field. field. Draw $\overline{BD}$ from home plate to dead center field and $\overline{AE}$ from the left field foul-ball pole to the right field foul-ball pole. Let C be the intersection point of $\overline{BD}$ and $\overline{AE}$. $\triangle ACB$ is a 45°-45°-90° triangle with an hypotenuse 330 feet long. Let $BC = x$ and $AC = x$ and use the Theorem 8–6 to solve for x, the distance from home plate to the the fielder.

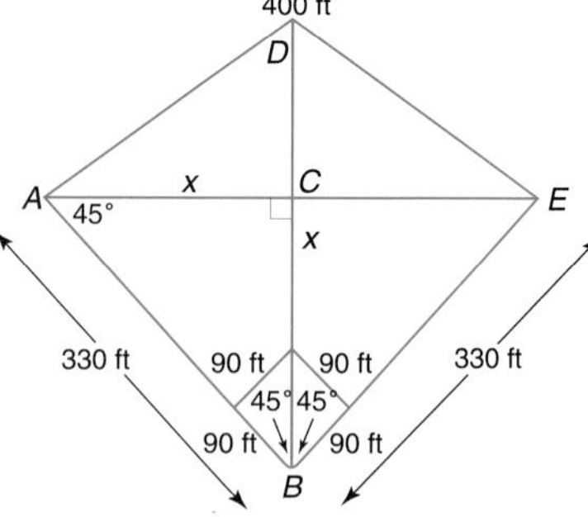

$$x\sqrt{2} = 330$$

$$x = \frac{330}{\sqrt{2}} \quad \textit{Divide each side by } \sqrt{2}.$$

$$x = \frac{330}{\sqrt{2}} \cdot \frac{\sqrt{2}}{\sqrt{2}} \quad \textit{Rationalize the denominator.}$$

$$x = \frac{330\sqrt{2}}{2}$$

$$x = 165\sqrt{2} \text{ or about } 233.35$$

The distance from home to the center fielder is about 233.35 feet. The distance from the center fielder to second base is about 233.35 − 127.28 or 106.07 feet. The distance from the center fielder to dead center field is about 400 − 233.35 or 166.65 feet. So the center fielder is closer to second base.

There is also a special relationship among the measures of the sides of a 30°-60°-90° triangle.

If an altitude is drawn from any vertex of an equilateral triangle, the triangle is separated into two congruent 30°-60°-90° triangles. Using the Pythagorean Theorem, it is possible to derive a formula relating the lengths of the sides to each other. Let each side of the equilateral triangle have a length of x units. $\overline{LM}$ and $\overline{MN}$ are congruent segments, so each measures $\frac{1}{2}x$ units. Let a represent the measure of the altitude.

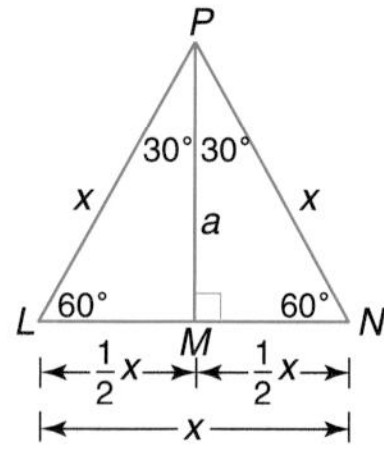

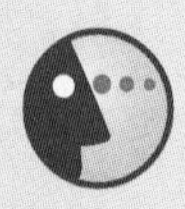

Alternative Learning Styles

Visual Have students draw an isosceles right triangle with legs measuring 1 inch. Have them use their rulers to find that the length of the hypotenuse is approximately $1\frac{7}{16}$ inches. Using a calculator, $1\frac{7}{16} \approx 1.4$, which is close to $\sqrt{2}$. Have them repeat the activity for a triangle with sides measuring 2 inches.

$$(PM)^2 + (LM)^2 = (PL)^2 \quad \textit{Pythagorean Theorem}$$

$$a^2 + \left(\frac{1}{2}x\right)^2 = x^2 \quad PM = a,\ LM = \frac{1}{2}x,\ PL = x$$

$$a^2 + \frac{1}{4}x^2 = x^2$$

$$a^2 = \frac{3}{4}x^2 \quad \textit{Subtraction Property (=)}$$

$$a = \sqrt{\frac{3}{4}x^2} \quad \textit{Take the square root of each side.}$$

$$a = \frac{\sqrt{3}}{2}x \quad \textit{Ignore the negative root since a is a measure.}$$

So, in the 30°-60°-90° triangle, the measures of the sides are x, $\frac{1}{2}x$, and $\frac{\sqrt{3}}{2}x$.

In Lesson 8–1, you learned about families of Pythagorean triples. Multiply the measure of each side by 2 to create a new set of measures for a 30°-60°-90° triangle.

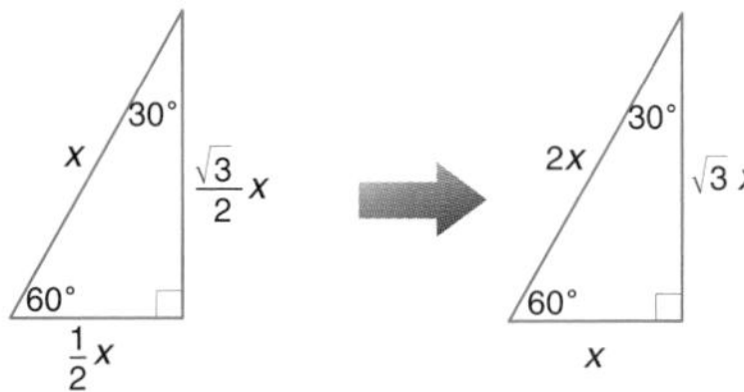

This relationship leads to Theorem 8–7.

Theorem 8–7 **In a 30°-60°-90° triangle, the hypotenuse is twice as long as the shorter leg, and the longer leg is $\sqrt{3}$ times as long as the shorter leg.**

The shorter leg is opposite the 30° angle and the longer leg is opposite the 60° angle.

You can use Theorem 8–7 to quickly find the measures of the sides of a 30°-60°-90° triangle if you know the measure of any side.

Example  **A *regular hexagon* is made up of six equilateral triangles. Find AC in the regular hexagon at the right.**

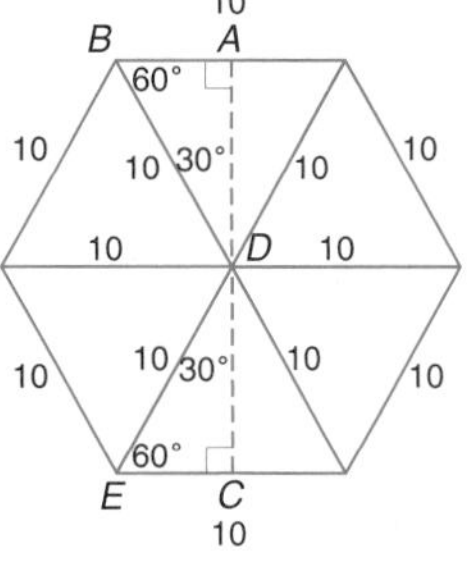

The shorter leg $\overline{AB}$ of $\triangle ABD$ is half as long as the hypotenuse $\overline{BD}$, which is 10 units long. Therefore, $AB = 5$. The longer leg is $\sqrt{3}$ times as long as the shorter leg. So, $AD = 5\sqrt{3}$. Likewise, in $\triangle DEC$, $DC = 5\sqrt{3}$. Since $AC = AD + DC$, $AC = 5\sqrt{3} + 5\sqrt{3}$ or $10\sqrt{3}$.

Special triangles can be graphed in a coordinate plane.

In-Class Example

For Example 2

Tiles in the shape of equilateral triangles are used to cover a countertop. The side of each tile measures 4 inches. Find the height of each tile.

$2\sqrt{3} \approx 3.5$ in.

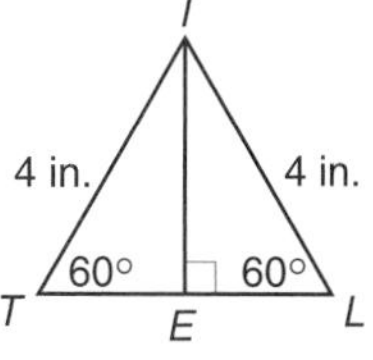

Teaching Tip After reading Theorem 8-7, point out that $\sqrt{3} \approx 1.732$. For estimating purposes, think about its value as being about $1\frac{3}{4}$.

In-Class Example

For Example 3

$\triangle EAT$ is a 30°-60°-90° triangle with right $\angle A$ and $\overline{AT}$ as the longer leg.

a. Find the coordinates of E in Quadrant I for $A(1, 2)$ and $T(5, 2)$. (1, 4.3)

b. Draw the triangle.

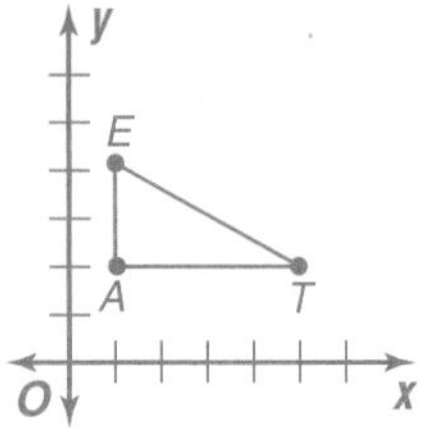

3 PRACTICE/APPLY

Check for Understanding

Exercises 1–10 are designed to help you assess your students' understanding through reading, writing, speaking, and modeling. You should work through Exercises 1–3 with your students and then monitor their work on Exercises 4–10.

Additional Answers

1.

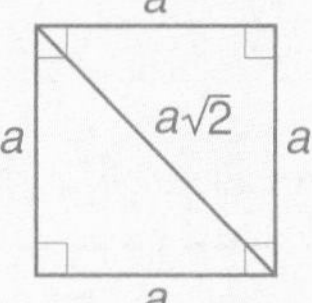

2.

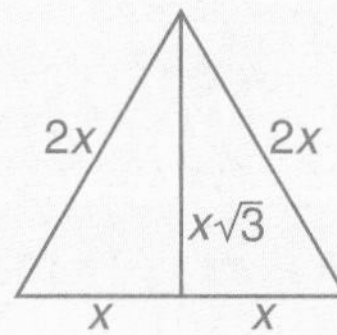

Example 3 INTEGRATION **Algebra**

$\triangle PCD$ is a 30°-60°-90° triangle with right $\angle C$ and $\overline{CD}$ as the longer leg.

a. Find the coordinates of P in Quadrant I for $C(3, 2)$ and $D(9, 2)$.

A graph of the given information can help you visualize the problem.

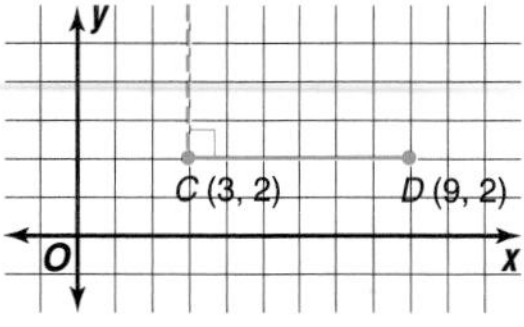

Notice that $\overline{CD}$ lies on a gridline of the coordinate plane and since $\overline{PC}$ will be perpendicular to $\overline{CD}$, it too lies on a gridline. Find the length of $\overline{CD}$.

$$CD = |9 - 3| = 6$$

$\overline{CD}$ is the longer leg, and $\overline{PC}$ is the shorter leg. So, $CD = \sqrt{3}(PC)$ by Theorem 8–7. Use CD to find PC.

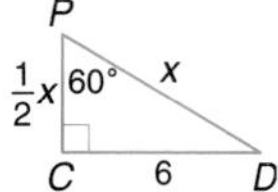

$$CD = \sqrt{3}(PC)$$

$$6 = \sqrt{3}(PC) \quad CD = 6$$

$$\frac{6}{\sqrt{3}} = PC \quad \text{Division Property } (=)$$

$$\frac{6}{\sqrt{3}} \cdot \frac{\sqrt{3}}{\sqrt{3}} = PC \quad \text{Rationalize the denominator.}$$

$$\frac{6\sqrt{3}}{3} = PC$$

$$2\sqrt{3} = PC$$

The x-coordinate of P is the same as the x-coordinate of C. P is located $2\sqrt{3}$ units above C. So, the coordinates of P are $(3, 2 + 2\sqrt{3})$ or about (3, 5.46)

b. Draw $\triangle PCD$.

Graph points C and D.

Draw $\overline{CD}$.

Graph P, and draw $\overline{PD}$ and $\overline{PC}$.

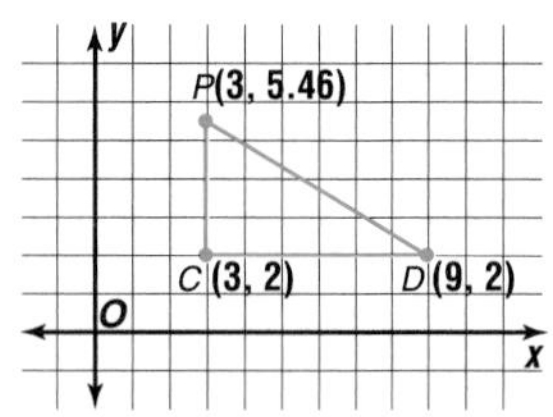

CHECK FOR UNDERSTANDING

Communicating Mathematics

Study the lesson. Then complete the following.

1. **Draw** a 45°-45°-90° triangle. Suppose the measure of one of the legs of the triangle is a. Label the measure of each segment in your figure. See margin.
2. **Draw** a 30°-60°-90° triangle. Suppose the measure of the hypotenuse of the triangle is $2x$. Label the measure of each segment in your figure. See margin.
3. **You Decide** The students in Ms. Esparza's geometry class have been given the assignment to draw a 30°-60°-90° triangle with $X(0, 0)$, $Y(-5, 0)$, and $\angle X$ as the right angle. Jamal claims there are two triangles that satisfy the conditions. Winona claims there are four such triangles and Mary claims there is only one. Who is correct? Explain.

3. Winona; The third vertex could be located above or below X, and $\overline{XY}$ could be the shorter or longer leg.

Reteaching

Using Competition Have class groups find the missing measures. Give points based on their rationale for each answer and time taken.

Find the missing measures in $\triangle ABC$.

1. $s = 6$
   $h = 6\sqrt{2} \approx 8.5$
2. $h = 8\sqrt{2}$
   $s = 8$

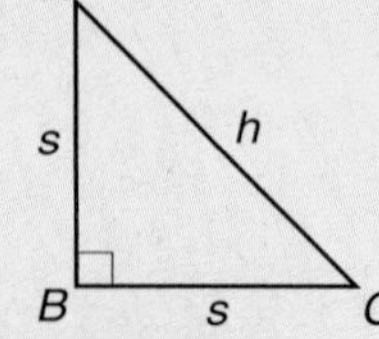

Find the missing measures in $\triangle DEF$.

3. $d = 4$
   $f = 4\sqrt{3} \approx 6.9$, $e = 8$
4. $e = 10$
   $d = 5$, $f = 5\sqrt{3} \approx 8.7$

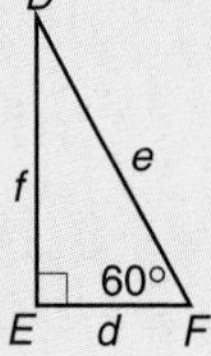

Guided Practice

4. 3; $3\sqrt{2} \approx 4.24$
5. 16; $8\sqrt{3} \approx 13.86$

Find the values of x and y.

4.

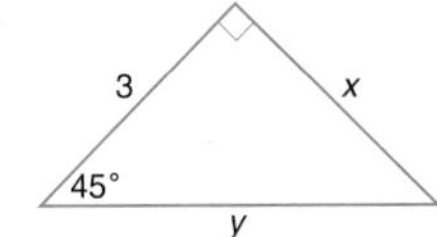

5. (triangle: 8, y, 60°, 30°, x)

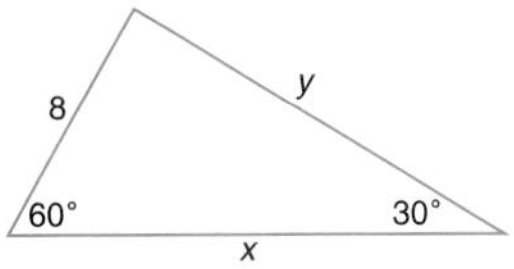

6. The length of a diagonal of a square is $10\sqrt{2}$ inches. Find the length of one side of the square. **10 in.**
7. The length of an altitude of an equilateral triangle is $\frac{\sqrt{3}}{2}$ feet. Find the length of a side of the triangle. **1 ft**
8. The perimeter of a square is 44 meters. Find the length of a diagonal of the square. **$11\sqrt{2}$ or about 15.56 m**

9. $(-4, -2 + 10\sqrt{3})$ or about $(-4, 15.32)$

Algebra

9. $\triangle PCD$ is a 30°-60°-90° triangle with right $\angle C$ and $\overline{CD}$ the shorter leg. Find the coordinates of P in Quadrant II for $C(-4, -2)$ and $D(6, -2)$.

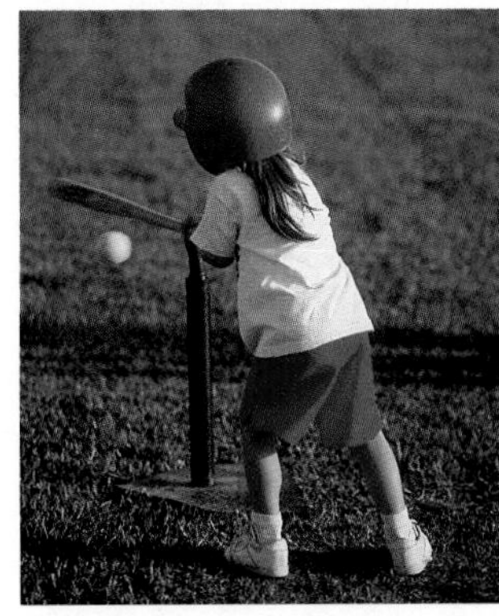

10. **Tee-Ball** Many younger children like to play a game similar to baseball called tee-ball. Instead of trying to hit a ball thrown by a pitcher, the batter hits the ball off a tee. To accommodate younger children, the bases are only 40 feet apart. Find the distance between home plate and second base in tee-ball. **$40\sqrt{2} \approx 56.57$**

EXERCISES

Practice

11. $4.5\sqrt{3} \approx 7.79$; 4.5
12. $6.5\sqrt{2} \approx 9.19$; 45
13. $\frac{3\sqrt{2}}{2} \approx 2.12$; $3\sqrt{2} \approx 4.24$
14. 45; $4\sqrt{2} \approx 5.66$
15. $9\sqrt{3} \approx 15.59$; 9
16. $\frac{31\sqrt{3}}{3} \approx 17.90$; $\frac{31\sqrt{3}}{3} \approx 17.90$

Find the values of x and y.

A

11.

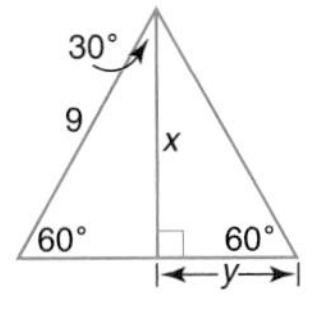

12.

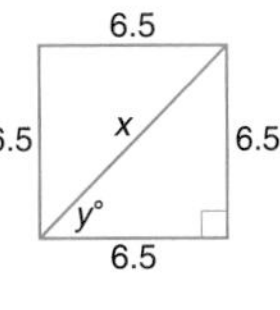

13.

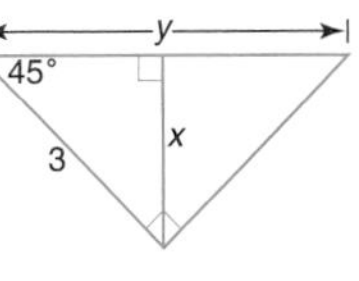

14.

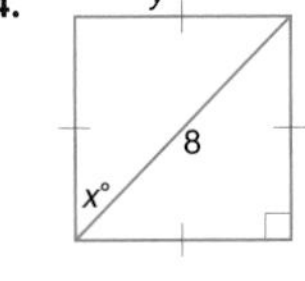

15.

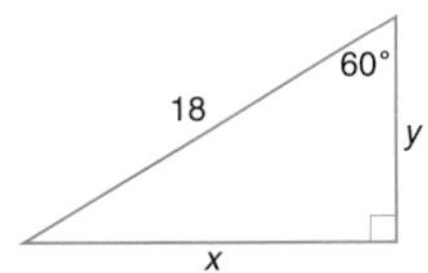

16.

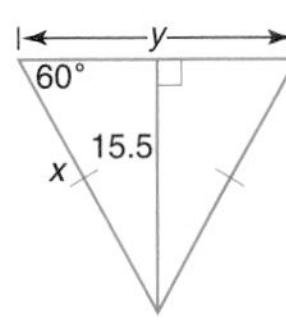

B

17. The length of one side of a square is 13.5 centimeters. Find the length of a diagonal of the square. **$13.5\sqrt{2}$ or about 19.09 cm**
18. The length of a diagonal of a square is 10 inches. Find the length of one side of the square. **$5\sqrt{2}$ or about 7.07 in.**
19. The length of one side of an equilateral triangle is $6\sqrt{3}$ meters. Find the length of one altitude of the triangle. **9 m**
20. The length of an altitude of an equilateral triangle is 12 feet. Find the length of a side of the triangle. **$8\sqrt{3}$ or about 13.86 ft**
21. The perimeter of an equilateral triangle is 39 centimeters. Find the length of an altitude of the triangle. **$6.5\sqrt{3}$ or about 11.26 cm**

Assignment Guide

Core (with proof): 11–33 odd, 34–41
Core (informal): 11–33 odd, 34–41
Enriched: 12–30 even, 31–41

For **Extra Practice**, see p. 778.

The red A, B, and C flags, printed only in the Teacher's Wraparound Edition, indicate the level of difficulty of the exercises.

Study Guide Masters, p. 44

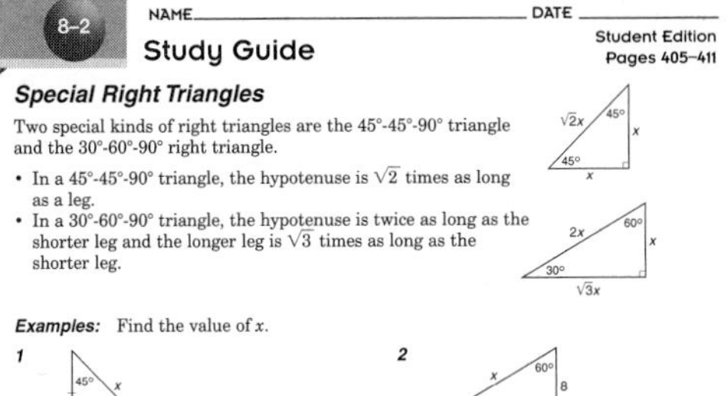

8–2 NAME ______ DATE ______

Study Guide — Student Edition Pages 405–411

Special Right Triangles

Two special kinds of right triangles are the 45°-45°-90° triangle and the 30°-60°-90° right triangle.

- In a 45°-45°-90° triangle, the hypotenuse is $\sqrt{2}$ times as long as a leg.
- In a 30°-60°-90° triangle, the hypotenuse is twice as long as the shorter leg and the longer leg is $\sqrt{3}$ times as long as the shorter leg.

Examples: Find the value of x.

1. Since the triangle is a 45°-45°-90° triangle, the hypotenuse is $\sqrt{2}$ times as long as the leg. So $x = 5\sqrt{2}$ or about 7.1.

2. Since the triangle is a 30°-60°-90° triangle, the hypotenuse is twice as long as the shorter leg. So $x = 2(8)$ or 16.

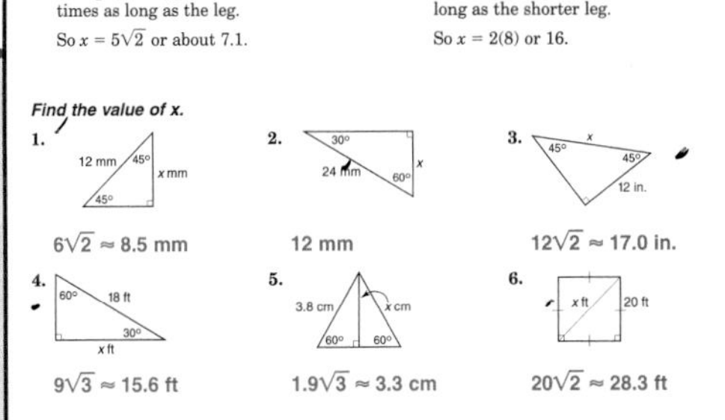

Find the value of x.

1. $6\sqrt{2} \approx 8.5$ mm
2. 12 mm
3. $12\sqrt{2} \approx 17.0$ in.
4. $9\sqrt{3} \approx 15.6$ ft
5. $1.9\sqrt{3} \approx 3.3$ cm
6. $20\sqrt{2} \approx 28.3$ ft
7. Find the perimeter of the triangle shown at the right. $40 + 20 + 20\sqrt{3} \approx 94.6$ *cm*

Additional Answers

31a.

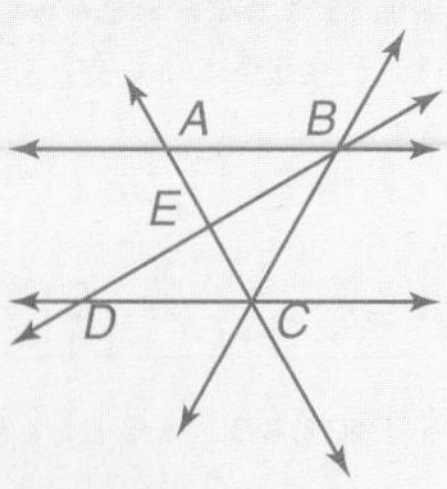

31b. $m\angle BCD = 120$, $m\angle BCE = 60$, $m\angle ABC = 180 - 120$ or 60, $m\angle EBC = 30$, $m\angle BEC = 180 - 60 - 30$ or 90, $\triangle BEC$ is a 30°-60°-90° triangle.

35. AA Similarity, SSS Similarity, SAS Similarity

36.

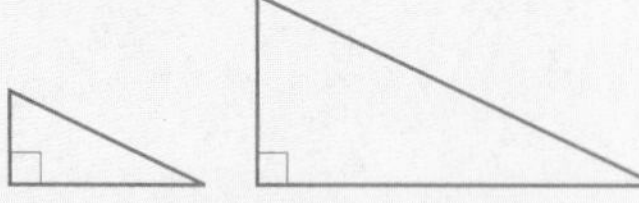

37. Yes; since $QS = 2\sqrt{26}$ and $RT = 2\sqrt{26}$, the diagonals are congruent.

Practice Masters, p. 44

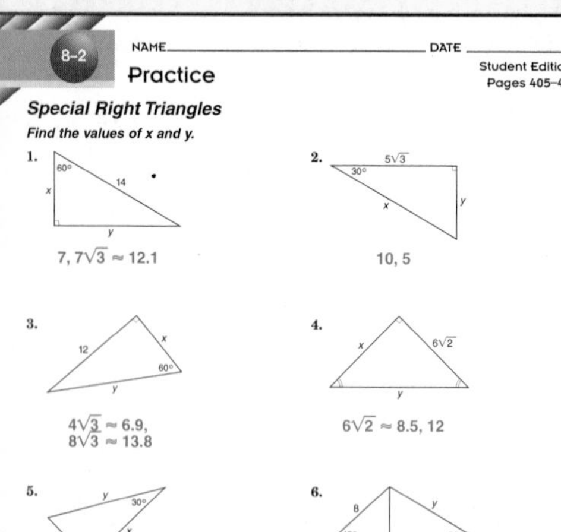

8-2 NAME________ DATE________

Practice — Student Edition Pages 405–411

Special Right Triangles

Find the values of x and y.

1. $7, 7\sqrt{3} \approx 12.1$
2. 10, 5
3. $4\sqrt{3} \approx 6.9$, $8\sqrt{3} \approx 13.8$
4. $6\sqrt{2} \approx 8.5$, 12
5. $4\sqrt{3} \approx 6.9$, 8
6. $4\sqrt{6} \approx 9.8$, $8\sqrt{2} \approx 11.3$
7. Find the length of a diagonal of a square with sides 10 in. long. $10\sqrt{2} \approx 14.1$ in.
8. Find the length of a side of a square whose diagonal is 4 cm. $2\sqrt{2} \approx 2.8$ cm
9. One side of an equilateral triangle measures 6 cm. Find the measure of an altitude of the triangle. $3\sqrt{3} \approx 5.2$ cm

22. The length of a diagonal of a square is $18\sqrt{2}$ millimeters. Find the perimeter of the square. **72 mm**

23. The altitude of an equilateral triangle is 5.2 meters long. Find the perimeter of the triangle. **$10.4\sqrt{3}$ or about 18.01 m**

24. The diagonals of a rectangle are 12 inches long and intersect at an angle of 60°. Find the perimeter of the rectangle. **$12 + 12\sqrt{3}$ or about 32.78 in.**

25. The sum of the squares of the measures of the sides of a square is 196. Find the measure of a diagonal of the square. **$7\sqrt{2} \approx 9.90$**

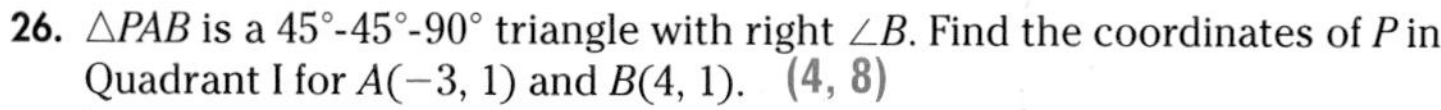

INTEGRATION Algebra

26. $\triangle PAB$ is a 45°-45°-90° triangle with right $\angle B$. Find the coordinates of P in Quadrant I for $A(-3, 1)$ and $B(4, 1)$. **(4, 8)**

27. $\triangle PCD$ is a 30°-60°-90° triangle with right $\angle C$ and $\overline{CD}$ the longer leg. Find the coordinates of P in Quadrant III for $C(-3, -6)$ and $D(-3, 7)$.

27. $\left(-3 - \frac{13\sqrt{3}}{3}, -6\right)$ or about $(-10.51, -6)$

28. $\triangle PCD$ is a 30°-60°-90° triangle with $m\angle C = 30$ and $\overline{CD}$ the hypotenuse. Find the coordinates of P for $C(2, -5)$ and $D(10, -5)$ if P lies above $\overline{CD}$.

28. $(8, -5 + 2\sqrt{3})$ or about $(8, -1.54)$

29. Each triangle in the figure at the right is a 30°-60°-90° triangle. Find the value of x. **2.25**

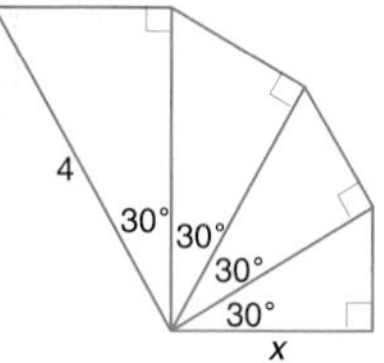

30. Each edge of the cube is s units long.
    a. Write an expression for AF in terms of s. **$s\sqrt{3}$**
    b. Find $m\angle BFD$. **60**

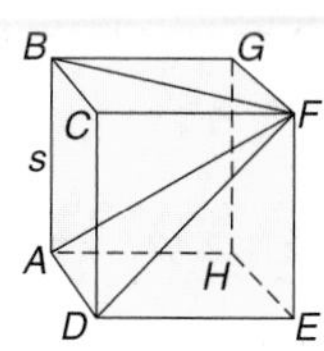

Critical Thinking

31. Two parallel lines are cut by a transversal. The transversal makes a 120° angle with one of the parallel lines. A line bisects the 120° angle. Another line bisects the consecutive interior angle.
    a. Draw the figure. **See margin.**
    b. Show that the triangle formed by the two angle bisectors and the transversal is a 30°-60°-90° triangle. **See margin.**
    c. If the measure of the longer leg is 22, find the measure of the shorter leg of the triangle and the measure of the transversal between the parallel lines. **$\frac{22\sqrt{3}}{3} \approx 12.70$; $\frac{44\sqrt{3}}{3} \approx 25.40$**

Applications and Problem Solving

32. **City Planning** Granmichele is a city with a population of about 15,000 people in northern Sicily. The city has a hexagonal design as shown by the aerial view at the right. Consider all sides of the hexagons to be congruent. Suppose the perpendicular distance from the center of the city to each side of the one hexagon is 0.5 mile. Find the perimeter of that hexagon.

32. $2\sqrt{3}$ or about 3.46 mi

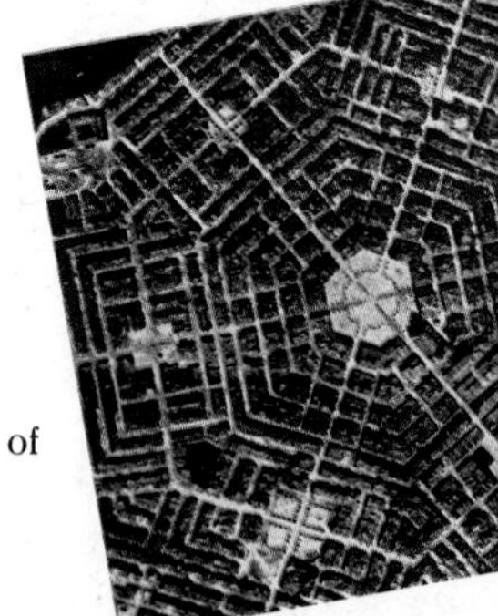

Extension

Problem Solving A regular hexagon is composed of six equilateral triangles. Find the perimeter of the hexagon if its apothem measures 8 cm. (The apothem is the altitude of one of the equilateral triangles.)

$32\sqrt{3} \approx 55.4$

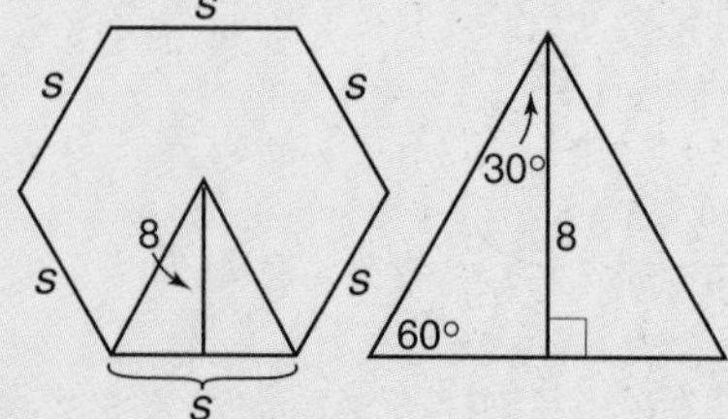

33. **Landscaping** Juana is a landscaper. She wishes to determine the height of a tree. Holding a drafter's 45° triangle so that one leg is horizontal, she sights the top of the tree along the hypotenuse as shown at the right. If she is 6 yards from the tree and her eyes are 5 feet from the ground, find the height of the tree. **23 ft**

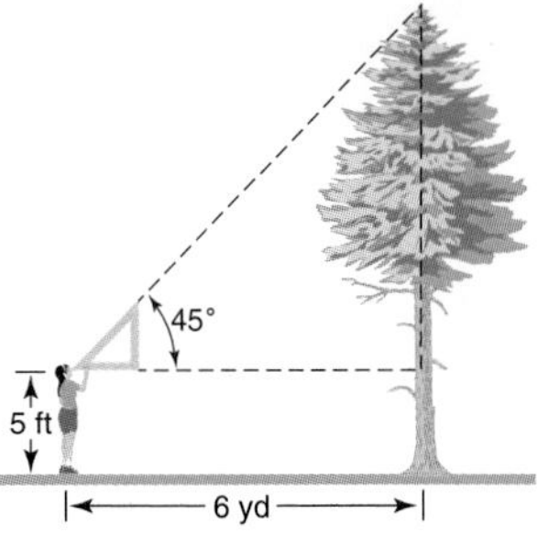

Mixed Review

34. **Sports** Digna is making a ramp to try out her car for a pinewood derby. The ramp support forms a right angle. The base is 12 feet long, and the height is 5 feet. What length of plywood does she need to complete the ramp? (Lesson 8–1) **13 ft**

35. Name three ways to prove two triangles are similar. (Lesson 7–3)

36. Draw two similar right, scalene triangles. (Lesson 7–2)

37. Is quadrilateral $QRST$ with vertices at $Q(1, 4)$, $R(5, 0)$, $S(-1, -6)$, and $T(-5, -2)$ a rectangle? Justify your answer. (Lesson 6–3)

38. Is it possible to have a triangle with vertices at $K(5, 8)$, $L(0, -4)$, and $M(-1, 1)$? Explain your answer. (Lesson 5–5) **35–38. See margin.**

39. Graph $x - 2y = 4$ and $J(-3, -1)$. Then construct a perpendicular segment and find the distance from point J to the line. (Lesson 3–5) **See margin for graph; $\sqrt{5}$ units.**

INTEGRATION Algebra

40. Simplify $\frac{8k^4}{2k}$. **$4k^3$**

41. Factor $8g^2 + 32g$. **$8g(g + 4)$**

WORKING ON THE In•ves•ti•ga•tion

Refer to the Investigation on pages 392–393.

On your mission to Mars, you will need to generate power for running equipment in the space station. One option for energy production is solar energy cells. Solar cells convert the energy of sunlight directly into electrical energy.

1. List the equipment that you think will be needed in the space station. Then estimate the energy needed by the space station each day.
2. If solar cells produce about 0.01 watt of electricity per square centimeter, what must be the total area of the solar cells?
3. Many solar cells are hexagonal because congruent hexagons can tessellate, or cover, a flat surface without leaving any gaps. The formula for the area A of a regular hexagon is $A = \frac{1}{2}Pa$, where P is the perimeter. If the solar cell below produces 15 watts, find the length of a side of the cell.

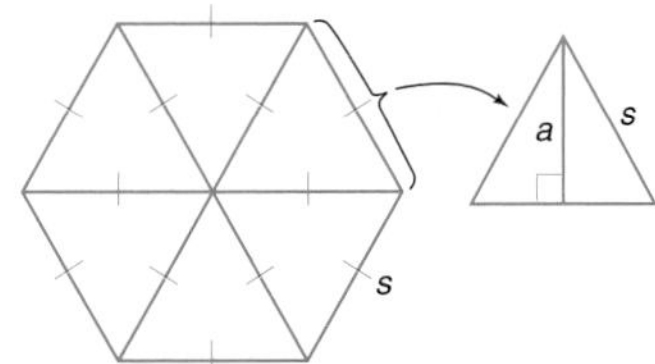

4. How many solar cells like the one above will be needed to power the space station? Is this a reasonable way to produce the energy?

Add the results of your work to your Investigation Folder.

4 ASSESS

Closing Activity

Writing Have students write a letter to an absent (or fictitious) classmate, explaining how to find the measures of both legs of a 30°-60°-90° triangle given the measure of its hypotenuse.

Chapter 8 Quiz A (Lessons 8-1 and 8-2) is available in the *Assessment and Evaluation Masters*, p. 212.

Additional Answers

38. Yes; the lengths of the segments satisfy the Triangle Inequality Theorem.

39.

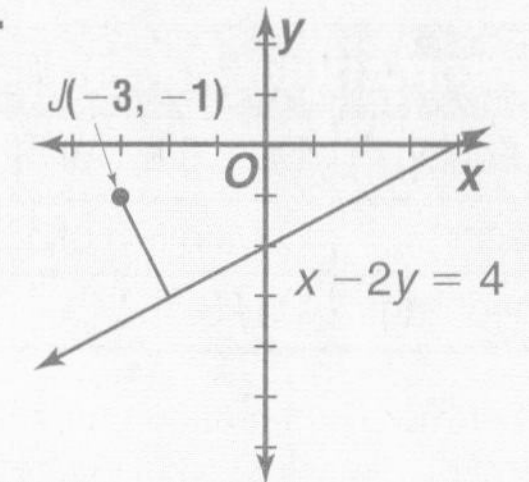

Enrichment Masters, p. 44

8-2 NAME______ DATE______

Enrichment

Student Edition Pages 405–411

Creating Pythagorean Puzzles

By drawing two squares and cutting them in a certain way, you can make a puzzle that demonstrates the Pythagorean Theorem. A sample puzzle is shown. You can create your own puzzle by following the instructions below.

Sample:

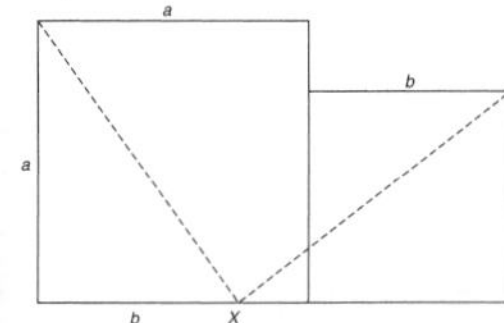

1. Carefully construct a square and label the length of a side as a. Then construct a smaller square to the right of it and label the length of a side as b, as shown in the figure above. The bases should be adjacent and collinear.
2. Mark a point X that is b units from the left edge of the larger square. Then draw the segments from the upper left corner of the larger square to point X, and from point X to the upper right corner of the smaller square.
3. Cut out and rearrange your five pieces to form a larger square. Draw a diagram to show your answer. See students' work. A sample answer is shown.

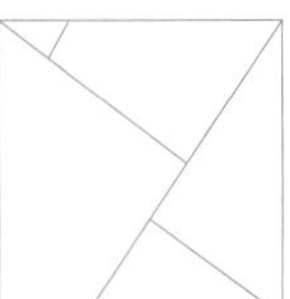

4. Verify that the length of each side is equal to $\sqrt{a^2 + b^2}$. See students' work.

In•ves•ti•ga•tion

Working on the Investigation

The Investigation on pages 392–393 is designed to be a long-term project that is completed over several days or weeks. Encourage students to keep their materials in their Investigation Folder as they work on the Investigation.

Answers for Working on the Investigation

1. See students' work.
2. See students' work.
3. Sample answer: 24 cm
4. See students' work.

8-3 LESSON NOTES

NCTM Standards: 1–5, 7, 9

Instructional Resources

- Study Guide Master 8-3
- Practice Master 8-3
- Enrichment Master 8-3
- Assessment and Evaluation Masters, pp. 211, 212
- Modeling Mathematics Masters, pp. 43–46
- Multicultural Activity Masters, p. 16

Transparency 8-3A contains the 5-Minute Check for this lesson; **Transparency 8-3B** contains a teaching aid for this lesson.

Recommended Pacing	
Standard Pacing	Days 6 & 7 of 12
Honors Pacing	Days 5 & 6 of 11
Block Scheduling*	Day 3 of 7

*For more information on pacing and possible lesson plans, refer to the *Block Scheduling Booklet*.

1 FOCUS

5-Minute Check
(over Lesson 8-2)

Find the measure of the diagonal of each square.

1. $s = 3$ $3\sqrt{2} \approx 4.2$
2. $s = 7$ $7\sqrt{2} \approx 9.9$
3. $s = 4.5$ $4.5\sqrt{2} \approx 6.4$

Solve.

4. The shorter leg of a 30°-60°-90° triangle measures 6 cm. What are the measurements of the other two sides? $6\sqrt{3}$ or about 10.4 cm, 12 cm
5. The hypotenuse of a 30°-60°-90° triangle measures 20 inches. Find the measurements of the other two sides. 10 in., $10\sqrt{3}$ or about 17.3 in.

8-3 Integration: Trigonometry Ratios in Right Triangles

What YOU'LL LEARN

- To find trigonometric ratios using right triangles, and
- to solve problems using trigonometric ratios.

Why IT'S IMPORTANT

You can use trigonometric ratios to find missing measures of right triangles involved in aviation, medicine, and astronomy.

APPLICATION

Golf

David Leadbetter is rapidly becoming one of the most well-known golf instructors in the world. His students include Nick Price and Debra McHaffie. He invented a golfing aid called the RIGHT ANGLE. The RIGHT ANGLE fits onto the arm and positions the arm correctly on the backswing to add consistency in the swing. The right triangle formed by the arm is modeled by the right triangle below.

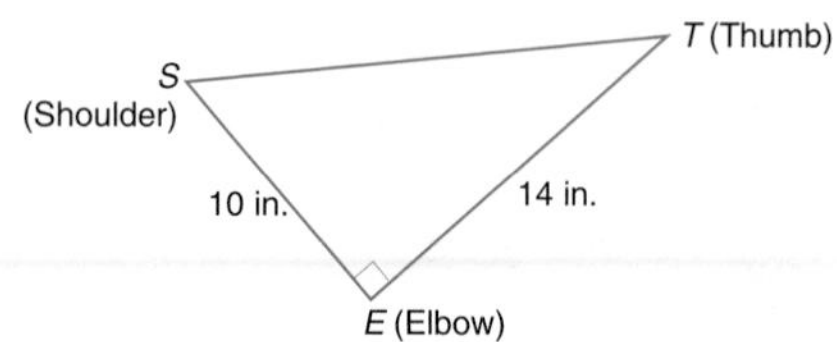

Since $\angle E$ is a right angle, $\triangle SET$ is a right triangle. What is the measure of the angle at the top of the swing represented by $\angle T$? **Trigonometry** will allow us to solve the problem. *You will be asked to find this measure in Exercise 51.*

The word *trigonometry* comes from two Greek terms, *trigon* meaning triangle and *metron* meaning measure. The study of trigonometry involves triangle measurement. A ratio of the lengths of sides of a right triangle is called a **trigonometric ratio**. The three most common ratios are **sine**, **cosine**, and **tangent**. Their abbreviations are *sin*, *cos*, and *tan*, respectively.

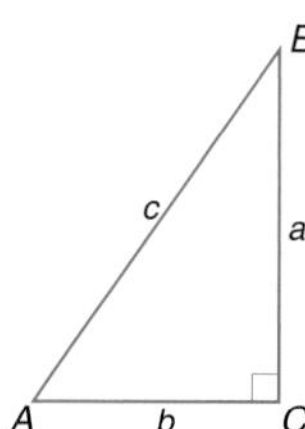

Trigonometric Ratio	Abbreviation	Definition
sine of A	sin A	$\frac{\text{leg opposite } \angle A}{\text{hypotenuse}} = \frac{a}{c}$
cosine of A	cos A	$\frac{\text{leg adjacent to } \angle A}{\text{hypotenuse}} = \frac{b}{c}$
tangent of A	tan A	$\frac{\text{leg opposite to } \angle A}{\text{leg adjacent to } \angle A} = \frac{a}{b}$

A represents the measurement of $\angle A$.

Trigonometric ratios are related to the acute angles of a right triangle, *not* the right angle.

You can use paper folding to explore trigonometric ratios.

MODELING MATHEMATICS — Trigonometric Ratios

Materials:  paper 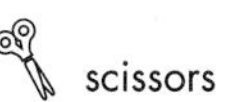 scissors  ruler

- Fold a rectangular piece of paper along a diagonal from *A* to *C*. Then cut along the fold to form a right $\triangle ABC$. Write the name of each angle on the inside of the triangle.

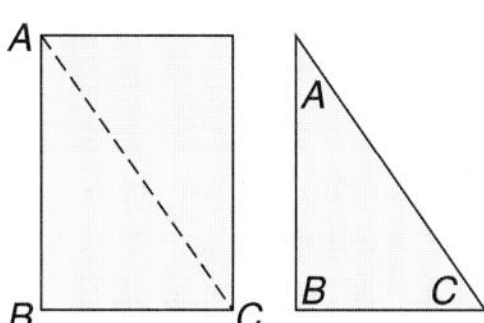

- Fold the triangle so that two segments perpendicular to $\overline{AB}$ result. Label points *D*, *E*, *F*, and *G* as shown. Use a ruler to find the measures *AC*, *AB*, *BC*, *AF*, *AG*, *FG*, *AD*, *AE*, and *DE* to the nearest millimeter.

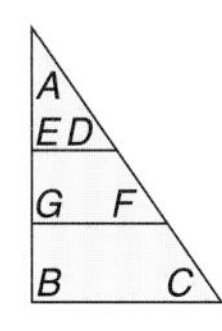

Your Turn

a. For $\angle A$, the sine could be expressed as $\frac{DE}{AD}$ for $\triangle ADE$, $\frac{FG}{AF}$ for $\triangle AFG$, and $\frac{BC}{AC}$ for $\triangle ACB$. How do these fractions compare? **equivalent**

b. For $\angle A$, the cosine could be expressed as $\frac{AE}{AD}$ for $\triangle ADE$, $\frac{AG}{AF}$ for $\triangle AFG$, and $\frac{AB}{AC}$ for $\triangle ACB$. How do these fractions compare? **equivalent**

c. For $\angle A$, the tangent could be expressed as $\frac{ED}{AE}$ for $\triangle ADE$, $\frac{GF}{AG}$ for $\triangle AFG$, and $\frac{BC}{AB}$ for $\triangle ACB$. How do these fractions compare?

d. Make a conjecture about the value of a trigonometric function based on your observations.

c. equivalent

d. See students' work.

As the Modeling Mathematics activity suggests, the value of a trigonometric ratio depends *only* on the measure of the angle. It does not depend on the size of the triangle.

Example 1 **Find the sin *S*, cos *S*, tan *S*, sin *E*, cos *E* and tan *E*. Express each ratio as a fraction and as a decimal.**

$\sin S = \frac{ME}{SE} = \frac{3}{5}$ or 0.6

$\cos S = \frac{SM}{SE} = \frac{4}{5}$ or 0.8

$\tan S = \frac{ME}{SM} = \frac{3}{4}$ or 0.75

$\sin E = \frac{SM}{SE} = \frac{4}{5}$ or 0.8

$\cos E = \frac{ME}{SE} = \frac{3}{5}$ or 0.6

$\tan E = \frac{SM}{ME} = \frac{4}{3}$ or $1.\overline{3}$

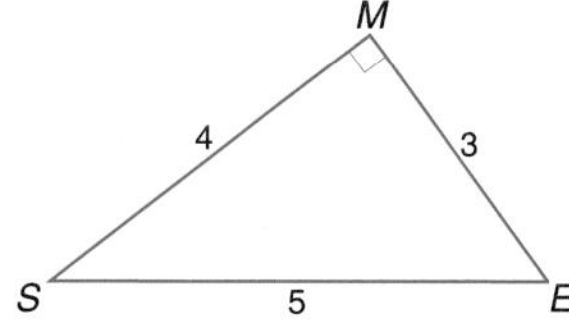

You can use a scientific calculator to evaluate expressions involving trigonometric ratios.

Motivating the Lesson

Situational Problem Ask students what they know about surveying. Discuss the need for more than a ruler or protractor to find distances across rivers and valleys and to the tops of mountains.

MODELING MATHEMATICS This modeling activity introduces students to the idea that the trigonometric ratios depend only on the measures of angles, and not on the lengths of sides.

2 TEACH

Teaching Tip After reading the trigonometric ratios for $\angle A$ on the first page of the lesson, have students use $\triangle ABC$ to name the sine, cosine, and tangent ratios for $\angle B$. $\frac{b}{c}, \frac{a}{c}, \frac{b}{a}$

In-Class Example

For Example 1

Find sin *A*, cos *A*, tan *A*, sin *B*, cos *B*, and tan *B*. Express each ratio as a fraction and as a decimal.

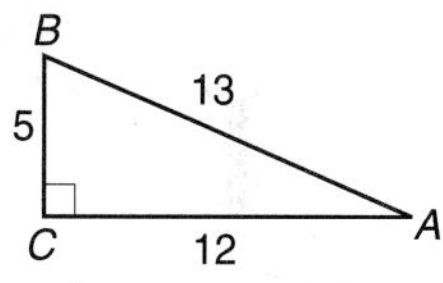

$\sin A = \frac{5}{13} \approx 0.385$

$\cos A = \frac{12}{13} \approx 0.923$

$\tan A = \frac{5}{12} \approx 0.417$

$\sin B = \frac{12}{13} \approx 0.923$

$\cos B = \frac{5}{13} \approx 0.385$

$\tan B = \frac{12}{5} \approx 2.4$

Alternative Learning Styles

Kinesthetic Use thumbtacks and string to model the following situation. A surveyor wants to measure the width of a river at point *A*. She has placed stakes 100 feet apart on the far side of the river and is standing at point *A*. Her instrument tells her that $m\angle BAC = 27$ and $m\angle ABC = 90$. How wide is the river? **196.3 ft**

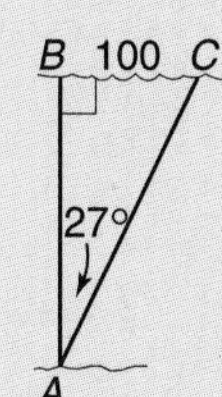

In-Class Examples

For Example 2
Find each value using a calculator. Round to the nearest ten thousandth.
a. sin 43° **0.6820**
b. cos 84° **0.1045**

For Example 3
A 16-foot ladder is propped against a building. The angle it forms with the ground measures 55°. How far up the side of the building does the ladder reach? **about 13.1 ft**

For Example 4
Find $m\angle A$ in right $\triangle ABC$ for $A(2, 3)$, $B(7, 3)$, and $C(6, 5)$. **26.6**

Teaching Tip On some scientific calculators, the ARC key or INV key takes the place of the 2nd key.

Example 2 **Find each value using a calculator. Round to the nearest ten thousandths.**
Make sure the calculator is in degree mode.

TECHNOLOGY Tip
If you are using a graphing calculator, the trigonometric ratio name is entered first and then the angle measure.

a. cos 41°

ENTER: 41 COS 0.75470958 *Your calculator display may differ slightly from the ones shown.*

cos 41° ≈ 0.7547

b. sin 78°

ENTER: 78 SIN 0.9781476

sin 78° ≈ 0.9781

You can use trigonometric ratios to find missing measures of a right triangle.

Example 3 **APPLICATION Aviation**

A plane is one mile above sea level when it begins to climb at a constant angle of 2° for the next 70 ground miles. How far above sea level is the plane after its climb?

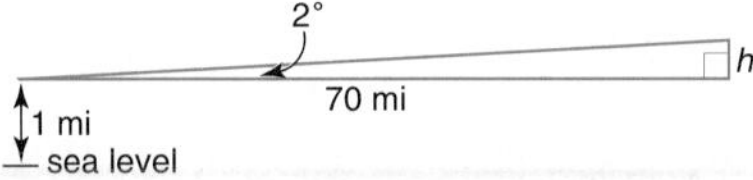

$\tan 2° = \frac{h}{70}$ $\quad$ *tan = $\frac{\text{opposite}}{\text{adjacent}}$*

$70 \tan 2° = h$ $\quad$ *Multiplication Property (=)*

Use your calculator to find h.

ENTER: 70 × 2 TAN = 2.444453864

The plane has climbed about 2.4 miles.
Therefore it is about 2.4 + 1 or 3.4 miles above sea level.

You can use trigonometric ratios and a calculator to find the measure of an angle if you know the measures of two sides of the right triangle. The inverse of each trigonometric ratio yields the angle measure. The inverse of the sine, cosine, and tangent ratios are indicated by $\sin^{-1}$, $\cos^{-1}$, and $\tan^{-1}$, respectively.

Example 4 **INTEGRATION Algebra**

Find $m\angle A$ in right triangle ABC for $A(1, 2)$, $B(6, 2)$, and $C(5, 4)$.

Explore The problem gives the coordinates of the vertices of a right triangle and asks for the measure of one of the angles. From the graph, $\angle C$ appears to be the right angle.

Plan Use the distance formula to find the measures of each side. Then find one of the trigonometric ratios for $\angle A$. Use the inverse of that ratio to find $m\angle A$.

Solve $AB = \sqrt{(6 - 1)^2 + (2 - 2)^2}$
$= \sqrt{25 + 0}$ or 5
$BC = \sqrt{(5 - 6)^2 + (4 - 2)^2}$
$= \sqrt{1 + 4}$ or $\sqrt{5}$
$AC = \sqrt{(5 - 1)^2 + (4 - 2)^2}$
$= \sqrt{16 + 4}$ or $\sqrt{20}$

Use the cosine ratio.

$\cos A = \frac{AC}{AB}$ $\cos = \frac{\text{adjacent}}{\text{hypotenuse}}$

$= \frac{\sqrt{20}}{5}$

Use a scientific calculator to find $m\angle A$.

ENTER: 20 [2nd] [$\sqrt{x}$] [÷] 5 [=] [2nd] [$\cos^{-1}$] 26.56505118

The measure of $\angle A$ is about 26.6.

Examine Use the sine ratio to check the answer.

$\sin A = \frac{BC}{AB}$ $\sin = \frac{\text{opposite}}{\text{hypotenuse}}$

$= \frac{\sqrt{5}}{5}$

ENTER: 5 [2nd] [$\sqrt{x}$] [÷] 5 [=] [2nd] [$\sin^{-1}$] 26.56505118

The answer is correct.

CAREER CHOICES

An **airline pilot** must have a commercial pilot's license with an instrument rating issued by the FAA. They must also pass strict physical examinations, including eyesight and hearing. While instruments help pilots to navigate in many situations, they also depend on their visual skills to evaluate angles and distances.

For more information, contact:

Airline Pilots Association
1625 Massachusetts Ave., NW
Washington, DC 20036

CHECK FOR UNDERSTANDING

Communicating Mathematics

Study the lesson. Then complete the following.

1. **State** the meaning of trigonometry. triangle measurement
2. **Compare and contrast** the sine, cosine, and tangent ratios. See margin.
3. **Explain** why trigonometric ratios do not depend on the size of the right triangle. The triangles are similar, so the ratios remain the same.
4. **Distinguish** between cos and $\cos^{-1}$. See margin.

MODELING MATHEMATICS

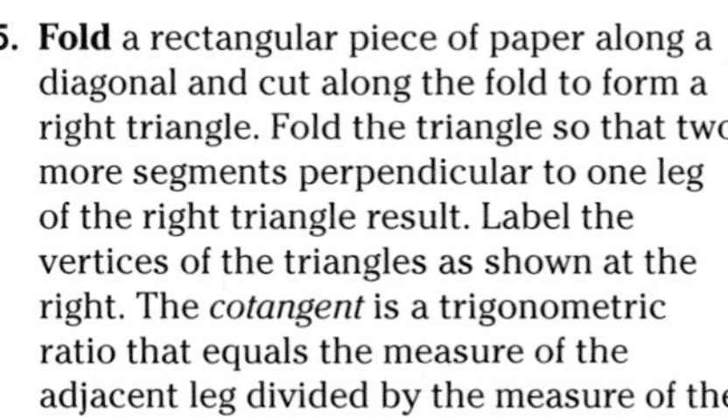

5. **Fold** a rectangular piece of paper along a diagonal and cut along the fold to form a right triangle. Fold the triangle so that two more segments perpendicular to one leg of the right triangle result. Label the vertices of the triangles as shown at the right. The *cotangent* is a trigonometric ratio that equals the measure of the adjacent leg divided by the measure of the opposite leg.

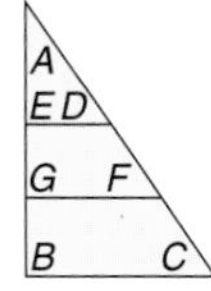

   a. **Write** an expression for the cotangent of A for $\triangle ADE$, $\triangle AFG$, and $\triangle ACB$.
   b. **Measure** the segments to the nearest tenth of a centimeter. Calculate the cotangent of A for each triangle. See students' work.
   c. Compare the results. The ratios are equal.

5a. $\frac{AE}{ED}, \frac{AG}{GF}, \frac{AB}{BC}$

3 PRACTICE/APPLY

Check for Understanding

Exercises 1–16 are designed to help you assess your students' understanding through reading, writing, speaking, and modeling. You should work through Exercises 1–5 with your students and then monitor their work on Exercises 6–16.

Additional Answers

2. All three ratios involve two sides of a right triangle. The sine ratio is the measure of the opposite side divided by the measure of the hypotenuse. The cosine ratio is the measure of the adjacent side divided by the measure of the hypotenuse. The tangent ratio is the measure of the opposite side divided by the measure of the adjacent side.
4. The cos is the ratio of the measure of the adjacent side divided by the measure of the hypotenuse for a given angle in a right triangle. The $\cos^{-1}$ is the measure of the angle with a certain cosine ratio.

Reteaching

Using Hands-On Activity Form groups of three or four students. Have each group use string and some chairs to outline a right triangle on the floor. Have them use a ruler and a protractor to find the measure of one side and one acute angle. Have two members use the appropriate trigonometric ratio to find the measure of another side. Have the other member(s) check by actually measuring the side. Repeat the activity for the third side of the triangle, with students switching roles within the group.

Assignment Guide

Core (with proof): 17–49 odd, 50, 51, 53–65
Core (informal): 17–49 odd, 50, 51, 53–61, 63–65
Enriched: 18–48 even, 50–65
All: Self Test, 1–10

For **Extra Practice**, see p. 778.

The red A, B, and C flags, printed only in the Teacher's Wraparound Edition, indicate the level of difficulty of the exercises.

Additional Answers

6. $\frac{20}{29} \approx 0.6897$
7. $\frac{20}{21} \approx 0.9524$
8. $\frac{20}{29} \approx 0.6897$

Guided Practice

Find the indicated trigonometic ratio as a fraction and as a decimal rounded to the nearest ten-thousandth. 6–8. See margin.

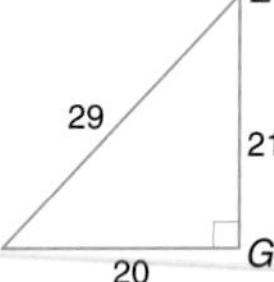

6. cos M 7. tan E 8. sin E

Find the value of each ratio to the nearest ten-thousandth.

9. sin 14° 0.2419
10. cos 51° 0.6293

Find the measure of each angle to the nearest tenth of a degree.

11. tan A = 2.0035 63.5
12. cos B = 0.7980 37.1

Find the value of x. Round to the nearest tenth.

13. 58.0

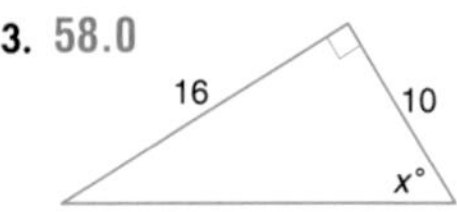

14.

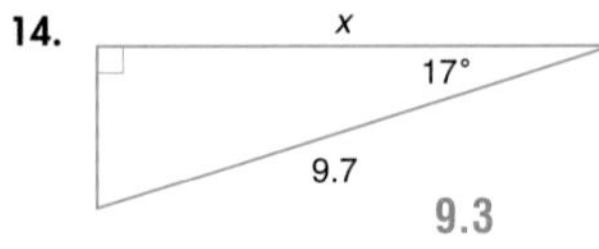

9.3

15. Find $m\angle A$ in right $\triangle ABC$ for $A(0, 8)$, $B(3, 0)$, and $C(0, 0)$.

16. **Safety** To guard against a fall, a ladder should make an angle of 75° or less with the ground. What is the maximum height that a ten-foot ladder can reach safely? about 9.66 ft

15. about 20.6

EXERCISES

Practice

Find the indicated trigonometric ratio as a fraction and as a decimal rounded to the nearest ten-thousandth.

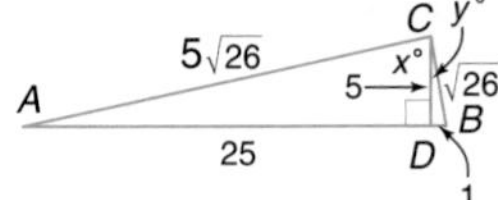

17. sin A 18. cos x 19. cos A
20. sin x 21. tan x 22. tan A
23. cos B 24. sin y 25. tan B

17. $\frac{\sqrt{26}}{26} \approx 0.1961$ A
18. $\frac{\sqrt{26}}{26} \approx 0.1961$
19. $\frac{5\sqrt{26}}{26} \approx 0.9806$
20. $\frac{5\sqrt{26}}{26} \approx 0.9806$ B
21. $\frac{5}{1} = 5.0000$
22. $\frac{1}{5} = 0.2000$
23. $\frac{\sqrt{26}}{26} \approx 0.1961$
24. $\frac{\sqrt{26}}{26} \approx 0.1961$
25. $\frac{5}{1} = 5.0000$

Find the value of each ratio to the nearest ten-thousandth.

26. sin 7° 0.1219
27. cos 24° 0.9135
28. tan 54° 1.3764
29. sin 72° 0.9511
30. cos 42° 0.7431
31. tan 20° 0.3640

Find the measure of each angle to the nearest tenth of a degree.

32. sin A = 0.7245 46.4
33. cos B = 0.2493 75.6
34. tan C = 9.4618 84.0
35. sin D = 0.4567 27.2
36. cos E = 0.1212 83.0
37. tan R = 0.4279 23.2

Find the value of x. Round to the nearest tenth. 38. 44.6

38.

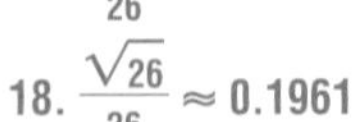

39. 32.6

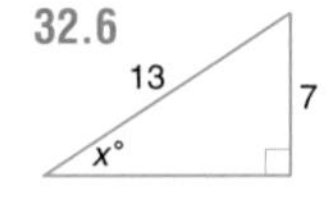

40. 7.0

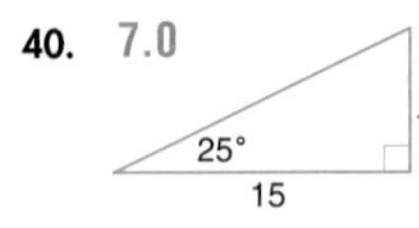

41. 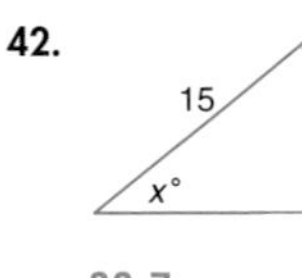

20.4

42. 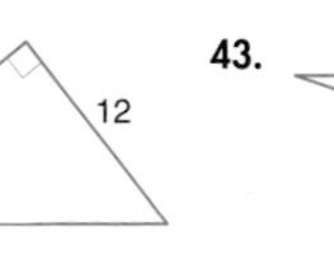

38.7

43. 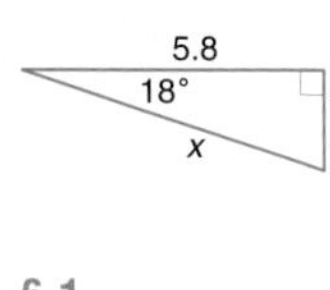

6.1

44. Find $m\angle J$ in right $\triangle JCL$ for $J(2, 2)$, $C(2, -2)$, and $L(7, -2)$. about 51.3

45. about 54.5

45. Find $m\angle C$ in right $\triangle BCD$ for $B(-1, -5)$, $C(-6, -5)$, and $D(-1, 2)$.

46. Find $m\angle X$ in right $\triangle XYZ$ with vertices $X(-5, 0)$, $Y(7, 0)$, and $Z(0, \sqrt{35})$. 49.8

Find the values of *x* and *y*. Round to the nearest tenth.

47. 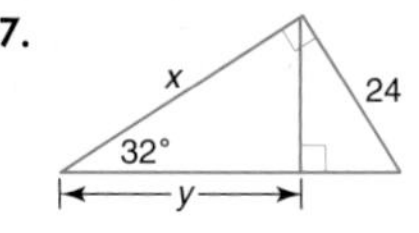

38.4; 32.6

48. 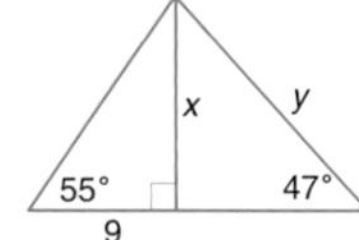

12.9; 17.6

49. 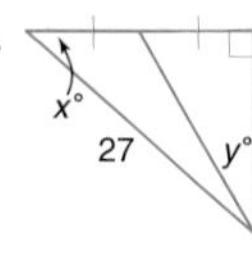

41.8; 29.2

Critical Thinking

50. The figure at the right is one-fourth of a circle drawn in Quadrant I of a coordinate plane. The radius of the circle is 1 unit and each division on the *x*- and *y*-axes is 0.2 unit. Right triangles are formed by drawing a vertical line from the point on the *x*-axis to the circle, then connecting that point with the origin.

a. Use the Pythagorean Theorem and trigonometry to complete the table of values for each triangle.

Length of Hypotenuse	x value	y value	sin O	cos O
1	0.2	0.98	0.98	0.2
1	0.4	0.92	0.92	0.4
1	0.6	0.80	0.80	0.6
1	0.8	0.60	0.60	0.8

b. Make a conjecture about the ordered pair (x, y) and the sine and cosine of $\angle O$. $x = \cos O$, $y = \sin O$

Applications and Problem Solving

51. about 35.5

51. **Golf** Refer to the application at the beginning of the lesson. Find $m\angle T$.

52. **Medicine** A patient is being treated with radiotherapy for a tumor that is behind a vital organ. In order to prevent damage to the organ, the radiologist must angle the rays to the tumor. If the tumor is 6.3 centimeters below the skin and the rays enter the body 9.8 centimeters to the right of the tumor, find the angle the rays should enter the body to hit the tumor. about 32.7°

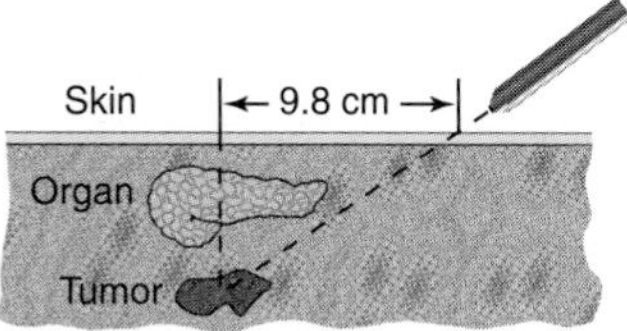

Study Guide Masters, p. 45

NAME________ DATE________

Study Guide

Student Edition Pages 412–419

Integration: Trigonometry Ratios in Right Triangles

A ratio of the lengths of two sides of a right triangle is called a **trigonometric ratio**. The three most common ratios are **sine**, **cosine**, and **tangent**. Their abbreviations are *sin*, *cos*, and *tan*, respectively. These ratios are defined for the acute angles of right triangles, though your calculator will give the values of sine, cosine, and tangent for angles of greater measure.

$\sin R = \frac{\text{leg opposite } \angle R}{\text{hypotenuse}} = \frac{r}{t}$

$\cos R = \frac{\text{leg adjacent to } \angle R}{\text{hypotenuse}} = \frac{s}{t}$

$\tan R = \frac{\text{leg opposite to } \angle R}{\text{leg adjacent to } \angle R} = \frac{r}{s}$

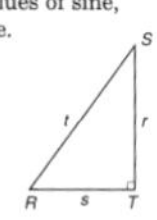

Example: Find sin *D*, cos *D*, and tan *D*. Express each ratio as a fraction and as a decimal rounded to the nearest thousandth.

$\sin D = \frac{5}{13} \approx 0.385$

$\cos D = \frac{12}{13} \approx 0.923$

$\tan D = \frac{5}{12} \approx 0.417$

Find the indicated trigonometric ratio as a fraction and as a decimal rounded to the nearest ten-thousandth.

1. $\sin M$ $\frac{8}{17} \approx 0.4706$
2. $\cos Z$ $\frac{3}{5} \approx 0.6000$
3. $\tan L$ $\frac{15}{8} \approx 1.8750$
4. $\sin X$ $\frac{3}{5} \approx 0.6000$
5. $\cos L$ $\frac{8}{17} \approx 0.4706$
6. $\tan Z$ $\frac{4}{3} \approx 1.3333$

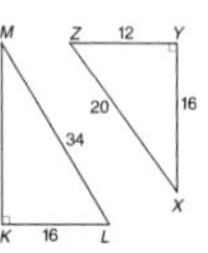

Find the value of each ratio to the nearest ten-thousandth.

7. sin 12° 0.2079
8. cos 32° 0.8480
9. tan 74° 3.4874
10. sin 55° 0.8192

Additional Answer

62. Given: $\overline{AB} \perp \overline{BD}$
$\overline{DE} \perp \overline{DB}$
$\overline{DB}$ bisects $\overline{AE}$.
Prove: $\angle A \cong \angle E$

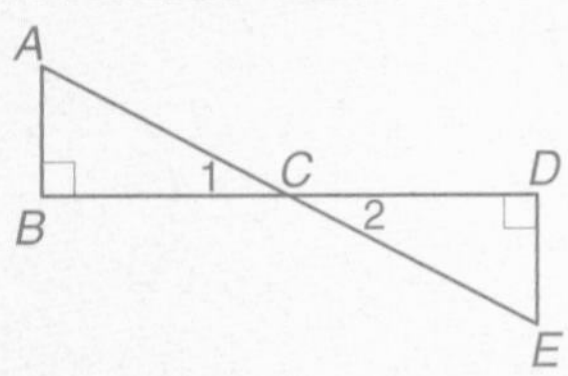

Proof:
Statements (Reasons)
1. $\overline{AB} \perp \overline{BD}$, $\overline{DE} \perp \overline{DB}$, $\overline{DB}$ bisects $\overline{AE}$. (Given)
2. $\angle B$ is a right angle, $\angle D$ is a right angle. (Def. $\perp$ lines)
3. $\angle B \cong \angle D$ (All rt. $\angle$s are $\cong$.)
4. $\overline{AC} \cong \overline{EC}$ (Def. segment bisector)
5. $\angle 1 \cong \angle 2$ (Vert. $\angle$s are $\cong$.)
6. $\triangle ABC \cong \triangle EDC$ (AAS)
7. $\angle A \cong \angle E$ (CPCTC)

Practice Masters, p. 45

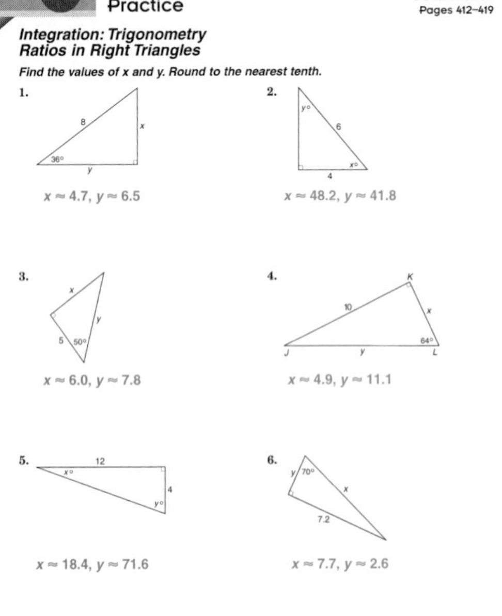

8-3 NAME ____________ DATE ______

Practice — Student Edition Pages 412–419

Integration: Trigonometry Ratios in Right Triangles

Find the values of x and y. Round to the nearest tenth.

1. $x \approx 4.7$, $y \approx 6.5$
2. $x \approx 48.2$, $y \approx 41.8$
3. $x \approx 6.0$, $y \approx 7.8$
4. $x \approx 4.9$, $y \approx 11.1$
5. $x \approx 18.4$, $y \approx 71.6$
6. $x \approx 7.7$, $y \approx 2.6$

53. Astronomy One way to find the distance between the Sun and a relatively close star is to determine the angles of sight for the star exactly six months apart. Half the measure formed by these two angles of sight is called the *stellar parallax*. Distances in space are sometimes measured in *astronomical units*. An astronomical unit is equal to the average distance between Earth and the Sun. The stellar parallax for the star Alpha Centauri is 0.00021°.

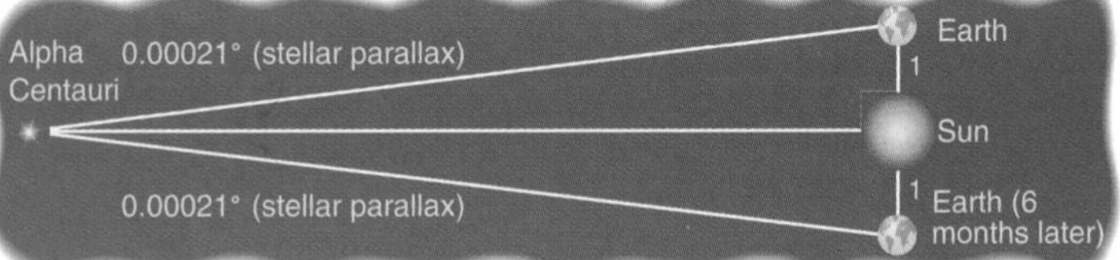

53a. about 272,837 astronomical units

a. Find the distance between Alpha Centauri and the Sun.

b. Make a conjecture as to why this method is used only for relatively close stars. **The stellar parallax would be too small.**

Mixed Review

54. The measure of the longer leg of a 30°-60°-90° triangle is 14. Write expressions for the measures of the shorter leg and the hypotenuse. (Lesson 8–2) $\frac{14\sqrt{3}}{3}, \frac{28\sqrt{3}}{3}$

55. Find the geometric mean between 8 and 9. (Lesson 8–1) $6\sqrt{2} \approx 8.49$

56. In the figure, $\triangle XYZ \sim \triangle XWV$. The perimeter of $\triangle XYZ$ is 34 units, and the perimeter of $\triangle XWV$ is 14 units. If $WV = 4$, find YZ. (Lesson 7–5) **about 9.7**

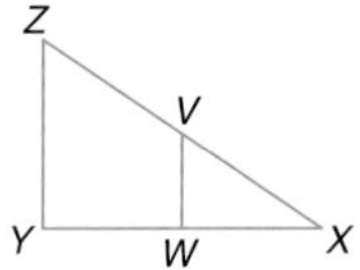

57. Find the value of j. (Lesson 7–2) **15**

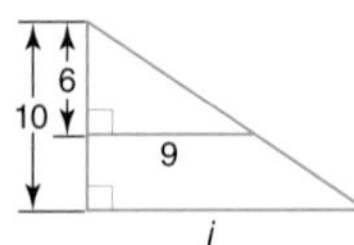

58. If $AC = 6g + 2$ and $AR = 4g - 2$, find RC. (Lesson 6–2) **10**

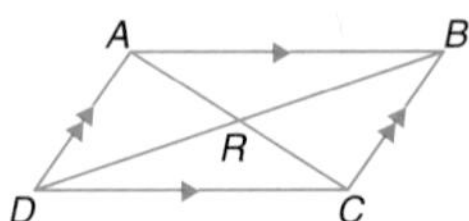

59. **Transportation** When driving from home to school, Isabel was unable to take her usual route straight down Dogwood Street because of road construction. The best options remaining were Walnut Street to Oak Street or Chestnut Street to Maple Street. Which would be the shortest route to school? (Lesson 5–6) **Walnut St. to Oak St.**

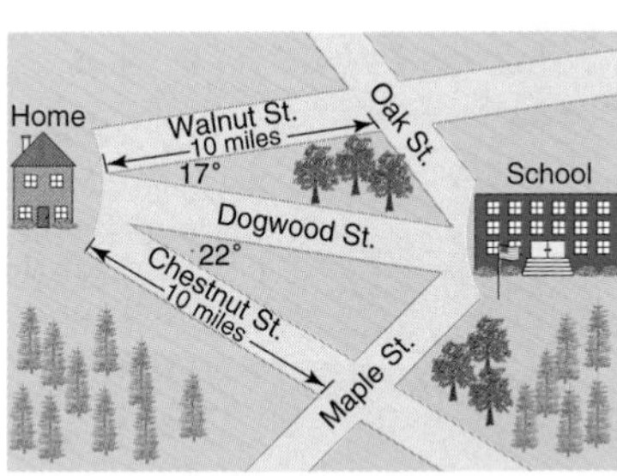

60. **Law** A man accused of shoplifting claims that he is innocent because he was at a movie with a friend when the crime took place. What needs to be established in order to prove that the man is guilty by indirect reasoning? (Lesson 5–3) **Show that he wasn't at the movies.**

61. Find the values of r and s so that $\triangle RST \cong \triangle XYZ$ by LA. (Lesson 5–2) **7.5; 12**

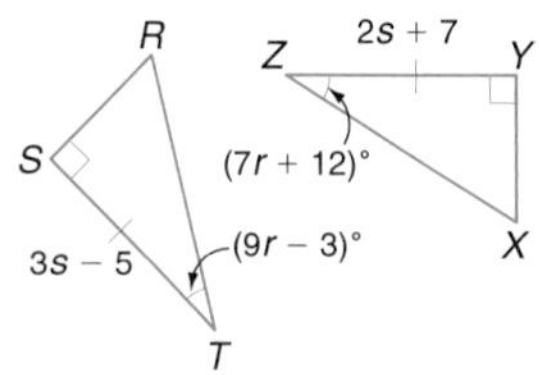

62. Write a two-column proof. (Lesson 4–5) **See margin.**

Given: $\overline{AB} \perp \overline{BD}$
$\overline{DE} \perp \overline{DB}$
$\overline{DB}$ bisects $\overline{AE}$.

Prove: $\angle A \cong \angle E$

63. If $\triangle GHI \cong \triangle RST$, which angle in $\triangle GHI$ is congruent to $\angle SRT$ in $\triangle RST$? (Lesson 4–3) **$\angle HGI$**

64. Define a variable, write an inequality, and solve the following problem. *The sum of a number and 11 is at least 25.*

64. Let x = the number; $x + 11 \geq 25$; $\{x \mid x \geq 14\}$.

65. Use substitution to solve the system of equations.
$y = 2x$ **(2, 4)**
$3x + y = 10$

SELF TEST

1. Find the geometric mean between 15 and 9. (Lesson 8–1) **$3\sqrt{15} \approx 11.62$**

Find the value of x. (Lesson 8–1)

2. **$2\sqrt{7} \approx 5.29$**

3. **4**

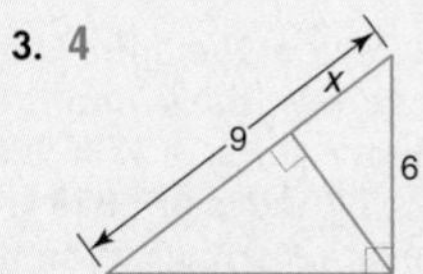

4. **1**

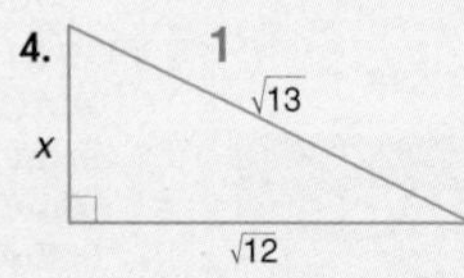

5. The length of one side of a square is 7.5 centimeters. Find the length of a diagonal of the square. (Lesson 8–2) **$7.5\sqrt{2} \approx 10.61$**

6. The length of one side of an equilateral triangle is 6 inches. Find the length of one altitude of the triangle. (Lesson 8–2) **$3\sqrt{3} \approx 5.20$**

Find the indicated trigonometry ratio as a fraction and as a decimal rounded to the nearest ten-thousandth. (Lesson 8–3)

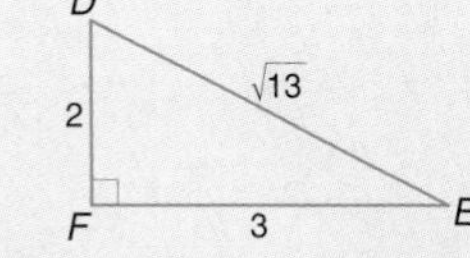

7. $\sin D$ **$\frac{3\sqrt{13}}{13} \approx 0.8321$**

8. $\cos D$ **8. $\frac{2\sqrt{13}}{13} \approx 0.5547$**

9. $\tan E$ **$\frac{2}{3} \approx 0.6667$**

10. **Road Construction** The 600 block of Powell Street in San Francisco has a steep rise. If it takes 66 feet along the horizontal for the street to rise 10 feet, find the angle the street makes with the horizontal. (Lesson 8–3) **about 8.6°**

Extension

Connections Determine how many different ratios can be written using the measures of the sides of a right triangle. What is the probability that a ratio chosen at random is the sine of an acute angle of the triangle?

$6; \frac{1}{3}$

SELF TEST

The Self Test provides students with a brief review of the concepts and skills in Lessons 8-1 through 8-3. Lesson numbers are given to the right of exercises or instruction lines so students can review concepts not yet mastered.

4 ASSESS

Closing Activity

Speaking Have students, one at a time, present an example of using a trigonometric ratio to find a missing measure. They may use the chalkboard or overhead.

Chapter 8 Quiz B (Lesson 8-3) is available in the *Assessment and Evaluation Masters*, p. 212.

Mid-Chapter Test (Lessons 8-1 through 8-3) is available in the *Assessment and Evaluation Masters*, p. 211.

Enrichment Masters, p. 45

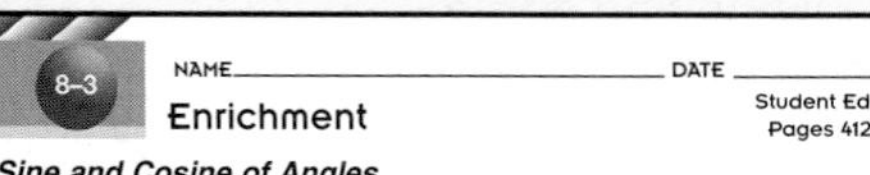

8–3 NAME ______ DATE ______

Enrichment

Student Edition Pages 412–419

Sine and Cosine of Angles

The following diagram can be used to obtain approximate values for the sine and cosine of angles from 0° to 90°. The radius of the circle is 1. So, the sine and cosine values can be read directly from the vertical and horizontal axes.

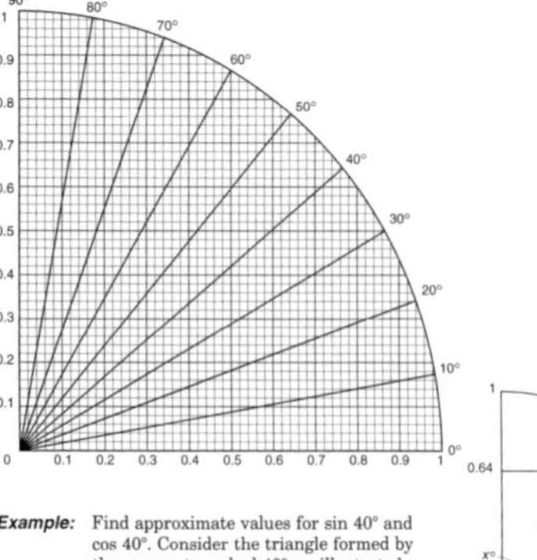

Example: Find approximate values for sin 40° and cos 40°. Consider the triangle formed by the segment marked 40°, as illustrated by the shaded triangle at right.

$\sin 40° = \frac{a}{c} \approx \frac{0.64}{1}$ or 0.64 $\sin 40° = \frac{b}{c} \approx \frac{0.77}{1}$ or 0.77

1. Use the diagram above to complete the chart of values.

$x°$	0°	10°	20°	30°	40°	50°	60°	70°	80°	90°
$\sin x°$	0	0.17	0.34	0.5	0.64	0.77	0.87	0.94	0.98	1
$\cos x°$	1	0.98	0.94	0.87	0.77	0.64	0.5	0.34	0.17	0

2. Compare the sine and cosine of two complementary angles (angles whose sum is 90°). What do you notice? The sine of an angle is equal to the cosine of the complement of the angle.

8-4 LESSON NOTES

NTCM Standards: 1–5, 7, 9

Instructional Resources

- Study Guide Master 8-4
- Practice Master 8-4
- Enrichment Master 8-4
- Real-World Applications, 16
- Tech Prep Applications Masters, p. 16

Transparency 8-4A contains the 5-Minute Check for this lesson; **Transparency 8-4B** contains a teaching aid for this lesson.

Recommended Pacing	
Standard Pacing	Day 8 of 12
Honors Pacing	Day 7 of 11
Block Scheduling*	Day 4 of 7

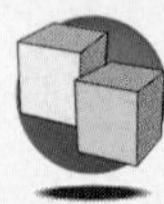

*For more information on pacing and possible lesson plans, refer to the *Block Scheduling Booklet*.

1 FOCUS

5-Minute Check
(over Lesson 8-3)

State the trigonometric ratio that corresponds to each value and angle.

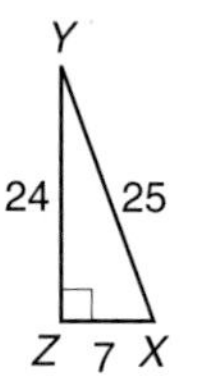

1. $\frac{7}{25}$, $\angle Y$ sin Y
2. $\frac{24}{7}$, $\angle X$ tan X
3. $\frac{24}{25}$, $\angle Y$ cos Y
4. $\frac{7}{25}$, $\angle X$ cos X
5. $\frac{7}{24}$, $\angle Y$ tan Y

Motivating the Lesson

Questioning Ask students how they might use trigonometry to find the height of the school building.

8-4 Angles of Elevation and Depression

What YOU'LL LEARN

- To use trigonometry to solve problems involving angles of elevation or depression.

Why IT'S IMPORTANT

You can use angles of elevation and depression to find missing measures involved in aerospace, architecture, and meteorology.

Tourism

Suppose Arnoldo is on the Skydeck of the Sears Tower looking through a telescope at Hanna who is on the Observation Deck of the John Hancock Center. Hanna is looking through a telescope at Arnoldo. An angle is formed by a horizontal line between the John Hancock Center and the Sears Tower and the line of sight from Hanna to Arnoldo. This angle is called the **angle of elevation**. Another angle is formed by a horizontal line between the Sears Tower and the John Hancock Center and the line of sight from Arnoldo to Hanna. This angle is called the **angle of depression**.

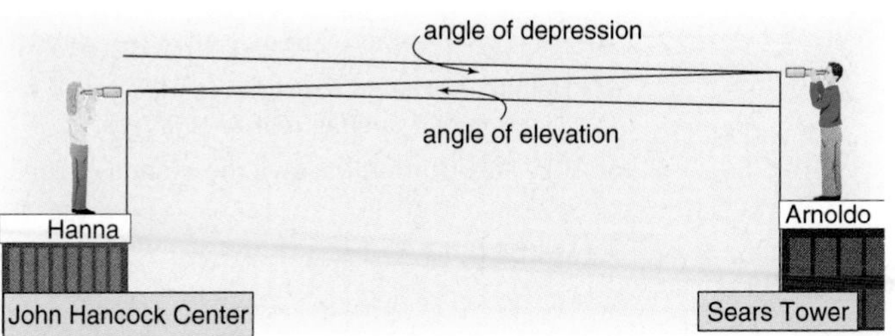

Sometimes drawing a diagram of the situation described in a problem can help you to solve problems involving angles of elevation and depression.

Example 1

Architecture

The Observation Deck of the John Hancock Center is on the 94th floor, which is 1030 feet above the ground. The Skydeck of the Sears Tower is on the 103rd floor, which is 1335 feet from the ground. The John Hancock Center is 1.7 miles or 8976 feet from the Sears Tower. What is the angle of elevation from Hanna to Arnoldo?

Explore The problem gives the height of the Observation Deck and the height of the Skydeck. It also gives the distance between the John Hancock Center and the Sears Tower. It asks for the angle of elevation from Hanna to Arnoldo.

Plan Draw a diagram.

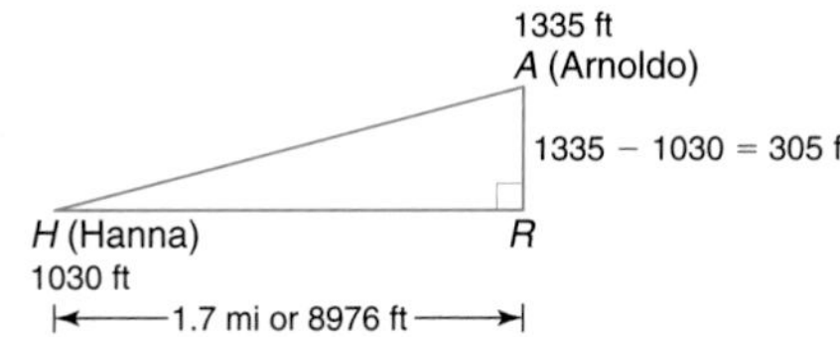

The length of one leg of the right triangle is 1.7 miles or 8976 feet. The length of the other leg is 1335 − 1030 or 305 feet. The angle of elevation can be found by using $\tan^{-1} H$.

Solve $\tan H = \frac{305}{8976}$

ENTER: 305 [÷] 8976 [=] [2nd] [tan^{-1}] [=] 1.9461332

The angle of elevation is about 1.9°.

Examine Since $\tan A = \frac{8976}{305}$, use a calculator to find $m\angle A$.

ENTER: 8976 [÷] 305 [=] [2nd] [tan^{-1}] [=] 88.053867

Since $1.9461332 + 88.053867 \approx 90$, the angles are complementary, and the answer is verified.

Angles of elevation or depression to two different objects can be used to find the distance between those objects.

Example 2

Aerospace

On July 20, 1969, Neil Armstrong became the first human to walk on the moon. During this mission, the lunar lander *Eagle* traveled aboard *Apollo 11*. Before sending *Eagle* to the surface of the moon, *Apollo 11* orbited the moon three miles above the surface. At one point in the orbit, the onboard guidance system measured the angles of depression to the far and near edges of a large crater. The angles measured 18° and 25°, respectively. Find the distance across the crater.

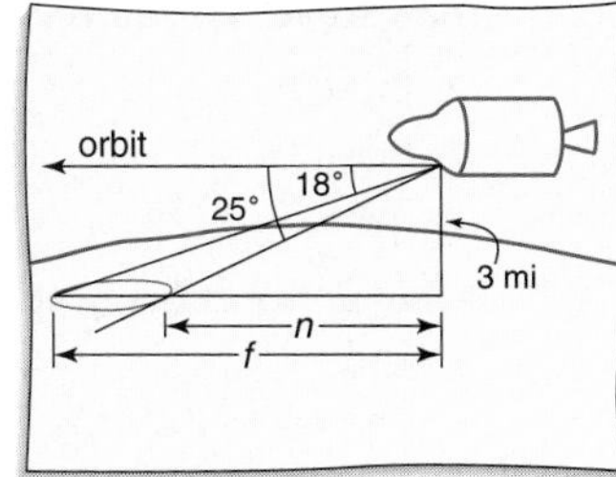

Let f be the ground distance from *Apollo 11* to the far edge of the crater and n be the ground distance to the near edge.

$\tan 18° = \frac{3}{f}$ *$\tan = \frac{\text{opposite}}{\text{adjacent}}$*

$f \tan 18° = 3$ *Cross multiply.*

$f = \frac{3}{\tan 18°}$ *Division Property (=)*

ENTER: 3 [÷] 18 [TAN] [=] 9.2330506 *Make sure your calculator is in degree mode.*

$\tan 25° = \frac{3}{n}$ *$\tan = \frac{\text{opposite}}{\text{adjacent}}$*

$n \tan 25° = 3$ *Cross multiply.*

$n = \frac{3}{\tan 25°}$ *Division Property (=)*

ENTER: 3 [÷] 25 [TAN] [=] 6.4335208

Since $f \approx 9.2$ and $n \approx 6.4$, the distance across the crater is about $9.2 - 6.4$ or 2.8 miles.

2 TEACH

In-Class Examples

For Example 1

A surveyor is 130 feet from a tower. The tower is 86 feet high. The surveyor's instrument is 4.75 feet above the ground. Find the angle of elevation.

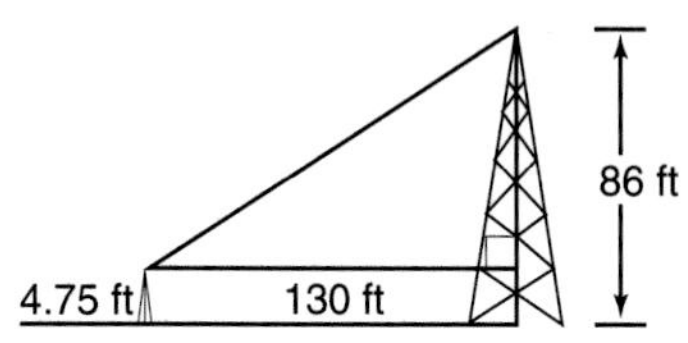

32°

For Example 2

A plane P is 3 miles above ground. The pilot sights the airport A at an angle of depression of 15°. He sights his house H at an angle of depression of 32°. What is the ground distance d between the pilot's house and the airport?

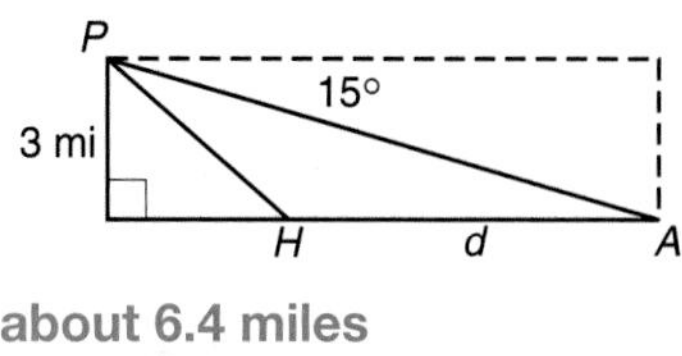

about 6.4 miles

Teaching Tip Before Example 2, ask students to demonstrate an angle of depression with their eyes by looking straight ahead and then down to a point on the floor.

Alternative Learning Styles

Auditory Ask students to brainstorm and list several situations in which a measurement can be found using an angle of elevation or an angle of depression.

3 PRACTICE/APPLY

Check for Understanding

Exercises 1–10 are designed to help you assess your students' understanding through reading, writing, speaking, and modeling. You should work through Exercises 1–4 with your students and then monitor their work on Exercises 5–10.

Error Analysis

In exercises about angles of depression, students may mistake the height they are given for the adjacent side of the angle of depression. Have them make diagrams as in Example 2 that clearly show which triangle is being used.

Additional Answers

1. Sample answer: the angle formed by the horizontal and the line of sight by looking at something below
2. Sample answer:

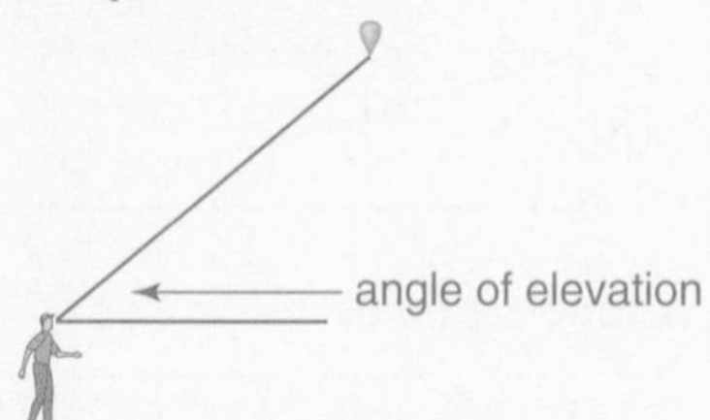

3. Use sin if the measures of the side opposite the angle and the hypotenuse are known. Use cos if the measures of the side adjacent to the angle and the hypotenuse are known. Use tan if the measures of the 2 legs are known.
4. $\sin = \frac{\text{opposite}}{\text{hypotenuse}}$

   $\cos = \frac{\text{adjacent}}{\text{hypotenuse}}$

   $\tan = \frac{\text{opposite}}{\text{adjacent}}$

   See students' work.

FYI

On March 26, 1937, the Popeye statue was unveiled in Popeye Park, Crystal City, Texas. It was the first monument to comic characters.

CHECK FOR UNDERSTANDING

Communicating Mathematics

Study the lesson. Then complete the following. 1. See margin.

1. **Describe** in your own words the meaning of angle of depression.
2. **Draw** an example showing an angle of elevation. Identify the angle of elevation. See margin.
3. **Explain** how you decide whether to use sin, cos, or tan when you are finding the measure of an acute angle in a right triangle. See margin.

4. **Assess Yourself** Name three common trigonometric ratios and describe each ratio. What is the meaning of each ratio? When you solve a problem involving trigonometric ratios, do you find a diagram helpful? What suggestions can you make to help your classmates solve these types of problems? See margin.

Guided Practice

5. Name the angles of elevation and depression in the figure at the right. $\angle OCP$; $\angle DPC$

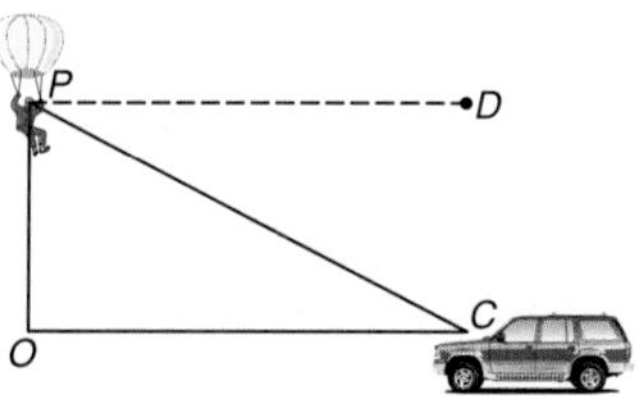

State an equation that would enable you to solve each problem. Then solve. Round answers to the nearest tenth.

6. sin 15° = $\frac{QR}{37}$; 9.6

6. Given $m\angle P = 15$ and $PQ = 37$, find QR.
7. Given $PR = 2.3$ and $PQ = 5.5$, find $m\angle P$. $\cos P = \frac{2.3}{5.5}$; 65.3

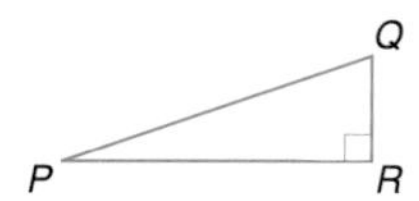

Refer to the chart at the right for Exercises 8–9.

8. Charo is 50 feet from the tallest totem pole. If Charo's eyes are 5 feet from the ground, find the angle of elevation for her line of sight to the top of the totem. about 73.4°
9. Derrick is visiting the San Jacinto State Park outside Houston, Texas. The angle of elevation for his line of sight to the top of the San Jacinto Column is 75°. If his eyes are 6 feet from the ground, how far is he from the base of the column? about 151.1 ft

FYI

The San Jacinto Column is the tallest monumental column in the world.

Monument Heights

Monument	Height
San Jacinto Column near Houston	570 feet
Gateway to the West Arch St. Louis	630 feet
Washington Monument Washington, D.C.	555 feet
Statue of Liberty New York City	305 feet
Tallest totem pole Alberta Bay, Canada	173 feet

Source: *Comparisons*

10b. Yes; 140 m is higher than 61 m.

10. **Aviation** The cloud ceiling is the lowest altitude at which solid cloud is present. If the cloud ceiling is below a certain level, usually about 61 meters, airplanes are not allowed to take off or land. One way that meteorologists can find the cloud ceiling at night is to shine a searchlight straight up and observe the spot of light on the clouds from a location away from the searchlight.
    a. If the searchlight is located 200 meters from the meteorologist and the angle of elevation to the spot of light on the clouds is 35°, how high is the cloud ceiling? about 140.0 m
    b. Can the airplanes land and take off under these conditions? Explain.

Reteaching

Using Word Problems Show students the two figures. For each, ask them to write a word problem that uses either angle of elevation or angle of depression. Have them write the solutions to their problems.

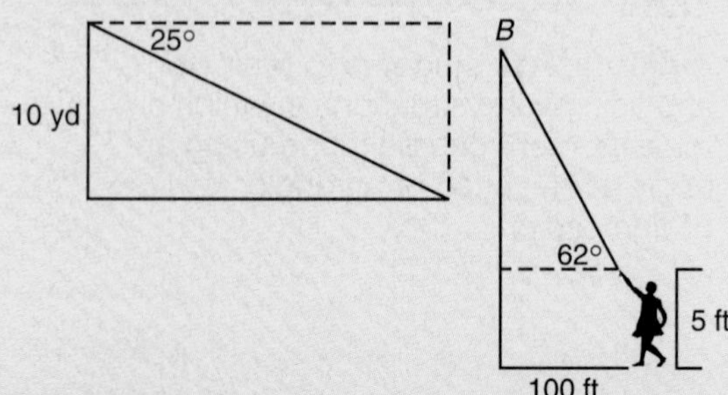

EXERCISES

Practice

Name the angles of elevation and depression in each figure.

11. $\angle GEF$; $\angle HFE$
12. $\angle BFT$; $\angle ATF$
13. $\angle KHJ$; $\angle IJH$
14. $\angle RUS$; $\angle TSU$

A

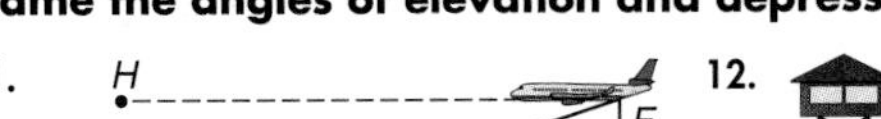

11.

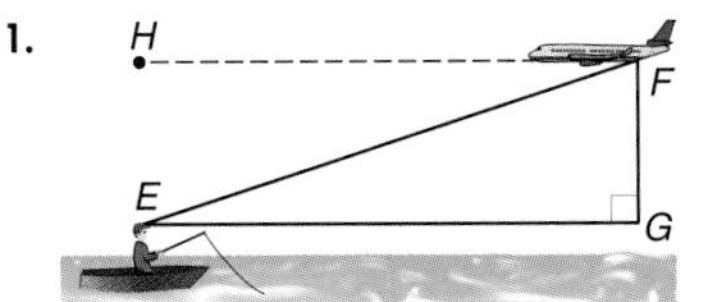

12.

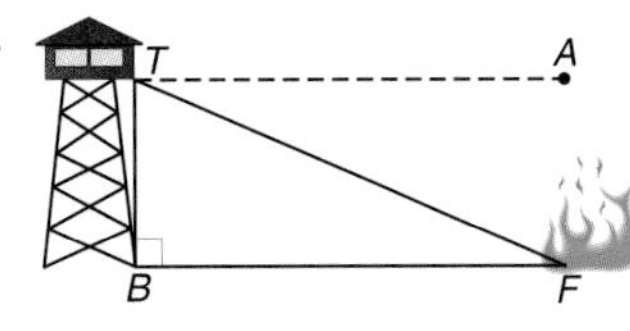

13.

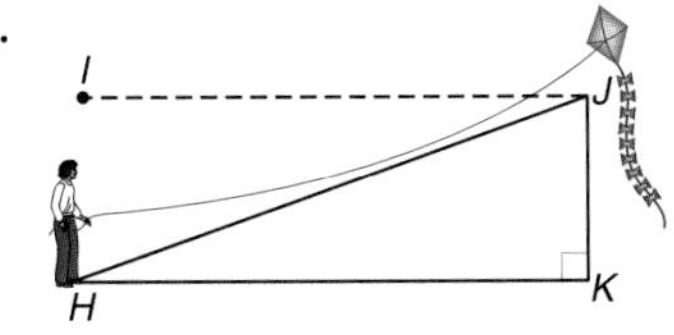

14.

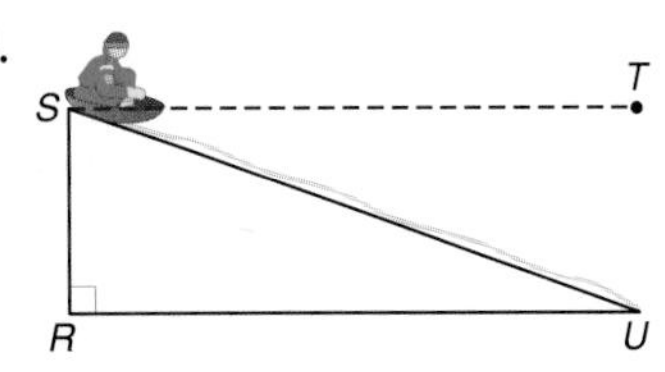

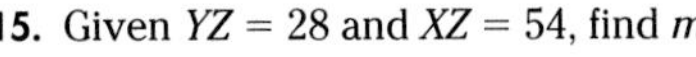

State an equation that would enable you to solve each problem. Then solve. Round answers to the nearest tenth.

15. $\tan Y = \frac{54}{28}$; 62.6
16. $\sin 28° = \frac{YZ}{15}$; 7.0
17. $\cos 66° = \frac{7}{XY}$; 17.2
18. $\cos Y = \frac{4}{15}$; 74.5
19. $\cos X = \frac{4.5}{6.6}$; 47.0
20. $\sin 65.5° = \frac{XZ}{22.4}$; 20.4

15. Given $YZ = 28$ and $XZ = 54$, find $m\angle Y$.
16. Given $XY = 15$ and $m\angle X = 28$, find YZ.
17. Given $m\angle Y = 66$ and $YZ = 7$, find XY.
18. Given $YZ = 4$ and $XY = 15$, find $m\angle Y$.
19. Given $XZ = 4.5$ and $XY = 6.6$, find $m\angle X$.
20. Given $XY = 22.4$ and $m\angle Y = 65.5$, find XZ.

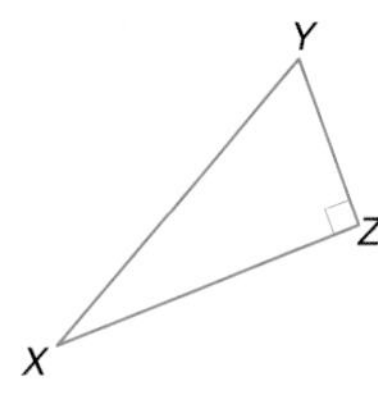

B

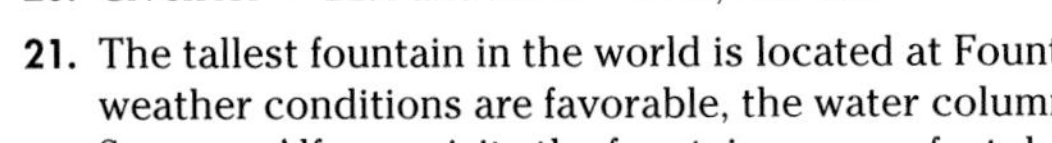

21. The tallest fountain in the world is located at Fountain Hills, Arizona. If weather conditions are favorable, the water column can reach 625 feet. Suppose Alfonso visits the fountain on a perfect day and his eyes are 5 feet from the ground. b. about 166.1 ft
    a. If Alfonso stands 40 feet from the fountain, find the angle of elevation for his line of sight to the top of the spray. about 86.3°
    b. If Alfonso moves so that the angle of elevation for his line of sight to the top of the spray is 75°, how far is he from the base of the spray?
22. After flying at an altitude of 9 kilometers, an airplane starts to descend when its ground distance from the landing field is 175 kilometers. What is the angle of depression for this portion of the flight? about 2.9°
23. A golfer is standing on a tee with the green in a valley below. If the tee is 43 yards higher than the green and the angle of depression from the tee to the hole is 14°, find the distance from the green to the hole. about 177.7 yd
24. Kierra is flying a kite. She has let out 55 feet of string. If the string makes a 35° angle with the ground, how high above the ground is the kite? about 31.5 ft
25. A trolley car track rises vertically 40 feet over a horizontal distance of 630 feet. What is the angle of elevation of the track? about 3.6°
26. A ski slope is 550 yards long with a vertical drop of 130 yards. Find the angle of depression of the slope. about 13.7°

Assignment Guide

Core (with proof): 11–29 odd, 30, 31, 33, 35–43
Core (informal): 11–29 odd, 30, 31, 33, 35–43
Enriched: 12–28 even, 30–43

For **Extra Practice**, see p. 779.

The red A, B, and C flags, printed only in the Teacher's Wraparound Edition, indicate the level of difficulty of the exercises.

Study Guide Masters, p. 46

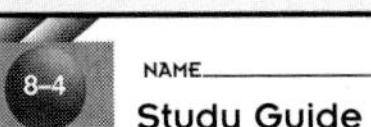

8-4 NAME__________ DATE________

Study Guide

Student Edition Pages 420–425

Angles of Elevation and Depression

Many problems in daily life can be solved by using trigonometry. Often such problems involve an **angle of elevation** or an **angle of depression**.

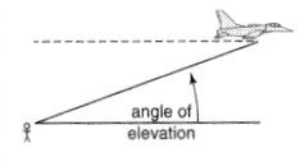

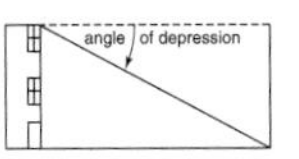

Example: The angle of elevation from point A to the top of a cliff is 38°. If point A is 80 feet from the base of the cliff, how high is the cliff?

Let x represent the height of the cliff. Then $\tan 38° = \frac{x}{80}$.

$80 \tan 38° = x$

Use a calculator set for the degree mode to find x.

ENTER: 80 [×] 38 [TAN] [=] 62.502850

The cliff is about 63 feet high.

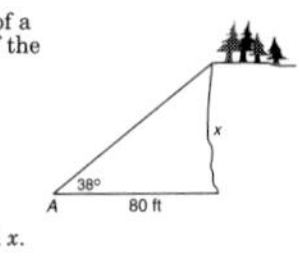

Solve each problem. Round measures of segments to the nearest hundredth and measures of angles to the nearest degree.

1. From the top of a tower, the angle of depression to a stake on the ground is 72°. The top of the tower is 80 feet above ground. How far is the stake from the foot of the tower? 25.99 ft
2. A tree 40 feet high casts a shadow 58 feet long. Find the measure of the angle of elevation of the sun. 35°
3. A ladder leaning against a house makes an angle of 60° with the ground. The foot of the ladder is 7 feet from the foundation of the house. How long is the ladder? 14 ft
4. A balloon on a 40-foot string makes an angle of 50° with the ground. How high above the ground is the balloon if the hand of the person holding the balloon is 6 feet above the ground? 36.64 ft

Classroom Vignette

"Have students be Egyptian rope stretchers. Give groups of students a length of rope and ask them to use it to form a perfect square. You may wish to suggest that they begin by using the Pythagorean Theorem to choose dimensions for a right triangle. Then after a right angle is formed, it can be drawn at each vertex to form the square."

Michael L. Berthusen

Michael L. Berthusen
Eddyville High School
Eddyville, Iowa

27. The waterway between Lake Huron and Lake Superior separates the United States and Canada at Sault Sainte Marie. The railroad drawbridge located at Sault Saint Marie is normally 13 feet above the water when it is closed. Each section of this drawbridge is 210 feet long. Suppose the angle of elevation of each section is 70°.

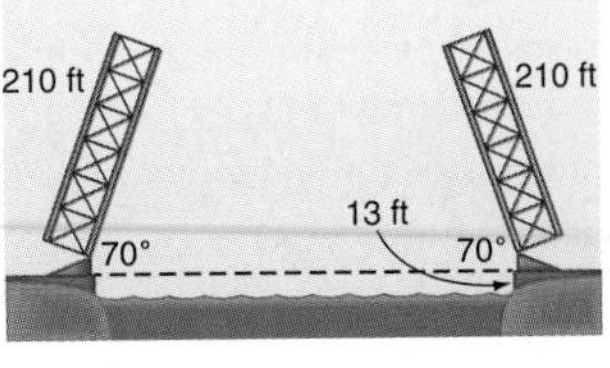

a. Find the distance from the top of a section of the drawbridge to the water. **about 210.3 ft**

b. Find the width of the gap created by the two sections of the bridge. **about 276.4 ft**

28. Carol is in the Skydeck of the Sears Tower overlooking Lake Michigan. She sights two sailboats going due east from the tower. The angles of depression to the two boats are 42° and 29°. If the Skydeck is 1335 feet high, how far apart are the boats? **about 925.7 ft**

29. Ulura or Ayers Rock is a sacred place for Aborigines of the western desert of Australia. Chun-Wei uses a surveying device to measure the angle of elevation to the top of the rock to be 11.5°. He walks half a mile closer and measures the angle of elevation to be 23.9°. How high is Ayers Rock in feet? **about 993.0 ft**

Critical Thinking

30. Imagine that a fly and an ant are in one corner of a rectangular box. The end of the box is 4 inches by 6 inches, and the diagonal across the bottom of the box makes an angle of 21.8° with the longer edge of the box. There is food in the corner opposite the insects.

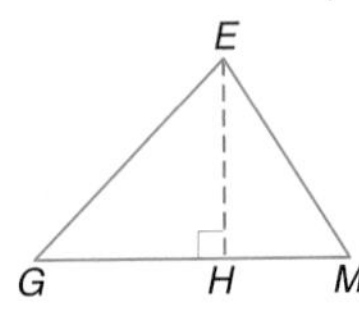

a. What is the shortest distance the fly must fly to get to the food? **about 16.64 in.**

b. What is the shortest distance the ant must crawl to get to the food? **about 20.16 in.**

31. Given acute $\triangle GME$ with altitude $\overline{EH}$, write a trigonometric expression for the ratio $\frac{GE}{EM}$. $\frac{\sin M}{\sin G}$

E G H M

Applications and Problem Solving

32. **Literature** *In The Adventures of Sherlock Holmes: The Adventures of the Musgrave Ritual*, Sherlock Holmes uses trigonometry to solve the mystery. To find a treasure, he must determine where the end of the shadow of an elm tree was located at a certain time of day. Unfortunately, the elm had been cut down, but Mr. Musgrave remembers that his tutor required him to calculate the height of the tree as part of his trigonometry class. Mr. Musgrave tells Sherlock Holmes that the tree was exactly 64 feet. Sherlock needs to find the length of the shadow at a time of day when the shadow from an oak tree is a certain length. The angle of elevation of the sun at this time of day is 33.7°. What was the length of the shadow of the elm? **about 95.96 ft**

Practice Masters, p. 46

NAME______________ DATE ________

8–4 Practice — Student Edition Pages 420–425

Angles of Elevation and Depression

Solve each problem. Round measures of segments to the nearest hundredth and measures of angles to the nearest degree.

1. A 20-foot ladder leans against a wall so that the base of the ladder is 8 feet from the base of the building. What angle does the ladder make with the ground? 66°

2. A 50-meter vertical tower is braced with a cable secured at the top of the tower and tied 30 meters from the base. What angles does the cable form with the vertical tower? 31°

3. At a point on the ground 50 feet from the foot of a tree, the angle of elevation to the top of the tree is 53°. Find the height of the tree. 66.35 ft

4. From the top of a lighthouse 210 feet high, the angle of depression of a boat is 27°. Find the distance from the boat to the food of the lighthouse. The lighthouse was built at sea level. 412.15 ft

5. Richard is flying a kite. The kite string makes an angle of 57° with the ground. If Richard is standing 100 feet from the point on the ground directly below the kite, find the length of the kite string. 183.61 ft

6. An airplane rises vertically 1000 feet over a horizontal distance of 1 mile. What is the angle of elevation of the airplane's path? 11°

33. **Architecture** Diana is an architect who designs houses so that the windows receive minimum sun in the summer and maximum sun in the winter. For Seattle, Washington, the angle of elevation of the sun at noon on the longest day is 66° and on the shortest day is 19°. Suppose a house is designed with a south-facing window that is 6 feet tall. The top of the window is to be installed 1 foot below the overhang.

x; 1 ft; Window 6 ft high; 66°

a. How long should Diana make the overhang so that the window gets no direct sunlight at noon on the longest day? about 3.12 ft

b. Using the overhang from part a, how much of the window will get direct sunlight at noon on the shortest day? the bottom 5.93 ft of the window

c. To find the angle of elevation of the sun on the longest day of the year where you live, subtract your latitude from 90° and add 23.5°. To find the elevation of the sun on the shortest day, subtract the latitude from 90° and then subtract 23.5°. Draw a solar design for a south-facing window and corresponding overhang for a home in your community. See students' work.

34. **Meteorology** Two weather observation stations are 7 miles apart. A weather balloon is located between the stations. From Station 1, the angle of elevation to the weather balloon is 35°. From Station 2, the angle of elevation to the balloon is 54°. Find the altitude of the balloon to the nearest tenth of a mile. (*Hint:* Find the distance from Station 2 to the point directly below the balloon.) about 3.25 mi

Mixed Review

35a. $\frac{4}{5} = 0.8000$

35b. $\frac{4}{5} = 0.8000$

35c. $\frac{4}{3} \approx 1.3333$

35. Find the indicated trigonometric ratio as a fraction and as a decimal rounded to the nearest ten-thousandth. (Lesson 8–3)

a. $\sin A$ b. $\cos B$ c. $\tan A$

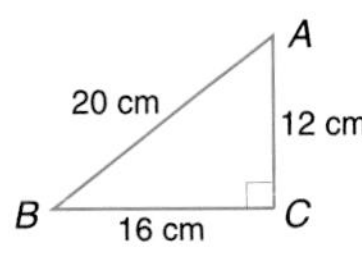

36. The perimeter of an equilateral triangle is 42 centimeters. Find the length of an altitude of the triangle. (Lesson 8–2) $7\sqrt{3} \approx 12.12$

37. **Construction** Find the length of a diagonal brace needed for a rectangular section of wall that is 6 feet wide and 8 feet high. (Lesson 8–1) 10 ft

38. **Photography** Chapa wants to enlarge a photograph that is currently 4 inches wide by 5 inches long so that the new photograph is 12 inches long. How wide will the new photograph be? (Lesson 7–2) 9.6 inches

39. See margin.

39. Draw a trapezoid with two right angles and one obtuse angle. (Lesson 6–5)

40. Quadrilateral $WXYZ$ is a parallelogram. If $WX = 3g + 7$, $XY = 7h - 1$, $YZ = 6g - 2$, and $WZ = 2h + 9$, find the perimeter of $WXYZ$. (Lesson 6–1) 58

41. The base of an isosceles triangle is 18 inches long. If the legs are $3y + 21$ and $10y$ inches long, find the perimeter of the triangle. (Lesson 4–6) 78 in.

Algebra

42. Find the slope and y-intercept of the graph of $2x - y = 16$. 2; −16

43. Solve the system of inequalities by graphing. See margin.
$y < 5$
$y > 2x + 1$

4 ASSESS

Closing Activity

Modeling Have each student make up one problem that uses angle of elevation and one problem that uses angle of depression. Each problem should be accompanied by a solution that includes a model.

Additional Answers

39.

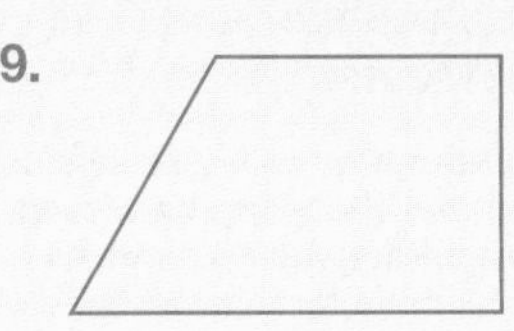

43.

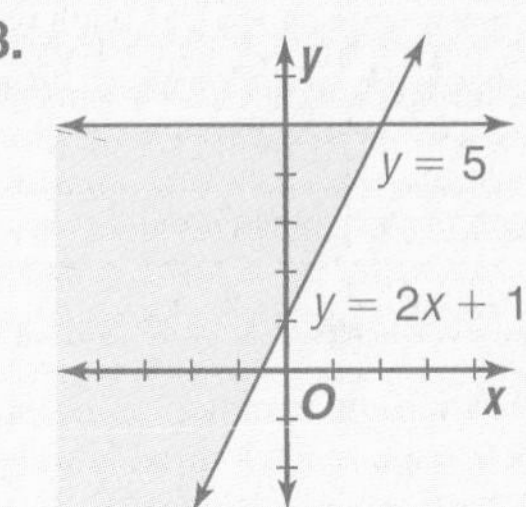

Enrichment Masters, p. 46

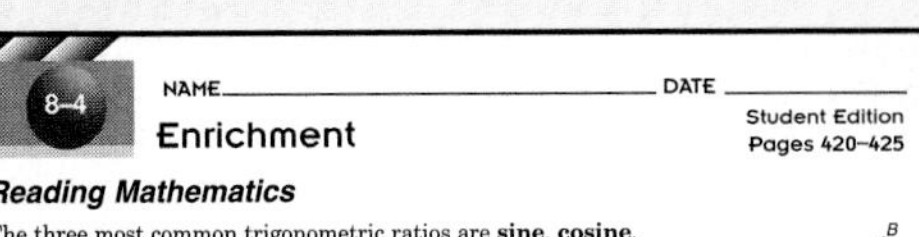

NAME__________ DATE______

Enrichment

Student Edition Pages 420–425

Reading Mathematics

The three most common trigonometric ratios are **sine**, **cosine**, and **tangent**. Three other ratios are the **cosecant**, **secant**, and **cotangent**. The chart below shows abbreviations and definitions for all six ratios. Refer to the triangle at the right.

Abbreviation	Read as:	Ratio
sin A	the sine of $\angle A$	$\frac{\text{leg opposite } \angle A}{\text{hypotenuse}} = \frac{a}{c}$
cos A	the cosine of $\angle A$	$\frac{\text{leg adjacent to } \angle A}{\text{hypotenuse}} = \frac{b}{c}$
tan A	the tangent of $\angle A$	$\frac{\text{leg opposite to } \angle A}{\text{leg adjacent } \angle A} = \frac{a}{b}$
csc A	the cosecant of $\angle A$	$\frac{\text{hypotenuse}}{} = \frac{c}{a}$
sec A	the secant of $\angle A$	$\frac{\text{hypotenuse}}{\text{leg adjacent to } \angle A} = \frac{c}{b}$
cot A	the cotangent of $\angle A$	$\frac{\text{leg adjacent to } \angle A}{\text{leg opposite } \angle A} = \frac{b}{a}$

Use the abbreviations to rewrite each statement as an equation.

1. The secant of angle A is equal to 1 divided by the cosine of angle A. $\sec A = \frac{1}{\cos A}$

2. The cosecant of angle A is equal to 1 divided by the sine of angle A. $\csc A = \frac{1}{\sin A}$

3. The cotangent of angle A is equal to 1 divided by the tangent of angle A. $\cot A = \frac{1}{\tan A}$

4. The cosecant of angle A multiplied by the sine of angle A is equal to 1. $\csc A \sin A = 1$

5. The secant of angle A multiplied by the cosine of angle A is equal to 1. $\sec A \cos A = 1$

6. The cotangent of angle A times the tangent of angle A is equal to 1. $\cot A \tan A = 1$

Use the triangle at right. Write each ratio.

7. $\sec R$ $\frac{t}{s}$ 8. $\csc R$ $\frac{t}{r}$ 9. $\cot R$ $\frac{s}{r}$

10. $\sec S$ $\frac{t}{r}$ 11. $\csc S$ $\frac{t}{s}$ 12. $\cot S$ $\frac{r}{s}$

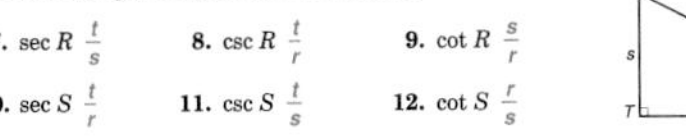

13. If $\sin x° = 0.289$, find the value of $\csc x°$. ≈ 3.46

14. If $\tan x° = 1.376$, find the value of $\cot x°$. ≈ 0.727

Extension

Problem Solving A surveyor finds the angle of elevation to the top of a tree to be 20°. She then moves directly toward the tree to a point 100 feet closer to the tree and finds the angle of elevation to be 30°. How far from the tree was she for the second reading? If her transit was 5 feet from the ground for both readings, how tall was the tree? about 171 ft; about 104 ft

8-5 LESSON NOTES

NCTM Standards: 1–5, 7, 9

Instructional Resources

- Study Guide Master 8-5
- Practice Master 8-5
- Enrichment Master 8-5
- Assessment and Evaluation Masters, p. 213
- Graphing Calculator and Computer Masters, p. 8

Transparency 8-5A contains the 5-Minute Check for this lesson; **Transparency 8-5B** contains a teaching aid for this lesson.

Recommended Pacing	
Standard Pacing	Day 9 of 12
Honors Pacing	Day 8 of 11
Block Scheduling*	Day 5 of 7

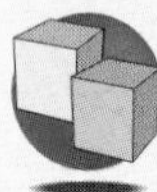

*For more information on pacing and possible lesson plans, refer to the *Block Scheduling Booklet*.

1 FOCUS

5-Minute Check
(over Lesson 8-4)

Refer to the figure below. Find each measure to the nearest whole number.

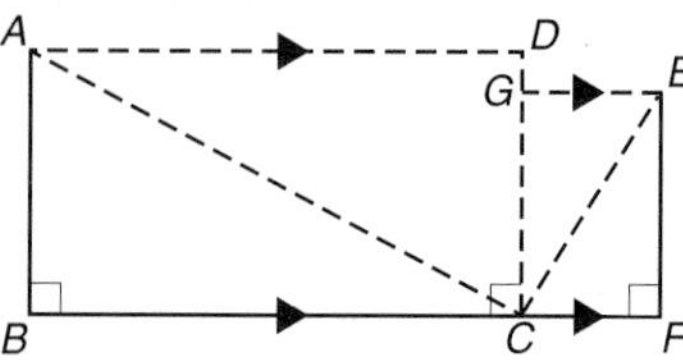

1. $AB = 10$ meters, $m\angle DAC = 27$. Find BC. **20 m**
2. $m\angle BCA = 35$, $BC = 25$ yards. Find AB. **18 yd**
3. $m\angle FCE = 54$, $CF = 6$ miles. Find EF. **8 mi**
4. $EF = 120$ feet, $m\angle GEC = 63$. Find CF. **61 ft**

8–5 Using the Law of Sines

What YOU'LL LEARN

- To use the Law of Sines to solve triangles.

Why IT'S IMPORTANT

You can use the Law of Sines to find missing measures of triangles involved in surveying, aviation, and fire fighting.

APPLICATION
Surveying

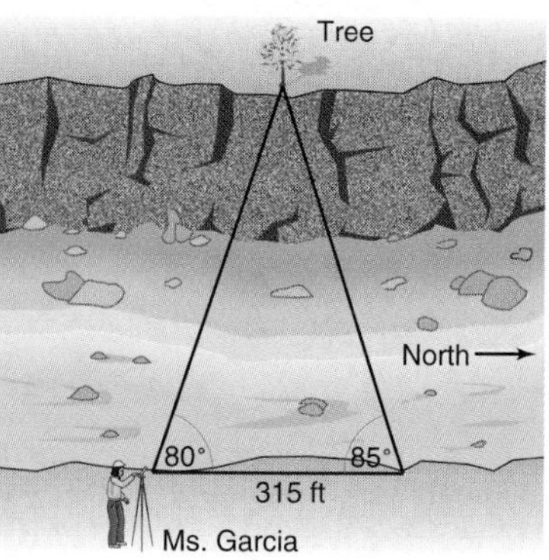

Anna Garcia is a surveyor who has the job of determining the distance across the Rio Grande Gorge in northern New Mexico. Standing at one side of the ridge, she measures the angle formed by the edge of the ridge and the line of sight to a tree on the other side of the ridge. She then walks along the ridge 315 feet north and measures the angle formed by the edge of the ridge and the new line of sight to the same tree. If the first angle is 80° and the second angle is 85°, find the distance across the gorge. *This problem will be solved in Example 2.*

You can use trigonometric functions to solve problems like this that involve triangles that are *not* right triangles. One such method is to use the **Law of Sines**.

Law of Sines	**Let $\triangle ABC$ be any triangle with a, b, and c representing the measures of sides opposite angles with measures A, B, and C, respectively. Then, $\frac{\sin A}{a} = \frac{\sin B}{b} = \frac{\sin C}{c}$.**

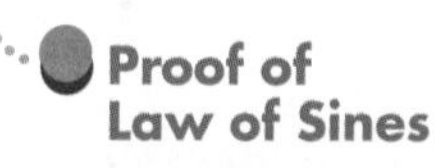

Proof of Law of Sines

$\triangle ABC$ is a triangle with an altitude from C that intersects $\overline{AB}$ at D. Let h represent the measure of $\overline{CD}$. Since $\triangle ACD$ and $\triangle BCD$ are right triangles, we can find $\sin A$ and $\sin B$.

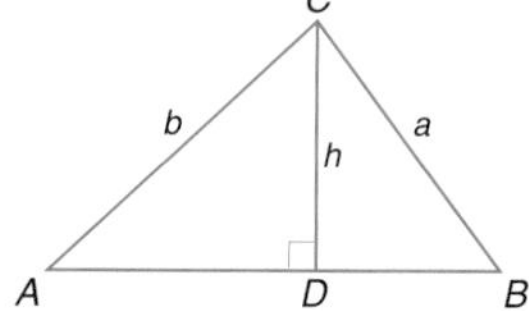

$\sin A = \frac{h}{b}$ $\qquad$ $\sin B = \frac{h}{a}$

$b \sin a = h$ $\qquad$ $a \sin B = h$

$b \sin A = a \sin B$ *Substitution Property (=)*

$\frac{\sin A}{a} = \frac{\sin B}{b}$ *Divide each side by ab.*

The proof can be completed by using a similar technique with another altitude to show that $\frac{\sin A}{a} = \frac{\sin B}{b} = \frac{\sin C}{c}$.

Finding the measures of all the angles and sides of a triangle is called **solving the triangle**. The Law of Sines can be used to solve a triangle in the following cases.

Case 1. You know the measures of two angles and any side of a triangle.
Remember that if you know the measures of two angles of a triangle, you can find the measure of the third angle by using the Angle Sum Theorem.

Case 2. You know the measures of two sides and an angle opposite one of these sides of the triangle.
Examples of this case:
(1) knowing the measures of $\angle A$ and $\angle B$ and length a or b;
(2) knowing the measures of $\angle A$ and $\angle C$ and length a or c;
(3) knowing the measures of $\angle B$ and $\angle C$ and length b or c;

Example  **Solve $\triangle XYZ$ if $m\angle X = 33$, $m\angle Z = 47$, and $z = 14$.**

Since the measures of two angles and a side are known, this is an example of Case 1. Consider the four parts $\angle X$, $\angle Z$, x, and z.

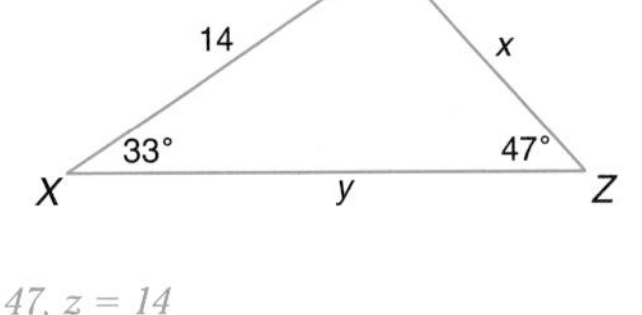

TECHNOLOGY Tip
If you round before finding the final result, your answers will differ slightly from answers in which there was no rounding.

$$\frac{\sin Z}{z} = \frac{\sin X}{x} \quad \textit{Law of Sines}$$
$$\frac{\sin 47°}{14} = \frac{\sin 33°}{x} \quad m\angle X = 33,\ m\angle Z = 47,\ z = 14$$
$$x \sin 47° = 14 \sin 33° \quad \textit{Cross multiply.}$$
$$x = \frac{14 \sin 33°}{\sin 47°} \quad \textit{Division Property (=)}$$
$$x \approx 10.4 \quad \textit{Use a calculator.}$$

Now we know the measures of two sides and two angles of the triangle. We can find the measure of the third angle by using the Angle Sum Theorem.

$$m\angle X + m\angle Y + m\angle Z = 180 \quad \textit{Angle Sum Theorem}$$
$$33 + m\angle Y + 47 = 180 \quad m\angle X = 33,\ m\angle Z = 47$$
$$m\angle Y = 100 \quad \textit{Subtraction Property (=)}$$

When finding y, use z instead of x since z is exact and x is approximate.

$$\frac{\sin Z}{z} = \frac{\sin Y}{y} \quad \textit{Law of Sines}$$
$$\frac{\sin 47°}{14} = \frac{\sin 100°}{y} \quad m\angle Z = 47,\ m\angle Y = 100,\ z = 14$$
$$y \sin 47° = 14 \sin 100° \quad \textit{Cross multiply.}$$
$$y = \frac{14 \sin 100°}{\sin 47°} \quad \textit{Division Property (=)}$$
$$y \approx 18.9 \quad \textit{Use a calculator.}$$

Therefore, $m\angle Y = 100$, $x \approx 10.4$, and $y \approx 18.9$, and the triangle is solved.

In the application at the beginning of the lesson, you know the measures of two angles and the side between the angles.

Motivating the Lesson

Hands-On Activity Ask students to find representations of triangles in the classroom that are not right triangles and to measure the sides and angles. Then have them think of examples of such triangles outside the classroom.

2 TEACH

In-Class Example

For Example 1
Solve $\triangle XYZ$ if $m\angle Z = 39$, $m\angle Y = 88$, and $y = 70$.

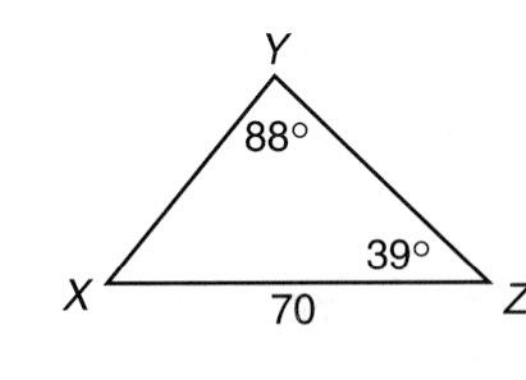

$m\angle X = 53$, $x \approx 56$, $z \approx 44$

Proof Pointer

After reading the plan for the proof of the Law of Sines as far as it has been given, develop with students the plan for the rest of the proof.

Teaching Tip Bear in mind that the ambiguous case in which the Law of Sines yields 0 or 2 solutions has not been presented in this text, and that such situations have been avoided in the exercises.

Cooperative Learning

Brainstorming Have students work in groups to write a paragraph explaining how the Law of Sines can be used to solve a triangle. They should mention what parts of the triangle must be given for the Law of Sines to apply. For more information on the brainstorming strategy, see *Cooperative Learning in the Mathematics Classroom*, one of the titles in the Glencoe Mathematics Professional Series, page 30.

In-Class Example

For Example 2
A triangular brace has one side 9 meters long. The angle opposite the 9-meter side measures 70° and $m\angle C = 31°$. What is the height of the brace? 4.8 m

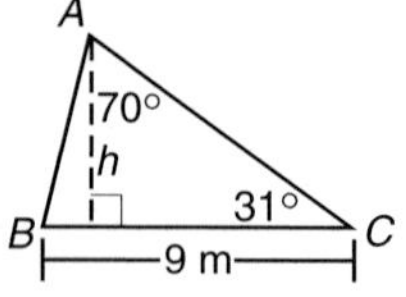

3 PRACTICE/APPLY

Check for Understanding

Exercises 1–9 are designed to help you assess your students' understanding through reading, writing, speaking, and modeling. You should work through Exercises 1–3 with your students and then monitor their work on Exercises 4–9.

Additional Answer

1. Law of Sines can be used with any triangle. However, $\sin = \frac{\text{opposite}}{\text{hypotenuse}}$ can only be used for right triangles.

Study Guide Masters, p. 47

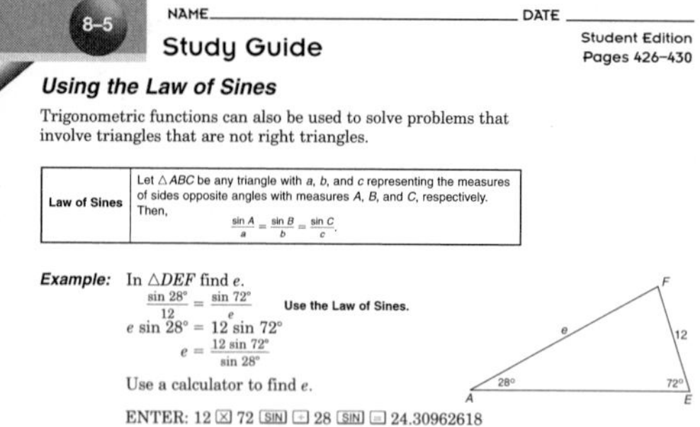

8–5 NAME ______ DATE ______

Study Guide — Student Edition Pages 426–430

Using the Law of Sines

Trigonometric functions can also be used to solve problems that involve triangles that are not right triangles.

Law of Sines	Let $\triangle ABC$ be any triangle with a, b, and c representing the measures of sides opposite angles with measures A, B, and C, respectively. Then, $\frac{\sin A}{a} = \frac{\sin B}{b} = \frac{\sin C}{c}$.

Example: In $\triangle DEF$ find e.

$\frac{\sin 28°}{12} = \frac{\sin 72°}{e}$ **Use the Law of Sines.**

$e \sin 28° = 12 \sin 72°$

$e = \frac{12 \sin 72°}{\sin 28°}$

Use a calculator to find e.

ENTER: 12 × 72 SIN ÷ 28 SIN = 24.30962618

So, $e \approx 24.3$.

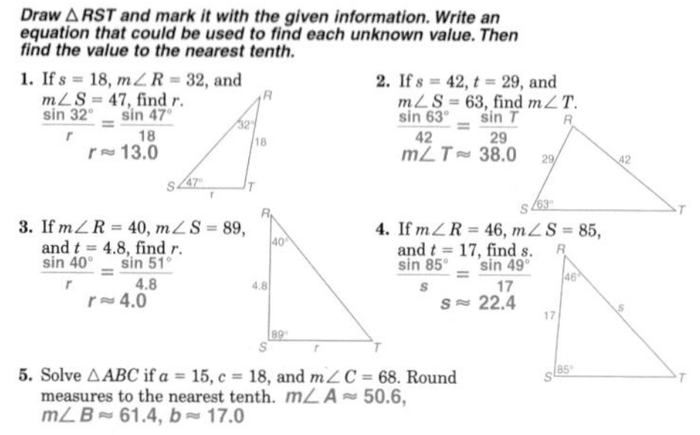

Draw △RST and mark it with the given information. Write an equation that could be used to find each unknown value. Then find the value to the nearest tenth.

1. If $s = 18$, $m\angle R = 32$, and $m\angle S = 47$, find r. $\frac{\sin 32°}{r} = \frac{\sin 47°}{18}$ $r \approx 13.0$
2. If $s = 42$, $t = 29$, and $m\angle S = 63$, find $m\angle T$. $\frac{\sin 63°}{42} = \frac{\sin T}{29}$ $m\angle T \approx 38.0$
3. If $m\angle R = 40$, $m\angle S = 89$, and $t = 4.8$, find r. $\frac{\sin 40°}{r} = \frac{\sin 51°}{4.8}$ $r \approx 4.0$
4. If $m\angle R = 46$, $m\angle S = 85$, and $t = 17$, find s. $\frac{\sin 85°}{s} = \frac{\sin 49°}{17}$ $s \approx 22.4$
5. Solve $\triangle ABC$ if $a = 15$, $c = 18$, and $m\angle C = 68$. Round measures to the nearest tenth. $m\angle A \approx 50.6$, $m\angle B \approx 61.4$, $b \approx 17.0$

Example 2 **Refer to the application at the beginning of the lesson. What is the approximate distance across the gorge?**

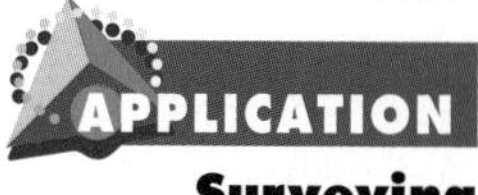

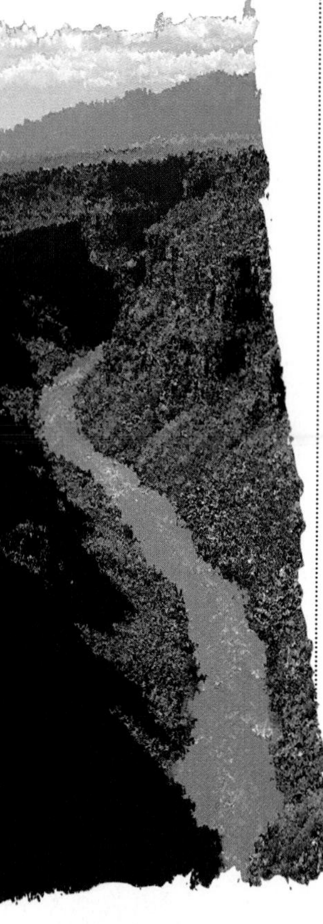

We do not have enough information to solve either right triangle for h using the trigonometric ratios. However, if we can find e using the Law of Sines, we can use sin 80° to find h.

We know the measures of two angles and a side, so this is an example of Case 1. However, we do not know the measure of a side opposite one of the given angles. We must first find $m\angle OGE$.

$m\angle O + m\angle E + m\angle OGE = 180$	*Angle Sum Theorem*
$80 + 85 + m\angle OGE = 180$	*$m\angle O = 80$, $m\angle E = 85$*
$m\angle OGE = 15$	*Subtraction Property (=)*

Choose the proportion that involves e and three known measures.

$\frac{\sin G}{g} = \frac{\sin E}{e}$	*Law of Sines*
$\frac{\sin 15°}{315} = \frac{\sin 85°}{e}$	*$m\angle G = 15$, $m\angle E = 85$, $g = 815$*
$e \sin 15° = 315 \sin 85°$	*Cross multiply.*
$e = \frac{315 \sin 85°}{\sin 15°}$	*Division Property (=)*
$e \approx 1212.4$	*Use a calculator.*

To find h, use right $\triangle GRO$ and sin O.

$\sin O = \frac{h}{e}$	*$\sin = \frac{\text{opposite}}{\text{hypotenuse}}$*
$\sin 80° \approx \frac{h}{1212.4}$	*$m\angle O = 80$, $e \approx 1212.4$*
$1212.4 \sin 80° \approx h$	*Multiplication Property (=)*
$1194.0 \approx h$	*Use a calculator.*

The distance across the gorge is about 1194 feet.

CHECK FOR UNDERSTANDING

Communicating Mathematics

Study the lesson. Then complete the following.

1. **Compare** when you use the Law of Sines rather than the sine ratio as it is defined. See margin.
2. **Explain** in your own words the meaning of solving a triangle.
3. **Describe** two cases when the Law of Sines can be used to solve a triangle.

2–3. See Solutions Manual.

Guided Practice

Draw △*REG* and mark it with the given information. Write an equation that could be used to find each unknown value. Then find the value to the nearest tenth. 4–5. See margin.

4. If $e = 8.5$, $m\angle G = 31$, and $m\angle E = 68$, find g.
5. If $e = 7$, $g = 11$, and $m\angle G = 37$, find $m\angle E$.

Reteaching

Using Diagrams Refer to the figure to complete each statement. Have students redraw the figure for each statement including only the necessary information.

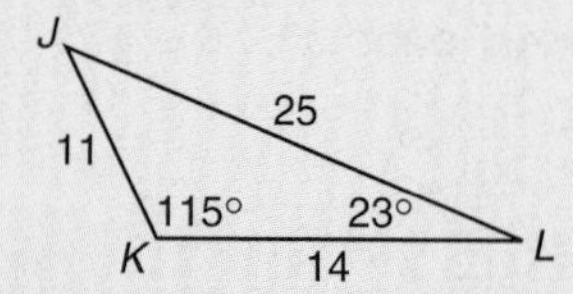

1. $\frac{\sin K}{JL} = \frac{\sin L}{?}$ JK
2. $\frac{\sin J}{KL} = \frac{?}{JK}$ sin L
3. $\frac{\sin 115°}{?} = \frac{\sin 23°}{?}$ 25; 11
4. $\frac{\sin ?}{14} = \frac{\sin ?}{25}$ 42°; 115°
5. $\frac{\sin 23°}{?} = \frac{\sin ?}{14}$ 11; 42°

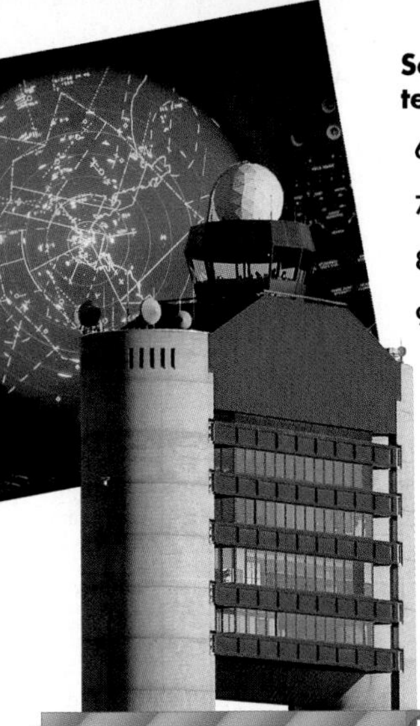

Solve each $\triangle BWY$ described below. Round measures to the nearest tenth.

6. $m\angle Y = 66, m\angle W = 59, b = 72$ $m\angle B = 55, w \approx 75.3, y \approx 80.3$
7. $b = 24, y = 18, m\angle B = 102$ $m\angle Y \approx 47.2, m\angle W \approx 30.8, w \approx 12.6$
8. $m\angle B = 33, m\angle Y = 58, w = 22$ $m\angle W = 89, b \approx 12.0, y \approx 18.7$
9. **Aviation** Two radar stations that are 20 miles apart located an unidentified plane that vanished from their screens at the same time. The first station indicated that the position of the plane made an angle of 43° with the line between the stations. The second station indicated that it made an angle of 48° with the same line.
   a. Draw and label a diagram of this situation. See margin.
   b. How far is each station from the point where they lost contact with the plane? about 14.9 mi, about 13.6 mi

EXERCISES

Practice

Draw $\triangle EPR$ and mark it with the given information. Write an equation that could be used to find each unknown value. Then find the value to the nearest tenth. 10–15. See margin.

10. If $m\angle P = 45, m\angle E = 63$, and $p = 22$, find e.
11. If $r = 9, e = 13$, and $m\angle E = 47$, find $m\angle R$.
12. If $e = 3.2, m\angle P = 52$, and $m\angle E = 70$, find p.
13. If $e = 48, r = 10$, and $m\angle E = 96$, find $m\angle R$.
14. If $m\angle P = 62, m\angle E = 26$, and $r = 2.6$, find p.
15. If $m\angle R = 59, p = 8.3$, and $r = 14.8$, find $m\angle P$.

Solve each $\triangle DFR$ described below. Round measures to the nearest tenth.

16. $m\angle R = 71, r = 7.4, m\angle F = 41$ $m\angle D = 68, d \approx 7.3, f \approx 5.1$
17. $f = 9.1, r = 20.1, m\angle R = 107$ $m\angle F \approx 25.7, m\angle D \approx 47.3, d \approx 15.4$
18. $m\angle F = 25, m\angle D = 52, r = 15.6$ $m\angle R = 103, d \approx 12.6, f \approx 6.8$
19. $m\angle R = 34, f = 9.1, r = 27$ $m\angle F \approx 10.9, m\angle D \approx 135.1, d \approx 34.1$
20. $m\angle D = 38, m\angle R = 115, d = 8.5$ $m\angle F = 27, f \approx 6.3, r \approx 12.5$
21. $m\angle D = 43, m\angle R = 77, d = 0.8$ $m\angle F = 60, f \approx 1.0, r \approx 1.1$
22. $d = 30, r = 9.5, m\angle D = 107$ $m\angle R \approx 17.6, m\angle F \approx 55.4, f \approx 25.8$
23. $f = 16, d = 21, m\angle D = 88$ $m\angle F \approx 49.6, m\angle R \approx 42.4, r \approx 14.2$
24. $f = 23, m\angle F = 45, m\angle D = 51$ $m\angle R = 84, d \approx 25.3, r \approx 32.3$

25. An isosceles triangle has a base of 22 centimeters and a vertex angle of 36°. Find the perimeter. about 93.2 cm

26. about 22.6 ft and 31.9 ft

26. The longest side of a triangle is 34 feet. The measures of two angles of the triangle are 40 and 65. Find the lengths of the other two sides.
27. Given parallelogram $PERA$, find the length of $\overline{PE}$ and $\overline{PR}$. about 11.3; about 9.8

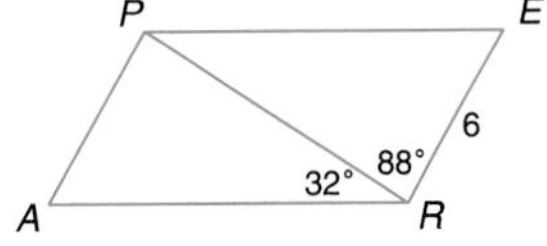

Assignment Guide

Core (with proof): 11–31 odd, 33–40
Core (informal): 11–31 odd, 33–40
Enriched: 10–26 even, 28–40

For **Extra Practice**, see p. 779.

The red A, B, and C flags, printed only in the Teacher's Wraparound Edition, indicate the level of difficulty of the exercises.

Additional Answers

2. Sample answer: Solving a triangle means finding all the missing measures.
3. The Law of Sines can be used if
   a. the measures of 2 angles and a side are known or
   b. the measures of 2 sides and an angle opposite one of the sides are known.

4–5. See Solutions Manual for drawings.

4. $\frac{\sin 68^\circ}{8.5} = \frac{\sin 31^\circ}{g}$; about 4.7
5. $\frac{\sin 37^\circ}{11} = \frac{\sin E}{7}$; about 22.5

9a. 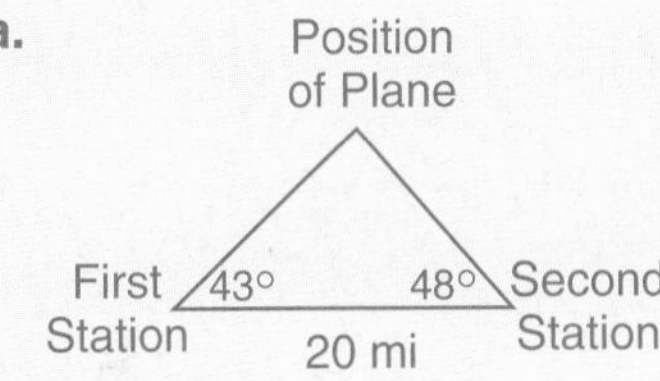

Practice Masters, p. 47

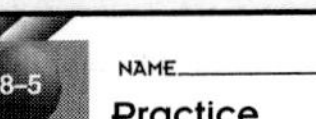

NAME ______ DATE ______

Practice — Student Edition Pages 426–430

Using the Law of Sines

Solve each $\triangle ABC$. Round measures to the nearest tenth.

1. $a = 12, m\angle B = 70, m\angle C = 15$ $m\angle A = 95, b \approx 11.3, c \approx 3.1$
2. $a = 12, b = 5, m\angle A = 110$ $m\angle B \approx 23.1, m\angle C \approx 46.9, c \approx 9.3$
3. $a = 8, m\angle A = 60, m\angle C = 40$ $m\angle B = 80, b \approx 9.1, c \approx 5.9$
4. $a = 5, c = 4, m\angle A = 65$ $m\angle C \approx 46.5, m\angle B \approx 68.5, b \approx 5.2$
5. $b = 6, m\angle A = 44, m\angle B = 68$ $m\angle C = 68, a \approx 4.5, c = 6$
6. $a = 7, m\angle A = 37, m\angle B = 76$ $m\angle C = 67, b \approx 11.3, c \approx 10.7$
7. $a = 9, b = 9, m\angle C = 20°$ $m\angle A = 80, m\angle B = 80, c \approx 3.1$
8. A ship is sighted from two radar stations 43 km apart. The angle between the line segment joining the two stations and the radar beam of the first station is 37°. The angle between the line segment joining the two stations and the beam from the second station is 113°. How far is the ship from the second station? 51.8 km

Additional Answers

10–15. See Solutions Manual for drawings.

10. $\frac{\sin 45^\circ}{22} = \frac{\sin 63^\circ}{e}$; about 27.7
11. $\frac{\sin 47^\circ}{13} = \frac{\sin R}{9}$; about 30.4
12. $\frac{\sin 70^\circ}{3.2} = \frac{\sin 52^\circ}{p}$; about 2.7
13. $\frac{\sin 96^\circ}{48} = \frac{\sin R}{10}$; about 12.0
14. $\frac{\sin 92^\circ}{2.6} = \frac{\sin 62^\circ}{p}$; about 2.3
15. $\frac{\sin 59^\circ}{14.8} = \frac{\sin P}{8.3}$; about 28.7

4 ASSESS

Closing Activity

Modeling Draw a triangle on the chalkboard or overhead and give measures for the two angles and one side of the triangle. Have volunteers explain how to solve the triangle. Repeat for a triangle for which you give the measures of two sides and one angle opposite one of these sides.

Chapter 8 Quiz C (Lessons 8-4 and 8-5) is available in the *Assessment and Evaluation Masters*, p. 213.

Additional Answer

29. Yes; by definition, $\sin A = \frac{a}{c}$ and $\sin B = \frac{b}{c}$. According to the Law of Sines, $\frac{\sin A}{a} = \frac{\sin B}{b}$.

$$\frac{\frac{a}{c}}{a} \stackrel{?}{=} \frac{\frac{b}{c}}{b}$$

$$\frac{1}{c} = \frac{1}{c}$$

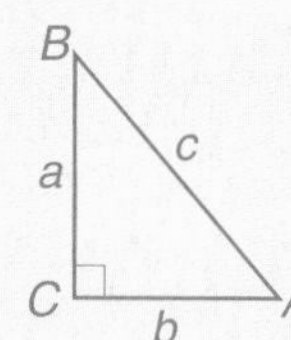

Enrichment Masters, p. 47

NAME________________________ DATE ____________

Enrichment

Student Edition Pages 426–430

Identities

An **identity** is an equation that is true for all values of the variable for which both sides are defined. One way to verify an identity is to use a right triangle and the definitions for trigonometric functions.

Example 1: Verify that $(\sin A)^2 + (\cos A)^2 = 1$ is an identity.

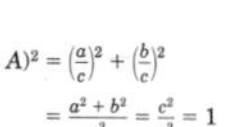

$$(\sin A)^2 + (\cos A)^2 = \left(\frac{a}{c}\right)^2 + \left(\frac{b}{c}\right)^2 = \frac{a^2+b^2}{c^2} = \frac{c^2}{c^2} = 1$$

To check whether an equation *may* be an identity, you can test several values. However, since you cannot test all values, you cannot be *certain* that the equation is an identity.

Example 2: Test $\sin 2x = 2 \sin x \cos x$ to see if it could be an identity.

Try $x = 20$. Use a calculator to evaluate each expression.

$\sin 2x = \sin 40 \approx 0.643$

$2 \sin x \cos x = 2 (\sin 20)(\cos 20) \approx 2(0.342)(0.940) \approx 0.643$

Since the left and right sides seem equal, the equation may be an identity.

Use triangle ABC shown above. Verify that each equation is an identity.

1. $\frac{\cos A}{\sin A} = \frac{1}{\tan A}$

$\frac{\cos A}{\sin A} = \frac{b}{c} \div \frac{a}{c} = \frac{b}{a} = \frac{1}{\tan A}$

2. $\frac{\tan B}{\sin B} = \frac{1}{\cos B}$

$\frac{\tan B}{\sin B} = \frac{b}{a} \div \frac{b}{c} = \frac{c}{a} = \frac{1}{\cos B}$

3. $\tan B \cos B = \sin B$

$\tan B \cos B = \frac{b}{a} \cdot \frac{a}{c} = \frac{b}{c} = \sin B$

4. $1 - (\cos B)^2 = (\sin B)^2$

$1(\cos B)^2 = 1 - \left(\frac{a}{c}\right)^2 = \frac{c^2}{c^2} - \frac{a^2}{c^2} = \frac{c^2 - a^2}{c^2} = \frac{b^2}{c^2} = (\sin B)^2$

Try several values for x to test whether each equation could be an identity.

5. $\cos 2x = (\cos x)^2 - (\sin x)^2$
Yes; see students' work.

6. $\cos (90 - x) = \sin x$
Yes; see students' work.

Critical Thinking

28. Refer to the figure at the right. Which segment looks longer, $\overline{AB}$ or $\overline{BC}$? Check your answer by using the Law of Sines to find AB and BC. **$\overline{AB}$; $AB \approx 7.2$, $BC \approx 7.5$**

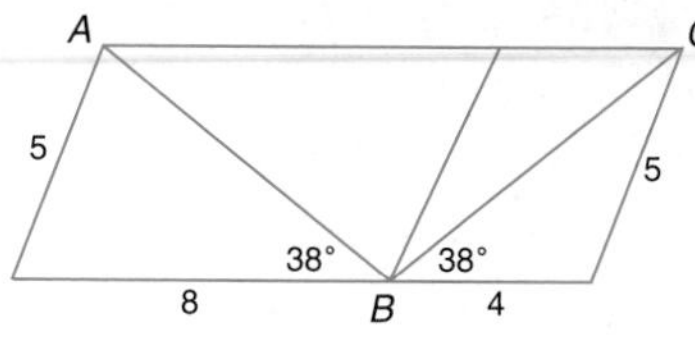

29. Does the Law of Sines hold true for the acute angles of right triangles? Justify your answer. **See margin.**

Applications and Problem Solving

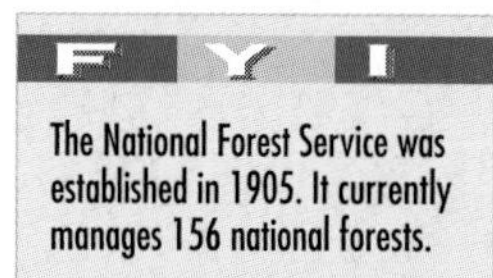

FYI
The National Forest Service was established in 1905. It currently manages 156 national forests.

30. Gardening Yana is planning a triangular garden. He wants to put a fence around it. The length of one side of the garden is 30 feet. If the angles at each end of this side are 44° and 58°, find the length of the fence needed to enclose the garden. **about 77.3 ft.**

31. Fire Fighting Two ranger stations 5 miles apart spot a forest fire. The Kips station determines that the angle formed by the line of sight to the fire and the line connecting the two stations is 37°. The Rips station determines that the angle formed by the line of sight to the fire and the line connecting the two stations is 52°. How far are the two stations from the fire?

31. about 3.0 mi and 3.9 mi

32. Aviation Jalisa Thompson is flying an airplane due east. To avoid a severe thunderstorm, she finds it necessary to change her course. She turns the plane 23° toward the north and flies 55 miles. Then she makes another turn of 120° and heads back to her original course.

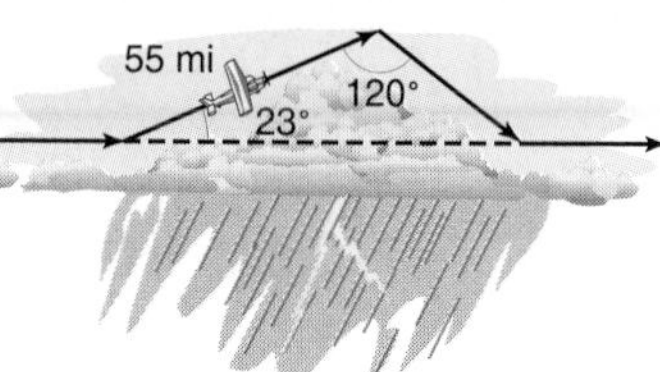

a. How far must Ms. Thompson fly after her second turn to return to her original course? **about 35.7 mi**

b. What angle must she turn to resume her course? **143°**

c. How many miles did she add to her flight by taking the detour? **about 11.6 mi**

Mixed Review

33. Surveying A surveyor is 100 meters from a building and finds that the angle of elevation to the top of the building is 23°. If the surveyor's eye level is 1.55 meters above the ground, find the height of the building. (Lesson 8–4)

33. about 44.0 ft

34. The length of a diagonal of a square is $7\sqrt{2}$. Find the perimeter of the square. (Lesson 8–2) **28 units**

35. The measures of the sides of a triangle are 5, 9, and 12. If the shortest side of a similar triangle measures 15, what are the measures of its other two sides? (Lesson 7–3) **27, 36**

36. The ratio of the measures of the angles of a triangle is 1:5:6. What is the measure of each angle in the triangle? (Lesson 7–1) **15, 75, 90**

37. Use a compass and straightedge to construct a square with sides 6 centimeters long. (Lesson 6–4)

37. See students' work.

38. The opposite angles of a parallelogram have measures of $9x + 12$ and $15x$. Find the measures of the angles of the parallelogram. (Lesson 6–1) **30, 150, 30, 150**

Algebra

39. Find $(5x^2 + 3xy - 2y) + (4x^2 - xy + 2y)$. **$9x^2 + 2xy$**

40. Solve $y(y - 12) = 0$. **{0, 12}**

Extension

Reasoning Refer to the figure. If $x = \frac{10 \sin 120°}{\sin 41°}$, find $m\angle Q$. **41**

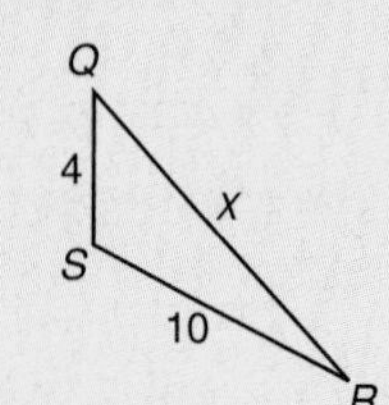

FYI

The U.S. National Park Service was established in 1916 to manage the growing number of national parks.

8-6 Using the Law of Cosines

What YOU'LL LEARN

- To use the Law of Cosines to solve triangles, and
- to choose the appropriate strategy for solving a problem.

Why IT'S IMPORTANT

You can use the Law of Cosines to find missing measures of triangles involved in technology, astronomy, and sports.

APPLICATION

Technology

In 1996, Allen Johnson ran the 110-meter hurdles in 12.92 seconds making him the fourth athlete to run this event in less than thirteen seconds. To help him improve his performance, Johnson used a computer model.

This problem will be solved in Example 2.

What was the measure of the angle formed by his takeoff point to the top of the hurdle and the top of the hurdle to his landing point?

The **Law of Cosines** allows us to solve a triangle when the Law of Sines cannot be used.

Law of Cosines

Let $\triangle ABC$ be any triangle with a, b, and c representing the measures of sides opposite angles with measures A, B, and C, respectively. Then, the following equations hold true.

$$a^2 = b^2 + c^2 - 2bc \cos A \qquad b^2 = a^2 + c^2 - 2ac \cos B$$
$$c^2 = a^2 + b^2 - 2ab \cos C$$

The Law of Cosines can be used to solve a triangle in the following cases.

Case 1. You know the measures of two sides and the included angle.

Case 2. You know the measures of three sides.

Example  **For $\triangle ABC$, find a if $m\angle A = 52$, $c = 7$, and $b = 4$.**

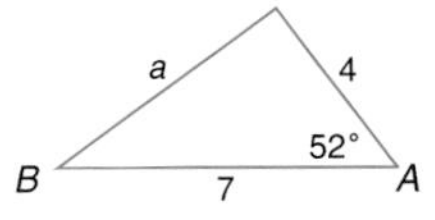

Since the measures of two sides and the included angle are known, this is an example of Case 1 for the Law of Cosines.

$a^2 = b^2 + c^2 - 2bc \cos A$ *Law of Cosines*

$a^2 = 4^2 + 7^2 - 2(4)(7)(\cos 52°)$ *$c = 7$, $b = 4$, $m\angle A = 52$*

$a = \sqrt{4^2 + 7^2 - 2(7)(4)(\cos 52°)}$ *Take the square root of each side.*

$a \approx 5.5$ *Use a calculator.*

8-6 LESSON NOTES

NCTM Standards: 1–5, 7, 9

Instructional Resources

- Study Guide Master 8-6
- Practice Master 8-6
- Enrichment Master 8-6
- Assessment and Evaluation Masters, p. 213
- Real-World Applications, 17

Transparency 8-6A contains the 5-Minute Check for this lesson; **Transparency 8-6B** contains a teaching aid for this lesson.

Recommended Pacing	
Standard Pacing	Day 10 of 12
Honors Pacing	Day 9 of 11
Block Scheduling*	Day 6 of 7

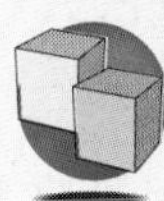

*For more information on pacing and possible lesson plans, refer to the *Block Scheduling Booklet*.

1 FOCUS

5-Minute Check
(over Lesson 8-5)

In $\triangle ABC$, $m\angle A = 52$, the side opposite $\angle A$ measures 15 inches, and the side opposite $\angle B$ measures 12 inches. Find each measure.

1. $m\angle B$ 39
2. $m\angle C$ 89
3. c 19

In $\triangle DEF$, $m\angle D = 64$, $m\angle E = 29$, and the side opposite $\angle F$ measures 20 centimeters. Find each measure.

4. $m\angle F$ 87
5. d 18
6. e 10

Motivating the Lesson

Questioning Ask students to recall what information about a triangle they need in order to solve it using the Law of Sines. Ask them to think of situations in which the Law of Sines could not be used.

2 TEACH

Teaching Tip Ask students to look for patterns in the Law of Cosines that will help them remember the equations. For example, the statement that begins "$a^2 =$" uses cos A.

In-Class Examples

For Example 1
For $\triangle ABC$ find b if $m\angle B = 48$, $a = 7$, and $c = 4$. $b \approx 5.25$

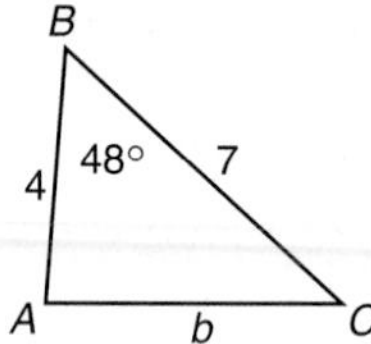

For Example 2
A triangular field is fenced on two sides. The third side of the field is formed by a river. If the fences measure 150 meters and 98 meters and the side along the river is 172 meters, what is the measure of the angle between the fences? $m\angle C \approx 85$

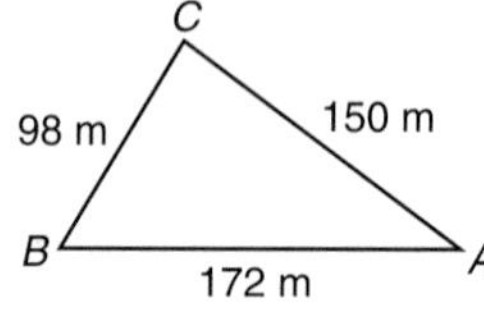

For Example 3
Solve $\triangle ABC$ if $m\angle A = 35$, $b = 16$, and $c = 19$. $a \approx 10.91$, $m\angle B \approx 57.3$, $m\angle C \approx 87.7$

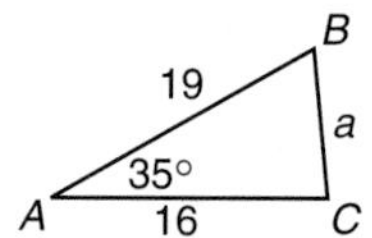

Teaching Tip When reading the examples, have students check the calculations with their own calculators to verify that they know the correct key sequence.

In the application at the beginning of the lesson, you know the measures of the three sides and need to find the measure of an angle of the triangle. This is an example of Case 2 of the Law of Cosines.

Example 2

Technology

Refer to the application at the beginning of the lesson. Find the measure of the angle formed by his takeoff point to the top of the hurdle and the top of the hurdle to his landing point.

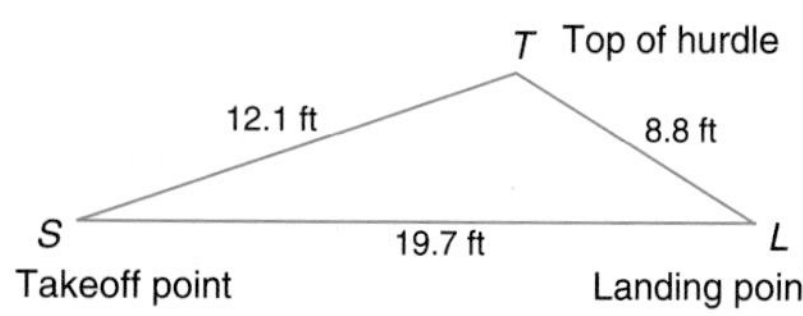

$t^2 = s^2 + \ell^2 - 2s\ell \cos T$	*Law of Cosines*
$19.7^2 = 8.8^2 + 12.1^2 - 2(8.8)(12.1)(\cos T)$	*$t = 19.7$, $s = 8.8$, $\ell = 12.1$*
$19.7^2 - 8.8^2 - 12.1^2 = -2(8.8)(12.1)(\cos T)$	*Subtraction Property*
$\dfrac{19.7^2 - 8.8^2 - 12.1^2}{-2(8.8)(12.1)} = \cos T$	*Division Property*
$-0.7712 \approx \cos T$	*Use a calculator.*
$140.5 \approx T$	

The measure of the angle is about 140.5.

Most problems can be solved in more than one way. Choosing the most efficient way to solve a problem is sometimes not obvious. **Decision making** is an important problem-solving strategy.

When solving right triangles, you can use the sine, cosine, and/or tangent ratios. When solving other triangles, you can use the Law of Sines and/or the Law of Cosines. You must decide how to solve each problem.

Example 3

PROBLEM SOLVING
Decision Making

Solve ΔUVW if $m\angle U = 41$, $v = 13$, and $w = 12$.

Since we do not know if this triangle is a right triangle, we must use the Law of Sines or the Law of Cosines. The Law of Cosines must be used to solve a triangle when the measures of two sides and the included angle are given (Case 1).

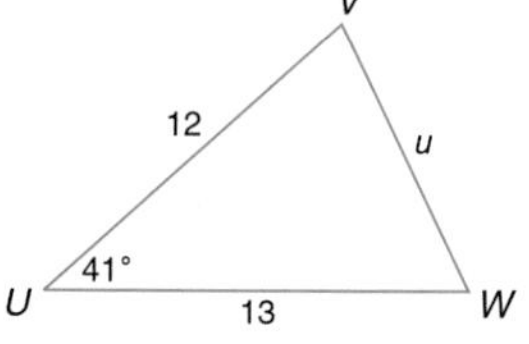

$u^2 = v^2 + w^2 - 2vw \cos U$	*Law of Cosines*
$u^2 = 13^2 + 12^2 - 2(13)(12)(\cos 41°)$	*$v = 13$, $w = 12$, $m\angle U = 41$*
$u = \sqrt{13^2 + 12^2 - 2(13)(12)(\cos 41°)}$	*Take the square root of each side.*
$u \approx 8.8$	*Use a calculator.*

Next, we can find $m\angle V$ or $m\angle W$. Suppose we decide to find $m\angle V$. We can use either the Law of Sines (Case 2) or the Law of Cosines (Case 2) to find this value. In this case, we will use the Law of Sines.

Additional Answers

1. The Law of Cosines can be used if
   a. the measures of 3 sides are known or
   b. the measures of 2 sides and the included angle are known.
2. Both the Law of Sines and the Law of Cosines are used to solve triangles. The Law of Sines can be used if the measures of 2 angles and a side are known or the measures of 2 sides and an angle opposite one of the sides are known. The Law of Cosines can be used if the measures of 3 sides are known or if the measures of 2 sides and the included angle are known.

$\frac{\sin V}{v} = \frac{\sin U}{u}$ *Law of Sines*

$\frac{\sin V}{13} \approx \frac{\sin 41°}{8.8}$ *$v = 13$, $u = 8.8$, $m\angle U = 41$*

$8.8 \sin V \approx 13 \sin 41°$ *Cross multiply.*

$\sin V \approx \frac{13 \sin 41°}{8.8}$ *Division Property (=)*

$\sin V \approx 0.9692$ *Use a calculator.*

$V \approx 75.7$ *Use a calculator.*

The last value that we must find is $m\angle W$. We can use the Law of Sines, the Law of Cosines, or the Angle Sum Theorem. We will use the Angle Sum Theorem.

$m\angle U + m\angle V + m\angle W = 180$ *Angle Sum Theorem*

$41 + 75.7 + m\angle W \approx 180$ *$m\angle U = 41$, $m\angle V \approx 75.5$*

$m\angle W \approx 63.3$ *Subtraction Property (=)*

Therefore, $u \approx 8.8$, $m\angle V \approx 75.7$, and $m\angle W \approx 63.3$, and the triangle is solved.

CHECK FOR UNDERSTANDING

Communicating Mathematics

Study the lesson. Then complete the following. 1, 2, 3a. See margin.

1. **Describe** two cases when the Law of Cosines can be used to solve a triangle.
2. **Compare** the Law of Sines and the Law of Cosines.
3. **Study** Example 3.
   a. Why must you find u first?
   b. Would you prefer to use the Law of Sines or the Law of Cosines to find $m\angle V$? Explain. See students' work.
   c. Would you prefer to use the Law of Sines, the Law of Cosines, or the Angle Sum Theorem to find $m\angle W$? Explain. See students' work.

4. a. Draw a right triangle and label its sides and angles. Write an equation for the sine, cosine, and tangent ratios for each acute angle. See margin.
   b. Draw an acute triangle and label its sides and angles. State the Law of Sines for the triangle. Write a summary of the cases that use the Law of Sines to solve the triangle. See margin.
   c. Draw another acute triangle and label the sides and angles differently than the previous triangle. State the Law of Cosines for the triangle. Write a summary of the cases that use the Law of Cosines to solve the triangle. See margin.

Guided Practice

5. Refer to $\triangle DEF$.
   a. Determine whether the Law of Sines or the Law of Cosines should be used first to solve the triangle. Law of Sines
   b. Solve the triangle. Round the measures to the nearest tenth. $m\angle E \approx 60.1$, $m\angle F \approx 47.9$, $f \approx 3.5$

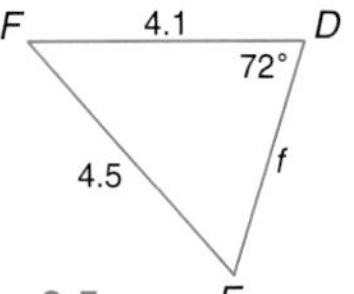

Reteaching

Using Cooperative Groups Have students work in small groups to create a problem such as Example 1. Each student should solve the triangle independently. The group members can compare, correct, and discuss their work together.

3 PRACTICE/APPLY

Check for Understanding

Exercises 1–12 are designed to help you assess your students' understanding through reading, writing, speaking, and modeling. You should work through Exercises 1–4 with your students and then monitor their work on Exercises 5–12.

Additional Answers

3a. With the information given, only the Law of Cosines can be used to solve for u.

4a.

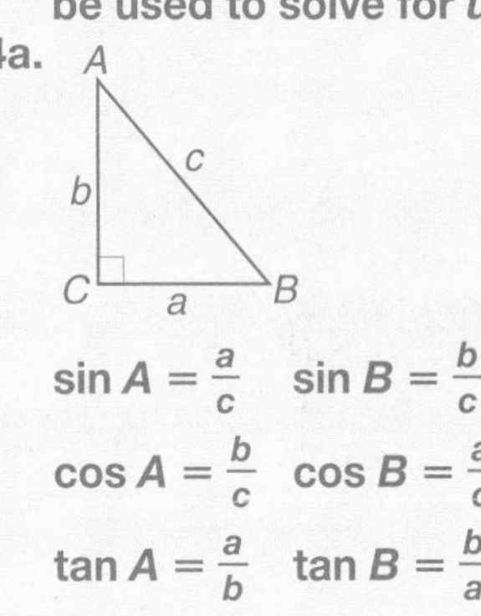

$\sin A = \frac{a}{c}$ $\sin B = \frac{b}{c}$

$\cos A = \frac{b}{c}$ $\cos B = \frac{a}{c}$

$\tan A = \frac{a}{b}$ $\tan B = \frac{b}{a}$

4b.

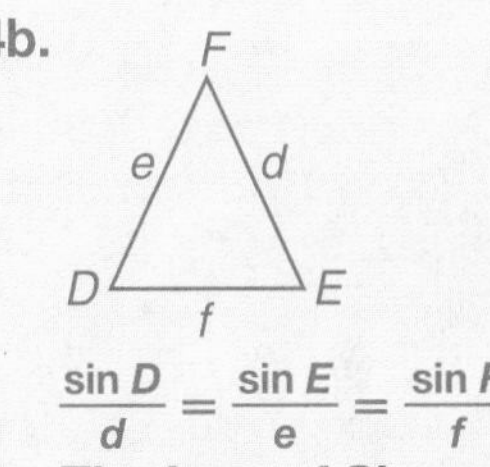

$\frac{\sin D}{d} = \frac{\sin E}{e} = \frac{\sin F}{f}$

The Law of Sines can be used if the measures of 2 angles and a side are known or the measures of 2 sides and an angle opposite one of the sides are known.

4c.

$g^2 = h^2 + i^2 - 2hi \cos G$

$h^2 = g^2 + i^2 - 2gi \cos H$

$i^2 = g^2 + h^2 - 2gh \cos I$

The Law of Cosines can be used if the measures of the 3 sides are known or the measures of 2 sides and the included angle are known.

Assignment Guide

Core (with proof): 13–35 odd, 36, 37, 39, 41–50
Core (informal): 13–35 odd, 36, 37, 39, 41–50
Enriched: 14–34 even, 36–50

For **Extra Practice**, see p. 779.

The red A, B, and C flags, printed only in the Teacher's Wraparound Edition, indicate the level of difficulty of the exercises.

Additional Answers

6a.

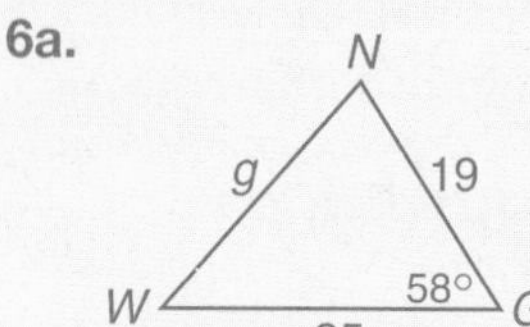

6b. Law of Cosines; two sides and the included angle require the Law of Cosines.

6c. $g \approx 22.0$, $m\angle N \approx 74.5$, $m\angle W \approx 47.5$

16–19. See Solutions Manual for drawings.

16. Law of Sines; $m\angle G = 61$, $g \approx 5.4$, $r \approx 4.1$

17. Law of Cosines; $d \approx 4.9$, $m\angle G \approx 63.8$, $m\angle R \approx 84.2$

18. Law of Cosines; $m\angle D \approx 98.2$, $m\angle G \approx 46.0$, $m\angle R \approx 35.8$

19. Law of Sines; $m\angle G = 99$, $g \approx 31.3$, $r \approx 25.3$

Study Guide Masters, p. 48

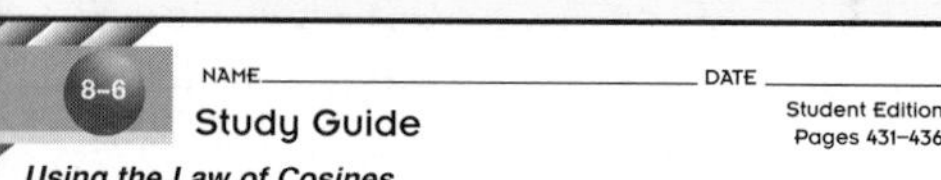

8-6 NAME ______ DATE ______

Study Guide — Student Edition Pages 431–436

Using the Law of Cosines

The Law of Cosines often allows you to solve a triangle when the Law of Sines cannot be used.

Law of Cosines	Let $\triangle ABC$ be any triangle with a, b, and c representing the measures of sides opposite angles with measures A, B, and C, respectively. Then, the following equations hold true. $a^2 = b^2 + c^2 - 2bc \cos A$; $b^2 = a^2 + c^2 - 2ac \cos B$; $c^2 = a^2 + b^2 - 2ab \cos C$

Example: For $\triangle ABC$, find a if $m\angle A = 28$, $c = 30$, and $b = 17$.

Since the measures of two sides and the included angle are known, this is an example of Case 1 for the Law of Cosines.

$a^2 = b^2 + c^2 - 2bc \cos A$
$a^2 = 17^2 + 30^2 - 2(17)(30) \cos 28$
$a^2 = 289 + 900 - 1020 \cos 28$
$a^2 \approx 288.393$
$a \approx 16.98$

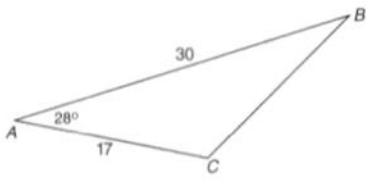

Use a calculator to solve each triangle ABC described below. Round measures to the nearest tenth.

1. $m\angle C = 60$, $a = 12$, $b = 15$ — $c \approx 13.7$, $m\angle A \approx 49.2$, $m\angle B \approx 70.8$
2. $a = 34$, $c = 27$, $m\angle B = 60$ — $b \approx 31.1$, $m\angle A \approx 71.2$, $m\angle C \approx 48.8$
3. $m\angle C = 65$, $a = 8.4$, $b = 9.6$ — $c \approx 9.7$, $m\angle A \approx 51.6$, $m\angle B \approx 63.4$
4. $a = 5$, $b = 9$, $c = 10$ — $m\angle C \approx 86.2$, $m\angle A \approx 29.9$, $m\angle B \approx 63.9$

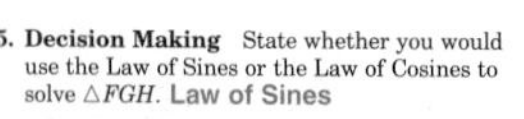

5. **Decision Making** State whether you would use the Law of Sines or the Law of Cosines to solve $\triangle FGH$. Law of Sines

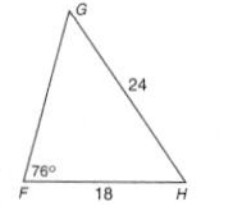

10. $b \approx 17.9$, $m\angle A \approx 54.8$, $m\angle C \approx 78.2$

11. about 35.7

6. In $\triangle WGN$, $w = 19$, $n = 25$, and $m\angle G = 58$.
   a. Sketch $\triangle WGN$. a–c. See margin.
   b. Determine whether the Law of Sines or the Law of Cosines should be used first to solve the triangle. Explain your choice.
   c. Solve the triangle. Round measures to the nearest tenth.

7. $m\angle C = 81$, $a \approx 9.1$, $b \approx 12.1$
8. $m\angle A \approx 20.2$, $m\angle B \approx 43.7$, $m\angle C \approx 116.1$

Solve each $\triangle ABC$ described below. Round measures to the nearest tenth. 9. $m\angle A \approx 15.8$, $m\angle C \approx 145.2$, $c \approx 106.9$

7. $m\angle A = 40$, $m\angle B = 59$, $c = 14$
8. $a = 5$, $b = 10$, $c = 13$
9. $a = 51$, $b = 61$, $m\angle B = 19$
10. $a = 20$, $c = 24$, $m\angle B = 47$
11. The measures of the sides of a triangle are 6.8, 8.4, and 4.9. Find the measure of the smallest angle to the nearest tenth.
12. **Golf** In golf, a *slice* is a shot that curves to the right of its intended path (for a right-handed player) and a *hook* curves to the left. Yolanda's shot from the third tee is a 180-yard slice 25° from the path straight to the cup. If the tee is 240 yards from the cup, how far does Yolanda's ball lie from the cup? about 108.1 yd

EXERCISES

Practice

Determine whether the Law of Sines or the Law of Cosines should be used first to solve each triangle. Then solve each triangle. Round measures to the nearest tenth.

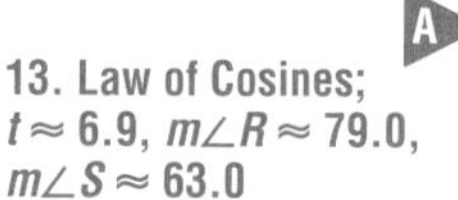

13. Law of Cosines; $t \approx 6.9$, $m\angle R \approx 79.0$, $m\angle S \approx 63.0$

14. Law of Sines; $m\angle C \approx 30.9$, $m\angle B \approx 109.1$, $b \approx 14.7$

15. Law of Cosines; $m\angle X \approx 78.5$, $m\angle Y \approx 44.4$, $m\angle Z \approx 57.1$

16–19. See margin.

13. (triangle TSR: $TS = 11$, $TR = 10$, $m\angle T = 38°$, side t)

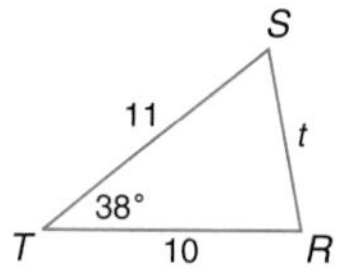

14. (triangle ABC: $AB = 8$, $BC = 10$, $m\angle A = 40°$, side b)

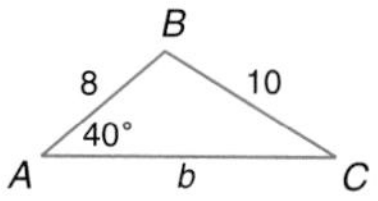

15.

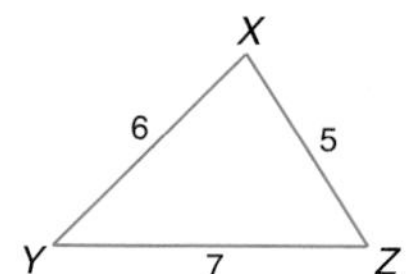

Sketch each $\triangle RGD$. Determine whether the Law of Sines or the Law of Cosines should be used first to solve each triangle. Then solve each triangle. Round measures to the nearest tenth.

16. $m\angle R = 42$, $m\angle D = 77$, $d = 6$
17. $r = 9.1$, $g = 8.3$, $m\angle D = 32$
18. $r = 13$, $g = 16$, $d = 22$
19. $m\angle R = 53$, $m\angle D = 28$, $d = 14.9$

Solve each $\triangle HJK$ described below. Round measures to the nearest tenth. 20–31. See margin.

20. $j = 44$, $h = 54$, $m\angle H = 23$
21. $j = 33$, $h = 56$, $k = 65$
22. $j = 19$, $k = 28$, $m\angle H = 49$
23. $m\angle J = 46$, $m\angle H = 55$, $k = 16$
24. $j = 364$, $h = 669$, $k = 436$
25. $m\angle J = 55$, $h = 6.3$, $k = 6.7$
26. $m\angle J = 27$, $h = 5$, $k = 10$
27. $k = 25$, $j = 27$, $h = 22$
28. $h = 14$, $k = 21$, $m\angle J = 60$
29. $h = 14$, $j = 15$, $k = 16$
30. $m\angle H = 51$, $h = 40$, $k = 35$
31. $h = 5$, $j = 6$, $k = 7$

Cooperative Learning

Group Discussion Have students work in small groups. Have them verify the Law of Cosines by measuring two sides and the included angle of a triangle. Measurements should be as accurate as possible. Have them compute the measure of the opposite side using the Law of Cosines. Have them check by measuring the side. They should compare their group's results with other groups. For more information on the group discussion strategy, see *Cooperative Learning in the Mathematics Classroom*, one of the titles in the Glencoe Mathematics Professional Series, page 31.

32. $m\angle A \approx 78.5$, $m\angle B \approx 101.5$, $m\angle C \approx 78.5$, $m\angle D \approx 101.5$

32. In parallelogram $ABCD$, $AB = 5$, $AD = 6$, and the measure of diagonal $\overline{BD}$ is 7. Find the measure of each angle of the parallelogram.
33. In parallelogram $EFGH$, $EH = 8$, $EF = 11$, and $m\angle E = 110$. Find the measure of each diagonal of the parallelogram. $HF \approx 15.7$, $EG \approx 11.2$
34. The sides of a triangle are 50 meters, 70 meters, and 85 meters long. Find the measure of the largest angle. 88.6
35. The lengths of the diagonals of a rhombus are 26 inches and 18 inches. Find the length of each side of the rhombus and the measure of each angle. 15.8, 15.8, 15.8, 15.8; 110.7, 69.3, 110.7, 69.3

Critical Thinking

36b. $(a - x)^2 = a^2 - 2ax + x^2$

36d. See margin.

36. Explain why each step of the derivation of the Law of Cosines is valid.
   a. $c^2 = (a - x)^2 + h^2$ See margin.
   b. $c^2 = a^2 - 2ax + x^2 + h^2$
   c. $c^2 = a^2 - 2ax + b^2$ See margin.
   d. $c^2 = a^2 - 2a(b \cos C) + b^2$
   e. $c^2 = a^2 + b^2 - 2ab \cos C$ Commutative Property (=)

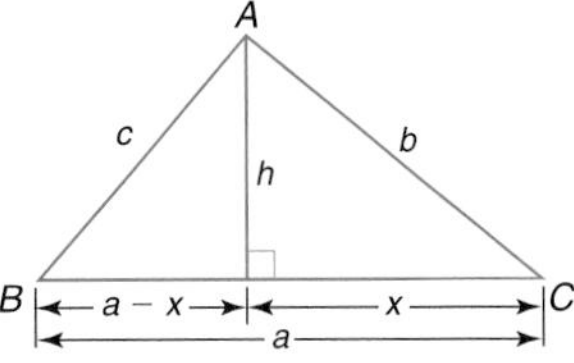

Applications and Problem Solving

37. **Astronomy** The angle formed by the line of sight from Earth to Sirius and the line of sight from Earth to Alpha Centauri is 44°. Find the distance between Sirius and Alpha Centauri. about 6.4 light-years

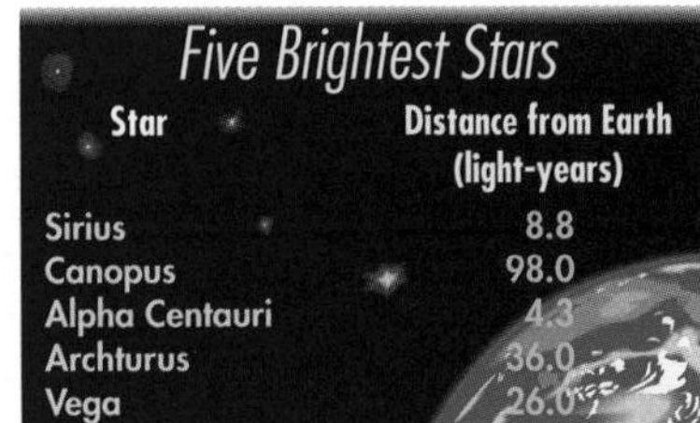

Five Brightest Stars

Star	Distance from Earth (light-years)
Sirius	8.8
Canopus	98.0
Alpha Centauri	4.3
Archturus	36.0
Vega	26.0

FYI

Structural design of the bicycle and the study of aerodynamics have resulted in new Olympic records for cycling events. Olympic bicycles are custom fit to the body type of the rider. The price of such a bike is about $10,000.

38. **Olympic Cycling** The front wheel of an Olympic-type bicycle shown at the left has 14-inch spokes. Each spoke forms a 30° angle with the spoke next to it. Find the distance between the points where the spokes attach to the wheel. about 7.2 in.

39. **Field Hockey** Juanita and Liz are playing field hockey. Juanita is standing 40 feet from one post of the goal and 45 feet from the other post. Liz is standing 30 feet from one post of the goal and 20 feet from the other post. If the goal is 12 feet wide, which player has a greater angle to make a shot on goal? Liz

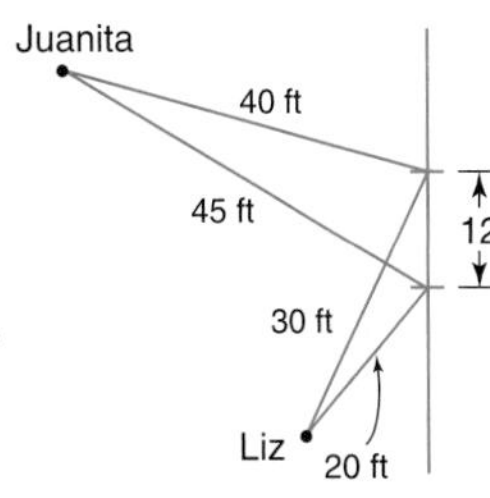

40. 9; 1 half-dollar, 1 quarter, 2 dimes, 1 nickel, 4 pennies

40. **Decision Making** What is the least number of coins you can have and be able to have exact change for any purchase that is less than a dollar?

FYI

China has the largest number of bicycles on the road, 300 million.

Additional Answers

20. $m\angle J \approx 18.6$, $m\angle K \approx 138.4$, $k \approx 91.8$
21. $m\angle H \approx 59.5$, $m\angle J \approx 30.5$, $m\angle K \approx 90.0$
22. $h \approx 21.1$, $m\angle J \approx 42.8$, $m\angle K \approx 88.2$
23. $m\angle K = 79$, $h \approx 13.4$, $j \approx 11.7$
24. $m\angle H \approx 113.2$, $m\angle J \approx 30.0$, $m\angle K \approx 36.8$
25. $j \approx 6.0$, $m\angle H \approx 59.3$, $m\angle K \approx 65.7$
26. $j \approx 6.0$, $m\angle H \approx 22.2$, $m\angle K \approx 130.8$
27. $m\angle H \approx 49.9$, $m\angle J \approx 69.8$, $m\angle K \approx 60.3$
28. $j \approx 18.5$, $m\angle H \approx 40.9$, $m\angle K \approx 79.1$
29. $m\angle H \approx 53.6$, $m\angle J \approx 59.6$, $m\angle K \approx 66.8$
30. $m\angle K \approx 42.8$, $m\angle J \approx 86.2$, $j \approx 51.4$
31. $m\angle H \approx 44.4$, $m\angle J \approx 57.1$, $m\angle K \approx 78.5$

Practice Masters, p. 48

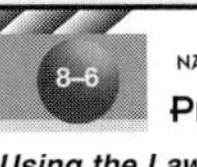

NAME________ DATE________

Practice

Student Edition Pages 431–436

Using the Law of Cosines

Solve each triangle △ABC described below. Round measures to the nearest tenth.

1. $a = 16$, $b = 20$, $m\angle B = 40$
   $m\angle A \approx 30.9$
   $m\angle C \approx 109.1$
   $c \approx 29.4$
2. $a = 10$, $b = 15$, $c = 12$
   $m\angle A \approx 41.6$
   $m\angle B \approx 84.8$
   $m\angle C \approx 53.6$
3. $a = 42$, $c = 60$, $m\angle B = 58$
   $m\angle A \approx 43.3$
   $m\angle C \approx 78.7$
   $b \approx 51.9$
4. $m\angle A = 60$, $m\angle B = 72$, $c = 9$
   $m\angle C \approx 48$
   $a \approx 10.5$
   $b \approx 11.5$
5. $a = 7$, $b = 12$, $c = 15$
   $m\angle A \approx 26.6$
   $m\angle B \approx 50.1$
   $m\angle C \approx 103.3$
6. $m\angle A = 43$, $b = 23$, $c = 26$
   $m\angle B \approx 59.5$
   $m\angle C \approx 77.5$
   $a \approx 18.2$
7. $a = 16$, $m\angle A = 23$, $m\angle B = 87$
   $m\angle C \approx 70$
   $b \approx 40.9$
   $c \approx 38.5$
8. $c = 15.6$, $a = 12.9$, $b = 18.4$
   $m\angle A \approx 43.6$
   $m\angle B \approx 79.6$
   $m\angle C \approx 56.8$
9. **Decision-Making** Complete the addition problem at the right. If a letter is used more than once, it represents the same digit each time.

SOME	1780
+MORE	+8730
SENSE	10510

Tech Prep

Welder Many two-year colleges offer programs in welding technology. Welders use geometry and trigonometry to weld at precise angles. For more information on tech prep, see the *Teacher's Handbook*.

Additional Answers

36a. The Pythagorean Theorem states that the sum of the squares of the measures of the legs ($a - x$ and h) is equal to the square of the measure of the hypotenuse.

36c. By the Pythagorean Theorem, $x^2 + h^2 = b^2$.

36d. Since $\cos C = \frac{x}{b}$, $b \cos C = x$.

4 ASSESS

Closing Activity

Speaking Have students in small groups make up problems about a triangle for which they must use the Law of Cosines. Have groups present their problems to the class and explain what information is required to solve the problems.

Chapter 8 Quiz D (Lesson 8-6) is available in the *Assessment and Evaluation Masters*, p. 213.

Additional Answer

44. Both pairs of opposite sides are congruent.
One pair of opposite sides are both parallel and congruent.
The diagonals bisect each other.
Both pairs of opposite angles are congruent.

Enrichment Masters, p. 48

NAME_________________ DATE_________

Enrichment

Student Edition Pages 431–436

Constructing Values of Square Roots

The diagram at the right shows a right isosceles triangle with two legs of length 1 inch. By the Pythagorean Theorem, the length of the hypotenuse is $\sqrt{2}$ inches. By constructing an adjacent right triangle with legs of $\sqrt{2}$ inches and 1 inch, you can create a segment of length $\sqrt{3}$.

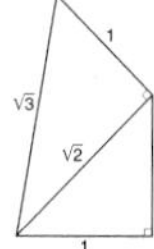

By continuing this process as shown below, you can construct a "wheel" of square roots. This wheel is called the "Wheel of Theodorus" after a Greek philosopher who lived about 400 B.C.

Continue constructing the wheel until you make a segment of length $\sqrt{18}$.

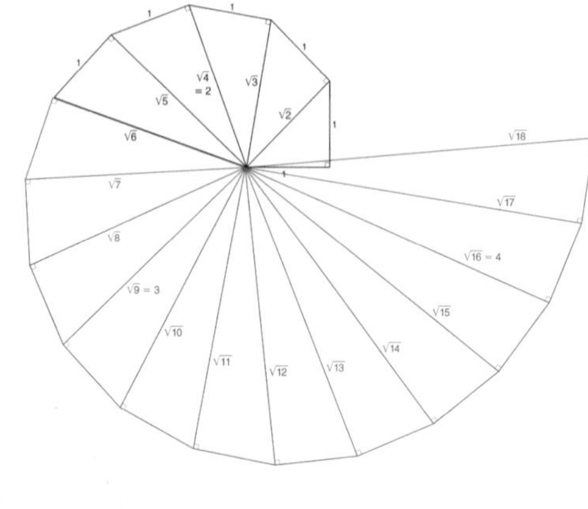

Mixed Review

41. Solve $\triangle RPQ$ if $m\angle R = 50$, $m\angle P = 75$, and $r = 10$. (Lesson 8–5) **$m\angle Q = 55$, $p \approx 12.6$, $q \approx 10.7$**

42. **Broadcasting** A guy wire is attached to a tower at a point 100 feet above the ground. If the tower is perpendicular to the ground and the wire makes an angle of 55° with the ground, find the length of the wire. (Lesson 8–3) **about 122.08 ft**

43. Find the values of x and y. (Lesson 8–1) **$2\sqrt{6} \approx 4.90$; $\sqrt{33} \approx 5.74$**

x, y, 8, 3

44. Explain the four methods used to determine if a quadrilateral is a parallelogram. (Lesson 6–2) **See margin.**

45. Find all possible ordered pairs for the fourth vertex of a parallelogram with vertices $G(-3, 1)$, $H(1, 5)$, and $I(0, -1)$. (Lesson 6–1) **(4, 3), (−2, 7), (−4, −5)**

46. **Quilting** Tim is making a quilt. The pieces of the quilt are made from equilateral triangles. What will he have to check to ensure all the pieces are the right shape and size? (Lesson 4–6) **He must check that all sides are the same length.**

47. Is the statement *A right triangle can be scalene* true or false? Explain. (Lesson 4–1) **true; Sample explanation: A 3-4-5 triangle is a right triangle with different length sides.**

48. Find the slope of the line that passes through points at (3, 0) and (8, −2). (Lesson 3–3) $-\frac{2}{5}$

INTEGRATION Algebra

49. Find $(x + 6)(x + 3)$. $x^2 + 9x + 18$

50. Find $\frac{z}{z+4} \div \frac{z+9}{z+4}$. $\frac{z}{z+9}$

WORKING ON THE

In•ves•ti•ga•tion

Refer to the Investigation on pages 392–393.

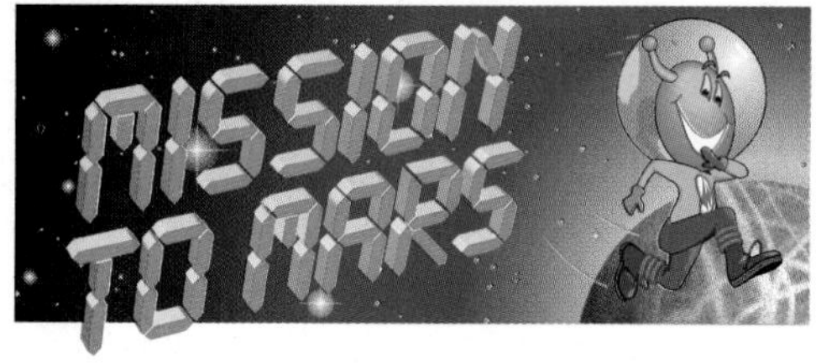

Some of the scientists who will be following you to Mars have expressed concern over the difference in the Martian year. They have asked you to develop a calendar for them to use on Mars.

1. Research the development of the calendar used on Earth. How were the lengths of the months determined? How does the calendar account for the length of time needed for Earth to complete an orbit of the Sun? How were the months named?
2. Refer to the information that you have gathered, along with the information on page 393 of the Investigation. Determine a reasonable number of Martian months and the length of each month.
3. Will you need to have a leap-year system in the Martian calendar? If so, determine how often.
4. Give the months of the Martian calendar names different from those in the Earth calendar. Note your rationale for choosing each name.
5. Complete your calendar and write a clear description of it for the scientists. Include a method for converting the Martian date to the Earth date.

1–5. See students' work.

Add the results of your work to your Investigation Folder.

Extension

Reasoning Have students name a distance or an angle in the vicinity of the school that can be found by the Law of Cosines. Have them describe the problem and how the necessary measurements could be obtained.

In•ves•ti•ga•tion

Working on the Investigation

The Investigation on pages 392–393 is designed to be a long-term project that is completed over several days or weeks. Encourage students to keep their materials in their Investigation Folder as they work on the Investigation.

CHAPTER 8

HIGHLIGHTS

VOCABULARY

After completing this chapter, you should be able to define each term, property, or phrase and give an example or two of each.

Geometry

Pythagorean Theorem (p. 399)

Pythagorean triple (p. 400)

Algebra

extremes (p.397)

geometric mean (p. 397)

means (p.397)

Trigonometry

angle of depression (p. 420)

angle of elevation (p. 420)

cosine (p. 412)

Law of Cosines (p. 431)

Law of Sines (p. 426)

sine (p. 412)

solving the triangle (p. 427)

tangent (p. 412)

trigonometric ratio (p. 412)

trigonometry (p. 412)

Problem Solving

decision making (p. 432)

4. angle of depression 7. Pythagorean triple 14. angle of elevation

UNDERSTANDING AND USING THE VOCABULARY

Choose the term from the list above that best completes each statement or phrase.

1. In a right triangle, the ratio of the measure of the opposite leg divided by the measure of the hypotenuse is called the _?_ of an angle. sine
2. In the proportion $\frac{3}{4} = \frac{9}{12}$, 3 and 12 are called the _?_. extremes
3. The _?_ states that $c^2 = a^2 + b^2 - 2ab \cos C$. Law of Cosines
4. The angle formed by a horizontal line and a line of sight down to an object is called the _?_.
5. The branch of mathematics that involves triangle measurement is called _?_. trigonometry
6. In a right triangle, the ratio of the measure of the opposite leg divided by the measure of the adjacent leg is called the _?_ of an angle. tangent
7. A group of three whole numbers that satisfies the equation $a^2 + b^2 = c^2$ is called a(n) _?_.
8. Six is the _?_ between 4 and 9. geometric mean
9. The proportion $\frac{\sin A}{a} = \frac{\sin B}{b} = \frac{\sin C}{c}$ is the _?_. Law of Sines
10. For a right triangle, the _?_ states that the sum of the squares of the measures of the legs equals the square of the measure of the hypotenuse. Pythagorean Theorem
11. The sine, cosine, and tangent ratios are three examples of _?_. trigonometric ratios
12. In a right triangle, the ratio of the measure of the adjacent leg divided by the measure of the hypotenuse is called the _?_ of an angle. cosine
13. In the proportion $\frac{a}{b} = \frac{c}{d}$, b and c are called the _?_. means
14. The angle formed by a horizontal line and a line of sight up to an object is called the _?_.

Using the CHAPTER HIGHLIGHTS

The Chapter Highlights begins with a listing of the new terms, properties, and phrases that were introduced in this chapter. Have students define each term and provide an example or two of it, if appropriate.

Instructional Resources

Three multiple-choice tests and three free-response tests are provided in the *Assessment and Evaluation Masters*. Forms 1A and 2A are for honors pacing, Forms 1B and 2B are for average pacing, and Forms 1C and 2C are for basic pacing. Chapter 8 Test, Form 1B, is shown at the right. Chapter 8 Test, Form 2B, is shown on the next page.

Postulates, Theorems, and Corollaries

A complete list of postulates, theorems, and corollaries begins on page 806.

Assessment and Evaluation Masters, pp. 199–200

8 NAME ____________ DATE ________

Chapter 8 Test, Form 1B

Write the letter for the correct answer in the blank at the right of each problem.

1. Find the geometric mean between 7 and 12.
   A. $\sqrt{84}$ B. 9.5 C. $\sqrt{12}$ D. $\sqrt{7}$ 1. A

For Questions 2–4, refer to the figure at the right.

2. If $a = 4$ and $x = 2$, find c.
   A. 6 B. 12
   C. 8 D. 16 2. C
3. If $c = 18$ and $y = 10$, find h.
   A. 8 B. 32.4
   C. 5.5 D. $\sqrt{80}$ 3. D
4. If $h = 5$ and $b = 10$, find y.
   A. $\sqrt{75}$ B. 15
   C. 75 D. $\sqrt{125}$ 4. A
5. Determine which set of numbers can be the measures of the sides of a right triangle.
   A. 7, 12, 13 B. 5, 6, 7 C. 8, 15, 17 D. 5, 10, 15 5. C
6. In a right triangle, the measures of the legs are 12 and $x - 5$, and the measure of the hypotenuse is $x + 3$. Find the value of x.
   A. 10 B. 5 C. 11 D. 13 6. A
7. The length of a rectangle is 6 cm, and the width is 3 cm. Find the length of a diagonal.
   A. 13 cm B. $\sqrt{45}$ cm C. $\sqrt{27}$ cm D. 9 cm 7. B
8. The perimeter of a square is 20 cm. Find the length of a diagonal.
   A. $\sqrt{10}$ cm B. $5\sqrt{2}$ cm C. $2\sqrt{5}$ cm D. 4 cm 8. B
9. The shortest side of a 30°-60°-90° triangle measures 8. What is the measure of the longest side?
   A. 16 B. 4 C. $8\sqrt{2}$ D. $8\sqrt{3}$ 9. A
10. Find the length of a diagonal of a rectangular box whose edges are 6 cm, 8 cm, and 10 cm long.
   A. $10\sqrt{3}$ cm B. $5\sqrt{3}$ cm C. $10\sqrt{2}$ cm D. $5\sqrt{2}$ cm 10. C

8 NAME ____________ DATE ________

Chapter 8 Test, Form 1B (continued)

For Questions 11 and 12, refer to the figure at the right.

11. Which trigonometric ratio equals $\frac{\sqrt{3}}{3}$?
   A. $\tan B$ B. $\sin B$
   C. $\cos B$ D. $\tan A$ 11. A
12. Find $m\angle A$ to the nearest degree.
   A. 63 B. 60
   C. 30 D. 27 12. B
13. In $\triangle PQR$, $\angle Q$ is the right angle. If $\tan R = \frac{8}{15}$, find $\cos P$.
   A. $\frac{15}{17}$ B. $\frac{15}{8}$ C. $\frac{8}{15}$ D. $\frac{8}{17}$ 13. D
14. A forest ranger spots a fire from the top of a lookout tower. The tower is 160 m tall, and the angle of depression to the fire is 12°. About how far is the fire from the foot of the tower?
   A. 164 m B. 753 m C. 770 m D. 610 m 14. B
15. In $\triangle DEF$, $d = 8$, $e = 12$, and $m\angle E = 34$. Find $m\angle D$ to the nearest degree.
   A. 68 B. 45 C. 22 D. 40 15. C
16. In $\triangle DEF$, $m\angle E = 75$, $m\angle F = 28$, and $d = 43$. Find f to the nearest whole unit.
   A. 21 B. 89 C. 17 D. 42 16. A
17. In $\triangle DEF$, $d = 12$, $e = 14$, and $m\angle F = 47$. Find f to the nearest tenth of a unit.
   A. 4.5 B. 9.7 C. 10.5 D. 23.9 17. C
18. The measures of the sides of a triangle are 13, 14, and 15. Find the measure of the smallest angle to the nearest degree.
   A. 37 B. 53 C. 59 D. 67 18. B
19. Choose the best strategy for solving the following problem: Jon had 22 points after he played four hands with Jan. In the second hand, Jon lost half of his points. In the third hand, he won 12 times what he had. In the fourth hand, Jon won 9 more points. How many points did Jon have after the first hand?
   A. look for a pattern B. draw a diagram
   C. work backward D. make a chart 19. C
20. Solve the problem in Question 19.
   A. 21 B. 30 C. 12 D. 2 20. D

Bonus

In $\triangle PQR$, $m\angle Q = 90$ and $\sin P = \frac{2}{5}$. Find $\cos P$.

A. $\frac{3}{5}$ B. $\frac{5}{2}$ C. $\frac{5}{\sqrt{21}}$ D. $\frac{\sqrt{21}}{5}$ Bonus D

Using the STUDY GUIDE AND ASSESSMENT

Skills and Concepts Encourage students to refer to the objectives and examples on the left as they complete the review exercises on the right.

Assessment and Evaluation Masters, pp. 205–206

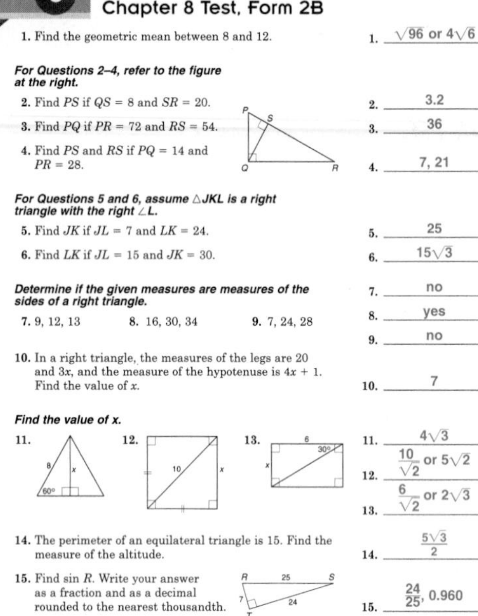

8 NAME ______ DATE ______

Chapter 8 Test, Form 2B

1. Find the geometric mean between 8 and 12. — 1. $\sqrt{96}$ or $4\sqrt{6}$

For Questions 2–4, refer to the figure at the right.

2. Find PS if $QS = 8$ and $SR = 20$. — 2. 3.2
3. Find PQ if $PR = 72$ and $RS = 54$. — 3. 36
4. Find PS and RS if $PQ = 14$ and $PR = 28$. — 4. 7, 21

For Questions 5 and 6, assume $\triangle JKL$ is a right triangle with the right $\angle L$.

5. Find JK if $JL = 7$ and $LK = 24$. — 5. 25
6. Find LK if $JL = 15$ and $JK = 30$. — 6. $15\sqrt{3}$

Determine if the given measures are measures of the sides of a right triangle.

7. 9, 12, 13 — 7. no
8. 16, 30, 34 — 8. yes
9. 7, 24, 28 — 9. no

10. In a right triangle, the measures of the legs are 20 and $3x$, and the measure of the hypotenuse is $4x + 1$. Find the value of x. — 10. 7

Find the value of x.

11. — 11. $4\sqrt{3}$
12. — 12. $\frac{10}{\sqrt{2}}$ or $5\sqrt{2}$
13. — 13. $\frac{6}{\sqrt{2}}$ or $2\sqrt{3}$

14. The perimeter of an equilateral triangle is 15. Find the measure of the altitude. — 14. $\frac{5\sqrt{3}}{2}$
15. Find $\sin R$. Write your answer as a fraction and as a decimal rounded to the nearest thousandth. — 15. $\frac{24}{25}$, 0.960

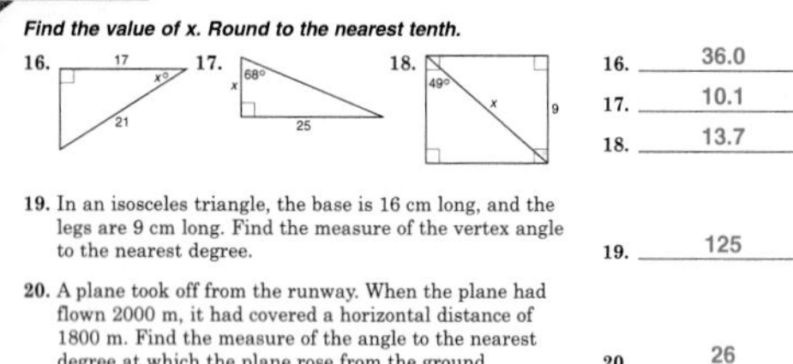

8 NAME ______ DATE ______

Chapter 8 Test, Form 2B (continued)

Find the value of x. Round to the nearest tenth.

16. — 16. 36.0
17. — 17. 10.1
18. — 18. 13.7

19. In an isosceles triangle, the base is 16 cm long, and the legs are 9 cm long. Find the measure of the vertex angle to the nearest degree. — 19. 125
20. A plane took off from the runway. When the plane had flown 2000 m, it had covered a horizontal distance of 1800 m. Find the measure of the angle to the nearest degree at which the plane rose from the ground. — 20. 26
21. A cliff is 80 m above the sea. From the cliff, the angle of depression of a boat is 35°. To the nearest meter, how far is the boat from the base of the cliff? — 21. 114 m

Find the indicated measure for each $\triangle JKL$. Round to the nearest tenth.

22. If $JK = 23$, $m\angle L = 28$, and $m\angle J = 74$, find JL. — 22. 47.9
23. If $JL = 52$, $LK = 34$, and $m\angle K = 85$, find $m\angle J$. — 23. 40.6
24. If $JK = 16$, $JL = 24$, and $m\angle J = 30$, find LK. — 24. 12.9
25. Identify the strategy you would use to solve the following problem and solve. There are 25 light bulbs on a square light board. There are five bulbs in each of five rows. The lights on the diagonals are yellow, and the rest are white. How many white light bulbs are there? — 25. draw a figure; 16

Bonus

If $\angle A$ and $\angle B$ are acute angles of right $\triangle ABC$, what is the value of $(\tan A)(\tan B)$? — Bonus 1

CHAPTER 8 STUDY GUIDE AND ASSESSMENT

SKILLS AND CONCEPTS

OBJECTIVES AND EXAMPLES

Upon completing this chapter, you should be able to:

- find the geometric mean between two numbers (Lesson 8–1)

Find the geometric mean between 16 and 77.

Let x represent the geometric mean.

$\frac{16}{x} = \frac{x}{77}$ *Definition of geometric mean*

$x^2 = 1232$ *Cross multiply.*

$x = \sqrt{1232}$

$x \approx 35.1$

The geometric mean is about 35.1.

REVIEW EXERCISES

Use these exercises to review and prepare for the chapter test.

Find the geometric mean between each pair of numbers.

15. 9 and 27 $9\sqrt{3} \approx 15.59$
16. 13 and 39 $13\sqrt{3} \approx 22.52$
17. 6 and $\frac{2}{3}$ 2
18. $\frac{3}{4}$ and $\frac{8}{3}$ $\sqrt{2} \approx 1.41$

- solve problems involving relationships between parts of a right triangle and the altitude to its hypotenuse (Lesson 8–1)

If $\triangle ABC$ is a right triangle with altitude $\overline{BD}$, then the following relationships hold true.

$\frac{AD}{BD} = \frac{BD}{DC}$

$\frac{AC}{BC} = \frac{BC}{DC}$

$\frac{AC}{AB} = \frac{AB}{AD}$

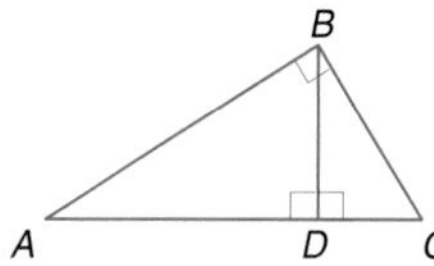

Use right triangle *KLM* and the given information to solve each problem.

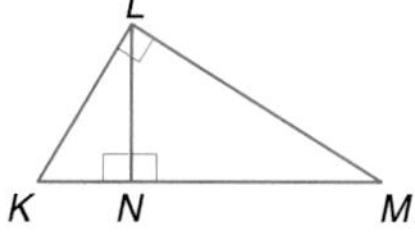

19. Find KL if $KM = 18$ and $KN = 4$. $6\sqrt{2} \approx 8.49$
20. Find LN if $KN = 4$ and $NM = 6$. $2\sqrt{6} \approx 4.90$
21. Find KM if $LM = 19$ and $NM = 14$. about 25.79
22. Find LK if $KN = \frac{1}{3}$ and $NM = \frac{1}{4}$. $\frac{\sqrt{7}}{6} \approx 0.44$
23. Find LM if $KN = 7$ and $NM = 3$. $\sqrt{30} \approx 5.48$
24. Find KN if $LK = 0.6$ and $KM = 1.5$. 0.24

- use the Pythagorean Theorem and its converse (Lesson 8–1)

A right triangle has a hypotenuse 61 inches long and a leg 11 inches long. Find the length of the other leg.

$a^2 + b^2 = c^2$ *Pythagorean Theorem*

$a^2 + 11^2 = 61^2$ *$b = 11$, $c = 61$*

$a^2 + 121 = 3721$

$a^2 = 3600$ *Subtraction Property (=)*

$a = 60$ *Take the square root of each side.*

The other leg of the triangle is 60 inches long.

Find the value of x.

25. 17

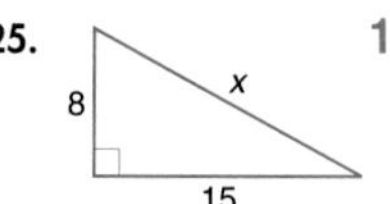

26. 4.94

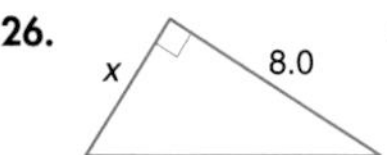

Determine if the given measures are measures of the sides of a right triangle.

27. 19, 24, 30 no
28. 4, 7.5, 8.5 yes

GLENCOE Technology

Test and Review Software

You may use this software, a combination of an item generator and item bank, to create your own tests or worksheets. Types of items include free response, multiple choice, short answer, and open ended.

For IBM & Macintosh

OBJECTIVES AND EXAMPLES

- use the properties of 45°-45°-90° and 30°-60°-90° triangles. (Lesson 8–2)

In a 45°-45°-90° triangle, the hypotenuse is $\sqrt{2}$ times as long as a leg.

In a 30°-60°-90° triangle, the hypotenuse is twice as long as the shorter leg, and the longer leg is $\sqrt{3}$ times as long as the shorter leg.

31. $1.55\sqrt{3} \approx 2.68$
32. $7.1\sqrt{2} \approx 10.04$

REVIEW EXERCISES

Find the value of x.

29. x in. 3.0 in.

$3\sqrt{2} \approx 4.24$

30. 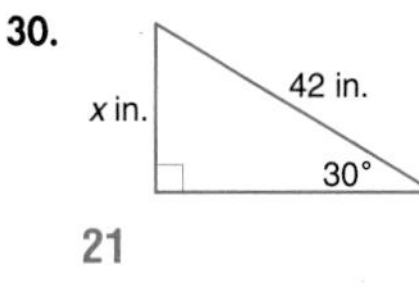

21

31. 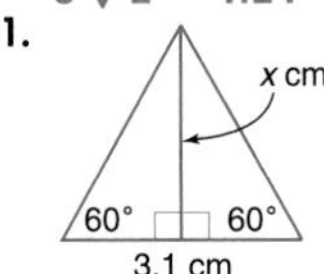

32. 
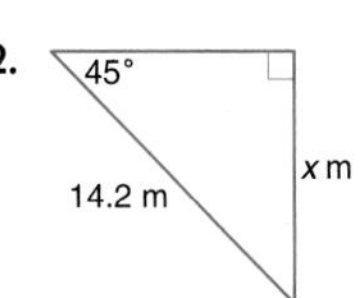

- find trigonometric ratios using right triangles (Lesson 8–3)

$\sin A = \frac{a}{c}$

$\cos A = \frac{b}{c}$

$\tan A = \frac{a}{b}$

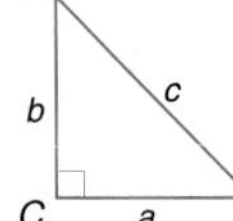

Find the indicated trigonometric ratio as a fraction and as a decimal rounded to the nearest ten-thousandth.

33. $\sin M$ $\frac{15}{17} \approx 0.8824$
34. $\tan M$ $\frac{15}{8} \approx 1.8750$
35. $\cos K$ $\frac{15}{17} \approx 0.8824$
36. $\tan K$ $\frac{8}{15} \approx 0.5333$

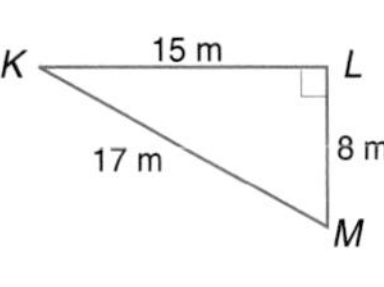

- use trigonometry to solve problems involving angles of elevation and depression (Lesson 8–4)

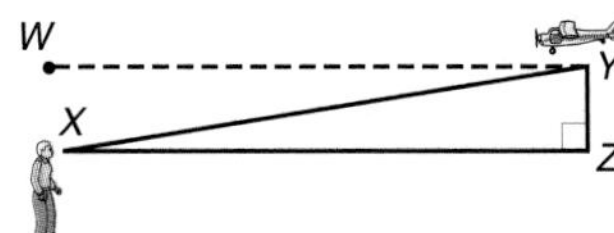

$\angle YXZ$ is the angle of elevation.

$\angle WYX$ is the angle of depression.

Solve each problem. Round answers to the nearest tenth.

37. The top of a lighthouse is 120 meters above sea level. The angle of depression from the top of the lighthouse to the ship is 23°. How far is the ship from the foot of the lighthouse? about 282.7 m
38. At a certain time of day, the angle of elevation of the sun is 44°. Find the length of a shadow cast by a building 30 meters high. about 31.1 m
39. A railroad track rises 30 feet for every 400 feet of track. What is the measure of the angle of elevation of the track? about 4.3°

Applications and Problem Solving Encourage students to work through the exercises in the Applications and Problem Solving section to strengthen their problem-solving skills.

OBJECTIVES AND EXAMPLES	REVIEW EXERCISES
use the Law of Sines to solve triangles (Lesson 8–5) According to the Law of Sines, $\frac{\sin A}{a} = \frac{\sin B}{b} = \frac{\sin C}{c}$. 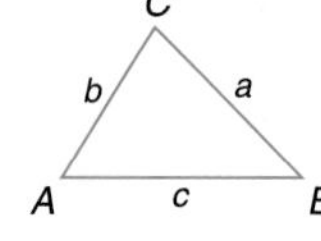 43. $m\angle B = 80$, $b \approx 14.8$, $c \approx 14.1$	**Solve each △ABC described below. Round measures to the nearest tenth.** 40. $a = 4.2$, $b = 6.8$, $m\angle B = 22$ $m\angle A \approx 13.4$, $m\angle C \approx 144.6$, $c \approx 10.5$ 41. $m\angle B = 46$, $m\angle C = 83$, $b = 65$ $m\angle A = 51$, $a \approx 70.2$, $c \approx 89.7$ 42. $a = 80$, $b = 10$, $m\angle A = 65$ $m\angle B \approx 6.5$, $m\angle C \approx 108.5$, $c \approx 83.7$ 43. $m\angle C = 70$, $a = 7.5$, $m\angle A = 30$
use the Law of Cosines to solve triangles (Lesson 8–6) According to the Law of Cosines, $a^2 = b^2 + c^2 - 2bc \cos A$, $b^2 = a^2 + c^2 - 2ac \cos B$, and $c^2 = a^2 + b^2 - 2ab \cos C$. 47. $m\angle A \approx 40.8$, $m\angle B \approx 78.6$, $m\angle C \approx 60.6$	**Solve each △ABC described below. Round measures to the nearest tenth.** 44. $m\angle C = 55$, $a = 8$, $b = 12$ $c \approx 9.9$, $m\angle A \approx 41.4$, $m\angle B \approx 83.6$ 45. $a = 44$, $c = 32$, $m\angle B = 44$ $b \approx 30.6$, $m\angle A \approx 87.3$, $m\angle C \approx 48.7$ 46. $m\angle C = 78$, $a = 4.5$, $b = 4.9$ $c \approx 5.9$, $m\angle A \approx 48.2$, $m\angle B \approx 53.8$ 47. $a = 6$, $b = 9$, $c = 8$

APPLICATIONS AND PROBLEM SOLVING

48. **Recreation** Jeremy is flying a kite whose string makes a 70°-angle with the ground. The kite string is 65 meters long. How far is the kite above the ground? (Lesson 8–3) **about 61.1 m**

49. **Aviation** A jet airplane begins a steady climb of 15° and flies for two ground miles. What was its change in altitude? (Lesson 8–3) **about 0.5 mi**

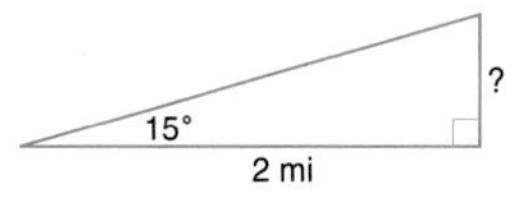

50. **Fire Fighting** Ranger Barojas sights a fire from a fire tower in Wayne National Forest. She finds that the angle of depression to the fire is 22°. If the tower is 75 meters tall, how far is the fire from the base of the tower? (Lesson 8–4) **about 185.6 m**

51. **Architecture** Bianca Jones is an architect designing a new parking garage for the city. The floors of the garage are to be 10 feet apart. The exit ramps between each pair of floors are to be 75 feet long. What is the measure of the angle of elevation of each ramp? (Lesson 8–4) **about 7.7°**

52. **Aviation** Sonia Ortega flew her airplane 1000 kilometers north before turning 20° clockwise and flying another 700 kilometers. How far is Ms. Ortega from her starting point? (Lesson 8–6) **about 1675.0 km**

53. **Decision Making** Which problem-solving strategies might you use to find the remainder for $5^{100} \div 7$? Find the value of the remainder for $5^{100} \div 7$. (Lesson 8–6) **Sample answer: look for a pattern; 2**

A practice test for Chapter 8 is available on page 800.

ALTERNATIVE ASSESSMENT

COOPERATIVE LEARNING PROJECT

Parallax In this chapter, you explored properties of right triangles and began the study of trigonometry. Your body is one of the best measuring tools available; you never leave home without it. In this project, you will learn to use your eyes and one finger to measure the size of an object if you know its distance, or the distance to an object if you know its size.

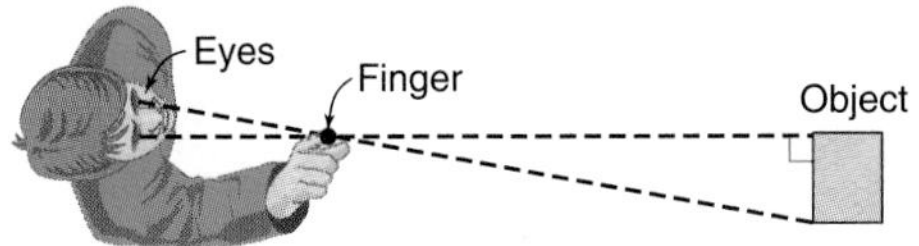

Follow these steps to conduct your measurement experiments.

- Place a chair or desk 10 feet away. As in the diagram above, place your index finger pointing up, about 6 inches in front of your right eye. Make sure your line of sight is perpendicular to the line connecting your eyes. Measure the distance between your eyes. Look with your right eye, then with your left eye.
- Does your finger move from one side of the chair to the other? Move your finger in or out so that through your right eye, your finger is at the left edge of the chair and through your left eye your finger is at the right edge of the chair.
- Use what you know about proportions of similar triangles or trigonometry to determine the size of the chair.
- Now measure the chair to confirm your indirect measurement.
- Now have another student select an object and measure its size. Have the student place the object at a distance from you without telling you the distance.
- Modify the above method to measure the distance to the object.
- As a group, complete the following lists. Make a list of common objects with known sizes in situations in which you could not tell the distance; for example, a car at an unknown distance. Make a list of situations where you would know the distance, but observe an object of unknown size; for example, an object across the street outside your bedroom window.

THINKING CRITICALLY

The angle of the sun's rays varies throughout the year. So, the shadow cast by a sundial at the same time of day will vary throughout the year. Can a sundial be useful all year? Explain.

PORTFOLIO

Trigonometry allows you to measure dimensions you cannot measure directly. Make a list, with photographs or drawings, of ten situations where trigonometry allows you to measure indirectly what cannot be measured directly. Keep this collection in your portfolio.

SELF EVALUATION

Relationships with friends and family can be represented geometrically. Ask other students in your group how they solve interpersonal conflicts. Write the solution methods in a step-by-step format. Then convert these into flow diagrams.

Assess yourself When you have an argument with a friend or family member, how do you resolve it? Which problem-solving methods do you think are most effective at creating harmonious relationships out of disharmony?

Assessment and Evaluation Masters, pp. 210, 221

NAME ______________________ DATE __________

Chapter 8 Performance Assessment

Instructions: ***Demonstrate your knowledge by giving a clear, concise solution to each problem. Be sure to include all relevant drawings and to justify your answers. You may show your solution in more than one way or investigate beyond the requirements of the problem.***

1. A bridge support is designed as shown below.

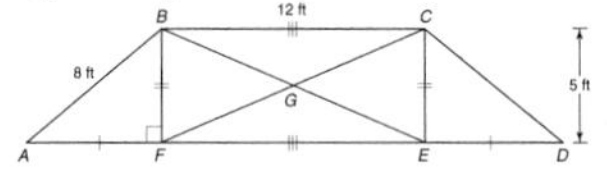

a. State the Pythagorean Theorem in your own words. Include a diagram if necessary.

b. Find the length of cross member $\overline{BE}$. Show your work.

c. Find the length of the bridge, AD. Explain each step.

d. Is $\angle BGC$ a right angle? Explain.

e. Find $m\angle A$ in two ways. Show your work.

2. **a.** State the Law of Sines and the cases in which it may be used to solve triangles. If necessary, include a diagram.

b. Give all three forms of the Law of Cosines and the cases in which they may be used to solve triangles.

c. A forest fire is sighted from two fire towers 5 miles apart. How far is the fire from each of the towers? Show your work.

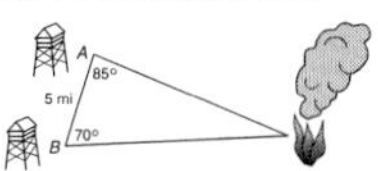

d. Jackson, Mississippi, (JM) is 212 miles south of Memphis, Tennessee, (MT). If it is 246 miles from Jackson to Montgomery, Alabama, (MA) and 328 miles from Memphis to Montgomery, find the measure of α, the flight bearing from Jackson to Montgomery. Find the measure of β, the flight bearing from Memphis to Montgomery.

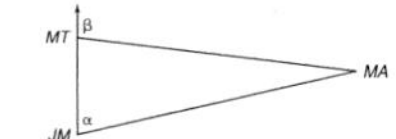

Scoring Guide
Chapter 8
Performance Assessment

Level	Specific Criteria
3 Superior	• Shows thorough understanding of the concepts of *the Pythagorean Theorem; sine, cosine, and tangent of an angle;* and *the laws of sines and cosines.* • Uses appropriate strategies to solve problems. • Computations are correct. • Written explanations are exemplary. • Goes beyond requirements of some or all problems.
2 Satisfactory, with Minor Flaws	• Shows understanding of the concepts of *the Pythagorean Theorem; sine, cosine, and tangent of an angle;* and *the laws of sines and cosines.* • Uses appropriate strategies to solve problems. • Computations are mostly correct. • Written explanations are effective. • Satisfies all requirements of some or all problems.
1 Nearly Satisfactory, with Serious Flaws	• Shows understanding of most of the concepts of *the Pythagorean Theorem; sine, cosine, and tangent of an angle;* and *the laws of sines and cosines.* • May not use appropriate strategies to solve problems. • Computations are mostly correct. • Written explanations are satisfactory. • Satisfies most requirements of some or all problems.
0 Unsatisfactory	• Shows little or no understanding of the concepts of *the Pythagorean Theorem; sine, cosine, and tangent of an angle;* and *the laws of sines and cosines.* • May not use appropriate strategies to solve problems. • Computations are incorrect. • Written explanations are not satisfactory. • Does not satisfy requirements of some or all problems.

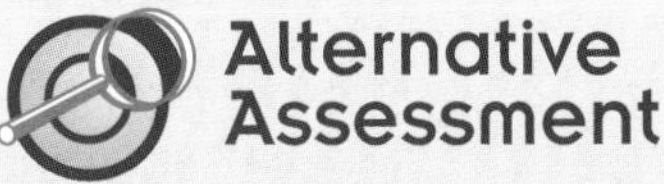

Alternative Assessment

The Alternative Assessment section provides students with the opportunity to assess their own work by thinking critically, working with others, keeping a portfolio, and honestly evaluating their own progress. For more information on alternative forms of assessment, see *Alternative Assessment in the Mathematics Classroom,* one of the titles in the Glencoe Mathematics Professional Series.

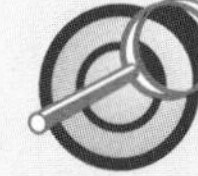

Performance Assessment

Performance Assessment tasks for this chapter are included in the *Assessment and Evaluation Masters.* A scoring guide is also provided.

Using the COLLEGE ENTRANCE EXAM PRACTICE

These two pages review the skills and concepts presented in Chapters 1–8. This review is formatted to reflect new trends in college entrance testing.

Assessment and Evaluation Masters, pp. 215–216

NAME ______________________ DATE ____________

Chapter 8 Cumulative Review

For Questions 1–4, refer to the figure at the right.

1. Find OP if $MP = 15$ and $PN = 60$. 1. 30
2. Find MO if $ON = 8$ and $MN = 17$. 2. 15
3. Find ON if $m\angle N = 30$ and $OM = 7$. 3. $7\sqrt{3}$
4. Write an equation using a trigonometric ratio that you could use to find MO if $m\angle M = 55$ and $ON = 9$. 4. $MO = \frac{9}{\tan 55°}$
5. A flagpole casts a 35-ft shadow. Nearby, a girl 5 ft tall casts a 7-ft shadow. Find the height of the flagpole. 5. 25 ft

For Questions 6 and 7, determine if each pair of triangles is similar. If they are similar, state the reason and find the missing measure.

6. 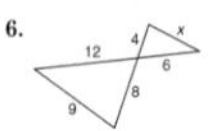
7.

6. yes, SAS, $x = 4.5$
7. yes, AA, $x = 9$

8. The lengths of two corresponding sides of a pair of similar polygons are 5 cm and 18 cm. The perimeter of the larger polygon is 126 cm. Find the perimeter of the smaller polygon. 8. 35 cm
9. Quadrilateral $ABCD$ is a parallelogram. Find the measure of $\angle B$ if $m\angle A = 7x + 5$ and $m\angle D = 9x - 1$. 9. 98
10. Identify quadrilateral $FGHI$ as a parallelogram, rectangle, rhombus, or square if the coordinates of the vertices are $F(-1, 2)$, $G(3, 2)$, $H(4, -1)$, and $I(0, -1)$. 10. parallelogram
11. Quadrilateral $JKLM$ is a rhombus. Find the measures of $\angle MJK$ and $\angle JKL$ if $m\angle LMK = 34.5$. 11. $m\angle MJK = 111$, $m\angle JKL = 69$
12. A segment that connects a vertex of a triangle to the midpoint of the opposite side is called a ____ of the triangle. 12. median

NAME ______________________ DATE ____________

Chapter 8 Cumulative Review (continued)

13. Use >, =, or < to complete the following: AD ____ BC. Name the theorem or postulate that justifies your conclusion. 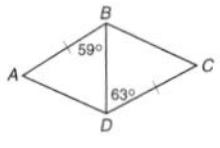 13. $AD < BC$; SAS Inequality
14. Is it possible to have a triangle with sides whose lengths are 6 cm, 15 cm, and 7 cm? Explain your answer. 14. no, $15 > 6 + 7$; it violates the △ inequality.

For Questions 15 and 16, refer to the figure below. Quadrilateral RSTU is an isosceles trapezoid with V as the midpoint of $\overline{RU}$.

15. What congruence theorem or postulate would you use to prove that $\triangle RSV \cong \triangle UTV$? 15. SAS
16. If $SV = 5x + 3$ and $TV = 7x - 9$, find the value of x. 16. 6
17. Find the value of x so that $p \parallel q$. 17. 37
18. State the slope of a line perpendicular to the line passing through the points $A(-1, 4)$ and $B(2, 2)$. 18. $\frac{3}{2}$
19. Write the converse of "If $x > 0$, then $x^2 > 0$". Determine whether the converse is true or false. 19. If $x^2 > 0$, then $x > 0$; false.
20. From an airplane that is flying at an altitude of 3000 ft, the angle of depression to a ground signal is 27°. Find the direct-line distance between the airplane and the ground signal. 20. 6608 ft

COLLEGE ENTRANCE EXAM PRACTICE

CHAPTERS 1–8

SECTION ONE: MULTIPLE CHOICE

There are nine multiple-choice questions in this section. After working each problem, write the letter of the correct answer on your paper.

1. $(8\sqrt{4})(2\sqrt{9})$ **C**

- **A.** 48
- **B.** 72
- **C.** 96
- **D.** 298

2. If the area of an isosceles right triangle is 36 square units, what is its perimeter? **A**

- **A.** $(12 + 12\sqrt{2})$ units
- **B.** $(12 + 6\sqrt{2})$ units
- **C.** $(6 + 12\sqrt{2})$ units
- **D.** $(12\sqrt{2})$ units

3. If $af = 6$, $fg = 1$, $ag = 24$, and $a > 0$, find afg. **C**

- **A.** 4
- **B.** 6
- **C.** 12
- **D.** 144

4. $\sqrt{25 \cdot 64}$ **D**

- **A.** 1600
- **B.** 400
- **C.** 80
- **D.** 40

5. If $x - \frac{2}{x - 3} = \frac{x - 1}{3 - x}$, find the value of x. **C**

- **A.** 3 or -1
- **B.** 3 or -3
- **C.** -1
- **D.** 3

6. In $\triangle ABC$ below, if $AB = BC$ and if $\angle ABC$ is a right angle, exactly how many 45° angles are formed by pairs of line segments? **D**

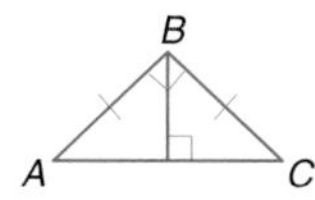

- **A.** none
- **B.** two
- **C.** three
- **D.** four

7. If a and b are negative numbers, which of the following must be negative? **C**

- **A.** ab
- **B.** $(ab)^2$
- **C.** $a + b$
- **D.** $a - b$

8. Find the value of x. **D**

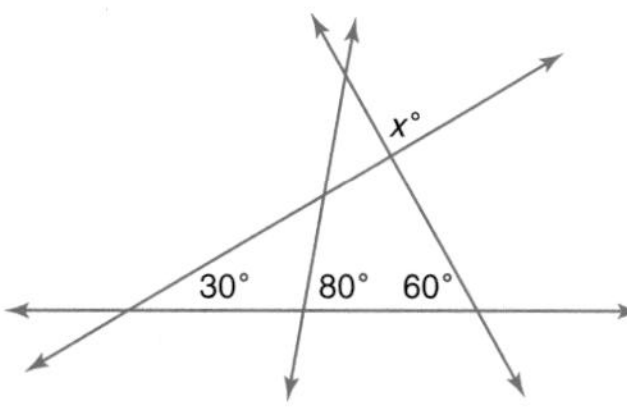

- **A.** 40
- **B.** 50
- **C.** 60
- **D.** 90

9. Find the value of x if $\overline{AE} \parallel \overline{BD}$, $AB = 10$, $BC = x$, $ED = x + 3$, and $DC = x + 6$. **C**

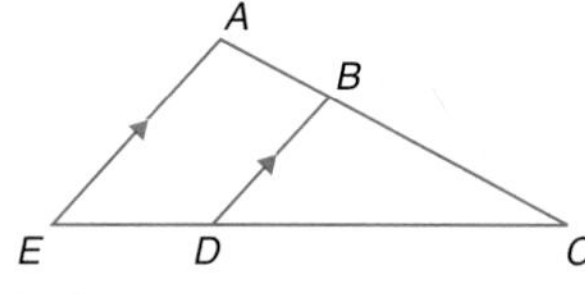

- **A.** 6
- **B.** 7
- **C.** 12
- **D.** 13

Standardized Test Practice Questions are also provided in the *Assessment and Evaluation Masters*, p. 214.

A more traditional cumulative review is provided in the *Assessment and Evaluation Masters*, pp. 215–216.

SECTION TWO: SHORT ANSWER

This section contains seven questions for which you will provide short answers. Write your answer on your paper.

10. If the area of $\triangle ABC$ is 30 square units, what is the value of x? **7.5 units**

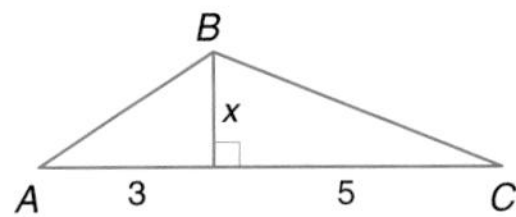

11. Find $\sin\theta$, $\cos\theta$, and $\tan\theta$.

$\sin\theta = \frac{3}{5}$, $\cos\theta = \frac{4}{5}$, $\tan\theta = \frac{3}{4}$

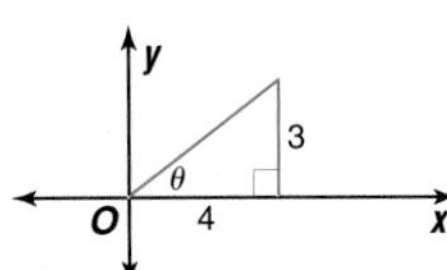

12. In the figure below, $\overline{AB}$ and $\overline{CD}$ are parallel line segments. Find the area of isosceles trapezoid $ABCD$. **76 units²**

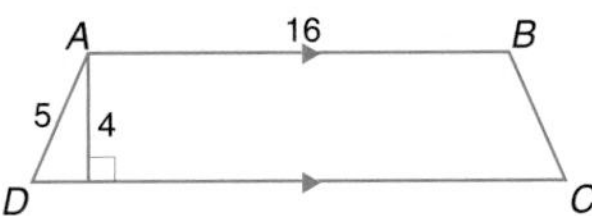

13. For all real numbers a and b, where $b \neq 0$, $a * b = \frac{a^2}{b^2}$, find $(3 * 4)(5 * 3)$. $\frac{25}{16}$

14. Find the resulting expression when x is subtracted from y and this difference is divided by the sum of x and y. $\left\{\frac{y-x}{x+y}\right\}$

15. If J, K, and L each represent a different digit in the multiplication problem below, what digit does J represent? **4**

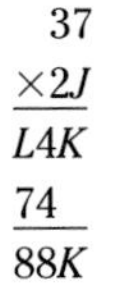

16. If the base and height of the triangle below are decreased by $2x$ units, what is the area of the resulting triangle? $30x^2$

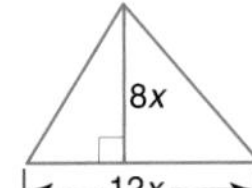

SECTION THREE: COMPARISON

This section contains five comparison problems that involve comparing two quantities, one in column A and one in column B. In certain questions, information related to one or both quantities is centered above them. All variables used represent real numbers.

Compare quantities A and B below.

- Write A if quantity A is greater.
- Write B if quantity B is greater.
- Write C if the two quantities are equal.
- Write D if there is not enough information to determine the relationship.

Column A	Column B
17. perimeter of $\triangle ABC$ (right triangle, legs 2 and 3)	perimeter of $\triangle DEF$ ($DE = 2$, $EF = 3$, $DF = \sqrt{3}$) **A**
Three angles of a triangle are $x°$, $x°$, and $\frac{x°}{2}$.	**A**
18. x	60
19. $(3 + 5)^2$	$3^2 + 5^2$ **A**
20. the perimeter of a triangle	the perimeter of a rectangle **20. D**
21. a	b **B**

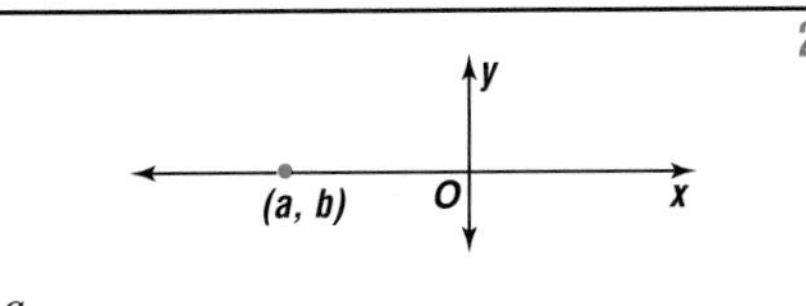

TEST-TAKING TIP

Geometric Figures

The drawings provided for test items are often not drawn to scale. Remind students what they can and cannot assume from a geometric figure.

What can be assumed from a figure:

- When points appear on a line or a line segment, they are collinear.
- Angles that appear to be adjacent or vertical are.
- When lines, line segments, or rays appear to intersect, they do.

What cannot be assumed from a figure:

- Line segments or angles that appear to be congruent are.
- Line segments or angles that appear not to be congruent are not.
- Lines or line segments that appear to be parallel or perpendicular are.
- A point that appears to be a midpoint of a segment is.

CHAPTER 9

Analyzing Circles

PREVIEWING THE CHAPTER

This chapter provides an in-depth study of circles beginning with definitions for radius, diameter, and circumference. Major and minor arcs and their relationships to central angles are explored. Chords, secants, and tangents and their properties are used to develop theorems involving inscribed angles and intercepted arcs. Emphasis on measurement of angles and arcs provides problem-solving experiences throughout. Compass and straightedge constructions are used occasionally to verify theorems. The chapter concludes with the application of these concepts to the coordinate plane including a study of equations for circles.

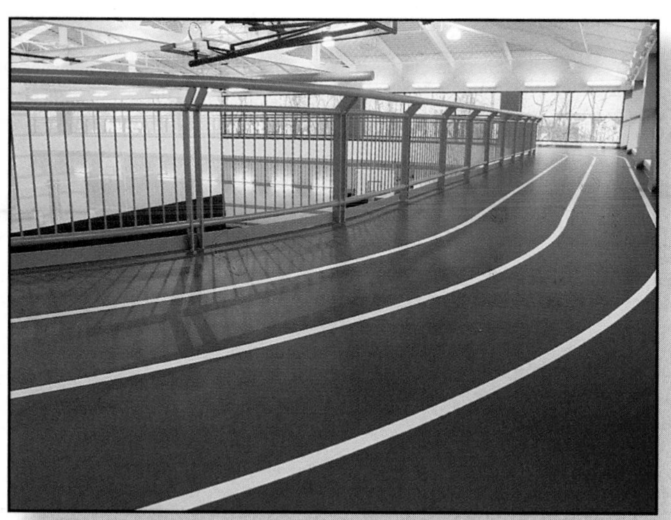

Lesson (Pages)	Lesson Objectives	NCTM Standards	State/Local Objectives
9-1 (446–451)	Identify and use parts of circles. Solve problems involving the circumference of a circle.	1–4, 7	
9-2 (452–458)	Recognize major arcs, minor arcs, semicircles, and central angles. Find measures of arcs and central angles. Solve problems by making circle graphs.	1–4, 7, 10	
9-3 (459–465)	Recognize and use relationships among arcs, chords, and diameters.	1–4, 7	
9-4 (466–473)	Recognize and find measures of inscribed angles. Apply properties of inscribed figures.	1–5, 7	
9-5A (474)	Use the TI-92 calculator to explore characteristics of tangents.	2, 3, 7	
9-5 (475–482)	Recognize tangents and use properties of tangents.	1–4, 7	
9-6 (483–490)	Find the measures of angles formed by intersecting secants and tangents in relation to intercepted arcs.	1–5, 7	
9-7 (491–497)	Use properties of chords, secants, and tangents to solve segment measure problems.	1–5, 7	
9-8 (498–503)	Write and use the equation of a circle in the coordinate plane.	1–5, 7, 8	

ORGANIZING THE CHAPTER

A complete, 1-page lesson plan is provided for each lesson in the ***Lesson Planning Guide***. Answer keys for each lesson are available in the ***Answer Key Masters***.

You may want to refer to the **Course Planning Calendar** on page T12 for detailed information on pacing.
PACING: Standard—15 days; **Honors**—13 days; **Block**—9 days

LESSON PLANNING CHART

Lesson (Pages)	Materials/ Manipulatives	Extra Practice (Student Edition)	Blackline Masters								Real-World Applications	Teaching Transparencies
			Study Guide	Practice	Enrichment	Assessment & Evaluation	Modeling Mathematics	Multicultural Activity	Tech Prep Applications	Graphing Calc. & Computer		
9-1 (446–451)	circular objects tape measure scientific calculator string chalk	p. 780	p. 49	p. 49	p. 49			p. 17	p. 17			9-1A 9-1B
9-2 (452–458)	compass* protractor*	p. 780	p. 50	p. 50	p. 50	p. 240		p. 18			18	9-2A 9-2B
9-3 (459–465)	compass* straightedge* patty paper	p. 780	p. 51	p. 51	p. 51		p. 87				19	9-3A 9-3B
9-4 (466–473)	protractor* straightedge*	p. 781	p. 52	p. 52	p. 52	pp. 239, 240	pp. 47–50					9-4A 9-4B
9-5A (474)	TI-92 calculator									p. 26		
9-5 (475–482)	compass* straightedge* patty paper	p. 781	p. 53	p. 53	p. 53				p. 18			9-5A 9-5B
9-6 (483–490)	compass* straightedge*	p. 781	p. 54	p. 54	p. 54	p. 241						9-6A 9-6B
9-7 (491–497)	TI-92 calculator or Cabri II software TI-82/83 graphing calculator	p. 782	p. 55	p. 55	p. 55					p. 9		9-7A 9-7B
9-8 (498–503)	TI-92 calculator	p. 782	p. 56	p. 56	p. 56	p. 241						9-8A 9-8B
Study Guide/ Assessment (505–509)						pp. 225–238, 242–244						

*Included in Glencoe's High School Manipulative Kit and Overhead Manipulative Resources.

ORGANIZING THE CHAPTER

Other Chapter Resources

Student Edition
Chapter Opener, pp. 444–445
Mathematics and Society, p. 458
Working on the Investigation, pp. 465, 497
Closing the Investigation, p. 504

Teacher's Classroom Resources
Investigations and Projects Masters, pp. 57–60
Block Scheduling Booklet

Technology
Test and Review Software (IBM and Macintosh)
CD-ROM Multimedia Applications (Windows and Macintosh)
Mindjogger Videoquizzes (VHS)

Professional Publications
Glencoe Mathematics Professional Series

Outside Resources

Books/Periodicals
The Circular Geoboard Book, ETA
Dancing Curves: A Dynamic Demonstration of Geometric Principle, NCTM
Circles, NASCO

Software
Geometry Toolkit, Ventura Educational Systems
Green Globs and Graphing Equations, Sunburst

Videos/CD-ROMs
The Story of π, Dale Seymour Publicatons, P.O. Box 10888, Palo Alto, CA 94303

Assessment Resources

Student Edition
Math Journal, pp. 449, 469, 486, 500
Mixed Review, pp. 451, 457, 465, 472, 482, 490, 497, 503
Self Test, p. 473
Chapter Highlights, p. 505
Chapter Study Guide and Assessment, pp. 506–508
Alternative Assessment, p. 509
Portfolio, p. 509
Chapter Test, p. 801

Teacher's Wraparound Edition
5-Minute Check, pp. 446, 452, 459, 466, 475, 483, 491, 498
Check for Understanding, pp. 449, 454, 461, 469, 478, 486, 494, 500
Closing Activity, pp. 451, 458, 465, 473, 482, 490, 497, 503
Cooperative Learning, pp. 460, 476

Assessment and Evaluation Masters
Multiple-Choice Tests, Forms 1A (Honors), 1B (Average), 1C (Basic), pp. 225–230
Free-Response Tests, Forms 2A (Honors), 2B (Average), 2C (Basic), pp. 231–236
Calculator-Based Test, p. 237
Performance Assessment, p. 238
Mid-Chapter Test, p. 239
Quizzes A–D, pp. 240–241
Standardized Test Practice, p. 242
Cumulative Review, pp. 243–244

ENHANCING THE CHAPTER

Examples of some of the materials for enhancing Chapter 9 are shown below.

DIVERSITY

Multicultural Activity Masters, pp. 17, 18

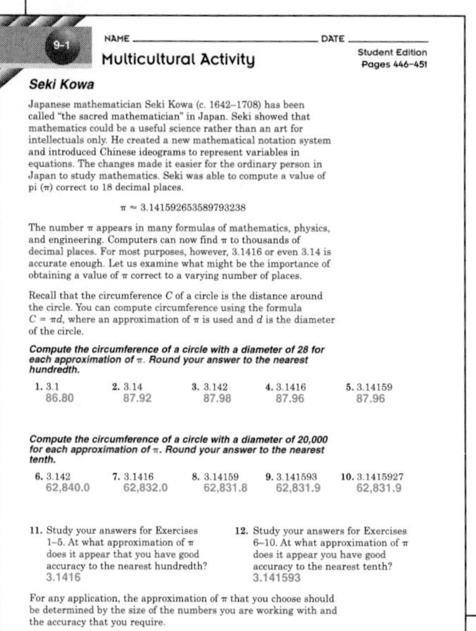

9-1 NAME ______________ DATE ______

Multicultural Activity

Student Edition Pages 446–451

Seki Kowa

Japanese mathematician Seki Kowa (c. 1642–1708) has been called "the sacred mathematician" in Japan. Seki showed that mathematics could be a useful science rather than an art for intellectuals only. He created a new mathematical notation system and introduced Chinese ideograms to represent variables in equations. The changes made it easier for the ordinary person in Japan to study mathematics. Seki was able to compute a value of pi (π) correct to 18 decimal places.

$$\pi \approx 3.141592653589793238$$

The number π appears in many formulas of mathematics, physics, and engineering. Computers can now find π to thousands of decimal places. For most purposes, however, 3.1416 or even 3.14 is accurate enough. Let us examine what might be the importance of obtaining a value of π correct to a varying number of places.

Recall that the circumference C of a circle is the distance around the circle. You can compute circumference using the formula $C = \pi d$, where an approximation of π is used and d is the diameter of the circle.

Compute the circumference of a circle with a diameter of 28 for each approximation of π. Round your answer to the nearest hundredth.

1. 3.1 86.80 **2.** 3.14 87.92 **3.** 3.142 87.98 **4.** 3.1416 87.96 **5.** 3.14159 87.96

Compute the circumference of a circle with a diameter of 20,000 for each approximation of π. Round your answer to the nearest tenth.

6. 3.142 62,840.0 **7.** 3.1416 62,832.0 **8.** 3.14159 62,831.8 **9.** 3.141593 62,831.9 **10.** 3.1415927 62,831.9

11. Study your answers for Exercises 1–5. At what approximation of π does it appear that you have good accuracy to the nearest hundredth? 3.1416

12. Study your answers for Exercises 6–10. At what approximation of π does it appear you have good accuracy to the nearest tenth? 3.141593

For any application, the approximation of π that you choose should be determined by the size of the numbers you are working with and the accuracy that you require.

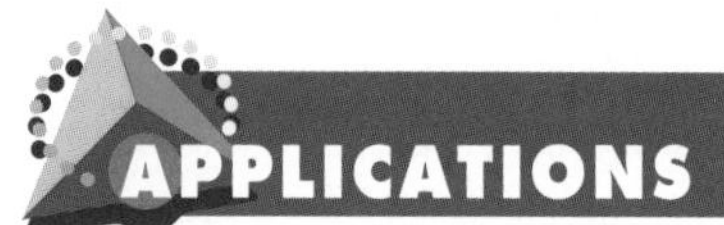

APPLICATIONS

Real-World Applications, 18, 19

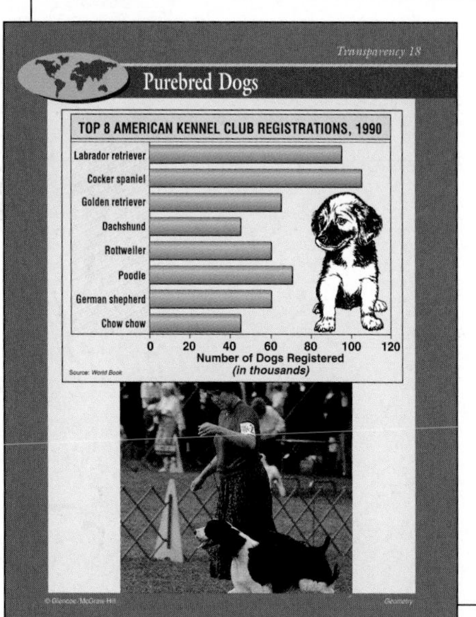

TECHNOLOGY

Graphing Calculator and Computer Masters, p. 9

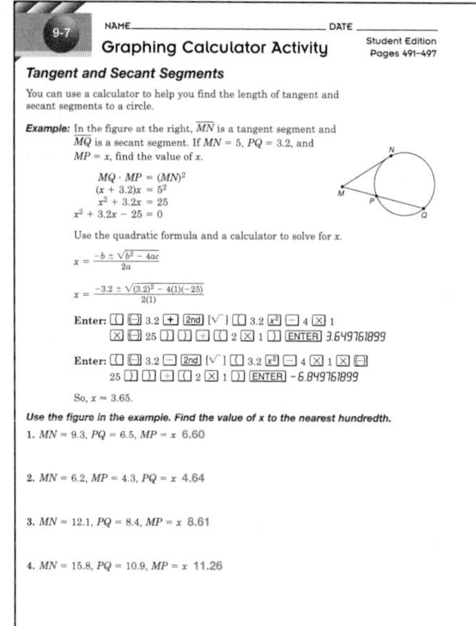

9-7 NAME ______________ DATE ______

Graphing Calculator Activity

Student Edition Pages 491–497

Tangent and Secant Segments

You can use a calculator to help you find the length of tangent and secant segments to a circle.

Example: In the figure at the right, $\overline{MN}$ is a tangent segment and $\overline{MQ}$ is a secant segment. If $MN = 5$, $PQ = 3.2$, and $MP = x$, find the value of x.

$$MQ \cdot MP = (MN)^2$$
$$(x + 3.2)x = 5^2$$
$$x^2 + 3.2x = 25$$
$$x^2 + 3.2x - 25 = 0$$

Use the quadratic formula and a calculator to solve for x.

$$x = \frac{-b \pm \sqrt{b^2 - 4ac}}{2a}$$

$$x = \frac{-3.2 \pm \sqrt{(3.2)^2 - 4(1)(-25)}}{2(1)}$$

Enter: ((-) 3.2 + 2nd [√] (3.2 x² − 4 × 1 × (-) 25)) ÷ (2 × 1) ENTER 3.649761899

Enter: ((-) 3.2 − 2nd [√] (3.2 x² − 4 × 1 × (-) 25)) ÷ (2 × 1) ENTER −6.849761899

So, $x \approx 3.65$.

Use the figure in the example. Find the value of x to the nearest hundredth.

1. $MN = 9.3$, $PQ = 6.5$, $MP = x$ 6.60
2. $MN = 6.2$, $MP = 4.3$, $PQ = x$ 4.64
3. $MN = 12.1$, $PQ = 8.4$, $MP = x$ 8.61
4. $MN = 15.8$, $PQ = 10.9$, $MP = x$ 11.26

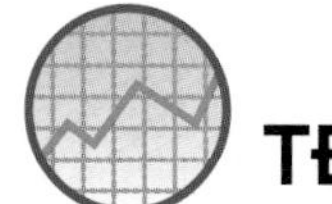

TECH PREP

Tech Prep Applications Masters, pp. 17, 18

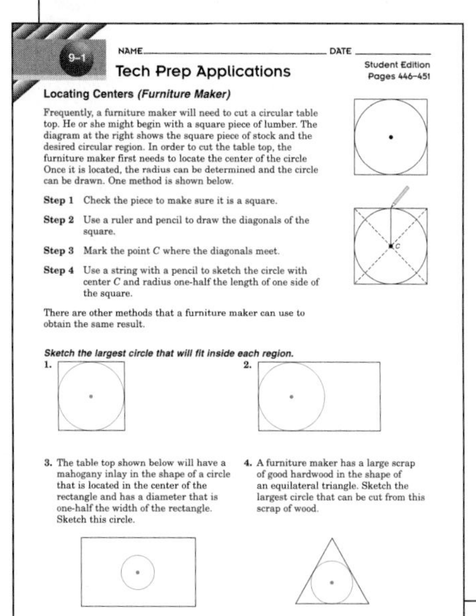

9-1 NAME ______________ DATE ______

Tech Prep Applications

Student Edition Pages 446–451

Locating Centers *(Furniture Maker)*

Frequently, a furniture maker will need to cut a circular table top. He or she might begin with a square piece of lumber. The diagram at the right shows the square piece of stock and the desired circular region. In order to cut the table top, the furniture maker first needs to locate the center of the circle Once it is located, the radius can be determined and the circle can be drawn. One method is shown below.

Step 1 Check the piece to make sure it is a square.

Step 2 Use a ruler and pencil to draw the diagonals of the square.

Step 3 Mark the point C where the diagonals meet.

Step 4 Use a string with a pencil to sketch the circle with center C and radius one-half the length of one side of the square.

There are other methods that a furniture maker can use to obtain the same result.

Sketch the largest circle that will fit inside each region.

1. **2.**

3. The table top shown below will have a mahogany inlay in the shape of a circle that is located in the center of the rectangle and has a diameter that is one-half the width of the rectangle. Sketch this circle.

4. A furniture maker has a large scrap of good hardwood in the shape of an equilateral triangle. Sketch the largest circle that can be cut from this scrap of wood.

PROBLEM SOLVING

Problem-of-the-Week Cards, 24, 25, 26

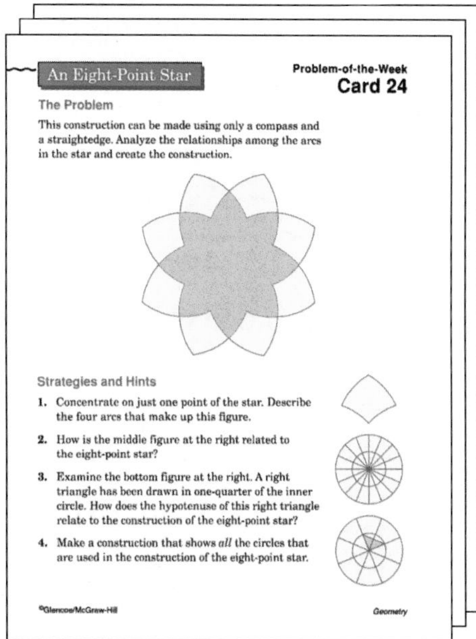

An Eight-Point Star

Problem-of-the-Week Card 24

The Problem

This construction can be made using only a compass and a straightedge. Analyze the relationships among the arcs in the star and create the construction.

Strategies and Hints

1. Concentrate on just one point of the star. Describe the four arcs that make up this figure.
2. How is the middle figure at the right related to the eight-point star?
3. Examine the bottom figure at the right. A right triangle has been drawn in one-quarter of the inner circle. How does the hypotenuse of this right triangle relate to the construction of the eight-point star?
4. Make a construction that shows *all* the circles that are used in the construction of the eight-point star.

©Glencoe/McGraw-Hill

Geometry

MAKING MATHEMATICS RELEVANT

This two-page introduction to the chapter provides students with an opportunity to explore an application of geometry that has historic roots as well as modern uses.

Background Information

The Popularity of Bicycling Here are some interesting facts about worldwide bicycle usage:

- In Beijing, China, roughly 75% of the workforce commutes by bicycle, but in the United States barely 1% of the workforce commutes by bicycle.
- More people commute to work by bicycle in China and India than the number of people who commute by car in all of the rest of the world.
- In the United States, half of the land in cities is used for highways. One lane can support 40,000 bicycles per hour, but 12 lanes are needed to support 40,000 cars per hour.
- A bicycle can support 10 times its own weight, which is something no car, truck, or airplane can equal.

Because bicycles are based on circles (wheels, sprockets, etc.), they provide a great way to talk about circles. Ask students to identify the various parts of a circle in the context of a bicycle or ask them to find the central angle for the arc defined by two adjacent spokes of a bicycle wheel.

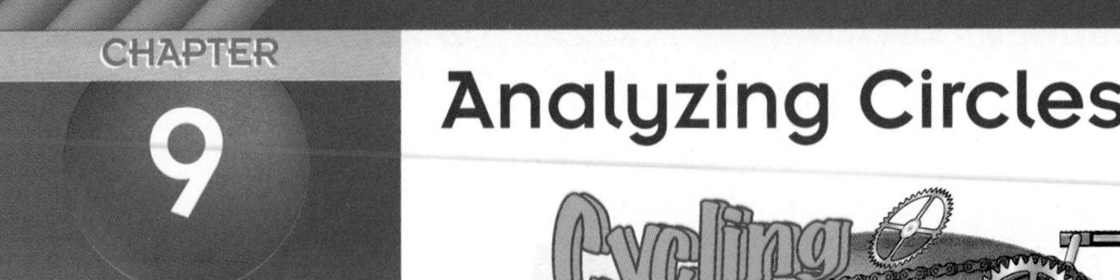
CHAPTER
9

Analyzing Circles

Cycling
Sensation

Objectives

In this chapter, you will:

- find the degree and linear measures of arcs,
- find the measures of angles in circles,
- solve problems by making circle graphs,
- use properties of chords, tangents, and secants to solve problems, and
- write equations of circles.

Geometry: Then and Now Do you have a bicycle? If so, do you ride it for recreation or is it your primary mode of transportation? In 1900, more than one million bicycles were being produced per year in the United States. Barely ten years later, automobiles and motorcycles became the vehicles of choice and nearly forced the bicycle industry out of existence. Bicycles became popular again in the United States later in the century, but mainly as recreation. However, much of the rest of the world, especially Asia and Africa, continues to rely heavily on the bicycle as a mode of transportation. Today, there are more than 800 million bicycles in use in the world—that's twice the number of automobiles!

TIME Line

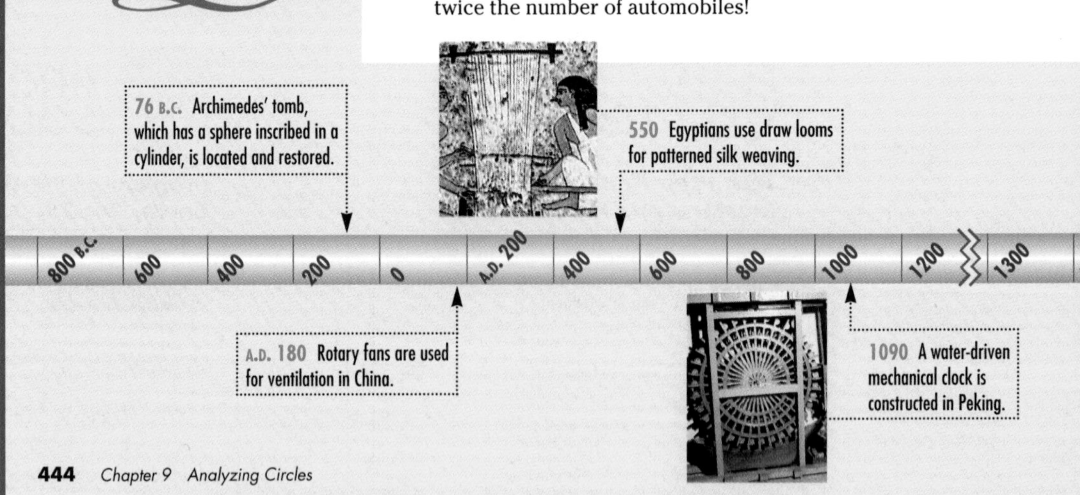
76 B.C. Archimedes' tomb, which has a sphere inscribed in a cylinder, is located and restored.
550 Egyptians use draw looms for patterned silk weaving.
800 B.C.
600
400
200
0
A.D. 200
400
600
800
1000
1200
1300
A.D. 180 Rotary fans are used for ventilation in China.
1090 A water-driven mechanical clock is constructed in Peking.

TIME Line

Students may find it interesting to look at the history of instruments used in map making and surveying. Tools used to measure angles have become increasingly precise.

interNET CONNECTION

Bicycle safety, mountain biking, and commuting are among the items you will find at this bicycling web site.

World Wide Web
http://www.cascade.org/links.html

PEOPLE IN THE NEWS

The Iditasport winter wilderness race is a unique competition for cross-country skiers, snowshoers, runners, and mountain bikers. **Fred Bull** of Anchorage, Alaska, was one of only two bicyclists under the age of 20 to complete in the 1995 Iditasport. Every February, professional and amateur athletes from around the world compete in this 170-mile race through the Alaskan wilderness. Fred finished the race in about two and a half days, having slept once for eight hours and once for nine hours. His goal for his next Iditasport is to finish the course in 25 hours. Fred says, "I won't be able to sleep this time."

Chapter Project

Work in groups of four to conduct a survey on bicycle ownership and usage.

- Each person in the group should survey 10 students to determine what kind of bicycles, if any, they own and how they use their bicycles.
- Combine your results. Make a circle graph for the responses to each survey question. Use computer software, if available.
- Compare your graphs with other groups. Make some conclusions about bicycle ownership and usage, and report your findings.
- How different do you think your results would be if you had surveyed adults instead? Explain your reasoning.
- Conduct the same survey with adults only. Make a circle graph for the responses to each survey. Are the results what you expected? Explain.

1492 Mathematician and cartographer Pedro Nunes, reportedly the first to develop an instrument for measuring angles, is born.

1861 The first practical bicycle, called the *vélocipède*, is built.

1400 1450 1500 1550 1600 1650 1700 1750 1800 1850 1900 1950 2000

1670 Isaac Barrow's book *Lectiones Geomericae* discusses drawing tangents to curves.

1996 Mountain biking becomes a new event in the Summer Olympics.

PEOPLE IN THE NEWS

The 170-mile Iditasport course through Alaska starts at Big Lake (north of Anchorage), crosses the Little Susitna River, continues north to Skwentna, and then heads south again along the Yentna River to finish at Big Lake. The course itself is just a snowmobile rut in the snow. Competitors must carry their own stove and fuel, sleeping bag, and one day's supply of food. Food is resupplied by dropping prepacked bags from the air.

Chapter Project

Cooperative Learning Students may work in pairs within their groups. They can compile the two pairs' results to create a larger sample for their circle graphs. Encourage students to ask the following:

1. What kind of bike do you own?
___ Mountain bike
___ 10-speed bike
___ Other type of bike
___ Don't own a bike

2. If you own a bike, is it used mostly for recreation or transportation?

Investigations and Projects Masters, p. 57

NAME__________ DATE ________

Chapter 9 Project A

Student Edition Pages 444–509

Hitting the Trail

1. Who is surveyed and how survey questions are written are two very important aspects of collecting data. Work in groups of four. Together you are trying to sell cycling products and need to conduct a survey about your products. Design two surveys: one with leading questions (such as, "Do you like cycling?" which assumes the person has ridden a bicycle before) and one with no leading questions. Discuss as a group how you think the responses will differ for each survey.
2. Ask 10 people the questions from one survey, and ask 10 different people the questions from the other survey. (Each student will survey 20 people altogether.) Record their responses.
3. Combine your results. Make a circle graph for each question in the survey.
4. Use the graphs to compare responses. Decide which graphs your company might use in its advertisement. Explain your thinking. Write a report about your survey to present to the company. Include some of the circle graphs in the report.

Alternative Chapter Projects

Two other chapter projects are included in the *Investigations and Projects Masters*. In Chapter 9 Project A, pp. 57–58, students extend the topic in the chapter opener. In Chapter 9 Project B, pp. 59–60, students examine the circular structure in buildings from other countries.

9-1 LESSON NOTES

NCTM Standards: 1–4, 7

Instructional Resources

- Study Guide Master 9-1
- Practice Master 9-1
- Enrichment Master 9-1
- Multicultural Activity Masters, p. 17
- Tech Prep Applications Masters, p. 17

Transparency 9-1A contains the 5-Minute Check for this lesson; **Transparency 9-1B** contains a teaching aid for this lesson.

Recommended Pacing	
Standard Pacing	Day 1 of 15
Honors Pacing	Day 1 of 13
Block Scheduling*	Day 1 of 9

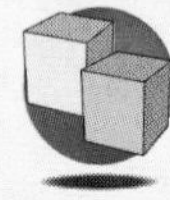

*For more information on pacing and possible lesson plans, refer to the *Block Scheduling Booklet*.

1 FOCUS

5-Minute Check
(over Chapter 8)

Solve. Round answers to the nearest tenth.

1. One leg of a right triangle measures 4 inches. The hypotenuse measures 8 inches. Find the length of the other leg. **6.9 in.**
2. Find the geometric mean between 8 and 25. $\sqrt{200} \approx 14.1$
3. In right $\triangle RED$, $m\angle R = 51$ and the hypotenuse measures 200 yards. Find the length of the side opposite $\angle R$. **155.4 yd**
4. In $\triangle ABC$, $m\angle A = 71$, $b = 20$, and $c = 15$.
   a. Find a. **20.7**
   b. Find $m\angle B$. **66.0**

9–1 Exploring Circles

What **YOU'LL LEARN**

- To identify and use parts of circles, and
- to solve problems involving the circumference of a circle.

Why **IT'S IMPORTANT**

You can use properties of circles to solve problems involving surveying, sports, and space travel.

APPLICATION

Surveying

When a house is sold, a land surveyor must construct and draw a *Plat of Survey* that verifies the house and land measurements of the property. In order to measure the dimensions of the property, surveyors use a *trundle wheel.*

Radii is the plural of radius.

A trundle wheel is a model of a **circle**. A circle is the set of all points in a plane that are a given distance from a given point in that plane. The given point is the **center** of the circle. Each spoke of the wheel represents the **radius** of a circle. A radius of a circle is a segment that has one endpoint at the center of the circle and the other endpoint on the circle. Notice that all the spokes of the wheel appear to be congruent. It follows from the definition of a circle that all radii are congruent.

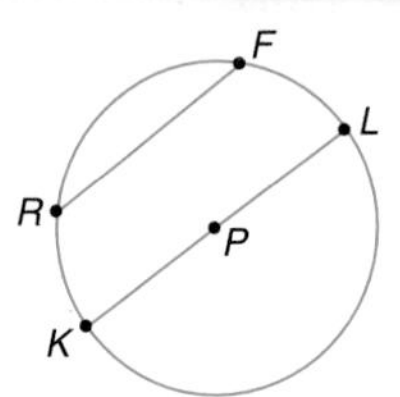

A circle is usually named by its center. The circle at the right is called circle P. This is symbolized as $\odot P$.

$\overline{RF}$ and $\overline{KL}$ are **chords** of $\odot P$. A chord of a circle is a segment that has its endpoints on the circle.

$\overline{KL}$ is also a **diameter** of $\odot P$. A diameter of a circle is a chord that contains the center of the circle. Notice that $\overline{PK}$ and $\overline{PL}$ are radii. By the Segment Addition Postulate, $KP + PL = KL$. Let r represent the measure of a radius and let d represent the measure of the diameter.

The terms diameter and radius can refer to segments in a circle or the measures of those segments.

$$KP + PL = KL$$
$$r + r = d$$
$$2r = d$$
$$r = \frac{1}{2}d$$

Thus, a diameter is twice as long as a radius, or a radius is half as long as a diameter.

Example 1

APPLICATION

Surveying

Refer to the application at the beginning of the lesson. The surveyor's wheel has a diameter of 1.59 feet. What is the approximate length of each spoke of the wheel?

Each spoke is a radius of the wheel.

$$d = 2r$$
$$1.59 = 2r$$
$$0.795 = r$$

The length of each spoke is 0.795 feet.

A trundle wheel is used to measure property lines by counting the number of revolutions. Each time the wheel makes one revolution, the distance it travels is the same as the **circumference** of the wheel. The circumference of a circle is the distance around the circle.

MODELING MATHEMATICS

Circumference of a Circle

Materials:  circular objects tape measure

- Choose three or four objects that have a circular shape such as different-sized paper plates, cookies, coins, or old records.

- Use a tape measure to measure the diameter and circumference of each object to the nearest millimeter.

Your Turn a. See students' work.

a. Record your results in a table like the one below. In the third column, find the ratio of the circumference C to the diameter d.

Object	d	C	$\frac{C}{d}$
1			
2			
3			

b. What seems to be true about the results for $\frac{C}{d}$ in the third column? It's about 3.14.

c. Use these results to make a conjecture about the relationship between the circumference and diameter of a circle. It is constant.

The Modeling Mathematics activity provides insight into the relationship between the radius and the circumference of a circle. By definition, the ratio of the circumference of a circle to its diameter is the irrational number called π (**pi**). If $\frac{C}{d} = \pi$, then $C = \pi d$. Since $d = 2r$, $C = 2\pi r$.

Circumference of a Circle	**If a circle has a circumference of C units and a radius of r units, then $C = 2\pi r$.**

In this book, we will use a calculator for evaluating expressions involving π. If no calculator is available, 3.14 is a good estimate for π.

Example  **Find the circumference of a circle with a radius of 6.8 centimeters.**

$C = 2\pi r$ *Formula for the circumference of a circle*

$C = 2\pi(6.8)$

$C = 13.6\pi$

The *exact* circumference is 13.6π centimeters. To *estimate* the circumference, use a calculator.

Enter: 13.6 [×] [2nd] [π] [=] 42.72566009

The circumference is about 42.7 centimeters.

Motivating the Lesson

Questioning Draw a circle and an ellipse or oval on the chalkboard or overhead. Ask students if the two figures are circles. What is the difference between an ellipse and a circle? All points on a circle are the same distance from the center.

2 TEACH

In-Class Examples

For Example 1
The size of a bicycle is determined by the diameter of the wheel. So, a 26-inch bicycle has a wheel with a 26-inch diameter. What is the length of a spoke of a 26-inch bicycle? 13 inches

For Example 2
Find the circumference of a circle with a radius of 2.3 miles. 14.4 miles

Teaching Tip Remind students that in this textbook the words *diameter* and *radius* are used to refer to the lengths of these segments as well as to the segments themselves.

 In this activity, students will discover approximations for π by measuring circular objects in the real world. A string can be used to measure large objects such as tree trunks when tape measures are inconvenient.

Alternative Learning Styles

Kinesthetic Have students tie a string around a piece of chalk and use it to draw circles on the chalkboard. Have them hold the string down with their fingers and pull the string tight with the chalk. Now have them use the chalk to draw a circle. Ask students to explain what the length of the string represents. radius of the circle

In-Class Examples

For Example 3
If the radius of a bicycle wheel is 17 inches, and the wheel turned 10,500 revolutions, how far did the bicycle travel? **$357{,}000\pi$ inches or about 17.87 miles**

For Example 4
Find the exact circumference of $\odot P$. **5π**

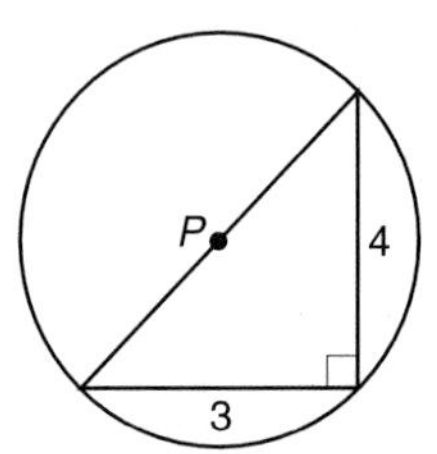

Teaching Tip In Example 4, remind students that the formula for circumference is $C = \pi d$, where d is the hypotenuse of the triangle.

fabulous **FIRSTS**

Born in Albany, Georgia, Coachman truly loved sports. During a time when few women competed in high level sports, she began her career as a sprinter at age 13. She won several national championships as a runner and high jumper while attending Tuskegee Institute and Albany State College. She was named to five All-American teams, including basketball.

Circumference can be used to measure distance.

Example 3

Marathons

In 1996, inspectors checked the accuracy of the Olympic marathon course by using the "calibrated bicycle method." The inspectors rode bicycles equipped with counters that tracked the turns of the wheels. If the radius of a bicycle wheel is 13 inches and the counter shows 20,300 revolutions at the end of a course, how long is the course?

The total distance in inches would equal the number of revolutions times the distance per revolution (circumference of the wheel).

$$\begin{aligned}\text{total distance} &= 20{,}300 \cdot (2\pi r)\\ &= 20{,}300 \cdot 2\pi \cdot 13\\ &= 527{,}800\pi \text{ inches}\end{aligned}$$

To find the distance in miles, divide by (12 inches per foot · 5280 feet per mile) or 63,360 inches/mile.

$$\begin{aligned}\text{total distance} &= \frac{527{,}800\pi \text{ inches}}{63{,}360 \text{ inches per mile}}\\ &\approx 26.17 \text{ miles} \quad \textit{Use a calculator.}\end{aligned}$$

The course is approximately 26 miles long, which is the length of a marathon.

fabulous **FIRSTS**

Alice Coachman (Davis) (1923–)

Alice Coachman became the first African American woman to win an Olympic gold medal. She set an Olympic record for the high jump in the 1948 Olympics in London.

In Lesson 9–3, you will learn about polygons whose vertices lie on a circle. You can use these kinds of polygons to help find the circumference of circles.

Example 4 **Find the exact circumference of $\odot P$ shown at the right.**

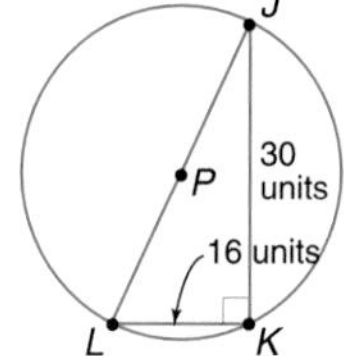

To find the circumference of the circle, we need to know the diameter. Since the triangle is a right triangle, we can use the Pythagorean Theorem to find the measure of the diameter $\overline{JL}$.

Let $c = JL$, $a = LK$, and $b = JK$.

$$\begin{aligned}a^2 + b^2 &= c^2 && \textit{Pythagorean Theorem}\\ 16^2 + 30^2 &= c^2 && a = 16,\ b = 30\\ 1156 &= c^2\\ 34 &= c && \textit{Take the square root of each side.}\end{aligned}$$

The diameter is 34 units. Now use that value to find the circumference.

$$\begin{aligned}C &= \pi d\\ &= \pi \cdot 34 \text{ or } 34\pi\end{aligned}$$

The exact circumference is 34π units.

Classroom Vignette

"At the beginning of the chapter, I give students what I call "Circle Theorem Sheets" for recording the circle theorems we learn. The students record each theorem on the sheets and draw a picture to illustrate it. At the beginning of the chapter, I provide the theorem and assist with the picture. Later in the chapter, I provide the theorem and have a student lead the class in drawing the picture."

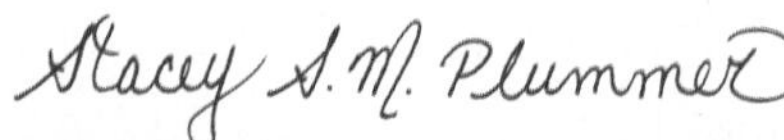

Stacey S. M. Plummer
Merrimack High School
Merrimack, New Hampshire

CHECK FOR UNDERSTANDING

Communicating Mathematics

Study the lesson. Then complete the following.

1. **Explain** why a compass can be used to draw a circle. See margin.
2. **Explain** why a diameter is the longest chord of a circle.
3. **You Decide** Fernando claims that "every diameter is a chord of a circle" and Koleka says, "every chord is a diameter of a circle." Who do you think is correct? Explain. See margin.
4. **Explain** why $C = \pi d$ and $C = 2\pi r$ are equivalent formulas. See margin.

2. The diameter contains the center of a circle.

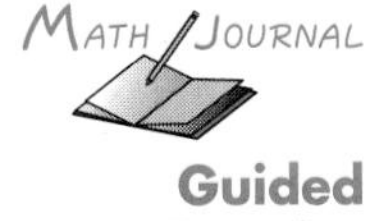

5. Use a dictionary to determine why the word *circle* is used to describe a set of points equidistant from a given point. Circle is derived from the Latin *circus*, which means ring.

Guided Practice

Refer to ⊙K for Exercises 6–8.

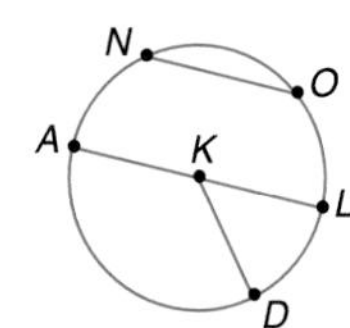

6. Name two chords of ⊙K. $\overline{AL}$, $\overline{NO}$
7. If $AL = 9.4$, find KD. 4.7
8. Is $NO < AL$? Explain.

8. Yes; $\overline{AL}$ is longer because it is a diameter.

In Exercises 9 and 10, the radius, diameter, or circumference of a circle is given. Find the other measures to the nearest tenth.

9. $r = 7$, $d = \underline{?}$, $C = \underline{?}$
10. $C = 76.4$, $d = \underline{?}$, $r = \underline{?}$

9. $d = 14$, $C = 44.0$
10. $d = 24.3$, $r = 12.2$

11. Find the exact circumference of ⊙T. 9π

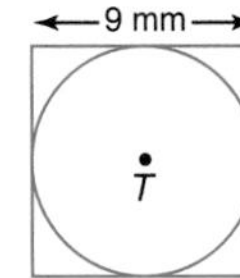

12. Circle P has a radius of 6 units, and ⊙T has a radius of 4 units.
    a. If $QR = 1$, find RT. 3
    b. If $QR = 1.2$, find AQ. 10.8

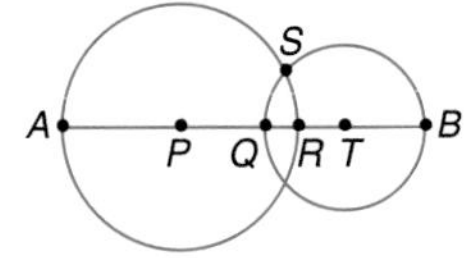

13. **Fencing** The Grabowskis wish to put a fence around their circular pool. The pool has a circumference of 60 feet. If they want the fence to be 6 feet from the pool all the way around, how many feet of fencing will be needed? (*Hint:* Draw a diagram.) 97.7 feet

EXERCISES

Practice

Refer to ⊙M for Exercises 14–22.

14. Name the center of ⊙M. M
15. Name a chord that is also a diameter. $\overline{RI}$
16. If $MD = 5$, find RI. 10
17. Is $\overline{MI}$ a chord of ⊙M? Explain. No; it is a radius.
18. Is $\overline{MA} \cong \overline{MI}$? Explain. Yes; they are both radii of ⊙M.
19. Name four radii of ⊙M. $\overline{RM}$, $\overline{AM}$, $\overline{DM}$, $\overline{IM}$
20. Is $RI > SU$? Explain. Yes; since $\overline{RI}$ is a diameter.
21. If $RI = 11.8$, find MA. 5.9
22. Draw $\overline{AR}$. What type of triangle is $\triangle MAR$? Explain.

22. It is isoceles, since $AM = RM$.

Reteaching

Using Generalization Draw five concentric circles with radii of 1", 2", 3", 4", and 5". With a tape measure, measure the circumference of each. Divide each circumference measurement by its corresponding diameter. What pattern results? The final results are all equal to π.

3 PRACTICE/APPLY

Check for Understanding

Exercises 1–13 are designed to help you assess your students' understanding through reading, writing, speaking, and modeling. You should work through Exercises 1–5 with your students and then monitor their work on Exercises 6–13.

Additional Answers

1. The point stays in one place and is the center of the circle. The tip of the pencil moves around the point, always at the same distance from the point.
3. Fernando is correct since not all chords pass through the center of a circle.
4. The measure of the diameter is twice the measure of the radius, $d = 2r$.

Study Guide Masters, p. 49

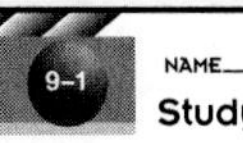

NAME________________ DATE________

Study Guide Student Edition Pages 446–451

Exploring Circles

A **circle** is the set of all points in a plane that are a given distance from a given point in the plane called the **center**. Various parts of a circle are labeled in the figure at the right.

The distance around a circle is called the **circumference**.

Circumference of a Circle	If a circle has a circumference of C units and a radius of r units, then $C = 2\pi r$.

Example: Find the circumference of the circle shown at the right.

$C = 2\pi r$
$C = 2\pi(13)$
$C = 26\pi$
$C \approx 81.7$

The circumference is about 81.7 cm.

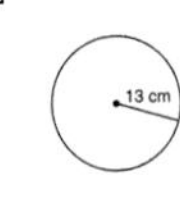

Refer to ⊙S for Exercises 1–6.

1. Name the center of ⊙S. S
2. Name three radii of ⊙S. $\overline{SR}$, $\overline{SM}$, $\overline{ST}$
3. Name a diameter. $\overline{RT}$
4. Name a chord. $\overline{XY}$, or $\overline{RT}$
5. If $RT = 8.2$, find SM. 4.1
6. Is $\overline{SR} \cong \overline{SM}$? Explain. Yes; they are both radii of ⊙S.

In Exercises 7–10, the radius, diameter, or circumference of a circle is given. Find the other measures to the nearest tenth.

7. $r = 7$, $d = \underline{?}$, $C = \underline{?}$ 14.0; 44.0
8. $d = 32.4$, $r = \underline{?}$, $C = \underline{?}$ 16.2; 101.8
9. $C = 116.5$, $d = \underline{?}$, $r = \underline{?}$ 37.1; 18.6
10. $r = 12$, $d = \underline{?}$, $C = \underline{?}$ 24; 75.4

Assignment Guide

Core (with proof): 15–45 odd, 47–56
Core (informal): 15–45 odd, 47–56
Enriched: 14–38 even, 39–56

For **Extra Practice**, see p. 780.

The red A, B, and C flags, printed only in the Teacher's Wraparound Edition, indicate the level of difficulty of the exercises.

Additional Answers

36. $AS = IA$ and $KS = KC$ since all radii of a circle are congruent. Therefore, the perimeter of $\triangle ASK = IA + AK + KC$ or IC.

39. Given: $\odot P$ with diameter $\overline{SA}$ and chord $\overline{KR}$
Prove: $SA > KR$
Proof:
By the Segment Addition Postulate, $SA = SP + PA$. Draw $\overline{PK}$ and $\overline{PR}$ since through any two points there is one line. Since all radii of a circle are congruent, $\overline{SP} \cong \overline{PK}$ and $\overline{PA} \cong \overline{PR}$. By substitution, $SA = PK + PR$. By the Triangle Inequality Theorem, $PK + PR > KR$. By substitution, $SA > KR$.

Practice Masters, p. 49

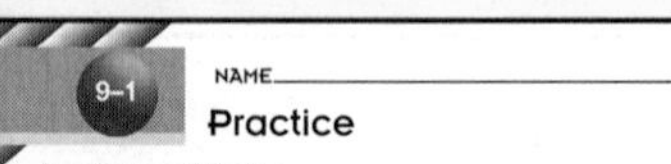

Exploring Circles

Refer to the figure at the right.

1. Name the center of $\odot P$. P
2. Name the three radii of the circle. PA, PD, PB
3. Name a diameter. $\overline{AB}$
4. Name two chords. $\overline{AD}$, $\overline{AB}$

Find the circumference of a circle with a radius of the given length. Round your answers to the nearest tenth.

5. 3 cm 18.8 cm
6. 2 ft 12.6 ft
7. 34 mm 213.6 mm
8. 4.5 m 28.3 m

Find the exact circumference of each circle.

9. 14π in.

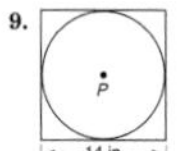

10. 26π cm

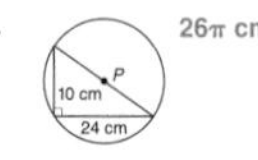

11. $5\sqrt{2}\pi$ in.

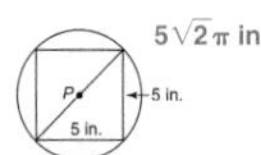

12. 16π cm

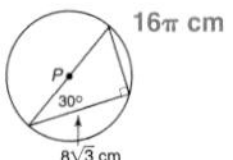

In Exercises 23–28, the radius, diameter, or circumference of a circle is given. Find the other measures to the nearest tenth.

23. $r = 5$, $d = \underline{?}$, $C = \underline{?}$ $d = 10$, $C = 31.4$
24. $d = 26.8$, $r = \underline{?}$, $C = \underline{?}$ $r = 13.4$, $C = 84.2$
25. $C = 136.9$, $d = \underline{?}$, $r = \underline{?}$ $d = 43.6$, $r = 21.8$
26. $r = \frac{x}{6}$, $d = \underline{?}$, $C = \underline{?}$ $d = 0.3x$, $C = 0.9x$ or $C = 1.0x$
27. $d = 2x$, $r = \underline{?}$, $C = \underline{?}$ $r = x$, $C = 6.3x$
28. $C = 2368$, $d = \underline{?}$, $r = \underline{?}$ $d = 753.8$, $r = 376.9$

Find the exact circumference of each circle.

29. 8π cm
30. 13π cm

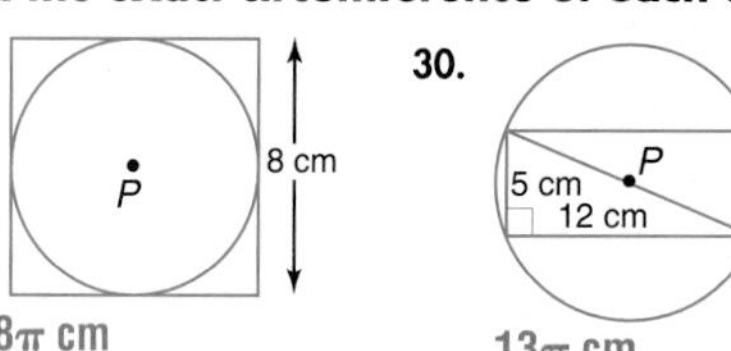

31. $6\sqrt{2}\pi$ cm

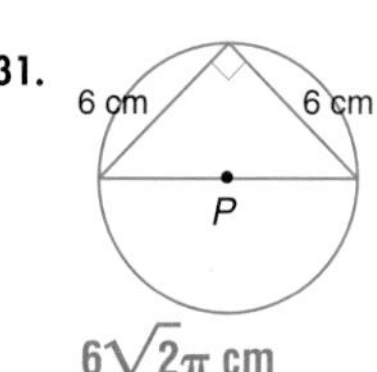

Circle A has a radius of 4 units, and $\odot K$ has a radius of 7 units. If $JE = 2$, find each measure.

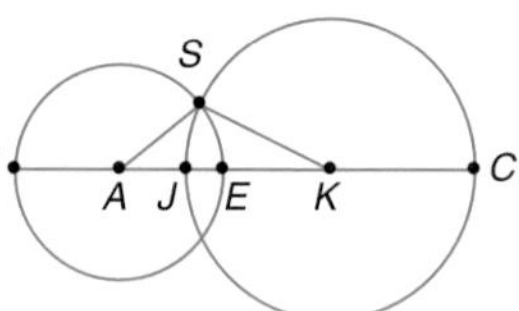

32. EK 5
33. IJ 6
34. IC 20
35. perimeter of $\triangle ASK$ 20
36. Explain why the perimeter of $\triangle ASK$ equals the length of $\overline{IC}$. See margin.
37. The diameter of a circle is 6 feet. If the diameter is doubled, what is the effect on the circumference? The circumference is doubled.
38. The radius of a circle is 18 centimeters. If the radius is divided by 3, what is the effect on the circumference? The circumference is divided by 3.

Critical Thinking

39. Use the Triangle Inequality Theorem to show diameter $\overline{SA}$ is the longest chord in $\odot P$. That is, write a paragraph proof that shows $SA > KR$. (*Hint:* Draw $\overline{PK}$ and $\overline{PR}$.) See margin.

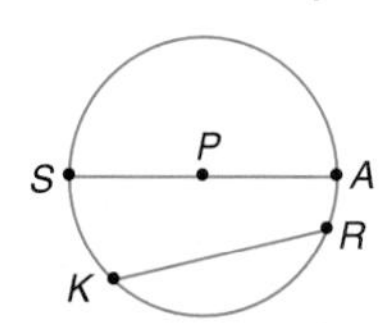

Applications and Problem Solving

40. **Sports** The new track, soccer, and football field for Prairie Ridge High School is shown below. To the nearest tenth of a yard, find how far a runner will go if she runs around the inside lane of the track four times. 1626.0 yards

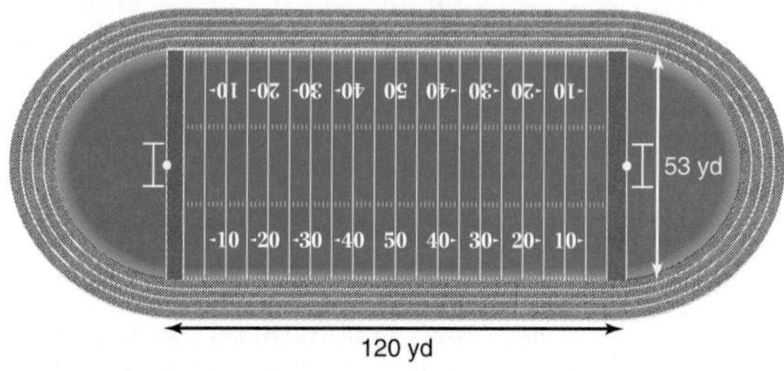

41. **Music** Using a centimeter tape measure, carefully measure the diameter and circumference of one of your favorite compact discs (CDs). Find the ratio of the circumference and the diameter. $d \approx 11.8$ cm, $C \approx 37.1$ cm, $\frac{C}{d} \approx 3.144$ or π
42. **Native American Culture** About one thousand years ago, wooden poles formed a gigantic circle 125 meters across in southern Illinois. The structure was a giant solar calendar that kept track of the seasons and the movement of the sun for Native North Americans. What was the circumference of this structure? 392.7 m

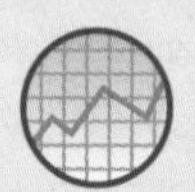

Tech Prep

Veterinary Technologist Many two-year colleges offer programs in veterinary related fields. A good background in biological science and basic mathematics is necessary. For more information on tech prep, see the *Teacher's Handbook*.

43. 3141.6 cm

44. 4822.9 revolutions

43. Pet A hamster wheel is 10 centimeters in diameter. If the hamster runs so that the wheel makes 100 revolutions, how far did the hamster run?

44. Surveying Refer to Example 3. An inspector uses a 33-centimeter radius bicycle wheel to determine that the total length of a racewalking course is 10 kilometers. How many revolutions did the wheel make?

45. Crops The following article appeared in the *Orlando Sentinel* in July, 1996.

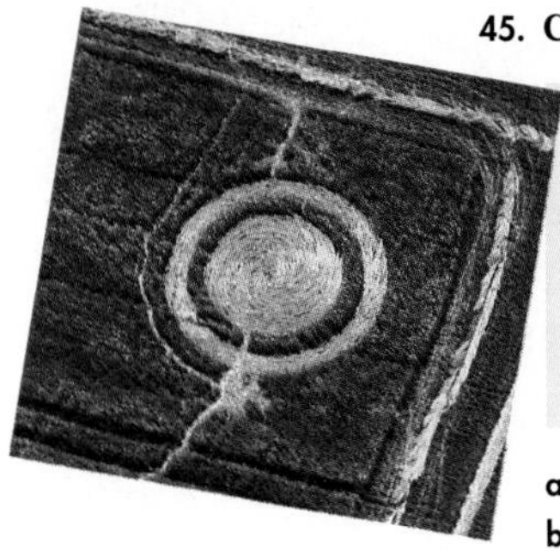

> Authorities were searching Wednesday for the origin of mysterious crop circles that appeared in a nearby barley field this week. "Nothing like this has ever happened around here before," Boyd County Sheriff Duane Pavel said...He said a circle about 25 feet in diameter was surrounded by a wide swath of barley and then another cleared ring...In all, the design was about 42 feet in diameter.

a. Find the circumference of the first circle to the nearest foot. **79 ft**

b. Find the circumference of the entire crop design to the nearest foot. **132 ft**

46. Space Travel Bernard A. Harris Jr., M.D. is the first African-American astronaut to walk in space. In 1993, he logged 4,164,183 miles in space on a mission aboard the *STS-55*. If the orbit was 250 miles above Earth and the diameter of Earth is about 8000 miles, find how many orbits around Earth Dr. Harris made on that mission. **about 156 orbits**

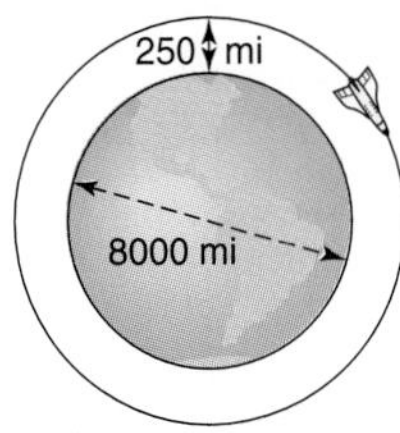

Mixed Review

47. Find the perimeter of $\triangle GEF$ to the nearest tenth of a centimeter. (Lesson 8–6) **273.9**

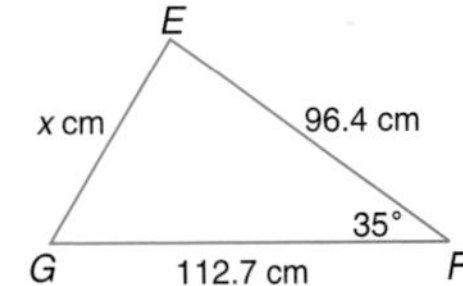

48. 6970 ft

48. Meteorology A searchlight 6500 feet from a weather station is turned on. If the angle of elevation to the spot of light on the clouds above the station is 47°, how high is the cloud ceiling? (Lesson 8–4)

49. Find the perimeter of a rhombus with diagonals 30 inches and 16 inches long. (Lesson 8–1) **68 in.**

50. yes; AA Similarity

50. One right triangle has an acute angle measuring 67°, and a second right triangle has an acute angle measuring 23°. Are the triangles similar? Explain. (Lesson 7–3)

51. 17.5

51. Solve the proportion $\frac{x}{5} = \frac{7}{2}$. (Lesson 7–1)

52. The median of a trapezoid is 18 inches long. If one base is 29 inches long, what is the length of the other base? (Lesson 6–5) **7 in.**

53. Yes; the sum of any two sides is greater than the length of the third side.

53. Can 7, 9, and 11 be the measures of the sides of a triangle? Explain. (Lesson 5–5)

54. Find the slope of a line that passes through the points $W(2, -9)$ and $C(0, 3)$. (Lesson 3–3) **−6**

Algebra

55. Use elimination to solve the system of equations. **(0, −4)**

$3x - 2y = 8$

$3x + 4y = -16$

56. Simplify $(a^3b)(a^2b^2)$. a^5b^3

Extension

Problem Solving The Greek symbol π is defined as the ratio $\frac{\text{circumference}}{\text{diameter}}$. Many people use the fraction $\frac{22}{7}$ as an approximation for the value of π. Is $\frac{22}{7}$ an appropriate approximation? Why or why not? **depends upon the accuracy required**

4 ASSESS

Closing Activity

Writing Have students write a few sentences describing the relationship between the radius, diameter, and circumference.

Teaching Tip Have students repeat Exercises 37 and 38 with other measures. Then ask if they come to the same conclusions.

Enrichment Masters, p. 49

9–1 NAME ______ DATE ______

Enrichment

Student Edition Pages 446–451

The Four Color Problem

Mapmakers have long believed that only four colors are necessary to distinguish among any number of different countries on a plane map. Countries that meet only at a point may have the same color provided they do not have an actual border. The conjecture that four colors are sufficient for every conceivable plane map eventually attracted the attention of mathematicians and became known as the "four-color problem." Despite extraordinary efforts over many years to solve the problem, no definite answer was obtained until the 1980s. Four colors are indeed sufficient, and the proof was accomplished by making ingenious use of computers.

The following problems will help you appreciate some of the complexities of the four-color problem. For these "maps," assume that each closed region is a different country.

1. What is the minimum number of colors necessary for each map?

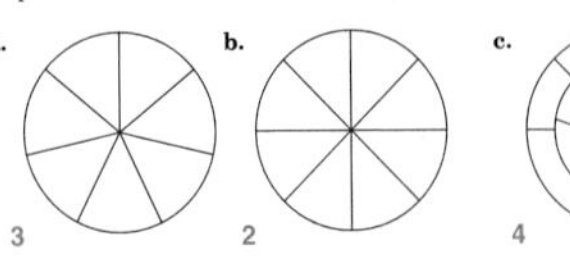

a. 3 b. 2 c. 4

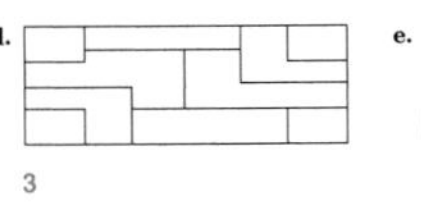

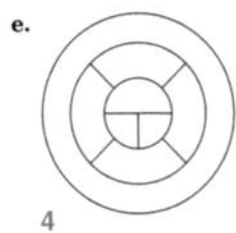

d. 3 e. 4

2. Draw some plane maps on separate sheets. Show how each can be colored using four colors. Then determine whether fewer colors would be enough. **See students' work.**

9-2 LESSON NOTES

NCTM Standards: 1–4, 7, 10

Instructional Resources

- Study Guide Master 9-2
- Practice Master 9-2
- Enrichment Master 9-2
- Assessment and Evaluation Masters, p. 240
- Multicultural Activity Masters, p. 18
- Real-World Applications, 18

Transparency 9-2A contains the 5-Minute Check for this lesson; **Transparency 9-2B** contains a teaching aid for this lesson.

Recommended Pacing	
Standard Pacing	Day 2 of 15
Honors Pacing	Day 2 of 13
Block Scheduling*	Day 2 of 9

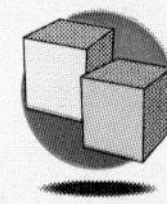

*For more information on pacing and possible lesson plans, refer to the *Block Scheduling Booklet*.

1 FOCUS

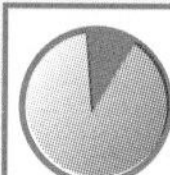

5-Minute Check
(over Lesson 9-1)

1. The circumference of Earth is about 25,000 miles. What is Earth's approximate diameter? **7957.75 miles**
2. If the circumference of a circle is 24 meters, find the radius. $\frac{24}{\pi}$ **meters**
3. If $d = 5\pi$, find C. $5\pi^2$
4. If a 24-inch bicycle makes 1000 revolutions, how far will the bicycle travel? (Remember that the size of the bicycle refers to the diameter of the wheels.) **about 75,360 inches or 2.4 miles**
5. Find the exact circumference of the circle below.

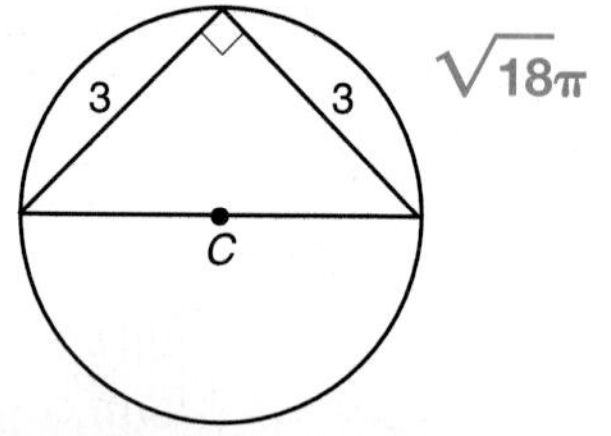

$\sqrt{18}\pi$

9-2 Angles and Arcs

What YOU'LL LEARN

- To recognize major arcs, minor arcs, semicircles, and central angles,
- to find measures of arcs and central angles, and
- to solve problems by making circle graphs.

Why IT'S IMPORTANT

Circle graphs are a useful way to display data. They allow you to see important characteristics of the data at a glance.

INTEGRATION

Statistics

The graph at the right shows how much families planned to spend on their vacations in 1996. How do you think the artist knew how large to make each section? *This will be solved in Example 1.*

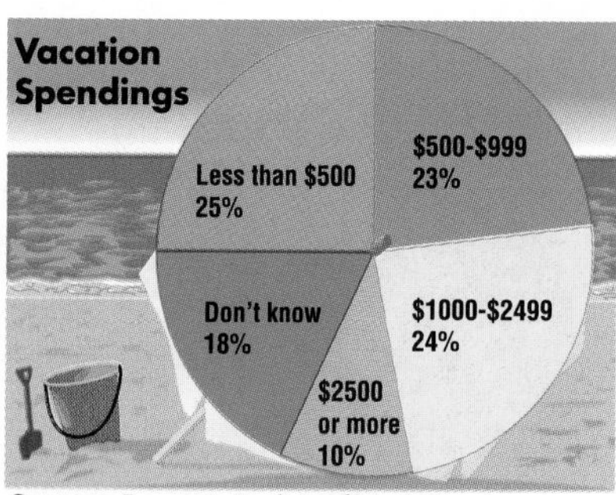

Source: *Economic Analysis of North American Ski Areas*

The graph above is a circle graph. Circle graphs are used to compare parts of a whole, usually as percents. Before drawing a circle graph, the artist needs to know that each section of the graph is a **central angle** of the circle. A central angle is an angle whose vertex is at the center of a circle.

Suppose we draw a diameter of $\odot C$ and name it $\overline{AB}$. We could then construct a diameter perpendicular to $\overline{AB}$ through C and name it $\overline{DE}$. $\angle ACD$, $\angle DCB$, $\angle BCE$, and $\angle ECA$ are central angles with no interior points in common. What is the sum of the measures of these angles?

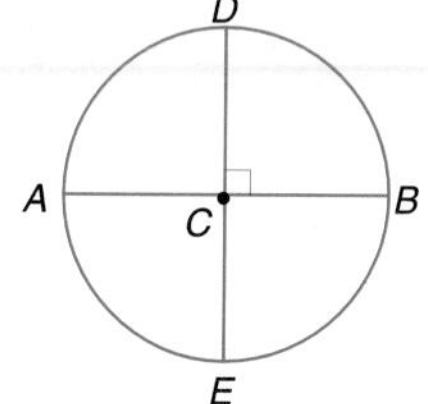

$$90 + 90 + 90 + 90 = 360$$

Sum of Central Angles — **The sum of the measures of the central angles of a circle with no interior points in common is 360.**

The central angles do not have to be right angles for the sum of the measures to be 360.

Example 1 Refer to the application at the beginning of the lesson. Determine the measure of each central angle used by the artist to draw the circle graph.

PROBLEM SOLVING

Make a Circle Graph

The artist knows that in a circle, the sum of the measures of the central angles should be 360. So the central angle that represents each different category of spending amounts for vacation should be proportional to the percent given in the statistics.

Central Angle	Category	%	Number of Degrees
1	Less than $500	25	25% of 360 = 90.0
2	$500 – $999	23	23% of 360 = 82.8
3	$1000 – $2499	24	24% of 360 = 86.4
4	$2500 or more	10	10% of 360 = 36.0
5	Don't know	18	18% of 360 = 64.8
	Total	100	360

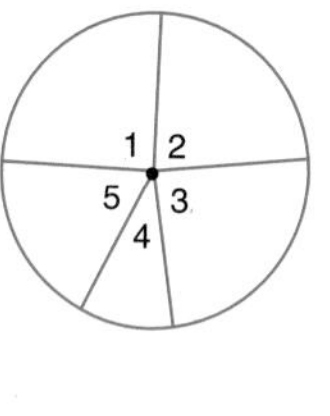

You can use a protractor to verify each angle measure in the graph.

TECHNOLOGY Tip

You can use software that has statistical graphing capabilities, like Microsoft Word® or Claris Works®, to make a circle graph.

From the circle graph, you may have noticed that a central angle separates a circle into two **arcs**.

Two letters are usually used to name a minor arc. However, three letters are often used when the measure of an arc is uncertain.

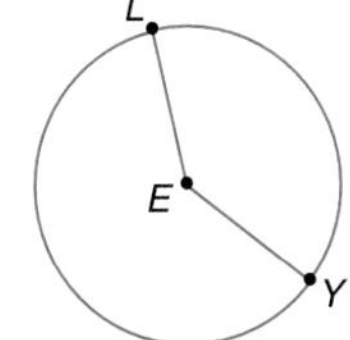

$\overset{\frown}{LY}$ is a **minor arc** of ⊙E.

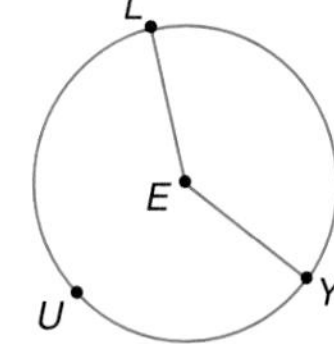

$\overset{\frown}{LUY}$ is a **major arc** of ⊙E.

Notice that minor arc LY, written $\overset{\frown}{LY}$, consists of its endpoints and all points on the circle interior to $\angle LEY$. Major arc LUY, written $\overset{\frown}{LUY}$, uses three letters to name the arc and consists of its endpoints and all points on the circle exterior to $\angle LEY$.

A semicircle is usually named by three letters.

Arcs are measured by their corresponding central angles. In ⊙C at the right, $m\angle PCM = 110$, so $m\overset{\frown}{PM} = 110$. The measure of a minor arc is always less than 180. The measure of a major arc is greater than 180. If the measure of an arc is 180, it is called a **semicircle**. Semicircles are congruent arcs formed when the diameter of a circle separates the circle into two arcs.

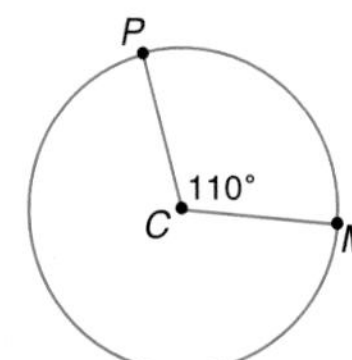

Definition of Arc Measure	**The measure of a minor arc is the measure of its central angle. The measure of a major arc is 360 minus the measure of its central angle. The measure of a semicircle is 180.**

Adjacent arcs are arcs of a circle that have exactly one point in common. As with adjacent angles, the measures of adjacent arcs can be added to find the measure of the arc formed by the adjacent arcs.

Postulate 9–1 Arc Addition Postulate	**The measure of an arc formed by two adjacent arcs is the sum of the measures of the two arcs. That is, if Q is a point on $\overset{\frown}{PR}$, then $m\overset{\frown}{PQ} + m\overset{\frown}{QR} = m\overset{\frown}{PQR}$.**

Example 2 **In ⊙E, $m\angle AEN = 18$, $\overline{JN}$ is a diameter, and $m\angle JES = 90$. Find each measure.**

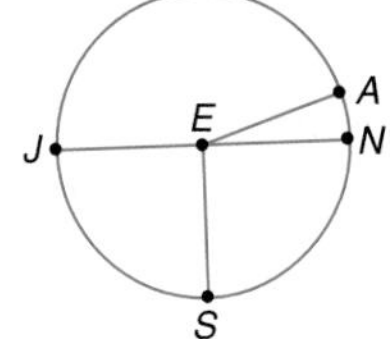

a. $m\overset{\frown}{AN}$

Since $\angle AEN$ is a central angle, $m\angle AEN = m\overset{\frown}{AN}$. Thus, $m\overset{\frown}{AN} = 18$.

b. $m\overset{\frown}{JA}$

By the Arc Addition Postulate,

$m\overset{\frown}{JAN} = m\overset{\frown}{JA} + m\overset{\frown}{AN}$

$180 = m\overset{\frown}{JA} + 18$ *$\overset{\frown}{JAN}$ is a semicircle.*

$m\overset{\frown}{JA} = 162$

c. $m\overset{\frown}{JAS}$

$\overset{\frown}{JAS}$ is the major arc for $\angle JES$. Therefore, $m\overset{\frown}{JAS} = 360 - m\overset{\frown}{JS}$. Since $m\overset{\frown}{JS} = 90$, $m\overset{\frown}{JAS} = 360 - 90$ or 270.

Motivating the Lesson

Questioning Ask students how they would cut a pie into five or six even pieces or how they would cut a pizza into eight even pieces. Relate this to central angles and arcs.

2 TEACH

In-Class Examples

For Example 1

A landfill in Jackson County was analyzed and it was found to contain 6% disposable diapers, 9% aluminum or metal, 15% plastic, 16% leaves and grass, and 54% paper. Draw a circle graph of the data.

Diapers: 6% of 360 = 21.6
Alum./metal: 9% of 360 = 32.4
Plastic: 15% of 360 = 54.0
Leaves/grass: 16% of 360 = 57.6
Paper: 54% of 360 = 194.4

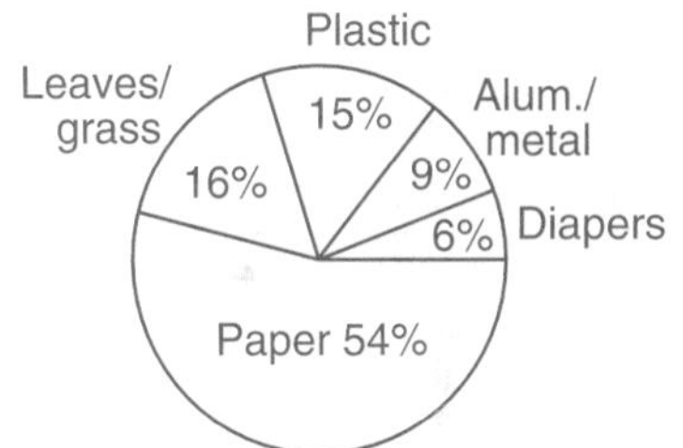

For Example 2

In ⊙Q, $\overline{AC}$ is a diameter and $m\angle CQD = 40$. Find $m\overset{\frown}{CD}$, $m\overset{\frown}{CAD}$, $m\overset{\frown}{AD}$, and $m\overset{\frown}{DCA}$.

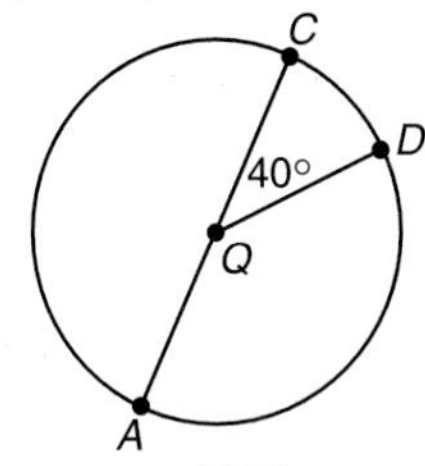

$m\overset{\frown}{CD} = 40$, $m\overset{\frown}{CAD} = 320$, $m\overset{\frown}{AD} = 140$, $m\overset{\frown}{DCA} = 220$

Teaching Tip If a section on a circle graph is greater than 50%, the angle will be greater than 180°. Also, it may be necessary to round the degree measures differently to get a total of 360°.

In-Class Example

For Example 3

In $\odot P$, $PR = 5$ and $m\angle QPR = 90$. Find the length of $\overset{\frown}{QR}$.

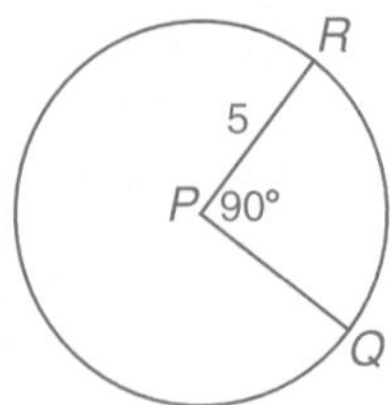

$\frac{10\pi}{4}$, or about 7.9 units

Teaching Tip In Example 3, point out to students that they can see that the area cut out by the two radii is proportional to the arc cut out by the two radii.

3 PRACTICE/APPLY

Check for Understanding

Exercises 1–18 are designed to help you assess your students' understanding through reading, writing, speaking, and modeling. You should work through Exercises 1–5 with your students and then monitor their work on Exercises 6–18.

Additional Answers

3. A diameter divides a circle into two semicircles. Each semicircle has an arc measure of 180.
4. The word *minor* denotes something smaller, and the word *major* denotes something greater, just as a minor arc is a smaller part of a circle and a major arc is a greater part of a circle.

You can also use the measure of the central angle to determine the **arc length**. The arc length (or length of an arc) is different from the degree measure of an arc. Suppose a circle was made of string. The length of the arc would be the linear distance of that piece of string representing the arc. The length of the arc is a part of the circumference proportional to the measure of the central angle when compared to the entire circle.

Example 3 **In $\odot P$, $PR = 9$ and $m\angle QPR = 120$. Find the length of $\overset{\frown}{QR}$.**

First, find what part of the circle is represented by $\angle QPR$.

$\frac{120}{360} = \frac{1}{3}$

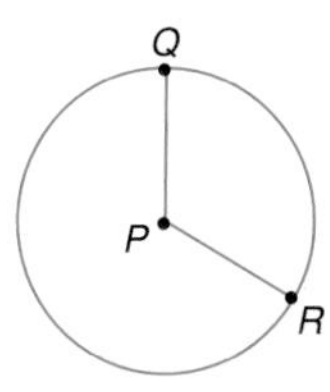

The angle is $\frac{1}{3}$ of the circle, so the length of $\overset{\frown}{QR}$ is $\frac{1}{3}$ of the circumference of $\odot P$.

length of $\overset{\frown}{QR} = \frac{1}{3}(2\pi r)$

$= \frac{1}{3}(2\pi)(9)$ $PR = 9$

$= 6\pi$ or about 18.8 units

The ripples after a stone is dropped into a pool of water are an example of concentric circles. **Concentric circles** lie in the same plane and have the same center, but have different radii. All circles are **similar circles**, so concentric circles are also similar.

Circles that have the same radius are **congruent circles**. Congruent circles are also similar circles. As with segments and angles, if two arcs of one circle have the same measure, then they are **congruent arcs**. Congruent arcs also have the same arc length.

CHECK FOR UNDERSTANDING

Communicating Mathematics

Study the lesson. Then complete the following. 3–4. See margin.

1. **Explain** why three letters are used to name a major arc of a circle. *1. to make it clear which points the arc goes through*
2. Refer to $\odot A$.
   a. **Explain** why $\overset{\frown}{BNR}$ and $\overset{\frown}{BRN}$ in $\odot A$ are not the same arc. *2a. $\overset{\frown}{BNR}$ has endpoints B and R and is a minor arc; $\overset{\frown}{BRN}$ has endpoints B and N and is a major arc.*
   b. **Explain** how to find $m\overset{\frown}{BRN}$ in $\odot A$ if $m\overset{\frown}{BN} = 25$. Subtract 360 − 25 to get 335.

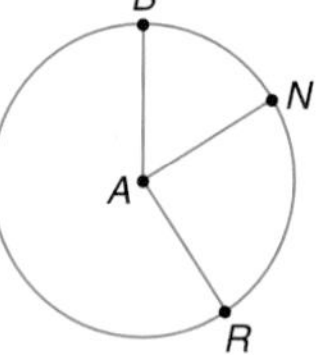

3. **Explain** why a diameter of a circle creates two arcs of degree measure 180.
4. **Discuss** the use of the words *minor* and *major* for arcs. Why are these terms used?

Alternative Teaching Strategies

Reading Geometry Find examples of circle graphs in magazines or newspapers. Have students analyze the data given in the graphs. Have them create examples and use the graphs to draw conclusions. Have students trade examples and answer questions based on the circle graphs.

Reteaching

Using Vocabulary Draw several circles on the chalkboard or overhead with central angles drawn and labeled. Ask each student to name a central angle, minor arc, or major arc of a circle.

Modeling Mathematics

5b. 150, 180; yes; arcs can be added similar to the way angles are added by the Angle Addition Postulate.

5. Draw a circle A with diameter $\overline{CT}$. Use your protractor to draw radius $\overline{AS}$ so that the measure of central angle SAT is 30.
   a. What is the measure of $\widehat{SCT}$? 330
   b. Arcs ST and SC have exactly one point in common. Use your protractor to measure $\widehat{SC}$ and $\widehat{TSC}$. Does $m\widehat{TSC} = m\widehat{CS} + m\widehat{ST}$? Compare this observation to the Angle Addition Postulate.

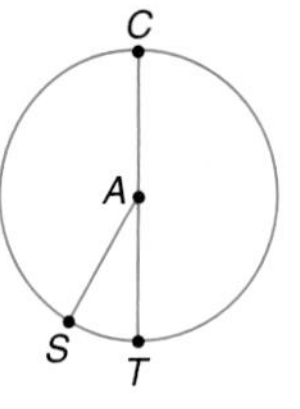

Guided Practice

In $\odot S$, $\overline{TE}$ and $\overline{KR}$ are diameters with $m\angle TSR = 42$. Determine whether each arc is a minor arc, a major arc, or a semicircle. Then find the degree measure of each arc. 6. semicircle; 180

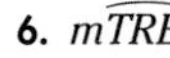

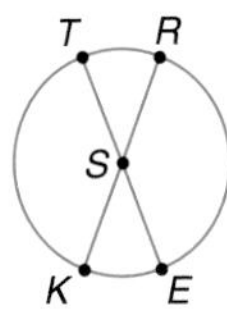

6. $m\widehat{TRE}$
7. $m\widehat{TK}$ minor; 138
8. $m\widehat{TRK}$ major; 222

In $\odot J$, $\overline{KM}$ is a diameter with $JL = 18$ and $m\angle KJL = 140$. Find the length of each arc. Round to the nearest tenth.

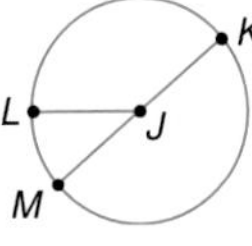

9. $\widehat{LM}$ 12.6
10. $\widehat{KL}$ 44.0
11. $\widehat{LMK}$ 69.1

INTEGRATION Algebra

In $\odot C$, $\overline{IL}$ is a diameter, $m\angle ICR = 3x + 5$, and $m\angle RCL = x - 1$.

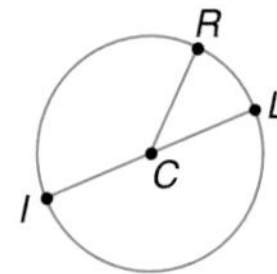

12. Find x. 44
13. Find $m\angle ICR$. 137
14. Find $m\widehat{ILR}$. 223

Determine whether each statement is *true* or *false*. Explain your reasoning. 15. true; definition of concentric circles

16. false; def. of concentric circles

17. True; concentric circles have the same center.

15. Two concentric circles have the same center.
16. Two concentric circles have the same radius.
17. Two concentric circles never intersect.
18. **Skiing** The graph below shows how people spend their money when they go skiing. b. The circle represents the sum of all parts of the whole.
    a. Draw a circle graph by hand or by using computer software. Label each part of the circle graph. See margin.
    b. Explain why the total percentage should add up to 100%.

Source: *Economic Analysis of North American Ski Areas*

Teaching Tip To begin a circle graph for Exercise 18, draw a horizontal line from the center to the edge of the circle. Use this as one side of the first angle measure. Then go around the circle counterclockwise and use the leg of the previous angle as the base of the angle you are drawing.

Additional Answer

18a.

Ski-lift tickets 54%
Miscellaneous 12%
Food/Beverage 11%
Non-skiing recreation 8%
Ski lessons 7%
Ski equipment/ clothing 4%
Equipment rental 4%

Assignment Guide

Core (with proof): 19–55 odd, 57–68
Core (informal): 19–55 odd, 57–68
Enriched: 20–50 even, 52–68

For **Extra Practice**, see p. 780.

The red A, B, and C flags, printed only in the Teacher's Wraparound Edition, indicate the level of difficulty of the exercises.

EXERCISES

Practice

Refer to $\odot E$ for Exercises 19–36. If $m\angle TEG = 21$ and $\overline{TR}$ is a diameter, determine whether each arc is a minor arc, a major arc, or a semicircle. Then find the degree measure of each arc.

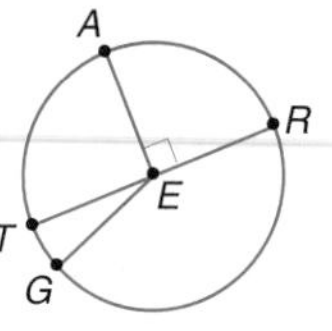

19. $m\widehat{TG}$ minor, 21
20. $m\widehat{ATR}$ major, 270
21. $m\widehat{AR}$ minor, 90
22. $m\widehat{TAR}$ — 22. semicircle, 180
23. $m\widehat{ATG}$ minor, 111
24. $m\widehat{ARG}$ major, 249
25. $m\widehat{RAG}$ major, 201
26. $m\widehat{TAG}$ major, 339
27. $m\widehat{GR}$ minor, 159

If $TR = 12$, find the length of each arc. Round to the nearest tenth.

28. $\widehat{TG}$ 2.2
29. $\widehat{ATR}$ 28.3
30. $\widehat{AR}$ 9.4
31. $\widehat{TAR}$ 18.8
32. $\widehat{ATG}$ 11.6
33. $\widehat{ARG}$ 26.1
34. $\widehat{RAG}$ 21.0
35. $\widehat{TAG}$ 35.5
36. $\widehat{GR}$ 16.7

Algebra

In $\odot G$, $\angle NGE \cong \angle EGT$, $m\angle AGJ = 4x$, $m\angle JGT = 2x + 24$, and $\overline{AT}$ and $\overline{JN}$ are diameters. Find each of the following.

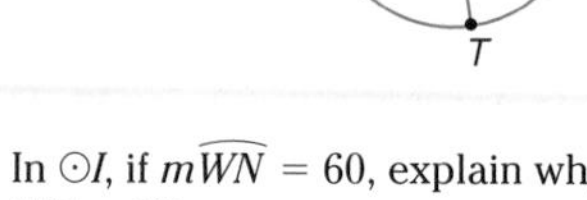

37. x 26
38. $m\angle AGJ$ 104
39. $m\angle JGT$ 76
40. $m\widehat{NE}$ 52
41. $m\widehat{NJT}$ 256
42. $m\widehat{JNE}$ 232

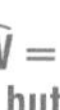

43. In $\odot I$, if $m\widehat{DG} = 132$, find $m\angle DGI$.

43. 24

44. In $\odot I$, if $m\widehat{WN} = 60$, explain why $WN = NI$.

44. Since $m\widehat{WN} = 60$, $m\angle WIN = 60$, but $\overline{WI} \cong \overline{IN}$ since they are radii, so $\angle IWN \cong \angle INW$.

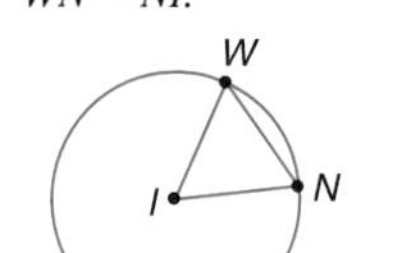

In the figure at the right, P is the center of two concentric circles. Determine whether each statement is *true* or *false*. Explain your reasoning.

45. If $m\widehat{RS} = 25$ and $m\widehat{CD} = 25$, then $\widehat{RS} \cong \widehat{CD}$.

45. False; arcs are not of the same circle.

46. If $m\angle CPD = 40$, then $m\widehat{RS}$ and $m\widehat{CD}$ are each 40. true
47. If $m\widehat{RS} = m\widehat{CD}$, then $\widehat{RS}$ must be congruent to $\widehat{CD}$.

47. False; arcs are not of the same circle.

48. $\overline{PC} \cong \overline{PD}$ true
49. If $m\widehat{RS} = 42$, then $m\widehat{CKD} = 318$. true

50. If $\overline{RT} \parallel \overline{AK}$ and $m\angle 3 = 2x$, find $m\widehat{AC}$. Justify your reasoning. $2x$

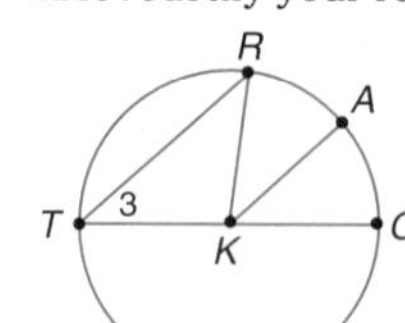

51. If $AE = 1$, find the measure of each side of square $ABCD$. $\frac{\sqrt{2}}{2}$

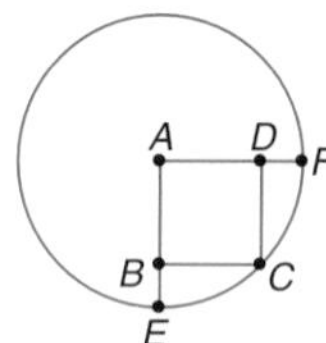

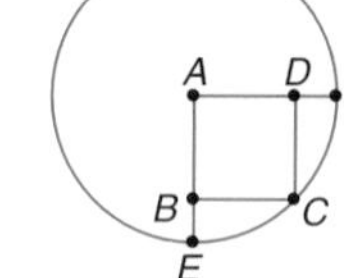

Study Guide Masters, p. 50

NAME__________ DATE __________

Study Guide

Student Edition
Pages 452–458

Angles and Arcs

An angle whose vertex is at the center of a circle is called a **central angle**. A central angle separates a circle into two arcs called a **major arc** and a **minor arc**. In the circle at the right, $\angle CEF$ is a central angle. Points C and F and all points of the circle interior to $\angle CEF$ form a minor arc called arc CF. This is written $\widehat{CF}$. Points C and F and all points of the circle exterior to $\angle CEF$ form a major arc called $\widehat{CGF}$.

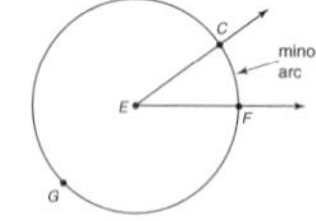

You can use central angles to find both the degree measure and the length of an arc. The arcs determined by a diameter are called semicircles and have measures of 180.

Examples: In $\odot R$, $m\angle ARB = 42$, $RB = 12$, and $\overline{AC}$ is a diameter.

1 Find $m\widehat{AB}$ and $m\widehat{ACB}$.
Since $\angle ARB$ is a central angle and $m\angle ARB = 42$, then $m\widehat{AB} = 42$.
$m\widehat{ACB} = 360 - m\widehat{AB} = 360 - 42$ or 318

2 Find the length of $\widehat{AB}$.
First, find what part of the circle is represented by $\angle ARB$.
$\frac{42}{360} = \frac{7}{60}$
So, the length of $\widehat{AB}$ is $\frac{7}{60}$ of the circumference of $\odot R$.
length of $\widehat{AB} = \frac{7}{60}(2\pi r)$
$= \frac{7}{60}(2\pi)(12)$ or about 8.8 units

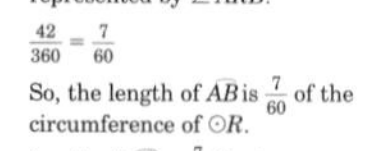

Refer to $\odot P$ for Exercises 1–8. If $\overline{SN}$ and $\overline{MT}$ are diameters with $m\angle SPT = 51$ and $m\angle NPR = 29$, determine whether each arc is a minor arc, a major arc, or a semicircle. Then find the degree measure of each arc.

1. $m\widehat{NR}$ minor; 29
2. $m\widehat{ST}$ minor; 51
3. $m\widehat{TSR}$ major; 260
4. $m\widehat{MST}$ semicircle; 180

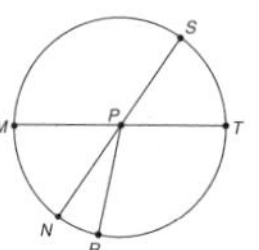

If $MT = 15$, find the length of each arc. Round to the nearest tenth.

5. $\widehat{NR}$ 3.8
6. $\widehat{ST}$ 6.7
7. $\widehat{TSR}$ 34.0
8. $\widehat{MST}$ 23.6

Critical Thinking

52–53. See margin.

52. Draw a diagram to explain how it is possible for two central angles to be congruent, yet their corresponding minor arcs are *not* congruent.

53. Draw a Venn diagram that illustrates the relationship among congruent, similar, and concentric circles.

Applications and Problem Solving

54. **Clocks** The Floral Clock in Frankfort, Kentucky, has a diameter of 34 feet. The hands on the clock form a central angle with the circular timepiece. Suppose the hour hand is on the 10 and the minute hand is on the 2.
    a. Find the measure of central angle ABC. 120
    b. Find the arc length of the minor arc. 35.6 ft

Where Do All the Apples Go?

Source: International Apple Institute

55. **Produce** Each year, the United States produces approximately 11 billion pounds of apples. The graph at the left shows where they go.
    a. Since $m\angle AEB = m\angle AED$, then is $m\widehat{AB} = m\widehat{AD}$? Explain.
    b. If diameter $AC = 16$, find the length of $\widehat{AB}$ to the nearest tenth. 20.6

a. Yes; if two central angles of one circle have equal measures, then their corresponding arcs have equal measures.

56. **Statistics** In 1995, a BKG Youth/Nintendo survey of students ages 9–18 was taken to find whom students would most like to trade places with. They were given five choices and the table at the right shows the results.

Person to Trade Places With	Number of Students
Head of a major company	700
NASA astronaut	350
President of the U.S.	1050
Professional athlete	1850
Teacher	1050

56a. 50.4, 25.2, 75.6, 133.2, 75.6

a. Determine the central angle measures to the nearest tenth degree needed to accurately construct a circle graph of the data.
b. Draw and label a circle graph that represents the data. Draw the graph by hand or by using computer software. See Solutions Manual.
c. Explain the advantages of displaying data in a graph rather than in a table. Use the graph you drew in part b to explain your answer. See margin.

Mixed Review

57. If the radius of a circle is 22 millimeters, find the circumference. Round to the nearest tenth. (Lesson 9–1) 138.2 mm

58. A square has a perimeter of 27 centimeters. Find the length of a diagonal of the square to the nearest tenth. (Lesson 8–2) 9.5 cm

59. Find the value of x if the triangles are similar. (Lesson 7–5) 4.5

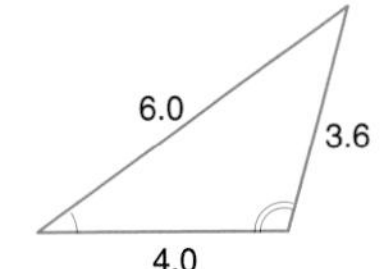

60. Determine whether the statement *If the corresponding sides of two polygons are proportional, the polygons are similar* is true or false. (Lesson 7–2) false

Additional Answers

52. Sample answer: $\angle CXD \cong \angle AXB$, but $\widehat{AB}$ is not congruent to $\widehat{CD}$.

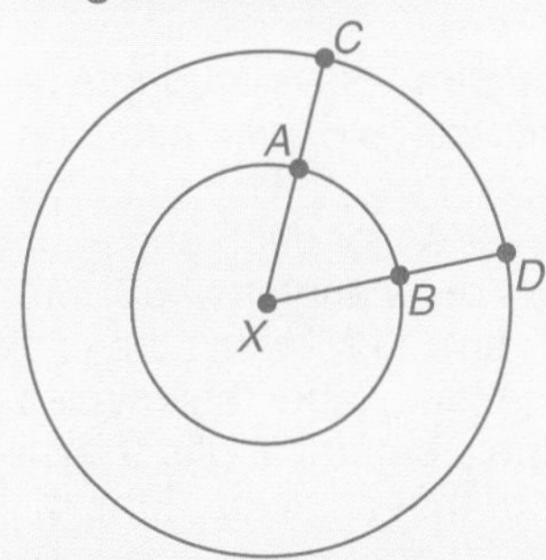

53. Sample answer:

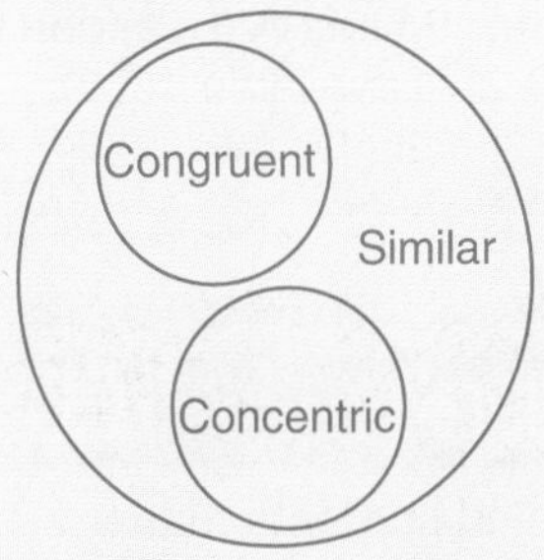

56c. It is helpful to compare data that are displayed in a graph. For example, it is easy to see that fewest students preferred trading places with an astronaut and the greatest number preferred trading places with a professional athlete.

Practice Masters, p. 50

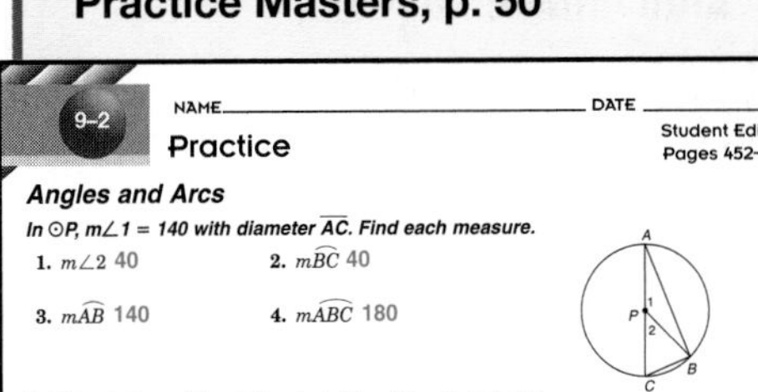

9–2 NAME______________ DATE ________

Practice — Student Edition Pages 452–458

Angles and Arcs

In ⊙P, $m\angle 1 = 140$ with diameter $\overline{AC}$. Find each measure.

1. $m\angle 2$ 40
2. $m\widehat{BC}$ 40
3. $m\widehat{AB}$ 140
4. $m\widehat{ABC}$ 180

In ⊙P, $m\angle 2 = m\angle 1$, $m\angle 2 = 4x + 35$, $m\angle 1 = 9x + 5$ with diameters $\overline{BD}$ and $\overline{AC}$. Find each of the following.

5. x 6
6. $m\widehat{AE}$ 59
7. $m\widehat{ED}$ 59
8. $m\angle 3$ 62
9. $m\widehat{AB}$ 62
10. $m\widehat{EC}$ 121
11. $m\widehat{EB}$ 121
12. $m\angle CPB$ 118
13. $m\widehat{CB}$ 118
14. $m\widehat{CEB}$ 242
15. $m\widehat{DC}$ 62
16. $m\widehat{CEA}$ 180

17. In ⊙A, $AB = 12$ and $m\angle BAC = 60$. Find the length of $\widehat{BC}$. 12.6

18. **Using Graphs** The table below shows how federal funds were spent on education in 1990.

1990 Federal Funds Spent for Education	
Elementary/Secondary	$ 7,945,177
Education for the Disabled	4,204,099
Post-Secondary Education	12,645,630
Public Library Services	145,367
Other	760,616
Total	$25,700,889

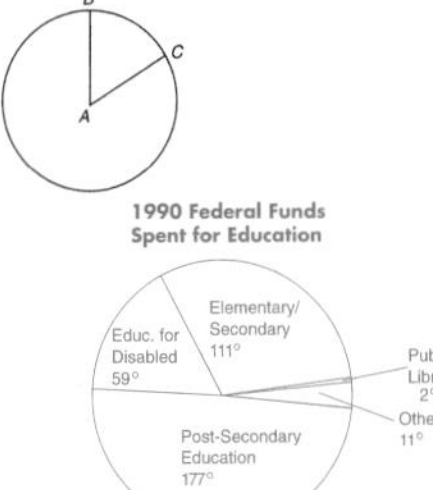

a. Use the information to make a circle graph.
b. Out of the $12,645,630 spent on post-secondary education, $10,801,185 went to post-secondary financial assistance. What percent is that of the $12,645,630? 85.4%

4 ASSESS

Closing Activity

Modeling Have students use a compass to draw a circle on a piece of paper. Using the center as the vertex for each angle, have them draw angles of 62°, 37°, 94°, 55°, and 112° within the circle. Have them label their circles and name a central angle, the minor arc of that angle, and the major arc of that angle.

Chapter 9 Quiz A (Lessons 9-1 and 9-2) is available in the *Assessment and Evaluation Masters*, p. 240.

Mathematics and SOCIETY

Remind students that any equation in two variables has a geometric representation on the plane. Point out that any equation in three variables has a geometric representation in space. So, any such equation is a plan for a sculpture.

Enrichment Masters, p. 50

9-2 NAME_____ DATE_____

Enrichment — Student Edition Pages 452–458

Measuring Angles

You measure an angle with a protractor. As you know, one unit for measuring angles is the degree. For any given circle, an arc of measure 1° is $\frac{1}{360}$ of the whole circle.

What if you had no protractor handy, but wanted to measure several angles in degrees? Could you use a straightedge and compass to construct an angle of 1° and then use that angle to make a protractor?

1. Suppose the only constructions you use are those of bisecting an angle (to get angles of lesser measure) and copying angles (to get angles of greater measure). If you start with an angle of measure 60 (from an equilateral triangle), will you be able to construct an angle of exactly 1°? Explain your answer.
No; bisecting and copying will always result in a measure of the form $\frac{15^m}{2^n}$, where m and n are whole numbers. (This is the measure you get by bisecting a 60° angle $n + 2$ times and putting m copies of the resulting small angle together to get a greater angle). $\frac{15^m}{2^n}$ cannot equal 1, since 3 is a factor of 15^m but is not a factor of 2^n.

2. Devise a practical method for making a fairly accurate protractor marked in degrees. Answers may vary. Possible answer: Mark off 360 equal segments on a long, straight strip of paper and join the ends to form a circular band.

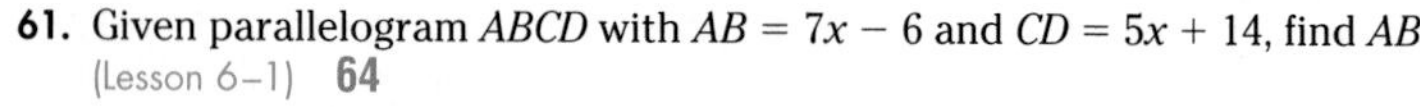

61. Given parallelogram $ABCD$ with $AB = 7x - 6$ and $CD = 5x + 14$, find AB. (Lesson 6–1) 64

62. State the assumption you would make to start an indirect proof of the statement *If two chords are congruent, they are equidistant from the center of the circle.* (Lesson 5–3)

62. If two chords are congruent, they are not equidistant from the center of the circle.

63. Find the value of x. (Lesson 4–6) 50

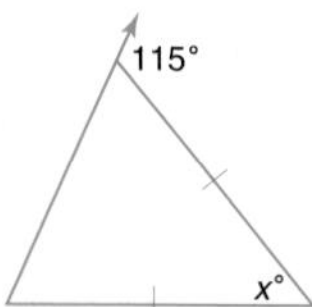

64. Describe the difference between obtuse and acute triangles. (Lesson 4–1)

64. Obtuse triangles have one obtuse angle and all the angles in an acute triangle are acute.

65. Determine whether the statement *If two parallel lines are cut by a transversal, then two consecutive interior angles are congruent* is true or false. (Lesson 3–2) false

66. **Sports** Write the converse of the statement *If a student plays a fall sport, then he or she must have a physical exam in the spring.* (Lesson 2–2) See margin.

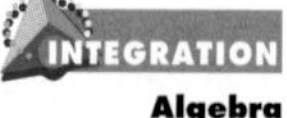

Algebra

67. Find $(3x^2y^4)(5x^4y)$. $15x^6y^5$

68. Find $\frac{9}{n-3} \cdot \frac{n^2-9}{12}$. $\frac{3n+9}{4}$

Math and the Sculptor

The excerpt below appeared in *Science News* on February 17, 1996.

From the outside, Helaman Ferguson's modest, two-car garage resembles just about any other garage in his suburban neighborhood....This is Ferguson's studio. Here, he carves stone and other materials to fashion works of art rooted in mathematical ideas.... a graceful, 3-foot-tall figure gradually emerges from a gleaming chunk of white Carrara marble. This sculpture is one of a series based on a type of mathematical form called a minimal surface. It's related to the shape of a soap film stretched across a bent ring.....Sculptors often make a small model...of what they want to carve....Ferguson has pioneered an alternative approach....he can translate geometric forms drawn on the computer screen directly into instructions on how much material the artist should remove at any point from an uncarved stone's surface to reveal the object. ■

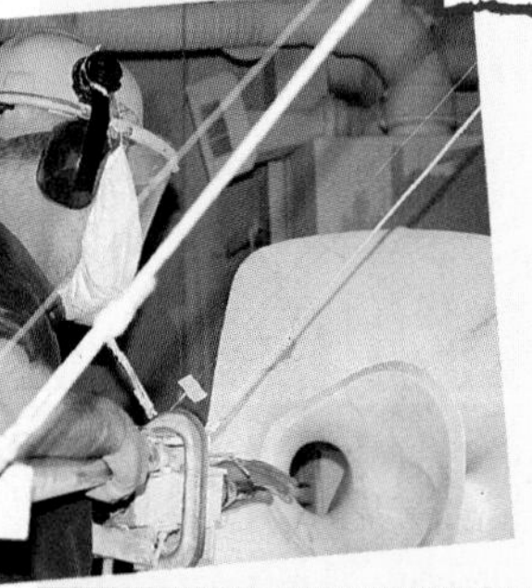

1. Which basic arithmetic operation could be used to describe the process of sculpting by stone carving? Which basic arithmetic operation could be used to describe the creation of a piece of art by combining pieces of aluminum, cloth, and wire? subtraction, addition
2. If you were viewing a three-dimensional model of a certain sculpture on a computer screen, it might be difficult to clearly see the curvature of all the surfaces. What would you suggest as a way to make the surface curvatures easier to see?
3. If you were creating a sculpture, would you like to work from a physical model, computer image, or an idea in your head? Why? Answers will vary.

2. Sample answer: Use different colors to represent areas of greater and lesser curvature.

Extension

Connections Have each student use a compass to draw a circle on a piece of paper. Have them draw five radii anywhere in the circle. Have them use a protractor to measure each angle created by the radii. What is the total measure of all the angles in each student's circle? 360 Have students form hypotheses based on this experiment.

Additional Answer

66. If a student must have a physical exam in the spring, then he or she plays a fall sport.

9-3 Arcs and Chords

What YOU'LL LEARN

- To recognize and use relationships among arcs, chords, and diameters.

Why IT'S IMPORTANT

You can use arcs and chords to solve problems involving geology.

Road Signs

The yellow railroad-crossing sign can be used to help explain special relationships that exist between arcs and chords. The sign is circular and has two perpendicular diameters.

If we label the center of the circle E and endpoints of the diameters A, B, C, and D, we can draw chords $\overline{AB}$, $\overline{BC}$, $\overline{CD}$, and $\overline{DA}$. When a minor arc and a chord share the same endpoints, we call the arc the **arc of the chord**. For example, in the railroad-crossing sign, $\widehat{AB}$ is the arc of $\overline{AB}$.

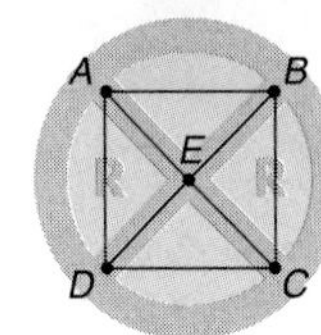

Since the diameters in the sign are perpendicular, we know the measure of each central angle is 90. Thus, $m\angle AEB = m\angle BEC = 90$, which implies $m\widehat{AB} = 90$. Since $m\widehat{AB} = m\widehat{BC}$, then $\widehat{AB} \cong \widehat{BC}$. How are $\overline{AB}$ and $\overline{BC}$ related? This relation is stated and proved below.

Theorem 9–1	In a circle or in congruent circles, two minor arcs are congruent if and only if their corresponding chords are congruent.

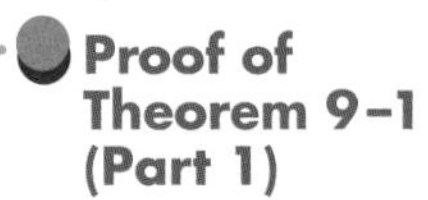

Proof of Theorem 9–1 (Part 1)

Prove that if two arcs of a circle are congruent, then their corresponding chords are congruent.

Given: $\odot E$

$\widehat{AB} \cong \widehat{DC}$

Prove: $\overline{AB} \cong \overline{DC}$

Proof:

Statements	Reasons
1. $\odot E$, $\widehat{AB} \cong \widehat{DC}$	1. Given
2. $\overline{AE} \cong \overline{CE}$ $\overline{BE} \cong \overline{DE}$	2. All radii of a $\odot$ are $\cong$.
3. $\angle AEB \cong \angle CED$	3. Vertical $\angle$s are $\cong$.
4. $\triangle AEB \cong \triangle CED$	4. SAS
5. $\overline{AB} \cong \overline{DC}$	5. CPCTC

You will be asked to prove the other part of Theorem 9–1 in Exercise 41. That is, prove that if two chords of a circle are congruent, then their corresponding arcs are congruent.

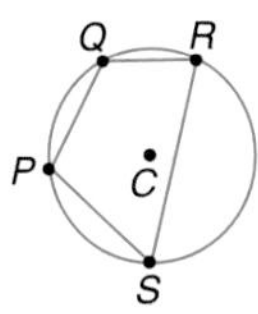

A polygon is an **inscribed polygon** if each of its vertices lies on a circle. The polygon is said to be inscribed in the circle. In the figure at the right, quadrilateral $PQRS$ is inscribed in $\odot C$.

9-3 LESSON NOTES

NCTM Standards: 1–4, 7

Instructional Resources

- Study Guide Master 9-3
- Practice Master 9-3
- Enrichment Master 9-3
- Modeling Mathematics Masters, p. 87
- Real-World Applications, 19

Transparency 9-3A contains the 5-Minute Check for this lesson; **Transparency 9-3B** contains a teaching aid for this lesson.

Recommended Pacing	
Standard Pacing	Days 3 & 4 of 15
Honors Pacing	Day 3 of 13
Block Scheduling*	Day 3 of 9

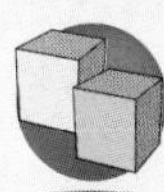

*For more information on pacing and possible lesson plans, refer to the *Block Scheduling Booklet*.

1 FOCUS

5-Minute Check
(over Lesson 9-2)

In the figure below, O is the center of two concentric circles. $\overline{AD}$ is a diameter, $m\angle AOB = 22$, and $\angle DOC$ is a right angle. Find each measure.

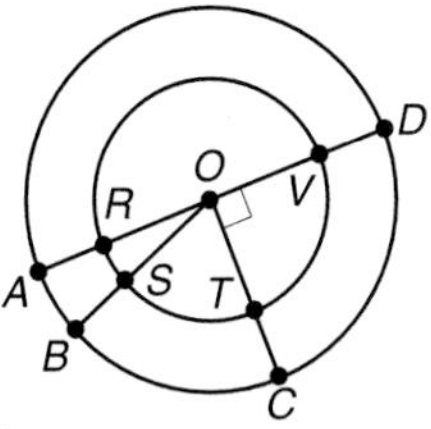

1. $m\widehat{AD}$ 180
2. $m\widehat{ST}$ 68
3. $m\angle DOS$ 158
4. $m\widehat{TVR}$ 270
5. $m\widehat{CDA}$ 270

Motivating the Lesson

Hands-On Activity Have students draw a circle and a central angle. Have them connect the two endpoints of the minor arc with a straight line. Point out that this is a chord of the circle. Have students draw a secant line through the circle. Point out that a chord is also formed here.

2 TEACH

In-Class Examples

For Example 1
Find the measure of each minor arc created when an equilateral triangle is inscribed in a circle. 120

For Example 2
In $\odot P$, $\overline{AB} \cong \overline{AC}$. Find the value of x to the nearest tenth.

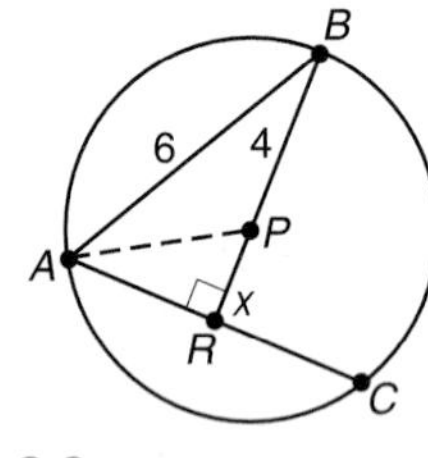

$\sqrt{7} \approx 2.6$

Teaching Tip When discussing Theorem 9-3, remind students that "if and only if" means that the theorem and its converse are true.

In this activity, use a chord and the center of the circle to create symmetry. The line of symmetry is perpendicular to the chord and passes through the center.

Answer for Modeling Mathematics

b. If a diameter is perpendicular to a chord, then it bisects the chord and its arc.

Example 1 **Road Signs**

A stop sign is an octagon with congruent sides. It can be inscribed in a circle by using the center of the sign and a vertex of the sign as endpoints for the radius of the circle, as in $\overline{QR}$. Find the measure of each of the eight corresponding arcs of a circle around the stop sign shown at the right.

Since all eight chords of the circle (or sides of the sign) are congruent, the eight corresponding arcs are congruent by Theorem 9–1. Therefore, each arc measures $\frac{360}{8}$ or 45.

The following activity provides insight into relationships between arcs and chords.

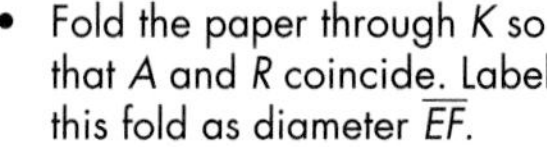

Chords and Diameters

Materials: 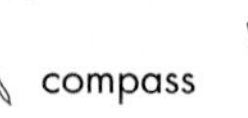 compass  patty paper 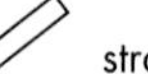 straightedge

- Use a compass to draw a circle on a piece of patty paper. Label the center K. Draw a chord that is not a diameter. Name it $\overline{AR}$.

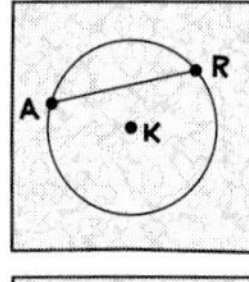

- Fold the paper through K so that A and R coincide. Label this fold as diameter $\overline{EF}$.

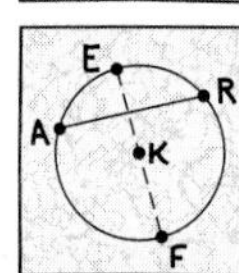

Your Turn

a. When the paper is folded, compare the lengths of $\widehat{AE}$ and $\widehat{ER}$. Then compare the lengths of $\widehat{AF}$ and $\widehat{FR}$. **They are equal; they are equal.**

b. Write a statement about the relationship between a diameter that is perpendicular to a chord and the chord and its arc. **See margin.**

Theorem 9–2	**In a circle, if a diameter is perpendicular to a chord, then it bisects the chord and its arc.**

You will be asked to prove this theorem in Exercise 28.

Example 2 **Chords $\overline{CH}$ and $\overline{IR}$ are equidistant from the center of $\odot S$. If $IR = 48$, find CH. SA is the distance from S to $\overline{CH}$, and SB is the distance from S to $\overline{IR}$.**

Remember that the distance from a point to a line is measured by a perpendicular segment.

Draw radii $\overline{SH}$ and $\overline{SR}$.
Since $\overline{SB} \perp \overline{IR}$, $BR = 24$ by Theorem 9–2.

$\triangle SAH$ and $\triangle SBR$ are right triangles.
Since $\overline{SB} \cong \overline{SA}$ (given) and $\overline{SR} \cong \overline{SH}$ (all radii are $\cong$), then $\triangle SAH \cong \triangle SBR$ by HL.

Since $\overline{AH} \cong \overline{BR}$ and $BR = 24$, then $AH = 24$.

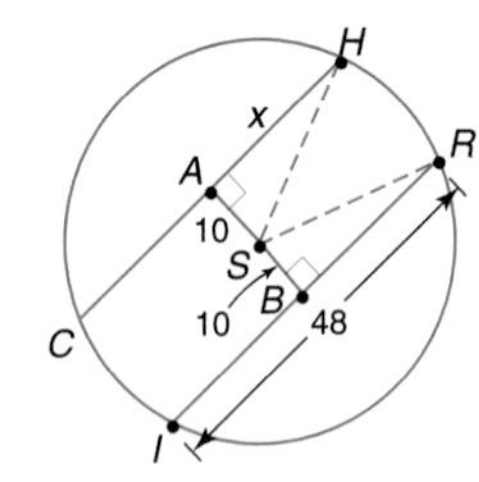

Since $\overline{AS}$ is a perpendicular bisector of $\overline{CH}$, by Theorem 9–2, $CH = 48$.

Cooperative Learning

Brainstorming Have students work in groups to explore regular polygons inscribed in a circle. They can inscribe regular n-gons for $n = 3, \ldots, 8$. They should use a ruler and protractor to measure angles and lengths of sides. Have them compile the results in a table and look for a pattern. For more information on the brainstorming strategy, see *Cooperative Learning in the Mathematics Classroom*, one of the titles in the Glencoe Mathematics Professional Series, page 30.

In Example 2, recall that $\overline{CH}$ and $\overline{IR}$ were equidistant from the center S. Also notice that $\overline{CH} \cong \overline{IR}$. This leads to the next theorem.

Theorem 9-3	**In a circle or in congruent circles, two chords are congruent if and only if they are equidistant from the center.**

You will be asked to prove this theorem in Exercises 42 and 43.

Example 3 **Suppose a chord of a circle is 10 inches long and 12 inches from the center of the circle. Find the length of the radius.**

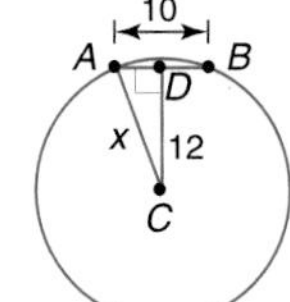

Draw a diagram and then solve.

In $\odot C$, $AB = 10$, $CD = 12$, and $\overline{AC}$ is the radius. Since $\triangle ADC$ is a right triangle, use the Pythagorean Theorem to find x, the length of the radius.

$(AD)^2 + 12^2 = x^2$ *Pythagorean Theorem*

$5^2 + 12^2 = x^2$ *AD = 5 Why?*

$25 + 144 = x^2$

$169 = x^2$

$x = 13$

Therefore, the length of the radius is 13.

CHECK FOR UNDERSTANDING

Communicating Mathematics

1. *ABCD* is a square. All arcs are congruent and all sides (chords) are congruent.

Study the lesson. Then complete the following.

1. **Examine** the labeled railroad-crossing sign at the beginning of the lesson. What type of quadrilateral is *ABCD*? Explain your reasoning.
2. **Extend** Example 1 by connecting every other vertex of the stop sign. What type of figure is formed? Explain how you could verify your conclusion using the theorems from this lesson. See margin.
3. **You Decide** Gabriella claims that $\overline{AB}$ bisects $\overline{PQ}$ since $\overline{AB} \perp \overline{PQ}$. Tiarri says $\overline{AB}$ doesn't bisect $\overline{PQ}$ because $\overline{AB}$ is not a diameter of K. Who is right and why?
Tiarri is right; to bisect $\overline{PQ}$, $\overline{AB}$ must be a diameter.

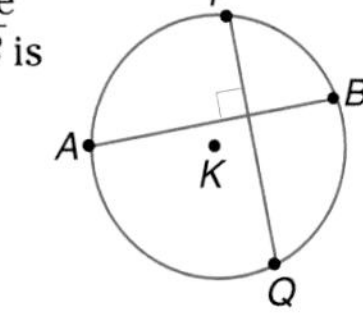

4. The intersecton of ℓ and m is the center of the circle because ℓ and m contain diameters.

MODELING MATHEMATICS

4. Use a compass to draw a circle on a piece of patty paper. Draw two nonparallel chords and label them $\overline{JK}$ and $\overline{RS}$. Fold the paper so that the endpoints of $\overline{JK}$ coincide. Label this fold as line ℓ. Fold the paper again so that the endpoints of $\overline{RS}$ overlap. Label this fold as line m. Write a conjecture about the intersection of lines ℓ and m. Explain.

Guided Practice

5. State the theorem that justifies the following statement for $\odot H$.
If $\overline{HG} \perp \overline{EF}$, then $\overline{EG} \cong \overline{GF}$.
See margin.

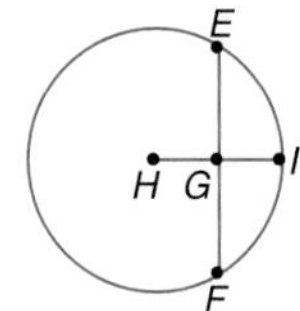

In-Class Example

For Example 3

You discovered a crop circle in a nearby farm. A chord of the circle is 500 feet long and 600 feet from the center of the circle. Find the length of the radius. 650 feet

3 PRACTICE/APPLY

Check for Understanding

Exercises 1–12 are designed to help you assess your students' understanding through reading, writing, speaking, and modeling. You should work through Exercises 1–4 with your students and then monitor their work on Exercises 5–12.

Error Analysis

In Theorem 9-2, make sure students understand that the segment must be a diameter in order to bisect a chord and its arc. Draw examples of circles that contain two perpendicular chords that are not congruent or bisected.

Additional Answers

2. Square; by Theorem 9-1, since the chords are congruent, the corresponding arcs are congruent. Then use the Arc Addition Postulate to show that the sides of the quadrilateral are congruent, forming the sides of the square.
5. In a circle, if a diameter is perpendicular to a chord, then it bisects the chord and its arc.

Reteaching

Using Rewriting Have students take the three theorems in this lesson and rewrite them as five theorems in if-then form. You may also want to ask them to write the converse of Theorem 9-2 in if-then form to make six theorems. Have them draw a picture for each theorem and write the given information as well as what the theorem is proving.

Assignment Guide

Core (with proof): 13–43 odd, 44, 45, 47–56
Core (informal): 13–27 odd, 29–39 odd, 44, 45, 47–56
Enriched: 14–42 even, 44–56

For **Extra Practice**, see p. 780.

The red A, B, and C flags, printed only in the Teacher's Wraparound Edition, indicate the level of difficulty of the exercises.

Additional Answers

11a. 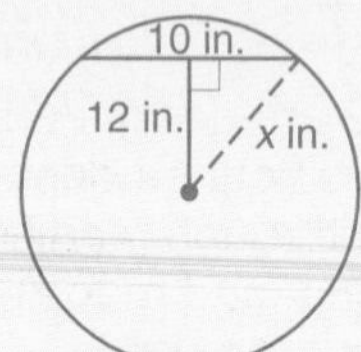

13. In a circle or in congruent circles, two minor arcs are congruent if and only if their corresponding chords are congruent.
14. In a circle, if a diameter is perpendicular to a chord, then it bisects the chord and its arc.
15. In a circle or in congruent circles, two chords are congruent if and only if they are equidistant from the center.

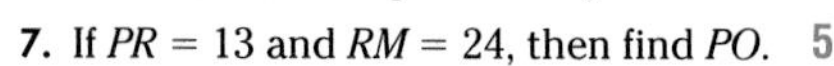
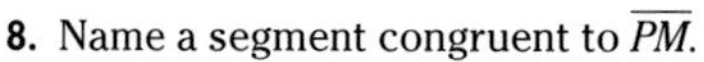

In ⊙P, $\overline{PQ} \perp \overline{RM}$.

6. Name an arc congruent to $\widehat{QR}$. $\widehat{QM}$
7. If $PR = 13$ and $RM = 24$, then find PO. 5
8. Name a segment congruent to $\overline{PM}$. 8. $\overline{PQ}$ or $\overline{PR}$

9. In ⊙R, $TR = 6.4$ and $EN = 10.8$. Find RO to the nearest tenth. 3.4

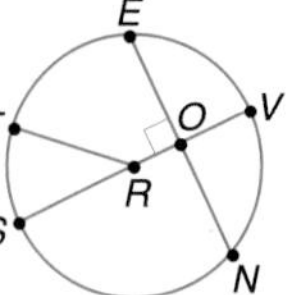

10. In ⊙O, $MO = 6$ and $LN = 16$. Find x. 10

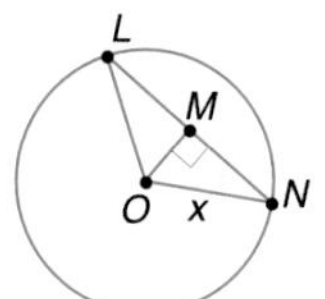

11. Suppose a chord of a circle is 10 inches long and is 12 inches from the center of the circle.
    a. Draw and label a figure. See margin.
    b. Find the length of the radius. 13 in.
12. **Food** Cecilia is barbecuing chicken. The grill on her barbecue is in the shape of a circle with a diameter of 54 centimeters. The horizontal wires are supported by 2 wires that are 12 centimeters apart as shown in the figure at the right. If the grill is symmetrical and the wires are evenly spaced, what is the length of each support wire? about 52.65 cm

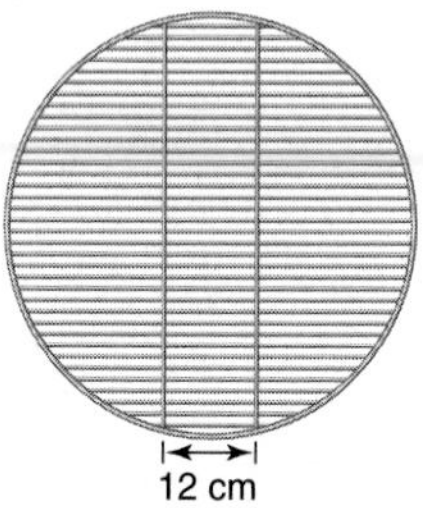

EXERCISES

Practice

State the theorem that justifies each statement. 13–15. See margin.

13. If $\overline{AC} \cong \overline{DG}$, then $\widehat{AC} \cong \widehat{DG}$.
14. If $\overline{JF} \perp \overline{DG}$, then $\overline{DE} \cong \overline{EG}$.
15. If $\overline{KP} \cong \overline{EP}$, then $\overline{AC} \cong \overline{DG}$.

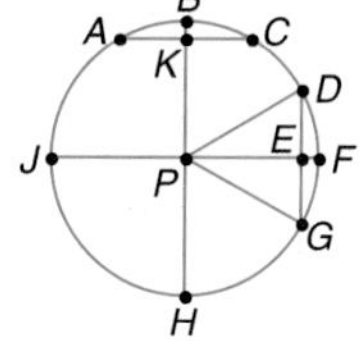

In ⊙D, $\overline{VR}$ and $\overline{QU}$ are diameters with $\overline{QU} \perp \overline{PR}$ and $\overline{QU} \perp \overline{VW}$.

16. Name the midpoint of $\overline{QU}$. D
17. Name the midpoint of $\overline{PR}$. S
18. If $VA = 9$, find VW. 18
19. Name two arcs congruent to $\widehat{QR}$. 19. $\widehat{PQ}$, $\widehat{VU}$, and $\widehat{UW}$
20. Name two segments congruent to $\overline{DQ}$. 20. $\overline{DU}$, $\overline{DV}$, $\overline{DT}$, $\overline{DR}$
21. Which chord is longer, $\overline{PR}$ or $\overline{VW}$? 21. Neither; they are ≅.
22. If $DS = 14$ and $PR = 32$, find DR to the nearest tenth. 21.3
23. Explain why $\overline{VW} \parallel \overline{PR}$. They are perpendicular to the same line.
24. Name a segment congruent to $\overline{QS}$. $\overline{AU}$

Alternative Learning Styles

Visual Have students draw two or three circles and two chords in each circle. Have them use a ruler to find the midpoints of the chords and make diameters in each circle by connecting the center with the midpoint of a chord. What relationship does the diameter have to the chord? They are perpendicular. Have students formulate conclusions from their discovery.

In each circle, M is the center. Find each measure.

25. $m\angle CAM$ 28
26. $m\widehat{SE}$ 100
27. CS 21

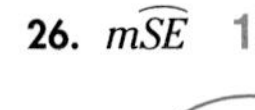

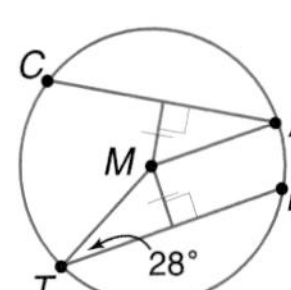

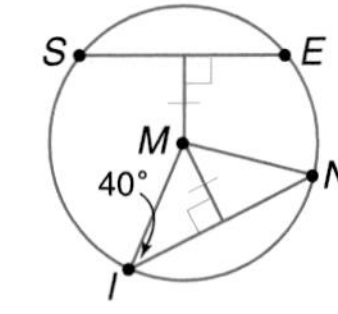

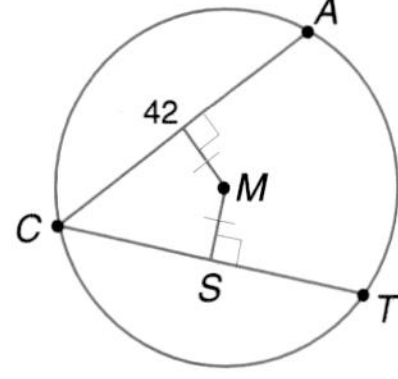

Proof

28. Complete the following proof of Theorem 9–2.

Given: $\odot P$
$\overline{AB} \perp \overline{TK}$

Prove: $\overline{AR} \cong \overline{BR}$
$\widehat{AK} \cong \widehat{BK}$

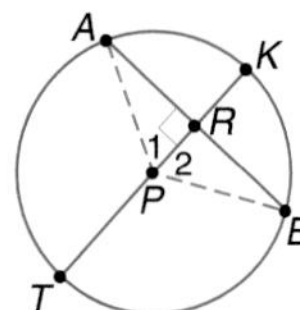

Proof:

Statements	Reasons
a. Draw radii $\overline{PA}$ and $\overline{PB}$.	a. ? Through any 2 pts. there is 1 line.
b. $\odot P$, $\overline{AB} \perp \overline{TK}$	b. ? Given
c. $\overline{PA} \cong \overline{PB}$	c. ? All radii of a circle are $\cong$.
d. $\overline{PR} \cong \overline{PR}$	d. ? Reflexive Prop. of $\cong$ Segments
e. ? $\triangle ARP \cong \triangle BRP$	e. HL
f. $\overline{AR} \cong \overline{BR}$ $\angle 1 \cong \angle 2$	f. ? CPCTC
g. ? $\widehat{AK} \cong \widehat{BK}$	g. Def. of $\cong$ arcs

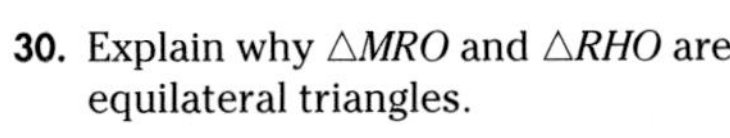

Refer to $\odot O$ with rhombus $RHOM$ for Exercises 29 and 30.

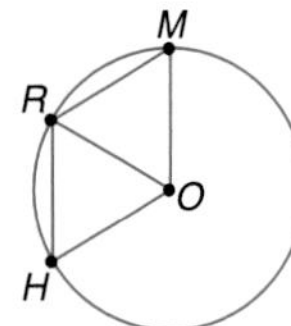

29. Explain why $\widehat{RH} \cong \widehat{MR}$. $\overline{RH} \cong \overline{MR}$

30. Explain why $\triangle MRO$ and $\triangle RHO$ are equilateral triangles.

30. Since $RHOM$ is a rhombus, $\overline{MR} \cong \overline{RH} \cong \overline{HO} \cong \overline{OM}$. Since all radii of a circle are congruent, $\overline{OR} \cong \overline{OM}$. Therefore, $\triangle MRO$ and $\triangle RHO$ are equilateral triangles.

31. In the figures at the right, $\overline{PC} \cong \overline{QT}$ and $\overline{AB} \cong \overline{RS}$. Is $\widehat{AB} \cong \widehat{RS}$? Explain your reasoning.
Yes; since radii are congruent, $\odot P \cong \odot Q$. Since the corresponding chords are congruent, by Theorem 9–1, $\widehat{AB} \cong \widehat{RS}$.

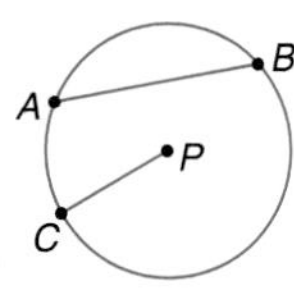

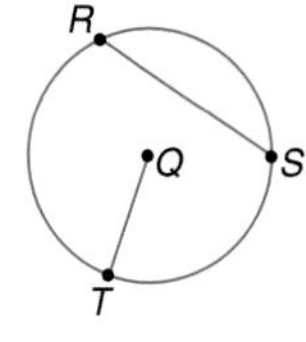

In each figure, J is the center of the circle. Find x to the nearest tenth.

32. 8.7
33. 24
34. 4

32.

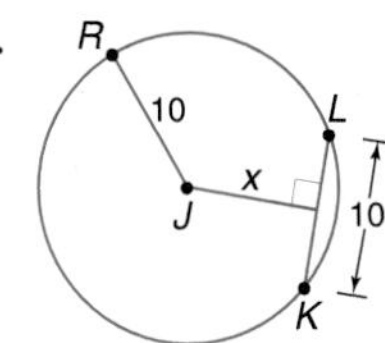

33.

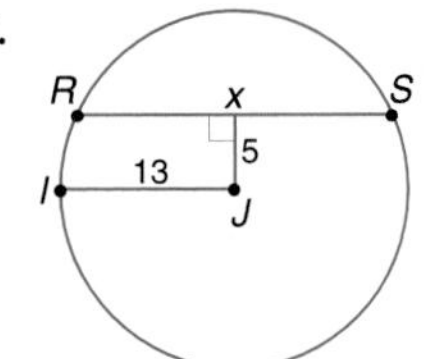

34.

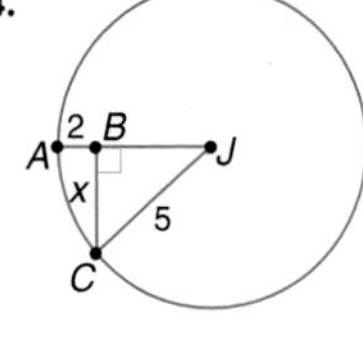

Food Service Worker Many two-year colleges offer programs in food service. Information is available from the Council on Hotel, Restaurant, and Institutional Education (202) 331-5990. For more information on tech prep, see the *Teacher's Handbook*.

Study Guide Masters, p. 51

NAME ______ DATE ______

9-3 Study Guide

Student Edition Pages 459–465

Arcs and Chords

The following theorems state relationships between arcs, chords, and diameters.

- In a circle or in congruent circles, two minor arcs are congruent if and only if their corresponding chords are congruent.
- In a circle, if a diameter is perpendicular to a chord, then it bisects the chord and its arc.
- In a circle or in congruent circles, two chords are congruent if and only if they are equidistant from the center.

Example: In the circle, O is the center, $OD = 15$, and $CD = 24$. Find x.

$ED = \frac{1}{2}CD$
$= \frac{1}{2}(24)$
$= 12$

$(OE)^2 + (ED)^2 = (OD)^2$
$x^2 + 12^2 = 15^2$
$x^2 + 144 = 225$
$x^2 = 81$
$x = 9$

In each circle, O is the center. Find each measure.

1. $m\widehat{NP}$ 80

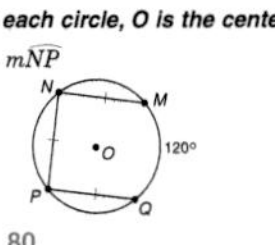

2. KM 24

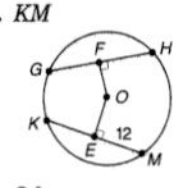

3. XY 32

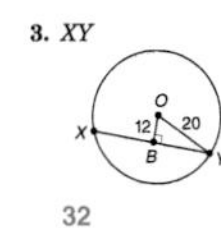

4. Suppose a chord is 20 inches long and is 24 inches from the center of the circle. Find the length of the radius. 26 in.

5. Suppose a chord of a circle is 5 inches from the center and is 24 inches long. Find the length of the radius. 13 in.

6. Suppose the diameter of a circle is 30 centimeters long and a chord is 24 centimeters long. Find the distance between the chord and the center of the circle. 9 cm

Additional Answers

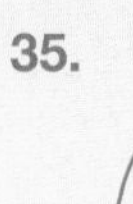

35.

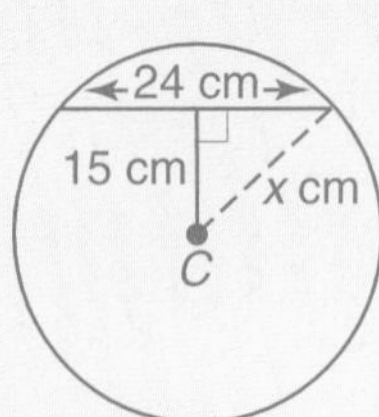

36.

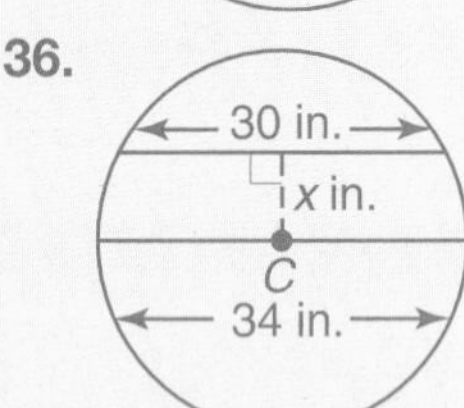

37.

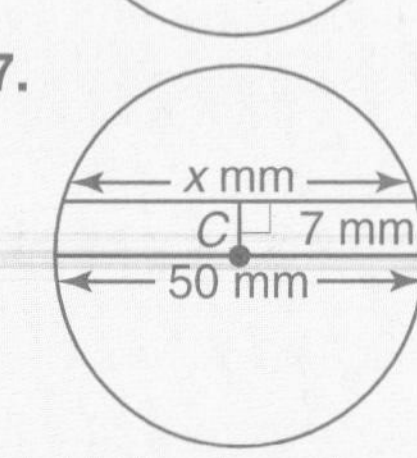

44. As the lengths of the chords increase, their distances to the center decrease.

Practice Masters, p. 51

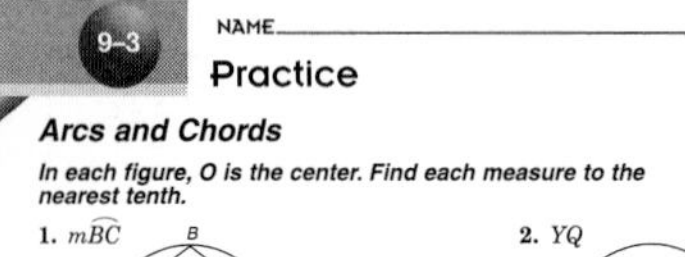

9-3 NAME______ DATE______

Practice

Student Edition Pages 459–465

Arcs and Chords

In each figure, O is the center. Find each measure to the nearest tenth.

1. $m\widehat{BC}$

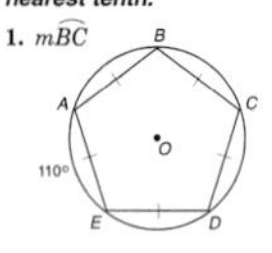

62.5

2. YQ

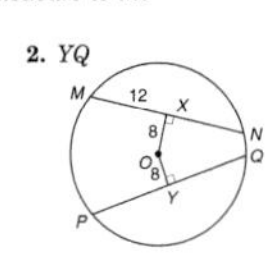

12

3. Suppose a chord of a circle is 16 inches long and is 6 inches from the center of the circle. Find the length of a radius. 10 in.

4. Find the length of a chord that is 5 inches from the center of a circle with a radius of 13 inches. 24 in.

5. Suppose a radius of a circle is 17 units and a chord is 30 units long. Find the distance from the center of the circle to the chord. 8 units

6. Find AB.

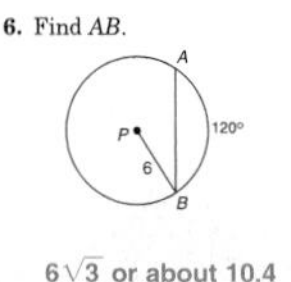

$6\sqrt{3}$ or about 10.4

7. Find AB

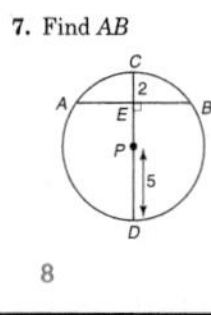

8

Draw a figure and then solve each problem.

35–37. See margin for figures.

36. 8 in.

35. Suppose a chord of a circle is 24 centimeters long and is 15 centimeters from the center of the circle. Find the length of the radius. 19.2 cm

36. Suppose the diameter of a circle is 34 inches long and a chord is 30 inches long. Find the distance between the chord and the center of the circle.

37. Suppose the diameter of a circle is 50 millimeters long and a chord is 7 millimeters from the center of the circle. Find the length of the chord. 48 mm

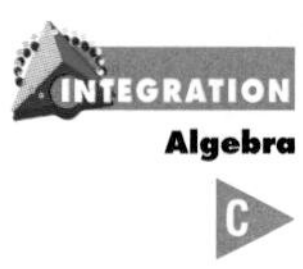

INTEGRATION Algebra C

In each figure, *O* is the center of the circle.

38. $\overline{MA} \cong \overline{TH}$, $MA = 8x + 4$, $TH = 12$, and $OQ = x^2$. Find OA. 6.1

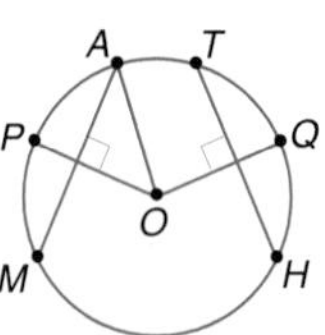
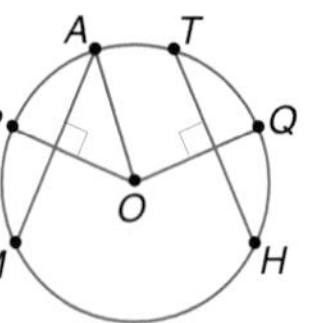

39. $\overline{PA}$ is perpendicular bisector for radius $\overline{OK}$ and $OK = 16$. Find AP. 27.7

40. Find the length of a chord that is the perpendicular bisector of a radius of length 30 units in a circle. 52.0 units

Proof

Draw a figure and write a proof for each theorem.

41–43. See Solutions Manual.

41. In a circle, if two chords are congruent, then their corresponding minor arcs are congruent. (second part of Theorem 9–1)

42. In a circle, if two chords are equidistant from the center, then they are congruent. (Theorem 9–3)

43. In a circle, if two chords are congruent, then they are equidistant from the center. (Theorem 9–3)

Critical Thinking

44. Draw a circle *R* and choose a point *P* on the circle. Now draw six chords with *P* as one endpoint. Label the other endpoints *A*, *B*, *C*, *D*, *E*, and *F*. Make a conjecture as to how the lengths of each chord are related to their distances from the center of the circle. Explain your reasoning. See margin.

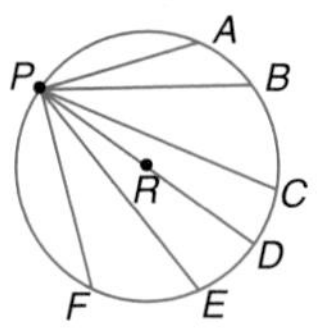

45a. See students' work.

45b. It is the only point equidistant from *T*, *S*, and *A*.

Applications and Problem Solving

45c. Minneapolis, Minnesota

45. **Geology** In order to locate the *epicenter* of an earthquake, geologists need data from three different seismograph stations. During an earthquake, suppose seismograph stations record the same earthquake intensity from Salt Lake City, Utah; Atlanta, Georgia; and Albuquerque, New Mexico.

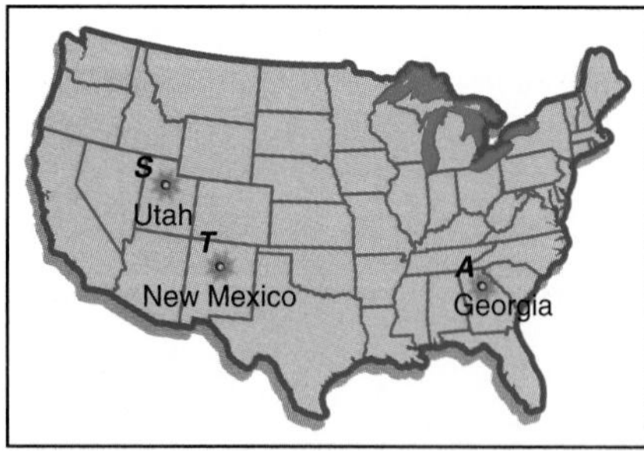

a. Copy the map at the left or use another map of the United States. Label each station as shown. (Salt Lake City–*S*, Atlanta–*A*, Albuquerque–*T*)

b. Since the stations recorded the same earthquake intensity, they are each equidistant from the epicenter or center of the circle that contains *T*, *S*, and *A*. Locate the epicenter by constructing the perpendicular bisectors of $\overline{ST}$ and $\overline{AT}$. Explain how you know the intersection of the two perpendicular bisectors is the center of the earthquake.

c. Use your map to locate a city near the epicenter.

Extension

Problem Solving Extend Theorem 9-2 to show that the major arcs of a chord are also bisected if a diameter is perpendicular to the chord. (See Exercise 28 for figure.) Using the figure for the proof of Theorem 9-2, draw $\overline{AT}$ and $\overline{BT}$. By Theorem 9-2, $\overline{AR} \cong \overline{BR}$. Also, $\overline{RT} \cong \overline{RT}$ since congruence of segments is reflexive. Since $\angle ART$ and $\angle BRT$ are right angles by Theorem 2-8, we can say $\triangle ART \cong \triangle BRT$ by LL. Therefore, $\overline{AT} \cong \overline{BT}$ by CPCTC. So, by Theorem 9-1, $\widehat{AT} \cong \widehat{BT}$.

46. See margin.

47. True; all points on a circle are equidistant from the center.

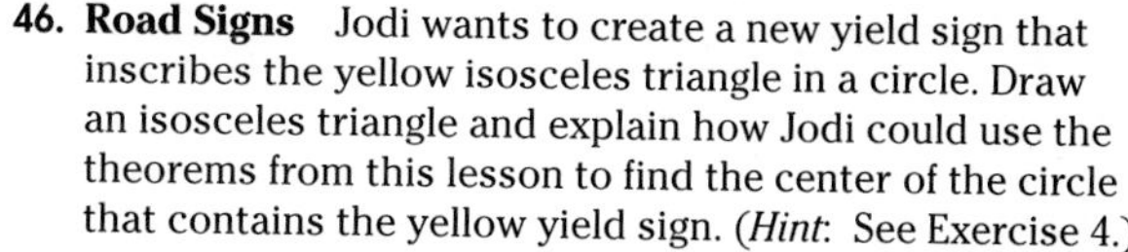

46. Road Signs Jodi wants to create a new yield sign that inscribes the yellow isosceles triangle in a circle. Draw an isosceles triangle and explain how Jodi could use the theorems from this lesson to find the center of the circle that contains the yellow yield sign. (*Hint*: See Exercise 4.)

Mixed Review

47. Determine whether the statement *All radii of a circle are congruent* is true or false. Explain your answer. (Lesson 9–2)

48. Native American Agriculture The *Hidatsa* are a people of the plains who live in what is now North Dakota. Their gardens, which are sometimes round, include corn, beans, squash, and sunflowers. If a typical round Hidatsa garden is 22 feet across, find the circumference of the garden to the nearest tenth of a foot. (Lesson 9–1) **69.1 ft**

49. Find the perimeter of $\triangle JKL$. (Lesson 8–2) **29.8**

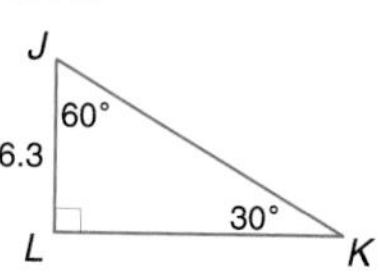

50. If $\triangle RAT \sim \triangle OFT$, name the proportional parts of the triangles. (Lesson 7–3) $\overline{RA}, \overline{OF}; \overline{AT}, \overline{FT}; \overline{RT}, \overline{OT}$

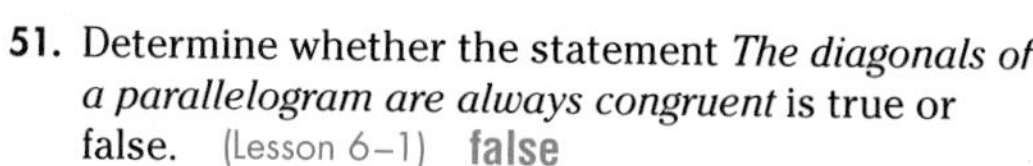

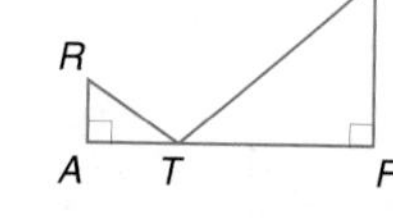

51. Determine whether the statement *The diagonals of a parallelogram are always congruent* is true or false. (Lesson 6–1) **false**

52. Can a median of a triangle also be an altitude? If so, what type of triangle is it? (Lesson 5–1) **yes; isosceles or equilateral**

53. If $\triangle PQR \cong \triangle HLM$, then $\angle R \cong$ _?_. (Lesson 4–3) $\angle M$

54. Identify the hypothesis and conclusion of the condition *If it is raining, then I will bring an umbrella.* (Lesson 2–2) **hypothesis: it is raining; conclusion: I will bring an umbrella**

Algebra

55. Find $-6rs(4r^2 + 1) + 8(rs + 2r)$. $-24r^3s + 2rs + 16r$

56. Find $\frac{21x^3y^5z}{7xy^6z}$. $\frac{3x^2}{y}$

WORKING ON THE In•ves•ti•ga•tion

Refer to the Investigation on pages 392–393.

After the space station is established on Mars, a team of scientists will begin studying its moons.

1. Refer to the information on page 393 to find the time it takes Mars to make one complete rotation. Given the amount of time that it takes for the planet to revolve 360°, how many degrees of rotation does Mars make in one hour?
2. Two outposts of the space station will be located so that each moon will pass directly over an outpost on each orbit. Use the information about the orbit times of each moon to determine how often each moon will pass over its outpost.
3. Using the distances from Mars to its moons, find the relative sizes of the moons in Mars' night sky.
4. Determine whether there will there be eclipses of the Sun by either of Mars' moons.

Add the results of your work to your Investigation Folder.

In•ves•ti•ga•tion

Working on the Investigation

The Investigation on pages 392–393 is designed to be a long-term project that is completed over several days or weeks. Encourage students to keep their materials in their Investigation Folder as they work on the Investigation.

Sample Answers for Working on the Investigation

1. about 14.6°
2. Phobos: about 10 hours, 47 minutes; Deimos: about 138 hours, 28 minutes
3. Phobos appears larger in the night sky of Mars.
4. Since the moons are not in the same plane of orbit as the Sun, eclipses of the Sun will occur.

4 ASSESS

Closing Activity

Writing Have students write the converse of Theorems 9-1, 9-2, and 9-3.

Additional Answer

46. Construct the perpendicular bisectors of two sides of the triangle. Their intersection is the center of the circle.

Enrichment Masters, p. 51

NAME__________ DATE______

Enrichment

Student Edition Pages 459–465

Patterns from Chords

Some beautiful and interesting patterns result if you draw chords to connect evenly spaced points on a circle. On the circle shown below, 24 points have been marked to divide the circle into 24 equal parts. Numbers from 1 to 48 have been placed beside the points. Study the diagram to see exactly how this was done.

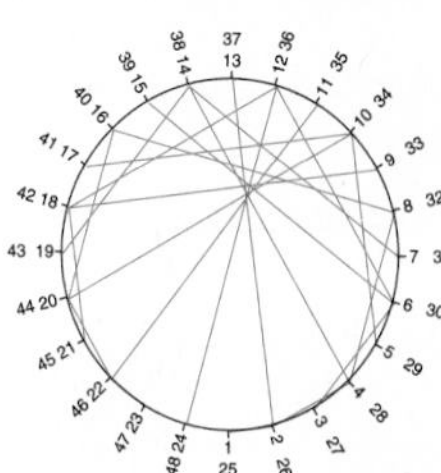

1. Use your ruler and pencil to draw chords to connect numbered points as follows: 1 to 2, 2 to 4, 3 to 6, 4 to 8, and so on. Keep doubling until you have gone all the way around the circle. What kind of pattern do you get? For figure, see above. The pattern is a heart-shaped figure.

2. Copy the original circle, points, and numbers. Try other patterns for connecting points. For example, you might try tripling the first number to get the number for the second endpoint of each chord. Keep special patterns for a possible class display. See students' work.

9-4 LESSON NOTES

NCTM Standards: 1–5, 7

Instructional Resources

- Study Guide Master 9-4
- Practice Master 9-4
- Enrichment Master 9-4
- Assessment and Evaluation Masters, pp. 239, 240
- Modeling Mathematics Masters, pp. 47–50

Transparency 9-4A contains the 5-Minute Check for this lesson; **Transparency 9-4B** contains a teaching aid for this lesson.

Recommended Pacing	
Standard Pacing	Days 5 & 6 of 15
Honors Pacing	Day 4 of 13
Block Scheduling*	Day 4 of 9

*For more information on pacing and possible lesson plans, refer to the *Block Scheduling Booklet*.

1 FOCUS

5-Minute Check
(over Lesson 9-3)

In ⊙B, $\overline{KN} \perp \overline{JL}$, and $\overline{KN}$ and $\overline{JP}$ are diameters.

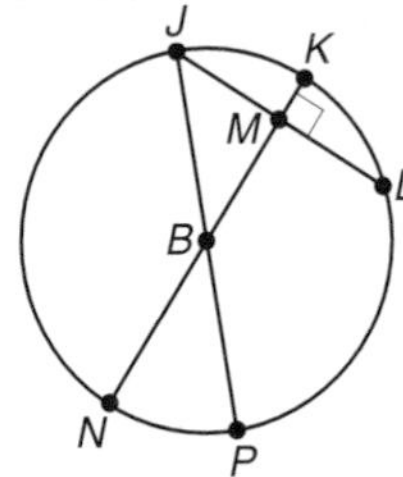

1. Name a segment congruent to $\overline{JM}$. $\overline{ML}$
2. Name an arc congruent to $\overset{\frown}{KL}$. $\overset{\frown}{JK}$
3. Name an arc congruent to $\overset{\frown}{JN}$. $\overset{\frown}{LN}$
4. Name a segment congruent to $\overline{JB}$. $\overline{KB}$, $\overline{BN}$, or $\overline{BP}$
5. Name the midpoint of $\overline{JL}$. M

9–4 Inscribed Angles

What YOU'LL LEARN

- To recognize and find measures of inscribed angles, and
- to apply properties of inscribed figures.

Why IT'S IMPORTANT

You can use inscribed angles to solve problems involving civil engineering and carpentry.

APPLICATION
Building Trades

Hardware stores often sell various nuts and bolts that are in a hexagonal shape. These fasteners have numerous uses in the building trades.

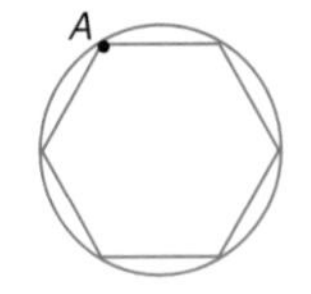

Machinists usually cut hexagonal shapes from round or circular stock. The machinist must trim or cut the stock just right to form a hexagon with all sides congruent. How could the machinist determine the appropriate cuts to make in the stock? *You will be asked to solve this problem in Exercise 16.*

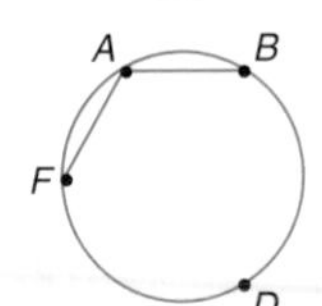

Each angle of the hexagon is an **inscribed angle**. An inscribed angle is an angle whose vertex is on the circle and whose sides each contain chords of the circle. We say that $\angle FAB$ intercepts $\overset{\frown}{FDB}$. $\overset{\frown}{FDB}$ is called the **intercepted arc** of $\angle FAB$. Notice that $\overset{\frown}{FDB}$ lies in the interior of $\angle FAB$.

The following activity explores the relationship between the measures of an inscribed angle and its intercepted arc.

MODELING MATHEMATICS — **Inscribed Angles**

Materials:  protractor  straightedge

- Trace ⊙P with inscribed hexagon $ABCDEF$ on a piece of paper. Note that all sides are congruent.

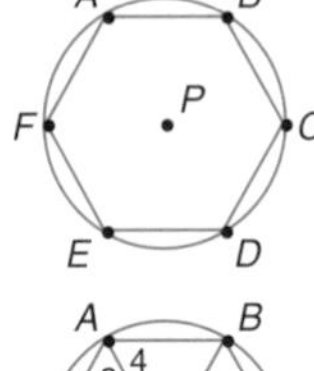

- Draw radii $\overline{PA}$, $\overline{PF}$, and $\overline{PB}$ and label angles 1, 2, 3, and 4.

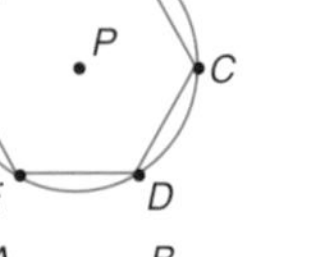
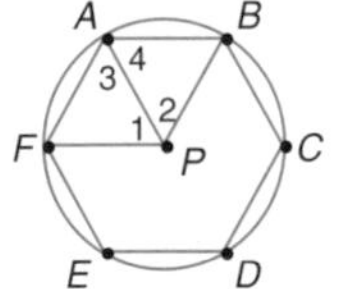

- Use a protractor to measure each numbered angle.

a. 60, 60; the measure of each minor arc equals the measure of its central angle.

Your Turn

a. Find $m\overset{\frown}{FA}$ and $m\overset{\frown}{AB}$. Explain your reasoning.
b. Find $m\angle FAB$ and $m\overset{\frown}{BDF}$. **120, 240**
c. Make a conjecture regarding the relationship between $m\angle FAB$ and $m\overset{\frown}{BDF}$. **The measure of $\angle FAB$ equals one-half the measure of $\overset{\frown}{BDF}$.**

The Modeling Math activity suggests the following theorem.

Theorem 9–4	**If an angle is inscribed in a circle, then the measure of the angle equals one-half the measure of its intercepted arc.**

This activity leads students to formulate a conjecture, which will be stated as Theorem 9-4. You may want to ask students to draw $\overline{PE}$ and make a conjecture about the relationship between $m\angle AFE$ and $m\overset{\frown}{ACE}$.

There are three cases to consider when writing a proof of this theorem.

Case 1: The center of the circle lies on one of the sides of the angle.

Case 2: The center of the circle is in the interior of the angle.

Case 3: The center lies in the exterior of the angle.

Proof of Theorem 9–4 (Case 1)

Given: $\angle PRQ$ inscribed in $\odot T$ and $\overline{PR}$ contains T.

Prove: $m\angle PRQ = \frac{1}{2}m\widehat{PQ}$

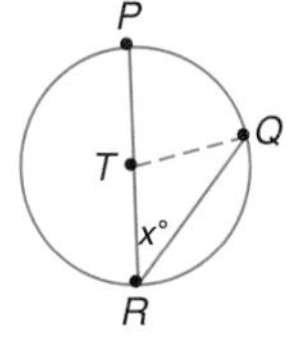

Paragraph Proof:

Draw radius $\overline{TQ}$ and let $m\angle PRQ = x$. Since $\angle PTQ$ is a central angle, $m\widehat{PQ} = m\angle PTQ$. Since $\overline{TQ}$ and $\overline{TR}$ are radii, $\triangle TQR$ is isosceles and $m\angle TQR = x$. By the Exterior Angle Theorem, $m\angle PTQ = 2x$. Therefore, $m\widehat{PQ} = 2x$ and $m\angle PRQ = \frac{1}{2}m\widehat{PQ}$.

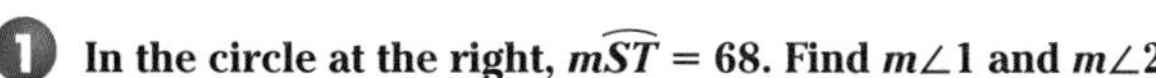

You will be asked to prove Cases 2 and 3 of Theorem 9–4 in Exercises 50 and 51, respectively.

Example 1

In the circle at the right, $m\widehat{ST} = 68$. Find $m\angle 1$ and $m\angle 2$.

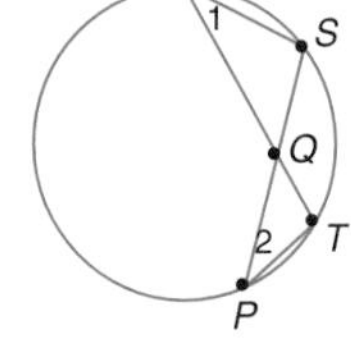

Since $\widehat{ST}$ is the intercepted arc for both $\angle 1$ and $\angle 2$, apply Theorem 9–4.

$m\angle 1 = \frac{1}{2}m\widehat{ST}$ $\quad$ $m\angle 2 = \frac{1}{2}m\widehat{ST}$

$= \frac{1}{2}(68)$ or 34 $\quad$ $= \frac{1}{2}(68)$ or 34

Therefore, $m\angle 1 = m\angle 2 = 34$.

Notice that Q is not the center of the circle.

Example 1 illustrates an important relationship between inscribed angles and intercepted arcs of equal measure.

Theorem 9–5	**If two inscribed angles of a circle or congruent circles intercept congruent arcs or the same arc, then the angles are congruent.**

Suppose an inscribed angle intercepts a semicircle on $\odot P$. The measure of semicircle FHG is 180, so $m\angle FEG = 90$. This is more formally stated in Theorem 9–6.

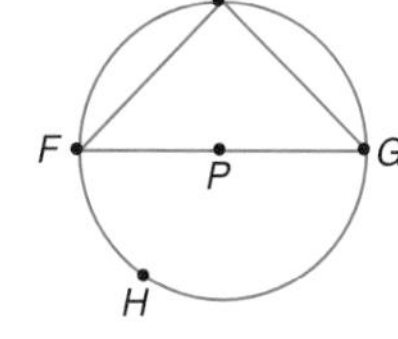

Notice that $\overline{FG}$ is a diameter of circle P.

Theorem 9–6	**If an inscribed angle of a circle intercepts a semicircle, then the angle is a right angle.**

You can use this theorem to find the center of a circle given any three points on the circle.

Motivating the Lesson

Situational Problem Ask students what an inscription is, or what it means to inscribe something. **Sample answer: writing on something, like a piece of jewelry.** Relate this to an inscribed angle, which is "an angle written on a circle" or an angle whose vertex lies on the circle.

2 TEACH

In-Class Example

For Example 1

In $\odot Q$, $\overline{AC}$ is a diameter, $m\widehat{CD} = 68$, and $m\widehat{BE} = 96$. Find $m\angle ABC$, $m\angle BDE$, $m\angle CED$, and $m\widehat{AD}$.

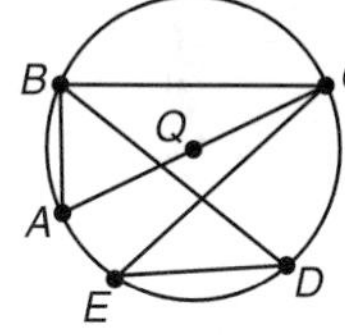

$m\angle ABC = 90$, $m\angle BDE = 48$, $m\angle CED = 34$, $m\widehat{AD} = 112$

Teaching Tip When discussing the three cases for Theorem 9-4, draw a figure for each case.

Teaching Tip Remind students that they can find the center of a circle by folding the circle in half at two different locations. The folds are diameters of the circle, and diameters meet at the center.

Classroom Vignette

"You may wish to have students construct circle P with inscribed hexagon $ABCDEF$ for the Modeling Mathematics activity. The instructions for this construction can be found on page 543 of the Student Edition."

Carol Malloy
Author

In-Class Example

For Example 2

In $\odot X$, $m\angle 4 = 7x + 3$, $m\angle 5 = 7x + 3$, and $m\angle 1 = 5x$. Find $m\angle 1$, $m\angle 2$, $m\angle 4$, and $m\angle 5$.

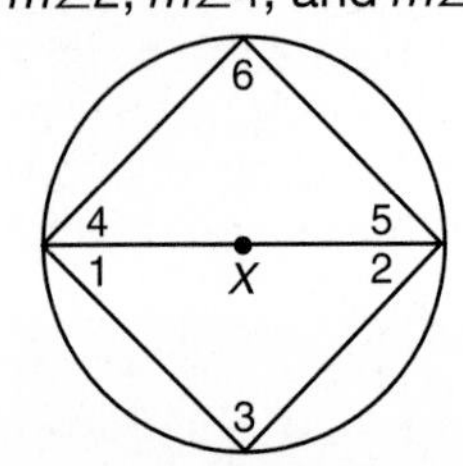

$m\angle 1 = 30$, $m\angle 2 = 60$, $m\angle 4 = m\angle 5 = 45$

Teaching Tip Not all polygons can be inscribed in a circle. Draw examples of polygons that cannot be inscribed in a circle.

CONSTRUCTION

Finding the Center of a Circle

Given three points on a circle, locate the center of the circle.

1. Draw a circle and locate three points A, B, and C on it.

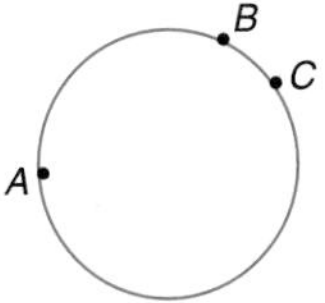

2. Draw $\overline{AB}$. Construct a segment through B perpendicular to $\overline{AB}$ and call it $\overline{BD}$.

   Draw $\overline{AD}$. Since $\overline{BD} \perp \overline{AB}$, $m\angle ABD = 90$ and the arc it intercepts has a measure of 180. This means that $\overline{AD}$ is a diameter.

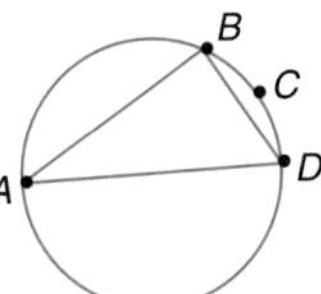

3. Draw $\overline{BC}$ and construct a segment through B perpendicular to $\overline{BC}$ and label it $\overline{BF}$. Draw $\overline{CF}$. Since $m\angle FBC = 90$, $\overline{CF}$ is also a diameter.

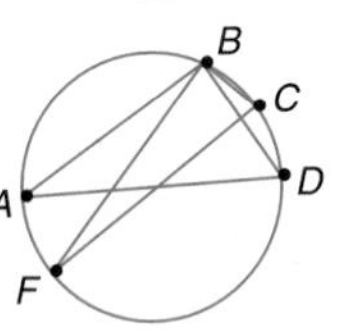

Conclusion: The point that all diameters have in common is the center of the circle. Thus, the intersection of $\overline{AD}$ and $\overline{CF}$ is the center of the circle.

You can also use Theorem 9–6 to find the measures of inscribed angles.

Example 2

INTEGRATION
Algebra

In $\odot A$, $m\angle 1 = 6x + 11$, $m\angle 2 = 9x + 19$, $m\angle 3 = 4y - 25$, $m\angle 4 = 3y - 9$, and $\widehat{PQ} \cong \widehat{RS}$. Find $m\angle 1$, $m\angle 2$, $m\angle 3$, and $m\angle 4$.

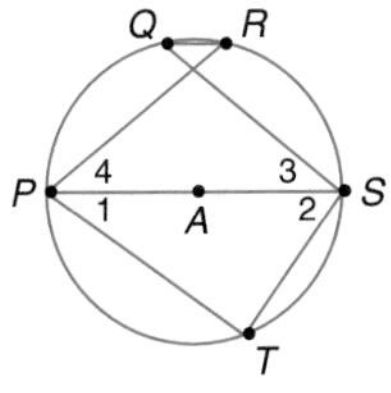

Since $\angle PTS$ is inscribed in a semicircle, by Theorem 9–6, $\angle T$ is a right angle. Therefore, $\triangle PTS$ is a right triangle, and $\angle 1$ and $\angle 2$ must be complementary.

$$m\angle 1 + m\angle 2 = 90 \quad \textit{Definition of complementary angles}$$
$$6x + 11 + 9x + 19 = 90$$
$$15x + 30 = 90$$
$$15x = 60$$
$$x = 4$$

Therefore, $m\angle 1 = 6(4) + 11 = 35$ $\quad m\angle 2 = 9(4) + 19 = 55$

Since $\widehat{PQ} \cong \widehat{RS}$, $\angle 3$ and $\angle 4$ intercept congruent arcs. Thus, $m\angle 3 = m\angle 4$.

$$4y - 25 = 3y - 9$$
$$y = 16$$

Therefore, $m\angle 3 = 4(16) - 25$ or 39

Check: $m\angle 4 = 3(16) - 9$ or 39 ✓

Alternative Learning Styles

Auditory Have students cut a rubber band and tape the ends to cardboard. Stretch the rubber band to form an angle, mark the vertex, and sketch the angle formed. Measure the angle and one of the legs of the angle. Stretch the rubber band until one leg is twice the measure of the original. Measure the angle. Compare the two measures. Have students relate their results to the class.

Example 3 **Quadrilateral *QRST* is inscribed in ⊙*C*. If $m\angle S = 28$ and $m\angle R = 110$, find $m\angle Q$ and $m\angle T$.**

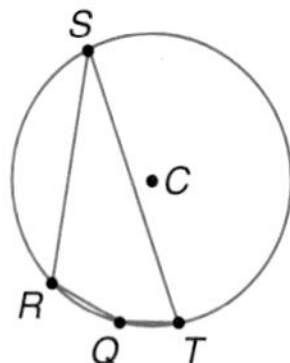

$m\angle S = \frac{1}{2}m\widehat{RQT}$ *Theorem 9–4*

$28 = \frac{1}{2}m\widehat{RQT}$ *$m\angle S = 28$*

$56 = m\widehat{RQT}$

$m\widehat{RST} = 360 - 56$ or 304

Since $\angle Q$ intercepts $\widehat{RST}$,

$m\angle Q = \frac{1}{2}m\widehat{RST}$

$= \frac{1}{2}(304)$ or 152

Use a similar strategy to find $m\angle T$.

$m\angle R = \frac{1}{2}m\widehat{STQ}$ *Theorem 9–4*

$110 = \frac{1}{2}m\widehat{STQ}$ *$m\angle R = 110$*

$220 = m\widehat{STQ}$

$m\widehat{QRS} = 360 - 220$ or 140

Since $\angle T$ intercepts $\widehat{QRS}$,

$m\angle T = \frac{1}{2}m\widehat{QRS}$

$= \frac{1}{2}(140)$ or 70

Notice that $\angle S$ and $\angle Q$ are supplementary and $\angle R$ and $\angle T$ are supplementary.

In Example 3, $\angle R$ and $\angle T$ are opposite angles and supplementary. $\angle Q$ and $\angle S$ are also opposite angles and supplementary. This leads to our next theorem.

Theorem 9–7	**If a quadrilateral is inscribed in a circle, then its opposite angles are supplementary.**

You will be asked to prove this theorem in Exercise 54.

CHECK FOR UNDERSTANDING

Communicating Mathematics

1. The measure of an inscribed angle equals one-half the measure of its intercepted arc.

Study the lesson. Then complete the following. 2–5. See margin.

1. **Explain** how an intercepted arc and an inscribed angle are related.
2. In Example 1, $\angle SQT$ appears to be obtuse, yet $m\widehat{ST} = 68$. Should $m\angle SQT$ be 68? Explain why or why not.
3. $\triangle ABC$ is inscribed in a circle so that $\overline{BC}$ is a diameter. What type of triangle is $\triangle ABC$? Explain your reasoning.
4. **Compare and contrast** inscribed angles and central angles of a circle. If they intercept the same arc, how are they related?

5. **Assess Yourself** Write about and describe three uses of hexagonal nuts, bolts or screws that appear at school or home. Why do you think the hexagon shape is used?

Guided Practice

In ⊙*P*, $\overline{AC}$ is a diameter.

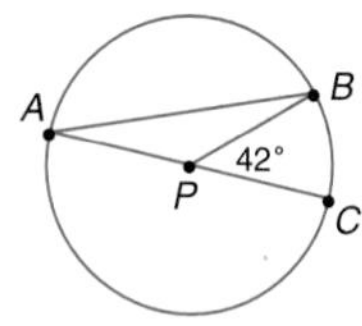

6. Name an inscribed angle. $\angle BAC$
7. Name the arc intercepted by $\angle BAC$. $\widehat{BC}$
8. If $m\angle BPC = 42$, find $m\angle BAC$. 21

In-Class Example

For Example 3

Quadrilateral *QRST* is inscribed in ⊙*C*. If $m\angle T = 95$ and $m\angle S = 100$, find $m\angle Q$ and $m\angle R$.
80, 85

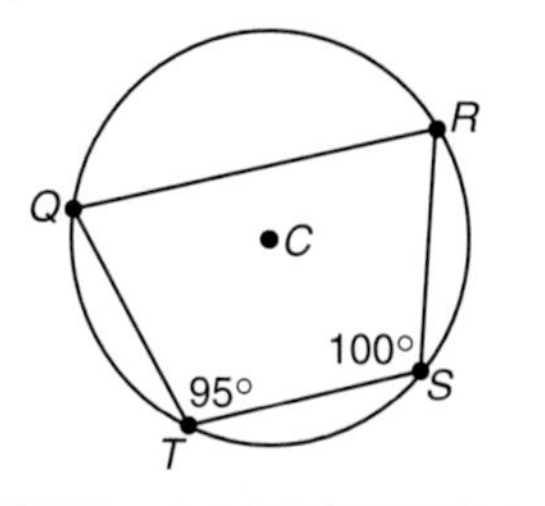

3 PRACTICE/APPLY

Check for Understanding

Exercises 1–16 are designed to help you assess your students' understanding through reading, writing, speaking, and modeling. You should work through Exercises 1–5 with your students and then monitor their work on Exercises 6–16.

Additional Answers

2. No; *Q* is not the center of the circle, so $\angle SQT$ is not a central angle.
3. Since the inscribed angle intercepts a semicircle, the angle is a right angle. Therefore, the triangle is a right triangle.
4. The measure of the inscribed angle equals half the measure of the central angle.
5. Sample answer: Keep tires on a wagon, drapery rods to walls, swingset, keep lug nuts on a car; used for easy alignment of tools and grip when fastening.

Reteaching

Using Diagrams Draw a large circle with a number of inscribed angles. Make sure some of them intercept the same arc. Label the points and ask students to name an inscribed angle. Then have them name an inscribed angle congruent to the first angle, if one exists. Assign measures to some of the arcs or angles and have students find the measures of some of the inscribed angles.

Assignment Guide

Core (with proof): 17–57 odd, 58–69
Core (informal): 17–45 odd, 49, 55, 57, 58–63, 65–69
Enriched: 18–54 even, 55–69
All: Self Test, 1–5

For **Extra Practice**, see p. 781.

The red A, B, and C flags, printed only in the Teacher's Wraparound Edition, indicate the level of difficulty of the exercises.

Additional Answers

47. Given: $\overline{BR} \parallel \overline{AC}$
Prove: $\widehat{RA} \cong \widehat{BC}$
Proof:
Statements (Reasons)
1. $\overline{BR} \parallel \overline{AC}$ (Given)
2. $\angle RBA \cong \angle BAC$ (Alt. Int. ∠ Theorem)
3. $m\angle RBA = m\angle BAC$ (≅ ∠s have = measures.)
4. $m\angle RBA = \frac{1}{2}m\widehat{RA}$, $m\angle BAC = \frac{1}{2}m\widehat{BC}$ (If an ∠ is inscribed in a ⊙, the measure of the ∠ = $\frac{1}{2}$ the measure of its intercepted arc.)
5. $\frac{1}{2}m\widehat{RA} = \frac{1}{2}m\widehat{BC}$ (Substitution Prop. (=))
6. $m\widehat{RA} = m\widehat{BC}$ (Mult. Prop. (=))
7. $\widehat{RA} \cong \widehat{BC}$ (Arcs that have = measures and are in the same ⊙ are ≅.)

48. Given: $\widehat{MHT}$ is a semicircle. $\overline{RH} \perp \overline{TM}$
Prove: $\frac{TR}{RH} = \frac{TH}{HM}$
Proof:
Statements (Reasons)
1. $\widehat{MHT}$ is a semicircle. $\overline{RH} \perp \overline{TM}$ (Given)
2. $\angle THM$ is a rt. ∠. (If an inscribed ∠ of a ⊙ intercepts a semicircle, then the ∠ is a rt. ∠.)
3. $\angle TRH$ is a rt. ∠. (Def. ⊥ lines)
4. $\angle THM \cong \angle TRH$ (All rt. ∠s are ≅.)
5. $\angle T \cong \angle T$ (Reflexive Prop. (=))
6. $\triangle TRH \sim \triangle THM$ (AA Similarity)
7. $\frac{TR}{RH} = \frac{TH}{HM}$ (Def. ~ △s)

9. Find x in ⊙N at the right. 40

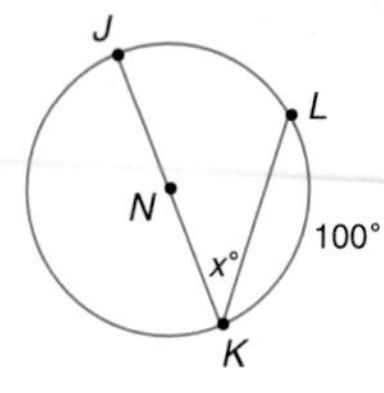

Pentagon *PENTA* is inscribed in ⊙*G*. All sides of *PENTA* are congruent. Find each measure.

10. $m\widehat{PE}$ 72
11. $m\angle PTN$ 72
12. $m\angle PEN$ 108

In ⊙*Z*, $\overline{AB} \parallel \overline{DC}$, $m\widehat{BC} = 94$, and $m\angle AZB = 104$. Find each measure.

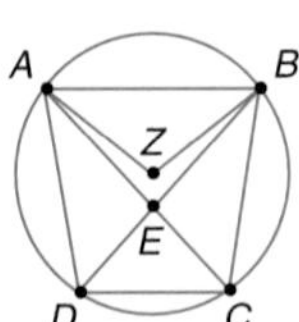

13. $m\widehat{AB}$ 104
14. $m\angle BAC$ 47
15. $m\angle ADB$ 52

16. **Building Trades** Refer to the application at the beginning of the lesson.
a. Given a round stock, how could a machinist determine the appropriate cuts to make in order to make a hexagonal shape with all sides congruent?
b. What diameter of round stock should be used to cut a hex nut one-half inch on each side? 1 inch

16a. Measure 60°, draw the chords, and cut.

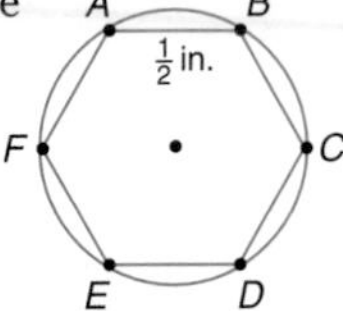

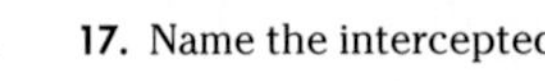

EXERCISES

Practice

A

In ⊙*A*, $\overline{HE}$ is a diameter.

17. Name the intercepted arc for $\angle HTC$. $\widehat{HC}$
18. If $m\angle HTC = 52$, find $m\widehat{CH}$. 104
19. Name the intercepted arc for $\angle TCH$.
20. Find $m\angle ECH$. 90
21. If $m\angle HTC = 52$, find $m\angle CEH$. 52
22. Name an inscribed angle.
23. Name a central angle. There are none shown in the figure.
24. If $m\angle HTC = 52$, find $m\widehat{CEH}$. 256

19. $\widehat{TCH}$

22. Sample answers: $\angle HCT$, $\angle HEU$, $\angle CEH$, $\angle CEU$, $\angle CTH$, $\angle THU$

Find the value of *x*.

25.

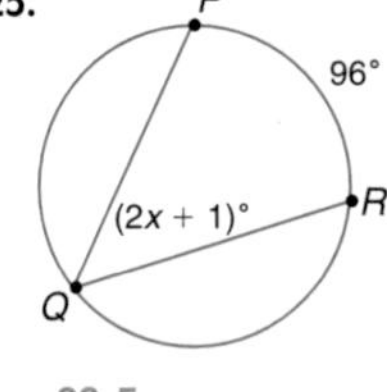

23.5

26.

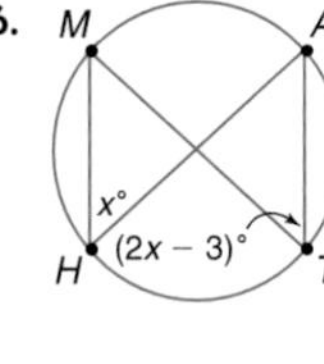

3

27. $\overline{KR}$ is a diameter of ⊙W.

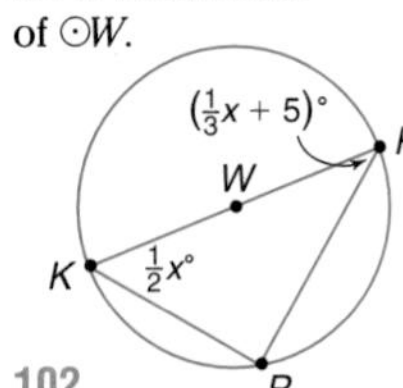

102

Quadrilateral *SPRT* is inscribed in ⊙*G*, and *PGTS* is a rhombus. If $m\angle PRS = -2x + 42$ and $m\angle SRT = 8x - 18$, find each measure.

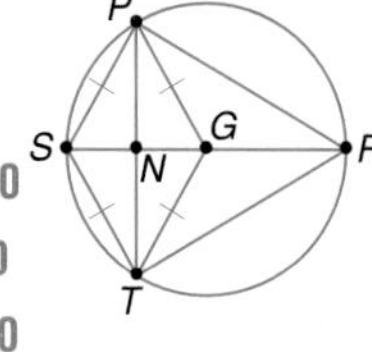

28. $m\widehat{PS}$ 60 29. $m\widehat{ST}$ 60 30. $m\angle RST$ 60
31. $m\angle PGR$ 120 32. $m\widehat{PR}$ 120 33. $m\widehat{PSR}$ 240
34. $m\angle PNG$ 90 35. $m\widehat{PRT}$ 240 36. $m\angle PSR$ 60

In ⊙*P*, $\widehat{EN} = 66$, $m\angle GPM = 89$, and $\overline{GN}$ is a diameter. Find each measure.

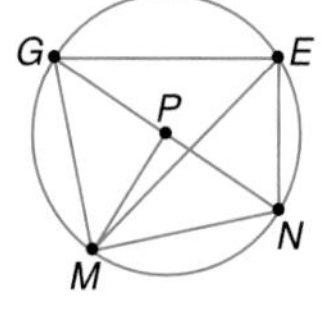

39. 114
42. 33
45. 78.5

37. $m\widehat{GM}$ 89 38. $m\widehat{NM}$ 91 39. $m\widehat{GE}$
40. $m\angle GEN$ 90 41. $m\angle EGN$ 33 42. $m\angle EMN$
43. $m\angle GNM$ 44.5 44. $m\angle GME$ 57 45. $m\angle EGM$

46. Equilateral triangle *PEQ* is inscribed in circle *R*. Point *A* bisects $\widehat{PE}$, point *B* bisects $\widehat{EQ}$, and point *C* bisects $\widehat{PQ}$. What must be true about $\triangle ABC$? Explain. $\triangle ABC$ is equilateral.

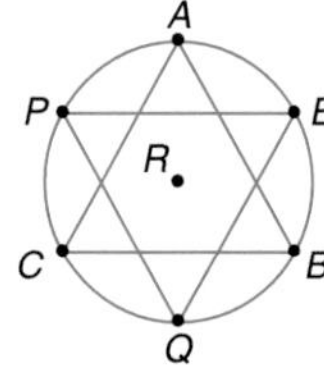

Write a two-column proof. 47–48. See margin.

47. **Given:** $\overline{BR} \parallel \overline{AC}$
Prove: $\widehat{RA} \cong \widehat{BC}$

48. **Given:** $\widehat{MHT}$ is a semicircle.
$\overline{RH} \perp \overline{TM}$
Prove: $\frac{TR}{RH} = \frac{TH}{HM}$

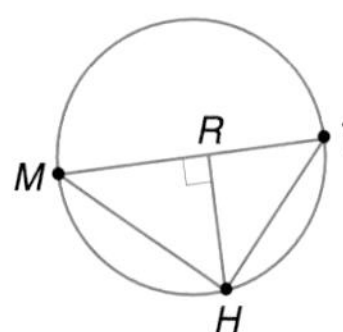

49. Is it possible for a parallelogram to be inscribed in a circle, yet not be a rectangle? Explain your reasoning. No; opposite angles must be congruent and supplementary. Therefore, they each must be 90°, or right angles.

Study the proof for Case 1 of Theorem 9–4. In Case 2 and Case 3, $\overline{PR}$ is not a diameter. To prove Cases 2 and 3, draw diameter $\overline{KR}$ and use Case 1 of Theorem 9–4, the Angle Addition Postulate, and the Arc Addition Postulate. 50–51. See margin.

50. Case 2
Given: *T* lies inside $\angle PRQ$.
Prove: $m\angle PRQ = \frac{1}{2}m\widehat{PQ}$

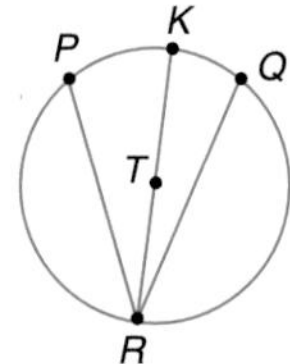

51. Case 3
Given: *T* lies outside $\angle PRQ$.
Prove: $m\angle PRQ = \frac{1}{2}m\widehat{PQ}$

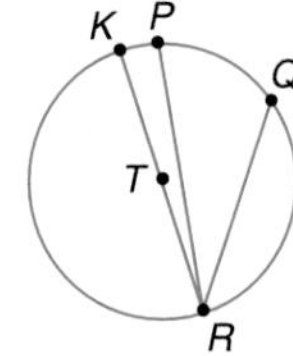

Additional Answers

50. $m\angle PRQ = m\angle PRK + m\angle KRQ$
$= \frac{1}{2}m\widehat{PK} + \frac{1}{2}m\widehat{KQ}$
$= \frac{1}{2}(m\widehat{PK} + m\widehat{KQ})$
$= \frac{1}{2}m\widehat{PQ}$

51. $m\angle PRQ = m\angle KRQ - m\angle KRP$
$= \frac{1}{2}m\widehat{KQ} - \frac{1}{2}m\widehat{KP}$
$= \frac{1}{2}(m\widehat{KQ} - m\widehat{KP})$
$= \frac{1}{2}m\widehat{PQ}$

Study Guide Masters, p. 52

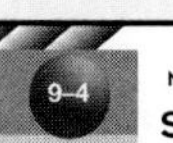

NAME ______ DATE ______

Study Guide

Student Edition Pages 466–473

Inscribed Angles

An **inscribed angle** of a circle is an angle whose vertex is on the circle and whose sides contain chords of the circle. We say that $\angle DEF$ *intercepts* $\widehat{DF}$. The following theorems involve inscribed angles.

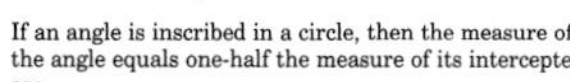

- If an angle is inscribed in a circle, then the measure of the angle equals one-half the measure of its intercepted arc.

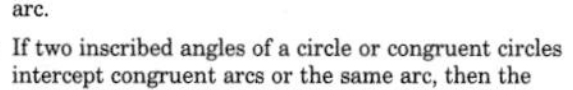

- If two inscribed angles of a circle or congruent circles intercept congruent arcs or the same arc, then the angles are congruent.
- If an inscribed angle of a circle intercepts a semicircle, then the angle is a right angle.
- If a quadrilateral is inscribed in a circle, then its opposite angles are supplementary.

Example: In the circle above, find $m\angle DEF$ if $m\widehat{DF} = 28$. Since $\angle DEF$ is an inscribed angle, $m\angle DEF = \frac{1}{2}m\widehat{DF} = \frac{1}{2}(28)$ or 14.

In ⊙P, $\overline{RS} \parallel \overline{TV}$.

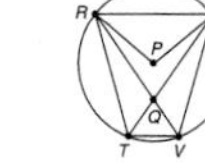

1. Name the intercepted arc for $\angle RTS$. $\widehat{RS}$
2. Name an inscribed angle. Sample answers: $\angle RTS$, $\angle RVS$, $\angle RTV$, $\angle TSR$, $\angle VSR$, $\angle TVS$, $\angle TVR$, $\angle VRS$
3. Name a central angle. $\angle RPS$

In ⊙P, $m\widehat{SV} = 86$ and $m\angle RPS = 110$. Find each measure.

4. $m\angle PRS$ 35 5. $m\widehat{RT}$ 86 6. $m\angle RVT$ 43 7. $m\angle SVT$ 98
8. $m\angle TQV$ 94 9. $m\angle RQT$ 86 10. $m\angle QRT$ 39 11. $m\widehat{RS}$ 110

Additional Answer

52. Given: inscribed $\angle MLN$, inscribed $\angle CED$, $\widehat{CD} \cong \widehat{MN}$
Prove: $\angle MLN \cong \angle CED$

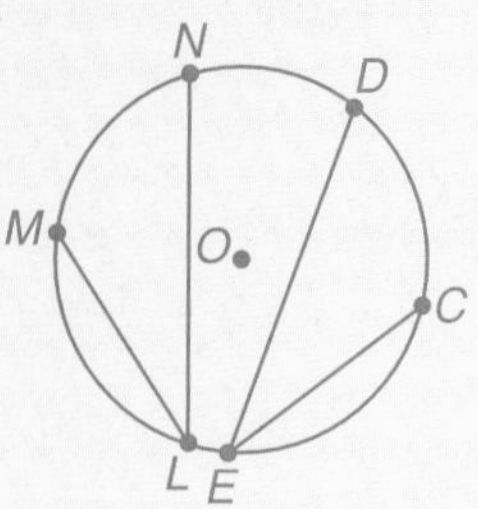

Proof:
Statements (Reasons)

1. $\angle MLN$ and $\angle CED$ are inscribed $\angle$s, $\widehat{CD} \cong \widehat{MN}$ (Given)
2. $m\angle MLN = \frac{1}{2}m\widehat{MN}$, $m\angle CED = \frac{1}{2}m\widehat{CD}$ (If an $\angle$ is inscribed in a $\odot$, the measure of the $\angle = \frac{1}{2}$ the measure of the intercepted arc.)
3. $m\angle MLN = \frac{1}{2}m\widehat{CD}$ (Substitution Prop. (=))
4. $m\angle MLN = m\angle CED$ (Trans. Prop. (=))
5. $\angle MLN \cong \angle CED$ (Def. $\cong$ $\angle$s)

Practice Masters, p. 52

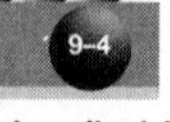

NAME________________ DATE________

Practice

Student Edition Pages 466–473

Inscribed Angles

In $\odot P$, $m\widehat{AB} = x$ and $m\widehat{BC} = 3x$. Find each measure.

1. $m\widehat{ADC}$ 180
2. $m\widehat{AB}$ 45
3. $m\widehat{BC}$ 135
4. $m\angle ABC$ 90
5. $m\angle A$ 67.5
6. $m\angle C$ 22.5

In $\odot Q$, $m\angle ABC = 72$ and $m\widehat{CD} = 46$. Find each measure.

7. $m\widehat{CA}$ 144
8. $m\widehat{AD}$ 98
9. $m\angle ABD$ 49
10. $m\widehat{BC}$ 134
11. $m\angle C$ 23
12. $m\angle A$ 49

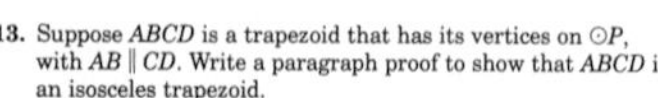

13. Suppose $ABCD$ is a trapezoid that has its vertices on $\odot P$, with $AB \parallel CD$. Write a paragraph proof to show that $ABCD$ is an isosceles trapezoid.
$\angle A$ and $\angle D$ are supplementary, since they are consecutive interior angles formed by two parallel lines and a transversal. $\angle B$ and $\angle D$ are supplementary, since they are opposite angles of a quadrilateral inscribed in a circle. It follows that $\widehat{ADC} \cong \widehat{DCB}$. By the arc addition postulate and the subtraction property of equality, $m\widehat{AD} = m\widehat{BC}$. Thus $\widehat{AD} \cong \widehat{BC}$ and, since if two arcs in a circle are congruent then their chords are congruent, $\overline{AD} \cong \overline{BC}$. Trapezoid $ABCD$ is isosceles by definition.

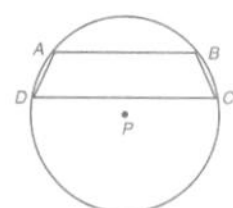

52–53. See margin.

52. Write a two-column proof to show that if two inscribed angles of a circle intercept congruent arcs, then the angles are congruent. (Theorem 9–5)

53. Write a paragraph proof to show that if an inscribed angle of a circle intercepts a semicircle, then the angle is a right angle. (Theorem 9–6)

54. Write a paragraph proof to show that if a quadrilateral is inscribed in a circle, then the opposite angles of the quadrilateral are supplementary. (Theorem 9–7) **See Solutions Manual.**

Critical Thinking

55. Can an isosceles trapezoid be inscribed in a circle? Write a brief paragraph explaining your reasoning. **Yes; see students' work.**

Applications and Problem Solving

56. Civil Engineering A civil engineer uses the formula $\ell = \frac{2\pi rm}{360}$ to calculate the length ℓ of a curve for a road, given a radius r and a central angle measure m. Find the length of the road from A to B in the diagram at the right. Round to the nearest foot. **410 ft**

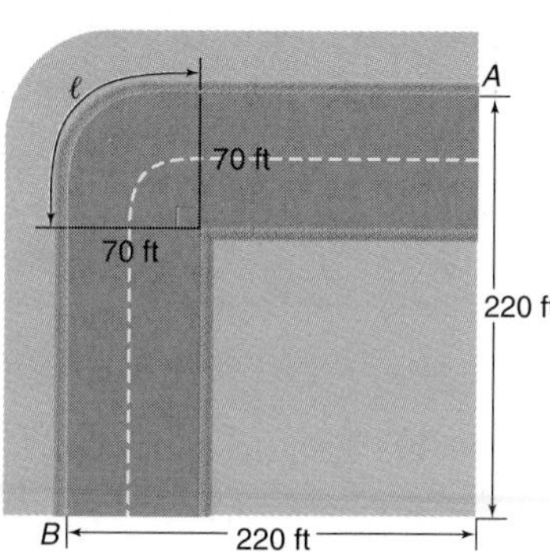

57. Carpentry A carpenter needs to find the center of a circular pattern of a parquet floor. How can the carpenter find the center of the circle using only a carpenter's square? Draw a diagram to explain your answer. **See Solutions Manual.**

Mixed Review

58. Food The graph at the right shows where Americans eat take-out food. Find each measure. (Lesson 9–2)

a. $m\angle ATD$ **50.4**

b. major arc AB **190.8**

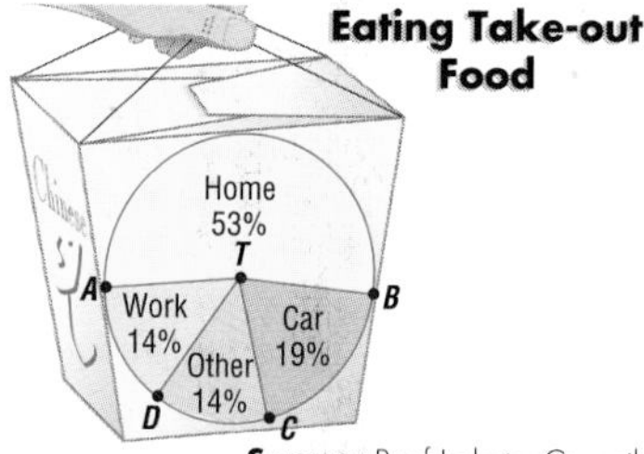

Source: Beef Industry Council

59. Botany *Fairy-ring mushrooms* form circles in grassy areas because their rootlike mycelia grow outward from a central point. If a ring of mushrooms is growing 8.5 feet from the central point, find the circumference of the mushroom circle. Round to the nearest tenth of a foot. (Lesson 9–1) **53.4 ft**

60. Find the geometric mean between 9 and 21. (Lesson 8–1) **13.7**

61. Draw a segment that is 11 centimeters long. Then by construction, separate the segment into three congruent parts. (Lesson 7–4) **See students' work.**

62. $JKLM$ is a rhombus with $m\angle 1 = 37$. Find $m\angle 2$. (Lesson 6–4) **53**

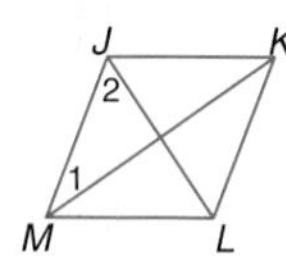

63. Gardening Sanequa is planting a rectangular garden. She has placed stakes at what she believes will be the four corners of the rectangle. Before she digs, what can she do to determine if the stakes are in the right places? (Lesson 6–3)

63. Sample answer: measure the diagonals to see if they are congruent.

Additional Answer

53. Given: $\widehat{PQR}$ is a semicircle of $\odot C$.
Prove: $\angle PQR$ is a right $\angle$.

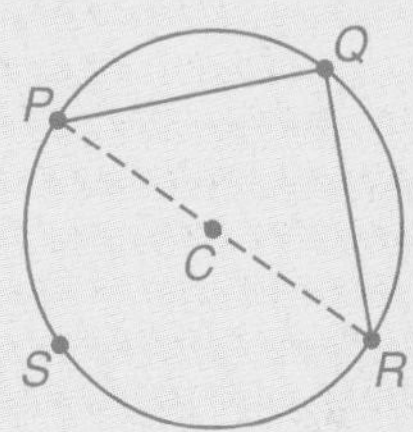

Proof:
Since $\widehat{PQR}$ is a semicircle, $\widehat{PSR}$ is also a semicircle and has a degree measure of 180. From the diagram, $\angle PQR$ is an inscribed angle, and $m\angle PQR = \frac{1}{2}(m\widehat{PSR})$ or 90. As a result, $\angle PQR$ is a right angle.

Proof 64. Complete the proof. (Lesson 4–4)

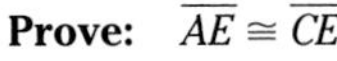

Given: $\overline{AB} \cong \overline{CD}$
$\overline{AB} \parallel \overline{CD}$
$\overline{AC}$ bisects $\overline{BD}$.

Prove: $\overline{AE} \cong \overline{CE}$

Proof:

Statements	Reasons
a. $\overline{AB} \cong \overline{CD}$, $\overline{AB} \parallel \overline{CD}$, $\overline{AC}$ bisects $\overline{BD}$.	a. ? Given
b. $\overline{BE} \cong \overline{DE}$	b. ? Definition of bisector
c. $\angle ABE \cong \angle CDE$	c. ? If $\parallel$, alt. int. $\angle$s are $\cong$.
d. $\triangle ABE \cong \triangle CDE$	d. ? SAS
e. $\overline{AE} \cong \overline{CE}$	e. ? CPCTC

65. Parallel lines ℓ and m are cut by a transversal t, and $\angle 1$ and $\angle 2$ are corresponding angles. If $m\angle 1 = 4x + 3$ and $m\angle 2 = 8x - 7$, what is the value of x? (Lesson 3–4) 2.5

66. Write a conjecture given points $A(-1, -3)$, $B(3, 0)$, and $C(3, -3)$. Draw a figure to illustrate your conjecture. (Lesson 2–1) See margin.

67. Given points $A(7, 4)$ and $B(-3, 1)$, find AB to the nearest tenth. (Lesson 1–4) 10.4

INTEGRATION Algebra

68. **Electronics** The total resistance in ohms R_T of a certain conductor can be given by the equation $\frac{1}{R_T} = \frac{1}{4} + \frac{1}{5}$. Find R_T. $2\frac{2}{9} \approx 2.22$ ohms

69. Use graphing to solve the system of equations. Check your solution algebraically.
$6a + 2b = 11$
$3a - 2b = 7$ $\left(2, -\frac{1}{2}\right)$

SELF TEST

Refer to ⊙P. (Lessons 9–1 and 9–2) 2. 106.8

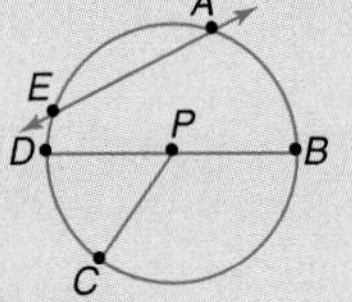

1. Name three radii of ⊙P. $\overline{PD}, \overline{PB}, \overline{PC}$
2. If $PD = 17$, find the circumference of ⊙P to the nearest tenth.
3. If $m\angle CPB = 125$ and $PB = 6$, find the length of $\widehat{CB}$. 13.1

Refer to ⊙O. Find each measure. (Lessons 9–3 and 9–4)

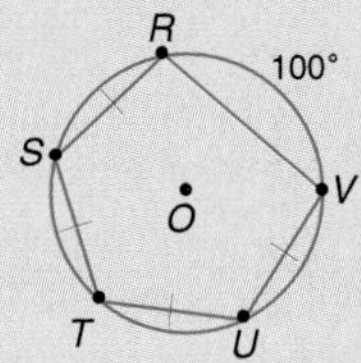

4. $m\widehat{ST}$ 65
5. $m\angle TUV$ 115

4 ASSESS

Closing Activity

Modeling Give each student a different arc measure. Have students use a protractor and a compass to draw a circle with three inscribed angles that intercept an arc of the given measure.

Chapter 9 Quiz B (Lessons 9-3 and 9-4) is available in the *Assessment and Evaluation Masters*, p. 240.

Mid-Chapter Test (Lessons 9-1 through 9-4) is available in the *Assessment and Evaluation Masters*, p. 239.

Additional Answer

66. $\triangle ABC$ is a right triangle.

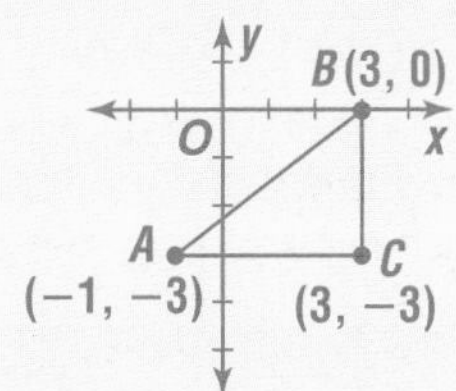

Enrichment Masters, p. 52

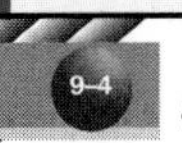

NAME ______ DATE ______

Enrichment

Student Edition Pages 466–473

Formulas for Regular Polygons

Suppose a regular polygon of n sides is inscribed in a circle of radius r. The figure shows one of the isosceles triangles formed by joining the endpoints of one side of the polygon to the center C of the circle. In the figure, s is the length of each side of the regular polygon, and a is the length of the segment from C perpendicular to $\overline{AB}$.

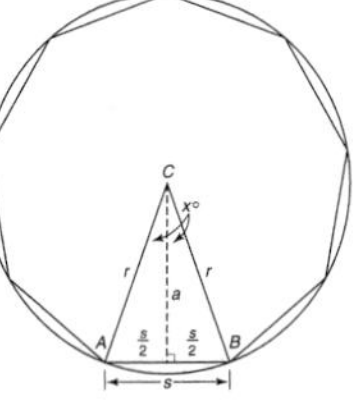

Use your knowledge of triangles and trigonometry to solve the following problems.

1. Find a formula for x in terms of the number of sides n of the polygon.
$x = \frac{180°}{n}$

2. Find a formula for s in terms of the number of n and r. Use trigonometry.
$s = 2r \sin\left(\frac{180°}{n}\right)$

3. Find a formula for a in terms of n and r. Use trigonometry.
$a = r \cos\left(\frac{180°}{n}\right)$

4. Find a formula for the *perimeter* of the regular polygon in terms of n and r.
perimeter $= 2nr \sin\left(\frac{180°}{n}\right)$

Extension

Reasoning Have students work alone or in groups to name some polygons that can always be inscribed in a circle and some polygons that cannot be inscribed in a circle. Sample answer: Any triangle and any equiangular polygon can always be inscribed; a non-square rhombus cannot be inscribed.

SELF TEST

The Self Test provides students with a brief review of the concepts and skills in Lessons 9-1 through 9-4. Lesson numbers are given to the right of the exercises or instruction lines so students can review concepts not yet mastered.

9-5A LESSON NOTES

NCTM Standards: 2, 3, 7

Objective

Use the TI-92 calculator to explore characteristics of tangents.

Recommended Time

20 minutes

Instructional Resources

Instructions for using the *Geometer's Sketchpad* for this activity are available in the *Graphing Calculator and Computer Masters*, p. 26. Answers may differ slightly from those given.

1 FOCUS

Motivating the Lesson

Ask students what it means when you *go off on a tangent*. Ask them how this might relate to geometry.

2 TEACH

Teaching Tip Make sure students understand why AT and AS touch circle C at only one point. Discuss how the measurements change if AT or AS is not tangent.

3 PRACTICE/APPLY

Assignment Guide
Core (with proof): 1–4
Core (informal): 1–4
Enriched: 1–4

4 ASSESS

Observing students working with technology is an excellent method of assessment.

9–5A Using Technology
Exploring Tangents

A Preview of Lesson 9–5

A line that intersects a circle in exactly one point is called a **tangent** to the circle. You can use a TI-92 calculator to explore some of the characteristics of tangents. Use the following steps to draw two lines that are tangent to a circle.

Step 1 Draw a circle by pressing F3 and choosing 1:Circle from the menu. Use the arrow key to position the cursor anywhere on the screen to locate the center of the circle. Press ENTER. Label the center C. Move to another place on the screen to designate the radius of the circle. Press ENTER to complete the circle.

Step 2 Next, press F2 and choose 1:Point from the menu to draw a point outside the circle. Label the point A.

Step 3 Press F2 and choose Line. Move the cursor to point A so that the prompt THRU THIS POINT occurs. Press ENTER. Move the cursor towards the circle until the line appears to touch the circle at one point. At the prompt ON THIS CIRCLE, press ENTER. Label this point T.

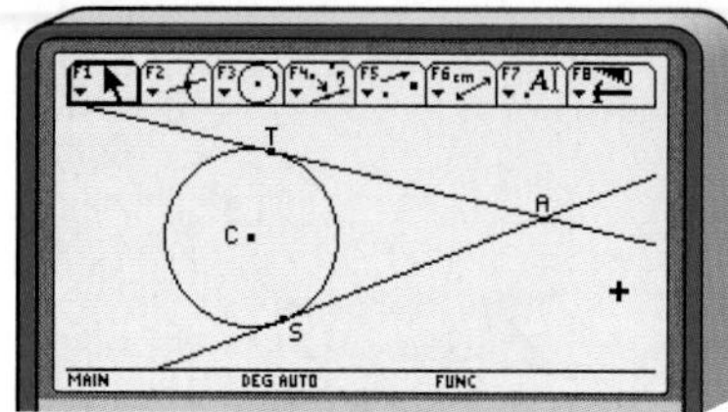

Step 4 Repeat the procedure in Step 3 to draw another line through point A that is tangent to $\odot C$ at point S.

The lines drawn to the circle are tangents to the circle. *Note that these tangents are approximate since it is difficult to find the exact point where the line touches the circle.*

EXERCISES

LOOK BACK Refer to Lesson 1-7A to review how to measure angles.

Use the measuring capabilities of the TI-92 to explore the characteristics of tangents. 2. Sample answer: the measures are equal.

1. Measure $\overline{AT}$ and $\overline{AS}$. See students' work.
2. Use the hand key to move point A closer to the circle. Make a conjecture about the measurements of $\overline{AT}$ and $\overline{AS}$.
3. Draw radii $\overline{CT}$ and $\overline{CS}$ by pressing F2 and choosing 5:Segment from the menu. Measure $\angle CTA$ and $\angle CSA$. See students' work.
4. Make a conjecture about the angles formed by a radius and a tangent to a circle. They are right angles.

Using Technology

This lesson offers an excellent opportunity for using technology in your geometry classroom. For more information on using technology, see *Graphing Calculators in the Mathematics Classroom*, one of the titles in the Glencoe Mathematics Professional Series.

9-5 Tangents

APPLICATION
Astronomy

What **YOU'LL LEARN**

- To recognize tangents and use properties of tangents.

Why **IT'S IMPORTANT**

You can use tangents to solve problems involving astronomy and aerospace.

A *solar eclipse* is an exciting natural phenomenon that occurs when the Moon passes in front of the Sun, blocking it from Earth. Some areas of the world experience a total eclipse, others, a partial eclipse, and some areas experience no eclipse at all.

In the diagram below, the Moon is between the Sun and Earth. The pink region indicates the portion of Earth that will experience a partial eclipse. The narrow blue region indicates a total eclipse. $\overline{AE}$ and $\overline{BF}$ are **tangent** to the circles that represent the Sun and Moon. A line is tangent to a circle if it intersects the circle in *exactly one* point. This point is called the **point of tangency**. *Parts of a tangent line may also be tangent to a circle.*

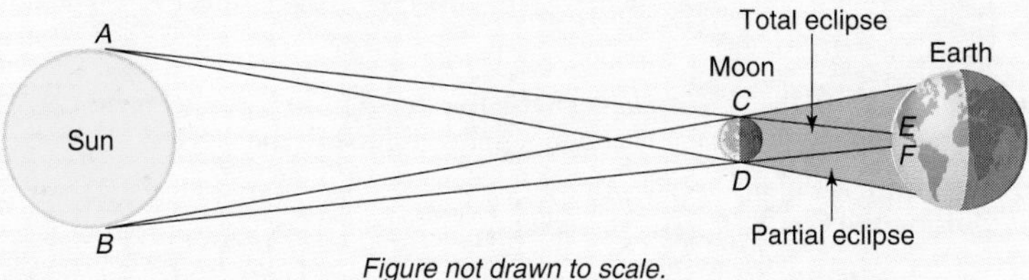

Figure not drawn to scale.

FYI

During the 20th century, 78 total solar eclipses will have taken place. Only 15 of these eclipses were visible from some part of the United States. The next total eclipse visible from the United States will be in 2017.

A circle separates a plane into three parts. The parts are the **interior**, the **exterior**, and the circle itself. In the figure at the right, T is a point of tangency, X is in the exterior of $\odot P$, and I is in the interior of $\odot P$. $\overline{PT}$ is a radius. Thus, $PX > PT$ and $PI < PT$.

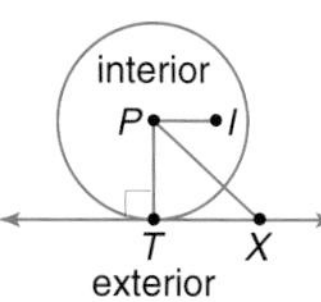

The shortest segment from a point to a line is a perpendicular segment. Thus, $\overline{PT} \perp \overleftrightarrow{TX}$. This leads to Theorem 9–8.

Theorem 9–8	**If a line is tangent to a circle, then it is perpendicular to the radius drawn to the point of tangency.**

You will be asked to prove this theorem in Exercise 47.

Example 1

INTEGRATION
Algebra

Refer to $\odot C$ with tangent $\overline{AB}$. Find x.

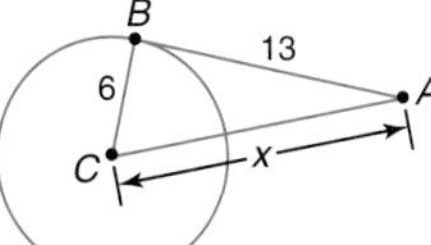

Since $\overline{AB}$ is tangent to $\odot C$, $\overline{AB} \perp \overline{BC}$ by Theorem 9–8. Therefore, $\triangle ABC$ is a right triangle.

By the Pythagorean Theorem,

$$(AB)^2 + (BC)^2 = (AC)^2$$
$$13^2 + 6^2 = x^2$$
$$169 + 36 = x^2$$
$$x^2 = 205$$
$$x = \sqrt{205} \text{ or about } 14.3$$

9-5 LESSON NOTES

NCTM Standards: 1–4, 7

Instructional Resources

- Study Guide Master 9-5
- Practice Master 9-5
- Enrichment Master 9-5
- Tech Prep Applications Masters, p. 18

Transparency 9-5A contains the 5-Minute Check for this lesson; **Transparency 9-5B** contains a teaching aid for this lesson.

Recommended Pacing	
Standard Pacing	Day 8 of 15
Honors Pacing	Day 6 of 13
Block Scheduling*	Day 5 of 9

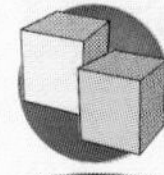

*For more information on pacing and possible lesson plans, refer to the *Block Scheduling Booklet*.

1 FOCUS

5-Minute Check
(over Lesson 9-4)

In $\odot T$, $m\widehat{AB} = 68$ and $m\angle DBC = 26$. Find each measure.

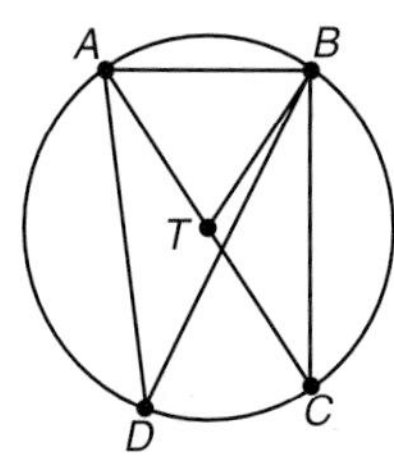

1. $m\angle BTC$ 112
2. $m\angle ADB$ 34
3. $m\widehat{DC}$ 52
4. $m\widehat{AD}$ 128
5. $m\angle BAD$ 82

FYI

In 1824, astronomer Friedrich Bessell invented mathematical functions that make it easier to predict eclipses.

Motivating the Lesson

Questioning Draw a circle and label the center. Draw a ray with its endpoint on the edge and the rest of it in the exterior of the circle. (Do not draw the ray tangent to the circle.) Ask students if the ray in the figure is tangent to the circle. Why or why not? **No, because if you extend the ray beyond the endpoint, the line formed would intersect the circle in two points.**

2 TEACH

In-Class Examples

For Example 1

If a spacecraft needs to eject its outer fuel tanks when it is 5000 km above Earth's surface, how far will it be from the horizon at the time it ejects its fuel tanks? The radius of Earth is about 6400 km. **The spacecraft will be about 9434 km from the horizon.**

For Example 2

Determine whether $\overline{RT}$ is tangent to $\odot S$. Explain. **Yes, since $\triangle RST$ is a right triangle with $m\angle R = 90$.**

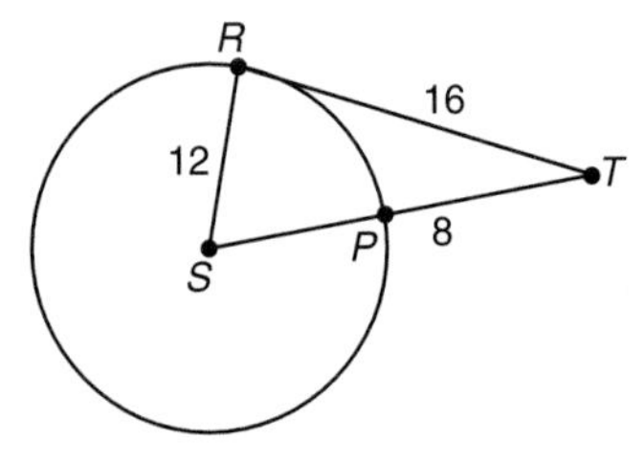

Teaching Tip For the constructions in this lesson, you may want to review how to construct perpendicular lines and angle bisectors.

Teaching Tip When discussing circumscribed and inscribed polygons, remind students that not all polygons can be circumscribed or have a circle inscribed.

The converse of Theorem 9–8 is also true and can be used to identify tangents to a circle.

Theorem 9–9 (Converse of Theorem 9–8)	In a plane, if a line is perpendicular to a radius of a circle at the endpoint on the circle, then the line is a tangent of the circle.

Example  **Refer to $\odot P$ with radius $\overline{PR}$. Show that $\overline{QR}$ is tangent to $\odot P$.**

Algebra

Since $\overline{PT}$ is also a radius of $\odot P$, then $PT = 5$.

$PQ = PT + TQ$

$= 5 + 8$ or 13

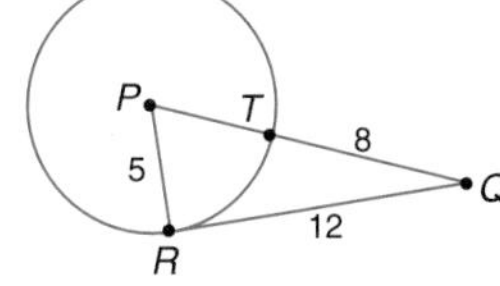

Notice that in $\triangle PQR$, $(PR)^2 + (QR)^2 = (PQ)^2$.

$$5^2 + 12^2 = 13^2$$
$$25 + 144 = 169$$
$$169 = 169$$

By the converse of the Pythagorean Theorem, $\triangle PQR$ is a right triangle with $\overline{PR} \perp \overline{QR}$. By Theorem 9–9, $\overline{QR}$ must be tangent to $\odot P$.

CONSTRUCTION

Tangent from a Point Outside the Circle

Construct a line tangent to a given circle C through a point A outside the circle.

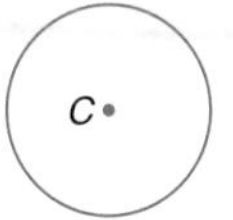

1. Draw $\overline{AC}$.
2. Construct the perpendicular bisector of $\overline{AC}$. Call this line ℓ. Call X the intersection of ℓ and $\overline{AC}$.
3. Using X as the center, draw a circle with radius measuring XC. Call D and E the intersection points of the two circles.
4. Draw $\overline{AD}$. Then $\overline{AD}$ is tangent to $\odot C$.

Conclusion: $\overline{AD}$ is a tangent to $\odot C$ if $\overline{AD} \perp \overline{DC}$. How do you know $\angle CDA$ is a right angle? **It is inscribed in a semicircle.**

A line or line segment that is tangent to two circles in the same plane is called a **common tangent** of the two circles. There are two types of common tangents.

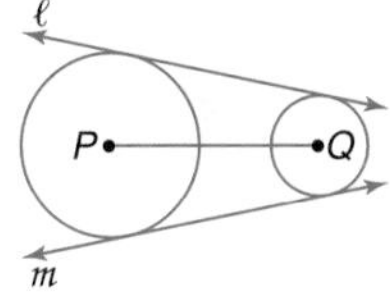

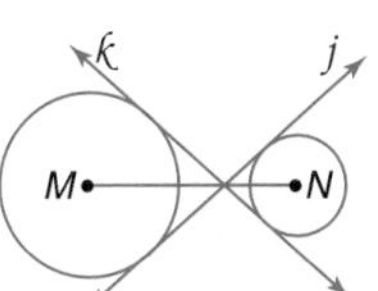

Common external tangents do not intersect the segment whose endpoints are the centers of the circles. In the figure above, lines ℓ and m are common external tangents to $\odot P$ and $\odot Q$.

Common internal tangents intersect the segment whose endpoints are the centers of the circles. In the figure above, lines j and k are common internal tangents to $\odot M$ and $\odot N$.

Cooperative Learning

Co-op Co-op After discussing Theorems 9-8 through 9-10, separate the class into three groups and assign each group one of the constructions in this lesson. Have each group follow the instructions and learn how to construct the desired figure. When every student understands how to construct the group's figure, have one member of each group explain the construction to the class. For more information on the co-op co-op strategy, see *Cooperative Learning in the Mathematics Classroom*, one of the titles in the Glencoe Mathematics Professional Series, page 32.

In the solar eclipse model, $\overline{EC}$ and $\overline{ED}$ are examples of two **tangent segments** drawn from a common point E outside the circle (the Moon). An important relationship exists between $\overline{EC}$ and $\overline{ED}$.

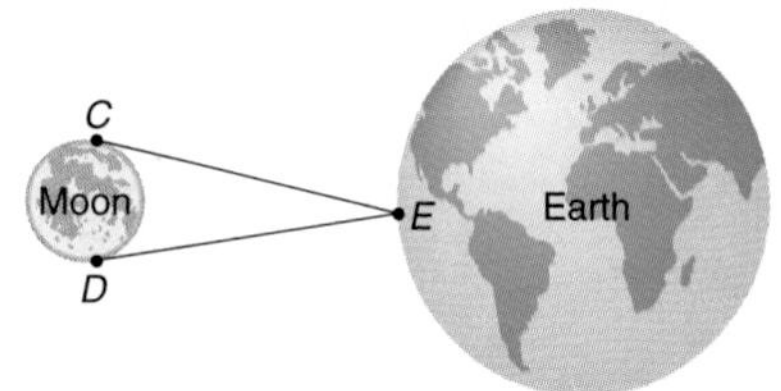

In the figure at the right, note that $\overrightarrow{PQ}$ and $\overrightarrow{PR}$ are both tangent to $\odot C$. Also, $\overline{PQ}$ and $\overline{PR}$ are tangent segments. By drawing $\overline{PC}$, $\overline{CR}$, and $\overline{CQ}$, two right triangles are formed. These triangles are congruent by HL, which leads us to the following theorem. *You will be asked to prove this theorem in Exercise 48.*

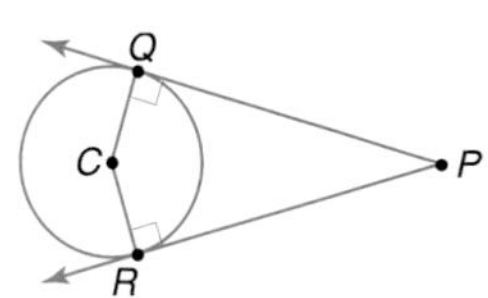

Theorem 9–10

If two segments from the same exterior point are tangent to a circle, then they are congruent.

We have worked with problems involving polygons inscribed in circles. We can use Theorem 9–10 to solve problems involving **circumscribed polygons**. A polygon is circumscribed *about* a circle if each side of the polygon is tangent to the circle. The following two statements are equivalent.

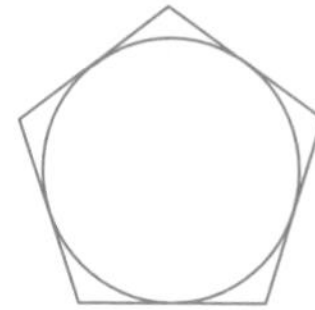

The polygons are circumscribed about the circles.

The circles are inscribed in the polygons.

CONSTRUCTION

Circle Inscribed in a Polygon

Construct a circle inscribed in a given triangle ABC.

1. Construct the angle bisectors of $\angle A$ and $\angle C$. Extend the bisectors to meet at point X.

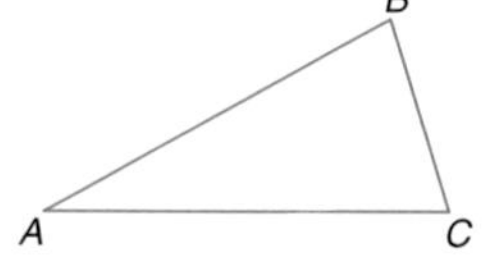

2. Construct a line from X perpendicular to $\overline{AC}$. Label the intersection of the perpendicular line and $\overline{AC}$, Y.

3. Setting the compass length equal to XY, draw $\odot X$.

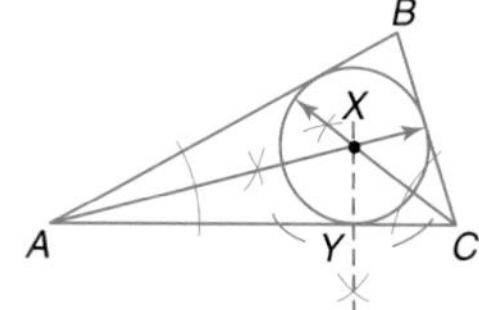

Conclusion: $\odot X$ is inscribed in $\triangle ABC$. $\overline{AB}$, $\overline{BC}$, and $\overline{AC}$ are tangent to $\odot X$.

You can use Theorem 9–10 to solve problems involving circumscribed polygons.

Teaching Tip To help students remember the difference between circumscribed and inscribed, discuss other words that begin with *circum-* and *in-*. For example: circumference and inside.

In-Class Example

For Example 3
$\triangle TRW$ is circumscribed about $\odot A$. If the perimeter of $\triangle TRW$ is 42, $MR = 6$, and $WM = 7$, find TR. 14

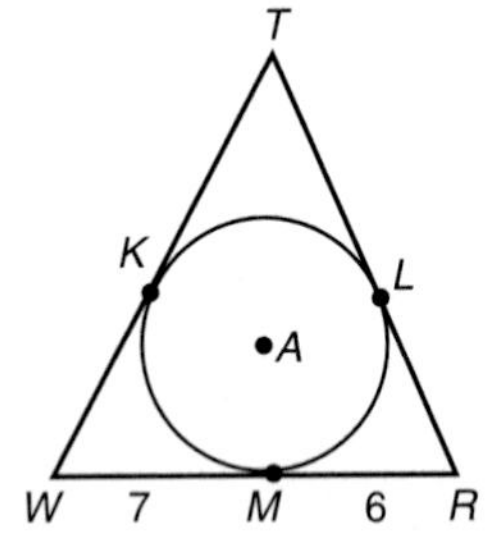

3 PRACTICE/APPLY

Check for Understanding

Exercises 1–16 are designed to help you assess your students' understanding through reading, writing, speaking, and modeling. You should work through Exercises 1–5 with your students and then monitor their work on Exercises 6–16.

Error Analysis

Two of the three constructions in this lesson may be confusing the first time through. Have students complete a number of constructions using different-size figures, and ask them to explain the importance of each step as they go along.

Additional Answers

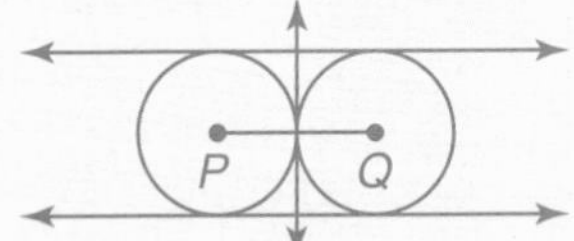

3c. 1; only one line can be drawn perpendicular to a radius of a circle at the endpoint on the circle.

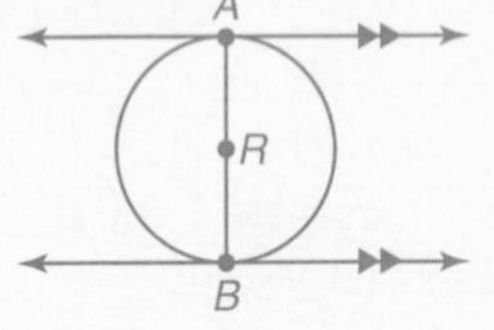

Example 3 **Triangle TRW is circumscribed about $\odot A$. If the perimeter of $\triangle TRW$ is 50, $TK = 3$, and $WM = 9.5$, find TR.**

Since $\odot A$ is inscribed in $\triangle TRW$ or circumscribed by $\triangle TRW$, K, L, and M are points of tangency. Thus, by Theorem 9–10, $WK = WM = 9.5$, $TK = TL = 3$, and $LR = RM$.

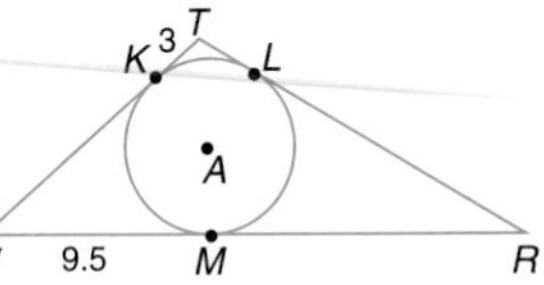

To find TR, let $\overline{LR}$ and $\overline{RM}$ have length x and use the known perimeter.

$$
\begin{aligned}
P &= WK + KT + TL + LR + RM + WM \quad \text{Perimeter of } \triangle TRW\\
50 &= 9.5 + 3 + 3 + x + x + 9.5\\
50 &= 25 + 2x\\
25 &= 2x\\
12.5 &= x
\end{aligned}
$$

The measures of $\overline{LR}$ and $\overline{RM}$ are each 12.5.

$$
\begin{aligned}
TR &= TL + LR \quad \text{Segment Addition Postulate}\\
&= 3 + 12.5\\
&= 15.5
\end{aligned}
$$

Therefore, the measure of $\overline{TR}$ is 15.5.

CHECK FOR UNDERSTANDING

Communicating Mathematics

Study the lesson. Then complete the following.

1. Refer to the diagram of a solar eclipse in the application at the beginning of the lesson. Name a pair of common external tangents and a pair of common internal tangents. $\overline{AC}$, $\overline{BD}$; $\overline{AD}$, $\overline{BC}$
2. **Draw** two circles having two common external tangents and one common internal tangent. See margin.
3. **State** the number of tangents that can be drawn to a circle for each case. Explain your answers.
   a. through a point outside the circle
   b. through a point inside the circle
   c. through a point on the circle See margin.
4. **You Decide** Sang Hee claims that two lines tangent to a circle must always intersect. Anjula says that she thinks it's possible for two lines tangent to a circle to be parallel. Who do you think is right? Explain. Draw a diagram to support your answer.

3a. 2; a tangent can be drawn to each "side" of the circle from the point outside of the circle.

3b. 0; a tangent cannot pass through a point inside the circle.

4. Anjula; two lines tangent to a circle could be parallel if the points of tangency were the endpoints of a diameter; see margin for diagram.

5. Draw circle P with radius $\overline{PQ}$ on patty paper. Fold the paper so the crease goes through Q and no other point of the circle. Put two points, R and S, on the line. Fold the paper along $\overline{PQ}$.
   a. What is the relationship of $\overline{PQ}$ to $\overleftrightarrow{RS}$? $\overline{PQ} \perp \overleftrightarrow{RS}$
   b. Draw a different radius and repeat the activity. What is true of any radius and the tangent through the endpoint of the radius?

5b. They are perpendicular to each other.

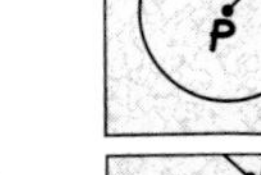

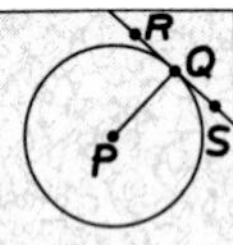

Reteaching

Using Diagrams Review Theorems 9-8 through 9-10 with students and draw a figure to represent each theorem. Write the given information and what is being proved. Have students investigate some of the logic behind the theorems and explain it in their own words.

Guided Practice

Refer to ⊙M for Exercises 6–13. $\overline{LK}$ and $\overline{LE}$ are tangent to ⊙M, m∠EML = 66, KM = 15, and LK = 36. Find each measure.

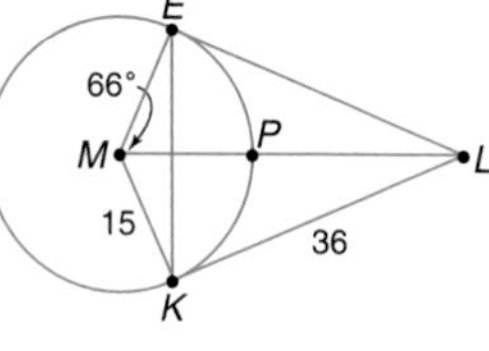

6. $m\angle MKL$ 90
7. $m\angle ELM$ 24
8. EL 36
9. $m\widehat{KPE}$ 132

Complete each statement.

10. $\triangle KML \cong$ __?__ $\triangle EML$
11. $\overline{KM} \perp$ __?__ $\overline{KL}$
12. $\triangle KLE$ is a(n) __?__ triangle. isosceles
13. Explain why ∠EMK and ∠ELK are supplementary.

13. In quadrilateral *KLEM*, ∠*K* and ∠*E* are right angles and the sum of their measures equals 180. Therefore, the sum of *m*∠*L* and *m*∠*M* must equal 180 and the angles must be supplementary.

For each ⊙C, find the value of x. Assume that segments that appear to be tangent are tangent.

14. 7

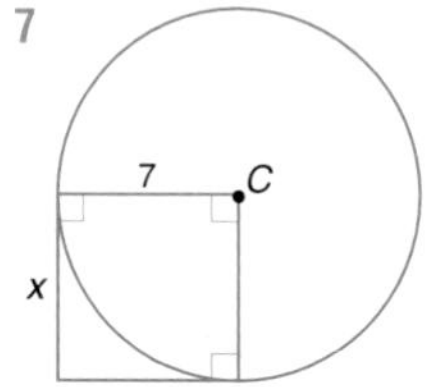

15. 8

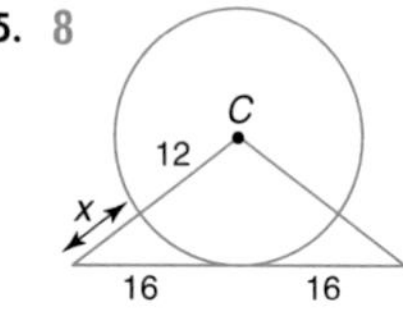

16. **Aerospace** A spacecraft is 4000 kilometers above Earth's surface. If the radius of Earth is about 6400 kilometers, find the distance between the spacecraft and the horizon. Round to the nearest tenth of a kilometer. 8197.6 km

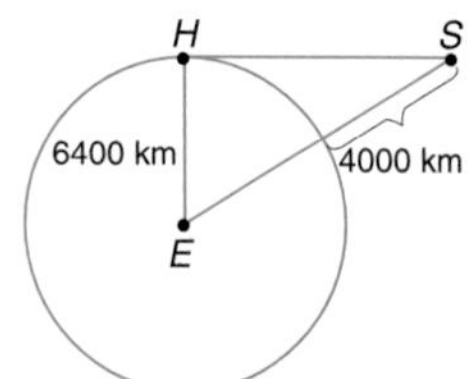

EXERCISES

Practice

Refer to the figure below for Exercises 17–30. $\overline{HL}$, $\overline{GF}$, and $\overline{AJ}$ are tangent to both ⊙P and ⊙Q, m∠APC = 60, PC = 5, and QE = 2. Find each measure.

17. AD $5\sqrt{3} \approx 8.7$
18. $m\angle DAP$ 90
19. PD 10
20. DF $2\sqrt{3} \approx 3.5$
21. $m\widehat{CA}$ 60
22. $m\angle ADP$ 30
23. $m\angle QJD$ 90
24. DQ 4
25. DJ $2\sqrt{3} \approx 3.5$
26. $m\widehat{FEJ}$ 120
27. $m\angle QDJ$ 30
28. AJ $7\sqrt{3} \approx 12.1$
29. Name two common internal tangents to ⊙P and ⊙Q. $\overline{GF}$ and $\overline{AJ}$
30. Explain why $\overline{GF} \cong \overline{AJ}$.

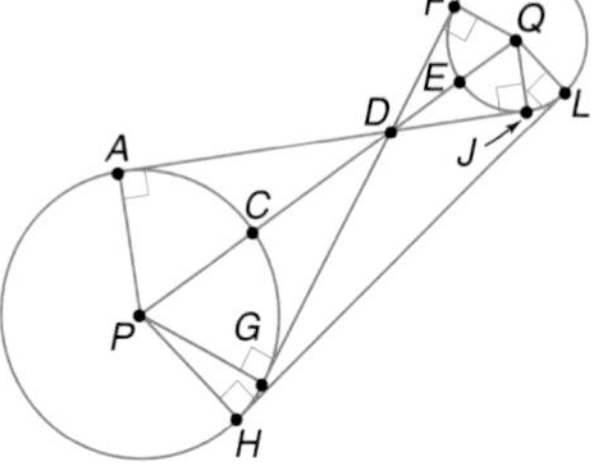

30. By Theorem 9–10, $\overline{GD} \cong \overline{AD}$ and $\overline{DF} \cong \overline{DJ}$. Since $GF = GD + DF$ and $AJ = AD + DJ$, then by substitution, $GF = AJ$ or $\overline{GF} \cong \overline{AJ}$.

Assignment Guide

Core (with proof): 17–51 odd, 52–62
Core (informal): 17–35 odd, 39–43 odd, 49, 51, 52–62
Enriched: 18–48 even, 49–62

For **Extra Practice**, see p. 781.

The red A, B, and C flags, printed only in the Teacher's Wraparound Edition, indicate the level of difficulty of the exercises.

Additional Answers

43. 4; $\triangle PQR$ is equilateral and $\overline{QN} \cong \overline{NR}$.

45. Given: $\overline{GR}$ is tangent to $\odot D$ at G. $\overline{AG} \cong \overline{DG}$

Prove: $\overline{GA}$ bisects $\overline{RD}$.

Proof:

Since $\overline{DA}$ is a radius, $\overline{DG} \cong \overline{DA}$. Since $\overline{AG} \cong \overline{DG} \cong \overline{DA}$, $\triangle GDA$ is equilateral. Therefore, each angle has a measure of 60. Since $\overline{GR}$ is tangent to $\odot D$, $m\angle RGD = 90$. Since $m\angle AGD = 60$, then by Angle Addition Post., $m\angle RGA = 30$. If $m\angle DAG = 60$, then $m\angle RAG = 120$. Then $m\angle R = 30$. Thus, $\triangle RAG$ is isosceles. By Trans. Prop. (=), $\overline{RA} \cong \overline{DA}$. Thus, $\overline{GA}$ bisects $\overline{RD}$.

46. Given: $\ell \perp \overline{AB}$

$\overline{AB}$ is a radius of $\odot A$.

Prove: ℓ is tangent to $\odot A$.

Proof:

Assume that ℓ is not tangent to $\odot A$. Since ℓ touches $\odot A$ at B, it must touch the circle in another place. Call this point C. Then $AB = AC$. But if $\overline{AB} \perp \ell$, AB must be the shortest distance between A and ℓ. There is a contradiction. Therefore, ℓ is tangent to $\odot A$.

Study Guide Masters, p. 53

9-5 NAME_______________ DATE_______

Study Guide Student Edition Pages 475–482

Tangents

Remember that a tangent is a line in the plane of a circle that intersects the circle in exactly one point. Three important theorems involving tangents are the following.

- If a line is a tangent to a circle, then it is perpendicular to the radius drawn to the point of tangency.
- In a plane, if a line is perpendicular to a radius of a circle at the endpoint on the circle, then the line is a tangent of the circle.
- If two segments from the same exterior point are tangent to a circle, then they are congruent.

Example: Find the value of x if $\overline{AB}$ is tangent to $\odot C$.

Tangent $\overline{AB}$ is perpendicular to radius $\overline{BC}$. Also, $AC = AD + BC = 17$.

$$(AB)^2 + (BC)^2 = (AC)^2$$
$$x^2 + 8^2 = 17^2$$
$$x^2 + 64 = 289$$
$$x^2 = 225$$
$$x = 15$$

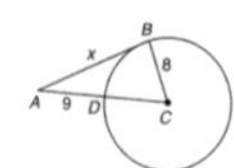

For each ⊙C, find the value of x. Assume that segments that appear to be tangent are tangent.

1. 19

2. 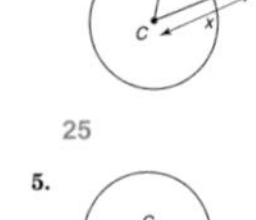 25

3. 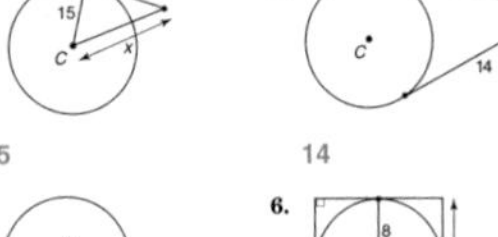 14

4. 20

5. 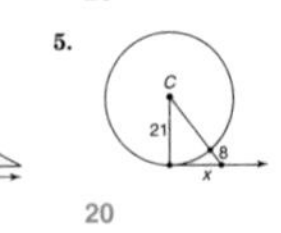 20

6. 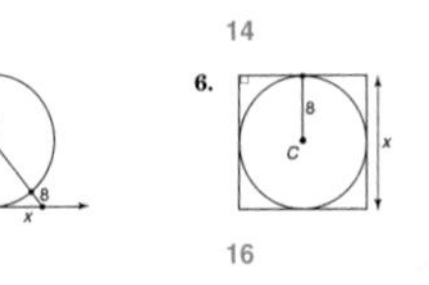 16

For each ⊙Q, find the value of x. Assume that segments that appear to be tangent are tangent. 31. 14 32. 10 33. 4

31. 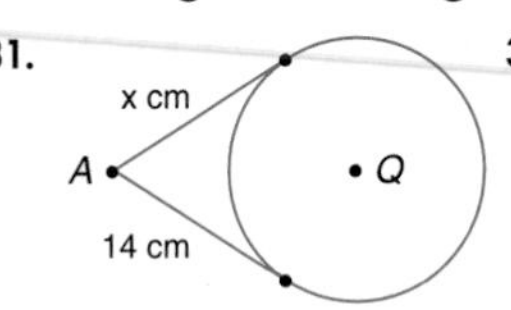

32. 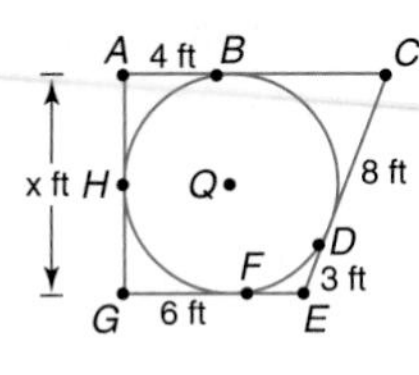

33. 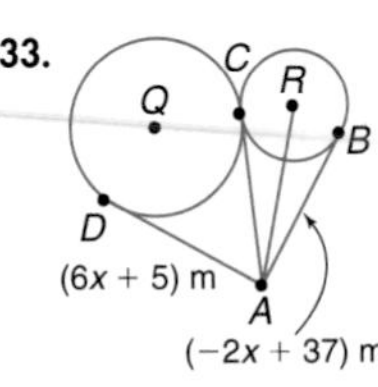

34. 16

35. 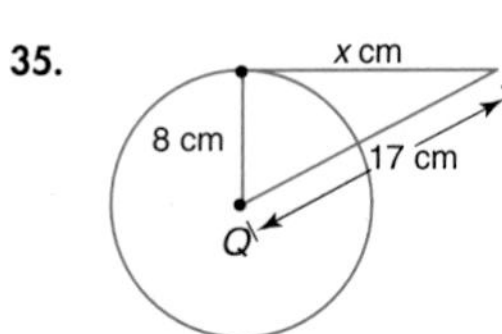

15

36. 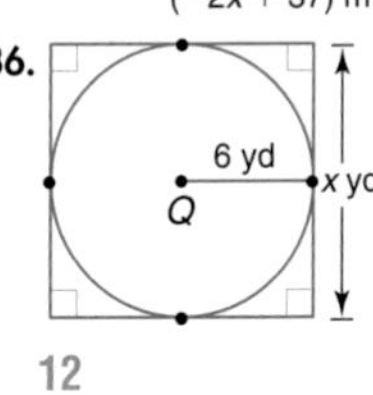

12

Proof

37. **Complete the following proof.**

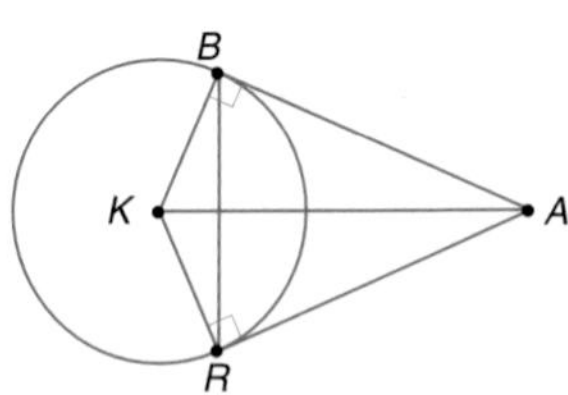

Given: $\overline{AB}$ and $\overline{AR}$ are tangent to $\odot K$.

Prove: $\angle BAK \cong \angle RAK$

Proof:

Statements	Reasons
a. $\overline{AB}$ and $\overline{AR}$ are tangent to $\odot K$.	a. __?__ Given
b. $\overline{AB} \cong \overline{AR}$	b. __?__
c. $\overline{BK} \cong \overline{RK}$	c. __?__
d. $\overline{AK} \cong \overline{AK}$	d. __?__ Reflexive Property (=)
e. $\triangle BAK \cong \triangle RAK$	e. __?__ SSS
f. $\angle BAK \cong \angle RAK$	f. __?__ CPCTC

37b. If 2 segments from the same exterior point are tangent to a ⊙, then they are ≅.

37c. Radii of the same circle are ≅.

INTEGRATION Algebra

Line ℓ is tangent to ⊙A at B.

38. Find the radius of $\odot A$. $\sqrt{13} \approx 3.6$

39. Find the slope of line ℓ. $-\frac{2}{3}$

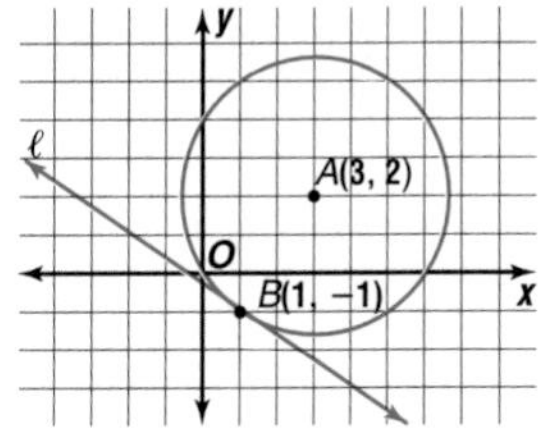

40. $\odot B$ is inscribed in quadrilateral $TANG$, $GN = 10$, $TG = 12$, and $AT = 18$. If the point of tangency bisects $\overline{AT}$, find AN. 16

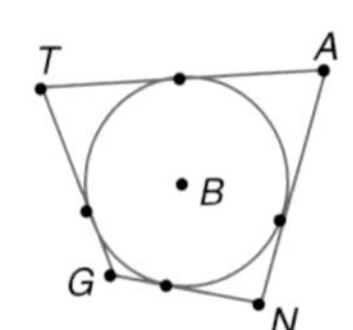

41–42. See students' work.

41. Draw a circle. Label the center P. Locate a point on the circle. Label it A. Construct a tangent to $\odot P$ at A.

42. Draw a circle. Label the center Q. Draw a point exterior to the circle. Label it B. Construct a tangent to $\odot Q$ containing B.

43. $\odot P$ is inscribed in hexagon $QRSTUV$. If the radius of $\odot P$ is 8 and all sides of $QRSTUV$ are congruent, find QN. Explain your reasoning. See margin.

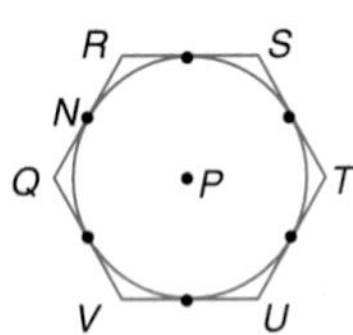

Additional Answer

47. Given: $\overleftrightarrow{CA}$ is tangent to $\odot X$ at A.

Prove: $\overline{XA} \perp \overleftrightarrow{CA}$

Proof:

Statements (Reasons)

1. $\overleftrightarrow{CA}$ is tangent to $\odot X$ at A. (Given)
2. Pick any point on $\overleftrightarrow{CA}$ other than A and call it B. Draw $\overline{XB}$. (Through any 2 pts. there is 1 line.)
3. $\overleftrightarrow{CA}$ intersects $\odot X$ at exactly one point, A, and B lies in the exterior of $\odot X$. (Def. tangent)
4. $XA < XB$ (The measure of a segment joining an exterior pt. to the center of a ⊙ is greater than the measure of a radius.)
5. $\overline{XA} \perp \overleftrightarrow{CA}$ ($\overline{XA}$ is the shortest segment from X to $\overleftrightarrow{CA}$.)

44. $\odot P$ is inscribed in right $\triangle CTA$. Find the perimeter of $\triangle CTA$ if the radius of $\odot P$ is 5 and $CT = 18$. 58.5

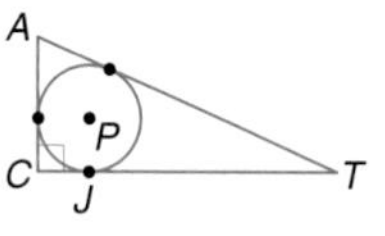

Proof

45. $\overline{GR}$ is tangent to $\odot D$ at G, and $\overline{AG} \cong \overline{DG}$. Write a paragraph proof to show that $\overline{GA}$ bisects $\overline{RD}$. See margin.

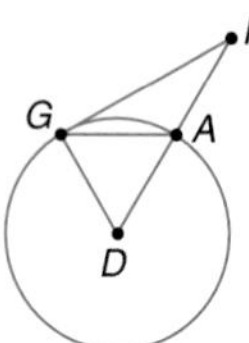

46. Write an indirect proof. See margin.

Given: $\ell \perp \overline{AB}$
$\overline{AB}$ is a radius of $\odot A$.

Prove: ℓ is tangent to $\odot A$. (Theorem 9–9)

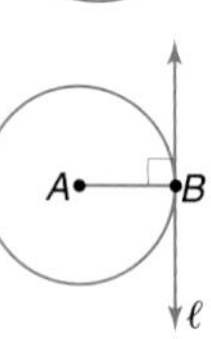

Write a two-column proof. 47–48. See margin.

47. If a line is tangent to a circle, then it is perpendicular to the radius drawn to the point of tangency. (Theorem 9–8)

48. If two segments from the same exterior point are tangent to a circle, then they are congruent. (Theorem 9–10)

Critical Thinking

49. A *unit circle* is a circle with a radius of 1. In the figure, $\odot O$ is a unit circle with $\overline{QR}$ tangent to $\odot O$ at R.

a. Use the right triangle trigonometry definition of tangent on page 412 to find $\tan \theta$. $\overline{QR}$

b. How do these two different definitions of tangent compare? See Solutions Manual.

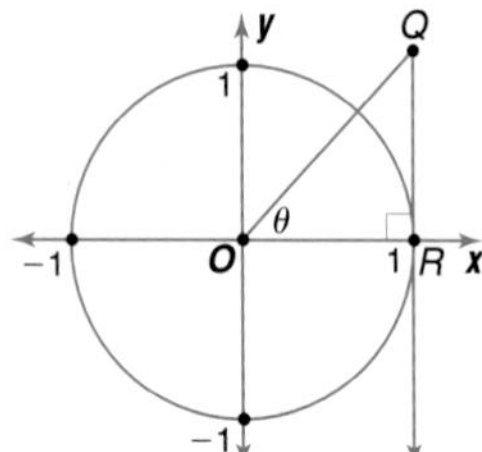

Applications and Problem Solving

50. **Astronomy** During an eclipse, the darkest portion of the shadow cast by the Moon is called the *umbra*. A *total* eclipse occurs when the umbra hits Earth's surface. The distance from S to E is about 93,000,000 miles. Round each answer to the nearest mile.

Celestial Body	Radius (miles)
Earth	3964
Moon	1080
Sun	432,000

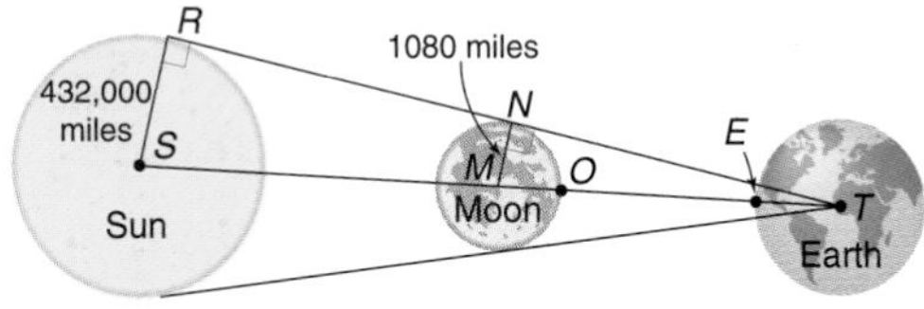

a. Find RT. 93,002,961 miles

b. Use similar triangles to find TN. 232,507 miles

c. Use $\overline{TN}$ and right $\triangle MNT$ to find MT. 232,510 miles

d. Find OE. 227,466 miles

e. Use your answer in part d to approximate how far apart Earth and the Moon must be during an eclipse in order to experience a total eclipse with the umbra casting a shadow on part of Earth.

50e. closer than 227,466 miles

Additional Answer

48. Given: $\overline{AB}$ is tangent to $\odot X$ at B.
$\overline{AC}$ is tangent to $\odot X$ at C.
Prove: $\overline{AB} \cong \overline{AC}$

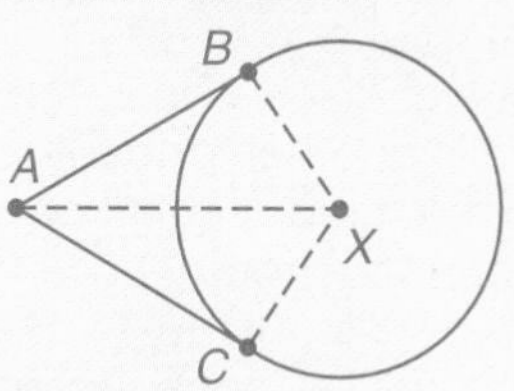

Proof:
Statements (Reasons)

1. $\overline{AB}$ is tangent to $\odot X$ at B; $\overline{AC}$ is tangent to $\odot X$ at C. (Given)
2. Draw $\overline{BX}$, $\overline{CX}$, and $\overline{AX}$. (Through any 2 pts. there is 1 line.)
3. $\angle ABX$ and $\angle ACX$ are rt. $\angle$s. (If a line is tangent to a $\odot$, then it is $\perp$ to the radius drawn to the pt. of tangency.)
4. $\overline{BX} \cong \overline{CX}$ (All radii of a $\odot$ are $\cong$.)
5. $\overline{AX} \cong \overline{AX}$ (Reflexive Prop. of $\cong$ segments)
6. $\triangle AXB \cong \triangle AXC$ (HL)
7. $\overline{AB} \cong \overline{AC}$ (CPCTC)

Practice Masters, p. 53

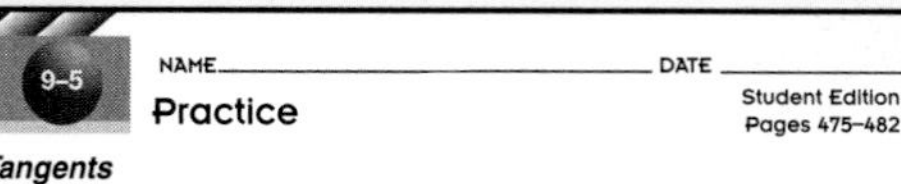
9-5 NAME ______ DATE ______
Practice
Student Edition Pages 475–482

Tangents

For each $\odot Q$, find the value of x. Assume that segments that appear to be tangent are tangent.

1.
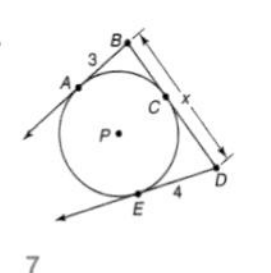

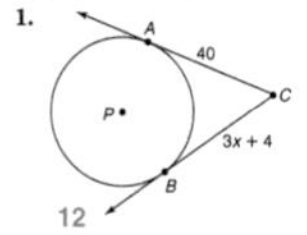

12

2.
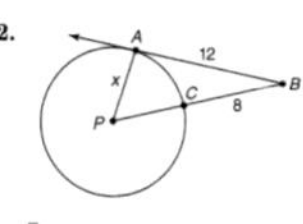

5

3.
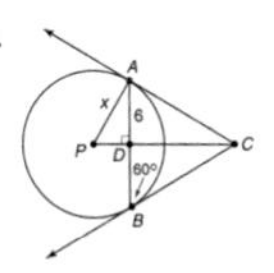

7

4.

$4\sqrt{3} \approx 6.9$

5.
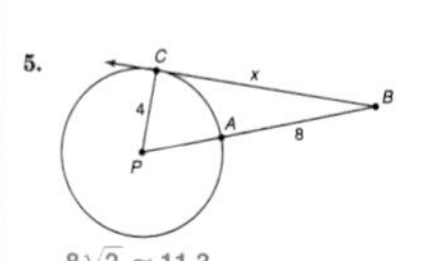

$8\sqrt{2} \approx 11.3$

6.
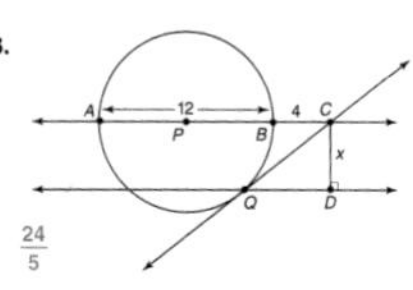

$\frac{24}{5}$

4 ASSESS

Closing Activity

Speaking Have a student give the first step of one of the constructions discussed in this lesson while another student performs that step on the board. Going around the room, have the next student give the next step, and so on until the construction has been completed. Do the same for the other two constructions.

Diskbolus, an ancient Greek sculpture of a discus thrower, is one of the most copied sculptures. It is attributed to Nankydes of Argos.

Enrichment Masters, p. 53

NAME______________________ DATE ________

Enrichment — Student Edition Pages 475–482

Magic Circles

Study the circles at the right. Their intersection points are labeled with numbers. The sum of the numbers for each circle equals the sum for each of the other circles. Such circles are called **magic circles**.

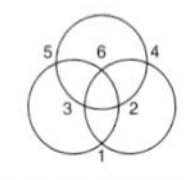

Each circle has a sum of 14.

1. Is the set of circles below a set of magic circles? If you answer yes, identify the sum. yes; 215

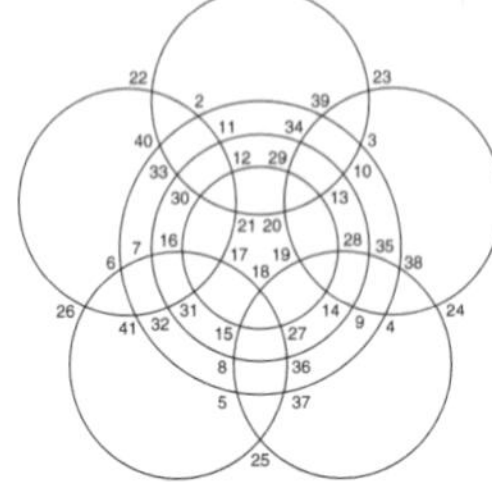

2. Label the intersection points for the circles at the right to obtain magic circles. Answers may vary. A possible answer is shown.

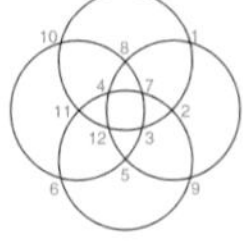

In 1988, Jurgen Schult of East Germany set a world record when he threw the discus 68.82 meters.

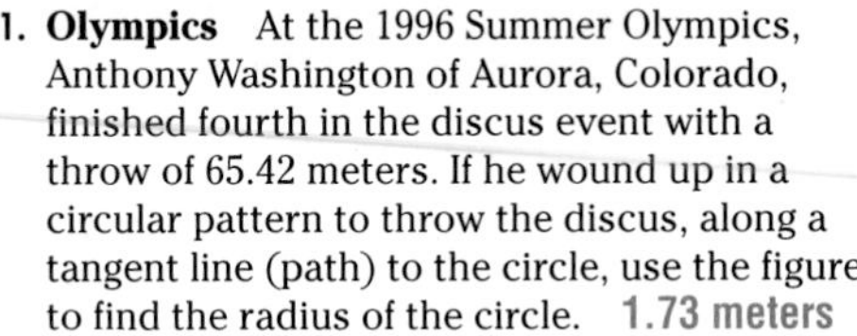

51. **Olympics** At the 1996 Summer Olympics, Anthony Washington of Aurora, Colorado, finished fourth in the discus event with a throw of 65.42 meters. If he wound up in a circular pattern to throw the discus, along a tangent line (path) to the circle, use the figure to find the radius of the circle. **1.73 meters**

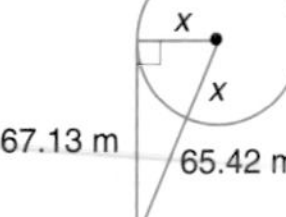

Mixed Review

52. Use the figure below to find $m\overset{\frown}{TZ}$. (Lesson 9–4) **110**

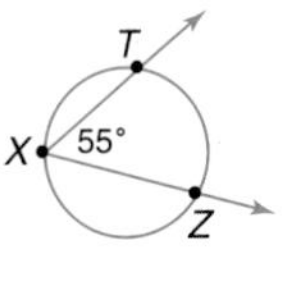

53. In ⊙O, find AC. (Lesson 9–3) **14**

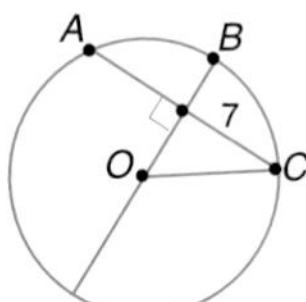

54. If the length of a side of an equilateral triangle is 6.5 meters, find the length of the altitude. (Lesson 8–4) **5.63 m**

55. Two triangles are similar, and their corresponding sides are in a ratio of 3:5. What is the measure of a side in the second triangle that corresponds to a side that measures 15 centimeters in the first triangle? (Lesson 7–3) **25 cm**

56. Construct a rectangle with length of 6 centimeters and width of 5 centimeters. (Lesson 6–3) **See students' work.**

57. Write an inequality to describe the possible values of x. (Lesson 5–6) $x > 5$

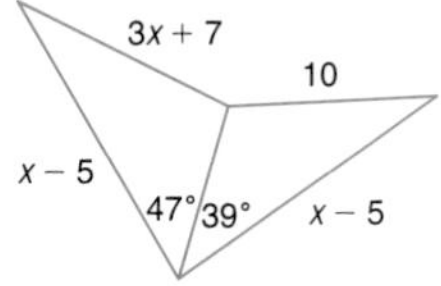

58. Can a triangle have two obtuse angles? Justify your answer. (Lesson 4–2)

58. No; the sum of the measures of the angles of a triangle must equal 180, and if two angles were obtuse, the sum of their measures would be greater than 180.

59. **Biology** Determine if a valid conclusion can be reached from the two true statements using the Law of Detachment or the Law of Syllogism. If a valid conclusion is possible, state it and the law that is used. If a valid conclusion does not follow, write *no conclusion*. (Lesson 2–3)
 (1) All species in the plant family *Maalvaceae* have five petals.
 (2) A wild rose has five petals. **no conclusion**

60. Draw opposite rays $\overrightarrow{QS}$ and $\overrightarrow{QR}$. (Lesson 1–6) **See students' work.**

INTEGRATION Algebra

61. Use a calculator to determine the approximate value of $30(4.2)^{-5.1}$ to the nearest hundredth. **0.02**

62. **Games** Monopoly® is the top-selling board game of all time. The graph at the right shows the retail prices of the game and the prices in inflation-adjusted dollars for the years 1935, 1962, and 1992.
 a. Find the percent of increase of the price in current dollars from 1935 to 1992. Round to the nearest whole percent. **186%**
 b. Find the percent of decrease of the retail price from 1935 to 1992. Round to the nearest whole percent. **72%**

Monopoly® for More or Less?

Retail price / In current dollars

Year	Retail price	In current dollars
1935	$3.50	$36.23
1962	$5.50	$25.82
1992	$10.00	$10.00

Source: *USA TODAY* research by Laurel Adams

Extension

Reasoning In two nonintersecting circles, will the points of tangency of a common internal tangent ever be the same as the points of tangency of a common external tangent? Why or why not? **No, because there is only one line tangent to a given point on a circle.**

9–6 Secants, Tangents, and Angle Measures

***What* YOU'LL LEARN**

- To find the measures of angles formed by intersecting secants and tangents in relation to intercepted arcs.

***Why* IT'S IMPORTANT**

A knowledge of secants, tangents, and angle measures will help you solve problems involving civil engineering, astronomy, and meteorology.

Civil Engineering

The *Luis I* bridge in Oporto, Portugal, was designed by T. Seyrig and completed in 1885. He worked closely with Gustav Eiffel (Eiffel tower) who built a railroad bridge of similar design in Gazabit, France.

The roads represent the two different ways that lines or segments can intersect a circle.

- The upper road appears to touch the circular support arch.
- The lower road appears to intersect the circular support arch in two places.

In Lesson 9–5, you learned that a line that touches or intersects a circle in exactly one point is a tangent. A line that intersects a circle in exactly *two* points is called a **secant** of the circle. A secant of a circle contains a chord of the circle.

You have calculated the measures of angles in circles whose vertices are either on the circle or are the center of the circle. You can also find the measures of angles formed by secants and tangents.

The *Luis I* bridge provides an example of an angle whose vertex is located *on* the circle at the point where tangent $\overleftrightarrow{TC}$ intersects the circle. $\overrightarrow{CA}$ is a secant ray with $\overline{CA}$ representing the chord from the point of tangency of the arch to the lower foot of the arch. Angle TCA intercepts $\widehat{CA}$ like the inscribed angle relationship studied in Lesson 9–4. The measure of $\angle TCA$ is one-half the measure of the intercepted arc or $\frac{1}{2}m\widehat{CA}$.

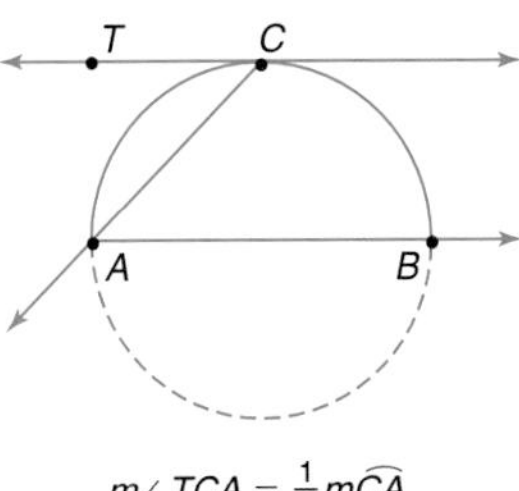

$m\angle TCA = \frac{1}{2}m\widehat{CA}$

Theorem 9–11	If a secant and a tangent intersect at the point of tangency, then the measure of each angle formed is one-half the measure of its intercepted arc.

You will be asked to prove this theorem in Exercise 47.

9-6 LESSON NOTES

NCTM Standards: 1–5, 7

Instructional Resources

- Study Guide Master 9-6
- Practice Master 9-6
- Enrichment Master 9-6
- Assessment and Evaluation Masters, p. 241

Transparency 9-6A contains the 5-Minute Check for this lesson; **Transparency 9-6B** contains a teaching aid for this lesson.

Recommended Pacing	
Standard Pacing	Days 9 & 10 of 15
Honors Pacing	Days 7 & 8 of 13
Block Scheduling*	Day 6 of 9

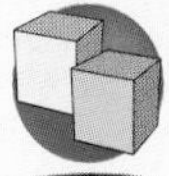

*For more information on pacing and possible lesson plans, refer to the *Block Scheduling Booklet*.

1 FOCUS

5-Minute Check (*over Lesson 9-5*)

$AF = 6$, $BJ = 3$, and $m\angle HAG = 60$. Find each measure.

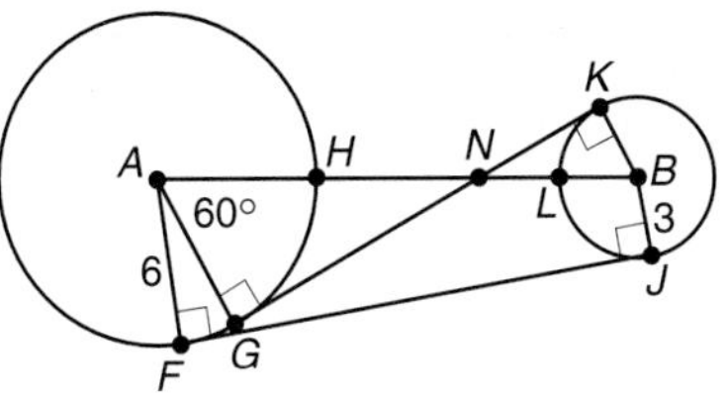

1. $m\angle ANG$ 30
2. $m\angle KBN$ 60
3. GN $6\sqrt{3} \approx 10.39$
4. $m\widehat{LK}$ 60

Motivating the Lesson

Questioning Have students draw a circle with a diameter and a tangent at one of the endpoints of the diameter. Ask if there is a relationship between the angle formed by the diameter and the tangent and the intercepted arc of that angle. **Yes; the angle measure is one-half the measure of the arc.** This can be said because the arc is a semicircle and the angle is a right angle. Suggest that this relationship can be extended to other angle measures.

2 TEACH

In-Class Example

For Example 1

$\overleftrightarrow{KD}$ is tangent to $\odot C$ at D. Find $m\angle KDE$. **90**

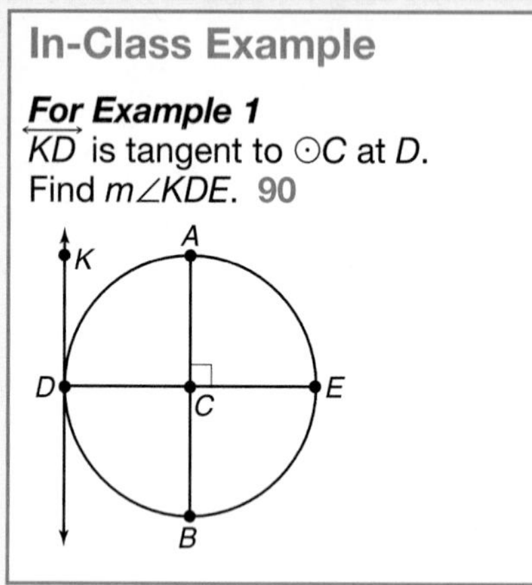

Teaching Tip For Theorem 9-12, review vertical angles and draw examples.

The TI-92 can be used to verify the results of each step of this activity. To do this, select the geometry application and utilize the measure feature of the program.

Answers for Modeling Mathematics

a. Exterior Angle Theorem

b. $m\angle 1 = \frac{1}{2}m\widehat{WX} + \frac{1}{2}m\widehat{ZY}$ or $\frac{1}{2}(m\widehat{WX} + m\widehat{ZY})$

c. Sample answer: The measure of an angle is half the sum of the measures of the arcs intercepted by the angle and its vertical angle.

Example 1 **The Darby bridge was built from 1777–1779 near Coalbrookdale, London. If the arch of the bridge is a semicircle and M, the midpoint of the semicircle, is the point of tangency, find $m\angle AMP$.**

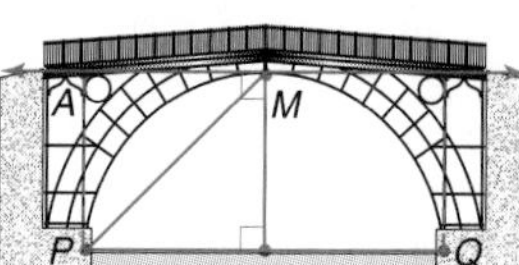

Since $\widehat{PMQ}$ is a semicircle with M as the midpoint, $m\widehat{PM} = 90$. By Theorem 9–11,

$$m\angle AMP = \frac{1}{2}m\widehat{PM}$$
$$= \frac{1}{2}(90) \text{ or } 45$$

Therefore, $m\angle AMP$ is 45.

FYI

The Darby bridge was the world's first iron bridge. The designer used circular supports to strengthen the bridge.

There is also a special relationship between the angles formed by two secants and the arcs they intercept.

MODELING MATHEMATICS — Angle Vertex in the Interior of a Circle

Materials: compass, straightedge

- Draw a circle and choose point V in the interior of the circle.

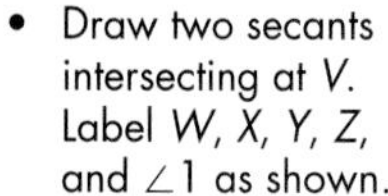

- Draw two secants intersecting at V. Label W, X, Y, Z, and $\angle 1$ as shown.

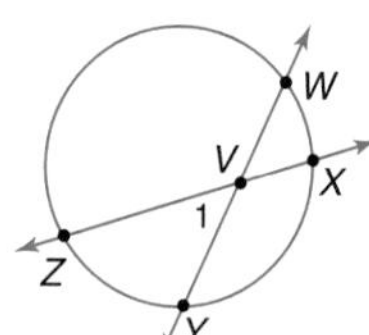

- Draw $\overline{XY}$ and label $\angle 2$ and $\angle 3$. Note that $m\angle 2 = \frac{1}{2}m\widehat{WX}$ and $m\angle 3 = \frac{1}{2}m\widehat{ZY}$, since $\angle 2$ and $\angle 3$ are inscribed angles.

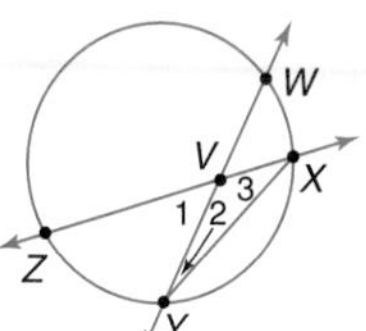

Your Turn

a. Explain why $m\angle 1 = m\angle 2 + m\angle 3$.

b. Use part a to write an equation for $m\angle 1$ in terms of the measures of $\widehat{WX}$ and $\widehat{ZY}$.

c. Make a conjecture about the measure of an angle formed by two secants in relation to the arcs of the circle. **a–c. See margin.**

The Modeling Mathematics activity illustrates Theorem 9–12. *You will be asked to prove this theorem in Exercise 45.*

Theorem 9–12	**If two secants intersect in the interior of a circle, then the measure of an angle formed is one-half the sum of the measures of the arcs intercepted by the angle and its vertical angle.**

In the circle at the right, two secants intersect inside the circle. From Theorem 9–12, we can conclude $m\angle 1 = \frac{1}{2}(m\widehat{RU} + m\widehat{ST})$ and $m\angle 2 = \frac{1}{2}(m\widehat{TU} + m\widehat{RS})$. Since $m\angle 1 = m\angle 3$, $m\angle 3 = \frac{1}{2}(m\widehat{RU} + m\widehat{ST})$ and, likewise, $m\angle 4 = \frac{1}{2}(m\widehat{TU} + m\widehat{RS})$.

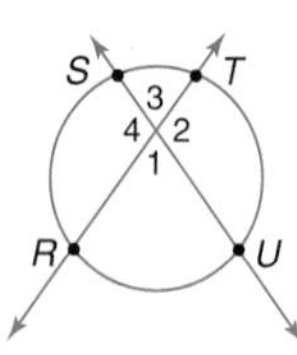

FYI

The longest steel arch bridge in the world is the New River Gorge Bridge in Fayetteville, West Virginia. It spans 1700 feet.

You can also use inscribed angles to find the measure of an angle formed by two secants that intersect outside the circle

Example  **If $m\widehat{BE} = 100$ and $m\widehat{AR} = 25$, find $m\angle S$.**

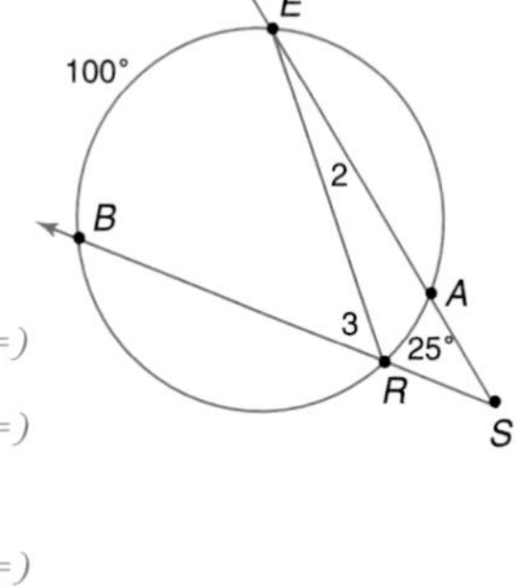

$\angle 3$ and $\angle 2$ are inscribed angles, so $m\angle 3 = \frac{1}{2}m\widehat{BE}$ and $m\angle 2 = \frac{1}{2}m\widehat{AR}$.

$m\angle 3 = m\angle S + m\angle 2$ *Exterior ∠ Th.*

$\frac{1}{2}m\widehat{BE} = m\angle S + \frac{1}{2}m\widehat{AR}$ *Substitution Prop. (=)*

$\frac{1}{2}(100) = m\angle S + \frac{1}{2}(25)$ *Substitution Prop. (=)*

$50 = m\angle S + 12.5$

$37.5 = m\angle S$ *Substitution Prop. (=)*

In Example 2, notice that $m\angle 1 = \frac{1}{2}(m\widehat{BE} - m\widehat{AR})$. This illustrates Theorem 9–13.

Theorem 9–13	**If two secants, a secant and a tangent, or two tangents intersect in the exterior of a circle, then the measure of the angle formed is one-half the positive difference of the measures of the intercepted arcs.**

There are three possible cases to illustrate Theorem 9–13.

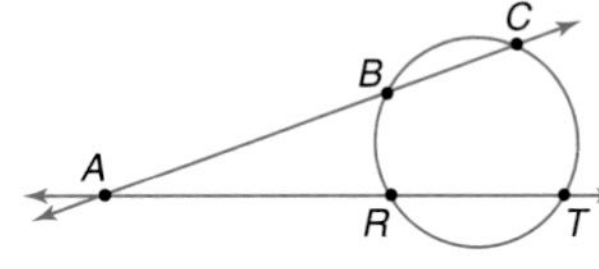

Case 1: two secants

$m\angle CAT = \frac{1}{2}(m\widehat{CT} - m\widehat{BR})$

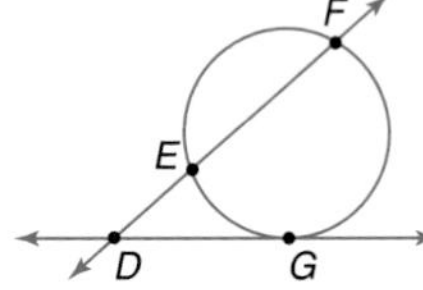

Case 2: a secant and a tangent

$m\angle FDG = \frac{1}{2}(\widehat{FG} - \widehat{EG})$

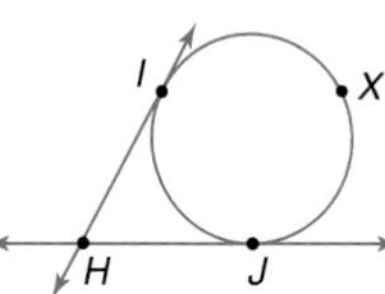

Case 3: two tangents

$m\angle IHJ = \frac{1}{2}(\widehat{IXJ} - \widehat{IJ})$

You will be asked to prove each of these cases in Exercise 46.

Example **3**

Algebra

a. Use ⊙K to find the value of x.

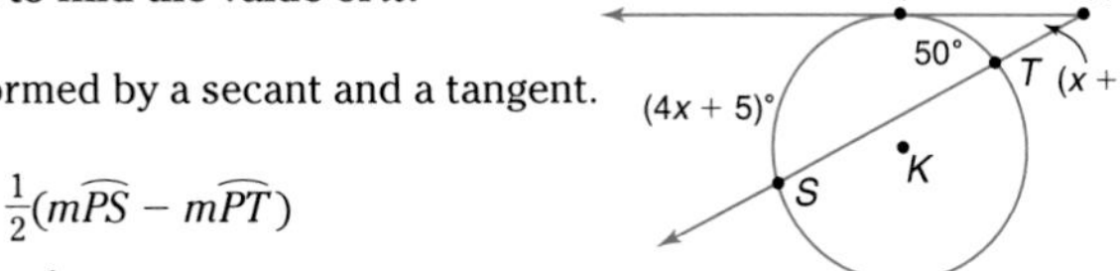

$\angle R$ is formed by a secant and a tangent.

$m\angle R = \frac{1}{2}(m\widehat{PS} - m\widehat{PT})$

$x + 2.5 = \frac{1}{2}[(4x + 5) - 50]$

$x + 2.5 = \frac{1}{2}(4x - 45)$

$x + 2.5 = 2x - 22.5$

$25 = x$

(continued on the next page)

In-Class Examples

For Example 2

If $m\widehat{AB} = 120$, $m\widehat{BC} = 30$, and $m\widehat{CD} = 160$, find $m\angle Q$. 10

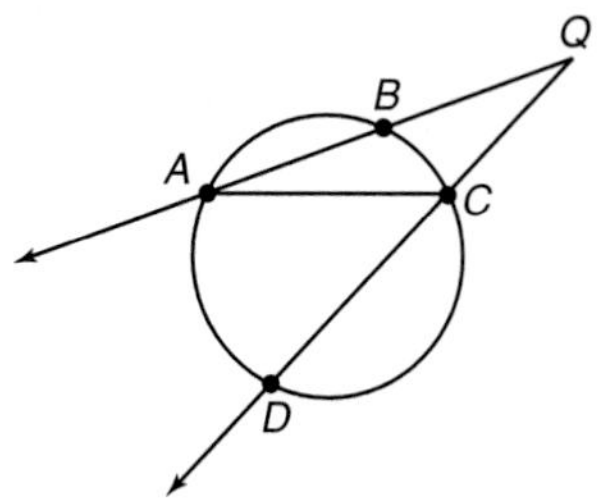

For Example 3

Alice is playing golf and her ball has landed behind a bush. If $m\widehat{AB} = 86$ and $m\widehat{AC} = 184$, find $m\angle AGB$. What does this tell Alice about her next shot?

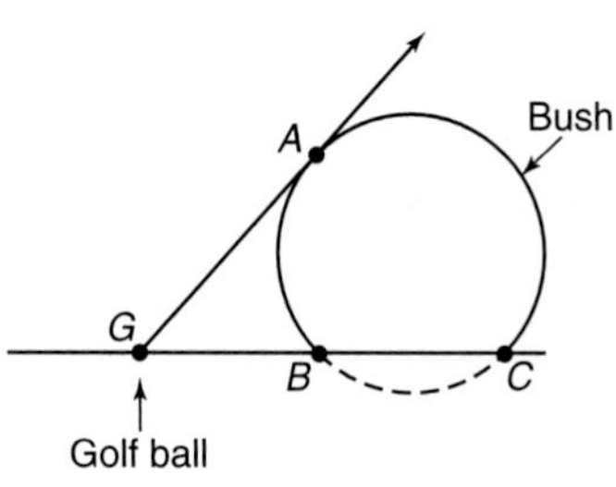

49; the angle of Alice's next shot must be greater than 49°.

Teaching Tip Example 3 illustrates that algebra can be applied to geometry in other ways than placing the figure on a coordinate system.

3 PRACTICE/APPLY

Check for Understanding

Exercises 1–13 are designed to help you assess your students' understanding through reading, writing, speaking, and modeling. You should work through Exercises 1–5 with your students and then monitor their work on Exercises 6–13.

Error Analysis

Students may have difficulty understanding when to subtract and when to add arc measures. Emphasize that, if the point of intersection is outside the circle, you subtract the arc measures. If the point of intersection is inside the circle, you add the arc measures.

Additional Answer

4. Since $\overrightarrow{DE}$ is tangent to ⊙P, then $\overline{ED} \perp \overline{BD}$, so $\angle BDE$ has a measure of 90. The intercepted arc is a semicircle, so by Theorem 9-11, $m\angle BDE = \frac{1}{2}(180)$ or 90.

b. Use ⊙S to find the value of y.

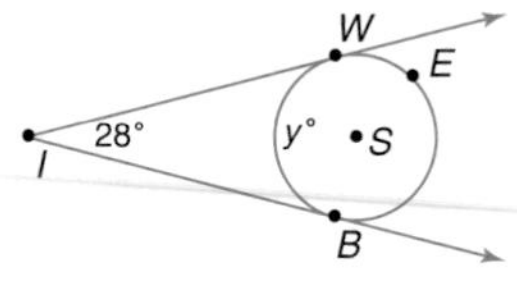

$\angle I$ is formed by two tangents. If $m\widehat{WB} = y$, then $m\widehat{WEB} = 360 - y$. *Why?*

$$m\angle I = \frac{1}{2}(m\widehat{WEB} - m\widehat{WB})$$
$$28 = \frac{1}{2}[(360 - y) - y]$$
$$28 = \frac{1}{2}(360 - 2y)$$
$$28 = 180 - y$$
$$y = 152$$

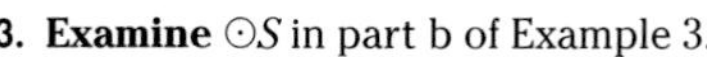

CHECK FOR UNDERSTANDING

Communicating Mathematics

Study the lesson. Then complete the following

1. **Explain** how to find the measure of an angle formed by two secants whose vertex is in the *exterior* of a circle. 1. Subtract the two intercepted arcs and take $\frac{1}{2}$.
2. **You Decide** Using ⊙P, Angelica argues that $m\angle 1 = \frac{1}{2}(m\widehat{JK} + m\widehat{ML})$ by Theorem 9–12. Sommer says $m\angle 1 = m\widehat{JK}$ since $\angle 1$ is a central angle. Who is right and why? 2. Both are right since $m\widehat{JK} = m\widehat{ML}$.

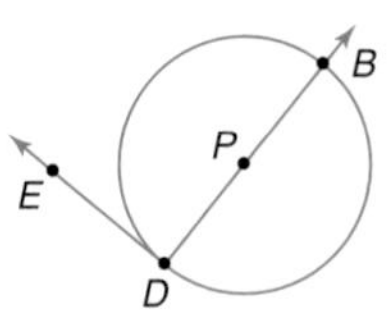

3. **Examine** ⊙S in part b of Example 3.
   a. What would be the value of y if $m\angle I$ were 100? 80
   b. Make a conjecture about the relationship between $m\angle I$ and the measure of the intercepted minor arc WB. They are supplementary.
4. $\overline{BD}$ is a diameter of ⊙P with tangent ray $\overrightarrow{DE}$. Provide two different reasons why $m\angle BDE$ is a right angle. See margin.

B
P
E
D

MATH JOURNAL

5. The word secant comes from the Latin word *secare*, meaning "to cut." Explain why secant is the word used for a line that intersects a circle in exactly two points. How is a secant different from a tangent? A secant "cuts" a circle into two parts or cuts away one part. A tangent only touches one point on the circle.

Guided Practice

6. Find the measure of $\angle 3$. 22

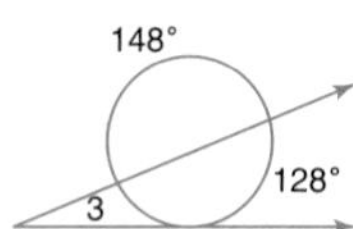

In ⊙K, $m\widehat{OB} = 98$, $m\widehat{OY} = 28$, $m\widehat{YD} = 62$, and $m\widehat{DA} = 38$. Find each measure.

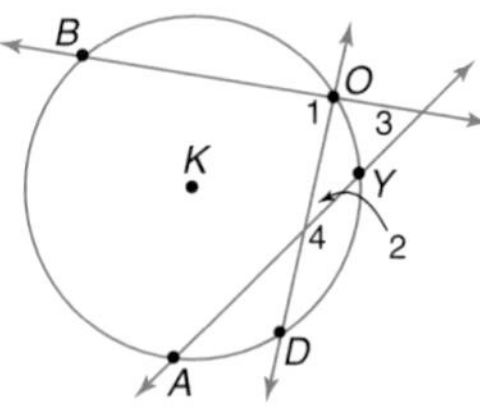

7. $m\widehat{AB}$ 134
8. $m\angle 1$ 86
9. $m\angle 2$ 33
10. $m\angle 3$ 53

Reteaching

Using Comparison Review the two ways of finding angle measures that previously have been discussed. If the angle is a central angle, then its measure is equal to the measure of its intercepted arc; if the angle is an inscribed angle, then it is one-half the measure of its intercepted arc. Have students draw other ways of intercepting a circle with an angle and apply the theorems in this lesson to their sketches.

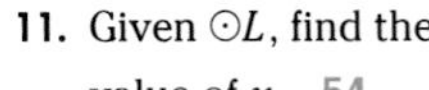

11. Given $\odot L$, find the value of x. 54

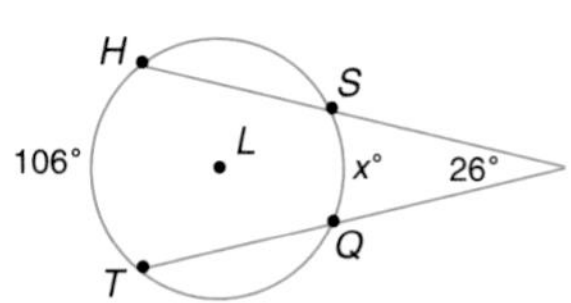

12. Refer to $\odot K$.

a. Find the value of x. 1

b. Find $m\angle AET$. 141

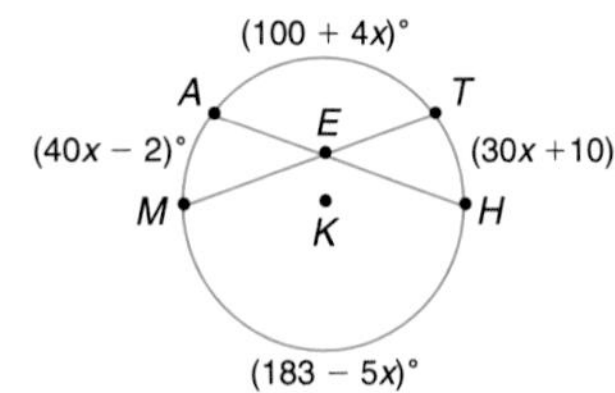

13. Astronomy A geostationary satellite that orbits about 33,000 miles above Earth rotates in the same direction as Earth and, as a result, appears to hover directly over a fixed point on the equator. Use the figure below to measure the arc on the equator of Earth visible to a geostationary satellite. 169

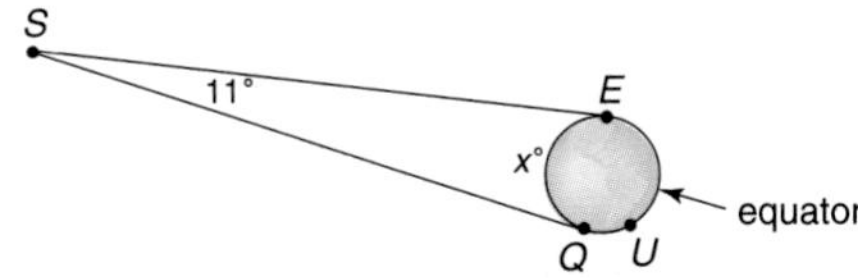

EXERCISES

Practice

Find the measure of each numbered angle.

14. 13

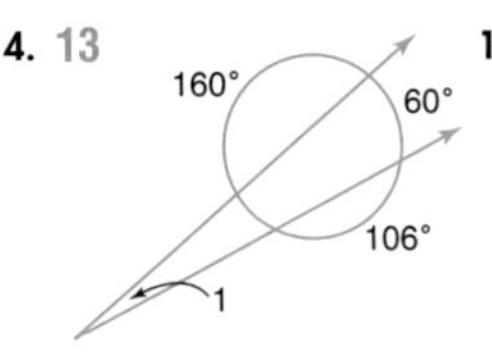

15. 157.5

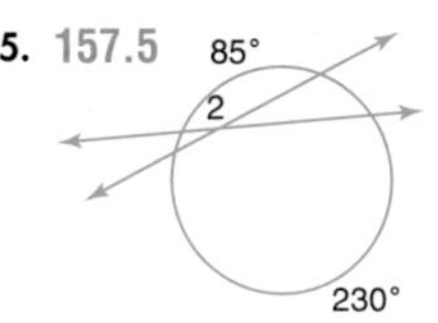

16. 129

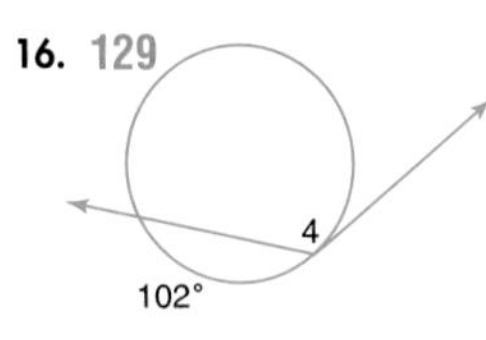

$\overleftrightarrow{CE}$ is tangent to $\odot S$ at E, and $m\widehat{IG} = 80$, $m\widehat{GE} = 140$, and $m\widehat{ER} = 130$. Find each measure.

17. $m\angle GRE$ 70

18. $m\widehat{IR}$ 10

19. $m\angle T$ 65

20. $m\angle GAE$ 75

21. $m\angle IEC$ 70

22. $m\angle IED$ 110

23. $m\angle IGA$ 5

24. $m\widehat{GIE}$ 220

25. $m\angle IAG$ 105

26. $m\angle TIE$ 110

27. $m\angle CER$ 65

28. $m\angle DEF$ 65

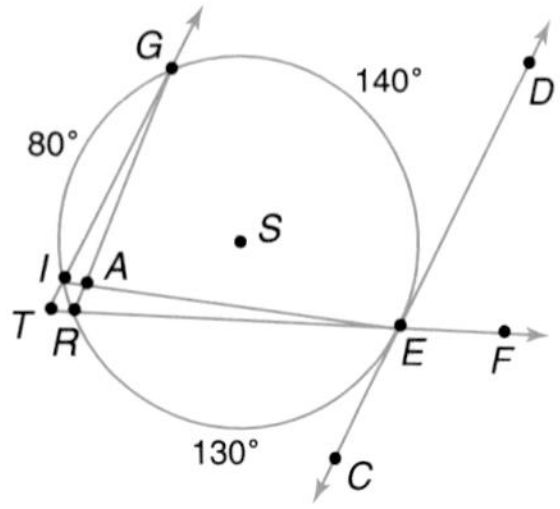

Given $\odot S$, find the value of x.

29. 20

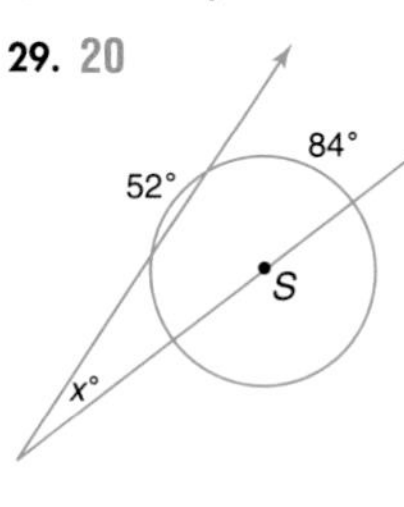

30. 40

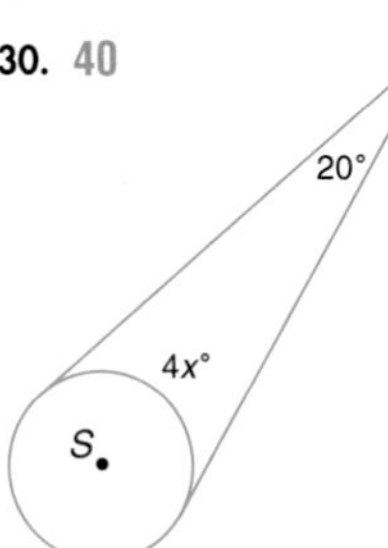

31. 26

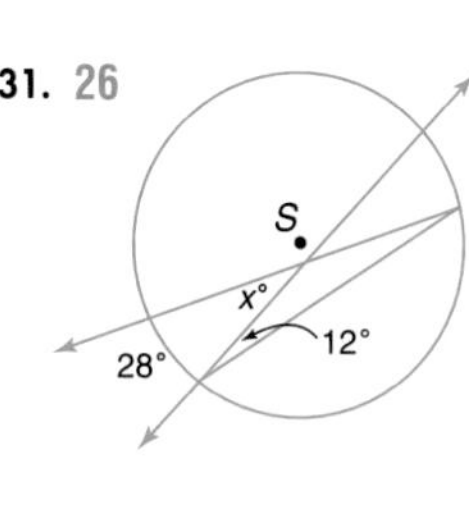

Assignment Guide

Core (with proof): 15–47 odd, 48, 49, 51–61
Core (informal): 15–43 odd, 48, 49, 51–61
Enriched: 14–46 even, 48–61

For **Extra Practice**, see p. 781.

The red A, B, and C flags, printed only in the Teacher's Wraparound Edition, indicate the level of difficulty of the exercises.

Additional Answer

46a. Proof:

Statements (Reasons)

1. $\overleftrightarrow{AC}$ and $\overleftrightarrow{AT}$ are secants to the circle. (Given)
2. Draw $\overline{CR}$. (Through any 2 pts. there is 1 line.)
3. $m\angle CRT = \frac{1}{2}m\widehat{CT}$, $m\angle ACR = \frac{1}{2}m\widehat{BR}$ (The measure of an inscribed $\angle = \frac{1}{2}$ the measure of its intercepted arc.)
4. $m\angle CRT = m\angle ACR + m\angle CAT$ (Exterior $\angle$ Th.)
5. $\frac{1}{2}m\widehat{CT} = \frac{1}{2}m\widehat{BR} + m\angle CAT$ (Substitution Prop. (=))
6. $\frac{1}{2}m\widehat{CT} - \frac{1}{2}m\widehat{BR} = m\angle CAT$ (Subtraction Prop. (=))
7. $\frac{1}{2}(m\widehat{CT} - m\widehat{BR}) = m\angle CAT$ (Distributive Prop.)

Study Guide Masters, p. 54

NAME____________________ DATE __________

Study Guide

Student Edition
Pages 483–490

Secants, Tangents, and Angle Measures

A line that intersects a circle in exactly two points is called a **secant** of the circle. You can find the measures of angles formed by secants and tangents by using the following theorems.

- If a secant and a tangent intersect at the point of tangency, then the measure of each angle formed is one-half the measure of its intercepted arc.
- If two secants intersect in the interior of a circle, then the measure of an angle formed is one-half the sum of the measures of the arcs intercepted by the angle and its vertical angle.
- If two secants, a secant and a tangent, or two tangents intersect in the exterior of a circle, then the measure of the angle formed is one-half the positive difference of the measures of the intercepted arcs.

Example: Find the measure of $\angle MPN$.

You can use the last theorem above.

$$m\angle MPN = \frac{1}{2}m(\widehat{MN} - \widehat{RS}) = \frac{1}{2}(34 - 18) = \frac{1}{2}(16) \text{ or } 8$$

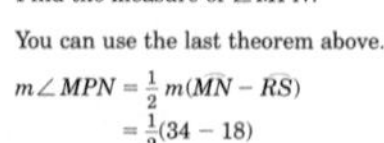

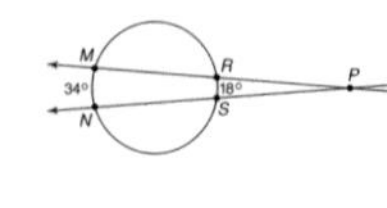

Find the measure of each numbered angle.

1. 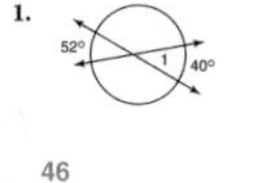

46

2. 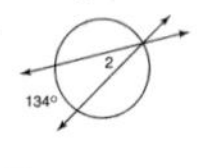

67

3. 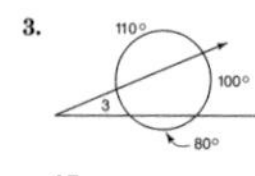

15

Given ⊙T, find the value of x.

4. 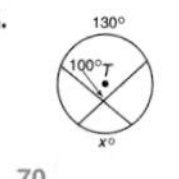

70

5. 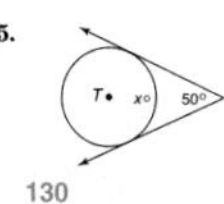

130

6. 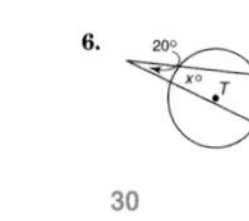

30

B

32. In the figure at the right, $m\widehat{CD} = 116$, $m\widehat{BE} = 38$, and $m\widehat{ATF} = 219$. Find $m\widehat{AF}$. **141**

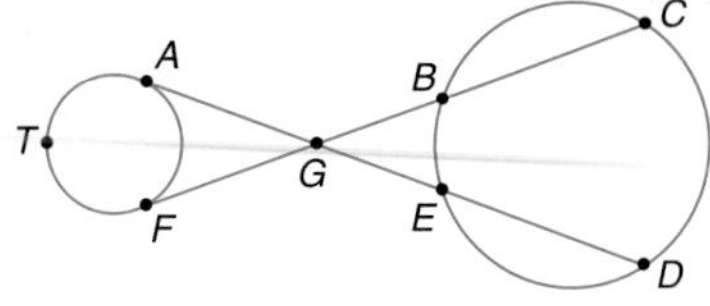

33. In $\odot X$, $m\widehat{AK} = 108$, $m\widehat{RE} = 118$, $m\angle KRE = 30$, and $m\angle KME = 52$. Find each measure.
    a. $m\widehat{SR}$ **44**
    b. $m\widehat{AS}$ **30**
    c. $m\angle KPE$ **15**

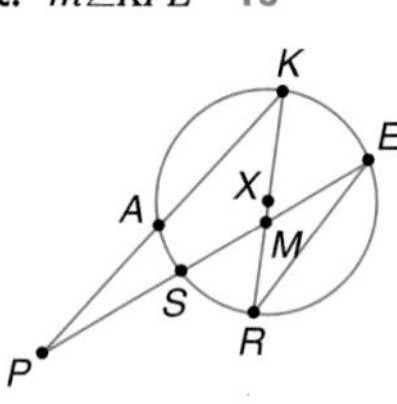

34. In the figure, $\overline{TA}$ and $\overline{TG}$ are tangent to $\odot P$. Find the values of x, y, and z. **$x = 85$, $y = 45$, $z = 40$**

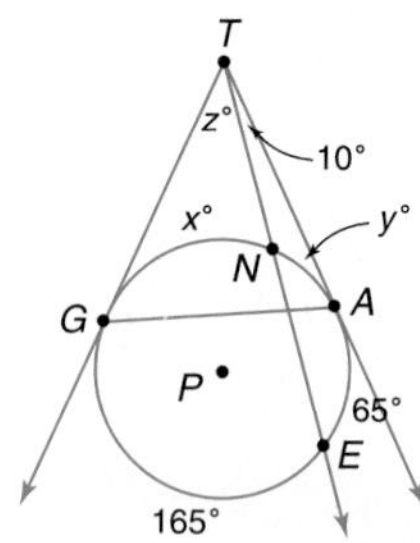

35. In $\odot T$, $m\widehat{UD} = 106$, $m\widehat{QU} = 96$, and $m\angle Q = 92$. Find $m\angle A$. **26**

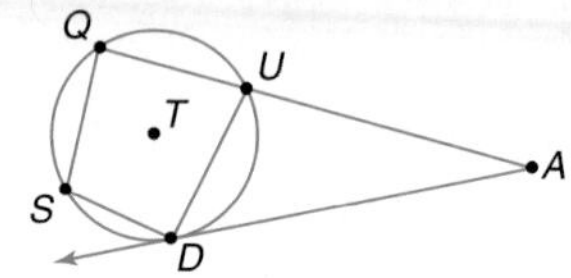

36. In the figure, $\overline{MR}$ and $\overline{MT}$ are tangent to $\odot B$ and $HTMR$ is a rhombus. Find $m\widehat{RT}$. **120**

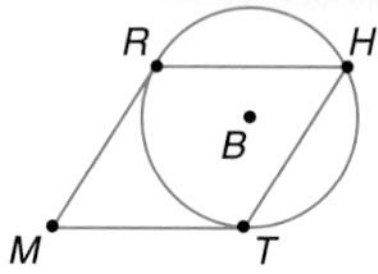

In the figure, quadrilateral *GERA* is inscribed in ⊙*P*, with $\overleftrightarrow{TA}$ tangent to ⊙*P* at *A*, $m\angle REG = 78$, $m\widehat{AR} = 46$, and $\overline{AR} \parallel \overline{GE}$. Find each measure.

37. $m\angle GAR$ **102**
38. $m\widehat{AG}$ **110**
39. $m\angle TAR$ **23**
40. $m\angle GAN$ **55**
41. $m\widehat{GE}$ **94**
42. $m\widehat{RE}$ **110**

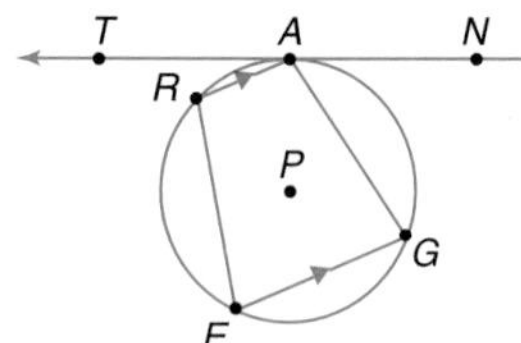

43. Refer to $\odot U$.
    a. Find the value of x. **$\sqrt{63} \approx 7.9$**
    b. Find $m\angle P$. **$2\sqrt{63} \approx 15.9$**

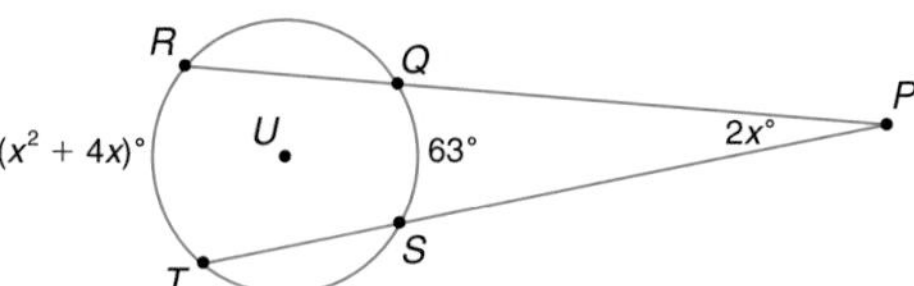

44. Refer to $\odot D$.
    a. Find the value of x. **5**
    b. Find $m\widehat{ABT}$. **310**

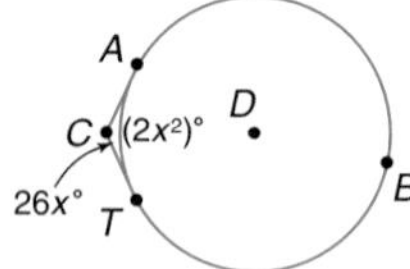

Write a proof.

45. If two secants intersect in the interior of a circle, then the measure of an angle formed is one-half the sum of the measures of the arcs intercepted by the angle and its vertical angle. (Theorem 9–12) See Solutions Manual.

46. If two secants, a secant and a tangent, or two tangents intersect in the exterior of a circle, then the measure of the angle formed is one-half the positive difference of the measures of the intercepted arcs. (Theorem 9–13)

a. Case 1 See margin.

Given: $\overleftrightarrow{AC}$ and $\overleftrightarrow{AT}$ are secants to the circle.

Prove: $m\angle CAT = \frac{1}{2}(m\widehat{CT} - m\widehat{BR})$

(*Hint:* Draw $\overline{CR}$.)

b. Case 2 See margin.

Given: $\overleftrightarrow{DG}$ is a tangent to the circle.
$\overleftrightarrow{DF}$ is a secant to the circle.

Prove: $m\angle FDG = \frac{1}{2}(m\widehat{FG} - m\widehat{GE})$

(*Hint:* Draw $\overline{FG}$.)

c. Case 3 See margin.

Given: $\overleftrightarrow{HI}$ and $\overleftrightarrow{HJ}$ are tangents to the circle.

Prove: $m\angle IHJ = \frac{1}{2}(m\widehat{IXJ} - m\widehat{IJ})$

(*Hint:* Draw $\overline{IJ}$.)

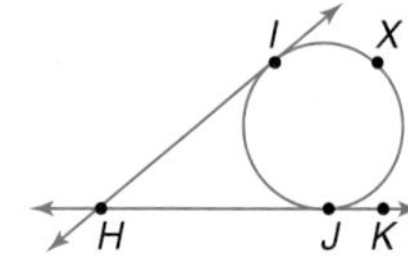

47. Write a paragraph proof for Theorem 9–11. a–b. See Solutions Manual.

a. Given: $\overleftrightarrow{AB}$ is a tangent of $\odot O$.
$\overrightarrow{AC}$ is a secant of $\odot O$.
$\angle CAB$ is acute.

Prove: $m\angle CAB = \frac{1}{2}m\widehat{CA}$

(*Hint:* Draw diameter $\overline{AD}$.)

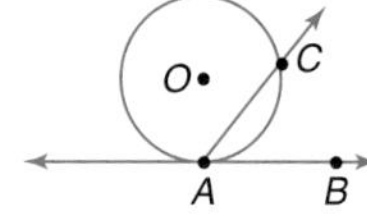

b. Prove Theorem 9–11 if the angle in part a is obtuse.

48. The diagonals of a rectangle are ≅ and bisect each other, thus creating the center of the circle that circumscribes the rectangle.

Critical Thinking

48. A set of points are *concyclic* if they all lie on the same circle. Explain why the vertices of any rectangle must always be concyclic.

Applications and Problem Solving

49. **Meteorology** Every rainbow is really a full circle whose center is at a point in the sky directly opposite the Sun. The position of a rainbow in the sky varies according to the viewer's position, but its angular size, $\angle ABC$, is always 42°, as shown in the diagram below. If $m\widehat{CD} = 160$, find the measure of the visible part of the rainbow, $m\widehat{AC}$. 76

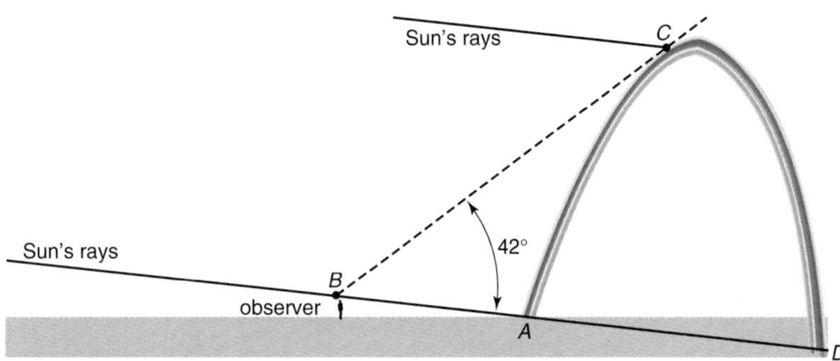

Additional Answers

46b. Proof:
Statements (Reasons)
1. $\overleftrightarrow{DG}$ is a tangent to the circle. $\overleftrightarrow{DF}$ is a secant to the circle. (Given)
2. Draw $\overline{FG}$. (Through any 2 pts. there is 1 line.)
3. $m\angle DFG = \frac{1}{2}m\widehat{GE}$, $m\angle FGH = \frac{1}{2}m\widehat{FG}$ (The measure of an inscribed $\angle = \frac{1}{2}$ the measure of its intercepted arc.)
4. $m\angle FGH = m\angle DFG + m\angle FDG$ (Exterior ∠ Th.)
5. $\frac{1}{2}m\widehat{FG} = \frac{1}{2}m\widehat{GE} + m\angle FDG$ (Substitution Prop. (=))
6. $\frac{1}{2}m\widehat{FG} - \frac{1}{2}m\widehat{GE} = m\angle FDG$ (Subtraction Prop. (=))
7. $\frac{1}{2}(m\widehat{FG} - m\widehat{GE}) = m\angle FDG$ (Distributive Prop.)

46c. Proof:
Statements (Reasons)
1. $\overleftrightarrow{HI}$ and $\overleftrightarrow{HJ}$ are tangents to the circle. (Given)
2. Draw $\overline{IJ}$. (Through any 2 pts. there is 1 line.)

(continued below)

46c. *(continued)*
3. $m\angle IJK = \frac{1}{2}m\widehat{IXJ}$, $m\angle HIJ = \frac{1}{2}m\widehat{IJ}$ (The measure of an inscribed $\angle = \frac{1}{2}$ the measure of its intercepted arc.)
4. $m\angle IJK = m\angle HIJ + m\angle IHJ$ (Exterior ∠ Th.)
5. $\frac{1}{2}m\widehat{IXJ} = \frac{1}{2}m\widehat{IJ} + m\angle IHJ$ (Subst. Prop. (=))
6. $\frac{1}{2}m\widehat{IXJ} - \frac{1}{2}m\widehat{IJ} = m\angle IHJ$ (Subtr. Prop. (=))
7. $\frac{1}{2}(m\widehat{IXJ} - m\widehat{IJ}) = m\angle IHJ$ (Dist. Prop.)

Practice Masters, p. 54

NAME____________ DATE ________

Practice

Student Edition Pages 483–490

Secants, Tangents, and Angle Measures

Assume that lines that appear to be tangents are tangents.

In $\odot Q$, $m\angle CQD = 120$, $m\widehat{BC} = 30$, and $m\angle BEC = 25$. Find each measure.

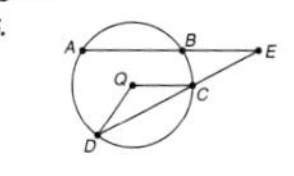

1. $m\widehat{DC}$ 120
2. $m\widehat{AD}$ 80
3. $m\widehat{AB}$ 130
4. $m\angle QDC$ 30

In $\odot Q$, $m\widehat{AE} = 140$, $m\widehat{BD} = y$, $m\widehat{AB} = 2y$, and $m\widehat{DE} = 2y$. Find each measure.

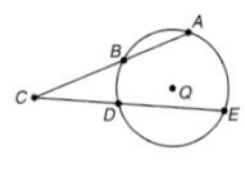

5. $m\widehat{BD}$ 44
6. $m\widehat{AB}$ 88
7. $m\widehat{DE}$ 88
8. $m\angle BCD$ 48

In $\odot P$, $m\widehat{BC} = 4x - 50$, $m\widehat{DE} = x + 25$, $m\widehat{EF} = x - 15$, $m\widehat{CD} = x$, and $m\widehat{FB} = 50$. Find each measure.

9. $m\angle A$ 50
10. $m\angle BCA$ 25
11. $m\angle ABC$ 105
12. $m\angle GBC$ 75
13. $m\angle FHE$ 42.5
14. $m\angle CFD$ 25

In $\odot P$, $m\angle A = 62$ and $m\widehat{BD} = 120$. Find each measure.

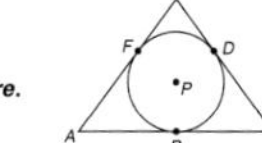

15. $m\angle C$ 60
16. $m\widehat{DF}$ 122
17. $m\angle E$ 58

4 ASSESS

Closing Activity

Writing For each of the three theorems in this lesson, have students draw a figure relating to the theorem and write an equation for the measure of the angle formed that demonstrates the theorem.

Chapter 9 Quiz C (Lessons 9-5 and 9-6) is available in the *Assessment and Evaluation Masters*, p. 241.

Some astronomers feel that Stonehenge was used to predict eclipses.

Enrichment Masters, p. 54

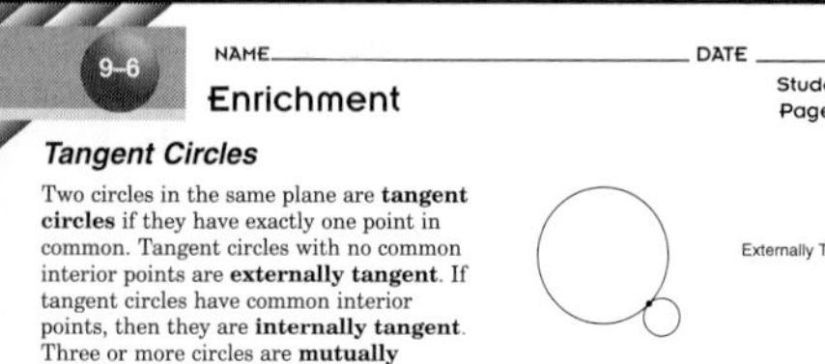

9-6 NAME__________ DATE________

Enrichment

Student Edition Pages 483–490

Tangent Circles

Two circles in the same plane are **tangent circles** if they have exactly one point in common. Tangent circles with no common interior points are **externally tangent**. If tangent circles have common interior points, then they are **internally tangent**. Three or more circles are **mutually tangent** if each pair of them are tangent.

Externally Tangent Circles

Internally Tangent Circles

1. Make sketches to show all possible positions of three mutually tangent circles.

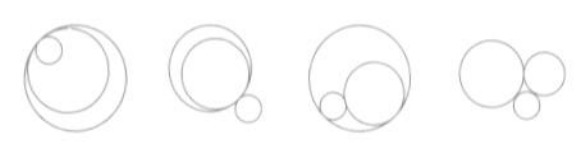

2. Make sketches to show all possible positions of four mutually tangent circles.

3. Make sketches to show all possible positions of five mutually tangent circles.

4. Write a conjecture about the number of possible positions for n mutually tangent circles if n is a whole number greater than four. Possible answer: For $n > 4$, there are $\frac{n}{2}$ positions if n is even and $\frac{1}{2}(n + 1)$ positions if n is odd.

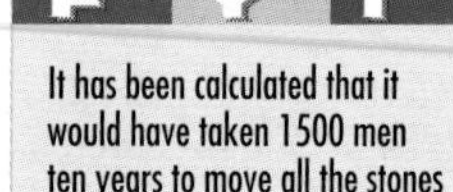

It has been calculated that it would have taken 1500 men ten years to move all the stones at Stonehenge, some of which weighed as much as 50 tons.

50. **Ancient Monuments** Stonehenge, built around 2100 B.C., is one of the most extraordinary and mysterious artifacts in Great Britain. The stones are arranged in a circular pattern according to the movements of Earth and the moon. If $m\widehat{AB} = 71$ and $m\widehat{BC} = 118$, find $m\angle AMB$, the angle that the north/south axis makes with the axis of the farthest north moonrise. 23.5°

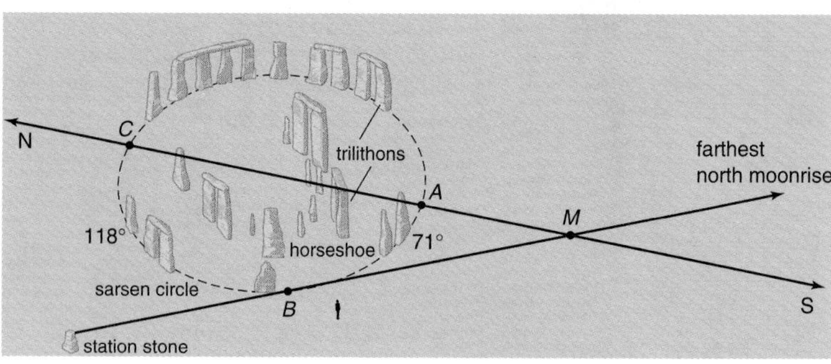

Mixed Review

52. No, one side is not a chord.

51. Find the value of x. (Lesson 9–5) 12

52. Determine whether $\angle A$ is an inscribed angle. Explain. (Lesson 9–4)

53. Find JL. (Lesson 9–3) 8

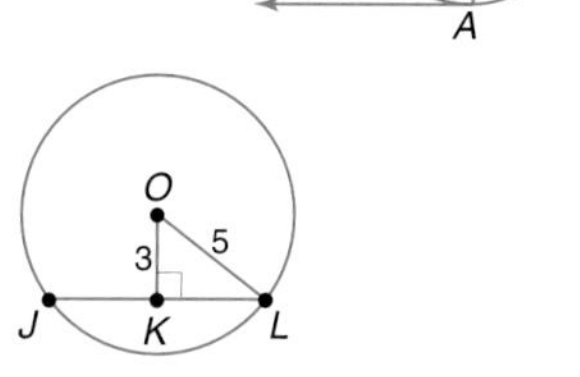

54. State the trigonometric ratio used to find the measure of $\angle Q$. Then find the measure to the nearest degree. (Lesson 8–3) tangent, 51

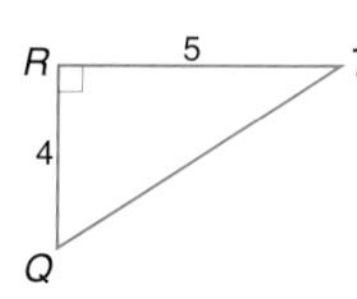

55. **Travel** A tour book costs 450 Taiwanese dollars. Solve the proportion $\frac{25.74}{1} = \frac{450}{x}$ to find the cost in U.S. dollars x. (Lesson 7–1) $17.48

56. In parallelogram $PQRS$, $m\angle P = 110$. Find $m\angle Q$. (Lesson 6–1) 70

57. Determine whether it is possible to draw a triangle with sides measuring 17 centimeters, 10 centimeters, and 29 centimeters. (Lesson 5–5) no

58. Determine whether the statement *Skew lines intersect in at least one point* is true or false. (Lesson 3–1) false

59. Name the property of equality that justifies the statement *If* $3x = 12$, *then* $x = 4$. (Lesson 2–4) Division Prop. (=) or Mult. Prop. (=)

INTEGRATION **Algebra**

60. Find $\frac{6w^2}{5} \cdot \frac{20}{12w}$. $2w$

61. Solve $x^2 + 2x - 3 = 0$. $-3, 1$

Extension

Communication With the students, make a list of the possible combinations of two intersecting lines (tangents and secants) in a circle. Then make two vertical angles out of sticks, pencils, or pipe cleaners. Draw a circle on the board or overhead and ask students to demonstrate an example of each situation.

9–7 Special Segments in a Circle

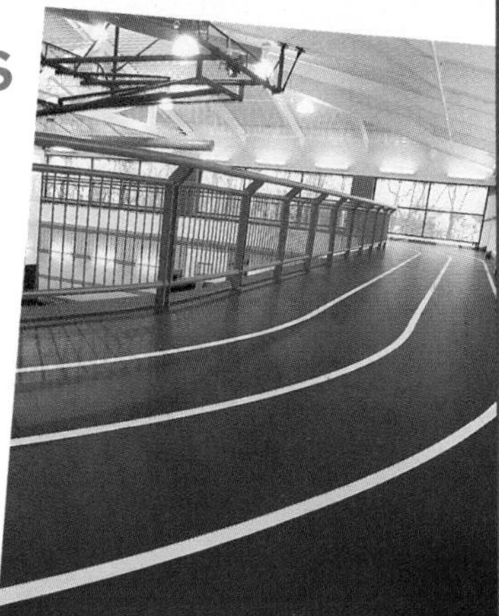

What YOU'LL LEARN

- To use properties of chords, secants, and tangents to solve segment measure problems.

Why IT'S IMPORTANT

A knowledge of special segments in a circle is helpful in solving problems involving carpentry and architecture.

Engineering

The indoor track being designed for the new Forest View Athletic Club must fit on an area 214 feet by 48 feet. The track design can be described as a rectangle with an arc at each end. The distance from the top of the arc to the side of the rectangle cannot exceed 12 feet.

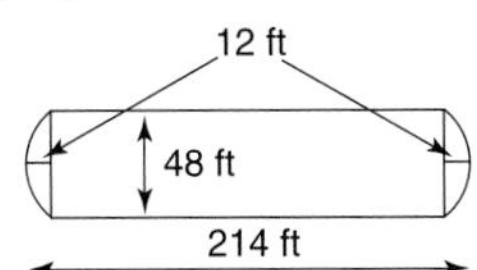

In order to draw the plans correctly, the architect needs to know the radius of the circle containing the arc. How could the architect find this radius? *This problem will be solved in Example 1.*

The side of the rectangle in the diagram above is a chord of the circle containing the arc. There are special relationships among the measures of chords in a circle. The following Exploration investigates such a relationship.

EXPLORATION CABRI GEOMETRY

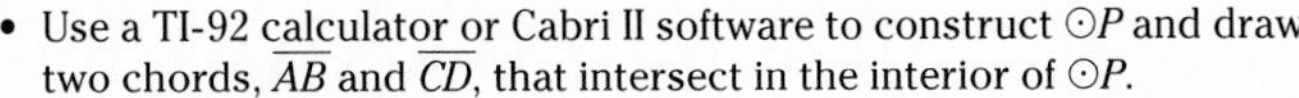

- Use a TI-92 calculator or Cabri II software to construct $\odot P$ and draw two chords, $\overline{AB}$ and $\overline{CD}$, that intersect in the interior of $\odot P$.
- Label the intersection of the chords E.
- Use the calculator or software to measure $\overline{AE}$, $\overline{EB}$, $\overline{EC}$ and $\overline{ED}$.
- Find the products $AE \cdot EB$ and $EC \cdot ED$.

Your Turn

a. Repeat the activity for two other circles. See students' work.

b. In each case, compare $AE \cdot EB$ with $EC \cdot ED$. Make a conjecture about these two products. The products are equal.

This Exploration suggests the following theorem. *You will be asked to prove this theorem in Exercise 28.*

Theorem 9–14	**If two chords intersect in a circle, then the products of the measures of the segments of the chords are equal.**

EXPLORATION

Students might also want to perform the Exploration with pencil, paper, compass, and ruler.

9-7 LESSON NOTES

NCTM Standards: 1–5, 7

Instructional Resources

- Study Guide Master 9-7
- Practice Master 9-7
- Enrichment Master 9-7
- Graphing Calculator and Computer Masters, p. 9

Transparency 9-7A contains the 5-Minute Check for this lesson; **Transparency 9-7B** contains a teaching aid for this lesson.

Recommended Pacing	
Standard Pacing	Days 11 & 12 of 15
Honors Pacing	Days 9 & 10 of 13
Block Scheduling*	Day 7 of 9

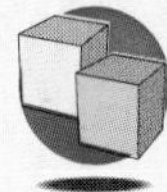

*For more information on pacing and possible lesson plans, refer to the *Block Scheduling Booklet*.

1 FOCUS

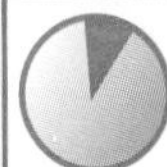

5-Minute Check
(over Lesson 9-6)

In the figure below, $\overleftrightarrow{NO}$ is tangent to the circle at N. $m\widehat{JK} = 5x - 5$, $m\widehat{KM} = 6x - 12$, $m\widehat{JN} = 13x + 11$, and $m\angle 2 = 9x + 6$. Find each value or measure.

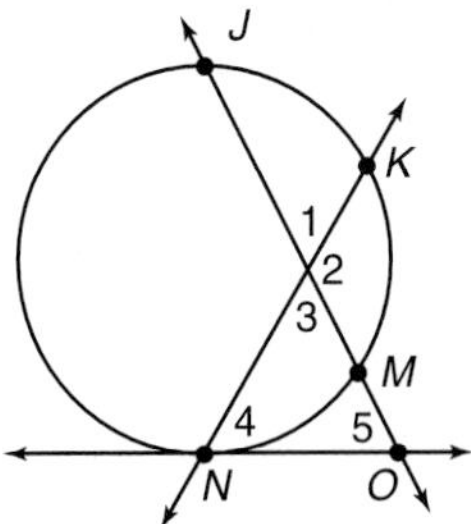

1. x 13
2. $m\widehat{NM}$ 54
3. $m\angle 3$ 57
4. $m\angle 4$ 60
5. $m\angle 5$ 63

Motivating the Lesson

Hands-On Activity Have students inscribe a triangle in a circle on patty paper. Give measures for each side of the triangle. Have them fold a median, an altitude, or an angle bisector and extend it until it becomes a chord of the circle. Ask them to look at the special segment folded and the side of the triangle that it intersects. How would they find the measure of each of the four segments formed?

2 TEACH

In-Class Example

For Example 1
A mechanic needs to know the radius of a rotor to find the correct size for a new part for a car. However, he can measure only part of the rotor without taking the engine apart. If he measures a segment of 6 inches and the distance from the midpoint to the arc is 1.5 inches, what is the radius of the rotor? 3.75 in.

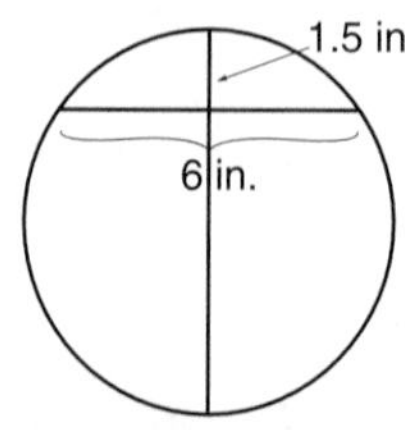

Teaching Tip Before discussing Theorem 9-15, take some time to make sure students understand what secant segments and external secant segments are.

We can use Theorem 9–14 to solve the problem at the beginning of the lesson.

Example 1

Engineering

Refer to the application at the beginning of the lesson. How can the engineer find the radius to accurately draw the arc for the track?

Explore To accurately draw the arc for the track, the engineer needs to know the radius of the circle.

Plan First, sketch the circle that contains the arc of the track for easier reference. Label the chords, which intersect at E, as $\overline{AB}$ and $\overline{CD}$. $\overline{DE}$ is 12 feet long and is part of the diameter of the circle because it is perpendicular to the chord and meets the chord at its center.

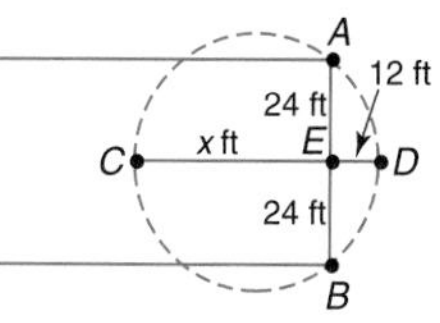

Solve We now have two intersecting chords, and we know the measures of three of the four segments. Let x represent the unknown measure.

$$DE \cdot EC = AE \cdot EB$$
$$12 \cdot x = 24 \cdot 24$$
$$12x = 576$$
$$x = 48$$

The radius of the circle is half the diameter.

$$DC = DE + EC$$
$$= 12 + 48 \text{ or } 60$$

The radius of the circle is $\frac{1}{2}(60)$ or 30 feet.

Examine Use the Pythagorean Theorem to check the triangle with vertices at A, E, and the center of the circle.

$$24^2 + (30 - 12)^2 \stackrel{?}{=} 30^2$$
$$576 + 324 \stackrel{?}{=} 900$$
$$900 = 900 \checkmark$$

The measures check.

In the figure at the right, $\overline{SC}$ and $\overline{SA}$ are **secant segments** and contain chords $\overline{EC}$ and $\overline{AN}$ of $\odot T$. The parts of $\overline{SC}$ and $\overline{SA}$ that are exterior to the circle are called **external secant segments**. $\overline{SE}$ and $\overline{SN}$ are examples of external secant segments.

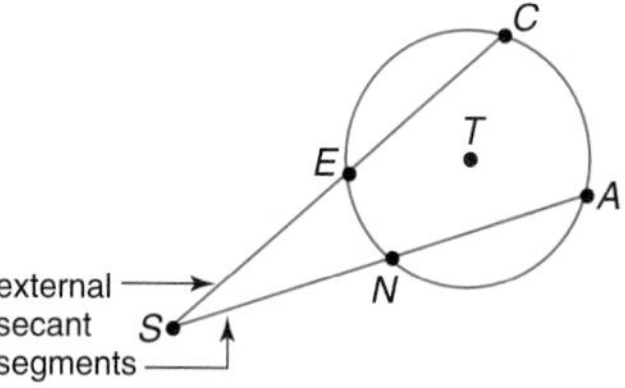

The segments formed by two secant segments also have a special relationship. Recall from Theorem 9–14, that for the two chords intersecting in $\odot E$, $AE \cdot EC = DE \cdot EB$. It is interesting that this relationship also holds true when the intersection of $\overline{AC}$ and $\overline{BD}$, E, occurs outside the circle.

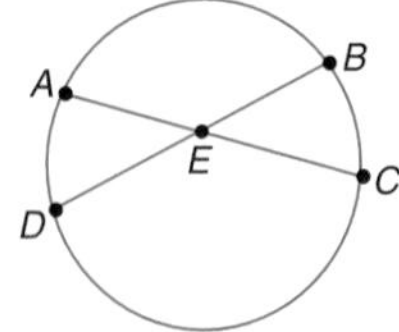

In the figure at the right, $\overline{AC}$ and $\overline{BD}$ are extended outside the circle to intersect at E as secant segments. It is true that $EA \cdot EC = ED \cdot EB$. Notice that $\overline{EA}$ and $\overline{ED}$ are external secant segments.

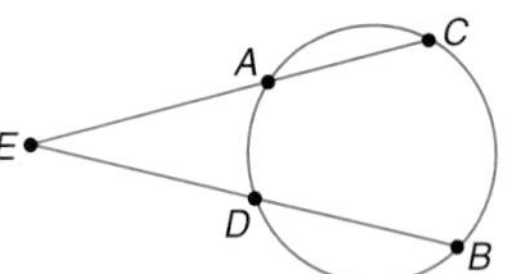

Theorem 9–15

If two secant segments are drawn to a circle from an exterior point, then the product of the measures of one secant segment and its external secant segment is equal to the product of the measures of the other secant segment and its external secant segment.

You will be asked to prove this theorem in Exercise 33.

Example  **Use the figure at the right to find OG.**

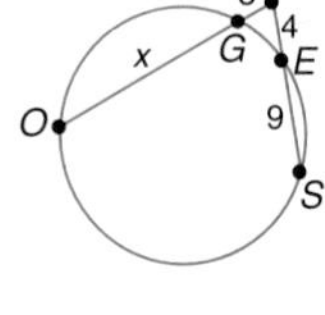

Let x represent OG.

$$LG \cdot LO = LE \cdot LS$$
$$3(x + 3) = 4 \cdot 13$$
$$3x + 9 = 52$$
$$3x = 43$$
$$x = 14\tfrac{1}{3}$$

Therefore, $OG = 14\frac{1}{3}$.

Suppose one of the secant segments from the exterior point is moved to become a tangent segment. In the figure at the right, $\overline{EC}$ is a tangent segment to the circle. $\overline{EC}$ now represents both the entire segment and the portion of the segment external to the circle. Thus, $EC \cdot EC = ED \cdot EB$. This suggests Theorem 9–16.

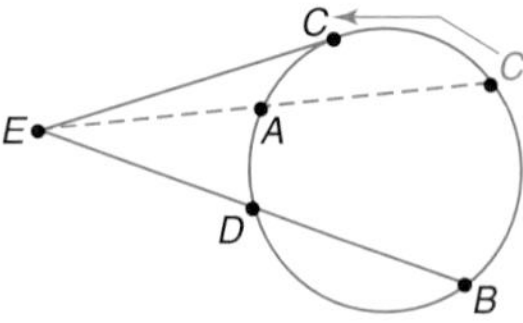

Theorem 9–16

If a tangent segment and a secant segment are drawn to a circle from an exterior point, then the square of the measure of the tangent segment is equal to the product of the measures of the secant segment and its external secant segment.

You will be asked to prove this theorem in Exercise 35.

Example  **In the figure at the right, $\overline{MA}$ is tangent to $\odot P$. Find the value of x.**

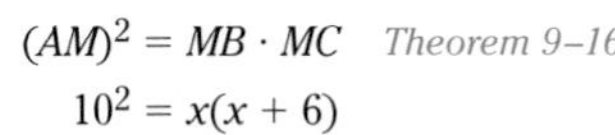

$$(AM)^2 = MB \cdot MC \quad \textit{Theorem 9–16}$$
$$10^2 = x(x + 6)$$
$$100 = x^2 + 6x$$
$$0 = x^2 + 6x - 100$$

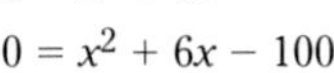

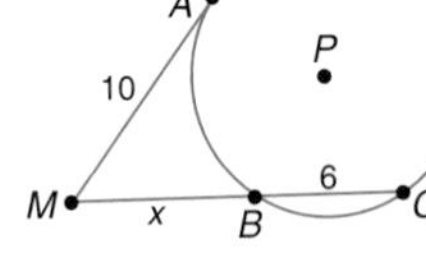

You can use the quadratic formula to solve an equation in the form $0 = ax^2 + bx + c$.

(continued on the next page)

In-Class Examples

For Example 2
Use the figure below to find OG.

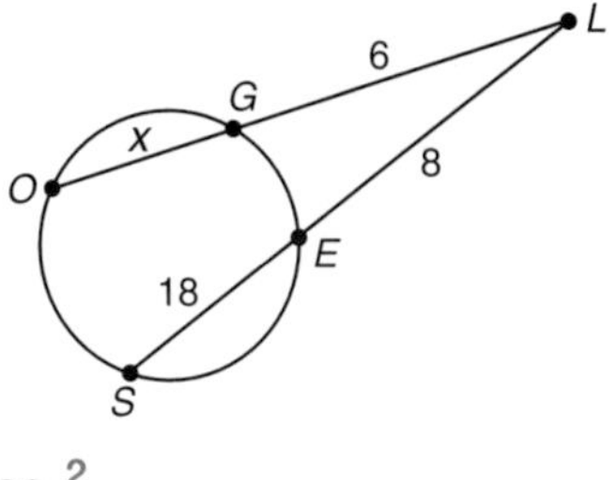

$28\frac{2}{3}$

For Example 3
In the figure below, $\overline{RT}$ is tangent to $\odot B$. Find the value of x.

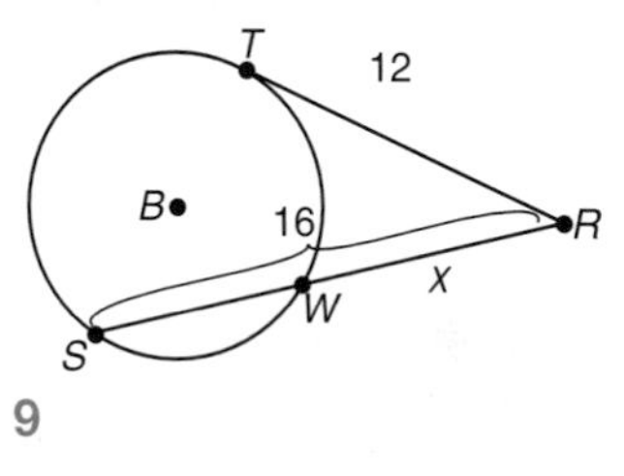

9

Teaching Tip Before discussing Theorem 9-16, have students work through the explanation at the bottom of page 492 and top of page 493, using a secant segment and a tangent segment.

3 PRACTICE/APPLY

Check for Understanding

Exercises 1–13 are designed to help you assess your students' understanding through reading, writing, speaking, and modeling. You should work through Exercises 1–5 with your students and then monitor their work on Exercises 6–13.

Error Analysis

It may be difficult for students to understand the differences among an external secant segment, a secant segment, and a secant, or between a tangent segment and a tangent. Point out that secants and tangents are lines, and that an external secant segment is part of the secant line but does not contain any points inside the circle. Also, a tangent segment, if it were to be extended, would be a tangent line.

Additional Answers

4. She is wrong. The measure of the exterior secant segment times the measure of the secant segment equals the measure of the other exterior secant segment times the measure of the other secant segment. The correct equation is $KA \cdot KN = KL \cdot KO$.
5. The distance from the center would have been 0, since a semicircle contains the diameter.

$$x = \frac{-b \pm \sqrt{b^2 - 4ac}}{2a}$$

$$x = \frac{-6 \pm \sqrt{6^2 - 4(1)(-100)}}{2(1)} \quad a = 1, b = 6, c = -100$$

$$x = \frac{-6 \pm \sqrt{436}}{2}$$

$$x = \frac{-6 + \sqrt{436}}{2} \qquad x = \frac{-6 - \sqrt{436}}{2}$$

$$x \approx 7.4 \qquad x \approx -13.4$$

Disregard the negative value. Thus, $x \approx 7.4$.

CHECK FOR UNDERSTANDING

Communicating Mathematics

Study the lesson. Then complete the following.

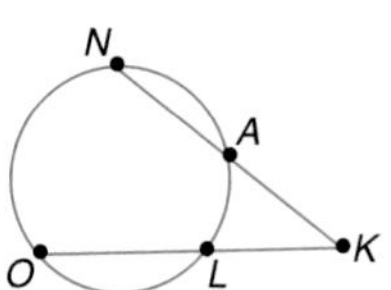

1. **Identify** an external secant segment in the figure at the right. Explain why you believe your choice is correct. $\overline{SC}$ or $\overline{ST}$
2. **Explain** why $CO \cdot OG = EO \cdot OM$.
3. **Define** tangent segment. State the relationship that exists between $\overline{SE}$, $\overline{ST}$, and $\overline{SN}$.
4. **You Decide** Ayashe uses the figure at the right to claim that $KA \cdot AN = KL \cdot LO$ by Theorem 9–15. Is she right? Explain why or why not. See margin.
5. **Examine** Example 1. The radius of the circle containing an arc 12 feet from the side of the rectangle was 30 feet, so the side was 30 − 12 or 18 feet from the center. Suppose the arc was to have been a semicircle. Would the "distance from the center" have been greater or less than 12 feet? Explain. See margin.

2. Chords CG and EM intersect in the circle at O, so the products of the measures of the segments of those chords are equal.

3. A tangent segment is a segment from an exterior point to the point of tangency; $SN \cdot ST = (SE)^2$.

Guided Practice

Find the value of x to the nearest tenth. Assume that segments that appear to be tangent are tangent.

6. 11.7

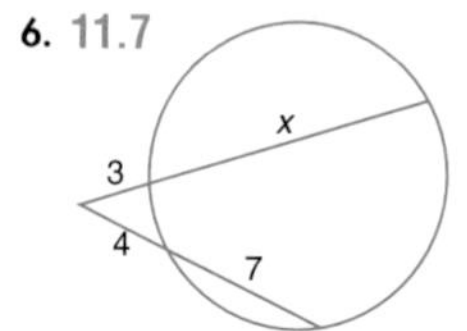

7. 2

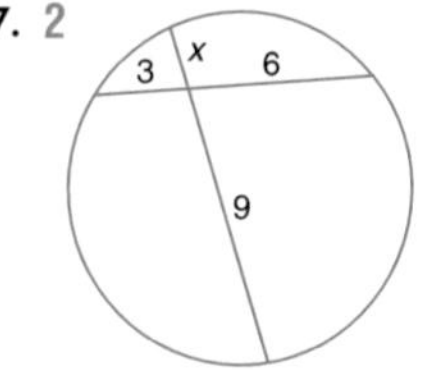

8. 10.8

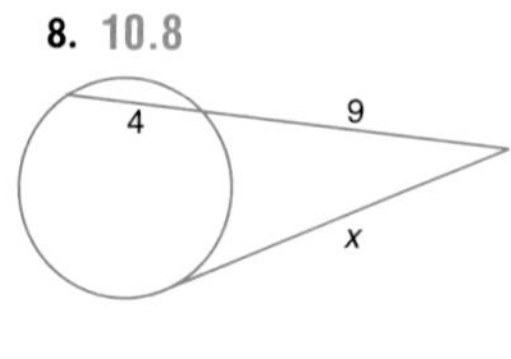

In ⊙A, $\overline{TS} \perp \overline{RE}$ with $TS = 10$ and $RE = 3$. Find each measure.

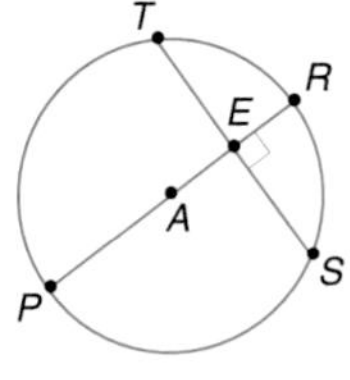

9. TE and ES 5, 5
10. PE $8\frac{1}{3}$
11. PR $11\frac{1}{3}$
12. radius of ⊙A $5\frac{2}{3}$

Reteaching

Using Synthesis Have students draw one figure for all three theorems discussed in this lesson. Have them write an equation for each theorem relating the measures of the segments in their pictures. Ask them to check each equation by measuring the segments involved in the equations and finding the products.

13. **Carpentry** An arch over a double-door entrance is 200 centimeters wide and 60 centimeters high. Find the radius of the circle that contains the arch. (*Hint:* Use the steps in Exercises 9–12.) $113\frac{1}{3}$ cm

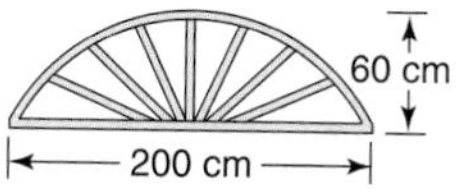

60 cm
200 cm

EXERCISES

Practice

Find the value of x to the nearest tenth. Assume that segments that appear to be tangent are tangent.

14. 12.5

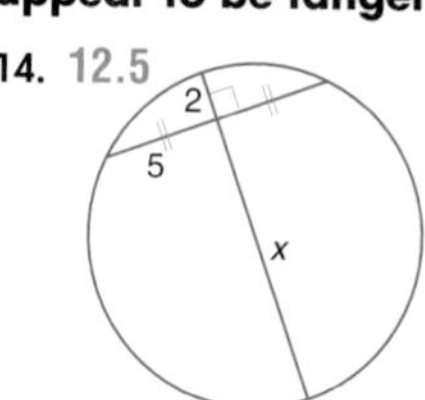

15. 28.1

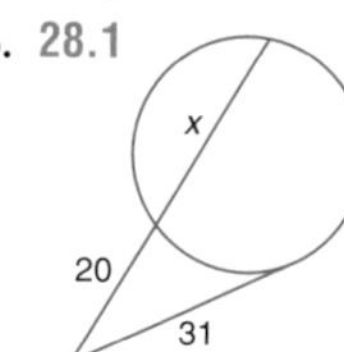

16. 7.4

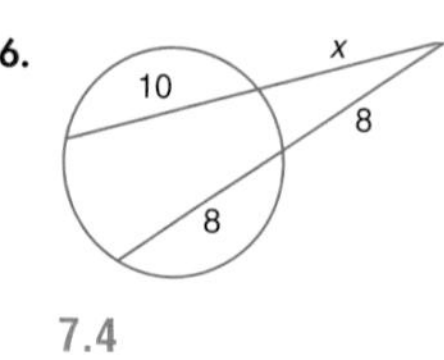

17. 3

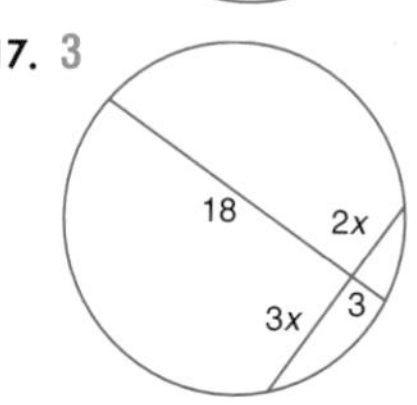

18. 4.7

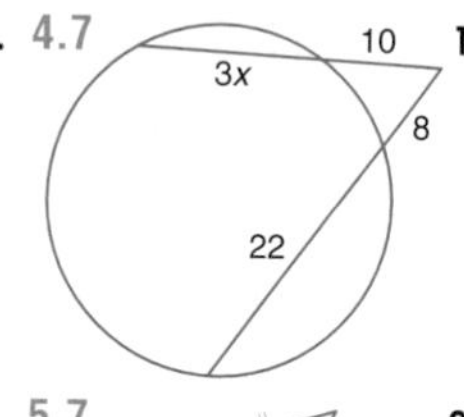

19. 6.4

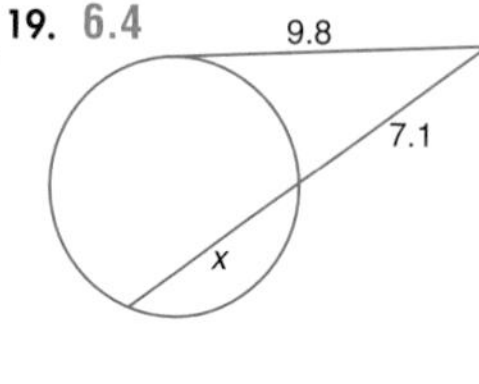

20. 5.6

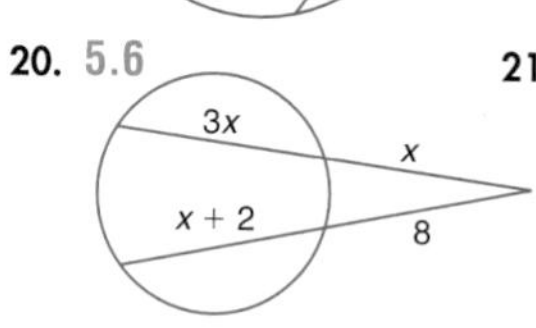

21. 5.7

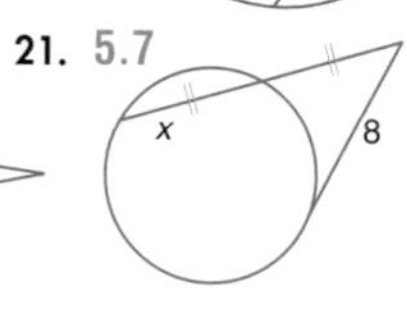

22. 4 or −5

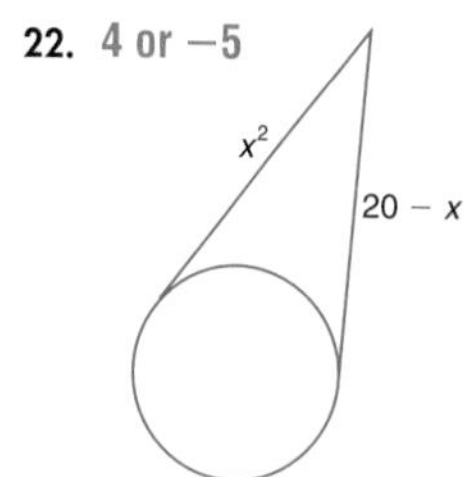

In ⊙D, $EX = 24$, $DE = 7$, $\overline{XT}$ is a secant segment, and $\overline{EX}$ and $\overline{AX}$ are tangent segments. Find each measure.

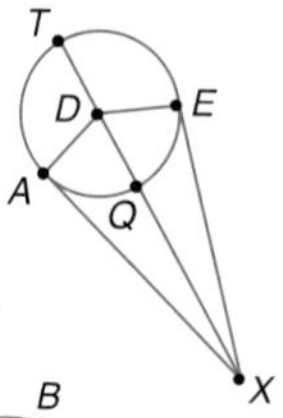

23. AX 24
24. DX 25
25. QX 18
26. TX 32
27. Show two different techniques for finding QX. See margin.

T
D
E
A
Q
X

Proof

28. Complete the reasons for each step in the proof of Theorem 9–14.

Given: $\overline{AC}$ and $\overline{BD}$ intersect at E.

Prove: $AE \cdot EC = BE \cdot ED$

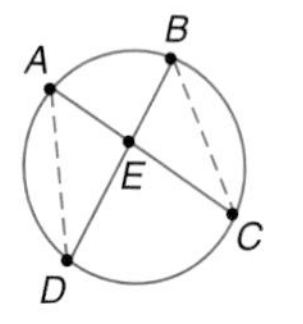

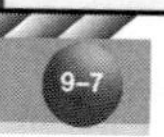

Statements	Reasons
a. Draw $\overline{AD}$ and $\overline{BC}$ forming $\triangle DAE$ and $\triangle CBE$.	a. ? Through any 2 pts. there is 1 line.
b. $\angle A \cong \angle B$, $\angle D \cong \angle C$	b. ?
c. $\triangle DAE \sim \triangle CBE$	c. ? AA Similarity
d. $\frac{AE}{BE} = \frac{ED}{EC}$	d. ? Definition of similar polygons
e. $AE \cdot EC = BE \cdot ED$	e. ? Cross products

28b. If 2 inscribed ∠s of a ⊙ intercept the same arc, then the ∠s are ≅.

Assignment Guide

Core (with proof): 15–39 odd, 40–48
Core (informal): 15–31 odd, 37, 39, 40–48
Enriched: 14–36 even, 37–48

For **Extra Practice**, see p. 782.

The red A, B, and C flags, printed only in the Teacher's Wraparound Edition, indicate the level of difficulty of the exercises.

Additional Answer

27. $DQ + QX = DX$ (Segment Add. Post.)
$QX = DX - DQ$
$= 25 - 7$ ($DQ = DE = 7$)
$= 18$
$(EX)^2 = QX \cdot TX$ (Th. 9-16)
$24^2 = x(x + 14)$
$0 = x^2 + 14x - 576$
$x = 18$ (quadratic formula)

Study Guide Masters, p. 55

9–7 NAME ______ DATE ______
Study Guide
Student Edition Pages 491–497

Special Segments in a Circle

The following theorems can be used to find the measure of special segments in a circle.

- If two chords intersect in a circle, then the products of the measures of the segments of the chords are equal.
- If two segments are drawn to a circle from an exterior point, then the product of the measures of one secant segment and its external secant segment is equal to the product of the measures of the other secant segment and its external secant segment.
- If a tangent segment and a secant segment are drawn to a circle from an exterior point, then the square of the measure of the tangent segment is equal to the product of the measures of the secant segment and its external secant segment.

Example: Find the value of x to the nearest tenth.

$(AB)^2 = BC \cdot BD$ Theorem 9–15
$(18)^2 = 15 \cdot (15 + x)$
$324 = 225 + 15x$
$99 = 15x$
$6.6 = x$

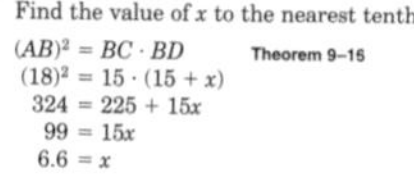

Find the value of x to the nearest tenth. Assume segments that appear tangent to be tangent.

1. 4
2. 7.7
3. 5.0

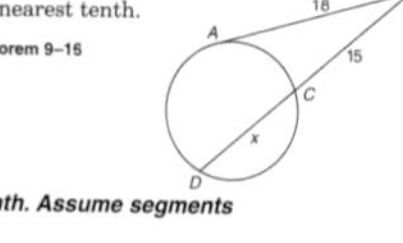

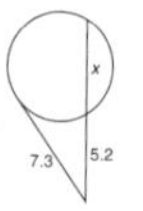

4. 4
5. 7.8
6. 2.3

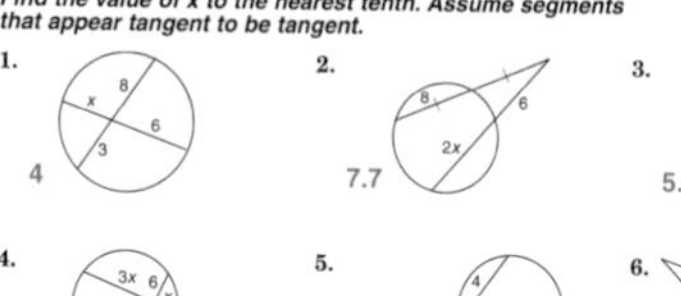

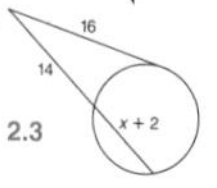

7. 1.9 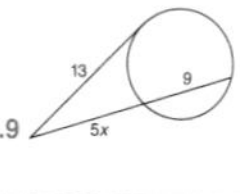
8. 1.7 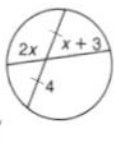
9. 3.9

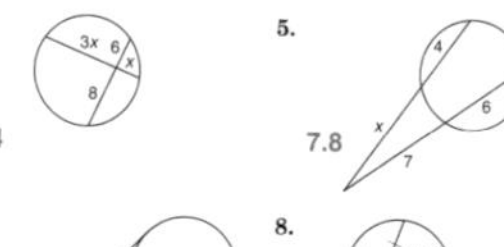

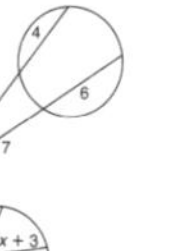

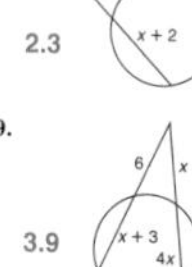

Using the Programming Exercises The program given in Exercise 36 is for use with the TI-82 or TI-83 graphing calculator. For other programmable calculators, have students consult their owner's manual for commands similar to those represented here.

Additional Answer

34. Proof:

Statements (Reasons)

1. Draw $\overline{DA}$ and $\overline{OM}$ forming $\triangle ADT$ and $\triangle MOT$. (Through any 2 pts. there is 1 line.)
2. $\odot H$, $\overline{AO} \perp \overline{DM}$ (Given)
3. $\overline{DT} \cong \overline{TM}$ (In a ⊙, if a diameter is ⊥ to a chord, then it bisects the chord and its arc.)
4. $\angle D \cong \angle O$, $\angle A \cong \angle M$ (If 2 inscribed ∠s of a ⊙ intercept the same arc, then the ∠s are ≅.)
5. $\triangle ADT \sim \triangle MOT$ (AA Similarity)
6. $\frac{DT}{OT} = \frac{TA}{TM}$ (Def. ~ polygons)
7. $DT \cdot TM = OT \cdot TA$ (Cross products)
8. $TM \cdot TM = OT \cdot TA$ (Substitution Prop. (=))
9. $OT \cdot TA = TM \cdot TM$ (Symmetric Prop. (=))
10. $OT \cdot TA = (TM)^2$ (Def. exponents)

Practice Masters, p. 55

9–7 NAME ______ DATE ______

Practice — Student Edition Pages 491–497

Special Segments in a Circle

Find the value of x to the nearest tenth. Assume segments that appear to be tangents are tangents.

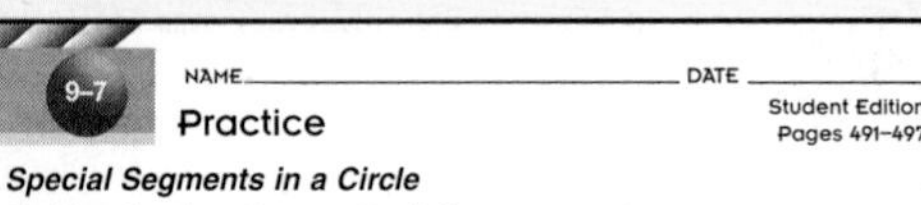

1. 6 2. 16

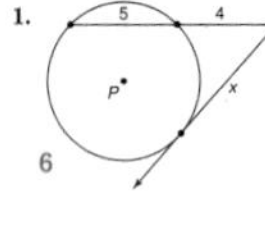

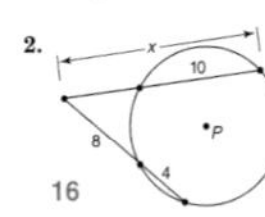

3. 8 4. 15.6

In ⊙P, CE = 6, CD = 16, and AB = 17. Find each measure.

5. EB 5

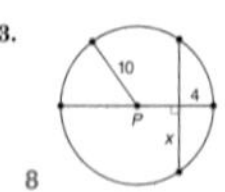

6. AE 12

In ⊙P, AC = 3, BC = 5, and AD = 2. Find each measure.

7. PD 5

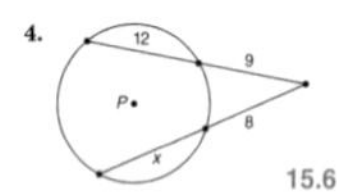

8. ED 10

9. PB 5

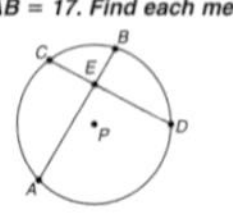

29. In ⊙K find KR. **5.3**

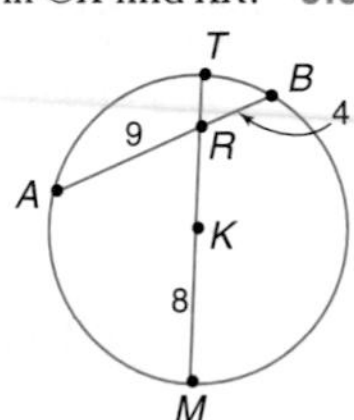

30. ⊙P and ⊙Q are tangent at S. If AB = 8, CD = 9, and BC = 10, find DE. **11**

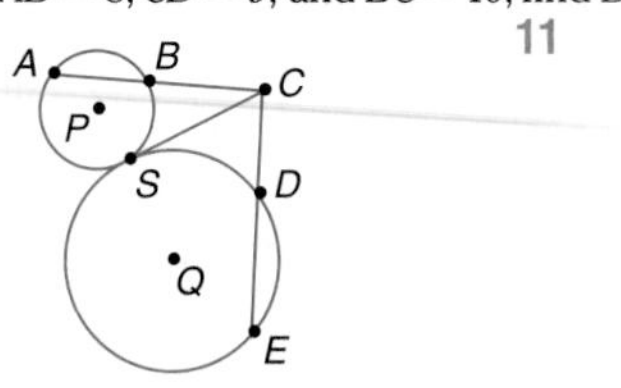

C

31. Find the values of x and y. **15; 22.5**

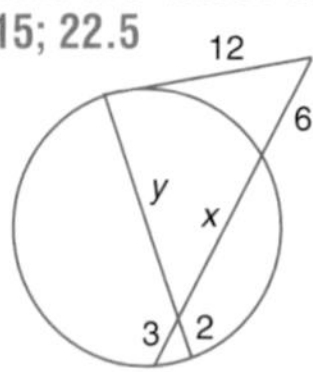

32. Find the values of x and y. **3.8; 4.2**

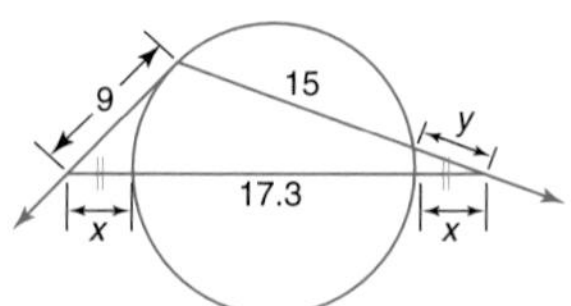

Proof

33. Prove Theorem 9–15. Start with a figure like the one shown at the right. Draw $\overline{AB}$ and $\overline{CD}$. **See Solutions Manual.**

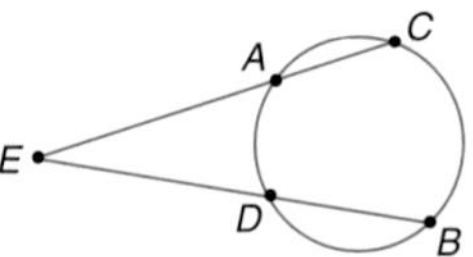

Write a two-column proof. **35. See Solutions Manual.**

34. Given: ⊙H
$\overline{AO} \perp \overline{DM}$
Prove: $OT \cdot TA = (TM)^2$

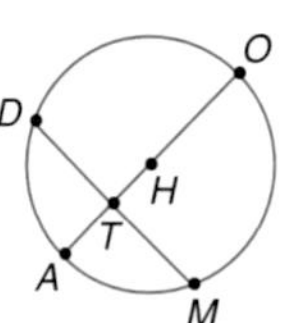

See margin.

35. Prove Theorem 9–16.
Given: $\overline{XY}$ is tangent to ⊙A. $\overline{WY}$ is a secant segment.
Prove: $(XY)^2 = WY \cdot ZY$
(*Hint:* Draw $\overline{WX}$ and $\overline{XZ}$.)

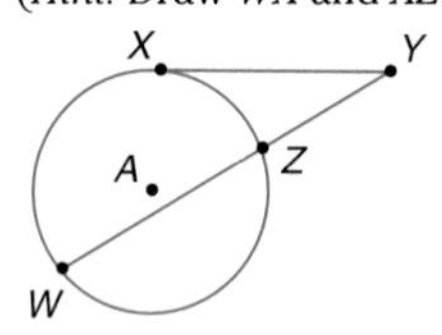

Programming

36. The TI-82/83 graphing calculator program at the right finds the measure of a segment in a circle. If you know the measures of $\overline{AE}$, $\overline{EB}$, and $\overline{CE}$, the program will find the measure of $\overline{ED}$.

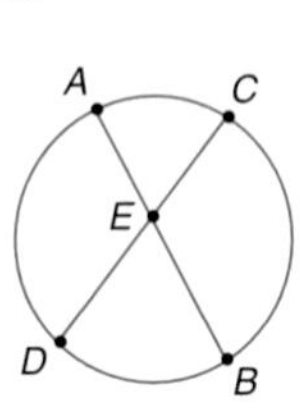

```
PROGRAM:SEGMENTS
:Disp "ENTER THE",
 "MEASURES"
:Input "AE = ", A
:Input "EB = ", E
:Input "CE =", C
:(A*E)/C→D
:Disp "ED = ", D
:Stop
```

Find the measure of $\overline{ED}$ for each of the following.

a. AE = 8, EB = 6, CE = 3 **16**
b. AE = 9, EB = 2, CE = 7 **about 2.6**
c. AE = 10, EB = 30, CE = 18 **about 16.7**

Critical Thinking

37. In ⊙P, $\overline{BC} \cong \overline{CD}$. Show that $AB = \sqrt{2} \cdot BC$. **See margin.**

Extension

Connections On the chalkboard or overhead, draw a circle that contains two chords that intersect at some point other than the center. Give measurements for three of the four segments and ask each student to find the measure of the fourth segment and to write a proportion using ratios of sides.

Applications and Problem Solving

38. 568 ft

38. **Architecture** The Louisiana Super Dome has a diameter that measures 680 feet. If the center of the dome is 113 feet higher than the sides of the stadium, how long is the radius of the sphere that forms the dome?

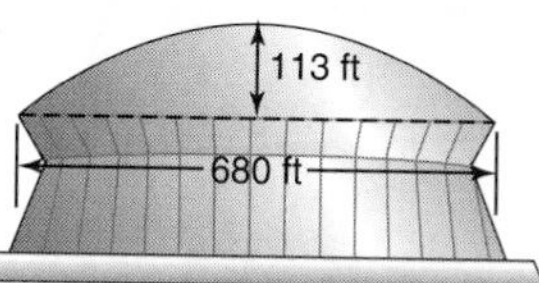

39. **Roman Architecture** The Roman Coliseum has many "entrances" in the shape of a door with an arch top. The ratio of the arch width to the arch height is 7:3. Find the ratio of the arch width to the radius of the circle that contains the arch. 7:3.5

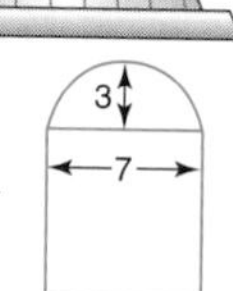

Mixed Review

40. Find $m\angle 1$. (Lesson 9–6) 63

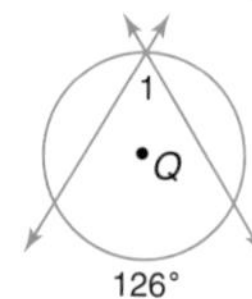

41. Find the value of x. (Lesson 9–5) 17

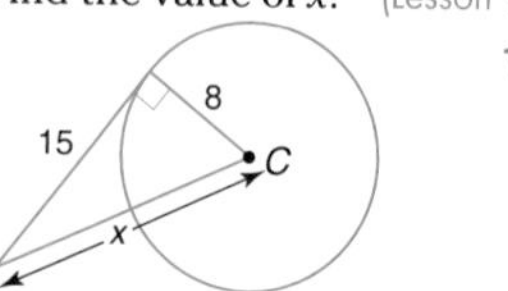

42. If the length of a side of a square is 30.4 yards, find the length of the diagonal to the nearest tenth of a yard. (Lesson 8–2) 43.0 yd

43. Solve the proportion $\frac{t}{18} = \frac{5}{6}$ by using cross products. (Lesson 7–1) 15

44. In $\square QRST$, if $m\angle Q = 76$, $m\angle QRT = 3x$, and $m\angle TRS = 2x$, find $m\angle TRS$. (Lesson 6–2) 41.6

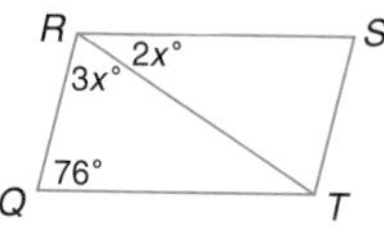

45. Name the property of inequality that justifies the statement, *If $XY + RM < EF + RM$, then $XY < EF$.* (Lesson 5–3) Subtraction Prop. ($\neq$)

46. Complete the statement with *always*, *sometimes*, or *never*. "Complementary angles are __?__ congruent." (Lesson 2–6) sometimes

INTEGRATION Algebra

47. Find $(2c - 1)(c + 7)$. $2c^2 + 13c - 7$

48. Factor $25a^2 - 4$. $(5a - 2)(5a + 2)$

WORKING ON THE In•ves•ti•ga•tion

Refer to the Investigation on pages 392–393.

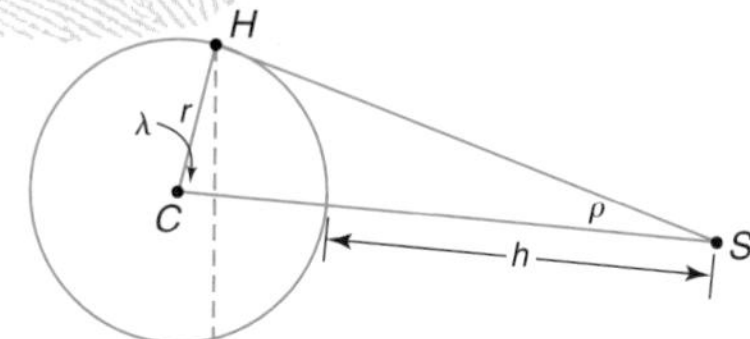

When a spacecraft is in orbit around a planet, only a portion of the surface of the planet can be seen at any one time. The circle that is the boundary of the visible area is called the *horizon circle.*

1 A satellite will be placed in orbit around Mars to relay information from the space station to Earth. The following diagram shows the position of the satellite S in relationship to Mars. C is the center of Mars, and H is a point on the horizon circle.

Find the relationships among ρ (rho), λ (lambda), h, and r.

2 If the satellite is in an orbit 160 miles above the surface of Mars, find the angular radius λ of the horizon circle seen by the satellite.

3 Approximately what percent of the surface of Mars will be visible by the satellite at any given moment?

Add the results of your work to your Investigation Folder.

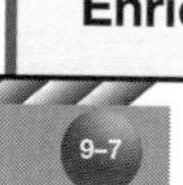

In•ves•ti•ga•tion

Working on the Investigation

The Investigation on pages 392–393 is designed to be a long-term project that is completed over several days or weeks. Encourage students to keep their materials in their Investigation Folder as they work on the Investigation.

Sample Answers for Working on the Investigation

1. $\angle H$ is a right angle. So $\sin \rho = \cos \lambda = \frac{r}{r+h}$.
2. about 21.6°
3. about 11.7%

4 ASSESS

Closing Activity

Speaking Have students draw and label figures for each of the three theorems. Have them discuss equations and how they relate to the theorem.

Additional Answer

37. $(AB)^2 = BC(BD)$ (Th. 9-16)
$= BC(BC + CD)$ (Seg. Add. Post.)
$= BC(BC + BC)$ $(BC = CD)$
$= BC(2BC)$
$= 2(BC)^2$
$AB = \sqrt{2}BC$

Enrichment Masters, p. 55

9–7 NAME ____________ DATE ______ Student Edition Pages 491–497

Enrichment

The Nine-Point Circle

The figure below illustrates a surprising fact about triangles and circles. Given any $\triangle ABC$, there is a circle that contains all of the following nine points:

(1) the midpoints K, L, and M of the sides of $\triangle ABC$

(2) the points X, Y, and Z, where $\overline{AX}$, $\overline{BY}$, and $\overline{CZ}$ are the attitudes of $\triangle ABC$

(3) the points R, S, and T which are the midpoints of the segments $\overline{AH}$, $\overline{BH}$, and $\overline{CH}$ that join the vertices of $\triangle ABC$ to the point H where the lines containing the altitudes intersect.

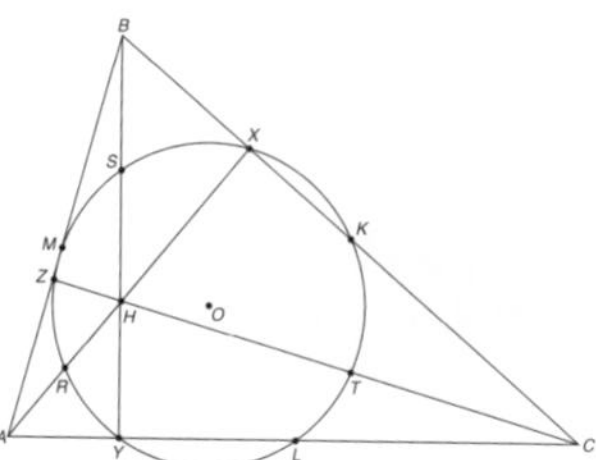

1. On a separate sheet of paper, draw an obtuse triangle ABC. Use your straightedge and compass to construct the circle passing through the midpoints of the sides. Be careful to make your construction as accurate as possible. Does your circle contain the other six points described above? For constructions, see students' work; yes.

2. In the figure you constructed for Exercise 1, draw $\overline{RK}$, $\overline{SL}$, and $\overline{TM}$. What do you observe? The segments intersect at the center of the nine-point circle.

9-8 LESSON NOTES

NCTM Standards: 1–5, 7, 8

Instructional Resources

- Study Guide Master 9-8
- Practice Master 9-8
- Enrichment Master 9-8
- Assessment and Evaluation Masters, p. 241

Transparency 9-8A contains the 5-Minute Check for this lesson; **Transparency 9-8B** contains a teaching aid for this lesson.

Recommended Pacing	
Standard Pacing	Day 13 of 15
Honors Pacing	Day 11 of 13
Block Scheduling*	Day 8 of 9

*For more information on pacing and possible lesson plans, refer to the *Block Scheduling Booklet*.

1 FOCUS

5-Minute Check
(over Lesson 9-7)

Refer to the figure below.

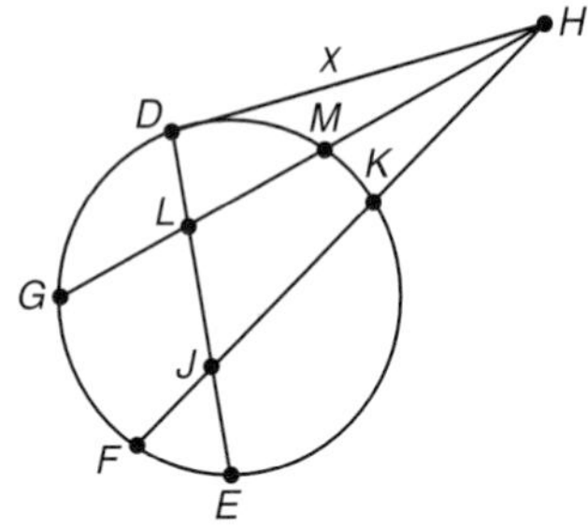

1. If $HK = 4$ and $KF = 10$, find the value of x. 7.48
2. If $DJ = 6$, $KJ = 5$, and $JE = 3$, find FJ. 3.6
3. If $DL = 3$, $LJ = 4$, $JE = 5$, and $LM = 12$, find GL. 2.25
4. If $GM = 9$, $GH = 15$, and $HK = 6$, find FK. 9

9–8 Integration: Algebra Equations of Circles

What YOU'LL LEARN

- To write and use the equation of a circle in the coordinate plane.

Why IT'S IMPORTANT

You can use equations of circles to solve problems involving aerodynamics and meteorology.

Aerodynamics

When a supersonic airplane's speed exceeds the speed of sound, cone-shaped shockwaves are produced as a result of overlapping waves of sound. The graph below shows the spherical sound waves produced by a supersonic airplane.

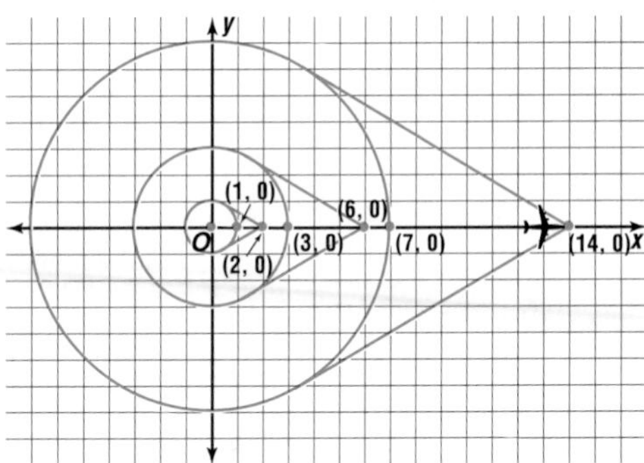

FYI

When Chuck Yeager broke the sound barrier in 1947 by flying at 700 mph, no one knew that he had two broken ribs and an immobile right arm from a riding accident the day before.

The wave is shown as an expanding circle in a two-dimensional coordinate plane. Notice that when the radius of the shockwave is 1 unit, the airplane is 2 units from the origin. A shockwave radius of 3 shows the airplane 6 units from the origin, and a shockwave radius of 7 shows the airplane 14 units from the origin. So, as sound travels from the center of each circle, the airplane is traveling twice as fast.

As supersonic aircraft travel through the sound barrier, it is sometimes useful to track the impact of the shockwaves being produced. To do this, it helps to know the equation for a circle that represents the size of the shockwave.

You can use the distance formula to write the equation for a circle given its center and radius. The circle at the right has its center at $C(-1, 4)$ and a radius of 5 units. Let $P(x, y)$ be any point on C. Then $\overline{CP}$ is a radius of the circle. The distance between $P(x, y)$ and $C(-1, 4)$ is 5 units.

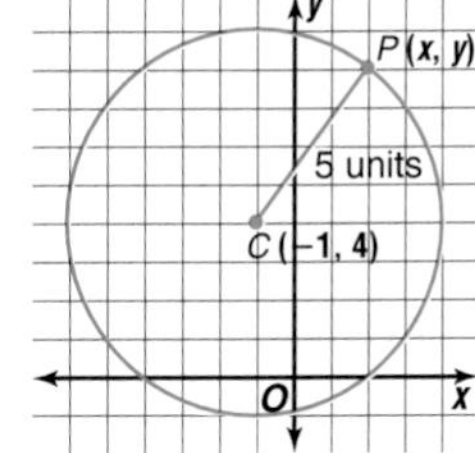

$$PC = 5$$

$$\sqrt{(x-(-1))^2 + (y-4)^2} = 5 \quad \textit{Distance formula}$$

$$(x+1)^2 + (y-4)^2 = 25 \quad \textit{Square each side.}$$

Thus, an equation for the circle with center at $C(-1, 4)$ and radius of 5 units is $(x + 1)^2 + (y - 4)^2 = 25$. Study the pattern in the chart to find the equation of a circle if the center of the circle is (h, k) and the radius is r.

FYI

Yeager was flying the experimental X-1 aircraft at an altitude of 40,000 feet when he broke the sound barrier.

Center	Radius	Equation of Circle
(1, 1)	$\sqrt{2}$	$(x-1)^2+(y-1)^2=2$
(2, 4)	5	$(x-2)^2+(y-4)^2=25$
(h, k)	r	$(x-h)^2+(y-k)^2=r^2$
(0, 0)	r	$x^2+y^2=r^2$

Standard Equation of a Circle

In general, an equation for a circle with center at (h, k) and a radius of r units is $(x-h)^2+(y-k)^2=r^2$.

The values of h and k tell how the circle is translated from the origin. That is, $x^2+y^2=r^2$ is moved h units horizontally and k units vertically.

Example  **Write an equation for a circle with center $C(-3, 6)$ and a diameter of 6 units.**

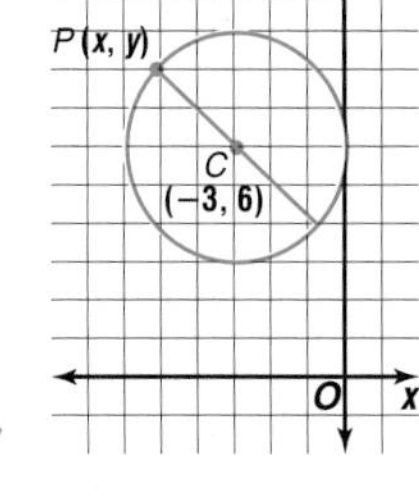

Since the diameter is 6, it follows that the radius must be 3.

Use the equation of a circle.

$$(x-h)^2+(y-k)^2=r^2$$
$$(x-(-3))^2+(y-6)^2=3^2 \quad (h, k) = (-3, 6), r = 3$$
$$(x+3)^2+(y-6)^2=9$$

An equation for the circle is $(x+3)^2+(y-6)^2=9$.

You can use the equation of a circle to draw its graph.

Example  **Graph the circle whose equation is $(x+2)^2+(y-3)^2=10$.**

Notice that the circle is translated 2 units to the left and 3 units up from the origin.

First, rewrite the equation in the form $(x-h)^2+(y-k)^2=r^2$.

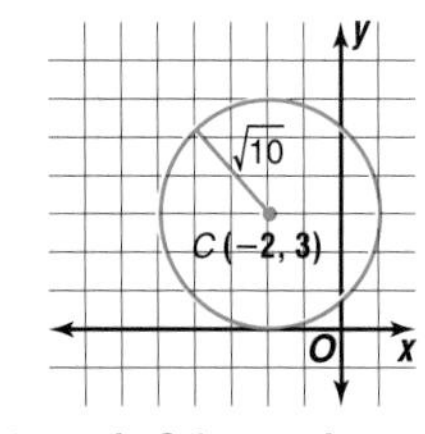

$$(x+2)^2+(y-3)^2=10$$
$$(x-(-2))^2+(y-3)^2=(\sqrt{10})^2$$

Therefore, $h=-2$, $k=3$, and $r=\sqrt{10}\approx 3.1$.
The center is at $(-2, 3)$ and the radius is approximately 3.1 units long.

Examples 1 and 2 explain how to write the equation of a circle and how to graph a circle given the center and radius. However, any three noncollinear points determine a circle. You can use the following steps to write the equation of a circle determined by three noncollinear points.

Step 1 Draw the triangle formed by the three points.

Step 2 Construct the perpendicular bisectors of two of the sides. The center of the circle is their point of intersection.

Step 3 Find the distance between the center and any of the three given points. This is the radius of the circle.

Step 4 Use the center and radius to write an equation of the circle.

Motivating the Lesson

Situational Problem
Superimpose a grid on a photograph of the moon. Place the center of the moon at the origin (0, 0) of the grid. Find the coordinates of several points along the circumference.

2 TEACH

In-Class Examples

For Example 1
Find the radius and diameter of the circle below. Then write an equation for the circle. $r=4$, $d=8$, $(x+2)^2+(y-4)^2=16$

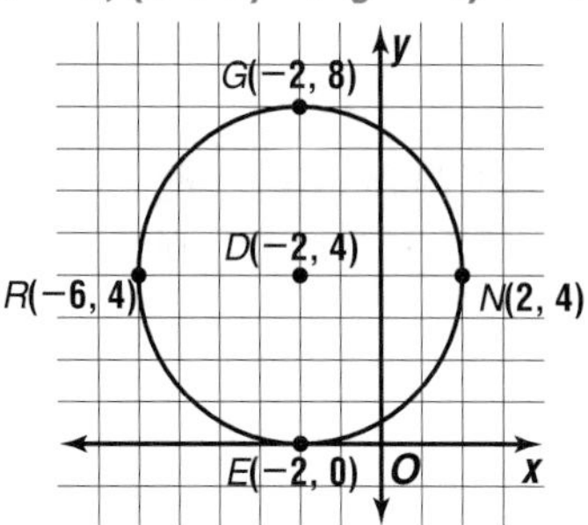

For Example 2
Graph the circle whose equation is $(x-3)^2+(y+2)^2=16$.

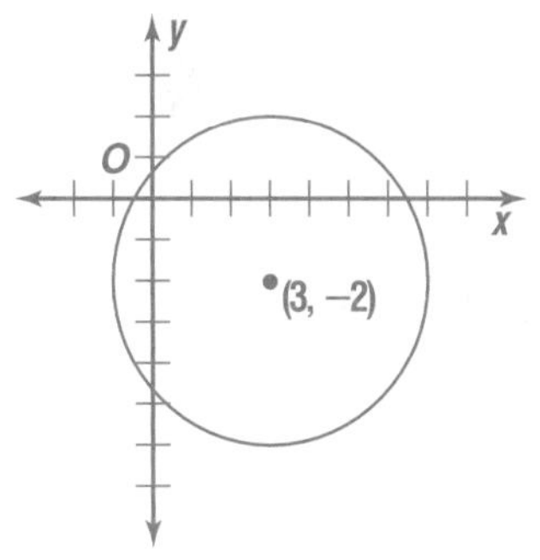

Teaching Tip Before stating the general equation of a circle, review the distance formula.

In-Class Example

For Example 3
Find an equation of the circle that passes through $A(1, 5)$, $B(7, 1)$, and $C(5, -3)$.
$(x - 2)^2 + (y - 1)^2 = 5^2$

3 PRACTICE/APPLY

Check for Understanding

Exercises 1–10 are designed to help you assess your students' understanding through reading, writing, speaking, and modeling. You should work through Exercises 1–3 with your students and then monitor their work on Exercises 4–10.

Error Analysis
When discussing the equation for a circle, $(x - h)^2 + (y - k)^2 = r^2$, students may have difficulty remembering that the center is (h, k) and not $(-h, -k)$. Remind them that to find the distance between two points (the center and a point on a circle), you must subtract the coordinates.

CAREER CHOICES

Salaries for radio announcers range from \$13,000 in small markets to \$45,000 in larger markets.

Additional Answers

2. Solve the system of equations for the two lines. The solution is the point of intersection, which is the center.

3a. Mach number is the ratio of the speed of an object to the speed of sound. Mach 1 is equal to the speed of sound, about 760 mph, Mach 2 is twice the speed of sound, or about 1520 mph.

3b. Austrian physicist and philosopher Ernst Mach developed the concept.

Example 3 A radio station needs to locate a broadcasting tower so that it reaches three cities. Find the equation of a circle that passes through the cities if their coordinates are $A(-2, 4)$, $B(10, 4)$, and $C(8, -3)$.

CAREER CHOICES

A **radio announcer** uses special skills in communication to relate news, weather, and entertainment to the listening audience. Formal training in broadcast journalism from a college or technical school is valuable; well-done audition tapes often get announcers their jobs.

For more information, contact:
National Association of Broadcasters
1771 N St., NW
Washington, DC 20036

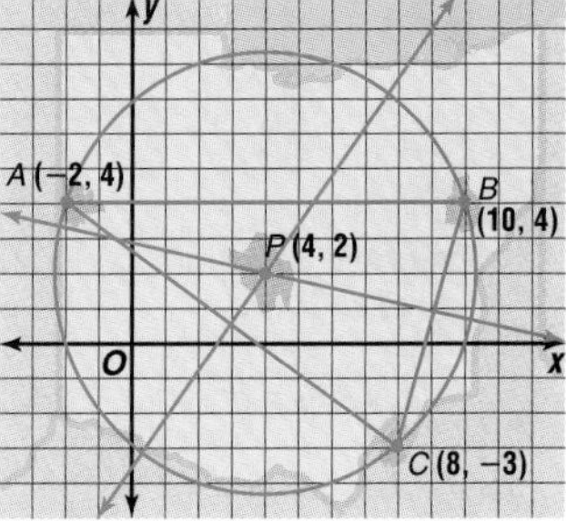

Step 1 First, graph the points and draw $\triangle ABC$.

Step 2 Construct the bisectors of $\overline{BC}$ and $\overline{AC}$. The coordinates of their intersection at P appear to be $(4, 2)$. Draw the circle by using $\overline{PA}$ as the radius.

Step 3 The radius of the circle is the distance between the center $P(4, 2)$ and any one of the three vertices.

$$d = \sqrt{(x_2 - x_1)^2 + (y_2 - y_1)^2}$$
$$= \sqrt{(10 - 4)^2 + (4 - 2)^2} \quad (x_1, y_1) = (4,2), (x_2, y_2) = (10, 4)$$
$$= \sqrt{6^2 + 2^2}$$
$$= \sqrt{40} \quad \text{The radius of } \odot P \text{ is } \sqrt{40}.$$

Step 4 Find an equation of the circle.

$$(x - h)^2 + (y - k)^2 = r^2$$
$$(x - 4)^2 + (y - 2)^2 = (\sqrt{40})^2 \quad (h, k) = (4, 2), r = \sqrt{40}$$
$$(x - 4)^2 + (y - 2)^2 = 40$$

An equation for the circle is $(x - 4)^2 + (y - 2)^2 = 40$.

CHECK FOR UNDERSTANDING

Communicating Mathematics

Study the lesson. Then complete the following.

1. **Explain** how the distance formula and the equation of a circle relate to each other. **The equation of a circle is derived using the distance formula.**

2. In Example 3, the center of the circle $P(4, 2)$ was conveniently located at a vertex of the graph paper. Explain how you could use the equations of the two perpendicular lines to determine the location of the center. **See margin.**

3. Aircraft traveling at approximately the speed of sound are assigned a *Mach number* of 1. **See margin.**
   a. What does the Mach number represent?
   b. Research to find the name and nationality of the physicist who developed the concept of the Mach number.

Reteaching

Using Vocabulary Review the terminology given in this lesson—circle, center, diameter, radius, chord, secant, tangent, common internal tangent, and common external tangent. Have students state the definition in their own words to see if they understand it.

Additional Answers

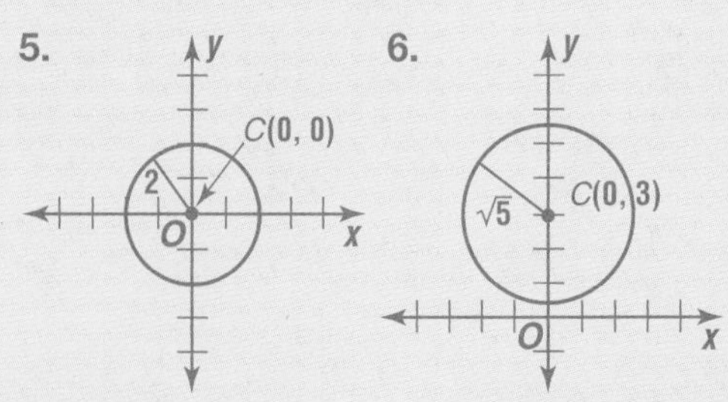

Guided Practice

4. Determine the coordinates of the center and the measure of the radius for the circle whose equation is $(x + 2)^2 + (y + 7)^2 = 81$. $(-2, -7), 9$

Graph each circle whose equation is given. Label the center and measure of the radius on each graph. 5–6. See margin.

5. $x^2 + y^2 = 4$ 6. $x^2 + (y - 3)^2 = 5$

Write an equation of circle *P* based on the given information.

7. center: $P(-2, 3)$
radius: $\sqrt{11}$
$(x + 2)^2 + (y - 3)^2 = 11$

8.

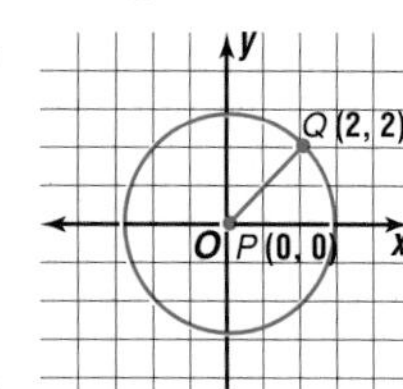

8. $x^2 + y^2 = 8$

9. $\odot P$ contains the three points $A(4, 0)$, $B(0, 4)$, and $C(-4, 0)$. $x^2 + y^2 = 16$

10. For any sample circle, the distance from the center of the circle to *N* will be 5 times the speed of sound.

10. **Supersonics** Refer to Example 1. Then copy the diagram shown below on graph paper.

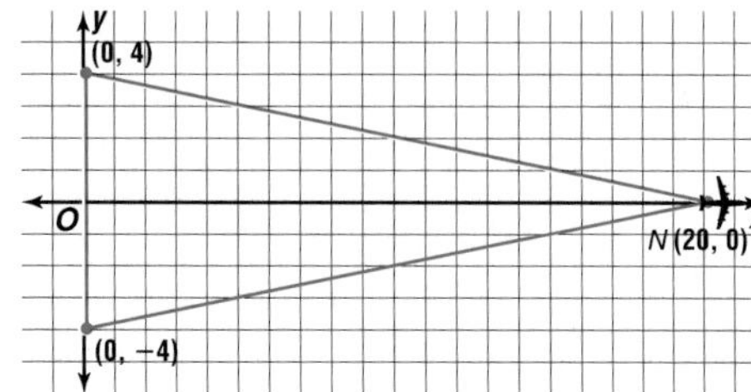

Draw your own circle tangent to the two segments and estimate the speed of the aircraft to produce the shockwave. The speed is about how many times the speed of sound?

EXERCISES

Practice

Determine the coordinates of the center and the measure of the radius for each circle whose equation is given.

11. $\left(\frac{3}{4}, -3\right), \frac{9}{2}$
12. $(-4, 0), 11$

11. $\left(x - \frac{3}{4}\right)^2 + (y + 3)^2 = \frac{81}{4}$ 12. $(x + 4)^2 + y^2 - 121 = 0$

13. $(x - 0.5)^2 + (y + 3.1)^2 = 17.64$ $(0.5, -3.1), 4.2$

Graph each circle whose equation is given. Label the center and measure of the radius on each graph. 14–19. See margin.

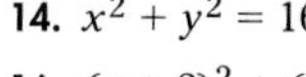

14. $x^2 + y^2 = 16$ 15. $(x + 3)^2 + y^2 = 9$

16. $(x + 2)^2 + (y - 3)^2 = 49$ 17. $x^2 + (y - 1)^2 = 8$

18. $\left(x - \frac{2}{5}\right)^2 + \left(y + \frac{1}{2}\right)^2 = \frac{1}{4}$ 19. $(x + 5)^2 + (y - 9)^2 = 20$

20. $x^2 + y^2 = 25$
21. $(x + 1)^2 + (y - 4)^2 = 15$
22. $x^2 + \left(y + \frac{3}{2}\right)^2 = \frac{16}{9}$

Write an equation of circle *P* based on the given information.

20. center: $P(0, 0)$
radius: 5

21. center: $P(-1, 4)$
radius: $\sqrt{15}$

22. center: $P\left(0, -\frac{3}{2}\right)$
radius: $\frac{4}{3}$

Additional Answers

16.

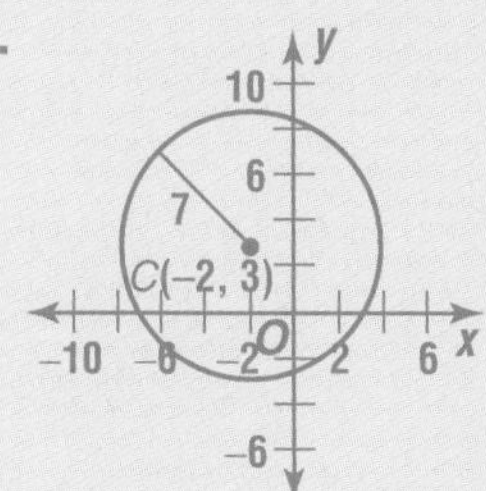

17.

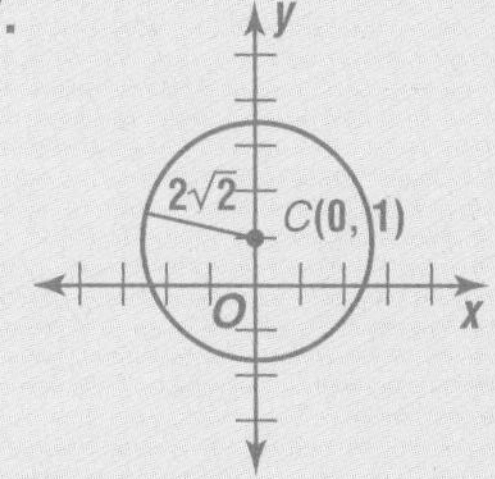

18.

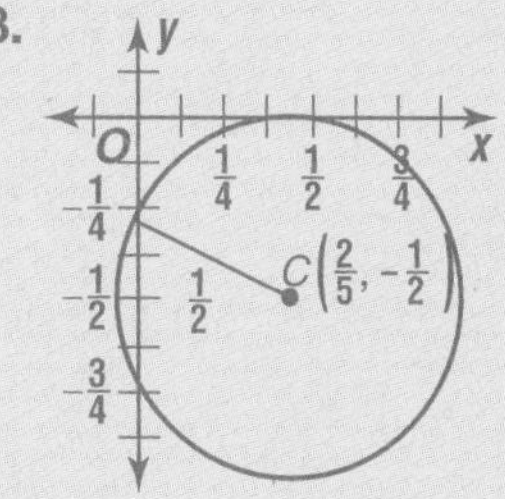

19. (see page 502)

Assignment Guide

Core (with proof): 11–37 odd, 38, 39, 41–49
Core (informal): 11–37 odd, 38, 39, 41–49
Enriched: 12–36 even, 38–49

For **Extra Practice**, see p. 782.

The red A, B, and C flags, printed only in the Teacher's Wraparound Edition, indicate the level of difficulty of the exercises.

Additional Answers

14.

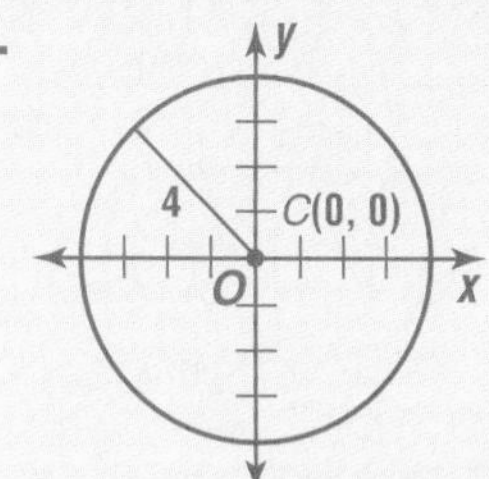

15.

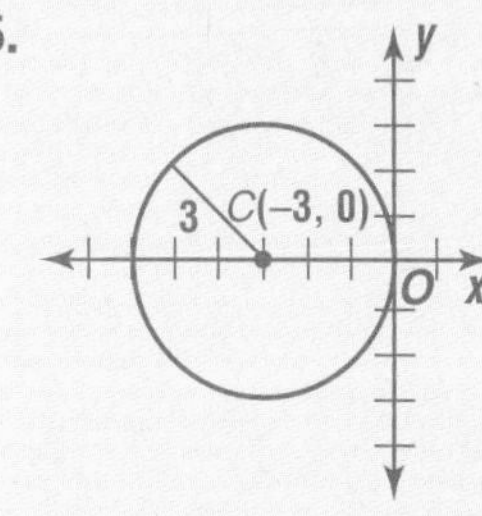

Study Guide Masters, p. 56

NAME________ DATE________

Study Guide Student Edition Pages 498–503

Integration: Algebra Equations of Circles

The **standard equation for a circle** is derived from using the distance formula given the coordinates of the center of the circle and the measure of its radius. An equation for a circle with center (h, k) and a radius of r units is $(x - h)^2 + (y - k)^2 = r^2$.

Example: Graph the circle whose equation is $(x + 3)^2 + (y - 1)^2 = 16$.

$(x - h)^2 + (y - k)^2 = r^2$ standard equation
$(x - (-3))^2 + (y - 1)^2 = (\sqrt{16})^2$ rewrite equation in standard form

Therefore, $h = -3$, $k = 1$, and $r = \sqrt{16} = 4$.
The center is at $(-3, 1)$ and the radius is 4 units.

Determine the coordinates of the center and the measure of the radius for each circle whose equation is given.

1. $(x - 7.2)^2 + (y + 3.4)^2 = 14.44$ $(7.2, -3.4), r = 3.8$

2. $\left(x + \frac{1}{2}\right)^2 + (y - 2)^2 = \frac{16}{25}$ $\left(-\frac{1}{2}, 2\right), r = \frac{4}{5}$

3. $(x - 6)^2 + (y - 3)^2 - 25 = 0$ $(6, 3), r = 5$

Graph each circle whose equation is given. Label the center and measure of the radius on each graph.

4. $(x - 2.5)^2 + (y + 1)^2 = 12.25$

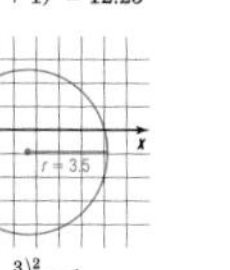

5. $(x + 3)^2 + (y - 4)^2 - 2.25 = 0$

6. $\left(x - \frac{1}{2}\right)^2 + \left(y - \frac{3}{4}\right)^2 = 1$

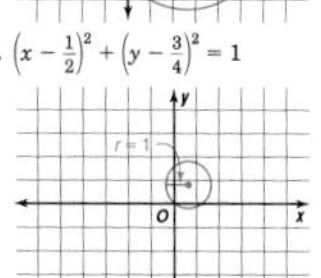

7. $x^2 + (y - 2)^2 = 9$

Additional Answers

19.

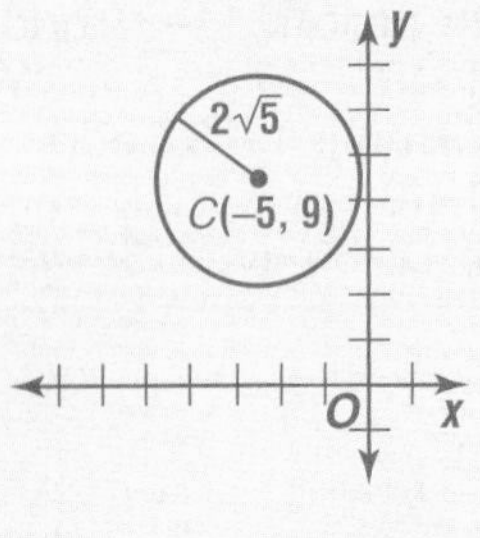

33a, c.

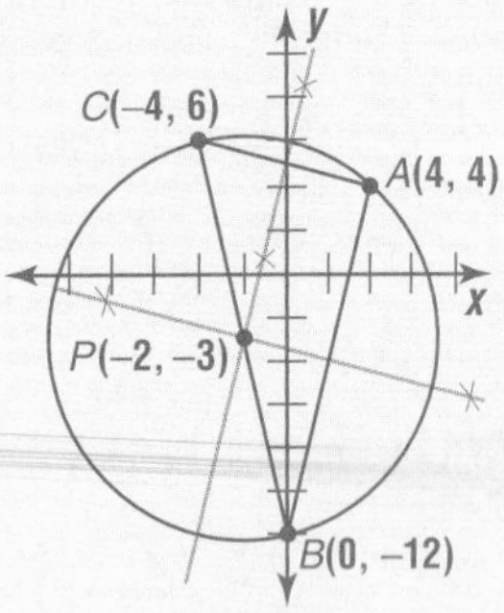

Practice Masters, p. 56

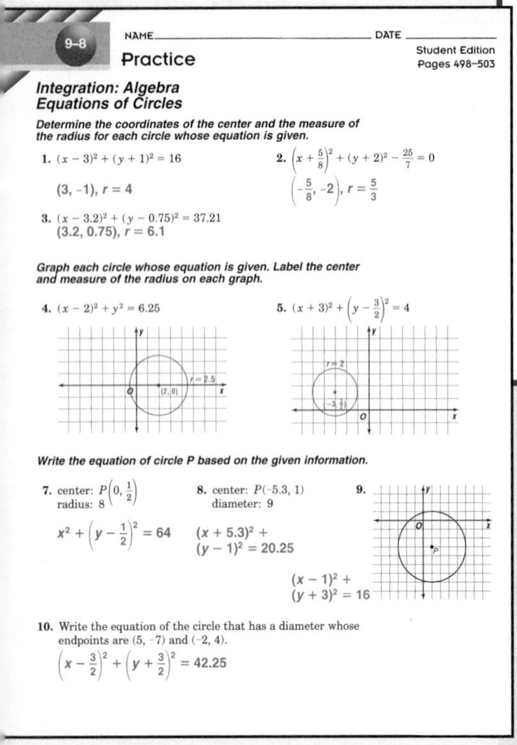
NAME ____ DATE ____

9-8 Practice

Student Edition Pages 498–503

Integration: Algebra
Equations of Circles

Determine the coordinates of the center and the measure of the radius for each circle whose equation is given.

1. $(x-3)^2+(y+1)^2=16$ — (3, −1), r = 4
2. $\left(x+\frac{5}{8}\right)^2+(y+2)^2-\frac{25}{7}=0$ — $\left(-\frac{5}{8}, -2\right)$, $r=\frac{5}{3}$
3. $(x-3.2)^2+(y-0.75)^2=37.21$ — (3.2, 0.75), r = 6.1

Graph each circle whose equation is given. Label the center and measure of the radius on each graph.

4. $(x-2)^2+y^2=6.25$
5. $(x+3)^2+\left(y-\frac{3}{2}\right)^2=4$

Write the equation of circle P based on the given information.

7. center: $P\left(0, \frac{1}{2}\right)$ radius: 8 — $x^2+\left(y-\frac{1}{2}\right)^2=64$
8. center: P(−5.3, 1) diameter: 9 — $(x+5.3)^2+(y-1)^2=20.25$
9. $(x-1)^2+(y+3)^2=16$

10. Write the equation of the circle that has a diameter whose endpoints are (5, −7) and (−2, 4). $\left(x-\frac{3}{2}\right)^2+\left(y+\frac{3}{2}\right)^2=42.25$

24. $(x+2)^2+(y-1)^2=10$

23.

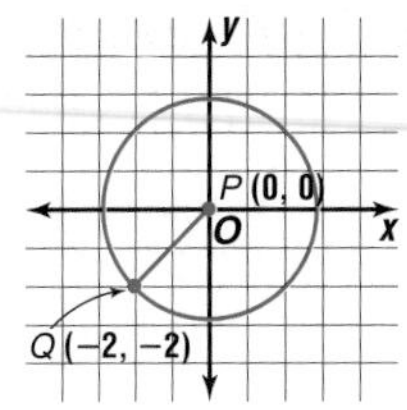

$x^2+y^2=8$

24.

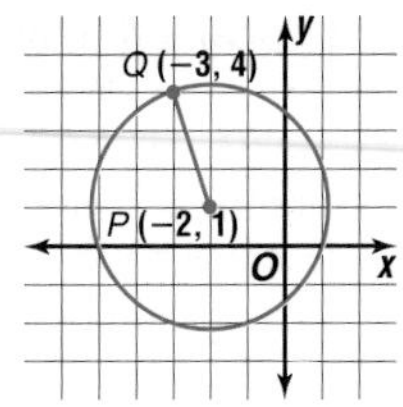

25.

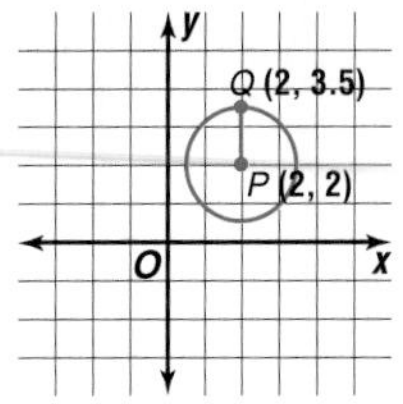

$(x-2)^2+(y-2)^2=2.25$

26. $(x+18)^2+(y+7)^2=36$

27. $(x+3)^2+(y+8)^2=49$

26. Write an equation of the circle that has a diameter of 12 units and whose center is translated 18 units to the left and 7 units down from the origin.

27. The graphs of $x=4$ and $y=-1$ are tangent to a circle that has its center in the third quadrant and a diameter of 14. Write an equation of the circle.

28. Write an equation of the circle that has a diameter whose endpoints are at (2, 7) and (−6, 15). $(x+2)^2+(y-11)^2=32$

29–32. See Solutions Manual for graphs.
29. secant
30. tangent
31. secant
32. secant

Graph the circle $(x-6)^2+(y+2)^2=36$ and the line having the given equation. Determine whether the line is a secant or tangent of the circle. Explain your reasoning.

29. $y=2x-2$ 30. $x=0$ 31. $y=-\frac{1}{6}x+5$ 32. $y=-x$

33. Use points $A(4, 4)$, $B(0, -12)$, and $C(-4, 6)$ to find each. a, c. See margin.
   a. Draw $\triangle ABC$ and construct the perpendicular bisectors of two sides.
   b. Determine the intersection of the two bisectors, P, which is the center of the circle. $P(-2, -3)$
   c. Draw the circle and use either A, B, or C with center P to find the radius. $\sqrt{85}\approx 9.2$
   d. Write an equation of the circle passing through points A, B, and C.

33d. $(x+2)^2+(y+3)^2=85$

34. Repeat the steps in Exercise 33 to write an equation of a circle passing through points $D(-2, 4)$, $E(4, 8)$, and $F(6, -8)$. $(x-5)^2+y^2=65$

C

35. a. Write an equation of the circle passing through points $A(0, 6)$, $B(6, 0)$, and $C(6, 6)$. $(x-3)^2+(y-3)^2=18$
   b. What type of triangle is $\triangle ABC$? Explain. right triangle
   c. What type of segment is $\overline{AB}$? diameter

36. Choose three noncollinear points that form an obtuse triangle in the coordinate plane. Would the center of the circle containing those points lie in the interior or exterior of the triangle? Justify your response. exterior

Cabri Geometry

37. You can use a TI-92 calculator to generate and display equations and coordinates of circles. a–d. See students' work.
   a. Display the x- and y-axes by pressing F8 and selecting 9:Format. Then select 2:RECTANGULAR from the Coordinate Axes option. Use F3 to draw a circle.
   b. Press F6 and select 5:Equation & Coordinates. Select the circle to find its equation.
   c. Select the center point of the circle to find the coordinates of the point.
   d. Move or change the circle and find the new equation and the new coordinates of the center point.

Critical Thinking

38b. $A(4, 2)$, $B(-4, -2)$

38. a. Draw the graphs of $x^2 + y^2 = 20$ and $x - 2y = 0$. See margin.
    b. Locate the two points of intersection and label them A and B.
    c. Find the distance $\overline{AB}$ between the two points. $4\sqrt{5} \approx 8.9$
    d. State the relationship between $\overline{AB}$ and the circle. $\overline{AB}$ is a diameter of the circle.

Applications and Problem Solving

39. **Aerodynamics** The diagram at the right shows a supersonic jet that travels at four times the speed of sound. Use the measurements to determine $m\angle BCD$. (*Hint*: Use trigonometry to find angle measures.) 29

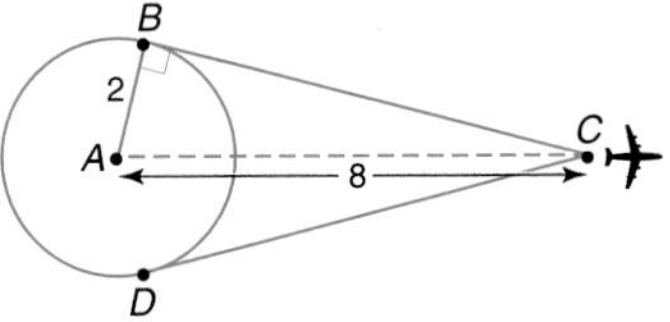

40. **Meteorology** Meteorologists often track the path of severe storms using radar. If the center of a storm is at $C(0, 0)$ and each concentric ring of the radar image has a width of 1 unit, determine the equation of the third concentric circle that encompasses the major part of the storm. $x^2 + y^2 = 9$

Mixed Review

41. Use the figure below to find the value of x. (Lesson 9–7) 5

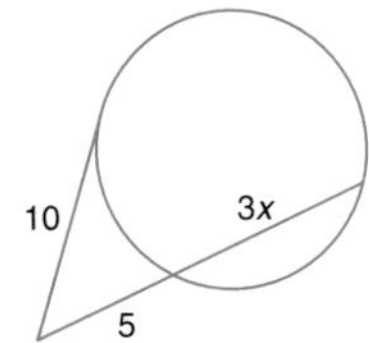

42. If the measure of inscribed angle ABC is 42, what is the measure of the intercepted arc AC? (Lesson 9–4) 84
43. In a circle with a radius of 7 inches, a chord is 5 inches from the center of the circle. What is the length of the chord? (Lesson 9–3) 9.8 in.
44. Solve $\triangle DHK$ if $m\angle K = 37$, $k = 8.1$, and $m\angle H = 62$. Round measures to the nearest tenth. (Lesson 8–5) $m\angle D = 81$, $d \approx 13.3$, $h \approx 11.9$
45. The measures of the sides of a triangle are 6, 9, and 11. If the shortest side of a similar triangle measures 12 units, what are the measures of its other two sides? (Lesson 7–3) 18 units, 22 units
46. In a right triangle, one angle measures 28°. Find the measures of the other two angles. (Lesson 4–2) 90, 62
47. Justify the statement $QT = QT$ with a property from algebra or a property of congruent segments. (Lesson 2–5) Reflexive Prop. (=)

Algebra

48. Solve the system of equations. (0, −3)
$y = x - 3$
$2x + y = -3$
49. **Savings** Jacqueline is investing $8000 in a savings certificate that matures in 5 years. The annual interest rate is 7% compounded quarterly. The equation $y = 8000(1 + 0.0175)^{4t}$ represents the balance after t years. Find the amount of Jacqueline's investment after 5 years. $10,377.82

4 ASSESS

Closing Activity

Speaking Go around the room and have each student state an important aspect relating to one of the terms or equations in this lesson.

Chapter 9 Quiz D (Lessons 9-7 and 9-8) is available in the *Assessment and Evaluation Masters*, p. 241.

Additional Answer

38a.

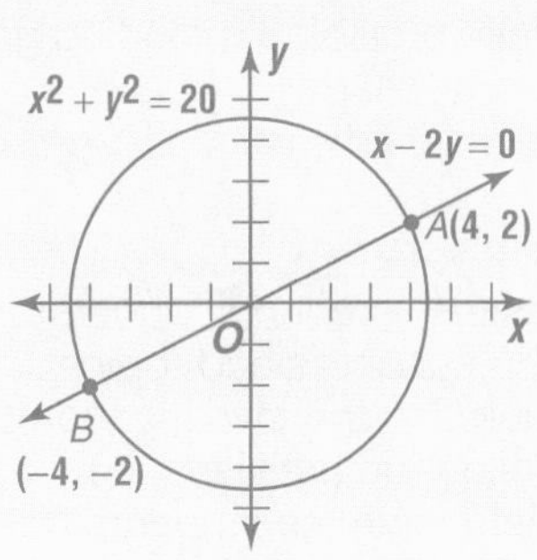

Enrichment Masters, p. 56

NAME ______ DATE ______

Enrichment

Student Edition Pages 498–503

The Unit Circle

A **unit circle** has a radius of 1. By using a unit circle with its center at the origin of a coordinate plane, the sine and cosine ratios can be defined for measures from 0° to 360°.

Consider a ray rotated through $n°$ from the positive x-axis and intersecting the circle at P, as shown at the right. Point $P(x, y)$ where the ray intersects the unit circle has coordinates ($\cos n°$, $\sin n°$).

$x = \cos n°$ $y = \sin n°$

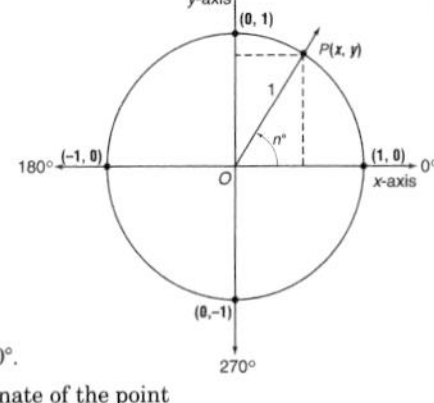

Example 1: Use the unit circle to find cos 180°.

The cosine of 180° is the x-coordinate of the point where the ray rotated 180° intersects the unit circle. So, cos 180° = −1.

Example 2: For measures between 270° and 360°, is the sine positive or negative?

The sine is the y-coordinate of a point on the unit circle in the fourth quadrant. So, the sine is negative.

Use the unit circle to find each value.

1. sin 180° 0	2. cos 270° 0	3. sin 270° −1	4. sin 90° 1
5. cos 90° 1	6. sin 360° 0	7. cos 360° 1	8. sin 0° 0

Use the unit circle to tell if each value is positive or negative. Write + or −.

9. sin 200° −	10. cos 100° −	11. sin 130° +	12. cos 310° +

13. What is the range of possible values for $\sin n°$? What is the range of possible values for $\cos n°$? Both sine and cosine can have values from −1 to 1, inclusive.

14. Consider a measure $n°$ between 0° and 180°. Compare $\sin n°$ and $\sin (360° - n°)$. Use a calculator to find several values. Then compare $\cos n°$ and $\cos (360 - n°)$. What do you notice? $\sin n° = -\sin (360° - n°)$; $\cos n° = \cos (360 - n°)$

Extension

Connections Is the graph of a circle a function? Why or why not? If not, how can you make it a function? No; it is not a function. Sample answer: There is more than one y value for some of the x values, and it fails the vertical line test. If you use the diameter parallel to the x-axis to divide the circle in half, each semicircle formed is a function by itself.

Closing the Investigation

This activity provides students an opportunity to bring their work on the Investigation to a close. For each Investigation, students should present their findings to the class. Here are some ways students can display their work.

- Conduct and report on an interview or survey.
- Write a letter, proposal, or report.
- Write an article for the school or local paper.
- Make a display, including graphs and/or charts.
- Plan an activity.

Assessment

To assess students' understanding of the concepts and topics explored in the Investigation and its follow-up activities, you may wish to examine students' Investigation Folders.

The scoring guide provided in the *Investigations and Projects Masters*, p. 15, provides a means for you to score students' work on the Investigation.

Investigations and Projects Masters, p. 15

Scoring Guide
Chapters 8 and 9
Investigation

Level	Specific Criteria
3 Superior	• Shows thorough understanding of the concepts of *planets, moon, hexagonal area, solar energy, calendar year, eclipse, horizon circle, orbit, angular radius,* and *terrain.* • Uses appropriate strategies to solve problems. • Computations are correct. • Written explanations are exemplary. • Calendar, report, and solutions are appropriate and sensible. • Goes beyond requirements of all or some problems.
2 Satisfactory, with Minor Flaws	• Shows understanding of the concepts of *planets, moon, hexagonal area, solar energy, calendar year, eclipse, horizon circle, orbit, angular radius,* and *terrain.* • Uses appropriate strategies to solve problems. • Computations are mostly correct. • Written explanations are effective. • Calendar, report, and solutions are appropriate and sensible. • Satisfies all requirements of problems.
1 Nearly Satisfactory, with Obvious Flaws	• Shows understanding of most of the concepts of *planets, moon, hexagonal area, solar energy, calendar year, eclipse, horizon circle, orbit, angular radius,* and *terrain.* • May not use appropriate strategies to solve problems. • Computations are mostly correct. • Written explanations are satisfactory. • Calendar, report, and solutions are appropriate and sensible. • Satisfies most requirements of problems.
0 Unsatisfactory	• Shows little or no understanding of the concepts of *planets, moon, hexagonal area, solar energy, calendar year, eclipse, horizon circle, orbit, angular radius,* and *terrain.* • Does not use appropriate strategies to solve problems. • Computations are incorrect. • Written explanations are not satisfactory. • Calendar, report, and solutions are not appropriate or sensible. • Does not satisfy requirements of problems.

Refer to the Investigation on pages 392–393.

Because it is Earth's closest neighbor, people have long been fascinated by Mars. Hundreds of movies have been made about fictional Martians, and many real-life explorations have been made there.

Analyze

You have prepared research and plans for a mission to Mars. It is now time to analyze your work and prepare your report.

PORTFOLIO ASSESSMENT

You may want to keep your work on this Investigation in your portfolio.

1 Look over your calculations regarding the energy needs of the space station. Verify your calculations. Then write a short description of the energy system you recommend for the station.

2 Review the Martian calendar you prepared. Make sure it is concise and complete. Then place the calendar aside to add to your report later.

3 Evaluate your calculations regarding the positions of Mars's moons. Rewrite your solutions and conclusions neatly to add to your report.

Write

Your report to NASA should explain your process for researching each of your recommendations. Also include materials that they could use to launch a campaign for presenting the project to the public and to Congress.

4 Include a neat, detailed description of the research you conducted to solve each problem that the scientists raised.

5 Review the list of questions that your team generated in your brainstorming session at the beginning of the Investigation. Are there any questions you have not answered that you could research and add to your report?

6 Research the terrain of Mars. Include any maps or photographs that you find in your report.

7 Investigate the history of the human study of Mars. Add a summary to your report.

8 Write a proposal for ways that NASA can promote the Mars space station project to Congress and the public. Include descriptions of the benefits of the project. You may wish to investigate products and innovations that were results of previous NASA projects.

9 Demonstrate how your research could be adapted to future expeditions to other planets.

CHAPTER 9 HIGHLIGHTS

VOCABULARY

After completing this chapter, you should be able to define each term, property, or phrase and give an example or two of each.

Geometry

- adjacent arc (p. 453)
- arc (p. 453)
- arc length (p. 454)
- arc of the chord (p. 459)
- center (p. 446)
- central angle (p. 452)
- chord (p. 446)
- circle (p. 446)
- circumference (p. 447)
- circumscribed polygon (p. 477)
- common external tangent (p. 476)
- common internal tangent (p. 476)
- common tangent (p. 476)
- concentric circles (p. 454)
- congruent arcs (p. 454)
- congruent circles (p. 454)
- diameter (p. 446)
- exterior (p. 475)
- external secant segment (p. 492)
- inscribed angle (p. 466)
- inscribed polygon (p. 459)
- intercepted arc (p. 466)
- interior (p. 475)
- major arc (p. 453)
- minor arc (p. 453)
- pi (p. 447)
- point of tangency (p. 475)
- radius (p. 446)
- secant (p. 483)
- secant segment (p. 492)
- semicircle (p. 453)
- similar circles (p. 454)
- tangent (p. 475)
- tangent segment (p. 477)

Problem Solving

- make a circle graph (p. 452)

UNDERSTANDING AND USING THE VOCABULARY

Choose the letter of the term that best matches each phrase.

1. an angle whose vertex is at the center of the circle i
2. circles lying in the same plane that have the same center but different radii j
3. a line that intersects a circle in exactly two points h
4. a segment that has its endpoints on the circle a
5. an angle whose vertex is on the circle and whose sides contain chords of the circle c
6. circles that have the same radius b
7. a line that intersects a circle in exactly one point f
8. the distance around a circle g
9. arcs of a circle that have exactly one point in common e
10. arcs that have the same measure d

a. chord
b. congruent circles
c. inscribed angle
d. congruent arcs
e. adjacent arcs
f. tangent
g. circumference
h. secant
i. central angle
j. concentric circles

Instructional Resources

Three multiple-choice tests and three free-response tests are provided in the *Assessment and Evaluation Masters*. Forms 1A and 2A are for honors pacing, Forms 1B and 2B are for average pacing, and Forms 1C and 2C are for basic pacing. Chapter 9 Test, Form 1B, is shown at the right. Chapter 9 Test, Form 2B, is shown on the next page.

Postulates, Theorems, and Corollaries

A complete list of postulates, theorems, and corollaries begins on page 806.

Using the CHAPTER HIGHLIGHTS

The Chapter Highlights begins with a listing of the new terms, properties, and phrases that were introduced in this chapter. Have students define each term and provide an example or two of it, if appropriate.

Assessment and Evaluation Masters, pp. 227–228

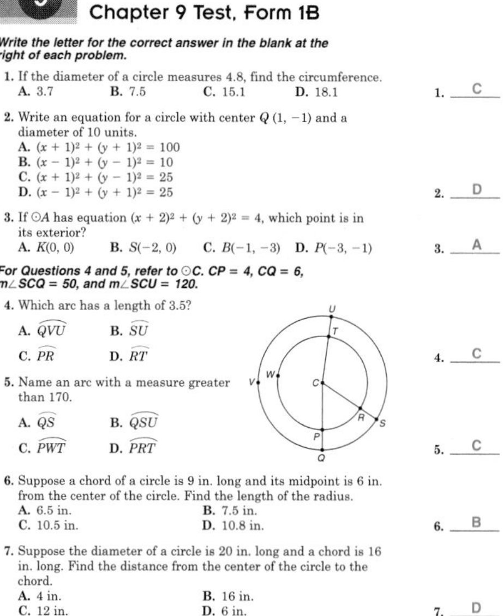

9 NAME_____ DATE_____

Chapter 9 Test, Form 1B

Write the letter for the correct answer in the blank at the right of each problem.

1. If the diameter of a circle measures 4.8, find the circumference.
A. 3.7 B. 7.5 C. 15.1 D. 18.1 — 1. C
2. Write an equation for a circle with center $Q(1, -1)$ and a diameter of 10 units.
A. $(x + 1)^2 + (y + 1)^2 = 100$
B. $(x - 1)^2 + (y - 1)^2 = 10$
C. $(x + 1)^2 + (y - 1)^2 = 25$
D. $(x - 1)^2 + (y + 1)^2 = 25$ — 2. D
3. If $\odot A$ has equation $(x + 2)^2 + (y + 2)^2 = 4$, which point is in its exterior?
A. $K(0, 0)$ B. $S(-2, 0)$ C. $B(-1, -3)$ D. $P(-3, -1)$ — 3. A

For Questions 4 and 5, refer to $\odot C$. $CP = 4$, $CQ = 6$, $m\angle SCQ = 50$, and $m\angle SCU = 120$.

4. Which arc has a length of 3.5?
A. $\widehat{QVU}$ B. $\widehat{SU}$ C. $\widehat{PR}$ D. $\widehat{RT}$ — 4. C
5. Name an arc with a measure greater than 170.
A. $\widehat{QS}$ B. $\widehat{QSU}$ C. $\widehat{PWT}$ D. $\widehat{PRT}$ — 5. C
6. Suppose a chord of a circle is 9 in. long and its midpoint is 6 in. from the center of the circle. Find the length of the radius.
A. 6.5 in. B. 7.5 in. C. 10.5 in. D. 10.8 in. — 6. B
7. Suppose the diameter of a circle is 20 in. long and a chord is 16 in. long. Find the distance from the center of the circle to the chord.
A. 4 in. B. 16 in. C. 12 in. D. 6 in. — 7. D

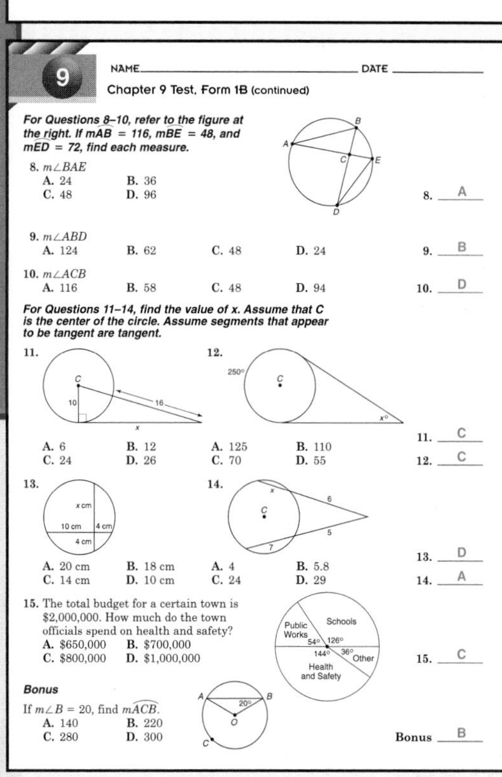

9 NAME_____ DATE_____

Chapter 9 Test, Form 1B (continued)

For Questions 8–10, refer to the figure at the right. If $m\widehat{AB} = 116$, $m\widehat{BE} = 48$, and $m\widehat{ED} = 72$, find each measure.

8. $m\angle BAE$
A. 24 B. 36 C. 48 D. 96 — 8. A
9. $m\angle ABD$
A. 124 B. 62 C. 48 D. 24 — 9. B
10. $m\angle ACB$
A. 116 B. 58 C. 48 D. 94 — 10. D

For Questions 11–14, find the value of x. Assume that C is the center of the circle. Assume segments that appear to be tangent are tangent.

11. A. 6 B. 12 C. 24 D. 26 — 11. C
12. A. 125 B. 110 C. 70 D. 55 — 12. C
13. A. 20 cm B. 18 cm C. 14 cm D. 10 cm — 13. D
14. A. 4 B. 5.8 C. 24 D. 29 — 14. A
15. The total budget for a certain town is \$2,000,000. How much do the town officials spend on health and safety?
A. \$650,000 B. \$700,000 C. \$800,000 D. \$1,000,000 — 15. C

Bonus

If $m\angle B = 20$, find $m\widehat{ACB}$.
A. 140 B. 220 C. 280 D. 300 — Bonus B

Using the STUDY GUIDE AND ASSESSMENT

Skills and Concepts Encourage students to refer to the objectives and examples on the left as they complete the review exercises on the right.

Assessment and Evaluation Masters, pp. 233–234

NAME ________ DATE ________

Chapter 9 Test, Form 2B

1. Write an equation for the circle with center $C(0, 0)$ and a diameter of 18 units. 1. $x^2 + y^2 = 81$
2. Find the radius and the coordinates of the center of the circle with equation $(x - 3)^2 + y^2 = 25$. 2. $r = 5$, center (3, 0)
3. Is the measure of a minor arc equal to the measure of its central angle? 3. yes
4. Points A, B, and D are on $\odot C$. If $m\angle ACB = 40$ and $\overline{AD}$ is a diameter, find $m\widehat{ADB}$. 4. 320
5. Suppose a chord of a circle is 48 cm long and is 7 cm from the center of the circle. Find the length of the radius. 5. 25 cm
6. Suppose the diameter of a circle is 50 ft long and a chord is 40 ft long. Find the distance from the center of the circle to the chord. 6. 15 ft

For Questions 7 and 8, refer to $\odot D$.

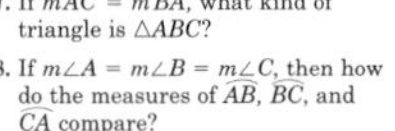

7. If $m\widehat{AC} = m\widehat{BA}$, what kind of triangle is $\triangle ABC$? 7. isosceles
8. If $m\angle A = m\angle B = m\angle C$, then how do the measures of $\widehat{AB}$, $\widehat{BC}$, and $\widehat{CA}$ compare? 8. They are =.

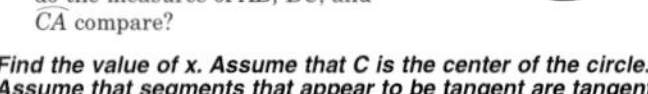

Find the value of x. Assume that C is the center of the circle. Assume that segments that appear to be tangent are tangent.

9. 9. 9

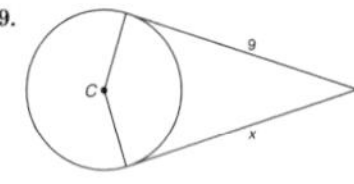

10. 10. 14

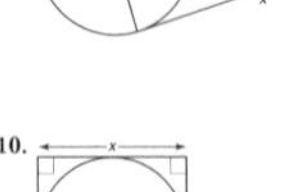

NAME ________ DATE ________

Chapter 9 Test, Form 2B (continued)

For Questions 11–14, find the value of x. Assume that segments that appear to be tangent are tangent.

11. 11. 40

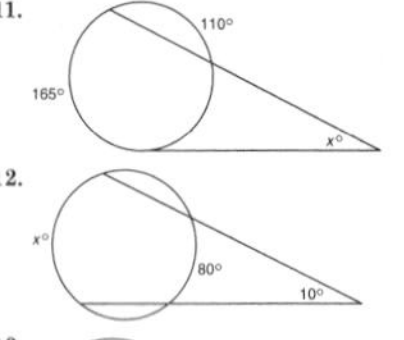

12. 12. 100

13. 13. 6.7 in.

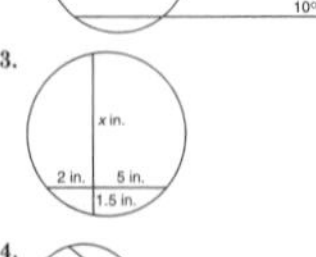

14. 14. 8.0 ft

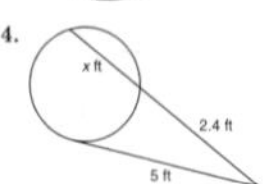

15. Complete the chart and the circle graph. Suppose the family earns $2,000 per month. How much is spent on rent?

Monthly Family Budget

Category	%	Number of Degrees
Rent	40	
Food	35	
Savings	15	
Other	10	

15. 144, 126, 54, 36; $800

Bonus

Two rays from an external point are tangent to $\odot C$ at A and B. The measure of the central angle ACB is 120. Find the measure of the angle formed by the two rays. Bonus 60

SKILLS AND CONCEPTS

OBJECTIVES AND EXAMPLES

Upon completing this chapter, you should be able to:

- identify and use parts of circles, and solve problems involving the circumference of circles (Lesson 9–1)

$\overline{JK}$, $\overline{LM}$, and $\overline{HI}$ are chords of $\odot C$. $\overline{LM}$ is the diameter, and $\overline{CQ}$, $\overline{CL}$, and $\overline{CM}$ are radii. If $CM = 9$, find the circumference.

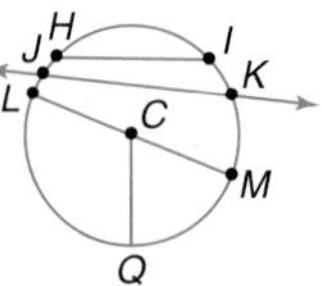

$C = 2\pi r$
$= 2\pi(9)$
$= 18\pi$ The circumference of circle P is 18π units.

- find the measures of arcs and central angles (Lesson 9–2)

$m\angle FOG = 36$

$m\widehat{FG} = 36$

If $FO = 4$, find the length of $\widehat{FG}$.

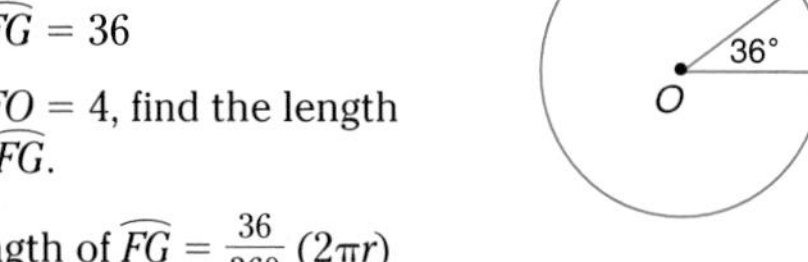

length of $\widehat{FG} = \frac{36}{360}(2\pi r)$
$= \frac{1}{10}(2\pi)(4)$ or 0.8π

- recognize and use relationships between arcs, chords, and diameters (Lesson 9–3)

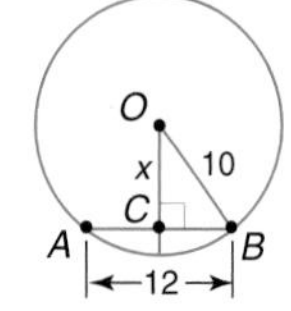

Find the value of x.

$CB = \frac{1}{2}AB$
$= \frac{1}{2}(12)$ or 6

$(OC)^2 + (CB)^2 = (OB)^2$
$x^2 + 6^2 = 10^2$
$x^2 + 36 = 100$
$x^2 = 64$
$x = 8$

REVIEW EXERCISES

Use these exercises to review and prepare for the chapter test.

Refer to $\odot T$ for Exercises 11–13.

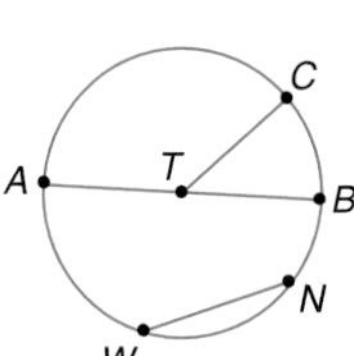

11. Name a chord that is also a diameter. $\overline{AB}$
12. If $TC = 6$, find AB. 12
13. If $AT = 14$, find the circumference to the nearest tenth. 88.0

Refer to $\odot P$ for Exercises 14–19. In $\odot P$, $\overline{XY}$ and $\overline{AB}$ are diameters. Find the degree measure of each arc.

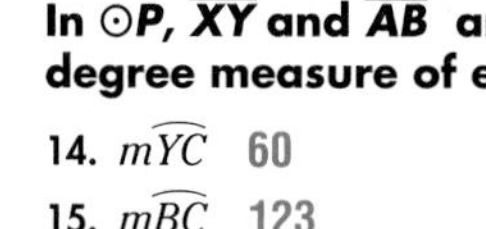

14. $m\widehat{YC}$ 60
15. $m\widehat{BC}$ 123
16. $m\widehat{BX}$ 117

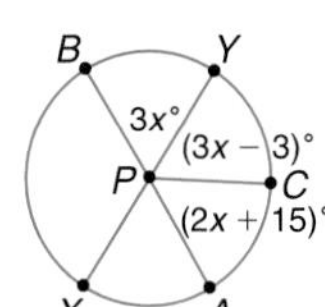

Find the length of each arc if $BA = 9$. Round to the nearest tenth.

17. $\widehat{CA}$ 4.5 18. $\widehat{YAX}$ 14.1 19. $\widehat{ABY}$ 19.1

20. A chord is 5 centimeters from the center of a circle with a radius of 13 centimeters. Find the length of the chord. 24 cm
21. Suppose a 24-centimeter chord of a circle is 32 centimeters from the center of the circle. Find the length of the radius. about 34.2 cm
22. Suppose the diameter of a circle is 20 centimeters long and a chord is 16 centimeters long. Find the distance between the chord and the center of the circle. 6 cm
23. Suppose the diameter of a circle is 10 inches long and a chord is 6 inches long. Find the distance between the chord and the center of the circle. 4 in.

GLENCOE Technology

Test and Review Software

You may use this software, a combination of an item generator and item bank, to create your own tests or worksheets. Types of items include free response, multiple choice, short answer, and open ended.

For IBM & Macintosh

OBJECTIVES AND EXAMPLES	REVIEW EXERCISES

- recognize and find measures of inscribed angles (Lesson 9–4)

$m\angle XYZ = \frac{1}{2}m\widehat{XZ}$

$= \frac{1}{2}(78)$

$= 39$

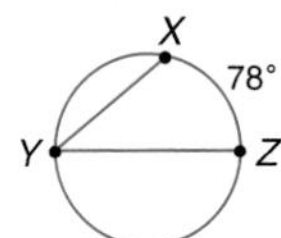

In ⊙P, $\overline{AB} \parallel \overline{CD}$, $m\widehat{BD} = 72$, and $m\angle CPD = 144$. Find each measure.

24. $m\angle DAB$ 36
25. $m\widehat{CD}$ 144
26. $m\widehat{CA}$ 72
27. $m\angle CDA$ 36
28. $m\widehat{AB}$ 72

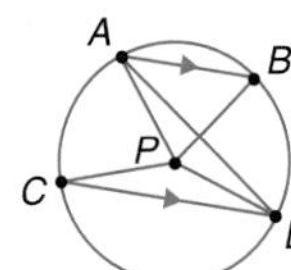

- recognize and use properties of tangents (Lesson 9–5)

Find the values of x and y.

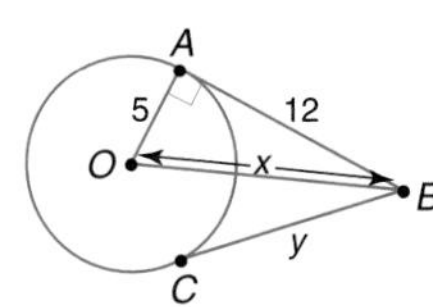

$5^2 + 12^2 = x^2$

$25 + 144 = x^2$

$169 = x^2$

$13 = x$

$CB = AB$

$y = 12$

For each ⊙C, find the value of x. Assume that segments that appear to be tangent are tangent.

29. 12

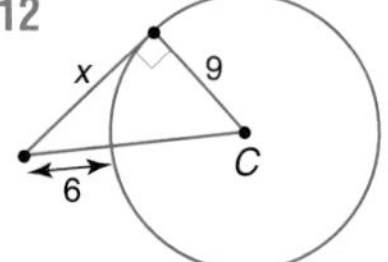

30. 6

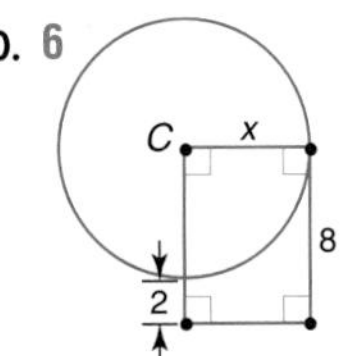

- find the measures of angles formed by intersecting secants and tangents in relation to intercepted arcs (Lesson 9–6)

$m\angle DXC = \frac{1}{2}(m\widehat{DC} + m\widehat{AB})$

$= \frac{1}{2}(60 + 20)$

$= \frac{1}{2}(80)$ or 40

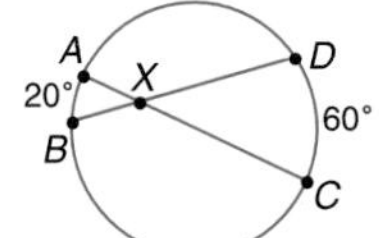

In ⊙P, $m\widehat{AB} = 29$, $m\angle AEB = 42$, $m\widehat{BG} = 18$, and $\overline{AC}$ is a diameter. Find each measure.

31. $m\angle DEC$ 42
32. $m\widehat{CD}$ 55
33. $m\angle GFD$ 18.5
34. $m\widehat{AD}$ 125
35. $m\angle AED$ 138
36. $m\widehat{GC}$ 133

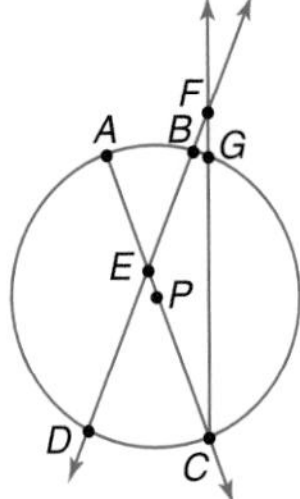

Applications and Problem Solving Encourage students to work through the exercises in the Applications and Problem Solving section to strengthen their problem-solving skills.

Additional Answer

43.

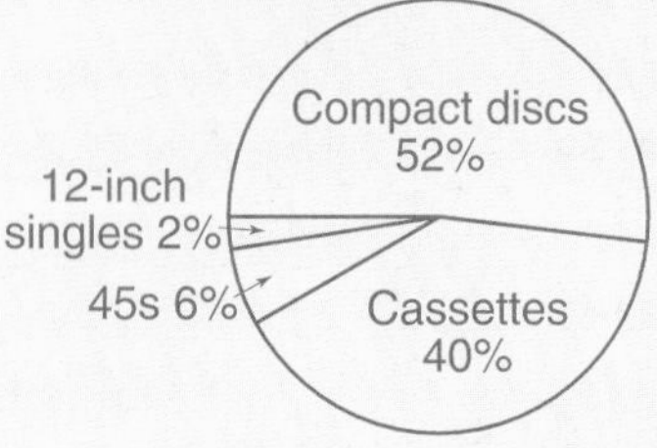

OBJECTIVES AND EXAMPLES

- use properties of chords, secants, and tangents to solve segment measure problems (Lesson 9–7)

Find the value of x.

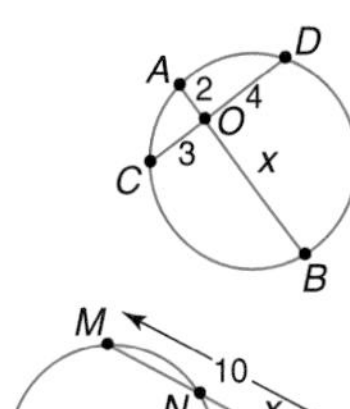

$(AO)(OB) = (CO)(OD)$
$2x = 3 \cdot 4$
$2x = 12$
$x = 6$

$(MP)(NP) = (QP)(RP)$
$10x = 5 \cdot 12$
$10x = 60$
$x = 6$

REVIEW EXERCISES

Find the value of x to the nearest tenth. Assume that segments that appear to be tangent are tangent.

37. **6.0 cm**

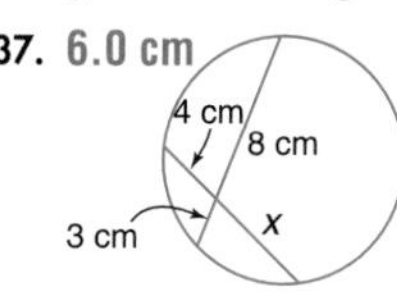

38. **0.8 ft**

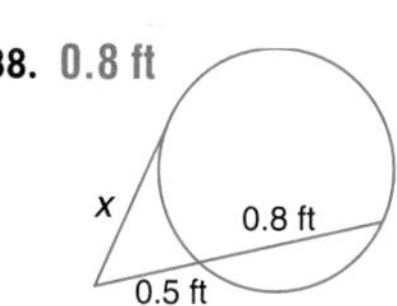

39. **15.3 cm** 40. **4.8 in.**

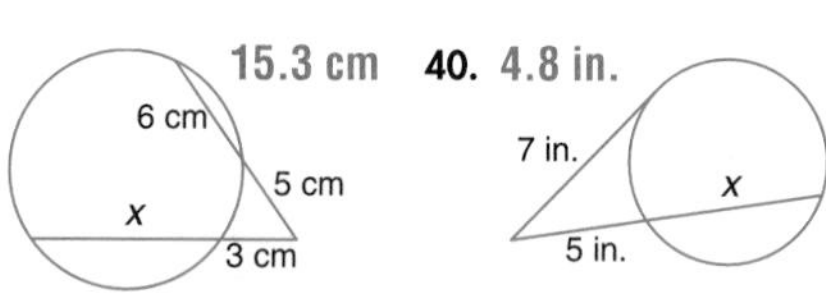

OBJECTIVES AND EXAMPLES

- write and use the equation of a circle in the coordinate plane (Lesson 9–8)

Write the equation of a circle with center at $(-1, 4)$ and radius 3.

$(x - h)^2 + (y - k)^2 = r^2$
$[x - (-1)]^2 + (y - 4)^2 = 3^2$
$(x + 1)^2 + (y - 4)^2 = 9$

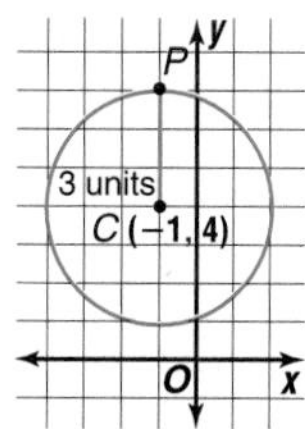

REVIEW EXERCISES

Write the equation of each circle based on the given information.

41. center: $P(-4, 3)$
radius: 6
$(x + 4)^2 + (y - 3)^2 = 36$

42.

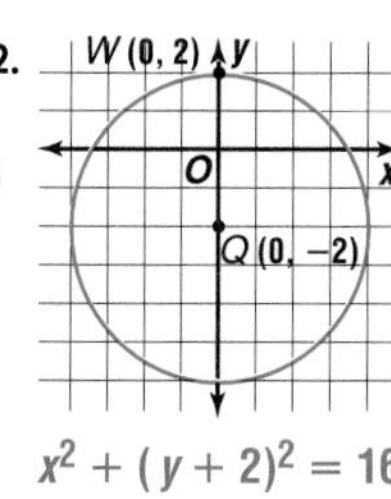

$x^2 + (y + 2)^2 = 16$

APPLICATIONS AND PROBLEM SOLVING

43. **Make a Circle Graph** Ms. Perez is opening a music store. She gathers information to help her decide what to sell. Use the information in the table below to make a circle graph showing the portion of sales for each type of recording. Draw the graph by hand or use graphing software. (Lesson 9–2) **See margin.**

Music Sales	
Compact discs	52%
Cassettes	40%
45s	6%
12-inch singles	2%

44. **Construction** The support of a bridge is in the shape of an arc. The span of the bridge (the length of the chord connecting the endpoints of the arc) is 28 meters. The highest point of the arc is 5 meters above the imaginary chord connecting the endpoints of the arc. What is the radius of the circle that forms the arc? (Lesson 9–3) **22.1 m**

45. **Crafts** Sara uses wooden spheres to make paperweights to sell at craft shows. She cuts off a flat surface for each base. If the original sphere has a radius of 4 centimeters and the diameter of the flat surface is 6 centimeters, what is the height of the paperweight? (Lesson 9–3) **about 6.6 cm**

A practice test for Chapter 9 is provided on page 801.

ALTERNATIVE ASSESSMENT

COOPERATIVE LEARNING PROJECT

Tilted circles Circular and spherical objects are everywhere throughout the universe. In this project, you will apply your knowledge of circles to planetary rings and spiral galaxies. Planetary rings and spiral galaxies are circular in shape, but because they tilt toward or away from us, they appear elliptical. The greater the tilt, the less circular they look.

An ellipse has a major and minor axis. You will measure these and use this information to determine the degree of tilt.

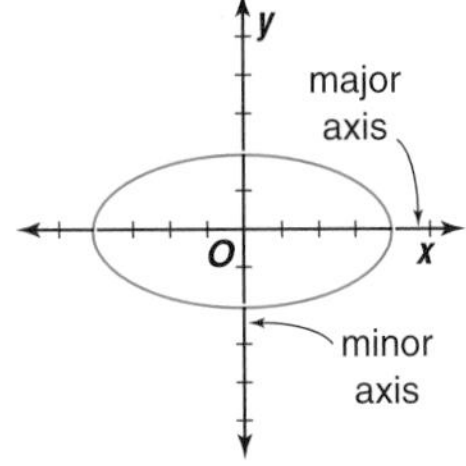

Follow these steps to determine the degree of tilt.

- Find five photographs of the planet Saturn with its rings at different degrees of tilt. Find five photographs of spiral galaxies with varying degrees of tilt.
- For the first photograph, measure the major and minor axes. Find the ratio of minor to major axis.
- Now cut out a 10-inch diameter circle from construction paper. Have one student hold it while another student views it edgewise, so that it appears paper thin. Slowly tilt it toward the observer until the proportions look the same as the photograph.

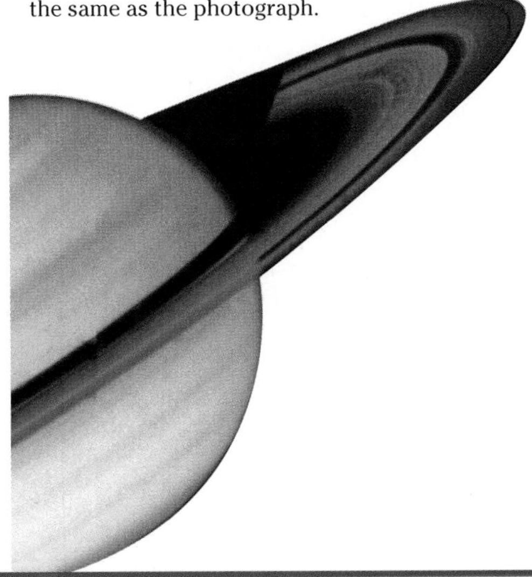

- Use a protractor to measure the angle of tilt of the circle. Use a ruler to measure the ratio of the minor to major axis as they appear to the student observing the tilted circle.
- Repeat the same procedure for each photograph and label each angle of tilt.

Follow these steps to discover how to use trigonometry to confirm the degree of tilt.

- Look at your tilted 10-inch diameter circle again. Do you see a right triangle that will allow you to apply trigonometry? Look at it from the side. Look at it from the observer's point of view.
- What information do you have? What information do you need to compute? Do you need to use sine, cosine, or tangent to draw your conclusion?
- Apply your trigonometric method to the measurements of major and minor axes from the photographs to compute the angle of tilt.

THINKING CRITICALLY

Circle graphs are geometric models used to represent nongeometric information. Think of other geometric models used to represent non-geometric information. Why do you think geometric models are useful? Are there ways in which they can be misleading?

PORTFOLIO

Before it was discovered that Earth revolved around the Sun, it was believed that the sun and all the planets revolved around Earth in perfect circular orbits. This model was known as the Ptolemaic system. Read about this model and draw a diagram of the solar system according to Ptolemy. Keep this diagram in your portfolio.

SELF EVALUATION

For one week, record the amount of time you spend doing each of the things you do, including class time, studying, leisure, and sleeping. Convert these to percents and draw a circle graph to represent these percentages.

Assess Yourself Many people are surprised to see how their time is spent. Is this how you want to spend your time? Make another circle graph of how you would prefer to spend your time.

Alternative Assessment

The Alternative Assessment section provides students with the opportunity to assess their own work by thinking critically, working with others, keeping a portfolio, and honestly evaluating their own progress. For more information on alternative forms of assessment, see *Alternative Assessment in the Mathematics Classroom,* one of the titles in the Glencoe Mathematics Professional Series.

Performance Assessment

Performance Assessment tasks for this chapter are included in the *Assessment and Evaluation Masters.* A scoring guide is also provided.

Sample Answer for Thinking Critically

Rectangles are used in bar graphs. Geometric models are useful since we have formulas to compute perimeter and area. Geometric models can be misleading if they are not drawn to scale.

Assessment and Evaluation Masters, pp. 238, 249

NAME ______________________ DATE ____________

Chapter 9 Performance Assessment

Instructions: ***Demonstrate your knowledge by giving a clear, concise solution to each problem. Be sure to include all relevant drawings and to justify your answers. You may show your solution in more than one way or investigate beyond the requirements of the problem.***

1. a. Draw a right triangle. Construct a circle circumscribing your triangle. Explain each step.

b. Draw a conclusion about inscribed right triangles and tell why you believe your conclusion to be true.

c. Draw an obtuse triangle. Construct a circle circumscribing your triangle.

d. Draw a conclusion about inscribed obtuse triangles. Justify your conclusion.

e. Make a conjecture about inscribed acute triangles. Tell why you believe your conjecture to be true.

f. Draw an acute triangle. Construct a circle circumscribing your triangle. Is your conjecture true for this triangle?

2. In the figure below, $\overline{CE}$ is tangent to both $\odot A$ and $\odot B$.

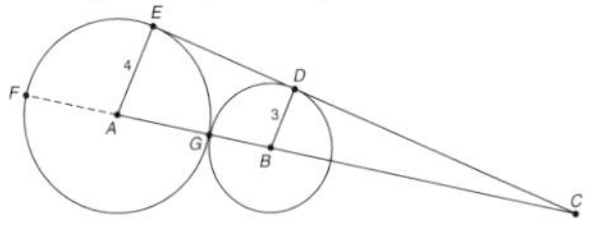

a. What can you conclude about $\overline{AE}$ and $\overline{BD}$? Why?

b. What can you conclude about $\triangle ACE$ and $\triangle BCD$? Explain.

c. Find BC. Show your work.

d. Find $m\angle EAC$. Show your work.

e. Find $m\widehat{FE}$. Justify your answer.

f. Find $m\angle C$ in at least two ways. Show your work.

Scoring Guide
Chapter 9
Performance Assessment

Level	Specific Criteria
3 Superior	• Shows thorough understanding of the concepts of *chord, tangent, secant, diameter, and radius of a circle; central and inscribed angles;* and *measure and length of an arc.* • Uses appropriate strategies to solve problems. • Computations are correct. • Written explanations are exemplary. • Diagrams are accurate and appropriate. • Goes beyond requirements of some or all problems.
2 Satisfactory, with Minor Flaws	• Shows understanding of the concepts of *chord, tangent, secant, diameter, and radius of a circle; central and inscribed angles;* and *measure and length of an arc.* • Uses appropriate strategies to solve problems. • Computations are mostly correct. • Written explanations are effective. • Diagrams are mostly accurate and appropriate. • Satisfies all requirements of some or all problems.
1 Nearly Satisfactory, with Serious Flaws	• Shows understanding of most of the concepts of *chord, tangent, secant, diameter, and radius of a circle; central and inscribed angles;* and *measure and length of an arc.* • May not use appropriate strategies to solve problems. • Computations are mostly correct. • Written explanations are satisfactory. • Diagrams are mostly accurate and appropriate. • Satisfies most requirements of some or all problems.
0 Unsatisfactory	• Shows little or no understanding of the concepts of *chord, tangent, secant, diameter, and radius of a circle; central and inscribed angles;* and *measure and length of an arc.* • May not use appropriate strategies to solve problems. • Computations are incorrect. • Written explanations are not satisfactory. • Diagrams are not accurate or appropriate. • Does not satisfy requirements of some or all problems.

In•ves•ti•ga•tion

TEACHER NOTES

NCTM Standards: 1–5, 7, 8

This Investigation is designed to be completed over several days or weeks. It may be considered optional. You may want to assign the Investigation and the follow-up activities to be completed at the same time.

Objective

Design a soccer ball that incorporates a variety of polygons.

Mathematical Overview

This Investigation will use the following mathematical skills and concepts from Chapters 10 and 11.

- summing measures of angles
- tessellating various polygons
- drawing nets for polygons
- drawing viewpoints of the soccer ball design
- finding volume, diameter, circumference, and surface area of the soccer ball
- finding length of sides, measures of interior angles, perimeter, and area of polygons

Recommended Time

Part	Pages	Time
Investigation	510–511	1 class period
Working on the Investigation	521, 550, 598	20 minutes each
Closing the Investigation	636	1 class period

Instructional Resources

Investigations and Projects Masters, pp. 17–20

A recording sheet, teacher notes, and scoring guide are provided for each Investigation in the *Investigations and Projects Masters*.

1 MOTIVATION

Bring in different soccer balls. Have students handle them and kick them around outside. Do different balls handle differently? Is a soccer ball a smooth sphere?

LONG-TERM PROJECT

In•ves•ti•ga•tion

Just For Kicks

MATERIALS NEEDED

ruler

protractor

calculator

posterboard

markers

Football, or soccer as it is called in the United States, is the most popular game in the world. It is enjoyed by millions of people in more than 150 countries. A game similar to soccer was played in China as early as 400 B.C. The rules that we use in a game of soccer today were established in England in the 1800s. Every four years, the best soccer teams in the world compete for the sport's top prize, the World Cup. The Fédération Internationale de Football Association (FIFA) is the world soccer authority. It establishes the rules for international play, including professional, semiprofessional, and interclub play.

Imagine that your group is a design team for a sporting goods manufacturer. Seeing how popular soccer is becoming in the United States, your company has decided to get involved in this growing market by producing a soccer ball. It is your team's job to design the ball that your company will produce.

CHECKING OUT THE COMPETITION

The marketing research division has produced design prints of the two best-selling soccer balls on the market. The top-selling soccer ball is made by Everkick. The design of this ball is comprised of pentagons and hexagons.

Below are three views of the Everkick soccer ball. Analyze the design pattern and determine the number of pentagons and hexagons it takes to make a ball. Explain your conclusion using mathematical reasoning.

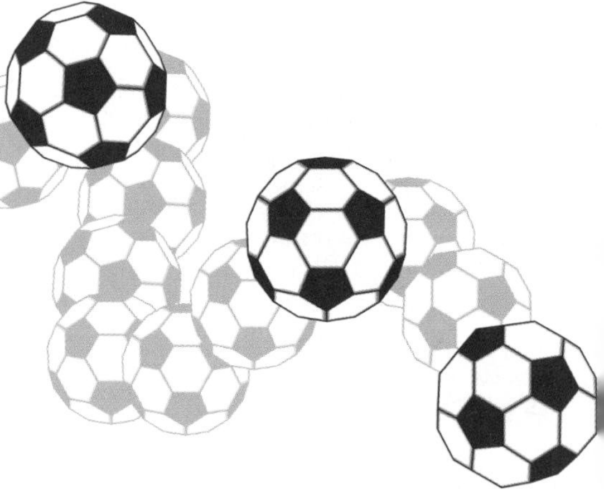

Cooperative Learning

This Investigation offers an excellent opportunity for using cooperative learning groups. For more information on cooperative learning strategies and group management, see *Cooperative Learning in the Mathematics Classroom*, one of the titles in the Glencoe Mathematics Professional Series.

The GoalScorer ball is second on the list of best-selling soccer balls. It has a different design, which is comprised of pentagons, triangles and squares. Above are three views of the GoalScorer ball. Analyze the design pattern and determine the number of pentagons, triangles and squares it takes to make a ball. Explain your conclusion using mathematical reasoning.

Discuss the advantages and disadvantages of each design. Brainstorm a list of ideas for your soccer ball design.

You will continue working on this Investigation throughout Chapters 10 and 11.

Be sure to keep your list of observations and materials in your Investigation Folder.

Just For Kicks Investigation

Working on the Investigation
Lesson 10–1, p. 521
Working on the Investigation
Lesson 10–5, p. 550
Working on the Investigation
Lesson 11–3, p. 598
Closing the Investigation
End of Chapter 11, p. 636

2 SETUP

Have a student read the first five paragraphs of the Investigation to provide background information for the soccer ball designing project. Read the remaining paragraphs, which explain what the class's tasks will be. Discuss the activity with students. Separate the class into groups of four or five.

3 MANAGEMENT

Each group member should be responsible for a specific task.

Recorder Keeps record of advantages and disadvantages of each design.

Measurer Measures angles and verifies validity of design.

Modeler Creates model of soccer ball.

Designer Sketches three views of each possible design.

At the end of the activity, each member should turn in his or her respective equipment.

Investigations and Projects Masters, p. 20

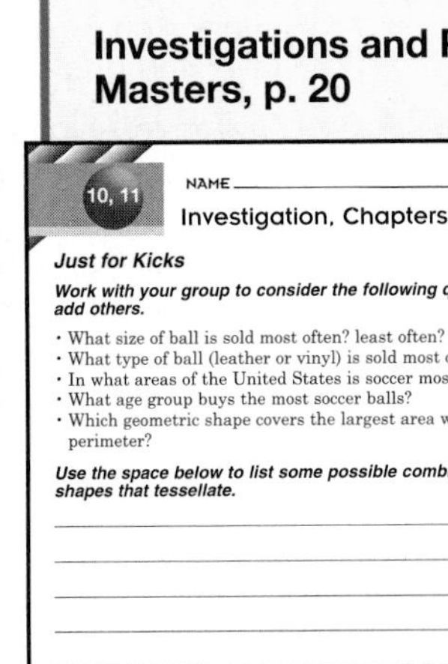

10, 11 NAME ______ DATE ______

Investigation, Chapters 10 and 11

Student Edition Pages 510–511, 521, 560, 598, 636

Just for Kicks

Work with your group to consider the following questions and to add others.

- What size of ball is sold most often? least often?
- What type of ball (leather or vinyl) is sold most often?
- In what areas of the United States is soccer most popular?
- What age group buys the most soccer balls?
- Which geometric shape covers the largest area with the least perimeter?

Use the space below to list some possible combinations of shapes that tessellate.

Use this chart to record your calculations for each soccer ball.

Calculation	Ball 1	Ball 2

Use this chart to record your calculations for each soccer ball.

Calculation	Size 3 ball	Size 4 ball	Size 5 ball

Please keep this page and any other research in your Investigation Folder.

Sample Answers

Answers will vary as they are based on each design.

CHAPTER

10 Exploring Polygons and Area

PREVIEWING THE CHAPTER

In this chapter, students learn to identify and name polygons. They investigate interior and exterior angle measures of convex and regular polygons. Students use paper and pencil to create tessellations by translating and rotating polygons. They learn to identify types of tessellations and create specific tessellations. They also learn to solve problems by using guess and check. Students use the TI-92 graphing calculator to compare the areas of parallelograms and rectangles. Then they use area formulas for parallelograms, triangles, rhombi, trapezoids, regular polygons, and circles. Students use area to investigate geometric probability. The chapter ends with students investigating polygon networks.

Lesson (Pages)	Lesson Objectives	NCTM Standards	State/Local Objectives
10-1 (514–521)	Identify and name polygons. Find the sum of the measures of interior and exterior angles of convex polygons and measures of interior and exterior angles of regular polygons. Solve problems involving angle measures of polygons.	1–5, 7, 8	
10-2A (522)	Create tessellations using translations and rotations.	1–3, 7, 8	
10-2 (523–527)	Identify regular and uniform (semi-regular) tessellations. Create tessellations with specific attributes. Solve problems by using guess and check.	1–5, 7, 8	
10-3A (528)	Compare the area of a rectangle with the area of a parallelogram having the same base length and the same parallel lines.	2–4, 7	
10-3 (529–534)	Find areas of parallelograms.	1–5, 7, 8	
10-4 (535–541)	Find areas of triangles, rhombi, and trapezoids.	1–5, 7, 8	
10-5A (542)	Create a regular *n*-gon.	2–4, 7	
10-5 (543–550)	Find areas of regular polygons. Find areas of circles.	1–5, 7, 8	
10-6 (551–558)	Use area to solve problems involving geometric probability.	1–5, 7, 8, 11	
10-7 (559–564)	Recognize nodes and edges as used in graph theory. Determine if a network is traceable. Determine if a network is complete.	1–5, 7, 8	

ORGANIZING THE CHAPTER

A complete, 1-page lesson plan is provided for each lesson in the ***Lesson Planning Guide***. Answer keys for each lesson are available in the ***Answer Key Masters***.

You may want to refer to the **Course Planning Calendar** on page T12 for detailed information on pacing.
PACING: Standard—15 days; **Honors**—14 days; **Block**—8 days

LESSON PLANNING CHART

Lesson (Pages)	Materials/ Manipulatives	Extra Practice (Student Edition)	Blackline Masters								Real-World Applications	Teaching Transparencies
			Study Guide	Practice	Enrichment	Assessment & Evaluation	Modeling Mathematics	Multicultural Activity	Tech Prep Applications	Graphing Calc. & Computer		
10-1 (514–521)	straightedge* protractor*	p. 782	p. 57	p. 57	p. 57							10-1A 10-1B
10-2A (522)							p. 97					
10-2 (523–527)	pattern blocks TI-92 calculator	p. 783	p. 58	p. 58	p. 58	p. 268						10-2A 10-2B
10-3A (528)	TI-92 calculator									p. 27		
10-3 (529–534)	TI-92 calculator grid paper	p. 783	p. 59	p. 59	p. 59					p. 10		10-3A 10-3B
10-4 (535–541)	grid paper scissors* straightedge* TI-82/83 graphing calculator	p. 783	p. 60	p. 60	p. 60	pp. 267, 268	pp. 51–54		p. 19		20	10-4A 10-4B
10-5A (542)	straightedge* hinged mirror* protractor*						p. 98					
10-5 (543–550)	compass* straightedge* scientific calculator grid paper	p. 784	p. 61	p. 61	p. 61		p. 88	p. 19	p. 20			10-5A 10-5B
10-6 (551–558)	TI-92 calculator	p. 784	p. 62	p. 62	p. 62	p. 269		p. 20				10-6A 10-6B
10-7 (559–564)		p. 784	p. 63	p. 63	p. 63	p. 269					21	10-7A 10-7B
Study Guide/ Assessment (565–569)						pp. 253–266, 270–272						

*Included in Glencoe's High School Manipulative Kit and Overhead Manipulative Resources.

ORGANIZING THE CHAPTER

Other Chapter Resources

Student Edition
Investigation, pp. 510–511
Chapter Opener, pp. 512–513
Mathematics and Society, p. 558
Working on the Investigation, pp. 521, 550

Teacher's Classroom Resources
Investigations and Projects Masters, pp. 61–64
Block Scheduling Booklet

Technology
Test and Review Software (IBM and Macintosh)
CD-ROM Multimedia Applications (Windows and Macintosh)
Mindjogger Videoquizzes (VHS)

Professional Publications
Glencoe Mathematics Professional Series

Outside Resources

Books/Periodicals
Introduction to Tessellations, NASCO
Exploring with Polydrons, Dale Seymour Publications
Geometry from Multiple Perspectives, NCTM

Software
Tesselmania, ETA
Shape Up, Sunburst
Geometry Connections: Tessellations, William K. Bradford Publishing Company

**Videos/CD-ROMs**
Mathematics: Making the Connection, NCTM, 1906 Association Drive, Reston, VA 20191
Learning About the TI-82, Venture Publishing, Bartlett Street, Suite 55, Andover, MA 01810

Assessment Resources

Student Edition
Math Journal, pp. 554, 561
Mixed Review, pp. 521, 527, 534, 540–541, 550, 557–558, 564
Self Test, p. 541
Chapter Highlights, p. 565
Chapter Study Guide and Assessment, pp. 566–568
Alternative Assessment, p. 569
Portfolio, p. 569
College Entrance Exam Practice, pp. 570–571
Chapter Test, p. 802

Teacher's Wraparound Edition
5-Minute Check, pp. 514, 523, 529, 535, 543, 551, 559
Check for Understanding, pp. 518, 525, 532, 538, 547, 554, 561
Closing Activity, pp. 521, 527, 534, 541, 550, 558, 564
Cooperative Learning, pp. 524, 560

Assessment and Evaluation Masters
Multiple-Choice Tests, Forms 1A (Honors), 1B (Average), 1C (Basic), pp. 253–258
Free-Response Tests, Forms 2A (Honors), 2B (Average), 2C (Basic), pp. 259–264
Calculator-Based Test, p. 265
Performance Assessment, p. 266
Mid-Chapter Test, p. 267
Quizzes A–D, pp. 268–269
Standardized Test Practice, p. 270
Cumulative Review, pp. 271–272

ENHANCING THE CHAPTER

Examples of some of the materials for enhancing Chapter 10 are shown below.

DIVERSITY

Multicultural Activity Masters, pp. 19, 20

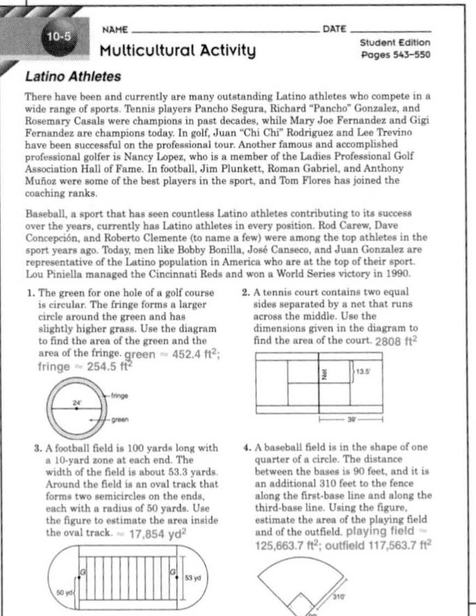

10-5 NAME ______________ DATE ______

Multicultural Activity

Student Edition Pages 543–550

Latino Athletes

There have been and currently are many outstanding Latino athletes who compete in a wide range of sports. Tennis players Pancho Segura, Richard "Pancho" Gonzalez, and Rosemary Casals were champions in past decades, while Mary Joe Fernandez and Gigi Fernandez are champions today. In golf, Juan "Chi Chi" Rodriguez and Lee Trevino have been successful on the professional tour. Another famous and accomplished professional golfer is Nancy Lopez, who is a member of the Ladies Professional Golf Association Hall of Fame. In football, Jim Plunkett, Roman Gabriel, and Anthony Muñoz were some of the best players in the sport, and Tom Flores has joined the coaching ranks.

Baseball, a sport that has seen countless Latino athletes contributing to its success over the years, currently has Latino athletes in every position. Rod Carew, Dave Concepción, and Roberto Clemente (to name a few) were among the top athletes in the sport years ago. Today, men like Bobby Bonilla, José Canseco, and Juan Gonzalez are representative of the Latino population in America who are at the top of their sport. Lou Piniella managed the Cincinnati Reds and won a World Series victory in 1990.

1. The green for one hole of a golf course is circular. The fringe forms a larger circle around the green and has slightly higher grass. Use the diagram to find the area of the green and the area of the fringe. green ≈ 452.4 ft^2; fringe ≈ 254.5 ft^2
2. A tennis court contains two equal sides separated by a net that runs across the middle. Use the dimensions given in the diagram to find the area of the court. 2808 ft^2
3. A football field is 100 yards long with a 10-yard zone at each end. The width of the field is about 53.3 yards. Around the field is an oval track that forms two semicircles on the ends, each with a radius of 50 yards. Use the figure to estimate the area inside the oval track. ≈ 17,854 yd^2
4. A baseball field is in the shape of one quarter of a circle. The distance between the bases is 90 feet, and it is an additional 310 feet to the fence along the first-base line and along the third-base line. Using the figure, estimate the area of the playing field and of the outfield. playing field ≈ 125,663.7 ft^2; outfield 117,563.7 ft^2

APPLICATIONS

Real-World Applications, 20, 21

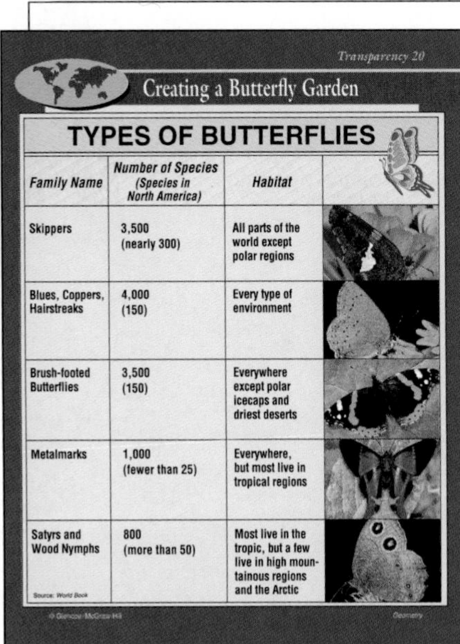

Transparency 20

Creating a Butterfly Garden

TYPES OF BUTTERFLIES

Family Name	Number of Species (Species in North America)	Habitat
Skippers	3,500 (nearly 300)	All parts of the world except polar regions
Blues, Coppers, Hairstreaks	4,000 (150)	Every type of environment
Brush-footed Butterflies	3,500 (150)	Everywhere except polar icecaps and driest deserts
Metalmarks	1,000 (fewer than 25)	Everywhere, but most live in tropical regions
Satyrs and Wood Nymphs	800 (more than 50)	Most live in the tropic, but a few live in high mountainous regions and the Arctic

Source: World Book

TECHNOLOGY

Graphing Calculator and Computer Masters, p. 10

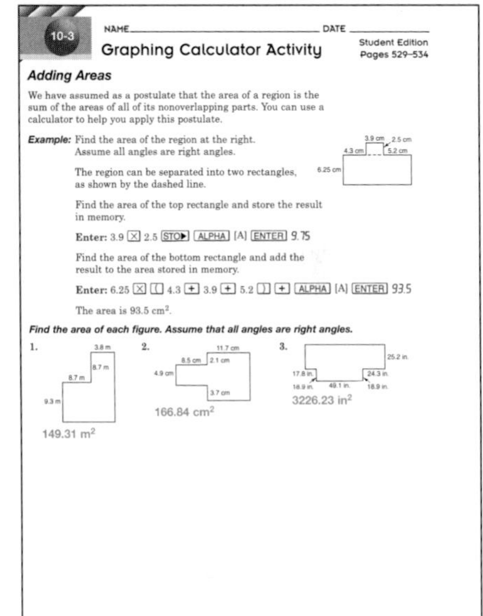

10-3 NAME ______________ DATE ______

Graphing Calculator Activity

Student Edition Pages 529–534

Adding Areas

We have assumed as a postulate that the area of a region is the sum of the areas of all of its nonoverlapping parts. You can use a calculator to help you apply this postulate.

Example: Find the area of the region at the right. Assume all angles are right angles.

The region can be separated into two rectangles, as shown by the dashed line.

Find the area of the top rectangle and store the result in memory.

Enter: 3.9 [×] 2.5 [STO▸] [ALPHA] [A] [ENTER] 9.75

Find the area of the bottom rectangle and add the result to the area stored in memory.

Enter: 6.25 [×] [(] 4.3 [+] 3.9 [+] 5.2 [)] [+] [ALPHA] [A] [ENTER] 93.5

The area is 93.5 cm^2.

Find the area of each figure. Assume that all angles are right angles.

1. 149.31 m^2
2. 166.84 cm^2
3. 3226.23 in^2

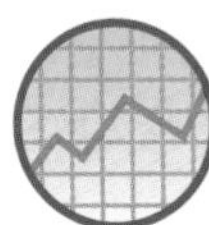

TECH PREP

Tech Prep Applications Masters, pp. 19, 20

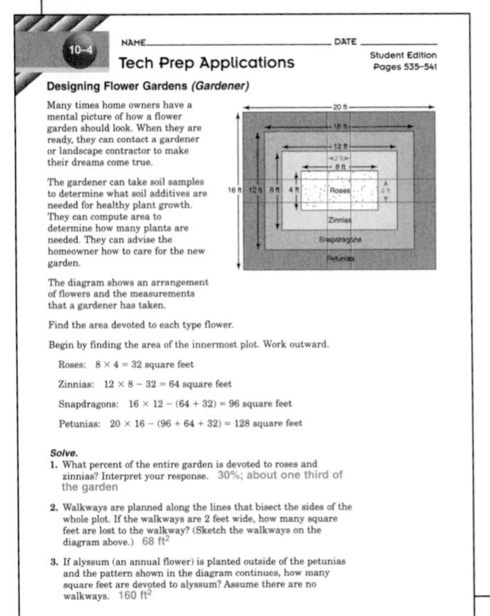

10-4 NAME ______________ DATE ______

Tech Prep Applications

Student Edition Pages 535–541

Designing Flower Gardens *(Gardener)*

Many times home owners have a mental picture of how a flower garden should look. When they are ready, they can contact a gardener or landscape contractor to make their dreams come true.

The gardener can take soil samples to determine what soil additives are needed for healthy plant growth. They can compute area to determine how many plants are needed. They can advise the homeowner how to care for the new garden.

The diagram shows an arrangement of flowers and the measurements that a gardener has taken.

Find the area devoted to each type flower.

Begin by finding the area of the innermost plot. Work outward.

Roses: $8 \times 4 = 32$ square feet

Zinnias: $12 \times 8 - 32 = 64$ square feet

Snapdragons: $16 \times 12 - (64 + 32) = 96$ square feet

Petunias: $20 \times 16 - (96 + 64 + 32) = 128$ square feet

Solve.

1. What percent of the entire garden is devoted to roses and zinnias? Interpret your response. 30%; about one third of the garden
2. Walkways are planned along the lines that bisect the sides of the whole plot. If the walkways are 2 feet wide, how many square feet are lost to the walkway? (Sketch the walkways on the diagram above.) 68 ft^2
3. If alyssum (an annual flower) is planted outside of the petunias and the pattern shown in the diagram continues, how many square feet are devoted to alyssum? Assume there are no walkways. 160 ft^2

PROBLEM SOLVING

Problem-of-the-Week Cards, 27, 28, 29

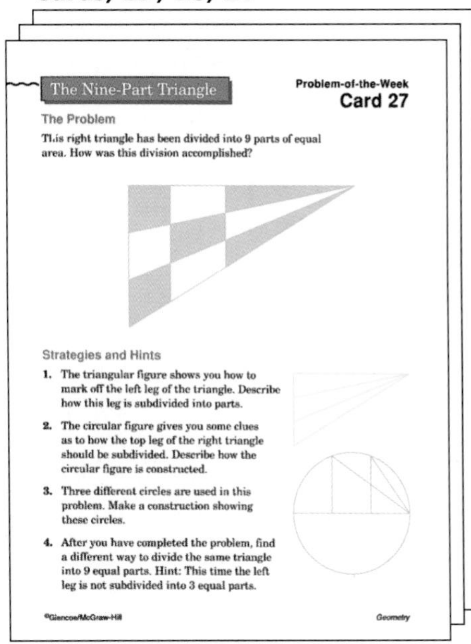

The Nine-Part Triangle

Problem-of-the-Week **Card 27**

The Problem

This right triangle has been divided into 9 parts of equal area. How was this division accomplished?

Strategies and Hints

1. The triangular figure shows you how to mark off the left leg of the triangle. Describe how this leg is subdivided into parts.
2. The circular figure gives you some clues as to how the top leg of the right triangle should be subdivided. Describe how the circular figure is constructed.
3. Three different circles are used in this problem. Make a construction showing these circles.
4. After you have completed the problem, find a different way to divide the same triangle into 9 equal parts. Hint: This time the left leg is not subdivided into 3 equal parts.

 Geometry

MAKING MATHEMATICS RELEVANT

This two-page introduction to the chapter provides students with an opportunity to explore an application of geometry that has historic roots as well as modern uses.

Background Information

Polygons in Architecture The design for the geodesic dome was patented in 1947 by American inventor R. Buckminster Fuller. Geodesic domes are usually constructed from interlocking tetrahedra or octahedra, or other polyhedra, that distribute weight evenly through the entire structure. The advantages of these domes are that they are strong, can be built with lightweight materials, and eliminate the need for any internal structural supports. Some other well-known geodesic domes include the Ford Rotunda Dome in Dearborn, Michigan; the Climatron in Saint Louis, Missouri; and the United States pavillion built for the 1967 World's Fair in Montreal, Canada.

CHAPTER 10

Exploring Polygons and Area

Objectives:

In this chapter, you will:

- find measures of interior and exterior angles of polygons,
- find areas of polygons and circles,
- solve problems involving geometric probability,
- determine characteristics of networks, and
- solve problems by using guess and check.

Polygons, Polygons
Everywhere

Geometry: Then and Now Polygons have always been used in structures. In the 1400s, the first structure that resembled a roller coaster was built in Russia. Its supporting framework was made of triangles and parallelograms, much like today's coasters. Interlocking polygons make up the structure of modern geodesic domes like Spaceship Earth at Walt Disney World's EPCOT Center. This spherical shell is 18 stories tall. How are polygons used in the designs of buildings in your community?

TIME Line

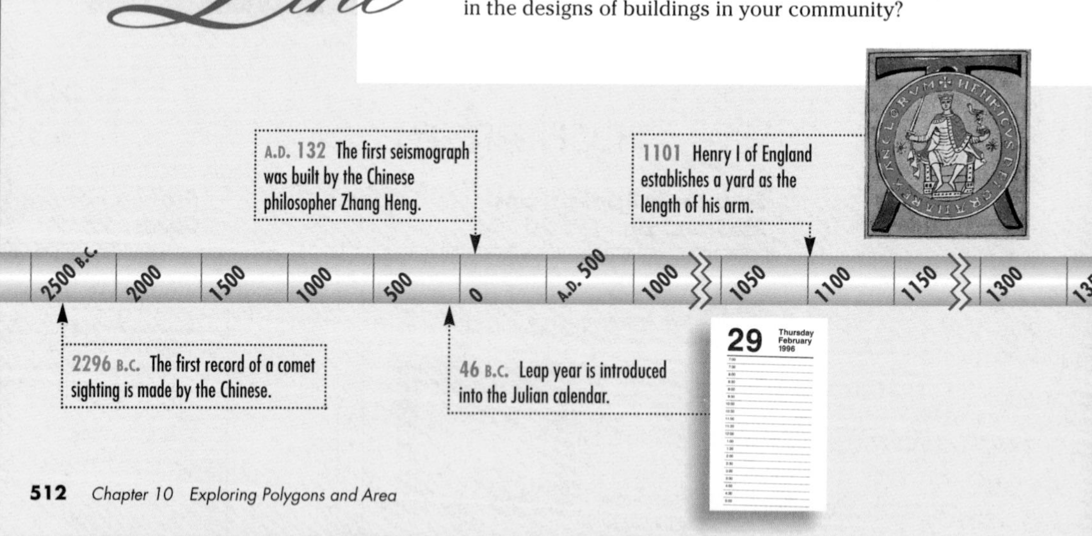
A.D. 132 The first seismograph was built by the Chinese philosopher Zhang Heng.
1101 Henry I of England establishes a yard as the length of his arm.
2500 B.C.
2000
1500
1000
500
0
A.D. 500
1000
1050
1100
1150
1300
1350
2296 B.C. The first record of a comet sighting is made by the Chinese.
46 B.C. Leap year is introduced into the Julian calendar.
29 Thursday February 1996

Students may find it interesting to research the origin of units of measure, such as inch, foot, mile, and so on.

interNET CONNECTION

Learn the history of domes and investigate recent developments in their design and construction.

World Wide Web
http://www.graphcomp.com/home/bfree/arch/domes.html

PEOPLE IN THE NEWS

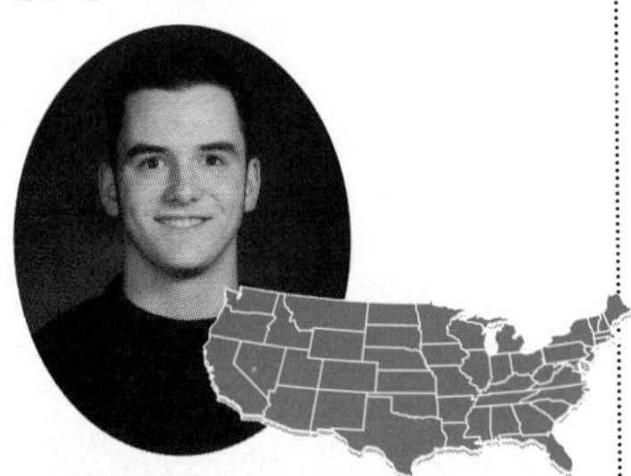

Seventeen-year-old **Ryan Morgan** of Baltimore, Maryland, has a geometry theorem named after him. As a high school freshman, Ryan investigated a theorem called (Marion) Walter's Theorem. This theorem states that if you divide the sides of a triangle into equal thirds and draw lines from each division point to the opposite vertex, the lines form a hexagon inside the triangle. Furthermore, the area of the hexagon is one tenth of the triangle's area. Ryan began experimenting with dividing the sides of triangles by other numbers. Ryan eventually found that dividing a triangle's sides by any odd number and connecting the endpoints of each middle segment to the opposite vertex also forms a hexagon. The area of the hexagon is always proportional to the area of the original triangle. Once his conjecture was proved by experts, it was officially named "Morgan's Theorem" in his honor.

Chapter Project

Work in small groups to explore Walter's Theorem and Morgan's Theorem. Morgan's Theorem states that if the sides of a triangle are divided n times, where n is an odd number greater than or equal to 3, and the endpoints of each middle segment are connected to the opposite vertex to form an interior hexagon, then the area of the triangle equals $\left(\frac{9}{8}n^2 - \frac{1}{8}\right) \times$ (area of the hexagon).

- Draw a sketch to demonstrate Walter's Theorem. Find the area of the original triangle and the area of the interior hexagon. Show that the area of the hexagon is one tenth of the triangle's area.
- Show that the area relationship given by Morgan's Theorem for $n = 3$ is equivalent to Walter's Theorem.
- Draw a sketch to demonstrate Morgan's Theorem for several values of n.
- Draw three triangles. Divide the sides of the first triangle into thirds, the second triangle into fifths, and the third triangle into sevenths.

In each triangle, connect *each division point* to the opposite vertex. How many interior hexagons are formed in each triangle? Do you see a pattern? Predict the number of interior hexagons formed in a triangle whose sides are divided into ninths. Verify your prediction with a sketch.

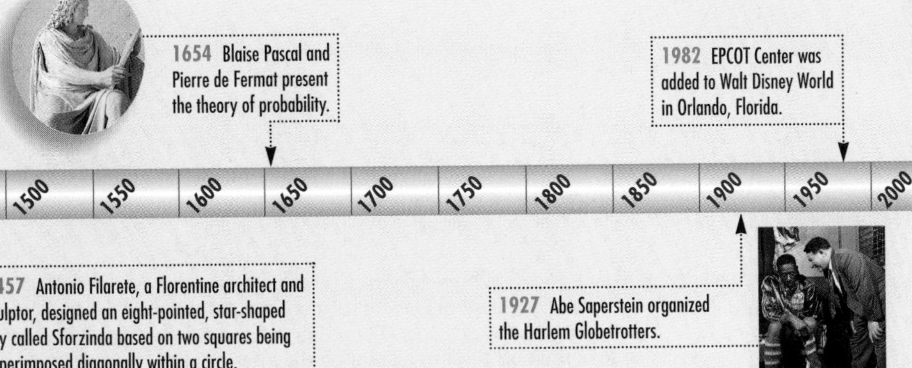
1400
1450
1500
1550
1600
1650
1700
1750
1800
1850
1900
1950
2000
1654 Blaise Pascal and Pierre de Fermat present the theory of probability.
1457 Antonio Filarete, a Florentine architect and sculptor, designed an eight-pointed, star-shaped city called Sforzinda based on two squares being superimposed diagonally within a circle.
1982 EPCOT Center was added to Walt Disney World in Orlando, Florida.
1927 Abe Saperstein organized the Harlem Globetrotters.

PEOPLE IN THE NEWS

Ryan Morgan discovered his theorem while experimenting with Walter's Theorem on the *Geometer's Sketchpad*. He tried to extend the theorem to polygons other than triangles but could not find a consistent relationship. He returned to exploring triangles. Once he found that dividing sides of a triangle into any odd number of segments also yielded an interior hexagon, he searched for a formula that related the hexagon's area to the area of the original triangle until he found one that worked. Ryan presented his conjecture to faculty and students at Towson State University in Maryland, and they later proved it.

Chapter Project

Cooperative Learning For the third and fourth parts of this project, students might prefer to work in pairs within the group. Each pair could work with a different value of n. The group should work together and consider everyone's results to make a prediction about $n = 9$.

Investigations and Projects Masters, p. 61

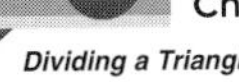

NAME__________ DATE __________

Chapter 10 Project A

Student Edition Pages 512–571

Dividing a Triangle

1. When you divide a triangle as described for Walter's Theorem and Morgan's Theorem, you also form pentagons, quadrilaterals, and triangles. Work in a small group. Draw fairly large sketches of scalene triangles with 3, 5, and 7 equal divisions on each side.
2. For each large triangle, record the number of hexagons, pentagons, quadrilaterals, and triangles formed inside. It may help to use color coding to count the polygons. For example, color all of the triangles blue, all of the quadrilaterals red, and so on, so it is easier to count them.
3. Using the recorded numbers, make conjectures about how many of each polygon will be formed with 9 divisions, 11 divisions, and n divisions on each side.
4. Think of other ways you could explore scalene triangles with divided sides. Make a list of other possible ways to divide triangles.

Alternative Chapter Projects

Two other chapter projects are included in the *Investigations and Projects Masters*. In Chapter 10 Project A, pp. 61–62, students extend the topic in the chapter opener. In Chapter 10 Project B, pp. 63–64, students design structures using shapes from this chapter.

Chapter 10 513

10-1 LESSON NOTES

NCTM Standards: 1–5, 7, 8

Instructional Resources

- Study Guide Master 10-1
- Practice Master 10-1
- Enrichment Master 10-1

Transparency 10-1A contains the 5-Minute Check for this lesson; **Transparency 10-1B** contains a teaching aid for this lesson.

Recommended Pacing	
Standard Pacing	Day 1 of 15
Honors Pacing	Day 1 of 14
Block Scheduling*	Day 1 of 8

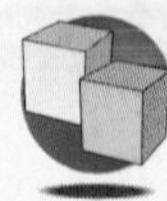

*For more information on pacing and possible lesson plans, refer to the *Block Scheduling Booklet*.

1 FOCUS

5-Minute Check
(over Chapter 9)

1. Draw a circle with center Q, diameter $\overline{AB}$, radius $\overline{QR}$, and tangent $\overleftrightarrow{ST}$. **Sample answer:**

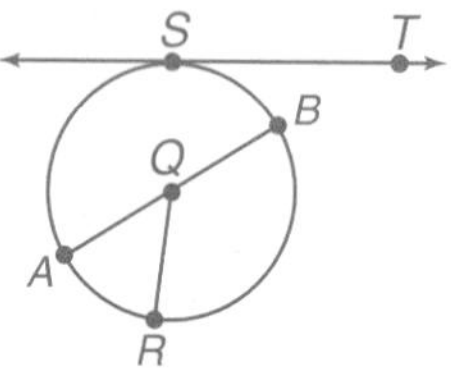

Refer to ⊙O.

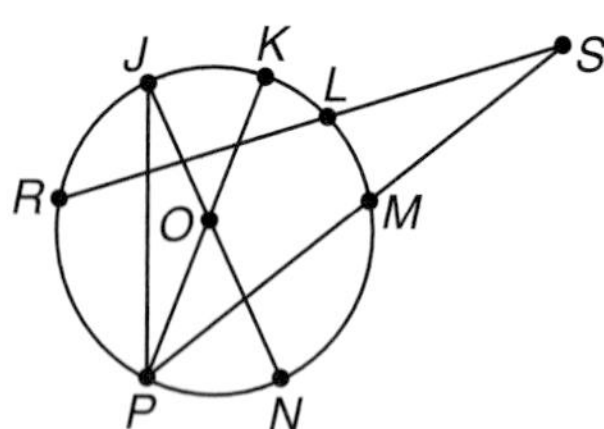

2. If $m\widehat{JK} = 53$, find $m\angle PON$. **53**
3. If $m\widehat{LM} = 18$ and $m\widehat{RP} = 69$, find $m\angle RSP$. **25.5**
4. If $SR = 18$, $SL = 4$, and $SM = 6$, find SP. **12**
5. If $m\angle NJP = 25$, find $m\angle NOP$. **50**

10-1 Polygons

What YOU'LL LEARN

- To identify and name polygons,
- to find the sum of the measures of interior and exterior angles of convex polygons and measures of interior and exterior angles of regular polygons, and
- to solve problems involving angle measures of polygons.

Why IT'S IMPORTANT

You can use polygons to solve problems involving architecture, landscaping, and recreation.

APPLICATION
Architecture

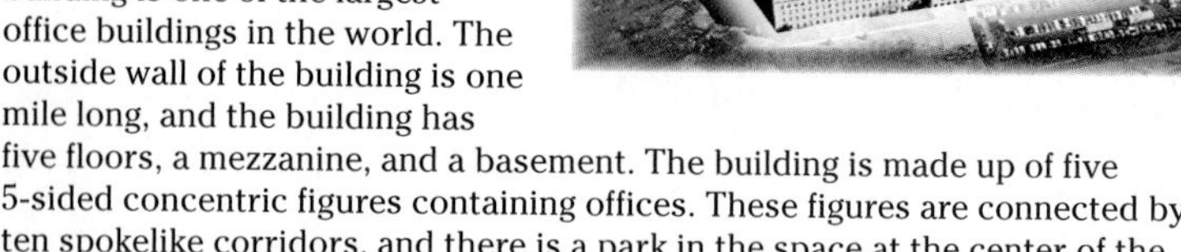

Completed in 1943, the Pentagon building is one of the largest office buildings in the world. The outside wall of the building is one mile long, and the building has five floors, a mezzanine, and a basement. The building is made up of five 5-sided concentric figures containing offices. These figures are connected by ten spokelike corridors, and there is a park in the space at the center of the innermost 5-sided figure. Each figure in the Pentagon building is a **polygon**.

The term *polygon* is derived from the Greek word meaning "many-angled." Look at the figures below. The figures on the left are polygons, and the figures on the right are not.

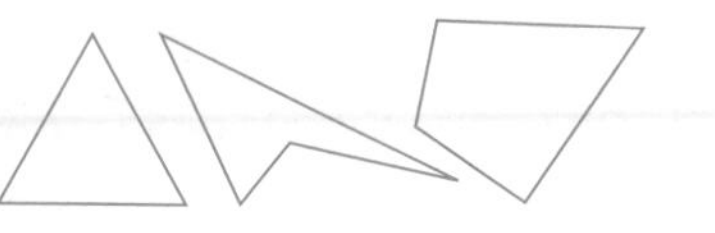

polygons

not polygons

You can determine the properties of polygons using these figures.

Definition of Polygon	**A polygon is a closed figure formed by a finite number of coplanar segments such that** **1. the sides that have a common endpoint are noncollinear, and** **2. each side intersects exactly two other sides, but only at their endpoints.**

A **convex** polygon is a polygon such that no line containing a side of the polygon contains a point in the interior of the polygon. A polygon that is not convex is *nonconvex* or **concave**.

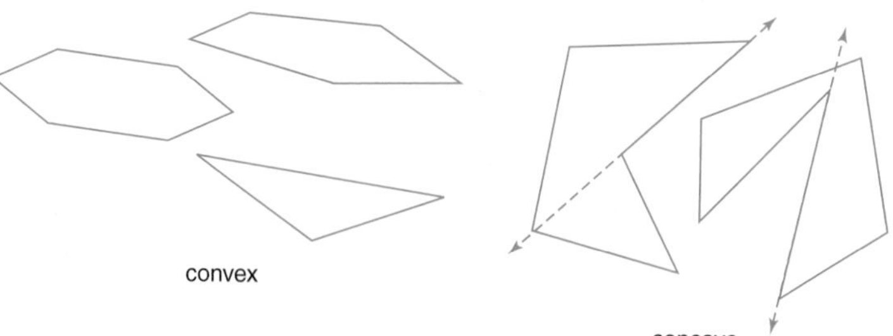

convex

concave

If a line segment in each of the concave polygons is extended, then the polygon fails to satisfy the definition for convex because a line containing a side of the polygon has a point on the interior of the polygon.

Polygons may be classified by the number of sides they have. The chart at the right gives some common names for polygons. In general, a polygon with n sides is called an ***n*-gon**. This means the *nonagon* can also be called a 9-gon.

Number of Sides	Polygon
3	triangle
4	quadrilateral
5	pentagon
6	hexagon
7	heptagon
8	octagon
9	nonagon
10	decagon
12	dodecagon
n	n-gon

When referring to a polygon, we use its name and list the vertices in consecutive order. Pentagon *RSTUV* and pentagon *TUVRS* are two possible correct names for the polygon at the right. In pentagon *RSTUV*, all the sides are congruent, and all the angles are congruent. When a polygon has these characteristics, it is called a **regular polygon**.

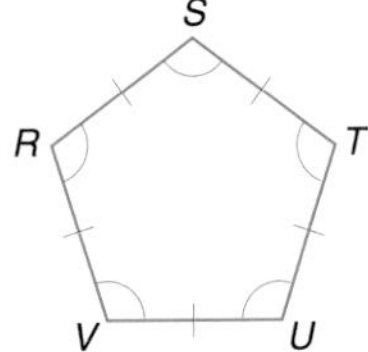

Definition of Regular Polygon	A regular polygon is a convex polygon with all sides congruent and all angles congruent.

An equilateral triangle is a regular polygon.

Example  **Name and classify each polygon**

a. by the number of sides,

b. as convex or concave, and

c. as regular or not regular.

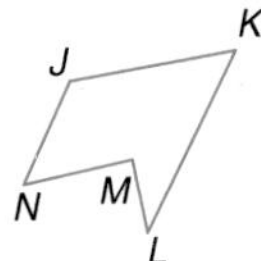

a. Polygon *JKLMN* has five sides. It is a pentagon.

b. If $\overline{ML}$ is extended, it passes through the interior of the pentagon. So, the pentagon is concave.

c. Since it is concave, pentagon *JKLMN* cannot be regular.

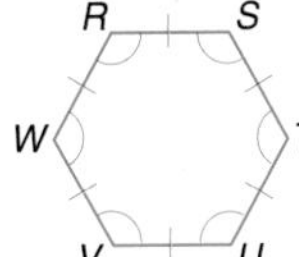

a. Polygon *RSTUVW* has six sides, so it is a hexagon.

b. No lines containing sides of the hexagon pass through the interior. Therefore, the polygon is convex.

c. Since all sides and all angles are congruent, hexagon *RSTUVW* is regular.

Consider each convex polygon below with all possible diagonals drawn from one vertex.

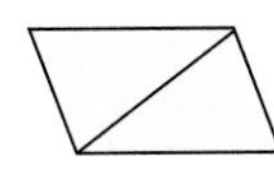
quadrilateral

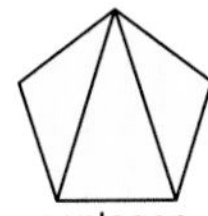
pentagon

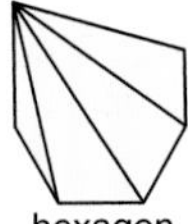
hexagon

heptagon

octagon

Motivating the Lesson

Situational Problem Have students make a list of naturally occurring polygons and of man-made polygons.

2 TEACH

In-Class Example

For Example 1
Classify each polygon

a. by the number of sides,
b. as convex or concave, and
c. as regular or not regular.

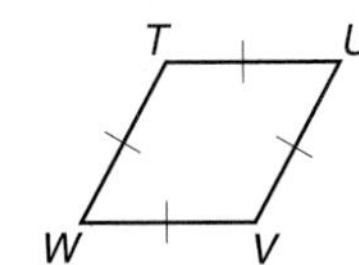

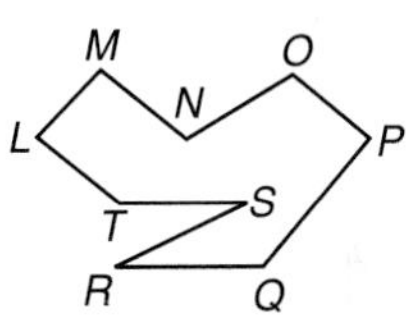

Polygon *LMNOPQRST*: nonagon; concave; not regular since its angles and sides are not congruent. Polygon *TUVW*: quadrilateral; convex; not regular

Teaching Tip After reading the definitions of convex and concave, ask which terms apply to a triangle and to a star. **triangle, convex; star, concave**

Teaching Tip In naming polygons, point out that the consecutive vertices can be listed in clockwise or counterclockwise order.

Alternative Teaching Strategies

Reading Geometry Ask students to look up the origins of these words and to find other words related to them: quadrilateral **Latin, four-sided—quadraphonic, quadruped;** geometry **Greek, earth measure—geography, geology;** perimeter **Greek, measure around—periscope, periphery;** bisector **Latin, cut in two—section, bicycle**

In-Class Example

For Example 2
Find the measure of each interior angle of a regular hexagon. **120**

Teaching Tip In drawing diagonals within polygons, point out that it does not matter which vertex is chosen.

LOOK BACK

Refer to Lesson 4-2 to review the Angle Sum Theorem.

Notice that in each case, the polygon is separated into triangles. The sum of the measures of the angles of each polygon can be found by adding the measures of the angles of the triangles. Since the sum of the measures of the angles in a triangle is 180, we can easily find this sum. Make a chart to find the sum of the angle measures for several convex polygons.

TECHNOLOGY Tip

A spreadsheet can be used to calculate each sum. If cell B2 contains the number of sides, cell B3 can contain the formula, B2 − 2, and the formula for B4 is 180 * B3.

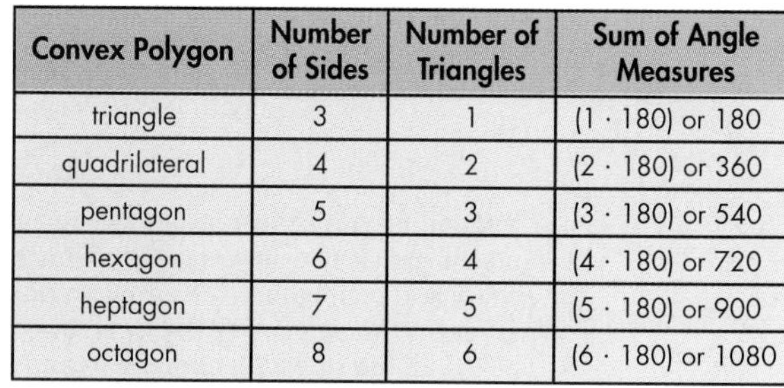

Convex Polygon	Number of Sides	Number of Triangles	Sum of Angle Measures
triangle	3	1	$(1 \cdot 180)$ or 180
quadrilateral	4	2	$(2 \cdot 180)$ or 360
pentagon	5	3	$(3 \cdot 180)$ or 540
hexagon	6	4	$(4 \cdot 180)$ or 720
heptagon	7	5	$(5 \cdot 180)$ or 900
octagon	8	6	$(6 \cdot 180)$ or 1080

Look for a pattern in the angle measures. In each case, the sum of the angle measures is 2 less than the number of sides in the polygon times 180. So in an n-gon, the sum of the angle measures will be $(n - 2)180$ or $180(n - 2)$. This conclusion, based on inductive reasoning, is stated formally in Theorem 10–1.

Theorem 10–1 Interior Angle Sum Theorem

If a convex polygon has *n* sides and *S* is the sum of the measures of its interior angles, then $S = 180(n - 2)$.

The Pentagon building, described in the application at the beginning of the lesson, has congruent sides and angles. Do you think that the angles formed in each of the five concentric pentagon-shaped buildings have the same measure, or are their measures different because the pentagons are of different sizes? Use Example 2 below to help you think about your answer. *You will be asked to write about this in Exercise 4.*

Example 2

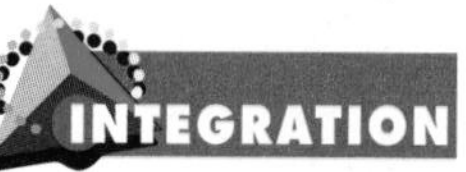

Algebra

Find the measure of each interior angle of the largest pentagon-shaped section of the Pentagon building.

Use the Interior Angle Sum Theorem to find the sum of the angle measures in a convex pentagon.

$S = 180(n - 2)$

$S = 180(5 - 2)$ *For a pentagon, $n = 5$.*

$S = 180(3)$ or 540

Because the outer pentagon of the Pentagon building is a regular pentagon, each of the interior angles will be congruent. Therefore, the measure of each angle is $\frac{540}{5}$ or 108.

Alternative Learning Styles

Kinesthetic Have students choose a polyhedron to make using cardboard and tape. They may design the piece beforehand or cut out several polygons and fit them together any way they can. Each student should prepare a sheet that tells the name of the polyhedron or describes it and gives the number of faces, edges, and vertices. Display students' work.

Example 3 INTEGRATION **Algebra**

Find the measure of each interior angle in quadrilateral *RSTU* if the measure of each consecutive angle is a consecutive multiple of x.

Draw a quadrilateral *RSTU* and label the measure of each angle. Let $m\angle R = x$, $m\angle S = 2x$, $m\angle T = 3x$, and $m\angle U = 4x$.

Since $n = 4$, the sum of the measures of the interior angles is $180(4 - 2)$ or 360. Write an equation to express the sum of the measures of the interior angles of the quadrilateral.

$$360 = m\angle R + m\angle S + m\angle T + m\angle U$$

$$360 = x + 2x + 3x + 4x \quad \textit{Substitution Property (=)}$$

$$360 = 10x \quad \textit{Combine like terms.}$$

$$36 = x \quad \textit{Divide each side by 10.}$$

Use the value of x to find the measure of each angle.

$m\angle R = 36$, $m\angle S = 2 \cdot 36$ or 72, $m\angle T = 3 \cdot 36$ or 108, and $m\angle U = 4 \cdot 36$ or 144.

The Interior Angle Sum Theorem identifies a relationship among the interior angles of a convex polygon. Is there a relationship among the exterior angles of a convex polygon?

MODELING MATHEMATICS: Sum of the Exterior Angles of a Polygon

Materials:  straightedge 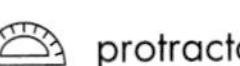 protractor

- Draw a triangle, a convex quadrilateral, a convex pentagon, a convex hexagon, and a convex heptagon.
- Extend the sides of each polygon to form exactly one exterior angle at each vertex.
- Use a protractor to measure each exterior angle of each polygon and record it on your drawing.

Your Turn a. See students' work.

a. Copy the table below and find the sum of the exterior angles for each of your polygons.

b. What conjecture can you make? **The sum of the measures of the exterior angles of any convex polygon is 360.**

Polygon	triangle	quadrilateral	pentagon	hexagon	heptagon
number of exterior angles					
sum of measures of exterior angles					

From the Modeling Mathematics activity, you should have discovered that the sum of the measures of the exterior angles of any convex polygon is 360. This is known as the Exterior Angle Sum Theorem.

Theorem 10-2 Exterior Angle Sum Theorem **If a polygon is convex, then the sum of the measures of the exterior angles, one at each vertex, is 360.**

You will be asked to prove this theorem in Exercise 53.

In-Class Example

For Example 3

Six angles of a convex octagon are congruent. Each of the other two angles has a measure 20 more than the measure of each of the other six angles. Find the measure of each angle. **Six angles measure 130° each, and the other two measure 150° each.**

Teaching Tip In Examples 2 and 3, point out that algebra is used to solve a problem in geometry.

Teaching Tip After reading Theorem 10-2, stress that it applies only to convex polygons.

MODELING MATHEMATICS Even though measurements will not be exact, students still get results close enough to 360° to be able to form a conjecture.

Classroom Vignette

"I have students build models of polyhedra using spaghetti joined with marshmallows or plastic straws joined together with paper clips. This project works well with cooperative groups."

Ingrid Hart
E.O. Smith High School
Storrs, Connecticut

In-Class Example

For Example 4
Use the Exterior Angle Sum Theorem to find the measure of an interior angle and an exterior angle of a regular hexagon.
exterior angle measure = 60
interior angle measure = 120

3 PRACTICE/APPLY

Check for Understanding

Exercises 1–18 are designed to help you assess your students' understanding through reading, writing, speaking, and modeling. You should work through Exercises 1–6 with your students and then monitor their work on Exercises 7–18.

Error Analysis

If students have trouble determining which figures are concave, encourage them to lay a straightedge along each side of the figure. If the straightedge passes through the figure, the figure is concave.

Additional Answers

3. Sample answer:

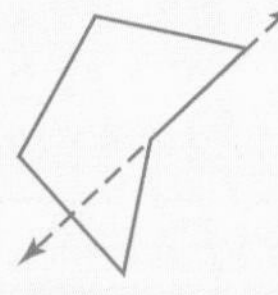

The pentagon is concave because when a line segment is extended, it contains a point in the interior of the pentagon.

4. The Interior Angle Sum Theorem is for a convex polygon, and it is only dependent on the number of sides the polygon has. Therefore, the size of the polygon makes no difference, and similar polygons all have the same interior angle sum. Furthermore, the pentagons are regular, and therefore each interior angle is congruent.
5. Subtract the measure of the known interior angle from 180. Divide 360 by that result. This answer is the number of sides in the polygon.

Example  **Use the Exterior Angle Sum Theorem to find the measure of each interior angle and exterior angle of a regular octagon.**

A regular octagon has eight congruent interior angles. Because each exterior angle is supplementary to an interior angle and all interior angles are congruent, all the exterior angles are also congruent. So the measure of each exterior angle y is $\frac{360}{8}$ or 45.

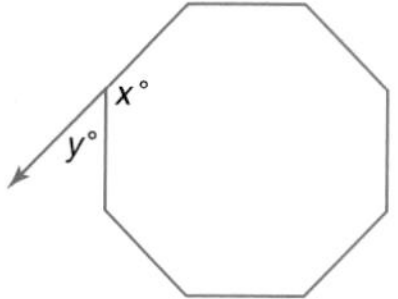

The measure of each interior angle x is $180 - 45$ or 135.

CHECK FOR UNDERSTANDING

Communicating Mathematics

Study the lesson. Then complete the following.

1. Determine if each figure below is a polygon. If the figure is not a polygon, explain why it is not. b. No; sides overlap.

a.  yes

b. 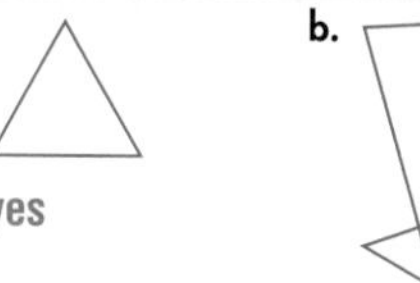

c. 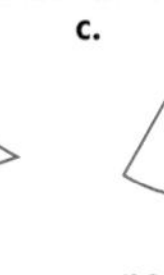 no; curved side

d. 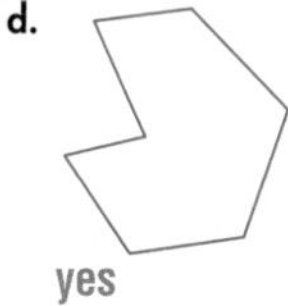 yes

2. Juanita is correct because $\frac{360}{15} = 24$, which is the exterior angle measure.

FYI
The Pentagon building has one of the world's largest private telephone systems. It handles more than 200,000 calls per day and utilizes about 100,000 miles of cable.

2. **You Decide** Mariana thinks that the measure of an exterior angle of a regular 15-gon is 156. Juanita says that the measure is 24. Who is correct and why?
3. **Draw** a concave pentagon and explain why it is concave. See margin.
4. **Describe** the relationship between the angles in each of the five concentric pentagon-shaped buildings in the Pentagon building described at the beginning of the lesson. See margin.
5. **Explain** how you would find the number of sides in a regular polygon if you know the measure of one of its interior angles. See margin.

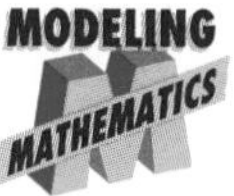

6. Demonstrate the Exterior Angle Sum Theorem with exterior angles of any polygon. b. The angles should form a circle or equal 360.
   a. Draw a polygon on a sheet of paper. Extend the sides of the polygon to form one exterior angle at each vertex. Label the angles by number and cut them out. See students' work.
   b. Arrange the angles so that all the vertices are at the same point and no angles share interior points. What do you notice?
   c. Will a different polygon have the same angle sum? yes

Guided Practice

Refer to polygon *ABCDEF* for Exercises 7 and 8.

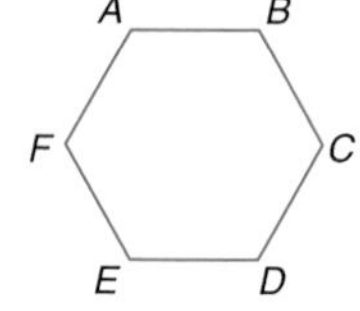

7. Name the sides of the polygon.
8. Is the polygon concave or convex? Explain. Convex; lines containing each side intersect only at the endpoints.

7. $\overline{AB}$, $\overline{BC}$, $\overline{CD}$, $\overline{DE}$, $\overline{EF}$, $\overline{FA}$

FYI
The Pentagon building houses the headquarters for the Army, Navy, Air Force, and the Department of Defense.

Reteaching

Using Drawing Ask students to draw each of the following:
1. five different polygons
2. a figure that is not a polygon
3. a polygon that is concave
4. a polyhedron with six faces
5. a polyhedron with six congruent faces

14. $\frac{180(a-2)}{a}, \frac{360}{a}$

17. $x = 34, 3x = 102, 2x - 1 = 67, 6x - 5 = 199, 4x + 2 = 138$

Find the sum of the measures of the interior angles of each convex polygon

9. decagon 1440
10. 21-gon 3420

The measure of an exterior angle of a regular polygon is given. Find the number of sides of the polygon.

11. 30 12
12. 8 45

The number of sides of a regular polygon is given. Find the measures of an interior angle and an exterior angle of the polygon.

13. 8 135, 45
14. a

The measure of an interior angle of a regular polygon is given. Find the number of sides in each polygon.

15. 120 6
16. 150 12
17. The measures of the interior angles of a pentagon are x, $3x$, $2x - 1$, $6x - 5$, and $4x + 2$. Find the measure of each angle.
18. **Landscaping** Anita Cruz is placing a large pentagon-shaped gazebo in her yard. The flower beds around the gazebo will be rectangular. What is the measure of the angle between two of the flower beds? 72

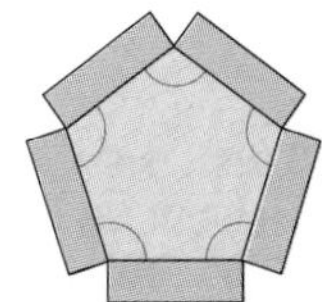

EXERCISES

Practice

Refer to polygon *ABCDEF* for Exercises 19–21.

19. Name the vertices of the polygon. *A, B, C, D, E, F*
20. Is the polygon regular? no
21. Name the polygon in two other ways. Sample answers: *BCDEFA* and *FABCDE*

Find the sum of the measures of the interior angles of each convex polygon.

22. 11-gon 1620
23. 26-gon 4320
24. 90-gon 15,840
25. 46-gon 7920
26. x-gon $180(x - 2)$
27. $3m$-gon $180(3m - 2)$

The measure of an exterior angle of a regular polygon is given. Find the number of sides of the polygon.

28. 72 5
29. 45 8
30. 18 20
31. 14.4 25
32. 20 18
33. x $\frac{360}{x}$

The number of sides of a regular polygon is given. Find the measures of an interior angle and an exterior angle for each polygon. Round to the nearest hundredth.

38. $\frac{60(3m-2)}{m}, \frac{120}{m}$

39. $\frac{180(x+2y-2)}{x+2y}, \frac{360}{x+2y}$

34. 30 168, 12
35. 16 157.5, 22.5
36. 22 163.64, 16.36
37. 14 154.29, 25.71
38. $3m$
39. $x + 2y$

The measure of an interior angle of a regular polygon is given. Find the number of sides in each polygon.

40. 135 8
41. 144 10
42. 157.5 16
43. 176.4 100
44. 165.6 25
45. $\frac{180(s-2)}{s}$ s

Assignment Guide

Core (with proof): 19–55 odd, 57–67
Core (informal): 19–55 odd, 57–67
Enriched: 20–52 even, 53–67

For **Extra Practice**, see p. 782.

The red A, B, and C flags, printed only in the Teacher's Wraparound Edition, indicate the level of difficulty of the exercises.

Study Guide Masters, p. 57

NAME________ DATE________

Study Guide Student Edition Pages 514–521

Polygons

A **polygon** is a plane figure formed by a finite number of segments such that (1) sides that have a common endpoint are noncollinear and (2) each side intersects exactly two other sides, but only at their endpoints. A **convex polygon** is a polygon such that no line containing a side of the polygon contains a point in the interior of the polygon. Convex polygons with all sides congruent and all angles congruent are called **regular**.

The following two theorems involve the interior and exterior angles of a convex polygon.

Interior Angle Sum Theorem	If a convex polygon has n sides and S is the sum of the measures of its interior angles, then $S = 180(n - 2)$.
Exterior Angle Sum Theorem	If a polygon is convex, then the sum of the measures of the exterior angles, one at each vertex, is 360.

Example: Find the sum of the measures of the interior angles of a convex polygon with 13 sides.

$S = 180(n - 2)$ **Interior Angle Sum Theorem**
$S = 180(13 - 2)$
$S = 180(11)$
$S = 1980$

Find the sum of the measures of the interior angles of each convex polygon.

1. 10-gon 1440
2. 16-gon 2520
3. 30-gon 5040

The measure of an exterior angle of a regular polygon is given. Find the number of sides of the polygon.

4. 30 12
5. 20 18
6. 5 72

The number of sides of a regular polygon is given. Find the measures of an interior angle and an exterior angle for each polygon.

7. 18 160; 20
8. 36 170; 10
9. 25 165.6; 14.4

10. The measure of the interior angle of a regular polygon is 157.5. Find the number of sides of the polygon. 16

Additional Answers

53. Consider the sum of the measures of the exterior angles for an n-gon.
sum of measures of exterior angles
N = sum of measures of linear pairs − sum of measures of interior angles
$= n \cdot 180 - 180(n - 2)$
$= 180n - 180n + 360$
$= 360$
So, the sum of the exterior angle measures is 360 for *any* convex polygon.

54. x, z, y; the sum of the exterior angles is 360, therefore, the larger the number of sides of a polygon, the smaller each exterior angle measure will be because there will be more angles to share the 360 degrees.

Find the measure of each angle.

47. $m\angle R = 125$, $m\angle S = 100$, $m\angle T = 100$, $m\angle U = 95$, $m\angle V = 120$

46. 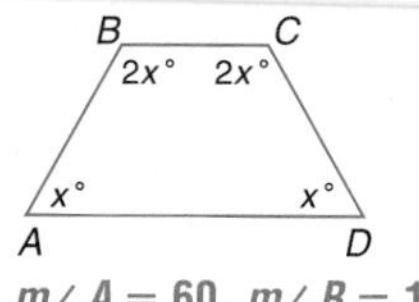

$m\angle A = 60$, $m\angle B = 120$, $m\angle C = 120$, $m\angle D = 60$

47. 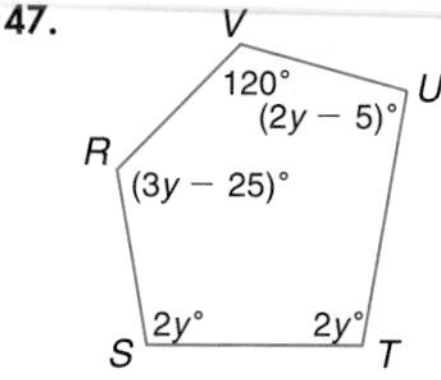

48. If the exterior angle of a regular polygon measures 36°, find the sum of the measures of the interior angles. 1440

49. If the exterior angle of a regular polygon measures 60°, find the sum of the measures of the interior angles. 720

50. Use the star at the right to find each of the following.
   a. the sum of the measures of the numbered angles in the exterior of the star 540
   b. the sum of the angles at each point of the star 180

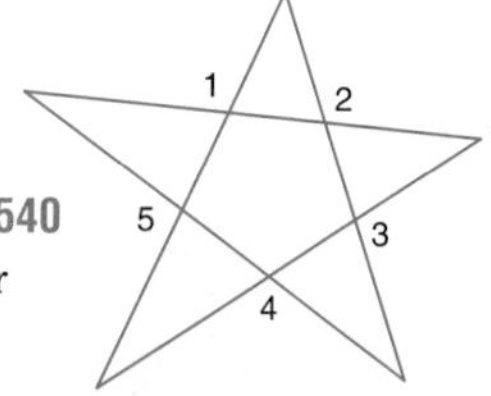

51. nonagon

51. If the measure of each of the interior angles of a regular polygon is 100 more than the measure of each of the exterior angles, name the polygon.

52. Each of the measures of the interior and exterior angles of a regular polygon are in a ratio of 5:1. Find the measures of an interior and an exterior angle and the name of the polygon. 150, 30, 12-gon

Critical Thinking

53. Use algebra to show that if an n-gon is convex, then the sum of the measures of the exterior angles, one at each vertex, is 360 (Exterior Angle Sum Theorem). See margin.

54. Let x, y, and z represent the measures of an exterior angle of a regular 20-gon, a regular decagon, and a regular dodecagon, respectively. Without calculating exact measures, order x, y, and z from least to greatest. Explain your reasoning. See margin.

Applications and Problem Solving

55. **Recreation** The soccer ball design at the right is made up of two types of regular polygons. Pentagons are surrounded and interlocked by regular hexagons. Could these polygons be interlocked in a plane? Explain. No; the sum of the measures of the angles at any vertex is less than 360, so the polygons would not interlock in a plane.

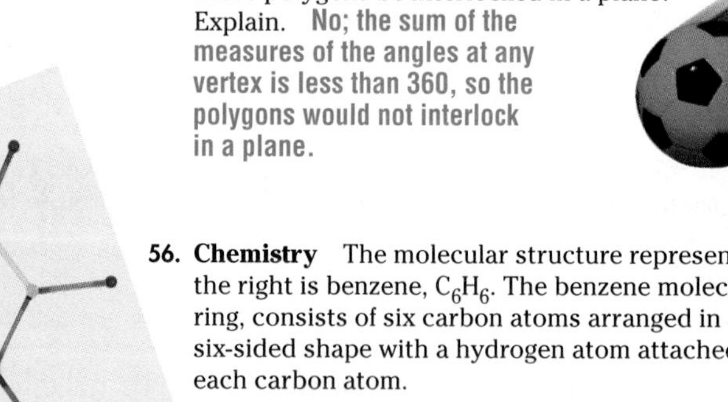

56. **Chemistry** The molecular structure represented at the right is benzene, C_6H_6. The benzene molecule, or ring, consists of six carbon atoms arranged in a flat, six-sided shape with a hydrogen atom attached to each carbon atom.
   a. What shape does the figure represent? hexagon
   b. Find the measure of each interior angle. 120
   c. Label all the measures of the exterior angles that are formed between the carbon and hydrogen atoms in the given representation.

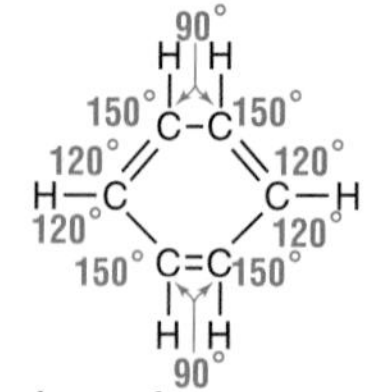

Practice Masters, p. 57

NAME ________________ DATE ________

Practice — Student Edition Pages 514–521

Polygons

State the number of sides for each convex polygon.

1. quadrilateral 4 2. octagon 8 3. 83-gon 83
4. heptagon 7 5. decagon 10 6. hexagon 6

Use polygon ABCDEFG to answer each question.

7. Name the vertices of the polygon. A, B, C, D, E, F, G
8. Name the angles of the polygon. ∠A, ∠B, ∠C, ∠D, ∠E, ∠F, ∠G
9. Name the sides of the polygon. $\overline{AB}$, $\overline{BC}$, $\overline{CD}$, $\overline{DE}$, $\overline{EF}$, $\overline{FG}$, $\overline{GA}$

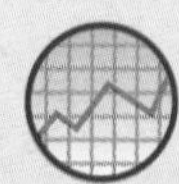

10. Is the polygon convex or concave. convex
11. Name the polygon according to the number of sides it has. heptagon
12. Is the polygon regular? Explain. Yes; its sides are all congruent, and its angles are all congruent, and it is convex.

Find the sum of the measures of the interior angles of each convex polygon.

13. heptagon 900 14. octagon 1080 15. 13-gon 1980

The number of sides of a regular polygon is given. Find the measure of an interior and an exterior angle of the polygon.

16. 5 105, 72 17. 9 140, 40 18. 10 144, 36

Tech Prep

Chemical Engineering Technician
Chemical engineering technicians apply chemical engineering principles to assist chemical engineers. They prepare charts, sketches, and diagrams, and record engineering data. Most two-year colleges offer programs in this area. A strong mathematics background is needed. For more information on tech prep, see the *Teacher's Handbook*.

Mixed Review

57. See margin.

57. Sketch the graph of the circle whose equation is $(x + 5)^2 + (y + 2)^2 = 16$. Label the center C and the radius measure on the graph. (Lesson 9–8)

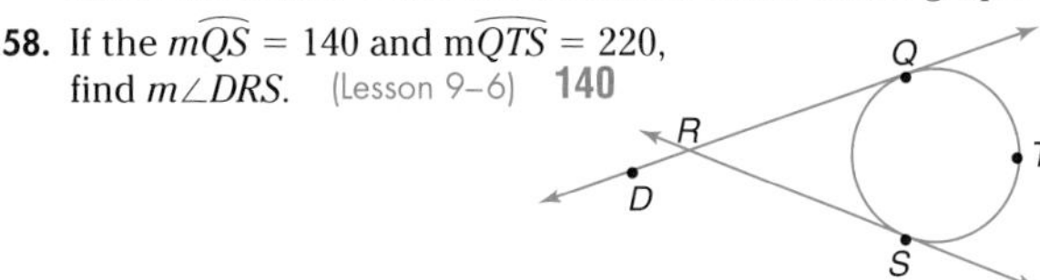

58. If the $m\widehat{QS} = 140$ and $m\widehat{QTS} = 220$, find $m\angle DRS$. (Lesson 9–6) **140**

59. **Cycling** A bicycle tire has a diameter of 16 inches. How far does the bike travel during one rotation of the wheel? (Lesson 9–1) **$16\pi \approx 50.27$ in.**

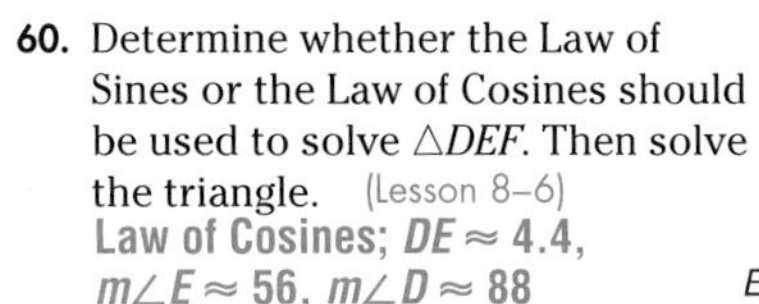

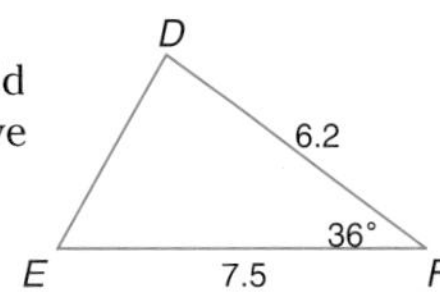

60. Determine whether the Law of Sines or the Law of Cosines should be used to solve $\triangle DEF$. Then solve the triangle. (Lesson 8–6) **Law of Cosines; $DE \approx 4.4$, $m\angle E \approx 56$, $m\angle D \approx 88$**

61. The measures of the legs of a right triangle are 4.0 and 5.6 centimeters. Find the measure of the hypotenuse. Round to the nearest tenth. (Lesson 8–1) **6.9 cm**

62. Use a compass and straightedge to construct a rhombus with sides equal to 8 centimeters. (Lesson 6–4) **See students' work.**

63. List the sides of $\triangle STV$ from largest to smallest if $m\angle S = 2x + 7$, $m\angle T = 5x - 4$, and $m\angle V = 4x + 12$. (Lesson 5–4) **ST, SV, TV**

64. Classify the statement *An isosceles triangle has three congruent sides* as *always*, *sometimes*, or *never* true. (Lesson 4–1) **sometimes**

65. What property justifies the statement $m\angle K = m\angle K$? (Lesson 2–4) **Reflexive Property (=)**

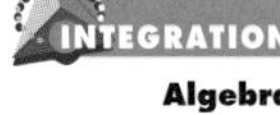

Algebra

66. Simplify $12\sqrt{15} + 7\sqrt{15}$. **$19\sqrt{15}$**

67. Solve $4^y = 4^{2y-3}$. **3**

WORKING ON THE

In•ves•ti•ga•tion

Refer to the Investigation on pages 510–511.

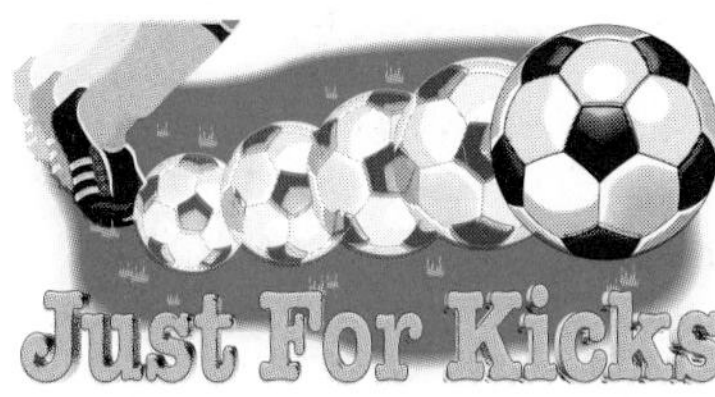

Study the patterns on the soccer balls shown on pages 510 and 511. Assume that all the polygons are regular.

1 Investigate the sum of the measures of the different combinations of angles that meet at a common vertex in each pattern.

2 Do the patterns fill the plane totally?

3 Explain your findings. How does this affect the design your team will propose for your company's soccer ball?

Add the results of your work to your Investigation Folder.

4 ASSESS

Closing Activity

Modeling Using pipe cleaners or straws, have students construct a polygon that is not regular and not convex.

Additional Answer

57.

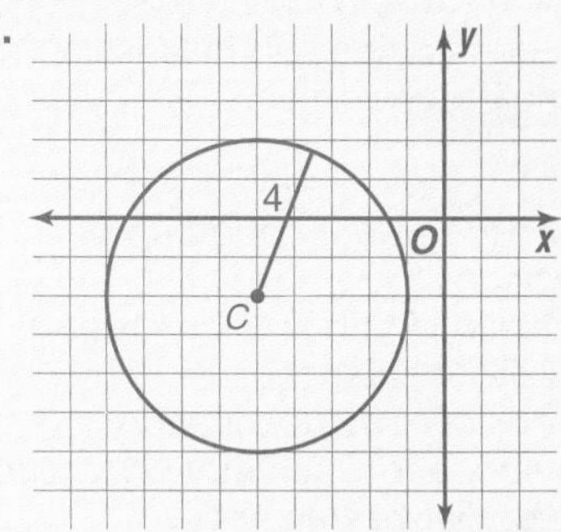

Extension

Communication Have students form small discussion groups to determine the number of diagonals they can draw in a 20-gon.

Working on the Investigation

The Investigation on pages 510–511 is designed to be a long-term project that is completed over several days or weeks. Encourage students to keep their materials in their Investigation Folder as they work on the Investigation.

Enrichment Masters, p. 57

NAME__________ DATE ______

Enrichment

Student Edition Pages 514–521

Constructing a Decagon

Consider a decagon inscribed in a circle with radius 1 unit, as shown below. Let x represent the length of each side. Since a decagon has 10 sides, each side is the base of an isosceles triangle that has a vertex angle of $\frac{360°}{10}$ or 36°. Each base angle measures $\frac{1}{2}(180° - 36°)$ or 72°. If one base angle is bisected, two isosceles triangles are formed as shown below at the right. Notice the relationships of the sides and angles.

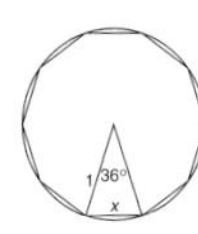

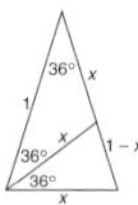

The value of x can be found by solving the proportion $\frac{x}{1-x} = \frac{1}{x}$. The solution is $x = \frac{\sqrt{5}-1}{2}$ units. You can construct a decagon by first constructing a segment of this length, then marking off arcs around a circle of radius 1 unit.

Follow these instructions to construct a decagon. Use separate paper.

See students' work.

1. Construct circle O of radius 1 unit, tangent to a line at point A.
2. Mark off 2 units from A to locate point P on the line.
3. Draw segment OP and label a point Q where the segment intersects the circle. By the Pythagorean Theorem, $OP = \sqrt{5}$. Also, $PQ = \sqrt{5} - 1$.
4. Bisect segment PQ and label the midpoint as R. Segment PR has length $\frac{\sqrt{5}-1}{2}$.

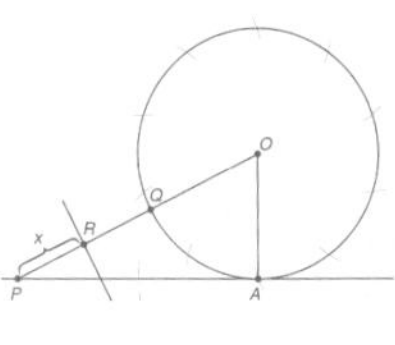

5. Set the compass to length PR. Starting at A, mark 10 points around the circle. Connect the 10 points to form a decagon.

10-2A LESSON NOTES

NCTM Standards: 1–3, 7, 8

Objective
Create tessellations using translations and rotations.

Recommended Time
Demonstration and discussion: 15 minutes; Exercises: 30 minutes

Instructional Resources
For each student or group of students
Modeling Mathematics Masters
• p. 97 (worksheet)
For teacher demonstration
Algebra and Geometry Overhead Manipulative Resources

1 FOCUS

Motivating the Lesson
Start this activity using simple shapes such as diamonds or rectangles. Show students the translations, reflections, and rotations with these simple shapes. As they work through the activity, have the figures get more complex.

2 TEACH

Teaching Tip Review the difference between reflections, translations, and rotations, and how these affect the tessellation.

3 PRACTICE/APPLY

Assignment Guide
Core (with proof): 1–5 **Core (informal):** 1–5 **Enriched:** 1–5

4 ASSESS

Observing students working in cooperative groups is an excellent method of assessment.

A Preview of Lesson 10–2

10–2A Tessellations and Transformations

Materials: plain paper  pencil

A **tessellation** is the result of covering a plane surface with a repeated pattern of figures without any spaces between the figures. Tessellations can be formed by transformations such as translations (slides), reflections (flips), and rotations (turns).

Activity 1 Make a tessellation using a translation.

Step 1 Start by drawing a square. Then draw a triangle inside the top and a curved figure with a dot inside the left of the square as shown.

Step 2 Translate the triangle on the top side to the bottom side.

Step 3 Translate the figure on the left side to the right side to complete the pattern unit.

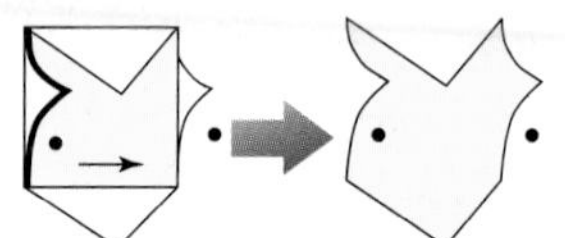

Step 4 Repeat this pattern unit on a tessellation of squares. *It is sometimes helpful to complete one pattern unit, cut it out, and trace it for the other units.*

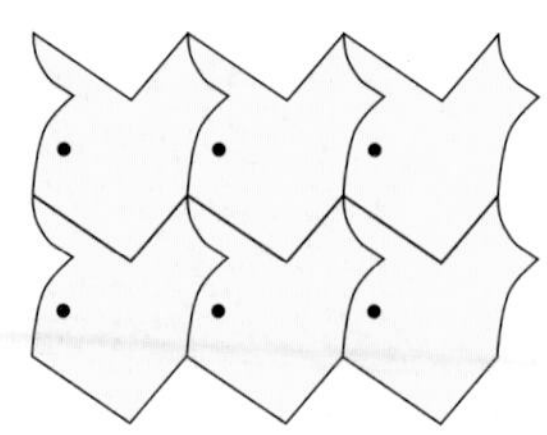

Activity 2 Make a tessellation using a rotation.

Step 1 Start by drawing an equilateral triangle. Then draw another triangle inside the right side of the triangle as shown below.

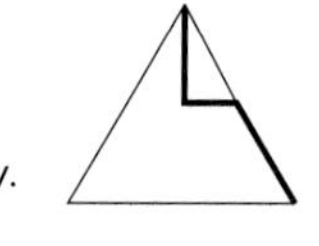

Step 2 Rotate the triangle so you can copy the change on the side indicated.

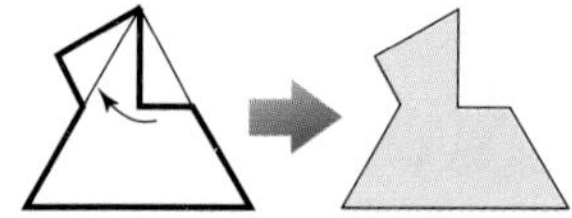

Step 3 Repeat this pattern unit on a tessellation of equilateral triangles. Alternating colors are used to best show the tessellation.

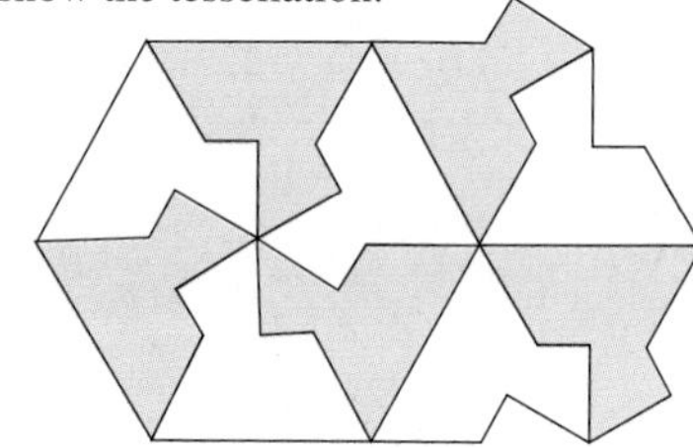

Write

1. Is the area of the original square in Step 1 of Activity 1 the same as the area of the new shape in Step 2? Explain. **1–2. See margin.**
2. Describe how you would create the unit for the pattern shown at the right.

Draw **Make a tessellation for each pattern described. Use a tessellation of two rows of three squares as your base.** **3–5. See Solutions Manual.**

3. 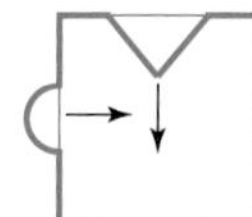
4. 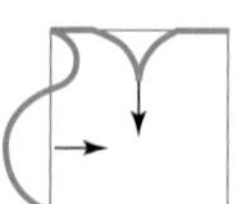
5.

Additional Answers

1. Yes; whatever space is taken out of the square is then added onto the outside of the square. The area does not change; only the shape changes.
2. Modify the bottom of the triangle to be like the right side of the triangle. Erase the bottom and right original sides of the triangle.

10-2 Tessellations

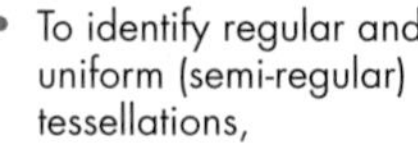
What YOU'LL LEARN

- To identify regular and uniform (semi-regular) tessellations,
- to create tessellations with specific attributes, and
- to solve problems by using guess and check.

Why IT'S IMPORTANT

You can use tessellations to solve problems involving interior design and quilting.

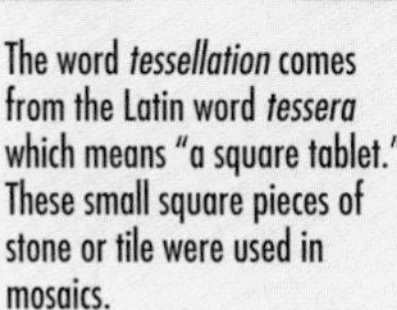
FYI

The word *tessellation* comes from the Latin word *tessera* which means "a square tablet." These small square pieces of stone or tile were used in mosaics.

Patterns that cover a plane with repeating figures so there is no overlapping or empty spaces are called **tessellations**. A **regular tessellation** uses only one type of regular polygon. In a tessellation, the sum of the measures of the angles of the polygons surrounding a point (at a vertex) is 360.

You can use what you know about angle measures in regular polygons to help determine which polygons tessellate.

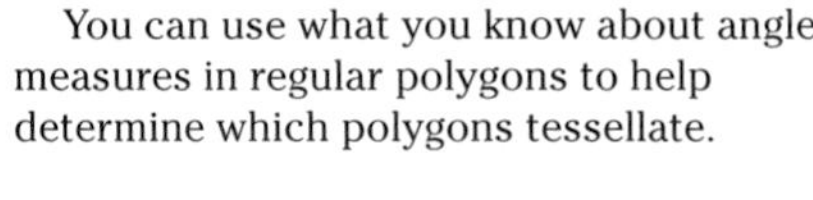

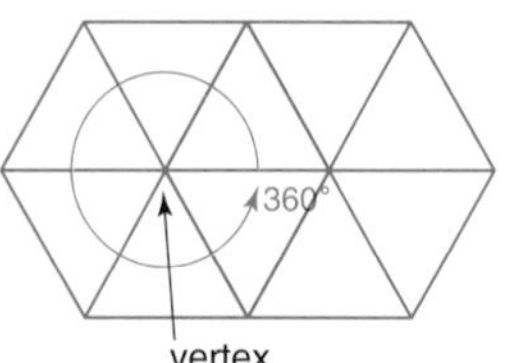

MODELING MATHEMATICS

Tessellations of Regular Polygons

Materials:  pattern blocks

- Use the pattern blocks to form regular tessellations of triangles, squares, and hexagons.
- Find the measure of one interior angle for each polygon.

Your Turn

a. Copy and complete the table below. **See students' work.**
b. Which regular polygons tessellate? **triangle, square, hexagon**
c. What conjecture can you make? **See margin.**

Polygon	triangle	square	pentagon	hexagon	heptagon	octagon
Does it tessellate the plane?						
measure of one interior angle						

From the Modeling Mathematics activity, you found that if a regular polygon has an interior angle with a measure that is a factor of 360, then the polygon will tessellate the plane.

Example 1 **Determine if a regular 20-gon will tessellate the plane.**

Find the measure of each interior angle of the regular 20-gon.

$$\frac{180\,(20-2)}{20} = 162 \qquad n = 20$$

Since 162 is not a factor of 360, the 20-gon will not tessellate the plane.

10-2 LESSON NOTES

NCTM Standards: 1–5, 7, 8

Instructional Resources

- Study Guide Master 10-2
- Practice Master 10-2
- Enrichment Master 10-2
- Assessment and Evaluation Masters, p. 268

Transparency 10-2A contains the 5-Minute Check for this lesson; **Transparency 10-2B** contains a teaching aid for this lesson.

Recommended Pacing	
Standard Pacing	Days 2 & 3 of 15
Honors Pacing	Day 2 of 14
Block Scheduling*	Day 2 of 8

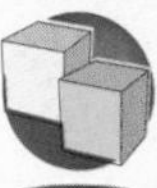
*For more information on pacing and possible lesson plans, refer to the *Block Scheduling Booklet.*

1 FOCUS

5-Minute Check
(over Lesson 10-1)

The number of sides of a regular polygon are given. Find the measure of an interior angle and an exterior angle of the polygon.

1. 6 **120, 60**
2. 9 **140, 40**
3. 10 **144, 36**

Draw each polygon.

4. a concave heptagon
5. a convex pentagon
6. a regular hexagon
See students' work.

Motivating the Lesson

Hands-On Activity Have students cut several polygons from a sheet of paper. They can be any shape, but roughly the same size. Tell students to try to fit them together with no overlapping and no empty space.

FYI

M.C. Escher is well-known for his artwork based on tessellations. He developed his ideas by studying Islamic art and architecture.

MODELING MATHEMATICS Check to make sure students don't leave gaps when testing for tessellations.

Answer for Modeling Mathematics

c. **If a regular polygon has an interior angle with a measure that is a factor of 360, then the polygon will tessellate the plane.**

2 TEACH

In-Class Examples

For Example 1
Determine if a regular 15-gon will tessellate. Explain. **Each interior angle of a regular 15-gon measures 156°. Since 156 is not a factor of 360, the 15-gon will not tessellate.**

For Example 2
Determine whether a semi-regular tessellation can be created from regular triangles and squares, all having sides 1 unit long. **yes**

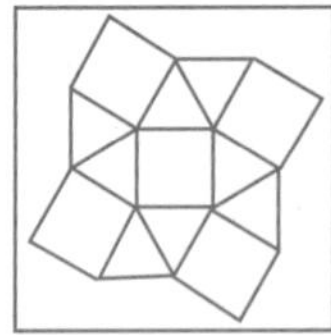

Teaching Tip Emphasize that it is not enough to avoid overlapping or gaps. The area must also be filled with no gaps or overhang at the borders.

A tessellation pattern can contain any number of figures. Tessellations containing the same combination of shapes and angles at each vertex are called **uniform**. The pattern below at the left will create a uniform tessellation. The pattern below at the right will form a tessellation, but it will not be uniform.

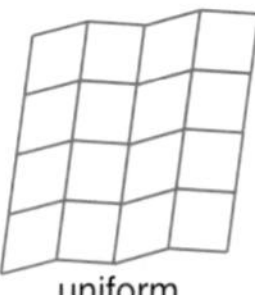
uniform

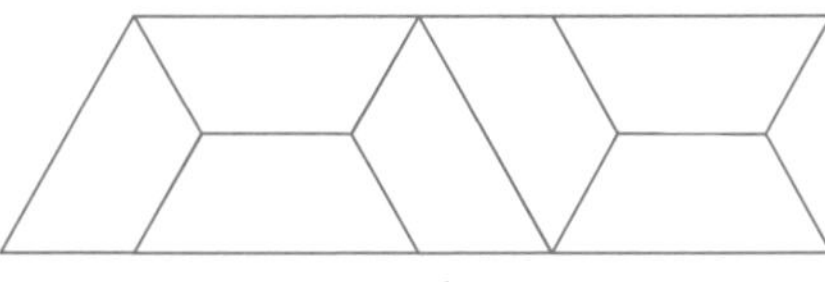
not uniform

Uniform tessellations containing two or more regular polygons are called **semi-regular**.

Example 2 **Determine whether a semi-regular tessellation can be created from regular octagons and squares, all having sides 1 unit long.**

There are two methods to solve this problem.

Method 1: Make a model

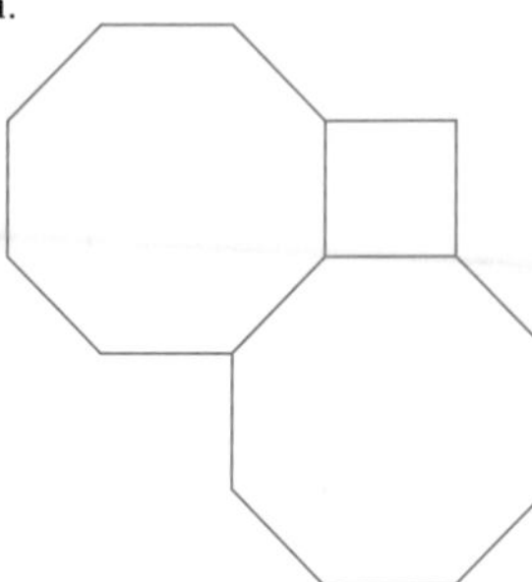

Either trace or place octagon pattern blocks side by side. You will notice that the space at each vertex could be filled in by a square. The uniform pattern at at each vertex is a square and two octagons.

Method 2: Solve algebraically

Calculate the measure of each interior angle of a regular octagon.

$$\frac{180(n-2)}{n} = \frac{180(8-2)}{8} \text{ or } 135 \quad n = 8$$

The measure of each interior angle is 135.

In the pattern, two octagons are adjacent, so there will be two angles with measures of 135 each.

$135 + 135 = 270$ *Find the sum.*

$360 - 270 = 90$ *Subtract the sum from 360.*

At each vertex, there is a 90° angle remaining. The sides will be congruent because the octagon is regular. Therefore, the figure could be a square.

The problem-solving strategy **guess and check**, or trial and error, is a powerful strategy. To use this strategy, first make an educated guess to answer a problem. Then use the conditions given in the problem to check if the answer is correct. If the first guess is not correct, use the information gathered from that guess to continue guessing until you find the correct answer.

Cooperative Learning

Co-op Co-op Separate the class into groups and have each group design a wall painting based on tessellations. Groups should select the polygons they will use and a color scheme. They should present their designs to the class and explain how they designed them. For more information on the co-op co-op strategy, see *Cooperative Learning in the Mathematics Classroom*, one of the titles in the Glencoe Mathematics Professional Series, page 32.

Example 3 Use the following three shapes and determine whether they can tessellate a row.

PROBLEM SOLVING
Guess and Check

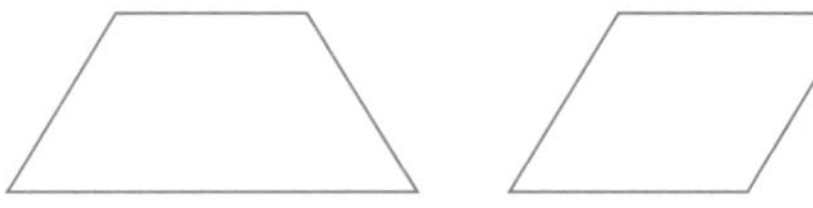

Use a model and the guess-and-check method to arrive at a possible answer.

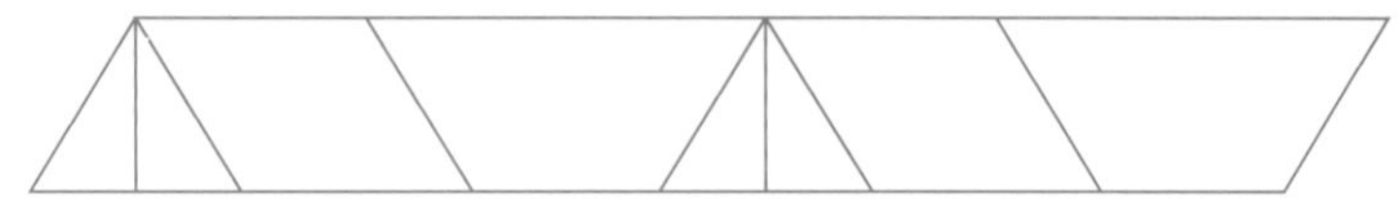

CHECK FOR UNDERSTANDING

Communicating Mathematics

1. Each angle is a factor of 360.

Study the lesson. Then complete the following. 2–4. See margin.

1. **Explain** why a rectangle will tessellate in a plane.
2. What is the difference between a regular and a semi-regular tessellation?
3. **Describe** the only three regular tessellations. Why are these the only ones?

4. Use the following blocks to create a tessellation.

5. Draw a tessellation that is semi-regular. See Solutions Manual.

Guided Practice

7. yes; interior angle = 60

Determine whether each regular polygon tessellates in a plane. If so, draw a sample figure. 6–7. See Solutions Manual for figures.

6. decagon no; interior angle = 144
7. equilateral triangle
8. Determine whether the pattern at the right will tesselate. yes

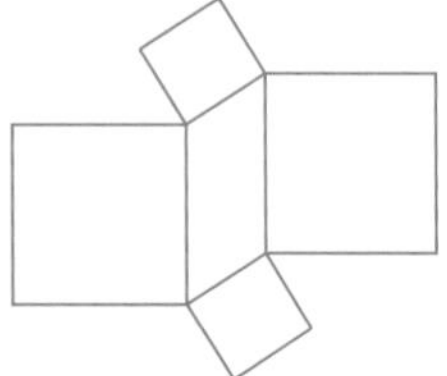

Determine whether each tessellation is *uniform, regular,* or *semi-regular*. Name all possibilities.

9. 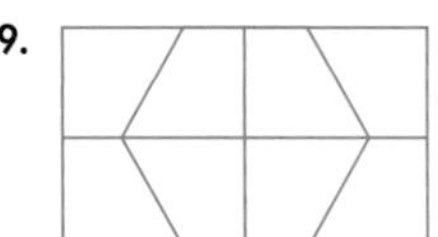 regular

10. 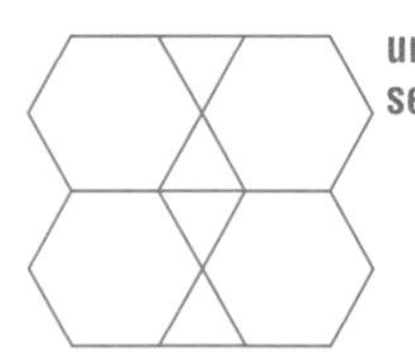 uniform, semi-regular

11. **Guess and Check** Replace each ? with an operation symbol and add parentheses in the following to make it true. $(8 - 4) \times 7 + 3 = 31$

$$8 \;?\; 4 \;?\; 7 \;?\; 3 = 31$$

In-Class Example

For Example 3
Use the following shape to determine if it can tessellate a row.

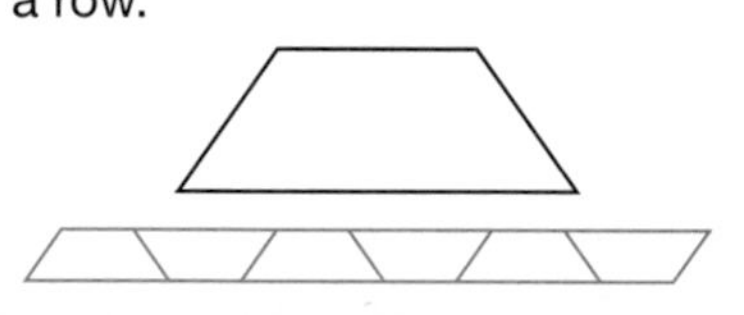

3 PRACTICE/APPLY

Check for Understanding

Exercises 1–11 are designed to help you assess your students' understanding through reading, writing, speaking, and modeling. You should work through Exercises 1–5 with your students and then monitor their work on Exercises 6–11.

Additional Answers

2. A regular tessellation uses one type of polygon and a semi-regular tessellation uses more than one type.
3. Equilateral triangles, squares, and regular hexagons; these are the only regular polygons that have an interior angle measure that is a factor of 360.

Study Guide Masters, p. 58

10-2 NAME__________ DATE ______
Study Guide Student Edition Pages 523–527

Tessellations

Tessellations are patterns that cover a plane with repeating polygons so that there are no overlapping or empty spaces. A **regular tessellation** uses only one type of regular polygon. **Uniform tessellations** contain the same combination of shapes and angles at each vertex. **Semi-regular tessellations** are uniform tessellations that contain two or more regular polygons.

regular, uniform

semi-regular

In a tessellation, the sum of the measures of the angles of the polygons surrounding a point (at a vertex) is 360. If a regular polygon has an interior angle with a measure that is a factor of 360, then the polygon will tessellate.

Example: Determine if a regular 16-gon will tessellate

$\frac{(180 - n)}{n}$ **Use this formula to find the measure of each interior angle.**

$\frac{(180 - 16)}{16} = 10.25$ $n = 16$

Since 10.25 is not a factor of 360, the 16-gon will not tessellate.

Determine whether each figure tessellates in a plane. If so, draw a sample figure.

1. scalene right triangle yes
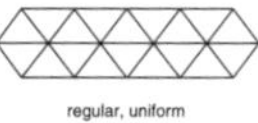
2. regular nonagon no
3. isosceles trapezoid yes

Determine if each pattern will tessellate.

4. square and isosceles triangle yes
5. rhombus, triangle, and octagon no
6. square and isosceles trapezoid yes

Reteaching

Using Models Bring in several examples of Islamic architecture using tessellations. In each tessellation, identify the repeating patterns. Have students model the patterns with pattern blocks.

Additional Answer

4. Sample answer:

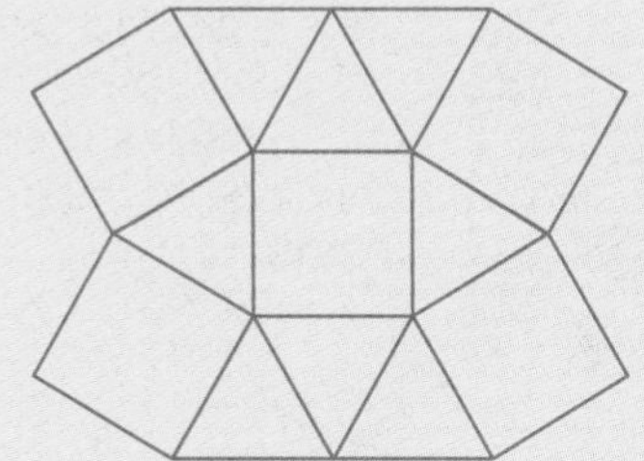

Assignment Guide

Core (with proof): 13–31 odd, 33–39
Core (informal): 13–31 odd, 33–39
Enriched: 12–24 even, 26–39

For **Extra Practice**, see p. 783.

The red A, B, and C flags, printed only in the Teacher's Wraparound Edition, indicate the level of difficulty of the exercises.

Additional Answers

12. Yes; sample answer:

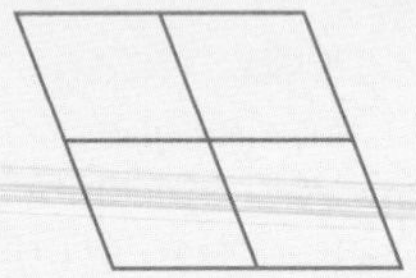

15. Yes; sample answer:

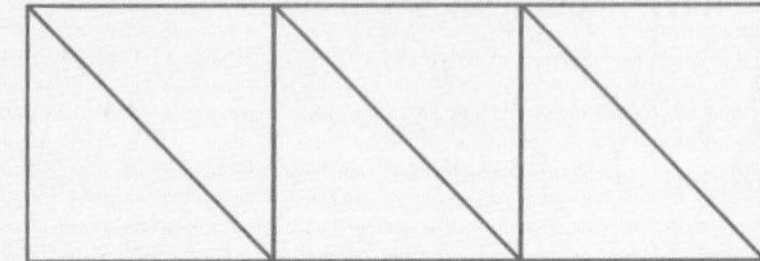

23. Sample answer:

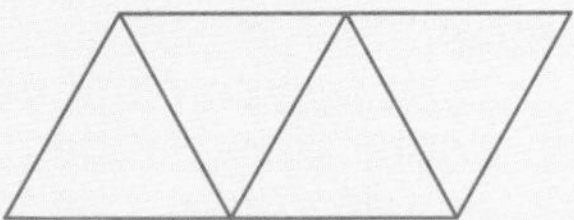

24. Sample answer:

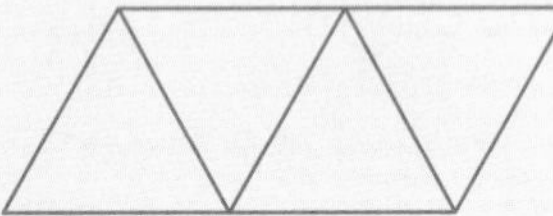

Practice Masters, p. 58

NAME __________ DATE ______

Practice

Student Edition Pages 523–527

Tessellations

Determine whether each figure tessellates in a plane. If so, draw a sample figure.

1. scalene triangle — Answers may vary.
2. regular 18-gon — Does not tessellate.
3. parallelogram — Answers may vary.

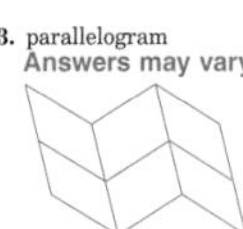

Determine if each pattern will tessellate.

4. octagon and isosceles triangle — No
5. three isosceles triangles — Yes
6. isosceles trapezoid and square — Yes

Determine whether each tessellation is regular, uniform, or semi-regular. Name all possibilities.

7. hexagon and triangle — uniform, semi-regular

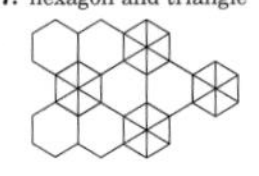

8. hexagon and triangle — semi-regular
9. obtuse triangle — not uniform

10. **Guess and check** Use the following three shapes and determine if they can tessellate a row. If so, draw a sample row.

   yes;  

EXERCISES

Practice

Determine whether each figure tessellates a plane. If so, draw a sample figure. 12–15. See margin for figures.

12. rhombus with no right angles yes
13. regular pentagon
14. regular 14-gon — no; interior angle = $154\frac{2}{7}$
15. nonequilateral triangle yes

13. no; interior angle = 108

Determine whether each pattern will tessellate a plane.

16. trapezoid and triangle yes

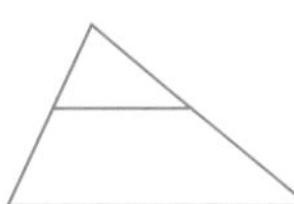

17. two rhombi yes

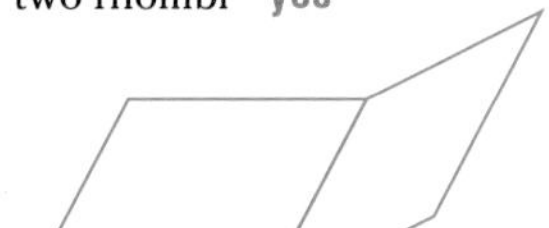

18. regular pentagon and triangle no

Determine whether each tessellation is *regular, uniform,* or *semi-regular*. Name all possibilities.

19. pentagon and triangle none

20. hexagon regular, uniform

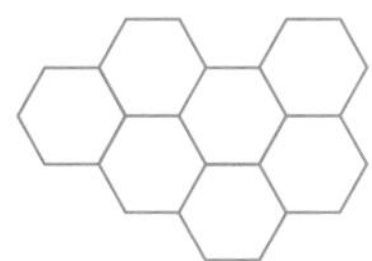

22. uniform, semi-regular

21. square and triangle

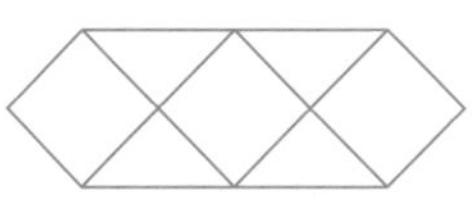

uniform, semi-regular

22. hexagon, square, and triangle

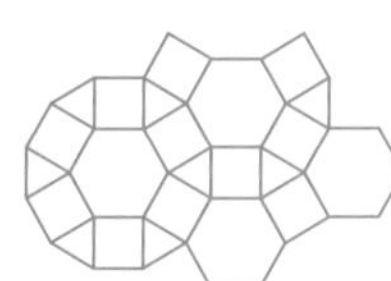

23. Draw a regular tessellation in a plane. 23–24. See margin for sample answers.

24. Draw a uniform tessellation in a plane.

Cabri Geometry

25. Use the reflection tool on the TI-92 to create a tessellation by reflecting a polygon over each of its sides. Draw your tessellation.
See students' work.

Critical Thinking

26. Sample answers: measures are 90, 90, 90, 135, 135.

26. What could be the measures of the interior angles in a pentagon that tessellate a plane? Is this tessellation regular? Is it uniform?

27–28. See Solutions Manual.

27. Can a tessellation be uniform but not semi-regular? Explain.
28. Can a tessellation be semi-regular but not uniform? Explain.

Classroom Vignette

"I bring examples of different polyhedra such as tetrahedrons, cubes, and octahedrons for students to observe. Then students use rulers, protractors, paper, and scissors, along with what they have learned about the measures of interior angles of polygons to construct the sides of a polyhedron and assemble a model."

Stan Williams

Stan Williams
Rockmart High School
Rockmart, Georgia

Applications and Problem Solving

29. **Guess and Check** Copy the figure at the right. Find the numbers that go in the squares whose sum on each side are the circled numbers between each pair of squares. 8, 15, 4

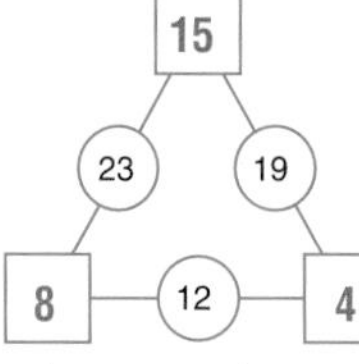

30. **Interior Design** Aiyana's family is tiling their rectangular recreation room in the basement with square tiles arranged so that the diagonal of the square is parallel to the wall.
    a. Draw a sketch of the pattern. See Solutions Manual.
    b. Describe the shape of the figure needed so that the entire floor is covered with tiles. See margin.

31. **Guess and Check** Use the number equation below to answer each of the following. a. See margin.
    a. If D = 5, find the unique digit represented by each letter.
    b. How does the problem change if more than one letter represents the same digit?

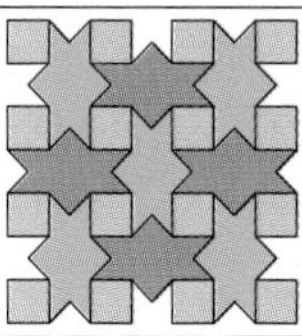

31b. More answers will result; sample answer: 561,535 + 207,535 = 769,070.

32. **Quilting** In the quilt at the right, what tessellation is shown? Describe the tessellation using the terms presented in this lesson. not regular, uniform, not semi-regular

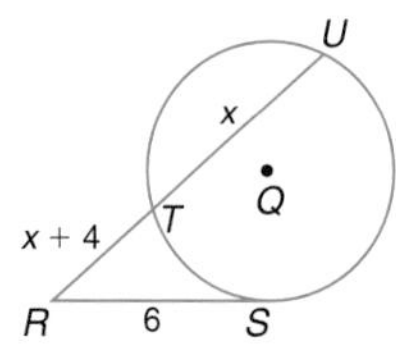

Mixed Review

33. The measure of an interior angle of a regular polygon is 160. Determine the number of sides of the polygon. (Lesson 10–1) 18

34. State the equation you would use to find the value of x if $\overline{RS}$ is tangent to $\odot Q$. Then find the value of x. (Lesson 9–7) $(2x + 4)(x + 4) = 6^2$; $-3 + \sqrt{19}$

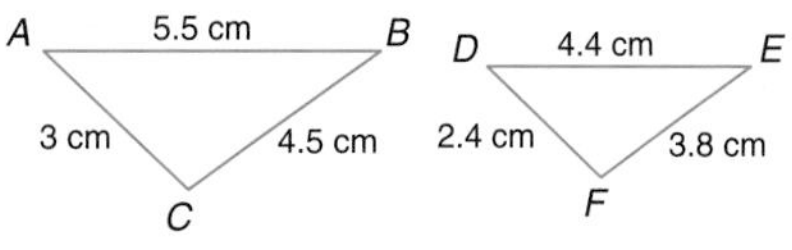

36. See margin.

35. A chord of a circle is 10 inches long, and is 12 inches from the center of the circle. What is the radius of the circle? (Lesson 9–3) 13 in.

36. **Make a Graph** The national parks in the United States are used for many recreational activities, which are recorded in visitor hours. In 1989, they were used for driving off-road vehicles 65,808 times, for camping 173,597 times, for hunting 46,760 times, for fishing 23,392 times, for boating 18,491 times, and for winter sports 3119 times. Make a circle graph of this data. (Lesson 9–2)

37. Determine whether $\triangle ABC \sim \triangle DEF$. Justify your answer. (Lesson 7–2) no; $\frac{5.5}{4.4} = \frac{3}{2.4} \neq \frac{4.5}{3.8}$

INTEGRATION Algebra

38. Determine the slope of the line that passes through points at (1, 8) and (−4, 7). $\frac{1}{5}$

39. Express the relation shown in the mapping at the right as a set of ordered pairs. Then state the domain, range, and inverse of the relation. {(6, 6), (0, 1), (−5, 6), (9, 2)}; D = {6, 0, −5, 9}; R = {2, 1, 6}; inverse = {(6, 6), (1, 0), (6, −5), (2, 9)}

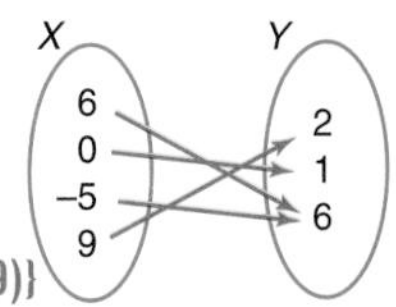

4 ASSESS

Closing Activity

Speaking Have students separate into small groups to discuss how tessellations are found in nature, art, and architecture.

Chapter 10 Quiz A (Lessons 10-1 and 10-2) is available in the *Assessment and Evaluation Masters*, p. 268.

Additional Answers

30b. a triangle with two sides equal to the side of the square and one side equal to the length of the diagonal

31a. $D = 5, Y = 2, N = 6, E = 9, A = 4, L = 8, S = 1, R = 7, C = 3$, and $T = 0$

Enrichment Masters, p. 58

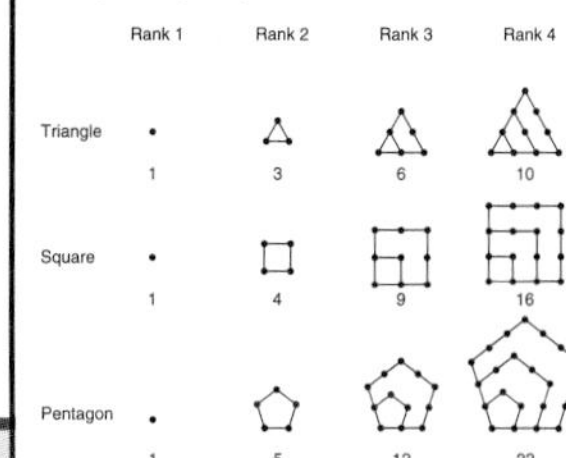
10-2 NAME ________ DATE ________
Enrichment
Student Edition Pages 523–527

Polygonal Numbers

Certain numbers related to regular polygons are called **polygonal numbers**. The chart shows several triangular, square, and pentagonal numbers. The **rank** of a polygon number is the number of dots on each "side" of the outer polygon. For example, the pentagonal number 22 has a rank of 4.

	Rank 1	Rank 2	Rank 3	Rank 4
Triangle	1	3	6	10
Square	1	4	9	16
Pentagon	1	5	12	22

Polygonal numbers can be described with formulas. For example, a triangular number T of rank r can be described by $T = \frac{r(r+1)}{2}$.

Answer each question.

1. Draw a diagram to find the triangular number of rank 5. 15
2. Draw a diagram to find the pentagonal number of rank 5. 35

3. Write a formula for a square number S of rank r. $S = r^2$
4. Write a formula for a pentagonal number P of rank r. $P = \frac{r(3r-1)}{2}$
5. What is the rank of the pentagonal number 70? 7
6. List the hexagonal numbers for ranks 1 to 5. (Hint: Draw a diagram.) 1, 6, 15, 28, 45

Extension

Connections Will a cube tessellate space? yes Will a 4-sided pyramid tessellate space? yes Will a cone tessellate space? no

Additional Answer

36.

Off-road vehicles 19.9%
Winter sports 0.9%
Boating 5.6%
Fishing 7.1%
Hunting 14.1%
Camping 52.4%

10-3A LESSON NOTES

NCTM Standards: 2–4, 7

Objective

Compare the area of a rectangle with the area of a parallelogram having the same base length and the same parallel lines.

Recommended Time

20 minutes

Instructional Resources

Instructions for using the *Geometer's Sketchpad* for this activity are available in the *Graphing Calculator and Computer Masters*, p. 27.

1 FOCUS

Motivating the Lesson

Draw a rectangle on grid paper. On the same paper, draw a parallelogram with the same base and height as the rectangle. Estimate the area of each by counting the squares inside each figure.

2 TEACH

Teaching Tip If students picture a parallelogram with angles of 170° and 10°, the relationship between the area of a rectangle with the same base may be more clear.

3 PRACTICE/APPLY

Assignment Guide
Core (with proof): 1–10 **Core (informal):** 1–10 **Enriched:** 1–10

4 ASSESS

Observing students working with technology is an excellent method of assessment.

10–3A Using Technology
Area of Parallelograms

A Preview of Lesson 10–3

You can use a TI-92 graphing calculator to draw a rectangle and find its area. Then using the same parallel lines and base length as the rectangle, you can draw a parallelogram and find its area. Compare the area of the rectangle to the area of the parallelogram.

Use these steps to draw a rectangle.

- Draw and label a horizontal line segment *AB*.
- At point *A*, draw a perpendicular line and label a point *C* on that line above $\overline{AB}$.
- Draw a line through *C* parallel to $\overline{AB}$.
- Construct point *D* on that line such that $CD = AB$.
- Use 4:Polygon on the F3 menu to draw a rectangle over *ABDC*.
- Then use 2:Area on the F6 menu to find the area of the polygon (rectangle *ABDC*).

Use the following steps to construct a parallelogram that is not a rectangle but has a base congruent to $\overline{AB}$ and is within the same parallel lines, $\overleftrightarrow{AB}$ and $\overleftrightarrow{CD}$, as rectangle *ABDC*.

- Using rectangle *ABDC* from above, label a point *E* between *C* and *D* that is on $\overleftrightarrow{CD}$.
- Label point *F* on $\overleftrightarrow{CD}$ such that $EF = AB$.
- Use 4:Polygon on the F3 menu to trace parallelogram *ABFE*.
- Then use 2:Area on the F6 menu to find the area of the polygon (parallelogram *ABFE*).

EXERCISES

Analyze your drawing. 1–5. See students' work.

1. What is the length of $\overline{AB}$?
2. What is the height of $\overline{AC}$?
3. What is the area of rectangle *ABDC*?
4. What is the height of parallelogram *ABFE*?
5. What is the area of parallelogram *ABFE*?
6. Compare the areas of the two figures. Explain what you found.
7. Use the rule for finding the area of a rectangle to make a conjecture about the rule for finding the area of a parallelogram. $A = bh$

6. The areas are the same.

Test your conjecture.

Use your TI-92 to draw and label each pair of figures so that they have the same area. 8–10. See Solutions Manual.

8. a square and a parallelogram that is not a square
9. a square and a rectangle that is not a square
10. two noncongruent rectangles

10-3 Area of Parallelograms

What YOU'LL LEARN

- To find areas of parallelograms.

Why IT'S IMPORTANT

You can use the area of parallelograms to solve problems involving housing, carpeting, and paving.

APPLICATION

Architecture

Both old and new buildings in New York City have geometric figures in their designs. One of the most prevalent figures in construction is a *parallelogram*.

Any side of a parallelogram can be called a **base**. For each base, there is a corresponding **altitude**. In ▱$ABCD$, $\overline{AB}$ is considered to be the base. A corresponding altitude is any perpendicular segment between the parallel lines $\overleftrightarrow{AB}$ and $\overleftrightarrow{DC}$. So in ▱$ABCD$, $\overline{CE}$ and $\overline{DB}$ are altitudes. The length of an altitude is called the **height** of the parallelogram.

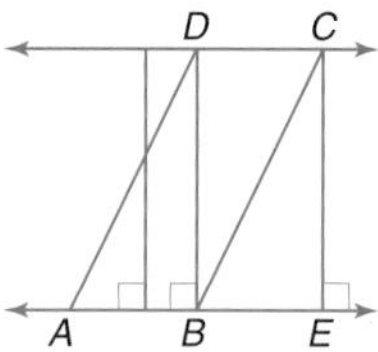

In Chapter 6, we learned that parallelograms have special properties. These properties will be used to investigate the **area** of a parallelogram. The *area* of a figure is the number of square units contained in the interior of the figure.

As you recall, the area of a rectangle, A square units, can be found by using the formula $A = \ell w$, where ℓ units is the length of the rectangle and w units is the width. The formula for the area of a parallelogram is closely related to the formula for the area of a rectangle.

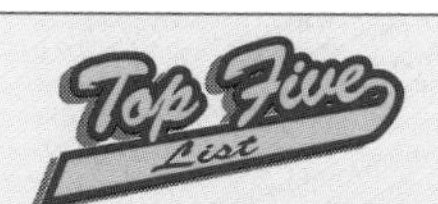

Top Five List

Tallest Buildings in New York City (feet, stories)

1. World Trade Center (1368, 110)
2. Empire State Building (1250, 102)
3. Chrysler Building (1046, 77)
4. American International (950, 67)
5. 40 Wall Tower (927, 71)

MODELING MATHEMATICS

Area of a Parallelogram

Materials: plain paper, pencil

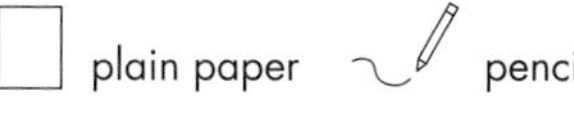

- Draw and cut out ▱$ABCD$.
- Fold ▱$ABCD$ so that A lies on B and C lies on D. A rectangle results.

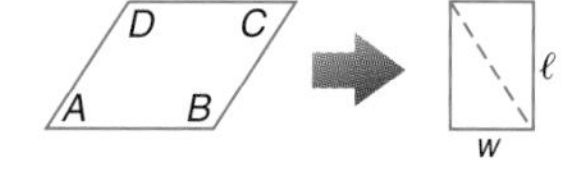

- This rectangle has an area of ℓw. Since there are 2 congruent layers of this rectangle, the area of the parallelogram is $2\ell w$.
- Unfolding the parallelogram, you see that $2w$ equals the base b of the parallelogram and that ℓ is its height h.

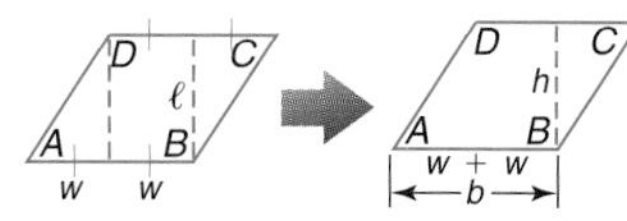

Your Turn a–b. See students' work.

a. Use a ruler to measure $\overline{AB}$ or b, w, and ℓ or h.

b. Compare these measures.

c. Summarize your findings in a formula. $A = bh$

10-3 LESSON NOTES

NCTM Standards: 1–5, 7, 8

Instructional Resources

- Study Guide Master 10-3
- Practice Master 10-3
- Enrichment Master 10-3
- Graphing Calculator and Computer Masters, p. 10

Transparency 10-3A contains the 5-Minute Check for this lesson; **Transparency 10-3B** contains a teaching aid for this lesson.

Recommended Pacing	
Standard Pacing	Day 5 of 15
Honors Pacing	Day 4 of 14
Block Scheduling*	Day 3 of 8

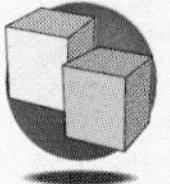

*For more information on pacing and possible lesson plans, refer to the *Block Scheduling Booklet*.

1 FOCUS

5-Minute Check
(over Lesson 10-2)

Determine whether each polygon will tessellate.

1. a regular 14-gon no
2. a regular 10-gon no
3. a regular 60-gon no
4. Find the numbers that go in the squares whose sum is the circled number between each pair of squares.

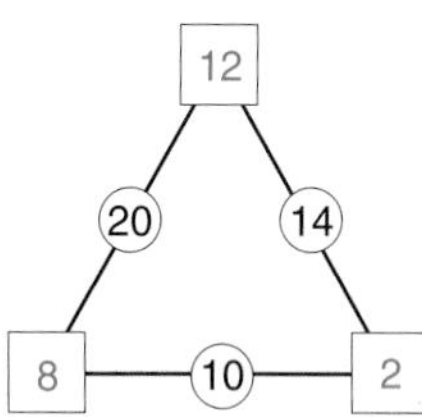

5. Draw a semi-regular tessellation in a plane. See students' work.

Top Five List

The 71st floor of the Chrysler Building is a visitor's center that displays Chrysler's first tool kit. The building project was purchased by Walter P. Chrysler in 1927 and completed in 1930.

MODELING MATHEMATICS Discuss other ways to divide the parallelogram, such as two triangles and a square. Find the area of each part and add them together. What does it equal? $A = bh$

Motivating the Lesson

Hands-On Activity Ask students to draw and cut out a square from a piece of paper. Have them cut the square into two pieces of any size. Ask students if the area of the whole square is equal to the area of these two parts. **yes**

2 TEACH

In-Class Examples

For Example 1

Find the area of ▱*STAR*.

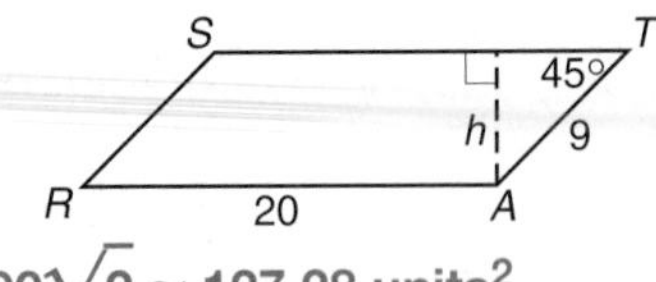

$90\sqrt{2} \approx 127.28$ units^2

For Example 2

The coordinates of the vertices of a quadrilateral are at (0, −5), (3, −4), (1, 2), and (−2, 1). Is it a square, a rectangle, or a parallelogram? Find its area.
rectangle, 20 units^2

Teaching Tip When reviewing the formulas for the area of a rectangle or square, ask students to explain what is meant by a square unit.

Area of a Parallelogram	**If a parallelogram has an area of *A* square units, a base of *b* units, and a height of *h* units, then *A* = *bh*.**

Example 1 **Find the area of parallelogram *RSTU*.**

Explore Study the diagram to determine what information is given, what information you will use, and what information you need.

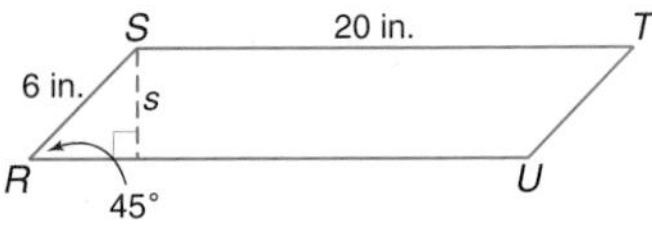

LOOK BACK
Refer to Lesson 8-2 to review relationships in 45°-45°-90° triangles.

Plan To find the area of the parallelogram, we must know the base and the height. The base, given in the figure, is 20 inches long. Since $m\angle R = 45$, the height can be found by using the 45°-45°-90° triangle that is formed when the altitude is drawn from *S*.

Solve Find the height by using the length of the hypotenuse. Recall that the length of the hypotenuse is $s\sqrt{2}$, where *s* is the length of the leg.

$6 = s\sqrt{2}$ *Substitute 6 for the hypotenuse.*

$\frac{6}{\sqrt{2}} = \frac{s\sqrt{2}}{\sqrt{2}}$ *Solve for s.*

$\frac{6\sqrt{2}}{\sqrt{2}\sqrt{2}} = s$ *Rationalize the denominator.*

$3\sqrt{2} = s$

The formula for the area of the parallelogram is $A = bh$, so $A = (20)(3\sqrt{2})$ or $60\sqrt{2}$. Therefore, the area of parallelogram *RSTU* is $60\sqrt{2}$ or about 84.85 square inches.

Examine Find the area of a 20-inch by 6-inch rectangle. The area of the parallelogram should be less than the area of the rectangle.

Example 2 **The coordinates of the vertices of a quadrilateral are at *A*(3, 7), *B*(7, 1), *C*(1, −3), and *D*(−3, 3).**

INTEGRATION
Algebra

a. Graph each point and draw the quadrilateral. Is it a *square*, a *rectangle*, or a *parallelogram*?

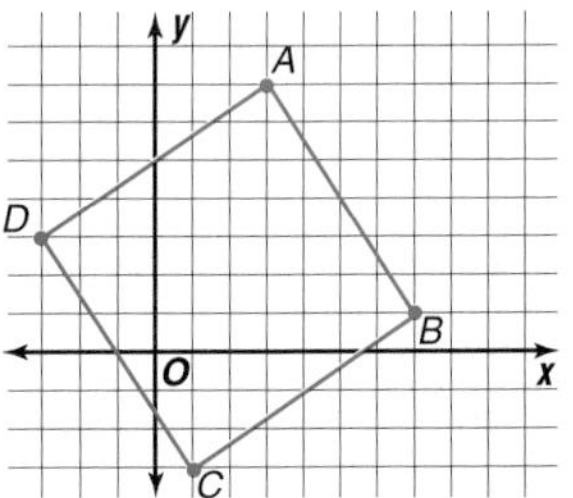

First, determine the slopes of each side.

slope of $\overline{AB} = \frac{7-1}{3-7}$ or $\frac{3}{-2}$

slope of $\overline{BC} = \frac{1-(-3)}{7-1}$ or $\frac{2}{3}$

slope of $\overline{CD} = \frac{-3-3}{1-(-3)}$ or $\frac{-3}{2}$

slope of $\overline{AD} = \frac{7-3}{3-(-3)}$ or $\frac{2}{3}$

The slopes of opposite sides are the same, so the opposite sides are parallel. The slopes of consecutive sides are negative reciprocals of each other, so those sides are perpendicular.

Alternative Learning Styles

Auditory Students should work in pairs. Have one student explain to another student how the area of a parallelogram is related to the area of a rectangle. Have the other student draw pictures to support the partner's explanation. Have two or three pairs present their explanations to the class.

Since consecutive sides are perpendicular, opposite sides are parallel, and all sides are congruent, the quadrilateral is a rectangle.

Next, we can use the distance formula to find the length of two adjacent sides.

$AB = \sqrt{(3-7)^2 + (7-1)^2}$ or $\sqrt{52}$ $s = \sqrt{(x_1 - x_2)2 + (y_1 - y_2)^2}$

$BC = \sqrt{(7-1)^2 + (1-(-3))^2}$ or $\sqrt{52}$

Since two adjacent sides are congruent, the rectangle is a square.

b. Find the area of quadrilateral *ABCD*.

Use the formula for the area of a square.

$A = s^2$

$= (\sqrt{52})^2$ *Substitute side length.*

$= 52$

The area of the square is 52 square units.

In-Class Example

For Example 3

Find the amount of carpet required to cover the hallway shown below.

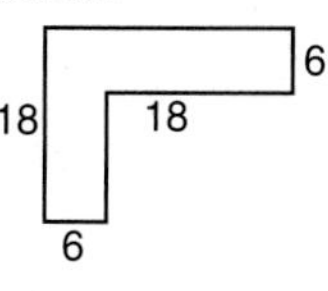

216 units^2

Teaching Tip After Example 3, ask students if they can see other ways of dividing the region into nonoverlapping parts.

Sometimes a figure can be divided into several figures. To find its area, you must find the area of each figure and then find the sum of those areas. This is stated in the postulate below.

Postulate 10–1	**The area of a region is the sum of the areas of all of its nonoverlapping parts.**

We can use this postulate to solve the following problem.

Example 3

Carpeting

Tyrone's grandparents are moving into a retirement home. Before they move in, they want to recarpet the two bedrooms and hall. Find the amount of carpeting in square yards that is needed to cover this floor space.

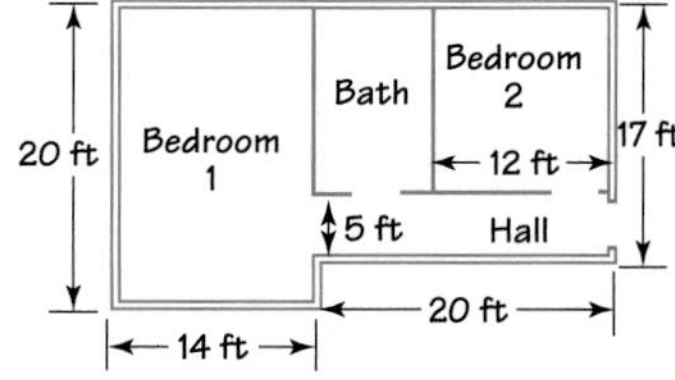

First, divide the carpeted area into regions whose areas you can find.

Area of large bedroom:	Area of small bedroom:	Area of hall:
$A = \ell w$	$A = \ell w$	$A = \ell w$
$A = (20)(14)$	$A = 12(17 - 5)$	$A = (20)(5)$
$A = 280$	$A = 144$	$A = 100$

The total area of the bedrooms and the hall is 280 + 144 + 100 or 524 square feet.

There are 9 square feet in one square yard, so divide the total number of square feet by 9 to find the number of square yards of carpeting needed.

$\frac{524}{9} \approx 58.2$

Therefore, Tyrone's grandparents will need 59 square yards of carpeting.

3 PRACTICE/APPLY

Check for Understanding

Exercises 1–9 are designed to help you assess your students' understanding through reading, writing, speaking, and modeling. You should work through Exercises 1–4 with your students and then monitor their work on Exercises 5–9.

Additional Answer

4. The two cuts are parallel to each other and at an angle less than 90 with the base. The two congruent shapes can be either right triangles or right trapezoids depending on how far in from the sides the cuts are made.

Study Guide Masters, p. 59

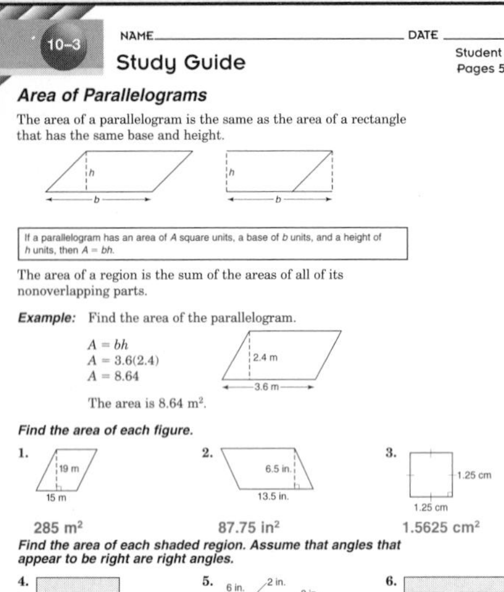

10-3 NAME__________ DATE________

Study Guide — Student Edition Pages 529–534

Area of Parallelograms

The area of a parallelogram is the same as the area of a rectangle that has the same base and height.

If a parallelogram has an area of A square units, a base of b units, and a height of h units, then $A = bh$.

The area of a region is the sum of the areas of all of its nonoverlapping parts.

Example: Find the area of the parallelogram.

$A = bh$
$A = 3.6(2.4)$
$A = 8.64$

The area is 8.64 m^2.

Find the area of each figure.

1. 285 m^2
2. 87.75 in^2
3. 1.5625 cm^2

Find the area of each shaded region. Assume that angles that appear to be right are right angles.

4. 972 m^2
5. 82 in^2
6. 147 m^2

7. The area of a parallelogram is 24.96 cm^2. The base is 6.4 cm. If the measures of the base and height are each doubled, find the area of the resulting parallelogram. 99.84 cm^2

8. A rectangle is 6 meters longer than it is wide. The area of the rectangle is 315 square meters. Find the length. 21 m

CHECK FOR UNDERSTANDING

Communicating Mathematics

1. Four meters is a linear measurement, while four square meters is an area measurement.

Study the lesson. Then complete the following.

1. **Describe** the difference between 4 meters and 4 square meters.
2. **Compare** the areas of rectangle *ABCD* and parallelogram *CDEF*.
   area of $CDEF = \frac{1}{2}$(area of $ABCD$)

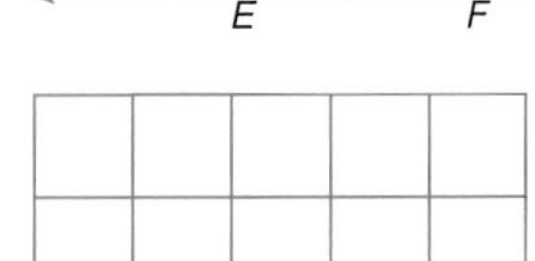

3. How can you add two squares to increase the area of the figure at the right by two square units, but not change its perimeter?

4. Draw and cut out a rectangle. Make two cuts on the rectangle to make it into a parallelogram. These two cuts must form two congruent shapes. Describe these two cuts and the shapes. Is there more than one way to make the cuts? Explain. See margin.

Guided Practice

Find the area of each figure. Assume the angles that appear to be right are right angles.

5. 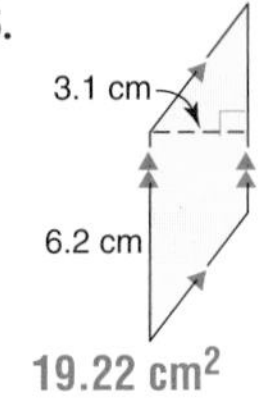

19.22 cm^2

6. 7. 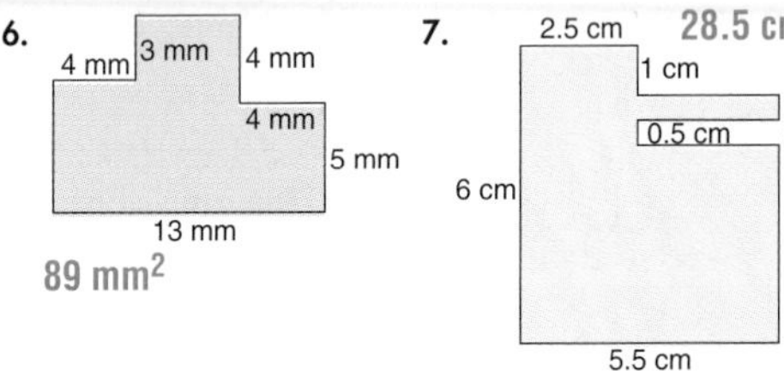

6. 89 mm^2
7. 28.5 cm^2

8. The coordinates of the vertices of a quadrilateral are (−1, −1), (1, 2), (6, 2), and (4, −1). Graph the points and draw the quadrilateral and an altitude. Identify the quadrilateral as a *square*, *rectangle*, or a *parallelogram*, and find its area. parallelogram, 15 $units^2$

9. **Maintenance** The University of North Carolina in Chapel Hill is resealing the parking lot around Peabody Hall. It costs \$0.52 per square yard to seal a parking lot. How much will it cost the University for their parking lot project? \$628.34

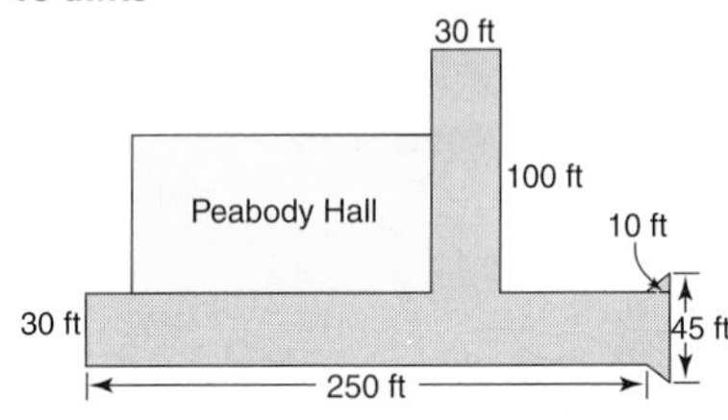

EXERCISES

Practice

Find the area of each figure or shaded region. Assume the angles that appear to be right are right angles.

10. 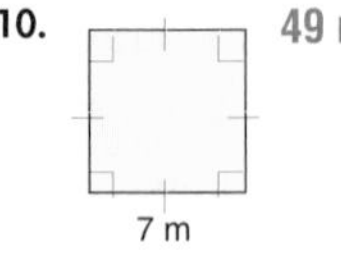

49 m^2

11. 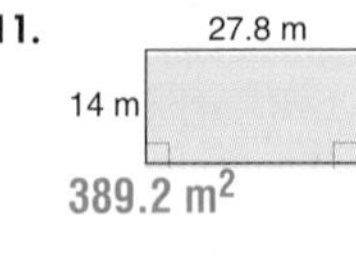

389.2 m^2

12. 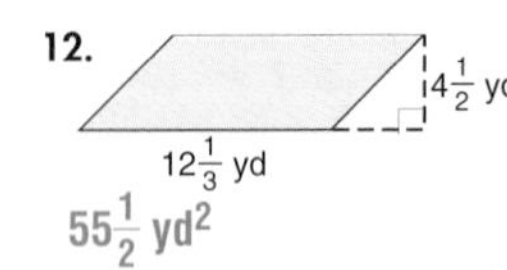

$55\frac{1}{2}$ yd^2

Reteaching

Using Alternative Strategies Show students how to find the area of the figure at the right by partitioning it into three rectangles: two 2 mm-by-6 mm rectangles and one 2 mm-by-4 mm rectangle. Have them find another way to partition the figure into rectangles. one 8 mm-by-2 mm rectangle and two 2 mm-by-4 mm rectangles Ask if they both give the same area. yes, 32 mm^2

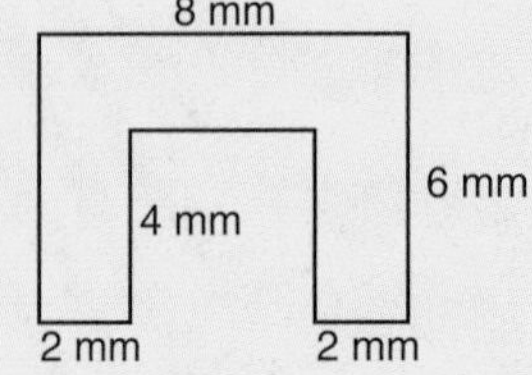

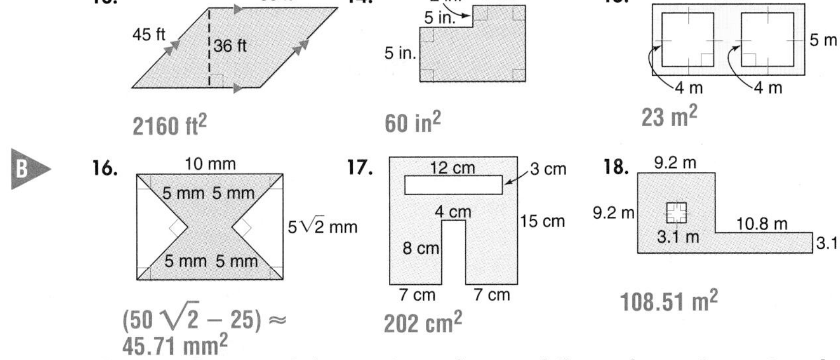

16. $(50\sqrt{2} - 25) \approx$ 45.71 mm^2

The coordinates of the vertices of a quadrilateral are given. Graph the points and draw the quadrilateral and an altitude.

a. Identify the quadrilateral as a *square*, a *rectangle*, or a *parallelogram*.

b. Find its area.

19. (4, −2), (10, −4), (10, −2), (4, −4) rectangle; 12 units^2

20. (−4, −5), (2, −5), (4, −8), (−2, −8) parallelogram; 18 units^2

21. (1, 10), (−1, 7), (2, 5), (4, 8) square; 13 units^2

22. How many 6-feet-by-8-feet rectangles can fit into a rectangle 18 feet long by 48 feet wide? Draw a figure to justify your answer. 18; see students' figures.

23. Find the area of the parallelogram at the right. $12\sqrt{3}$ mm^2

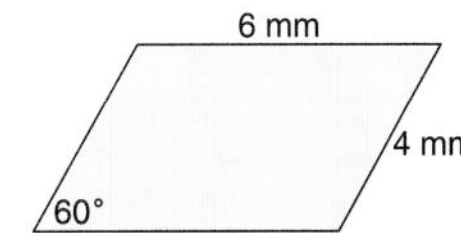

24. When the length of each side of a square is increased by 5 inches, the area of the resulting square is 2.25 times the area of the original square. What is the area of the original square? 100 in^2

25. Find the area of the shaded region of the congruent overlapping squares if E is the intersection of the diagonals of the square $ABDC$. Explain your solution.
25 cm^2; $\frac{1}{4}$ area of square

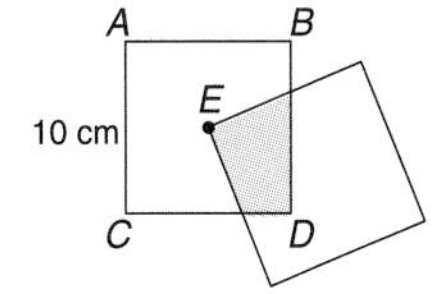

Cabri Geometry

26. Use the TI-92 graphing calculator to draw several parallelograms that have bases of the same length, but different heights. Compare the areas of the parallelograms to find the shape of the parallelogram that has a maximum area and the shape of the parallelogram that has a minimum area. See margin.

Critical Thinking

27. Given that the area of parallelogram $MNOP$ is 48 square units and $MX = 3$, find the length of the diagonal $\overline{NP}$. $NP = 16$ units

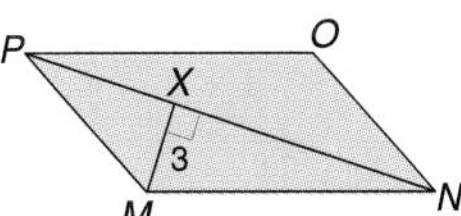

Assignment Guide

Core (with proof): 11–29 odd, 30–37
Core (informal): 11–29 odd, 30–32, 34–37
Enriched: 10–26 even, 27–37

For **Extra Practice**, see p. 783.

The red A, B, and C flags, printed only in the Teacher's Wraparound Edition, indicate the level of difficulty of the exercises.

Additional Answer

26. Maximum area would be a rectangle while minimum area would be approaching a line.

Practice Masters, p. 59

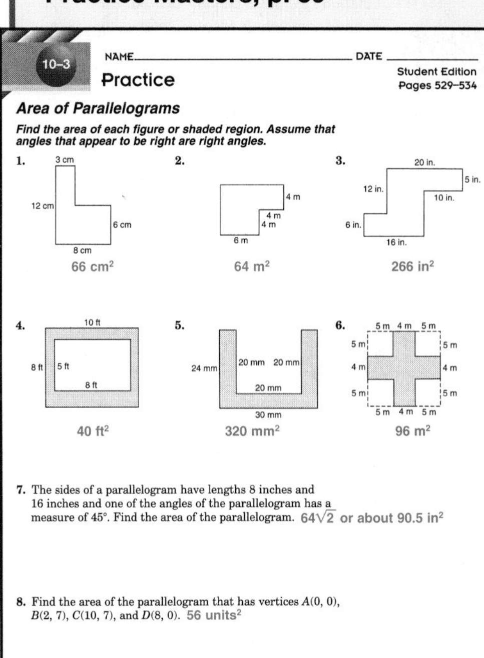

10–3 NAME______ DATE______
Practice
Student Edition Pages 529–534

Area of Parallelograms

Find the area of each figure or shaded region. Assume that angles that appear to be right are right angles.

1. 3 cm; 12 cm; 6 cm; 8 cm — 66 cm^2
2. 4 m; 4 m; 4 m; 6 m — 64 m^2
3. 20 in.; 5 in.; 12 in.; 10 in.; 6 in.; 16 in. — 266 in^2
4. 10 ft; 8 ft; 5 ft; 8 ft — 40 ft^2
5. 24 mm; 20 mm; 20 mm; 20 mm; 30 mm — 320 mm^2
6. 5 m 4 m 5 m; 5 m; 4 m; 5 m — 96 m^2

7. The sides of a parallelogram have lengths 8 inches and 16 inches and one of the angles of the parallelogram has a measure of 45°. Find the area of the parallelogram. $64\sqrt{2}$ or about 90.5 in^2

8. Find the area of the parallelogram that has vertices $A(0, 0)$, $B(2, 7)$, $C(10, 7)$, and $D(8, 0)$. 56 units^2

9. Find the area of the parallelogram that has vertices $W(-4, 15)$, $X(1, 15)$, $Y(4, 10)$, and $Z(-1, 10)$. 25 units^2

Tech Prep

Textile Scientist Some two-year colleges offer programs in textile sciences. Textile sciences involve the design and fabrication of textile products such as carpets, draperies, and so on. For more information on tech prep, see the *Teacher's Handbook*.

4 ASSESS

Closing Activity

Writing Ask each student to draw a figure with an area that can be found by adding the areas of nonoverlapping regions. Have students write the process they would use to find the area.

Additional Answers

28. going →, need 1 length 4 yd by 12 ft, 1 length 4 yd by 14 ft, and 1 length 4 yd by 34 ft; 80 yd^2

33. Statements (Reasons)
1. trapezoid *ABCD*, $\overline{AB} \parallel \overline{DC}$ (Given)
2. $\angle A$ and $\angle D$ are supplementary. (Consecutive Interior Angle Theorem)

35.

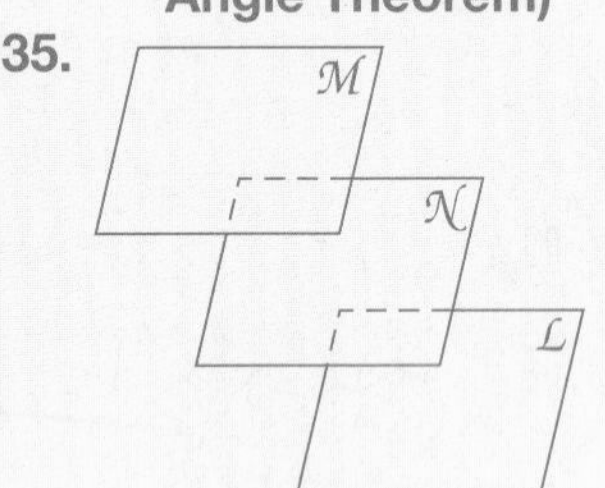

Enrichment Masters, p. 59

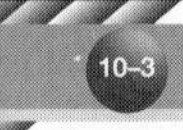

NAME______ DATE______

Enrichment

Student Edition Pages 529–534

Area of a Parallelogram

You can prove some interesting results using the formula you have proved for the area of a parallelogram by drawing auxiliary lines to form congruent regions. Consider the top parallelogram shown at the right. In the figure, *d* is the length of the diagonal $\overline{BD}$, and *k* is the length of the perpendicular segment from *A* to $\overline{BD}$. Now consider the second figure, which shows the same parallelogram with a number of auxiliary perpendiculars added. Use what you know about perpendicular lines, parallel lines, and congruent triangles to answer the following.

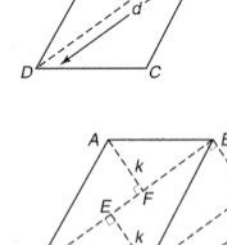

1. What kind of figure is *DBHG*? rectangle

2. If you moved $\triangle AFB$ to the lower-left end of figure *DBHG*, would it fit perfectly on top of $\triangle DGC$? Explain your answer. Yes; $\triangle AFB \cong \triangle CED$ (by HA) and $\triangle CED \cong \triangle DGC$ (since $\overline{DC}$ is a diagonal of rectangle *DECD*). So $\triangle AFB \cong \triangle DGC$.

3. Which two triangular pieces of ▱*ABCD* are congruent to $\triangle CBH$? $\triangle DAF$ and $\triangle BCE$

4. The area of ▱*ABCD* is the same as that of figure *DBHG*, since the pieces of ▱*ABCD* can be rearranged to form *DBHG*. Express the area of ▱*ABCD* in terms of the measurements *k* and *d*. Area of ▱*ABCD* = *dk*

Applications and Problem Solving

28. **Carpeting** Refer to Example 3. If carpeting comes in widths of 4 yards and the direction the carpet is laid must be the same throughout the house, how much carpeting must be purchased to cover the bedrooms and the hall in the diagram? Justify your calculations with a drawing. See margin.

29. **Housing** At the right is the floor plan of the home that the Summers are buying. They want to add a sunroom in place of the patio. They also want to increase their living space by one-fourth.

    a. How many square feet are in the existing house? 828 ft^2

    b. What is the area that will be added to the existing house? 207 ft^2

    c. What could be the dimensions of the new sunroom?

    d. Make a sketch of your design for the addition. See students' work.

    c. Sample answer: 12 ft by 17.25 ft

Mixed Review

30. **Guess and Check** Each letter in the multiplication problem at the right represents a digit from 0 to 9 each time it occurs. Find the two numbers that satisfy the conditions. (Lesson 10–2)

	RECAP
×	4
	PACER

30. 21,978 and 87,912

31. Use polygon *MNOPQ* to answer the following. (Lesson 10–1)

    a. Name the vertices of the polygon. *M, N, O, P, Q*

    b. Is the polygon convex or concave? convex

    c. Is the polygon regular? Explain.

    d. Name the polygon according to its sides. pentagon

31c. No; its sides are not all congruent.

32. In ⊙*R*, $\overline{AD}$ and $\overline{BE}$ are diameters, $m\angle ARB = 24$, $m\angle CRD = 44$, and $\angle CRD \cong \angle FRE$. Find each measure. (Lesson 9–2)

    a. $m\widehat{DB}$ 156

    b. $m\angle DRF$ 68

    c. $m\widehat{AFD}$ 180

    d. $m\widehat{BEC}$ 248

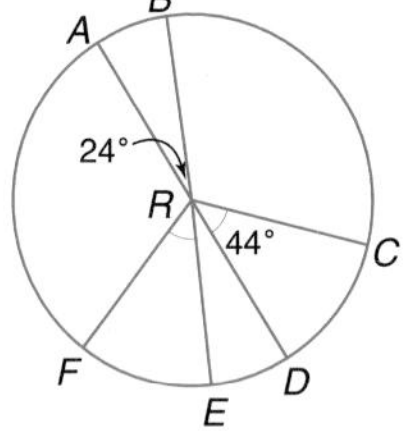

33. Write a two-column proof. (Lesson 6–5) See margin.

    Given: trapezoid *ABCD*, $\overline{AB} \parallel \overline{DC}$

    Prove: $\angle A$ and $\angle D$ are supplementary.

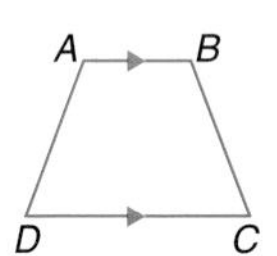

34. In the figure, $\overline{EF} \perp \overline{EH}$, $\overline{EH} \perp \overline{HG}$, and $\overline{HF} \perp \overline{JG}$. Name the segment whose length represents the distance between the following points and lines. (Lesson 3–5)

    a. *G* to $\overleftrightarrow{HF}$ $\overline{GJ}$

    b. *E* to $\overleftrightarrow{HG}$ $\overline{EH}$

    c. *F* to $\overleftrightarrow{EH}$ $\overline{FE}$

35. Draw and label a figure showing planes $\mathcal{M}$, $\mathcal{N}$, and $\mathcal{L}$ that do not intersect. (Lesson 1–2) See margin.

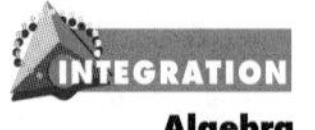

Algebra

36. Solve $a^2 - 5a + 6 = 0$ (3, 2)

37. Determine which ordered pairs are solutions to $y > 4x - 3$. a, d

    a. (0, 2) b. (3, 0) c. (1, −1) d. (−4, 2)

Extension

Problem Solving Find the area of a parallelogram with vertices *A*(−2, 0), *B*(0, 2), *C*(5, 2), and *D*(3, 0). 10 units^2

10-4 Area of Triangles, Rhombi, and Trapezoids

What YOU'LL LEARN

- To find areas of triangles, rhombi, and trapezoids.

Why IT'S IMPORTANT

You can use area of triangles, rhombi, and trapezoids to solve problems involving real estate, home repair, and design.

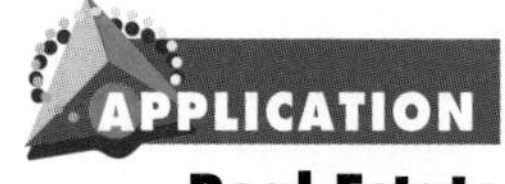

APPLICATION

Real Estate

When a new development is planned, it is divided into lot parcels. These lots sell for varying amounts and there are several factors that determine the price: area, location (corner, cul-de-sac, etc.), trees and terrain, and so on. The map at the right shows the layout of the streets and lots in the Linworth Village development in Columbus, Ohio.

The developers made the lots into shapes that were not just rectangular or square. Since the area of a lot is a major factor in its cost, the developer must use geometry to find the area and understand the property of area described in Postulate 10–2.

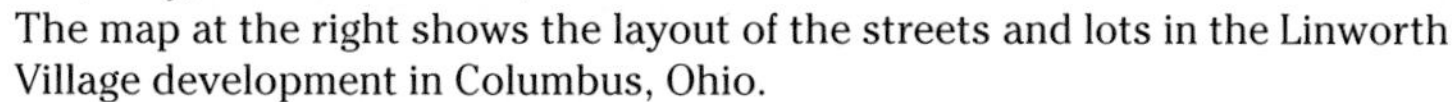

Postulate 10–2	**Congruent figures have equal areas.**

You know formulas for the areas of squares and rectangles. Now, use the following activity to discover the formula for the area of a triangle.

MODELING MATHEMATICS

Area of a Triangle

Materials:  grid paper  scissors  straightedge

You can determine a formula for the area of a triangle by using the area of a rectangle.

- Draw a triangle on grid paper so that one edge is along a horizontal line as shown at the right. Label the vertices of the triangle *A*, *B*, and *C*.
- Draw a line perpendicular to $\overline{AC}$ through *A*.
- Draw a line perpendicular to $\overline{AC}$ through *C*.
- Draw a line parallel to $\overline{AC}$ through *B*.
- Label the points of intersection of the lines drawn as *D* and *E* as shown.
- Find the area of rectangle *ACDE* in square units.
- Cut out rectangle *ACDE*. Then cut out $\triangle ABC$. Place the two smaller pieces over $\triangle ABC$ to completely cover the triangle.

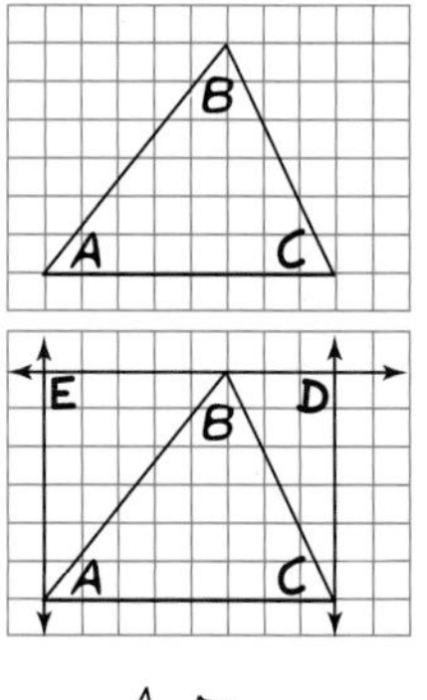

Your Turn a. areas of $\triangle BEA + \triangle DBC = \triangle ABC$

a. What do you observe about the two smaller triangles and $\triangle ABC$?

b. What portion of rectangle *ACDE* is $\triangle ABC$? half

c. What is a formula that could be used to find the area of $\triangle ABC$? $A = \frac{1}{2}bh$

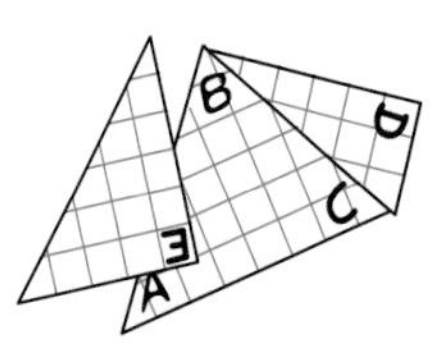

MODELING MATHEMATICS Students will use grid paper to estimate the area of a triangle. Help students develop a general formula for the area of a triangle.

10-4 LESSON NOTES

NCTM Standards: 1–5, 7, 8

Instructional Resources

- Study Guide Master 10-4
- Practice Master 10-4
- Enrichment Master 10-4
- Assessment and Evaluation Masters, pp. 267, 268
- Modeling Mathematics Masters, pp. 51–54
- Real-World Applications, 20
- Tech Prep Applications Masters, p. 19

Transparency 10-4A contains the 5-Minute Check for this lesson; **Transparency 10-4B** contains a teaching aid for this lesson.

Recommended Pacing	
Standard Pacing	Days 6 & 7 of 15
Honors Pacing	Days 5 & 6 of 14
Block Scheduling*	Day 4 of 8

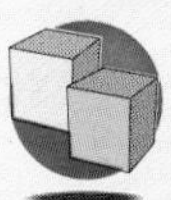

*For more information on pacing and possible lesson plans, refer to the *Block Scheduling Booklet*.

1 FOCUS

5-Minute Check *(over Lesson 10-3)*

Refer to the figure below. Find each area.

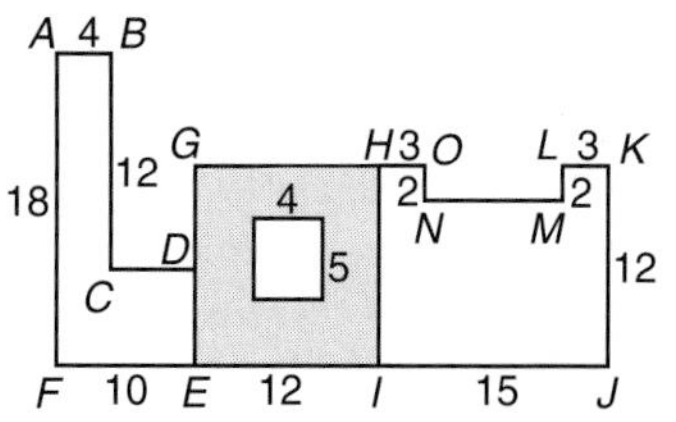

1. *ABCDEF* 108 units^2
2. shaded part of *GHIE* 124 units^2
3. *HONMLKJI* 162 units^2
4. the entire figure including the unshaded part of *GHIE* 414 units^2

Motivating the Lesson

Hands-On Activity Have students draw any type of triangle on graph paper and cut out two copies of it. Ask them to fit the two triangles together so that they form a parallelogram. Ask them how the base and height of the triangle relate to the base and height of the parallelogram. They are the same.

2 TEACH

In-Class Example

For Example 1
Find each area.

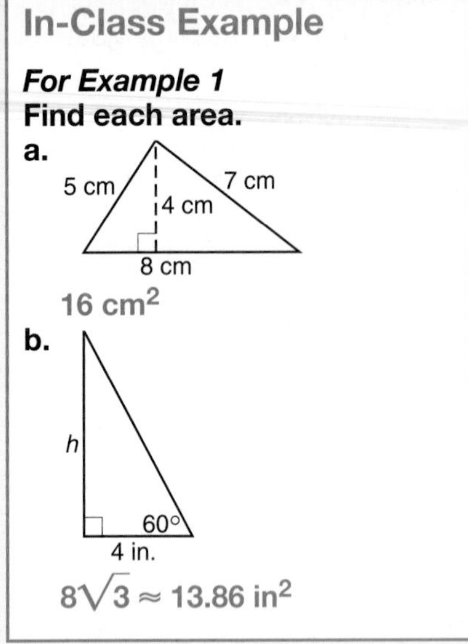

$8\sqrt{3} \approx 13.86$ in^2

Teaching Tip In the statement of the area of a triangle, point out the phrase "corresponding height," and be sure students understand what this means.

Since the combined area of the two smaller triangles *ABE* and *CBD* is the same as the area of the larger triangle *ABC,* we can see that the area of *ABC* is one-half the area of the rectangle. In Lesson 10–3, we found the area of a parallelogram by multiplying the measures of the base and the height. This leads us to the formula for the area of a triangle.

Area of a Triangle — **If a triangle has an area of *A* square units, a base of *b* units, and a corresponding height of *h* units, then $A = \frac{1}{2}bh$.**

You can find the areas of some shapes by finding the total areas of the figures that make up that shape.

Example 1 **Find the area of quadrilateral *ABCD* if *BE* = 12, *ED* = 5, and *AC* = 20.**

area of quadrilateral *ABCD* = area of △*ABC* + area of △*ADC*

Find the areas of △*ABC* and △*ACD*.

area of △*ABC*:
$A = \frac{1}{2}bh$
$= \frac{1}{2}(20)12$
$= 120$

area of △*ACD*:
$A = \frac{1}{2}bh$
$= \frac{1}{2}(20)(5)$
$= 50$

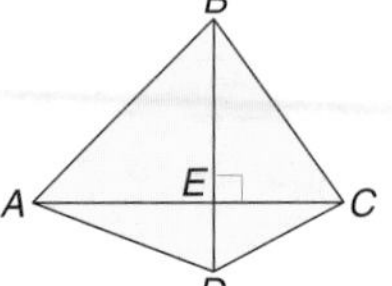

The area of quadrilateral *ABCD* is 120 + 50 or 170 square units.

Now, let's use the formula for the area of a triangle to derive the formula for the area of a trapezoid.

Suppose we are given trapezoid *RSTU* with diagonal $\overline{SU}$. The two bases of the trapezoid are parallel. Therefore, the altitude *h* from vertex *U* to the extension of base $\overline{RS}$ is the same length as the altitude from vertex *S* to the base $\overline{UT}$. Since the area of the trapezoid is the area of the two nonoverlapping parts, we can write the equation below.

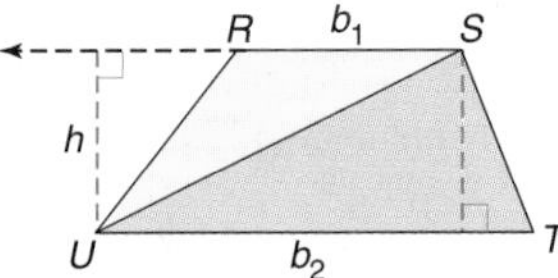

area of trapezoid *RSTU* = area of △*RSU* + area of △*STU*

area of trapezoid *RSTU* = $\frac{1}{2}(b_1)h + \frac{1}{2}(b_2)h$ *Let RS be b_1 and UT be b_2.*

$= \frac{1}{2}(b_1 + b_2)h$ *Factor.*

$= \frac{1}{2}h(b_1 + b_2)$ *Commutative Property (=)*

This leads us to the formula for the area of a trapezoid.

Area of a Trapezoid — **If a trapezoid has an area of *A* square units, bases of b_1 units and b_2 units, and height of *h* units, then $A = \frac{1}{2}h(b_1 + b_2)$.**

Alternative Learning Styles

Visual Have students design a chart that lists all the area formulas covered so far and that includes a diagram of each figure. Have them update this chart with all new formulas.

If you know the area of a figure and some of its measures, you can find missing measures.

Example 2 **Find the height of a trapezoid that has an area of 287 square inches and bases of 38 inches and 44 inches.**

Use the formula for the area of a trapezoid and solve for h.

$$A = \frac{1}{2}h(b_1 + b_2)$$

$$287 = \frac{1}{2}h(38 + 44) \quad A = 287,\ b_1 = 38,\ b_2 = 44$$

$$287 = 41h$$

$$7 = h \quad \text{Divide each side by 41.}$$

The height of the trapezoid is 7 inches.

The formula for the area of a triangle can also be used to derive the formula for the area of a rhombus. *You will be asked to derive this formula in Exercise 3.*

Area of a Rhombus	If a rhombus has an area of A square units and diagonals of d_1 and d_2 units, then $A = \frac{1}{2}d_1d_2$.

Example 3

Landscaping

Sonja Washington wants to place decorative brick edging around a flower garden that is in the shape of a rhombus. One diagonal is 12 feet long, and the area is 264 square feet. How much brick must she purchase?

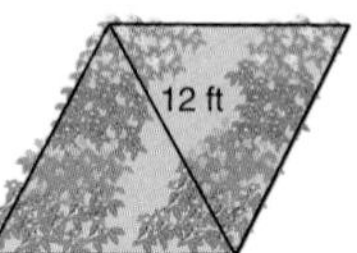

To solve the problem, we need to find the perimeter of the rhombus. First, use the area formula to find the length of the other diagonal.

$$A = \frac{1}{2}d_1d_2$$

$$264 = \frac{1}{2}(12)d_2$$

$$264 = 6d_2$$

$$44 = d_2$$

The diagonals are 12 feet and 44 feet long.

The diagonals of a rhombus are perpendicular and they bisect each other. We can use the Pythagorean Theorem to find the length of a side.

$$s^2 = 6^2 + 22^2$$

$$s^2 = 36 + 484$$

$$s^2 = 520$$

$$s = \sqrt{520} \approx 22.8$$

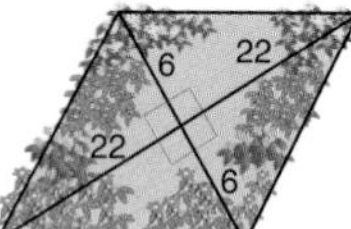

Since each side is about 22.8 feet long, the perimeter is about 4(22.8) or 91.2 feet.

Therefore, Ms. Washington needs to buy 92 feet of brick edging.

In-Class Examples

For Example 2
Find the height of a trapezoid that has an area of 50 square yards and bases of 8 yards and 12 yards. **5 yards**

For Example 3
A rhombus with a 20-centimeter diagonal has an area of 300 square centimeters. Find the perimeter of the rhombus. **72.11 cm**

Teaching Tip You may want to have students copy these formulas into a math notebook.

3 PRACTICE/APPLY

Check for Understanding

Exercises 1–10 are designed to help you assess your students' understanding through reading, writing, speaking, and modeling. You should work through Exercises 1–5 with your students and then monitor their work on Exercises 6–10.

Error Analysis

Students may forget to multiply by $\frac{1}{2}$ when using the formulas for triangles, trapezoids, or rhombi. Encourage them to write the appropriate area formula first and then substitute the given data.

Additional Answers

3. A rhombus is made up of two congruent triangles and using d_1 and d_2 instead of b and h, its area in reference to $A = \frac{1}{2}bh$ is $2\left[\frac{1}{2}(d_1)\left(\frac{1}{2}d_2\right)\right]$ or $\frac{1}{2}d_1d_2$.
4. See students' work. Sample answer: use trapezoid formula, use median of trapezoid, and split trapezoid up into shape of other figures and then add areas together.

5a. See students' work. Sample answer:

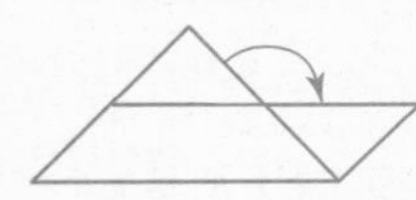

5b. Each new figure has the same area as its original triangle.

The following chart summarizes the area formulas you have learned thus far.

Summary of Area Formulas			
Square	$A = s^2$	Triangle	$A = \frac{1}{2}bh$
Rectangle	$A = \ell w$	Trapezoid	$A = \frac{1}{2}h(b_1 + b_2)$
Parallelogram	$A = bh$	Rhombus	$A = \frac{1}{2}d_1d_2$

CHECK FOR UNDERSTANDING

Communicating Mathematics

Study the lesson. Then complete the following.

1. **Explain** whether $\triangle ABC$ and $\triangle DEF$ have the same area.
 Yes; $\frac{1}{2}(3)(7) = \frac{1}{2}(7)(3)$ by the Commutative Property (=).

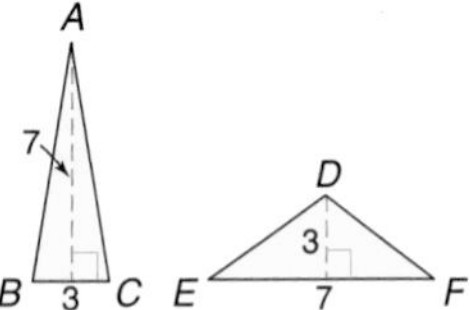

2. **Analyze** the following quilting patterns and determine the formula(s) you would use to find the area of each section in each pattern.

2d. triangle and rhombus

a. 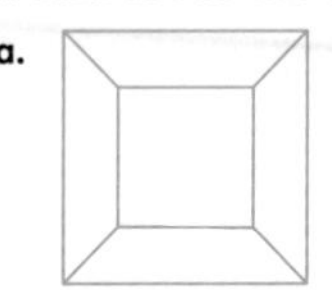 trapezoid and square
b.  triangle
c. 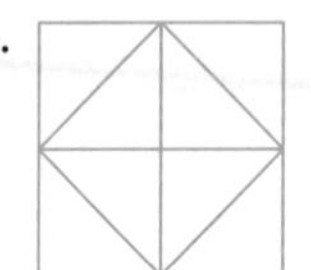 triangle and square
d. 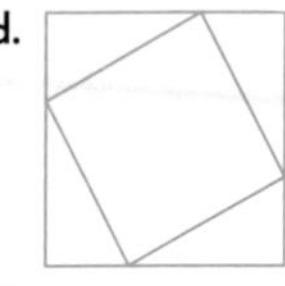

3. **Describe** how the formula for the area of a rhombus is related to the formula for the area of a triangle. See margin.
4. **Draw** a trapezoid and measure the height and each base. Determine three different ways to find the area of the trapezoid. Show and explain your work. See margin.

5. Draw a right triangle, an acute triangle, and an obtuse triangle on a piece of paper. Cut out each of them. Cut one line through each of the triangles and position the two pieces to form another figure. a–b. See margin.
 a. Describe the figures formed by each pair of pieces.
 b. What is true of the areas of the original triangles and the new figures?

Guided Practice

Find the area of each figure.

6. 3.4 in., 7.3 in. — 12.41 in²

7. 60 cm²

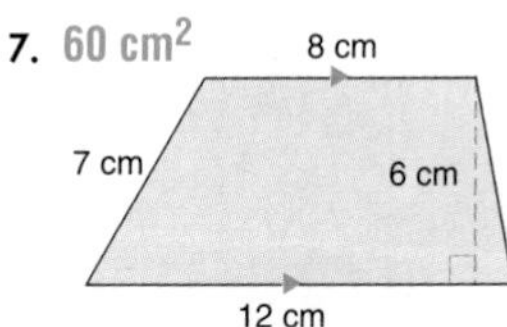

8. If the area of $\triangle ABC$ is 24 square units, find the value of x. 6 units

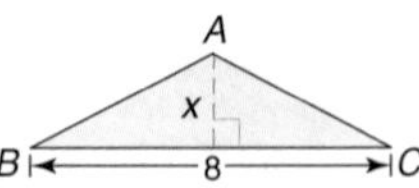

Reteaching

Using Word Problems Use these three exercises to review how to find the area of a triangle, a trapezoid, and a rhombus. Draw a picture of each figure on the chalkboard, and explain the formulas used.

1. triangle with base measuring 12 inches and height of 5 inches 30 in²
2. trapezoid with bases measuring 10 centimeters and 13 centimeters and height of 7 centimeters 80.5 cm²
3. rhombus with diagonals measuring 8 meters and 11 meters 44 m²

9. A rhombus has side lengths of 12 inches each and an angle measure of 120. Find the area of the rhombus. $72\sqrt{3} \approx 124.7$ in^2

10. **Cleaning** City Carpeting will clean up to 325 square feet of carpeting for $78. Yolanda Alvarez has three rectangular bedrooms which all have carpeting. Their measures are 8 feet by 13 feet, 8 feet by 10 feet, and 12 feet by 12 feet. Will Mrs. Alvarez have to pay more than $78 to have the three bedrooms cleaned? Explain. Yes; the total area is 328 ft^2.

EXERCISES

Practice **Find the area of each figure.**

11.
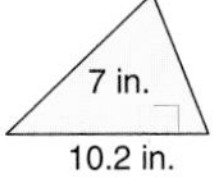

35.7 in^2

12.
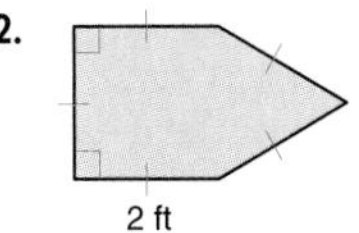

$(4 + \sqrt{3}) \approx 5.7$ ft^2

13.
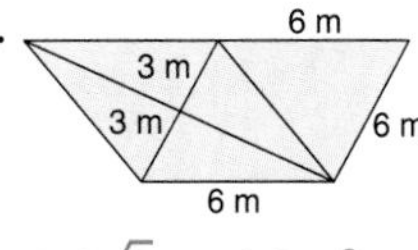

$27\sqrt{3} \approx 46.8$ m^2

14.
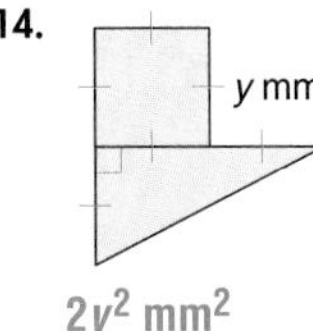

$2y^2$ mm^2

15.

3 m 12 m 20 m 4 m

88 m^2

16.
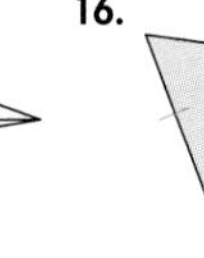

$(144 + 72\sqrt{3}) \approx 268.7$ yd^2

Find the value of *x* in each figure.

17. $A = 12$ 4
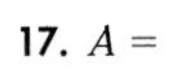
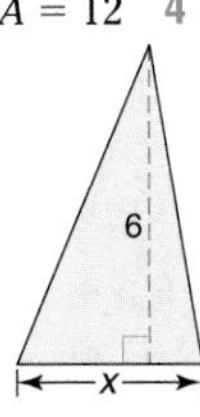

18. $A = 56$ 7
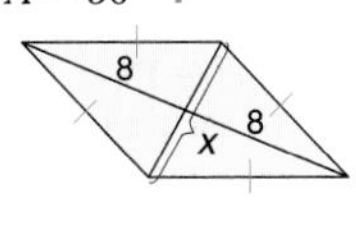

19. $A = 95$ 10
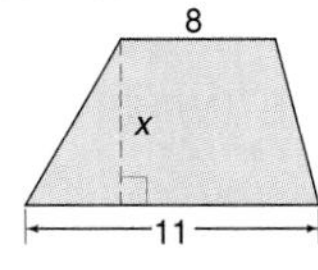

20. A trapezoid has an area of 75 square inches, and its two bases are 8 and 17 inches long. Find the height of the trapezoid. 6 in.

21. A rhombus has an angle measure of 120, and its longer diagonal has a length of 10 inches. Find the area of the rhombus. about 28.9 in^2

22. The shorter diagonal of a rhombus is 10 inches long, and the angle the diagonal forms with the side measures 60°. Find the area of the rhombus. $50\sqrt{3} \approx 86.6$ in^2

23. $16\sqrt{3} \approx 27.7$ cm^2

23. Find the area of an equilateral triangle whose perimeter is 24 centimeters.

C

24. A trapezoid has an area of 126 square feet, a height of 9 feet, and one base of 13 feet. Find the length of the second base. 15 ft

25. The area of an isosceles trapezoid is 77 square inches, the height is 4 inches, and the congruent sides of the trapezoid are each 5 inches long. Find the lengths of the bases. 16.25 in., 22.25 in.

Assignment Guide

Core (with proof): 11–31 odd, 33–40
Core (informal): 11–31 odd, 33–40
Enriched: 12–26 even, 27–40
All: Self Test, 1–10

For **Extra Practice**, see p. 783.

The red A, B, and C flags, printed only in the Teacher's Wraparound Edition, indicate the level of difficulty of the exercises.

Study Guide Masters, p. 60

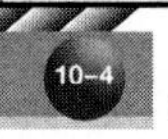

NAME ____________ DATE ________

Study Guide

Student Edition Pages 535–541

Area of Triangles, Rhombi, and Trapezoids

Formulas for the areas of triangles, trapezoids, and rhombi can be obtained from the formula for the area of a parallelogram.

triangle
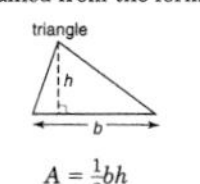

$A = \frac{1}{2}bh$

trapezoid
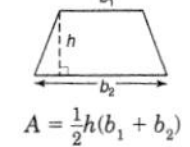

$A = \frac{1}{2}h(b_1 + b_2)$

rhombus
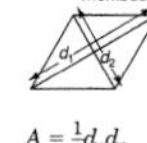

$A = \frac{1}{2}d_1d_2$

Example: Find the area of the trapezoid.

$A = \frac{1}{2}h(b_1 + b_2)$
$A = \frac{1}{2}(10)(12 + 19)$
$A = 155$

The area is 155 square meters

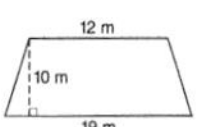

Find the area of each figure.

1.
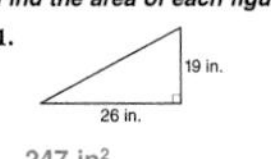

247 in^2

2.
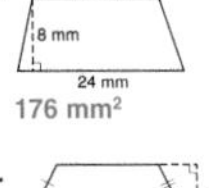

176 mm^2

3.
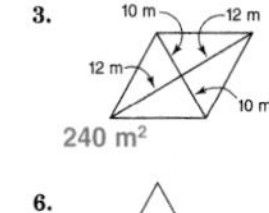

240 m^2

4.
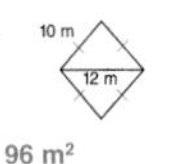

96 m^2

5. 24 m 9 m

216 m^2

6. 10 cm

$25\sqrt{3}$ cm^2 ≈ 43.3 cm^2

7. The area of a triangle is 150 square inches. If the height is 20 inches, find the length of the base. 15 in.

8. A rhombus has a perimeter of 100 meters and a diagonal 30 meters long. Find the area of the rhombus. 600 m^2

Using the Programming Exercises The program given in Exercise 29 is for use with a TI-82 or TI-83 graphing calculator. For other programmable calculators, have students consult their owner's manual for commands similar to those represented here.

Additional Answer

32. Sample answer:

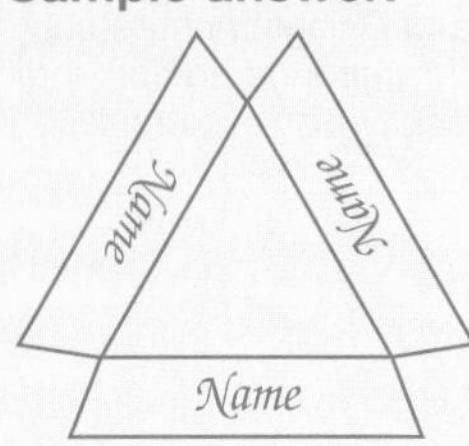

CAREER CHOICES

An industrial designer not only has to plan the structure, operation, and appearance of a product, but also has to keep in mind efficient production, distribution, and selling procedures. The designer balances mass appeal with efficiency and utility.

Practice Masters, p. 60

10–4 NAME__________ DATE________

Practice — Student Edition Pages 535–541

Area of Triangles, Rhombi, and Trapezoids

Find each missing measure.

1. The area of a triangle is 216 square units. If the height is 18 units, what is the length of the base? 24 units

2. The diagonals of a rhombus are 21 and 16 centimeters long. Find the area of the rhombus. 168 cm^2

3. The area of a trapezoid is 80 square units. If its height is 8 units, find the length of its median. 10 units

4. The height of a trapezoid is 9 cm. The bases are 8 cm and 12 cm long. Find the area. 90 cm^2

5. A trapezoid has an area of 908.5 cm^2. If the altitude measures 23 cm and one base measures 36 cm, find the length of the other base. 43 cm

6. The measure of the consecutive sides of an isosceles trapezoid are in the ratio 8:5:2:5. The perimeter of the trapezoid is 140 inches. If its height is 28 inches, find the area of the trapezoid. 980 in^2

26. What is the relationship between the base of a triangle and the base of a parallelogram that both have an area of 96 square centimeters and a height of 12 centimeters? The base of the triangle is twice as long as the base of the parallelogram.

Critical Thinking

27. Use $\triangle BCD$ and $\square ABDE$ at the right to derive the formula for the area of a trapezoid. See Solutions Manual.

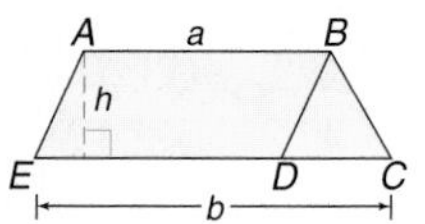

28. In the figure, the vertices of quadrilateral *HBDF* intersect the square *ACEG* and divides its sides into segments whose measures have a ratio of 1:2.

a. Find the area of quadrilateral *HBDF*. 45 m^2

b. What type of figure is *HBDF*? square

c. What is the relationship between the areas of quadrilateral *HBDF* and square *ACEG*? It will always have a ratio of 5:9.

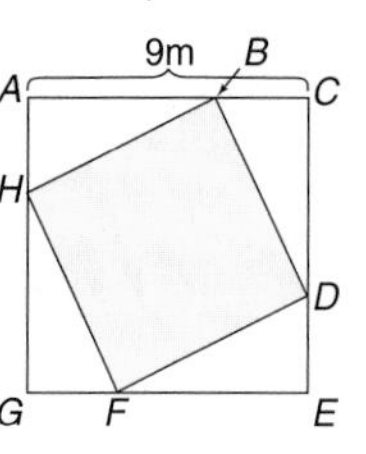

Programming

29. The TI-82/83 program at the right finds the area of a trapezoid given the height and the measures of the bases.

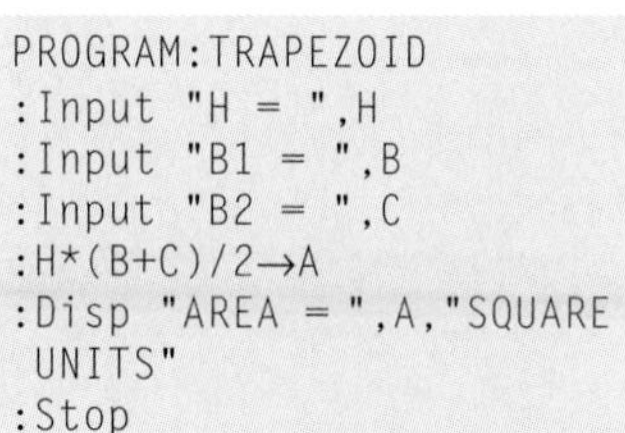

```
PROGRAM:TRAPEZOID
:Input "H = ",H
:Input "B1 = ",B
:Input "B2 = ",C
:H*(B+C)/2→A
:Disp "AREA = ",A,"SQUARE
 UNITS"
:Stop
```

Use the program to find the area of each trapezoid.

a. $h = 5, b_1 = 8, b_2 = 6$ 35 $units^2$

b. $h = 3.5, b_1 = 7.1, b_2 = 8.4$ 27.125 $units^2$

Applications and Problem Solving

30. Real Estate Kamali Narula planned to sell a trapezoid-shaped portion of her property that has parallel sides measuring 147.08 feet and 57.62 feet. The other sides measure 144.34 feet and 90.54 feet while the depth of the property is 50.74 feet.

a. What is the area of this property? 5193.239 ft^2

b. If an acre is 43,560 square feet, what percent of an acre is this property? b. 11.9%

CAREER CHOICES

An **industrial designer** develops and designs manufactured products and materials. Creativity, persistence, problem-solving skills, and communication of ideas, both visually and verbally, are required for this career.

For more information, contact:

The Industrial Designers Society of America
1142-E Walker Road
Great Falls, VA 22066

31. Home Repair Rosa Llarena needed to replace the roof on her home. The roof had two large trapezoids, one in the front and one in the back, two smaller trapezoids, one on either side of the roof, and a rectangle on the top. If each of the trapezoids has the same height and a package of shingles covers 100 square feet, how many packages of shingles should Ms. Llarena buy? 30 packages

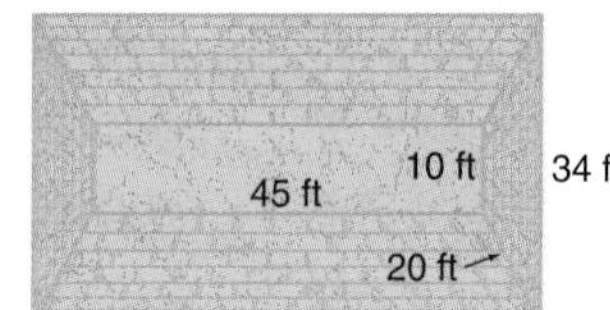

32. Design Zephia Watson was planning a meeting and was not satisfied with the usual place marker that could only be seen from two directions. She wanted to develop place markers that could be seen from three directions. Use what you know about triangles, rhombi, and trapezoids to design a creative pattern for a place marker that Zephia could use. See margin for sample answer.

Mixed Review

33. The area of a rectangle is 520 square units, and its perimeter is 106 units. What are its dimensions? (Lesson 10–3) 13 units by 40 units

34. Is the tessellation shown at the right *uniform*, *regular*, or *semi-regular*? Name all possibilities. (Lesson 10–2) uniform

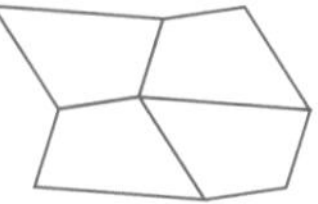

Extension

Problem Solving David Laing wants to fertilize his lawn. A diagram of his lot, which is in the shape of a trapezoid, is shown. Before he purchases the fertilizer, he needs to know the area of his lawn. Calculate the number of square feet the fertilizer must cover. 7850 ft^2

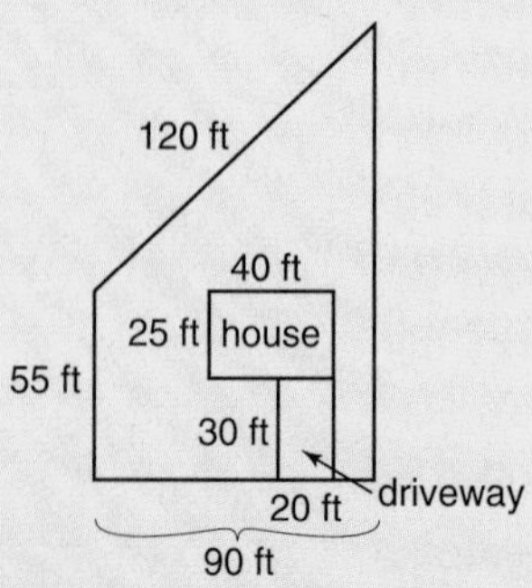

35. Find the value of h and j in the two concentric circles below. (Lesson 9–7) $h = 45, j = 60$

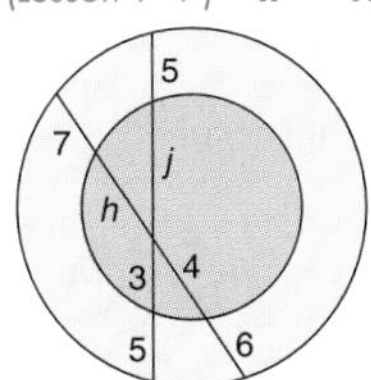

36. Find the value of r to the nearest tenth. (Lesson 8–3) 13.5 m

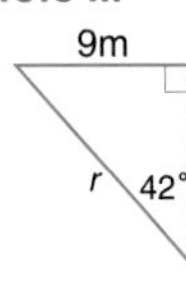

37. No; consecutive angles in a parallelogram must be supplementary.

37. In quadrilateral $JULY$, $m\angle U = 120$ and $m\angle L = 50$. Is $JULY$ a parallelogram? Justify your answer. (Lesson 6–1)

38. Find the value of g if $\overline{QT}$ is an altitude. (Lesson 5–1) $g = \frac{25}{3}$

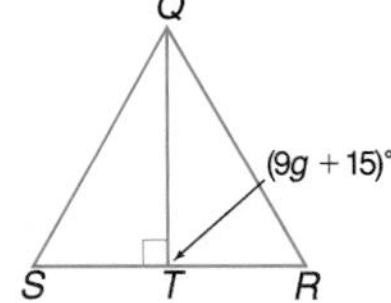

INTEGRATION Algebra

39. Find $\frac{k^3\ell m^2}{x^2y^2} \div \frac{k^3m}{x^3y^2}$. ℓmx

40. Find the supplement of a 64° angle. 116

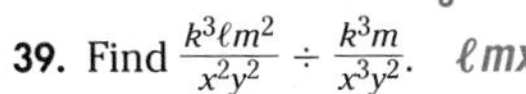

SELF TEST

1. **Refer to polygon *ABCDEFG*.** (Lesson 10–1)
   a. Identify the type of polygon. **heptagon**
   b. Is it regular? **no**
   c. Is it convex or concave? **convex**

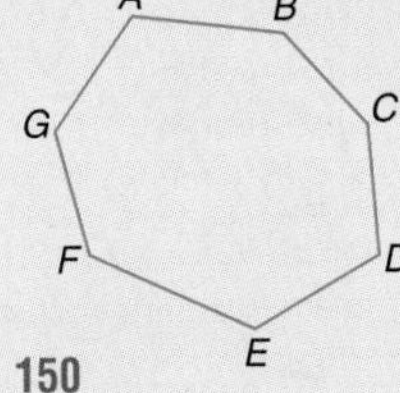

Find the measure of an interior angle of each regular polygon. (Lesson 10–1)

2. 20-gon **162**
3. dodecagon **150**

The measure of an interior angle of a regular polygon is given. Find the measure of an exterior angle and the number of sides of each regular polygon. (Lesson 10–1)

4. 160 **20, 18**
5. $168\frac{3}{4}$ **$11\frac{1}{4}$, 32**

6. Is the tessellation shown below *regular*, *uniform*, or *semi-regular*? Name all possibilities. (Lesson 10–2)

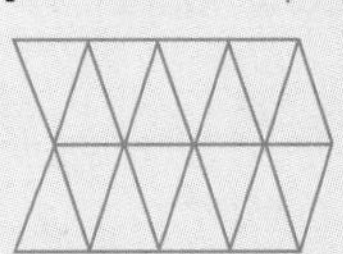

not regular, uniform, not semi-regular

7. Find the area of the green region below. (Lesson 10–3) **9200 ft²**

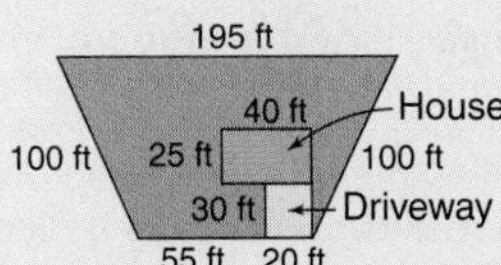

Find the area of each region. (Lesson 10–4)

8. **3.78 m²**

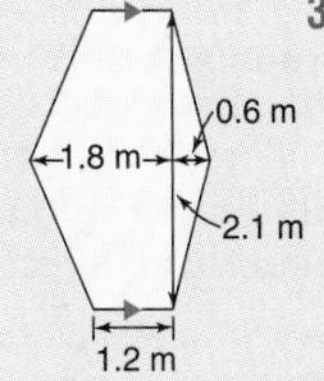

9. **150 cm²**

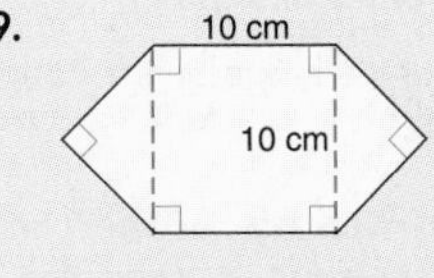

10. **Guess and Check** Place two addition symbols and two subtraction symbols in the expression below to make it a true equation. Note that digits can be put together to form larger numbers, but not rearranged. (Lesson 10–2) **Sample answer: 35 − 64 + 752 + 6 − 17 = 712**

3 5 6 4 7 5 2 6 1 7 = 712

SELF TEST

The Self Test provides students with a brief review of the concepts and skills in Lessons 10-1 through 10-4. Lesson numbers are given to the right of exercises or instruction lines so students can review concepts not yet mastered.

4 ASSESS

Closing Activity

Speaking Have students come one at a time to the chalkboard to state an area formula and give an example of its use. Continue until all six formulas stated at the top of page 538 have been demonstrated.

Chapter 10 Quiz B (Lessons 10-3 and 10-4) is available in the *Assessment and Evaluation Masters*, p. 268.

Mid-Chapter Test (Lessons 10-1 through 10-4) is available in the *Assessment and Evaluation Masters*, p. 267.

Enrichment Masters, p. 60

10-4 NAME______ DATE______

Enrichment — Student Edition Pages 535–541

Heron's Formula

If you know the lengths of the sides of a triangle, you can use Heron's formula to find the area.

Heron's Formula:
The area, A, of a triangle with sides measuring a, b, and c is $A = \sqrt{s(s-a)(s-b)(s-c)}$, where $s = \frac{a+b+c}{2}$.

Example: Use Heron's formula to find the area of the triangle.

$s = \frac{2.7 + 3.8 + 5.3}{2} = 5.9$

$A = \sqrt{5.9(5.9-2.7)(5.9-3.8)(5.9-5.3)}$

$= \sqrt{5.9(3.2)(2.1)(0.6)}$

$= \sqrt{23.8} \approx 4.9$

The area is about 4.9 cm².

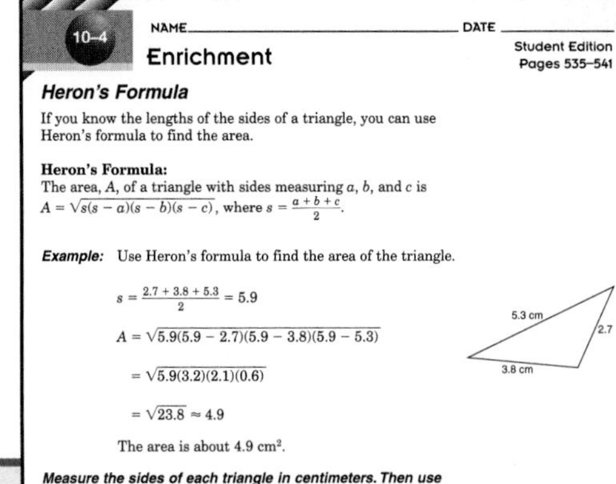

Measure the sides of each triangle in centimeters. Then use Heron's formula to find the area. Round your answers to the nearest tenth.

1. 8.4 cm²
2. 12.0 cm²
3. 9.2 cm²
4. Construct an altitude for the triangle in problem 2. Then measure the altitude in centimeters and compute the area using the formula $A = \frac{1}{2}bh$. Did you get the same result as with Heron's formula? See students' work. Results should be approximately the same.

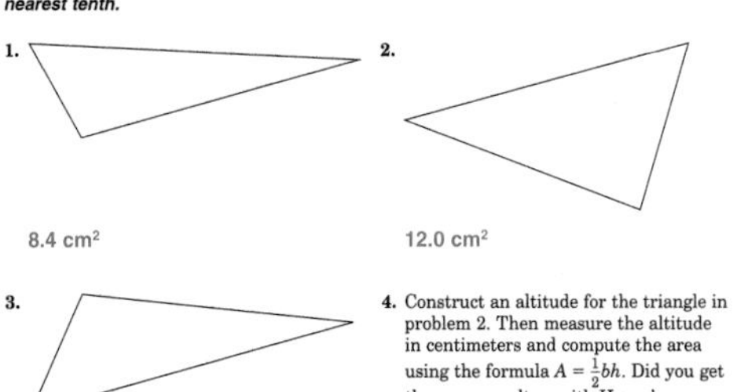

10-5A LESSON NOTES

NCTM Standards: 2–4, 7

Objective

Create a regular n-gon.

Recommended Time

Demonstration and discussion: 15 minutes; Exercises: 30 minutes

Instructional Resources

For each student or group of students
Student Manipulative Kit
- hinged mirrors
- protractors
- straightedges

Modeling Mathematics Masters
- p. 98 (worksheet)

For teacher demonstration
Algebra and Geometry Overhead Manipulative Resources

1 FOCUS

Motivating the Lesson

Students will use a hinged mirror to investigate regular n-gons. Give students time to experiment with the hinged mirror, noting how it works, perhaps using simple drawings or text.

2 TEACH

Teaching Tip Caution students to keep the plane of the mirror perpendicular to the plane of the paper.

3 PRACTICE/APPLY

Assignment Guide

Core (with proof): 1–9
Core (informal): 1–9
Enriched: 1–9

MODELING MATHEMATICS

A Preview of Lesson 10–5

10–5A Regular Polygons

Materials:  straightedge  hinged mirror  protractor  plain paper

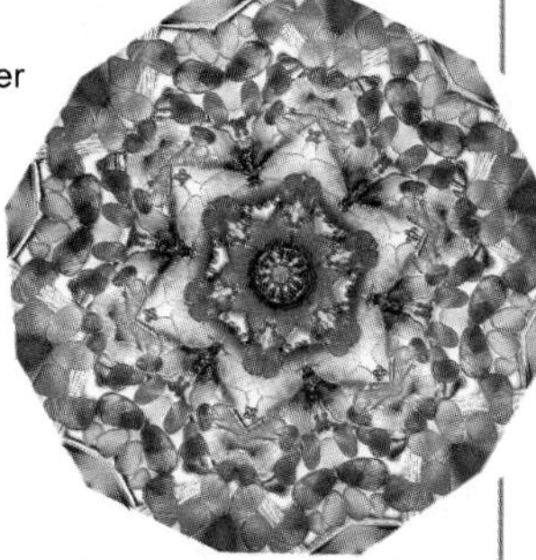

Mirrors are often used to create images in kaleidoscopes and telescopes as well as in interior design. You can use a hinged mirror to create any n-gon that is regular. The mirror uses multiple reflections to create the number of sides you desire.

Activity **Use a hinged mirror to create a regular hexagon.**

Step 1 Use a straightedge to draw a line on a piece of plain paper. Position the hinged mirror so that the sides of the mirror intersect the line at approximately the same distance from the hinge of the mirror.

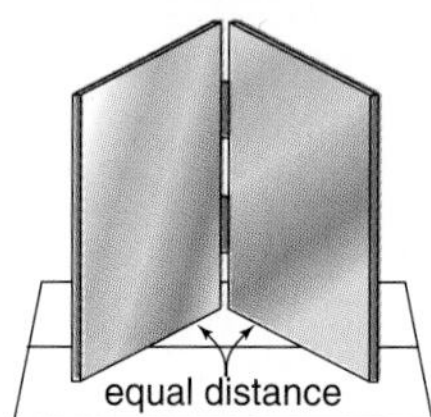

Step 2 Sketch what you see. Place a protractor on top of the mirror so that you can measure the angle of the mirror. Record this measurement.

Step 3 Open the mirror wider. Make sure the line still intersects the mirror's edges at equal distances from the hinge. What changes occur in the figure you see? Sketch the figure and record the angle of the mirror.

Step 4 Open or close the mirror until the image you see has six sides. Sketch the figure and record the angle of the mirror.

Write **Use the six-sided image you formed in Step 4 above to answer these questions.** 2a. same number of angles as sides

1. Find the sum of the angle measures for those angles whose vertex is the mirror's hinge. (*Hint:* Each of these angles is a central angle.) 360
2. a. How many angles are there in the image for each number of sides?
   b. What is the measure of each angle? 60
3. How does the measure of the hinged mirror angle compare with the measure you calculated in Exercise 2b? They are the same.

Draw **Use what you have learned about the central angles to create each regular polygon with a hinged mirror. Record the central angle measure and sketch the polygon.** 4–9. See students' work for sketch of polygon.

4. triangle 120
5. square 90
6. pentagon 72
7. octagon 45
8. decagon 36
9. dodecagon 30

4 ASSESS

Observing students working in cooperative groups is an excellent method of assessment.

Using Cooperative Learning

This lesson offers an excellent opportunity for using cooperative learning groups. For more information on cooperative learning strategies and group management, see *Cooperative Learning in the Mathematics Classroom*, one of the titles in the Glencoe Mathematics Professional Series.

10-5 Area of Regular Polygons and Circles

What YOU'LL LEARN
- To find areas of regular polygons, and
- to find areas of circles.

Why IT'S IMPORTANT

You can use the area of polygons and circles to solve problems involving landscape design and architecture.

APPLICATION
Gardening

The central feature of Elizabeth Park in Hartford, Connecticut, is the Rose Garden. Over 15,000 rosebushes of approximately 800 varieties are planted in the garden. In 1896, Theodore Wirth designed the garden with a rustic summer house encircled with concentric circles of rosebushes. Wide paths radiate from the center and divide the garden into square and circular sections.

The polygon and the circle are closely related. Both the circumference and the area of circles can be approximated using the perimeter and area of regular polygons. All regular polygons can be inscribed in a circle.

CONSTRUCTION
Regular Hexagon

Use a compass and straightedge to construct a regular hexagon.

1. Use your compass to draw a $\odot P$. Point P will also be the **center** of the hexagon. The radius of $\odot P$ is the **radius** of the hexagon and is congruent to a side of the hexagon.

2. Using the same compass setting, place the compass point on the circle and draw an arc, labeling the point of intersection with the circle, A.

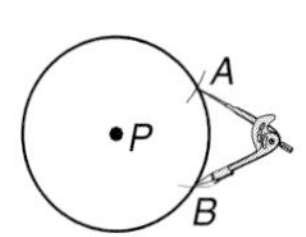

3. Place the compass point on A and draw another arc. Label the point of intersection with the circle, B.

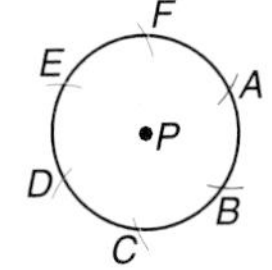

4. Continue this process, labeling points C, D, E, and F, on the circle. Point F should be where the point of the compass was placed to locate point A.

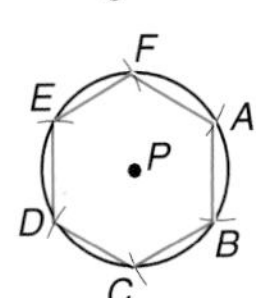

5. Use a straightedge to connect A, B, C, D, E, and F, consecutively.

Conclusion: $ABCDEF$ is a regular hexagon inscribed in $\odot P$ because in a circle, if minor arcs are congruent, then their corresponding chords are congruent.

GLOBAL CONNECTIONS

Circles or regular multisided polygons have often been used as plans for cities or monuments. Stonehenge, built about 1800–1400 B.C., is a circular monument in England. Baghdad, the capital of Iraq, began as a circular village in the early 600s.

10-5 LESSON NOTES

NCTM Standards: 1–5, 7, 8

Instructional Resources

- Study Guide Master 10-5
- Practice Master 10-5
- Enrichment Master 10-5
- Modeling Mathematics Masters, p. 88
- Multicultural Activity Masters, p. 19
- Tech Prep Applications Masters, p. 20

Transparency 10-5A contains the 5-Minute Check for this lesson; **Transparency 10-5B** contains a teaching aid for this lesson.

Recommended Pacing	
Standard Pacing	Days 9 & 10 of 15
Honors Pacing	Days 8 & 9 of 14
Block Scheduling*	Day 5 of 8

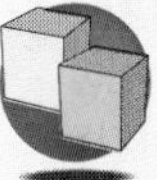

*For more information on pacing and possible lesson plans, refer to the *Block Scheduling Booklet*.

1 FOCUS

5-Minute Check
(over Lesson 10-4)

Find each area.

1. triangle with base measuring 50 millimeters and height of 100 millimeters 2500 mm^2
2. trapezoid with bases measuring 15 inches and 21 inches and a height of 10 inches 180 in^2
3. isosceles trapezoid with bases measuring 14 centimeters and 18 centimeters and leg measuring 9 centimeters 140.4 cm^2
4. rhombus with diagonals measuring 24 feet and 20 feet 240 ft^2
5. right triangle with one leg measuring 5 yards and a hypotenuse measuring 13 yards 30 yd^2

GLOBAL CONNECTIONS

Stonehenge was actually built in three separate phases. The second phase was built to align with the summer solstice but was dismantled during the third phase. It is assumed that Stonehenge was constructed as a place of worship, but it hasn't been aligned to any religion yet.

Motivating the Lesson

Questioning Ask students to name some real objects in the shape of regular polygons. Sample answers: tiles, faces of geodesic domes, road signs, pieces of leather on a soccer ball

2 TEACH

Teaching Tip Note that $\triangle PED$ in the figure at the top of the page is actually an equilateral triangle. When altitude $\overline{PT}$ is drawn, two 30°-60°-90° triangles are formed.

In-Class Example

For Example 1
Find the area of a regular pentagon whose perimeter is 60 centimeters.
247.7 cm^2

Teaching Tip After reading about the apothem, ask students how many different apothems can be drawn in the figure in Example 1. 8

In regular hexagon *ABCDEF*, suppose we draw $\overline{PE}$, $\overline{PD}$, and $\overline{PT}$ so that $\overline{PT}$ is perpendicular to $\overline{ED}$. A segment like $\overline{PT}$ that is drawn from the center of a regular polygon perpendicular to a side of the polygon is called an **apothem**.

$\triangle PED$ is an isosceles triangle, since sides $\overline{PE}$ and $\overline{PD}$ are radii of $\odot P$. If all of the radii of hexagon *ABCDEF* were drawn, they would separate the hexagon into six triangles all congruent to $\triangle PED$.

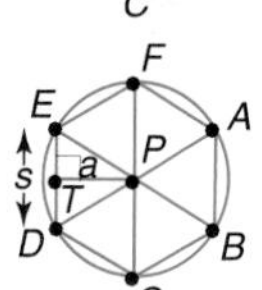

Now, since the area of a region is the sum of the areas of its nonoverlapping parts, you can find the area of the hexagon by adding the areas of the triangles. Since $\overline{PT}$ is perpendicular to $\overline{ED}$, it is an altitude of $\triangle PED$ as well as an apothem of hexagon *ABCDEF*. Let a represent the measure of $\overline{PT}$ and s represent the length of a side of the hexagon.

$$\text{area of } \triangle PED = \frac{1}{2}bh \quad \textit{Formula for area of a triangle}$$
$$= \frac{1}{2}sa$$

The area of one of the six triangles is $\frac{1}{2}sa$ units^2. So the area of the hexagon is $6\left(\frac{1}{2}sa\right)$ units^2. Notice that the perimeter of hexagon *ABCDEF* is $6s$ units. Therefore, if the perimeter of the hexagon is P units, the area will be $\frac{1}{2}Pa$ square units. This area formula can be used for *any* regular polygon.

Area of a Regular Polygon  **If a regular polygon has an area of *A* square units, a perimeter of *P* units, and an apothem of *a* units, then $A = \frac{1}{2}Pa$.**

Example 1 **Find the area of a regular octagon that has a perimeter of 72 inches.**

To use the formula for the area of a regular polygon, we must find the length of the apothem $\overline{OP}$.

The central angles of *ABCDEFGH* are all congruent. Therefore, the measure of each angle is $\frac{360}{8}$ or 45. Since $\overline{OP}$ is an apothem of the octagon, it is perpendicular to and bisects $\overline{AB}$. It also bisects central angle *AOB*. Therefore, $m\angle BPO = 90$ and $m\angle BOP = 22.5$. Since the perimeter is 72 inches, each side of the octagon is $\frac{72}{8}$ or 9 inches. Therefore, $PB = \frac{1}{2}(9)$ or 4.5.

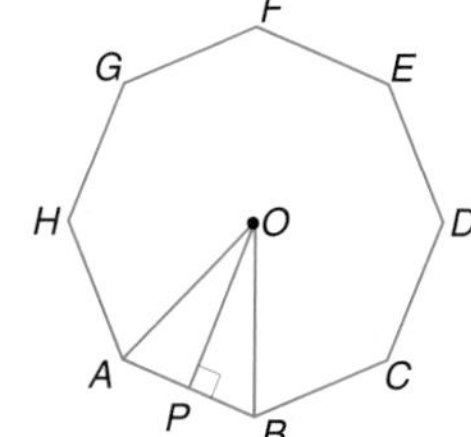

Use a trigonometric ratio to find the length of $\overline{OP}$.

$$\tan \angle BOP = \frac{PB}{OP}$$
$$\tan 22.5 = \frac{4.5}{OP} \quad \textit{m}\angle BOP = 22.5,\ PB = 4.5$$
$$OP (\tan 22.5) = 4.5 \quad \textit{Multiply each side by OP.}$$
$$OP = \frac{4.5}{\tan 22.5} \quad \textit{Divide each side by tan 22.5.}$$
$$OP \approx 10.9 \quad \textit{Use a calculator.}$$

Now use the formula for the area of a regular polygon.

$$A = \frac{1}{2}Pa$$
$$A = \frac{1}{2}(72)(10.9)$$
$$A \approx 392.4$$

The area of the regular octagon is about 392.4 square inches.

You can use a calculator to help derive the formulas for the area and circumference of a circle from the areas and perimeters of regular polygons.

EXPLORATION SCIENTIFIC CALCULATOR

Suppose each regular polygon is inscribed in a circle of radius r. Copy and complete the following chart. Round to the nearest hundredth.

	(triangle)	(pentagon)	(octagon)	(decagon)	(20-gon)	(50-gon)
Number of Sides	3	5	8	10	20	50
Measure of a Side	$1.73r$	$1.18r$	$0.77r$	$0.62r$	$0.31r$	$0.126r$
Perimeter	$5.20r$	$5.90r$	$6.16r$	$6.20r$	$6.20r$	$6.30r$
Measure of Apothem	$0.5r$	$0.81r$	$0.92r$	$0.95r$	$0.99r$	$0.998r$
Area	$1.30r^2$	$2.39r^2$	$2.83r^2$	$2.95r^2$	$3.07r^2$	$3.14r^2$

a. What happens to the perimeters as the number of sides increases?
b. What happens to the areas as the number of sides increases?

a. Perimeters of polygons approach the circumference of circle.
b. Areas of polygons approach the area of circle.

LOOK BACK
Refer to Lesson 9-1 to review circumference of a circle.

From this Exploration, you found that as the number of sides increases, the perimeter approaches $6.28r$, and the area approaches $3.14r^2$. Notice that $6.28 = 2 \times 3.14$ and 3.14 is an approximation of the irrational number called π (pi). So $6.28r \approx 2\pi r$ and $3.14r^2 \approx \pi r^2$. Therefore, the perimeters approach the circumference of the circle, and the areas approach the area of the circle.

Area of a Circle	**If a circle has an area of A square units and a radius of r units, then $A = \pi r^2$.**

Example 2

Design

The area of a circular pool is approximately 7850 square feet. The owner wants to replace the tiling at the edge of the pool.

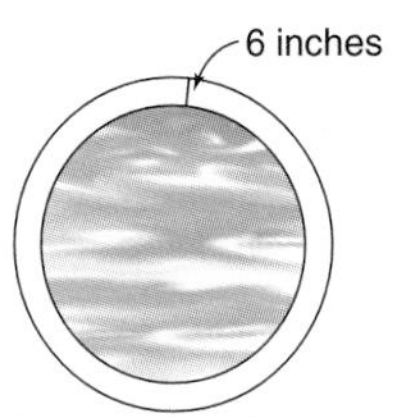

a. He plans to use tiles that are each 6-inch squares since the edging is 6 inches wide. How many tiles should he purchase?

(continued on the next page)

EXPLORATION

In this Exploration, students will discover that a circle is the limit of a regular polygon as the number of sides increases.

In-Class Example

For Example 2

The Rosenthals have built a gazebo in the shape of a regular hexagon. The length of each side of the gazebo is 8 yards. They want to buy outdoor carpeting for it.

a. Find the area of the floor of the gazebo. $96\sqrt{3} \approx 166.28$ yd^2

b. Find the cost of carpeting the gazebo if carpeting costs \$18 per square yard. \$2993.04

In-Class Example

For Example 3

Find the area of the shaded region.

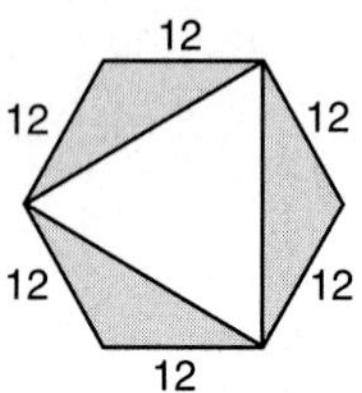

$108\sqrt{3} \approx 187.1$ units^2

Use the area formula to find the radius of the pool.

$$
\begin{aligned}
A &= \pi r^2 && \textit{Formula for area of circle} \\
7850 &= \pi r^2 \\
\frac{7850}{\pi} &= r^2 && \textit{Divide each side by } \pi. \\
50.0 &\approx r && \textit{Use a calculator.}
\end{aligned}
$$

Now use the circumference formula to find the distance around the pool.

$$
\begin{aligned}
C &= 2\pi r && \textit{Formula for circumference} \\
C &\approx 2\pi(50) \\
C &\approx 100\pi \\
C &\approx 314.2 && \textit{Use a calculator.}
\end{aligned}
$$

Divide the circumference by the length of each tile to find the number of tiles to go around the edge of the pool.

$\frac{314.2}{6} \approx 52.4$

The owner should purchase 52 tiles and space evenly.

b. Once the square tiles are placed around the circular pool, there will be extra space between each square. What shape of tile will fill this space and how many tiles of this shape will he need to go around the pool?

The best shape would be a triangle. The owner will need the same number of triangular tiles as 6-inch square tiles.

In Lesson 10–3, you learned how to find the area of a specific region by adding the areas of various shapes.

Example 3 **Find the area of the shaded region if $m\angle BAC = m\angle BCA = 60$.**

The area of the shaded region is the area of $\odot O$ minus the area of $\triangle ABC$.

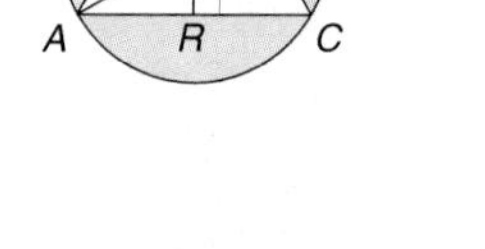

Find the area of the circle.

$$
\begin{aligned}
A &= \pi r^2 \\
A &= \pi(8^2) && r = 8 \\
A &= 64\pi \\
A &\approx 201.1 && \textit{Use a calculator.}
\end{aligned}
$$

Now find the area of $\triangle ABC$. $\triangle AOR$ is a 30°-60°-90° triangle, so $OR = 4$ and $AR = 4\sqrt{3}$. Since $AC = 2(AR)$, $AC = 2(4\sqrt{3})$ or $8\sqrt{3}$. The perimeter of $\triangle ABC$ is $3(8\sqrt{3})$ or $24\sqrt{3}$. Now use the formula for the area of a regular polygon.

$$
\begin{aligned}
A &= \frac{1}{2}Pa \\
&= \frac{1}{2}(24\sqrt{3})(4) \\
A &\approx 83.1
\end{aligned}
$$

The area of the shaded region is about 201.1 − 83.1 or 118.0 square units.

CHECK FOR UNDERSTANDING

Communicating Mathematics

Study the lesson. Then complete the following. 1–3. See margin.

1. **Explain** how you would find the apothem of a regular hexagon if you know its perimeter.
2. **Describe** the relationship between the radius and the apothem of a regular polygon.
3. The circumference of a circle is sometimes expressed as $C = \pi d$, where d is the measure of the diameter of the circle. Why is this formula equivalent to $C = 2\pi r$?

4. Using graph paper, draw several regular polygons. Inscribe a circle in each polygon and then circumscribe a circle about each polygon. Describe when a radius and an apothem will be the same and when they will be different. They will be the same when the circle is inscribed in the polygon and different when the circle is circumscribed about the polygon.

Guided Practice

For Exercises 5–7, use the figure at the right. 5. $8\sqrt{2} \approx 11.31$ cm

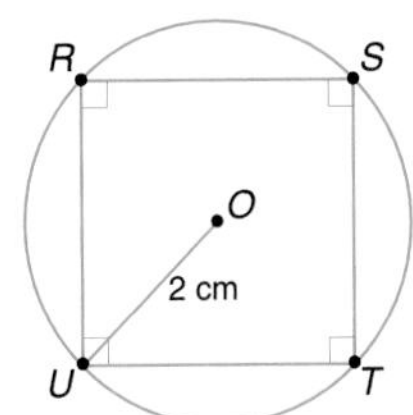

5. Find the perimeter of $RSTU$.
6. Find the area and the circumference of $\odot O$.

6. $4\pi \approx 12.57$ cm^2, $4\pi \approx 12.57$ cm

7. How do the perimeter of $RSTU$ and the circumference of $\odot O$ compare? They are close in length.

Find the apothem, the area, and the perimeter of each regular polygon. Round to the nearest tenth.

8. 5.5 cm, 121 cm^2, 44 cm

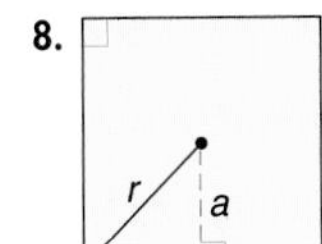

9. 2.3 in., 27.7 in^2, 24 in.

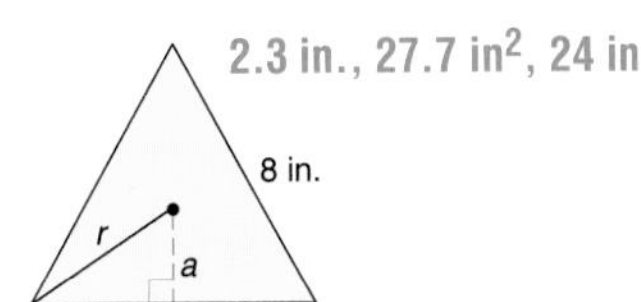

10. Find the circumference of a circle with an area of 10π square feet. Round to the nearest tenth. 19.9 ft

13. $(456 + 72\pi) \approx 682.19$ ft^2

Find the area of each shaded region. Assume that all polygons are regular. Round to the nearest tenth.

11. 56.7 cm^2

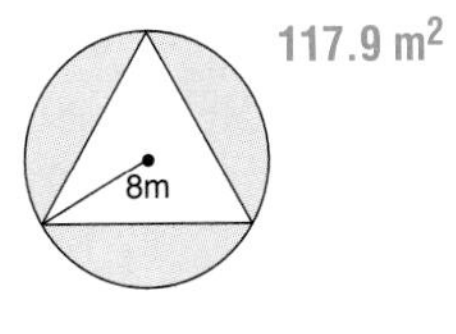

12. 117.9 m^2

13. **Sports** Shorem High School colors are blue and white. On the basketball court, the entire free throw areas and the entire large center circle of the basketball court are painted blue. What is the area of the court that is painted blue?

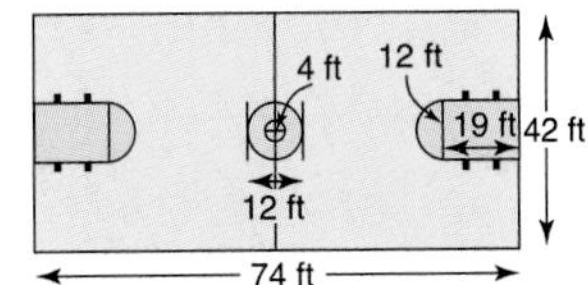

3 PRACTICE/APPLY

Check for Understanding

Exercises 1–13 are designed to help you assess your students' understanding through reading, writing, speaking, and modeling. You should work through Exercises 1–4 with your students and then monitor their work on Exercises 5–13.

Additional Answers

1. The central angle of a regular hexagon is 60. Therefore, a 30-60-90 triangle is formed with the apothem, half of a side, and the radius. Working with the perimeter, you can find the measure of the apothem.
2. The radius is the distance from the center to a vertex, and the apothem is the distance from the center to the middle of a side. Together they form a triangle along with half of a side.
3. d represents diameter and the diameter of a circle is equal to twice the radius. Therefore, $d = 2r$ and can be substituted into the formula.

Reteaching

Using Drawing Use these exercises to review how to find the area of regular polygons. Draw a picture of each polygon on the chalkboard. Then find the area of each polygon.

1. a pentagon with an apothem 6.9 meters long and a side 10 meters long 172.5 m^2
2. a hexagon with a side that measures 11 inches and an apothem that measures 9.5 inches 313.5 in^2
3. an octagon that measures 20 millimeters on a side and has an apothem of 24.1 millimeters 1928 mm^2

Assignment Guide

Core (with proof): 15–41 odd, 42–51
Core (informal): 15–41 odd, 42–51
Enriched: 14–36 even, 37–51

For **Extra Practice**, see p. 784.

The red A, B, and C flags, printed only in the Teacher's Wraparound Edition, indicate the level of difficulty of the exercises.

Study Guide Masters, p. 61

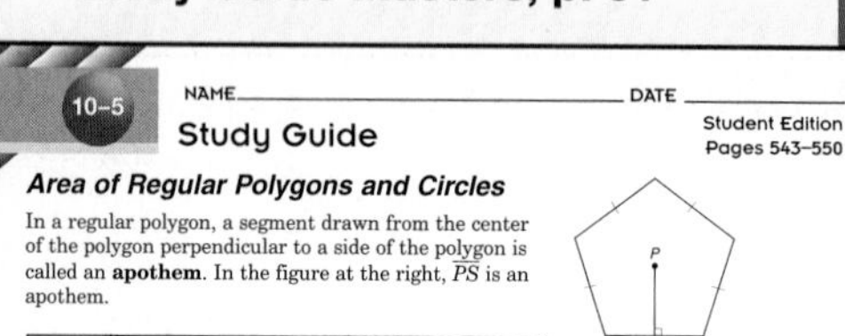

NAME ______ DATE ______

10-5 **Study Guide**

Student Edition Pages 543–550

Area of Regular Polygons and Circles

In a regular polygon, a segment drawn from the center of the polygon perpendicular to a side of the polygon is called an **apothem**. In the figure at the right, $\overline{PS}$ is an apothem.

Area of a Regular Polygon	If a regular polygon has an area of *A* square units, a perimeter of *P* units, and an apothem of *a* units, then $A = \frac{1}{2}Pa$.
Area of a Circle	If a circle has an area of *A* square units and a radius of *r* units, then $A = \pi r^2$.

Example: Find the area of a regular pentagon with an apothem of 2.8 cm and a perimeter of 20.34 cm.

$A = \frac{1}{2}Pa$ area of a regular polygon

$A = \frac{1}{2}(20.34)(2.8)$

$A = 28.476$

The area is 28.476 square centimeters.

Find the apothem, area, and perimeter of each regular polygon. Round your answers to the nearest tenth.

1. 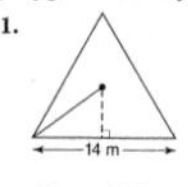

42 m, 120, 4.0, 84 m²

2. 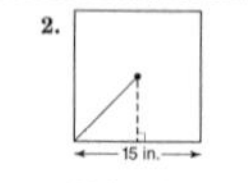

60 in., 90, 7.5 in., 225 in²

3. 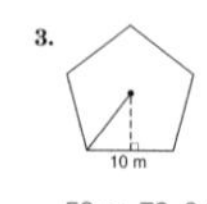

50 m, 72, 6.9 m, 172.5 m²

Find the area of each regular polygon described. Round your answers to the nearest tenth.

4. a hexagon with an apothem of $5\sqrt{3}$ cm 259.8 cm²

5. a pentagon with a perimeter of 54.49 m and an apothem of 7.5 m 204.3 m²

Find the area of each shaded region. Assume that all polygons are regular. Round your answers to the nearest tenth.

6. 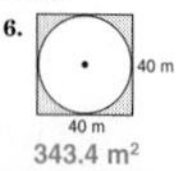

343.4 m²

7. 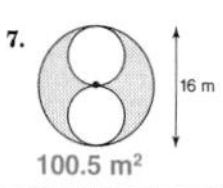

100.5 m²

EXERCISES

Practice

Find the area of each regular polygon described. Round to the nearest tenth.

14. square with apothem length of 12 centimeters 576 cm²
15. triangle with side length of 15.5 inches 104.0 in²
16. square with perimeter of $84\sqrt{2}$ meters 882 m²
17. hexagon with perimeter of 60 feet 259.8 ft²
18. octagon with side length of 10 kilometers 482.8 km²
19. hexagon with apothem length 24 inches 1995.3 in²

Find the circumference and the area of circles with the given radius. Round to the nearest tenth.

20. 34 meters 213.6m, 3631.7 m²
21. 8.5 centimeters 53.4 cm, 227.0 cm²
22. 15 millimeters 94.2 mm, 706.9 mm²
23. 10.25 inches 64.4 in., 330.1 in²
24. $5\frac{1}{3}$ feet 33.5 ft, 89.4 ft²
25. $21\frac{3}{4}$ centimeters 136.7 cm, 1486.2 cm²

Find the area of each shaded region. Assume that all polygons are regular. Round to the nearest tenth.

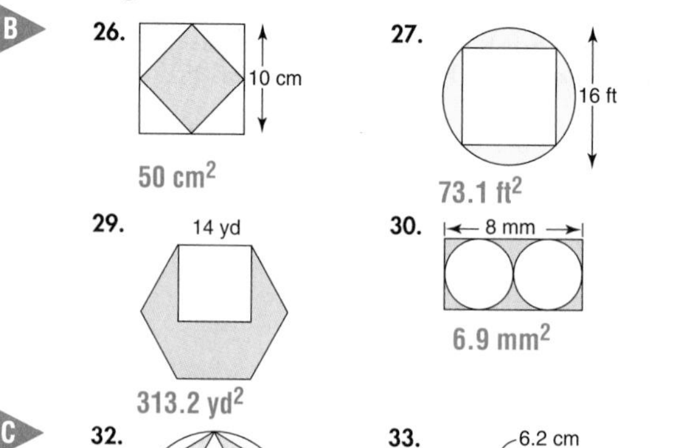

26. 50 cm²
27. 73.1 ft²
28. (4 m) 32.9 m²
29. 313.2 yd²
30. 6.9 mm²
31. (3.5 in.) 15.8 in²

32. 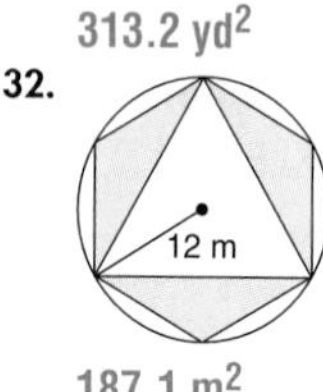

187.1 m²

33. 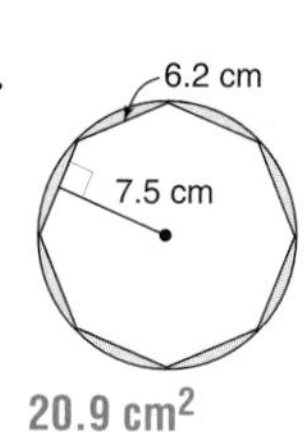

20.9 cm²

34. 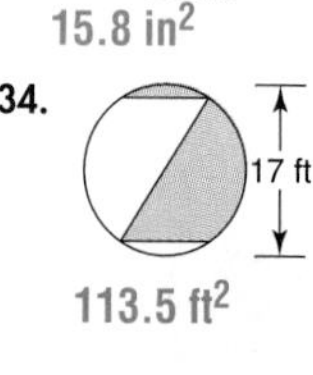

113.5 ft²

35. A circle is inscribed in a rhombus whose diagonals are 24 and 32 feet long.
    a. Find the area of the circle. 92.16π ft²
    b. What is the area of the region that is inside the rhombus but outside the circle? $(384 - 92.16\pi)$ ft²

36. A circle inscribes a regular hexagon and circumscribes another. If the radius of the circle is 10 units long, find the ratio of the area of the smaller hexagon to the area of the larger hexagon. 3 to 4

Critical Thinking

37. The number 3.14 is often used to approximate π. Use the findings in the Exploration on page 545 to explain why 3.14 is a good approximation of π. See margin.

38. Find the ratio of the area of $\triangle ABC$ to square $BCDE$. $\sqrt{3}:4$

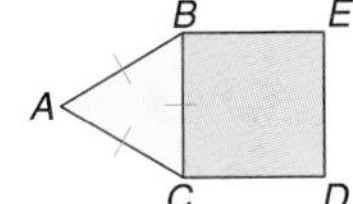

Applications and Problem Solving

39. **Shopping** The Pizza Shoppe sells a 12-inch cheese pizza for \$6.98 and a 16-inch cheese pizza for \$9.98.
    a. Which will give Kwam the most pizza, buying two 12-inch pizzas or one 16-inch pizza? two 12-inch pizzas
    b. Will Kwam get the best deal if he chooses to buy the option that will give him the most pizza? Explain. No; the unit cost of two 12-inch pizzas is more expensive than the unit cost of one 16-inch pizza.

40. **Gardening** Refer to the application about Elizabeth Park at the beginning of the lesson. Use the diameters of the gazebo and the two concentric rose garden plots to answer the following questions.

FYI

Elizabeth Park began in 1894 when a tract of land was bequeathed to the City of Hartford, Connecticut, by the will of Charles H. Pond to be used as a park and named for his wife, Elizabeth.

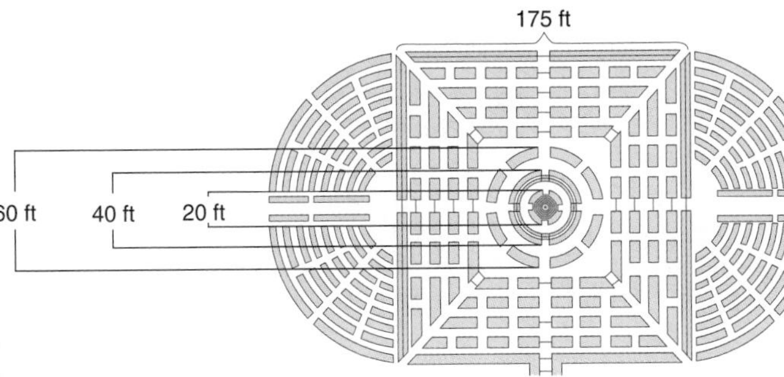

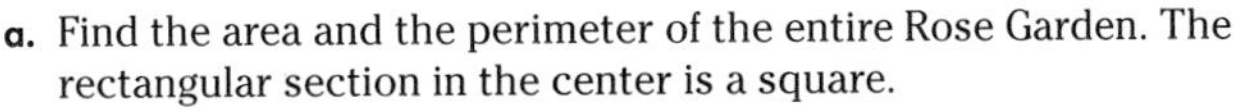

a. Find the area and the perimeter of the entire Rose Garden. The rectangular section in the center is a square.
b. What is the total of the circumferences of the three concentric circles formed by the gazebo and the two circular rose garden plots?
c. Each rose plot has a width of 5 feet. What is the area of the path between the outer two complete circles of rose garden plots? $225\pi \approx 706.9$ ft^2

41. **Architecture** The Anraku Temple in Japan is composed of four octagonal floors of different sizes that are separated by four octagonal roofs of different sizes. If the sides of the octagonal floors are in a ratio of 1:3:5:7, are the areas of each of the four octagonal floors in the same ratio? Explain your answer. See margin.

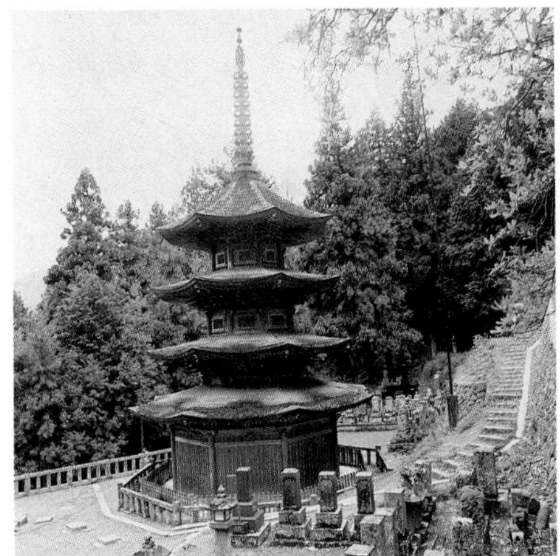

40a. $(30{,}625 + 7656.25\pi) \approx 54{,}677.82$ ft^2, $(350 + 175\pi) \approx 899.78$ ft

40b. $120\pi \approx 377.0$ ft

Additional Answers

37. As the polygon increases in the number of sides, the length of the apothem has a limit of the radius of the circle, and the measure of the side of a polygon becomes increasingly small and has as factors of the limit (2)(3.14).

41. No; they will increase by the squares of 1, 3, 5, and 7, or in other words, 1, 9, 25, and 49. This is because the apothem has a factor of the side and so does the perimeter.

Practice Masters, p. 61

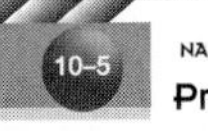

10–5 NAME________ DATE ________

Practice — Student Edition Pages 543–550

Area of Regular Polygons and Circles

Find the area of each regular polygon. Round your answers to the nearest tenth.

1. an octagon with an apothem 4.8 centimeters long and a side 4 centimeters long 76.8 cm^2
2. a square with a side 24 inches long and an apothem 12 inches long 576 in^2
3. a hexagon with a side 23.1 meters long and an apothem 20.0 meters long 1386 m^2
4. a pentagon with an apothem 316.6 millimeters long and a side 460 millimeters long 364,090 mm^2

Find the apothem, area, and perimeter of each regular polygon. Round your answers to the nearest tenth.

5. 8 in. — 4 in., 83.2 in^2, 41.6 in.

6. 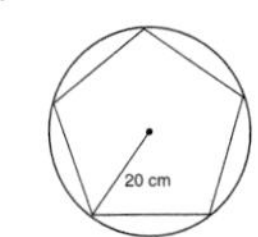

16.2 cm, 952.6 cm^2, 117.6 cm

7. 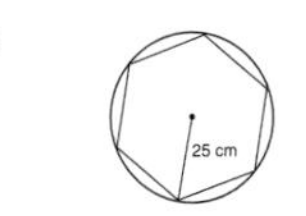

21.7 cm, 1627.5 cm^2, 150 cm

8. 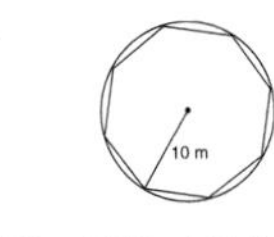

9.2 m, 281.5 m^2, 61.2 m

FYI

Landscape architects who design gardens have to balance plants and pathways to maximize aesthetic appeal. Have students think of local parks or gardens that use geometry in the design.

4 ASSESS

Closing Activity

Modeling Have students cut out a regular polygon from a sheet of paper. Have them count the number of sides and measure one of its sides. Then find the area of the polygon.

Additional Answer

48. 
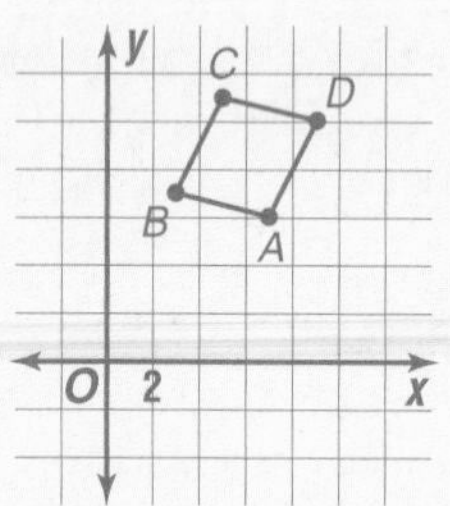

Enrichment Masters, p. 61

NAME__________ DATE ________

Enrichment

Student Edition Pages 543–550

Aerial Surveyors and Area

Many land regions have irregular shapes. Aerial surveyors often use coordinates when finding areas of such regions. The coordinate method described in the steps below can be used to find the area of *any* polygonal region. Study how this method is used to find the area of the region at the right.

Step 1 List the ordered pairs for the vertices in counter-clockwise order, repeating the first ordered pair at the bottom of the list.

Step 2 Find D, the sum of the downward diagonal products (from left to right).
$D = (5 \cdot 5) + (2 \cdot 1) + (2 \cdot 3) + (6 \cdot 7)$
$= 25 + 2 + 6 + 42$ or 75

Step 3 Find U, the sum of the upward diagonal products (from left to right).
$U = (2 \cdot 7) + (2 \cdot 5) + (6 \cdot 1) + (5 \cdot 3)$
$= 14 + 10 + 6 + 15$ or 45

(5, 7)
(2, 5)
(2, 1)
(6, 3)
(5, 7)

Step 4 Use the formula $A = \frac{1}{2}(D - U)$ to find the area.

$A = \frac{1}{2}(D - U)$

$= \frac{1}{2}(75 - 45)$

$= \frac{1}{2}(30)$ or 15

The area is 15 square units. Count the number of square units enclosed by the polygon. Does this result seem reasonable?

Use the coordinate method to find the area of each region in square units.

1. 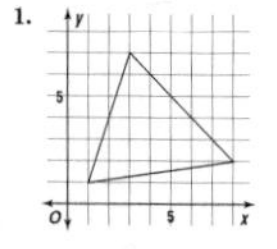
20 sq. units

2. 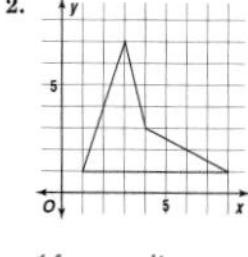
14 sq. units

3. 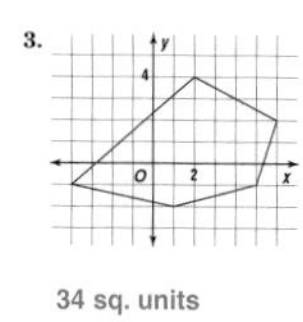
34 sq. units

Mixed Review

42. The perimeter of the triangle at the right is 18 units. Find the area of the triangle. (Lesson 10–4) **12 units^2**

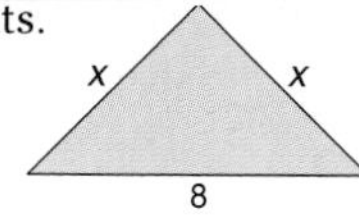

43. Find the area of the parallelogram at the right. (Lesson 10–3)
$12\sqrt{3}$ cm^2

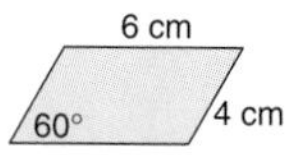

44. Will a regular dodecagon tessellate in a plane? If so, draw a sample figure. (Lesson 10–2) **no, interior angle = 162**

45. Write an equation for the circle whose diameter has endpoints at $(-1, 6)$ and $(5, 2)$. (Lesson 9–8) **$(x - 2)^2 + (y - 4)^2 = 13$**

46. The length of a side of an equilateral triangle is $\frac{8}{5}$ feet. Find the length of an altitude of the triangle. (Lesson 8–2) **$\frac{4\sqrt{3}}{5} \approx 1.4$ ft**

47. Identify the similar triangles in the figure at the right. Explain your answer. (Lesson 7–3)

$\triangle QRT \sim \triangle QTS$ by AA, $\triangle QTS \sim \triangle TRS$ by AA, $\triangle QRT \sim \triangle TRS$ by Transitive Property

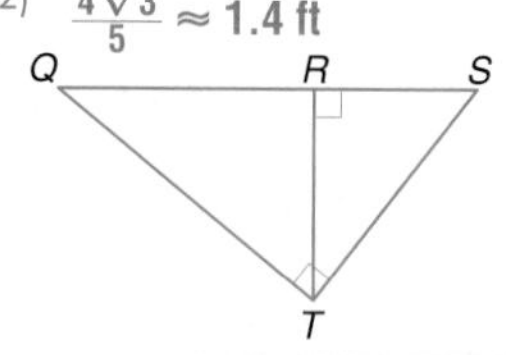

48. Graph $A(7, 6)$, $B(3, 7)$, $C(5, 11)$, and $D(9, 10)$. Draw quadrilateral $ABCD$. Determine if $ABCD$ is a parallelogram. Justify your answer. (Lesson 6–2)

48. Yes; both pairs of opposite sides are congruent. See margin for graph.

49. The measures of the angles in a triangle are $x + 16$, $8x + 7$, and $11x - 3$. Is the triangle *acute*, *obtuse*, *right*, or *equiangular*? (Lesson 4–2) **acute**

Algebra

50. Factor $21xy^3 + 3y^5$. **$3y^3(7x + y^2)$**

51. Simplify $(2a^3)^2$. **$4a^6$**

WORKING ON THE

In•ves•ti•ga•tion

Refer to the Investigation on pages 510–511.

Most adult soccer teams, including Major League and World Cup teams, use a size 5 soccer ball. A size 5 ball must have a circumference of 27 to 28 inches and weigh 14 to 16 ounces.

1. About how many polygons of each type appear on each soccer ball design shown on pages 510 and 511?
2. Estimate the length of an edge of a pentagon in each design. Then use that estimate to find the area of each type of polygon.
3. What is the total area of the figures on each ball design?
4. Do you think one of the two designs would be less expensive to manufacture than the other? Explain.
5. Use your results to describe some things you will consider when you determine a proposed design for your company's soccer ball.

Add the results of your work to your Investigation Folder.

Extension

Reasoning Find the area of the shaded region. The figure is a regular hexagon. **110.4 units^2**

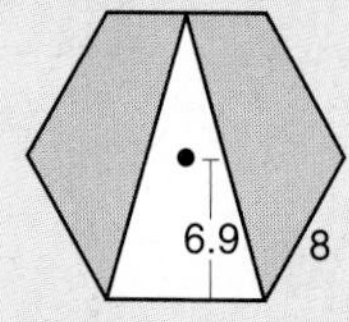

In•ves•ti•ga•tion

Working on the Investigation

The Investigation on pages 510–511 is designed to be a long-term project that is completed over several days or weeks. Encourage students to keep their materials in their Investigation Folder as they work on the Investigation.

10-6 Integration: Probability Geometric Probability

What YOU'LL LEARN

- To use area to solve problems involving geometric probability.

Why IT'S IMPORTANT

You can use geometric probability to solve problems involving navigation and entertainment.

Radio

WGLN radio station is having a "Listen and Call" contest. A song of the day is announced and played at 7:00 A.M. each day. Then, once during each hour of the day, that song is played. Since WGLN is at 91.1 on the radio dial, the ninety-first person to call the station when the song of the day begins to play wins \$100. If you turn on your radio at 2:35 P.M., what is the probability that you have missed the start of the song of the day when it is played in the 2:00 P.M. to 3:00 P.M. hour?

This problem can be solved using **geometric probability**. Geometric probability involves using principles of length and area to find the probability of an event. One of the principles of geometric probability is stated in Postulate 10–3.

Postulate 10–3 Length Probability Postulate	**If a point on $\overline{AB}$ is chosen at random and C is between A and B, then the probability that the point is on $\overline{AC}$ is $\frac{\text{length of } \overline{AC}}{\text{length of } \overline{AB}}$.**

To find the probability that you have missed the start of the song of the day, draw a line segment to represent the time from 2:00 to 3:00. Let point C represent the time that you started listening.

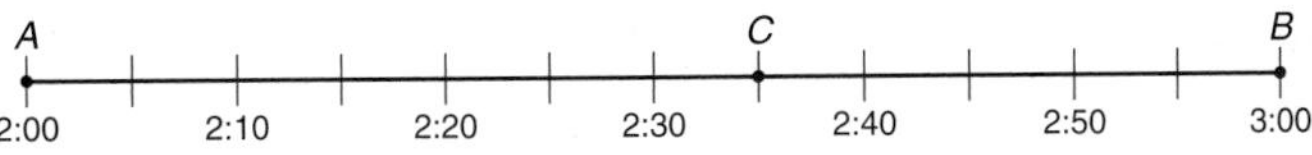

The entire hour is represented by the length of $\overline{AB}$. The length of $\overline{AC}$ represents the time from 2:00 to 2:35, and the length of $\overline{CB}$ represents the time from 2:35 to 3:00.

You have missed the song of the day if the station does not play the song after 2:35. So, the probability of having missed the song is $\frac{\text{length of } \overline{AC}}{\text{length of } \overline{AB}}$.

$$P(\text{missed the song}) = \frac{\text{length of } \overline{AC}}{\text{length of } \overline{AB}}$$

$$= \frac{35 \text{ minutes}}{60 \text{ minutes}}$$

$$= \frac{7}{12}$$

The probability that you missed the song is $\frac{7}{12}$ or about 58%.

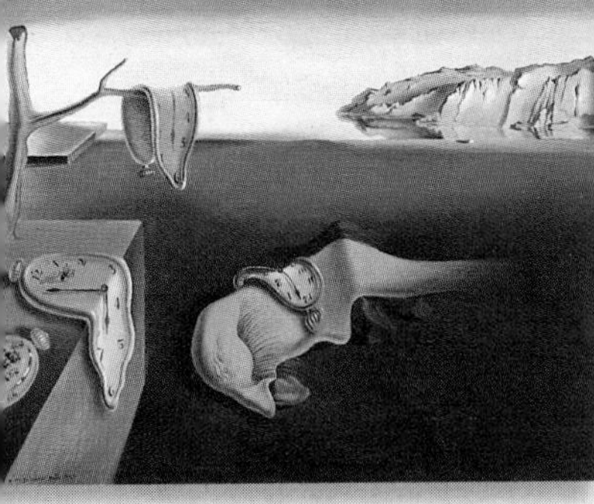

Persistence of Memory by Salvador Dali (1931)

10-6 LESSON NOTES

NCTM Standards: 1–5, 7, 8, 11

Instructional Resources

- Study Guide Master 10-6
- Practice Master 10-6
- Enrichment Master 10-6
- Assessment and Evaluation Masters, p. 269
- Multicultural Activity Masters, p. 20

Transparency 10-6A contains the 5-Minute Check for this lesson; **Transparency 10-6B** contains a teaching aid for this lesson.

Recommended Pacing	
Standard Pacing	Day 11 of 15
Honors Pacing	Day 10 of 14
Block Scheduling*	Day 6 of 8

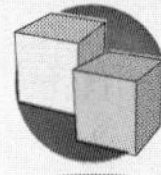

*For more information on pacing and possible lesson plans, refer to the *Block Scheduling Booklet*.

1 FOCUS

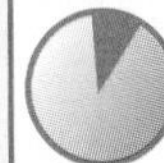

5-Minute Check *(over Lesson 10-5)*

Find the area of each polygon.

1. square with a side of 4 centimeters and an apothem of 2 centimeters **16 cm²**
2. regular pentagon with a side of 6 yards and an apothem of 4.13 yards **61.95 yd²**
3. regular octagon with a side of 10 inches and an apothem of 12.07 inches **482.8 in²**
4. regular decagon with a side of 7 feet **377.02 ft²**

Motivating the Lesson

Situational Problem Ask students if they have ever tried to win a radio contest. What strategies did they use to try to be the caller chosen?

2 TEACH

Teaching Tip After reading Postulate 10-3, discuss with students why the line segment is a good model for time in this problem.

In-Class Example

For Example 1

Joanna designed a new dart game. A dart in section A earns 10 points; a dart in section B earns 5 points; a dart in section C earns 2 points. Find the probability of earning each score. Round to the nearest hundredth.

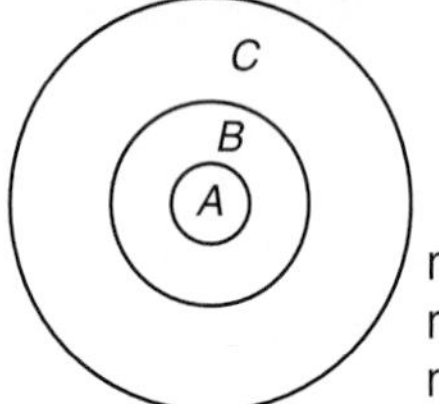

radius $\odot A = 2$ in.
radius $\odot B = 5$ in.
radius $\odot C = 10$ in.

10 points: 0.04; 5 points: 0.21; 2 points: 0.75

Teaching Tip After Example 1, point out that the sum of all the probabilities is 1.

Postulate 10–4 states a second principle of geometric probability.

Postulate 10–4 Area Probability Postulate	**If a point in region A is chosen at random, then the probability that the point is in region B, which is in the interior of region A, is $\frac{\text{area of region } B}{\text{area of region } A}$.**

Example 1

Festivals

Every summer the Hayti Heritage Center in Durham, North Carolina, has a festival at Eno River Park. The Center invites entertainers and vendors to celebrate the diversity of the community.

The rectangular area of the park where vendors are located has shaded and sunny areas as shown in the diagram. Seventy-five percent of the park is used by vendors. The other 25% is for walkways. Each vendor space has exactly the same area.

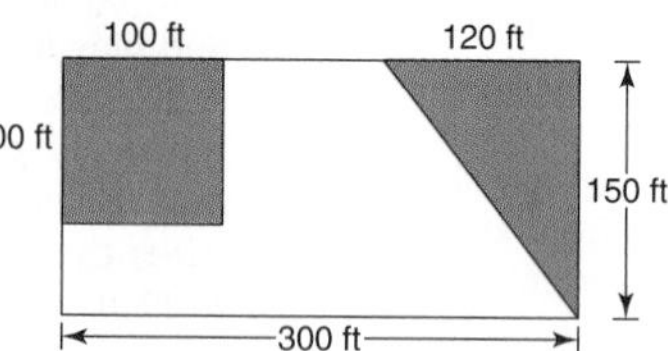

Because it is usually very hot during the festival, the shaded spots are most desired. If vendors are randomly placed in the park, what is the probability that a vendor will be placed in a shaded spot?

Explore The dimensions of the entire park and also the dimensions of the shaded areas are given. We need to subtract the area of the walkways away from these areas. Then we can use the Area Probability Postulate to find the probability.

Plan To find the probability that a vendor will be in a shaded area, we must find the area of the entire region without the walkways and the area of the shaded regions without the walkways.

Solve

- Find the area of the entire region without the walkways.

  area of entire region = (150)(300) or 45,000
  area of entire region without walkways = 0.75(45,000) or 33,750

  The vendor area of the entire region is 33,750 square feet.

- Find the area of the shaded regions without the walkways.

  area of square region = (100)(100) or 10,000
  area of square region without walkways = 0.75(10,000) or 7500

  area of triangular region = $\frac{1}{2}(120)(150)$ or 9000

  area of triangular region without walkways = 0.75(9000) or 6750

  The area of the shaded regions without the walkways is 7500 + 6750 or 14,250 square feet.

- Find the probability of being in a shaded spot.

$$P(\text{shaded region}) = \frac{\text{shaded area}}{\text{total area}}$$
$$= \frac{14{,}250}{33{,}750} \text{ or } \frac{19}{45}$$

The probability that the vendor would be placed in a shaded region is $\frac{19}{45}$ or about 42%.

Examine Look at the diagram. Is about 40% of the total area shaded?

Sometimes when you are finding a geometric probability, you need to find the area of a **sector of a circle**. A sector of a circle is a region of a circle bounded by a central angle and its intercepted arc.

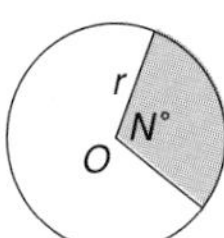

Area of a Sector of a Circle	If a sector of a circle has an area of A square units, a central angle measuring $N°$, and a radius of r units, then $A = \frac{N}{360}\pi r^2$.

Example 2 **Nuela has to spin either a 3 or a 5 to stay in the game. What is the probability that she will spin either one of these two numbers?**

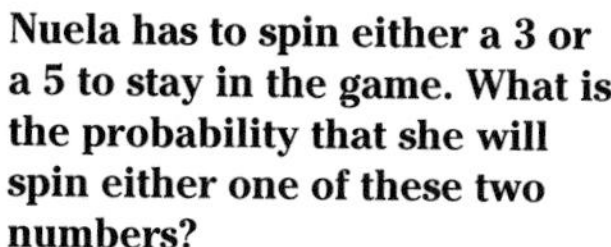

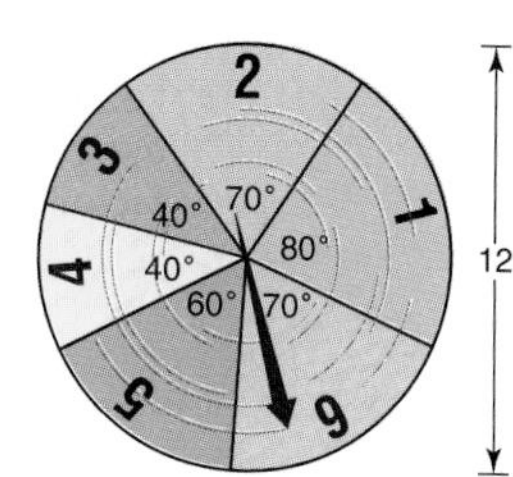

Find the area of the circle and the areas of the appropriate sectors of the circle.

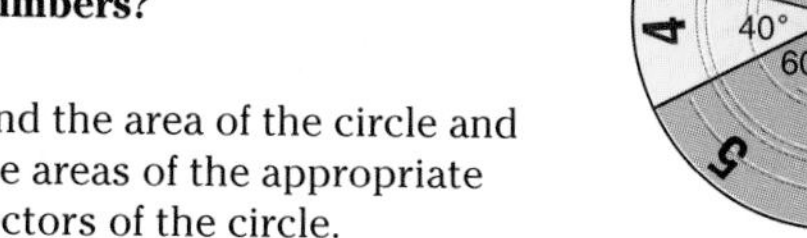

area of the 3 sector	area of the 5 sector	area of circle
$A = \frac{N}{360}\pi r^2$	$A = \frac{N}{360}\pi r^2$	$A = \pi r^2$
$= \frac{40}{360}\pi(6)^2$	$= \frac{60}{360}\pi(6)^2$	$= \pi(6)^2$
$= \frac{1440\pi}{360}$	$= \frac{2160\pi}{360}$	$= 36\pi$
$= 4\pi$	$= 6\pi$	

The total area of the two sectors is 10π.

Now find the geometric probability.

$$P(3 \text{ or } 5) = \frac{\text{area of sectors}}{\text{area of circle}}$$
$$= \frac{4\pi + 6\pi}{36\pi}$$
$$= \frac{10\pi}{36\pi} \text{ or } \frac{5}{18}$$

The probability that Nuela will spin a 3 or a 5 is $\frac{5}{18}$ or about 28%.

In-Class Example

For Example 2
Scott is planning a circular flower garden with a radius of 2 meters. He plans to plant tulips in the sectors marked with a T and daffodils in the sectors marked with a D. What is the area that will be planted with tulips?

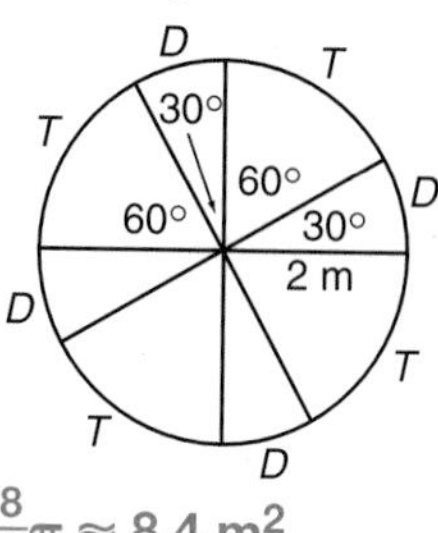

$\frac{8}{3}\pi \approx 8.4 \text{ m}^2$

Classroom Vignette

"In order to integrate trigonometry with area and make use of calculators, I develop the formula for the area of a regular polygon in terms of sine and cosine. The area of a regular n-gon is $n\left(\frac{s^2 \cdot \cos\theta}{4 \sin\theta}\right)$, where s is the measure of a side and $\theta = \frac{360}{2n}$. A figure like the one at the right can be used to help prove the theorem."

George May
Salem High School
Salem, Virginia

Figure for Vignette

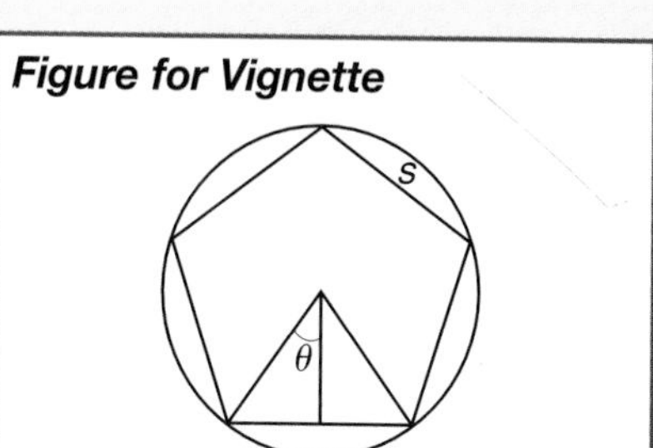

3 PRACTICE/APPLY

Check for Understanding

Exercises 1–9 are designed to help you assess your students' understanding through reading, writing, speaking, and modeling. You should work through Exercises 1–4 with your students and then monitor their work on Exercises 5–9.

Additional Answers

2. Adina is right. Yes; you can use the probability of an event occurring to estimate the area of a region.

3a. Area Probability Postulate

3b. Length Probability Postulate

CHECK FOR UNDERSTANDING

Communicating Mathematics

1. The Length Probability Postulate is one-dimensional and the Area Probability Postulate is two-dimensional. They are similar otherwise.

Study the lesson. Then complete the following.

1. **Compare and contrast** the Length Probability Postulate with the Area Probability Postulate.

2. **You Decide** Terrence needs to find the area of a region that has an irregular shape. Adina said that he could place the region in a square dartboard, throw darts randomly at the square, and then use the number of times the darts land in that region to find the area of the region. Terrence said this was impossible. Who is right? Can the probability of an event occurring be used to estimate the area of a region? See margin.

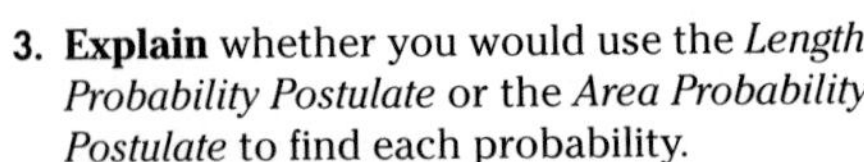

3. **Explain** whether you would use the *Length Probability Postulate* or the *Area Probability Postulate* to find each probability.
   a. the probability that a point is in a particular region See margin.
   b. the probability that a point is on a particular segment See margin.

4. **Draw** a target made up of three concentric circles so that the probability of a dart landing in each ring, as the circumferences get larger, increases by a factor of 24. Justify your drawing. Sample answer: $r = 1$, $r = 5$, and $r = 7$.

Guided Practice

Find the probability that a point chosen at random on $\overline{DI}$ is also a part of each of the following segments.

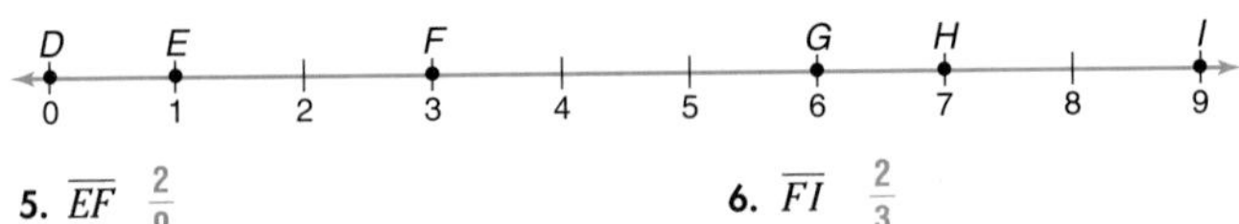

5. $\overline{EF}$ $\frac{2}{9}$

6. $\overline{FI}$ $\frac{2}{3}$

7. Find the probability that a point chosen at random in the square at the right lies in the shaded region. $\frac{1}{2}$

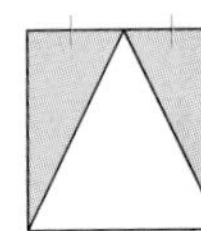

8a. See students' work.

8c. No; as long as the dart is randomly thrown and the cards do not overlap, the area of the board and the cards is always the same.

8. To win at a carnival game, you must throw a dart at a board that is 6-feet by 3-feet and hit one of the 25 playing cards on the board. The playing cards are each $2\frac{1}{2}$-inches by $3\frac{1}{2}$-inches.
   a. Draw a diagram of the dartboard.
   b. What is the probability that a randomly thrown dart that hits the board hits a playing card? Round to the nearest hundredth. 0.08
   c. Does the arrangement of the cards on the board affect the probability? Explain.

Reteaching

Using Diagrams Find the probability that a point on $\overline{AE}$, chosen at random, will be on the given segment.

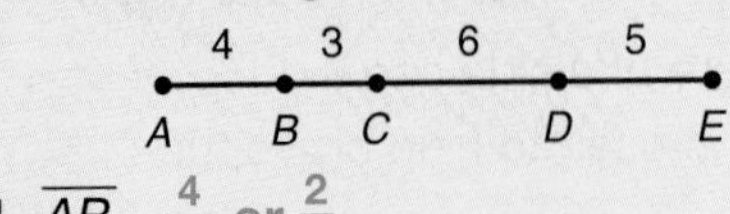

1. $\overline{AB}$ $\frac{4}{18}$ or $\frac{2}{9}$
2. $\overline{BD}$ $\frac{1}{2}$
3. $\overline{DE}$ $\frac{5}{18}$

Find the probability that a point chosen at random in this circle will be in the given section.

4. A $\frac{5}{36}$
5. C $\frac{1}{6}$
6. D $\frac{1}{9}$

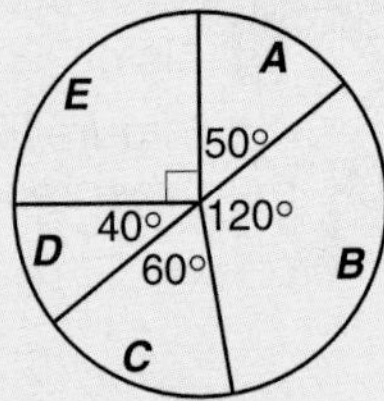

9. Entertainment The children at Jasmine's birthday party are playing a game for prizes. Each child tosses a beanbag at the target on the floor. Depending on where one marked corner of the beanbag lands, the following prizes are given:

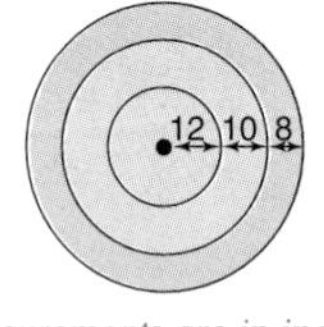

Measurements are in inches.

red: jumbo squirt gun
blue: yo-yo
green: candy bar

If the marked corner of the beanbag lands on the target, find the probability of a child winning a jumbo squirt gun. Round to the nearest hundredth. $\frac{4}{25}$ or 0.16

EXERCISES

Practice

10. What is the probability that a random point on $\overline{AB}$ will be closer to point A than to point X? $\frac{1}{3}$

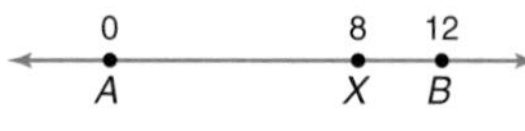

11. A point is chosen at random on $\overline{XY}$. If Z is the midpoint of $\overline{XY}$, W is the midpoint of $\overline{XZ}$, and V is the midpoint of $\overline{WY}$, what is the probability that the point is on $\overline{XV}$? $\frac{5}{8}$ or 0.625

Find the probability that a point chosen at random in each figure lies in the shaded region.

12. $\frac{3}{4}$ or 75%

12.
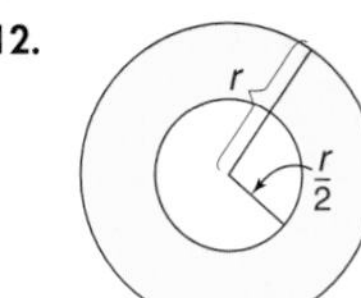

13.
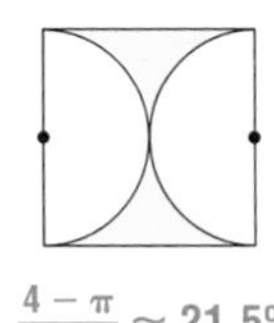
$\frac{4-\pi}{4} \approx 21.5\%$

14.
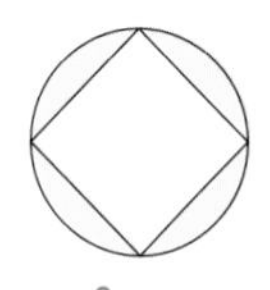
$\frac{\pi - 2}{\pi} \approx 36.3\%$

Find the probability for each outcome on the game spinner shown at the right.

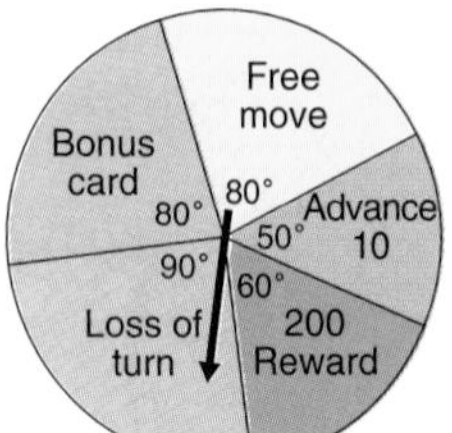

15. free move $\frac{2}{9}$
16. loss of turn $\frac{1}{4}$
17. draw bonus card $\frac{2}{9}$
18. get $200 reward or a free move $\frac{7}{18}$

Find the probability for each outcome on the target shown at the right. The center ring has a radius of 1 unit. Each successive ring has a radius 1 unit greater than the previous one.

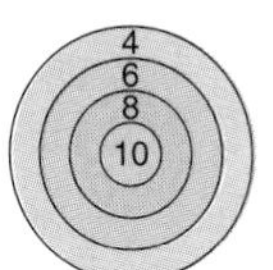

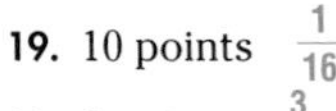
19. 10 points $\frac{1}{16}$
20. 8 points $\frac{3}{16}$
21. 6 points $\frac{5}{16}$

22. A toy car moves on tracks that consist of the sides of a regular hexagon and all of its diagonals. What is the probability that at a random moment the car will:

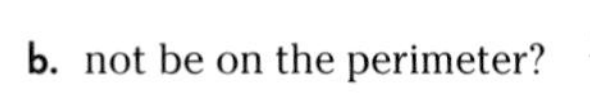
a. be on the longest diagonal? $\frac{1}{5}$ b. not be on the perimeter? $\frac{3}{5}$

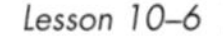

Assignment Guide

Core (with proof): 11–31 odd, 32, 33, 35, 36–43
Core (informal): 11–31 odd, 32, 33, 35, 36–43
Enriched: 10–30 even, 32–43

For **Extra Practice**, see p. 784.

The red A, B, and C flags, printed only in the Teacher's Wraparound Edition, indicate the level of difficulty of the exercises.

Additional Answers

23a. 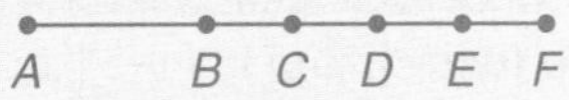 A, B, C, D, E, F (points on a line)

32. Sample answer: $m\angle 1 = 90$, $m\angle 2 = 90$, $m\angle 3 = 45$, $m\angle 4 = 45$, and $m\angle 5 = 90$.

23b. $\frac{1}{3}$

23c. $\frac{1}{6}$

23d. $\frac{1}{2}$

24. $\frac{5}{16}$

25. $\frac{1}{3}$

26. $\frac{2}{3}$

27. $\frac{\pi}{4} \approx 78.5\%$

28. $\frac{3\sqrt{3}}{2\pi} \approx 82.7\%$

LOOK BACK

Refer to Lesson 5-1 to review median of a triangle.

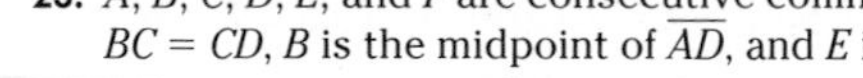

23. A, B, C, D, E, and F are consecutive collinear points such that $AD = BF$, $BC = CD$, B is the midpoint of $\overline{AD}$, and E is the midpoint of $\overline{DF}$.
   a. Draw the figure. See margin.
   b. What is the probability that a random point will fall on $\overline{AB}$?
   c. What is the probability that a random point will fall on $\overline{DE}$?
   d. What is the probability that a random point will fall on $\overline{BE}$?

24. Ms. Perez often gives a short one-problem quiz during her 48-minute geometry class. If a student is out of the room at the time she gives the quiz, he or she will miss that grade. Natalia was fifteen minutes late for class. What is the probability that she missed the start of the quiz?

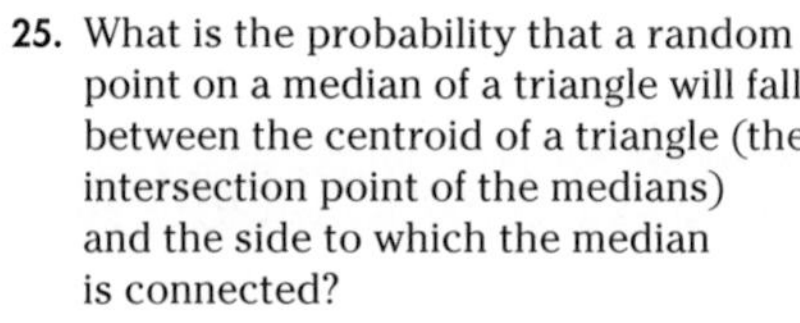

25. What is the probability that a random point on a median of a triangle will fall between the centroid of a triangle (the intersection point of the medians) and the side to which the median is connected?

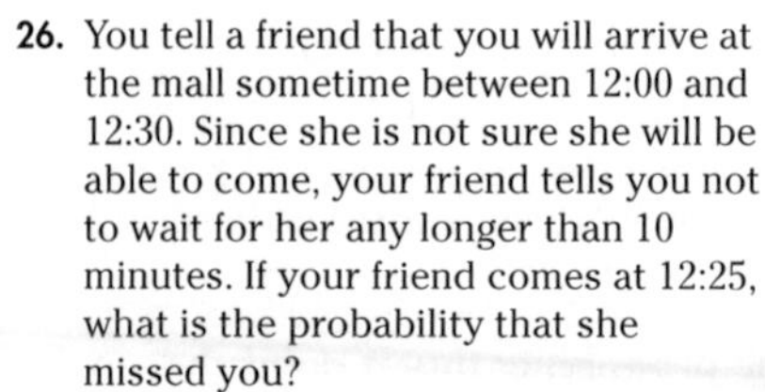

26. You tell a friend that you will arrive at the mall sometime between 12:00 and 12:30. Since she is not sure she will be able to come, your friend tells you not to wait for her any longer than 10 minutes. If your friend comes at 12:25, what is the probability that she missed you?

27. If a unit circle is inscribed in a unit square, what is the probability of a random point being in the circle?

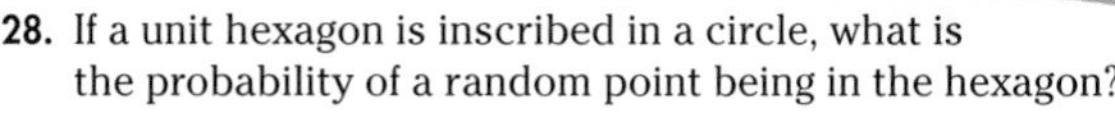

28. If a unit hexagon is inscribed in a circle, what is the probability of a random point being in the hexagon?

29. The figure at the right shows two squares. Determine the value of s if the shaded part is 75% of the area of the figure. 6 cm

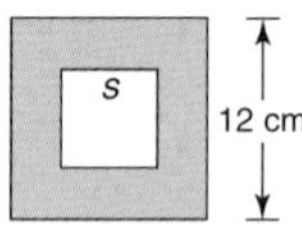

30. A circle is circumscribed about a square with side lengths of 7 centimeters. If a dart is thrown at random into the circle, what is the probability that it lands in the circle, but outside the square? $\frac{\pi - 2}{\pi} \approx 36.3\%$

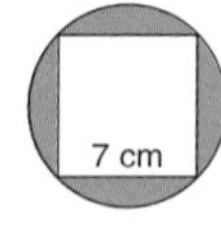

Cabri Geometry

31. Use a TI-92 and the steps below to find the probability that a point on the median of a triangle is between the centroid and the midpoint of the side.
   a. Construct $\triangle ABC$. a–e. See students' work.
   b. Construct midpoints on each side of $\triangle ABC$.
   c. Construct two medians, $\overline{AD}$ and $\overline{BF}$, and their point of intersection G.
   d. Construct the third median.
   e. Measure the distance from B to F and from F to G.
   f. Find the ratio of $\overline{FG}$ to $\overline{BF}$. $\frac{1}{3}$

Critical Thinking

32. How would you design a spinner having 5 sectors, numbered 1 through 5, if you want to have a probability of spinning a one, a two, or a five to be twice the probability of spinning a three or a four? See margin.

Study Guide Masters, p. 62

10-6 NAME______ DATE______

Study Guide — Student Edition Pages 551–558

Integration: Probability
Geometric Probability

Geometric probability involves using length and area to find the probability of an event.

- If a point on $\overline{AB}$ is chosen at random and C is between A and B, then the probability that the point is on AC is $\frac{\text{length of } \overline{AC}}{\text{length of } \overline{AB}}$.
- If a point in region A is chosen at random, then the probability that the point is in region B, which is in the interior of region A, is $\frac{\text{area of region } B}{\text{area of region } A}$.

Example: Suppose a dart is thrown at random at a circular dartboard like the one shown at the right and hits the dartboard. What is the probability that the dart will land in the bullseye?

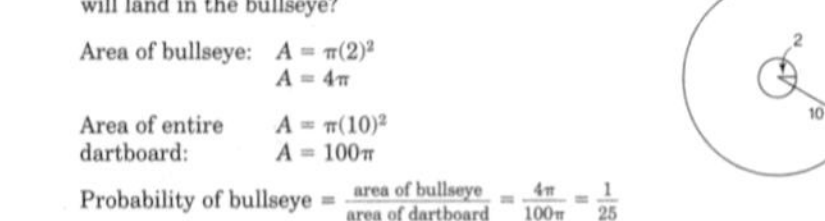

Area of bullseye: $A = \pi(2)^2$, $A = 4\pi$

Area of entire dartboard: $A = \pi(10)^2$, $A = 100\pi$

Probability of bullseye $= \frac{\text{area of bullseye}}{\text{area of dartboard}} = \frac{4\pi}{100\pi} = \frac{1}{25}$

Find the probability that a point chosen at random on $\overline{AH}$ is also a part of each of the following segments. Assume the twelve shortest segments are all congruent.

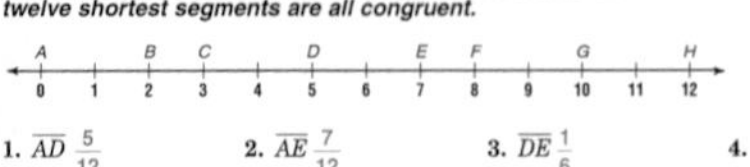

1. $\overline{AD}$ $\frac{5}{12}$ 2. $\overline{AE}$ $\frac{7}{12}$ 3. $\overline{DE}$ $\frac{1}{6}$ 4. $\overline{CG}$ $\frac{7}{12}$

Find the probability that a point chosen at random in each figure lies in the shaded region. Round your answers to the nearest hundredth.

5. 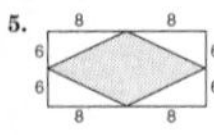 0.50

6. 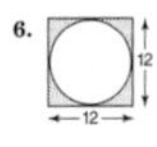 0.21

7. 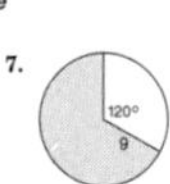  0.67

Applications and Problem Solving

33. Accommodations The convention center in Washington, D.C., lies in the northwest sector of the city between New York and Massachusetts Avenues. These streets intersect at a 130° angle. If the amount of hotel space is evenly distributed over an area with that intersection as the center and a radius of 1.5 miles, what is the probability that a visitor, randomly assigned to a hotel in the city, will be housed in the sector that contains the convention center? **$\frac{13}{36} \approx 36.1\%$**

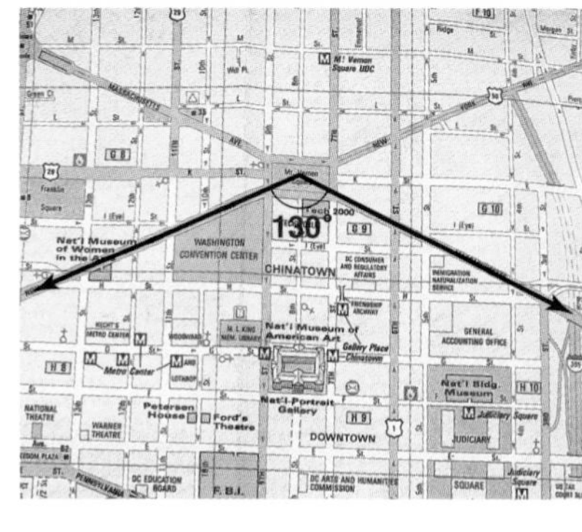

34. Conventions The Annual Meeting of the National Council of Teachers of Mathematics will be held in Orlando, Florida, in 2001. If there are 1100 sessions evenly distributed over the three full days of the convention, what is the probability that a particular session will fall on the first day of the meeting? **$\frac{1}{3}$**

35. Navigation As part of a scuba diving exercise, a 12-foot by 3-foot rectangular-shaped rowboat was sunk in a quarry. A motorboat takes a scuba diver to a random spot in the triangular section of the quarry and anchors there so that the diver can search for the rowboat.

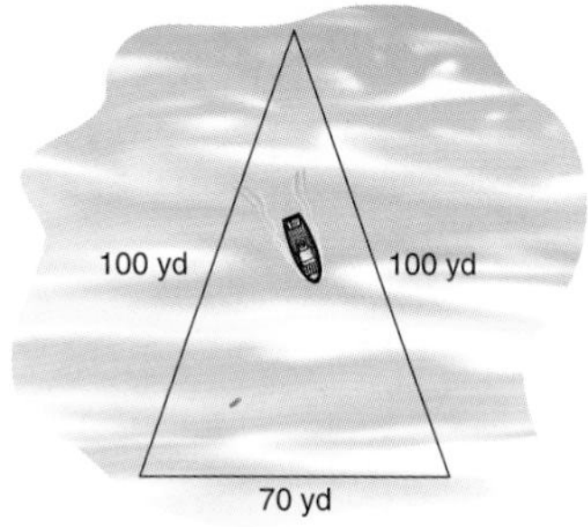

a. What is the approximate area of the triangular section of the quarry? **3279 yd^2**

b. What is the area of the rowboat? **36 ft^2**

c. What is the probability that the motorboat will anchor over the sunken rowboat? **$\frac{36}{29{,}511} \approx 0.12\%$**

Mixed Review

36. Find the area of an octagon with an apothem 7.5 feet long and a side 6.2 feet long. (Lesson 10–5) **186 ft^2**

37. If the lengths of the sides of a triangle are multiplied by 3, what is the ratio of the area of the new triangle to the area of the old triangle? (Lesson 10–4) **9:1**

38. The measure of an interior angle of a regular polygon is 160. How many sides does the polygon have? (Lesson 10–1) **18**

39. Find $m\widehat{IJK}$. (Lesson 9–4) **192**

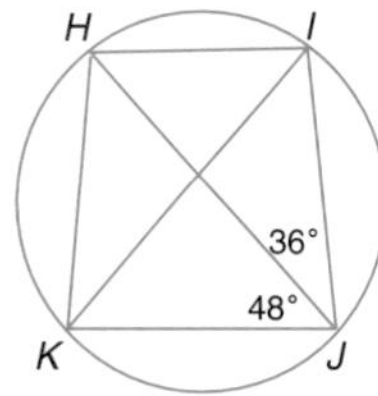

40. Gardening A flower bed is in the shape of an obtuse triangle with the shortest side measuring 7.5 feet. One angle measures 103°, and the side opposite that angle measures 14 feet. Find the measures of the remaining sides and angles. (Lesson 8–5) **10.3 ft; 31; 46**

Extension

Communication Separate students into groups to discuss whether the following statement is true or false. Have each group give a reason for its answer. *To find the geometric probability that a point chosen in a circle is in a particular sector of the circle, it is not necessary to know the radius of the circle.* **True; the probability will be the ratio of the central angle to 360° and is not affected by the radius.**

Practice Masters, p. 62

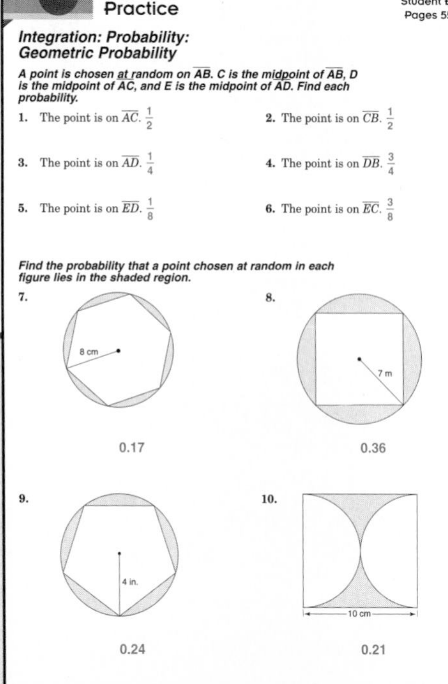
10–6 NAME________ DATE ________

Practice

Student Edition Pages 551–558

Integration: Probability: Geometric Probability

A point is chosen at random on $\overline{AB}$. C is the midpoint of $\overline{AB}$, D is the midpoint of $\overline{AC}$, and E is the midpoint of $\overline{AD}$. Find each probability.

1. The point is on $\overline{AC}$. $\frac{1}{2}$
2. The point is on $\overline{CB}$. $\frac{1}{2}$
3. The point is on $\overline{AD}$. $\frac{1}{4}$
4. The point is on $\overline{DB}$. $\frac{3}{4}$
5. The point is on $\overline{ED}$. $\frac{1}{8}$
6. The point is on $\overline{EC}$. $\frac{3}{8}$

Find the probability that a point chosen at random in each figure lies in the shaded region.

7. 8 cm — 0.17
8. 7 m — 0.36
9. 4 in. — 0.24
10. 10 cm — 0.21

4 ASSESS

Closing Activity

Speaking Have students describe a situation that can be solved by finding a geometric probability and explain how to find the probability in that case, without actually computing it.

Chapter 10 Quiz C (Lessons 10-5 and 10-6) is available in the *Assessment and Evaluation Masters*, p. 269.

Enrichment Masters, p. 62

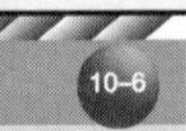

NAME______________ DATE______ Student Edition Pages 551–558

10-6 Enrichment

Polygon Probability

Each problem on this page involves one or more regular polygons. To find the probability of a point chosen at random being in the shaded region, you need to find the ratio of the shaded area to the total area. If you wish, you may substitute numbers for the variables.

Find the probability that a point chosen at random in each figure is in the shaded region. Assume polygons that appear to be regular are regular. Round your answer to the nearest hundredth.

1. 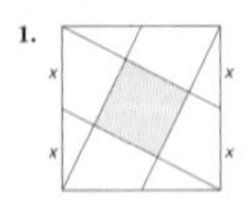

$\frac{1}{5}$ or 0.20

2. 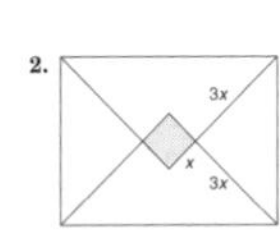

$\frac{1}{24} \approx 0.04$

3. 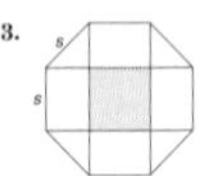

$\frac{1}{2+2\sqrt{2}} \approx 0.21$

4. 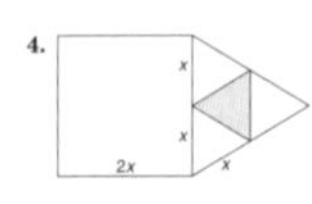

$\frac{\sqrt{3}}{16+4\sqrt{3}} \approx 0.08$

5. 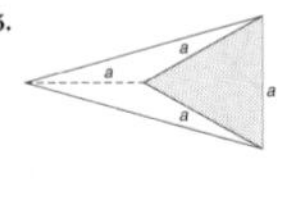

$\frac{\sqrt{3}}{2+\sqrt{3}} \approx 0.46$

6. 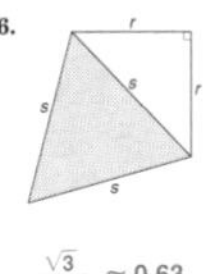

$\frac{\sqrt{3}}{\sqrt{3}+1} \approx 0.63$

41. Determine whether the pair of triangles at the right is congruent. If so, state the postulate or theorem that you used to come to that conclusion. (Lesson 5–2) **not enough information**

INTEGRATION Algebra

42. Solve $2x^2 - 6x + 1 = 0$ by using the quadratic formula. Approximate irrational roots to the nearest hundredth. **0.18, 2.82**

43. **Percent** In general, the more education a person has, the higher the earnings for that person. The graph below shows the mean annual income in 1992 for people 18 years or older, by level of education. Use the graph to find the percent of increase in each level of education. **46.3%, 5.0%, 24.1%, 33.7%, 23.7%, 36.0%, 35.8%**

Higher Education, Higher Earnings

Mean annual income in 1992 for people 18 or older, by level of education:

Level of education	Income
Professional	$74,560
Doctorate	$54,904
Master	$40,368
Bachelor	$32,629
Associate	$24,398
Some college, no degree	$19,666
High school graduate	$18,737
Not high school graduate	$12,809

1. to improve telescope operation by getting away from city lights, air pollution, and atmospheric disturbances

2. 40-meter telescope, a great improvement over a 5-meter telescope

Mathematics and SOCIETY

Seeing Double with Telescopes

The excerpt below appeared in *Science News* in June, 1996.

THE DOMES OF EIGHT TELESCOPES DOT the barren landscape atop...an extinct Hawaiian volcano. Reigning supreme among them...has been the 10-meter W.M. Keck Telescope. Now, the world's largest optical telescope has a partner....Like the original, Keck II uses a mirror composed of 36 [hexagonal] glass tiles to form a parabolic reflecting surface equivalent to that of a single 10-meter mirror....By adjusting individual tiles of the segmented mirror 100 times a second, the telescope is expected to produce images with a resolution as sharp as 0.04 arc second....That's like distinguishing between two barely touching pennies viewed from a distance of 600 kilometers. Spaced 85 meters apart, Keck I and Keck II can make simultaneous observations of the same heavenly body. In that way, the paired instruments would act as a single, 85-meter telescope. ■

1. Why are optical telescopes usually located on mountains or volcanoes in remote areas?
2. If two 5-meter telescopes were paired 40 meters apart, they would have the same capability as what size single-mirror telescope?
3. If a parabolic mirror is a curved surface, are the hexagons regular? Explain your answer. **no**

Mathematics and SOCIETY

Placing two telescopes at a distance from each other forms a long base for a triangle. The third vertex is the astronomical object the telescopes are observing. The longer the base, the greater the angle at the astronomical object.

10-7 Integration: Graph Theory Polygons as Networks

What YOU'LL LEARN

- To recognize nodes and edges as used in graph theory,
- to determine if a network is traceable, and
- to determine if a network is complete.

Why IT'S IMPORTANT

You can use graph theory to solve games and problems involving routing.

Internet

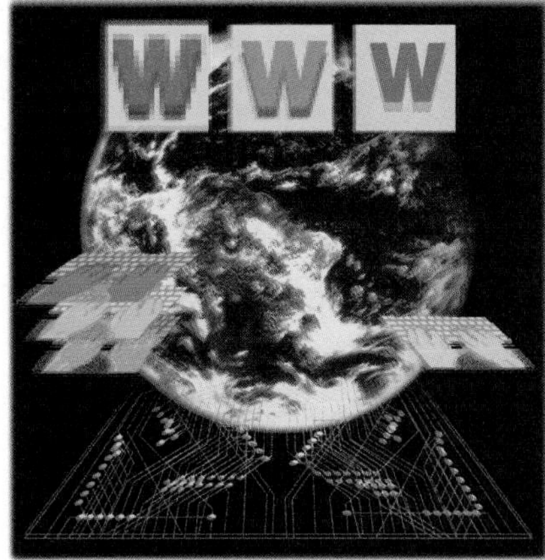

The international computer information system, Internet, was called the information highway in the early 1990s. It is the networking of computers, connections, and information. Internet is composed of nodes that allow for information to be transferred from one computer site to another.

The diagram at the right represents a **network**. Such a diagram illustrates a branch of mathematics called **graph theory**. In networking, the points are called **nodes**, and the paths connecting the nodes are called **edges**. Edges can be straight or curved.

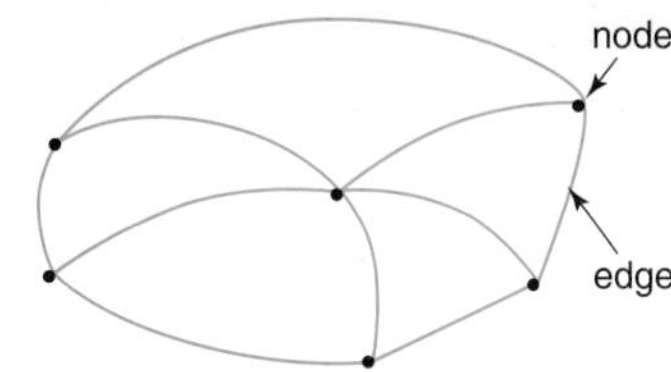

Straight edges can be used to form *closed* or *open* graphs. If an edge of a closed graph intersects exactly two other edges only at the endpoints, then the graph forms a polygon.

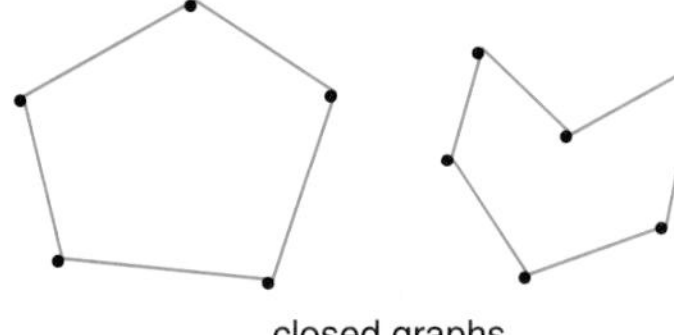

closed graphs

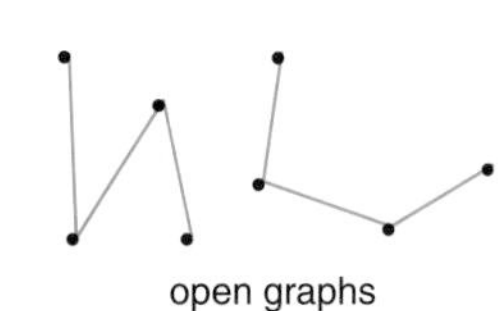

open graphs

However, not all pairs of nodes are connected by an edge in some networks. A network like this is called **incomplete**. Therefore, a **complete network** has at least one path between each pair of nodes.

Intersection points of edges are not nodes unless the network is shown with a solid vertex point.

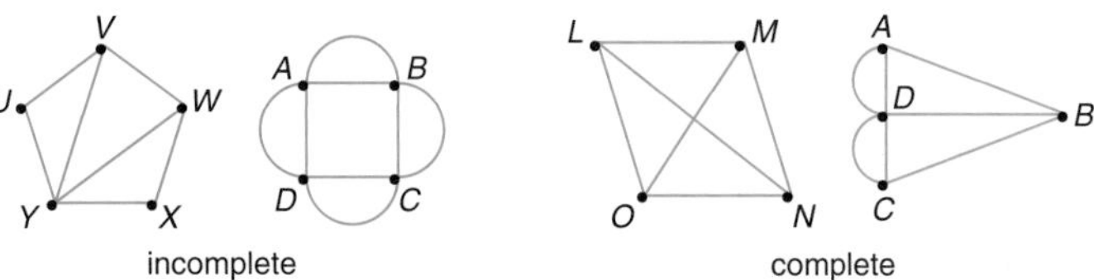

incomplete   complete

You have probably seen puzzles that ask you to trace over a figure without lifting your pencil and without tracing any lines more than once. In graph theory, if all nodes can be connected and each edge of a network can be covered exactly once, then a network is said to be **traceable**.

10-7 LESSON NOTES

NCTM Standards: 1–5, 7, 8

Instructional Resources

- Study Guide Master 10-7
- Practice Master 10-7
- Enrichment Master 10-7
- Assessment and Evaluation Masters, p. 269
- Real-World Applications, 21

Transparency 10-7A contains the 5-Minute Check for this lesson; **Transparency 10-7B** contains a teaching aid for this lesson.

Recommended Pacing	
Standard Pacing	Days 12 & 13 of 15
Honors Pacing	Days 11 & 12 of 14
Block Scheduling*	Day 7 of 8

*For more information on pacing and possible lesson plans, refer to the *Block Scheduling Booklet*.

1 FOCUS

5-Minute Check
(over Lesson 10-6)

A point on $\overline{UZ}$ is chosen at random. What is the probability that the point will be on each segment?

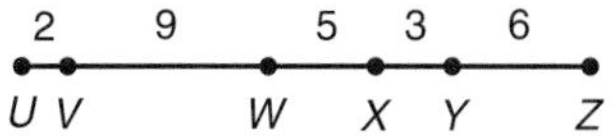

Number line: U, V, W, X, Y, Z with segment lengths 2, 9, 5, 3, 6

1. $\overline{UV}$ $\frac{2}{25}$ 2. $\overline{WX}$ $\frac{1}{5}$
3. $\overline{VW}$ $\frac{9}{25}$ 4. $\overline{WZ}$ $\frac{14}{25}$

A point in a circle is chosen at random. What is the probability that it will be in the sector whose central angle is given?

5. 75° $\frac{5}{24}$ 6. 100° $\frac{5}{18}$
7. 125° $\frac{25}{72}$ 8. 10° $\frac{1}{36}$

Motivating the Lesson

Questioning Ask students to define *graph* in their own words. Tell them that their definition must be broad enough to include every kind of graph.

2 TEACH

In-Class Examples

For Example 1
Determine if each network is traceable.

a.

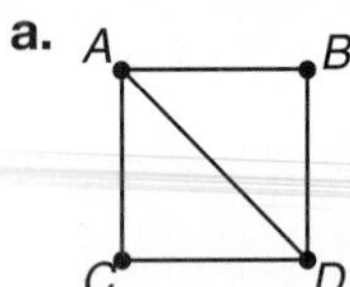

Yes; exactly two nodes have odd degrees; the others are even.

b.

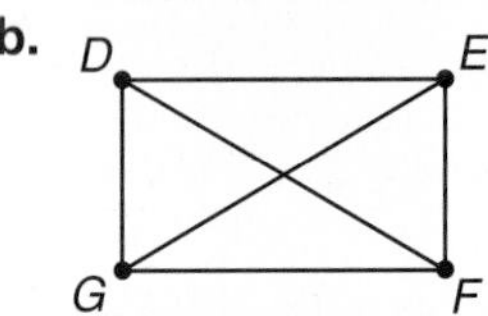

No; more than two nodes have odd degrees.

For Example 2
Determine if each network is complete. If it is not complete, name the edges that must be added to make it complete.

a.

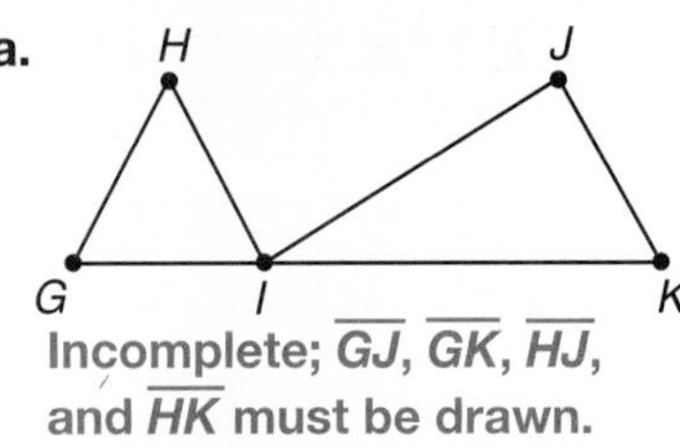

Incomplete; $\overline{GJ}$, $\overline{GK}$, $\overline{HJ}$, and $\overline{HK}$ must be drawn.

b.

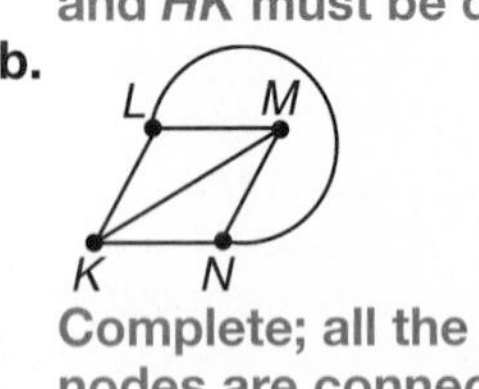

Complete; all the pairs of nodes are connected by an edge.

Teaching Tip To trace a figure, have students copy it on a sheet of paper and use a different color to trace it.

Teaching Tip After reading Example 2, you may want students to actually try to trace the networks.

Example 1

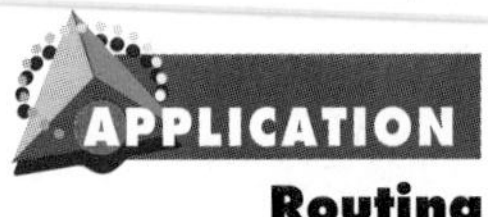

Routing

County highway inspectors need to visually inspect roads that connect several small towns. They must travel a route that takes them over each section of road exactly once. Use the networks of towns shown below to determine if this is possible for each set of towns. That is, determine if each network is traceable.

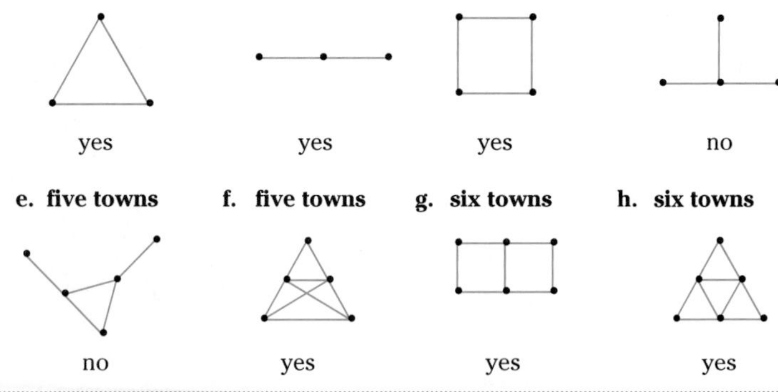

The **degree of a node** is the number of edges that are connected to that node. The traceability of a network is related to the degrees of the nodes in the network. What conjecture can you make about the design of the networks in Example 1 and their traceability?

Network Traceability Tests	**A network is traceable if and only if one of the following is true.** **1. All of the nodes in the network have even degrees.** **2. Exactly two nodes in the network have odd degrees.**

Example 2 **Determine if each network is traceable and complete. If not complete, name the edges that need to be added to make it complete.**

a.

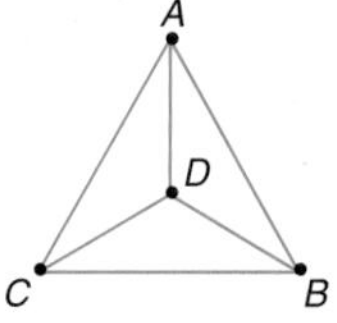

The network is not traceable because all of the nodes have odd degrees.

Since all nodes are connected, the network is complete.

b.

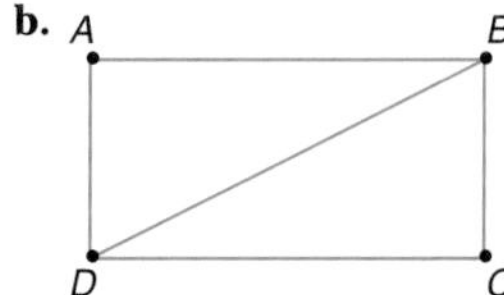

Exactly two nodes, B and D, have odd degrees. So the network is traceable.

The network is incomplete because nodes A and C are not connected.

Cooperative Learning

Brainstorming Give each small group of students a map. Ask them to draw a network representing a messenger service route. The map should include five nodes, be traceable, and be complete. For more information on the brainstorming strategy, see *Cooperative Learning in the Mathematics Classroom*, one of the titles in the Glencoe Mathematics Professional Series, page 30.

Example 3 a. **Find the degree of each node in the network shown at the right.**

b. **Is the network traceable?**

c. **If so, list a sequence of segments that demonstrates traceability.**

a. *A*: degree 2, *B*: degree 4, *C*: degree 3, *D*: degree 3, *E*: degree 4

b. Exactly two nodes, *C* and *D*, have odd degrees. The network is traceable.

c. $\overline{CB}$ to $\overline{BA}$ to $\overline{AE}$ to $\overline{ED}$ to $\overline{DB}$ to $\overline{BE}$ to $\overline{EC}$ to $\overline{CD}$

Look at the network in Example 3. Did it matter where you started in order to determine the sequence of segments to show its traceability? When a path goes through a node, it uses two edges. When a traceable network has an odd node, it must be a starting or finishing point of the traceability path. Therefore, you could start at vertex *C* or *D* and end at the one in which you didn't start. What would happen if there were three vertices with odd degree? *It would not be traceable.*

When you determined if a network was traceable, you probably had to try different strategies on some of the figures. The following patterns may have become apparent as you worked.

Starting Points for Network Traceability	**If a node has an odd degree, then the tracing must start or end at that node.** **If there are more than two nodes of odd degree, then the network is not traceable.** **If the network has no node of odd degree, then the tracing can start at any node and will end at the starting point.**

CHECK FOR UNDERSTANDING

Communicating Mathematics

Study the lesson. Then complete the following.

1. **Compare** a network to a convex polygon. See margin.
2. **You Decide** Amiri thought that the traceability of a network was related to whether the network was open or closed. Terri did not think that open or closed had anything to do with traceability. Who was correct? Explain. 2. See margin.
3. **Draw** an example of a network that is traceable and incomplete. 3. Sample answer: see margin.
4. **Explain** why only two nodes can have an odd degree. See margin.

5. **Assess Yourself** Describe in your journal why this network is or is not traceable. Can you determine visually whether a network is traceable? Draw a traceable and a not traceable network and show how the pattern works. See margin.

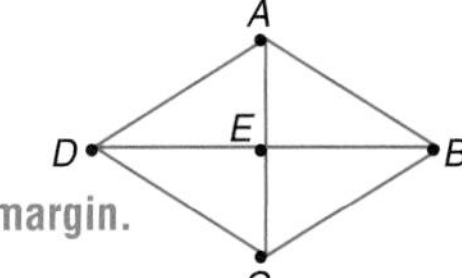

Guided Practice

6. Find the degree of each node in the network at the right.
A: degree 2, *B*: degree 3, *C*: degree 2, *D*: degree 2, *E*: degree 3, *F*: degree 2

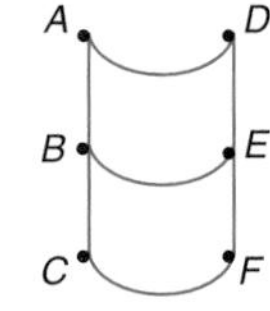

Reteaching

Using Cooperative Learning Have each student draw a picture of a network. Some might choose to produce traceable networks, some complete networks, and some incomplete networks. Have each student in turn present his or her network to the class. As a class, decide which of the networks are traceable, which are complete, and why.

In-Class Example

For Example 3

a. Find the degree of each node in the network below.
A: degree 2, *B*: degree 3, *C*: degree 4, *D*: degree 3, *E*: degree 2, *F*: degree 4

b. Is the network traceable? Explain. Yes; exactly two nodes have odd degrees; the others are even.

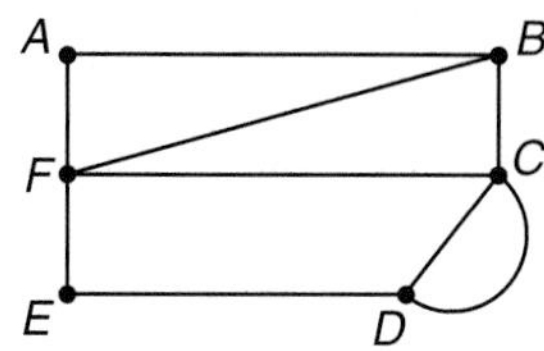

c. If so, list a sequence of segments that demonstrates traceability.
$\overline{BA}$ to $\overline{AF}$ to $\overline{FB}$ to $\overline{BC}$ to $\overline{CF}$ to $\overline{FE}$ to $\overline{ED}$ to $\overline{DC}$ to $\overline{CD}$

3 PRACTICE/APPLY

Check for Understanding

Exercises 1–9 are designed to help you assess your students' understanding through reading, writing, speaking, and modeling. You should work through Exercises 1–5 with your students and then monitor their work on Exercises 6–9.

Additional Answers

1. Polygons must be closed and made of line segments that do not overlap, while networks can be open, curved, and overlap.
2. Terri was correct. Traceability has to do with degree of nodes.
3. Sample answer: 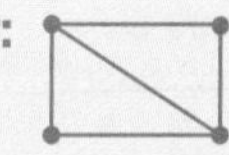
4. When a node has an odd degree, it must be a starting or finishing point because there is only one path and therefore must only exit or only enter. When a path goes through a node, it uses two edges.
5. No; it is not traceable. It has more than two nodes that have odd degrees.

Assignment Guide

Core (with proof): 11–29 odd, 30–37
Core (informal): 11–29 odd, 30–37
Enriched: 10–24 even, 25–37

For **Extra Practice**, see p. 784.

The red A, B, and C flags, printed only in the Teacher's Wraparound Edition, indicate the level of difficulty of the exercises.

Additional Answers

13a. traceable, complete
13b. $\overline{AB}$, $\overline{BC}$, $\overline{CA}$
13c. none
14a. traceable, incomplete
14b. $\overline{ED}$, $\overline{DC}$, $\overline{CB}$, $\overline{BA}$, $\overline{AF}$, $\overline{FC}$, $\overline{CE}$, $\overline{EF}$
14c. $\overline{AC}$, $\overline{AD}$, $\overline{AE}$, $\overline{BD}$, $\overline{BE}$, $\overline{BF}$, $\overline{DF}$
15a. not traceable, incomplete
15b. none
15c. $\overline{AC}$, $\overline{BD}$, $\overline{CE}$
16a. traceable, incomplete
16b. $\overline{CD}$, $\overline{DE}$, $\overline{EF}$, $\overline{FA}$, $\overline{AB}$, $\overline{BC}$, $\overline{CE}$, $\overline{EB}$, $\overline{BF}$
16c. $\overline{AC}$, $\overline{AD}$, $\overline{AE}$, $\overline{BD}$, $\overline{CF}$, $\overline{DF}$
17a. traceable, complete
17b. $\overline{AB}$, $\overline{BC}$, $\overline{CA}$, $\overline{AC}$, $\overline{CB}$, $\overline{BA}$
17c. none
18a. not traceable, incomplete
18b. none
18c. $\overline{AC}$, $\overline{AF}$, $\overline{AG}$, $\overline{AH}$, $\overline{BD}$, $\overline{BE}$, $\overline{BH}$, $\overline{BG}$, $\overline{CF}$, $\overline{CE}$, $\overline{CH}$, $\overline{DE}$, $\overline{DF}$, $\overline{DG}$, $\overline{EG}$, $\overline{FH}$

Study Guide Masters, p. 63

NAME ______ DATE ______

10-7 **Study Guide** — Student Edition Pages 559–564

Integration: Graph Theory
Polygons As Networks

A network is a set of points called **nodes** that are connected by paths called **edges**. Often the nodes represent locations on a map and the edges represent connecting routes. The edges can be straight or curved. A network is **traceable** if you can trace it with your finger without lifting your finger or retracing an edge. A **complete** network is one with at least one edge between each pair of nodes.

The **degree** of a node is the number of edges connected to it. A network is traceable if and only if one of the following is true.

- All of the nodes in the network have even degrees.
- Exactly two nodes in the network have odd degrees.

Example: Determine if the network at the right is traceable.

Node	Degree
A	3
B	4
C	3

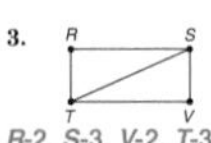

The network is traceable.

Find the degree of each node in the network.

1. 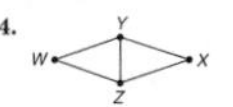 A-3, B-3, C-2
2. 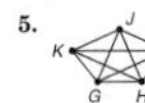 D-3, E-1, F-2
3. 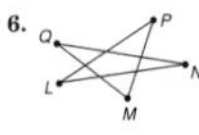 R-2, S-3, V-2, T-3

Determine if each network is traceable and complete. If a network is not complete, name the edges that need to be added to make the network complete.

4. not complete; edge WX
5. complete
6. not complete; edges LM, MN, NP, PQ, QL

Determine whether each network is traceable.

7. 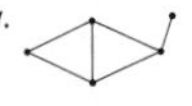 not traceable
8. 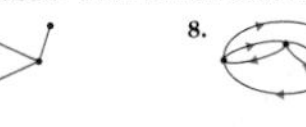 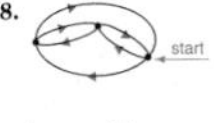 traceable
9. 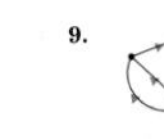 traceable

7a. traceable, incomplete
7b. $\overline{AE}$, $\overline{ED}$, $\overline{DC}$, $\overline{CB}$, $\overline{BD}$, $\overline{DA}$, $\overline{AB}$
7c. $\overline{AC}$, $\overline{BE}$, $\overline{CE}$
8a. traceable, incomplete
8b. $\overline{AB}$, $\overline{BC}$, $\overline{CD}$, $\overline{DB}$
8c. $\overline{AC}$, $\overline{AD}$

Consider each network.
a. Determine if it is traceable and complete.
b. If traceable, list a sequence of segments that demonstrates traceability.
c. If not complete, name the edges that need to be added to make it complete.

7. 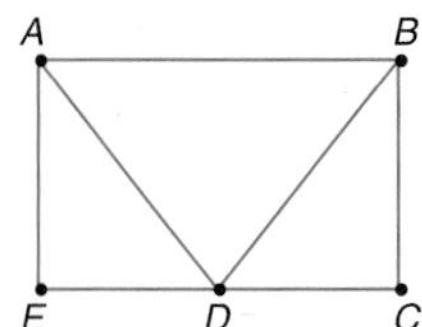
8. 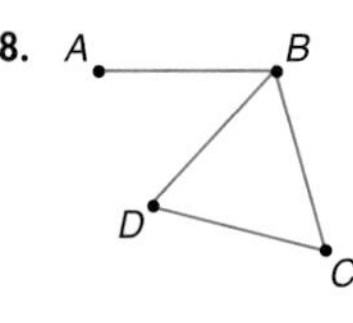

9. **Neighborhoods** Three neighbors decided to share the cost of groundskeeping equipment for the summer and the winter. One purchased a lawn mower, another purchased a snowblower, and the third purchased various gardening, pruning, and trimming tools. All the equipment purchased was stored in three separate sheds at the back of the three properties.
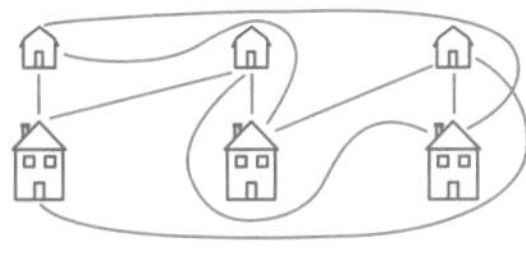
   a. Draw a network that shows possible paths for each of the neighbors.
   b. Do the paths have to cross? Explain. yes; at least once

EXERCISES

Practice

Find the degree of each node in each network.

10. 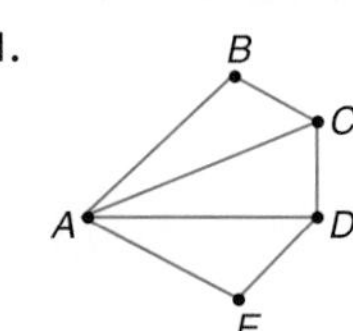
11. 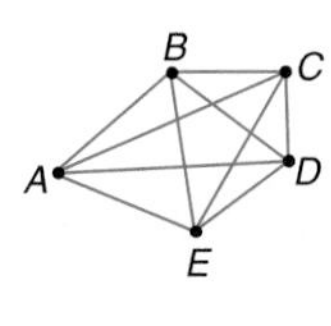
12.

10. A: degree 2, B: degree 2, C: degree 2, D: degree 2, E: degree 2
11. A: degree 4, B: degree 2, C: degree 3, D: degree 3, E: degree 2
12. A: degree 4, B: degree 4, C: degree 4, D: degree 4, E: degree 4

Consider each network.
a. Determine if it is traceable and complete.
b. If traceable, list a sequence of segments that demonstrates traceability.
c. If not complete, name the edges that need to be added to make it complete. 13–18. See margin.

13. 
14. 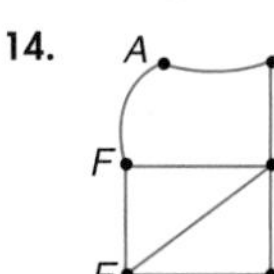
15. 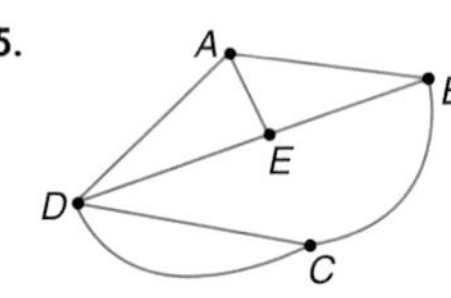
16. 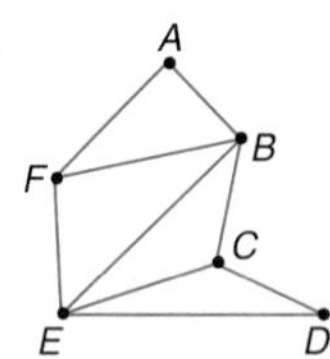
17. 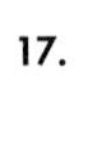 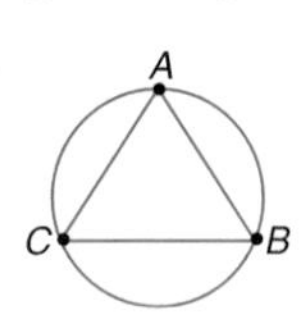
18.

Determine whether the network puzzles below *can't be traced*, *can be traced and started at one odd vertex and finished at the other*, or *can be traced and started and finished at the same vertex*. If it can be traced, show how.

20. Can be traced; start at one odd vertex and finish at the other.

19. can't be traced

20. 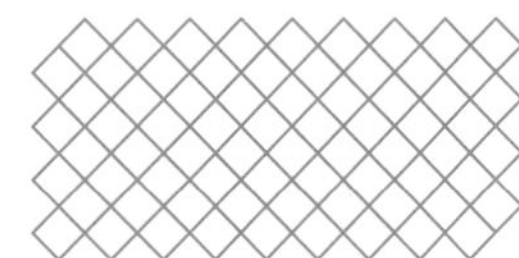

If it is possible, draw a network that satisfies the following conditions. 21–23. See margin for sample drawings.

21. both traceable and complete

22. complete, but not traceable

23. not traceable and incomplete

24. If you can use your pencil to draw a path through a network so that the path starts and ends at two different nodes and no node is passed more than once, the path is called an *Euler* (pronounced Ol-ler) *path*. If the path can be drawn so that the path starts and ends at the same node and no node is passed more than once, the path is called an *Euler circuit*.
    a. Does the network at the right have an Euler path? yes
    b. Does the network contain an Euler circuit? no
    c. Do you think a network could contain both an Euler path and an Euler circuit? Explain. No; if a network has an Euler circuit, each node has an even degree, so any path will return to its starting node.

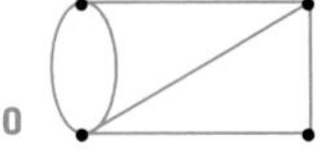

Critical Thinking

25. Show why this statement is true: *If a network has no node of odd degree, then the tracing can start at any node and will end at the starting point.* See margin.

26a. *A*: degree 2, *B*: degree 3, *C*: degree 3, *D*: degree 2, *E*: degree 3, *F*: degree 3

26b. 16

26c. The sum of the degrees of the nodes is twice the number of edges.

26. a. Find the degree of each node in the network at the right.
    b. Find the sum of the degrees of all the nodes.
    c. Compare this sum to the number of edges in the network. Is there a relationship between the number of edges and the sum of the degrees of the nodes?
    d. Draw several networks to verify your answer. See students' work.

Applications and Problem Solving

27. **Math History** In 1735, Leonhard Euler used networks to solve a famous problem about the bridges in the city of Köenigsberg, which is now Kaliningrad. Köenigsberg had seven bridges connecting both sides of the Predgel River to two islands in the river. Try to find a path that will take you over all seven bridges without crossing the same bridge twice.

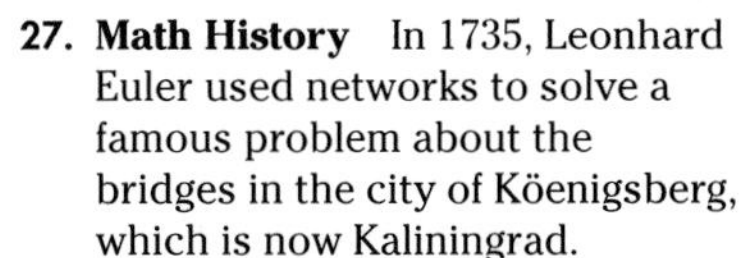

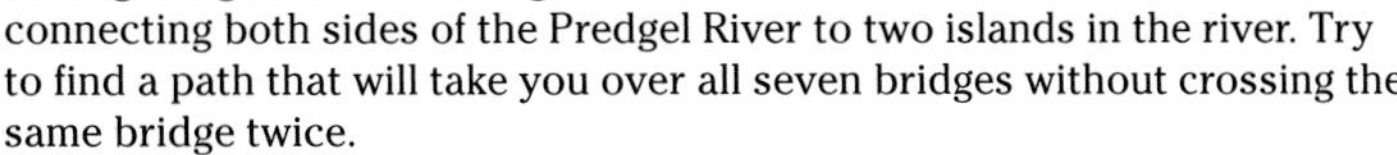

27b. No; all four nodes have odd degree.

    a. Draw the network. See margin.
    b. Is the network traceable? Explain your answer.
    c. If the bridge is not traceable, could you add one bridge to make it traceable? See margin.

Additional Answers

27a.

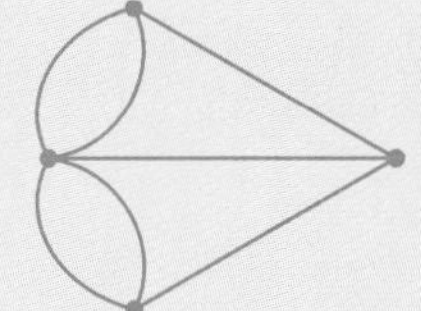

27c.

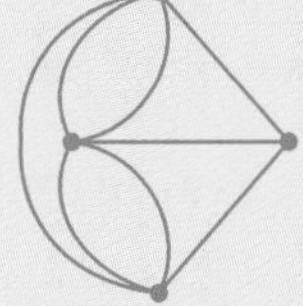

Additional Answers

21.

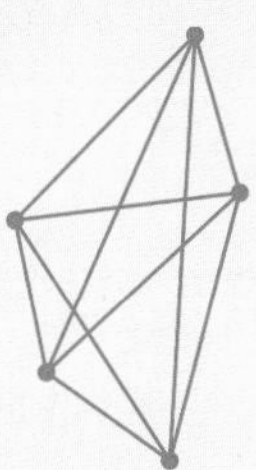

22. Sample answer:

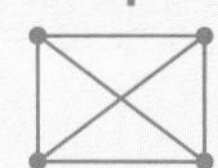

23. Sample answer:

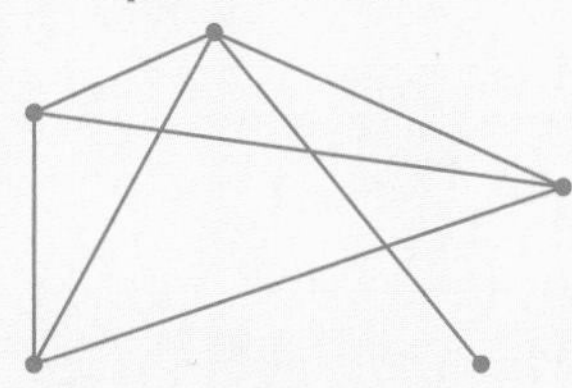

25. When a path goes through a node, it uses two edges, so wherever you start you use one edge from that node, and the edges of every subsequent node are both used, one for entering and one for exiting. Therefore, the starting node is the only node left to go to because only one edge was used to start.

Practice Masters, p. 63

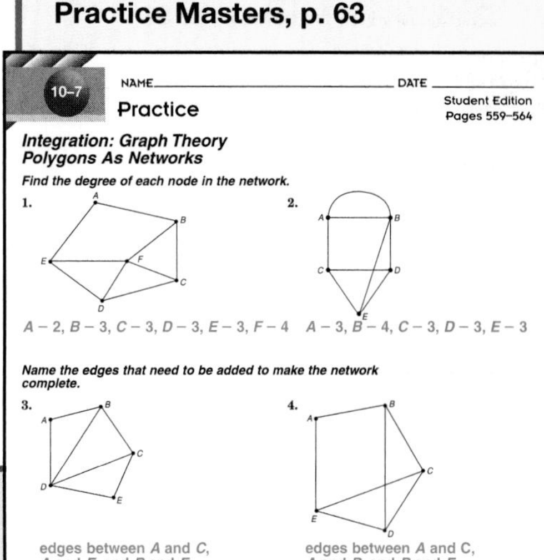

10-7 NAME ______ DATE ______

Practice — Student Edition Pages 559–564

Integration: Graph Theory Polygons As Networks

Find the degree of each node in the network.

1. A – 2, B – 3, C – 3, D – 3, E – 3, F – 4

2. A – 3, B – 4, C – 3, D – 3, E – 3

Name the edges that need to be added to make the network complete.

3. edges between *A* and *C*, *A* and *E*, and *B* and *E*

4. edges between *A* and C, *A* and *D*, and *B* and *E*

Determine whether each network is traceable. If a network is traceable, use arrows to trace the network.

5. traceable

6. not traceable

7. Draw two network of your own that are traceable. Then draw two networks that are not traceable. See students' work.

4 ASSESS

Closing Activity

Writing Have students write a brief paragraph explaining networks, nodes, and traceability.

Chapter 10 Quiz D (Lesson 10-7) is available in the *Assessment and Evaluation Masters*, p. 269.

Additional Answers

28a.

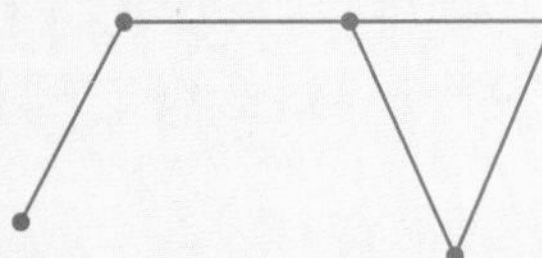

34. Given: Lines ℓ and m intersect at P.
Prove: Plane $\mathcal{R}$ contains both ℓ and m.
Assumption: Plane $\mathcal{R}$ does not contain both ℓ and m.

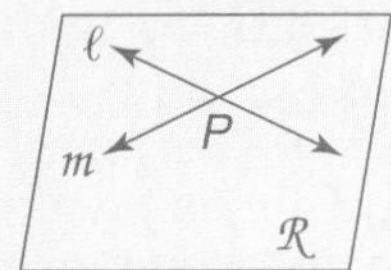

Enrichment Masters, p. 63

NAME________________ DATE ________

Enrichment

Student Edition Pages 559–564

Defining Numbers

A network may be defined by a table of its edges and nodes. Each edge connects one or two nodes. If an edge connects only one node, the edge is a **loop**. Study the network and its corresponding table.

Edge	Nodes
1	{A, B}
2	{B, C}
3	{C}
4	{C, D}
5	{B, D}
6	{A, B}
7	{A, D}

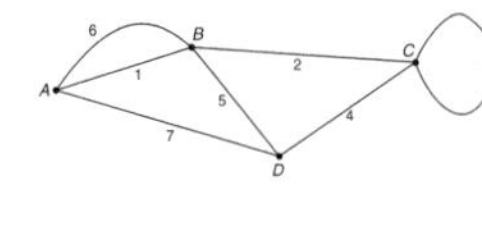

Notice that the network has a loop at node C. Also, notice that two edges, 1 and 6, connect A and B.

Complete the table of edges and nodes for each network.

1.

Edge	Nodes
1	{F, G}
2	{G, H}
3	{H, K}
4	{G, K}
5	{F, J}
6	{F, H}
7	{H, J}
8	{J, K}

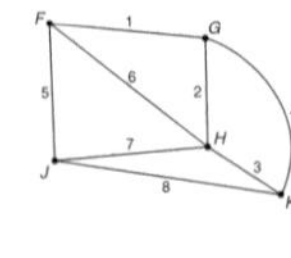

2.

Edge	Nodes
1	{W}
2	{W, X}
3	{X}
4	{X, Z}
5	{X, Y}
6	{W, Y}

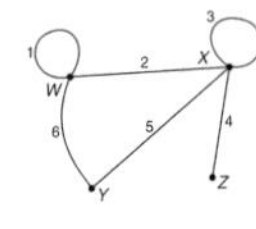

3. Draw a network with five nodes. Make a table of edges and nodes. See students' work.

28. **Safety** The fire department suggests that families establish exit paths in the case of an emergency. Use the floor plan at the right to establish a path that can be used to check to make sure that everyone is out of the house. Because children often hide behind doors in an emergency, the path should go through each door once. **a. See margin.**
   a. Draw a network that represents a plan for checking the rooms.
   b. Determine if the floor plan is traceable and if so, show how. **yes**

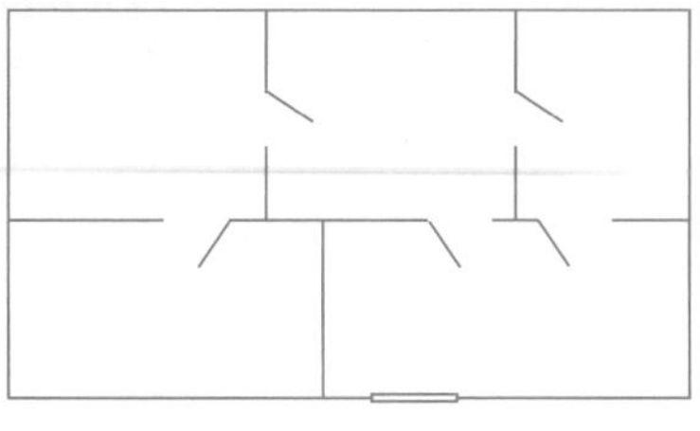

29. **Games** Shongo children from Zaire, a country in central Africa, spend time tracing patterns in the sand near their homes. The designs often look like those found on the clothing and woodcarvings their parents make. These patterns are actually traceable networks. Design your own network and, like the Shongo children, pattern it after a design on your clothes, the wallpaper in your bedroom, or another design. Exchange with a classmate when completed to see if they can trace your design. **See students' work.**

Mixed Review

30. If a point is chosen at random in the interior of $\odot O$ with radius 5, what is the probability that the point is in the interior of right triangle AOB where A and B lie on $\odot O$ and the right angle is at O? (Lesson 10–6) $\frac{1}{2\pi} \approx 15.9\%$

31. If $QRSTV$ is a regular pentagon and $ABVT$ is a square, find the area of the shaded region. (Lesson 10–5) **31.5 units²**

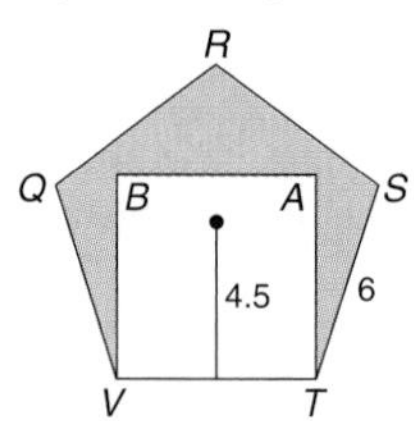

32. What is the area of the figure below? (Lesson 10–4) $1.5\sqrt{3} \approx 2.6$ **units²**

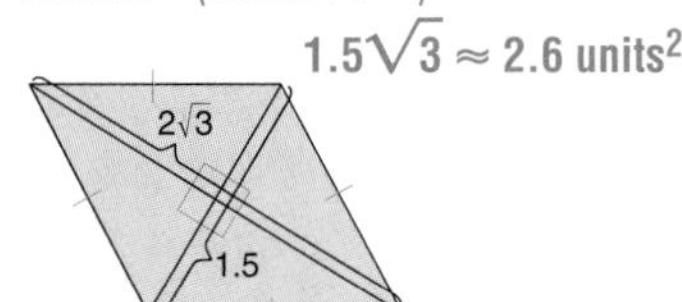

33. In $\triangle RST$, X lies on $\overline{RT}$ and Y lies on $\overline{RS}$. Determine the value of g that would make $\overline{YX} \parallel \overline{ST}$ if $RS = 12$, $RY = 8$, $RX = 10$, and $XT = g$. (Lesson 7–4) **5**

34. Write the given statement and draw the figure for a proof of the statement *If two lines intersect, then at least one plane contains both lines.* Then write the assumption you would make to write an indirect proof. (Lesson 5–3) **See margin.**

35. Given $\triangle UVW \cong \triangle HJK$, $UV = 9x - 12$, $VW = 12x - 30$, $UW = 2x + 6$, and $JK = 7x - 5$, find the length of $\overline{JK}$. (Lesson 4–3) **30 units**

Algebra

36. Simplify $6\frac{1}{4} \div 1\frac{1}{4}$. **5**

37. Use substitution to solve the system of equations. If the system does not have exactly one solution, state whether it has *no* solution, or *infinitely many* solutions. **no solution**

$2s = 10r + 6$

$s = 5r - 1$

Extension

Reasoning If all the possible diagonals of an octagon are drawn, will the figure be a traceable network? Explain your answer. **No; from each node there will be seven edges.**

VOCABULARY

After completing this chapter, you should be able to define each term, property, or phrase and give an example or two of each.

Geometry

altitude (p. 529)
apothem (p. 544)
area (p. 529)
base (p. 529)
center (p. 543)
complete network (p. 559)
concave (p. 514)
convex (p. 514)
degree of a node (p. 560)
edge (p. 559)
geometric probability (p. 551)
graph theory (p. 559)
height (p. 529)
incomplete network (p. 559)
n-gon (p. 515)
network (p. 559)
nodes (p. 559)
polygon (p. 514)
radius (p. 543)
regular polygon (p. 515)
regular tessellation (p. 523)
sector of a circle (p. 553)
semi-regular (p. 524)
tessellation (p. 522, 523)
traceable (p. 559)
uniform (p. 524)

Problem Solving

guess and check (p. 524)

UNDERSTANDING AND USING THE VOCABULARY

Determine whether each statement is *true* or *false*. If it is false, change the underlined word to make it true.

1. An <u>apothem</u> connects the center of a polygon with one of its vertices. false; radius
2. An <u>equilateral triangle</u> is a regular polygon. true
3. Figure 3 is <u>convex</u>. true
4. Figure 2 is a <u>regular polygon</u>. false; network
5. The degree of node A in Figure 1 is <u>4</u>. false; 5
6. Figure 3 is a regular <u>apothem</u>. false; polygon or octagon
7. Figure 1 is <u>traceable</u>. true
8. Figure 2 is a <u>complete</u> network. true
9. The segment drawn in Figure 3 is an <u>altitude</u>. false; apothem
10. Figure 4 is a <u>uniform</u> tessellation. true

Figure 1

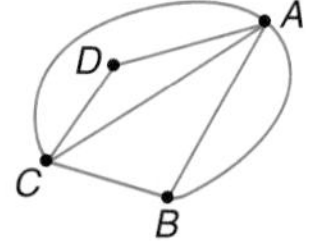

Figure 2

Figure 3

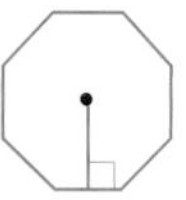

Figure 4

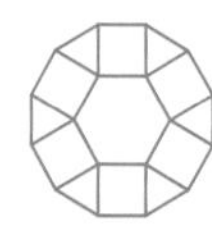

Using the CHAPTER HIGHLIGHTS

The Chapter Highlights begins with a listing of the new terms, properties, and phrases that were introduced in this chapter. Have students define each term and provide an example or two of it, if appropriate.

Assessment and Evaluation Masters, pp. 255–256

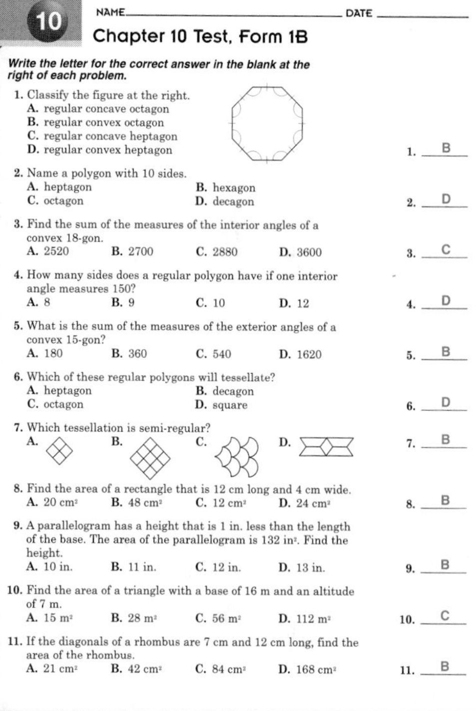

10 NAME________ DATE________

Chapter 10 Test, Form 1B

Write the letter for the correct answer in the blank at the right of each problem.

1. Classify the figure at the right.
A. regular concave octagon
B. regular convex octagon
C. regular concave heptagon
D. regular convex heptagon — 1. B
2. Name a polygon with 10 sides.
A. heptagon B. hexagon
C. octagon D. decagon — 2. D
3. Find the sum of the measures of the interior angles of a convex 18-gon.
A. 2520 B. 2700 C. 2880 D. 3600 — 3. C
4. How many sides does a regular polygon have if one interior angle measures 150?
A. 8 B. 9 C. 10 D. 12 — 4. D
5. What is the sum of the measures of the exterior angles of a convex 15-gon?
A. 180 B. 360 C. 540 D. 1620 — 5. B
6. Which of these regular polygons will tessellate?
A. heptagon B. decagon
C. octagon D. square — 6. D
7. Which tessellation is semi-regular?
A. B. C. D. — 7. B
8. Find the area of a rectangle that is 12 cm long and 4 cm wide.
A. 20 cm² B. 48 cm² C. 12 cm² D. 24 cm² — 8. B
9. A parallelogram has a height that is 1 in. less than the length of the base. The area of the parallelogram is 132 in². Find the height.
A. 10 in. B. 11 in. C. 12 in. D. 13 in. — 9. B
10. Find the area of a triangle with a base of 16 m and an altitude of 7 m.
A. 15 m² B. 28 m² C. 56 m² D. 112 m² — 10. C
11. If the diagonals of a rhombus are 7 cm and 12 cm long, find the area of the rhombus.
A. 21 cm² B. 42 cm² C. 84 cm² D. 168 cm² — 11. B

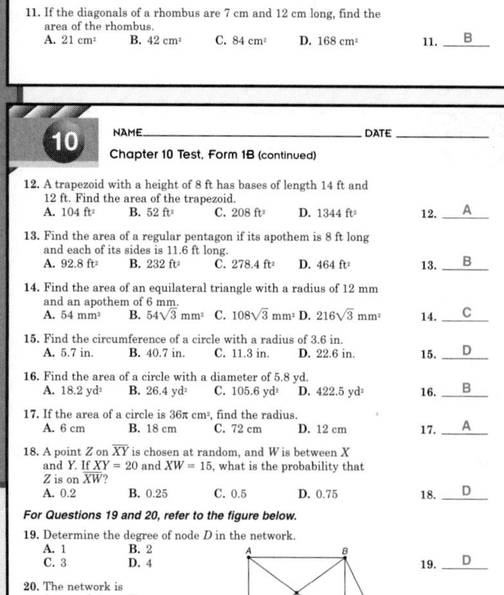

10 NAME________ DATE________

Chapter 10 Test, Form 1B (continued)

12. A trapezoid with a height of 8 ft has bases of length 14 ft and 12 ft. Find the area of the trapezoid.
A. 104 ft² B. 52 ft² C. 208 ft² D. 1344 ft² — 12. A
13. Find the area of a regular pentagon if its apothem is 8 ft long and each of its sides is 11.6 ft long.
A. 92.8 ft² B. 232 ft² C. 278.4 ft² D. 464 ft² — 13. B
14. Find the area of an equilateral triangle with a radius of 12 mm and an apothem of 6 mm.
A. 54 mm² B. $54\sqrt{3}$ mm² C. $108\sqrt{3}$ mm² D. $216\sqrt{3}$ mm² — 14. C
15. Find the circumference of a circle with a radius of 3.6 in.
A. 5.7 in. B. 40.7 in. C. 11.3 in. D. 22.6 in. — 15. D
16. Find the area of a circle with a diameter of 5.8 yd.
A. 18.2 yd² B. 26.4 yd² C. 105.6 yd² D. 422.5 yd² — 16. B
17. If the area of a circle is 36π cm², find the radius.
A. 6 cm B. 18 cm C. 72 cm D. 12 cm — 17. A
18. A point Z on $\overline{XY}$ is chosen at random, and W is between X and Y. If $XY = 20$ and $XW = 15$, what is the probability that Z is on $\overline{XW}$?
A. 0.2 B. 0.25 C. 0.5 D. 0.75 — 18. D

For Questions 19 and 20, refer to the figure below.

19. Determine the degree of node D in the network.
A. 1 B. 2
C. 3 D. 4 — 19. D
20. The network is
A. complete. B. open.
C. traceable. D. not traceable. — 20. C

Bonus

The area of circle P is 176.625 in². If $CB = 12$ in., find the area of $\triangle ABC$.
A. 90 in² B. 108 in²
C. 54 in² D. 180 in² — Bonus C

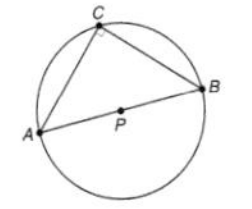

Instructional Resources

Three multiple-choice tests and three free-response tests are provided in the *Assessment and Evaluation Masters*. Forms 1A and 2A are for honors pacing, Forms 1B and 2B are for average pacing, and Forms 1C and 2C are for basic pacing. Chapter 10 Test, Form 1B, is shown at the right. Chapter 10 Test, Form 2B, is shown on the next page.

Postulates, Theorems, and Corollaries

A complete list of postulates, theorems, and corollaries begins on page 806.

Using the STUDY GUIDE AND ASSESSMENT

Skills and Concepts Encourage students to refer to the objectives and examples on the left as they complete the review exercises on the right.

Assessment and Evaluation Masters, pp. 261–262

NAME__________ DATE __________

Chapter 10 Test, Form 2B

For Questions 1–3, determine whether each statement is always, sometimes, or never true.

1. A regular polygon has all its angles congruent. — 1. always
2. If an octagor. is convex, then it is regular. — 2. sometimes
3. If the circumference of a circle is 16π, then the area is 64π. — 3. always
4. Is the polygon at the right convex or concave? — 4. convex
5. The sum of the measures of the interior angles of a convex polygon is 1980. How many vertices does the polygon contain? — 5. 13
6. The sum of the measures of three of the interior angles of a pentagon is 280. The measure of the fourth angle is three times the measure of the fifth. Find the measure of the fourth angle. — 6. 195
7. Find the sum of the measures of the exterior angles of a convex 19-gon. — 7. 360
8. Draw a tessellation, using a square and an equilateral triangle. — 8. Sample answer:
9. The coordinates of the vertices of a parallelogram are $A(-2, 3)$, $B(2, 3)$, $C(3, 0)$, and $D(-1, 0)$. Find the area of the parallelogram. — 9. 12 units²
10. The length of a rectangle is 4 in. more than the width. If the perimeter is 56 in., find the area of the rectangle. — 10. 192 in²
11. A square has an area of 49 yd². What is the perimeter of the square? — 11. 28 yd
12. The area of a triangle is 120 m². If the length of the base is 24 m, what is the height? — 12. 10 m
13. The diagonals of a rhombus are 32 cm and 10 cm long. Find the area of the rhombus. — 13. 160 cm²

NAME__________ DATE __________

Chapter 10 Test, Form 2B (continued)

14. One base of a trapezoid is 12 in. long and the altitude is 16 in. long. If the area is 248 in², find the length of the other base. — 14. 19 in.
15. Find the area of an equilateral triangle with an apothem 3.5 cm long and a side 12 cm long. — 15. 63 cm²
16. Find the area of a regular hexagon with an apothem 8 m long. Round to the nearest tenth. — 16. 221.7 m²
17. Find the area of a circle with a circumference of 32 ft. Round to the nearest tenth. — 17. 81.5 ft²
18. Maria told Missy that she would arrive at her house sometime between 9:30 A.M. and 10:00 A.M. Missy says she must leave at 9:50 A.M. What is the probability that Maria will arrive by 9:50 A.M.? — 18. $\frac{2}{3}$

For Questions 19 and 20, refer to the figure below.

X Y W V Z

19. Name the additional edges that must be drawn to make the network complete. — 19. edges between *X* and *Z*, *Y* and *V*, *W* and *Y*, and *W* and *Z*.
20. When the network is complete, is it traceable? Explain why or why not. — 20. Yes; all nodes have even degrees.

Bonus

A square has sides 20 in. long. If the square is inscribed in a circle, find the area of the circle. — Bonus 628.3 in²

SKILLS AND CONCEPTS

OBJECTIVES AND EXAMPLES

Upon completing this chapter, you should be able to:

- find measures of angles in polygons (Lesson 10–1)

Interior Angle Sum Theorem (Theorem 10–1)

If a convex polygon has n sides and S is the sum of the measures of its interior angles, then $S = 180(n - 2)$.

Exterior Angle Sum Theorem (Theorem 10–2)

If a polygon is convex, then the sum of the measures of the exterior angles, one at each vertex, is 360.

REVIEW EXERCISES

Use these exercises to review and prepare for the chapter test.

11. Find the measure of an interior angle of a regular decagon. **144**
12. The sum of the measures of the interior angles of a convex polygon is 1980. How many sides does the polygon have? **13**
13. A regular polygon has 20 sides. Find the measure of an interior angle and an exterior angle of the polygon. **162; 18**
14. The measure of each interior angle of a regular polygon is eleven times that of an exterior angle. How many sides are in the polygon? **24**

- identify regular and uniform (semi-regular) tessellations (Lesson 10–2)

A *regular tessellation* uses only one type of regular polygon.

A *uniform tessellation* contains the same combination of shapes and angles at each vertex.

Uniform tessellations that contain two or more regular polygons are *semi-regular.*

Determine whether the following tessellations are *regular, uniform,* or *semi-regular.* Name all possibilites.

15.

regular, uniform

16.

regular

- find the areas of parallelograms (Lesson 10–3)

Find the area of the parallelogram.

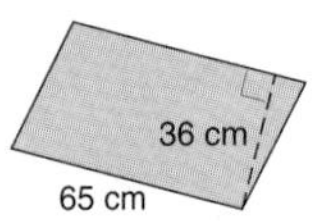

$A = bh$ *Formula for area of a parallelogram*

$A = (65)(36)$ $b = 65, h = 36$

$A = 2340$ square centimeters

Find the area of each shaded region.

17.

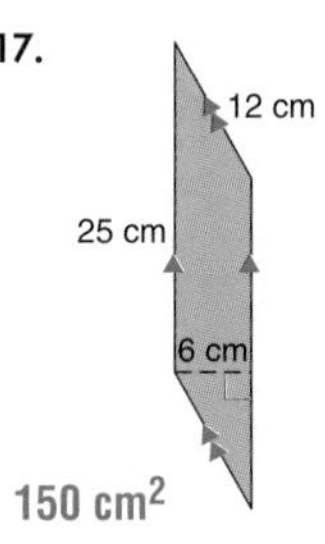

150 cm²

18.

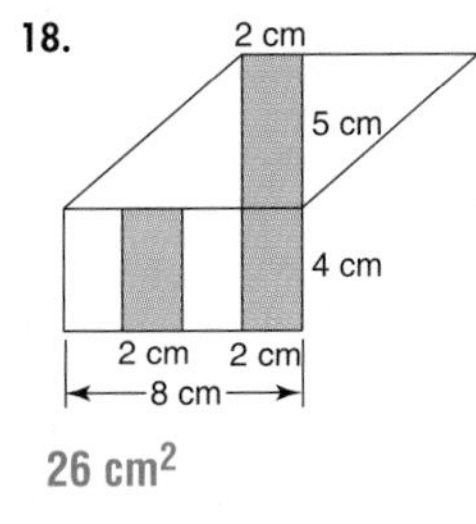

26 cm²

19. The area of parallelogram *ABCD* is 134.19 square inches. If the base is 18.9 inches long, find the height. **7.1 in.**

GLENCOE Technology

Test and Review Software

You may use this software, a combination of an item generator and item bank, to create your own tests or worksheets. Types of items include free response, multiple choice, short answer, and open ended.

For IBM & Macintosh

OBJECTIVES AND EXAMPLES

- find the areas of triangles, rhombi, and trapezoids (Lesson 10–4)

Area Formulas	
Triangle	$A = \frac{1}{2}bh$
Rhombus	$A = \frac{1}{2}d_1d_2$
Trapezoid	$A = \frac{1}{2}h(b_1 + b_2)$

OBJECTIVES AND EXAMPLES

Find the area of each figure. 20. 42.42 m^2

20.

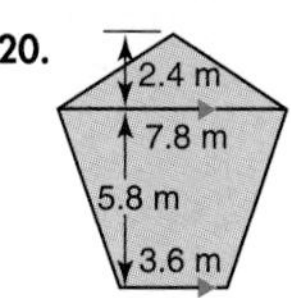

21.

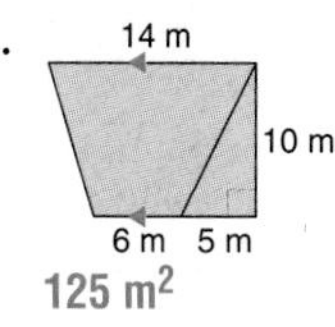

125 m^2

22. A rhombus has a diagonal 8.6 centimeters long and an area of 54.18 square centimeters. What is the length of each side? 7.6 cm

23. The area of an isosceles trapezoid is 7.2 square feet. The perimeter is 6.8 feet. If a leg is 1.9 feet long, find the height. 4.8 ft

- find the areas of regular polygons (Lesson 10–5)

Find the area of a regular hexagon with an apothem 12 centimeters long.

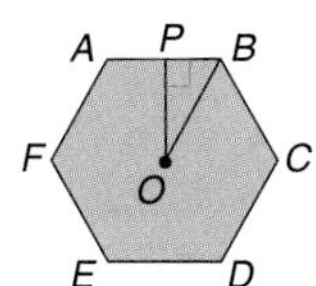

Since $\triangle POB$ is a 30°-60°-90° triangle, $PB = \frac{12}{\sqrt{3}}$ or $4\sqrt{3}$.

So, $AB = 8\sqrt{3}$ and $P = 48\sqrt{3}$.

$A = \frac{1}{2}Pa$

$A = \frac{1}{2}(48\sqrt{3})(12)$

$A = 288\sqrt{3}$ or about 498.8 cm^2

26. 10,641.7 mm^2

27. $4n^2$ cm^2

Find the area of each regular polygon. Round to the nearest tenth. 25. 0.6 ft^2

24. an equilateral triangle with an apothem 8.9 inches long 411.6 in^2

25. a pentagon with an apothem 0.4 feet long

26. a hexagon with sides 64 millimeters long

27. a square with an apothem n centimeters long

28. Find the area of the regular pentagon shown at the right with a perimeter of 45 inches. 139.4 in^2

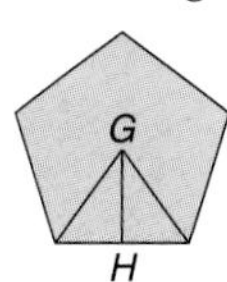

- find the area of a circle (Lesson 10–5)

If a circle has a radius of r units, then its area is πr^2.

Find the area of a circle with a diameter of 14 centimeters.

$A = \pi r^2$

$A = \pi(7)^2$ *d = 14, so r = 7.*

$A = 49\pi$

The area is 49π or about 153.9 square centimeters.

Find the circumference and area of a circle with the given radius. Round to the nearest tenth.

29. 7 mm 44.0 mm; 153.9 mm^2

30. 19 in. 119.4 in.; 1134.1 in^2

31. 0.9 ft 5.7 ft; 2.5 ft^2

32. $3\frac{1}{3}$ cm 20.9 cm; 34.9 cm^2

33. The diagonal of a square is 8 feet. Find the circumference and the area of a circle inscribed in the square. $4\pi\sqrt{2} \approx 17.8$ ft, $8\pi \approx 25.1$ ft^2

Applications and Problem Solving Encourage students to work through the exercises in the Applications and Problem Solving section to strengthen their problem-solving skills.

OBJECTIVES AND EXAMPLES

- use area to solve problems involving geometric probability (Lesson 10–6)

Length Probability Postulate (*Postulate 10–3*)

If a point on $\overline{AB}$ is chosen at random and C is between A and B, then the probability that the point is on $\overline{AC}$ is $\frac{\text{length of } \overline{AC}}{\text{length of } \overline{AB}}$.

Area Probability Postulate (*Postulate 10–4*)

If a point in region A is chosen at random, then the probability that the point is in region B, which is in region A, is $\frac{\text{area of region } B}{\text{area of region } A}$.

OBJECTIVES AND EXAMPLES

34. During the morning rush hour, a bus arrives at the bus stop at Indianola and Morse Roads every seven minutes. A bus will wait for thirty seconds before departing. If you arrive at a random time, what is the probability that there will be a bus waiting? $\frac{1}{14}$

35. Circle O has a radius of 4.5 centimeters. What is the probability that a point chosen at random will lie in the shaded region? 9.08%

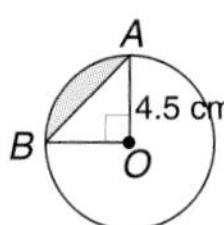

- determine if a network is traceable or complete (Lesson 10–7)

The network at the right is traceable, since exactly two of the nodes have odd degrees. However, it is not complete since there is no edge between M and A.

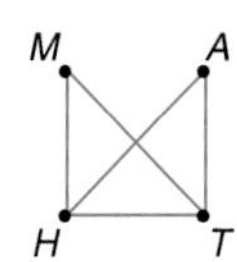

36. nodes: A, B, C, D, E; edges: $\overline{AB}, \overline{BE}, \overline{BD}, \overline{BC}, \overline{CD}$

Use the network at the right to answer each question.

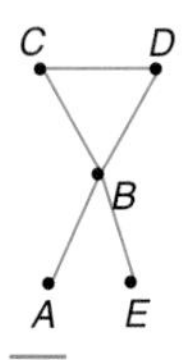

36. Name the nodes and edges in the network.

37. Is the network traceable? yes

38. Is the network complete? If not, what edges need to be added for the network to be complete? no; $\overline{CA}, \overline{CE}, \overline{DA}, \overline{DE}, \overline{AE}$

APPLICATIONS AND PROBLEM SOLVING

39. Guess and Check Using the digits 1, 2, 3, 4, 5, and 6 only once, find two whole numbers whose product is as great as possible. (Lesson 10–2) 631 and 542

40. Interior Design John and Cynthia are buying paint for the walls of their kitchen. Three walls are each 8 feet long, and one is 10 feet long. The walls are all 8 feet high. If a gallon of paint covers 400 square feet, how many gallons of paint will they need for two coats of paint? (Lesson 10–3) 2 gallons

41. Manufacturing A wooden planter has a square base and sides that are shaped like trapezoids. An edge of the base is 8 inches long, and the top edge of each side is 10 inches long. If the height of a side is 12 inches, how much wood does it take to make a planter? (Lesson 10–4) 496 in^2

A practice test for Chapter 10 is provided on page 802.

ALTERNATIVE ASSESSMENT

COOPERATIVE LEARNING PROJECT

Networks In this chapter, you saw that polygons can be described by networks. Networks have many applications. In this project, you will apply networks to several areas.

Follow these steps to explore applications of networks.

- Each step of a proof follows from a previous step in the proof or a proved theorem or postulate. Select a proof from any chapter in this text and draw a network for the proof.
- Your daily schedule can be represented by a network. Each activity occurs in a sequence. Draw a network for your schedule for tomorrow.
- Making a decision can be difficult because there are so many choices. Each choice can be represented as a branch on a tree. Each choice can lead to several possible outcomes. Each outcome can be represented by a branch from the choice branch. Select a decision you will be faced with in the near future and draw a network for the possible choices and their outcomes.
- As a group, discuss other applications of networks.

THINKING CRITICALLY

It is possible to find the area of a region by adding the areas of its nonoverlapping parts. How would you determine the area of a region if you know the areas of regions that do overlap? **Subtract the areas of the regions formed by the overlapping parts.**

PORTFOLIO

Make plans for a gazebo you might build for your yard or a nearby park. Its base should be a regular polygon. Its sides should consist of rectangles, and the roof should consist of triangles joined together. How might you use what you've learned in this chapter to determine how much paint would be needed for the gazebo? Include a drawing of the gazebo in your portfolio.

SELF EVALUATION

People make decisions in a variety of ways. Some people work alone; others seek assistance. Some people follow their feelings; others approach problems rationally. Think of decisions you have faced.

Assess Yourself How do you make decisions? Do you lay out all the alternatives and study their outcomes? Do you follow your feelings? Do you seek advice from a friend or advisor?

Alternative Assessment

The Alternative Assessment section provides students with the opportunity to assess their own work by thinking critically, working with others, keeping a portfolio, and honestly evaluating their own progress. For more information on alternative forms of assessment, see *Alternative Assessment in the Mathematics Classroom,* one of the titles in the Glencoe Mathematics Professional Series.

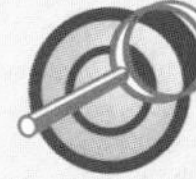

Performance Assessment

Performance Assessment tasks for this chapter are included in the *Assessment and Evaluation Masters.* A scoring guide is also provided.

Assessment and Evaluation Masters, pp. 266, 277

NAME________________ DATE ________

Chapter 10 Performance Assessment

Instructions: ***Demonstrate your knowledge by giving a clear, concise solution to each problem. Be sure to include all relevant drawings and to justify your answers. You may show your solution in more than one way or investigate beyond the requirements of the problem.***

1. Myrtle Beach High School wishes to refinish its basketball court. The shaded area will be painted. The rest will have a hardwood finish.
   a. Give a plan for finding the area to be painted.
   b. Give a plan for finding the area to be given a hardwood finish.
   c. Find the area to be painted. Show your work.
   d. Find the area to be given a hardwood finish.
   e. Find the circumference of the center circle. If eight players stand around the center circle for a jump ball, what is the arc length available to each player?

50'
94'
6'
6'
19'
3'
4'

2. A Franklin County subdivision has nine streets arranged as shown on the network below.

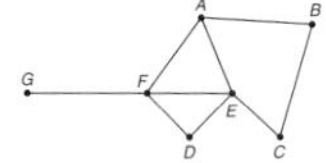

   a. Tell what is meant by a node and an edge.
   b. Define the degree of a node. Name a node in the network above and give its degree.
   c. Is the network shown above complete? Why or why not?
   d. Tell whether the network is traceable. If so, list at least two sequences, if possible, that demonstrate traceability.
   e. If a mail carrier wants to travel each street only once, where would you suggest that he or she start the mail deliveries? Why?

Scoring Guide
Chapter 10
Performance Assessment

Level	Specific Criteria
3 Superior	• Shows thorough understanding of the concepts of *the area of a polygon, area and circumference of a circle, network, node, degree of a node,* and *complete and traceable networks.* • Uses appropriate strategies to solve problems. • Computations are correct. • Written explanations are exemplary. • Goes beyond requirements of some or all problems.
2 Satisfactory, with Minor Flaws	• Shows understanding of the concepts of *the area of a polygon, area and circumference of a circle, network, node, degree of a node,* and *complete and traceable networks.* • Uses appropriate strategies to solve problems. • Computations are mostly correct. • Written explanations are effective. • Satisfies all requirements of some or all problems.
1 Nearly Satisfactory, with Serious Flaws	• Shows understanding of most of the concepts of *the area of a polygon, area and circumference of a circle, network, node, degree of a node,* and *complete and traceable networks.* • May not use appropriate strategies to solve problems. • Computations are mostly correct. • Written explanations are satisfactory. • Satisfies most requirements of some or all problems.
0 Unsatisfactory	• Shows little or no understanding of the concepts of *the area of a polygon, area and circumference of a circle, network, node, degree of a node,* and *complete and traceable networks.* • May not use appropriate strategies to solve problems. • Computations are incorrect. • Written explanations are not satisfactory. • Does not satisfy requirements of some or all problems.

Using the
COLLEGE ENTRANCE EXAM PRACTICE

These two pages review the skills and concepts presented in Chapters 1–10. This review is formatted to reflect new trends in college entrance testing.

Assessment and Evaluation Masters, pp. 271–272

NAME________________ DATE ________

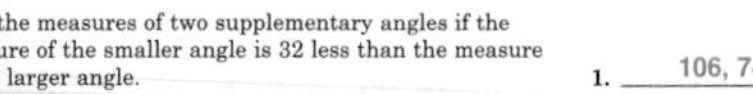

Chapter 10 Cumulative Review

1. Find the measures of two supplementary angles if the measure of the smaller angle is 32 less than the measure of the larger angle. 1. 106, 74
2. Identify the conclusion of the conditional statement "If $y = 9$, then $10y = 90$." 2. $10y = 90$
3. Determine the value of r so that a line through the points with the coordinates $(-3, 2)$ and $(r, 6)$ has a slope of 1. 3. 1
4. Given vertices $X(-1, 4)$, $Y(5, 4)$, and $Z(5, -2)$, describe $\triangle XYZ$ in terms of its angles and sides. 4. right isosceles

For Questions 5 and 6, refer to the figure at the right. Compare each statement by using <, =, or >.

5. $m\angle 1$ ___ $m\angle 3$ 5. <
6. $m\angle 4$ ___ $m\angle 6$ 6. >
7. List the sides of $\triangle ABC$ in order from the longest to shortest if $m\angle A = 2x + 3$, $m\angle B = 2x - 4$, and $m\angle C = x + 16$. 7. $\overline{BC}, \overline{AC}, \overline{AB}$
8. Given parallelogram $GHIJ$ with $m\angle G = x$ and $m\angle J = 7x + 4$. Find $m\angle H$. 8. 158
9. $RSTU$ is a parallelogram. If $RS = 2x + 3$ and $ST = 3x - 4$, for what value of x is $RSTU$ a rhombus? 9. 7

For Questions 10 and 11, use rhombus MNOP.

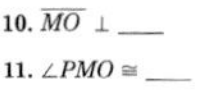

10. $\overline{MO} \perp$ ___ 10. $\overline{NP}$
11. $\angle PMO \cong$ ___ 11. $\angle OMN$, $\angle NOM$, or $\angle POM$

Solve.

12. $\frac{4}{10} = \frac{8}{x}$ 12. 20
13. $\frac{a}{a-2} = \frac{2}{5}$ 13. $-\frac{4}{3}$
14. One angle in a right triangle has a measure of 42. A second right triangle has an angle with a measure of 48. Are the two triangles similar? Explain. 14. yes; AA Similarity

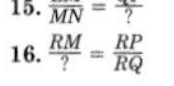

NAME________________ DATE ________

Chapter 10 Cumulative Review (continued)

Refer to the figure at the right. $NO \parallel MP \parallel LQ$.

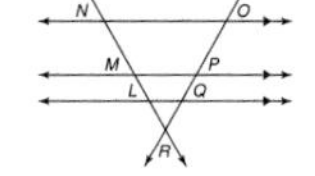

15. $\frac{LM}{MN} = \frac{QP}{?}$ 15. PO
16. $\frac{RM}{?} = \frac{RP}{RQ}$ 16. RL

Refer to the figure at the right.

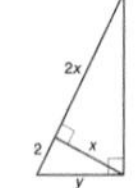

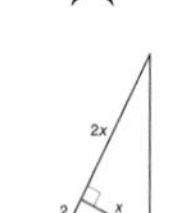

17. Find the value of x. 17. 8
18. Find the value of y. 18. $\sqrt{80}$

For Questions 19 and 20, determine whether each statement is true or false.

19. A tangent of a circle is a secant of the circle. 19. false
20. A concave polygon can be regular. 20. false
21. Find the measure of an interior angle of a regular octagon. 21. 135
22. Find the base of a parallelogram whose area is 180 cm² and whose height is 12 cm. 22. 15 cm
23. Find the area of a rhombus whose diagonals measure 17 cm and 23 cm. 23. 195.5 cm²
24. What is the probability that a point chosen at random on $\overline{XY}$ will lie on $\overline{XZ}$? 24. 0.25
25. What is the degree of node A? 25. 4

COLLEGE ENTRANCE EXAM PRACTICE

CHAPTERS 1–10

SECTION ONE: MULTIPLE CHOICE

There are eight multiple-choice questions in this section. After working each problem, write the letter of the correct answer on your paper.

1. The circle below with center P is divided into six equal sectors. If the area of the circle is 144π, find the perimeter of the shaded sector. B

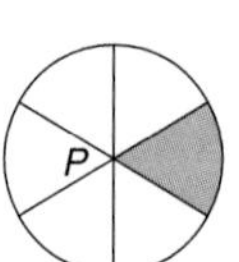

A. $12 + 3\pi$
B. $24 + 4\pi$
C. $24 + 24\pi$
D. $12 + 4\pi$

2. If $3x = \pm\sqrt{25} - \sqrt{16}$, then find the solution set for x. A

A. $\left\{\frac{1}{3}, -3\right\}$
B. $\left\{-\frac{1}{3}, 9\right\}$
C. $\left\{\frac{1}{3}, \frac{1}{9}\right\}$
D. $\left\{-\frac{1}{3}, -\frac{1}{9}\right\}$

3. If $2x + 4 = 9$, then find $x - \frac{1}{2}$. B

A. 1
B. 2
C. 12
D. 14

4. In the circle below, C is the center of the circle with radius r, and $ABCD$ is a rectangle. Find the length of diagonal $\overline{BD}$. A

A. r
B. $r\frac{\sqrt{2}}{2}$
C. $r\sqrt{3}$
D. $r\frac{\sqrt{3}}{2}$

5. In the circle below, the shaded region is what fractional part of the area of the circle? D

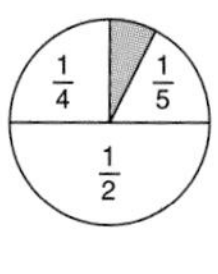

A. $\frac{1}{40}$
B. $\frac{1}{60}$
C. $\frac{1}{10}$
D. $\frac{1}{20}$

6. In the figure below, isosceles right triangle BEF overlaps square $ABCD$. If $AB = 2$ and $EB = 4$, find the area of $CDEF$. A

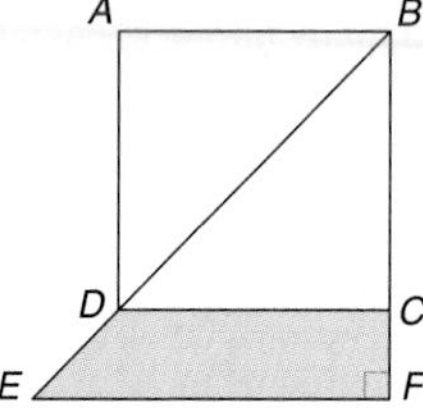

A. 2
B. $\sqrt{2}$
C. $\frac{\sqrt{2}}{2}$
D. 1

7. What are the values of x for which $\frac{x(x+1)}{(x-4)} \cdot \frac{1}{(x+3)}$ is undefined? A

A. $-3, 4$
B. $3, -4$
C. $1, 3, -4$
D. $-1, -3, 4$

8. The graph of $y = 4x + 8$ is which of the following? C

A. a horizontal line
B. a vertical line
C. a line that rises to the right
D. a line that falls to the right

Standardized Test Practice Questions are also provided in the *Assessment and Evaluation Masters*, p. 270.

A more traditional cumulative review is provided in the *Assessment and Evaluation Masters*, pp. 271–272.

SECTION TWO: SHORT ANSWER

This section contains seven questions for which you will provide short answers. Write your answer on your paper.

9. If $x^2 - 11 < x^2 + 2x - 5 < x^2 + 25$, find an inequality that represents x. **$-3 < x < 15$**

10. The figure below is formed from a semicircle and a rectangle. The diameter of the semicircle is the length of the rectangle. Write a formula for the area of the figure if the length of the rectangle is $3x$ and the width is x. **$\frac{9\pi x^2}{8} + 3x^2$ units^2**

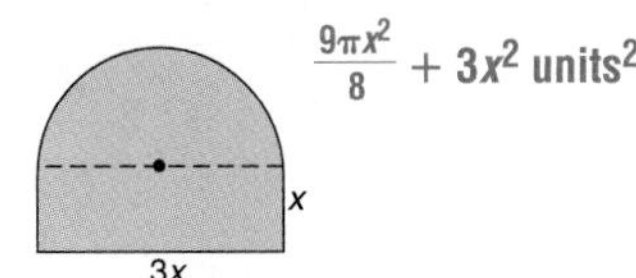

11. The average of six numbers is 10, and the average of ten other numbers is 6. Find the average of all sixteen numbers. **7.5**

12. In the figure, P and Q lie on $\odot O$. $\overline{OP}$ is 8 units long, and $\angle POQ$ measures 80°. Find the length of $\widehat{PQ}$.

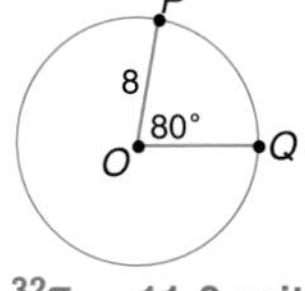

$\frac{32\pi}{9} \approx 11.2$ units

13. Simplify $\frac{8x^5y^{-2}z}{16x^{-2}yz^2}$. **$\frac{x^7}{2y^3z}$**

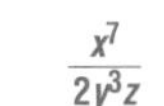

14. Find the value of x in the figure at the right.

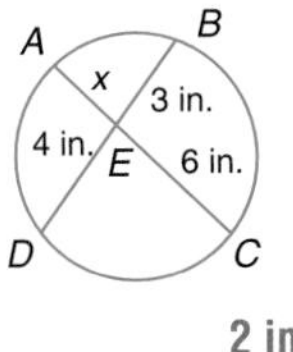

2 in.

15. The figure below is composed of two squares with areas shown. What is the value of x? **140 cm**

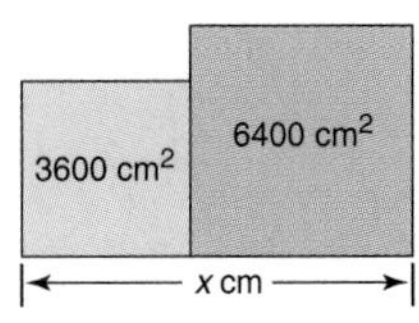

SECTION THREE: COMPARISON

This section contains five comparison problems that involve comparing two quantities, one in column A and one in column B. In certain questions, information related to one or both quantities is centered above them. All variables used represent real numbers.

Compare quantities A and B below.

- Write A if quantity A is greater.
- Write B if quantity B is greater.
- Write C if the two quantities are equal.
- Write D if there is not enough information to determine the relationship.

Column A	Column B
Parallelogram P and rectangle R have equal bases and equal areas. P is not a rectangle.	
16. perimeter of R	perimeter of P **B**
17. the length of the hypotenuse of right triangle MNP	the length of side $\overline{PQ}$ of equilateral triangle PQR **D**
18. the largest prime factor of 858	the largest prime factor of 2310 **A**
19. perimeter of $ABCD$	circumference of the circle with diameter $\overline{DC}$ **B**
20. the length of the altitude of $\triangle XYZ$ drawn to side $\overline{YZ}$	the length of the altitude of $\triangle PQR$ drawn to side $\overline{QR}$ **C**

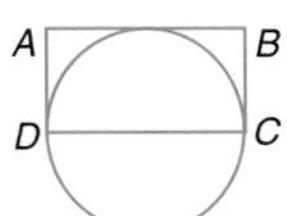

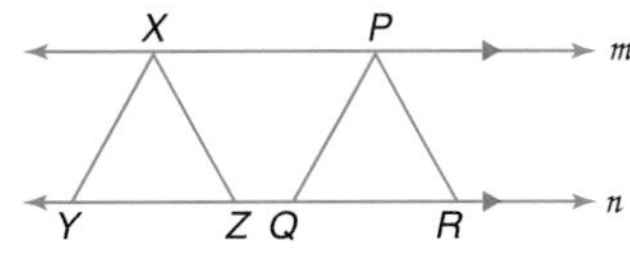

TEST-TAKING TIP

Quantitative Comparisons

Have students substitute values for variables to make a quantitative comparison. Remind them that a variable may represent a positive number greater than 1, a negative number less than -1, a number between 0 and 1, or a number between -1 and 0. Numbers from these different groups often behave differently, so it is important to examine a comparison for each case.

Example

Column A	Column B
$x^2 - 1$	$x^3 + 1$

Try $x = 2$.

$(2)^2 - 1 = 3$ $\quad$ $(2)^3 + 1 = 9$

Column B is greater.

Try $x = -2$.

$(-2)^2 - 1 = 3$ $\quad$ $(-2)^3 + 1 = -7$

Column A is greater.

Try $x = \frac{1}{2}$.

$\left(\frac{1}{2}\right)^2 - 1 = -\frac{3}{4}$ $\quad$ $\left(\frac{1}{2}\right)^3 + 1 = 1\frac{1}{8}$

Column B is greater.

Try $x = -\frac{1}{2}$.

$\left(-\frac{1}{2}\right)^2 - 1 = -\frac{3}{4}$ $\quad$ $\left(-\frac{1}{2}\right)^3 + 1 = \frac{7}{8}$

Column B is greater.

Since the actual value of x is not known and no pattern seems apparent, there is not enough information to determine the relationship between the quantities.

CHAPTER 11

Investigating Surface Area and Volume

PREVIEWING THE CHAPTER

In this chapter, students make models of three-dimensional figures and look at cross sections and slices. Students learn to recognize many three-dimensional figures and learn to draw three-dimensional figures. Students apply their learning to make tetrahedron kites. Students flatten three-dimensional figures and look at nets and surface area. Students learn how to determine surface area and volume for prisms, cylinders, pyramids, cones, and spheres. Students finish the chapter by using comparison skills to determine congruent and similar solids.

Lesson (Pages)	Lesson Objectives	NCTM Standards	State/Local Objectives
11-1A (574)	Create cross sections and other slices of solids.	2, 3, 7	
11-1 (575–581)	Use top, front, side, and corner views of three-dimensional solids to make models. Describe and draw cross sections and other slices of three-dimensional figures.	1–3, 7	
11-1B (582–583)	Construct a tetrahedron kite.	1–4, 7	
11-2 (584–589)	Draw three-dimensional figures on isometric dot paper. Make two-dimensional nets for three-dimensional solids. Find surface areas.	1–5, 7	
11-2B (590)	Investigate Plateau's problem using soap film.	2, 3, 7	
11-3 (591–598)	Find the lateral area and surface area of a right prism. Find the lateral area and surface area of a right cylinder.	1–5, 7	
11-4 (599–606)	Find the lateral area and surface area of a regular pyramid. Find the lateral area and surface area of a right circular cone.	1–5, 7	
11-5 (607–613)	Find the volume of a right prism. Find the volume of a right cylinder.	1–5, 7	
11-6A (614)	Compare the volumes of prisms and pyramids and the volumes of cylinders and cones.	2–5, 7	
11-6 (615–620)	Find the volume of a pyramid. Find the volume of a circular cone.	1–5, 7	
11-7 (621–628)	Recognize and define basic properties of spheres. Find the surface area and the volume of a sphere.	1–5, 7	
11-8 (629–635)	Identify congruent or similar solids. State properties of congruent solids.	1–5, 7	

ORGANIZING THE CHAPTER

A complete, 1-page lesson plan is provided for each lesson in the ***Lesson Planning Guide***. Answer keys for each lesson are available in the ***Answer Key Masters***.

You may want to refer to the **Course Planning Calendar** on page T12 for detailed information on pacing.
PACING: Standard—18 days; **Honors**—16 days; **Block**—10 days

LESSON PLANNING CHART

			Blackline Masters									
Lesson (Pages)	**Materials/ Manipulatives**	**Extra Practice (Student Edition)**	**Study Guide**	**Practice**	**Enrichment**	**Assessment & Evaluation**	**Modeling Mathematics**	**Multicultural Activity**	**Tech Prep Applications**	**Graphing Calc. & Computer**	**Real-World Applications**	**Teaching Transparencies**
11-1A (574)	modeling clay dental floss						p. 99					
11-1 (575–581)	isometric dot paper	p. 785	p. 64	p. 64	p. 64							11-1A 11-1B
11-1B (582–583)	straws, string, glue, colored tissue paper						p. 100					
11-2 (584–589)	isometric dot paper	p. 785	p. 65	p. 65	p. 65	p. 296		p. 21		p. 11		11-2A 11-2B
11-2B (590)	toothpicks, gumdrops, thread, dishwashing detergent, deep bowl, water						p. 101					
11-3 (591–598)		p. 785	p. 66	p. 66	p. 66							11-3A 11-3B
11-4 (599–606)	envelopes straightedge* scissors*	p. 786	p. 67	p. 67	p. 67	pp. 295, 296	pp. 55–58, 89					11-4A 11-4B
11-5 (607–613)	small cubes spreadsheet software	p. 786	p. 68	p. 68	p. 68			p. 22	p. 21			11-5A 11-5B
11-6A (614)	rectangular dot paper, scissors*, tape, rice, compass*						p. 102					
11-6 (615–620)	TI-82/83 graphing calculator	p. 786	p. 69	p. 69	p. 69	p. 297			p. 22		22	11-6A 11-6B
11-7 (621–628)	Styrofoam™ ball, scissors*, tape, straight pins	p. 787	p. 70	p. 70	p. 70						23	11-7A 11-7B
11-8 (629–635)	spreadsheet software	p. 787	p. 71	p. 71	p. 71	p. 297						11-8A 11-8B
Study Guide/ Assessment (637–641)						pp. 281–294, 298–300						

*Included in Glencoe's High School Manipulative Kit and Overhead Manipulative Resources.

ORGANIZING THE CHAPTER

Other Chapter Resources

Student Edition
Chapter Opener, pp. 572–573
Mathematics and Society, p. 628
Working on the Investigation, p. 598
Closing the Investigation, p. 636

Teacher's Classroom Resources
Investigations and Projects Masters, pp. 63–68
Block Scheduling Booklet

Technology
Test and Review Software (IBM and Macintosh)
CD-ROM Multimedia Applications (Windows and Macintosh)
Mindjogger Videoquizzes (VHS)

Professional Publications
Glencoe Mathematics Professional Series

Outside Resources

Books/Periodicals
Cut and Assemble 3-D Geometric Shapes, NASCO
D.I.M.E. 3-D Sketching Project, Dale Seymour Publications
Paper Engineering, Dale Seymour Publications
Boxes, Squares, and Other Things, NCTM

Software
Perimeter, Area, Volume—Gamco, NASCO
Geometry Concepts, Ventura Educational Systems

Videos/CD-ROMs
The Story of Pi, NCTM, 1906 Association Drive, Reston, VA 20191
Curse of the Tomb of King Tut Tut Cubit, ETA, 620 Lakeview Pkwy, Vernon Hills, IL 60061

Assessment Resources

Student Edition
Math Journal, pp. 578, 595, 625
Mixed Review, pp. 581, 589, 597, 606, 613, 620, 628, 635
Self Test, p. 606
Chapter Highlights, p. 637
Chapter Study Guide and Assessment, pp. 638–640
Alternative Assessment, p. 641

Portfolio, p. 641
Chapter Test, p. 803

Teacher's Wraparound Edition
5-Minute Check, pp. 575, 584, 591, 599, 607, 615, 621, 629
Check for Understanding, pp. 578, 586, 595, 603, 610, 617, 625, 632
Closing Activity, pp. 581, 589, 598, 606, 613, 620, 628, 635
Cooperative Learning, pp. 587, 601

Assessment and Evaluation Masters
Multiple-Choice Tests, Forms 1A (Honors), 1B (Average), 1C (Basic), pp. 281–286
Free-Response Tests, Forms 2A (Honors), 2B (Average), 2C (Basic), pp. 287–292
Calculator-Based Test, p. 293
Performance Assessment, p. 294
Mid-Chapter Test, p. 295
Quizzes A–D, pp. 296–297
Standardized Test Practice, p. 298
Cumulative Review, pp. 299–300

ENHANCING THE CHAPTER

Examples of some of the materials for enhancing Chapter 11 are shown below.

DIVERSITY

Multicultural Activity Masters, pp. 21, 22

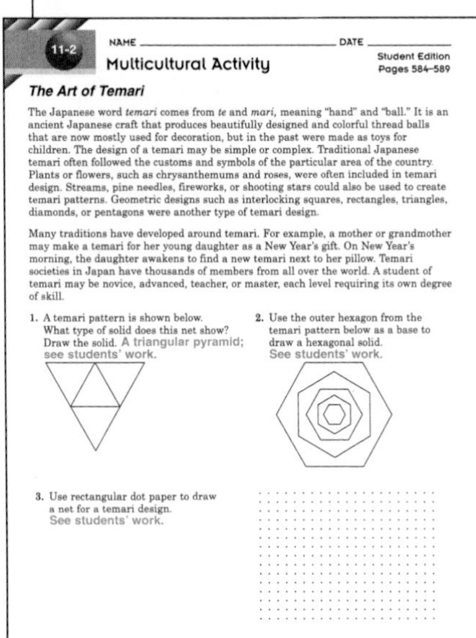

11-2 NAME ____________ DATE ____________

Multicultural Activity — Student Edition Pages 584–589

The Art of Temari

The Japanese word *temari* comes from *te* and *mari*, meaning "hand" and "ball." It is an ancient Japanese craft that produces beautifully designed and colorful thread balls that are now mostly used for decoration, but in the past were made as toys for children. The design of a temari may be simple or complex. Traditional Japanese temari often followed the customs and symbols of the particular area of the country. Plants or flowers, such as chrysanthemums and roses, were often included in temari design. Streams, pine needles, fireworks, or shooting stars could also be used to create temari patterns. Geometric designs such as interlocking squares, rectangles, triangles, diamonds, or pentagons were another type of temari design.

Many traditions have developed around temari. For example, a mother or grandmother may make a temari for her young daughter as a New Year's gift. On New Year's morning, the daughter awakens to find a new temari next to her pillow. Temari societies in Japan have thousands of members from all over the world. A student of temari may be novice, advanced, teacher, or master, each level requiring its own degree of skill.

1. A temari pattern is shown below. What type of solid does this net show? Draw the solid. A triangular pyramid; see students' work.
2. Use the outer hexagon from the temari pattern below as a base to draw a hexagonal solid. See students' work.
3. Use rectangular dot paper to draw a net for a temari design. See students' work.

APPLICATIONS

Real-World Applications, 22, 23

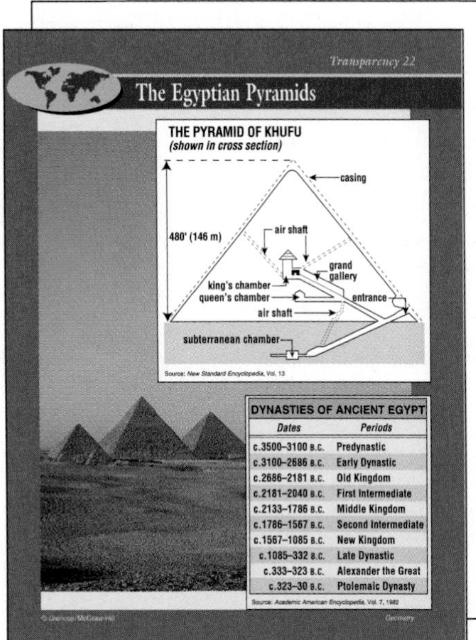

Transparency 22

The Egyptian Pyramids

THE PYRAMID OF KHUFU *(shown in cross section)*

Source: *New Standard Encyclopedia*, Vol. 13

DYNASTIES OF ANCIENT EGYPT

Dates	Periods
c.3500–3100 B.C.	Predynastic
c.3100–2686 B.C.	Early Dynastic
c.2686–2181 B.C.	Old Kingdom
c.2181–2040 B.C.	First Intermediate
c.2133–1786 B.C.	Middle Kingdom
c.1786–1567 B.C.	Second Intermediate
c.1567–1085 B.C.	New Kingdom
c.1085–332 B.C.	Late Dynastic
c.333–323 B.C.	Alexander the Great
c.323–30 B.C.	Ptolemaic Dynasty

Source: *Academic American Encyclopedia*, Vol. 7, 1982

TECHNOLOGY

Graphing Calculator and Computer Masters, p. 11

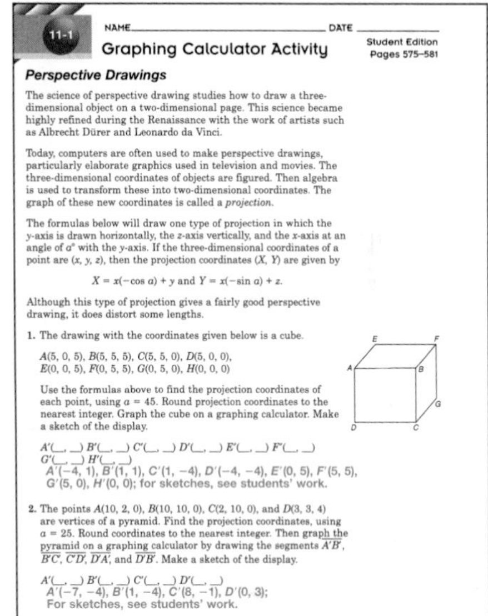

11-1 NAME ____________ DATE ____________

Graphing Calculator Activity — Student Edition Pages 575–581

Perspective Drawings

The science of perspective drawing studies how to draw a three-dimensional object on a two-dimensional page. This science became highly refined during the Renaissance with the work of artists such as Albrecht Dürer and Leonardo da Vinci.

Today, computers are often used to make perspective drawings, particularly elaborate graphics used in television and movies. The three-dimensional coordinates of objects are figured. Then algebra is used to transform these into two-dimensional coordinates. The graph of these new coordinates is called a *projection*.

The formulas below will draw one type of projection in which the y-axis is drawn horizontally, the z-axis vertically, and the x-axis at an angle of $a°$ with the y-axis. If the three-dimensional coordinates of a point are (x, y, z), then the projection coordinates (X, Y) are given by

$$X = x(-\cos a) + y \text{ and } Y = x(-\sin a) + z.$$

Although this type of projection gives a fairly good perspective drawing, it does distort some lengths.

1. The drawing with the coordinates given below is a cube.

   $A(5, 0, 5)$, $B(5, 5, 5)$, $C(5, 5, 0)$, $D(5, 0, 0)$, $E(0, 0, 5)$, $F(0, 5, 5)$, $G(0, 5, 0)$, $H(0, 0, 0)$

   Use the formulas above to find the projection coordinates of each point, using $a = 45$. Round projection coordinates to the nearest integer. Graph the cube on a graphing calculator. Make a sketch of the display.

   $A'(\_,\_)$ $B'(\_,\_)$ $C'(\_,\_)$ $D'(\_,\_)$ $E'(\_,\_)$ $F'(\_,\_)$ $G'(\_,\_)$ $H'(\_,\_)$

   $A'(-4, 1)$, $B'(1, 1)$, $C'(1, -4)$, $D'(-4, -4)$, $E'(0, 5)$, $F'(5, 5)$, $G'(5, 0)$, $H'(0, 0)$; for sketches, see students' work.

2. The points $A(10, 2, 0)$, $B(10, 10, 0)$, $C(2, 10, 0)$, and $D(3, 3, 4)$ are vertices of a pyramid. Find the projection coordinates, using $a = 25$. Round coordinates to the nearest integer. Then graph the pyramid on a graphing calculator by drawing the segments $\overline{A'B'}$, $\overline{B'C'}$, $\overline{C'D'}$, $\overline{D'A'}$, and $\overline{D'B'}$. Make a sketch of the display.

   $A'(\_,\_)$ $B'(\_,\_)$ $C'(\_,\_)$ $D'(\_,\_)$

   $A'(-7, -4)$, $B'(1, -4)$, $C'(8, -1)$, $D'(0, 3)$; For sketches, see students' work.

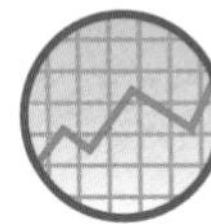

TECH PREP

Tech Prep Applications Masters, pp. 21, 22

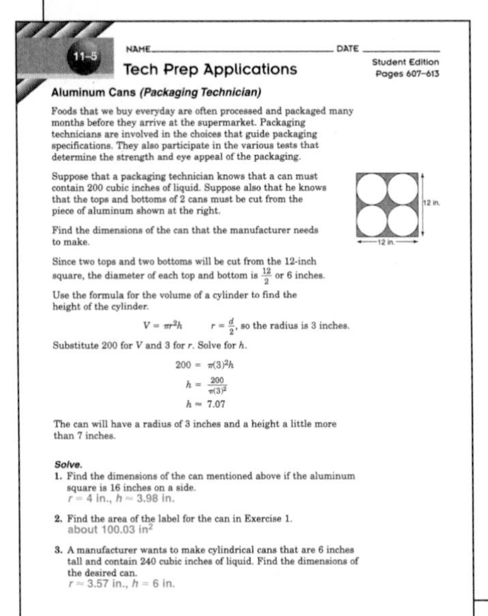

11-5 NAME ____________ DATE ____________

Tech Prep Applications — Student Edition Pages 607–613

Aluminum Cans *(Packaging Technician)*

Foods that we buy everyday are often processed and packaged many months before they arrive at the supermarket. Packaging technicians are involved in the choices that guide packaging specifications. They also participate in the various tests that determine the strength and eye appeal of the packaging.

Suppose that a packaging technician knows that a can must contain 200 cubic inches of liquid. Suppose also that he knows that the tops and bottoms of 2 cans must be cut from the piece of aluminum shown at the right.

Find the dimensions of the can that the manufacturer needs to make.

Since two tops and two bottoms will be cut from the 12-inch square, the diameter of each top and bottom is $\frac{12}{2}$ or 6 inches.

Use the formula for the volume of a cylinder to find the height of the cylinder.

$$V = \pi r^2 h \qquad r = \frac{d}{2}, \text{ so the radius is 3 inches.}$$

Substitute 200 for V and 3 for r. Solve for h.

$$200 = \pi(3)^2 h$$
$$h = \frac{200}{\pi(3)^2}$$
$$h \approx 7.07$$

The can will have a radius of 3 inches and a height a little more than 7 inches.

Solve.

1. Find the dimensions of the can mentioned above if the aluminum square is 16 inches on a side. $r = 4$ in., $h \approx 3.98$ in.
2. Find the area of the label for the can in Exercise 1. about 100.03 in^2
3. A manufacturer wants to make cylindrical cans that are 6 inches tall and contain 240 cubic inches of liquid. Find the dimensions of the desired can. $r \approx 3.57$ in., $h = 6$ in.

PROBLEM SOLVING

Problem-of-the-Week Cards, 30, 31, 32

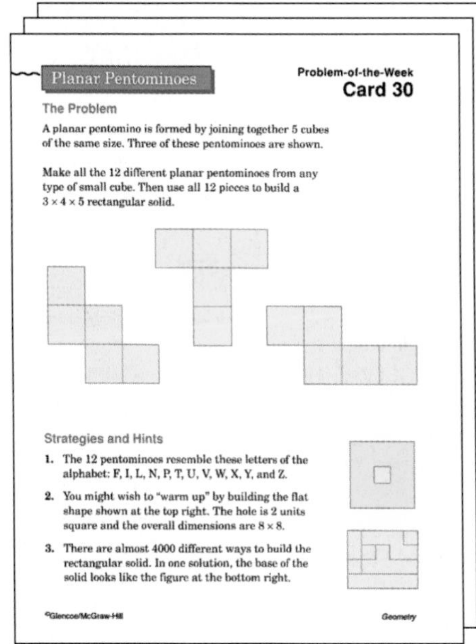

Planar Pentominoes — Problem-of-the-Week Card 30

The Problem

A planar pentomino is formed by joining together 5 cubes of the same size. Three of these pentominoes are shown.

Make all the 12 different planar pentominoes from any type of small cube. Then use all 12 pieces to build a $3 \times 4 \times 5$ rectangular solid.

Strategies and Hints

1. The 12 pentominoes resemble these letters of the alphabet: F, I, L, N, P, T, U, V, W, X, Y, and Z.
2. You might wish to "warm up" by building the flat shape shown at the top right. The hole is 2 units square and the overall dimensions are 8×8.
3. There are almost 4000 different ways to build the rectangular solid. In one solution, the base of the solid looks like the figure at the bottom right.

©Glencoe/McGraw-Hill — Geometry

MAKING MATHEMATICS RELEVANT

This two-page introduction to the chapter provides students with an opportunity to explore an application of geometry that has historic roots as well as modern uses.

Background Information

Watching the Big Parade The Tournament of Roses Parade is the longest floral parade in the world. The football game, the Rose Bowl, was added to the celebration in 1902. In that first game, Michigan beat Stanford 49-0. Football was then suspended until 1916, due to World War I. Here are some interesting facts about the Rose Parade:

- The Rose Parade route along Pasadena's Colorado Avenue is $5\frac{1}{2}$ miles long. The parade route takes 2 hours to complete.
- Decorating each of the 60 floral floats requires an average of 7000 person-hours. Because so many fresh flowers are used, float decoration normally doesn't begin until December 29 and must be completed by 3 A.M. on January 1.
- During the month of December, over 30,000 volunteers arrive at Pasadena float construction sites to help with final decorations.

Ask students to use this information to calculate:

- the average speed of a float in the parade.
- the number of days a single person would need to decorate an average float, working around the clock.
- the number of days a single person would need to decorate an average float, working 10 hours each day.
- the number of people needed to work on a float per hour to finish decorating the float by the 3 A.M. deadline.
- the average number of decorating hours donated by each volunteer.

For more information, you can contact the Tournament of Roses Parade offices at:
391 South Orange Grove Blvd.
Pasadena, CA 91184
(818) 449-7673 (24-hour recording)

CHAPTER 11

Investigating Surface Area and Volume

Objectives

In this chapter, you will:

- describe and draw cross sections and other slices of three-dimensional figures,
- draw three-dimensional figures,
- make two-dimensional nets for three-dimensional solids,
- find lateral areas, surface areas, and volumes of solids, and
- identify and state properties of congruent or similar solids.

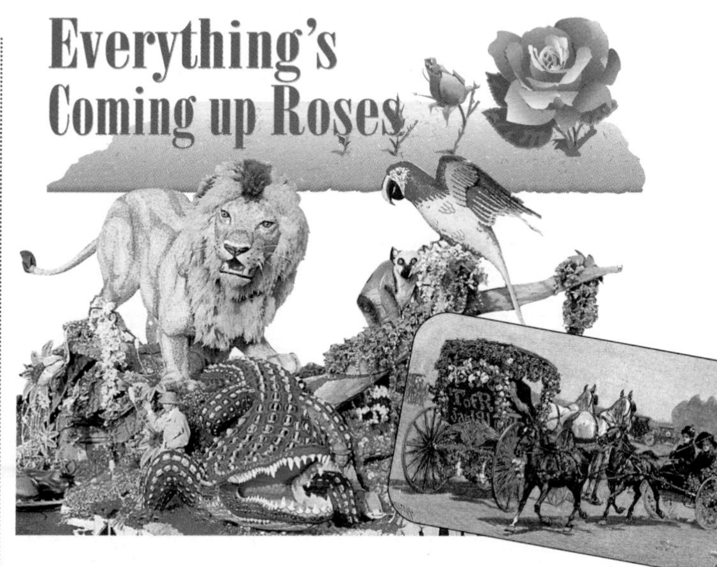

Geometry: Then and Now Is watching the Tournament of Roses Parade on January 1 a tradition in your family? This favorite New Year's Day pastime first took place in 1890. Residents of Pasadena, California were eager to show off their mild winter weather to the rest of the country, which may be buried under snow in January. Two thousand people watched that first parade of carriages decorated with fresh flowers and fruit from the Pasadena area. Today, the annual two-hour parade features 60 floats, each entirely decorated with flowers, leaves, bark, mosses, seeds, or other living or dried plant parts. The Tournament of Roses Parade is watched by over a million spectators along the streets of Pasadena and by another 425 million television viewers around the world.

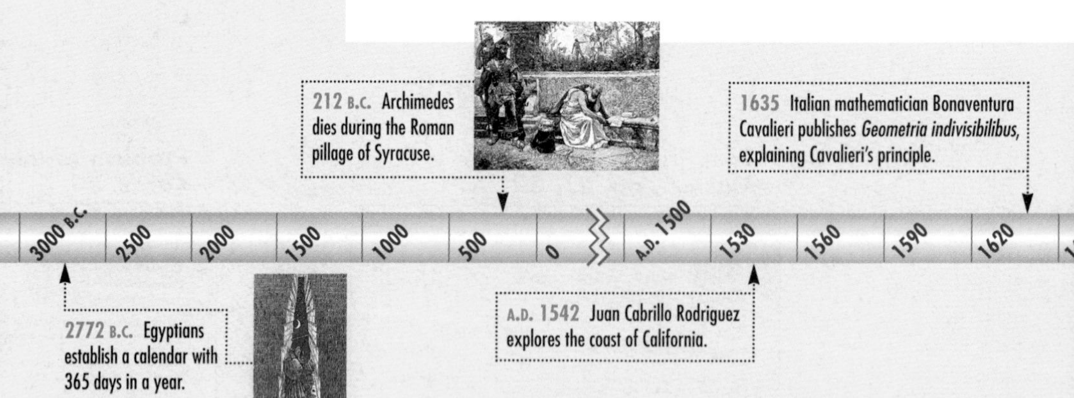

212 B.C. Archimedes dies during the Roman pillage of Syracuse.

1635 Italian mathematician Bonaventura Cavalieri publishes *Geometria indivisibilibus*, explaining Cavalieri's principle.

2772 B.C. Egyptians establish a calendar with 365 days in a year.

A.D. 1542 Juan Cabrillo Rodriguez explores the coast of California.

Students might find it interesting to write a report on calendars. Many cultures have different calendars. Students could study the reasons for the differences.

interNET CONNECTION

Read about floats, marching bands, equestrians, and other entries in the Tournament of Roses Parade.

World Wide Web
http://www.citycent.com/tournamentroses/roseflot.htm

PEOPLE IN THE NEWS

When International House of Pancakes sponsored a "Dream-Up-Our-Float" design contest for its entry in Pasadena's annual Tournament of Roses Parade, fourteen-year-old **Adalee Velasquez** of Arcadia, California, jumped at the chance to enter. She chose a horse-riding theme and sketched her idea in about two hours. Her design for the 55-foot-long float won the contest. The horses in her float were made with mosses, seeds, and spices, and other areas of the float were decorated with carnations, daisies, roses, and orchids. Adalee was honored with a ride on the completed float in the parade.

Chapter Project

Decorating a float for the Tournament of Roses Parade takes roughly 5 tons of flowers. The flower used most often is the chrysanthemum for its wide variety of colors and its ability to last. Twenty chrysanthemums cover about one square foot.

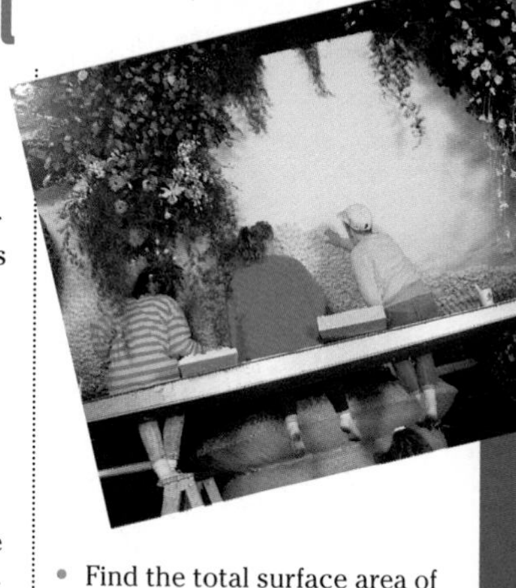

- The average dimensions of a Tournament of Roses Parade floral float is 55 feet long by 18 feet wide by 17 feet tall. Estimate the volume of space occupied by an average float.
- Choose a theme for a float for the Tournament of Roses Parade. Then design your float.
- Find the total surface area of your float. Next, calculate the number of chrysanthemums needed to decorate it. If you would like to use flowers other than chrysanthemums, conduct research to help you determine how many will be needed to cover your float.
- Finally, research the cost of flowers and estimate the cost of constructing your float.

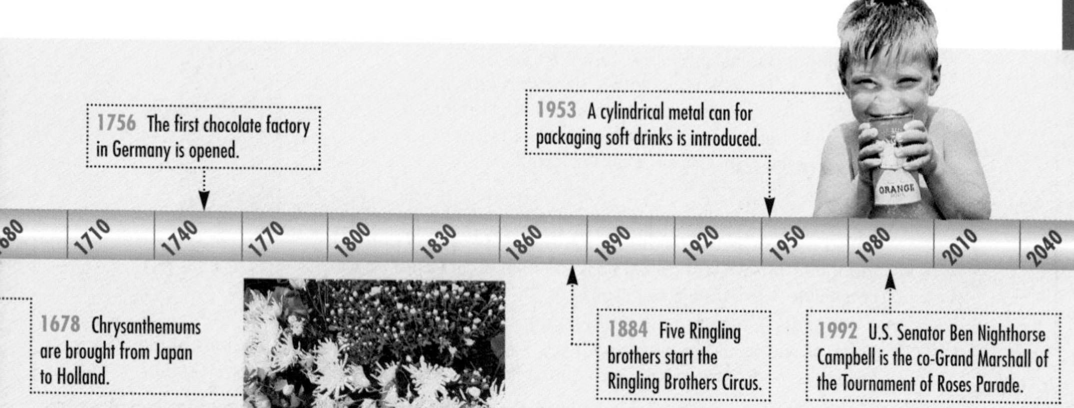
1756 The first chocolate factory in Germany is opened.
1953 A cylindrical metal can for packaging soft drinks is introduced.
1680
1710
1740
1770
1800
1830
1860
1890
1920
1950
1980
2010
2040
1678 Chrysanthemums are brought from Japan to Holland.
1884 Five Ringling brothers start the Ringling Brothers Circus.
1992 U.S. Senator Ben Nighthorse Campbell is the co-Grand Marshall of the Tournament of Roses Parade.

ORANGE

PEOPLE IN THE NEWS

Building a float for the Rose Parade is a year-long process. Planning starts shortly after January 1. By April, construction of the chassis begins. It is completed by September. Props assembly, float painting, and decoration with nonperishables take up the rest of the fall. Perishable decorations are attached during the last few days of December. A float can cost anywhere from $30,000 to $300,000.

Chapter Project

Cooperative Learning After deciding as a group on the theme and basic plan for a float, each group member could design one portion of the float. Students can find the surface areas of their portion, then combine their surface area calculations and work together to calculate the number of flowers needed. Set up teams so that each member has specific responsibilities. A student who likes literature might develop the theme. One who likes art may develop the design. The mathematician may estimate the surface area. A student with a flair for business may compute the cost.

Investigations and Projects Masters, p. 65

11 NAME ______ DATE ______

Chapter 11 Project A

Student Edition Pages 572–641

Full of Hot Air

1. Parades, such as the Macy's Thanksgiving Day Parade, include helium-filled balloons that are moved along the parade route by a team of people walking on the ground. Select a theme for a balloon and design it.
2. Use the maximum measurements of 55 feet long by 18 feet wide by 17 feet tall. Indicate on the drawing the exact measurements of the parts of the balloon. Estimate the surface area for the entire balloon, then use this to estimate how much fabric of each color the balloon will need.
3. Use your measurements to estimate the volume of air that is inside the balloon when it is filled.
4. Helium comes in 80 ft^3 tanks. Use the volume estimate to find the minimum number of helium tanks needed to fill the balloon.

Alternative Chapter Projects

Two other chapter projects are included in the *Investigations and Projects Masters*. In Chapter 11 Project A, pp. 65–66, students extend the topic in the chapter opener. In Chapter 11 Project B, pp. 67–68, students investigate the Seven Wonders of the Ancient World.

11-1A LESSON NOTES

NCTM Standards: 2, 3, 7

Objective

Create cross sections and other slices of solids.

Recommended Time

Demonstration and discussion: 15 minutes; Exercises: 30 minutes

Instructional Resources

For each student or group of students
Modeling Mathematics Masters
• p. 99 (worksheet)
For teacher demonstration
Algebra and Geometry Overhead Manipulative Resources

1 FOCUS

Motivating the Lesson

Have students make a list of solid objects with which they are familiar. Then have them speculate about what a cross section of each would look like.

2 TEACH

Teaching Tip Point out to students that they will not be able to form a perfect circle or triangle.

3 PRACTICE/APPLY

Assignment Guide
Core (with proof): 1–3
Core (informal): 1–3
Enriched: 1–3

4 ASSESS

Observing students working in cooperative groups is an excellent method of assessment.

11-1A Cross Sections and Slices of Solids

Materials:  modeling clay dental floss

A Preview of Lesson 11–1

Every day you deal with three-dimensional objects, or **solids**. A **slice** of a solid is the figure formed by making a straight cut across the solid. We can investigate solids and their slices using modeling clay and dental floss.

Activity Make a cylinder.

Step 1 Roll a piece of modeling clay on your desk to form a thick tube. Then use dental floss to cut off the ends. Make your cuts perpendicular to the sides of the tube. The figure you have formed is a **cylinder**. The ends of your cylinder should be circles. Sketch a diagram of the cylinder.

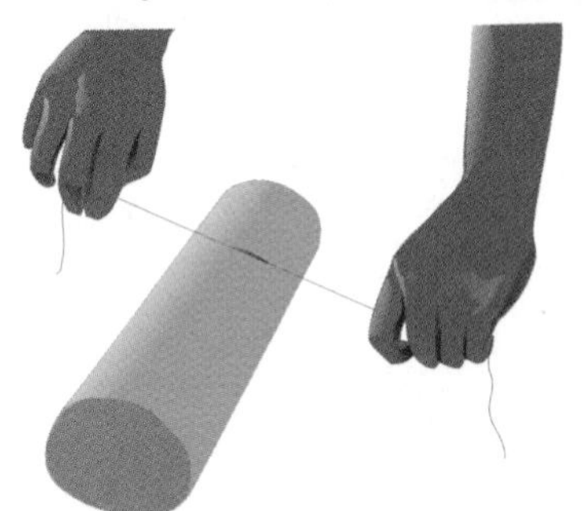

Step 2 Set your cylinder on the table on one of its flat surfaces. Make a cut parallel to the table. What shape is the new surface that is exposed by making this slice? **circle**

Step 3 Make a cut along a diameter of a base of your cylinder and perpendicular to the table. A rectangular surface will be exposed. What are the dimensions of the rectangle?

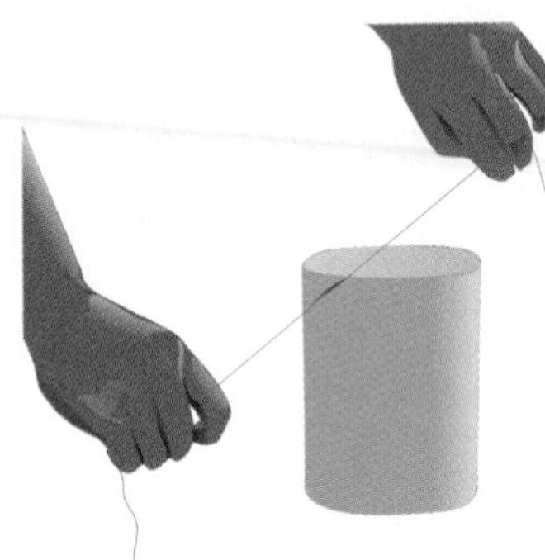

Model

1. Make a **triangular prism**. First roll a tube of modeling clay. Then flatten one side of the tube by pressing it against the table. Turn the tube and flatten another side, then flatten the rest to form a tube that is triangular instead of round. Cut off the ends with dental floss to form a solid like the one shown at the right. **See students' work.**

Write

2. A slice made parallel to the two parallel sides of a prism is called a **cross section**. What shape is a cross section of a triangular prism? Use dental floss to make a cut to test your conjecture. **a triangle congruent to the bases**

3. **a rectangle whose length is the length of the prism**

3. What shape would be revealed if you made a cut along one of the medians of the triangular end of the prism? Use dental floss to make a cut to test your conjecture.

Using Cooperative Learning

This lesson offers an excellent opportunity for using cooperative learning groups. For more information on cooperative learning strategies and group management, see *Cooperative Learning in the Mathematics Classroom*, one of the titles in the Glencoe Mathematics Professional Series.

11–1 Exploring Three-Dimensional Figures

What YOU'LL LEARN

- To use top, front, side, and corner views of three-dimensional solids to make models, and
- to describe and draw cross sections and other slices of three-dimensional figures.

Why IT'S IMPORTANT

Different perspectives of three-dimensional figures are used in art and architecture.

History

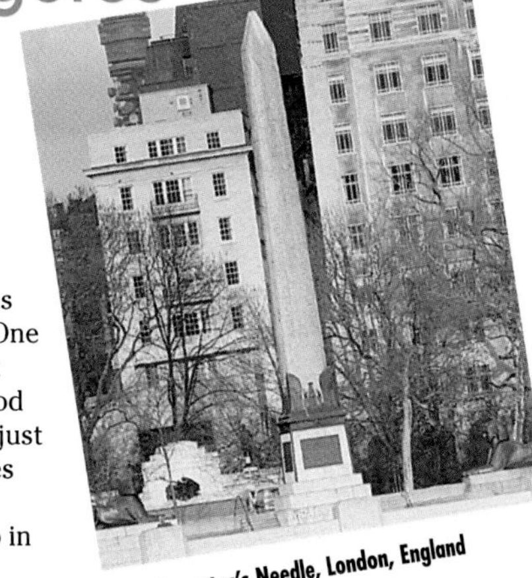

Cleopatra's Needle, London, England

The ancient Egyptians left many remnants of their civilization for us to appreciate. One of these is Cleopatra's Needles, two giant granite obelisks, or pillars, that once stood at the entrance to the Temple of the Sun just north of Cairo. One of Cleopatra's Needles was given to the United Kingdom by the Turkish people in gratitude for their help in resisting Napoleon. The other obelisk is displayed in Central Park in New York City.

Looking at a photograph of just one view of Cleopatra's Needle does not allow you to determine its three-dimensional shape. Front, top, and side views give you more information.

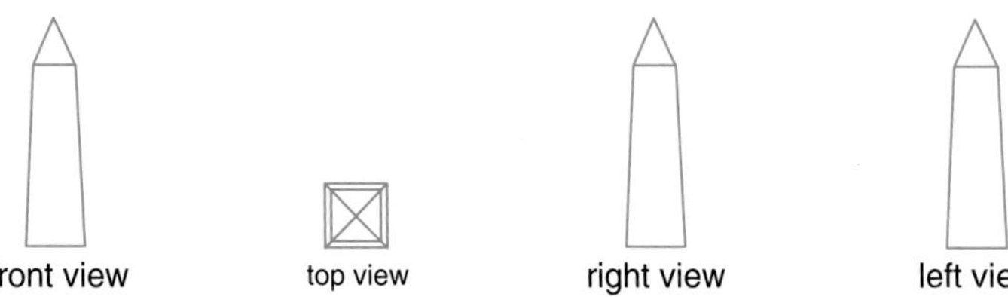

If you are given top, front, and side views of a three-dimensional object, you can **make a model** of the object. Making a model is often helpful in solving problems.

Example 1

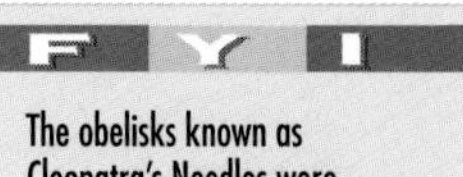

Make a Model

Various views of a solid figure are shown below. The edge of one block represents one unit of length. A dark segment indicates a break in the surface.

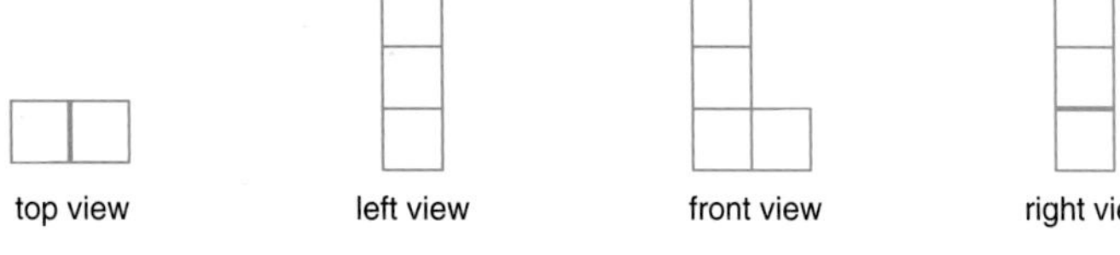

a. Make a model of the figure.

b. Draw the back view of the figure.

a. Use each view to determine the number of blocks needed.

- The top view tells us that there are two columns of blocks. The heavy segment tells us that one column is taller than the other. Start with two blocks as the base of the figure.
- The left view tells us that the figure is 3 blocks high. Since there are no heavy segments in this view, you know that the left column is 3 blocks high. Add blocks to match this view. (continued on the next page)

FYI

The obelisks known as Cleopatra's Needles were commissioned by Pharaoh Tuthmosis III. They were built about 1500–1450 B.C.

FYI

When you hear the name Cleopatra, you're probably thinking of Cleopatra VII. There were in fact seven Cleopatras—queens of ancient Egypt's Ptolemaic dynasty.

11-1 LESSON NOTES

NCTM Standards: 1–3, 7

Instructional Resources

- Study Guide Master 11-1
- Practice Master 11-1
- Enrichment Master 11-1

Transparency 11-1A contains the 5-Minute Check for this lesson; **Transparency 11-1B** contains a teaching aid for this lesson.

Recommended Pacing	
Standard Pacing	Day 2 of 18
Honors Pacing	Day 2 of 16
Block Scheduling*	Days 1 & 2 of 10

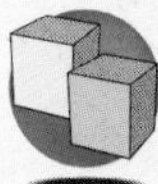

*For more information on pacing and possible lesson plans, refer to the *Block Scheduling Booklet.*

1 FOCUS

5-Minute Check
(over Chapter 10)

1. Draw a polygon that is not convex. Answers will vary.
2. Find the measure of an exterior angle of a regular decagon. 36

Find the area of each polygon. Round your answers to the nearest tenth.

3. 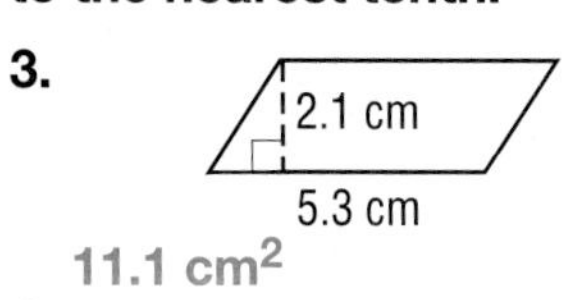

11.1 cm^2

4. 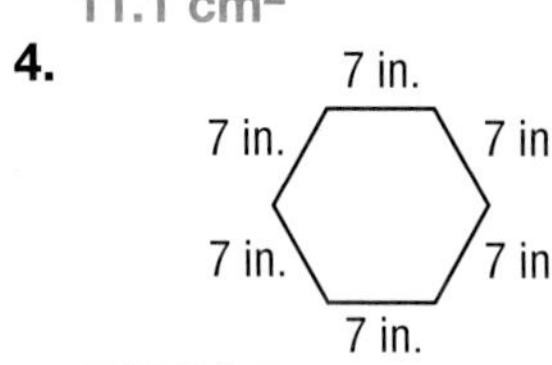

127.3 in^2

5. 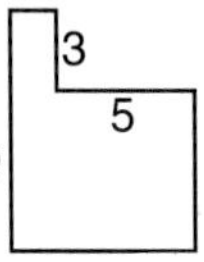 (figure: 2, 3, 5, 9)

48 square units

6. Find the circumference and area of a circle with a radius of 4 inches. 25.1 in.; 50.3 in^2

Motivating the Lesson

Hands-On Activity Have students find five three-dimensional objects from the natural world or in the classroom. They should sketch each object from the top, bottom, and four sides.

2 TEACH

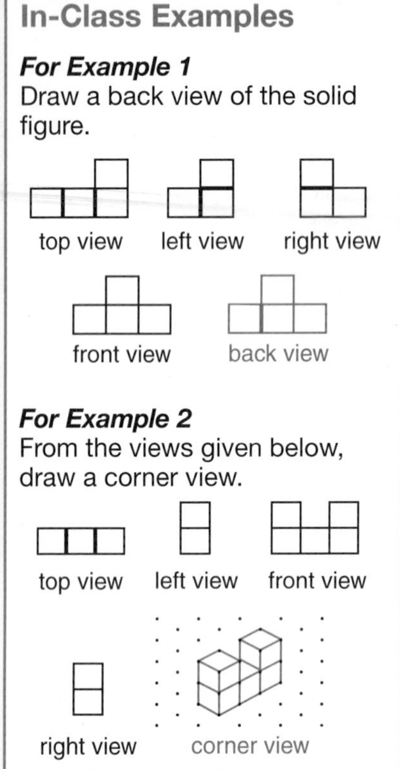

In-Class Examples

For Example 1
Draw a back view of the solid figure.

For Example 2
From the views given below, draw a corner view.

Teaching Tip Students may find it helpful to construct a paper model of the solid figure in order to determine how it looks from each view.

CAREER CHOICES

Graphic artists can work in a variety of fields. They do advertising layouts, window displays, product design, and much more.

CAREER CHOICES

Graphic artists use various perspectives to portray ideas using a variety of media. They may use hand drawn artwork, computer generated art, or film to create designs. Art classes in drawing and painting in high school will help prepare a student for a BA degree in fine arts, visual communications, or graphic design.

For more information, contact:

The American Institute of Graphic Arts
164 Fifth Ave.
New York, NY 10010

- The front view shows the figure has 3 blocks in the left column and 1 in the right. Our figure meets this requirement.
- The right view says that the figure is 3 blocks high, but there is a heavy segment at the 1-block level. This means that the right column is exactly 1 block high. Our model still works!
- After completing your model, check to see that all views correspond to the model.

b. Now turn the figure to the back view and draw what you see. No heavy segments are needed since all of the surfaces are flush at the back.

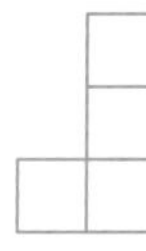

back view

A **corner view** or **perspective view** is the view from a corner of the figure. You can use isometric dot paper to draw a corner view. The figure at the right shows a corner view of a cube.

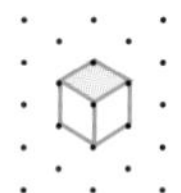

The portion of a drawing that represents the top surface of a three-dimensional figure is often shaded to give a three-dimensional look.

Example 2 **From the views given below, make a model and then draw a corner view.**

top view　　left view

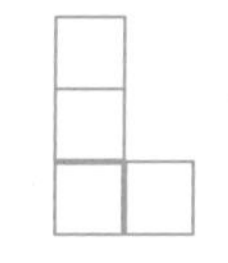

front view

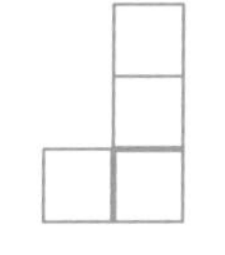

right view

Use cubes to make a model that satisfies each view.

- The top view tells us that the base has 3 blocks and that the figure has columns of different heights.
- The left view shows blocks flush with the surface. Add blocks to the left back column so that the figure is 3 blocks tall.
- The front view and right view show us that the other columns are 1 block tall. Your model should look like the one shown at the left.

To make a corner drawing, turn your model so that you are looking at the edge of the highest column and the other columns as well.

Connect dots on the isometric dot paper to represent the edges you see. Shade the tops of each column.

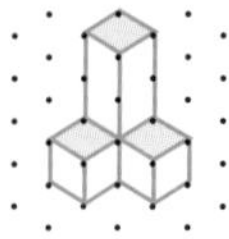

Classroom Vignette

"I photocopy the top portions of the Study Guide Masters at a reduced size and arrange them so that the entire chapter fits on two pages. Then I give them to my students as a review tool. Often students will cut them out and put them on index cards for studying purposes."

Patricia McDaniel
Pinellas Park High School
Largo, Florida

All of the surfaces of the solids in Examples 1 and 2 are flat surfaces or **faces**. Solids with all flat surfaces that enclose a single region of space are called **polyhedrons** or **polyhedra**. All of the faces are polygons, and the line segments where the faces intersect are called **edges**.

A **prism** is a polyhedron with two congruent faces that are polygons contained in parallel planes. These two faces are called the **bases** of the prism. The other faces, called **lateral faces**, are shaped like parallelograms. In the prism at the right, pentagons *ABCDE* and *FGHJK* are the bases, and ▱*AEKF* and ▱*DJHC* are two of the lateral faces.

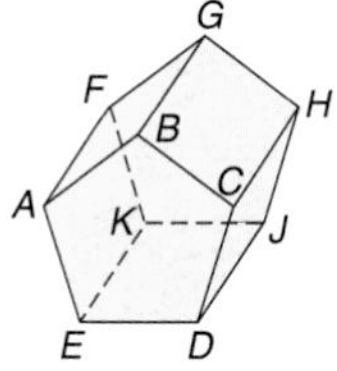

regular prism

Prisms are named by the shape of their bases. A **regular prism** is a prism whose bases are regular polygons. A **cube** is a prism in which all the faces are squares.

A polyhedron that has all faces except one intersecting at one point is a **pyramid**. A **cylinder** is a solid with congruent bases in a pair of parallel planes. However, since its bases are circular regions, it is not a polyhedron. A **cone** has a circular base and a vertex. A **sphere** is a set of points in space that are a given distance from a given point.

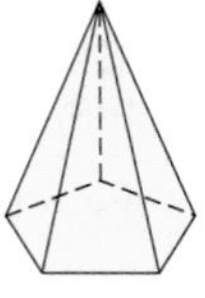
pyramid

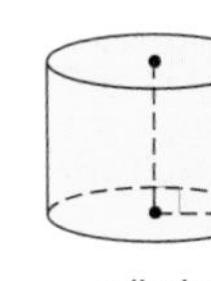
cylinder

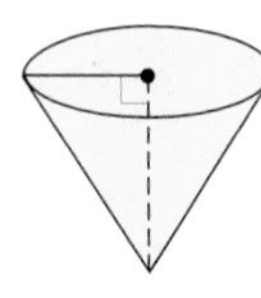
cone

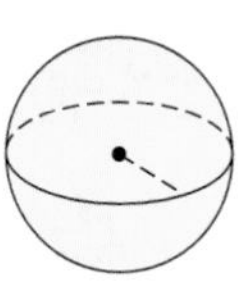
sphere

A polyhedron is **regular** if all of its faces are shaped like congruent regular polygons. Since all of the faces of a regular polyhedron are regular and congruent, all of the edges of a regular polyhedron are congruent. There are exactly five types of regular polyhedra. These are called the **Platonic solids**, because Plato described them so fully in his writings.

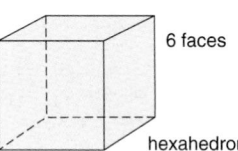

Plato with his students

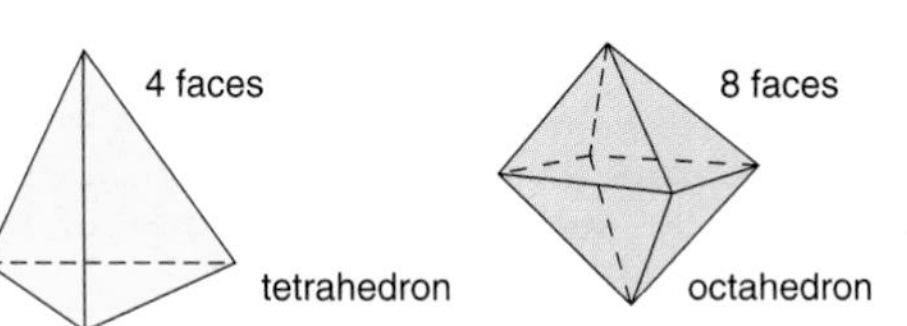

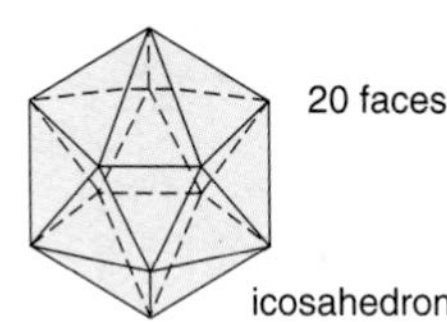

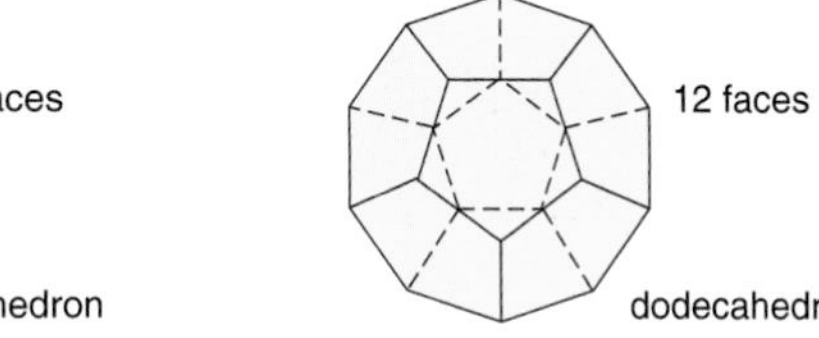

Interesting shapes occur when a plane intersects, or **slices**, a solid figure. If the plane is parallel to the base or bases of the solid, then the intersection of the plane and the solid is called a **cross section** of the solid.

Teaching Tip Encourage students to make models of each 3-D figure and practice drawing them.

Alternative Teaching Strategies

Reading Geometry Ask students what they think of when they read the term *3-D*. Explain that something that is 3-D has three dimensions—length, width, and depth.

In-Class Example

For Example 3

Imagine vertically cutting a cone whose base is horizontal. What would each slice look like? a parabola, or a triangle, if you cut through the vertex If the cone were cut horizontally, what would each slice look like? a circle

3 PRACTICE/APPLY

Check for Understanding

Exercises 1–13 are designed to help you assess your students' understanding through reading, writing, speaking, and modeling. You should work through Exercises 1–4 with your students and then monitor their work on Exercises 5–13.

Additional Answers

2. A prism and a cylinder both have two parallel bases, but a cylinder's bases are circles and a prism's bases are polygons. A cone and a pyramid both have one base and a vertex, but the base of a cone is a circle and the base of a pyramid is a polygon.
3. Rachel is correct. A sphere is not a polyhedron because a polyhedron has sides that are polygons and a sphere has no flat sides.
4. Sample answers: prism: cereal box; cube: die; pyramid: Egyptian pyramids; cylinder: plumbing pipes; cone: ice cream cone; sphere: basketball
5. 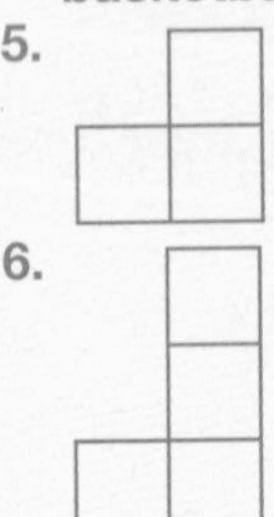
6.

Example 3

Foods

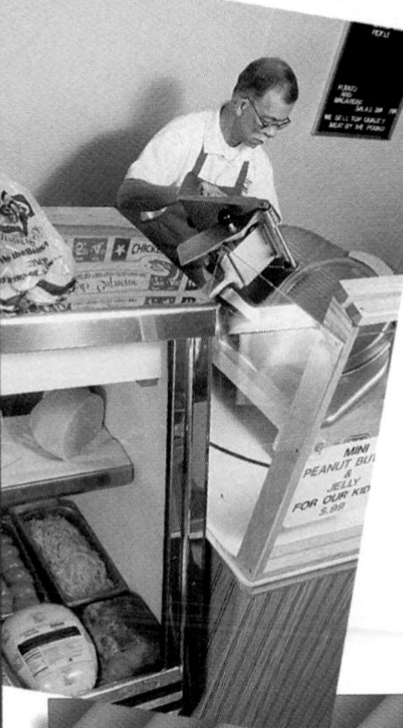

At Jersey Mike's Delicatessen, a slicer is used to slice whole pieces of meat and cheese for sandwiches. Suppose Dan had a cylindrical piece of cheddar cheese.

a. What type of slices would you get if the cheese is placed in the slicer so that the long edge of the cheese is perpendicular to the blade?

b. Suppose Dan placed the cheese with the long edge parallel to the blade. Describe the first ten slices of cheese he would cut.

c. Which slice is a cross section of the cylinder?

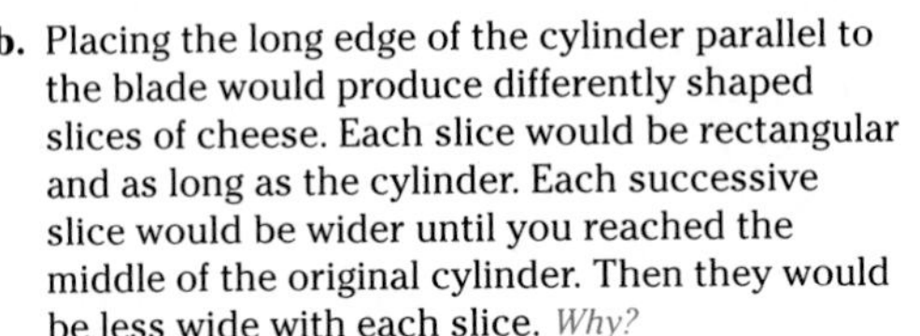

a. If the cylinder is placed so that the long edge is perpendicular to the blade, then the slices would be circles congruent to the bases of the cylinder.

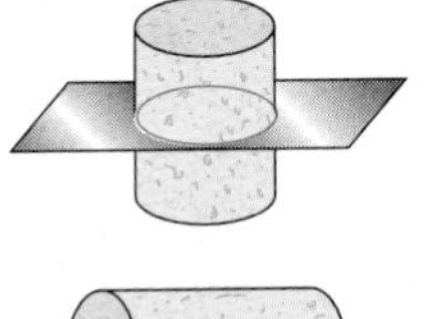

b. Placing the long edge of the cylinder parallel to the blade would produce differently shaped slices of cheese. Each slice would be rectangular and as long as the cylinder. Each successive slice would be wider until you reached the middle of the original cylinder. Then they would be less wide with each slice. *Why?*

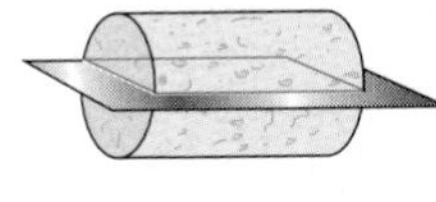

c. The circular slice is the cross section because the slice was made parallel to the bases.

CHECK FOR UNDERSTANDING

Communicating Mathematics

Study the lesson. Then complete the following.

1. Refer to the application at the beginning of the lesson. Is an obelisk a polyhedron? Explain. Yes; all of the surfaces of an obelisk are polygons.
2. **Compare and contrast** a cylinder with a prism and a pyramid with a cone. How are they alike and how are they different? See margin.
3. **You Decide** Evita says that a sphere is a polyhedron. Rachel disagrees. Who is correct and why? See margin.

4. Study the types of solids shown on page 7. Identify a real-life example of an object of each shape. See margin.

Guided Practice

Various views of a solid figure are shown below. The edge of one block represents one unit of length. A dark segment indicates a break in the surface. Make a model of each figure. Then draw the back view of the figure.

5–6. See students' models; see margin.

5. top view | left view | front view | right view

6. top view | left view | front view | right view

Reteaching

Using Models Collect three-dimensional models of a cube, cone, cylinder, and pyramid. Have students sketch the models from front, back, top, right, and left views.

7–8. See margin.

From the views of a solid figure given below, draw a corner view.

7.

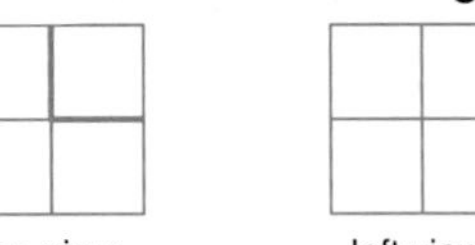
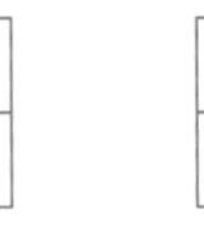
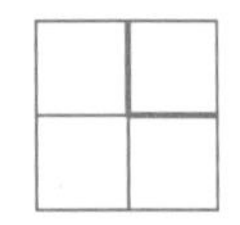

top view   left view   front view   right view

8.

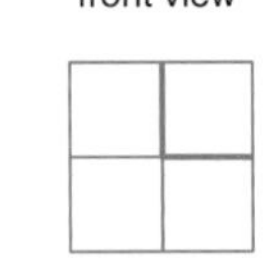
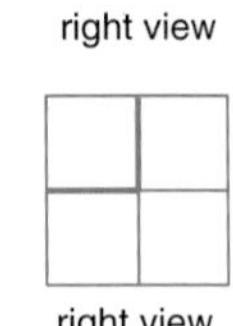

top view   left view   front view   right view

9. The corner view of a figure is given at the right. Draw the top, left, front, right, and back views of the figure. See margin.

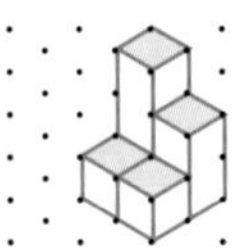

Determine if each figure is a polyhedron. Name the edges, faces, and vertices of each polyhedron.

10.

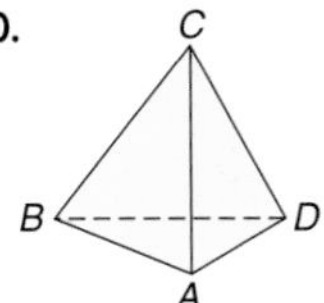

11. no

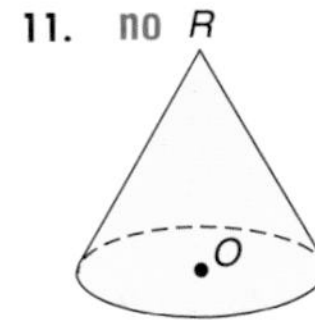

10. yes; edges: $\overline{AB}$, $\overline{BD}$, $\overline{AD}$, $\overline{AC}$, $\overline{BC}$, $\overline{DC}$; faces: $\triangle ABC$, $\triangle ADC$, $\triangle BCD$, $\triangle ABD$; vertices: A, B, C, D

12. Suppose pyramid $ABCDE$ is sliced by a plane. What is the shape of each slice? b. isosceles trapezoid

a. The plane is parallel to the base and does not contain E. square

b. The plane is perpendicular to the base and parallel to $\overline{AB}$ and $\overline{CD}$ and does not contain E.

c. The plane is perpendicular to the base and passes through E. isosceles triangle

13. Art The photograph at the left shows a front view of *Cubi XIX* by David Smith. Choose one element of the sculpture and draw a left, right, and back view. See students' work.

EXERCISES

Practice

Various views of a solid figure are shown below. The edge of one block represents one unit of length. A dark segment indicates a break in the surface. Make a model of each figure. Then draw the back view of the figure. See students' models; see margin for drawings.

14.

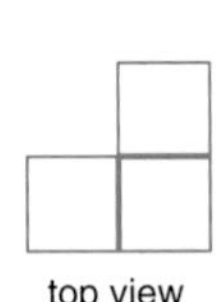
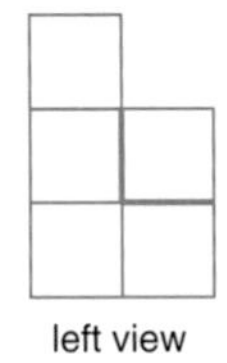
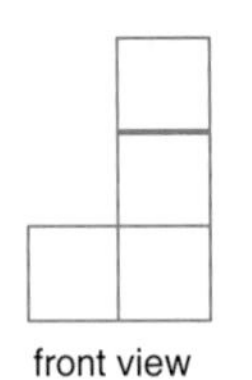
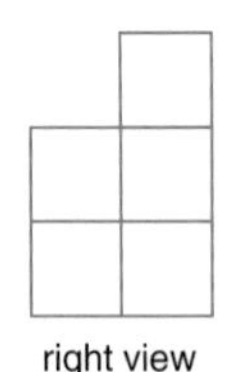

top view   left view   front view   right view

Additional Answer

14.

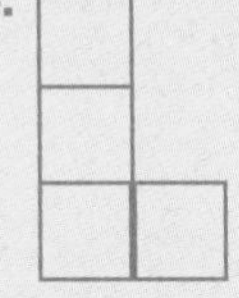

Assignment Guide

Core (with proof): 15–35 odd, 37–44
Core (informal): 15–35 odd, 37–44
Enriched: 14–30 even, 32–44

For **Extra Practice**, see p. 785.

The red A, B, and C flags, printed only in the Teacher's Wraparound Edition, indicate the level of difficulty of the exercises.

Additional Answers

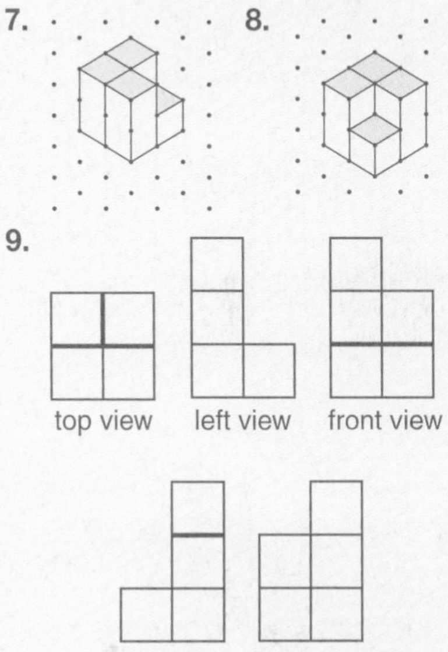

Study Guide Masters, p. 64

NAME______ DATE______

11-1 Study Guide

Student Edition Pages 575–581

Exploring Three-Dimensional Figures

Looking at a picture of a building does not allow you to determine its three-dimensional shape. Depth and the complete configuration of an object can be determined if the front, top, and side views of the object are provided.

Example: Various views of a solid figure are shown below. The edge of one block represents one unit of length. A dark segment indicates a break in the surface. To draw the back view of the figure, make a model based on the information given in the views below.

top view   left view   front view   right view

- The top view indicates that one column is taller than the other.
- The left view indicates the taller column is 3 blocks high while the shorter column is 2 blocks high, and is the closest column to you in this view.

This is the back view based on the model built from the information given.

Make a model of each figure. Then draw the back view of the figure.

top view   left view   front view   right view   back view

1.

2.

3.

4.

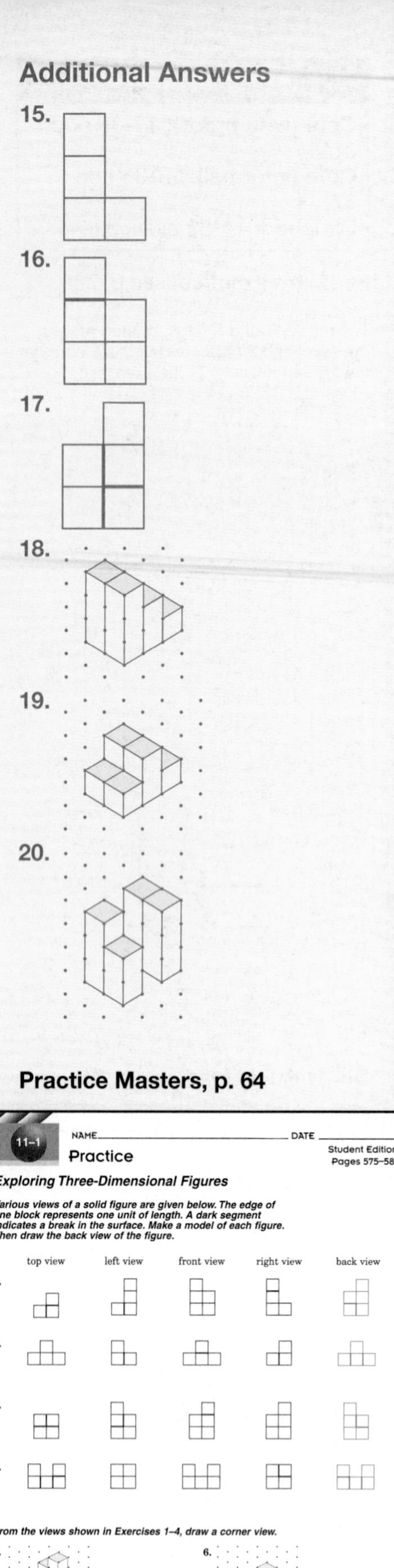

15–20. See students' models; see margin for drawings.

Various views of a solid figure are shown below. The edge of one block represents one unit of length. A dark segment indicates a break in the surface. Make a model of each figure. Then draw the back view of the figure.

15. top view, left view, 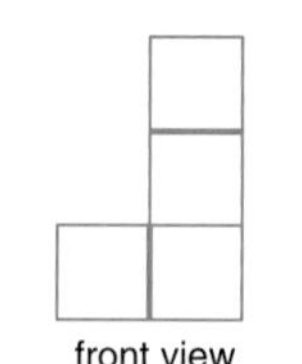 front view, 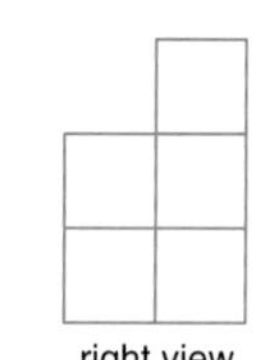 right view

16. top view, left view, front view, right view

17. 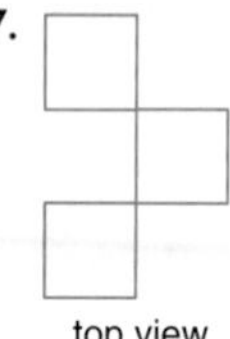 top view, left view, 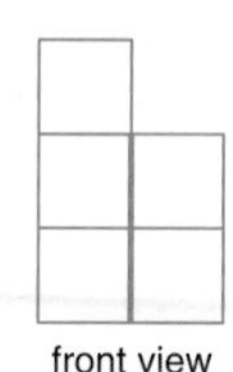 front view, 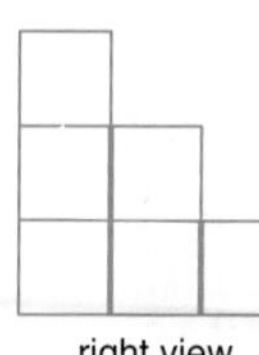 right view

From the views of a solid figure given below, draw a corner view.

18. 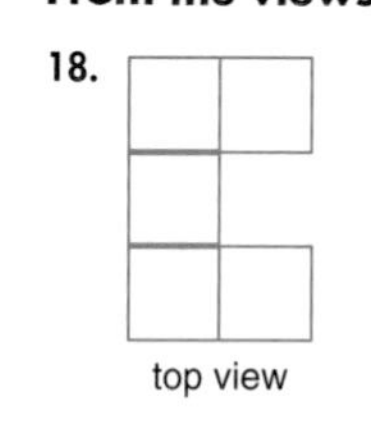 top view, 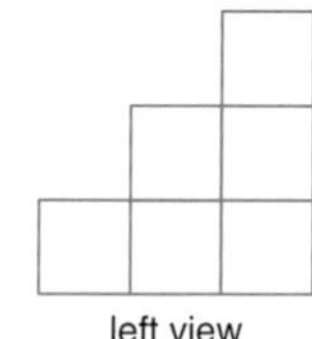 left view, 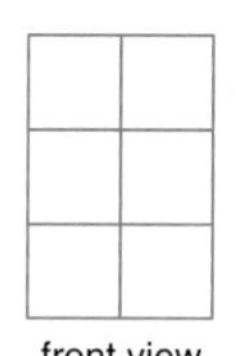 front view, 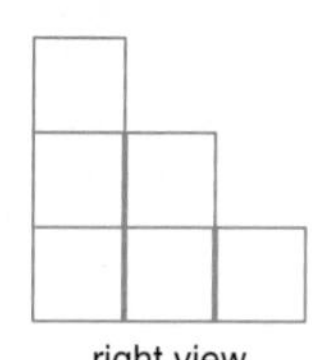 right view

19. 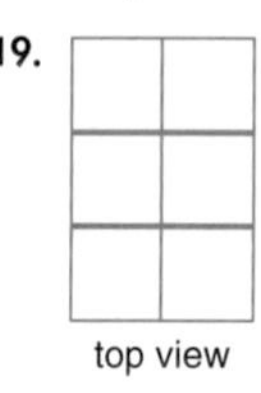 top view, 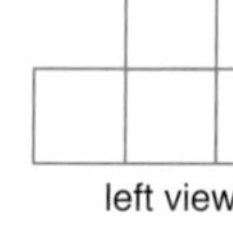 left view, front view, right view

20. top view, left view,  front view, 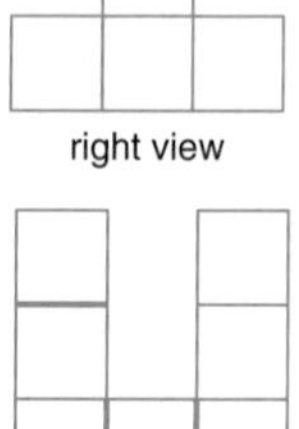 right view

Describe the slice resulting from each cut of the cone.

21. 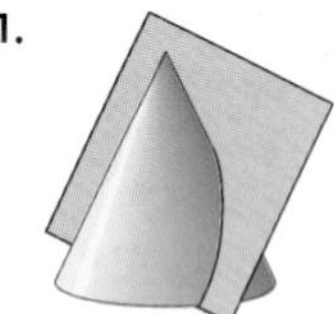

22.

23. 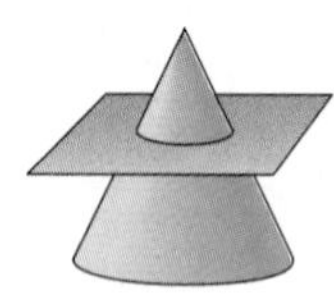

21. upside down U
22. triangle
23. circle

Additional Answers

32. If the plane passes through the cube so that the intersection is a quadrilateral whose vertices are two opposite vertices of the top face of the cube and two points on consecutive sides of the bottom face of the cube, then the intersection is a trapezoid.

33. The only polygon that can be formed by the intersection of a plane and a cylinder is a rectangle.

34. A cross section is the intersection of a solid with a plane parallel to the base(s). This intersection is always congruent to the base(s).

The corner view of a figure is given. Draw the top, left, front, right, and back views of the figure. 24–26. See Solutions Manual.

24. 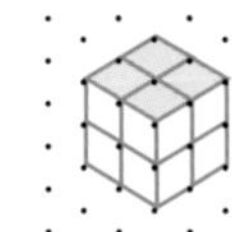

25. 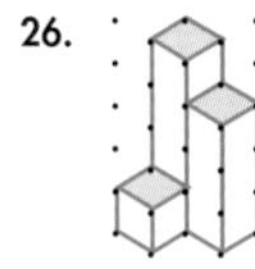

26.

27. yes; edges: $\overline{AB}$, $\overline{BC}$, $\overline{CD}$, $\overline{DE}$, $\overline{EA}$, $\overline{AF}$, $\overline{BG}$, $\overline{CH}$, $\overline{DI}$, $\overline{EJ}$, $\overline{FG}$, $\overline{GH}$, $\overline{HI}$, $\overline{IJ}$, $\overline{JF}$; faces: *ABCDE, ABGF, BCHG, CDIH, DEJI, EAFJ, FGHIJ*; vertices: *A, B, C, D, E, F, G, H, I, J*

28. yes; edges: $\overline{KL}$, $\overline{LM}$, $\overline{MN}$, $\overline{NK}$, $\overline{KO}$, $\overline{LP}$, $\overline{MQ}$, $\overline{NR}$, $\overline{OP}$, $\overline{PQ}$, $\overline{QR}$, $\overline{RO}$; faces: *KLMN, KLPO, LMQP, MNRQ, KNRO, OPQR*; vertices: *K, L, M, N, O, P, Q, R*

Determine if each figure is a polyhedron. Name the edges, faces, and vertices of each polyhedron.

27. 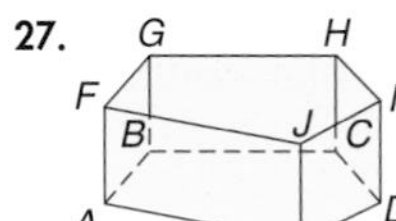

28. 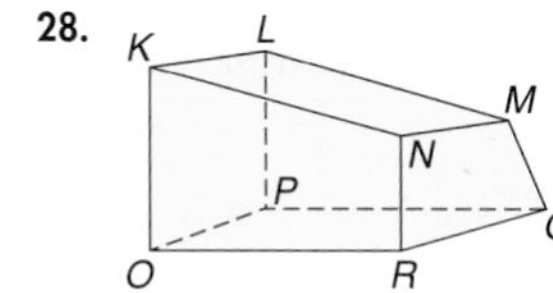

Describe the shape of the intersection in each situation. 30. circle

29. A cube is intersected by a plane that is parallel to the base. square

30. A sphere is intersected by a plane that contains the center of the sphere.

31. The plane perpendicular to the base of a cone and containing the vertex intersects the cone. triangle

Critical Thinking

32–34. See margin for explanations.

Determine whether each statement is *true* or *false*. Explain.

32. A plane can intersect a cube in a trapezoid. true

33. The intersection of a plane and a cylinder can be a triangle. false

34. The cross section of a prism is always congruent to the base. true

Applications and Problem Solving

36. An MRI scan creates images of cross sections of the patient's body.

35. **Architecture** When a building is being designed and constructed, an architect provides a set of elevation drawings. Elevation drawings show how the buildings appear from each side. Draw a set of elevation drawings for your home or school. See students' work.

36. **Medicine** Physicians use magnetic resonance imaging, called MRI scans, to examine their patients for injuries or disease. Research MRI scans in the library or on the Internet. How does an MRI scanner use the concepts discussed in this lesson?

Mixed Review

37. $A = 5$, $B = 5$, $C = 5$, $D = 5$, not traceable

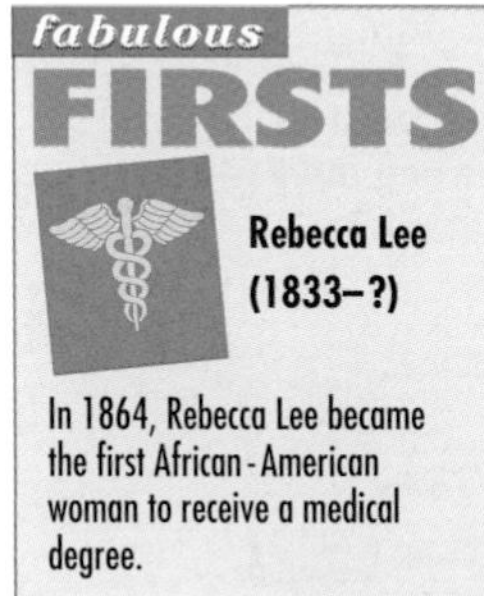

fabulous **FIRSTS**

Rebecca Lee (1833–?)

In 1864, Rebecca Lee became the first African-American woman to receive a medical degree.

37. Find the degree of each node in the network at the right. Is the network traceable? (Lesson 10–7)

38. Find the sum of the measures of the interior angles of a 22-gon. (Lesson 10–1) 3600

39. Classify the statement *A radius of a circle is perpendicular to a line tangent to the circle* as always, sometimes, or never true. (Lesson 9–5) sometimes

40. **Engineering** A machinist is given a diagram of a right triangular brace that has side measures of 2.7, 3.0, and 5.3. Is the diagram correct? Explain. (Lesson 8–1) No; the measures do not satisfy the Pythagorean Theorem.

41. Determine whether *ABCD* is a parallelogram for $A(-3, 3)$, $B(3, 4)$, $C(1, -1)$, and $D(-4, -2)$. Explain your answer. (Lesson 6–2) See margin.

42. Graph the line that passes through the point at $(2, -3)$ and is parallel to the line through points at $(-2, 6)$ and $(4, -6)$. (Lesson 3–3) See Solutions Manual.

INTEGRATION Algebra

43. Evaluate $\sqrt{t}$ if $t = 324$. 18

44. State whether $\{(7, 7), (8, 12), (12, 1), (0, 7)\}$ is a function. yes

4 ASSESS

Closing Activity

Modeling Have students work in pairs with one student building figures with blocks and the other student drawing the figure. Trade off frequently.

fabulous **FIRSTS**

Rebecca Lee received her degree from the New England Female Medical College in Boston.

Additional Answer

41. No; the slope of $\overline{AB} = \frac{1}{6}$, but the slope of $\overline{CD} = \frac{1}{5}$ and $\overline{AB} \parallel \overline{CD}$ if *ABCD* is a parallelogram.

Enrichment Masters, p. 64

NAME ______ DATE ______

11-1 Enrichment

Student Edition Pages 575–581

Drawing Solids on Isometric Dot Paper

Isometric dot paper is helpful for drawing solids. Remember to use dashed lines for hidden edges.

For each solid shown, draw another solid whose dimensions are twice as large.

1. 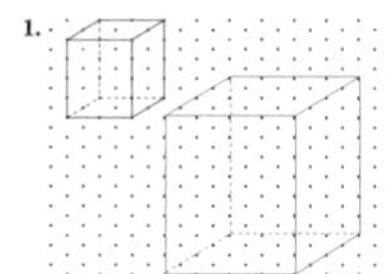

2. 

3. 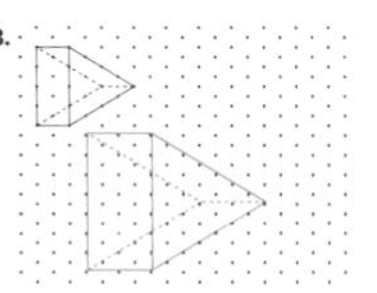

4. 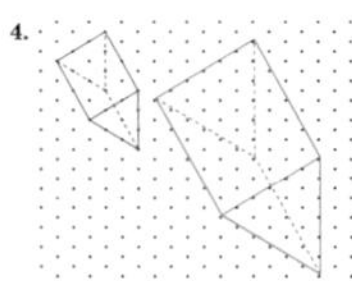

5. 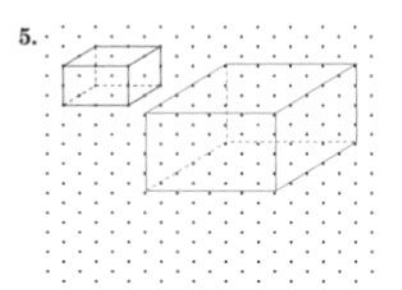

6. 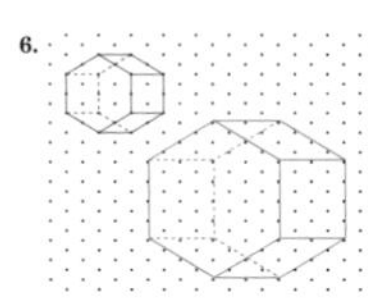

Extension

Reasoning Have students think of solids that have the same view from the top, front, back, right, and left.

Tech Prep

Medical Technologist Many two-year colleges offer degrees in medical technology and related fields. Graduates of these programs operate X-ray and MRI equipment. For more information on tech prep, see the *Teacher's Handbook*.

11-1B LESSON NOTES

NCTM Standards: 1–4, 7

Objective

Construct a tetrahedron kite.

Recommended Time

Demonstration and discussion: 30 minutes; Exercises: 30 minutes

Instructional Resources

For each student or group of students
Modeling Mathematics Masters
• p. 100 (worksheet)
For teacher demonstration
Algebra and Geometry Overhead Manipulative Resources

1 FOCUS

Motivating the Lesson

Ask students what qualities a kite must have in order to fly. Discuss different shapes and the possible advantages and disadvantages of each shape.

2 TEACH

Teaching Tip Point out to students that the pattern is created by folding so that sides that should be congruent are congruent.

Teaching Tip Mylar® can be substituted for tissue paper. It is lightweight, inexpensive, and sturdier for actually flying a kite.

MODELING MATHEMATICS

11–1B Tetrahedron Kites

An Extension of Lesson 11–1

Materials:  straws  string  glue 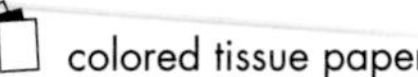 colored tissue paper

You can use straws and string to make a tetrahedron. Then you can combine several tetrahedra to form a type of box kite that really flies.

Activity 1 Use strings and straws to make a tetrahedron.

Step 1 For each tetrahedron, you will need 6 straws, a 40-inch piece of string, and a 20-inch piece of string.

Step 2 Thread the 40-inch string through three of the straws. Tie the string so the straws form a triangle with one end of the string very short and the other end very long.

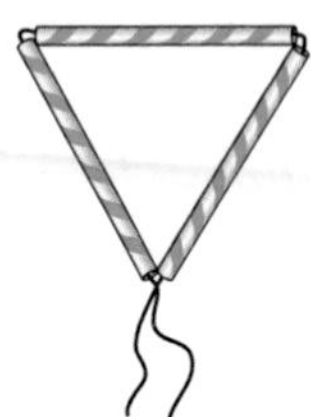

Step 3 Place two straws on the 20-inch string and tie them to the vertices of the triangle that do not contain the original knot.

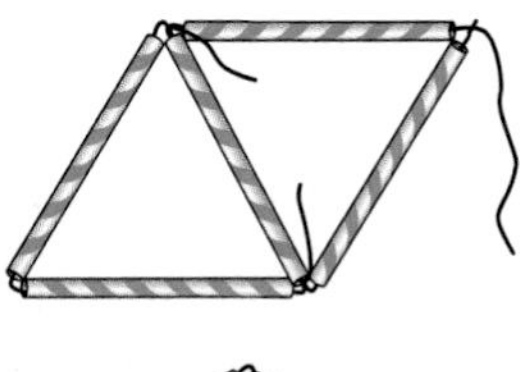

Step 4 Place one straw on the long end of the original 40-inch string and tie it (*A*) to the free vertex (*B*) of the equilateral triangle to form the tetrahedron.

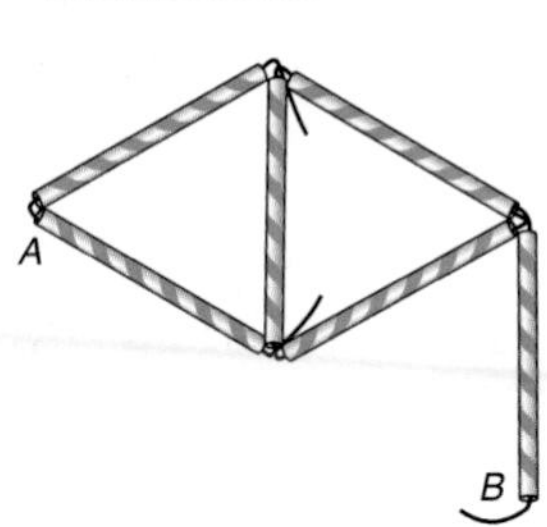

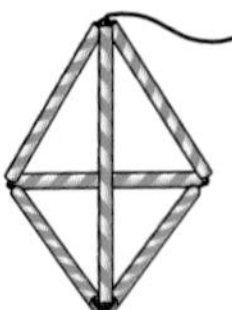

To make a tetrahedron kite, you will need at least four tetrahedra. In order to make the kite fly, you will have to add material to each tetrahedron to catch the wind. The following activity tells how to make the pattern for each tetrahedron's surface and how to assemble the tetrahedra to form a kite.

Activity 2 Make a tetrahedron kite.

Step 1 Draw a template similar to the one below as a pattern for cutting tissue paper.

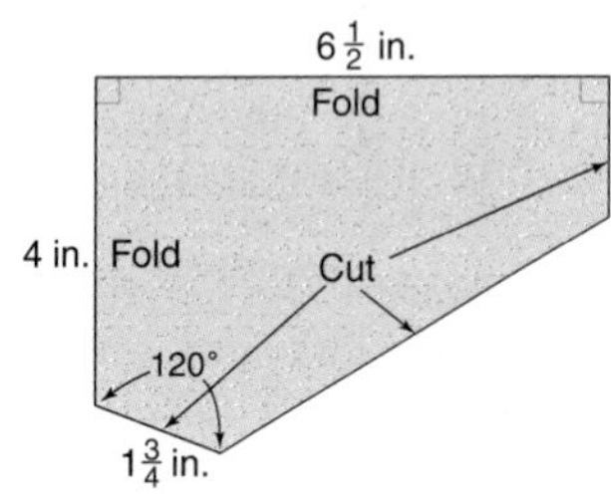

Note which sides say "Fold" and which ones say "Cut."

Step 2 Fold a 20-inch by 26-inch piece of tissue paper into fourths. Then fold the resulting 10-inch by 13-inch paper into fourths. The end result is a rectangle that is 5-inches by 6.5-inches.

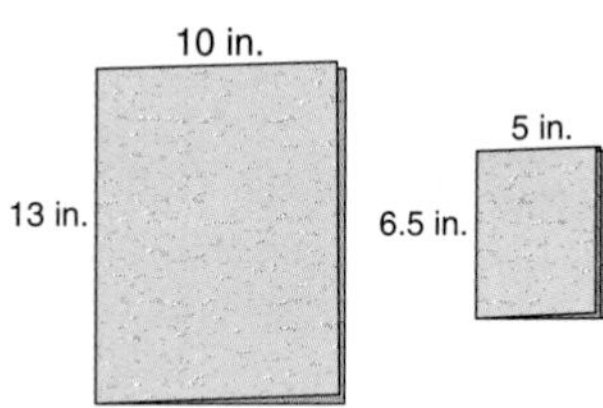

Using Cooperative Learning

This lesson offers an excellent opportunity for using cooperative learning groups. For more information on cooperative learning strategies and group management, see *Cooperative Learning in the Mathematics Classroom*, one of the titles in the Glencoe Mathematics Professional Series.

Step 3 Place the pattern template on the paper matching the pattern edges with the folds of the paper. Cut along the three lines. Separate the pieces of paper. You should have four shapes like the one shown at the right.

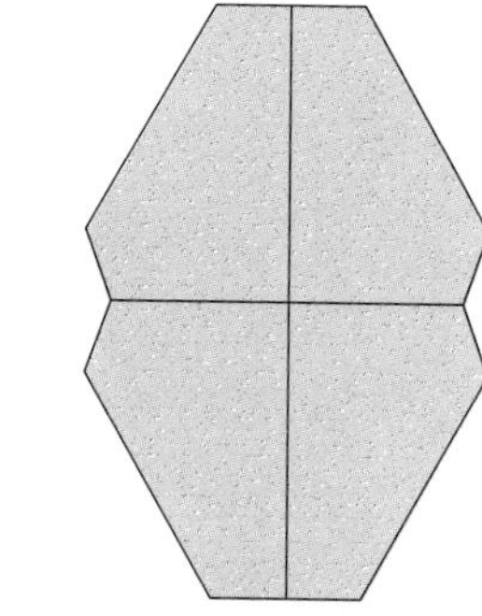

Step 4 Lay one edge of the tetrahedron on the horizontal crease of the tissue paper. Glue the other edges around the four straws that attach to that edge.

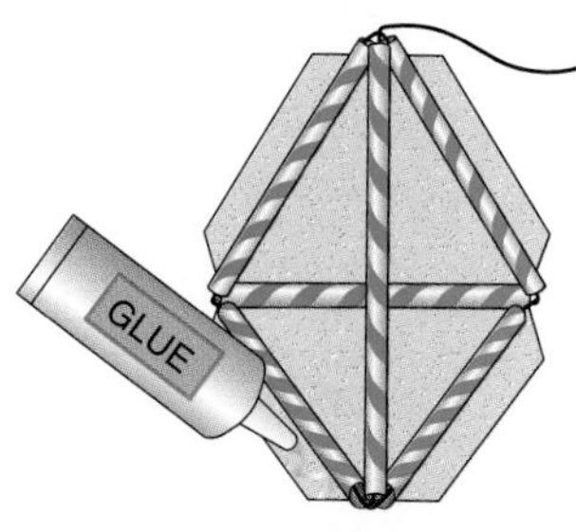

Step 5 Use the long ends of the strings to connect four tetrahedra to form a larger tetrahedron. This is your box kite.

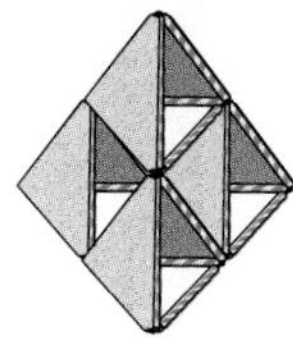

Write

1. Describe the types of triangles used to form the tetrahedron. **equilateral**

2–3. See students' work.

2. Explain how you determined how each tetrahedron should be connected to the others in order to form a tetrahedron that would catch the wind.
3. Estimate what percent of the outer surface of the tetrahedron is covered with tissue paper. Explain how you arrived at your estimate.
4. What is the ratio of the size of the smaller tetrahedron to the larger one? $\frac{1}{2}$

Model

5a–c. See students' work.

5. **a.** Attach a 16-inch string to the top (point *A*) and bottom (point *B*) of the front edge of the top tetrahedron. Put a loop in the middle of the string.

   b. Attach a 24-inch string to the back bottom vertex (point *C*) of the tetrahedron, then thread it through the loop, and tie it to the other back bottom vertex (point *D*) of the tetrahedron.

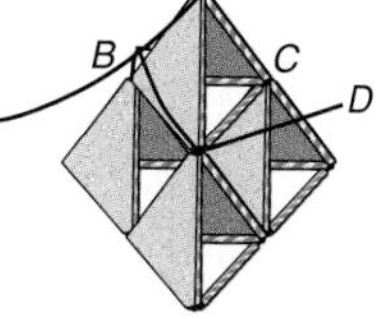

   c. Attach the flying string to the loop and find a gentle breeze to test your kite.

6. Suppose you wanted to build a larger tetrahedron kite. How many small tetrahedra would you need for a kite that is three layers high? four layers high? **10; 20**

3 PRACTICE/APPLY

Assignment Guide

Core (with proof): 1–6
Core (informal): 1–6
Enriched: 1–6

4 ASSESS

Observing students working in cooperative groups is an excellent method of assessment. You may wish to ask a student from each group to explain how to model and solve a problem. Also watch for and acknowledge students who are helping others understand the concept being taught.

11-2 LESSON NOTES

NCTM Standards: 1–5, 7

Instructional Resources

- Study Guide Master 11-2
- Practice Master 11-2
- Enrichment Master 11-2
- Assessment and Evaluation Masters, p. 296
- Graphing Calculator and Computer Masters, p. 11
- Multicultural Activity Masters, p. 21

Transparency 11-2A contains the 5-Minute Check for this lesson; **Transparency 11-2B** contains a teaching aid for this lesson.

Recommended Pacing	
Standard Pacing	Day 4 of 18
Honors Pacing	Day 4 of 16
Block Scheduling*	Day 3 of 10

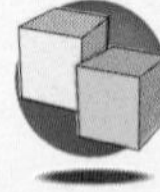

*For more information on pacing and possible lesson plans, refer to the *Block Scheduling Booklet.*

1 FOCUS

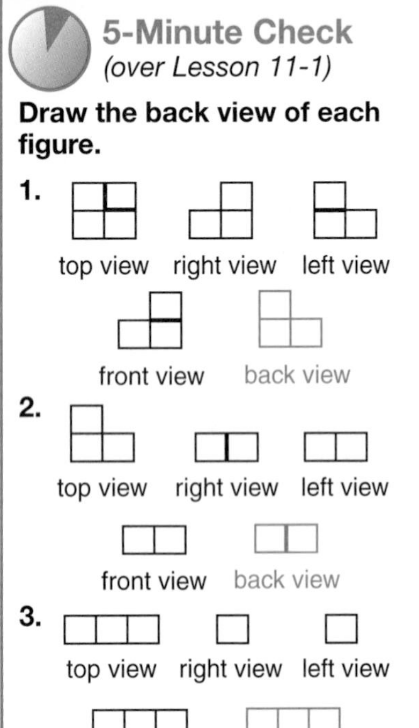

5-Minute Check *(over Lesson 11-1)*

Draw the back view of each figure.

1. top view right view left view front view back view
2. top view right view left view front view back view
3. top view right view left view front view back view

Determine whether each statement is *true* or *false*.

4. A sphere is a polyhedron. false
5. A pentagon is a polyhedron. false

11-2 Nets and Surface Area

What YOU'LL LEARN

- To draw three-dimensional figures on isometric dot paper,
- to make two-dimensional nets for three-dimensional solids, and
- to find surface areas.

Why IT'S IMPORTANT

You can use nets to make models of three-dimensional figures and find their surface area.

APPLICATION
Sports

The Cherokee National Forest of southern Tennessee was the site of the whitewater competitions of the 1996 Summer Olympics. Designers enhanced the natural features of the Ocoee River to make it a challenge for the world's best kayakers and canoers. After two years of construction and a $25 million investment, the 1725-foot course was completed.

In sports such as kayaking where speed is important, designers are conscious of the **surface area** of the equipment. The surface area of a three-dimensional object or solid is the sum of the areas of its outer surfaces. If a piece of sporting equipment has less surface area, it will have less friction with anything it contacts. The reduced friction allows it to travel faster.

In this lesson, we will consider the surfaces of polyhedrons. Recall that the surfaces that make up a polyhedron are polygons. Some common polyhedra are shown below.

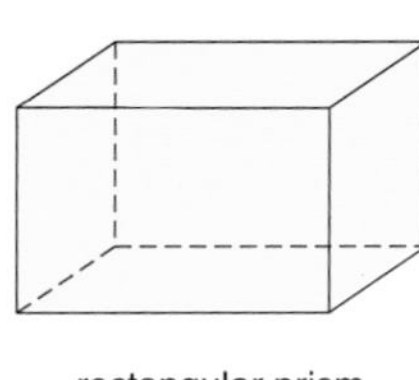

rectangular prism

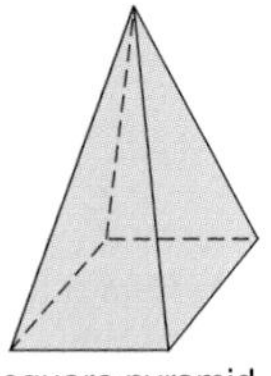

square pyramid

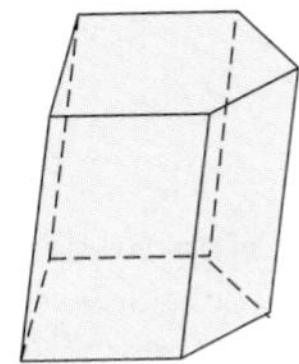

pentagonal prism

In Lesson 11–1, you drew corner views of three-dimensional figures using isometric dot paper. You can also use isometric dot paper to draw geometric solids.

Example  **Sketch a rectangular solid 3 units high, 4 units long, and 5 units wide using isometric dot paper.**

- Draw the top of the solid 4 by 5 units.
- Draw a segment 3 units down from each vertex. Hidden edges are shown by dashed lines.
- Connect the corresponding vertices.

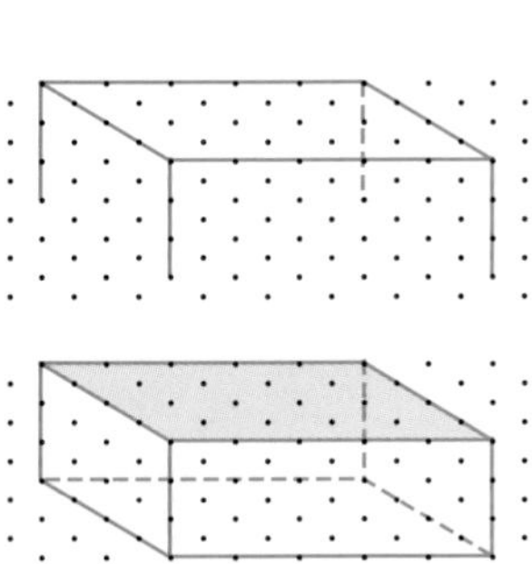

Imagine that you cut a cardboard box along its edges and laid it out flat. The result is a two-dimensional figure known as **net**. Nets can be made for any solid figure. A net for triangular pyramid $ABCD$ is shown at the right. Nets are very useful in seeing the polygons whose areas must be computed to find the surface area of a polyhedron. For the pyramid, the surface area is the sum of the areas of $\triangle ABC$, $\triangle ACD$, $\triangle ABD$, and $\triangle BCD$.

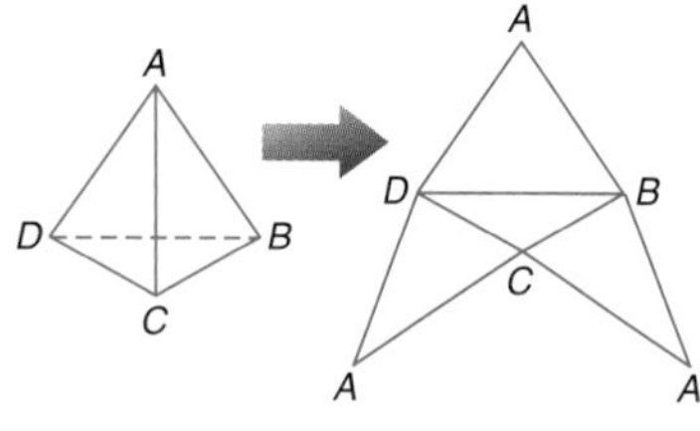

Example 2 **Refer to the rectangular prism in Example 1.**

a. Use rectangular dot paper to draw a net for the prism.

b. Find the surface area of the prism.

This is only one of several possible nets that you could draw for the prism. However, the surface area should be the same regardless of which net is used.

a. Choose one surface to begin your drawing. Then visualize unfolding the solid along the edges so that it is a two-dimensional figure.

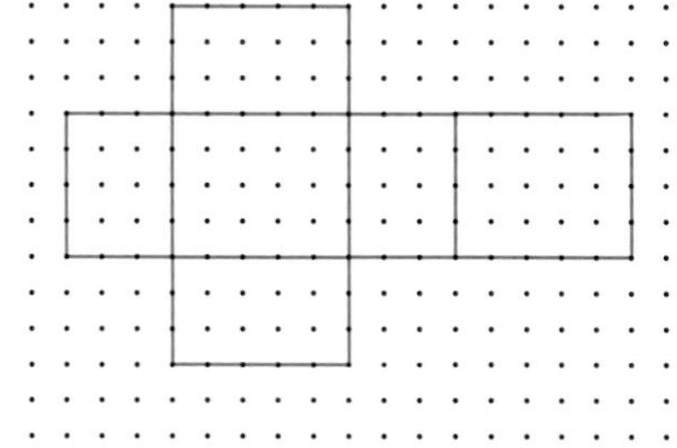

b. Find the area of each rectangle in the net. The surface area is the sum of these areas.

$(3 \cdot 5) + (5 \cdot 4) + (3 \cdot 4) + (5 \cdot 4) + (3 \cdot 4) + (3 \cdot 5)$

$= 15 + 20 + 12 + 20 + 12 + 15$

$= 94$ square units

Often times there are several different nets that could be used to produce the same solid. Could you design a different one for the rectangular prism drawn in Example 1?

Example 3 **APPLICATION Set Design**

Alegria is covering a box 40 centimeters high, 30 centimeters long, and 20 centimeters wide with expensive fabric to use as a prop in a play. She wants to cover the box using the smallest amount of fabric possible. What size of fabric is needed for a net to cover the box?

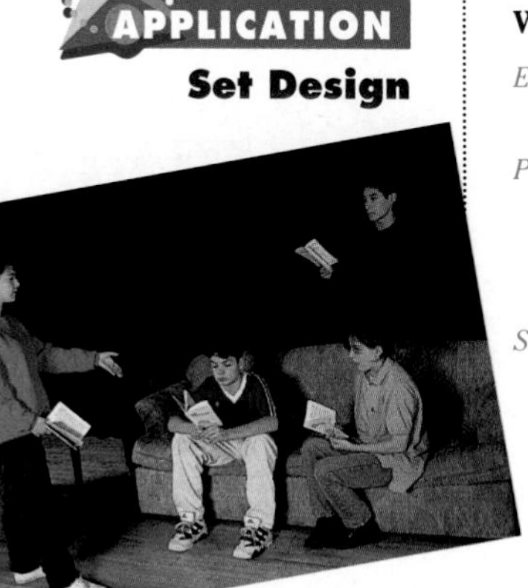

Explore We know the dimensions of the box and need to find the size of the smallest piece of fabric that could be used to cover the box.

Plan First, we can sketch the box using the given dimensions. Next, we can find the dimensions of each face of the box. Then we can draw nets for the box to find the smallest piece of fabric that can be used.

Solve Draw a corner view of the box using isometric dot paper. Let each dot represent 10 centimeters.

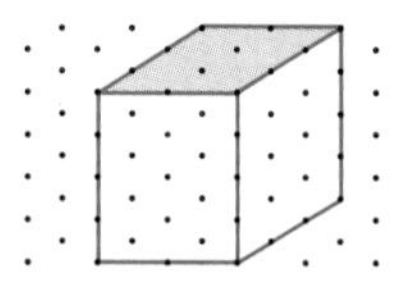

(continued on the next page)

Motivating the Lesson

Situational Problem Have students think of several sports where speed depends on reducing friction. For each piece of sports equipment, how does reduced surface area contribute to reduced friction?

2 TEACH

In-Class Examples

For Example 1
Sketch a rectangular solid 1 unit high, 3 units long, and 2 units wide using isometric dot paper.

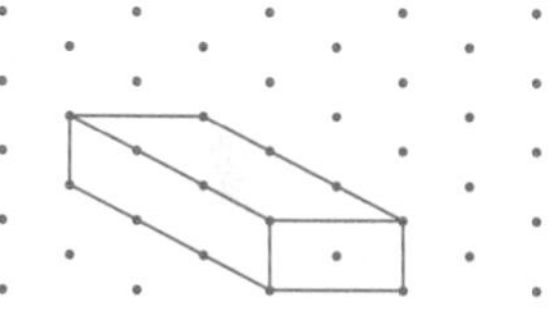

For Example 2
Use rectangular dot paper to draw a net for a rectangular solid 5 units high, 6 units long, and 3 units wide.

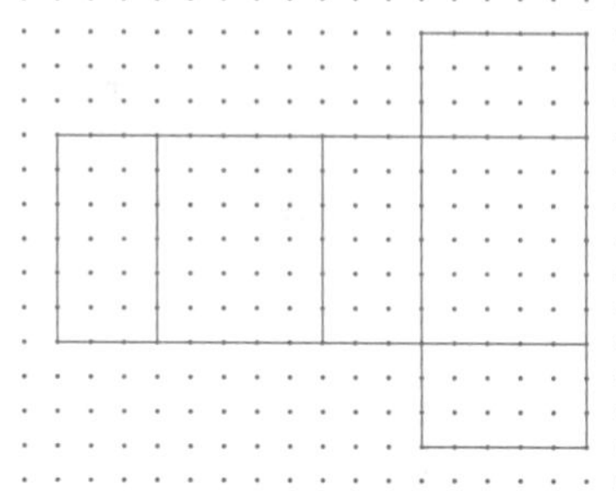

For Example 3
Ruth is covering a box 12 feet high, 8 feet long, and 3 feet wide with fabric. What size fabric is needed for a net to cover the box? **896 ft²**

Teaching Tip When drawing a net, emphasize that other nets can be drawn for the same figure, and have students draw an alternative.

Alternative Learning Styles

Kinesthetic Have each student bring an empty cardboard box to class. Have them cut open the box along the edges until it lies flat in a single layer. Explain that this is a net of the box. If two students have the same box, have them compare nets. They may not be the same.

3 PRACTICE/APPLY

Check for Understanding

Exercises 1–10 are designed to help you assess your students' understanding through reading, writing, speaking, and modeling. You should work through Exercises 1–4 with your students and then monitor their work on Exercises 5–10.

Error Analysis
If students are having difficulty with nets, have them copy some of the nets in this lesson, cut them out, and see if they will fold into a three-dimensional object. Or have students label the faces of a box, open it up to make a net, and study where each of the faces lies in the net.

Additional Answers

1. On isometric dot paper, the dots are arranged in triangles, which aid in drawing three-dimensional objects and corner views. On rectangular dot paper, the dots are arranged in squares, which aid in drawing two-dimensional objects such as nets, and right, left, top, front, and back views.
2. A net is the unfolded two-dimensional pattern of a three-dimensional object. The sum of the areas of the polygonal shapes in the net is the surface area of the polyhedron.
3. If a piece of sporting equipment has less surface area, the reduced friction allows it to travel faster.
4. No; there is no way to cover both the top and the bottom of the solid.
5. 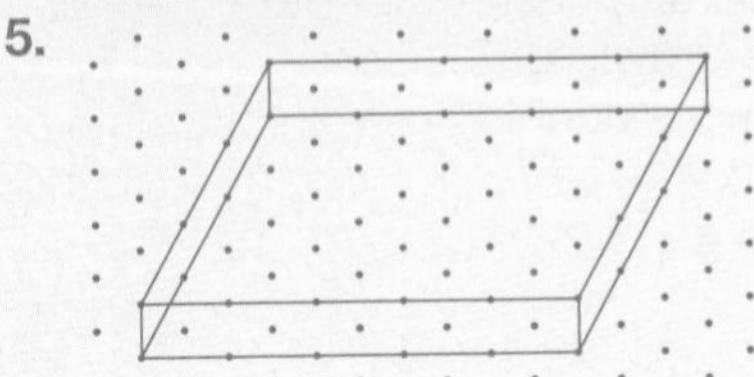

Now, determine the dimensions of each face of the box.

top and bottom	3×2
small ends	4×2
sides	3×4

Use grid paper and these dimensions to draw each face of the box. Then arrange the rectangles to form a net so that the total width and the total length of the net uses the minimum amount of fabric. Two of the nets are shown below.

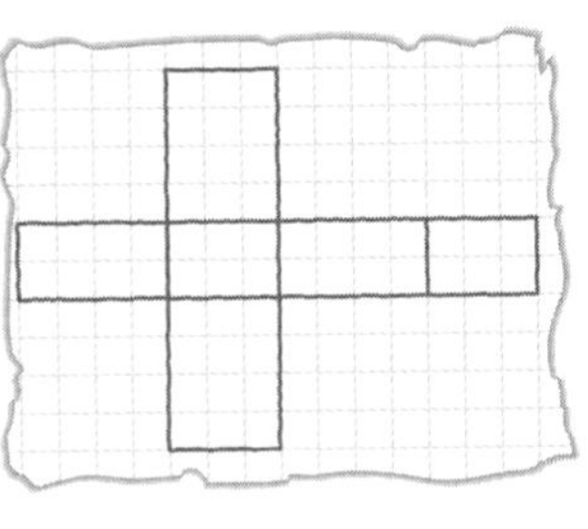

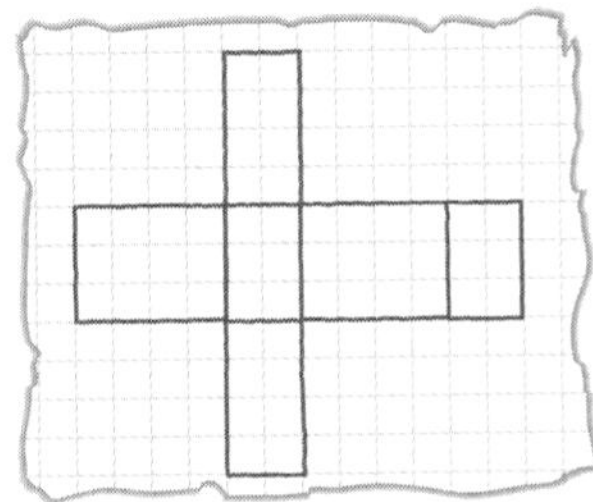

total length = 140 cm	total length = 110 cm
total width = 100 cm	total width = 120 cm
area of fabric = 14,000 cm^2	area of fabric = 13,200 cm^2

The net on the right uses less fabric. Alegria needs an 110-cm by 120-cm piece of fabric.

Examine Copy one of the nets on rectangular grid paper and fold it into a box. Try covering the box in different-sized pieces of paper to verify the solution.

CHECK FOR UNDERSTANDING

Communicating Mathematics

Study the lesson. Then complete the following. 1–4. See margin.

1. **Compare and contrast** isometric dot paper and rectangular dot paper. When is each type of paper useful?
2. What is the relationship between a net and surface area?
3. Refer to the application at the beginning of the lesson. Explain how kayak design is related to the geometric concept of surface area.
4. **Determine** whether the figure at the right produces a cube when folded. Why or why not? Copy the figure on rectangular dot paper, cut it out, and fold to check your answer.

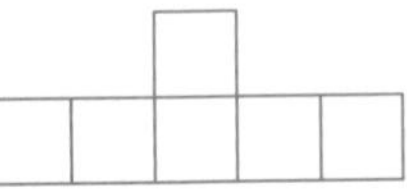

Guided Practice

Use isometric dot paper to draw each polyhedron. 5–6. See margin.

5. a rectangular solid 1 unit high, 5 units long, and 3 units wide
6. a cube 4 units on each edge
7. Given the polyhedron at the right, copy its net and label the remaining vertices.

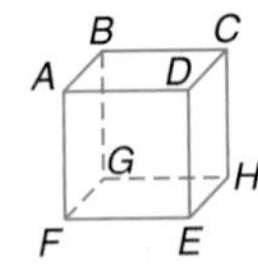

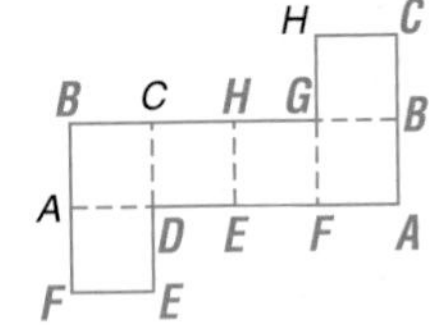

Reteaching

Using Vocabulary Have students think about the words *surface* and *area*. They can also look the terms up in the dictionary. Putting two words together for "surface area" means putting together the meaning of the two words, or the amount of area on the surface.

Additional Answer

6. 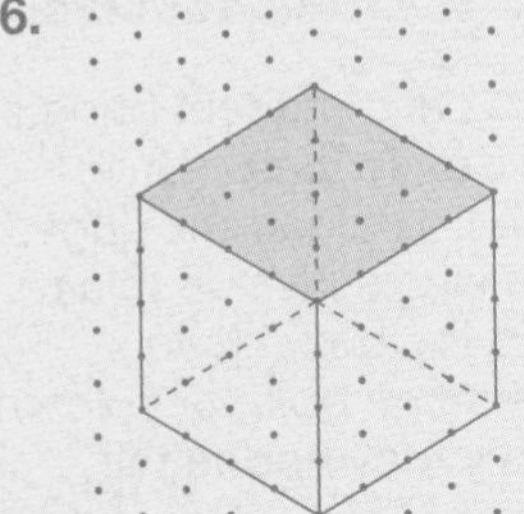

Use rectangular dot paper to draw a net for each solid. Then find the surface area of the solid. See Solutions Manual for sample nets.

Wright Brothers' plane at Kittyhawk, 1903

8. 168 units^2

9. 80.8 units^2

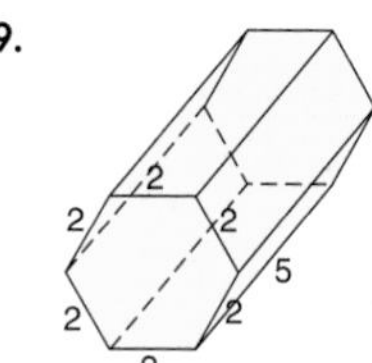

10. **Aeronautical Engineering** The surface area of the wing on an aircraft is used to determine a design factor known as wing loading. If the total weight of the aircraft and its load is w and the total surface area of its wings is s, then the formula for the wing-loading factor l is $l = \frac{w}{s}$. If the wing-loading factor is exceeded, the pilot must either reduce the fuel load or remove passengers or cargo. Find the wing-loading factor for the Wright Brothers' plane if it had a maximum take-off weight of 750 pounds and the surface area of the wings was 532 square feet. 1.41

EXERCISES

Practice

Use isometric dot paper to draw each polyhedron. 11–12. See margin.

13–16. See Solutions Manual.

11. a rectangular prism 3 units high, 6 units long, and 4 units wide
12. a cube 6 units on each edge
13. a rectangular prism 9 units high, 3 units long, and 2 units wide
14. a triangular prism 2 units high, whose bases are equilateral triangles with sides 5 units long
15. a cube with a surface area of 150 square units
16. a triangular prism 5 units high, whose bases are triangles with sides 3, 4, and 5 units long

Given each polyhedron, copy its net and label the remaining vertices.

17.

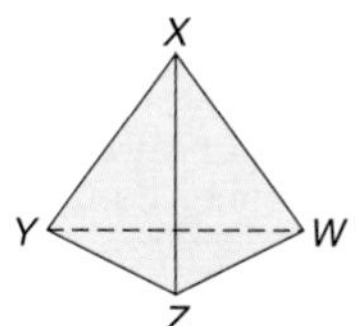

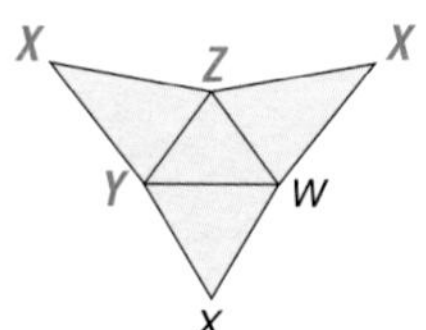

18.

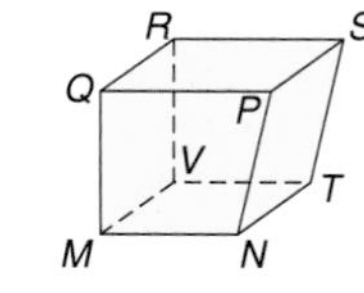

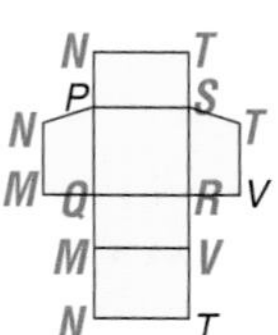

19.

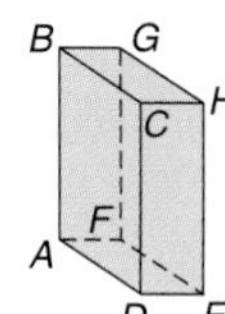

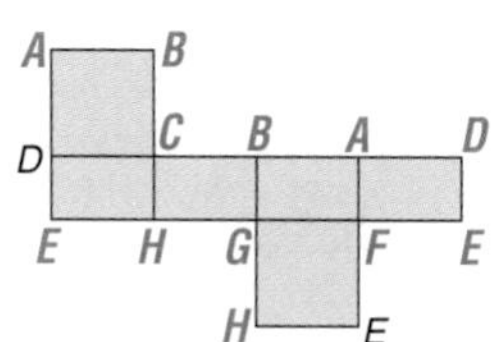

Assignment Guide

Core (with proof): 11–29 odd, 30, 31, 33, 34–41
Core (informal): 11–29 odd, 30, 31, 33, 34–41
Enriched: 12–28 even, 30–41

For **Extra Practice**, see p. 785.

The red A, B, and C flags, printed only in the Teacher's Wraparound Edition, indicate the level of difficulty of the exercises.

Additional Answers

11.

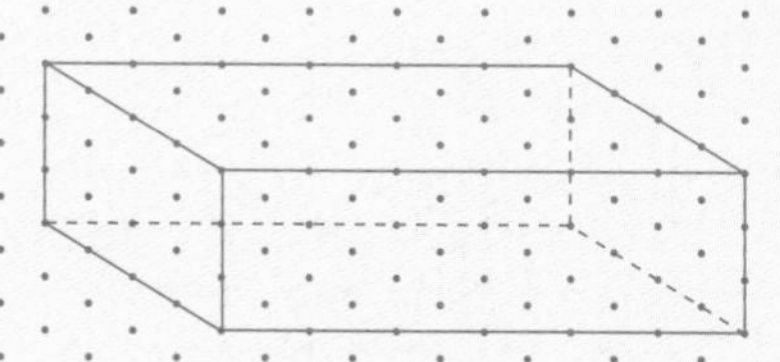

12.

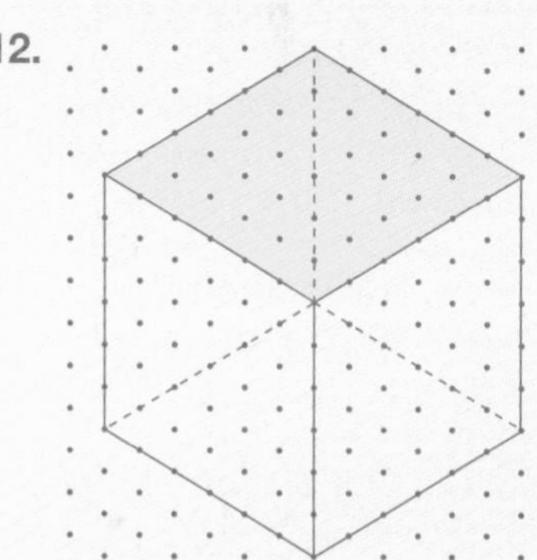

Study Guide Masters, p. 65

11-2 NAME ______ DATE ______ **Study Guide** Student Edition Pages 584–589

Nets and Surface Area

The **surface area** of a three-dimensional object or solid is the sum of the areas of its outer surfaces. The surfaces that make up a polyhedron are polygons.

pyramid

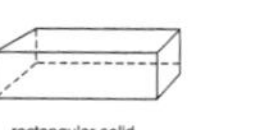
rectangular solid

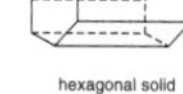
hexagonal solid

When a polyhedron is unfolded, the result is a two-dimensional figure called a **net**.

Example: Match the net at the right with one of the solids below.

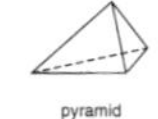

a.  b. 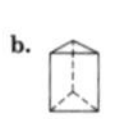 c. 

The net represents solid c.

Identify the number and type of polygons that are faces of each polyhedron.

1. 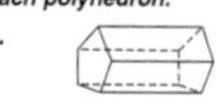
2 pentagons, 5 rectangles

2. 
1 rectangle, 4 triangles

3. 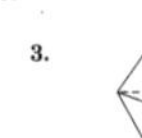
6 triangles

Given each polyhedron, copy its net and label the remaining vertices.

4. 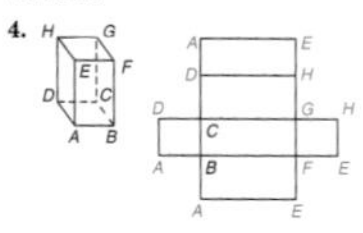

5. 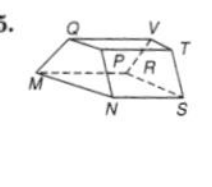 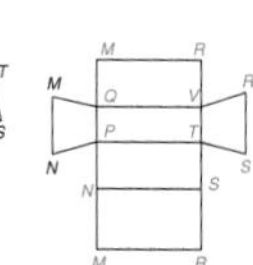

Cooperative Learning

Co-op Co-op Separate the class into teams. Have each team use stiff paper or cardboard to create a polyhedron such as a cube or pyramid. Each team should measure the surface area of the polyhedron and then explain the process to the class. For more information on the co-op co-op strategy, see *Cooperative Learning in the Mathematics Classroom*, one of the titles in the Glencoe Mathematics Professional Series, page 32.

Additional Answers

26.

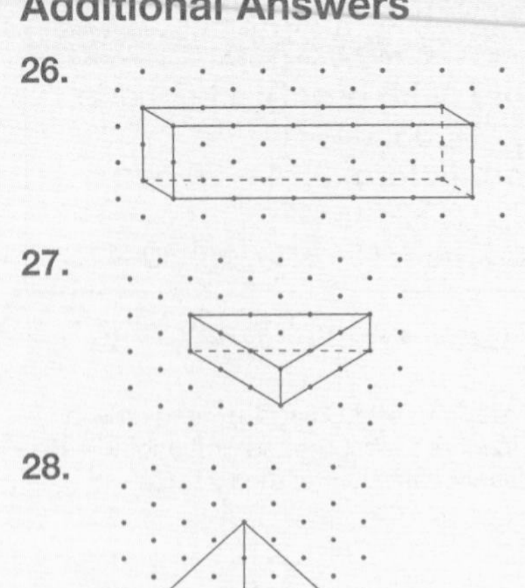

27.

28.

29c. See students' conjectures. The surface area of a solid whose dimensions have been tripled is nine times the surface area of the original solid.

30. 4; Since the integer on each side of the cube will be used in the vertex number for exactly four of the vertices, then $S = 4(a + b + c + d + e + f + g + h)$.

Practice Masters, p. 65

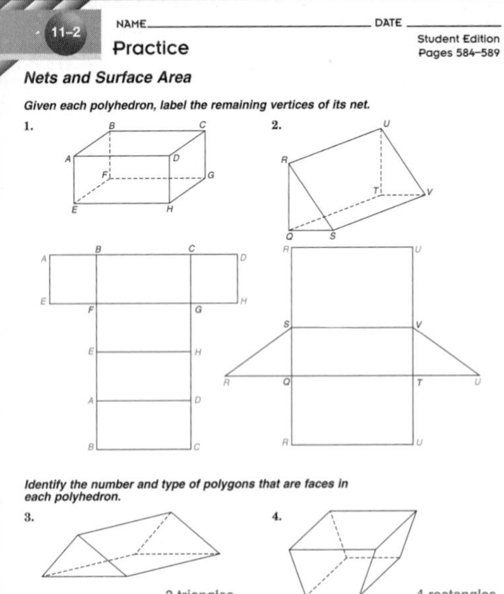

NAME
DATE
11-2
Practice
Student Edition Pages 584–589
Nets and Surface Area
Given each polyhedron, label the remaining vertices of its net.
Identify the number and type of polygons that are faces in each polyhedron.
2 triangles
3 rectangles
4 rectangles
2 trapezoids
6 squares
5 rectangles
2 pentagons

20–25. See Solutions Manual for sample nets.

20. 104 units^2

21. $\approx$ 116.3 units^2

22. $\approx$ 294.6 units^2

23. 120 units^2

24. $\approx$ 92.8 units^2

25. $\approx$ 6.9 units^2

Use rectangular dot paper to draw a net for each solid. Then find the surface area of the solid.

20. 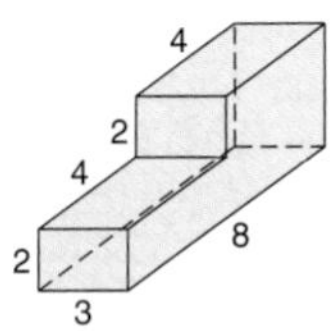

21. 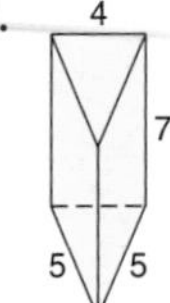

22. 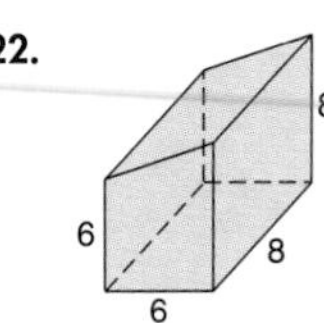

23. 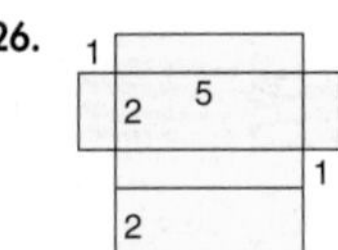

24. 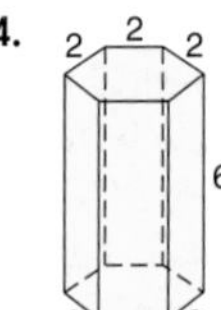

25. 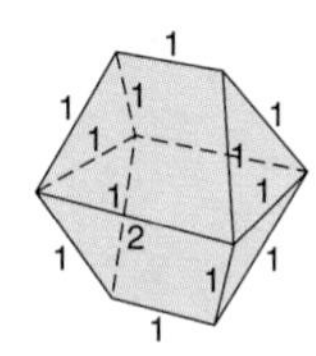

Use isometric dot paper to draw the polyhedron represented by each net. 26–28. See margin for drawings.

26. 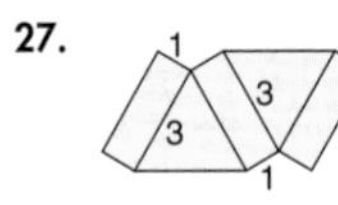

27.

28. 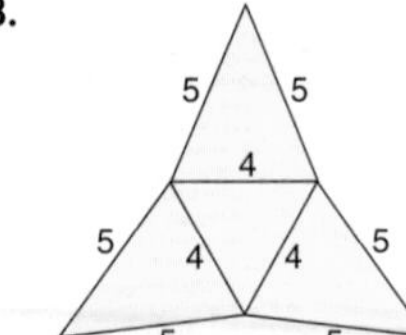

29a–b. See Solutions Manual for drawings.

29a. cube surface area = 6 units^2; triangular prism surface area $\approx$ 9.9 units^2; rectangular prism surface area = 76 units^2

29b. cube surface area = 24 units^2; triangular prism surface area $\approx$ 39.5 units^2; rectangular prism surface area = 304 units^2

29. Investigate the change in surface area that results from a change in the dimensions of a solid.
    a. Draw a net for each solid and find its surface area.
    b. Double each dimension of each solid. Then draw a net for each new solid. Find the surface area of the new solid.
    c. Study the results of part b. Write a conjecture about the surface area of a solid whose dimensions have been tripled. Check your conjecture by drawing a net and finding the surface area of one solid whose dimensions have been tripled. See margin.

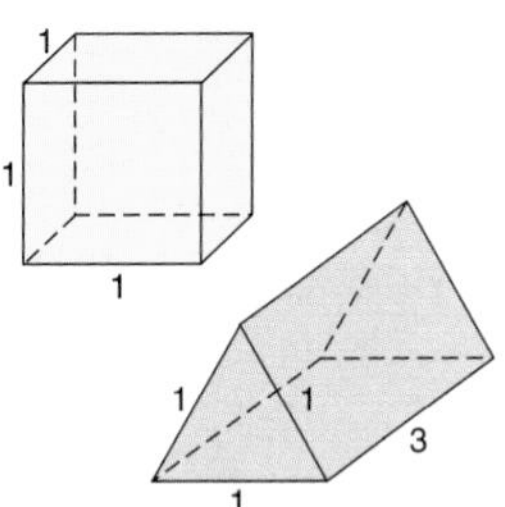

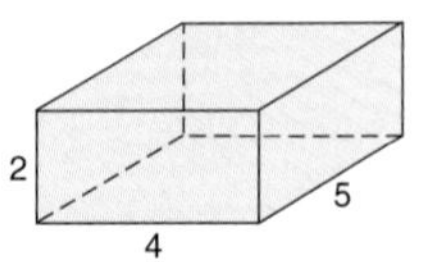

Critical Thinking

30. Suppose you wrote a different integer on each face of cube *ABCDEFGH*. Then a is the sum of the integers on the faces that meet at vertex A, b is the sum of the integers on the faces that meet at vertex B, and so on. Suppose $N = a + b + c + d + e + f + g + h$. What is the greatest number by which N must be divisible? See margin.

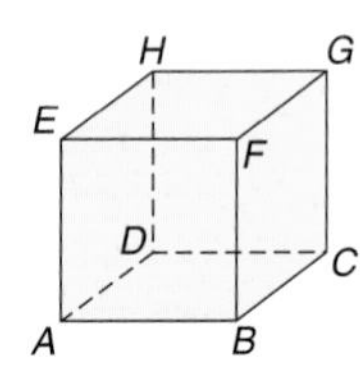

Applications and Problem Solving

31. **Construction** The roof shown at the right is a hip-and-valley style. Use the dimensions given to find the area of the roof that would need to be shingled. 2344.8 ft^2

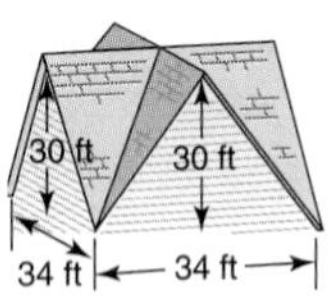

Extension

Problem Solving Have students think of closed solid figures that are not polyhedrons. Have them sketch two.

Sample answers:

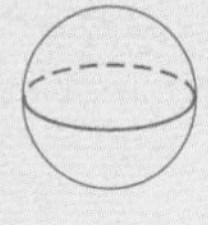

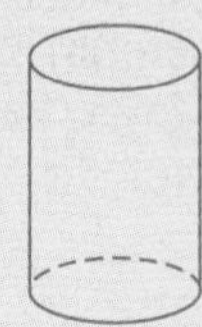

32. **Chemistry** Mineral crystals form as solids with flat faces. Mineralogists can determine what mineral a crystal is by the shape of the crystal. The forms of three common minerals are shown below. Draw a net of each crystal. **See Solutions Manual.**

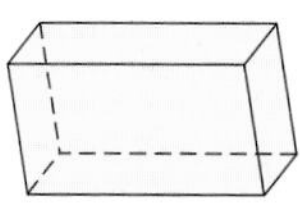
borax

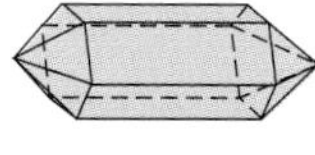
quartz

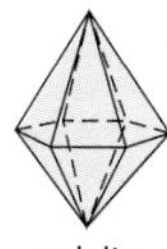
calcite

33. **Games** Many board games use a standard die like the one shown at the right. The opposite faces of a standard die are marked with numbers of dots so that their sum is 7. Determine whether each numbered net can be folded to result in a standard die.

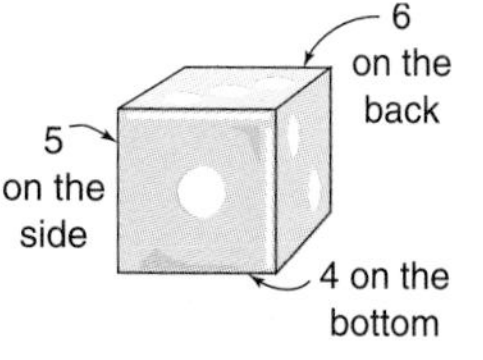

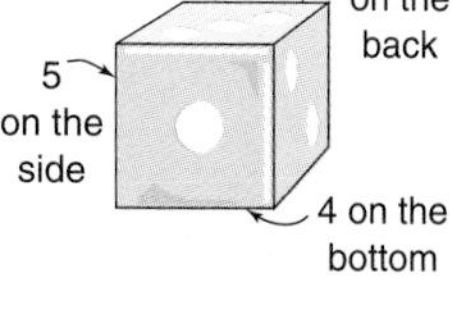

a. **yes** b. **no** c. **yes**

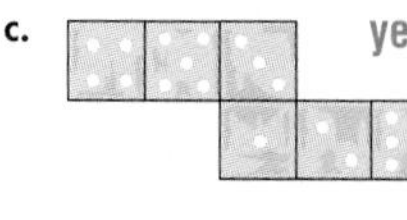

Mixed Review

34. The corner view of a figure is given at the right. Draw the top, left, front, right, and back views of the figure. (Lesson 11–1) **See margin.**

35. Find the area of a trapezoid whose median is 8.5 feet long and whose altitude is 7.1 feet long. (Lesson 10–4) **60.35 ft^2**

36. Circle L has equation $(x - 12)^2 + (y - 3)^2 = 20$. Determine whether point $M(7, -5)$ is on the circle, in the interior, or in the exterior. (Lesson 9–8) **exterior**

37. If $b = 6.4$, $m\angle A = 38$, and $m\angle B = 79$ in $\triangle ABC$, find a. (Lesson 8–5) **4.0**

38. What is the length of a diagonal of a square that has sides 53.7 meters long? (Lesson 8–2) **75.9 m**

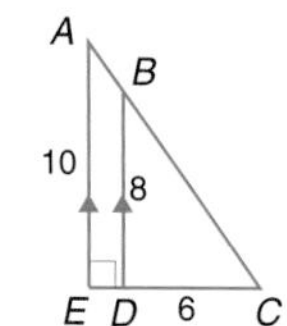

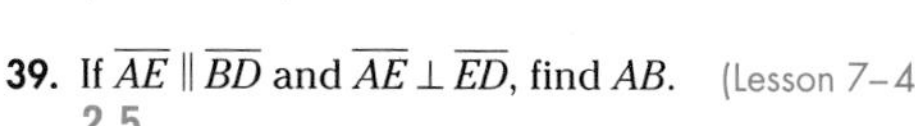

39. If $\overline{AE} \parallel \overline{BD}$ and $\overline{AE} \perp \overline{ED}$, find AB. (Lesson 7–4) **2.5**

Algebra

40. **Nutrition** The table at the right shows the number of Calories in two slices of a medium Pizza Hut™ pizza. Find the range and median for the set of data. **236; 526**

PIZZA HUT™

Type	Calories
Pan, Cheese	522
Pan, Pepperoni	530
Pan, Supreme	622
Pan, Super Supreme	646
Thin 'n Crispy, Cheese	410
Thin 'n Crispy, Pepperoni	430
Thin 'n Crispy, Supreme	514
Thin 'n Crispy, Super Supreme	540

Source: Fast Food Facts

41. Solve $-0.15 \geq y + (-0.03)$. **$\{y \mid y \leq -0.12\}$**

Additional Answer

34.

top view | left view | front view | right view | back view

4 ASSESS

Closing Activity

Modeling Have each student select a three-dimensional object in the classroom and draw a net for the solid.

Chapter 11 Quiz A (Lessons 11-1 and 11-2) is available in the *Assessment and Evaluation Masters*, p. 296.

Enrichment Masters, p. 65

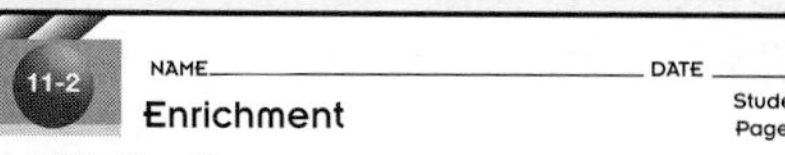

11-2 NAME________________ DATE________

Enrichment Student Edition Pages 584–589

Can-didly Speaking

Beverage companies have mathematically determined the ideal shape for a 12-ounce soft-drink can. Why won't any shape do? Some shapes are more economical than others. In order to hold 12 ounces of soft drink, an extremely skinny can would have to be very tall. The total amount of aluminum used for such a shape would be greater than the amount used for the conventional shape, thus costing the company more to make the skinny can. The radius r chosen determines the height h needed to hold 12 ounces of liquid. This also determines the amount of aluminum needed to make the can. Companies also have to keep in mind that the top of the can is three times thicker than the bottom and sides. Why? So you won't tear off the entire top when you open the can! The following formulas can be used to find the height and amount of aluminum a needed for a 12-ounce soft-drink can.

$$h = \frac{17.89}{\pi r^2}$$

$$a = 0.02\pi\left[\frac{4r^3 + 35.78}{r}\right]$$ The values of r and h are measured in inches.

Find the height needed for a 12-ounce can for each radius. Round to the nearest tenth.

1. $2\frac{2}{3}$ in. 0.8 in. **2.** 1 in. 5.7 in. **3.** 3 in. 0.6 in. **4.** $1\frac{3}{4}$ in. 1.9 in.

Sketch the shape of each can in Exercises 1–4.

5. Exercise 1 **6.** Exercise 2 **7.** Exercise 3 **8.** Exercise 4

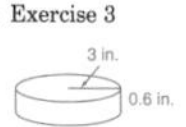

9a. Measure the radius of a soft-drink can. $1\frac{1}{8}$ in.

b. Use the formula to find the height of the can. $4\frac{1}{2}$ in.

c. Measure the height of a soft-drink can. $\approx 4\frac{3}{4}$ in.

d. How does this measure compare to your findings in part a? See students' work.

10. Find the amount of aluminum used in making a soft-drink can. $\approx$ 2.3 oz

11-2B LESSON NOTES

NCTM Standards: 2, 3, 7

Objective

Investigate Plateau's problem using soap film.

Recommended Time

Demonstration and discussion: 15 minutes; Exercises: 30 minutes

Instructional Resources

For each student or group of students
Modeling Mathematics Masters
• p. 101 (worksheet)
For teacher demonstration
Algebra and Geometry Overhead Manipulative Resources

1 FOCUS

Motivating the Lesson

Imagine a cube-shaped frame with a pliable rubber covering. Think of ways to decrease the volume enclosed by the rubber covering.

2 TEACH

Teaching Tip Dawn® dishwashing liquid works best for longer lasting bubbles.

3 PRACTICE/APPLY

Assignment Guide

Core (with proof): 1–5
Core (informal): 1–5
Enriched: 1–5

4 ASSESS

Observing students working in cooperative groups is an excellent method of assessment.

An Extension of Lesson 11–2

11–2B Plateau's Problem

Materials: toothpicks gumdrops thread  water  dishwashing detergent deep bowl

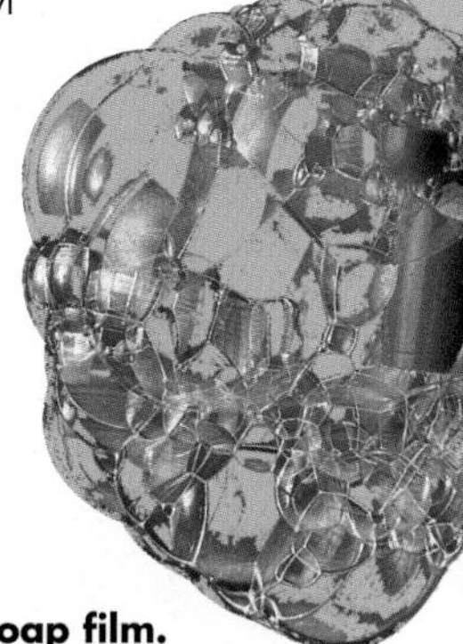

Children and adults alike have long been fascinated with bubbles. Joseph Antoine Ferdinand Plateau (1801–1883) spent much of his life studying the surface properties of fluids. The problem of finding the surfaces of three-dimensional figures that have the least surface area is now called Plateau's problem in his honor. You can investigate Plateau's problem with soap film even though it was not solved mathematically until 1976.

Activity **Make a model and find the surface area of the model using soap film.**

Step 1 Make a soap mixture from a gallon of water and 6 ounces of dishwashing detergent.

Step 2 Build a triangular pyramid, a cube, and a triangular prism. Use toothpicks to represent the edges of the solids and gumdrops to represent the vertices of the solids.

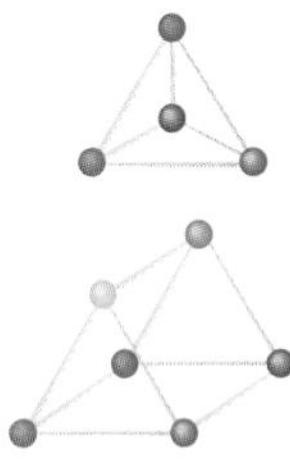

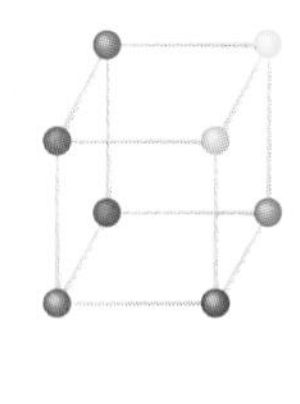

Step 3 Cut a piece of thread and loop it around an edge of your pyramid. Hold onto both ends of the thread and dip the pyramid into the soap mixture. Remove the pyramid from the soap, and observe the shape made by the soap bubble. Dip the cube and the prism in the soap and observe the shapes made by the bubbles.

Write

1. Make a conjecture about the way that the soap will cling to the toothpicks to make the minimum surface area on a cube. **See students' work.**

Model

2. Make solids of other shapes using toothpicks and gumdrops. Dip in the soap mixture and observe the results. **See students' work.**
3. Does a soap bubble cling to the sides of a solid? Explain. **3. No; it forms a shape inside of the frame.**
4. Measure the angle made by three soap surfaces that meet in a line. What does the angle made by four soap surfaces that meet in a line measure? **See margin.**
5. Why do you think the soap behaves as it does? Was the surface made by the soap what you expected? Explain. **See students' work.**

Using Cooperative Learning

This lesson offers an excellent opportunity for using cooperative learning groups. For more information on cooperative learning strategies and group management, see *Cooperative Learning in the Mathematics Classroom*, one of the titles in the Glencoe Mathematics Professional Series.

Additional Answer

4. Three soap surfaces meet at an angle of 120°. Four soap surfaces meet at an angle of about 110°.

11–3

Surface Area of Prisms and Cylinders

What YOU'LL LEARN

- To find the lateral area and surface area of a right prism, and
- to find the lateral area and surface area of a right cylinder.

Why IT'S IMPORTANT

You can solve surface area problems involving manufacturing, gardening, and camping.

APPLICATION

Architecture

You have probably seen buildings covered in brick, stucco, aluminum siding, and wood. But have you ever seen a building covered in corn? In 1892, the citizens of Mitchell, South Dakota, erected the first Corn Palace. The Corn Palace that stands in Mitchell now was built in 1919 and is covered in colored corn and grain each September. The locally grown decorations now cost about $40,000 annually and feed the birds through the winter.

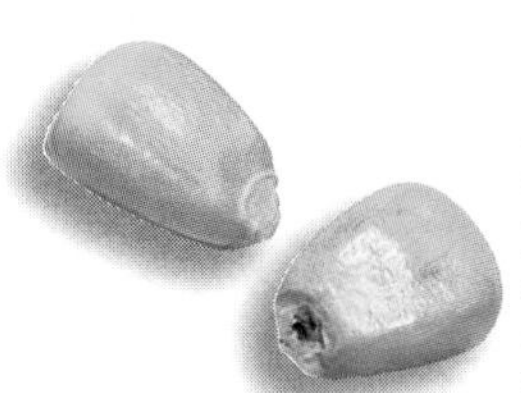

How do you think the citizens of Mitchell go about covering the Corn Palace each year? The first step is to determine the amount of grain to purchase. The amount of grain needed is a function of the surface area of the building. The surface area of the Corn Palace can be estimated by finding the surface area of a rectangular prism. Prisms have the following characteristics.

- Two faces, called **bases**, are formed by congruent polygons that lie in parallel planes.
- The **lateral faces**, the faces that are not bases, are formed by parallelograms.

altitude
base
lateral edge
base
lateral face

- The intersections of two adjacent lateral faces are called **lateral edges** and are parallel segments.

A segment perpendicular to the planes containing the two bases, with an endpoint in each plane is called an **altitude** of the prism. The length of an altitude of a prism is called the **height** of the prism. A prism whose lateral edges are also altitudes is called a **right prism**. If a prism is not right, then it is an **oblique prism**.

11-3 LESSON NOTES

NCTM Standards: 1–5, 7

Instructional Resources

- Study Guide Master 11-3
- Practice Master 11-3
- Enrichment Master 11-3

Transparency 11-3A contains the 5-Minute Check for this lesson; **Transparency 11-3B** contains a teaching aid for this lesson.

Recommended Pacing	
Standard Pacing	Day 6 of 18
Honors Pacing	Day 6 of 16
Block Scheduling*	Day 4 of 10

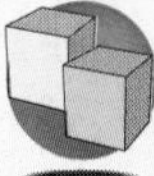

*For more information on pacing and possible lesson plans, refer to the *Block Scheduling Booklet.*

1 FOCUS

5-Minute Check
(over Lesson 11-2)

Refer to the figure below.

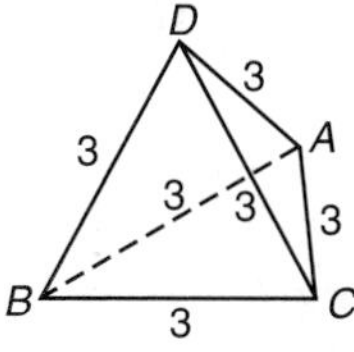

1. Identify the number and type of polygons that are faces. 4 equilateral triangles
2. Draw a net for the solid and label each vertex.

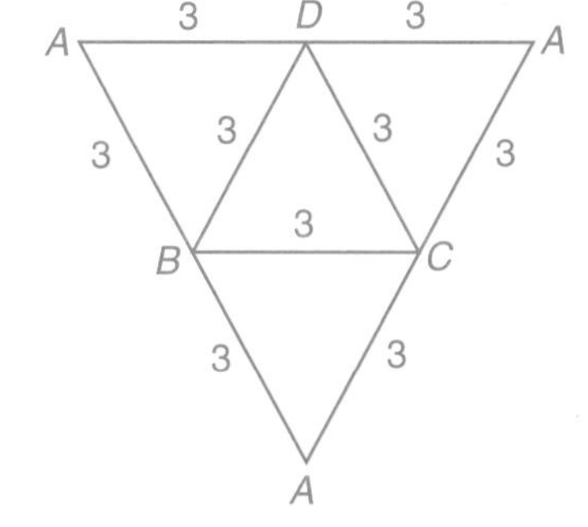

Motivating the Lesson

Questioning Ask students to name the parts of a prism before giving them any of the terminology in this lesson. Then explain that the words *face*, *base*, *vertex*, and *lateral edge* can be used to describe different parts.

2 TEACH

In-Class Example

For Example 1
Find the lateral area and surface area of the triangular prism.

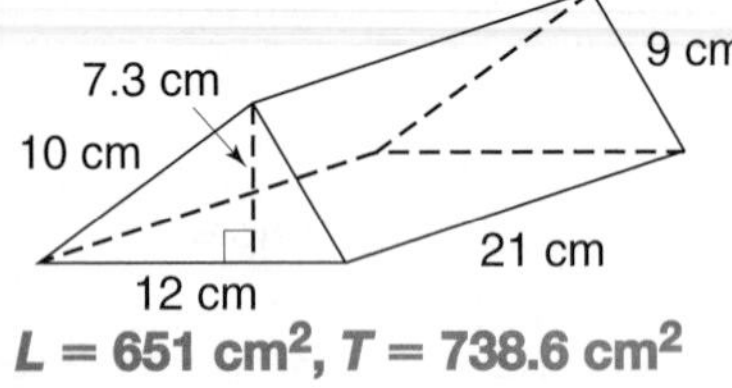

***L* = 651 cm², *T* = 738.6 cm²**

Teaching Tip When defining surface area of a right cylinder, point out that πr^2 is the area of a circle, so the area of both bases is $2\pi r^2$.

You can classify a prism by the shape of its bases.

The lateral faces of right prisms are rectangles.

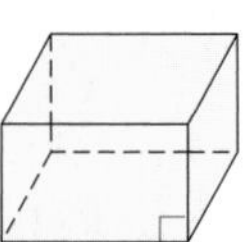
Right rectangular prism

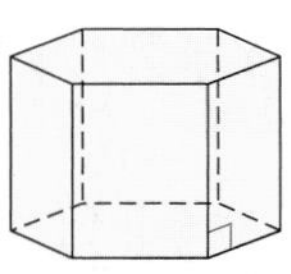
Right hexagonal prism

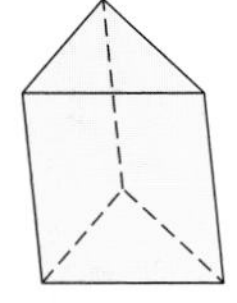
Oblique triangular prism

The lateral edges of oblique prisms are not altitudes.

You can find the **surface area** T of a prism by adding the areas of each of the faces. However, if you can use a formula, the process is much faster. The first step in finding a formula for the surface area of a prism is to find a formula for the area of the lateral faces.

The area of all the lateral faces of a prism is called the **lateral area** L. As a result of the Distributive Property, the lateral area of a right prism can be found by multiplying the height by the perimeter P of the base as shown in its net. For example in the hexagonal prism at the right, a, b, c, d, e, and f are the measures of the sides of the base.

$$L = ah + bh + ch + dh + eh + fh$$
$$= (a + b + c + d + e + f)h$$
$$= Ph \qquad P = a + b + c + d + e + f$$

This formula, $L = Ph$, can be used to find the lateral area of any right prism.

Lateral Area of a Right Prism	If a right prism has a lateral area of L square units, a height of h units, and each base has a perimeter of P units, then $L = Ph$.

The bases of a right prism are congruent, so they have the same area B. Thus, the surface area T of a right prism is found by adding the lateral area to the area of both bases $2B$.

Surface Area of a Right Prism	If the total surface area of a right prism is T square units, its height is h units, and each base has an area of B square units and a perimeter of P units, then $T = Ph + 2B$.

Example 1 **Find the lateral area and the surface area of a right triangular prism with a height of 20 inches and a right triangular base with legs of 8 and 6 inches.**

First, use the Pythagorean Theorem to find the measure of the hypotenuse, c.

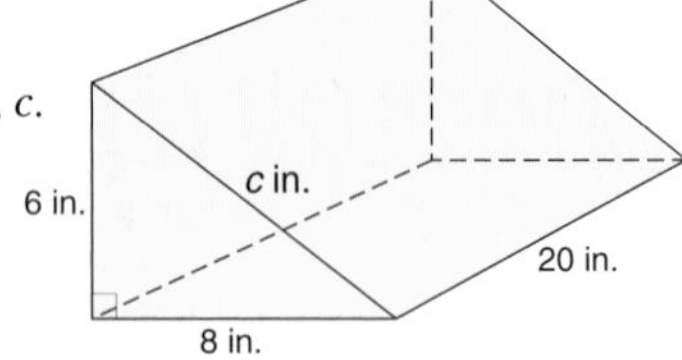

$c^2 = 6^2 + 8^2$
$c^2 = 100$
$c = 10$

Alternative Learning Styles

Auditory Have students draw nets as you describe objects. Start simple and get more challenging. To verify the correctness of nets, students could cut them out and tape them together. Discuss what information is needed for drawing a net from a verbal description.

Next, use the value of c to find the perimeter.

$P = 6 + 8 + 10$ or 24

Now find the area of a base.

$B = \frac{1}{2}bh$

$B = \frac{1}{2}(6)(8)$ or 24 in^2 $b = 6, h = 8$

Finally, find the surface area.

$T = Ph + 2B$

$T = 24 \cdot 20 + 2 \cdot 24$ or 528 in^2 $P = 24, h = 20, B = 24$

The formulas for the lateral area and the surface area of a right prism can be used to estimate areas of other three-dimensional objects.

Example 2 Refer to the application at the beginning of the lesson.

APPLICATION
Architecture

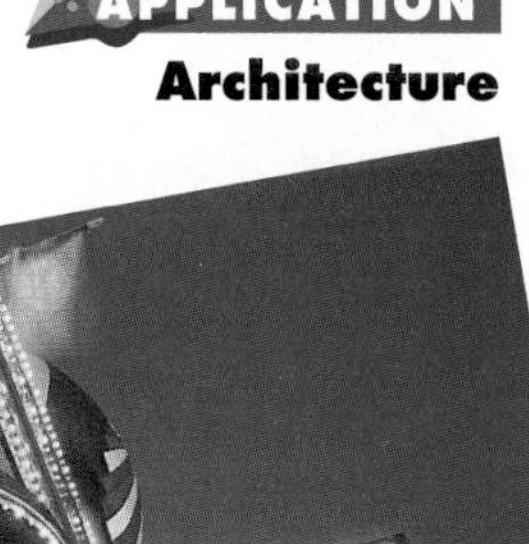

a. Estimate the area of the Corn Palace to be covered if its base is 152 by 196 feet and it is 40 feet tall, not including the turrets.

b. Suppose a bushel of grain can cover 15 square feet. How many bushels of grain will it take to cover the Corn Palace?

c. Will the actual amount of grain needed be higher or lower than the estimate?

a. Assume that the Corn Palace is a right rectangular prism. Then since neither the floor nor roof of the Palace will be covered, the area to consider is the lateral area.

$$L = Ph$$
$$= [2(152 + 196)][40]$$
$$= 27{,}840 \text{ ft}^2$$

b. It will take $27{,}840 \div 15$ or about 1856 bushels of grain to cover the Corn Palace.

c. The actual amount of grain needed will be higher because the area of the turrets was not accounted for in the estimate.

FYI

Actually, the walls of the Corn Palace are not covered entirely by grain. Fourteen huge grain murals cover a large portion of the surface area.

A **cylinder** is another common type of solid. Like a prism, a cylinder has parallel, congruent **bases**. However, the bases of a cylinder are circles. The **axis** of the cylinder is the segment whose endpoints are centers of these circles.

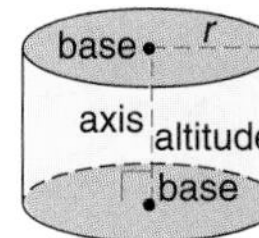

right cylinder

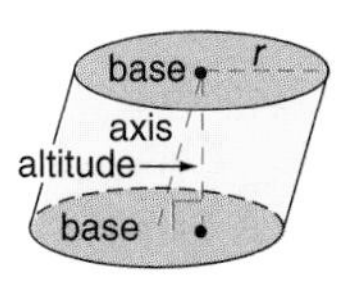

oblique cylinder

An **altitude** of a cylinder is a segment perpendicular to the planes containing the bases with an endpoint in each plane. If the axis of a cylinder is also an altitude of the cylinder, then the cylinder is called a **right cylinder**. Otherwise, the cylinder is an **oblique cylinder**.

In-Class Example

For Example 2

Find the number of square feet of cardboard needed to make the sides of 30,000 frozen concentrate orange juice cans if each can is 6 inches tall and 3 inches in diameter.

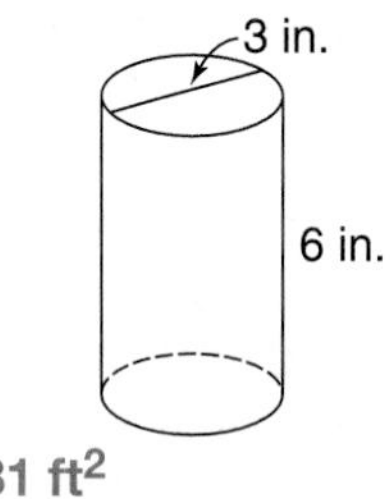

11,781 ft^2

FYI

In July, Mitchell inhabitants hold the Corn Palace Stampede, which is a rodeo, and in September, they have the Corn Palace Festival. For more information contact:
The Corn Palace
604 N. Main St.
Mitchell, South Dakota 57301
Or phone (605) 996-7311

Classroom Vignette

"I have students work at the chalkboard breaking down surface area problems into steps with each step of the problem being completed by a different student. For example, to find the surface area of a prism, the first student draws the prism, the second finds the perimeter of the base, the third finds the lateral area, a fourth finds the area of the base, and the last student finds the total surface area."

Roberta L. Cook

Roberta L. Cook
Waterford Union High School
Waterford, Wisconsin

In-Class Example

For Example 3

Find the number of square feet needed to make the lids of 30,000 frozen orange juice cans. Each lid is 3 inches in diameter and the lids are cut from a 3.5 inch wide roll of metal. How much metal will be left? **There will be 805 ft^2 of metal left.**

Teaching Tip A calculator has been used to calculate all answers involving π with rounding occurring at the final stage. If students use 3.14 for π or round at an earlier stage, their answers will differ from those given in this book.

The figure below is a net for a right cylinder.

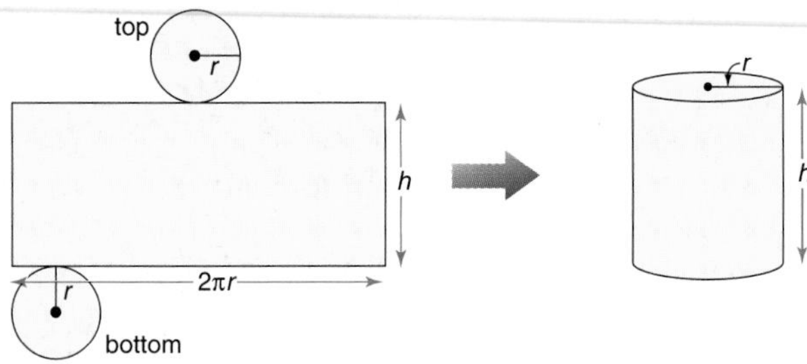

The lateral area of a cylinder is the area of the curved surface. Like a prism, the lateral area of a cylinder can be found by using the formula $L = Ph$. However, since the base is a circle, the perimeter is the circumference $(2\pi r)$ of the circle. Thus, the curved surface is a rectangle whose width is the height of the cylinder, h units, and whose length is the circumference of one of its bases, $2\pi r$ units.

Lateral Area of a Right Cylinder	**If a right cylinder has a lateral area of L square units, a height of h units, and the bases have radii of r units, then $L = 2\pi rh$.**

The surface area of a cylinder is the sum of the lateral area and the areas of the bases. Each base is a circle whose area is πr^2.

Surface Area of a Right Cylinder	**If a right cylinder has a total surface area of T square units, a height of h units, and the bases have radii of r units, then $T = 2\pi rh + 2\pi r^2$.**

Example 3

Find the number of square feet of aluminum used to make 50,000 cylindrical vegetable cans if each can is 5 inches tall and 2.5 inches in diameter.

If the diameter of the base is 2.5 inches, then the radius is 1.25 inches.

$T = 2\pi rh + 2\pi r^2$

$T = 2\pi(1.25)(5) + 2\pi(1.25)^2$

$T = 12.5\pi + 3.125\pi$

$T = 15.625\pi$

$T \approx 49.1 \text{ in}^2$

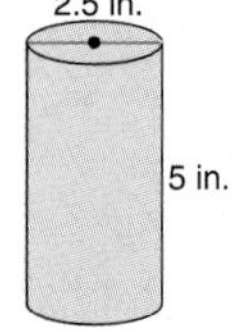

The surface area of each can is about 49.1 square inches, so the area of aluminum used for 50,000 cans is about 50,000 × 49.1 or about 2,455,000 square inches. There are 144 square inches in a square foot. Therefore, about 2,455,000 ÷ 144 or 17,049 square feet of aluminum are used.

You can use the formulas for lateral and surface area to find missing measures.

Example 4 The surface area of a right cylinder is 301.6 cm². If the height is 8 centimeters, find the radius of the base.

Remember that for $ax^2 + bx + c = 0$, the value of x can be found using $x = \frac{-b \pm \sqrt{b^2 - 4ac}}{2a}$. This is the ***quadratic formula****.*

$$T = 2\pi rh + 2\pi r^2$$
$$301.6 = 2\pi r(8) + 2\pi r^2$$
$$301.6 = 50.3r + 6.3r^2$$
$$0 = 6.3r^2 + 50.3r - 301.6$$

Now use the quadratic formula to solve this equation. Let r replace x.

$$r = \frac{-b \pm \sqrt{b^2 - 4ac}}{2a}$$
$$= \frac{-50.3 \pm \sqrt{50.3^2 - 4(6.3)(-301.6)}}{2(6.3)} \quad a = 6.3,\ b = 50.3, \text{ and } c = -301.6.$$
$$\approx 4.0 \text{ or } -12.0 \quad \textit{Use a calculator.}$$

Since a circle cannot have a negative radius, the radius of the base must be 4.0 centimeters.

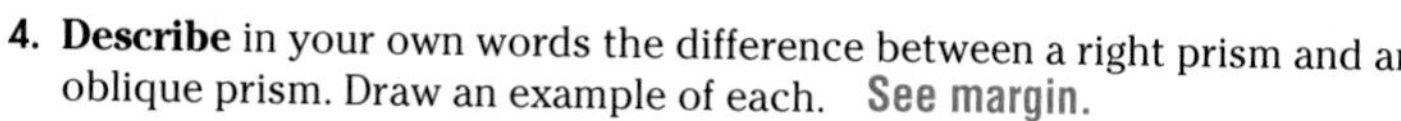

CHECK FOR UNDERSTANDING

Communicating Mathematics

Read and study the lesson to answer these questions.

1. Refer to the triangular prism in Example 1. **b–c. See Solutions Manual.**
   a. What are the shapes of the polygonal faces that make up the solid? Be specific. **2 congruent triangles and 3 rectangles**
   b. Draw a net of the solid on rectangular dot paper.
   c. Draw a corner view of the solid on isometric dot paper.
2. What dimensions of a soft-drink can would be important for you to know if you are designing a new label for the can? Explain. **See margin.**
3. **Compare and contrast** lateral area and surface area. **See margin.**

4. **Describe** in your own words the difference between a right prism and an oblique prism. Draw an example of each. **See margin.**

Guided Practice

Refer to the prism at the right to answer each of the following.

5. Is the prism a right prism or an oblique prism? Explain.
6. What is the shape of its bases and lateral faces?
7. If the bases are regular polygons with sides 6 units long, find the perimeter of the base. **30 units**
8. If the perimeter of the base is 60 units and the length of a lateral edge is 15 units, find the lateral area of the prism. **900 units²**

5. Right prism; the lateral edges are perpendicular to the bases.

6. bases: pentagons; lateral faces: rectangles

Find the surface area of the right rectangular prism for each set of measures.

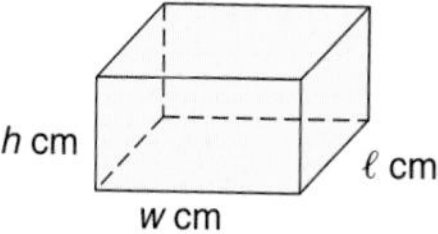

9. $\ell = 8$, $w = 4$, and $h = 2$ **112 cm²**
10. $\ell = 6.5$, $w = 6.5$, and $h = 6.5$ **253.5 cm²**

Find the surface area of a cylinder with each set of measures. Round to the nearest tenth.

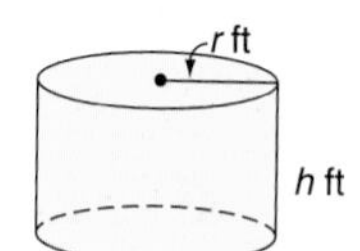

11. $r = 4$ and $h = 6$ **251.3 ft²**
12. $r = 8.3$ and $h = 6.6$ **777.0 ft²**

Reteaching

Using Vocabulary Draw a prism and a cylinder on the chalkboard or overhead. Ask students about the parts of the solids and what type of solid they are. For example, ask them to identify the bases, faces, lateral edges, altitudes, and axis of the cylinder; ask them if the solid is right or oblique and why.

In-Class Example

For Example 4
The surface area of a right cylinder is 100 cm². If the height is 5 centimeters, find the radius of the base. **2.2 cm**

3 PRACTICE/APPLY

Check for Understanding

Exercises 1–14 are designed to help you assess your students' understanding through reading, writing, speaking, and modeling. You should work through Exercises 1–4 with your students and then monitor their work on Exercises 5–14.

Error Analysis

Finding the bases of prisms may pose difficulty for students. Emphasize that the bases of prisms (and cylinders) are parallel and congruent, and that the bases may not be the sides that are horizontal. Also, the altitude of the prism connects the two bases. In some prisms, more than one pair of sides could be used as the bases.

Additional Answers

2. To make the label, the designer needs to know the circumference of the cylindrical base and the height of the cylinder. The area of the label is closely associated with the lateral area of the cylinder.
3. Lateral area is the area of all the lateral surfaces. Surface area is the area of the entire surface including the bases. The surface area is the sum of the areas of the bases and the lateral area.
4. In a right prism, the lateral edges of the prism are also altitudes of the prism. If this is not the case, then the prism is an oblique prism. See students' drawings.

Assignment Guide

Core (with proof): 15–39 odd, 40–49
Core (informal): 15–39 odd, 40–49
Enriched: 16–34 even, 35–49

For **Extra Practice**, see p. 785.

The red A, B, and C flags, printed only in the Teacher's Wraparound Edition, indicate the level of difficulty of the exercises.

Study Guide Masters, p. 66

NAME_______________ DATE_______

Study Guide — Student Edition Pages 591–598

Surface Area of Prisms and Cylinders

Prisms are polyhedrons with congruent polygonal bases in parallel planes. **Cylinders** have congruent and parallel circular bases. An **altitude** is a perpendicular segment joining the planes of the bases. The length of an altitude is the **height** of the figure. **Right prisms** have lateral edges that are altitudes. A right cylinder is one whose **axis** is an altitude.

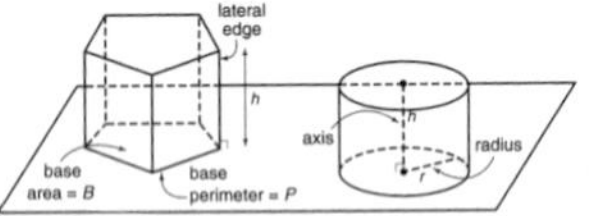

In the following formulas, L is lateral and T is total surface area.

Prisms $L = Ph$, $T = Ph + 2B$ **Cylinders** $L = 2\pi rh$, $T = 2\pi rh + 2\pi r^2$

Example: Find the surface area of the right cylinder.

$T = 2\pi rh + 2\pi r^2$
$T = 2\pi(3.5)(6) + 2\pi(3.5)^2$
$T = 66.5\pi$ or about 208.92 cm²

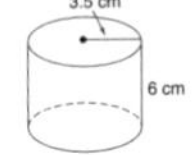

Find the lateral area and the surface area of each right prism or right cylinder. Round your answers to the nearest tenth.

1. 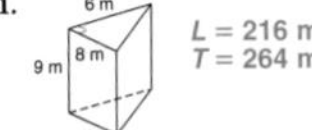

$L = 216$ m², $T = 264$ m²

2. 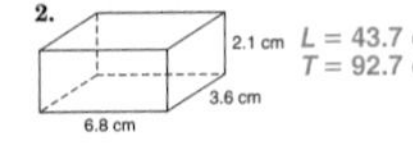

$L = 43.7$ cm², $T = 92.7$ cm²

3. 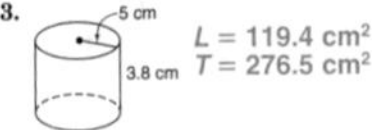

$L = 119.4$ cm², $T = 276.5$ cm²

4. 

$L = 62.6$ m², $T = 71.6$ m²

13. The height of a right cylinder is 28 inches, and its surface area is 1977.7 square inches. Find the radius of the base of the cylinder. **8.6 in.**

14. **Manufacturing** Many baking pans are given a special coating to make food stick to the surface less.
 a. A rectangular cake pan is 9 inches by 13 inches and 2 inches deep. What is the area of the surface to be coated? **205 in²**
 b. Round cake pans have a diameter of 9 inches and a height of 2 inches. Find the area of the surface to be coated. **120.2 in²**

EXERCISES

Practice

Use the rectangular prism at the right to answer each of the following.

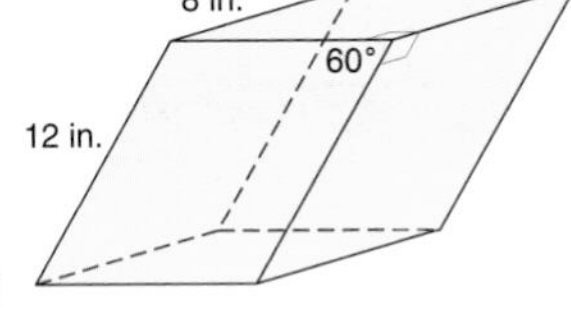

15. Is the prism a right prism or an oblique prism? **oblique prism**
16. What shape are its bases and lateral faces? **16. rectangles, parallelograms**
17. Find the perimeter of the base. **36 in.**

B

18. What is the lateral area of the prism? **18. ≈ 399.8 in²**
19. Find the surface area of the prism. **559.8 in²**

Find the surface area of the right triangular prism for each set of measures. Round to the nearest tenth.

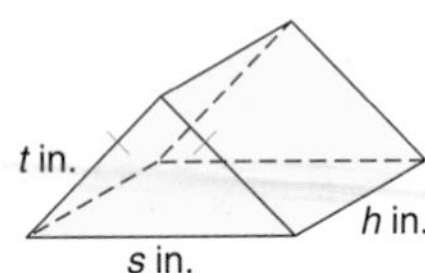

20. $t = 8, s = 14$, and $h = 7$ **264.2 in²**
21. $t = 10, s = 8$, and $h = 20.4$ **644.5 in²**
22. $t = 14, s = 18$, and $h = 30.5$ **1596.0 in²**

Find the lateral area and the surface area of each right prism. Round to the nearest tenth. **24. 473.2 yd²; 646.4 yd²**

23. 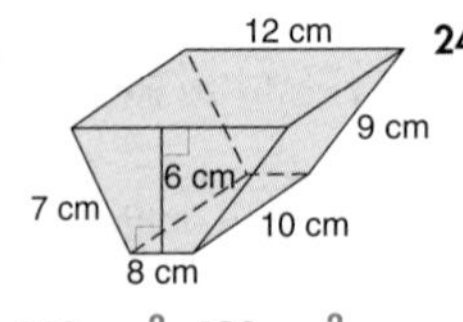

360 cm²; 480 cm²

24. 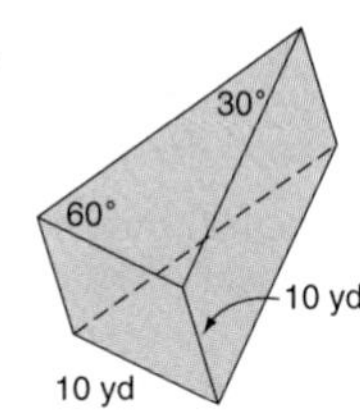

25. 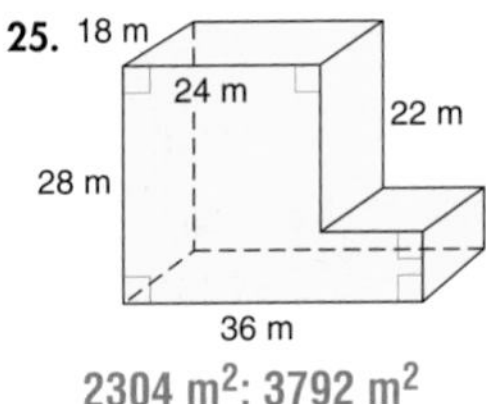

2304 m²; 3792 m²

Find the surface area of a cylinder with each set of measures. Round to the nearest tenth.

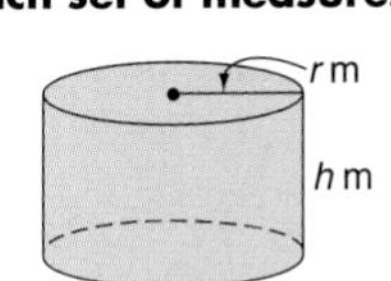

26. $r = 11$ and $h = 11$ **1520.5 m²**
27. $r = 13$ and $h = 15.8$ **2352.4 m²**
28. $r = 6.8$ and $h = 1.9$ **371.7 m²**

Find the lateral area and the surface area of each right cylinder. Round to the nearest tenth. **30. 1231.5 mm²; 2463.0 mm²**

29. 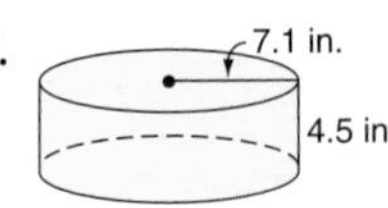

200.7 in²; 517.5 in²

30. 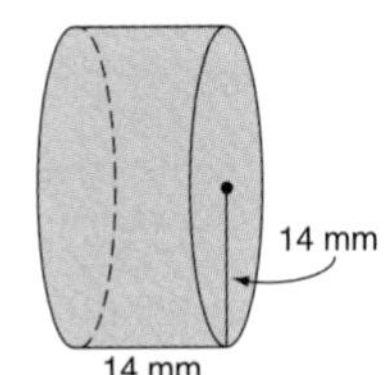

31. 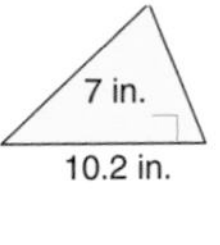

24.9 ft²; 30.0 ft²

Tech Prep

Landscape Designer Many two-year colleges offer degrees in landscape design. Landscape designers plan gardens and yards for private residences and public corporate offices. For more information on tech prep, see the *Teacher's Handbook*.

32. The surface area of a cube is 864 square units. What is the length of a lateral edge of the cube? **12 units**

33. Suppose the lateral area of a right rectangular prism is 144 square centimeters. If its length is three times its width and its height is twice its width, find its surface area. **198 cm^2**

34. A cylinder has a surface area of 301.6 square meters. Find the diameter of the base if the cylinder is 8 meters tall. **8 m**

Critical Thinking

35. Suppose the height of a right prism is doubled. What effect does it have on the surface area? **See margin.**

36. Suppose the height of a right cylinder is tripled. Is the surface area tripled? Explain. **See margin.**

Additional Answers

35. If the height of the prism is doubled, the *lateral area* is doubled. The surface area is increased, but it is not doubled.

36. No; if the height of a cylinder is tripled, the *lateral area* is tripled. The surface area is increased, but it is not tripled.

Applications and Problem Solving

37. 1680 ft^2

37. **Gardening** A greenhouse is designed to allow gardeners to grow plants in areas where they would not normally grow. The surface of a greenhouse is covered with plastic or glass. Find the amount of plastic that would be needed to construct a greenhouse like the one shown at the right.

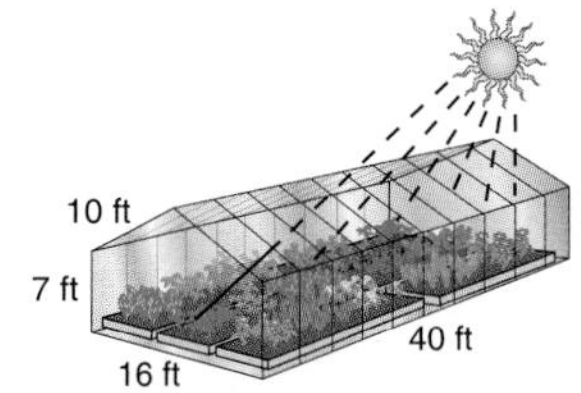

38. **Agriculture** The acid associated with filling a silo can weaken the cement walls and seriously damage the silo's structure. So the inside of the silo must occasionally be resurfaced. The cost of the resurfacing is a function of the lateral area of the inside of the silo. Find the lateral area of a silo 13 meters tall with an interior diameter of 5 meters. **about 204.2 m^2**

39. **Camping** Campers can use a solar cooker to use the energy of the Sun to prepare food. You can make a solar cooker from supplies you have on hand. The reflector in the cooker shown at the right is half of a cardboard cylinder covered with aluminum foil. The reflector is 18 inches long and has a diameter of $5\frac{1}{2}$ inches. How much aluminum foil was needed to cover the inside of the reflector? **179.3 in^2**

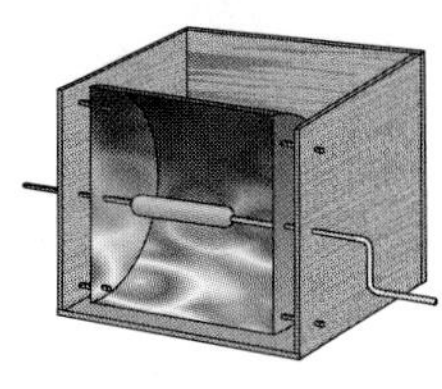

Mixed Review

40. Draw a net for the solid shown at the right. (Lesson 11–2) **See Solutions Manual.**

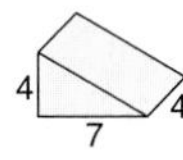

41. **SHOE**

a. What view of the Washington Monument is shown in the photograph in the comic—corner, top, left, right, front, or back? (Lesson 11–1) **top**

b. Find a photograph of the Washington Monument in a reference book or on the Internet. Draw the top, left, right, front, back, and corner views of the monument. (Lesson 11–1) **See margin.**

Additional Answer

41b.

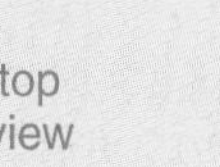

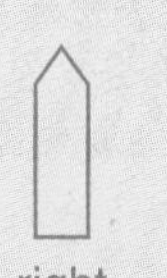

top view | left view | right view | front view | back view

Practice Masters, p. 66

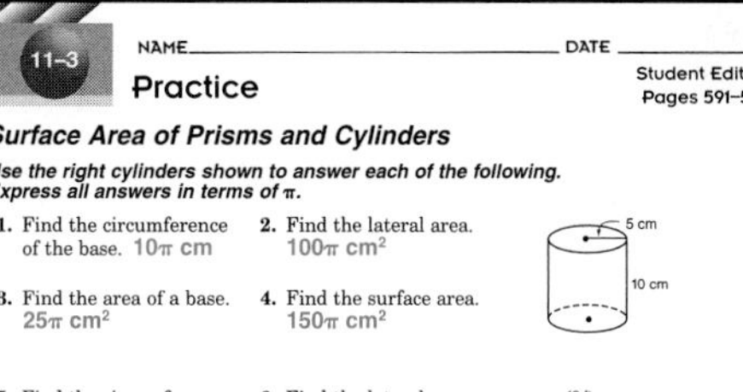

11-3 NAME ______ DATE ______

Practice Student Edition Pages 591–598

Surface Area of Prisms and Cylinders

Use the right cylinders shown to answer each of the following. Express all answers in terms of π.

1. Find the circumference of the base. 10π cm
2. Find the lateral area. 100π cm^2
3. Find the area of a base. 25π cm^2
4. Find the surface area. 150π cm^2

5. Find the circumference of the base. 20π ft
6. Find the lateral area. 200π ft^2
7. Find the area of a base. 100π ft^2
8. Find the surface area. 400π ft^2

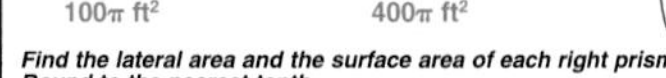

Find the lateral area and the surface area of each right prism. Round to the nearest tenth.

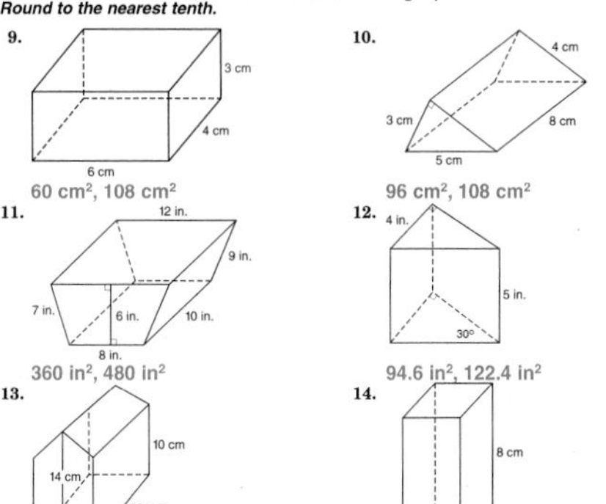

9. 60 cm^2, 108 cm^2
10. 96 cm^2, 108 cm^2
11. 360 in^2, 480 in^2
12. 94.6 in^2, 122.4 in^2
13. 786.3 cm^2, 978.3 cm^2
14. 128 cm^2, 160 cm^2

4 ASSESS

Closing Activity

Writing Have students write a description of a prism and a cylinder. Make sure they include all the parts of each and name the type of solid it is.

Enrichment Masters, p. 66

11-3 NAME ____________ DATE ____________

Enrichment Student Edition Pages 591–598

Cross Sections of Prisms

When a plane intersects a solid figure to form a two-dimensional figure, the results is called a **cross section**. The figure at the right shows a plane intersecting a cube. The cross section is a hexagon.

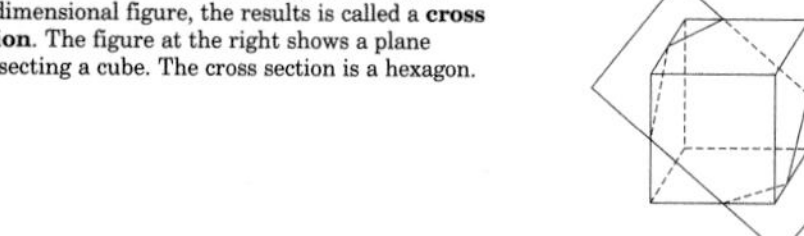

For each right prism, connect the labeled points in alphabetical order to show a cross section. Then identify the polygon.

1. rectangle
2. triangle
3. trapezoid

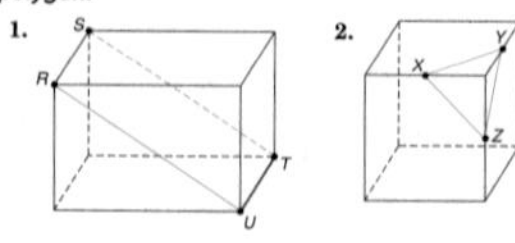

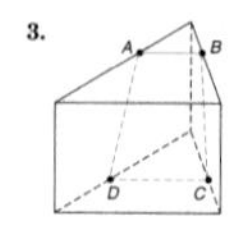

Refer to the right prisms shown below. In the rectangular prism, A and C are midpoints. Identify the cross-section polygon formed by a plane containing the given points.

4. *A, C, H* rectangle
5. *C, E, G* triangle
6. *H, C, E, F* trapezoid
7. *H, A, E* pentagon
8. *B, D, F* rectangle
9. *V, X, R* triangle
10. *R, T, Y* rectangle
11. *R, S, W* trapezoid

43. 64 cm
44. The measure of the inscribed angle is half the measure of the central angle.
45. $\frac{1}{5}$
46. acute; isosceles; $m\angle J = 63$, $m\angle K = 63$, and $m\angle L = 54$, thus, no angle is right or obtuse, so the triangle is acute. $m\angle J = m\angle K$ so the sides opposite these angles are congruent and $\triangle JKL$ is isosceles.
47. A triangle is equiangular if each angle measures 60°; Law of Syllogism.

INTEGRATION **Algebra**

42. **Probability** Find the probability that a point chosen at random in the figure is in the shaded region. Round to the nearest hundredth. (Lesson 10–6) **0.82**

43. One diagonal of a rhombus with an area of 1664 square centimeters is 52 centimeters long. Find the length of the other diagonal. (Lesson 10–4)

44. How are the measures of a central angle and an inscribed angle that intercept the same arc related? (Lesson 9–4)

45. The vertices of $\triangle ABC$ are $A(9, 5)$, $B(6, 2)$, and $C(7, -3)$. Find the slope of the altitude to $\overline{BC}$. (Lesson 5–1)

46. In $\triangle JKL$, $m\angle J = 10x - 7$, $m\angle K = 63$, and $m\angle L = 7x + 5$. Classify $\triangle JKL$ by its angles and sides. Justify your answer. (Lesson 4–6)

47. If possible, write a valid conclusion. State the law of logic that you used. (Lesson 2–3)
    (1) A triangle is equiangular if it is equilateral.
    (2) Each angle of an equilateral triangle measures 60°.

48. Determine whether $\frac{8}{5}x + \frac{4}{y} = 10$ is a linear equation. If it is linear, rewrite it in standard form, $Ax + By = C$. **not linear**

49. **Statistics** Explain whether a scatter plot showing cities' annual per capita consumption of coffee and the population of cities would probably show a *positive*, *negative*, or *no* correlation. **positive**

WORKING ON THE **In·ves·ti·ga·tion**

Refer to the Investigation on pages 510–511.

Soccer balls that are made of leather or vinyl are stitched together by following a pattern. The pattern can be created by first drawing a net of the surface of the ball.

1 Choose a length for the edge of a pentagon in each design shown on pages 510 and 511. Then draw a net for each design.

2 Which design will require more stitching to complete?

3 Which ball would be easier to manufacture?

4 Note your observations to consider when completing your soccer ball design.

Add the results of your work to your Investigation Folder.

Extension

Connections Healthy Food Company sells vegetables in cans that are 4 inches high and have a diameter of 3 inches. How many square feet of paper are required to make the labels for 15,000 cans if there is a 1-inch overlap of the label on each can? **4343.7 ft²**

Working on the Investigation

The Investigation on pages 510–511 is designed to be a long-term project that is completed over several days or weeks. Encourage students to keep their materials in their Investigation Folder as they work on the Investigation.

11-4 Surface Area of Pyramids and Cones

What YOU'LL LEARN

- To find the lateral area and surface area of a regular pyramid, and
- to find the lateral area and surface area of a right circular cone.

Why IT'S IMPORTANT

Pyramids and cones have been used in structures since ancient times.

You can use an envelope to make a *triangular pyramid.*

MODELING MATHEMATICS **Triangular Pyramids**

Materials: 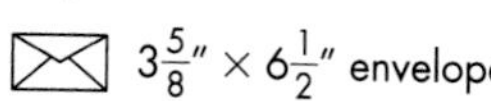 $3\frac{5}{8}'' \times 6\frac{1}{2}''$ envelope

 straightedge  scissors

- Seal a $3\frac{5}{8}'' \times 6\frac{1}{2}''$ envelope. Then draw both diagonals of the envelope.

- Fold and crease along each diagonal. Then fold and crease along the perpendicular bisector of the long edge of the envelope. Label the point of intersection of the diagonals as point *A*.

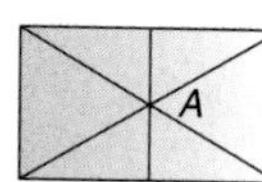

- Carefully cut from the top of the envelope along each diagonal to point *A*. Remove the triangular piece.

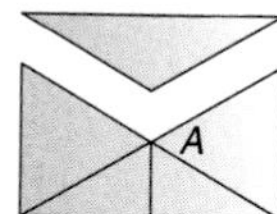

- Open the envelope. Fold along the perpendicular bisector so that the top corners come together. Tuck one corner inside the other and push until the corner meets the bottom edge. The solid formed is a triangular pyramid.

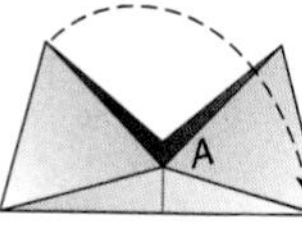

Your Turn d. See margin.

a. What shape is each face of the pyramid? triangle

b. Is each face of the pyramid congruent? Explain.

c. Suppose that the face with the letter *A* on it is the base of the pyramid. What do you observe about the other faces?

d. Repeat the activity using a $4\frac{1}{8}'' \times 9\frac{1}{2}''$ envelope and describe your results.

b. Yes; see students' explanations.

c. Sample answer: They all intersect in one point.

11-4 LESSON NOTES

NCTM Standards: 1–5, 7

Instructional Resources

- Study Guide Master 11-4
- Practice Master 11-4
- Enrichment Master 11-4
- Assessment and Evaluation Masters, pp. 295, 296
- Modeling Mathematics Masters, pp. 55–58, 89

Transparency 11-4A contains the 5-Minute Check for this lesson; **Transparency 11-4B** contains a teaching aid for this lesson.

Recommended Pacing	
Standard Pacing	Days 7 & 8 of 18
Honors Pacing	Days 7 & 8 of 16
Block Scheduling*	Day 5 of 10

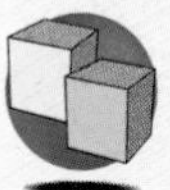

*For more information on pacing and possible lesson plans, refer to the *Block Scheduling Booklet.*

1 FOCUS

5-Minute Check
(over Lesson 11-3)

Refer to the figure below.

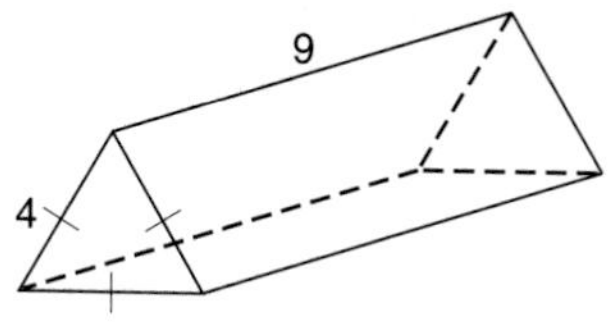

1. Find the lateral area. 108 units²
2. Find the surface area. about 121.9 units²

Refer to the figure below.

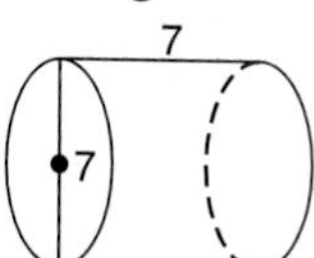

3. Find the lateral area. about 153.9 units²
4. Find the surface area. about 230.9 units²

In this modeling activity, students will construct a triangular pyramid. Have a finished example ready for students who have problems following the directions.

Answer for Modeling Mathematics

d. The faces of the pyramid are congruent isosceles triangles that have a smaller vertex angle than those in the pyramid made from the first envelope.

Motivating the Lesson

Questioning Discuss the pyramids of ancient Egypt. Ask students what they know about them. Ask how they think the pyramids were built. Have students compare building methods used now with ancient Egyptian methods.

2 TEACH

Teaching Tip When defining a regular pyramid, put the definition into if-then language: if the perpendicular segment from the vertex of a pyramid to the base intersects the center of the base, then it is a regular pyramid.

A triangular pyramid is just one type of pyramid. A pyramid has the following characteristics.

- All the faces except one intersect at a point called the **vertex**.
- The **base** is the face that does not intersect the other faces at the vertex. The base is always a polygon.
- The faces that intersect at the vertex are called **lateral faces** and form triangles. The edges of the lateral faces that have the vertex as an endpoint are called **lateral edges**.

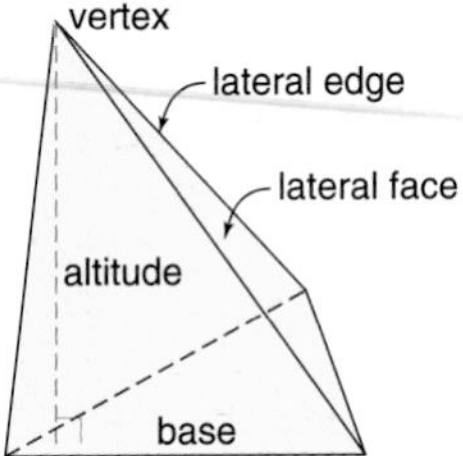

- The **altitude** is the segment from the vertex perpendicular to the base.

If the base of a pyramid is a regular polygon and the segment whose endpoints are the center of the base and the vertex is perpendicular to the base, then the pyramid is called a **regular pyramid**. In a regular pyramid, the segment whose endpoints are the center of the base and the vertex is also the altitude.

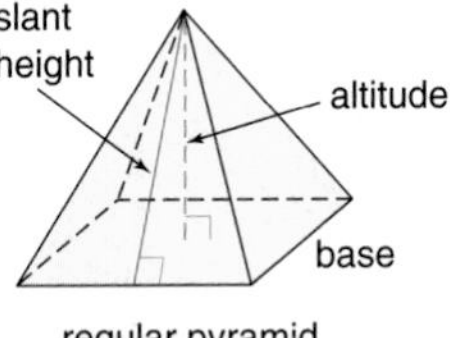

regular pyramid

In a regular pyramid, all of the lateral faces are congruent isosceles triangles. The height of each lateral face is called the **slant height** ℓ of the pyramid.

The figure below is a regular hexagonal pyramid. Its lateral area L can be found by adding the areas of all its congruent triangular faces as shown in its net.

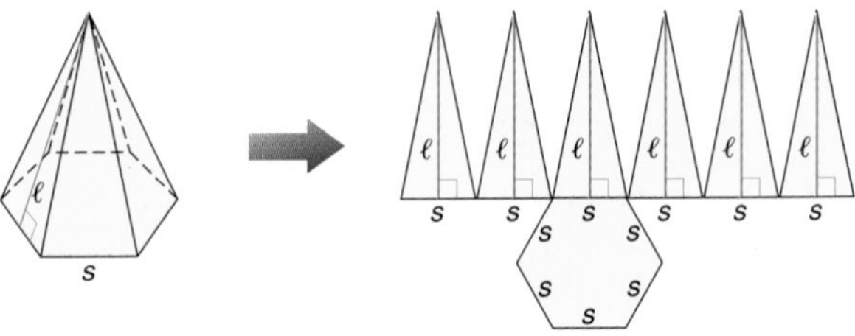

$$L = \frac{1}{2}s\ell + \frac{1}{2}s\ell + \frac{1}{2}s\ell + \frac{1}{2}s\ell + \frac{1}{2}s\ell + \frac{1}{2}s\ell$$
$$= \frac{1}{2}(s + s + s + s + s + s)\ell$$
$$= \frac{1}{2}P\ell \quad P = (s + s + s + s + s + s)$$

This suggests the following formula.

Lateral Area of a Regular Pyramid	**If a regular pyramid has a lateral area of L square units, a slant height of ℓ units, and its base has a perimeter of P units, then $L = \frac{1}{2}P\ell$.**

Example 1 **APPLICATION** **Architecture**

The Luxor Hotel in Las Vegas, Nevada, is shaped like a gigantic black glass pyramid. The base of the pyramid is a square with edges 646 feet long. The hotel is 350 feet tall. Find the area of the glass on the Luxor.

The area of the glass is the lateral area of the pyramid.

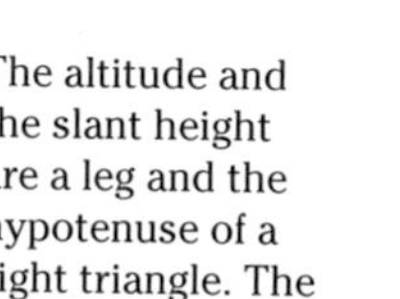

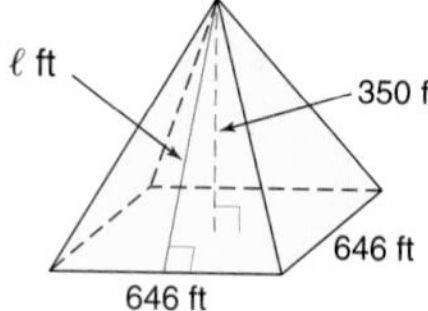

The altitude and the slant height are a leg and the hypotenuse of a right triangle. The other leg is half of the measure of a side of the base. Use the Pythagorean Theorem, $c^2 = a^2 + b^2$, to find the slant height of the pyramid.

$$\ell^2 = 350^2 + \left(\frac{1}{2} \cdot 646\right)^2$$
$$\ell^2 = 226{,}829$$
$$\ell = \sqrt{226{,}829} \text{ or about } 476.3 \text{ ft}$$

$$L = \frac{1}{2}P\ell$$
$$\approx \frac{1}{2}(4 \cdot 646)(476.3)$$
$$\approx 615{,}379.6$$

The area of the glass on the Luxor is about 615,380 ft^2.

The surface area of a regular pyramid is the sum of its lateral area and the area of its base.

Surface Area of a Regular Pyramid	**If a regular pyramid has a surface area of *T* square units, a slant height of *ℓ* units, and its base has a perimeter of *P* units and an area of *B* square units, then $T = \frac{1}{2}P\ell + B$.**

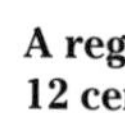

Example 2 **A regular pyramid has a slant height of 13 centimeters and a height of 12 centimeters. If the base is a regular pentagon, find the surface area of the pyramid.**

The slant height, the altitude, and the apothem form a right triangle. Use the Pythagorean Theorem to find the length of the apothem.

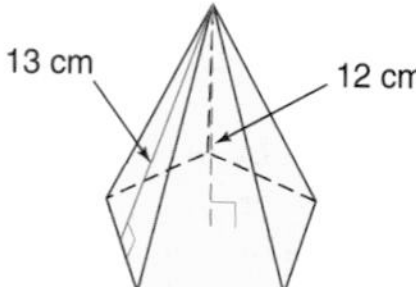

$$13^2 = 12^2 + a^2$$
$$25 = a^2$$
$$5 = a$$

(continued on the next page)

In-Class Examples

For Example 1
A regular triangular pyramid has a slant height of 12 units and a lateral edge of 13 units. Find its lateral area. $L = 180$ units^2

For Example 2
A regular pyramid has a slant height of 17 units and a height of 15 units. If the base is a square, find its surface area. 800 units^2

Teaching Tip Encourage students to break each problem down into smaller parts to ease solving.

Cooperative Learning

Brainstorming Have students work in teams to draw a pyramid on a piece of paper. Have teams critique each other's drawings to determine which ones are correct and why. They should then try to formulate a definition of a pyramid. For more information on the brainstorming strategy, see *Cooperative Learning in the Mathematics Classroom*, one of the titles in the Glencoe Mathematics Professional Series, page 30.

Teaching Tip Have students try to determine where these figures would be found or used.

Additional Answer

1. The lateral edge of a pyramid is the side of a triangle in which one endpoint is on the base, while the other is the vertex. The lateral edge of a prism is the side of a parallelogram, and its endpoints are both on bases.

LOOK BACK

You can review the trigonometric functions in Lesson 8-3.

Now find the length of the sides of the base of the pyramid. The central angle of the pentagon measures $\frac{360°}{5}$ or 72°. Let a represent the measure of the angle formed by a radius and the apothem. Thus, $a = \frac{72}{2}$ or 36.

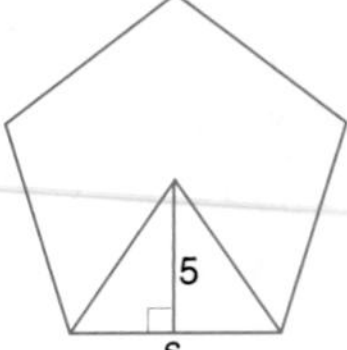

Use trigonometry to find the length of the sides.

$$\tan 36° = \frac{\frac{1}{2}s}{5}$$
$$5 \tan 36° = \frac{1}{2}s$$
$$10 \tan 36° = s$$
$$7.3 \approx s$$

Now use the formula to find the surface area of the pyramid.

$$T = \frac{1}{2}P\ell + B$$
$$\approx \frac{1}{2}(5 \cdot 7.3)(13) + \frac{1}{2}(5 \cdot 7.3)(5) \quad B = \frac{1}{2}Pa$$
$$\approx 328.5$$

The surface area of the pyramid is about 328.5 cm^2.

The figure at the right is a **circular cone**. Its **base** is a circle, and its **vertex** is at V. Its **axis**, $\overline{VX}$, is the segment whose endpoints are the vertex and the center of the base. The segment that has the vertex as one endpoint and is perpendicular to the base is called the **altitude** of the cone.

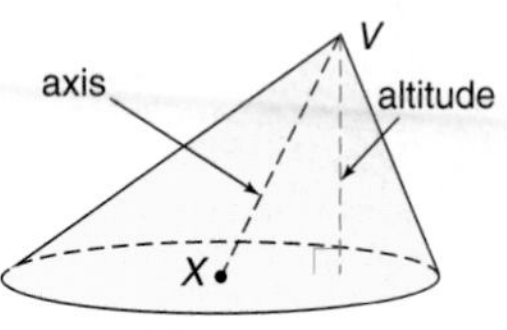

oblique cone

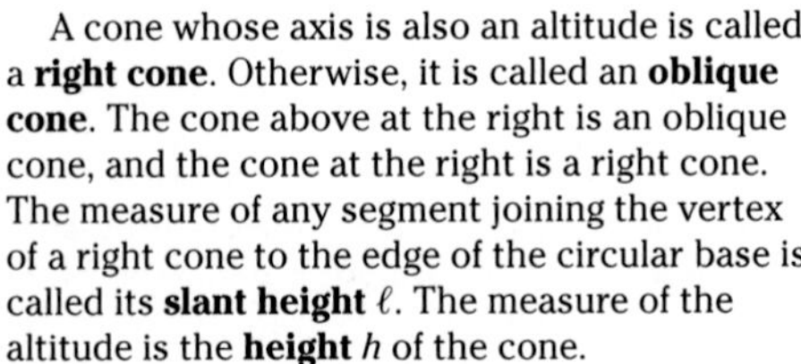

A cone whose axis is also an altitude is called a **right cone**. Otherwise, it is called an **oblique cone**. The cone above at the right is an oblique cone, and the cone at the right is a right cone. The measure of any segment joining the vertex of a right cone to the edge of the circular base is called its **slant height** ℓ. The measure of the altitude is the **height** h of the cone.

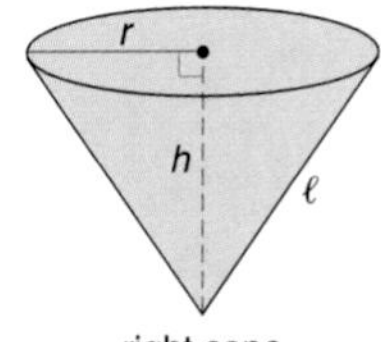

right cone

You will use formulas similar to the formulas for finding the lateral area and the surface area of a regular pyramid to find those same measures for a right cone. In the net for a cone, the region of the cone that is not the base is a sector of a circle whose radius is the slant height ℓ.

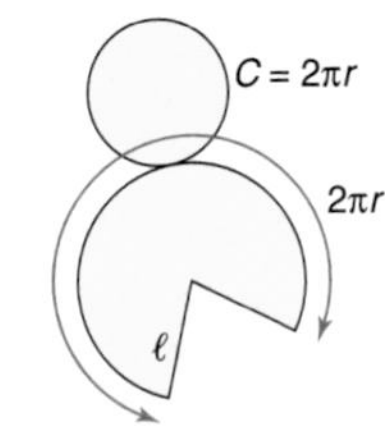

The area of the sector is proportional to the area of the circle. Notice that the arc length of this sector is equal to the circumference of the base of the original cone, $2\pi r$.

$$\frac{\text{area of sector}}{\text{area of circle}} = \frac{\text{measure of arc}}{\text{circumference of circle}}$$

$$\frac{\text{area of sector}}{\pi\ell^2} = \frac{2\pi r}{2\pi\ell}$$

$$\text{area of sector} = \pi r\ell \quad \textit{Multiply each side by } \pi\ell^2.$$

Lateral Area and Surface Area of a Right Circular Cone	If a right circular cone has a lateral area of L square units, a surface area of T square units, a slant height of ℓ units, and the radius of the base is r units, then $L = \pi r\ell$ and $T = \pi r\ell + \pi r^2$.

Example 3 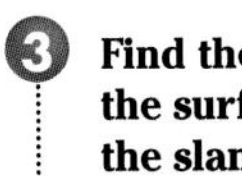

Find the lateral area and the surface area of a cone if the slant height is 13 feet and the diameter of the base is 10 feet.

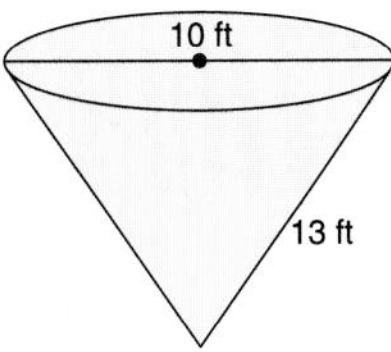

If the diameter is 10 feet, then the radius is $\frac{1}{2} \times 10$ or 5 feet.

$$\begin{aligned} L &= \pi r\ell \\ &= \pi(5)(13) \\ &= 65\pi \\ &\approx 204.2 \text{ ft}^2 \end{aligned}$$

$$\begin{aligned} T &= \pi r\ell + \pi r^2 \\ &= \pi(5)(13) + \pi(5)^2 \\ &= 90\pi \\ &\approx 282.7 \text{ ft}^2 \end{aligned}$$

CHECK FOR UNDERSTANDING

Communicating Mathematics

Study the lesson. Then complete the following. 1. See margin.

1. **Compare and contrast** the lateral edges of a pyramid and those of a prism.
2. **Analyze** the change in the shape of the base of a regular pyramid as the number of sides increases.
   a. What shape is it?
   b. As this transformation takes place, what happens to the shape of the pyramid? See margin.

2a. As the number of sides increases, the base resembles a circle.

3. **Sketch** a regular pyramid. Which is longer, a lateral edge or the slant height? Explain. See margin.
4. **Describe** a cone in which the axis is not also the altitude of the cone. See margin.

MODELING MATHEMATICS

5. Cut the side of a cone-shaped drinking cup from the brim to the vertex. Flatten out the cup.
   a. What is the shape of the surface? a circle with a sector missing
   b. Measure the radius and the angle formed by the cut edges.
   c. Find the area of the surface. How does the area relate to the formula for the lateral area of a cone? See students' work. The area of the surface is the same as the lateral area of the cone.

5b. See students' work.

Guided Practice

Determine whether the condition given is characteristic of a *pyramid* or *prism, both,* or *neither.*

6. The lateral faces are parallelograms. prism
7. It can have as few as five faces. both

In-Class Example

For Example 3

The So-Good Ice Cream Company makes Cluster Cones. For packaging, they must cover each cone with paper. If the diameter of the top of each cone is 6 cm and its slant height is 15 cm, what is the area of the paper necessary to cover one cone? about 141.4 cm²

3 PRACTICE/APPLY

Check for Understanding

Exercises 1–13 are designed to help you assess your students' understanding through reading, writing, speaking, and modeling. You should work through Exercises 1–5 with your students and then monitor their work on Exercises 6–13.

Error Analysis

Understanding the formulas in this lesson may be difficult. Reinforce them by using pictures and three-dimensional objects to develop the area formulas for each solid. Have students draw diagrams and use them to try to understand what is happening in this lesson.

Additional Answers

2b. As this takes place, the lateral faces become narrower so the lateral surface looks rounder.

3. The lateral edge is longer because it is the hypotenuse of the right triangle formed by the slant height, the lateral edge, and half of a side of the base.

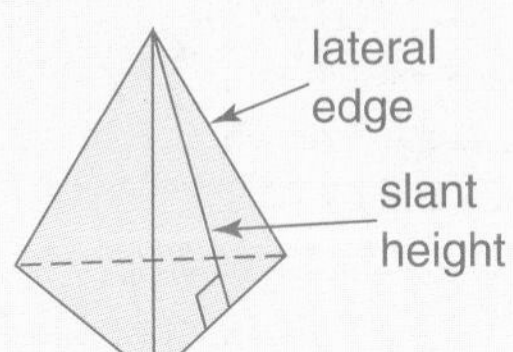

4. a cone that is oblique and not a right cone

Reteaching

Using Comparison Go through the terminology in this lesson and compare it with the terminology in Lessons 11-2 and 11-3. Similar ideas occur, such as regular or right solids as opposed to oblique solids, altitudes of the segments, vertices, etc. Select common terms and make sure students understand the connections.

Assignment Guide

Core (with proof): 15–35 odd, 37–44
Core (informal): 15–35 odd, 37–44
Enriched: 14–32 even, 33–44
All: Self Test, 1–5

For **Extra Practice**, see p. 786.

The red A, B, and C flags, printed only in the Teacher's Wraparound Edition, indicate the level of difficulty of the exercises.

Study Guide Masters, p. 67

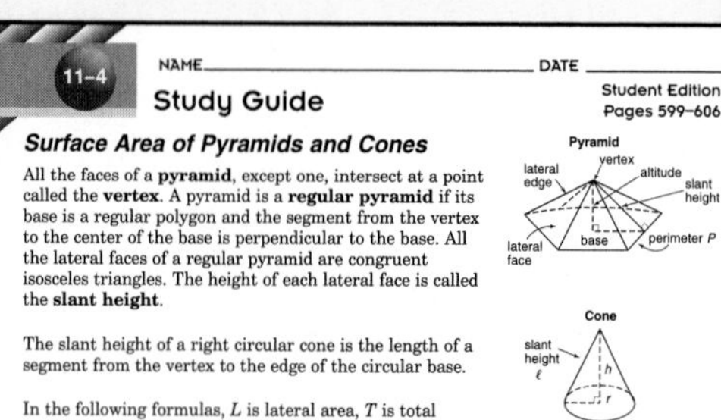

11-4 NAME__________ DATE ________

Study Guide — Student Edition Pages 599–606

Surface Area of Pyramids and Cones

All the faces of a **pyramid**, except one, intersect at a point called the **vertex**. A pyramid is a **regular pyramid** if its base is a regular polygon and the segment from the vertex to the center of the base is perpendicular to the base. All the lateral faces of a regular pyramid are congruent isosceles triangles. The height of each lateral face is called the **slant height**.

The slant height of a right circular cone is the length of a segment from the vertex to the edge of the circular base.

In the following formulas, L is lateral area, T is total surface area, P is perimeter, and ℓ is slant height.

Regular Pyramids $L = \frac{1}{2}P\ell$; $T =$ Lateral Area + Area of Base

Cones $L = \pi r\ell$; $T = \pi r\ell + \pi r^2$

Example: Find the surface area of the cone.

$T = \pi r\ell + \pi r^2$
$T = \pi(6)(10) + \pi(6)^2$
$T = 60\pi + 36\pi$
$T = 96\pi$ or about 301.6 cm

Find the lateral area and the surface area of each regular pyramid or right cone. Round your answers to the nearest tenth.

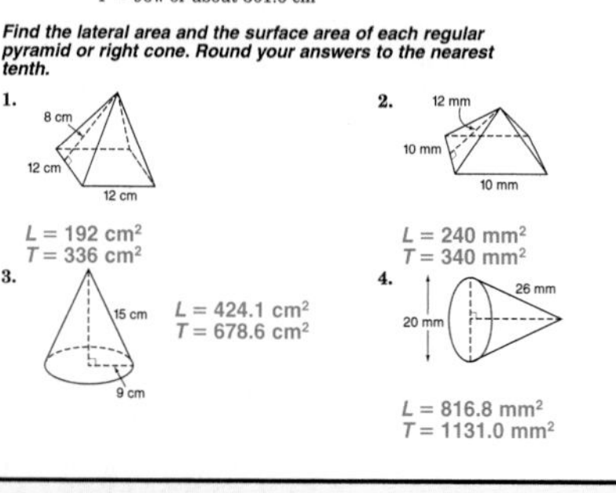

1. $L = 192$ cm^2, $T = 336$ cm^2
2. $L = 240$ mm^2, $T = 340$ mm^2
3. $L = 424.1$ cm^2, $T = 678.6$ cm^2
4. $L = 816.8$ mm^2, $T = 1131.0$ mm^2

Find the lateral area and the surface area of each regular pyramid or right cone. Round to the nearest tenth.

8. 70 cm^2; 119 cm^2
9. 47.1 m^2; 75.4 m^2
10. 120 ft^2; 181.9 ft^2

8. **9.**

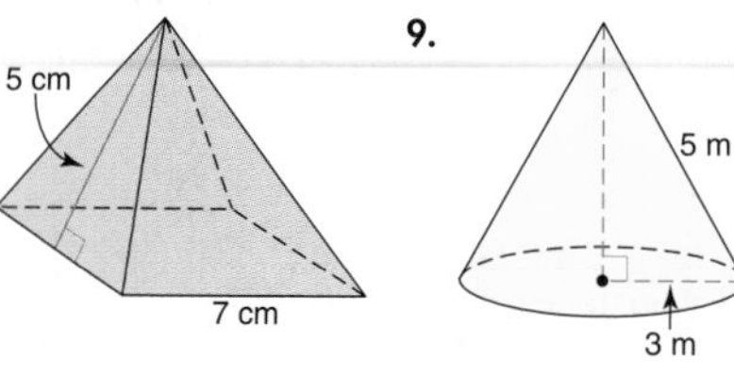

10.

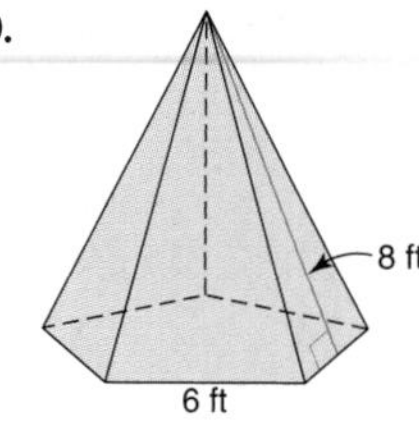

11. Find the surface area of the solid at the right. Round to the nearest tenth. 475.2 in^2

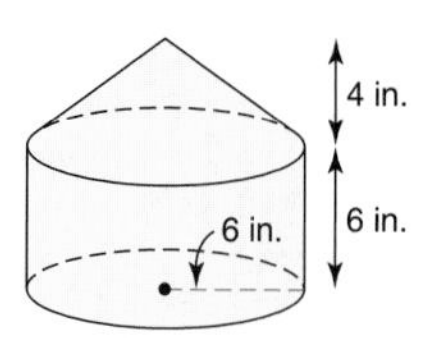

12. The base of a rectangular pyramid is 15 inches long and 8 inches wide. The height is 4 inches. Find the surface area of the pyramid if all of the lateral edges are congruent. 272.9 in^2

13. **History** Historians believe that the Great Pyramids of Egypt were once covered with gold or white stones that have worn away or have been removed to be used for other purposes. The diagram at the right shows the approximate dimensions of the Great Pyramid of Khufu. Find the surface area that would have to have been covered. 924,974.6 ft^2

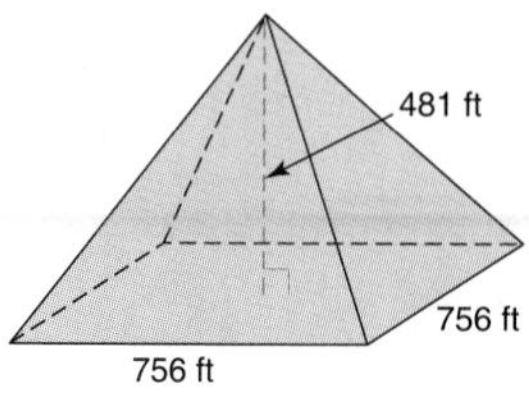

EXERCISES

Practice

Determine whether the condition given is characteristic of a *pyramid* or *prism*, *both*, or *neither*.

14. There is exactly one base. pyramid
15. It always has an even number of faces. neither
16. There are two bases. prism
17. The lateral faces are triangles. pyramid
18. It has the same number of lateral faces as vertices. neither
19. There can be as few as four faces. pyramid
20. The number of edges is always even. pyramid

Find the lateral area and the surface area of each regular pyramid or right cone. Round to the nearest tenth.

21. 284.3 in^2; 485.2 in^2
22. 240 cm^2; 340 cm^2
23. 81 cm^2; 133.6 cm^2

B

21.

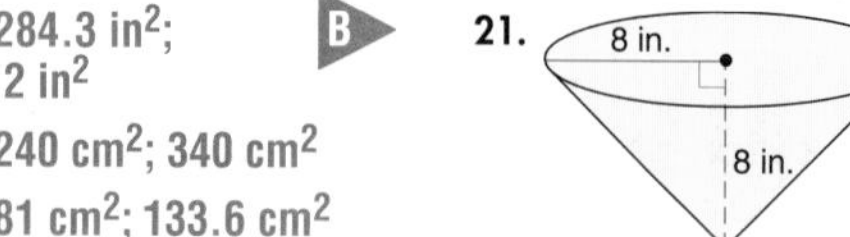

22.

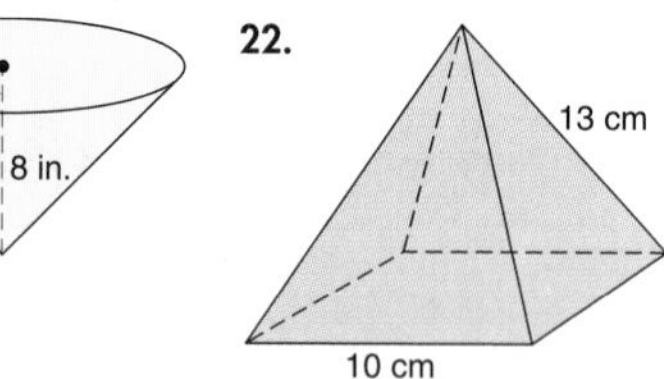

23.

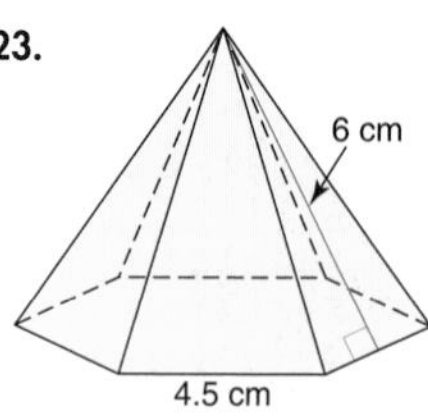

24. 95.2 cm²; 157.6 cm²
25. 188.5 ft²; 301.6 ft²
26. 204.2 m²; 282.7 m²

24.
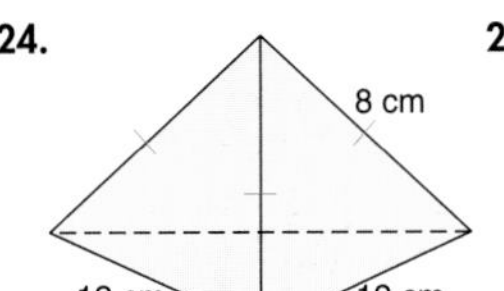

25.
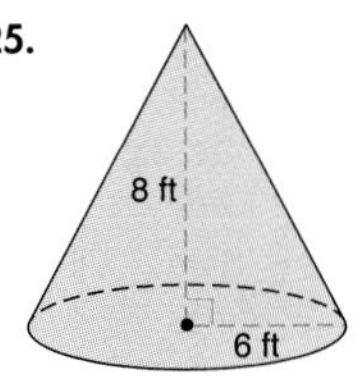

26.
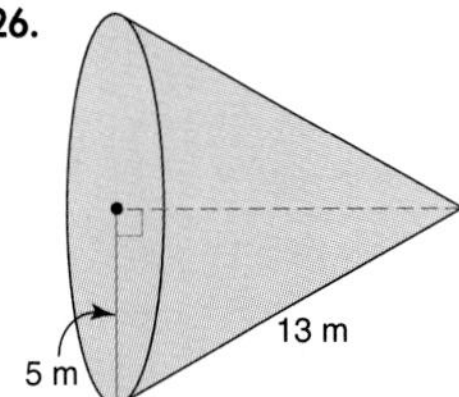

Find the surface area of each solid. Round to the nearest tenth.

27. 169.6 ft²
28. 112.9 in²
29. 3696 yd²

27.
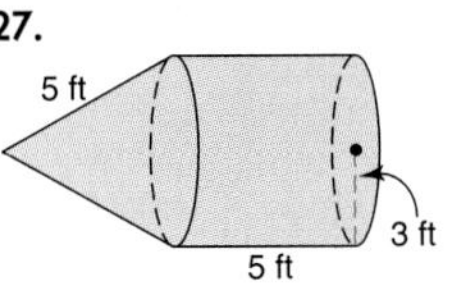

28.
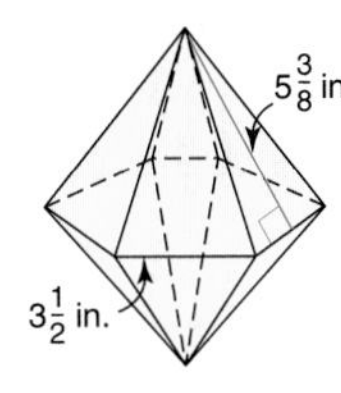

29.
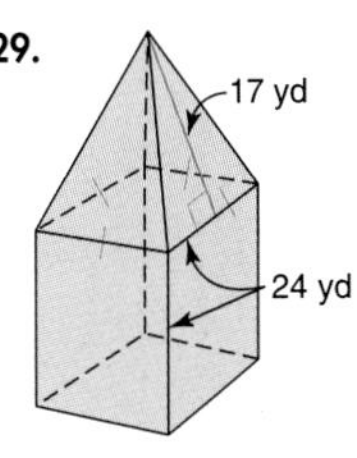

C

30. A regular pyramid has a slant height of 13 feet. The area of its square base is 100 square feet. Find its surface area. 360 ft²

31. In the given cube, A, B, and C are the vertices of the base of the pyramid with vertex D. If the edge of the cube is 8 units long, find the lateral area and the surface area of the pyramid. $L = 96$ units²; $T = 151.4$ units²

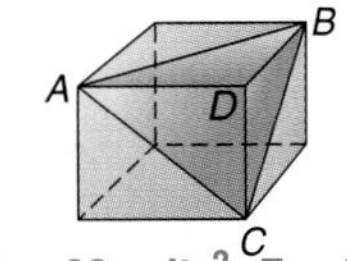

32. A *frustum* is the part of a solid that remains after the top portion of the solid has been cut off by a plane parallel to the base.

a. The figure below is a frustum of a regular pyramid. Find its lateral area. 36 yd²

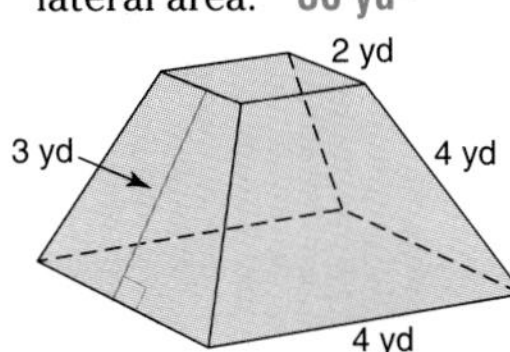

b. Find the surface area of the frustum of a cone shown below. 395.8 mm²

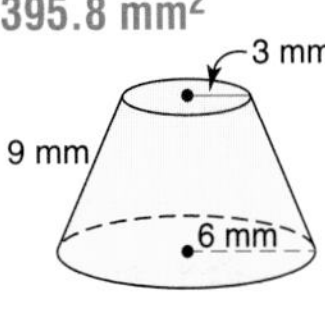

Critical Thinking

33. If you were to move the vertex of a right cone down the axis toward the center of the base, what would happen to the lateral area of the cone? Be as specific as possible and demonstrate the validity of your answer with a series of diagrams. See margin.

Applications and Problem Solving

34. **Dwellings** The largest tepee in the United States belongs to Dr. Michael Doss of Washington, D.C. Dr. Doss is a member of Montana's Crow Tribe. The tepee is a right cone with a diameter of 42 feet and a slant height of about 47.9 feet. How much canvas was used to cover the tepee? 3160.1 ft²

35. **Art** In 1921, Italian immigrant Simon Rodia bought a home in Los Angeles, California, and began building conical towers in his backyard. The structures are made of steel mesh and cement mortar with no rivets, bolts, or welds. The first tower completed was the East Tower, which stands 55 feet high. The diameter of the base of the East Tower is $8\frac{1}{2}$ feet. Find the lateral area of the tower. 736.5 ft²

Additional Answer

33. The lateral area approaches the area of the base. This can be seen by showing a series of cones cut on their slant heights and folded out into sectors. As the altitude approaches zero, the slant height approaches the radius of the base, and the sector narrows, approaching a complete circle. See students' diagrams.

Practice Masters, p. 67

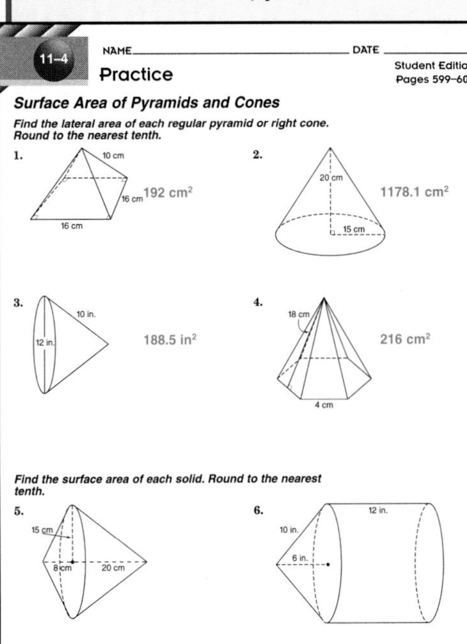

11-4 NAME____ DATE____

Practice — Student Edition Pages 599–606

Surface Area of Pyramids and Cones

Find the lateral area of each regular pyramid or right cone. Round to the nearest tenth.

1. 192 cm² (10 cm, 16 cm, 16 cm)
2. 1178.1 cm² (20 cm, 15 cm)
3. 188.5 in² (10 in., 12 in.)
4. 216 cm² (18 cm, 4 cm)

Find the surface area of each solid. Round to the nearest tenth.

5. 1979.2 cm² (15 cm, 8 cm, 20 cm)
6. 1055.6 in² (12 in., 10 in., 6 in.)

Extension

Connections How can you relate the formulas for the lateral area of a pyramid and for the lateral area of a cone? Both lateral areas are found by multiplying $\frac{1}{2}$ (perimeter or circumference) by the slant height ℓ.

4 ASSESS

Closing Activity

Speaking Have each student define one of the terms in this lesson in his or her own words or give one of the formulas in this lesson using words instead of letters.

Chapter 11 Quiz B (Lessons 11-3 and 11-4) is available in the *Assessment and Evaluation Masters*, p. 296.

Mid-Chapter Test (Lessons 11-1 through 11-4) is available in the *Assessment and Evaluation Masters*, p. 295.

Additional Answer

38.

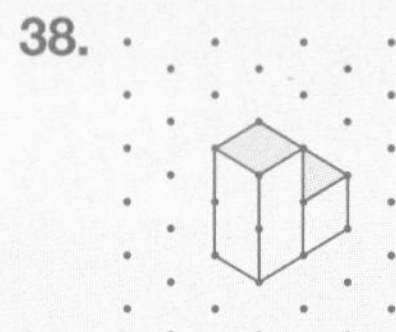

Enrichment Masters, p. 67

11-4 NAME_______________ DATE_______ Student Edition Pages 599–606

Enrichment

Conic Sections

A double conical surface is formed by all the lines ℓ that pass through point P (not in the plane of $\odot C$) and that intersect $\odot C$. Each such line is an **element** of the surface. Cross sections formed by planes that cut the surface are called **conic sections**.

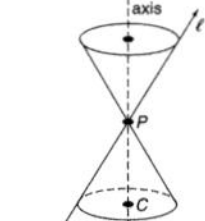

A circle or ellipse is formed by a plane that intersects one cone and is not parallel to an element. A **parabola** is formed by a plane parallel to an element. A **hyperbola** is formed by a plane that intersects both cones. Study the figures below.

Circle Ellipse Parabola Hyperbola

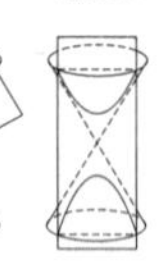

Answer each question.

1. Points E and F are the same distance from the vertex, G. Which conic section would be formed by a plane through points E and F that is perpendicular to the axis? circle
2. Which conic section would be formed by a plane through points H and K that is parallel to edge FG. parabola
3. Which conic section would be formed by a plane through points D and F that intersects only the top cone? ellipse
4. Which conic section would be formed by a plane through points B, C, I, and J? hyperbola
5. What type of figure would be formed by a plane that contains the axis? intersecting lines

36. **History** The Cahokia Mounds stand close to East St. Louis, Illinois, where there was once the largest ancient city in America, north of Mexico. The inhabitants began building 120 earthen pyramids there around A.D. 600. The largest mound, Monk's Mound, has a height of 30.5 meters and a rectangular base with sides 216.6 and 329.4 meters long. Find the lateral area of Monk's Mound. (*Hint:* Is Monk's Mound a regular pyramid?)

36. 73,342.3 m^2

Mixed Review

37. If the lateral area of a right rectangular prism is 784 square centimeters, its length is three times its width, and its height is twice its width, find its surface area. (Lesson 11–3) 1078 cm^2

38. From the views given below, draw a corner view. (Lesson 11–1)

38. See margin.

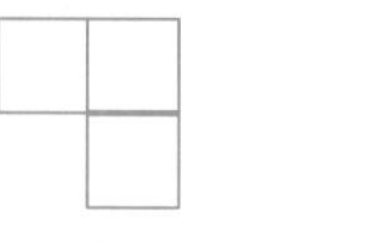
top view

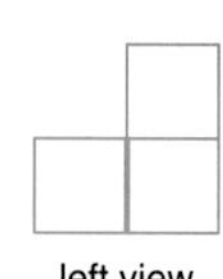
left view

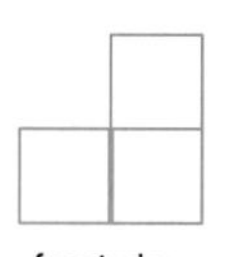
front view

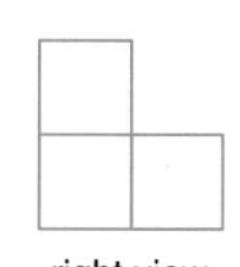
right view

39. Find the area of a regular nonagon with an apothem 12.2 centimeters long and a side 10.4 centimeters long. (Lesson 10–5) 570.96 cm^2
40. The diameter of a circle measures 10 centimeters, and the length of a chord is 8 centimeters. Find the distance from the chord to the center of the circle. (Lesson 9–3) 3 cm
41. **Photography** Kandhi is taking pictures at the zoo. If a gnu is 4.3 feet tall, the film is 1 inch from the camera lens, and the camera lens is standing 5 feet from the gnu, how tall will the gnu's image be on the film? (Lesson 7–5)
42. Graph $(-6, -3)$ and $(1, 8)$. Then draw the line that passes through the points. Through which quadrants does the line pass? (Lesson 1–1)

41. 0.86 inches

42. See Solutions Manual for graph. The line passes through quadrants I, II, and III.

INTEGRATION Algebra

43. Translate the sentence *The quantity x is equal to 18 more than the square of b* into an equation. $x = b^2 + 18$
44. Use elimination to solve the system of equations. $(1, -4)$

$2x + 4y = -14$

$3x - 5y = 23$

SELF TEST

1. The corner view of a figure is given at the right. Draw the top, left, front, right, and back views of the figure. (Lesson 11–1) 1–2. See Solutions Manual.
2. Use isometric dot paper to draw a corner view of a rectangular solid 2 units high, 4 units long, and 2 units wide. Then draw a net of the solid on rectangular dot paper. (Lesson 11–2)

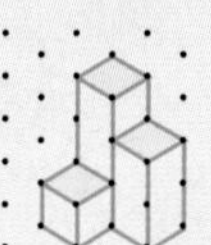

Find the lateral area and the surface area of each solid below. Round to the nearest tenth. (Lesson 11–3)

3.

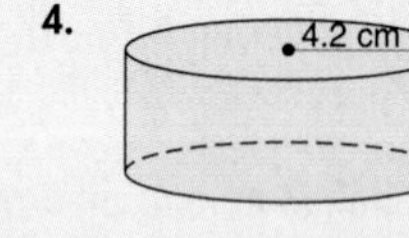

$L = 432$ in^2; $T = 619.1$ in^2

4.
4.2 cm
3.1 cm

$L = 81.8$ cm^2; $T = 192.6$ cm^2

5. **Architecture** The Transamerica Tower in San Francisco is a regular pyramid with a square base that is 149 feet on each side and a height of 853 feet. Find its lateral area. (Lesson 11–4)

5. about 255,161.7 ft^2

SELF TEST

The Self Test provides students with a brief review of the concepts and skills in Lessons 11-1 through 11-4. Lesson numbers are given to the right of exercises or instruction lines so students can review concepts not yet mastered.

11-5 Volume of Prisms and Cylinders

What YOU'LL LEARN

- To find the volume of a right prism, and
- to find the volume of a right cylinder.

Why IT'S IMPORTANT

Prisms and cylinders are common shapes used for containers.

The measure of the amount of space that a figure encloses is the volume of the figure. Volume is measured in cubic units. You can use the skills you learned about creating solid figures from different views of the figure to investigate the volume of a right rectangular prism.

MODELING MATHEMATICS — Volume of a Right Rectangular Prism

Materials:  small cubes

- Make a model of a prism for the left, front, and top views given below.

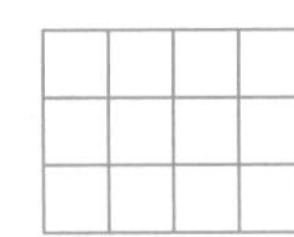
left view

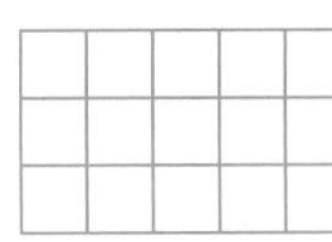
front view

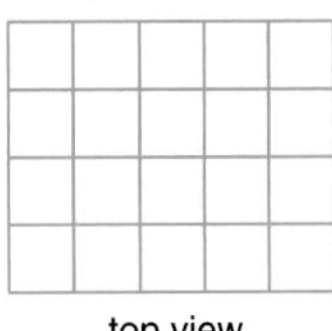
top view

- Disassemble the prism and count the number of cubes that make up the prism.
- Multiply the length, width, and height of the prism.

Your Turn b. See students' work.

a. How do the number of cubes used to make the prism and the product of the length, width, and height of the prism compare? They are the same.

b. Repeat the activity using a different prism and describe your results.

c. Write a formula for the volume of a right rectangular prism. $V = \ell wh$

The Modeling Mathematics activity leads us to the following formula for the volume of a right prism.

Volume of a Right Prism	If a right prism has a volume of V cubic units, a base with an area of B square units, and a height of h units, then $V = Bh$.

Example  **Find the volume of the right triangular prism shown at the right.**

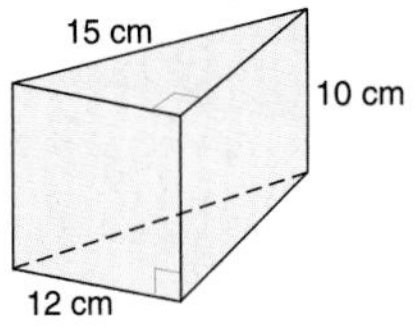

First, use the Pythagorean Theorem to find the height of the base of the prism.

$$a^2 + 12^2 = 15^2$$
$$a^2 = 15^2 - 12^2$$
$$a^2 = 81$$
$$a = 9$$

(continued on the next page)

MODELING MATHEMATICS Have students determine the perimeter and surface area of the right rectangular prism to emphasize the difference between these measurements and volume.

11-5 LESSON NOTES

NCTM Standards: 1–5, 7

Instructional Resources

- Study Guide Master 11-5
- Practice Master 11-5
- Enrichment Master 11-5
- Multicultural Activity Masters, p. 22
- Tech Prep Applications Masters, p. 21

Transparency 11-5A contains the 5-Minute Check for this lesson; **Transparency 11-5B** contains a teaching aid for this lesson.

Recommended Pacing	
Standard Pacing	Days 9 & 10 of 18
Honors Pacing	Day 9 of 16
Block Scheduling*	Day 6 of 10

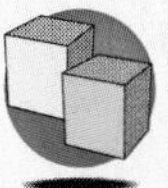

*For more information on pacing and possible lesson plans, refer to the *Block Scheduling Booklet.*

1 FOCUS

5-Minute Check *(over Lesson 11-4)*

Refer to the figures below.

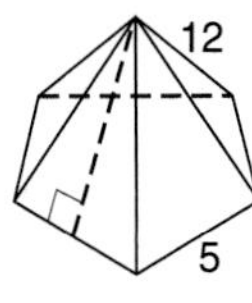

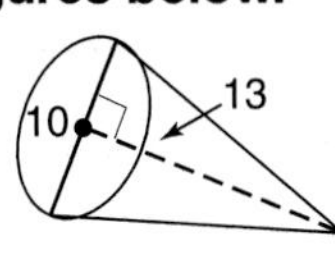

1. Find the slant height of the regular pyramid. about 11.7 units
2. Find the lateral area of the regular pyramid. about 146.7 units^2
3. Find the slant height of the right circular cone. about 13.9 units
4. Find the lateral area of the right circular cone. about 218.8 units^2
5. Find the surface area of the right circular cone. about 297.3 units^2

Motivating the Lesson

Questioning Ask students what they think the word *volume* means. What is the difference between area and volume? **Sample answer: Area is two-dimensional and volume is three-dimensional.** Point out that volume is the amount of space that a solid takes up.

2 TEACH

In-Class Examples

For Example 1
Find the cubic yards of cement that are required for a 60-foot-long driveway that is 8 inches thick and 20 feet wide.
29.6 yd^3

For Example 2
To resurface a football field, it was dug 6 inches deep. How many cubic feet of dirt are necessary to fill the field? A football field is 65 yards wide and 120 yards long.
35,100 ft^3

Teaching Tip Explain the relationship between linear units cubed and volume units in the metric system. For example, 1 cm^3 = 1 milliliter.

Now find the volume of the prism.

$V = Bh$

$V = \left(\frac{1}{2} \cdot 12 \cdot 9\right) \cdot 10$

$V = 540 \text{ cm}^3$

The volume formula can be used to solve real-life problems that involve objects shaped like prisms.

Example 2

Before the athletes could hit the water for the 1996 Olympic Games, the new main competition pool at the Georgia Tech Aquatic Center had to be filled. The pool is 50 meters long and 25 meters wide. The adjustable bottom of the pool can be up to 3 meters deep for competition and as shallow as 0.3 meter deep for recreation. A liter of water has a volume of 0.001 cubic meter. Suppose the pool was filled to the recreational level and then the floor was lowered to the competition level. How much water had to be added to fill the pool?

Explore We know the dimensions of the pool when the floor is set for competition and when it is set for recreation. We also know the amount of water needed for a cubic meter of volume. We need to find the amount of water needed to fill the pool.

Plan First find the volume of water needed to fill the pool for competition and the volume of water needed to fill the pool for recreation. Find the difference in the two volumes. Then determine the number of liters of water needed to fill the difference in volume.

Solve When the pool is set for competition, it is a right rectangular prism 50 meters long, 25 meters wide, and 3 meters deep. The pool is 50 meters long, 25 meters wide, and 0.3 meters deep when it is set for recreation.

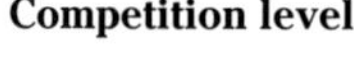

Competition level

$V = Bh$

$V = (50 \cdot 25)(3)$

$V = 3750 \text{ m}^3$

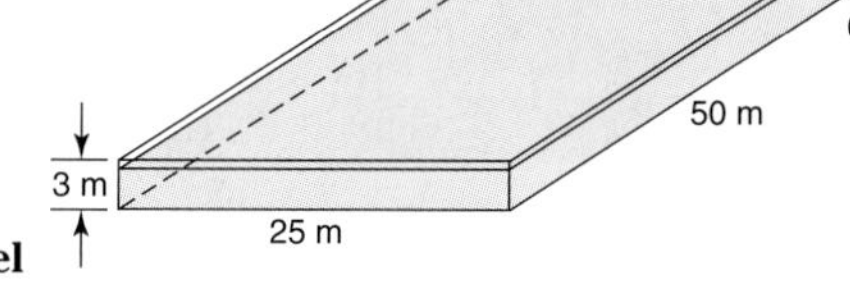

Recreation level

$V = Bh$

$V = (50 \cdot 25)(0.3)$

$V = 375 \text{ m}^3$

The difference in volumes is 3750 − 375 or 3375 m^3.

A liter of water occupies 0.001 cubic meter of volume. Thus, 3375 ÷ 0.001 or 3,375,000 liters of water were needed to fill the pool.

Examine Draw a model of the prism of water that needs to be added. Then find its volume and amount of water needed to confirm the solution.

Changing the dimensions of a prism affects the surface area and the volume of the prism. You can investigate the changes by using a spreadsheet.

EXPLORATION — SPREADSHEETS

Remember that a spreadsheet is like a table in which you enter data and formulas to calculate desired quantities.

- Use column A to enter the length, column B to enter the width, and column C to enter the height of the prism.
- Enter the formula (2A1+ 2B1)*C1 + 2*(A1*B1) in cell D1. This formula will find the surface area of the prism. Copy the formula into the other cells in column D.
- Write a formula to find the volume of the prism. Enter the formula in the cells in column E.
- Use a spreadsheet to find the surface areas and volumes of prisms with the dimensions given below in columns A, B, and C.

	A	B	C	D	E
1	1	2	3	22	6
2	2	4	6	88	48
3	3	6	9	198	162
4	4	8	12	352	384
5	8	16	24	1408	3072

- Print the spreadsheet. If a printer is not available, copy the information into a table.

Your Turn

a. Compare the dimensions of prisms 1 and 2, prisms 2 and 4, and prisms 4 and 5. The dimensions are doubled.

b. Compare the surface areas of prisms 1 and 2, prisms 2 and 4, and prisms 4 and 5.

c. Compare the volumes of prisms 1 and 2, prisms 2 and 4, and prisms 4 and 5.

d. Write a statement about the change in the surface area and volume of a prism when the dimensions are doubled. See margin.

b. The surface area of the larger prism is four times the surface area of the smaller prism.

c. The volume of the larger prism is eight times the volume of the smaller prism.

Like a right prism, the formula for the volume of a right cylinder is $V = Bh$. The area of the base is the area of a circle, πr^2.

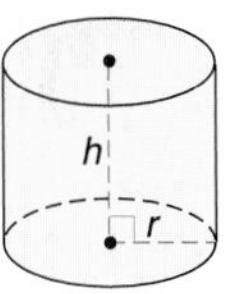

$$\text{volume} = \text{area of base} \times \text{height}$$
$$V = Bh$$
$$V = \pi r^2 \cdot h$$
$$\text{or } \pi r^2 h$$

Volume of a Right Cylinder	**If a right cylinder has a volume of *V* cubic units, a height of *h* units, and a radius of *r* units, then $V = \pi r^2 h$.**

EXPLORATION

This activity encourages students to use a spreadsheet to discover a pattern in the surface area and volume of a prism. Encourage students to expand the spreadsheet if they need more examples.

Answer for the Exploration

d. The surface area is multiplied by four and the volume is multiplied by eight.

In-Class Example

For Example 3

There are 150 1-inch washers in a box. When the washers are stacked, they measure 9 inches in height. If the inside hole of each washer has a diameter of $\frac{3}{4}$ inch, find the volume of metal in one washer. **Each washer has about 0.02 in^3 of metal.**

Teaching Tip In In-Class Example 3, point out that you do not multiply the number of washers, just the area represented by one washer multiplied by the height of the stack.

3 PRACTICE/APPLY

Check for Understanding

Exercises 1–13 are designed to help you assess your students' understanding through reading, writing, speaking, and modeling. You should work through Exercises 1–5 with your students and then monitor their work on Exercises 6–13.

Error Analysis

Students may have difficulty understanding the concept of volume. If this is the case, explain that the surface area of a figure surrounds the volume of the figure.

Additional Answers

2. The volumes of two congruent geometric solids are equal.
3. You find the volume of both a prism and a cylinder by using the formula $V = Bh$. However, the base of a cylinder is always a circle, so its formula can be written as $V = \pi r^2h$.
4. Kiki is correct. A cubic yard is 3 feet long, 3 feet wide, and 3 feet high, so the volume is $3(3 \cdot 3)$ or 27 cubic feet.

Example 3 **Find the volume of each right cylinder. Round to the nearest tenth.**

a.

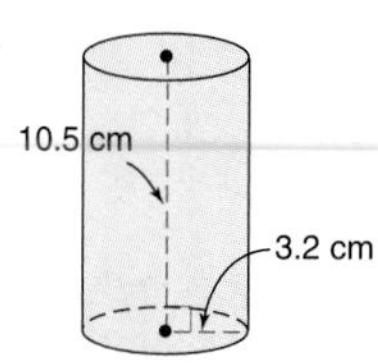

In this cylinder, the height h is 10.5 cm and the radius r is 3.2 cm.

$$V = \pi r^2h$$
$$= \pi(3.2)^2(10.5)$$
$$= 107.52\pi$$
$$\approx 337.8 \text{ cm}^3$$

b.

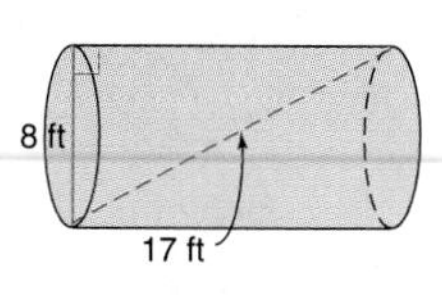

The diameter of the base, the diagonal, and the lateral edge of the cylinder form a right triangle. Use the Pythagorean Theorem to find the height.

$$h^2 + 8^2 = 17^2$$
$$h^2 = 17^2 - 8^2$$
$$h^2 = 225$$
$$h = 15$$

Now find the volume.

$$V = \pi r^2h$$
$$= \pi(4)^2(15)$$
$$= 240\pi$$
$$\approx 754.0 \text{ ft}^3$$

CHECK FOR UNDERSTANDING

Communicating Mathematics

Study the lesson. Then complete the following.

1. **Research** to find three items that are sold by volume. State the units of measure used in their sale. **1. Sample answers: soft drinks are sold by the liter, cement is sold by the cubic yard, and gasoline is sold by the gallon.**
2. **Describe** the relationship between the volumes of two geometric solids that are the same size and shape. **See margin.**
3. **Compare and contrast** the procedures for finding the volume of a prism and finding the volume of a cylinder. **See margin.**
4. **You Decide** Shalena says there are 3 cubic feet in a cubic yard. Kiki thinks there are 27 cubic feet in a cubic yard. Who is correct and why? **See margin.**

5. Make a model of a prism for the top, front, and right views given below.

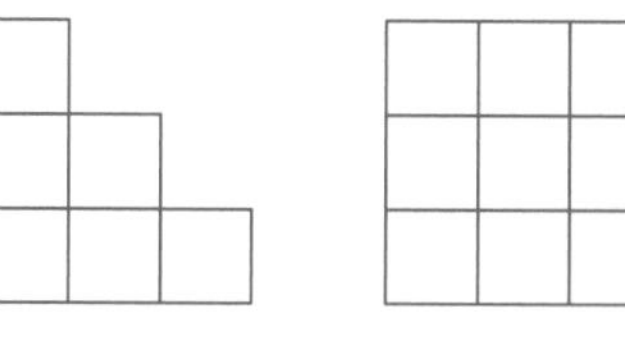

top view front view

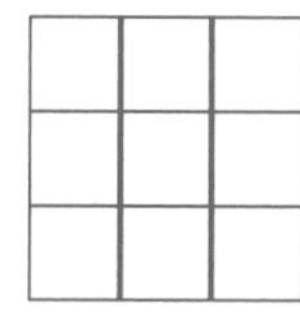

right view

a. Count the number of cubes used to build the figure. **18 cubes**
b. Compute the volume using the formula for the volume of a prism. **5b. 6(3) or 18 cubic units**
c. Does the number of cubes you counted verify the formula? If not, why not? **Yes, counting verifies the formula.**

Reteaching

Using Hands-On Activity Bring examples of prisms and cylinders to class and have students measure the solids and find their volumes.

Guided Practice **Find the volume of each right prism or right cylinder. Round to the nearest tenth.**

6.
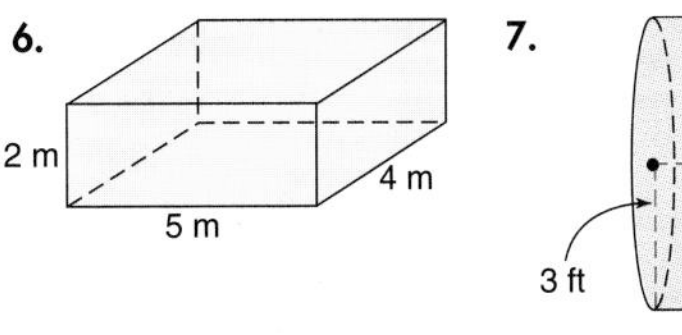

40 m^3

7.
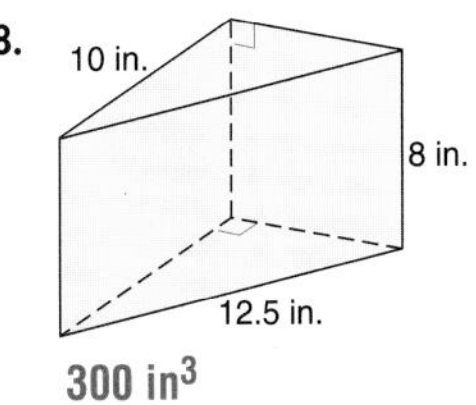

127.2 ft^3

8.
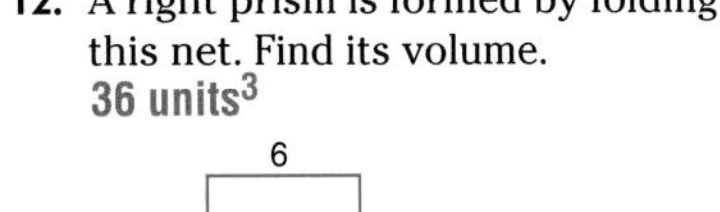

300 in^3

9. What is the volume of a right cylinder whose diameter is 12 yards long and has a height of 15 yards? Round to the nearest tenth. about 1696.5 yd^3

10. Find the volume of a right hexagonal prism that has a height of 10 centimeters and whose base is a regular hexagon with sides of 6 centimeters. Round to the nearest cubic centimeter. about 935 cm^3

11. Find the volume of the partial right cylinder shown below. Round to the nearest tenth.

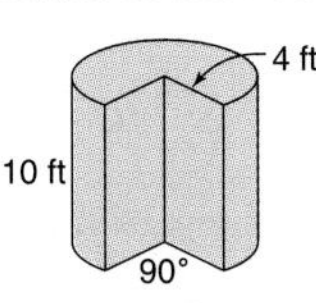

377.0 ft^3

12. A right prism is formed by folding this net. Find its volume. 36 $units^3$

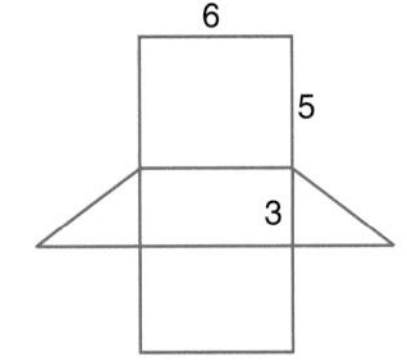

13. **Meteorology** On April 1, 1960, the world's first weather satellite, *TIROS 1*, was launched. *TIROS 1* orbited Earth taking pictures as it passed over sites. The weather satellites in service today are often placed in orbits so that they are over the same area all of the time. One of these satellites is cylindrical with a diameter of 7 feet and a height of 12 feet. What is the volume available for carrying weather instruments and other hardware? 461.8 ft^3

EXERCISES

Practice **Find the volume of each right prism or right cylinder. Round to the nearest tenth.**

16. 2752 cm^3

14.
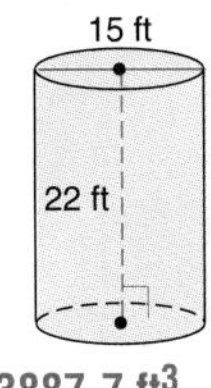

3887.7 ft^3

15.
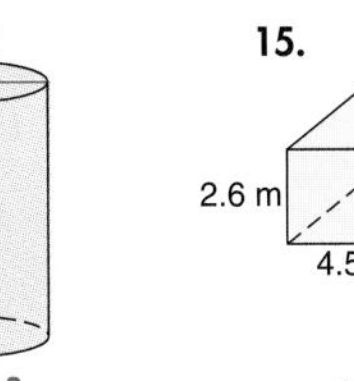

98.3 m^3

16.
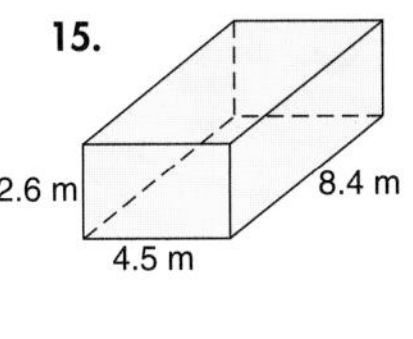
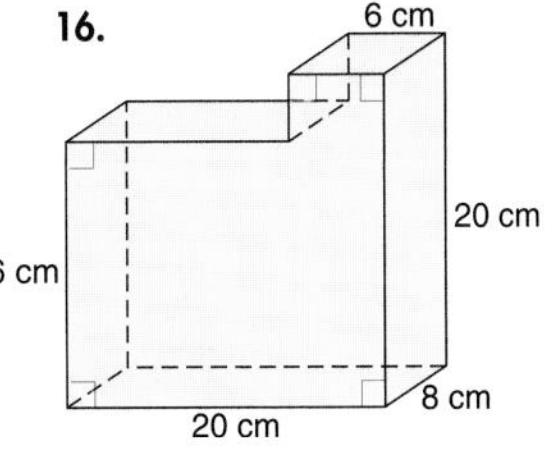

17.
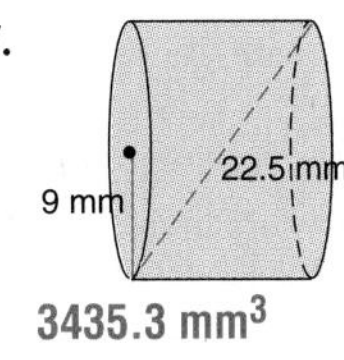

3435.3 mm^3

18.
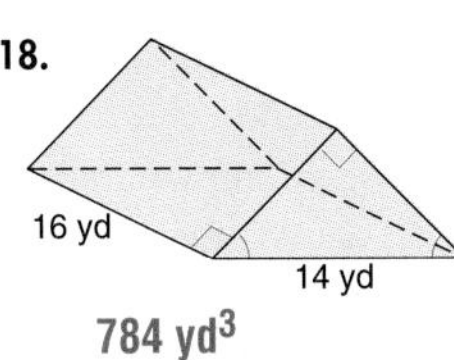

784 yd^3

19.
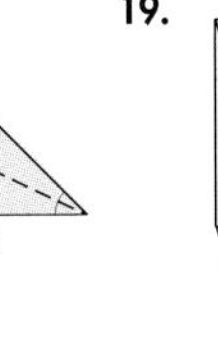

1800 in^3

Assignment Guide

Core (with proof): 15–37 odd, 38–46
Core (informal): 15–37 odd, 38–46
Enriched: 14–32 even, 33–46

For **Extra Practice**, see p. 786.

The red A, B, and C flags, printed only in the Teacher's Wraparound Edition, indicate the level of difficulty of the exercises.

Study Guide Masters, p. 68

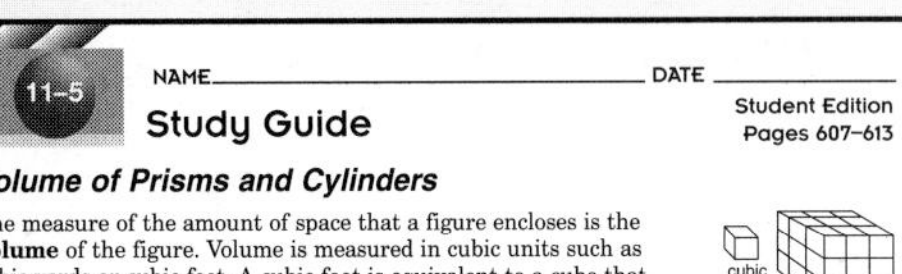

NAME______ DATE______

11-5 Study Guide

Student Edition Pages 607–613

Volume of Prisms and Cylinders

The measure of the amount of space that a figure encloses is the **volume** of the figure. Volume is measured in cubic units such as cubic yards or cubic feet. A cubic foot is equivalent to a cube that is 1 foot long on each side. A cubic yard is equivalent to 27 cubic feet.

Volume of a Right Prism	If a right prism has a volume of V cubic units, a base of B square units, and a height of h units, then $V = Bh$.
Volume of a Right Cylinder	If a right cylinder has a volume of V cubic units, a height of h units, and a radius of r units, then $V = \pi r^2 h$.

Examples: Find the volume of each right prism or right cylinder.

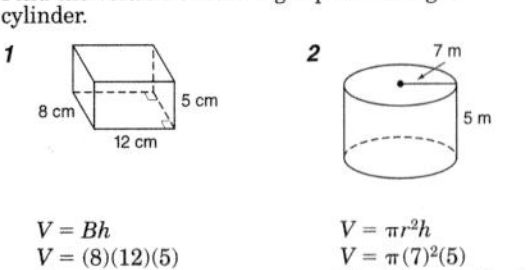

1 $V = Bh$
$V = (8)(12)(5)$
$V = 480\ cm^3$

2 $V = \pi r^2 h$
$V = \pi(7)^2(5)$
$V = 245\pi$ or about 769.7 m^3

Find the volume of each right prism or right cylinder. Round your answers to the nearest tenth.

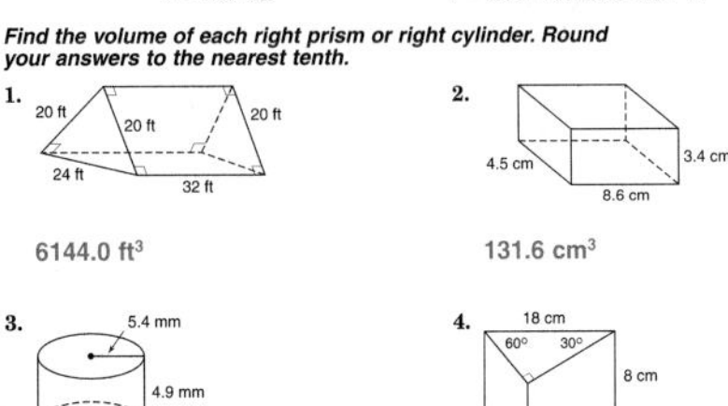

1. 6144.0 ft^3

2. 131.6 cm^3

3. 448.9 mm^3

4. 561.2 cm^3

Additional Answers

32a. For cylinders A and B, and B and C: the height of the larger cylinder is twice the height of the smaller cylinder. For cylinders A and D, and D and E: the radius of the larger cylinder is twice the radius of the smaller cylinder.

32c. For cylinders A and B, and B and C: the volume of the larger cylinder is twice the volume of the smaller cylinder. For cylinders A and D, and D and E: the volume of the larger cylinder is four times the volume of the smaller cylinder.

32d. If the height of a cylinder is doubled, the surface area and volume are both doubled. If the radius of a cylinder is doubled, the surface area is doubled and the volume is quadrupled.

Practice Masters, p. 68

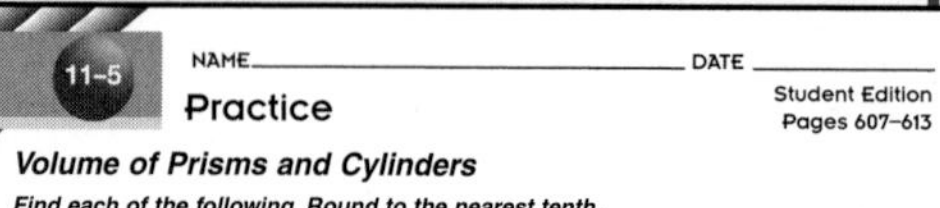

11-5 NAME__________ DATE ______

Practice — Student Edition Pages 607–613

Volume of Prisms and Cylinders

Find each of the following. Round to the nearest tenth.

1. the volume of a right prism whose square base has sides of 4 feet and whose height is 9 feet 144.0 ft^3

2. the volume of a cylinder with a height of 2 meters and a radius of 0.5 meters 1.6 m^3

Find the volume of each solid. Round to the nearest tenth.

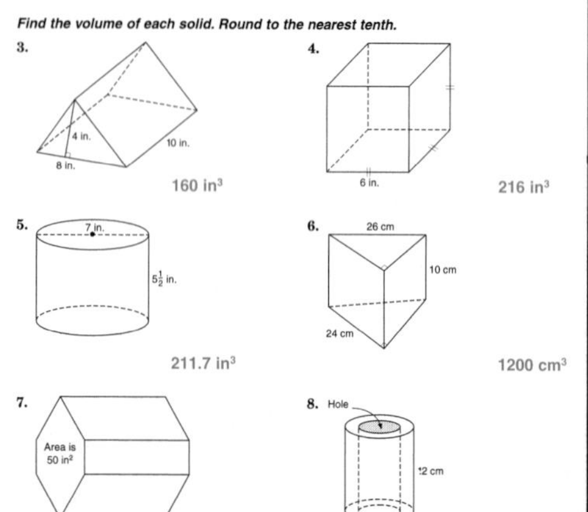

20. What is the volume of a cube that has an edge of 6 meters long? 216 m^3
21. The base of a right prism has an area of 25 square meters. The prism is 4.2 meters high. Find the volume of the prism. 105 m^3
22. Find the volume of a right cylinder whose diameter is 12 centimeters and height is 8 centimeters. about 904.8 cm^3
23. A regular hexagonal prism has a length of 40 feet and a base that is a regular hexagon with sides 5 feet long. Find the volume of the prism. 2598.1 ft^3
24. A cylinder has a height of 4 meters and a volume of about 452.4 cubic meters. What is the diameter of the base of the cylinder? 12 m
25. Find the length of a lateral edge of a right prism with a volume of 648 cubic inches and a base whose area is 36 square inches. 18 in.

Find the volume of each solid.

C

26. 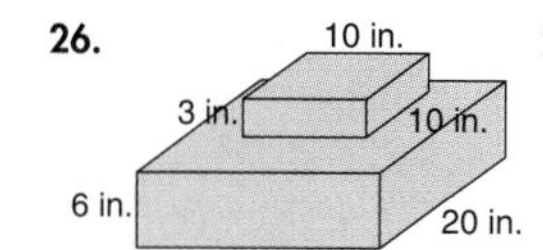

2700 in^3

27. 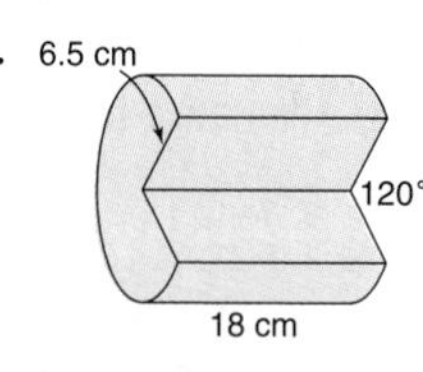

1592.8 cm^3

28. 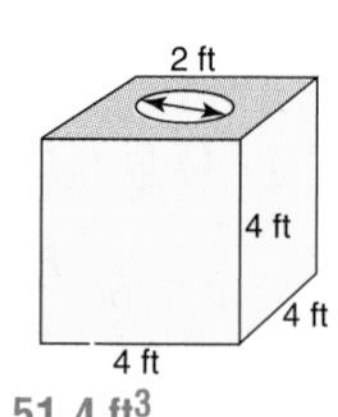

51.4 ft^3

Find the volume of the solid formed by each net.

29. 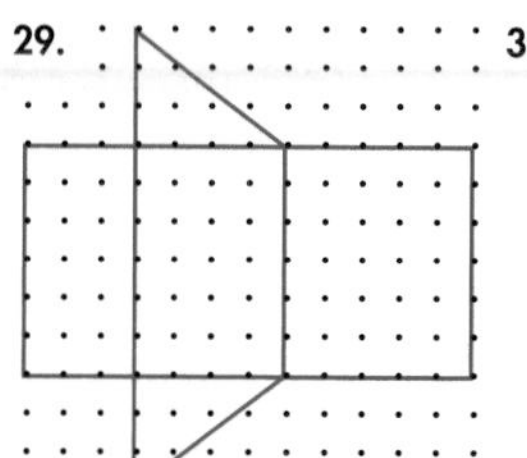

29. 36 $units^3$

30. 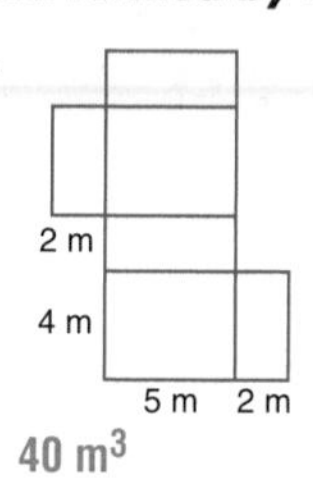

40 m^3

31. 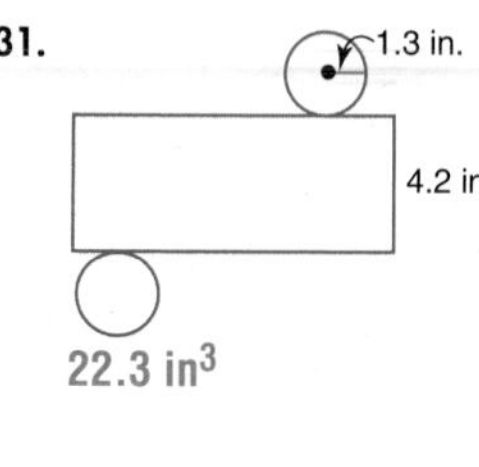

22.3 in^3

32. **Technology** Use a spreadsheet to investigate the change in the surface area and volume of a right cylinder when the dimensions are changed. Use column A to name cylinders A, B, C, D, and E. Use column B to enter the radius of the base, column C for the height, column D for the lateral area, and column E for the volume. Complete the spreadsheet for the values given below.

Cylinder	Radius	Height	Lateral Area	Volume
A	1	2	12.6	6.3
B	1	4	25.1	12.6
C	1	8	50.3	25.1
D	2	2	25.1	25.1
E	4	2	50.3	100.5

a. How do the dimensions of cylinders A and B, B and C, A and D, and D and E compare? See margin.
b. Compare the lateral areas of cylinders A and B, B and C, A and D, and D and E.
c. Compare the volumes of cylinders A and B, B and C, A and D, and D and E.
d. Write a statement about the change in the lateral area and volume of a cylinder when its height is doubled, and a statement about the change in the lateral area and volume of a cylinder when its radius is doubled.

32b. The lateral area of the larger cylinder is twice the lateral area of the smaller cylinder.

32c–d. See margin.

Critical Thinking

33. *True* or *false*: The volume of every oblique prism or oblique cylinder is equal to the area of its base times it height. Justify your answer. See margin.

Applications and Problem Solving

34. **Engineering** Machinists make parts for complicated pieces of equipment for manufacturing. While making a part for a weaving machine, a machinist drilled a hole in a block of copper as shown at the right.

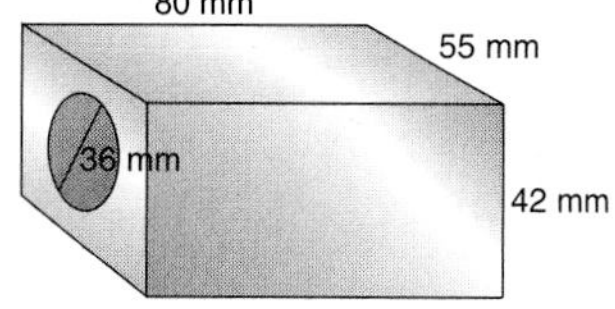

a. Find the volume of the resulting solid.

b. The *density* of a substance is its mass per unit of volume. At room temperature, the density of copper is 8.9 grams per cubic centimeter. What is the mass of this block? 920 grams

34a. 103369.9 mm^3 or 103.4 cm^3

35. **Music** To play a concertina, you push and pull the end plates and press the keys. Then the air pressure causes vibrations of the metal reeds that make the notes. When fully expanded, the concertina at the right is 36 inches from one endplate to the other. If the concertina is compressed, it is 7 inches from one endplate to the other. Find the volume of air in the instrument when it is fully expanded and when it is compressed. (*Hint*: There is no air in the endplates.)

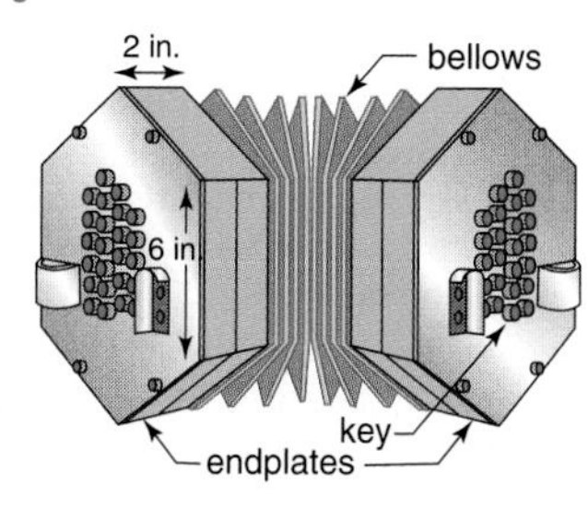

35. fully expanded: 2993.0 in^3; fully compressed: 280.6 in^3

36. 775,710 gallons

36. **Sports** The diving pool at the Georgia Tech Aquatic Center was used for the springboard and platform diving competitions of the 1996 Olympic Games. The pool measures 78 feet wide, 78 feet long, and 17 feet deep. If it takes about $7\frac{1}{2}$ gallons of water to fill one cubic foot of space, approximately how many gallons of water are needed to fill the diving pool?

37. **Automotive Engineering** A car muffler reduces the noise that the exhaust makes when it is released. Study the diagram of the muffler at the right. If each of the louver tubes has a diameter of 3 inches, find the total volume of the louver tubes. 381.7 in^3

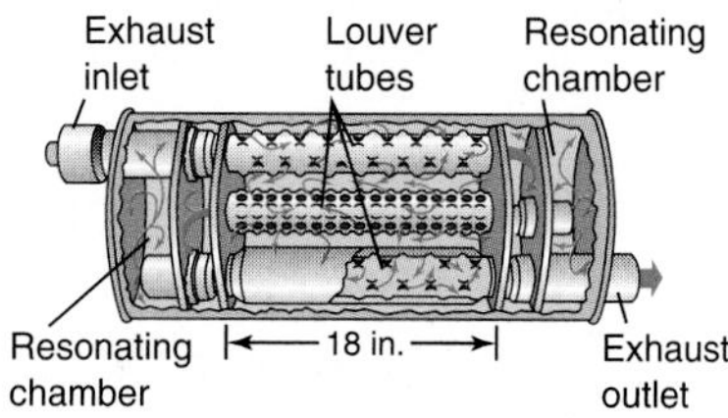

Mixed Review

38. **Highway Management** A building used to store salt is pyramid-shaped with height 32 feet and slant height 56 feet. If the salt pile is cone-shaped, what is the greatest radius of the base of the pile? (Lesson 11–4) about 46 ft

39. 384 in^2

39. Find the surface area of a cube whose edges are 8 inches long. (Lesson 11–3)

40. A circular area rug has a diameter of 4 yards. What is the area of the rug? (Lesson 10–5) 4π or about 12.6 yd^2

41. The sum of the measure of 7 interior angles of a convex decagon is 1290. The other three angles are in the ratio 1:1:3. Find the measures of the other three angles. (Lesson 10–1) 30°, 30°, 90°

42. Is the statement *All diameters of a circle are congruent* true or false? (Lesson 9–1) true

43. $\overline{RS}, \overline{MN}; \overline{RT}, \overline{MO}; \overline{ST}, \overline{NO}$

44. parallel

43. If $\triangle RST \sim \triangle MNO$, name the proportional parts of the triangles. (Lesson 7–3)

44. Determine if $\overleftrightarrow{AB} \parallel \overleftrightarrow{CD}$ for $A(6, 0)$, $B(8, 4)$, $C(0, -3)$, and $D(2, 1)$. (Lesson 3–3)

INTEGRATION Algebra

45. Factor $54c^2d$ completely. Do not use exponents. $2 \cdot 3 \cdot 3 \cdot 3 \cdot c \cdot c \cdot d$

46. Find the degree of $x^3y + 5x^4y^2$. 6

4 ASSESS

Closing Activity

Modeling Have students make a prism or cylinder. Have them measure the solid and find its volume.

Additional Answer

33. Sample answer: True; imagine you sliced the oblique cylinder or prism parallel to its base to form a large number of pieces. Then you could slide the pieces so that they stack as a right cylinder or prism. This sliding will not affect the volume or the height. Since the volume of the right cylinder or prism is the product of the area of the base and the height, then the volume of the oblique cylinder or prism is the product of the area of the base and the height.

Extension

Problem Solving For a given cylinder, the height is twice the radius, and the total surface area is 54π. Find the volume of the cylinder in terms of π. 54π units^3

Enrichment Masters, p. 68

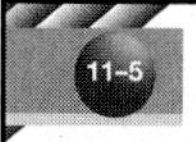

NAME ______ DATE ______

Student Edition Pages 607–613

Enrichment

Doubling Sizes

Consider what happens to surface area when the sides of a figure are doubled.

The sides of the large cube are twice the size of the sides of the small cube.

3 in.

1. How long are the edges of the large cube? 6 in.
2. What is the surface area of the small cube? 54 in^2
3. What is the surface area of the large cube? 216 in^2
4. The surface area of the large cube is how many times greater than that of the small cube? 4 times

The radius of the large sphere at the right is twice the radius of the small sphere.

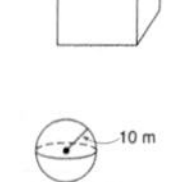

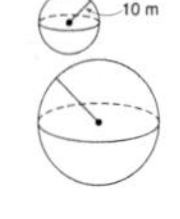

5. What is the surface area of the small sphere? 400π m^2
6. What is the surface area of the large sphere? 1600π m^2
7. The surface area of the large sphere is how many times greater than the surface area of the small sphere? 4 times
8. It appears that if the dimensions of a solid are doubled, the surface area is multiplied by ______. 4

Now consider how doubling the dimensions affects the volume of a cube.

The sides of the large cube are twice the size of the small cube.

5 in.

9. How long are the edges of the large cube? 10 in.
10. What is the volume of the small cube? 125 in^3
11. What is the volume of the large cube? 1000 in^3
12. The volume of the large cube is how many times greater than that of the small cube? 8 times

The large sphere at the right has twice the radius of the small sphere.

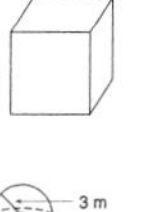

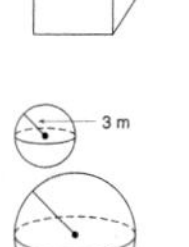

13. What is the volume of the small sphere? 36π m^3
14. What is the volume of the large sphere? 288π m^3
15. The volume of the large sphere is how many times greater than the volume of the small sphere? 8 times
16. It appears that if the dimensions of a solid are doubled, the volume is multiplied by ______. 8

11-6A LESSON NOTES

NCTM Standards: 2–5, 7

Objective

Compare the volumes of prisms and pyramids and the volumes of cylinders and cones.

Recommended Time

Demonstration and discussion: 15 minutes; Exercises: 30 minutes

Instructional Resources

For each student or group of students

Modeling Mathematics Masters

- p. 3 (square dot paper)
- p. 102 (worksheet)

For teacher demonstration

Algebra and Geometry Overhead Manipulative Resources

1 FOCUS

Motivating the Lesson

Encourage students to think of examples of cones and pyramids in the real world and how they would determine the volume.

2 TEACH

Teaching Tip Point out to students that if it were possible, water might give a more accurate measure.

3 PRACTICE/APPLY

Assignment Guide

Core (with proof): 1–9
Core (informal): 1–9
Enriched: 1–9

4 ASSESS

Observing students working in cooperative groups is an excellent method of assessment.

MODELING MATHEMATICS

11–6A Investigating Volumes of Pyramids and Cones

A Preview of Lesson 11–6

Materials:  plain paper 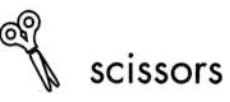 scissors  masking tape  rice 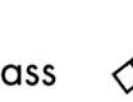 compass centimeter ruler

You can compare the volumes of prisms and pyramids and the volumes of cylinders and cones by making a model of each solid and filling them with rice.

Activity **Compare the volumes of a prism and pyramid with the same height and base.**

Step 1 The nets at the right will fold into a prism and a pyramid with open tops. Draw the two nets on plain paper using a centimeter ruler.

Step 2 Cut out the nets and fold on the dashed lines.

Step 3 Fully tape the edges together to form models of the solids.

Step 4 Estimate how much larger the volume of the prism is than the volume of the pyramid.

Step 5 Fill the pyramid with rice. Then pour this rice into the prism. Repeat until the prism is filled.

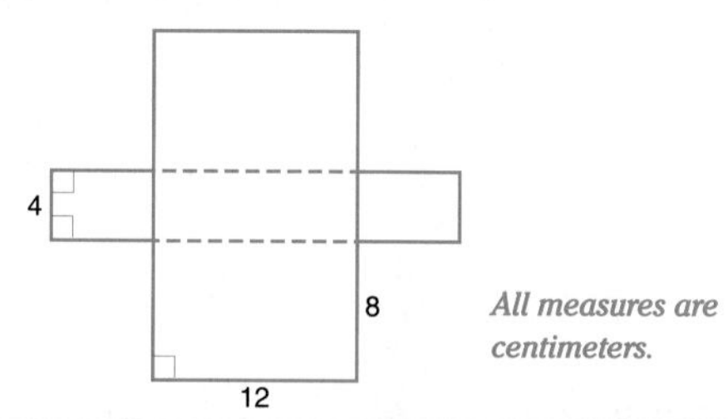

All measures are centimeters.

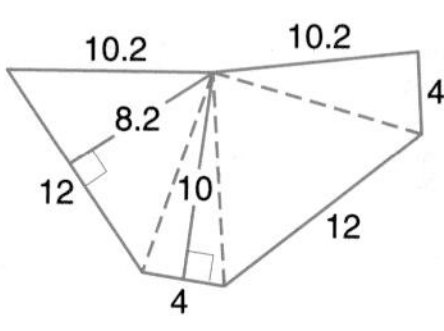

Write

1. How many pyramids of rice did it take to fill the prism? **3**
2. What do you know about the areas of the bases of the prism and the pyramid? **2. The areas of the bases of the prism and the pyramid are the same.**
3. Compare the heights of the prism and the pyramid. **See margin.**
4. Write a formula for the volume of a pyramid. $V = \frac{1}{3}Bh$

Model

5. The nets below fold into a cylinder and a cone. Repeat the activity above using these nets. **See students' work.**

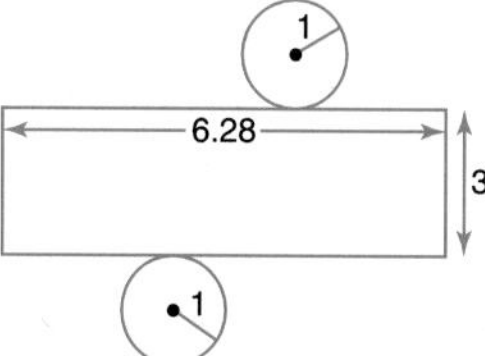

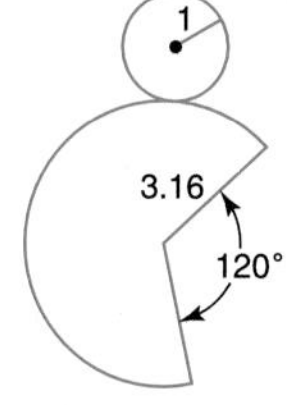

Write

6. How many cones of rice did it take to fill the cylinder? **3**
7. What do you know about the areas of the bases of the cone and the cylinder? **7. The areas of the bases of the cone and the cylinder are the same.**
8. Compare the heights of the cone and the cylinder. **8. The heights of the cone and the cylinder are the same.**
9. Write a formula for the volume of a cone. $V = \frac{1}{3}Bh$

Additional Answer

3. The heights of the prism and the pyramid are the same.

Using Cooperative Learning

This lesson offers an excellent opportunity for using cooperative learning groups. For more information on cooperative learning strategies and group management, see *Cooperative Learning in the Mathematics Classroom*, one of the titles in the Glencoe Mathematics Professional Series.

11-6 Volume of Pyramids and Cones

***What* YOU'LL LEARN**
- To find the volume of a pyramid, and
- to find the volume of a circular cone.

***Why* IT'S IMPORTANT**

You can use pyramids and cones in solving problems about history, geology, and engineering.

APPLICATION
History

The American Indians who lived in the plains followed herds of buffalo, which they hunted for food and hides for making shelters and clothes. When hunting, they lived in tepees made by stretching buffalo skins over poles. The tepees were many-sided pyramids that resembled cones. The American Heritage Center at the University of Wyoming was designed with a conical shape to represent a celebration of the culture of the American Indians.

The heating and cooling systems chosen for the American Heritage Center had to be able to handle the amount of space that had to be heated or cooled. A formula for the volume of a cone was used since the building is shaped like a cone. *You will find the volume of the American Heritage Center in Example 1.*

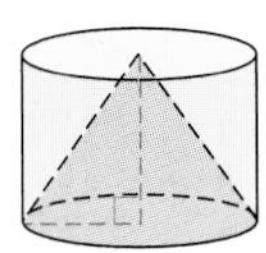

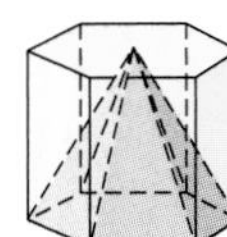

Study the figures at the right. The cone and cylinder have the same base and height, and the pyramid and prism have the same base and height. As you can see, the volume of the cone is less than the volume of the cylinder, and the volume of the pyramid is less than the volume of the prism. In the Modeling Mathematics on page 614, you discovered that the ratio of the volumes in each case is 1:3. This relationship is stated in the formulas below.

Volume of a Right Circular Cone	**If a right circular cone has a volume of V cubic units, a height of h units, and the area of the base is B square units, then $V = \frac{1}{3}Bh$.**

Volume of a Right Pyramid	**If a right pyramid has a volume of V cubic units, a height of h units, and the area of the base is B square units, then $V = \frac{1}{3}Bh$.**

11-6 LESSON NOTES

NCTM Standards: 1–5, 7

Instructional Resources

- Study Guide Master 11-6
- Practice Master 11-6
- Enrichment Master 11-6
- Assessment and Evaluation Masters, p. 297
- Real-World Applications, 22
- Tech Prep Applications Masters, p. 22

Transparency 11-6A contains the 5-Minute Check for this lesson; **Transparency 11-6B** contains a teaching aid for this lesson.

Recommended Pacing	
Standard Pacing	Days 12 & 13 of 18
Honors Pacing	Days 11 & 12 of 16
Block Scheduling*	Day 7 of 10

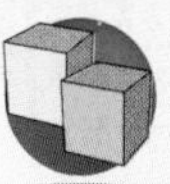

*For more information on pacing and possible lesson plans, refer to the *Block Scheduling Booklet.*

1 FOCUS

5-Minute Check
(over Lesson 11-5)

Find the volume of each solid.

1.

802.5 ft^3

2.

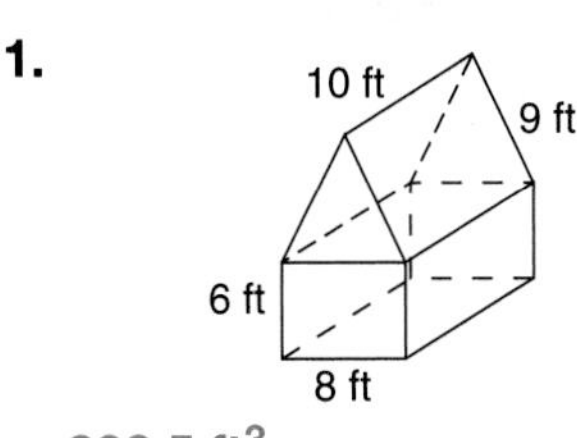

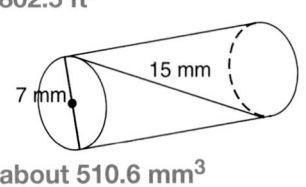

about 510.6 mm^3

Motivating the Lesson

Situational Problem Ask how much ice cream an ice-cream cone will hold. How can students find out without filling the cone with ice cream?

2 TEACH

In-Class Examples

For Example 1
Find the volume of a right circular cone with a radius of 5 centimeters and a height of 9 centimeters. Round your answer to the nearest tenth.
$V \approx 235.6$ cm^3

For Example 2
Find the volume of a regular square pyramid if the height is 11 centimeters and the diagonal of the base is 14 centimeters.
$V \approx 359.3$ cm^3

Teaching Tip When defining the volume of a right circular cone, substitute the area of a circle, πr^2, for B in the equation.

Teaching Tip Comparing cross-sectional areas means using a plane to cut two solids with the same height (right or oblique) and looking at the areas of the figures that the solids make in the plane. The plane should be parallel to the base of the solid.

Example 1

Mechanical Engineering

Refer to the application at the beginning of the lesson. The American Heritage Center has a height of 77 feet. The area of the base is about 38,000 square feet. Find the volume of air that the heating and cooling systems would have to be able to accommodate. Round to the nearest tenth.

$$V = \frac{1}{3}Bh$$
$$\approx \frac{1}{3}(38{,}000)(77)$$
$$\approx 975{,}333.3 \text{ ft}^3$$

The heating and cooling systems would have to accommodate about 975,333 cubic feet of air.

You can use trigonometry to help solve many problems involving the volume of solids.

Example 

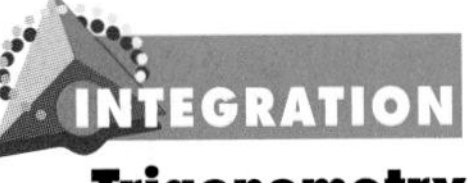

Trigonometry

Find the volume of each solid. Round to the nearest tenth.

a.

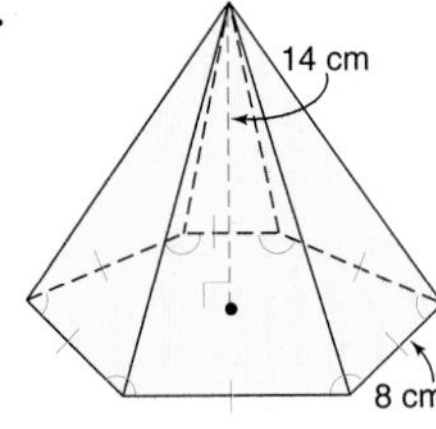

$$V = \frac{1}{3}Bh$$
$$= \frac{1}{3}\left(\frac{1}{2}Pa\right)h$$
$$= \frac{1}{3}\left(\frac{1}{2} \cdot 48 \cdot 4\sqrt{3}\right)14$$
$$= 448\sqrt{3}$$
$$\approx 776.0 \text{ cm}^3$$

b.

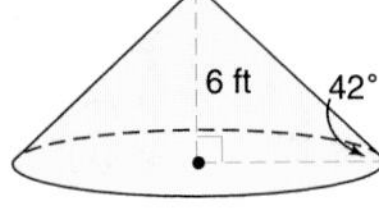

Use trigonometry to find the radius of the base.

$$\tan 42° = \frac{6}{r}$$
$$r = \frac{6}{\tan 42°}$$
$$r \approx 6.7$$

Now find the volume.

$$V = \frac{1}{3}Bh$$
$$= \frac{1}{3}\pi(6.7)^2(6)$$
$$= 89.78\pi$$
$$\approx 282.1 \text{ ft}^3$$

If you solve problems using calculators without rounding, your answers may vary slightly.

In this chapter so far, only the formulas for the volume of right prisms, cylinders, pyramids, and cones have been presented. Do you think the same formulas can be applied to all oblique solids?

Study the photograph at the right. It shows two matching stacks of index cards. The stack on the left represents a right prism and the stack on the right represents an oblique prism. Since the stacks have the same number of cards, with all cards the same size and shape, the two prisms represented by the stacks have the same volume. This observation was first made by Cavalieri, an Italian mathematician of the seventeenth century. It is known as **Cavalieri's Principle**.

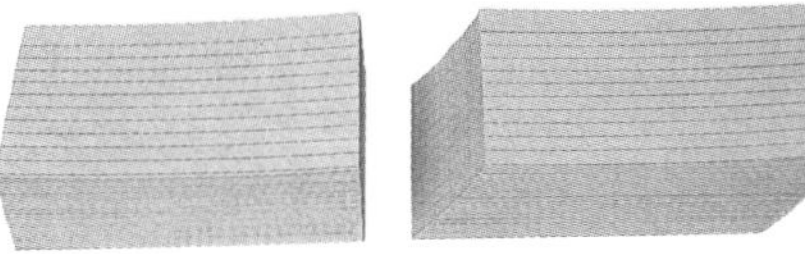

Cavalieri's Principle	**If two solids have the same height and the same cross-sectional area at every level, then they have the same volume.**

As a result of Cavalieri's Principle, if a prism has a base with an area of B square units and a height of h units, then its volume is Bh cubic units, whether it is right or oblique. The volume formulas for cylinders, cones, and pyramids hold whether they are right or oblique as well.

Example 3 **Find the volume of the oblique pyramid at the right.**

First find the height of the pyramid.

$$h^2 + 15^2 = 17^2$$
$$h^2 = 64$$
$$h = 8$$

Now find the area of the base.

$$a^2 + 4^2 = 15^2$$
$$a^2 = 209$$
$$a = \sqrt{209}$$

Finally, find the volume.

$$V = \frac{1}{3}Bh$$
$$= \frac{1}{3}\left(\frac{1}{2} \cdot 8 \cdot \sqrt{209}\right)8$$
$$\approx 154.2$$

The volume is about 154.2 cubic meters.

CHECK FOR UNDERSTANDING

Communicating Mathematics

Study the lesson. Then complete the following.

1. **Explain** how the volume of a cone is related to that of a cylinder with the same altitude and a base congruent to that of the cone. See margin.
2. Suppose you found the cross section of an oblique prism and a right prism that had the same length, width, and height. How would the cross sections compare? Explain. See margin.

In-Class Example

For Example 3
Using Cavalieri's Principle, find the volume of the cylinder below.

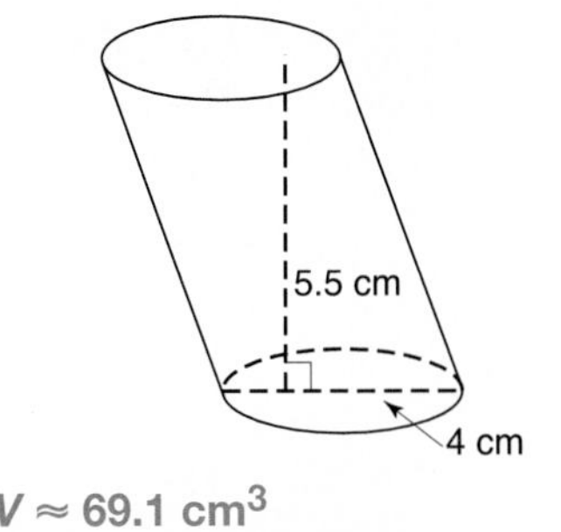

$V \approx 69.1 \text{ cm}^3$

3 PRACTICE/APPLY

Check for Understanding

Exercises 1–9 are designed to help you assess your students' understanding through reading, writing, speaking, and modeling. You should work through Exercises 1–3 with your students and then monitor their work on Exercises 4–9.

Additional Answers

1. The volume of a cone is one-third the volume of a cylinder of the same height as the cone and with bases congruent to the base of the cone.
2. The cross sections taken at the same height in both prisms would be congruent. Each cross section is like one of the cards in the stack described in the lesson. At each level, the prisms would be the same size and shape.

Reteaching

Using Diagrams Have a student draw a picture of a pyramid or cone. Have another student give a measure for the height, and a third give a measure for either the radius or the length of an edge. Have students continue to add data until they are able to compute the volume of the solid. Continue in this fashion with several problems until all students have participated.

Additional Answer

3. Sample answer: Measure the radius and the height and use the formula for the volume of a cone.

Assignment Guide

Core (with proof): 11–25 odd, 26, 27, 29–35
Core (informal): 11–25 odd, 26, 27, 29–35
Enriched: 10–24 even, 26–35

For **Extra Practice**, see p. 786.

The red A, B, and C flags, printed only in the Teacher's Wraparound Edition, indicate the level of difficulty of the exercises.

Study Guide Masters, p. 69

11-6 NAME________________ DATE ________

Study Guide Student Edition Pages 615–620

Volume of Pyramids and Cones

Volume of a Right Circular Cone	If a right circular cone has a volume of V cubic units, a height of h units, and the area of the base is B units, then $V = \frac{1}{3}Bh$
Volume of a Right Pyramid	If a right pyramid has a volume of V cubic units, a height of h units, and the area of the base is B square units, then $V = \frac{1}{3}Bh$.

Examples: Find the volume of each solid.

1 (9 m, 5 m)

$V = \frac{1}{3}Bh$
$V = \frac{1}{3}\pi(5^2)(9)$
$V = 75\pi$ or about 235.6 m³

2 (10 ft, 7 cm, 7 cm)

$V = \frac{1}{3}Bh$
$V = \frac{1}{3}(49)\ 10$
$V = \frac{490}{3}$ or about 163.3 cm³

Find the volume of each solid. Round your answers to the nearest tenth.

1. 255.5 m³

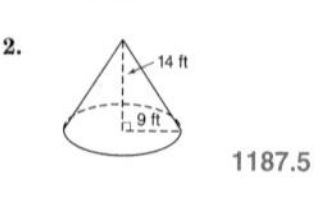

2. 1187.5 ft³

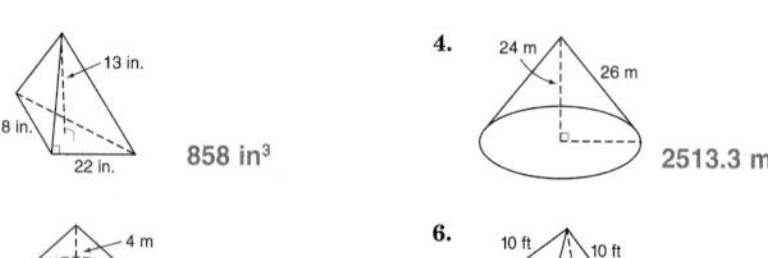

3. 858 in³

4. 2513.3 m³

5. 235.6 m³

6. 1982.0 ft³

3. **Describe** how you would find the volume of an empty ice-cream cone. See margin.

Guided Practice

4. 320 in³
5. 301.6 m³
6. 490.1 cm³

Find the volume of each pyramid or cone. Round your answer to the nearest tenth.

4.

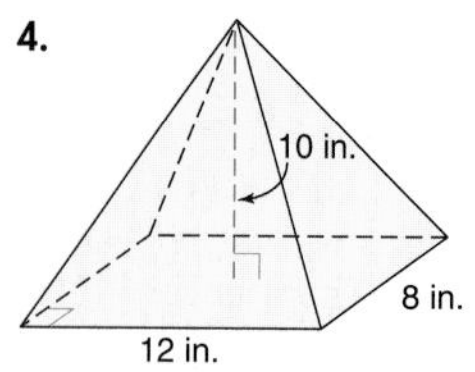

5. 6.

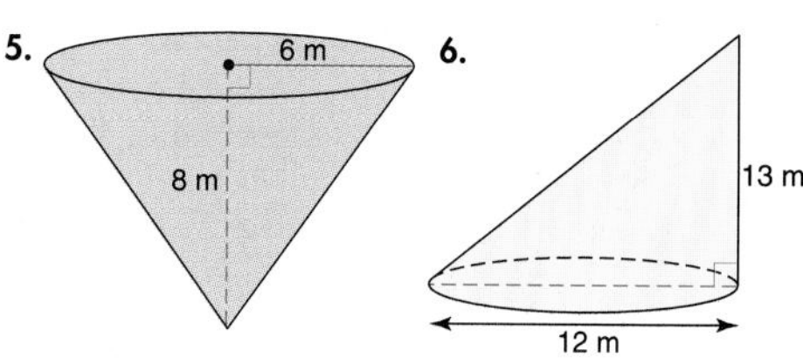

7. Find the volume of the solid shown at the right. about 5178.8 mm³

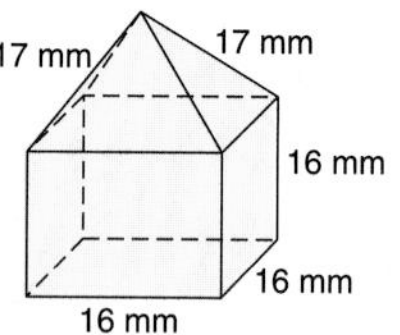

8. What is the volume of a right cone with a slant height of 20 feet and a 60° angle at the vertex of the cone? 1813.8 ft³

9. Mauna Loa: 22,415.8 km³; Fuji: 169.1 km³; Paricutín: 171,137,610 m³; Vesuvius: 4,912,194 m³

9. **Geology** Geologists describe volcanoes as one of three major types: cinder cone, shield dome, and composite. The slope of a volcano is the angle made by the side of the cone and a horizontal line. The table below shows the name, location, type, and characteristics of several volcanoes. Find the volume of material in each volcano.

Volcano	Location	Type	Characteristics
Mauna Loa	Hawaii, U.S.A.	shield dome	9100 m tall, 97 km across at base
Fuji	Honshu, Japan	composite	3776 m tall, slope of 30°
Paricutín	Michoacán, Mexico	cinder cone	410 m tall, 33° slope
Vesuvius	Campania, Italy	composite	120 across at base, 1303 m tall

EXERCISES

Find the volume of each pyramid or cone. Round to the nearest tenth.

10. 796.4 m³
11. 382.5 in³
12. 471.5 cm³

A

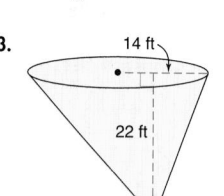

10. (13 m, 18 m) 11. ($8\frac{1}{2}$ in., 9 in., 15 in.) 12.

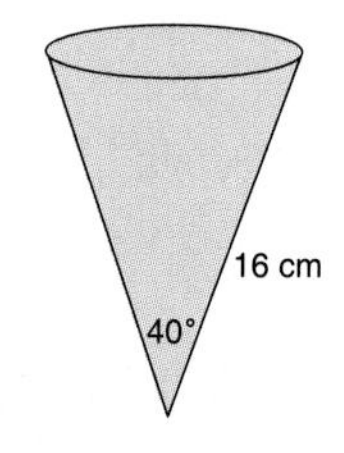

13. 4515.5 ft³
14. 609.7 cm³
15. 1082.8 mm³

B

13. (14 ft, 22 ft) 14.

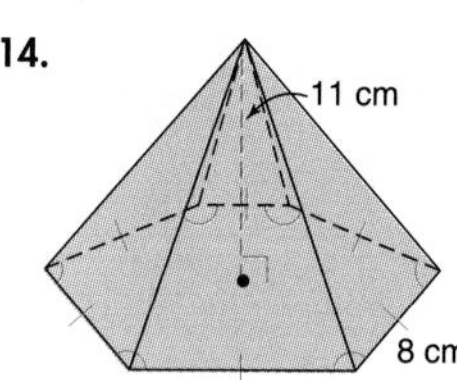

15.

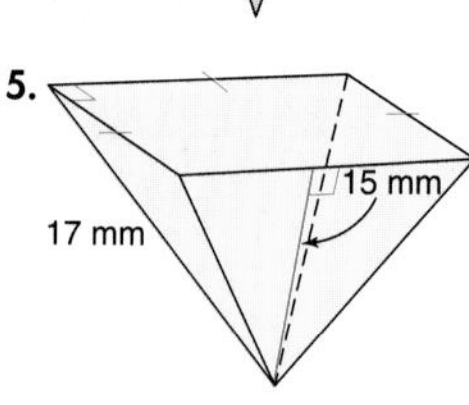

Alternative Learning Styles

Visual Have students make two cones with the same height and base, one oblique and one right. Have them draw a circle around each cone halfway between the top and bottom. Then have them measure the circumferences of the circles. Since the circumferences are the same, the areas must be the same. This demonstrates Cavalieri's Principle.

16. 1631.3 ft^3
17. 277.1 m^3
18. 192 in^3

16.

17.

18.

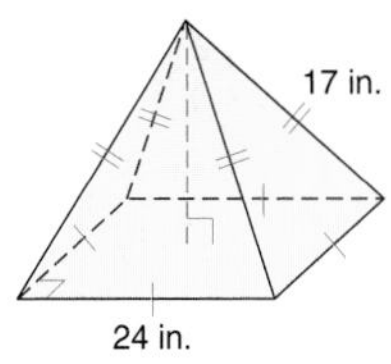

Find the volume of each solid. Round to the nearest tenth.

19. 5730.3 units^3
20. 469,144.5 cm^3
21. 814.6 yd^3

C

19.

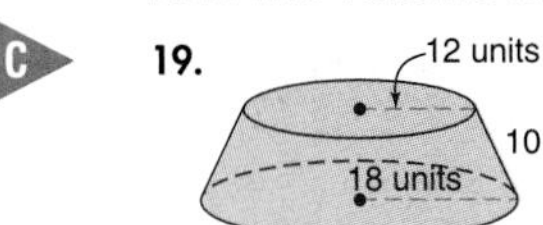

20.

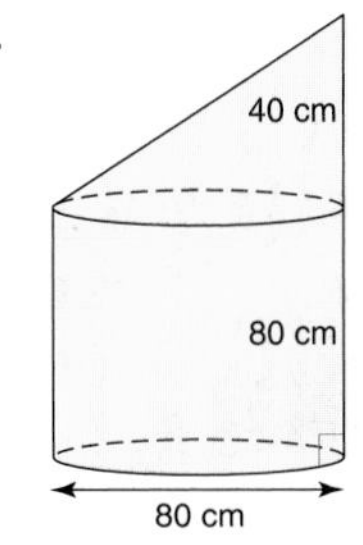

21.

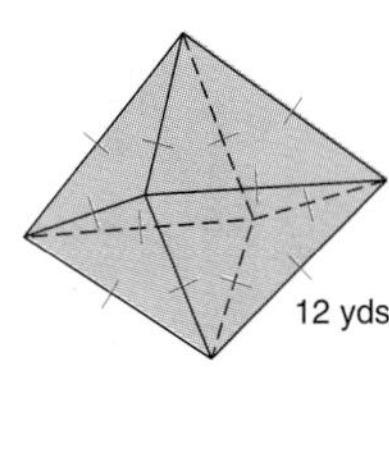

22. A pyramid has a volume of 729 cubic units. If the area of the base is 243 square units, what is the height of the pyramid? **9 units**

23. 58.9 in^3

23. One right circular cone is inside a larger right circular cone. The two cones have the same axis, the same vertex, and the same height. Find the volume of the space between the cones if the diameter of one cone is 6 inches, the diameter of the other is 9 inches, and the height of both is 5 inches.

24. A regular tetrahedron is a pyramid with four congruent equilateral triangles as its only faces. If the length of one edge of a tetrahedron measures 12 units, find the volume of the tetrahedron. **about 203.6 units^3**

Programming

25. The TI-82/83 program at the right will find the volume of a rectangular pyramid or of a circular cone rounded to the nearest cubic unit.

```
PROGRAM:VOLUME
:Disp "CHOOSE:"
:Disp "1: PYRAMID"
:Input "2: CONE", C
:If C=2
:Goto 2
:Input "LENGTH = ", L
:Input "WIDTH = ", W
:Input "HEIGHT = ", H
:round(L*W*H/3,0)→V
:Disp "VOLUME = ", V
:Stop
:Lbl 2
:Input "RADIUS = ", R
:Input "HEIGHT = ", H
:round(π*R²*H/3,0)→V
:Disp "VOLUME = ", V
:Stop
```

Use the program to find the volume of each solid. **a. 1493 ft^3**

a. a pyramid that is 20 feet tall with a base 16 feet long and 14 feet wide

b. a cone with a diameter of 44 inches and a height of 54 inches **27,370 in^3**

c. a pyramid with a base 3 meters by 5 meters and a height of 7 meters **35 m^3**

d. How could you change the program to find the volume of a prism or a cylinder?

e. Would the program have to be altered to find the volume of an oblique pyramid or cone? Explain.

25d. Delete the "/3" in each of the lines that contains the volume formulas.

25e. No; the formulas are the same for a right or oblique solid.

Critical Thinking

26. Write a ratio comparing the volumes of the solids shown at the right. Justify your answer. **See margin.**

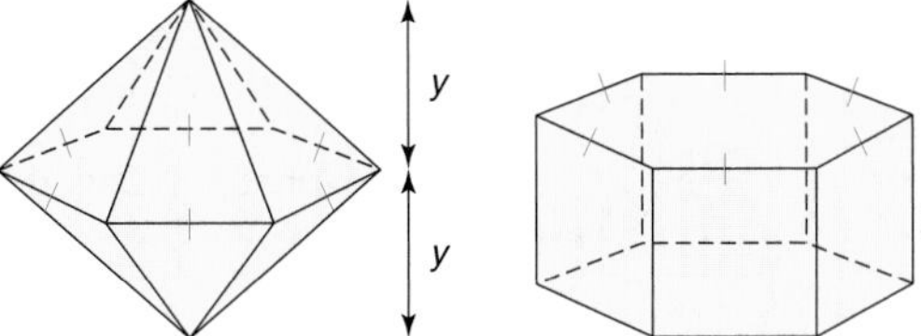

Using the Programming Exercises The program given in Exercise 25 is for use with a TI-82 or TI-83 graphing calculator. For other programmable calculators, have students consult their owner's manual for commands similar to those represented here.

Additional Answer

26. $\frac{2}{3}$; The volume of each pyramid that makes up the solid on the left is $\frac{1}{3}$ the volume of the prism, so the total volume of the solid on the left is $\frac{1}{3} + \frac{1}{3}$ or $\frac{2}{3}$ the volume of the prism.

Practice Masters, p. 69

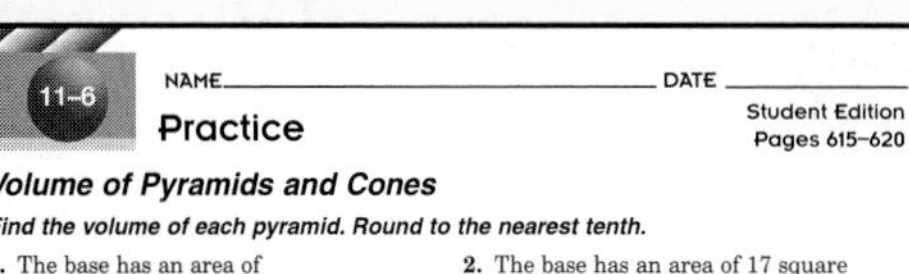

11–6
NAME
DATE
Practice
Student Edition
Pages 615–620

Volume of Pyramids and Cones

Find the volume of each pyramid. Round to the nearest tenth.

1. The base has an area of 84.3 square centimeters, and the height is 16.4 centimeters 460.8 cm^3

2. The base has an area of 17 square feet, and the height is 3 feet. 17.0 ft^3

Find the volume of each cone. Round to the nearest tenth.

3. The base has a radius of 16 centimeters, and the height is 12 centimeters 3217.0 cm^3

4. The base has a diameter of 24 meters, and the height is 15.3 meters 2307.2 m^3

Find the volume of each solid. Round to the nearest tenth.

5.

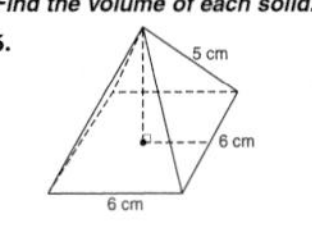

31.7 cm^3

6.

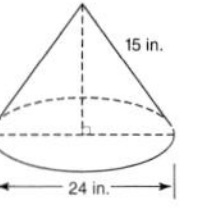

1357.2 in^3

7.

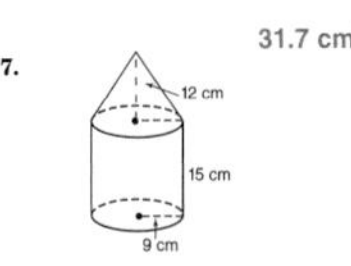

4834.9 cm^3

8.

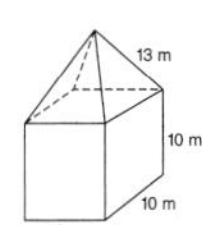

1363.6 m^3

4 ASSESS

Closing Activity

Writing Instruct each student to draw a cone or pyramid and label the measure of the height as well as the measure of the radius or an edge of the base. Have students exchange pictures and find the volume of the solids.

Chapter 11 Quiz C (Lessons 11-5 and 11-6) is available in the *Assessment and Evaluation Masters*, p. 297.

Enrichment Masters, p. 69

NAME________________ DATE ________

Enrichment

Student Edition Pages 615–620

Prism Puzzles

1. The surface area of a cube is 384 cm². Another cube is one unit longer on each edge. What is the surface area of the second cube? 486 cm²

2. The area of each end of a rectangular prism is 42 in². The area of the top is 77 in², and the area of the front is 66 in². What is the volume of the prism? 462 in³

3. Place the digits 1 through 8 at the corners of this cube so the sum of the four numbers for each face is 18.

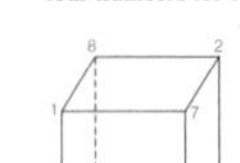

4. Find the area of the cross section of this right prism that contains points *A*, *B*, *C*, and *D*.

A, D, B, C, 9 cm, 8 cm, 15 cm

153 cm²

5. Find the value of x for the prism below so the surface area of the prism is 2050 square units. 5

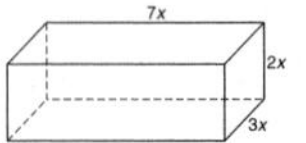

6. Find the value of x for the prism below so the volume of the prism is 4116 cubic units. 7

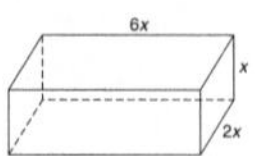

7. Find the value of x for the prism below so the surface area of the prism is 768 square units. 4

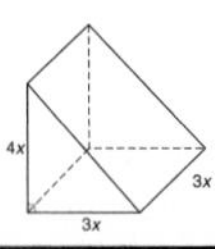

8. Find the value of x for the prism in Problem 7 so the volume of the prism is 486 cubic units. 3

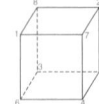

Applications and Problem Solving

27. Geology The way that the ice inside glaciers moves depends on the depth of the ice. The changes in different areas of a glacier are shown in the diagram below. How does the volume of the ice change as each of the movements takes place? Explain.

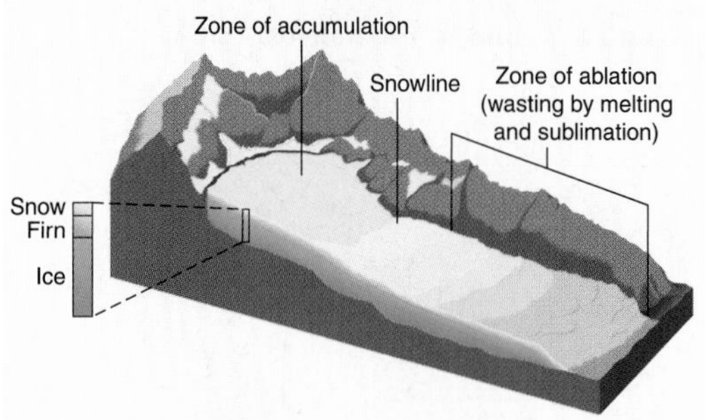

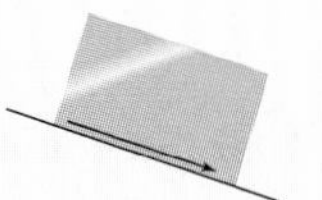

In the top layers of a glacier, ice slides along the base.

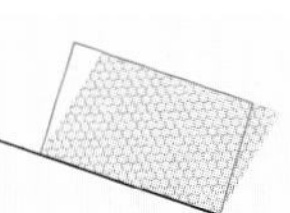

At lower levels, grains of snow and ice move within the glacier.

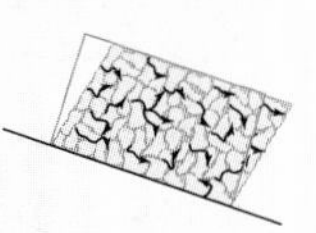

Pressure can cause ice to move by melting and refreezing.

Sheets of ice can slip in planes to move a glacier.

27. According to Cavalieri's Principle, the volume of each solid stays the same.

28. Architecture In an attempt to rid Florida's Lower Sugarloaf Key of mosquitoes, Richter Perky built a tower to attract bats. The Perky Bat Tower is a frustum of a pyramid with a square base. Each side of the base of the tower is 15 feet long, the top is a square with sides 8 feet long, and the tower is 35 feet tall. How many cubic feet of space does the tower supply for bats? **4771.7 ft³**

Mixed Review

29. Engineering The oil drilling platform called the Statfjord B is located off the coast of Norway in the North Sea. The base of the platform is made up of 24 concrete cylinders or cells. Twenty of the cells are used for oil storage. The pillars that support the platform deck rest on the four other cells. Find the total volume of the storage cells. (Lesson 11–5) **18,555,031.6 ft³**

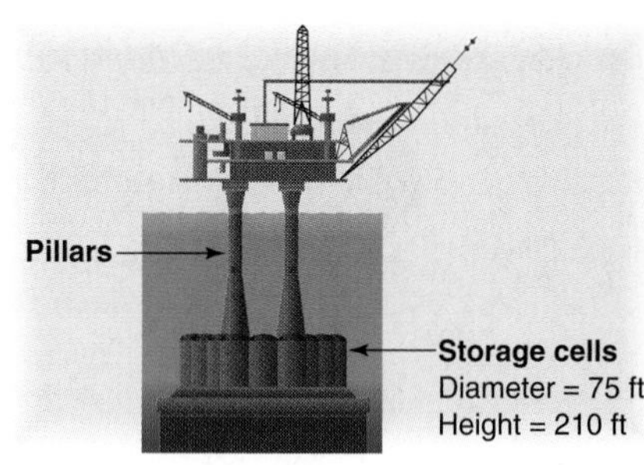

30. Find the surface area of a cylindrical water tank that is 8 meters tall and has a diameter of 8 meters. (Lesson 11–3) **301.6 m²**

31. A rhombus with a 56-inch diagonal has an area of 1344 in². Find the length of a side of the rhombus. (Lesson 10–4) **36.9 in.**

32. Cycling Hermosa's little brother's tricycle tire has a diameter of 12 inches. About how far does the tricycle travel in one turn of the wheel? (Lesson 9–1) **12π or about 37.7 inches**

33. The sides of a right triangle have measures of $x + 8$, $x + 1$, and $x + 9$ units. Find the value of x. (Lesson 8–1) **4**

Algebra

34. Write the standard form of an equation of the line that passes through the point at $(9, -3)$ and has a slope of -1. $x + y = 6$

35. Define a variable, write an inequality, and solve *if nine times a number is at most 108.* **Let x = the number; $9x \leq 108$; $\{x \mid x \leq 12\}$**

Extension

Problem Solving Joshua is planning to sell popcorn at the fair. He is either going to use a 2 in. by 6 in. by 8 in. box for the popcorn or he is going to use a cone with a radius of 3 inches and a slant height of 10 inches. Which will be more economical for him to use and why?

The cone; both shapes hold about the same amount of popcorn (box, 96 in³; cone, 90 in³), but the box has a much larger surface area (152 in² vs. 94 in²), making it likely to be more expensive to produce.

11–7 Surface Area and Volume of Spheres

What YOU'LL LEARN

- To recognize and define basic properties of spheres,
- to find the surface area and the volume of a sphere.

Why IT'S IMPORTANT

You can use spheres to solve problems involving geography, astronomy, and housing.

History

According to legend, Christopher Columbus set out to prove that Earth was round when he discovered America. But in reality, Greek mathematicians not only knew that Earth was round sixteen centuries earlier, they had calculated its circumference!

Eratosthenes of Cyrene (275–194 B.C.) was director of the Alexandrian Library, which was the academic center of the ancient world. Eratosthenes used the position of the Sun and the distance between two cities to calculate that the circumference of Earth was 250,000 stades or about 39,375 km. This is remarkably close to the current accepted value of 40,075 km.

Earth is roughly a **sphere**. To visualize a sphere, consider infinitely many congruent circles in space, all with the same point for their center. Considered together, all these circles form a sphere. In space, a sphere is the set of all points that are a given distance from a given point called its **center**.

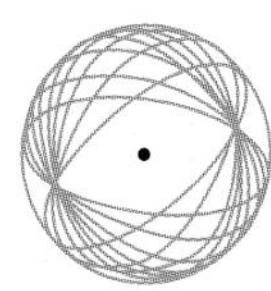

There are several special segments and lines related to spheres.

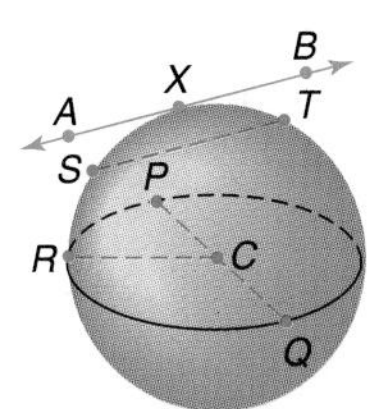

- A segment whose endpoints are the center of the sphere and a point on the sphere is a **radius** of the sphere. In the figure, $\overline{CR}$, $\overline{CP}$, and $\overline{CQ}$ are radii.
- A **chord** of a sphere is a segment whose endpoints are points on the sphere. In the figure, $\overline{TS}$ and $\overline{PQ}$ are chords.
- A chord that contains the sphere's center is a **diameter** of the sphere. In the figure, $\overline{PQ}$ is a diameter.
- A **tangent** to a sphere is a line that intersects the sphere in exactly one point. In the figure, $\overleftrightarrow{AB}$ is tangent to the sphere at X.

11-7 LESSON NOTES

NCTM Standards: 1–5, 7

Instructional Resources

- Study Guide Master 11-7
- Practice Master 11-7
- Enrichment Master 11-7
- Real-World Applications, 23

Transparency 11-7A contains the 5-Minute Check for this lesson; **Transparency 11-7B** contains a teaching aid for this lesson.

Recommended Pacing	
Standard Pacing	Days 14 & 15 of 18
Honors Pacing	Day 13 of 16
Block Scheduling*	Day 8 of 10

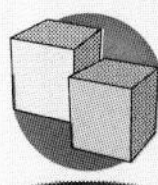

*For more information on pacing and possible lesson plans, refer to the *Block Scheduling Booklet*.

1 FOCUS

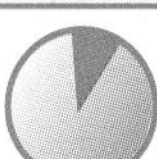

5-Minute Check
(over Lesson 11-6)

1. Using Cavalieri's Principle, find the volume of the cone below. **about 139.8 units^3**

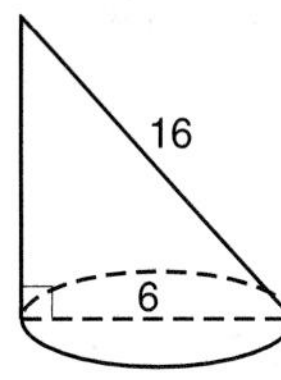

2. A regular square pyramid has a slant height of 16 meters and a lateral edge of 22 meters. Find the volume of the pyramid. **about 1608.6 m^3**

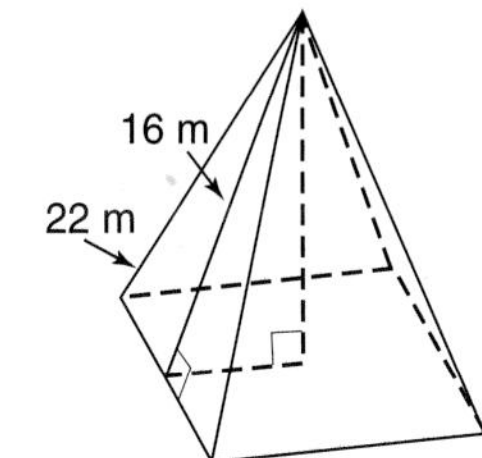

Motivating the Lesson

Hands-On Activity Bring in and discuss examples of spheres. Ask students what criteria must be met for the solid to be a sphere. You can also use a soft rubber ball to demonstrate a sphere and then squash the ball to form circular solids that are not spheres. This may help students define the criteria for a sphere.

MODELING MATHEMATICS Point out to students that all methods for discovering a formula for the surface area of a sphere are approximations. The formula is an attempt to get the exact measure.

Teaching Tip When discussing great circles, emphasize that the lines of longitude and the equator are great circles because they have the same center as the sphere.

A plane can intersect a sphere in a point or in a circle.

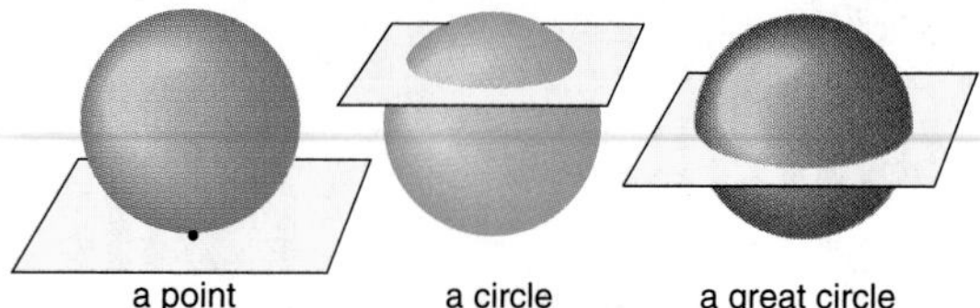

LOOK BACK
You can review great circles of spheres in Lesson 3-6.

When a plane intersects a sphere so that it contains the center of the sphere, the intersection is called a **great circle**. A great circle has the same center as the sphere, and its radii are also radii of the sphere. On the surface of a sphere, the shortest distance between any two points is the length of the arc of a great circle passing through those two points. Each great circle separates a sphere into two congruent halves called **hemispheres**.

You can use a model to investigate the surface area of a sphere.

MODELING MATHEMATICS — Surface Area of a Sphere

Materials: Styrofoam ball, scissors, tape, straight pins

- Cut the Styrofoam ball along a great circle. Trace around the edge to draw a circle. Then cut out the paper circle.

- Fold the circle into eighths. Then unfold and cut the eight pieces apart. Tape the pieces back together in the arrangement shown at the right.

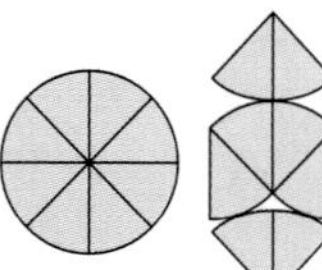

- Use tape or a straight pin to put the two pieces of the Styrofoam ball back together. Then use pins to attach the pattern to the sphere.

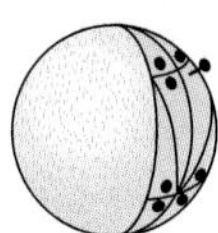

Your Turn a. $\frac{1}{4}$

a. Estimate how much of the total surface of the sphere was covered by the pattern.

b. What is the area of the pattern in terms of r, the radius of the sphere? πr^2

c. Write a formula for the surface area of the sphere. $4\pi r^2$

The Modeling Mathematics activity leads us to the formula for the surface area of a sphere.

Surface Area of a Sphere	**If a sphere has a surface area of T square units and a radius of r units, then $T = 4\pi r^2$.**

Many sports use balls that are shaped like spheres. When the balls are made, the manufacturer needs to find its surface area to determine the amount of material needed.

Example 1

APPLICATION

Sports

Find the surface area of each sports ball described.

a. An NCAA basketball has a radius of $4\frac{3}{4}$ inches.

b. An Olympic-sized volleyball has a circumference of 27 inches.

a. $T = 4\pi r^2$

$= 4\pi\left(4\frac{3}{4}\right)^2$

$\approx 283.5 \text{ in}^2$ *Use a calculator.*

The surface area of an NCAA basketball is about 283.5 square inches.

b. Use the formula for circumference to find the radius of a volleyball.

$C = 2\pi r$

$27 = 2\pi r$

$4.30 \approx r$ *Use a calculator.*

Now find the surface area.

$T = 4\pi r^2$

$\approx 4\pi(4.30)^2$

≈ 232.4 *Use a calculator*

The surface area of an Olympic-sized volleyball is about 232.4 square inches.

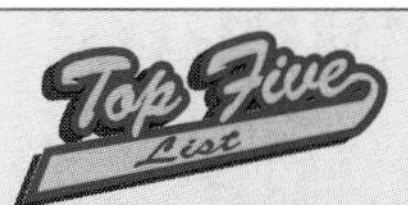

The *Sports Almanac*'s Most Outstanding Players in Women's College Basketball

1995, Rebecca Lobo (Connecticut)

1994, Charlotte Smith (North Carolina)

1993, Sheryl Swoopes (Texas Tech)

1992, Molly Goodenbour (Stanford)

1991, Dawn Staley (Virginia)

You can relate finding a formula for the volume of a sphere to finding the volume of a right pyramid and the surface area of a sphere.

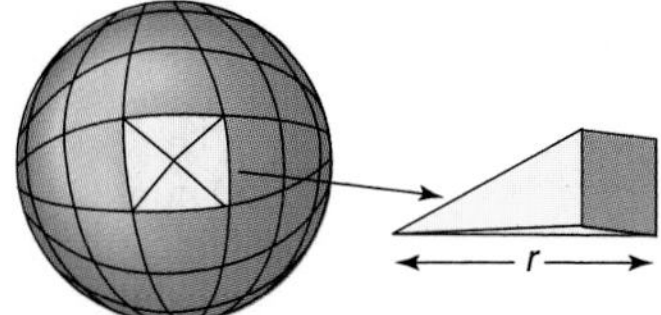

Picture separating the space inside a sphere into infinitely many small pyramids, all with their vertices located at the center of the sphere. This is shown above. Observe that the height of these very small pyramids is equal to the radius r of the sphere. The sum of the areas of all the pyramid bases equals the surface of the sphere.

Each pyramid has a volume of $\frac{1}{3}Bh$, where B is the area of its base and h is its height. The volume of the sphere is equal to the sum of the volumes of all the infinitely many small pyramids. Thus, the volume V of the sphere can be represented as follows.

In-Class Example

For Example 1
The diameter of a softball is about 3.5 inches. Find its surface area.
$A \approx 38.5 \text{ in}^2$

Teaching Tip When calculating a value for r^3, use the y^x key on your calculator or the ^ key on your graphing calculator.

Basketball was invented in 1891 by James Naismith at the International YMCA Training School in Springfield, Mass. The first game was governed by 13 axioms formulated by Naismith. These axioms are still part of the game.

In-Class Example

For Example 2

Find the volume of the empty space in a box containing three golf balls. The diameter of each is about 1.5 inches. The box is 4.5 inches by 1.5 inches by 1.5 inches.

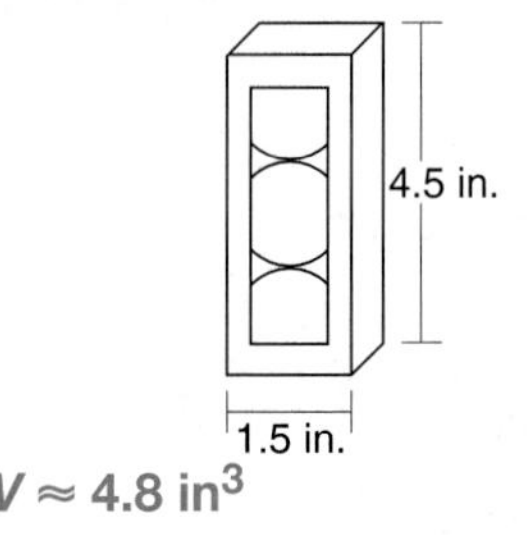

$V \approx 4.8 \text{ in}^3$

Teaching Tip Encourage students to make a reference table that groups all of the formulas that they have learned in this chapter.

Archimedes is also known for his approximation for π, between $3\frac{10}{71}$ and $3\frac{1}{7}$.

$$
\begin{aligned}
V &= \tfrac{1}{3}B_1h_1 + \tfrac{1}{3}B_2h_2 + \tfrac{1}{3}B_3h_3 + \dots + \tfrac{1}{3}B_nh_n \\
&= \tfrac{1}{3}B_1r + \tfrac{1}{3}B_2r + \tfrac{1}{3}B_3r + \dots + \tfrac{1}{3}B_nr \\
&= \tfrac{1}{3}r(B_1 + B_2 + B_3 + \dots + B_n) \\
&= \tfrac{1}{3}r(4\pi r^2) \quad \text{Replace } (B_1 + B_2 + B_3 + \dots + B_n)\text{, which is the surface area of the sphere, with } 4\pi r^2. \\
&= \tfrac{4}{3}\pi r^3
\end{aligned}
$$

Volume of a Sphere	**If a sphere has a volume of V cubic units and a radius of r units, then $V = \frac{4}{3}\pi r^3$.**

Example 2 **Find the volume of each sphere. Round to the nearest tenth.**

a.

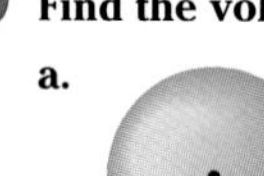

$$
\begin{aligned}
V &= \tfrac{4}{3}\pi r^3 \\
&= \tfrac{4}{3}\pi(18)^3 \\
&\approx 24{,}429.0 \text{ cm}^3
\end{aligned}
$$

b.

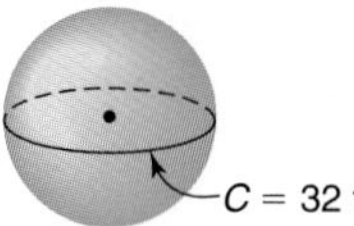

Find the radius of the sphere.

$$
\begin{aligned}
C &= 2\pi r \\
32 &= 2\pi r \\
\tfrac{16}{\pi} \text{ ft} &= r
\end{aligned}
$$

Now find the volume.

$$
\begin{aligned}
V &= \tfrac{4}{3}\pi r^3 \\
&= \tfrac{4}{3}\pi\left(\tfrac{16}{\pi}\right)^3 \\
&\approx 553.3 \text{ ft}^3
\end{aligned}
$$

F Y I

Archimedes thought this relationship was so important that he asked to have the figure shown engraved on his tombstone.

Over 2200 years ago, the mathematician Archimedes discovered a relationship between the volume of a cylinder and an inscribed sphere.

The radius of the sphere is r units. So the radius of the cylinder is also r units. The height of the cylinder is the diameter of the sphere, $2r$.

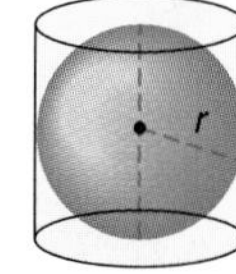

$$
\begin{aligned}
\frac{\text{volume of sphere}}{\text{volume of cylinder}} &= \frac{\frac{4}{3}\pi r^3}{\pi r^2 h} \quad \text{Substitute the volume formulas.} \\
&= \frac{\frac{4}{3}\pi r^3}{\pi r^2(2r)} \quad h = 2r \\
&= \frac{\frac{4}{3}\pi r^3}{2\pi r^3} \\
&= \frac{\frac{4}{3}}{2} \text{ or } \frac{2}{3} \quad \text{Simplify.}
\end{aligned}
$$

The ratio of the volume of a sphere to that of a cylinder in which it is inscribed is $\frac{2}{3}$.

CHECK FOR UNDERSTANDING

Communicating Mathematics

Study the lesson. Then complete the following.

1. **Draw and label a diagram** of a sphere with a great circle. See margin.
2. Is a sphere a polyhedron? Justify your answer.
3. **Explain** how the formula for the volume of a sphere was developed in this lesson. See margin.
4. **Compare and contrast** squares and cubes, and circles and spheres. See margin.
5. **Assess Yourself** Describe some models of spheres that you encounter in your life. See students' work.

2. No, a polyhedron has flat faces and a sphere has no flat surfaces.

Guided Practice

Describe each object as a model of a *circle*, a *sphere*, or *neither*.

6. baseball sphere
7. jelly jar neither

Determine whether each statement is *true* or *false*.

8. All chords of a sphere are diameters. false
9. All diameters of a sphere are chords. true
10. If a great circle of a sphere is congruent to a great circle of another sphere, then the spheres are congruent. true

In the figure, *C* is the center of the sphere, and plane $\mathcal{B}$ intersects the sphere in circle *R*.

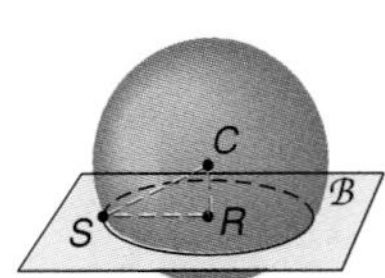

11. Suppose $CR = 4$ and $SR = 3$. What is the length of a radius of the sphere? 5
12. If the radius of the sphere is 13 units and the radius of $\odot R$ is 12 units, find CR. 5

13. drizzle: $T < 0.005$ in^2, $V < 0.000034$ in^3; rain: $T > 0.005$ in^2, $V > 0.000034$ in^3

13. **Meteorology** The figure below shows the differences meteorologists say there are between rain and drizzle. Assume that rain and drizzle fall in spherical drops. Write a statement or two about the surface area and volume of drops of rain and drizzle.

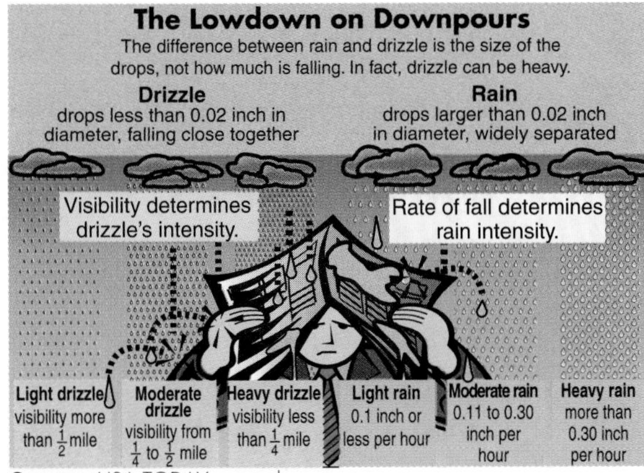

The Lowdown on Downpours
The difference between rain and drizzle is the size of the drops, not how much is falling. In fact, drizzle can be heavy.
Drizzle
drops less than 0.02 inch in diameter, falling close together
Rain
drops larger than 0.02 inch in diameter, widely separated
Visibility determines drizzle's intensity.
Rate of fall determines rain intensity.
Light drizzle visibility more than 1/2 mile
Moderate drizzle visibility from 1/4 to 1/2 mile
Heavy drizzle visibility less than 1/4 mile
Light rain 0.1 inch or less per hour
Moderate rain 0.11 to 0.30 inch per hour
Heavy rain more than 0.30 inch per hour

Source: USA TODAY research

3 PRACTICE/APPLY

Check for Understanding

Exercises 1–16 are designed to help you assess your students' understanding through reading, writing, speaking, and modeling. You should work through Exercises 1–5 with your students and then monitor their work on Exercises 6–16.

Additional Answers

1. [sphere diagram with great circle]

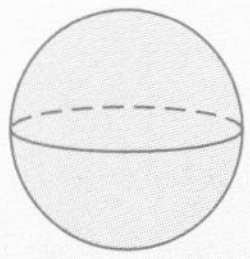

3. The volume of a sphere was generated by adding the volumes of an infinite number of small pyramids. Each pyramid has its base on the surface of the sphere and its height from the base to the center of the sphere.
4. Sample answer: Squares and circles are two-dimensional and cubes and spheres are three-dimensional. A cross section of a cube is a square, and a cross section of a sphere is a circle.

Reteaching

Using Vocabulary Draw a sphere on the chalkboard or overhead. Do not label any of its parts. Then go around the room and ask students to draw and label parts of the sphere. Use the terminology defined in this lesson.

Assignment Guide

Core (with proof): 17–51 odd, 52–62
Core (informal): 17–51 odd, 52–62
Enriched: 18–44 even, 46–62

For **Extra Practice**, see p. 787.

The red A, B, and C flags, printed only in the Teacher's Wraparound Edition, indicate the level of difficulty of the exercises.

Study Guide Masters, p. 70

11-7 NAME______ DATE______

Study Guide — Student Edition Pages 621–628

Surface Area and Volume of Spheres

The following is a list of definitions related to the study of spheres.

Sphere	the set of all points that are a given distance from a given point (center)
Radius	a segment whose endpoints are the center of the sphere and a point on the sphere
Chord	a segment whose endpoints are points on the sphere
Diameter	a chord that contains the sphere's center
Tangent	a line that intersects the sphere in exactly one point
Great Circle	the intersection of a plane and a sphere so that the center of the sphere is included
Hemispheres	two congruent halves of a sphere separated by a great circle

Describe each object as a model of a circle, sphere, or neither.

1. tennis ball can neither
2. pancake circle
3. sun sphere
4. basketball rim circle
5. globe sphere
6. lipstick container neither

Determine whether each statement is true or false.

7. All lines intersecting a sphere are tangent to the sphere. false
8. Every plane that intersects a sphere creates a great circle. false
9. The eastern hemisphere of Earth is congruent to the western hemisphere of Earth. true
10. The diameter of a sphere is congruent to the diameter of its great circle. true

Find the surface area and volume of each sphere described below. Round to the nearest tenth.

14. A radius is 12 centimeters long. $T = 1809.6$ cm^2; $V = 7238.2$ cm^3
15. One of its great circles has an area of 113.04 square feet.
16. The volume of a sphere is $\frac{32}{3}\pi$ cubic meters. Find the surface area of the sphere. 50.3 m^2

15. $T = 452.4$ ft^2; $V = 904.8$ ft^3

EXERCISES

Practice

Describe each object as a model of a *circle*, a *sphere*, or *neither*.

17. compact disc
18. the Moon sphere
19. shot put sphere
20. football neither
21. hot-air balloon neither
22. soda can neither

17. circle

Determine whether each statement is *true* or *false*.

23. In a sphere, all the radii are congruent. true
24. A sphere is contained in a plane. false
25. A sphere's longest chord will always pass through the sphere's center.
26. A radius of a great circle of a sphere is also a radius of the sphere. true
27. All of the chords of a sphere are congruent. false
28. Two spheres may intersect in exactly one point. true
29. In a sphere, two different great circles may intersect in exactly one point.
30. If two spheres intersect, their intersection may be a circle. true
31. The intersection of two spheres with congruent radii may be a great circle. false

25. true
29. false

In the figure, *O* is the center of the sphere, and plane *C* intersects the sphere in ⊙*R*.

32. If $OR = 9$ and $SR = 12$, find OS. 15
33. Given $OS = 16$ and $RS = 12.8$, find OR. 9.6
34. If the radius of the sphere is 15 units and the radius of the circle is 10 units, what is OR? 11.2
35. If O and R are distinct points, is ⊙R a great circle? no
36. If M is a point on ⊙R and $OS = 18$, what is OM? 18

Find the surface area and volume of each sphere described below. Round to the nearest tenth. 38. $T = 2642.1$ cm^2; $V = 12{,}770.1$ cm^3

37. The radius is 25 inches long. $T = 7854.0$ in^2; $V = 65{,}449.8$ in^3
38. The radius of a great circle is 14.5 centimeters.
39. The diameter is 450 meters. $T = 636{,}172.5$ m^2; $V = 47{,}712{,}938.4$ m^3
40. A radius is $6\frac{1}{2}$ inches long. $T = 530.9$ in^2; $V = 1150.3$ in^3
41. The diameter of a great circle is 3.4 meters. $T = 36.3$ m^2; $V = 20.6$ m^3
42. A great circle has a circumference 43.96 centimeters.

42. $T = 615.8$ cm^2; $V = 1436.8$ cm^3

43. $\frac{32}{3}\pi$ cm^3 or about 33.5 cm^3

C **43.** What is the volume of a sphere if its surface area is 16π cm^2?

44. A sphere is circumscribed about a cube with a volume of 1728 cubic centimeters. What is the volume of the sphere? 4701 cm^3

45. Find the ratio of the radii of two spheres if the surface area of one is 4 times the surface area of the other. 2:1

Critical Thinking

46. Desta, a pilot, is leaving for a flight. Her friend Mei needs to fly to another city. Desta offers to take her. Mei says that she doesn't want to make Desta go out of her way. Desta replies that no matter where Mei is going it will not take her out of her way. Explain how this could be true. See margin.

47. A plane slices a sphere 5 centimeters from its center. The sphere has a radius of 13 centimeters. What is the area of the slice to the nearest hundredth? 452.39 cm^2

Applications and Problem Solving

48. Astronomy Imagine that the lines of longitude and latitude are projected on the sky as if Earth was surrounded by a giant sphere with the stars on it. This sphere is called the *celestial sphere*.

a. Is a longitude line in the celestial sphere a great circle? yes

b. Is a latitude line in the celestial sphere a great circle? no

c. Can you see an entire hemisphere of the celestial sphere? no

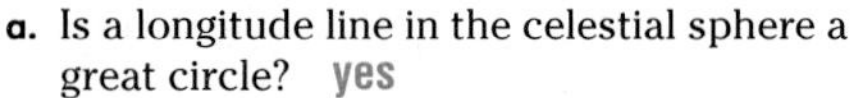

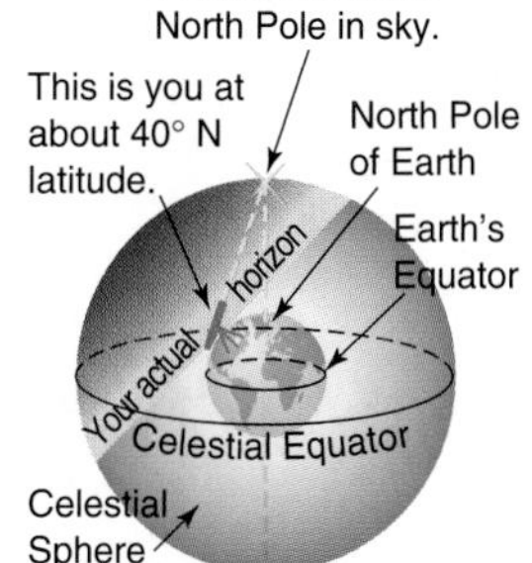
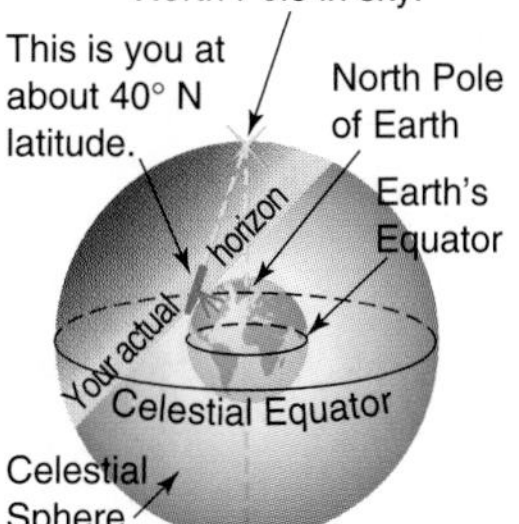

FYI

British colonists probably brought recipes for ice cream to America in the early 1700s. Ice cream was made at home until the first ice-cream plant was established in 1851.

49. Food The first ice-cream cone was made at the World's Fair in St. Louis in 1904 when an ice-cream seller ran out of cups. Suppose a sugar cone for ice cream is 10 centimeters deep and has a diameter of 4 centimeters. A scoop of ice cream with a diameter of 4 centimeters rests on the top of the cone.

a. If all the ice cream melts into the cone, will the cone overflow? no

b. If the cone does not overflow, what percent of the cone will be filled? 80%

50. $T = 157.1$ ft^2; $V = 261.8$ ft^3

50. Dwellings The traditional shelter of the Algonquian people of northern Canada, Greenland, Alaska, and eastern Siberia was the igloo. Igloos were made of hard-packed snow cut into blocks 2 to 3 feet long and 1 to 2 feet wide. The blocks were put together in a spiral that grew smaller at the top to form a hemispherical dome. An air vent at the top allowed fresh air to enter and the long entrance trapped cold air to keep the interior warm. Find the surface area and volume of the living area in the igloo shown above.

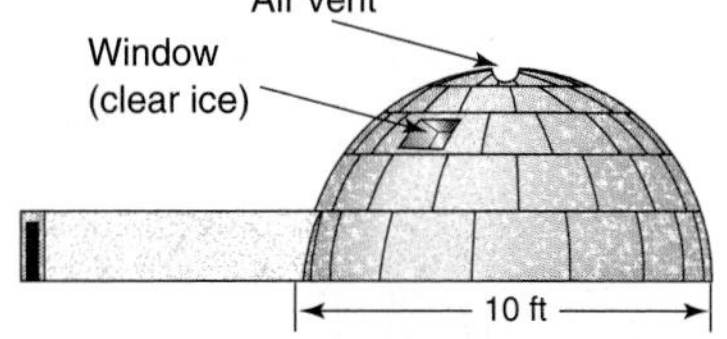

51. See margin.

51. Architecture Buckminster Fuller was an engineer and inventor who worked to create new home designs that maximized space with a minimum amount of materials. Research Fuller's designs in a reference book or on the Internet. How does his work relate to the information in this lesson?

Extension

Communication Suppose a sphere and a cube have equal surface area. Let r represent the measure of the radius of the sphere and s represent the measure of an edge of the cube. Write an equation to show the relationship between r and s.

FYI

Ice cream, by FDA standards, must contain at least 10% milk fat. A high quality ice cream contains between 16% and 20% milk fat. Ice milk contains between 2% and 7% milk fat, while sherbet has only 1% to 2% milk fat.

Additional Answers

46. Desta's destination is the polar opposite of the point of departure. Any city in the world would be on a great circle through these two points. So Mei's destination would be on the way.

51. Sample answer: Buckminster Fuller designed geodesic domes. The domes are portions of spheres.

Practice Masters, p. 70

NAME ______ DATE ______

Practice

Student Edition Pages 621–628

Surface Area and Volume of Spheres

Find the surface area and volume of each sphere described below. Round to the nearest tenth.

1. The diameter is 100 centimeters. 31,415.9 cm^2; 523,598.8 cm^3

2. A great circle has a circumference 83.92 meters. 2241.7 m^2; 9980.3 m^3

3. The radius is 12 inches long. 1809.6 in^2; 7238.2 in^3

4. A great circle has an area of 70.58 square feet. 282.3 ft^2; 446.1 ft^3

Find the surface area and volume of each solid. Round to the nearest tenth.

5. 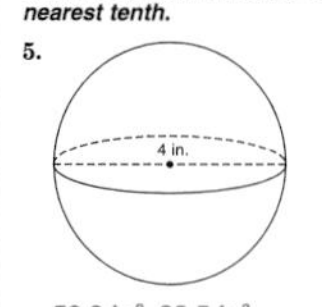

50.3 in^2, 33.5 in^3

6. 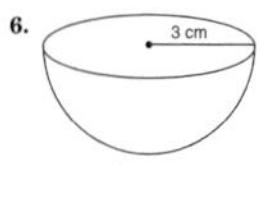

84.8 cm^2, 56.5 cm^3

7. 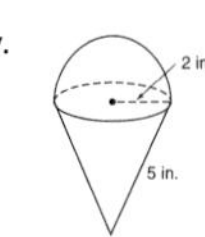

56.5 in^2, 36.0 in^3

8. 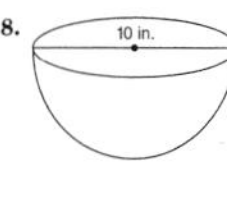

235.6 in^2, 261.8 in^3

Mixed Review

52. Refer to the figure at the right. (Lesson 11–6)
    a. Find the volume of the pyramid cut from the rectangular solid shown. **108 units^3**
    b. What is the ratio of the volume of the pyramid to the volume of the rectangular solid? **1 to 6**

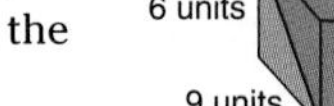

53. **Consumerism** Lucita is comparing two sports bags. One is cylindrical with a diameter of 8 inches and a length of 20 inches. The other is a rectangular prism with a base that is 8 by 8 inches and a length of 18 inches. Which bag has the greater volume? (Lesson 11–5)
54. *True* or *false*: The axis in an oblique cylinder is also an altitude. (Lesson 11–3)
55. **Probability** Nien's dartboard has a radius of 8 inches. The radius of the bull's-eye is 1.5 inches. Find the probability that a randomly thrown dart will not hit the bull's-eye. (Lesson 10–6) **0.96**
56. What are the measures of an interior angle and an exterior angle of a regular 24-gon? (Lesson 10–1) **interior: 165; exterior: 15**
57. Quadrilateral $ABCD$ is inscribed in a circle. If $m\angle A = 63$, find $m\angle C$. (Lesson 9–4) **117**
58. *True* or *false*: A radius of a circle is a chord of the circle. (Lesson 9–3) **false**
59. The bases of a trapezoid are $3x + 4$ and $5x - 2$ units long. If the median measures 21 units, what is the value of x? (Lesson 6–5) **5**
60. **Work Backward** Decrease a number by 52, multiply by 12, add 20, divide by 4, and the result is 32. What was the original number? (Lesson 5–3) **61**

53. the bag shaped like a rectangular prism
54. false

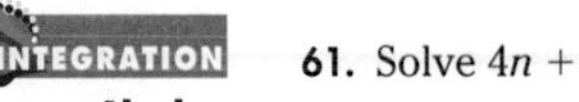

Algebra

61. Solve $4n + 7 < 15$. **$\{n \mid n < 2\}$**
62. Express 6.14×10^5 in standard notation. **614,000**

Dimples for Distance

The excerpt below appeared in an article in *Popular Science* in February, 1995.

A GOLF BALL'S PERFORMANCE IS LARGELY determined by its dimple pattern....Dimple patterns work with the ball's spin to move the ball through the air....A ball with no dimples will travel 130 yards and behave much like a bullet, with a straight trajectory. A dimpled ball, on the other hand, can travel 280 yards, rising through the air because of its lift. Dimple design has become quite sophisticated, and patterns are now complex geometric constructions, such as icosahedrons, octahelixes, and cuboctahedrons. Symmetry is a key element because it ensures that no matter how the ball is spinning, it will fly straight. Designers have become obsessed with fitting more and more of these impressions on the spheroids. Ball designers...hit on the concept that one way to fit more dimples was to increase the surface area....the solution was obvious: make a bigger ball. ■

1. A standard golf ball cannot be less than 1.68 inches in diameter. Compared to a 1.68-inch ball, how much more surface area does an oversize ball with a diameter of 1.74 inches have? (Give your answer in square inches and as a percentage.) **1–3. See margin.**
2. Why might a larger ball not travel as far as designers expect? Explain.
3. Think of another sport or game that uses a ball. Explain why the geometric characteristics of the ball, such as size, shape, and surface features, are necessary or helpful.

4 ASSESS

Closing Activity

Speaking Go around the room and have each student state one fact about spheres, their surface area, or their volume.

Mathematics and SOCIETY

On February 9, 1980, Lang Martin set a record by balancing seven golf balls vertically on a flat surface. Students might speculate about the role of the dimple in this feat. Is there a dimple design that might work better for this?

Enrichment Masters, p. 70

NAME ______________________ DATE ____________

Student Edition Pages 621–628

Enrichment

Spheres and Density

The **density** of a metal is a ratio of its mass to its volume. For example, the mass of aluminum is 2.7 grams per cubic centimeter. Here is a list of several metals and their densities.

Aluminum	2.7 g/cm^3	Copper	8.96 g/cm^3
Gold	19.32 g/cm^3	Iron	7.874 g/cm^3
Lead	11.35 g/cm^3	Platinum	21.45 g/cm^3
Silver	10.50 g/cm^3		

To calculate the mass of a piece of metal, multiply volume by density.

Example: Find the mass of a silver ball that is 0.8 cm in diameter.

$$M = D \cdot V$$
$$= 10.5 \cdot \tfrac{4}{3}\pi(0.4)^3$$
$$\approx 10.5\,(0.27)$$
$$\approx 2.83$$

The mass is about 2.83 grams.

Find the mass of each metal ball described. Assume the balls are spherical. Round your answers to the nearest tenth.

1. a copper ball 1.2 cm in diameter **8.1 g**
2. a gold ball 0.6 cm in diameter **2.2 g**
3. an aluminum ball with radius 3 cm **305.4 g**
4. a platinum ball with radius 0.7 cm **30.8 g**

Solve. Assume the balls are spherical. Round your answers to the nearest tenth.

5. A lead ball weighs 326 g. Find the radius of the ball to the nearest tenth of a centimeter. **1.9 cm**
6. An iron ball weighs 804 g. Find the diameter of the ball to the nearest tenth of a centimeter. **5.8 cm**
7. A silver ball and a copper ball each have a diameter of 3.5 cm. Which weighs more? How much more? **silver; 34.6 g**
8. An aluminum ball and a lead ball each have a radius of 1.2 cm. Which weighs more? How much more? **lead; 62.6 g**

Answers for Mathematics and Society

1. Larger ball has 0.65 square inches more surface area, an increase of 7.3%.
2. Larger ball will encounter more air resistance (or drag) that will oppose its forward motion. Larger ball could have greater mass, thus not traveling as far when the same hitting force is applied.
3. See students' work. Sports or games could include football, basketball, tennis, baseball, softball, bowling, pool, billiards, and many others.

11-8 Congruent And Similar Solids

What YOU'LL LEARN

- To identify congruent or similar solids, and
- to state properties of congruent solids.

Why IT'S IMPORTANT

You can apply what you learned about similar and congruent polygons to similar and congruent solids.

Folk Art

Artists have been making traditional matryoshka dolls in Russia and the Ukraine for centuries. Each doll in a set opens to reveal a smaller doll of the same shape. Sets of dolls usually contain 3, 6, or 12 dolls and are often painted to allow telling a story as the dolls are opened. Old dolls were hand carved, but now the dolls are made using a wood lathe. Matryoshka dolls are **similar solids**.

Similar solids are solids that have exactly the same shape but not necessarily the same size. You can determine if two solids are similar by comparing the ratios of corresponding linear measurements. For example, in the similar solids at the right, $\frac{1}{4} = \frac{3}{12} = \frac{5}{20}$. The ratio of the measures is called the **scale factor**. In two similar polyhedra, all of the corresponding faces are similar and all of the corresponding edges are proportional. Below are examples of pairs of similar and non-similar solids.

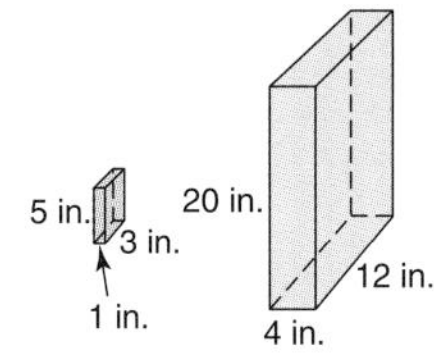

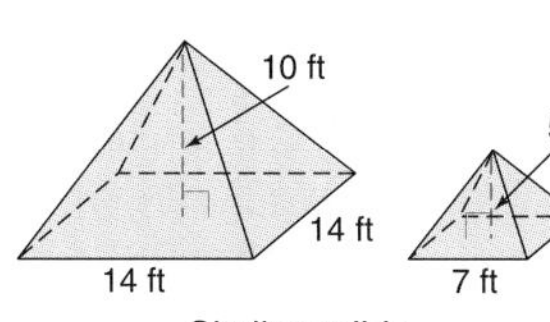

Similar solids

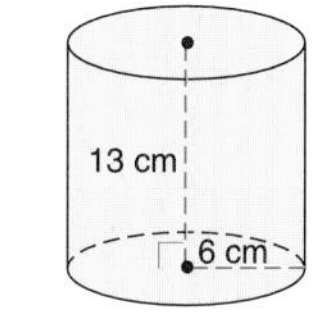

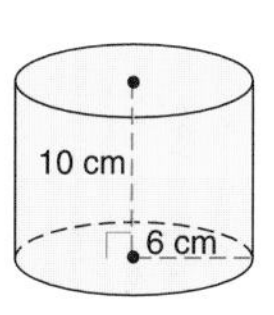

Non-similar solids

If the ratio of corresponding measurements of two solids is 1:1, then the solids are **congruent**. For two solids to be congruent, all of the following conditions must be met.

Two solids are congruent if:

- the corresponding angles are congruent,
- corresponding edges are congruent,
- areas of corresponding faces are congruent, and
- the volumes are congruent.

Congruent polyhedra are exactly the same shape and exactly the same size.

11-8 LESSON NOTES

NCTM Standards: 1–5, 7

Instructional Resources

- Study Guide Master 11-8
- Practice Master 11-8
- Enrichment Master 11-8
- Assessment and Evaluation Masters, p. 297

Transparency 11-8A contains the 5-Minute Check for this lesson; **Transparency 11-8B** contains a teaching aid for this lesson.

Recommended Pacing	
Standard Pacing	Day 16 of 18
Honors Pacing	Day 14 of 16
Block Scheduling*	Day 9 of 10

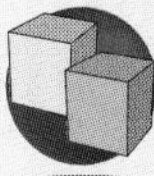

*For more information on pacing and possible lesson plans, refer to the *Block Scheduling Booklet.*

1 FOCUS

5-Minute Check
(over Lesson 11-7)

Find the surface area of each sphere to the nearest unit.

1. The radius is 10 inches. 1257 in^2
2. The diameter is 5 inches. 79 in^2
3. The circumference of a great circle is 62.8 meters. 1255 m^2

Find the volume of each sphere to the nearest hundredth.

4. The radius is 1 mile. 4.19 mi^3
5. The diameter is 10 inches. 523.60 in^3

Motivating the Lesson

Situational Problem Have students think of an example of a pair of solids in the real world that are congruent, and a pair of solids in the real world that are similar.

2 TEACH

In-Class Example

For Example 1

Determine if the pair of solids are *similar*, *congruent*, or *neither*. All measures are in inches. Explain your answer.

a. Cone A: radius = 4.3, height = 12, slant height = 14.3
Cone B: radius = 8.6, height = 25, slant height = 28.6
Neither; the ratios of the measures are not equal, so the cones are not similar and, therefore, are also not congruent.

b. Cylinder A: radius = 5.5, height = 7.3
Cylinder B: radius = 5.5, height = 7.3 **congruent**

EXPLORATION

This activity leads students to discover ratios between scale, surface area, and volume. Include more examples if students have difficulty seeing the relationships.

Example 1 **Determine if each pair of solids are *similar*, *congruent*, or *neither*.**

a.

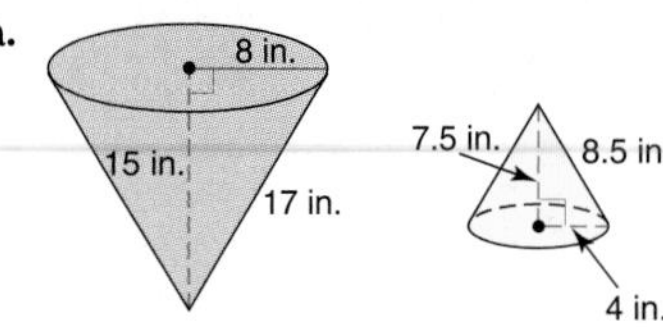

Find the ratios between corresponding parts of the cones.

$\frac{\text{radius of larger cone}}{\text{radius of smaller cone}} = \frac{8}{4} \text{ or } 2$

$\frac{\text{height of larger cone}}{\text{height of smaller cone}} = \frac{15}{7.5} \text{ or } 2$

$\frac{\text{slant height of larger cone}}{\text{slant height of smaller cone}} = \frac{17}{8.5} \text{ or } 2$

The ratios of the measures are equal, so we can conclude that the cones are similar. Since the scale factor is not 1, the solids are not congruent.

b.

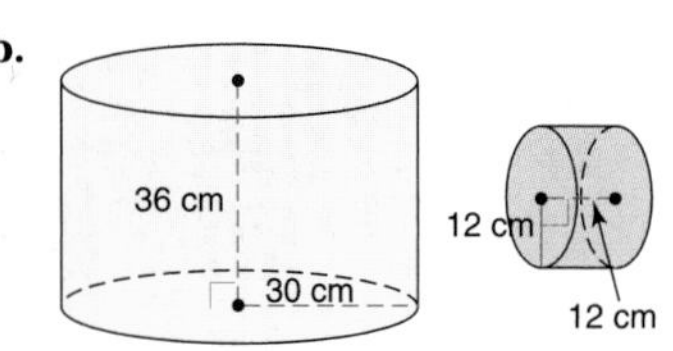

Compare the ratios between corresponding parts of the cylinders.

$\frac{\text{radius of larger cylinder}}{\text{radius of smaller cylinder}} = \frac{36}{12} \text{ or } \frac{3}{1}$

$\frac{\text{height of larger cylinder}}{\text{height of smaller cylinder}} = \frac{30}{12} \text{ or } \frac{5}{2}$

Since the ratios are not the same, the cylinders are not similar.

You can investigate the relationships between similar solids using spreadsheets.

EXPLORATION

- In column A, enter the labels *length, width, height, surface area, volume, scale factor, ratio of surface areas,* and *ratio of volumes* in the first eight cells.
- You will be using columns B, C, D, and E for four similar prisms.
- Type the formula (2B1+ 2B2)*B3 + 2*(B1*B2) in cell B4. This formula will find the surface area of a prism B. Copy the formula into the other cells in row 4.
- Write a similar formula to find the volume of a prism. Enter the formula in the cells in row 5.
- Enter the formula C1/B1 in cell C6, enter D1/B1 in cell D6, and so on. These formulas find the scale factor of prism B and each other solid.
- Type the formula C4/B4 in cell C7, type D4/B4 in cell D7, and so on. This formula will find the ratio of the surface area of prism B to the surface area of each other prism.
- Write a formula for the ratio of the volume of prism C to the volume of prism B. Enter the formula in cell C8. Enter similar formulas in cell 8 of each other column.
- Use the spreadsheet to find the surface areas, volumes, and ratios for prisms with the dimensions given on the next page.

	A	B	C	D	E
1	length	1	2	3	4
2	width	4	8	12	16
3	height	6	12	18	24
4	surface area	68	272	612	1088
5	volume	24	192	648	1536
6	scale factor		2	3	4
7	ratio of surface areas		4	9	16
8	ratio of volumes		8	27	64

Your Turn

a. Compare the ratios in cells 6, 7, and 8 of columns C, D, and E. What do you observe? **If the number in row 6 is a, then row 7 contains a^2, and row 8 contains a^3.**

b. Write a statement about the ratio between the surface areas of two solids if the scale factor is a:b. **The ratio between the surface areas is a^2:b^2.**

c. Write a statement about the ratio between the volumes of two solids if the scale factor is a:b. **The ratio between the volumes is a^3:b^3.**

The Exploration leads us to the following theorem.

Theorem 11-1 **If two solids are similar with a scale factor of a:b, then the surface areas have a ratio of a^2:b^2 and the volumes have a ratio of a^3:b^3.**

Example 2

Tourism

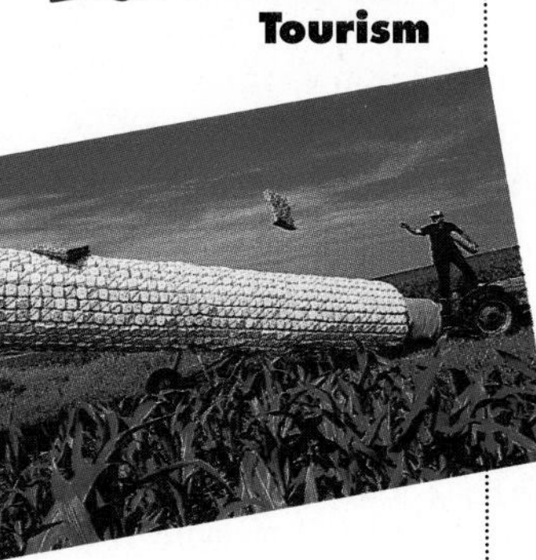

Dale Ungerer, a farmer in Hawkeye, Iowa, constructed a gigantic ear of corn to attract tourists to his farm. The ear of corn is 32 feet long and has a radius of 12 feet. Each "kernel" is a one-gallon milk jug with a volume of 231 cubic inches.

a. What is the scale factor between the gigantic ear of corn and a similar real ear of corn that is 14 inches long?

b. Estimate the volume of a kernel of the 14-inch real ear of corn.

a. Write the ratio between corresponding measures of the ears of corn.

$$\frac{\text{length of gigantic ear}}{\text{length of real ear}} = \frac{384}{14} \quad \textit{32 feet is } 32 \cdot 12 \textit{ or 384 inches.}$$
$$= \frac{192}{7}$$

The scale factor is 192:7.

b. According to Theorem 11–1 if the scale factor is a:b, then the ratio of the volumes is a^3:b^3.

$$\frac{\text{volume of gigantic ear}}{\text{volume of real ear}} = \frac{a^3}{b^3}$$
$$= \frac{192^3}{7^3} \quad a = 192,\ b = 7$$
$$= \frac{7{,}077{,}888}{343}$$

(continued on the next page)

In-Class Example

For Example 2

Jupiter has a diameter 11 times the diameter of Earth. How much greater are Jupiter's surface area and volume? **surface area: 121 times; volume: 1331 times**

Teaching Tip Remind students that Theorem 11-1 applies only if the two solids are similar.

3 PRACTICE/APPLY

Check for Understanding

Exercises 1–13 are designed to help you assess your students' understanding through reading, writing, speaking, and modeling. You should work through Exercises 1–3 with your students and then monitor their work on Exercises 4–13.

Additional Answers

1. Find the ratio between each pair of corresponding linear measures. If the ratios are all the same, then the solids are similar.
2. If two solids are similar with a scale factor of $a:b$, then the surface areas have a ratio of $a^2:b^2$ and the volumes have a ratio of $a^3:b^3$.
3. The corresponding measures of similar solids are proportional. The corresponding measures of congruent solids are congruent. All congruent solids are also similar, but not all similar solids are congruent.

Use a proportion to find the volume of a kernel on the real ear of corn.

$$\frac{7{,}077{,}888}{343} = \frac{231}{x}$$

$7{,}077{,}888x = 79{,}233$ *Multiply cross products.*

$x \approx 0.011$

The volume of a kernel on the real ear of corn is about 0.011 in^3.

CHECK FOR UNDERSTANDING

Communicating Mathematics

Study the lesson. Then complete the following. 1–3. See margin.

1. **Describe** how you can determine if two polyhedra are similar.
2. **Explain** how the surface areas and volumes of similar solids are related.
3. **Compare and contrast** similar solids and congruent solids.

Guided Practice

Determine if each pair of solids is *similar, congruent,* or *neither.*

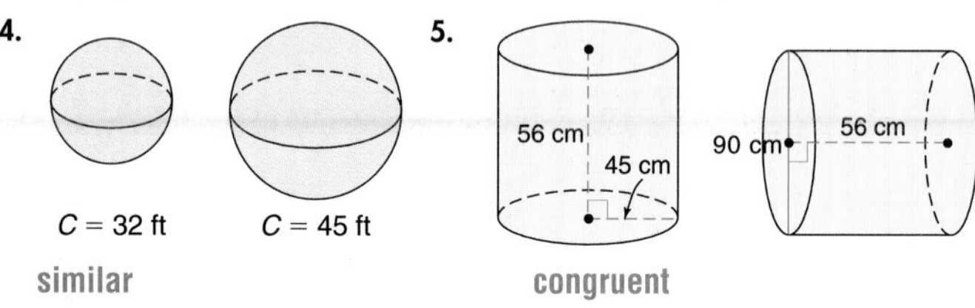

4. similar
5. congruent

Refer to the cones at the right.

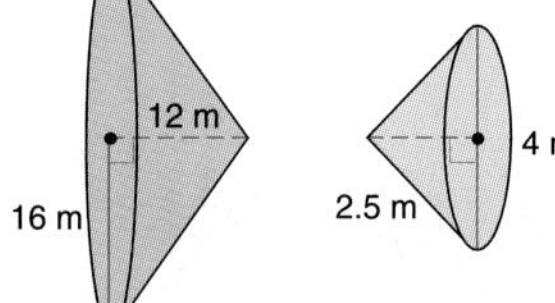

6. What is the ratio of the height of the larger cone to the height of the smaller cone? 8:1
7. Find the ratio of the surface areas. 64:1
8. What is the ratio of the circumferences?
9. If the volume of the larger cone is x cubic meters, what is the volume of the smaller cone? $\frac{x}{512}$ m^3

8. 8:1

11. False; if two pyramids have square bases and all the linear measurements are proportional, then they must be similar.

12. False; if the edge of one cube is twice that of another cube, then its surface area is four times that of the smaller cube.

Determine if each statement is *true* or *false*. If the statement is false, rewrite it so that it is true.

10. All spheres are similar. true
11. If two pyramids have square bases, then they must be similar.
12. If the edge of one cube is twice that of another cube, then its surface area is twice that of the smaller cube.
13. **World Records** According to the Guinness Book of World Records, Nippondenso's Micro-Car is the world's smallest car. It is a miniature version of Toyota's 1936 Model AA sedan. The scale factor between the full-sized car and the Micro-Car is 1000:1.
    a. If the door handle of the full-sized car is 15 centimeters long, how long is the door handle on the Micro-Car? 0.015 cm or 0.15 mm
    b. If the surface area of the Micro-Car is x square centimeters, what is the surface area of the full-sized car? $1{,}000{,}000x$ cm^2

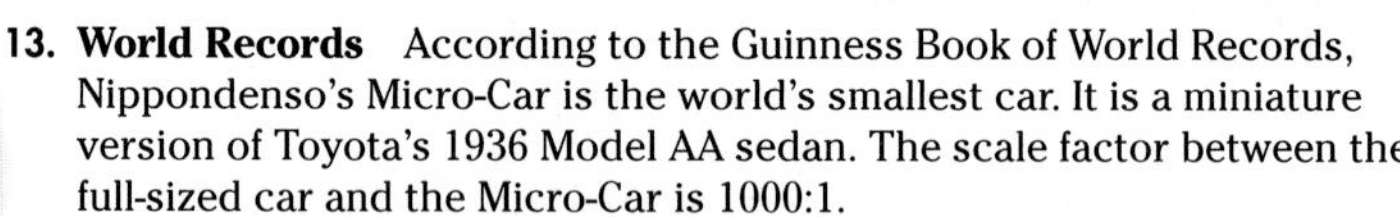

Reteaching

Using Cooperative Learning Have students make up five true/false questions about similar and congruent solids. Trade questions and have students critique them. Discuss as many of the questions as possible, especially incorrect ones.

EXERCISES

Practice **Determine if each pair of solids is *similar, congruent,* or *neither.***

14.

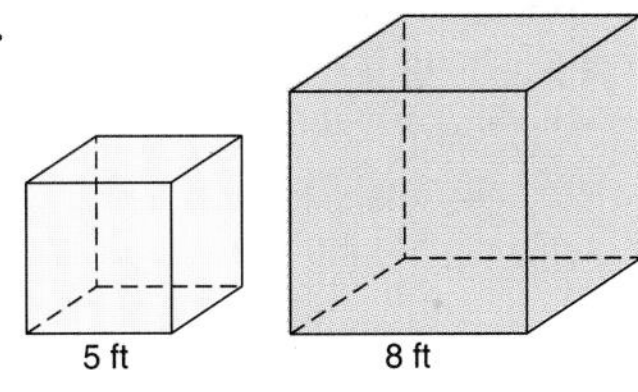

similar

15.

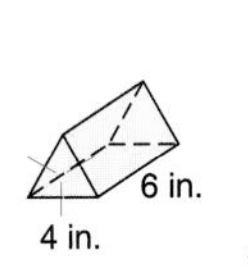

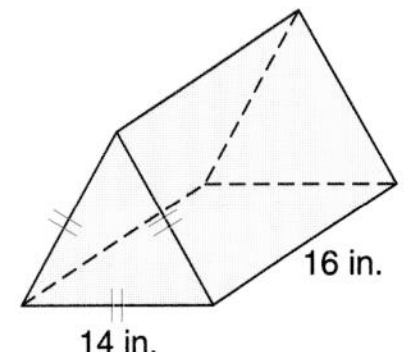

neither

16.

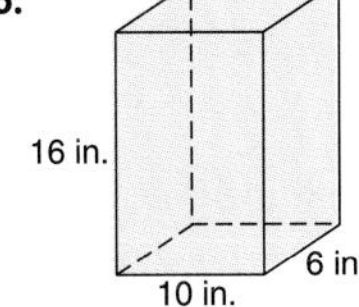

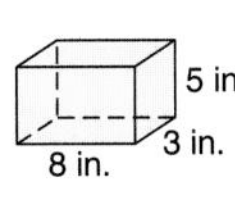

similar

17.

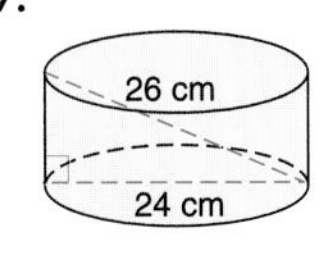

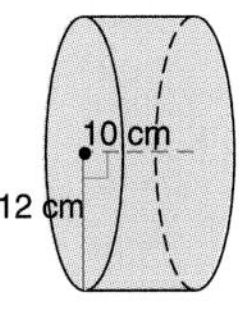

congruent

18.

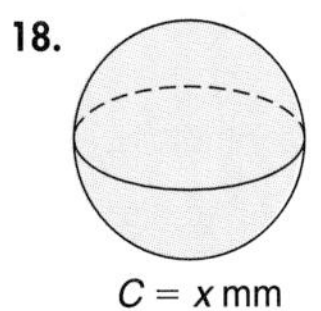

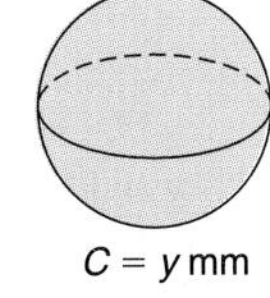

similar

19.

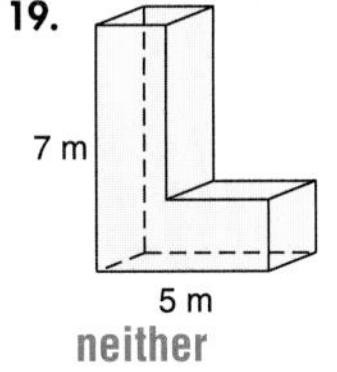

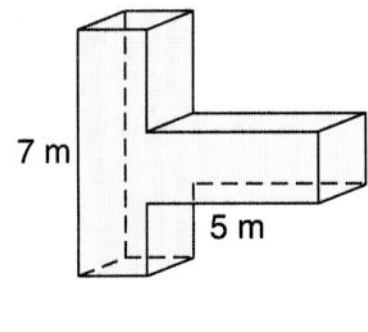

neither

The two right rectangular prisms shown at the right are similar.

2 in. 3 in.

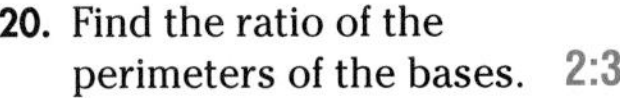

20. Find the ratio of the perimeters of the bases. 2:3

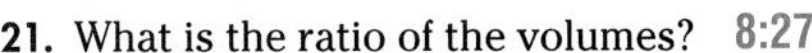

21. What is the ratio of the volumes? 8:27
22. Suppose the volume of the larger prism is 54 cm^3. Find the volume of the smaller prism. 16 cm^3
23. The diameters of two similar cylinders are in the ratio of 4 to 5. If the volume of the smaller cylinder is 48π cubic units, what is the height of the larger cylinder? $3\frac{3}{4}$ in.

Determine if each statement is *true* or *false*. If the statement is false, rewrite it so that it is true.

24. All cubes are similar. true
25. If an edge length of a cube is tripled, then its volume is nine times greater.
26. If the surface area of one of two similar pyramids is one-fourth the surface area of the other, then the volume of the smaller is $\frac{1}{8}$ of the volume of the larger. true
27. Doubling the radius of a sphere doubles the surface area.
28. If two solids are congruent, then their volumes are equal. true

25. False; if an edge length of a cube is tripled, then its volume is twenty-seven times greater.

27. False; doubling the radius of a sphere quadruples the surface area.

Assignment Guide

Core (with proof): 15–29 odd, 30, 31, 33, 35–42
Core (informal): 15–29 odd, 30, 31, 33, 35–39, 41, 42
Enriched: 14–28 even, 30–42

For **Extra Practice**, see p. 787.

The red A, B, and C flags, printed only in the Teacher's Wraparound Edition, indicate the level of difficulty of the exercises.

Study Guide Masters, p. 71

NAME________ DATE________

Study Guide

Student Edition Pages 629–635

Congruent and Similar Solids

Solids that have the same shape but are different in size are said to be **similar**. You can determine if two solids are similar by comparing the ratios **(scale factors)** of corresponding linear measurements. If the scale factor is 1:1, then the solids are **congruent**.

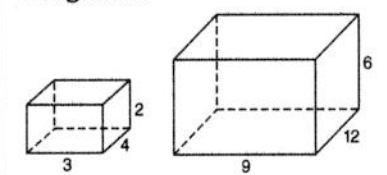

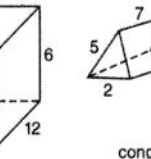

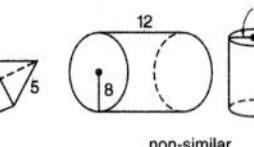

Determine if each pair of solids is similar, congruent, or neither.

1.

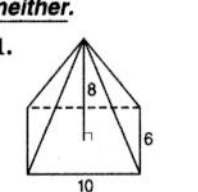

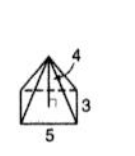

similar

2.

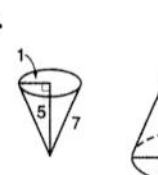

neither

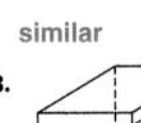

3.

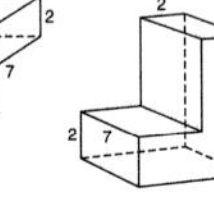

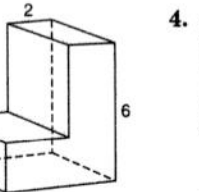

congruent

4.

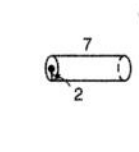

similar

5.

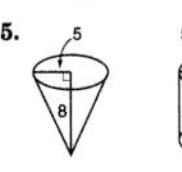

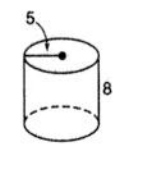

neither

6.

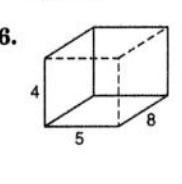

congruent

29. When a cone is cut by a plane parallel to its base, a cone similar to the original is formed. Suppose the slant height of the smaller cone is half that of the original.
   a. What is the ratio of the volume of the frustum to that of the original cone? to the smaller cone? 7:8; 7:1
   b. What is the ratio of the surface area of the frustum to that of the original cone? to the smaller cone? 3:4; 3:1

Critical Thinking

30. Is there a value for x for which the rectangular prisms at the right would be similar? Explain. See margin.

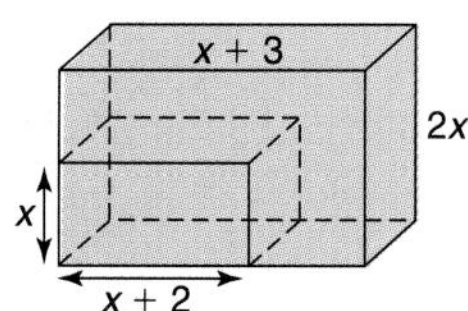

31. Write a convincing argument to show that all spheres are similar. See margin.

Applications and Problem Solving

32. **Literature** Read the poem *One Inch Tall* by Shel Silverstein.

ONE INCH TALL

If you were only one inch tall, you'd ride a worm to school.
The teardrop of a crying ant would be your swimming pool.
A crumb of cake would be a feast
And last you seven days at least,
A flea would be a frightening beast
If you were one inch tall.

If you were one inch tall, you'd walk beneath the door,
And it would take about a month to get down to the store.
A bit of fluff would be your bed,
You'd swing upon a spider's thread,
And wear a thimble on your head
If you were one inch tall.

You'd surf across the kitchen sink upon a stick of gum.
You couldn't hug your mama, you'd just have to hug her thumb.
You'd run from people's feet in fright,
To move a pen would take all night,
(This poem took fourteen years to write —
'Cause I'm just one inch tall.)

   a. Choose one of the statements in the poem and determine whether it is correct. For example, could you swing upon a spider's thread if you were one inch tall? See students' work.
   b. A person who is 6 feet tall has lungs that have a surface area of 810 ft^2 and a volume of $\frac{1}{2}$ liter. Estimate the surface area and volume of your lungs if you were one inch tall.

32b. $\frac{810}{5184}$ ft^2 or 22.5 in^2; $\frac{\frac{1}{2}}{72^3}$ liters or about 0.0013 mL

33. **Architecture** Citizens of Padova, Italy built a model of the Basilica di Sant' Antonio di Padova from empty beverage cans. The model was in a scale of 1:4. The model measured 96 feet by 75 feet by 56 feet. Find the dimensions of the actual Basilica di Sant' Antonio di Padova. 384 ft × 300 ft × 224 ft

FYI

The smallest living adult is Madge Bester of Johannesburg, South Africa. She is $25\frac{1}{2}$ inches tall.

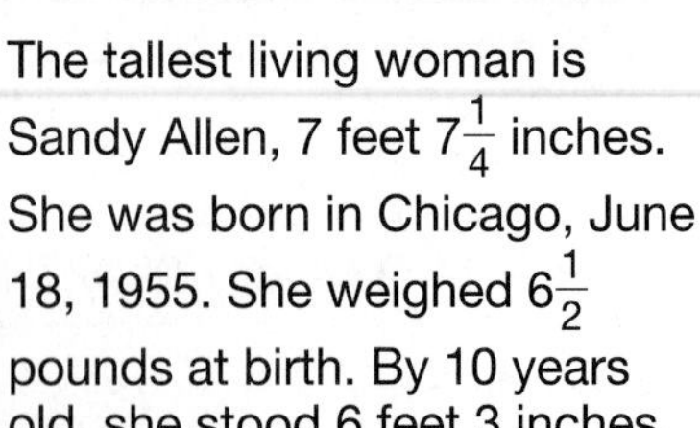

FYI

The tallest living woman is Sandy Allen, 7 feet $7\frac{1}{4}$ inches. She was born in Chicago, June 18, 1955. She weighed $6\frac{1}{2}$ pounds at birth. By 10 years old, she stood 6 feet 3 inches. In 1977 she had a pituitary gland operation to inhibit further growth.

Additional Answers

30. No; the values for which corresponding lengths would be proportional are 0 and −1. Since neither of these answers gives a length that is possible geometrically, the situation is impossible.
31. Since the only linear measure involved in a sphere is the radius, the linear measures of two spheres are always proportional.

Practice Masters, p. 71

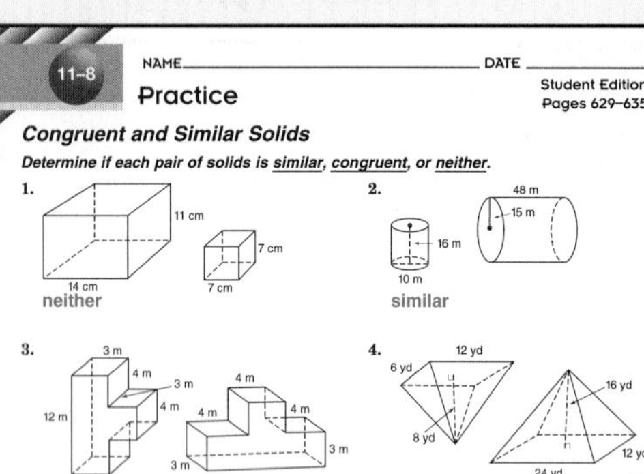

11-8 NAME____ DATE____

Practice — Student Edition Pages 629–635

Congruent and Similar Solids

Determine if each pair of solids is similar, congruent, or neither.

1. neither
2. similar
3. congruent
4. similar

The two right rectangular prisms shown at the right are similar.

5. Find the ratio of the perimeters of the bases. $\frac{7}{5}$
6. What is the ratio of the surface areas? $\frac{49}{25}$
7. Suppose the volume of the smaller prism is 60 in^3. Find the volume of the larger prism. 84 in^3

Determine if each statement is true or false. If the statement is false, rewrite it so that it is true.

8. If two cylinders are similar, then their volumes are equal. False; if two cylinders are congruent, then their volumes are equal.
9. Doubling the height of a cylinder doubles the volume. True
10. Two solids are congruent if they have the same shape. False; two solids are congruent if they have the same shape and size.

34. Food The world's largest cherry pie was made by the Oliver Rotary Club of Oliver, British Columbia, Canada. It measured 20 feet in diameter and was completed on July 14, 1990. Most pies are 8 inches in diameter. If the largest pie was similar to a standard pie, how much greater was the volume of the largest pie?
27,000 times as large

Mixed Review

35. World Records The world's largest ball of string is 13 feet $2\frac{1}{2}$ inches in diameter. It took 2 years to complete and is now on display in Valley View, Texas. Find the surface area and volume of the ball of string. (Lesson 11–7)
$T = 548.08$ ft^2; $V = 1206.5$ ft^3

36. Find the volume of the solid formed by the net shown at the right. (Lesson 11–5)
360 cm^3

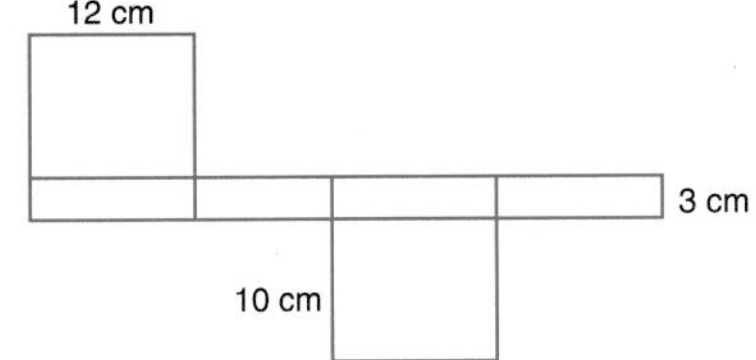

37. Make a Model Inez gave her little brother a set of blocks for his third birthday. The cubical blocks all have edges 5 centimeters long, and the blocks fit perfectly in the cubical box in which they came. Inez and her brother built a tower consisting of three cubes of descending size made up of blocks. They used every block in the set to make the tower, which Inez's brother could just see over. How high was the tower? (Lesson 11–1) **60 cm**

38. The coordinates of the vertices of a quadrilateral are $(-2, 3)$, $(3, 5)$, $(-2, -6)$, and $(3, -4)$. Graph the quadrilateral and determine whether it is a *square*, a *rectangle*, or a *parallelogram*. Then find its area. (Lesson 10–3)
parallelogram; 45 units^2

39. Solve $\triangle DFG$ if $m\angle D = 45$, $m\angle G = 37$, and $DG = 15$. (Lesson 8–5)
$m\angle F = 98$, $DF \approx 9.1$, $FG \approx 10.7$

40. Given: $\angle S \cong \angle W$
$\overline{SY} \cong \overline{WY}$

Prove: $\overline{ST} \cong \overline{WV}$ (Lesson 4–4)
See margin.

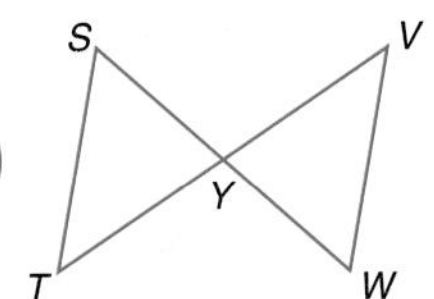

Algebra

41. Determine which ordered pairs are solutions of $y \geq 8x + 2$. **a, d**

a. $(0, 2)$ b. $(3, 24)$ c. $(1, -1)$ d. $(-4, -30)$

42. Probability The graph at the right shows the chances of an amateur bowling a perfect 300 game.

a. What is the probability of a man bowling a perfect game? $\frac{1}{12,500}$

b. What are the odds of a woman bowling a perfect game? $\frac{1}{643,999}$

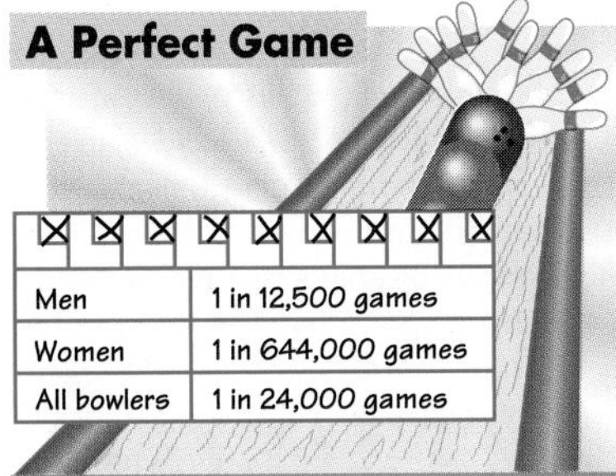
A Perfect Game

Men	1 in 12,500 games
Women	1 in 644,000 games
All bowlers	1 in 24,000 games

Source: American Bowling Congress

4 ASSESS

Closing Activity

Writing Have students write a few sentences explaining similar solids to someone who has never taken geometry. They should use their knowledge of similar polygons.

Chapter 11 Quiz D (Lessons 11-7 and 11-8) is available in the *Assessment and Evaluation Masters*, p. 297.

Additional Answer

40. Given: $\angle S \cong \angle W$
$\overline{SY} \cong \overline{YW}$
Prove: $\overline{ST} \cong \overline{WV}$
Proof:
Statements (Reasons)
1. $\angle S \cong \angle W$, $\overline{SY} \cong \overline{YW}$ (Given)
2. $\angle SYT \cong \angle WYV$ (Vert. ∠s are ≅.)
3. $\triangle SYT \cong \triangle WYV$ (ASA)
4. $\overline{ST} \cong \overline{WV}$ (CPCTC)

Enrichment Masters, p. 71

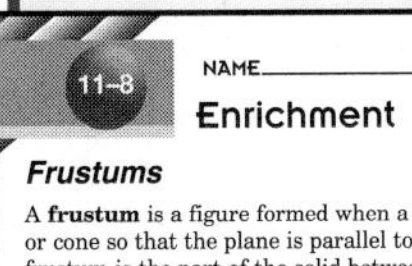
11-8 NAME______ DATE______
Enrichment
Student Edition Pages 629–635

Frustums

A **frustum** is a figure formed when a plane intersects a pyramid or cone so that the plane is parallel to the solid's base. The frustum is the part of the solid between the plane and the base. To find the volume of a frustum, the areas of both bases must be calculated and used in the formula,

$V = \frac{1}{3}h(B_1 + B_2 + \sqrt{B_1B_2})$,

where h = height (perpendicular distance between the bases), B_1 = area of top base, and B_2 = area of bottom base

Describe the shape of the bases of each frustum. Then find the volume. Round to the nearest tenth.

1.
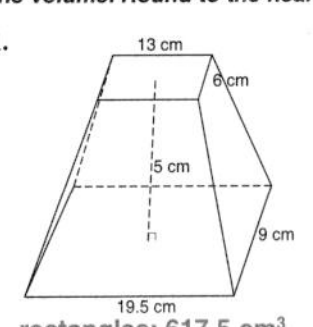

rectangles; 617.5 cm^3

2.
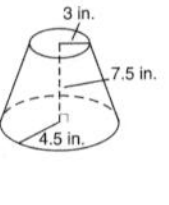

circles; 335.8 in^3

3.
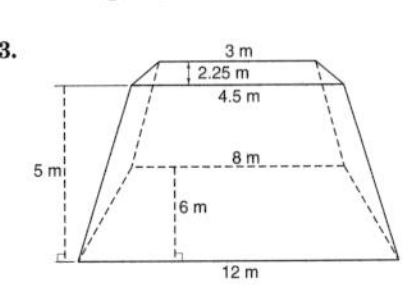

trapezoids; 174.2 m^3

4.
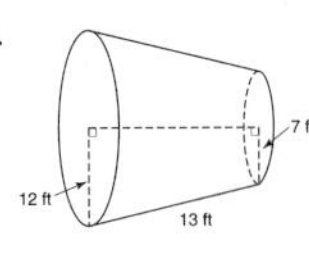

circles; 3543.7 ft^3

Extension

Communication Have students explain how to construct a solid that is congruent to a given solid.

In•ves•ti•ga•tion

TEACHER NOTES

Closing the Investigation

This activity provides students an opportunity to bring their work on the Investigation to a close. For each Investigation, students should present their findings to the class. Here are some ways students can display their work.

- Conduct and report on an interview or survey.
- Write a letter, proposal, or report.
- Write an article for the school or local paper.
- Make a display, including graphs and/or charts.
- Plan an activity.

Assessment

To assess students' understanding of the concepts and topics explored in the Investigation and its follow-up activities, you may wish to examine students' Investigation Folders.

The scoring guide provided in the *Investigations and Projects Masters*, p. 19, provides a means for you to score students' work on the Investigation.

Investigations and Projects Masters, p. 19

Scoring Guide
Chapters 10 and 11
Investigation

Level	Specific Criteria
3 Superior	• Shows thorough understanding of the concepts of *design pattern, tessellation, angle measures, polygon, area, volume, surface area, perimeter,* and *diameter.* • Uses appropriate strategies to solve problems. • Computations are correct. • Written explanations are exemplary. • Models, blueprints, design report, cost analysis, and presentation are appropriate and sensible. • Goes beyond requirements of all or some problems.
2 Satisfactory, with Minor Flaws	• Shows understanding of the concepts of *design pattern, tessellation, angle measures, polygon, area, volume, surface area, perimeter,* and *diameter.* • Uses appropriate strategies to solve problems. • Computations are mostly correct. • Written explanations are effective. • Models, blueprints, design report, cost analysis, and presentation are appropriate and sensible. • Satisfies all requirements of problems.
1 Nearly Satisfactory, with Obvious Flaws	• Shows understanding of most of the concepts of *design pattern, tessellation, angle measures, polygon, area, volume, surface area, perimeter,* and *diameter.* • May not use appropriate strategies to solve problems. • Computations are mostly correct. • Written explanations are satisfactory. • Models, blueprints, design report, cost analysis, and presentation are appropriate and sensible. • Satisfies most requirements of problems.
0 Unsatisfactory	• Shows little or no understanding of the concepts of *design pattern, tessellation, angle measures, polygon, area, volume, surface area, perimeter,* and *diameter.* • Does not use appropriate strategies to solve problems. • Computations are incorrect. • Written explanations are not satisfactory. • Models, blueprints, design report, cost analysis, and presentation are not appropriate or sensible. • Does not satisfy requirements of problems.

CLOSING THE

In•ves•ti•ga•tion

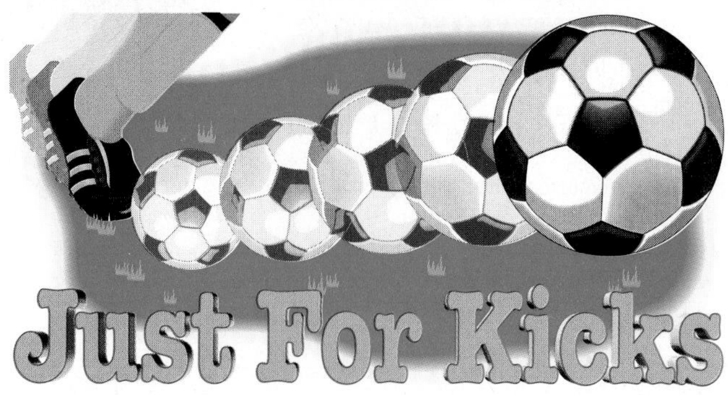

Refer to the Investigation on pages 510–511.

Soccer players must master a variety of skills to be effective. They include *kicking* the ball, *passing* the ball to another player, *heading* or hitting the ball with the head, *dribbling* or moving the ball while running, and *tackling* or taking the ball from an opponent.

Analyze

You have investigated the designs for the two most popular soccer balls. It is now time to analyze your findings and complete your soccer ball design.

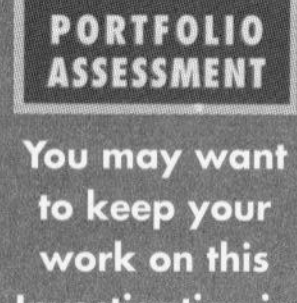

PORTFOLIO ASSESSMENT

You may want to keep your work on this Investigation in your portfolio.

1 Look over the observations you made about the two most popular soccer ball designs. Note the characteristics that are the same and those that are different. List the characteristics you will incorporate in your ball design.

2 Use various polygons to create a spherical shaped object. Build a design model using construction paper and tape. Draw a blueprint design of the ball showing at least three different views of the soccer ball. List the types of polygons used in the design. For each polygon, state the quantity needed to construct a ball.

3 Your company will produce soccer balls in sizes 3, 4, and 5. The circumference of each type of ball is shown in the table.

Size	Circumference (inches)
3	23
4	25
5	27–28

Prepare a design specification report for all three sizes of soccer balls. For each size of ball, state the volume, diameter, circumference, and surface area. Also, list the dimensions of each polygon used in your design, including the length of each side, the measure of an interior angle, the perimeter, and the area of each polygon.

4 Research the cost of leather and vinyl. Then estimate the cost of materials for a ball of each size. If possible, interview a sporting goods manufacturer regarding the cost of labor for constructing a product and the selling price. Complete a cost analysis and suggest a retail price for each size of soccer ball.

Present

The class represents the board of directors of your company. Present your proposed design model and a written report.

5 Summarize the process you used for creating your design and write a justification to the executive board explaining the advantages of your design.

6 Select a name for your ball. Explain a plan for marketing your soccer ball to attract sales to children, youth, and adult players.

HIGHLIGHTS

VOCABULARY

After completing this chapter, you should be able to define each term, property, or phrase and give an example or two of each.

Geometry

altitude (pp. 591, 593, 600, 602)
axis (pp. 593, 602)
base (pp. 577, 591, 593, 600, 602)
Cavalieri's Principle (p. 617)
center (p. 621)
chord (p. 621)
circular cone (p. 602)
cone (p. 577)
congruent solids (p. 629)
corner view (p. 576)
cross section (pp. 574, 577)
cube (p. 577)
cylinder (pp. 574, 577, 593)
diameter (p. 621)
edge (p. 577)
face (p. 577)
great circle (p. 622)
height (pp. 591, 602)
hemisphere (p. 622)
lateral area (p. 592)
lateral edge (pp. 591, 600)
lateral face (pp. 577, 591, 600)
net (p. 585)
oblique cone (p. 602)
oblique cylinder (p. 593)
oblique prism (p. 591)
perspective view (p. 576)
Platonic solids (p. 577)
polyhedron (p. 577)
prism (p. 577)
pyramid (p. 577)
radius (p. 621)
regular polyhedron (p. 577)
regular prism (p. 577)
regular pyramid (p. 600)
right cone (p. 602)
right cylinder (p. 593)
right prism (p. 591)
scale factor (p. 629)
similar solids (p. 629)
slant height (pp. 600, 602)
slice (pp. 574, 577)
solid (p. 574)
sphere (pp. 577, 621)
surface area (pp. 584, 592)
tangent (p. 621)
triangular prism (p. 574)
vertex (pp. 600, 602)
volume (p. 607)

Problem Solving

make a model (p. 575)

UNDERSTANDING AND USING THE VOCABULARY

Choose the correct best term to complete each statement.

1. If a plane that is parallel to the base or bases of a solid slices the solid, then the intersection of the plane and the solid is a (cross section, corner view).
2. A (cylinder, prism) is a polyhedron with two congruent faces that are polygons in parallel planes.
3. If the axis of a cone is perpendicular to the base of the cone, then the cone is a(n) (right cone, oblique cone).
4. A (net, slice) is a two-dimensional figure that shows the shapes that could be folded to form a three-dimensional figure.
5. The lateral area of a solid is the area of the (lateral faces, lateral edges).
6. In a regular pyramid, the height of a lateral face is the (altitude, slant height) of the pyramid.
7. In a (right, regular) prism, the bases are regular polygons, and the lateral faces are perpendicular to the bases.
8. A ball bearing is a model of a (sphere, great circle).
9. If two solids are (similar, congruent), then they have the same shape but are different sizes.
10. The height of one solid is a units and the height of a similar solid is b units. Then the scale factor is (a:b, a^2:b^2).

Using the CHAPTER HIGHLIGHTS

The Chapter Highlights begins with a listing of the new terms, properties, and phrases that were introduced in this chapter. Have students define each term and provide an example or two of it, if appropriate.

Assessment and Evaluation Masters, pp. 283–284

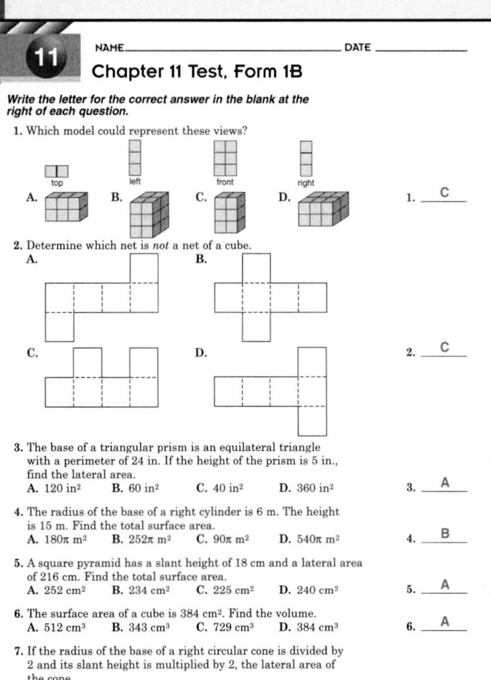
11 NAME ______ DATE ______

Chapter 11 Test, Form 1B

Write the letter for the correct answer in the blank at the right of each question.

1. Which model could represent these views?

A. B. C. D. — 1. C

2. Determine which net is *not* a net of a cube.
A. B. C. D. — 2. C

3. The base of a triangular prism is an equilateral triangle with a perimeter of 24 in. If the height of the prism is 5 in., find the lateral area.
A. 120 in² B. 60 in² C. 40 in² D. 360 in² — 3. A

4. The radius of the base of a right cylinder is 6 m. The height is 15 m. Find the total surface area.
A. 180π m² B. 252π m² C. 90π m² D. 540π m² — 4. B

5. A square pyramid has a slant height of 18 cm and a lateral area of 216 cm. Find the total surface area.
A. 252 cm² B. 234 cm² C. 225 cm² D. 240 cm² — 5. A

6. The surface area of a cube is 384 cm². Find the volume.
A. 512 cm³ B. 343 cm³ C. 729 cm³ D. 384 cm³ — 6. A

7. If the radius of the base of a right circular cone is divided by 2 and its slant height is multiplied by 2, the lateral area of the cone
A. is decreased. B. remains the same.
C. is increased. D. cannot be determined. — 7. B

NAME ______ DATE ______

Chapter 11 Test, Form 1B (continued)

For Questions 8–10, refer to the figure at the right.

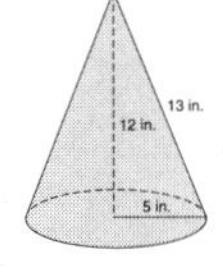

8. Find the lateral area.
A. 60π cm² B. 325π cm²
C. 130π cm² D. 65π cm² — 8. D

9. Find the total surface area.
A. 75π cm² B. 70π cm²
C. 90π cm² D. 85π cm² — 9. C

10. Find the volume.
A. 108π cm³ B. 100π cm³
C. 40π cm³ D. 260π cm³ — 10. B

11. The volume of a sphere is 36π cm³. Find the radius.
A. 9 cm B. 3 cm C. 6 cm D. 27 cm — 11. B

12. Find the volume of the space between the cylinder and the rectangular prism to the nearest cubic inch.
A. 1728 in³ B. 1239 in³
C. 2688 in³ D. 1432 in³ — 12. B

5 in. 16 in. 12 in. 13 in.

13. The radius of the base of a soup can is 5 centimeters, and the height is 20 centimeters. Two dozen cans are to be packed in a cardboard box. Which of the following are the dimensions of a box in which they can be packed using the least amount of cardboard?
A. 30 cm × 40 cm × 40 cm B. 30 cm × 80 cm × 20 cm
C. 60 cm × 40 cm × 20 cm D. 240 cm × 10 cm × 20 cm — 13. A

14. A sphere with radius of 6 cm is inscribed in a cube. Find the volume of the cube.
A. 216 cm³ B. 864 cm³ C. 1728 cm³ D. 904 cm³ — 14. C

15. The volume of the larger prism is 108 cm³. If the prisms are similar, what is the volume of the smaller prism?
3 cm 2 cm
A. 364.5 cm³ B. 48 cm³ C. 72 cm³ D. 32 cm³ — 15. D

Bonus

A sphere is inscribed in a cube. Find the ratio of the volume of the cube to the volume of the sphere.

A. $2:\frac{1}{3}\pi$ B. $2:3\pi$ C. $1:\frac{4}{3}\pi$ D. $1:\frac{2}{3}\pi$ — Bonus A

Instructional Resources

Three multiple-choice tests and three free-response tests are provided in the *Assessment and Evaluation Masters*. Forms 1A and 2A are for honors pacing, Forms 1B and 2B are for average pacing, and Forms 1C and 2C are for basic pacing. Chapter 11 Test, Form 1B, is shown at the right. Chapter 11 Test, Form 2B, is shown on the next page.

Postulates, Theorems, and Corollaries

A complete list of postulates, theorems, and corollaries begins on page 806.

Using the STUDY GUIDE AND ASSESSMENT

Skills and Concepts Encourage students to refer to the objectives and examples on the left as they complete the review exercises on the right.

Assessment and Evaluation Masters, pp. 289–290

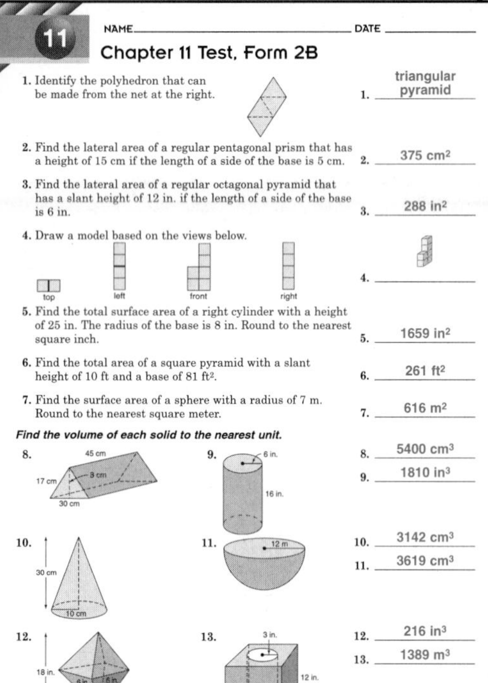

11 NAME________________ DATE ________

Chapter 11 Test, Form 2B

1. Identify the polyhedron that can be made from the net at the right. 1. triangular pyramid
2. Find the lateral area of a regular pentagonal prism that has a height of 15 cm if the length of a side of the base is 5 cm. 2. 375 cm²
3. Find the lateral area of a regular octagonal pyramid that has a slant height of 12 in. if the length of a side of the base is 6 in. 3. 288 in²
4. Draw a model based on the views below. 4.

top left front right

5. Find the total surface area of a right cylinder with a height of 25 in. The radius of the base is 8 in. Round to the nearest square inch. 5. 1659 in²
6. Find the total area of a square pyramid with a slant height of 10 ft and a base of 81 ft². 6. 261 ft²
7. Find the surface area of a sphere with a radius of 7 m. Round to the nearest square meter. 7. 616 m²

Find the volume of each solid to the nearest unit.

8. 45 cm, 17 cm, 8 cm, 30 cm 8. 5400 cm³
9. 6 in., 16 in. 9. 1810 in³
10. 30 cm, 10 cm 10. 3142 cm³
11. 12 m 11. 3619 cm³
12. 18 in., 6 in., 6 in. 12. 216 in³
13. 3 in., 12 in., 12 in. 13. 1389 m³

11 NAME________________ DATE ________

Chapter 11 Test, Form 2B (continued)

For Questions 14 and 15, refer to the partial right cylinder at the right.

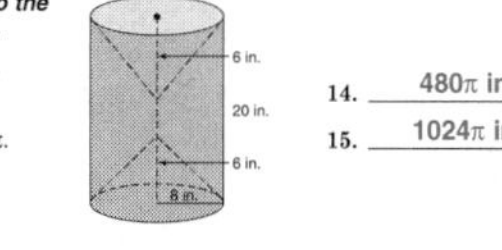

14. Find the total surface area in terms of π. 14. 480π in²
15. Find the volume in terms of π. 15. 1024π in³
16. Find the volume of a cone with a height of 18 ft if the radius of the base is 8 ft. Round to the nearest cubic foot. 16. 1206 ft³
17. The surface area of a cube is 486 in². Find the volume. 17. 729 in³
18. Two rectangular prisms are similar. The measures of two corresponding sides are 4 cm and 5 cm. What is the ratio of the volumes of the prisms? 18. 64 : 125
19. To the nearest tenth of a pound, find the weight of a cast-iron water pipe 4 ft long, with an outside diameter of 2 in. and a thickness of $\frac{1}{4}$ in. Iron weighs 0.26 lb/in³. 19. 17.2 lb
20. A set of eight game pieces contains four cubes that measure 3 cm on an edge and four triangular prisms with a height of 3 cm and bases with sides that are 3 cm, 3 cm, and $3\sqrt{2}$ cm. Make a model to find the dimensions of a box in which they can be packed with no extra space. 20. Sample answer: 6 cm by 9 cm by 3 cm

Bonus

A sphere is circumscribed about a regular octahedron. Find the ratio of the volume of the octahedron to the volume of the sphere. Bonus 1 : π

SKILLS AND CONCEPTS

OBJECTIVES AND EXAMPLES

Upon completing this chapter, you should be able to:

- use top, front, side, and corner views of three-dimensional solids to make models (Lesson 11–1)

From the views given, draw a perspective view.

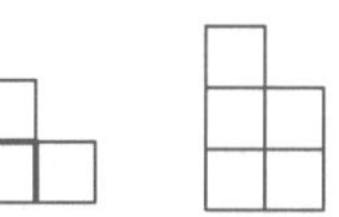
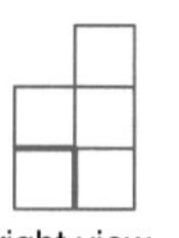

top view left view front view right view

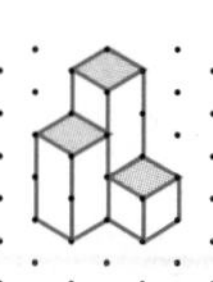

The top view tells us that the base has 3 blocks and that the columns have different heights. The other views allow us to determine the height of each column.

REVIEW EXERCISES

Use these exercises to review and prepare for the chapter test. **11–12. See margin.**

11. From the views of a solid figure given, draw a corner view.

top view left view front view right view

12. The corner view of a figure is given. Draw the top, left, front, right, and back views of the solid.

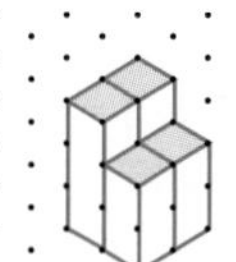

- make two-dimensional nets for three-dimensional solids (Lesson 11–2)

A polyhedron and one of its two-dimensional nets are shown below.

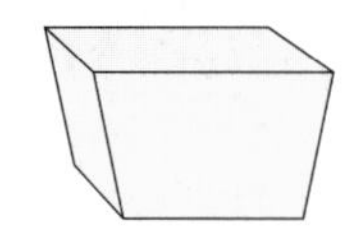
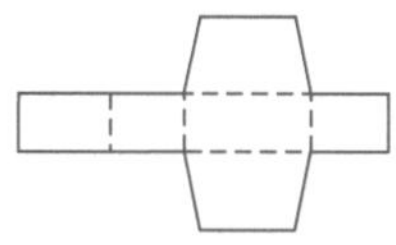

13. Use isometric dot paper to draw a rectangular prism that is 4 units high, 7 units long, and 2 units wide. **See margin.**
14. Use rectangular dot paper to draw a net for the solid shown at the right. Then find the surface area of the solid. **See margin.**

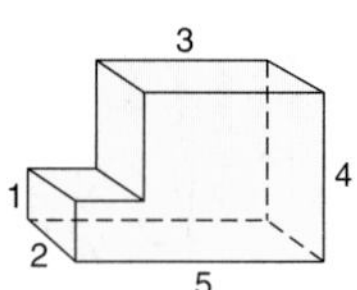

- find the lateral areas and surface areas of right prisms and right cylinders (Lesson 11–3)

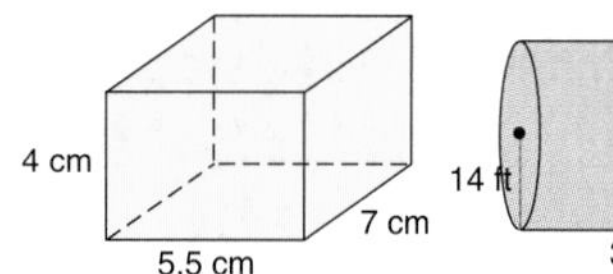

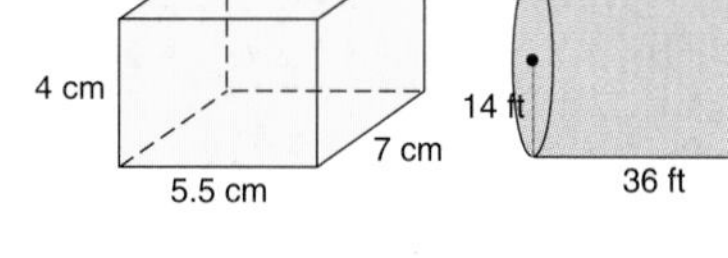

$L = Ph$
$= (25)(4)$
$= 100 \text{ cm}^2$
$T = Ph + 2B$
$= 100 + 2(5.5 \cdot 7)$
$= 177 \text{ cm}^2$

$L = 2\pi rh$
$= 2\pi(14)(36)$
$\approx 3166.7 \text{ ft}^2$
$T = 2\pi rh + 2\pi r^2$
$\approx 3166.7 + 2\pi(14)^2$
$\approx 4398.2 \text{ ft}^2$

Find the lateral area and the surface area of each right prism or right cylinder. Round to the nearest tenth. 15–18. See margin.

15.
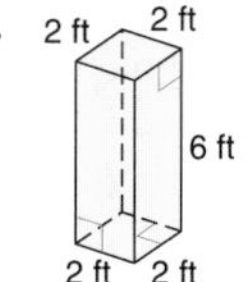

16.
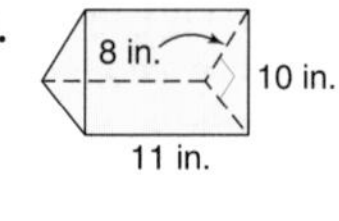

17.
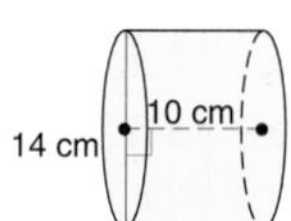

18.
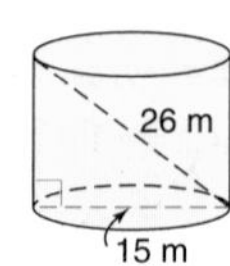

GLENCOE Technology

Test and Review Software

You may use this software, a combination of an item generator and item bank, to create your own tests or worksheets. Types of items include free response, multiple choice, short answer, and open ended.

For IBM & Macintosh

Additional Answer

11.

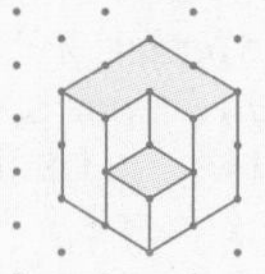

OBJECTIVES AND EXAMPLES

- find the lateral areas and surface areas of regular pyramids and right circular cones (Lesson 11–4)

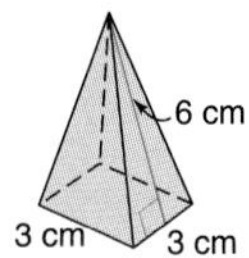

$L = \frac{1}{2}P\ell$

$= \frac{1}{2}(12)(6)$ or 36 cm^2

$T = L + B$

$= 36 + 9$

$= 45 \text{ cm}^2$

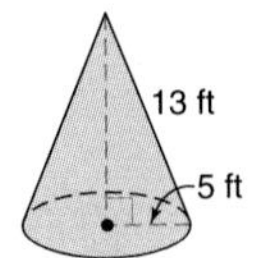

$L = \pi r\ell$

$= \pi(5)(13)$

≈ 204.2 ft

$T = \pi r\ell + \pi r^2$

$= 204.2 + \pi(5^2)$

$\approx 282.7 \text{ ft}^2$

REVIEW EXERCISES

Find the lateral area and the surface area of each regular prism or right cone. Round to the nearest tenth. 20. $L = 47.1 \text{ ft}^2$; $T = 75.4 \text{ ft}^2$

19. $L = 20 \text{ in}^2$ $T = 24 \text{ in}^2$

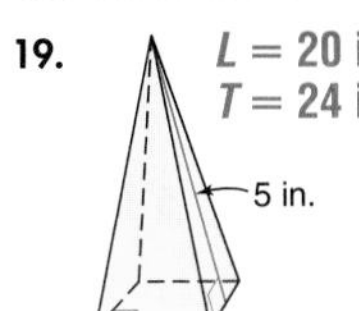

20.

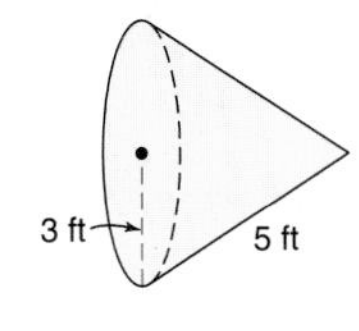

21.

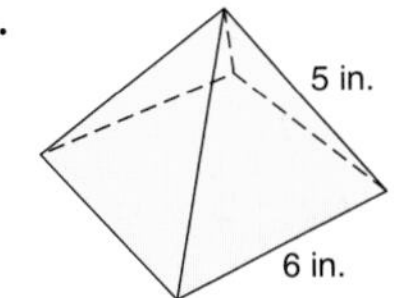

22.

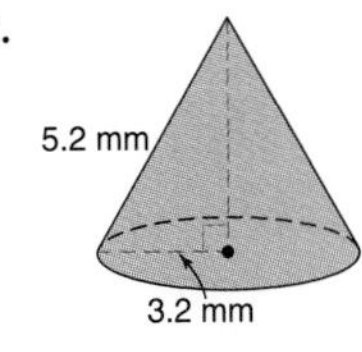

21. $L = 48 \text{ in}^2$, $T = 84 \text{ in}^2$

22. $L = 52.3 \text{ mm}^2$; $T = 84.4 \text{ mm}^2$

- find the volumes of right prisms and right cylinders (Lesson 11–5)

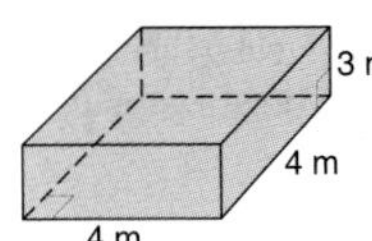

$V = Bh$

$= (16)(3)$

$= 48 \text{ m}^3$

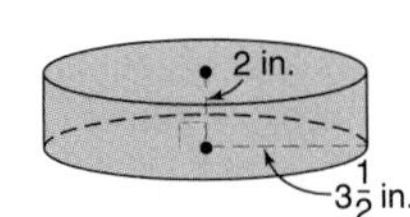

$V = \pi r^2 h$

$= \pi\left(3\frac{1}{2}\right)^2 2$

$\approx 77.0 \text{ in}^3$

Find each of the following. Round to the nearest tenth.

23. What is the volume of a hexagonal prism if its radius is 10 centimeters and its height is 20 centimeters? 5196.2 cm^3

24. A right cylinder has a radius of 10 centimeters and a height of 20 centimeters. Find the volume of the cylinder. 6283.2 cm^3

25. Find the volume of a right cylinder if its diameter is 10 feet and its height is 13 feet. 1021.0 ft^3

- find the volumes of circular cones and pyramids (Lesson 11–6)

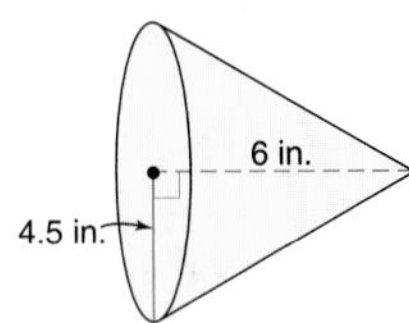

$V = \frac{1}{3}Bh$

$= \frac{1}{3}\pi(4.5^2)(6)$

$\approx 127.2 \text{ in}^3$

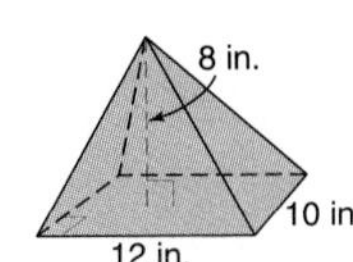

$V = \frac{1}{3}Bh$

$= \frac{1}{3}(12 \cdot 10)(8)$

$= 320 \text{ in}^3$

Find each of the following. Round to the nearest tenth.

26. The base of a triangular pyramid is an equilateral triangle with sides 9 centimeters long. The pyramid's height is 15 centimeters. Find the volume of the pyramid. 175.4 cm^3

27. What is the volume of a right circular cone if its height is 22 centimeters and its radius is 11 centimeters? 2787.6 cm^3

28. The circumference of the base of a right circular cone is 62.8 millimeters. The cone has a height of 15 millimeters. Find the volume of the cone. 1569.2 mm^3

Additional Answers

12.

top view | left view | right view

front view | back view

13.

14.

15. $L = 48 \text{ ft}^2$; $T = 56 \text{ ft}^2$

16. $L = 264 \text{ in}^2$; $T = 312 \text{ in}^2$

17. $L = 439.8 \text{ cm}^2$; $T = 747.7 \text{ cm}^2$

18. $L = 1000.8 \text{ m}^2$; $T = 1354.2 \text{ m}^2$

Classroom Vignette

"As a final review for this chapter, I give students a surface area measurement and have them design a prism that has the surface area and the greatest possible volume."

Kathy A. Folske

Kathy A. Folske
Harrison High School
Farmington Hills, Michigan

Applications and Problem Solving Encourage students to work through the exercises in the Applications and Problem Solving section to strengthen their problem-solving skills.

Additional Answer

30. about 14,657,415 mi^2

OBJECTIVES AND EXAMPLES

- find the surface area and volume of a sphere (Lesson 11–7)

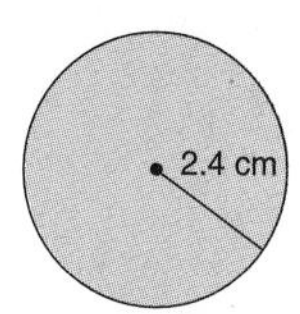

$T = 4\pi r^2$ $\quad V = \frac{4}{3}\pi r^3$

$= 4\pi(2.4)^2$ $\quad = \frac{4}{3}\pi(2.4)^3$

≈ 72.4 cm^2 $\quad \approx 57.9$ cm^3

REVIEW EXERCISES

Answer each of the following.

29. *True* or *false*: In a sphere, all great circles are congruent. **true**
30. The radius of the moon is approximately 1080 miles. Find its surface area. **See margin.**
31. The area of a great circle of a sphere is 50.24 square centimeters. Find the volume of the sphere. **267.9 cm^3**

- identify congruent or similar solids (Lesson 11–8)

Determine if the cylinders are *congruent, similar,* or *neither.*

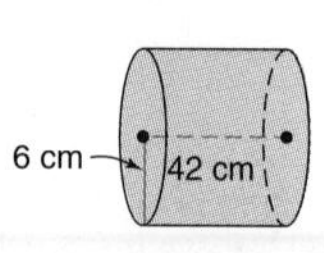

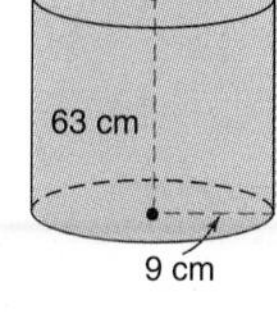

$\frac{\text{radius of larger cylinder}}{\text{radius of smaller cylinder}} = \frac{9}{6} \text{ or } \frac{3}{2}$

$\frac{\text{height of larger cylinder}}{\text{height of smaller cylinder}} = \frac{63}{42} \text{ or } \frac{3}{2}$

Since the ratios are the same, the cylinders are similar. However since the ratio is not 1, the cylinders are not congruent.

32. Determine if the solids below are *similar, congruent,* or *neither.* **neither**

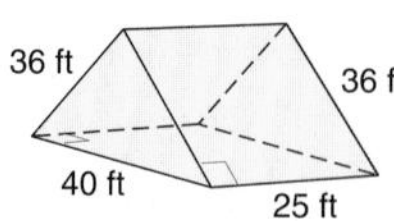

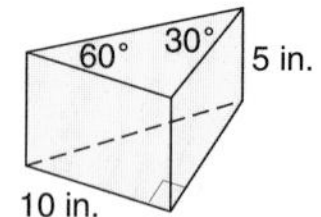

Determine if each statement is *true* or *false*. If the statement is false, rewrite it so that it is true. **35. False; A solid is always similar to itself.**

33. Two congruent prisms are also similar. **true**
34. The radius of a sphere is twice the radius of another sphere. Thus, the surface area of the larger sphere is four times the surface area of the smaller sphere. **true**
35. A solid is never similar to itself.

APPLICATIONS AND PROBLEM SOLVING

36. **Home Maintenance** Different types of paint are used for different purposes. A wall with a rough texture will diffuse more light than a smooth-surfaced one. A smooth wall can withstand more traffic and cleaning. When minute differences are considered, a wall with a rough texture will have a greater surface area than the same wall with a smooth texture. Study the magnified paint surfaces below. Which finish has the most surface area? (Lesson 11–2)

Ceiling, Flat, Eggshell, Semi-Gloss, Gloss **ceiling**

37. **Engineering** In 1986, the Water Spheroid was completed in Edmond, Oklahoma. This water tank is the largest in the world, with a diameter of 218 feet. Find its surface area and volume. (Lesson 11–7) **149,301.0 ft^2; 5,424,604.8 ft^3**

38. **Communications** Coaxial cable is used to transmit long-distance telephone calls, cable television programming, and other communications. A typical coaxial cable contains 22 coaxials, or copper tubes, with copper wire and plastic insulation. A 22-coaxial cable is about 3 inches in diameter. What is the lateral area and the volume of a coaxial cable that is 500 feet long? (Lessons 11–3, 11–5) **392.7 ft^2; 24.5 ft^3**

39. **Music** The largest guitar in the world was built by students in the Shakamak High School in Jasonville, Indiana. If a folk guitar is 18 inches wide and the largest guitar is 38 feet 2 inches wide, find the scale factor between the guitars. (Lesson 11–8) **18:458 or about 1:25.4**

A practice test for Chapter 11 is provided on page 803.

ALTERNATIVE ASSESSMENT

COOPERATIVE LEARNING PROJECT

Chartography In this chapter, you learned about the surface area of figures that have smooth surfaces. An acre of land has an area of 43,560 square feet, but this applies to a perfectly flat surface, which is rare on Earth. In this project, you will estimate the surface area of a plot of land with a contour that is not flat.

The U.S.G.S. (United States Geological Survey) provides contour maps for every region of the United States. You can obtain these from the library, federal government office, or some bookstores. Obtain the map for your region, or the nearest region which contains hills and valleys. You will select a rectangular region of the map and estimate its surface area.

WEST KENYA: Contour Map

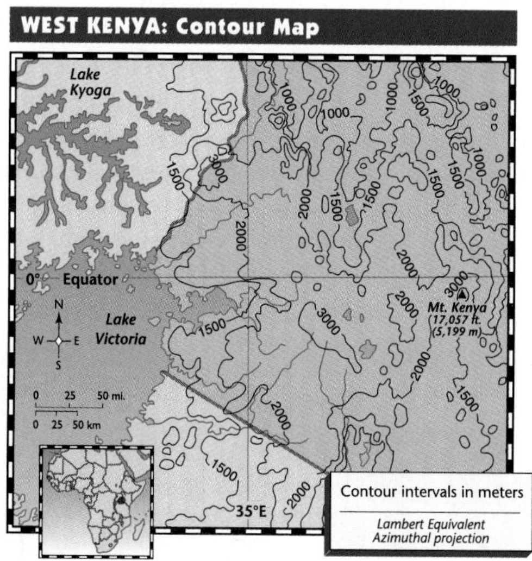

Follow these steps to estimate the surface area of the selected region.

- Draw a border around the rectangular region you will measure.
- Subdivide the rectangle into smaller regions so that the boundary of each region is a polygon.
- The vertices of a polygonal region may be at different elevations. Select regions where the vertices all lie in the same plane.
- Use formulas for area to determine the area of each polygon.
- Sum up the areas to obtain an estimate of the surface area for the entire plot of land.

Now imagine that the area is to be flooded to create a recreational lake. You need to know how much water is needed to fill the area so that the deepest point is six feet. As a group, discuss the method you would use to estimate the volume of water needed. Compute the volume.

THINKING CRITICALLY

Discuss the sentence *Length is to area as area is to volume*. Include examples or drawings to suggest your conclusions. **It could mean one dimension is to two dimensions as two dimensions are to three dimensions. The analogy is valid because in each case you extend it by one dimension to get the next term. But the analogy does not work if actual values for a problem are substituted.**

PORTFOLIO

Select a problem from this chapter that you found challenging. Write a paragraph about how you solved it, and why it was so challenging. Keep this problem and your essay in your portfolio.

SELF EVALUATION

Creativity requires connecting things that do not seem connected. Many factors aid the creative process. Some are knowledge of unrelated areas, the ability to scan many ideas quickly, and the freedom from conventional ways of thinking.

Assess Yourself Think back to a time when you thought up a new idea. Are you able to describe the process? Did the idea come from an area unrelated to the topic you were thinking about? Review the problems in this chapter. Were you better at the problems that required creativity or those that did not?

Assessment and Evaluation Masters, pp. 294, 305

NAME ______ DATE ______

Chapter 11 Performance Assessment

***Instructions:** Demonstrate your knowledge by giving a clear, concise solution to each problem. Be sure to include all relevant drawings and to justify your answer. You may show your solution in more than one way or investigate beyond the requirements of the problem.*

1. a. Tell what is meant by the term *polyhedron*.
   b. Draw a polyhedron. Use dashed lines for hidden edges.
   c. Give at least two nets for the polyhedron you drew in part b.
   d. Explain the surface area of a polyhedron.
   e. Assign dimensions to the figure you drew in part b and find the surface area.
2. The city of Anderson is installing a new water tower like the one shown below.

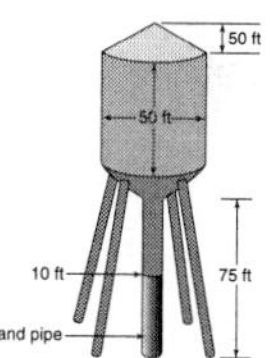

   a. Give a plan for finding the volume of the water tank, including the standpipe and conical roof.
   b. Use your plan to find the volume of the water tank. Show your work.
   c. If the surface area of each leg is 125 ft², find the total surface area of the water tank.
   d. If the name of the city is to be painted on two opposite sides of the tower with 10 ft between the words on each side, about how wide should each letter be?

Scoring Guide
Chapter 11
Performance Assessment

Level	Specific Criteria
3 Superior	• Shows thorough understanding of the concepts of *drawing three-dimensional figures and nets, surface area,* and *volume of a solid.* • Uses appropriate strategies to solve problems. • Computations are correct. • Written explanations are exemplary. • Diagrams are accurate and appropriate. • Goes beyond requirements of some or all problems.
2 Satisfactory, with Minor Flaws	• Shows understanding of the concepts of *drawing three-dimensional figures and nets, surface area,* and *volume of a solid.* • Uses appropriate strategies to solve problems. • Computations are mostly correct. • Written explanations are effective. • Diagrams are mostly accurate and appropriate. • Satisfies all requirements of some or all problems.
1 Nearly Satisfactory, with Serious Flaws	• Shows understanding of most of the concepts of *drawing three-dimensional figures and nets, surface area,* and *volume of a solid.* • May not use appropriate strategies to solve problems. • Computations are mostly correct. • Written explanations are satisfactory. • Diagrams are mostly accurate and appropriate. • Satisfies most requirements of some or all problems.
0 Unsatisfactory	• Shows little or no understanding of the concepts of *drawing three-dimensional figures and nets, surface area,* and *volume of a solid.* • May not use appropriate strategies to solve problems. • Computations are incorrect. • Written explanations are not satisfactory. • Diagrams are not accurate or appropriate. • Does not satisfy requirements of some or all problems.

Alternative Assessment

The Alternative Assessment section provides students with the opportunity to assess their own work by thinking critically, working with others, keeping a portfolio, and honestly evaluating their own progress. For more information on alternative forms of assessment, see *Alternative Assessment in the Mathematics Classroom,* one of the titles in the Glencoe Mathematics Professional Series.

Performance Assessment

Performance Assessment tasks for this chapter are included in the *Assessment and Evaluation Masters.* A scoring guide is also provided.

NCTM Standards: 1–5, 7, 8

This Investigation is designed to be completed over several days or weeks. It may be considered optional. You may want to assign the Investigation and the follow-up activities to be completed at the same time.

Objective

Make a movie flip book using coordinate geometry.

Mathematical Overview

This Investigation will use the following mathematical skills and concepts from Chapters 12 and 13.

- identifying locations on a grid using coordinates
- using trigonometry to draw a three-dimensional figure in two dimensions
- using reflections to create movie special effects
- using dilations to create specific effects with graphic images

Recommended Time		
Part	**Pages**	**Time**
Investigation	642–643	1 class period
Working on the Investigation	651, 686, 737, 753	20 minutes each
Closing the Investigation	754	1 class period

Instructional Resources

Investigations and Projects Masters, pp. 21–24

A recording sheet, teacher notes, and scoring guide are provided for each Investigation in the *Investigations and Projects Masters*.

1 MOTIVATION

Hold a magnifying glass up to a working TV screen. Look at the individual colors that make up the picture. Discuss how these individual parts combine to give a whole picture.

LONG-TERM PROJECT

In•ves•ti•ga•tion

MATERIALS NEEDED

- grid paper
- ruler
- markers
- calculator

Were you amazed when Forrest Gump shook hands with President Kennedy? Or astounded when dinosaurs came to life in Jurassic Park? Then you have experienced the magic that digital computers can create at the movies.

The field of computer special effects for movies began when George Lucas couldn't find a company to create the effects he envisioned for the 1976 movie *Star Wars*. At the time, filmmakers considered Lucas' ideas experimental and not commercially viable. But by 1996, roughly half of the movies released used digital visuals of some kind!

Computer graphics are created using *coordinate geometry*. The computer screen is comprised of thousands of little lights, called pixels, arranged in rows and columns across and down the screen. An image is created by assigning each pixel a color. Pixels are identified by a coordinate position on the screen. For example, (34, 78) may refer to a pixel that is 34 pixels to the right and 78 pixels above the center of the screen. When several pictures that differ only slightly are shown very quickly, the subject appears to move across the screen. The different pictures used to create moving images can often be created by performing transformations on the objects in the picture.

Cooperative Learning

This Investigation offers an excellent opportunity for using cooperative learning groups. For more information on cooperative learning strategies and group management, see *Cooperative Learning in the Mathematics Classroom*, one of the titles in the Glencoe Mathematics Professional Series.

THE SCREEN

To develop your skill in creating computer graphics, begin by creating a simple design on a "screen." Use graph paper to create a model of a computer screen that measures 50 pixels horizontally and 30 pixels vertically. Let the center of the screen be at (0, 0). What are the coordinates of each corner of the screen? Lay the model screen over an existing picture or draw a picture on the screen. Your picture could be a geometric figure or a simple illustration. What are the coordinates of ten different points used to create the picture? What are the colors of the pixels you identified? Could you give someone else the coordinates and instructions so that they could recreate your picture without seeing the finished picture?

THE MOVIE

Work with the members of your group to choose a subject and plot for a "movie." You will be drawing the images to make the objects or characters move. Then the movie images can be bound as a small book. When the pages of the book are flipped quickly, the images of your movie will appear to come to life.

You will continue working on this Investigation throughout Chapters 12 and 13.

Be sure to keep your computer screen model, movie ideas, and other materials in your Investigation Folder.

Movie Magic Investigation

Working on the Investigation
Lesson 12–1, p. 651

Working on the Investigation
Lesson 12–6, p. 686

Working on the Investigation
Lesson 13–6, p. 736

Working on the Investigation
Lesson 13–8, p. 753

Closing the Investigation
End of Chapter 13, p. 754

2 SETUP

Have a student read the first four paragraphs of the Investigation to provide background information for the movie-making project. Read the remaining paragraphs that explain the class's tasks. Discuss the activity with students. Separate the class into groups of four or five.

3 MANAGEMENT

Each group member should be responsible for a specific task.

Recorder Keeps track of the progress of the pictures.

Artist Draws the more complex pictures.

Writer Writes the plot for the movie.

At the end of the activity, each member should turn in his or her respective equipment.

Sample Answers

Answers will vary based on movie designs and plots.

Investigations and Projects Masters, p. 24

12, 13 NAME ________________ DATE ________

Investigation, Chapters 12 and 13

Student Edition Pages 642–643, 651, 686, 736, 753, 754

Movie Magic

Use this chart to record the colors for each pixel of your computer screen.

Pixel	Color	Pixel	Color	Pixel	Color

Use the space below to calculate the coordinates for your cube.

Use the space below to list ways reflections can be used in your movie.

Use the space below to list ways dilations can be used in your movie.

Please keep this page and any other research in your Investigation Folder.

CHAPTER 12

Continuing Coordinate Geometry

PREVIEWING THE CHAPTER

In this chapter, students use two methods to graph linear equations. They determine the equation of a line given information about its graph and solve problems by using equations. They use statistics and equations of lines to expand upon geometric concepts. Students use coordinate proofs to prove theorems. They integrate physics to look at a geometric model of motion. Students investigate vectors by finding the magnitude and direction and by comparing vectors. They perform vector operations. The chapter ends with students investigating three-dimensional space by locating points, using the distance and midpoint formulas, and finding the center and radius of a sphere.

Lesson (Pages)	Lesson Objectives	NCTM Standards	State/Local Objectives
12-1 (646–651)	Graph linear equations using the intercepts method and the slope-intercept method.	1–6	
12-2A (652)	Use a graphing calculator to determine an equation of the line that passes through plotted points.	2, 4, 5	
12-2 (653–659)	Write an equation of a line given information about its graph. Solve problems by using equations.	1–6	
12-3 (660–665)	Relate statistics and equations of lines to geometric concepts.	1–6, 10	
12-4 (666–671)	Prove theorems using coordinate proofs.	1–5, 8	
12-5A (672)	Use a geometric model of motion.	2, 3	
12-5 (673–679)	Find the magnitude and direction of a vector. Determine if two vectors are equal. Perform operations with vectors.	1–6, 8	
12-6 (680–686)	Locate a point in space. Use the distance and midpoint formulas for points in space. Determine the center and radius of a sphere.	1–8	

ORGANIZING THE CHAPTER

A complete, 1-page lesson plan is provided for each lesson in the ***Lesson Planning Guide***. Answer keys for each lesson are available in the ***Answer Key Masters***.

You may want to refer to the **Course Planning Calendar** on page T12 for detailed information on pacing.
PACING: Standard—12 days; **Honors**—11 days

LESSON PLANNING CHART

			Blackline Masters									
Lesson (Pages)	**Materials/ Manipulatives**	**Extra Practice (Student Edition)**	**Study Guide**	**Practice**	**Enrichment**	**Assessment & Evaluation**	**Modeling Mathematics**	**Multicultural Activity**	**Tech Prep Applications**	**Graphing Calc. & Computer**	**Real-World Applications**	**Teaching Transparencies**
12-1 (646–651)	square tiles grid paper straightedge* TI-82/83 graphing calculator	p. 787	p. 72	p. 72	p. 72					p. 12	24	12-1A 12-1B
12-2A (652)	TI-82/83 graphing calculator									pp. 28, 29		
12-2 (653–659)		p. 788	p. 73	p. 73	p. 73	p. 324	pp. 59–62					12-2A 12-2B
12-3 (660–665)	grid paper	p. 788	p. 74	p. 74	p. 74	pp. 323,324		p. 23	p. 23		25	12-3A 12-3B
12-4 (666–671)	scissors* ruler* grid paper	p. 788	p. 75	p. 75	p. 75		p. 90					12-4A 12-4B
12-5A (672)	battery-powered car grid paper masking tape meter stick straightedge* butcher paper						p. 103					
12-5 (673–679)	TI-92 calculator grid paper	p. 789	p. 76	p. 76	p. 76	p. 325			p. 24			12-5A 12-5B
12-6 (680–686)		p. 789	p. 77	p. 77	p. 77	p. 325		p. 24			26	12-6A 12-6B
Study Guide/ Assessment (687–691)						pp. 309–322, 326–328						

*Included in Glencoe's High School Manipulative Kit and Overhead Manipulative Resources.

ORGANIZING THE CHAPTER

OTHER CHAPTER RESOURCES

Student Edition
Investigation, pp. 642–643
Chapter Opener, pp. 644–645
Mathematics and Society, p. 659
Working on the Investigation, pp. 651, 686

Teacher's Classroom Resources
Investigations and Projects Masters, pp. 69–72
Block Scheduling Booklet

Technology
Test and Review Software (IBM and Macintosh)
CD-ROM Multimedia Applications (Windows and Macintosh)
Mindjogger Videoquizzes (VHS)

Professional Publications
Glencoe Mathematics Professional Series

OUTSIDE RESOURCES

Books/Periodicals
Shulte, Albert P. and Stuart A. Choate, *What Are My Chances?*, Dale Seymour Publications
Modelling With Force and Motion, The School Mathematics Project, Cambridge University
How to Draw a Straight Line, NCTM

Software
CampOS Math: Numbers & Number Theory, Pierian Spring
MacStat, MECC

Videos/CD-ROMs
Probability and Statistics, ETA, 620 Lakeview Parkway, Vernon Hills, IL 60061
The TI-82 Video, Venture Publishing, Bartlett Street, Suite 55, Andover, MA 01810

ASSESSMENT RESOURCES

Student Edition
Math Journal, pp. 668, 676
Mixed Review, pp. 651, 658–659, 665, 671, 679, 686
Self Test, p. 665
Chapter Highlights, p. 687
Chapter Study Guide and Assessment, pp. 688–690
Alternative Assessment, p. 691
Portfolio, p. 691
College Entrance Exam Practice, pp. 692–693
Chapter Test, p. 804

Teacher's Wraparound Edition
5-Minute Check, pp. 646, 653, 660, 666, 673, 680
Check for Understanding, pp. 649, 656, 662, 668, 676, 683
Closing Activity, pp. 651, 659, 665, 671, 679, 686
Cooperative Learning, pp. 647, 661

Assessment and Evaluation Masters
Multiple-Choice Tests, Forms 1A (Honors), 1B (Average), 1C (Basic), pp. 309–314
Free-Response Tests, Forms 2A (Honors), 2B (Average), 2C (Basic), pp. 315–320
Calculator-Based Test, p. 321
Performance Assessment, p. 322
Mid-Chapter Test, p. 323
Quizzes A–D, pp. 324–325
Standardized Test Practice, p. 326
Cumulative Review, pp. 327–328

ENHANCING THE CHAPTER

Examples of some of the materials for enhancing Chapter 12 are shown below.

Multicultural Activity Masters, pp. 23, 24

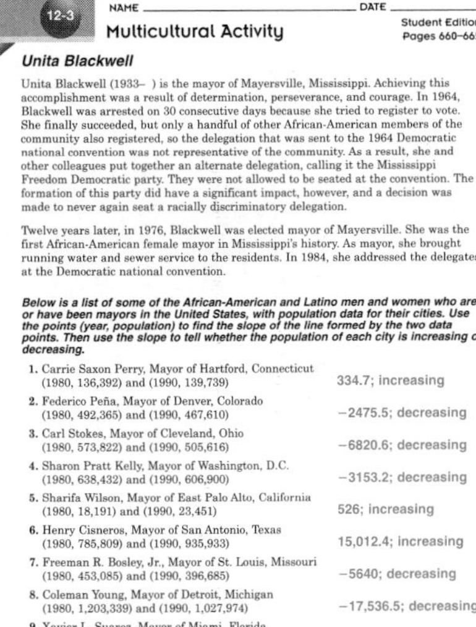

12-3 NAME ______ DATE ______

Multicultural Activity — Student Edition Pages 660–665

Unita Blackwell

Unita Blackwell (1933–) is the mayor of Mayersville, Mississippi. Achieving this accomplishment was a result of determination, perseverance, and courage. In 1964, Blackwell was arrested on 30 consecutive days because she tried to register to vote. She finally succeeded, but only a handful of other African-American members of the community also registered, so the delegation that was sent to the 1964 Democratic national convention was not representative of the community. As a result, she and other colleagues put together an alternate delegation, calling it the Mississippi Freedom Democratic party. They were not allowed to be seated at the convention. The formation of this party did have a significant impact, however, and a decision was made to never again seat a racially discriminatory delegation.

Twelve years later, in 1976, Blackwell was elected mayor of Mayersville. She was the first African-American female mayor in Mississippi's history. As mayor, she brought running water and sewer service to the residents. In 1984, she addressed the delegates at the Democratic national convention.

Below is a list of some of the African-American and Latino men and women who are or have been mayors in the United States, with population data for their cities. Use the points (year, population) to find the slope of the line formed by the two data points. Then use the slope to tell whether the population of each city is increasing or decreasing.

1. Carrie Saxon Perry, Mayor of Hartford, Connecticut (1980, 136,392) and (1990, 139,739) — 334.7; increasing
2. Federico Peña, Mayor of Denver, Colorado (1980, 492,365) and (1990, 467,610) — −2475.5; decreasing
3. Carl Stokes, Mayor of Cleveland, Ohio (1980, 573,822) and (1990, 505,616) — −6820.6; decreasing
4. Sharon Pratt Kelly, Mayor of Washington, D.C. (1980, 638,432) and (1990, 606,900) — −3153.2; decreasing
5. Sharifa Wilson, Mayor of East Palo Alto, California (1980, 18,191) and (1990, 23,451) — 526; increasing
6. Henry Cisneros, Mayor of San Antonio, Texas (1980, 785,809) and (1990, 935,933) — 15,012.4; increasing
7. Freeman R. Bosley, Jr., Mayor of St. Louis, Missouri (1980, 453,085) and (1990, 396,685) — −5640; decreasing
8. Coleman Young, Mayor of Detroit, Michigan (1980, 1,203,339) and (1990, 1,027,974) — −17,536.5; decreasing
9. Xavier L. Suarez, Mayor of Miami, Florida (1980, 346,865) and (1990, 358,548) — 1168.3; increasing
10. Louis E. Saavedra, Mayor of Albuquerque, New Mexico (1980, 331,767) and (1990, 384,736) — 5296.9; increasing

Real-World Applications, 24, 25, 26

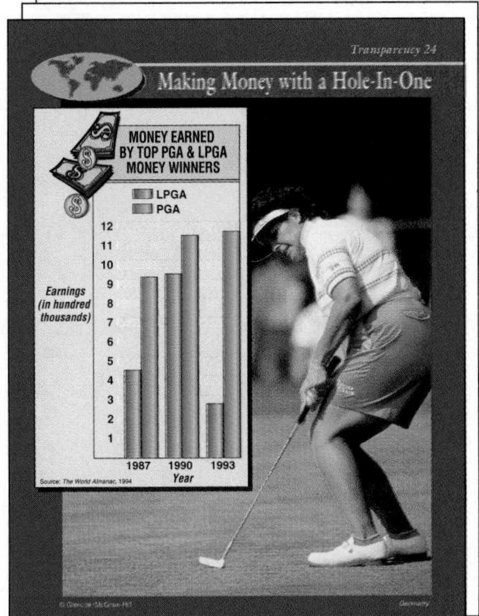

TECHNOLOGY

Graphing Calculator and Computer Masters, p. 12

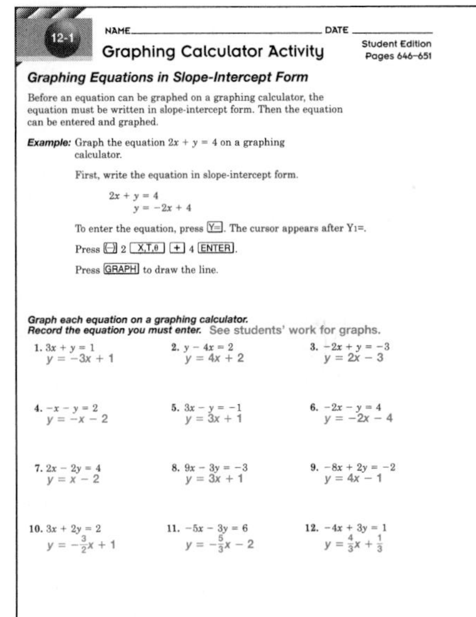

12-1 NAME ______ DATE ______

Graphing Calculator Activity — Student Edition Pages 646–651

Graphing Equations in Slope-Intercept Form

Before an equation can be graphed on a graphing calculator, the equation must be written in slope-intercept form. Then the equation can be entered and graphed.

Example: Graph the equation $2x + y = 4$ on a graphing calculator.

First, write the equation in slope-intercept form.

$$2x + y = 4$$
$$y = -2x + 4$$

To enter the equation, press [Y=]. The cursor appears after Y1=.

Press [(−)] 2 [X,T,θ] [+] 4 [ENTER].

Press [GRAPH] to draw the line.

Graph each equation on a graphing calculator. Record the equation you must enter. See students' work for graphs.

1. $3x + y = 1$ $y = -3x + 1$	2. $y - 4x = 2$ $y = 4x + 2$	3. $-2x + y = -3$ $y = 2x - 3$
4. $-x - y = 2$ $y = -x - 2$	5. $3x - y = -1$ $y = 3x + 1$	6. $-2x - y = 4$ $y = -2x - 4$
7. $2x - 2y = 4$ $y = x - 2$	8. $9x - 3y = -3$ $y = 3x + 1$	9. $-8x + 2y = -2$ $y = 4x - 1$
10. $3x + 2y = 2$ $y = -\frac{3}{2}x + 1$	11. $-5x - 3y = 6$ $y = -\frac{5}{3}x - 2$	12. $-4x + 3y = 1$ $y = \frac{4}{3}x + \frac{1}{3}$

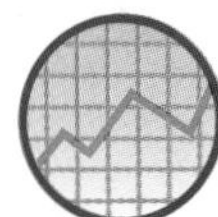

TECH PREP

Tech Prep Applications Masters, pp. 23, 24

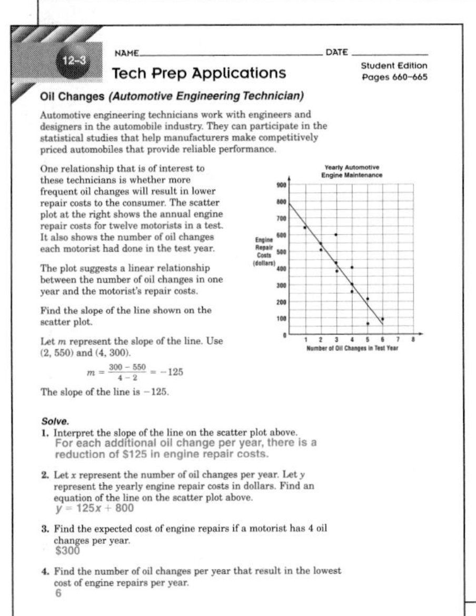

12-3 NAME ______ DATE ______

Tech Prep Applications — Student Edition Pages 660–665

Oil Changes *(Automotive Engineering Technician)*

Automotive engineering technicians work with engineers and designers in the automobile industry. They can participate in the statistical studies that help manufacturers make competitively priced automobiles that provide reliable performance.

One relationship that is of interest to these technicians is whether more frequent oil changes will result in lower repair costs to the consumer. The scatter plot at the right shows the annual engine repair costs for twelve motorists in a test. It also shows the number of oil changes each motorist had done in the test year.

The plot suggests a linear relationship between the number of oil changes in one year and the motorist's repair costs.

Find the slope of the line shown on the scatter plot.

Let m represent the slope of the line. Use (2, 550) and (4, 300).

$$m = \frac{300 - 550}{4 - 2} = -125$$

The slope of the line is −125.

Solve.

1. Interpret the slope of the line on the scatter plot above.
   For each additional oil change per year, there is a reduction of $125 in engine repair costs.
2. Let x represent the number of oil changes per year. Let y represent the yearly engine repair costs in dollars. Find an equation of the line on the scatter plot above.
   $y = 125x + 800$
3. Find the expected cost of engine repairs if a motorist has 4 oil changes per year.
   $300
4. Find the number of oil changes per year that result in the lowest cost of engine repairs per year.
   6

Problem-of-the-Week Cards, 33, 34

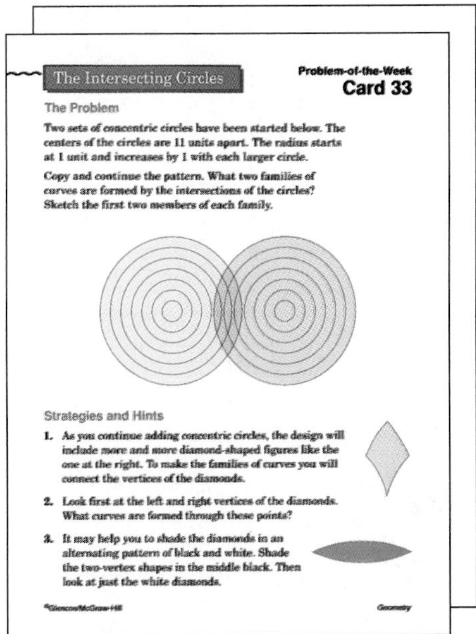

The Intersecting Circles — Problem-of-the-Week Card 33

The Problem

Two sets of concentric circles have been started below. The centers of the circles are 11 units apart. The radius starts at 1 unit and increases by 1 with each larger circle.

Copy and continue the pattern. What two families of curves are formed by the intersections of the circles? Sketch the first two members of each family.

Strategies and Hints

1. **As you continue adding concentric circles, the design will include more and more diamond-shaped figures like the one at the right. To make the families of curves you will connect the vertices of the diamonds.**
2. **Look first at the left and right vertices of the diamonds. What curves are formed through these points?**
3. **It may help you to shade the diamonds in an alternating pattern of black and white. Shade the two-vertex shapes in the middle black. Then look at just the white diamonds.**

©Glencoe/McGraw-Hill — Geometry

MAKING MATHEMATICS RELEVANT

This two-page introduction to the chapter provides students with an opportunity to explore an application of geometry that has historic roots as well as modern uses.

Background Information

A Funny Business DC Comics has published over 500 comic book titles since it began. Well-known DC Comics superheroes include Superman, Batman, Flash, Green Lantern, and Wonder Woman. Over its history, Marvel Comics has published more than 900 titles featuring such well-known characters as Spider-Man, Captain America, the Incredible Hulk, Thor the Thunder God, and the Thing. These two industry giants shared 80% of the comic book market in 1991, but independent publishers have flourished in the last decade and drastically reduced these leaders' market share.

CHAPTER 12

Continuing Coordinate Geometry

Objectives

In this chapter, you will:

- write and graph linear equations,
- relate statistics and equations of lines to geometric concepts,
- prove theorems using coordinate proofs,
- perform operations with vectors, and
- locate points in space.

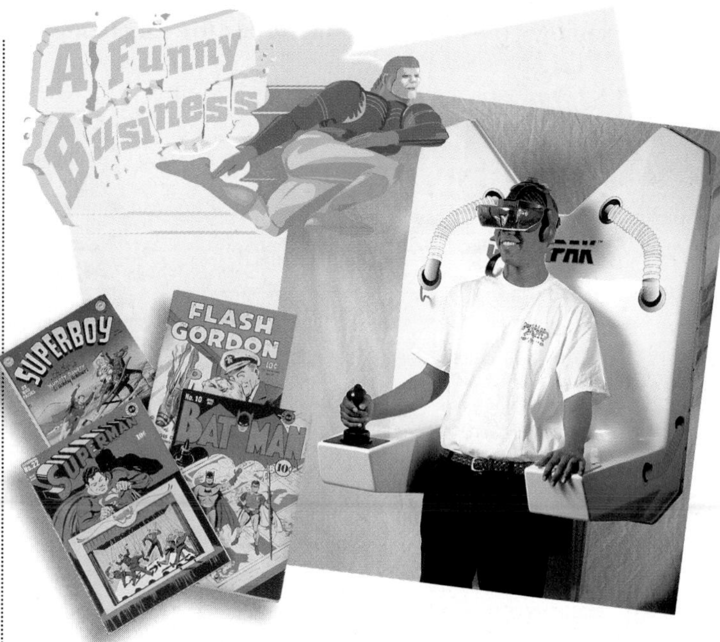

A Funny Business
SUPERBOY
FLASH GORDON
BATMAN
SUPERMAN

Geometry: Then and Now It's a bird, it's a plane, it's an American phenomenon! Superheros have captured the imagination of generations. The first modern comic book, *Funnies on Parade,* was printed in 1933 by Eastern Color Printing Company of Waterbury, Connecticut. Within weeks, all 10,000 copies were sold, and a new industry was born. Now that computers can create a virtual reality, comic book characters have entered a new dimension. Three-dimensional displays and data-input sensors allow interaction with characters that appear to be there in the flesh.

TIME Line

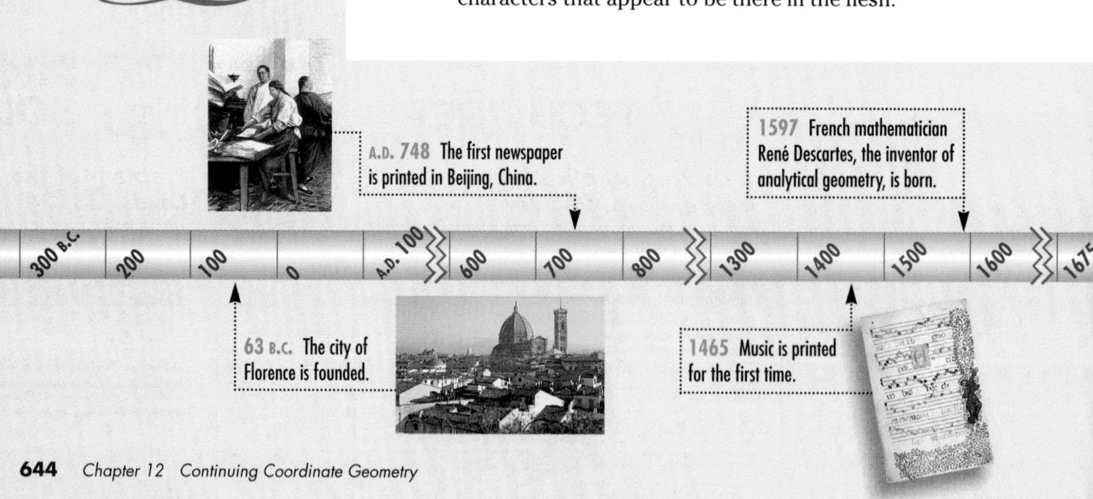

A.D. 748 The first newspaper is printed in Beijing, China.
1597 French mathematician René Descartes, the inventor of analytical geometry, is born.
300 B.C.
200
100
0
A.D. 100
600
700
800
1300
1400
1500
1600
1675
63 B.C. The city of Florence is founded.
1465 Music is printed for the first time.

TIME Line

Students might want to investigate music notation. Time signatures utilize basic mathematical concepts, fractions, and ratios.

interNET CONNECTION

Check out some all time favorite comic books, reviews of recent releases, and the Comic Book Message Board.

World Wide Web
http://envisionww.com/jonahw/comics/indexframe.shtml

PEOPLE IN THE NEWS

Sixteen-year-old **Leander Morgan** of Albuquerque, New Mexico, creates his own comic books. Once interested only in violent comics, Leander has now found a way to make a difference in the world with his creations. He wrote a comic book emphasizing respect and individuality that is being used in elementary schools. Before starting his most recent project, he met with a diverse group of students from his school to discuss what they had in common. Working together, they wrote a comic book titled *Defaced.* Leander explains its message as, "Stand up for yourself, even if you're standing alone, because you'll be amazed at how many are standing with you."

Chapter Project

Marvel Entertainment Group, Inc., is a leading creator of youth entertainment products including Marvel comic books, toys, trading cards, stickers, and books. It also licenses the use of the characters it creates. The table below shows Marvel Entertainment Group, Inc. sales (in millions of dollars) for the years 1991–1995.

Year	Sales (millions of dollars)
1991	115.1
1992	223.8
1993	415.2
1994	514.8
1995	829.3

Source: *Hoover's Company Profile Database*

- Draw a graph of the data in the table, using each year as the x-coordinate of an ordered pair and the corresponding sales as the y-coordinate. Is there a pattern in the graph? If so, describe it.
- If you were a potential investor in Marvel Entertainment Group, Inc., what does this graph tell you? What other types of information would you want to know about the company before making an investment decision?
- Draw a line through two of the points that you graphed. Then find the equation of the line. What does the equation's slope mean in relation to Marvel's sales?
- Use the equation to predict Marvel's sales for the year 2001. Do you think this prediction is reasonable? Why or why not?
- Discuss potential dangers of making predictions with the equation of a line drawn in this way.

1738 The first cuckoo clocks are made in the Black Forest of Germany.

1993 Annual sales of comic books topped $1 billion for the first time.

1700 | 1725 | 1750 | 1775 | 1800 | 1825 | 1850 | 1875 | 1900 | 1925 | 1950 | 1975 | 2000

1881 *Vector Analysis*, the manual for calculations using vectors, is published.

1938 The first *Superman* comic is published.

Alternative Chapter Projects

Two other chapter projects are included in the *Investigations and Projects Masters*. In Chapter 12 Project A, pp. 69–70, students extend the topic in the chapter opener. In Chapter 12 Project B, pp. 71–72, students develop a three-dimensional coordinate system for their rooms.

PEOPLE IN THE NEWS

Encourage students to talk about how they can promote mutual respect among different groups of students at their school. How could a collaborative project like Leander's comic book increase tolerance and understanding?

Chapter Project

Cooperative Learning Encourage students to research other data that could be represented by a linear equation. For instance, students might be interested in comparing Marvel's sales to those of another youth entertainment products company or researching the attendance trends at the events of their favorite sports teams. Students might also want to try collecting their own data through surveys to analyze.

Investigations and Projects Masters, p. 69

12 NAME__________ DATE ________

Chapter 12 Project A — Student Edition Pages 644–691

Up or Down?

1. Oshkosh B'Gosh® produces various types of clothing. This company is probably most well known for their high-quality children's apparel. The table shows the sales for Oshkosh B'Gosh (in millions of dollars) for the years 1991–1995.

Year	Sales (millions of dollars)
1991	365.2
1992	346.2
1993	340.2
1994	363.4
1995	432..3

Source: *Standard & Poor's NASDAQ Stock Reports*

2. Graph the data in the table, using each year as the x-coordinate of an ordered pair and the corresponding sales as the y-coordinate.
3. Draw four lines, each using (1991, 365.2) as one point. Relate the direction that each line is heading with the change in sales. Describe how the two are related. Write the equation of each line and relate its slope to the change in sales.
4. Do you think any of the lines give a good indication of the company's sales? Explain. Draw a line that represents the company's average sales over the five-year period. (It does not have to contain any of the five points.)
5. Make a list of some positive and negative things that the data indicate. If you were an investor, would you invest in this company? Why or why not?

12-1 LESSON NOTES

NCTM Standards: 1–6

Instructional Resources

- Study Guide Master 12-1
- Practice Master 12-1
- Enrichment Master 12-1
- Graphing Calculator and Computer Masters, p. 12
- Real-World Applications, 24

Transparency 12-1A contains the 5-Minute Check for this lesson; **Transparency 12-1B** contains a teaching aid for this lesson.

Recommended Pacing	
Standard Pacing	Day 1 of 12
Honors Pacing	Day 1 of 11

1 FOCUS

5-Minute Check
(over Chapter 11)

1. Find the volume of a right circular cone with radius 4 meters and height 7 meters. **about 117.3 m^3**
2. Find the surface area of a right cylinder with a height of 3 feet and a radius of 5 feet. **about 251.3 ft^2**
3. Find the surface area of a regular square pyramid with a slant height of 18 inches and a base with sides measuring 20 inches. **1120 in^2**
4. Find the volume of a right rectangular prism with a length of 19 units, a height of 3 units, and a width of 6 units. **342 $units^3$**
5. Find the volume of a sphere with diameter 15 centimeters. **about 1767.1 cm^3**

Motivating the Lesson

Situational Problem Ask students to think of a steep road they know. Ask them how they know that a road is steep. **Sample answers: It is harder to walk up; balls roll down faster.**

12-1 Integration: Algebra
Graphing Linear Equations

***What* YOU'LL LEARN**

- To graph linear equations using the intercepts method and the slope-intercept method.

***Why* IT'S IMPORTANT**

You can use linear equations to model relationships and solve real-world problems.

Many relationships can be represented using linear equations.

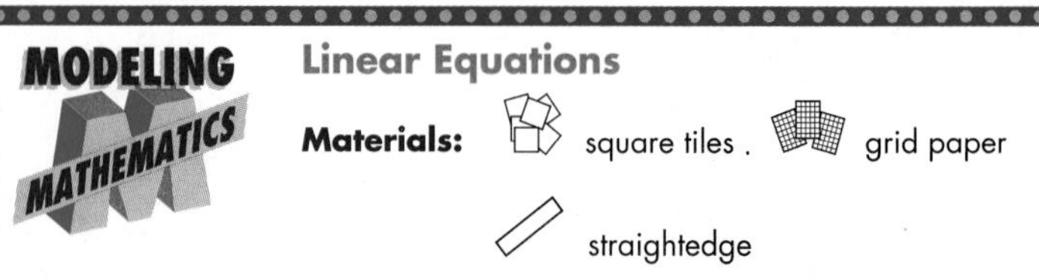

MODELING MATHEMATICS **Linear Equations**

Materials: square tiles, grid paper, straightedge

- The figures below represent the first three figures in a sequence. The perimeter of each figure is noted in the table below at the right.

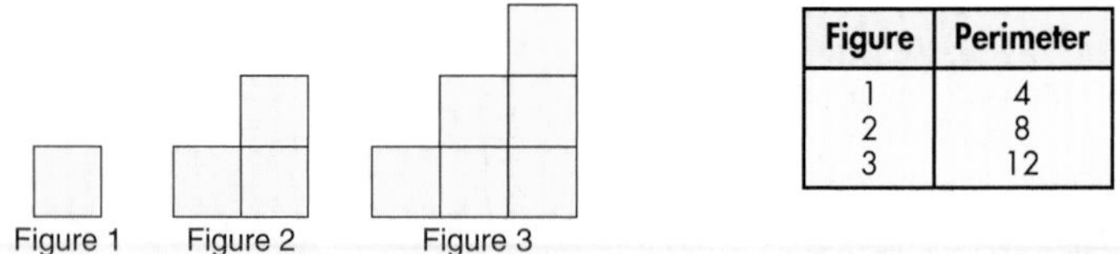

Figure	Perimeter
1	4
2	8
3	12

- Arrange square tiles in the patterns shown. Then extend the pattern to make the fourth, fifth, and sixth figures.
- Copy the table shown above. Then find the perimeter of figures 4, 5, and 6 and record them in your table.

Your Turn

a. Suppose that the number of the figure is x and its perimeter is y. Verify that for each figure you drew, $y = 4x$. **See students' work.**

b. Graph each ordered pair (x, y) on a coordinate plane. **See students' work.**

c. Draw a line through the points. **See students' work.**

d. Use the graph to estimate the perimeter of figure 40 in this sequence. **160**

The line that you graphed in the Modeling Mathematics activity has the equation $y = 4x$. Every line drawn in a coordinate plane has a corresponding **linear equation** that describes the line algebraically. A linear equation can be written in the form $Ax + By = C$, where A, B, and C are any real numbers and A and B are not both zero. A linear equation in the form $Ax + By = C$ is said to be in **standard form**.

The equations $2x + y = 10$, $3x = 5y - 8$, $y = 2$, and $\frac{1}{2}x + \frac{3}{4}y = 0$ are all linear equations. Each can be written in the form $Ax + By = C$. The equations $y = 2x^2 + 3$ and $x + \frac{1}{y} = 0$ are not linear equations. *Why not?*

The graph of a linear equation is the set of all points with coordinates (x, y) that satisfy the equation $Ax + By = C$. One technique for graphing a linear equation is called the **intercepts method**. This method uses two special values, the *x-intercept* and the *y-intercept*. The x-intercept is the value of x when y equals zero. Likewise, the y-intercept is the value of y when x equals zero. These values can be used as long as neither A nor B is zero.

MODELING MATHEMATICS Students will look for a pattern to find the perimeter of a figure. Ask students how the perimeter varies based on the arrangement of the tiles.

Example **1**

Finance

In this application, for any ordered pair, (x, y), x represents the number of months, and y represents the amount owed to Aunt Guillerma after x months.

Maita is buying a used car from her Aunt Guillerma for $4500. She agreed to pay her $150 each month. The equation $y = 4500 - 150x$ represents the amount owed, y, after x payments. Graph the equation by finding the intercepts.

First write the equation in standard form.

$$y = 4500 - 150x$$
$$150x + y = 4500$$

Let $x = 0$ to find the y-intercept.

$$150(0) + y = 4500$$
$$y = 4500$$

Let $y = 0$ to find the x-intercept.

$$150x + 0 = 4500$$
$$150x = 4500$$
$$x = 30$$

The x-intercept is 30, and the y-intercept is 4500. To graph $150x + y = 4500$, plot (30, 0) and (0, 4500) and draw a line through the points.

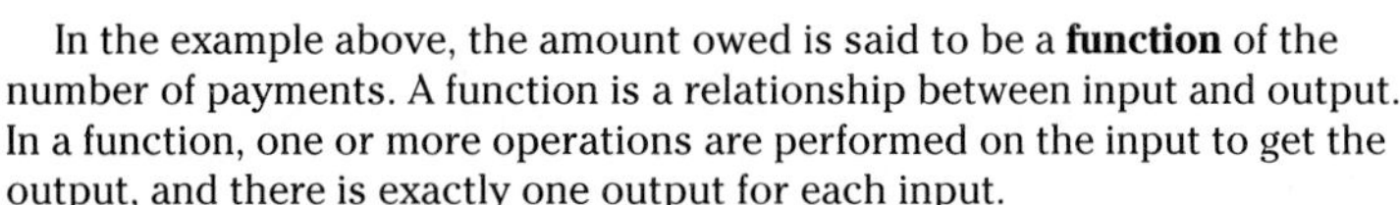

In the example above, the amount owed is said to be a **function** of the number of payments. A function is a relationship between input and output. In a function, one or more operations are performed on the input to get the output, and there is exactly one output for each input.

When an equation is rewritten so that one side has y by itself, another method for graphing is more convenient to use. Use a graphing calculator, graphing software, or graph paper to graph the *family of equations* that appears below.

$$y = \frac{2}{3}x + 2 \qquad y = \frac{2}{3}x + 0 \qquad y = \frac{2}{3}x - 3$$

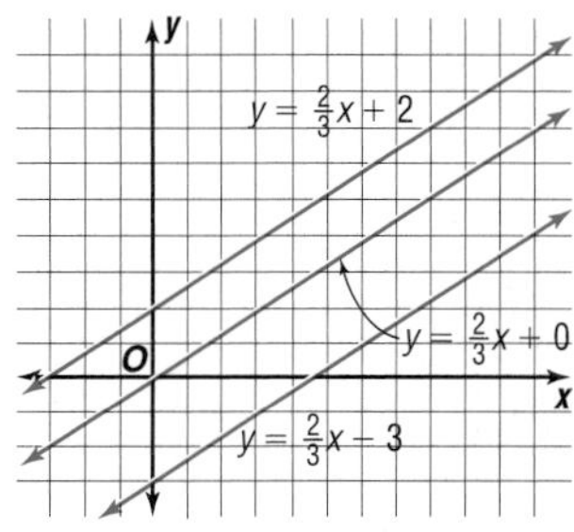

The graphs are parallel lines, and the slope of each line is $\frac{2}{3}$. How does the slope of each line compare to its equation? In each case, the slope of each line is equal to the coefficient of x in its equation. Also the y-intercept of each line is equal to the constant term of the equation. A linear equation written in the form $y = mx + b$ is called the **slope-intercept form** of the equation.

LOOK BACK

Refer to Lesson 3-3 to review how to find the slope of a line.

Theorem 12–1 Slope-Intercept Form	**If the equation of a line is written in the form $y = mx + b$, m is the slope of the line, and b is the y-intercept.**

In Theorem 12–1, notice that if $x = 0$, $y = b$. Therefore, b is the y-intercept. If $x = 1$, then $y = m + b$. Since the ordered pairs for two points on the line have coordinates $(0, b)$ and $(1, m + b)$, the slope is $\frac{(m + b) - b}{1 - 0}$ or m.

Horizontal and vertical lines are special cases. The graph of an equation of the form $x = a$ is a vertical line and has an undefined slope. If the equation of a vertical line is written in the standard form, $Ax + By = C$, then $B = 0$. The graph of an equation of the form $y = b$ is a horizontal line and has a slope of 0. When the equation of a horizontal line is written in the standard form, $A = 0$.

2 TEACH

Teaching Tip In the definition of a linear equation in standard form, point out that *A and B are not both zero* means that one or the other but not both may be zero.

In-Class Example

For Example 1

Golden Sporting Goods makes tennis and squash racquets. The equation $3x + 4y = 36{,}000$ represents the cost of making x tennis racquets and y squash racquets. Graph the equation by finding the intercepts.

(0, 9000); (12,000, 0)

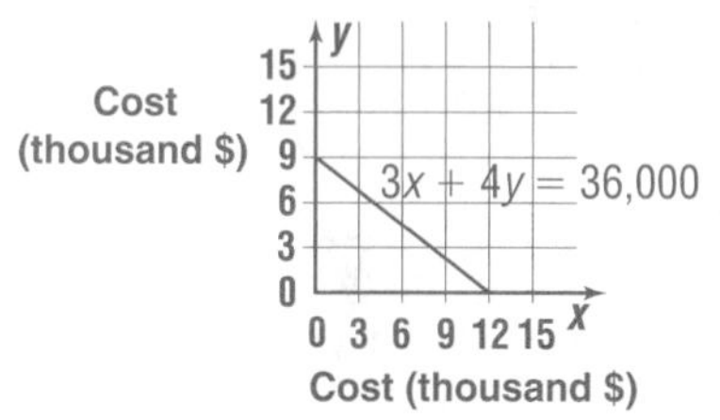

Cooperative Learning

Co-op Co-op Have students work in pairs. One student will write an equation in standard form. The other student will change it into an equation in slope-intercept form. Pairs will then exchange problems. For more information on the co-op co-op strategy, see *Cooperative Learning in the Mathematics Classroom*, one of the titles in the Glencoe Mathematics Professional Series, page 32.

In-Class Examples

For Example 2

Graph $y = -3x + 1$ using the slope and the y-intercept. **The slope is -3, the y-intercept is 1.**

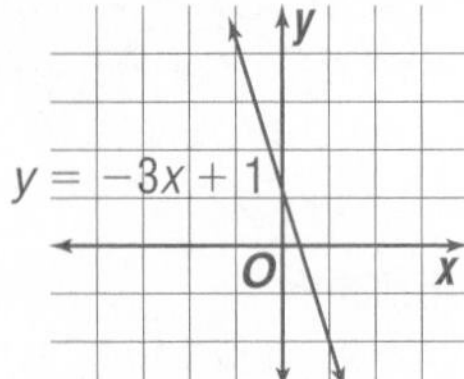

For Example 3

Use the description to graph each line.

a. the line parallel to the graph of $y = 2x + 1$ through the point at $(1, 4)$

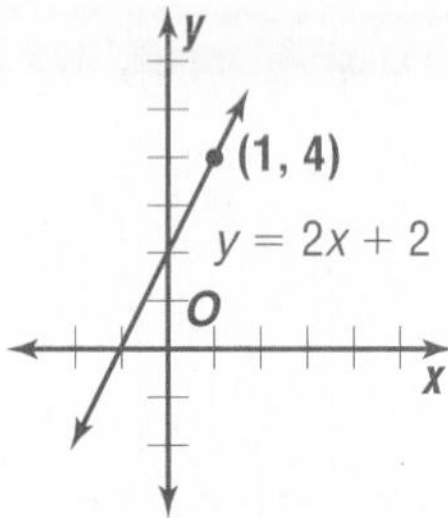

b. the line perpendicular to the graph of $y = -3x + 9$ through the point at $(3, 0)$

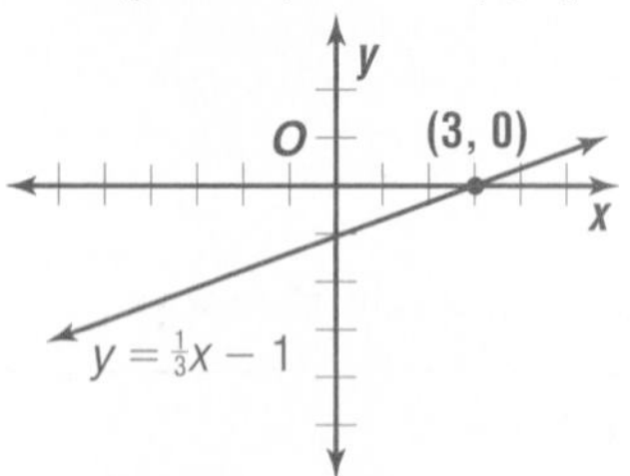

Additional Answers

1. Sample answer: Substitute the coordinates of the point into the equation and see if the result is a true sentence.
2. Sample answer: The x-coordinates of points on a vertical line are the same. When finding the slope, the denominator is zero and division by zero is undefined.
3. intercepts, slope-intercept; Answers may vary.

You can graph a line if you know its slope and y-intercept.

Example  **Graph $y = -\frac{1}{3}x + 4$ using the slope and the y-intercept.**

Since the equation is in slope-intercept form, the slope is $-\frac{1}{3}$, and the y-intercept is 4.

The ordered pair for the y-intercept, 4, is $(0, 4)$. Graph $(0, 4)$. From this point, move 1 unit down and 3 units right. This point, which has coordinates $(3, 3)$, must also lie on the line. Draw the line containing the two points.

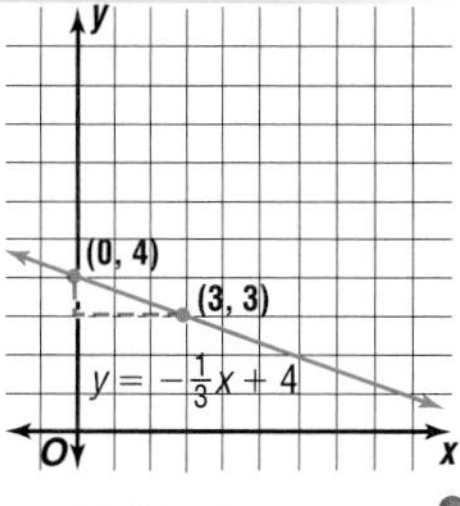

The slope-intercept form of a linear equation can also be used to determine if two lines are parallel or perpendicular without graphing the equations. Remember that if two nonvertical lines are parallel, they have the same slope. If two lines are perpendicular and neither is vertical, the product of their slopes is -1.

Example  **Use the description to graph each line.**

a. the line parallel to the graph of $y = 3x + 8$ through the point at $(1, -2)$

Graph $(1, -2)$. By Postulate 3–2, the line must have the same slope as the graph of $y = 3x + 8$. Its slope is 3. So, from the point at $(1, -2)$, move up 3 units and right 1 unit. This point, which has coordinates $(2, 1)$, must also lie on the line. Draw the line containing the two points.

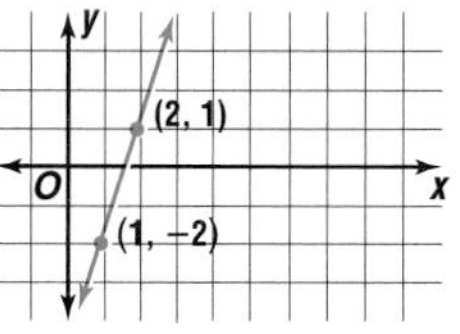

b. the line perpendicular to the graph of $y = 2x$ through the point at $(2, 4)$

Graph $(2, 4)$. By Postulate 3–3, the product of the slopes of two perpendicular lines is -1. Since the slope of the line for $y = 2x$ is 2, the slope of the perpendicular must be $-\frac{1}{2}$. From the point at $(2, 4)$, move down 1 unit and right 2 units. This point, which has coordinates $(4, 3)$, must also lie on the line. Draw the line containing the two points.

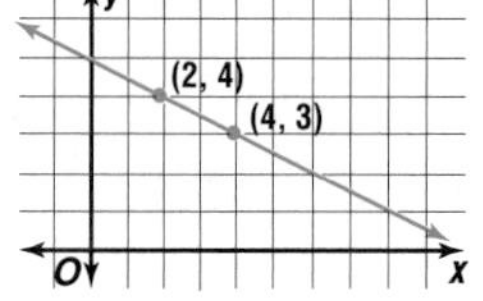

CHECK FOR UNDERSTANDING

Communicating Mathematics

Study the lesson. Then complete the following. **1–3. See margin.**

1. **Describe** how you would determine if a given point is on a given line.
2. **Explain** why the slope of a vertical line is undefined.
3. **Describe** two methods of graphing a linear equation. Which method is more convenient for graphing $5x - 7y = 16$? Explain.

Alternative Teaching Strategies

Reading Geometry Have students look up the meaning of *slope* in the dictionary. Then have them think of examples of slopes that fit the dictionary definition.

4. **Determine** whether $y = 3x + 4$ is a linear equation. Explain your reasoning. Yes, it can be written in the form $Ax + By = C$.
5. The figures at the right form a sequence. Figure number x has a perimeter of y units. a–b. See students' work.

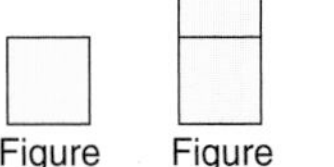
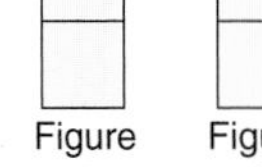

Figure 1 Figure 2 Figure 3

   a. Construct the first eight figures of the sequence.
   b. Verify that the equation $y = 2x + 2$ holds true for the first eight figures.
   c. Then graph the equation on a coordinate grid.
   d. What is the perimeter of figure 40? 82

5c. See Solutions Manual.

Guided Practice

Find the *x*- and *y*-intercepts and slope of the graph of each equation. 8–9. See Solutions Manual.

6. $y = x$ 0; 0; 1
7. $10x + 35y = 280$ 28; 8; $-\frac{2}{7}$
8. Graph $10x - 15y = 120$ by using the intercepts method.
9. Graph $y = -0.5x + 20$ by using the slope and the y-intercept.

Use the description to graph each line. 10–11. See Solutions Manual.

10. $m = -1$; passes through $A(0,1)$
11. line parallel to the line $y = 2x - 5$; passes through $B(0, -4)$
12. **Physics** The velocity of an object v in feet per second, that is dropped from a given height can be expressed by the equation $v = 32t$ where t is the time in seconds since the object was dropped.
    a. Graph $v = 32t$. See margin.
    b. What does the slope of the line represent?
    c. Why is the v-intercept 0? When the object is dropped, the beginning velocity is 0.

    b. the acceleration, or the change in the velocity that occurs in each second that the object drops

EXERCISES

Practice

Find the *x*- and *y*-intercepts and slope of the graph of each equation.

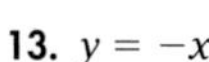

13. $y = -x$
14. $x + 3y = 0$
15. $x = 2$
16. $y = 0$ none, 0; 0
17. $x = 0$ 0; none; undefined
18. $18x - 42y = 210$ $11\frac{2}{3}$; -5; $\frac{3}{7}$

13. 0; 0; −1
14. 0; 0; $-\frac{1}{3}$
15. 2; none; undefined

Graph each equation. Explain the method you used.

19. $y = -4x + 3$
20. $x + y = 8$
21. $3x + 7y = 0$
22. $y = \frac{1}{4}x + 6$
23. $10x + 25y = 100$
24. $21x - 7y = 14$

19–29. See Solutions Manual.

Use the description to graph each line.

25. $m = 0$; $b = 2$
26. line perpendicular to the line $y = 2x - 5$; passes through $C(0, -4)$
27. line parallel to the y-axis through $D(-8, 15)$
28. line perpendicular to the y-axis through $E(-8, 15)$
29. $m = \frac{3}{10}$; passes through $F(0, 32)$

3 PRACTICE/APPLY

Check for Understanding

Exercises 1–12 are designed to help you assess your students' understanding through reading, writing, speaking, and modeling. You should work through Exercises 1–5 with your students and then monitor their work on Exercises 6–12.

Assignment Guide

Core (with proof): 13–41 odd, 42–50
Core (informal): 13–41 odd, 42–50
Enriched: 14–36 even, 37–50

For **Extra Practice**, see p. 787.

The red A, B, and C flags, printed only in the Teacher's Wraparound Edition, indicate the level of difficulty of the exercises.

Study Guide Masters, p. 72

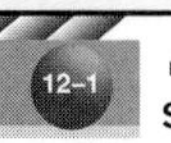

NAME ______ DATE ______

Study Guide — Student Edition Pages 646–651

Integration: Algebra
Graphing Linear Equations

One method for graphing a linear equation is called the **intercepts method**. The ***x*-intercept** is the value of x when $y = 0$. The ***y*-intercept** is the value of y when $x = 0$. By plotting the points on the axes that correspond to the intercepts, you can then draw the line for the equation. (This method will not work if the line is parallel to either axis.)

When an equation is written in the form $y = mx + b$, the equation is said to be in **slope-intercept form**, where m is the slope of the line and b is the y-intercept. The graph of an equation of the form $x = a$ is a vertical line and has an undefined slope. The graph of an equation of the form $y = b$ is a horizontal line and has a slope of 0.

Example: Graph $y = \frac{2}{3}x + 4$ using the slope and the y-intercept.

Since the equation is in slope-intercept form, the slope is $\frac{2}{3}$ and the y-intercept is 4.

Plot the point with coordinates (0, 4). From this point move up 2 units and to the right 3 units. The point at (3, 6) must also lie on the line.

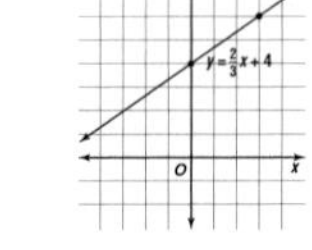

Find the x- and y-intercepts and slope of the graph of each equation.

1. $6x + 4y = -24$ −4, −6, $m = -\frac{3}{2}$
2. $y + 5 = x$ 4, −5, $m = 1$
3. $x = 5$ 5, no y-intercept, m is undefined

Graph each equation.

4. $3x - 6y = 12$
5. $2x + y = 3$
6. $y + 4x = 2$

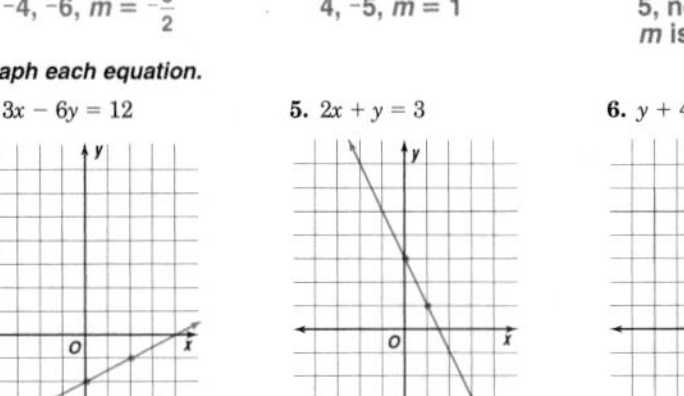

Reteaching

Using Drawing Draw a variety of lines tilting different ways. Show which lines have positive or negative slope. Move lines up and down to show varying y-intercepts.

Additional Answer

12a.

Additional Answers

30. See students' graphs. Sample answer: All are of the form $y = -4x + b$, but all have a different value for b.
31. See students' graphs. Sample answer: All are of the form $y = mx - 1$, but all have a different value for m.
32. Sample answer: The graphs of the first and third equations are parallel and the graph of the second equation is perpendicular to the other two.
33. The slope is undefined. Sample answer: The x-coordinates of points on a vertical line are the same. When finding the slope, the denominator (the change in x) is zero and division by zero is undefined.
37. $y = -2.5x = 3.5$; Sample answer: The distance between two parallel lines is the length of any perpendicular segment that connects points on the two lines. The distance between the y-intercepts is 5. The y-intercept midway between the two lines is 3.5. Since the line is parallel to the other two, the slope must be the same.

Practice Masters, p. 72

12-1 NAME____________ DATE________

Practice — Student Edition Pages 646–651

Integration: Algebra
Graphing Linear Equations

Graph each pair of linear equations on the same coordinate plane. Determine if the lines are parallel, perpendicular, or neither by finding the slope of each line.

1. $y = x + 2$ parallel; 1, 1
   $y - x = 4$

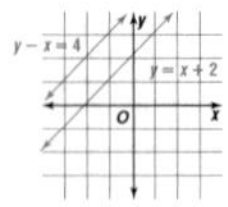

2. $2x - y = 3$ parallel; 2, 2
   $y = 2x + 4$

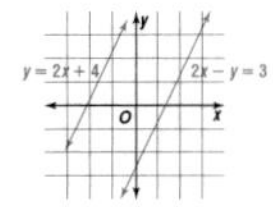

3. $3y = x + 1$ perpendicular; $\frac{1}{3}$, −3
   $y = -3x + 2$

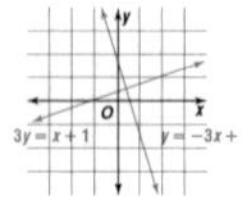

4. $y = 2x + 1$ neither; 2, 3
   $y = 3x - 1$

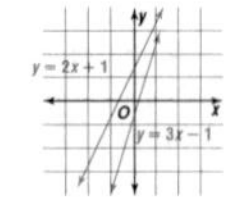

Find the slope and y-intercept of the graph of each equation.

5. $x + y = 8$ −1, 8
6. $2x - y = 4$ 2, −4
7. $2x - 3y = 10$ $\frac{2}{3}$, $-\frac{10}{3}$

Determine the x- and y-intercepts of each line. Then graph the equation.

8. $y = 2x - 3$

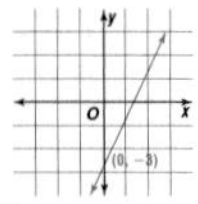

$\frac{3}{2}$, −3

9. $y = 5$

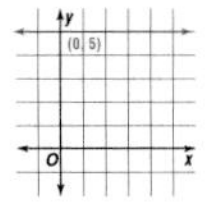

no x-intercept, 5

30. Graph several lines having a slope of −4. Describe how the equations of these lines are alike and how they are different. See margin.
31. Graph several lines passing through the point $G(0, -1)$. Describe how the equations of these lines are alike and how they are different. See margin.
32. Describe the relationship among the graphs of the three lines whose equations are $3x - y = 10$, $3y + x = 6$, and $y = 3x - 2$. See margin.
33. What is the slope of any line that is parallel to the y-axis? Explain why this is true. See margin.
34. The equations for two lines are $3x + 4y = 12$ and $Ax + By = 10$. Find an A and B that will satisfy each condition.
    a. The two lines are parallel.
    b. The two lines are perpendicular.

34a. $A = 3c$ and $B = 4c$ for any real number c, $c \neq 0$.
34b. $A = 4c$ and $B = -3c$ for any real number c, $c \neq 0$.

35. The equation of a line is $y = -2x + 6$. Suppose you added 3 to the x value and 2 to the y value of every ordered pair of points on the line.
    a. How would the graph of the line change?
    b. How would the equation of the line change? m would still be −2, b would change to 14.

35a. The new line would be parallel to the first line with y-intercept 14.

Graphing Calculator

36. Enter the equation $y = -2x + 4$ into the Y= list of a TI-82/83 graphing calculator and then press ZOOM 6. Determine the equation of a line that is perpendicular to it, enter the equation in the Y= list, and press ZOOM 5 to visually check your answer. Sample answer: $y = \frac{1}{2}x + 4$

Critical Thinking

37. The graphs of $5x + 2y = 12$ and $5x + 2y = 2$ are parallel lines. Find the equation of the line that is parallel to both lines and lies midway between them. Explain why your answer is correct. See margin.

Applications and Problem Solving

38. **Transportation** According to research by *Travel & Leisure* magazine, the equations $y = 1.80x + 1.70$, $y = 1.80x + 1.80$, and $y = 1.40x + 1.50$ represent the costs of a taxi in San Francisco, Philadelphia, and Atlanta, respectively. The slope is the cost per mile, and the constant term is the initial fee. Graph the equations on the same coordinate axes and compare the cost of taking a taxi in these three cities. See Solutions Manual.
39. **Nutrition** The equation $C = 12f + 180$ describes the relationship between the amount of fat f in grams, and the number of Calories C in some food items.
    a. What is the slope of the line? What does it indicate about the relationship between fat and Calories in the food items?
    b. What is the y-intercept? 180
    c. Graph the equation. See margin.
    d. If a food item contains 30 grams of fat, how many Calories would you expect to be in that item? 540 Calories

39a. 12; for every gram of fat in the food, there are 12 times more Calories.

FYI

According to *The Guiness Book of World Records*, the steepest street in the world is Baldwin Street, Dunedin, New Zealand. It has a maximum road grade of 79%.

40. **Civil Engineering** The maximum road grade recommended by the Federal Highway Commission is 12%. This means that the slope of a road should be no more than $\frac{12}{100}$ or 0.12. At the maximum grade, a road would change 12 feet vertically for every 100 feet horizontally.
    a. How many horizontal feet will it take for a road with the maximum grade to drop 240 feet? 2000 feet
    b. If the maximum grade for railroad tracks is 2%, how many more miles will it take a train to climb 500 feet than it will take a car on a road that has the maximum grade? $20{,}833\frac{1}{3}$ feet or about 3.9 miles

FYI

The steepest street in San Francisco is Filbert Street between Leavenworth and Hyde. It has a maximum road grade of 31.5%.

Additional Answer

39c.

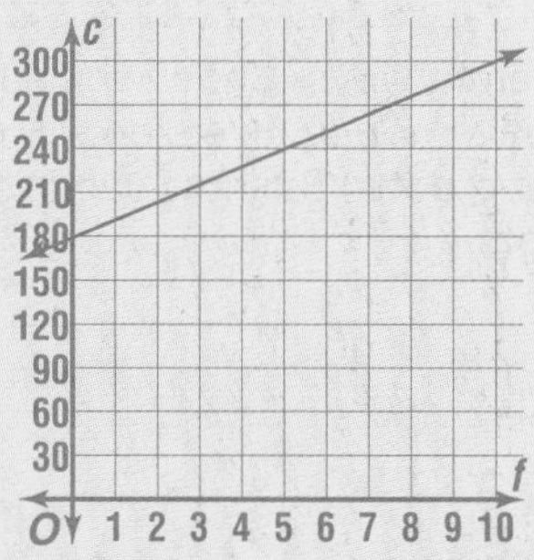

41. **Transportation** A small van can carry 10 people and a large van can carry 15 people. Let x represent the number of small vans and y the number of large vans. If 120 people need transportation to a camp in the mountains, an equation describing the transportation of the people is $10x + 15y = 120$.
   a. Explain what $10x$ represents.
   b. What are the intercepts? What does each represent?
   c. Graph the equation using the intercepts.
   d. Is it possible to have an arrangement of 6 small vans and 4 large vans? Explain why or why not.
   e. Does the line representing the equation pass through the point at (3, 6)? Explain why or why not. **yes; $10(3) + 15(6) = 120$**

Mixed Review

41a. the number of people that can be carried in x small vans

41b–c. See margin.

41d. yes; $10(6) + 15(4) = 120$

42. 26 ft $\frac{1}{2}$ in.

43. 18 m, 6 m

42. **World Records** The world's smallest model train was built by Jean Damery of Paris, France. Built in a 1:1000 scale, the engine on Damery's train measures $\frac{5}{16}$ inches. How long is the engine on a life-sized train similar to Damery's? (Lesson 11–8)
43. If the height of a parallelogram is 3 times the base, and the area of the parallelogram is 108 square meters, find the height and the base. (Lesson 10–3)
44. Write an equation for the circle with a center at (6, −5) and radius $7\sqrt{3}$ units long. (Lesson 9–8) **$(x - 6)^2 + (y + 5)^2 = 147$**
45. **Aviation** A jet takes off from Columbus International Airport at an angle of 22° with the ground. The jet covers 6.9 miles of ground distance before leveling off. At what altitude does the jet level off? (Lesson 8–3) **2.8 miles**
46. Determine whether the statement *The opposite angles of a parallelogram are supplementary* is true or false. (Lesson 6–1) **false**
47. Graph $\triangle ABC$ and list the angles from least to greatest measure for $A(1, 3)$, $B(5, -2)$, and $C(-2, -4)$. (Lesson 5–4) **$\angle C$, $\angle A$, $\angle B$**
48. Find the measure of $\overline{AB}$ for $A(5, -2)$, and $B(-3, -7)$. (Lesson 3–5) **$\sqrt{89} \approx 9.4$**

Algebra

49. Find the factors of 22. **1, 2, 11, 22**
50. Find $5y(8y^3 + 7y^2 - 3y)$. **$40y^4 + 35y^3 - 15y^2$**

Refer to the Investigation on pages 642–643.

1 Begin creating your movie by drawing a central object or character.
2 Draw the image on a computer screen model.
3 Then create a list of pixels to be colored in different colors. You may wish to use lines to enclose the figure and list the colors as regions instead of individual pixels.

Add the results of your work to your Investigation Folder.

4 ASSESS

Closing Activity

Writing Have students write a brief paragraph explaining two ways of graphing linear equations. Have them compare the methods.

Additional Answers

41b. x: 12; the number of small vans if all of the people were to be transported in small vans and none in the large vans; y: 8; the number of large vans if all of the people were to be transported in the large vans and none in the small vans.

41c.

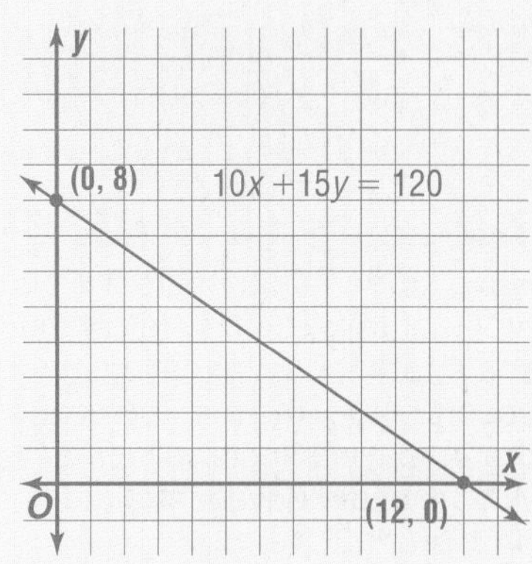

Extension

Problem Solving Write the equation in slope-intercept form of a line that has a y-intercept of −3 and passes through the point (2, 2).
$y = \frac{5}{2}x - 3$

In·ves·ti·ga·tion

Working on the Investigation

The Investigation on pages 642–643 is designed to be a long-term project that is completed over several days or weeks. Encourage students to keep their materials in their Investigation Folder as they work on the Investigation.

Enrichment Masters, p. 72

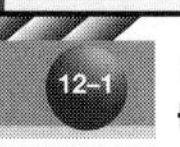

NAME________ DATE ________

Enrichment

Student Edition Pages 646–651

Graphing Linear Inequalities

The graph of $y = 2x + 5$ separates the coordinate plane into two **half-planes**. The half-plane below the line for $y = 2x + 5$ consists of all points (x, y) whose coordinates satisfy the inequality $y < 2x + 5$. The half-plane above the line consists of all points that satisfy $y > 2x + 5$. The graph of $y = 2x + 5$ is the **boundary** of the half-planes.

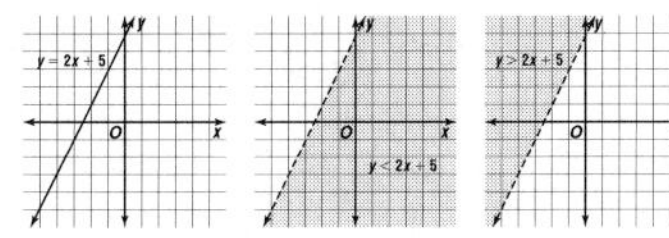

Notice the dashed line used for $y = 2x + 5$ in the graphs of the inequalities. This boundary line is not part of either of the half-planes.

Graph each inequality. Use shading to show half-planes.

1. $y = \frac{1}{2}x + 3$

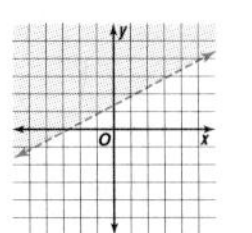

2. $y < -x$
3. $y < -2x + 4$
4. $y > x + 1$

12-2A LESSON NOTES

NCTM Standards: 2, 4, 5

Objective

Use a graphing calculator to determine an equation of the line that passes through plotted points.

Recommended Time

20 minutes

Instructional Resources

Instructions for using the TI-81 and Casio calculators for this activity are available in the *Graphing Calculator and Computer Masters*, pp. 28, 29.

1 FOCUS

Motivating the Lesson

Have students compare the two types of tax. The first tax is the same rate for all taxpayers. In the second type, the rate increases as the taxpayer's income increases. Which would have a straight line on a graph? the first type

2 TEACH

Teaching Tip Point out to students that the y-intercept is zero only if all people who earn an income pay tax.

3 PRACTICE/APPLY

Assignment Guide

Core (with proof): 1–4
Core (informal): 1–4
Enriched: 1–4

4 ASSESS

Observing students working with technology is an excellent method of assessment.

12–2A Using Technology Writing Equations

A Preview of Lesson 12–2

You can use a graphing calculator to plot points and determine an equation of the line that passes through the points.

The table at the right shows the amount of Indiana state income tax owed by several single taxpayers with no dependents. Find an equation for the amount of tax owed, y, if a taxpayer has an income of x dollars.

Indiana State Income Tax

Income	Tax Owed
1000	0.00
5500	153.00
10,800	333.20
16,900	540.60
27,600	904.40
34,700	1145.80
51,100	1703.40

Source: U.S. Advisory Commission on Intergovernmental Relations

Step 1 Clear the statistical lists by pressing STAT 4 2nd [L1] , 2nd [L2] ENTER .

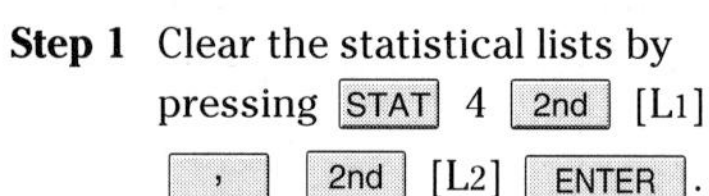

Step 2 Next, enter the data into the statistical lists.To access the lists, press STAT ENTER . Enter the income amounts into list L1. Then press the right arrow button and enter the tax amounts into list L2.

Step 3 To have the calculator find the equation of the line, press STAT ▶ then choose LinReg($ax + b$) and press ENTER .

The calculator will display values for a and b for the equation $y = ax + b$. In this case, $a = 0.034$ and $b = -34$. So an equation for the amount of Indiana state tax owed by a single taxpayer with no dependents is $y = 0.034x - 34$. *The value of $r = 1$ displayed on the TI-82 indicates that the equation fits the data entered in the lists perfectly.*

EXERCISES

Study the results of the activity.

1. Choose any two ordered pairs from the list and find the slope between them. **The slope between any two points is 0.034.**
2. Use the equation $y = mx + b$. Substitute the x and y values of one data point and the slope you found in Exercise 1 for x, y, and m, respectively. Then solve for b. $b = -34$
3. Write the equation for the relationship in slope-intercept form. How does the equation you found compare to the equation that the calculator found?
4. Use the calculator to find an equation that represents the data in the table at the right. Then follow the steps in Exercises 1–3 to find the equation. Compare your results. $y = 0.25x + 6$

x	y
−5	4.75
−3	5.25
0	6.0
2	6.5
6	7.5

3. $y = 0.034x - 34$; see students' work.

Using Technology

This lesson offers an excellent opportunity for using technology in your geometry classroom. For more information on using technology, see *Graphing Calculators in the Mathematics Classroom*, one of the titles in the Glencoe Mathematics Professional Series.

12-2 Integration: Algebra
Writing Equations of Lines

What YOU'LL LEARN

- To write an equation of a line given information about its graph, and
- to solve problems by using equations.

Why IT'S IMPORTANT

You can write equations to solve problems involving sports and computers.

APPLICATION
Tourism

Do you dream of visiting far away places? One of the world's most popular vacation spots is Rio de Janeiro, Brazil. The harbor there has been called one of the seven wonders of the natural world because of its beautiful, low mountain ranges.

Like most countries other than the United States, Brazil uses the Celsius scale to report their temperatures. In the United States, the Fahrenheit scale is used. According to U.S. Department of Commerce data, the high temperature in Rio de Janeiro averages 25° Celsius during the month of January. What is the equivalent Fahrenheit temperature?

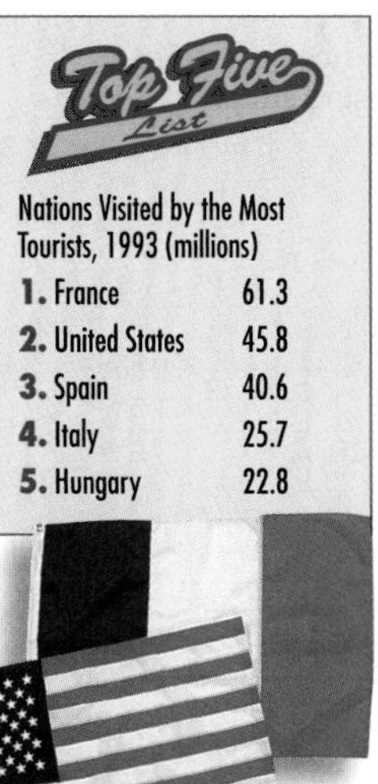

Top Five List

Nations Visited by the Most Tourists, 1993 (millions)

1. France	61.3
2. United States	45.8
3. Spain	40.6
4. Italy	25.7
5. Hungary	22.8

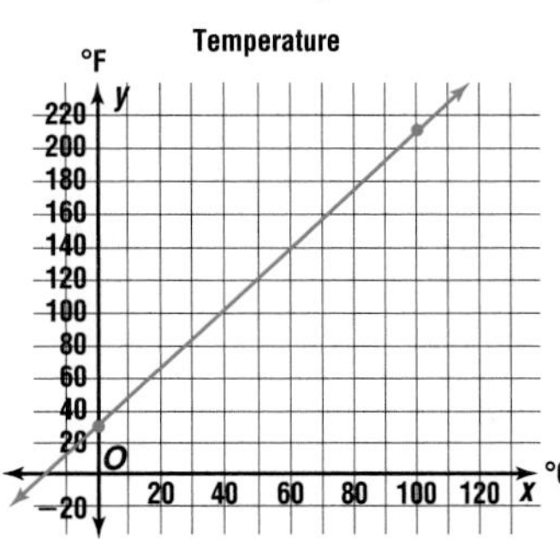

The graph at the left shows the relationship between temperatures in Celsius and temperatures in Fahrenheit. The y-intercept, 32, represents the temperature where water freezes. Let $A(100, 212)$ represent the temperature at which water boils. We can use the ordered pairs (0, 32) and (100, 212) to find the slope of the line.

$$\frac{y_2 - y_1}{x_2 - x_1} = \frac{212 - 32}{100 - 0} = \frac{180}{100} \text{ or } \frac{9}{5}$$

Since the graph of this relationship is a line, it can be represented by a linear equation. If we substitute $\frac{9}{5}$ for m and 32 for b in $y = mx + b$, we can find the slope-intercept form of the equation for this line. Thus, the equation is $y = \frac{9}{5}x + 32$, where y represents the temperature in degrees Fahrenheit and x represents the temperature in degrees Celsius.

In general, you can write an equation of a line if you are given:

- **Case 1:** the slope and the y-intercept,
- **Case 2:** the slope and the coordinates of a point on the line, or
- **Case 3:** the coordinates of two points on the line.

The application above illustrates Case 1.

Example 1 illustrates Case 2. You can find an equation of a line given the slope and the coordinates of a point on the line using the **point-slope form** of a linear equation. The point-slope form is $y - y_1 = m(x - x_1)$, where (x_1, y_1) are the coordinates of a point on the line and m is the slope of the line.

12-2 LESSON NOTES

NCTM Standards: 1–6

Instructional Resources

- Study Guide Master 12-2
- Practice Master 12-2
- Enrichment Master 12-2
- Assessment and Evaluation Masters, p. 324
- Modeling Mathematics Masters, pp. 59–62

Transparency 12-2A contains the 5-Minute Check for this lesson; **Transparency 12-2B** contains a teaching aid for this lesson.

Recommended Pacing	
Standard Pacing	Days 2 & 3 of 12
Honors Pacing	Day 2 of 11

1 FOCUS

5-Minute Check
(over Lesson 12-1)

Find the x- and y-intercepts and the slope of the graph of each equation.

1. $2x + 3y = -12$ **$-4, -6, -\frac{2}{3}$**
2. $x - 5y = 15$ **$-3, 15, \frac{1}{5}$**
3. $4x - 3y = -6$ **$2, -\frac{3}{2}, \frac{4}{3}$**

Find the slope and y-intercept of the graph of each equation.

4. $4x - 2y = 5$ **$2, -\frac{5}{2}$**
5. $x + 3y = 9$ **$-\frac{1}{3}, 3$**
6. $5x + y = -7$ **$-5, -7$**

Top Five List

There are 3142 counties in the U.S. In 1991 Allen F. Zondlak completed his attempt to visit every county in the U.S.

Motivating the Lesson

Situational Problem Tell students that Cole has $10 in savings and gets an allowance of $4 per week. Write ordered pairs that show how much money Cole will have in 2 weeks and in 4 weeks if he saves all his money. Graph the points on a coordinate grid on the chalkboard and ask students to use the graph to write the equation of the line. (2, 18), (4, 26); $y = 4x + 10$

2 TEACH

In-Class Examples

For Example 1
Write an equation of the line whose slope is -2 and x-intercept is 6.
$y = -2x + 12$

For Example 2
Write an equation for each line.

a. line b

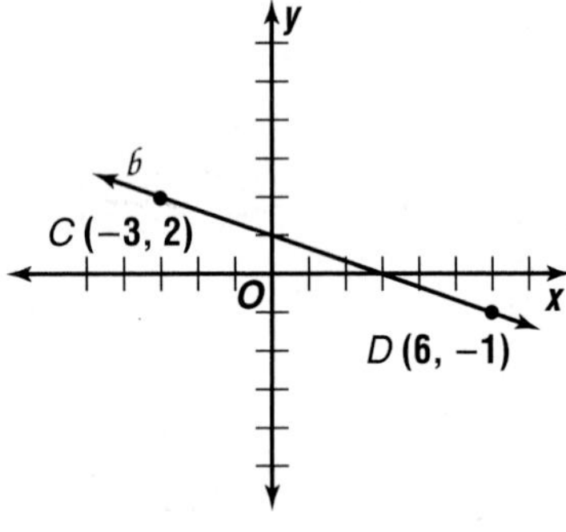

$y = -\frac{1}{3}x + 1$

b. a line that is parallel to line b and passes through the point at (4, 4) $y = -\frac{1}{3}x + 5\frac{1}{3}$

Teaching Tip After reading Example 2, point out that the equation in part a could have also been found using the other point.

Example 1

Algebra

Write an equation of the line whose slope is -2 and passes through the point at (4, 10).

Method 1: Point-Slope Form

$y - y_1 = m(x - x_1)$ *Use point-slope form.*
$y - 10 = -2(x - 4)$ *Substitute -2 for m and (4, 10) for (x_1, y_1).*

The point-slope form of the equation is $y - 10 = -2(x - 4)$.

Method 2: Slope-Intercept Form

Since we know that the slope is -2, we can substitute -2 for m in $y = mx + b$.

$$y = -2x + b$$

The point at (4, 10) is on the line. Substitute the coordinates of this point into the equation to find the value of b.

$y = -2x + b$
$10 = -2(4) + b$ *$y = 10$ and $x = 4$*
$10 = -8 + b$
$18 = b$ *Add 8 to each side.*

The slope-intercept form of the equation of the line is $y = -2x + 18$.

The first step in Case 3 is to find the slope by using the two ordered pairs. Then you can write the equation in point-slope form by using either of the two ordered pairs and simplifying.

Example 2 **Write an equation for each line.**

a. line ℓ

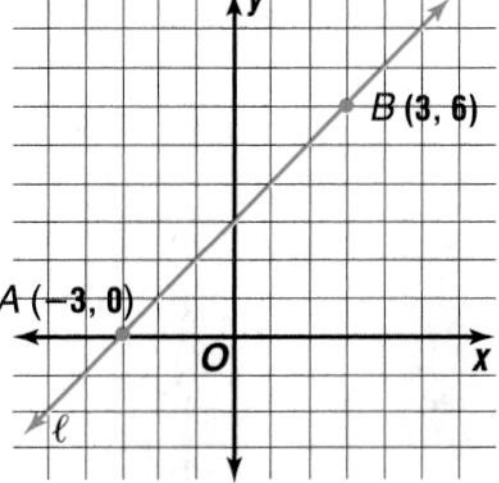

Find the slope of line ℓ by using the points $A(-3, 0)$ and $B(3, 6)$.

$$m = \frac{y_2 - y_1}{x_2 - x_1} = \frac{6 - 0}{3 - (-3)} = \frac{6}{6} \text{ or } 1$$

Now use the point-slope form to write the linear equation.

$y - y_1 = m(x - x_1)$ *Use point-slope form.*
$y - 0 = 1(x - (-3))$ *The slope is 1, and the point with coordinates $(-3, 0)$ is on the line.*
$y = x + 3$

The slope-intercept form of the equation of the line is $y = x + 3$.

b. a line that is parallel to line ℓ and passes through the point at (4, 3)

Since the slope of line ℓ is 1, the slope of a line parallel to it is 1. *Why?*

$y - y_1 = m(x - x_1)$ *Use point-slope form.*
$y - 3 = 1(x - 4)$ *The slope is 1 and the point with coordinates (4, 3) is on the line.*
$y - 3 = x - 4$
$y = x - 1$

The slope-intercept form of the equation of the line is $y = x - 1$.

Alternative Learning Styles

Auditory Separate students into pairs. Have one student describe a line without mentioning its slope. Have the other student determine what kind of slope the line has and give a possible equation for the line.

One strategy for solving problems is to write and solve an equation.

Example 3

PROBLEM SOLVING

Write an Equation

Wired Ventures, Inc. is the San Francisco-based publisher of *Wired* magazine and other computer publications. In 1993, the first year of operation, Wired had revenues of $2.9 million. By 1995, revenues were $25.2 million. If the increase in earnings remains constant, how much will the company earn in its eighth year of operation?

Explore You know how much the company made in the third year and how much it made in the first year.

Plan Define the variables and write a linear equation that describes the company's earnings.

Let d represent dollars and let t represent time in years.

Solve The line passes through the points at (1, 2.9) and (3, 25.2). Find the slope of the line.

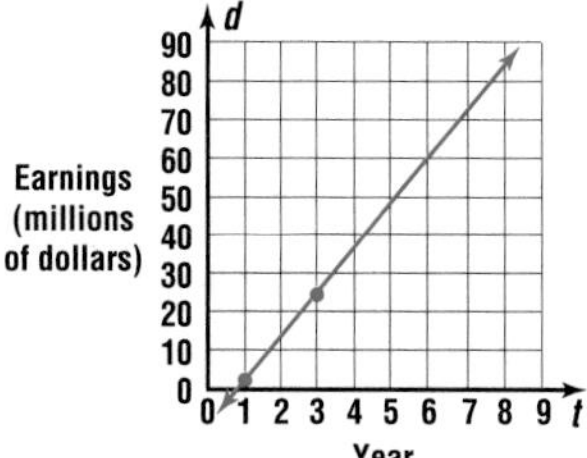

$$m = \frac{d_2 - d_1}{t_2 - t_1}$$

$$= \frac{25.2 - 2.9}{3 - 1}$$

$$= \frac{22.3}{2} \text{ or } 11.15$$

Now use the point-slope form to write the linear equation.

$d - d_1 = m(t - t_1)$ *Use point-slope form.*

$d - 25.2 = 11.15(t - 3)$ *Use the coordinates of either point for (t_1, d_1).*

$d - 25.2 = 11.15t - 33.45$ *We chose (3, 25.2).*

$d = 11.15t - 8.25$

Use the equation to find how much money the company is predicted to earn in its eighth year.

$$d = 11.15t - 8.25$$

$$d = 11.15(8) - 8.25$$

$$d = 89.2 - 8.25$$

$$d = 80.95$$

If the rate of increase remains the same, the company will earn $80.95 million in its eighth year. This is represented by the ordered pair (8, 80.95).

Examine Check by looking at the x-coordinate of the graph of the line when y is about 80.95.

CHECK FOR UNDERSTANDING

Communicating Mathematics

Study the lesson. Then complete the following. 1–3. See margin.

1. **Explain** how you would write the equation of a line whose slope is 3 and y-intercept is -10.
2. **Write** equations for two different lines that pass through the point at (7, 2).
3. **Explain** why it is important to define a variable before you write an equation to solve a problem.

In-Class Example

For Example 3

A hotel charges $2.50 to fax 3 pages and $3.50 to fax 5 pages.

a. Write a linear equation that describes the cost of faxing x pages. $y = \frac{1}{2}x + 1.00$

b. How much would it cost to fax 10 pages? Faxing 10 pages would cost $6.00.

Check for Understanding

Exercises 1–12 are designed to help you assess your students' understanding through reading, writing, speaking, and modeling. You should work through Exercises 1–5 with your students and then monitor their work on Exercises 6–12.

Error Analysis

Students may make mistakes putting equations in slope-intercept form. They may have a wrong sign or forget to divide both terms on one side of an equation by the number that divides the other side. Encourage students who make such mistakes to write out every step in the solution of an equation.

Additional Answers

1. Sample answer: Substitute 3 for m and -10 for b into $y = mx + b$.
2. Sample answers: $y = x - 5$ and $y = 2x - 12$
3. Sample answer: You need to know what the variable represents in order to write the equation.

Reteaching

Using Examples Show students how they can use graphing to determine the equation of a line, given its slope and a point on the graph. For example, given a slope of 4 and a point (1, −3), have them plot the point, and from that point draw a line with a slope of 4. By extending the line to where it crosses the y-axis, they can obtain the y-intercept and write the equation in slope-intercept form. $y = 4x - 7$

3 PRACTICE/APPLY

Assignment Guide

Core (with proof): 13–45 odd, 47–56
Core (informal): 13–45 odd, 47–53, 55–56
Enriched: 14–40 even, 42–56

For **Extra Practice**, see p. 788.

The red A, B, and C flags, printed only in the Teacher's Wraparound Edition, indicate the level of difficulty of the exercises.

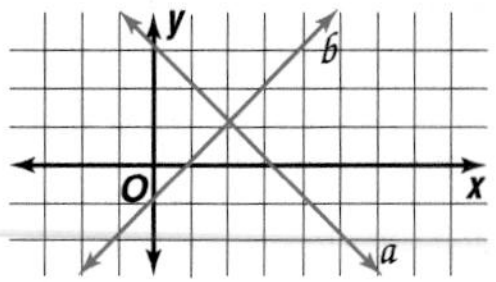

4. **State** the slope and y-intercept of line a. Then write its equation in slope-intercept form. $-1; 3; y = -x + 3$
5. **Explain** why the slope of line b is 1 if line a is perpendicular to line b. Sample answer: If the slope of line a is -1, the slope of line b must be 1 by Postulate 3–3.

Guided Practice

Write the equation in slope-intercept form of the line having the given slope and passing through the point with the given coordinates.

6. 4, (1, 5) $y = 4x + 1$
7. $-\frac{1}{3}$, (−2, 4) $y = -\frac{1}{3}x + \frac{10}{3}$

Write the equation in slope-intercept form of the line that satisfies the given conditions.

8. $m = -3$; y-intercept $= 5$ $y = -3x + 5$
9. $m = \frac{1}{3}$; passes through the point at (6, −1) $y = \frac{1}{3}x - 3$
10. $m = 5$; x-intercept is -2 $y = 5x + 10$
11. perpendicular to $y = \frac{1}{2}x - 5$; through the point at (4, −1) $y = -2x + 7$
12. **Business** A news retrieval service allows a customer to have news delivered to his or her computer via the Internet. The Dow Jones News Retrieval Service, which specializes in stock market news, advertises a \$29.95 start-up fee and a \$19.95 yearly fee with the first year's fee waived.
    a. Write an equation that will show the cost of using the service over the years. What assumptions do you have to make?
    b. Explain how you would know if the point at (3, 89.8) is on the graph of the equation. Sample answer: See if $89.8 = 29.95 + 19.95(3 - 1)$ is a true sentence.

12a. $c = 29.95 + 19.95(x - 1)$; Sample answer: You must assume the fees remain constant.

EXERCISES

Practice

Refer to the graph for Exercises 13–15. State the slope and y-intercept for each line. Then write the equation of the line in slope-intercept form.

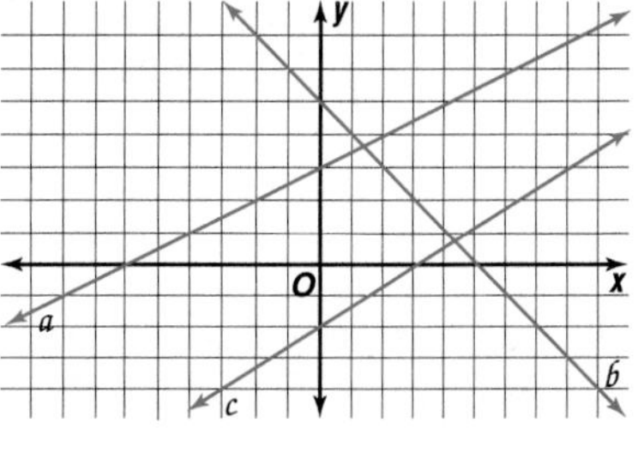

13. a 14. b 15. c

13. $\frac{1}{2}$, 3; $y = \frac{1}{2}x + 3$
14. −1, 5; $y = -x + 5$
15. $\frac{2}{3}$, −2; $y = \frac{2}{3}x - 2$

Write the equation in slope-intercept form of the line having the given slope and passing through the point with the given coordinates.

16. 0.3, (0, −6)
17. 0.5, (40, 19)
18. $\frac{1}{3}$, (−3, −15)
19. 1, (2, 4) $y = x + 2$
20. $-\frac{3}{4}$, (24, 3)
21. 0, (2.5, 8) $y = 8$

16. $y = 0.3x - 6$
17. $y = 0.5x - 1$
18. $y = \frac{1}{3}x - 14$
20. $y = -\frac{3}{4}x + 21$

Write the equation in slope-intercept form of the line that satisfies the given conditions.

22. $m = 3$; y-intercept $= -4$ $y = 3x - 4$
23. $m = 0$; y-intercept $= 6$ $y = 6$
24. x-intercept is 5; y-intercept is 3
25. passes through points at $(2, -3)$ and $(-1, -9)$ $y = 2x - 7$
26. parallel to $y = \frac{1}{2}x - 5$; passes through the point at $(4, -1)$ $y = \frac{1}{2}x - 3$
27. $m = -0.1$; passes through the point at $(20, 17)$ $y = -0.1x + 19$
28. $m = 5$; passes through the point at $(-8, -7)$ $y = 5x + 33$
29. parallel to $y = 5x - 2$; passes through the point at $(4, 0)$ $y = 5x - 20$
30. parallel to the y-axis; passes through the point at $(4, -3)$ $x = 4$
31. perpendicular to the y-axis; passes through the point at $(-5, 10)$ $y = 10$
32. x-intercept is 5; y-intercept is -1 $y = \frac{1}{5}x - 1$
33. passes through points at $(4, -1)$ and $(-2, -1)$ $y = -1$
34. passes through points at $(5, 7)$ and $(-3, 11)$

B

24. $y = -\frac{3}{5}x + 3$

34. $y = -\frac{1}{2}x + 9\frac{1}{2}$

Use the information to write an equation representing the cost in relation to time.

35. A lawn mower costs \$340, and the price has increased an average of \$20 per year. $c = 20t + 340$
36. The cost of a local call from a pay phone was 25¢ in 1985. The cost has not changed since then. $c = 0.25$

C

37. Line a has slope $-\frac{3}{5}$ and passes through the point at $(6, 3)$. Line b is perpendicular to line a. Write an equation for line b in slope-intercept form if lines a and b have the same x-intercept.
38. Write the equation in slope-intercept form of the line that is the perpendicular bisector of the segment whose endpoints have the coordinates $(5, -3)$ and $(11, 7)$.
39. $\angle ABC$ is a right angle. If the coordinates of points A and B are $(4, 6)$ and $(5, 9)$, respectively, write an equation for line BC.
40. Verify that $\triangle DEF$ is a right triangle for $D(0, -4)$, $E(2, -3)$, and $F(-4, 9)$.

37. $y = \frac{5}{3}x - \frac{55}{3}$

38. $y = -\frac{3}{5}x + \frac{34}{5}$

39. $y = -\frac{1}{3}x + \frac{32}{3}$

40. See margin.

Programming

41. The TI-82/83 graphing calculator program below finds the equation of a line containing two points at (x_1, y_1) and (x_2, y_2). Then it displays the equation in slope-intercept form.

```
PROGRAM:WRITEQN
:ClrDraw:AxesOff
:Input  "X1=", Q
:Input  "Y1=", R
:Input  "X2=", S
:Input  "Y2=", T
:Text (8, 5, "EQUATION IS")
:If S-Q = 0
:Goto  1
:(T-R)/(S-Q)→M
:(R-M*Q)→B
:If  M=0
:Then
:Text (14, 5, "Y=", R)
:Else
:If  B<0
:Then
:abs(B)→B
:Text (14, 5, "Y= ", M,
 "X-", B)
:Else
:Text (14, 5, "Y= ", M,
 "X+", B)
:Stop
:Lbl  1
:Text (14, 5, "X= ", S)
```

a. Use the program to find the equation of the line that passes through $A(6, -3)$ and $B(3, -2)$.

b. Use the program to check your equations in Exercises 33–34.

41a. $y = -\frac{1}{3}x - 1$

41b. See students' work.

Using the Programming Exercises The program given in Exercise 41 is for use with a TI-82 or TI-83 graphing calculator. For other programmable calculators, have students consult their owner's manual for commands similar to those represented here.

Additional Answer

40. slope of $\overline{DE} = \frac{1}{2}$ and slope of $\overline{DF} = -2$. Since $\frac{1}{2} \cdot -2 = -1$, $\overline{DE} \perp \overline{DF}$ and $\triangle DEF$ is a right triangle.

Study Guide Masters, p. 73

12-2 NAME__________ DATE______

Study Guide Student Edition Pages 653–659

Integration: Algebra
Writing Equations of Lines

You can write an equation of a line if you are given
- the slope and the coordinates of a point on the line, or
- the coordinates of two points on the line.

Example: Write the equation in slope-intercept form of the line that has slope 5 and an x-intercept of 3.

Since the slope is 5, you can substitute 5 for m in
$y = mx + b$.
$y = 5x + b$

Since the x-intercept is 3, the point (3, 0) is on the line.
$y = 5x + b$
$0 = 5(3) + b$ $y = 0$ and $x = 3$
$0 = 15 + b$
$b = -15$ Solve for b.
So the equation is $y = 5x - 15$.

If you know two points on a line, you will need to use the point-slope form of the equation, that is, $y - y_1 = m(x - x_1)$.

Write the equation in slope-intercept form of the line that satisfies the given conditions.

1. $m = 3$, y-intercept $= -4$ $y = 3x - 4$
2. $m = -\frac{2}{5}$, x-intercept $= 6$ $y = -\frac{2}{5}x + \frac{12}{5}$
3. passes through $(-5, 10)$ and $(2, 4)$ $y = -\frac{6}{7}x + \frac{40}{7}$
4. passes through $(8, 6)$ and $(-3, -3)$ $y = \frac{9}{11}x - \frac{6}{11}$
5. perpendicular to the y-axis, passes through $(-6, 4)$ $y = 4$
6. parallel to the y-axis, passes through $(-7, 3)$ $x = -7$
7. $m = 3$ and passes through $(-4, 6)$ $y = 3x + 18$
8. perpendicular to the graph of $y = 4x - 1$ and passes through $(6, -3)$ $y = -\frac{1}{4}x - \frac{3}{2}$

Additional Answers

42c. Sample answer: The graphs of the inverse equation is a reflection about the line $y = x$. The line $y = x$ passes through the intersection of the two inverse equations.

45b. Sample answer: Ameritech ISDN service is always cheaper for the first customer.

47. $m = \frac{3}{2}$; $b = -\frac{3}{4}$

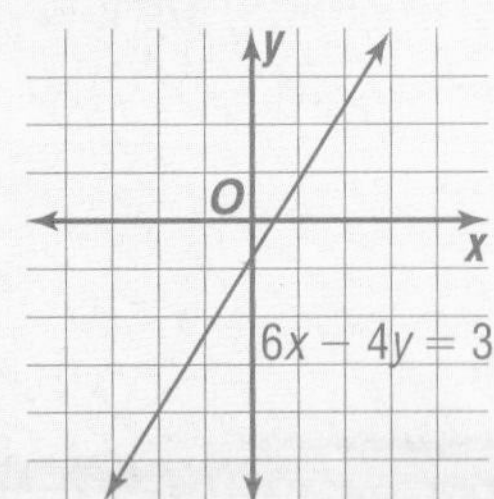

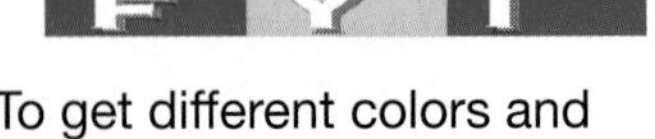

To get different colors and textures of cosmetics and dyes, early Egyptians, Greeks, and Romans had to use proportional reasoning.

Practice Masters, p. 73

NAME__________ DATE ________

Practice

Student Edition Pages 653–659

Integration: Algebra
Writing Equations of Lines

State the slope and y-intercept for each line. Then write the equation of the line in slope-intercept form.

1. $\overleftrightarrow{AB}$ 1, 4, $y = x + 4$

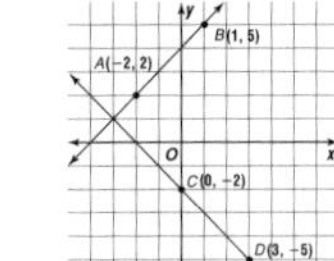

2. $\overleftrightarrow{CD}$ −1, −2, $y = -x - 2$

Write the equation in slope-intercept form of the line satisfying the given conditions.

3. parallel to $y = 3x - 2$; y-intercept = 1 $y = 3x + 1$
4. perpendicular to $y = \frac{1}{2}x + 8$; passes through the point at (0, 3) $y = -2x + 3$
5. perpendicular to the line passing through points at (−2, 3) and (2, 1); y-intercept = −3 $y = 2x - 3$
6. passes through the points at (1, 5) and (−3, 5) $y = 5$
7. x-intercept = 5; y-intercept = −2 $y = \frac{2}{5}x - 2$
8. $m = -\frac{2}{3}$; passes through the point at (−1, 2) $y = -\frac{2}{3}x + \frac{4}{3}$
9. $m = -4$; x-intercept = 0 $y = -4x$
10. The length of a rectangular garden is 30 feet more than 2 times its width. Its perimeter is 300 feet. Find its length and width. 110 feet, 40 feet

Critical Thinking

42. The inverse of any equation is found by interchanging the variables and solving the equation for y. **a.** $y = 8x - 88$

a. Determine the equation that is the inverse of $y = 0.125x + 11$.

b. Write an equation that is its own inverse. $y = x$

c. Compare the graphs of any two inverse equations with the graph of the equation that is its own inverse. Describe the relationship. See margin.

43. Study the design at the right. **a.** See students' work.

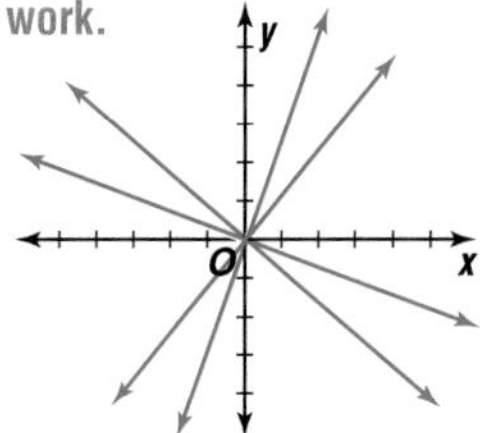

a. Create a similar design by writing four equations and graphing them.

b. If you moved the center of the design from (0, 0) to (1, 3) how would your equations change? The y-intercepts will all change.

Applications and Problem Solving

44. 66 minutes

44. Write an Equation Sprint® long distance phone company advertises that with its Sprint Sense™ program, calls during the prime daytime hours cost \$0.22 per minute and calls during evening hours are \$0.10 per minute. Zina Harrison's bill for long distance calls in September was \$28.52 before taxes. If she was billed for eight more than twice as many minutes during the evening as during the day, how many minutes was she billed for use during the day?

45. Computers An Integrated Services Digital Network, or ISDN, uses telephone lines to transmit digital signals. The charges for an ISDN depend on the distance from the company to your home. Ameritech charges one ISDN customer \$192 for 6 months of service and \$576 for 18 months of service. The bill for a second customer was \$456 for one year and \$570 for 15 months of service.

a. Write an equation that describes the cost of using each ISDN in terms of months of service. $y = 32x$; $y = 38x$

b. Describe the difference between the bills for the two customers based on what you know from the equations. See margin.

46. Sports It costs about \$1200 to equip a typical Olympic runner in 1996. The costs are expected to increase \$82 per year. Write the linear equation in slope-intercept form that represents the approximate cost of equipping a runner in the future. Assume that the rate of increase remains constant. Let t = number of years after 1996; $c = 82t + 1200$.

Mixed Review

47. See margin for graph. $m = \frac{3}{2}$; $b = -\frac{3}{4}$

47. Graph $6x - 4y = 3$. State the slope and y-intercept of the line. (Lesson 12–1)

48. Find the volume of a cylinder with a radius of 5 meters and a height of 2 meters. (Lesson 11–5) $157.1\ m^3$

49. Entertainment One midway game at the Virginia State Fair is a dart game involving a board that is 6 feet by 8 feet with forty 8-inch-square targets attached. If you throw a dart at random and it hits the board, what is the probability that you will hit a target? (Lesson 10–6) 0.37

Refer to the graph at the right for Exercises 50 and 51.

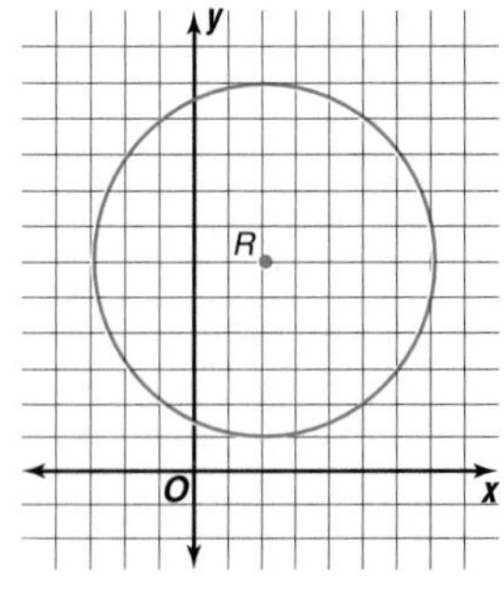

50. Write the equation of circle R. (Lesson 9–8) $(x - 2)^2 + (y - 6)^2 = 25$

51. Describe a circle S that has no common tangents with circle R. (Lesson 9–1)
Sample answer: A circle S that has all of its points in the interior of circle R would have no common tangents with circle R.

Extension

Reasoning Is it possible for a line to have a negative slope, a positive y-intercept, and a negative x-intercept? Explain your answer. No; sample answer: Look at the signs in the equation. Since b is positive, $-b$ is negative; m is negative; therefore, x must be positive.

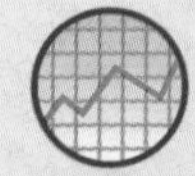

Tech Prep

Computer Programming Associate Two-year colleges offer a variety of degrees related to computers. These schools train workers to write software, repair hardware, and design and build robots. Each requires a background in mathematics. For more information on tech prep, see the *Teacher's Handbook*.

52. **Engineering** The recommended grade for a highway in a non-mountainous region is 7%. In other words, the road rises 7 feet vertically for every 100 feet of distance horizontally. What is the angle of elevation of a road with this grade? (Lesson 8–4) **4°**

53. Suppose $\triangle EFG \sim \triangle JKL$. If $EG = 3.6$, $JL = 2.4$, and the perimeter of $\triangle JKL$ is 6.8, find the perimeter of $\triangle EFG$. (Lesson 7–5) **10.2**

Proof

54. **Given:** $\overline{MN} \cong \overline{QP}$, $\overline{MQ} \cong \overline{NP}$

Prove: $\overline{MN} \parallel \overline{QP}$

(Lesson 4–4) **See margin.**

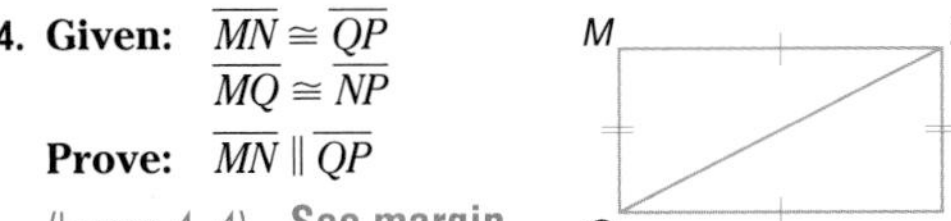

INTEGRATION Algebra

55. Simplify $\frac{8a^2bc}{28abc^2}$. State the excluded values of a, b, and c.

55. $\frac{2a}{7c}$; $a \neq 0$, $b \neq 0$, $c \neq 0$

56. **Cosmetics** The graph at the right shows the percent of teenage girls who use each different type of cosmetic.
   a. Do you think there are girls who use more than one type of cosmetic listed? Explain.
   b. Would a circle graph be appropriate for displaying this data? Why or why not? **No; a circle graph compares parts of a whole. These statistics do not represent one whole where each person is represented in just one category.**

56a. Yes; if each girl used only one type of cosmetic, the total of the percents would be 100%.

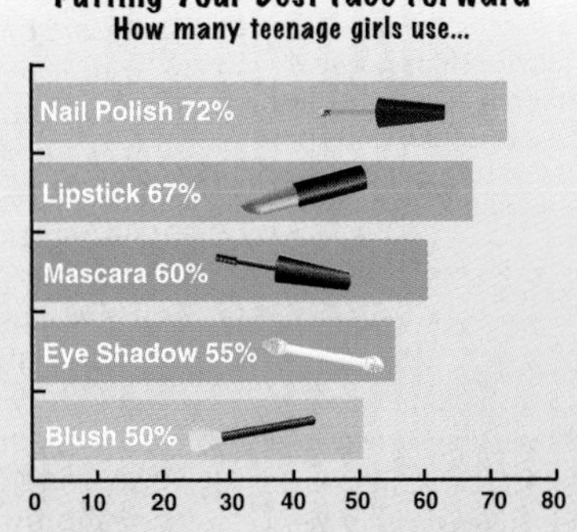

Source: Simmons Market Research Bureau

F Y I

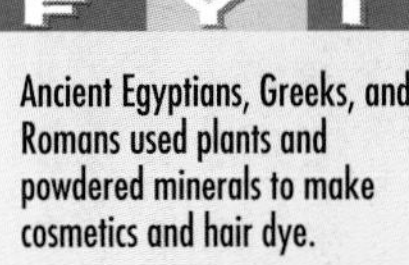

Ancient Egyptians, Greeks, and Romans used plants and powdered minerals to make cosmetics and hair dye.

Mathematics and SOCIETY

Mapping the Undersea Landscape

The excerpt below appeared in an article in *Earth* in June, 1996.

> Top-secret data recently declassified by the Navy has enabled scientists to view the seafloor almost as if the oceans had been drained completely of water....scientists knew...that much of the seafloor's topography is mirrored on the surface by slight variations in sea level. This is due to the subtle gravitational pull exerted by seamounts and other hidden features of the deep....a satellite called Geosat...carried out its mission for the Navy, spinning a dense web of tracks over the oceans while conducting a detailed global survey of the height of the sea surface....The gravity data reveals all discrete seafloor features larger than about 3,000 feet high and six miles across....a vast improvement over prior global maps. ■

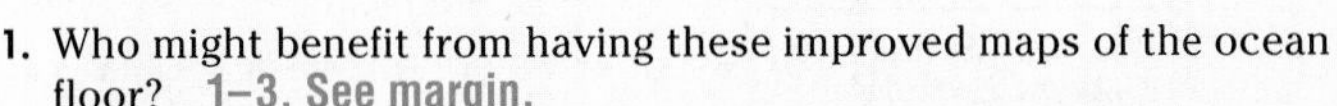

1. Who might benefit from having these improved maps of the ocean floor? **1–3. See margin.**
2. Which type of orbit did the Geosat satellite have to follow in order to map Earth's oceans, north-to-south between the poles or east-to-west along the equator? Explain.
3. Ships using sonar could provide even more detailed maps of the ocean floor. What might be the main advantage of using a satellite for this task?

4 ASSESS

Closing Activity

Modeling Have students select two points in the classroom. Measure the height of each point from the floor. They should determine the slope of the line joining the two points.

Chapter 12 Quiz A (Lessons 12-1 and 12-2) is available in the *Assessment and Evaluation Masters*, p. 324.

Additional Answer

54. Statements (Reasons)
1. $\overline{MN} \cong \overline{PQ}$, $\overline{MQ} \cong \overline{NP}$ (Given)
2. $\overline{QN} \cong \overline{QN}$ (Congruence of segments is reflexive.)
3. $\triangle MNQ \cong \triangle PQN$ (SSS)
4. $\angle MNQ \cong \angle PQN$ (CPCTC)
5. $\overline{MN} \parallel \overline{QP}$ (If ≅ and alt. int. ∠s are ≅, then the lines are ∥.)

Mathematics and SOCIETY

Have students discuss how geometry might be used in mapping.

Enrichment Masters, p. 73

12-2 NAME________ DATE________

Enrichment — Student Edition Pages 653–659

Equations of Circles and Tangents

Recall that the circle whose radius is r and whose center has coordinates (h, k) is the graph of $(x - h)^2 + (y - k)^2 = r^2$. You can use this idea and what you know about circles and tangents to find an equation of the circle that has a given center and is tangent to a given line.

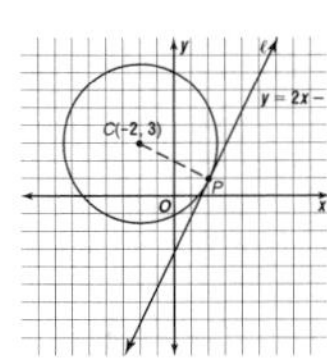

Use the following steps to find an equation for the circle that has center C(−2, 3) and is tangent to the graph y = 2x − 3. Refer to the figure.

1. State the slope of the line ℓ that has equation $y = 2x - 3$. 2
2. Suppose $\odot C$ with center $C(-2, 3)$ is tangent to line ℓ at point P. What is the slope of radius $\overline{CP}$? $-\frac{1}{2}$
3. Find an equation for the line that contains $\overline{CP}$. $y = -\frac{1}{2}x + 2$
4. Use your equation from Exercise 3 and the equation $y = 2x - 3$. At what point do the lines for these equations intersect? What are its coordinates? P; $(2, 1)$
5. Find the measure of radius $\overline{CP}$. $\sqrt{20}$
6. Use the coordinate pair $C(-2, 3)$ and your answer for Exercise 5 to write an equation for $\odot C$. $(x - (-2))^2 + (y - 3)^2 = 20$ or $(x + 2)^2 + (y - 3)^2 = 20$

Answers for Mathematics and Society

1. commercial fisherman, oil companies, geologists, and the Navy
2. North-south orbit; all of Earth's surface would eventually pass beneath the satellite's path as Earth rotates on its axis.
3. Satellites can do this job faster and cheaper than ships. It would cost several billion dollars and require a hundred years for a ship to completely map Earth's oceans. By contrast, the Geosat satellite cost $80 million and did the job in 18 months.

12-3 LESSON NOTES

NCTM Standards: 1–6, 10

Instructional Resources

- Study Guide Master 12-3
- Practice Master 12-3
- Enrichment Master 12-3
- Assessment and Evaluation Masters, pp. 323, 324
- Multicultural Activity Masters, p. 23
- Real-World Applications, 25
- Tech Prep Applications Masters, p. 23

Transparency 12-3A contains the 5-Minute Check for this lesson; **Transparency 12-3B** contains a teaching aid for this lesson.

Recommended Pacing	
Standard Pacing	Day 4 of 12
Honors Pacing	Day 3 of 11

1 FOCUS

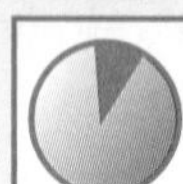

5-Minute Check (over Lesson 12-2)

Find the equation of the line with the given slope and passing through the point with the given coordinates.

1. 3, (2, −1) $y = 3x - 7$
2. $-\frac{3}{4}$, (4, −2) $y = -\frac{3}{4}x + 1$
3. −4, (2, −5) $y = -4x + 3$

Find the equation of the line passing through points with the given coordinates.

4. (3, 2) and (4, 1) $y = -x + 5$
5. (4, −2) and (7, −4) $y = -\frac{2}{3}x + \frac{2}{3}$
6. (−2, 5) and (6, 5) $y = 5$

12-3 Integration: Algebra and Statistics
Scatter Plots and Slopes

What YOU'LL LEARN
- To relate statistics and equations of lines to geometric concepts.

Why IT'S IMPORTANT

Statistics are used to find relationships between quantities in fields such as education and sports.

Aviation

Many proponents of the 55 mile per hour speed limit on U.S. highways say that it is less expensive to run a car at lower speeds. The average speed at which a car is operated is just one of many factors that affect its operating costs. Do you think that the speed of an airplane affects its cost of operation? The table at the right shows the average airborne speed and average operating cost for the most commonly used models.

Plane	Speed (mph)	Cost per hr
B747-100	478	$5571
DC-10-10	439	$4504
B747-400	507	$6592
DC-10-40	444	$3904
B727-200	357	$2241
L-1011-100/200	429	$3783
DC-10-30	479	$4349
A300-600	393	$3648
MD11	484	$5024
L-1011-500	486	$3958
B757-200	397	$2362

Source: *The World Almanac*, 1996

Because an airplane must maintain a minimum speed to remain airborne, the reasonable values of the domain and range are restricted.

A **scatter plot**, like the one below, shows the relationship between speed and cost by plotting a set of data as ordered pairs in the coordinate plane. In the graph below, we can fit a line as a way to summarize the data, and find an equation that will express the approximate cost in terms of speed.

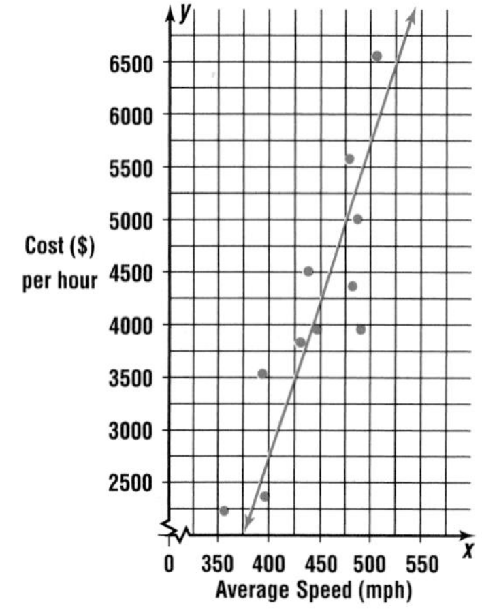

Notice that the points do not lie in a straight line, but they do suggest a linear pattern. You can determine the equation for this line using techniques from algebra. The line drawn in the graph at the left passes through points with coordinates (525, 6500) and (450, 4250).

First, find the slope of the line.

$$m = \frac{y_2 - y_1}{x_2 - x_1}$$
$$= \frac{6500 - 4250}{525 - 450}$$
$$= \frac{2250}{75} \text{ or } 30$$

Substitute the slope and the coordinates of one of the points into the point-slope form to find the equation.

$$y - y_1 = m(x - x_1)$$
$$y - 4250 = 30(x - 450)$$ *The slope is 30, and one point has coordinates (450, 4250).*
$$y - 4250 = 30x - 13{,}500$$
$$y = 30x - 9250$$

An equation that relates the airborne speed and the cost is $y = 30x - 9250$.
Since the scatter plot shows a pattern, it is reasonable to say that the average speed at which an airplane flies may be a factor in its cost of operation.

You can use the equation to predict the cost per hour for a given speed or find the speed that predicts a given cost per hour. Find the approximate cost of traveling at a speed of 500 miles per hour by substituting 500 for x in the equation.

$$y = 30x - 9250$$
$$y = 30(500) - 9250$$
$$y = 15{,}000 - 9250 \text{ or } 5750$$

The cost per hour will be approximately $5750.

Whenever you work with coordinates, it is a good idea to begin by graphing the information you are given.

Example  **Determine whether $K(-8, 12)$, $E(4, -6)$, and $B(6, -9)$ are collinear.**

One way to approach the problem is to find the equation of $\overleftrightarrow{KE}$ and see if the coordinates of B satisfy the equation. That is, does B lie on $\overleftrightarrow{KE}$? First find the slope of $\overleftrightarrow{KE}$.

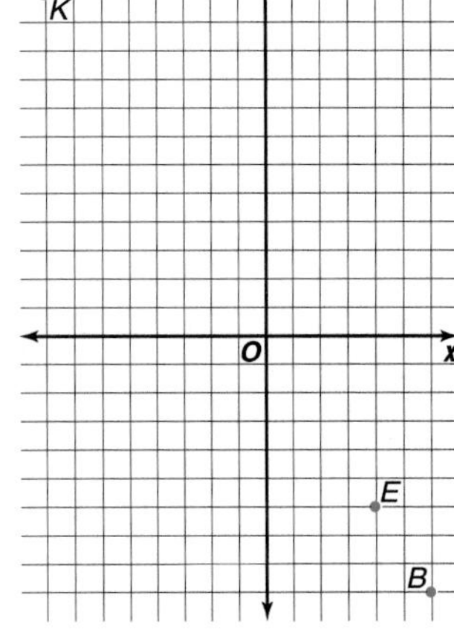

$$m = \frac{y_2 - y_1}{x_2 - x_1}$$
$$= \frac{12 - (-6)}{-8 - 4}$$
$$= \frac{18}{-12} \text{ or } -\frac{3}{2}$$

$$y - y_1 = m(x - x_1)$$
$$y - 12 = -\frac{3}{2}(x - (-8))$$
$$y - 12 = -\frac{3}{2}x - 12$$
$$y = -\frac{3}{2}x$$

The equation of $\overleftrightarrow{KE}$ is $y = -\frac{3}{2}x$. Since $-9 = -\frac{3}{2}(6)$, $B(6, -9)$ satisfies the equation, and the points are collinear.

You can also use algebraic techniques to find the coordinates of the point of intersection of two lines.

Example 2 **Find the coordinates of the point of intersection of the medians $\overline{CD}$ and $\overline{BE}$ for $\triangle ABC$ if the vertices are $A(4, -5)$, $B(8, 11)$, and $C(0, 15)$.**

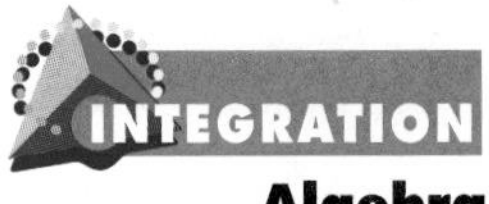

Algebra

The median is a segment whose endpoints are the vertex of an angle and the midpoint of the side opposite that angle. To find the coordinates of D and E, first find the midpoints of $\overline{AB}$ and $\overline{AC}$. Use the midpoint formula.

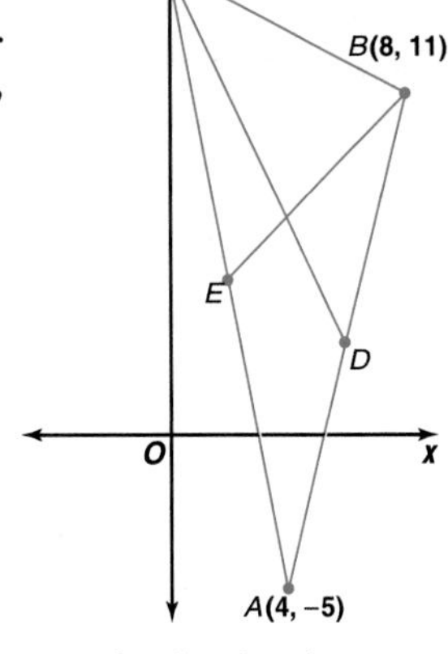

(continued on the next page)

Motivating the Lesson

Questioning Ask students to give examples of kinds of predictions that are based on gathered data. **Sample answers: teams that are favored to win in sports, accident frequency on a road**

2 TEACH

In-Class Examples

For Example 1
Determine whether $P(2, 5)$, $Q(4, 6)$, and $R(1, 3)$ are collinear. **The points are not collinear.**

For Example 2
Given $\triangle PQR$ with $P(-3, 2)$, $Q(1, 6)$, and $R(5, -3)$, write the equation of the line containing the altitude from P.
$y = \frac{4}{9}x + \frac{10}{3}$

Teaching Tip The line on the graph of the scatter plot is sometimes called the *line of best fit*. In higher mathematics, various methods are used to better determine this line.

Teaching Tip After Example 1, ask students what figure would be formed if the points were not collinear. **a triangle**

Cooperative Learning

Group Discussion Have groups of students perform the following experiment. Roll a single die six times, $r_1, r_2, r_3, r_4, r_5, r_6$. Plot the points $(1, r_1)$, $(2, r_2)$, $(3, r_3)$, $(4, r_4)$, $(5, r_5)$, $(6, r_6)$. Have students discuss how likely it is that the scatter plot is linear. For more information on the group discussion strategy, see *Cooperative Learning in the Mathematics Classroom*, one of the titles in the Glencoe Mathematics Professional Series, page 31.

3 PRACTICE/APPLY

Check for Understanding

Exercises 1–10 are designed to help you assess your students' understanding through reading, writing, speaking, and modeling. You should work through Exercises 1–3 with your students and then monitor their work on Exercises 4–10.

Error Analysis
When writing equations of lines that are perpendicular to a given line, students may forget that the slope must be the *opposite* reciprocal. Tell students to be sure to check that the signs of the two slopes are opposites.

Additional Answers

2. Sample answer: Use the equation of the line determined by two of the points and check to see if the third was on the line; check to see if the slope determined by the points was the same.
3. Sample answer: Neither; both procedures may have flaws and may be good or may not.

10b. Answers will vary. The points (1912, 84.6) and (1996, 54.10) produce slope of about −0.363 and the equation $y = -0.363x + 778.656$.

10d. Using the equation $y = -0.363x + 778.656$, the predicted time would be 45.396 seconds. Answers will vary. Sample answer: Yes; the time would be about 9 seconds faster than the winning 1996 time. That was accomplished in the 8 years between 1912 and 1920. Sample answer: No; the times have stayed around 55 seconds for the last 20 years.

midpoint of $\overline{AB}$

$$D(x, y) = \left(\frac{x_1 + x_2}{2}, \frac{y_1 + y_2}{2}\right) = \left(\frac{4+8}{2}, \frac{-5+11}{2}\right) = (6, 3)$$

midpoint of $\overline{AC}$

$$E(x, y) = \left(\frac{x_1 + x_2}{2}, \frac{y_1 + y_2}{2}\right) = \left(\frac{4+0}{2}, \frac{-5+15}{2}\right) = (2, 5)$$

Now find the equation of the line containing each median. Find the slope of each segment.

slope of $\overline{CD}$

$$m = \frac{y_2 - y_1}{x_2 - x_1} = \frac{15-3}{0-6} = \frac{12}{-6} \text{ or } -2$$

Since the point at (0, 15) is the *y*-intercept, use the slope-intercept form.

$$y = mx + b$$
$$y = -2x + 15$$

slope of $\overline{BE}$

$$m = \frac{y_2 - y_1}{x_2 - x_1} = \frac{11-5}{8-2} = \frac{6}{6} \text{ or } 1$$

Use the point-slope form.

$$y - y_1 = m(x - x_1)$$
$$y - 5 = 1(x - 2)$$
$$y - 5 = x - 2$$
$$y = x + 3$$

Find the coordinates of the intersection of the lines containing the medians, $y = -2x + 15$ and $y = x + 3$.

$$y = x + 3$$
$$-2x + 15 = x + 3 \quad \textit{Substitution Property of Equality}$$
$$-3x = -12$$
$$x = 4$$

Substitute 4 into one of the equations to find *y*.

$$y = x + 3$$
$$y = 4 + 3$$
$$y = 7$$

The two medians of the triangle intersect at (4, 7).

CHECK FOR UNDERSTANDING

Communicating Mathematics

Study the lesson. Then complete the following. 2–3. See margin.

1. **Explain** a way in which a scatter plot would be useful. 1. Sample answer: A scatter plot can be useful in analyzing relationships between data.
2. **Describe** two ways to determine whether three points are collinear.
3. **You Decide** Conchita and Dalila were analyzing a scatter plot of data. Conchita decided to draw a line to summarize the data by connecting the first and the last data point. Dalila thought she had a better procedure because she drew a line that had the same number of data points on either side. Who is correct? Explain your reasoning.

Guided Practice

Determine whether the three points listed are collinear.

4. $B(4, 9)$, $C(7, 18)$, $D(-2, -9)$ yes
5. $R(-6, 10)$, $S(-1, -1)$, $T(5, 1)$ no

Reteaching

Using a Graphing Calculator Explain how to use a graphing calculator to find an equation of a line when given the coordinates of two points on the line. Then give students the coordinates of three points and ask how they can use the calculator to decide if the three points are collinear. Sample answer: Find the equation of the line between two of the points. Then substitute the coordinates of the third point in the equation to see if the coordinates satisfy the equation. Then have them use that method to test these points: $D(3, 5)$, $E(5, 9)$, $F(4, 8)$. no

Side $\overline{RS}$ of $\triangle RST$ lies on line a. The equation of line a is $y = -\frac{2}{5}x + 8$. The coordinates of T are $(-10, 15)$. Identify each of the following as *true* or *false*. If the statement is false, explain why.

6\. False; the product of the slopes is not -1.

6. The altitude from T to $\overline{RS}$ lies on the line $y = -\frac{5}{2}x - 4$.
7. Point R could have coordinates $(10, 4)$. false; $4 \neq -10 + 8$
8. Point T could be at the intersection of line a and the line whose equation is $y = 4x - 20$. False; T is not on the line.
9. Point S could be at the intersection of line a and the line whose equation is $x = 5$. true
10. **Sports** The table below contains the winning times in selected Olympics for the men's 100-meter backstroke. a. See Solutions Manual.

YEAR	WINNER, COUNTRY	TIME
1912	Arno Bieberstein, Ger	84.6
1920	Warren Kealoha, USA	75.2
1928	George Koiac, USA	68.2
1936	Adolph Kiefer, USA	65.9
1952	Yoshinobu Oyakawa, USA	65.4
1960	David Thiele, Australia	61.9
1972	Roland Matthes, E Ger	56.58
1980	Bengt Baron, Sweden	56.33
1988	Daichi Suzuki, Japan	55.05
1996	Jeff Rouse, USA	54.10

Source: *The World Almanac, 1996*

Jeff Rouse

b, d. See margin.

a. Draw a scatter plot to show how year x and time y are related.
b. Write an equation that relates the year to the approximate winning time.
c. Predict the winning time for the 2000 Olympics Games.
d. If you used this equation to predict for the year 2020, what time would you get? Is this a reasonable prediction? Explain.

c. Using the equation $y = -0.363x + 778.656$, the predicted time would be 52.656 seconds.

EXERCISES

Practice

Determine whether the three points listed are collinear.

11. $A(0, 0), B(6, 4), C(-3, -2)$ yes
12. $D(19, -13), E(14, 18), F(10, 10)$ no
13. $M(6, 10), N(5, 10), P(-2, -6)$ no
14. $G(0, 0), H(4, 12), J(7, 21)$ yes
15. $R(1, -3), S(4, -18), T(-3, 17)$ yes
16. $U(4, 3), V(-2, -9), W(-7, 9)$ no
17. The table at the right lists United States birthrates. The rates are per 1000 people living in the United States. a–d. See margin.

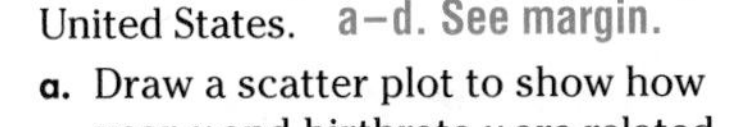

Year	Rate
1940	19.4
1950	24.1
1960	23.7
1970	18.4
1980	15.9
1990	16.7

   a. Draw a scatter plot to show how year x and birthrate y are related.
   b. Write an equation that relates the year to the approximate birthrate.
   c. What does the slope tell you about the birthrate every ten years?
   d. Predict the birthrate for 2000. How reliable do you think your prediction is? Explain your reasoning.

Source: *National Center for Health Statistics*

18\. $y = \frac{4}{3}x + 4$; $y = -\frac{3}{4}x + 10\frac{1}{4}$

18. Find the equations of the lines containing the legs of an isosceles triangle if they meet at $(3, 8)$ and a vertex of a base angle is at the x-intercept of the graph of $-x + 7y = 3$, which contains the base.

Alternative Learning Styles

Kinesthetic Find 10 people and record their heights and the span of their arms when outstretched. Make a graph on the floor using masking tape. Graph the data with the height on the x-axis and the arm span on the y-axis and have each person stand at their point on the graph. Have two students hold a rope to represent a line that best fits the data and write an equation for it. Answers will vary.

Assignment Guide

Core (with proof): 11–31 odd, 32–39
Core (informal): 11–31 odd, 32–39
Enriched: 12–28 even, 29–39
All: Self Test, 1–10

For **Extra Practice**, see p. 788.

The red A, B, and C flags, printed only in the Teacher's Wraparound Edition, indicate the level of difficulty of the exercises.

Additional Answers

17a.

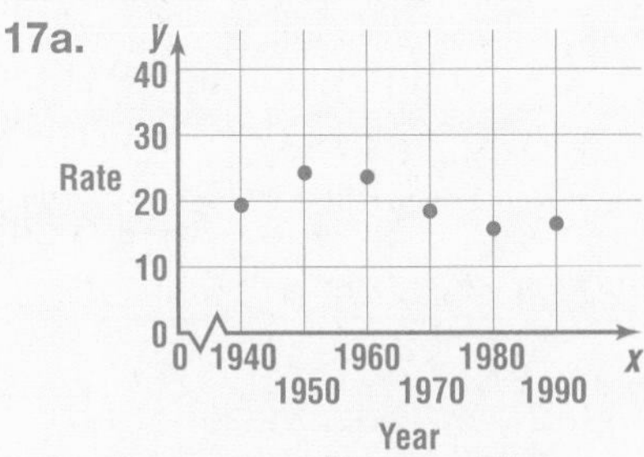

17b. Answers will vary. The points (1950, 24.1) and (1990, 16.7) produce slope of -0.185 and the equation $y = -0.185x + 384.85$.

17c. Sample answer: Each year the birth rate decreases by 0.185 so every ten years about 2 fewer children are born per 1000 people.

17d. Sample answer: 14.85; It will be reliable only if the trend continues as it has in the past 50 years.

Study Guide Masters, p. 74

NAME__________ DATE__________

Study Guide — Student Edition Pages 660–665

Integration: Algebra and Statistics
Scatter Plots and Slope

The methods you have learned for writing equations of lines are helpful when applied to geometric concepts. For example, when you are given the coordinates of the vertices for a triangle, you can use these coordinates to find the equations of various lines, such as the lines that contain the altitudes, medians, or perpendicular bisectors.

Example: The vertices of $\triangle ABC$ are $A(0, 12)$, $B(^-6, 6)$, and $C(4, 8)$. Write the equation of the line containing the perpendicular bisector of $\overline{BC}$.

The midpoint of $\overline{BC}$ is $\left(\frac{^-6+4}{2}, \frac{6+8}{2}\right)$ or $(^-1, 7)$.

Slope of $\overline{BC} = \frac{8-6}{4-(^-6)} = \frac{2}{10} = \frac{1}{5}$

So the perpendicular bisector of $\overline{BC}$ must have a slope of $^-5$.

Since you know the slope of the perpendicular bisector and one point on it, you can write the equation.

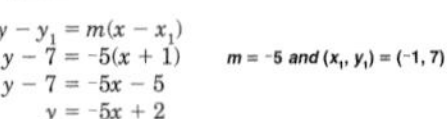

$$y - y_1 = m(x - x_1)$$
$$y - 7 = {^-5}(x + 1) \quad m = {^-5} \text{ and } (x_1, y_1) = (^-1, 7)$$
$$y - 7 = {^-5}x - 5$$
$$y = {^-5}x + 2$$

The vertices of $\triangle RST$ are $R(0, 10)$, $S(^-5, 0)$ and $T(5, 5)$. Write in slope-intercept form the equations of the lines that contain the segments described.

1. the sides of $\triangle RST$
   $y = 2x + 10$
   $y = {^-x} + 10$
   $y = \frac{1}{2}x + \frac{5}{2}$
2. the medians of $\triangle RST$
   $x = 0$
   $y = x + 5$
   $y = 5$
3. the altitudes of $\triangle RST$
   $y = {^-2}x + 10$
   $y = {^-\frac{1}{2}}x + \frac{15}{2}$
   $y = x + 5$
4. the perpendicular bisectors of $\triangle RST$
   $y = {^-2}x + \frac{5}{2}$
   $y = {^-\frac{1}{2}}x + \frac{15}{4}$
   $y = x + 5$

Additional Answers

23a. $y = -\frac{1}{2}x - \frac{1}{2}$

23b. The line through the mdpt. of the two sides of a triangle is ∥ to the third side and equal to half the length of the third side. Let D be the mdpt. of $\overline{AB}$ and let E be the mdpt. of $\overline{AC}$. Then D will have coordinates $(-3, 1)$ and $E(1, -1)$. The slope of $\overline{BC} = -\frac{1}{2}$. The slope of $\overline{DE} = -\frac{1}{2}$. The lines do not share a common point but have the same slope so they are parallel. $BC = \sqrt{(2-6)^2 + (8-4)^2} = \sqrt{80}$ or $4\sqrt{5}$. $DE = \sqrt{(-3-1)^2 + [1-(-1)]^2} = 2\sqrt{5}$. Thus, the length of the segment connecting the midpoints is one half the length of the third side.

23c. $(0, 7)$ is a point on $\overline{BC}$ and on the perpendicular from $\overline{BC}$ to $D(-3, 1)$. The distance from $\overline{DE}$ to $\overline{BC}$ is $\sqrt{45} = 3\sqrt{5}$ or 6.71.

28c. Sample answer: The coordinates of D can be found by finding the averages of the x- and y-coordinates.

Practice Masters, p. 74

12-3 NAME__________ DATE ________

Practice

Student Edition Pages 660–665

Integration: Algebra and Statistics Scatter Plots and Slope

1. **Travel** The table below lists the number of gallons of gasoline used during 5 different trips.

Length of trip (in miles)	374	506	2020	144	1034
Gallons of gasoline	17	22	101	6	47

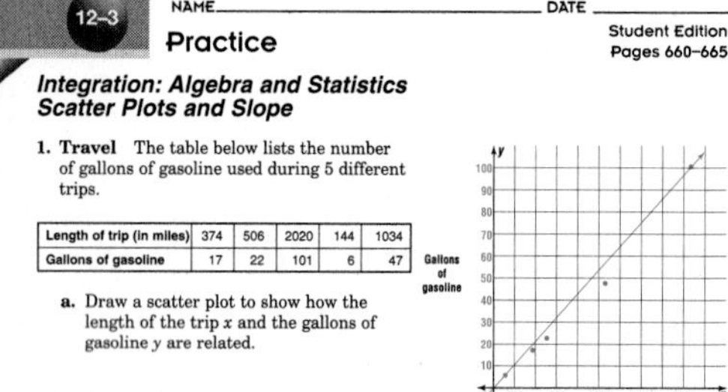

a. Draw a scatter plot to show how the length of the trip x and the gallons of gasoline y are related.

b. Write an equation that relates the trip length to the gallons of gasoline used. Sample answer: $y = \frac{1}{22}x$

c. About how many gallons of gasoline will be used on an 800-mile trip? 36 gallons

2. Find the equation of the line that is the perpendicular bisector of the segment whose endpoints are (3, 7) and (−3, −1). $y = -\frac{3}{4}x + 3$

3. Find the equations of the lines that contain sides of an isosceles triangle if the vertex of the vertex angle is at the y-intercept of $y = -2x + 3$ and the vertex of a base angle is at (6, 1). $y = 1, y = \frac{1}{3}x + 3, y = -\frac{1}{3}x + 3$ or $x = 6, y = \frac{1}{3}x + 3, y = -\frac{1}{3}x + 3$

The vertices of △DEF are D(0, 12), E(−2, 10), and F(4, 6).

4. Write the equation of the line that contains the altitude to $\overline{EF}$. $y = \frac{3}{2}x + 12$

5. Write the equation of the line line that contains the perpendicular bisector to $\overline{ED}$. $y = -x + 10$

The vertices of △ABC are A(2, 18), B(−2, −4), and C(6, 12).

19. Write the equation of the line containing the altitude to $\overline{BC}$. $y = -\frac{1}{2}x + 19$

20. Write the equation of the line containing the perpendicular bisector of $\overline{BC}$. $y = -\frac{1}{2}x + 5$

21. Write the equation of the line containing the median from A to $\overline{BC}$. $x = 2$

B 22. The vertices of △RST are $R(-10, 2)$, $S(-2, -4)$, and $T(-3, 3)$. Find the equation of the altitude to $\overline{RS}$ and of the median to $\overline{RS}$. What conclusion can you make?

22. △RST is an isosceles triangle; the equations of both lines are $y = \frac{4}{3}x + 7$; the altitude and median are the same in an isosceles triangle.

23. The vertices of △ABC are $A(-4, -6)$, $B(-2, 8)$, and $C(6, 4)$.

a. Write the equation of the line through D and E, the midpoints of $\overline{AB}$ and $\overline{AC}$, respectively.

b. What is the relationship between the line through D and E and the line containing $\overline{BC}$? Explain your reasoning.

c. Find the distance between $\overline{BC}$ and $\overline{DE}$.

23a–c. See margin.

24. Find the coordinates of D, E, and F in △DEF if the vertices lie on the lines $y = 4x + 7$, $y = x - 2$, and $y = \frac{2}{5}x + 3\frac{2}{5}$. (−1, 3), (9, 7), (−3, −5)

An equation of a circle is $(x-3)^2 + (y-5)^2 = 100$.

C 25. Show that the point with coordinates $(-3, 13)$ is on the circle. Find another point you know will be on the circle.

25. Sample answer: $(-3-3)^2 + (13-5)^2 = (-6)^2 + (8)^2 = 36 + 64$ or 100; (9, −3), (11, 11), (11, −1) are some of the possible points.

26. Is the point at $(-2, 5)$ inside the circle or outside the circle? Explain.

26. Sample answer: inside; when you substitute $(-2, 5)$ for (x, y), you get 25, which is less than 100.

27. Write the equation of the tangent to the circle at $(-3, 13)$. $y = 0.75x + 15.25$

28. Consider the points $A(4, 15)$, $B(-2, 7)$, and $C(13, 2)$.

a. Find D such that D is $\frac{1}{3}$ the distance from $\overline{AB}$ to C. $D(5, 8)$

b. The median fit line is the line that passes through D and is parallel to $\overline{AB}$. Write the equation of the median fit line. $y = 8 + \frac{4}{3}(x - 5)$

c. Point D could be considered an "average" point for the triangle. Explain. See margin.

Critical Thinking

29. Plot the points $A(2, 14)$, $B(0, 10)$, $C(12, 34)$ and $D(6, 8)$. Which of the two equations, $y = 1.8x + 7.3$ or $y = 2x + 10$, seems like a better fit for summarizing the linear pattern? Explain your reasoning.

29. See Solutions Manual.

Applications and Problem Solving

30a–b. See Solutions Manual. 30c. Sample answer using (482, 413) and (469, 424): $V = -0.85M + 822.7$ 30d. 429

30. **Education** The SAT is a college entrance examination given in two parts, mathematics and verbal. The 1994 mean math and verbal scores for the twelve most populous states are shown in the table below.

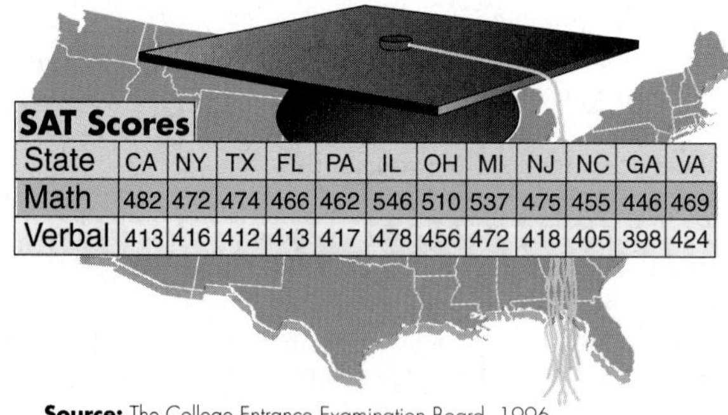

SAT Scores

State	CA	NY	TX	FL	PA	IL	OH	MI	NJ	NC	GA	VA
Math	482	472	474	466	462	546	510	537	475	455	446	469
Verbal	413	416	412	413	417	478	456	472	418	405	398	424

Source: The College Entrance Examination Board, 1996

a. Draw a scatter plot to show how the scores are related.

b. Does it make a difference if you plot (verbal, math) or (math, verbal)? Explain why or why not.

c. Write an equation that relates a state's mean math score to its mean verbal score.

d. The mean SAT score on the math portion in the state of Maine was 463 in 1994. What prediction might you make for its mean verbal score?

Classroom Vignette

"I provide a piece of cardboard of a given size to each student. The students cut a square from each corner and fold up the sides to form an open box. We predict which square will produce the box with the greatest volume. After the students calculate and graph some volume, we write an equation and graph on a TI-82. The trace function allows us to find the maximum volume."

Rose Peters

Rose M. Peters
Marshfield Senior High School
Marshfield, Wisconsin

Car Costs

Car	Original price	Operating cost per year	Fixed cost per year
Buick Le Sabre Ltd	\$24,087	\$2771	\$7978
Cadillac Deville	\$42,638	\$2910	\$11,383
Chevrolet Caprice CL	\$21,047	\$2585	\$7183
Chevrolet Cavalier LS	\$12,543	\$2006	\$5210
Chrysler Intrepid ES	\$20,286	\$2597	\$7430
Ford Excort LX	\$10,402	\$1913	\$4902
Ford Taurus GL	\$16,228	\$2365	\$6499
Honda Accord LX	\$15,141	\$2099	\$5734
Lincoln Town Car	\$32,895	\$2899	\$11,830
Nissan Maxima GSE	\$20,288	\$2214	\$7872
Toyota Camry DX	\$15,470	\$1948	\$6246

Source: *Wilson's New Car Price Guide, 1996*

31. **Statistics** The base price for buying a new car in 1996 is given in the table along with the annual operating costs and fixed costs. Operating costs include items such as gas, oil, maintenance, and repairs. Fixed costs include insurance, license, taxes, and finance charges. **a–b. See margin.**
    a. Make a scatter plot using any two columns containing dollar amounts.
    b. Draw a line that represents the data. Find the equation of the line.

Mixed Review

32. *True* or *false:* To find an equation for the line that passes through the points at (21, 4) and (5, 0), first find the y-intercept and then find the slope. (Lesson 12–2) **false**

33. $6, -3; \frac{1}{2}$

33. Find the x- and y-intercepts and the slope of the graph of $3x - 6y = 18$. (Lesson 12–1)

34. about 523.6 in^3

34. Find the volume of a sphere with a radius of 5 inches. (Lesson 11–7)
35. *True* or *false:* An equilateral polygon must also be an equiangular polygon. (Lesson 10–1) **false**
36. **Food** A round pizza is cut into 12 congruent pieces. What is the measure of the central angle of each piece? (Lesson 9–2) **30**
37. *True* or *false:* In $\odot C$, if D is a point on the circle, then $\overline{CD}$ is a radius of the circle. (Lesson 9–1) **true**

Algebra

38. Find $(m + 3n)^2$. $m^2 + 6mn + 9n^2$
39. Find $(2a^2 + 3a - 7) \div (2a + 1)$. $a + 1 - \frac{8}{2a+1}$

SELF TEST

Find the x- and y-intercepts and slope of the graph of each equation. (Lesson 12–1)

1. $y = 12$ **no x-intercept; 12; 0**
2. $3x - 4y = 24$ **$8; -6; \frac{3}{4}$**

Use the description to graph each line. (Lesson 12–1) **3–4. See Solutions Manual.**

3. $m = -2; b = 3$
4. perpendicular to the line $y = 3x + 4$; passes through $A(1, 2)$

Write the equation in slope-intercept form of the line that satisfies the given conditions. (Lesson 12–2) **5. $y = 5x + 15$** **6. $y = 0.35x + 0.5$**

5. $m = 5$, y-intercept = 15
6. $m = 0.35$; passes through the point at (4, 1.9)
7. passes through points at $(-10, -1)$ and $(8, 1)$ $y = \frac{1}{9}x + \frac{1}{9}$
8. **Sports** Nascha is lining a soccer field. The length should be 75 yards shorter than 3 times the width. If the perimeter is 370 yards, what are the length and the width of the field? (Lesson 12–2) **length: 120 yd; width: 65 yd**

Determine whether the three points listed are collinear. (Lesson 12–3)

9. $A(9, 0), B(4, 2), C(2, -1)$ **no**
10. $L(0, 4), M(2, 3), N(-4, 6)$ **yes**

4 ASSESS

Closing Activity

Writing Have students write a brief paragraph explaining how to find an equation that relates two variables, given data points for the variables that form a linear pattern when graphed.

Chapter 12 Quiz B (Lesson 12-3) is available in the *Assessment and Evaluation Masters*, p. 324.

Mid-Chapter Test (Lessons 12-1 through 12-3) is available in the *Assessment and Evaluation Masters*, p. 323.

Additional Answers

31a. See students' work.
31b. See students' work. Sample answer: $y = 0.264x + 4339.5$ where x is the original cost and y is the sum of the operating and fixed costs or the total cost of operating the car after it has been purchased.

Enrichment Masters, p. 74

NAME ______ DATE ______

Enrichment

Student Edition Pages 660–665

Absolute Zero

All matter is made up of atoms and molecules that are in constant motion. Temperature is one measure of this motion. Absolute zero is the theoretical temperature limit at which the motion of the molecules and atoms of a substance is the least possible.

Experiments with gaseous substances yield data that allow you to estimate just how cold absolute zero is. For any gas of a constant volume, the pressure, expressed in a unit called atmospheres, varies linearly as the temperature. That is, the pressure P and the temperature t are related by an equation of the form $P = mt + b$, where m and b are real numbers.

1. Sketch a graph for the data in the table. Use the axis shown below.

t (in °C)	P (in atmospheres)
-25	0.91
0	1.00
25	1.09
100	1.36

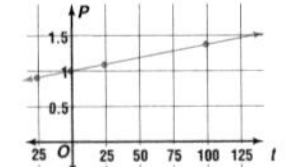

2. Use the data and your graph to find values for m and b in the equation $P = mt + b$ which relates temperature to pressure. $m = 0.0036$, $b = 1.00$

3. Estimate absolute zero in degrees Celsius by setting P equal to 0 in the equation above and using the values m and b that you obtained in Exercise 2. about −278°C (The actual value is slightly higher.)

Extension

Communication Have students work in small groups to discuss how they would write a general equation for a line that is perpendicular to a line with slope r that passes through the point $T(a, b)$.
$y = \frac{-1}{r}x + \frac{a + br}{r}$

SELF TEST

The Self Test provides students with a brief review of the concepts and skills in Lessons 12-1 through 12-3. Lesson numbers given to the right of exercises or instruction lines so students can review concepts not yet mastered.

12-4 LESSON NOTES

NCTM Standards: 1–5, 8

Instructional Resources

- Study Guide Master 12-4
- Practice Master 12-4
- Enrichment Master 12-4
- Modeling Mathematics Masters, p. 90

Transparency 12-4A contains the 5-Minute Check for this lesson; **Transparency 12-4B** contains a teaching aid for this lesson.

Recommended Pacing	
Standard Pacing	Days 5 & 6 of 12
Honors Pacing	Days 4 & 5 of 11

1 FOCUS

5-Minute Check
(over Lesson 12-3)

Determine whether the three points are collinear.

1. $A(2, -3)$, $B(4, -7)$, $C(-1, 3)$ **yes**
2. $D(-4, -2)$, $E(5, 7)$, $F(0, 2)$ **yes**
3. $G(-1, -2)$, $H(3, 6)$, $I(2, -6)$ **no**

$\triangle QRS$ has vertices with coordinates $Q(-1, 4)$, $R(3, 1)$, and $S(-5, 1)$.

4. Find the slope of $\overline{QS}$. $\frac{3}{4}$
5. Find the slope of a line perpendicular to $\overline{QS}$. $-\frac{4}{3}$
6. Find the equation of the line containing the altitude from R to $\overline{QS}$. $y = -\frac{4}{3}x + 5$

Motivating the Lesson

Questioning Ask students to use variables to represent each of the following.

1. any point in the coordinate plane **Sample answer: (a, b)**
2. any point on the x-axis **Sample answer: $(c, 0)$**
3. any point on the y-axis **Sample answer: $(0, d)$**

12-4 Coordinate Proof

What YOU'LL LEARN

- To prove theorems using coordinate proofs.

Why IT'S IMPORTANT

You can use coordinate proofs to prove theorems in geometry and algebra.

You can use paper folding to discover geometric relationships that can be proved using a coordinate proof.

MODELING MATHEMATICS — **Linear Equations**

Materials:  scissors, ruler, paper

- Cut a quadrilateral $ABCD$ with no two sides parallel or congruent from a piece of paper.
- Fold to find the midpoint of $\overline{AB}$ and pinch to make a crease at the point. Mark the point with a pencil.
- Fold to find the midpoints of $\overline{BC}$, $\overline{CD}$, and $\overline{DA}$. Mark the points.
- Draw the quadrilateral formed by the midpoints of the segments.

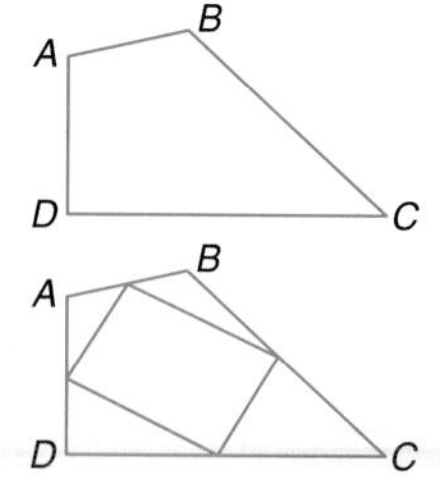

Your Turn

a. Measure each side of the quadrilateral formed by the midpoints. Record your measurements on the sides. **See students' work.**

b. What type of quadrilateral is formed by the midpoints? Justify your answer.

b. Parallelogram; The opposite sides are congruent.

In the Modeling Mathematics activity, you discovered that the quadrilateral formed by connecting the midpoints of the sides of any quadrilateral is a parallelogram. You can prove this using a **coordinate proof**. In geometry, we can assign coordinates to figures and use the coordinates to prove theorems. An important part of planning a coordinate proof is the placement of the figure on the coordinate plane. *This proof will be completed in Example 2.*

Guidelines for Placing Figures on a Coordinate Plane

1. **Use the origin as a vertex or center.**
2. **Place at least one side of a polygon on an axis.**
3. **Keep the figure within the first quadrant if possible.**
4. **Use coordinates that make computations as simple as possible.**

Example  **Position and label a parallelogram with one side a units long on the coordinate plane.**

- Use the origin as vertex A of the parallelogram.
- Place the side of length a on the positive x-axis. Label the vertices A, B, C, and D.
- Since B is on the x-axis, its y-coordinate is 0. Its x-coordinate is a because $\overline{AB}$ is a units.

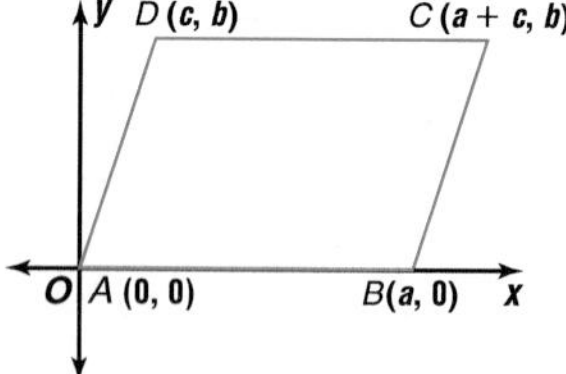

MODELING MATHEMATICS In this modeling activity, students use paper folding to discover geometric relationships. Make sure students use a convex quadrilateral.

- Let D have coordinates (c, b).
- Since the altitude in a parallelogram from D is equal to the altitude from C, the y-coordinate at C will be the same as the y-coordinate at D. The x-coordinate at C will be the sum of the x-coordinates at B and D, so C will have coordinates $(a + c, b)$.

Some examples of figures placed on the coordinate plane are given below. The figures have been placed so that the coordinates of the vertices are as simple to use as possible.

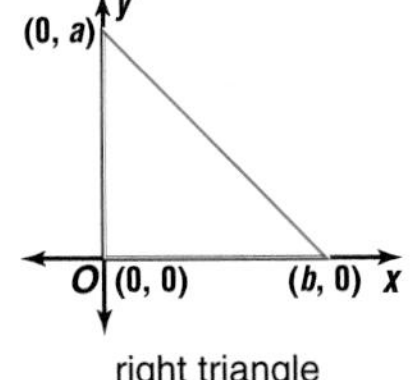

right triangle

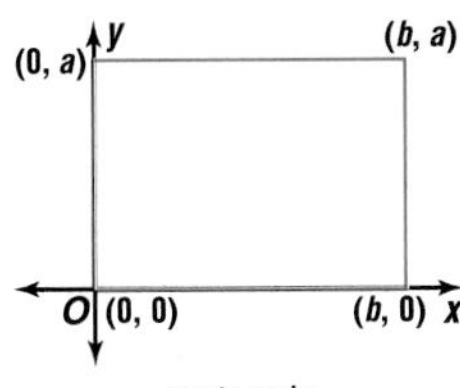

rectangle

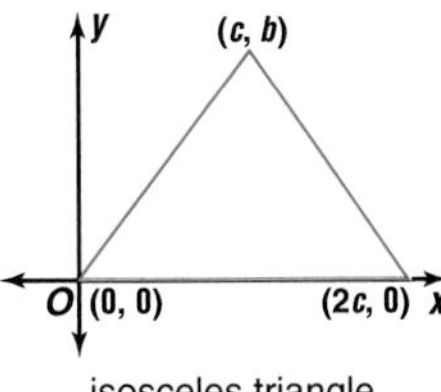

isosceles triangle

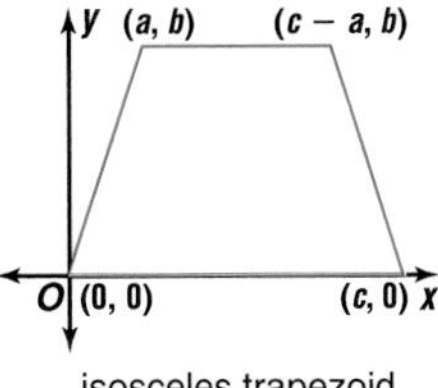

isosceles trapezoid

Example 2 Use a coordinate proof to prove that the segments joining the midpoints of the sides of a quadrilateral form a parallelogram.

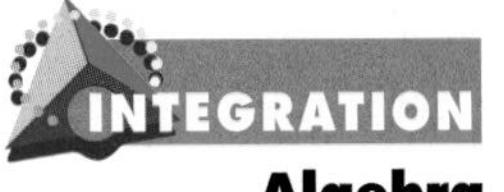

Given: *RSTV* is a quadrilateral.
A, B, C, and D are midpoints of sides $\overline{RS}$, $\overline{ST}$, $\overline{TV}$, and $\overline{VR}$, respectively.

Prove: *ABCD* is a parallelogram.

Proof:

Place quadrilateral *RSTV* on the coordinate plane and label coordinates as shown. (Using coordinates that are multiples of 2 will make the computation easier.) By the midpoint formula, the coordinates of A, B, C, and D are

$A\left(\frac{2a}{2}, \frac{2e}{2}\right) = (a, e)$; $B\left(\frac{2d + 2a}{2}, \frac{2e + 2b}{2}\right) = (d + a, e + b)$; $C\left(\frac{2d + 2c}{2}, \frac{2b}{2}\right) = (d + c, b)$; and $D\left(\frac{2c}{2}, \frac{0}{2}\right) = (c, 0)$.

Find the slopes of $\overline{AB}$ and $\overline{DC}$.

slope of $\overline{AB}$

$$m = \frac{y_2 - y_1}{x_2 - x_1} = \frac{(e + b) - e}{(d + a) - a} = \frac{b}{d}$$

slope of $\overline{DC}$

$$m = \frac{y_2 - y_1}{x_2 - x_1} = \frac{0 - b}{c - (d + c)} = \frac{-b}{-d} \text{ or } \frac{b}{d}$$

The slopes of $\overline{AB}$ and $\overline{DC}$ are the same so the segments are parallel.

Use the distance formula to find AB and DC.

(continued on the next page)

Classroom Vignette

"You can have students discover methods for placing a figure on a coordinate plane so that the coordinates are convenient for calculations. Cut different geometric figures out of cardboard. Have students work in pairs to place the figures on a plane in different ways and write the coordinates of the vertices for each placement. Have them choose the "best" position of the figure and explain their choice."

Jerry Cummins
Author

2 TEACH

Teaching Tip When reading the guidelines for placing figures on a coordinate plane, ask students to explain how each guideline helps simplify proof-writing.

In-Class Examples

For Example 1
Position and label an isosceles right triangle with legs of a units on the coordinate plane.

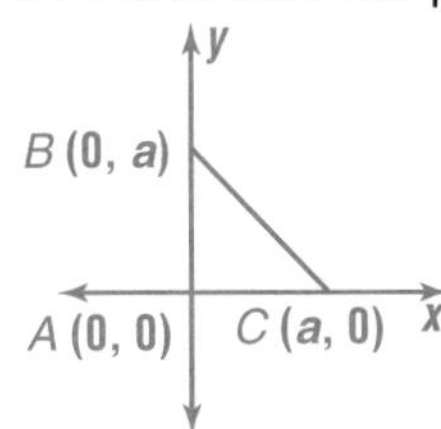

For Example 2
Use a coordinate proof to prove that if a segment from a vertex in a triangle is the perpendicular bisector of the opposite side, the triangle is isosceles.

Given: $\triangle MIT$, $\overline{MC}$ is the $\perp$ bisector of $\overline{IT}$.
Prove: $\triangle MIT$ is isosceles.
Proof:

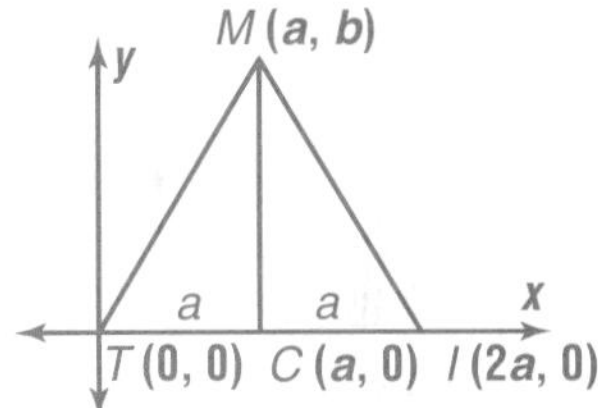

Because $\overline{MC} \perp \overline{TI}$, $\overline{MC}$ is a vertical segment and $\overline{TI}$ is a horizontal segment. Since C is on the perpendicular bisector and on $\overline{TI}$, it has the same x-coordinate as M, and its y-coordinate is halfway between that of T and I. The Distance Formula can be used to show that $\overline{MT} \cong \overline{MI}$.

$$MI = \sqrt{(2a - a)^2 + (0 - b)^2} = \sqrt{a^2 + b^2}$$

$$MT = \sqrt{(a - 0)^2 + (b - 0)^2} = \sqrt{a^2 + b^2}$$

Since $MI = MT$, $\overline{MI} \cong \overline{MT}$ and $\triangle MIT$ is isosceles.

3 PRACTICE/APPLY

Check for Understanding

Exercises 1–11 are designed to help you assess your students' understanding through reading, writing, speaking, and modeling. You should work through Exercises 1–4 with your students and then monitor their work on Exercises 5–11.

Error Analysis
Some students may use coordinates that imply more than the given information. For example, they may draw and label an isosceles or right triangle when the proof calls for a statement about any triangle. Remind them that their selection of points must include all possible cases.

Additional Answers

2. Answers will vary. Many problems that involve finding the midpoint of a segment are simpler if the sum of the coordinates has a factor of two because the result will not have any fractions.
3. Sample answer: Dashiki is correct. RJ is $a\sqrt{2}$; JK is $2a$. The sides do not have the same measures.
8. The coordinates of M, the midpoint of $\overline{BC}$, will be $\left(\frac{2c}{2}, \frac{2b}{2}\right) = (c, b)$. The distance from M to each of the vertices can be found using the distance formula.
$MB = \sqrt{(c-0)^2 + (b-2b)^2}$
$= \sqrt{c^2 + b^2}$
$MC = \sqrt{(c-2c)^2 + (b-0)^2}$
$= \sqrt{c^2 + b^2}$
$MA = \sqrt{(c-0)^2 + (b-0)^2}$
$= \sqrt{c^2 + b^2}$
Thus, $MB = MC = MA$, and M is equidistant from the vertices.

$AB = \sqrt{((d+a)-a)^2 + ((e+b)-e)^2}$ $\quad DC = \sqrt{((d+c)-c)^2\ (b-0)^2}$
$= \sqrt{d^2 + b^2}$ $\quad = \sqrt{d^2 + b^2}$

Thus, $AB = DC$. Therefore, $ABCD$ is a parallelogram because if one pair of opposite sides of a quadrilateral are both parallel and congruent, then the quadrilateral is a parallelogram.

CHECK FOR UNDERSTANDING

Communicating Mathematics

Study the lesson. Then complete the following.

1. **Explain** why it is often helpful to place as many vertices of a polygon as possible on an axis when planning a coordinate proof. *1. Sample answer: When the coordinates are 0, the computations are easier.*
2. **Write** a sample problem in which it would be helpful to place a figure on a coordinate grid so that the coordinates are multiples of 2. **See margin.**
3. **You Decide** Lorenza positioned a triangle on a coordinate plane shown at the right. She claimed it was an equilateral triangle. Dashiki disagreed. Who is correct? Explain your reasoning. **See margin.**

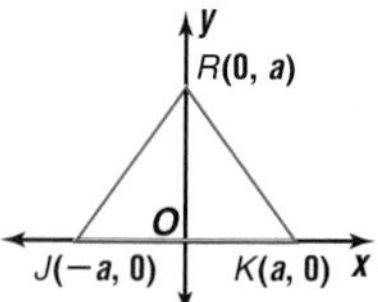

4. Select an area from your community or region and place it on a coordinate plane. Identify the coordinates of at least one point of interest in the area. Write a paragraph explaining what you did and how you selected the coordinates. **See students' work.**

Guided Practice

Name the missing coordinates in terms of the given variables.

5. CBD is an isosceles right triangle. *5. $D(0, 2c)$*
6. $PQRS$ is a rectangle. *6. $R(a, b)$, $S(-a, b)$*

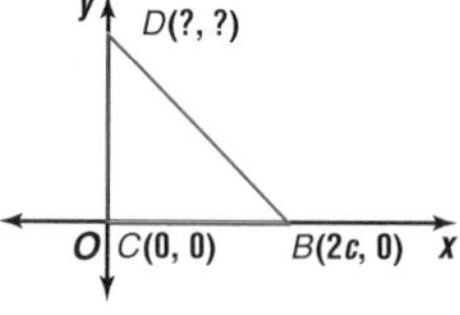

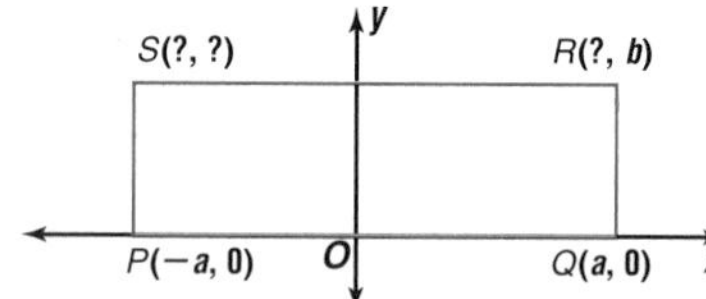

7. Use quadrilateral $ABCD$ with coordinates as indicated in the figure to answer each of the following.
   a. What kind of quadrilateral is $ABCD$? Explain your answer. *7a. Square; $AB = BC = CD = DA = a$, and $\angle DAB$ is right.*
   b. Find the midpoints of $\overline{AC}$ and $\overline{DB}$. What conclusion can you make? *7b. Both midpoints are $\left(\frac{a}{2}, \frac{a}{2}\right)$; $\overline{AC}$ and $\overline{BD}$ bisect each other.*
   c. Find the slope of $\overline{AC}$. **1**
   d. Find slope of $\overline{DB}$. **−1**
   e. What conclusion can you make about $\overline{AC}$ and $\overline{DB}$? **$\overline{AC}$ and $\overline{DB}$ are perpendicular.**

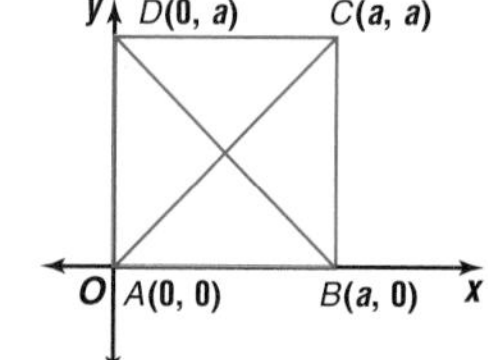

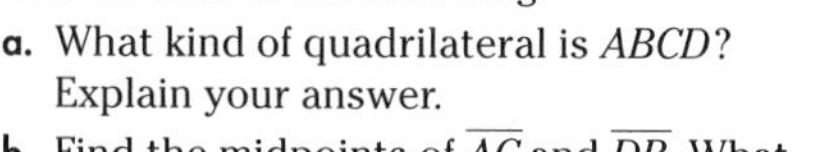

Use the triangle at the right to prove each of the following. **8–9. See margin.**

8. The midpoint of the hypotenuse is equidistant from each of the vertices.
9. The measure of the median to the hypotenuse is one-half the measure of the hypotenuse.

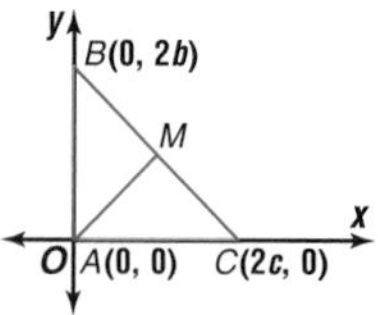

Reteaching

Using Cooperative Groups Have groups of students draw and label figures on the coordinate plane for each of the following: square, rectangle, parallelogram, right triangle, triangle. Remind them to use as few variables as possible. Have groups compare their work and then go back and alter it if they choose.

10. Position and label a parallelogram on the coordinate plane. Then write a coordinate proof to prove that the diagonals of a parallelogram bisect each other. **See margin.**

11. Recreation Refer to the map of Yellowstone National Park shown below. If a helicopter flew directly from Morning Glory Pool to Inspiration Point, what distance would it fly? **about 29.2 miles**

Yellowstone National Park

Old Faithful, the most famous geyser in the park, thrills visitors with streams of water shot 100 feet into the air.

Another popular attraction is *Morning Glory Pool*. The hot water of this pool brings minerals to the surface that make it resemble a morning glory flower in shape and color.

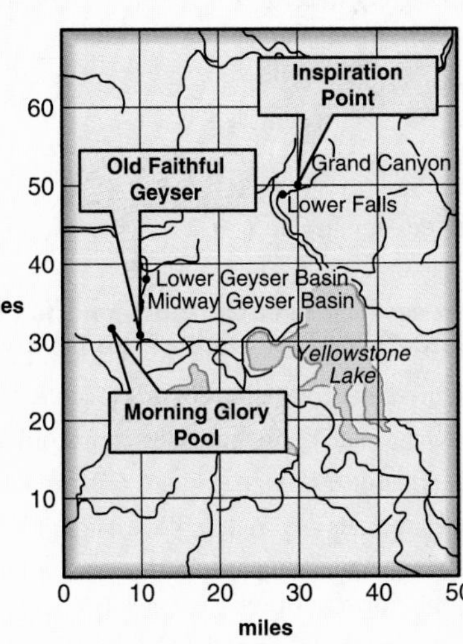

Two waterfalls grace the Yellowstone River that flows through the park. *Inspiration Point*, near the Lower Falls, offers an especially beautiful view.

Yellowstone is home to more than 275 species of birds and nearly 50 other kinds of animals. Trumpeter swans, bald eagles, elk, grizzly bears, moose, bighorn sheep, and cougars are just some of the residents.

EXERCISES

Practice

12. $K(-b, 0)$, $N(0, c)$ A
13. $D(-a, 0)$, $F(0, b)$

Name the missing coordinates in terms of the given variables.

12. *KLMN* is an isosceles trapezoid.

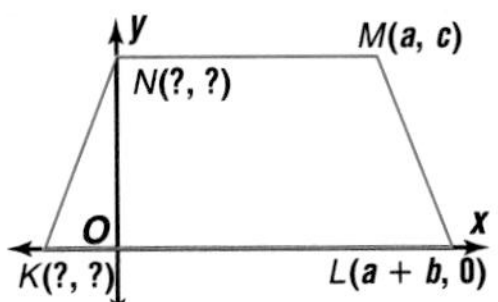

13. $\triangle DEF$ is isosceles.

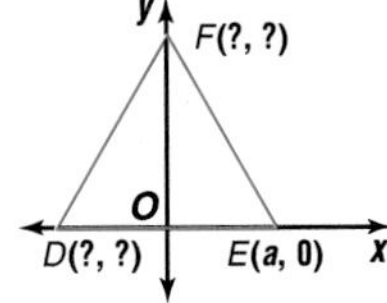

14. $Q(a, 0)$, $S(-b, c)$
15. $B(a, 0)$, $D(a, d)$, $F(-b, c)$

14. *PQRS* is a parallelogram.

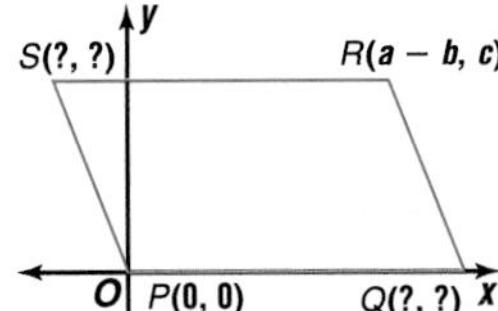

15. *ABCDEF* is a regular hexagon.

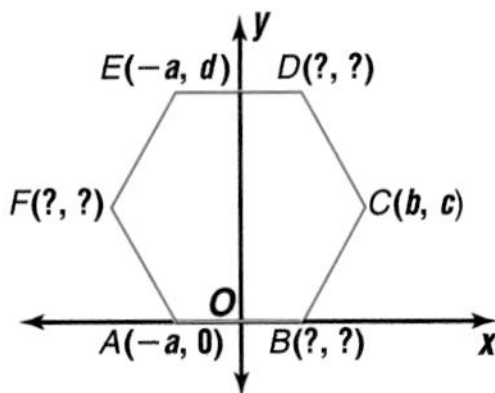

Additional Answer

10.

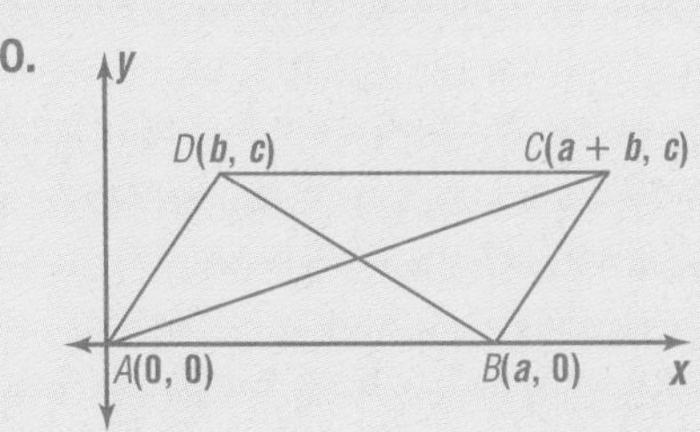

Midpoint of $\overline{AC}$ is $\left(\frac{(a+b)+0}{2}, \frac{c+0}{2}\right)$ or $\left(\frac{a+b}{2}, \frac{c}{2}\right)$.

Midpoint of $\overline{BD}$ is $\left(\frac{a+b}{2}, \frac{c+0}{2}\right)$ or $\left(\frac{a+b}{2}, \frac{c}{2}\right)$.

$\overline{AC}$ and $\overline{DB}$ bisect each other.

Assignment Guide

Core (with proof): 13–33 odd, 35–42
Core (informal): 13–15 odd, 31, 33, 35–42
Enriched: 12–30 even, 31–42

For **Extra Practice**, see p. 788.

The red A, B, and C flags, printed only in the Teacher's Wraparound Edition, indicate the level of difficulty of the exercises.

Additional Answer

9. Let M be the midpoint of $\overline{BC}$. The coordinates of M will be $\left(\frac{2c}{2}, \frac{2b}{2}\right) = (c, b)$.

$$MA = \sqrt{(c-0)^2 + (b-0)^2} = \sqrt{c^2 + b^2}$$

$$BC = \sqrt{(2c-0)^2 + (0-2b)^2} = 2\sqrt{c^2 + b^2}$$

Since $2MA = BC$, $MA = \frac{1}{2}BC$.

Study Guide Masters, p. 75

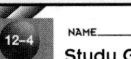

12–4 NAME______ DATE______

Study Guide

Student Edition Pages 666–671

Coordinate Proof

You can place figures in the coordinate plane and use algebra to prove theorems. The following guidelines for positioning figures can help keep the algebra simple.

- Use the origin as a vertex or center.
- Place at least one side of a polygon on an axis.
- Keep the figure within the first quadrant if possible.
- Use coordinates that make computations simple.

The distance formula, the midpoint formula, and your knowledge of slopes are useful tools for coordinate proofs.

Example: Use a coordinate proof to prove that the diagonals of a rectangle are congruent.

Use (0, 0) as one vertex. Place another vertex on the x-axis at $(a, 0)$ and another on the y-axis at $(0, b)$. The fourth vertex must then be (a, b).

Use the distance formula to find OB and AC.

$OB = \sqrt{(a-0)^2 + (b-0)^2} = \sqrt{a^2 + b^2}$
$AC = \sqrt{(0-a)^2 + (b-0)^2} = \sqrt{a^2 + b^2}$

Since $OB = AC$, the diagonals are congruent.

Name the missing coordinates in terms of the given variables.

1. *ABCD* is a rectangle. $D(-a, c)$, $A(-a, 0)$

2. *HIJK* is a parallelogram. $H(0, 0)$, $K(0, c)$, $J(a, b + c)$

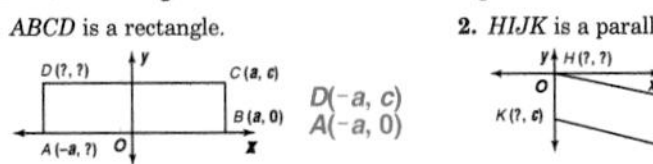

3. Use a coordinate proof to show that the opposite sides of any parallelogram are congruent. Label the vertices $A(0, 0)$, $B(a, 0)$, $C(b, c)$, and $D(a + b, c)$. Then use the distance formula to find AB, CD, AC, and BD.

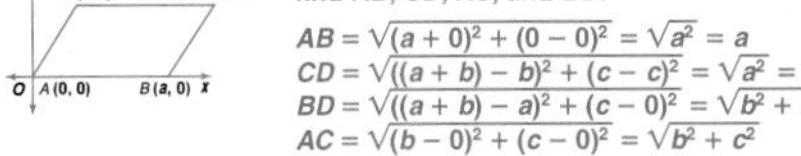

$AB = \sqrt{(a+0)^2 + (0-0)^2} = \sqrt{a^2} = a$
$CD = \sqrt{((a+b)-b)^2 + (c-c)^2} = \sqrt{a^2} = a$
$BD = \sqrt{((a+b)-a)^2 + (c-0)^2} = \sqrt{b^2 + c^2}$
$AC = \sqrt{(b-0)^2 + (c-0)^2} = \sqrt{b^2 + c^2}$

So $AB = CD$ and $AC = BD$. Therefore, the opposite sides of a parallelogram are congruent.

Additional Answers

16. $FJ =$
$\sqrt{[-b - (-a)]^2 + (c - 0)^2} =$
$\sqrt{b^2 - 2ab + a^2 + c^2}$
$HK =$
$\sqrt{(b - a)^2 + (c - 0)^2} =$
$\sqrt{b^2 - 2ab + a^2 + c^2}$
Slope of $\overline{FH}$ is $\frac{c - c}{b - (-b)} = \frac{0}{2b}$ or 0
Slope of $\overline{JK}$ is $\frac{0 - 0}{a - (-a)} = \frac{0}{2a}$ or 0
$FJ = HK$ and $\overline{FH} \parallel \overline{JK}$
$FHKJ$ is an isosceles trapezoid.

17. Slope of $\overline{LM}$ is $\frac{b - 0}{b - 0} = \frac{b}{b}$ or 1
Slope of $\overline{MN}$ is $\frac{b - 0}{b - 2b} = \frac{b}{-b}$ or -1
$\overline{LM}$ and $\overline{MN}$ are perpendicular and $\triangle LMN$ is a right triangle.

18. $RS = \sqrt{(0 - 0)^2 + (2b - 0)^2}$
$= \sqrt{4b^2}$ or $2b$
$ST =$
$\sqrt{(b\sqrt{3} - 0)^2 + (b - 2b)^2} =$
$\sqrt{3b^2 + b^2} = \sqrt{4b^2}$ or $2b$
$RT =$
$\sqrt{(b\sqrt{3} - 0)^2 + (b - 0)^2} =$
$\sqrt{3b^2 + b^2} = \sqrt{4b^2}$ or $2b$
$RS = ST = RT$ and $\overline{RS} \cong \overline{ST} \cong \overline{RT}$; $\triangle RST$ is equilateral.

Practice Masters, p. 75

NAME______________________ DATE__________

Practice

Student Edition Pages 666–671

Coordinate Proof

Name the missing coordinates in terms of the given variables.

1. $\triangle XYZ$ is isosceles and right.

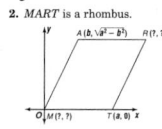

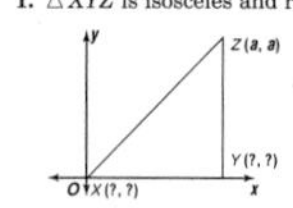

$X(0, 0)$ $Y(a, 0)$

2. $MART$ is a rhombus.

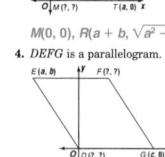

$M(0, 0)$, $R(a + b, \sqrt{a^2 - b^2})$

3. $RECT$ is a rectangle.

$R(0, 0)$ $C(a, b)$

4. $DEFG$ is a parallelogram.

$D(0, 0)$, $F(a + c, b)$

5. Use a coordinate proof to prove that the diagonals of a rhombus are perpendicular. Draw the diagram at the right.

Typical proof:

slope of $\overline{AC} = \frac{\sqrt{a^2 - b^2} - 0}{a + b - 0} = \frac{\sqrt{a^2 - b^2}}{a + b}$

slope of $\overline{BD} = \frac{\sqrt{a^2 - b^2} - 0}{b - a} = \frac{\sqrt{a^2 - b^2}}{b - a}$

$\frac{\sqrt{a^2 - b^2}}{a + b} \cdot \frac{\sqrt{a^2 - b^2}}{b - a} = \frac{a^2 - b^2}{b^2 - a^2} = -1$

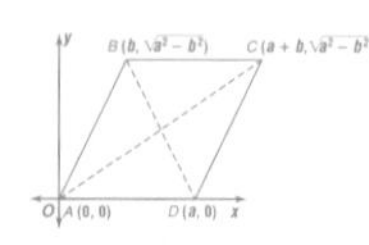

Proof **Prove using a coordinate proof.** 16–19. See margin.

16. $JFHK$ is an isosceles trapezoid.

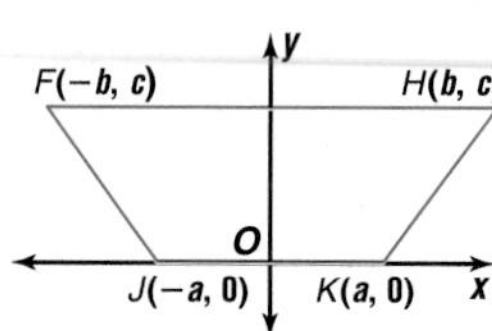

17. $\triangle LMN$ is a right triangle.

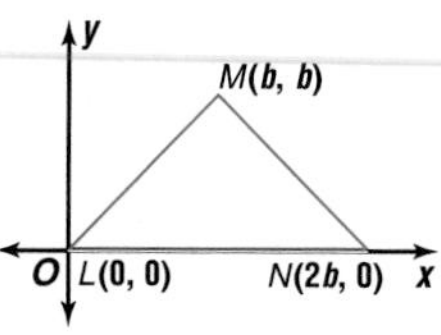

B

18. $\triangle RST$ is equilateral.

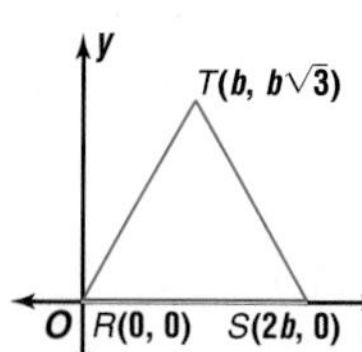

19. $EFGH$ is a rhombus.

y F($a\sqrt{2}$, $a\sqrt{2}$) G($2a + a\sqrt{2}$, $a\sqrt{2}$)
O E(0, 0) H(2a, 0) x

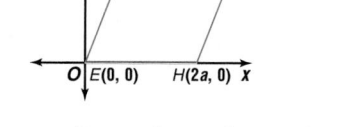

Position and label each figure on the coordinate plane. Then write a coordinate proof for each of the following.

20–30. See Solutions Manual.

20. The medians to the legs of an isosceles triangle are congruent.
21. The diagonals of an isosceles trapezoid are congruent.
22. If the diagonals of a parallelogram are congruent, then it is a rectangle.
23. The diagonals of a rectangle are congruent.

C

24. The three segments joining the midpoints of the sides of an isosceles triangle form another isosceles triangle.
25. The segments joining the midpoints of the opposite sides of a quadrilateral bisect each other.
26. If the diagonals of a parallelogram are perpendicular, then the parallelogram is a rhombus.
27. The segments joining the midpoints of the sides of an isosceles trapezoid form a rhombus.
28. If a line segment joins the midpoints of two sides of a triangle, then it is parallel to the third side.
29. If a line segment joins the midpoints of two sides of a triangle, then its length is equal to one-half the length of the third side.
30. The line segments joining the midpoints of the sides of a rectangle form a rhombus.

31a. Answers will vary. Two possible coordinates for C are $(a, 0)$ or $(0, b)$.

31b. Answers will vary. Two possible coordinates for C are $(2a, 0)$ or $(0, 2b)$.

31c. Answers will vary. Two possible solutions are $C(a - b, a + b)$, $D(-b, a)$, and $C(a + b, b - a)$, $D(b, -a)$.

32. Sample answer: The line would be a vertical line that passes through (22, 0). The equation for that line is $x = 22$.

Critical Thinking

31. Point A has coordinates (0, 0), and B has coordinates (a, b).
    a. Find the coordinates of point C so $\triangle ABC$ is a right triangle.
    b. Find the coordinates of point C so $\triangle ABC$ is isosceles.
    c. Find the coordinates of C and D so $ABCD$ is a rectangle.

Applications and Problem Solving

32. **Theater** An orchestra pit is in the shape of a semicircle with a 44-foot diameter. If a scale drawing of the orchestra pit is assigned coordinates as shown at the right, find the equation of the line that bisects the orchestra pit into two congruent parts.

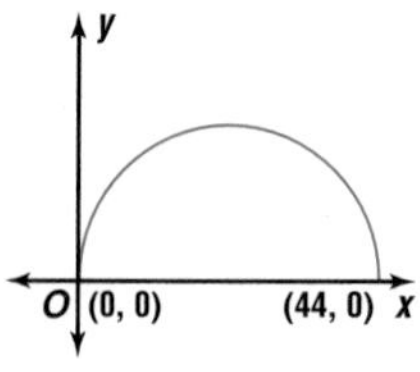

Additional Answer

19. $EF = \sqrt{(a\sqrt{2} - 0)^2 + (a\sqrt{2} - 0)^2} = \sqrt{2a^2 + 2a^2} = \sqrt{4a^2}$ or $2a$
$FG = \sqrt{[(2a + a\sqrt{2}) - a\sqrt{2}]^2 + (a\sqrt{2} - a\sqrt{2})^2} = \sqrt{(2a)^2 + 0^2} = \sqrt{4a^2}$ or $2a$
$GH = \sqrt{[(2a + a\sqrt{2}) - 2a]^2 + (a\sqrt{2} - 0)^2} = \sqrt{2a^2 + 2a^2} = \sqrt{4a^2}$ or $2a$
$EH = \sqrt{(0 - 0)^2 + (2a - 0)^2} = \sqrt{0^2 + (2a)^2} = \sqrt{4a^2}$ or $2a$
$EF = FG = GH = EH$; $\overline{EF} \cong \overline{FG} \cong \overline{GH} \cong \overline{EH}$; $EFGH$ is a rhombus.

33. Air Traffic Control An airplane is 4 kilometers east and 5 kilometers north of the airport while a second airplane is 3 kilometers west and 8 kilometers south. Assuming the planes are flying at the same altitude, find the distance between the airplanes. $\sqrt{218} \approx 14.8$ km

34. Statistics Consider scalene triangle ABC.

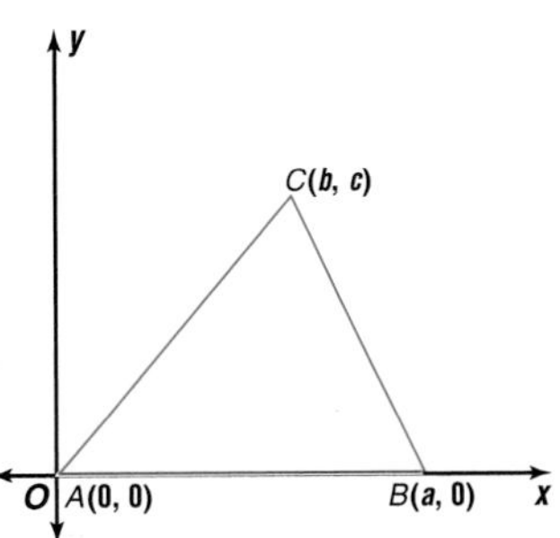

a. Find the coordinates of point D if the x-coordinate of D is the mean of the x-coordinates of the vertices of $\triangle ABC$ and the y-coordinate is the mean of the y-coordinates of the vertices of $\triangle ABC$.

b. Prove that D is the intersection of the medians of $\triangle ABC$. See margin.

34a. $D\left(\frac{a+b}{3}, \frac{c}{3}\right)$

Mixed Review

FYI

The Census Bureau has been taking a census of the United States every ten years since 1790. The primary purpose is to distribute seats for the U.S. House of Representatives.

35a. See Solutions Manual.

36. $y = \frac{1}{6}x + \frac{7}{3}$

38. 262 units^2; 431 units^2

35. Politics The table at the right shows the number of millions of Hispanic-American citizens of voting age in several congressional election years. (Lesson 12–3) b–c. See margin.

a. Draw a scatter plot to show how the year and the number of voters are related.

b. Write an equation that relates the year and the number of millions of Hispanic-Americans that are eligible voters.

c. Predict how many Hispanic-Americans will be eligible to vote in the congressional elections of 2010.

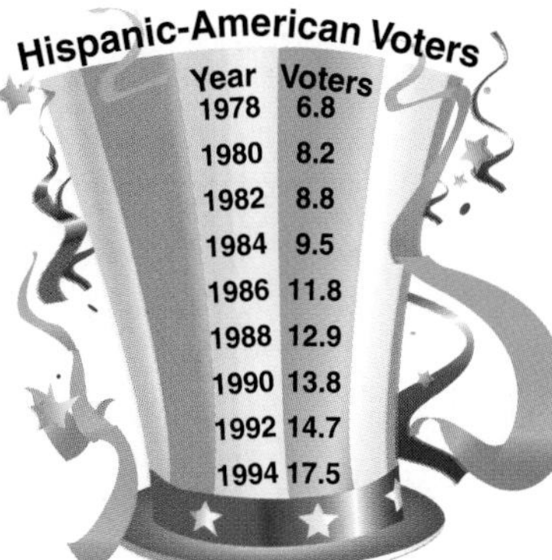

Hispanic-American Voters

Year	Voters
1978	6.8
1980	8.2
1982	8.8
1984	9.5
1986	11.8
1988	12.9
1990	13.8
1992	14.7
1994	17.5

Source: U.S. Bureau of the Census

36. The vertices of $\triangle TUV$ are $T(1, 5)$, $U(7, 1)$, and $V(-2, 2)$. Find the equation of the line containing $\overline{VX}$ if $\overline{VX}$ is a median of $\triangle TUV$. (Lesson 12–2)

Refer to the regular square pyramid for Exercises 37–38.

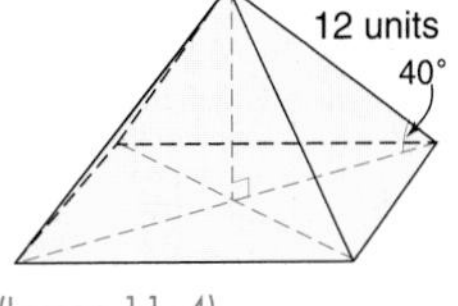

37. Find its volume. Round to the nearest cubic unit. (Lesson 11–6) 435 units^3

38. What are the lateral area and the surface area of the pyramid? Round to the nearest square unit. (Lesson 11–4)

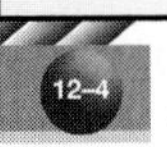

39. A car tire has a radius of 8 inches. How far does the car travel in one revolution of the tire? (Lesson 9–1) about 50.3 inches

40. The sides of a rhombus are each 25 units long. If a diagonal makes an angle of 30° with a side, what is the length of each diagonal to the nearest tenth? (Lesson 8–3) 43.3 units and 25 units

INTEGRATION **Algebra**

41. State whether $7b^3 - 4ab$ is a polynomial. If it is a polynomial, identify it as a *monomial*, a *binomial*, or a *trinomial*. yes; binomial

42. Factor $b^2 + 6b + 12$, if possible. If the trinomial cannot be factored using integers, write *prime*. prime

4 ASSESS

Closing Activity

Speaking Have students come to the chalkboard and draw a figure on a coordinate plane. Ask students to explain their choice of position and labels.

Additional Answers

34b. Label the midpoints of $\overline{AB}$, $\overline{BC}$, and $\overline{CA}$ as E, F, and G respectively. Then E, F, and G are at $\left(\frac{a}{2}, 0\right)$, $\left(\frac{a+b}{2}, \frac{c}{2}\right)$, and $\left(\frac{b}{2}, \frac{c}{2}\right)$ respectively. Slope of $\overline{AF} = \frac{c}{a+b}$ and slope of $\overline{AD} = \frac{c}{a+b}$ so D is on $\overline{AF}$; slope of $\overline{BG} = \frac{c}{b-2a}$ and slope of $\overline{BD} = \frac{c}{b-2a}$ so D is on $\overline{BG}$; slope of $\overline{CE} = \frac{2c}{2b-a}$ and slope of $\overline{CD} = \frac{2c}{2b-a}$ so D is on $\overline{AF}$. Since D is on $\overline{AF}$, $\overline{BG}$, and $\overline{CE}$, it is the intersection point of the three lines.

35b. Sample answer: $y = 0.66875x - 1315.9875$

35c. Sample answer: 28.2 million

Enrichment Masters, p. 75

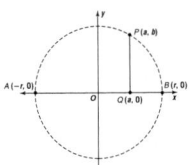

12-4 NAME______ DATE______

Enrichment Student Edition Pages 666–671

Coordinate Proofs with Circles

You can prove many theorems about circles by using coordinate geometry. Whenever possible locate the circle so that its center is at the origin.

1. Prove that an angle inscribed in a semicircle is a right angle. Use the figure at right. (Hint: Write an equation for the circle. Use your equation to help show that (slope of $\overline{AP}$) · (slope of $\overline{PB}$) = −1).

slope of $\overline{AP} = \frac{b-0}{a-(-r)} = \frac{b}{a+r}$

slope of $\overline{PB} = \frac{b-0}{a-r} = \frac{b}{a-r}$

(slope of $\overline{AP}$) · (slope of $\overline{PB}$)

$= \frac{b}{a+r} \cdot \frac{b}{a-r} = \frac{b^2}{a^2-r^2}$

$a^2 + b^2 = r^2$, since (a, b) is on the graph of $x^2 + y^2 = r^2$. Therefore

$b^2 = r^2 - a^2$, and $\frac{b^2}{a^2-r^2} = \frac{r^2-a^2}{a^2-r^2} = -1$.

2. Suppose $\overline{PQ} \perp \overline{AB}$, Q is between A and B, and $\overline{PQ}$ is the geometric mean between $\overline{AQ}$ and $\overline{QB}$. Prove that P is on the circle that has $\overline{AB}$ as a diameter. Use the figure at the right.

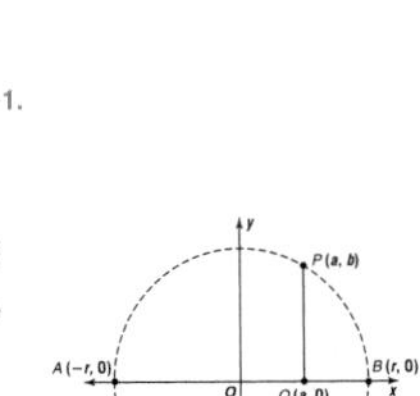

$(PQ)^2 = (AQ) \cdot (QB)$

$(b)^2 = (a + r) \cdot (r - a)$

$b^2 = (r + a) \cdot (r - a)$

$b^2 = r^2 - a^2$

Therefore $a^2 + b^2 = r^2$, which means that (a, b) is on the circle with the equation $x^2 + y^2 = r^2$. This is the circle that has $\overline{AB}$ as a diameter.

Extension

Connections A circle with radius r is placed on a coordinate plane with its center at the origin. Segment $\overline{AB}$ is drawn from the origin to a point (x, y) on the circle and in the first quadrant. What are the sine, cosine, and tangent of the angle, C, formed by $\overline{AB}$ and the x-axis?

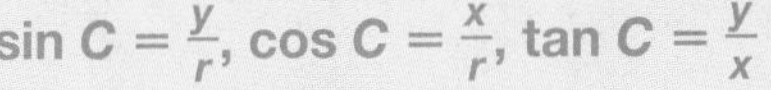

$\sin C = \frac{y}{r}$, $\cos C = \frac{x}{r}$, $\tan C = \frac{y}{x}$

FYI

Social scientists have evidence that on a census, people tend to exaggerate the amount of education they have.

12-5A LESSON NOTES

NCTM Standards: 2, 3

Objective

Use a geometric model of motion.

Recommended Time

Demonstration and discussion: 15 minutes; Exercises: 30 minutes

Instructional Resources

For each student or group of students
Modeling Mathematics Masters
• p. 103 (worksheet)
For teacher demonstration
Algebra and Geometry Overhead Manipulative Resources

1 FOCUS

Motivating the Lesson

Ask students what effect wind has on the flight of an airplane. Have them imagine the differences between a head-wind, a tail-wind, and no wind.

2 TEACH

Teaching Tip Before students draw an exact coordinate grid, they should draw a rough diagram to help them visualize the situation.

3 PRACTICE/APPLY

Assignment Guide
Core (with proof): 1–4
Core (informal): 1–4
Enriched: 1–4

4 ASSESS

Observing students working in cooperative groups is an excellent method of assessment.

A Preview of Lesson 12–5

12–5A Representing Motion

Materials:  battery-powered car  meterstick 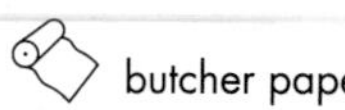 butcher paper  graph paper 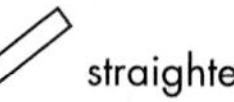 straightedge 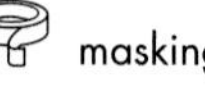 masking tape

When a power boat travels up or down a river, the speed of the river current affects how far the boat travels in a given amount of time. How do you think the current affects the motion of a boat that is traveling across the river? You can use a model car to investigate this situation.

Activity 1 **Investigate the motion of the boat if there is no current in the river.**

Step 1 The butcher paper represents the river and the car represents the boat. Place a piece of masking tape on the floor and lay the butcher paper over it so that the tape shows past each edge of the paper.

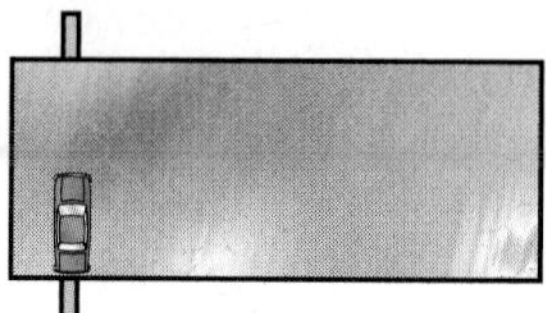

Step 2 Set your car at the edge of the paper so that its left wheels align with the tape. Make a mark on the edge of the paper where the left rear tire of the car is placed. Then allow the car to travel across the width of the paper and stop it when its front wheels reach the edge.

Step 3 Suppose that the starting position of the car is the origin and the tape is the y-axis. Measure the distance that the car traveled vertically and horizontally from the origin.

Step 4 Graph the ending position of the car as point A on a coordinate plane. Then draw segment OA.

Activity 2 **Now investigate the motion of the boat in a river with a current.**

Step 1 Place the car at the origin again. When the car is released, have one of your group members pull the paper at a constant speed to simulate the current. Stop the river motion and the car when the car reaches the opposite side of the river. Mark the ending position of the car on the floor with a piece of tape.

Step 2 Measure the distance that the car traveled vertically and horizontally from the origin.

Step 3 Graph the ending position of the car as point B on the coordinate plane. Then draw segment OB.

Step 4 Write an ordered pair to represent the ending position of the mark from which the car was released. Graph this as point C on the coordinate plane. Then draw segment OC.

Segment $\overline{OA}$ represents the displacement of the boat without the effect of the current, $\overline{OC}$ represents the current, and $\overline{OB}$ represents the displacement of the boat with the current.

Draw

1. Draw $\overline{BC}$ and $\overline{AB}$ as dashed segments on your graph. See students' work.

Write

2. What type of figure is quadrilateral $OABC$? rectangle
3. How is $\overline{OB}$ related to quadrilateral $OABC$? It is a diagonal.
4. How does the current affect the motion of the boat? It makes the boat travel downstream as it travels across the river.

Using Cooperative Learning

This lesson offers an excellent opportunity for using cooperative learning groups. For more information on cooperative learning strategies and group management, see *Cooperative Learning in the Mathematics Classroom*, one of the titles in the Glencoe Mathematics Professional Series.

12-5 Vectors

What YOU'LL LEARN

- To find the magnitude and direction of a vector,
- to determine if two vectors are equal, and
- to perform operations with vectors.

Why IT'S IMPORTANT

You can use vectors to solve problems in physics and transportation.

Benjamin Oliver Davis, Jr. (1912–)

The leader of the first squadron to graduate from the Tuskegee Institute was Benjamin Oliver Davis, Jr. He went on to become the first Air Force general and the first African-American to command an air base.

APPLICATION Aviation

In 1942, the United States Army established the first and only training facility for African-American pilots in Tuskegee, Alabama. About six hundred men were trained at this facility during World War II. Known as The Tuskegee Airmen, these pilots distinguished themselves in battles in Europe. Their efforts opened the door for the integration of the United States armed forces.

In addition to flying skills, pilots at the Tuskegee Institute were taught to plot a course. Even today, pilots must file a flight plan with the Federal Aviation Agency before taking off. Often a pilot plans a course for a trip involving flying to one airport, refueling, and then flying to another airport. A course like this may be represented by the diagram at the right. In this case, the course shows that the pilot travels 200 miles at a direction of 22° east of north and then travels 100 miles at a direction of 65° east of north.

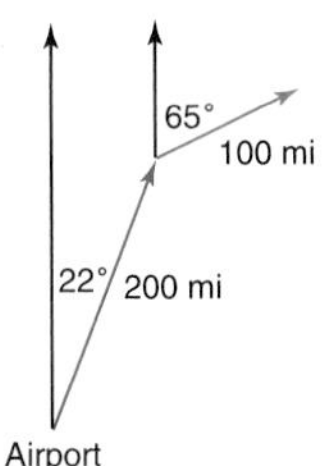

The distance and direction of the flight can be represented by a directed segment called a **vector**. A vector is any quantity that has both **magnitude** (length) and **direction**. In this case, the length of the segment represents the distance flown by the plane. Aviators use the angle that the vector makes with north to indicate direction.

In symbols, a vector is written as $\vec{v}$ or $\overrightarrow{AB}$. A vector can be either in **standard position**, with the initial point at the origin, or it can be drawn anywhere in the coordinate plane. A vector can be represented by an ordered pair (change in x, change in y). In the diagram below, vector OB ($\overrightarrow{OB}$) can be represented by the ordered pair (3, 4). To represent $\overrightarrow{CD}$ as an ordered pair, find the change in x and the corresponding change in y and write it as an ordered pair.

$$\overrightarrow{CD} = (x_2 - x_1, y_2 - y_1)$$
$$= (8 - 4, -2 - 6)$$
$$= (4, -8)$$

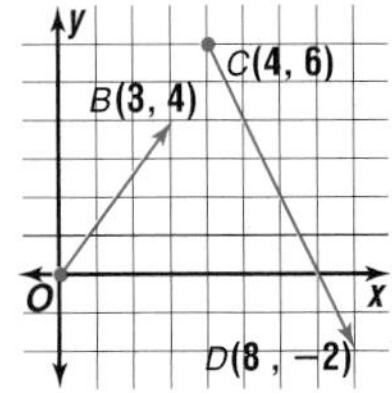

Because the magnitude and direction are not changed by moving, or translating, the vector (4, −8) represents the same vector as $\overrightarrow{CD}$.

You can use the distance formula to find the magnitude, or length, of a vector. The symbol for the magnitude of $\overrightarrow{AB}$ is $|\overrightarrow{AB}|$. In geometry, the direction of a vector is the measure of the angle that the vector, or an equal vector in standard position, forms with the positive x-axis. You can use the trigonometric ratios to find the direction of a vector.

fabulous FIRSTS

The Tuskegee Institute was founded in 1881 by Booker T. Washington. It was originally an elementary and secondary school for African Americans.

12-5 LESSON NOTES

NCTM Standards: 1–6, 8

Instructional Resources

- Study Guide Master 12-5
- Practice Master 12-5
- Enrichment Master 12-5
- Assessment and Evaluation Masters, p. 325
- Tech Prep Applications Masters, p. 24

Transparency 12-5A contains the 5-Minute Check for this lesson; **Transparency 12-5B** contains a teaching aid for this lesson.

Recommended Pacing	
Standard Pacing	Days 7 & 8 of 12
Honors Pacing	Days 6 & 7 of 11

1 FOCUS

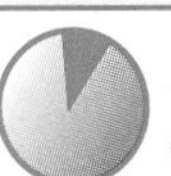

5-Minute Check *(over Lesson 12-4)*

Find the coordinates of each.

1. A of square $SQAR$ with vertices $S(0, 0)$, $Q(0, a)$, and $R(a, 0)$ **(a, a)**
2. E of parallelogram $TILE$ with vertices $T(0, 0)$, $I(a, b)$, and $L(c + a, b)$ **$(c, 0)$**
3. Z of equilateral triangle XYZ with vertices $X(0, 0)$ and $Y(2w, 0)$ **$(w, w\sqrt{3})$**
4. Prove that $\overline{PQ} \cong \overline{QR}$ for $P(0, 0)$, $Q(p, q)$, and $R(2p, 0)$.
   $PQ = \sqrt{p^2 + q^2}$
   $QR = \sqrt{p^2 + q^2}$
   Since $PQ = QR$, $\overline{PQ} \cong \overline{QR}$

Motivating the Lesson

Hands-On Activity Have students sketch a diagram on grid paper of their route to school. Ask how they could show the difference between going to school and returning home. **Sample answer: Use directed line segments.**

2 TEACH

In-Class Example

For Example 1
Given $P(2, 4)$ and $Q(5, 1)$, find the magnitude and direction of $\overrightarrow{PQ}$.

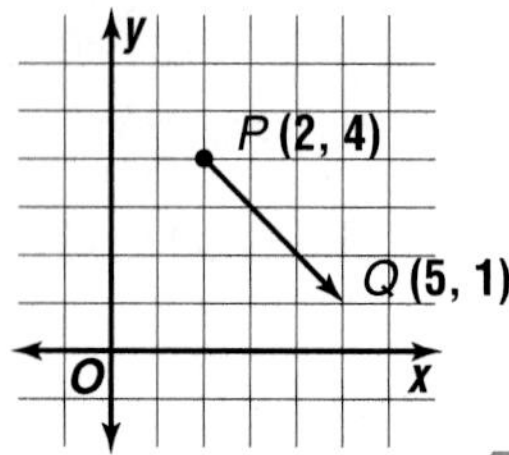

The magnitude is $\sqrt{18}$; the direction is $-45°$.

Teaching Tip After reading the first two lines below Example 1, point out that $\overrightarrow{AB}$ is not the same as $\overrightarrow{BA}$.

Example 1 **Given $C(8, -2)$ and $D(4, 6)$, find the magnitude and direction of $\overrightarrow{CD}$.**

magnitude

$$|\overrightarrow{CD}| = \sqrt{(8-4)^2 + (-2-6)^2}$$
$$= \sqrt{80} \text{ or about 8.9 units}$$

direction

$$\tan C = \frac{8}{4}$$
$$m\angle C \approx 63.4 \quad \textit{Use a calculator.}$$

A vector in standard position that is equal to $\overrightarrow{CD}$ forms a 63.4° angle with the negative x-axis, so it forms a $180 - 63.4$ or 116.6° angle with the positive x-axis. Thus, $\overrightarrow{CD}$ has a magnitude of about 8.9 units and a direction of about 116.6°.

Two vectors are equal only if they have the same magnitude and direction. They are parallel if they have the same direction or slope.

$\overrightarrow{RS} \parallel \overrightarrow{AB}$	Both have a slope of $\frac{4}{3}$, but have different lengths.
$\overrightarrow{RS} = \overrightarrow{KL}$	Both have a length of 5 and a slope of $\frac{4}{3}$.
$\overrightarrow{RS} \neq \overrightarrow{EC}$	Both have a length of 5, but have different slopes.

$\overrightarrow{RS}$ is *not* considered to be parallel to $\overrightarrow{ET}$ because they have opposite directions.

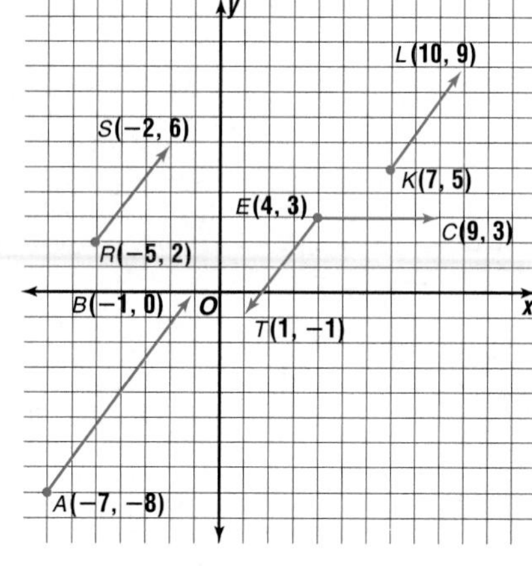

A vector can be multiplied by a constant that will change the magnitude of the vector but not affect the direction. If $\vec{v} = (1, 5)$, then $3\vec{v} = (3 \times 1, 3 \times 5)$ or $(3, 15)$. Now compare the magnitudes of $\vec{v}$ and $3\vec{v}$.

$$|\vec{v}| = \sqrt{1^2 + 5^2} = \sqrt{1 + 25} = \sqrt{26}$$

$$|3\vec{v}| = \sqrt{3^2 + 15^2} = \sqrt{9 + 225} = \sqrt{234} = 3\sqrt{26}$$

Notice that multiplying the vector by 3 tripled its magnitude. Multiplying a vector by a constant is called **scalar multiplication**.

It is also possible to add vectors. Suppose a plane flew from Minneapolis (M) to St. Louis (S) and then from St. Louis to New York City (N). This has the same result as flying directly from Minneapolis to New York City. In terms of vectors, $\overrightarrow{MS} + \overrightarrow{SN} = \overrightarrow{MN}$.

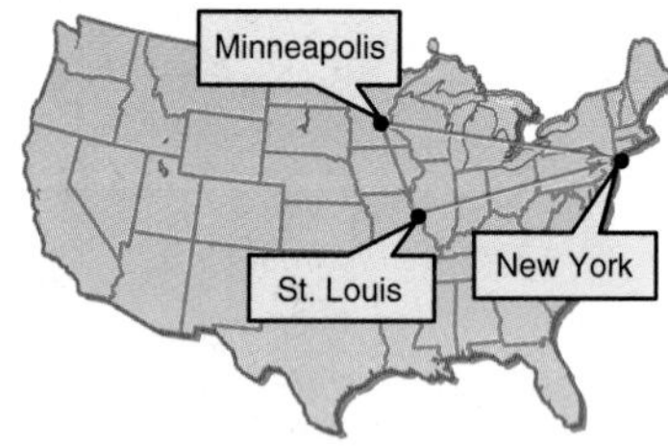

Alternative Learning Styles

Visual Separate the class into small groups. Each group will need grid paper, a stop watch, and a small ball. Have students practice rolling the object in a straight line down one side of the grid paper. Then, have one member blow on the object in a direction perpendicular to its path. After 2 seconds of rolling and blowing, stop the ball. That point will be the end of the resultant vector. Have students trace these paths and determine the magnitude and direction of each vector.

To add vectors, you can use the **parallelogram law**. The **resultant** or sum of two vectors is the diagonal of the parallelogram made by using the given vectors as sides. To add $\overrightarrow{QR}$ and $\overrightarrow{QT}$, draw parallelogram $QRST$ using the magnitude and direction of $\overrightarrow{QR}$ for side $\overrightarrow{TS}$ and that of $\overrightarrow{QT}$ for side $\overrightarrow{RS}$. The sum of the vectors is the diagonal of the parallelogram from Q to S.

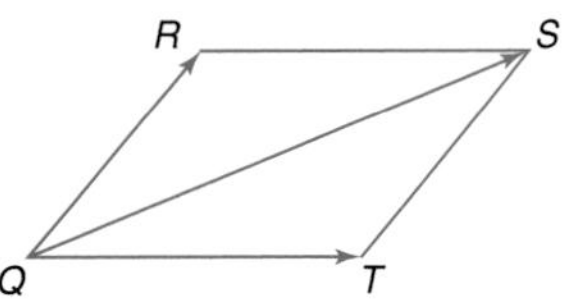

Teaching Tip The procedure shown at the top of the page is called the parallelogram method of adding vectors.

Teaching Tip Note that the method of adding vectors by adding their coordinates is somewhat like finding a translation of a point.

EXPLORATION CABRI GEOMETRY

The TI-92 will create a resultant vector that is the sum of two vectors.

- Use 9: Format on F8 to turn on the coordinate axes of a TI-92 calculator.
- Create a vector with its initial point at the origin.
- Create a vector with a different magnitude and direction having its initial point at the origin.
- Create the resultant vector that is the sum of the two vectors using the Vector Sum tool in the Construction menu F4 .

Your Turn

a. Use 5:Equation & Coordinates on the F6 menu to find the coordinates of the endpoints of all three vectors. See students' work.

b. How do the coordinates of the three endpoints compare?

c. Write a conjecture about the coordinates of two vectors and the coordinates of their sum. See students' work.

b. Sample answer: The coordinates of the endpoint of the resultant are the sums of the coordinates of the endpoints of the addends.

EXPLORATION

Have students repeat the activity for different vectors to emphasize the relationships.

In-Class Example

For Example 2

A boat is traveling due north at a speed of 8 miles per hour across a river that is flowing east at 3 miles per hour. How does the current of the river affect the speed and direction of the boat? The river moves the boat off course by 20.6° and increases its speed to 8.5 miles per hour.

Vectors can also be added by adding their coordinates.

$$(a, b) + (c, d) = (a + c, b + d)$$
$$\overrightarrow{AB} + \overrightarrow{BC} = \overrightarrow{AC}$$
$$(6, 2) + (1, 2) = (6 + 1, 2 + 2)$$
$$= (7, 4)$$

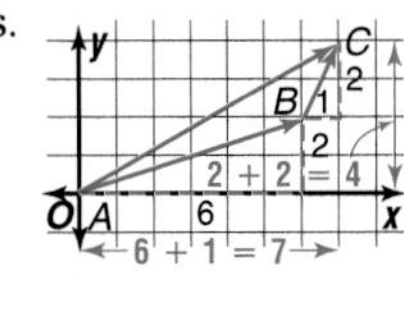

Vectors are used in physics to represent motion or forces acting upon objects.

Example 2

Physics

A river has a current of 2 kilometers per hour. A swimmer can swim at a rate of 3.5 kilometers per hour. How does the current affect the speed and direction of the swimmer as she swims across the river?

Use coordinates to make a model. Let each unit represent 0.5 kilometer. $\overrightarrow{OS}$ is the vector that represents the speed and direction of the swimmer. $\overrightarrow{OC}$ is the vector that represents the speed and direction of the current.

$\overrightarrow{OS} = (3.5, 0)$ $\overrightarrow{OC} = (0, 2)$

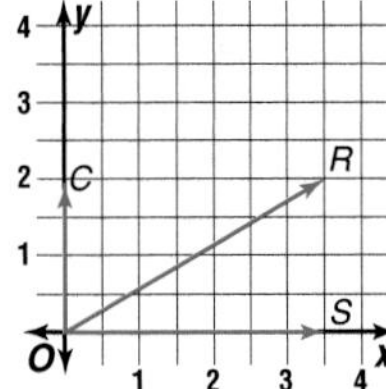

(continued on the next page)

$\overrightarrow{OR}$ is the resultant of $\overrightarrow{OS} + \overrightarrow{OC}$.

$$\begin{aligned}\overrightarrow{OR} &= \overrightarrow{OS} + \overrightarrow{OC}\\ &= (3.5, 0) + (0, 2)\\ &= (3.5 + 0, 0 + 2) \text{ or } (3.5, 2)\end{aligned}$$

Find the magnitude of $\overrightarrow{OR}$.

$$\begin{aligned}|\overrightarrow{OR}| &= \sqrt{3.5^2 + 2^2}\\ &= \sqrt{12.25 + 4}\\ &= \sqrt{16.25}\\ &\approx 4.03\end{aligned}$$

Find the direction of $\overrightarrow{OR}$.

$$\begin{aligned}\tan x &= \frac{2}{3.5} \quad \tan x = \frac{RS}{OS}\\ \tan x &\approx 0.57\\ x &\approx 29.7\end{aligned}$$

The current pushes the swimmer off course by about 29.7° and increases her speed to about 4.03 kilometers per hour.

Vectors can also be represented in an array called a *matrix*. The vector (4, 2) can be written as a **column matrix** $\begin{bmatrix}4\\2\end{bmatrix}$. You can add column matrices by adding corresponding entries.

Example 3 Given $\vec{v} = \begin{bmatrix}3\\7\end{bmatrix}$ and $\vec{u} = \begin{bmatrix}4\\-1\end{bmatrix}$, **find each sum.**

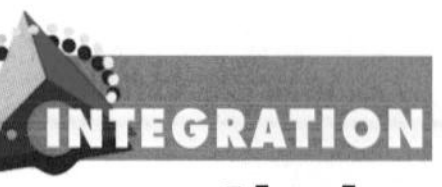

Algebra

a. $\vec{v} + \vec{u}$

$$\vec{v} + \vec{u} = \begin{bmatrix}3\\7\end{bmatrix} + \begin{bmatrix}4\\-1\end{bmatrix} = \begin{bmatrix}3+4\\7+(-1)\end{bmatrix} = \begin{bmatrix}7\\6\end{bmatrix}$$

b. $\vec{v} + 2\vec{u}$

$$\vec{v} + 2\vec{u} = \begin{bmatrix}3\\7\end{bmatrix} + 2\begin{bmatrix}4\\-1\end{bmatrix} = \begin{bmatrix}3+2(4)\\7+2(-1)\end{bmatrix} = \begin{bmatrix}3+8\\7+(-2)\end{bmatrix} = \begin{bmatrix}11\\5\end{bmatrix}$$

CHECK FOR UNDERSTANDING

Communicating Mathematics

Study the lesson. Then complete the following. 1–4. See margin.

1. **Describe** the difference between two parallel vectors and two equal vectors.
2. **Explain** how two vectors can have the same slope but are not considered parallel vectors.
3. **Draw** two vectors that meet the conditions.
   a. They have the same magnitude but different directions.
   b. They have the same direction but different magnitudes.

4. **Assess Yourself** List three different methods for adding vectors. Which of these methods do you prefer and why?

5. See students' work; $\sqrt{125} \approx 11.2$; 80°.
6. See students' work; $\sqrt{50} \approx 7.1$; 45°.

Guided Practice

Sketch each vector. Then find the magnitude to the nearest tenth and the direction to the nearest degree.

5. $\overrightarrow{AB} = (2, 11)$
6. $\overrightarrow{RS}$ if $R(-1, 3)$ and $S(4, 8)$

In-Class Example

For Example 3

Given $\vec{x} = \begin{bmatrix}2\\5\end{bmatrix}$ and $\vec{y} = \begin{bmatrix}1\\6\end{bmatrix}$, find each sum.

a. $\vec{x} + \vec{y}$ $\begin{bmatrix}3\\11\end{bmatrix}$

b. $\vec{x} + 3\vec{y}$ $\begin{bmatrix}5\\23\end{bmatrix}$

3 PRACTICE/APPLY

Check for Understanding

Exercises 1–16 are designed to help you assess your students' understanding through reading, writing, speaking, and modeling. You should work through Exercises 1–4 with your students and then monitor their work on Exercises 5–16.

Error Analysis

When finding a scalar multiple of a vector, some students may multiply the constant by the *x*-coordinate only. Remind them to multiply both coordinates by the scalar.

Additional Answers

1. Sample answer: Parallel vectors have the same direction but not necessarily the same magnitude. Equal vectors have both the same direction and same magnitude.
2. Sample answer: Two vectors can have the same slope and are not considered parallel when they are going in opposite directions.

3a. Sample answer:

3b. Sample answer:

4. Sample answer: by adding the coordinates, by writing as column matrices, and by finding the resultant of the parallelogram created by the two vectors; preferences will vary.

Reteaching

Using Hands-On Activity Find the magnitude to the nearest tenth and the direction to the nearest degree for each vector.

1. $\overrightarrow{AB} = (2, 6)$ 6.3, 72°
2. $\overrightarrow{CD} = (3, 4)$ 5, 53°
3. $\overrightarrow{EF} = (7, 2)$ 7.3, 16°
4. $\overrightarrow{GH} = (10, 5)$ 11.2, 27°

Have students draw each of these vectors on grid paper and make a ruler from a strip of the grid paper. Have them use the grid-paper ruler to check their calculations of the magnitude of the vector and a protractor to check the direction. Measurements will be approximate but reasonable estimates for each exercise.

7. $\overrightarrow{AB} \parallel \overrightarrow{KM} \parallel \overrightarrow{TR}$

8. $\overrightarrow{AB} = \overrightarrow{KM}$

Use the coordinate plane to answer each question. Explain your answers.

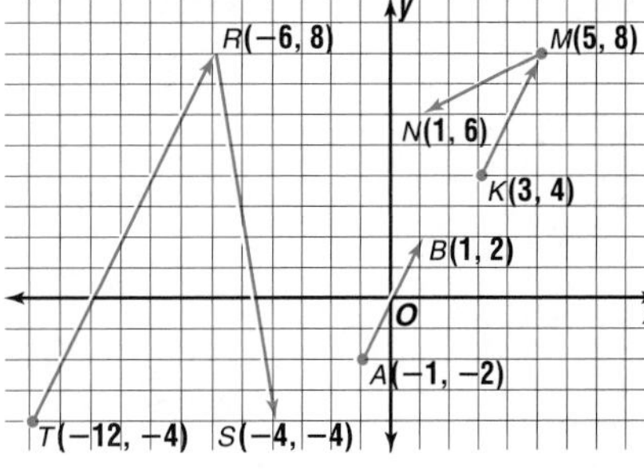

7. Which vectors are parallel?
8. Which vectors are equal?
9. Name a pair of vectors that have the same direction but different magnitudes.
$\overrightarrow{TR}$ and $\overrightarrow{AB}$ or $\overrightarrow{TR}$ and $\overrightarrow{KM}$

Given $\vec{u} = (-4, 10)$, $\vec{v} = (0, 5)$, and $\vec{w} = (6, 2)$, represent each of the following as an ordered pair.

10. $\vec{u} + 2\vec{v}$ (−4, 20)

11. $3\vec{w} - (\vec{u} + \vec{v})$ (22, −9)

Find each sum or difference.

12. $\begin{bmatrix} 1 \\ 3 \end{bmatrix} + \begin{bmatrix} -5 \\ 2 \end{bmatrix}$ $\begin{bmatrix} -4 \\ 5 \end{bmatrix}$

13. $\begin{bmatrix} 2 \\ -6 \end{bmatrix} + \begin{bmatrix} 0 \\ 3 \end{bmatrix} - \begin{bmatrix} -4 \\ -1 \end{bmatrix}$ $\begin{bmatrix} 6 \\ -2 \end{bmatrix}$

Copy each pair of vectors and draw a resultant vector.

14.

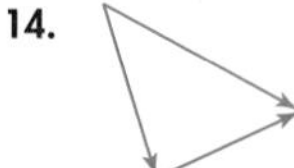

15.

16a. 0.25 hour or 15 minutes

16. **Physics** Suppose the river in Example 2 is 1 kilometer wide.
   a. How long will it take the swimmer to cross the river?
   b. How far downstream from the point perpendicular across from her starting point will she land? She will land at a point 0.57 kilometer downstream from her starting point.

EXERCISES

Practice

Sketch each vector. Then find the magnitude to the nearest tenth and the direction to the nearest degree. 19–22. See margin.

A

17. See students' work; $\sqrt{17} \approx 4.1$ units, 76°.

18. See students' work; $\sqrt{97} \approx 9.8$ units, −66°.

17. $\overrightarrow{AB} = (1, 4)$
18. $\vec{v} = (4, -9)$
19. $\overrightarrow{AB}$ if $A(4, 2)$ and $B(7, 22)$
20. $\overrightarrow{CD}$ if $C(0, -20)$ and $D(40, 0)$
21. $\overrightarrow{EF}$ if $E(0, 6)$ and $F(-6, 0)$
22. $\overrightarrow{GH}$ if $G(12, -4)$ and $H(19, 1)$

Given a path from *A* south 25 units to *B*, then east 10 units to *C*, answer each question.

23. What is the total length of the path? 35 units
24. What is the magnitude of $\overrightarrow{AC}$? $\sqrt{725} \approx 26.9$ units
25. What is the direction from *A* to *C*? −68°

Find each sum or difference.

26. $\begin{bmatrix} 9 \\ 21 \end{bmatrix} + \begin{bmatrix} 8 \\ -2 \end{bmatrix}$ $\begin{bmatrix} 17 \\ 19 \end{bmatrix}$

27. $\begin{bmatrix} 2 \\ 4 \end{bmatrix} + \begin{bmatrix} -2 \\ -3 \end{bmatrix}$ $\begin{bmatrix} 0 \\ 1 \end{bmatrix}$

28. $\begin{bmatrix} -5 \\ -8 \end{bmatrix} + \begin{bmatrix} 6 \\ -1 \end{bmatrix}$ $\begin{bmatrix} 1 \\ -9 \end{bmatrix}$

29. $\begin{bmatrix} 12 \\ 4 \end{bmatrix} - \begin{bmatrix} 4 \\ -1 \end{bmatrix}$ $\begin{bmatrix} 8 \\ 5 \end{bmatrix}$

30. $\begin{bmatrix} 5 \\ 4 \end{bmatrix} - \begin{bmatrix} 7 \\ -3 \end{bmatrix}$ $\begin{bmatrix} -2 \\ 7 \end{bmatrix}$

31. $\begin{bmatrix} 3 \\ -3 \end{bmatrix} + \begin{bmatrix} -3 \\ 4 \end{bmatrix} - \begin{bmatrix} 2 \\ -5 \end{bmatrix}$

31. $\begin{bmatrix} -2 \\ 6 \end{bmatrix}$

Assignment Guide

Core (with proof): 17–57 odd, 58–69
Core (informal): 17–43 odd, 47–57 odd, 59–69
Enriched: 18–52 even, 53–69

For **Extra Practice**, see p. 789.

The red A, B, and C flags, printed only in the Teacher's Wraparound Edition, indicate the level of difficulty of the exercises.

Additional Answers

19. See students' work. $\sqrt{409} \approx 20.2$ units, 81°
20. See students' work. $\sqrt{2000} \approx 44.7$ units, 27°
21. See students' work. $\sqrt{72} \approx 8.5$ units, −45°
22. See students' work. $\sqrt{74} \approx 8.6$ units, 36°

Study Guide Masters, p. 76

NAME ______ DATE ______

12-5 **Study Guide**

Student Edition Pages 673–679

Vectors

A **vector** is any quantity that has both **magnitude** (length) and **direction**. In symbols, a vector is written as $\vec{v}$ or $\overrightarrow{AB}$. Vectors with the initial point at (0, 0) can be represented by the ordered pair for point *B*. You can write $\vec{v} = (5, 3)$.

You can use the distance formula to find the magnitude of a vector. The symbol for the magnitude of $\overrightarrow{AB}$ is $|\overrightarrow{AB}|$. The direction of a vector is the measure of the angle that the vector forms with the positive *x*-axis or any other horizontal line. Equal vectors are vectors with the same magnitude and direction.

Example: Given *A*(5, 2) and *B*(8, 7), find the magnitude and direction of $\overrightarrow{AB}$.

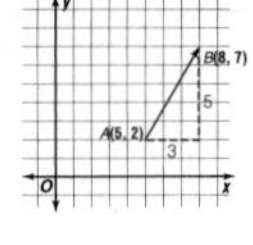

Magnitude
$|\overrightarrow{AB}| = \sqrt{(8-5)^2 + (7-2)^2}$
$= \sqrt{34}$ or about 5.8 units

Direction
$\tan A = \frac{5}{3}$ $\tan A = \frac{\text{opposite}}{\text{adjacent}}$
$m\angle A \approx 59.0$

The magnitude is about 5.8 units, and the direction is about 59°.

To find the **resultant** or sum of two vectors, you can represent them by equal vectors with initial points at (0, 0) and then add corresponding coordinates.

For each vector, find the magnitude to the nearest tenth and the direction to the nearest degree.

1. $\vec{v} = (8, 2)$ $\sqrt{68} \approx 8.2$, 14°
2. $\overrightarrow{AB}$ if $A(3, 0)$ and $B(5, 7)$ $\sqrt{53} \approx 7.3$, 74°

Given $\vec{a} = (4, 6)$ and $\vec{b} = (9, -4)$, represent each of the following as an ordered pair.

3. $\vec{a} + 2\vec{b}$ (22, −2)
4. $\vec{a} - \vec{b}$ (−5, 10)
5. $2\vec{a} - \vec{b}$ (−1, 16)
6. $3\vec{a} + (\vec{a} - \vec{b})$ (7, 28)

Additional Answers

46. Let $a = (a_1, a_2)$ and $b = (b_1, b_2)$ and $c = (c_1, c_2)$.
$(a + b) + c = [(a_1, a_2) + (b_1, b_2)] + (c_1, c_2)$
$= (a_1 + b_1, a_2 + b_2) + (c_1, c_2)$ Vector addition
$= [(a_1 + b_1) + c_1, (a_2 + b_2) + c_2]$ Vector addition
$= [a_1 + (b_1 + c_1), a_2 + (b_2 + c_2)]$ Addition of real numbers is associative
$= (a_1, a_2) + [(b_1, b_2) + (c_1, c_2)]$
$= a + (b + c)$

49. A line through the origin will have equation $y = cx$ for some constant c. If (k, m) lies on the line, then $m = ck$. If (r, s) is a scalar multiple of (k, m), then $(r, s) = c(k, m) = (ck, cm)$. Substituting into $y = cx$, $cm = c(ck)$. Dividing by c, $m = ck$ which is true because (k, m) lies on the line. Therefore, (r, s) lies on the line also.

53a. They have the same magnitude but opposite directions.

Practice Masters, p. 76

NAME ________ DATE ________

Practice — Student Edition Pages 673–679

Vectors

Given $\vec{a} = (8, 6)$, $\vec{b} = (4, 3)$, $\vec{c} = (1, 2)$, and $\vec{d} = (-3, -6)$, answer each of the following.

1. Find the magnitude of $\vec{a}$. 10
2. Find the magnitude $\vec{c}$. $\sqrt{5} \approx 2.2$
3. Determine if $\vec{b}$ and $\vec{d}$ are equal. no
4. Determine if $\vec{c}$ and $\vec{d}$ are equal. no
5. Find the coordinates of $\vec{a} + \vec{b}$. (12, 9)
6. Find the coordinates of $(\vec{b} + \vec{c}) + \vec{d}$. (2, −1)
7. Given $A(2, 5)$ and $B(7, 10)$, find the magnitude and direction of $\overline{AB}$. $\sqrt{50} \approx 7.1$, 45°
8. Given $C(0, 1)$ and $D(8, 12)$, find the magnitude and direction of $\overline{CD}$. $\sqrt{185} \approx 13.6$, about 54°

Given path from A south 5 units to B, then east 12 units to C, answer each question.

9. What is the total length of the path? 17 units
10. What is the magnitude of $\overline{AC}$? 13 units

Given $A(2, 4)$, $B(1, 7)$, $C(-1, 0)$, $D(-3, 6)$, $E(4, 6)$, $F(7, 7)$, $G(-3, -1)$, $H(-2, -4)$, and $I(1, -3)$, draw $\overrightarrow{AB}$, $\overrightarrow{CD}$, $\overrightarrow{EF}$, $\overrightarrow{HG}$, and $\overrightarrow{HI}$. Use your diagram to answer each question. Explain your answers.

32. $\overrightarrow{AB} \parallel \overrightarrow{CD} \parallel \overrightarrow{HG}$; $\overrightarrow{EF} \parallel \overrightarrow{HI}$
33. $\overrightarrow{AB} = \overrightarrow{HG}$; $\overrightarrow{EF} = \overrightarrow{HI}$

B

32. Which vectors are parallel?
33. Which vectors are equal?
34. Which vectors have the same magnitude? $|\overrightarrow{AB}| = |\overrightarrow{EF}| = |\overrightarrow{HG}| = |\overrightarrow{HI}|$

Given the quadrilateral *ACDE*, complete each statement.

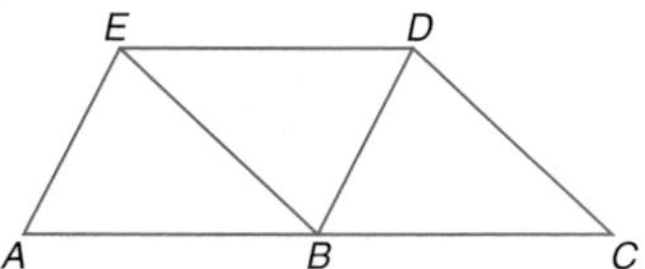

35. $\overrightarrow{AE} + \overrightarrow{AB} =$ _?_ $\overrightarrow{AD}$
36. $\overrightarrow{BC} + \overrightarrow{CD} =$ _?_ $\overrightarrow{BD}$
37. $\overrightarrow{BA} + \overrightarrow{AE} =$ _?_ $\overrightarrow{BE}$
38. $\overrightarrow{EB} + \overrightarrow{ED} =$ _?_ $\overrightarrow{EC}$

Given $\vec{v} = (3, 1)$ and $\vec{u} = (5, 8)$, represent $\vec{w}$ as an ordered pair.

40. (−1, −3.5)

39. $\vec{u} + 2\vec{v} = \vec{w}$ (11, 10)
40. $\vec{u} + 2\vec{w} = \vec{v}$
41. $\vec{w} = 4\vec{u}$ (20, 32)
42. $\vec{w} = 2\vec{u} + 3\vec{u}$ (25, 40)
43. $\vec{w} = \vec{v} + \vec{u}$ (8, 9)
44. $3\vec{v} - \vec{w} = \vec{u}$ (4, −5)

C

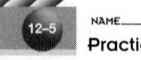

45. Complete the proof.

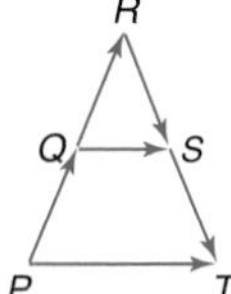

Given: $\overrightarrow{PQ} = \overrightarrow{QR}$ and $\overrightarrow{RS} = \overrightarrow{ST}$

Prove: $\overrightarrow{QS} = \frac{1}{2}\overrightarrow{PT}$

Proof:

Statements	Reasons
1. $\overrightarrow{QR} + \overrightarrow{RS} = \overrightarrow{QS}$	1. _?_ Definition of vector addition
2. $\overrightarrow{PR} + \overrightarrow{RT} = \overrightarrow{PT}$	2. _?_ Definition of vector addition
3. $\overrightarrow{PQ} = \overrightarrow{QR}$, $\overrightarrow{RS} = \overrightarrow{ST}$	3. _?_ Given
4. $2\overrightarrow{QR} + 2\overrightarrow{RS} = \overrightarrow{PT}$	4. _?_ Substitution Property (=)
5. $2(\overrightarrow{QR} + \overrightarrow{RS}) = \overrightarrow{PT}$	5. _?_ Distributive Property (=)
6. $2\overrightarrow{QS} = \overrightarrow{PT}$	6. _?_ Substitution Property (=)
7. $\overrightarrow{QS} = \frac{1}{2}\overrightarrow{PT}$	7. _?_ Multiplication Property (=)

46. Prove that addition of vectors is associative. See margin.

47. $\begin{bmatrix} ac \\ ab \end{bmatrix}$; the process has to model the way scalar multiplication of vectors written as ordered pairs is written.

47. How would you define $a \cdot \begin{bmatrix} c \\ b \end{bmatrix}$? Explain your reasoning.
48. If $R(4, 12)$, $S(9, 2)$, $T(x, 8)$, and $M(2, 7)$, find x if $\overrightarrow{RS} \parallel \overrightarrow{TM}$. 1.5
49. Vector (k, m) has its initial point at the origin and is on a line that passes through the origin. Prove that if (r, s) is a scalar multiple of (k, m), then (r, s) lies on that line. See margin.

If two vectors are represented by a single ordered pair, it is possible to determine whether they are perpendicular by using the dot product test. If $\vec{a} = (x_1, y_1)$ and $\vec{b} = (x_2, y_2)$, their dot product, $\vec{a} \cdot \vec{b}$, is $x_1 \cdot x_2 + y_1 \cdot y_2$. If the dot product is 0, the vectors are perpendicular. Find each dot product and determine whether the vectors are perpendicular. Write *yes* or *no*.

50. $(3, -5) \cdot (-6, 10)$ no
51. $(3, -5) \cdot (5, 3)$ yes
52. $(3, -5) \cdot (-5, 3)$ no

Critical Thinking

53. Points C, D, E, F, G, and H are noncollinear, and $\overrightarrow{CD} + \overrightarrow{FG} = 0$.
    a. What is the relationship between $\overrightarrow{CD}$ and $\overrightarrow{FG}$? See margin.
    b. If $\overrightarrow{DE} + \overrightarrow{GH} = 0$ and $\overrightarrow{EF} + \overrightarrow{HC} = 0$, what is true about the polygon with vertices C, D, E, F, G, and H? It is a hexagon.

Extension

Connections Have students write vector problems that model everyday situations. Use examples such as pulling a wagon, pushing a lawn mower, and walking in a strong wind. Have students identify the vectors in each situation. For example, if the lawn mower handle is the resultant vector, what are the other two vectors? the weight of the mower pulling down and the person pushing forward

Applications and Problem Solving

54. **Transportation** A truck driver drives 40 miles due north and then 75 miles due west. What is the direction and distance of the truck from the starting point? **85 miles at 28°**

55. **Sailing** A boat is traveling north at 20 miles per hour. A wind from the west is blowing the boat eastward at 5 miles per hour. Find the speed of the boat and the direction in which it is moving. **See margin.**

56. **Physics** A force of 40 pounds is acting on an object at a 30° angle and a force of 20 pounds is acting on the same object at a 60° angle. In what direction and with what force will the object move? **See margin.**

57. **Sports** Two soccer players kick the ball at the same time. One exerts a force of 72 newtons east. The other exerts a force of 45 newtons north. What is the magnitude and direction of the resultant force on the ball? **84.9 newtons, 32° northeast**

58. See margin.

Mixed Review

58. $\triangle ABC$ is a right isosceles triangle. M is the midpoint of $\overline{AB}$. Use a coordinate proof to show that $\overline{CM}$ is perpendicular to $\overline{AB}$. (Lesson 12–4)

59. **Transportation** The table below shows the number of thousands of people who travel to work alone or in car pools in different southern states. (Lesson 12–3)

Source: U.S. Bureau of the Census

a. Write an equation that relates the number of people who drive alone to the number of people who carpool. You may wish to use a scatter plot.

b. The Census Bureau found that there are 2281 thousand workers who drove to work alone in the state of Virginia. Predict how many people carpool in Virginia. **441 thousand**

59a. Sample answer: Using points (1195, 229) and (5821, 1134) the equation is $y = 0.2x - 4.8$.

60. about 341.3 units^3

60. Find the volume of a regular square pyramid if its lateral edges are 12 centimeters and the base is 16 centimeters on a side. (Lesson 11–6)

61. Triangle ABC has an area of 88 square meters. If the height is 8 meters, what is the length of the base? (Lesson 10–4) **22 m**

62. Find the measure of an interior angle of a regular polygon with 36 sides. (Lesson 10–1) **170°**

63. $\angle NMP$ is inscribed in a circle. If $m\angle NMP = 62$, what is the measure of the arc intercepted by $\angle NMP$? (Lesson 9–4) **124**

64. *True* or *false*: Concentric circles must have the same center. (Lesson 9–2) **true**

65. The perimeter of an equilateral triangle is 72 inches. What is the length of an altitude of the triangle? (Lesson 8–2) $12\sqrt{3} \approx 20.8$ **in.**

66. The number 10 has exactly four factors, 1, 2, 5, and 10. Find the least number with exactly five factors. (Lesson 7–6) **16**

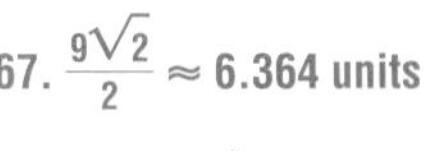

67. $\frac{9\sqrt{2}}{2} \approx 6.364$ units

67. Find the length of the median of a trapezoid whose vertices are at (2, 1), (4, 0), (7, 3), and (8, 7). (Lesson 6–5)

INTEGRATION Algebra

68. Express $2m(3x - y) + 7n(3x - y)$ in factored form. $(2m + 7n)(3x - y)$

69. Find $(3u^2 - 4) \div 2u$. $\frac{3}{2}u - \frac{2}{u}$

Tech Prep

Civil Engineering Assistant Some two-year colleges train students to assist civil engineers. Civil engineers design solutions to problems regarding a city's transportation and energy problems. For more information on tech prep, see the *Teacher's Handbook*.

4 ASSESS

Closing Activity

Modeling Ask each student to think of a situation that can be modeled by vectors. Have students model the situation by drawing pictures and minimizing use of words. Have students present their models to the class.

Chapter 12 Quiz C (Lessons 12-4 and 12-5) is available in the *Assessment and Evaluation Masters*, p. 325.

Additional Answers

55. $\sqrt{425} \approx 20.62$ miles per hour at a direction of 76°

56. The object will move with a force of 58.14 pounds in the direction of 40°.

58. Midpoint M is $\left(\frac{0+a}{2}, \frac{a+0}{2}\right) = \left(\frac{a}{2}, \frac{a}{2}\right)$. Slope of $\overline{AB}$ is $\frac{0-a}{a-0} = \frac{-a}{a}$ or -1. Slope of $\overline{CM}$ is $\frac{\frac{a}{2}-0}{\frac{a}{2}-0} = \frac{\frac{a}{2}}{\frac{a}{2}}$ or 1. Since $-1 \cdot 1 = -1$, $\overline{CM}$ is perpendicular to $\overline{AB}$.

Enrichment Masters, p. 76

NAME ______ DATE ______

Enrichment

Student Edition Pages 673–679

Reading Mathematics

Many quantities in nature can be thought of as vectors. The science of physics involves many vector quantities. In reading about applications of mathematics, ask yourself whether the quantities involve only magnitude or both magnitude and direction. The first kind of quantity is called **scalar**. The second kind is a **vector**.

Classify each of the following. Write scalar or vector.

1. the mass of a book scalar
2. a car traveling north at 55 mph vector
3. a balloon rising 24 feet per minute vector
4. the size of a shoe scalar
5. a room temperature of 22 degrees Celsius scalar
6. a west wind of 15 mph vector
7. the batting average of a baseball player scalar
8. a car traveling 60 mph vector
9. a rock falling at 10 mph vector
10. your age scalar
11. the force of Earth's gravity acting on a moving satellite vector
12. the area of a record rotating on a turntable scalar
13. the length of a vector in the coordinate plane scalar

12-6 LESSON NOTES

NCTM Standards: 1–8

Instructional Resources

- Study Guide Master 12-6
- Practice Master 12-6
- Enrichment Master 12-6
- Assessment and Evaluation Masters, p. 325
- Multicultural Activity Masters, p. 24
- Real-World Applications, 26

Transparency 12-6A contains the 5-Minute Check for this lesson; **Transparency 12-6B** contains a teaching aid for this lesson.

Recommended Pacing	
Standard Pacing	Day 10 of 12
Honors Pacing	Day 9 of 11

1 FOCUS

5-Minute Check
(over Lesson 12-5)

Find the magnitude to the nearest tenth and the direction to the nearest degree for each vector.

1. $\overrightarrow{AB} = (1, 4)$ 4.1, 76°
2. $\overrightarrow{CD} = (2, 2)$ 2.8, 45°

Use the vectors above to solve.

3. $3\overrightarrow{AB}$ 3, 12
4. $\overrightarrow{AB} + \overrightarrow{CD}$ 3, 6

Motivating the Lesson

Hands-On Activity Have each student hold a pencil in the air. Have them try to describe its position. Ask them if they could describe its position using coordinates.

12-6 Coordinates in Space

What YOU'LL LEARN

- To locate a point in space,
- to use the distance and midpoint formulas for points in space, and
- to determine the center and radius of a sphere.

Why IT'S IMPORTANT

A three-dimensional coordinate system is a good way to represent locations in space.

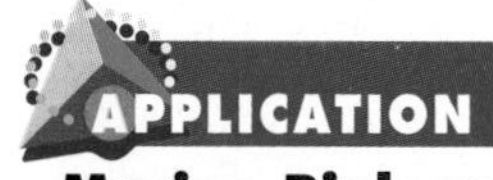

APPLICATION
Marine Biology

Hal Whitehead, his wife Lindy Weilgart, and their three children have an unusual home away from home. Every two years, the family spends months sailing the waters near the Galápagos Islands in search of sperm whales. Hal, Lindy, and the other scientists and students who join the expeditions study the social habits and communications of the whales.

Sperm whales make clicking noises that biologists can use to locate them. A special microphone, called a hydrophone, is used to listen underwater. When whales are located, their location is recorded in a chart. One way they can record an underwater location is by using the latitude, longitude, and depth. They would write an ordered triple like (2°N, 150°W, −1500 ft) to record the location.

In a coordinate plane, the ordered pair for each point has two numbers, or coordinates, to describe its location because a plane has two dimensions. In space, each point requires three numbers, or coordinates, to describe its location because space has three dimensions. In space, the x-, y-, and z-axes are perpendicular to each other.

A point in space is represented by an **ordered triple** of real numbers (x, y, z). In the figure at the right, the ordered triple $(-1, 2, 4)$ locates point P. Notice that a parallelogram is used to help show perspective and convey the idea of the third dimension.

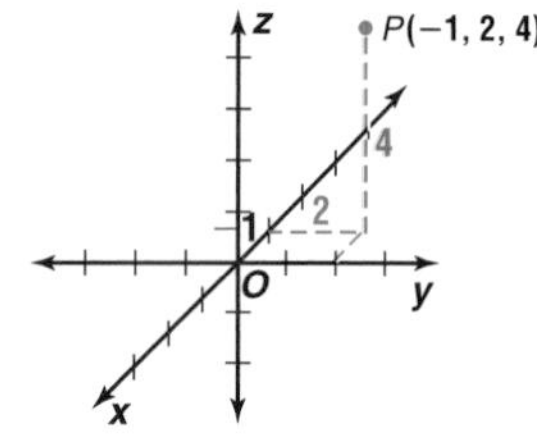

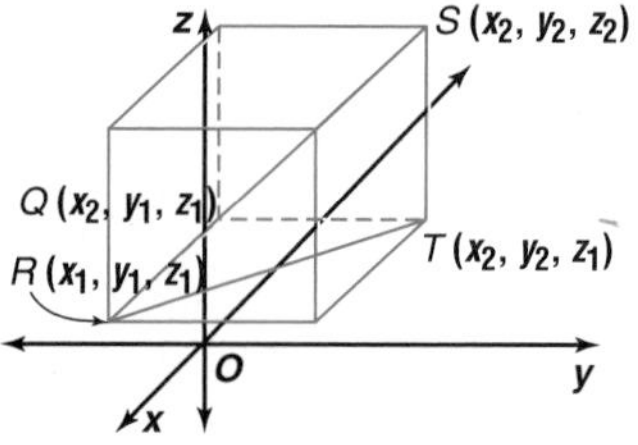

Just as the Pythagorean Theorem can be used to find the formula for the distance between two points in a plane, it can also be used to find the formula for the distance between two points in space.

Since R and T are both in the same plane that is parallel to the xy-plane, the z-coordinates for both points are the same. Since $\triangle QRT$ is a right triangle, the distance between points R and T can be found as follows.

$RT = \sqrt{(x_2 - x_1)^2 + (y_2 - y_1)^2 + (z_1 - z_1)^2}$ or $\sqrt{(x_2 - x_1)^2 + (y_2 - y_1)^2}$

Therefore, $(RT)^2 = (x_2 - x_1)^2 + (y_2 - y_1)^2$.

Likewise, S and T are in the same plane that is parallel to the yz-plane, so the x-coordinates for both points are the same. Since $\triangle RTS$ is a right triangle, the distance between S and T can be found as follows.

CAREER CHOICES

The oceans are the largest unexplored territories left on Earth. Each year **marine biologists** learn more about the plants and animals that live there.

For more information, contact:

National Ocean Industries Association
1050 17th St., NW, Suite 700
Washington, DC 20036

CAREER CHOICES

The deepest ocean trench is the Marianas Trench in the western Pacific Ocean. It is 11,034 feet deep.

$ST = \sqrt{(x_2 - x_2)^2 + (y_2 - y_2)^2 + (z_2 - z_1)^2}$ or $\sqrt{(z_2 - z_1)^2}$
Therefore, $(ST)^2 = (z_2 - z_1)^2$.

$(RS)^2 = (RT)^2 + (TS)^2$	*Pythagorean Theorem*
$(RS)^2 = ((x_2 - x_1)^2 + (y_2 - y_1)^2) + ((z_2 - z_1)^2)$	*Substitution Property (=)*
$RS = \sqrt{(x_2 - x_1)^2 + (y_2 - y_1)^2 + (z_2 - z_1)^2}$	*Take the square root of each side.*

Theorem 12-2

Given two points $A(x_1, y_1, z_1)$ and $B(x_2, y_2, z_2)$ in space, the distance between A and B is given by the following equation.

$$AB = \sqrt{(x_2 - x_1)^2 + (y_2 - y_1)^2 + (z_2 - z_1)^2}$$

This formula is an extension of the distance formula in the two-dimensional coordinate system. The midpoint formula can also be extended to the three-dimensional system.

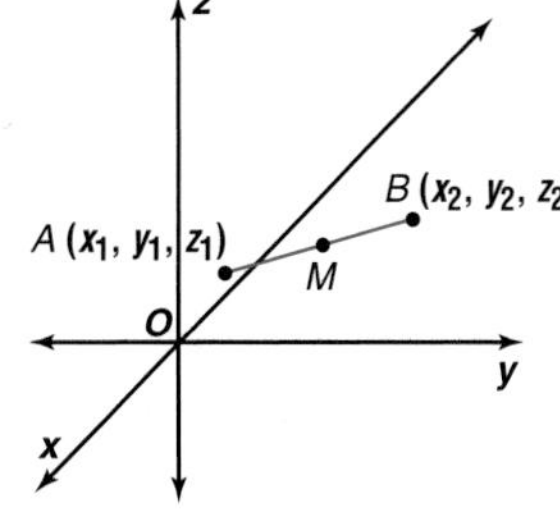

Suppose M is the midpoint of $\overline{AB}$, a segment in space. The midpoint has the following coordinates.

$$\left(\frac{x_1 + x_2}{2}, \frac{y_1 + y_2}{2}, \frac{z_1 + z_2}{2}\right)$$

Example 1 **Consider $F(4, 14, 8)$, $G(1, 2, -1)$, and $H(2, 6, 2)$.**

a. Use the distance formula to prove that the points are collinear.

b. Determine whether H is the midpoint of $\overline{FG}$.

a. Use the distance formula to find the lengths of $\overline{GH}$, $\overline{FH}$, and $\overline{GF}$.

$GH = \sqrt{(1 - 2)^2 + (2 - 6)^2 + (-1 - 2)^2}$
$= \sqrt{26}$

$HF = \sqrt{(4 - 2)^2 + (14 - 6)^2 + (8 - 2)^2}$
$= \sqrt{104}$ or $2\sqrt{26}$

$GF = \sqrt{(1 - 4)^2 + (2 - 14)^2 + (-1 - 8)^2}$
$= \sqrt{234}$ or $3\sqrt{26}$

Since $3\sqrt{26} = \sqrt{26} + 2\sqrt{26}$, $GF = GH + HF$.

By the Segment Addition Postulate, if $GF = GH + HF$, then H is between G and F. Thus, G, F, and H are collinear points.

b. Use the midpoint formula to find the coordinates of the midpoint of $\overline{FG}$. Let $(4, 14, 8)$ be (x_1, y_1, z_1) and $(1, 2, -1)$ be (x_2, y_2, z_2).

$$\left(\frac{4 + 1}{2}, \frac{14 + 2}{2}, \frac{8 + (-1)}{2}\right) = \left(\frac{5}{2}, \frac{16}{2}, \frac{7}{2}\right)$$
$$= (2.5, 8, 3.5)$$

The coordinates of the midpoint are $(2.5, 8, 3.5)$. Thus, H is not the midpoint of $\overline{FG}$.

2 TEACH

Teaching Tip It may be difficult for some students to follow the steps given for finding RS. It may help to have students copy the diagram and use colored pencils to outline the two different right triangles being used.

In-Class Example

For Example 1
Consider $R(-3, 4, 2)$, $S(1, 1, -1)$, and $T(5, -2, -4)$.

a. Use the distance formula to prove that the points are collinear. **$RS = ST = \sqrt{34}$; $RT = 2\sqrt{34}$; by the Segment Addition Postulate, if $RT = RS + ST$, then S is between R and T. Thus, R, S, and T are collinear points.**

b. Determine whether S is the midpoint of $\overline{RT}$. **The coordinates of the midpoint are $(1, 1, -1)$. Thus S is the midpoint of $\overline{RT}$.**

In-Class Examples

For Example 2
Write the equation of a sphere that has a diameter with endpoints at (1, 3, −4) and (7, 9, 6). $(x - 4)^2 + (y - 6)^2 + (z - 1)^2 = 43$

For Example 3
With reference to an airport, an airplane is 4 miles west, 2 miles north, and 5 miles up. Another airplane is 2 miles east, 5 miles south, and 2 miles up. Find the distance between the two airplanes.

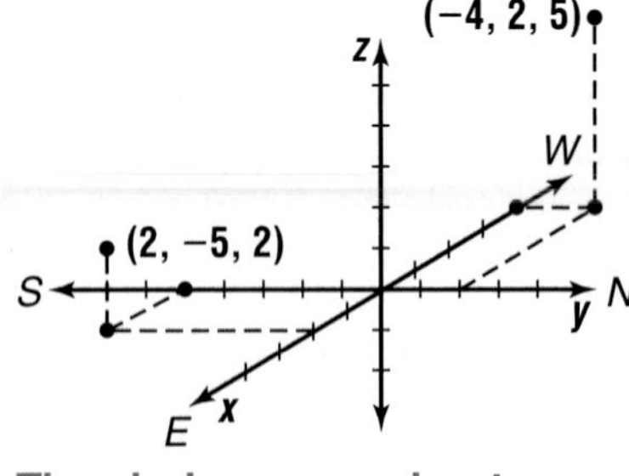

The airplanes are about 9.7 miles apart.

Teaching Tip Remind students that, when finding the radius in Example 2, they can use either given point.

The formula for the equation of a sphere is an extension of the formula for the equation of a circle. The equation of a sphere whose center is at (0, 0, 0) and whose radius is r units long is as follows.

$$x^2 + y^2 + z^2 = r^2$$

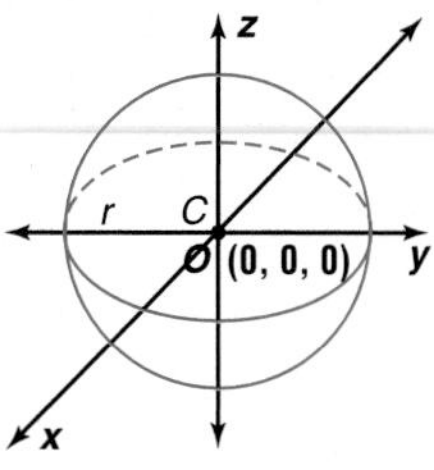

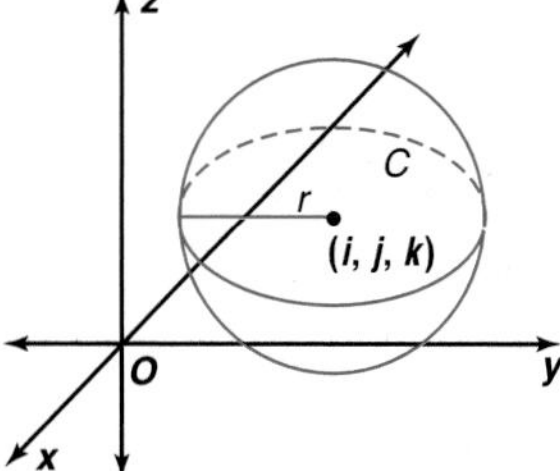

The equation of a sphere whose center is at (i, j, k) and whose radius is r units long is as follows.

$$(x - i)^2 + (y - j)^2 + (z - k)^2 = r^2$$

Example 2

Algebra

Write an equation for the sphere that has a diameter with endpoints at (−5, 10, −2) and (11, 22, 14).

The center of the sphere is the midpoint of the diameter.
First, find the midpoint of the diameter.
Let (−5, 10, −2) be (x_1, y_1, z_1) and (11, 22, 14) be (x_2, y_2, z_2).

$$\left(\frac{-5 + 11}{2}, \frac{10 + 22}{2}, \frac{-2 + 14}{2}\right) = \left(\frac{6}{2}, \frac{32}{2}, \frac{12}{2}\right)$$
$$= (3, 16, 6) \quad \textit{Center of sphere}$$

Next, find the radius.
Let $i = 3$, $j = 16$, and $k = 6$.

$r^2 = (x - i)^2 + (y - j)^2 + (z - k)^2$

$r^2 = (-5 - 3)^2 + (10 - 16)^2 + (-2 - 6)^2$ *Substitute 3 for i, 16 for j, and 6 for k. Substitute −5 for x, 10 for y, and −2 for z.*

$r^2 = (-8)^2 + (-6)^2 + (-8)^2$

$r^2 = 164$

The equation of the sphere is $(x - 3)^2 + (y - 16)^2 + (z - 6)^2 = 164$.

You can solve problems using a three-dimensional coordinate system.

Example 3

Aviation

An airplane at an elevation of 2 miles is 50 miles east and 100 miles north of an airport. Another airplane at an elevation of 2.5 miles is 240 miles west and 140 miles north of the airport. Find the distance between the two planes.

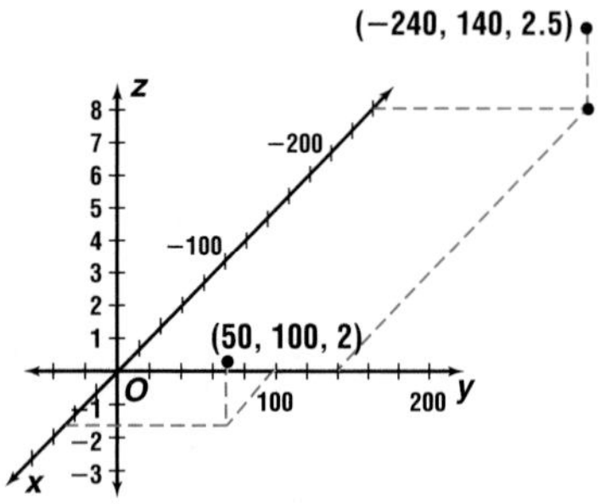

Explore Since the distances are relative to the same airport, the airport can be represented by the origin. The coordinates of the first airplane are (50, 100, 2). The other plane has coordinates (−240, 140, 2.5).

Plan Use Theorem 12–2 to find the distance between the two airplanes.

Solve $\sqrt{(x_2 - x_1)^2 + (y_2 - y_1)^2 + (z_2 - z_1)^2}$

$= \sqrt{(50 - (-240))^2 + (100 - 140)^2 + (2 - 2.5)^2}$

$= \sqrt{290^2 + (-40)^2 + (-0.5)^2}$

$= \sqrt{85{,}700.25} \approx 292.75$

The airplanes are about 292.75 miles apart.

Examine Model the problem on a three-dimensional coordinate system. Draw a rectangular solid having the coordinates of the two planes as vertices. Check the solution by using the Pythagorean Theorem.

CHECK FOR UNDERSTANDING

Communicating Mathematics

Study the lesson. Then complete the following. 1–3. See margin.

1. **Explain** how to locate the point at $(-3, 1, -2)$ in space. Then sketch the point on a three-dimensional coordinate system.
2. **Compare** the midpoint and distance formulas in space to the midpoint and distance formulas in a plane.
3. **Describe** the sphere whose equation is $(x + 2)^2 + (y + 4)^2 + (z - 4)^2 = 25$.

Guided Practice

Plot each point in a three-dimensional coordinate system.

4. $J(2, 5, 6)$ See margin.
5. $L(4, 2, -3)$ See margin.

Determine the distance between each pair of points and determine the coordinates of the midpoint of the segment containing them.

6. $R(0, 0, 4)$ and $S(0, -1, 6)$ $\sqrt{5} \approx 2.24$; $(0, -0.5, 5)$
7. $A(19, -8, 40)$ and $B(20, -2, 18)$ $\sqrt{521} \approx 22.83$; $(19.5, -5, 29)$

Identify each of the following as *true* or *false*. If the statement is false, explain why.

8. The point at $(0, 0, 5)$ lies on the z-axis. true
9. The distance between the points at $(0, 0, 0)$ and (x, y, z) can be expressed as $\sqrt{x^2 + y^2 + z^2}$. true
10. The point at $(2, 1, 5)$ lies on the sphere whose equation is $(x + 2)^2 + (y - 4)^2 + (z - 1)^2 = 25$. False; the point does not satisfy the equation.
11. Determine the coordinates of the center and the measure of the radius for the sphere with equation $(x - 5)^2 + (y + 2)^2 + z^2 = 36$. $(5, -2, 0)$, 6

Write the equation of the sphere using the given information.

12. center at $(-1, 2, 5)$, radius of 10 $(x + 1)^2 + (y - 2)^2 + (z - 5)^2 = 100$
13. $\overline{AB}$ is a diameter where $A(3, -5, 18)$ and $B(-7, 11, 32)$

13. $(x + 2)^2 + (y - 3)^2 + (z - 25)^2 = 138$

Reteaching

Using Comparison Review the formulas developed for finding the distance between two points in the coordinate plane and the equation for a circle in the coordinate plane. Then present the formulas developed for the same general concepts in this lesson for 3-space. Compare the formulas. Discuss how they are similar and how they differ.

3 PRACTICE/APPLY

Check for Understanding

Exercises 1–14 are designed to help you assess your students' understanding through reading, writing, speaking, and modeling. You should work through Exercises 1–3 with your students and then monitor their work on Exercises 4–14.

Additional Answers

1. Sample answer: First find -3 on the x-axis, 1 on the y-axis, and -2 on the z-axis. Then imagine a plane perpendicular to the x-axis at -3 and planes perpendicular to the y- and z-axes at 1 and -2. The three planes will intersect at $(-3, 1, -2)$. See students' work.
2. Sample answer: The midpoint formula in a plane is one-half the sum of the x and y values. The midpoint formula in space is one-half the sum of the x, y, and z values. The distance formula in a plane is the square root of the sum of the squares of the differences of the x and y values. The distance formula in space is the square root of the sum of the squares of the differences of the x, y, and z values.
3. The sphere has its center at $(-2, -4, 4)$ with a radius of 5.
4.

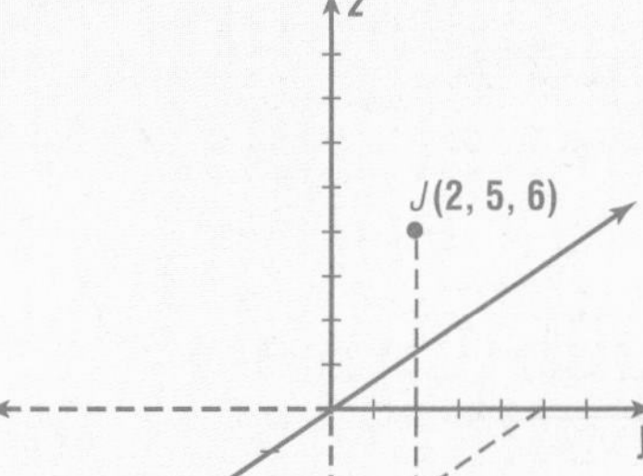

5.

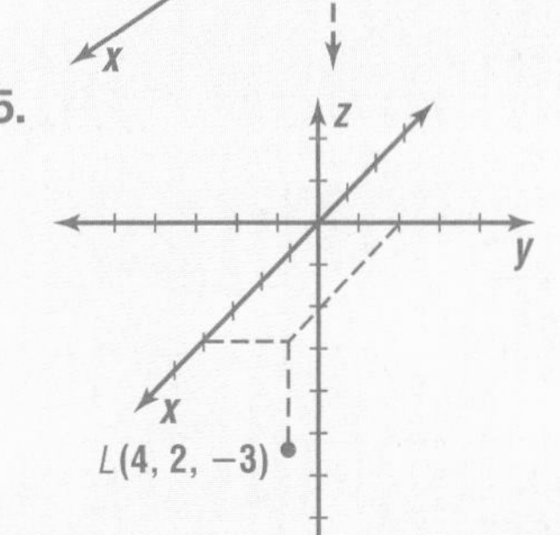

Assignment Guide

Core (with proof): 15–51 odd, 53–63
Core (informal): 15–45 odd, 49, 51, 53, 55–63
Enriched: 16–48 even, 49–63

For **Extra Practice**, see p. 789.

The red A, B, and C flags, printed only in the Teacher's Wraparound Edition, indicate the level of difficulty of the exercises.

Additional Answers

38. $(x-1)^2 + (y-1)^2 + (z-22)^2 = 350$

42. $BA = \sqrt{142}$, $AC = \sqrt{142}$, $BC = \sqrt{568}$ or $2\sqrt{142}$, so $BA + AC = BC$

43. $AB = 11$, $AC = 11$ so $\triangle ABC$ is isosceles. If $AB^2 + AC^2 = BC^2$, $\triangle ABC$ will be a right triangle. $BC = \sqrt{242}$ and $11^2 + 11^2 = 242$, so $\triangle ABC$ is a right triangle.

Study Guide Masters, p. 77

NAME____________ DATE________

Study Guide

Student Edition Pages 680–686

Coordinates in Space

In space, you need an **ordered triple** of real numbers to describe the location of a point. In space, the *x*-, *y*-, and *z*-axes are perpendicular to each other. In the figure at the right, the ordered triple (-2, -3, 1) locates point *P*.

Given two points $A(x_1, y_1, z_1)$ and $B(x_2, y_2, z_2)$ in space, the distance between *A* and *B* is given by the following equation.

$$AB = \sqrt{(x_2 - x_1)^2 + (y_2 - y_1)^2 + (z_2 - z_1)^2}$$

The midpoint of $\overline{AB}$ has coordinates $\left(\frac{x_1 + x_2}{2}, \frac{y_1 + y_2}{2}, \frac{z_1 + z_2}{2}\right)$.

Example: Find *AB* given *A* (3, 2, -5) and *B*(-4, 6, 9). Then find the coordinates of the midpoint of $\overline{AB}$.

$$AB = \sqrt{(-4-3)^2 + (6-2)^2 + (9-(-5))^2}$$
$$AB = \sqrt{(-7)^2 + (4)^2 + (14)^2}$$
$$AB = \sqrt{49 + 16 + 196}$$
$$AB = \sqrt{261} \approx 16.2$$
$$\text{midpoint of } \overline{AB} = \left(\frac{3 + (-4)}{2}, \frac{2+6}{2}, \frac{-5+9}{2}\right) = \left(-\frac{1}{2}, 4, 2\right)$$

Plot each point in a three-dimensional coordinate system.

1. *S*(0, 4, 2)

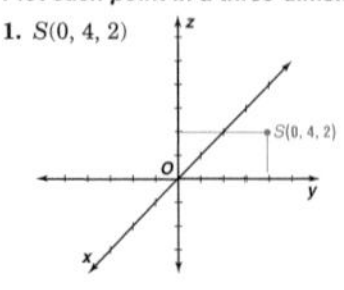

2. *T*(5, -3, 6)

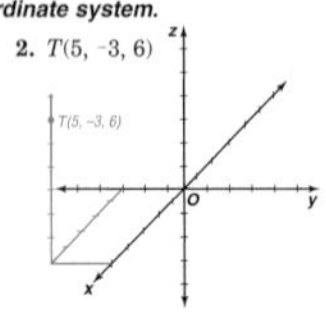

Determine the distance between each pair of points.

3. *A*(0, 7, -4) and *B*(-2, 8, 3) $\sqrt{54} \approx 7.3$
4. *C*(-7, 6, 5) and *D*(10, 2, -5) $\sqrt{405} \approx 20.1$

Determine the coordinates of the midpoint of each line segment whose endpoints are given.

5. *E*(6, 10, -8) and *F*(1, 20, 8) $\left(\frac{7}{2}, 15, 0\right)$
6. *G*(0, -8, 4) and *H*(8, 4, 3) $\left(4, -2, \frac{7}{2}\right)$

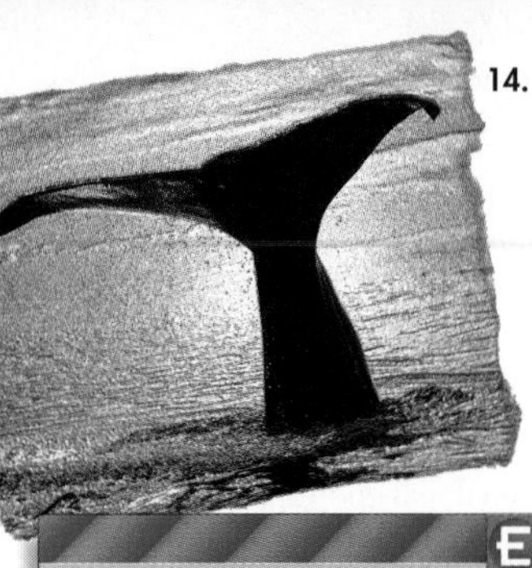

14. **Marine Biology** A group of sperm whales was found feeding at a location of (3°N, 152°W, −1500 ft). The next day a group is found at (2°N, 154°W, −1300 ft). The distance between latitude lines that differ by one degree is about 60 nautical miles. Near the equator, the distance between longitude lines that differ by one degree is about 69 nautical miles. A nautical mile is 6076.115 feet.
    a. Write an ordered pair for each group sighted in terms of nautical miles from (0, 0, 0). **(180, 10,488, −0.25); (120, 10,626, −0.21)**
    b. Find the distance between the locations of the groups. **about 150.5**

EXERCISES

Practice

15–20. See Solutions Manual.

A **Plot each point in a three-dimensional coordinate system.**

15. $A(2, 1, 5)$
16. $B(3, 5, 4)$
17. $C(-6, 0, 0)$
18. $D(-1, 2, -2)$
19. $E(4, -2, 8)$
20. $F(-3, -4, -5)$

Determine the distance between each pair of points, and determine the coordinates of the midpoint of the segment connecting them.

21. $J(3, -7, 0)$ and $K(5, 1, 7)$ **$\sqrt{117} \approx 10.8$ units; $(4, -3, 3.5)$**
22. $L(17, -22, -41)$ and $M(-19, 34, -53)$ **$\sqrt{4576} \approx 67.6$; $(-1, 6, -47)$**
23. $A(6, 1, 1)$ and $B(9, 0, 1)$ **$\sqrt{10} \approx 3.16$ units; $(7.5, 0.5, 1)$**
24. $C(4, -8, 12)$ and $D(7, 20, 18)$ **$\sqrt{829} \approx 28.8$ units; $(5.5, 6, 15)$**
25. $E(3, 7, -1)$ and $F(5, 7, 2)$ **$\sqrt{13} \approx 3.6$ units; $(4, 7, 0.5)$**
26. $G(2, 2, 2)$ and $H(-25, 4, 18)$ **$\sqrt{989} \approx 31.4$ units; $(-11.5, 3, 10)$**

Identify each of the following as *true* or *false*. If the statement is false, explain why. 28. False, it is outside of the sphere.

27. Every point on the *yz*-plane has coordinates (c, y, z) for any real number *c*. **False; $c = 0$; the points will be of the form $(0, y, z)$.**
28. The point at $(1, 8, -12)$ is inside the sphere $(x-3)^2 + (y-5)^2 + (z+2)^2 = 9$.
29. The intersection of the *xy*-plane, the *yz*-plane, and the *xz*-plane is the point at $(0, 0, 0)$. **true**
30. The set of points in space 5 units from the point at $(1, -1, 3)$ can be described by the equation $(x-1)^2 + (y+1)^2 + (z-3)^2 = 25$. **true**
31. The set of points equidistant from $A(2, 5, 8)$ and $B(-3, 4, 7)$ is a line that is the perpendicular bisector of $\overline{AB}$. **False; it is a plane that contains that line.**

Determine the coordinates of the center and the measure of the radius for each sphere whose equation is given.

32. $x^2 + (y-3)^2 + (z+8)^2 = 81$ **(0, 3, −8), 9**
33. $(x-5)^2 + (y+4)^2 + (z-10)^2 = 9$ **(5, −4, 10), 3**
34. $x^2 + y^2 + (z-3)^2 = 49$ **(0, 0, 3), 7**
35. $(x+4)^2 + (y-2)^2 + (z+12)^2 = 18$ **$(-4, 2, -12)$, $\sqrt{18} \approx 4.24$**

Write the equation of the sphere using the given information.

36. The center is at $(-5, 11, -3)$, and the radius is 4. **$(x+5)^2 + (y-11)^2 + (z+3)^2 = 16$**
37. The center is at $(-2, 3, -4)$, and it contains the point at $(5, -1, -1)$. **$(x+2)^2 + (y-3)^2 + (z+4)^2 = 74$**
38. The diameter has endpoints at $(14, -8, 32)$ and $(-12, 10, 12)$. **See margin.**
39. It is concentric with the sphere with equation $(x+5)^2 + (y-4)^2 + (z-19)^2 = 9$, and it has a radius of 6 units. **$(x+5)^2 + (y-4)^2 + (z-19)^2 = 36$**

40. It is inscribed in a cube determined by the points at (0, 0, 0), (4, 0, 0), (0, 4, 0), and (4, 4, 4). **$(x-2)^2 + (y-2)^2 + (z-2)^2 = 4$**
41. Find the perimeter of a triangle with vertices $A(-1, 3, 2)$, $B(0, 2, 4)$, and $C(-2, 0, 3)$. **$\sqrt{6} + \sqrt{11} + 3$ or about 8.77 units**
42. Show that $A(2, -1, 3)$, $B(-3, 5, -6)$, and $C(7, -7, 12)$ are collinear. **See margin.**

44a. Each side measures $\sqrt{24}$ units.

45. 150 units^2, 125 units^3

43. Show that $\triangle ABC$ is an isosceles right triangle if the vertices are $A(3, 2, -3)$, $B(5, 8, 6)$, and $C(-3, -5, 3)$. See margin.

44. Consider $R(6, 1, 3)$, $S(4, 5, 5)$, and $T(2, 3, 1)$.
   a. Determine the measures of $\overline{RS}$, $\overline{ST}$, and $\overline{RT}$.
   b. If $\overline{RS}$, $\overline{ST}$, and $\overline{RT}$ are the sides of a triangle, what type of triangle is $\triangle RST$? $\triangle RST$ is an equilateral triangle.

45. Find the surface area and volume of the rectangular prism at the right.

46. Find z if the distance between $R(5, 4, -1)$ and $S(3, -2, z)$ is 7. 2 or -4

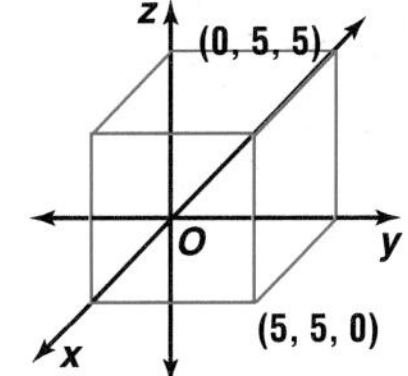

47. Write a coordinate proof to prove that the diagonals of a rectangular prism are congruent and bisect each other. See margin.

48. The center of a sphere is at $(4, -2, 6)$, and the endpoint of a diameter is at $(8, 10, -2)$. What are the coordinates of the other endpoint of the diameter? $(0, -14, 14)$

Critical Thinking

49. The sphere whose equation is $(x - 1)^2 + (y - 4)^2 + (z - 8)^2 = 25$ is inscribed in a cube. Determine the coordinates of the cube and make a drawing of the figure. $(6, -1, 3)$, $(6, 9, 3)$, $(-4, 9, 3)$, $(-4, -1, 3)$, $(6, -1, 13)$, $(6, 9, 13)$, $(-4, 9, 13)$, $(-4, -1, 13)$; see Solutions Manual for drawing.

Applications and Problem Solving

50. **Statistics** Each year *Money* magazine collects data on cities and uses that data to rate the cities in terms of the "best" places to live. The data collected on the unemployment rate, the property taxes, and the number of robberies per 100,000 people for four cities are given in the table. Suppose the "best" city was the city that was the closest to 0 in each of the categories.

City	Unemployment Rate	Average Property Tax	Violent Crimes per 100,000 people
Austin, TX	3.5%	2800	539
Gainesville, FL	3.8%	1875	1377
Rochester, MN	2.5%	1300	176
Seattle, WA	4.7%	2250	542

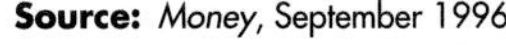

Source: *Money*, September 1996

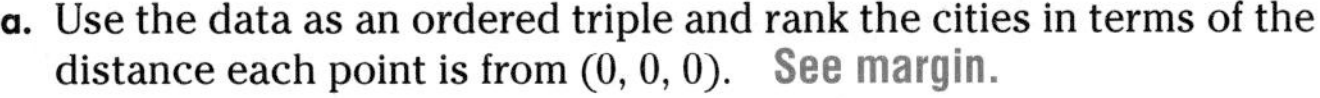

   a. Use the data as an ordered triple and rank the cities in terms of the distance each point is from $(0, 0, 0)$. See margin.
   b. Do you think this method of ranking cities is fair? Explain your reasoning. See margin.

51. **Recreation** Two children are playing a three-dimensional tic-tac-toe game. Three Xs or three Os in any row wins the game. The positions of the Xs are at $(2, 1, 1)$ and $(2, 3, 3)$, and the position of the O is at $(1, 2, 3)$. Where should the next O be placed? Explain your answer. See margin.

52. **Sports** Two hot-air balloons take off from the same location. One hot-air balloon is 8 miles west and 10 miles south of the take-off point and 0.3 mile above the ground. The other hot-air balloon is 4 miles west and 8 miles south of the take-off point and 0.4 mile above the ground. Find the distance between the hot-air balloons. $\sqrt{20.01} \approx 4.5$ miles

Additional Answers

50a. Rochester: ≈ 1311.9; Seattle: ≈ 2314.4; Gainesville: ≈ 2326.3; Austin: ≈ 2851.4

50b. The largest contributor to the rank is the numbers in the thousands, so the percent and the tax don't really have much effect on the outcomes.

51. $(2, 2, 2)$ to block a row of Xs

Additional Answer

47. The vertices are $A(a, 0, 0)$, $B(a, b, 0)$, $C(0, b, 0)$, $D(0, 0, 0)$, $E(a, 0, c)$, $F(a, b, c)$, $G(0, b, c)$, and $H(0, 0, c)$.

$AG = \sqrt{(a - 0)^2 + (0 - b)^2 + (0 - c)^2} = \sqrt{a^2 + b^2 + c^2}$

$BH = \sqrt{(a - 0)^2 + (b - 0)^2 + (0 - c)^2} = \sqrt{a^2 + b^2 + c^2}$

$CE = \sqrt{(0 - a)^2 + (b - 0)^2 + (0 - c)^2} = \sqrt{a^2 + b^2 + c^2}$

$DF = \sqrt{(0 - a)^2 + (0 - b)^2 + (0 - c)^2} = \sqrt{a^2 + b^2 + c^2}$

$AG = BH = CE = DF$ and $\overline{AG} \cong \overline{BH} \cong \overline{CE} \cong \overline{DF}$

Midpoint of $\overline{AG}$ is $\left(\frac{a+0}{2}, \frac{0+b}{2}, \frac{0+c}{2}\right)$ or $\left(\frac{a}{2}, \frac{b}{2}, \frac{c}{2}\right)$

Midpoint of $\overline{CE}$ is $\left(\frac{0+a}{2}, \frac{b+0}{2}, \frac{0+c}{2}\right)$ or $\left(\frac{a}{2}, \frac{b}{2}, \frac{c}{2}\right)$

Midpoint of $\overline{BH}$ is $\left(\frac{a+0}{2}, \frac{b+0}{2}, \frac{0+c}{2}\right)$ or $\left(\frac{a}{2}, \frac{b}{2}, \frac{c}{2}\right)$

Midpoint of $\overline{DF}$ is $\left(\frac{0+a}{2}, \frac{0+b}{2}, \frac{0+c}{2}\right)$ or $\left(\frac{a}{2}, \frac{b}{2}, \frac{c}{2}\right)$

$\overline{AG}$, $\overline{CE}$, $\overline{BH}$, and $\overline{DF}$ all bisect each other.

Practice Masters, p. 77

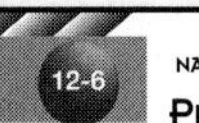

12-6 NAME________ DATE________

Practice — Student Edition Pages 680–686

Coordinates in Space

Determine the distance between each pair of points.

1. $A(0, 0, 0)$ and $B(1, 2, 3)$ $\sqrt{14} \approx 3.7$
2. $C(4, -2, 3)$ and $D(0, 2, 0)$ $\sqrt{41} \approx 6.4$
3. $E(1, -2, 5)$ and $F(1, 2, 5)$ 4
4. $P(0, 1, 0)$ and $Q(-1, 0, 1)$ $\sqrt{3} \approx 1.7$

Determine the coordinates of the midpoint of each line segment whose endpoints are given.

5. $A(0, 0, 4)$, $B(4, -6, 6)$ $(2, -3, 5)$
6. $C(-1, 2, 4)$, $D(3, -6, 8)$ $(1, -2, 6)$
7. $E(-2, -3, 6)$, $F(0, -6, 8)$ $\left(-1, -\frac{9}{2}, 7\right)$
8. $G\left(-1, 5, \frac{3}{2}\right)$, $H\left(1, -5, \frac{1}{2}\right)$ $(0, 0, 1)$

Write an equation of the sphere given the coordinates of the center and the measure of the radius.

9. $C(0, -3, 1)$, $r = 3$ $x^2 + (y + 3)^2 + (z - 1)^2 = 9$
10. $C(-2, 1, 3)$, $r = 1\frac{1}{2}$ $(x + 2)^2 + (y - 1)^2 + (z - 3)^2 = \frac{9}{4}$
11. $C(3, 0, -1)$, $r = \frac{\sqrt{2}}{2}$ $(x - 3)^2 + y^2 + (z + 1)^2 = \frac{1}{2}$
12. $C(5, 5, 5)$, $r = 7$ $(x - 5)^2 + (y - 5)^2 + (z - 5)^2 = 49$
13. Find the perimeter of a triangle whose vertices are $A(0, 2, 1)$, $B(-2, 2, 6)$, and $C(4, 2, -2)$. $15 + \sqrt{29}$ units

The diameter of a sphere has endpoints $A(-3, 2, 4)$ and $B(1, -6, 5)$.

14. Determine the center of the sphere. $(-1, -2, 4.5)$
15. Determine the radius of the sphere. 4.5 units
16. Write the equation of the sphere. $(x + 1)^2 + (y + 2)^2 + (z - 4.5)^2 = 20.25$
17. Find the surface area of the sphere. 81π square units or about 254.5 square units

4 ASSESS

Closing Activity

Speaking Have students explain in their own words how to find the equation of a sphere, given the coordinates of its center and its radius, and why that procedure works. **Every point on a sphere is the same distance, *r*, from the center. Therefore, use the distance formula for three dimensions to find the equation of the sphere.**

Chapter 12 Quiz D (Lesson 12-6) is available in the *Assessment and Evaluation Masters*, p. 325.

Enrichment Masters, p. 77

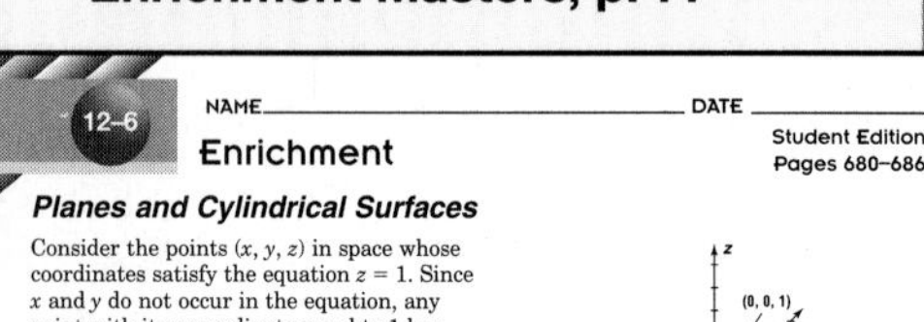

12-6 NAME________ DATE________

Enrichment

Student Edition Pages 680–686

Planes and Cylindrical Surfaces

Consider the points (x, y, z) in space whose coordinates satisfy the equation $z = 1$. Since x and y do not occur in the equation, any point with its z-coordinate equal to 1 has coordinates that satisfy the equation. These are the points in the plane 1 unit above the xy-plane. This plane is perpendicular to the z-axis at (0, 0, 1).

Next consider the points (x, y, z) whose coordinates satisfy $x^2 + y^2 = 16$. In the xy-plane, all points on the circle with center (0, 0, 0) and radius 4 have coordinates that satisfy the equation. In the plane perpendicular to the z-axis at $(0, 0, k)$, the points that satisfy the equation are those on the circle with center $(0, 0, k)$ and radius 4. The graph in space of $x^2 + y^2 = 16$ is an infinite cylindrical surface whose axis is the z-axis and whose radius is 4.

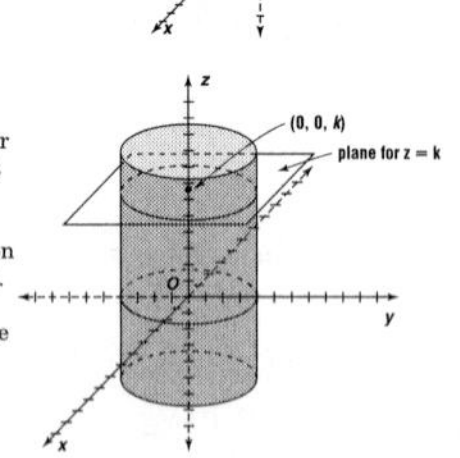

Describe the graph in space of each equation. You may find it helpful to make sketches on a separate sheet.

1. $x = 5$ the plane perpendicular to the x-axis at (5, 0, 0)
2. $y = -2$ the plane perpendicular to the y-axis at (0, −2, 0)
3. $x + y = 7$ the plane parallel to the z-axis and containing the line through (0, 7, 0) and (7, 0, 0)
4. $z^2 + y^2 = 25$ the infinite cylindrical surface whose axis is the x-axis and whose radius is 5
5. $(x - 2)^2 + (y - 5)^2 = 1$ the infinite cylindrical surface whose axis is the line parallel to the z-axis and passing through (2, 5, 0) and whose radius is 1
6. $x^2 + y^2 + z^2 = 0$ the point (0, 0, 0)

Mixed Review

53. $|\overleftrightarrow{LN}| = \sqrt{85} \approx 9.2$; 13°

54. See Solutions Manual.

PBS

Source: Courtesy, WOSU

59. $288\sqrt{2}$ or about 407.3 m²

53. Sketch $\overleftrightarrow{LN}$ if L is at (−3, 5) and N is at (6, 7). Then find the magnitude of $\overleftrightarrow{LN}$ to the nearest tenth and the direction to the nearest degree. (Lesson 12–5)

54. Position and label a right triangle on the coordinate plane. Then write a coordinate proof to show that the measure of the median to the hypotenuse of a right triangle is half of the measure of the hypotenuse. (Lesson 12–4)

55. Communication There are currently 349 public television stations in the United States. About 11 new stations begin broadcasting each year. (Lesson 12–2) **a. $y = 11x + 349$**

a. Write an equation that represents the total number of public televisions stations x years from now if the rate of increase stays the same.

b. Approximately how many public television stations will there be in fifteen years if the rate of increase stays the same? **514**

56. If the volume of a sphere is 972π cubic centimeters, what is the diameter of the sphere? (Lesson 11–7) **18 cm**

57. The diameter of the base of a right circular cone measures 6.4 millimeters, and the slant height measures 5.2 millimeters. Find the lateral area. (Lesson 11–4) **52.3 mm²**

58. A circle is inscribed in a square whose diagonal is 18 feet long. Find the circumference and the area of the circle. (Lesson 10–5) **$9\sqrt{2}\pi$ ft; $\frac{81}{2}\pi$ ft²**

59. The sides of a parallelogram are 18 and 32 meters long. One angle of the parallelogram is 135°. What is the area of the parallelogram? (Lesson 10–4)

60. The sum of the measures of eight interior angles of a convex nonagon is 1210. What is the measure of the ninth angle? (Lesson 10–1) **50**

61. Suppose $\triangle ABC \sim \triangle DEF$. If $AC = 30$, $DF = 18$, and $AB = 7.5$, find DE. (Lesson 7–3) **4.5**

INTEGRATION Algebra

62. Factor $v^2 - 16$. **$(v - 4)(v + 4)$**

63. Solve $y + 3 = -15$. **−18**

WORKING ON THE In·ves·ti·ga·tion

Refer to the Investigation on pages 642–643.

Making believable movies with computer graphics requires drawing three-dimensional objects on a two-dimensional surface. Computers begin perspective drawings by finding the three-dimensional coordinates of the object being drawn. Then algebra is used to transform these to two-dimensional coordinates. The image created is called a projection. The formulas $X = x(-\cos a°) + y$ and $Y = x(-\sin a°) + z$ will draw a projection in which the x-axis is horizontal, the y-axis is vertical, and the z-axis is drawn at an angle of $a°$ with the x-axis. If the three-dimensional coordinates of a point are (x, y, z), then the projected coordinates are (X, Y).

A cube has vertices $A(5, 0, 5)$, $B(5, 5, 5)$, $C(5, 5, 0)$, $D(5, 0, 0)$, $E(0, 0, 5)$, $F(0, 5, 5)$, $G(0, 5, 0)$, and $H(0, 0, 0)$.

1. Draw the cube using a projection with $a = 45°$.
2. Then create a drawing of an image for your movie using a projection.

Add the results of your work to your Investigation Folder.

Extension

Problem Solving Find the center and the radius of the sphere whose equation is $x^2 + y^2 - 4y + z^2 + 2z = 11$. **(0, 2, −1), $r = 4$**

In·ves·ti·ga·tion

Working on the Investigation

The Investigation on pages 642–643 is designed to be a long-term project that is completed over several days or weeks. Encourage students to keep their materials in their Investigation Folder as they work on the Investigation.

CHAPTER 12 HIGHLIGHTS

VOCABULARY

After completing this chapter, you should be able to define each term, property, or phrase and give an example or two of each.

Geometry

- coordinate proof (p. 666)
- direction of a vector (p. 673)
- magnitude of a vector (p. 673)
- ordered triple (p. 680)
- parallelogram law (p. 675)
- resultant (p. 675)
- standard position (p. 673)
- vector (p. 673)

Algebra

- column matrix (p. 676)
- function (p. 647)
- intercepts method (p. 646)
- linear equation (p. 646)
- point-slope form (p. 653)
- scalar multiplication (p. 674)
- slope-intercept form (p. 647)
- standard form (p. 646)

Statistics

- scatter plot (p. 660)

Problem Solving

- write an equation (p. 655)

UNDERSTANDING AND USING THE VOCABULARY

Choose the term from the list above that best completes each statement. 7. scatter plot

1. An important part of _?_ is the placement of the figure on the coordinate plane. coordinate proof
2. Multiplying a vector by a constant is called _?_. scalar multiplication
3. The equation $y = -3x + 5$ is written in _?_. slope-intercept form
4. The _?_ can be found by using the distance formula. magnitude of a vector
5. The equation $4x + 5y = -5$ is written in _?_. standard form
6. The _?_ can be used to add vectors. column matrix
7. The equation of the line suggested by the points on a(n) _?_ can be used to make predictions.
8. A point in space is represented by a(n) _?_ of real numbers. ordered triple
9. An equation whose graph is a straight line is called a(n) _?_. linear equation
10. The equation $y + 6 = -3(x - 1)$ is written in _?_. point-slope form

Using the CHAPTER HIGHLIGHTS

The Chapter Highlights begins with a listing of the new terms, properties, and phrases that were introduced in this chapter. Have students define each term and provide an example or two of it, if appropriate.

Assessment and Evaluation Masters, pp. 311–312

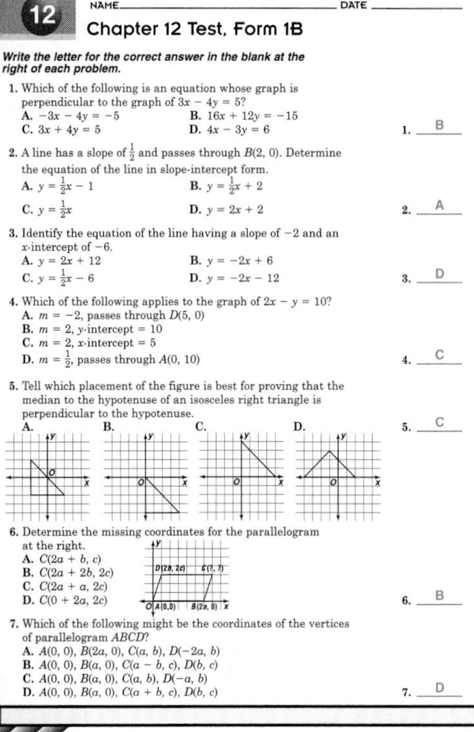

12 NAME ______ DATE ______

Chapter 12 Test, Form 1B

Write the letter for the correct answer in the blank at the right of each problem.

1. Which of the following is an equation whose graph is perpendicular to the graph of $3x - 4y = 5$?
   A. $-3x - 4y = -5$ B. $16x + 12y = -15$
   C. $3x + 4y = 5$ D. $4x - 3y = 6$ 1. B
2. A line has a slope of $\frac{1}{2}$ and passes through $B(2, 0)$. Determine the equation of the line in slope-intercept form.
   A. $y = \frac{1}{2}x - 1$ B. $y = \frac{1}{2}x + 2$
   C. $y = \frac{1}{2}x$ D. $y = 2x + 2$ 2. A
3. Identify the equation of the line having a slope of -2 and an x-intercept of -6.
   A. $y = 2x + 12$ B. $y = -2x + 6$
   C. $y = \frac{1}{2}x - 6$ D. $y = -2x - 12$ 3. D
4. Which of the following applies to the graph of $2x - y = 10$?
   A. $m = -2$, passes through $D(5, 0)$
   B. $m = 2$, y-intercept $= 10$
   C. $m = 2$, x-intercept $= 5$
   D. $m = \frac{1}{2}$, passes through $A(0, 10)$ 4. C
5. Tell which placement of the figure is best for proving that the median to the hypotenuse of an isosceles right triangle is perpendicular to the hypotenuse.
   A. B. C. D. 5. C
6. Determine the missing coordinates for the parallelogram at the right.
   A. $C(2a + b, c)$
   B. $C(2a + 2b, 2c)$
   C. $C(2a + a, 2c)$
   D. $C(0 + 2a, 2c)$ 6. B
7. Which of the following might be the coordinates of the vertices of parallelogram $ABCD$?
   A. $A(0, 0)$, $B(2a, 0)$, $C(a, b)$, $D(-2a, b)$
   B. $A(0, 0)$, $B(a, 0)$, $C(a - b, c)$, $D(b, c)$
   C. $A(0, 0)$, $B(a, 0)$, $C(a, b)$, $D(-a, b)$
   D. $A(0, 0)$, $B(a, 0)$, $C(a + b, c)$, $D(b, c)$ 7. D

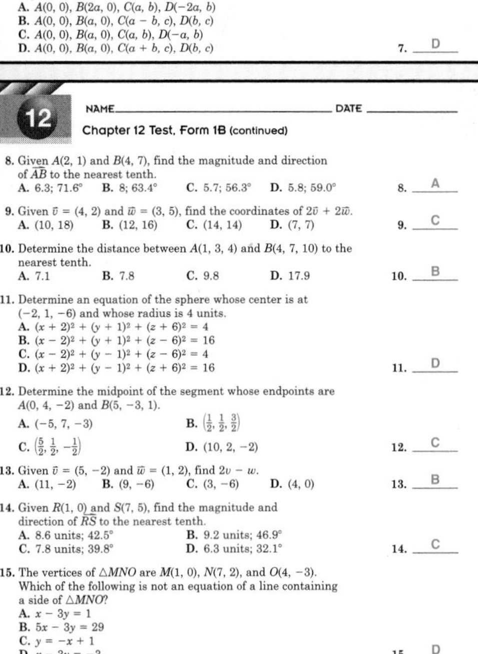

12 NAME ______ DATE ______

Chapter 12 Test, Form 1B (continued)

8. Given $A(2, 1)$ and $B(4, 7)$, find the magnitude and direction of $\overline{AB}$ to the nearest tenth.
   A. 6.3; 71.6° B. 8; 63.4° C. 5.7; 56.3° D. 5.8; 59.0° 8. A
9. Given $\vec{v} = (4, 2)$ and $\vec{w} = (3, 5)$, find the coordinates of $2\vec{v} + 2\vec{w}$.
   A. (10, 18) B. (12, 16) C. (14, 14) D. (7, 7) 9. C
10. Determine the distance between $A(1, 3, 4)$ and $B(4, 7, 10)$ to the nearest tenth.
    A. 7.1 B. 7.8 C. 9.8 D. 17.9 10. B
11. Determine an equation of the sphere whose center is at $(-2, 1, -6)$ and whose radius is 4 units.
    A. $(x + 2)^2 + (y + 1)^2 + (z + 6)^2 = 4$
    B. $(x - 2)^2 + (y + 1)^2 + (z - 6)^2 = 16$
    C. $(x - 2)^2 + (y - 1)^2 + (z - 6)^2 = 4$
    D. $(x + 2)^2 + (y - 1)^2 + (z + 6)^2 = 16$ 11. D
12. Determine the midpoint of the segment whose endpoints are $A(0, 4, -2)$ and $B(5, -3, 1)$.
    A. $(-5, 7, -3)$ B. $\left(\frac{1}{2}, \frac{1}{2}, \frac{3}{2}\right)$
    C. $\left(\frac{5}{2}, \frac{1}{2}, -\frac{1}{2}\right)$ D. $(10, 2, -2)$ 12. C
13. Given $\vec{v} = (5, -2)$ and $\vec{w} = (1, 2)$, find $2v - w$.
    A. (11, −2) B. (9, −6) C. (3, −6) D. (4, 0) 13. B
14. Given $R(1, 0)$ and $S(7, 5)$, find the magnitude and direction of $\overline{RS}$ to the nearest tenth.
    A. 8.6 units; 42.5° B. 9.2 units; 46.9°
    C. 7.8 units; 39.8° D. 6.3 units; 32.1° 14. C
15. The vertices of $\triangle MNO$ are $M(1, 0)$, $N(7, 2)$, and $O(4, -3)$. Which of the following is not an equation of a line containing a side of $\triangle MNO$?
    A. $x - 3y = 1$
    B. $5x - 3y = 29$
    C. $y = -x + 1$
    D. $x - 3y = -3$ 15. D

Bonus

$\triangle WXY$ has vertices $W(-3, -5)$, $X(1, 7)$, and $Y(7, 3)$. Find the measure of median $\overline{YT}$.
A. $\sqrt{40}$ B. $\sqrt{52}$ C. $\sqrt{68}$ D. $\sqrt{80}$ Bonus C

Instructional Resources

Three multiple-choice tests and three free-response tests are provided in the *Assessment and Evaluation Masters*. Forms 1A and 2A are for honors pacing, Forms 1B and 2B are for average pacing, and Forms 1C and 2C are for basic pacing. Chapter 12 Test, Form 1B, is shown at the right. Chapter 12 Test, Form 2B, is shown on the next page.

Postulates, Theorems, and Corollaries

A complete list of postulates, theorems, and corollaries begins on page 806.

Using the STUDY GUIDE AND ASSESSMENT

Skills and Concepts Encourage students to refer to the objectives and examples on the left as they complete the review exercises on the right.

Assessment and Evaluation Masters, pp. 317–318

12 NAME ______________________ DATE __________

Chapter 12 Test, Form 2B

1. Graph $y = \frac{1}{2}x + 1$ by using the slope and the y-intercept. Write the slope and the y-intercept.

For Questions 2–5, write the equation in slope-intercept form of the line satisfying the given conditions.

1. $m = \frac{1}{2}$, y-int. $= 1$

2. $m = 5$, y-intercept $= 0$ — 2. $y = 5x$

3. $m = -1$, passes through $C(-2, 1)$ — 3. $y = -x - 1$

4. passes through $F(0, -2)$ and $C(-4, 4)$ — 4. $y = -\frac{3}{2}x - 2$

5. perpendicular to $y = -\frac{1}{4}x - 3$, passes through $H(0, 2)$ — 5. $y = 4x + 2$

6. Name the missing coordinates in terms of the given variables if $EFGH$ is a rectangle. — 6. (a, b)

7. Draw the resultant of the pair of vectors. What law did you use? — 7. Parallelogram Law

8. Given $A(0, -1)$ and $B(6, 2)$, find the magnitude of $\overrightarrow{AB}$ to the nearest tenth. Then find the direction of $\overrightarrow{AB}$ to the nearest degree. — 8. 6.7, 27°

For Questions 9 and 10, given $\vec{w} = (4, -3)$ and $\vec{v} = (6, 1)$, find the coordinates.

9. $\vec{w} + \vec{v}$ — 9. $(10, -2)$

10. $2\vec{w} + 3\vec{v}$ — 10. $(26, -3)$

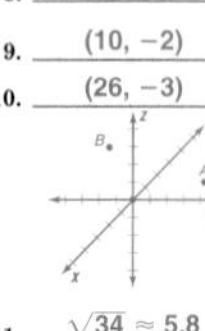

11. Plot points $A(1, 5, 2)$ and $B(4, 2, 6)$ in the three-dimensional coordinate system at the right. Then determine the distance between the points to the nearest tenth. — 11. $\sqrt{34} \approx 5.8$

12 NAME ______________________ DATE __________

Chapter 12 Test, Form 2B (continued)

12. Determine the midpoint of the segment whose endpoints are $C(1, 5, 1)$ and $D(5, 1, 4)$. — 12. $(3, 3, 2.5)$

13. Determine the center and the measure of the radius for the sphere whose equation is $(x + 1)^2 + (y - 3)^2 + (z - 4)^2 = 4$. — 13. $(-1, 3, 4)$, 2

14. Write the equation of a sphere whose center is at $(4, 4, 1)$ and whose radius is 5 units. — 14. $(x - 4)^2 + (y - 4)^2 + (z - 1)^2 = 25$

15. Lawrence and Dublin are 750 miles apart by plane. A jet leaves Lawrence at the same time an airplane leaves Dublin. The jet averages 150 miles per hour faster than the airplane. The two aircraft pass each other in 2.5 hours. How fast is each plane traveling? — 15. plane, 75 mph; jet, 225 mph

16. The length of a rectangular house is 10 feet more than twice its width. Its perimeter is 170 feet. Find its length and width. — 16. $l = 60$ ft, $w = 25$ ft

17. Determine whether $A(-3, 1)$, $B(1, 2)$ and $C(4, 6)$ are collinear. — 17. no

For Questions 18 and 19, assume that the vertices of $\triangle RST$ are $R(1, 1)$, $S(7, 1)$, and $T(4, 6)$.

18. Write the equation in slope-intercept form of the line containing side $\overline{RT}$. — 18. $y = \frac{5}{3}x - \frac{2}{3}$

19. Write the equation of the line containing the median to side $\overline{RS}$. — 19. $x = 4$

20. On a separate sheet of paper, write a coordinate proof to show that the length of the median of a trapezoid is one-half the sum of the lengths of the bases. Use the figure at the right.

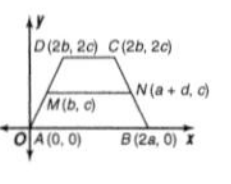

20. Sample answer:
a. $MN = a + d - b$
b. $\frac{1}{2}(AB + CD) = \frac{1}{2}(2a + 2d - 2b) = a + d - b$
c. $MN = \frac{1}{2}(AB + CD)$

Bonus

Find the area of the triangle formed by connecting the points $A(10, 0, 0)$, $B(0, 10, 0)$, and $C(0, 0, 10)$. — Bonus $25\sqrt{12} \approx 86.6$ square units

CHAPTER 12 STUDY GUIDE AND ASSESSMENT

SKILLS AND CONCEPTS

OBJECTIVES AND EXAMPLES

Upon completing this chapter, you should be able to:

- graph linear equations using the intercepts method (Lesson 12–1)

$x + 4y = 4$

***x*-intercept**

$x + 4(0) = 4$
$x = 4$
$(4, 0)$

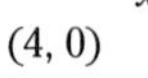

***y*-intercept**

$0 + 4y = 4$
$y = 1$
$(0, 1)$

REVIEW EXERCISES

Use these exercises to review and prepare for the chapter test.

Graph each equation using the intercepts method. 11–12. See margin.

11. $y = 3x + 3$
12. $5x - y = 10$

- graph linear equations using the slope and y-intercept (Lesson 12–1)

$y = 4x - 1$

slope

$m = 4$

***y*-intercept**

$y = 4(0) - 1$
$y = -1$
$(0, -1)$

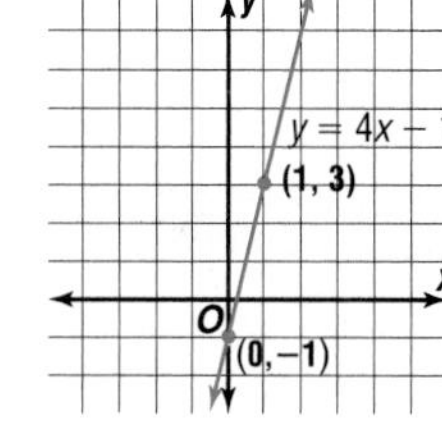

Graph each equation using the slope and *y*-intercept. 13–14. See margin.

13. $y = 5x - 3$
14. $x + 2y = 4$

- write an equation of a line given information about its graph (Lesson 12–2)

The slope of the line that passes through points at $(4, -3)$ and $(2, 1)$ is $\frac{1-(-3)}{2-4}$ or -2. To find an equation of the line, use the point-slope form.

$y - y_1 = m(x - x_1)$ *Point-slope form*

$y - 1 = -2(x - 2)$ *$m = -2$, $(x_1, y_1) = (2, 1)$*

$y - 1 = -2x + 4$

$y = -2x + 5$

16. $y = 5x + 8$ 17. $y = -5x - 31$
18. $y = -\frac{3}{5}x + 3$ 20. $y = -\frac{1}{3}x + 2$

Write the equation in slope-intercept form of the line that satisfies the given conditions.

15. $m = 4$; y-intercept $= 2$ $y = 4x + 2$
16. $m = 5$; passes through the point at $(-1, 3)$
17. passes through points at $(-7, 4)$ and $(-5, -6)$
18. x-intercept $= 5$; y-intercept $= 3$
19. line parallel to the graph of $y = x - 5$; passes through the point at $(0, 8)$ $y = x + 8$
20. line perpendicular to the graph of $y = 3x - 1$; passes through the point at $(6, 0)$

GLENCOE Technology

Test and Review Software

You may use this software, a combination of an item generator and item bank, to create your own tests or worksheets. Types of items include free response, multiple choice, short answer, and open ended.

For IBM & Macintosh

Additional Answer

11.

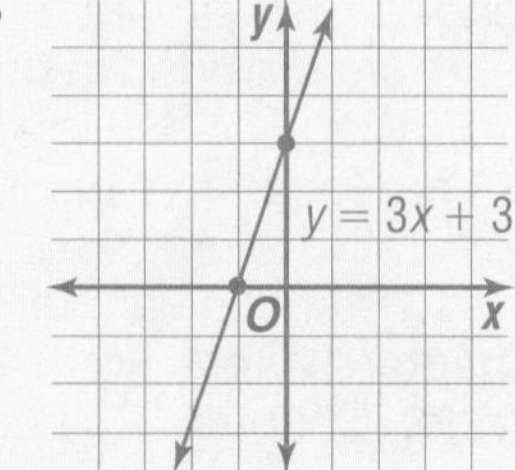

OBJECTIVES AND EXAMPLES

- relate equations of lines to geometric concepts (Lesson 12–3)

To find the equation of the perpendicular bisector of a segment whose endpoints have coordinates (5, −3) and (−1, 1), first find the coordinates of the midpoint.

$\left(\frac{5-1}{2}, \frac{-3+1}{2}\right) = \left(\frac{4}{2}, \frac{-2}{2}\right)$ or (2, −1)

Then find the slope of the segment.

$\frac{-3-1}{5-(-1)} = \frac{-4}{6}$ or $-\frac{2}{3}$

The perpendicular bisector will pass through (2, −1) and have a slope of $\frac{3}{2}$.

$y - y_1 = m(x - x_1)$

$y - (-1) = \frac{3}{2}(x - 2)$

$y = \frac{3}{2}x - 4$

The equation is $y = \frac{3}{2}x - 4$.

REVIEW EXERCISES

The vertices of △*XYZ* are at *X*(2, −1), *Y*(6, 1), and *Z*(0, −3).

21. Find the equations of the lines containing the sides of △*XYZ*.
22. Find the equations of the lines containing the medians of △*XYZ*.
23. Find the equations of the lines containing the altitudes of △*XYZ*.
24. Find the equations of the lines containing perpendicular bisectors of the sides of △*XYZ*. $y = -2x + 8,\ y = -\frac{3}{2}x + \frac{7}{2},\ y = -x - 1$

21. $y = \frac{1}{2}x - 2,\ y = \frac{2}{3}x - 3,\ y = x - 3$
22. $y = \frac{3}{4}x - 3,\ y = -1,\ y = \frac{3}{5}x - \frac{13}{5}$
23. $y = -2x - 3,\ y = -\frac{3}{2}x + 2,\ y = -x + 7$

- prove theorems using coordinate proofs (Lesson 12–4)

When planning a coordinate proof, use the following guidelines to position the figure on a coordinate plane.

- Use the origin as a vertex or center.
- Position at least one side of a polygon on a coordinate axis.
- Keep the figure in the first quadrant if possible.
- Use coordinates that make computations simple.

Prove using a coordinate proof.

25. The segment through the midpoints of the nonparallel sides of a trapezoid is parallel to the bases. See margin.
26. The length of the segment through the midpoints of the nonparallel sides of a trapezoid is one-half the sum of the lengths of the bases. See margin.

- find the magnitude and direction of a vector (Lesson 12–5)

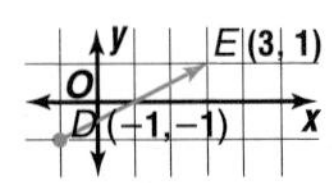

$\tan \angle D = \frac{2}{4}$ or 0.5

$m\angle D \approx 26.6$

$|\overrightarrow{DE}| = \sqrt{(3-(-1))^2 + (1-(-1))^2}$

$= \sqrt{4^2 + 2^2}$

$= 20$ or about 4.5

$\overrightarrow{DE}$ has a magnitude of about 4.5 units and a direction of about 26.6°.

Find the magnitude and direction of each vector. Round to the nearest tenth.

27. $\vec{v} = (7, 1)$ $\sqrt{50} \approx 7.1$ units; about 8.1°
28. $\overrightarrow{AB}$ with $A(1, 0)$ and $B(7, 5)$ $\sqrt{61} \approx 7.8$ units; about 39.8°

12.

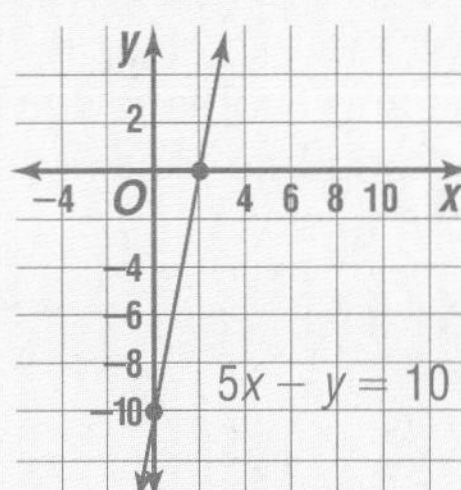

13.

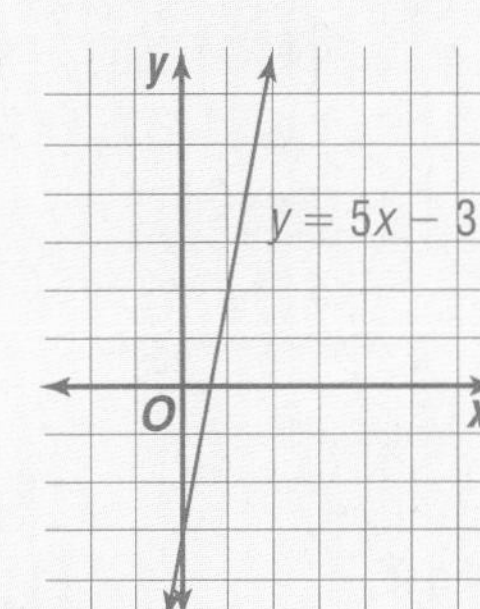

14.

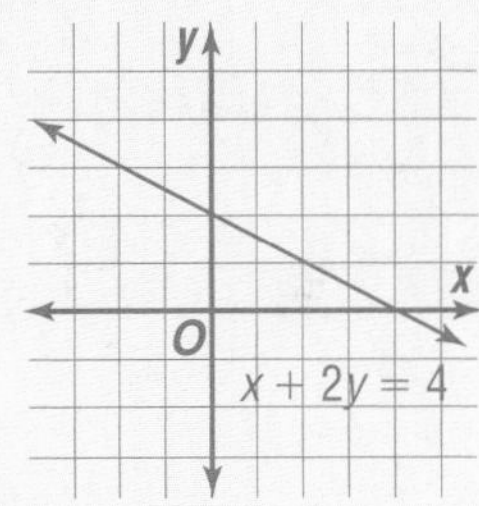

25.

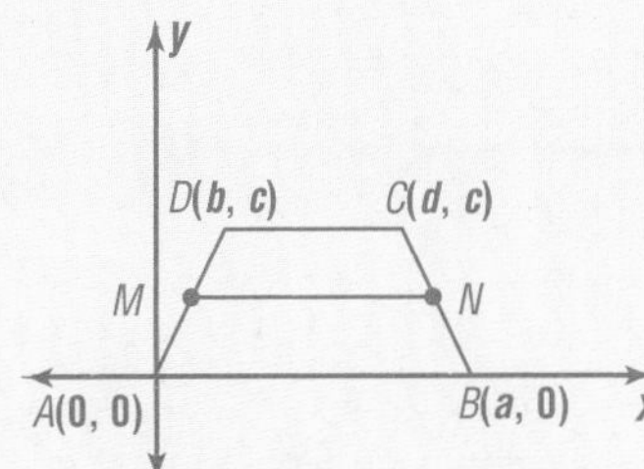

Midpoint *M* is $\left(\frac{b+0}{2}, \frac{c+0}{2}\right)$ or $\left(\frac{b}{2}, \frac{c}{2}\right)$.

Midpoint *N* is $\left(\frac{a+d}{2}, \frac{c+0}{2}\right)$ or $\left(\frac{a+d}{2}, \frac{c}{2}\right)$.

Slope of $\overline{DC}$ is $\frac{c-c}{d-b}$ or $\frac{0}{d-b}$ or 0.

Slope of $\overline{MN}$ is $\frac{\frac{c}{2}-\frac{c}{2}}{\frac{a+d}{2}-\frac{b}{2}}$ or $\frac{0}{\frac{a+d-b}{2}}$ or 0.

Slope of $\overline{AB}$ is $\frac{0-0}{a-0}$ or $\frac{0}{a}$ or 0.

$\overline{DC} \parallel \overline{MN} \parallel \overline{AB}$

Additional Answer

26.

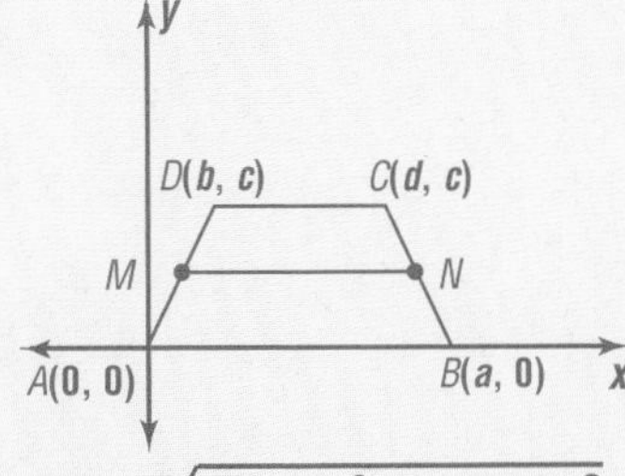

$DC = \sqrt{(d-b)^2 + (c-c)^2}$

$= \sqrt{(d-b)^2 + 0^2} = d - b$

$AB = \sqrt{(a-0)^2 + (0-0)^2}$

$= \sqrt{a^2 + 0^2} = a$

$MN = \sqrt{\left(\frac{a+d}{2} - \frac{b}{2}\right)^2 + \left(\frac{c}{2} - \frac{c}{2}\right)^2}$

$= \sqrt{\left(\frac{a+d-b}{2}\right)^2 + 0} = \frac{a+d-b}{2}$

$\frac{1}{2}(DC + AB) = \frac{1}{2}[(d-b) + a]$

$= \frac{d-b-a}{2}$ or $\frac{a+d-b}{2} = MN$

Applications and Problem Solving Encourage students to work through the exercises in the Applications and Problem Solving section to strengthen their problem-solving skills.

Additional Answer

37.

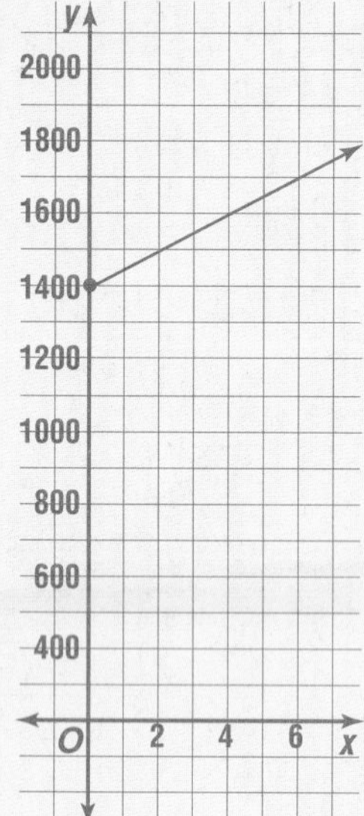

OBJECTIVES AND EXAMPLES	REVIEW EXERCISES
perform operations with vectors (Lesson 12–5)	Given $\vec{v} = (0, 8)$ and $\vec{s} = (4, 0)$, represent each of the following as an ordered pair.

Find the sum of $\vec{a}$ and $\vec{b}$.

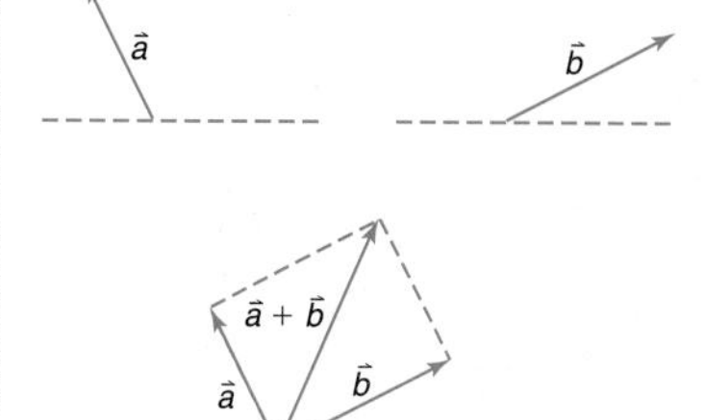

Given $\vec{v} = (0, 8)$ and $\vec{s} = (4, 0)$, represent each of the following as an ordered pair.

29. $\vec{v} + \vec{s}$ (4, 8)

30. $3\vec{v} + \vec{s}$ (4, 24)

Find each sum.

31. $\begin{bmatrix} 4 \\ 6 \end{bmatrix} + \begin{bmatrix} 3 \\ 2 \end{bmatrix}$ $\begin{bmatrix} 7 \\ 8 \end{bmatrix}$

32. $\begin{bmatrix} 2 \\ -3 \end{bmatrix} + \begin{bmatrix} 0 \\ 9 \end{bmatrix}$ $\begin{bmatrix} 2 \\ 6 \end{bmatrix}$

33. $4\sqrt{2} \approx 5.7$ units; (5, −3, 3)

34. $2\sqrt{3} \approx 3.5$ units; (1, 3, 5)

---

use the distance and midpoint formulas for points in space (Lesson 12–6)

The distance between the points at (2, 2, 2) and (−6, 0, 5) is $\sqrt{(-6-2)^2 + (0-2)^2 + (5-2)^2}$ or $\sqrt{77}$ units. The coordinates of the midpoint of a line segment whose endpoints are (2, 2, 2) and (−6, 0, 5) are $\left(\frac{2-6}{2}, \frac{2+0}{2}, \frac{2+5}{2}\right)$ or (−2, 1, 3.5).

36. $(x + 1)^2 + (y - 2)^2 + (z + 3)^2 = 16$

Determine the distance between each pair of points and the coordinates of the midpoint of the segment whose endpoints are given.

33. $A(3, -3, 1), B(7, -3, 5)$ 34. $C(2, 4, 6), D(0, 2, 4)$

Write the equations of the sphere given the coordinates of the center and measure of the radius. 35. $x^2 + y^2 + z^2 = 25$

35. (0, 0, 0), 5

36. (−1, 2, −3), 4

39a. See Soutions Manual.

APPLICATIONS AND PROBLEM SOLVING

37. **Business** Just Like Grandma's Bake Shop spends \$1400 a month for rent and utilities. For each day of operation, they spend \$500 for employees' wages, benefits, and baking supplies. If x represents the number of days of operation in a month, $y = 500x + 1400$ represents the cost of operations for the month. Draw a graph that represents the cost of operation for one week. What does the y-intercept represent? (Lesson 12–1) \$1400 a month for rent and utilities

38. **Economics** In 1960, a candy bar cost 10¢. In 1996, the same candy bar cost 60¢. Write an equation describing the cost of the candy bar over time. If the rate of change in price has been constant, how much will the candy bar cost in the year 2010? (Lesson 12–2) $c = 0.1 + 0.0139(t - 1960)$; Sample answer: \$0.80

A practice test for Chapter 12 is provided on page 804.

A practice test for Chapter 12 is provided on page 804.

39. **Architecture** The table below shows the height and number of stories of several notable structures around the world. (Lesson 12–3)
   a. Draw a scatter plot to show how height x and number of stories y are related.
   b. Write an equation that relates the height to the number of stories.
   c. The Governor Philip Tower in Sydney, Australia, is 745 feet tall. How many stories do you think the tower has?

39b. Sample answer: $y = 0.09x - 15$ 39c. Sample answer: 52 stories. The tower actually has 54 stories.

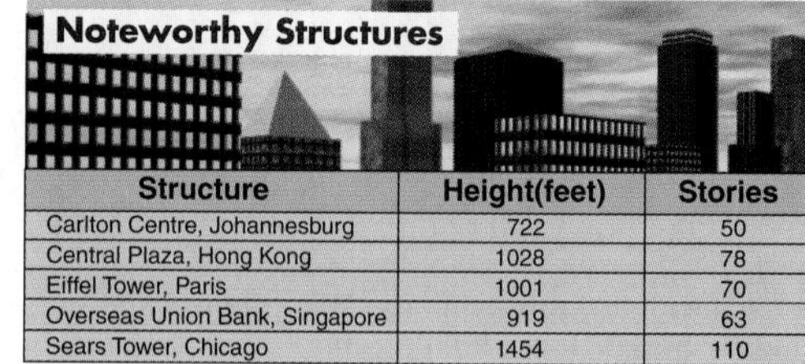

Noteworthy Structures

Structure	Height(feet)	Stories
Carlton Centre, Johannesburg	722	50
Central Plaza, Hong Kong	1028	78
Eiffel Tower, Paris	1001	70
Overseas Union Bank, Singapore	919	63
Sears Tower, Chicago	1454	110

Source: *World Almanac, 1996*

ALTERNATIVE ASSESSMENT

COOPERATIVE LEARNING PROJECT

Architectural Design In this chapter, you learned how to write coordinates for points in space. In this project, you will apply this know-how to describe a simple building you plan to construct.

Design a simple building with surfaces that are polygons. Follow these steps to describe your building mathematically.

- Construct a scale model of your design. Only the frame of the building is needed.
- Remove the top and two adjoining sides of a cardboard box. Mark the edges of the remaining sides, x, y, and z.
- Position the model so that one of its corners is at the origin.
- Using a ruler, determine the coordinates of each point on the model that is the intersection of two line segments.
- Record the ordered triples for these points in a table.
- Use three pieces of graph paper to model the positive quadrants in the plane. Label the edges of each plane with appropriate axes.
- Plot the points from the table on each graph.
- Connect the appropriate points. The result is a view of your building from three perspectives.

THINKING CRITICALLY

Is vector addition commutative? Justify your answer with a proof. **See Solutions Manual.**

PORTFOLIO

Find a photograph of a building with simple geometric structure. The photograph should view the building from one side. Position the lower left corner of the building at the origin of a two-dimensional coordinate system. Determine values of points that are the intersection of line segments. From this, determine linear equations for each line. Keep the photograph and the equations in your portfolio.

SELF EVALUATION

The right hemisphere of the brain processes visual and spatial information. So, geometric figures are processed in the right hemisphere. The left hemisphere processes linear information. An equation is a linear sequence of symbols. So, an equation of a geometric figure is processed in the left hemisphere.

Assess Yourself Based on your experience with the concepts in this chapter, do you believe you are more left-brained or more right-brained?

Alternative Assessment

The Alternative Assessment section provides students with the opportunity to assess their own work by thinking critically, working with others, keeping a portfolio, and honestly evaluating their own progress. For more information on alternative forms of assessment, see *Alternative Assessment in the Mathematics Classroom,* one of the titles in the Glencoe Mathematics Professional Series.

Performance Assessment

Performance Assessment tasks for this chapter are included in the *Assessment and Evaluation Masters.* A scoring guide is also provided.

Assessment and Evaluation Masters, pp. 322, 333

NAME________________ DATE ________

Chapter 12 Performance Assessment

Instructions: ***Demonstrate your knowledge by giving a clear, concise solution to each problem. Be sure to include all relevant drawings and to justify your answers. You may show your solution in more than one way or investigate beyond the requirements of the problem.***

1. Suppose $\triangle ABC$ has vertices $A(0, 0)$, $B(3, 4)$, and $C\left(0, \frac{17}{3}\right)$.
   a. Draw $\triangle ABC$ on the coordinate plane.
   b. Write the equations of the lines containing each side of $\triangle ABC$.
   c. Explain how to find the equation of a line perpendicular to a given line through a point not on the line.
   d. Write the equations of the lines containing the altitudes of $\triangle ABC$.
   e. Write a coordinate proof to show that $\triangle ABC$ is scalene.

2. The following questions refer to the figure below.

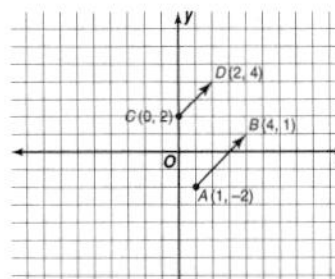

   a. Represent $\overrightarrow{AB}$ and $\overrightarrow{CD}$ by ordered pairs.
   b. Find the magnitude and direction of $\overrightarrow{AB}$.
   c. Tell what it means for two vectors to be equivalent.
   d. Draw and label a $\overrightarrow{EF}$ equivalent to $\overrightarrow{AB}$ on the coordinate plane.
   e. Write a word problem that involves finding the sum of two vectors.
   f. Solve your word problem. Explain each step and give the meaning of the answer.

Scoring Guide
Chapter 12
Performance Assessment

Level	Specific Criteria
3 Superior	• Shows thorough understanding of the concepts of *writing the equation of a line, using slope to draw a line perpendicular to a line, vector, magnitude and direction of a vector,* and *equivalent vectors.* • Uses appropriate strategies to solve the problem and write the proof. • Computations are correct. • Written explanations are exemplary. • Word problem concerning vectors is appropriate and makes sense. • Graphs are accurate and appropriate. • Goes beyond requirements of some or all problems.
2 Satisfactory, with Minor Flaws	• Shows understanding of the concepts of *writing the equation of a line, using slope to draw a line perpendicular to a line, vector, magnitude and direction of a vector,* and *equivalent vectors.* • Uses appropriate strategies to solve the problem and write the proof. • Computations are mostly correct. • Written explanations are effective. • Word problem concerning vectors is appropriate and makes sense. • Graphs are mostly accurate and appropriate. • Satisfies all requirements of some or all problems.
1 Nearly Satisfactory, with Serious Flaws	• Shows thorough understanding of the concepts of *writing the equation of a line, using slope to draw a line perpendicular to a line, vector, magnitude and direction of a vector,* and *equivalent vectors.* • May not use appropriate strategies to solve the problem and write the proof. • Computations are mostly correct. • Written explanations are satisfactory. • Word problem concerning vectors is mostly appropriate and sensible. • Graphs are mostly accurate and appropriate. • Satisfies most requirements of some or all problems.
0 Unsatisfactory	• Shows little or no understanding of the concepts of *writing the equation of a line, using slope to draw a line perpendicular to a line, vector, magnitude and direction of a vector,* and *equivalent vectors.* • May not use appropriate strategies to solve the problem and write the proof. • Computations are incorrect. • Written explanations are not satisfactory. • Word problem concerning vectors is not appropriate and sensible. • Graphs are not accurate or appropriate. • Does not satisfy requirements of some or all problems.

Using the COLLEGE ENTRANCE EXAM PRACTICE

These two pages review the skills and concepts presented in Chapters 1–12. This review is formatted to reflect new trends in college entrance testing.

Assessment and Evaluation Masters, pp. 327–328

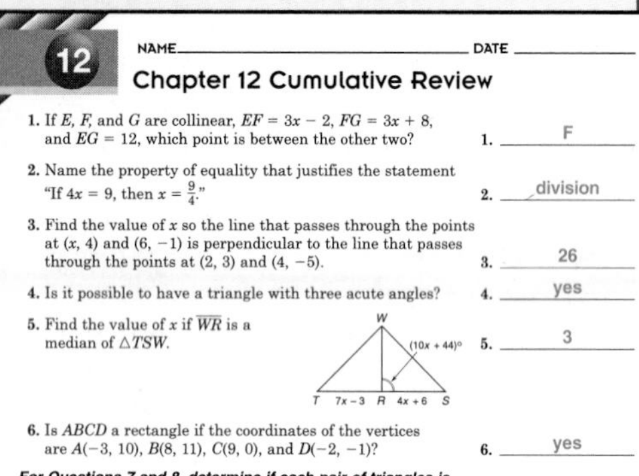

12 NAME______ DATE ______

Chapter 12 Cumulative Review

1. If E, F, and G are collinear, $EF = 3x - 2$, $FG = 3x + 8$, and $EG = 12$, which point is between the other two? 1. F
2. Name the property of equality that justifies the statement "If $4x = 9$, then $x = \frac{9}{4}$." 2. division
3. Find the value of x so the line that passes through the points at $(x, 4)$ and $(6, -1)$ is perpendicular to the line that passes through the points at $(2, 3)$ and $(4, -5)$. 3. 26
4. Is it possible to have a triangle with three acute angles? 4. yes
5. Find the value of x if $\overline{WR}$ is a median of $\triangle TSW$. 5. 3
6. Is $ABCD$ a rectangle if the coordinates of the vertices are $A(-3, 10)$, $B(8, 11)$, $C(9, 0)$, and $D(-2, -1)$? 6. yes

For Questions 7 and 8, determine if each pair of triangles is similar. Write yes or no. If they are similar, give a reason.

7\. 7. yes, SSS Similarity

8\. 8. no

9. Find the geometric mean between 5 and 20. 9. 10
10. Find sin A as a fraction and as a decimal rounded to the nearest thousandth. 10. $\frac{36}{39}$ or $\frac{12}{13}$; 0.923

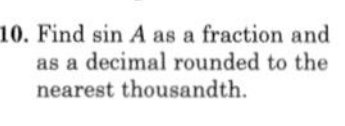

Find the value of x for each ⊙C. Assume that segments that appear to be tangent are tangent.

11\. 11. 8

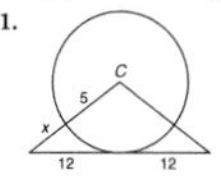

12\. 12. 20

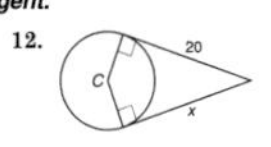

12 NAME______ DATE ______

Chapter 12 Cumulative Review (continued)

13. Find the sum of the measures of the interior angles of a convex 19-gon. 13. 3060
14. Find the area of the shaded region. 14. 264 m²

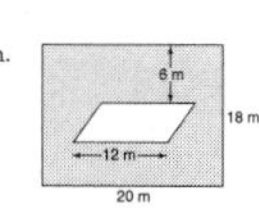

15. Find the surface area of the cylinder. Round to the nearest tenth. 15. 131.9 ft²

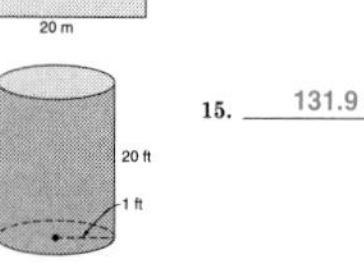

16. Find the volume of the prism. Round to the nearest cubic unit. 16. 346 cm³

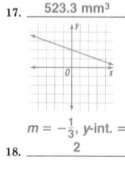

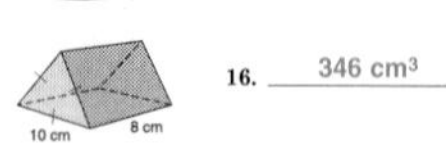

17. Find the volume of a sphere with a radius of 5 millimeters. Round your answer to the nearest tenth. 17. 523.3 mm³
18. Graph the equation $x + 3y = 6$. State the slope and y-intercept. 18. $m = -\frac{1}{3}$, y-int. = 2

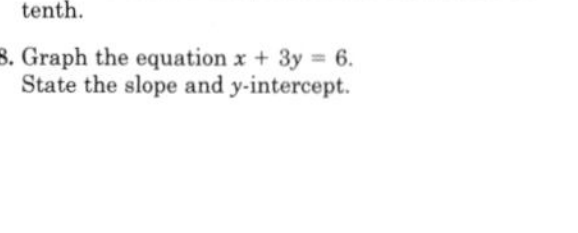

19. Write the equation in slope-intercept form of the line whose slope is 3 and that passes through the point at $(2, -8)$. 19. $y = 3x - 14$
20. Determine the center and the measure of the radius for a sphere with equation $(x + 1)^2 + (y - 4)^2 + (z + 5)^2 = 49$. 20. $(-1, 4, -5)$, 7

COLLEGE ENTRANCE EXAM PRACTICE

CHAPTERS 1–12

SECTION ONE: MULTIPLE CHOICE

There are nine multiple-choice questions in this section. After working each problem, write the letter of the correct answer on your paper.

1. If $4^a = 16$, then find $2^a \times 2$. B

A. 4
B. 8
C. 16
D. 32

2. If the volume of a cube is 125 cm³, find its surface area. B

A. 250 cm²
B. 150 cm²
C. 100 cm²
D. 125 cm²

3. If the operation # is defined by the equation $x \# y = 2x + y$, what is the value of a in the equation $2 \# a = a \# 3$? C

A. -1
B. 0
C. 1
D. 4

4. If x is between 0 and 1, which of the following increases as x increases? A

I. $x + 1$
II. $1 - x^2$
III. $\frac{1}{x}$

A. I only
B. III only
C. I and II only
D. I, II, and III

5. If $x + 2 = 0$, then find x^3. D

A. 4
B. -4
C. 8
D. -8

6. In the figure below, if $\overline{PQ} \perp \overline{QT}$, $\overline{QR} \perp \overline{RT}$, and $\overline{QS} \perp \overline{SV}$, what is the value of $y - x$? C

A. 60
B. 45
C. 30
D. 20

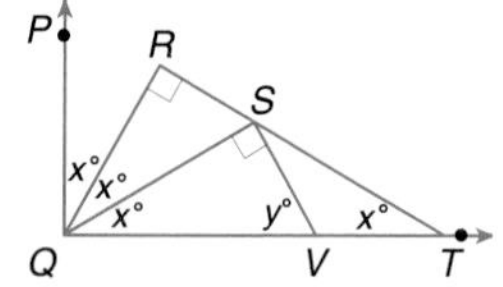

7. If $\frac{x}{3}$ is the measure of the perimeter of a square, then find the measure of the area of the square. A

A. $\frac{x^2}{144}$
B. $\frac{x^2}{36}$
C. $\frac{x^2}{16}$
D. $\frac{x^2}{9}$

8. A cylindrical can has a diameter of 12 inches and a height of 8 inches. If one gallon of liquid occupies 231 cubic inches, what is the capacity of the can in gallons rounded to the nearest tenth? A

A. 3.9 gallons
B. 4.2 gallons
C. 2.3 gallons
D. 4.4 gallons

9. What is the mean of $a + 5$, $2a - 4$, and $3a + 8$? C

A. $2a$
B. $6a + 9$
C. $2a + 3$
D. $3a + 3$

Standardized Test Practice Questions are also provided in the *Assessment and Evaluation Masters*, p. 326.

A more traditional cumulative review is provided in the *Assessment and Evaluation Masters*, pp. 327–328.

SECTION TWO: SHORT ANSWER

This section contains seven questions for which you will provide short answers. Write your answer on your paper.

10. Right triangles ABC and DEF are similar. If $BC = 9$, $AC = 21$, and $EF = 24$, find DF. **56 units**

11. Find the volume of a right cylinder with a radius of $3x$ meters and a height of $6x$ meters. **$54\pi x^3$ meters^3**

12. In the figure below, find x if O is the center of the circle. **76°**

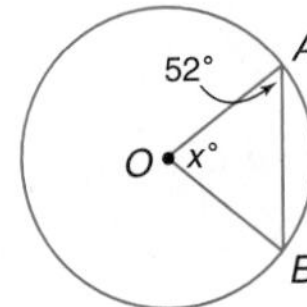

13. The volume of a sphere S is $\frac{4}{3}\pi r^3$, and the volume of sphere Q is $2\pi r^3$. The volume of S is what percent of the volume of Q? **$66\frac{2}{3}\%$**

14. Find the distance in three-dimensional space between points at $(-1, 2, -3)$ and $(2, 4, -1)$ **$\sqrt{17}$ units**

15. A cone and a cylinder both have a height of 48 units and a radius of 15 units. Find the ratios of their volumes. **$\frac{1}{3}$**

16. The corner is cut from a rectangular piece of cardboard as shown below. Find the area of the remaining cardboard. **114 ft^2**

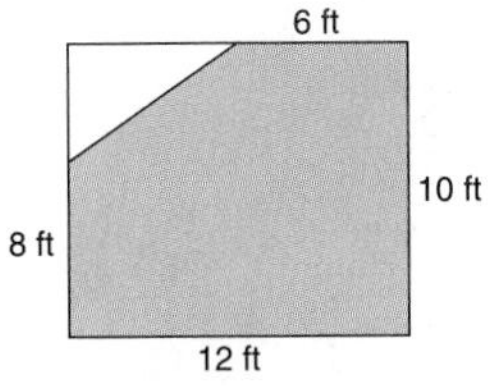

SECTION THREE: COMPARISON

This section contains five comparison problems that involve comparing two quantities, one in column A and one in column B. In certain questions, information related to one or both quantities is centered above them. All variables used represent real numbers.

Compare quantities A and B below.

- Write A if quantity A is greater.
- Write B if quantity B is greater.
- Write C if the two quantities are equal.
- Write D if there is not enough information to determine the relationship.

19. B 21. D

Column A		Column B
	$x > 0$	**C**
17. the perimeter of a square with side of length $9x$		the perimeter of an equilateral triangle with side of length $12x$
	$3y - x = 8$ $y + x = 2$	**B**
18. x		y
19. the volume of a sphere with a radius of 2		the volume of a hemisphere with a radius of 4
	$0 < x < 1$	**A**
20. $2x$		x^2
21. the volume of a cylinder with a radius of 2		the volume of a cylinder with a radius of 4

TEST-TAKING TIP

Time Management

Tell students that if they finish a college entrance exam before the allotted time is up, they should use the remaining time wisely.

Remind students to check to be sure that they have marked all of their answers in the correct place on the answer sheet.

Tell them to be certain that they have answered all of the questions. If guessing is not penalized, students should answer the questions they can't complete by eliminating as many answers as they can and then guessing.

Tell students to make sure that they have marked only one answer for each question. If two answers are marked, the question will be marked wrong even if one of the answers was the correct choice.

CHAPTER

13

Investigating Loci and Coordinate Transformations

PREVIEWING THE CHAPTER

In this chapter, students learn what a locus is and how to locate and draw it. Students find loci for given situations and solve problems using loci. Students recognize different types of transformations and draw the resulting images. They use a geomirror to investigate translations and to draw translations using two reflections. Students determine scale factors for dilations and draw dilations based on scale factors.

Lesson (Pages)	Lesson Objectives	NCTM Standards	State/Local Objectives
13-1 (696–701)	Locate, draw, and describe a locus in a plane or in space.	1–4, 7, 8	
13-2 (702–708)	Find the locus of points that are solutions of a system of equations by graphing, by substitution, or by elimination.	1–5, 7, 8	
13-3 (709–714)	Solve locus problems that satisfy more than one condition.	1–5, 7, 8	
13-4 (715–721)	Recognize an isometry or congruence transformation. Solve problems by making a table.	1–5, 7, 8	
13-5 (722–729)	Name, recognize, and draw reflection images, lines of symmetry, and points of symmetry.	1–5, 7, 8	
13-6A (730)	Use a geomirror to investigate translations.	1–4, 7, 8	
13-6 (731–737)	Name and draw translation images of figures with respect to parallel lines.	1–5, 7, 8	
13-6B (738)	Use a TI-92 calculator to translate a triangle in the direction and magnitude of a given vector.	1–4, 7, 8	
13-7 (739–745)	Name and draw rotation images of figures with respect to intersecting lines.	1–5, 7, 8	
13-8 (746–753)	Use scale factors to determine if a dilation is an enlargement, a reduction, or a congruence transformation. Find the center and scale factor for a given dilation, and vice versa.	1–5, 7, 8	

ORGANIZING THE CHAPTER

A complete, 1-page lesson plan is provided for each lesson in the ***Lesson Planning Guide***. Answer keys for each lesson are available in the ***Answer Key Masters***.

You may want to refer to the **Course Planning Calendar** on page T12 for detailed information on pacing.
PACING: Honors—13 days

LESSON PLANNING CHART

			Blackline Masters									
Lesson (Pages)	**Materials/ Manipulatives**	**Extra Practice (Student Edition)**	**Study Guide**	**Practice**	**Enrichment**	**Assessment & Evaluation**	**Modeling Mathematics**	**Multicultural Activity**	**Tech Prep Applications**	**Graphing Calc. & Computer**	**Real-World Applications**	**Teaching Transparencies**
13-1 (696–701)	ruler* TI-82/83 graphing calculator	p. 789	p. 78	p. 78	p. 78							13-1A 13-1B
13-2 (702–708)	TI-82/83 graphing calculator	p. 790	p. 79	p. 79	p. 79	p. 352		p. 25	p. 25	p. 13	27	13-2A 13-2B
13-3 (709–714)	tracing paper ruler* protractor* TI-82/83 graphing calculator	p. 790	p. 80	p. 80	p. 80							13-3A 13-3B
13-4 (715–721)		p. 790	p. 81	p. 81	p. 81	pp. 351,352					28	13-4A 13-4B
13-5 (722–729)	grid paper	p. 791	p. 82	p. 82	p. 82		pp. 63–66	p. 26	p. 26			13-5A 13-5B
13-6A (730)	geomirror* straightedge*						p. 104					
13-6 (731–737)	thin cardboard grid paper straightedge*	p. 791	p. 83	p. 83	p. 83	p. 353	p. 91					13-6A 13-6B
13-6B (738)	TI-92 calculator									p. 30		
13-7 (739–745)	grid paper tracing paper	p. 792	p. 84	p. 84	p. 84							13-7A 13-7B
13-8 (746–753)		p. 792	p. 85	p. 85	p. 85	p. 353						13-8A 13-8B
Study Guide/ Assessment (755–759)						pp. 337–350, 354–356						

*Included in Glencoe's High School Manipulative Kit and Overhead Manipulative Resources.

ORGANIZING THE CHAPTER

OTHER CHAPTER RESOURCES

Student Edition
Chapter Opener, pp. 694–695
Mathematics and Society, p. 708
Working on the Investigation, pp. 737, 753
Closing the Investigation, p. 754

Teacher's Classroom Resources
Investigations and Projects Masters, pp. 73–76
Block Scheduling Booklet

Technology
Test and Review Software (IBM and Macintosh)
CD-ROM Multimedia Applications (Windows and Macintosh)
Mindjogger Videoquizzes (VHS)

Professional Publications
Glencoe Mathematics Professional Series

OUTSIDE RESOURCES

Books/Periodicals
Geometry: Constructions and Transformations, Dale Seymour Publications
Transformational Geometry, Dale Seymour Publications
Informal Geometry Explorations, NASCO

Software
Geometry Connections: Tessellations, William K. Bradford Publishing Company
Geometric Golfer, ETA

Videos/CD-ROMs
Fermat's Last Theorem, NCTM, 1906 Association Drive, Reston, VA 20191
Mathematics: Making the Connection, NCTM, 1906 Association Drive, Reston, VA 20191

ASSESSMENT RESOURCES

Student Edition
Math Journal, pp. 706, 749
Mixed Review, pp. 701, 707–708, 714, 721, 729, 736–737, 745, 752–753
Self Test, p. 721
Chapter Highlights, p. 755
Chapter Study Guide and Assessment, pp. 756–758
Alternative Assessment, p. 759
 Portfolio, p. 759
College Entrance Exam Practice, pp. 760–761
Chapter Test, p. 805

Teacher's Wraparound Edition
5-Minute Check, pp. 696, 702, 709, 715, 722, 731, 739, 746
Check for Understanding, pp. 699, 706, 712, 718, 725, 733, 741, 749
Closing Activity, pp. 701, 708, 714, 721, 729, 737, 745, 753
Cooperative Learning, pp. 698, 723

Assessment and Evaluation Masters
Multiple-Choice Tests, Forms 1A (Honors), 1B (Average), 1C (Basic), pp. 337–342
Free-Response Tests, Forms 2A (Honors), 2B (Average), 2C (Basic), pp. 343–348
Calculator-Based Test, p. 349
Performance Assessment, p. 350
Mid-Chapter Test, p. 351
Quizzes A–D, pp. 352–353
Standardized Test Practice, p. 354
Cumulative Review, pp. 355–356

ENHANCING THE CHAPTER

Examples of some of the materials for enhancing Chapter 13 are shown below.

DIVERSITY

Multicultural Activity Masters, pp. 25, 26

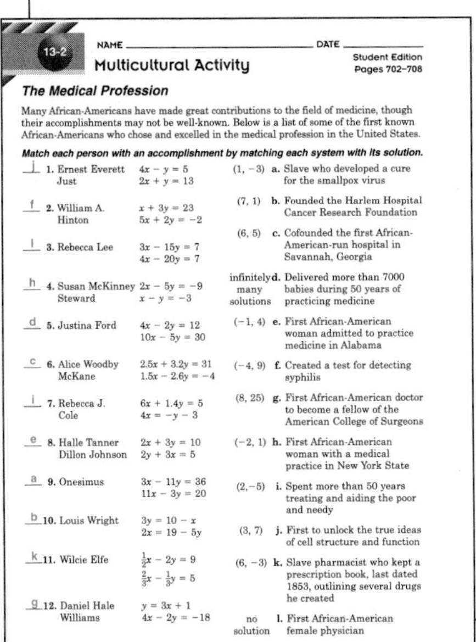

13-2 NAME ______ DATE ______

Multicultural Activity — Student Edition Pages 702–708

The Medical Profession

Many African-Americans have made great contributions to the field of medicine, though their accomplishments may not be well-known. Below is a list of some of the first known African-Americans who chose and excelled in the medical profession in the United States.

Match each person with an accomplishment by matching each system with its solution.

Answer	Person	System	Solution	Accomplishment
j	1. Ernest Everett Just	$4x - y = 5$, $2x + y = 13$	(1, −3)	a. Slave who developed a cure for the smallpox virus
f	2. William A. Hinton	$x + 3y = 23$, $5x + 2y = -2$	(7, 1)	b. Founded the Harlem Hospital Cancer Research Foundation
l	3. Rebecca Lee	$3x - 15y = 7$, $4x - 20y = 7$	(6, 5)	c. Cofounded the first African-American-run hospital in Savannah, Georgia
h	4. Susan McKinney Steward	$2x - 5y = -9$, $x - y = -3$	infinitely many solutions	d. Delivered more than 7000 babies during 50 years of practicing medicine
d	5. Justina Ford	$4x - 2y = 12$, $10x - 5y = 30$	(−1, 4)	e. First African-American woman admitted to practice medicine in Alabama
c	6. Alice Woodby McKane	$2.5x + 3.2y = 31$, $1.5x - 2.6y = -4$	(−4, 9)	f. Created a test for detecting syphilis
i	7. Rebecca J. Cole	$6x + 1.4y = 5$, $4x = -y - 3$	(8, 25)	g. First African-American doctor to become a fellow of the American College of Surgeons
e	8. Halle Tanner Dillon Johnson	$2x + 3y = 10$, $2y + 3x = 5$	(−2, 1)	h. First African-American woman with a medical practice in New York State
a	9. Onesimus	$3x - 11y = 36$, $11x - 3y = 20$	(2, −5)	i. Spent more than 50 years treating and aiding the poor and needy
b	10. Louis Wright	$3y = 10 - x$, $2x = 19 - 5y$	(3, 7)	j. First to unlock the true ideas of cell structure and function
k	11. Wilcie Elfe	$\frac{1}{2}x - 2y = 9$, $\frac{2}{3}x - \frac{1}{3}y = 5$	(6, −3)	k. Slave pharmacist who kept a prescription book, last dated 1853, outlining several drugs he created
g	12. Daniel Hale Williams	$y = 3x + 1$, $4x - 2y = -18$	no solution	l. First African-American female physician

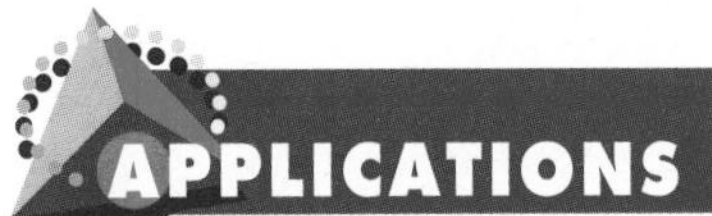

APPLICATIONS

Real-World Applications, 27, 28

TECHNOLOGY

Graphing Calculator and Computer Masters, p. 13

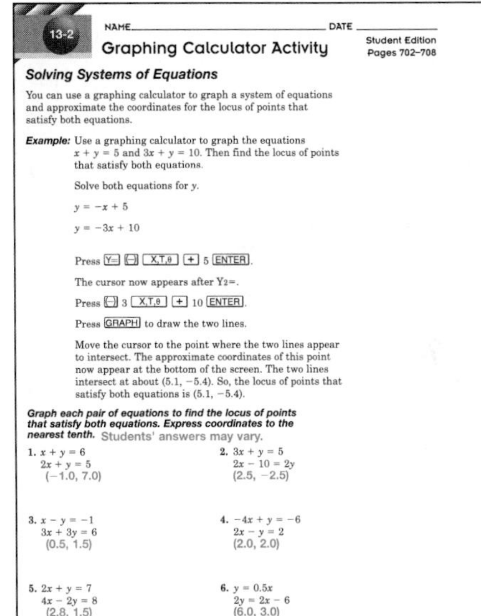

13-2 NAME ______ DATE ______

Graphing Calculator Activity — Student Edition Pages 702–708

Solving Systems of Equations

You can use a graphing calculator to graph a system of equations and approximate the coordinates for the locus of points that satisfy both equations.

Example: Use a graphing calculator to graph the equations $x + y = 5$ and $3x + y = 10$. Then find the locus of points that satisfy both equations.

Solve both equations for y.

$y = -x + 5$

$y = -3x + 10$

Press [Y=] [(−)] [X,T,θ] [+] 5 [ENTER].

The cursor now appears after Y_2=.

Press [(−)] 3 [X,T,θ] [+] 10 [ENTER].

Press [GRAPH] to draw the two lines.

Move the cursor to the point where the two lines appear to intersect. The approximate coordinates of this point now appear at the bottom of the screen. The two lines intersect at about (5.1, −5.4). So, the locus of points that satisfy both equations is (5.1, −5.4).

Graph each pair of equations to find the locus of points that satisfy both equations. Express coordinates to the nearest tenth. Students' answers may vary.

1. $x + y = 6$, $2x + y = 5$ (−1.0, 7.0)
2. $3x + y = 5$, $2x - 10 = 2y$ (2.5, −2.5)
3. $x - y = -1$, $3x + 3y = 6$ (0.5, 1.5)
4. $-4x + y = -6$, $2x - y = 2$ (2.0, 2.0)
5. $2x + y = 7$, $4x - 2y = 8$ (2.8, 1.5)
6. $y = 0.5x$, $2y = 2x - 6$ (6.0, 3.0)

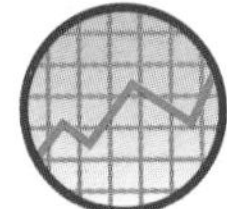

TECH PREP

Tech Prep Applications Masters, 25, 26

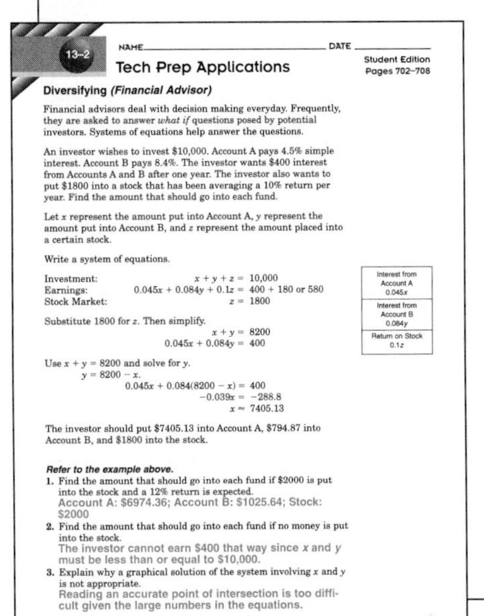

13-2 NAME ______ DATE ______

Tech Prep Applications — Student Edition Pages 702–708

Diversifying *(Financial Advisor)*

Financial advisors deal with decision making everyday. Frequently, they are asked to answer *what if* questions posed by potential investors. Systems of equations help answer the questions.

An investor wishes to invest $10,000. Account A pays 4.5% simple interest. Account B pays 8.4%. The investor wants $400 interest from Accounts A and B after one year. The investor also wants to put $1800 into a stock that has been averaging a 10% return per year. Find the amount that should go into each fund.

Let x represent the amount put into Account A, y represent the amount put into Account B, and z represent the amount placed into a certain stock.

Write a system of equations.

Investment: $x + y + z = 10{,}000$
Earnings: $0.045x + 0.084y + 0.1z = 400 + 180$ or 580
Stock Market: $z = 1800$

Interest from Account A $0.045x$
Interest from Account B $0.084y$
Return on Stock $0.1z$

Substitute 1800 for z. Then simplify.

$x + y = 8200$
$0.045x + 0.084y = 400$

Use $x + y = 8200$ and solve for y.

$y = 8200 - x$.
$0.045x + 0.084(8200 - x) = 400$
$-0.039x = -288.8$
$x \approx 7405.13$

The investor should put $7405.13 into Account A, $794.87 into Account B, and $1800 into the stock.

Refer to the example above.

1. Find the amount that should go into each fund if $2000 is put into the stock and a 12% return is expected. Account A: $6974.36; Account B: $1025.64; Stock: $2000
2. Find the amount that should go into each fund if no money is put into the stock. The investor cannot earn $400 that way since x and y must be less than or equal to $10,000.
3. Explain why a graphical solution of the system involving x and y is not appropriate. Reading an accurate point of intersection is too difficult given the large numbers in the equations.

PROBLEM SOLVING

Problem-of-the-Week Cards, 35, 36

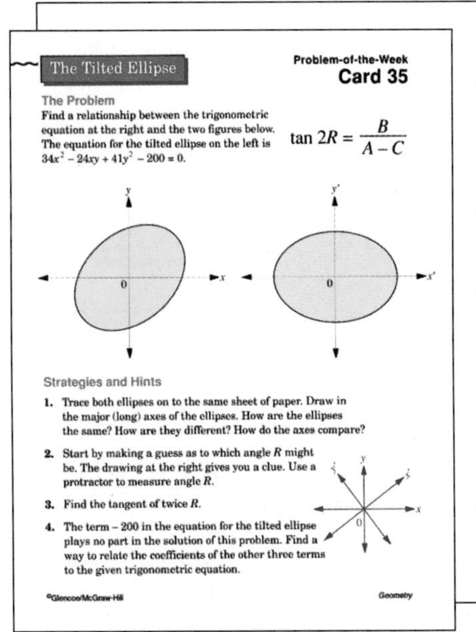

The Tilted Ellipse — Problem-of-the-Week Card 35

The Problem

Find a relationship between the trigonometric equation at the right and the two figures below. The equation for the tilted ellipse on the left is $34x^2 - 24xy + 41y^2 - 200 = 0$.

$$\tan 2R = \frac{B}{A - C}$$

Strategies and Hints

1. Trace both ellipses on to the same sheet of paper. Draw in the major (long) axes of the ellipses. How are the ellipses the same? How are they different? How do the axes compare?
2. Start by making a guess as to which angle R might be. The drawing at the right gives you a clue. Use a protractor to measure angle R.
3. Find the tangent of twice R.
4. The term − 200 in the equation for the tilted ellipse plays no part in the solution of this problem. Find a way to relate the coefficients of the other three terms to the given trigonometric equation.

©Glencoe/McGraw-Hill — Geometry

MAKING MATHEMATICS RELEVANT

This two-page introduction to the chapter provides students with an opportunity to explore an application of geometry that has historic roots as well as modern uses.

Background Information

All Dressed Up Adult butterflies and moths have no means of protection from their predators other than hiding from them or scaring them away. The coloring of butterflies and moths is their primary defense. Many species have developed a type of camouflage that allows them to blend in with tree bark or dead leaves. Others have developed pairs of eyespots on their wings that, when suddenly flashed, can help to startle off predators. Still other species are poisonous to their predators. These species tend to use bright colors, such as reds, yellows, and oranges, to advertise their inedibility. This type of protection is so effective that it isn't uncommon for nearby nonpoisonous species to mimic the color patterns of the poisonous species to also be avoided by predators.

CHAPTER 13

Investigating Loci and Coordinate Transformations

Objectives

In this chapter, you will:

- draw, locate, or describe a locus in a plane or in space,
- determine line symmetry and point symmetry,
- draw reflections, translations, rotations, and dilations on the coordinate plane, and
- solve problems by making a table.

Geometry: Then and Now Brightly colored butterflies have fascinated people for centuries. Ancient Egyptians, Chinese, Japanese, and Native Americans included images of butterflies in their art and pottery. Today over 170,000 butterflies and moths have been identified in the insect order, called Lepidoptera. This Latin name, meaning "scaly wings," fits because the fragile wings of butterflies and moths are covered with thousands of tiny scales. These colored scales make unique patterns that may have specific purposes, such as camouflage or frightening would-be predators. The wing markings on butterflies and moths are usually symmetrical.

TIME Line

776 B.C. The first recorded Olympic games are played.

870 Johannes Scotus Erigena compiles information for an encyclopedia on nature.

800 B.C. | 600 | 400 | 200 | 0 | A.D. 200 | 400 | 600 | 800 | 1000 | 1420 | 1460 | 1500

A.D. 580 An iron-chain suspension bridge is built in China.

1434 Leone Battista publishes a book on drawing that is considered the forerunner of projective geometry.

TIME Line

Students might find it interesting to write an essay on projective geometry and its applications to computer graphics. Computer graphics is becoming more sophisticated as 3-D features are added.

interNET CONNECTION

Spread your wings and explore links related to butterflies and moths from around the world.

World Wide Web
http://www.chebucto.ns.ca/Environment/NHR/lepidoptera.html

PEOPLE IN THE
NEWS

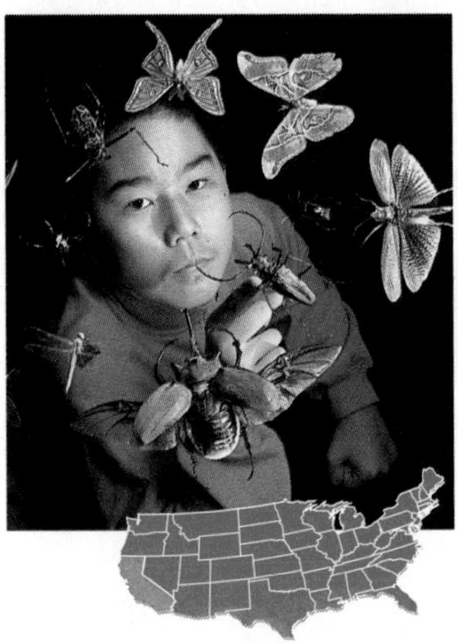

James Fujita of Oxnard, California, likes bugs. He has been studying and collecting insects ever since he was bitten by a cricket at age 3. He now has hundreds of species of insects in his collection. His hobby has taken him as far away as Costa Rica and Japan to study and collect insects. James also teaches workshops on insects and has donated specimens to zoos and museums. His knowledge of insects really paid off when he identified a previously undiscovered species of cricket.

Chapter Project

Work in small groups to explore the use of transformations in the world around you.

- Examine photos of butterflies in books, electronic encyclopedias, or the Internet.
- Name the four types of transformations. Which transformations are used in the photos?
- For each type of transformation, give an example of the transformation's use. Can you think of any situation in nature that is related to the transformation?

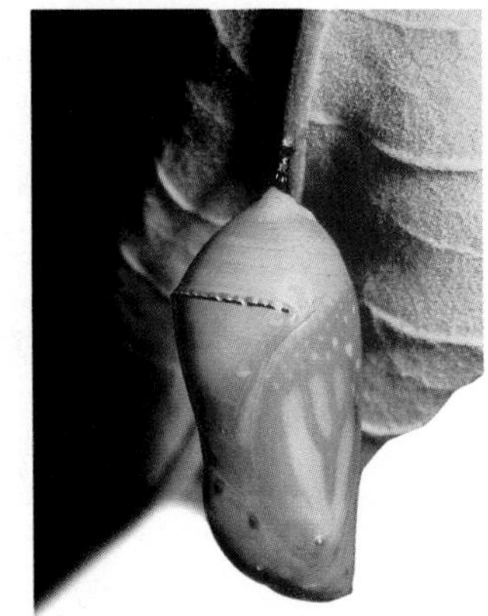

- Choose any two transformations. Discuss how you could write a computer or calculator program that would move a shape like trapezoid $ABCD$ according to the transformations you chose.

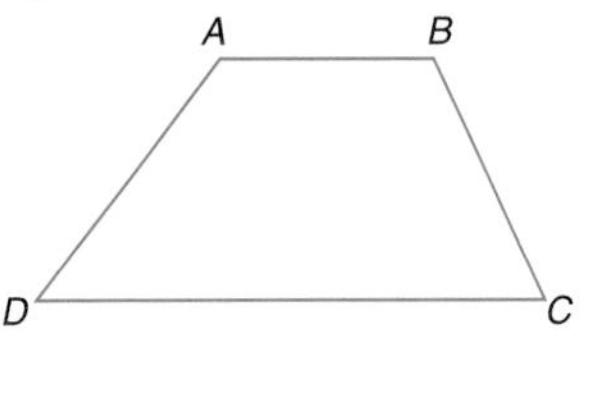

1647 A Massachusetts law requiring every town of 50 or more families to provide a school where children could learn to read and write is written.

1912 Prizes are added to each package of Cracker Jacks®.

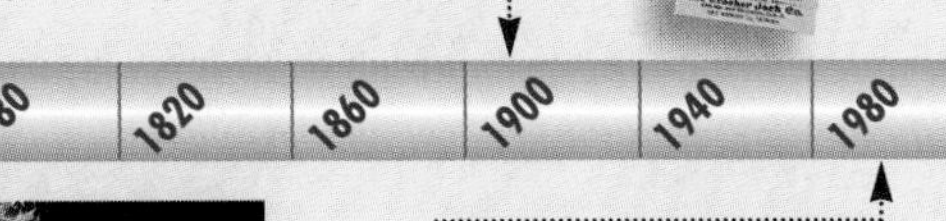

1758 Carl von Linne, a Swedish naturalist, gives the Tiger Swallowtail, a common North American butterfly, its genus and species name, *papilio glaucus*.

1994 NASA astronauts repair the Hubble Telescope to improve reflective images of its mirrors.

PEOPLE IN THE
NEWS

It's easy to get started in studying insects because, as James Fujita says, they are everywhere. Patience and some simple equipment, including a magnifying glass, a notebook, field guides, and a camera, will help. Students can find out more information on the study of insects by contacting the Young Entomologists' Society, International Headquarters, 1915 Peggy Place, Lansing, MI 48910.

Chapter Project

Cooperative Learning You might want to structure groups so that each group has one member with an interest in nature, one member with an interest in programming, and one member with management skills to plan and direct the project. Each student could be responsible for researching or listing instances of the translation, rotation, and dilation transformations. Then the members should discuss these transformations together as a group.

Investigations and Projects Masters, p. 73

NAME ______ DATE ______

13 **Chapter 13 Project A** — Student Edition Pages 694–759

Finding Transformations

1. Collect from around school, home, or other places objects or pictures that can be described as involving at least one type of transformation. You may draw pictures if you cannot collect them. Include at least one object or picture for each type of transformation.
2. Put the pictures and the objects together as a collage or display. You may want to group them by transformation or choose another arrangement.
3. Indicate what type of transformation each object or picture contains.
4. As a class, make a list of all of the objects and pictures that you found along with the different places that you found them. Discuss the extent to which transformations appear in nature and are used in society.

Alternative Chapter Projects

Two other chapter projects are included in the *Investigations and Projects Masters*. In Chapter 13 Project A, pp. 73–74, students extend the topic in the chapter opener. In Chapter 13 Project B, pp. 75–76, students investigate the skills needed to make computerized movies.

13-1 LESSON NOTES

NCTM Standards: 1–4, 7, 8

Instructional Resources

- Study Guide Master 13-1
- Practice Master 13-1
- Enrichment Master 13-1

Transparency 13-1A contains the 5-Minute Check for this lesson; **Transparency 13-1B** contains a teaching aid for this lesson.

Recommended Pacing	
Honors Pacing	Days 1 & 2 of 13

1 FOCUS

5-Minute Check
(over Chapter 12)

1. Write an equation of the line parallel to the graph of $y = -3x + 5$ and passing through the point at (4, 1). $y = -3x + 13$
2. Determine whether the points $A(2, 5)$, $B(-2, 3)$, and $C(0, 9)$ are collinear. **no**
3. Find the magnitude and direction of $\overrightarrow{AB}$, with $A(2, 1)$ and $B(6, 4)$. **5, 37°**
4. Determine the distance between $A(0, -1, 4)$ and $B(-2, 6, 3)$. $\sqrt{54} \approx 7.3$
5. Determine the coordinates of the midpoint of the segment whose endpoints have coordinates (0, 4, 3) and (−6, 2, 1). **(−3, 3, 2)**

Motivating the Lesson

Questioning Show students the orbit of Earth around the sun. Ask them to describe the orbit.

13-1 What Is Locus?

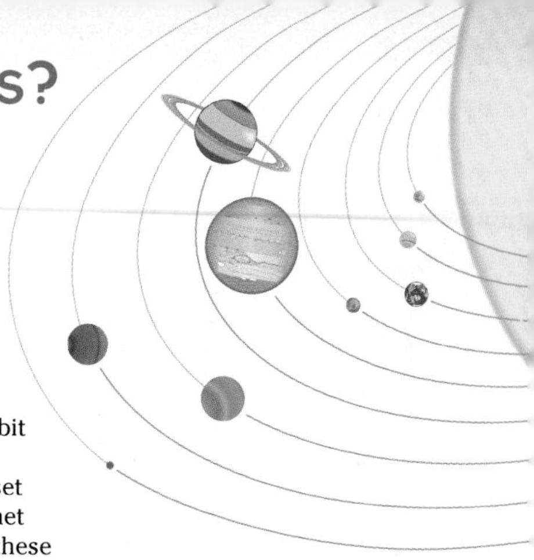

What YOU'LL LEARN

- To locate, draw, and describe a locus in a plane or in space.

Why IT'S IMPORTANT

Using loci is a way to describe many sets of points that have unique characteristics.

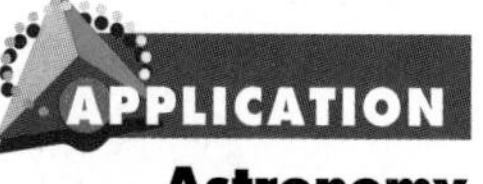

APPLICATION
Astronomy

In our solar system, nine planets orbit the Sun. The path of each planet is different, and each path satisfies a set of conditions. The path of each planet can be defined and predicted from these conditions. A path can be thought of as a **locus** of points. The word locus comes from a Latin word meaning "location" or "place." The plural of locus is *loci* (pronounced *LOW-sigh*).

FYI

The Solar system consists of the Sun, which is a star, and all of the objects that travel around it. These include nine planets and their satellites (moons), asteroids, meteoroids, comets, interplanetary dust, and electrically charged gas called plasma.

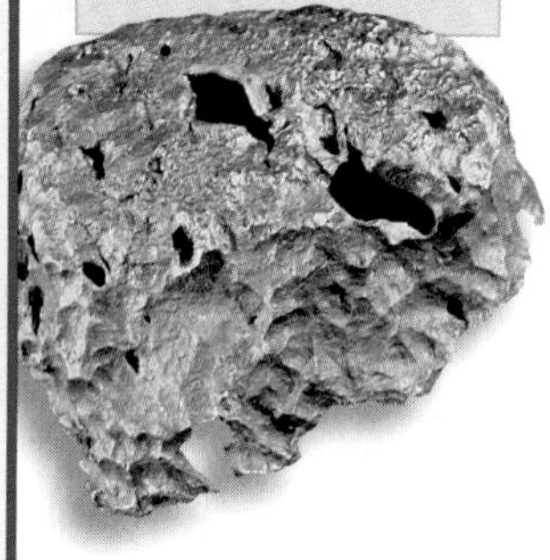

In geometry, a figure is a locus if it is the set of all the points that satisfy a given condition or set of conditions. A locus may also be defined as the path of a moving point satisfying a given set of conditions. A locus may be one or more points, lines, planes, surfaces, or any combination of these. A locus can occur in a plane or in space.

In order to describe a locus for a given problem, you should follow these steps.

Problem Find the locus of all points in a plane that are 5 centimeters from a given point in that plane.

Step 1 Draw the given figure. •*C* The given figure is point *C*.

Step 2 Locate points that satisfy the conditions. (5 cm, •*C*) Draw points that are 5 centimeters from point *C*. Locate enough points to suggest the shape of the locus.

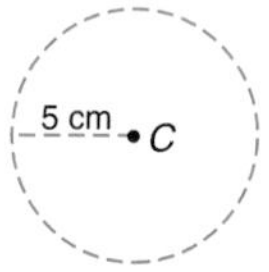

Step 3 Draw a curve or line.

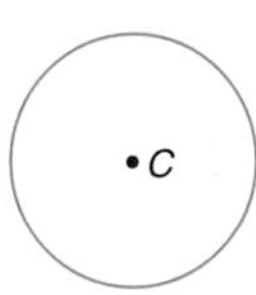

The points suggest a circle.

Step 4 Describe the locus. The locus of the points in the plane that are 5 centimeters from given point *C* is a circle with a radius of 5 centimeters.

FYI

Solar activity goes in cycles. The best known cycle is the 11-year sunspot cycle. The number of sunspots visible on the sun's surface varies over this period.

The steps below summarize the procedure for determining a locus.

Procedure for Determining Locus	**After reading the problem carefully, follow these steps.** **1. Draw the given figure.** **2. Locate the points that satisfy the given conditions.** **3. Draw a smooth geometric figure.** **4. Describe the locus.**

A locus in a plane may be different from a locus in space that has the same description.

Locus of All Points Equidistant from Two Parallel Lines	
In a Plane	**In Space**
a, ℓ, b Line ℓ is the locus.	a, $\mathcal{M}$, b Plane $\mathcal{M}$ is the locus.
The locus of points equidistant from two parallel lines in a plane is a line located halfway between the two lines.	The locus of points equidistant from two parallel lines in space is a plane located halfway between the two lines.

Example  **Find the locus of all points in space that are 20 millimeters from a given point *P*.**

Step 1 *Draw the given figure.* • P The given figure is point P.

Step 2 *Locate the points that satisfy the given conditions.* Draw enough points to determine the shape.

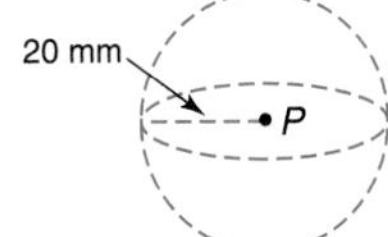

Step 3 *Draw a smooth geometric figure.* The points describe a sphere.

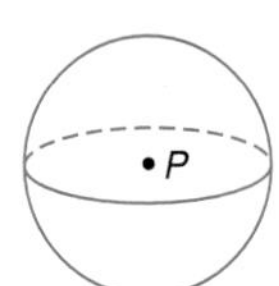

Step 4 *Describe the locus.* The locus of the points in space that are 20 millimeters from given point P is a sphere with a radius of 20 millimeters.

Many loci can be determined by theorems you have learned in previous chapters.

2 TEACH

In-Class Example

For Example 1
Find the locus of all points in space that are 7 centimeters from a point P.

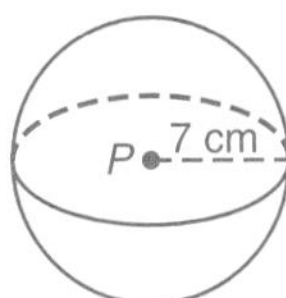

The locus of points in space that are 7 cm from point *P* is a sphere with point *P* as its center and a radius of 7 cm.

Teaching Tip When defining *locus*, point out that a locus in a plane is limited to points that lie in that plane, while a locus in space is not limited in any way.

Teaching Tip Point out that there are some indefinite steps in the process of finding a locus. For example, the number of points necessary to get an idea of the shape of the figure varies from problem to problem. Students must decide when they have enough points.

Alternative Teaching Strategies

Reading Geometry Have students look up the word *locus* in a dictionary and explain how they think it may apply to geometry.

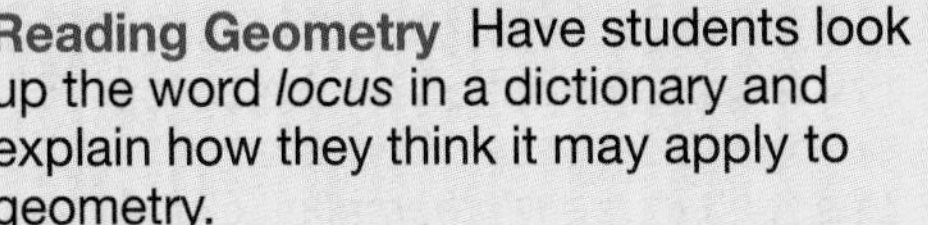

Alternative Learning Styles

Kinesthetic Cut a circle out of paper and draw a thick line around its edge. Also, bring in a tennis ball or other type of hollow ball, an empty can of food with both the bottom and top removed, or other types of objects that can be used to demonstrate loci.

In-Class Examples

For Example 2
Determine the locus of all points in a plane that are equidistant from two intersecting lines. **The locus is two lines passing through the intersection point that are bisectors of the vertical angles formed by the lines.**

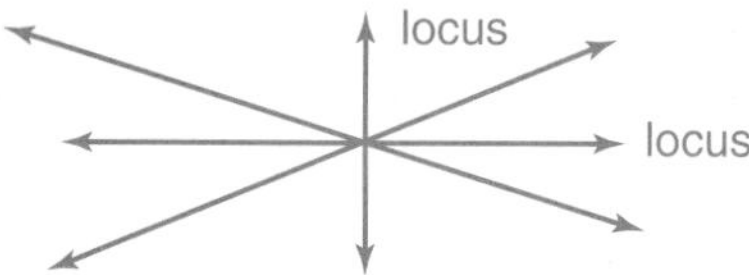

For Example 3
Draw a diagram and describe the locus of points in space that are 1 inch from a given line.

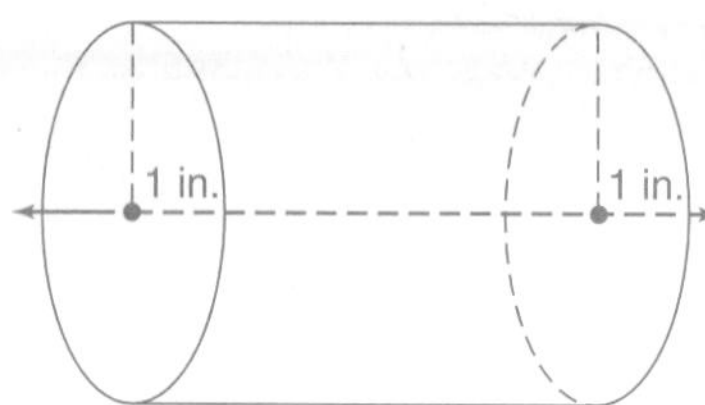

an infinite cylinder of radius 1 inch

For Example 4
A dog is tied to a clothesline with a 13-foot leash so that the leash can move along the clothesline, but stops at each end. If the clothesline is 39 feet long and is 5 feet above the dog's collar, find the locus of the boundary of the dog's running area.

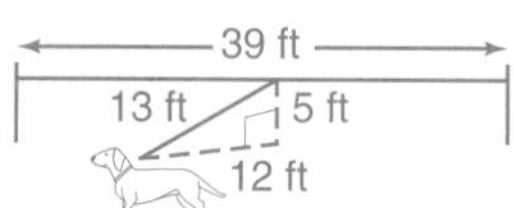

The locus is a path composed of a semicircle with a radius of 12 feet on each end of the clothesline and two parallel lines, one on each side of the clothesline, 39 feet long and 12 feet away from the line on the ground directly below the

Teaching Tip In Example 4, students may understand the situation more easily if they actually act out the example.

Example  **Determine the locus of all points in a plane that are equidistant from the sides of a given angle and are in the interior of that angle.**

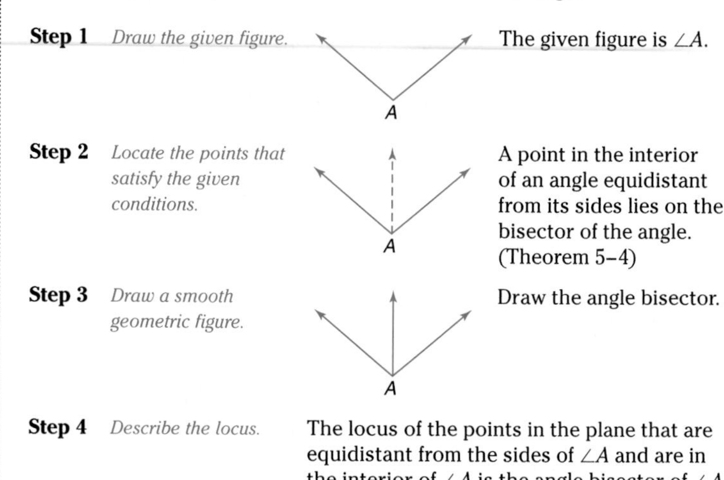

Step 1 *Draw the given figure.* The given figure is $\angle A$.

Step 2 *Locate the points that satisfy the given conditions.* A point in the interior of an angle equidistant from its sides lies on the bisector of the angle. (Theorem 5–4)

Step 3 *Draw a smooth geometric figure.* Draw the angle bisector.

Step 4 *Describe the locus.* The locus of the points in the plane that are equidistant from the sides of $\angle A$ and are in the interior of $\angle A$ is the angle bisector of $\angle A$.

A locus can be composed of more than one figure.

Example  **Draw a diagram and describe the locus of points in space that are 0.5 inch from the endpoints of a given line segment.**

The locus of all points in space at a specific distance from a given point is a sphere. Thus, for this problem, the locus of points is two spheres each with a radius of 0.5 inch whose centers are the endpoints of the given line segment.

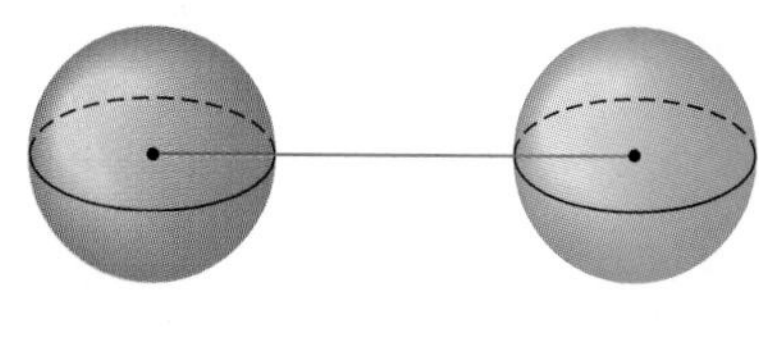

A locus can also describe a region in a real-life situation.

Example 4 **Paloma designed a type of sprinkler that looks like a small tractor. It moves along a straight line as it sprays water in a circular motion. Suppose this sprinkler can cast a spray 30 feet in diameter and travels 10 feet in an hour.**

a. Draw a figure showing the boundary of the sprinkler's spray in an hour.

When not moving, the sprinkler makes a circular spray. As it moves, the circle moves leaving a wet path.

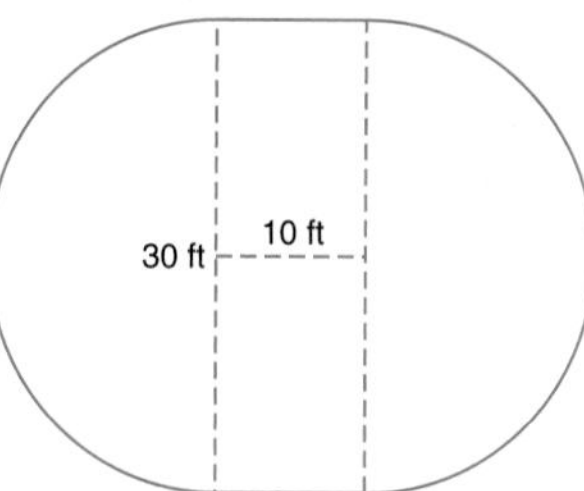

Cooperative Learning

Co-op Co-op Separate the class into small groups and give each group a few problems that involve finding the locus of a figure. They may use any method to help them find each locus. When each group is finished, have students explain what the locus of each problem is and how they found it. Every student in each group should understand the solutions. For more information on the co-op co-op strategy, see *Cooperative Learning in the Mathematics Classroom*, one of the titles in the Glencoe Mathematics Professional Series, page 32.

b. Describe the locus.

The locus of points is a rectangle having dimensions 10 feet by 30 feet with a semicircle of radius 15 feet attached to each end.

c. How large is the area that this sprinkler will water?

$$\begin{aligned}\text{total area} &= \text{area of rectangle} + \text{area of circle}\\ &= \ell w + \pi r^2\\ &= (10)(30) + \pi(15)^2\\ &= 300 + 225\pi\\ &\approx 1006.86 \quad \textit{Use a calculator.}\end{aligned}$$

The sprinkler will water about 1007 square feet.

LOOK BACK

Refer to Lesson 10-3 to review area of several regions and Lesson 10-5 to review area of circles.

CHECK FOR UNDERSTANDING

Communicating Mathematics

1–3. See margin.

6. a cylindrical surface of radius 10 mm with line m as the axis

Study the lesson. Then complete the following.

1. **Explain** why a locus is a geometric figure.
2. **Describe** the locus defined by *a circle with a radius r.*
3. **Explain** the difference between a locus in a plane and a locus in space.

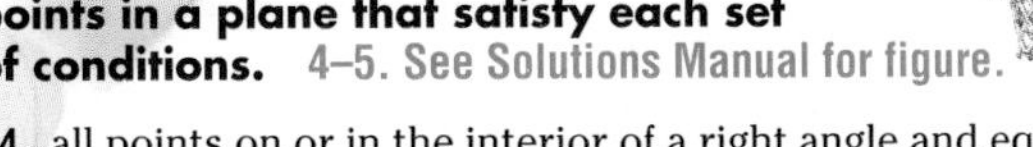

Guided Practice

7. a line perpendicular to the plane of the square through the point of intersection of the diagonals

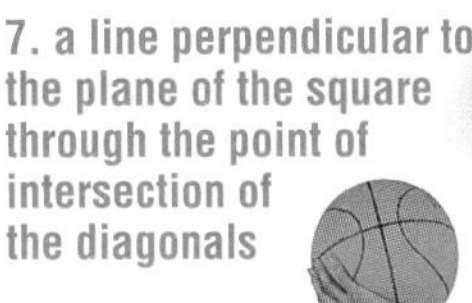

Draw a figure and describe the locus of points in a plane that satisfy each set of conditions. 4–5. See Solutions Manual for figure.

4. all points on or in the interior of a right angle and equidistant from the rays that form the angle   the angle bisector
5. all points equidistant from two given points   the ⊥ bisector of the segment that joins the two points

Draw a figure and describe the locus of points in space that satisfy each set of conditions. 6–7. See Solutions Manual for figure.

6. all points 10 millimeters from a given line m
7. all points equidistant from the vertices of a given square
8. Write a locus description of the drawing at the right. a pair of parallel lines, one on each side of m, and each 10 mm from m

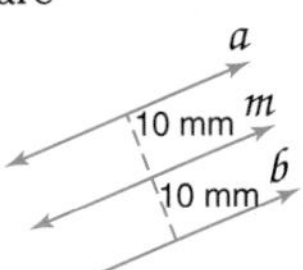

9. Use isosceles trapezoid $ABCD$ in plane $\mathcal{R}$ to describe the locus of points in plane $\mathcal{R}$ equidistant from the vertices of $ABCD$.   the perpendicular bisector of $\overline{AB}$ and $\overline{DC}$ in plane $\mathcal{R}$

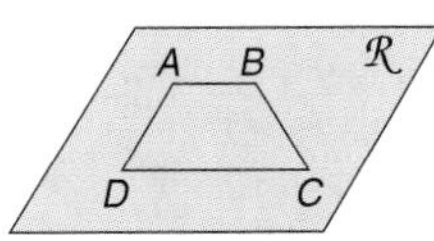

10. **Sports** In high school basketball, three points are awarded for a shot that is made from a distance greater than 19 feet 9 inches from the basket.
    a. Draw a figure showing the locus of points on a basketball court from which a 3-point shot can be made.   See margin.
    b. Describe the locus.   See margin.

3 PRACTICE/APPLY

Check for Understanding

Exercises 1–10 are designed to help you assess your students' understanding through reading, writing, speaking, and modeling. You should work through Exercises 1–3 with your students and then monitor their work on Exercises 4–10.

Error Analysis

Emphasize that a locus is a set of points. Even a locus that is a path can be thought of as a set of points because the locus would contain all points on the path.

Additional Answers

1. Both a locus and a geometric figure are a set of points that satisfy given conditions.
2. all points in a plane a given distance r from a point
3. In a plane, the locus could be 1- or 2-dimensional. In space, the locus could be 1-, 2-, or 3-dimensional.

10a.

Study Guide Masters, p. 78

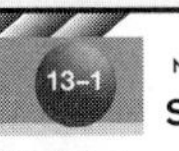

NAME ________ DATE ________

Study Guide

Student Edition Pages 696–701

What Is Locus?

A **locus** is the set of all points that satisfy a given condition or set of conditions. In determining a locus, it is often helpful to draw a figure showing several points that meet the given conditions. Remember that the locus is the set of *all* points meeting the conditions. Be sure that your final figure includes all possibilities. Also be careful to note whether you are considering all points in space or only points in a plane.

Example: Describe the locus of points in a plane that are 1 inch from a circle with a radius of 5 inches.

The locus consists of two circles having the same center as the 5-inch circle. One has a radius of 4 inches and the other has a radius of 6 inches.

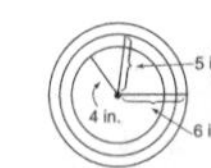

Draw a figure and describe the locus of points that satisfy each set of conditions.

1. all points in a plane equidistant from two parallel lines that are 3 cm apart
   3 cm
   a line parallel to the two parallel lines and halfway between them
2. all points in a plane that are 18 mm from a given point

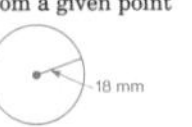

   a circle with an 18 mm radius and the given point as center
3. all points in a plane that are less than 5 centimeters from a given point P

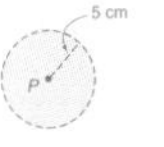

   all interior points for the circle of radius 5 cm and center P
4. all points in a plane that are equidistant from the sides of the angle and not outside the angle

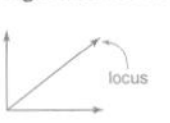

   the angle bisector
5. all points in space 2 inches from a given point
   2 in.
   a sphere with a radius of 2 inches and having the given point as center
6. all points in space 2 inches from a given plane
   2 in.
   2 in.
   two planes parallel to the given plane and 2 inches from either side of it

Reteaching

Using Manipulatives As a group project, have the class find the loci of several problems. Have students give examples of representations of each locus. For example, have them use a thumbtack and a string with a piece of chalk attached to draw loci that are circles. They can vary the length of the string to produce different size circles.

Additional Answer

10b. the interior of a rectangle that makes up the basketball court, minus two semicircles with centers at the midpoint of the ends with the basket and a radius of 19 feet 9 inches

Assignment Guide

Core (with proof): 11–37 odd, 38–46
Core (informal): 11–37 odd, 38–46
Enriched: 12–32 even, 34–46

For **Extra Practice**, see p. 789.

The red A, B, and C flags, printed only in the Teacher's Wraparound Edition, indicate the level of difficulty of the exercises.

Additional Answers

18. a plane parallel to the floor and 3 feet from the floor which contains a part of the desks

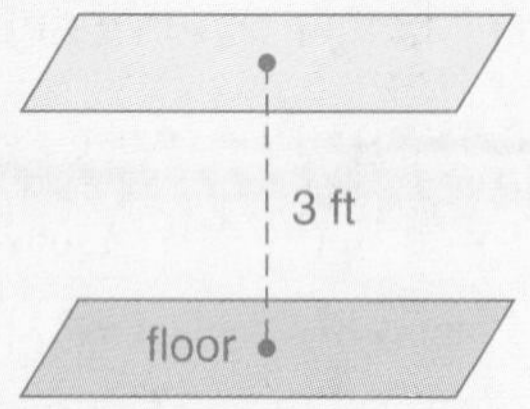

19. a pair of parallel planes, one on each side of $\mathcal{M}$, and each 4 meters from $\mathcal{M}$

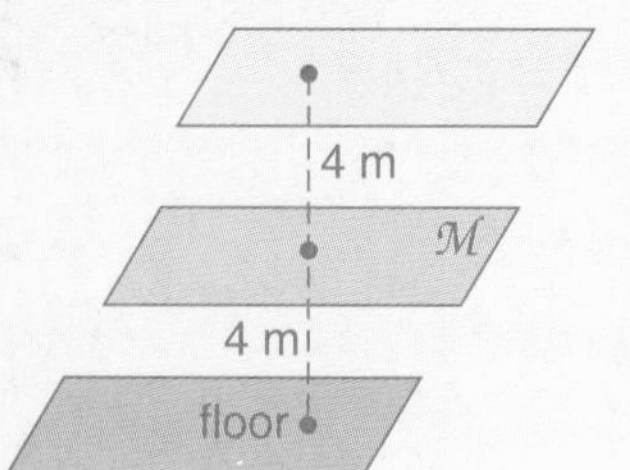

Practice Masters, p. 78

13-1 NAME______ DATE______
Practice
Student Edition Pages 696–701

What Is Locus?

Draw a figure and describe the locus of points that satisfy each set of conditions.

1. all points in a plane that are midpoints of the radii of a given circle

a circle whose radius is half the length of the radius of the given circle

2. all points in a plane that are equidistant from the endpoints of a given segment
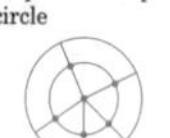
the perpendicular bisector of the segment

3. all points in a plane that are equidistant from two parallel lines 12 centimeters apart
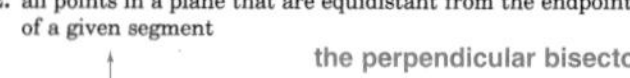
a line that is between the two given lines and 6 cm from each

4. all points in a plane that are 4 centimeters away from $\overline{AB}$
a pair of lines parallel to $\overline{AB}$, one on each side of $\overline{AB}$ and 4 cm from $\overline{AB}$

5. all points in a plane that are equidistant from two concentric circles whose radii are 8 inches and 12 inches
a circle of radius 10, concentric with the given circles

6. all points in a plane that are centers of circles having a given line segment as a chord
the perpendicular bisector of the line segment

7. all points in a plane that belong to a given angle or its interior and are equidistant from the sides of the given angle
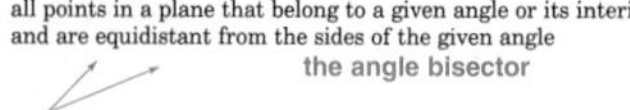
the angle bisector

EXERCISES

Practice

Draw a figure and describe the locus of points in a plane that satisfy each set of conditions. 11–17. See Solutions Manual for drawings.

11. all points equidistant from the centers of two given intersecting circles of the same radius   the line that joins the two points of intersection
12. all points on the midpoint of the radii of given circle C whose radius is 6 centimeters   concentric circle C with radius 3 cm
13. all points r units from a given line ℓ
14. all points that are the third vertices of triangles having a given base with length b and a given altitude length a
15. all points equidistant from the vertices of a given regular pentagon
16. all points on the perpendicular bisector of all chords in a given circle
17. all points that are equidistant from two concentric circles with larger radius r and smaller radius s concentric circle of radius $s + \frac{r-s}{2}$

13. a pair of parallel lines, one on each side of ℓ, and each r units from ℓ
14. a pair of parallel lines a units from the base
15. center of the pentagon
16. center of the circle

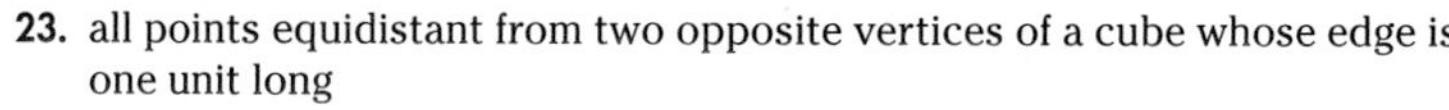

Draw a figure and describe the locus of points in space that satisfy each set of conditions. 18–20. See margin.

18. all points 3 feet from the floor in your classroom
19. all points 4 meters from a given plane $\mathcal{M}$
20. all points that are the centers of the wheels of a car moving in a straight line
21. all points that are third vertices of equilateral triangle ABC with side length s and with $\overline{BC}$ in plane $\mathcal{M}$
22. all points equidistant to three noncollinear points in a given plane $\mathcal{P}$
23. all points equidistant from two opposite vertices of a cube whose edge is one unit long
24. all points that are the centers of spheres with radius 7 units and tangent to a sphere of radius r units with center C
25. Write a locus description of the figure at the right.   a plane midway between two ∥ lines, perpendicular to the plane containing the two ∥ lines

21–24. See Solutions Manual.

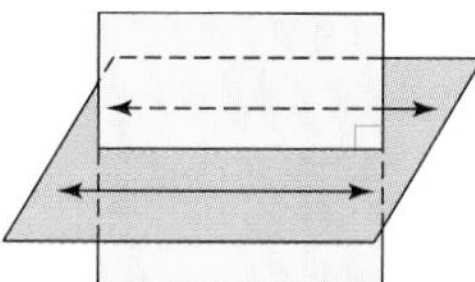

26–28. See Solutions Manual for figure.

Refer to ⊙O and inscribed △ABC in plane $\mathcal{R}$. Draw a figure and describe each locus in Exercises 26–28.

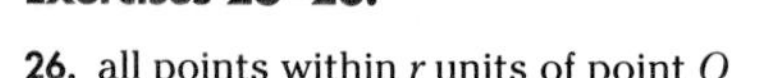

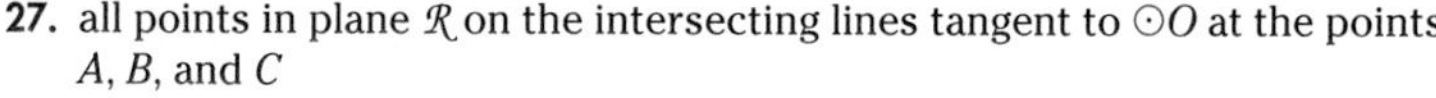

26. all points within r units of point O
27. all points in plane $\mathcal{R}$ on the intersecting lines tangent to ⊙O at the points A, B, and C
28. all points that are the midpoints of all chords of ⊙O of a given length
29. Given ∠ABC with bisector $\overline{BD}$ in a plane, describe the locus of the centers of circles that are tangent to $\overline{BD}$ and the outside ray of the angle.
30. Describe the locus of points in space that are all the centers of circles that contain two given points.
31. Given a hypotenuse of length h and an altitude of length a from the right angle of a right triangle, what is the locus of points in space formed by the right angle vertex of all possible right triangles?   See Solutions Manual.

26. sphere with center O and radius r
27. a triangle similar to △ABC containing points A, B, and C
28. a concentric circle with radius of given length and center O
29. the rays that bisect ∠ABD and ∠BDC
30. concentric circles in the plane ⊥ to the segment joining the two points with center at the midpoint of that segment

Additional Answer

20. set of two parallel lines parallel to plane of road and through the centers of the wheels

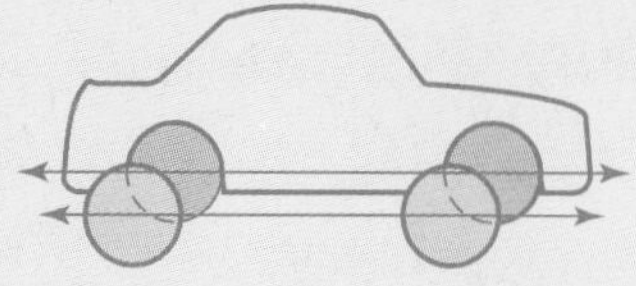

Tech Prep

Mine Technician Mining Technology degrees are offered at 14 two-year colleges in the United States. General mathematics courses are part of the requirements for the degree. For more information on tech prep, see the *Teacher's Handbook*.

32. Suppose a semicircle has center $(a, 0)$. What is the result of rotating the semicircle in space about the x-axis? **a sphere with center $(a, 0)$**

Graphing Calculator

33. Graph the functions $y = 0$ and $y = x^2$ on your calculator.
   a. What is the locus of points in the plane equidistant from the two graphs? **a curve, parabola with ordinate one-half that of $y = x^2$**
   b. Write an equation that represents this locus. $y = \frac{x^2}{2}$

Critical Thinking

34. Draw two perpendicular lines a and b. Describe the locus of points in a plane equidistant from these two lines.

34. the angle bisectors of the right angles

35. Mrs. Richter is part of a team selected to plan the placement of new fire stations in a town where a fire station is required to be within two miles of each building. Some of the members of the team thought that some buildings would be within two miles of more than one fire station; others thought this would never happen. Who was correct. Draw a diagram to support your answer. **See margin.**

Applications and Problem Solving

36. **Sports** Two friends who live on opposite sides of town usually ride bicycles together three days a week. Use the map to determine the locus of points where they can meet that are equidistant from both homes. **points along the ⊥ bisector of the segment connecting the two homes**

37. **Copper Mining** What is the locus of points that describes how particles will disperse in an explosion above ground if the expected distance a particle could travel is 300 feet? **half of a sphere with a radius of 300 ft**

Mixed Review

38. Draw a network to represent a regular octagon and all of its diagonals. (Lesson 10–7) **See Solutions Manual.**

39. **Space** A satellite is orbiting 220 miles above Earth. If the diameter of Earth is about 8000 miles, what is the length of the longest line of sight from the satellite to Earth? (Lesson 9–7) **1345 miles**

40. Find the value of x in $\odot P$. (Lesson 9–2) **75**

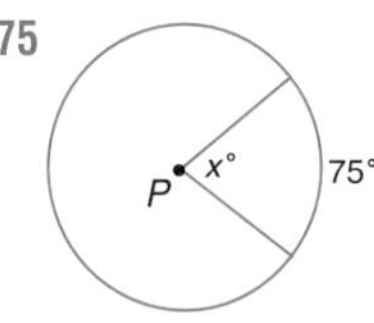

41. The shortest side of a triangle is 18 centimeters long. If two angles of the triangle measure 108° and 56°, find the lengths of the other two sides. (Lesson 8–5) **about 54.1 cm and 62.1 cm**

42. Classify the statement *The diagonals of a rhombus are congruent* as always, sometimes, or never true. (Lesson 6–4) **sometimes**

43. **Algebra** Given $\triangle YLO \cong \triangle GRN$, $m\angle Y = 60$, $m\angle L = 78$, $m\angle O = 42$, and $m\angle N = \frac{2}{3}x + 6$, find x. (Lesson 4–3) **$x = 54$**

44. If $\angle 1$ and $\angle 2$ are congruent alternate interior angles and are supplementary, describe the lines that form them. (Lesson 3–4) **∥ lines with ⊥ transversal**

INTEGRATION Algebra

45. Graph $y = 2x - 4$ by using the x- and y-intercepts. **See margin.**

46. Determine whether $9x = 5 - 3y$ is a linear equation. If it is linear, rewrite it in standard form, $Ax + By = C$. **yes; $9x + 3y = 5$**

Extension

Problem Solving Describe the locus of points in a plane that are equidistant from the three vertices of a triangle. **the point where the perpendicular bisectors of the sides intersect**

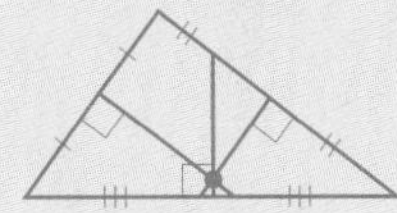

Additional Answer

45.

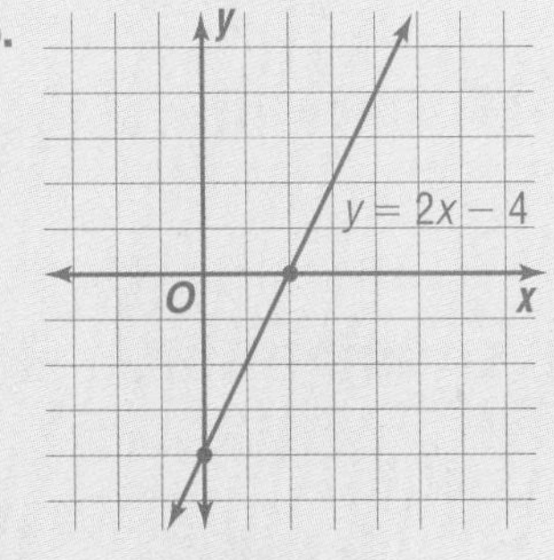

4 ASSESS

Closing Activity

Speaking Have students make a locus for a particular problem. Have them tell why it represents the locus of that problem.

Additional Answer

35. When three or more circles are tangent, there is an area among them that is not covered; therefore, the circles must overlap to cover all the area. This would put some of the buildings within two miles of more than one fire station.

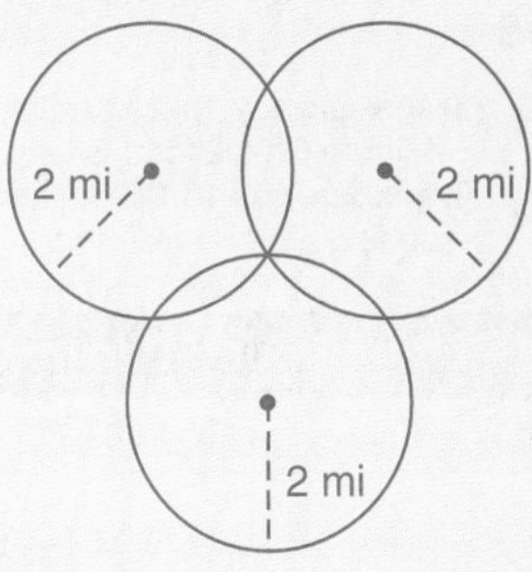

Enrichment Masters, p. 78

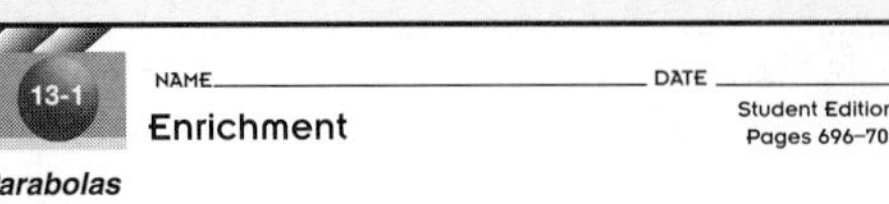

13-1 NAME__________ DATE ________

Enrichment

Student Edition Pages 696–701

Parabolas

A **parabola** is the set of all points equidistant from a given line and a given point not on the line. Coordinate geometry can help you discover what such a locus looks like. Consider the figure at the right. A point $P(x, y)$ is equidistant from the line $\frac{1}{4}$ unit below the x-axis and the point $F\left(0, \frac{1}{4}\right)$ if $y + \frac{1}{4}$ (the distance from the given point to the given line) equals PF. You can write this as an equation in terms of x and y and then simplify.

$$y + \frac{1}{4} = \sqrt{(x - 0)^2 + \left(y - \frac{1}{4}\right)}$$

$$y^2 + 2\left(\frac{1}{4}\right)y + \frac{1}{16} = x^2 + y^2 - 2\left(\frac{1}{4}\right)y + \frac{1}{16} \quad \text{Square each side.}$$

$$y^2 + \frac{1}{2}y + \frac{1}{16} = x^2 + y^2 - \frac{1}{2}y + \frac{1}{16} \quad \text{Simplify.}$$

$$y = x^2 \quad \text{Addition and subtraction properties of equality}$$

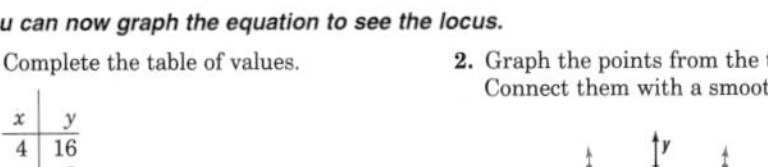

You can now graph the equation to see the locus.

1. Complete the table of values.

x	y
4	16
3	9
2	4
1	1
0	0
-1	1
-2	4
-3	9
-4	16

2. Graph the points from the table. Connect them with a smooth curve.

13-2 LESSON NOTES

NCTM Standards: 1–5, 7, 8

Instructional Resources

- Study Guide Master 13-2
- Practice Master 13-2
- Enrichment Master 13-2
- Assessment and Evaluation Masters, p. 352
- Graphing Calculator and Computer Masters, p. 13
- Multicultural Activity Masters, p. 25
- Real-World Applications, 27
- Tech Prep Applications Masters, p. 25

Transparency 13-2A contains the 5-Minute Check for this lesson; **Transparency 13-2B** contains a teaching aid for this lesson.

Recommended Pacing	
Honors Pacing	Day 3 of 13

1 FOCUS

5-Minute Check
(over Lesson 13-1)

Draw a figure showing the locus of points. Describe the locus.

1. all points in a plane that are equidistant from the endpoints of a semicircular arc

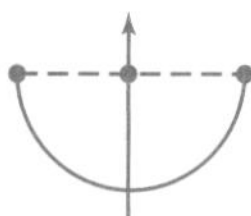

the perpendicular bisector of the segment joining the two endpoints of the arc

2. all points in space that are equidistant from two opposite corners of a cube

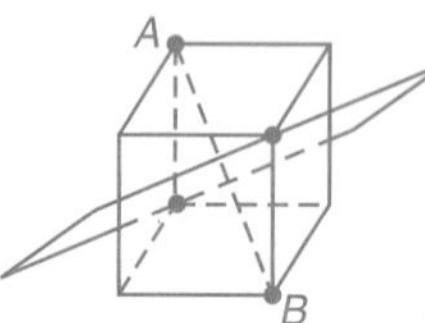

the plane that is the perpendicular bisector of the segment connecting the two opposite corners

13-2

Integration: Algebra
Locus and Systems of Linear Equations

What **YOU'LL LEARN**

- To find the locus of points that are solutions of a system of equations by graphing, substitution, or elimination.

Why **IT'S IMPORTANT**

You can use systems of equations to solve problems involving health, sports, and demographics.

APPLICATION
Physical Fitness

Millions of people exercise regularly to maintain or improve their cardiovascular health. During exercise many people monitor their heart rate to make sure that they reach a pulse rate that exercises but does not overtax their heart. The recommended maximum heart rate is a function of the age of the person and can be represented in a graph. This graph represents a locus of points that satisfy a certain set of conditions that can also be given in a table and represented by a linear equation.

Graph

Rate (beats per minute): 130, 140, 150, 160, 170, 180
Age (years): 20, 30, 40, 50, 60, 70

Linear Equation

$y = -0.8x + 190$

Table

Guide to Reasonable Heart Rate (rate taken immediately after exercise)

Age	Highest Rate (for 85% working level)
20	174
25	170
30	166
35	162
40	157
45	153
50	149
55	145
60	140
65	136
70	132

Source: *Longevity* by Kenneth R. Pelletier

Example 1

APPLICATION
Travel

A to Z Travel offered a vacation package plan to the Krivaneks that was based on a plane fare of \$350 per person and a rate of \$125 per day per person and included a hotel room and a rental car.

a. Write an equation that represents the total cost of this plan per person based on the number of days that the Krivaneks stay.

Let x = the number of days stayed and let y = the total cost of the plan.

$$\underbrace{\text{total cost}}_{y} = \underbrace{\text{(rate)}}_{125} \underbrace{\times}_{\cdot} \underbrace{\text{(number of days)}}_{x} \underbrace{+}_{+} \underbrace{\text{plane fare}}_{350}$$

$$y = 125x + 350$$

b. Draw a graph of the locus of points that represents this plan.

$y = 125x + 350$		
x	y	(x, y)
2	600	(2, 600)
6	1100	(6, 1100)
10	1600	(10, 1600)

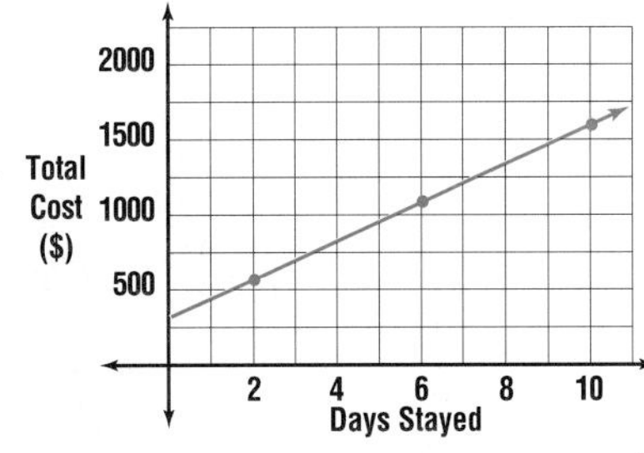

Relationships exist between pairs of lines. To find the locus of points that are common to two lines, you can use graphing to solve the system of equations.

Example 2

Consumerism

The Krivaneks then went to Northwest Travel to see what plan they could offer. Northwest Travel presented a plan that had a plane fare of $470 per person and a rate of $95 per day per person and included a hotel room and a rental car. Describe the locus of points in which the total cost of this plan and the total cost of the plan described in Example 1 are the same.

Explore You have two travel-plan options. One plan has a plane fare of $350 per person and a rate of $125 per day per person. The other plan has a plane fare of $470 per person and a rate of $95 per day per person.

Plan Write an equation for each of these plans and graph them on the same coordinate grid. Construct a table of values for these two equations. Then describe the locus of points in which the total cost of each plan is equal.

Solve Let x = the number of days stayed and y = the total cost of the plan.

$$\underbrace{\text{total cost}} = \underbrace{\text{(rate)}} \times \underbrace{\text{(number of days)}} + \underbrace{\text{plane fare}}$$

$$y \quad = \quad 95 \quad \cdot \quad x \quad + \quad 470$$

$$y = 95x + 470$$

The equation for the A to Z plan in Example 1 is $y = 125x + 350$.

The graph of these two equations is shown at the right. The graphs intersect at $P(4, 850)$. Therefore, P is the locus of points that satisfy the graph of both equations, $y = 95x + 470$ and $y = 125x + 350$. Therefore, if the Krivaneks stay for 4 days, the total cost of each plan will be $850.

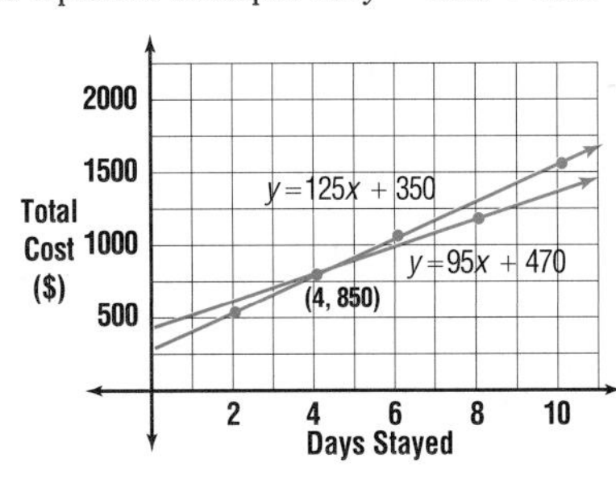

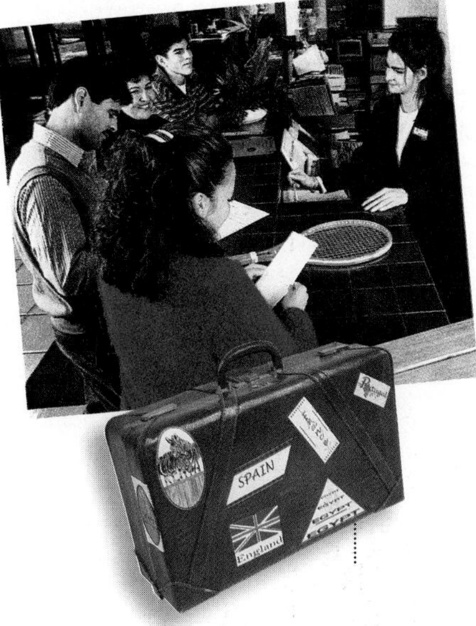

(continued on the next page)

Motivating the Lesson

Situational Problem Ask students to describe geographic or architectural features in which straight lines are used.

2 TEACH

In-Class Examples

For Example 1

In 1990, the population of Springfield, Illinois, was approximately 105,000 and was growing at the rate of 500 people per year.

a. Write an equation that represents the total population of Springfield. **$y = 500x + 105{,}000$**

b. Draw a graph of the locus of points that represents the population of Springfield in x number of years assuming the growth continues at the same rate.

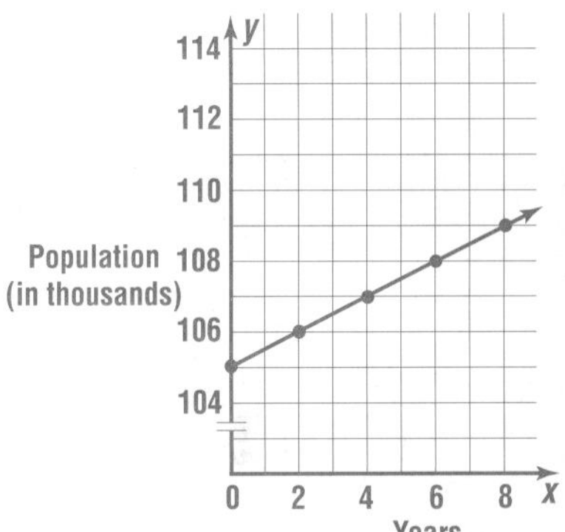

For Example 2

In 1990, Peoria, Illinois, had a population of about 113,000, which was decreasing at the rate of 1000 people per year. Describe the locus of points that represents the time at which the populations of Springfield and Peoria will be equal. **In about $5\frac{1}{3}$ years the populations will each be about 107,667 people. This could be represented on the graph by $\left(5\frac{1}{3}, 107{,}667\right)$.**

In-Class Example

For Example 3

a. Use substitution to find the locus of points that satisfy both equations $y = 2x + 1$ and $3x + y = 11$. $(2, 5)$

b. Use elimination to find the locus of points that satisfy both equations $2x + 3y = 5$ and $5x + 6y = 13$. $\left(3, -\frac{1}{3}\right)$

Teaching Tip Have students check their work by substituting the solution back into both equations.

Examine Study the table at the right. The table confirms the point of intersection is (4, 850).

x	95x + 470	125x + 350
1	565	475
2	660	600
3	755	725
4	**850**	**850**
5	945	975
6	1040	1100
7	1135	1225
8	1230	1350
9	1325	1475
10	1420	1600

Notice that the graphs in Example 2 intersect at the point with coordinates (4, 850). Since this point lies on both graphs, its coordinates satisfy both $y = 95x + 470$ and $y = 125x + 350$. The equations $y = 95x + 470$ and $y = 125x + 350$ are called a **system of equations**. The solution of this system is (4, 850).

You may remember from algebra that a system of equations can be solved by algebraic methods as well as by graphing. Two algebraic methods are the *substitution method* and the *elimination method*. The substitution method involves solving one equation for one of the variables and then substituting that value for the variable in the second equation. The elimination method involves combining two equations by addition or subtraction to eliminate one of the variables.

Example 3

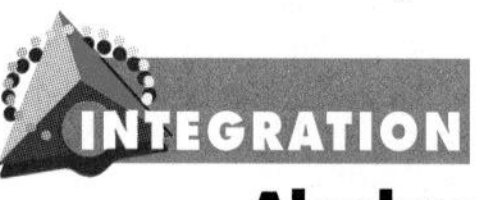

Algebra

Find the locus of points that satisfy the graphs of both equations $5x - 2y = 11$ and $y = x - 1$.

Method 1: Substitution

The second equation states that $y = x - 1$. Therefore, $x - 1$ can be substituted for y in the first equation.

$$5x - 2y = 11$$
$$5x - 2(x - 1) = 11 \quad \textit{Substitute } x - 1 \textit{ for } y \textit{ and solve.}$$
$$5x - 2x + 2 = 11$$
$$3x + 2 = 11$$
$$3x = 9$$
$$x = 3 \quad \textit{The x-coordinate is 3.}$$

Find y by substituting 3 for x in the second equation.

$$y = x - 1$$
$$y = 3 - 1$$
$$y = 2 \quad \textit{The y-coordinate is 2.}$$

The point with coordinates (3, 2) is the locus of points that satisfy the graphs of both equations $5x - 2y = 11$ and $y = x - 1$.

Method 2: Elimination

Rewrite the second equation in standard form, $Ax + By = C$.

$$y = x - 1$$
$$-x + y = -1 \quad \textit{Subtract x from each side.}$$

Sometimes adding or subtracting two equations will eliminate a variable. In this case, adding or subtracting the two equations will not eliminate a variable. If both sides of the second equation are multiplied by 5, the system can be solved by adding the two equations.

$5x - 2y = 11$

$-x + y = -1$  Multiply by 5.

$$5x - 2y = 11$$
$$\underline{-5x + 5y = -5}$$
$$3y = 6 \quad \textit{Add to eliminate x.}$$
$$y = 2 \quad \textit{The y-coordinate is 2.}$$

Substitute 2 for y in the second equation. Then solve for x.

$$y = x - 1$$
$$2 = x - 1$$
$$3 = x \quad \textit{The x-coordinate is 3.}$$

The point with coordinates (3, 2) is the locus of points that satisfy the graphs of both equations $5x - 2y = 11$ and $y = x - 1$.

Note in Example 3 that the solution to the system of equations was the same regardless of the method used. However, one method of solution may be simpler than another.

A graphing calculator can also help you find the locus of points that satisfy the graphs of two linear equations.

EXPLORATION GRAPHING CALCULATOR

Use a graphing calculator to find the locus of points that satisfy the graphs of $y = 3.4x + 2.1$ and $y = -5.1x + 8.3$.

- Graph both functions.
- Use the CALC menu and select 5:Intersect from the menu.
- As the calculator prompts you, press ENTER to select each line and determine the intersection point.

The approximate coordinates of the locus will appear at the bottom of the screen.

Your Turn c. when an approximation will do; fairly accurate

a. According to your exploration, what are the coordinates of the point of intersection to the nearest hundredth? (0.73, 4.58)

b. Solve the system of equations algebraically. Is your solution close to the results found on the graphing calculator? yes

c. When do you think finding a solution to a system of equations with a graphing calculator is appropriate? Does this method of solution always give an accurate solution?

EXPLORATION

Discuss with students the advantages and disadvantages of using a graphing calculator. Is it faster? More accurate? Easier? What are the limitations of the calculator? What are the limitations of pencil and paper? What are some other options?

3 PRACTICE/APPLY

Check for Understanding

Exercises 1–11 are designed to help you assess your students' understanding through reading, writing, speaking, and modeling. You should work through Exercises 1–4 with your students and then monitor their work on Exercises 5–11.

Assignment Guide

Core (with proof): 13–33 odd, 34, 35, 37–45
Core (informal): 13–33 odd, 34, 35, 37–45
Enriched: 12–32 even, 34–45

For **Extra Practice**, see p. 790.

The red A, B, and C flags, printed only in the Teacher's Wraparound Edition, indicate the level of difficulty of the exercises.

Additional Answers

2. no points when the lines are parallel, one point when the lines intersect, all points when the lines are the same line
4. Sample answer: substitution, solve for x in the second equation.

Study Guide Masters, p. 79

NAME ______ DATE ______

Study Guide — Student Edition Pages 702–708

Locus and Systems of Linear Equations

The graph of a linear equation in x and y is a straight line. Two linear equations form a **system of equations**. The locus of points that satisfy both equations of a system can be found graphically or algebraically. Two frequently used algebraic methods are the **substitution method** and the **elimination method**.

Example: Find the locus of points that satisfy both equations at the right by using both the substitution method and the elimination method.
$2x - y = 3$
$3x + 2y = 22$

Substitution Method
Solve the first equation for y in terms of x.

$2x - y = 3$
$y = 2x - 3$

Substitute $2x - 3$ for y in the second equation and solve for x.

$3x + 2(2x - 3) = 22$
$3x + 4x - 6 = 22$
$7x = 28$
$x = 4$

Find y by substituting 4 for x in the first equation.

$2(4) - y = 3$
$8 - y = 3$
$y = 5$

Elimination Method
$2x - y = 3$ ◄ Multiply by 2.
$3x + 2y = 22$

$4x - 2y = 6$
$3x + 2y = 22$ ◄ Add to eliminate y.
$7x = 28$
$x = 4$

Substitute 4 for x in the first equation. Solve for y.

$2(4) - y = 3$
$y = 5$

The point with coordinates (4, 5) is the locus of points that satisfy both equations.

State the locus of points that satisfy the intersection of each pair of lines.

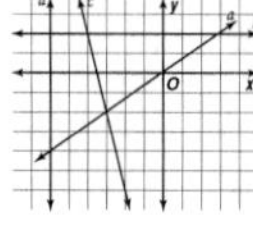

1. a and b (3, 2)
2. a and c (−3, −2)
3. b and c (−4, 2)
4. a and d (−6, −4)

Use either substitution or elimination to find the locus of points that satisfy both equations.

5. $y = 7x + 1$, $6x + y = 27$ (2, 15)
6. $3x + y = 7$, $2x - 3y = 12$ (3, −2)
7. $3x + 4y = 36$, $2x - 3y = -10$ (4, 6)

CHECK FOR UNDERSTANDING

Communicating Mathematics

Study the lesson. Then complete the following.

1. **Explain** whether a point could possibly satisfy all of a certain set of conditions in the definition of a locus if the point is not in the locus. 1. No; a point that is outside of the locus does not satisfy the conditions. This is an if and only if situation.
2. **Describe** the possible loci of points that satisfy the graphs of two linear equations. See margin.
3. **Health** The suggested maximum pulse rate for a person 15 years old is 160 beats per minute. The suggested maximum pulse rate for a person 40 years old is 140 beats per minute. If the pulse rate decreases at a constant rate, what is the equation whose graph models the locus of points that represent the maximum heart rate for any age a? $y = -\frac{4}{5}a + 172$

4. Determine whether the most efficient way to find the locus of points that satisfy the graphs of $3x + 2y = 9$ and $-x + 3y = 8$ is graphing, substitution, or elimination. Explain your answer. Sample answer: see margin.

Guided Practice

Graph each pair of equations to find the locus of points that satisfy the graphs of both equations.

5. $y = 2x - 2$, $x + y = 4$ (2, 2)
6. $y = 3x + 1$, $y = -3x + 1$ (0, 1)

Use either substitution or elimination to find the locus of points that satisfy the graphs of both equations.

7. $2x - y = 4$, $x - y = 5$ (−1, −6)
8. $y = \frac{1}{3}x - 4$, $y = -x + 1$ $\left(\frac{15}{4}, \frac{-11}{4}\right)$
9. $y = 4x$, $x + y = 10$ (2, 8)
10. $12 - 3y = 4x$, $40 + 4x = 10y$ (0, 4)

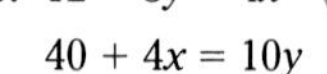

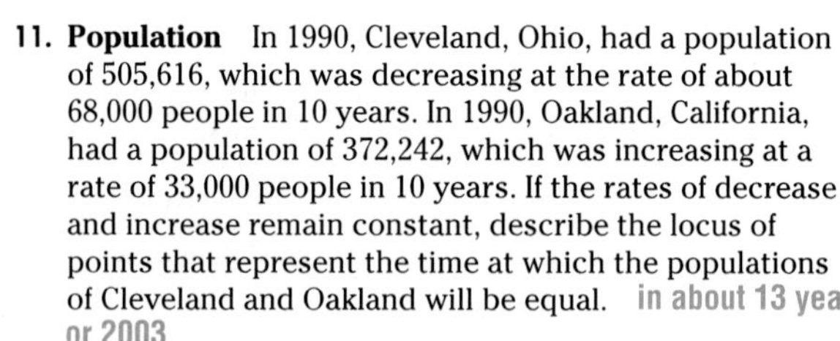

11. **Population** In 1990, Cleveland, Ohio, had a population of 505,616, which was decreasing at the rate of about 68,000 people in 10 years. In 1990, Oakland, California, had a population of 372,242, which was increasing at a rate of 33,000 people in 10 years. If the rates of decrease and increase remain constant, describe the locus of points that represent the time at which the populations of Cleveland and Oakland will be equal. in about 13 years or 2003

Cleveland

Oakland

EXERCISES

Practice

Graph each pair of equations to find the locus of points that satisfy the graphs of both equations. 12–17. See margin for graphs.

12. $x + 3y = 5$, $y = 2x - 10$ (5, 0)
13. $y = 4x + 7$, $y = -x - 3$ (−2, −1)
14. $4x - 3y = 6$, $2x - 3y = 12$ (−3, −6)
15. $y = -3x$, $6y - x = -38$ (2, −6)
16. $2x + y = 4$, $x - y = 2$ (2, 0)
17. $x - 2y = 0$, $y = 2x - 3$ (2, 1)

Reteaching

Using Comparison Have students compare the graphing, elimination, and substitution methods of solving a system of equations.

Additional Answer

12.

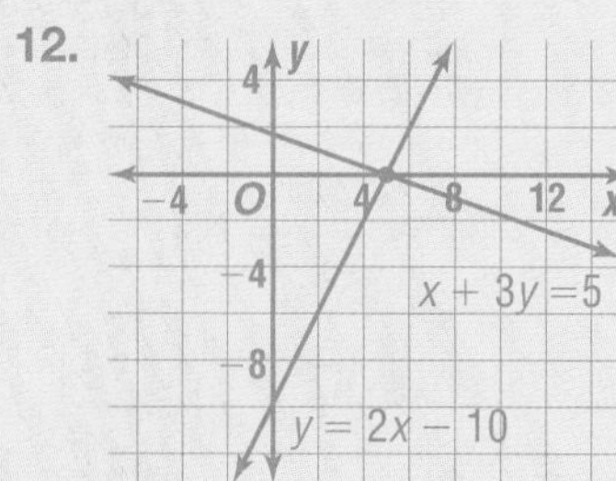

Use either substitution or elimination to find the locus of points that satisfy the graphs of both equations.

18. $x + 2y = 10$, $y = x - 4$ (6, 2)
19. $3x + y = 8$, $y = x$ (2, 2)
20. $x + y = 4$, $y = x + 6$ (−1, 5)

21. $2x + y = 9$, $x = 4y$ (4, 1)
22. $y = -x + 1$, $y = 6$ (−5, 6)
23. $y = -2x + 2$, $y = \frac{x}{2} + 2$ (0, 2)
24. $3x - y = 11$, $x + y = 5$ (4, 1)
25. $x - 4y = 9$, $2x - 3y = 8$ (1, −2)
26. $x - 2y = 7$, $2x + 3y = 0$ (3, −2)
27. $x + 4y = 27$, $x + 2y = 21$ (15, 3)
28. $y = x + 7$, $3x + y = 8$ $\left(\frac{1}{4}, \frac{29}{4}\right)$
29. $y = \frac{1}{3}x$, $x + 2y = -3$ $\left(-\frac{9}{5}, -\frac{3}{5}\right)$
30. The graphs of $y = 3$, $y = 7$, $y = 2x$, and $y = 2x - 13$ intersect to form a parallelogram.
    a. Find the coordinates of the vertices. (1.5, 3); (3.5, 7); (8, 3); (10, 7)
    b. Find the area of the parallelogram. 26
    c. Find the locus of points that satisfy the graphs of all four equations.

30c. There is no point that satisfies all equations.

31. Find the locus of points that satisfy the graphs of $(x + 3)^2 - (y + 4)^2 = 9$ and $x = 2$. (2, 0) and (2, 8)
32. Given $\overline{AG}$ with endpoints $A(-1, -3)$ and $G(9, 0)$, find the equation that represents the locus of points that is the perpendicular bisector of $\overline{AG}$.

32. $20x + 6y = 71$

33. Find the locus of points that satisfy the graphs of $y = (x - 1)^2$ and $y = x + 1$. (0, 1) and (3, 4)

Critical Thinking

34. Write three pairs of linear equations so that one pair would best be solved by using substitution, another by using elimination, and the third by graphing. Explain your reasoning for each pair of equations. See students' work.

Applications and Problem Solving

35. **Sports** For an opening-game promotion, the Orange Hill Volleyball Boosters purchased caps for all children under 12 years at \$1.50 each and computer mouse pads for everyone 12 years and over at \$2.50 each. The ticket price for the game for children under 12 was \$1.00 and \$4.00 for everyone else. What is the locus of points that represents the distribution of tickets sold if merchandise costs were \$405 and ticket sales were \$550? (70, 120)

36. **Population** From 1993 to 2020, the number of people age 65 or older will surge as baby boomers head into old age. Pennsylvania's population of persons age 65 or older was 1,908,000 in 1993 and is expected to be 2,303,000 by 2020. In 1993, the population in Texas of persons age 65 or older was 1,835,000, and in 2020, it will be 3,640,000. Will there be a time when Pennsylvania and Texas have the same population of people age 65 or older? If so, describe the locus of points that will represent that time.

36. Yes; in 1994, Texas and Pennsylvania will both have about 1,928,000 people age 65 or older; the intersection of the graphs of two equations representing the number of people age 65 or older.

Mixed Review

37. Draw and describe the locus of points in a plane formed by the vertex of the right triangles with hypotenuse $\overline{DC}$. (Lesson 13–1)

37. See margin for drawing; a circle with radius $\frac{DC}{2}$ and the midpoint of $\overline{DC}$ as its center.

Additional Answers

13.

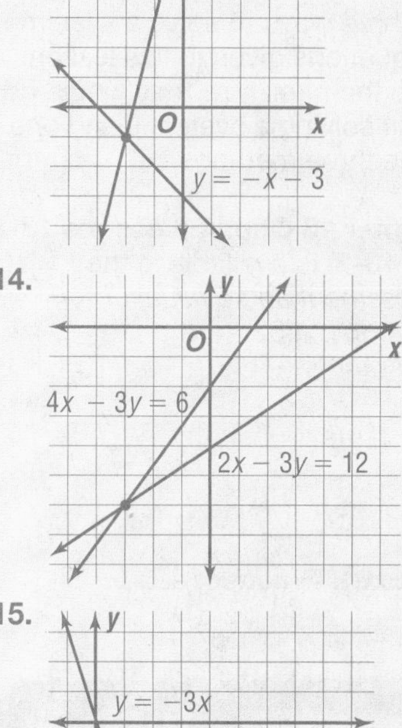

y = 4x + 7

y = −x − 3

Practice Masters, p. 79

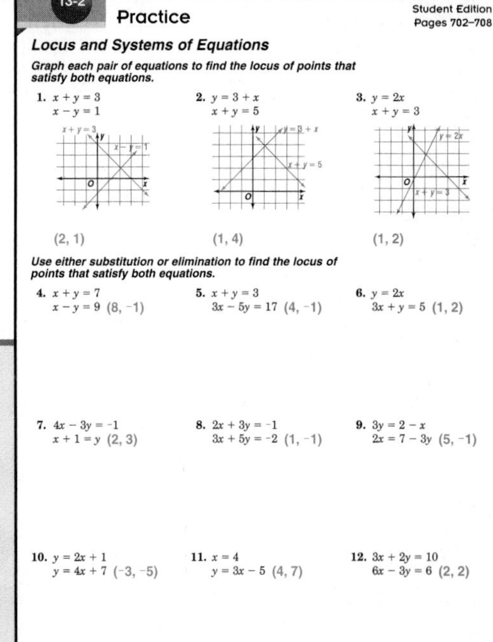

13-2 NAME_____ DATE_____

Practice — Student Edition Pages 702–708

Locus and Systems of Equations

Graph each pair of equations to find the locus of points that satisfy both equations.

1. $x + y = 3$, $x - y = 1$ (2, 1)
2. $y = 3 + x$, $x + y = 5$ (1, 4)
3. $y = 2x$, $x + y = 3$ (1, 2)

Use either substitution or elimination to find the locus of points that satisfy both equations.

4. $x + y = 7$, $x - y = 9$ (8, −1)
5. $x + y = 3$, $3x - 5y = 17$ (4, −1)
6. $y = 2x$, $3x + y = 5$ (1, 2)
7. $4x - 3y = -1$, $x + 1 = y$ (2, 3)
8. $2x + 3y = -1$, $3x + 5y = -2$ (1, −1)
9. $3y = 2 - x$, $2x = 7 - 3y$ (5, −1)
10. $y = 2x + 1$, $y = 4x + 7$ (−3, −5)
11. $x = 4$, $y = 3x - 5$ (4, 7)
12. $3x + 2y = 10$, $6x - 3y = 6$ (2, 2)

Additional Answers

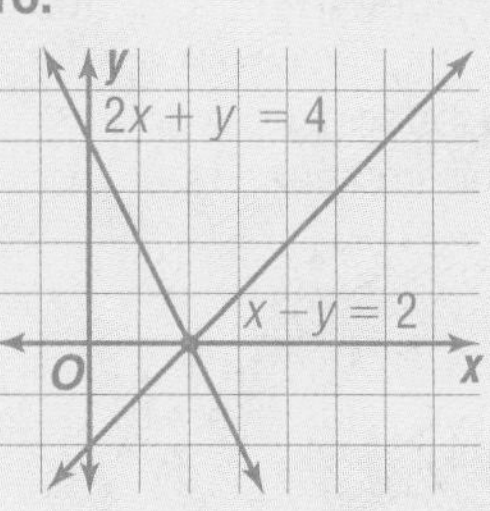

17.

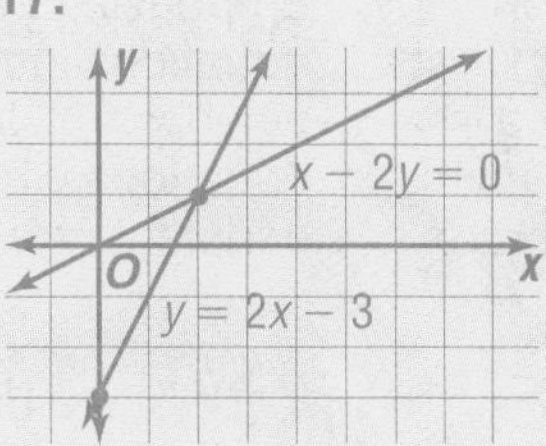

37.

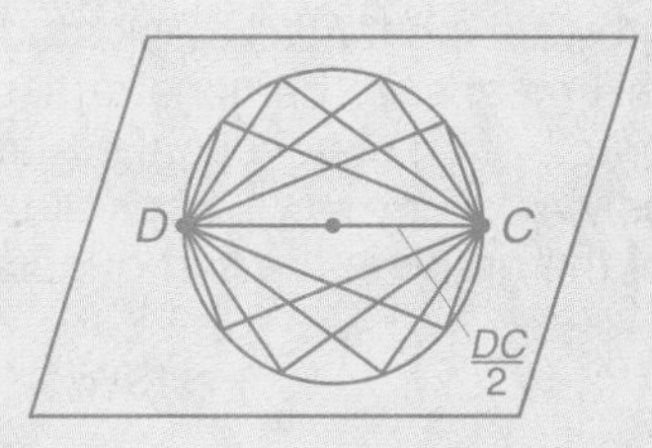

4 ASSESS

Closing Activity

Speaking Have students identify the three ways to solve a system of equations given in this lesson. Have them tell how they would go about solving a system using one of the three methods.

Chapter 13 Quiz A (Lessons 13-1 and 13-2) is available in the *Assessment and Evaluation Masters*, p. 352.

Additional Answer

41.

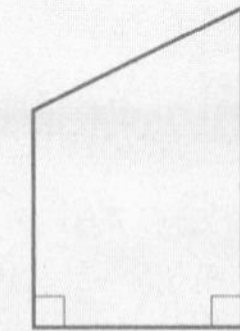

Mathematics and SOCIETY

Discuss with students other mathematical models that are used for people, such as supply and demand curves, exponential growth, and the bell curve.

Enrichment Masters, p. 79

NAME______ DATE ______

13-2 Enrichment — Student Edition Pages 702–708

Equations for Angles

Recall that the absolute value of x, written $|x|$, is equal to x when $x \geq 0$ and $-x$ when $x < 0$. You can write equations for angles by using absolute value expressions.

Graph each equation.

1. $y = |2x|$
2. $y = |x - 3|$

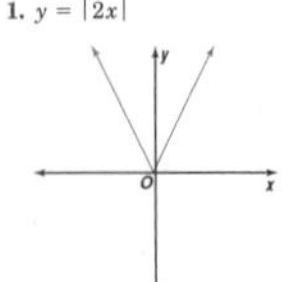

3. $y = -|x - 3|$
4. $y = -2x - 2 + 3|x - 2|$

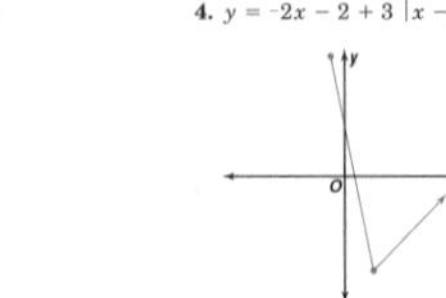

5. Write an equation for an angle that opens to the left or right. Then graph your equation in the coordinate plane. Answers will vary. A possible equation and its graph are shown.

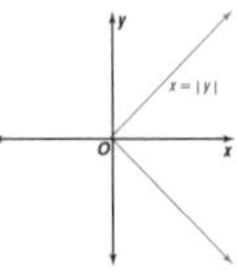

38. The measures of the exterior angles of a convex pentagon are g, $2g$, $1.5g$, $0.5g$, and g. Find the value of g and the measure of each exterior angle. (Lesson 10–1) **$g = 60$; 60, 120, 90, 30, 60**
39. If the equation of $\odot O$ is $(x - 2)^2 + (y - 3)^2 = 25$, find the value of j so that the point at $(6, j)$ lies on the circle. (Lesson 9–8) **0 or 6**
40. Use the information in the figure at the right to find the values of x, y, and z. *(Lesson 8–1)* **$x = \sqrt{76}$ or 8.7, $y = \sqrt{92}$ or 9.6, $z = \sqrt{437}$ or 20.9**

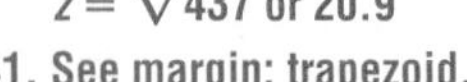

41. See margin; trapezoid.

41. Draw a quadrilateral with exactly one obtuse angle and one pair of parallel sides. What kind of quadrilateral is it? (Lesson 6–5)
42. Find $m\angle 1$ and $m\angle 2$. Then classify the triangle by its angles and by its sides. (Lesson 4–2) **60, 60; equiangular and equilateral**

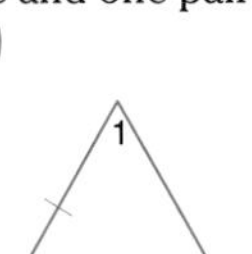

43. Parallel lines m and n are cut by transversal p forming angles 1 through 8. If all the even-numbered angles have measures of 35, draw a figure labeling the lines and the angles and determine the measures of all odd-numbered angles. (Lesson 3–2) **See Solutions Manual; all odd-numbered angles have measure 145.**

44. −15, −13, −11

INTEGRATION Algebra

44. The sum of three consecutive odd integers is −39. What are the integers?
45. Determine the value of r so the line that passes through points at $(2, r)$ and $(0, 3)$ has a slope of $\frac{3}{2}$. **6**

Fractal Models for City Growth

The excerpt below appeared in *Science News* on January 6, 1996.

During the last decade, researchers have begun to examine the possibility of using mathematical forms called fractals to capture the irregular shapes of developing cities....A magnified portion of a fractal looks very similar, if not identical, to the overall structure. Further magnification reveals details that again resemble the full pattern. Hence, fractal objects look the same whatever the magnification. Zooming in for a closer view doesn't smooth out the irregularities or end the branching....The recent focus on fractal models of urban development represents one aspect of a renewed interest in the importance of local actions, individual decisions, and self-organization in shaping cities....realistic models of how local activities lead to large-scale patterns and order may eventually serve as important tools for planning and predicting urban development. ■

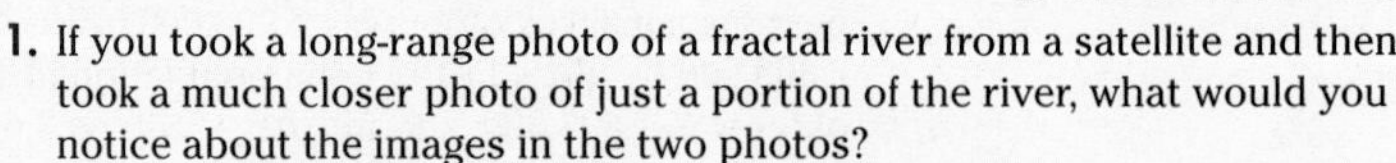

1–2. See Solutions Manual.

1. If you took a long-range photo of a fractal river from a satellite and then took a much closer photo of just a portion of the river, what would you notice about the images in the two photos?
2. How does the concept of locus play a part in city planning? What factors might be considered to predict and plan growth?
3. Do you think cities should grow according to a master plan or according to decisions that its citizens make about where they want to live? Explain.

3. See students' work.

Extension

Connections The world population grows exponentially. Ask students what this means. To demonstrate, draw a graph of an exponential equation, such as $y = x^2$ or $y = x^4$. Look at the portion of the graph in the first quadrant. Emphasize that y values in exponential equations increase much more rapidly when $x > 1$, as opposed to linear equations where the increase is a constant rate.

13-3 Intersection of Loci

What YOU'LL LEARN

- To solve locus problems that satisfy more than one condition.

Why IT'S IMPORTANT

You can use the intersection of loci to solve problems involving agriculture and sports.

You can use tracing paper to create a triangle that has certain specifications.

MODELING MATHEMATICS **Creating a Triangle**

Materials: tracing paper, ruler, protractor, plain paper

Use tracing paper to create a triangle ABC that contains $\angle A$, a side AC units long, and an altitude BD units long.

- Trace $\angle A$ on a piece of tracing paper. Extend the rays to the edges of the paper.
- Lay the paper over $\overline{AC}$, aligning both As and a ray of $\angle A$ with $\overline{AC}$. Mark the position of C on a ray of $\angle A$.

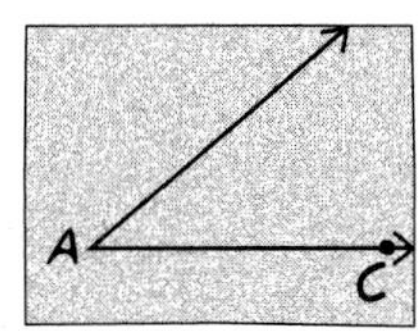

- On another piece of paper, draw a line ℓ and draw a segment perpendicular to ℓ with an endpoint on ℓ that is the same length as $\overline{BD}$.

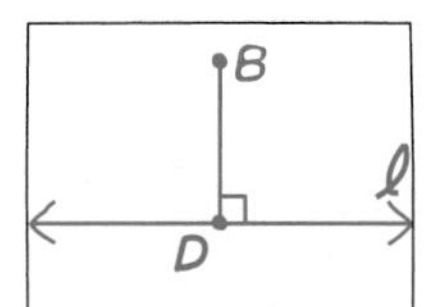

- Place the tracing paper over your drawing of ℓ and $\overline{BD}$ so that ℓ and $\overline{AC}$ coincide. Move the tracing paper right or left, keeping the two lines aligned, until point B appears to lie on the other ray of $\angle A$.

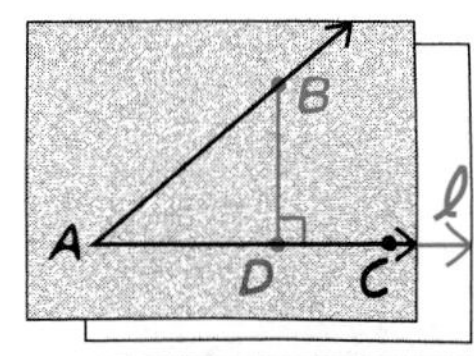

- Mark the position of B on your angle and draw $\triangle ABC$.

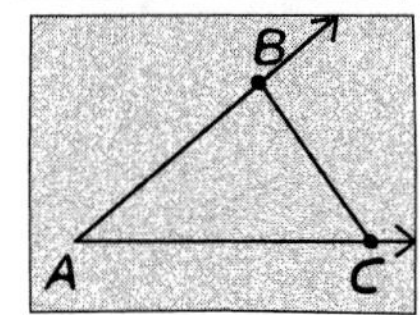

Your Turn

a. Measure $\angle BAC$, $\overline{AC}$, and $\overline{BD}$ in $\triangle ABC$. Are they the same measures as the original measures of $\angle A$, $\overline{AC}$, and $\overline{BD}$? yes

b. Repeat the activity putting C on the other ray. How does your result compare to $\triangle ABC$ in the activity? the same result

c. Are there other possible ways to draw $\triangle ABC$? no

13-3 LESSON NOTES

NCTM Standards: 1–5, 7, 8

Instructional Resources

- Study Guide Master 13-3
- Practice Master 13-3
- Enrichment Master 13-3

Transparency 13-3A contains the 5-Minute Check for this lesson; **Transparency 13-3B** contains a teaching aid for this lesson.

Recommended Pacing	
Honors Pacing	Day 4 of 13

1 FOCUS

5-Minute Check *(over Lesson 13-2)*

1. Graph the system of equations to find the locus of points that satisfy the equations $y = -\frac{3}{2}x - 2$ and $y = \frac{3}{2}x + 10$.

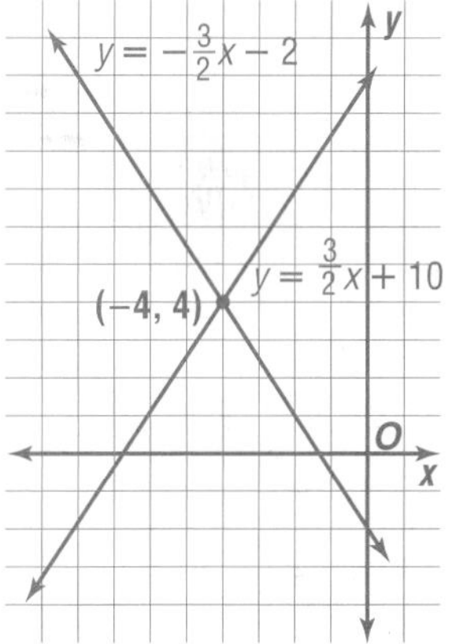

(−4, 4)

2. Use substitution to find the locus of points that satisfy the equations $x = 2y - 1$ and $5x - 6y = 3$. (3, 2)
3. Use elimination to find the locus of points that satisfy the equations $3x + y = -3$ and $6x - 4y = 12$. (0, −3)

MODELING MATHEMATICS Have students discuss why a triangle should be constructed in this way. Try different methods of creating a triangle with certain specifications and compare them.

Motivating the Lesson

Situational Problem Draw a small map of three city blocks on the chalkboard or overhead. Ask students to find the intersections of the streets. Point out that, if the streets represent loci, the intersections are where the loci intersect.

2 TEACH

Teaching Tip Before beginning the construction in Example 1, point out that $\overline{BD}$ must be perpendicular to $\overleftrightarrow{AC}$ to be an altitude.

In-Class Examples

For Example 1

Keshia is looking out the north window of a house, and Keiko is looking out the west window. Both windows are the same distance from the northwest corner of the room. If both Keshia and Keiko have an angle of sight of 160°, determine the locus of the area outside the house that both girls can see.

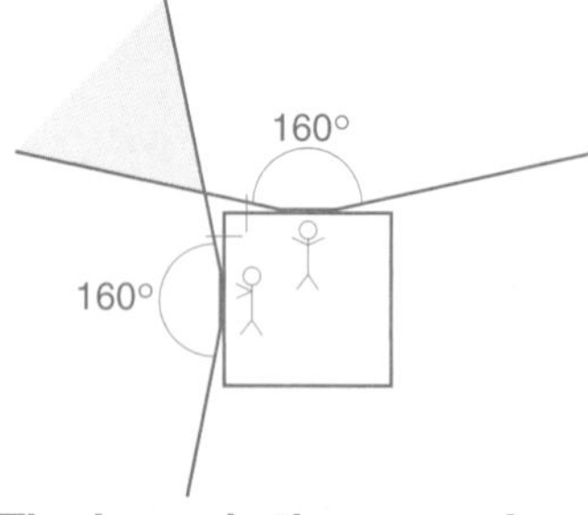

The locus is the area where the angles of sight intersect.

For Example 2

If the irrigation system is expanded to a radius of 150 meters, find the locus of points in the 200-meter by 200-meter field that is within reach of the irrigation system. What is the area in that field that is being irrigated? **≈ 17,671.5 m^2**

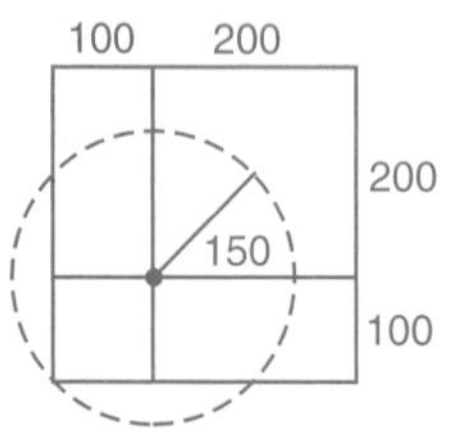

In Chapters 4 and 5, you learned that different sets of angles and sides could define unique triangles. Let's investigate the relationship between one given angle, one given altitude, and one given side of a triangle and how they can define one or more triangles.

Example 1 **Determine the ways in which the locus of points that satisfy the conditions in the Modeling Mathematics activity on the previous page can be drawn.**

Draw line ℓ. Construct $\overline{AC}$ on line ℓ.

The locus of points for B, BD units from $\overline{AC}$, is two parallel lines on either side of line ℓ. Construct the locus of points BD units from ℓ.

Construct $\angle A$ on the upper and lower sides of the base and mark the points where the rays intersect the parallel lines.

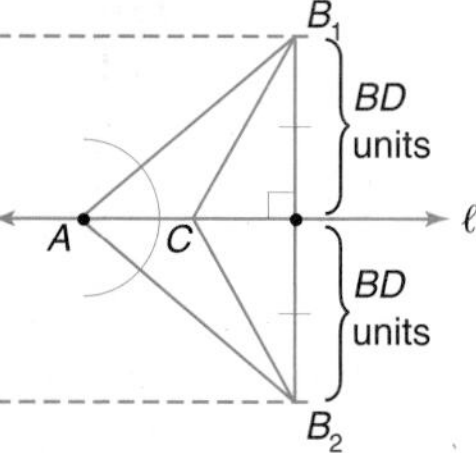

Note: $\triangle AB_1C$ and $\triangle AB_2C$ are congruent.

Label the points B_1 and B_2. Draw $\overline{B_1C}$ and $\overline{B_2C}$.

There are two possible ways to draw $\triangle ABC$.

Often loci satisfy several conditions. Such loci can usually be determined by finding the intersection of the loci that meet each separate condition.

Example 2

Agriculture

An above-ground irrigation system with a radius of 100 meters is positioned at the intersection of four fields as shown at the right.

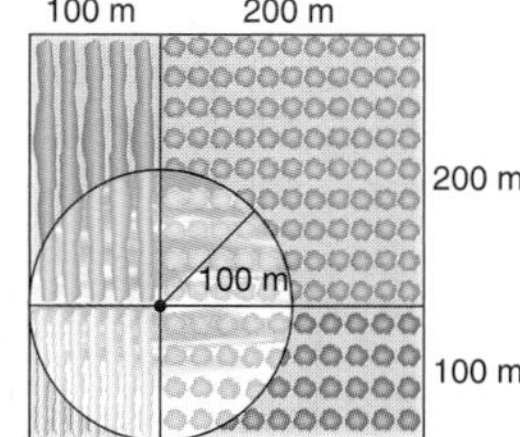

a. Find the locus of points in the 200-meter-by-200-meter soybean field that are within the reach of the irrigation system.

b. What percent of the soybean field is in the locus?

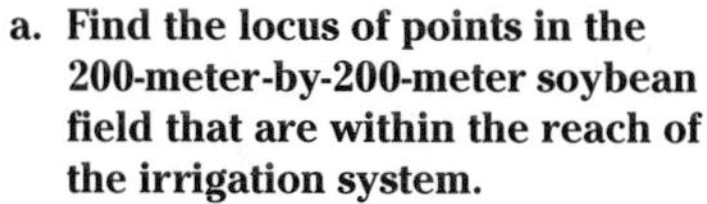

FYI

The soybean is one of the world's most inexpensive and useful sources of protein. Soybeans grown on an acre of land can provide about 10 times as much protein as beef cattle raised on the same land.

a. The locus of the irrigation system arm is the set of points on and within a circle with a radius of 100 meters. Only one fourth of this locus is in the 200-meter-by-200-meter soy bean field. Therefore, the locus of points in the soybean field is a sector that covers one fourth of the circle.

b. First, find the area of the sector that is irrigated.

$$A = \frac{1}{4}\pi r^2$$
$$= \frac{1}{4}\pi(100)^2 \text{ or about 7854 square meters}$$ *Use a calculator.*

The area of the soybean field is 200^2 or 40,000 square meters.

To find the percent of the soybean field that is in the locus, divide the area of the sector by the area of the field and write as a percent.

$$\frac{7854}{40{,}000} = 0.19635 \text{ or about } 19.6\%$$

About 19.6% of the field is in the locus.

FYI

Soybeans contain seven of the eight essential amino acids. If wheat or corn is part of one's diet, complete protein is supplied.

Alternative Learning Styles

Visual Collect or make examples of loci and use them to demonstrate how loci can intersect. For example, you may use a pencil or pen as an example of a line, and a tennis ball as a sphere. Have students experiment with the intersection of these objects. Have them use these models to visualize a homework exercise.

Example 3 **Find the locus of points in the coordinate plane that satisfy the graphs of $y = (x - 1)^2$ and $y = 5x - 11$.**

Algebra

The locus of points that satisfy $y = (x - 1)^2$ can be drawn by first making a table of some of the values and then sketching the graph.

x	$(x - 1)^2$	y	(x, y)
−2	$(-2 - 1)^2$	9	(−2, 9)
−1	$(-1 - 1)^2$	4	(−1, 4)
0	$(0 - 1)^2$	1	(0, 1)
1	$(1 - 1)^2$	0	(1, 0)
2	$(2 - 1)^2$	1	(2, 1)

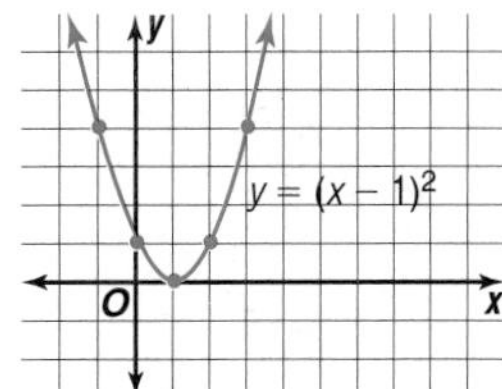

Therefore, from the graph, the locus of points that satisfy $y = (x - 1)^2$ is a parabola with the vertex at (1, 0).

The locus of points that satisfy the graph of $y = 5x - 11$ is a line with slope 5 that passes through (2, −1).

Method 1: Solve this system of equations graphically.

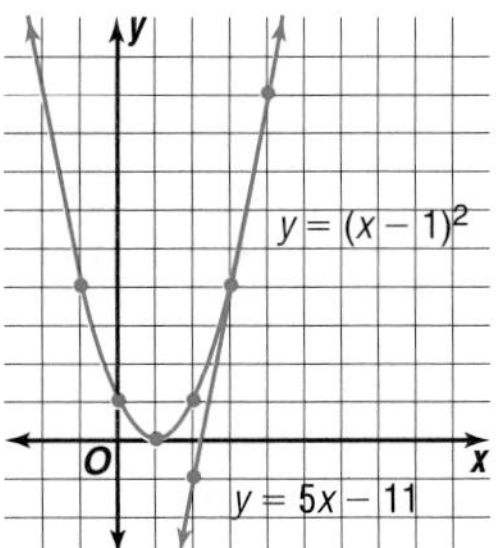

The graphs intersect at (3, 4) and (4, 9).

Method 2: Solve this system of equations algebraically. Substitute $5x - 11$ for y in the first equation.

$$y = (x - 1)^2$$
$$5x - 11 = (x - 1)^2$$
$$5x - 11 = x^2 - 2x + 1$$
$$0 = x^2 - 7x + 12$$
$$0 = (x - 3)(x - 4)$$
$$x = 3 \text{ or } x = 4$$

Using substitution, $y = 4$ for $x = 3$ and $y = 9$ for $x = 4$. The solutions are (3, 4) and (4, 9).

The locus of points that satisfy both conditions are points at (3, 4) and (4, 9).

The conditions of loci can also be in space. These loci conditions will be drawn and described differently than those in a plane.

Example 4 **Describe the locus of points in space that satisfy the graph of $y = 2x - 10$.**

The locus of points is a plane perpendicular to the xy-plane whose intersection with the xy-plane is the graph of $y = 2x - 10$.

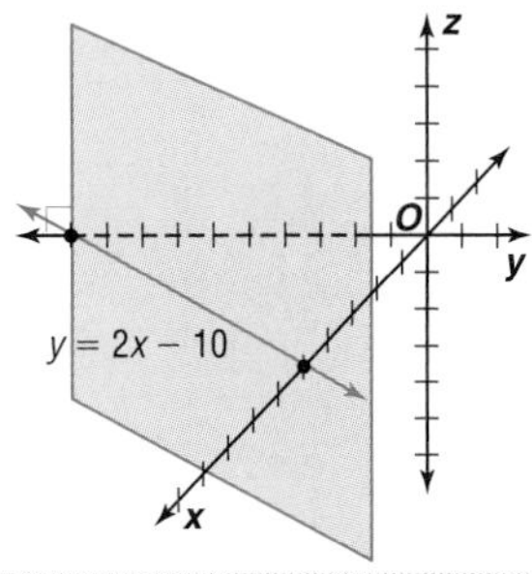

In-Class Examples

For Example 3
Find the locus of points in the coordinate plane that satisfy the graphs of $y = (x + 1)^2$ and $y = x + 7$. **(−3, 4), (2, 9)**

For Example 4
Describe the geometric figure whose locus in space satisfies the graph of $y = -3x + 2$.
a plane perpendicular to the xy-plane whose intersection with the xy-plane is the line $y = -3x + 2$

Teaching Tip Remind students that there are different methods for solving some questions, all of which are correct. Encourage them to be familiar with more than one method.

3 PRACTICE/APPLY

Check for Understanding

Exercises 1–11 are designed to help you assess your students' understanding through reading, writing, speaking, and modeling. You should work through Exercises 1–4 with your students and then monitor their work on Exercises 5–11.

Error Analysis

Since some loci can intersect in several ways, students may not see all possible intersections. To help, have them use or make examples of the loci so they can visualize the solutions.

Additional Answers

6.

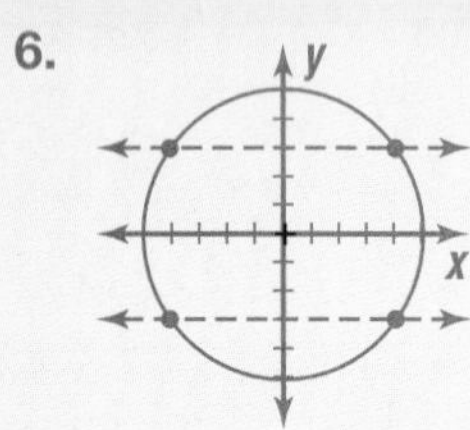

7.

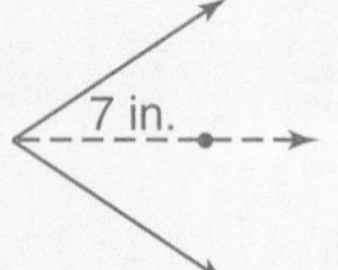

Study Guide Masters, p. 80

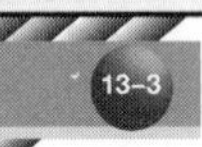

NAME______________________ DATE________

Study Guide

Student Edition Pages 709–714

Intersection of Loci

Sometimes locus problems involve sets of points that satisfy more than one set of conditions. In these cases, it is helpful to draw a figure for each set of conditions and then find the point or points where the figures intersect.

Example: Describe the locus of points in a plane 4 cm from a line and 6 cm from a point P on the line.

The locus of points 4 cm from the line is two parallel lines, each one 4 cm from the given line. The locus of points 6 cm from point P is a circle with a radius 6 cm, with P its center. The required locus consists of the 4 points where the two parallel lines intersect the circle.

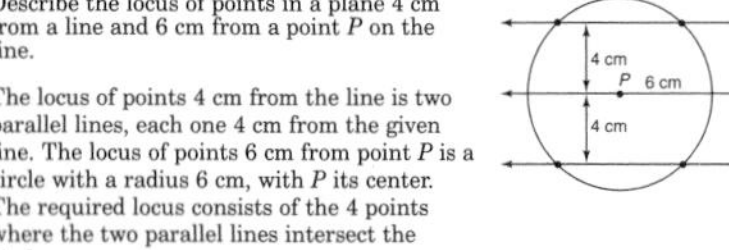

Draw a diagram to find the locus of points that satisfy the conditions. Then describe the locus.

1. all points in the coordinate plane that are 3 units from the graph of $x = 2$ and equidistant from the graphs $y = 1$ and $y = 5$

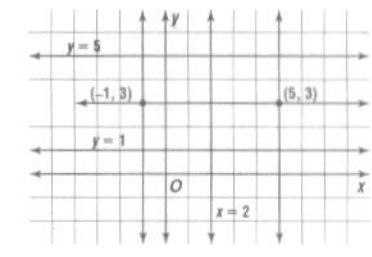

points (5, 3) and (−1, 3)

2. all the points in a plane that are 10 cm from a given line and 12 cm from a given point P on the line

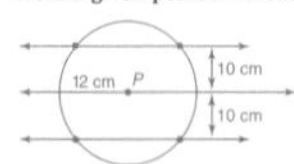

4 points at the intersection of a circle with a radius of 12 cm and 2 parallel lines each 10 cm from the given line

3. all interior points of an angle that are equidistant from the sides of the angle and 2 cm from the vertex of the angle

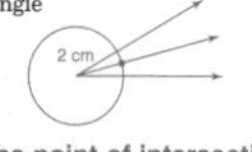

the point of intersection of the angle bisector and the circle of radius 2 cm with center at the vertex

4. all points in space that are 5 inches from a given line and 4 inches from a given point on the line

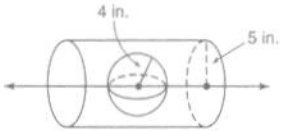

empty set

CHECK FOR UNDERSTANDING

Communicating Mathematics

Study the lesson. Then complete the following.

1. **Explain** two ways to determine the locus of points that satisfy the graphs of two equations in a coordinate plane. by graphing or algebraically
2. **Draw** and describe all possible loci involving a line and a parabola in a coordinate plane.
3. How many $\triangle ABC$s can be drawn with a given $\angle A$ and given sides of length AB and BC? at most two

2. See Solutions Manual for figure; intersection could be no points, one point, or two points.

4. Use the same $\overline{AC}$ and $\overline{BD}$ from the Modeling Mathematics activity on page 709, but draw $\angle A$ so that it is obtuse. Determine $\triangle ABC$ with these new specifications. What did you find? Did you get the same or different results? Explain. See students' work.

Guided Practice

5. Describe the different possible intersections of two circles. none, a point, two points, or a circle

Draw a diagram to find the locus of points that satisfy the conditions. Then describe the locus. 6–8. See margin for diagrams.

6. all the points in the coordinate plane 5 units from the origin and 3 units from the x-axis
7. all points in a plane 7 inches from the vertex of an angle and equidistant to the sides of the angle a point on the angle bisector 7 in. from the vertex
8. the graphs of $y = (x + 4)^2$ and $y = -x - 2$ two points, (−3, 1) and (−6, 4)

6. four points in the intersection of the circle with radius 5 units and center (0, 0) and the parallel lines 3 units on either side of the x-axis

Describe the locus of points in space that satisfy the graph of each equation. 9–10. See Solutions Manual.

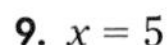
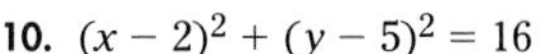

9. $x = 5$
10. $(x - 2)^2 + (y - 5)^2 = 16$

11. **Travel** The Flynn family is planning a family reunion. Members of the family live in Madison, Wisconsin; Philadelphia, Pennsylvania; and Durham, North Carolina. Use the map to draw a diagram showing what area of the United States is equidistant from the three cities and should be considered for the reunion. See margin.

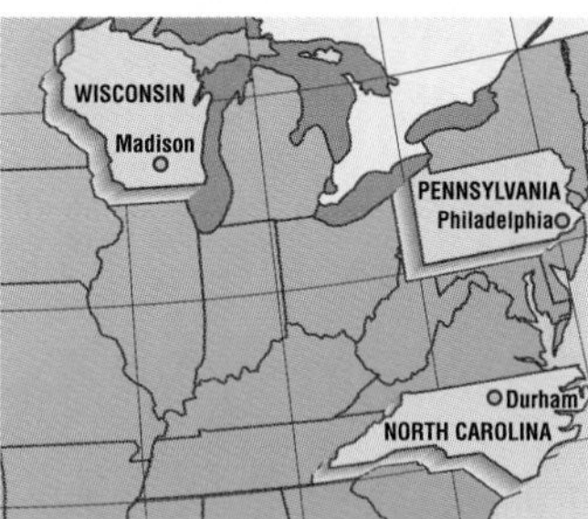

EXERCISES

A Practice

12. Determine the least and the greatest number of points in the intersection of a circle and a line MN that satisfy these conditions:
    a. M and N are in the exterior of the circle. none; two
    b. M and N are in the interior of the circle. two; two
    c. M and N are on the circle. two; two
    d. M is in the exterior and N is on the circle. one; two

Describe the different possible intersections of each pair of figures.

13. a point and a line none or a point
14. a line and a plane
15. a cube and a line
16. two spheres

14. none, a point, or a line
15. none, a point, two points, or a line segment
16. none, a point, a circle, or a sphere

Additional Answer

11. Construct the perpendicular bisectors of each side of the triangle formed by these three cities. This intersection area is the area that should be considered; Columbus, Ohio, area.

Reteaching

Using Examples Use examples to help students find the intersection of two loci. For example, have students find the locus of all points in a plane equidistant from two given intersecting lines and also equidistant from two given parallel lines. The intersection can be one point or two points.

17–23. See Solutions Manual for drawings.

17. four points whose coordinates are $(\pm 3\sqrt{2}, \pm 3\sqrt{2})$

18. the intersection of the ⊥ bis. of the segment connecting 2 pts. and the line halfway between 2 ∥ lines

20. four points that are the intersection of the two parallel lines 2 cm from the given line and a circle of radius 3 cm whose center is the given point on the line

21. two circles with a radius of 5 cm that each lie in the parallel planes and whose centers are the intersection of the ⊥ line with each plane

22. the intersection of the ⊥ bis. of the segment whose endpoints are the two points and the angle bisector

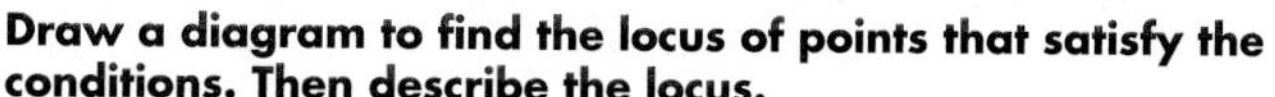

Draw a diagram to find the locus of points that satisfy the conditions. Then describe the locus.

17. all the points in the coordinate plane 6 units from the origin and equidistant from the x-and y-axes

18. all the points in a plane equidistant from two given points and equidistant from two parallel lines

19. all the points in space equidistant from a cylinder the axis of the cylinder

20. all points in a plane that are 2 centimeters from a given line and 3 centimeters from a given point on the line

21. all points in space 5 centimeters from a line perpendicular to two parallel planes that lie in the two planes

22. all points in the plane that are equidistant from the rays of an angle and equidistant from two points on one side of the angle

23. all points in space equidistant from two intersecting planes a set of two intersecting planes

Describe the geometric figure whose locus in space satisfies the graph of each equation. 24–29. See margin.

24. $y = -3x + 11$

25. $y = (x - 9)^2$

26. $y + x = 0$

27. $(x - 4)^2 + (y - 1)^2 = 25$

28. $(x + 6)^2 + (y - 3)^2 + (z - 1)^2 = 25$

29. $(x - 3)^2 + (y - 4)^2 + (z - 5)^2 = 16$

30. Construct an isosceles triangle with base $\overline{RT}$ and vertex angle S. How many triangles can be drawn that satisfy these conditions? one

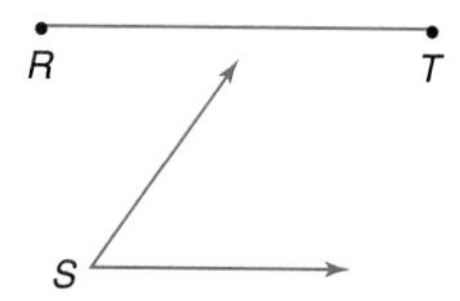

Find the locus of points in the coordinate plane that satisfy the graphs of both equations or inequalities.

31. $(x + 3)^2 + (y - 1)^2 = 25$ and $y = -3x - 13$ $(-6, 5), (-3, -4)$

32. $x + y = 6$ and $(x - 1)^2 + y^2 = 17$ $(5, 1), (2, 4)$

33. $x^2 + y^2 = 4$ and $y = x + 6$ no points

34. $y > 0$ and $x^2 + y^2 \leq 9$ points on and in the interior of the circle that lie above the x-axis

35–36. See Solutions Manual.

35. Circumscribe an equilateral triangle around a circle.

36. Point M is on line ℓ and point N is not on the line. Construct a circle that is tangent to line ℓ, and contains both points M and N.

Critical Thinking

37. How would the construction in Example 1 be different if the side given were $\overline{BC}$? There would be many solutions.

38. Place a point in a plane with two intersecting lines so that the locus of points equidistant from the two lines and a distance t from the point are at a maximum. Place the point at the intersection of the two lines.

Assignment Guide

Core (with proof): 13–41 odd, 42–51

Core (informal): 13–41 odd, 42–51

Enriched: 12–36 even, 37–51

For **Extra Practice**, see p. 790.

The red A, B, and C flags, printed only in the Teacher's Wraparound Edition, indicate the level of difficulty of the exercises.

Additional Answers

8.

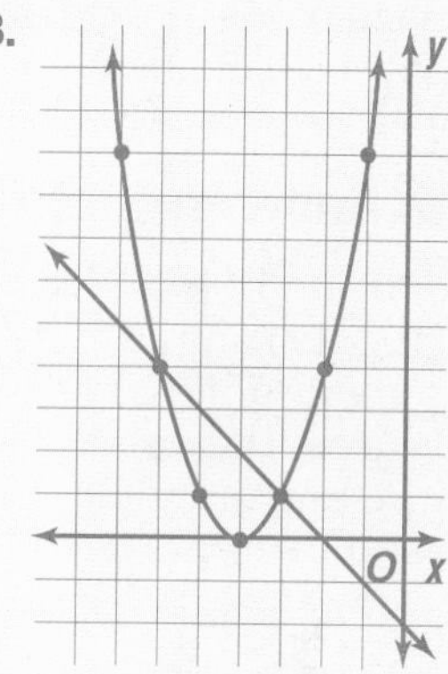

24. a plane perpendicular to the xy-plane whose intersection with the xy-plane in the line $y = -3x + 11$

25. a parabola that opens up with vertex at (9, 0)

26. a plane perpendicular to the xy-plane whose intersection with the xy-plane is the line $y = -x$

Practice Masters, p. 80

NAME________ DATE ________

Practice

Student Edition Pages 709–714

Intersection of Loci

Describe the locus of points in a plane that satisfy each condition.

1. $x + y = 4$ a line with slope −1 that passes through (0, 4)

2. $x = y$ a line with slope 1 that passes through (0, 0)

3. $y = 5$ a line with slope 0 that passes through (0, 5)

4. $x^2 + (y - 2)^2 = 25$ a circle with center (0, 2) and radius 5

Describe the geometric figure whose locus in space satisfies each condition.

5. $(x + 2)^2 + (y - 3)^2 + z^2 = 64$ a sphere with center (−2, 3, 0) and radius 8

6. $(x - 2)^2 (y + 3)^2 + (z - 1)^2 = 81$ a sphere with center (2, −3, 1) and radius 9

Draw a diagram and describe the locus of points.

7. all points in a plane that are 3 inches from a given segment and equidistant from the two endpoints two points on the perpendicular bisector of the segment, one on each side, whose perpendicular distance to the segment is 3 in.

8. all points in a plane that are 3 inches from a given line and 3 inches from a given point on the line two points, one on each side of the given point on the line, whose perpendicular distance from the point is 3 in.

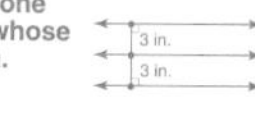

9. all points in a plane that are equidistant from the vertices of a given square the center of the square; the point of intersection of the diagonals of the square

10. all points in a plane that are 5 centimeters from a given point A and equidistant from points A and B that are 8 centimeters apart two points, the intersection of the perpendicular bisector of $\overline{AB}$ and the circle with center A and radius 5

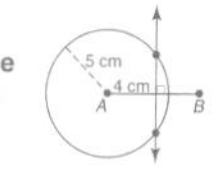

Tech Prep

Business Manager Business Management degrees are offered at most two-year colleges across the United States. Graphs are used at many points in planning, a vital part of management. For more information on tech prep, see the *Teacher's Handbook*.

Additional Answers

27. a circle with center (4, 1) and radius 5

28. a sphere with center (−6, 3, 1) and radius 5

29. a sphere with center (3, 4, 5) and radius 4

4 ASSESS

Closing Activity

Writing Have each student write two examples of loci. Have students exchange papers and write the intersection of the two loci.

A regulation soccer ball has a circumference of 27–28 inches and weighs 14–16 ounces.

Additional Answers

40. on the midline of the field so that the circle formed is at least tangent to the short side of the rectangle
41. a semicircle with radius 80 feet and center at the center of the goal
43.

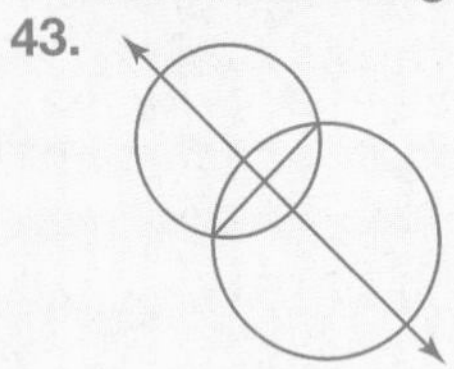

Enrichment Masters, p. 80

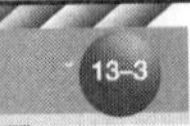

NAME__________ DATE________

Enrichment — Student Edition Pages 709–714

Points Equidistant from Segments

The distance from a point to a segment is zero if the point is on the segment. Otherwise, it is the length of the shortest segment from the point to the segment.

The locus of all points that are $\frac{1}{4}$ inch from the segment AB is shown by two dashed segments with semicircles at both ends.

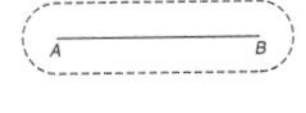

1. Suppose A, B, C, and D are four different points, and consider the locus of all points x units from $\overline{AB}$ *and* x units from $\overline{CD}$. Use any unit you find convenient. The locus can take different forms. Sketch at least three possibilities. List some of the things that seem to affect the form of the locus.
Answers will vary. Some possible answers are shown.

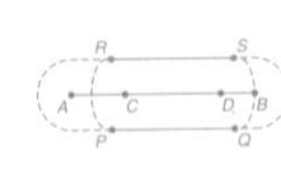

locus is 1 segment XY midway between $\overline{AB}$ and $\overline{CD}$

locus is a set of 2 points, X and Y

locus is a pair of line segments, $\overline{RS}$ and $\overline{PQ}$

The locus of points x units from $\overline{AB}$ and $\overline{CD}$ depends on the distance x and how $\overline{AB}$ and $\overline{CD}$ are situated relative to one another.

2. Conduct your own investigation of the locus of points *equidistant* from two segments. Describe your results on a separate sheet of paper. See students' work.

Applications and Problem Solving

FYI

In 1993, Volkhart Caro from Canada set a record for juggling a regulation soccer ball without it ever touching the ground. He juggled for 18 hours nonstop with his feet, legs, and head.

39. **Management** The Mathematics Club from Northwood High School was planning a trip to Washington, D.C. On one day, three groups of students wanted to visit three attractions, the Capitol, Union Station, and the Air and Space Museum. Use the map to locate a meeting point equidistant from these three buildings. **the corner of E St. and 3rd St.**

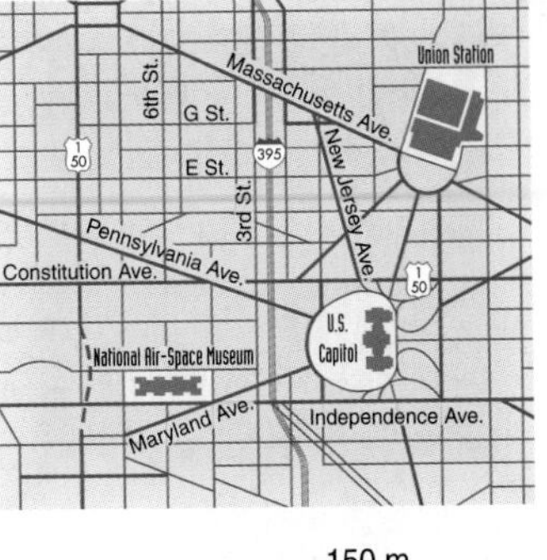

40. **Agriculture** On the field at the right, where would you place a circular irrigation system with a radius of 50 meters, so that the maximum area of the field is irrigated? **See margin.**

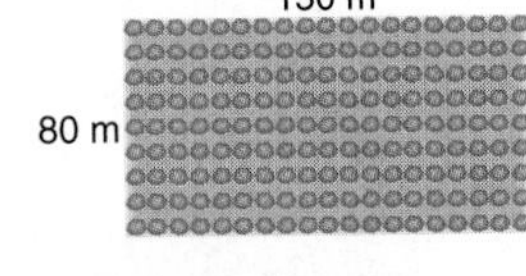

41. **Sports** Josh is the goalie for a 16-year-and-under soccer team. His kicks from the goal area average 160 feet in length before they hit the ground. What is the locus of points representing the area in which most of his kicks from the goal area land? **See margin.**

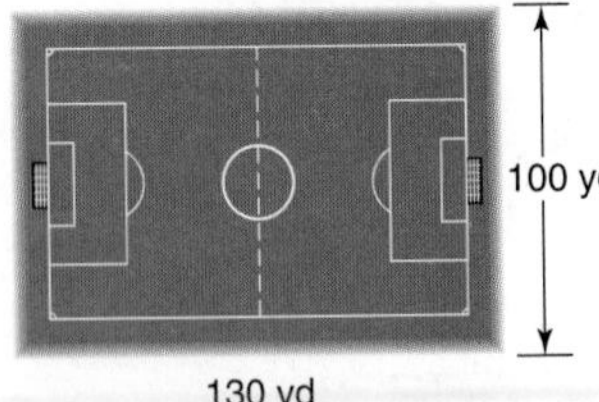

Mixed Review

42. Use an algebraic method to find the locus of points that satisfy the graphs of $9x + y = 20$ and $3x + 3y = 12$. (Lesson 13–2) **(2, 2)**
43. Draw and describe the locus of points in a plane that are equidistant from the points of intersection of two given circles. (Lesson 13–1)
43. ⊥ bisector of segment connecting the two points of intersection; see margin for drawing.
44. Write the equation of the sphere with center at $(4, -3, -2)$ and the point at $(-2, -5, 7)$ on the sphere. (Lesson 12–6)
44. $(x - 4)^2 + (y + 3)^2 + (z + 2)^2 = 121$
45. Find the surface area and volume of a sphere with a great circle that has a circumference of 15.7 meters. (Lesson 11–7) **$T \approx 78.5$ m²; $V \approx 65.4$ m³**
46. A trapezoid has a median 17.5 centimeters long. If its height is 18 centimeters, find the area of the trapezoid. (Lesson 10–4) **315 cm²**
47. **Sports** Two cross-country skiers left from a ski cabin at 8:00 A.M. The first skier traveled in a path 17° east of north, and the second skier traveled 22° west of north. By 9:00 A.M., the first skier had gone 15 miles, and the second skier had gone 17 miles. How far apart are the skiers? (Lesson 8–6)
47. 10.8 miles
48. If A, B, and C are the midpoints of the sides of $\triangle FGE$, find the perimeter of $\triangle FGE$. (Lesson 7–4) **23.8**

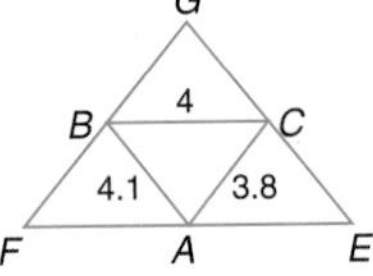

49. If $\triangle ABC \cong \triangle CDE$, name the corresponding congruent parts of the two triangles. (Lesson 4–3) **$\angle A \cong \angle C$, $\angle B \cong \angle D$, $\angle C \cong \angle E$, $\overline{AB} \cong \overline{CD}$, $\overline{BC} \cong \overline{DE}$, $\overline{AC} \cong \overline{CE}$**

50. Solve $x^2 + 8x - 5 = 0$ by completing the square. Leave irrational roots in simplest radical form. **$-4 \pm \sqrt{21}$**
51. Solve $x^2 + 5x - 3 = 0$ by using the quadratic formula. Approximate irrational roots to the nearest hundredth. **−5.54, 0.54**

Extension

Communication Discuss the locus of points P in a plane such that the sum of the distances from P to two given points is a constant $c > 0$. Students will discover that it is an ellipse.

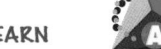

13-4 Mappings

What YOU'LL LEARN

- To recognize an isometry or congruence transformation, and
- to solve problems by making a table.

Why IT'S IMPORTANT

Mappings can be used to describe the movement of a figure in a plane.

APPLICATION
Engineering

The Golden Gate Bridge in San Francisco is tied down to an anchor block with many cables that support the bridge. The cables are arranged like the roots of a tree causing the braces on the cables to be different sizes.

Each cable is bound by each of the four braces labeled A, B, C, and D. Every contact point for a cable in one brace can be associated with a contact point in another corresponding brace. These correspondences are called **mappings** or **transformations**. A transformation maps a **preimage** onto an **image**.

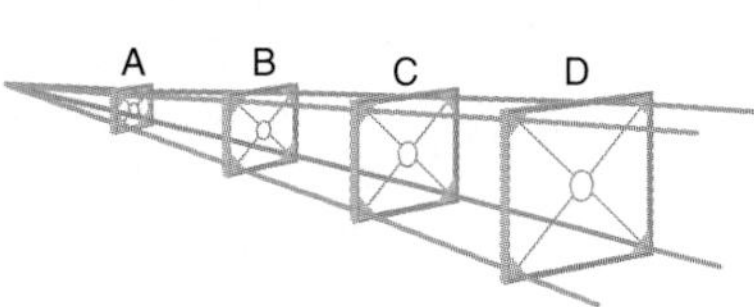

Definition of Transformation	In a plane, a mapping is a transformation if each preimage point has exactly one image point, and each image point has exactly one preimage point.

Below are some of the types of transformations.

dilation

A figure can be enlarged or reduced.

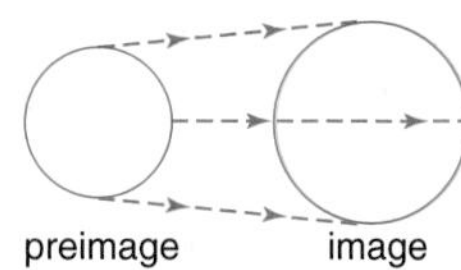

rotation

A figure can be turned.

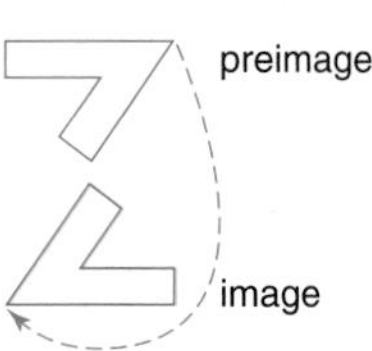

translation

A figure can be slid.

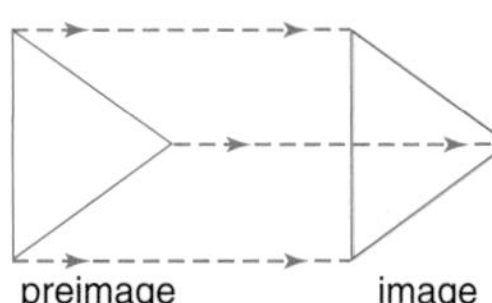

reflection

A figure can be flipped.

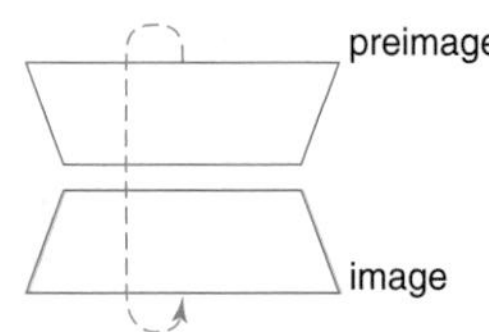

13-4 LESSON NOTES

NCTM Standards: 1–5, 7, 8

Instructional Resources

- Study Guide Master 13-4
- Practice Master 13-4
- Enrichment Master 13-4
- Assessment and Evaluation Masters, pp. 351, 352
- Real-World Applications, 28

Transparency 13-4A contains the 5-Minute Check for this lesson; **Transparency 13-4B** contains a teaching aid for this lesson.

Recommended Pacing	
Honors Pacing	Day 5 of 13

1 FOCUS

5-Minute Check
(over Lesson 13-3)

1. Describe a geometric figure whose locus in a plane satisfies the graph of $(x - 3)^2 + (y + 4)^2 = 121$. **a circle with radius 11 and center at (3, −4)**
2. Write an equation for the locus of all points in space that are 1 unit from the point at (7, 5, −1). $(x - 7)^2 + (y - 5)^2 + (z + 1)^2 = 1$
3. Draw a diagram to find the locus of points in a plane that are 3 centimeters from a given line and 3 centimeters from a given point on the line. Describe the locus.

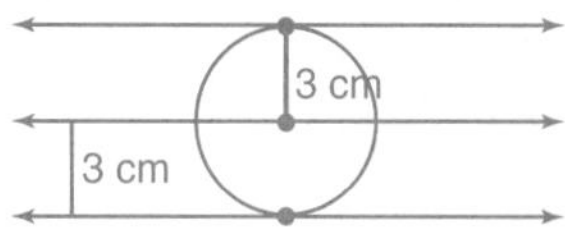

two points that lie on the circle with radius 3 centimeters and on the line perpendicular to the given line at the given point

Motivating the Lesson

Situational Problem Bring in a book that contains some of Escher's work. Have students use art to look for patterns and mappings.

2 TEACH

Teaching Tip Point out that the symbol "→" is also used in mapping functions.

In-Class Examples

For Example 1
What type of transformation is used in inflating a round balloon? **dilation**

For Example 2
Suppose $\triangle DEF \rightarrow \triangle GJH$. Show that this mapping is an isometry.

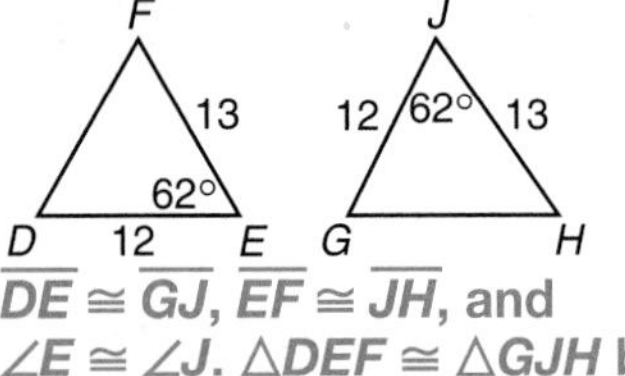

$\overline{DE} \cong \overline{GJ}$, $\overline{EF} \cong \overline{JH}$, and $\angle E \cong \angle J$. $\triangle DEF \cong \triangle GJH$ by SAS. $\triangle DEF$ is the preimage. $\triangle GJH$ is the image.

Teaching Tip In the mapping of two triangles, the order of letters indicates how the sides of the triangles correspond, as well as the vertices.

Example 1 **Refer to the application at the beginning of the lesson.**

a. What type of transformation is used in the braces of the Golden Gate Bridge?

The brace transformation on the Golden Gate Bridge is a dilation.

b. Determine whether *A*, *B*, and *C* are preimages of *D*.

Yes, each previous image is a preimage of the enlargement.

The symbol → is used to indicate a mapping. For example, $\triangle ABC \rightarrow \triangle PQR$ means each point in $\triangle ABC$ is mapped onto exactly one point of $\triangle PQR$. $\triangle ABC$ is the preimage, and $\triangle PQR$ is the image. The order of the letters indicates the correspondence of the preimage to the image; that is, *A* and *P* are corresponding vertices, *B* and *Q* are corresponding vertices, and *C* and *R* are corresponding vertices.

When a geometric figure and its transformation image are congruent, the mapping is called an **isometry** or a **congruence transformation**.

Definition of Isometry (Congruence Transformation)	**In a plane, an isometry is a transformation that maps every segment to a congruent segment.**

Transformations can occur in the coordinate plane.

Example 2 **Show that $\triangle WNR \rightarrow \triangle KAP$ in the coordinate plane at the right is an isometry.**

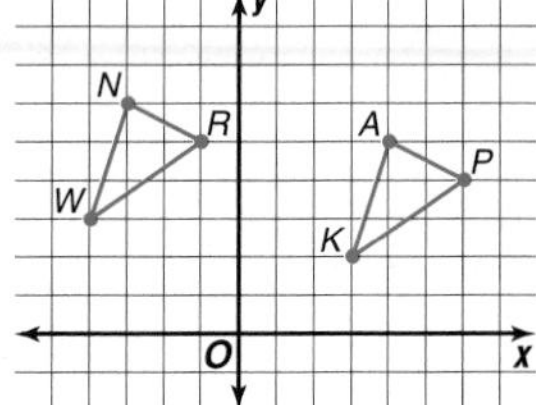

Use the distance formula to show that the sides of $\triangle WNR$ are congruent to the sides of $\triangle KAP$.

The coordinates are $W(-4, 3)$, $N(-3, 6)$, $R(-1, 5)$, $K(3, 2)$, $A(4, 5)$, and $P(6, 4)$.

Use the distance formula to find the measure of each side.

$WN = \sqrt{(-4+3)^2 + (3-6)^2}$ or $\sqrt{10}$

$NR = \sqrt{(-3+1)^2 + (6-5)^2}$ or $\sqrt{5}$

$WR = \sqrt{(-4+1)^2 + (3-5)^2}$ or $\sqrt{13}$

$KA = \sqrt{(3-4)^2 + (2-5)^2}$ or $\sqrt{10}$

$AP = \sqrt{(4-6)^2 + (5-4)^2}$ or $\sqrt{5}$

$KP = \sqrt{(3-6)^2 + (2-4)^2}$ or $\sqrt{13}$

LOOK BACK

You can review the distance formula in Lesson 1-4.

Since the measures of the corresponding sides are equal, $WN = KA$, $NR = AP$, and $WR = KP$. So, the corresponding sides are congruent, and $\triangle WNR \cong \triangle KAP$ by SSS.

Therefore, the mapping $\triangle WNR \rightarrow \triangle KAP$ is a congruence transformation or an isometry.

Alternative Learning Styles

Kinesthetic Use tiles to help students understand the concept of mappings. Explain that if you flip them over, turn them, and slide them, you are actually performing a geometric transformation. Have students use tiles to make a design, and have them explain how one tile is mapped to another tile that is the same or similar.

When a figure and its transformation image are similar, the mapping is called a **similarity transformation**.

Example  **In the figure at the right, $\overline{AC} \rightarrow \overline{DF}$.**

a. Is this a transformation?

b. If so, what type?

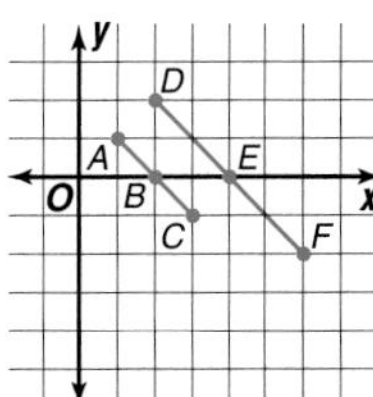

a. A and D are corresponding points, C and F are corresponding points, and B and E are corresponding points. Every point on $\overline{AC}$ corresponds to a point on $\overline{DF}$, and every point on $\overline{DF}$ corresponds to a point on $\overline{AC}$. Thus, the mapping is a transformation.

b. Use the Pythagorean Theorem to determine the length of each segment.

$AB = \sqrt{1^2 + 1^2}$ or $\sqrt{2}$ | $DE = \sqrt{2^2 + 2^2}$ or $2\sqrt{2}$

$BC = \sqrt{1^2 + 1^2}$ or $\sqrt{2}$ | $EF = \sqrt{2^2 + 2^2}$ or $2\sqrt{2}$

$AC = \sqrt{2} + \sqrt{2}$ or $2\sqrt{2}$ | $DF = 2\sqrt{2} + 2\sqrt{2}$ or $4\sqrt{2}$

LOOK BACK

You can review the Pythagorean Theorem by referring to Lesson 1-4.

Since $\overline{AC} \not\cong \overline{DF}$, this is *not* a congruence transformation. But, since $\frac{AB}{DE} = \frac{BC}{EF} = \frac{\sqrt{2}}{2\sqrt{2}}$ or $\frac{1}{2}$ and $\frac{AC}{DF} = \frac{2\sqrt{2}}{4\sqrt{2}}$ or $\frac{1}{2}$, then $\overline{AC}$ and $\overline{DF}$ are proportional in every respect. Therefore, $\overline{AC} \rightarrow \overline{DF}$ is a similarity transformation.

Sometimes you can observe patterns in mappings by **making a table**.

Example  **The figure at the right shows a reflection of $\triangle ABC$ over the x-axis.**

PROBLEM SOLVING

Make a Table

a. Make a table to compare the coordinates of the vertices of the preimage and the image.

△ABC		△DEF	
x	y	x	y
−3	2	−3	−2
2	5	2	−5
7	3	7	−3

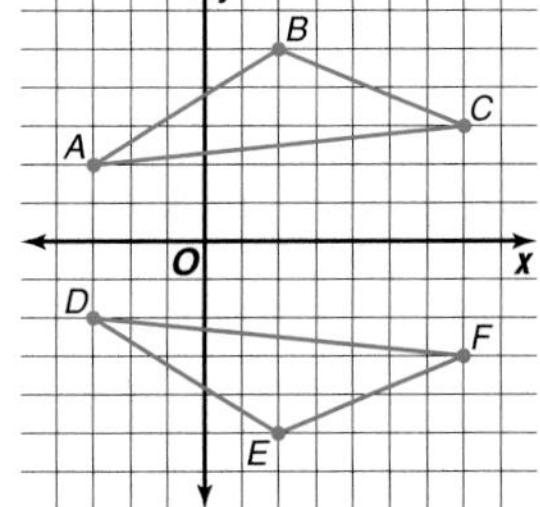

b. Describe any pattern shown in the table.

The x-coordinates are exactly the same, and the y-coordinates are the opposite of each other.

c. Use the pattern to predict the image of (a, b) reflected over the x-axis.

For (a, b), the x-coordinate of the image would be a, and the y-coordinate would be $-b$. The image of (a, b) would be $(a, -b)$.

In-Class Examples

For Example 3

In the figure, $\triangle ABC \rightarrow \triangle DEF$.

a. Is this a transformation?

b. If so, what type? yes, similarity transformation

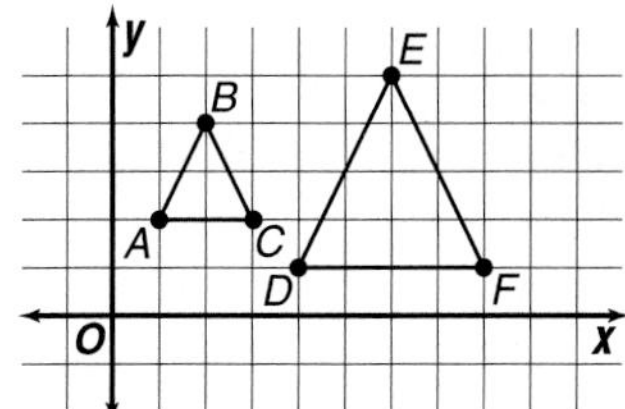

For Example 4

a. Complete the table below.

ABCD		reflection of ABCD over the y-axis	
x	y	x	y
2	2	−2	2
4	2	−4	2
2	4	−2	4
4	4	−4	4

b. Describe any pattern implied by the table.
The y-coordinates of the image are the same as those of the preimage, and the x-coordinates of the image are the opposite of the x-coordinates of the preimage.

c. Predict the image of (a, b) reflected over the y-axis.
The image of (a, b) is $(-a, b)$.

3 PRACTICE/APPLY

Check for Understanding

Exercises 1–12 are designed to help you assess your students' understanding through reading, writing, speaking, and modeling. You should work through Exercises 1–4 with your students and then monitor their work on Exercises 5–12.

Assignment Guide

Core (with proof): 13–37 odd, 38–45
Core (informal): 13–37 odd, 38–45
Enriched: 14–32 even, 33–45
All: Self Test, 1–10

For **Extra Practice**, see p. 790.

The red A, B, and C flags, printed only in the Teacher's Wraparound Edition, indicate the level of difficulty of the exercises.

Additional Answers

2. The preimage and the image of both isometries and enlargements are similar figures. Those that have isometry preserve congruence and those that are enlargements don't preserve congruence.
4. They organize data, which makes analysis more efficient.

CHECK FOR UNDERSTANDING

Communicating Mathematics

1. because there are actually three transformations
3. See students' work.

Study the lesson. Then complete the following.

1. **Explain** why in Example 1 there is more than one preimage.
2. Compare transformations that have isometry to those that are enlargements. How are they similar and how are they different? See margin.
3. **Describe** the four different transformations in your own words.
4. **Explain** how tables are helpful in problem solving. See margin.

Guided Practice

9. two reflections or a rotation

Suppose trapezoid *RSTU* → trapezoid *ABCD*.

5. Name the image of $\overline{TU}$. $\overline{CD}$
6. Name the preimage of $\angle ADC$. $\angle RUT$
7. Name the preimage of $\angle BCD$. $\angle STU$
8. Name the image of $\overline{ST}$. $\overline{BC}$
9. What transformations were necessary in the mapping?
10. Describe the transformation(s) that occurs in the mapping of $\triangle ABC \to \triangle DEC$. reduction, rotation

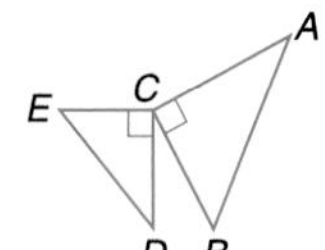

11. Determine if pentagon $QRSTU \to$ pentagon $ABCDE$ is an isometry. Yes; all angles and sides are congruent.

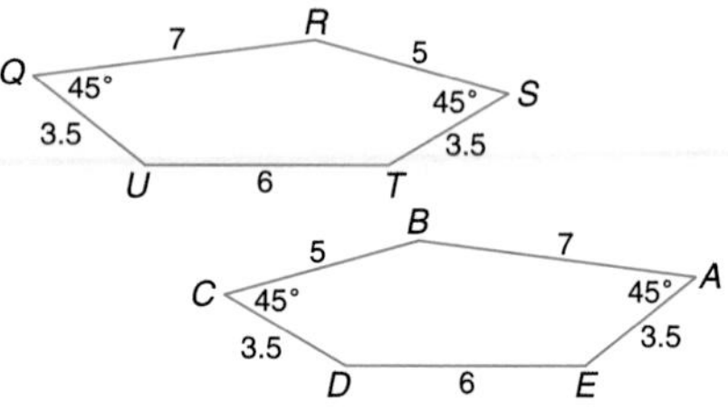

12. **Make a Table** Refer to the arrangement of circles at the right. Use a table to find how to place the digits from 1 to 6 in the circles so that every side of the triangle has the same sum.

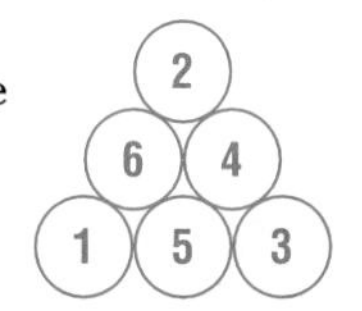

EXERCISES

Practice

Suppose △*ABC* → △*EFD*.

13. Name the preimage of $\angle FED$. $\angle BAC$
14. Name the image of $\angle C$. $\angle D$
15. Name the image of $\overline{AC}$. $\overline{ED}$
16. Name the preimage of $\overline{DE}$. $\overline{CA}$
17. What transformations were necessary in the mapping? enlargement and reflection

Describe the transformations that occurred in the mappings.

18. rotation or reflection
19. enlargement, reflection
20. translation

18. 
19. 
20. 

Reteaching

Using Manipulatives Have each student draw and cut out a two-dimensional figure. Have them use that figure to draw examples of the four transformations discussed in this lesson: a reflection, a rotation, a slide, and an enlargement or reduction.

Determine if each transformation is an isometry.

22. no; $GH \neq AF$ and $DE \neq JI$

21. $\triangle RST \to \triangle WUV$

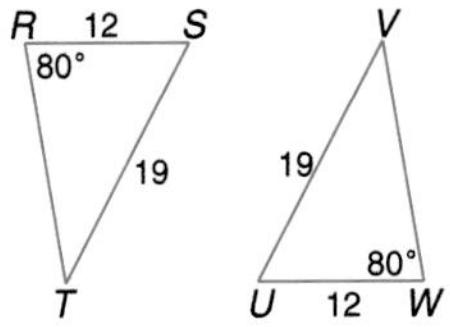

Yes; all angles and sides are congruent.

22. hexagon $ABCDEF \to$ hexagon $GLKJIH$

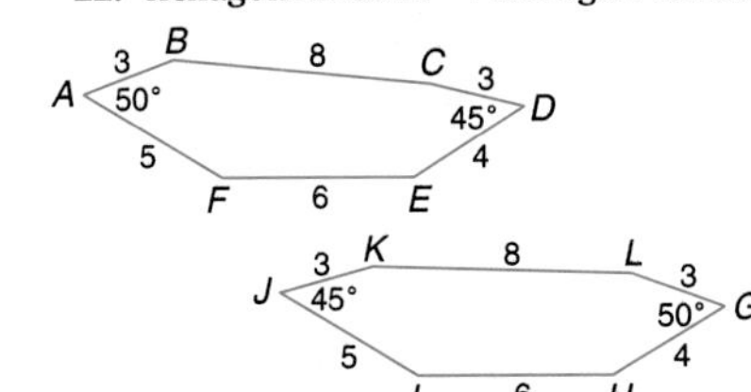

23. In a coordinate plane, quadrilateral $ABCD \to$ quadrilateral $EFGH$ for $A(-4, 0)$, $B(0, 3)$, $C(1, -2)$, $D(-3, -4)$, $E(0, -4)$, $F(-3, 0)$, $G(2, 1)$, and $H(4, -3)$. a. See margin for graph; rotation.

a. Graph the quadrilaterals and describe the transformation.

b. Is the transformation an isometry? yes

Each figure below has a preimage that is an isometry. Write the image of each given preimage listed in Exercises 24–29.

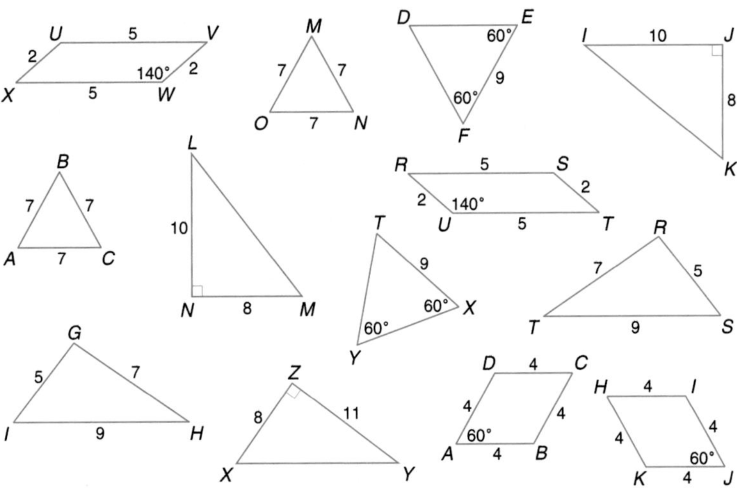

25. quadrilateral $SRUT$

24. $\triangle ABC$ $\triangle MNO$

25. quadrilateral $UVWX$

26. $\triangle DEF$ $\triangle TXY$

27. $\triangle RST$ $\triangle GIH$

28. $\triangle LMN$ $\triangle IKJ$

29. quadrilateral $ABCD$ quadrilateral $HIJK$

30. **Make a Table** Use the figure at the right and a table of coordinates to determine the coordinates of the image after a reflection of the given preimage over the x-axis.

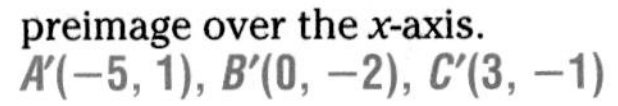

$A'(-5, 1)$, $B'(0, -2)$, $C'(3, -1)$

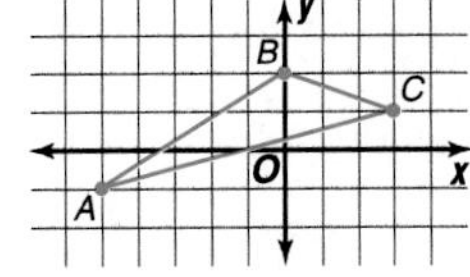

31. $A'(3, -3)$, $B'(1, -2)$, $C'(2, -4)$, $D'(4, -5)$

31. **Make a Table** Use a table of coordinates and the given figure to determine the coordinates of the image after a slide 4 units down and a reflection across the y-axis.

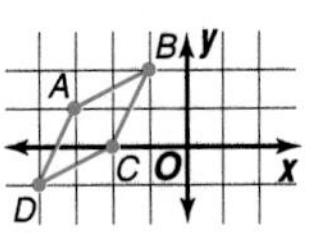

Additional Answer

23a.

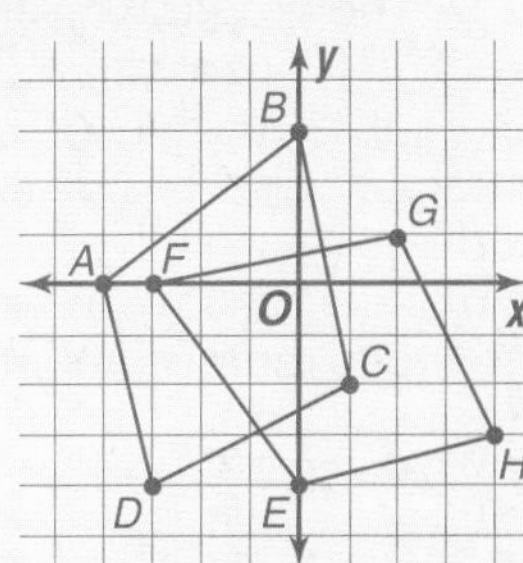

Study Guide Masters, p. 81

13-4 NAME________ DATE________

Study Guide

Student Edition Pages 715–721

Mappings

A figure can be moved by any of the following methods:

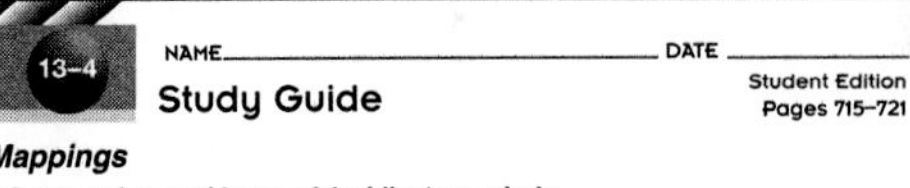

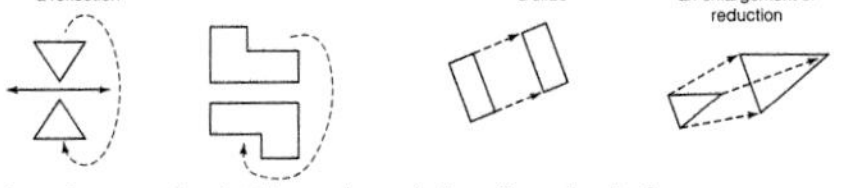

In each case, each point (the **preimage**) of one figure is paired with exactly one point (the **image**) of the corresponding figure. This kind of one-to-one mapping is called a **transformation**. The symbol $\to$ is used to indicate a mapping.

Example: If quadrilateral $ABCD \to$ quadrilateral $EFGH$, name the preimage of each vertex of $EFGH$.

The preimages of E, F, G, and H are A, B, C, and D.

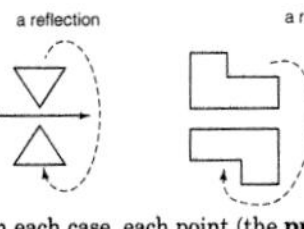

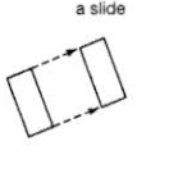

Each figure below has a preimage or image of isometry. Write the image of each given preimage.

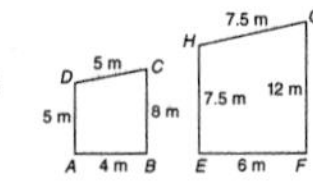

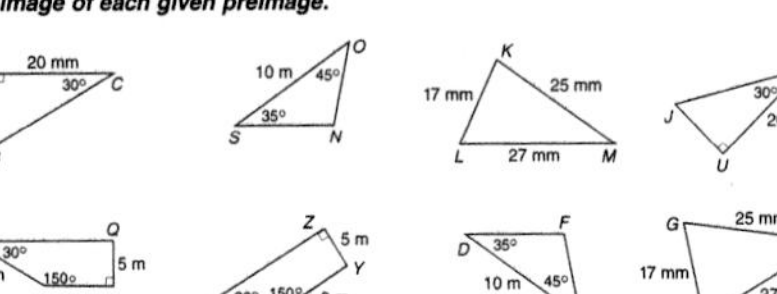

1. $\triangle ABC$ $\triangle UJV$
2. $\triangle DEF$ $\triangle SON$
3. quadrilateral $XYZW$ quadrilateral $RTQP$
4. $\triangle LMK$ $\triangle HIG$
5. $\triangle MKL$ $\triangle IGH$
6. quadrilateral $TRPQ$ quadrilateral $YXWZ$

7. Use a table of coordinates and the given figure to determine the coordinates of the image after a slide of 3 units up and a reflection across the y-axis. $(-2, 7)$, $(-4, 4)$, $(0, 1)$

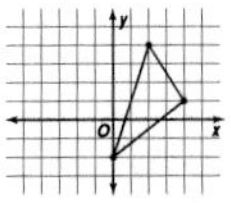

Using the Programming Exercises The program given in Exercise 32 is for use with a TI-82 or TI-83 graphing calculator. For other programmable calculators, have students consult their owner's manual for commands similar to those represented here.

Additional Answers

32a.

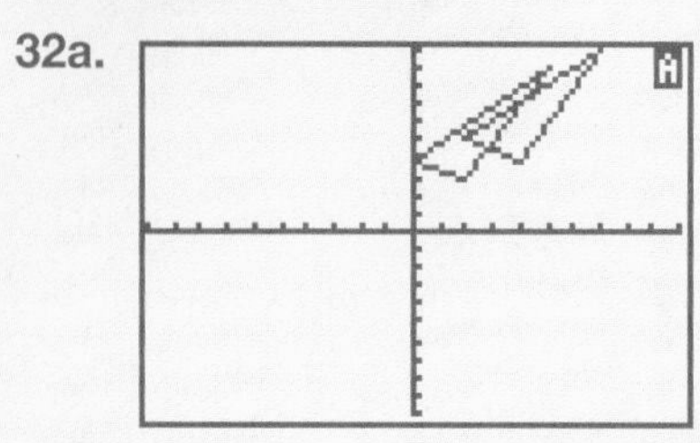

32b.

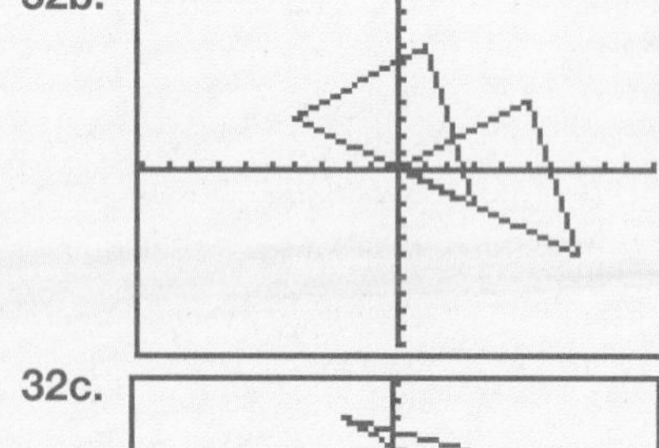

32c.

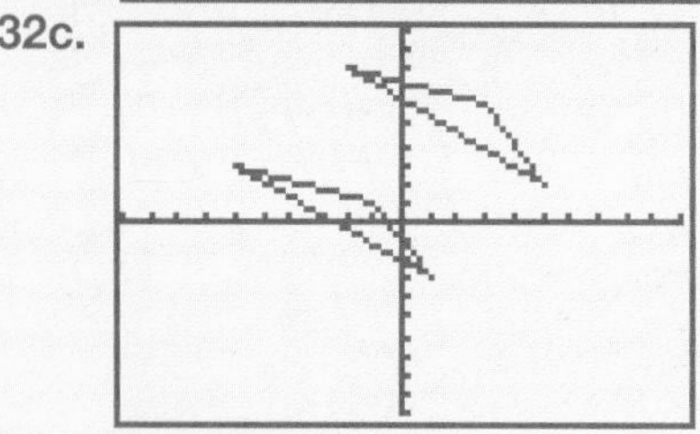

33. two reflections or a rotation
34. rotation

Practice Masters, p. 81

13-4 NAME______________ DATE______

Practice Student Edition Pages 715–721

Mappings

Each figure below has a preimage or is the image of isometry. Write the image of each given preimage listed in Exercise 1–6.

1. $\triangle ABC$ *$\triangle MNO$*
2. $\triangle DEF$ *$\triangle JKL$*
3. $\triangle GHI$ *$\triangle RPQ$ or $\triangle QPR$*
4. $\triangle ONM$ *$\triangle CBA$*
5. Quadrilateral *STUV* *quadrilateral ZWXY or quadrilateral WZYX*
6. Quadrilateral *XYZW* *quadrilateral UVST or quadrilateral VUTS*
7. **Make a Table** Benjamin has seven employees who work in his store: Janell, Pedro, David, Elyse, Faye, Gordon, and Irene. Janell and Gordon work full time, 5 days in a row per week. Pedro and Faye each work 4 days per week. Pedro cannot work on Thursdays. David and Elyse also work 4 days per week each. David cannot work on weekends, and Elyse cannot work on Wednesdays. Irene works any three days assigned each week. Benjamin needs three people to work on Monday through Thursday, five to work on Friday, and six to work on weekends. Make a possible schedule for Benjamin to use.
Possible answer:

	M	T	W	T	F	S	S
Janell	X	X	X			X	X
Pedro	X				X	X	X
David		X	X	X	X		
Elyse				X	X	X	X
Faye				X	X	X	X
Gordon	X	X	X			X	X
Irene					X	X	X

Programming

32. The TI-82/83 program below draws a triangle and its image when it slides a certain horizontal and vertical distance. You enter the coordinates of its vertices and the distances to slide. Then the mapping will be drawn.

TECHNOLOGY Tip

After the graph of $\triangle ABC$ is shown, press ENTER to continue with the program.

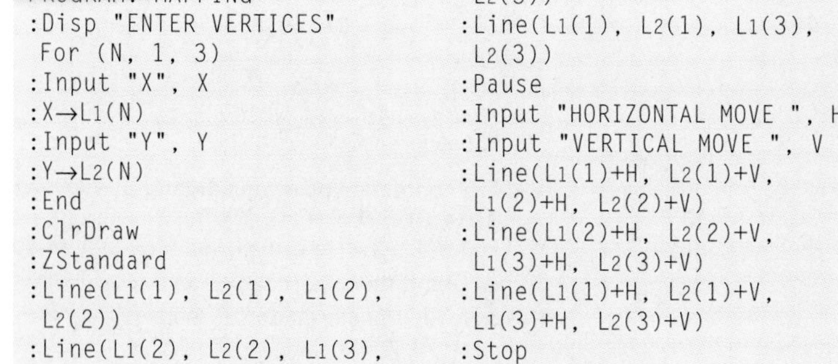

```
PROGRAM: MAPPING
:Disp "ENTER VERTICES"
 For (N, 1, 3)
:Input "X", X
:X→L1(N)
:Input "Y", Y
:Y→L2(N)
:End
:ClrDraw
:ZStandard
:Line(L1(1), L2(1), L1(2),
 L2(2))
:Line(L1(2), L2(2), L1(3),
 L2(3))
:Line(L1(1), L2(1), L1(3),
 L2(3))
:Pause
:Input "HORIZONTAL MOVE ", H
:Input "VERTICAL MOVE ", V
:Line(L1(1)+H, L2(1)+V,
 L1(2)+H, L2(2)+V)
:Line(L1(2)+H, L2(2)+V,
 L1(3)+H, L2(3)+V)
:Line(L1(1)+H, L2(1)+V,
 L1(3)+H, L2(3)+V)
:Stop
```

Use the program to draw $\triangle ABC$ and its translated image.

32a–c. See margin.

a. $A(2, 3)$, $B(5, 9)$, $C(0, 4)$; horizontal move: 2; vertical move: 1
b. $A(-4, 3)$, $B(1, 7)$, $C(3, -2)$; horizontal move: 4; vertical move: -3
c. $A(3, 6)$, $B(5, 2)$, $C(-2, 8)$; horizontal move: -4; vertical move: -5
d. How are the triangles and their images related? 32d. They are congruent triangles.
e. Is the slide performed by the program an isometry? Justify your answer. Yes; when a figure and its image are congruent, the mapping is an isometry.

Critical Thinking

Separate each figure into two congruent figures. Explain what transformation is used on one of the two congruent figures to form the given original figure. 33–34. See margin for explanations.

33.

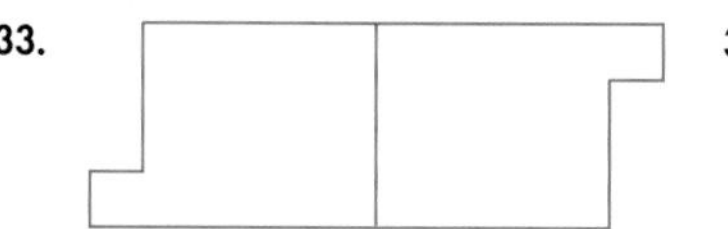

34.

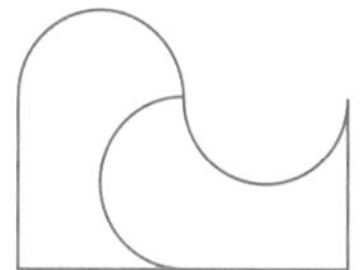

Applications and Problem Solving

35. **Writing** Small children often write letters backward. What transformations did a child use in writing her name? reflection of order and reflection of L and J

AIJUL

36. **Art** What transformations were used in the pattern on the Incan tunic shown at the left? See students' work.
37. **Fractals** What transformations are apparent in the development of the fractal below called the Networked-Growing Twig? reflections

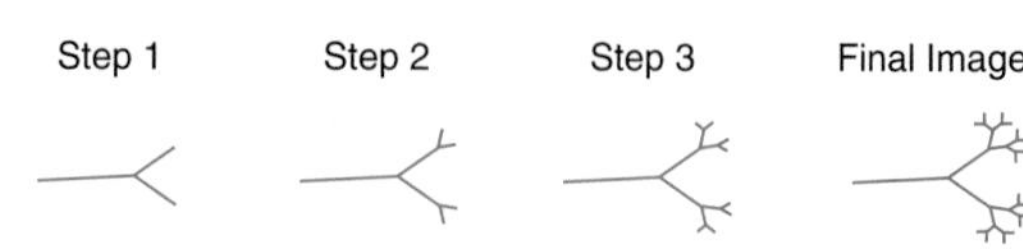

Extension

Communication Graph $y = x^2$ and $y = (x - 2)^2 + 4$ on the same graph. Students should discuss the similarities and differences and make conjectures.

Mixed Review

38. Describe and draw a diagram of the locus of all points in a plane equidistant from two given parallel lines and a given distance from an intersecting line. (Lesson 13–3) **Two points; see Solutions Manual for drawing.**

39. Graph the equations $y = x + 3$ and $3y + x = 5$ to find the locus of points that satisfy the graphs of both equations. (Lesson 13–2) **(−1, 2)**

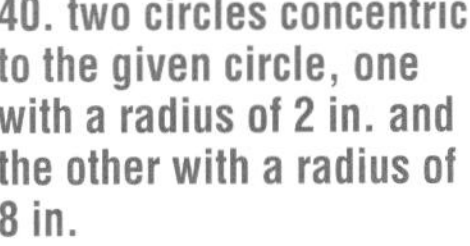

40. two circles concentric to the given circle, one with a radius of 2 in. and the other with a radius of 8 in.

40. Describe the locus of all points in a plane that are 3 inches from a circle with radius measuring 5 inches. (Lesson 13–1)

41. Write the equation in slope-intercept form of the line that passes through points at (7, 6) and (3, −2). (Lesson 12–2) $y = 2x - 8$

42. Classify the statement *Chords that are equidistant from the center of a circle are congruent* as always, sometimes, or never true. (Lesson 9–3) **always**

43. In right triangle XYZ, $m\angle X = 30$ and $m\angle Y = 60$. If $XY = 10$, find YZ and XZ. (Lesson 8–2) $5, 5\sqrt{3}$

INTEGRATION Algebra

44. Write an equation in functional notation for the relation shown in the table at the right. $f(x) = 3x + 1$

x	1	2	3	4	5
f(x)	4	7	10	13	16

45. **Manufacturing** The graph at the right shows the number of cans produced from a pound of aluminum. Find the percent of change from 1972 to 1992. Round to the nearest whole percent. 35%

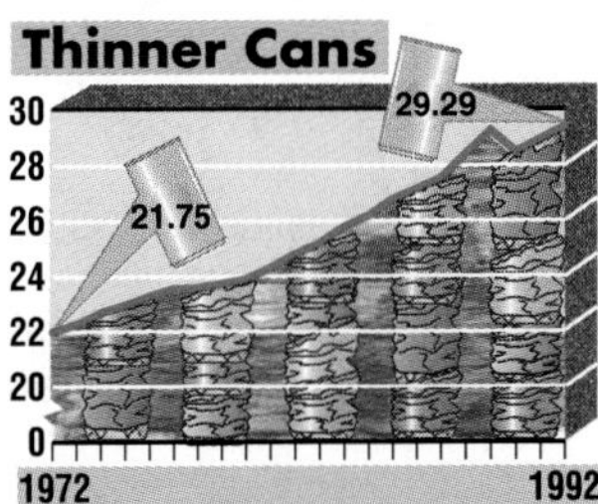

Source: The Aluminum Association

SELF TEST

1. Point R is on line ℓ. What is the locus of points in space that are 8 centimeters from ℓ and 8 centimeters from R? (Lesson 13–1) **See margin.**

2. Points P and Q are 6 centimeters apart. What is the locus of points in a plane that are equidistant from P and Q and are 8 centimeters from P? Sketch the locus. (Lesson 13–1) **See Solutions Manual.**

Use either substitution or elimination to find the locus of points that satisfy the graphs of both equations. (Lesson 13–2)

3. $x - 7y = 13$ **(6, −1)**
   $3x - 5y = 23$

4. $x = 8 + 3y$ **(−16, −8)**
   $2x - 5y = 8$

5. Sketch the circle for $x^2 + y^2 = 10$ and the line for $y = 3x + 10$. Describe the locus of points that satisfy both graphs. (Lesson 13–3) **One point; (−3, 1); see Solutions Manual for graph.**

Describe the different possible intersections of each pair of figures. (Lesson 13–3)

6. a plane and a sphere in space
   none, a point, or a circle

7. two perpendicular lines and a circle in a plane
   0, 1, 2, 3, or 4 points

8. Describe and draw the locus of points in the plane of $\angle DEF$, equidistant from the sides of the angle, and 4 centimeters from $\overline{EF}$. (Lesson 13–3) **The two points that are the intersection of the bisector of $\angle DEF$ and a pair of rays parallel to $\overline{EF}$, and each 4 cm from $\overline{EF}$; see margin.**

Suppose $\triangle ABC \rightarrow \triangle EBD$. (Lesson 13–4)

9. Name the image of $\overline{CB}$. $\overline{DB}$

10. Name the preimage of $\angle BDE$. $\angle BCA$

Answers for the Self Test

1. the circle that is the intersection of the cylindrical surface with axis ℓ and radius 8 cm, and the sphere with center R and radius 8 cm

8.

4 ASSESS

Closing Activity

Modeling Have each student select a type of transformation and make a model to represent it. They can use any available materials. Have them explain their mappings.

Chapter 13 Quiz B (Lessons 13-3 and 13-4) is available in the *Assessment and Evaluation Masters*, p. 352.

Mid-Chapter Test (Lessons 13-1 through 13-4) is available in the *Assessment and Evaluation Masters*, p. 351.

SELF TEST

The Self Test provides students with a brief review of the concepts and skills in Lessons 13-1 through 13-4. Lesson numbers are given to the right of exercises or instruction lines so students can review concepts not yet mastered.

Enrichment Masters, p. 81

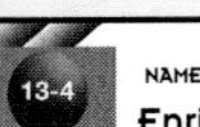

13-4 NAME________ DATE________

Enrichment

Student Edition Pages 715–721

Transformations in The Coordinate Plane

The following statement tells one way to map preimage points to image points in the coordinate plane:

$(x, y) \rightarrow (x + 6, y - 3)$

This can be read, "The point with coordinates (x, y) is mapped to the point with coordinates $(x + 6, y - 3)$." With this transformation, for example, (3, 5) is mapped to $(3 + 6, 5 - 3)$ or (9, 2). The figure shows how the triangle ABC is mapped to triangle XYZ.

1. Does the transformation above appear to be an isometry? Explain your answer. Yes; the transformation slides the figure to the lower right without changing its size or shape.

Draw the transformation image for each figure. Then tell whether the transformation is or is not an isometry.

2. $(x, y) \rightarrow (x - 4, y)$

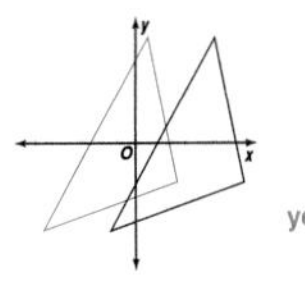

yes

3. $(x, y) \rightarrow (x + 8, y + 7)$

yes

4. $(x, y) \rightarrow (-x, -y)$

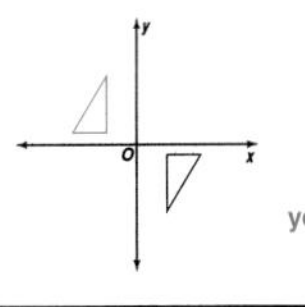

yes

5. $(x, y) \rightarrow \left(-\frac{1}{2}x, y\right)$

no

13-5 LESSON NOTES

NCTM Standards: 1–5, 7, 8

Instructional Resources

- Study Guide Master 13-5
- Practice Master 13-5
- Enrichment Master 13-5
- Modeling Mathematics Masters, pp. 63–66
- Multicultural Activity Masters, p. 26
- Tech Prep Applications Masters, p. 26

Transparency 13-5A contains the 5-Minute Check for this lesson; **Transparency 13-5B** contains a teaching aid for this lesson.

Recommended Pacing	
Honors Pacing	Day 6 of 13

1 FOCUS

5-Minute Check
(over Lesson 13-4)

Suppose ▱ABCD → ▱MLKN. Write the name of the image that corresponds to each preimage.

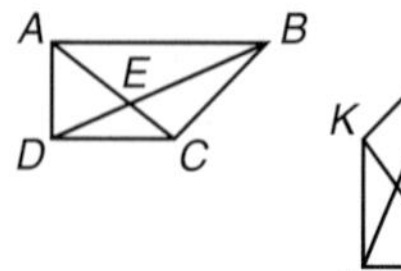

1. ∠ABC ∠MLK
2. $\overline{DC}$ $\overline{NK}$
3. BE LO
4. △BDC △LNK
5. ∠AEB ∠MOL

Motivating the Lesson

Hands-On Activity Bring a mirror to class and use it to demonstrate reflection. Emphasize that the mirror does not change the shape of a person or object, and that the parts of the person or object visible to the mirror are reflected so that the reflection is identical to the original.

13-5 Reflections

What YOU'LL LEARN

- To name, recognize, and draw reflected images, lines of symmetry, and points of symmetry.

Why IT'S IMPORTANT

Reflections can be used to solve problems involving mirrors, billiards, and golf.

APPLICATION
Mirrors

As light strikes the full-length mirror in the photo at the right, it reflects an image that appears to be behind the mirror, but it is reversed from left to right. The image is the same size as the woman it reflects. It also appears to be the same distance from the mirror as the woman.

A reflection is sometimes called a flip of the figure.

A **reflection** is a type of transformation. Just as the mirror is a plane of reflection for the woman and her image, line ℓ is a **line of reflection** for P and its image R. If this page were folded along line ℓ, P and R would coincide. This means that ℓ is the perpendicular bisector of $\overline{PR}$. Therefore, point Q is the midpoint of $\overline{PR}$. Since point Q is already on the line of reflection, Q is its own image.

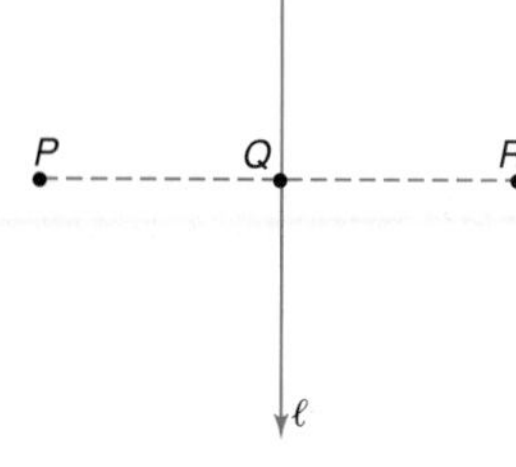

It is also possible to reflect an image with respect to a point. Consider the figure above without line ℓ. Point P is the reflection of R with respect to Q. Therefore, Q is a **point of reflection**.

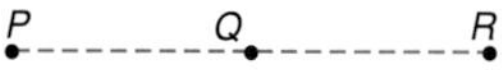

What happens with the reflected images of collinear points S, T, and U? The reflected points X, Y, and Z are also collinear. Thus, it is said that reflections preserve collinearity.

Reflections also preserve betweenness of points. Point T is between points S and U. Likewise, the reflection of T, point Y, is between reflected points X and Z.

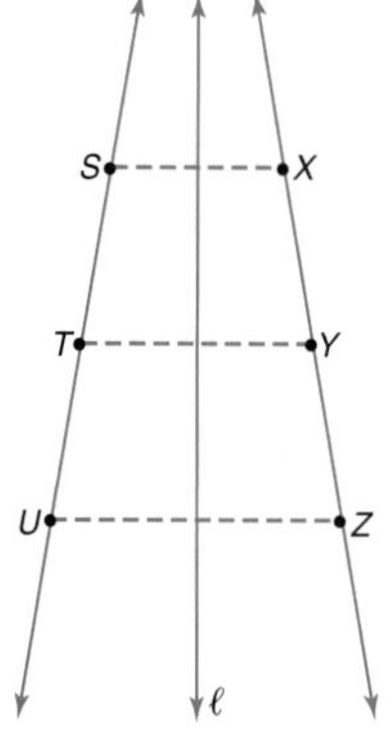

In the figure below, $\triangle ABC$ is reflected over the y-axis. $\triangle XYZ$ is the reflected image of $\triangle ABC$. Are there any other relationships between the two triangles? By measuring the corresponding parts of the triangles, it appears that the two triangles are also congruent.

In addition to preserving collinearity and betweenness of points, reflections also preserve angle measure and distance measure.

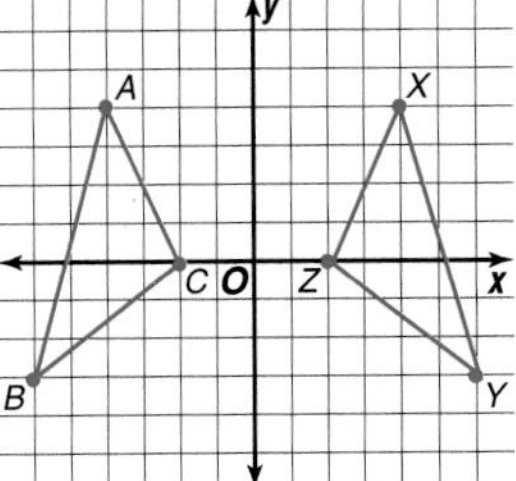

Points A, B, and C can be read in a counterclockwise order. $\triangle ABC$ is said to have a counterclockwise orientation.

Corresponding points X, Y, and Z are then in a clockwise order. $\triangle XYZ$ is said to have a clockwise orientation.

2 TEACH

Teaching Tip Emphasize that reflections preserve collinearity, betweenness of points, angle measure, and distance measure.

Teaching Tip Instead of a formal construction, you may wish to have students draw $\triangle ABC$ and line ℓ on tracing paper. Have them fold the paper along ℓ and place the paper on a window or other light source to trace the reflection.

CONSTRUCTION

Reflecting a Triangle

Construct the reflected image of $\triangle ABC$ over line ℓ.

1. Since A is on line ℓ, A is its own reflection. Construct perpendiculars from B and C through line ℓ.
2. Locate Y and Z so that line ℓ is the perpendicular bisector of $\overline{BY}$ and $\overline{CZ}$. Points Y and Z are the reflected points of B and C.
3. Connect vertices A, Y, and Z.

Conclusion: Since points A, Y, and Z are reflected images of points A, B, and C, $\triangle AYZ$ is the reflection of $\triangle ABC$.

Reflections are also used in playing many sports.

In-Class Example

For Example 1

You are at a miniature golf course. The first hole is a straight hole with a windmill in the center. There is no room to go under the windmill. How can you make a hole in one?

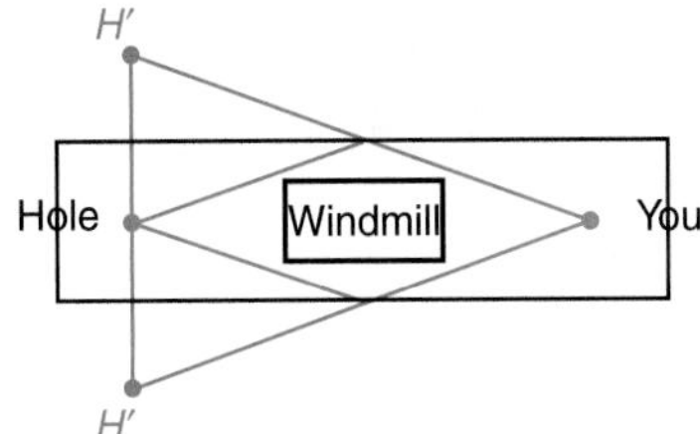

You visualize the reflection image H' of the hole H with respect to either the left or the right side of the hole. Then aim for H'.

Example 1

APPLICATION

Recreation

A top or bottom spin of the cue ball will affect the angle at which the cue ball rebounds off another ball, while a side spin of the cue ball will affect the angle at which the cue ball rebounds off a cushion.

Jenelle is playing billiards. She wants to use the white cue ball to hit the black 8 ball into the lower right pocket. How can she use a reflection to make the shot?

She can mentally reflect the pocket over the upper side of the pool table. Then she can align the eight ball and the reflected pocket to find point P on the side of the table. If the ball is hit without a spin, she can aim the cue ball to hit the 8 ball so that it will hit point P on the table and then bounce off the table side at an angle that will take it to the pocket.

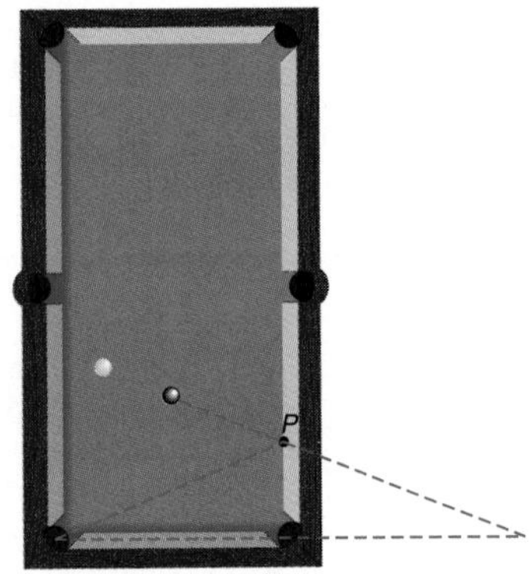

Teaching Tip When defining reflections, emphasize that a point that lies on the line of reflection *does* have an image even though it is itself.

Cooperative Learning

Group Discussion Have small groups of students plan a music video. They should select the music, one song, and a variety of images that will be dilated, rotated, translated, and reflected in a patterned sequence to match the music. For more information on the group discussion strategy, see *Cooperative Learning in the Mathematics Classroom*, one of the titles in the Glencoe Mathematics Professional Series, page 31.

In-Class Examples

For Example 2

Determine and show all lines of symmetry for each figure below.

a.

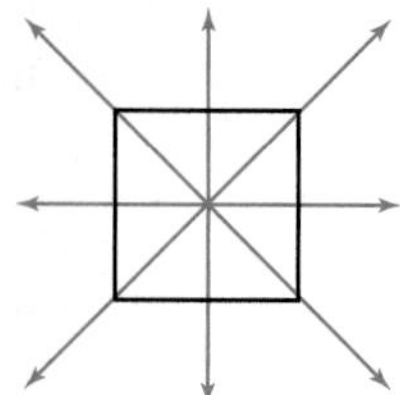

4 lines of symmetry

b.

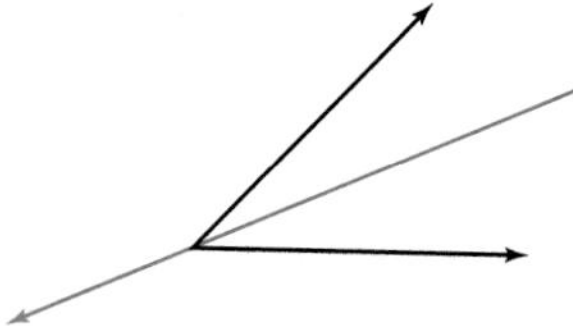

1 line of symmetry

For Example 3

a. Draw the reflected image of $\triangle SEH$ over the y-axis.

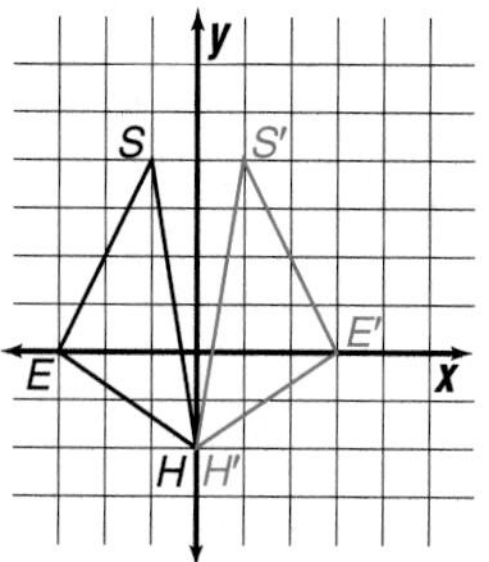

b. Describe any pattern you see in the coordinates of the triangle and its reflected image.

$S(-1, 4) \rightarrow S'(1, 4)$
$E(-3, 0) \rightarrow E'(3, 0)$
$H(0, -2) \rightarrow H'(0, -2)$

In a reflection over the y-axis, the y-coordinate remains the same and the x-coordinate is the opposite of the x-coordinate of the preimage.

Teaching Tip Remind students that to verify a line of symmetry they can fold the paper in half and match the sides.

The union of any figure and its reflected image is always a figure that has a line of symmetry.

A line of reflection can be drawn through some plane figures so that one side is a reflected image of the other side. This line of reflection is called a **line of symmetry**. This line of symmetry can be determined and drawn in certain figures.

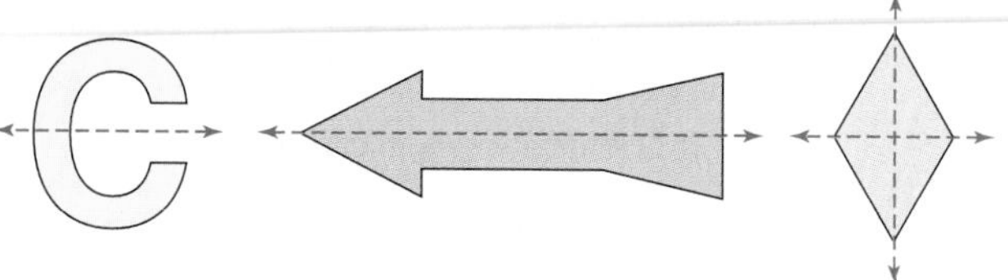

Example 2 **Determine and show all the lines of symmetry for each figure below.**

a. equilateral triangle

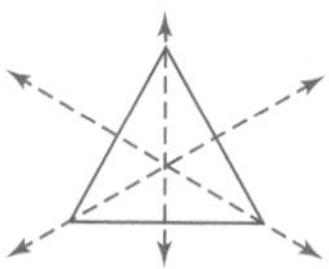

3 lines of symmetry

b. line segment

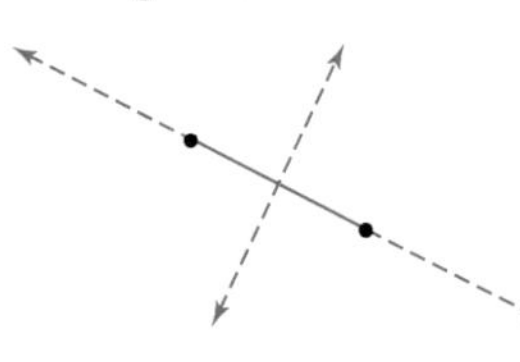

2 lines of symmetry

In part b of Example 2, note that one of the lines of symmetry is a perpendicular bisector of the segment. You can use perpendicular bisectors to find images of reflections.

Example  **a. Draw the reflected image of $\triangle PQR$ over the x-axis.**

Reflect each point over the x-axis so that the x-axis is the perpendicular bisector of the segment formed by each point and its image.

Connect the reflected image points to form $\triangle P'Q'R'$.

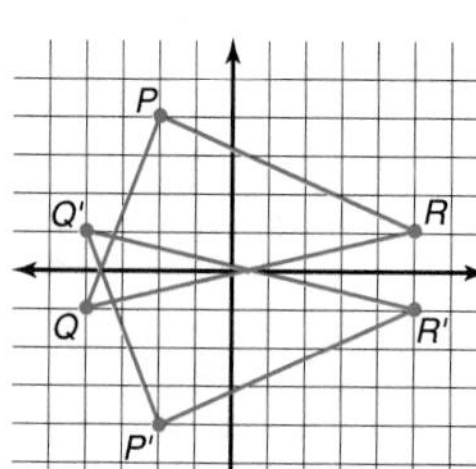

b. Describe any pattern you see in the coordinates of the triangle and its reflected image.

$Q(-4, -1) \rightarrow Q'(-4, 1)$

$P(-2, 4) \rightarrow P'(-2, -4)$

$R(5, 1) \rightarrow R'(5, -1)$

In a reflection over the x-axis, the x-coordinate remains the same, and the y-coordinate is the opposite of the y-coordinate of the preimage.

A pattern also exists when you reflect a figure over the y-axis.

Example 4 **The vertices of quadrilateral $PQRS$ are $P(-5, 4)$, $Q(-1, -1)$, $R(-3, -6)$, and $S(-7, -3)$, and the vertices of quadrilateral $P'Q'R'S'$ are $P'(5, 4)$, $Q'(1, -1)$, $R'(3, -6)$, and $S'(7, -3)$.**

a. Construct a table of these vertices and look for a pattern in the x- and y-coordinates. What type of transformation occurred?

The x-coordinates of quadrilateral $P'Q'R'S'$ are the opposites of the x-coordinates of quadrilateral $PQRS$. The y-coordinates stayed the same. Thus, the pattern indicates that the vertices must have been reflected over the y-axis.

quadrilateral PQRS	quadrilateral P'Q'R'S'
P(−5, 4),	P'(5, 4)
Q(−1, −1)	Q'(1, −1)
R(−3, −6)	R'(3, −6)
S(−7, −3)	S'(7, −3)

b. Graph the preimage, the image, and the line of reflection to verify the transformation.

Each point is an equal distance from the y-axis, which is the line of reflection.

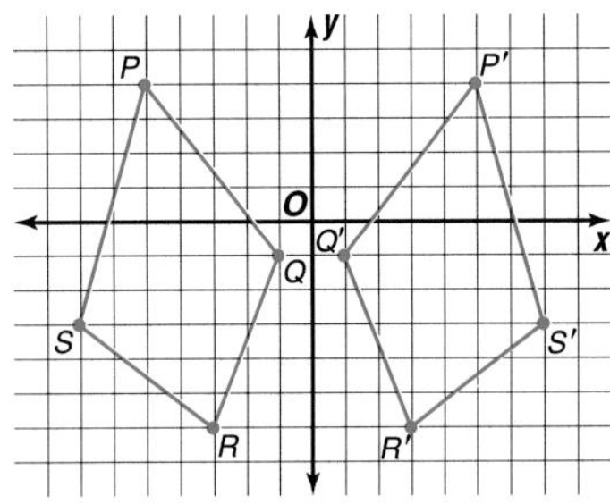

For many figures, a point can be found that is a point of reflection for all points on the figure. This point of reflection is called a **point of symmetry**. A point of symmetry must be a midpoint for all segments that pass through it and have endpoints on the figure. In the two figures below, P and Q are midpoints of the segments drawn. In the third figure, R is not a point of symmetry because R is not the midpoint of $\overline{XZ}$.

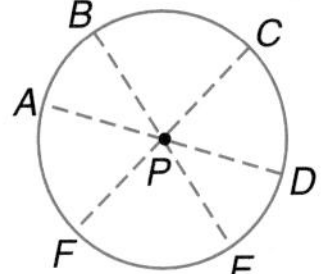

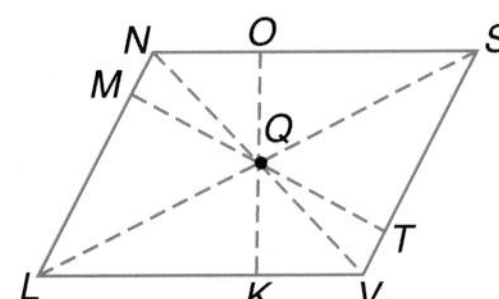

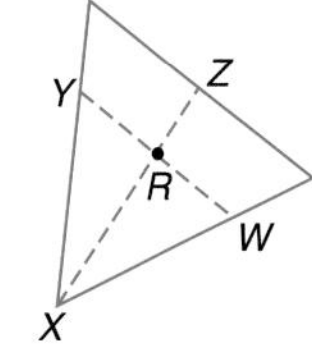

P and Q are points of symmetry. R is not a point of symmetry.

CHECK FOR UNDERSTANDING

Communicating Mathematics

1. The line of reflection is the ⊥ bisector of the line segment that joins a point and its reflected image.

Study the lesson. Then complete the following.

1. **Describe** the relationship between the line of reflection and the line segment that joins a point and its reflected image.
2. **Name** four properties that are preserved in reflections. See margin.
3. **You Decide** Damaris was asked to name the geometric plane figure that has the most lines of symmetry. Damaris thinks it is the square. Sarah disagrees. She thinks there is another geometric plane figure that has more lines of symmetry than a square. Who is right? See margin.

In-Class Example

For Example 4

▱$ABCD$ has vertices $A(1, 3)$, $B(4, 4)$, $C(5, 2)$, and $D(2, 1)$. ▱$EFGH$ has vertices $E(-1, 3)$, $F(-4, 4)$, $G(-5, 2)$, and $H(-2, 1)$.

a. Construct a table of these vertices and look for a pattern in the x- and y-coordinates. What type of transformations occurred? **reflection over the y-axis**

b. Graph the preimage, the image, and the line of reflection to verify the transformation.

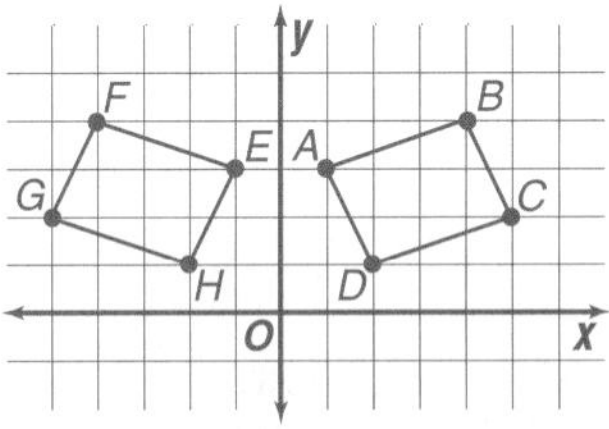

3 PRACTICE/APPLY

Check for Understanding

Exercises 1–15 are designed to help you assess your students' understanding through reading, writing, speaking, and modeling. You should work through Exercises 1–3 with your students and then monitor their work on Exercises 4–15.

Error Analysis

It may be hard to find a point of symmetry. Point out that students can test to see if a point is a point of symmetry by turning the figure 180°. If the rotated figure overlaps the original figure, then the point is a point of symmetry.

Additional Answers

2. collinearity, betweenness of points, distance measure, angle measure
3. Sarah is right. A circle is a plane figure and has an infinite number of lines of symmetry. A square only has 4 lines of symmetry.

Reteaching

Using Constructions Have each student draw a triangle and a line on a piece of paper. Work through the construction on page 723 with students, using the triangles and lines that they just drew.

Assignment Guide

Core (with proof): 17–57 odd, 59–70
Core (informal): 17–57 odd, 59–70
Enriched: 16–52 even, 53–70

For **Extra Practice**, see p. 791.

The red A, B, and C flags, printed only in the Teacher's Wraparound Edition, indicate the level of difficulty of the exercises.

Additional Answer

14.

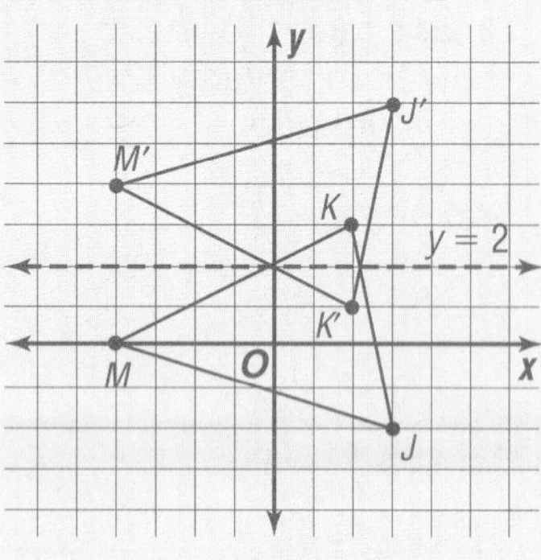

Guided Practice

For the figure at the right, name the reflected image of each of the following over line ℓ.

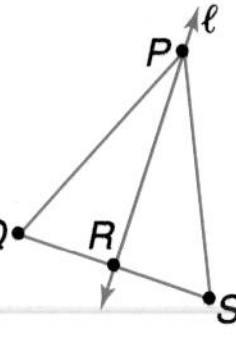

4. P **P**
5. $\overline{PQ}$ **$\overline{PS}$**
6. $\angle PQS$ **$\angle PSQ$**
7. $\triangle QSP$ **$\triangle SQP$**

Copy each figure. Draw the reflected image of each figure over line p.

8.

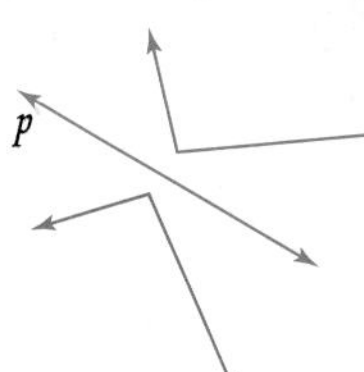

9.

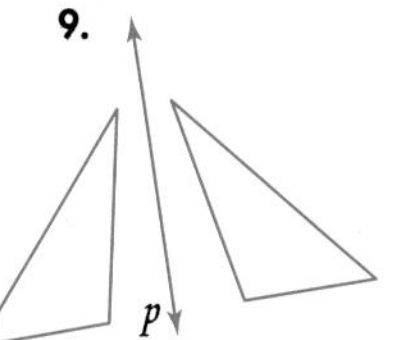

Determine how many lines of symmetry each figure has. Then, identify those figures that have point symmetry.

10.

10. two lines of symmetry; point symmetry

11.

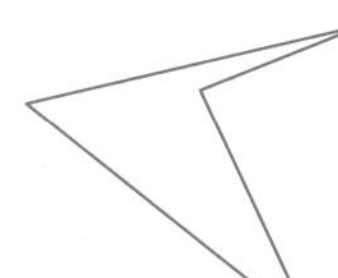

11. no lines of symmetry; no point symmetry

12.

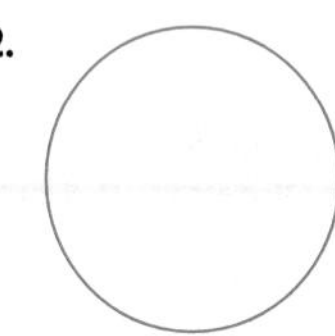

12. infinite number of lines of symmetry; point symmetry

13.

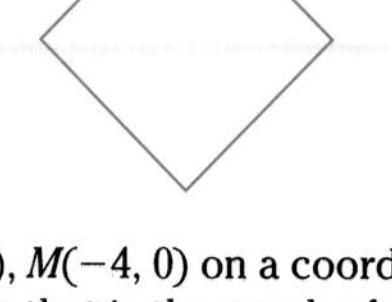

13. four lines of symmetry; point symmetry

14. Graph $\triangle JKM$ with vertices $J(3, -2)$, $K(2, 3)$, $M(-4, 0)$ on a coordinate plane. Then draw its reflected image over the line that is the graph of $y = 2$. **14. See margin.**

15. **Grooming** Jamal uses two mirrors to check the appearance of his hair from the back. Using the measurements as indicated in the figure, how far away does the image of the back of his head appear to Jamal? **168 cm**

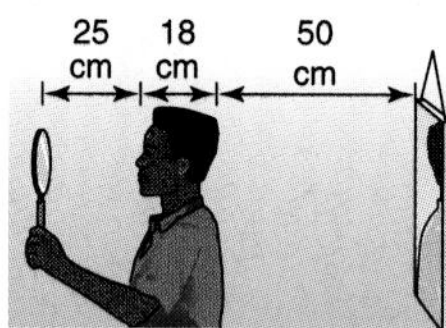

EXERCISES

Practice

For the figure below, name the reflected image of each figure over line m.

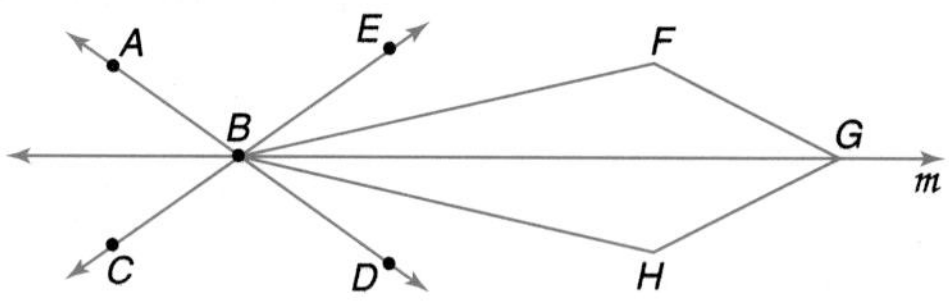

16. A **C**
17. $\overline{FG}$ **$\overline{HG}$**
18. $\angle ABE$ **$\angle CBD$**
19. $\angle ABC$ **$\angle CBA$**
20. $\overline{BG}$ **$\overline{BG}$**
21. $\triangle BFG$ **$\triangle BHG$**
22. $\angle GBF$ **$\angle GBH$**
23. quadrilateral $FGHB$ **quadrilateral $HGFB$**

Copy each figure. Draw all possible lines of symmetry. If none exist, write *none*.

24. N none

25. T

26. 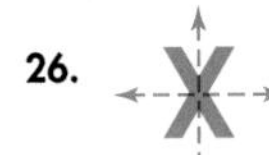

27.  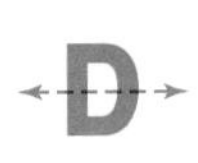

Copy each figure. Use a straightedge to draw the reflected image of each figure over line p.

28.  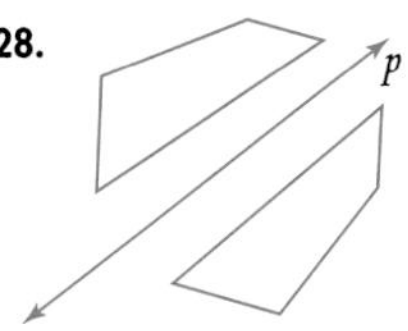

29. 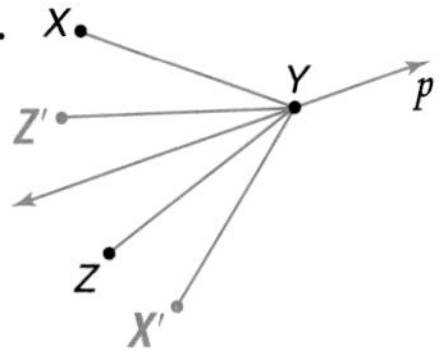

30. 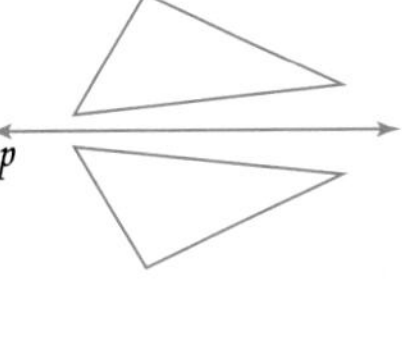

31. 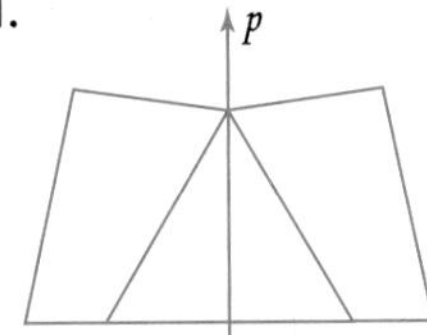

32. 

33. 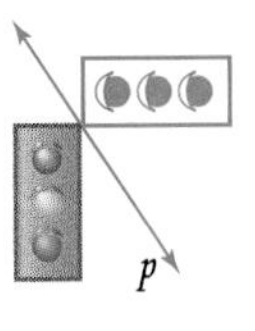

For each figure, determine whether ℓ is a line of symmetry. Write *yes* or *no*. Explain your answer. 36–37. See margin.

34. 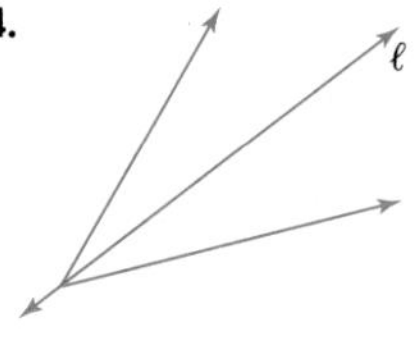

35. 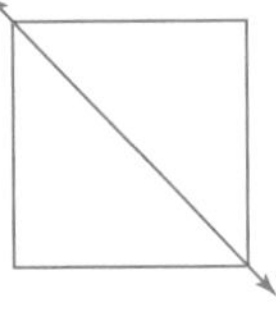

36. 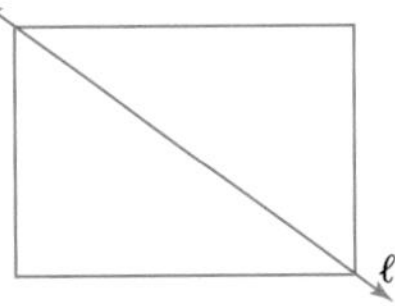

37. 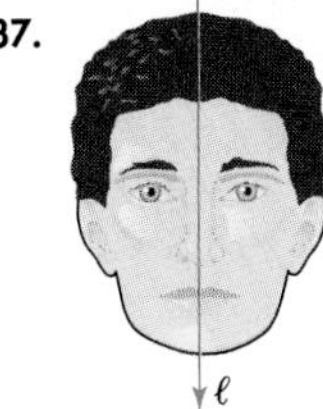

38. 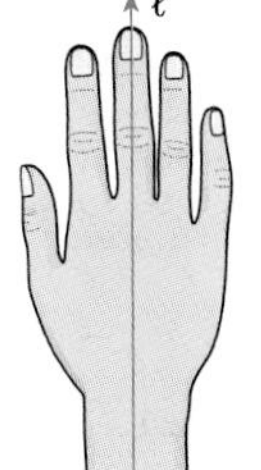

39. 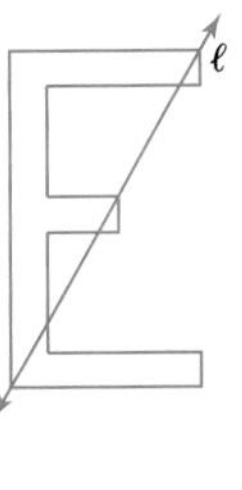

34. Yes; ℓ is the line of symmetry because it is possible to find, for every point A, another point B so that ℓ is the perpendicular bisector of $\overline{AB}$.

35. Yes; ℓ is the line of symmetry because it is possible to find, for every point A, another point B so that ℓ is the perpendicular bisector of $\overline{AB}$.

38. No; not all points are the same distance from ℓ.

39. No; not all points are the same distance from ℓ.

40. In Example 3, what pattern do you notice when there is a reflection over the x-axis? See margin.

Graph each figure on a coordinate plane. Then draw its reflected image over the indicated line of reflection.

41–43. See Solutions Manual.

41. $\overline{AB}$ for $A(4, 3)$ and $B(0, 3)$ over the x-axis.

42. Trapezoid $ABCD$ for $A(2, 2)$, $B(2, -1)$, $C(-1, -2)$, $D(-1, 3)$ over the graph of $y = 3$.

43. $\triangle PQR$ for $P(0, 6)$, $Q(-4, 0)$, $R(-4, -6)$ over the graph of $y = x$.

Additional Answers

36. No; not all points are the same distance from ℓ.

37. Yes; ℓ is the line of symmetry because it is possible to find, for every point A, another point B so that ℓ is the perpendicular bisector of $\overline{AB}$.

40. The x-coordinates of the preimage and the image stay the same. However, the y-coordinate of the preimage is negated to get the y-coordinate of the image.

Study Guide Masters, p. 82

NAME__________ DATE______

Study Guide

Student Edition Pages 722–729

Reflections

A **reflection** is a type of transformation that can be described by folding. If the diagram at the right is traced and folded along line ℓ, each point will coincide with its image. That is, A is mapped to D, B is mapped to E, and C is mapped to itself.

Note that each point and its image are on lines which are perpendicular to ℓ, the **line of reflection**. This line of reflection is also called the **line of symmetry**.

It is also possible to have a reflection image with respect to a point. In the figure at the right, Z is the reflection of X with respect to Y. The point of reflection, Y, is the midpoint of the segment drawn from X to Z.

For some figures, a point can be found that is a point of reflection for all points on the figure. This point of reflection is called a **point of symmetry**. A point of symmetry must be a midpoint for all segments that pass through it and have endpoints on the figure.

Examples: 1 Is the dashed line a line of symmetry for the figure?

No, the two parts do not coincide if the figure is folded along the dashed line.

2 Is point Q a point of symmetry for the rhombus?

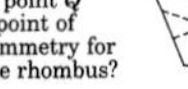

Yes, Q is a midpoint for all segments that pass through it and have endpoints on the figure.

Copy each figure. Use a straightedge to draw the reflection image of each figure over line m.

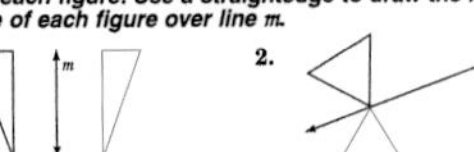

Determine if each figure has line symmetry, point symmetry, or both.

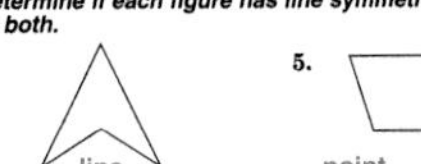

4. line 5. point 6. both

7. Given $A(-1, 5)$ and $B(4, -2)$, graph $\overline{AB}$ on a coordinate plane. Then draw the reflection image over the line of reflection, the x-axis.

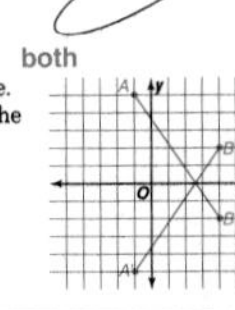

Copy each figure below. Indicate any points of symmetry, If none exist, write *none*.

44. 

45.  none

46. 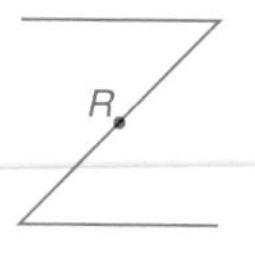

47.  none

48. none

49. 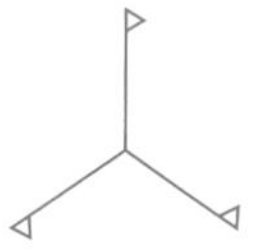 none

For each figure, indicate if the figure has *line symmetry, point symmetry*, or *both*.

50. 
point symmetry

51. 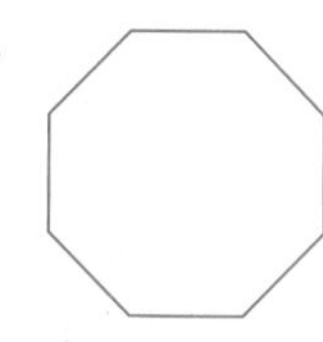
both

52. 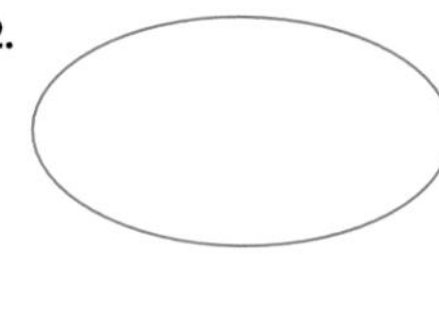
both

Critical Thinking

53. Study Example 3, Example 4, and Exercise 40. Make a conjecture for reflecting over each of the following.
    a. y-axis $(x, y) \rightarrow (-x, y)$ when reflected over the y-axis
    b. x-axis $(x, y) \rightarrow (x, -y)$ when reflected over the x-axis
    c. graph of $y = x$ $(x, y) \rightarrow (y, x)$ when reflected over line $y = x$

54. Copy the figure at the right. Draw the reflected image over line m of the reflected image of pentagon $ABCDE$ over line ℓ. Label the vertices A', B', C', D', and E' to correspond to A, B, C, D, and E, respectively.

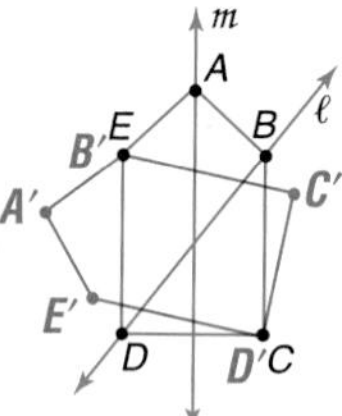

Applications and Problem Solving

55. **Algebra** For the parabola shown, give an example of a line of reflection and a point of reflection.
    Line m is a line of reflection; the vertex is a point of reflection.

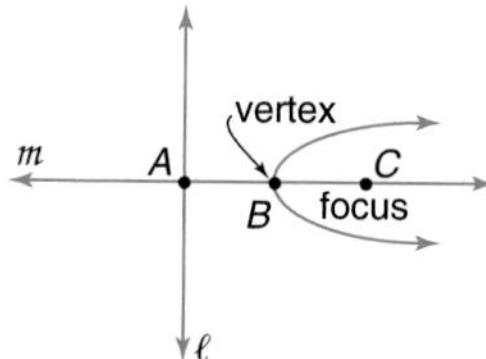

56. **Tessellations** Draw a regular hexagon and reflect it over each of its sides.
    a. Describe the pattern that you made. Do any of the areas of the figures overlap? interlocking hexagons; no
    b. After the first set of reflections, suppose you perform the same reflections on the resulting hexagons. Describe the pattern you see.
    c. What would happen if you reflected a regular octagon in the same way?

56b. A honeycomb-like pattern develops.

56c. Figures would overlap and not tessellate correctly.

Practice Masters, p. 82

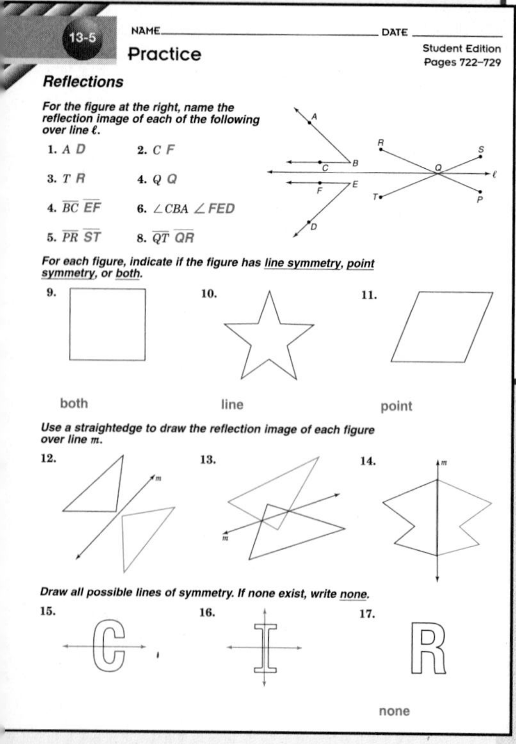

13-5 NAME______________ DATE ______

Practice

Student Edition Pages 722–729

Reflections

For the figure at the right, name the reflection image of each of the following over line ℓ.

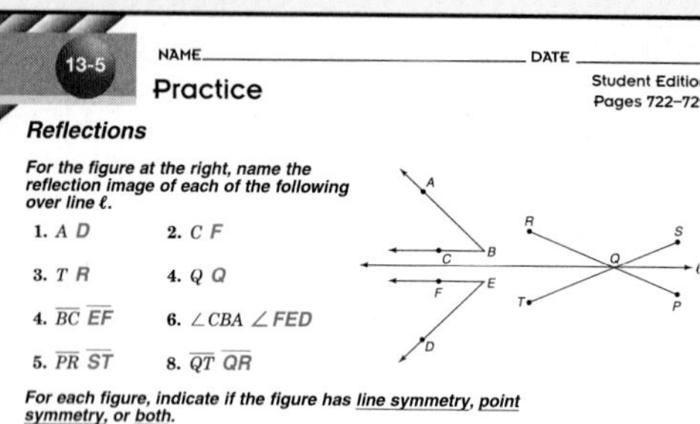

1. A D
2. C F
3. T R
4. Q Q
5. $\overline{BC}$ $\overline{EF}$
6. ∠CBA ∠FED
7. $\overline{PR}$ $\overline{ST}$
8. $\overline{QT}$ $\overline{QR}$

For each figure, indicate if the figure has line symmetry, point symmetry, or both.

9. both

10. 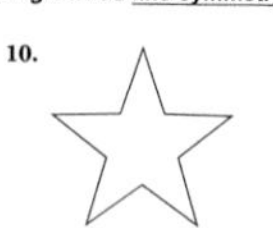
line

11. 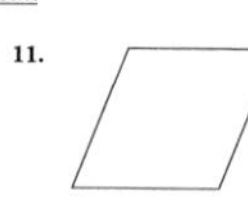
point

Use a straightedge to draw the reflection image of each figure over line m.

12. 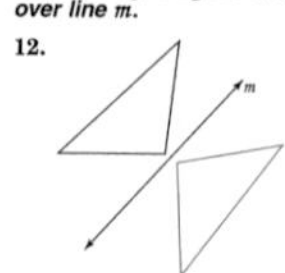

13.

14. 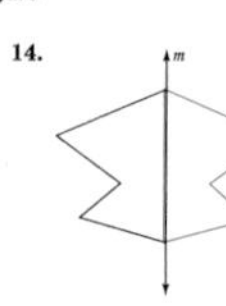

Draw all possible lines of symmetry. If none exist, write none.

15. 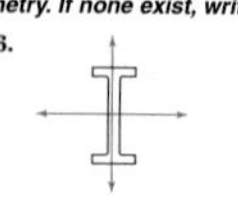

16. 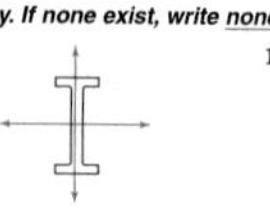

17. 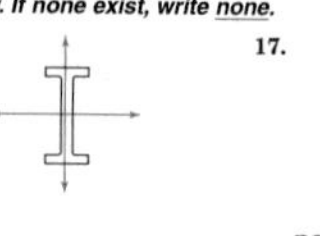
none

57a. The y-coordinates are negated and the x-coordinates are unchanged. It is reflected over the x-axis.

57. Make a Table Use the conjectures you formed in Exercise 53. Describe each transformation without graphing.

a. The vertices of $\triangle XYZ$ on a coordinate plane are $X(-2, 1)$, $Y(0, 5)$, and $Z(6, 2)$, and the vertices of $\triangle X'Y'Z'$ are $X'(-2, -1)$, $Y'(0, -5)$, and $Z'(6, -2)$.

b. The vertices of $\triangle ABC$ on a coordinate plane are $A(3, -2)$, $B(4, 1)$, and $C(-3, 2)$, and the vertices of $\triangle A'B'C'$ are $A'(-3, -2)$, $B'(-4, 1)$, and $C'(3, 2)$.

58. Recreation Moira is playing miniature golf. If the ball is at point B, how can she make a hole-in-one at H for the situation shown at the right?

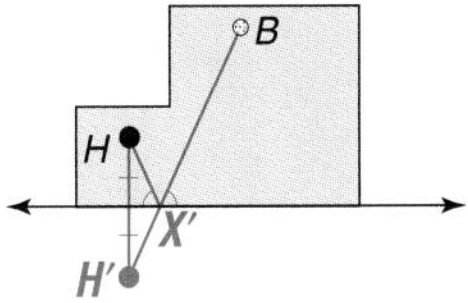

Mixed Review

59. Suppose quadrilateral $ABCD \rightarrow$ quadrilateral $RSTU$. (Lesson 13–4)

a. Name the image of $\angle B$. $\angle S$

b. Name the preimage of $\overline{TU}$. $\overline{CD}$

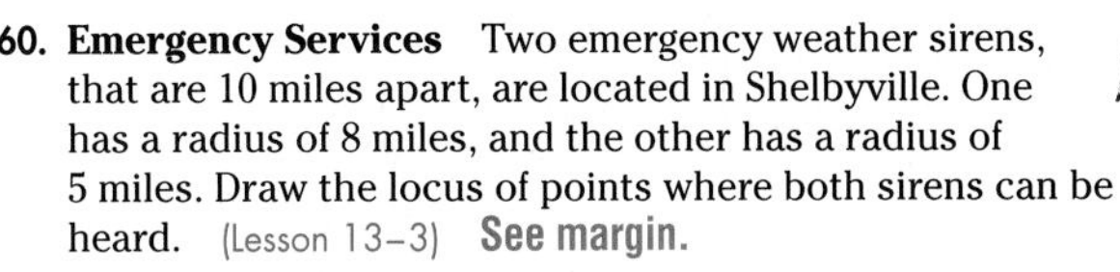

60. Emergency Services Two emergency weather sirens, that are 10 miles apart, are located in Shelbyville. One has a radius of 8 miles, and the other has a radius of 5 miles. Draw the locus of points where both sirens can be heard. (Lesson 13–3) See margin.

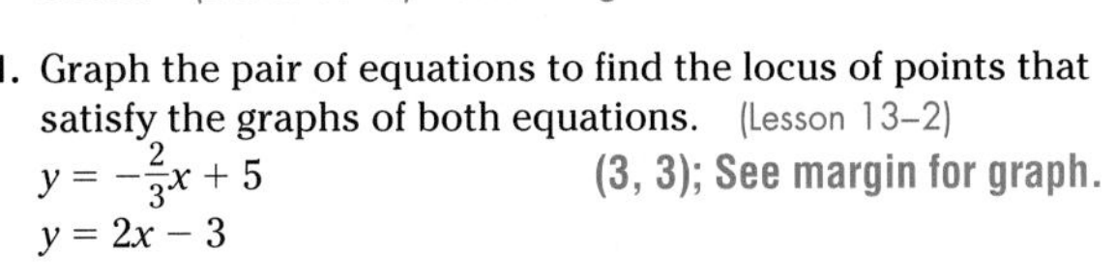

61. Graph the pair of equations to find the locus of points that satisfy the graphs of both equations. (Lesson 13–2)

$y = -\frac{2}{3}x + 5$ (3, 3); See margin for graph.

$y = 2x - 3$

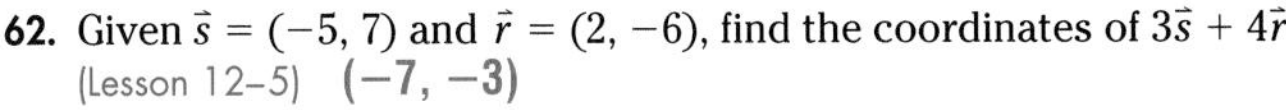

62. Given $\vec{s} = (-5, 7)$ and $\vec{r} = (2, -6)$, find the coordinates of $3\vec{s} + 4\vec{r}$. (Lesson 12–5) $(-7, -3)$

57b. The x-coordinates are negated and the y-coordinates are unchanged. It is reflected over the y-axis.

63. $y = -\frac{3}{4}x + \frac{7}{4}$

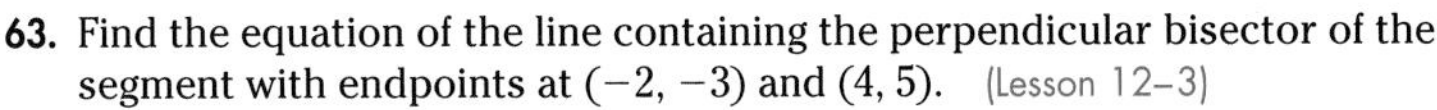

63. Find the equation of the line containing the perpendicular bisector of the segment with endpoints at $(-2, -3)$ and $(4, 5)$. (Lesson 12–3)

64. Complete: In the linear equation $y = mx + b$, m represents the _?_ of the line. (Lesson 12–1) slope

65. Find the lateral area of the triangular pyramid. (Lesson 11–4) 62.89 units^2

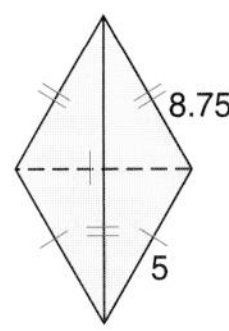

66. Find the area of the parallelogram. (Lesson 10–3) 72 ft^2

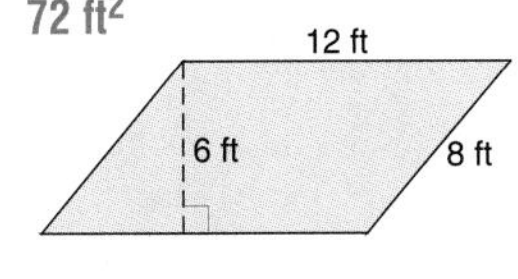

67. Complete: The longest chords of a circle are _?_. (Lesson 9–1) diameters

68. Use the figure to complete each statement with < or >. (Lesson 5–3)

a. $m\angle 1$ _?_ $m\angle 4$ $>$

b. $m\angle 2$ _?_ $m\angle 6$ $<$

c. $m\angle 7$ _?_ $m\angle 3$ $<$

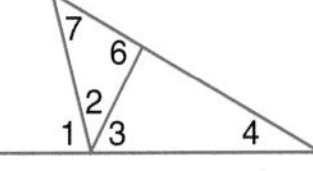

INTEGRATION Algebra

69. Solve $|8n - 5| < 0$. $\varnothing$

70. Solve $8b \geq 64$. $\{b \mid b \geq 8\}$

4 ASSESS

Closing Activity

Speaking Draw a line on the chalkboard or overhead and use it as a line of reflection. Draw several points, lines, line segments, and their reflections also. Have students describe the reflection image of each point or line.

Additional Answers

60.

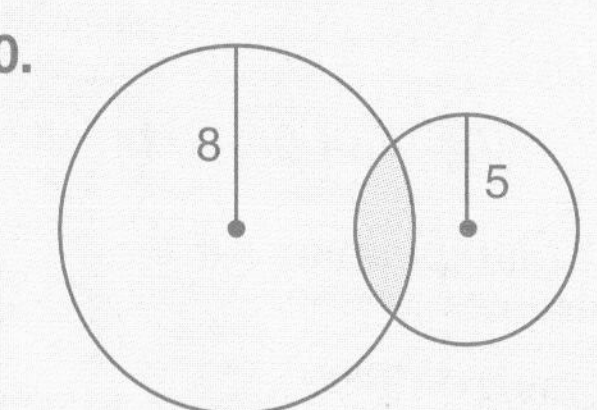

61.

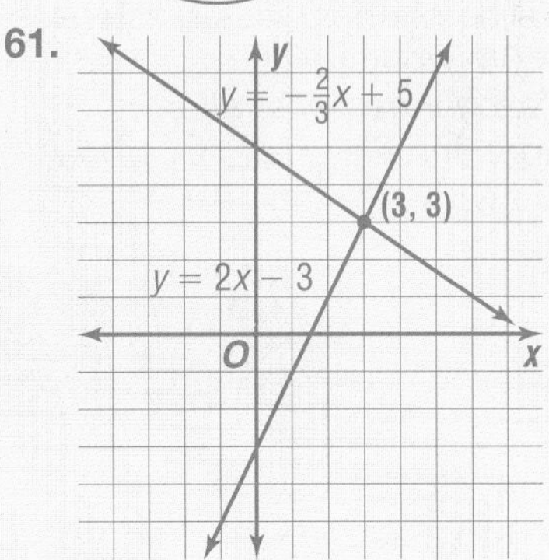

Extension

Reasoning Draw segments, points, or lines on the coordinate plane. Have students reflect each segment, point, or line with respect to the x-axis, y-axis, origin, and line $y = x$.

Enrichment Masters, p. 82

NAME________ DATE________

Enrichment

Student Edition Pages 722–729

Reflections in the Coordinate Plane

Study the diagram at the right. It shows how the triangle ABC is mapped onto triangle XYZ by the transformation $(x, y) \rightarrow (-x + 6, y)$. Notice that $\triangle XYZ$ is the reflection image with respect to the vertical line with equation $x = 3$.

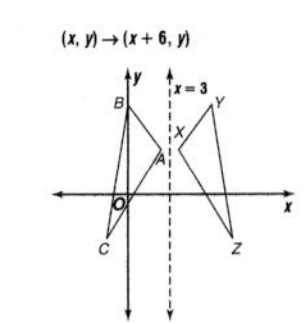

1. Prove that the vertical line with equation $x = 3$ is the perpendicular bisector of the segment with endpoints (x, y) and $(-x + 6, y)$. (Hint: Use the midpoint formula.)

Midpoint $= \left(\frac{x + (-x + 6)}{2}, \frac{y + y}{2}\right)$ or $(3, y)$

The segment joining (x, y) and $(-x + 6, y)$ is horizontal and hence is perpendicular to the vertical line.

2. Every transformation of the form $(x, y) \rightarrow (-x + 2h, y)$ is a reflection with respect to the vertical line with equation $x = h$. What kind of transformation is $(x, y) \rightarrow (x, -y + 2k)$?
reflection with respect to the horizontal line with equation $y = k$

Draw the transformation image for each figure and the given transformation. Is it a reflection transformation? If so, with respect to what line?

3. $(x, y) \rightarrow (-x - 4, y)$

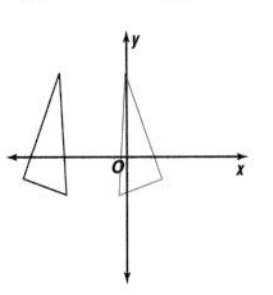

yes; $x = -2$

4. $(x, y) \rightarrow (x, -y + 8)$

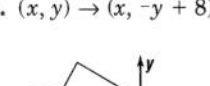

yes; $y = 4$

13-6A LESSON NOTES

NCTM Standards: 1–4, 7, 8

Objective

Use a geomirror to investigate translations.

Recommended Time

Demonstration and discussion: 15 minutes; Exercises: 30 minutes

Instructional Resources

For each student or group of students
Student Manipulative Kit
- geomirror
- ruler

Modeling Mathematics Masters
- p. 104 (worksheet)

For teacher demonstration
Algebra and Geometry Overhead Manipulative Resources

1 FOCUS

Motivating the Lesson

Look into a mirror. Using a dry-erase marker to write on the mirror, label your right and left ear. Label your right and left eye.

2 TEACH

Teaching Tip Students might want to try the activity with tracing paper instead of the geomirror.

3 PRACTICE/APPLY

Assignment Guide

Core (with proof): 1–5
Core (informal): 1–5
Enriched: 1–5

4 ASSESS

Observing students working in cooperative groups is an excellent method of assessment.

A Preview of Lesson 13–6

13–6A Reflections and Translations

Materials: paper  straightedge  geomirror

A geomirror is a construction instrument that allows you to find the reflection image of a figure. Use a geomirror with the following activity to investigate translations.

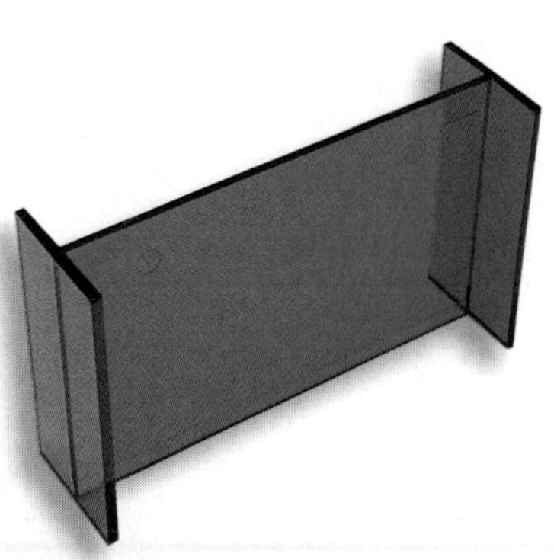

- Draw two parallel lines ℓ and m and a triangle ABC.
- Place the geomirror so that the edge is aligned with line ℓ. Look into the geomirror to see the reflection image of $\triangle ABC$.
- Use a straightedge to draw the image of $\triangle ABC$. Label the vertices A', B', and C'.
- Align the edge of the geomirror with line m. Look into the geomirror to see the reflection image of $\triangle A'B'C'$.
- Draw the image of $\triangle A'B'C'$. Label the vertices A'', B'', and C''.

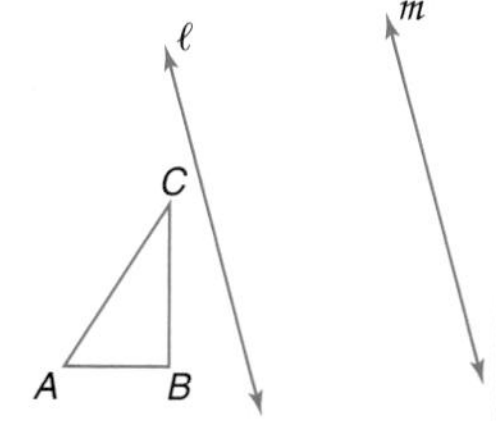

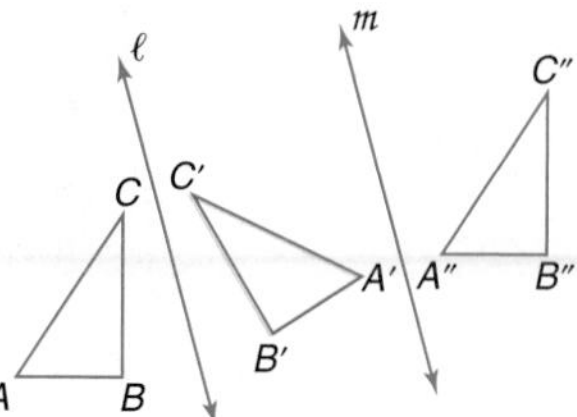

Write **Examine your drawings.**

1. You know that a reflection changes the orientation of its image. What happens when a figure is reflected twice? **The orientation remains unchanged.**
2. Are the points A, A', and A'' collinear? How about B, B', and B'' and C, C', and C''? Use a straightedge to verify your answer. **yes**
3. Compare the lengths of $\overline{AA''}$, $\overline{BB''}$, and $\overline{CC''}$, and the distance between ℓ and m. $AA'' = BB'' = CC'' = 2(\text{distance from } \ell \text{ to } m)$
4. Describe how you could map $\triangle ABC$ onto $\triangle A''B''C''$ in one motion instead of two reflections. **Use a translation.**

Model

5. **Use the geomirror to test whether two reflections will translate a regular polygon.** **b. The image is a translation of the preimage.**
   a. Draw two parallel lines and a regular polygon. **See students' work.**
   b. Reflect the regular polygon twice and compare the preimage and the image.
   c. Does your conjecture in Exercise 4 about mapping in one motion instead of two reflections hold true? **yes**

Using Cooperative Learning

This lesson offers an excellent opportunity for using cooperative learning groups. For more information on cooperative learning strategies and group management, see *Cooperative Learning in the Mathematics Classroom*, one of the titles in the Glencoe Mathematics Professional Series.

13-6 Translations

What YOU'LL LEARN

- To name and draw translation images of figures with respect to parallel lines.

Why IT'S IMPORTANT

Translations are used to describe linear movement in the real-world as well as on a coordinate plane.

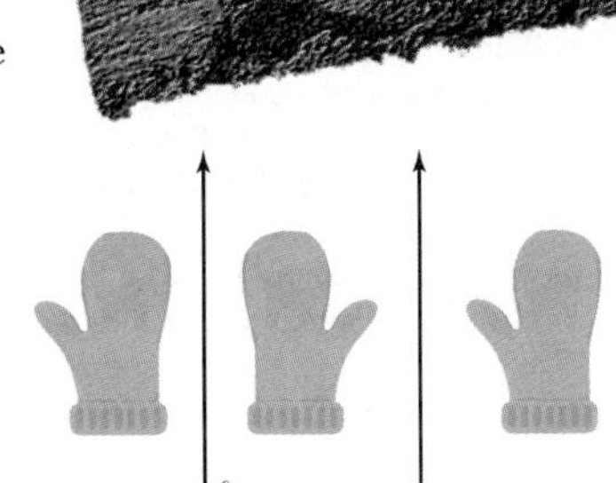

APPLICATION
Transportation

In order to preserve an old building that was going to be demolished to widen a road, the Reynoldsburg (Ohio) Historical Society had the building moved to another location. The result of a movement in one direction is a transformation called a **translation**. In a translation, all points are moved the same distance in the same direction.

One way to find a translation image is to perform one reflection after another with respect to two parallel lines. For example, the mitten at the right is first reflected over line ℓ. Then that image is reflected over line m. Two successive reflections such as this are called a **composite of reflections**.

Do you see a relationship between the first and the last images? A translation is often referred to as a glide.

Since translations are composites of two reflections, all translations are isometries. As a result, all properties preserved by reflections are preserved by translations. These properties include collinearity, betweenness of points, and angle and distance measure.

Example 1

INTEGRATION
Algebra

Find the translation image of $\square PQRS$ with respect to the parallel lines s and t.

Using the coordinate plane, reflect $\square PQRS$ over line s. Then draw the reflection image of $\square P'Q'R'S'$ with respect to line t.

The image reflected over line t is the translation image of $\square PQRS$.

Is there a way that $\square PQRS$ can be directly translated to $\square P''Q''R''S''$?

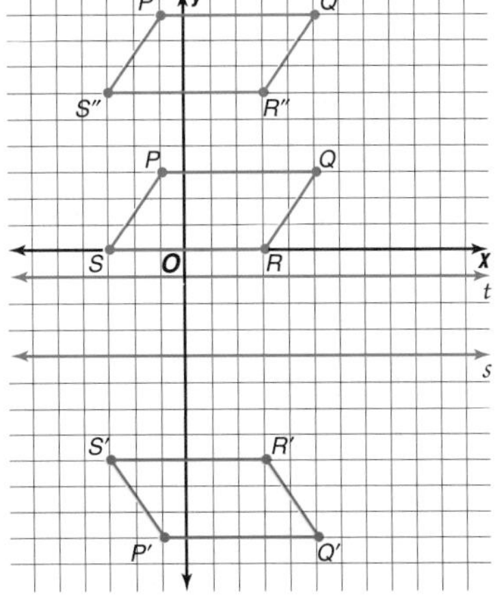

13-6 LESSON NOTES

NCTM Standards: 1–5, 7, 8

Instructional Resources

- Study Guide Master 13-6
- Practice Master 13-6
- Enrichment Master 13-6
- Assessment and Evaluation Masters, p. 353
- Modeling Mathematics Masters, p. 91

Transparency 13-6A contains the 5-Minute Check for this lesson; **Transparency 13-6B** contains a teaching aid for this lesson.

Recommended Pacing	
Honors Pacing	Day 8 of 13

1 FOCUS

5-Minute Check
(over Lesson 13-5)

Refer to the figure below. Name the reflection image of each of the following over line ℓ.

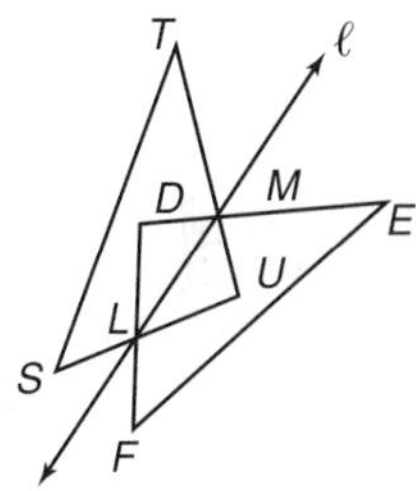

1. $\overline{ST}$ $\overline{FE}$
2. U D
3. $\overline{DM}$ $\overline{UM}$
4. $\overline{LM}$ $\overline{LM}$
5. $\triangle MDL$ $\triangle MUL$

Motivating the Lesson

Questioning Discuss with students what a translator does. Emphasize that a translator helps two other people communicate without changing their message. Relate this to geometric translations.

2 TEACH

Teaching Tip Emphasize that for a translation, the lines of reflection must be parallel.

In-Class Example

For Example 1

Find the translation image of $\triangle ABC$ with respect to the parallel lines s and t.

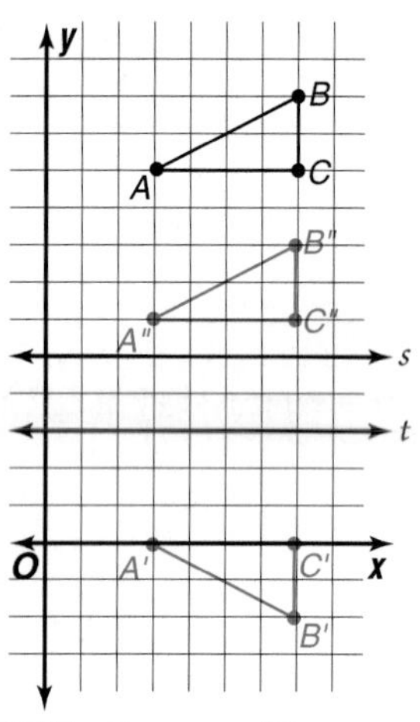

***A″B″C″* is the resulting translation image.**

Teaching Tip When discussing translations, emphasize that the same properties preserved by reflections are preserved by translations.

MODELING MATHEMATICS Have students compare their drawn result with an actual prism. Discuss the difficulties of drawing 3-D figures in 2-D. As an extension, have them use other figures as bases and draw a prism.

Since a translation is an isometry, congruence of shapes is preserved. You can use a translation to draw a prism by drawing the base and translating it to form a congruent base.

MODELING MATHEMATICS **Drawing Prisms Using Translations**

Materials: thin cardboard, plain paper, straightedge

Use a translation to draw a triangular prism.

- Cut a triangle out of a piece of thin cardboard. Label its angles 1, 2, and 3.

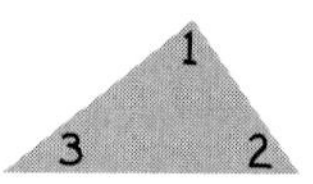

- Place the triangle on a piece of notebook paper, aligning the common side of $\angle 2$ and $\angle 3$ with a horizontal rule. Trace the triangle on the paper. Label the vertices so that the vertex of $\angle 1$ is X, the vertex of $\angle 2$ is Y, and the third vertex is Z.
- Now slide the cutout triangle to another place on the paper, making sure that the common side of $\angle 2$ and $\angle 3$ is still aligned with a horizontal rule. Trace the cutout triangle again. Label the vertices of this triangle X', Y', and Z' so that they correspond to the vertices of the first triangle you drew.

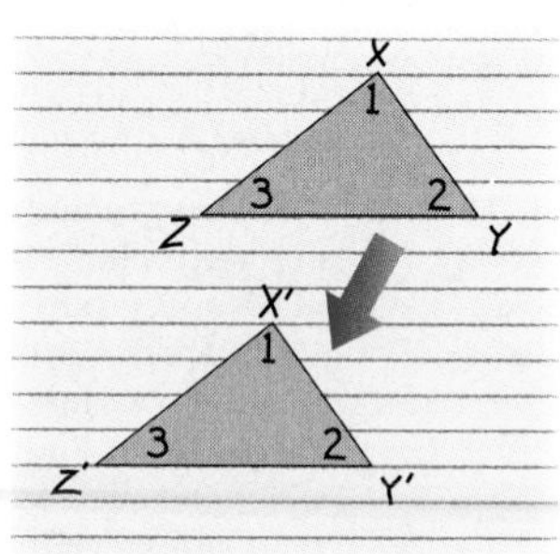

- Use a straightedge to draw $\overline{XX'}$, $\overline{YY'}$, and $\overline{ZZ'}$.

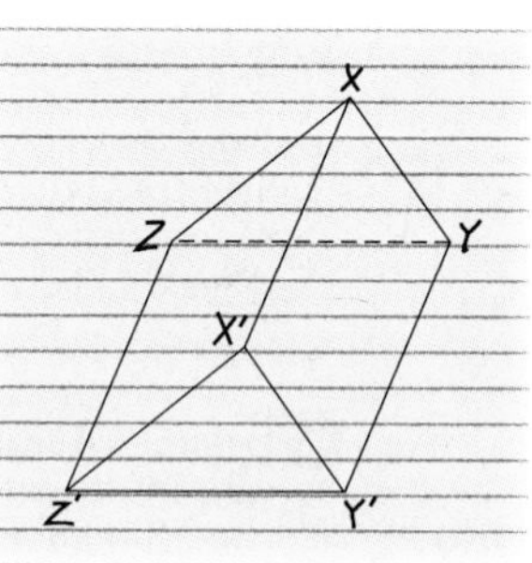

Your Turn

a. Why is $\overline{ZY}$ a dashed segment? **It is hidden from view.**

b. When you started with two triangular bases, how many more faces evolved when the prism was completed? **3**

c. Describe the shapes of the faces that were formed from this translation. How many of each shape are there? **3 rectangles**

Translations on a coordinate plane are easily drawn if you know the direction in which the figure is moving horizontally and vertically. Let's use a table to look at the three images from Example 1.

▱PQRS		▱P′Q′R′S′		▱P″Q″R″S″
P(−1, 3)	→	P′(−1, −11)	→	P″(−1, 9)
Q(5, 3)	→	Q′(5, −11)	→	Q″(5, 9)
R(3, 0)	→	R′(3, −8)	→	R″(3, 6)
S(−3, 0)	→	S′(−3, −8)	→	S″(−3, 6)

From ▱$PQRS$ to ▱$P''Q''R''S''$, the x-coordinates stay the same, but the y-coordinates have been translated 6 units up. So, (x, y) maps to $(x, y + 6)$. In general, if (a, b) describes the translation horizontally a units and vertically b units, then the image of (x, y) is $(x + a, y + b)$.

Example 2 **Translate $\triangle ABC$ with vertices $A(5, 4)$, $B(3, -1)$, and $C(0, 2)$, so that A' is located at (3, 1). Graph both triangles, and state the coordinates of B' and C'.**

Graph $\triangle ABC$ and A'.

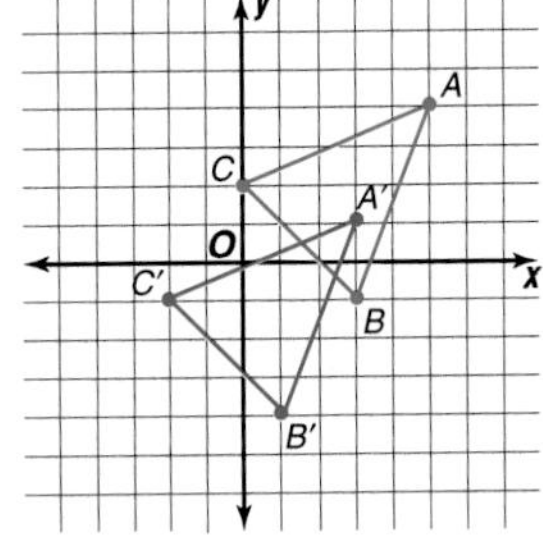

Point A was translated 2 units left and 3 units down to become A'. Perform this same translation on B and C to determine the coordinates of B' and C'.

B' has coordinates of $(1, -4)$, and C' has coordinates of $(-2, -1)$.

CHECK FOR UNDERSTANDING

Communicating Mathematics

Study the lesson. Then complete the following. 2–3. See margin.

1. **Name** the properties that are preserved in a reflection. Since a translation consists of two successful reflections, what properties will be preserved in a translation? Collinearity, betweenness of points, distance measure, angle measure; each of these properties are also preserved in a translation.
2. **Explain** how to translate a figure using a method other than two reflections.
3. **You Decide** Johnna says that the designs of wallpaper are usually designs of translations. Do you agree with her? Why or why not? You may use drawings to support your answer.
4. Suppose $\triangle BCD$ is slid along the y-axis until D has coordinates $(0, -8)$. What are the new coordinates of point B? (2, −5)

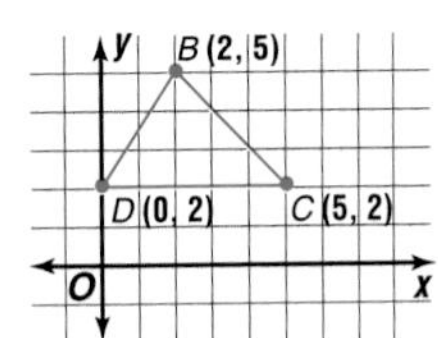

In-Class Example

For Example 2
$\triangle ABC$ has vertices $A(4, 2)$, $B(0, 6)$, and $C(-2, 2)$. State the coordinates of B' and C' where A' is $(-1, -1)$ and $\triangle A'B'C'$ is a translation of $\triangle ABC$.
$B'(-5, 3)$, $C'(-7, -1)$

3 PRACTICE/APPLY

Check for Understanding

Exercises 1–11 are designed to help you assess your students' understanding through reading, writing, speaking, and modeling. You should work through Exercises 1–5 with your students and then monitor their work on Exercises 6–11.

Additional Answers

2. A translation is like sliding a figure in one direction such that all points move the same number of units.
3. Sample answer: Yes; the designs on wallpaper keep repeating. The pictures are usually translations of each other. While hanging paper, it is important to match parts of the repeating pattern.

Reteaching

Using Diagrams Draw two parallel lines on the chalkboard or overhead and use a tile or a picture of a geometric object to demonstrate translations. Move the tile or object around and have students translate the figure at each location.

Assignment Guide

Core (with proof): 13–35 odd, 36–44
Core (informal): 13–35 odd, 36–44
Enriched: 12–32 even, 33–44

For **Extra Practice**, see p. 791.

The red A, B, and C flags, printed only in the Teacher's Wraparound Edition, indicate the level of difficulty of the exercises.

Additional Answer

5b. There will be an additional x number of rectangular faces after the translation, where x is the number of sides on the polygon that is the base. There will also be 2 base faces of a given polygonal shape.

5. Refer to the Modeling Mathematics activity. Use a translation to draw a pentagonal prism.
   a. How many additional faces were formed from these bases in the translation? 5
   b. Write a generalization about the number of faces and their shapes when a prism is formed from a translation. See margin.

Guided Practice

For each of the following, lines ℓ and m are parallel. Determine whether each red figure is a translation image of the blue figure. Write *yes* or *no*. Explain your answer.

6.

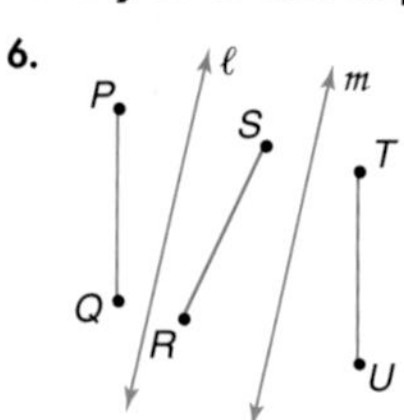

Yes; $\overline{TU}$ is a glide of $\overline{PQ}$.

7.

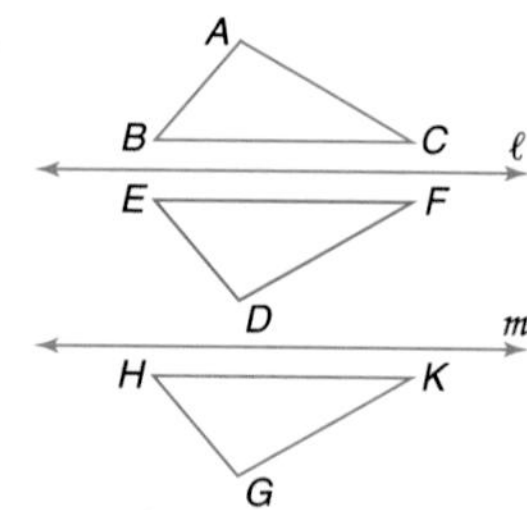

No; $\triangle HGK$ is not turned correctly.

10. $H'(1, 1)$, $I'(5, 3)$, $J'(5, 1)$, $K'(1, -1)$

Graph each preimage on a coordinate plane. Then graph the image after performing the translation given, and list the coordinates of that image. 8–10. See Solutions Manual for graphs.

8. $S(-3, 5)$; slide 3 down, 4 right $S'(1, 2)$
9. $\triangle EFG$, $E(-4, 1)$, $F(-1, 3)$, $G(-1, 1)$; reflection over the graph of $x = -2$ and then the graph of $x = 2$ $E'(4, 1)$, $F'(7, 3)$, $G'(7, 1)$
10. $\square HIJK$, $H(-2, 7)$, $I(2, 9)$, $J(2, 7)$, $K(-2, 5)$; slide 3 units right, 6 units down
11. **Design** The diagram at the right shows the floor plan of Eleanor's kitchen. She recently had her kitchen remodeled and her refrigerator (square A) was moved. The new position is at square B. Describe the move. *Each square on the diagram represents 3 × 3 or 9 square feet.* It is a move of 15 ft away from the wall and 21 ft to the left and rotated.

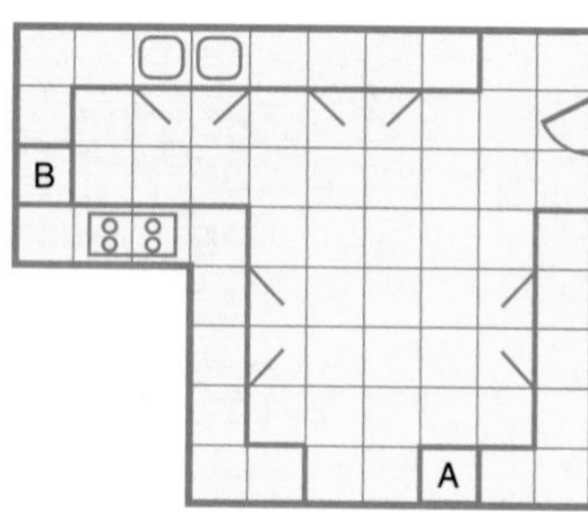

EXERCISES

Practice

In the figure at the right, $a \parallel b$. Name the translation image for each point with respect to line a, then line b.

12. K R	13. M T
14. J Q	15. P U
16. L S	17. N none

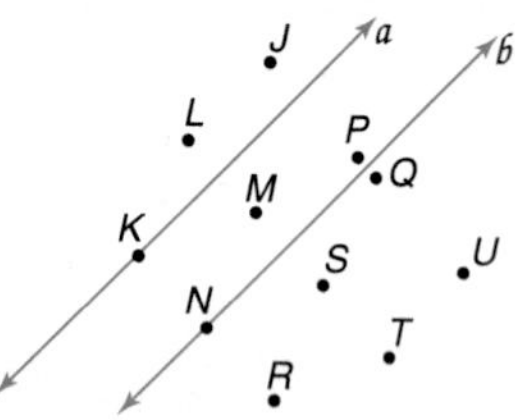

In each figure for Exercises 18–21, $\ell \parallel m$. Determine whether each red figure is a translation image of the blue figure. Write *yes* or *no*. Explain your answer. 18–21. See margin for explanations.

18. yes

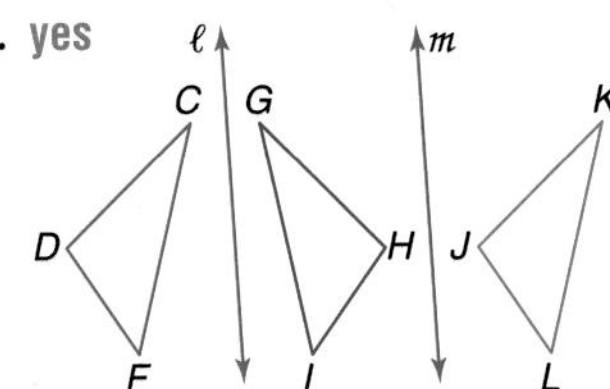

19. yes

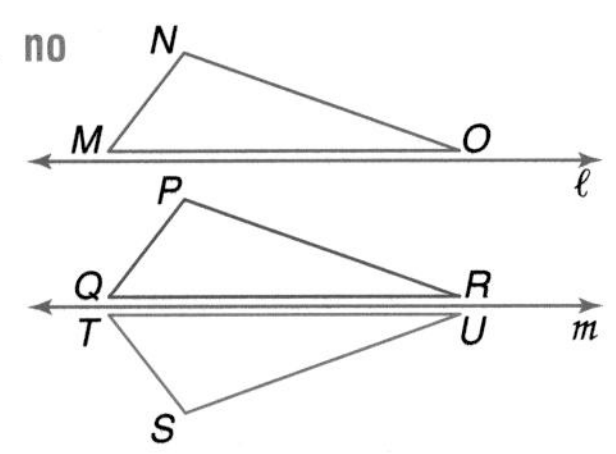

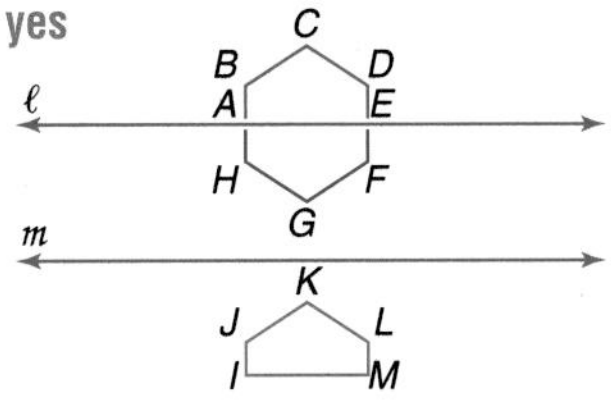

20.

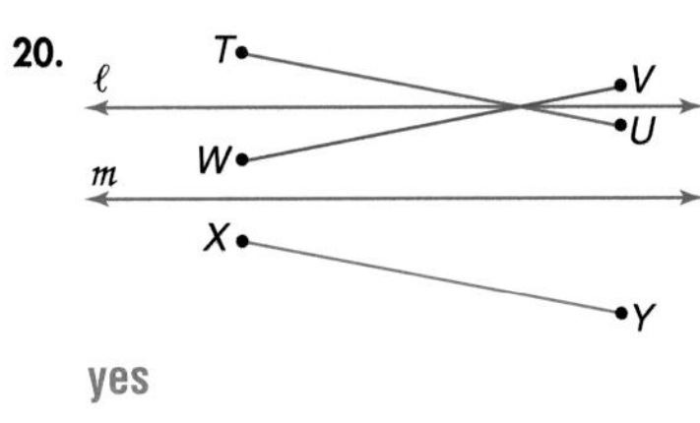

yes

21. no

Graph each preimage on a coordinate plane. Then graph the image after performing the translation given, and list the coordinates of that image. 23. $A'(-8, -4)$, $B'(-3, -2)$, $C'(-4, -5)$, $D'(-7, -6)$

22–25. See Solutions Manual for graphs.

22. $\overline{WX}$, $W(4, 8)$, $X(7, 5)$; slide 5 left, 3 down $W'(-1, 5)$, $X'(2, 2)$

23. Quadrilateral $ABCD$, $A(-8, 4)$, $B(-3, 6)$, $C(-4, 3)$, $D(-7, 2)$; reflect over the graph of $y = 2$ and then the graph of $y = -2$

24. $\triangle CAT$, $C(-2, -1)$, $A(0, 2)$, $T(2, -2)$; reflect over the graph of $x = 1$ and then the graph of $x = -3$ $C'(-10, -1)$, $A'(-8, 2)$, $T'(-6, -2)$

25. Trapezoid $NICK$, $N(2, 3)$, $I(6, 2)$, $C(6, -1)$, $K(-2, 1)$; slide N to $N'(-4, 1)$ $N'(-4, 1)$, $I'(0, 0)$, $C'(0, -3)$, $K'(-8, -1)$

Copy each figure on a coordinate plane. Then find the translation image of each geometric figure with respect to the parallel lines n and ℓ. 26–29. See margin.

26.

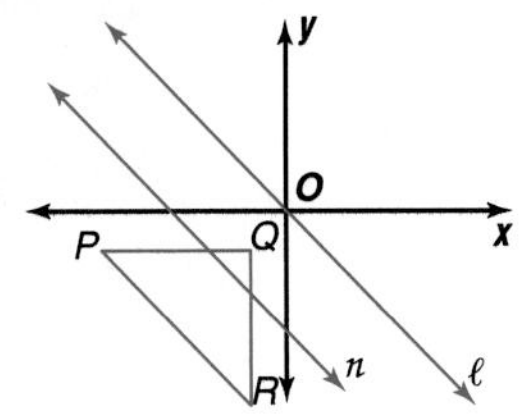

27.

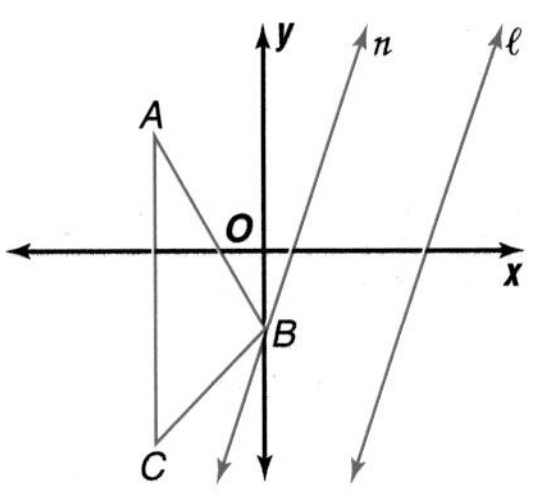

28.

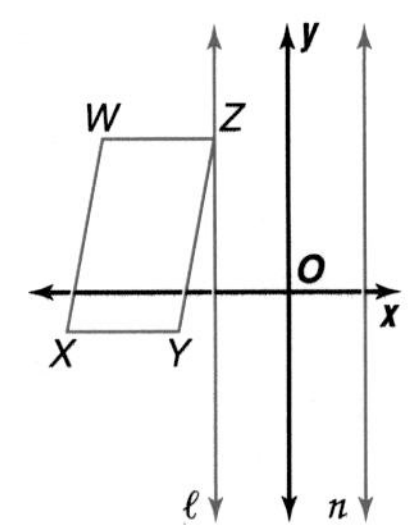

29.

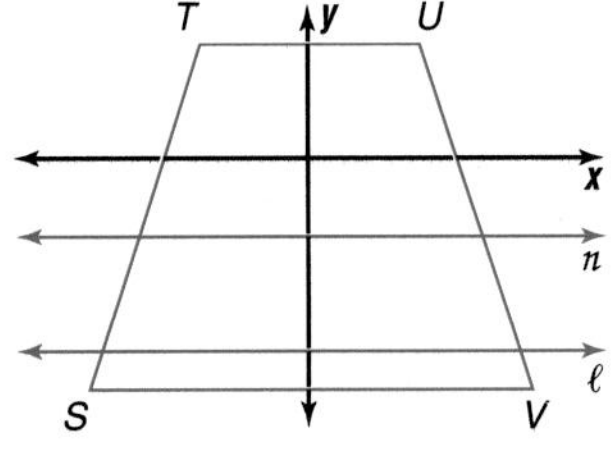

Additional Answers

18. It is one reflection after another with respect to two parallel lines.
19. It is one reflection after another with respect to two parallel lines.
20. It is one reflection after another with respect to two parallel lines.
21. It is a translation and then a reflection with respect to a line.

26.

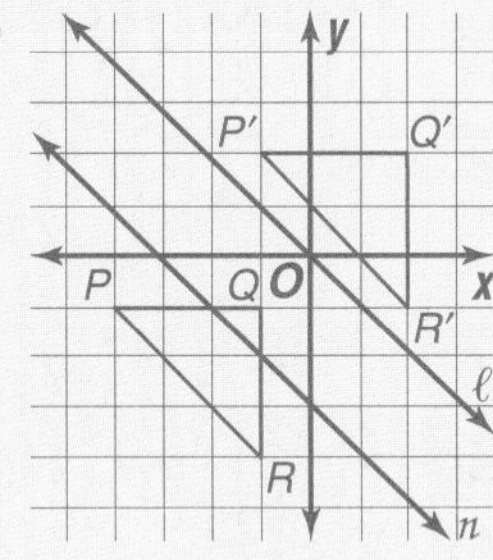

27.

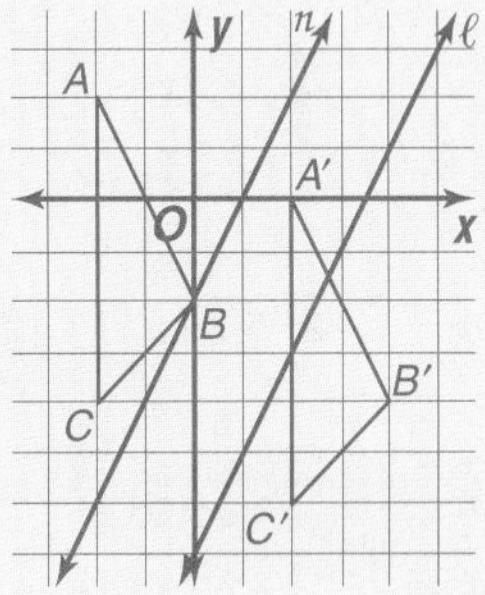

Additional Answers

28.

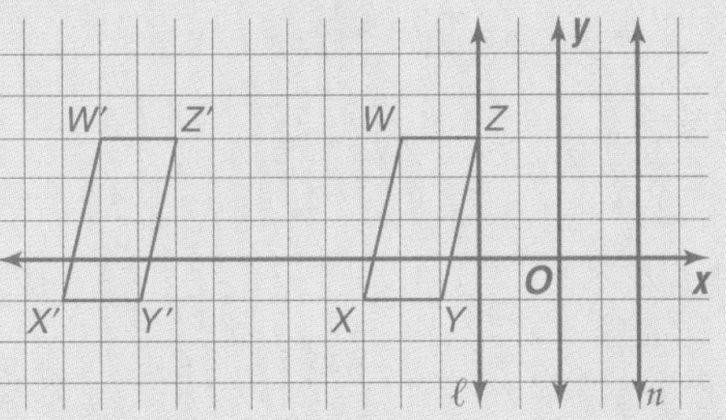

29.

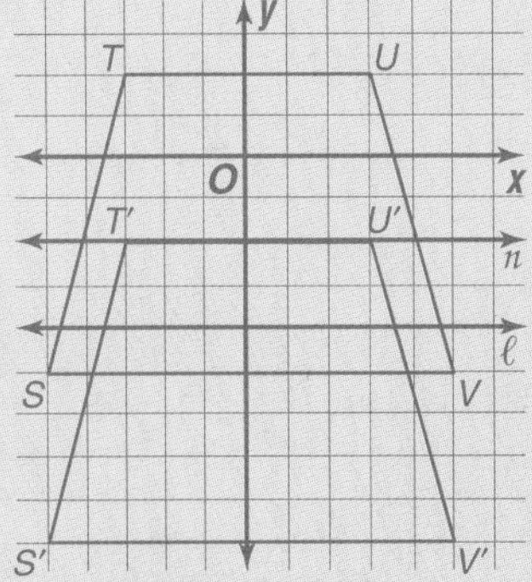

Study Guide Masters, p. 83

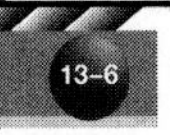

NAME________________ DATE________

Study Guide

Student Edition Pages 731–737

Translations

A **composite of reflections** is the transformation that results from performing one reflection after another. A **translation** is the composite of two reflections over parallel lines.

Example: Draw the translation image of $\triangle ABC$ with respect to parallel lines p and q.

First draw the reflection image of $\triangle ABC$ with respect to line p. Then draw the reflection image of that figure with respect to line q. The translation image of $\triangle ABC$ is $\triangle FGH$.

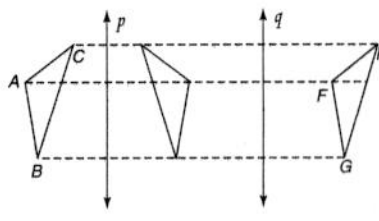

Notice that you can also think of a translation as a slide.

Draw the translation image of each figure with respect to ℓ and m. Assume $\ell \parallel m$. Draw an X inside the translation image to identify it.

1.

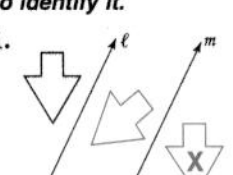

2.

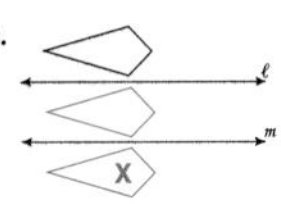

Use the figures below to name each triangle. Assume $d \parallel e$.

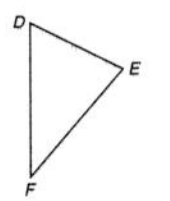

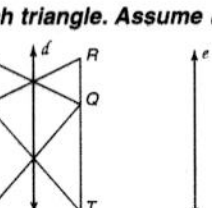

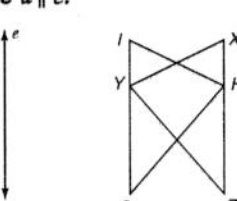

3. reflection image of $\triangle RST$ with respect to d $\triangle PQM$

4. reflection image of $\triangle DEF$ with respect to d $\triangle XYZ$

5. translation image of $\triangle PQM$ with respect to d and e $\triangle IHG$

6. Find the translation image of the figure with respect to the parallel lines n and t.

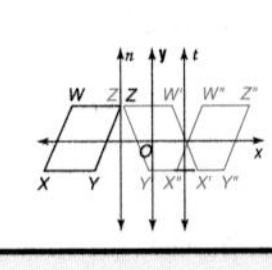

Additional Answers

32. A translation is composed of two successive reflections over parallel lines. The first reflection with respect to line ℓ preserves angle and distance measure. The second reflection with respect to line m preserves angle and distance measure. Therefore, by transitivity, angle and distance measure is preserved from preimage to image.

34.

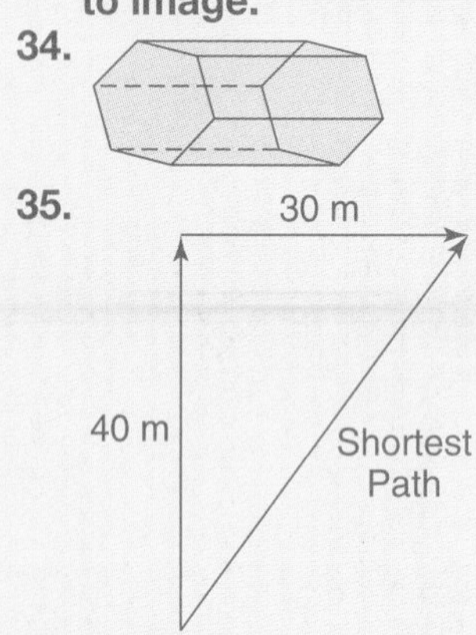

35.

30 m

40 m

Shortest Path

Practice Masters, p. 83

13-6 NAME_______________ DATE_______

Practice

Student Edition Pages 731–737

Translations

For each of the following, lines ℓ and m are parallel. Determine whether Figure 3 is a translation image of Figure 1. Write yes or no. Explain your answer.

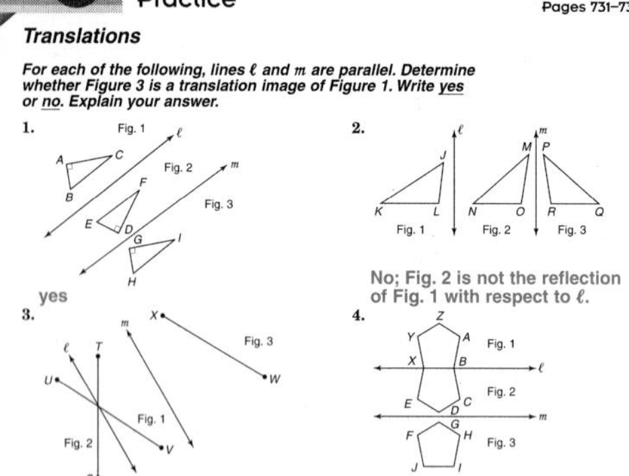

1. yes

2. No; Fig. 2 is not the reflection of Fig. 1 with respect to ℓ.

3. yes

4. yes

For Exercises 5–8, refer to the figures above.

5. Name the reflection of $\overline{ST}$ with respect to line ℓ. If none is drawn, write *none*. VU

6. Name the reflection of $\triangle JKL$ with respect to line ℓ. If none is drawn, write *none*. none

7. Name the reflection of pentagon $XYZAB$ with respect to line ℓ. If none is drawn, write *none*. pentagon $XEDCB$

8. Name the reflection of $\overline{UV}$ with respect to line m. If none is drawn, write *none*. none

Find the translation image of each geometric figure with respect to the parallel lines ℓ and m.

9.

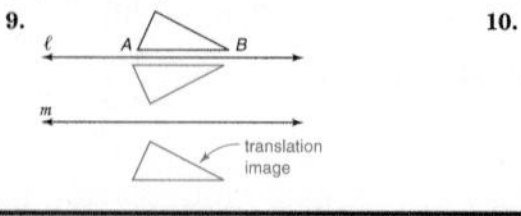

10.

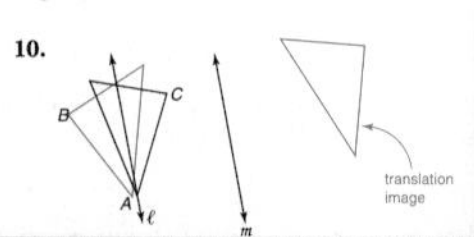

Use the figures to name each triangle. Assume $n \parallel t$.

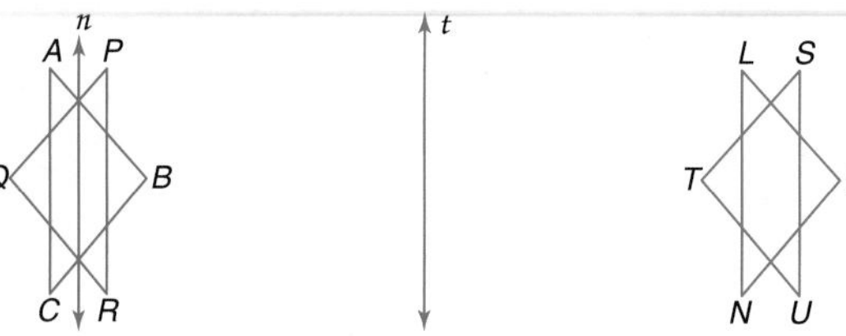

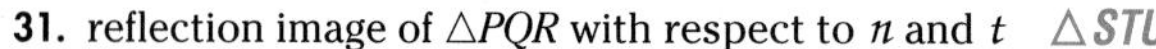

30. reflection image of $\triangle ABC$ with respect to n and t $\triangle LMN$

31. reflection image of $\triangle PQR$ with respect to n and t $\triangle STU$

32. Plan a proof to show that the translation image of $\triangle ABC$ with respect to parallel lines ℓ and m preserves angle measure and distance measure. See margin.

Critical Thinking

33. Triangle RST has vertices $R(3, -7)$, $S(7, -4)$, and $T(9, -8)$. Triangle ABC has vertices $A(3, 3)$, $B(7, 6)$, and $C(9, 2)$. If $\triangle ABC$ is the translation image of $\triangle RST$ with respect to two parallel lines, find the equations that represent two possible parallel lines. Sample answer: $y = 1$, $y = -4$

Applications and Problem Solving

34. **Art** Mbwane is an illustrator working on a gemology book. One section is about quartz crystals that are six-sided prisms. Using a translation, draw a prism with hexagonal bases that could be used as a sketch for the quartz crystals. See margin.

35. **Environment** In the photograph, a cloud of dense gas and dust pour out of Surtsey, a volcanic island off the south coast of Iceland. If the cloud blows 40 miles north and then 30 miles east, make a sketch to show the translation of the smoke particles. Then indicate the shortest path that would take the particles to the same position. See margin.

Mixed Review

36. Copy quadrilateral $QRST$. Draw the reflection image of quadrilateral $QRST$ with respect to line ℓ. (Lesson 13–5)

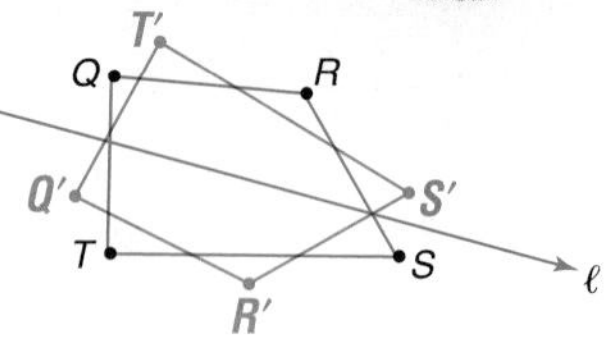

37. Describe the locus of all points in space that are a given distance from a line segment and equidistant from the endpoints of the line segment. (Lesson 13–3) a circle

38. Use an algebraic method to find the locus of points that satisfy the graphs of both equations. (Lesson 13–2) $\left(\frac{4}{13}, -\frac{16}{13}\right)$

$2x - 6y = 8$
$3x + 4y = -4$

39. Given ▱$QRST$, complete each statement. (Lesson 12–5)
a. $\overrightarrow{SR} + \overrightarrow{RQ} = \underline{?}$ $\overrightarrow{SQ}$
b. $\overrightarrow{QR} + \overrightarrow{RT} = \underline{?}$ $\overrightarrow{QT}$
c. $\overrightarrow{TQ} + \overrightarrow{TS} = \underline{?}$ $\overrightarrow{TR}$

40. $y = \frac{3}{4}x - 3$; see margin for graph.

40. Graph the line with slope equal to $\frac{3}{4}$ and that passes through the point at (0, −3). Then write the equation for the line. (Lesson 12–2)

41. Determine whether the pair of triangles at the right is congruent. If so, state the postulate or theorem used. (Lesson 5–2) yes; HL

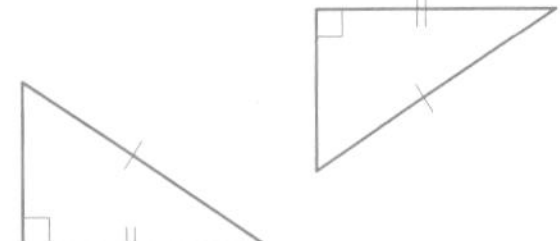
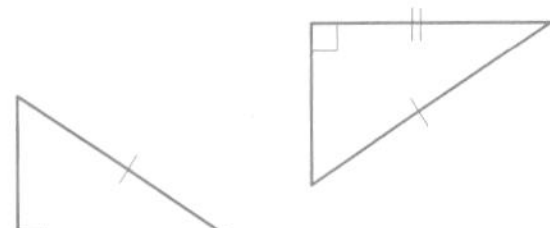

44a. 42, 66, 66, 75, 83, 84, 84, 156, 158

44b. $\frac{2}{9}$

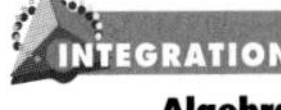

Algebra

42. Describe how an equilateral triangle is also an isosceles triangle. (Lesson 4–1) An isosceles triangle has at least two sides congruent, and an equilateral triangle has three congruent sides.

43. Factor $8x^2 + 48xy + 72y^2$. $8(x + 3y)^2$

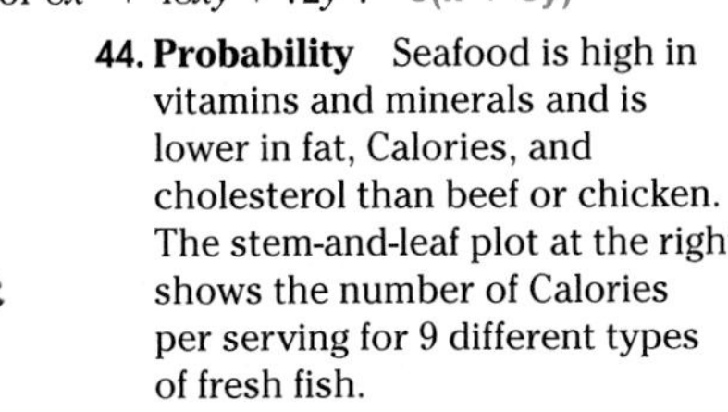

44. **Probability** Seafood is high in vitamins and minerals and is lower in fat, Calories, and cholesterol than beef or chicken. The stem-and-leaf plot at the right shows the number of Calories per serving for 9 different types of fresh fish.

Stem	Leaf
4	2
6	6 6
7	5
8	3 4 4
15	6 8 7∣5 = 75

a. What numbers are represented by the stem-and-leaf plot?
b. What is the probability that a type of fish chosen at random will contain 84 Calories per serving?
c. What is the probability that a type of fish chosen at random will contain less than 100 Calories per serving? $\frac{7}{9}$

WORKING ON THE In·ves·ti·ga·tion

Refer to the Investigation on pages 642–643.

Computer movie images can use reflections to create a variety of effects. For example, a bouncing ball, a reflection in a mirror, and any type of symmetric object could be made by reflecting an image.

1 Brainstorm with your group about ways that reflections can be used to make your movie.
2 Incorporate one or more reflections into your movie.
3 Write a sentence describing how each reflection affects the coordinates of the object being reflected in your movie.

Add the results of your work to your Investigation Folder.

4 ASSESS

Closing Activity

Writing Have students write the steps needed to draw a translation.

Chapter 13 Quiz C (Lessons 13-5 and 13-6) is available in the *Assessment and Evaluation Masters*, p. 353.

Additional Answer

40.

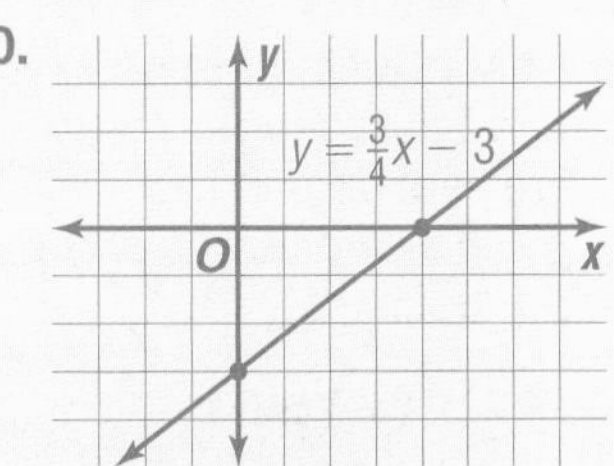

Enrichment Masters, p. 83

NAME ____ DATE ____

13-6 Enrichment

Student Edition Pages 731–737

Translations in The Coordinate Plane

You can use algebraic descriptions of reflections to show that the composite of two reflections with respect to parallel lines is a translation (that is, a slide).

1. Suppose a and b are two different real numbers. Let S and T be the following reflections.
S: $(x, y) \to (-x + 2a, y)$
T: $(x, y) \to (-x + 2b, y)$
S is reflection with respect to the line with equation $x = a$, and T is reflection with respect to the line with equation $x = b$.

a. Find an algebraic description (similar to those above for S and T) to describe the composite transformation "S followed by T." $(x, y) \to (x + 2(b - a), y)$

b. Find an algebraic description for the composite transformation "T followed by S." $(x, y) \to (x + 2(a - b), y)$

2. Think about the results you obtained in Exercise 1. What do they tell you about how the distance between two parallel lines is related to the distance between a preimage and image point for a composite of reflections with respect to these lines? The distance between a preimage and its image point is twice the distance between the parallel lines.

3. Illustrate your answers to Exercises 1 and 2 with sketches. Use a separate sheet if necessary. See students' work.

Extension

Problem Solving Translate an object twice, once through one pair of parallel lines and then through a different pair of parallel lines intersecting the first pair. What do you notice about the object? It was slid and rotated.

In·ves·ti·ga·tion

Working on the Investigation

The Investigation on pages 642–643 is designed to be a long-term project that is completed over several days or weeks. Encourage students to keep their materials in their Investigation Folder as they work on the Investigation.

13-6B LESSON NOTES

NCTM Standards: 1–4, 7, 8

Objective
Use a TI-92 calculator to translate a triangle in the direction and magnitude of a given vector.

Recommended Time
20 minutes

Instructional Resources
Instructions for using the *Geometer's Sketchpad* for this activity are available in the *Graphing Calculator and Computer Masters*, p. 30.

1 FOCUS

Motivating the Lesson
Cut out a triangle from a piece of paper. Move it around on a flat surface. Trace the triangle in different locations. Draw the path the triangle follows using straight lines. Do not tilt the triangle. Compare this to translation of triangles on a calculator.

2 TEACH

Teaching Tip Point out to students that if a polygon tilts while it is being moved around the plane, the motion is not a translation.

3 PRACTICE/APPLY

Assignment Guide

Core (with proof): 1–4
Core (informal): 1–4
Enriched: 1–4

4 ASSESS

Observing students working with technology is an excellent method of assessment.

13–6B Using Technology
Translating Polygons with Vectors

An Extension of Lesson 13–6

LOOK BACK
Refer to Lesson 12-5 to review vectors.

TECHNOLOGY Tip
To scroll the whole screen so that a figure is visible, hold the 2nd key down and press the direction pad to move the screen in the desired direction.

You can use a TI-92 calculator to translate a triangle in the direction and magnitude (length) of a given vector.

- Draw a triangle by pressing F3 and selecting 3:Triangle.
- A vector is needed beside the triangle in order to determine how you want the triangle translated—direction and magnitude of the translation. To draw a vector on your screen, press F2 and then select 7:Vector.
- Now translate the triangle by placing the cursor on the triangle (The message will say, "THIS TRIANGLE.") and press F5 and select 1:Translation.

  The message will say, "TRANSLATE THIS TRIANGLE." Press ENTER. Then move the cursor to the vector. The message will say, "BY THIS VECTOR." Press ENTER. The translated triangle will appear.
- Hide the vector by pressing F7 and selecting 1:Hide/Show. Then place the cursor on the vector and press ENTER.
- Use the button F6 and select 5:Equation & Coordinates to find the coordinates of the six vertices. A sample screen is shown at the right.

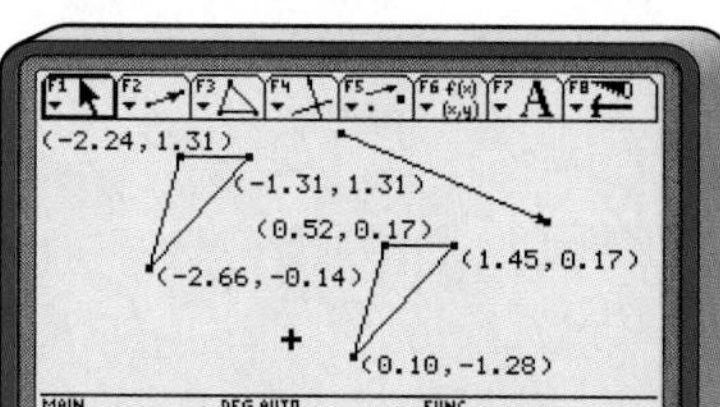

EXERCISES

1–4. See students' work.

Examine your drawings.

1. Were you able to see all the parts of the translated triangle?
2. Compare the coordinates of the corresponding vertices. What do you notice?
3. Describe the translation in terms of horizontal and vertical movement. Make a conjecture.

Test your conjecture.

4. Use the TI-92 to test a translation of a regular polygon.
   a. Draw a regular polygon.
   b. Draw a vector in a different direction.
   c. Translate the regular polygon and compare the preimage and the image coordinates.
   d. Does your conjecture in Exercise 3 about describing translations hold true?

Using Technology
This lesson offers an excellent opportunity for using technology in your geometry classroom. For more information on using technology, see *Graphing Calculators in the Mathematics Classroom*, one of the titles in the Glencoe Mathematics Professional Series.

13–7 Rotations

What YOU'LL LEARN

- To name and draw rotation images of figures with respect to intersecting lines.

Why IT'S IMPORTANT

Rotations are an important part of motion, which affects every aspect of your life.

APPLICATION

Industry

Modern windmills called wind turbines are used to generate electric current. These turbines are made with two or three blades that rotate on a shaft. As a windmill rotates, each blade moves in a circular motion to a new position. This is a **rotation**, which is another type of transformation.

Turning motions such as a doorknob being turned or the movement of the paddles on a paddle wheel involve rotation around a fixed point called the **center of rotation**.

Remember, a translation is a composite of two reflections over parallel lines.

MODELING MATHEMATICS — **Rotating a Triangle**

Materials:  grid paper  tracing paper

- Use grid paper to draw two intersecting lines p and q on a coordinate plane.
- Draw $\triangle ABC$ so that it does not intersect either line.
- Reflect $\triangle ABC$ over line p. Label the image $A'B'C'$.
- Reflect $A'B'C'$ over line q. Label this reflection $A''B''C''$. This is the rotated triangle.
- Trace the original triangle on a separate sheet of paper.
- Place the traced triangle over $\triangle ABC$ on the grid paper. Using your pencil as the center of rotation, place it on the point of intersection for lines p and q. Turn your paper until the traced triangle coincides with the image of $\triangle A''B''C''$.

Your Turn

a. This rotation consists of two reflections. What properties were preserved in this rotation? **congruence of sides and angles**

b. Is a rotation an isometry? Explain. **Yes; it is a congruence transformation.**

This Modeling Mathematics activity shows that a rotation is a composite of two reflections with respect to two intersecting lines.

Point out to students that if the rotation is not a 180° rotation, it is not equal to a reflection.

13-7 LESSON NOTES

NCTM Standards: 1–5, 7, 8

Instructional Resources

- Study Guide Master 13-7
- Practice Master 13-7
- Enrichment Master 13-7

Transparency 13-7A contains the 5-Minute Check for this lesson; **Transparency 13-7B** contains a teaching aid for this lesson.

Recommended Pacing	
Honors Pacing	Day 10 of 13

1 FOCUS

5-Minute Check *(over Lesson 13-6)*

List the coordinates of the image after performing the translation given.

1. $A(2, 6)$; slide 2 down, 3 right **$A'(5, 4)$**
2. $\triangle ABC$, $A(1, 1)$, $B(2, 5)$, $C(6, 3)$; slide 2 up, 2 left **$A'(-1, 3)$, $B'(0, 7)$, $C'(4, 5)$**
3. $\square EFGH$, $E(0, 2)$, $F(3, 2)$, $G(1, 5)$, $H(4, 5)$; slide 3 down **$E'(0, -1)$, $F'(3, -1)$, $G'(1, 2)$, $H'(4, 2)$**

In the figure below, $x \parallel y$. For each of the following, name the translation image of each segment with respect to x and y.

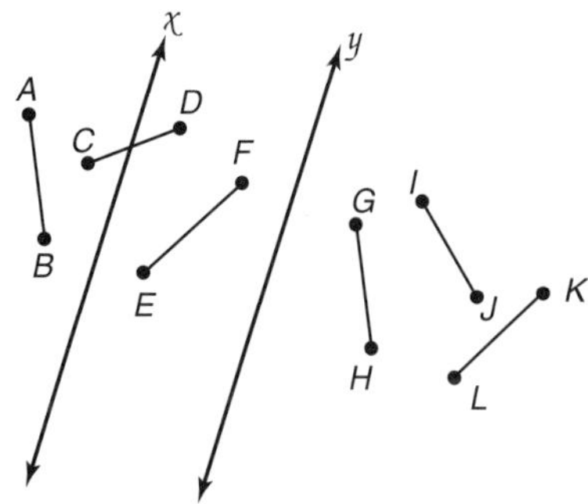

4. translation of $\overline{FE}$ **$\overline{KL}$**
5. translation of $\overline{AB}$ **$\overline{GH}$**

Motivating the Lesson

Hands-On Activity Draw a two-dimensional figure on a stiff piece of paper. Poke the tip of a pencil through the paper at the center of the given figure. Spin the paper while holding the pencil. Explain that you are rotating the figure.

2 TEACH

Teaching Tip When defining rotation, rotating with respect to t and then s is not the same as rotating with respect to s and then t.

In-Class Examples

For Example 1

Find the rotation image of $\triangle ABC$ with respect to t and then s using reflections.

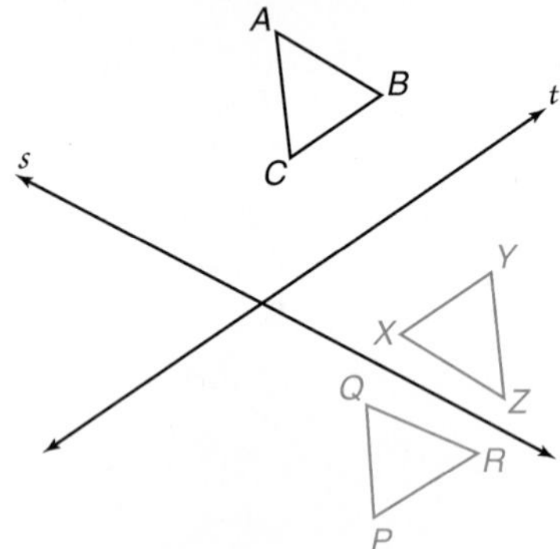

For Example 2

Lines a and b intersect at Q to form a 75° angle. Use the angle of rotation to find the rotation image of $\overline{TZ}$.

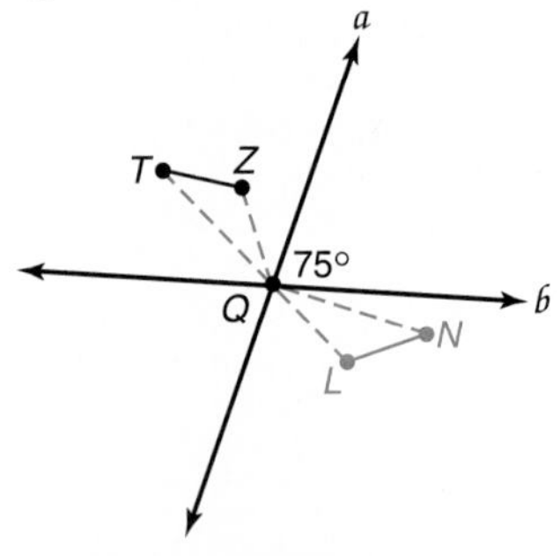

The angle of rotation has a measure of 2(75) or 150. Construct $\angle TQN$ such that its measure is 150 and $\overline{TQ} \cong \overline{QN}$. Then construct $\angle ZQL$ so that its measure is 150 and $\overline{ZQ} \cong \overline{QL}$. The rotation image of $\overline{TZ}$ is $\overline{NL}$.

Teaching Tip Emphasize that for a rotation the two lines must intersect.

Example 1 **Find the rotation image of rhombus $ABCD$ over lines m and n using reflections.**

First reflect rhombus $ABCD$ over line m. Label the image $A'B'C'D'$.

Next, reflect the image over line n.

Label this reflection $A'B''C''D''$.

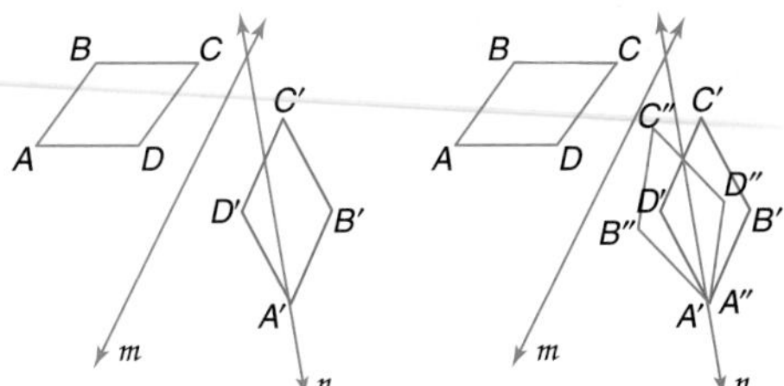

Rhombus $A'B''C''D''$ is a rotation image of rhombus $ABCD$.

There are two methods by which a rotation can be performed.

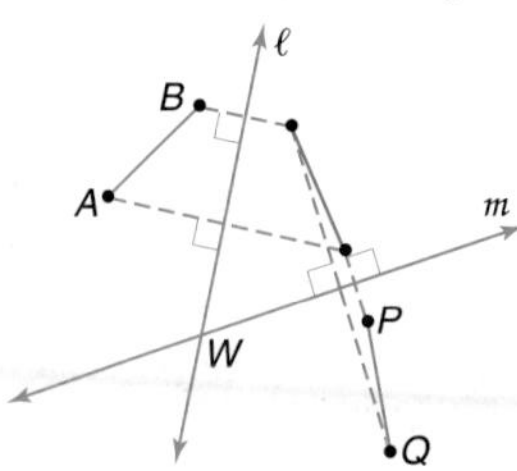

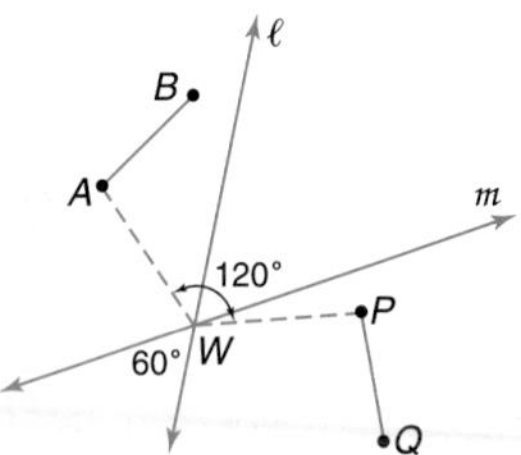

A rotation image can be determined by a composite of two successive reflections over two intersecting lines. $\overline{AB}$ is reflected over ℓ and then m. The rotation image of $\overline{AB}$ is $\overline{PQ}$.

A rotation image can also be determined by using the angle formed by the intersecting lines. This is called the **angle of rotation**. The point at which the two lines intersect is the center of rotation.

There is a relationship between the angle formed by the intersecting lines and the angle of rotation.

Postulate 13–1

In a given rotation, if A is the preimage, P is the image, and W is the center of rotation, then the measure of the angle of rotation $\angle AWP$ is twice the measure of the angle formed by the intersecting lines of reflection.

Example 2 **Find the rotation image of $\overline{XY}$ over intersecting lines ℓ and m, which form a 40° angle, by using the angle of rotation.**

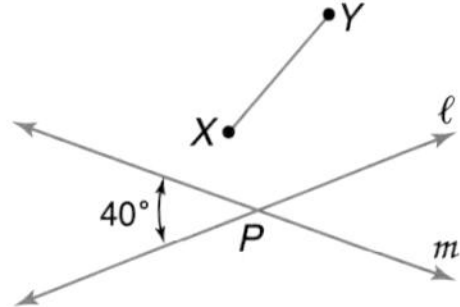

Explore Since the angle formed by lines ℓ and m measures 40°, the angle of rotation measures 2(40°) or 80°.

Plan Draw the angles of rotation and rotate each point.

Classroom Vignette

"I enhance this lesson by having students use software such as *Geometer's Sketchpad* to create regular polygons by rotating a given segment. Then to extend the lesson, they can rotate the polygon to create tessellations."

Loraine M. Duclos
Loraine M. Duclos
Fairfield High School
Fairfield, Connecticut

Solve Draw $\angle XPR$ so that its measure is 80 and $\overline{XP} \cong \overline{PR}$. Draw $\angle YPQ$ so that its measure is 80 and $\overline{YP} \cong \overline{PQ}$. Connect R and Q to form the rotation image of $\overline{XY}$.

Examine Reflect $\overline{XY}$ over the two lines to verify that $\overline{RQ}$ is the rotation image of $\overline{XY}$.

Example 3

PROBLEM SOLVING

Look for a Pattern

$\triangle XYZ$ is rotated 180° on the coordinate plane to form $\triangle X'Y'Z'$.

a. Make a table of the coordinates of the vertices of $\triangle XYZ$ and $\triangle X'Y'Z'$.

$\triangle XYZ$	$\triangle X'Y'Z'$
X(−2, 1) Y(2, −3) Z(3, 5)	X′(2, −1) Y′(−2, 3) Z′(−3, −5)

b. Use the table to find a pattern in the coordinates of the vertices of $\triangle XYZ$ and $\triangle X'Y'Z'$. Make a conjecture about what happens in a coordinate plane for a 180° rotation.

The x- and y-coordinates of the image are the opposites of the preimage coordinates.

$P(x, y) \rightarrow P(-x, -y)$ for a rotation of 180°.

Example 4

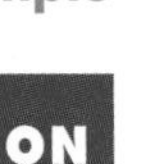

APPLICATION

Design

One of the two screws securing a stop sign has come out, and the sign is hanging upside down. Assuming the remaining screw can be considered the center of rotation, find the angle of rotation needed to straighten the sign.

A stop sign is the shape of a regular octagon. From the center of rotation, imagine lines to each of the corners of the sign. The angles formed by the intersecting lines form 45° angles, because 360 ÷ 8 is 45. The measure of the angle of rotation is 2(45) or 90. The bottom side of the sign would have to be rotated two times or 180°.

CHECK FOR UNDERSTANDING

Communicating Mathematics

Study the lesson. Then complete the following.

1. **Compare and contrast** a translation and a rotation. See margin.
2. **Describe** two techniques that can be used to locate a rotation image with respect to two intersecting lines. See margin.
3. **Determine** if a regular hexagon has 60° rotational symmetry. Explain.

3. Yes; it is made of 6 equilateral triangles that repeat every 60°.

In-Class Examples

For Example 3

Square *ABCD* has vertices *A*(2, 3), *B*(6, 3), *C*(6, 7), and *D*(2, 7). It is rotated 180° around point *A* to form the square *A′B′C′D′*.

a. Make a table of the coordinates of the vertices of square *ABCD* and square *A′B′C′D′*.

□*ABCD*	□*A′B′C′D′*
A(2, 3)	*A′*(−2, −3)
B(6, 3)	*B′*(−6, −3)
C(6, 7)	*C′*(−6, −7)
D(2, 7)	*D′*(−2, −7)

b. What pattern do you see from the table? $P(x, y) \rightarrow P'(-x, -y)$ for a rotation of 180°

For Example 4

Zina makes artificial flowers for the Floral Fantasia Company. She designed one flower to look like a dahlia with 8 petals. What is the measure of the angle of rotation if petal A moves to the position of petal B? 45

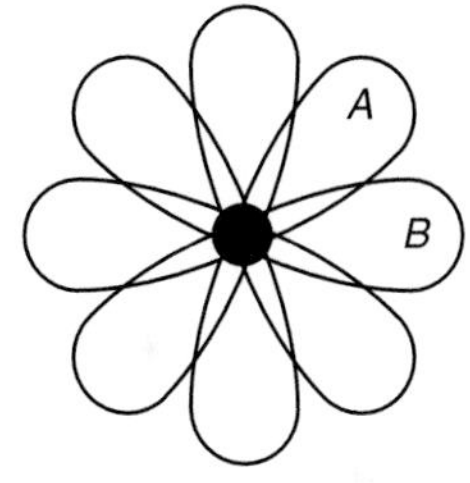

3 PRACTICE/APPLY

Check for Understanding

Exercises 1–14 are designed to help you assess your students' understanding through reading, writing, speaking, and modeling. You should work through Exercises 1–4 with your students and then monitor their work on Exercises 5–14.

Error Analysis

Students may find it difficult to perceive the first reflection without trying to relate it to the second line. If this is the case, have them draw the first line of reflection and complete the reflection, then add the second line of reflection and complete the second reflection.

Additional Answers

1. Both a translation and a rotation are made up of two reflections. The difference is that a translation reflects across parallel lines and a rotation reflects across intersecting lines.
2. A rotation image can be found by reflecting the image over a line, then reflecting that image over the second of the two intersecting lines. Another method is to rotate each point using the angle of reflection.

Additional Answers

4b. Sample answer: a point on the perpendicular bisector of $\overline{AY}$.

4c. Any point on the perpendicular bisector of $\overline{AY}$ could be a center of rotation.

4. Point Y is the image of A under a rotation.
   a. Copy the figure and construct a point that could be the center of the rotation. a. Sample answer: midpoint of $\overline{AY}$
   b. Construct a second point that could be the center of rotation. Sample answer: See margin.
   c. Generalize your conclusions for parts a and b. See margin.

5. Two lines intersect to form an angle of 37°. What is the angle of rotation? 74°

Guided Practice

Refer to the figure at the right for Exercises 6–11.

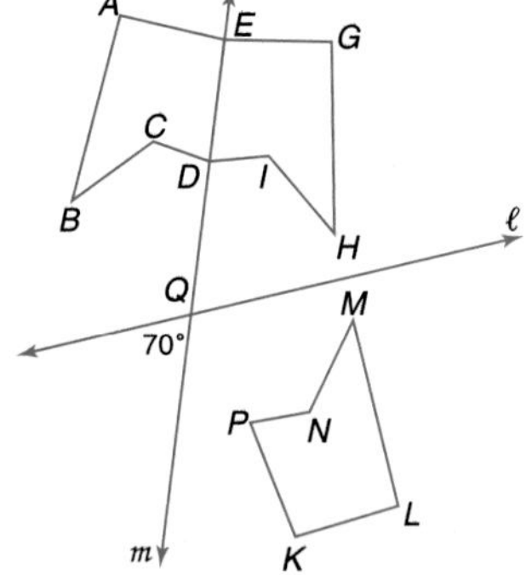

6. Find the reflection image of pentagon $ABCDE$ with respect to line m. 6. pentagon $GHIDE$
7. Find the reflection image of pentagon $DEGHI$ with respect to line ℓ. 7. pentagon $PKLMN$
8. Find the rotation image of pentagon $ABCDE$ with respect to lines ℓ and m. 8. pentagon $LMNPK$
9. Name the angle of rotation used to rotate C. $\angle CQN$
10. Find the angle measure of $\angle AQL$. 140
11. Find the rotation image of $\overline{CE}$ with respect to lines m and ℓ. $\overline{NK}$
12. Copy the figure below. Then use a composite of reflections to find the rotation image with respect to lines n and t.

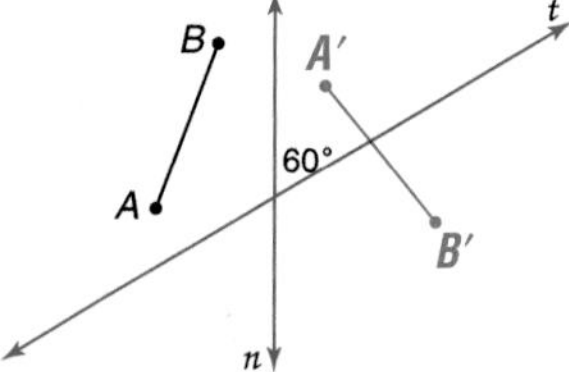

13. Copy the figure below. Use the angle of rotation to find the rotation image with respect to lines n and t.

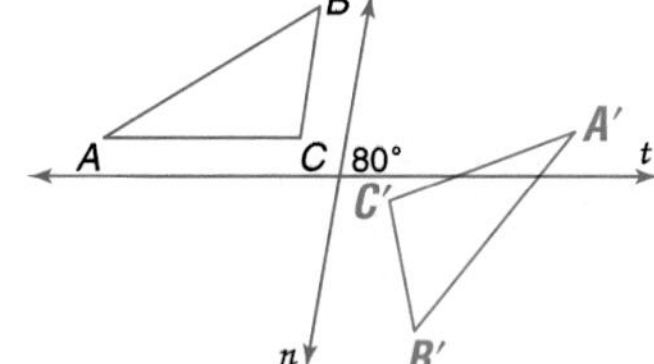

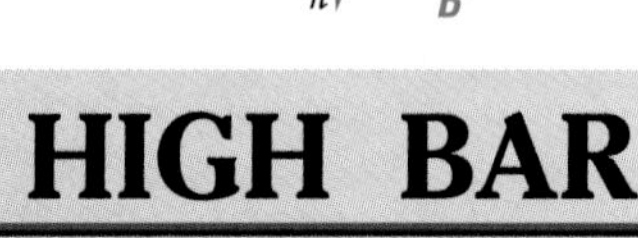

14. **Gymnastics** The high bar is a men's gymnastics event. If the drawing at the right is put on a coordinate grid with the hand grips at (0, 0), then the upper left body figure has toe coordinates of (−5, 6), and the lower right body figure has toe coordinates of (5, −6).
    a. What is the change that occurred in the coordinates? 14a. The x- and y-coordinates of the preimage were negated to get the coordinates of the image.
    b. What kind of a rotation is this? This is a 180° rotation.

HIGH BAR

A routine with continuous flow to quick changes in body position.

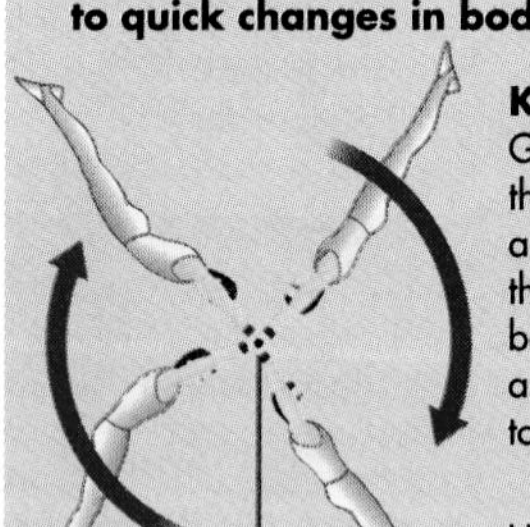

Key move: Giant swing. As the body swings around the bar the body should be straight with a slight hollow to the chest.

Height: $8\frac{1}{2}$ feet
Length: 8 feet

Reteaching

Using Reasoning Draw two intersecting lines on the chalkboard or overhead, and draw a geometric figure in the area where the angle formed by the intersecting lines is greater than 90°. Have students tell you how to rotate the figure, and follow their instructions as you complete the exercise.

EXERCISES

Practice

15. Yes; it is a proper successive reflection with respect to two intersecting lines.

16. Yes; it is a proper successive reflection with respect to two intersecting lines.

A

For each of the following, determine whether the indicated composition of reflections is a rotation. Explain your answer.

15.

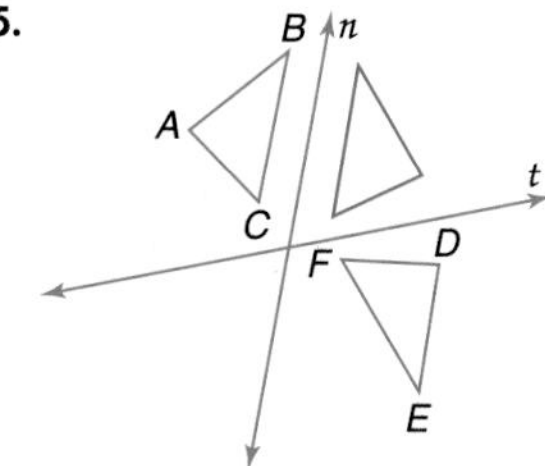

16.

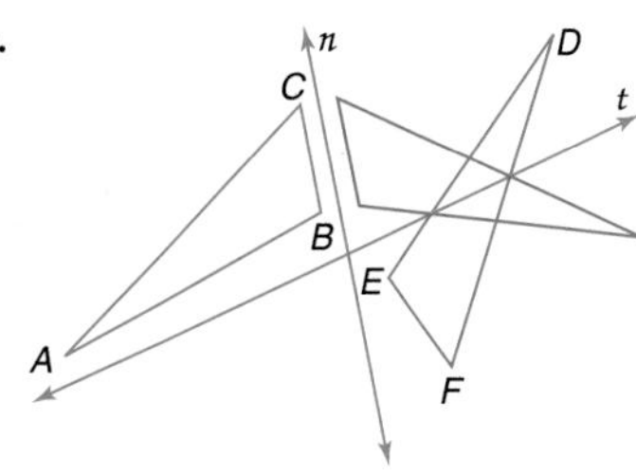

Two lines intersect to form an angle with the following measure. Find the angle of rotation for each.

17. 55° 110 **18.** 28.5° 57 **19.** 74° 148

Copy each figure. Then use a composite of reflections to find the rotation image with respect to lines ℓ and t.

20.

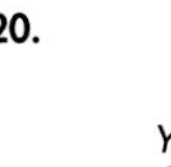

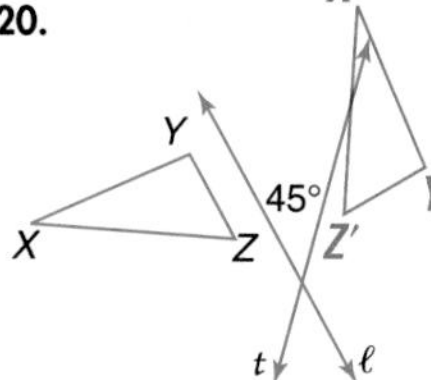

21.

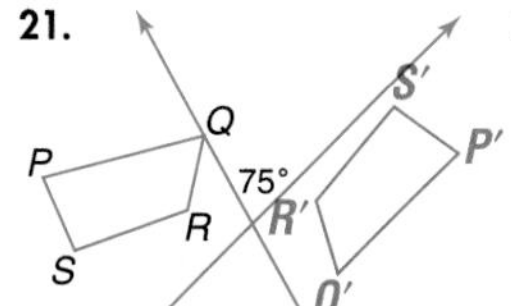

22.

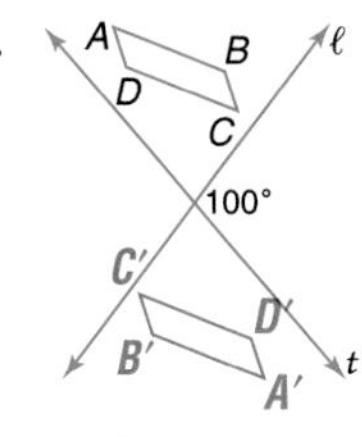

Copy each figure. Then use the angle of rotation to find the rotation image with respect to lines ℓ and t.

B

23. **24.** **25.**

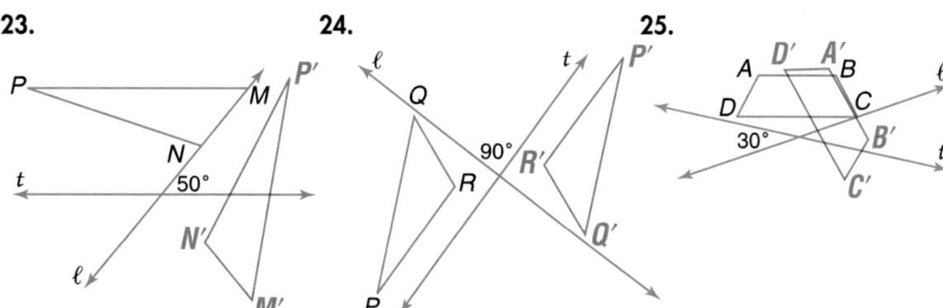

In the following figures, a rotation of $\triangle ABC$ has been performed with respect to lines p and q. Use a protractor to find the angle of rotation.

26.

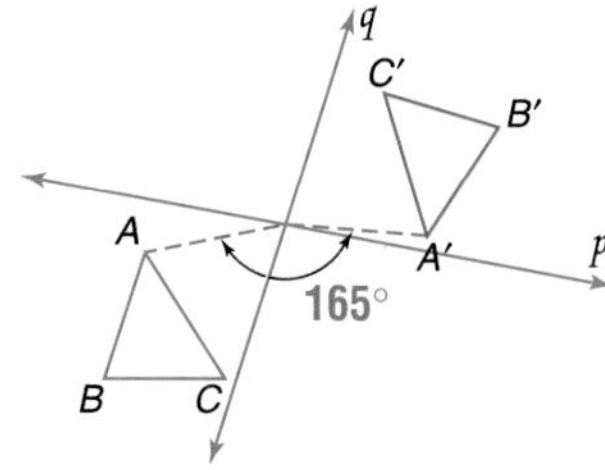

27.

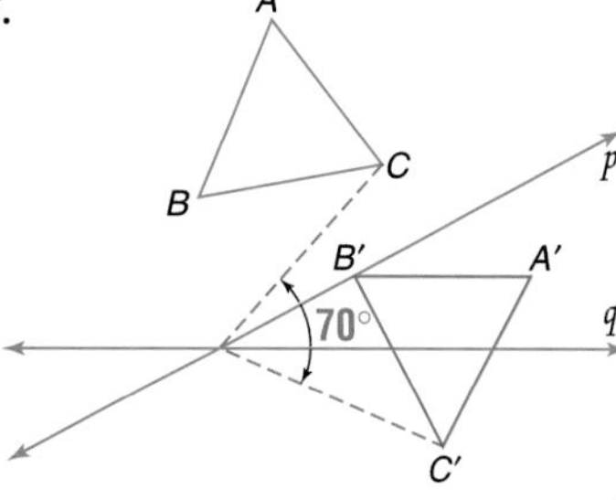

Assignment Guide

Core (with proof): 15–43 odd, 45–54
Core (informal): 15–43 odd, 45–47, 49–54
Enriched: 16–40 even, 41–54

For **Extra Practice**, see p. 792.

The red A, B, and C flags, printed only in the Teacher's Wraparound Edition, indicate the level of difficulty of the exercises.

Study Guide Masters, p. 84

NAME__________ DATE ______

Study Guide

Student Edition Pages 739–745

Rotations

The composite of reflections with respect to two intersecting lines is a transformation called a **rotation**.

Example: Draw the rotation image of *ABCD* with respect to lines *m* and *n*.

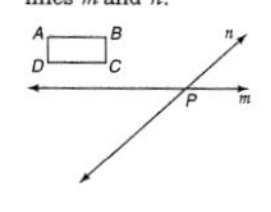

ABCD is on one side of *m* and *n*.

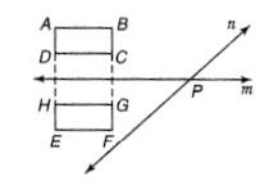

First reflect *ABCD* with respect to *m*. The image is *EFGH*.

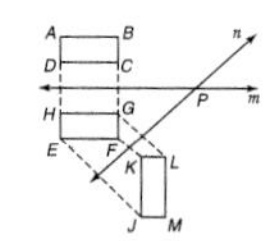

Then reflect *EFGH* with respect to *n*. The image is *JKLM*.

Thus, JKLM is the rotation image of *ABCD* with respect to *m* and *n*.

Notice that *P* is the **center of rotation**. You can think of a rotation as a turn around point *P*.

The following postulate involves the measure of the angle of rotation.

In a given rotation, if *A* is the preimage, *P* is the image, and *W* is the center of rotation, then the measure of the angle of rotation, ∠ *AWP*, equals twice the measure of the angle formed by intersecting lines of reflection.

Use a composite of reflections to find the rotation image with respect to lines s and t. Then state the measure of the angle of rotation.

1.

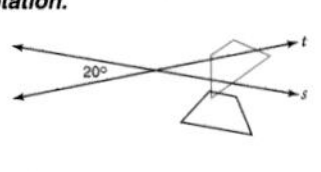

40°

2.

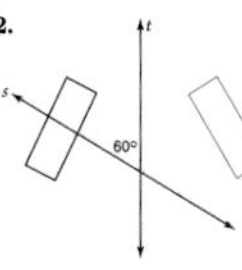

120°

Additional Answer

41. Angles of rotation with measures of 90 or 180 would be easier on a coordinate plane because of the grids used in graphing.

Practice Masters, p. 84

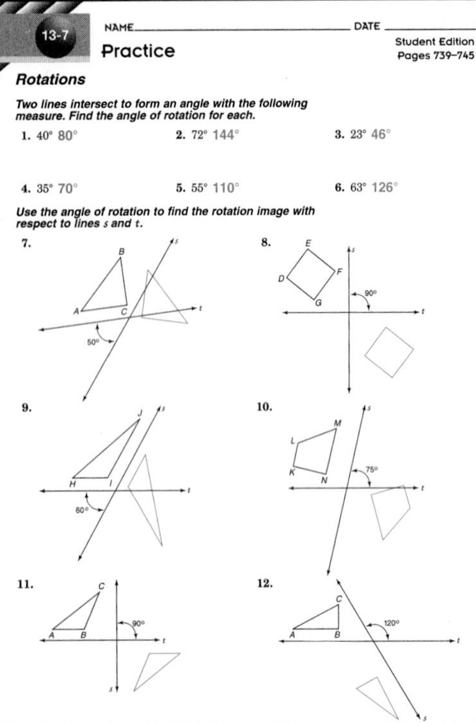
13-7 NAME ____________ DATE ________

Practice — Student Edition Pages 739–745

Rotations

Two lines intersect to form an angle with the following measure. Find the angle of rotation for each.

1. 40° 80°
2. 72° 144°
3. 23° 46°
4. 35° 70°
5. 55° 110°
6. 63° 126°

Use the angle of rotation to find the rotation image with respect to lines s and t.

7. 8. 9. 10. 11. 12.

Copy and complete the table below. Determine whether each translation preserves the given property. Write *yes* or *no*.

		angle measure	betweenness of points	orientation	collinearity	distance measure
28.	reflection	yes	yes	no	yes	yes
29.	translation	yes	yes	yes	yes	yes
30.	rotation	yes	yes	no	yes	yes

31. Draw a segment and two intersecting lines. Find the rotation image of the segment with respect to the two intersecting lines. See students' work.

32. Draw an isosceles triangle and two intersecting lines. Find the rotation image of the triangle with respect to the two intersecting lines. See students' work.

33. Kite *ABCD* in the figure at the right is soaring upward. If kite *ABCD* goes through a 180° rotation, it will form kite *STUV*. Draw kite *STUV* on the coordinate grid.

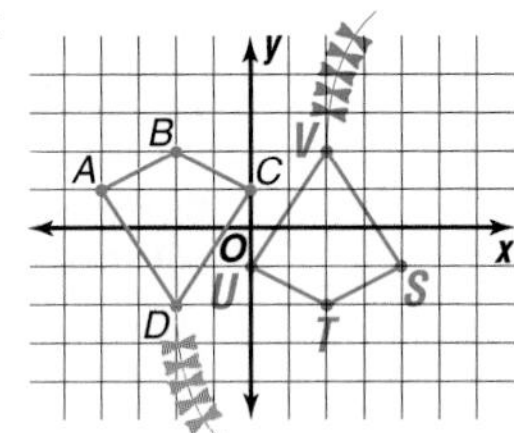

Hexagon *ABCDEF* is a regular hexagon with diagonals as drawn. Point *C* is the rotation image of point *A*.

34. What is the measure of the angle formed by each set of intersecting lines? Explain your answer. $360 \div 6 = 60$

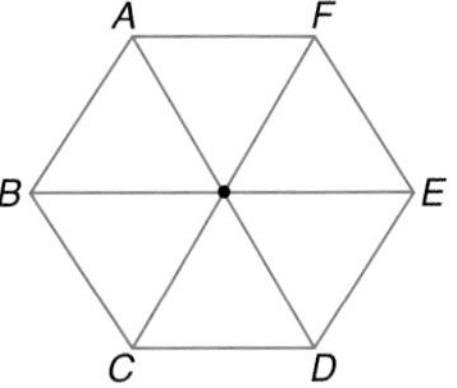

35. What is the measure of the angle of rotation? 120

36. Determine the number of rotations needed to rotate *C* to *A*'s original position if the rotation is counterclockwise. two rotations

In the figure at the right, $\ell \parallel m$. Name the type of transformation represented by the mappings shown.

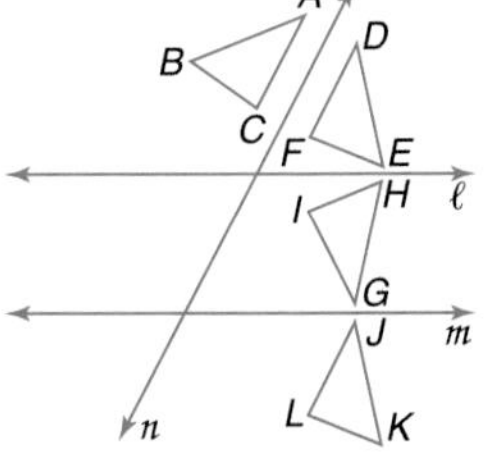

37. $\triangle GHI \to \triangle DEF$ reflection

38. $\triangle ABC \to \triangle GHI$ rotation

39. $\triangle DEF \to \triangle JKL$ translation

40. $\triangle ABC \to \triangle DEF$ reflection

Critical Thinking

41. If a rotation is performed on a coordinate plane, what angles of rotation would make the rotations easier? Explain your answer. See margin.

42. What capital letters produce the same letter after being rotated 180°? 42. H, I, N, O, S, X, Z

Applications and Problem Solving

FYI

The tallest Ferris wheel in the United States is the Texas Star at Fair Park in Dallas, Texas. It is 212 feet 6 inches high.

43. **Recreation** A Ferris wheel's motion is an example of a rotation.
    a. What is the measure of the angle of rotation if seat 1 of a 16-seat Ferris wheel is moved to the seat 5 position? **90**
    b. If seat 1 of a 16-seat Ferris wheel is rotated 135°, find the seat whose position it now occupies. **seat 7**

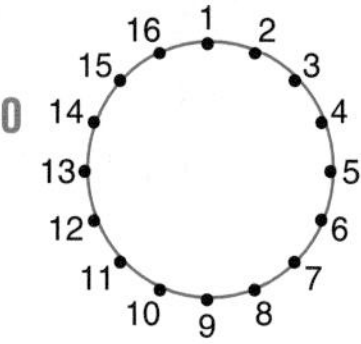

44. **Technology** A computer designer can form various solids by rotating a 2-dimensional geometric figure about a line.
    a. Describe the solid that would be formed by a rotation of $\triangle PQR$ about the y-axis.
    b. Describe the geometric figure that should be rotated about the y-axis to form a cube. **square**

Mixed Review

44a. right circular cone

45. yes; the upper left, upper right, and center quadrilaterals

45. In the tessellation shown at the right, it is possible to slide, or glide, the lower left quadrilateral so it fits exactly on top of the lower right quadrilateral. Is it possible to glide the lower left quadrilateral so it fits exactly on any of the other quadrilaterals? (Lesson 13–6)

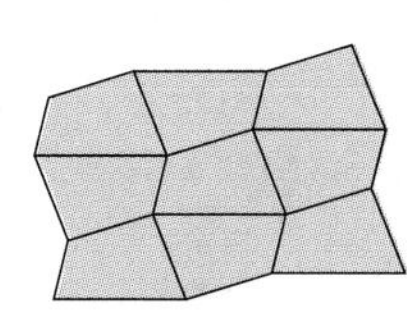

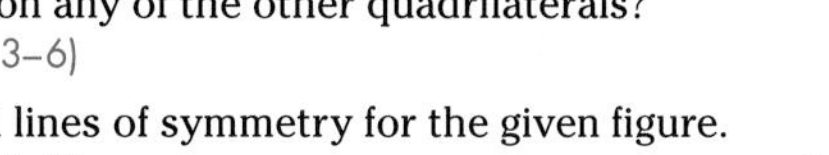

46. Draw all lines of symmetry for the given figure. (Lesson 13–5)

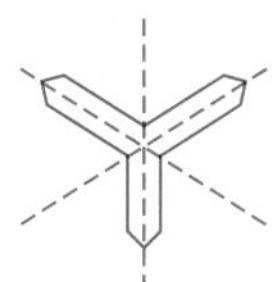

47. two circles concentric to the given circle, one with a radius of 2 in. and the other with a radius of 8 in.

47. Describe the locus of points in a plane that are 3 inches from a circle with a radius measuring 5 inches. (Lesson 13–1)

48. Prove the statement *If both pairs of opposite sides of a quadrilateral are congruent, then the quadrilateral is a parallelogram* using a coordinate proof. (Lesson 12–4) **See margin.**

50. See Solutions Manual.

49. Find the volume of the cone below. (Lesson 11–6) **$768\pi \approx 2412.7$ ft^3**

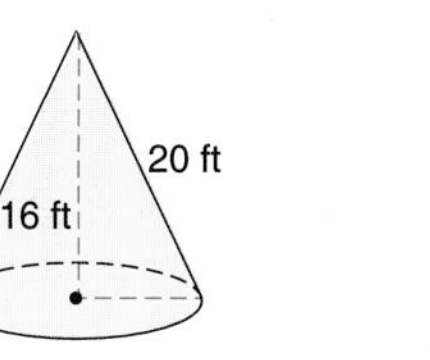

50. Draw a net for the solid below. (Lesson 11–2)

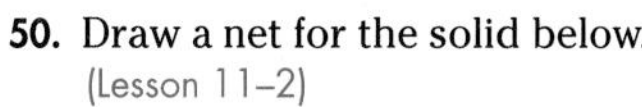

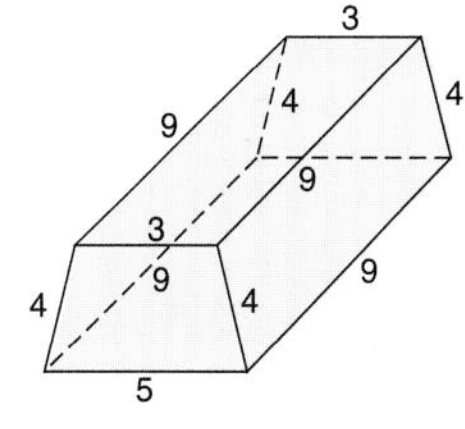

52. yes; $m\angle S = 53$, $m\angle C = 37$, $AB = 4.5$, $QS = 10.0$

51. Find the value of x in $\odot P$. (Lesson 9–4) **37.5**

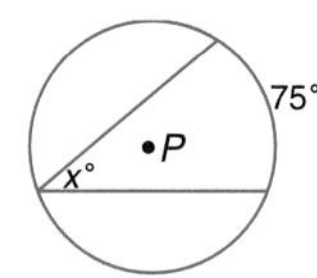

52. Is $\triangle QRS \sim \triangle CAB$? If so, find the missing measures. (Lesson 7–3)

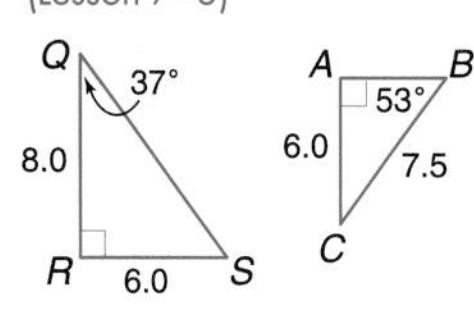

53. Find $\frac{4}{x+9} - \frac{3}{x+9}$. $\frac{1}{x+9}$

54. Factor $3z^2 - 75$. $3(z-5)(z+5)$

4 ASSESS

Closing Activity

Modeling Have students complete a rotation of a two-dimensional object and demonstrate it by using the center of rotation.

Additional Answer

48. Use the distance formula to find AB, DC, AD, and BC.

$$AB = \sqrt{(a-0)^2 + (0-0)^2} = \sqrt{a^2 + 0^2} = \sqrt{a^2} \text{ or } a$$

$$DC = \sqrt{(a+c-c)^2 + (b-b)^2} = \sqrt{a^2 + 0^2} = \sqrt{a^2} \text{ or } a$$

$$AD = \sqrt{(c-0)^2 + (b-0)^2} = \sqrt{c^2 + b^2}$$

$$BC = \sqrt{(a+c-a)^2 + (b-0)^2} = \sqrt{c^2 + b^2}$$

Thus, $AB = DC$ and $AD = BC$. By the definition of congruence, $\overline{AB} \cong \overline{DC}$ and $\overline{AD} \cong \overline{BC}$. Therefore, $ABCD$ is a parallelogram.

Enrichment Masters, p. 84

13-7 NAME______ DATE______

Enrichment Student Edition Pages 739–745

Finding the Center of Rotation

Suppose you are told that $\triangle X'Y'Z'$ is the rotation image of $\triangle XYZ$, but you are not told where the center of rotation is nor the measure of the angle of rotation. Can you find them? Yes, you can. Connect two pairs of corresponding vertices with segments. In the figure, the segments YY' and ZZ' are used. Draw the perpendicular bisectors, ℓ and m, of these segments. The point C where ℓ and m intersect is the center of rotation.

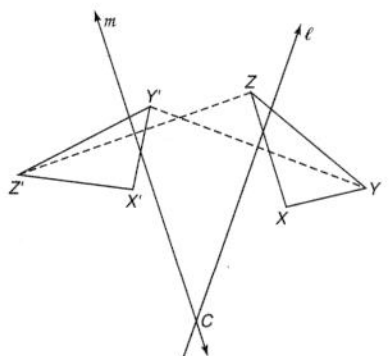

1. How can you find the measure of the angle of rotation in the figure above? Draw $\angle XCX'$ ($\angle YCY'$ or $\angle ZCZ'$ would also do) and measure it with a protractor.

Locate the center of rotation for the rotation that maps WXYZ onto W'X'Y'Z'. Then find the measure of the angle of rotation.

2. 3.

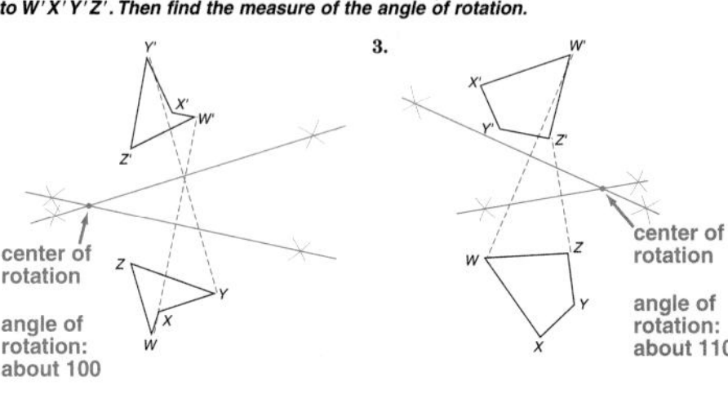

The segments students choose to bisect may vary. See students' work.

FYI

The Ferris wheel was invented by an American engineer, G.W. Ferris. His original ride could carry more than 2000 people.

Extension

Connections Describe the similarities and differences among reflections, translations, and rotations. Sample answer: They are all isometries; reflections involve one line, translations involve two parallel lines, and rotations involve two intersecting lines.

13-8 LESSON NOTES

NCTM Standards: 1–5, 7, 8

Instructional Resources

- Study Guide Master 13-8
- Practice Master 13-8
- Enrichment Master 13-8
- Assessment and Evaluation Masters, p. 353

Transparency 13-8A contains the 5-Minute Check for this lesson; **Transparency 13-8B** contains a teaching aid for this lesson.

Recommended Pacing	
Honors Pacing	Day 11 of 13

1 FOCUS

5-Minute Check
(over Lesson 13-7)

Refer to the figure below.

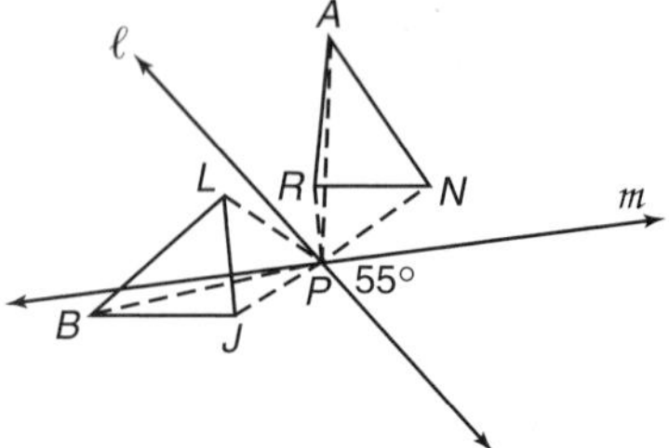

1. Name the lines of reflection. **line ℓ and line *m***
2. Name the rotation image of △*ARN*. **△*BJL***
3. Find the angle of rotation. **110°**
4. Name two angles of rotation. **∠*APB*, ∠*NPL*, ∠*RPJ***

Motivating the Lesson

Hands-On Activity Bring a magnifying glass to class and ask what it does. Have students relate the image that the glass produces to a dilation.

13-8 Dilations

What **YOU'LL LEARN**

- To use scale factors to determine if a dilation is an enlargement, a reduction, or a congruence transformation, and
- to find the center and scale factor for a given dilation, and vice versa.

Why **IT'S IMPORTANT**

You can use dilations to solve problems in photography, drafting, and art.

Video Cameras

One night, while experimenting with his video camera, Mick hooked up the camera to play on the television. His son Troy was very amused to see himself on television. He also noticed that he looked smaller on television than in real life. This is an example of yet another transformation called a **dilation**.

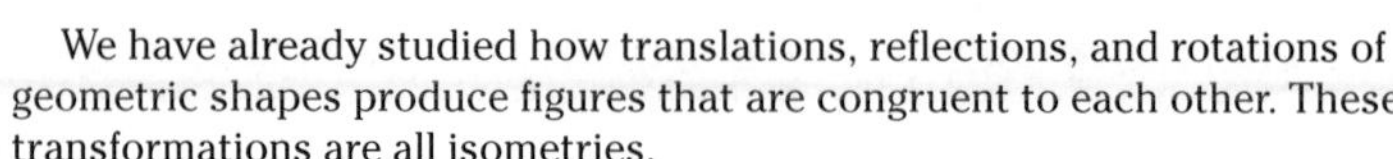

We have already studied how translations, reflections, and rotations of geometric shapes produce figures that are congruent to each other. These transformations are all isometries.

A dilation is a transformation that alters the size of the geometric figure, but does not change its shape. So, a dilation is a *similarity transformation*.

CAREER CHOICES

A **cinematographer** is a photographer of motion pictures. An artistic eye and hands-on experiences are the best qualifications for this job. Courses in cinematography are often offered by photography institutes and universities or through special seminars.

For more information, contact:

American Society of Media Photographers
Washington Rd., Suite 502
Princeton Junction, NJ 08550

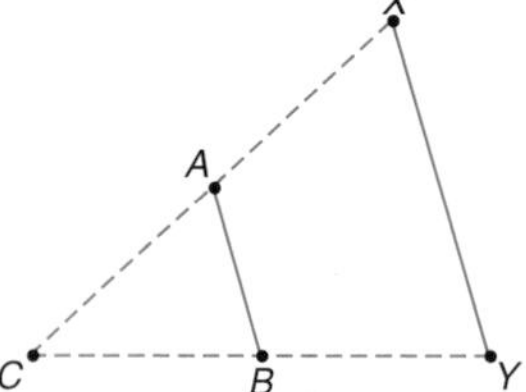

In the figure above, the measure of the distance from C to a point on $\overline{XY}$ is twice the distance from C to a corresponding point on $\overline{AB}$.

$$CX = 2(CA)$$
$$CY = 2(CB)$$

In this transformation, $\overline{AB}$ with **center** C and a **scale factor** of 2 is enlarged to $\overline{XY}$.

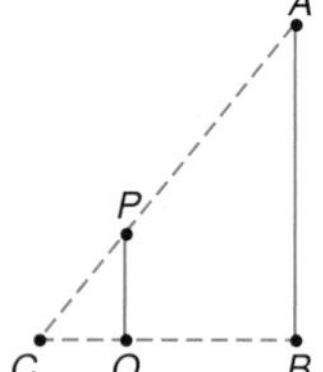

In the figure above, the measure of the distance from C to a point on $\overline{PQ}$ is one-third the distance from C to a corresponding point on $\overline{AB}$.

$$CP = \frac{1}{3}(CA)$$
$$CQ = \frac{1}{3}(CB)$$

In this transformation, $\overline{AB}$ with **center** C and a **scale factor** of $\frac{1}{3}$ is reduced to $\overline{PQ}$.

Theorem 13–1	If a dilation with center C and a scale factor k maps A onto E and B onto D, then $ED = k(AB)$.

CAREER CHOICES

If students are interested in cinematography, have them research some major contributors to film, such as D.W. Griffith, Charles Chaplin, Cecil B. DeMille, Louis Lumière, Alfred Hitchcock, Walt Disney, Orson Welles, George Lucas, and Steven Spielberg.

Notice that when the scale factor is 2, the figure is enlarged. When the scale factor is $\frac{1}{3}$, the figure is reduced.

In general, if k is the scale factor for a dilation with center C, then the following is true.

> **If $k > 0$, P', the image of point P, lies on $\overrightarrow{CP}$, and $CP' = k \cdot CP$.**
> **If $k < 0$, P', the image of point P, lies on the ray opposite $\overrightarrow{CP}$, and $CP' = |k| \cdot CP$.** *The center of a dilation is always its own image.*
>
> **If $|k| > 1$, the dilation is an enlargement.**
> **If $0 < |k| < 1$, the dilation is a reduction.**
> **If $|k| = 1$, the dilation is a congruence transformation.**

Example 1 **Given center C and each scale factor k, find the dilation image of $\overline{XY}$.**

a. $k = \frac{3}{4}$

Since $k < 1$, the dilation is a reduction.

Draw $\overrightarrow{CX}$ and $\overrightarrow{CY}$.

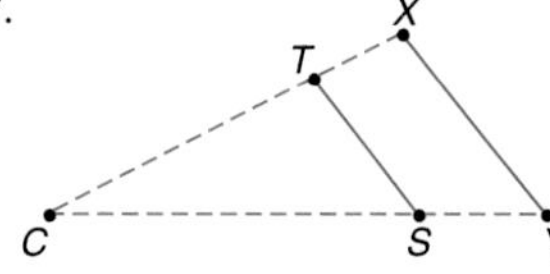

$CT = \frac{3}{4}(CX)$

$CS = \frac{3}{4}(CY)$

$\overline{TS}$ is the dilation of $\overline{XY}$ with scale factor $\frac{3}{4}$.

b. $k = 5$

Since $k > 1$, the dilation is an enlargement.

Draw $\overrightarrow{CX}$ and $\overrightarrow{CY}$.

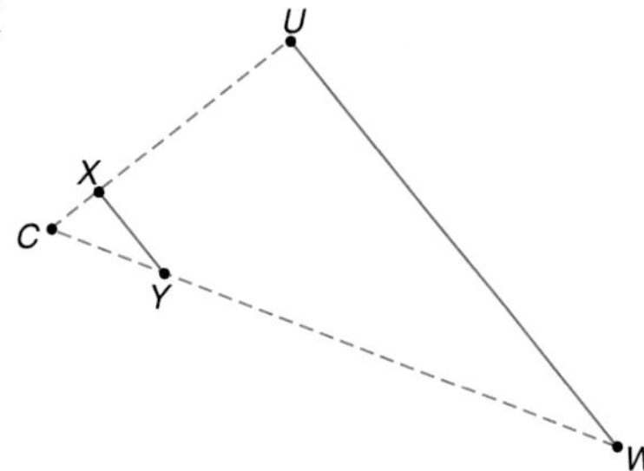

$CU = 5(CX)$

$CW = 5(CY)$

$\overline{UW}$ is the dilation of $\overline{XY}$ with scale factor 5.

In Chapter 7, you learned about scale factors of similar figures. Dilations form similar figures. If you know the measurements of an image and its dilation, you can determine the scale factor.

2 TEACH

In-Class Example

For Example 1

a. Given center C and a scale factor of $\frac{3}{2}$, find the dilation image of $\triangle PQR$.

b. Given center C and a scale factor of $\frac{1}{2}$, find the dilation image of $\triangle PQR$.

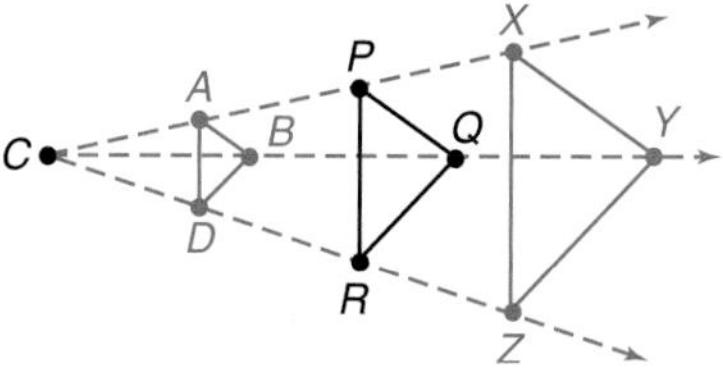

a. Since the absolute value of the scale factor is greater than 1, the dilation is an enlargement. Draw $\overrightarrow{CX}$, $\overrightarrow{CY}$, and $\overrightarrow{CZ}$ so that $CX = \frac{3}{2}(CP)$, $CY = \frac{3}{2}(CQ)$, and $CZ = \frac{3}{2}(CR)$. $\triangle XYZ$ is the dilation image of $\triangle PQR$ with scale factor $\frac{3}{2}$.

b. Since $0 < |k| < 1$, the dilation is a reduction. Draw $\overrightarrow{CP}$, $\overrightarrow{CQ}$, and $\overrightarrow{CR}$ so that $CA = \frac{1}{2}(CP)$, $CB = \frac{1}{2}(CQ)$, and $CD = \frac{1}{2}(CR)$. $\triangle ABC$ is the dilation image of $\triangle PQR$ with scale factor $\frac{1}{2}$.

Teaching Tip Point out that all aspects of similar triangles also hold true for dilation triangles.

Alternative Learning Styles

Auditory Have the class finish the following statement by calling out responses. "If a dilation is an enlargement, then it is absolutely true that . . . ." Write out all of the responses on the chalkboard or overhead. Discuss which are correct and why. Repeat the process for each type of dilation.

Teaching Tip Have students use a ruler to help determine scale factors. Compare actual measures.

In-Class Examples

For Example 2
For each figure, a dilation with center C produced the dotted line figure. What is the scale factor for each transformation?

a.

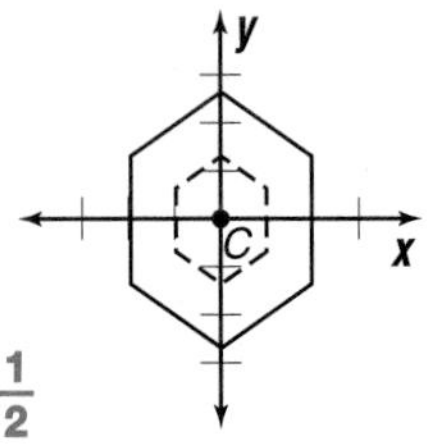

$\frac{1}{2}$

b.

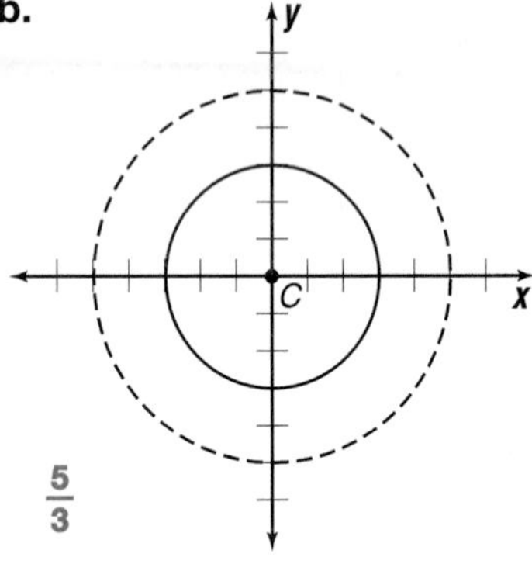

$\frac{5}{3}$

For Example 3
The line segment with endpoints $(-1, -2)$ and $(3, 6)$ has been dilated. The result is the line segment with endpoints $(-11, -22)$ and $(13, 26)$. What is the scale factor? 6

Example  **Determine the scale factor used for each dilation with center *C*. The dilation image in the figure is red.**

a. 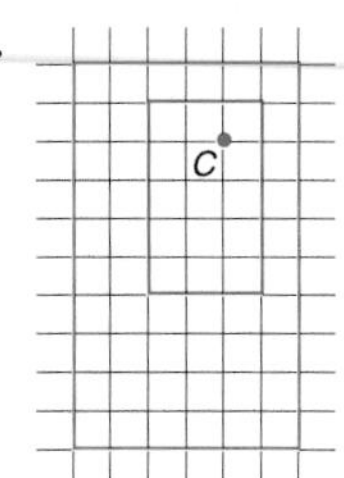 **b.** 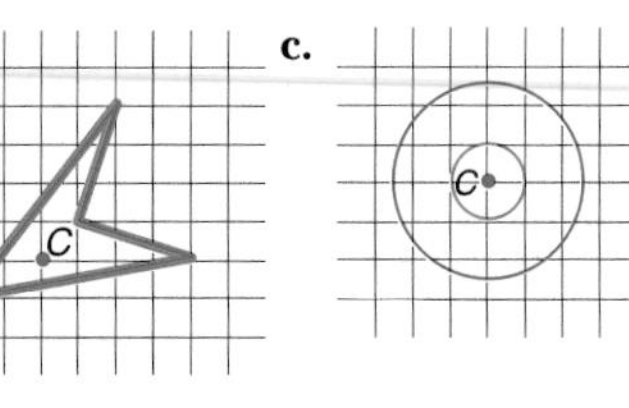 **c.**

Count the grids to determine sizes of each figure and its dilation image. Compare these values to obtain a scale factor.

a. The dilation is an enlargement of the preimage. The ratio of the measures of the dilation to the original is 6:3 so the scale factor is 2.

b. The dilation is the same size as the preimage so it is a congruence transformation. The scale factor is 1.

c. The dilation is a reduction of the preimage. The ratio of the diameters of the dilation to the original is 1:2.5. The scale factor is $\frac{1}{2.5}$ or $\frac{2}{5}$.

You can also use algebraic skills to find the scale factor of a dilation.

Example 3 **The dilation image for trapezoid *ABCD* with center *G* is shown in the coordinate plane below. Compare the length of $\overline{AB}$ with the length of its image to find the scale factor of the dilation.**

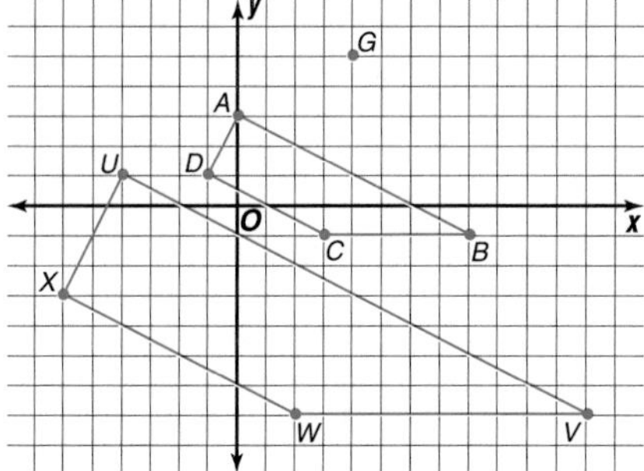

Use the distance formula to find the lengths.

$AB = \sqrt{(3 + 1)^2 + (0 - 8)^2}$ or $4\sqrt{5}$

$UV = \sqrt{(1 + 7)^2 + (-4 - 12)^2}$ or $8\sqrt{5}$

$\overline{UV}$, the image, is 2 times as long as the preimage $\overline{AB}$. The scale factor is $\frac{8\sqrt{5}}{4\sqrt{5}}$ or 2.

Classroom Vignette

"I have students draw an enlargement of a figure by using a grid. Students have a lot of fun with this activity. They lay an acetate grid over their favorite cartoon character. Then they copy the markings in each grid square onto a poster that has been marked with larger grid squares."

Karyn S. Cummins

Karyn S. Cummins
Franklin Central High School
Indianapolis, Indiana

Example 4 **Scale Drawing**

Maria had to make a scale drawing of her bedroom for a class project. If her bedroom is 16 feet long and 12 feet wide, what scale factor can she use to produce her drawing on an $8\frac{1}{2}$-inch by 11-inch piece of construction paper? What would be the size of this scale drawing?

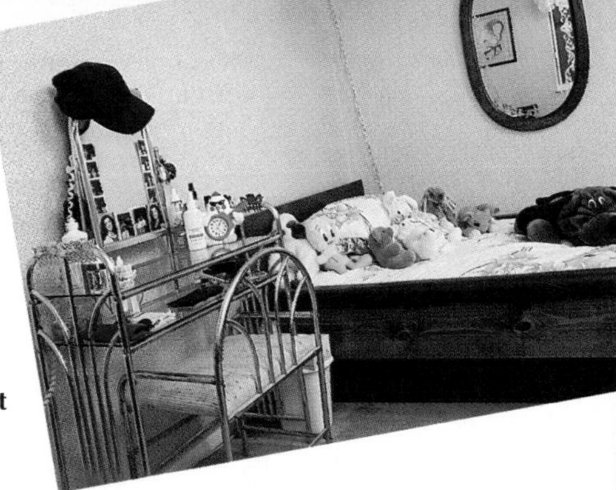

The scale factor Maria uses for the reduction must allow the drawing to fit on the paper. Even though there are many scales she could use, Maria decides to let 1 inch represent 2 feet. So, the scale factor is $\frac{1 \text{ inch}}{2 \text{ feet}}$. The size of her drawing will be 16 feet $\times \frac{1 \text{ inch}}{2 \text{ feet}}$ or 8 inches in length and 12 feet $\times \frac{1 \text{ inch}}{2 \text{ feet}}$ or 6 inches in width.

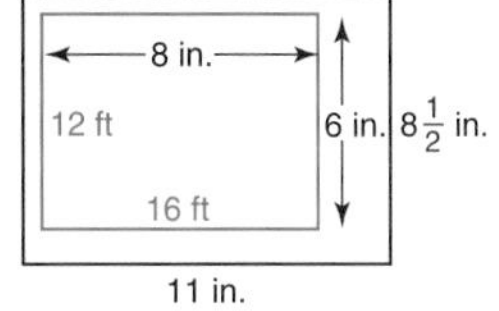

CHECK FOR UNDERSTANDING

Communicating Mathematics

Study the lesson. Then complete the following. 1–3. See margin.

1. **Explain** the difference between a dilation and other transformations that you have studied.
2. **Identify** how you can determine if a dilation is a reduction or an enlargement.
3. **Explain** how a negative value for k affects the dilation image.
4. **Discuss** why a dilation is also called a similarity transformation. A dilation produces figures that are similar.
5. **Assess Yourself** Which of the transformations studied do you find the easiest to perform? Which do you find are the most difficult? Explain why. See students' work.

MATH JOURNAL

Guided Practice

In the figure, $\triangle XYZ$ is a dilation image of $\triangle ABC$ with a scale factor of 8. Complete.

6. If $QB = 6$, then $QY =$ _?_. 48
7. BC _?_ YZ $<$
8. If $XY = 32$, then $AB =$ _?_. 4
9. If $m\angle BCA = 62$, then $m\angle YZX =$ _?_. 62
10. $\triangle ABC$ is _?_ to $\triangle XYZ$. similar

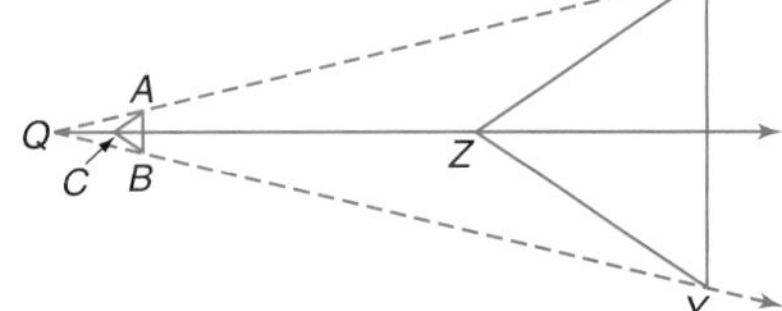

For each scale factor, determine if the dilation is an *enlargement*, a *reduction*, or a *congruence transformation*.

11. 1.00
12. $5\frac{2}{3}$
13. -0.75
14. $\frac{3}{5}$ reduction

11. congruence transformation
12. enlargement
13. reduction

Reteaching

Using Comparison Review the four properties that are preserved by isometries: collinearity, betweenness of points, angle measure, and distance measure. Ask if any of these are not preserved by dilations. Distance measure is not preserved except in the case where $|k| = 1$. Then discuss the other methods of finding the dilation, using such things as scale factor.

In-Class Example

For Example 4
The back wheels of a tricycle have a diameter of 6 inches. The front wheel has a diameter of 10 inches. Find the scale factor used to reduce the size of the front wheel to the size of the back wheels.
The scale factor is $\frac{6}{10}$ or $\frac{3}{5}$.

3 PRACTICE/APPLY

Check for Understanding

Exercises 1–18 are designed to help you assess your students' understanding through reading, writing, speaking, and modeling. You should work through Exercises 1–5 with your students and then monitor their work on Exercises 6–18.

Error Analysis
Figuring out whether the dilation is a reduction or an enlargement when no picture is involved may be difficult. Emphasize the importance of knowing which figure is the preimage and which is the image. Review Theorem 13-1 and point out that the scale factor, k, is multiplied by the measures of the preimage to find the measure of the image.

Additional Answers

1. All transformations, except dilations, produce congruent figures. The dilation image may be congruent or may be similar.
2. A dilation is an enlargement if the absolute value of the scale factor is greater than one. It is a reduction if the absolute value of the scale factor is a value between 0 and 1.
3. If the scale factor is a negative value, a dilation of P with center C would have its image point on the ray opposite $\overrightarrow{CP}$.

Assignment Guide

Core (with proof): 19–49 odd, 51–61
Core (informal): 19–49 odd, 51–61
Enriched: 20–46 even, 47–61

For **Extra Practice**, see p. 792.

The red A, B, and C flags, printed only in the Teacher's Wraparound Edition, indicate the level of difficulty of the exercises.

Additional Answers

15.

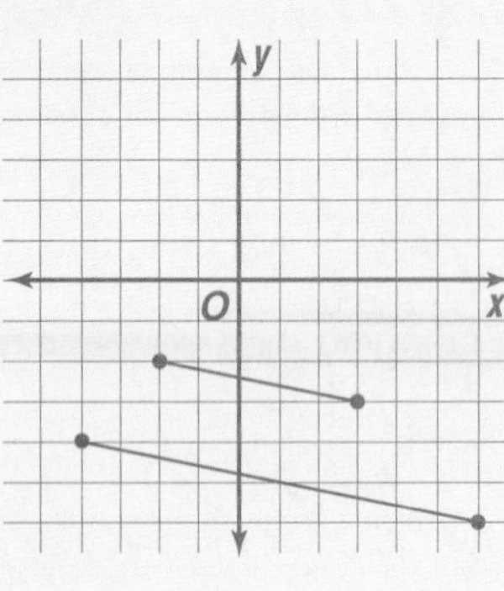

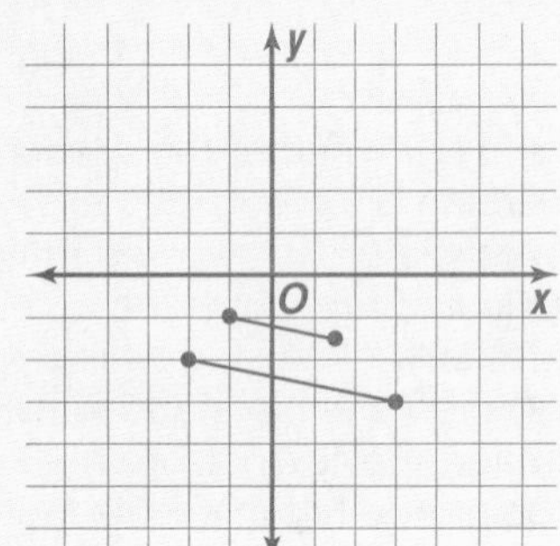

16.

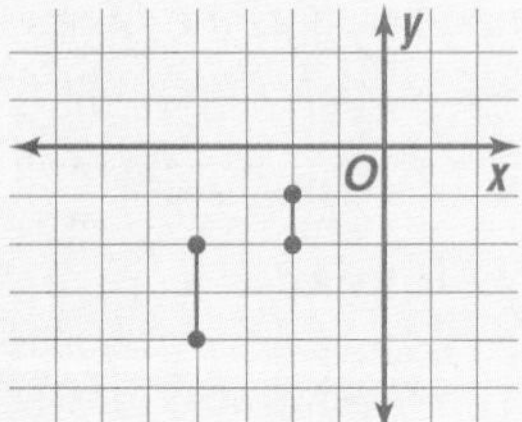

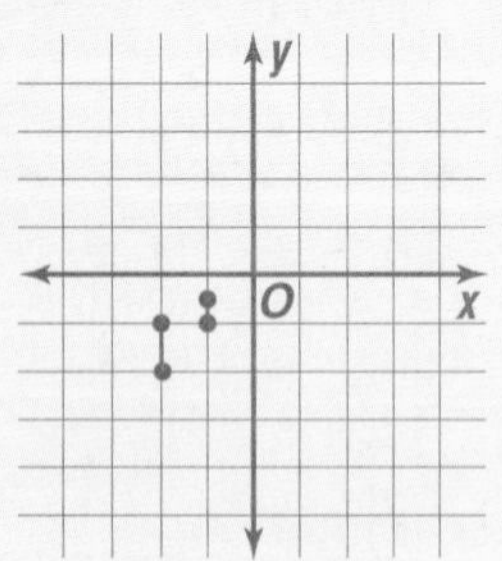

On a coordinate plane, graph the segment whose endpoints are given. Using the origin as the center of dilation and a scale factor of 2, draw the dilation image. Repeat using a scale factor of $\frac{1}{2}$.

15–16. See margin.

15. $A(3, -3), B(-2, -2)$

16. $C(-2, -1), D(-2, -2)$

17. For the given figure at the right, a dilation with center C produced the figure in red. What is the scale factor for this transformation? $\frac{1}{2}$

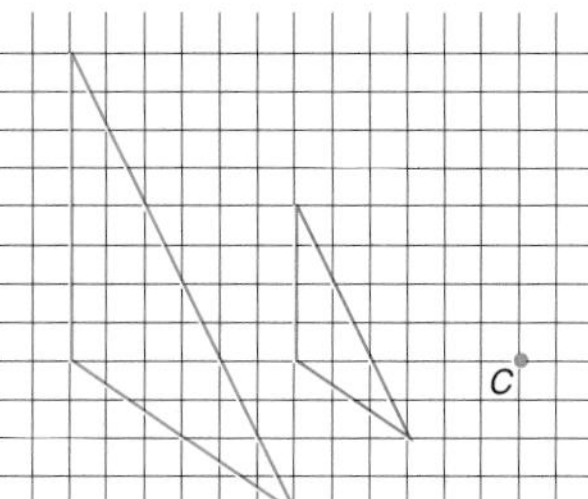

18. **Publishing** Tara is doing a layout for a new book. She has a drawing that is 12 centimeters by 9 centimeters, but the maximum space available for the drawing is 8 centimeters by 6 centimeters. If she wants the drawing to be as large as possible in the book, what scale factor should she use to reduce the original drawing? $\frac{2}{3}$

EXERCISES

Practice

For each scale factor, find the image of *A* with respect to a dilation with center *P*.

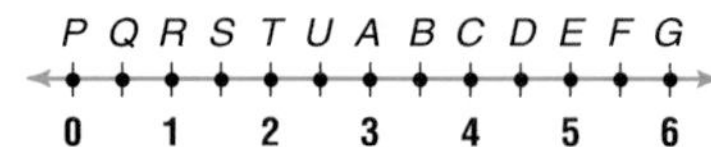

19. $1\frac{1}{6}$ *B*

20. $\frac{2}{3}$ *T*

21. $\frac{1}{2}$ *S*

22. $1\frac{5}{6}$ *F*

Find the measure of the dilation image of $\overline{AB}$ with the given scale factor.

23. $AB = 5, k = -6$ 30

24. $AB = \frac{2}{3}, k = \frac{1}{2}$ $\frac{1}{3}$

25. $AB = 16, k = 1.5$ 24

26. $AB = 12, k = \frac{1}{4}$ 3

27. $AB = 3, k = -1$ 3

28. $AB = 3.1, k = -5$ 15.5

A dilation with center *C* and a scale factor of *k* maps *A* onto *D* and *B* onto *E*. Find $|k|$ for each dilation. Then determine whether each dilation is an *enlargement*, a *reduction*, or a *congruence transformation*.

29. $CD = 10, CA = 5$

30. $CB = 6, CE = 4$

31. $AB = 7, DE = 7$

32. $DE = 12, AB = 4$

33. $CE = 7, CB = 28$

34. $CE = 18, CB = 27$

29. 2, enlargement
30. $\frac{2}{3}$, reduction
31. 1, congruence transformation
32. 3, enlargement
33. $\frac{1}{4}$, reduction
34. $\frac{2}{3}$, reduction

For each figure, a dilation with center C produced the figure in red. What is the scale factor for each transformation?

35. 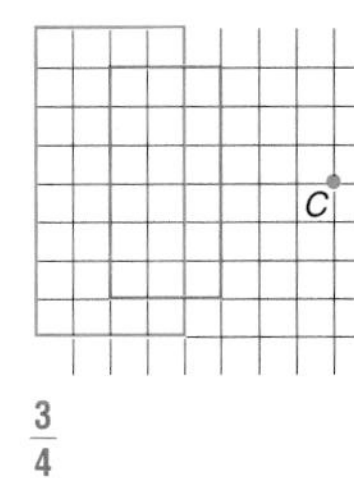

$\frac{3}{4}$

36. 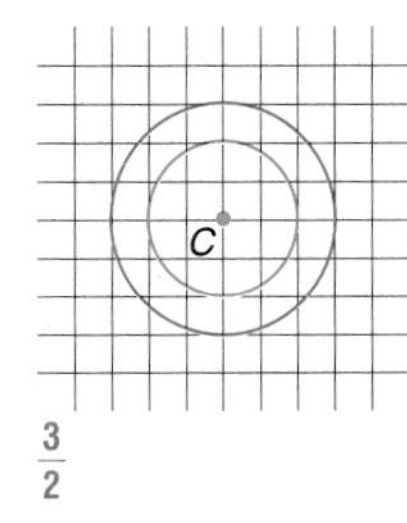

$\frac{3}{2}$

37. 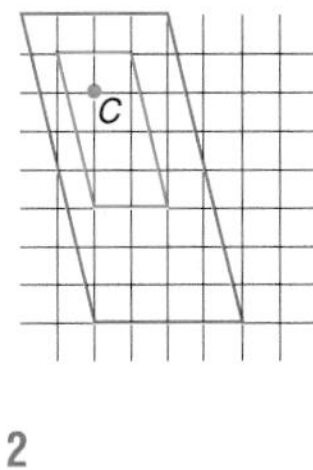

2

On a coordinate plane, graph the polygons whose vertices are given. Using the origin as the center of dilation and a scale factor of 2, draw the dilation image. Repeat using a scale factor of $\frac{1}{2}$.

38–43. See Solutions  Manual.

38. $A(3, 4)$, $B(6, 10)$, $C(-3, 5)$
39. $D(6, 5)$, $E(4, 5)$, $F(3, 7)$
40. $G(-1, 4)$, $H(0, 1)$, $I(2, 3)$
41. $J(1, -2)$, $K(4, -3)$, $L(6, -1)$
42. $M(1, 2)$, $N(3, 3)$, $O(3, 5)$, $P(1, 4)$
43. $Q(4, 2)$, $R(-4, 6)$, $S(-6, -8)$, $T(6, -10)$

44a. The perimeter of the image will be four times the perimeter of the preimage.

44b. The area of the image will be sixteen times the area of the preimage.

44. A dilation on a rectangle has a scale factor of 4.

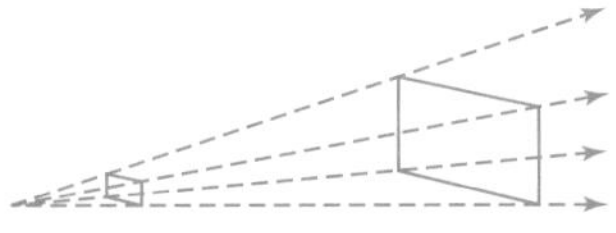

a. What is the effect of the dilation on the perimeter of the rectangle?
b. What is the effect of the dilation on the area of the rectangle?

45a. The surface area of the image will be nine times the surface area of the preimage.

45. A dilation on a cube has a scale factor of 3.

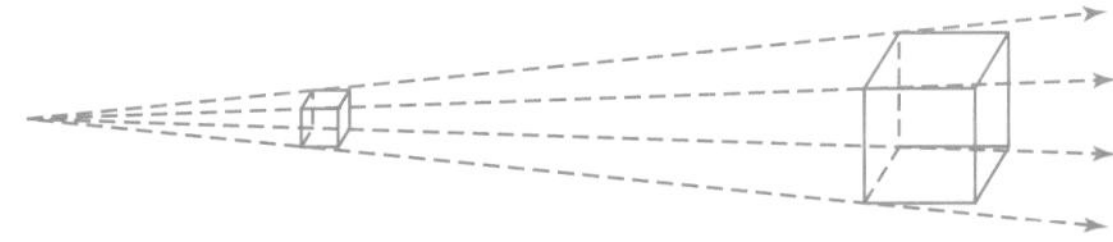

a. What is the effect of the dilation on the surface area of the cube?
b. What is the effect of the dilation on the volume of the cube? See margin.

46. Write a paragraph proof for *If a dilation with center C and a scale factor k maps A onto E and B onto D, then ED = k(AB).* (Theorem 13–1) See margin.

Critical Thinking

47. Graph $\triangle ABC$ with vertices $A(3, 4)$, $B(4, 3)$, and $C(2, 1)$ and $\triangle RST$ with vertices $R(7.5, 2.5)$, $S(10, 0)$, and $T(5, -5)$. If $\triangle RST$ is the dilation image of $\triangle ABC$, find the coordinates of the center and the scale factor. $(0, 5)$; $\frac{5}{2}$

Applications and Problem Solving

48. 6 in. by 7.5 in.

48. **Photography** An 8-inch by 10-inch photograph is being reduced by a scale factor of $\frac{3}{4}$. What are the dimensions of the new photograph?

Additional Answers

45b. The volume of the image will be 27 times the volume of the preimage.

46. Given: Dilation with center C and scale factor k
Prove: $ED = k(AB)$

Proof:
$CE = k(CA)$ and $CD = k(CB)$ by the definition of a dilation. $\frac{CE}{CA} = k$ and $\frac{CD}{CB} = k$. So, $\frac{CE}{CA} = \frac{CD}{CB}$ by substitution. $\angle ACB \cong \angle ECD$ since congruence of angles is reflexive. Therefore, by SAS Similarity, $\triangle ACB \sim \triangle ECD$. The corresponding sides of similar triangles are proportional, so $\frac{ED}{AB} = \frac{CE}{CA}$. We know that $\frac{CE}{CA} = k$, so $\frac{ED}{AB} = k$ by substitution. Therefore, $ED = k(AB)$ by the Multiplication Property of Equality.

Study Guide Masters, p. 85

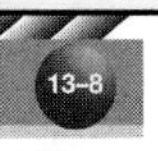

NAME ______ DATE ______

Study Guide — Student Edition Pages 746–753

Dilations

A **dilation** is a similarity transformation that alters the size of a geometric figure, but does not change its shape. In the figure below, the measure from the center C of dilation to a corresponding point on $\overline{RT}$ is three times the distance from C to a point on $\overline{AB}$.

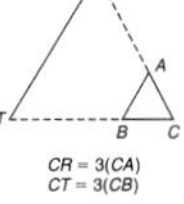

Thus $\overline{AB}$ with center C and a scale factor of 3 is enlarged to $\overline{RT}$. The following theorem applies to the above information.

Theorem 13-1	If a dilation with center C and a scale factor k maps A onto E and B onto D, then $ED = k(AB)$.

If $k > 0$, P', the image of point P, lies on $\overrightarrow{CP}$, and $CP' = k \cdot CP$.
If $k < 0$, P', the image of point P, lies on the ray opposite $\overrightarrow{CP}$, and $CP' = |k| \cdot CP$. The center of a dilation is always its own image.
If $|k| > 1$, the dilation is an enlargement.
If $0 < |k| < 1$, the dilation is a reduction.
If $|k| = 1$, the dilation is a congruence transformation.

Example: Given center C, find the measure of the dilation image XY if $AB = 4$ with a scale factor of $\frac{5}{4}$.
$XY = k(AB)$
$XY = \frac{5}{4}(4)$ Since $|k| > 1$, the dilation is an enlargement.
$XY = 5$

For each scale factor, find the image of A with respect to a dilation with center P.

P Q R S T U A B C D E F G (number line 1 2 3 4 5 6 7)

1. $1\frac{1}{2}$ E
2. $\frac{1}{4}$ P
3. $\frac{3}{8}$ Q
4. $\frac{5}{8}$ F

Find the measure of the dilation image of AB with the given scale factor. Then determine whether each dilation is an enlargement, a reduction, or a congruence transformation.

5. $AB = 3$, $k = 4$ 12, enlargement
6. $AB = 1.3$, $k = 0.6$ 0.78, reduction
7. $AB = 12$, $k = -\frac{1}{4}$ −3, reduction
8. $AB = 1\frac{5}{7}$, $k = \frac{3}{4}$ $\frac{9}{7}$, reduction
9. $AB = 9$, $k = 1.0$ 9, congruence transformation
10. $AB = \frac{5}{6}$, $k = -3$ $-2\frac{1}{2}$, enlargement

The top tourist attractions in California include Disneyland, the San Diego Zoo, Sea World, the state's six national parks, and Death Valley National Monument.

Practice Masters, p. 85

NAME________________ DATE ________

Practice

Student Edition Pages 746–753

Dilations

A dilation with center C and a scale factor k maps X onto Y. Find |k| for each dilation. Then determine whether each dilation is an enlargement, a reduction, or a congruence transformation.

1. $CY = 15$, $CX = 10$ $\frac{3}{2}$; enlargement
2. $CY = 2$, $CX = 2$ 1; congruence
3. $CX = 5$, $CY = 2$ $\frac{2}{5}$; reduction
4. $CY = 20$, $CX = \frac{1}{2}$ 40; enlargement

Find the measure of the dilation image of $\overline{AB}$ with the given scale factor.

5. $AB = 6$ in., $k = -\frac{2}{3}$ 4 in.
6. $AB = 4$ in., $k = 1$ 4 in.
7. $AB = 1\frac{1}{2}$ in., $k = \frac{1}{2}$ $\frac{3}{4}$ in.
8. $AB = 20$ in., $k = -2\frac{1}{2}$ 50 in.

Find each scale factor, find the image of A with respect to a dilation with center C.

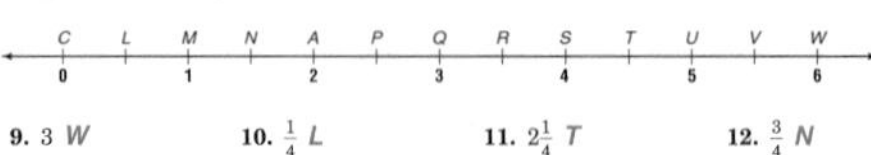

9. 3 *W*
10. $\frac{1}{4}$ *L*
11. $2\frac{1}{4}$ *T*
12. $\frac{3}{4}$ *N*

Graph each set of ordered pairs. Then connect the points in order. Using (0, 0) as the center of dilation and a scale factor of 2, draw the dilation image. Repeat this using a scale factor of $\frac{1}{2}$.

13. (2, 2), (4, 6), (6, −2)

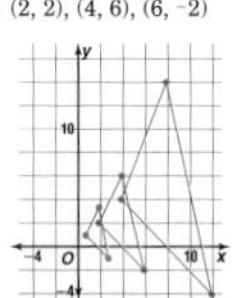

14. (0, 2), (−4, 2), (−4, −2)

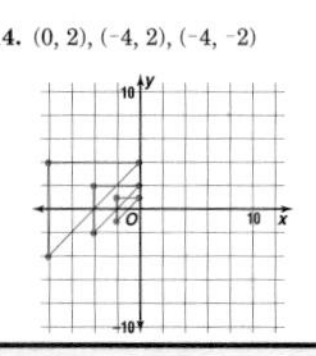

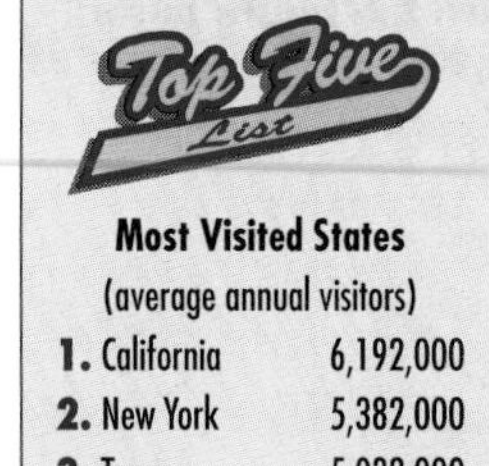

Most Visited States
(average annual visitors)

1.	California	6,192,000
2.	New York	5,382,000
3.	Texas	5,032,000
4.	Florida	3,853,000
5.	Hawaii	2,203,000

49. Travel In trying to calculate how far she must travel for an appointment, Gunja measured the distance between Richmond, Virginia, and Charlotte, North Carolina, on a map. The distance on the map was 2.25 inches, and the scale factor was 1 inch equals 150 miles. How far must she travel? **337.5 miles**

50. Art The picture below is Mauritz C. Escher's *Fish and Scales*.

a. Describe how Escher has used dilations in this piece of art.

b. How are tessellation images used in this piece of art?
The various fishes are tessellated so that the entire surface is covered.

50a. Sample answer: the same fish are used at different sizes.

Mixed Review

51. Draw and label $\triangle EFG$. Draw the dilation image of $\triangle EFG$ with a scale factor of $\frac{1}{2}$ and center *C*. (Lesson 13–8)

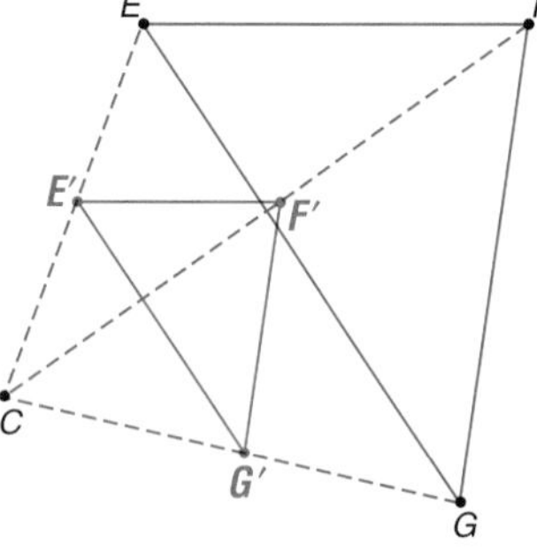

52. What is the measure of the angle of rotation if two lines intersect to form a 63° angle? (Lesson 13–7) **126**

53. Copy the figure at the right. Then find the translation image of the geometric figure with respect to parallel lines ℓ and m. (Lesson 13–6)

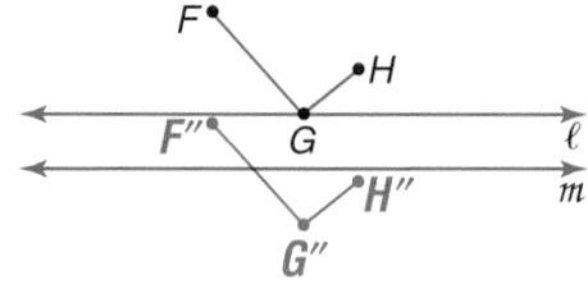

54. Read the problem carefully. Draw the given figure. Locate the points that satisfy the given conditions. Draw a smooth geometric figure. Describe the locus.

54. Describe the procedure for determining locus. (Lesson 13–1)

55. Determine the distance between points $L(-2, 3, 7)$ and $M(5, 6, -4)$. (Lesson 12–6) $\sqrt{179} \approx 13.4$ **units**

56. **Water Management** A water tank holds 17,650 gallons of water. (Lesson 12–2)

a. If it is losing 895 gallons per hour, write an equation to describe the number of gallons left after *x* hours. $y = 17{,}650 - 895x$

b. After how many hours will the tank be empty? **about 19 h 45 min**

57. **Manufacturing** A piece of bubble gum is shaped like a right hexagonal prism. Each edge of the hexagon measures 8 millimeters, and the gum is 60 millimeters tall. If the gum wrapper overlaps 3 millimeters along the lateral surface area and needs to be 40 millimeters longer than the gum, what is the area of the wrapper? (Lesson 11–3) **5100 mm²**

58. What is the measure of $\angle B$? Justify your answer. (Lesson 8–2)

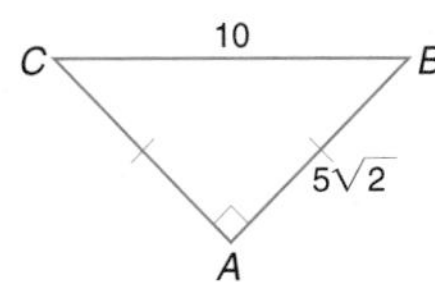

59. If $\overline{GJ}$ bisects $\angle HGK$, $GH = 9$, $GK = 12$, and $JK = 10$, find HJ. (Lesson 7–5) **7.5**

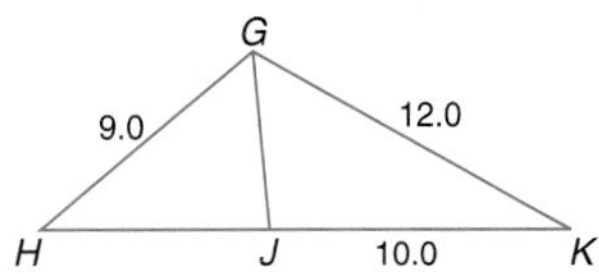

INTEGRATION Algebra

60. Use elimination to solve the system of equations. **(−3, 1)**
$3p + 4q = -5$
$5p - 12q = -27$

58. 45; The hypotenuse is $\sqrt{2}$ times as long as the leg, so it is a 45°-45°-90° triangle.

61. **Probability** For the dinner special at the Silver Inn, you can select one item from each of the categories at the right.

Entree	Potato	Vegetable
chicken fish	baked sweet fried	broccoli peas

a. See Solutions Manual.

a. Draw a tree diagram showing all of the dinner special combinations.
b. What is the probability that a customer will select fish with a baked potato? $\frac{2}{12}$ or $\frac{1}{6}$

WORKING ON THE **In•ves•ti•ga•tion**

Refer to the Investigation on pages 642–643.

In the making of *Jurassic Park*, small models of dinosaurs were used to produce the images of dinosaurs. Then, those images were enlarged and added to the scenes with the actors. Dilations can be used to create similar effects with graphic images.

1 Use a dilation to change the size and position of a drawing you created for your movie.

2 Write a sentence describing how the dilation affects the coordinates of the object being dilated in your movie.

3 Is there any part of the story that could be accomplished by using a dilation?

Add the results of your work to your Investigation Folder.

4 ASSESS

Closing Activity

Writing Have students write an essay using 5 scale factors that will produce an enlargement and 5 scale factors that will produce a reduction. Then have them draw an example using each scale factor.

Chapter 13 Quiz D (Lessons 13-7 and 13-8) is available in the *Assessment and Evaluation Masters*, p. 353.

Enrichment Masters, p. 85

NAME ______ DATE ______
13–8 **Enrichment** Student Edition Pages 746–753

Similar Circles

You may be surprised to learn that two noncongruent circles that lie in the same plane and have no common interior points can be mapped one onto the other by more than one dilation.

1. Here is diagram that suggests one way to map a smaller circle onto a larger one using a dilation. The circles are given. The lines suggest how to find the center for the dilation. Describe how the center is found. Use segments in the diagram to name the scale factor. Draw the two common external tangents. The point where they intersect is the center of a dilation that maps ⊙A onto ⊙A′; scale factor = $\frac{A'M'}{AM}$.

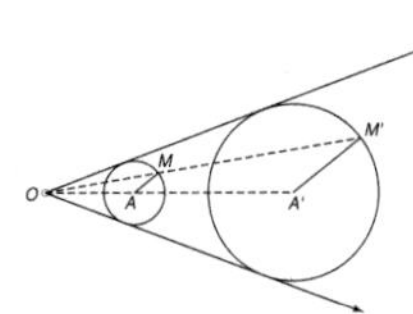

2. Here is another pair of noncongruent circles with no common interior point. From Exercise 1, you know you can locate a point off to the left of the smaller circle that is the center for a dilation mapping ⊙C onto ⊙C′. Find another center for another dilation that maps ⊙C onto ⊙C′. Mark and label segments to name the scale factor. Find the intersection of the two common internal tangents; scale factor: $-\frac{C'X'}{CX}$.

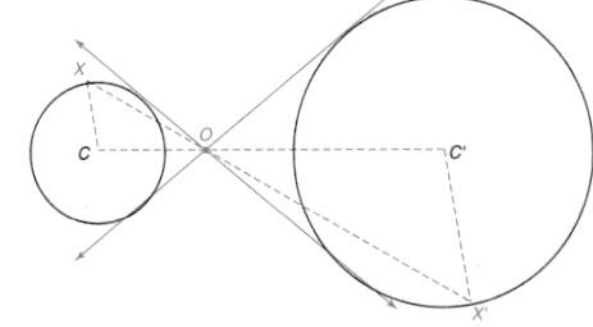

Extension

Reasoning A figure undergoes a series of dilations with center at C. The first scale factor applied is −1. Then a scale factor of 2 is applied to the image. What scale factor must be applied to the second image to produce the original figure in its original position?
$-\frac{1}{2}$

In•ves•ti•ga•tion

Working on the Investigation

The Investigation on pages 642–643 is designed to be a long-term project that is completed over several days or weeks. Encourage students to keep their materials in their Investigation Folder as they work on the Investigation.

Closing the Investigation

This activity provides students an opportunity to bring their work on the Investigation to a close. For each Investigation, students should present their findings to the class. Here are some ways students can display their work.

- Conduct and report on an interview or survey.
- Write a letter, proposal, or report.
- Write an article for the school or local paper.
- Make a display, including graphs and/or charts.
- Plan an activity.

Assessment

To assess students' understanding of the concepts and topics explored in the Investigation and its follow-up activities, you may wish to examine students' Investigation Folders.

The scoring guide provided in the *Investigations and Projects Masters*, p. 23, provides a means for you to score students' work on the Investigation.

Investigations and Projects Masters, p. 23

Scoring Guide
Chapters 12 and 13
Investigation

Level	Specific Criteria
3 Superior	• Shows thorough understanding of the concepts of *pixel, coordinates, two-dimensional, three-dimensional, image, trigonometric values, reflection,* and *dilation.* • Uses appropriate strategies to solve problems. • Computations are correct. • Written explanations are exemplary. • Movie books, titles, introductions, and promotional items are appropriate and sensible. • Goes beyond requirements of all or some problems.
2 Satisfactory, with Minor Flaws	• Shows understanding of the concepts of *pixel, coordinates, two-dimensional, three-dimensional, image, trigonometric values, reflection,* and *dilation.* • Uses appropriate strategies to solve problems. • Computations are mostly correct. • Written explanations are effective. • Movie books, titles, introductions, and promotional items are appropriate and sensible. • Satisfies all requirements of problems.
1 Nearly Satisfactory, with Obvious Flaws	• Shows understanding of most of the concepts of *pixel, coordinates, two-dimensional, three-dimensional, image, trigonometric values, reflection,* and *dilation.* • May not use appropriate strategies to solve problems. • Computations are mostly correct. • Written explanations are satisfactory. • Movie books, titles, introductions, and promotional items are appropriate and sensible. • Satisfies most requirements of problems.
0 Unsatisfactory	• Shows little or no understanding of the concepts of *pixel, coordinates, two-dimensional, three-dimensional, image, trigonometric values, reflection,* and *dilation.* • Does not use appropriate strategies to solve problems. • Computations are incorrect. • Written explanations are not satisfactory. • Movie books, titles, introductions, and promotional items are not appropriate or sensible. • Does not satisfy requirements of problems.

CLOSING THE In·ves·ti·ga·tion

Refer to the Investigation on pages 642–643.

In 1995, the movie *Casper* became the first movie to feature a character that was completely a digital image. In the near future, filmmakers believe we may see a movie produced with a digital human character. As the power of computers grows, making amazing special effects that were once impossible will become more and more simple.

Production

PORTFOLIO ASSESSMENT

You may want to keep your work on this Investigation in your portfolio.

1 Complete the story line for your movie.
2 If your movie will need sound, write a script or choose music.
3 Make a list of characters and background sets that must be designed. Then complete each design.
4 Use the images of the sets and characters to create at least fifteen images to make the movie. Bind the images into a book.
5 Flip the movie book to test the flow of the movie. Add or change images as needed to complete the production.

Screening

6 Choose another group of students in your class to represent the executives of your production company.
7 Write an introduction to the movie, choose a title, and create promotional posters and previews.
8 Show your movie to the executives. Present the promotional materials and previews.
9 Explain your technique for making the movie. Include any sketches or story research.

VOCABULARY

After completing this chapter, you should be able to define each term, property, or phrase and give an example or two of each.

Geometry

- angle of rotation (p. 740)
- center of dilation (p. 746)
- center of rotation (p. 739)
- composite of reflections (p. 731)
- congruence transformation (p. 716)
- dilation (p. 746)
- image (p. 715)
- isometry (p. 716)
- line of reflection (p. 722)
- line of symmetry (p. 724)
- locus (p. 696)
- mapping (p. 715)
- point of reflection (p. 722)
- point of symmetry (p. 725)
- preimage (p. 715)
- reflection (p. 722)
- rotation (p. 739)
- scale factor (p. 746)
- similarity transformation (p. 717)
- transformation (p. 715)
- translation (p. 731)

Algebra

- system of equations (p. 704)

Problem Solving

- look for a pattern (p. 741)
- make a table (p. 717)

UNDERSTANDING AND USING THE VOCABULARY

State whether each sentence is *true* or *false*. If false, replace the underlined word to make a true sentence.

1. The scale factor in figure e is 2. true
2. An image is a set of all points that satisfy a given set of conditions. false; locus
3. In figure c, $\triangle JKL$ demonstrates a translation image of $\triangle ABC$. true
4. Figure b is an example of an isometry. 4. false; dilation
5. If quadrilateral $KLMN \rightarrow$ quadrilateral $UVWX$, the preimage of $\overline{LM}$ is $\overline{VW}$. false; image
6. Figure d represents a rotation. true
7. A transformation maps a preimage onto an image. true
8. Figure a represents a dilation. false; reflection
9. In figure c, $\triangle ABC$ is the image of $\triangle JKL$. false; preimage
10. In figure c, $\triangle XYZ$ is a reflection over line ℓ of $\triangle ABC$. true

a.

b.

c.

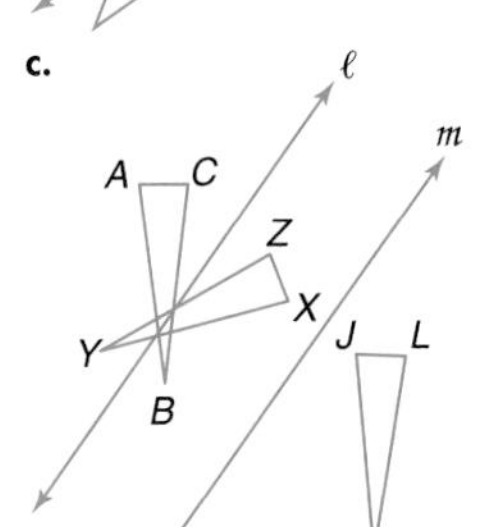

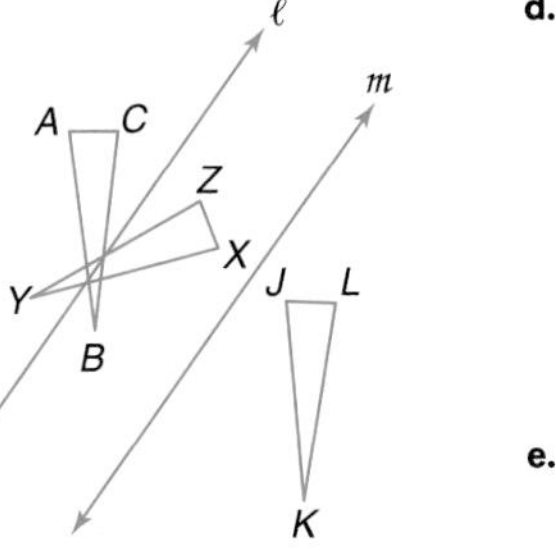

d.

e.

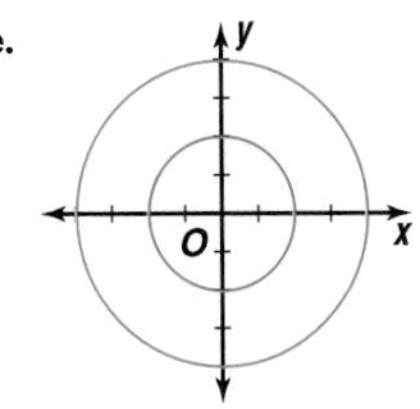

Using the CHAPTER HIGHLIGHTS

The Chapter Highlights begins with a listing of the new terms, properties, and phrases that were introduced in this chapter. Have students define each term and provide an example or two of it, if appropriate.

Assessment and Evaluation Masters, pp. 339–340

13 NAME______ DATE______

Chapter 13 Test, Form 1B

Write the letter for the correct answer in the blank at the right of each problem.

1. Describe the locus of all points in a plane that are inside or on a given square and equidistant from at least two sides of the square.
   A. the center of the square
   B. the lines containing the diagonals
   C. the perpendicular bisectors of the sides
   D. the two diagonals and the segments joining the midpoints of opposite sides — 1. D
2. How many points in a plane are in the locus of all points equidistant from three parallel lines in the plane?
   A. 0 B. 1 C. 2 D. infinitely many — 2. A
3. Find the locus of points that satisfy both of the equations $3x + 2y = -2$ and $2x - 3y = 16$.
   A. (2, −2) B. (14, 4) C. (2, −4) D. (−14, −4) — 3. C
4. Find the area of the figure enclosed by the lines $y = -2$, $x = 2$, and $2y = x - 4$.
   A. 3 B. 2 C. 1 D. 5 — 4. C
5. Which does *not* represent a possible intersection of a line and a right cone with height 4 cm?
   A. a segment B. 3 points C. 2 points D. 1 point — 5. B

For Questions 6–9, refer to the figure at the right.

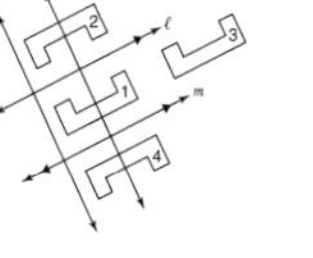

6. The reflection image of figure 1 with respect to line m is
   A. figure 2. B. figure 3. C. figure 4. D. none of these. — 6. C
7. The rotation image of figure 2 with respect to lines k and ℓ is
   A. figure 1. B. figure 3. C. figure 4. D. none of these. — 7. A
8. Which figure is a translation image of figure 3?
   A. figure 1 B. figure 2 C. figure 4 D. none of these — 8. A
9. Which figure is the rotation image of figure 1 with respect to lines j and ℓ?
   A. figure 2 B. figure 3 C. figure 4 D. none of these — 9. D

13 NAME______ DATE______

Chapter 13 Test, Form 1B (continued)

10. Which does *not* represent a possible intersection of a sphere and two parallel lines?
    A. 3 points B. 5 points C. 2 points D. 4 points — 10. B
11. Refer to figures 1 and 2 on the right. If figures 1 and 2 are congruent, then figure 1 is an image of figure 2 with respect to
    A. a reflection. B. a dilation. C. a rotation. D. all of these. — 11. D
12. Which kind of transformation may *not* preserve segment lengths?
    A. reflection B. dilation C. rotation D. translation — 12. B
13. Which of the following has only one line of symmetry?
    A. square B. regular pentagon C. isosceles trapezoid D. equilateral triangle — 13. C
14. If the angle of rotation with respect to lines ℓ and m measures 20°, then ℓ and m intersect to form an angle of
    A. 10°. B. 20°. C. 40°. D. 160°. — 14. A

For Questions 15–19, find the coordinates of the image of point A(−1, 3) for each transformation.

15. translation with respect to the line $y = 1$ and the x-axis
    A. (1, 3) B. (−1, −3) C. (1, −3) D. (−1, 1) — 15. D
16. reflection with respect to the y-axis
    A. (3, 1) B. (1, 3) C. (−3, −1) D. (−1, 3) — 16. B
17. rotation with respect to the x-axis and the y-axis
    A. (1, 3) B. (3, 1) C. (1, −3) D. (−1, 3) — 17. C
18. dilation with center at the origin, scale factor 2
    A. (2, 6) B. (−2, −6) C. (−2, 6) D. (6, −2) — 18. C
19. reflection with respect to the line $y = -2$
    A. (−1, −7) B. (−1, −2) C. (−3, 3) D. (−1, 3) — 19. A
20. In a coordinate plane, (2, 3) is the reflection image of (3, 2) with respect to which of the following lines?
    A. x-axis B. y-axis C. $y = -x$ D. $y = x$ — 20. D

Bonus

In a coordinate plane, $\overline{AB}$ is translated up 2 units and then reflected over the y-axis. For $A(-1, 3)$ and $B(3, -3)$, find the midpoint of the final image of $\overline{AB}$.
A. (1, 2) B. (−1, 2) C. (−3, −2) D. (3, 2) — Bonus B

Instructional Resources

Three multiple-choice tests and three free-response tests are provided in the *Assessment and Evaluation Masters*. Forms 1A and 2A are for honors pacing, Forms 1B and 2B are for average pacing, and Forms 1C and 2C are for basic pacing. Chapter 13 Test, Form 1B, is shown at the right. Chapter 13 Test, Form 2B, is shown on the next page.

Postulates, Theorems, and Corollaries

A complete list of postulates, theorems, and corollaries begins on page 806.

Using the STUDY GUIDE AND ASSESSMENT

Skills and Concepts Encourage students to refer to the objectives and examples on the left as they complete the review exercises on the right.

Assessment and Evaluation Masters, pp. 345–346

13 NAME ______ DATE ______

Chapter 13 Test, Form 2B

Describe the locus of points.

1. all points in space that are 2 in. from the center C of a sphere that has a radius of 3 in. — 1. The sphere with $r = 2$ in. and center C
2. all points in a plane that are 3 cm from a given vertex V of a square with a side of 6 cm — 2. the circle with $r = 3$ cm and center V
3. all points in space that are equidistant from the vertices of a given square — 3. the line ⊥ to the plane of the square through the intersection point of the diagonals of the square
4. all points in a plane that are represented by the center of a circle if the circle rolls along the diameter of circle M and the diameter is always tangent to the circle — 4. a segment ∥ to the diameter of circle M
5. all points in space equidistant from two intersecting lines — 5. 2 planes that bisect the ∠s formed by the given lines and ⊥ to their plane

For Questions 6–8, use an algebraic method to find the locus of points that satisfy both equations.

6. $3x - 2y = -12$, $2x - y = -7$ — 6. $(-2, 3)$
7. $2x - 4y = -2$, $4x + 8y = 0$ — 7. $\left(-\frac{1}{2}, \frac{1}{4}\right)$
8. $x - 2y = 9$, $3x + 2y = -5$ — 8. $(1, -4)$
9. Find the area of the figure bounded by $y = -1$, $x = 5$, and $y - x = -2$. — 9. 8 units^2
10. List the number of possible points of intersection for a sphere and two parallel lines. — 10. 0, 1, 2, 3, or 4 points
11. Describe all possible ways two circles with diameters of different lengths can intersect. — 11. 0, 1, or 2 points

Refer to quadrilateral ABCD → quadrilateral EFGH.

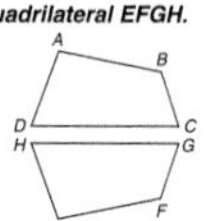

12. Name the image of $\overline{BC}$. — 12. $\overline{FG}$
13. Name the image of $\angle A$. — 13. $\angle E$
14. Name the preimage of $\overline{EF}$. — 14. $\overline{AB}$

13 NAME ______ DATE ______

Chapter 13 Test, Form 2B (continued)

15. Name the kind of isometry that maps $\triangle XYZ$ onto $\triangle MNP$. — 15. translation
16. Does a rectangle have line symmetry, point symmetry, or both? — 16. line symmetry

For Questions 17–19, refer to the figure below. Name the translation image with respect to line j, then line k.

17. A — 17. D
18. B — 18. R
19. C — 19. E
20. If two lines intersect to form a 56° angle, find the measure of the angle of rotation. — 20. 112
21. The measure of the angle of rotation formed by two intersecting lines is 132. Find the measure of the angle formed by the intersecting lines. — 21. 66
22. Is the dilation with a scale factor of $2\frac{2}{3}$ *an enlargement, a reduction, or a congruence transformation?* — 22. an enlargement
23. Find the measure of the dilation image of $\overline{ZW}$ if $ZW = 12$ and scale factor $k = -3$. — 23. 36
24. A dilation with center C and scale factor k maps B onto D and C onto E. Find $|k|$ if $BC = 8$ and $DE = 20$. — 24. $\frac{5}{2}$
25. In a coordinate plane, $\overline{AB}$ is rotated 180° about the origin and then reflected over the y-axis. For $A(0, 3)$ and $B(-1, -5)$, find the midpoint of the final image of $\overline{AB}$. — 25. $\left(-\frac{1}{2}, 1\right)$

Bonus

Trapezoid $ABCD$ has coordinates $A(0, 0)$, $B(4, 0)$, $C(3, -2)$, and $D(1, -2)$. If this trapezoid is reflected over the y-axis and then dilated by a scale factor of $k = \frac{1}{2}$ with the origin as the center, what are the coordinates of the vertices of the resulting image? — Bonus $A''(0, 0)$, $B''(-2, 0)$, $C''\left(-1\frac{1}{2}, -1\right)$, $D''\left(-\frac{1}{2}, -1\right)$

CHAPTER 13 STUDY GUIDE AND ASSESSMENT

SKILLS AND CONCEPTS

OBJECTIVES AND EXAMPLES

Upon completing this chapter, you should be able to:

- locate, draw, and describe a locus in a plane or in space (Lesson 13–1)

The locus of all points in a plane that are equidistant from the endpoints of a given line segment is the perpendicular bisector of the line segment.

The locus of all points in space that are equidistant from the endpoints of a given line segment is the plane that is the perpendicular bisector of the line segment.

REVIEW EXERCISES

Use these exercises to review and prepare for the chapter test.

Describe the locus of points for each set of conditions.

11. all points in a plane that are less than 6 centimeters from a given point
12. all points in space that are equidistant from two given points
13. all points in a plane that are equidistant from three noncollinear points A, B, and C

11–13. See margin.

- find the locus of points that are solutions of a system of equations by graphing, by substitution, or by elimination (Lesson 13–2)

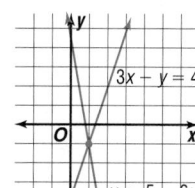

$3x - y = 4$

$y = 5 - 6x$

The locus of points that satisfy both graphs is $(1, -1)$.

14–15. See Solutions Manual for graphs.

Graph each pair of equations to find the locus of points that satisfy the graphs of both equations.

14. $y = x - 2$, $2x + y = 13$ **(5, 3)**
15. $3x - 4y = -1$, $-2x + y = -1$ **(1, 1)**

Use either substitution or elimination to find the locus of points that satisfy the graphs of both equations.

16. $y = 2x - 1$, $x + y = 7$ $\left(\frac{8}{3}, \frac{13}{3}\right)$
17. $3x + y = 5$, $2x + 3y = 8$ **(1, 2)**

- solve locus problems that satisfy more than one condition (Lesson 13–3)

The locus of points in a plane 3 inches from a line and 4 inches from a point on the line is the four points where the parallel lines of one locus intersect the circle of the other.

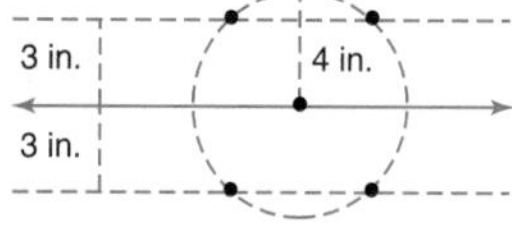

Draw a diagram to find the locus of points that satisfy each set of conditions. Then describe the locus. **18–20. See margin.**

18. all points in a plane equidistant from lines j and k, which intersect at point P, and are 2 centimeters from point P
19. all points in space that are 3 centimeters from the given points P and Q, which are 5 centimeters apart
20. all points in a plane equidistant from two parallel lines and also equidistant from two points on one of the lines

GLENCOE Technology

Test and Review Software

You may use this software, a combination of an item generator and item bank, to create your own tests or worksheets. Types of items include free response, multiple choice, short answer, and open ended.

For IBM & Macintosh

OBJECTIVES AND EXAMPLES	REVIEW EXERCISES

- name the image and preimage of a mapping (Lesson 13–4)

$\triangle RST \rightarrow \triangle R'S'T'$

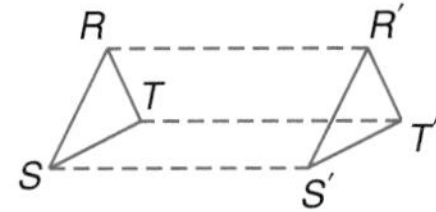

The image of $\angle S$ is $\angle S'$. The preimage of $\overline{R'T'}$ is $\overline{RT}$.

Suppose $\triangle ABE \rightarrow \triangle CBD$. 22. $\angle CBD$

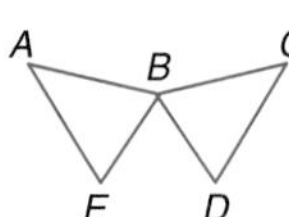

21. Name the preimage of D. E
22. Name the image of $\angle ABE$.
23. Name the image of $\overline{AE}$. $\overline{CD}$
24. Name the preimage of $\angle D$. $\angle E$
25. Describe the transformation that occurred in the mapping. reflection

- draw reflection images, lines of symmetry, and points of symmetry (Lesson 13–5)

M is the reflection of *X* over line ℓ.

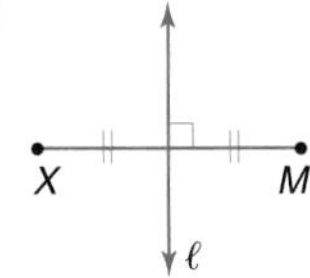

Copy each figure. Then draw the reflection image over line ℓ.

26.

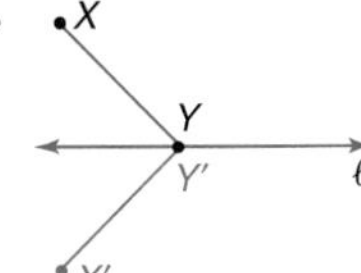

27.

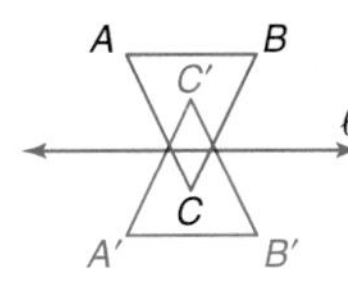

- name and draw translation images of figures with respect to parallel lines (Lesson 13–6)

$\triangle XYZ$ is the translation image of $\triangle ABC$ with respect to parallel lines ℓ and m.

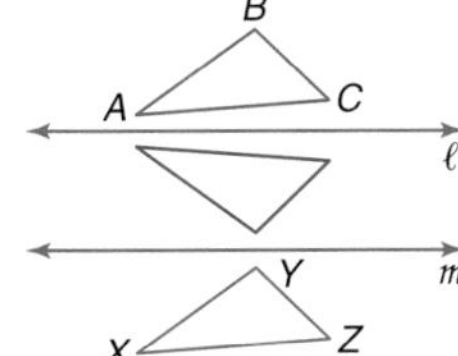

Copy each figure. Then find the translation image of each geometric figure with respect to parallel lines ℓ and m.

28.

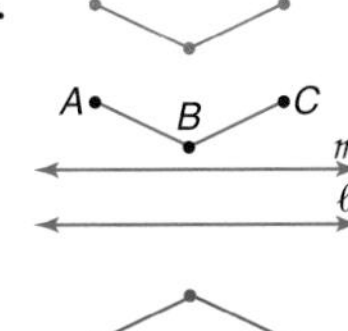

29.

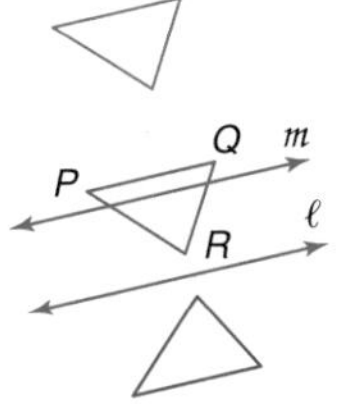

- name and draw rotation images of figures with respect to intersecting lines (Lesson 13–7)

$\triangle RST$ is the rotation image of $\triangle DEF$ with respect to lines r and t.

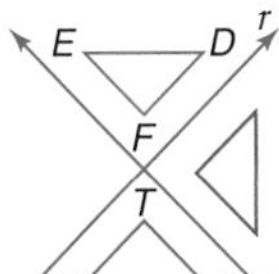

Refer to the figure below for Exercises 30–31. 30. 140

30. Find the measure of the angle of rotation of the rotation image with respect to lines t and m.
31. Copy the figure at the right. Then draw the rotation image with respect to lines t and m.

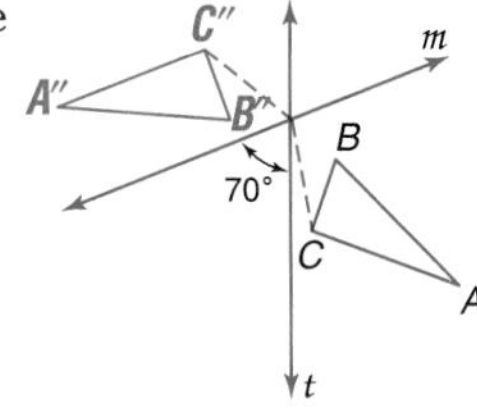

Additional Answers

11. the interior of a circle with center at the given point and radius 6 cm
12. a plane that is the perpendicular bisector of the segment joining the two points
13. the circumcenter; the point where the three perpendicular bisectors of the segments joining *A*, *B*, and *C* intersect
18. the four points that are the intersection of ⊙*P* with radius 2 cm and the bisector of the angles formed by *j* and *k*

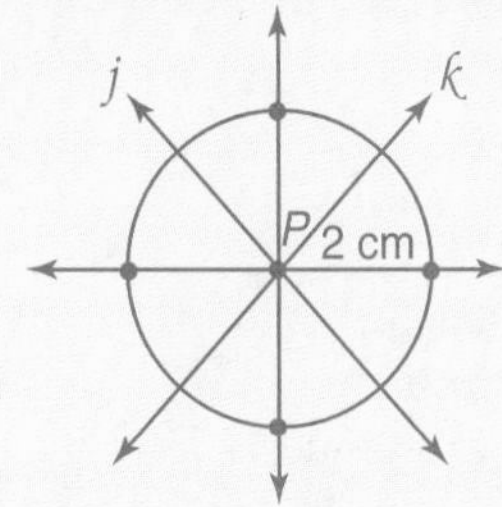

19. the circle that is the intersection of spheres *P* and *Q* with radii 3 cm

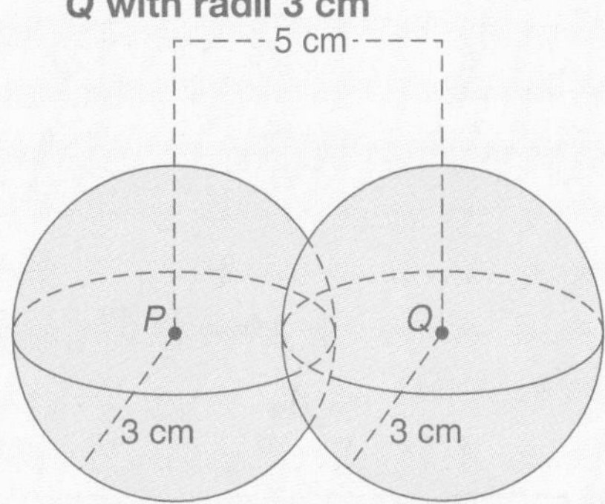

20. the point of intersection between the perpendicular bisector of the two points and the line parallel to and half-way between the two parallel lines

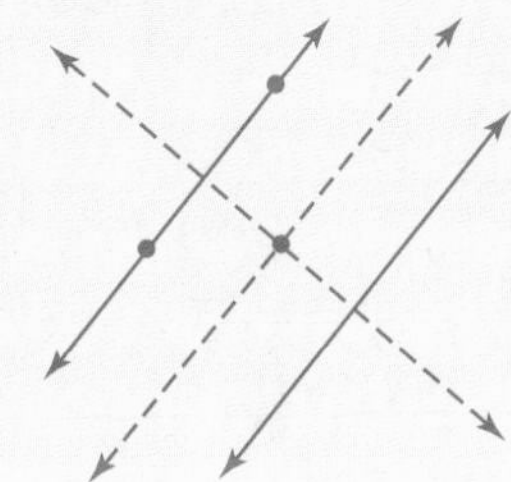

Applications and Problem Solving Encourage students to work through the exercises in the Applications and Problem Solving section to strengthen their problem-solving skills.

Additional Answers

34. The stake should be a perpendicular distance of at least 18 ft from, and all the way around, the horizontal pole if it were lying on the ground.
35. Imagine the reflection of the 5 ball with respect to the line formed by one side of the billiard table, and aim for the reflection.

OBJECTIVES AND EXAMPLES

- find the dilation image for a given center and scale factor (Lesson 13–8)

$\triangle XYZ$ is the dilation image of $\triangle PQR$ with center A and scale factor of $\frac{1}{2}$. Since the absolute value of the scale factor is less than 1, the dilation is a reduction.

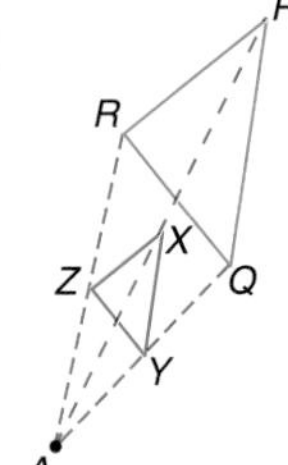

REVIEW EXERCISES

Refer to the figure at the right for Exercises 32 and 33.

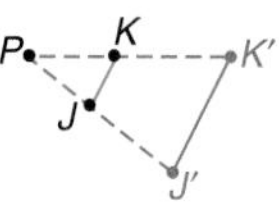

32. Copy the figure at the right. Then draw the dilation image with a scale factor of 3 and center at P.
33. Is the image an *enlargement*, a *reduction*, or a *congruence transformation*? Explain your answer. **enlargement; $|3| > 1$**

APPLICATIONS AND PROBLEM SOLVING

34. **Pet care** Perez has 2 dogs, Toby and Gretel. Part of each day, he ties the dogs in the backyard. Toby's 5-foot leash is tied to a long horizontal pole 4 feet above the ground so that it can slide along the pole. Gretel is tied to a stake with a 15-foot leash that can pivot around the stake. Explain how Perez should place the stake so that Toby and Gretel cannot tangle their leashes or get into a fight. (Lesson 13–3) **See margin.**

35. **Recreation** Dominique is playing billiards. How can he hit the 5 ball with the cue ball (all white) if the cue ball must first hit a side of the billiard table? Draw a diagram. (Lesson 13–5) **See margin; see Solutions Manual for diagram.**

36. **Photography** Use the photo below to describe the transformation that it depicts. (Lesson 13–5) **reflection**

37. **Manufacturing** Mr. Rex designs paddle fans for Functional Fans Inc. He designs one fan with 5 paddles. What is the measure of the angle of rotation if paddle A moves to the position of paddle B? (Lesson 13–6) **144**

A practice test for Chapter 13 is provided on page 805.

ALTERNATIVE ASSESSMENT

COOPERATIVE LEARNING PROJECT

Motion pictures In this chapter, you became familiar with several types of transformations. Imagine you are out for a walk. With your eyes looking directly ahead, you notice that each point in your visual field is moving. In fact, each point is moving from the vanishing point straight ahead on the road, toward the periphery of your visual field. You take a turn to the right. What just happened to the points in your visual field? In this project, you will write transformations for this situation.

Follow these steps to write the transformations.

- Take a walk down the hallway. Take a photograph of the view ahead at intervals of five steps. (Drawings can be used instead of photos.) Walk in a straight line until you reach a turn in the hallway. Walk through the turn slowly, taking another photograph every two steps.
- Have the photographs developed and printed. On each photograph, place the origin of the coordinates at the vanishing point.
- For the sequence of photographs prior to reaching the turn, select one point in each quadrant and assign coordinates to the points. Write an equation for the transformation that describes the motion of the points away from the origin and toward the periphery.
- For the sequence of photographs of the walk through the turn, select a few points in the field of view. Write transformations for the turning sequence.

THINKING CRITICALLY

How many points are contained in a locus?

There may be none, one, a finite number of points, or an infinite number of points.

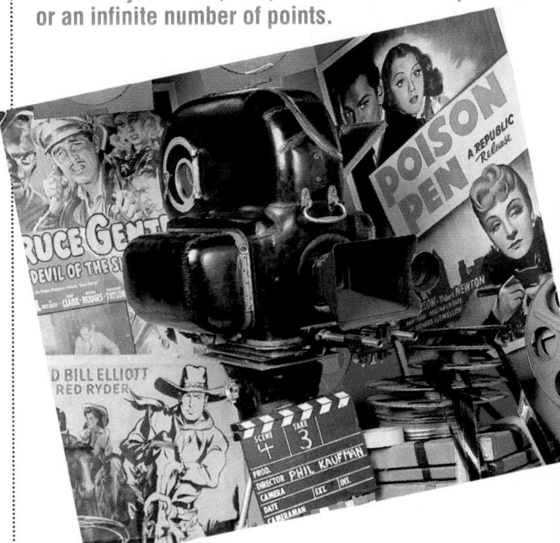

PORTFOLIO

Rent one of your favorite movies and view it. Find a sequence in the film where there are interesting camera angles and changes of camera angles. Write a short essay describing the sequence and the types of transformations for each camera movement. Keep this essay in your portfolio.

SELF EVALUATION

In this text, you have seen that there are many connections between geometry and art. Geometry is used in perspective drawing, landscape design, architecture, sculpture, animation, and other applications. Some people find they are a natural at art, while others find that they are a natural at geometry.

Assess Yourself At which are you better, geometry or art? Did your skill at one help you learn and appreciate the other? What skills did you use from one to help you learn the other?

Assessment and Evaluation Masters, pp. 350, 361

13 NAME__________ DATE ________

Chapter 13 Performance Assessment

Instructions: ***Demonstrate your knowledge by giving a clear, concise solution to each problem. Be sure to include all relevant drawings and to justify your answers. You may show your solution in more than one way or investigate beyond the requirements of the problem.***

1. A treasure map indicates that a treasure chest is buried 10 ft from a fence and 30 ft from an oak tree. There is more than one oak tree in the vicinity.
   - **a.** Describe the locus of points 10 ft from a straight fence.
   - **b.** Describe the locus of points 30 ft from a tree.
   - **c.** Draw the loci that satisfy the conditions given on the map for different locations of the tree.
2. **a.** Draw a triangle ABC and a line ℓ intersecting the triangle. Draw the reflection image of $\triangle ABC$ with respect to ℓ. Label the reflection as $\triangle A'B'C'$.
   - **b.** Draw a triangle DEF. Use a composite of reflections to produce a translation image $\triangle D''E''F''$.
   - **c.** Draw a triangle GHI and two lines m and n that intersect at an angle of 30°. Use a composite of reflections with respect to m and n to produce a rotation image of $\triangle GHI$.
   - **d.** What is the angle of rotation for $\triangle GHI$? Explain your method of solution.
   - **e.** Draw a triangle JKL. Choose a scale factor k and a point B outside of $\triangle JKL$. With center B and the scale factor k, draw the dilation image of $\triangle JKL$.

Scoring Guide
Chapter 13
Performance Assessment

Level	Specific Criteria
3 Superior	• Shows thorough understanding of the concepts of *a locus in a plane* and *reflection, translation, rotation, and dilation images.* • Uses appropriate strategies to solve problems. • Computations are correct. • Written explanations are exemplary. • Diagrams are accurate and appropriate. • Goes beyond requirements of some or all problems.
2 Satisfactory, with Minor Flaws	• Shows understanding of the concepts of *a locus in a plane* and *reflection, translation, rotation, and dilation images.* • Uses appropriate strategies to solve problems. • Computations are mostly correct. • Written explanations are effective. • Diagrams are mostly accurate and appropriate. • Satisfies all requirements of some or all problems.
1 Nearly Satisfactory, with Serious Flaws	• Shows understanding of most of the concepts of *a locus in a plane* and *reflection, translation, rotation, and dilation images.* • May not use appropriate strategies to solve problems. • Computations are mostly correct. • Written explanations are satisfactory. • Diagrams are mostly accurate and appropriate. • Satisfies most requirements of some or all problems.
0 Unsatisfactory	• Shows little or no understanding of the concepts of *a locus in a plane* and *reflection, translation, rotation, and dilation images.* • May not use appropriate strategies to solve problems. • Computations are incorrect. • Written explanations are not satisfactory. • Diagrams are not accurate and appropriate. • Does not satisfy requirements of some or all problems.

Alternative Assessment

The Alternative Assessment section provides students with the opportunity to assess their own work by thinking critically, working with others, keeping a portfolio, and honestly evaluating their own progress. For more information on alternative forms of assessment, see *Alternative Assessment in the Mathematics Classroom,* one of the titles in the Glencoe Mathematics Professional Series.

Performance Assessment

Performance Assessment tasks for this chapter are included in the *Assessment and Evaluation Masters.* A scoring guide is also provided.

Using the COLLEGE ENTRANCE EXAM PRACTICE

These two pages review the skills and concepts presented in Chapters 1–13. This review is formatted to reflect new trends in college entrance testing.

Assessment and Evaluation Masters, pp. 355–356

13 NAME______________________ DATE ____________

Chapter 13 Cumulative Review

For Questions 1–6, determine whether the statement is true or false.

1. An obtuse triangle has exactly one obtuse angle. 1. true
2. A segment can have two distinct midpoints. 2. false
3. Two triangles are congruent if two sides of one triangle are congruent to two corresponding sides of the other triangle. 3. false
4. If two rectangles have equal perimeters, they must have equal areas. 4. false
5. The translation image of a triangle is congruent to its preimage. 5. true
6. Two lines are skew if and only if they do not intersect and are coplanar. 6. false
7. In $\triangle ABC$, $m\angle A = 50$ and $m\angle B = 55$. List the sides of $\triangle ABC$ in order from the shortest to the longest. 7. $\overline{BC}, \overline{AC}, \overline{AB}$

For Questions 8 and 9, refer to right $\triangle ABC$.

8. Find AB. 8. $10\sqrt{3}$
9. Find AC. 9. 20

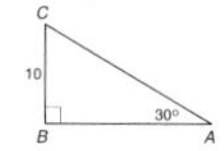

For Questions 10 and 11, refer to right $\triangle LMN$. Round to the nearest hundredth.

10. Find $\tan \angle L$. 10. 0.42
11. Find $\cos \angle N$. 11. 0.38
12. At a certain time of day, the angle of elevation of the sun is 35. To the nearest tenth, find the length of the shadow cast by a building that is 115 ft tall. 12. 164.2
13. In $\triangle ABC$, $a = 10$, $m\angle A = 51$, and $m\angle B = 63$. Use the Law of Sines to solve the triangle. Round to the nearest tenth. 13. $m\angle C = 66$, $c \approx 11.8$, $b \approx 11.5$
14. Determine the center and the radius of the circle whose equation is $x^2 + (y + 2)^2 = \frac{9}{4}$. 14. $(0, -2), \frac{3}{2}$

13 NAME______________________ DATE ____________

Chapter 13 Cumulative Review (continued)

For Questions 15–17, find the value of x. Assume that all lines that appear to be tangent are tangent.

15.

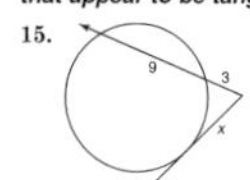

16.

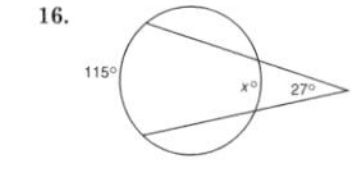

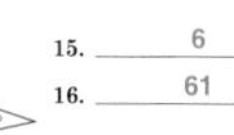

15. 6
16. 61

17.

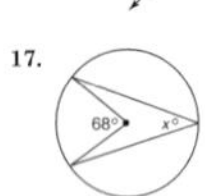

17. 34

For Questions 18 and 19, round to the nearest tenth.

18. Find the area and circumference of a circle with a diameter of 8 cm. 18. 50.3 cm², 25.1 cm
19. Find the volume of a right circular cone with radius of 6 in. and slant height of 10 in. 19. 301.6 in³

For Questions 20 and 21, find the volume of the figure. Round to the nearest tenth.

20.

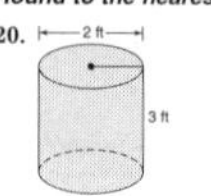

21.

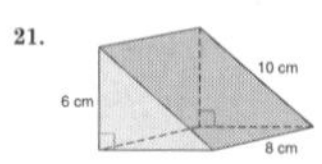

20. 9.4 ft³
21. 192 cm³

22. Find the total surface area of the figure in Question 21. 22. 240 cm²
23. The endpoints of a line segment are at $(-3, 5, 7)$ and $(5, -1, 5)$. Determine the length and the midpoint of the segment. 23. $2\sqrt{26}$ or ≈ 10.2, $(1, 2, 6)$
24. Two lines ℓ and m intersect to form an angle with a measure of 76. Find the measure of the angle of rotation for a rotation with respect to ℓ and m. 24. 152
25. Describe the locus of all points in a plane that are equidistant from two given points. 25. ⊥ bisector of segment connecting points

COLLEGE ENTRANCE EXAM PRACTICE

CHAPTERS 1–13

SECTION ONE: MULTIPLE CHOICE

There are eight multiple-choice questions in this section. After working each problem, write the letter of the correct answer on your paper.

1. Ten segments of equal length form the figure below. If the perimeter of $\triangle XYZ$ is eight inches, what is the perimeter of octagon $RSTUVWYZ$? B

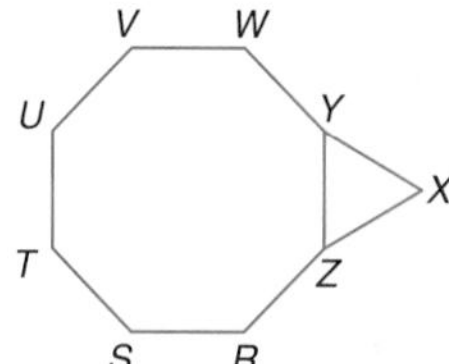

A. $16\frac{2}{3}$ inches

B. $21\frac{1}{3}$ inches

C. 16 inches

D. $18\frac{2}{3}$ inches

2. How many hours equal 282 minutes? D

A. $4\frac{1}{4}$ hours

B. $4\frac{2}{3}$ hours

C. $4\frac{4}{5}$ hours

D. $4\frac{7}{10}$ hours

3. Three vertices of a parallelogram are at (2, 1), (−1, −3), and (6, 4). Which of the following could be the coordinates of the remaining vertex? B

A. (0, 3)

B. (3, 0)

C. (−1, 4)

D. (6, 1)

4. Find the value of p if the ratio of p to q is 2 to 3 and the ratio of q to r is 6 to 11. A

A. 4

B. −4

C. 8

D. −8

5. 8(624) + 624 is equivalent to which of the following? B

A. 5(624) + 3(624)

B. 5(624) + 4(624)

C. 4(624) + 6(624)

D. 3(624) + 9(624)

6. Which of the following represents the solution set of the inequality $-3 - x < 2x < 9 - x$? A

A. $\{x \mid -1 < x < 3\}$

B. $\{x \mid 1 < x < 3\}$

C. $\{x \mid -3 < x < 1\}$

D. $\{x \mid -3 < x < -1\}$

7. Solve $\frac{a}{3a + 6} - \frac{a}{5a + 10} = \frac{2}{5}$. B

A. −2

B. −3

C. −3 and −2

D. 3 and 2

8. Find the decimal form of $\frac{2}{1000} + \frac{3}{10} + \frac{4}{100}$. A

A. 0.342

B. 0.432

C. 0.234

D. 0.243

Standardized Test Practice Questions are also provided in the *Assessment and Evaluation Masters*, p. 354.

A more traditional cumulative review is provided in the *Assessment and Evaluation Masters*, pp. 355–356.

SECTION TWO: SHORT ANSWER

This section contains seven questions for which you will provide short answers. Write your answer on your paper.

9. One angle of a triangle measures 68°. The measures of the other two angles are in a 1 to 3 ratio. Find the measures of the other two angles. **28, 84**

10. If the center of square *ABCD* is at the origin and the sides of the square are parallel to the *x*- and *y*-axes, find the position of the four vertices if the square is rotated 180° about the *y*-axis and then rotated 180° about the *x*-axis. **same as *ABCD***

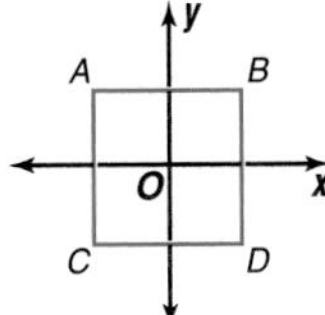

11. Simplify $\frac{3}{4-x} \div \frac{x}{x^2-16}$. **$\frac{-3(x+4)}{x}$**

12. Find the maximum number of common tangents that can be drawn to any two circles. **4**

13. Find the solution of $\sqrt{y^2 - 4y + 9} = y - 1$. **{4}**

14. Find *x* if *m* is parallel to *n*. **72°**

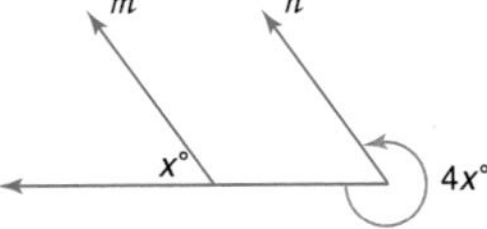

15. In the following diagram, $MN = NO = OP = PQ = QR = 1$. Find the length of $\overline{MR}$. **$\sqrt{5}$**

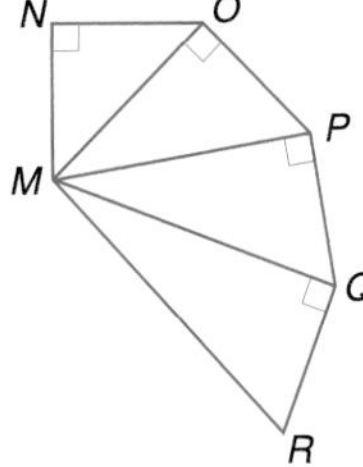

SECTION THREE: COMPARISON

This section contains five comparison problems that involve comparing two quantities, one in column A and one in column B. In certain questions, information related to one or both quantities is centered above them. All variables used represent real numbers.

Compare quantities A and B below.

- Write A if quantity A is greater.
- Write B if quantity B is greater.
- Write C if the two quantities are equal.
- Write D if there is not enough information to determine the relationship.

	Column A	Column B	
16.	$\sqrt{3^4}$	9	**C**
17.	a given chord in a given circle	the radius of the same circle	**D**

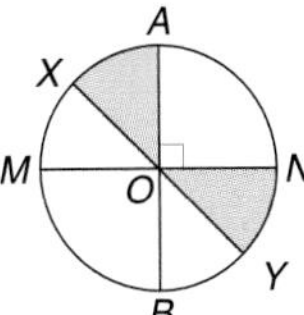

O is the center of a circle with $\overline{MN}$, $\overline{AB}$, and $\overline{XY}$ as diameters, **A**

	Column A	Column B
18.	30% of the circle's area	shaded part of the circle

$a @ b$ is defined to be $\frac{b}{a-b}$ and $a \neq b$. **D**

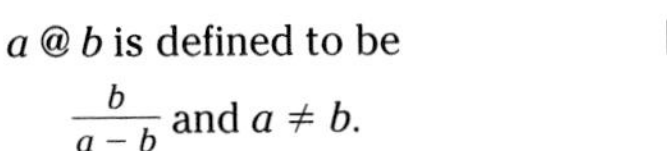

	Column A	Column B
19.	$a @ b$	$b @ a$

$Q = \{1, 2, 3, 4, 5, 6, 7, 8, 9\}$ **A**

	Column A	Column B
20.	the product of the odd integers in *Q*	the product of the even integers in *Q*

TEST-TAKING TIP

Remind students that there is more to taking a test than knowing the subject matter. Present students with this summary of the Test-Taking Tips presented in this book.

- **Getting Ready** Before the test, familiarize yourself with its form and content. Find out if you are penalized for guessing.
- **Managing Your Time** Pace yourself during the test. Develop a plan of attack and stick to it. If you finish early, check your work.
- **Quantitative Comparisons** Use the rules of inequalities. Substitute values for variables and evaluate. Examine a comparison from different sets of numbers such as positive, negative, and zero.
- **Geometric Figures** Do not assume the picture is accurate. Remember what you can and cannot assume.

STUDENT HANDBOOK

For Additional Assessment

For Reference

Additional Answers for Lesson 1-1

5–7.

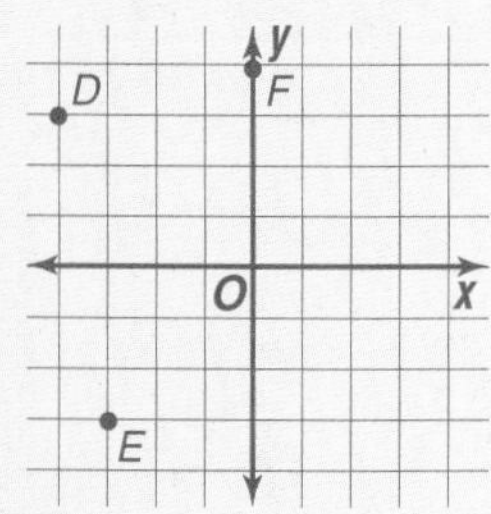

Additional Answers for Lesson 1-2

4.

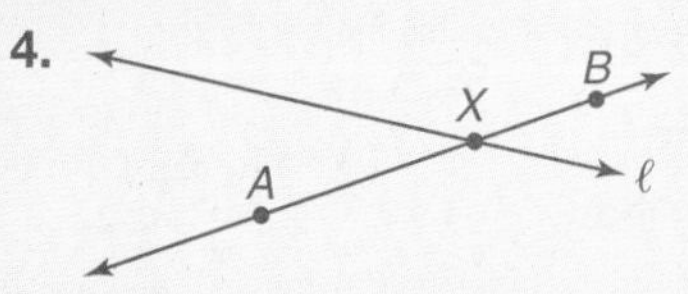

5.

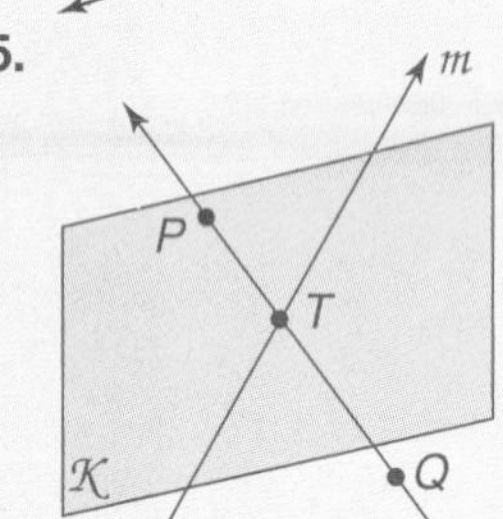

EXTRA PRACTICE

EXTRA PRACTICE

Lesson 1-1 **Write the ordered pair for each point shown at the right.**

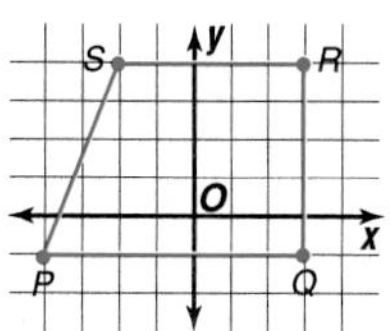

1. P (−4, −1)
2. Q (3, −1)
3. R (3, 4)
4. S (−2, 4)

Graph each point on the same coordinate plane. 5–7. See margin.

5. $D(-4, 3)$
6. $E(-3, -3)$
7. $F(0, 4)$

Points $R(-1, -5)$ and $S(3, 3)$ lie on the graph of $y = 2x - 3$. Determine whether each point is collinear with R and S.

8. $T(0, -3)$ collinear
9. $U(-0.5, -4)$ collinear
10. $V(2, -1)$ noncollinear

Lesson 1-2 **Refer to the figure at the right to name each of the following.**

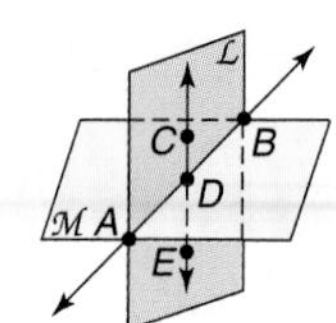

1. the intersection of planes $\mathcal{M}$ and $\mathcal{L}$ $\overleftrightarrow{AB}$
2. a point collinear with points C and E D
3. a plane containing line $\overleftrightarrow{CE}$ ℒ

Draw and label a figure for each relationship. 4–5. See margin.

4. Lines $\overleftrightarrow{AB}$ and ℓ both contain point X.
5. Lines $\overleftrightarrow{PQ}$ and m, and plane $\mathcal{K}$ intersect at T.

Refer to figure at the right to answer each question.

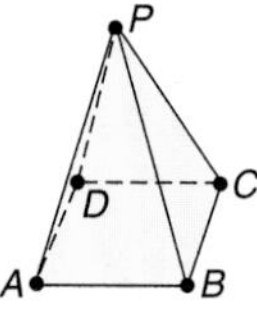

6. The flat surfaces of the figure are called *faces*. Name the five planes that contain the faces of the pyramid. planes *PAB, PBC, PCD, PAD, ABC*
7. Name three lines that intersect at B. $\overleftrightarrow{AB}$, $\overleftrightarrow{CB}$, $\overleftrightarrow{PB}$

Lesson 1-3 **Find the perimeter and area of each rectangle.**

1. 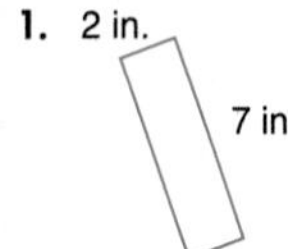

18 in., 14 in^2

2. 3.4 m by 3.4 m — 13.6 m, 11.56 m^2

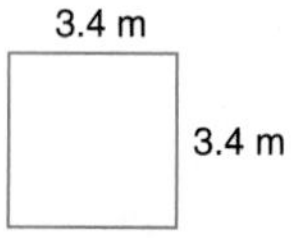

3. 12 cm by 3 cm — 30 cm, 36 cm^2

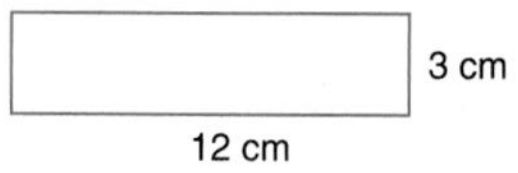

Find the missing measure in each formula if $P = 2\ell + 2w$ and $A = \ell w$.

4. $\ell = 6, w = 4.5, A = ?$ 27
5. $\ell = 12, w = 8, P = ?$ 40
6. $P = 17, \ell = 5.5, w = ?$ 3
7. $A = 91, w = 7, \ell = ?$ 13

Find the maximum area for the given perimeter of a rectangle. State the length and width of the rectangle. For Exercises 8–10, $\ell = w$.

8. 36 cm 81 cm^2, 9 cm
9. 68 in. 289 in^2, 17 in.
10. 118 yd 870.25 yd^2, 29.5 yd

Lesson 1–4 **Refer to the number line below to find each measure.**

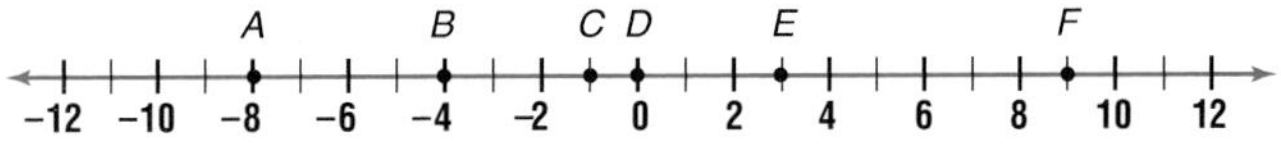

1. CD 1
2. BF 13
3. CF 10
4. EB 7

Refer to the coordinate plane at the right to find each measure. Round your answers to the nearest hundredth.

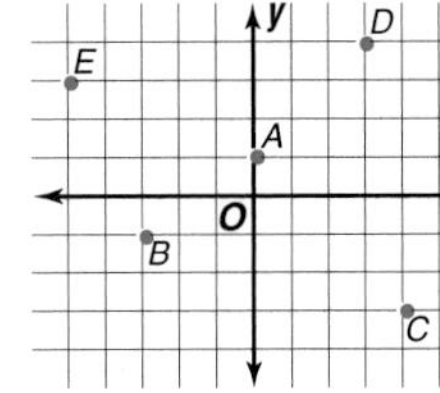

5. BA 3.61
6. ED 8.06
7. AC 5.66
8. CD 7.07

Use the Pythagorean Theorem to find the missing length x in each right triangle.

9.
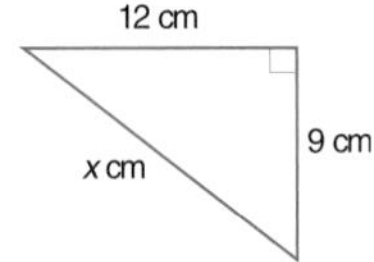

15 cm

10.
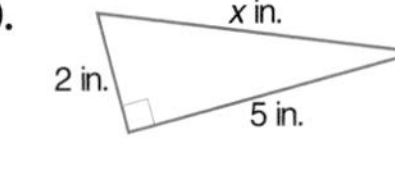

$\sqrt{29} \approx 5.39$ in.

11.
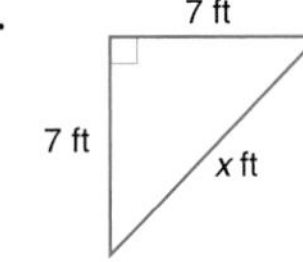

$7\sqrt{2} \approx 9.90$ ft

Lesson 1–5 **Refer to the number line at the right to find the coordinate of the midpoint of each segment.**

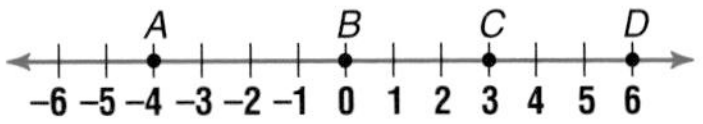

1. $\overline{BD}$ 3
2. $\overline{AB}$ −2
3. $\overline{BC}$ $1\frac{1}{2}$

Given three points A, B, and M, and M is the midpoint of $\overline{AB}$, find the coordinates of the third point.

4. $A(-9, 4), B(3, -2)$ $M(-3, 1)$
5. $M(-1, 8), B(2, 6)$ $A(-4, 10)$
6. $A(-5, -4), M(1, -5)$ $B(7, -6)$
7. $B(7, 3), A(2, 4)$ $M(4.5, 3.5)$

In the figure at the right, $\overline{CX}$ bisects $\overline{AB}$ at X, and $\overleftrightarrow{CD}$ bisects $\overline{XB}$ at Y. For each of the following, find the value of x and the measure of the indicated segment.

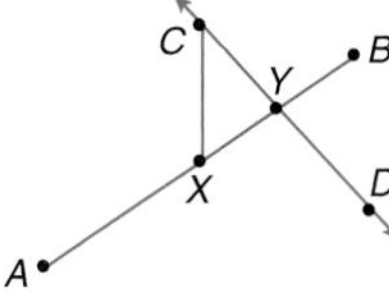

8. $AX = 2x + 11, XB = 4x - 5; \overline{AB}$ 8; 54
9. $YB = 23 - 2x, XY = 2x + 3; \overline{AB}$ 5; 52
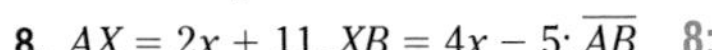
10. $AB = 5x - 4, XY = x + 1; \overline{AX}$ 8; 18

Lesson 1–6 **In the figure at the right, $\overrightarrow{RQ}$ and $\overrightarrow{RS}$ are opposite rays, and $\overrightarrow{RU}$ bisects $\angle QRT$.**

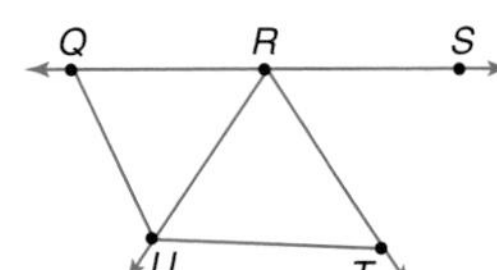

1. Name two congruent angles. $\angle QRU, \angle TRU$
2. Name three angles that have U as a vertex. $\angle QUT, \angle QUR, \angle TUR$
3. List all the angles that have $\overrightarrow{RT}$ as a side. $\angle SRT, \angle TRU, \angle TRQ$
4. If $m\angle SRT = 17$ and $m\angle TRU = 80$, what is the measure of $\angle SRU$? 97
5. If $m\angle QRU = 4x - 3$ and $m\angle TRU = 2x + 23$, find $m\angle QRU$. 49
6. Given that $m\angle SRU = 3y + 8$ and $m\angle URQ = 2y + 7$, find $m\angle URQ$. 73
7. Find $m\angle URT$ if $m\angle URQ = 3n + 11$ and $m\angle QRT = 9n - 14$. 47

Extra Practice 765

Additional Answers for Lesson 2-1

1. Sample example:

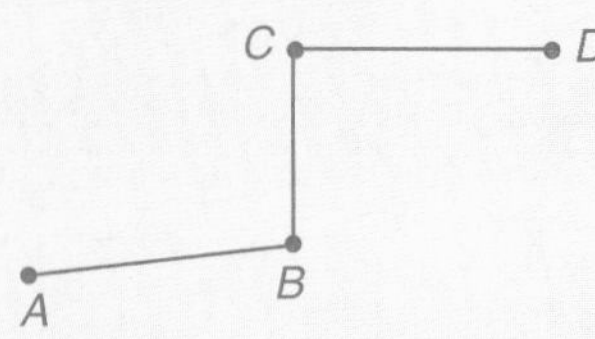

2. The square of a real number is always a nonnegative number.
3.

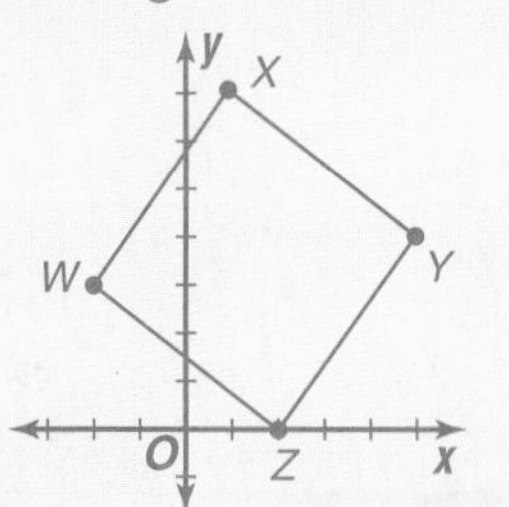

4.

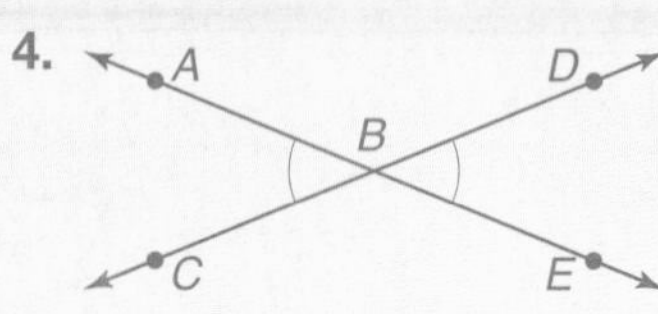

5.

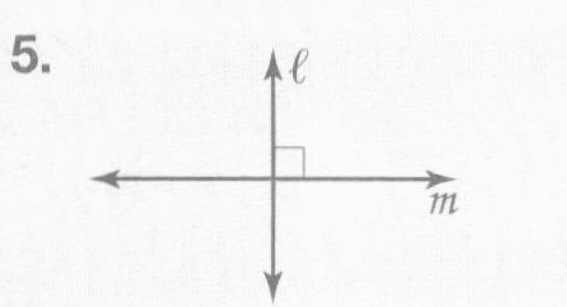

6.

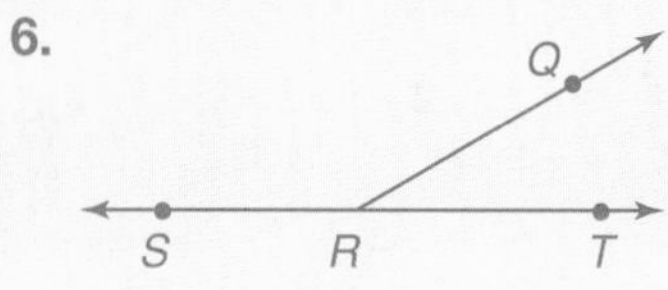

Additional Answers for Lesson 2-2

1. Hypothesis: a container holds 32 oz
   Conclusion: it holds a qt
2. Hypothesis: a candy bar is a Milky Way®
   Conclusion: it contains caramel
6. Converse: If the distance of a race is about 6.2 mi, then it is 10 km; true.
   Inverse: If the distance of a race is not 10 km, then it is not about 6.2 mi; true.
   Contrapositive: If the distance of a race is not about 6.2 mi, then it is not 10 km; true.

EXTRA PRACTICE

Lesson 1-7 **Refer to the figure at the right for Exercises 1–6.**

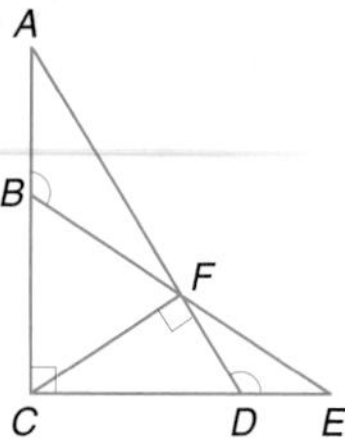

1. Name a pair of vertical angles. $\angle AFB$ and $\angle EFD$
2. Which angle is complementary to $\angle ECF$? $\angle ACF$
3. Name a pair of angles that are congruent and supplementary. $\angle ACE$ and $\angle CFD$
4. Identify $\angle ABF$ and $\angle CBF$ as adjacent, vertical, complementary, supplementary, or as a linear pair. List all possibilities.
5. Can you assume that $\angle ABF \cong \angle EDF$? Explain.
6. Does $\angle EDF$ appear to be obtuse, acute, or right? obtuse
7. The measure of an angle is one-third the measure of its supplement. Find the measure of the angle. 45
8. Suppose $\angle C$ and $\angle D$ are complementary. Find $m\angle C$ and $m\angle D$ if $m\angle C = 3x + 5$ and $m\angle D = 4x - 6$. 44, 46

4. adjacent, supplementary, linear pair
5. Yes; the markings on the angles indicate that they are congruent.

Lesson 2-1 **Determine if each conjecture is *true* or *false* based on the given information. Explain your answer.** 1–3. See margin for explanation.

1. **Given:** $\overline{AB}$, $\overline{BC}$, and $\overline{CD}$
   Conjecture: *A*, *B*, *C*, and *D* are collinear. false
2. **Given:** *x* is a real number.
   Conjecture: x^2 is a nonnegative number. true
3. **Given:** $W(-2, 3)$, $X(1, 7)$, $Y(5, 4)$, and $Z(2, 0)$
   Conjecture: *WXYZ* is a square. true

Write a conjecture based on the given information. If appropriate, draw a figure to illustrate your conjecture. 4–6. See margin for drawings.

4. **Given:** $\angle ABC$ and $\angle DBE$ are vertical angles. $\angle ABC \cong \angle DBE$
5. **Given:** Lines ℓ and m intersect to form right angles. $\ell \perp m$
6. **Given:** $\angle QRS$ and $\angle QRT$ form a linear pair. $\angle QRS$ and $\angle QRT$ are supplementary.

Lesson 2-2 **Identify the hypothesis and conclusion of each conditional statement.**

1. If a container holds 32 ounces, then it holds a quart. 1–2. See margin.
2. If a candy bar is a Milky Way®, then it contains caramel.

Write each conditional in if-then form.

3. Right angles are congruent. If two angles are right angles, then they are congruent.
4. A car has four wheels. If a vehicle is a car, then it has four wheels.
5. A triangle contains exactly three angles. If a figure is a triangle, then it contains exactly three angles.

Write the converse, inverse, and contrapositive of each conditional. Determine if the converse, inverse, and contrapositive are *true* or *false*. If false, give a counterexample. 6–7. See margin.

6. If the distance of a race is 10 kilometers, then it is about 6.2 miles.
7. An apple is a fruit.

Additional Answer for Lesson 2-2

7. Converse: If a food is a fruit, then it is an apple; false. Sample counterexample: Pears are fruit, but they are not apples.
   Inverse: If a food is not an apple, then it is not a fruit; false. Sample counterexample: Pears are not apples, but they are fruit.
   Contrapositive: If a food is not a fruit, then it is not an apple; true.

Lesson 2-3 **Determine if statement (3) follows from statements (1) and (2) by the Law of Detachment or the Law of Syllogism. If it does, state which law was used. If it does not, write *invalid*.**

1. (1) All pilots must pass a physical examination.
   (2) Kris Thomas must pass a physical examination.
   (3) Kris Thomas is a pilot. invalid
2. (1) If a student is enrolled at Lyons High, then the student has an ID number.
   (2) Joel Nathan is enrolled at Lyons High.
   (3) Joel Nathan has an ID number. yes; Law of Detachment

Determine if a valid conclusion can be reached from the two true statements using the Law of Detachment or the Law of Syllogism. If a valid conclusion is possible, state it and the law that is used. If a valid conclusion does not follow, write *no conclusion*. 3. Basalt was formed by volcanoes; Law of Syllogism

3. (1) Basalt is an igneous rock.
   (2) Igneous rocks were formed by volcanoes.
4. (1) If a quadrilateral is a rectangle, then it has four right angles.
   (2) A rectangle has diagonals that are congruent. no conclusion

Lesson 2-4 **Name the property of equality that justifies each statement.**

1. If $AB + BC = AC$ and $AC = EF + GH$, then $AB + BC = EF + GH$. Transitive Property (=)
2. If $x + y = 9$ and $x - y = 12$, then $2x = 21$. Addition Property (=)

Copy the proof. Then name the property that justifies each statement.

3. **Given:** $x - 1 = \frac{x - 10}{-2}$

   Prove: $x = 4$

Statements	Reasons
1. $x - 1 = \frac{x - 10}{-2}$	1. ? Given
2. $-2(x - 1) = x - 10$	2. ? Multiplication Property (=)
3. $-2x + 2 = x - 10$	3. ? Distributive Property (=)
4. $12 = 3x$	4. ? Addition Property (=)
5. $4 = x$	5. ? Division Property (=)
6. $x = 4$	6. ? Symmetric Property (=)

Lesson 2-5 **Justify each statement with a property from algebra or a property of congruent segments.**

1. If $2MN = TS$, then $MN = \frac{1}{2}TS$. Division Property (=)
2. If $AN - 8 = IN - 8$, then $AN = IN$. Addition Property (=)
3. If $EF = GH$ and $GH = JK$, then $EF = JK$. Transitive Property (=) or Substitution Property (=)

Write the given and prove statements you would use to prove the theorem. Draw a figure if applicable. 4–5. See margin.

4. If two angles are vertical, then they are congruent.
5. The diagonals of a rectangle are congruent.

Additional Answers for Lesson 2-5

4. Given: $\angle 1$ and $\angle 2$ are vertical angles.
   Prove: $\angle 1 \cong \angle 2$

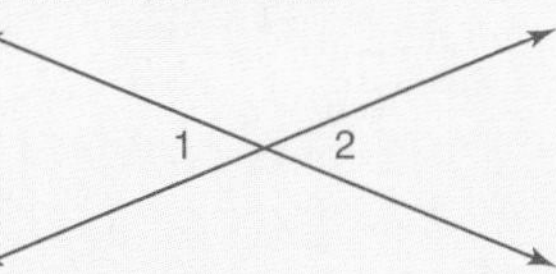

5. Given: $ABCD$ is a rectangle.
   Prove: $\overline{AC} \cong \overline{BD}$

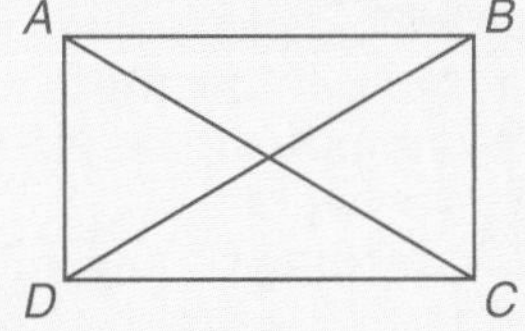

Extra Practice **767**

Additional Answer for Lesson 2-6

5. $\angle VWQ$ and $\angle QWS$, $\angle VWR$ and $\angle RWS$, $\angle QWR$ and $\angle RWT$, $\angle QWS$ and $\angle SWT$, $\angle SWT$ and $\angle TWV$, $\angle SWU$ and $\angle UWV$, $\angle TWU$ and $\angle UWQ$, $\angle TWV$ and $\angle VWQ$

Additional Answers for Lesson 3-1

5. Using n as the transversal, $\angle 6$ and $\angle 11$ are on opposite sides of the transversal and between lines ℓ and m.
6. $\angle 4$ and $\angle 9$ are corresponding angles, not alternate exterior angles.
7. Using m as the transversal, $\angle 7$ and $\angle 11$ are in the same relative position and therefore corresponding.

Lesson 2-6 **Complete each statement with *always*, *sometimes*, or *never*.**

1. Congruent angles are __?__ vertical angles. sometimes
2. If two angles form a linear pair, then they are __?__ supplementary. always
3. If two lines are perpendicular, they __?__ form acute angles. never

Refer to the figure at the right to answer each question.

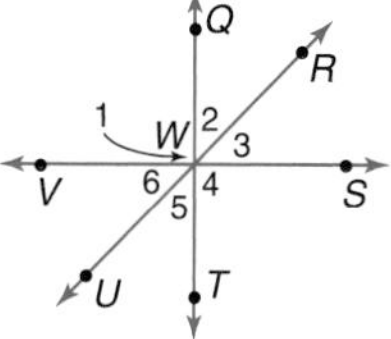

4. Name three pairs of vertical angles. $\angle 1$ and $\angle 4$, $\angle 2$ and $\angle 5$, $\angle 3$ and $\angle 6$
5. Name eight pairs of angles that form a linear pair. See margin.
6. If $m\angle 1 = 4x + 14$ and $m\angle 4 = 3x + 33$, find the value of x and $m\angle 4$. 19; 90
7. $\angle 2$ and $\angle 3$ are complementary. If $m\angle 2 = 2x - 14$ and $m\angle 3 = x + 17$, find $m\angle 3$. 46
8. If $m\angle 5 = 3x - 7$ and $m\angle UWQ = 8x$, find $m\angle UWQ$. 136

Lesson 3-1 **Describe each of the following as *intersecting*, *parallel*, or *skew*.**

1. a floor and a ceiling parallel
2. plaid fabric intersecting and parallel
3. flag pole in a football stadium and the 50-yard line on the field skew
4. train tracks parallel

Determine whether each statement is *true* or *false*, Explain Your reasoning 5–7. See margin for explanations.

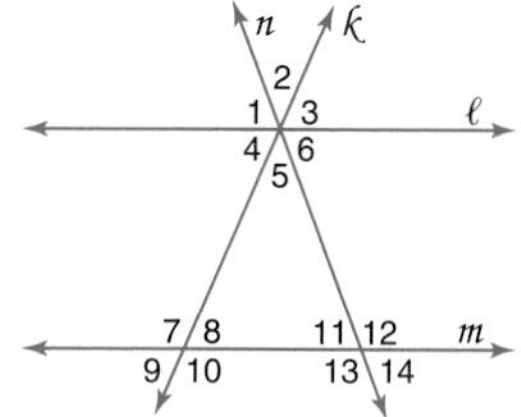

5. $\angle 6$ and $\angle 11$ are alternate interior angles. true
6. $\angle 4$ and $\angle 9$ are alternate exterior angles. false
7. $\angle 7$ and $\angle 11$ are corresponding angles. true

State the transversal that forms each pair of angles. Then identify the special angle name for each pair of angles.

8. $\angle 3$ and $\angle 8$ k; corresponding angles
9. $\angle 14$ and $\angle 7$ m; alternate exterior angles
10. $\angle 8$ and $\angle 5$ k; consecutive interior angles

Lesson 3-2 **In the figure, $m\angle 2 = 62$, $m\angle 1 = 41$, $\overline{XS} \parallel \overline{YT}$, and $\overline{SY} \parallel \overline{TZ}$. Find the measure of each angle.**

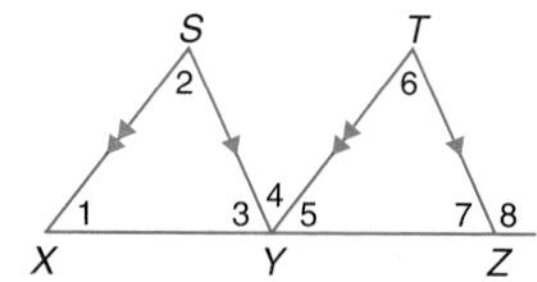

1. $\angle 4$ 62
2. $\angle 3$ 77
3. $\angle 5$ 41
4. $\angle 6$ 62
5. $\angle 7$ 77
6. $\angle 8$ 103

Find the values of x, y, and z in each figure.

7.

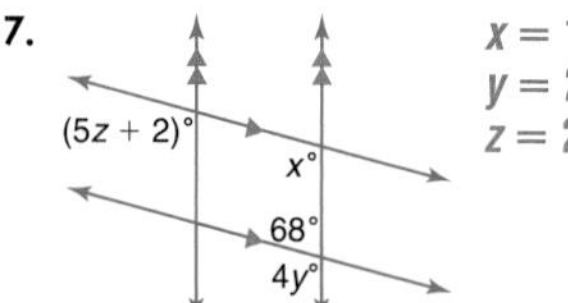

8.

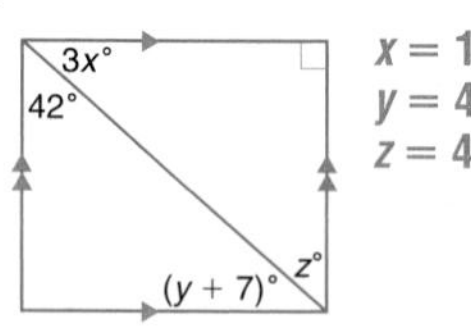

$x = 16$
$y = 41$
$z = 42$

9. $x = 33$, $y = 19$, $z = 19$

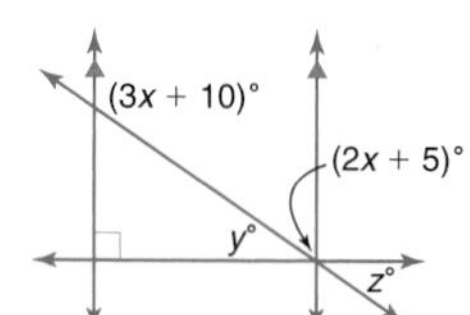

768 Extra Practice

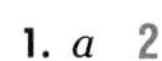
Lesson 3-3 **Determine the slope of each line named below.**

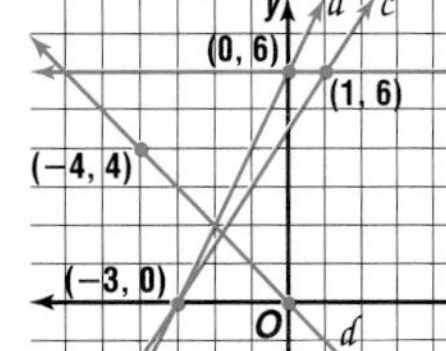

1. a 2
2. b 0
3. c $\frac{3}{2}$
4. d −1
5. any line perpendicular to b undefined
6. any line parallel to a 2
7. any line perpendicular to c $-\frac{2}{3}$

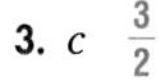

Graph the line that satisfies each description. 8–10. See margin.

8. slope = 0, passes through $P(2, 6)$
9. slope = $\frac{2}{5}$, passes through $P(0, -2)$
10. passes through $P(2, 1)$ and is parallel to $\overline{AB}$ with $A(-2, 5)$ and $B(1, 8)$

Lesson 3-4 **Given the following information, determine which lines, if any, are parallel. State the postulate or theorem that justifies your answer.** 1–4. See margin for explanation.

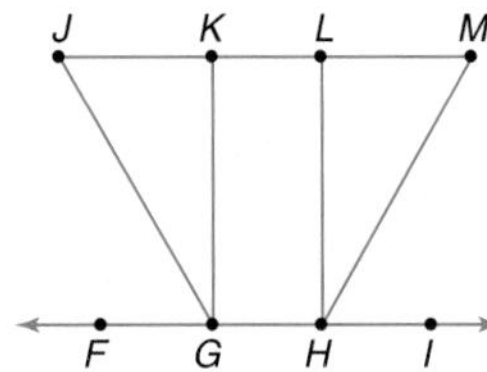

1. $\angle HLK \cong \angle GKJ$ $\overline{GK} \parallel \overline{HL}$
2. $\angle IHL \cong \angle HLK$ $\overline{HI} \parallel \overline{KL}$
3. $\angle FGJ \cong \angle KJG$ $\overline{FG} \parallel \overline{JK}$
4. $m\angle GJK + m\angle HLK = 180$ $\overline{GJ} \parallel \overline{HL}$

Find the value of x so that $\ell \parallel m$.

5.

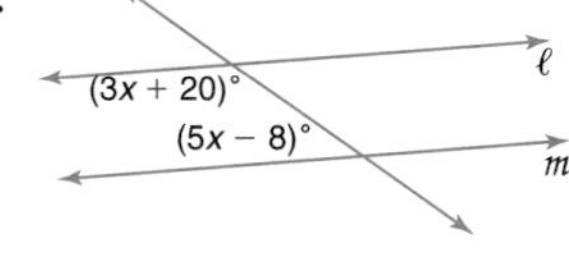

21

6.

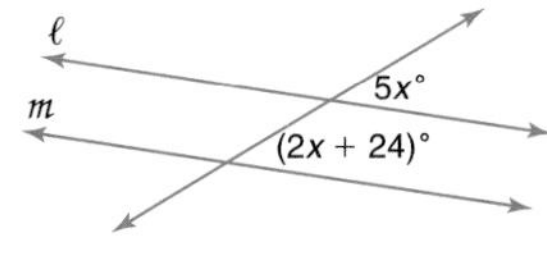

8

7.

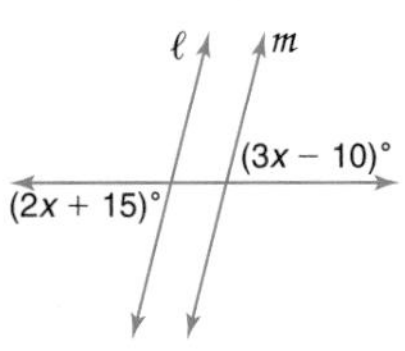

25

Lesson 3-5 **Copy each figure and draw the segment that represents the distance indicated.**

1. C to $\overline{DE}$

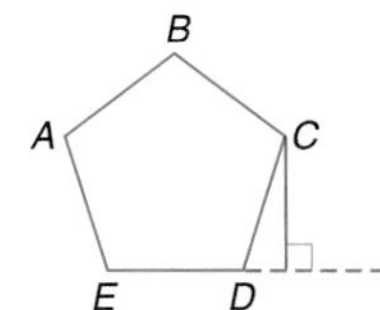

2. N to $\overline{MP}$

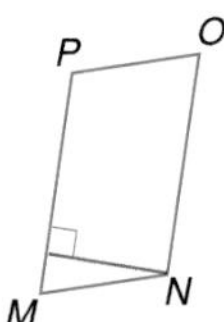

3. R to $\overline{ST}$

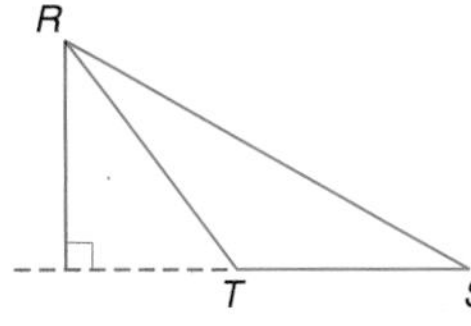

Graph each equation and plot the given ordered pair. Then construct a perpendicular segment and find the distance from the point to the line. 4–6. See margin for graphs.

4. $x = 1, (4, 5)$ 3
5. $y = -3, (-2, 4)$ 7
6. $y = -2x + 3, (-2, -3)$ $2\sqrt{5} \approx 4.47$

In the figure at the right, $\overline{QR}$, $\perp \overline{RT}$, $\overline{QR} \perp \overline{QU}$, and $\overline{VT} \perp \overline{RU}$. Name the segment whose length represents the distance between the following points and lines.

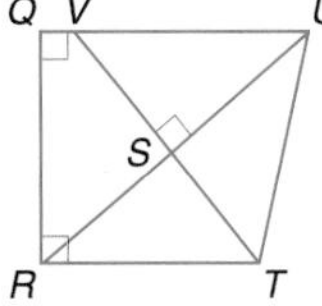

7. U to $\overline{QR}$ $\overline{QU}$
8. Q to $\overline{RT}$ $\overline{QR}$
9. V to $\overline{RU}$ $\overline{VS}$
10. T to $\overline{US}$ $\overline{TS}$

Additional Answer for Lesson 3-5

6.

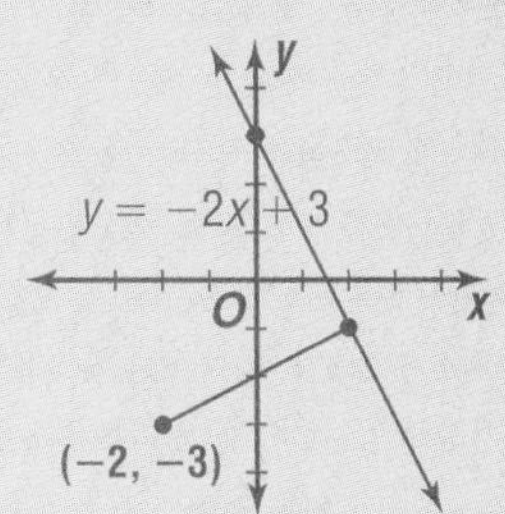

Additional Answers for Lesson 3-3

8.

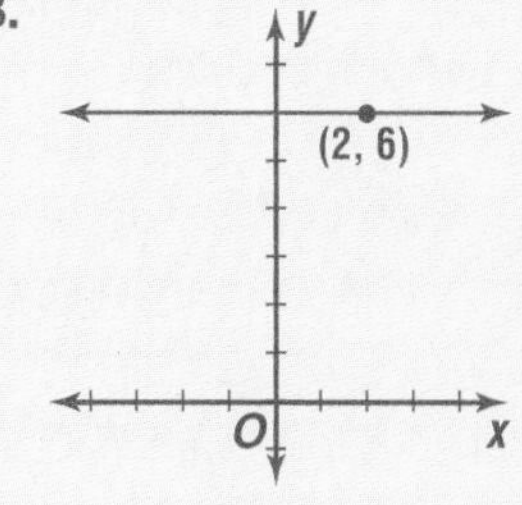

9.

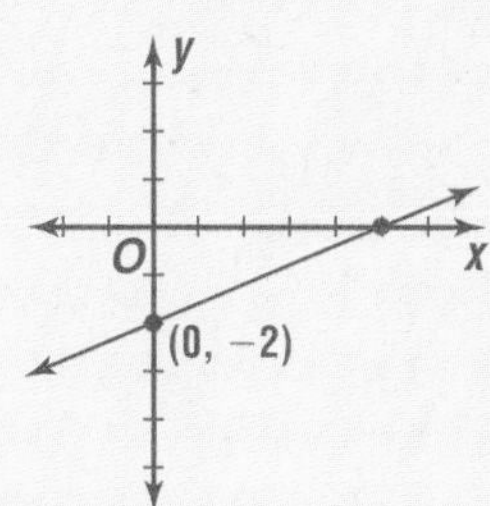

10.

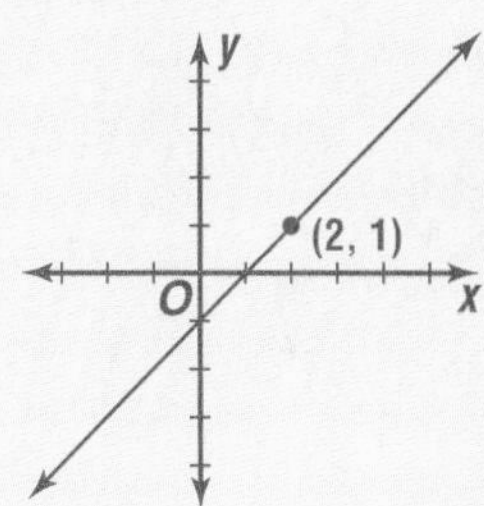

Additional Answers for Lesson 3-4

1. If ⫽̸ and corr. ∡s are ≅, then the lines are ∥.
2. If ⫽̸ and a pair of alt. int. ∡s is ≅, then the lines are ∥.
3. If ⫽̸ and a pair of alt. int. ∡s is ≅, then the lines are ∥.
4. If ⫽̸ and a pair of consecutive int. ∡s is supplementary, then the lines are ∥.

Additional Answers for Lesson 3-5

4.

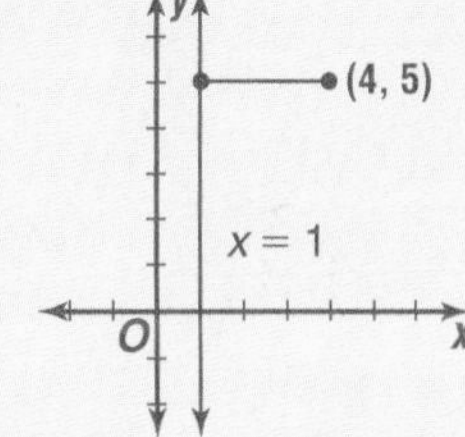

5.

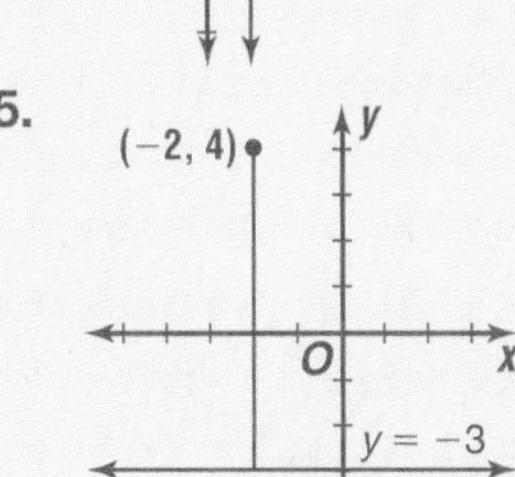
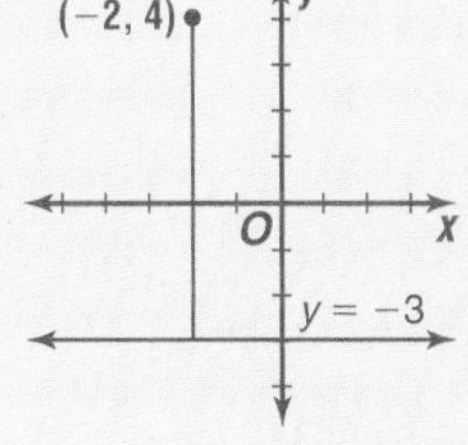

Additional Answer for Lesson 3-6

4. An arc of a great circle is the shortest path between 2 points.

Lesson 3-6 **If spherical points are restricted to be nonpolar points, decide which statements from Euclidean geometry are true in spherical geometry. If false, explain your reasoning.**

1. Given points A and B, there is a unique straight line containing A and B. true
2. Perpendicular lines divide a plane into four infinite regions.
3. Parallel lines have no point of intersection. False; in spherical geometry, there are no parallel lines.

2. False; perpendicular great circles divide a sphere into 4 finite regions.

For each property listed from plane Euclidean geometry, write a corresponding statement for non-Euclidean spherical geometry.

4. A line segment is the shortest path between two points. See margin.
5. A line has infinite length. A great circle is finite.
6. The intersecting lines intersect at one point. Intersecting great circles intersect at 2 points.
7. If three points are collinear, exactly one is between the other two.
   If 3 points are collinear, any one of the 3 points is between the other two.

Lesson 4-1 **Refer to the figure for Exercises 1–5.**

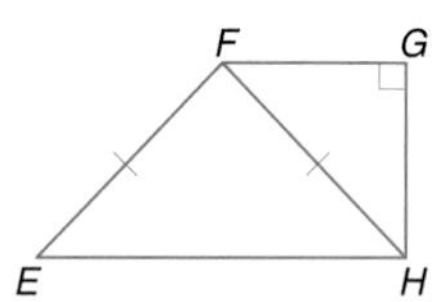

1. Name the hypotenuse. $\overline{FH}$
2. Name the base angles. $\angle FEH$, $\angle FHE$
3. Name the side opposite $\angle EFH$. $\overline{EH}$
4. Name the vertex angle. $\angle EFH$
5. Name the angle opposite $\overline{FG}$. $\angle GHF$

Use the distance formula to classify each triangle by the measures of its sides.

6. $\triangle ABC$ with vertices $A(6, 4)$, $B(-2, 4)$, and $C(2, 7)$. isosceles
7. $\triangle PQR$ with vertices $P(-3, 4)$, $Q(0, 1)$, and $R(2, 3)$. scalene

Copy each sentence. Fill in the blank with *sometimes*, *always*, or *never*.

8. Scalene triangles are _?_ obtuse. sometimes
9. Equilateral triangles are _?_ acute. always
10. Right triangles are _?_ isosceles. sometimes

Lesson 4-2 **Find the value of *x*.**

1. 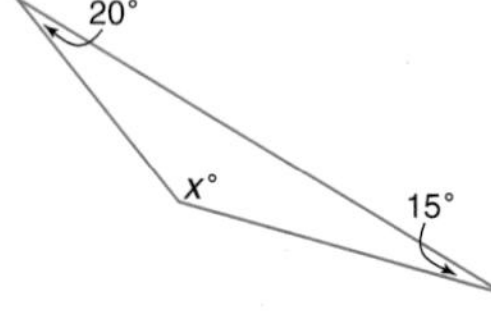

145

2. 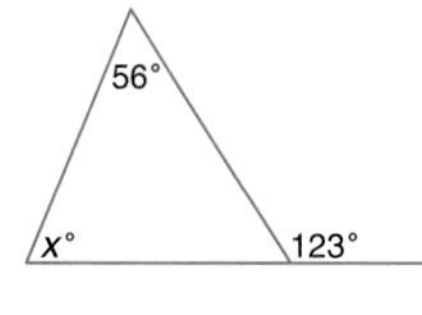

67

3. 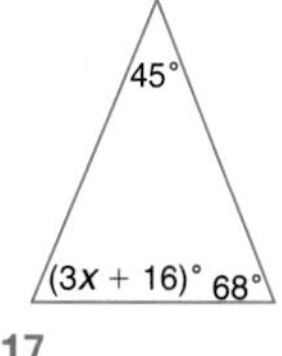

17

If $\overline{PQ}$ is perpendicular to $\overline{QR}$, find the measure of each angle in the figure at the right.

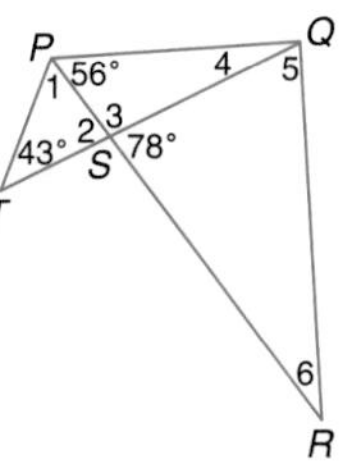

4. $m\angle 2$ 78
5. $m\angle 1$ 59
6. $m\angle 3$ 102
7. $m\angle 4$ 22
8. $m\angle 5$ 68
9. $m\angle 6$ 34

Lesson 4–3 **For each pair of congruent triangles, name all the corresponding sides and angles. Use $\leftrightarrow$ to indicate each correspondence. Draw a figure for each pair of triangles, and mark the corresponding parts.** 1–3. See margin.

1. $\triangle MNO \cong \triangle JKL$
2. $\triangle XYZ \cong \triangle ZPR$
3. $\triangle DEF \cong \triangle ABC$

Complete each congruence statement.

4. $\triangle MST \cong \triangle$ __?__ *NST*

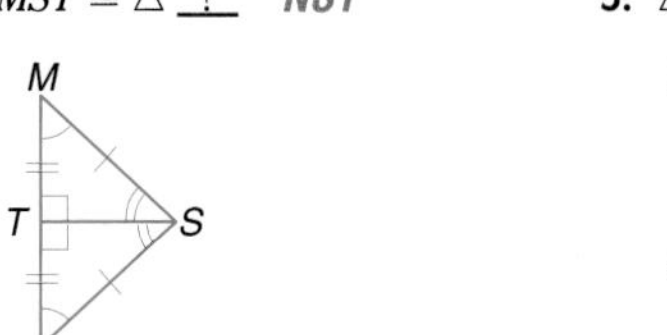

5. $\triangle ABX \cong \triangle$ __?__ *DCX*

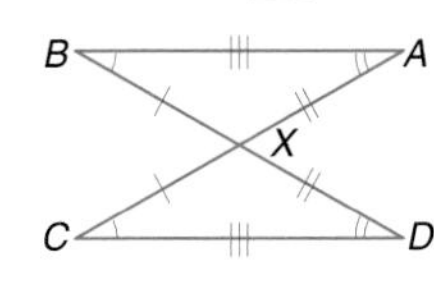

6. $\triangle WYX \cong \triangle$ __?__ *YWZ*

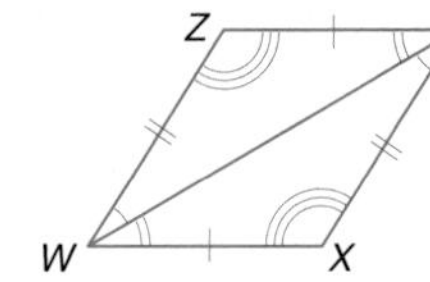

7. Given $\triangle QRS \cong \triangle LMN$, $RS = 8$, $QR = 14$, $QS = 10$, and $MN = 3x - 25$.
   a. Draw and label a figure to show the congruent triangles. See margin.
   b. Find the value of x. 11

Lesson 4–4 **Determine which postulate can be used to prove the triangles congruent. If it is not possible to prove them congruent, write *not possible*.**

1. ASA

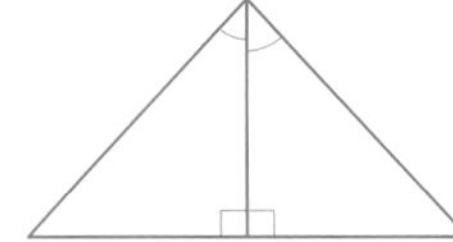

2. SSS

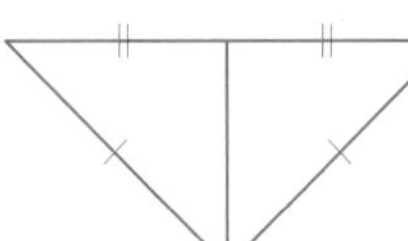

3. SAS

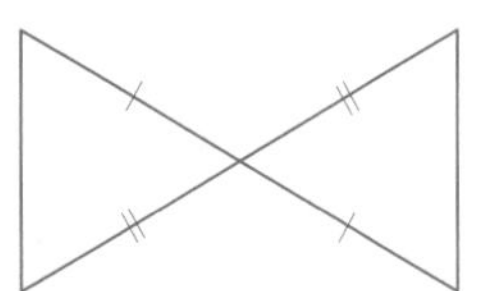

4. $\triangle JKL \cong \triangle STU$; $JK = ST = \sqrt{41}$, $JL = SU = \sqrt{53}$, $KL = TU = \sqrt{58}$

Determine if each pair of triangles is congruent. Justify your answer.

4. The vertices of $\triangle JKL$ are $J(-3, 1)$, $K(-8, 5)$, and $L(-1, 8)$, and the vertices of $\triangle STU$ are $S(0, 1)$, $T(4, 6)$, and $U(7, -1)$.
5. The vertices of $\triangle BCD$ are $B(-4, 1)$, $C(-1, 2)$, and $D(-1, 4)$, and the vertices of $\triangle NMR$ are $N(1, -1)$, $M(0, -3)$, and $R(4, -4)$.
   not congruent; $BD = NR = 3\sqrt{2}$, $BC = \sqrt{10}$, $CD = 2$, $NM = \sqrt{5}$, $MR = \sqrt{17}$
6. Write a proof.

   Given: $\angle A \cong \angle D$
   $\overline{AO} \cong \overline{OD}$

   Prove: $\triangle AOB \cong \triangle DOC$

   See margin.

Lesson 4–5 **Refer to the figure at the right for Exercises 1–5.**

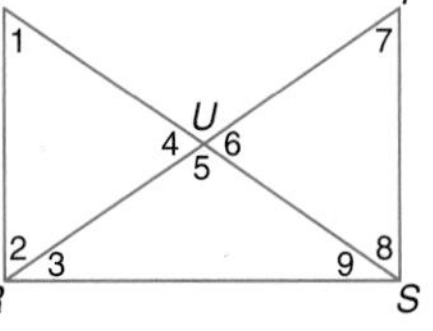

1. Name the included side for $\angle 2$ and $\angle 4$. $\overline{RU}$
2. Name a nonincluded side for $\angle 6$ and $\angle 8$. See margin.
3. $\overline{TS}$ is included between what two angles? $\angle 7$ and $\angle 8$
4. Name a pair of angles so that $\overline{RS}$ is not included. See margin.
5. If $\angle 1 \cong \angle 8$, $\angle 4 \cong \angle 6$, and $\overline{QR} \cong \overline{ST}$, then $\triangle QUR \cong \triangle$ __?__ by __?__. *SUT; AAS*

Additional Answers for Lesson 4-3

1. $\angle M \leftrightarrow \angle J$, $\angle N \leftrightarrow \angle K$, $\angle O \leftrightarrow \angle L$, $\overline{MN} \leftrightarrow \overline{JK}$, $\overline{NO} \leftrightarrow \overline{KL}$, $\overline{MO} \leftrightarrow \overline{JL}$

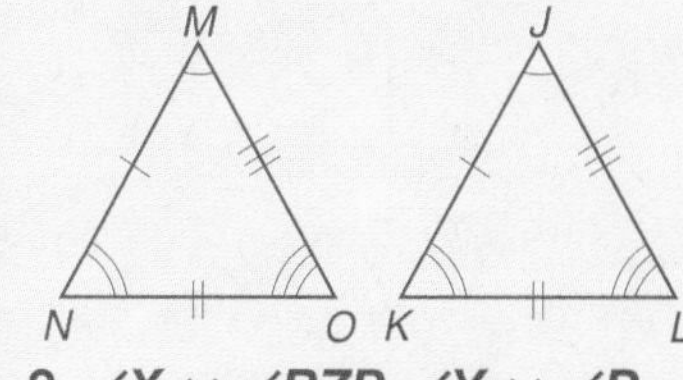

2. $\angle X \leftrightarrow \angle RZP$, $\angle Y \leftrightarrow \angle P$, $\angle XZY \leftrightarrow \angle R$, $\overline{XY} \leftrightarrow \overline{ZP}$, $\overline{YZ} \leftrightarrow \overline{PR}$, $\overline{XZ} \leftrightarrow \overline{ZR}$

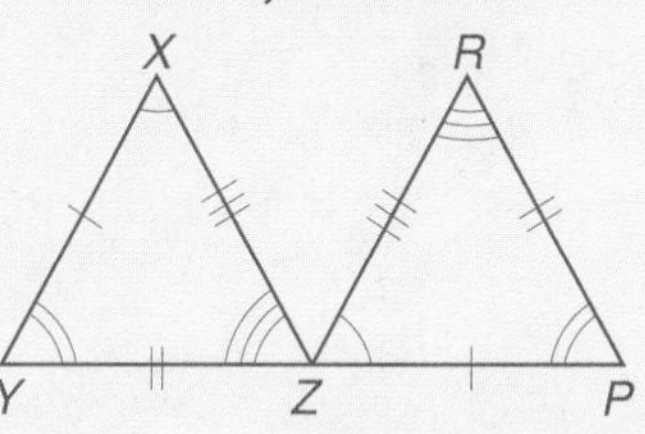

3. $\angle D \leftrightarrow \angle A$, $\angle E \leftrightarrow \angle B$, $\angle F \leftrightarrow \angle C$, $\overline{DE} \leftrightarrow \overline{AB}$, $\overline{EF} \leftrightarrow \overline{BC}$, $\overline{DF} \leftrightarrow \overline{AC}$

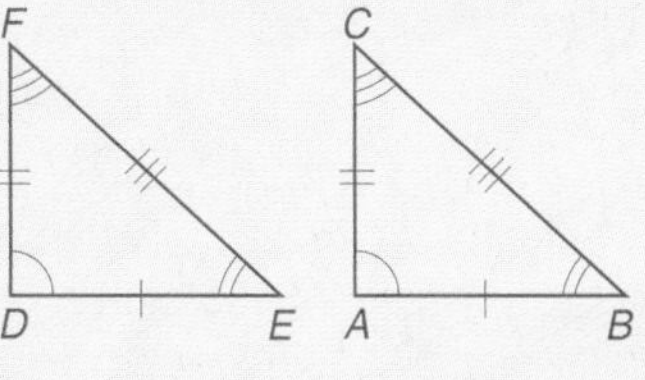

7a.

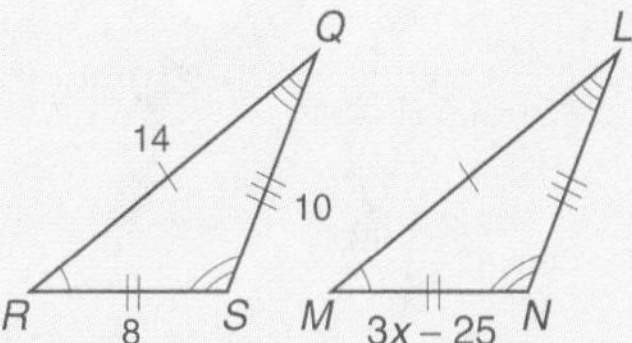

Additional Answer for Lesson 4-4

6. Given: $\angle A \cong \angle D$
   $\overline{AO} \cong \overline{OD}$
   Prove: $\triangle AOB \cong \triangle DOC$
   Proof:
   Statements (Reasons)
   1. $\angle A \cong \angle D$, $\overline{AO} \cong \overline{OD}$ (Given)
   2. $\angle AOB \cong \angle DOC$ (Vert. $\angle$s are $\cong$.)
   3. $\triangle AOB \cong \triangle DOC$ (ASA)

Additional Answers for Lesson 4-5

2. any side except $\overline{US}$; Sample answer: $\overline{ST}$
4. Sample answer: $\angle 3$ and $\angle 5$

Additional Answers for Lesson 5-1

1.

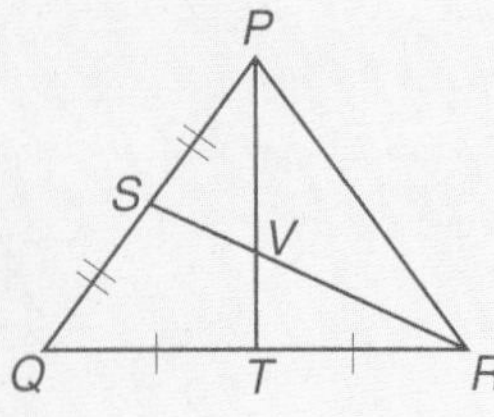

2.

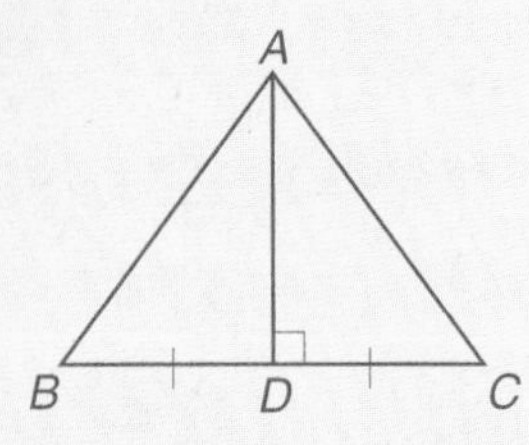

3. F

G E

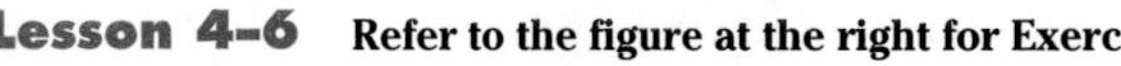

Lesson 4-6 **Refer to the figure at the right for Exercises 1–5.**

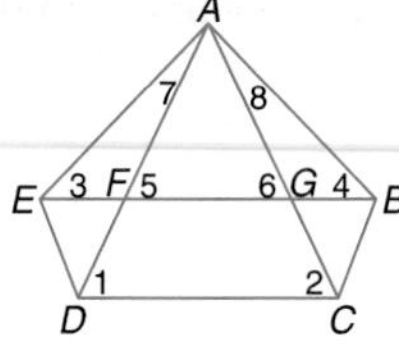

1. If $\overline{FA} \cong \overline{GA}$, name two congruent angles. $\angle 5 \cong \angle 6$
2. If $\overline{AE} \cong \overline{AB}$, name two congruent angles. $\angle 3 \cong \angle 4$
3. If $\overline{AD} \cong \overline{AC}$, name two congruent angles. $\angle 1 \cong \angle 2$
4. If $\angle 6 \cong \angle 3$, name two congruent segments. $\overline{AE} \cong \overline{AG}$
5. If $\angle 5 \cong \angle 4$, name two congruent segments. $\overline{FA} \cong \overline{AB}$

Find the value of x.

6.

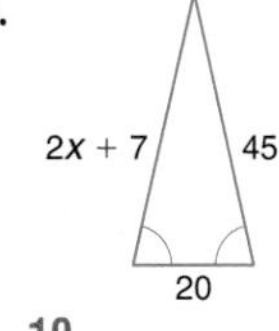

19

7.

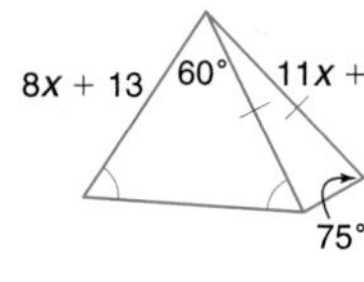

3

8.

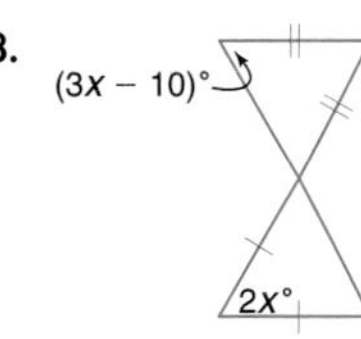

25

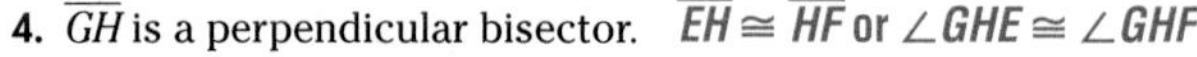

Lesson 5-1 **Draw and label a figure to illustrate each situation.** 1–3. See margin.

1. $\overline{PT}$ and $\overline{RS}$ are medians of $\triangle PQR$ and intersect at V.
2. $\overline{AD}$ is a median and an altitude of $\triangle ABC$.
3. $\overline{EG}$ and $\overline{FG}$ are altitudes of $\triangle EFG$.

Refer to $\triangle EFG$. Write at least one conclusion you can make from each statement.

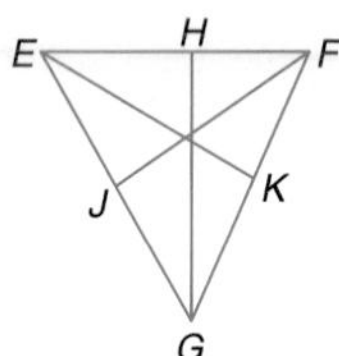

4. $\overline{GH}$ is a perpendicular bisector. $\overline{EH} \cong \overline{HF}$ or $\angle GHE \cong \angle GHF$
5. $\overline{FJ}$ is a median. $\overline{EJ} \cong \overline{JG}$
6. $\angle FEK \cong \angle GEK$ $\overline{EK}$ is an angle bisector.

If possible, describe a triangle for which each statement is true. If no triangle exists, write *no such triangle*.

7. The three medians of a triangle intersect at a point inside the triangle. any triangle
8. The three angle bisectors of a triangle intersect at a point outside the triangle. no such triangle
9. The three altitudes of a triangle intersect at a vertex of the triangle. right triangle

Lesson 5-2 **State the additional information needed to prove each pair of triangles congruent by the given theorem or postulate.**

1. LL

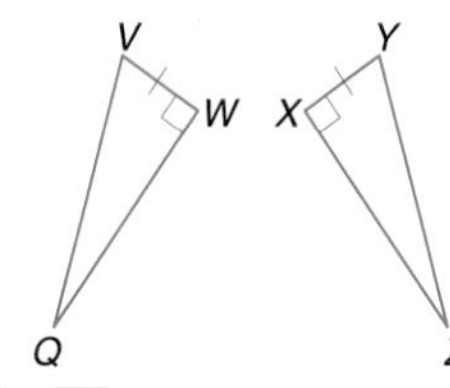

$\overline{QW} \cong \overline{ZX}$

2. HA

$\angle ABD \cong \angle CDB$ or $\angle ADB \cong \angle CBD$

3. HL

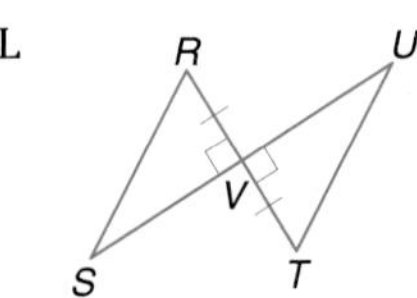

$\overline{RS} \cong \overline{TU}$

Find the value of x so that $\triangle DEF \cong \triangle PQR$ by the indicated theorem or postulate.

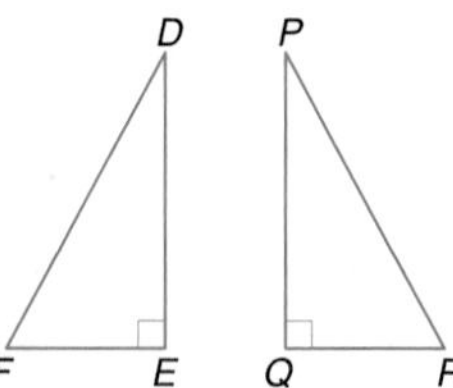

4. $DF = 6x + 1$, $FE = 8$, $PR = 10x - 19$, $RQ = 8$; HL 5
5. $FE = 9$, $m\angle F = 8x - 3$, $RQ = 9$, $m\angle R = 7x + 4$; LA 7

772 Extra Practice

Lesson 5-3 **State the assumption you would make to start an indirect proof of each statement.**

1. Points M, N, and P are collinear Points M, N, and P are not collinear.
2. Triangle ABC is an acute triangle. $\triangle ABC$ is a right or obtuse triangle.
3. The angle bisector of the vertex angle of an isosceles triangle is also an altitude of the triangle. The angle bisector of the vertex angle of an isosceles triangle is not an altitude of the triangle.

Refer to the figure at the right for Exercises 4–6.

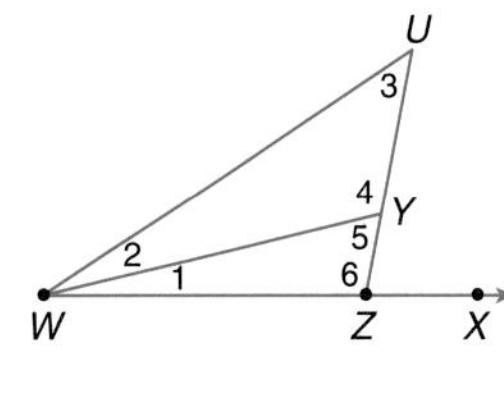

4. Which is greater, $m\angle 4$ or $m\angle 5$? $m\angle 4$
5. If $m\angle 1 < m\angle 3$, which is greater $m\angle 5$ or $m\angle 1$? $m\angle 5$
6. Name an angle whose measure is less than $m\angle UWZ$. $\angle 1$ or $\angle 2$
7. Write an indirect proof. See margin.

   Given: $\overline{PQ} \cong \overline{PR}$

   $m\angle 1 \neq m\angle 2$

   Prove: $\overline{PZ}$ is not a median of $\triangle PQR$.

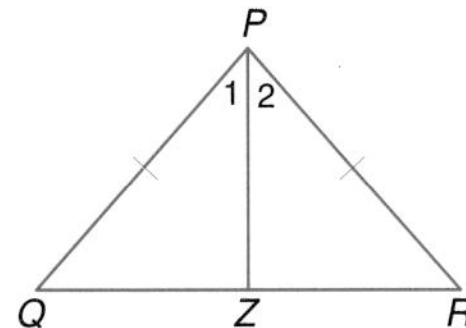

Lesson 5-4 **Refer to the figure at the right for Exercises 1–2.**

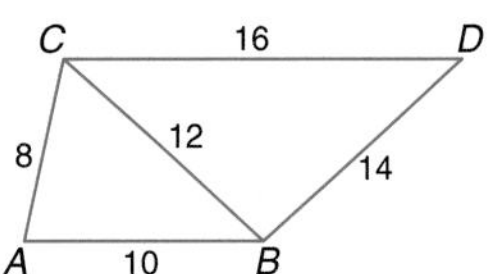

1. Name the angles with the least and greatest measures in $\triangle ABC$.
2. Name the angles with the least and greatest measures in $\triangle BCD$.

1. $\angle CBA$; $\angle A$ 2. $\angle D$; $\angle CBD$

Find the value of x and list the sides of $\triangle PQR$ in order from shortest to longest if the angles have the indicated measures.

3. $m\angle P = 7x + 8$, $m\angle Q = 8x - 10$, $m\angle R = 7x + 6$ 8; $\overline{PR}$, $\overline{PQ}$, $\overline{QR}$
4. $m\angle P = 3x + 44$, $m\angle Q = 68 - 3x$, $m\angle R = x + 61$ 7; $\overline{PR}$, $\overline{QR}$, $\overline{PQ}$
5. Write a two-column proof.

   Given: $QR > QP$

   $\overline{PR} \cong \overline{PQ}$

   Prove: $m\angle P > m\angle Q$

   See margin.

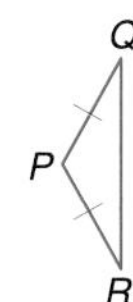

Lesson 5-5 **Determine whether it is possible to draw a triangle with sides of the given measures. Write *yes* or *no*. If yes, draw the triangle.**

1. 12, 11, 17 Yes; see students' work.
2. 5, 100, 100 Yes; see students' work.
3. 4.7, 9, 4.1 no
4. 2.3, 12, 12.2 Yes; see students' work.

The measures of two sides of a triangle are given. Between what two numbers must the measure of the third side fall?

5. 12 and 15 3 and 27
6. 4 and 13 9 and 17
7. 21 and 17 4 and 38

Determine whether it is possible to have a triangle with the given vertices. Write *yes* or *no*, and then explain your answer. 8–9. See margin for explanations.

8. $A(4, -3)$, $B(0, 0)$, $C(-4, 3)$ no
9. $G(-2, 4)$, $H(-6, 5)$, $I(-3, 3)$ yes

Additional Answer for Lesson 5-3

7. Given: $\overline{PQ} \cong \overline{PR}$
   $m\angle 1 \neq m\angle 2$
   Prove: $\overline{PZ}$ is not a median of $\triangle PQR$.
   Proof: Assume that $\overline{PZ}$ is a median of $\triangle PQR$. $\overline{QZ} \cong \overline{ZR}$ since Z is the midpoint of $\overline{QR}$ by definition of median. We are given that $\overline{PQ} \cong \overline{PR}$ and we know that $\overline{PZ} \cong \overline{PZ}$ since the congruence of segments is reflexive. Therefore, $\triangle PZQ \cong \triangle PZR$ by SSS and $\angle 1 \cong \angle 2$ by CPCTC. If $\angle 1 \cong \angle 2$, then $m\angle 1 = m\angle 2$ by definition of congruence which is a contradiction of the given $m\angle 1 \neq m\angle 2$. Therefore, the assumption is incorrect and $\overline{PZ}$ is not a median of $\triangle PQR$.

Additional Answer for Lesson 5-4

5. Given: $\overline{QR} > \overline{QP}$
   $\overline{PR} \cong \overline{PQ}$
   Prove: $m\angle P > m\angle Q$
   Proof:
   Statements (Reasons)
   1. $QR > QP$ (Given)
   2. $m\angle P > m\angle R$ (If one side of a $\triangle$ is longer than another side, then the $\angle$ opp. the longer side $>$ the $\angle$ opp. the shorter side.)
   3. $\overline{PR} \cong \overline{PQ}$ (Given)
   4. $\angle Q \cong \angle R$ (Isos. $\triangle$ Th.)
   5. $m\angle Q = m\angle R$ (Def. $\cong$ $\angle$s)
   6. $m\angle P > m\angle Q$ (Subst. Prop. (=))

Additional Answers for Lesson 5-5

8. $AB = 5$, $BC = 5$, $AC = 10$; $5 + 5 = 10$
9. $GH = \sqrt{17} \approx 4.12$, $HI = \sqrt{13} \approx 3.61$, $GI = \sqrt{2} \approx 1.41$; $4.12 + 3.61 > 1.14$, $3.61 + 1.41 > 4.12$, $4.12 + 1.41 > 3.16$

Additional Answer for Lesson 5-6

7. Given: $\overline{TR} \cong \overline{EU}$
   Prove: $TE > RU$
   Proof:
   Statements (Reasons)
   1. $\overline{TR} \cong \overline{EU}$ (Given)
   2. $\overline{TU} \cong \overline{TU}$ (≅ of segments is reflexive.)
   3. $m\angle 1 > m\angle 2$ (If an ∠ is an ext. ∠ of a △, then its measure > the measure of either of its corr. remote int. ∠s.)
   4. $TE > RU$ (SAS Inequality)

Additional Answers for Lesson 6-1

1. Definition of a parallelogram
2. Opposite sides of a parallelogram are ≅.
3. SAS or SSS
4. Opp. ∠s of a parallelogram are ≅.
5. Diagonals of a parallelogram bisect each other.
6. Alt. Int. ∠s Th.

Additional Answers for Lesson 6-2

1. Opp. sides are ∥.
2. We know that 2 sides are ∥ and another 2 sides are ≅. This is insufficient information to determine if the quadrilateral is a parallelogram.
3. The diagonals bisect each other.

EXTRA PRACTICE

Lesson 5-6 Refer to the figure at the right to write an inequality relating the given pair of angle or segment measures.

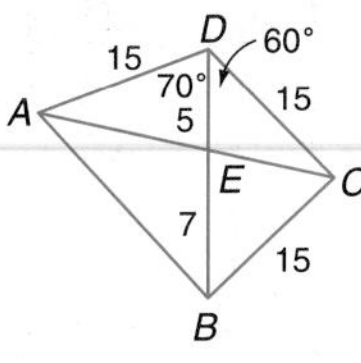

1. AE, CE $AE > CE$
2. $m\angle DCE$, $m\angle ECB$ $m\angle DCE < m\angle ECB$
3. AB, BC $AB > BC$

Write an inequality or pair of inequalities to describe the possible values of x.

4.
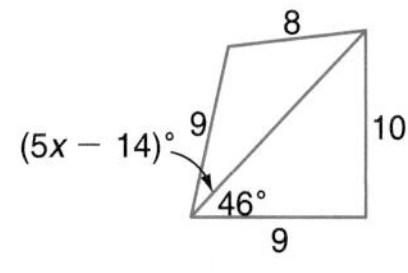

$2.8 < x$ and $x < 12$

5.
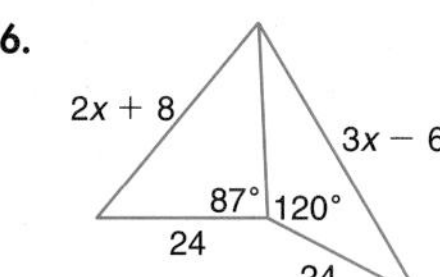

$4 < x$ and $x < 10$

6.

$x > 14$

7. Write a proof.
   Given: $\overline{TR} \cong \overline{EU}$
   Prove: $TE > RU$ See margin.

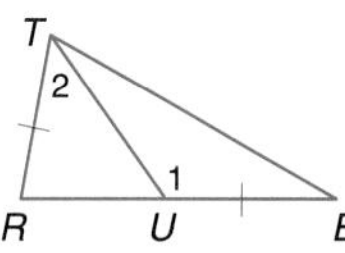

Lesson 6-1 Complete each statement about ▱$ABCD$. Then name the theorem or definition that justifies your answer. 1–6. See margin for reasons.

1. $\overline{AB} \parallel$ __?__ $\overline{DC}$
2. $\overline{DA} \cong$ __?__ $\overline{CB}$
3. $\triangle ADC \cong$ __?__ $\triangle CBA$
4. $\angle CDA \cong$ __?__ $\angle ABC$
5. $\overline{DE} \cong$ __?__ $\overline{EB}$
6. $\angle BAC \cong$ __?__ $\angle ACD$

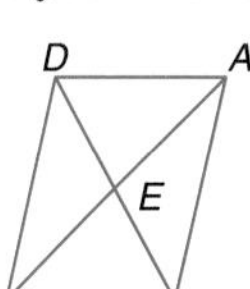

For each parallelogram, find the values of x, y, and z.

7.
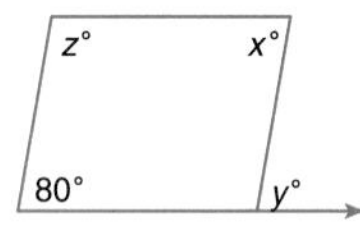

$x = 80$, $y = 80$, $z = 100$

8.

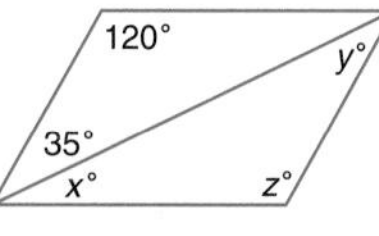

$x = 25$, $y = 35$, $z = 120$

9.

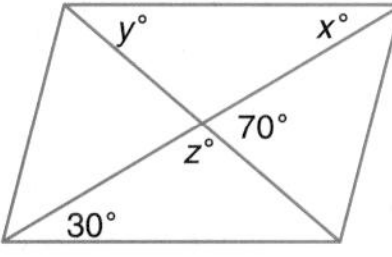

$x = 30$, $y = 40$, $z = 110$

Lesson 6-2 Determine if each quadrilateral is a parallelogram. Justify your answer.

1.
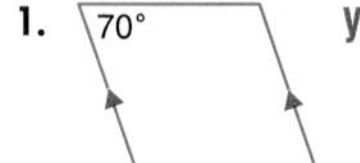

yes

2.

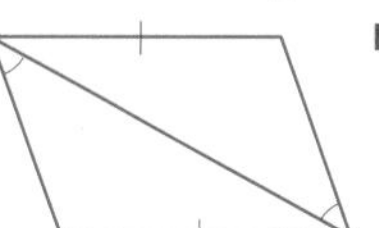

no

3.
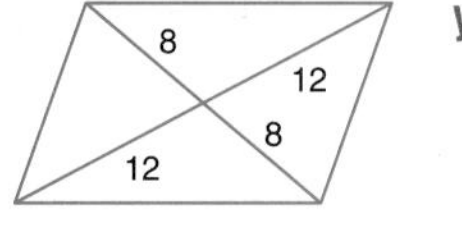

yes

1–3. See margin for reasons.

Find the values of x and y that ensure each quadrilateral is a parallelogram. 4. $x = 2$, $y = -3$

4.
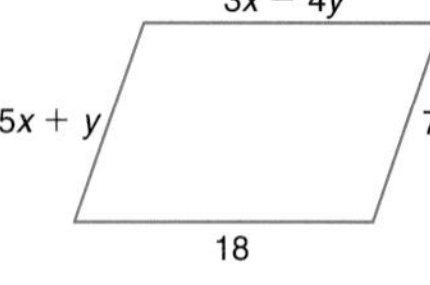

5.
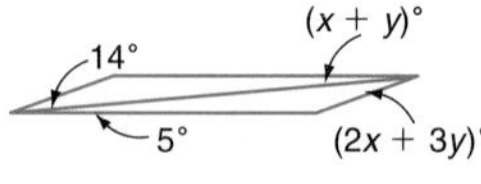

$x = 1$, $y = 4$

6.
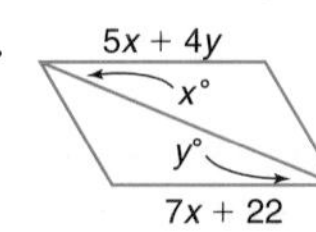

$x = 11$, $y = 11$

Lesson 6-3 **Use rectangle *QRST* and the given information to solve each problem.**

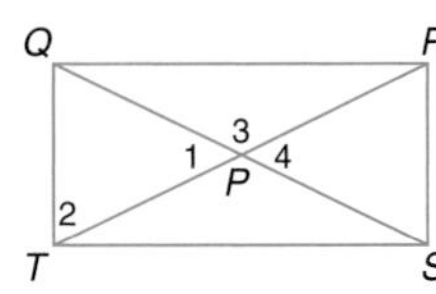

1. $QP = 6$, find RT. 12
2. $QT = 8$, find RS. 8
3. $PT = 3x$ and $PS = 18$, find x. 6
4. $m\angle 1 = 55$, find $m\angle 2$. 62.5
5. $m\angle 3 = 110$, find $m\angle 4$. 70

Determine whether each statement is *true* or *false*. Justify your answer. 6–8. See margin for reasons

6. If a parallelogram has congruent diagonals, then it is a rectangle. true
7. If the diagonals of a quadrilateral bisect each other, then it is a rectangle. false
8. If a quadrilateral is a rectangle, then it has four right angles. true

Determine whether *PQRS* is a rectangle. Justify your answer.

9. $P(12, 2)$, $Q(12, 8)$, $R(-3, 8)$, $S(-3, 2)$ yes; opposite sides parallel and all right angles
10. $P(0, -3)$, $Q(4, 8)$, $R(11, 7)$, $S(7, -4)$ no; not all right angles

Lesson 6-4 **Use rhombus *IJKL* and the given information to find each value.**

1. If $m\angle 3 = 62$, find $m\angle 1$. 28
2. If $m\angle 4 = 3x - 1$ and $m\angle 3 = 2x + 30$, find the value of x. 31
3. If $m\angle 5 = 2(x + 1)$ and $m\angle 3 = 4(x + 1)$, find the value of x. 14
4. If $m\angle 6 = 7x + 13$, find the value of x. 11
5. If $m\angle LKJ = x^2 - 17$ and $m\angle 2 = x + 23$, find the value of x. 9 or -7

Name all the quadrilaterals—*parallelogram*, *rectangle*, *rhombus*, or *square*—that have each property.

6. The opposite sides are parallel. parallelogram, rectangle, rhombus, square
7. The opposite sides are congruent. parallelogram, rectangle, rhombus, square
8. All angles congruent. rectangle, square

Lesson 6-5 ***PQRS* is an isosceles trapezoid with bases $\overline{PS}$ and $\overline{QR}$ and median $\overline{TV}$. Use the given information to solve each problem.**

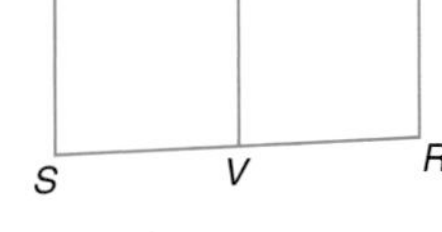

1. If $PS = 20$ and $QR = 14$, find TV. 17
2. If $QR = 14.3$ and $TV = 23.2$, find PS. 32.1
3. If $TV = x + 7$ and $PS + QR = 5x + 2$, find x. 4
4. If $m\angle RVT = 57$, find $m\angle QTV$. 57
5. If $m\angle VTP = a$, find $m\angle TPS$ in terms of a. $180 - a$

6. trapezoid 7. quadrilateral 8. parallelogram

Determine whether each figure is a *trapezoid*, a *parallelogram*, a *rectangle*, or a *quadrilateral*.

6. 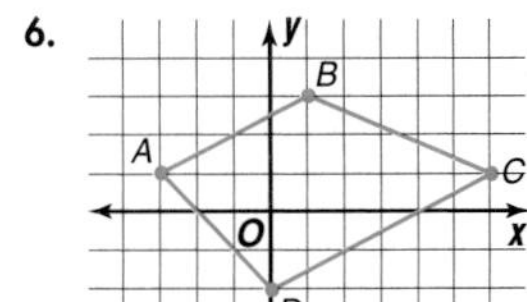

7. 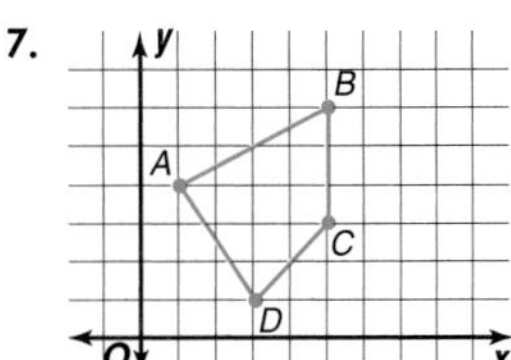

8. 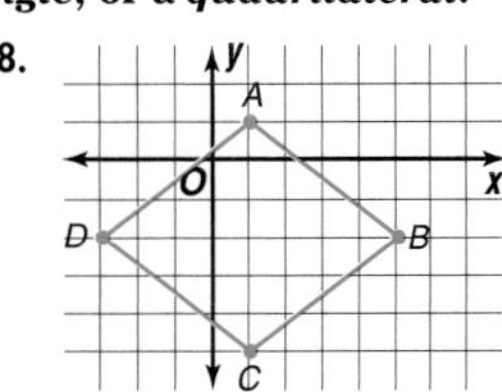

Additional Answers for Lesson 6-3

6. A parallelogram with ≅ diagonals is a rectangle.
7. If the diagonals of a quadrilateral bisect each other, then it is a parallelogram.
8. A rectangle is a quadrilateral with 4 rt. ⦣s.

Additional Answers for Lesson 7-3

1. yes; AA Similarity; $\triangle GHI \sim \triangle OMN$
2. yes; SAS Similarity; $\triangle JKL \sim \triangle QRP$
3. yes; AA Similarity; $\triangle RSX \sim \triangle TUX$

Extra Practice

EXTRA PRACTICE

Lesson 7-1 **Find each ratio and express it as a fraction in simplest form.**

1. There are 96 girls and 74 boys in the senior class. Find the ratio of girls to boys. $\frac{48}{37}$
2. A 5-pound can of cashews cost \$15.00. Find the ratio of pounds to cost. $\frac{1}{3}$
3. A designated hitter made 6 hits in 9 games. Find the ratio of hits to games. $\frac{2}{3}$

Solve each proportion by using cross products.

4. $\frac{4}{n} = \frac{7}{8}$ $\frac{32}{7}$
5. $\frac{x+1}{x} = \frac{7}{2}$ 0.4
6. $\frac{5}{17} = \frac{2x}{51}$ 7.5

Corresponding sides of polygon *PQRS* are proportional to the sides of polygon *ABCD*.

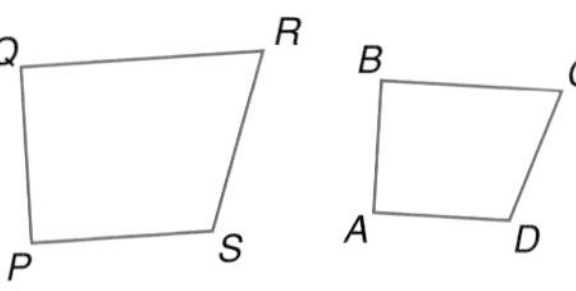

7. If $AB = 4$, $AD = 8$, and $PQ = 6$, find PS. 12
8. If $RS = 4.5$, $CD = 6.3$, and $BC = 7$, find QR. 5
9. If $CD = 21.8$, $DA = 43.6$, and $SR = 33$, find SP. 66

Lesson 7-2 **Each pair of figures is similar. Find the values of *x* and *y*.**

1.

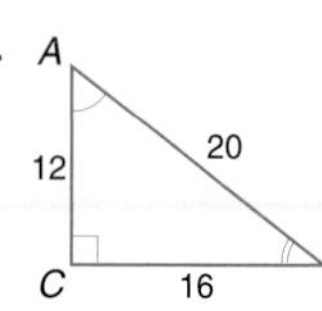

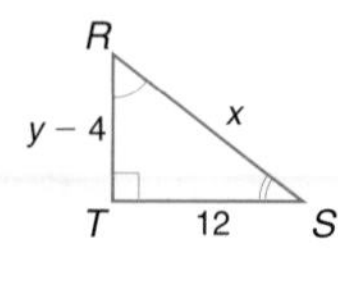

15; 13

2.

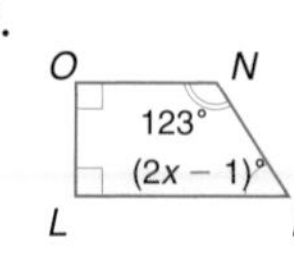

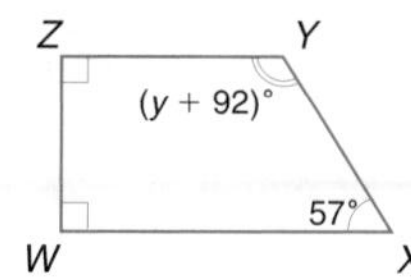

29; 31

Given quadrilateral *ABCD* ~ quadrilateral *EFGH*, find each of the following.

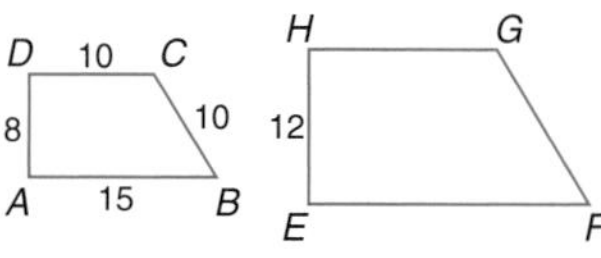

3. scale factor of $ABCD$ to $EFGH$ $\frac{2}{3}$
4. **a.** GH 15 **b.** FG 15 **c.** EF 22.5
5. **a.** perimeter of $ABCD$ 43
   b. perimeter of $EFGH$ 64.5
   c. ratio of the perimeters of $ABCD$ and $EFGH$ $\frac{43}{64.5} = \frac{2}{3}$

Lesson 7-3 **Determine whether each pair of triangles is similar. Give a reason for your answer. If similarity exists, write a mathematical sentence relating the two triangles.** 1–3. See margin.

1.

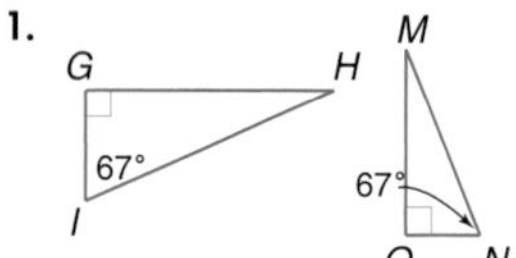

2.

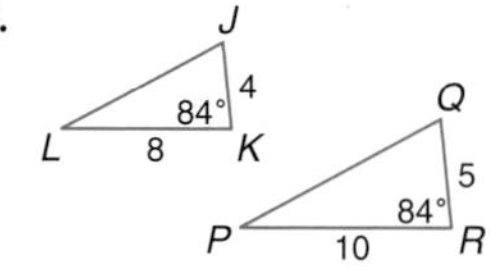

3.

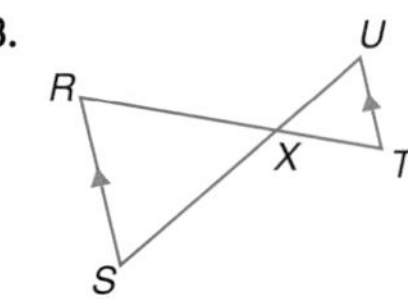

Use the given information to find each measure.

4. $\overline{BE} \parallel \overline{CD}$, find CD, AC, and BC. 12, 22.5, 7.5

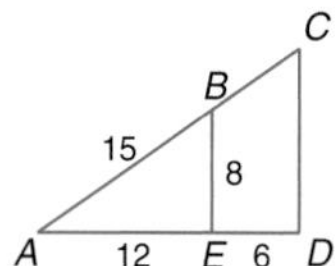

5. If $RSTU$ is a parallelogram, find SV, WS, and RW. 5, 4, 16

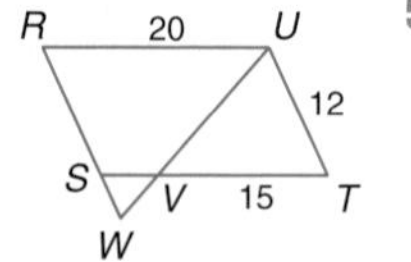

Lesson 7-4 **In the figure at the right, $AB \parallel CD \parallel EF$. Complete each statement.**

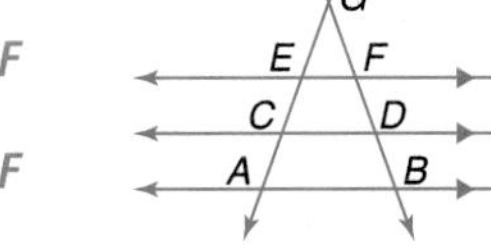

1. $\frac{CE}{AC} = \frac{DF}{?}$ BD
2. $\frac{AE}{AC} = \frac{?}{BD}$ BF
3. $\frac{?}{DF} = \frac{AE}{CE}$ BF
4. $\frac{EG}{CE} = \frac{FG}{?}$ DF

In $\triangle ACE$, $\overline{BD} \parallel \overline{AE}$. Determine whether each statement is *true* or *false*. If false, explain why.

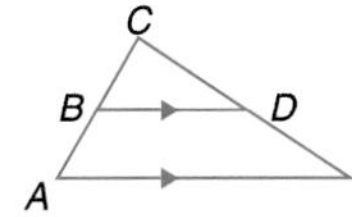

5. $\frac{BC}{ED} = \frac{AB}{CD}$ false; $\frac{BC}{AB} = \frac{CD}{ED}$
6. $\frac{AB}{BC} = \frac{DE}{CD}$ true

In $\triangle XYZ$, find x so that $\overline{YZ} \parallel \overline{MN}$.

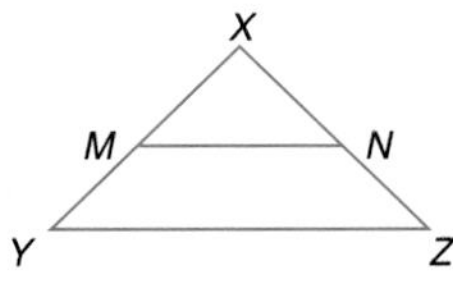

7. $YM = 6$, $MX = 9$, $NX = 12$, $ZN = x$. 8
8. $XN = 20$, $NZ = 16$, $XM = x - 6$, $MY = 20$. 31
9. $MX = 5$, $MY = x - 2$, $NX = 9$, $NZ = x + 6$. 12

Lesson 7-5 **In the figure at the right, $\triangle ABC \sim \triangle DEF$.**

1. If $\overline{AR} \cong \overline{RC}$, $\overline{DS} \cong \overline{SF}$, $AC = 20$, $DF = 12$, and $ES = 5$, $BR =$ __?__. $8\frac{1}{3}$
2. If $\overline{BR} \perp \overline{AC}$ and $\overline{ES} \perp \overline{DF}$, $\frac{BR}{ES} = \frac{?}{EF}$. BC
3. If $\overline{BR}$ and $\overline{ES}$ are angle bisectors, $AB = 12$, $DE = 5$, and $ES = 4$, $BR =$ __?__. 9.6
4. If $\overline{BR}$ bisects $\angle ABC$, $AB = 15$, $AR = 6$, and $RC = 8$, $CB =$ __?__. 20
5. If the perimeter of $\triangle DEF$ is 30, $AC = 12$, and $DF = 8$, find the perimeter of $\triangle ABC$. 45

Lesson 7-6 **Find the value of each expression. Then, use that value as the next x in the expression. Repeat the process until you can make some observations. Describe what happens in each of the iterations.**

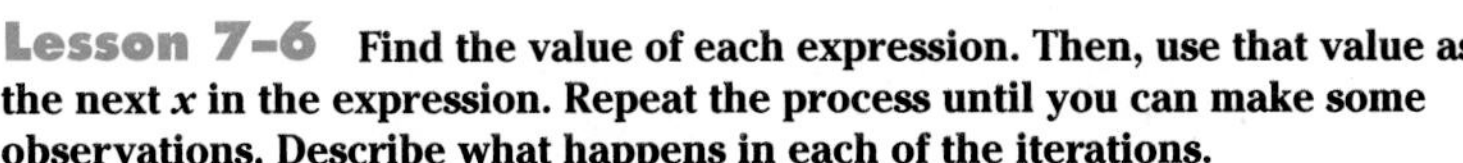

1. $\sqrt[3]{x}$, where x initially equals 28 $\sqrt[3]{x}$ approaches 1
2. $\frac{1}{x^2}$, where x initially equals 2 $\frac{1}{x^2}$ alternates between approaching infinity and approaching 0.
3. $x^{\frac{1}{2}}$, where x initially equals 0.4 $x^{\frac{1}{2}}$ approaches 1

Use grid paper to draw a square that is 20 units long on each side. Connect the midpoints on each side to make four squares. Shade the bottom right square. This is stage 1. Repeat this process in the the unshaded squares.

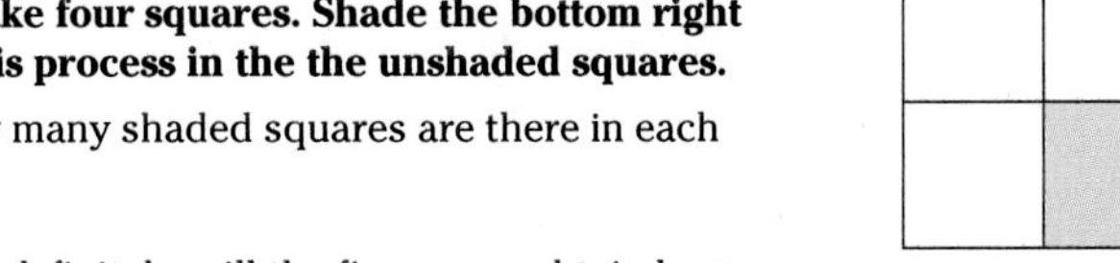

4. Draw stage 2 and stage 3. How many shaded squares are there in each stage? See margin.
5. If your continue the process indefinitely, will the figure you obtain be strictly self similar? Explain. yes

Additional Answer for Lesson 7-6

4.

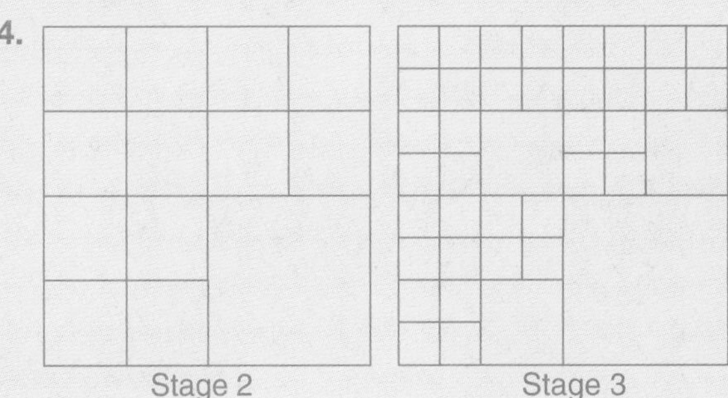

stage 1, 1
stage 2, 4
stage 3, 13

Extra Practice 777

Additional Answers for Lesson 8-3

1. $\frac{21}{29} \approx 0.7241$
2. $\frac{20}{29} \approx 0.6897$
3. $\frac{21}{20} = 1.0500$
4. $\frac{21}{29} \approx 0.7241$
5. $\frac{3}{5} = 0.6000$
6. $\frac{3}{4} = 0.7500$

Additional Answers for Lesson 8-4

1. $\sin 15° = \frac{LN}{37}$
2. $\sin 47° = \frac{10}{AL}$
3. $\cos 16° = \frac{13.4}{AL}$
4. $\tan 72° = \frac{13}{LN}$
5. $\tan 74° = \frac{AN}{33.6}$

Additional Answers for Lesson 8-5

1. 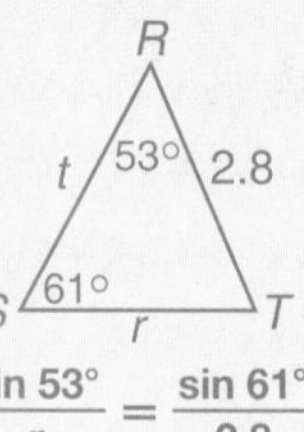

$\frac{\sin 53°}{r} = \frac{\sin 61°}{2.8}$

2. 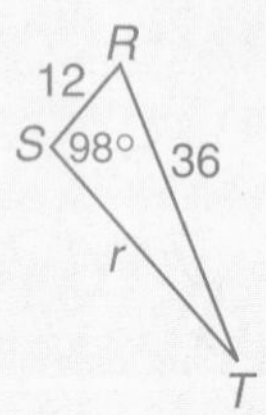

$\frac{\sin 98°}{36} = \frac{\sin T}{12}$

3. 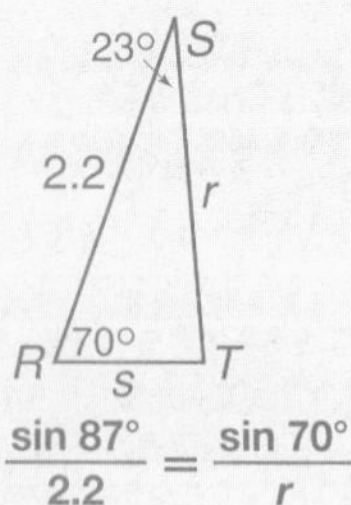

$\frac{\sin 87°}{2.2} = \frac{\sin 70°}{r}$

4. 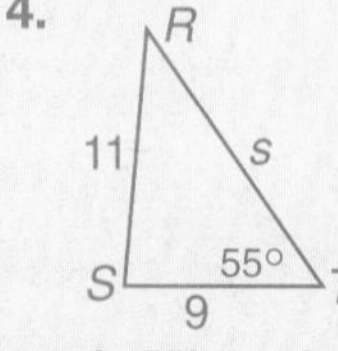

$\frac{\sin 55°}{11} = \frac{\sin R}{9}$

Lesson 8-1 **Find the geometric mean between each pair of numbers.**

1. 3 and 5 $\sqrt{15} \approx 3.9$
2. $\frac{1}{4}$ and 9 $\frac{3}{2}$
3. $\frac{3}{8}$ and $\frac{8}{3}$ 1

Find the values of x and y.

4. 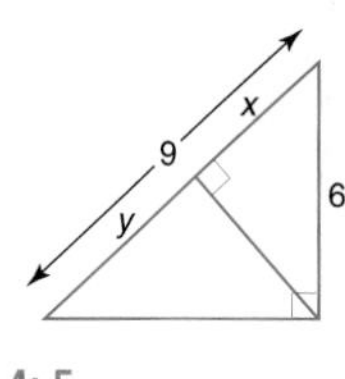

4; 5

5. 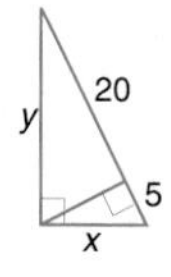

$\sqrt{125} \approx 11.2$; $\sqrt{500} \approx 22.4$

6. 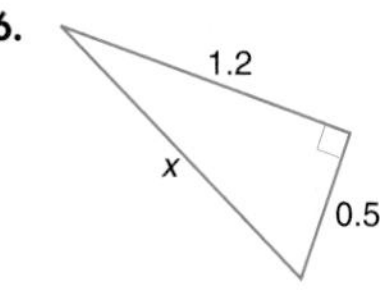

1.3

Determine if the given measures are measures of the sides of a right triangle.

7. 25, 20, 15 yes
8. 1.6, 3.0, 3.4 yes
9. 18, 34, 39 no

Lesson 8-2 **Find the values of x and y.** 3. $10.4\sqrt{3} \approx 18.01$; $20.8\sqrt{3} \approx 36.03$

1. 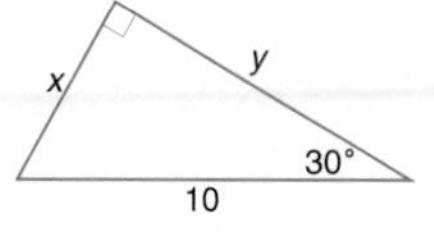

5; $5\sqrt{3} \approx 8.66$

2. 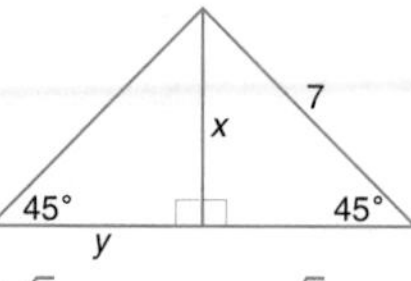

$\frac{7\sqrt{2}}{2} \approx 4.95$, $\frac{7\sqrt{2}}{2} \approx 4.95$

3. 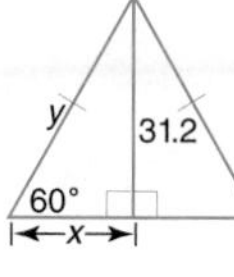

4. The length of one side of a square is 17 meters. Find the length of a diagonal of the square. $17\sqrt{2} \approx 24.04$ m
5. The length of an altitude of an equilateral triangle is $\frac{\sqrt{3}}{3}$ yards. Find the length of one side of the triangle. $\frac{2}{3}$ yd
6. The length of a diagonal of a square is 6 inches. Find the length of one side of the square. $3\sqrt{2} \approx 4.24$ in.
7. The perimeter of an equilateral triangle is 42 centimeters. Find the length of an altitude of the triangle. $7\sqrt{3} \approx 12.12$ cm

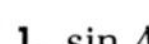

Lesson 8-3 **Find the indicated trigonometric ratio as a fraction and as a decimal rounded to the nearest ten-thousandth.**

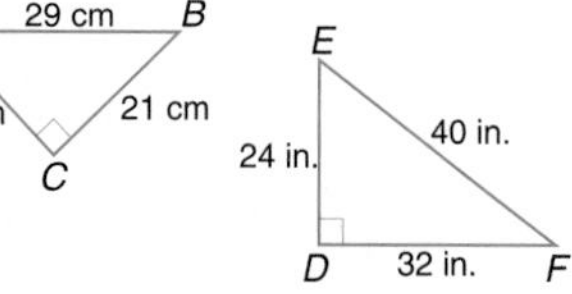

1. $\sin A$
2. $\sin B$
3. $\tan A$
4. $\cos B$
5. $\cos E$
6. $\tan F$

1–6. See margin.

Find the value of x. Round to the nearest tenth.

7. 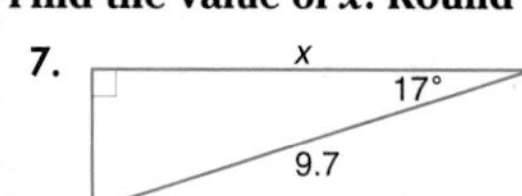

9.3

8. 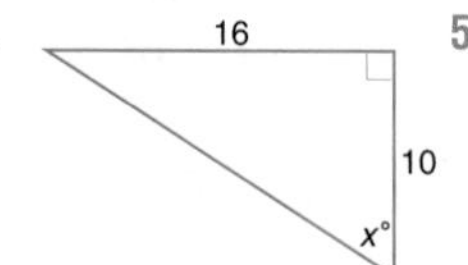

58.0

9. 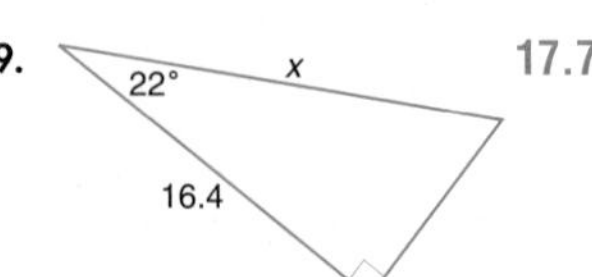

17.7

Lesson 8-4 **State an equation that would enable you to solve each problem. Then solve. Round answers to the nearest tenth.** 1–5. See margin for equations.

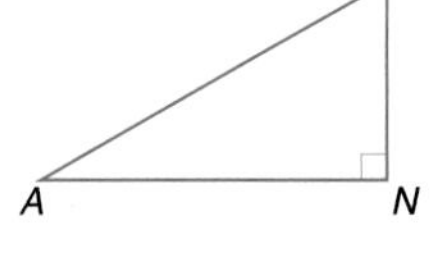

1. Given $m\angle A = 15$ and $AL = 37$, find LN. 9.6
2. Given $m\angle A = 47$ and $LN = 10$, find AL. 13.7
3. Given $AN = 13.4$ and $m\angle A = 16$, find AL. 13.9
4. Given $m\angle L = 72$ and $AN = 13$, find LN. 4.2
5. Given $LN = 33.6$ and $m\angle L = 74$, find AN. 117.2
6. A surveyor is standing 100 meters from a bridge. She determines that the angle of elevation to the top of the bridge is 35°. The surveyor's eye level is 1.5 meters above the ground. Find the height of the bridge. about 71.5 m
7. A ladder leaning against the side of a house forms an angle of 65° with the ground. The foot of the ladder is 8 feet from the building. Find the length of the ladder. about 18.9 ft
8. From the top of a lighthouse, the angle of depression to a buoy is 25°. If the top of the lighthouse is 150 feet above sea level, find the distance from the buoy to the foot of the lighthouse. about 321.7 ft
9. Bill Owens is an architect designing a new parking garage for the city. The floors of the garage are to be 10 feet apart. The exit ramps between the pair of floors are to be 75 feet long. What is the measurement of the angle of elevation of each ramp? about 7.7°
10. A jet takes off and rises at an angle of 19° with the ground until it hits 28,000 feet. How much ground distance is covered in miles? (*Hint*: 5280 feet = 1 mile) about 15.4 mi

Lesson 8-5 **Draw $\triangle RST$ and mark it with the given information. Write an equation that could be used to find each unknown value. Then find the value to the nearest tenth.** 1–4. See margin for diagrams and equations

1. If $s = 2.8$, $m\angle R = 53$, and $m\angle S = 61$, find r. 2.6
2. If $s = 36$, $t = 12$, and $m\angle S = 98$, find $m\angle T$. 19.3
3. If $m\angle R = 70$, $m\angle S = 23$, and $t = 2.2$, find r. 2.1
4. If $m\angle T = 55$, $r = 9$, and $t = 11$, find $m\angle R$. 42.1

Solve each $\triangle EFG$ described below. Round measures to the nearest tenth. 5–10. See margin.

5. $m\angle G = 70, g = 8, m\angle E = 30$
6. $e = 10, g = 25, m\angle G = 124$
7. $m\angle E = 29, m\angle F = 62, g = 11.5$
8. $m\angle G = 35, e = 7.5, g = 24$
9. $m\angle F = 36, m\angle G = 119, f = 8$
10. $m\angle F = 47, m\angle G = 73, e = 0.9$

Lesson 8-6 **Sketch each $\triangle ERQ$. Determine whether the Law of Sines or the Law of Cosines should be used first to solve each triangle. Then solve each triangle. Round measures to the nearest tenth.** 1–4. See margin.

1. $m\angle E = 40, m\angle Q = 70, q = 4$
2. $e = 11, r = 10.5, m\angle Q = 35$
3. $e = 11, r = 17, m\angle R = 42$
4. $m\angle E = 56, m\angle Q = 26, q = 12.2$

Solve each $\triangle PSV$ described below. Round measures to the nearest tenth. 5–10. See margin.

5. $p = 51, s = 61, m\angle S = 19$
6. $p = 5, s = 12, v = 13$
7. $p = 20, v = 24, m\angle S = 47$
8. $m\angle P = 40, m\angle V = 59, s = 14$
9. $p = 345, v = 648, s = 442$
10. $m\angle P = 29, v = 5, s = 4.9$

Additional Answers for Lesson 8-5

5. $m\angle F = 80$, $e \approx 4.3$, $f \approx 8.4$
6. $m\angle E \approx 19.4$, $m\angle F \approx 36.6$, $f \approx 18.0$
7. $m\angle G = 89$, $e \approx 5.6$, $f \approx 10.2$
8. $m\angle E \approx 10.3$, $m\angle F \approx 134.7$, $f \approx 29.7$
9. $m\angle E = 25$, $e \approx 5.8$, $g \approx 11.9$
10. $m\angle E = 60$, $f \approx 0.8$, $g \approx 1.0$

Additional Answers for Lesson 8-6

1.

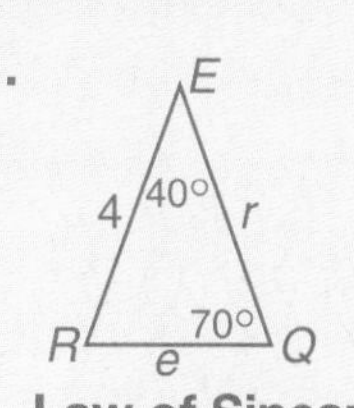

Law of Sines; $e \approx 2.7$, $m\angle R = 70$, $r = 4$

2.

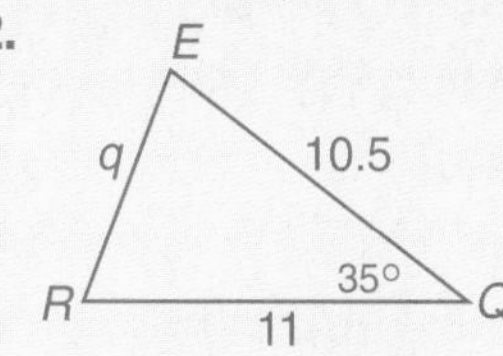

Law of Cosines; $q \approx 6.5$, $m\angle E \approx 76.1$, $m\angle R \approx 68.9$

3.

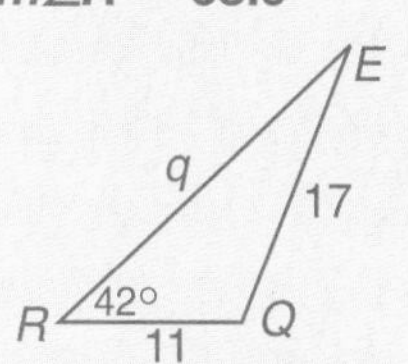

Law of Sines; $m\angle E \approx 25.7$, $m\angle Q \approx 112.3$, $q \approx 23.5$

4.

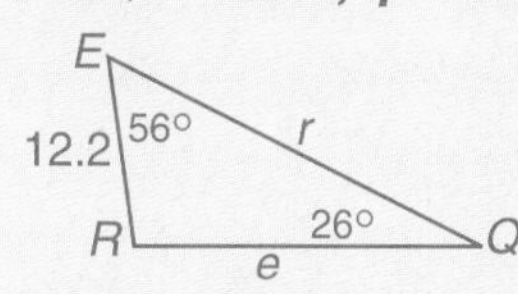

Law of Sines; $e \approx 23.1$, $m\angle R = 98$, $r = 27.6$

5. $m\angle P \approx 15.8$, $m\angle V \approx 145.2$, $v \approx 106.9$
6. $m\angle P \approx 22.6$, $m\angle S \approx 67.3$, $m\angle V \approx 90.1$
7. $s \approx 17.9$, $m\angle P \approx 54.8$, $m\angle V \approx 78.2$
8. $m\angle S = 81$, $p \approx 9.1$, $v \approx 12.1$
9. $m\angle P \approx 30.0$, $m\angle S \approx 39.8$, $m\angle V \approx 110.2$
10. $p \approx 2.5$, $m\angle S \approx 71.8$, $m\angle V \approx 79.2$

Lesson 9–1 Refer to the circle at the right.

1. Name the center of $\odot P$. P
2. Is $\overline{PD}$ a chord of $\odot P$? Explain.
3. If $PC = 6$, find DB. 12
4. Name a chord that is not a diameter. $\overline{EA}$
5. Is $\overline{PC} \cong \overline{PB}$? Explain.
6. If $DB = 17$, find PG. 8.5

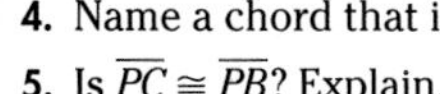

2. No; both endpoints are not on the circle.
5. Yes; all radii of a circle are congruent.

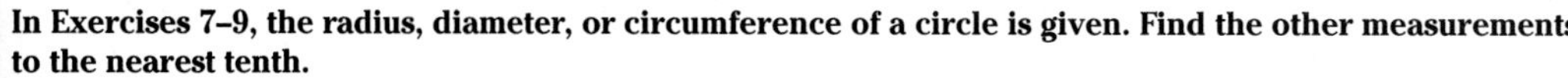

In Exercises 7–9, the radius, diameter, or circumference of a circle is given. Find the other measurements to the nearest tenth.

7. $r = 3.8$, $d =$ _?_ , $C =$ _?_ 7.6; 23.9
8. $C = 11$, $d =$ _?_ , $r =$ _?_ 3.5; 1.8
9. $d = x$, $r =$ _?_ , $C =$ _?_ $0.5x$; πx

Find the exact circumference of each circle.

10.
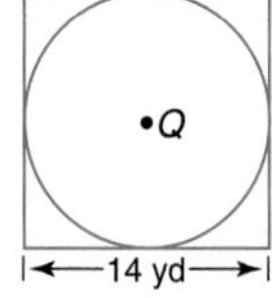

14π yd

11.
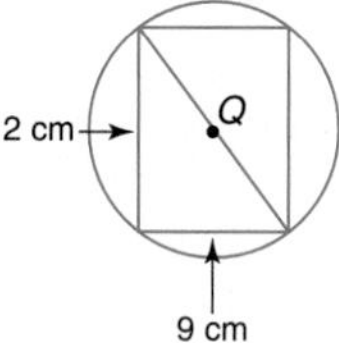

15π cm

12.
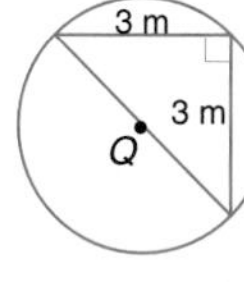

$3\sqrt{2}\,\pi$ m

Lesson 9–2 Refer to $\odot R$ for Exercises 1–8. In $\odot R$, $m\angle QRS = 40$ and $m\angle SRT = 90$ with diameters $\overline{XS}$ and $\overline{QV}$. Determine whether each arc is a minor arc, a major arc, or a semicircle. Then find the degree measure of each arc.

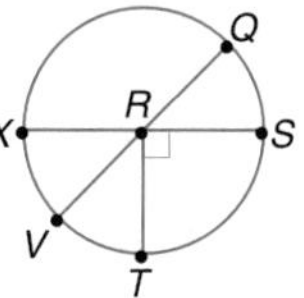

1. $m\widehat{XQ}$ minor; 140
2. $m\widehat{VSX}$ major; 320
3. $m\widehat{QXV}$ semicircle; 180
4. $m\widehat{TSV}$ major; 310

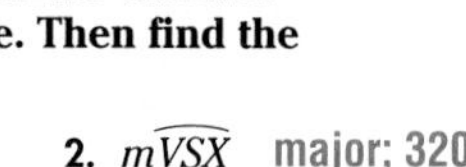

If $XS = 16$, find the length of each arc. Round to the nearest tenth.

5. $\widehat{XQ}$ 19.5
6. $\widehat{VSX}$ 44.7
7. $\widehat{QXV}$ 25.1
8. $\widehat{TSV}$ 43.3

Lesson 9–3 In $\odot A$, $\overline{YS}$ and $\overline{ZR}$ are diameters, and $\overline{SY} \perp \overline{QT}$.

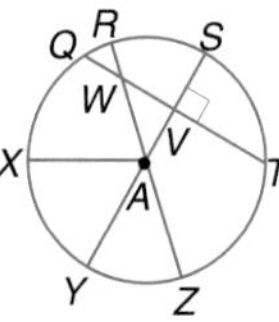

1. Name the midpoint of $\overline{RZ}$. A
2. Name an arc congruent to $\widehat{ST}$. $\widehat{QS}$
3. Name a segment congruent to $\overline{VT}$. $\overline{QV}$
4. Which segment is longer, $\overline{WA}$ or $\overline{VA}$? $\overline{WA}$
5. Name a segment congruent to $\overline{RZ}$. $\overline{SY}$

In each circle, O is the center. Find each measure.

6. AC 14
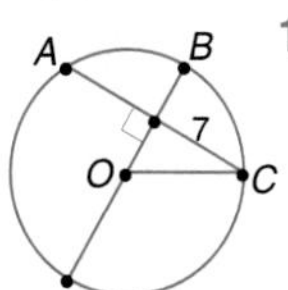

7. $m\widehat{JK}$ 75
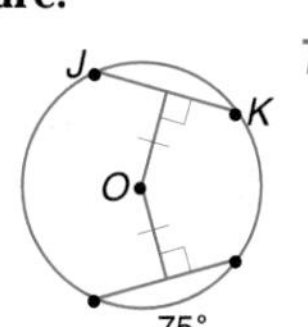

8. ON 10
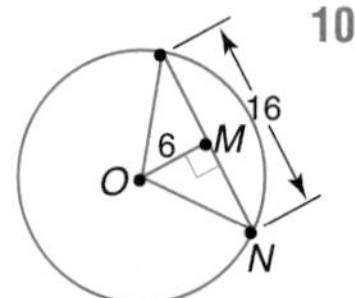

780 Extra Practice

Lesson 9-4 In $\odot O$, $\overline{GM}$ is a diameter.

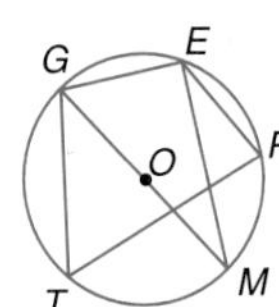

1. Name the intercepted arc for $\angle GTR$. $\widehat{GR}$
2. If $m\angle GTR = 62$, find $m\widehat{GR}$. 124
3. Find $m\angle MEG$. 90
4. If $m\angle GTR = 62$, find $m\widehat{GTR}$. 236
5. Name an inscribed angle. Sample answer: $\angle GTR$

Quadrilateral $GHIJ$ is inscribed in $\odot Q$. $m\angle GHI = 2x$, $m\angle HGJ = 2x - 10$, and $m\angle IJG = 2x + 10$. Find each measure.

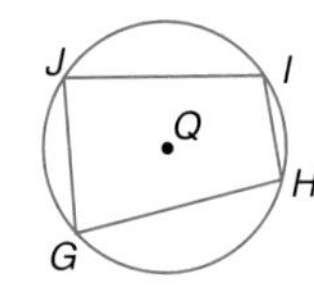

6. $m\angle GHI$ 85
7. $m\widehat{GJI}$ 170
8. $m\widehat{JIH}$ 150
9. $m\widehat{JGH}$ 210

Lesson 9-5 $\overline{AB}$ and $\overline{CD}$ are both tangent to $\odot P$ and $\odot Q$, $AP = 8$, $BQ = 5$, and $m\angle CPE = 45$. Find each measure.

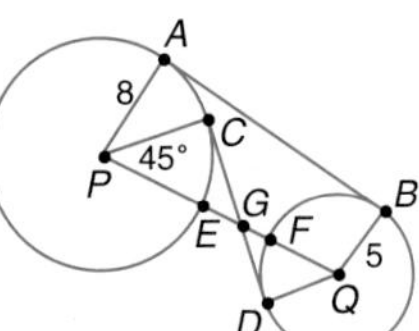

1. $m\widehat{CE}$ 45
2. $m\angle PCG$ 90
3. $m\angle CGP$ 45
4. CG 8
5. $m\angle FQD$ 45
6. $m\widehat{DF}$ 45
7. DQ 5
8. DG 5

For each circle C, find the value of x. Assume that segments that appear to be tangent are tangent.

9. 12

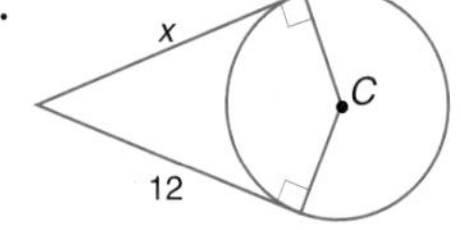

10. 8

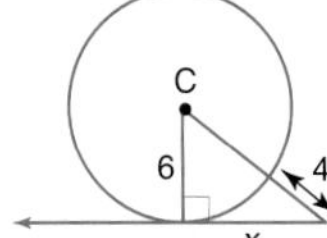

11. 8

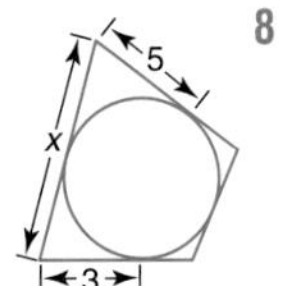

Lesson 9-6 $\overline{AF}$ and $\overline{AB}$ are tangent to $\odot Y$, and $m\widehat{BC} = 84$, $m\widehat{CD} = 38$, $m\widehat{DE} = 64$, and $m\widehat{EF} = 60$. Find each measure.

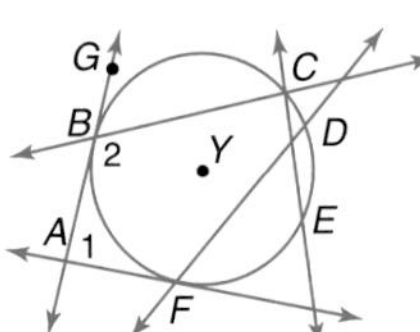

1. $m\widehat{BF}$ 114
2. $m\widehat{BDF}$ 246
3. $m\angle 1$ 66
4. $m\widehat{BFC}$ 276
5. $m\angle 2$ 138
6. $m\angle GBC$ 42

Find the value of x.

7. 145

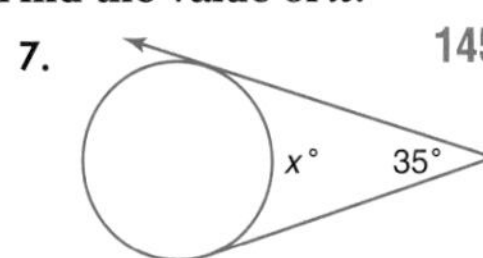

8. 50

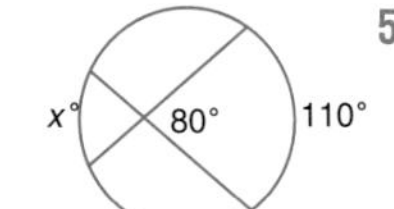

9. 10

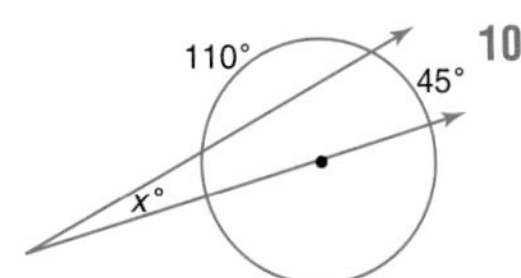

Extra Practice

Extra Practice 781

Additional Answers for Lesson 9-8

3.

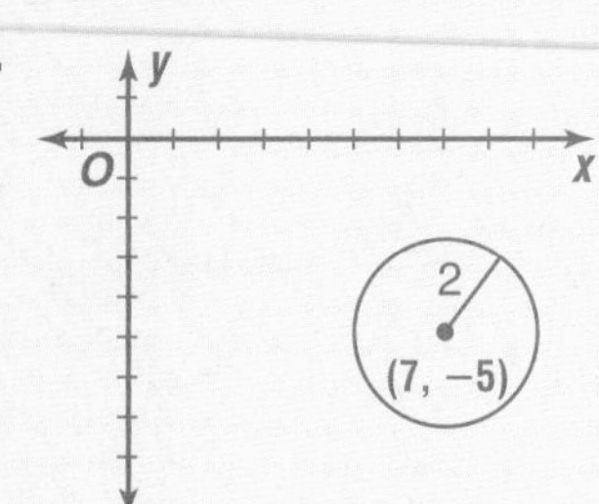

4.

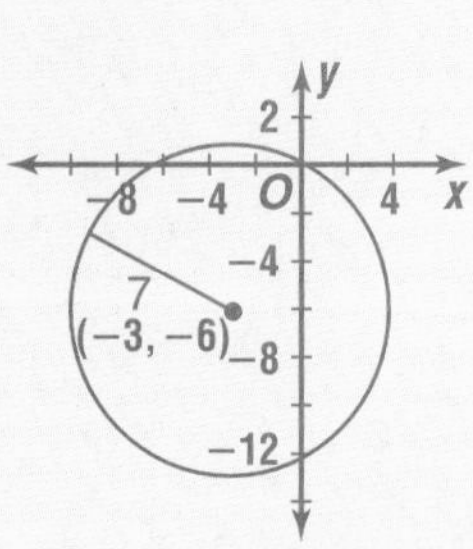

EXTRA PRACTICE

Lesson 9-7 **Find the value of x to the nearest tenth. Assume segments that appear tangent to be tangent.**

1. 1.8

2.

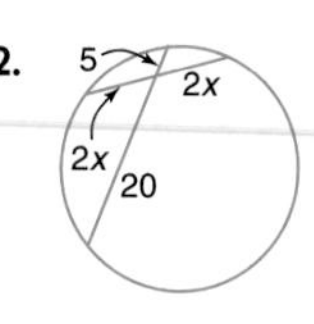

5

3. 2.1

In ⊙L, $\overline{NM}$ and $\overline{NO}$ are tangent segments, $LO = 8$, $NM = 13$, and $\overline{QN}$ is a secant segment. Find each measure.

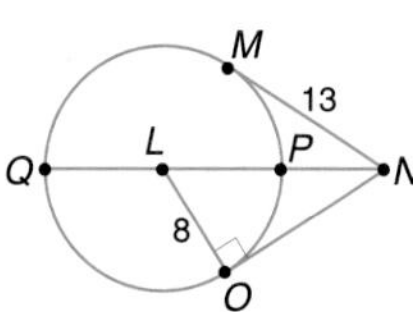

4. NO 13 **5.** NL 15.3 **6.** PN 7.3

Lesson 9-8 **Determine the coordinates of the center and the measure of the radius for each circle whose equation is given.**

1. $(x + 2)^2 + (y - 7)^2 = 81$ (−2, 7); 9

2. $\left(x - \frac{2}{3}\right)^2 + (y + 4)^2 - 169 = 0$ $\left(\frac{2}{3}, -4\right)$; 13

Graph each circle whose equation is given. Label the center and measure of the radius on each graph.

3. $(x - 7)^2 + (y + 5)^2 = 4$

4. $(x + 3)^2 + (y + 6)^2 = 49$ 3–4. See margin.

Write an equation of circle P based on the given information.

5. center: $P(3, 4)$
radius: 6
$(x - 3)^2 + (y - 4)^2 = 36$

6. center: $P(-2, 7)$
radius: $\sqrt{17}$
$(x + 2)^2 + (y - 7)^2 = 17$

Lesson 10-1 **Find the sum of the measures of the interior angles of each convex polygon.**

1. 15-gon 2340 **2.** 59-gon 10,260 **3.** $2t$-gon $360t - 360$

The measure of an exterior angle of a regular polygon is given. Find the number of sides of the polygon.

4. 24 15 **5.** 60 6 **6.** $51\frac{3}{7}$ 7

The number of sides of a regular polygon is given. Find the measures of an interior angle and an exterior angle for each polygon.

7. 4 90, 90 **8.** 16 157.5, 22.5 **9.** s $\frac{180(s-2)}{s}, \frac{360}{s}$

The measure of an interior angle of a regular polygon is given. Find the number of sides in each polygon.

10. 160 18 **11.** 177.6 150 **12.** 120 6

Lesson 10-2 **Determine whether each figure tessellates in a plane. If so, draw a sample figure.**

1. isosceles triangle Yes; see margin.
2. quadrilateral with no two sides congruent no

Determine if each pattern will tessellate.

3. regular octagon and equilateral triangle yes

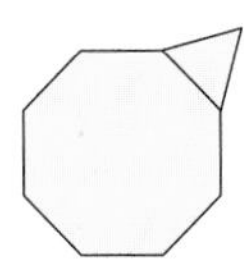

4. regular pentagon and square no

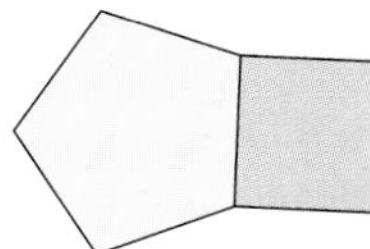

Determine whether each tessellation is *regular*, *uniform*, or *semi-regular*. Name all possibilities.

5. quadrilateral none

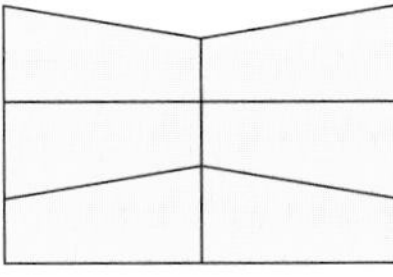

6. rhombus uniform

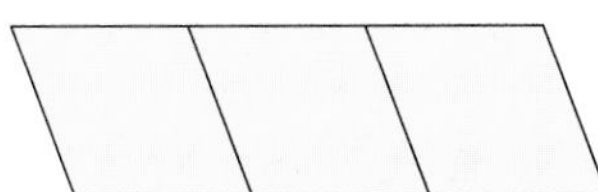

Lesson 10-3 **Find the area of each figure. Assume that angles that appear to be right are right angles.**

1. 56in^2

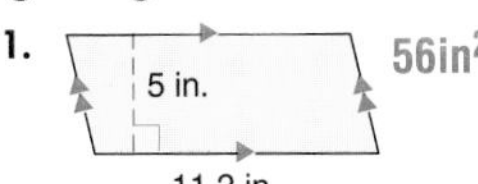

2. 144 yd^2

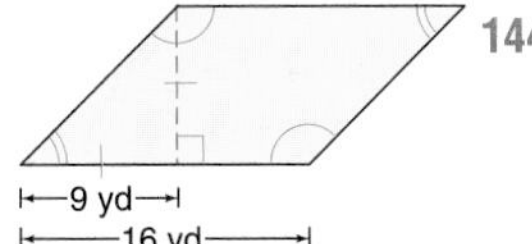

3. 100 mm^2

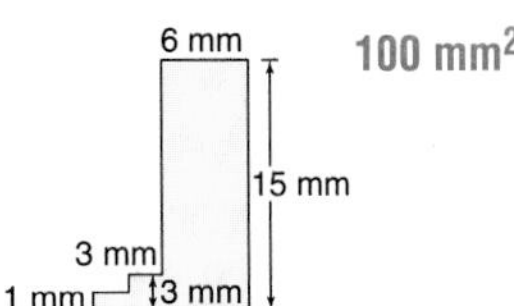

4–5. See margin for graphs.

The coordinates of the vertices of a quadrilateral are given. Graph the points and draw the quadrilateral and an altitude. Identify the quadrilateral as a *square, rectangle*, or a *parallelogram*, and find its area.

4. (3, 4), (2, 1), (8, 4), (9, 7)
parallelogram; $A = 15$ units^2
5. (0, −5), (3, −4), (1, 2), (−2, 1)
rectangle; $A = 20$ units^2

Lesson 10-4 **Find the area of each figure.**

1. 52.5 ft^2

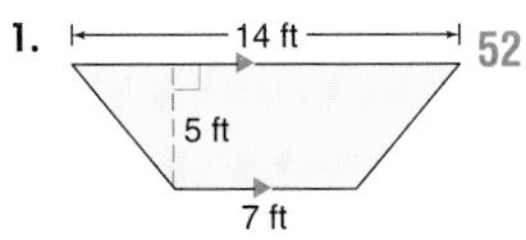

2. 56 in^2

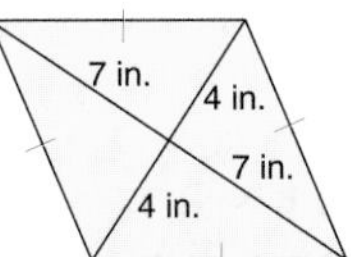

3. $9\sqrt{3} \approx 15.6$ m^2

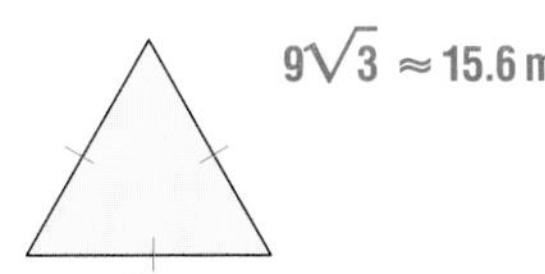

4. The perimeter of a trapezoid is 29 inches. Its nonparallel sides are 4 inches and 5 inches long. If the height of the trapezoid is 3 inches, find its area. 30 in^2
5. A rhombus has a perimeter of 52 units and a diagonal 24 units long. Find the area of the rhombus. 120 units^2
6. The area of an isosceles trapezoid is 36 square centimeters. The perimeter is 28 centimeters. If a leg is 5 centimeters long, find the height of the trapezoid. 4 cm

Additional Answer for Lesson 10-2

1.

Additional Answers for Lesson 10-3

4. (9, 7) (3, 4) (8, 4) (2, 1)

5.

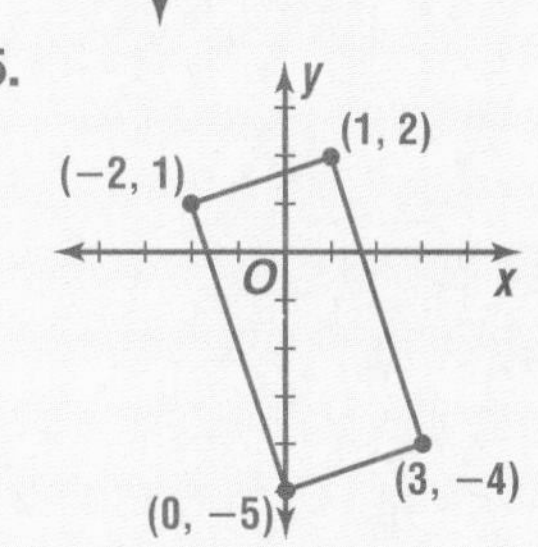

Additional Answer for Lesson 10-6

4a.

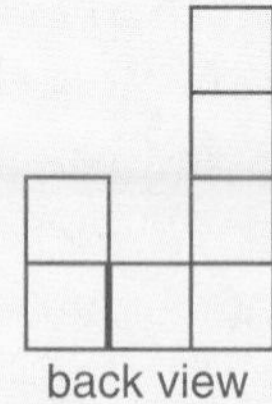

Additional Answers for Lesson 10-7

1. A, 3; B, 4; C, 2; D, 4; E, 3; F, 4
2. A, 3; B, 2; C, 3; D, 2; E, 5; F, 2; G, 3
3. A, 3; B, 6; C, 3

Additional Answers for Lesson 11-1

1. See students' models.

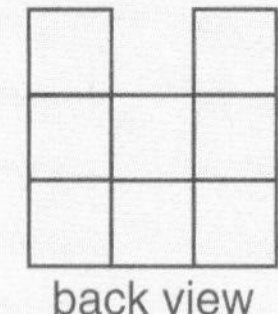

2. See students' models.

back view

3.

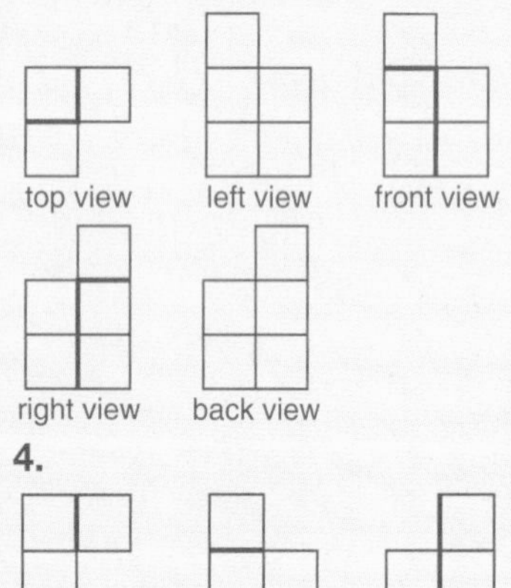

4.

top view left view front view

right view back view

5.

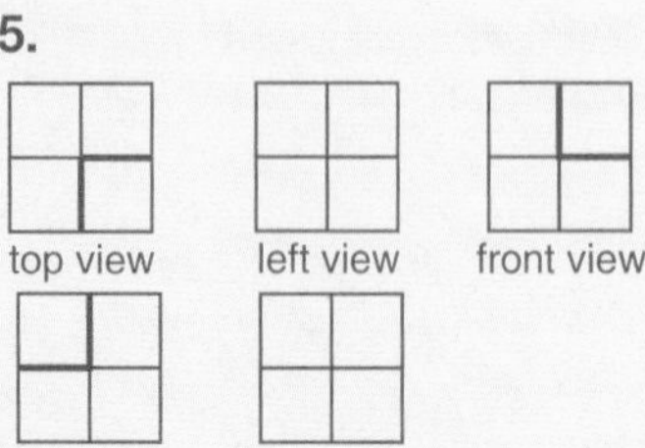

Lesson 10-5 **Find the area of each regular polygon described. Round to the nearest tenth.**

1. triangle with an apothem length of 5.8 centimeters 174.8 cm^2
2. square with apothem length of 8 inches 256 in^2
3. a hexagon with a side length of 19.1 millimeters 947.8 mm^2
4. a pentagon with side length of 13.0 miles 290.8 mi^2
5. Find the circumference and the area of a circle with radius of 18 inches. Round to the nearest tenth. 113.1 in.; 1017.9 in^2

Find the area of each shaded region. Assume that all polygons are regular. Round to the nearest tenth.

6.

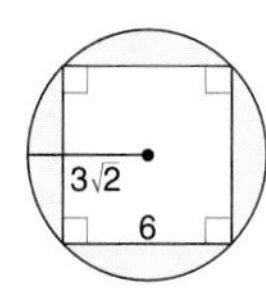

7.

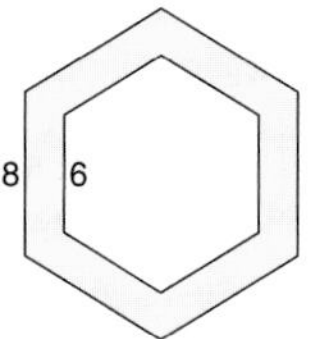

8.

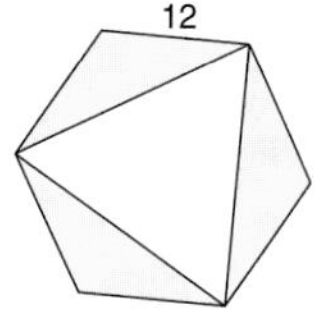

$18\pi - 36 \approx 20.5$ units^2 $42\sqrt{3} \approx 72.7$ units^2 $108\sqrt{3} \approx 187.1$ units^2

Lesson 10-6 **Find the probability that a point chosen at random in each figure lies in the shaded region.**

1. 0.50

2. 0.57

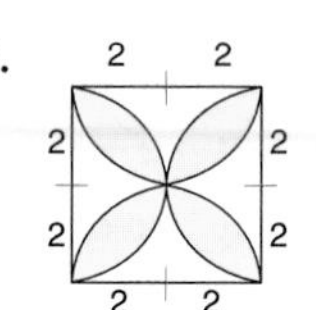

3. 0.19

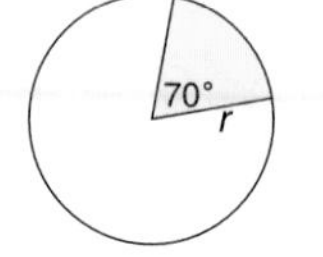

4. A, B, C, D, and E are consecutive collinear points such that $\overline{AB} \cong \overline{BD}$, $\overline{BC} \cong \overline{CD}$, and $\overline{AB} \cong \overline{DE}$.
   a. Draw a diagram of the figure. See margin.
   b. What is the probability that a random point will fall on $\overline{AC}$? $\frac{1}{2}$
   c. What is the probability that a random point will fall on $\overline{AD}$? $\frac{2}{3}$

Lesson 10-7 **Find the degree of each node in each network.** 1–3. See margin.

1.

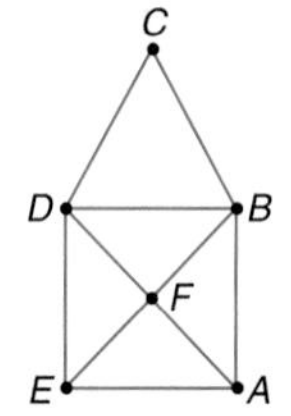

2.

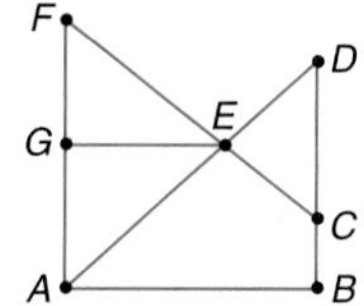

3.

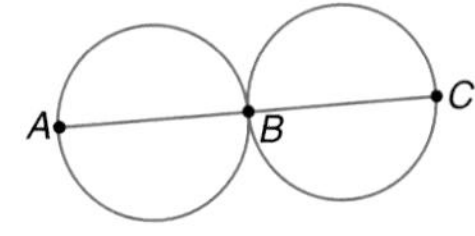

Determine if each network is traceable and complete. If a network is not complete, name the edges that need to be added to make it complete.

4. no; yes

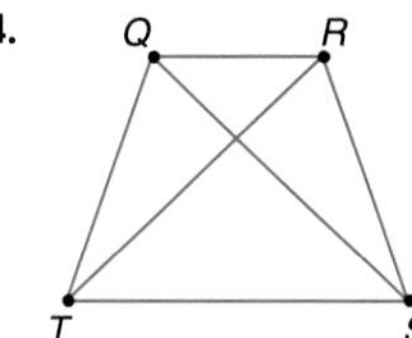

5. yes; yes

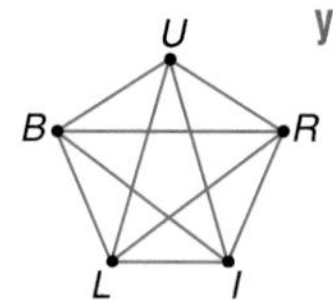

6. yes; no $\overline{AR}$, $\overline{JY}$, $\overline{JR}$, $\overline{JG}$, $\overline{GY}$

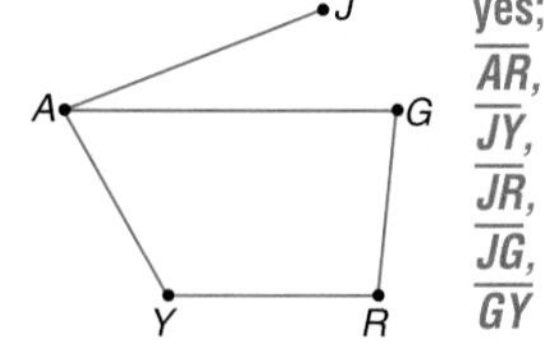

Lesson 11-1 Various views of a solid figure are given. The edge of one block represents one unit of length. A dark segment indicates a break in the surface. Make a model of each figure. Then draw the back view of the figure. 1–2. See margin.

1.

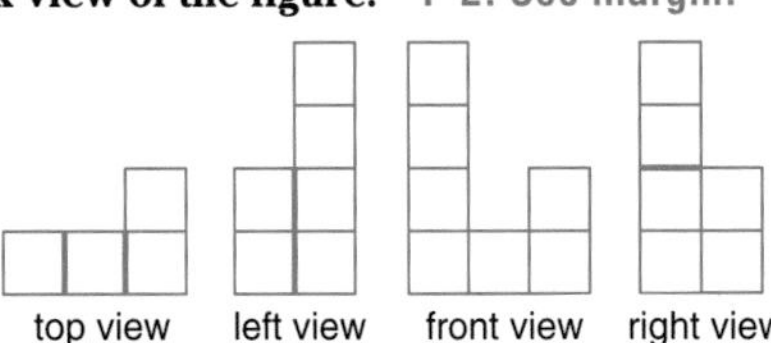

2.

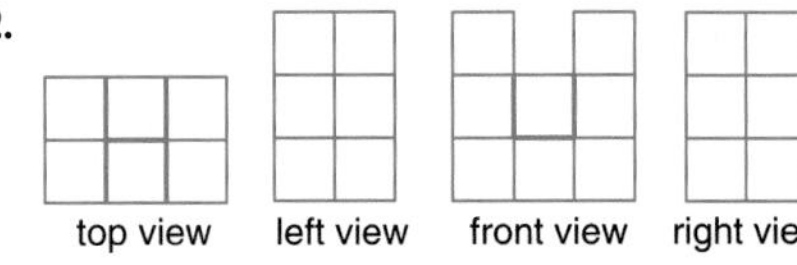

The corner view of a figure is given. Draw the top, left, front, right, and back views of the figure.

3. 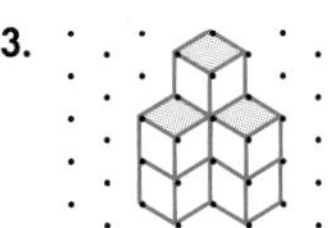

4. 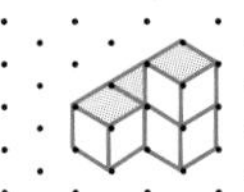

5. 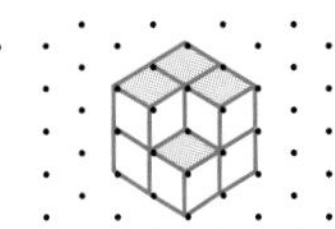

3–5. See margin.

Lesson 11-2 Use isometric dot paper to draw each polyhedron. 1–3. See margin.

1. a rectangular prism 4 units high, 7 units long, and 2 units wide
2. a cube 3 units on each edge
3. a triangular prism 4 units high, whose bases are equilateral triangles with sides 2 units long

Use rectangular dot paper to draw a net for each solid. Then find the surface area of the solid.

4. 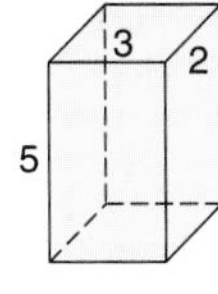

62 units^2

5. 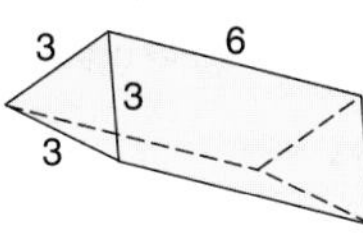

61.8 units^2

6. 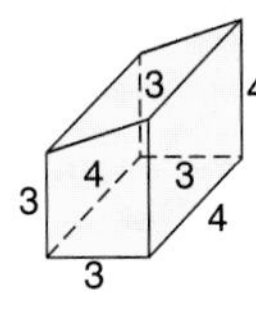

73.6 units^2

4–6. See margin for nets.

Lesson 11-3 Find the surface area of each right prism. Round to the nearest tenth.

1. 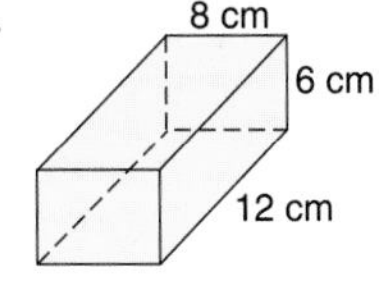

432 cm^2

2. 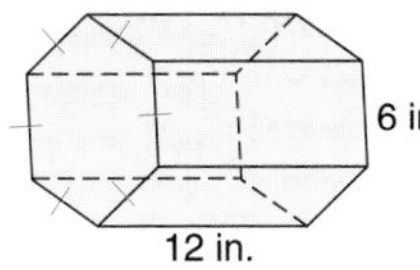
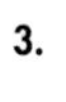

619.1 in^2

3. 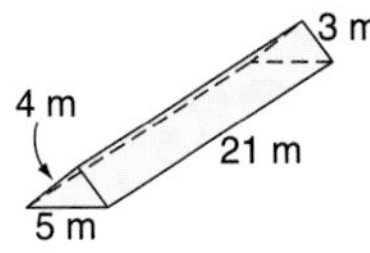

264 m^2

Find the lateral area and surface area of each right cylinder. Round to the nearest tenth.

4. 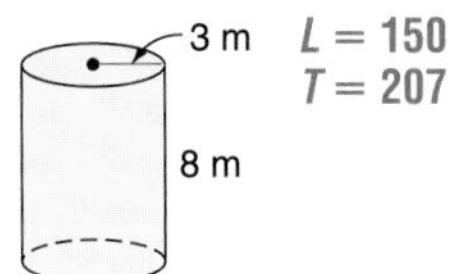

$L = 150.8$ m^2, $T = 207.3$ m^2

5. 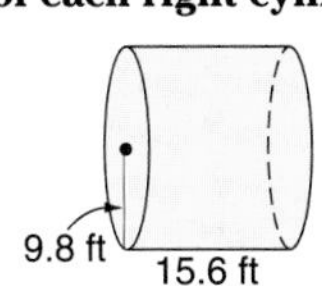

$L = 960.6$ ft^2, $T = 1564.0$ ft^2

6. 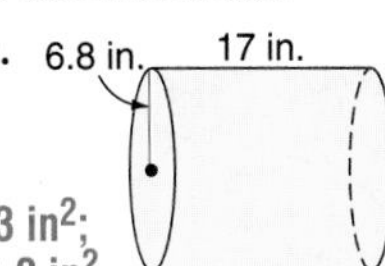

6. $L = 726.3$ in^2; $T = 1016.9$ in^2

Additional Answers for Lesson 11-2

1. 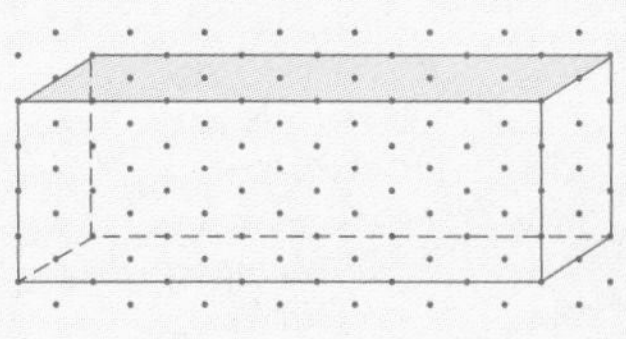

2. 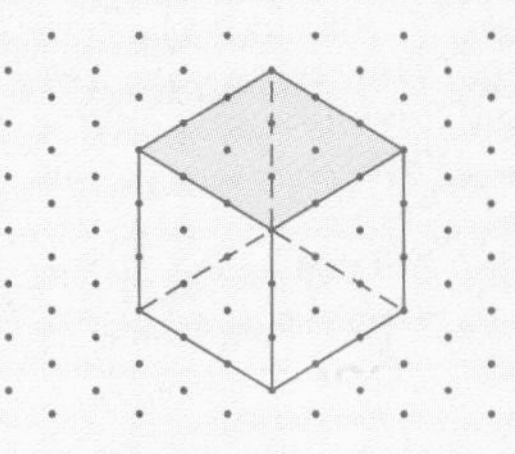

3. 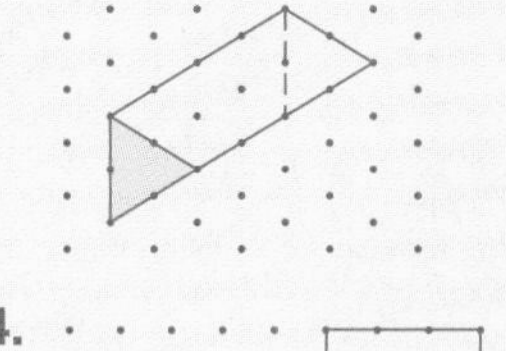

4. 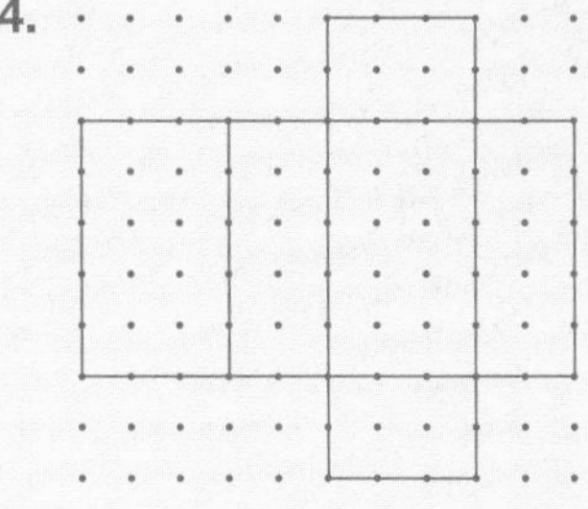

Additional Answers for Lesson 11-2

5. 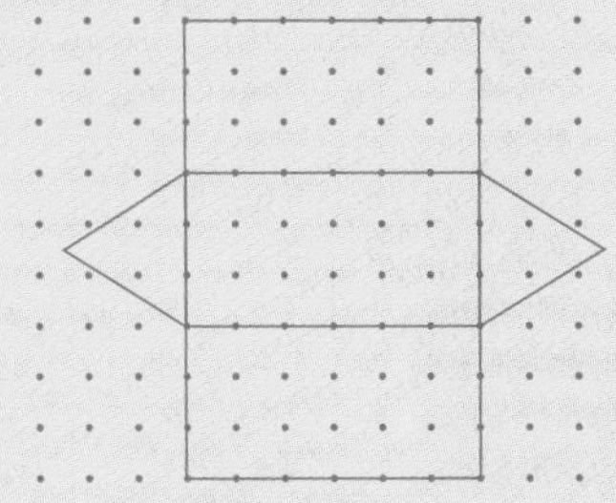

6. 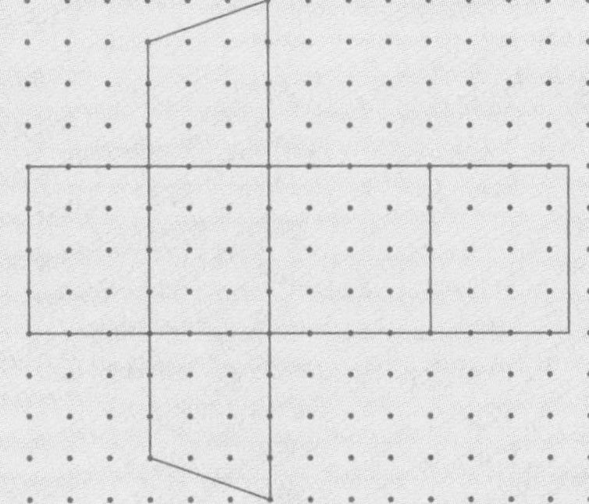

Extra Practice

EXTRA PRACTICE

Lesson 11-4 Find the surface area of each solid. Round to the nearest tenth.

1. 15 in., 16 in. 628.3 in^2

2. 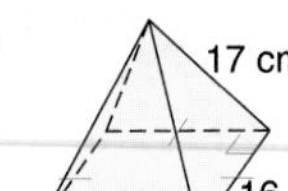

736 cm^2

3. 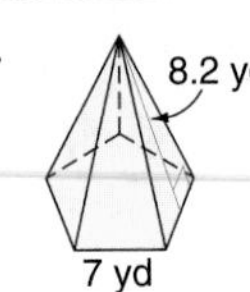

227.8 yd^2

4. 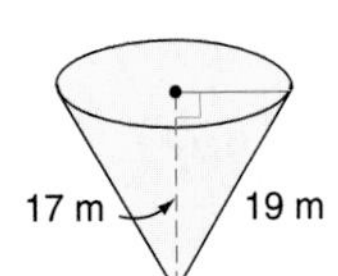

732.7 m^2

5. 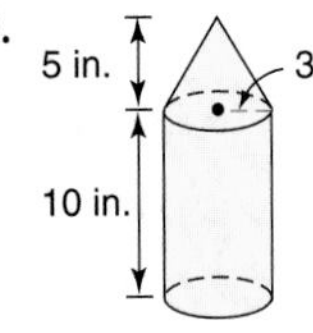

271.7 in^2

6. 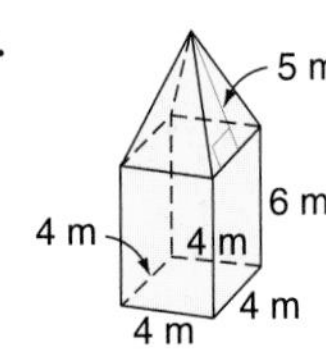

152 m^2

Lesson 11-5 Find the volume of each prism or right cylinder. Round to the nearest tenth.

1. 36 ft, 36 ft, 40 ft, 25 ft

14,966.6 ft^3

2. 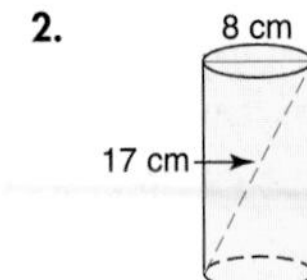

754.0 cm^3

3. 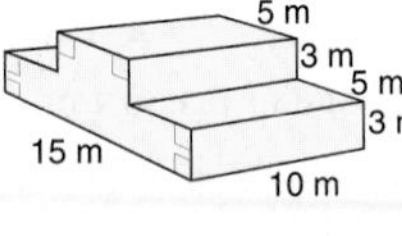

600 m^3

4. The base of a right prism has an area of 17.5 square centimeters. The prism is 14 centimeters high. Find the volume of the prism. 245 cm^3

5. Find the volume of a right cylinder whose radius is 3.2 centimeters and height is 10.5 centimeters. 337.8 cm^3

6. Find the volume of a right prism whose base has an area of 16 square feet and a height of 4.2 feet. 67.2 ft^3

7. Find the volume of a cube for which a diagonal of one of its faces measures 12 meters. 610.9 cm^3

Lesson 11-6 Find the volume of each pyramid or cone. Round to the nearest tenth.

1. 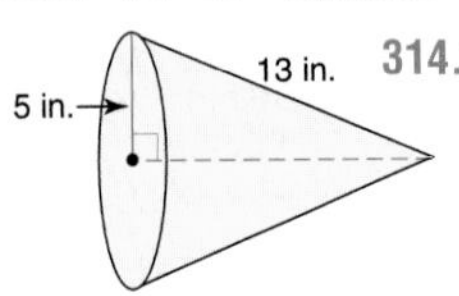

314.2 in^3

2. 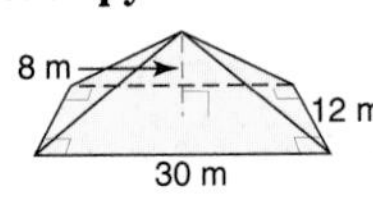

960 m^3

3. 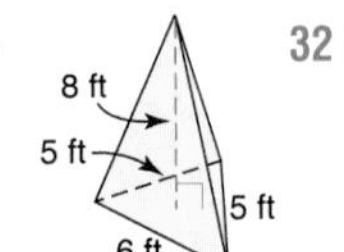

32 ft^3

4. 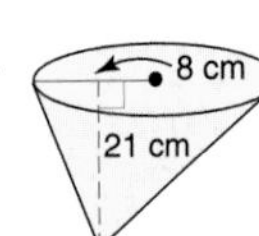

1407.4 cm^3

5. 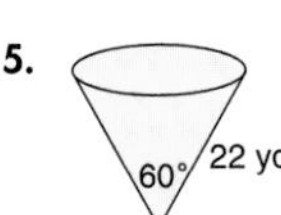

2414.2 yd^3

6. 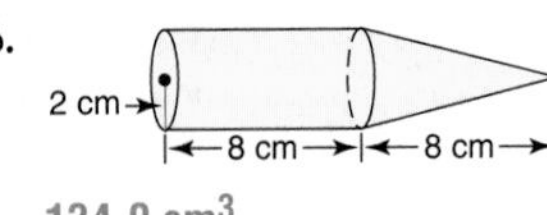

134.0 cm^3

786 Extra Practice

Lesson 11-7 **Determine whether each statement is *true* or *false*.**

1. All diameters of a sphere are congruent. true
2. A plane and a sphere may intersect in exactly two points. false
3. A diameter of a great circle is a diameter of the sphere. true
4. All great circles of the same sphere are congruent. true

In the figure, P is the center of the sphere, and plane $\mathcal{B}$ intersects the sphere in $\odot R$.

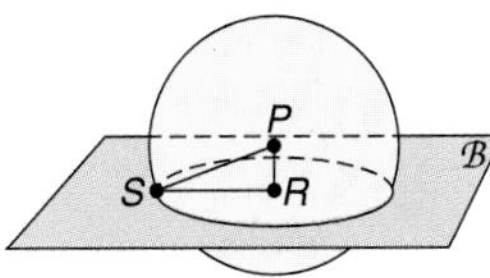

5. Suppose $PS = 15$ and $PR = 9$, find RS. 12
6. Suppose $PS = 26$ and $RS = 24$, find PR. 10

Find the surface area and volume of each sphere described below. Round to the nearest tenth.

7. The diameter is 400 feet long. $T \approx 502{,}654.8$ ft^2, $V \approx 33{,}510{,}321.6$ ft^3
8. A great circle has a circumference of 18.84 meters. $T \approx 113.0$ m^2, $V \approx 112.9$ m^3

Lesson 11-8 **Determine if each pair of solids is *similar*, *congruent*, or *neither*.**

1.
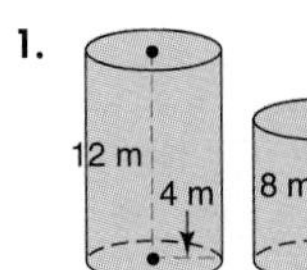

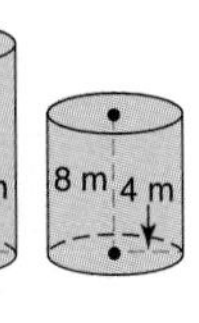

neither

2.
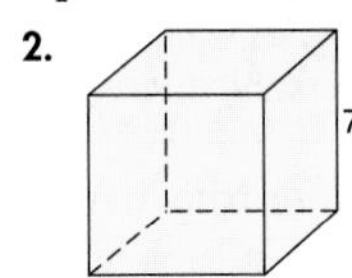

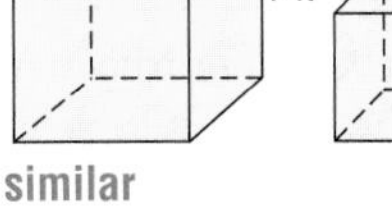

similar

3.
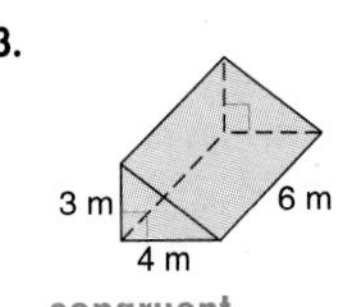

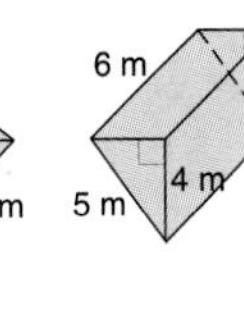

congruent

The two right rectangular prisms shown at the right are similar.

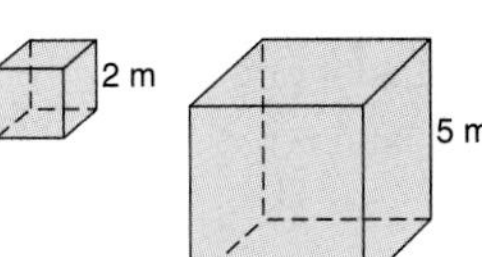

4. Find the ratio of the surface areas. 4:25
5. Suppose the volume of the smaller prism is 18 cubic meters. Find the volume of the larger prism. 281.25 m^3

Lesson 12-1 **Find the x- and y-intercepts and slope of the graph of each equation.**

1. $4x - y = 4$ 1; -4; 4
2. $x = 4$ 4; no y-intercept; slope is undefined.
3. $x + 2y = 6$ 6; 3; $-\frac{1}{2}$
4. $3x - 6y = 6$ 2; -1; $\frac{1}{2}$

Graph each equation. Explain the method you used to draw the graph. 5–8. See margin.

5. $y = 4x - 2$
6. $5x + 2y = 10$
7. $y = 2x - 10$
8. $3x - y = 6$

Use the description to graph each line. 9–12. See margin.

9. $m = -\frac{1}{3}$; $b = -2$
10. parallel to y-axis through $A(2, -1)$
11. perpendicular to the line $y = 2x - 4$ through $G(0, 3)$
12. $m = \frac{2}{3}$; passes through $D(-2, 1)$

Additional Answers for Lesson 12-1

5.
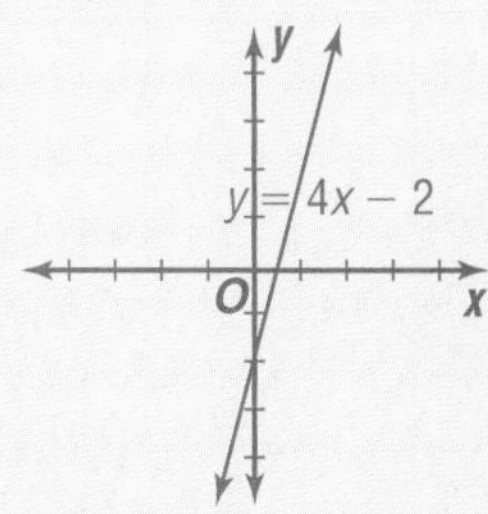

Sample answer: slope-intercept

6.
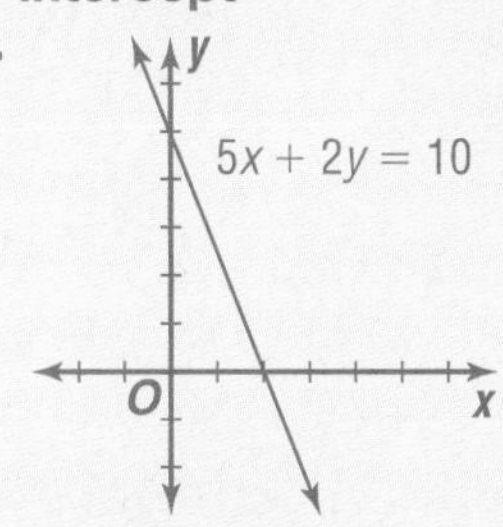

Sample answer: x- and y-intercepts

7.
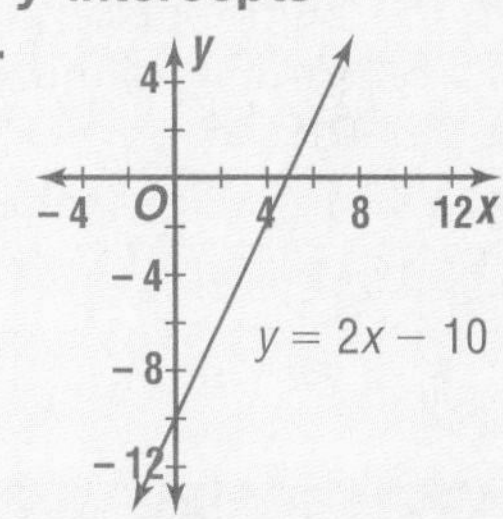

Sample answer: slope-intercept

8.
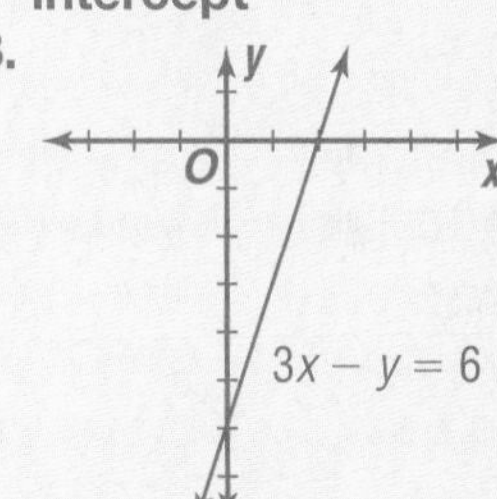

Sample answer: x- and y-intercepts

9.
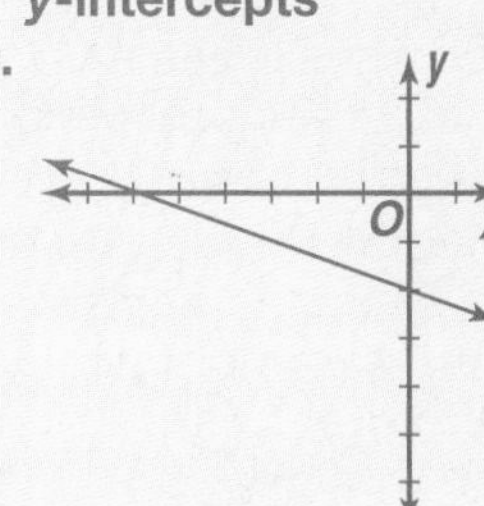

10.
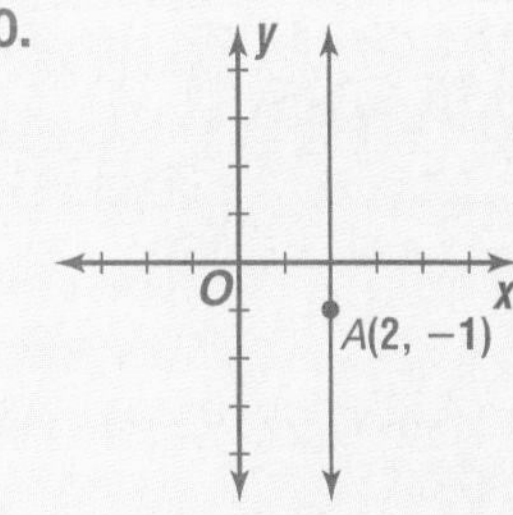

Additional Answers for Lesson 12-1

11.

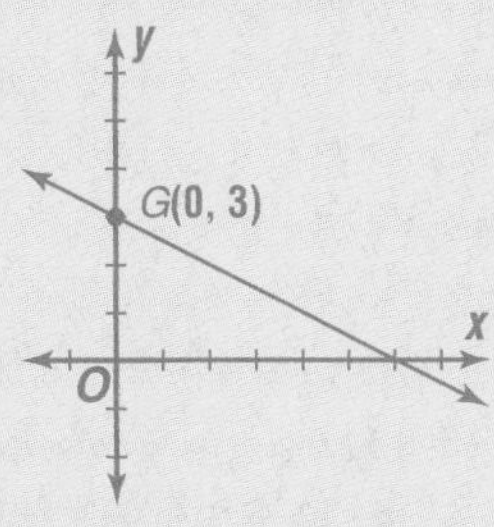

12.
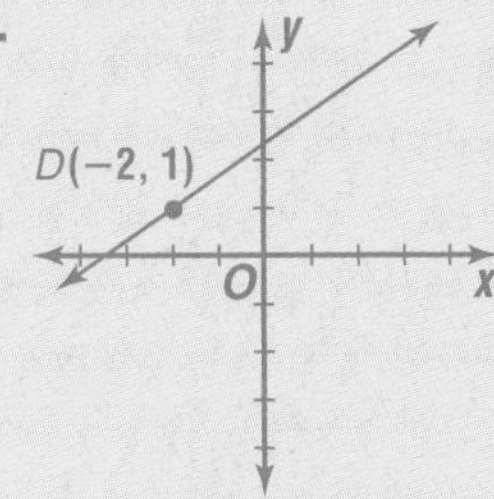

Additional Answers for Lesson 12-3

4a.

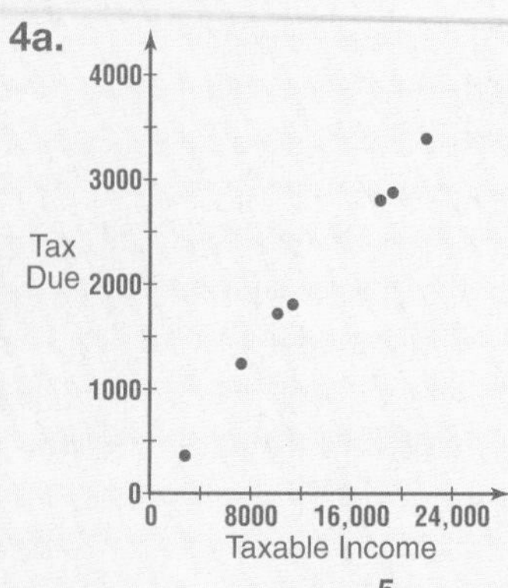

5. $y = -2x - 2$, $y = \frac{5}{2}x + 7$, $y = \frac{1}{4}x + \frac{5}{2}$
6. $y = -x + 4$, $y = 2x + 4$, $y = 4$

Additional Answers for Lesson 12-4

3. Proof: Use the distance formula to find AB and CB.

$AB = \sqrt{(a-0)^2 + (b-0)^2} = \sqrt{a^2 + b^2}$

$CB = \sqrt{(a-2a)^2 + (b-0)^2} = \sqrt{a^2 + b^2}$

Thus, $AB = CB$ and $\overline{AB} \cong \overline{CB}$. By the definition of isosceles, $\triangle ABC$ is an isosceles $\triangle$.

4. Proof:
Find the slope of $\overline{PQ}$ and $\overline{RQ}$.

$m(\overline{PQ}) = \frac{a-0}{a-0} = \frac{a}{a} = 1$

$m(\overline{RQ}) = \frac{a-0}{a-2a} = \frac{a}{-a} = -1$

Thus, the slopes are negative reciprocals of each other. So, $\overline{PQ}$ and $\overline{RQ}$ are $\perp$. By def. of $\perp$ lines, the angles formed are right angles. Therefore, $\triangle PQR$ is a right triangle.

Additional Answers for Lesson 12-5

1.

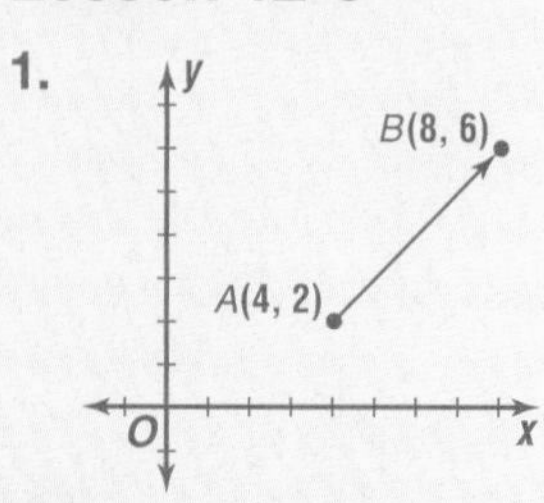

Lesson 12-2 Write the equation in slope-intercept form of the line having the given slope and passing through the point with the given coordinates.

1. -4, $(-3, -2)$ $y = -4x - 14$
2. $\frac{1}{6}$, $(12, -3)$ $y = \frac{1}{6}x - 5$
3. 0, $(6, 7)$ $y = 7$

Write the equation in slope-intercept form of the line that satisfies the given conditions.

4. $m = \frac{3}{4}$; y-intercept $= 8$ $y = \frac{3}{4}x + 8$
5. parallel to $y = -4x + 1$; passes through the point at $(-3, 1)$ $y = -4x - 11$
6. perpendicular to the y-axis; passes through the point at $(-8, 2)$ $y = 2$
7. passes through points at $(0, 3)$ and $(4, -3)$ $y = -\frac{3}{2}x + 3$

Lesson 12-3 Determine whether the three points listed are collinear.

1. $A(9, 0)$, $B(4, 2)$, $C(2, -1)$ no
2. $X(6, 9)$, $Y(3, -1)$, $Z(4, 0)$ no
3. $L(0, 4)$, $M(2, 3)$, $N(-4, 6)$ yes
4. The table below lists the Federal Income Tax due from a single person with the given taxable income for 1995.

Taxable Income	11,905	7412	22,898	19,054	10,995	3268	18,753
Tax Due	1789	1114	3491	2861	1646	491	2816

a. Draw a scatter plot to show how taxable income x and Federal Income Tax due y are related. See margin.
b. Write an equation that relates a singe person's taxable income and their approximate Federal Income Tax due. Sample answer: $y = 0.15x + 3$
c. Angela's taxable income for 1995 was $12,982. Approximately how much did she owe in Federal Income Tax? Sample answer: $1950.30

The vertices of $\triangle RST$ are $R(-6, -8)$, $S(6, 4)$, and $T(-6, 10)$. 5–6. See margin.

5. Write the equations of the lines containing the medians of $\triangle RST$.
6. Write the equations of the lines containing the altitudes of $\triangle RST$.

Lesson 12-4 Name the missing coordinates in terms of the given variables.

1. $\triangle RST$ is isosceles and right. $R(-b, 2b)$

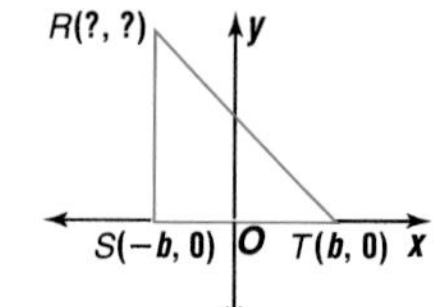

2. $ABCD$ is an isosceles trapezoid. $C(a - b, c)$, $D(0, 0)$

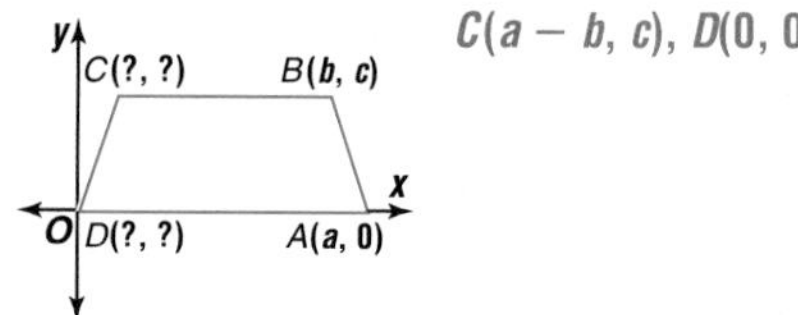

Prove using a coordinate proof. 3–4. See margin.

3. $\triangle ABC$ is an isosceles triangle.

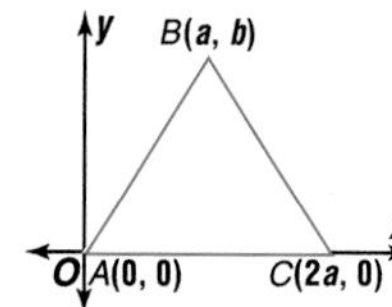

4. $\triangle PQR$ is a right triangle.

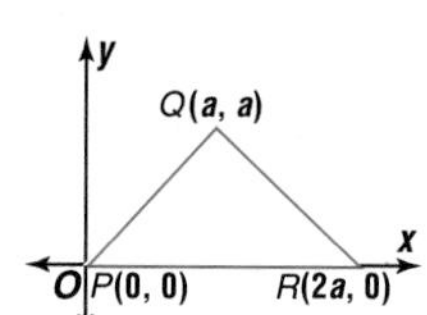

2.

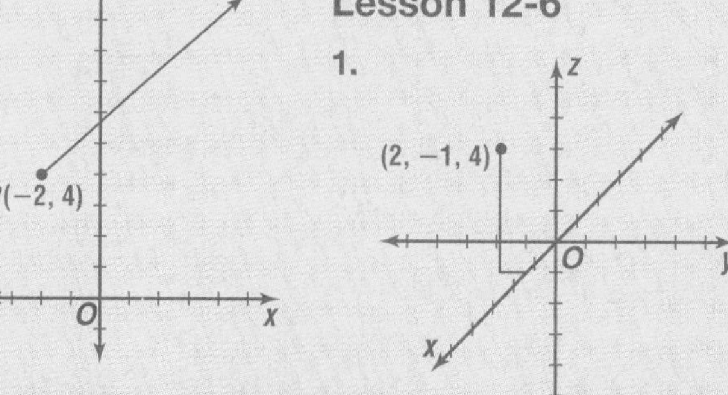

Additional Answers for Lesson 12-6

1.

2.

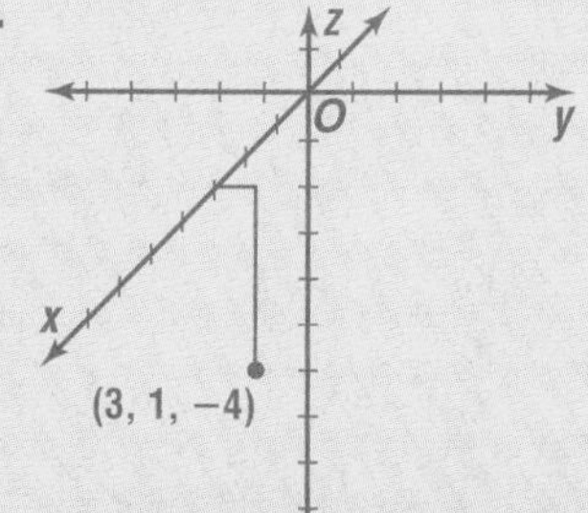

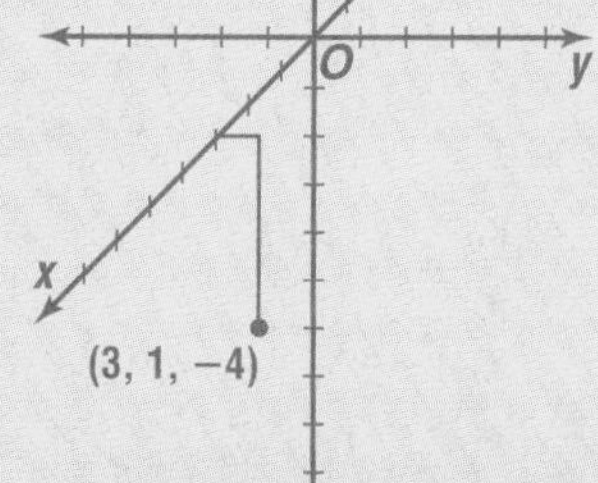

Lesson 12-5 **Sketch each vector. Then find the magnitude to the nearest tenth and the direction to the nearest degree.** 1–2. See margin for drawings.

1. $\overrightarrow{AB}$ if $A(4, 2)$ and $B(8, 6)$ $4\sqrt{2} \approx 5.7$ units, 45°
2. $\overrightarrow{RS}$ if $R(-2, 4)$ and $S(5, 10)$ $\sqrt{85} \approx 9.2$ units, 41°

Find the sum or difference of the given vectors.

3. $\begin{bmatrix} 2 \\ -1 \end{bmatrix} + \begin{bmatrix} -6 \\ 3 \end{bmatrix}$ $\begin{bmatrix} -4 \\ 2 \end{bmatrix}$
4. $\begin{bmatrix} 3 \\ 5 \end{bmatrix} - \begin{bmatrix} 1 \\ 7 \end{bmatrix}$ $\begin{bmatrix} 2 \\ -2 \end{bmatrix}$
5. $\begin{bmatrix} 7 \\ 9 \end{bmatrix} + \begin{bmatrix} 6 \\ -4 \end{bmatrix} - \begin{bmatrix} 12 \\ 3 \end{bmatrix}$ $\begin{bmatrix} 1 \\ 2 \end{bmatrix}$

Given the quadrilateral *RPTQ*, complete each statement.

6. $\overrightarrow{RP} + \overrightarrow{PT} = \underline{\ ?\ }$ $\overrightarrow{RT}$
7. $\overrightarrow{PH} + \overrightarrow{HR} = \underline{\ ?\ }$ $\overrightarrow{PR}$

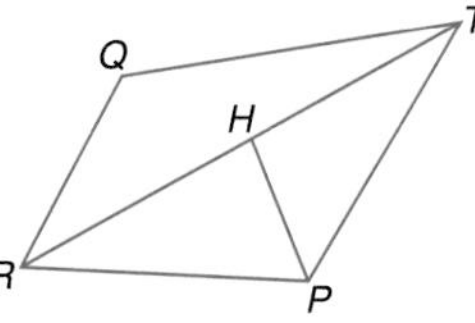

Given $\vec{v} = (2, 5)$ and $\vec{u} = (7, 1)$, represent each of the following as an ordered pair.

8. $\vec{v} + \vec{u}$ (9, 6)
9. $\vec{v} + 2\vec{u}$ (16, 7)
10. $2\vec{u} + 3\vec{v}$ (20, 17)

Lesson 12-6 **Plot each point in a three-dimensional coordinate system.** 1–2. See margin.

1. (2, −1, 4)
2. (3, 1, −4)

Determine the distance between each pair of points, and determine the coordinates of the midpoint of the segment connecting them.

3. $A(0, -2, 5)$ and $B(-3, 4, -2)$ $\sqrt{94} \approx 9.7$; (−1.5, 1, 1.5)
4. $J(9, 1, 0)$ and $K(5, -7, 4)$ $\sqrt{96} \approx 9.8$; (7, −3, 2)

Determine the coordinates of the center and the measure of the radius for each sphere whose equation is given.

5. $(x - 6)^2 + (y + 5)^2 + (z - 1)^2 = 81$ (6, −5, 1); 9
6. $x^2 + (y - 2)^2 + (z - 4)^2 = 4$ (0, 2, 4); 2

Write the equation of the sphere using the given information.

7. The center is at (6, −1, 3), and the radius is 12. $(x - 6)^2 + (y + 1)^2 + (z - 3)^2 = 144$
8. A diameter has endpoints at (−3, 5, 7) and (5, −1, 5). $(x - 1)^2 + (y - 2)^2 + (z - 6)^2 = 26$
9. The center is at (0, −2, 1), and the diameter is 16. $x^2 + (y + 2)^2 + (z - 1)^2 = 64$
10. The center is at (2, −2, −1), and one endpoint of a radius is located at (5, −5, 2). $(x - 2)^2 + (y + 2)^2 + (z + 1)^2 = 27$

Lesson 13-1 **Draw a figure and describe the locus of points in a plane that satisfy each set of conditions.** 1–4. See margin.

1. all points that are 3 inches from a circle with a radius of 6 inches
2. all points equidistant from the endpoints of a given line segment
3. all points that are equidistant from two intersecting lines
4. all points that are the midpoints of parallel chords of a given circle

Draw a figure and describe the locus of points in space that satisfy each set of conditions. 5–8. See margin.

5. all points 4 meters from a given line *m*
6. all points that are equidistant from two intersecting planes
7. all points that are 6 inches from a plane
8. all points equidistant from all points on a circle

Additional Answers for Lesson 13-1

1.

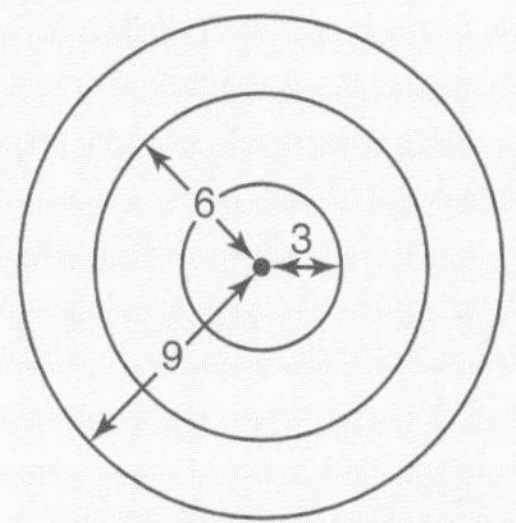

two concentric circles, one with a radius of 3 in. and one with a radius of 9 in.

2.

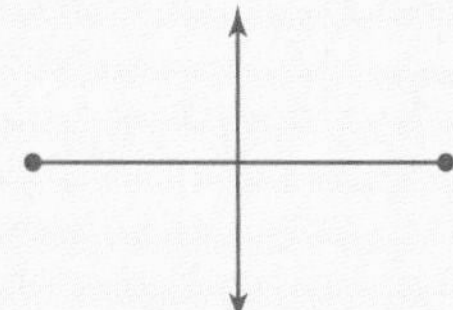

the perpendicular bisector of the line segment

3.

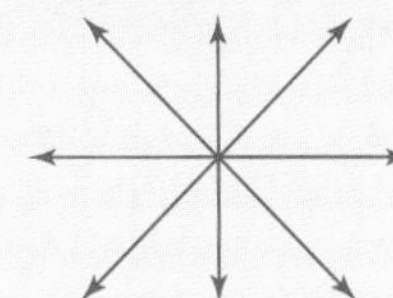

two intersecting lines that are bisectors of the angles formed by the given lines

4.

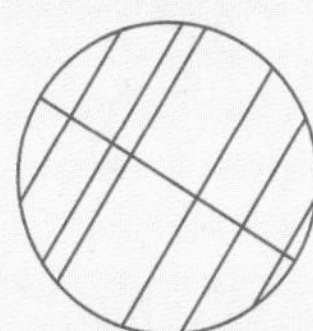

a diameter that is the perpendicular bisector of the parallel chords

5.

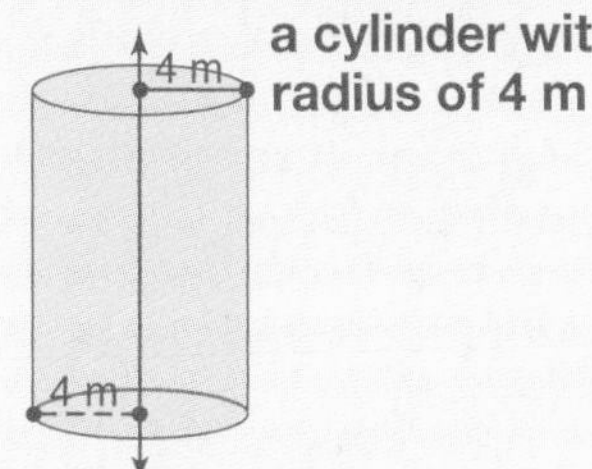

a cylinder with radius of 4 m

6.

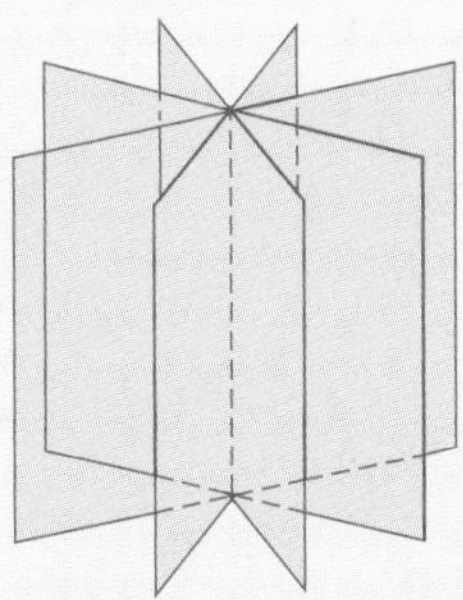

a pair of intersecting planes

Additional Answers for Lesson 13-1

7.

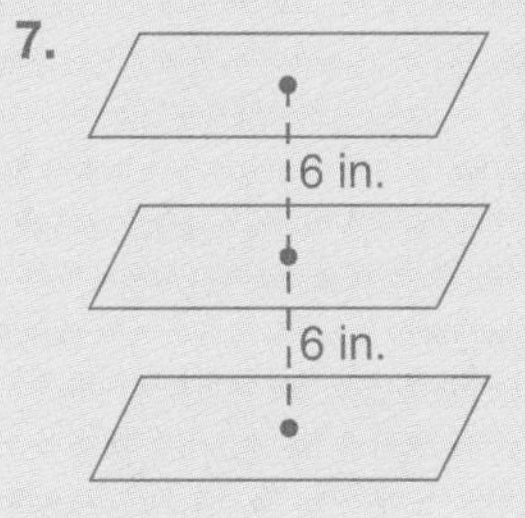

two parallel planes on either side of the given plane

8.

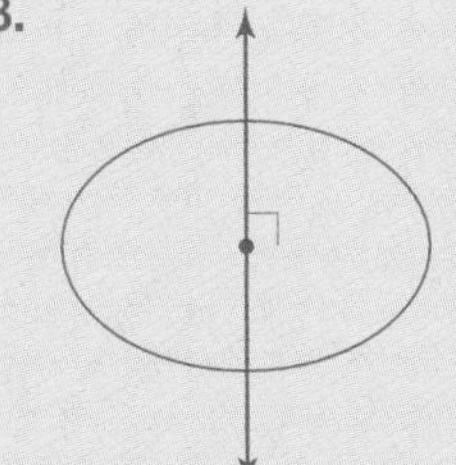

a line ⊥ to the plane of the circle at the center of the circle

Lesson 13-2 **Graph each pair of equations to find the locus of points that satisfy the graphs of both equations.** 1–4. See Solutions Manual for graphs.

1. $3x - 2y = 10$
   $x + y = 0$ (2, −2)
2. $x + 2y = 7$
   $y = 2x + 1$ (1, 3)
3. $x + y = 6$
   $x - y = 2$ (4, 2)
4. $y = x - 1$
   $x + y = 11$ (6, 5)

Use either substitution or elimination to find the locus of points that satisfy the graphs of both equations.

5. $y = 3x$
   $x + 2y = -21$ (−3, −9)
6. $x - y = 5$
   $x + y = 25$ (15, 10)
7. $x - y = 6$
   $x + y = 5$ (5.5, −0.5)
8. $y = x - 1$
   $4x - y = 19$ (6, 5)
9. $x - 2y = 5$
   $3x - 5y = 8$ (−9, −7)
10. $9x + 7y = 4$
    $6x - 3y = 18$ (2, −2)

Lesson 13-3 **Describe the different possible intersections of the figures.** 1–4. See margin.

1. a sphere and a plane
2. two concentric circles and a line
3. a line and a sphere
4. a circle and a plane

Draw a diagram to find the locus of points that statisfy the conditions. Then describe the locus of points. 5–7. See margin.

5. all points in a coordinate plane that are 5 units from the graph of $x = 6$ and equidistant from the graphs of $y = 1$ and $y = 7$
6. all points in a plane that are 2 inches from a given line and 5 inches from a point on the line
7. all points in a plane that are equidistant from two intersecting lines and 3 units from the point of intersection

Describe the geometric figure whose locus in space satisfies the graph of each equation.

8. $(x - 2)^2 + (y + 4)^2 = 36$ 8–9. See margin.
9. $(x - 3)^2 + (y - 4)^2 + (z - 5)^2 = 16$

Lesson 13-4 **Describe the transformations that occurred in the mappings.**

1.  rotation, translation
2. dilation, translation

Each figure below has a preimage that is an isometry. Write the image of each preimage given in Exercises 3–8.

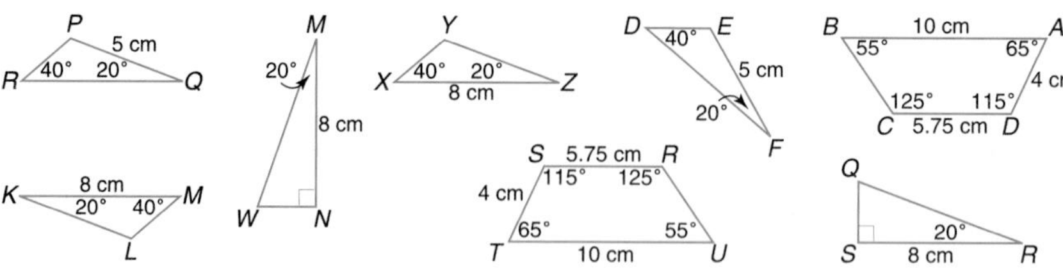

3. $\triangle MWN$ $\triangle RQS$
4. $\triangle PQR$ $\triangle EFD$
5. $\triangle SRQ$ $\triangle NMW$
6. $\triangle LMK$ $\triangle YXZ$
7. quadrilateral $RSTU$ $CDAB$
8. $\triangle ZYX$ $\triangle KLM$

Additional Answers for Lesson 13-3

1. no points, the point of tangency, a circle
2. no points, 1, 2, 3, or 4 points
3. no points, 1 or 2 points
4. no points, 1 or 2 points, a circle
5. 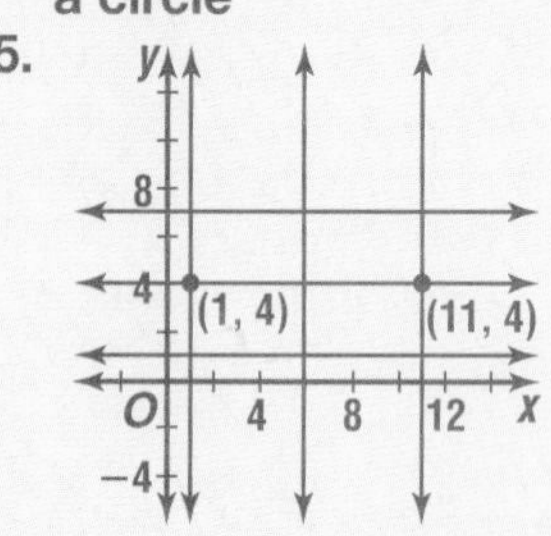

   (1, 4) and (11, 4)
6. 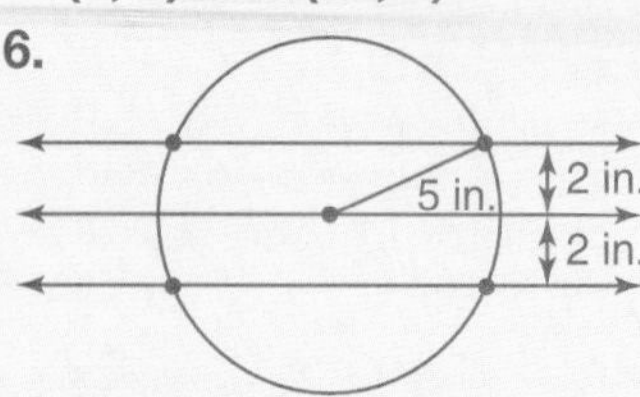

   4 points 5 in. from the given point that are located on 2 parallel lines 2 in. above and below the given line
7. 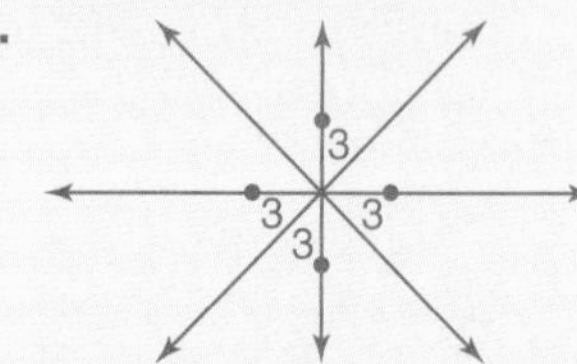

   4 points 3 units from the intersection point on the 2 intersecting lines that bisect the angles formed by the given lines
8. a cylinder with radius 6 units whose axis is perpendicular to the xy-plane through the point at (2, −4, 0)
9. a sphere with radius 4 units and center at (3, 4, 5)

Additional Answers for Lesson 13-5

7. 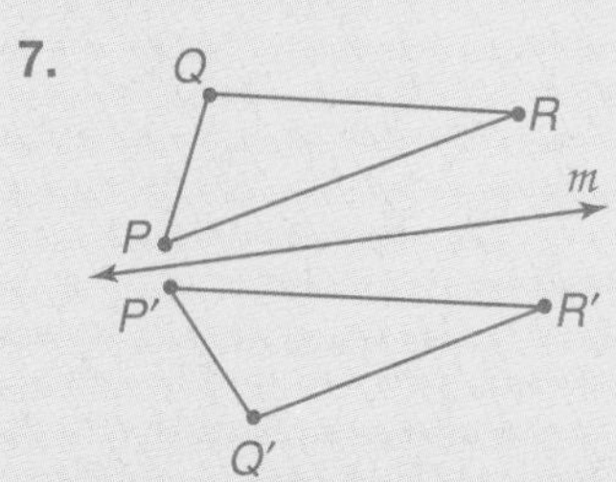

8. 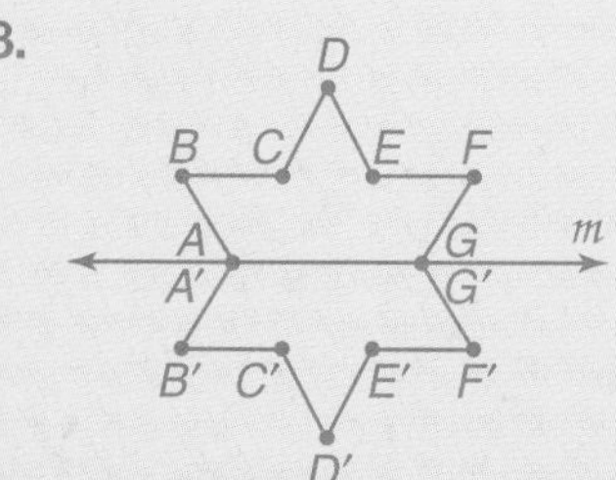

EXTRA PRACTICE

Lesson 13-5 For the figure at the right, name the reflection image of each of the following over line ℓ.

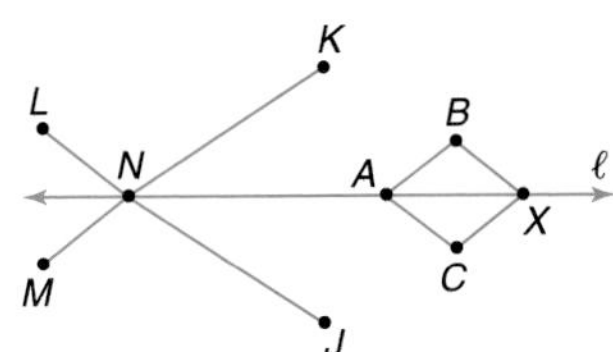

1. M L
2. K J
3. $\overline{KM}$ $\overline{JL}$
4. $\triangle BXA$ $\triangle CXA$
5. N N
6. $\angle LNK$ $\angle MNJ$

Copy each figure. Use a straightedge to draw the reflection image of each figure over line m.

7. 

8. 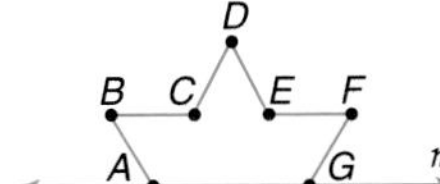

9. 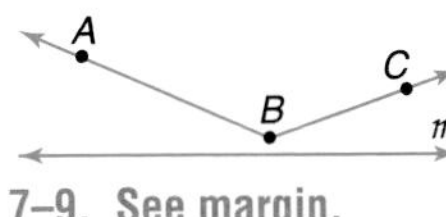

7–9. See margin.

For each figure, determine whether p is a line of symmetry. Write *yes* or *no*. Explain your answer.

10. 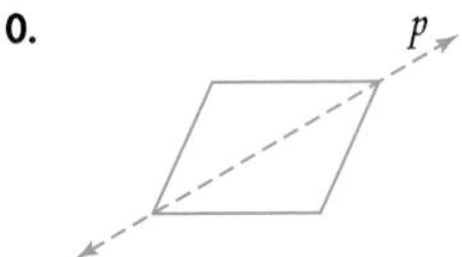

11. 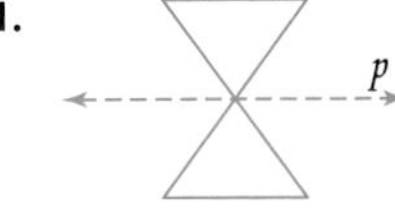

12. 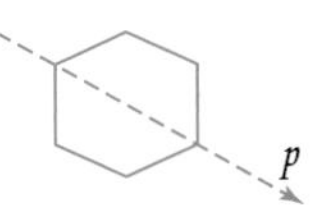

10–12. See margin.

Graph each figure on a coordinate plane. Draw the reflection image for each over the indicated line of reflection. 13–14. See margin.

13. $\overline{ST}$: $S(-1, 2)$, $T(3, 4)$
line of reflection: y-axis

14. $\triangle XYZ$: $X(1, 5)$, $Y(0, -2)$, $Z(-1, 1)$
line of reflection: $x = 1$

Lesson 13-6 For each of the following, lines ℓ and m are parallel. Determine whether each red figure is a translation image of the blue figure. Write *yes* or *no*. Explain your answer.

1. 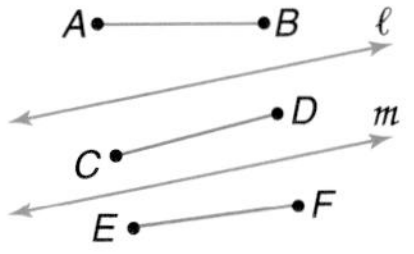 yes

2. 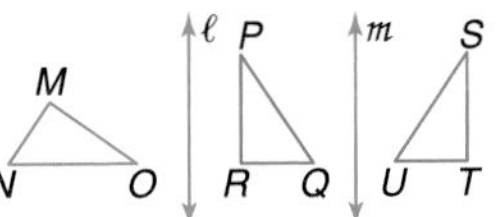

No; the green image is incorrect.

3. 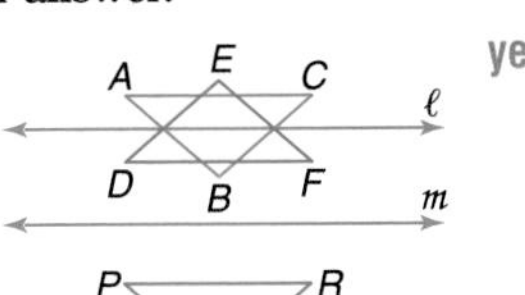 yes

Graph each of the following preimages on a coordinate plane. Then graph the image after performing the translation given, and list the coordinates of the vertices of that image. 4–7. See margin for graphs.

4. $\overline{RS}$, $R(2, 5)$, $S(-4, -2)$; slide 3 right, 4 down $R'(5, 1)$, $S'(-1, -6)$
5. $\overline{AB}$, $A(2, -1)$, $B(0, 2)$; reflect over the graph of $y = 1$ and then over the graph of $y = 4$ $A'(2, 5)$, $B'(0, 8)$
6. $\triangle DFG$, $D(-2, 5)$, $F(3, 6)$, $G(2, 1)$; slide D to $D'(0, 2)$ $F'(5, 3)$, $G'(4, -2)$
7. $\triangle XYZ$, $X(-1, -2)$, $Y(0, 3)$, $Z(1, 1)$; reflect over the graph of $y = -2$ and then over the graph of $y = 0$ $X'(-1, 2)$, $Y'(0, 7)$, $Z'(1, 5)$

Copy each figure on a coordinate plane. Then find the translation image of each geometric figure with respect to the parallel lines m and ℓ.

8. 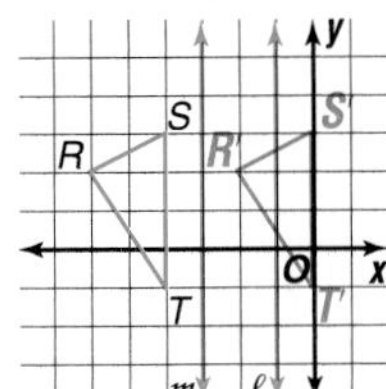

9. 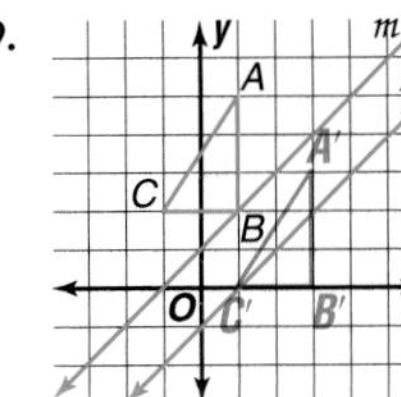

10. 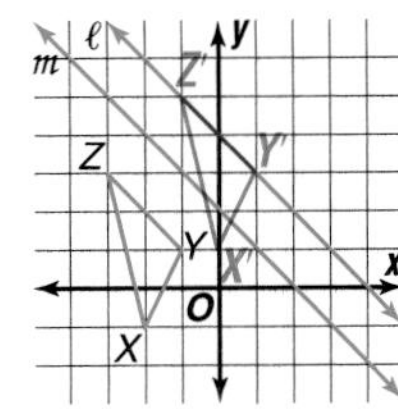

Additional Answers for Lesson 13-5

9. 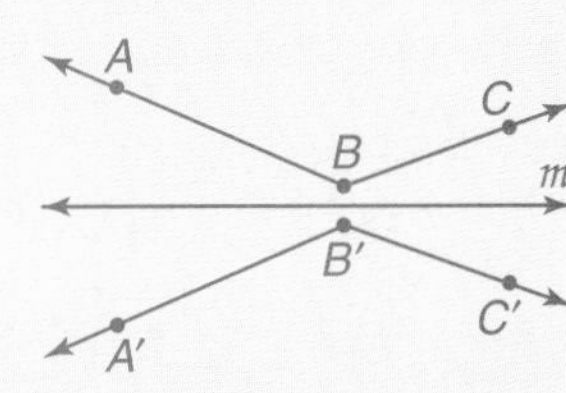

10. No; not all points are the same distance from p.
11. Yes; for any point A on the figure on one side of p, it is possible to find another point B on the figure on the other side of p so that p is the perpendicular bisector of $\overline{AB}$.
12. Yes; for any point A on the figure on one side of p, it is possible to find another point B on the figure on the other side of p so that p is the perpendicular bisector of $\overline{AB}$.

13. 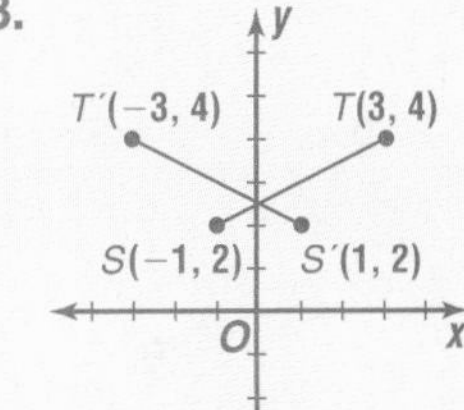

14. 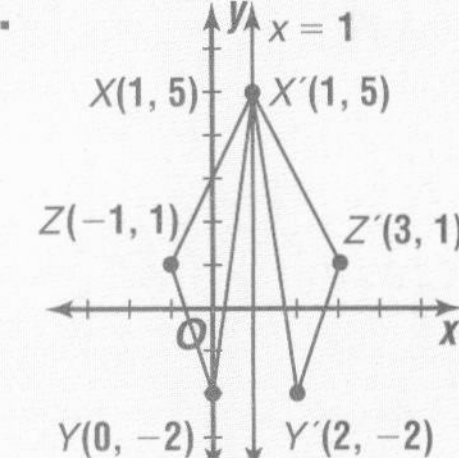

Additional Answers for Lesson 13-6

4. 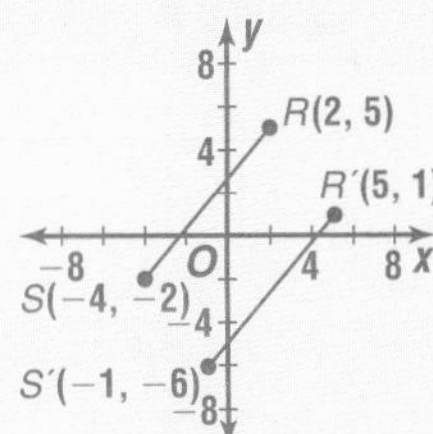

5. 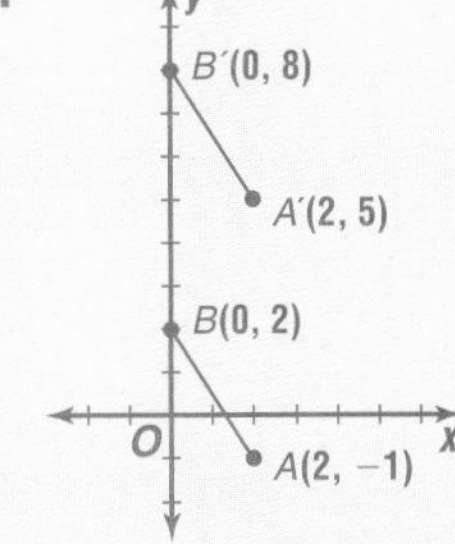

Additional Answers for Lesson 13-6

6. 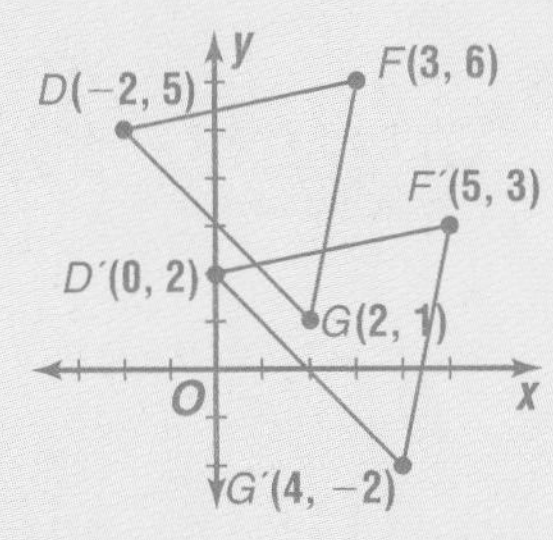

7. 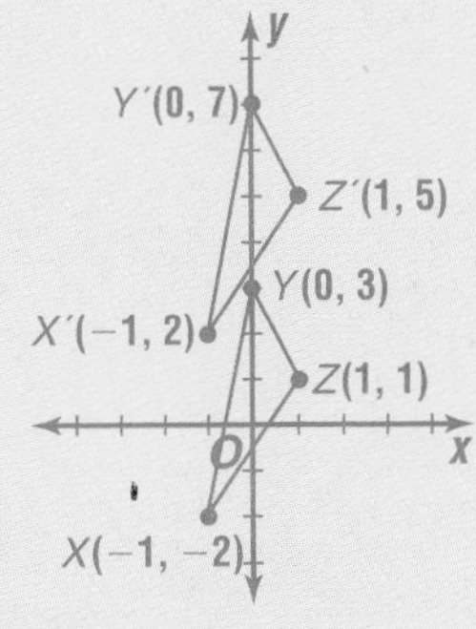

Additional Answers for Lesson 13-7

4.
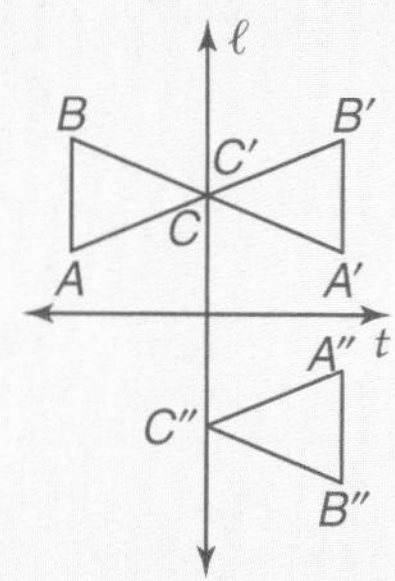

5.
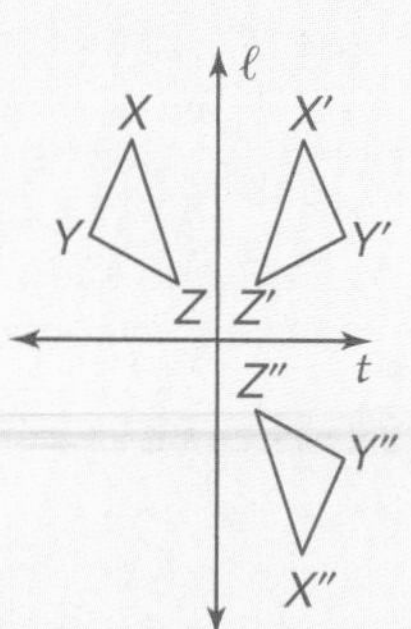

6.
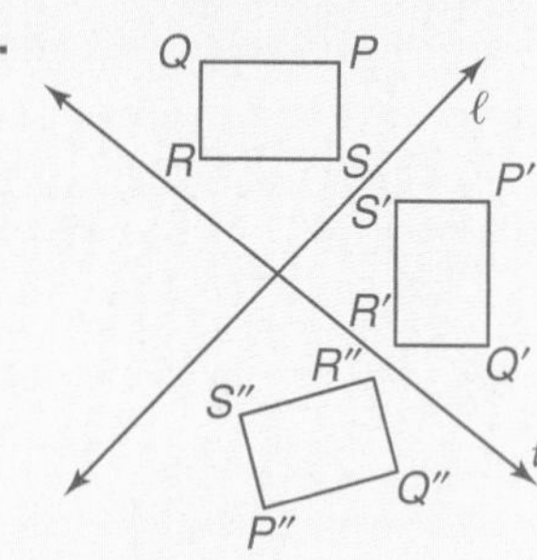

7.
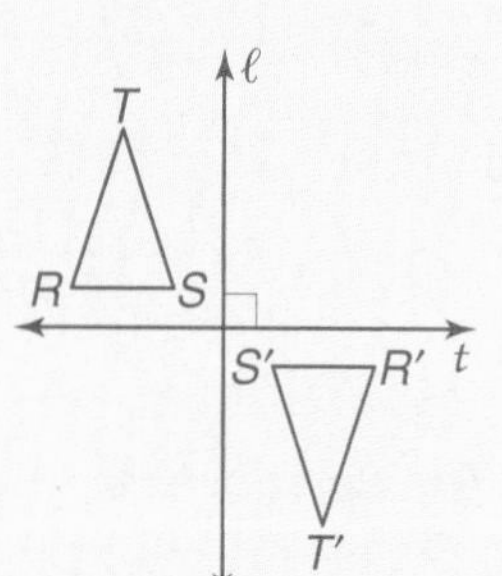

8.
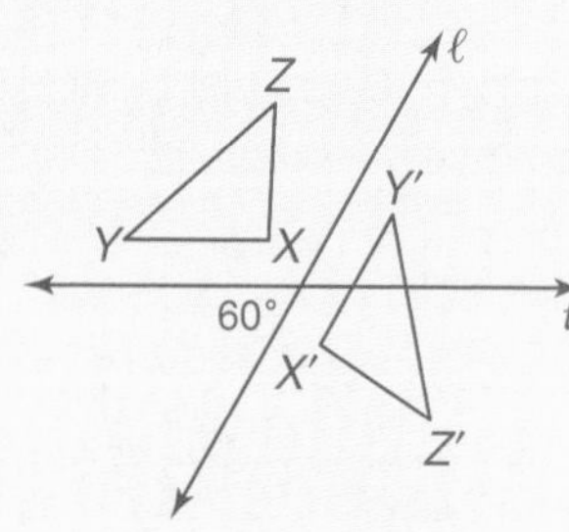

9.
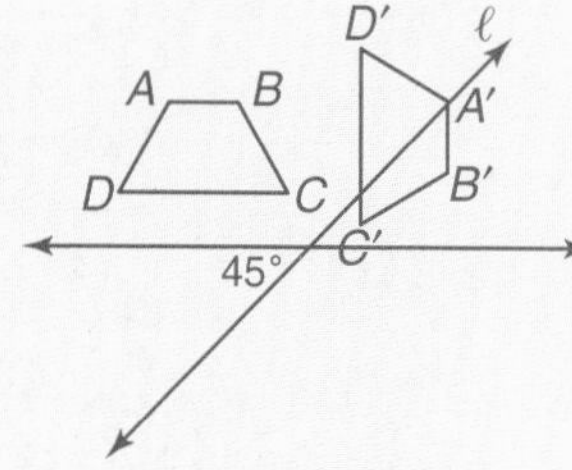

EXTRA PRACTICE

Lesson 13-7 **For each of the following, determine whether the indicated composition of reflections is a rotation. Explain your answer.**

1. 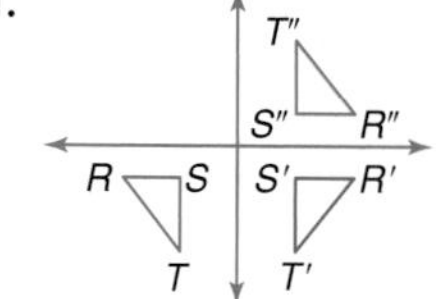

yes

2. 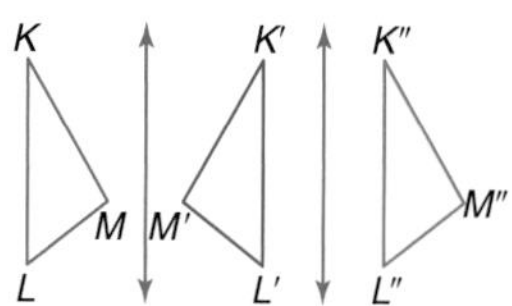

No, the lines of reflection are parallel.

3. 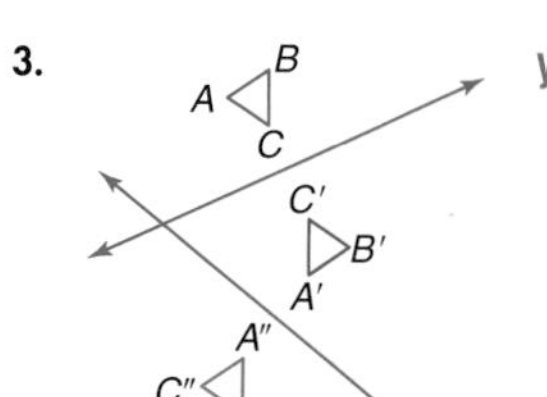

yes

Copy each figure. Then use a composite of reflections to find the rotation image with respect to lines ℓ and t. 4–6. See margin.

4. 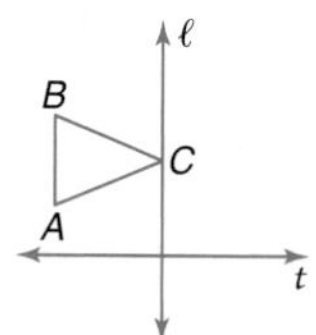

5. 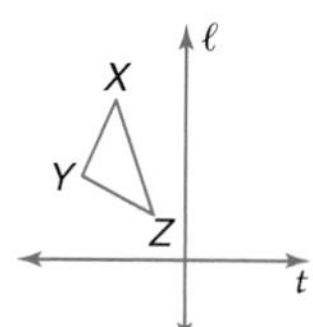

6. 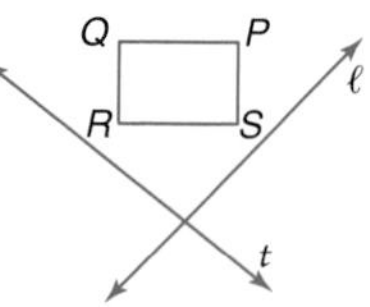

Copy each figure. Then use the angle of rotation to find the image with respect to lines ℓ and t.

7. 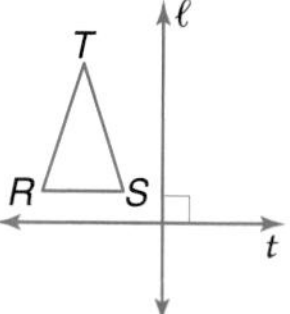

8. 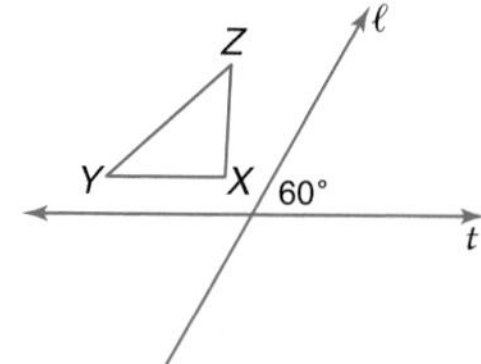

9. 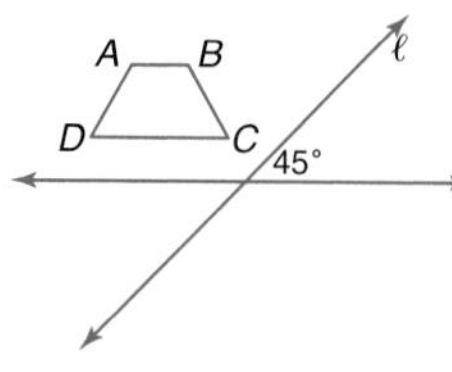

7–9. See margin.

Lesson 13-8 **Find the measure of the dilation image of $\overline{AB}$ with the given scale factor.**

1. $AB = 4, k = -3$ 12
2. $AB = \frac{3}{5}, k = 25$ 15
3. $AB = 18, k = \frac{2}{3}$ 12

A dilation with center C and a scale factor of k maps A onto D and B onto E. Find $|k|$ for each dilation. Then determine whether each dilation is an enlargement, a reduction, or a congruence transformation.

4. $CE = 18, CB = 9$
2; enlargement
5. $AB = 3, DE = 1$
$\frac{1}{3}$; reduction
6. $AB = 3, DE = 4$
$\frac{4}{3}$; enlargement

For each figure, a dilation with center C produced the figure in red. What is the scale factor for each transformation?

7. 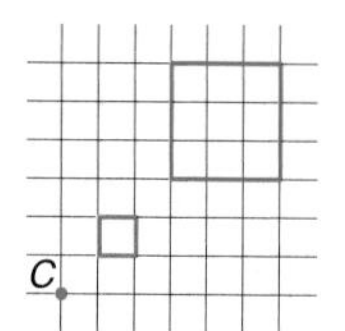

$\frac{1}{3}$

8. 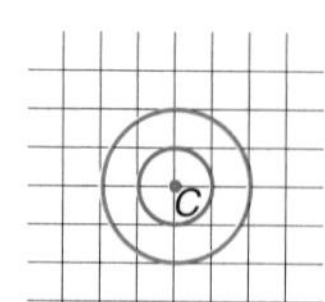

2

9. 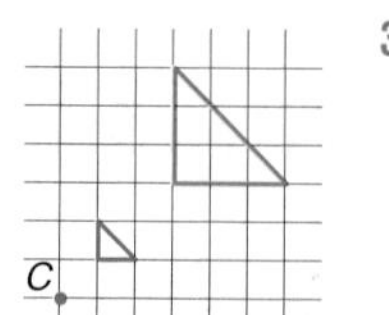

3

Graph each set of ordered pairs. Then connect the points in order. Using the origin as the center of dilation and a scale factor of 2, draw the dilation image. Repeat using a scale factor of $\frac{1}{2}$.

10. (4, 2), (7, −2), (−1, −3) 10–11. See margin
11. (2, 0), (3, 4), (−1, 4), (−4, 2)

Additional Answers for Lesson 13-8

10.
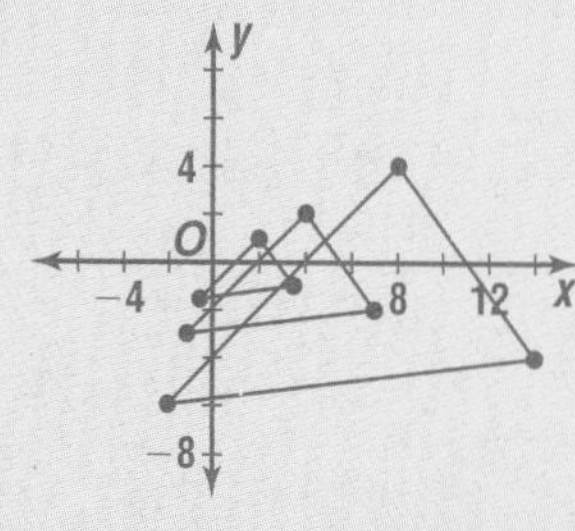

11.
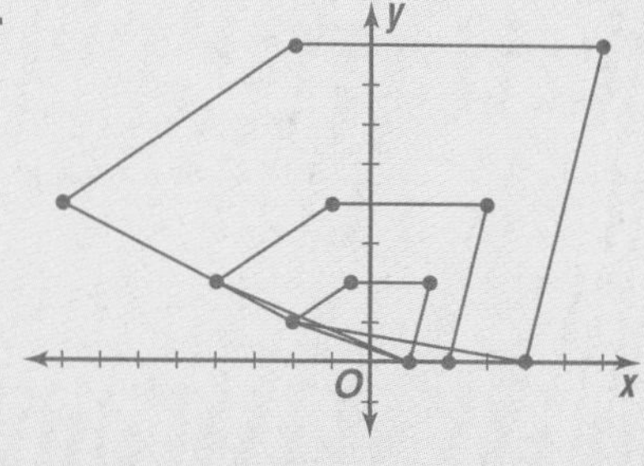

CHAPTER 1 TEST

Draw and label a figure for each relationship. See margin.

1. Lines ℓ, m, and n all intersect at point X.
2. Planes $\mathcal{Q}$ and $\mathcal{R}$ do not intersect.
3. Plane $\mathcal{P}$ contains point A but does not contain $\overline{BC}$.
4. $A(4, 3)$, $B(-2, -4)$, $C(-3, 4)$, and $D(0, -1)$ lie on a coordinate plane.

Refer to the coordinate grid at the right to answer each question.

5. What ordered pair names point S? $(-2, 3)$
6. What is the length of $\overline{PQ}$? $4\sqrt{2}$ units
7. What are the coordinates of the midpoint of $\overline{QR}$? $\left(-\frac{3}{2}, -1\right)$
8. What is the y-coordinate of any point collinear to Q and S? 3

Refer to the number line to answer each question.

9. What is the measure of $\overline{AC}$? 6 units
10. What is the coordinate of the midpoint of $\overline{CF}$? 3
11. What segment is congruent to $\overline{BF}$? $\overline{AD}$
12. What is the coordinate of G if C is between D and G and $DG = 14$? -10

Refer to the figure at the right to answer each question. 14, 15, 20. See margin.

13. Which of the numbered angles appears to be obtuse? $\angle 7$
14. Name a pair of congruent supplementary angles.
15. Name a pair of adjacent angles that do not form a linear pair.
16. If $\angle 1 \cong \angle 5$, does $\overrightarrow{VG}$ bisect $\angle BVF$? yes
17. *True* or *false*: $\overleftrightarrow{CG}$ bisects $\overline{AE}$. True
18. If $AC = 4x + 1$ and $CE = 16 - x$, find AE. 26
19. If $m\angle BVF = 7x - 1$ and $m\angle FVA = 6x + 12$, is $\overleftrightarrow{AB} \perp \overleftrightarrow{VF}$? yes
20. Which two angles must be complementary if $\angle AVF$ is a right angle?
21. If $m\angle 5 = 3x + 14$, $m\angle 6 = x + 30$, and $m\angle FVB = 9x - 11$, find $m\angle FVB$. 88

Use the city block at the right to answer each question.

22. How far will Kelli walk her dog if they walk around the block? 2380 ft
23. The entire block is a park. What is the area of the park? 23. 319,800 ft^2
24. The Department of Parks and Recreation wants to pave a path that goes through the park at a diagonal from corner to corner. How long will the path be? about 881.2 ft
25. **Sports** Three darts are thrown at the target shown at the right. If we assume that each dart lands within a ring, how many different point totals are possible? List them. See margin.

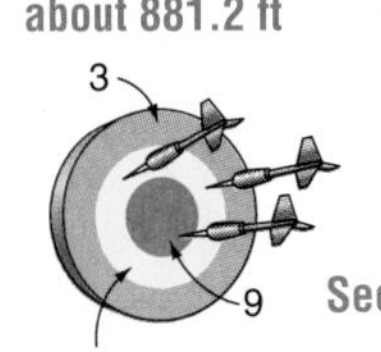

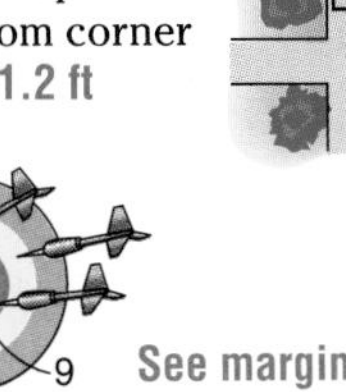

Chapter 1 Test Answers

1. Sample answer:

2. Sample answer:

3. Sample answer:

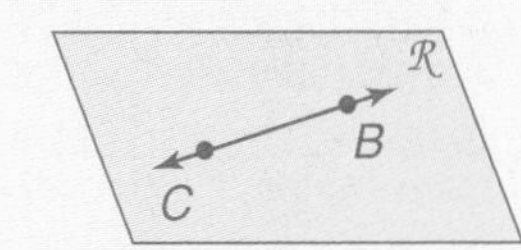

4. Sample answer:

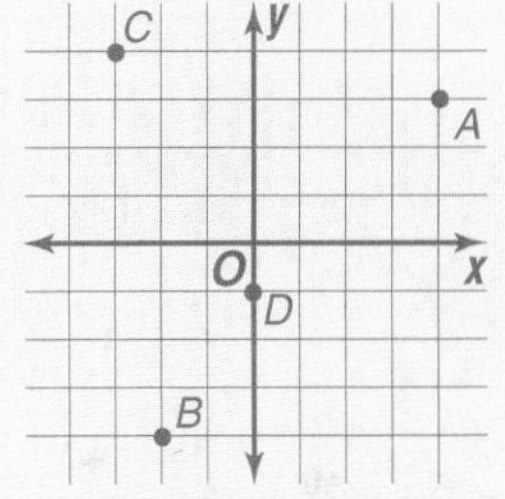

14. Sample answers: $\angle 3$ and $\angle 4$, $\angle 4$ and $\angle VCE$, $\angle VCE$ and $\angle VCA$, $\angle VCA$ and $\angle 3$
15. Sample answers: $\angle 6$ and $\angle 5$, $\angle 5$ and $\angle 2$, $\angle 2$ and $\angle 1$
20. $\angle 5$ and $\angle 6$ or $\angle 5$ and $\angle 1$
25. 10 totals; 3, 3, 3; 3, 3, 4; 3, 3, 9; 4, 4, 4; 4, 4, 9; 4, 9, 9; 3, 4, 4; 3, 4, 9; 9, 9, 9; 9, 9, 3

Chapter 2 Test Answers

1. False; counterexample: $x = -1$; -1 is real, but $-(-1) > 0$.
2. True; congruent angles have the same measures, and the measures of the angles are real numbers; therefore, the Symmetric Property of Equality holds true.
3. True; the angles form a straight angle which measures 180°.
4. False; it is possible that $x = -4$.
5. If something is a rolling stone, then it gathers no moss.
   Hypothesis: something is a rolling stone
   Conclusion: it gathers no moss
   Converse: If something gathers no moss, then it is a rolling stone.
   Inverse: If something is not a rolling stone, then it gathers moss.
   Contrapositive: If something gathers moss, then it is not a rolling stone.
6. If you eat an apple a day, then you won't need to go to the doctor.
   Hypothesis: you eat an apple a day
   Conclusion: you won't need to go to the doctor
   Converse: If you don't need to go to the doctor, then you ate an apple each day.
   Inverse: If you don't eat an apple a day, then you will need to go to the doctor.
   Contrapositive: If you do need to go to the doctor, then you didn't eat an apple each day.

CHAPTER 2 TEST

Determine if each conjecture is *true* or *false*. Explain your answer and give a counterexample for any false conjecture. 1–4. See margin.

1. **Given:** x is a real number.
   Conjecture: $-x < 0$
2. **Given:** $\angle 1 \cong \angle 2$
   Conjecture: $\angle 2 \cong \angle 1$
3. **Given:** $\angle 1$ and $\angle 2$ form a linear pair.
   Conjecture: $m\angle 1 + m\angle 2 = 180$
4. **Given:** $3x^2 = 48$
   Conjecture: $x = 4$

Write each conditional statement in if-then form. Identify the hypothesis and conclusion of each conditional. Then write the converse, inverse, and contrapositive of each conditional. 5–9. See margin.

5. A rolling stone gathers no moss.
6. An apple a day keeps the doctor away.
7. Two parallel planes do not intersect.
8. Two points make a line.
9. Through any two points there is exactly one line.

10. Wise investments with Petty-Bates build for the future; syllogism.

Determine if a valid conclusion can be reached from the two true statements using the Law of Detachment or the Law of Syllogism. If a valid conclusion is possible, state it and the law that is used. If a valid conclusion does not follow, write *no conclusion*.

10. (1) Wise investments with Petty-Bates pay off.
    (2) Investments that pay off build for the future.
11. (1) Perpendicular lines intersect.
    (2) Lines ℓ and m are perpendicular.
    Lines ℓ and m intersect; detachment.
12. (1) Vertical angles are congruent.
    (2) $\angle 1$ is congruent to $\angle 2$. no conclusion
13. (1) All integers are real numbers.
    (2) 7 is an integer.
    7 is a real number; detachment.

Name the property of equality that justifies each statement. 14. Symmetric 17. Substitution

14. If $m\angle A = m\angle B$, then $m\angle B = m\angle A$.
15. If $x + 9 = 12$, then $x = 3$. Subtraction
16. If $2ST = 4UV$, then $ST = 2UV$. Division
17. If $AB = 7$ and $CD = 7$, then $AB = CD$.
18. **Transportation** An empty commuter train picked up passengers on Monday morning at a rate of one passenger at the first stop, three at the second stop, five at the third stop, seven at the fourth, and so on.
    a. How many passengers got on the train at the 15th stop? 29
    b. What was the total number of passengers on the train after 10 stops? 100 passengers

20b. Reflex. Prop.

Copy and complete the following proofs. 20d. Def. linear pair 20g. Subst. Prop.

19. **Given:** $\overline{AC} \cong \overline{BD}$
    Prove: $\overline{AB} \cong \overline{CD}$

A B C D

Statements	Reasons
a. $\overline{AC} \cong \overline{BD}$	a. ? Given
b. $AC = BD$	b. ? Def. ≅ seg.
c. ? $AB + BC = AC$; ? $BC + CD = BD$	c. Segment Addition Postulate
d. $AB + BC = BC + CD$	d. ? Subst. Prop.
e. $AB = CD$	e. ? Subtr. Prop.
f. ? $\overline{AB} \cong \overline{CD}$	f. Def. ≅ seg.

20. **Given:** $m\angle 1 = m\angle 3 + m\angle 4$
    Prove: $m\angle 3 + m\angle 4 + m\angle 2 = 180$

1 2 3 4

Statements	Reasons
a. ? $m\angle 1 = m\angle 3 + m\angle 4$	a. Given
b. $m\angle 2 = m\angle 2$	b. ?
c. $m\angle 1 + m\angle 2 = m\angle 3 + m\angle 4 + m\angle 2$	c. ? Add. Prop.
d. $\angle 1$ and $\angle 2$ form a linear pair.	d. ?
e. $\angle 1$ and $\angle 2$ are supplementary.	e. ? Suppl. Th.
f. ? $m\angle 1 + m\angle 2 = 180$	f. Def. suppl. ∡
g. $m\angle 3 + m\angle 4 + m\angle 2 = 180$	g. ?

7. If two planes are parallel, then they do not intersect.
   Hypothesis: two planes are parallel
   Conclusion: they do not intersect
   Converse: If two planes do not intersect, then they are parallel.
   Inverse: If two planes are not parallel, then they intersect.
   Contrapositive: If two planes intersect, then they are not parallel.
8. If there are two points, then they make a line.
   Hypothesis: there are two points
   Conclusion: they make a line
   Converse: If there is a line, then there are two points on it.
   Inverse: If there are not two points, then there is not a line.
   Contrapositive: If there is not a line, then there are not two points on that line.

CHAPTER 3 TEST

In the figure, $\ell \parallel m$. Determine whether each statement is *true* or *false*. Justify your answer. 1–7. See margin for justifications.

1. $\angle 1$ and $\angle 14$ are alternate exterior angles. true
2. $\angle 5$ and $\angle 11$ are consecutive interior angles. false
3. $\angle 2$ and $\angle 6$ are vertical angles. false
4. $\angle 6$ and $\angle 12$ are supplementary angles. true
5. $\angle 3 \cong \angle 8$ true
6. $m\angle 7 + m\angle 10 = 180$ false
7. $m\angle 4 + m\angle 5 + m\angle 11 = 180$ true

Use the figure at the right to answer each question.

8. If $\angle 5 \cong \angle 4$, which lines are parallel and why? $a \parallel b$
9. If $\angle 3 \cong \angle 9$, which lines are parallel and why? $m \parallel n$
10. If $m\angle 4 + m\angle 9 + m\angle 6 = 180$, which lines are parallel and why? $a \parallel b$

8–10. See margin for justifications.

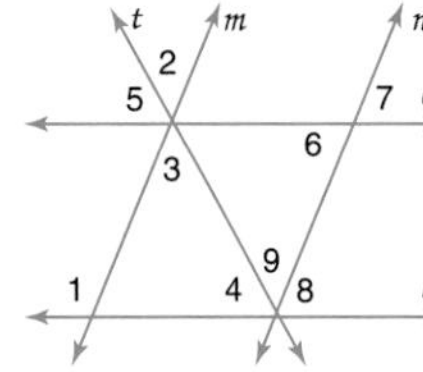

Determine the slope of each line.

11. $\overleftrightarrow{EF}$ $\frac{1}{2}$
12. $\overleftrightarrow{AB}$ $-\frac{8}{3}$
13. any line parallel to $\overleftrightarrow{CD}$ $-\frac{4}{7}$
14. any line perpendicular to $\overleftrightarrow{EF}$ -2

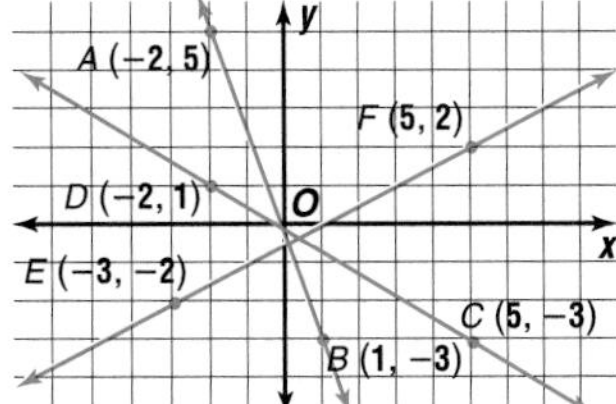

Name the segment in the figure at the right whose length represents the distance between the following points and lines.

15. from F to $\overleftrightarrow{CE}$ $\overline{FD}$
16. from A to $\overleftrightarrow{CE}$ $\overline{AC}$
17. from D to $\overleftrightarrow{BC}$ $\overline{DC}$

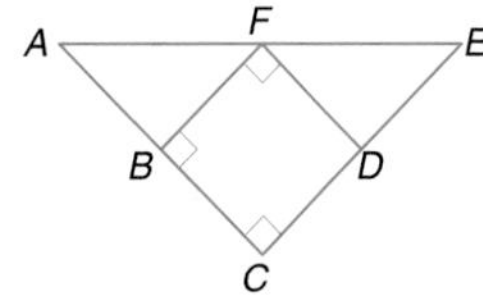

18. Write a two-column proof. See margin.

    Given: $\angle 1 \cong \angle 2$

    $\overline{ST} \parallel \overline{PR}$

    Prove: $\angle P \cong \angle R$

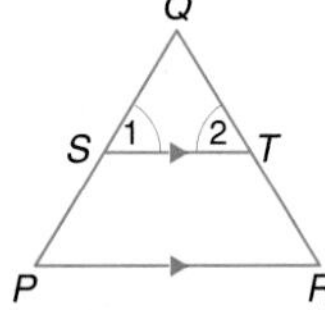

19. For the property *Two perpendicular lines form four right angles* from plane Euclidean geometry, write a corresponding statement for non-Euclidean spherical geometry. Two perpendicular lines form eight right angles.
20. **Business** The Carpet Experts cleaning team can clean the carpet in a room that is 10 feet by 10 feet in 20 minutes. Draw a diagram and find how long it would take them to clean a walk-in closet that is 5 feet by 5 feet. 5 minutes

Chapter 3 Test Answers

1. def. alt. ext. ⦞s
2. def. consec. int. ⦞s
3. def. vertical ⦞s
4. Consecutive Interior Angles Theorem
5. Corresponding Angles Postulate
6. def. supp. ⦞s
7. Consecutive Interior Angles Theorem
8. If ⧣ so that corr. ⦞s are ≅, then the lines are ∥.
9. If ⧣ so that a pair of alt. int. ⦞s is ≅, then the lines are ∥.
10. If ⧣ so that a pair of consec. int. ⦞s is supp., then the lines are ∥.

18. Given: $\angle 1 \cong \angle 2$
    $\overline{ST} \parallel \overline{PR}$
    Prove: $\angle P \cong \angle R$
    Proof:
    Statements (Reasons)
    1. $\angle 1 \cong \angle 2$; $\overline{ST} \parallel \overline{PR}$ (Given)
    2. $\angle 1 \cong \angle P$; $\angle 2 \cong \angle R$ (Corresponding Angles Postulate)
    3. $\angle P \cong \angle R$ (Congruence of ⦞s is transitive.)

Chapter 2 Test Answer

9. If there are any two points, then there is exactly one straight line through them.
   Hypothesis: there are any two points
   Conclusion: there is exactly one straight line through them
   Converse: If there is exactly one line, then there are two points on it.
   Inverse: If there are not two points, then there is not exactly one straight line through them.
   Contrapositive: If there is not exactly one line, then there are not two points on it.

19. Statements (Reasons)
1. $\overline{AC} \perp \overline{BD}$; $\angle B \cong \angle D$ (Given)
2. $\overline{CA} \cong \overline{CA}$ ($\cong$ of segments is reflexive.)
3. $\angle BCA$ and $\angle DCA$ are rt. $\angle$s. ($\perp$ lines form 4 rt. $\angle$s.)
4. $\angle 3 \cong \angle 4$ (All rt. $\angle$s are $\cong$.)
5. $\triangle BCA \cong \triangle DCA$ (AAS)
6. $\overline{BC} \cong \overline{DC}$ (CPCTC)
7. C is the midpoint of $\overline{BD}$. (Def. midpoint)

20. Statements (Reasons)
1. $m\angle ACB = 110$; $m\angle DAC = 40$ (Given)
2. $m\angle D + m\angle DAC = m\angle ACB$ (Ext. $\angle$ Th.)
3. $m\angle D + 40 = 110$ (Subst. Prop. (=))
4. $m\angle D = 70$ (Subtr. Prop. (=))
5. $m\angle ACD + m\angle D + m\angle DAC = 180$ ($\angle$ Sum Th.)
6. $m\angle ACD + 70 + 40 = 180$ (Subst. Prop. (=))
7. $m\angle ACD = 70$ (Subtr. Prop. (=))
8. $m\angle ACD = m\angle D$ (Subst. Prop. (=))
9. $\angle ACD \cong \angle D$ (Def. $\cong$ $\angle$s)
10. $\overline{AD} \cong \overline{AC}$ (If 2 $\angle$s of a $\triangle$ are $\cong$, the sides opp. the $\angle$s are $\cong$.)
11. $\triangle ACD$ is isosceles. (Def. isosceles $\triangle$)

Chapter Tests

CHAPTER 4 TEST

Use figure *PQRST* to answer each of the following.

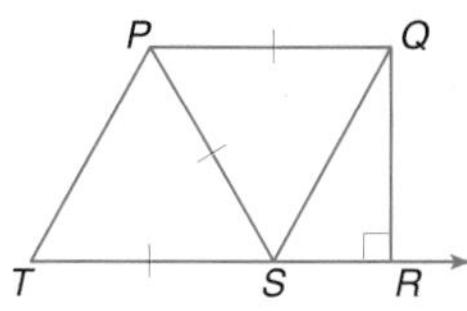

1. Name the isosceles triangle(s). **$\triangle PSQ$, $\triangle PST$**
2. Which triangle is a right triangle? **$\triangle QRS$**
3. Which side of $\triangle PTS$ is opposite $\angle T$? **$\overline{PS}$**
4. Name the triangle(s) that appear to be scalene. **$\triangle QRS$**

In the figure, $\overline{GH} \cong \overline{GL}$, $\overline{GI} \cong \overline{GK}$, $\overline{GJ} \perp \overline{HL}$, $m\angle 3 = 30$, and $m\angle 4 = 20$. Find each measure.

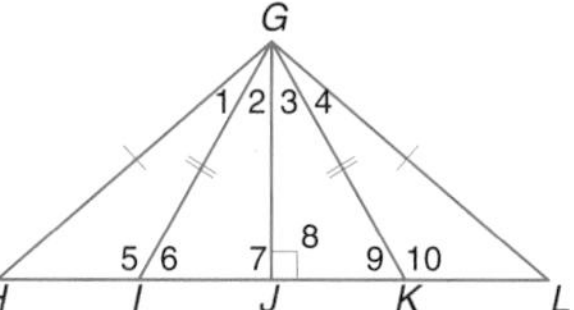

5. $m\angle 8$ **90**
6. $m\angle 6$ **60**
7. $m\angle 2$ **30**
8. $m\angle 10$ **120**
9. $m\angle H$ **40**
10. $m\angle 1$ **20**

Use the figure at the right to answer each of the following. **11. $\angle 10$ and $\angle 5$ or $\angle 11$ and $\angle 4$**

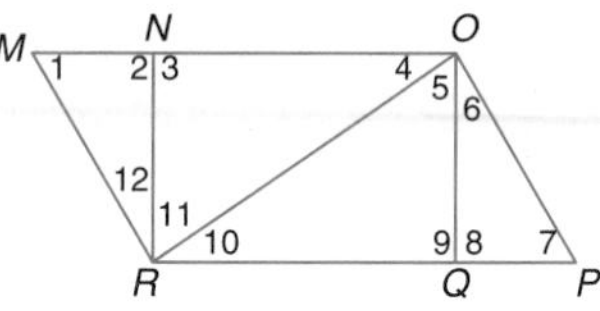

11. $\overline{RO}$ is included between what two angles?
12. If $\angle 1 \cong \angle 7$ and $\overline{MO} \parallel \overline{RP}$, then $\triangle MOR \cong$ _?_ by _?_. **$\triangle PRO$; AAS**
13. If $\overline{NO} \cong \overline{RQ}$ and $\overline{NR} \cong \overline{OQ}$, then $\triangle ONR \cong$ _?_ by _?_. **$\triangle RQO$; SSS**
14. If $\overline{MO} \parallel \overline{RP}$ and $\overline{QO} \parallel \overline{RN}$, then $\triangle NRO \cong$ _?_ by _?_. **$\triangle QOR$; ASA**
15. If $\overline{NR} \perp \overline{MO}$, $\overline{RP} \perp \overline{OQ}$, $\overline{MN} \cong \overline{QP}$, and $\overline{NR} \cong \overline{OQ}$, then $\triangle MNR \cong$ _?_ by _?_. **$\triangle PQO$; SAS**

For each triangle, find the value of *x*.

16. **46**

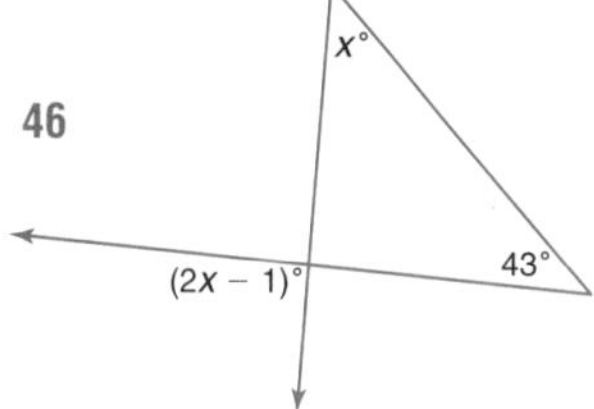

17. **4**

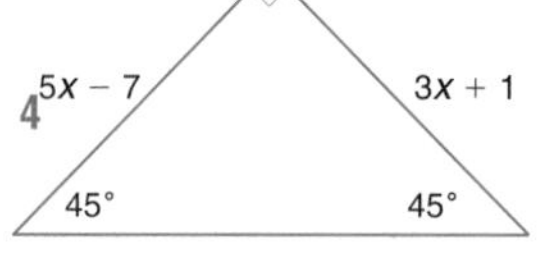

18. **Eliminate the Possibilities** Umeko, Jim, and Gwen each participate in an extracurricular activity and have an after-school job. One of them is in the Spanish Club, one is in the Drama Club, and one is in the marching band. One of them is a pizza delivery person, one is a math tutor, and one is a lifeguard at the community pool. Jim is tutoring the Drama Club member's brother in Algebra. The Spanish Club member cannot swim or drive a car. The lifeguard is teaching Umeko to read music. Who does what? **Umeko — Drama Club, delivery person; Jim — Spanish Club, tutor; Gwen — marching band, lifeguard**

Write a proof.

19. **Given:** $\overline{AC} \perp \overline{BD}$
$\angle B \cong \angle D$
Prove: C is the midpoint of $\overline{BD}$. **See margin.**

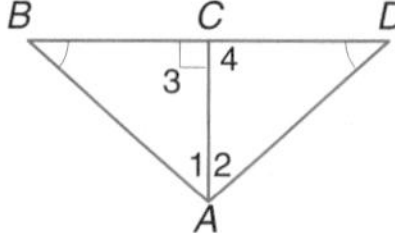

20. **Given:** $m\angle ACB = 110$
$m\angle DAC = 40$
Prove: $\triangle ACD$ is isosceles. **See margin.**

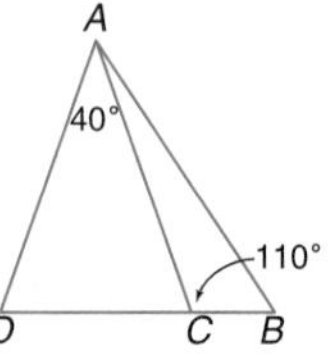

In $\triangle AHW$, $m\angle A = 64$ and $m\angle AWH = 36$. If $\overline{WP}$ is an angle bisector and $\overline{HQ}$ is an altitude, find each measure.

1. $m\angle AQH$ **90**
2. $m\angle AHQ$ **26**
3. $m\angle APW$ **98**
4. $m\angle HXW$ **108**
5. If $\overline{WP}$ is a median, $AP = 3y + 11$, and $PH = 7y - 5$, find AH. **46**

Refer to the figure at the right. Complete each statement with < or >.

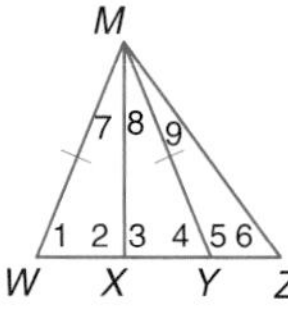

6. $m\angle 2$? $m\angle 9$ **>**
7. $m\angle 6$? $m\angle 1$ **<**
8. If $m\angle 8 < m\angle 7$, then WX ? XY. **>**
9. If $MX < MZ$, then $m\angle 4$? $m\angle 5$. **<**

Find the value of x so that $\triangle ABE \cong \triangle DBC$ by the indicated theorem or postulate.

10. $AB = 2x$, $BE = 3x - 1$, $BC = 23$, $BD = 5x - 24$; LL **8**
11. $AB = x + 6$, $AE = 2x + 13$, $BD = 15$, $CD = 4x - 5$; HL **9**

12. List the least and the greatest angle measures in $\triangle LAN$. **$m\angle NAL$, $m\angle NLA$**

13. List all the angles whose measures are greater than $m\angle 6$. **$\angle 4$, $\angle 5$, $\angle 8$**

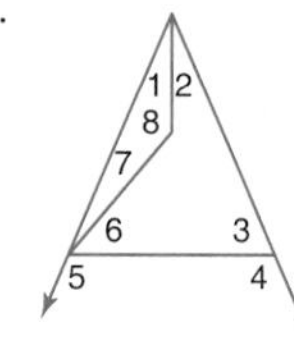

14. Find the longest segment in $\triangle ABC$ if $m\angle A = 5x + 31$, $m\angle B = 74 - 3x$, and $m\angle C = 4x + 9$. **$\overline{BC}$**

15. Determine whether it is possible to have a triangle with vertices $A(1, -1)$, $B(7, 7)$, and $C(2, -5)$. Write *yes* or *no*, and explain.
Yes; the measures satisfy the Triangle Inequality Theorem.

16. Write an inequality to describe the possible values of x. **$x < 30$**

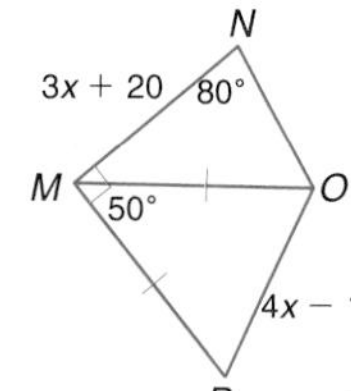

Refer to the figure at the right for Questions 17 and 18.

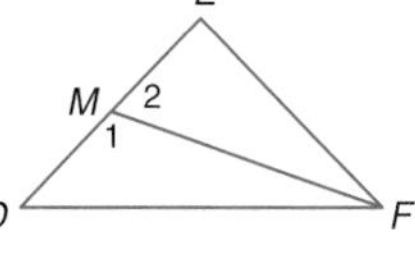

17. If M is the midpoint of $\overline{DE}$, $m\angle 1 > m\angle 2$, $DF = 13x - 5$, and $EF = 7x + 25$, find all possible values of x. **$x > 5$**

18. Write an indirect proof. **See margin.**

Given: $\overline{FM}$ is a median of $\triangle DEF$.
$m\angle 1 < m\angle 2$

Prove: $DF \neq EF$

Write a two-column proof. **19–20. See margin.**

19. **Given:** $NO = QP$
$PN > OQ$
Prove: $MP > MO$

20. **Given:** $\overline{AD} \perp \overline{DC}$
$\overline{AB} \perp \overline{BC}$
$AB = DC$
Prove: $\overline{DC} \perp \overline{BC}$

Chapter 5 Test Answers

18. Proof: Assume $DF = EF$. M is the midpoint of DE and $\overline{DM} \cong \overline{EM}$ since it is given that $\overline{FM}$ is a median of $\triangle DEF$. $\overline{FM} \cong \overline{FM}$ by the Reflexive Prop. (=) and $m\angle 1 > m\angle 2$ is given, so $DF > EF$ by SAS Inequality. But $DF = EF$ by assumption. This contradicts the Comparison Prop. Therefore, the assumption is false, and $DF \neq EF$.

19. Statements (Reasons)
1. $NO = QP$ (Given)
2. $\overline{NO} \cong \overline{QP}$ (Def. of $\cong$)
3. $\overline{OP} \cong \overline{OP}$ ($\cong$ of segments is reflexive.)
4. $PN > OQ$ (Given)
5. $m\angle MOP > m\angle MPO$ (SSS Inequality)
6. $MP > MO$ (If one $\angle$ of a $\triangle$ is greater than another, the side opp. the greater $\angle$ is longer than the side opp. the lesser $\angle$.)

20. Statements (Reasons)
1. $\overline{AD} \perp \overline{DC}$; $\overline{AB} \perp \overline{BC}$; $AB = DC$ (Given)
2. $\overline{AB} \cong \overline{DC}$ (Def. of $\cong$)
3. $\overline{AC} \cong \overline{AC}$ ($\cong$ of segments is reflexive.)
4. $\angle D$ and $\angle B$ are rt. $\angle$s. ($\perp$ lines form 4 rt. $\angle$s.)
5. $\triangle ADC$ and $\triangle CBA$ are rt. $\triangle$s. (Def. rt. $\triangle$)
6. $\triangle ADC \cong \triangle CBA$ (HL)
7. $\angle ACD \cong \angle CAB$ (CPCTC)
8. $\overline{AB} \parallel \overline{CD}$ (If $\nparallel$ and alt. int. $\angle$s are $\cong$, lines are $\parallel$.)
9. $\overline{DC} \perp \overline{BC}$ (Perpendicular Transversal Theorem)

Chapter 6 Test Answers

1. The diagonals of a parallelogram bisect each other.
2. Opposite angles of a parallelogram are congruent.
3. Def. of parallelogram
4. SAS
5. The diagonals of a rhombus are perpendicular.
6. Each diagonal of a rhombus bisects each pair of opposite angles.
8. slope of $\overline{AB}$ = $\frac{6-11}{-2-2} = \frac{5}{4}$
   slope of $\overline{BC}$ = $\frac{11-8}{2-3} = -3$
   slope of $\overline{CD}$ = $\frac{8-3}{3-(-1)} = \frac{5}{4}$
   slope of $\overline{DA}$ = $\frac{3-6}{-1-(-2)} = -3$
   ABCD is a parallelogram because the slopes of the opposite sides are equal.
9. slope of $\overline{AB}$ = $\frac{-3-(-2)}{7-4} = -\frac{1}{3}$
   slope of $\overline{BC}$ = $\frac{-2-4}{4-6} = 3$
   slope of $\overline{CD}$ = $\frac{4-2}{6-12} = -\frac{1}{3}$
   slope of $\overline{DA}$ = $\frac{2-(-3)}{12-7} = 1$
   ABCD is not a parallelogram because the slopes of both pairs of opposite sides are not equal.
11. Sample answer:

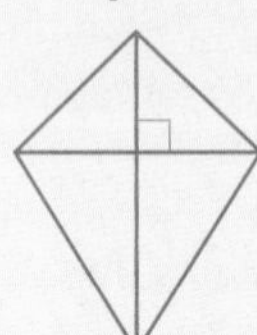

12. Sample answer:

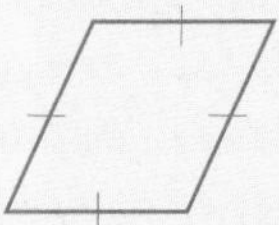

CHAPTER 6 TEST

Complete each statement about ▱*DHGF*. Then name the theorem or definition that justifies your answer. 1–6. See margin for justifications.

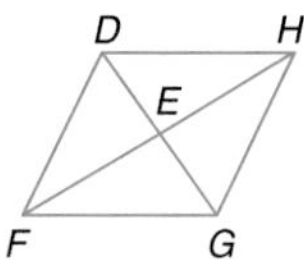

1. $\overline{DE} \cong$ __?__ $\overline{EG}$
2. $\angle FDH \cong$ __?__ $\angle FGH$
3. $\overline{FD} \parallel$ __?__ $\overline{GH}$
4. $\triangle FDG \cong$ __?__ $\triangle HGD$
5. If *DHGF* is a rhombus, then $\overline{DG} \perp$ __?__. $\overline{FH}$
6. If *DHGF* is a rhombus, then $\overline{DG}$ bisects __?__ and __?__. $\angle FDH$, $\angle FGH$
7. Find a three-digit number that is a perfect square and a perfect cube. 729

Determine whether *ABCD* is a parallelogram for each set of vertices. Justify your answer.

8. $A(-2, 6), B(2, 11), C(3, 8), D(-1, 3)$
9. $A(7, -3), B(4, -2), C(6, 4), D(12, 2)$

8–9. See margin.

10–13. See margin for drawings.

Determine whether each conditional is *true* or false. If false, draw a counterexample.

10. If a quadrilateral has four right angles, then it is a rectangle. true
11. If the diagonals of a quadrilateral are perpendicular, then it is a rhombus. false
12. If a quadrilateral has all four sides congruent, then it is a square. false
13. If a quadrilateral has opposite sides congruent and one right angle, then it is a rectangle. true

Find the values of *x* and *y* that ensure each quadrilateral is a parallelogram.

14.  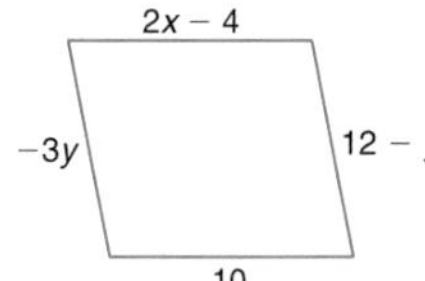

$x = 7, y = -6$

15. 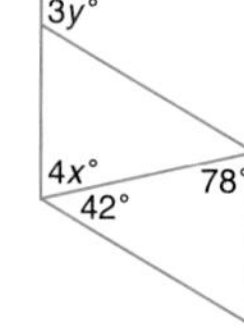

$x = 19.5, y = 40$

***PQRS* is an isosceles trapezoid with bases $\overline{QR}$ and $\overline{PS}$. Use the figure and the given information to find each measure.**

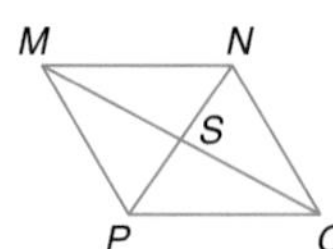

16. If $PS = 32$ and $TV = 26$, find QR. 20
17. If $m\angle QTV = 79$, find $m\angle TVS$. 101
18. If $QR = 9x$ and $PS = 13x$, find TV. $11x$
19. If $QR = x - 3$, $PS = 2x + 4$, and $TV = 3x - 10$, find QR, PS, and TV. $QR = 4, PS = 18, TV = 11$
20. Use rhombus *MNOP* to find the value of *x* if $m\angle MPS = 4x + 7$ and $m\angle OPS = 7x - 38$. $x = 15$

M N S P O

Solve each proportion.

1. $\frac{x}{28} = \frac{60}{16}$ 105
2. $\frac{21}{1-x} = \frac{7}{x}$ 0.25
3. A recent survey showed that 14 out of 25 teens prefer orange juice over apple juice. How many of the 1600 students at Dublin High School would you expect to prefer orange juice? 896 students

Identify each statement as *true* or *false*. If false, state why. 5. See margin for reasons.

4. All equilateral triangles are similar. true
5. All isosceles triangles are similar. false

Determine whether each pair of triangles is similar. Justify with a reason. If similarity exists, write a mathematical sentence relating the two triangles.

6.

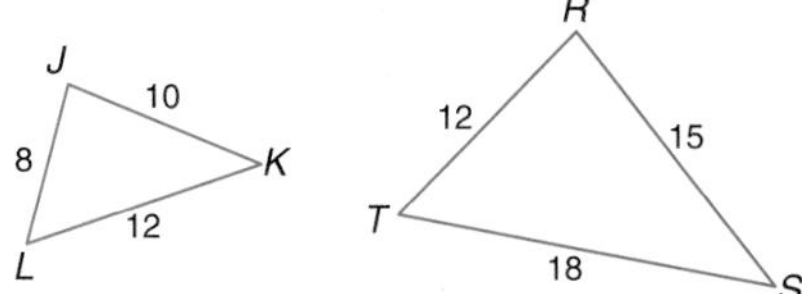

Yes, SSS Similarity; $\triangle JKL \sim \triangle RST$

7.

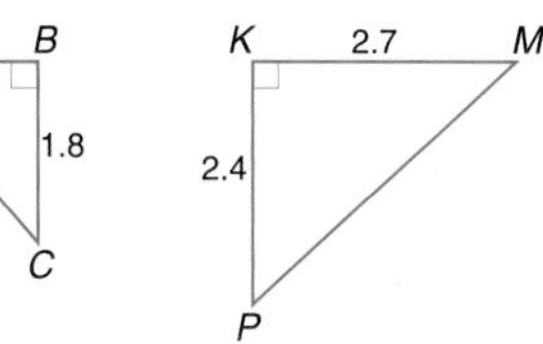

Yes, SAS Similarity; $\triangle ABC \sim \triangle PKM$

Use $\triangle ACE$ to find the value of x.

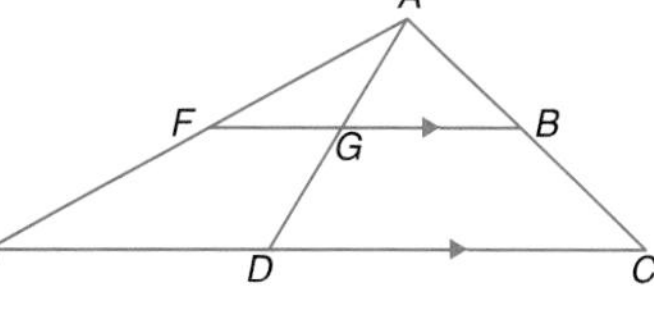

8. $AF = 3x + 1$, $FE = 18$, $AG = 8$, $GD = 9$ 5
9. $AB = BC = AF = FE$, $ED = 2x + 5$, $DC = 4x - 7$, $FB = 17$ 6
10. Find the value of $a^{\frac{1}{2}}$, where a initially equals 15. Then, use that value as the next a in the expression. Repeat the process until you can make some observations. Describe what happens in the iteration. converges to 1
11. If $\overline{BD} \parallel \overline{AE}$ and $3(BD) = AE$, find the coordinates of A and E.

$A(-6, 1)$, $E(0, -2)$

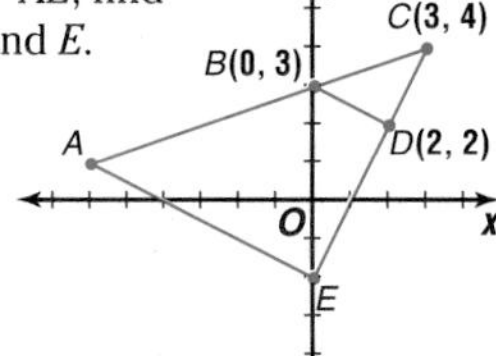

12. $\triangle JKL \sim \triangle BKA$, and the perimeter of $\triangle BKA$ is 42. Find the value of x. 3

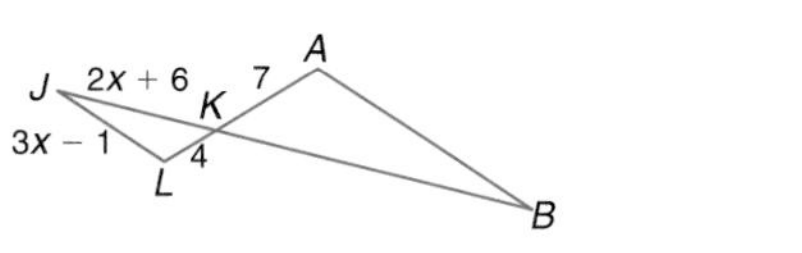

Use the figure at the right and the given information to find the value of x.

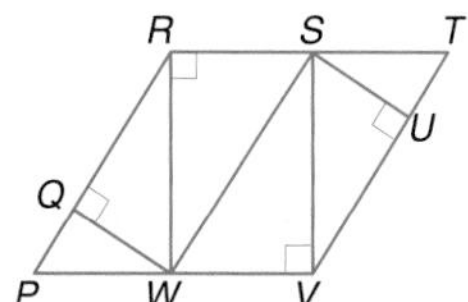

13. $\triangle RWP \sim \triangle VST$, $PW = x$, $SU = 5$, $QW = 1$, $ST = x + 5$ $1\frac{1}{4}$
14. $\triangle WRS \sim \triangle VUS$, $SW = 3$, $SV = 2$, $RS = 1\frac{1}{3}$, $SU = x + \frac{2}{3}$ $\frac{2}{9}$

Complete each of the following.

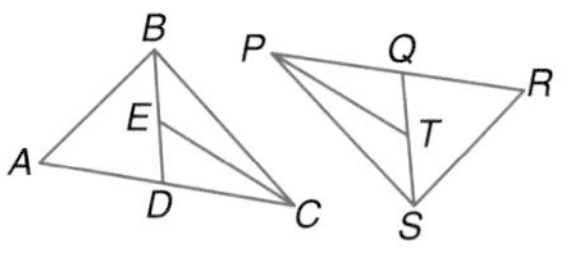

15. **Given:** $\triangle ABC \sim \triangle RSP$ See margin.
D is the midpoint of $\overline{AC}$.
Q is the midpoint of $\overline{PR}$.
Prove: $\triangle SPQ \sim \triangle BCD$

Chapter 7 Test Answers

5. All angles in isosceles $\triangle$s are not the same measure.

15. Statements (Reasons)
    1. $\triangle ABC \sim \triangle RSP$; D is the midpoint of $\overline{AC}$; Q is the midpoint of $\overline{PR}$. (Given)
    2. $\angle RSP \cong \angle ABC$ (Def. $\sim$ polygons)
    3. $\overline{BD}$ is a median of $\triangle ABC$. $\overline{QS}$ is a median of $\triangle RSP$. (Def. median)
    4. $\frac{SQ}{BD} = \frac{SP}{BC} = \frac{PR}{AC}$ (If 2 $\triangle$s are $\sim$, then the measures of corr. medians are proportional to the measures of corr. sides.)
    5. $PQ = QR$; $DC = DA$ (Def. midpoint)
    6. $AD + DC = AC$; $PQ + QR = PR$ (Seg. Add. Post.)
    7. $2(DC) = AC$; $2(PQ) = PR$ (Subst. Prop. (=))
    8. $\frac{2(PQ)}{2(DC)} = \frac{SQ}{BD}$ (Subst. Prop. (=))
    9. $\frac{PQ}{DC} = \frac{SQ}{BD}$ (Subst. Prop. (=))
    10. $\triangle SPQ \sim \triangle BCD$ (SSS Similarity)

CHAPTER 8 TEST

CHAPTER TEST

Find the geometric mean between each pair of numbers.

1. 3 and 12 6
2. 5 and 4 $\sqrt{20} \approx 4.5$
3. 28 and 56 $28\sqrt{2} \approx 39.6$

Use the figure below and the given information to find each measure. Round answers to the nearest tenth.

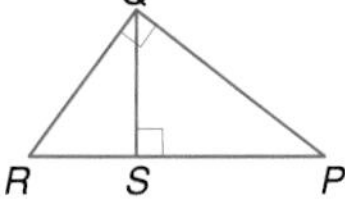

4. Find QS if $PS = 8$ and $SR = 5$. 6.3
5. Find QP if $SP = 9.5$ and $SR = 3$. 10.9

Determine if the given measures are measures of the sides of a right triangle.

6. 10, 26, 24 yes
7. 12, 27, 19 no
8. 39, 80, 89 yes

Find the value of x.

9. 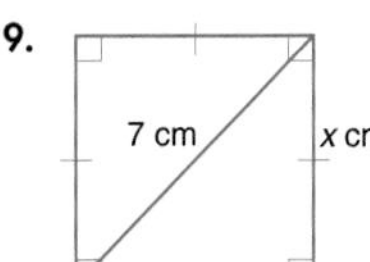

$\frac{7\sqrt{2}}{2} \approx 4.9$

10. 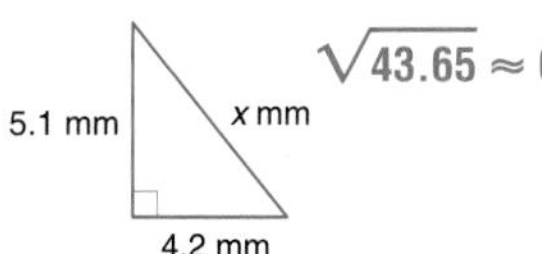

$\sqrt{43.65} \approx 6.6$

11. (Figure: x ft, 6.8 ft, 60°, 60°) $3.4\sqrt{3} \approx 5.9$

12. The perimeter of an equilateral triangle is 51 inches. Find the length of an altitude of the triangle. $8.5\sqrt{3} \approx 14.72$ in.

Find the indicated trigonometric ratio as a fraction and as a decimal rounded to the nearest ten-thousandth.

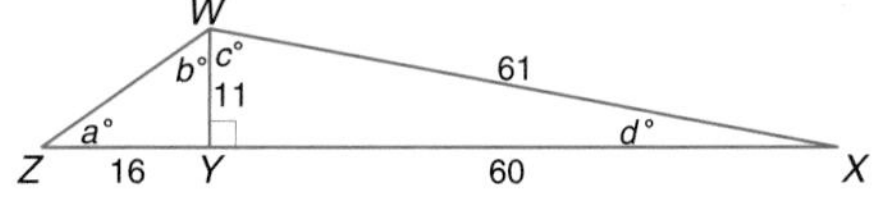

13. $\sin d°$ $\frac{11}{61} \approx 0.1803$
14. $\tan c°$ $\frac{60}{11} \approx 5.4545$
15. $\cos a°$ $\frac{16\sqrt{377}}{377} \approx 0.8240$
16. $\tan b°$ $\frac{16}{11} \approx 1.4545$

17. **Fire Safety** A firefighter's 36-foot ladder leans against a building. The top of the ladder touches the building 28 feet above the ground. What is the measure of the angle the ladder forms with the ground? about 51

18. The longest side of a triangle is 30 centimeters long. Two of the angles have measures of 45 and 79. Find the measures of the other two sides and the remaining angle. sides: 21.6, 25.3; angle: 56

19. **Gemology** A jeweler is making a sapphire earring in the shape of an isosceles triangle with sides 22, 22, and 28 millimeters long. What is the measure of the vertex angle? about 79

20. Sample answer: making a list; $\frac{5}{50}$ or $\frac{1}{10}$.

20. **Algebra** Which problem-solving strategy might you use to find the fractional part of the odd whole numbers less than 100 that are perfect squares? Find the fraction.

Chapter Tests

800 Chapter 8 Test

Refer to ⊙N for Questions 1–6.

1. Name a radius of ⊙N. $\overline{NA}$, $\overline{NB}$, $\overline{NC}$, or $\overline{ND}$
2. What kind of a triangle is $\triangle BNC$? isosceles
3. Is $ED > AD$? Explain. No; $\overline{AD}$ is a diameter.
4. If $AD = 21$, what is the radius? $10\frac{1}{2}$
5. If $NC = 5$, what is the circumference? 10π or 31.4
6. If $C = 126.5$, what is the diameter? 40.3

Find the exact circumference of each circle.

7. 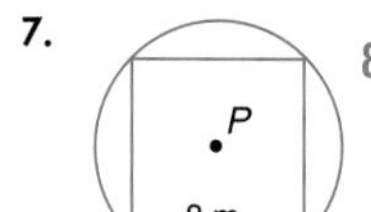 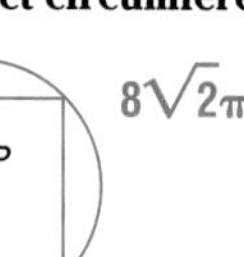 $8\sqrt{2}\pi$

8. 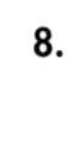 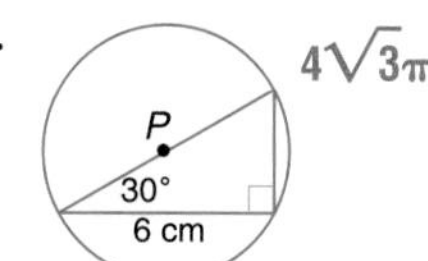 $4\sqrt{3}\pi$

For each ⊙C, find the value of x. Assume that segments that appear to be tangent are tangent.

9. 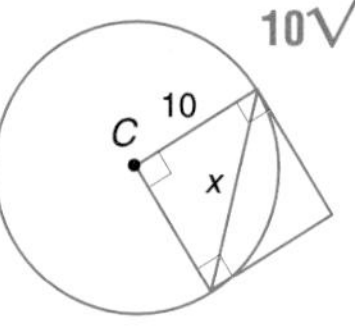 $10\sqrt{2}$

10. 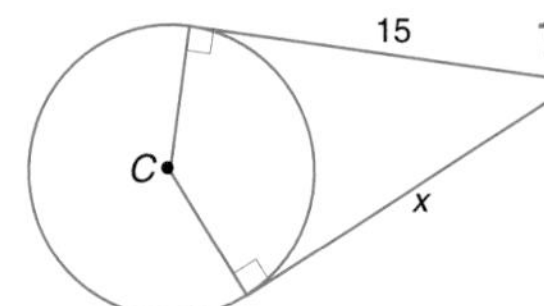 15

11. 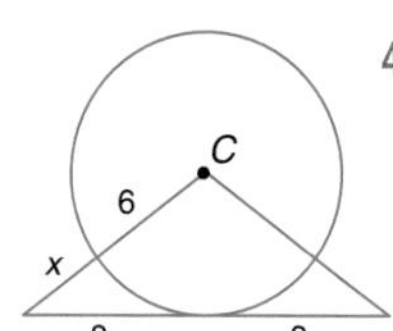 4

12. 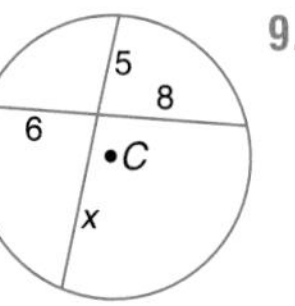 9.6

13. 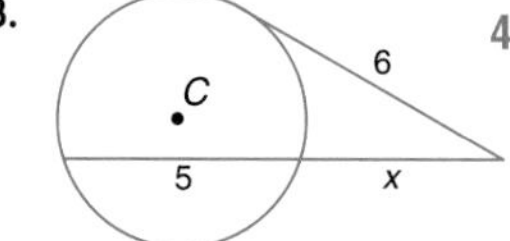 4

14. 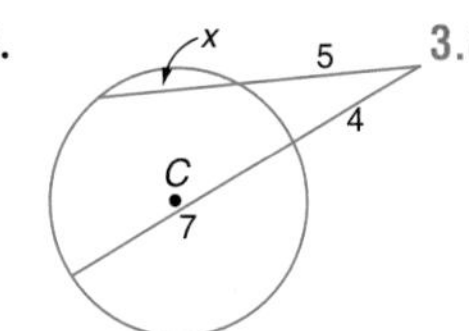 3.8

In ⊙P, $\overline{AB} \parallel \overline{CD}$, $m\widehat{BD} = 42$, $m\widehat{BE} = 12$, and $\overline{CF}$ and $\overline{AB}$ are diameters. Find each measure.

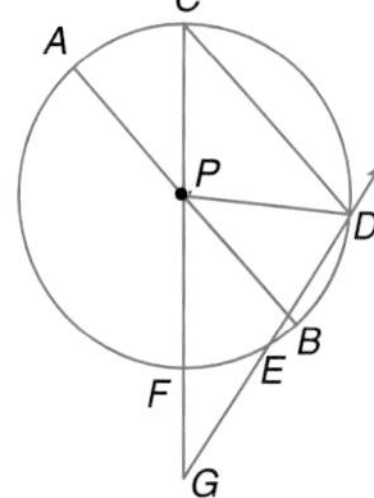

15. $m\widehat{AC}$ 42
16. $m\widehat{CD}$ 96
17. $m\angle BPF$ 42
18. $m\angle CPD$ 96
19. $m\widehat{AF}$ 138
20. $m\angle G$ 33
21. $m\angle FCD$ 42
22. $m\angle EDC$ 105
23. Suppose the diameter of a circle is 10 inches long and a chord is 6 inches long. Find the distance between the chord and the center of the circle. 4 in.
24. **Make a Graph** In 1995, a survey of parents of children under the age of 18 was taken to find how parents wake up their children. Of the parents surveyed, 64 said their kids wake on their own, 88 said their kids wake with an alarm, 172 said their kids have them call to wake them, and 76 said their kids have other ways of waking. Draw and label a circle graph that represents the data. See margin.
25. **Crafts** Teri plans to paint a jack-o-lantern face on a circular piece of wood. She wants it to be able to stand freely on a shelf so she cuts along a chord of the circle to form a flat bottom. If the radius of the circle is 4 inches and the chord is 5 inches, find the height of the final product. about 7.1 in.

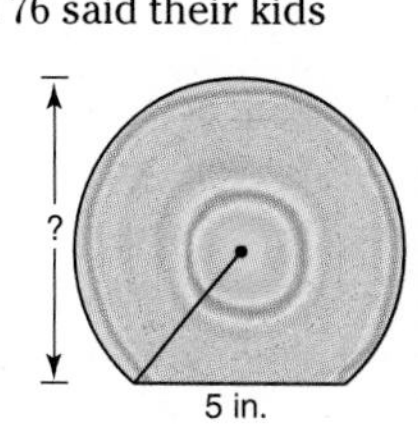

Chapter 9 Test Answer

24. How parents wake up their children:

Call to them 43%
Other 19%
Alarm clock 22%
Kids wake on own 16%

CHAPTER 10 TEST

Refer to the polygon at the right for Questions 1–3.

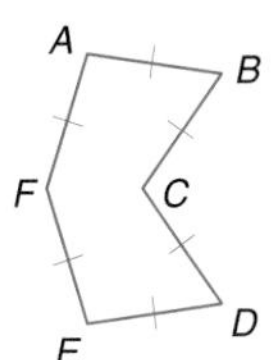

1. Classify the polygon by the number of sides. hexagon
2. Classify the polygon as convex or concave. concave
3. Classify the polygon as regular or not regular. Explain.
4. Find the sum of the measures of the interior angles of a convex dodecagon. 1800
5. Find the measure of one interior angle and one exterior angle of a regular 15-gon. 156, 24
6. Draw a regular tessellation. See students' work.

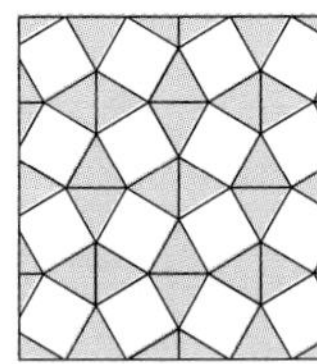

7. Determine whether the tessellation at the right is *regular, uniform,* or *semi-regular.* Name all possibilities. uniform, semi-regular
8. Use the strategy of guess and check to find the least prime number greater than 720. 727
9. Find the area of a parallelogram with a base of 4.95 feet and a height of 0.51 feet. 2.5245 ft^2
10. A square has a perimeter of 258 inches. Find the length of one side and the area of the square. 64.5 in.; 4160.25 in^2
11. A triangle has a base of 16 feet and a height of 30.6 feet. Find its area. 244.8 ft^2
12. The area of $\triangle ABC$ is $(3x^2 + 6x)$ square units. If the height is $3x + 6$ units, find the measure of the base. $2x$ units
13. The median of a trapezoid is 13 feet 6 inches long. If the height is 10 feet, what is the area of the trapezoid? 135 ft^2
14. A regular hexagon has sides 10 centimeters long. What is its area? 259.8 cm^2

3. Not regular; polygon is not convex and all angles are not congruent.

$\odot O$ has a diameter of 4.5 inches. Find the missing measures. Round to the nearest tenth.

15. What is the area of $\odot O$? 15.9 in^2
16. What is the circumference of $\odot O$? 14.1 in.
17. A sector of $\odot O$ has a central angle measuring 30°. Find the area of the sector. 1.3 in^2

A circular dartboard has a diameter of 18 inches. The bull's-eye has a diameter of 3 inches, and the blue ring around the bull's-eye is 4 inches wide.

18. What is the probability that a randomly-thrown dart that hits the dartboard will land in the bull's-eye? $\frac{1}{36}$ or about 2.78%
19. What is the probability that a randomly-thrown dart that hits the dartboard will land in the blue ring? $\frac{28}{81}$ or about 34.57%
20. Draw a complete network with 6 nodes. a–c. See students' work.
    a. How many edges does it have?
    b. Find the degree of each node.
    c. Is it traceable? If so, list a sequence of segments for traceability.

802 Chapter 10 Test

CHAPTER 11 TEST

1. From the views of the given solid figure, draw a corner view. 1–3. See margin.

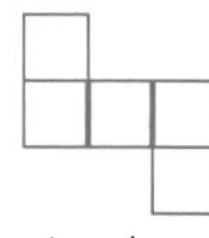
top view

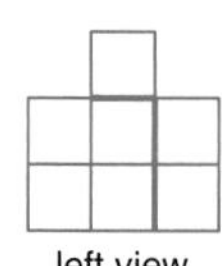
left view

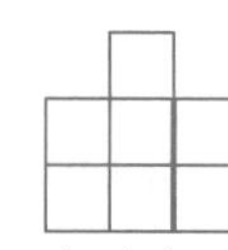
front view

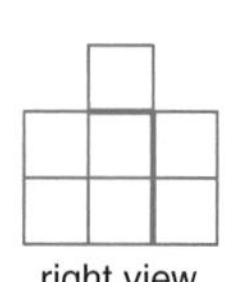
right view

2. Use isometric dot paper to draw a rectangular prism with a square base that is 5 units on each side and a height that is 10 units.
3. Use rectangular dot paper to sketch a net of a right square pyramid.

Find the surface area of each solid figure. Round to the nearest tenth.

4. 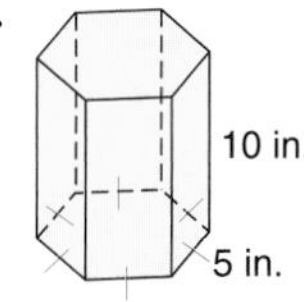

429.9 in^2

5. 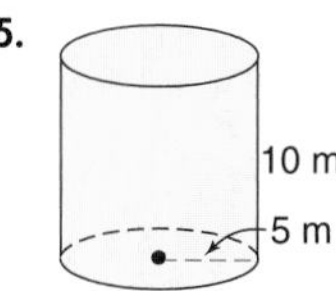

471.2 m^2

6. 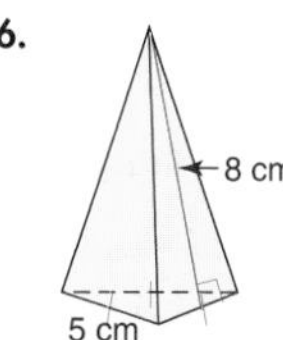

70.8 cm^2

7. a right circular cone with a radius of 2.7 millimeters and a height of 30 millimeters 278.4 mm^2
8. a sphere with a diameter of 6 inches 113.1 in^2

Find the volume of each solid figure. Round to the nearest tenth.

9. 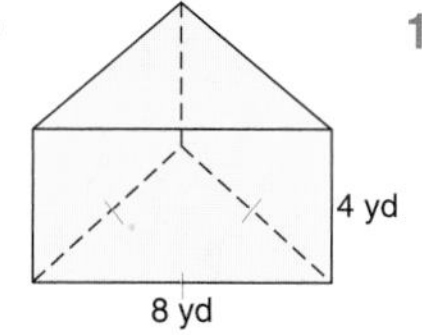

110.9 yd^3

10. 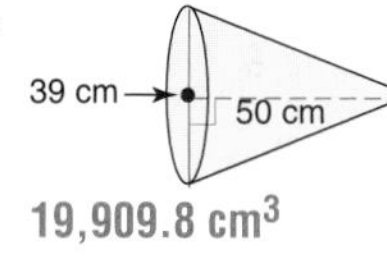

19,909.8 cm^3

11. 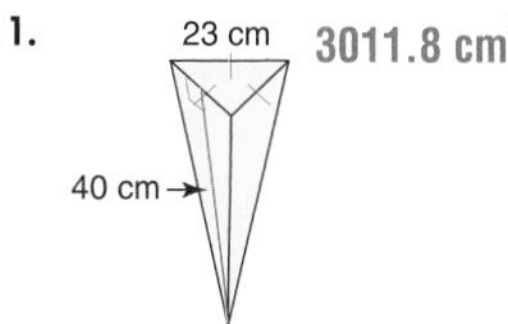

3011.8 cm^3

12. a right cylinder with a diameter of 39 centimeters and a height of 50 centimeters 59,729.5 cm^3
13. a sphere with a radius of 18 millimeters 24,429.0 mm^3
14. **Sports** A rectangular swimming pool is 4 meters wide and 10 meters long. A concrete walkway is poured around the pool. The walkway is 1 meter wide and 0.1 meter deep. What is the volume of the concrete? 3.2 m^3
15. The base of a right prism is a right triangle with legs 6 inches and 8 inches. If the height of the prism is 16 inches, find the lateral area of the prism. 384 in^2

Chapter 11 Test Answers

1. 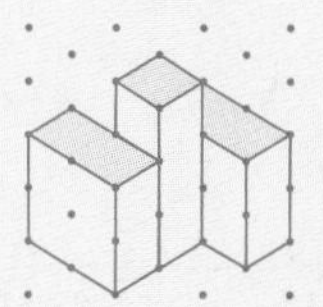

2. Sample answer:

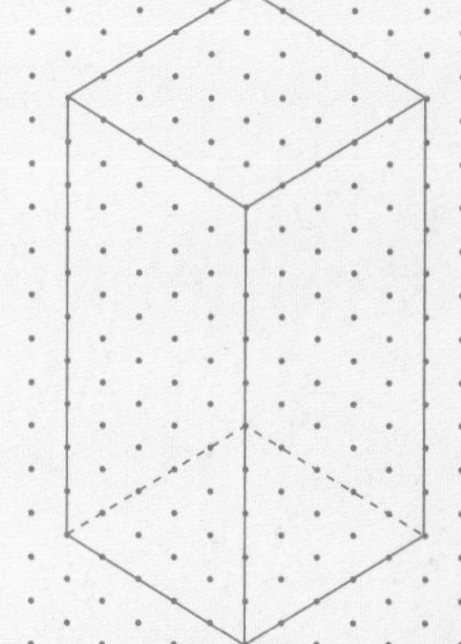

3. Sample answer:

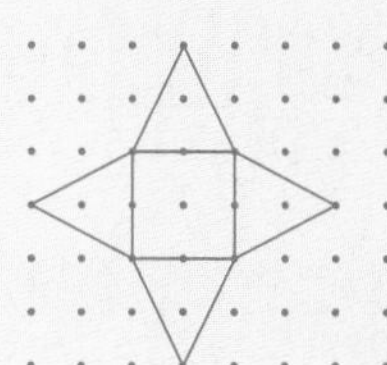

Chapter 12 Test Answers

1.

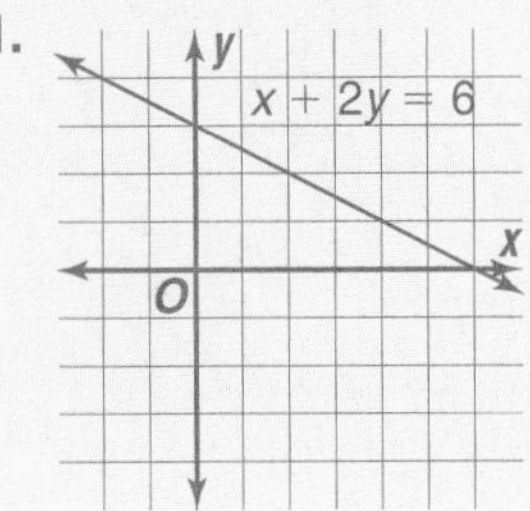

2.

y
x = 3
O
x

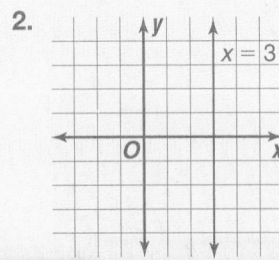

3.

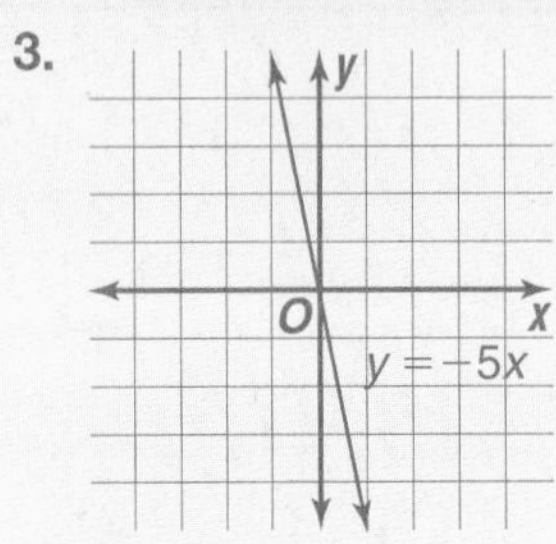

19. **Proof:**
slope of $\overleftrightarrow{NK} = \frac{d-0}{a-a} = \frac{d}{0}$; undefined; slope of $\overline{OM} = \frac{0-0}{0-2a} = \frac{0}{-2a} = 0$; $\overleftrightarrow{NK} \perp \overline{OM}$ because vertical and horizontal lines are perpendicular.
$OK = \sqrt{(a-0)^2 + (0-0)^2} = a$
$KM = \sqrt{(2a-a)^2 + (0-0)^2} = a$
Since $OK = KM$ and $\overleftrightarrow{NK} \perp \overline{OM}$, $\overleftrightarrow{NK}$ is a perpendicular bisector of $\overline{OM}$.

CHAPTER 12 TEST

Graph each equation. State the slope and *y*-intercept. 1–3. See margin for graphs.

1. $x + 2y = 6$ $\frac{-1}{2}$; 3
2. $x = 3$ none; none
3. $y = -5x$ −5, 0

Write the equation in slope-intercept form of the line that satisfies the given conditions.

4. $m = -4$; passes through the point at $(3, -2)$ $y = -4x + 10$
5. passes through points at $(-4, 11)$ and $(-6, 3)$ $y = 4x + 27$
6. parallel to the graph of $y = 2x - 5$; passes through the point at $(-1, -4)$ $y = 2x - 2$
7. parallel to the *y*-axis; passes through the point at $(-4, -2)$ $x = -4$

Given $\vec{a} = (-3, 5)$ and $\vec{b} = (0, 7)$, represent each of the following as an ordered pair.

8. $\vec{a} + \vec{b}$ $(-3, 12)$
9. $\vec{b} - 3\vec{a}$ $(9, -8)$

Determine the coordinates of the center and the measure of the radius for each sphere whose equation is given.

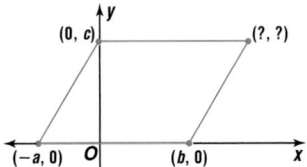

10. $(x-4)^2 + (y-5)^2 + (z+2)^2 = 81$ $(4, 5, -2)$; 9
11. $x^2 + y^2 + z^2 = 7$ $(0,0,0)$; $\sqrt{7} \approx 2.6$
12. Given $\vec{v} = (-5, -3)$, find the magnitude of $\vec{v}$. $\sqrt{34} \approx 5.8$ units
13. Given $A(3, 7)$ and $B(-2, 5)$, find the magnitude of $\overline{AB}$. $\sqrt{29} \approx 5.4$ units
14. Determine the distance between points at $(2, 4, 5)$ and $(2, 4, 7)$. 2 units
15. Determine the midpoint of the segment whose endpoints are $X(0, -4, 2)$ and $Y(3, 0, 2)$. $\left(\frac{3}{2}, -2, 2\right)$
16. Write the equation of the sphere whose diameter has endpoints at $P(-3, 5, 7)$ and $Q(5, -1, 5)$. $(x-1)^2 + (y-2)^2 + (z-6)^2 = 26$
17. Write an equation for the perpendicular bisector of the segment whose endpoints are at $(5, 2)$ and $(1, -4)$. $y = \frac{-2}{3}x + 1$
18. Name the missing coordinates for the parallelogram in terms of the given variables. $(a + b, c)$
19. Use a coordinate proof to show that $\overline{NK}$ is a perpendicular bisector of a side of $\triangle LOM$. See margin.

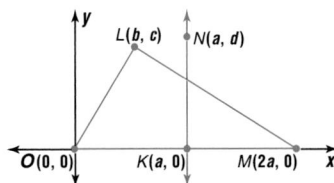

20. The table below lists radio advertising revenue in billions of dollars.

Year	1980	1982	1984	1986	1988	1990	1992	1994
Revenue	4	4.8	5.9	7	7.8	8.7	8.5	10.3

a. Draw a scatter plot to show how year *x* and advertising revenue *y* are related. See margin.
b. Write an equation that relates the year to the dollar revenue. Sample answer: $y - 4{,}000{,}000{,}000 = 0.475(x - 1980)$
c. Graph the equation. See margin.

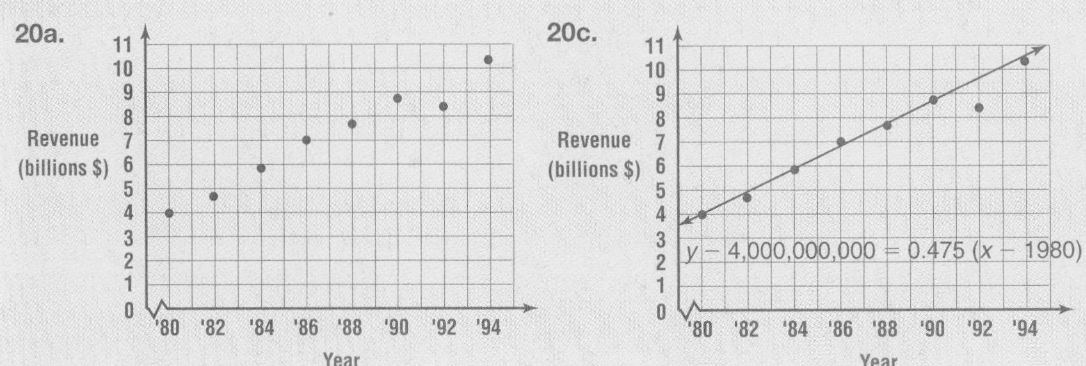

CHAPTER 13 TEST

Describe the locus of points that satisfy each set of conditions. See margin.

1. all the points in a plane that are 5 inches from a given line ℓ
2. all the points in a plane that are 3.5 centimeters from a vertex of a given rhombus
3. all the points in space that are 4 meters from a given corner of a room
4. all the points in space that are 10 feet from a given line n

Describe all the possible ways the figures can intersect.

5. two concentric spheres and a line
   0, 1, 2, 3, or 4 points
6. two parallel planes and a sphere
   0 points, a point, 2 points, a circle, 2 circles

Graph the following pair of equations to find the locus of points that satisfy the graphs of both equations.

7. $y = 4x$
   $x + y = 5$ (1, 4); see margin for graph.

Use either substitution or elimination to find the locus of points that satisfy the graphs of both equations.

8. $3x - 2y = 10$
   $x + y = 0$ (2, −2)
9. $x - 4y = 7$
   $2x + y = -4$ (−1, −2)

Determine if each of the following is an isometry. Write *yes* or *no*.

10. reflection yes
11. translation yes
12. rotation yes
13. dilation no

Describe each of the following as a *reflection*, a *rotation*, a *translation*, or a *dilation*.

14. 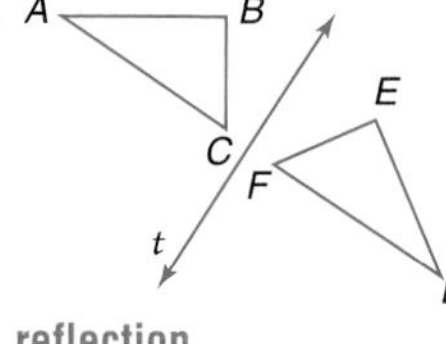

reflection

15. 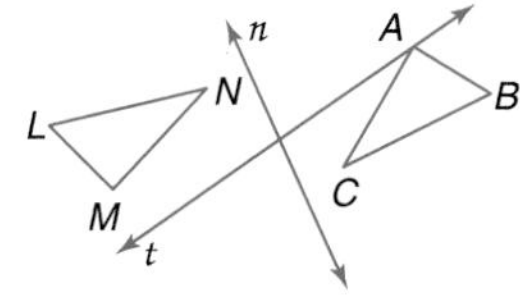

rotation

16. 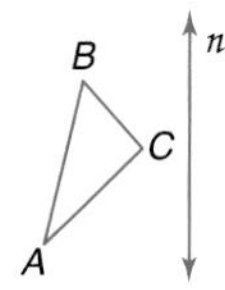

t Q R P
translation

17. Copy the figure at the right. Then draw all the possible lines of symmetry.

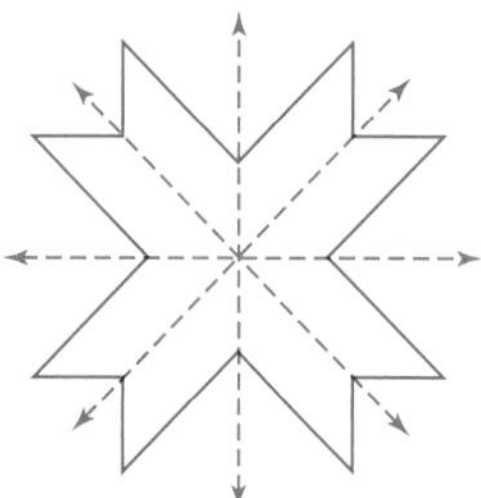

For each scale factor, determine whether the image is an *enlargement*, a *reduction*, or a *congruence transformation*.

18. $k = \frac{2}{3}$ reduction
19. $k = 1$ congruence transformation
20. **Entertainment** On a TV game show called "The Big 24", a contestant is paid \$7 for each correct answer, but must pay back \$5 for each wrong answer. After answering 24 questions on the show, Sam broke even. How many questions did he answer correctly? 10

Chapter 13 Test Answers

1. 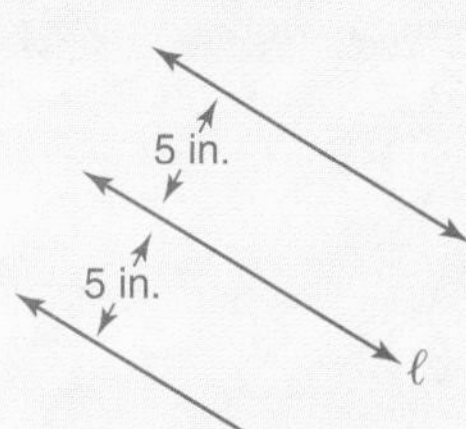

2. 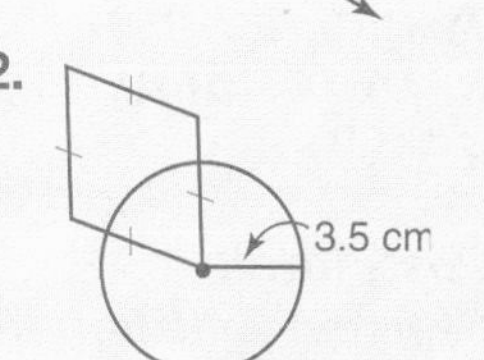

3. 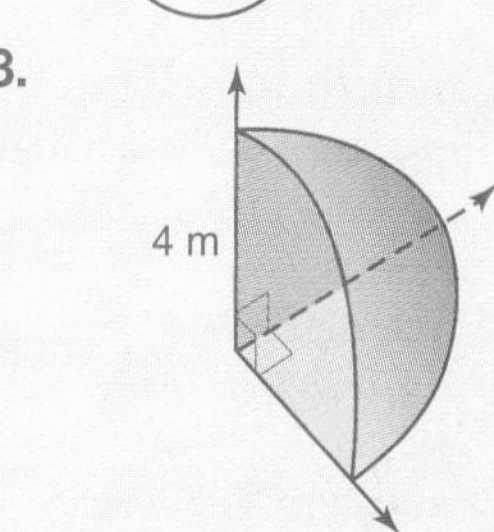

This is a quarter of a half sphere.

4. 
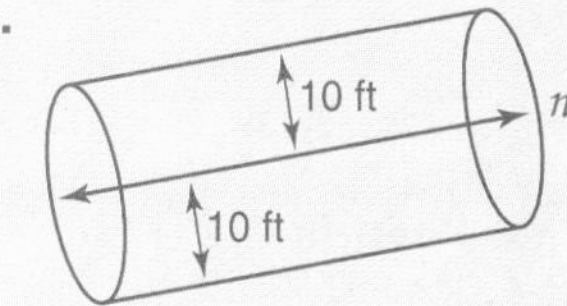

7. 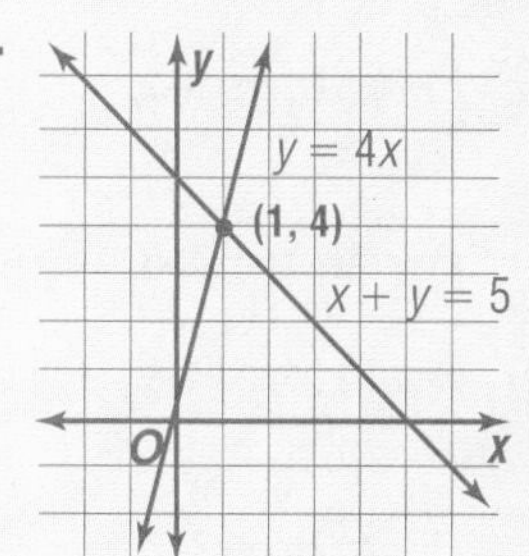

Chapter 13 Test 805

POSTULATES, THEOREMS, AND COROLLARIES

Chapter 1 Discovering Points, Lines, Planes, and Angles

Postulate 1–1
Ruler Postulate

The points on any line can be paired with the real numbers so that, given any two points P and Q on the line, P corresponds to zero, and Q corresponds to a positive number. (29)

Postulate 1–2
Segment Addition Postulate

If Q is between P and R, then $PQ + QR = PR$. If $PQ + QR = PR$, then Q is between P and R. (29)

Postulate 1–3
Protractor Postulate

Given $\overrightarrow{AB}$ and a number r between 0 and 180, there is exactly one ray with endpoint A, extending on each side of $\overrightarrow{AB}$, such that the measure of the angle formed is r. (46)

Postulate 1–4
Angle Addition Postulate

If R is in the interior of $\angle PQS$, then $m\angle PQR + m\angle RQS = m\angle PQS$. If $m\angle PQR + m\angle RQS = m\angle PQS$, then R is in the interior of $\angle PQS$. (46)

Theorem 1–1
Midpoint Theorem

If M is the midpoint of $\overline{AB}$, then $\overline{AM} \cong \overline{AB}$. (39)

Chapter 2 Connecting Reasoning and Proof

Postulate 2–1

Through any two points there is exactly one line. (79) *(Through any 2 pts. there is 1 line.)*

Postulate 2–2

Through any three points not on the same line there is exactly one plane. (79) *(Through any 3 noncollinear pts. there is 1 plane.)*

Postulate 2–3

A line contains at least two points. (79) *(A line contains at least 2 pts.)*

Postulate 2–4

A plane contains at least three points not on the same line. (79) *(A plane contains at least 3 noncollinear pts.)*

Postulate 2–5

If two points lie in a plane, then the entire line containing those two points lies in that plane. (79) *(If 2 pts. are in a plane, then the line that contains them is in the plane.)*

Postulate 2–6

If two planes intersect, then their intersection is a line. (79) *(The intersection of 2 planes is a line.)*

Theorem 2–1

Congruence of segments is reflexive, symmetric, and transitive. (101) *(Reflexive Prop. of ≅ Segments, Symmetric Prop. of ≅ Segments, Transitive Prop. of ≅ Segments.)*

Theorem 2–2
Supplement Theorem

If two angles form a linear pair, then they are supplementary angles. (107)

Theorem 2–3

Congruence of angles is reflexive, symmetric, and transitive. (107) *(Reflexive Prop. of ≅ ∠s, Symmetric Prop. of ≅ ∠s, Transitive Prop. of ≅ ∠s.)*

Theorem 2–4

Angles supplementary to the same angle or to congruent angles are congruent. (108) *(∠s supp. to same ∠ or ≅ ∠s are ≅.)*

Theorem 2–5

Angles complementary to the same angle or to congruent angles are congruent. (109) *(∠s compl. to same ∠ or ≅ ∠s are ≅.)*

Theorem 2–6 All right angles are congruent. (109) *(All rt. $\angle$s are $\cong$.)*

Theorem 2–7 Vertical angles are congruent. (109) *(Vert. $\angle$s are $\cong$.)*

Theorem 2–8 Perpendicular lines intersect to form four right angles. (110) *($\perp$ lines form 4 rt. $\angle$s.)*

Chapter 3 Using Perpendicular and Parallel Lines

Postulate 3–1 Corresponding Angles Postulate It two parallel lines are cut by a transversal, then each pair of corresponding angles is congruent. (132)

Theorem 3–1 Alternate Interior Angles Theorem If two parallel lines are cut by a transversal, then each pair of alternate interior angles is congruent. (132)

Theorem 3–2 Consecutive Interior Angles Theorem If two parallel lines are cut by a transversal, then each pair of consecutive interior angles is supplementary. (132)

Theorem 3–3 Alternate Exterior Angles Theorem If two parallel lines are cut by a transversal, then each pair of alternate exterior angles is congruent. (132)

Theorem 3–4 Perpendicular Transversal Theorem In a plane, if a line is perpendicular to one of two parallel lines, then it is perpendicular to the other. (133)

Postulate 3–2 Two nonvertical lines have the same slope if and only if they are parallel. (139)

Postulate 3–3 Two nonvertical lines are perpendicular if and only if the product of their slopes is -1. (139)

Postulate 3–4 If two lines in a plane are cut by a transversal so that the corresponding angles are congruent, then the lines are parallel. (147) *(If ⇸ and corr. $\angle$s are $\cong$, then the lines are $\parallel$.)*

Postulate 3–5 Parallel Postulate If there is a line and a point not on the line, then there exists exactly one line through the point that is parallel to the given line. (147)

Theorem 3–5 If two lines in a plane are cut by a transversal so that a pair of alternate exterior angles is congruent, then the two lines are parallel. (147) *(If ⇸ and alt. ext. $\angle$s are $\cong$, then the lines are $\parallel$.)*

Theorem 3–6 If two lines in a plane are cut by a transversal so that a pair of consecutive interior angles is supplementary, then the lines are parallel. (147) *(If ⇸ and consec. int. $\angle$s are supp., then the lines are $\parallel$.)*

Theorem 3–7 If two lines in a plane are cut by a transversal so that a pair of alternate interior angles is congruent, then the lines are parallel. (147) *(If ⇸ and alt. int. $\angle$s are $\cong$, then the lines are $\parallel$.)*

Theorem 3–8 In a plane, if two lines are perpendicular to the same line, then they are parallel. (147) *(In a plane, if 2 lines are $\perp$ to the same line, they are $\parallel$.)*

Postulates, Theorems, and Corollaries

Chapter 4 Identifying Congruent Triangles

Theorem 4–1 ***Angle Sum Theorem***	The sum of the measures of the angles of a triangle is 180. (189)
Theorem 4–2 ***Third Angle Theorem***	If two angles of one triangle are congruent to two angles of a second triangle, then the third angles of the triangles are congruent. (190)
Theorem 4–3 ***Exterior Angle Theorem***	The measure of an exterior angle of a triangle is equal to the sum of the measures of the two remote interior angles. (190)
Corollary 4–1	The acute angles of a right triangle are complementary. (192) *(The acute ∡ of a rt. △ are comp.)*
Corollary 4–2	There can be at most one right or obtuse angle in a triangle. (192) *(There can be at most 1 rt. or obtuse ∠in a △.)*
Theorem 4–4	Congruence of triangles is reflexive, symmetric, and transitive. (198) *(Congruence of triangles is (reflexive/symmetric/transitive).)*
Postulate 4–1 ***SSS Postulate***	If the sides of one triangle are congruent to the sides of a second triangle, then the triangles are congruent. (206)
Postulate 4–2 ***SAS Postulate***	If two sides and the included angle of one triangle are congruent to two sides and the included angle of another triangle, then the triangles are congruent. (207)
Postulate 4–3 ***ASA Postulate***	If two angles and the included side of one triangle are congruent to two angles and the included side of another triangle, the triangles are congruent. (207)
Theorem 4–5 ***AAS***	If two angles and a nonincluded side of one triangle are congruent to the corresponding two angles and side of a second triangle, the two triangles are congruent. (214)
Theorem 4–6 ***Isosceles Triangle Theorem***	If two sides of a triangle are congruent, then the angles opposite those sides are congruent. (222)
Theorem 4–7	If two angles of a triangle are congruent, then the sides opposite those angles are congruent. (223) *(If 2 ∡ of a △ are ≅, the sides opp. the ∡ are ≅.)*
Corollary 4–3	A triangle is equilateral if and only if it is equiangular. (224) *(An (equilateral/equiangular) △ is (equiangular/equilateral).)*
Corollary 4–4	Each angle of an equilateral triangle measures 60°. (224) *(Each ∠ of an equilateral △ measures 60°.)*

Chapter 5 Applying Congruent Triangles

Theorem 5–1	A point on the perpendicular bisector of a segment is equidistant from the endpoints of the segment. (238) *(A pt. on the ⊥ bisector of a segment is equidistant from the endpts. of the segment.)*
Theorem 5–2	A point equidistant from the endpoints of a segment lies on the perpendicular bisector of the segment. (238) *(A pt. equidistant from the endpts. of a segment lies on the ⊥ bisector of the segment.)*
Theorem 5–3	A point on the bisector of an angle is equidistant from the sides of the angle. (240) *(A pt. on the bisector of an ∠ is equidistant from the sides of the ∠.)*

Theorem 5–4	A point on or in the interior of an angle and equidistant from the sides of an angle lies on the bisector of the angle. (240) *(A pt. on or in the int. of an ∠ and equidistant from the sides of an ∠ lies on the bisector of the ∠.)*
Theorem 5–5 LL	If the legs of one right triangle are congruent to the corresponding legs of another right triangle, then the triangles are congruent. (245)
Theorem 5–6 HA	If the hypotenuse and an acute angle of one right triangle are congruent to the hypotenuse and corresponding acute angle of another right triangle, then the two triangles are congruent. (246)
Theorem 5–7 LA	If one leg and an acute angle of one right triangle are congruent to the corresponding leg and acute angle of another right triangle, then the triangles are congruent. (247)
Postulate 5–1 HL	If the hypotenuse and a leg of one right triangle are congruent to the hypotenuse and corresponding leg of another right triangle, then the triangles are congruent. (247)
Theorem 5–8 Exterior Angle Inequality Theorem	If an angle is an exterior angle of a triangle, then its measure is greater than the measure of either of its corresponding remote interior angles. (253)
Theorem 5–9	If one side of a triangle is longer than another side, then the angle opposite the longer side has a greater measure than the angle opposite the shorter side. (259) *(If one side of a △ is longer than another, the ∠ opp. the longer side > the ∠ opp. the shorter side.)*
Theorem 5–10	If one angle of a triangle has a greater measure than another angle, then the side opposite the greater angle is longer than the side opposite the lesser angle. (259) *(If an ∠ of a △ > another, the side opp. the greater ∠ is longer than the side opp. the lesser ∠.)*
Theorem 5–11	The perpendicular segment from a point to a line is the shortest segment from the point to the line. (261) *(The ⊥ segment from a pt. to a line is the shortest segment from the pt. to the line.)*
Corollary 5–1	The perpendicular segment from a point to a plane is the shortest segment from the point to the plane. (262) *(The ⊥ segment from a pt. to a plane is the shortest segment from the pt. to the plane.)*
Theorem 5–12 Triangle Inequality Theorem	The sum of the lengths of any two sides of a triangle is greater than the length of the third side. (267)
Theorem 5–13 SAS Inequality (Hinge Theorem)	If two sides of one triangle are congruent to two sides of another triangle, and the included angle in one triangle is greater than the included angle in the other, then the third side of the first triangle is longer than the third side in the second triangle. (273)
Theorem 5–14 SSS Inequality	If two sides of one triangle are congruent to two sides of another triangle and the third side in one triangle is longer than the third side in the other, then the angle between the pair of congruent sides in the first triangle is greater than the corresponding angle in the second triangle. (274)

Chapter 6 Exploring Quadrilaterals

Theorem 6–1	Opposite sides of a parallelogram are congruent. (292) *(Opp. sides of a ▱ are ≅.)*

Theorem 6–2	Opposite angles of a parallelogram are congruent. (292) *(Opp. ∠s of a ▱ are ≅ .)*
Theorem 6–3	Consecutive angles in a parallelogram are supplementary. (292) *(Consec. ∠s of a ▱ are supp.)*
Theorem 6–4	The diagonals of a parallelogram bisect each other. (293) *(Diagonals of a ▱ bisect each other.)*
Theorem 6–5	If both pairs of opposite sides of a quadrilateral are congruent, then the quadrilateral is a parallelogram. (298) *(If opp. sides of a quad. are ≅, it is a ▱.)*
Theorem 6–6	If both pairs of opposite angles of a quadrilateral are congruent, then the quadrilateral is a parallelogram. (298) *(If both pairs of opp. ∠s of a quad. are ≅, it is a ▱.)*
Theorem 6–7	If the diagonals of a quadrilateral bisect each other, then the quadrilateral is a parallelogram. (298) *(If the diagonals of quad. bisect, it is a ▱.)*
Theorem 6–8	If one pair of opposite sides of a quadrilateral are both parallel and congruent, then the quadrilateral is a parallelogram. (299) *(If a pair of opp. sides of a quad. are ≅ and ∥, it is a ▱.)*
Theorem 6–9	If a parallelogram is a rectangle, then its diagonals are congruent. (306) *(If a ▱ is a rect., then its diagonals are ≅.)*
Theorem 6–10	If the diagonals of a parallelogram are congruent, then the parallelogram is a rectangle. (308) *(If the diagonals of a ▱ are ≅, then the ▱ is a rect.)*
Theorem 6–11	The diagonals of a rhombus are perpendicular. (313) *(Diagonals of a rhom. are ⊥.)*
Theorem 6–12	If the diagonals of a parallelogram are perpendicular, then the parallelogram is a rhombus. (313) *(If the diagonals of a ▱ are ⊥, then the ▱ is a rhombus.)*
Theorem 6–13	Each diagonal of a rhombus bisects a pair of opposite angles. (313) *(Each diagonal of a rhom. bisects opp. ∠s.)*
Theorem 6–14	Both pairs of base angles of an isosceles trapezoid are congruent. (321) *(Base ∠s of an iso. trap. are ≅ .)*
Theorem 6–15	The diagonals of an isosceles trapezoid are congruent. (321) *(The diagonals of an isos. trap. are ≅ .)*
Theorem 6–16	The median of a trapezoid is parallel to the bases, and its measure is one-half the sum of the measures of the bases. (323) *(Median of a trap. is ∥ to the bases. Length of median of a trap. = $\frac{1}{2}$(sum of the lengths of bases).)*

Chapter 7 Connecting Proportion and Similarity

Postulate 7–1 ***AA Similarity***	If two angles of one triangle are congruent to two angles of another triangle, then the triangles are similar. (355)
Theorem 7–1 ***SSS Similarity***	If the measures of the corresponding sides of two triangles are proportional, then the triangles are similar. (355)
Theorem 7–2 ***SAS Similarity***	If the measures of two sides of a triangle are proportional to the measures of two corresponding sides of another triangle and the included angles are congruent, then the triangles are similar. (356)

Theorem 7–3	Similarity of triangles is reflexive, symmetric, and transitive. (357) *(Reflexive Prop. of ~ △s, Symmetric Prop. of ~ △s, Transitive Prop. of ~ △s.)*
Theorem 7–4 Triangle Proportionality	If a line is parallel to one side of a triangle and intersects the other two sides in two distinct points, then it separates these sides into segments of proportional lengths. (362)
Theorem 7–5	If a line intersects two sides of a triangle and separates the sides into corresponding segments of proportional lengths, then the line is parallel to the third side. (363)
Theorem 7–6	A segment whose endpoints are the midpoints of two sides of a triangle is parallel to the third side of the triangle and its length is one-half the length of the third side. (363)
Corollary 7–1	If three or more parallel lines intersect two transversals, then they cut off the transversals proportionally. (364)
Corollary 7–2	If three or more parallel lines cut off congruent segments on one transversal, then they cut off congruent segments on every transversal. (364)
Theorem 7–7 Proportional Perimeters	If two triangles are similar, then the perimeters are proportional to the measures of corresponding sides. (370)
Theorem 7–8	If two triangles are similar, then the measures of the corresponding altitudes are proportional to the measures of the corresponding sides. (371)
Theorem 7–9	If two triangles are similar, then the measures of the corresponding angle bisectors are proportional to the measures of the corresponding sides. (371)
Theorem 7–10	If two triangles are similar, then the measures of the corresponding medians are proportional to the measures of the corresponding sides. (371)
Theorem 7–11 Angle Bisector Theorem	An angle bisector in a triangle separates the opposite side into segments that have the same ratio as the other two sides. (372)

Chapter 8 Applying Right Triangles and Trigonometry

Theorem 8–1	If the altitude is drawn from the vertex of the right angle of a right triangle to its hypotenuse, then the two triangles formed are similar to the given triangle and to each other. (398)
Theorem 8–2	The measure of the altitude drawn from the vertex of the right angle of a right triangle to its hypotenuse is the geometric mean between the measures of the two segments of the hypotenuse. (398)
Theorem 8–3	If the altitude is drawn to the hypotenuse of a right triangle, then the measure of a leg of the triangle is the geometric mean between the measures of the hypotenuse and the segment of the hypotenuse adjacent to the leg. (399)
Theorem 8–4 Pythagorean Theorem	In a right triangle, the sum of the squares of the measures of the legs equals the square of the measure of the hypotenuse. (399)
Theorem 8–5 Converse of the Pythagorean Theorem	If the sum of the squares of the measures of two sides of a triangle equals the square of the measure of the longest side, then the triangle is a right triangle. (395)

Theorem 8–6	In a 45°-45°-90° triangle, the hypotenuse is $\sqrt{2}$ times as long as a leg. (405)
Theorem 8–7	In a 30°-60°-90° triangle, the hypotenuse is twice as long as the shorter leg, and the longer leg is $\sqrt{3}$ times as long as the shorter leg. (407)

Chapter 9 Analyzing Circles

Postulate 9–1 Arc Addition Postulate	The measure of an arc formed by two adjacent arcs is the sum of the measures of the two arcs. That is, if Q is a point on $\overset{\frown}{PR}$, then $m\overset{\frown}{PQ} + m\overset{\frown}{QR} = m\overset{\frown}{PQR}$. (453)
Theorem 9–1	In a circle or in congruent circles, two minor arcs are congruent if and only if their corresponding chords are congruent. (459) *(In a ⊙ or in ≅ ⊙s, 2 minor arcs are ≅ if and only if their corr. chords are ≅.)*
Theorem 9–2	In a circle, if a diameter is perpendicular to a chord, then it bisects the chord and its arc. (460) *(In a ⊙, if a diameter is ⊥ to a chord, then it bisects the chord and its arc.)*
Theorem 9–3	In a circle or in congruent circles, two chords are congruent if and only if they are equidistant from the center. (461) *(In a ⊙ or in ≅ ⊙s, 2 chords are ≅ if and only if they are equidistant from the center.)*
Theorem 9–4	If an angle is inscribed in a circle, then the measure of the angle equals one-half the measure of the intercepted arc. (466) *(If an ∠ is inscribed in a ⊙, then the measure of the ∠ = $\frac{1}{2}$ the measure of the intercepted arc.)*
Theorem 9–5	If two inscribed angles of a circle or congruent circles intercept congruent arcs or the same arc, then the angles are congruent. (467) *(If 2 inscribed ∠s of a ⊙ or ≅ ⊙s intercept ≅ arcs or the same arc, then the ∠s are ≅.)*
Theorem 9–6	If an inscribed angle of a circle intercepts a semicircle, then the angle is a right angle. (467) *(If an inscribed ∠ of a ⊙ intercepts a semicircle, then the ∠ is a rt. ∠.)*
Theorem 9–7	If a quadrilateral is inscribed in a circle, then its opposite angles are supplementary. (469) *(If a quad. is inscribed in a ⊙, then its opp. ∠s are supp.)*
Theorem 9–8	If a line is tangent to a circle, then it is perpendicular to the radius drawn to the point of tangency. (475)
Theorem 9–9	In a plane, if a line is perpendicular to a radius of a circle at the endpoint on the circle, then the line is a tangent of the circle. (476)
Theorem 9–10	If two segments from the same exterior point are tangent to a circle, then they are congruent. (477)
Theorem 9–11	If a secant and a tangent intersect at the point of tangency, then the measure of each angle formed is one-half the measure of its intercepted arc. (483)
Theorem 9–12	If two secants intersect in the interior of a circle, then the measure of an angle formed is one-half the sum of the measure of the arcs intercepted by the angle and its vertical angle. (484)
Theorem 9–13	If two secants, a secant and a tangent, or two tangents intersect in the exterior of a circle, then the measure of the angle formed is one-half the positive difference of the measures of the intercepted arcs. (485)
Theorem 9–14	If two chords intersect in a circle, then the products of the measures of the segments of the chords are equal. (491)

Theorem 9–15 If two secant segments are drawn to a circle from an exterior point, then the product of the measures of one secant segment and its external secant segment is equal to the product of the measures of the other secant segment and its external secant segment. (493)

Theorem 9–16 If a tangent segment and a secant segment are drawn to a circle from an exterior point, then the square of the measure of the tangent segment is equal to the product of the measures of the secant segment and its external secant segment. (493)

Chapter 10 Exploring Polygons and Area

Theorem 10–1 Interior Angle Sum Theorem If a convex polygon has n sides and S is the sum of the measures of its interior angles, then $S = 180(n - 2)$. (516)

Theorem 10–2 Exterior Angle Sum Theorem If a polygon is convex, then the sum of the measures of the exterior angles, one at each vertex, is 360. (517)

Postulate 10–1 The area of a region is the sum of the areas of all of its nonoverlapping parts. (531)

Postulate 10–2 Congruent figures have equal areas. (535)

Postulate 10–3 Length Probability Postulate If a point on $\overline{AB}$ is chosen at random and C is between A and B, then the probability that the point is on $\overline{AC}$ is $\frac{\text{length of } \overline{AC}}{\text{length of } \overline{AB}}$. (551)

Postulate 10–4 Area Probability Postulate If a point in region A is chosen at random, then the probability that the point is in region B, which is in the interior of region A, is $\frac{\text{area of region } B}{\text{area of region } A}$. (552)

Chapter 11 Investigating Surface Area and Volume

Theorem 11–1 If two solids are similar with a scale factor of $a{:}b$, then the surface areas have a ratio of $a^2{:}b^2$ and the volumes have a ratio of $a^3{:}b^3$. (631)

Chapter 12 Continuing Coordinate Geometry

Theorem 12–1 Slope-Intercept Form If the equation of a line is written in the form $y = mx + b$, m is the slope of the line and b is the y-intercept. (647)

Theorem 12–2 Given two points $A(x_1, y_1, z_1)$ and $B(x_2, y_2, z_2)$ in space, the distance between A and B is given by the following equation. (681)

$$AB = \sqrt{(x_2 - x_1)^2 + (y_2 - y_1)^2 + (z_2 - z_1)^2}$$

Chapter 13 Investigating Loci and Coordinate Transformations

Postulate 13–1 In a given rotation, if A is the preimage, P is the image, and W is the center of rotation, then the measure of the angle of rotation, $\angle AWP$, equals twice the measure of the angle formed by intersecting lines of reflection. (740)

Theorem 13–1 If a dilation with center C and a scale factor k maps A onto E and B onto D, then $ED = k(AB)$. (746)

GLOSSARY

A

absolute value (29) For any real number, the number of units the number is from zero on the number line.

acute angle (46) An angle whose degree measure is less than 90.

acute triangle (180) A triangle all of whose angles are acute angles.

adjacent angles (53) Two angles in the same plane that have a common vertex and a common side, but no common interior points.

adjacent arcs (453) Two arcs of a circle that have exactly one point in common.

alternate exterior angles (126) In the figure, transversal t intersects lines ℓ and m. $\angle 5$ and $\angle 3$, and $\angle 6$ and $\angle 4$ are alternate exterior angles.

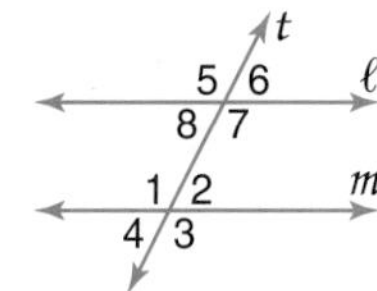

alternate interior angles (126) In the figure above, transversal t intersects lines ℓ and m. $\angle 1$ and $\angle 7$, and $\angle 2$ and $\angle 8$ are alternate interior angles.

altitude of a cone (602) The segment from the vertex to the plane of the base and perpendicular to the plane of the base.

altitude of a cylinder (593) A segment perpendicular to the base planes and having an endpoint in each plane.

altitude of a parallelogram (529) Any perpendicular segment between the lines containing two of the parallel sides.

altitude of a prism (591) A segment perpendicular to the base planes of the prism, with an endpoint in each plane. The length of an altitude is called the height of the prism.

altitude of a pyramid (600) The segment from the vertex to the plane of the base and perpendicular to the plane of the base.

altitude of a triangle (238) A segment from a vertex of the triangle to the line containing the opposite side and perpendicular to the line containing that side.

angle (44) A figure consisting of two noncollinear rays with a common endpoint. The rays are the sides of the angle. The common endpoint of the rays is the vertex of the angle. An angle separates a plane into three parts, the interior of the angle, the exterior of the angle, and the angle itself.

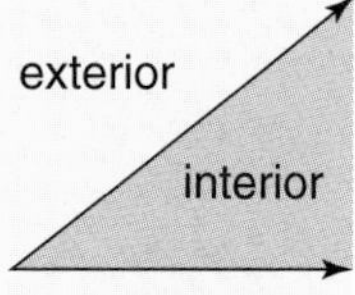

angle bisector (48) The ray, QS, is the bisector of $\angle PQR$ if S is in the interior of the angle and $\angle PQS \cong \angle RQS$.

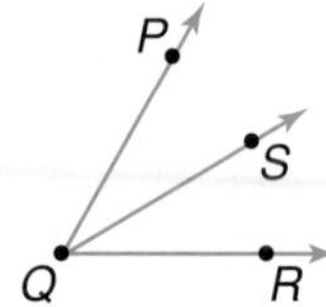

angle bisector of a triangle (240) A segment that bisects an angle of the triangle and has one endpoint at a vertex of the triangle and the other endpoint at another point on the triangle.

angle of depression (420) An angle formed by a horizontal line and the line of sight to an object below the level of the horizontal.

angle of elevation (400) An angle formed by a horizontal line and the line of sight to an object above the level of the horizontal.

angle of rotation (740) The angle of rotation, $\angle ABC$, is determined by the preimage A, the center of rotation B, and the rotation image C.

apothem (544) A segment that is drawn from the center of a regular polygon perpendicular to a side of the polygon.

arc (453) An unbroken part of a circle.

arc length (454) The linear distance representing the arc. The length of the arc is a part of the circumference proportional to the measure of the central angle when compared to the entire circle.

arc measure (453) The degree measure of a minor arc is the degree measure of its central angle. The degree measure of a major arc is 360 minus the degree measure of its central angle. The degree measure of a semicircle is 180.

arc of a chord (459) A minor arc that has the same endpoints as a chord is called an arc of the chord.

area (19, 529) The number of square units contained in the interior of a figure.

auxiliary line (189) A line or line segment added to a given figure to help in proving a result.

axis **1.** (6) In a coordinate plane, the *x*-axis is the horizontal number line and the *y*-axis is the vertical number line. **2.** (593) The axis of a cylinder is the segment whose endpoints are the centers of the bases. **3.** (602) The axis of a cone is the segment whose endpoints are the vertex and the center of the base.

B

base **1.** (182) In an isosceles triangle, the side opposite the vertex angle is called the base. **2.** (321) In a trapezoid, the parallel sides are called bases. **3.** (529) Any side of a parallelogram can be called a base. **4.** (577, 591) In a prism, the bases are the two faces formed by congruent polygons that lie in parallel planes, all of the other faces being parallelograms. **5.** (593) In a cylinder, the bases are the two congruent and parallel circular regions that form the ends of the cylinder. **6.** (600) In a pyramid, the base is the face that does not intersect the other faces at the vertex. The base is a polygonal region. **7.** (602) In a cone, the base is the flat, circular portion of the cone.

base angle **1.** (182) In an isosceles triangle, either angle formed by the base and one of the legs is called a base angle. **2.** (321) In the trapezoid at the right, $\angle A$ and $\angle D$, and $\angle B$ and $\angle C$ are pairs of base angles.

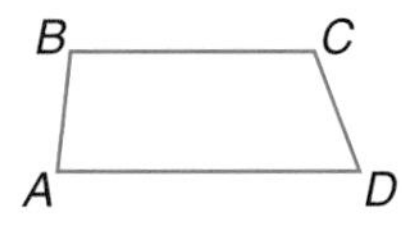

between (28) In general, *B* is between *A* and *C* if and only if *A*, *B*, and *C* are collinear and $AB + BC = AC$.

C

Cavalieri's Principle (617) If two solids have the same height and the same cross-sectional area at every level, then they have the same volume.

center of a circle (446) A point from which all given points in a plane are equidistant.

center of a regular polygon (543) The common center of the inscribed and circumscribed circles of the regular polygon.

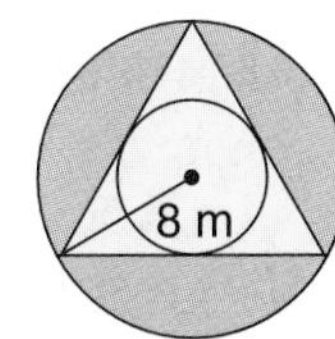

center of a sphere (621) The point from which a set of all points in space are a given distance.

center of dilation (747) A fixed point used for measurement when altering the size of a geometric figure without changing its shape.

center of rotation (739) A fixed point around which shapes move in a circular motion to a new position.

central angle **1.** (452) For a given circle, an angle that intersects the circle in two points and has its vertex at the center of the circle. **2.** (544) An angle formed by two segments drawn to consecutive vertices of a regular polygon from its center.

chord **1.** (446) For a given circle, a segment whose endpoints are points on the circle. **2.** (621) For a given sphere, a segment whose endpoints are on the sphere.

circle (446) A set of points that consists of all points in a plane that are a given distance from a given point in the plane, called the center.

circumference (447) The limit of the perimeters of the inscribed regular polygons as the number of sides increases.

circumscribed polygon (477) A polygon is circumscribed about a circle if each side of the polygon is tangent to the circle.

collinear points (8) Points that lie on the same line.

column matrix (676) A matrix that has only one column.

common external tangent (476) A common tangent that does not intersect the segment whose endpoints are the centers of the two circles.

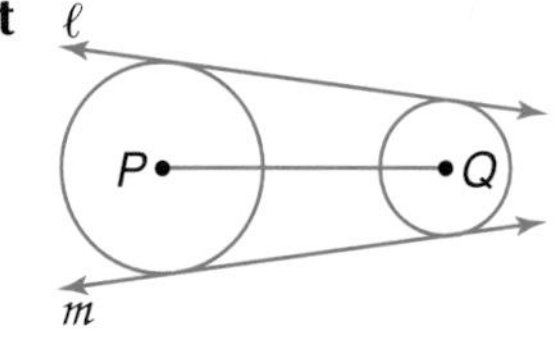

common internal tangent (476) A common tangent that intersects the segment whose endpoints are the centers of the two circles.

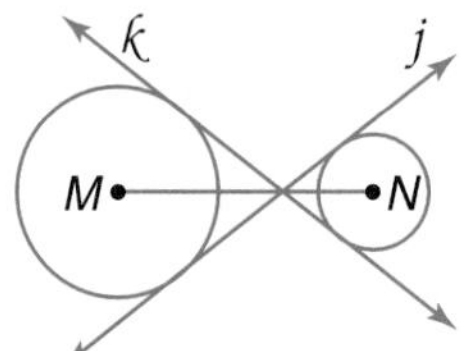

common tangent (476) A line that is tangent to two circles that are in the same plane.

compass (38) An instrument used to draw circles and arcs of circles.

complementary angles (55) Two angles whose degree measures have a sum of 90.

complete network (559) In graph theory, a network that has at least one path between each pair of nodes.

composite of reflections (731) Two successive reflections.

concave polygon (514) A polygon for which there is a line containing a side of the polygon and a point in the interior of the polygon.

concentric circles (454) Circles that lie in the same plane and have the same center but have different radii.

conclusion (76) In a conditional statement, the statement that immediately follows the word *then*.

conditional statement (76) A statement of the form "If A, then B." that can be written in if-then form. The part following *if* is called the hypothesis. The part following *then* is called the conclusion.

cone (577, 602) A solid with a circular base, a vertex V not contained in the same plane as the base, and a lateral surface area composed of all points in the segments connecting the vertex to the edge of the base.

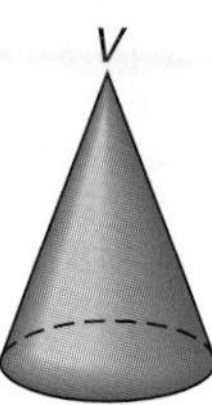

congruence transformation (196, 716) When a geometric figure and its transformation image are congruent, the mapping is called a congruence transformation or isometry.

congruent angles (47) Angles that have the same measure.

congruent arcs (454) Two arcs that have the same measure.

congruent circles (454) Circles that have the same radius.

congruent segments (31) Segments that have the same length.

congruent solids (629) Two solids are congruent if all of the following conditions are met.

1. The corresponding angles are congruent.
2. Corresponding edges are congruent.
3. Areas of corresponding faces are congruent.
4. The volumes are congruent.

congruent triangles (196) Triangles that have their corresponding parts congruent.

conjecture (70) An educated guess.

consecutive interior angles (126) In the figure, transversal t intersects lines ℓ and m. There are two pairs of consecutive interior angles: $\angle 8$ and $\angle 1$, and $\angle 7$ and $\angle 2$.

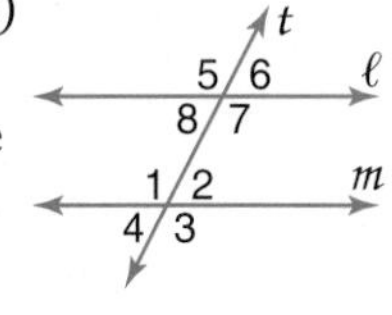

contrapositive (78) Given a conditional statement, the negation of the hypothesis and conclusion of the converse of the given condition.

converse (77) For a given conditional statement, the statement formed by interchanging the hypothesis and conclusion of the original conditional.

convex polygon (514) A polygon for which there is no line that contains both a side of the polygon and a point in the interior of the polygon.

coordinate (7) The numbers in an ordered pair of numbers. The first number is called the x-coordinate and the second number is called the y-coordinate.

coordinate plane (6) A plane for which two perpendicular number lines that intersect at their zero points have been used to match the points of the plane one-to-one with ordered pairs of numbers.

coordinate proof (666) A proof that uses figures in a coordinate plane so that geometric results can be proved by means of algebra.

coplanar points (13) Points that lie in the same plane.

corner view (576) The view from a corner of a figure (also called perspective view).

corollary (192) A statement that can be easily proven using a theorem is called a corollary of that theorem.

corresponding angles (126) In the figure, transversal t intersects lines ℓ and m. There are four pairs of corresponding angles: $\angle 5$ and $\angle 1$, $\angle 8$ and $\angle 4$, $\angle 6$ and $\angle 2$, and $\angle 7$ and $\angle 3$.

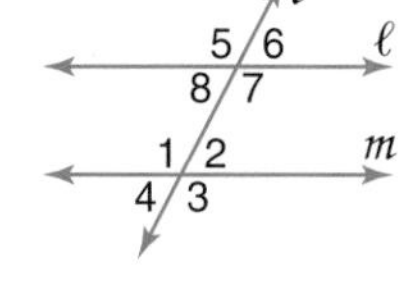

cosine (412) For an acute angle of a right triangle, the ratio of the length of the leg adjacent to the acute angle to the length of the hypotenuse.

counterexample (72) An example used to show that a given general statement is not always true.

cross products (339) In the proportion $\frac{a}{b} = \frac{c}{d}$, where $b \neq 0$ and $d \neq 0$, the cross products are ad and bc. The proportion is true if and only if the cross products are equal.

cross section (574, 577) The intersection of a plane parallel to the base or bases of a solid.

cube (577) A prism in which all the faces are squares.

cylinder (577) A figure whose bases are formed by congruent circles in parallel planes.

D

deductive reasoning (86) A system of reasoning used to reach conclusions that must be true whenever the assumptions on which the reasoning is based are true.

degree (45) A unit of measure used in measuring angles and arcs of circles. An arc of a circle with a measure of $1°$ is $\frac{1}{360}$ of the entire circle.

degree of a node (560) In a network, the number of edges meeting at a given node.

diagonal (293) In a polygon, a segment joining nonconsecutive vertices of the polygon.

diameter **1.** (446) In a circle, a chord that contains the center of the circle. **2.** (621) In a sphere, a segment that contains the center of the sphere, and whose endpoints are on the sphere.

dilation (348, 746) A transformation determined by a center point C and a scale factor $k > 0$. For any point P in the plane, the image P' of P is the point on $\overrightarrow{CP}$ such that $CP' = k \cdot CP$.

direction of a vector (673) The measure of the angle that the vector forms with the positive x-axis or any other horizontal line.

distance between a point and a line (154) For a point not on a given line, the length of the segment perpendicular to the line from the point. If the point is on the line, then the distance between the point and the line is zero.

distance between two parallel lines (156) The distance between one of the lines and any point on the other line.

E

edge **1.** (577) In a polyhedron, a line segment in which a pair of faces intersect. **2.** (559) In graph theory, a path connecting two nodes.

equiangular triangle (181) A triangle with all angles congruent.

equidistant (156) The distance between two lines measured along a perpendicular line to the line is always the same.

equilateral triangle (181) A triangle with all sides congruent.

exterior angle of a polygon (190) An angle that forms a linear pair with one of the angles of the polygon. In the figure, $\angle 2$ is an exterior angle.

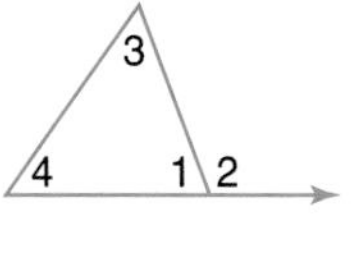

exterior angles (126) In the figure, transversal t intersects lines ℓ and m. The exterior angles are $\angle 3$, $\angle 4$, $\angle 5$, and $\angle 6$.

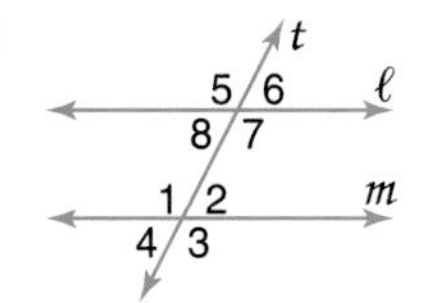

exterior point **1.** (45) For a given angle, a point that is neither on the angle or in the interior of the angle. **2.** (475) For a circle, a point whose distance from the center of the circle is greater than the radius of the circle.

external secant segment (492) The part of a secant segment that is exterior to the circle. *See secant segment.*

extremes (339, 397) In the proportion, $\frac{a}{b} = \frac{c}{d}$, the numbers a and d.

F

face (577) In a polyhedron, the flat polygonal surfaces that intersect to form the edges of the polyhedron.

flow proof (191) A proof which organizes a series of statements in logical order, starting with the given statements. Each statement along with its reasons is written in a box. Arrows are used to show how each statement leads to another.

fractals (378) A figure generated by repeating a special sequence of steps infinitely often. Fractals often exhibit self-similarity.

function (647) A relationship between input and output. In a function, one or more operations are performed on the input to get the output, and there is exactly one output for each input.

G

geometric mean (397) For any positive numbers a and b, the positive number x such that $\frac{a}{x} = \frac{x}{b}$.

geometric probability (551) Involves using the principles of length and area to find the probability of an event.

graph theory (559) The study of properties of figures consisting of points, called nodes, and paths that connect the nodes in various ways.

great circle (622) For a given sphere, the intersection of the sphere and a plane that contains the center of the sphere.

H

height (529, 591, 602) The length of an altitude of a figure.

hemisphere (622) One of the two congruent parts into which a great circle separates a given sphere.

hypotenuse (181) In a right triangle, the side opposite the right angle.

hypothesis (76) In a conditional statement, the statement that immediately follows the word *if.*

I

if and only if (139) When both a conditional and its converse are true.

if-then statement (76) A compound statement of the form "if A, then B", where A and B are statements.

image (715) The result of a transformation. If A is mapped onto A', then A' is called the image of A. The preimage of A' is A.

included angle (207) In a triangle, the angle formed by two sides is the included angle for these two sides.

included side (207) The side of a triangle that forms a side of two given angles.

incomplete network (559) In graph theory, a network with at least one pair of nodes not connected by an edge.

indirect proof (252) Proof by contradiction. In an indirect proof, one assumes that the statement to be proved is false. One then uses logical reasoning to deduce a statement that contradicts a postulate, theorem, or one of the assumptions. Once a contradiction is obtained, one concludes that the statement assumed false must in fact be true.

indirect reasoning (252) Reasoning that assumes the conclusion is false and then shows that this assumption leads to a contradiction of the hypothesis or some other accepted fact, like a postulate, theorem, or corollary. Then, since the assumption has been proved false, the conclusion must be true.

inductive reasoning (70) Reasoning that uses a number of specific examples to arrive at a plausible generalization or prediction. Conclusions arrived at by inductive reasoning lack the logical certainty of those arrived at by deductive reasoning.

inequality (254) For any real numbers a and b, $a > b$ if there is a positive number c such that $a = b + c$.

inscribed angle (466) An angle having its vertex lie on a given circle and containing two chords of the circle.

inscribed polygon (459) A polygon is inscribed in a circle if each of its vertices lies on the circle.

intercepted arc (466) An angle intercepts an arc if and only if each of the following conditions holds.

1. The endpoints of the arc lie on the angle.
2. All points of the arc, except the endpoints, are in the interior of the circle.
3. Each side of the angle contains an endpoint of the arc.

intercepts method (646) A technique for graphing a linear equation by locating the x- and y-intercepts.

interior **1.** (45) A point is in the interior of an angle if it does not lie on the angle itself and it lies on a segment whose endpoints are on the sides of the angle. **2.** (475) A point is in the interior of a circle if the measure of the segment joining the point to the center of the circle is less that the measure of the radius.

interior angles (126) In the figure, transversal t intersects lines ℓ and m. The interior angles are $\angle 1$, $\angle 2$, $\angle 7$, and $\angle 8$.

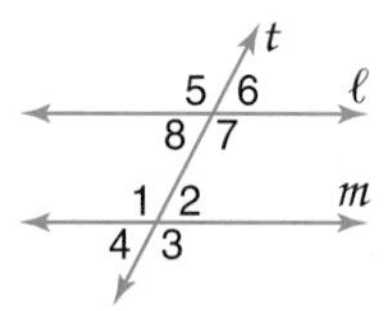

intersection (14) The intersection of two figures is the set of points that are in both figures.

inverse (78) The denial of a statement.

isometry (716) A mapping for which the original figure and its image are congruent.

isosceles trapezoid (321) A trapezoid in which the legs are congruent. Both pairs of base angles are congruent and the diagonals are congruent.

isosceles triangle (181) A triangle with at least two sides congruent. The congruent sides are called legs. The angles opposite the legs are base angles. The angle formed by two legs is the vertex angle. The third side is the base.

iteration (378) A process of repeating the same procedure over and over again.

kite (320) A quadrilateral with exactly two distinct pairs of adjacent congruent sides.

lateral area (592) For prisms, pyramids, cylinders, and cones, the area of the figure not including the bases.

lateral edge **1.** (591) In a prism, the intersection of two adjacent lateral faces. Lateral edges are parallel segments that join corresponding vertices of the bases. **2.** (600) In a pyramid, lateral edges are the edges of the lateral faces that join the vertex to vertices of the base.

lateral faces **1.** (591) In a prism, a face that is not a base of the figure. **2.** (600) In a pyramid, faces that intersect at the vertex.

Law of Cosines (431) Let $\triangle ABC$ be any triangle with a, b, and c representing the measures of sides opposite the angles with measures A, B, and C, respectively. Then the following equations hold true.

$$a^2 = b^2 + c^2 - 2bc \cos A$$
$$b^2 = a^2 + c^2 - 2ac \cos B$$
$$c^2 = a^2 + b^2 - 2ab \cos C$$

Law of Detachment (86) If $p \rightarrow q$ is a true conditional and p is true, then q is true.

Law of Sines (426) Let $\triangle ABC$ be any triangle with a, b, and c representing the measures of sides opposite the angles with measures A, B, and C, respectively. Then, $\frac{\sin A}{a} = \frac{\sin B}{b} = \frac{\sin C}{c}$.

Law of Syllogism (87) If $p \rightarrow q$ and $q \rightarrow r$ are true conditionals, then $p \rightarrow r$ is also true.

leg **1.** (181) In a right triangle, the sides opposite the acute angles. **2.** (182) In an isosceles triangle, the congruent sides. **3.** (321) In a trapezoid, the nonparallel sides.

line (12) A basic undefined term of geometry. Lines extend indefinitely and have no thickness or width. In a figure, a line is shown with arrows at each end. Lines are usually named by lower case script letters or by writing capital letters for two points on the line, with a double arrow over the pair of letters.

linear equation (646) An equation that can be written in the form $Ax + By = C$, where A, B, and C are real numbers, with A and B not both 0.

linear pair (53) A pair of adjacent angles whose noncommon sides are opposite rays.

line of reflection (722) Line ℓ is a line of reflection for a figure if, for every point A of the figure not on ℓ, there is a point A' of the figure such that ℓ is the perpendicular bisector of $\overline{AA'}$. The points A and A' are reflection images of each other.

line of symmetry (724) A line that can be drawn through a plane figure so that the figure on one side is the reflection image of the figure on the opposite side.

locus (696) In geometry, a figure is a locus if it is the set of all points and only those points that satisfy a given condition.

M

magnitude of a vector (673) The length of a vector.

major arc (453) If $\angle APB$ is a central angle of circle P, and C is any point on the circle and in the exterior of the angle, then points A and B and all points of the circle exterior to $\angle APB$ form a major arc called $\overset{\frown}{ACB}$ Three letters are needed to name a major arc.

mapping (715) A one-to-one correspondence between points of two figures.

means (339, 397) In the proportion, $\frac{a}{b} = \frac{c}{d}$, the numbers b and c.

measure **1.** (28) The length of $\overline{AB}$, written AB, is the distance between A and B. **2.** (45) A protractor can be used to find the measure of an angle.

median **1.** (238) In a triangle, a segment that joins a vertex of the triangle and the midpoint of the opposite side. **2.** (322) In a trapezoid, the segment joining the midpoints of the legs.

midpoint (36) A point M is the midpoint of segment PQ if M is between P and Q, and $PM = MQ$.

minor arc (453) If $\angle APB$ is a central angle of circle P, then points A and B and all points on the circle interior to the angle form a minor arc called $\overset{\frown}{AB}$.

negation (78) The denial of a statement.

net (585) A two-dimensional figure that, when folded, forms the surfaces of a three-dimensional object.

network (559) A figure consisting of points, called nodes, and edges that join various nodes to one another.

***n*-gon** (515) A polygon with n sides.

node (559) In graph theory, the points of a network.

noncollinear points (8) Points that do not lie on the same line.

non-Euclidean geometry (164) The study of geometrical systems which are not in accordance with the Parallel Postulate of Euclidean geometry.

oblique cone (602) A cone that is not a right cone.

oblique cylinder (593) A cylinder that is not a right cylinder.

oblique prism (591) A prism that is not a right prism.

obtuse angle (46) An angle with degree measure greater than 90 and less than 180.

obtuse triangle (180) A triangle with an obtuse angle.

opposite rays (44) Two rays $\overrightarrow{BA}$ and $\overrightarrow{BC}$ such that B is between A and C.

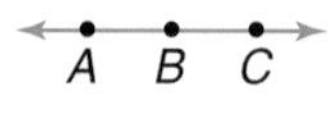

ordered pair (6) A pair of numbers given in a specific order. Ordered pairs are used to locate points in a plane.

ordered triple (680) Three numbers given in a specific order. Ordered triples are used to locate points in space.

origin (6) In a coordinate plane or a three-dimensional coordinate system, the point where the coordinate axes intersect (usually designated by O).

P

paragraph proof (39) A proof written in the form of a paragraph (as opposed to a two-column proof).

parallel lines (124) Lines in the same plane that do not intersect.

parallelogram (290, 291) A quadrilateral in which both pairs of opposite sides are parallel. Any side of a parallelogram may be called a base. For each base, a segment called an altitude is a segment perpendicular to the base and having its endpoints on the lines containing the base and the opposite side.

parallelogram law (675) The parallelogram law is the basis for a method for adding two vectors. The two vectors with the same initial point form part of a parallelogram. The resultant or sum of the two vectors is the diagonal of the parallelogram.

perpendicular bisector (238) A bisector of a segment that is perpendicular to the segment.

perpendicular bisector of a triangle (238) A line or line segment that passes through the midpoint of a side of a triangle and is perpendicular to that side.

perpendicular lines (53) Two lines that intersect to form a right angle.

perspective view (576) The view from a corner of a figure (also called corner view).

pi (447) The ratio of the circumference of a circle to its diameter.

plane (12) A basic undefined term of geometry. Planes can be thought of as flat surfaces that extend indefinitely in all directions and have no thickness. In a figure, a plane is often represented by a parallelogram. Planes are usually named by a capital script letter or by three noncollinear points on the plane.

plane Euclidean geometry (164) Geometry based on Euclid's axioms dealing with a system of points, lines, and planes.

Platonic solid (577) Any one of the five regular polyhedrons: tetrahedron, hexahedron, octahedron, dodecahedron, or icosohedron.

point (12) A basic undefined term of geometry. Points have no size. In a figure, a point is represented by a dot. Points are named by capital letters.

point of reflection (722) Point S is the reflection of point R with respect to point Q, the point of reflection, if Q is the midpoint of the segment drawn from R to S.

point of symmetry (725) The point of reflection for all points of a figure.

point of tangency (475) For a line that intersects a circle in only one point, the point in which they intersect.

point-slope form (654) An equation of the form $y - y_1 = m(x - x_1)$ for the line passing through a point whose coordinates are (x_1, y_1) and having a slope of m.

polygon (180, 514) A figure in a plane that meets the following conditions.

1. It is a closed figure formed by three or more coplanar segments called sides.
2. Sides that have a common endpoint are noncollinear.
3. Each side intersects exactly two other sides, but only at their endpoints.

polyhedron (577) A closed three-dimensional figure made up of flat polygonal regions. The flat regions formed by the polygons and their interiors are called faces. Pairs of faces intersect in line segments called edges. Points where three or more edges intersect are called vertices.

postulate (28) A statement that describes a fundamental relationship between the basic terms of geometry. Postulates are accepted as true without proof.

preimage (715) For a transformation, if A is mapped onto A', then A is the preimage of A'.

prism (577) A solid with the following characteristics.

1. Two faces, called bases, are formed by congruent polygons that lie in parallel planes.
2. The faces that are not bases, called lateral faces, are formed by parallelograms.
3. The intersections of two adjacent lateral faces are called lateral edges and are parallel segments.

probability (294) The ratio that tells how likely it is that an event will occur.

$$P(\text{event}) = \frac{\text{number of successful outcomes}}{\text{total number of possible outcomes}}$$

proof (39, 94) A logical argument showing that the truth of a hypothesis guarantees the truth of the conclusion.

proportion (339) An equation of the form $\frac{a}{b} = \frac{c}{d}$ that states that two ratios are equivalent.

protractor (45) A tool used to find the degree measure of angles.

pyramid (577) A solid with the following characteristics.

1. All the faces, except one face, intersect at a point called the vertex.
2. The face that does not contain the vertex is called the base and is a polygonal region.
3. The faces meeting at the vertex are called lateral faces and are triangular regions.

Pythagorean Theorem (30, 399) In a right triangle, the sum of the squares of the measures of the legs equals the sum of the square of the measure of the hypotenuse.

Pythagorean triple (400) A set of numbers, a, b, and c, that satisfy the equation $a^2 + b^2 = c^2$.

Q

quadrant (6) One of the four regions into which the two perpendicular axes of a coordinate plane divide the plane.

quadrilateral (291) A four-sided polygon.

R

radius **1.** (446) A radius of a circle is a segment whose endpoints are the center of the circle and a point on the circle. **2.** (543) A segment is a radius of a regular polygon if it is a radius of a circle circumscribed about the polygon. **3.** (621) A radius of a sphere is a segment whose endpoints are the center and a point on the sphere.

ratio (338) A comparison of two numbers using division.

ray (44) $\overrightarrow{PQ}$ is a ray if it is the set of points consisting of $\overline{PQ}$ and all points S for which Q is between P and S.

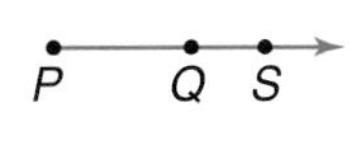

rectangle (305, 306) A quadrilateral with four right angles.

reflection (722) A transformation that flips a figure over a line called the line of reflection. *Also see line of reflection and point of reflection.*

regular polygon (515) A convex polygon with all sides congruent and all angles congruent.

regular polyhedron (577) A polyhedron in which all faces are congruent regular polygons.

regular prism (577) A right prism whose bases are regular polygons.

regular pyramid (600) A pyramid whose base is a regular polygon and in which the segment from the vertex to the center of the base is perpendicular to the base. This segment is called the altitude of the pyramid.

regular tessellation (523) A tessellation consisting entirely of regular polygons.

remote interior angles (190) The angles of a triangle that are not adjacent to a given exterior angle.

resultant (675) The sum of two or more vectors.

rhombus (313) A quadrilateral with all four sides congruent.

right angle (46) An angle whose degree measure is 90.

right circular cone (602) A cone that has a circular base and whose axis (the segment from the vertex to the center of the base) is perpendicular to the base. The axis is also the altitude of the cone.

right cylinder (593) A cylinder whose axis is also an altitude.

right prism (591) A prism in which the lateral edges are also altitudes.

right triangle (180) A triangle with a right angle. The side opposite the right angle is called the hypotenuse. The other two sides are called legs.

rotation (739) A transformation that is the composite of two reflections with respect to two intersecting lines. The intersection of the two lines is called the center of rotation.

Ruler Postulate (28) The points on any line can be paired with real numbers so that, given any two points *P* and *Q* on the line, *P* corresponds to zero, and *Q* corresponds to a positive number.

S

scalar multiplication (674) Multiplication of a vector by a real number.

scale factor **1.** (346) The ratio of the lengths of two corresponding sides of two similar polygons. **2.** (629) The ratio of the lengths of two corresponding sides of two similar solids. **3.** (746) For a dilation transformation with center *C*, the number *k* such that $ED = k(AB)$, where *E* is the image of *A* and *D* is the image of *B*.

scalene triangle (181) A triangle with no two sides congruent.

scatter plot (660) Two sets of data plotted as ordered pairs in a coordinate plane.

secant (483) For a circle, a line that intersects the circle in exactly two points.

secant segment (492) A segment from a point exterior to a circle to a point on the circle and containing a chord of the circle. The part of a secant segment that is exterior to the circle is called an external secant segment.

sector (553) A region bounded by a central angle and the intercepted arc.

segment (12) A part of a line that consists of two points, called endpoints, and all the points between them.

segment bisector (38, 48) A segment, line, or plane that intersects a segment at its midpoint.

self-similar (378) If any parts of a fractal image are replicas of the entire image, the image is self-similar.

semicircle (453) Either of the two parts into which a circle is separated by a line containing a diameter of the circle.

semi-regular (524) Uniform tessellations containing two or more regular polygons.

side **1.** (44) For an angle, the two rays that form the sides of the angle. **2.** (180) For a polygon, a segment joining two consecutive vertices of the polygon.

Sierpinski triangle (378) A fractal triangle created by connecting midpoints of an equilateral triangle. The resulting fractal is named after the Polish mathematician, Waclaw Sierpinski.

similar circles (454) Circles that have different radii.

similar figures (346) Figures that have the same shape but that may differ in size.

similar polygons (346) Two polygons are similar if there is a correspondence between their vertices such that corresponding angles are congruent and the measures of corresponding sides are proportional.

similar solids (629) Solids that have exactly the same shape but not necessarily the same size.

similarity transformation (717) When a figure and its transformation image are similar.

sine (412) For an acute angle of a right triangle, the ratio of the length of the leg opposite the acute angle to the length of the hypotenuse.

skew lines (125) Lines that do not intersect and are not in the same plane.

slant height **1.** (600) For a regular pyramid, the height of lateral face. **2.** (602) For a right circular cone, the length of any segment joining the vertex to the edge of the circular base.

slice of a solid (574, 577) The figure formed by making a straight cut across the solid.

slope (138) For a (nonvertical) line containing two points (x_1, y_1) and (x_2, y_2), the number *m* given by $m = \frac{y_2 - y_1}{x_2 - y_1}$ where $x_2 \neq x_1$.

slope-intercept form (647) A linear equation of the form $y = mx + b$. The graph of such an equation has slope m and y-intercept b.

solid (574) A three-dimensional figure consisting of all of its surface points and all of its interior points.

solving the triangle (427) Finding the measures of all the angles and sides of a triangle.

space (12) A boundless three-dimensional set of all points.

sphere (577, 621) In space, the set of all points that are a given distance from a given point, called the center.

spherical geometry (164) The branch of geometry which deals with a system of points, great circles (lines), and spheres (planes) (also known as Riemannian geometry).

spreadsheets (21) Computer programs designed especially for creating charts involving many calculations.

square (315) A quadrilateral with four right angles and four congruent sides.

standard form (646) For linear equations, an equation of the form $Ax + By = C$, where A and B are not both zero.

straight angle (44) A figure formed by two opposite rays.

straightedge (31) An instrument used to draw lines.

strictly self-similar (379) A figure is strictly self-similar if any of its parts, no matter where they are located or what size is selected, contain the same figure as the whole.

supplementary angles (55) Two angles whose degree measures have a sum of 180.

surface area (584) The sum of the areas of all faces and side surfaces of a three-dimensional figure.

system of equations (704) A group of two or more equations with the same variables.

tangent **1.** (412) For an acute angle of a right triangle, the ratio of the length of the leg opposite the acute angle to the length of the leg adjacent to the acute angle. **2.** (474) A tangent to a circle is a line in the plane of the circle that intersects the circle in exactly one point. The point of intersection is called the point of tangency. **3.** (621) A tangent to a sphere is a line or plane that intersects the sphere in exactly one point.

tangent segment (477) A segment AB such that one endpoint is on a circle, the other is outside the circle, and the line AB is tangent to the circle.

tessellation (522, 523) Tile-like patterns formed by repeating shapes to fill a plane without gaps or overlaps.

theorem (39, 101) A statement, usually of a general nature, that can be proved by appeal to postulates, definitions, algebraic properties, and rules of logic.

traceable network (559) A network that can be traced in one continuous path without retracing any edge. A network is traceable if its nodes all have even degrees or if exactly two nodes have odd degrees.

transformation (715) In a plane, a mapping for which each point has exactly one image point and each image point has exactly one preimage point.

translation (731) A composite of two reflections over two parallel lines. A translation slides figures the same distance in the same direction.

transversal (126) A line that intersects two or more lines in a plane at different points.

trapezoid (321) A quadrilateral that has exactly one pair of parallel sides. The parallel sides of a trapezoid are called bases. The nonparallel sides are called legs. The pairs of angles with their vertices at the endpoints of the same base are called base angles. The line segment joining the midpoints of the legs of a trapezoid is called the median. An altitude is a segment perpendicular to the lines containing the bases and having its endpoints on these lines.

triangle (180) A figure formed by the segments determined by three noncollinear points. The three segments are called sides. The endpoints are called the vertices of the triangle. A triangle separates a plane into three parts, the triangle, its interior, and its exterior.

trigonometric ratio (412) A ratio of the measures of two sides of a right triangle is called a trigonometric ratio.

trigonometry (412) The study of the properties of triangles and trigonometric functions and their applications.

two-column proof (94) A formal proof in which statements are listed in one column and the reasons for each statement are listed in a second column.

U

undefined term (13) A word, usually readily understood, that is not formally explained by means of more basic words and concepts. The basic undefined terms of geometry are point, line, and plane.

uniform (524) Tessellations containing the same combination of shapes and angles at each vertex.

vector (673) A directed segment. Vectors possess both magnitude (length) and direction.

vertex **1.** (44) For an angle, the common endpoint of the two rays that form the angle. **2.** (180) In a polygon, the endpoints of the sides is called a vertex. **3.** (577) In a polyhedron, where three or more edges intersect. **4.** (600) In a pyramid, the vertex that is not contained in the base of the pyramid. **5.** (602) In the figure, the vertex of the cone is point *V*.

vertex angle (182) In an isosceles triangle, the angle formed by the congruent sides (legs).

vertical angles (53) Two nonadjacent angles formed by two intersecting lines.

volume (607) The measure of the amount of space enclosed by a three dimensional figure.

***x*-axis** (6) The horizontal number line in a coordinate plane.

***x*-coordinate** (6) The first number in an ordered pair of numbers.

***y*-axis** (6) The vertical number line in a coordinate plane.

***y*-coordinate** (6) The second number in an ordered pair of numbers.

SPANISH GLOSSARY

absolute value/valor absoluto (29) El número de unidades que un número real dista de cero en una recta numérica.

acute angle/acutángulo (46) Ángulo que mide menos de 90 grados.

acute triangle/triángulo acutángulo (180) Triángulo en el que todos los ángulos son agudos.

adjacent angles/ángulos adyacentes (53) Dos ángulos en el mismo plano que tienen el vértice y un lado en común, pero sin puntos interiores en común.

adjacent arcs/arcos adyacentes (453) Dos arcos de un círculo que tienen solo un punto en común.

alternate exterior angles/ángulos alternos externos (126) En la figura, la transversal t interseca las rectas ℓ y m. $\angle 5$ y $\angle 3$, y $\angle 6$ y $\angle 4$ se conocen como ángulos alternos externos.

alternate interior angles/ángulos alternos internos (126) En la figura anterior, la transversal t interseca las rectas ℓ y m. $\angle 1$ y $\angle 7$, y $\angle 2$ y $\angle 8$ son ángulos alternos internos.

altitude of a cone/altitud de un cono (602) Segmento perpendicular trazado desde el vértice del cono al plano que contiene la base.

altitude of a cylinder/altitud de un cilindro (593) Segmento perpendicular a los planos que contienen las bases del cilindro y que tiene un extremo en cada uno de los planos.

altitude of a parallelogram/altitud de un paralelogramo (529) Cualquier segmento perpendicular entre las rectas que contienen dos de los lados paralelos del paralelogramo.

altitude of a prism/altitud de un prisma (591) Segmento perpendicular a los planos que contienen las bases del prisma y con un extremo en cada uno de ellos. A la medida de la altitud se le llama la altura del prisma.

altitude of a pyramid/altitud de una pirámide (600) Segmento perpendicular a la base trazado desde el vértice de la pirámide hasta el plano que contiene la base.

altitude of a triangle/altitud de un triángulo (238) Segmento perpendicular trazado del vértice de un triángulo a la recta que contiene el lado opuesto.

angle/ángulo (44) Figura que con-consiste de dos rayos no colineales con un extremo común. Los rayos son los lados del ángulo y el extremo común es su vértice. Un ángulo separa el plano en tres partes: el interior y el exterior del ángulo y el ángulo mismo.

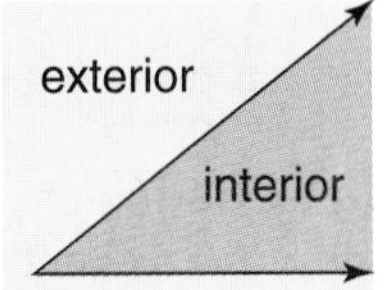

angle bisector/bisectriz de un ángulo (48) El rayo, QS, es la bisectriz del $\angle PQR$ si S está en el interior del ángulo y $\angle PQS \cong \angle RQS$.

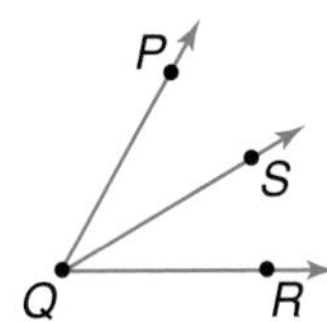

angle bisector of a triangle/bisectriz del ángulo de un triángulo (240) Segmento que biseca el ángulo de un triángulo y que tiene un extremo en un vértice del triángulo y el otro extremo en otro punto del triángulo.

angle of depression/ángulo de depresión (420) Ángulo formado por una recta horizontal y la recta de visión a un objeto, debajo del nivel de la recta horizontal.

angle of elevation/ángulo de elevación (400) Ángulo formado por una recta horizontal y la recta de visión a un objeto, encima del nivel de la recta horizontal.

angle of rotation/ángulo de rotación (740) El ángulo de rotación, $\angle ABC$, está determinado por la preimagen A, el centro de rotación B y la imagen de rotación C.

apothem/apotema (544) Segmento perpendicular a un lado trazado desde el centro de un polígono regular hasta uno de sus lados.

arc/arco (453) Parte continua de un círculo.

arc length/longitud de arco (454) Distancia lineal que representa el arco. La longitud de un arco es una parte de la circunferencia proporcional a la medida del ángulo central correspondiente cuando se compara con el círculo completo.

arc measure/medida de arco (453) La medida en grados de un arco menor es la medida de su ángulo central. La de un arco mayor es igual a 360°, menos la medida de su ángulo central. La de un semicírculo es 180°.

arc of a chord/arco de cuerda (459) Un arco menor que tiene los mismos extremos que una cuerda dada.

area/área (19, 529) El número de unidades cuadradas contenidas en el interior de una figura.

auxiliary line/recta auxiliar (189) Recta o segmento de recta que se agrega a una figura para ayudar a demostrar un resultado.

axis/eje **1.** (6) En un plano de coordenadas, el eje x es la recta numérica horizontal y el eje y es es la recta numérica vertical. **2.** (593) El eje de un cilindro es el segmento cuyos extremos son los centros de las bases. **3.** (602) El eje de un cono es el segmento cuyos extremos son el vértice y el centro de la base.

B

base/base **1.** (182) En un triángulo isósceles, el lado opuesto al ángulo del vértice. **2.** (321) En un trapecio, las bases son los lados paralelos. **3.** (529) Cualquier lado de un paralelogramo recibe el nombre de base. **4.** (577, 591) En un prisma, las bases son las dos caras formadas por polígonos congruentes que yacen en planos paralelos, mientras que todas las otras caras son paralelogramos. **5.** (593) En un cilindro, las bases son las regiones circulares, congruentes y paralelas que forman los extremos del cilindro. **6.** (600) En una pirámide, la base es la cara que no interseca las otras caras en el vértice. La base es una región poligonal. **7.** (602) En un cono, la base es la porción plana y circular del cono.

base angle/ángulo basal **1.** (182) Cualquiera de los ángulos de un triángulo isósceles formado por la base y uno de los catetos. **2.** (321) En el trapecio de la derecha, $\angle A$ y $\angle D$, y $\angle B$ y $\angle C$ are son pares de ángulos basales.

between/estar entre (28) B está y entre A y C si y solo si A, B y C son colineales y $AB + BC = AC$.

C

Cavalieri's Principle/Principio de Cavalieri (617) Si dos sólidos tienen la misma altura y la misma área transversal en cada nivel, entonces tienen el mismo volumen.

center of a circle/centro de un círculo (446) Punto desde el cual equidistan todos los puntos en un plano.

center of a regular polygon/centro de un polígono regular (543) El centro común de los círculos inscritos y circunscritos del polígono regular.

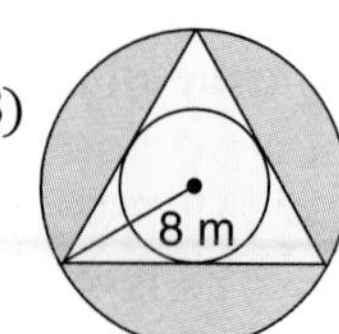

center of a sphere/centro de una esfera (621) Un punto en el espacio del cual equidistan todos los puntos de una esfera.

center of dilation/centro de dilatación (747) Un punto fijo que se usa para la medición cuando se altera el tamaño de una figura geométrica sin cambiar su forma.

center of rotation/centro de rotación (739) Un punto fijo alrededor del cual las figuras se mueven circularmente a una nueva posición.

central angle/ángulo central **1.** (452) Ángulo que interseca un círculo en dos puntos y que tiene su vértice en el centro del círculo. **2.** (544) Ángulo formado por dos segmentos trazados desde el centro de un polígono regular hasta los vértices consecutivos del mismo.

chord/cuerda **1.** (446) Segmento de recta cuyos extremos están en un círculo. **2.** (621) Segmento de recta cuyos extremos están en una esfera.

circle/círculo (446) Conjunto de puntos del plano que están a una distancia dada de un punto dado del plano, llamado centro.

circumference/circunferencia (447) El límite de los perímetros de los polígonos regulares inscritos a medida que aumenta el número de lados.

circumscribed polygon/polígono circunscrito (477) Un polígono está circunscrito a un círculo si cada lado del polígono es tangente al círculo.

collinear points/puntos colineales (8) Puntos que yacen sobre la misma recta.

column matrix/matriz columna (676) Matriz que solo tiene una columna.

common external tangent/ tangente externa común (476) Tangente común que no interseca el segmento cuyos extremos son los centros de los dos círculos.

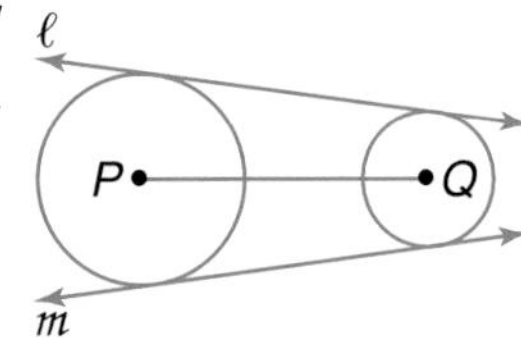

common internal tangent/ tangente interna común (476) Tangente común que interseca el segmento cuyos extremos son los centros de dos círculos.

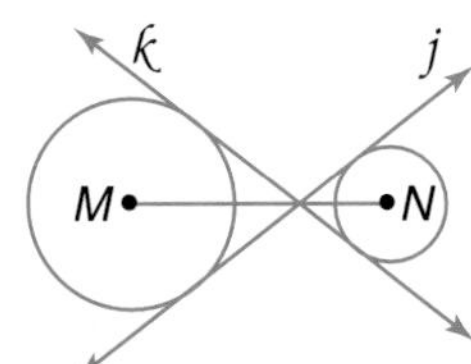

common tangent/tangente común (476) Recta tangente a dos círculos que están en el mismo plano.

compass/compás (38) Instrumento usado para trazar círculos y arcos de círculos.

complementary angles/ángulos complementarios (55) Dos ángulos cuyas medidas suman 90°.

complete network/red completa (559) En teoría de grafos, un circuito que tiene al menos un camino entre cada par de nodos.

composite of reflections/composición de reflexiones (731) Dos reflexiones sucesivas.

concave polygon/polígono cóncavo (514) Polígono para el cual existe una recta que contiene un lado del polígono y un punto interior del polígono.

concentric circles/círculos concéntricos (454) Círculos que yacen sobre el mismo plano y que tienen el mismo centro pero distintos radios.

conclusion/conclusión (76) La afirmación que le sigue a *entonces* en un enunciado condicional.

conditional statement/enunciado condicional (76) Enunciado de la forma "Si A, entonces B", que puede escribirse en la forma *si-entonces.* La parte que sigue a *si* se llama hipótesis y la que sigue a *entonces* se llama conclusión.

cone/cono (577, 602) Sólido de base circular, un vértice *V* fuera del plano de la base y un área de superficie lateral compuesta de todos los puntos de los segmentos que conectan al vértice con las aristas de la base.

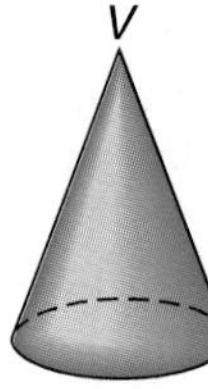

congruence transformation/transformación de congruencia (196, 716) Cuando una figura geométrica y su imagen son congruentes, la transformación se llama transformación de congruencia o isometría.

congruent angles/ángulos congruentes (47) Ángulos que tienen la misma medida.

congruent arcs/arcos congruentes (454) Dos arcos que tienen la misma medida.

congruent circles/círculos congruentes (454) Círculos que tienen el mismo radio.

congruent segments/segmentos congruentes (31) Segmentos que tienen la misma longitud.

congruent solids/sólidos congruentes (629) Dos sólidos son congruentes si se cumplen todas las siguientes condiciones:

1. Los ángulos correspondientes son congruentes.
2. Las aristas correspondientes son congruentes.
3. Las caras correspondientes son congruentes.
4. Los volúmenes son congruentes.

congruent triangles/triángulos congruentes (196) Triángulos cuyas partes correspondientes son congruentes.

conjecture/conjetura (70) Suposición informada.

consecutive interior angles/ángulos consecutivos internos (126) En la figura, la transversal t interseca las rectas ℓ y m. Hay dos pares de ángulos consecutivos internos: $\angle 8$ y $\angle 1$, y $\angle 7$ y $\angle 2$.

contrapositive/antítesis (78) Negación de la hipótesis y de la conclusión del recíproco de un enunciado condicional dado.

converse/recíproco (77) En un enunciado condicional dado, el enunciado formado al intercambiar la hipótesis y la conclusión del enunciado original.

convex polygon/polígono complejo (514) Polígono para el que no existe recta alguna que contenga un lado del polígono y un punto interior del polígono.

coordinate/coordenada (7) Los números en un par ordenado de números. El primer número se llama la coordenada *x* y el segundo, la coordenada *y*.

coordinate plane/plano de coordenadas (6) Plano en el que se han usado dos rectas numéricas perpendiculares que se intersecan en sus puntos cero para aparear los puntos del plano con pares ordenados de números.

coordinate proof/prueba de coordenada (666) Demostración que usa figuras en un plano coordenado para demostrar resultados algebraicos.

coplanar points/puntos coplanares (13) Puntos que están en el mismo plano.

corner view/vista de esquina (576) También llamada vista de perspectiva, es la vista desde una esquina de una figura.

corollary/corolario (192) La afirmación que puede demostrarse fácilmente mediante un teorema es el corolario de dicho teorema.

corresponding angles/ángulos correspondientes (126) En la figura, la transversal t interseca las rectas ℓ y m. Hay cuatro pares de ángulos correspondientes: $\angle 5$ y $\angle 1$; $\angle 8$ y $\angle 4$; $\angle 6$ y $\angle 2$; y $\angle 7$ y $\angle 3$.

cosine/coseno (412) La razón de la longitud del cateto adyacente al ángulo agudo, a la longitud de la hipotenusa, en un ángulo agudo de un triángulo rectángulo.

counterexample/contraejemplo (72) Un ejemplo que se usa para demostrar que un enunciado dado no es siempre cierto.

cross products/productos cruzados (339) En la proporción $\frac{a}{b} = \frac{c}{d}$, donde $b \neq 0$ y $d \neq 0$, los productos cruzados son ad bc. La proporción es verdadera si y solo si los productos cruzados son iguales.

cross section/corte transversal (574, 577) La intersección de un sólido con un plano paralelo a la base o bases del sólido.

cube/cubo (577) Prisma en el que todas las caras son cuadradas.

cylinder/cilidro (577) Figura cuyas bases están formadas por círculos congruentes que yacen en planos paralelos.

D

deductive reasoning/razonamiento deductivo (86) Sistema de razonamiento usado para llegar a conclusiones que deben ser verdaderas cada vez que las suposiciones en las que está basado el razonamiento son verdaderas.

degree/grado (45) Unidad de medida para medir ángulos y arcos de círculo. El arco de un círculo con una medida de 1° es $\frac{1}{360}$ del círculo completo.

degree of a node/grado de un nodo (560) Número de aristas que convergen en un nodo de una red.

diagonal/diagonal (293) Segmento que une vértices no consecutivos de un polígono.

diameter/diámetro **1.** (446) Cuerda que contiene el centro de un círculo. **2.** (621) Segmento que contiene el centro de una esfera y cuyos extremos están en la esfera.

dilation/dilatación (348, 746) Transformación determinada por un punto central C y un factor de escala $k > 0$. Para cada punto P del plano, la imagen P' de P es el punto en $\overrightarrow{CP}$ que satisface $CP' = k \cdot CP$.

direction of a vector/dirección de un vector (673) Medida del ángulo entre un vector y el eje x positivo o cualquier otra recta horizontal.

distance between a point and a line/distancia entre un punto y una recta (154) Para un punto fuera de una recta, la longitud del segmento perpendicular trazado desde el punto a la recta. Si el punto está sobre la recta, entonces la distancia entre el punto y la recta es cero.

distance between two parallel lines/distancia entre dos paralelas (156) Distancia entre una de las rectas y cualquier punto en la otra recta.

edge/arista **1.** (577) Segmento de recta en el que se intersecan dos caras de un poliedro. **2.** (559) En teoría de grafos, un camino que conecta dos nodos.

equiangular triangle/triángulo equiangular (181) Aquél en que todos los ángulos son congruentes.

equidistant/equidistante (156) La distancia entre dos rectas medida a lo largo de una perpendicular a la recta es siempre la misma.

equilateral triangle/triángulo equilátero (181) Aquél en que todos los lados son congruentes.

exterior angle of a polygon/ángulo exterior de un polígono (190) Ángulo que forma un par lineal con uno de los ángulos del polígono. En la figura, $\angle 2$ es un ángulo exterior.

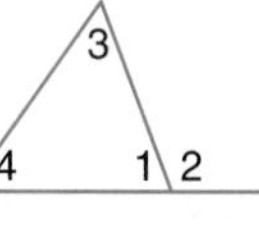

exterior angles/ángulos exteriores (126) En la figura, la transversal t interseca las rectas ℓ y m. Los ángulos exteriores son $\angle 3$, $\angle 4$, $\angle 5$ y $\angle 6$.

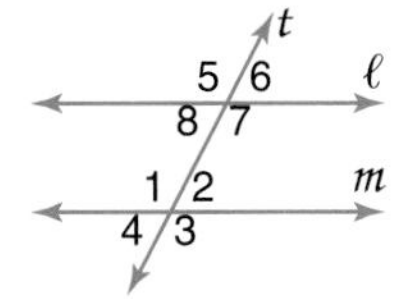

exterior point/punto exterior **1.** (45) Para un ángulo dado, un punto que no está ni en el ángulo ni en el interior del ángulo. **2.** (475) En un círculo, un punto cuya distancia al centro del círculo es mayor que el radio del mismo.

external secant segment/segmento externo de la secante (492) La parte de un segmento de secante exterior al círculo. *Ver segmento de secante.*

extremes/extremos (339, 397) Los números a y d, en la proporción, $\frac{a}{b} = \frac{c}{d}$.

face/cara (577) Superficies poligonales planas de un poliedro que al intersecarse forman las aristas del poliedro.

flow proof/demostración de flujo (191) La que organiza una serie de enunciados en orden lógico, comenzando con los enunciados dados. Cada enunciado junto con sus razones se escribe en un rectángulo. Se usan flechas para demostrar cómo cada enunciado conduce a otro.

fractals/fractales (378) Figura generada por la repetición infinita de una sucesión especial de pasos. A menudo exhiben autosemejanza.

function/función (647) Relación entre las entradas y las salidas. Una o más operaciones son ejecutadas en las entradas para obtener las salidas y hay una única salida para cada entrada.

geometric mean/media geométrica (397) Para números positivos a y b cualesquiera, el número positivo x tal que $\frac{a}{x} = \frac{x}{b}$.

geometric probability/probabilidad geométrica (551) El uso de los principios de longitud y área para calcular la probabilidad de un evento.

graph theory/teoría de grafos (559) Estudio de las propiedades de figuras que consisten de puntos llamados nodos y de caminos que conectan los nodos de varias formas.

great circle/círculo máximo (622) Intersección de una esfera con un plano que contiene el centro de la esfera.

height/altura (529, 591, 602) La longitud de la altitud de una figura.

hemisphere/semiesfera (622) Una de las dos partes congruentes en las que un círculo máximo divide una esfera.

hypotenuse/hipotenusa (181) Lado opuesto al ángulo recto en un triángulo rectángulo.

hypothesis/hipótesis (76) La afirmación que sigue inmediatamente a la palabra *si* en un enunciado condicional.

if and only if/si y solo si (139) Cuando un enunciado condicional y su recíproco son verdaderos.

if-then statement/enunciado si-entonces (76) Enunciado compuesto que tiene la forma "si A, entonces B", donde A y B son enunciados.

image/imagen (715) El resultado de una transformación. Si A es mapeado en A', A' se llama la imagen de A. La preimagen de A' es A.

included angle/ángulo incluido (207) El ángulo formado por dos lados de un triángulo recibe el nombre de ángulo incluido de esos lados.

included side/lado incluido (207) Lado de un triángulo, común a dos de sus ángulos.

incomplete network/red incompleta (559) En teoría de grafos, una red que tiene al menos un par de nodos no conectados por una arista.

indirect proof/demostración indirecta (252) Demostración por contradicción. Uno asume que el enunciado por demostrar es falso, en una demostración indirecta. Usando razonamiento lógico, deduce un enunciado que contradice un postulado, un teorema o una de las suposiciones. Al obtener una contradicción, concluye que el enunciado que se supuso falso debe ser cierto.

indirect reasoning/razonamiento indirecto (252) Razonamiento que supone que la conclusión es falsa y demuestra que esa suposición contradice la hipótesis o algún hecho aceptado, como un postulado, teorema o corolario. Luego, como se ha demostrado que la suposición es falsa, la conclusión debe ser verdadera.

inductive reasoning/razonamiento inductivo (70) Razonamiento que usa ejemplos específicos para llegar a una generalización o predicción plausible. Las conclusiones obtenidas mediante este tipo de razonamiento carecen de la certeza lógica de aquellas a las que se ha llegado mediante el razonamiento deductivo.

inequality/desigualdad (254) Para números reales a y b cualesquiera, $a > b$ si existe un número positivo c tal que $a = b + c$.

inscribed angle/ángulo inscrito (466) Un ángulo que tiene su vértice en un círculo dado y cuyos lados son dos cuerdas del mismo.

inscribed polygon/polígono inscrito (459) Un polígono está inscrito en un círculo, si y solo si, todos sus vértices están en el círculo.

intercepted arc/arco interceptado (466) Un ángulo interseca un arco si se cumplen las condiciones siguientes:

1. Los extremos del arco yacen en el ángulo.
2. Todos los puntos del arco, excepto los extremos, están en el interior del ángulo.
3. Cada lado del ángulo contiene un extremo del arco.

intercepts method/método de las intersecciones axiales (646) Técnica para trazar la gráfica de una ecuación lineal usando las intersecciones de esta con los ejes de coordenadas.

interior/interior **1.** (45) Un punto está en el interior de un ángulo si no está en el ángulo mismo, pero está en un segmento cuyos extremos yacen en los lados del ángulo. **2.** (475) Un punto está en el interior de un círculo si la medida del segmento que une el punto con el centro del círculo es menor que la medida del radio del círculo.

interior angles/ángulos interiores (126) En la figura, la transversal t interseca las rectas ℓ y m. Los ángulos interiores son $\angle 1$, $\angle 2$, $\angle 7$ y $\angle 8$.

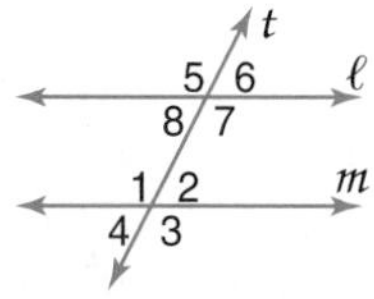

intersection/intersección (14) La intersección de dos figuras es el conjunto de puntos que están en ambas figuras.

inverse/inversa (78) Negación de un enunciado.

isometry/isometría (716) Una relación en que la figura original y su imagen son congruentes.

isosceles trapezoid/trapecio isósceles (321) Trapecio en que los catetos son congruentes. Ambas bases son congruentes y las diagonales también son congruentes.

isosceles triangle/triángulo isósceles (181) El que tiene al menos dos lados congruentes. Los lados congruentes se llaman catetos. Los ángulos opuestos a los catetos se llaman ángulos basales. El ángulo formado por los catetos es el ángulo del vértice y el tercer lado es la base.

iteration/iteración (378) Proceso en que se repite el mismo procedimiento una y otra vez.

kite/cometa (320) Cuadrilátero que tiene exactamente dos pares distintos de lados congruentes adyacentes.

lateral area/área lateral (592) En prismas, pirámides, cilindros y conos, el área de la figura excluyendo las bases.

lateral edge/arista lateral **1.** (591) La intersección de dos caras laterales adyacentes de un prisma. Las aristas laterales son segmentos paralelos que unen los vértices correspondientes de las bases. **2.** (600) Las aristas laterales de una pirámide son las aristas de las caras laterales que unen el vértice a los vértices de la base.

lateral faces/caras laterales **1.** (591) En un prisma, una cara que no es la base del prisma. **2.** (600) En una pirámide, las caras que se intersecan en el vértice de la pirámide.

Law of Cosines/Ley de los cosenos (431) Sea $\triangle ABC$ un triángulo cualquiera y sean a, b y c los lados opuestos a los ángulos con medidas A, B y C, respectivamente. Entonces las siguientes ecuaciones son verdaderas:

$$a^2 = b^2 + c^2 - 2bc \cos A$$
$$b^2 = a^2 + c^2 - 2ac \cos B$$
$$c^2 = a^2 + b^2 - 2ab \cos C$$

Law of Detachment/Ley de indiferencia (86) Si $p \rightarrow q$ es un enunciado condicional verdadero y p es verdadero, entonces q es verdadero.

Law of Sines/Ley de los senos (426) Sea $\triangle ABC$ un triángulo cualquiera y sean a, b y c los lados opuestos a los ángulos con medidas A, B y C, respectivamente. Entonces, $\frac{\sin A}{a} = \frac{\sin B}{b} = \frac{\sin C}{c}$.

Law of Syllogism/Ley del silogismo (87) Si $p \rightarrow q$ y $q \rightarrow r$ son enunciados condicionales verdaderos, entonces $p \rightarrow r$ es también verdadero.

leg/cateto **1.** (181) Los lados opuestos a los ángulos agudos en un triángulo rectángulo. **2.** (182) Los lados congruentes en un triángulo isósceles. **3.** (321) Los lados no paralelos en un trapecio.

line/recta (12) Término primitivo en geometría. Las rectas se extienden indefinidamente en ambos sentidos y no tienen grosor ni ancho. Las rectas se muestran con flechas en cada extremo y se designan mediante letras caligráficas minúsculas o mediante letras mayúsculas que designan dos puntos en la recta, con una flecha doble trazada encima del par de letras.

linear equation/ecuación lineal (646) Ecuación que tiene la forma $Ax + By = C$, donde A, B y C son números reales, con al menos uno de A o B distinto de 0.

linear pair/par lineal (53) Par de ángulos adyacentes cuyos lados no comunes son rayos opuestos.

line of reflection/línea de reflexión (722) La recta ℓ es una línea de reflexión de una figura si para cada punto A de la figura que no está en ℓ, existe un punto A' de la figura de modo que ℓ es la bisectriz perpendicular de $\overline{AA'}$. Los puntos A y A' son imagenes reflexivas uno del otro.

line of symmetry/línea de simetría (724) Línea que puede trazarse a través de una figura plana de modo que la figura en un lado de la recta sea la imagen reflexiva de la figura en el otro lado.

locus/lugar geométrico (696) Una figura es un lugar geométrico si es el conjunto de todos los puntos y, solo esos puntos que satisfacen una condición dada.

magnitude of a vector/magnitud de un vector (673) La longitud del vector.

major arc/arco mayor (453) Si $\angle APB$ es un ángulo central del círculo P y C es un punto cualquiera en el círculo y en el exterior del ángulo, entonces los puntos A y B y todos los puntos del círculo exteriores al $\angle APB$ forman el arco mayor $\overset{\frown}{ACB}$. Se necesitan tres letras para identificar un arco mayor.

mapping/relación (715) Correspondencia uno-a-uno entre los puntos de dos figuras.

means/medios (339, 397) Los números b y c en la proporción, $\frac{a}{b} = \frac{c}{d}$.

measure/medida **1.** (28) La longitud de $\overline{AB}$, designada por AB, es la distancia entre A y B. **2.** (45) Para calcular la medida de un ángulo puede usarse un transportador.

median/mediana **1.** (238) Segmento que une un vértice de un triángulo con el punto medio del lado opuesto. **2.** (322) El segmento que une los puntos medios de los catetos de un trapecio.

midpoint/punto medio (36) El punto M es el punto medio del segmento PQ si M está entre P y Q y $PM = MQ$.

minor arc/arco menor (453) Si $\angle APB$ es un ángulo central del círculo P, entonces los puntos A y B y todos los puntos del círculo, interiores al $\angle APB$, forman un arco que recibe el nombre de arco menor y que se designa por $\overset{\frown}{AB}$.

negation/negación (78) Negación de un enunciado.

net/red (585) Figura bidimensional que una vez plegada forma la superficie de un objeto tridimensional.

network/red (559) Figura que consiste de puntos llamados nodos y aristas que unen los nodos.

***n*-gon/*en*ágono** (515) Polígono de n lados.

node/nodo (559) En teoría de grafos, los puntos de una red.

noncollinear points/puntos no colineales (8) Puntos que no están en la misma recta.

non-Euclidean geometry/geometría no Euclidiana (164) El estudio de los sistemas geométricos que no satisfacen el Postulado de las Paralelas de la geometría Euclidiana.

oblique cone/cono oblicuo (602) Cono que no es un cono recto.

oblique cylinder/cilindro oblicuo (593) Cilindro que no es un cilindro recto.

oblique prism/prisma oblicuo (591) Prisma que no es un prisma recto.

obtuse angle (46) Ángulo que mide más de 90° y menos de 180°.

obtuse triangle/triángulo obtuso (180) Triángulo que tiene un ángulo obtuso.

opposite rays/rayos opuestos (44) Dos rayos $\overrightarrow{BA}$ y $\overrightarrow{BC}$ tales que B está entre A y C.

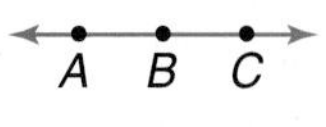

ordered pair/par ordenado (6) Un par de números dados en un orden específico. Se s se usan para ubicar puntos en un plano.

ordered triple/triple ordenado (680) Tres números dados en un orden específico. Se usan para ubicar puntos en el espacio.

origin/origen (6) En un plano de coordenadas o en un sistema de coordenadas tridimensional, el punto en el que se intersecan los ejes (designado generalmente con *O*).

P

paragraph proof/demostración de párrafo (39) Demostración escrita en forma de párrafo (en contraste con la demostración de dos columnas).

parallel lines/rectas paralelas (124) Rectas en el mismo plano que no se intersecan.

parallelogram/paralelogramo (290, 291) Cuadrilátero en el que ambos pares de lados opuestos son paralelos. Cualquier lado de un paralelogramo puede llamarse base. Para cada base, el segmento denominado altitud es un segmento perpendicular a la base. Dicho segmento tiene sus extremos en las rectas que contienen la base y el lado opuesto.

parallelogram law/ley del paralelogramo (675) La ley del paralelogramo se usa para sumar dos vectores. Los dos vectores con el mismo punto inicial forman parte de un paralelogramo. La resultante o suma de los dos vectores es la diagonal del paralelogramo.

perpendicular bisector/bisectriz perpendicular (238) Bisectriz de un segmento que es perpendicular al segmento.

perpendicular bisector of a triangle/bisectriz perpendicular de un triángulo (238) Recta o segmento de recta que pasa por el punto medio del lado de un triángulo y que es perpendicular al lado.

perpendicular lines/rectas perpendiculares (53) Dos rectas que se intersecan para formar un ángulo recto.

perspective view/vista de perspectiva (576) Vista desde una esquina de una figura, también llamada vista de esquina.

pi/pi (447) La razón de la circunferencia de un círculo a su diámetro.

plane/plano (12) Término primitivo en geometría. Los planos pueden visualizarse como superficies planas que se extienden indefinidamente en todas direcciones y que no tienen grosor. En una figura, los planos se representan generalmente mediante paralelogramos y se designan, por lo general, con una letra caligráfica mayúscula o mediante tres puntos no colineales del plano.

plane Euclidean geometry/geometría del plano euclidiano (164) Geometría basada en los axiomas de Euclides, un sistema de puntos, rectas y planos.

Platonic solid/sólido platónico (577) Cualquiera de los cinco poliedros regulares: tetraedro, hexaedro, octaedro, dodecaedro o icosaedro.

point/punto (12) Término primitivo indefinido en geometría. Los puntos no tienen tamaño. En una figura, los puntos se representan con puntos dibujados y son designados mediante letras mayúsculas.

point of reflection/punto de reflexión (722) El punto *S* es la reflexión del punto *R* con respecto al punto *Q*, si *Q* es el punto medio del segmento que une *R* con *S*.

point of symmetry/punto de simetría (725) El punto de reflexión de todos los puntos de una figura.

point of tangency/punto de tangencia (475) Para una recta que interseca un círculo en un único punto, el punto en el que se intersecan.

point-slope form/forma punto-pendiente (654) Ecuación de la forma $y - y_1 = m(x - x_1)$ de la recta que pasa por el punto (x_1, y_1) y que tiene pendiente m.

polygon/polígono (180, 514) Una figura en el plano que cumple con las siguientes condiciones:

1. Es una figura cerrada formada por tres o más segmentos coplanares llamados lados.
2. Los lados que tienen un extremo común no son colineales.
3. Cada lado interseca exactamente dos lados, pero solo en sus extremos.

polyhedron/poliedro (577) Figura cerrada tridimensional que consiste de regiones poligonales planas. Las regiones planas formadas por los polígonos y sus interiores se llaman caras. Dos caras se intersecan en segmentos llamados aristas. Los puntos en que se intersecan tres o más aristas se llaman vértices.

postulate/postulado (28) Enunciado que describe una relación fundamental entre los términos geométricos primitivos. Los postulados se aceptan como verdaderos sin necesidad de demostración.

preimage/preimagen (715) Para una transformación, si A está relacionado con A', entonces A es la preimagen de A'.

prism/prisma (577) Sólido con las siguientes características:

1. Dos caras, llamadas bases, formadas por polígonos congruentes que yacen en planos paralelos.
2. Las caras que no son bases, llamadas caras laterales, están formadas por paralelogramos.
3. Las intersecciones de dos caras adyacentes laterales se llaman aristas laterales y son segmentos paralelos.

probability/probabilidad (294) Razón que indica el grado de certeza de que un evento ocurra.

$$P\,(\text{evento}) = \frac{\text{número de resultados exitosos}}{\text{número total de posibles resultados}}$$

proof/demostración (39, 94) Argumento lógico que muestra que la verdad de una hipótesis garantiza la verdad de la conclusión.

proportion/proporción (339) Ecuación de la forma $\frac{a}{b} = \frac{c}{d}$ que afirma que dos razones son equivalentes.

protractor/transportador (45) Instrumento que se usa para calcular la medida de los ángulos.

pyramid/pirámide (577) Sólido con las siguientes características:

1. Todas las caras, excepto una, se intersecan en un punto llamado vértice.
2. La cara que no contiene el vértice se llama base y es una región poligonal.
3. Las caras que se encuentran en el vértice se llaman caras laterales y son regiones triangulares.

Pythagorean Theorem/teorema de Pitágoras (30, 399) En un triángulo rectángulo, la suma de los cuadrados de las medidas de los catetos es igual al cuadrado de la medida de la hipotenusa.

Pythagorean triple/triplete de Pitágoras (400) Tres números, a, b y c que satisfacen la ecuación $a^2 + b^2 = c^2$.

Q

quadrant/cuadrante (6) Una de las cuatro regiones en que los ejes perpendiculares de un plano de coordenadas dividen dicho plano.

quadrilateral/cuadrilátero (291) Polígono de cuatro lados.

R

radius/radio **1.** (446) El radio de un círculo es un segmento cuyos extremos son el centro del círculo y un punto en el mismo. **2.** (543) El radio de un polígono regular es el radio del círculo circunscrito al polígono. **3.** (621) El radio de una esfera es un segmento cuyos extremos son el centro de la esfera y un punto en la misma.

ratio/razón (338) Comparación de dos números usando división.

ray/rayo (44) $\overrightarrow{PQ}$ es un rayo si es el conjunto de puntos que consisten de P, Q y todos los puntos S para los cuales Q se encuentra entre P y S.

P Q S

rectangle/rectángulo (305, 306) Cuadrilátero que tiene cuatro ángulos rectos.

reflection/reflexión (722) Transformación que relaciona simétricamente una figura a través de una línea llamada línea de reflexión. *Véase también línea de reflexión y punto de reflexión.*

regular polygon/polígono regular (515) Polígono convexo cuyos lados son todos congruentes y cuyos ángulos son también congruentes.

regular polyhedron/poliedro regular (577) Poliedro en el que todas las caras son polígonos regulares congruentes.

regular prism/prisma regular (577) Prisma recto cuyas bases son polígonos regulares.

regular pyramid/pirámide regular (600) Pirámide cuya base es un polígono regular y en la cual el segmento desde el vértice al centro de la base es perpendicular a la base. Este segmento se llama la altitud de la pirámide.

regular tessellation/teselado regular (523) Teselado que consiste únicamente de polígonos regulares.

remote interior angles/ángulos internos no adyacentes (190) Los ángulos de un triángulo que no son adyacentes a un ángulo exterior dado.

resultant/resultante (675) La suma de dos o más vectores.

rhombus/rombo (313) Cuadrilátero cuyos lados son todos congruentes.

right angle/ángulo rectángulo (46) Un ángulo que mide 90°.

right circular cone/cono circular recto (602) Cono cuyo eje (el segmento trazado desde el vértice al centro de la base) es perpendicular a la base. El eje es asimismo la altitud del cono.

right cylinder/cilindro recto (593) Cilindro cuyo eje es también una altitud.

right prism/prisma recto (591) Prisma en el que las aristas laterales son también altitudes.

right triangle/triángulo rectángulo (180) Triángulo que tiene un ángulo recto. El lado opuesto al ángulo recto se llama hipotenusa. Los otros dos lados se llaman catetos.

rotation/rotación (739) Transformación que es la composición de dos reflexiones con respecto a dos rectas que se intersecan. La intersección de las rectas se llama el centro de rotación.

Ruler Postulate/postulado de la regla (28) Establece que los puntos de una recta cualquiera pueden aparearse con números reales de modo que, dados dos puntos cualesquiera P y Q en la recta, P corresponda a cero y Q corresponda a un número positivo.

S

scalar multiplication/multiplicación escalar (674) Multiplicación de un vector por un número real.

scale factor/factor de escala **1.** (346) La razón de las longitudes de dos lados correspondientes de dos polígonos similares. **2.** (629) La razón de las longitudes de dos lados correspondientes de dos sólidos similares. **3.** (746) Para una transformación de dilatación de centro C, el número k tal que $ED = k(AB)$, donde E es la imagen de A y D es la imagen de B.

scalene triangle/triángulo escaleno (181) Aquél que no tiene ningún par de lados congruentes.

scatter plot/diagrama de dispersión (660) Dos conjuntos de datos graficados como pares ordenados en un plano de coordenadas.

secant/secante (483) Recta que interseca un círculo exactamente en dos puntos.

secant segment/segmento de secante (492) Segmento trazado desde un punto exterior a un círculo hasta un punto en el círculo y que contiene una cuerda del círculo. La parte de un segmento de secante que es exterior al círculo se llama segmento externo de la secante.

sector/sector (553) Región acotada por un ángulo central y el arco interceptado.

segment/segmento (12) Parte de una recta que consiste de dos puntos, llamados extremos, y de todos los puntos que están entre ellos.

segment bisector/bisectriz de segmento (38, 48) Segmento, recta o plano que interseca un segmento en su punto medio.

self-similar/autosemejanza (378) Si cualquier parte de una imagen fractal es una réplica de la imagen total, la imagen es autosemejante.

semicircle/semicírculo (453) Cualquiera de las dos partes en que un círculo es dividido por una recta que contiene un diámetro del círculo.

semi-regular/semi-regular (524) Teselados uniformes que contienen dos o más polígonos regulares.

side/lado **1.** (44) Los dos rayos que forman los lados de un ángulo. **2.** (180) Segmento que une dos vértices consecutivos de un polígono.

Sierpinski triangle/triángulo de Sierpinski (378) Triángulo fractal creado al conectar los puntos medios de los lados de un triángulo equilátero. El fractal resultante lleva el nombre del matemático polaco Waclau Sierpinsk.

similar circles/círculos semejantes (454) Círculos que tienen distintos radios.

similar figures/figuras semejantes (346) Figuras que tienen la misma forma, pero que pueden diferir en tamaño.

similar polygons/polígonos semejantes (346) Dos polígonos son semejantes si existe una correspondencia entre sus vértices de modo que los ángulos correspondientes sean congruentes y las medidas de los lados correspondientes sean proporcionales.

similar solids/sólidos semejantes (629) Sólidos que tienen exactamente la misma forma, pero no necesariamente el mismo tamaño.

similarity transformation/transformación de semejanza (717) Cuando una figura y su imagen transformada son semejantes.

sine/seno (412) Para un ángulo agudo de un triángulo rectángulo, la razón de la longitud del cateto opuesto al ángulo a la longitud de la hipotenusa.

skew lines/rectas alabeadas (125) Rectas que no se intersecan y que no están en el mismo plano.

slant height/altura oblicua **1.** (600) La longitud de la altitud de una cara lateral en una pirámide. **2.** (602) La longitud de cualquier segmento que une el vértice de un cono recto circular con la arista de la base circular.

slice of a solid/corte de un sólido (574, 577) Figura que se forma al hacer un corte recto a lo largo de un sólido.

slope/pendiente (138) Para una recta no vertical que contiene los puntos (x_1, y_1) y (x_2, y_2), el número m dado por $m = \frac{y_2 - y_1}{x_2 - y_1}$ donde $x_2 \neq x_1$.

slope-intercept form/forma pendiente-intersección (647) Ecuación lineal de la forma $y = mx + b$. La gráfica de tal ecuación tiene pendiente m e intersección y igual a b.

solid/sólido (574) Figura tridimensional que consiste de todos los puntos de su superficie más todos sus puntos interiores.

solving the triangle/resolviendo un triángulo (427) Hallar todas las medidas de los ángulos y lados de un triángulo.

space/espacio (12) Conjunto tridimensional no acotado de todos los puntos.

sphere/esfera (577, 621) Conjunto de puntos en el espacio que están a una distancia dada de un punto dado llamado centro.

spherical geometry/geometría esférica (164) También conocida con el nombre de geometría riemaniana, es la rama de la geometría que estudia los sistemas de puntos, círculos máximos (rectas) y esferas (planos).

spreadsheets/hojas de cálculo (21) Programas de computación diseñados para crear tablas que involucran cálculos.

square/cuadrado (315) Cuadrilátero que tiene cuatro ángulos rectos y cuyos lados son todos congruentes entre sí.

standard form/forma estándar (646) Para las ecuaciones lineales, una ecuación de la forma $Ax + By = C$, donde A y B no son ambos cero.

straight angle/ángulo extendido (44) Figura formada por dos rayos opuestos.

straightedge/regla (31) Instrumento usado para trazar rectas.

strictly self-similar/estrictamente autosemejanza (379) Una figura es estrictamente autosemejante si cualquiera de sus partes, sin importar su ubicación o tamaño, contiene la figura completa.

supplementary angles/ángulos suplementarios (55) Dos ángulos cuyas medidas suman 180°.

surface area/área de superficie (584) La suma de las áreas de todas las caras y superficies laterales de una figura tridimensional.

system of equations/sistema de ecuaciones (704) Un grupo de dos o más ecuaciones con las mismas variables.

tangent/tangente **1.** (412) Para un ángulo agudo de un triángulo rectángulo, la razón de la longitud del cateto opuesto al ángulo a la longitud del cateto adyacente al ángulo. **2.** (474) Una tangente a un círculo es una recta en el plano del círculo que lo interseca en un único punto. El punto de intersección se llama punto de tangencia. **3.** (621) Una tangente a una esfera es una recta o plano que interseca la esfera en un único punto.

tangent segment/segmento de tangente (477) Un segmento AB tal que un extremo está en un círculo, el otro está fuera de él y la recta AB es tangente al círculo.

tessellation/teselado (522, 523) Patrones que semejan mosaicos que se forman repitiendo figuras para así cubrir completamente un plano sin traslapación.

theorem/teorema (39, 101) Un enunciado, a menudo de carácter general, que puede ser demostrado apelando a postulados, definiciones, propiedades algebraicas y reglas de lógica.

traceable network/red trazable (559) Red que puede recorrerse completamente sin pasar dos veces por una misma arista. Una red es trazable si todos sus nodos tienen grado par o si hay exactamente dos nodos de grado impar.

transformation/transformación (715) En un plano, una relación en que cada punto tiene un único punto imagen y cada punto imagen tiene un único punto preimagen.

translation/traslación (731) Composición de dos reflexiones a través de rectas paralelas. Una traslación desliza las figuras la misma distancia y en la misma dirección.

transversal/transversal (126) Recta que interseca dos o más rectas en el plano en puntos distintos.

trapezoid/trapecio (321) Cuadrilátero con sólo un par de lados paralelos llamados bases. Los lados no paralelos son los catetos. Los pares de ángulos con sus vértices en los extremos de la misma base se llaman ángulos basales. El segmento de recta que une los puntos medios de los catetos es la mediana. Una altitud es un segmento perpendicular a las rectas que contienen las bases y que tiene sus extremos en estas rectas.

triangle/triángulo (180) Figura formada por los segmentos determinados por tres puntos no colineales. Los tres segmentos se llaman lados. Los extremos de los segmentos se llaman vértices. Un triángulo divide el plano en tres partes: el triángulo, su interior y su exterior.

trigonometric ratio/razón trigonométrica (412) Razón de las medidas de dos lados de un triángulo rectángulo.

trigonometry/trigonometría (412) Estudio de las propiedades de los triángulos y de las funciones trigonométricas y sus aplicaciones.

two-column proof/demostración a dos columnas (94) Demostración formal en la que los enunciados aparecen en una columna y las razones de cada enunciado en la otra columna.

undefined term/término primitivo (13) Una palabra, habitualmente fácil de entender, que no se puede explicar mediante palabras y conceptos más básicos. Los términos primitivos de la geometría son el punto, la recta y el plano.

uniform/uniforme (524) Teselados que contienen la misma combinación de formas y ángulos en cada vértice.

vector/vector (673) Segmento dirigido. Los vectores tienen magnitud (longitud) y dirección.

vertex/vértice **1.** (44) El extremo común de los dos rayos que forman un ángulo. **2.** (180) El extremo de un lado cualquiera en un polígono. **3.** (577) El lugar de intersección de tres o más aristas de un poliedro. **4.** (600) En una pirámide, el vértice que no está contenido en la base de la pirámide. **5.** (602) En la figura, el vértice del cono es el punto *V*.

vertex angle/ángulo del vértice (182) El ángulo formado por los lados congruentes (catetos) de un triángulo isósceles.

vertical angles/ángulos opuestos por el vértice (53) Dos ángulos no adyacentes formados por dos rectas que se intersecan.

volume/volumen (607) Medida de la cantidad de espacio encerrado por una figura tridimensional.

x*-axis/eje *x (6) La recta numérica horizontal de un plano de coordenadas.

x*-coordinate/coordenada *x (6) El primer número en un par ordenado de números.

y*-axis/eje *y (6) La recta numérica vertical de un plano de coordenadas.

y*-coordinate/coordenada *y (6) El segundo número en un par ordenado de números.

CHAPTER 1 DISCOVERING POINTS, LINES, PLANES, AND ANGLES

Pages 9–11 Lesson 1-1

5. $(-4, -3)$ **6–7.**

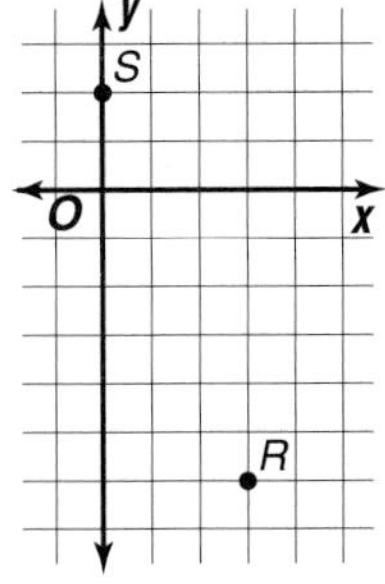

9. noncollinear **11.** $(4, 3)$ **13.** $(-3, 2)$ **15.** $(1, -4)$

17–22.

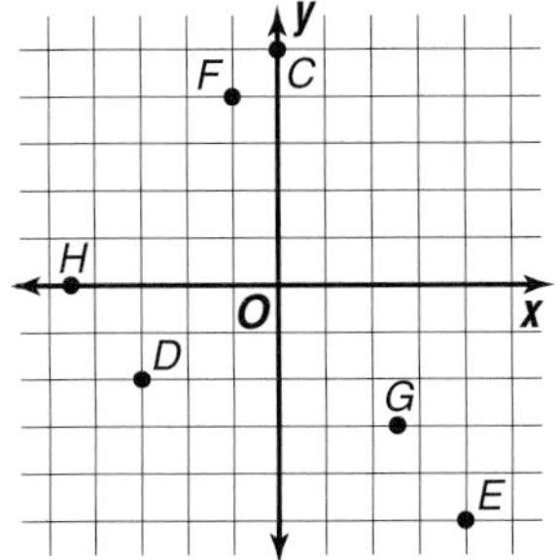

23. collinear **25.** noncollinear **27.** $(0, 4)$ **29.** $(3, -2)$ **31.** $(-5, 0)$ **33.** Sample points: $(0, -9)$, $(1, -3)$, $(2, 3)$

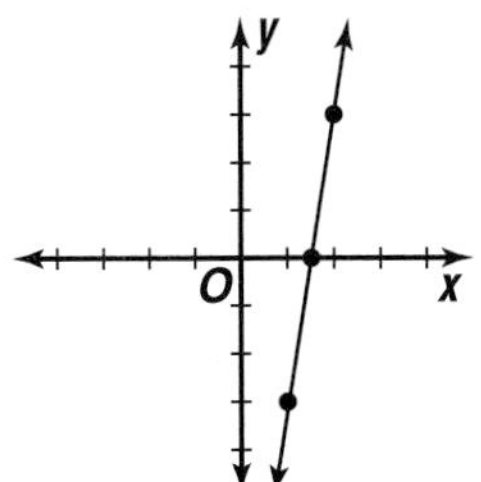

sample point not on line: $(0, 0)$ **35a.** 2; I, IV **35b.** $(2, 0)$; $(6, 0)$ **35c.** Sample answer: $(4, -3)$ **35d.** -3 **35e.** 2

37. $(-3, -4)$; Substitute the -3 for x and -4 for y in each of the equations to make sure these coordinates satisfy both equations. Since it satisfies both equations, these are the coordinates of a point that lies on the graphs of both equations. In order for the point to lie on two distinct lines, it must be the point they have in common, or their intersection.

39.

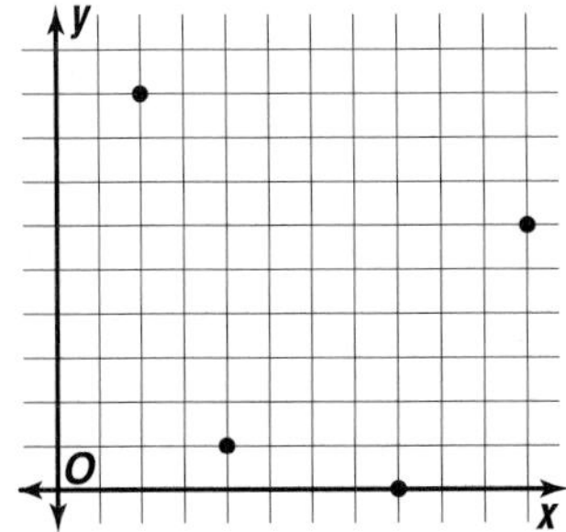

41. 1 **43.** 4 **45.** 6 **47a.** 5 or more hours

47b. 1920 parents **47c.** 1–4 hours.

Pages 16–18 Lesson 1-2

7. point **9.** line **11.** Sample answer: plane $\mathcal{A}$

13.

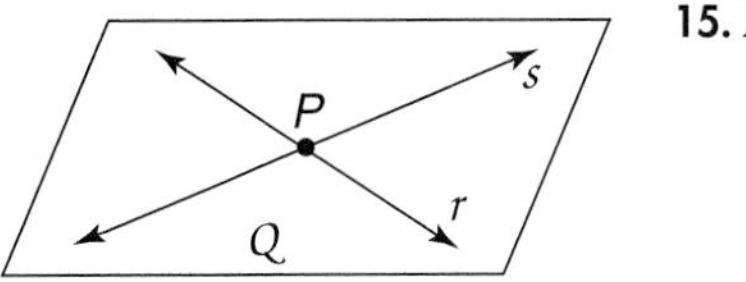

15. $\overrightarrow{PR}$ **17.** $\overleftrightarrow{PR}$

19. line **21.** plane **23.** lines **25.** point **27.** line

29–33. Sample names for figures are given. **29.** $\overrightarrow{BF}$

31. F, C, E, or D **33.** $\mathcal{R}$

35. sample answer:

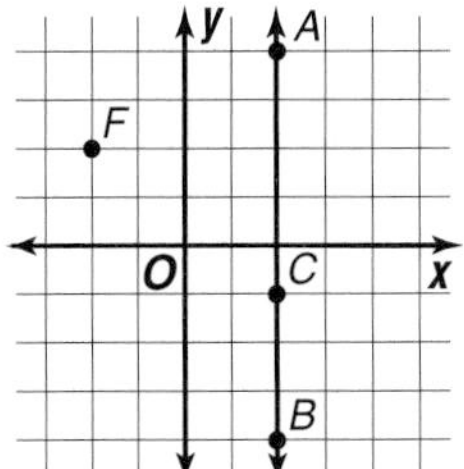

37. sample answer:

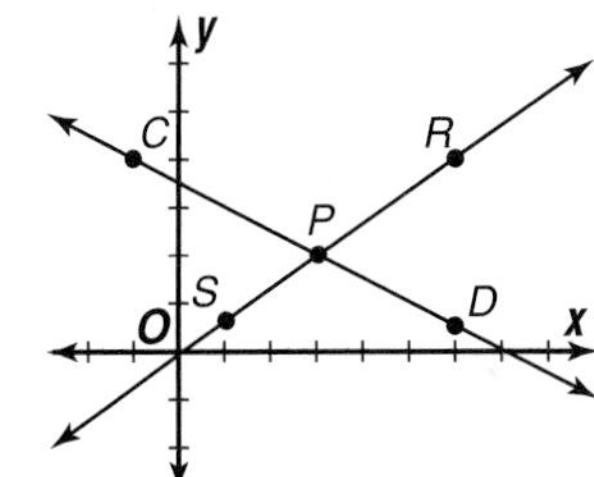

39.

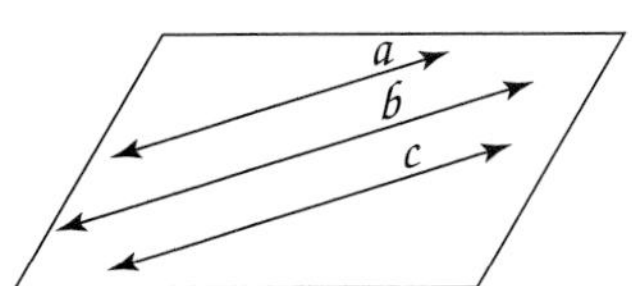

41.

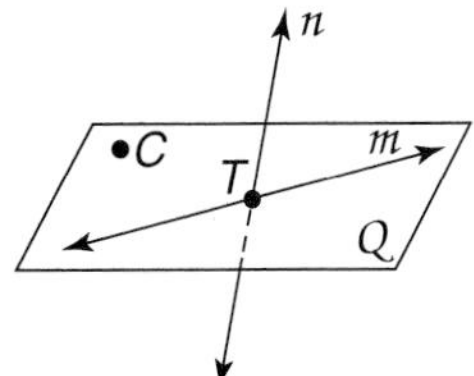

43–51. Sample names for figures are given. **43.** AFH, BHD, CDE, EGF, AGC, FED **45.** $\overleftrightarrow{GE}$, $\overleftrightarrow{DE}$, $\overleftrightarrow{FE}$ **47.** $\overleftrightarrow{GE}$, $\overleftrightarrow{GC}$ **49.** G

51. Yes; you can draw $\overleftrightarrow{BG}$ and $\overleftrightarrow{HE}$ to form a plane that fits diagonally through the figure.

53.

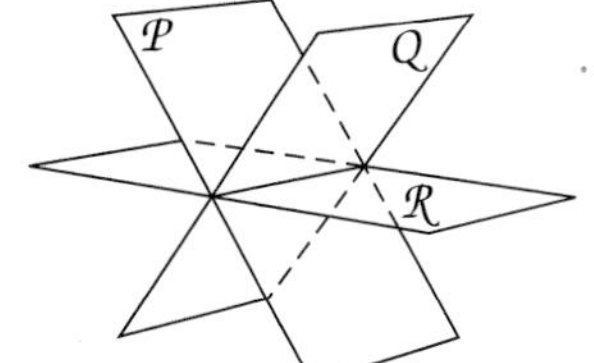

55.

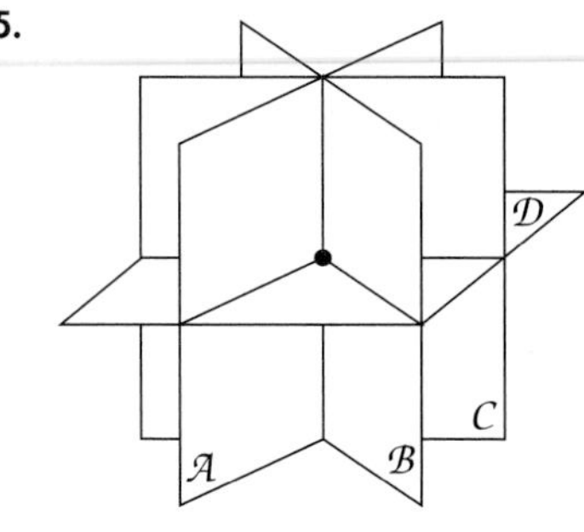

57. maximum: 6, minimum: 0

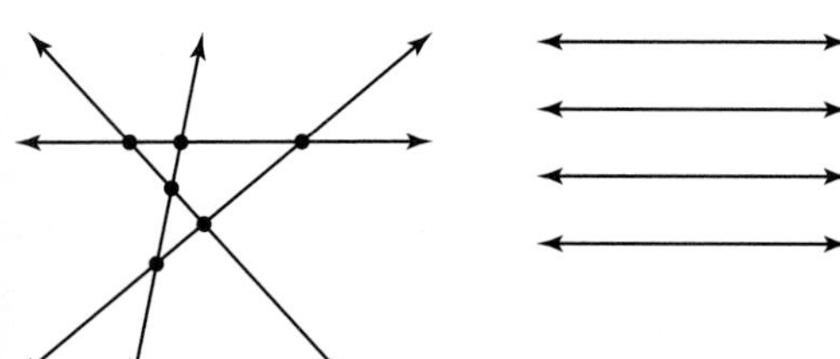

61.

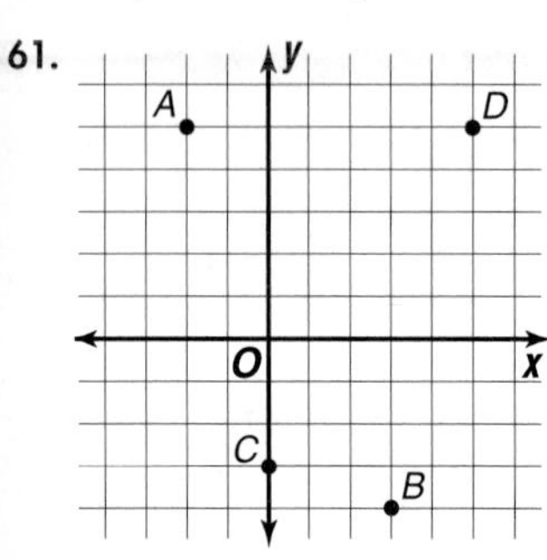

63a. (3, 0) **63b.** (−2, 3) **63c.** (−2, −3) **65.** −11 **67.** −4 **69.** −4 **71.** 7

Pages 22–27 Lesson 1–3

5. 28 cm, 40 cm^2 **7.** 28 **9.** 2 **11.** $T = 5d + t$ **13.** 81 mi^2; $\ell = w = 9$ mi **15.** 26 m, 30 m^2 **17.** 10 cm, 6.25 cm^2 **19.** 14 mi, 8.25 mi^2 **21.** 20 **23.** 32 **25.** 6 **27.** 17 **29.** 30 **31.** 49 in^2; $\ell = w = 7$ in. **33.** 64 ft^2; $\ell = w = 8$ ft **35.** $14\frac{1}{16}$ yd^2; $\ell = w = 3\frac{3}{4}$ yd **37.** 60 $units^2$ **39.** 17.5 $units^3$ **41.** 4225 ft^2 **43a.** B **43b.** C **45.** Sample answers: $\overrightarrow{ST}$, $\overrightarrow{SV}$, $\overrightarrow{SR}$ **47a.** (1, 3) **47b.** (−1, 1) **47c.** (4, −2) **49.** $7x^2 + 7x$ **51.** $8c - 3d$

Page 26 Self Test

1a. (4, 0) **1b.** (1, −3) **1c.** (−2, 1) **3.** $\overleftrightarrow{MN}$, $\overleftrightarrow{NM}$, $\overleftrightarrow{OM}$, $\overleftrightarrow{MO}$, $\overleftrightarrow{ON}$, $\overleftrightarrow{NO}$, line q **5.** No; any three of the points are coplanar, but the fourth does not lie in the same plane as the other three.

7.

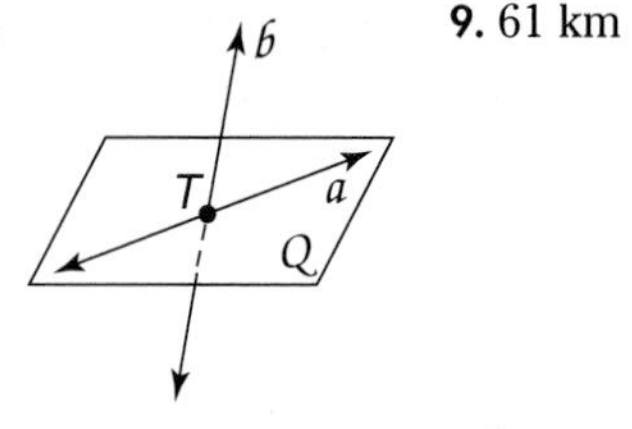

9. 61 km

Page 27 Lesson 1–4A

1. any number **3.** The calculator shows the measurement of the entire segment containing $\overline{EA}$, not just $\overline{EA}$.

5. Sample answer:

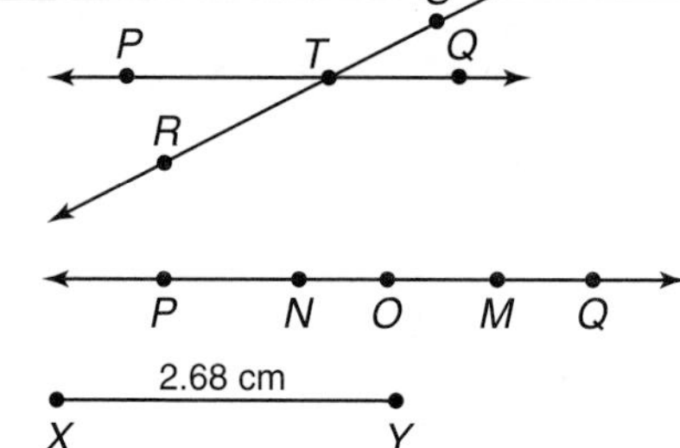

7. Sample answer:

P N O M Q

9. Sample answer:

2.68 cm
X Y

Pages 32–35 Lesson 1–4

5. 8 units **7.** 23 **9.** 3.61 **11.** $NP < QR$ **13.** 3, 20

15a.

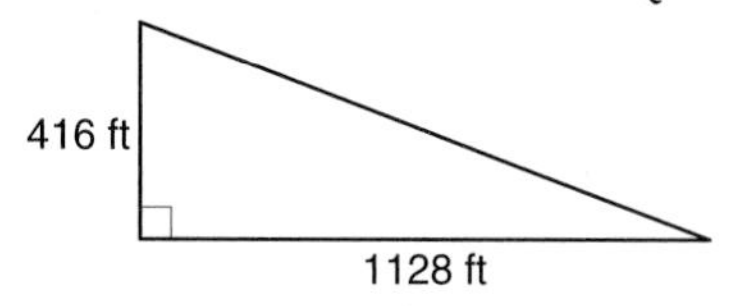

15b. about 1202 feet **17.** 4 **19.** 5 **21.** 4 **23.** $5\frac{1}{3}$ **25.** 8.35 **27.** 9 **29.** 1.41 **31.** 5 **33.** $FJ = JL$ **35.** 10 **37.** $\sqrt{2} \approx 1.414$ **39.** 2, 7

41.

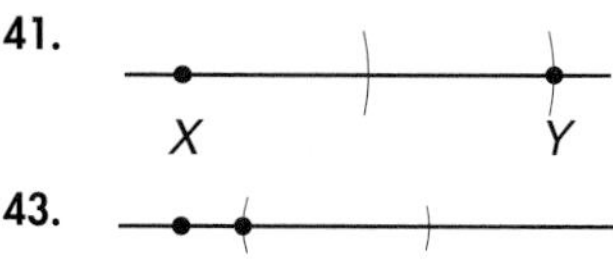

43.

X Y

45a.

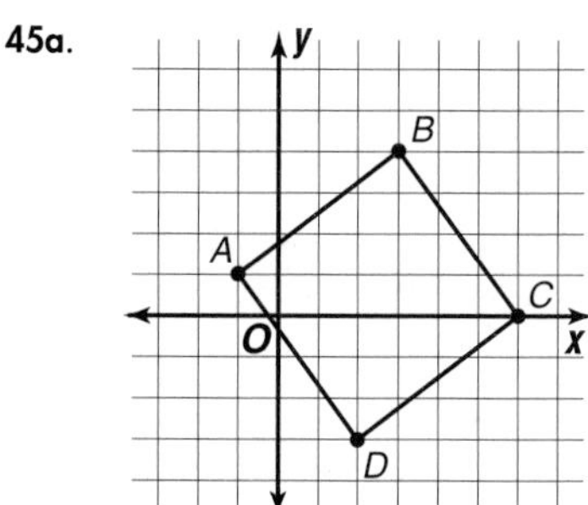

45b. $A = 25$ $units^2$, $P = 20$ units **47a.** $E(6, 4)$, $F(4, 6)$ **47b.** The x-coordinate will be the same as the x-coordinate of F and the y-coordinate will be the same as the y-coordinate of E. Thus, $G(4, 4)$. **47c.** $DG = GB$; use the distance formula to find DG and GB, which yields about 2.83. **49.** The distance is measured along a number line. The distance from 6 to 9 is 3 units. **51.** 6 cm

53.

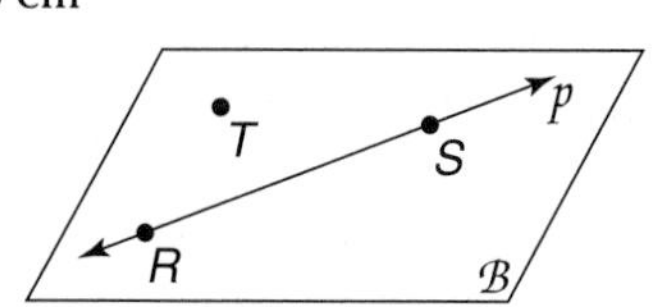

55.

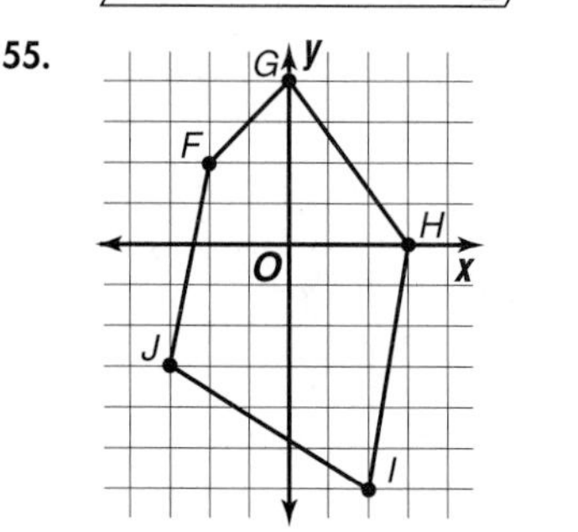

57. $3x^2 + 18x$ **59.** $x^2 - x - 12$

Pages 40–43 Lesson 1-5

7. true **9.** (1.5, −2) **13a.** 2; 18 **13b.** 1; 2 **15.** −2 **17.** −3.5 **19.** False; sample answer: E is the midpoint of $\overline{HJ}$. **21.** true **23.** true **25.** $S = -18$ **27.** $Y(2, 5)$ **29.** $X(-6, -4)$ **31.** $Z\left(\frac{8}{3}, 11\right)$ **33.** 8; 60 **35.** 1; 1 **37.** 2; 6 **39.** (scale 2:5), $AB = 5$ in., $AC = \frac{1}{8}AB$

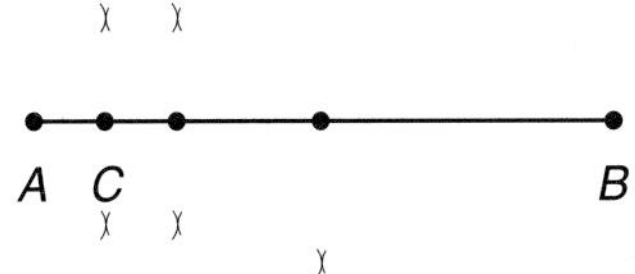

41. $R(4, -1)$, $S(6, -5)$

43.

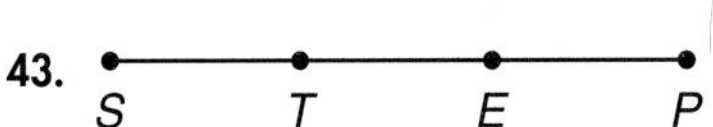

By the Segment Addition Property, we know that $SP = SE + EP$ and $SE = ST + TE$. By substituting for SE in the first equation, $SP = ST + TE + EP$.

45a–b.

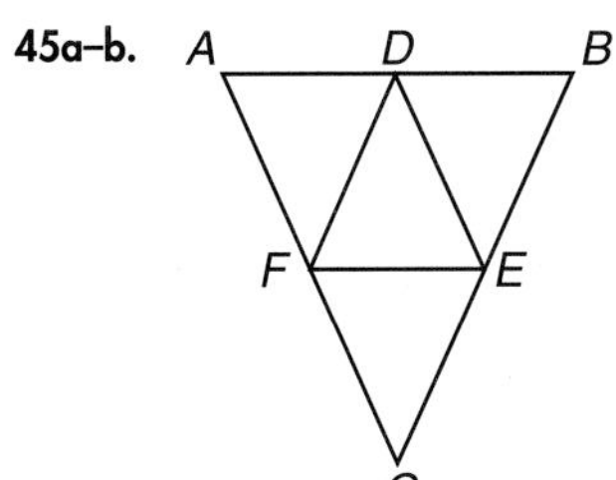

45d. The perimeter of the larger triangle is twice that of the smaller triangle. **45e.** The area of the larger triangle is 4 times that of the smaller triangle. Sample answer: If you make four copies of the small triangle, you can fit them on the surface of the large rectangles. **47.** 159 **49a.** path 1: 8 units; path 2: 8 units; path 3: 5.83 units **49b.** path 3, because you can't drive diagonally through city blocks.

51a.

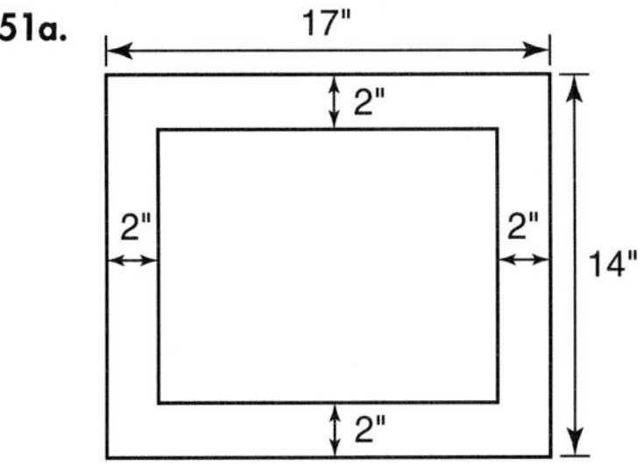

51b. 130 in^2

53.

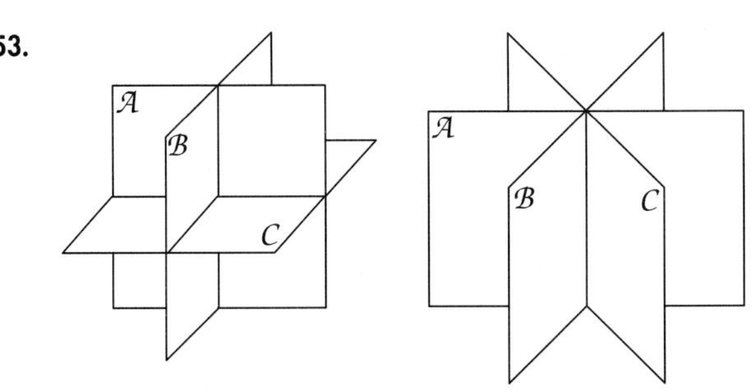

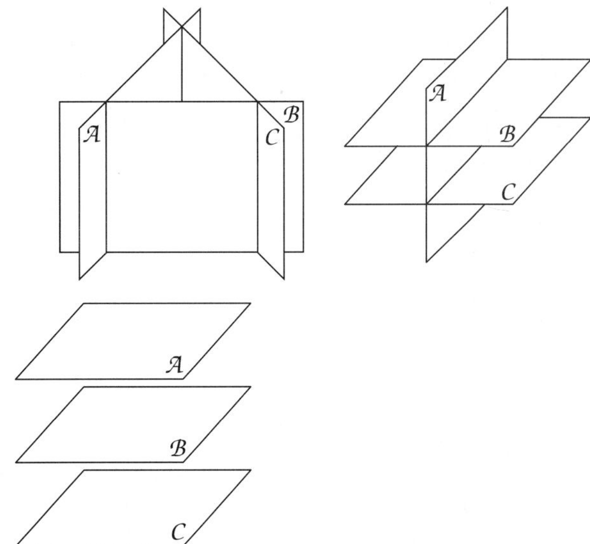

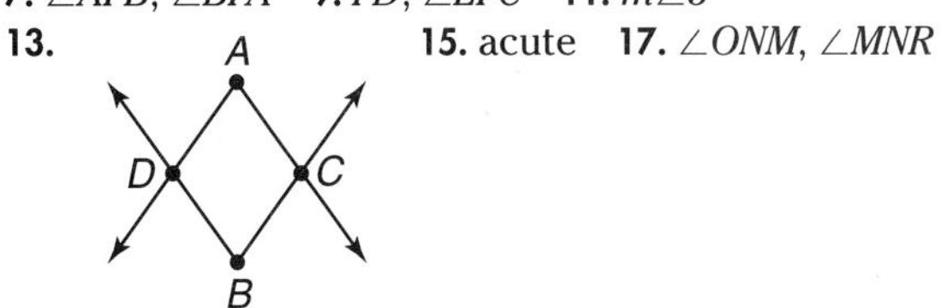

55. $x = 4$

Pages 49–51 Lesson 1-6

7. $\angle AFB$, $\angle BFA$ **9.** $\overrightarrow{FD}$; $\angle EFC$ **11.** $m\angle 3$

13. A, D, C, B

15. acute **17.** $\angle ONM$, $\angle MNR$

19. I or P **21.** obtuse **23.** $\angle NML$, $\angle NMK$, $\angle NMJ$, $\angle NMP$, $\angle NMO$ **25.** 120 **27.** $\angle JMQ$ and $\angle NMK$

29. **31.**

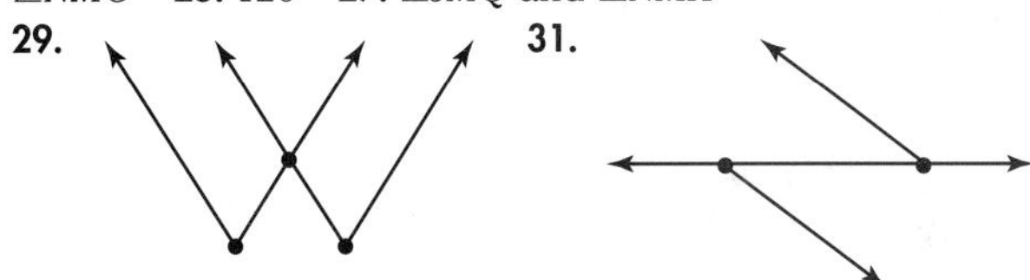

33. 50 **35.** 22 **37.** 2.8 **39a.** 1, 3, 6, 10, 15 **39b.** For 4 rays, there are $(4 \cdot 3) \div 2$ or 6 angles. For 5 rays, there are $(5 \cdot 4) \div 2$ or 10 angles. For 6 rays, there are $(6 \cdot 5) \div 2$ or 15 angles. **39c.** 21, 45 **39d.** $a = \frac{n(n-1)}{2}$, for a = number of angles and n = number of rays **41a.** acute **41b.** 15 **43.** 10 **45.** 5.5 mm

47.

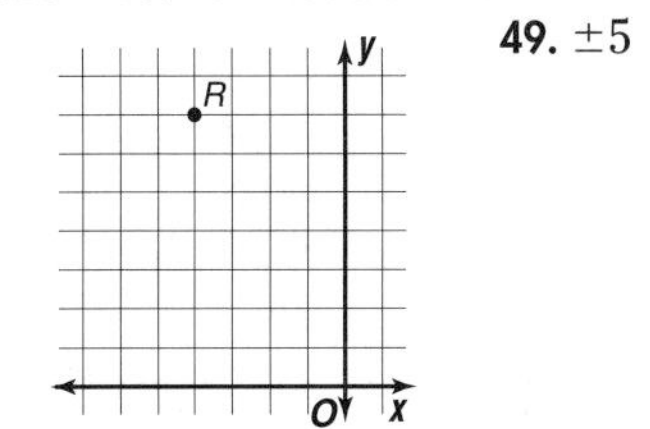

49. ± 5

Page 52 Lesson 1-7A

1a. $\angle CBA$ and $\angle DBE$; $\angle CBD$ and $\angle EBA$ **1b.** The measures of the vertical angles are equal. **5.** Regardless of the position of the lines, vertical angles have equal measures.

Pages 58–60 Lesson 1-7

7. $\overline{AD}$ **9.** $\angle AFB$ and $\angle BFC$, $\angle BFC$ and $\angle CFD$, or $\angle CFD$ and $\angle DFE$ **11.** No, because there are no markings or measures given to indicate that the segments are congruent. **13.** 148 **15.** $\angle YUW$ and $\angle XUV$ or $\angle YUX$ and $\angle VUW$ **17.** $\angle TWU$ **19.** No, there are no measures or markings to indicate that right angles are present.

21. congruent, adjacent supplementary, linear pair **23.** No, there is no indication that $\angle SRT$ is a 90° angle. **25.** right **27a.** perpendicular **27c.** The angle bisectors are perpendicular, because the angles formed by the two lines measure 45° + 45° or 90°. **29.** 112, 68 **31.** 67.8; 22.2 **33.** 36, 17, 75, 15, 55, 35

35. Given: $\angle PQR$ and $\angle RQS$ are complementary angles; $m\angle PQR = 45$.

Prove: $\angle PQR \cong \angle RQS$

Sample paragraph proof: Since $\angle PQR$ and $\angle RQS$ are complementary, $m\angle PQR + m\angle RQS = 90$. Substitute 45 for $m\angle PQR$ in the equation. This results in, 45 + $m\angle RQS = 90$. By subtracting 45 from each side, $m\angle PQR = 45$. Since $\angle PQR$ and $\angle QRS$ have the same measure, $\angle PQR \cong \angle QRS$.

37. Sample answer: Complementary is defined as something that completes, and complementary angles complete a right angle. Supplementary is defined as something that completes or makes an addition, and supplementary angles add up to be a straight angle. **39.** obtuse **41.** about 14 feet 2 inches

43.

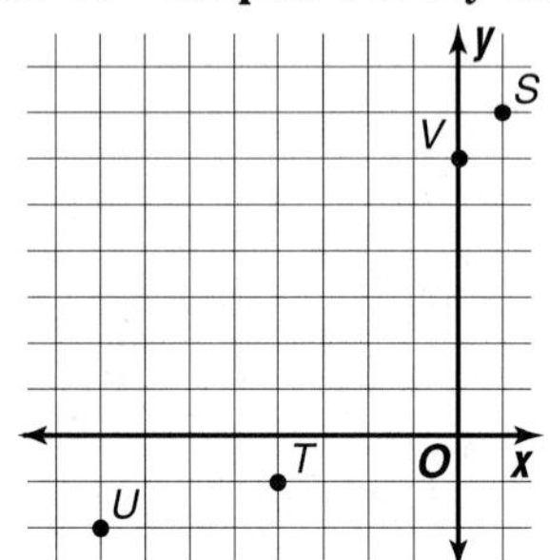

45. $4ab$

Page 61 Chapter 1 Highlights

1. quadrants **3.** supplementary **5.** line **7.** origin **9.** Vertical

Pages 62–64 Chapter 1 Study Guide and Assessment

11–14.

15. Quadrant III **17.** point A **19.** 43 mph **21.** 56.25 cm^2 **23.** $3\frac{1}{2}$ **25.** 5 **27.** 12.21 **29.** (5.5, −0.5) **31.** 12 **33.** yes **35.** 148 **37.** 40 **39.** $\angle PTN$ **41.** 120, 60

43.

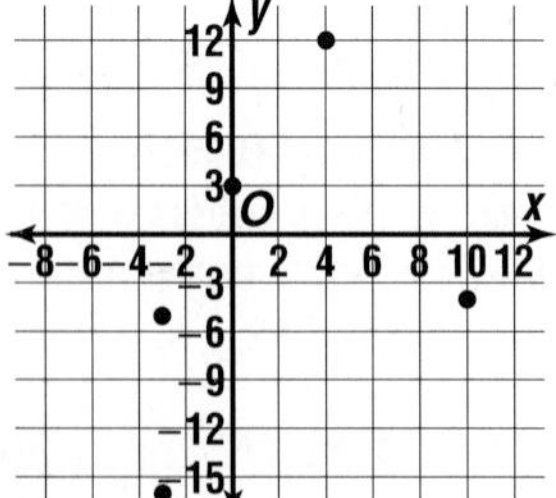

45a. As the loft of the club increases, the ball will travel higher in the air and for a shorter distance, as long as the ball is struck with the same amount of force.

45b.

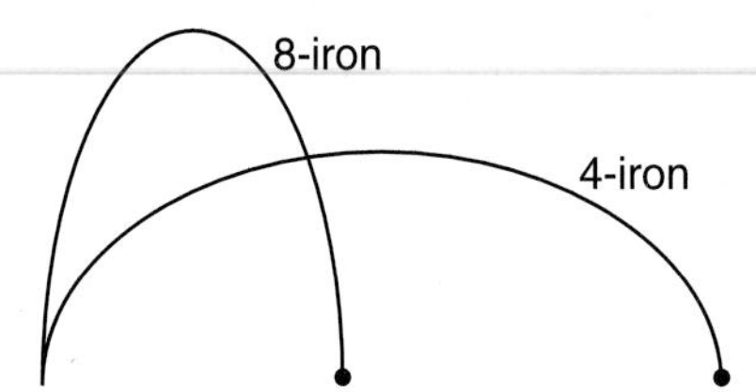

CHAPTER 2 CONNECTING REASONING AND PROOF

Pages 72–75 Lesson 2–1

7. If ℓ and m are perpendicular, then they form a right angle. **9.** Points H, I, and J are noncollinear. **11.** Sample answer: Earth is flat, Earth is the center of the universe. **13.** False; $XZ + YZ = XY$ by the Segment Addition Post.

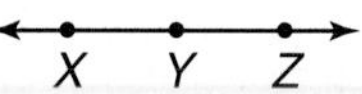

15. Points A, B, and C do not lie on a line.

A•

•C

B•

17. X, Y, Z, and W are noncollinear

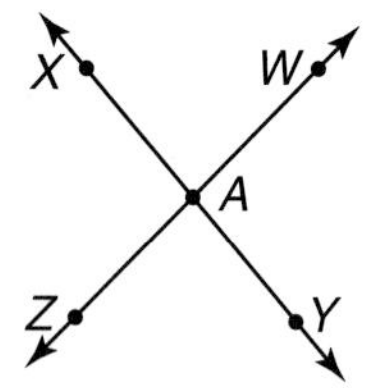

19. Points R, S, and T are collinear.

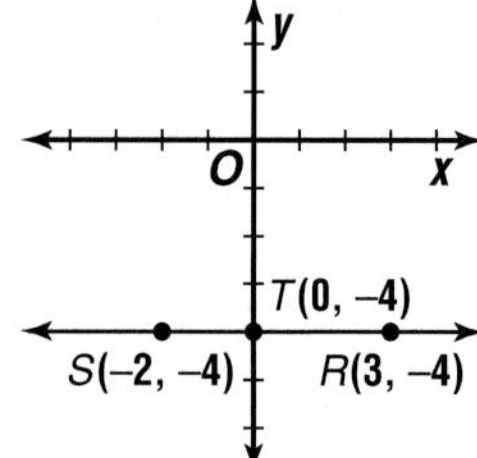

21. $m\angle ABD = m\angle CBD$

23. false; counterexample:

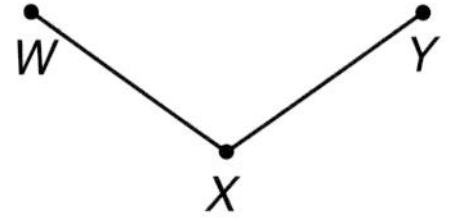

25. False; K, L, and M are collinear.

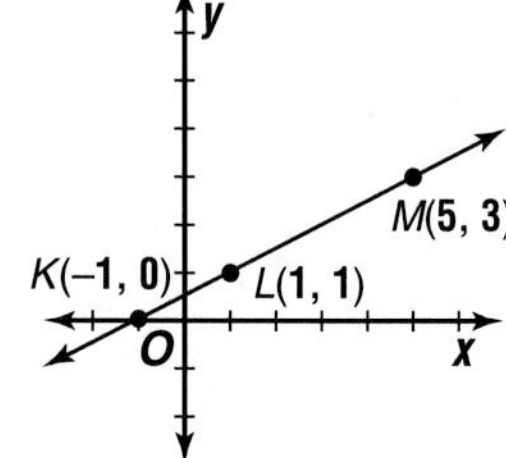

27. *PQRST* is a pentagon.

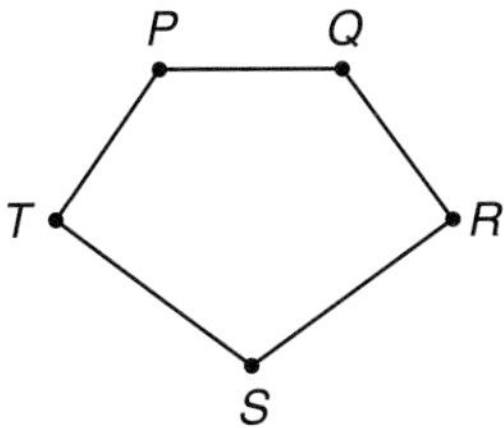

29a. They are congruent. **29b.** Sample answer: Use the Angle option on the F6 menu to find the measures of $\angle PQS$ and $\angle QSR$. **31.** The ball will strike two rails and then continue on a path to the corner on the opposite side of the table at the opposite end. **33.** 20, no **35.** (10, 12) **37.** 3; 13 cm by 9 cm **39.** $\frac{3}{20}$

Pages 80–83 Lesson 2–2
7. Hypothesis: three points lie on a line; Conclusion: they are collinear **9.** If angles have the same measure, then they are congruent. **11.** Four points are coplanar. **13. Converse:** If an angle is a right angle, then it measures 90°; true. **Inverse:** If an angle does not measure 90°, then it is not a right angle; true. **Contrapositive:** If an angle is not a right angle, then it does not measure 90°; true. **15.** false **17.** false **19.** Hypothesis: a man hasn't discovered something that he will die for; Conclusion: he isn't fit to live **21.** Hypothesis: we would have new knowledge; Conclusion: we must get a whole world of new questions **23.** Hypothesis: $3x - 5 = -11$; Conclusion: $x = -2$ **25.** If people are happy, then they rarely correct their faults. **27.** If angles are adjacent, then they have a common vertex. **29.** If angles have measures between 90 and 180, then they are obtuse. **31.** A book is not a mirror. **33.** Rectangles are squares. **35.** You do not live in Dallas. **37. Converse:** If there are three points that are noncollinear, then they are not on the same line; true. **Inverse:** If there are three points on the same line, then they are collinear; true. **Contrapositive:** If there are three points that are collinear, then they are on the same line; true. **39. Converse:** If an angle has a measure less than 90, then it is acute; true. **Inverse:** If an angle is not acute, then it does not have a measure less than 90; true. **Contrapositive:** If an angle does not have a measure less than 90, then it is not acute; true. **41. Converse:** If you don't live in Illinois, then you don't live in Chicago; true. **Inverse:** If you live in Chicago, then you live in Illinois; true. **Contrapositive:** If you live in Illinois, then you live in Chicago; false. Counterexample: you may live in another city in Illinois. **43.** true **45.** true
47. three

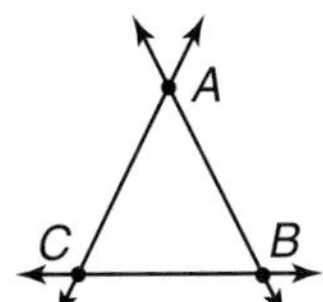

49. one **51a.** doubles **51b.** quadruples **51c.** is multiplied by 8 **53a.** The Hatter is right; Alice exchanged the hypothesis and conclusion. **53b.** These statements are converses of each other. **55a.** If you try Casa Fiesta, then you're looking for a fast, easy way to add some fun to your family's menu. **55b.** They are a fast, easy way to add fun to your family's menu. **55c.** No; the conclusion is implied. **57.** Sample answer: $\angle C \cong \angle D$ **59.** Sample answer: *RSTU* has four congruent sides. **61.** acute **63.** 12 in. by 8 in., 15 in. by 10 in., 18 in. by 12 in., 21 in. by 14 in. **65.** no **67.** false

Page 84 Lesson 2–2B
1. The result is the same, $x < -3$. **3.** true **5.** False; if $12 - 3x > 23 - 14x$, then $x > 1$.

Pages 88–91 Lesson 2–3
7. yes; syllogism **9.** Patricia Gorman should get 8 hours of sleep each day; detachment. **11.** If the measure of an angle is less than 90, then it is not obtuse; syllogism. **13a.** You'll become a true Beatles fan. **13b.** no conclusion **13c.** no conclusion **13d.** You'll love this album. **15.** yes; detachment **17.** invalid **19.** invalid **21.** invalid **23.** Odina lives to eat; detachment. **25.** If *M* is the midpoint of $\overline{AB}$, then $\overline{AM} \cong \overline{MB}$; syllogism. **27.** no conclusion **29.** Planes $\mathcal{M}$ and $\mathcal{N}$ intersect in a line; detachment. **31.** Line *t* is perpendicular to line *q*; detachment. **33.** $\overline{XY}$ lies in plane $\mathcal{P}$; detachment. **35a.** (2) I like pizza with everything. (3) I'll like Jimmy's pizza. **35b.** (2) If you like Jimmy's pizza, then you are a pizza connoisseur. (3) If you like pizza with everything, then you are a pizza connoisseur. **37.** (1) If a person is a baby, then the person is not logical. (statement 1) (2) If a person is not logical, then the person is despised. (statement 3) (3) If a person is a baby, then the person is despised. (Law of Syllogism) (4) If a person is despised, then the person cannot manage a crocodile. (contrapositive of statement 2) (5) If a person is a baby, then the person cannot manage a crocodile. (Law of Syllogism) **39.** 1
41. If two lines intersect, then they are perpendicular; false. Counterexample:

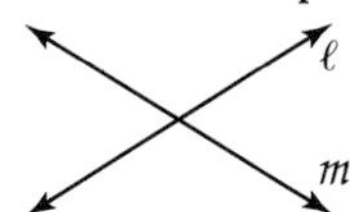

43. Sample answer: *A*, *B*, *C*, and *D* are collinear.

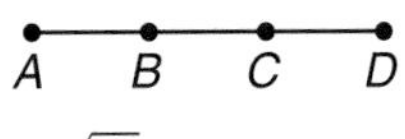

47. $2\sqrt{10}$ **49.** $3.22

Page 91 Self Test
1. False; counterexample:

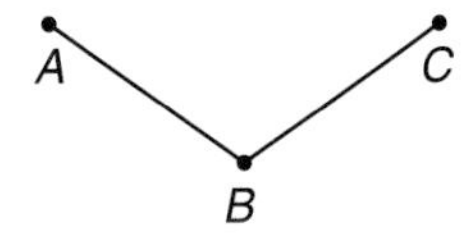

3a. Sample answer: insufficient light or water **3b.** Yes; the fungus is killing the plants. **3c.** Introduce the fungus to some healthy plants and see if they droop. **5.** If there are clouds, then it is raining. **7.** If a figure does not have four sides, then it is not a square. **9.** invalid

Pages 95–99 Lesson 2–4
7. Addition Property (=) **9a.** 1 **9b.** 3 **9c.** 4 or 5 **9d.** 5 or 4 **9e.** 2

11. Given: $k = \frac{\Delta\ell}{\ell(T - t)}$

Prove: $T = \frac{\Delta\ell}{k\ell} + t$

Proof:

Statements	Reasons
1. $k = \frac{\Delta\ell}{\ell(T - t)}$	1. Given
2. $k(T - t) = \frac{\Delta\ell}{\ell}$	2. Multiplication Property (=)
3. $T - t = \frac{\Delta\ell}{k\ell}$	3. Division Property (=)
4. $T = \frac{\Delta\ell}{k\ell} + t$	4. Add. Property (=)

13. Symmetric Property (=) **15.** Multiplication Property (=) or Division Property (=) **17.** Division Property (=) **19.** Transitive Property (=) **21.** Subtraction Property (=) **23a.** 2 **23b.** 1 **23c.** 4 **23d.** 3 **23e.** 5 **23f.** 6 **25a.** Given **25b.** Multiplication Property (=) **25c.** Distributive Property **25d.** Subtraction Property (=) **25e.** Division Property (=) **27a.** $m\angle TUV = 90$, $m\angle XWV = 90$, $m\angle 1 = m\angle 3$ **27b.** Substitution Property (=) **27c.** Angle Addition Postulate **27d.** $m\angle 1 + m\angle 2 = m\angle 3 + m\angle 4$ **27e.** Substitution Property (=) **27f.** $m\angle 2 = m\angle 4$

29. Given: $2x + 6 = 3 + \frac{5}{3}x$

Prove: $x = -9$

Proof:

Statements	Reasons
1. $2x + 6 = 3 + \frac{5}{3}x$	1. Given
2. $3(2x + 6) = 3(3 + \frac{5}{3}x)$	2. Multiplication Property (=)
3. $6x + 18 = 9 + 5x$	3. Distributive Property
4. $x + 18 = 9$	4. Subtraction Property (=)
5. $x = -9$	5. Subtraction Property (=)

31. Sample answer: Both properties are transitive; equality relates numbers, congruence relates sets of points.

33. Given: $AC = BD$

Prove: $AB = CD$

Proof:

Statements	Reasons
1. $AC = BD$	1. Given
2. $AB + BC = AC$ $BC + CD = BD$	2. Segment Addition Postulate
3. $AB + BC = BC + CD$	3. Substitution Property (=)
4. $AB = CD$	4. Subtraction Property (=)

35. If $m\angle 1 \neq 27$, then $\angle 1$ is not acute; false, if $m\angle 1 = 32$, then $\angle 1$ is acute. **37.** Complement does not exist; 21. **41a.** $\frac{100 \text{ lb}}{\$300} = \frac{390 \text{ lb}}{x}$ **41b.** \$1170 **41c.** \$269,100

Pages 103–106 Lesson 2–5

5. Symmetric Property of ≅ Segments **7.** Addition Property (=) **9.** Given: x is a whole number. Prove: x is an integer. **11a.** 3 **11b.** 1 or 4 **11c.** 2 **11d.** 1 or 4

13. Given: $WX = XY$

Prove: $WY = 2XY$

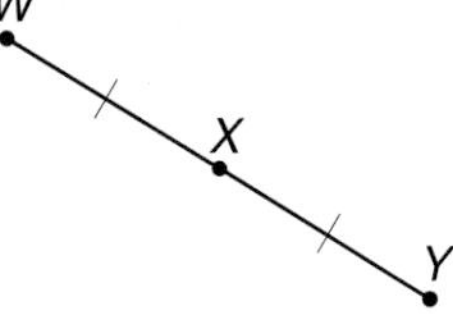

Proof:

Statements	Reasons
1. $WX = XY$	1. Given
2. $WY = WX + XY$	2. Segment Addition Post.
3. $WY = XY + XY$	3. Substitution Property (=)
4. $WY = 2XY$	4. Substitution Property (=)

15. Reflexive Property of ≅ Segments **17.** Addition Property (=) **19.** Transitive Property of ≅ Segments **21.** Distributive Property **23.** Given: $x > 2$ and x is prime. Prove: x is odd.

25. Given: $AB = CD$, $EF = CD$

Prove: $AB = EF$

27. Given: A, B are on ℓ.

Prove: $\overline{AB}$ is on ℓ.

29. Given: x is a rational number.

Prove: x is a real number.

31a. Given **31b.** Definition ≅ Segments **31c.** $PM = MS$, $RM = MQ$ **31d.** Segment Addition Postulate **31e.** Substitution Property (=) **31f.** Substitution Property (=) **31h.** Division Property (=) **31i.** Definition ≅ Segments

33. Given: $NL = NM$
$AL = BM$

Prove: $NA = NB$

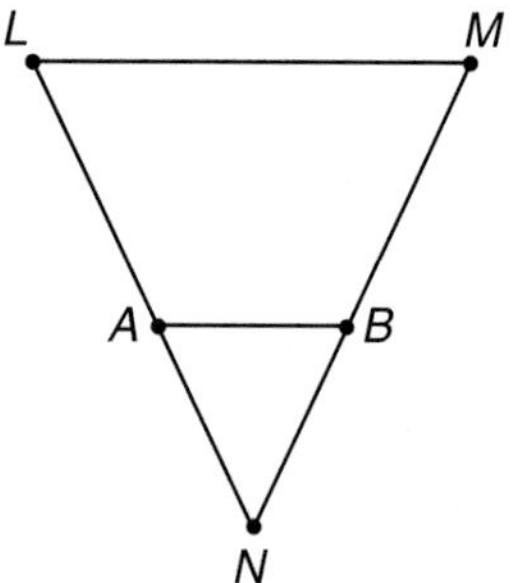

Proof:

Statements	Reasons
1. $NL = NM$ $AL = BM$	1. Given
2. $NL = NA + AL$ $NM = NB + BM$	2. Segment Addition Postulate
3. $NA + AL = NB + BM$	3. Substitution Property (=)
4. $NA + BM = NB + BM$	4. Substitution Property (=)
5. $NA = NB$	5. Subtraction Property (=)

35. Sample answers: $\overline{LN} \cong \overline{QO}$ and $\overline{LM} \cong \overline{MN} \cong \overline{RS} \cong \overline{ST} \cong \overline{QP} \cong \overline{PO}$ **37a.** about 15.2 cm **37b.** Yes; the scales are different. **39.** Law of Syllogism; if lines are parallel, then they have no point in common. **41.** (1, 4) **43.** 136 **45.** $2\frac{1}{2}$ yd

Pages 111–114 Lesson 2-6

11. $m\angle 1 = 121$, $m\angle 2 = 59$ **13d.** $m\angle XYZ = m\angle 1 + m\angle 2$ **13e.** Substitution Property (=) **13f.** Definition of supp. ∠s **15.** sometimes **17.** always **19.** sometimes **21.** sometimes **23.** never **25.** sometimes **27.** $m\angle 5 = m\angle 6 = 58$ **29.** $m\angle 1 = 124$, $m\angle 2 = 56$ **31.** $m\angle 5 = m\angle 6 = 62$ **33a.** ∠1 and ∠8, ∠2 and ∠6, ∠3 and ∠5, ∠4 and ∠7 **33b.** ∠*AXB* and ∠*BXE*, ∠*BXC* and ∠*CXF*, ∠*CXD* and ∠*DXH*, ∠*DXE* and ∠*EXG*, ∠*EXF* and ∠*FXA*, ∠*FXH* and ∠*HXB*, ∠*HXG* and ∠*GXC*, ∠*GXA* and ∠*AXD*

35. Given: ∠*ABD* and ∠*CBD* form a linear pair.
∠*YXZ* and ∠*WXZ* form a linear pair.
∠*ABD* ≅ ∠*YXZ*

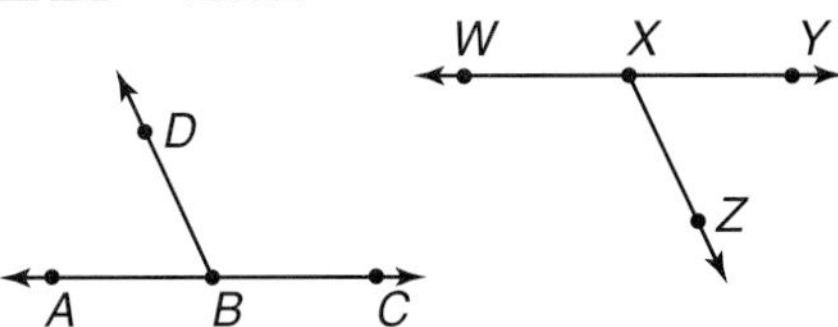

Prove: ∠*CBD* ≅ ∠*WXZ*

Proof:

Statements	Reasons
1. ∠*ABD* and ∠*CBD* form a linear pair, ∠*YXZ* and ∠*WXZ* form a linear pair, ∠*ABD* ≅ ∠*YXZ*.	1. Given
2. ∠*ABD* and ∠*CBD* are supplementary, ∠*YXZ* and ∠*WXZ* are supplementary.	2. Supp. Theorem
3. ∠*CBD* ≅ ∠*WXZ*	3. ∠s supp. to ≅ ∠s are ≅.

37. Given: ∠*A* ≅ ∠*B*

Prove: ∠*B* ≅ ∠*A*

Proof:

Statements	Reasons
1. ∠*A* ≅ ∠*B*	1. Given
2. $m\angle A = m\angle B$	2. Definition of ≅ ∠s
3. $m\angle B = m\angle A$	3. Symmetric Property (=)
4. ∠*B* ≅ ∠*A*	4. Definition of ≅ ∠s

39. Given: ∠1 and ∠2 form a linear pair.
∠1 is a right angle.

Prove: ∠2 is a right angle.

Proof:

Statements	Reasons
1. ∠1 and ∠2 form a linear pair. ∠1 is a right angle.	1. Given
2. $m\angle 1 + m\angle 2 = 180$	2. Definition of linear pair
3. $m\angle 1 = 90$	3. Definition of rt. ∠
4. $90 + m\angle 2 = 180$	4. Substitution Property (=)
5. $m\angle 2 = 90$	5. Subtraction Property (=)
6. ∠2 is a right angle.	6. Definition of rt. ∠

41. No, they are parallel; hold at an angle and look at it. **43a.** The first pair of vertical lines appear to curve inward, the second pair appear to curve outward. **43b.** They appear parallel.

43c. Given: ∠4 ≅ ∠2

Prove: ∠3 ≅ ∠1

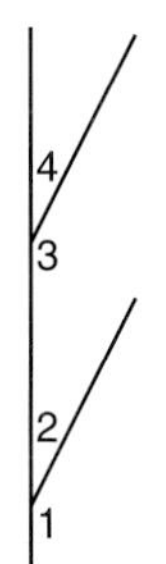

Proof:

Statements	Reasons
1. ∠4 ≅ ∠2	1. Given
2. ∠4 and ∠3 form a linear pair. ∠2 and ∠1 form a linear pair.	2. Definition of linear pair
3. ∠4 and ∠3 are supplementary. ∠2 and ∠1 are supplementary.	3. If 2 ∠s form a linear pair, then they are supplementary.
4. ∠3 ≅ ∠1	4. ∠s supplementary to ≅ ∠s are ≅.

45. Substitution Property (=) **47.** 5, 15 **49.** 15 ft by 5 ft, 13 ft by 6 ft, 11 ft by 7 ft, 9 ft by 8 ft **51.** Thursday; the largest audience is available.

Page 115 Chapter 2 Highlights

1. k **3.** b **5.** c **7.** a **9.** h **11.** j

Pages 116–118 Chapter 2 Study Guide and Assessment

13. true; definition of midpoint **15.** If something is a cloud, then it has a silver lining. **17.** If a rock is obsidian, then it is a glassy rock produced by a volcano. **19. Converse:** If a rectangle is a square, then it has four congruent sides. **Inverse:** If a rectangle does not have four congruent sides, then it is not a square. **Contrapositive:** If a rectangle is not a square, then it does not have four congruent sides. **21. Converse:** If a month has 31 days, then it is January. **Inverse:** If the month is not January, then it does not have 31 days. **Contrapositive:** If a month does not have 31 days, then it is not January. **23.** ∠*A* and ∠*B* have measures with a sum of 180; Law of Detachment. **25.** The sun is in constant motion; Law of Syllogism. **27.** Reflexive Property (=) **29.** Subtraction Property (=) **31.** Substitution Property (=) **33.** never **35.** never **37.** 56 pairs **39.** Substitution Property (=) or Transitive Property (=)

CHAPTER 3 USING PERPENDICULAR AND PARALLEL LINES

Pages 127–129 Lesson 3-1

5. intersecting **7.** False; a transversal inters… in a plane. **9.** $\overline{AH}$; alternate interior **11.** p… plane *DEF*

13.

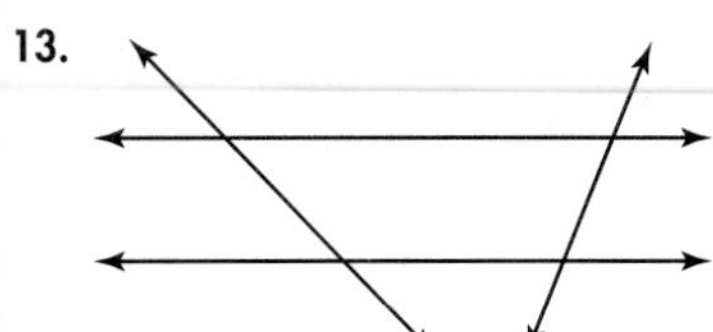

15. Skew lines; the planes are flying in different directions and at different altitudes. **17.** intersecting **19.** parallel **21.** parallel **23.** False; the angles are not formed by 2 lines and a transversal. **25.** True; the angles are in corresponding positions when the transversal *m* intersects *r* and *s*. **27.** False; the angles are formed by lines ℓ and *m* and transversal *s*. **29.** ℓ; alternate exterior **31.** ℓ; alternate interior **33.** *q*; alternate exterior **35.** $\overline{AE}$ and $\overline{DR}$, $\overline{AD}$ and $\overline{ER}$ **37.** none **39.** plane *ADR*, plane *DRM*, plane *ERM*, plane *AEM*

41. **43.**

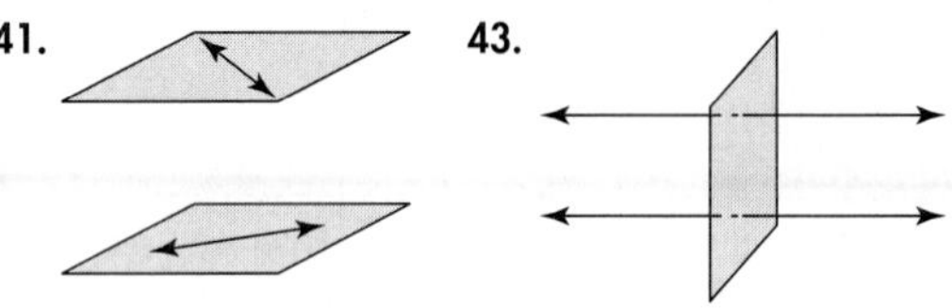

45a. $m\angle 1 = m\angle 2 = 160$ **45b.** The measures of alternate interior angles are equal. **47.** Sample answer: parallel circuits in electronics, parallel story lines in literature

49. 24 handshakes

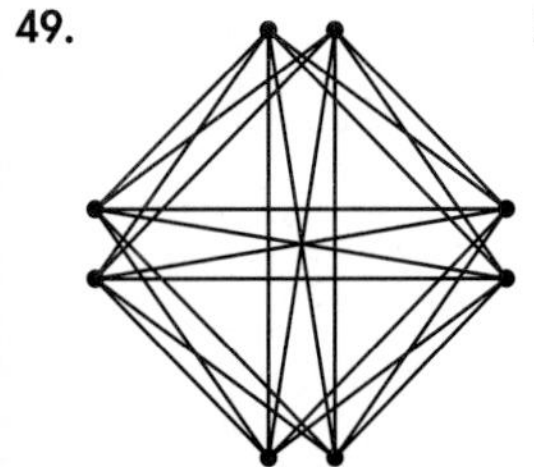

51. Given: *PARL* is a parallelogram.

Prove: $\angle 1 \cong \angle 2$
$\angle 3 \cong \angle 4$

P L 1 4 3 2 A R

53. If something is a cloud, then it is composed of millions of water droplets. **55.** (12, 18) **57.** Explore the problem; plan the solution; solve the problem; and examine the solution. **59.** $\frac{2}{5}$

Page 130 Lesson 3–2A

1. consecutive interior angles: $\angle CEF$ and $\angle AFE$, $\angle DEF$ and $\angle BFE$; alternate exterior angles: $\angle CEG$ and $\angle BFH$, $\angle DEG$ and $\angle AFH$; alternate interior angles: $\angle CEF$ and $\angle BFE$, $\angle DEF$ and $\angle AFE$; corresponding angles: $\angle CEG$ and $\angle AFE$, $\angle DEG$ and $\angle BFE$, $\angle CEF$ and $\angle AFH$, $\angle DEF$ and $\angle BFH$ **3a.** Corresponding angles are congruent. **3b.** Alternate interior angles are congruent. **3c.** Alternate exterior angles are congruent. **5a.** The measure of each of the eight angles is 90. **5b.** A transversal that is perpendicular to one of two parallel lines is perpendicular to the other.

Pages 134–137 Lesson 3–2

7. Alternate Interior Angles Theorem **9.** 107 **11.** 48 **13.** 59 **15.** $x = 12, y = 10$ **17.** Alternate Interior Angles Theorem **19.** Corresponding Angles Postulate **21.** 49 **23.** 49 **25.** 131 **27.** 47 **29.** 107 **31.** 49 **33.** 42 **35.** 42 **37.** 120 **39.** $x = 90, y = 15, z = 13.5$ **41.** Since $\angle 4$ and $\angle 8$ are alternate exterior angles formed when 2 parallel lines are cut by a transversal, $m\angle 4 = m\angle 8$ or $2x - 25 = x + 26$. Solving this equation, $x = 51$. Therefore, $m\angle 8 = 51 + 26$ or 77. Since $\angle 2$ and $\angle 8$ are vertical angles, $m\angle 2 = m\angle 8$ or $m\angle 2 = 77$.

43.
1. Given
2. Perpendicular lines form 4 right angles.
3. Definition of right angle
4. Corresponding Angles Postulate
5. Definition of congruent angles
6. Substitution Property (=)
7. Definition of right angle
8. Definition of $\perp$ lines

45. Given: $\ell \parallel m$

Prove: $\angle 3$ and $\angle 5$ are supplementary. $\angle 4$ and $\angle 6$ are supplementary.

p ℓ m 1 2 3 4 6 5 8 7

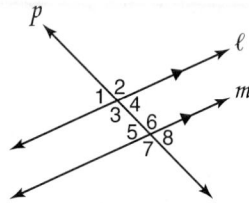

Proof:
We are given that $\ell \parallel m$. If two parallel lines are cut by a transversal, corresponding angles are congruent. So, $\angle 1 \cong \angle 5$ and $\angle 2 \cong \angle 6$ or $m\angle 1 = m\angle 5$ and $m\angle 2 = m\angle 6$. Since $\angle 1$ and $\angle 3$ form a linear pair and $\angle 2$ and $\angle 4$ form a linear pair, $\angle 1$ and $\angle 3$ are supplementary and $\angle 2$ and $\angle 4$ are supplementary. By the definition of supplementary angles, $m\angle 1 + m\angle 3 = 180$ and $m\angle 2 + m\angle 4 = 180$. By the Substitution Property (=), $m\angle 5 + m\angle 3 = 180$ and $m\angle 6 + m\angle 4 = 180$. Therefore, $\angle 3$ and $\angle 5$ are supplementary and $\angle 4$ and $\angle 6$ are supplementary by the definition of supplementary.

47. Given: $\overline{MQ} \parallel \overline{NP}$
$\angle 4 \cong \angle 3$

Prove: $\angle 1 \cong \angle 5$

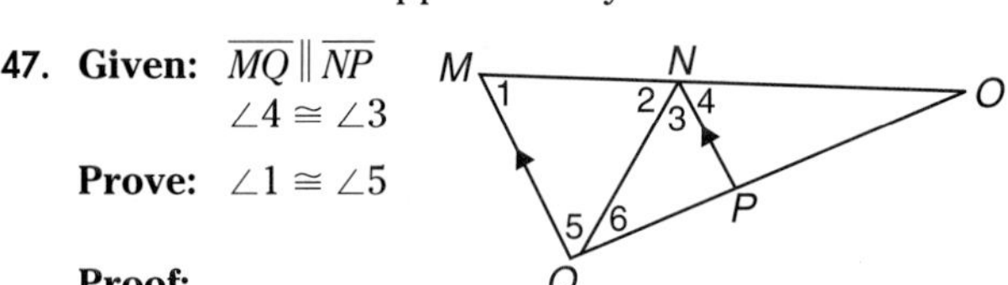

Proof:

Statements	Reasons
1. $\overline{MQ} \parallel \overline{NP}$ $\angle 4 \cong \angle 3$	1. Given
2. $\angle 3 \cong \angle 5$	2. Alternate Interior Angles Theorem
3. $\angle 4 \cong \angle 5$	3. Congruence of angles is transitive.
4. $\angle 4 \cong \angle 1$	4. Corresponding Angles Postulate
5. $\angle 1 \cong \angle 5$	5. Congruence of angles is transitive.

49. Since $\overline{AB} \parallel \overline{DC}$, we know $\angle 1 \cong \angle 4$ by the Alternate Interior Angle Theorem. However, $\angle 3$ and $\angle 2$ are not alternate interior angles for a transversal and sides $\overline{AB}$ and $\overline{DC}$. They are alternate interior angles for a transversal and sides $\overline{AD}$ and $\overline{BC}$ which may or may not be parallel. **51a.** Exclude; the transversal was used to show the number of teams not making the playoffs. **51b.** Sample answers: to make money, to create excitement during the season

53. yes; Transitive Property **55.** 43 **57a.** yes **57b.** Law of Detachment **59.** right angles **61.** 12 **63.** 20

Pages 141–145 Lesson 3–3

7. -5; falling **9.** $\frac{3}{4}$ **11.** 0 **13.**

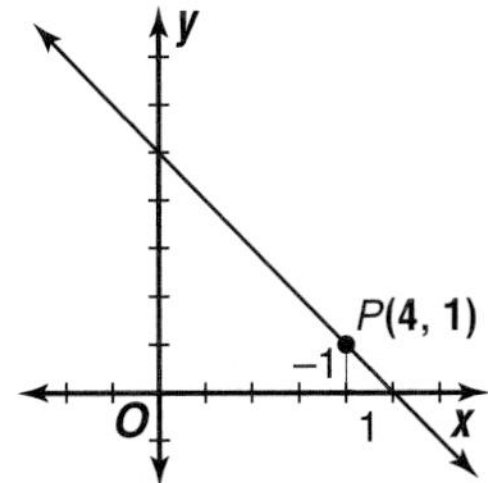

15. 8 **17.** $-\frac{3}{2}$; falling **19.** 0; horizontal **21.** $\frac{1}{2}$; rising **23.** $-\frac{5}{4}$ **25.** 1 **27.** undefined **29.** $\frac{4}{5}$ **31.** 0

33.

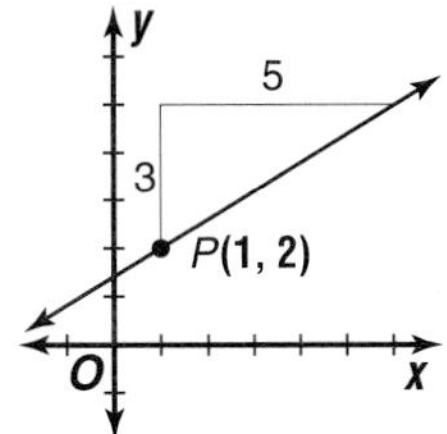

35.

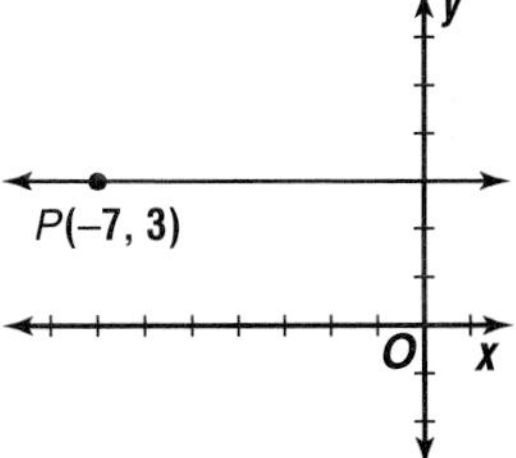

37.

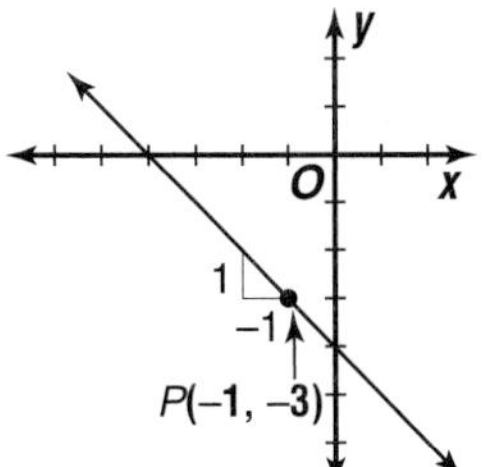

39. yes; slope of $\overleftrightarrow{MA} = -\frac{3}{5}$, slope of $\overleftrightarrow{TH} = -\frac{3}{5}$ **41.** yes; slope of $\overleftrightarrow{PQ} = \frac{1}{2}$, slope of $\overleftrightarrow{RS} = -2$, $\frac{1}{2}(-2) = -1$

43. 13

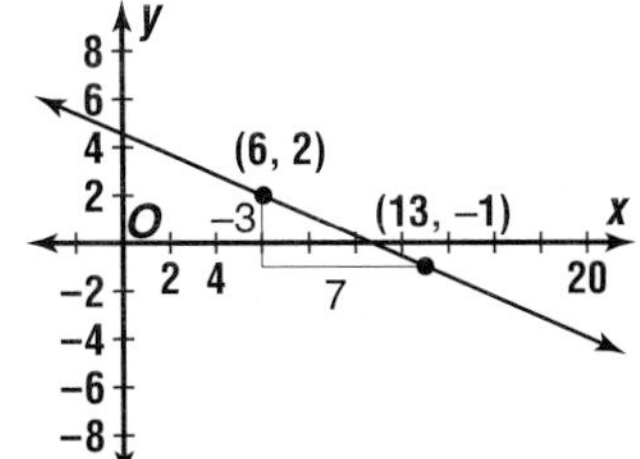

45a. slope of $\overline{PQ} = -1$, slope of $\overline{RS} = -1$, slope of $\overline{QR} = 1$, slope of $\overline{PS} = 1$ **45b.** $-1(1) = -1$ and $1(-1) = -1$

45c. $PQ = \sqrt{(5-1)^2 + (2-6)^2} = \sqrt{32}$

$QR = \sqrt{(1-(-3))^2 + (6-2)^2} = \sqrt{32}$

$RS = \sqrt{(-3-1)^2 + (2-(-2))^2} = \sqrt{32}$

$PS = \sqrt{(5-1)^2 + (2-(-2))^2} = \sqrt{32}$

45d. square **47a.** $-\frac{5}{6}$ **47b.** $-\frac{5}{2}$ **47c.** $\frac{5}{2}$ **47d.** $-\frac{3}{4}$

49. yes; $0.64 < 0.88$ **51a.** $\frac{1}{18}$ **51b.** 1.5 feet **53.** $x = 16$, $y = 11$

55. Given: $\angle 1 \cong \angle 2$

Prove: $\angle 1 \cong \angle 3$

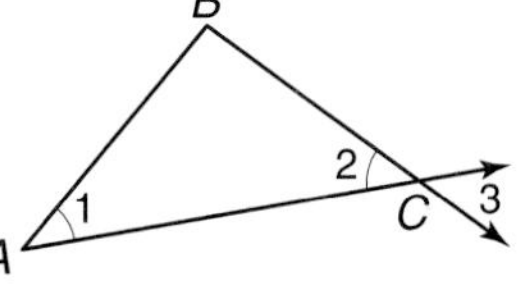

Proof:

Statements	Reasons
1. $\angle 1 \cong \angle 2$	1. Given
2. $\angle 2 \cong \angle 3$	2. Vertical ∡ are ≅.
3. $\angle 1 \cong \angle 3$	3. Congruence of angles is transitive.

57.

$A = 2\pi r^2 + 2\pi rh$ *Given*

$A - 2\pi r^2 = 2\pi rh$ *Subtraction Property (=)*

$\frac{A - 2\pi r^2}{2\pi r} = h$ *Division Property (=)*

59. Sample answer: $\left(\frac{1}{2}\right)^2 = \frac{1}{4}, \frac{1}{2} > \frac{1}{4}$ **61.** 18 area codes **63.** 22

Page 145 Self Test

1. $\overline{AE}, \overline{BD}$ **3.** plane ABC and plane EDF **5.** 40 **7.** $-\frac{1}{3}$ **9.** $-\frac{4}{3}$

Pages 149–153 Lesson 3–4

7. $a \parallel b$; If ⇄ cut by transversal, and a pair of alt. int. ∡ is ≅, then the lines are ∥. **9.** $\ell \parallel m$; If ⇄ cut by transversal, and corr. ∡ are ≅, then the lines are ∥. **11.** 7

13. **1.** Given
2. Vertical ∡ are ≅.
3. Congruence of ∡ is transitive.
4. If ⇄ cut by transversal, and corr. ∡ are ≅, then the lines are ∥.

15. $\overleftrightarrow{EC} \parallel \overleftrightarrow{HF}$; If ⇄ cut by transversal, and corr. ∡ are ≅, then the lines are ∥.
17. $\overline{JK} \parallel \overline{BL}$; If ⇄ cut by transversal, and a pair of consecutive int. ∡ is supplementary, then the lines are ∥. **19.** $p \parallel q$; If ⇄ cut by transversal, and a pair of alt. ext. ∡ is congruent, then the lines are ∥. **21.** $p \parallel q$; If ⇄ cut by transversal, and a pair of consecutive int. ∡ is supplementary, then the lines are ∥. **23.** $\ell \parallel m$; If ⇄ cut by transversal, and a pair of consecutive int. ∡ is supplementary, then the lines are ∥. **25.** 13 **27.** 9 **29.** $x = 10, y = 3$
31. $\ell \parallel m$; If ⇄ cut by transversal, and a pair of consecutive int. ∡ is supplementary, then the lines are ∥.

33. **1.** Given
2. Definition of linear pair
3. $\angle 2$ and $\angle 3$ are supplementary.
4. 2 ∡ supplementary to the same ∠ are ≅.
5. If ⇄ cut by transversal, and corr. ∡ are ≅, then the lines are ∥.

35. Given: $\angle 2 \cong \angle 1$
$\angle 1 \cong \angle 3$
Prove: $\overline{ST} \parallel \overline{YZ}$

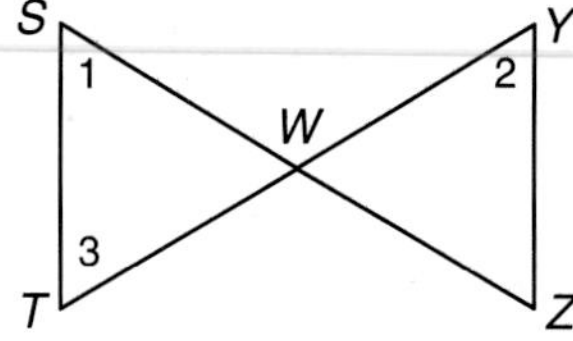

Proof:

Statements	Reasons
1. $\angle 2 \cong \angle 1$ $\angle 1 \cong \angle 3$	1. Given
2. $\angle 2 \cong \angle 3$	2. Congruence of ∠s is transitive.
3. $\overline{ST} \parallel \overline{YZ}$	3. If two lines are cut by a transversal, and a pair of alt. int. ∠s are ≅, then the lines are ∥.

37. Given: $\overline{AU} \perp \overline{QU}$
$\angle 1 \cong \angle 2$
Prove: $\overline{DQ} \perp \overline{QU}$

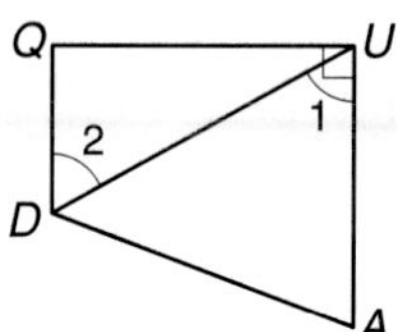

Proof:

Statements	Reasons
1. $\overline{AU} \perp \overline{QU}$ $\angle 1 \cong \angle 2$	1. Given
2. $\overline{AU} \parallel \overline{DQ}$	2. If two lines are cut by a transversal, and a pair of alt. int. ∠s is ≅, then the lines are ∥.
3. $\overline{DQ} \perp \overline{QU}$	3. Perpendicular Transversal Theorem

39a. Sample drawing:

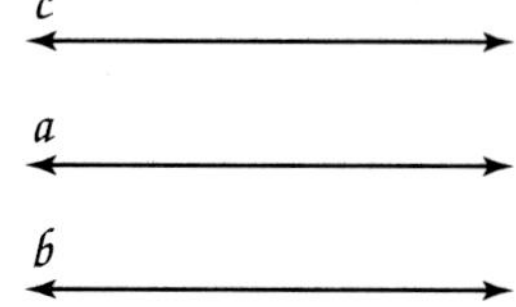

39b. Draw a transversal intersecting all three lines and number the angles formed.

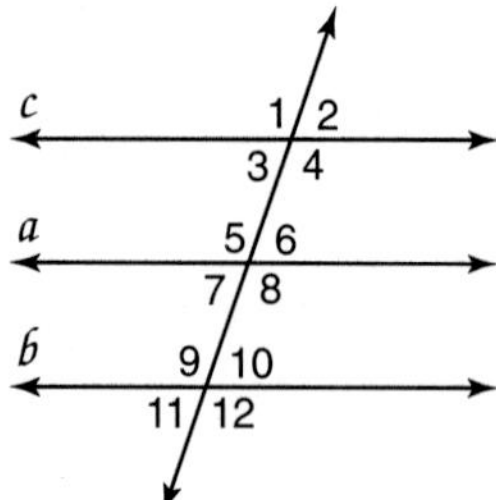

Sample plan for proof: Use corresponding angles to show $\angle 6 \cong \angle 10$ and $\angle 2 \cong \angle 6$. Then $\angle 2 \cong \angle 10$ which is enough to show that $b \parallel c$. **41.** yes **43.** $\frac{4}{5}$; rising
45. Intersecting; they all meet at the hub.
47. Lines p and m never meet; Law of Detachment.
49. (2, 9) **51.** $(2x - 3y)(8x + y)$

Pages 158–161 Lesson 3–5

7.

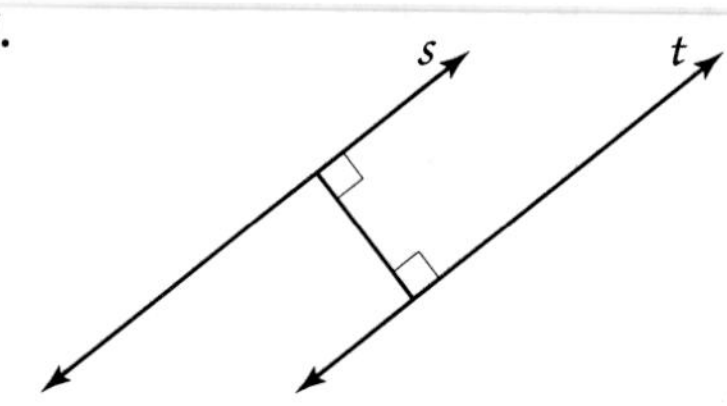

9. 1

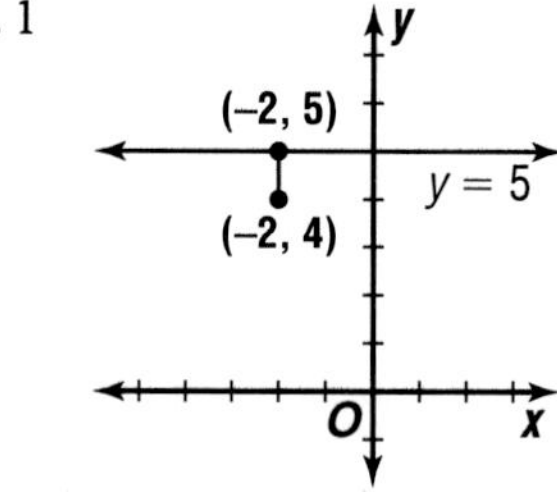

11. $\overline{PS}$ **13.** It is everywhere equidistant.

15.

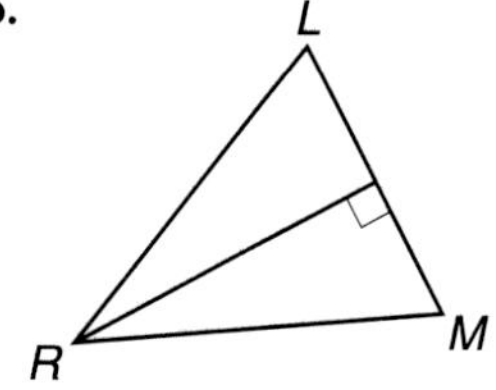

17.

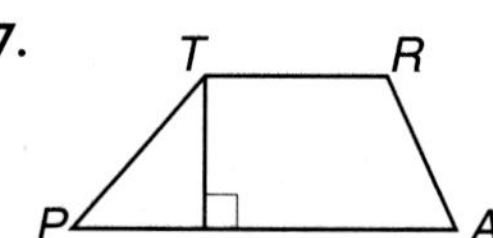

19.

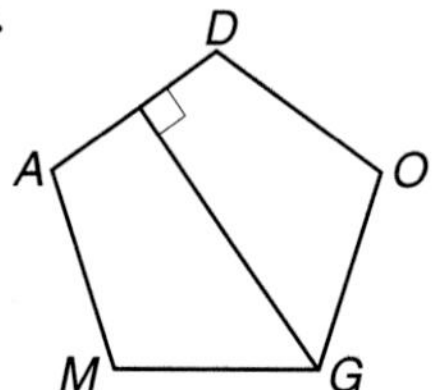

21. Yes; the lines are everywhere equidistant.

23. 5

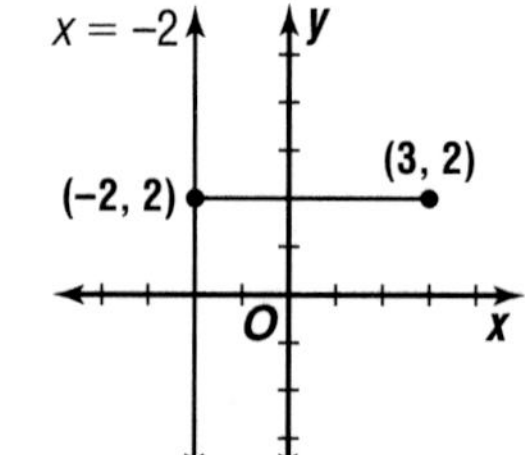

25. $\sqrt{5} \approx 2.24$

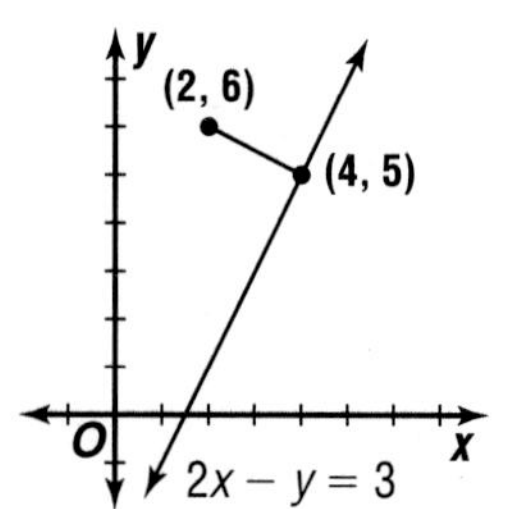

27. $\sqrt{13} \approx 3.61$

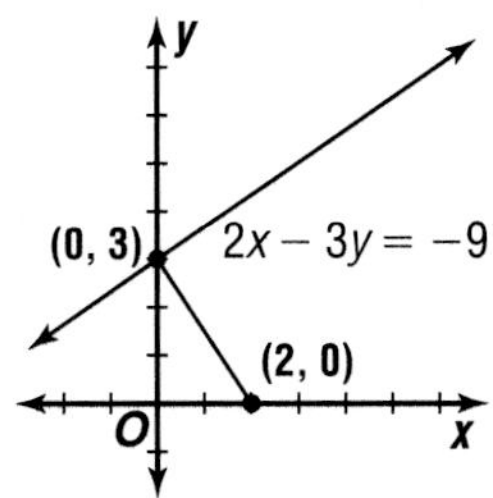

29. $\overline{MT}$ 31. $\overline{GM}$ 33. $\overline{HM}$ 35. $\sqrt{24.5} \approx 4.95$

37. $d = \frac{|3 \cdot 2 + 4 \cdot 5 - 1|}{\sqrt{3^2 + 4^2}} = 5$; Yes; both equal 5. 39a. yes

39b. no 41. $p \not\parallel q$; cons. int. ∠s are not supp. 43. $m\angle 2 = m\angle 4$; They are corresponding angles. 45. They are right angles.

47.

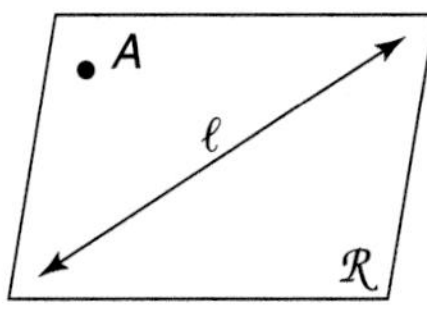

49. 8.1

Page 162 Lesson 3–5B

1. 1.41 3. 3.16

Pages 167–169 Lesson 3–6

5. true 7. 41° N, 29° E 9. The great circle is finite.
13. False; in spherical geometry, a line has finite length.
15. true 17. true 19. 30° N, 85° W 21. 19° N, 99° W
23. Reno, Nevada 25. There exists no parallel lines.
27. A pair of perpendicular great circles divides the sphere into 4 finite congruent regions. 29. There exists no parallel lines.
31. 1

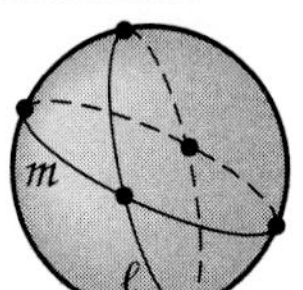

33.

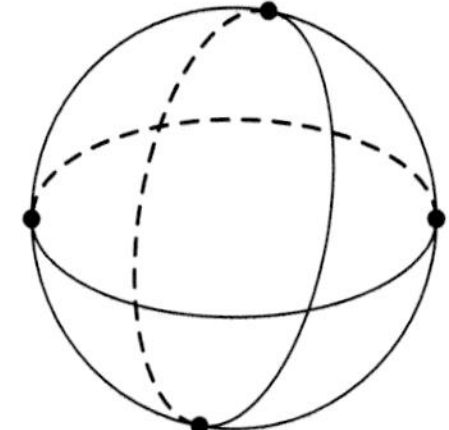

33a. They are each perpendicular to each other.
33b. They lie at the intersection of the other 2 circles.
35a. In a plane, if 2 lines are perpendicular to the same line, then they are parallel. 35b. Yes, 2 intersecting great circles can both be perpendicular to another great circle.

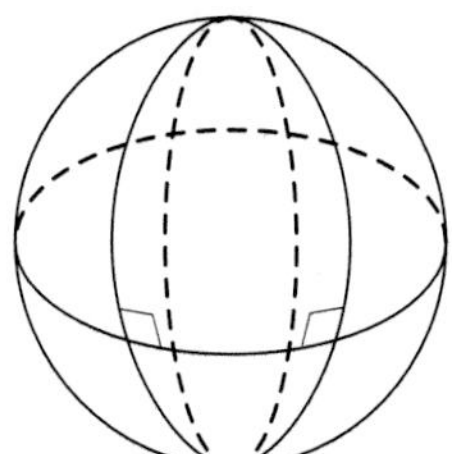

37. Sample answer: The spaceship will eventually return to its starting point. 39. The 2 lines must be perpendicular to the transversal.

41. 21

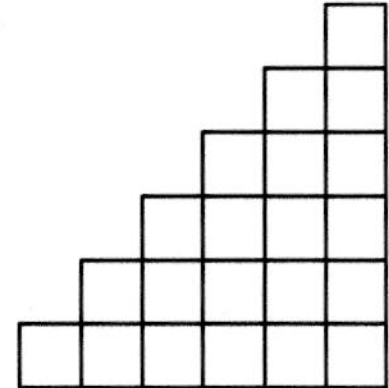

43. Sample answer: If the measure of an angle is 90, the angle is a right angle. Hypothesis: if the measure of an angle is 90; Conclusion: the angle is a right angle
45. Commutative (+)

Page 171 Chapter 3 Highlights

1. f 3. d 5. g 7. c 9. b 11. j

Pages 172–174 Chapter 3 Study Guide and Assessment

13. t and m 15. Sample answer: ℓ and n 17. $\angle 3 \cong \angle 5$, $\angle 3 \cong \angle 6$, $\angle 1 \cong \angle 4$, $\angle 5 \cong \angle 6$, $\angle GAB \cong \angle 10$, $\angle GAB \cong \angle ABH$, $\angle FAE \cong \angle 9$ 19. $\angle 3$, $\angle 8$ 21. 6; rising 23. $-\frac{1}{3}$; falling 25. $-\frac{9}{7}$; $\frac{7}{9}$ 27. $\frac{1}{6}$; -6 29. $\overline{AD} \parallel \overline{EF}$; $\angle 1 \cong \angle 4$ since vertical angles are congruent. So, $\angle 4$ and $\angle 2$ are supplementary. The lines are parallel because consecutive interior angles are supplementary.
31. $\overline{AD} \parallel \overline{EF}$; corr. ∠s are ≅. 33. $\overline{RQ}$ 35. $\overline{PQ}$ or $\overline{MR}$
37. The intersection of 2 great circles creates 8 angles.
39. The sortest path between two points is an arc on the great circle passing through the points.
41. 36 students

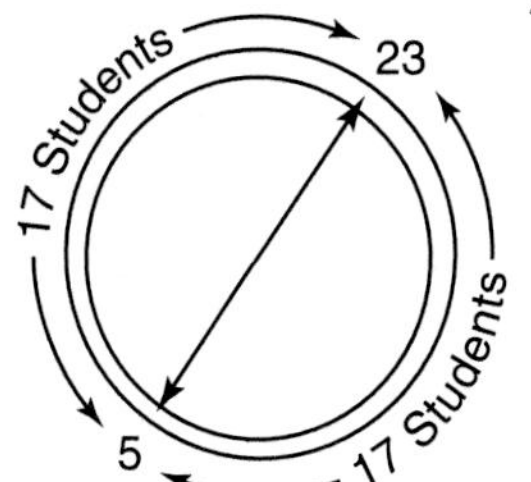

43. 65 mph

CHAPTER 4 IDENTIFYING CONGRUENT TRIANGLES

Pages 183–187 Lesson 4–1

7. $\overline{SM}$ 9. Sample answer: right, isosceles

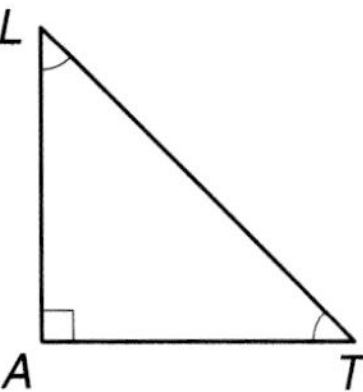

11. $x = 7$; $EQ = 25$, $QU = 25$, $EU = 25$ 13. sometimes
15a. right, scalene; obtuse, scalene

15b. Sample answer:

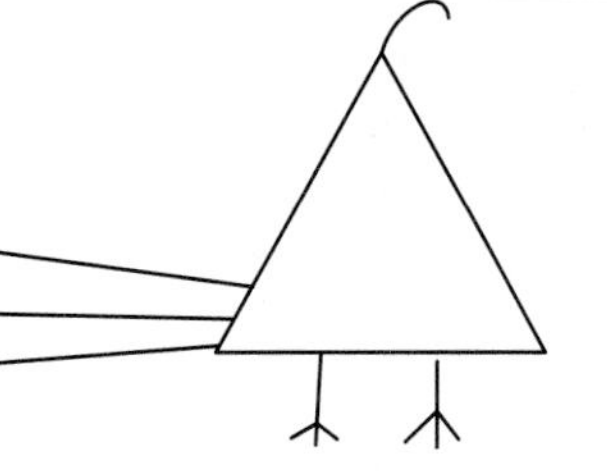

15c. Sample answer:

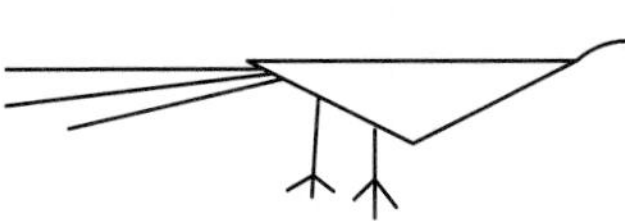

17. $\overline{LM}$ **19.** $\overline{LM}$ **21.** $\angle BLM$, $\angle BML$ **23.** $\overline{BL}$, $\overline{BM}$
25. Sample answer: right, isosceles

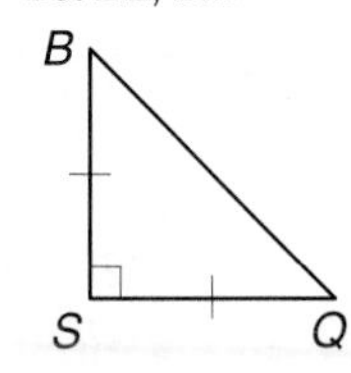

27. Sample answer: scalene

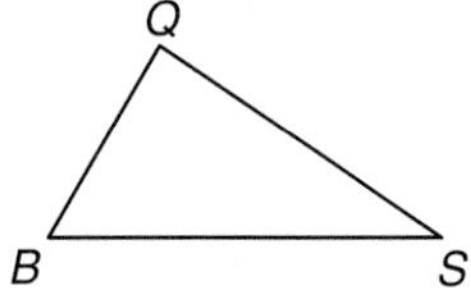

29. Sample answer: obtuse, isosceles

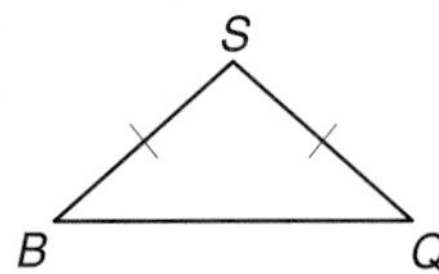

31. $x = 5$; $HK = 12$, $HT = 12$, $KT = 12$

33. $PQ = \sqrt{(3-0)^2 + (6-6)^2} = \sqrt{9+0} = \sqrt{9} = 3$
$QR = \sqrt{(3-3)^2 + (0-6)^2} = \sqrt{0+36} = \sqrt{36} = 6$
$PR = \sqrt{(3-0)^2 + (0-6)^2} = \sqrt{9+36} = \sqrt{45} =$
$3\sqrt{5}$; scalene

35. $KL = \sqrt{(-2-4)^2 + (0-0)^2} = \sqrt{36+0} = \sqrt{36} = 6$
$LM = \sqrt{(1-(-2))^2 + (5-0)^2} = \sqrt{9+25} = \sqrt{34}$
$KM = \sqrt{(1-4)^2 + (5-0)^2} = \sqrt{9+25} = \sqrt{34}$
isosceles

37. never **39.** sometimes **41.** always

43. **2.** Definition of perpendicular
3. Perpendicular Transversal Theorem
4. $\angle RST$ is a right angle.
5. $\triangle RST$ is a right triangle.

45. $\angle E$; Since $23 < 2x + 2 + 10 + x + 4 < 32$, $\frac{7}{3} < x < \frac{16}{3}$. If $2x + 2 = 10$, $x = 4$. If $2x + 2 = x + 4$, $x = 2$. If $x + 4 = 10$, $x = 6$. The only value of x that satisfies the conditions is $x = 4$, so $\overline{EF}$ and $\overline{ED}$ are the legs of the isosceles triangle and $\angle E$ is the vertex. **47.** If a triangle is an acute triangle, the square of the length of the longest side is less than the sum of the squares of the lengths of the other two sides. If a triangle is an obtuse triangle, the square of the length of the longest side is greater than the sum of the squares of the lengths of the other two sides. **49a.** $\triangle ABC$, $\triangle ADG$, $\triangle AHM$, $\triangle ANS$ **49b.** none **49c.** $\triangle BED$, $\triangle CFG$, $\triangle BJH$, $\triangle CKM$, $\triangle DIH$, $\triangle GLM$, $\triangle BPN$, $\triangle CQS$, $\triangle DON$, $\triangle GRS$ **51.** infinite number **53.** (1) If two lines in a plane are cut by a transversal so that corresponding angles are congruent, then the lines are parallel. (2) If two lines in a plane are cut by a transversal so that a pair of alternate interior angles is congruent, then the lines are parallel. (3) If two lines in a plane are cut by a transversal so that a pair of consecutive interior angles is supplementary, then the lines are parallel. (4) If two lines in a plane are cut by a transversal so that a pair of alternate exterior angles is congruent, then the lines are parallel. (5) In a plane, if two lines are perpendicular to the same line, then they are parallel. **55.** Sample answer: the line formed by the intersection of the floor and a wall of a room and the line formed by two walls on the opposite side of the room **57.** Transitive Property of Equality; Sample answer: The Law of Syllogism states if $p \to q$ and $q \to r$, then $p \to r$. The Transitive Property of Equality states if $p = q$ and $q = r$, then $p = r$. **59.** 45° **61.** −3

Page 188 Lesson 4–2A
3. The sum of the measures of the angles of a triangle is 180.

Pages 192–195 Lesson 4–2
7. 139 **9.** 54 **11a.** 45 **11b.** 90 **11c.** right triangle **13.** 30 **15.** 91 **17.** 115 **19.** 75 **21.** 68 **23.** 40 **25.** 26 **27.** 14 **29.** 51 **31.** scalene; none of the angles are congruent so none of the sides are congruent.

33.

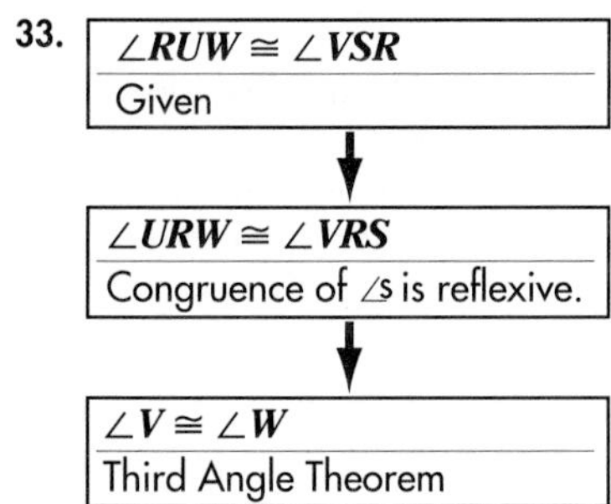

∠RUW ≅ ∠VSR
Given
∠URW ≅ ∠VRS
Congruence of ∠s is reflexive.
∠V ≅ ∠W
Third Angle Theorem

35. Given: $\triangle RED$ is equiangular.
Prove: $m\angle R = m\angle E = m\angle D = 60$

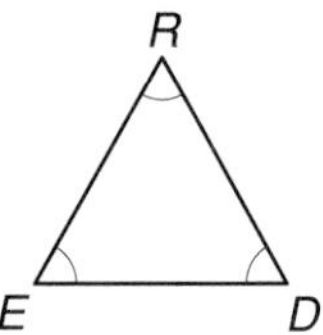

Proof:

Statements	Reasons
1. $\triangle RED$ is equiangular.	1. Given
2. $\angle R \cong \angle E \cong \angle D$	2. Definition of equiangular triangle
3. $m\angle R = m\angle E = m\angle D$	3. Definition of congruent angles
4. $m\angle R + m\angle E + m\angle D = 180$	4. Angle Sum Theorem
5. $m\angle R + m\angle R + m\angle R = 180$	5. Substitution Property (=)
6. $3m\angle R = 180$	6. Substitution Property (=)
7. $m\angle R = 60$	7. Division Property (=)
8. $m\angle R = m\angle E = m\angle D = 60$	8. Substitution Property (=)

37. Given: $\triangle RST$
$\angle R$ is a right angle.

Prove: $\angle S$ and $\angle T$ are complementary.

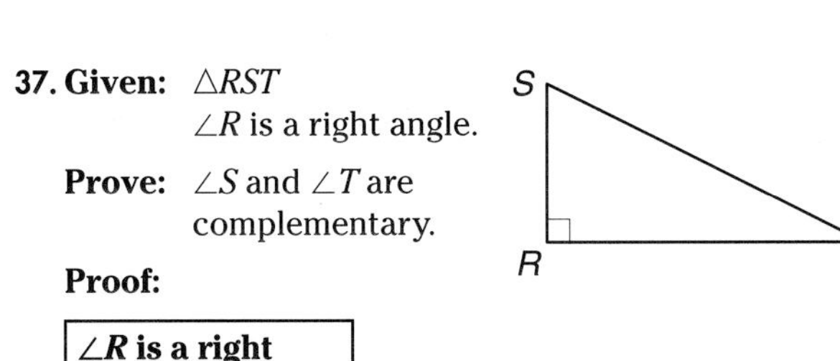

Proof:

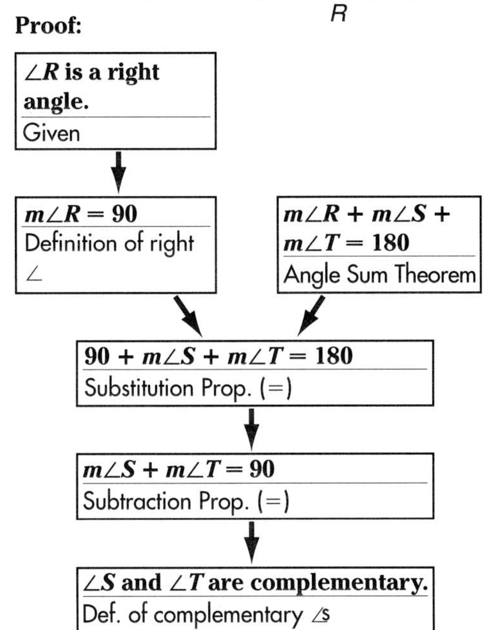

39. 360 **41a.** No; two of the angles are right angles, so the sum of their measures is 180. **41b.** The sum of the measures is greater than 180 and less than or equal to 360. The sum of the measures of two angles is 180. The measure of the third angle is greater than 0 and less than or equal to 180. **41c.** If the measure of the angle at the pole is less than 90, the triangle is an acute triangle. **41d.** If the measure of the angle at the pole is greater than 90, the triangle is an obtuse triangle. **41e.** If the measure of the angle at the pole is 90, the triangle is a right triangle. **43.** 130 **45.** No; If they are parallel, $m\angle 1 = m\angle 5$.

47. 28 games

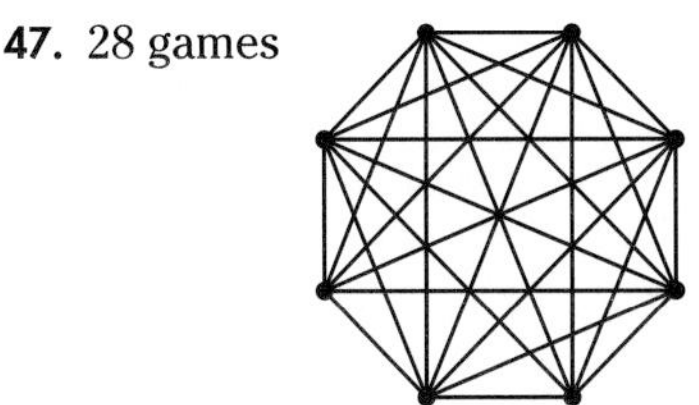

49a. Yes; Derrick will receive an A. **49b.** Law of Detachment **51.** III **53a.** 260 women **53b.** 395 women

Pages 199–203 Lesson 4–3

7. $\angle A \longleftrightarrow \angle X$, $\angle B \longleftrightarrow \angle Y$, $\angle C \longleftrightarrow \angle Z$; $\overline{AB} \longleftrightarrow \overline{XY}$, $\overline{AC} \longleftrightarrow \overline{XZ}$, $\overline{BC} \longleftrightarrow \overline{YZ}$

A B C X Y Z

9. 1. Given
2. Congruence of segments is reflexive.
3. Congruence of angles is reflexive.
4. Definition of congruent triangles

11a.

C, A, T: 14, 21, 18; D, O, G: 2x + 7

11b. 7

13. Given: $\triangle MXR$ is a right isosceles triangle with $\angle X$ the vertex angle.
$\overline{XY} \perp \overline{MR}$
Y is the midpoint of $\overline{MR}$.
$\angle M \cong \angle R$
$\overline{YX}$ bisects $\angle MXR$

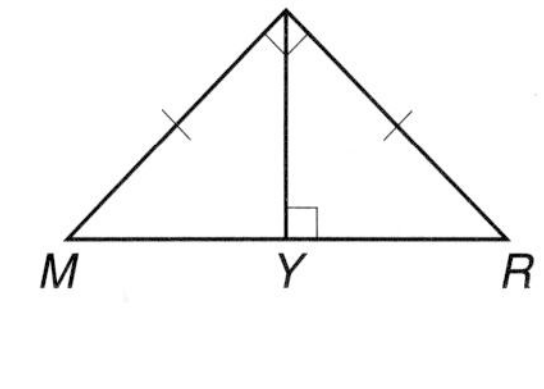

Prove: $\triangle MXY \cong \triangle RXY$

Proof:

Statements	Reasons
1. $\triangle MXR$ is isosceles with vertex $\angle X$.	1. Given
2. $\overline{XM} \cong \overline{XR}$	2. Definition of isosceles $\triangle$
3. $\overline{XY} \perp \overline{MR}$	3. Given
4. $\angle XYR$ and $\angle XYM$ are right ∠s.	4. Definition of $\perp$ lines
5. $\angle XYR \cong \angle XYM$	5. All right ∠s are $\cong$.
6. Y is the midpoint of $\overline{MR}$.	6. Given
7. $\overline{MY} \cong \overline{RY}$	7. Definition of midpoint
8. $\angle M \cong \angle R$	8. Given
9. $\overline{YX}$ bisects $\angle MXR$.	9. Given
10. $\angle MXY \cong \angle RXY$	10. Definition of angle bisector
11. $\overline{XY} \cong \overline{XY}$	11. Congruence of segments is reflexive.
12. $\triangle MXY \cong \triangle RXY$	12. Definition of $\cong$ triangles

15. $\triangle HGA$, $\triangle KLA$, $\triangle JBA$, $\triangle IEA$, $\triangle HEA$, $\triangle KGA$, $\triangle JLA$, $\triangle IBA$; $\triangle HDE$, $\triangle KFG$, $\triangle JML$, $\triangle ICB$; $\triangle EDA$, $\triangle GFA$, $\triangle LMA$, $\triangle BCA$ **17.** $\overline{PQ} \longleftrightarrow \overline{RS}$, $\overline{PR} \longleftrightarrow \overline{RT}$, $\overline{QR} \longleftrightarrow \overline{ST}$; $\angle P \longleftrightarrow \angle R$, $\angle Q \longleftrightarrow \angle S$, $\angle R \longleftrightarrow \angle T$

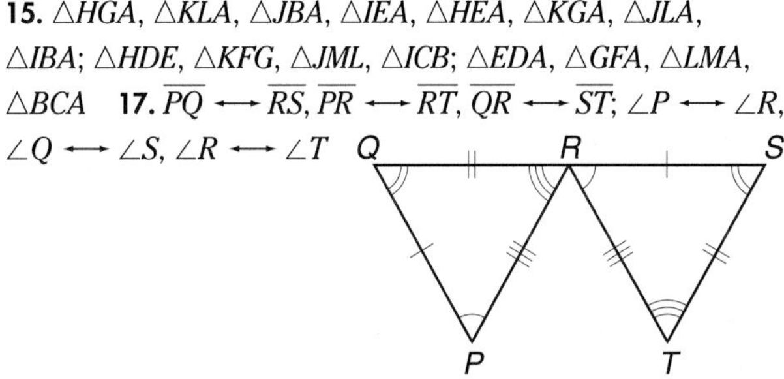

19. WXY **21.** ERG **23a.** Given **23b.** Given **23c.** Congruence of segments is reflexive. **23d.** Given **23e.** Definition of $\perp$ lines **23f.** Given **23g.** Definition of $\perp$ lines **23h.** All right ∠s are $\cong$. **23i.** Given **23j.** Alternate Interior Angle Theorem **23k.** Given **23l.** Alternate Interior Angle Theorem **23m.** Definition of congruent triangles

25a.

25b. $\frac{28}{3}$ **27.** Sample answer:

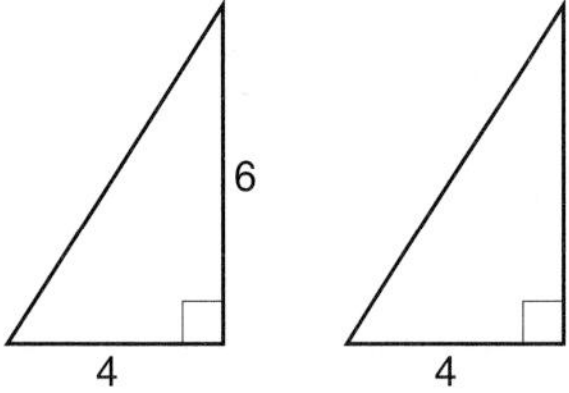

29. Sample answer:

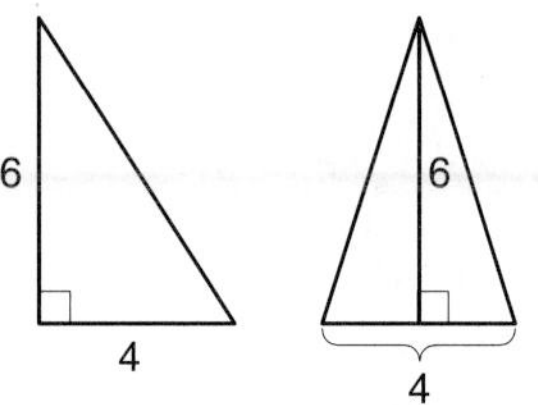

31. true; Since $AD = AB + BD$ and $AD = AC + CD$, $AB + BD = AC + CD$. By CPCTC $\overline{BD} \cong \overline{AC}$, so $BD = AC$. By Subtraction Property (=), $AB = CD$ and $\overline{AB} \cong \overline{CD}$ or $\overline{DC} \cong \overline{AB}$. **33.** true; CPCTC **35.** true; Since $m\angle AND = m\angle ANB + m\angle BND$ and $m\angle AND = m\angle ANC + m\angle CND$, $m\angle ANB + m\angle BND = m\angle ANC + m\angle CND$. By CPCTC, $\angle BND \cong \angle ANC$, so $m\angle BND = m\angle ANC$. By Subtraction Property (=), $m\angle ANB = m\angle CND$ and $\angle ANB \cong \angle CND$.

37. Given: $\triangle SIO \cong \triangle OIS$

Prove: $\triangle SIO$ is an isosceles triangle.

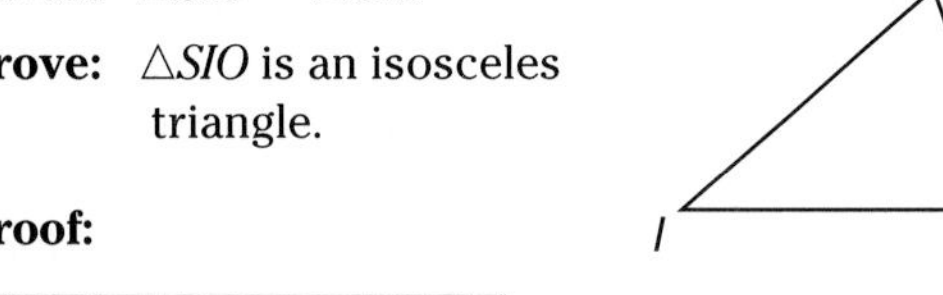

Proof:

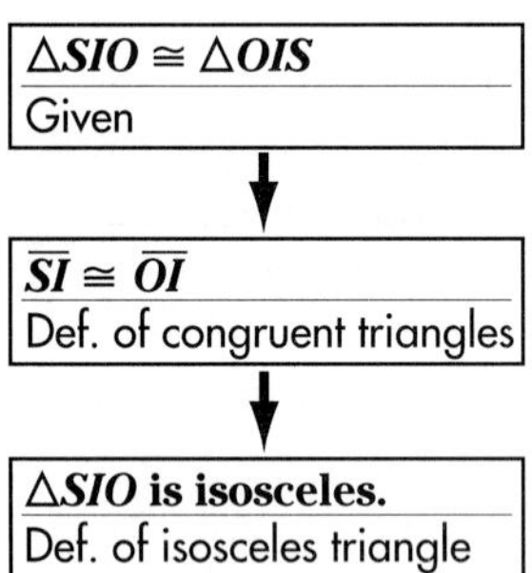

39a. 4 **39b.** 0 **41.** $\triangle RQU \cong \triangle RSU$, $\triangle QUP \cong \triangle SUT$, $\triangle RUP \cong \triangle RUT$ **43.** 33 **45.** $\angle 1 \cong \angle 7 \cong \angle 9$, $\angle 10 \cong \angle 8$, $\angle 3 \cong \angle 5 \cong \angle 6$, $\angle 2 \cong \angle 4$; $\angle 1$ and $\angle 10$, $\angle 7$ and $\angle 10$, $\angle 7$ and $\angle 8$, $\angle 8$ and $\angle 9$, $\angle 9$ and $\angle 10$, $\angle 1$ and $\angle 8$, $\angle 2$ and $\angle 6$, $\angle 2$ and $\angle 5$, $\angle 5$ and $\angle 4$, $\angle 4$ and $\angle 3$, $\angle 3$ and $\angle 2$, $\angle 6$ and $\angle 4$ are supplementary. **47.** Substitution Property (=) **49.** a straight angle **51.** $\{x \mid -3 < x \leq 2\}$

Page 203 Self Test

1. $\triangle DBC$, $\triangle ABD$ **3.** $\overline{AD}$, $\overline{AC}$ **5.** 107 **7.** 125 **9.** $\triangle YZX$

Page 205 Lesson 4–4A

1. It is congruent to $\triangle ABC$. **3.** It is congruent to $\triangle ABC$. **5.** If the sides of one triangle are congruent to the sides of another triangle, the triangles are congruent. **7.** If two angles and the included side of one triangle are congruent to two angles and the included side of another triangle, the triangles are congruent.

Pages 209–213 Lesson 4–4

7. SSS

9.
1. Given
2. Definition of segment bisector
3. Given
4. Alternate Interior Angle Theorem
5. Vertical angles are ≅.
6. ASA

11. Given: $\overline{MO} \cong \overline{PO}$
$\overline{NO}$ bisects $\overline{MP}$.

Prove: $\triangle MNO \cong \triangle PNO$

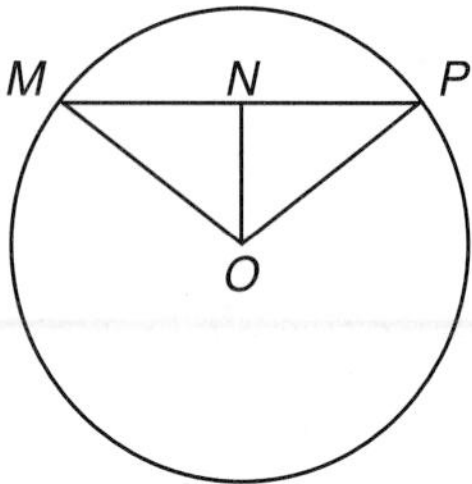

Proof:

Statements	Reasons
1. $\overline{MO} \cong \overline{PO}$; $\overline{NO}$ bisects $\overline{MP}$	1. Given
2. $\overline{MN} \cong \overline{PN}$	2. Definition of segment bisector
3. $\overline{NO} \cong \overline{NO}$	3. Congruence of segments is reflexive.
4. $\triangle MNO \cong \triangle PNO$	4. SSS

13a. the two middle frameworks **15.** SAS **17.** SAS **19.** SSS

21. $PQ = \sqrt{(0-(-1))^2 + (6-(-1))^2} = \sqrt{50} = 5\sqrt{2}$

$QR = \sqrt{(2-0)^2 + (3-6)^2} = \sqrt{13}$

$PR = \sqrt{(2-(-1))^2 + (3-(-1))^2} = \sqrt{25} = 5$

$XY = \sqrt{(5-3)^2 + (3-1)^2} = \sqrt{8} = 2\sqrt{2}$

$YZ = \sqrt{(8-5)^2 + (1-3)^2} = \sqrt{13}$

$XZ = \sqrt{(8-3)^2 + (1-1)^2} = \sqrt{25} = 5$

No; $\overline{QR} \cong \overline{YZ}$ and $\overline{PR} \cong \overline{XZ}$, but $\overline{PQ}$ is not congruent to $\overline{XY}$.

23. $\angle SRU \cong \angle TRU$

25. Given: $\angle J \cong \angle L$
B is the midpoint of $\overline{JL}$.

Prove: $\triangle JHB \cong \triangle LCB$

Proof:

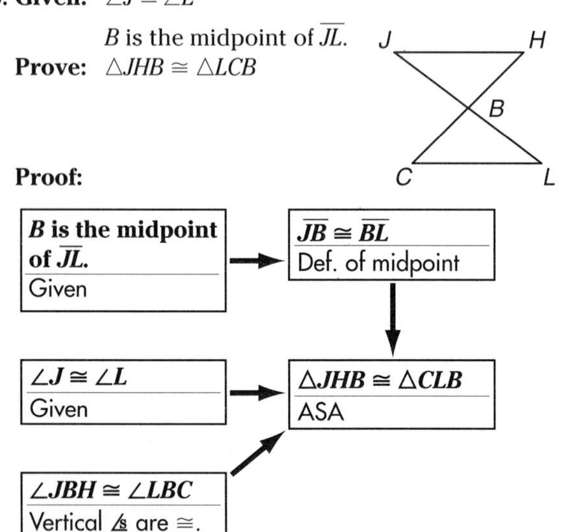

27. Given: $\triangle MGR$ is an isosceles triangle with vertex $\angle MGR$.
K is the midpoint of $\overline{MR}$.

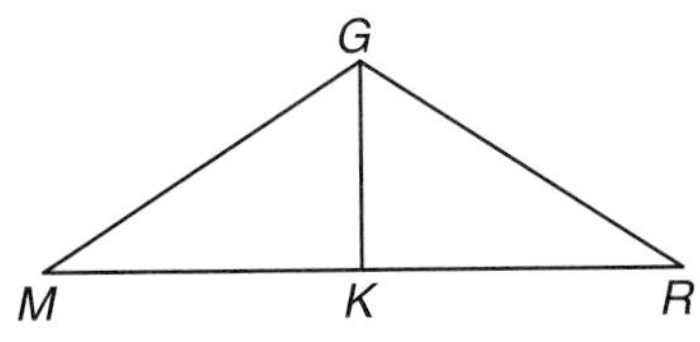

Prove: $\triangle MGK \cong \triangle RGK$

Proof:

Statements	Reasons
1. $\triangle MGR$ is an isosceles triangle with vertex $\angle MGR$.	1. Given
2. $\overline{GM} \cong \overline{GR}$	2. Definition of isosceles triangle
3. K is the midpoint of $\overline{MR}$.	3. Given
4. $\overline{MK} \cong \overline{RK}$	4. Definition of midpoint
5. $\overline{GK} \cong \overline{GK}$	5. Congruence of segments is reflexive.
6. $\triangle MGK \cong \triangle RGK$	6. SSS

29. Given: $\overline{RL} \parallel \overline{DC}$
$\overline{LC} \parallel \overline{RD}$

Prove: $\angle R \cong \angle C$

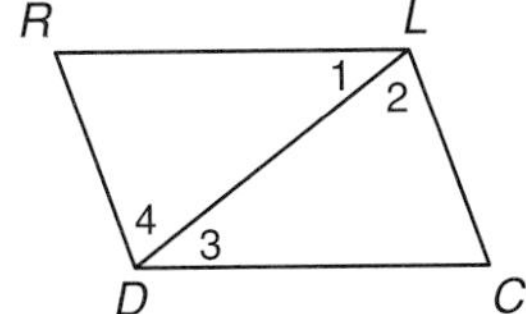

Proof:

Statements	Reasons
1. $RL \parallel \overline{DC}$	1. Given
2. $\angle 1 \cong \angle 3$	2. Alternate Interior Angle Theorem
3. $\overline{LC} \parallel \overline{RD}$	3. Given
4. $\angle 2 \cong \angle 4$	4. Alternate Interior Angle Theorem
5. $\overline{LD} \cong \overline{DL}$	5. Congruence of segments is reflexive.
6. $\triangle RLD \cong \triangle CDL$	6. ASA
7. $\angle R \cong \angle C$	7. CPCTC

31. Given: $\angle 1 \cong \angle 2$
$\angle 3 \cong \angle 4$
$\overline{LA} \cong \overline{RU}$

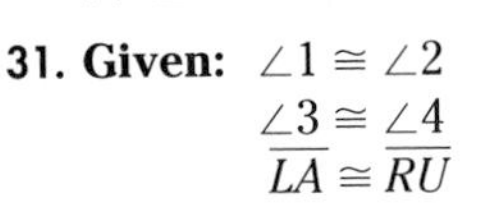

Prove: $\triangle WLU \cong \triangle WRA$

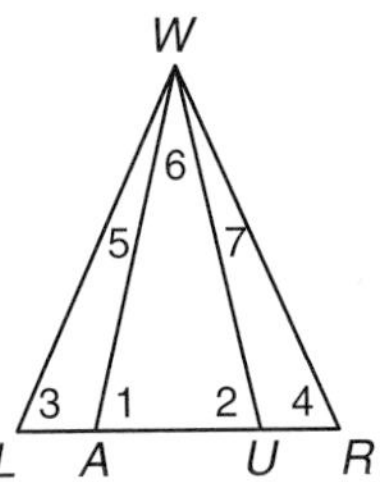

Proof:

Statements	Reasons
1. $\angle 1 \cong \angle 2$, $\angle 3 \cong \angle 4$, $\overline{LA} \cong \overline{RU}$	1. Given
2. $LA = RU$	2. Definition of congruent segments
3. $LU = LA + AU$, $RA = RU + AU$	3. Segment Addition Postulate
4. $RA = LA + AU$	4. Substitution Property (=)
5. $LU = RA$	5. Substitution Property (=)
6. $\overline{LU} \cong \overline{RA}$	6. Definition of congruent segments
7. $\triangle WLU \cong \triangle WRA$	7. ASA

33. Given: $\angle 5 \cong \angle 7$
$\angle 3 \cong \angle 4$
$\overline{LW} \cong \overline{RW}$

Prove: $\overline{LU} \cong \overline{RA}$

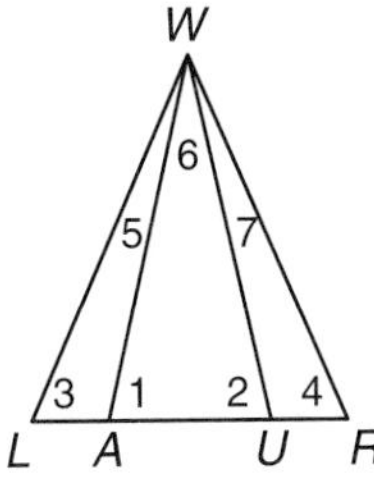

Proof:

Statements	Reasons
1. $\angle 5 \cong \angle 7$	1. Given
2. $m\angle 5 = m\angle 7$	2. Definition of congruent angles
3. $m\angle LWU = m\angle 5 + m\angle 6$, $m\angle RWA = m\angle 7 + m\angle 6$	3. Angle addition
4. $m\angle RWA = m\angle 5 + m\angle 6$	4. Substitution Property (=)
5. $m\angle LWU = m\angle RWA$	5. Substitution Property (=)
6. $\angle LWU \cong \angle RWA$	6. Definition of congruent angles
7. $\angle 3 \cong \angle 4$, $\overline{LW} \cong \overline{RW}$	7. Given
8. $\triangle LWU \cong \triangle RWA$	8. ASA
9. $\overline{LU} \cong \overline{RA}$	9. CPCTC

35. $\angle A$ is not the included angle. **37.** Since Jamal is perpendicular to the ground, two right triangles are formed and the two right angles are congruent. The angles of sight are the same and his height is the same, so the triangles are congruent by ASA. By CPCTC, the distances are the same and the method is valid. **39.** 30 **41.** $\frac{1}{10}$ **43.** Sample answer: The plants need more sunlight to live. Place them somewhere with more sun and see if they survive. **45.** 1.5 hours

Pages 217–221 Lesson 4–5

7. Sample answer: $\angle 1$ and $\angle 5$ **9.** Yes; Since $\angle 7 \cong \angle 11$, $\overline{AD} \parallel \overline{EV}$. Then $\angle 6 \cong \angle 10$, by the Alt. Interior Angle Theorem. Then you have two sides and the included angle congruent.

11. 1. Given
2. Vertical ∡ are ≅.
3. AAS
4. CPCTC

13. Given: $\overline{NM}$ bisects $\overline{RD}$.
$\angle 7 \cong \angle 8$

Prove: $\overline{MD} \cong \overline{NR}$

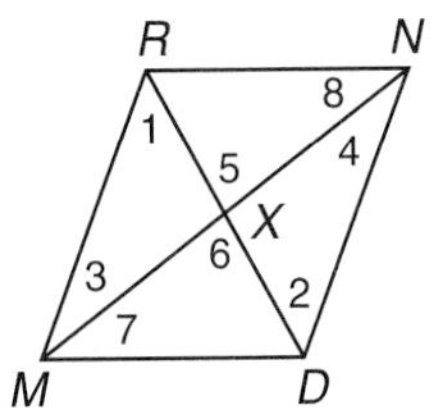

Proof:

Statements	Reasons
1. $\overline{NM}$ bisects $\overline{RD}$	1. Given
2. $\overline{RX} \cong \overline{XD}$	2. Definition of bisector
3. $\angle 7 \cong \angle 8$	3. Given
4. $\angle 6 \cong \angle 5$	4. Vertical ∡ are ≅.
5. $\triangle DXM \cong \triangle RXN$	5. AAS
6. $\overline{MD} \cong \overline{NR}$	6. CPCTC

15. $\overline{FD}$ **17.** $\overline{DR}$ or $\overline{RS}$ **19.** $\angle 10$ and $\angle 11$ **21.** $\angle 1$ and $\angle 4$ or $\angle 2$ and $\angle 4$ **23.** *SRD*; AAS **25.** $\overline{FD} \cong \overline{CD}$ and $\overline{DR} \cong \overline{DT}$ **27a.** Given **27b.** Given **27c.** Given **27d.** $\perp$ lines form 4 rt. $\angle$s. **27e.** $\perp$ lines form 4 rt. $\angle$s. **27f.** All rt. $\angle$s are $\cong$. **27g.** Given **27h.** AAS **27i.** CPCTC

29. Given: $\angle A \cong \angle D$
$\angle EBC \cong \angle ECB$
$\overline{AE} \cong \overline{DE}$

Prove: $\triangle ABE \cong \triangle DCE$

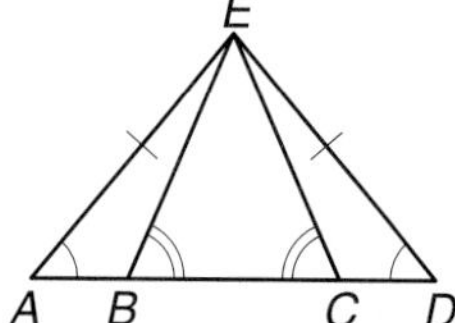

Proof:

Since $\angle EBC \cong \angle ECB$, by definition of congruent angles $m\angle EBC = m\angle ECB$. Since $\angle ABE$ and $\angle EBC$ are linear pairs, the angles are supplementary and $m\angle ABE + m\angle EBC = 180$. Likewise, $m\angle DCE + m\angle ECB = 180$. By substitution, $m\angle ABE + m\angle EBC = m\angle DCE + m\angle ECB$ and $m\angle ABE + m\angle ECB = m\angle DCE + m\angle ECB$. Using the Subtraction Property of Equality, $m\angle ABE = m\angle DCE$. By the definition of congruent angles, $\angle ABE \cong \angle DCE$. Since we are given that $\angle A \cong \angle D$ and $\overline{AE} \cong \overline{DE}$, $\triangle ABE \cong \triangle DCE$ by AAS.

31. Given: $\angle 3 \cong \angle 2$
$\angle T \cong \angle N$
$\overline{TC} \cong \overline{NA}$

Prove: $\triangle TEA \cong \triangle NEC$

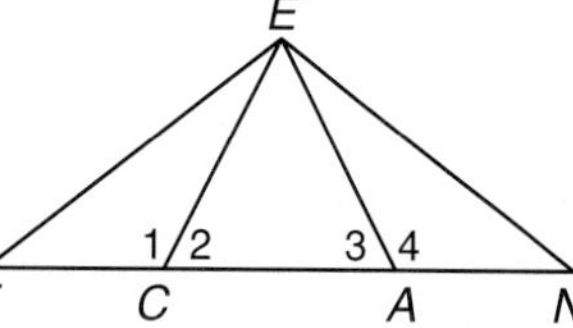

Proof:

Statements	Reasons
1. $\angle 3 \cong \angle 2$ $\angle T \cong \angle N$ $\overline{TC} \cong \overline{NA}$	1. Given
2. $TC = NA$	2. Definition of $\cong$ segments
3. $TA = TC + CA$ $NC = NA + AC$	3. Segment Addition Postulate
4. $TA = NA + AC$	4. Substitution Property (=)
5. $TA = NC$	5. Substitution Property (=)
6. $\overline{TA} \cong \overline{NC}$	6. Definition of $\cong$ segments
7. $\triangle TEA \cong \triangle NEC$	7. ASA

33. Given: $\overline{FP} \parallel \overline{ML}$
$\overline{FL} \parallel \overline{MP}$

Prove: $\overline{PM} \cong \overline{LF}$

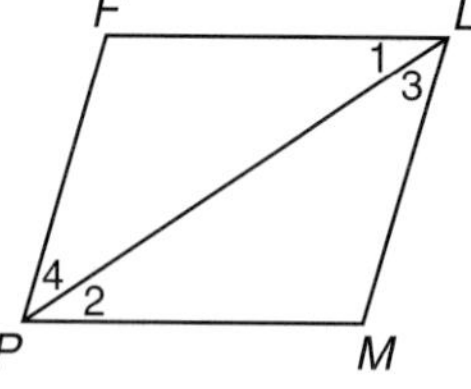

Proof:

Statements	Reasons
1. $\overline{FP} \parallel \overline{ML}$	1. Given
2. $\angle 3 \cong \angle 4$	2. Alternate Interior Angle Theorem
3. $\overline{FL} \parallel \overline{MP}$	3. Given
4. $\angle 1 \cong \angle 2$	4. Alternate Interior Angle Theorem
5. $\overline{PL} \cong \overline{LP}$	5. Congruence of segments is reflexive.
6. $\triangle FLP \cong \triangle MPL$	6. ASA
7. $\overline{PM} \cong \overline{LF}$	7. CPCTC

35. Given: $\triangle GVR$ is an isosceles triangle with base $\overline{GR}$.
$\triangle TVS$ is an isosceles triangle with base $\overline{TS}$.
$\overline{GV} \cong \overline{TV}$
$\angle 5 \cong \angle 6$

Prove: $\overline{GR} \cong \overline{TS}$

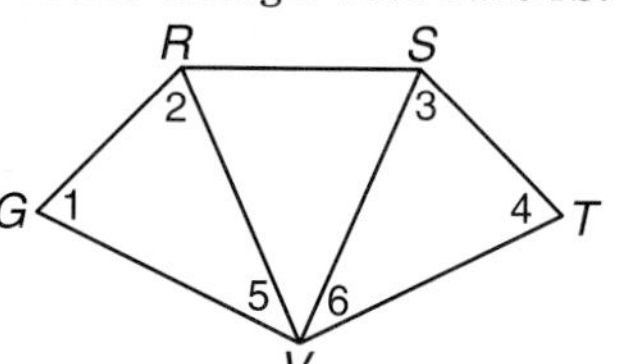

Proof:

Statements	Reasons
1. $\triangle GVR$ is an isosceles triangle with base $\overline{GR}$.	1. Given
2. $\overline{RV} \cong \overline{GV}$	2. Definition of isoceles triangle
3. $\triangle TVS$ is an isosceles triangle with base $\overline{TS}$.	3. Given
4. $\overline{TV} \cong \overline{SV}$	4. Definition of isoceles triangle
5. $\overline{GV} \cong \overline{TV}$	5. Given
6. $\overline{RV} \cong \overline{TV}$	6. Congruence of segments is transitive.
7. $\overline{RV} \cong \overline{SV}$	7. Congruence of segments is transitive.
8. $\angle 5 \cong \angle 6$	8. Given
9. $\triangle GRV \cong \triangle TSV$	9. SAS
10. $\overline{GR} \cong \overline{TS}$	10. CPCTC

37. Given: $\overline{PX} \cong \overline{LT}$
$\triangle PRL$ is an isosceles triangle with base $\overline{PL}$.

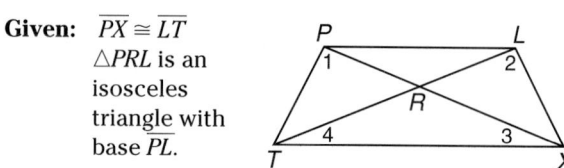

Prove: $\triangle TRX$ is an isosceles triangle.

Proof:

Statements	Reasons
1. $\overline{PX} \cong \overline{LT}$	1. Given
2. $PX = LT$	2. Definition of $\cong$ segments
3. $PX = PR + RX$ $LT = LR + RT$	3. Segment Addition Postulate
4. $PR + RX = LR + RT$	4. Substitution Property (=)
5. $\triangle PRL$ is an isosceles triangle with base $\overline{PL}$	5. Given
6. $\overline{PR} \cong \overline{LR}$	6. Definition of isosceles triangle
7. $PR = LR$	7. Definition of $\cong$ segments
8. $PR + RX = PR + RT$	8. Substitution Property (=)
9. $RX = RT$	9. Subtraction Property (=)
10. $\overline{RX} \cong \overline{RT}$	10. Definition of $\cong$ segments
11. $\triangle TRX$ is an isosceles triangle.	11. Definition of isosceles triangle

39. No; two equiangular triangles are an example of AAA, but the sides are not necessarily congruent. **41.** The guy wires are all the same length because the triangles are congruent by AAS. **43.** Neither player has a greater angle. The triangles formed are congruent by SSS and therefore the angles are congruent. **45.** no **47.** $-\frac{1}{9}$ **49.** a straight angle **51.** inverse; 18

Pages 224–228 Lesson 4–6

5. $\angle 5 \cong \angle 11$ **7.** $\overline{MT} \cong \overline{MR}$ **9.** 56

11a. $AB = \sqrt{(5-2)^2 + (2-5)^2} = \sqrt{18} = 3\sqrt{2}$

$BC = \sqrt{(2-5)^2 + (-1-2)^2} = \sqrt{18} = 3\sqrt{2}$

$AC = \sqrt{(2-2)^2 + (-1-5)^2} = \sqrt{36} = 6$

$\overline{AB} \cong \overline{BC}$

11b. $\angle A \cong \angle C$

13. Given: $\triangle ABC$
$\angle A \cong \angle C$

Prove: $\overline{AB} \cong \overline{CB}$

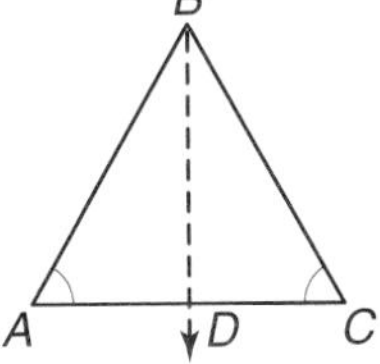

Proof:

Statements	Reasons
1. Let $\overrightarrow{BD}$ bisect $\angle ABC$.	1. Protractor Postulate
2. $\angle ABD \cong \angle CBD$	2. Definition of angle bisector
3. $\angle A \cong \angle C$	3. Given
4. $\overline{BD} \cong \overline{BD}$	4. Congruence of segments is reflexive.
5. $\triangle ABD \cong \triangle CBD$	5. AAS
6. $\overline{AB} \cong \overline{CB}$	6. CPCTC

15. $\angle 10 \cong \angle 6$ **17.** $\angle 11 \cong \angle 9$ **19.** $\angle OEF \cong \angle OFE$
21. $\overline{EF} \cong \overline{EO}$ **23.** $\overline{FB} \cong \overline{FA}$

25. Sample answer:

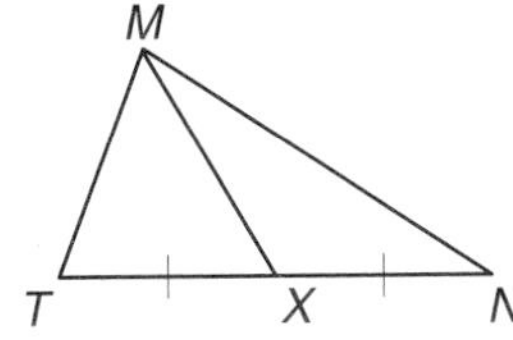

27. 60 **29.** 45 **31.** 18

33. $DE = \sqrt{(-2-(-4))^2 + (-1-(-3))^2} = \sqrt{8} = 2\sqrt{2}$

$EF = \sqrt{(0-(-2))^2 + (-3-(-1))^2} = \sqrt{8} = 2\sqrt{2}$

$DF = \sqrt{(0-(-4))^2 + (-3-(-3))^2} = \sqrt{16} = 4$

$\overline{DE} \cong \overline{EF}$, so the triangle is isosceles.

slope of $\overline{DE} = \frac{-1-(-3)}{-2-(-4)} = \frac{2}{2} = 1$

slope of $\overline{EF} = \frac{-3-(-1)}{0-(-2)} = \frac{-2}{2} = -1$

Since $1(-1) = -1$, $\overline{DE} \perp \overline{EF}$ and $\triangle DEF$ is right isosceles.

35. $m\angle 2 = 120$, $m\angle 3 = 30$, $m\angle 4 = 60$, $m\angle 5 = 60$, $m\angle 6 = 60$, $m\angle 7 = 120$, $m\angle 8 = 90$

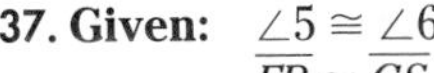

37. Given: $\angle 5 \cong \angle 6$
$\overline{FR} \cong \overline{GS}$

Prove: $\angle 4 \cong \angle 3$

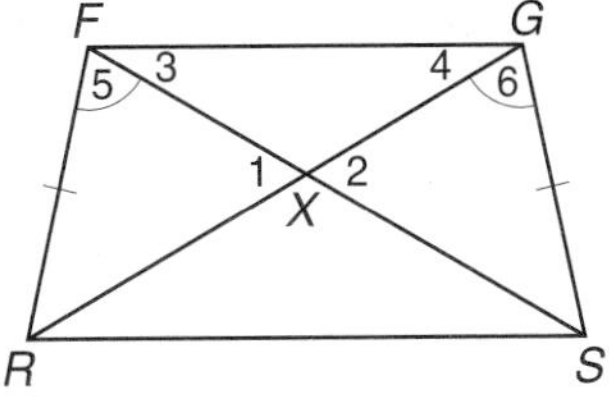

Proof:

Statements	Reasons
1. $\angle 5 \cong \angle 6$ $\overline{FR} \cong \overline{GS}$	1. Given
2. $\angle 1 \cong \angle 2$	2. Vertical ∡ are ≅.
3. $\triangle FXR \cong \triangle GXS$	3. AAS
4. $\overline{FX} \cong \overline{GX}$	4. CPCTC
5. $\angle 4 \cong \angle 3$	5. Isosceles Triangle Theorem

39. Given: $\triangle CAN$ is an isosceles triangle with vertex $\angle N$.
$\overline{CA} \parallel \overline{BE}$

Prove: $\triangle NEB$ is an isosceles triangle.

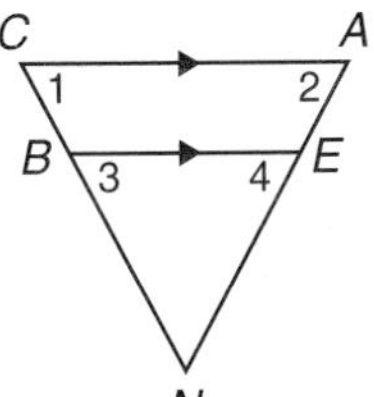

Proof:

Statements	Reasons
1. $\triangle CAN$ is an isosceles triangle with vertex $\angle N$.	1. Given
2. $\overline{NC} \cong \overline{NA}$	2. Definition of isosceles triangle
3. $\angle 2 \cong \angle 1$	3. Isosceles Triangle Theorem
4. $\overline{CA} \parallel \overline{BE}$	4. Given
5. $\angle 1 \cong \angle 3$ $\angle 4 \cong \angle 2$	5. Corresponding Angles Postulate
6. $\angle 2 \cong \angle 3$	6. Congruence of angles is transitive.
7. $\angle 4 \cong \angle 3$	7. Congruence of angles is transitive.
8. $\overline{BN} \cong \overline{NE}$	8. If 2 ∡ of a △ are ≅, then the sides opposite those ∡ are ≅.
9. $\triangle NEB$ is an isosceles triangle.	9. Definition of isosceles triangle

41. Given: $\triangle IOE$ is an isosceles triangle with base $\overline{OE}$.
$\overline{AO}$ bisects $\angle IOE$.
$\overline{AE}$ bisects $\angle$IEO.

Prove: $\triangle EAO$ is an isosceles triangle.

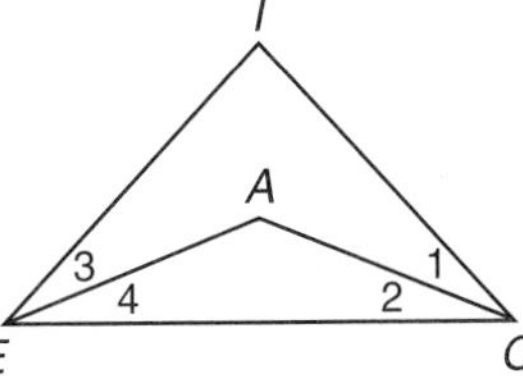

Proof:

Statements	Reasons
1. $\overline{AO}$ bisects $\angle IOE$. $\overline{AE}$ bisects $\angle IEO$.	1. Given
2. $\angle 1 \cong \angle 2$ $\angle 3 \cong \angle 4$	2. Definition of angle bisector
3. $m\angle 1 = m\angle 2$ $m\angle 3 = m\angle 4$	3. Definition of congruent angles

Statements	Reasons
4. $m\angle IOE = m\angle 1 + m\angle 2$ $m\angle IEO = m\angle 3 + m\angle 4$	4. Angle Addition Postulate
5. $m\angle IOE = 2m\angle 2$ $m\angle IEO = 2m\angle 4$	5. Substitution Property (=)
6. $\triangle IOE$ is an isosceles triangle with base $\overline{OE}$.	6. Given
7. $\overline{IE} \cong \overline{IO}$	7. Definition of isosceles triangle
8. $\angle IOE \cong \angle IEO$	8. Isosceles Triangle Theorem
9. $m\angle IOE = m\angle IEO$	9. Definition of congruent angles
10. $2m\angle 2 = 2m\angle 4$	10. Substitution Property (=)
11. $m\angle 2 = m\angle 4$	11. Division Property (=)
12. $\angle 2 \cong \angle 4$	12. Definition of ≅ angles
13. $\overline{AE} \cong \overline{AO}$	13. If 2 ∡ of a △ are ≅, then the sides opposite those ∡ are ≅.
14. $\triangle AEO$ is an isosceles triangle.	14. Definition of isosceles triangle

43. Given: $\triangle MNO$ is an equilateral triangle.

Prove: $m\angle M = m\angle N = m\angle O = 60$

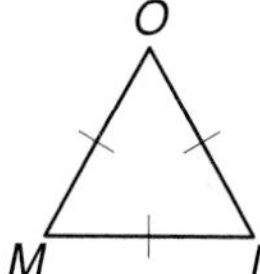

Proof:

Statements	Reasons
1. $\triangle MNO$ is an equilateral triangle.	1. Given
2. $\overline{MN} \cong \overline{MO} \cong \overline{NO}$	2. Definition of equilateral triangle
3. $\angle M \cong \angle N \cong \angle O$	3. Isosceles Triangle Theorem
4. $m\angle M = m\angle N = m\angle O$	4. Definition of ≅ angles
5. $m\angle M + m\angle N + m\angle O = 180$	5. Angle Sum Theorem
6. $3m\angle M = 180$	6. Substitution Property (=)
7. $m\angle M = 60$	7. Division Property (=)
8. $m\angle M = m\angle N = m\angle O = 60$	8. Substitution Property (=)

45.

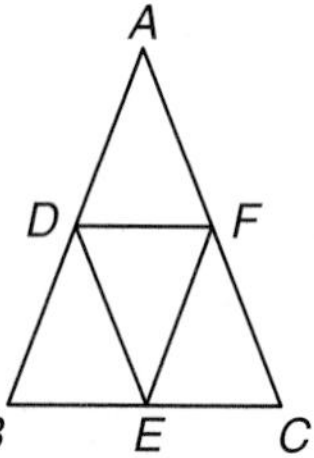

45a. $\triangle DEB \cong \triangle FEC$; Since E is the midpoint of $\overline{BC}$, $\overline{BE} \cong \overline{CE}$. Since $\overline{AC} \cong \overline{AB}$, $\angle B \cong \angle C$. Since D is the midpoint of $\overline{AB}$, F is the midpoint of $\overline{AC}$, and $\overline{AB} \cong \overline{AC}$; $\overline{AD} \cong \overline{AF}$. Therefore, $\triangle DEB \cong \triangle FEC$ by SAS. **45b.** $\triangle ABC$, $\triangle ADF$, $\triangle DEF$; We are given $\triangle ABC$ is an isosceles triangle. Since D is the midpoint of $\overline{AB}$, F is the midpoint of $\overline{AC}$, and $\overline{AB} \cong \overline{AC}$; $\overline{AD} \cong \overline{AF}$ and $\triangle ADF$ is an isosceles triangle. Since $\triangle DEB \cong \triangle FEC$, $\overline{DE} \cong \overline{FE}$ and $\triangle DEF$ is an isosceles triangle. **47.** point M; If the surface is level then it is horizontal. The plumb line will be vertical so it is perpendicular to the surface. A perpendicular dropped from the vertex angle of an isosceles triangle will pass through the midpoint of the base. **49.** SSA means 2 sides and a nonincluded angle are congruent to the corresponding sides and angle. It cannot be used as a proof for congruent triangles. SAS means 2 sides and the included angle are congruent to 2 sides and the included angle. It can be used as a proof for congruent triangles. **51.** $\angle C \longleftrightarrow \angle P$, $\angle D \longleftrightarrow \angle Q$, $\angle E \longleftrightarrow \angle R$, $\overline{CD} \longleftrightarrow \overline{PQ}$, $\overline{DE} \longleftrightarrow \overline{QR}$, $\overline{CE} \longleftrightarrow \overline{PR}$ **53.** 12 units

55. Given: $\overline{AB} \cong \overline{EF}$
$\overline{EF} \cong \overline{JK}$
$\overline{BC} \cong \overline{HJ}$

A B C
D E F G
H J K

Prove: $\overline{AC} \cong \overline{HK}$

Proof:

Statements	Reasons
1. $\overline{AB} \cong \overline{EF}$ $\overline{EF} \cong \overline{JK}$	1. Given
2. $\overline{AB} \cong \overline{JK}$	2. Congruence of segments is transitive.
3. $AB = JK$	3. Definition of ≅ segments
4. $\overline{BC} \cong \overline{HJ}$	4. Given
5. $BC = HJ$	5. Congruence of segments is transitive.
6. $AB + BC = JK + HJ$	6. Addition Property (=)
7. $AC = AB + BC$ $HK = JK + HJ$	7. Segment Addition Property
8. $AC = HK$	8. Substitution Property (=)
9. $\overline{AC} \cong \overline{HK}$	9. Definition of ≅ segments

57. $2\sqrt{29} \approx 10.77$; (2, 6) **59.** 2; −9

Page 229 Chapter 4 Highlights

1. true **3.** false; base angles **5.** false; nonincluded **7.** true **9.** false; flow proof **11.** true **13.** false; base **15.** true

Pages 230–232 Chapter 4 Study Guide and Assessment

17. $\triangle SUV$, $\triangle STU$ **19.** $\triangle SWT$, $\triangle VWU$ **21.** $\overline{SV}$ **23.** 90 **25.** 65 **27.** 40 **29.** 55 **31.** 25 **33.** 55 **35.** 65 **37.** $\overline{RT}$ **39.** $\angle ONM$ **41.** $\angle STR$

43. Given: $\overline{AM} \parallel \overline{CR}$
B is the midpoint of $\overline{AR}$.

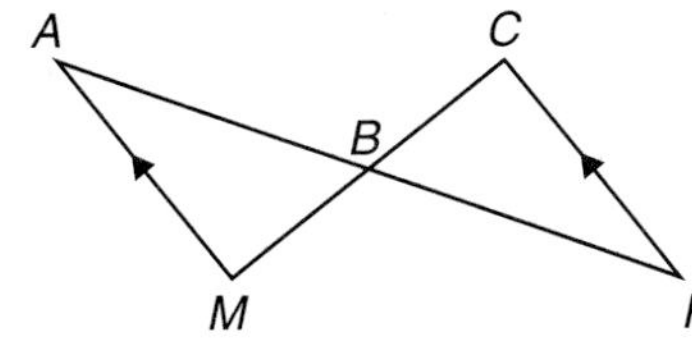

Prove: $\overline{AM} \cong \overline{RC}$

Proof:

Statements	Reasons
1. $\overline{AM} \parallel \overline{CR}$	1. Given
2. $\angle A \cong \angle R$	2. Alternate Interior Angle Theorem
3. *B* is the midpoint of $\overline{AR}$.	3. Given
4. $\overline{AB} \cong \overline{RB}$	4. Definition of midpoint
5. $\angle ABM \cong \angle RBC$	5. Vertical ∠s are ≅.
6. $\triangle ABM \cong \triangle RBC$	6. ASA
7. $\overline{AM} \cong \overline{RC}$	7. CPCTC

45. Given: $\overline{AC} \cong \overline{EC}$
$\angle 1 \cong \angle 2$
$\overline{BC} \cong \overline{DC}$

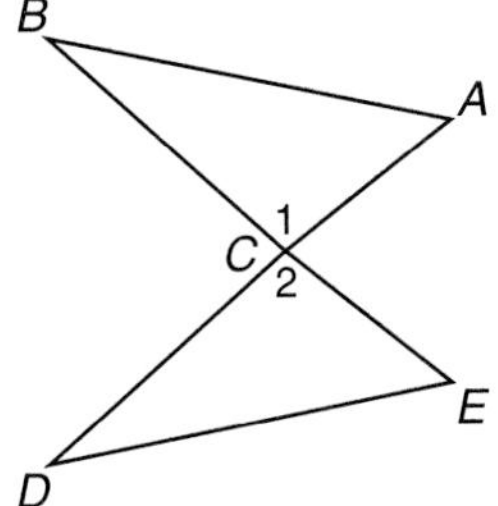

Prove: $\angle B \cong \angle D$

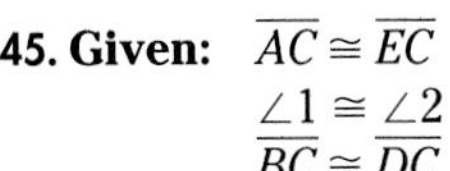

Proof:

Statements	Reasons
1. $\overline{AC} \cong \overline{EC}$, $\angle 1 \cong \angle 2$, $\overline{BC} \cong \overline{DC}$	1. Given
2. $\triangle ABC \cong \triangle EDC$	2. SAS
3. $\angle B \cong \angle D$	3. CPCTC

47. 32 **49.** 36 **51.** 135

CHAPTER 5 APPLYING CONGRUENT TRIANGLES

Pages 242–244 Lesson 5–1

7.

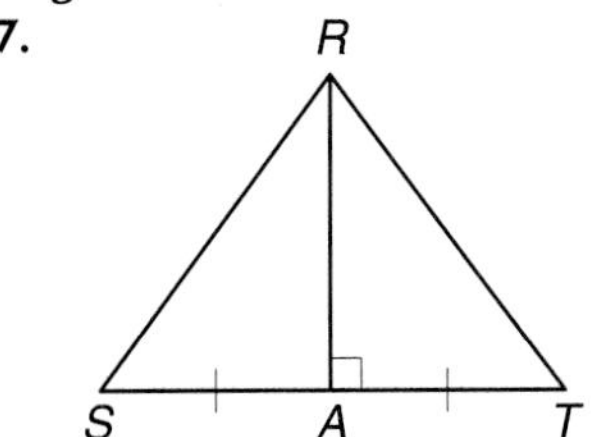

9.

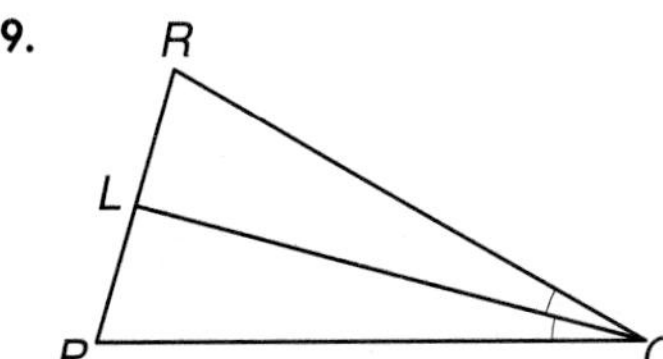

Sample answers given for Exercises 11–13. **11.** *N* is the midpoint of $\overline{SE}$; $\overline{RN}$ is a median of $\triangle RES$ **13.** $\overline{NR}$ is the angle bisector of $\angle SRE$ **15a.** (7, 5) **15b.** $\frac{7}{5}$ **15c.** No, because $\frac{7}{5} \cdot -3 \neq -1$. **15d.** No, $AT \approx 17.2$ and $AB \approx 21.5$ by using the distance formula. **17.** The flag is located at the intersection of the angle bisector between the west and shore roads and the perpendicular bisector of the segment joining the lookout tower to the entrance.

19.

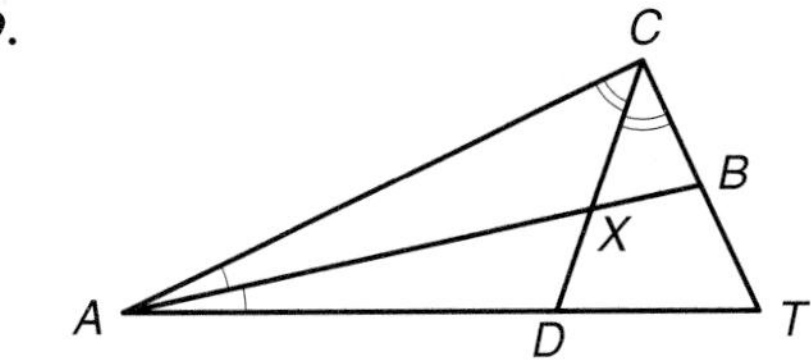

21.

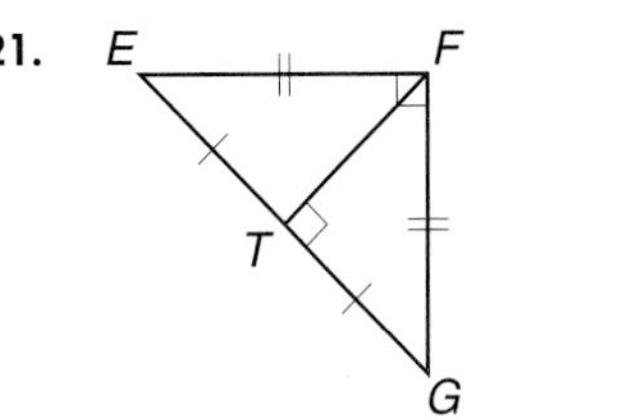

23.

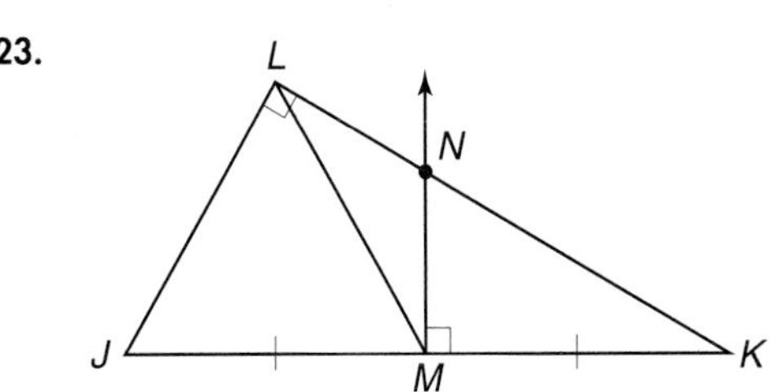

25. $\overline{AE} \cong \overline{EC}$; $\overline{BE} \perp \overline{AC}$; $\overline{AB} \cong \overline{BC}$ **27.** $\angle CAD \cong \angle DAB$ **29.** In a right triangle the altitudes intersect at the vertex of the right angle. **31.** an obtuse triangle **33a.** 40 **33b.** 35 **33c.** 18

35. Given: $\overline{UW}$ is the perpendicular bisector of $\overline{XZ}$.

Prove: For any point *V* on $\overline{UW}$, $VX = VZ$.

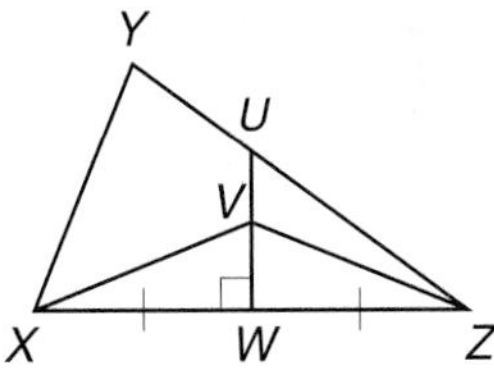

Proof:

Statements	Reasons
1. $\overline{UW}$ is the ⊥ bisector of $\overline{XZ}$.	1. Given
2. *W* is the midpoint of $\overline{XZ}$.	2. Def. ⊥ bisector
3. $\overline{XW} \cong \overline{WZ}$	3. Def. midpoint
4. $\overline{UW} \perp \overline{XZ}$	4. Def. ⊥ bisector
5. $\angle XWV$, $\angle ZWV$ are right ∠s.	5. ⊥ lines form four rt. ∠s.
6. $\angle XWV \cong \angle ZWV$	6. All rt. ∠s are ≅.
7. $\overline{VW} \cong \overline{VW}$	7. Congruence of segments is reflexive.
8. $\triangle XWV \cong \triangle ZWV$	8. SAS
9. $\overline{VX} \cong \overline{VZ}$	9. CPCTC
10. $VX = VZ$	10. Def. ≅ segments

37. Given: $\overline{BD}$ bisects $\angle ABC$.

Prove: $DE = DF$

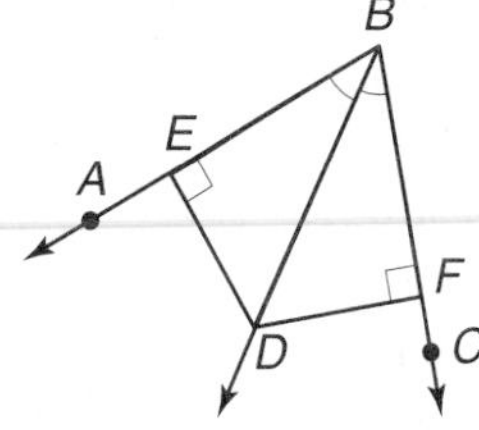

Proof:

Statements	Reasons
1. $\overline{BD}$ bisects $\angle ABC$.	1. Given
2. $\angle ABD \cong \angle CBD$	2. Def. $\angle$ bisector
3. Let DE = distance from D to $\overline{AB}$, and DF = distance from D to $\overline{BC}$.	3. Def. distance from a point to a line
4. $\overline{DE} \perp \overline{AB}$, $\overline{DF} \perp \overline{BC}$	4. Def. distance from a point to a line
5. $\angle DEB$, $\angle DFB$ are rt. $\angle$s.	5. $\perp$ lines form four rt. $\angle$s.
6. $\angle DEB \cong \angle DFB$	6. All rt. $\angle$s are $\cong$.
7. $\overline{BD} \cong \overline{BD}$	7. Congruence of segments is reflexive.
8. $\triangle DEB \cong \triangle DFB$	8. AAS
9. $\overline{DE} \cong \overline{DF}$	9. CPCTC
10. $DE = DF$	10. Def. $\cong$ segments

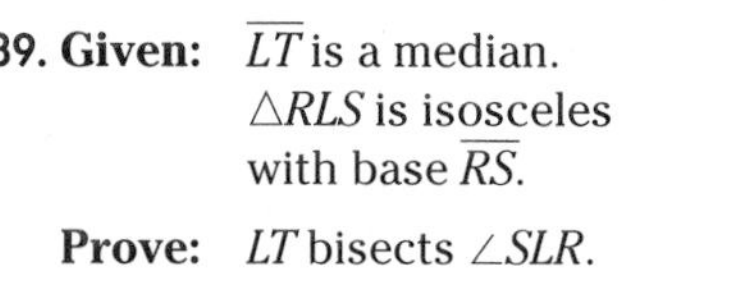

39. Given: $\overline{LT}$ is a median. $\triangle RLS$ is isosceles with base $\overline{RS}$.

Prove: $\overline{LT}$ bisects $\angle SLR$.

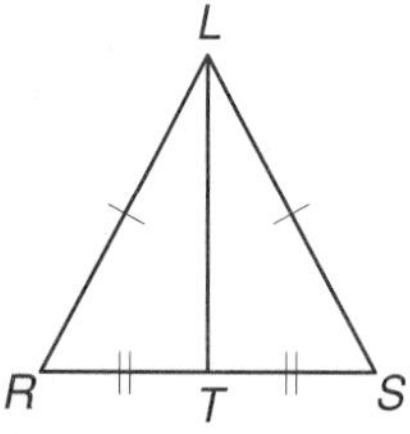

Proof:

Statements	Reasons
1. $\overline{LT}$ is a median. $\triangle RLS$ is isosceles with base $\overline{RS}$.	1. Given
2. $\overline{RT} \cong \overline{TS}$	2. Def. median
3. $\overline{RL} \cong \overline{LS}$	3. Def. isosceles $\triangle$
4. $\overline{LT} \cong \overline{LT}$	4. Congruence of segments is reflexive.
5. $\triangle RLT \cong \triangle SLT$	5. SSS
6. $\angle TLR \cong \angle TLS$	6. CPCTC
7. $\overline{LT}$ bisects $\angle SLR$.	7. Def. $\angle$ bisector

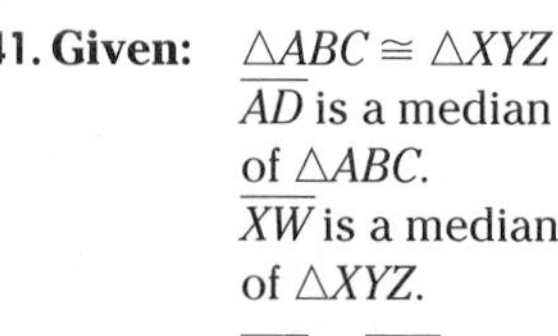

41. Given: $\triangle ABC \cong \triangle XYZ$ $\overline{AD}$ is a median of $\triangle ABC$. $\overline{XW}$ is a median of $\triangle XYZ$.

Prove: $\overline{AD} \cong \overline{XW}$

Proof:

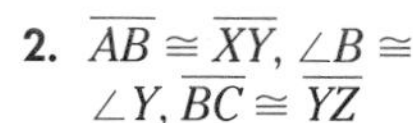

Statements	Reasons
1. $\triangle ABC \cong \triangle XYZ$ $\overline{AD}$ is a median of $\triangle ABC$. $\overline{XW}$ is a median of $\triangle XYZ$.	1. Given
2. $\overline{AB} \cong \overline{XY}$, $\angle B \cong \angle Y$, $\overline{BC} \cong \overline{YZ}$	2. CPCTC
3. $BC = YZ$	3. Def. $\cong$ segments
4. D is the midpoint of $\overline{BC}$. W is the midpoint of $\overline{YZ}$.	4. Def. median
5. $\overline{BD} \cong \overline{DC}$, $\overline{YW} \cong \overline{WZ}$	5. Def. midpoint
6. $BD = DC$, $YW = WZ$	6. Def. $\cong$ segments
7. $BC = BD + DC$, $YZ = YW + WZ$	7. Seg. Addition Post.
8. $BC = BD + BD$, $YZ = YW + YW$	8. Substitution Prop. (=)
9. $2BD = 2YW$	9. Substitution Prop. (=)
10. $BD = YW$	10. Division Prop. (=)
11. $\overline{BD} \cong \overline{YW}$	11. Def. $\cong$ segments
12. $\triangle ABD \cong \triangle XYW$	12. SAS
13. $\overline{AD} \cong \overline{XW}$	13. CPCTC

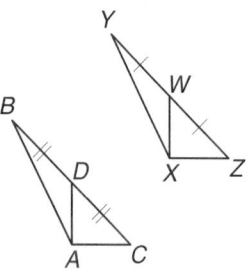

43. Acute triangle: all segments meet inside the triangle; right triangle: altitudes meet at the vertex of the right angle, other segments meet inside the triangle; obtuse triangle: altitudes meet outside the triangle, rest meet inside **45a.** 4 **45b.** 8

45c.

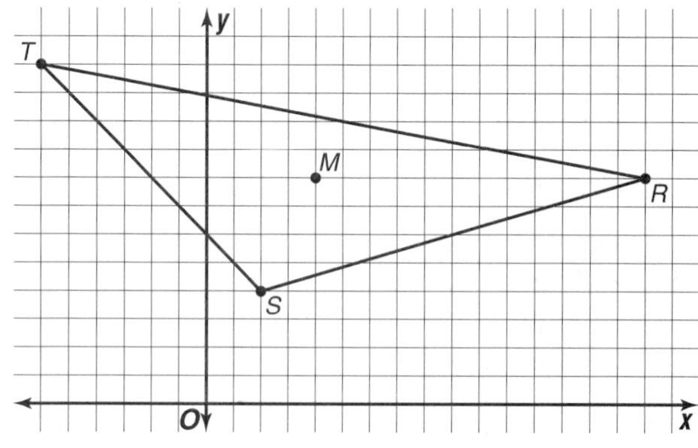

45d. The medians intersect at M. **47.** $BA = 5$, $AY = 8$, $BY = 5$ **49.** yes; both have slope $-\frac{2}{3}$

51. Given: $\overline{PQ}$ bisects $\overline{AB}$ at point M.

Prove: $\overline{AM} \cong \overline{MB}$

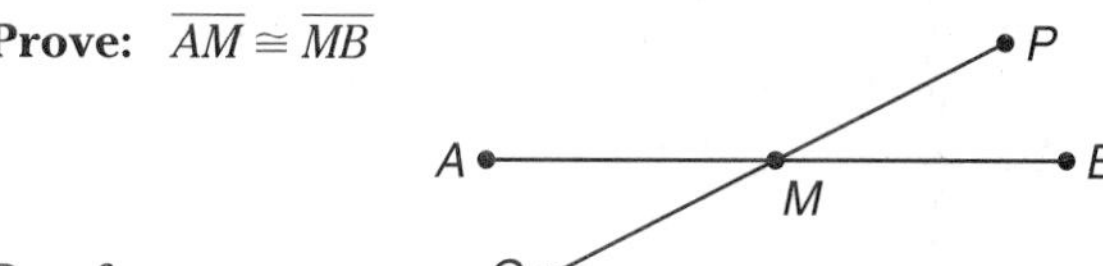

Proof:

Statements	Reasons
1. $\overline{PQ}$ bisects $\overline{AB}$ at point M.	1. Given
2. M is the midpoint of $\overline{AB}$.	2. Def. bisector
3. $AM = MB$	3. Def. midpoint
4. $\overline{AM} \cong \overline{MB}$	4. Def. $\cong$ segments

53. S; $RS + ST = RT$ **55.** 45 meters

Pages 248–51 Lesson 5–2

7. $\overline{ST} \cong \overline{TU}$ **9.** 6 **11.** 9 **13.** yes, LL **15.** $\overline{JT} \cong \overline{MR}$ and $\angle J \cong \angle M$ or $\overline{JT} \cong \overline{MR}$ and $\angle T \cong \angle R$ **17.** $\overline{MN} \cong \overline{OP}$ **19.** $\overline{VX} \cong \overline{XY}$ or $\overline{WX} \cong \overline{XZ}$ **21.** 1 **23.** 3

25. Given: $\angle M$ and $\angle P$ are right angles.
$\overline{MN} \parallel \overline{OP}$

Prove: $\overline{MN} \cong \overline{OP}$

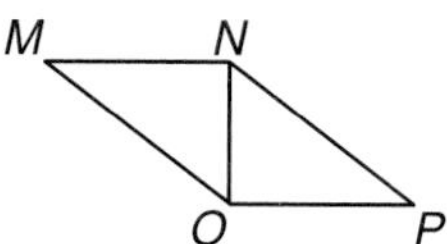

Proof:

Statements	Reasons
1. $\angle M$ and $\angle P$ are right angles. $\overline{MN} \parallel \overline{OP}$	1. Given
2. $\triangle MNO$ and $\triangle PON$ are rt. $\triangle$s.	2. Def. rt. $\triangle$
3. $\angle MNO \cong \angle PON$	3. If $\parallel$ lines, alt. int. $\angle$s are $\cong$.
4. $\overline{NO} \cong \overline{NO}$	4. Congruence of segments is reflexive.
5. $\triangle MNO \cong \triangle PON$	5. HA
6. $\overline{MN} \cong \overline{OP}$	6. CPCTC

27. Given: $\angle PBC \cong \angle PCB$
$\overline{AP} \perp$ plane $\mathcal{M}$

Prove: $\angle ABC \cong \angle ACB$

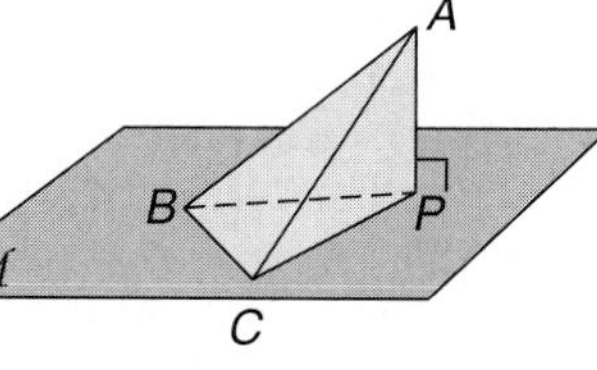

Proof:

Statements	Reasons
1. $\angle PBC \cong \angle PCB$ $\overline{AP} \perp$ plane $\mathcal{M}$	1. Given
2. $\angle APB$ and $\angle APC$ are rt. $\angle$s.	2. $\perp$ lines form four rt. $\angle$s.
3. $\triangle APB$ and $\triangle APC$ are rt. $\triangle$s.	3. Def. rt. $\triangle$
4. $\overline{BP} \cong \overline{CP}$	4. Isosceles Triangle Th.
5. $\overline{AP} \cong \overline{AP}$	5. Congruence of segments is reflexive.
6. $\triangle APC \cong \triangle APB$	6. LL
7. $\overline{AC} \cong \overline{AB}$	7. CPCTC
8. $\angle ABC \cong \angle ACB$	8. Isosceles Triangle Th.

29. Case 1:

Given: $\triangle ABC$ and $\triangle DEF$ are right triangles.
$\overline{AC} \cong \overline{DF}$
$\angle C \cong \angle F$

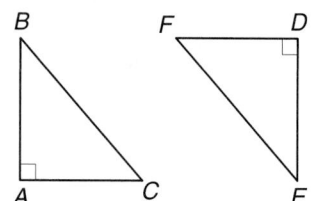

Prove: $\triangle ABC \cong \triangle DEF$

Proof:
It is given that $\triangle ABC$ and $\triangle DEF$ are right triangles, $\overline{AC} \cong \overline{DF}$, and $\angle C \cong \angle F$. By the definition of right triangle, $\angle A$ and $\angle D$ are right angles. Thus, $\angle A \cong \angle D$ since all right angles are congruent. Therefore, $\triangle ABC \cong \triangle DEF$ by Angle-Side-Angle.

Case 2:

Given: $\triangle ABC \cong \triangle DEF$ are right triangles.
$\overline{AC} \cong \overline{DF}$
$\angle B \cong \angle E$

Prove: $\triangle ABC \cong \triangle DEF$

Proof:
It is given that $\triangle ABC$ and $\triangle DEF$ are right triangles, $\overline{AC} \cong \overline{DF}$, and $\angle B \cong \angle E$. By the definition of right triangle, $\angle A$ and $\angle D$ are right angles. Thus, $\angle A \cong \angle D$ since all right angles are congruent. Therefore, $\triangle ABC$ and $\triangle DEF$ by Angle-Angle-Side.

31. Given: $\triangle LMN$ is isosceles with base $\overline{LN}$.
O is the midpoint of $\overline{LM}$.
P is the midpoint of $\overline{NM}$.
$\overline{OQ} \perp \overline{LN}$,
$\overline{PR} \perp \overline{LN}$

Prove: $\overline{OQ} \cong \overline{PR}$

Proof:
It is given that $\triangle LMN$ is isosceles. By the definition of isosceles triangle, $\overline{LM} \cong \overline{NM}$. Thus, $LM = NM$ by the definition of congruent segments. $\angle L \cong \angle N$ because if two sides of a triangle are congruent, then the angles opposite the sides are congruent also. We were given that O is the midpoint of $\overline{LM}$ and P is the midpoint of $\overline{NM}$. The definition of midpoint lets us say that $\overline{LO} \cong \overline{OM}$ and $\overline{NP} \cong \overline{PM}$. Thus by the definition of congruent segments, $LO = OM$ and $NP = PM$. By the segment addition postulate, $LM = LO + OM$ and $NM = NP + PM$. Thus, $LO + OM = NP + PM$ by substitution. Then $2LO = 2NP$ by substitution. By the division property of equality, $LO = NP$. Then $\overline{LO} \cong \overline{NP}$ by the definition of congruent segments. It was given that $\overline{OQ} \perp \overline{LN}$ and $\overline{PR} \perp \overline{LN}$. Then $\angle OQL$ and $\angle PRN$ are right angles since perpendicular lines form four right angles. Then $\triangle OQL$ and $\triangle PRN$ are right triangles by the definition of right triangle. By Hypotenuse-Angle, $\triangle OQL \cong \triangle PRN$. Therefore, $\overline{OQ} \cong \overline{PR}$ by CPCTC.

33. The corresponding sides are proportional, but not necessarily congruent. **35.** legs: 48 in. and 60 in.; hypotenuse and an angle: 76.8 in. and 51° or 76.8 in. and 39°; leg and an angle: 60 in. and 51°, or 60 in. and 39°, 48 in. and 51°, or 48 in. and 39°; hypotenuse and a leg: 76.8 in. and 60 in. or 78.6 in. and 48 in.

37.

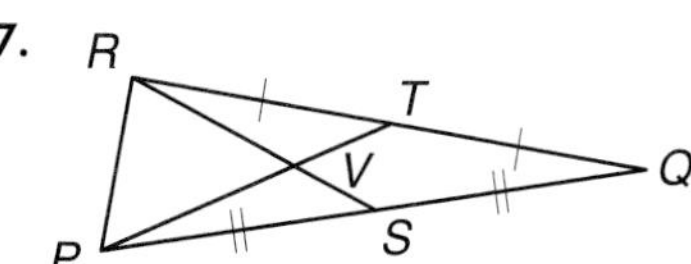

39. obtuse, isosceles

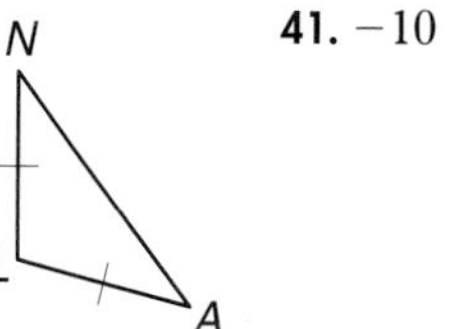

41. -10

43. converse: interchange the hypothesis and conclusion; **inverse:** negate the hypothesis and the conclusion; **contrapositive:** negate and interchange the hypothesis and conclusion **45.** {(1938, 50,000), (1939, 40,000), (1938, 5200), (1938, 4200), (1939, 4020)}; D = {1938, 1939}; R = {50,000, 40,000, 5200, 4200, 4020}

Pages 255–258 Lesson 5-3

5. Lines ℓ and m do not intersect at point X. **7.** Sabrina did not eat the leftover pizza. **9.** $\angle 1$, $\angle 4$, $\angle 8$ **11.** Division Property of Inequality **13.** Transitive Property of Inequality

15. Given: m is not parallel to n.
Prove: $m\angle 3 \neq m\angle 2$

Proof:

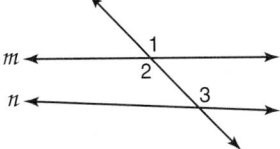

Step 1: Assume $m\angle 3 = m\angle 2$.

Step 2: $\angle 3$ and $\angle 2$ are alternate interior angles. If 2 lines are cut by a transversal so that alternate interior angles are congruent, then the lines are parallel. This means that $m \parallel n$. However, that contradicts the given statement.

Step 3: Therefore, since the assumption leads to a contradiction, the assumption must be false. Thus, $m\angle 3 \neq m\angle 2$.

17. $\overline{AB}$ is not congruent to $\overline{CD}$. **19.** The disk is not defective. **21.** If two altitudes of a triangle are congruent, then the triangle is not isosceles. **23.** $m\angle 7$ **25.** Sample answer: $\angle 1$ **27.** $\angle 2$, $\angle 3$, $\angle 4$, $\angle 5$, $\angle 6$ **29.** Division Property of Inequality **31.** Comparison Property of Inequality **33.** Division Property of Inequality

35. Given: $\triangle XYV$, $\triangle XZV$
Prove: $m\angle 7 > m\angle 6$

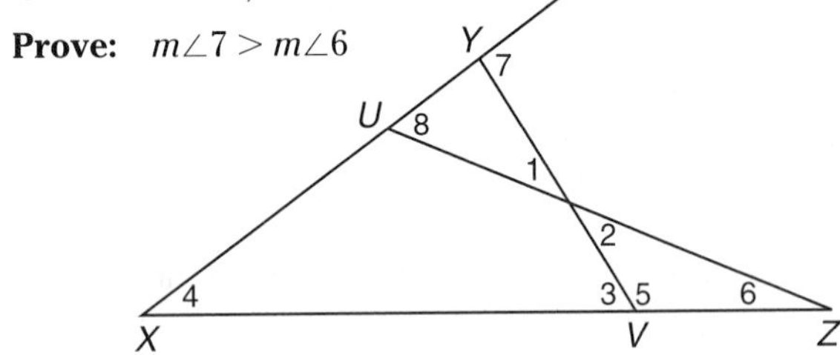

Proof:
Since if an angle is an exterior angle of a triangle its measure is greater than the measure of either corresponding interior angle, we can say that $m\angle 7 > m\angle 8$ and $m\angle 8 > m\angle 6$. Thus, by the Transitive Property of Inequality, $m\angle 7 > m\angle 6$.

37. Given: $\overline{FD} \perp \overline{AB}$
$\overline{GE} \perp \overline{AB}$
Prove: $m\angle 5 > m\angle 2$

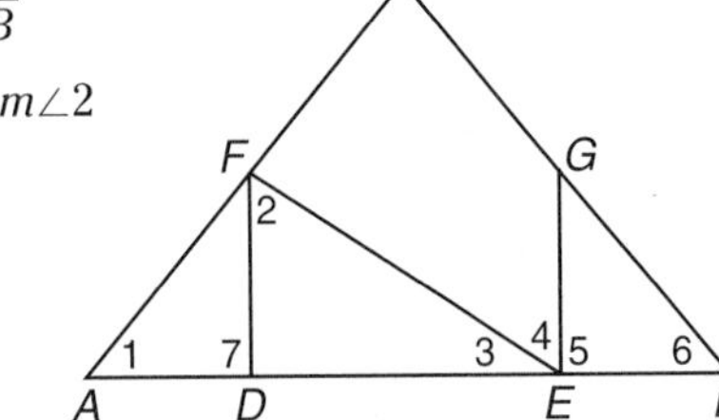

Proof:

Statements	Reasons
1. $\overline{FD} \perp \overline{AB}$, $\overline{GE} \perp \overline{AB}$	1. Given
2. $\angle 7$ and $\angle 5$ are rt. $\angle$s.	2. $\perp$ lines form four rt. $\angle$s.
3. $m\angle 7 > m\angle 2$	3. Exterior Angle Inequality Th.
4. $\angle 7 \cong \angle 5$	4. All rt. $\angle$s are $\cong$.
5. $m\angle 7 = m\angle 5$	5. Def. $\cong$ $\angle$s
6. $m\angle 5 > m\angle 2$	6. Substitution Prop. of Inequality

39. Given: $\overline{MO} \cong \overline{ON}$
$\overline{MP}$ is not congruent to $\overline{NP}$.
Prove: $\angle MOP$ is not congruent to $\angle NOP$.

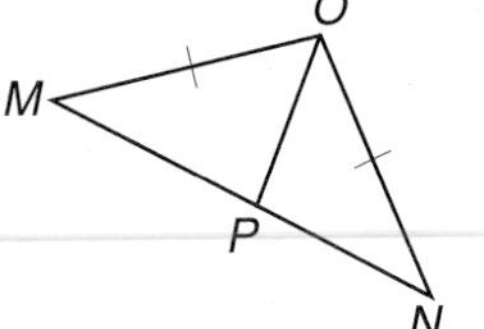

Proof:
Assume that $\angle MOP \cong \angle NOP$. We are given that $\overline{MO} \cong \overline{ON}$. If two sides of a triangle are congruent, then the angles opposite the sides are congruent also. Thus, $\angle M \cong \angle N$. Therefore, $\triangle MOP \cong \triangle NOP$ by ASA. We can then conclude that $\overline{MP} \cong \overline{NP}$. But this contradicts the given information. Therefore, the assumption is incorrect and $\angle MOP$ is not congruent to $\angle NOP$.

41. Given: $\triangle XYZ$
$m\angle X \neq m\angle Y \neq m\angle Z$
Prove: $XY \neq YZ \neq XZ$

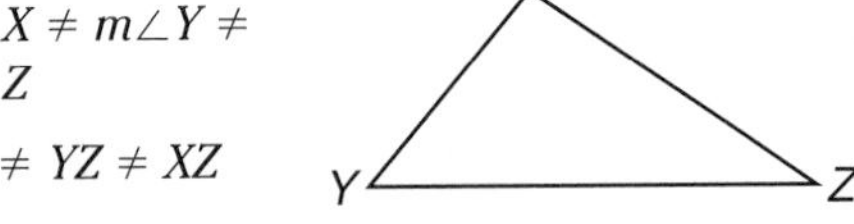

Proof:
Assume that $XY = YZ$. Then since if two sides of a triangle are congruent, then the angles opposite the sides are congruent, $\angle Z \cong \angle X$. Then by the definition of congruent angles, $m\angle Z = m\angle X$. However, we are given that $m\angle X \neq m\angle Y \neq m\angle Z$. Therefore the assumption is incorrect and $XY \neq YZ$. A similar argument could prove that $YZ \neq XZ$ and $XY \neq XZ$. Thus, if a triangle has no two angles congruent, then it has no two sides congruent.

43. The door on the left. If the sign on the door on the right were true, then both signs would be true. But one sign is false, so the sign the on the door on the right must be false. **45.** Yes. If you assume the client was at the scene of the crime, it is contradicted by his presence at the meeting in New York City. Thus, the assumption he was present at the crime is wrong.

47.

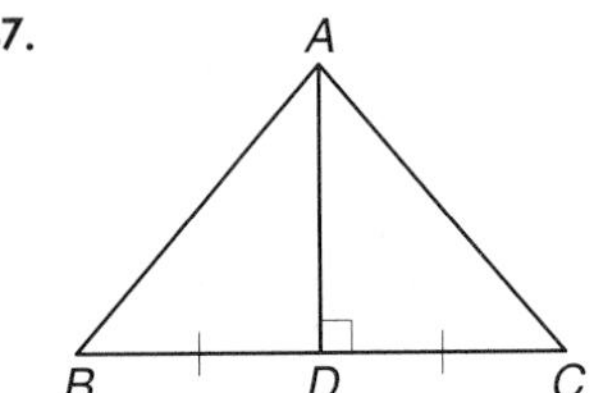

49. obtuse, isosceles
51. parallel **53.** 36 lunch specials **55.** 0.3728

Page 258 Self Test

1. 5 **3.** yes, by LL or SAS

5. Given: $\overleftrightarrow{AB}$ and $\overleftrightarrow{PQ}$ are skew.
Prove: $\overleftrightarrow{AP}$ and $\overleftrightarrow{BQ}$ are skew.

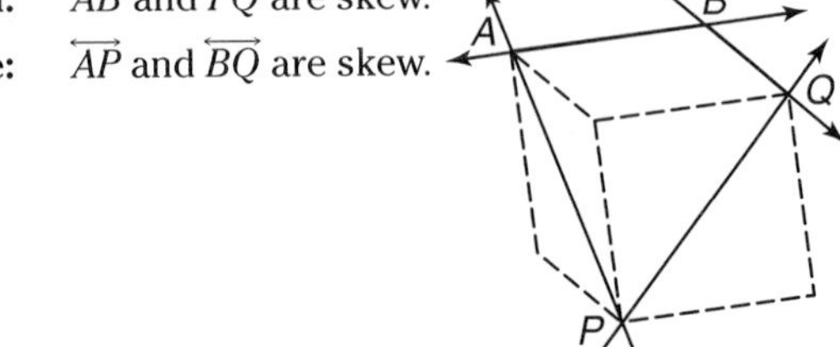

Proof:
Assume that $\overleftrightarrow{AP}$ and $\overleftrightarrow{BQ}$ are not skew. Then either $\overleftrightarrow{AP} \parallel \overleftrightarrow{BQ}$ or $\overleftrightarrow{AP}$ and $\overleftrightarrow{BQ}$ intersect. In either case, A, B, P, and Q are coplanar. It is given that $\overleftrightarrow{AB}$ and $\overleftrightarrow{PQ}$ are skew, so $\overleftrightarrow{AB}$ and $\overleftrightarrow{PQ}$ do not lie in the same plane, and A and B

are not coplanar with P and Q. This is a contradiction. Therefore, $\overleftrightarrow{AP}$ and $\overleftrightarrow{BQ}$ must be skew.

Lesson 262–265 Lesson 5–4

7. yes **9.** $\overline{TU}$ **11.** $\overline{TS}$

13. Given: $\angle B$ is a right angle.
$\overline{BD}$ is an altitude.

Prove: $BD < AB$ and $BD < BC$

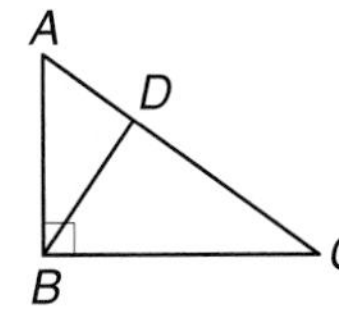

Proof:

We are given that $\overline{BD}$ is an altitude. Therefore by the definition of altitude, $\overline{BD} \perp \overline{AC}$. Then since perpendicular lines form four right angles, $\angle ADB$ and $\angle CDB$ are right angles. Then $\triangle ADB$ and $\triangle CDB$ are right triangles by the definition of right triangle. $\angle BAD$ and $\angle DBA$, and $\angle BCD$ and $\angle DBC$ are complementary since acute angles of right triangles are complementary. By the definition of complementary angles, $m\angle BAD + m\angle DBA = 90$ and $m\angle BCD + m\angle DBC = 90$. The definition of inequality allows us to say that $m\angle BAD < 90$ and $m\angle BCD < 90$. $m\angle ADB = 90$ and $m\angle CDB = 90$ by the definition of a right angle. Thus by substitution, $m\angle BAD < m\angle ADB$ and $m\angle BCD < m\angle CDB$. Therefore in $\triangle BCD$, $BD < AB$; and in $\triangle ABD$, $BD < BC$.

15. $\angle N$ **17.** $\angle T$ **19.** $\angle WYX$ **21.** $\overline{CE}$ **23.** $\overline{BE}$, In $\triangle EDC$, $\overline{EC}$ is longest. In $\triangle BEC$, $\overline{BE}$ is longest. In $\triangle ABE$, $\overline{BE}$ is longest. Thus, $\overline{BE}$ is the longest segment. **25.** No, the segments do not have the correct relative lengths.
27. $x = 12$; $m\angle A = 104$, $m\angle B = 32$, $m\angle C = 44$; so $AC < AB < BC$.

29. Given: $\overline{RQ}$ bisects $\angle SRT$.

Prove: $m\angle SQR > m\angle SRQ$

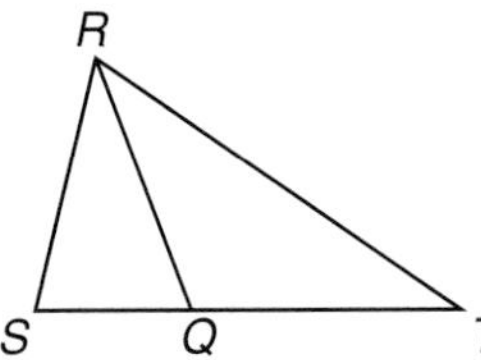

Proof:

Statements	Reasons
1. $\overline{RQ}$ bisects $\angle SRT$.	1. Given
2. $\angle SRQ \cong \angle QRT$	2. Def. $\angle$ bisector
3. $m\angle SRQ = m\angle QRT$	3. Def. $\cong$ $\angle$s
4. $m\angle SQR > m\angle QRT$	4. Exterior Angle Inequality Th.
5. $m\angle SQR > m\angle SRQ$	5. Substitution Prop. (=)

31. Given: D is on $\overline{AC}$.
$\angle BDC$ is acute.

Prove: $BA > BD$

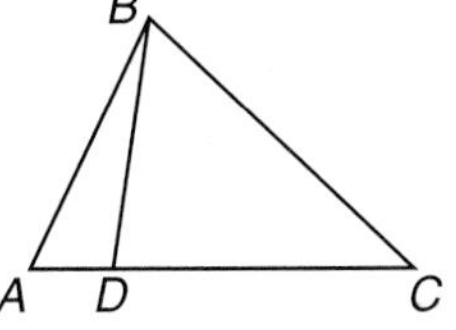

Proof:

Statements	Reasons
1. D is on $\overline{AC}$. $\angle BDC$ is acute.	1. Given
2. $m\angle BDC < 90$	2. Def. acute $\angle$
3. $m\angle BDC + x = 90$	3. Def. inequality
4. $m\angle BDC = 90 - x$	4. Subtraction Prop. (=)
5. $\angle BDC$ and $\angle BDA$ are a linear pair.	5. Def. linear pair
6. $m\angle BDC + m\angle BDA = 180$	6. Supplement Th.
7. $(90 - x) + m\angle BDA = 180$	7. Substitution Prop. (=)
8. $m\angle BDA = 90 + x$	8. Subtraction Prop. (=)
9. $m\angle BDA > 90$	9. Def. inequality
10. $\angle BDA$ is obtuse.	10. Def. obtuse
11. $\angle BAD$ and $\angle ABD$ are acute angles.	11. There can be at most 1 rt. or obtuse $\angle$ in a $\triangle$.
12. $m\angle BAD < 90$	12. Def. acute $\angle$
13. $m\angle BDA > m\angle BAD$	13. Transitive Prop. of Inequality
14. $BA > BD$	14. If an $\angle$ of a $\triangle >$ another $\angle$, the side opp. the greater $\angle$ is longer than the side opp. the lesser $\angle$.

33. Given: $\overline{AC}$ is an altitude of $\triangle AEB$.
$\overline{AF}$ is a median.
$\overline{AE} \not\cong \overline{AB}$

Prove: $AF > AC$

Proof:

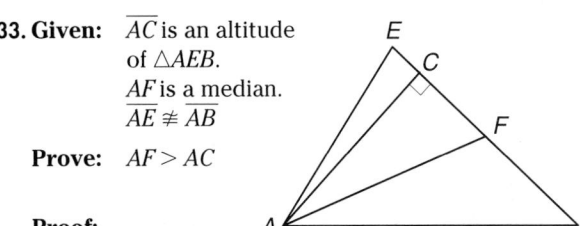

We are given that $\overline{AC}$ is an altitude of $\triangle AEB$. $\overline{AC} \perp \overline{EB}$ by the definition of an altitude. The perpendicular segment from a point to a line is the shortest segment from the point to the line, so $\overline{AC}$ is the shortest segment from A to $\overline{EB}$. The only way that $\overline{AF}$ is not longer than $\overline{AC}$ is if they are the same segment. If they are the same segment, then C and F are the same point and $EC = CB$ by the definition of median. In that case, $\angle ACE$ and $\angle ACB$ are right angles since perpendicular lines form four right angles. Then $\triangle ACE$ and $\triangle ACB$ are right triangles by the definition of right triangle. $\overline{AC} \cong \overline{AC}$ because congruence of segments is reflexive. Then $\triangle ACE \cong \triangle ACB$ by HL. Then $\overline{AE} \cong \overline{AB}$ by CPCTC. But this contradicts the given information. Thus, C and F do not correspond and $AF > AC$.

35. Given: $\overline{PQ} \perp$ plane $\mathcal{M}$

Prove: $\overline{PQ}$ is the shortest segment from P to plane $\mathcal{M}$.

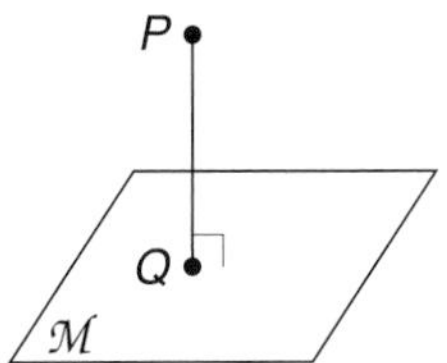

Proof:

By definition, $\overline{PQ}$ is perpendicular to plane $\mathcal{M}$ if it is perpendicular to every line in $\mathcal{M}$ that intersects it. But since the perpendicular segment from a point to a line is the shortest segment from the point to the line, that perpendicular segment is the shortest segment from the point to each of these lines. Therefore, $\overline{PQ}$ is the shortest segment from P to $\mathcal{M}$.

37a.

37b. The equation is the best predictor for Henry Rono's record in 1978. It is the worst predictor for Ron Clarke's record in 1965. **39.** Songan **43.** False; the hypotenuse must be longer than the other two sides.

45.

x	2x − 6	(x, y)
0	−6	(0, −6)
2	−2	(2, −2)
9	2	(4, 2)

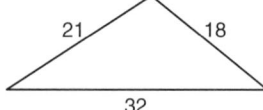

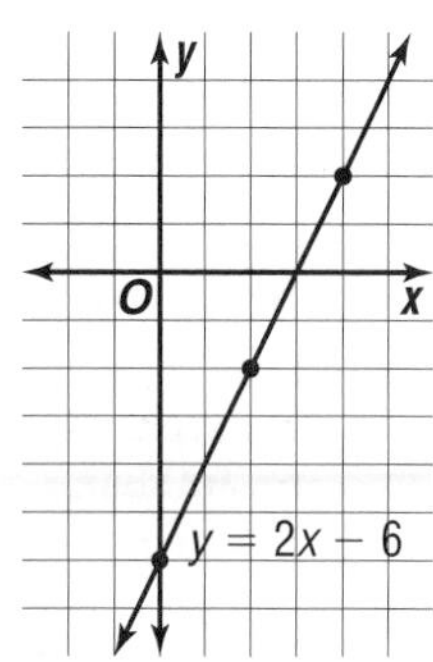

47. 2.3

Page 266 Lesson 5–5A

3. In the column of sets that do not make a triangle, either $S1 + S2 \leq S3$, $S2 + S3 \leq S1$, or $S1 + S3 \leq S2$. In the column of sets that do make a triangle, $S1 + S2 > S3$, $S2 + S3 > S1$, and $S1 + S3 > S2$.

Pages 269–272 Lesson 5–5

5. yes

7. 6 and 48 **9.** 0 and 60 **11.** always true; $DB < 7$ and $BC < 2 + DB$ **13.** Yes. The measures of the sides are 5, $\sqrt{51}$ and $\sqrt{90}$ and the Triangle Inequality Theorem applies. **15.** yes

17. yes **19.** yes

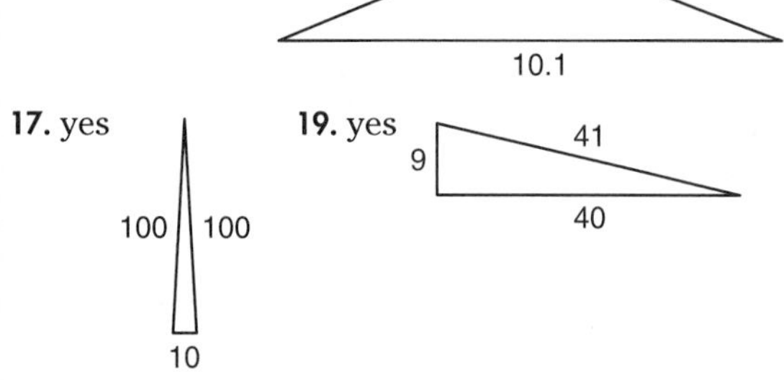

21. no **23.** 3 and 33 **25.** 12 and 56 **27.** 24 and 152 **29.** 24 and 118 **31.** $|a - b|$ and $a + b$ **33.** sometimes true; $BD + DC < BC$ and $BC + DC > BC$, so $10 < BC < 38$. 12 is one of the acceptable values for BC. **35.** always true; $AB + AC > BC$ and $BC + DC > BD$, so $AB + AC > BD - DC$, then $AB + AC > BD$. **37.** never true; $\overline{BA}$ is the perpendicular, so it is the shortest distance from B to $\overline{AC}$. **39.** yes; $AB = \sqrt{245} \approx 15.7$, $AC = \sqrt{244} \approx 15.6$, $BC = \sqrt{685} \approx 26.2$ These measures satisfy the Triangle Inequality Theorem. **41.** yes; $DE = \sqrt{41} \approx 6.4$, $DF = 8$, $EF = 5$ These measures satisfy the Triangle Inequality Theorem.

43. Given: quadrilateral $ABCD$

Prove: $AD + CD + AB > BC$

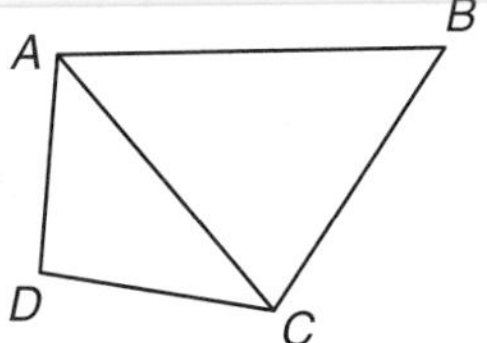

Proof:

Statements	Reasons
1. quadrilateral $ABCD$	1. Given
2. Draw $\overline{AC}$.	2. Through any 2 pts. there is 1 line.
3. $AD + CD > AC$ $AB + AC > BC$	3. Triangle Inequality
4. $AC > BC - AB$	4. Subtraction Prop. of Inequality
5. $AD + CD > BC - AB$	5. Transitive Prop. of Inequality
6. $AD + CD + AB > BC$	6. Addition Prop. of Inequality

45a. no triangle **45b.** Triangle exists. **45c.** no triangle **45d.** Triangle exists. **45e.** Triangle exists. **47a.** a is either 15 cm, or 16 cm; z is 14 cm, 15 cm, or 16 cm. The possible triangles that can be made from sides with those measures are (2 cm, 15 cm, 14 cm), (2 cm, 15 cm, 15 cm), (2 cm, 15 cm, 16 cm), (2 cm, 16 cm, 15 cm), (2 cm, 16cm, 16 cm) **47b.** $\frac{2}{5}$ **49.** $\overline{AC}$ **51.** no triangle **53.** 12 units

55. Given: $\ell \parallel m$
$s \parallel t$
$t \perp \ell$

Prove: $\angle 3$ is a right angle.

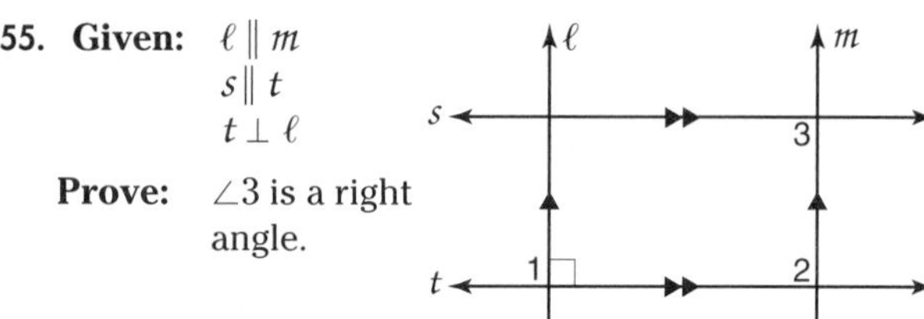

Proof:

Statements	Reasons
1. $\ell \parallel m$, $s \parallel t$, $t \perp \ell$	1. Given
2. $\angle 1$ is a right angle.	2. $\perp$ lines form four rt. $\angle$s.
3. $m\angle 1 = 90$	3. Def. rt. $\angle$
4. $\angle 1 \cong \angle 2$	4. Corresponding Angles Postulate
5. $m\angle 2 = 90$	5. Substitution Prop. (=)
6. $\angle 2$ and $\angle 3$ are supplementary.	6. Consecutive Interior Angle Theorem
7. $m\angle 2 + m\angle 3 = 180$	7. Def. supp.
8. $90 + m\angle 3 = 180$	8. Substitution Prop. (=)
9. $m\angle 3 = 90$	9. Subtraction Prop. (=)
10. $\angle 3$ is a right angle.	10. Def. rt. $\angle$

57. 27; 117 **59a.** the percentage of classical music listeners who are in different age groups **59b.** 880,000 people

Pages 276–279 Lesson 5–6

5. $m\angle DBE < m\angle BFA$ **7.** $m\angle FDB > m\angle BDC$ **9.** never **11.** $2.8 < x$ and $x < 12$ **13.** right arm **15.** $TM < RS$ **17.** $FH > GE$ **19.** $m\angle AOD > m\angle AOB$ **21.** always **23.** always **25.** never **27.** $x < 14$ **29.** $x > 4$

31. Given: $\overline{MN} \cong \overline{QR}$
$\overline{MN} \parallel \overline{QR}$
$m\angle MPQ >$
$m\angle QPR$

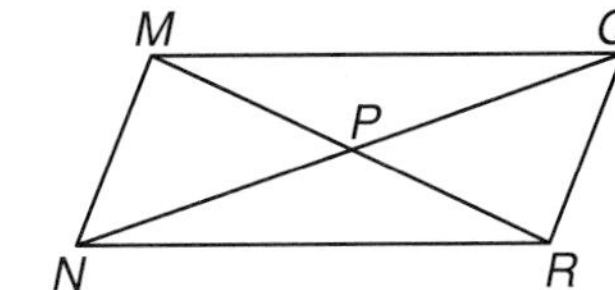

Prove: $MQ > QR$

Proof:

Statements	Reasons
1. $\overline{MN} \cong \overline{QR}$ $\overline{MN} \parallel \overline{QR}$ $m\angle MPQ > m\angle QPR$	1. Given
2. $\angle MNQ \cong \angle NQR$	2. If $\parallel$, alt. int. $\angle$s are $\cong$.
3. $\angle MPN \cong \angle QPR$	3. Vertical $\angle$s are $\cong$.
4. $\triangle MPN \cong \triangle RPQ$	4. AAS
5. $\overline{MP} \cong \overline{RP}$	5. CPCTC
6. $\overline{PQ} \cong \overline{PQ}$	6. Congruence of segments is reflexive.
7. $MQ > QR$	7. SAS Inequality

33. Given: $\triangle ABC$ is equilateral.
$\triangle ABD \cong \triangle CBD$
$BD > AD$

Prove: $m\angle BCD >$
$m\angle DAC$

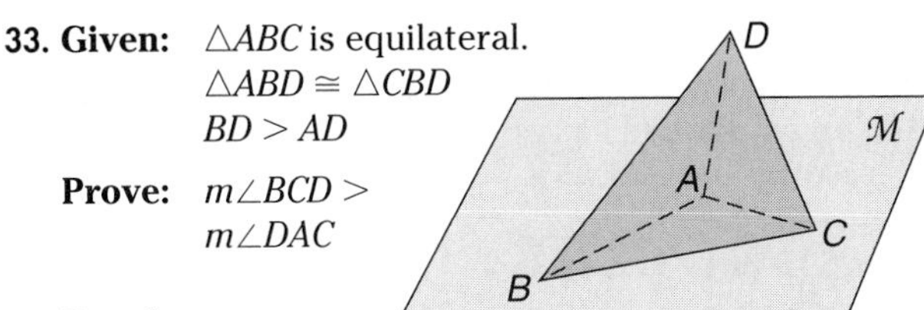

Proof:
It is given that $\triangle ABC$ is equilateral, $\triangle ABD \cong \triangle CBD$, and $BD > AD$. By the definition of an equilateral triangle, $\overline{AC} \cong \overline{BC}$. $\overline{DC} \cong \overline{DC}$ because congruence of segments is reflexive. The SSS Inequality allows us to say that $m\angle BCD > m\angle DCA$. $\overline{BC} \cong \overline{BA}$ by the definition of an equilateral triangle. $\overline{DB} \cong \overline{DB}$ because congruence of segments is reflexive. Thus, $\triangle DBC \cong \triangle DBA$ by SAS. Then by CPCTC, $\overline{DC} \cong \overline{DA}$. Therefore since if two sides of a triangle are congruent then the angles opposite those sides are congruent, $\angle DAC \cong \angle DCA$. $m\angle DAC = m\angle DCA$ by the definition of congruent angles. Then by substitution, $m\angle BCD > m\angle DAC$.

35. $\overline{AC}$ is not perpendicular to plane $\mathcal{F}$. **37.** The nutcracker is an example of an application of the SAS inequality. As force is applied to the arms of the lever, the distance between the ends of the arms of the lever is decreased. According to the SSS inequality, this makes the angle between the arms of the lever get smaller. This is the same angle as the one in the triangle formed by the arms of the lever and the segment between the points where the nut meets the arms of the lever. As this angle get smaller, the segment between the points where the nut meets the arms of the lever gets shorter, thereby crushing the nut. **39.** 6; $\overline{FG}, \overline{GH}, \overline{FH}$ **41.** $\left(3\frac{1}{2}, 3\frac{1}{2}\right)$ **43.** acute, isosceles **45.** $32y - 20$

Page 281 Chapter 5 Highlights
1. d **3.** b and l **5.** h **7.** m **9.** k

Pages 282–284 Chapter 5 Study Guide and Assessment
11. $m\angle NQO = m\angle NOQ = 57$, $m\angle ONQ = 66$ **13.** No; $m\angle MSQ \neq 90$. **15.** $x = 6, y = 26$ **17.** $x = 2, y = 1$

19. Given: $\triangle MJN \cong \triangle MJL$.
$\overline{MJ}$ does not bisect $\angle NML$.

Prove: $\overline{MJ}$ is an altitude of $\triangle NML$.

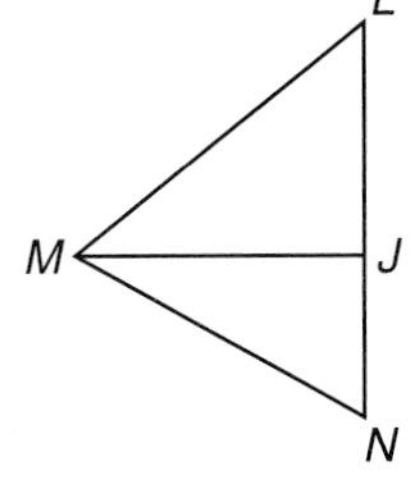

Proof:
We assume that $\overline{MJ}$ is not an altitude of $\triangle LMN$. This means that $\overline{MJ}$ is not perpendicular to $\overline{LN}$ by the definition of altitude. So, by the definition of perpendicular, this means $\angle MJN$ is not a right angle. Therefore, $\angle MJN$ is either acute or obtuse. Since $\angle MJL$ and $\angle MJN$ form a linear pair, these two angles are supplementary. Hence, if $\angle MJL$ is acute, $\angle MJN$ is obtuse (or vice versa), by the definition of supplementary. In either case, $\angle MJL \not\cong \angle MJN$. This is a contradiction, and hence our assumption must be false. Therefore $\overline{MJ}$ is an altitude of $\triangle LMN$.

21. Sample answer: $\angle 7$ **23.** $\angle YXB, \angle ACB, \angle 9, \angle 4, \angle 11$ **25.** $\overline{SP}$

27. Given: $FG < FH$

Prove: $m\angle 1 <$
$m\angle 2$

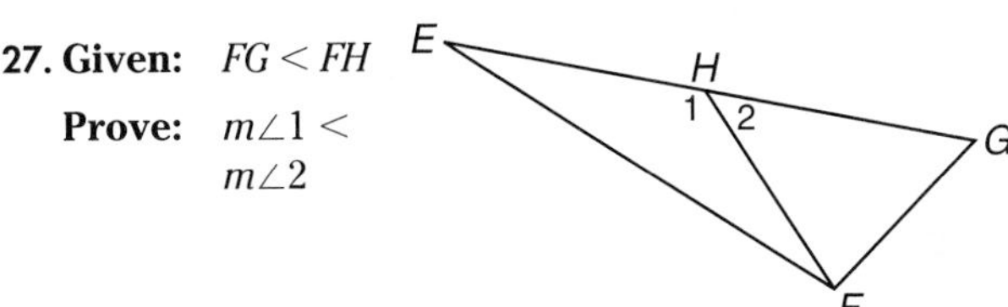

Proof:

Statements	Reasons
1. $FG < FH$	1. Given
2. $m\angle FGH > m\angle 2$	2. If one side of a $\triangle$ is longer than another, the $\angle$ opp. the longer side > the $\angle$ opp. the shorter side.
3. $m\angle 1 > m\angle FGH$	3. Exterior Angle Inequality Th.
4. $m\angle 1 > m\angle 2$	4. Transitive Property ($\neq$)

29. 17 and 31 **31.** No, the distances between the points do not satisfy the triangle inequality. **33.** $m\angle ALK < m\angle ALN$ **35.** $m\angle OLK > m\angle NLO$

37. Given: $AD = BC$

Prove: $AC > DB$

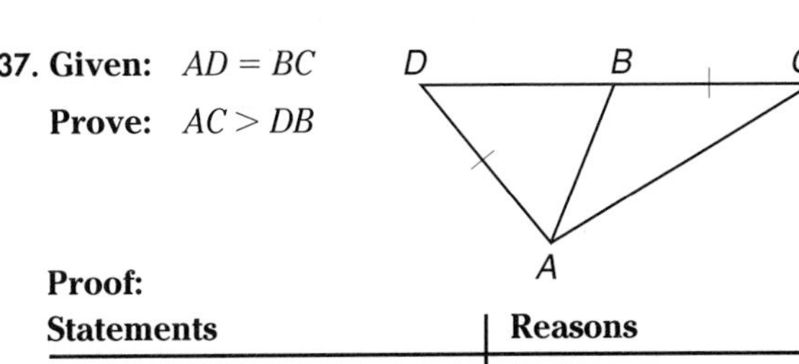

Proof:

Statements	Reasons
1. $AD = BC$	1. Given
2. $\overline{AD} \cong \overline{BC}$	2. Def. $\cong$ segments
3. $\overline{BA} \cong \overline{BA}$	3. Congruence of segments is reflexive.
4. $m\angle CBA > m\angle DAB$	4. Exterior Angle Inequality Th.
5. $AC > DB$	5. SAS Inequality Th.

39. 200 points

CHAPTER 6 EXPLORING QUADRILATERALS

Page 290 Lesson 6–1A

1. Sample answer: The opposite sides are congruent.
3. Sample answer: Yes, the same relationships hold true.

Pages 295–297 Lesson 6–1

7. $\overline{DC}$; opposite sides of a parallelogram are parallel.
9. $\overline{GF}$; opposite sides of a parallelogram are congruent.
11. $\triangle HDF$; SSS. Since the diagonals of a parallelogram bisect each other, $\overline{HC} \cong \overline{HF}$ and $\overline{HG} \cong \overline{HD}$ and since the opposite sides of a parallelogram are congruent, $\overline{GC} \cong \overline{DF}$. **13.** $m\angle Y = 47$, $m\angle X = 133 = m\angle Z$

15. Given: $PRSV$ is a parallelogram.
$\overline{PT} \perp \overline{SV}$
$\overline{QS} \perp \overline{PR}$

Prove: $\triangle PTV \cong \triangle SQR$

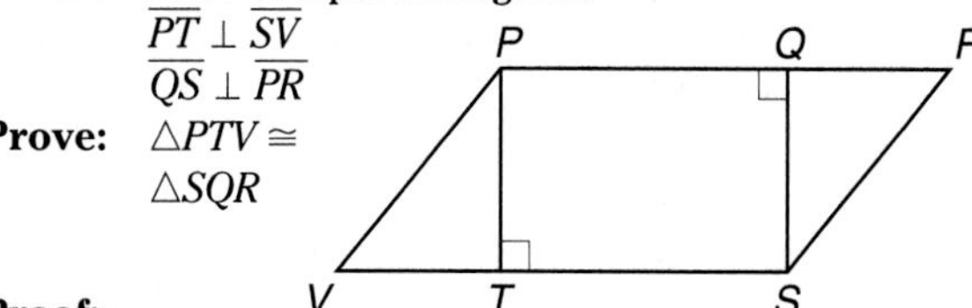

Proof:

Statements	Reasons
1. $PRSV$ is a parallelogram. $\overline{PT} \perp \overline{SV}$ $\overline{QS} \perp \overline{PR}$	1. Given
2. $\angle V \cong \angle R$	2. Opp. ∡ of a ▱ are ≅.
3. $\overline{PV} \cong \overline{RS}$	3. Opp. sides of a ▱ are ≅.
4. $\angle PTV$ and $\angle SQR$ are rt. ∡.	4. ⊥ lines form four rt. ∡.
5. $\triangle PTV$ and $\triangle SQR$ are rt. △s.	5. Def. rt. △
6. $\triangle PTV$ and $\triangle SQR$	6. HA

17. $\overline{AR}$; opposite sides of a parallelogram are parallel.
19. $\angle MAR$; opposite angles of a parallelogram are congruent. **21.** $\triangle RKM$; SSS. $\overline{MA} \cong \overline{RK}$ and $\overline{KM} \cong \overline{RA}$ because opposite sides of a parallelogram are congruent. $\overline{RM} \cong \overline{RM}$ because congruence of segments is reflexive.
23. $\overline{RK}$; opposite sides of a parallelogram are parallel.
25. $\triangle SKR$; SAS. Diagonals bisect each other, so $\overline{SA} \cong \overline{SK}$ and $\overline{SM} \cong \overline{SR}$. $\angle ASM \cong \angle KSR$ because they are vertical angles. **27.** $x = 118$; $y = 62$; $z = 118$ **29.** $x = 53$; $y = 15$; $z = 53$ **31.** The diagonals are not congruent.
33. $(0, -8)$, $(8, 6)$, or $(-8, 10)$ **35.** 19

37. Given: ▱$TEAM$
$MS = FS$

Prove: $\angle F \cong \angle TEA$

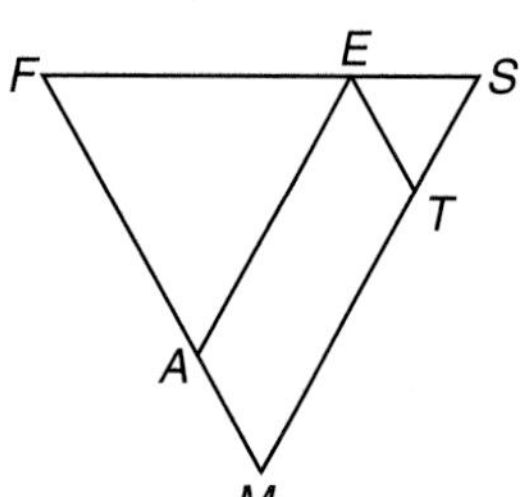

Proof:

Statements	Reasons
1. ▱$TEAM$ $MS = FS$	1. Given
2. $\angle F \cong \angle M$	2. Isosceles Triangle Th.
3. $\angle M \cong \angle TEA$	3. Opp. ∡ of a ▱ are ≅.
4. $\angle F \cong \angle TEA$	4. Congruence of angles is transitive.

39. Given: $ABCD$ is a parallelogram.

Prove: $\angle BAD \cong \angle DCB$
$\angle ABC \cong \angle CDA$

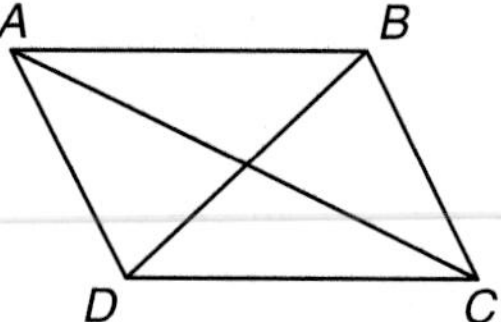

Proof:

We are given that $ABCD$ is a parallelogram. Since the opposite sides of a parallelogram are congruent, $\overline{AD} \cong \overline{BC}$, $\overline{AB} \cong \overline{CD}$. $\overline{BD} \cong \overline{BD}$ and $\overline{AC} \cong \overline{AC}$ because congruence of segments is reflexive. By Side-Side-Side, $\triangle BAD \cong \triangle DCB$ and $\triangle ABC \cong \triangle CDA$. Therefore we can conclude that $\angle BAD \cong \angle DCB$ and $\angle ABC \cong \angle CDA$ by CPCTC.

41. Sum = 360°; yes, two triangles are always formed by any quadrilateral and a diagonal.

45.

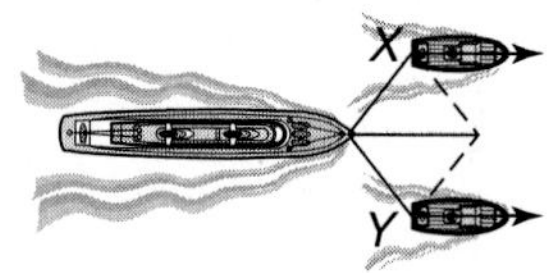

47. $m\angle ADC < m\angle ADB$ **49.** $\angle M$, $\angle K$, $\angle L$ **51.** obtuse; yes, $m\angle G$ and $m\angle H$ could be 30 each; no, $m\angle F \neq 60$.
53. If a quadrilateral has opposite sides parallel, then it is a parallelogram. **55.** 15%

Pages 301–304 Lesson 6–2

7. Yes; the triangles are congruent by SAS, so both pairs of opposite sides are congruent. **9.** $x = 4$, $y = 1$
11. False; the diagonals being congruent does not ensure that opposite sides will be congruent or parallel.

13. Given: $\overline{HD} \cong \overline{DN}$
$\angle DHM \cong \angle DNT$

Prove: Quadrilateral $MNTH$ is a parallelogram.

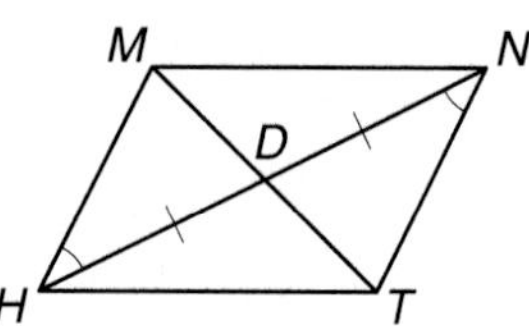

Proof:

Statements	Reasons
1. $\overline{HD} \cong \overline{DN}$ $\angle DHM \cong \angle DNT$	1. Given
2. $\angle MDH \cong \angle TDN$	2. Vert ∡ are ≅.
3. $\triangle MDH \cong \triangle TDN$	3. ASA
4. $\overline{MH} \cong \overline{TN}$	4. CPCTC
5. $\overline{MH} \parallel \overline{TN}$	5. If two lines are cut by a transversal, and alt. int. ∡ are ≅, then the lines are ∥.
6. $MNTH$ is a parallelogram.	6. If one pair of opp. sides of a quad. are both ∥ and ≅, then the quad. is a ▱.

15. Yes, both pairs of opposite angles are congruent.
17. Yes, both pairs of opposite angles are congruent.
19. No, none of the tests for parallelograms is met.
21. $x = 5$, $y = 16$ **23.** $x = 11$, $y = 12$ **25.** $x = 64$, $y = 23.5$
27. false, counterexample:

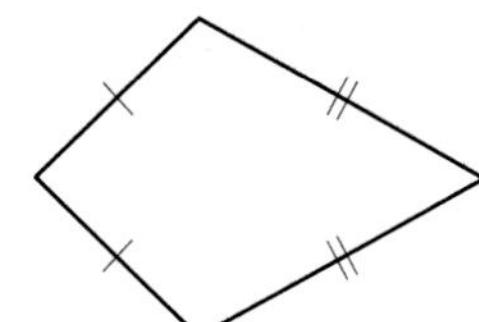

29. false, counterexample:

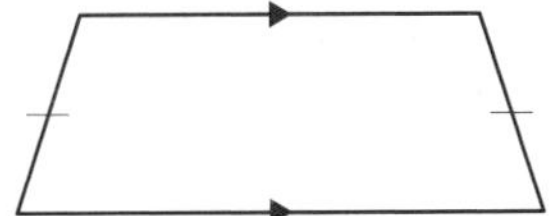

31. a parallelogram; slope of $\overline{FG} = -\frac{1}{4}$, slope of $\overline{HJ} = -\frac{1}{4}$, $\overline{FG} = \sqrt{17}$, $\overline{HJ} = \sqrt{17}$ since one pair of opposite sides is both parallel and congruent, *FGHJ* is a parallelogram.

33.

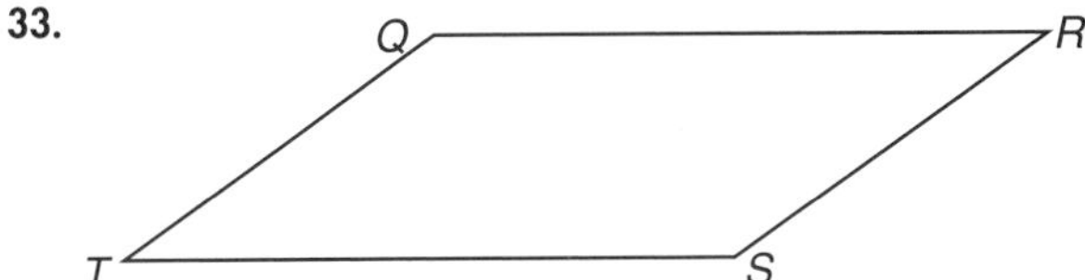

35. Given: $\overline{BD}$ bisects $\overline{AC}$.
$\overline{AC}$ bisects $\overline{BD}$.

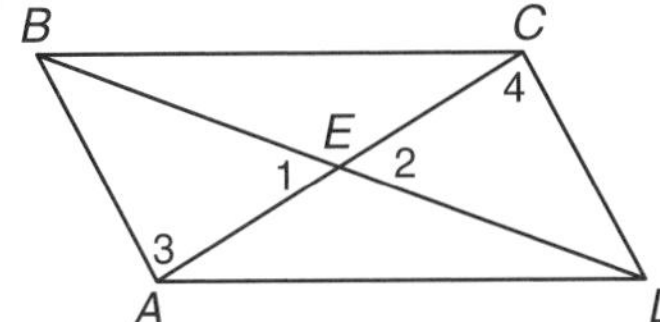

Prove: *ABCD* is a parallelogram.

Proof:

Statements	Reasons
1. $\overline{BD}$ bisects $\overline{AC}$. $\overline{AC}$ bisects $\overline{BD}$.	1. Given
2. $\overline{AE} \cong \overline{CE}$ $\overline{BE} \cong \overline{DE}$	2. Def. segment bisector
3. $\angle 1 \cong \angle 2$	3. Vert. ∠s are ≅.
4. $\triangle BEA \cong \triangle DEC$	4. SAS
5. $\angle 3 \cong \angle 4$ $\overline{AB} \cong \overline{CD}$	5. CPCTC
6. $\overline{AB} \parallel \overline{CD}$	6. If ⇸, and alt. int. ∠s are ≅, then the lines are ∥.
7. *ABCD* is a parallelogram.	7. If a pair of opp. sides of a quad. are ≅ and ∥, it is a ▱.

37. Subgoals:
1. Prove that $\triangle ABC \cong \triangle DEF$.
2. Use CPCTC to say that $\overline{AC} \cong \overline{FD}$.
3. State that $\overline{FA} \cong \overline{DC}$ by the definition of a regular hexagon.
4. Use both pair of opposite sides congruent to show *FDCA* is a parallelogram.

Given: *ABCDEF* is a regular hexagon.
Prove: *FDCA* is a ▱.

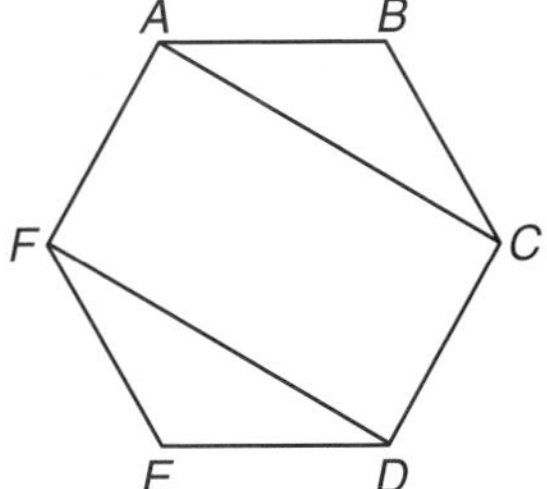

Proof:

Statements	Reasons
1. *ABCDEF* is a regular hexagon.	1. Given
2. $\overline{AB} \cong \overline{DE}$ $\overline{BC} \cong \overline{EF}$ $\angle E \cong \angle B$	2. Def. reg. hexagon
3. $\triangle ABC \cong \triangle DEF$	3. SAS
4. $\overline{AC} \cong \overline{DF}$	4. CPCTC
5. $\overline{FA} \cong \overline{CD}$	5. Def. reg. hexagon
6. *FDCA* is a ▱.	6. If opp. sides of a quad. are ≅, it is a ▱.

39a. Sample answer: Measure each pair of opposite segments. **39b.** Sample answer: Yes. The pattern yields quadrilaterals with guaranteed parallel opposite sides. **41.** 47 **43.** no solution **45.** $\overline{TA}$ **47.** The quadrilateral is a parallelogram. **49a.** 36% **49b.** 13,241,800

Page 305 Lesson 6–3A

1. The opposite sides of a rectangle are congruent. **3.** Yes; the opposite sides are congruent and the diagonals are congruent.

Pages 309–312 Lesson 6–3

5. $x = 15.5$ **7.** $x = 13.5$ **9.** *ABCD* is a rectangle. slope of $\overline{AB} = 1$; slope of $\overline{BC} = -1$; slope of $\overline{CD} = 1$; slope of $\overline{DA} = -1$ **11.** 26; 26; 64 **13.** Left; When stakes *F* and *G* are moved left an appropriate distance, $\overline{MK}$ lengthens and its midpoints will coincide with the midpoint of $\overline{JL}$, which becomes shorter after the move. **15.** 3 **17.** 12 **19.** 2 **21.** false; A counterexample:

23. True; The sum of the measures of the angles of a quadrilateral is 360. If all four angles are congruent, then the measure of each angle is $\frac{360}{4}$ or 90. Thus, all four angles are right and the definition of a rectangle is a quadrilateral with four right angles. **25.** 140 **27.** $-\frac{2}{3}$ or 4 **29.** 90 **31.** 33 **33.** 74 **35.** no; not a quadrilateral

37.

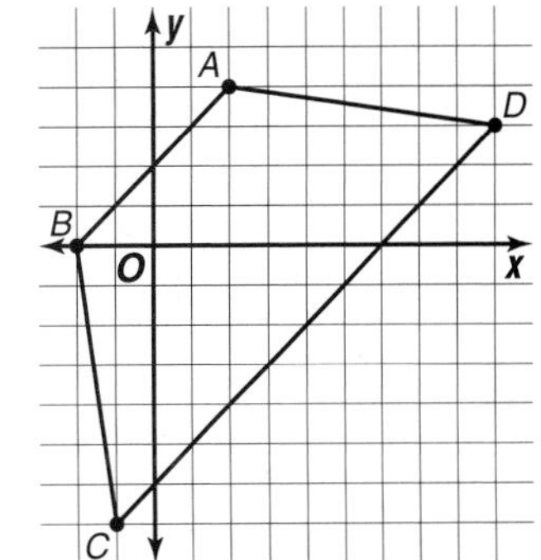

37a. $AC = \sqrt{130}$; $BD = \sqrt{130}$ **37b.** midpoint of $\overline{AC} = \left(\frac{1}{2}, -1\frac{1}{2}\right)$; midpoint of $\overline{BD} = \left(3\frac{1}{2}, 1\frac{1}{2}\right)$ **37c.** *ABCD* is not a rectangle because it is not a parallelogram.

39. Given: *ACDE* is a rectangle.
ABCE is a parallelogram.

Prove: $\triangle ABD$ is isosceles.

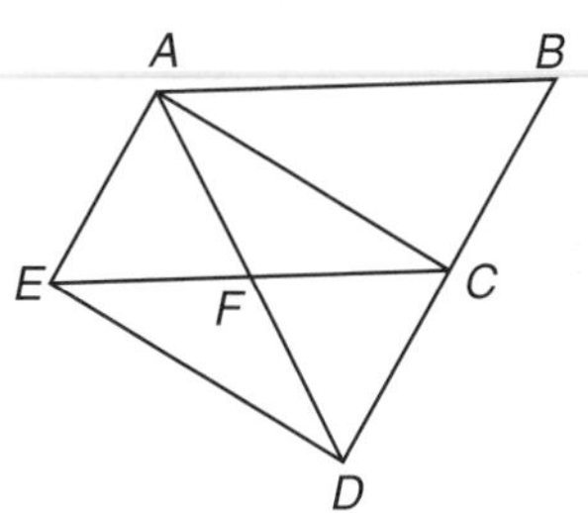

Proof:

Statements	Reasons
1. *ACDE* is a rectangle. *ABCE* is a parallelogram.	1. Given
2. $\overline{AD} \cong \overline{CE}$	2. If a ▱ is a rect. then its diagonals are ≅.
3. $\overline{AB} \cong \overline{CE}$	3. Opp. sides of a ▱ are ≅.
4. $\overline{AD} \cong \overline{AB}$	4. Congruence of segments is transitive.
5. $\triangle ABD$ is isosceles.	5. Def. isosceles △

41. Given: *PQMO* and *RQMN* are rectangles.
$\angle SVT \cong \angle UTV$

Prove: *STUV* is a parallelogram.

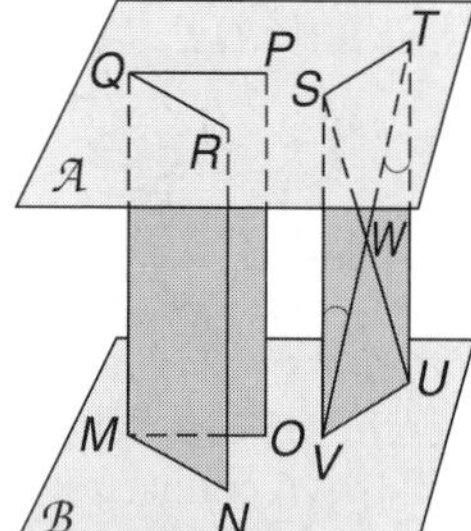

Proof:

Statements	Reasons
1. *PQMO* and *RQMN* are rectangles. $\angle SVT \cong \angle UTV$ $\overline{SU}$ and $\overline{TV}$ intersect at *W*.	1. Given
2. $\overline{PQ} \parallel \overline{MO}$ and $\overline{RQ} \parallel \overline{MN}$	2. Def. ▱
3. plane $\mathcal{N} \parallel$ plane $\mathcal{M}$	3. Def. ∥ planes
4. *S*, *U T*, *V* and *W* are in the same plane.	4. Def. intersecting lines
5. $\overline{ST} \parallel \overline{VU}$	5. Def. ∥ lines
6. $\overline{SV} \parallel \overline{TU}$	6. If ⟷, and alt. int. ∠s are ≅, then the lines are ∥.
7. *STUV* is a ▱.	7. Def. ▱

43. Sample answer: A rectangular package fits better in a box or display unit than a circular package would. **45.** no **45a.** Answers will vary. Sample answers: In architecture the golden rectangle is used in many well-known structures such as the Parthenon. In marketing: Many consumer products such as credit cards are approximately golden rectangles. **45b.** The diagonal is about 1.96 units long. **47.** HA, LL, and HL **49.** $\angle T \cong \angle X$ **51.** Transitive Property (=) **53.** −3

Page 312 Self Test

1. $\overline{PN}$; opposite sides of a parallelogram are parallel by definition. **3.** $\overline{LP}$; opposite sides of a parallelogram are congruent. **5.** yes **7.** yes

9. Given: ▱*WXZY*
$\angle 1$ and $\angle 2$ are complementary.

Prove: *WXZY* is a rectangle.

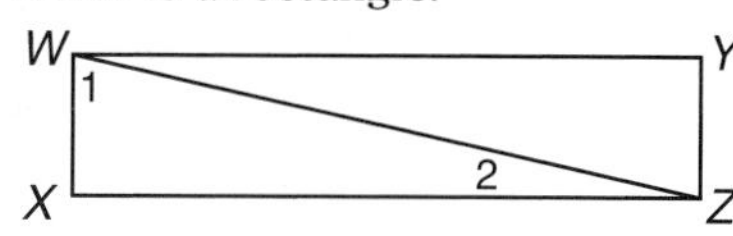

Proof:

Statements	Reasons
1. ▱*WXZY* $\angle 1$ and $\angle 2$ are complementary.	1. Given
2. $m\angle 1 + m\angle 2 = 90$	2. Def. complementary ∠s
3. $m\angle 1 + m\angle 2 + m\angle X = 180$	3. Angle Sum Theorem
4. $90 + m\angle X = 180$	4. Substitution Prop. (=)
5. $m\angle X = 90$	5. Subtraction Prop. (=)
6. $\angle X \cong \angle Y$	6. Opp. ∠s of a ▱ are ≅.
7. $m\angle Y = 90$	7. Substitution Prop. (=)
8. $\angle X$ and $\angle XWY$ are supp. $\angle X$ and $\angle XZY$ are supp.	8. Consec. ∠s of a ▱ are supp.
9. $m\angle X + m\angle XWY = 180$ $m\angle X + m\angle XZY = 180$	9. Def. Supp. ∠s
10. $90 + m\angle XWY = 180$ $90 + m\angle XZY = 180$	10. Substitution Prop. (=)
11. $m\angle XWY = 90$ $m\angle XZY = 90$	11. Subtraction Prop. (=)
12. $\angle Y$, $\angle XWY$, and $\angle XZY$ are rt. ∠s.	12. Def. rt. ∠s
13. *WXZY* is a rect.	13. Def. rect.

Pages 316–319 Lesson 6–4

7. Right; the diagonals of a rhombus are perpendicular, so $\angle PEN$ is right. **9.** No, the diagonals of a rhombus are not congruent unless the rhombus is a square. **11.** 22.5 **13.** 12 **15.** parallelogram, a rectangle **17.** rectangle, square

19. Given: *ABCD* is a rhombus.

Prove: $\overline{AC} \perp \overline{BD}$

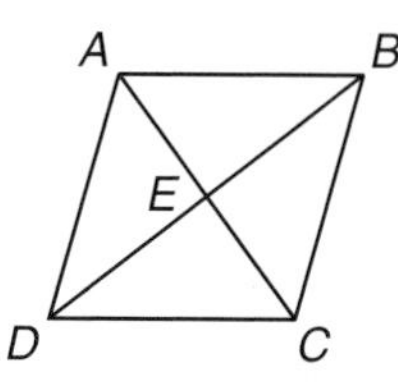

Proof:
The definition of a rhombus states that the four sides of the quadrilateral are congruent. Therefore, $\overline{AB} \cong \overline{BC} \cong \overline{CD} \cong \overline{AD}$. A rhombus is a parallelogram and the diagonals of a parallelogram bisect each other, so $\overline{BD}$ bisects $\overline{AC}$ at *E*. $\overline{AE} \cong \overline{CE}$ by the definition of a bisector. $\overline{BE} \cong \overline{BE}$ because congruence of segments is reflexive. Thus, $\triangle ABE \cong \triangle CBE$ by SSS. $\angle BEA \cong \angle BEC$ by CPCTC. $\angle BEA$ and $\angle BEC$ form a linear pair and if

two angles form a linear pair they are supplementary. Thus, $\angle BEA$ and $\angle BEC$ are supplementary. By the definition of supplementary angles, $m\angle BEA + m\angle BEC = 180$. Also, $m\angle BEA = m\angle BEC$ by the definition of congruent angles. So by substitution, $2m\angle BEA = 180$. Then $m\angle BEA = 90$ by the division property of equality. $\angle BEA$ is right by the definition of a right angle. Therefore, $\overline{AC} \perp \overline{BD}$ by the definition of perpendicular lines.

21. Scalene; the consecutive sides of *MNOP* are not congruent. **23.** Yes; the sides of *MNOP* are congruent and the diagonals are perpendicular, so the triangles are congruent by SAS. **25.** No; all squares are rhombi, but not all rhombi are squares. **27.** No; *MNOP* could be a rhombus if $\overline{PO} \cong \overline{ON}$, but the conditions are not sufficient for it to be a square. **29.** 5.25 **31.** 7.5 **33.** 12 **35.** parallelogram, rhombus **37.** parallelogram, rectangle, rhombus, square **39.** parallelogram, rectangle, rhombus, square **41.** rhombus, square **43.** False; some rhombi are not squares. **45.** True; the set of squares is a subset of the set of rectangles. **47.** False; some parallelograms are not rectangles.

49. Given: *JKLM* is a square.
Prove: $\overline{JL} \cong \overline{KM}$

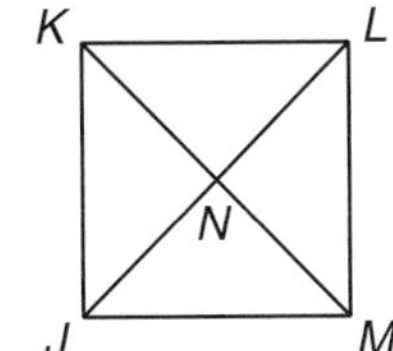

Proof:

Statements	Reasons
1. *JKLM* is a square.	1. Given
2. *JKLM* is a rectangle.	2. Def. square
3. $\overline{JL} \cong \overline{KM}$	3. If a ▱ is a rect. then its diagonals are ≅.

51. Given: *ABCD* is a parallelogram. $\overline{AC} \perp \overline{BD}$
Prove: *ABCD* is a rhombus.

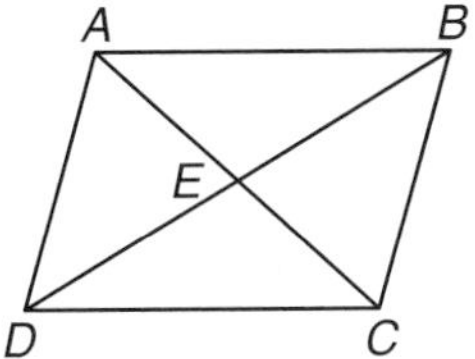

Proof:

We are given that *ABCD* is a parallelogram. The diagonals of a parallelogram bisect each other, so $\overline{AE} \cong \overline{EC}$. $\overline{BE} \cong \overline{BE}$ because congruence of segments is reflexive. We are also given that $\overline{AC} \perp \overline{BD}$. Thus, $\angle AEB$ and $\angle BEC$ are right angles by the definition of perpendicular lines. Then $\angle AEB \cong \angle BEC$ because all right angles are congruent. Therefore, $\triangle AEB \cong \triangle BEC$ by SAS. $\overline{AB} \cong \overline{BC}$ by CPCTC. Opposite sides of parallelograms are congruent, so $\overline{AB} \cong \overline{CD}$ and $\overline{BC} \cong \overline{AD}$. Then since congruence of segments is transitive, $\overline{AB} \cong \overline{CD} \cong \overline{BC} \cong \overline{AD}$. All four sides of *ABCD* are congruent, so *ABCD* is a rhombus by definition.

53. The flag of Denmark contains two red squares and two rectangles. The flag of St. Vincent and The Grenadines contains a blue rectangle, a green rectangle, a yellow rectangle, a blue and yellow rectangle, a yellow and green rectangle, and three green rhombi. The flag of Trinidad and Tobago contains two white parallelograms, and one black parallelogram. **55.** The quadrilateral formed by the back wall, the two side walls, and the short line is a square because all the sides are congruent and the angles are right. The quadrilateral formed by the front wall, the two side walls, and the short line is a square because all the sides are congruent and the angles are right. The back wall, the two side walls and the service line form a rectangle because the angles are all right, but the sides are not all congruent. The front wall, the two side walls and the service line form a rectangle because the angles are all right, but the sides are not all congruent. The two side walls, the short line, and the service line form a rectangle because the angles are all right, but the sides are not all congruent.
57. The legs are made so that they will bisect each other, so the quadrilateral formed by the ends of the legs is always a parallelogram. Therefore, the table top is parallel to the floor. **59.** Yes; If *V* is (12, 3), *TU* = *UV*. **61.** true **63.** false **65.** 4; 9 **67.** 217%

Page 320 Lesson 6–4B

5. Sample answer: For kite *ABCD*, $\angle B \cong \angle D$; $\overline{AC} \perp \overline{BD}$; $\overline{AC}$ bisects $\overline{BD}$, $\overline{AB} \cong \overline{AD}$; $\overline{BC} \cong \overline{CD}$; $\triangle ABC \cong \triangle ADC$.

Pages 324–328 Lesson 6–5

5. True; the diagonals of an isosceles trapezoid are congruent. **7.** False; if the diagonals bisected each other, *ABCD* would be a parallelogram not a trapezoid. **9.** 17 **11.** (5, 3), $\left(2\frac{1}{2}, 8\right)$ **13.** $RS \neq \frac{1}{2}(NP + MQ)$ because *MNPQ* is not a trapezoid. **15.** Sides are trapezoids; front, back & tops are rectangles. **17.** True; $\overline{TV} \cong \overline{TV}$ because congruence of segments is reflexive, $\angle TVR \cong \angle VTS$ because base angles of an isosceles trapezoid are congruent, and $\overline{TR} \cong \overline{VS}$ because diagonals of an isosceles trapezoid are congruent; thus $\triangle TRV \cong \triangle VST$ by SAS and $\angle TRV \cong \angle VST$ by CPCTC. **19.** True; $\overline{TV} \parallel \overline{SR}$ and consecutive interior angles are supplementary. **21.** 78 **23.** 5 **25.** 35 **27.** 2 **29.** slope of $\overline{RS} = 0$, slope of $\overline{TV} = 0$ **31.** midpoint of $\overline{RV} = \left(\frac{-1+(-2)}{2}, \frac{2+(-2)}{2}\right)$ or $\left(-1\frac{1}{2}, 0\right)$; midpoint of $\overline{ST} = \left(\frac{1+3}{2}, \frac{2+(-2)}{2}\right)$ or (2, 0); slope of $\overline{AB} = 0$; slope of $\overline{RS} = 0$ **33.** parallelogram **35.** trapezoid **37.** $180 - x$

39. Given: *DEFH* is a trapezoid with bases $\overline{DE}$ and $\overline{FH}$. $\overline{DE} \cong \overline{FG}$
Prove: *DEFG* is a parallelogram.

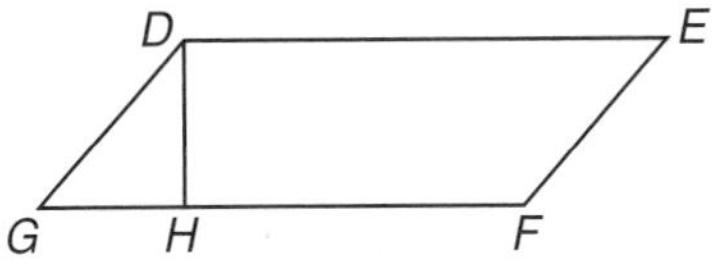

Proof:

Statements	Reasons
1. *DEFH* is a trapezoid with bases $\overline{DE}$ and $\overline{FH}$. $\overline{DE} \cong \overline{FG}$	1. Given
2. $\overline{DE} \parallel \overline{FH}$	2. Def. base of a trapezoid
3. *DEFG* is a parallelogram.	3. If a pair of opp. sides of a quad. are ≅ and ∥, it is a ▱.

41a. 6.18 **41b.** 4.74 **41c.** 26.6 **43a.** Sample answer: 10 and 30 **43b.** A measure cannot be negative and the average of the measures of the bases is 20. So the possible measures of the bases are 1 and 39, 2 and 38, ... and 19 and 21. The measures of the bases cannot be the same or the figure would be a parallelogram. Thus, there are 19 possible combinations. **43c.** There are an infinite number of combinations of possible measures if the integer restriction is removed because there is an infinite number of positive values with an average of 20. **45.** Top is a square, sides are isosceles trapezoids. **47.** 112 **49.** true **51.** yes; LA or AAS **53.** yes; detachment **55.** 123,454,321

Page 329 Chapter 6 Highlights

1. False; Every parallelogram is a quadrilateral. **3.** true **5.** False; You can prove that a parallelogram is a rectangle by proving that the diagonals are congruent. **7.** False; A square has all of the characteristics of a parallelogram, a rectangle, and a rhombus, but not the characteristics of a trapezoid. **9.** False; The legs of an isosceles trapezoid are congruent. **11.** If *QUAD* is a square, then it is also a parallelogram, a rectangle, a rhombus, and a quadrilateral. But *QUAD* is not a trapezoid.

Pages 330–332 Chapter 6 Study Guide and Assessment

13. $\angle WXY$; Opposite $\measuredangle$s in a $\square$ are $\cong$. **15.** $\overline{EZ}$; Diagonals of a $\square$ bisect each other. **17.** $\triangle ZWX$; SAS, or SSS **19.** $\overline{EW}$; Diagonals of a $\square$ bisect each other. **21.** $x = 28$, $y = 5$

23. Given: $\square ABCD$, $\overline{AE} \cong \overline{CF}$

Prove: Quadrilateral *EBFD* is a parallelogram.

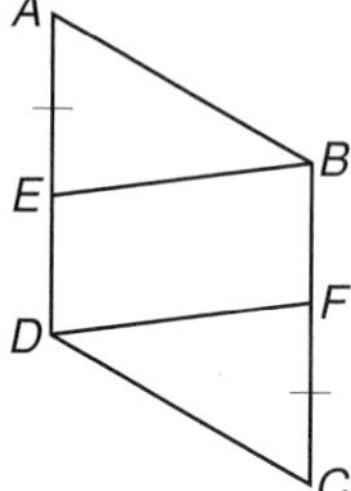

Proof:

Statements	Reasons
1. $\square ABCD$, $\overline{AE} \cong \overline{CF}$	1. Given
2. $\overline{BA} \cong \overline{DC}$	2. Opp. sides of a $\square$ are $\cong$.
3. $\angle A \cong \angle C$	3. Opp. $\measuredangle$s of a $\square$ are $\cong$.
4. $\triangle BAE \cong \triangle DCF$	4. SAS
5. $\overline{BE} \cong \overline{DF}$, $\angle BEA \cong \angle DFC$	5. CPCTC
6. $\overline{BC} \parallel \overline{AD}$	6. Def. $\square$
7. $\angle DFC \cong \angle FDE$	7. Alt. Int. $\angle$ Th.
8. $\angle BEA \cong \angle FDE$	8. Congruence of angles is transitive.
9. $\overline{BE} \parallel \overline{DF}$	9. If $\overleftrightarrow{}$, and corr. $\measuredangle$s are $\cong$, the lines are $\parallel$.
10. Quadrilateral *EBFD* is a parallelogram.	10. If a pair of opp. sides of a quad. are $\cong$ and $\parallel$, it is a $\square$.

25. 6 **27.** 1 or 2 **29.** rectangle, parallelogram **31.** parallelogram **33.** 28 **35.** 90 **37.** 38 **39.** 45 **41.** 47 **43.** 30 **45.** the kite flaps and the back are all trapezoids **47.** the deck top and the road surface

CHAPTER 7 CONNECTING PROPORTION AND SIMILARITY

Pages 341–345 Lesson 7–1

7. $\frac{170}{9}$ **9.** 14 **11.** yes **13.** no **15.** 18 in., 24 in., 30 in. **17.** $\frac{1000}{1}$ **19.** $\frac{4}{5}$ **21.** 1.295 **23.** $\frac{121}{7} \approx 17.29$ **25.** -11 **27.** 18 ft, 24 ft **29.** 3.9 **31.** $\frac{4}{9}$ **33.** 40 in., 60 in., 90 in. **35.** 49.92 **37.** 25%

39.

$$\begin{aligned} \frac{a+b}{b} &= \frac{c+d}{d} \\ (a+b)d &= (c+d)b \\ ad + bd &= cb + db \\ ad + bd &= cb + bd \\ ad + bd &= cb + bd \\ ad + bd - bd &= cb + bd - bd \\ ad &= cb \\ \frac{a}{b} &= \frac{c}{d} \end{aligned}$$

41. $x = 24$, $y = 12$ **43.** This is true because one of the angles measures 90° and the other two measure 45°. The ratio of 45:90 is 1:2. **45a.** The ratios (miles per gallon) are 27, 25, 29, 31, 20.3125, 29.44. **45b.** You cannot find the average number of miles per gallon by adding the number of miles per gallon at each stop and dividing by 6 because Keshia drove a different distance between each stop. To find the average, you must divide the total distance by the total amount of gas for all six stops. The average is about 26.86 miles per gallon. **47a.** 11,049; 10,862; 19,238; 12,985; 13,595 **47b.** You could find the ratio of people per movie screen to see which is smallest; Kansas City is the lowest at 10,862 people per screen, followed by Ann Arbor at 11,049 people per screen. **49a.** 6,658,400 lb **49b.** 16.4 lb

51a.

Currency	Cost of Making (cents)	Value (cents)	Ratio (cost/value)
penny	0.8	1	$\frac{0.8}{1} = 0.8$
nickel	2.9	5	$\frac{2.9}{5} = 0.58$
dime	1.7	10	$\frac{1.7}{10} = 0.17$
quarter	3.7	25	$\frac{3.7}{25} = 0.148$
half dollar	7.8	50	$\frac{7.8}{50} = 0.156$
dollar bill	3	100	$\frac{3}{100} = 0.03$

51b. Sample answer: No, it costs the most to make a penny since the ratio is the greatest at 0.8 and it costs the least to make a dollar bill since the ratio is the smallest at 0.03. **51c.** \$2.90, \$0.74 **53.** trapezoid **55.** It is less than 13 and greater than 3. **57.** 90, 52 **59.** 75, 105 **61.** False; *A*, *B*, and *P* are not necessarily collinear. **63.** -8, 2

Pages 350–353 Lesson 7-2

7. 12, 12 **9.** A **11.** Yes; the corresponding angles are congruent and the corresponding sides are proportional with a scale factor of $\frac{1}{3}$. **13.** No; the side on the second triangle measuring 18.4 should be 21.6. **15.** 71.05, 48.45 **17.** 91, 30 **19.** S **21.** N **23.** S **25.** A **27.** $\frac{4}{3}$ **29a.** 7.5 **29b.** 22.4 **29c.** 108 **29d.** 12.8 **31.** $\triangle ABC \sim \triangle IHG \sim \triangle JLK$ and $\triangle NMO \sim \triangle RPS$ **33.** Students should draw a rectangle $5\frac{1}{4}$ in. by $\frac{31}{8}$ in. **35.** Students should draw a rectangle $4\frac{1}{2}$ in. by $9\frac{3}{4}$ in. **37.** $L(16, 8)$ and $O(8, 8)$ or $L(16, -8)$ and $O(8, -8)$

39. In triangles ABC and DEF, $\overline{AC}$ and $\overline{DF}$ both have slope $\frac{3}{2}$ and $\overline{BC}$ and $\overline{EF}$ both have slope $-\frac{3}{2}$. Thus, $\overline{AC} \parallel \overline{DF}$ and $\overline{BC} \parallel \overline{EF}$. The horizontal axis is a transversal for both pairs of parallel lines. Thus, corresponding angles BAC and EDF are $\cong$. In the same way, corresponding angles B and E are $\cong$. $\angle C \cong \angle F$ because if two $\angle$s in a $\triangle$ are $\cong$ to two $\angle$s in another $\triangle$, the third $\angle$s are also $\cong$. Using the distance formula, $\frac{AC}{DF} = \frac{AB}{DE} = \frac{BC}{EF} = \frac{3}{2}$. Corresponding sides have the same ratio, and the angles are congruent, so the figures are similar. **41a.** 13.5 ft by 11.25 ft **41b.** 22.5 ft by 6.75 ft **43.** 12 **45.** No; one pair of sides parallel and the other pair of sides congruent is not a test for a parallelogram. **47.** 176 **49.** 84 **51.** $5x + 8y = -43$

Pages 357–361 Lesson 7-3

7. no **9.** $\triangle AEC \sim \triangle BDC$; AA Similarity **11.** SAS Similarity; $x = 4.5$ **13.** yes; $\triangle MNO \sim \triangle PQR$; SSS Similarity **15.** no **17.** yes; $\triangle RST \sim \triangle JKL$; AA Similarity **19.** False; this is not true for equilateral or isosceles triangles. **21.** False; the proportions are not the same. **23.** $\triangle QRS \sim \triangle QTR$ (AA Similarity); $\triangle QRS \sim \triangle RTS$ (AA Similarity); $\triangle QTR \sim \triangle RTS$ (Transitive Prop. of $\sim \triangle$s) **25.** $AE = 5$, $AC = 15$, $CF = 13\frac{1}{3}$, $CB = 20$, $CD = 12$, $EF = 16\frac{2}{3}$ **27.** $m\angle CDE = 43$, $m\angle A = 43$, $m\angle ABD = 47$, $m\angle DBE = 43$, $m\angle BDE = 47$

29. Given: $\overline{LP} \parallel \overline{MN}$

Prove: $\frac{LJ}{JN} = \frac{PJ}{JM}$

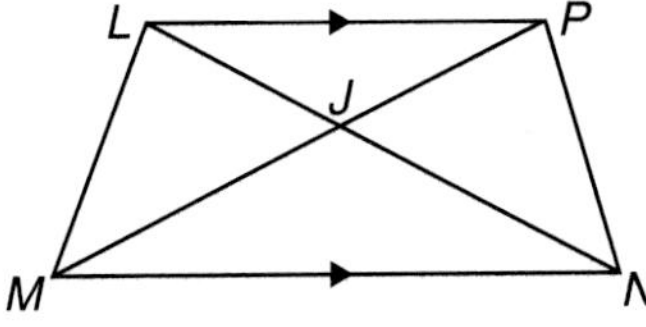

Proof:

Statements	Reasons
1. $\overline{LP} \parallel \overline{MN}$	1. Given
2. $\angle PLN \cong \angle LNM$, $\angle LPM \cong \angle PMN$	2. Alternate Interior $\angle$ Theorem
3. $\triangle LPJ \sim \triangle NMJ$	3. AA Similarity
4. $\frac{LJ}{JN} = \frac{PJ}{JM}$	4. Corr. sides of $\sim \triangle$s are proportional.

31. Given: $\triangle JFM \sim \triangle EFB$, $\triangle LFM \sim \triangle GFB$

Prove: $\triangle JFL \sim \triangle EFG$

Proof:

Since $\triangle JFM \sim \triangle EFB$ and $\triangle LFM \sim \triangle GFB$, then by the definition of similar triangles, $\frac{JF}{EF} = \frac{MF}{BF}$ and $\frac{MF}{BF} = \frac{LF}{GF}$. By the Transitive Property of Equality, $\frac{JF}{EF} = \frac{LF}{GF}$. $\angle F \cong \angle F$, by the Reflexive Property of congruent angles. Then, by SAS Similarity, $\triangle JFL \sim \triangle EFG$.

33. Given: $\angle B \cong \angle E$

$\frac{AB}{DE} = \frac{BC}{EF}$

Prove: $\triangle ABC \sim \triangle DEF$

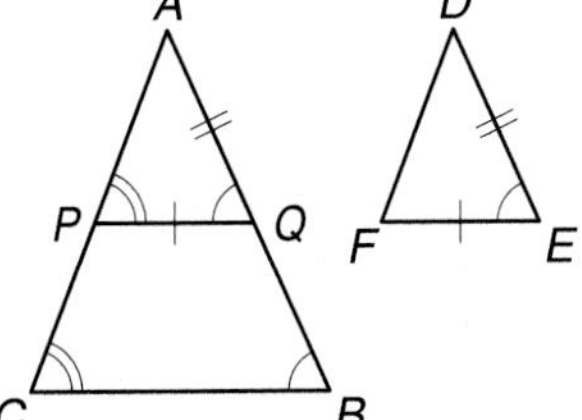

Proof:

Statements	Reasons
1. Draw $\overline{QP} \parallel \overline{BC}$ so that $\overline{QP} \cong \overline{EF}$	1. Parallel Postulate
2. $\angle APQ \cong \angle C$, $\angle AQP \cong \angle B$	2. Corresponding $\angle$s Postulate
3. $\angle B \cong \angle E$	3. Given
4. $\angle AQP \cong \angle E$	4. Transitive Prop. of $\cong \angle$s
5. $\triangle ABC \sim \triangle AQP$	5. AA Similarity
6. $\frac{AB}{AQ} = \frac{BC}{QP}$	6. Def. of $\sim \triangle$s
7. $\frac{AB}{DE} = \frac{BC}{EF}$	7. Given
8. $AB \cdot QP = AQ \cdot BC$, $AB \cdot EF = DE \cdot BC$	8. Equality of cross products
9. $QP = EF$	9. Def. of $\cong$ segments
10. $AB \cdot ERF = AQ \cdot BC$	10. Substitution Prop. (=)
11. $AQ \cdot BC = DE \cdot BC$	11. Substitution Prop. (=)
12. $AQ = DE$	12. Div. Prop. (=)
13. $\overline{AQ} \cong \overline{DE}$	13. Def. of $\cong$ segments
14. $\triangle AQP \cong \triangle DEF$	14. SAS
15. $\angle APQ \cong \angle F$	15. CPCTC
16. $\angle C \cong \angle F$	16. Transitive Prop. of $\cong \angle$s
17. $\triangle ABC \sim \triangle DEF$	17. AA Similarity

35. Proof of Reflexive Property

Given: $\triangle ABC$

Prove: $\triangle ABC \sim \triangle ABC$

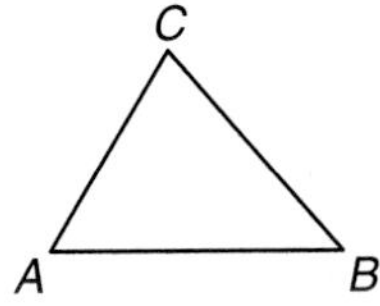

Proof:

Statements	Reasons
1. $\triangle ABC$	1. Given
2. $\angle A \cong \angle A$, $\angle B \cong \angle B$	2. Reflexive Prop. of $\cong \angle$s
3. $\triangle ABC \sim \triangle ABC$	3. AA Similarity

Proof of Symmetric Property

Given: $\triangle ABC \sim \triangle DEF$

Prove: $\triangle DEF \sim \triangle ABC$

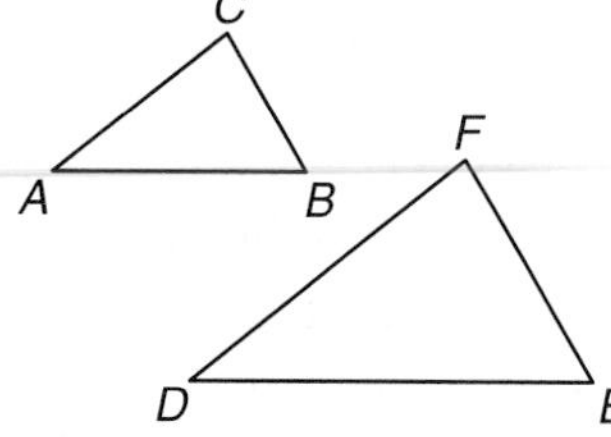

Proof:

Statements	Reasons
1. $\triangle ABC \sim \triangle DEF$	1. Given
2. $\angle A \cong \angle D$ $\angle B \cong \angle E$	2. Def. of ~ polygons
3. $\angle D \cong \angle A$ $\angle E \cong \angle B$	3. Symmetric Prop. of ≅ ∠s
4. $\triangle DEF \sim \triangle ABC$	4. AA Similarity

Proof of Transitive Property

Given: $\triangle ABC \sim \triangle DEF$
$\triangle DEF \sim \triangle GHI$

Prove: $\triangle ABC \sim \triangle GHI$

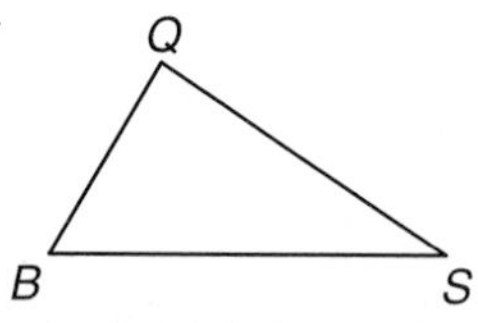

Proof:

Statements	Reasons
1. $\triangle ABC \sim \triangle DEF$ $\triangle DEF \sim \triangle GHI$	1. Given
2. $\angle A \cong \angle D$ $\angle B \cong \angle E$ $\angle D \cong \angle G$ $\angle E \cong \angle H$	2. Def. of ~ polygons
3. $\angle A \cong \angle G$ $\angle B \cong \angle H$	3. Transitive Prop. of ≅ ∠s
4. $\triangle ABC \sim \triangle GHI$	4. AA Similarity

37. 12.1 feet **39.** Yes; corresp. ∠s are ≅ and $\frac{2.0}{3.0} = \frac{1.6}{2.4} = \frac{1.8}{2.7}$. **41.** true

43.

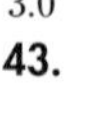

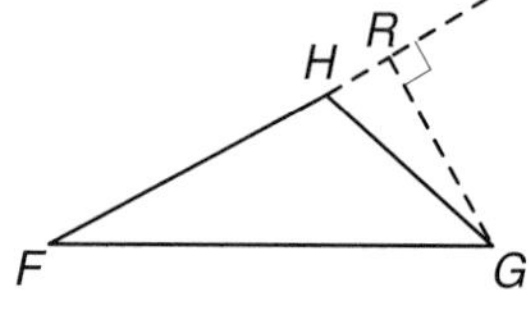

45. Reflexive Prop. of ≅ Segments **47.** $\{t \mid t \geq 5\}$

Page 361 Self Test

1. 2 **3.** about 487 soft drinks **5.** yes; SAS Similarity

Pages 366–369 Lesson 7–4

7a. true **7b.** false; $RS = 16$ **9.** Yes; if 3 or more ∥ lines cut off ≅ segments on one transversal, then they cut off ≅ segments on every transversal. **11.** Yes; if 3 or more ∥ lines intersect 2 transversals, then they cut off the transversals proportionately. **13.** DF **15.** GB **17.** true **19.** false; $\frac{BD}{DC} = \frac{2}{3}$ **21.** 21, 15 **23.** 10 **25.** 10 **27.** $B(5, -2)$; $C(11, 2)$ **29.** $AB = 7$, $BC = 10$, $AC = 9$

31.

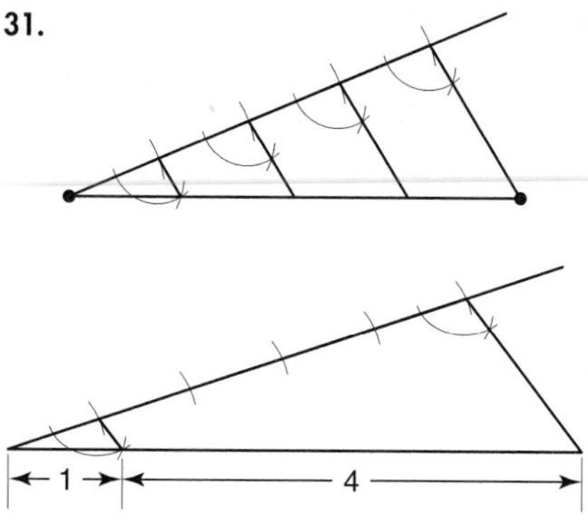

33. Given: D is the midpoint of $\overline{AB}$.
E is the midpoint of $\overline{AC}$.

Prove: $\overline{DE} \parallel \overline{BC}$
$DE = \frac{1}{2}BC$

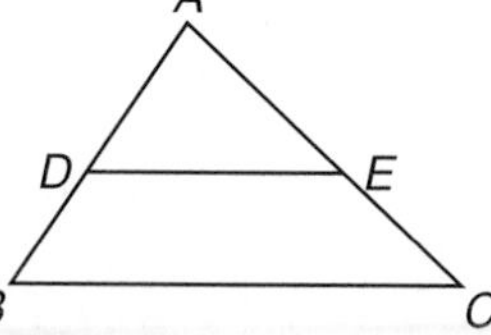

Proof:

Statements	Reasons
1. D is the midpoint of $\overline{AB}$. E is the midpoint of $\overline{AC}$.	1. Given
2. $\overline{AD} \cong \overline{DB}$ $\overline{AE} \cong \overline{EC}$	2. Def.of midpoint
3. $AD = DB$ $AE = EC$	3. Def. of ≅ segments
4. $AB = AD + DB$ $AC = AE + EC$	4. Segment Addition Postulate
5. $AB = AD + AD$ $AC = AE + AE$	5. Substitution Prop. (=)
6. $AB = 2AD$ $AC = 2AE$	6. Substitution Prop. (=)
7. $\frac{AB}{AD} = 2$ $\frac{AC}{AE} = 2$	7. Division Prop. (=)
8. $\frac{AB}{AD} = \frac{AC}{AE}$	8. Transitive Prop. (=)
9. $\angle A \cong \angle A$	9. Reflexive prop. of ≅ ∠s
10. $\triangle ADE \sim \triangle ABC$	10. SAS Similarity
11. $\angle ADE \cong \angle ABC$	11. Def. of ~ polygons
12. $\overline{DE} \parallel \overline{BC}$	12. If ⇸, and corr. ∠s are ≅, the lines are ∥.
13. $\frac{BC}{DE} = \frac{AB}{AD}$	13. Def. of ~ polygons
14. $\frac{BC}{DE} = 2$	14. Substitution Prop. (=)
15. $2DE = BC$	15. Mult. Prop. (=)
16. $DE = \frac{1}{2}BC$	16. Division Prop. (=)

35. $L(17, 8)$; $M(-7, -16)$ **37.** $x = \frac{14}{9}$ cm, $y = \frac{7}{3}$ cm, $z = \frac{35}{9}$ cm **39.** $\triangle ABC \sim \triangle ADE$ by SAS Similarity, so $\frac{AD}{AB} = \frac{DE}{BC}$. $AD = 60$ and $AB = 100$ so $\frac{60}{100} = \frac{DE}{BC}$. $\frac{3}{5} = \frac{DE}{BC}$ and $\frac{3}{5}BC = DE$. **41.** $1\frac{2}{3}$, $1\frac{1}{3}$ **43.** true **45.** $\angle RST \cong \angle ABC$ **47.** $(k - 3)(k + 6)$

Selected Answers

Pages 373–377 Lesson 7–5

5. AB; Angle Bisector Theorem **7.** 15 **9.** 5 **11.** Sample answer: 6, 8, 10 and 9, 12, 15 **13.** 8.6 **15.** 18 **17.** DB **19.** $11\frac{1}{5}$ **21.** 36.68 **23.** 6 **25.** 29.25 **27.** 36 **29.** 5, 13.5 **31.** $8\frac{8}{9}$, $11\frac{1}{9}$ **33.** 26.14

35. Given: $\overline{JF}$ bisects $\angle EFG$.
$\overline{EH} \parallel \overline{FG}$
$\overline{EF} \parallel \overline{HG}$

Prove: $\frac{EK}{KF} = \frac{GJ}{JF}$

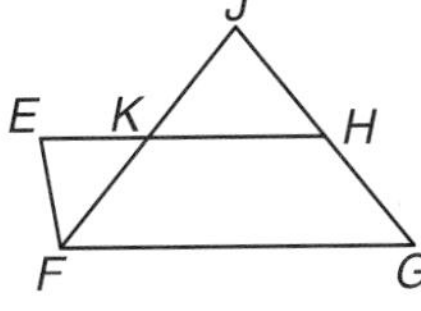

Proof:

Statement	Reasons
1. $\overline{EH} \parallel \overline{FG}$, $\overline{EF} \parallel \overline{HG}$, $\overline{JF}$ bisects $\angle EFG$.	1. Given
2. $\angle EFK \cong \angle KFG$	2. Def. of $\angle$ bisector
3. $\angle KFG \cong \angle JKH$	3. Corresponding $\angle$ Postulate
4. $\angle JKH \cong \angle EKF$	4. Vertical $\angle$s are $\cong$.
5. $\angle EFK \cong \angle EKF$	5. Transitive Prop. (=)
6. $\angle EFK \cong \angle FJH$	6. Alternate Interior $\angle$ Theorem
7. $\angle EKF \cong \angle FJH$	7. Transitive Prop. (=)
8. $\triangle EKF \sim \triangle GJF$	8. AA Similarity
9. $\frac{EK}{KF} = \frac{GJ}{JF}$	9. Def. of $\sim \triangle$s

37. Given: $\angle C \cong \angle BDA$

Prove: $\frac{AC}{DA} = \frac{AD}{BA}$

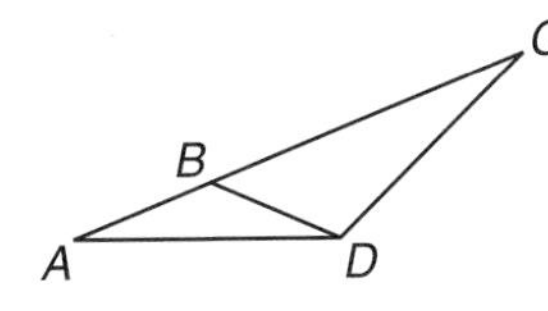

Proof:

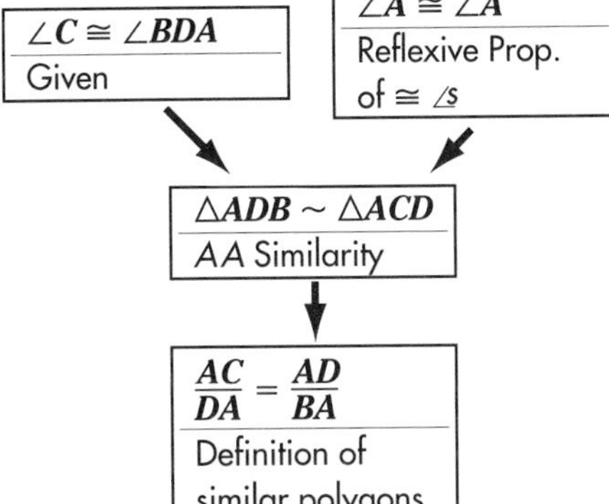

39. Given: $\triangle ABC \sim \triangle RST$
$\overline{AD}$ is a median of $\triangle ABC$.
$\overline{RU}$ is a median of $\triangle RST$.

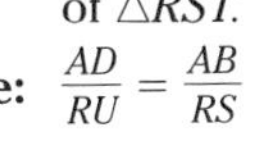
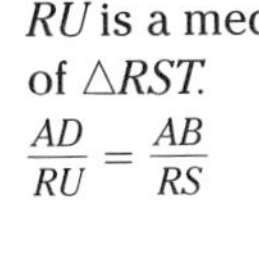

Prove: $\frac{AD}{RU} = \frac{AB}{RS}$

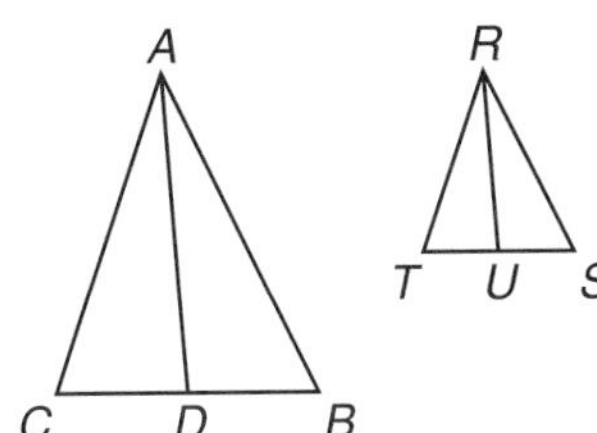

Proof:

We are given that $\triangle ABC \sim \triangle RST$, $\overline{AD}$ is median of $\triangle ABC$, and $\overline{RU}$ is a median of $\triangle RST$. So, by the definition of median, $CD = DB$ and $TU = US$. According to the definition of similar polygons, $\frac{AB}{RS} = \frac{CB}{TS}$. $CB = CD + DB$ and $TS = TU + US$ by the Segment Addition Postulate.

Substituting, $\frac{AB}{RS} = \frac{CD + DB}{TU + US}$

$\frac{AB}{RS} = \frac{DB + DB}{US + US}$

$\frac{AB}{RS} = \frac{DB}{US}$

$\angle B \cong \angle S$ by the definition of similar polygons and $\triangle ABD \sim \triangle RSU$ by SAS Similarity. Therefore, $\frac{AD}{RU} = \frac{AB}{RS}$ by the defintion of similar polygons.

41a. 6; 24; 1:2; 1:4 **41b.** 35.5; 319.5; 1:3; 1:9 **41c.** 16.1; 578.3; 1:6; 1:36; Sample answer: If the ratio of the measures of the sides of two similar triangles is a:b, then the ratio of their areas is a^2:b^2. **43.** Yes; the enlarged picture will take approximately 109.2 cm of piping. **45.** yes; proportional perimeters **47.** 8.4 **49.** 14 seniors **51.** Yes; diagonals bisect each other. **53.** No; the measures of corresponding sides are not equal. **55.** If a geometry test score is 89, then it is above average. **57.** 600%

Pages 381–383 Lesson 7–6

7. 1, 3, 6, 10, 15, ...; each difference is 1 more than the preceding difference. The triangular numbers are the numbers in the diagonal.

9. Given: $\triangle ABC$ is equilateral.
$CD = \frac{1}{3}CB$
$CE = \frac{1}{3}CA$

Prove: $\triangle CED \sim \triangle CAB$

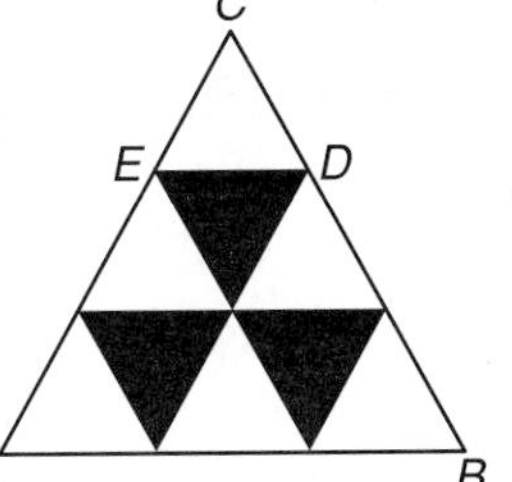

Proof:

Statements	Reasons
1. $\triangle ABC$ is equilateral. $CD = \frac{1}{3}CB$, $CE = \frac{1}{3}CA$	1. Given
2. $\overline{AC} \cong \overline{BC}$	2. Def. of equilateral $\triangle$
3. $AC = BC$	3. Def of $\cong$ segments
4. $\frac{1}{3}AC = \frac{1}{3}CB$	4. Mult. Prop. (=)
5. $CD = CE$	5. Substitution Prop. (=)
6. $\frac{CD}{CB} = \frac{CE}{CB}$	6. Division Prop. (=)
7. $\frac{CD}{CB} = \frac{CE}{CA}$	7. Substitution Prop. (=)
8. $\angle C \cong \angle C$	8. Reflexive Prop. of $\cong$ $\angle$s
9. $\triangle CED \sim \triangle CAB$	9. AA Similarity

11a. converges to 1 **11b.** alternates between 0.2 and 5.0 **11c.** converges to 1

13a.

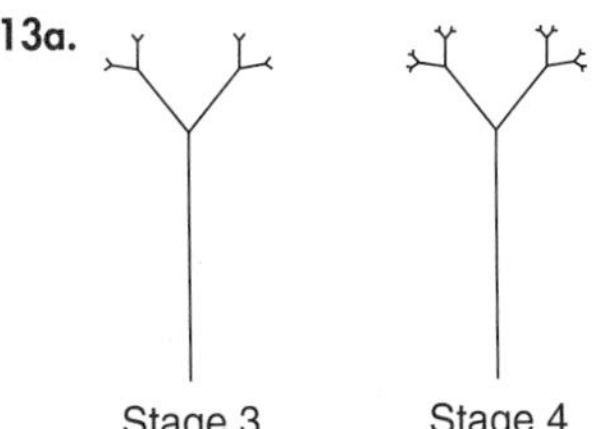

stage 1: 2, stage 2: 6, stage 3: 14, stage 4: 30 **13b.** Sample answer: at stage n, the number of branches would be $S_n = 2(1 - 2^n)$. **15.** Similar to stage 1 of the Sierpinski triangle variation from Exercise 10.

17a. stage 1: 3 units
stage 2: $3 \cdot \frac{4}{3}$ or 4 units
stage 3: $3 \cdot \frac{4}{3} \cdot \frac{4}{3} = 3\left(\frac{4}{3}\right)^2$ or $5\frac{1}{3}$ units
stage 4: $3 \cdot \frac{4}{3} \cdot \frac{4}{3} \cdot \frac{4}{3} = 3\left(\frac{4}{3}\right)^3$ or $7\frac{1}{9}$ units
17b. $P = 3\left(\frac{4}{3}\right)^{n-1}$; as the stages increase, the perimeter continues to increase. The perimeter will approach infinity. **19a.** Trisect each of the three edges; replace the middle section on the center edge with three segments of length equal to the length removed; replace the first section on each of the outside edges with three segments of length equal to that removed. Repeat. **21a.** 25 **21b.** 1275 **23.** 2400 units **25.** $-\frac{2}{7}$ **27.** no conclusion **29.** $\{(-1, -1), (0, 2), (2, 8), (4, 14)\}$

Page 385 Lesson 7–6B

1a. The midpoint of $\overline{PA}$ is (0.1, 0.2) in $\triangle L$; the midpoint of $\overline{PB}$ is (0.6, 0.2) in $\triangle R$; the midpoint of $\overline{PC}$ is (0.35, 0.65) in $\triangle T$. **1b.** The midpoint of $\overline{PA}$ is (0.25, 0.1) in $\triangle L$; the midpoint of $\overline{PB}$ is (0.75, 0.1) in $\triangle R$; the midpoint of $\overline{PC}$ is (0.5, 0.55) in $\triangle T$. **1c.** $\frac{1}{27}$ **1d.** $\frac{1}{9}$ **1e.** $\frac{1}{9}$ **3a.** $\frac{1}{9}$ **3b.** $\frac{1}{3}$ **3c.** $\frac{1}{27}$

Page 387 Chapter 7 Highlights

1. false; proportional **3.** false; two, two **5.** true **7.** true **9.** true **11.** false; $ad = bc$

Pages 388 390 Chapter 7 Study Guide and Assessment

13. $\frac{1}{2}$ **15.** 3.2 **17.** 1.25 **19.** $9\frac{1}{3}$ **21.** false **23.** 10 **25.** no; not enough information **27.** 2.5 **29.** $9\frac{1}{3}$
31.

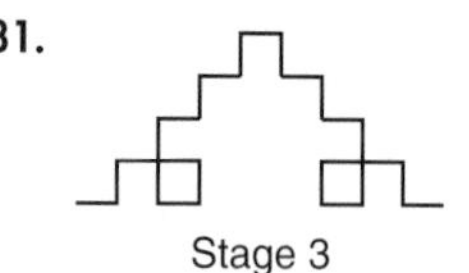

33. 29 **35.** 31.2 cm **37.** 2

CHAPTER 8 APPLYING RIGHT TRIANGLES AND TRIGONOMETRY

Page 396 Lesson 8–1A

1. yes. **3.** The sum of the areas of the two smaller squares ($a^2 + b^2$) is equal to the area of the larger square (c^2).

Pages 401–404 Lesson 8–1

7. 10 **9.** $\sqrt{10} \approx 3.16$; $\sqrt{14} \approx 3.74$ **11.** $\sqrt{51} \approx 7.14$
13.

a	b	c
3	4	5
6	8	10
9	12	15
12	16	20

13a. yes; 3, 4, 5 **13b.** Each triple is a multiple of the triple 3, 4, and 5. **13c.** Sample answer: The triples are all multiples of the triple 3, 4, and 5. **13d.** Yes; the measures of the sides are always multiples of 3, 4, and 5. **15.** 6 **17.** $\frac{\sqrt{2}}{3} \approx 0.47$ **19a.** $\overline{AH}, \overline{SH}$ **19b.** $\overline{AS}, \overline{SH}$ **19c.** $\overline{SH}, \overline{LS}$ **21.** $2\sqrt{6} \approx 4.90$; $\sqrt{33} \approx 5.74$ **23.** $7\frac{1}{2}$; $2\frac{1}{2}$ **25.** $5\sqrt{5} \approx 11.18$; $5\sqrt{30} \approx 27.39$ **27.** 4.5 **29.** 20 **31.** no

33. Sample answers: 16, 30, 34; 24, 45, 51 **35.** Sample answers: 18, 80, 82; 27, 120, 123

37. Given: $\angle ADC$ is a right angle. $\overline{DB}$ is an altitude of $\triangle ADC$.
Prove: $\frac{AB}{DB} = \frac{DB}{CB}$
Proof:
It is given that $\triangle ADC$ is a right triangle and $\overline{DB}$ is an altitude of $\triangle ADC$. $\triangle ADC$ is a right triangle by the definition of a right triangle. Therefore, $\triangle ADB \sim \triangle DCB$, because if the altitude is drawn from the vertex of the right angle to the hypotenuse of a right triangle, then the two triangles formed are similar to the given triangle and to each other. So $\frac{AB}{DB} = \frac{DB}{CB}$ by definition of similar polygons.

39. Given: $\triangle ABC$ with sides of measure a, b, and c, where $c^2 = a^2 + b^2$
Prove: $\triangle ABC$ is a right triangle.

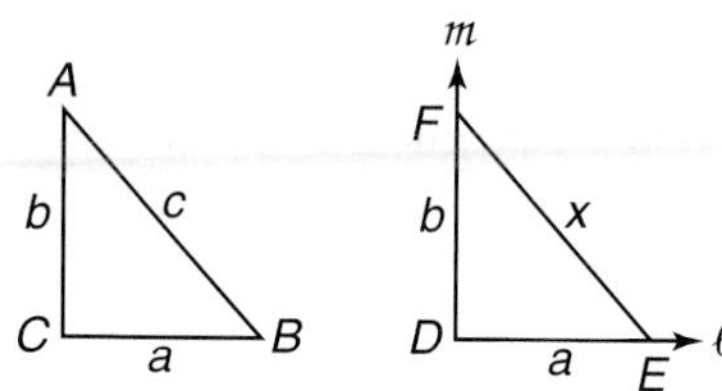

Proof:
Draw $\overline{DE}$ on line ℓ with measure equal to a. At D, draw line $m \perp \overline{DE}$. Locate point F on m so that $DF = b$. Draw $\overline{FE}$ and call its measure x. Because $\triangle FED$ is a right triangle, $a^2 + b^2 = x^2$. But $a^2 + b^2 = c^2$, so $x^2 = c^2$ or $x = c$. Thus, $\triangle ABC \cong \triangle FED$ by SSS. This means $\angle C \cong \angle D$. Therefore, $\angle C$ must be a right angle, making $\triangle ABC$ a right triangle. **41a.** 3, 4, 5; 6, 8, 10; 12, 16, 20; 24, 32, 40; 27, 36, 45 **41b.** yes **43.** 2.4 yd **45.** no; Sample counterexample: isosceles triangle with sides measuring 4, 4, and 7

$$\sqrt{4} + \sqrt{4} \stackrel{?}{=} \sqrt{7}$$
$$2 + 2 \stackrel{?}{=} 2.65$$
$$4 \neq 2.65$$

47. 254 mi **49.** 13

51. Given: $\overline{GH} \cong \overline{HJ}$, $GL < JL$
Prove: $m\angle 1 < m\angle 2$

Proof:
$\triangle GHL$ and $\triangle JHL$ satisfy the SSS Inequality. That is, $\overline{GL} \cong \overline{HJ}$ and $\overline{HL} \cong \overline{HL}$, and $GL < JL$. It follows that $m\angle 1 < m\angle 2$. **53.** 30, 60, 90

55.

x	y
0	1
1	4
−1	−2

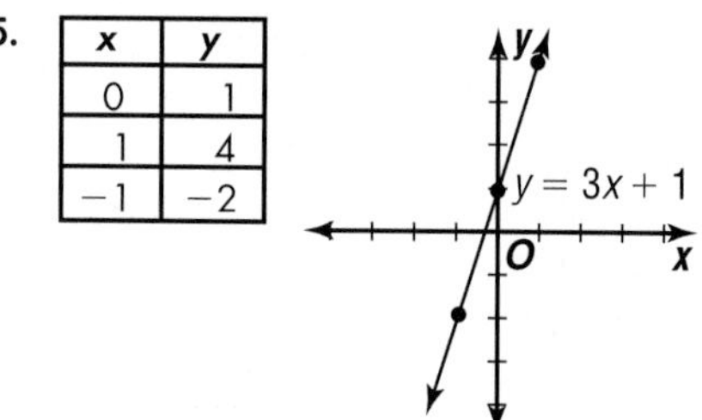

Pages 409–411 Lesson 8–2

5. 16; $8\sqrt{3} \approx 13.86$ **7.** 1 ft **9.** $(-4, -2 + 10\sqrt{3})$ or about $(-4, 15.32)$ **11.** $4.5\sqrt{3} \approx 7.79$; 4.5 **13.** $\frac{3\sqrt{2}}{2} \approx 2.12$; $3\sqrt{2} \approx 4.24$ **15.** $9\sqrt{3} \approx 15.59$; 9 **17.** $13.5\sqrt{2}$ or about 19.09 cm **19.** 9 m **21.** $6.5\sqrt{3}$ or about 11.26 cm **23.** $10.4\sqrt{3}$ or about 18.01 m **25.** $7\sqrt{2} \approx 9.90$ **27.** $\left(-3 - \frac{13\sqrt{3}}{3}, -6\right)$ or about $(-10.51, -6)$ **29.** 2.25

31a.

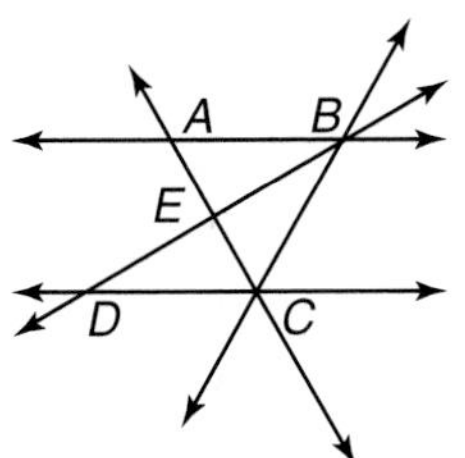

31b. $m\angle BCD = 120$, $m\angle BCE = 60$, $m\angle ABC = 180 - 120 = 60$, $m\angle EBC = 30$, $m\angle BEC = 180 - 60 - 30 = 90$; $\triangle BEC$ is a 30°-60°-90° triangle. **31c.** $\frac{22\sqrt{3}}{3} \approx 12.70$; $\frac{44\sqrt{3}}{3} \approx 25.40$ **33.** 23 ft **35.** AA Similarity, SSS Similarity, SAS Similarity **37.** Yes; since $QS = 2\sqrt{26}$ and $RT = 2\sqrt{26}$, the diagonals are congruent.

39. $\sqrt{5}$

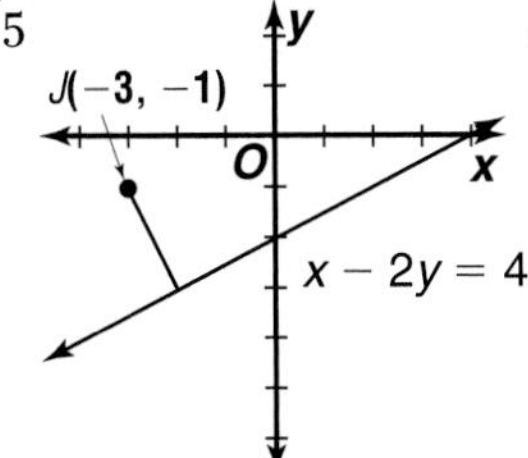

41. $8g(g + 4)$

Pages 416–420 Lesson 8–3

7. $\frac{20}{21} \approx 0.9524$ **9.** 0.2419 **11.** 63.5 **13.** 58.0 **15.** about 20.6 **17.** $\frac{\sqrt{26}}{26} \approx 0.1961$ **19.** $\frac{5\sqrt{26}}{26} \approx 0.9806$ **21.** $\frac{5}{1} = 5.0000$ **23.** $\frac{\sqrt{26}}{26} \approx 0.1961$ **25.** $\frac{5}{1} = 5.0000$ **27.** 0.9135 **29.** 0.9511 **31.** 0.3640 **33.** 75.6 **35.** 27.2 **37.** 23.2 **39.** 32.6 **41.** 20.4 **43.** 6.1 **45.** about 54.5 **47.** 38.4; 32.6 **49.** 41.8; 29.2 **51.** about 35.5 **53a.** about 272,837 astronomical units **53b.** The stellar parallax would be too small. **55.** $6\sqrt{2} \approx 8.49$ **57.** 15 **59.** Walnut to Oak **61.** 7.5; 12 **63.** $\angle HGI$ **65.** (2, 4)

Page 420 Self Test

1. $3\sqrt{15} \approx 11.62$ **3.** 4 **5.** $7.5\sqrt{2} \approx 10.61$ **7.** $\frac{3\sqrt{13}}{13} \approx 0.8321$ **9.** $\frac{2}{3} \approx 0.6667$

Pages 422–425 Lesson 8–4

5. $\angle OCP$; $\angle DPC$ **7.** $\cos P = \frac{2.3}{5.5}$; 65.3 **9.** about 151.1 ft **11.** $\angle GEF$; $\angle HFE$ **13.** $\angle KHJ$; $\angle IJH$ **15.** $\tan Y = \frac{54}{28}$; 62.6 **17.** $\cos 66° = \frac{7}{XY}$; 17.2 **19.** $\cos X = \frac{4.5}{6.6}$; 47.0 **21a.** about 86.3° **21b.** about 166.1 ft **23.** about 177.7 yd **25.** about 3.6° **27a.** about 210.3 ft **27b.** about 276.4 ft **29.** about 993.0 ft **31.** $\frac{\sin M}{\sin G}$ **33a.** about 3.12 ft **33b.** the bottom 5.93 ft of the window **35a.** $\frac{4}{5} = 0.8000$ **35b.** $\frac{4}{5} = 0.8000$ **35c.** $\frac{4}{3} \approx 1.3333$ **37.** 10 ft

39.

41. 78 in.

43.

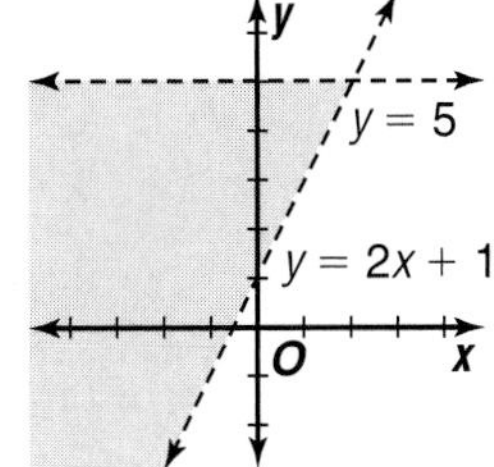

Pages 429–430 Lesson 8–5

5. $\sin \frac{37°}{11} = \frac{\sin E}{7}$; about 22.5

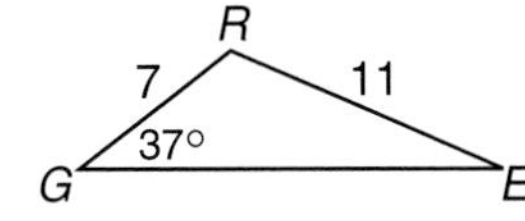

7. $m\angle Y \approx 47.2$, $m\angle W \approx 30.8$, $w = 12.6$

9a.

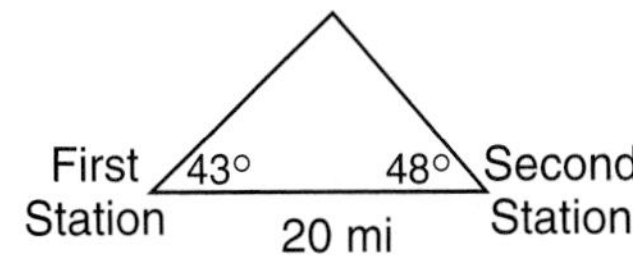

9b. about 14.9 mi, about 13.6 mi

11. $\sin \frac{47°}{13} = \frac{\sin R}{9}$; about 30.4

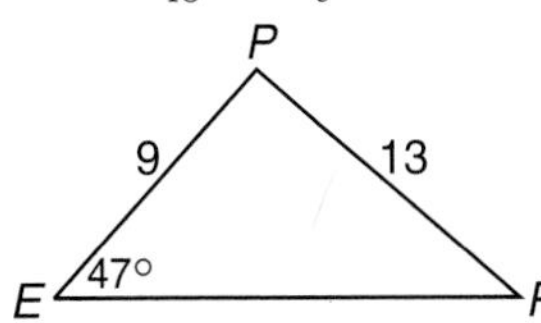

13. $\sin \frac{96°}{48} = \frac{\sin R}{10}$; about 12.0

15. $\frac{\sin 59°}{14.8} = \frac{\sin P}{8.3}$; about 28.7

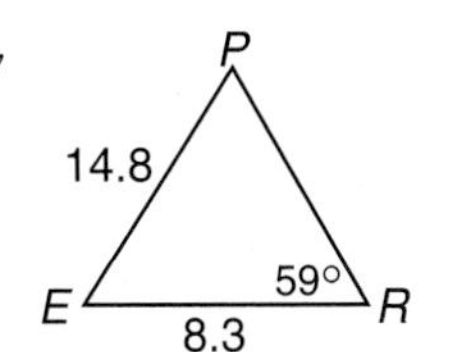

17. $m\angle F \approx 25.7$, $m\angle D \approx 47.3$, $d \approx 15.4$ **19.** $m\angle F \approx 10.9$, $m\angle D \approx 135.1$, $d \approx 34.1$ **21.** $m\angle F = 60$, $f \approx 1.0$, $r \approx 1.1$ **23.** $m\angle F \approx 49.6$, $m\angle R \approx 42.4$, $r \approx 14.2$ **25.** about 93.2 cm **27.** about 11.3; about 9.8 **29.** Yes; by definition, $\sin A = \frac{a}{c}$ and $\sin B = \frac{b}{c}$. According to the Law of Sines, $\frac{\sin A}{a} = \frac{\sin B}{b}$.

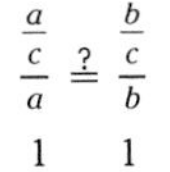

$$\frac{\frac{a}{c}}{a} \stackrel{?}{=} \frac{\frac{b}{c}}{b}$$

$$\frac{1}{c} = \frac{1}{c}$$

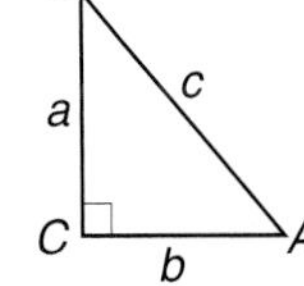

So the Law of Sines holds true for the acute angles of a right triangle. **31.** about 3.0 mi and 3.9 mi **33.** about 44.0 ft **35.** 27, 36 **39.** $9x^2 + 2xy$

Pages 433–436 Lesson 8–6

5a. Law of Sines **5b.** $m\angle E \approx 60.1$, $m\angle F \approx 47.9$, $f \approx 3.5$ **7.** $m\angle C = 81$, $a \approx 9.1$, $b \approx 12.1$ **9.** $m\angle A \approx 15.8$, $m\angle C \approx 145.2$, $c \approx 106.9$ **11.** about 35.7 **13.** Law of Cosines; $t \approx 6.9$, $m\angle R \approx 79.0$, $m\angle S \approx 63.0$ **15.** Law of Cosines; $m\angle X \approx 78.5$, $m\angle Y \approx 44.4$, $m\angle Z \approx 57.1$

17.

R
8.3
32°
D 9.1 G

Law of Cosines; $d \approx 4.9$, $m\angle G \approx 63.8$, $m\angle R \approx 84.2$

19.

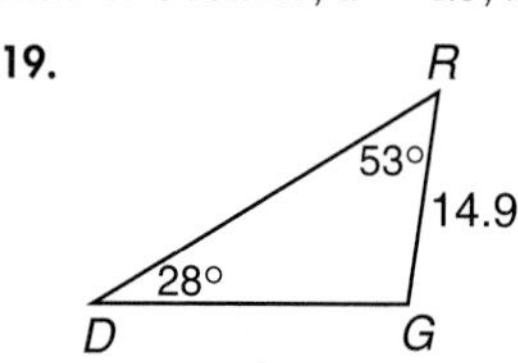

Law of Sines; $m\angle G = 99$, $g \approx 31.3$, $r \approx 25.3$ **21.** $m\angle H \approx 59.5$, $m\angle J \approx 30.5$, $m\angle K \approx 90.0$ **23.** $m\angle K = 79$, $h \approx 13.4$, $j \approx 11.7$ **25.** $j \approx 6.0$, $m\angle H \approx 59.3$, $m\angle K \approx 65.7$ **27.** $m\angle H \approx 49.9$, $m\angle J \approx 69.8$, $m\angle K \approx 60.3$ **29.** $m\angle H \approx 53.6$, $m\angle J \approx 59.6$, $m\angle K \approx 66.8$ **31.** $m\angle H \approx 44.4$, $m\angle J \approx 57.1$, $m\angle K \approx 78.5$ **33.** $HF \approx 15.7$, $EG \approx 11.2$ **35.** 15.8, 15.8, 15.8, 15.8; 110.7, 69.3, 110.7, 69.3 **37.** about 6.4 light-years **39.** Liz **41.** $m\angle Q = 55$, $p \approx 12.6$, $q \approx 10.7$ **43.** $2\sqrt{6} \approx 4.90$; $\sqrt{33} \approx 5.74$ **45.** (4, 3), (−2, 7), (−4, −5) **47.** True; Sample explanation: A 3-4-5 triangle is a right triangle with sides of different lengths. **49.** $x^2 + 9x + 18$

Page 437 Chapter 8 Highlights

1. sine **3.** Law of Cosines **5.** trigonometry **7.** Pythagorean triple **9.** Law of Sines **11.** trigonometric ratios **13.** means

Pages 438–440 Chapter 8 Study Guide and Assessment

15. $9\sqrt{3} \approx 15.59$ **17.** 2 **19.** $6\sqrt{2} \approx 8.49$ **21.** about 25.79 **23.** $\sqrt{30} \approx 5.48$ **25.** 17 **27.** no **29.** $3\sqrt{2} \approx 4.24$ **31.** $1.55\sqrt{3} \approx 2.68$ **33.** $\frac{15}{17} \approx 0.8824$ **35.** $\frac{15}{17} \approx 0.8824$ **37.** about 282.7 m **39.** about 4.3° **41.** $m\angle A = 51$, $a \approx 70.2$, $c \approx 89.7$ **43.** $m\angle B = 80$, $b \approx 14.8$, $c \approx 14.1$ **45.** $b \approx 30.6$, $m\angle A \approx 87.3$, $m\angle C \approx 48.7$ **47.** $m\angle A \approx 40.8$, $m\angle B \approx 78.6$, $m\angle C \approx 60.6$ **49.** about 0.5 mi **51.** about 7.7° **53.** Sample answer: Look for a pattern; 2.

CHAPTER 9 ANALYZING CIRCLES

Pages 449–451 Lesson 9–1

7. 4.7 **9.** $d = 14$, $C = 44.0$ **11.** 9π **13.** 97.7 feet **15.** $\overline{RI}$ **17.** No; it is a radius. **19.** $\overline{RM}$, $\overline{AM}$, $\overline{DM}$, $\overline{IM}$ **21.** 5.9 **23.** $d = 10$, $C = 31.4$ **25.** $d = 43.6$, $r = 21.8$ **27.** $r = x$, $C = 6.3x$ **29.** 8π cm **31.** $6\sqrt{2}\pi$ cm **33.** 6 **35.** 20 **37.** The circumference is doubled.

39. Given: $\odot P$ with diameter $\overline{SA}$ and chord $\overline{KR}$

Prove: $SA > KR$

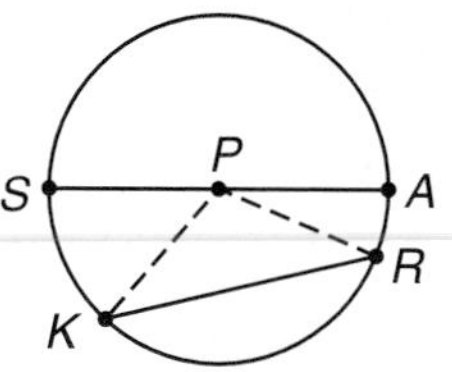

Proof:

By the Segment Addition Postulate, $SA = SP + PA$. Draw $\overline{PK}$ and $\overline{PR}$ since through any two points there is one line. Since all radii of a circle are congruent, $\overline{SP} \cong \overline{PK}$ and $\overline{PA} \cong \overline{PR}$. By substitution, $SA = PK + PR$. By the Triangle Inequality Theorem, $PK + PR > KR$. By substitution, $SA > KR$.

41. $d \approx 11.8$ cm, $C \approx 37.1$ cm, $\frac{C}{d} \approx 3.144$ or π **43.** 3141.6 cm **45a.** 79 ft **45b.** 132 ft **47.** 273.9 cm **49.** 68 in. **51.** 17.5 **53.** Yes; the sum of any two sides is greater than the length of the third side. **55.** (0, −4)

Pages 455–458 Lesson 9–2

7. minor; 138 **9.** 12.6 **11.** 69.1 **13.** 137 **15.** true; def. of concentric circles **17.** True; concentric circles have the same center. **19.** minor, 21 **21.** minor, 90 **23.** minor, 111 **25.** major, 201 **27.** minor, 159 **29.** 28.3 **31.** 18.8 **33.** 26.1 **35.** 35.5 **37.** 26 **39.** 76 **41.** 256 **43.** 24 **45.** False; arcs are not of the same circle. **47.** False; arcs are not of same circle. **49.** true **51.** $\frac{\sqrt{2}}{2}$

53. Sample answer:

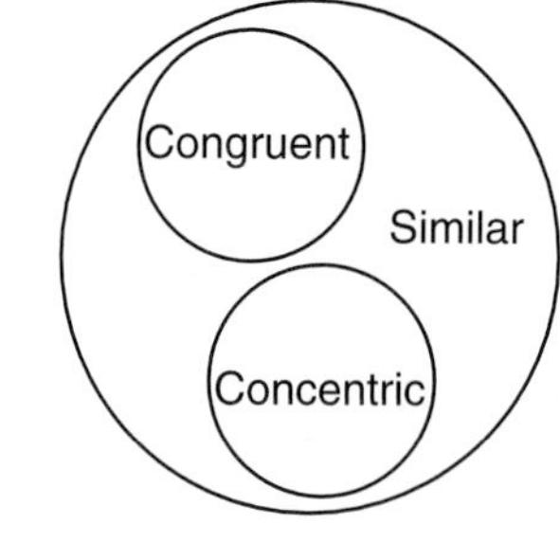

55a. Yes; if two central angles of one circle have equal measures, then their corresponding arcs have equal measures. **55b.** 20.6 **57.** 138.2 mm **59.** 4.5 **61.** 64 **63.** 50 **65.** false **67.** $15x^6y^5$

Pages 461–465 Lesson 9–3

5. In a circle, if a diameter is perpendicular to a chord, then it bisects the chord and its arc. **7.** 5 **9.** 3.4

11a.

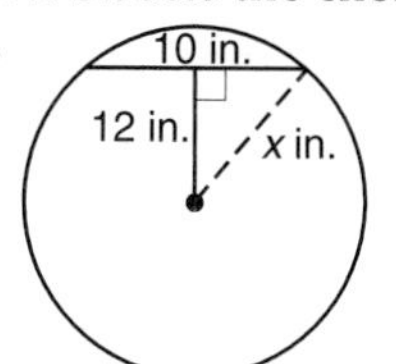

11b. 13 in.

13. In a circle or in congruent circles, two minor arcs are congruent if and only if their corresponding chords are congruent. **15.** In a circle or in congruent circles, two chords are congruent if and only if they are equidistant from the center. **17.** S **19.** $\widehat{PQ}$, $\widehat{VU}$, and $\widehat{UW}$ **21.** Neither; they are $\cong$. **23.** They are perpendicular to the same line. **25.** 28 **27.** 21 **29.** $\overline{RH} \cong \overline{MR}$ **31.** Yes; since radii are congruent, $\odot P \cong \odot Q$. Since the corresponding chords are congruent, by Theorem 9–1, $\widehat{AB} \cong \widehat{RS}$. **33.** 24

35. 19.2 cm **37.** 48 mm

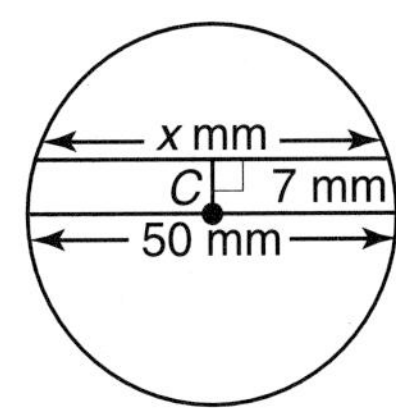

39. 27.7

41. Given: $\odot O$
$\overline{AB} \cong \overline{CD}$

Prove: $\widehat{AB} \cong \widehat{CD}$

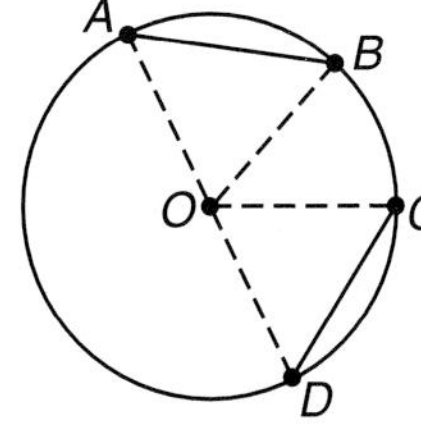

Proof:

Statements	**Reasons**
1. Draw radii $\overline{OA}$, $\overline{OB}$, $\overline{OC}$, and $\overline{OD}$.	**1.** Through any 2 pts. there is 1 line.
2. $\overline{OA} \cong \overline{OC}$ $\overline{OB} \cong \overline{OD}$	**2.** All radii of a $\odot$ are $\cong$.
3. $\overline{AB} \cong \overline{CD}$	**3.** Given
4. $\triangle ABO \cong \triangle CDO$	**4.** SSS
5. $\angle AOB \cong \angle COD$	**5.** CPCTC
6. $\widehat{AB} \cong \widehat{CD}$	**6.** In a $\odot$, 2 minor arcs are $\cong$ if and only if their corr. central $\angle$s are $\cong$.

43. Given: $\odot O$

$\overline{MN} \cong \overline{PQ}$

Prove: $\overline{OA} \cong \overline{OB}$

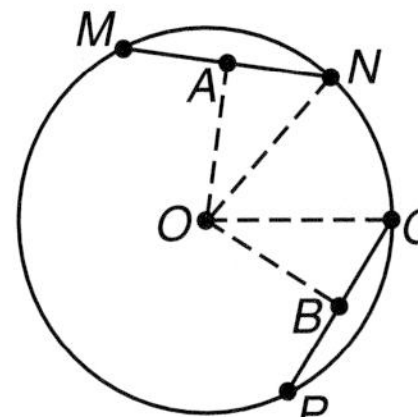

Proof:

Statements	**Reasons**
1. Draw radii $\overline{ON}$ and $\overline{OQ}$. Draw $\overline{OA}$ so that $\overline{OA} \perp \overline{MN}$ and draw $\overline{OB}$ so that $\overline{OB} \perp \overline{PQ}$.	**1.** Through any 2 pts. there is 1 line.
2. $\overline{OA}$ bisects $\overline{MN}$. $\overline{OB}$ bisects $\overline{PQ}$.	**2.** $\overline{OA}$ and $\overline{OB}$ can be extended to form radii, and a radius $\perp$ to a chord bisects the chord.
3. $AN = \frac{1}{2}MN$ $BQ = \frac{1}{2}PQ$	**3.** Def. of $\perp$ bisector
4. $\overline{MN} \cong \overline{PQ}$	**4.** Given
5. $MN = PQ$ $AN = BQ$	**5.** Def. of $\cong$ segments
6. $\overline{ON} \cong \overline{OQ}$	**6.** All radii of a $\odot$ are $\cong$.
7. $\triangle AON \cong \triangle BOQ$	**7.** HL
8. $\overline{OA} \cong \overline{OB}$	**8.** CPCTC

45b. It is the only point equidistant from T, S, and A. **45c.** Minneapolis, Minnesota **47.** True; all points on a circle are equidistant from the center. **49.** 29.8 **51.** false **53.** $\angle M$ **55.** $-24r^3s + 2rs + 16r$

Pages 469–473 Lesson 9–4

7. $\widehat{BC}$ **9.** 40 **11.** 72 **13.** 104 **15.** 52 **17.** $\widehat{HC}$ **19.** $\widehat{TCH}$ **21.** 52 **23.** There are none shown in the figure. **25.** 23.5 **27.** 102 **29.** 60 **31.** 120 **33.** 240 **35.** 240 **37.** 89 **39.** 114 **41.** 33 **43.** 44.5 **45.** 78.5

47. Given: $\overline{BR} \parallel \overline{AC}$

Prove: $\widehat{RA} \cong \widehat{BC}$

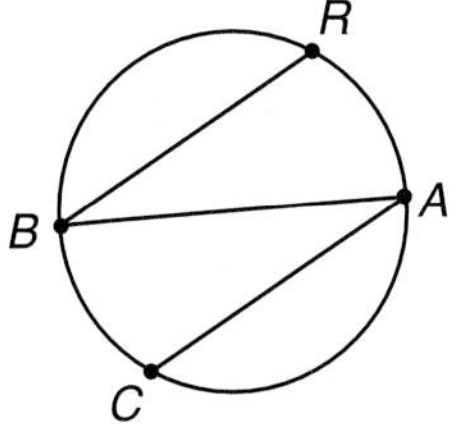

Proof:

Statements	**Reasons**
1. $\overline{BR} \parallel \overline{AC}$	**1.** Given
2. $\angle RBA \cong \angle BAC$	**2.** Alt. Int. $\angle$ Theorem
3. $m\angle RBA = m\angle BAC$	**3.** $\cong$ $\angle$s have = measures.
4. $m\angle RBA = \frac{1}{2}m\widehat{RA}$ $m\angle BAC = \frac{1}{2}m\widehat{BC}$	**4.** If an $\angle$ is inscribe in a $\odot$, the measure of the $\angle$ = $\frac{1}{2}$ the measure of its intercepted arc.
5. $\frac{1}{2}m\widehat{RA} = \frac{1}{2}m\widehat{BC}$	**5.** Substitution Prop. (=)
6. $m\widehat{RA} = m\widehat{BC}$	**6.** Mult. Prop. (=)
7. $\widehat{RA} \cong \widehat{BC}$	**7.** Arcs that have = measures and are in the same $\odot$ are $\cong$.

49. No; opposite angles must be congruent and supplementary. Therefore, they each must be 90° or right angles.

51. $m\angle PRQ = m\angle KRQ - m\angle KRP$
$= \frac{1}{2}m\widehat{KQ} - \frac{1}{2}m\widehat{KP}$
$= \frac{1}{2}(m\widehat{KQ} - m\widehat{KP})$
$= \frac{1}{2}m\widehat{PQ}$

53. Given: $\widehat{PQR}$ is a semicircle of $\odot C$.

Prove: $\angle PQR$ is a right $\angle$.

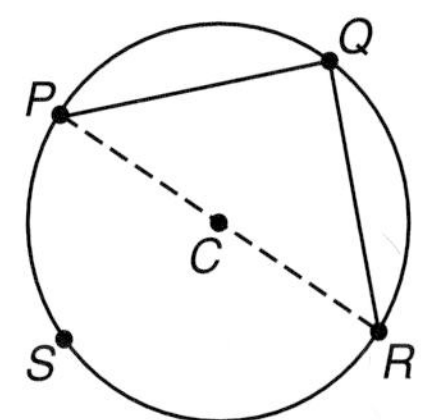

Proof:

Since $\widehat{PQR}$ is a semicircle, $\widehat{PSR}$ is also a semicircle and has a degree measure of 180. From the diagram, $\angle PQR$ is an inscribed angle, and $m\angle PQR = \frac{1}{2}(m\widehat{PSR})$ or 90. As a result, $\angle PQR$ is a right angle.

55. yes **57.** Position the carpenter's square so that the vertex of the right angle of the square is on the circle and the square forms an inscribed angle. Since the inscribed angle is a right angle, the measure of the intercepted arc is 180. So, the points where the square crosses the circle are the ends of a diameter (Theorem 9–6). Mark these points and draw the diameter. Draw another diameter using the same method. The point where the two diameters intersect is the center of the circle.

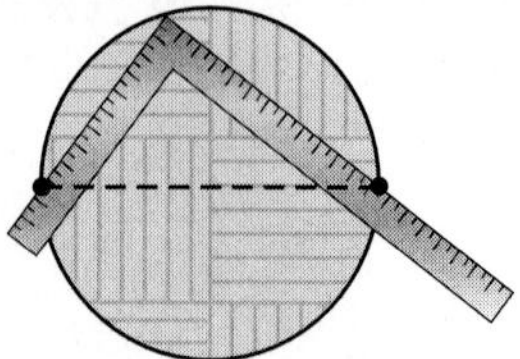

59. 53.4 ft **63.** Sample answer: measure the diagonals to see if they are congruent. **65.** 2.5 **67.** 10.4 **69.** $\left(2, -\frac{1}{2}\right)$

Page 473 Self Test

1. $\overline{PD}, \overline{PB}, \overline{PC}$ **3.** 13.1 **5.** 115

Pages 479–482 Lesson 9–5

7. 24 **9.** 132 **11.** $\overline{KL}$ **13.** In quadrilateral *KLEM*, $\angle K$ and $\angle E$ are right angles and the sum of their measures equals 180. Therefore, the sum of $m\angle L$ and $m\angle M$ must equal 180 and the angles must be supplementary. **15.** 8 **17.** $5\sqrt{3} \approx 8.7$ **19.** 10 **21.** 60 **23.** 90 **25.** $2\sqrt{3} \approx 3.5$ **27.** 30 **29.** $\overline{GF}$ and $\overline{AJ}$ **31.** 14 **33.** 4 **35.** 15 **37a.** Given **37b.** If 2 segments from the same exterior point are tangent to a ⊙, then they are ≅. **37c.** Radii of the same circle are ≅. **37d.** Reflexive Property (=) **37e.** SSS **37f.** CPCTC **39.** $-\frac{2}{3}$

41.

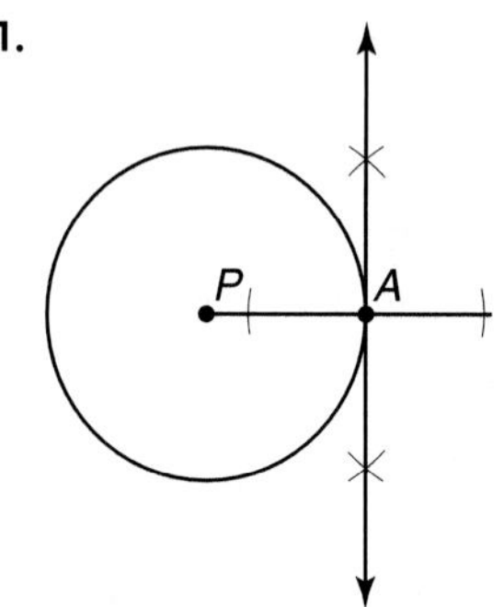

43. 4; $\triangle PQR$ is equilateral and $\overline{QN} \cong \overline{NR}$.

45. Given: $\overline{GR}$ is tangent to $\odot D$ at G. $\overline{AG} \cong \overline{DG}$

Prove: $\overline{GA}$ bisects $\overline{RD}$.

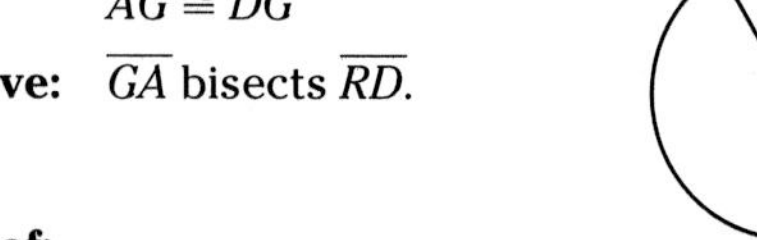

Proof:

Since $\overline{DA}$ is a radius, $\overline{DG} \cong \overline{DA}$. Since $\overline{AG} \cong \overline{DG} \cong \overline{DA}$, $\triangle GDA$ is equilateral. Therefore, each angle has a measure of 60. Since $\overline{GR}$ is tangent to $\odot D$, $m\angle RGD = 90$. Since $m\angle AGD = 60$, then by Angle Addition Post., $m\angle RGA = 30$. If $m\angle DAG = 60$, then $m\angle RAG = 120$. Then $m\angle R = 30$. Thus, $\triangle RAG$ is isosceles. By Transitive Prop. (=), $\overline{RA} \cong \overline{DA}$. Thus, $\overline{GA}$ bisects $\overline{RD}$.

47. Given: $\overleftrightarrow{CA}$ is tangent to the circle at A.

Prove: $\overline{XA} \perp \overleftrightarrow{CA}$

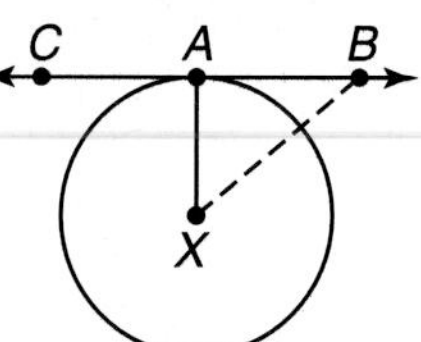

Proof:

Statements	Reasons
1. $\overleftrightarrow{CA}$ is tangent to the circle at *A*.	1. Given
2. Pick any point on $\overleftrightarrow{CA}$ other than *A* and call it *B*. Draw $\overline{XB}$.	2. Through any 2 pts. there is 1 line.
3. $\overleftrightarrow{CA}$ intersects $\odot X$ at exactly one point, *A*, and *B* lies in the exterior of $\odot X$.	3. Def. of tangent
4. $XA < XB$	4. The measure of a segment joining an exterior pt. to the center of a ⊙ is greater than the measure of a radius.
5. $\overline{XA} \perp \overleftrightarrow{CA}$	5. $\overline{XA}$ is the shortest segment from *X* to $\overleftrightarrow{CA}$.

49a. $\overline{QR}$ **49b.** One definition is a ratio of the measure of the leg opposite the acute angle to the measure of the leg adjacent to the acute angle. Another definition is a line in a plane that intersects a circle in the plane in exactly one point. **51.** 1.73 meters **53.** 14 **55.** 25 cm **57.** $x > 5$ **59.** no conclusion **61.** 0.02

Pages 486–490 Lesson 9–6

7. 134 **9.** 33 **11.** 54 **13.** 169 **15.** 157.5 **17.** 70 **19.** 65 **21.** 70 **23.** 5 **25.** 105 **27.** 65 **29.** 20 **31.** 26 **33a.** 44 **33b.** 30 **33c.** 15 **35.** 26 **37.** 102 **39.** 23 **41.** 94 **43a.** $\sqrt{63} \approx 7.9$ **43b.** $2\sqrt{63} \approx 15.9$

45. Given: Secants $\overleftrightarrow{AC}$ and $\overleftrightarrow{BD}$ intersect at X inside $\odot P$.

Prove: $m\angle AXB = \frac{1}{2}(m\widehat{AB} + m\widehat{CD})$

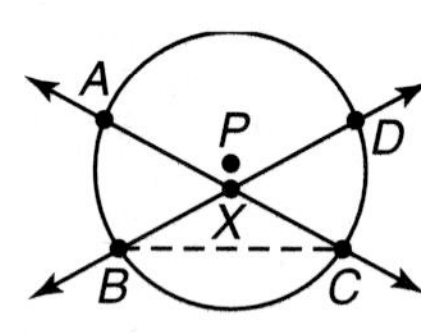

Proof:

Statements	Reasons
1. Secants $\overleftrightarrow{AC}$ and $\overleftrightarrow{BD}$. intersect at *X* inside $\odot P$	1. Given
2. Draw $\overline{BC}$.	2. Through any 2 pts. there is 1 line.
3. $m\angle XBC = \frac{1}{2}m\widehat{CD}$ $m\angle XCB = \frac{1}{2}m\widehat{AB}$	3. An ∠ inscribed in a ⊙ has the measure of $\frac{1}{2}$ the measure of its intercepted arc.

Statements	Reasons
4. $m\angle AXB = m\angle XCB + m\angle XBC$	4. Exterior Angle Theorem
5. $m\angle AXB = \frac{1}{2}m\widehat{AB} + \frac{1}{2}m\widehat{CD}$	5. Substitution Prop. (=)
6. $m\angle AXB = \frac{1}{2}(m\widehat{AB} + m\widehat{CD})$	6. Distributive Prop.

47a. Given: $\overleftrightarrow{AB}$ is a tangent to $\odot O$. $\overline{AC}$ is a secant to $\odot O$. $\angle CAB$ is acute.

Prove: $m\angle CAB = \frac{1}{2}m\widehat{CA}$

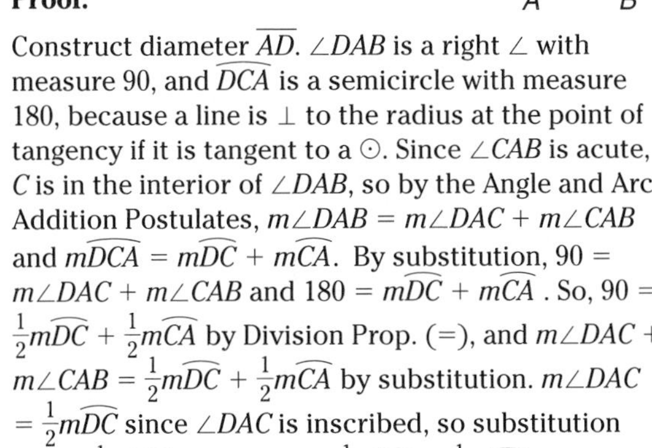

Proof:

Construct diameter $\overline{AD}$. $\angle DAB$ is a right $\angle$ with measure 90, and $\widehat{DCA}$ is a semicircle with measure 180, because a line is $\perp$ to the radius at the point of tangency if it is tangent to a $\odot$. Since $\angle CAB$ is acute, C is in the interior of $\angle DAB$, so by the Angle and Arc Addition Postulates, $m\angle DAB = m\angle DAC + m\angle CAB$ and $m\widehat{DCA} = m\widehat{DC} + m\widehat{CA}$. By substitution, $90 = m\angle DAC + m\angle CAB$ and $180 = m\widehat{DC} + m\widehat{CA}$. So, $90 = \frac{1}{2}m\widehat{DC} + \frac{1}{2}m\widehat{CA}$ by Division Prop. (=), and $m\angle DAC + m\angle CAB = \frac{1}{2}m\widehat{DC} + \frac{1}{2}m\widehat{CA}$ by substitution. $m\angle DAC = \frac{1}{2}m\widehat{DC}$ since $\angle DAC$ is inscribed, so substitution yields $\frac{1}{2}m\widehat{DC} + m\angle CAB = \frac{1}{2}m\widehat{DC} + \frac{1}{2}m\widehat{CA}$. By Subtraction Prop. (=), $m\angle CAB = \frac{1}{2}m\widehat{CA}$.

47b. Given: $\overleftrightarrow{AB}$ is a tangent to $\odot O$. $\overline{AC}$ is a secant to $\odot O$.

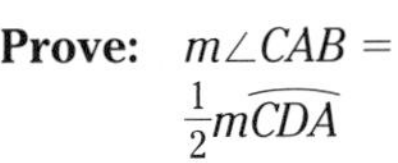

Prove: $m\angle CAB = \frac{1}{2}m\widehat{CDA}$

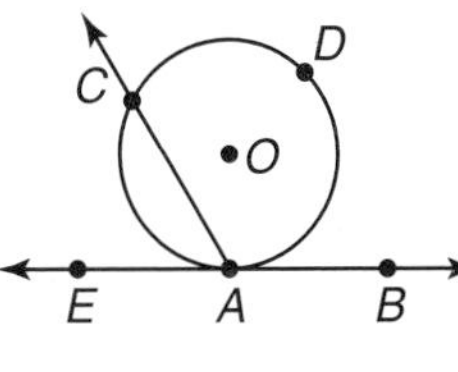

Proof:

$\angle CAB$ and $\angle CAE$ form a linear pair, so $m\angle CAB + m\angle CAE = 180$. Since $\angle CAB$ is obtuse, $\angle CAE$ is acute and Case 1 applies, so $m\angle CAE = \frac{1}{2}m\widehat{CA}$. $m\widehat{CA} + m\widehat{CDA} = 360$, so $\frac{1}{2}m\widehat{CA} + \frac{1}{2}m\widehat{CDA} = 180$ by Division Prop. (=), and $m\angle CAE + \frac{1}{2}m\widehat{CDA} = 180$ by substitution. By the Transitive Prop. (=), $m\angle CAB + m\angle CAE = m\angle CAE + \frac{1}{2}m\widehat{CDA}$, so by Subtraction Prop. (=), $m\angle CAB = \frac{1}{2}m\widehat{CDA}$.

49. 76 **51.** 12 **53.** 8 **55.** \$17.48 **57.** no **59.** Division Prop. (=) or Mult. Prop. (=) **61.** −3, 1

Pages 494–497 Lesson 9–7

7. 2 **9.** 5, 5 **11.** $11\frac{1}{3}$ **13.** $113\frac{1}{3}$ cm **15.** 28.1 **17.** 3 **19.** 6.4 **21.** 5.7 **23.** 24 **25.** 18

27.

$DQ + QX = DX$	*Segment Add. Post.*
$QX = DX - DQ$	
$= 25 - 7$	*DQ = DE = 7*
$= 18$	
$(EX)^2 = QX \cdot TX$	*Theorem 9–16*
$24^2 = x(x + 14)$	
$0 = x^2 + 14x - 576$	
$x = 18$	*Quadratic formula*

29. 5.3 **31.** 15; 22.5

33. Given: $\overline{EC}$ and $\overline{EB}$ are secant segments.

Prove: $EA \cdot EC = ED \cdot EB$

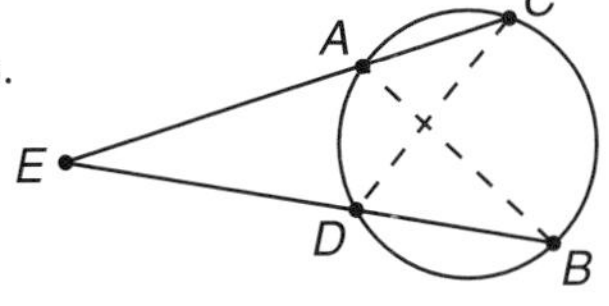

Proof:

Statements	Reasons
1. $\overline{EC}$ and $\overline{EB}$ are secant segments.	1. Given
2. Draw $\overline{AB}$ and $\overline{CD}$.	2. Through any 2 pts. there is 1 line.
3. $\angle DEC \cong \angle AEB$	3. Reflexive Prop. of $\cong$ ∠s
4. $\angle ECD \cong \angle EBA$	4. If 2 inscribed ∠s of a $\odot$ or $\cong$ $\odot$s intercept $\cong$ arcs or the same arc, then the ∠s are $\cong$.
5. $\triangle ABE \sim \triangle DCE$	5. AA Similarity
6. $\frac{EA}{ED} = \frac{EB}{EC}$	6. Def. of $\sim$ polygons
7. $EA \cdot EC = ED \cdot EB$	7. Cross products

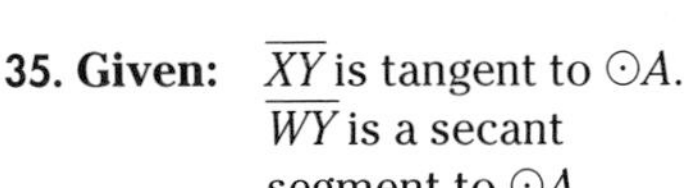

35. Given: $\overline{XY}$ is tangent to $\odot A$. $\overline{WY}$ is a secant segment to $\odot A$.

Prove: $(XY)^2 = WY \cdot ZY$

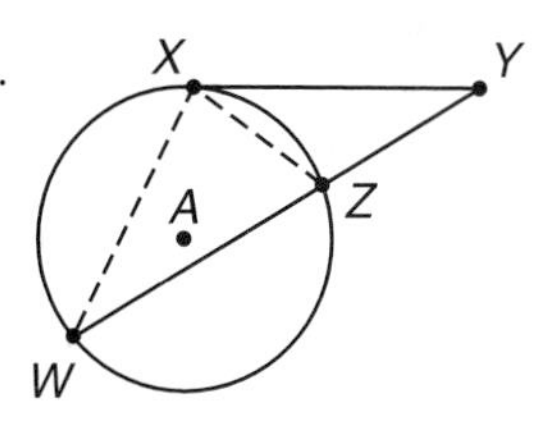

Proof:

Statements	Reasons
1. $\overline{XY}$ is tangent to $\odot A$. $\overline{WY}$ is a secant segment to $\odot A$.	1. Given
2. Draw $\overline{XZ}$ and $\overline{XW}$.	2. Through any 2 pts. there is 1 line.
3. $m\angle XWZ = \frac{1}{2}m\widehat{XZ}$	3. If an $\angle$ is inscribed in a $\odot$, the measure of the $\angle = \frac{1}{2}$ the measure of its intercepted arc.
4. $m\angle YXZ = \frac{1}{2}m\widehat{XZ}$	4. If a secant and a tangent intersect at the pt. of tangency, then the measure of the $\angle$ formed $= \frac{1}{2}$ the measure of its intercepted arc.
5. $m\angle YXZ = m\angle XWZ$	5. Substitution Prop. (=)
6. $\angle YXZ \cong \angle XWZ$	6. Def. of $\cong$ ∠s
7. $\angle Y \cong \angle Y$	7. Reflexive Prop. of $\cong$ ∠s
8. $\triangle YXZ \sim \triangle YWX$	8. AA Similarity
9. $\frac{XY}{WY} = \frac{ZY}{XY}$	9. Def. of $\sim$ polygons
10. $(XY)^2 = WY \cdot ZY$	10. Cross products

37. $(AB)^2 = BC(BD)$ *Th. 9–16*
$= BC(BC + CD)$ *Seg. Add. Post.*
$= BC(BC + BC)$ *$BC = CD$*
$= BC(2BC)$
$= 2(BC)^2$
$AB = \sqrt{2}BC$

39. 7:3.5 **41.** 17 **43.** 15 **45.** Subtraction Prop. (≠)
47. $2c^2 + 13c - 7$

Pages 501–503 Lesson 9–8

5.

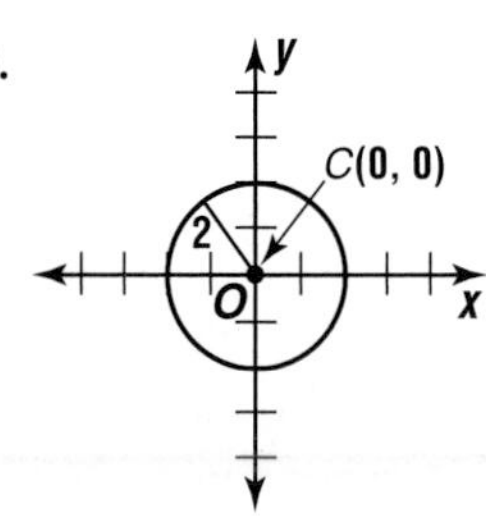

7. $(x + 2)^2 + (y - 3)^2 = 11$
9. $x^2 + y^2 = 16$ **11.** $\left(\frac{3}{4}, -3\right), \frac{9}{2}$
13. (0.5, −3.1), 4.2

15.

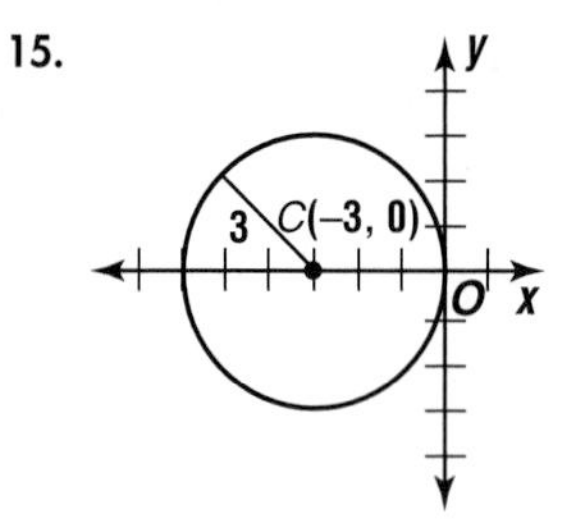

17.

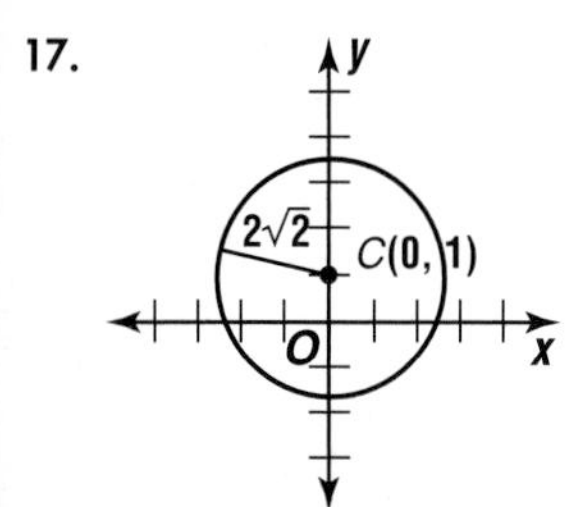

19.

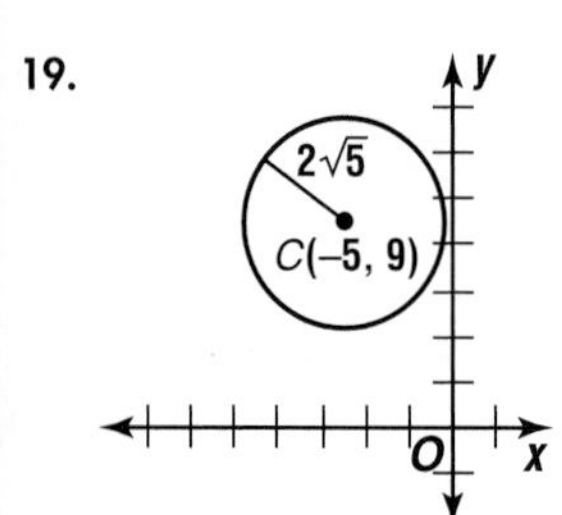

21. $(x + 1)^2 + (y - 4)^2 = 15$
23. $x^2 + y^2 = 8$
25. $(x - 2)^2 + (y - 2)^2 = 2.25$
27. $(x + 3)^2 + (y + 8)^2 = 49$

29–32.

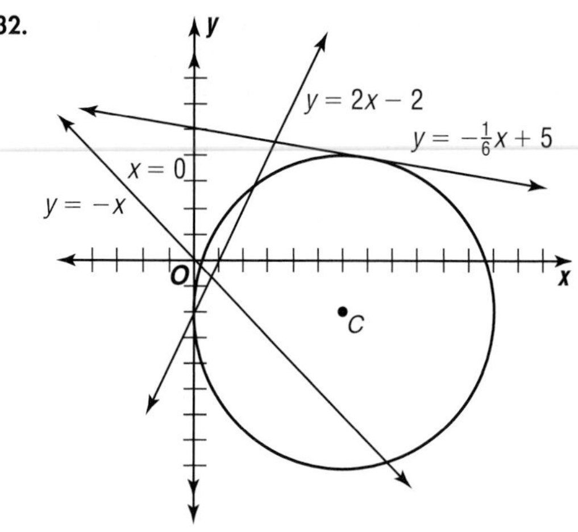

29. secant **31.** tangent

33a, c.

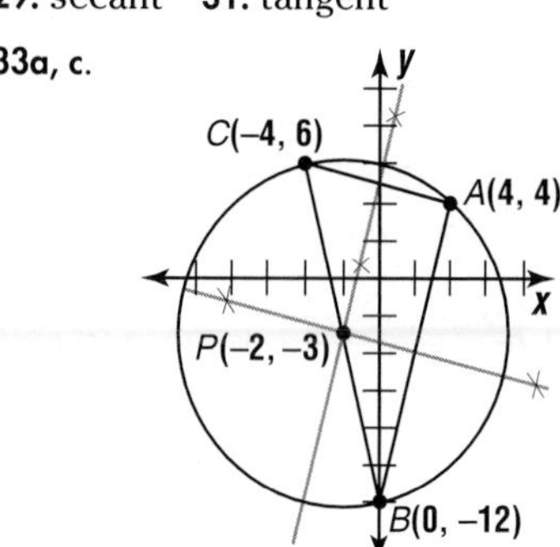

33b. $P(-2, -3)$ **33c.** $\sqrt{85} \approx 9.2$ **33d.** $(x + 2)^2 + (y + 3)^2 = 85$ **35a.** $(x - 3)^2 + (y - 3)^2 = 18$ **35b.** right triangle **35c.** diameter **39.** 29 **41.** 5 **43.** 9.8 in. **45.** 18 units, 22 units **47.** Reflexive Prop. (=) **49.** $10,377.82

Page 505 Chapter 9 Highlights

1. i **3.** h **5.** c **7.** f **9.** e

Pages 506–508 Chapter 9 Study Guide and Assessment

11. $\overline{AB}$ **13.** 88.0 **15.** 123 **17.** 4.5 **19.** 19.1 **21.** about 34.2 cm **23.** 4 in. **25.** 144 **27.** 36 **29.** 12 **31.** 42 **33.** 18.5 **35.** 138 **37.** 6.0 cm **39.** 15.3 cm **41.** $(x + 4)^2 + (y - 3)^2 = 36$

43. Portion of Sales per Recording Type

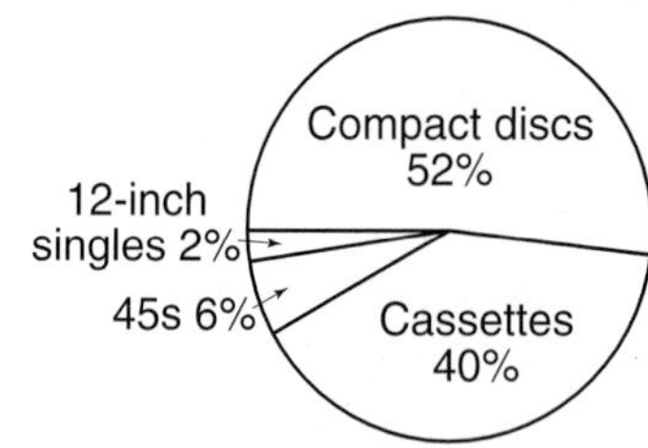

45. about 6.6 cm

CHAPTER 10 EXPLORING POLYGONS AND AREA

Pages 518–521 Lesson 10–1

7. $\overline{AB}, \overline{BC}, \overline{CD}, \overline{DE}, \overline{EF}, \overline{FA}$ **9.** 1440 **11.** 12 **13.** 135, 45 **15.** 6 **17.** $x = 34$, $3x = 102$, $2x - 1 = 67$, $6x - 5 = 199$, $4x + 2 = 138$ **19.** A, B, C, D, E, F **21.** Sample answers: *BCDEFA* and *FABCDE* **23.** 4320 **25.** 7920

27. $180(3m-2)$ **29.** 8 **31.** 25 **33.** $\frac{360}{x}$ **35.** 157.5, 22.5
37. 154.29, 25.71 **39.** $\frac{180(x+2y-2)}{x+2y}, \frac{360}{x+2y}$ **41.** 10
43. 100 **45.** s **47.** $m\angle R = 125$, $m\angle S = 100$, $m\angle T = 100$, $m\angle U = 95$, $m\angle V = 120$ **49.** 720 **51.** nonagon

53. Consider the sum of the measures of the exterior angles for an n-gon.

sum of measures of exterior angles	=	sum of measures of linear pairs	−	sum of measures of interior angles
	=	$n \cdot 180$	−	$180(n-2)$
	=	$180n$	−	$180n + 360$
	=	360		

So, the sum of the exterior angle measures is 360 for *any* convex polygon. **55.** No; the sum of the measures of the angles at any vertex is less than 360, so the polygons would not interlock in a plane.

57.

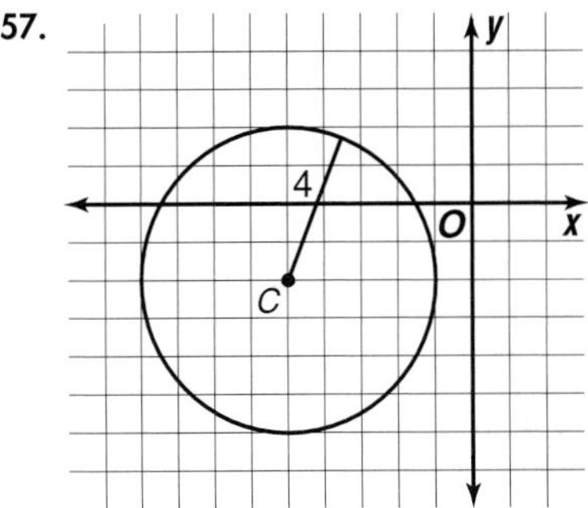

59. $16\pi \approx 50.27$ in. **61.** 6.9 cm **63.** $\overline{ST}, \overline{SV}, \overline{TV}$
65. Reflexive Property (=) **67.** 3

Page 522 Lesson 10–2A

1. Yes; whatever space is taken out of the square is then added onto the outside of the square. The area does not change, only the shape changes.

3.

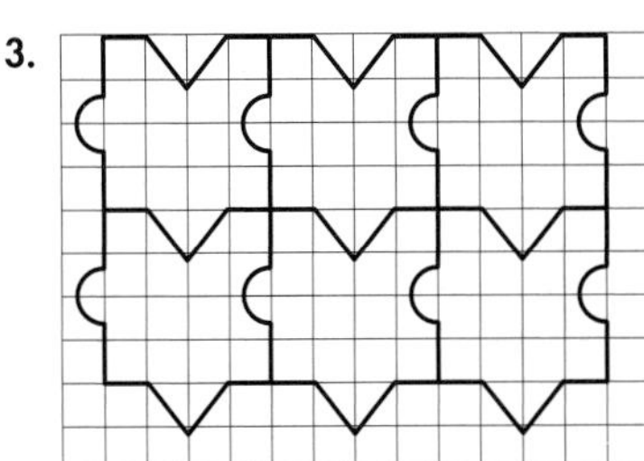

5.

Pages 525–527 Lesson 10–2

7. yes; interior angle = 60 **9.** regular **11.** $(8-4) \times 7 + 3 = 31$ **13.** no; interior angle = 108

15. yes; sample answer:

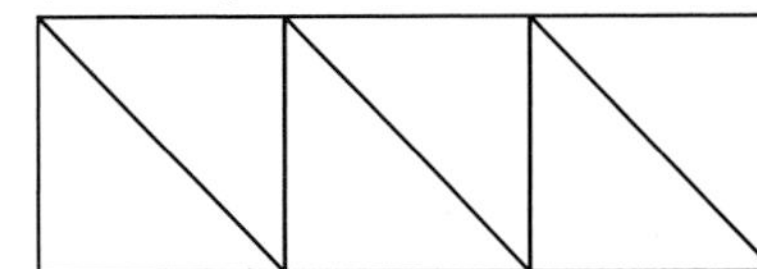

17. yes **19.** none **21.** uniform, semi-regular

23. Sample answer:

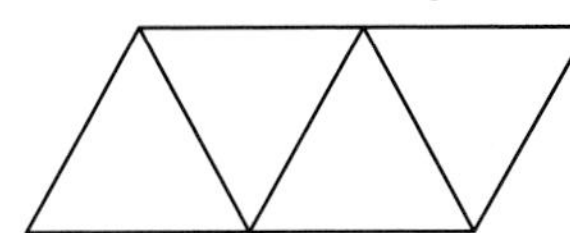

27. Yes; a tessellation can be uniform with only one regular polygon. Therefore, it would not be semi-regular.

29.

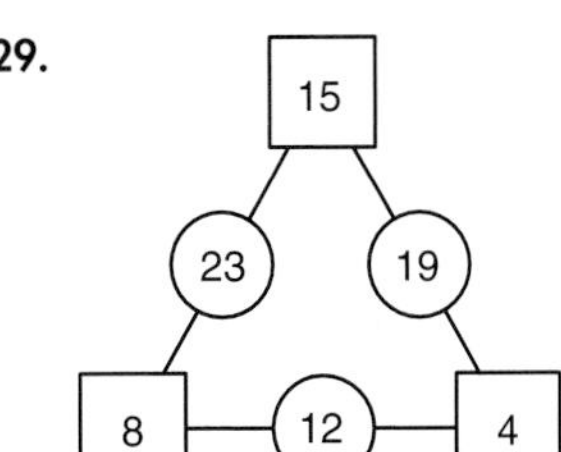

31a. $D = 5$, $Y = 2$, $N = 6$, $E = 9$, $A = 4$, $L = 8$, $S = 1$, $R = 7$, $C = 3$, and $T = 0$ **31b.** More answers will result; sample answers: 561,535 + 207,535 = 769,070. **33.** 18
35. 13 in. **37.** no; $\frac{5.5}{4.4} = \frac{3}{2.4} \neq \frac{4.5}{3.8}$ **39.** {(6, 6), (0, 1), (−5, 6), (9, 2)}; D = {6, 0, −5, 9}; R = {2, 1, 6}; inverse = {(6, 6), (1, 0), (6, −5), (2, 9)}

Page 528 Lesson 10–3A

7. $A = bh$

9. Sample answer:

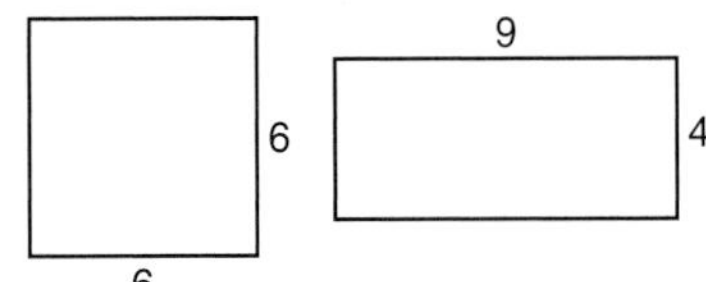

Pages 532–534 Lesson 10–3

5. 19.22 cm^2 **7.** 28.5 cm^2 **9.** \$628.33 **11.** 389.2 m^2 **13.** 2160 ft^2 **15.** 23 m^2 **17.** 202 cm^2 **19.** rectangle; 12 units^2 **21.** square; 13 units^2 **23.** $12\sqrt{3}$ mm^2 **25.** 25 cm^2; $\frac{1}{4}$ area of square **27.** NP = 16 units **29a.** 828 ft^2 **29b.** 207 ft^2 **29c.** Sample answer: 12 ft by 17.25 ft **31a.** M, N, O, P, Q **31b.** convex **31c.** No; its sides are not all congruent. **31d.** pentagon

33. Given: trapezoid $ABCD$, $\overline{AB} \parallel \overline{DC}$
Prove: $\angle A$ and $\angle D$ are supplementary.
Proof:

Statements	Reasons
1. trapezoid $ABCD$, $\overline{AB} \parallel \overline{DC}$	1. Given
2. $\angle A$ and $\angle D$ are supplementary.	2. Consecutive Interior Angle Theorem

35.

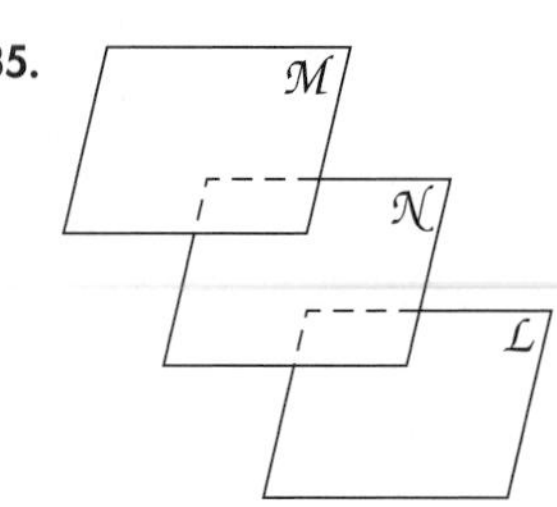

37. a, d

Pages 538–541 Lesson 10–4

7. 60 cm^2 **9.** $72\sqrt{3} \approx 124.7$ in^2 **11.** 35.7 in^2 **13.** $27\sqrt{3} \approx 46.8$ m^2 **15.** 88 m^2 **17.** 4 **19.** 10 **21.** about 28.9 in^2 **23.** $16\sqrt{3} \approx 27.7$ cm^2 **25.** 16.25 in., 22.25 in. **27.** total area = area of parallelogram + area of triangle

$$\begin{aligned} &= bh + \tfrac{1}{2}bh \\ &= ah + \tfrac{1}{2}(b - a)h \\ &= ah + \left(\tfrac{1}{2}b - \tfrac{1}{2}a\right)h \\ &= ah + \tfrac{1}{2}bh - \tfrac{1}{2}ah \\ &= \tfrac{1}{2}ah + \tfrac{1}{2}bh \\ &= \tfrac{1}{2}h(a + b) \end{aligned}$$

29a. 35 units^2 **29b.** 27.125 units^2 **31.** 30 packages **33.** 13 units by 40 units **35.** $h = 45, j = 60$ **37.** No; consecutive angles in a parallelogram must be supplementary. **39.** ℓmx

Page 541 Self Test

1a. heptagon **1b.** no **1c.** convex **3.** 150 **5.** $11\frac{1}{4}$, 32 **7.** 9200 ft^2 **9.** 150 cm^2

Page 542 Lesson 10–5A

1. 360 **3.** They are the same. **5.** 90 **7.** 45 **9.** 30

Pages 547–550 Lesson 10–5

5. $8\sqrt{2} \approx 11.31$ cm **7.** They are close in length. **9.** 2.3 in., 27.7 in^2, 24 in. **11.** 56.7 cm^2 **13.** $(456 + 72\pi) \approx$ 682.19 ft^2 **15.** 104.0 in^2 **17.** 259.8 ft^2 **19.** 1995.3 in^2 **21.** 53.4 cm, 227.0 cm^2 **23.** 64.4 in., 330.1 in^2 **25.** 136.7 cm, 1486.2 cm^2 **27.** 73.1 ft^2 **29.** 313.2 yd^2 **31.** 15.8 in^2 **33.** 20.9 cm^2 **35a.** 92.16π ft^2 **35b.** $(384 - 92.16\pi)$ ft^2 **37.** As the polygon increases in the number of sides, the length of the apothem has a limit of the radius of the circle, and the measure of the side of a polygon becomes increasingly small and has as factors of the limit (2)(3.14). **39a.** two 12-inch pizzas **39b.** No; the unit cost of two 12-inch pizzas is more expensive than the unit cost of one 16-inch pizza. **41.** No; they will increase by the squares of 1, 3, 5, and 7, or in other words, 1, 9, 25, and 49. This is because the apothem has a factor of the side and so does the perimeter. **43.** $12\sqrt{3}$ cm^2 **45.** $(x - 2)^2 + (y - 4)^2 = 13$ **47.** $\triangle QRT \sim \triangle QTS$ by AA, $\triangle QTS \sim \triangle TRS$ by AA, $\triangle QRT \sim \triangle TRS$ by Transitive Property **49.** acute **51.** $4a^6$

Pages 554–558 Lesson 10–6

5. $\frac{2}{9}$ **7.** $\frac{1}{2}$ **9.** $\frac{4}{25}$ or 0.16 **11.** $\frac{5}{8}$ or 0.625 **13.** $\frac{4 - \pi}{4} \approx 21.5\%$ **15.** $\frac{2}{9}$ **17.** $\frac{2}{9}$ **19.** $\frac{1}{16}$ **21.** $\frac{5}{16}$

23a.

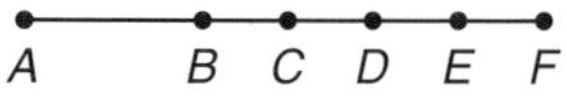

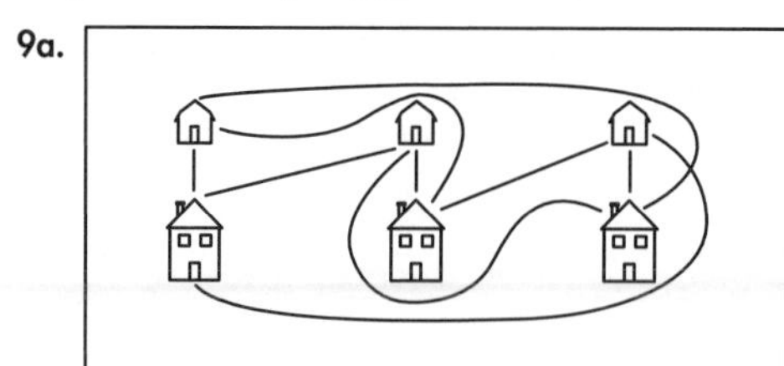

23b. $\frac{1}{3}$ **23c.** $\frac{1}{6}$ **23d.** $\frac{1}{2}$ **25.** $\frac{1}{3}$ **27.** $\frac{\pi}{4} \approx 78.5\%$ **29.** 6 cm **31f.** $\frac{1}{3}$ **33.** $\frac{13}{36} \approx 36.1\%$ **35a.** 3279 yd^2 **35b.** 36 ft^2 **35c.** $\frac{36}{29{,}508} \approx 0.12\%$ **37.** 9:1 **39.** 192 **41.** not enough information **43.** 46.3%, 5.0%, 24.1%, 33.7%, 23.7%, 36.0%, 35.8%

Pages 562–564 Lesson 10–7

7a. traceable, incomplete **7b.** $\overline{AE}, \overline{ED}, \overline{DC}, \overline{CB}, \overline{BD}, \overline{DA}, \overline{AB}$ **7c.** $\overline{AC}, \overline{BE}, \overline{CE}$

9a.

9b. yes; at least once **11.** A: degree 4, B: degree 2, C: degree 3, D: degree 3, E: degree 2 **13a.** traceable, complete **13b.** $\overline{AB}, \overline{BC}, \overline{CA}$ **13c.** none **15a.** not traceable, incomplete **15b.** none **15c.** $\overline{AC}, \overline{BD}, \overline{CE}$ **17a.** traceable, complete **17b.** $\overline{AB}, \overline{BC}, \overline{CA}, \overline{AC}, \overline{CB}, \overline{BA}$ **17c.** none **19.** can't be traced

21. Sample answer:

23. Sample answer:

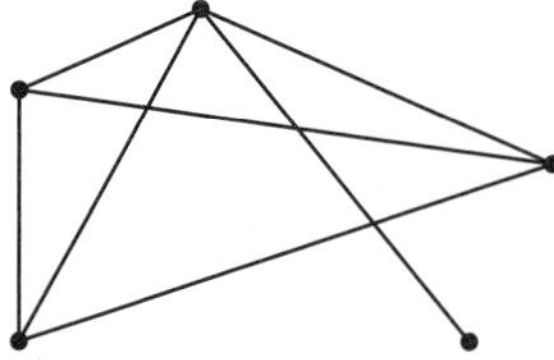

25. When a path goes through a node, it uses two edges, so wherever you start you use one edge from that node and the edges of every subsequent node are both used, one for entering and one for exiting. Therefore, the starting node is the only node left to go to because only one edge was used to start.

27a.

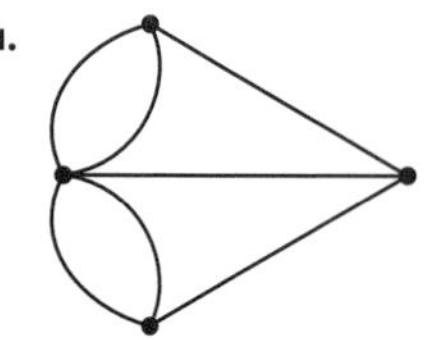

27b. No; all four nodes have odd degree.

27c. 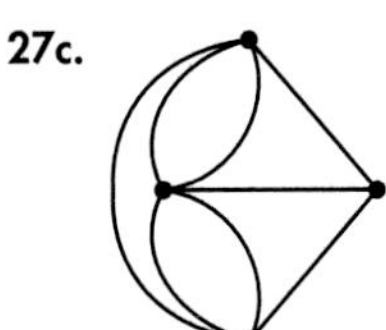

31. 31.5 units^2 33. 5 35. 30 units 37. no solution

Page 565 Chapter 10 Highlights

1. false; radius 3. true 5. false; 5 7. true 9. false; apothem

Pages 566–568 Chapter 10 Study Guide and Assessment

11. 144 13. 162; 18 15. regular, uniform 17. 150 cm^2 19. 7.1 in. 21. 125 m^2 23. 4.8 ft 25. 0.6 ft^2 27. $4n^2$ cm^2 29. 44.0 mm; 153.9 mm^2 31. 5.7 ft; 2.5 ft^2 33. $4\pi\sqrt{2} \approx$ 17.8 ft, $8\pi \approx 25.1$ ft^2 35. 9.08% 37. yes 39. 631 and 542 41. 496 in^2

CHAPTER 11 INVESTIGATING SURFACE AREA AND VOLUME

Page 574 Lesson 11–1A

3. a rectangle whose length is the length of the prism

Pages 578–581 Lesson 11–1

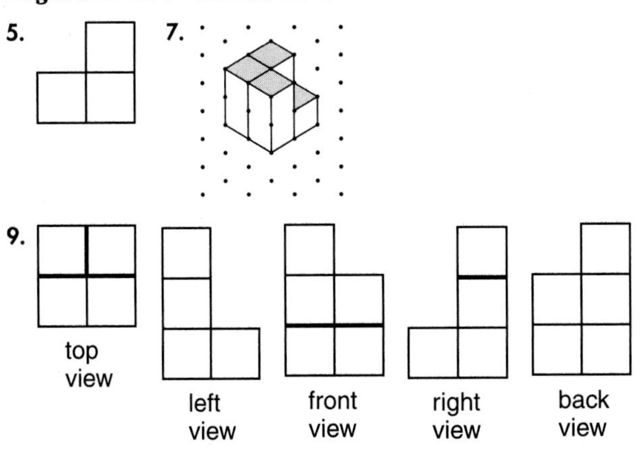

11. no

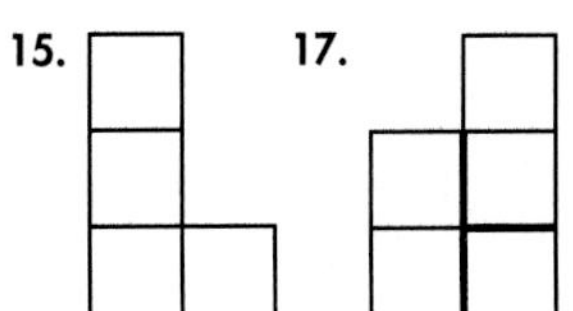

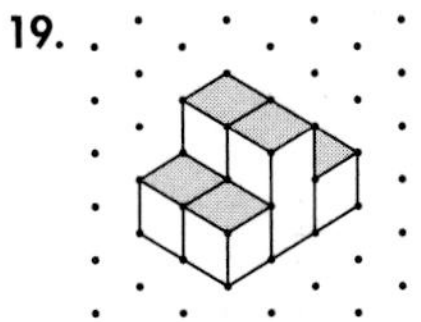

21. upside down U 23. circle

25. 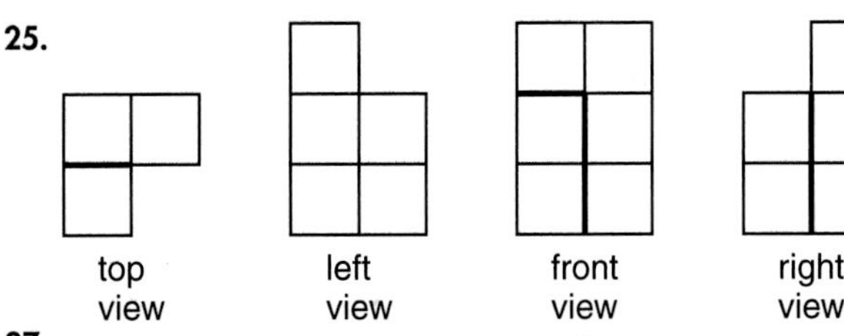

27. yes; edges: $\overline{AB}$, $\overline{BC}$, $\overline{CD}$, $\overline{DE}$, $\overline{EA}$, $\overline{AF}$, $\overline{BG}$, $\overline{CH}$, $\overline{DI}$, $\overline{EJ}$, $\overline{FG}$, $\overline{GH}$, $\overline{HI}$, $\overline{IJ}$, $\overline{JF}$; faces: $ABCDE$, $ABGF$, $BCHG$, $CDIH$, $DEJI$, $EAFJ$, $FGHIJ$; vertices: A, B, C, D, E, F, G, H, I, J 29. square 31. triangle 33. False; The only polygon that can be formed by the intersection of a plane and a cylinder is a rectangle. 37. $A = 5$, $B = 5$, $C = 5$, $D = 5$, not traceable 39. sometimes 41. No; the slope of $\overline{AB} = \frac{1}{6}$, but the slope of $\overline{CD} = \frac{1}{5}$ and $\overline{AB} \parallel \overline{CD}$ if $ABCD$ is a parallelogram. 43. 18

Pages 582–583 Lesson 11–1B

1. equilateral

Pages 586–589 Lesson 11–2

5. 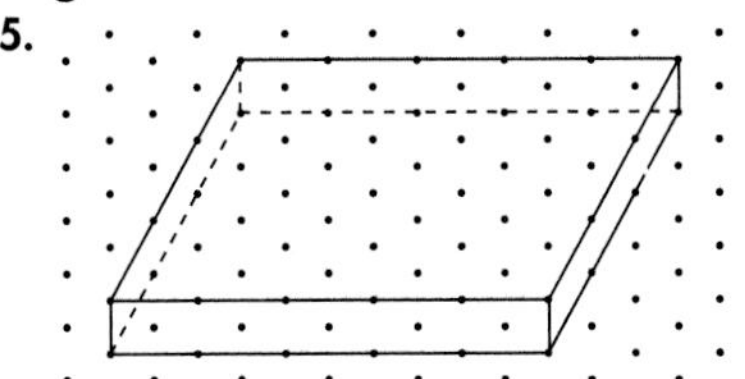

7. 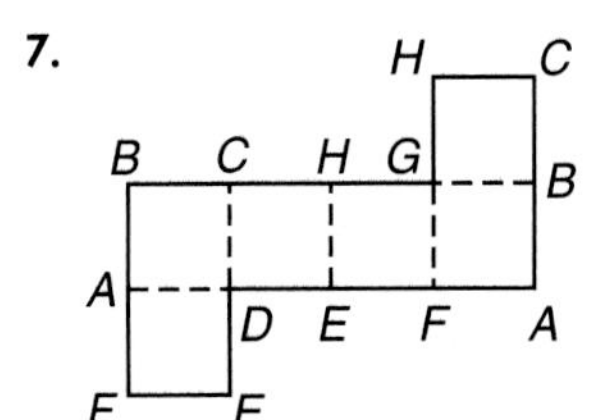

9. surface area = 80.8 units^2; sample net:

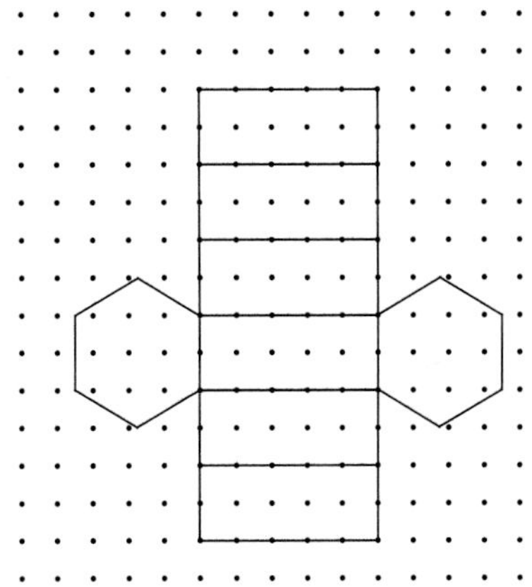

11. 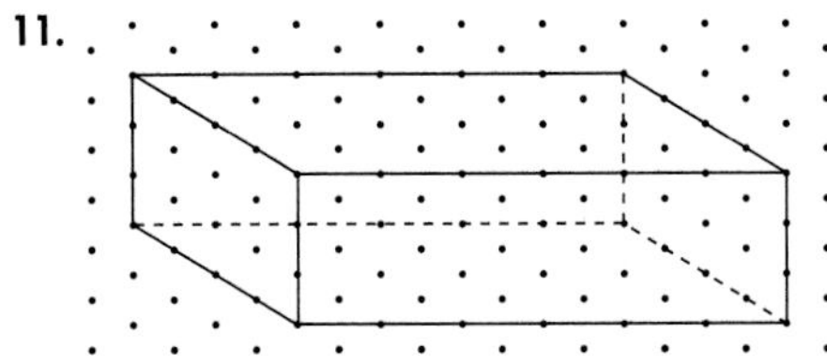

13.

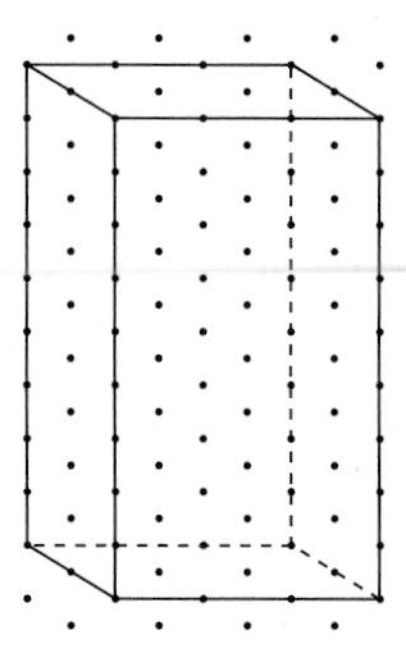

15.

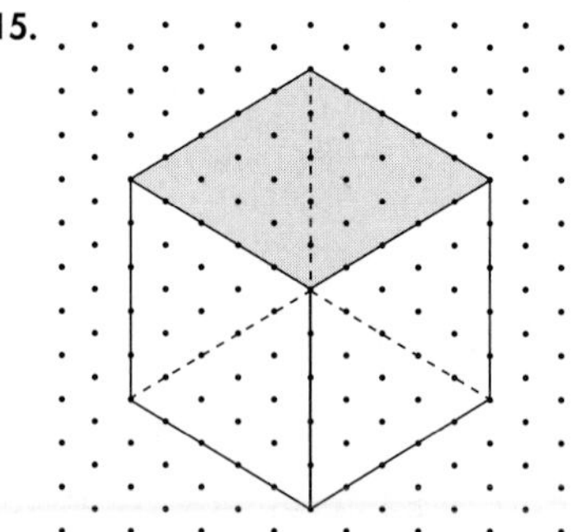

17.

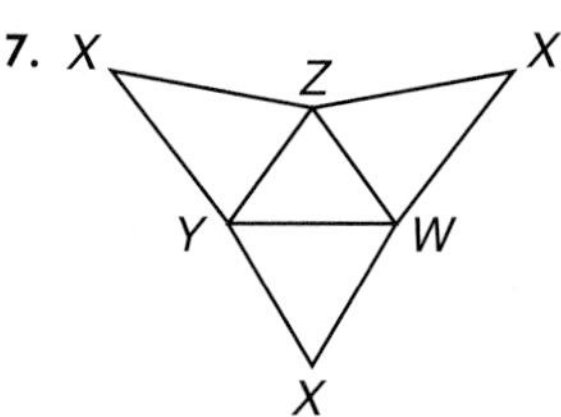

19.

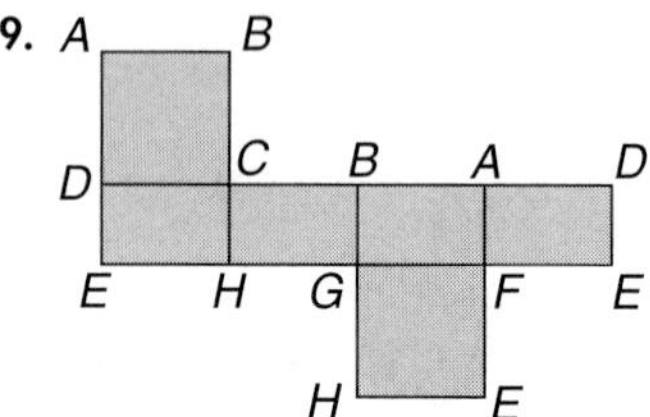

21. surface area ≈ 116.3 units^2;
sample net:

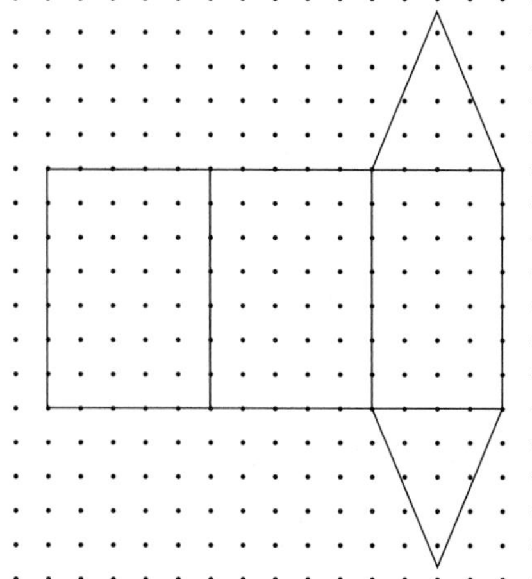

23. surface area = 120 units^2;
sample net:

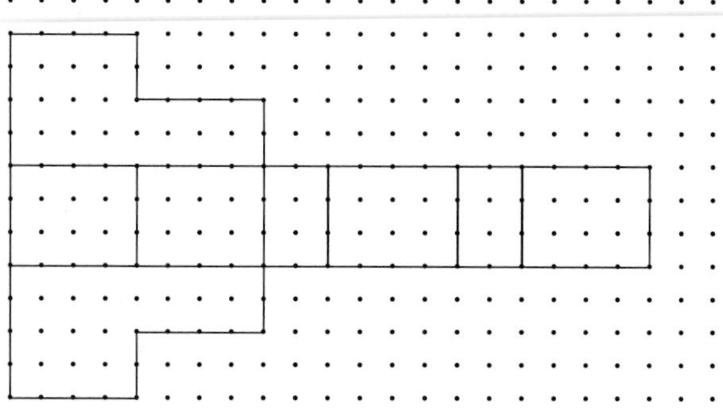

25. surface area = 6.9 units^2;
sample net:

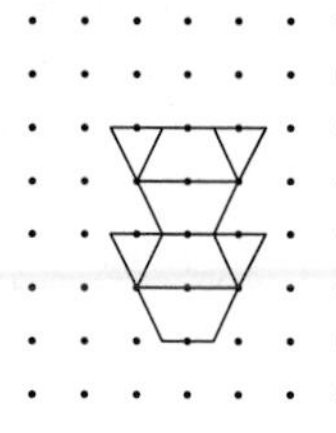

27.

29a.

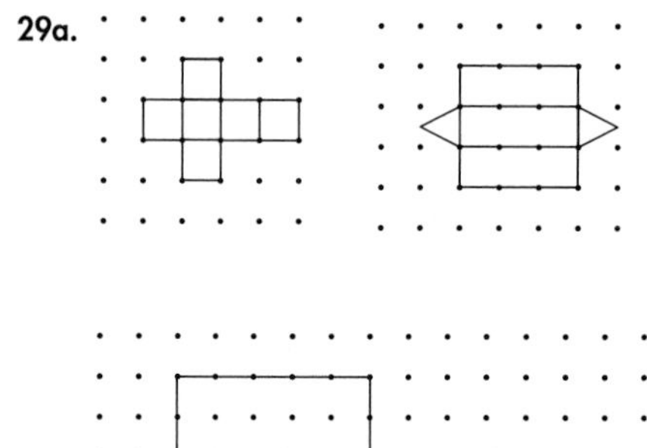

cube surface area = 6 units^2; triangular prism surface area = 9.9 units^2; rectangular prism surface area = 76 units^2

29b.

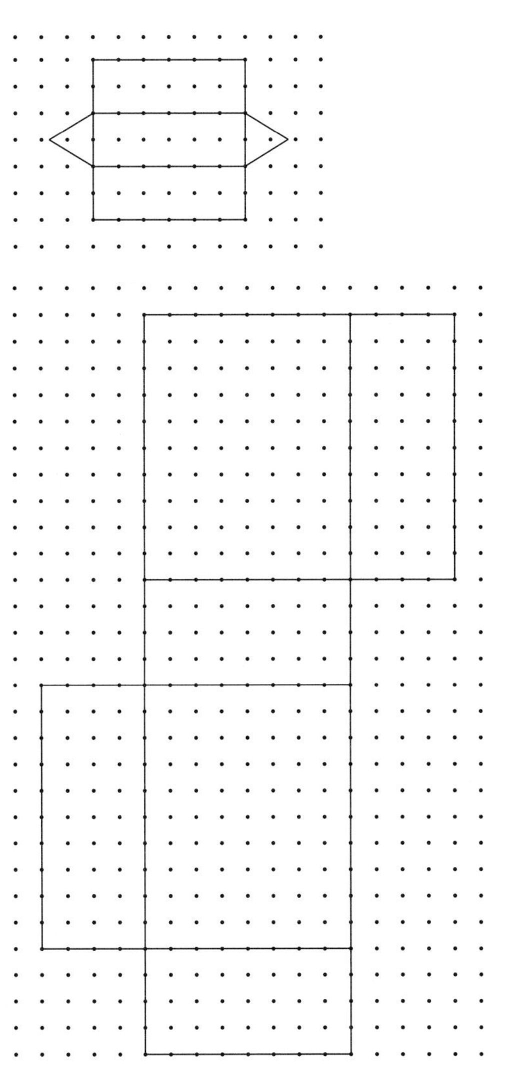

cube surface area = 24 units^2; triangular prism surface area = 39.5 units^2; rectangular prism surface area = 304 units^2 **29c.** The surface area of a solid whose dimensions have been tripled is nine times the surface area of the original solid. **31.** 2344.8 ft^2 **33a.** yes **33b.** no **33c.** yes **35.** 60.35 ft^2 **37.** 4.0 **39.** 2.5 **41.** $\{y \mid y \leq -0.12\}$

Page 590 Lesson 11–2B
3. No, it forms a shape inside of the frame.

Pages 595–598 Lesson 11–3
5. Right prism; the lateral edges are perpendicular to the bases. **7.** 30 units **9.** 112 cm^2 **11.** 251.3 ft^2 **13.** 8.6 in. **15.** oblique prism **17.** 36 in. **19.** 534.4 in^2 **21.** 644.5 in^2 **23.** 360 cm^2; 480 cm^2 **25.** 2304 m^2; 3792 m^2 **27.** 2352.4 m^2 **29.** 200.7 in^2; 517.5 in^2 **31.** 24.9 ft^2; 30.0 ft^2 **33.** 198 cm^2 **35.** If the height of the prism is doubled, the *lateral area* is doubled. The surface area is increased, but it is not doubled. **37.** 1680 ft^2 **39.** 179.3 in^2 **41a.** top
41b.

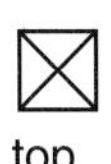
top view

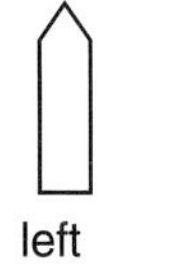
left view

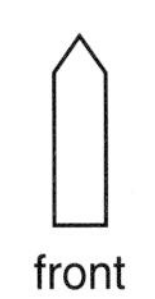
front view

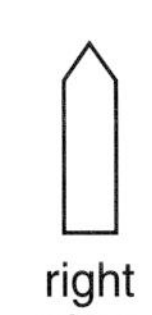
right view

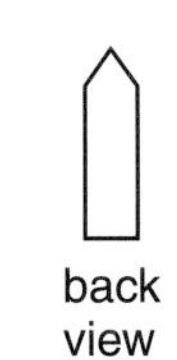
back view

43. 64 cm **45.** $\frac{1}{5}$ **47.** A triangle is equiangular if each angle measures 60°; Law of Syllogism **49.** positive

Pages 603–606 Lesson 11–4
7. both **9.** 47.1 m^2; 75.4 m^2 **11.** 475.2 in^2 **13.** 924,974.6 ft^2 **15.** neither **17.** pyramid **19.** pyramid **21.** 284.3 in^2; 485.2 in^2 **23.** 81 cm^2; 133.6 cm^2 **25.** 188.5 ft^2; 301.6 ft^2 **27.** 169.6 ft^2 **29.** 3696 yd^2 **31.** L = 96 units^2; T = 151.4 units^2 **33.** The lateral area approaches the area of the base. This can be seen by showing a series of cones cut on their slant heights and folded out into sectors. As the altitude approaches zero, the slant height approaches the radius of the base, and the sector narrows, approaching a complete circle. **35.** 736.5 ft^2 **37.** 1078 cm^2 **39.** 570.96 cm^2 **41.** 0.86 inches **43.** $x = b^2 + 18$

Page 606 Self Test
1.

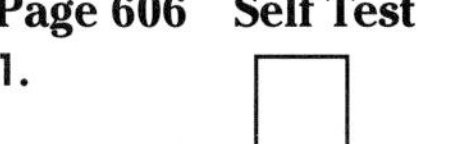

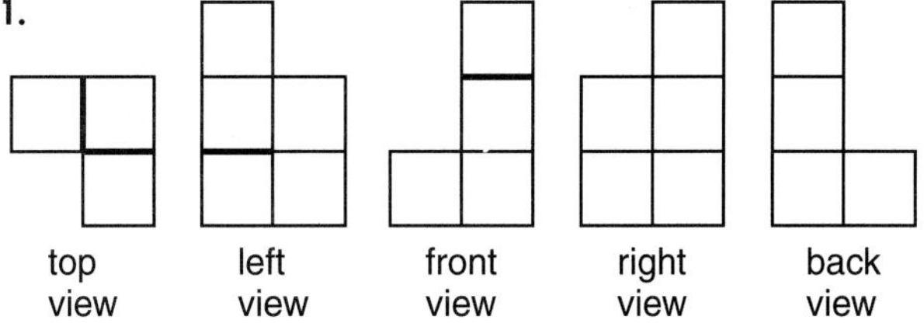

3. L = 432 in^2; T = 619.1 in^2 **5.** about 255,161.7 ft^2

Pages 610–613 Lesson 11–5
7. 127.2 ft^3 **9.** about 1696.5 yd^3 **11.** about 377.0 ft^3 **13.** 461.8 ft^3 **15.** 98.3 m^3 **17.** 3435.3 mm^3 **19.** 1800 in^3 **21.** 105 m^3 **23.** 2598.1 ft^3 **25.** 18 in. **27.** 1592.8 cm^3 **29.** 36 units^3 **31.** 22.3 in^3 **33.** Sample answer: true; Imagine you sliced the oblique cylinder or prism parallel to its base to form a large number of pieces. Then you could slide the pieces so that they stack as a right cylinder or prism. This sliding will not affect the volume or the height. Since the volume of the right cylinder or prism is the product of the area of the base and the height, then the volume of the oblique cylinder or prism is the product of the area of the base and the height. **35.** fully expanded: 2993.0 in^3; fully compressed: 280.6 in^3 **37.** 381.7 in^3 **39.** 384 in^2 **41.** 30°, 30°, 90° **43.** $\overline{RS}$, $\overline{MN}$; $\overline{RT}$, $\overline{MO}$; $\overline{ST}$, $\overline{NO}$ **45.** $2 \cdot 3 \cdot 3 \cdot 3 \cdot c \cdot c \cdot d$

Page 614 Lesson 11–6A
1. 3 **3.** The heights of the prism and the pyramid are the same. **7.** The areas of the bases of the cone and the cylinder are the same. **9.** $V = \frac{1}{3}Bh$

Pages 617–620 Lesson 11–6
5. 301.6 m^3 **7.** about 5178.8 mm^3 **9.** Mauna Loa: 22,415.8 km^3; Fuji: 169.1 km^3; Paricutín: 171,137,610 m^3; Vesuvius: 4,912,194 m^3 **11.** 382.5 in^3 **13.** 4515.5 ft^3 **15.** 1082.8 m^3 **17.** 277.1 m^3 **19.** 5730.3 units^3 **21.** 814.6 yd^3 **23.** 58.9 in^3 **25a.** 1493 ft^3 **25b.** 27,370 in^3 **25c.** 35 m^3 **25d.** Delete the "/3" in each of the lines that contains the volume formulas. **25d.** No, the formulas are the same for a right or oblique solid. **27.** According to Cavalieri's Principle, the volume of each solid stays the same. **29.** 18,555,031.6 ft^3 **31.** 36.9 in. **33.** 4 **35.** Let x = the number; $9x \leq 108$; $\{x \mid x \leq 12\}$

Pages 625–628 Lesson 11–7
7. neither **9.** true **11.** 5 **13.** drizzle: $T < 0.005$ in^2, $V < 0.000034$ in^3; rain: $T > 0.005$ in^2, $V > 0.000034$ in^3 **15.** $T = 452.4$ ft^2; $V = 904.8$ ft^3 **17.** circle **19.** sphere **21.** neither **23.** true **25.** true **27.** false **29.** false **31.** false **33.** 9.6 **35.** no **37.** $T = 7854.0$ in^2; $V = 65{,}449.8$ in^3 **39.** $T = 636{,}172.5$ m^2; $V = 47{,}712{,}938.4$ m^3 **41.** $T = 36.3$ m^2; $V = 20.6$ m^3 **43.** $\frac{32}{3}\pi$ cm^3 or about 33.5 cm^3 **45.** 2:1 **47.** 452.39 cm^2 **49a.** no **49b.** 80% **51.** Sample answer: Buckminster Fuller designed geodesic domes. The domes are portions of spheres. **53.** the bag shaped like a rectangular prism **55.** 0.96 **57.** 117 **59.** 5 **61.** $\{n \mid n < 2\}$

Pages 632–635 Lesson 11–8
5. congruent **7.** 64:1 **9.** $\frac{x}{512}$ m^3 **11.** False; if two pyramids have square bases and all the linear measurements are proportional, then they must be similar. **13a.** 0.015 cm or 0.15 mm **13b.** $1{,}000{,}000x$ cm^2 **15.** neither **17.** congruent **19.** neither **21.** 8:27 **23.** $3\frac{3}{4}$ in. **25.** False; if an edge length of a cube is tripled, then its volume is twenty-seven times greater. **27.** False; doubling the radius of a sphere quadruples the surface area. **29a.** 7:8; 7:1 **29b.** 3:4; 3:1 **31.** Since the only linear measure involved in a sphere is the radius, the linear measures of two spheres are always proportional. **33.** 384 ft × 300 ft × 224 ft **35.** $T = 548.08$ ft^2; $V = 1206.5$ ft^3 **37.** 60 cm **39.** $m\angle F = 98$, $DF \approx 9.1$, $FG \approx 10.7$ **41.** a and d

Page 637 Chapter 11 Highlights
1. cross section **3.** right cone **5.** lateral faces **7.** regular **9.** similar

Pages 638–640 Chapter 11 Study Guide and Assessment
11. **13.**

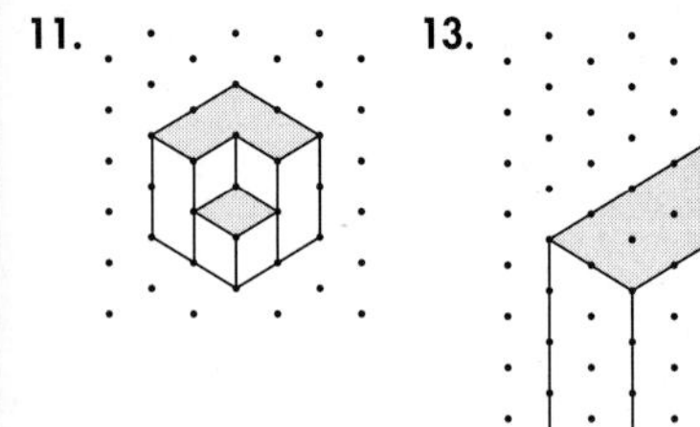

15. $L = 48$ ft^2; $T = 56$ ft^2 **17.** $L = 439.8$ cm^2; $T = 747.7$ cm^2 **19.** $L = 20$ in^2; $T = 24$ in^2 **21.** $L = 48$ in^2; $T = 84$ in^2 **23.** 5196.2 cm^3 **25.** 1021.0 ft^3 **27.** 2787.6 cm^3 **29.** true **31.** 267.9 cm^3 **33.** true **35.** False; a solid is always similar to itself. **37.** 149,301.0 ft^2; 5,424,604.8 ft^3 **39.** 18:458 or about 1:25.4

CHAPTER 12 CONTINUING COORDINATE GEOMETRY

Pages 649–651 Lesson 12–1
7. 28; 8; $-\frac{2}{7}$

9.
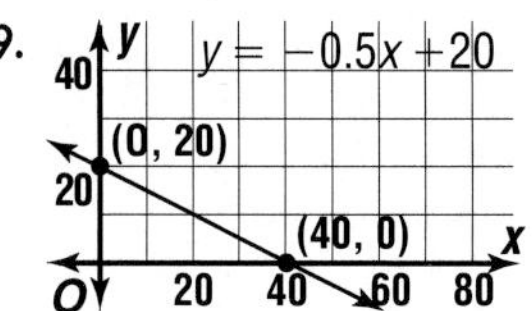

11.
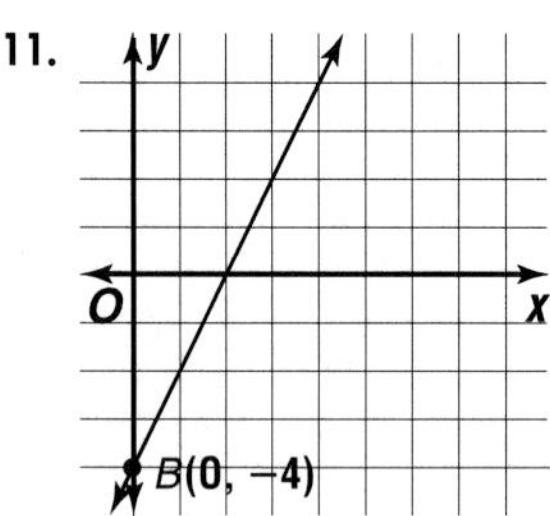

13. 0; 0; −1 **15.** 2; none; undefined **17.** 0; none; undefined

19.
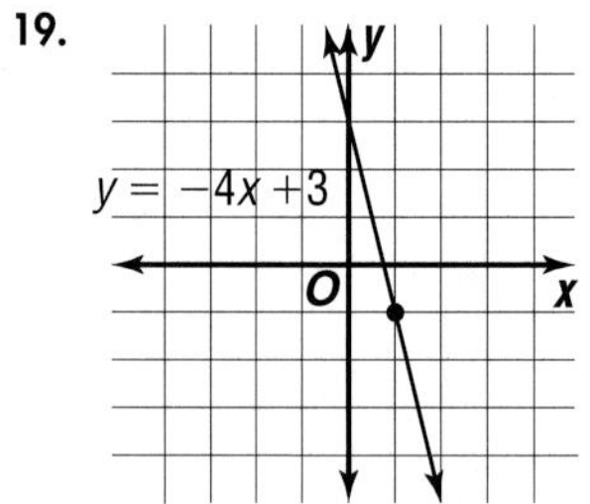

21.
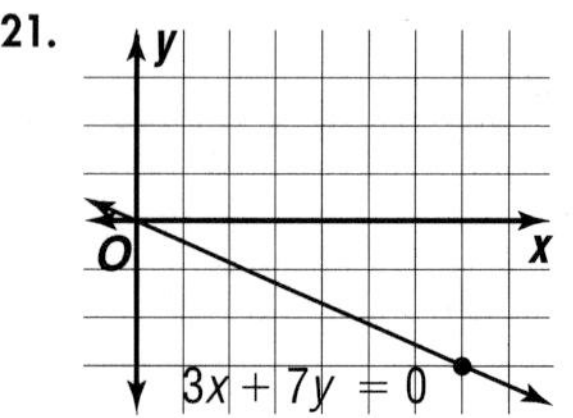

23.
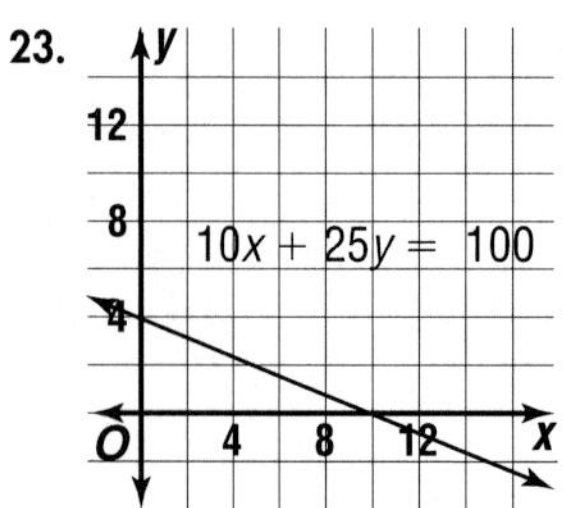

25.
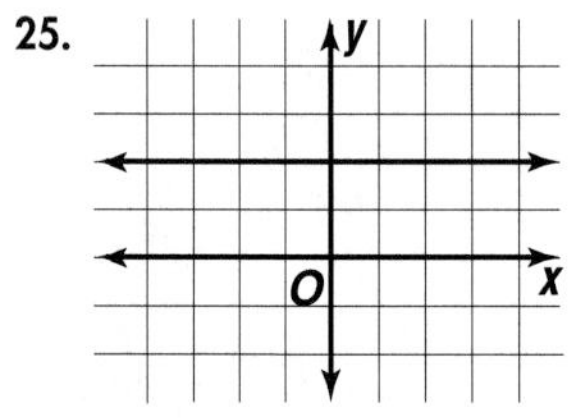

27. 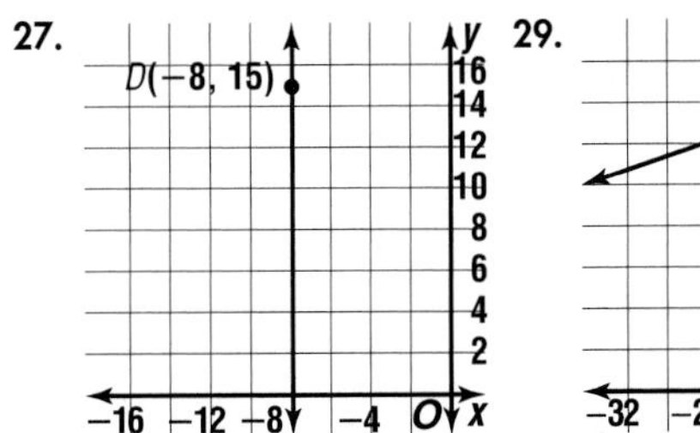

29. 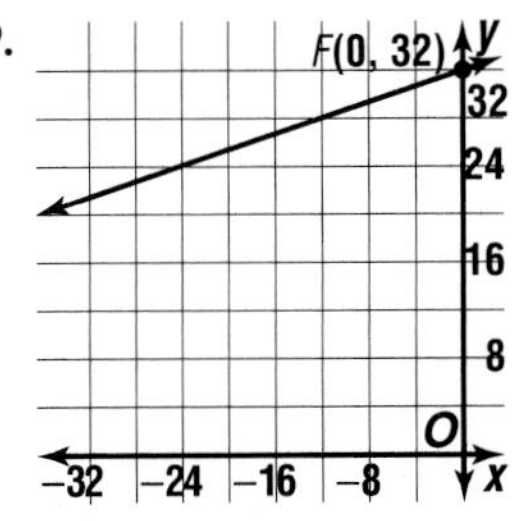

31. Sample answer: All are of the form $y = mx - 1$, but all have a different value for m. **33.** The slope is undefined. Sample answer: The x-coordinates of points on a vertical line are the same. When finding the slope, the denominator (the change in x) is zero and division by zero is undefined. **35a.** The new line would be parallel to the first line with y-intercept 14. **35b.** m would still be -2, b would change to 14 **37.** $y = -2.5x + 3.5$; Sample answer: The distance between two parallel lines is the length of any perpendicular segment that connects points on the two lines. The distance between the y-intercepts is 5. The y–intercept midway between the two lines is 3.5. Since the line is parallel to the other two, the slope must be the same. **39a.** 12; for every gram of fat in the food, there are 12 times more Calories **39b.** 180

39c. 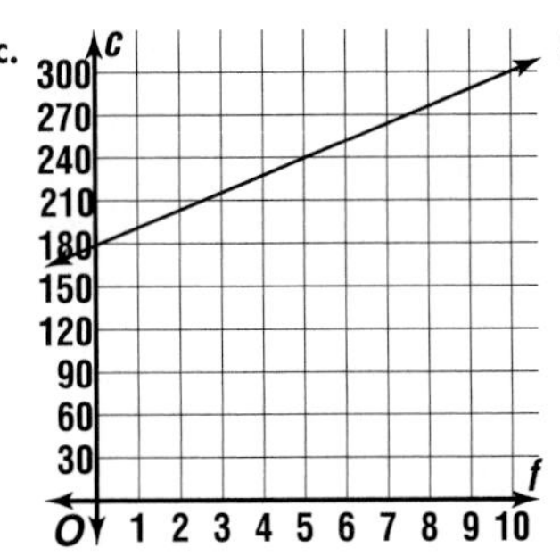

39d. 540 Calories

41a. the number of people that can be carried in x small vans **41b.** x: 12; the number of small vans if all of the people were to be transported in small vans and none in the large vans; y: 8, the number of large vans if all of the people were to be transported in the large vans and none in the small vans

41c. 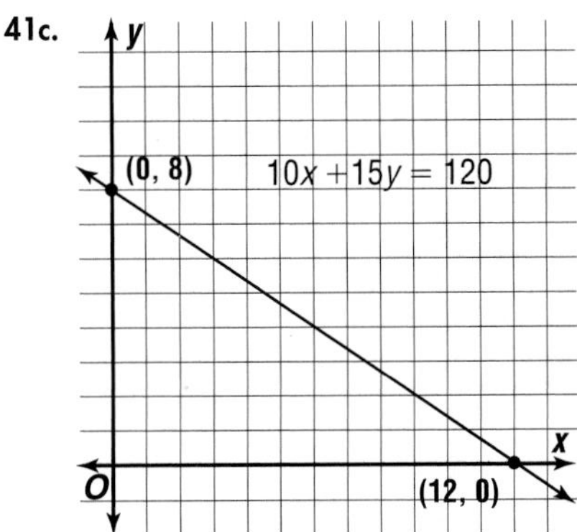

41d. yes; $10(6) + 15(4) = 120$ **41e.** yes; $10(3) + 15(6) = 120$ **43.** 18 m, 6 m **45.** 2.8 miles **47.** $\angle C$, $\angle A$, $\angle B$ **49.** 1, 2, 11, 22

Page 652 Lesson 12–2A

1. The slope between any two points is 0.034. **3.** $y = 0.034x - 34$

Pages 656–659 Lesson 12–2

7. $y = -\frac{1}{3}x + \frac{10}{3}$ **9.** $y = \frac{1}{3}x - 3$ **11.** $y = -2x + 7$ **13.** $\frac{1}{2}$, 3; $y = \frac{1}{2}x + 3$ **15.** $\frac{2}{3}$, -2; $y = \frac{2}{3}x - 2$ **17.** $y = 0.5x - 1$ **19.** $y = x + 2$ **21.** $y = 8$ **23.** $y = 6$ **25.** $y = 2x - 7$ **27.** $y = -0.1x + 19$ **29.** $y = 5x - 20$ **31.** $y = 10$ **33.** $y = -1$ **35.** $c = 20t + 340$ **37.** $y = \frac{5}{3}x - \frac{55}{3}$ **39.** $y = -\frac{1}{3}x + \frac{32}{3}$ **41a.** $y = -\frac{1}{3}x - 1$ **43b.** The y-intercepts will all change. **45a.** $y = 32x$; $y = 38x$ **45b.** Sample answer: Ameritech ISDN service is always cheaper for the first customer.

47. 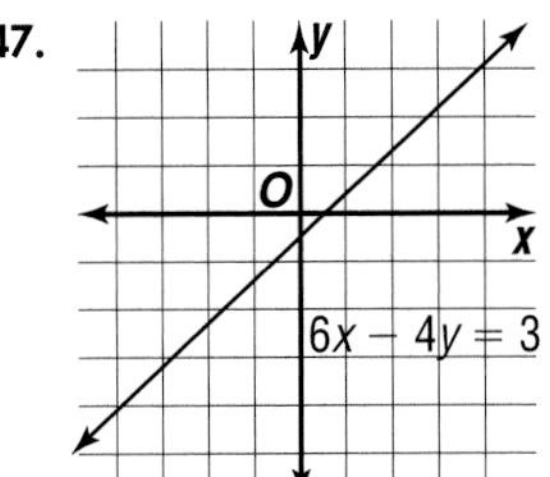

$m = \frac{3}{2}$; $b = -\frac{3}{4}$

49. 0.37 **51.** Sample answer: A circle S that has all of its points in the interior of circle R would have no common tangents with circle R. **53.** 10.2 **55.** $\frac{2a}{7c}$; $a \neq 0$, $b \neq 0$, $c \neq 0$

Pages 662–665 Lesson 12–3

5. no **7.** False; $4 \neq -10 + 8$ **9.** true **11.** yes **13.** no **15.** yes **17a.**

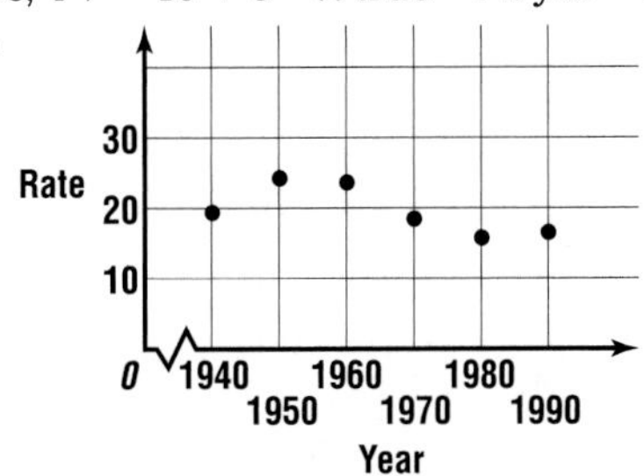

17b. Answers will vary. The points (1950, 24.1) and (1990, 16.7) produce slope of -0.185 and the equation $y = -0.185x + 384.85$. **17c.** Sample answer: Each year the birthrate decreases by 0.185 so every ten years about 2 fewer children are born per 1000 people. **17d.** Sample answer: 14.85; It will be reliable only if the trend continues as it has in the past 50 years. **19.** $y = -\frac{1}{2}x + 19$ **21.** $x = 2$ **23a.** $y = -\frac{1}{2}x - \frac{1}{2}$ **23b.** The line through the midpoint of the two sides of a triangle is parallel to the third side and equal to half of the length of the third side. Point D will have coordinates $(-3, 1)$ and $E(1, -1)$. The slope of the line containing $\overline{BC}$ is $-\frac{1}{2}$. The slope of $\overline{DE} = -\frac{1}{2}$. The lines do not share a common point but have the same slope so they are parallel. $BC = \sqrt{(-2-6)^2 + (8-4)^2} = \sqrt{80}$ or $4\sqrt{5}$. $DE = \sqrt{(-3-1)^2 + (1-(-1))^2} = 2\sqrt{5}$. Thus, the length of the segment connecting the midpoints is one half the length of the third side. **23c.** (0, 7) is a point on $\overline{BC}$ and on the perpendicular from $\overline{BC}$ to $D(-3, 1)$. The distance from $\overline{DE}$ to $\overline{BC}$ is $\sqrt{45} = 3\sqrt{5}$ or 6.71. **25.** Sample answer: $(-3-3)^2 + (13-5)^2 =$

$(-6)^2 + (8)^2 = 36 + 64$ or 100; (9, −3), (11, 11), (11, −1) are some of the possible points. **27.** $y = 0.75x + 5.25$ **29.** Sample answer: One approach might be to use the vertical distance as a measure of error. A, B, and C are on the line $y = 2x + 10$, but D is not. The error in predicting y for $x = 6$ at point D is 14. The error in predicting using $y = 1.8x + 7.3$ for point A is 3.1, B is 2.7, C is 5.1 and D is 10.1 for a total of 21. The better line is $y = 2x + 10$. **31b.** Sample answer: $y = 0.264x + 4339.5$ where x is the original cost and y is the sum of the operating and fixed costs or the total cost of operating the car after it has been purchased. **33.** $6, -3; \frac{1}{2}$ **35.** false **37.** true **39.** $a + 1 - \frac{8}{2a + 1}$

Page 665 Self Test

1. no x-intercept; 12; 0

3.

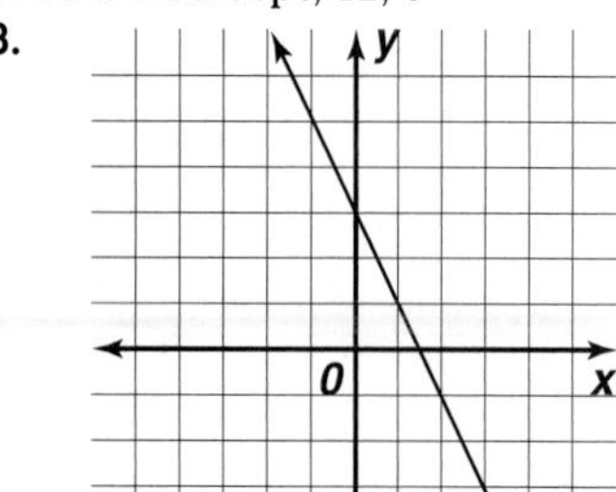

5. $y = 5x + 15$ **7.** $y = \frac{1}{9}x + \frac{1}{9}$ **9.** no

Pages 668–671 Lesson 12–4

5. $D(0, 2c)$ **7a.** square; $AB = BC = CD = DA$; and $\angle DAB$ is right. **7b.** Both midpoints are $\left(\frac{a}{2}, \frac{a}{2}\right)$; $\overline{AC}$ and $\overline{BD}$ bisect each other. **7c.** The slope of $\overline{AC}$ is 1. **7d.** The slope of $\overline{DB}$ is −1. **7e.** $\overline{AC}$ and $\overline{DB}$ are perpendicular. **9.** Let M be the midpoint of $\overline{BC}$. The coordinates of M will be $\left(\frac{2c}{2}, \frac{2b}{2}\right) = (c, b)$. $MA = \sqrt{(c-0)^2 + (b-0)^2} = \sqrt{c^2 + b^2}$. $BC = \sqrt{(2c-0)^2 + (0-2b)^2} = 2\sqrt{c^2 + b^2}$. Since $2MA = BC$, $MA = \frac{1}{2}BC$. **11.** about 29.2 miles **13.** $D(-a, 0)$, $F(0, b)$ **15.** $B(a, 0)$, $D(a, d)$; $F(-b, c)$ **17.** Slope of $\overline{LM}$ is $\frac{b-0}{b-0} = \frac{b}{b}$ or 1. Slope of $\overline{MN}$ is $\frac{b-0}{b-2b} = \frac{b}{-b}$ or −1. $\overline{LM}$ and $\overline{MN}$ are perpendicular and $\triangle LMN$ is a right triangle.

19. $EF = \sqrt{(a\sqrt{2} - 0)^2 + (a\sqrt{2} - 0)^2} = \sqrt{2a^2 + 2a^2}$
$= \sqrt{4a^2}$

$FG = \sqrt{((2a + a\sqrt{2}) - a\sqrt{2})^2 + (a\sqrt{2} - a\sqrt{2})^2}$
$= \sqrt{(2a)^2 + 0^2} = \sqrt{4a^2}$

$GH = \sqrt{((2a + a\sqrt{2}) - 2a)^2 + (a\sqrt{2} - 0)^2}$
$= \sqrt{2a^2 + 2a^2} = \sqrt{4a^2}$

$EH = \sqrt{(0-0)^2 + (2a-0)^2} = \sqrt{0^2 + (2a)^2} = \sqrt{4a^2}$

$EF = FG = GH = EH$

$\overline{EF} \cong \overline{FG} \cong \overline{GH} \cong \overline{EH}$ $EFGH$ is a rhombus.

21.

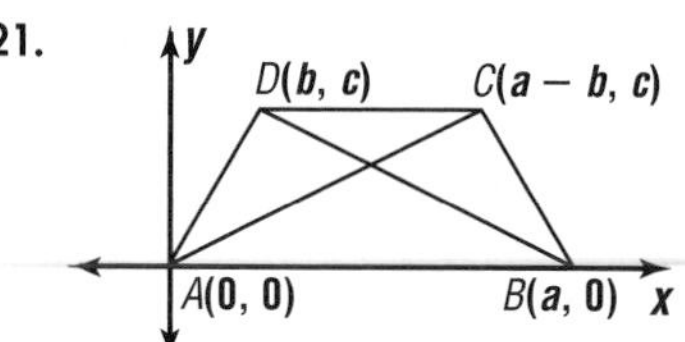

$DB = \sqrt{(a-b)^2 + (0-c)^2} = \sqrt{(a-b)^2 + c^2}$
$AC = \sqrt{((a-b) - 0)^2 + (c-0)^2} = \sqrt{(a-b)^2 + c^2}$
$DB = AC$ and $\overline{DB} \cong \overline{AC}$
$AC = \sqrt{(a + b - 0)^2 + (c - 0)^2}$
$BD = \sqrt{(b-a)^2 + (c-0)^2}$
But $AC = BD$ and $\overline{DB} \cong \overline{AC}$.

23.

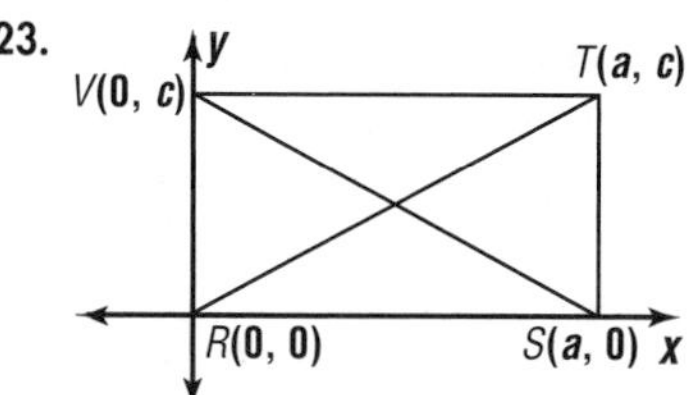

$VS = \sqrt{(0 - a)^2 + (c-0)^2} = \sqrt{a^2 + c^2}$
$RT = \sqrt{(a-0)^2 + (c-0)^2} = \sqrt{a^2 + c^2}$
$VS = RT$ and $\overline{VS} \cong \overline{RT}$

25.

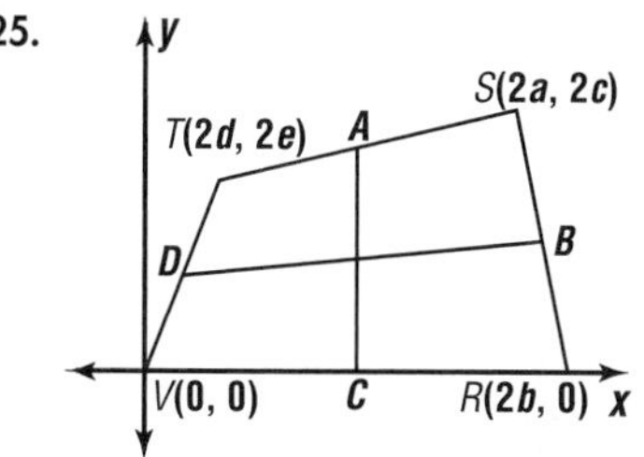

Midpoint A of $\overline{TS}$ is $\left(\frac{2d + 2a}{2}, \frac{2e + 2c}{2}\right)$ or $(d + a, e + c)$. Midpoint B of $\overline{SR}$ is $\left(\frac{2a + 2b}{2}, \frac{2c + 0}{2}\right)$ or $(a + b, c)$. Midpoint C of $\overline{VR}$ is $\left(\frac{0 + 2b}{2}, \frac{0 + 0}{2}\right)$ or $(b, 0)$. Midpoint D of $\overline{TV}$ is $\left(\frac{0 + 2d}{2}, \frac{0 + 2e}{2}\right)$ or (d, e). Midpoint of $\overline{AC}$ is $\left(\frac{d + a + b}{2}, \frac{e + c + 0}{2}\right)$ or $\left(\frac{a + b + d}{2}, \frac{c + e}{2}\right)$. Midpoint of $\overline{DB}$ is $\left(\frac{d + a + b}{2}, \frac{e + c}{2}\right)$ or $\left(\frac{a + b + d}{2}, \frac{c + e}{2}\right)$. $\overline{AC}$ and $\overline{DB}$ bisect each other.

27.

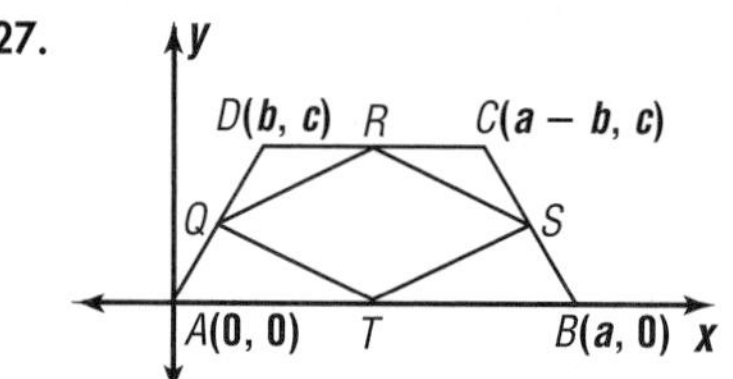

$QR = \sqrt{\left(\frac{b}{2} - \frac{a}{2}\right)^2 + \left(\frac{c}{2} - c\right)^2} = \frac{\sqrt{b^2 - 2ab + a^2 + c^2}}{2}$;

$TS = \left(\frac{2a - b}{2} - \frac{a}{2}\right)^2 + \left(\frac{c}{2} - 0\right)^2 = \frac{\sqrt{b^2 - 2ab + a^2 + c^2}}{2}$;

$QT = \left(\frac{b}{2} - \frac{a}{2}\right)^2 + \left(\frac{c}{2} - 0\right)^2 = \frac{\sqrt{b^2 - 2ab + a^2 + c^2}}{2}$;

$RS = \sqrt{\left(\frac{a}{2} - \frac{2a - b}{2}\right)^2 + \left(c - \frac{c}{2}\right)^2} = \frac{\sqrt{b^2 - 2ab + a^2 + c^2}}{2}$;

$QR = TS = QT = RS$, so $\overline{QR} \cong \overline{TS} \cong \overline{QT} \cong \overline{RS}$ and $QRST$ is a rhombus.

29.

$ST = \sqrt{\left(\frac{a+b}{2} - \frac{b}{2}\right)^2 + \left(\frac{c}{2} - \frac{c}{2}\right)^2} = \sqrt{\left(\frac{a}{2}\right)^2 + 0^2}$

$= \sqrt{\left(\frac{a}{2}\right)^2} = \frac{a}{2}$

$AB = \sqrt{(a-0)^2 + (0-0)^2} = \sqrt{a^2 + 0^2} = \sqrt{a^2} = a$

$ST = \frac{1}{2}AB$

31a. Two possible coordinates for C are $(a, 0)$ or $(0, b)$. **31b.** Two possible coordinates for C are $(2a, 0)$ or $(0, 2b)$. **31c.** Two possible solutions are $C(a - b, a + b)$, $D(-b, a)$ and $C(a + b, b - a)$, $D(b, -a)$. **33.** $\sqrt{218} \approx 14.8$ km

35a.

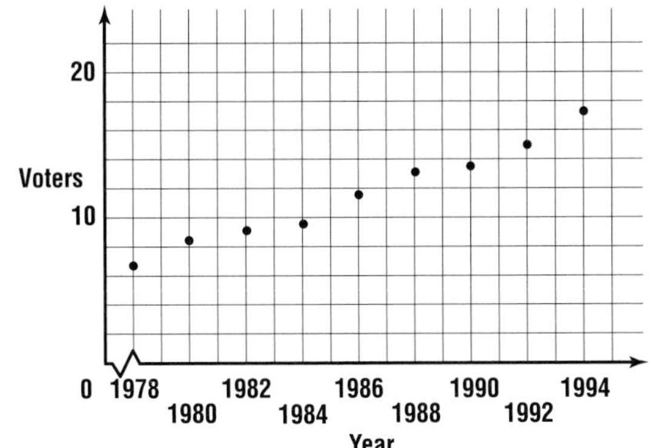

35b. Sample answer: $y = 0.66875x - 1315.9875$ **35c.** Sample answer: 28.2 million **37.** 435 units^3 **39.** about 50.3 inches **41.** yes; binomial

Page 672 Lesson 12–5A

3. It is a diagonal.

Pages 676–679 Lesson 12–5

5. $\sqrt{125} \approx 11.2$; 80° **7.** $\overrightarrow{AB} \parallel \overrightarrow{KM} \parallel \overrightarrow{TR}$ **9.** $\overrightarrow{TR}$ and $\overrightarrow{AB}$ or $\overrightarrow{TR}$ and $\overrightarrow{KM}$ **11.** $(22, -9)$ **13.** $\begin{bmatrix} 6 \\ -2 \end{bmatrix}$

15.

17. $\sqrt{17} \approx 4.1$ units, 76°

19. $\sqrt{409} \approx 20.2$ units, 81° **21.** $\sqrt{72} \approx 8.5$ units, $-135°$

23. 35 units **25.** $-68°$ **27.** $\begin{bmatrix} 0 \\ 1 \end{bmatrix}$ **29.** $\begin{bmatrix} 8 \\ 5 \end{bmatrix}$ **31.** $\begin{bmatrix} -2 \\ 6 \end{bmatrix}$

33. $\overrightarrow{AB} = \overrightarrow{HG}$ **35.** $\overrightarrow{AD}$ **37.** $\overrightarrow{BE}$ **39.** $(11, 10)$ **41.** $(20, 32)$; $\overrightarrow{EF} = \overrightarrow{HI}$ **43.** $(8, 9)$

45.
1. Definition of vector addition
2. Definition of vector addition
3. Given
4. Substitution Property (=)
5. Distributive Property (=)
6. Substitution Property (=)
7. Multiplication Property (=)

47. $\begin{bmatrix} ac \\ ab \end{bmatrix}$; The process has to model the way scalar multiplication of vectors written as ordered pairs is written. **49.** A line through the origin will have equation $y = cx$ for some constant c. If (k, m) lies on the line, then $m = ck$. If (r, s) is a scalar multiple of (k, m), then $(r, s) = c(k, m) = (ck, cm)$. Substituting into $y = cx$, $cm = c(ck)$. Dividing by c, $m = ck$ which is true because (k, m) lies on the line. Therefore, (r, s) lies on the line also. **51.** yes **53a.** They have the same magnitude but opposite directions. **53b.** It is a hexagon. **55.** $\sqrt{425} \approx 20.62$ miles per hour at a direction of 76° **57.** 84.9 newtons, 32° northeast **59a.** Sample answer: Using points (1195, 229) and (5821, 1134) the equation is $y = 0.2x - 4$. **59b.** 441 thousand **61.** 22 m **63.** 124 **65.** $12\sqrt{3} \approx 20.8$ in. **67.** $\frac{9\sqrt{2}}{2} \approx 6.364$ units **69.** $\frac{3}{2}u - \frac{2}{u}$

Pages 683–686 Lesson 12–6

5.

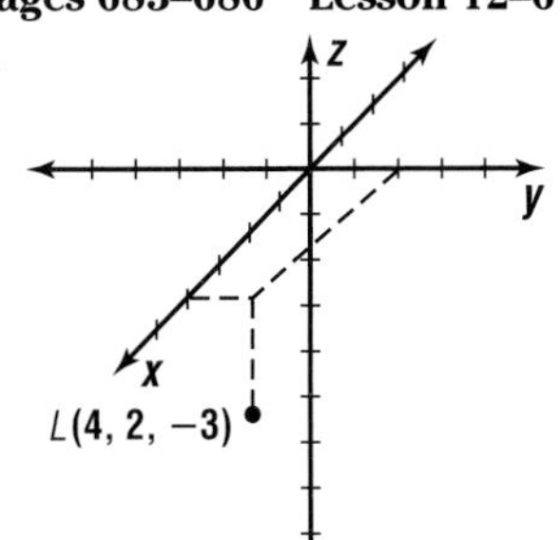

7. $\sqrt{521} \approx 22.83$; $(19.5, -5, 29)$ **9.** true **11.** $(5, -2, 0)$, 6 **13.** $(x + 2)^2 + (y - 3)^2 + (z - 25)^2 = 138$

15.

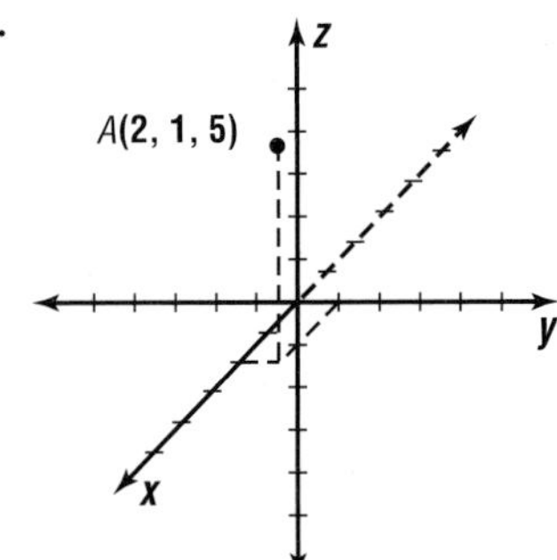

17.

19.

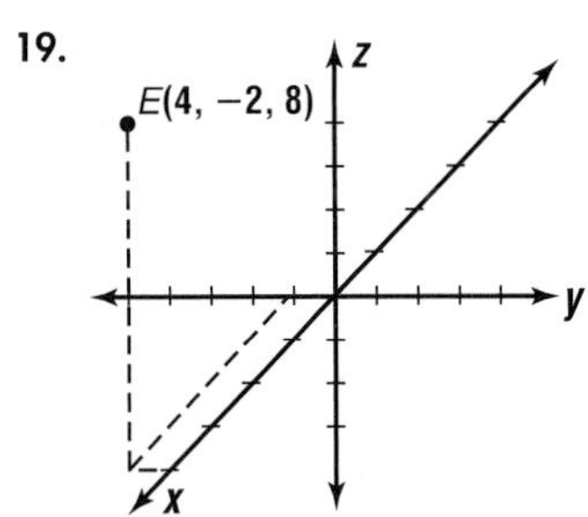

21. $\sqrt{117} \approx 10.8$ units; $(4, -3, 3.5)$ **23.** $\sqrt{10} \approx 3.16$ units; $(7.5, 0.5, 1)$ **25.** $\sqrt{13} \approx 3.6$ units; $(4, 7, 0.5)$ **27.** False; $c = 0$; the points will be of the form $(0, y, z)$. **29.** true **31.** False; it is a plane that contains that line. **33.** $(5, -4, 10), 3$ **35.** $(-4, 2, -12), \sqrt{18} \approx 4.24$ **37.** $(x + 2)^2 + (y - 3)^2 + (z + 4)^2 = 74$ **39.** $(x + 5)^2 + (y - 4)^2 + (z - 19)^2 = 36$ **41.** $\sqrt{6} + \sqrt{11} + 3$ or about 8.77 units **43.** $AB = 11$, $AC = 11$ so $\triangle ABC$ is isosceles. If $AB^2 + AC^2 = BC^2$, $\triangle ABC$ will be a right triangle. $BC = \sqrt{242}$ and $11^2 + 11^2 = 242$, so $\triangle ABC$ is a right triangle. **45.** 150 units^2, 125 units^3

47.

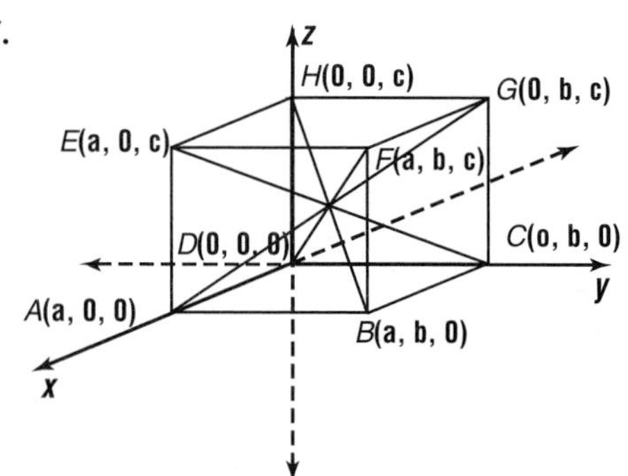

$AG = \sqrt{(a - 0)^2 + (0 - b)^2 + (0 - c)^2} = \sqrt{a^2 + b^2 + c^2}$

$BH = \sqrt{(a - 0)^2 + (b - 0)^2 + (0 - c)^2} = \sqrt{a^2 + b^2 + c^2}$

$CE = \sqrt{(0 - a)^2 + (b - 0)^2 + (0 - c)^2} = \sqrt{a^2 + b^2 + c^2}$

$DF = \sqrt{(0 - a)^2 + (0 - b)^2 + (0 - c)^2} = \sqrt{a^2 + b^2 + c^2}$

$AG = BH = CE = DF$ and $\overline{AG} \cong \overline{BH} \cong \overline{CE} \cong \overline{DF}$

Midpoint of $\overline{AG}$ is $\left(\frac{a+0}{2}, \frac{0+b}{2}, \frac{0+c}{2}\right)$ or $\left(\frac{a}{2}, \frac{b}{2}, \frac{c}{2}\right)$.

Midpoint of $\overline{CE}$ is $\left(\frac{0+a}{2}, \frac{b+0}{2}, \frac{0+c}{2}\right)$ or $\left(\frac{a}{2}, \frac{b}{2}, \frac{c}{2}\right)$.

Midpoint of $\overline{BH}$ is $\left(\frac{a+0}{2}, \frac{b+0}{2}, \frac{0+c}{2}\right)$ or $\left(\frac{a}{2}, \frac{b}{2}, \frac{c}{2}\right)$.

Midpoint of $\overline{DF}$ is $\left(\frac{0+a}{2}, \frac{0+b}{2}, \frac{0+c}{2}\right)$ or $\left(\frac{a}{2}, \frac{b}{2}, \frac{c}{2}\right)$.

$\overline{AG}$, $\overline{CE}$, $\overline{BH}$, and $\overline{DF}$ all bisect each other.

49. $(6, -1, 3), (6, 9, 3), (-4, 9, 3), (-4, -1, 3), (6, -1, 13), (6, 9, 13), (-4, 9, 13), (-4, -1, 13)$ **51.** $(2, 2, 2)$ to block a row of Xs **53.** $|\overline{LN}| = \sqrt{85} \approx 9.2$; 13° **55a.** $y = 11x + 349$ **55b.** 514 **57.** 52.3 mm^2 **59.** $288\sqrt{2}$ or about 407.3 m^2 **61.** 4.5 **63.** -18

Page 687 Chapter 12 Highlights

1. coordinate proof **3.** slope-intercept form **5.** standard form **7.** scatter plot **9.** linear equation

Page 688–690 Chapter 12 Study Guide and Assessment

11.

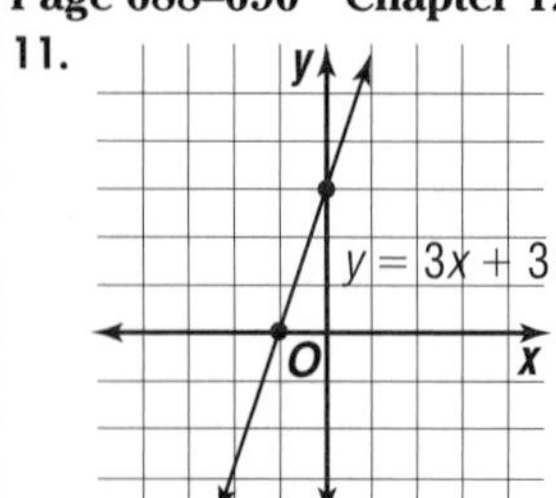

13.

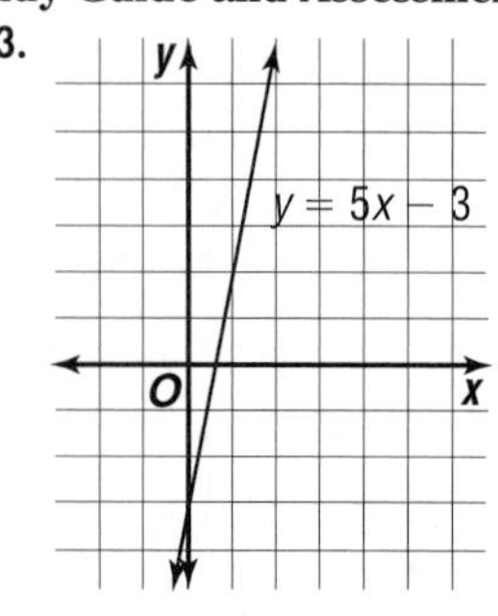

15. $y = 4x + 2$ **17.** $y = -5x - 31$ **19.** $y = x + 8$

21. $y = \frac{1}{2}x - 2, y = \frac{2}{3}x - 3, y = x - 3$

23. $y = -2x - 3, y = -\frac{3}{2}x + 2, y = -x + 7$

25.

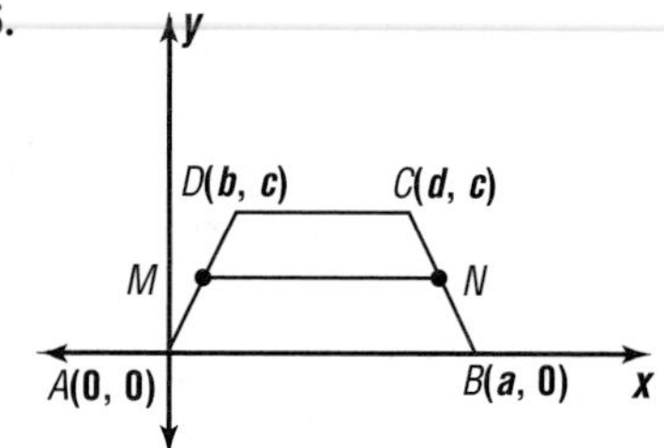

Midpoint M is $\left(\frac{b+0}{2}, \frac{c+0}{2}\right)$ or $\left(\frac{b}{2}, \frac{c}{2}\right)$.

Midpoint N is $\left(\frac{a+d}{2}, \frac{c+0}{2}\right)$ or $\left(\frac{a+d}{2}, \frac{c}{2}\right)$.

Slope of $\overline{DC}$ is $\frac{c-c}{d-b}$ or $\frac{0}{d-b}$ or 0.

Slope of $\overline{MN}$ is $\frac{\frac{c}{2} - \frac{c}{2}}{\frac{a+d}{2} - \frac{b}{2}}$ or $\frac{0}{\frac{a+d-b}{2}}$ or 0.

Slope of $\overline{AB}$ is $\frac{0-0}{a-0}$ or $\frac{0}{a}$ or 0.

$\overline{DC} \parallel \overline{MN} \parallel \overline{AB}$

27. $\sqrt{50} \approx 7.1$ units; about 8.1° **29.** $(4, 8)$ **31.** $\begin{bmatrix} 7 \\ 8 \end{bmatrix}$

33. $4\sqrt{2} \approx 5.7$ units; $(5, -3, 3)$ **35.** $x^2 + y^2 + z^2 = 25$

37. \$1400 a month for rent and utilities

39a.

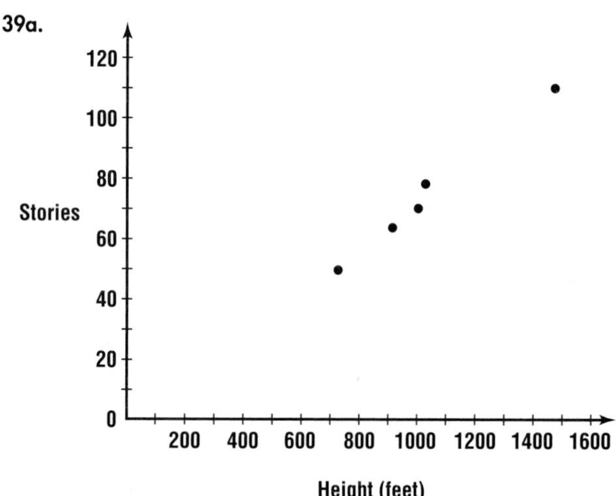

39b. Sample answer: $y = 0.09x - 15$ **39c.** Sample answer: 52 stories. The tower actually has 54 stories.

CHAPTER 13 INVESTIGATING LOCI AND COORDINATE TRANSFORMATIONS

Pages 699–701 Lesson 13–1

5. the bisector of the segment that joins the two points;

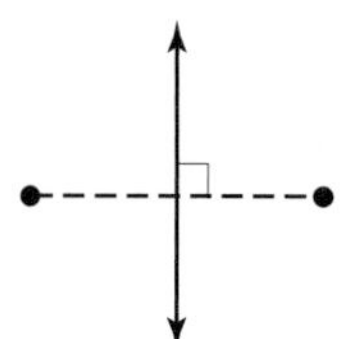

7. a line perpendicular to the plane of the square through the point of intersection of the diagonals;

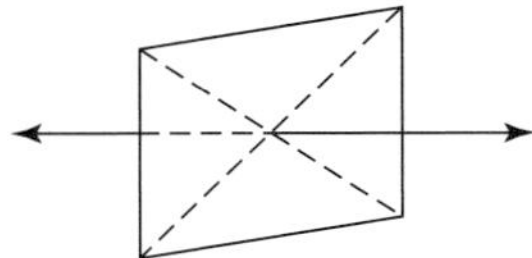

9. the perpendicular bisector of $\overline{AB}$ and $\overline{DC}$ in plane $\mathcal{R}$; **11.** the line that joins the two points of intersection;

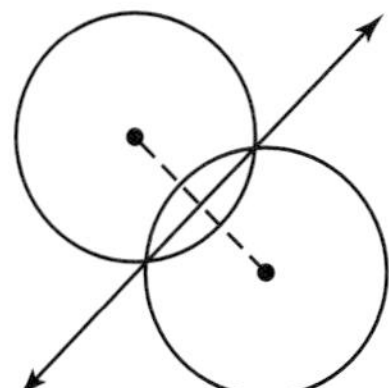

13. a pair of parallel lines, one on each side of ℓ, and each r units from ℓ;

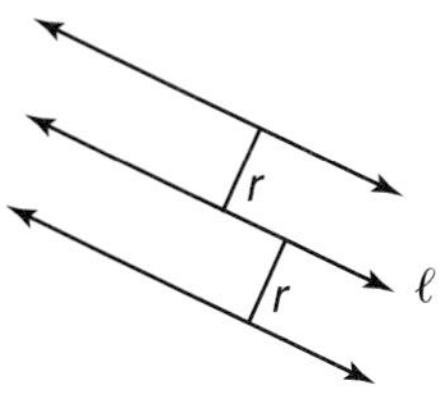

15. center of the pentagon;

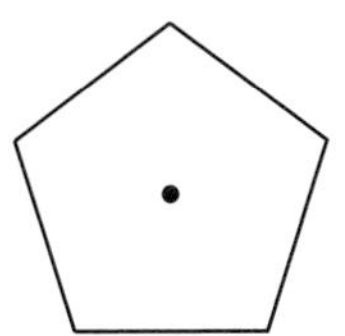

17. concentric circle of radius $s + \frac{(r-s)}{2}$;

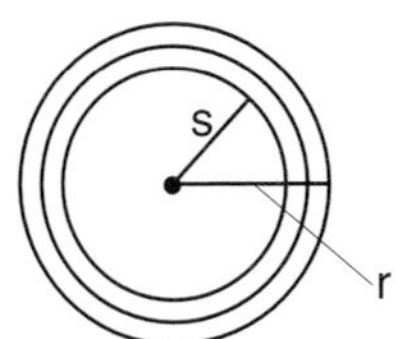

19. a pair of parallel planes, one on each side of plane $\mathcal{M}$, and each 4 m from $\mathcal{M}$;

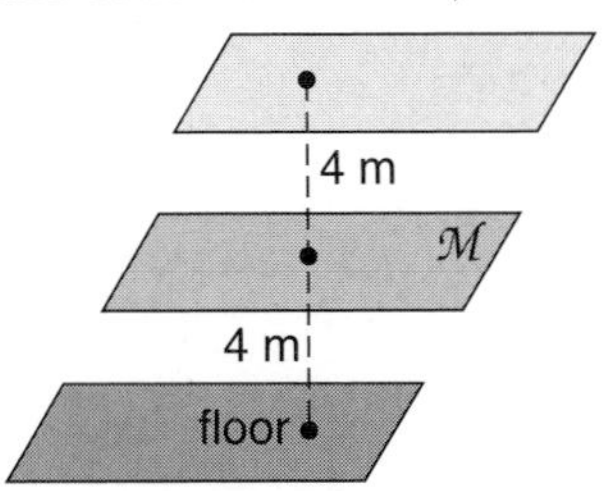

21. a circle with radius $\frac{\sqrt{3s}}{2}$ units in a plane perpendicular to the base with center on the midpoint of the base;

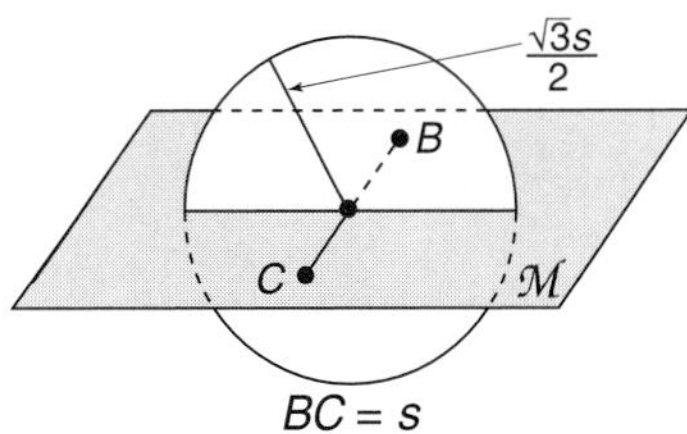

23. a plane perpendicular to the line containing the opposite vertices of the cube;

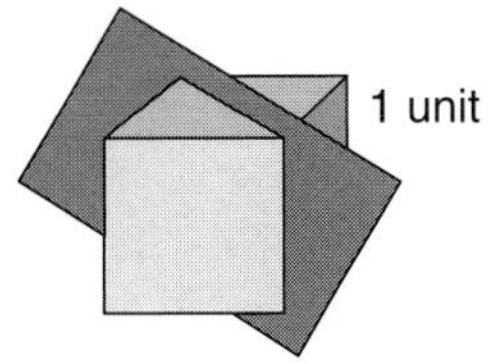

25. a plane midway between two parallel lines, perpendicular to the plane containing the two parallel lines **27.** a triangle similar to $\triangle ABC$ containing points A, B, and C;

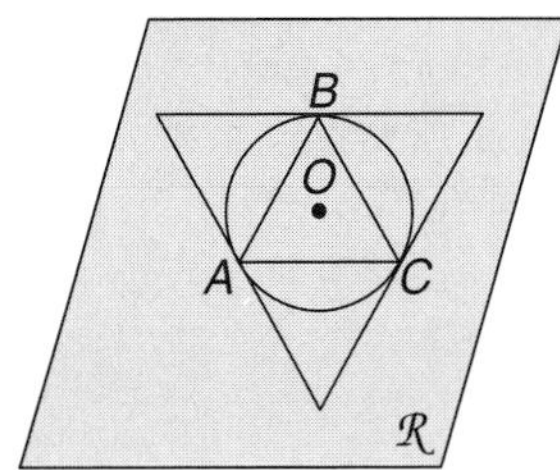

29. the rays that bisect $\angle ABD$ and $\angle BDC$ **31.** a pair of circles with radius a and center on h in two parallel planes perpendicular to the hypotenuse **33a.** a curve, parabola with ordinate one-half that of $y = x^2$ **33b.** $y = \frac{x^2}{2}$ **35.** When three or more circles are tangent there is an area among them that is not covered, therefore the circles must overlap in order to cover all of the area. This would then put some of the buildings within two miles of more than one fire station;

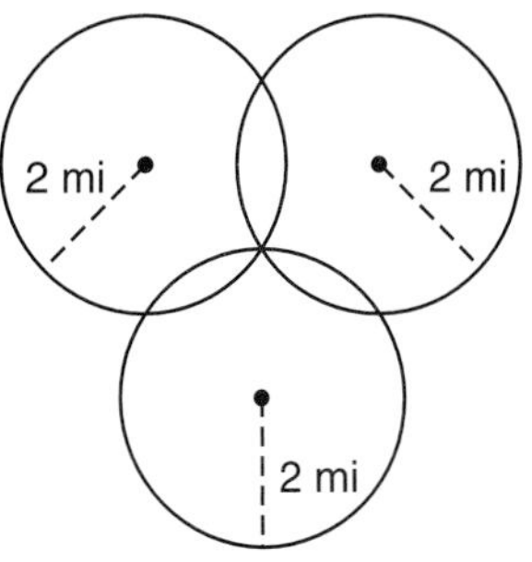

37. half of a sphere with a radius of 300 ft **39.** 1345 miles **41.** about 54.1 cm and 62.1 cm **43.** $x = 54$

45.

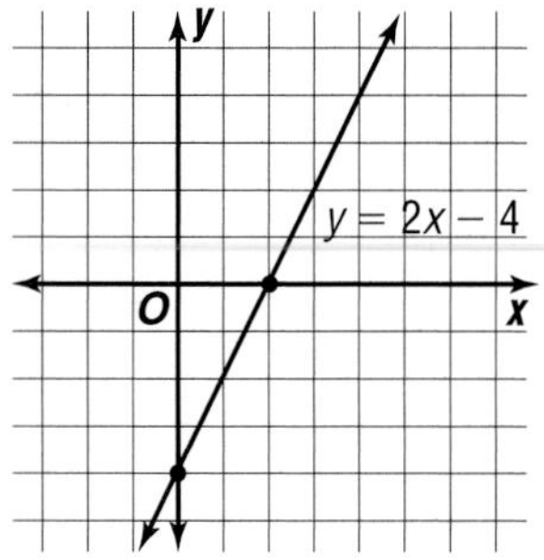

Pages 706–708 Lesson 13-2

5. (2, 2) **7.** (−1, −6) **9.** (2, 8) **11.** in about 13 years, 2003 **13.** (−2, −1);

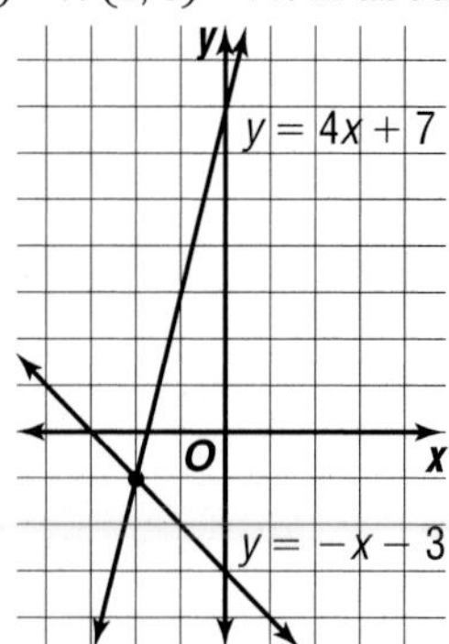

15. (2, −6);

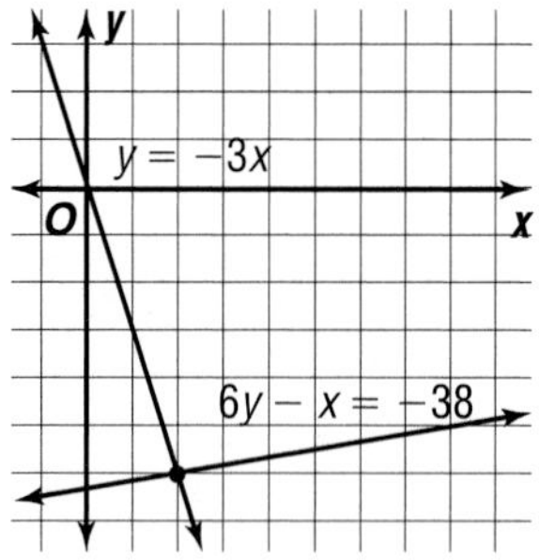

17. (2, 1);

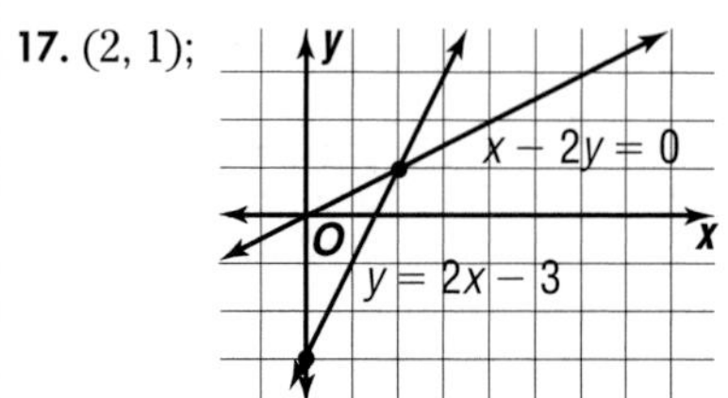

19. (2, 2) **21.** (4, 1) **23.** (0, 2) **25.** (1, −2) **27.** (15, 3) **29.** $\left(-\frac{9}{5}, -\frac{3}{5}\right)$ **31.** (2, 0) and (2, 8) **33.** (0, 1) and (3, 4) **35.** (70, 120) **37.** a circle with radius $\frac{DC}{2}$ and the midpoint of $\overline{DC}$ as its center;

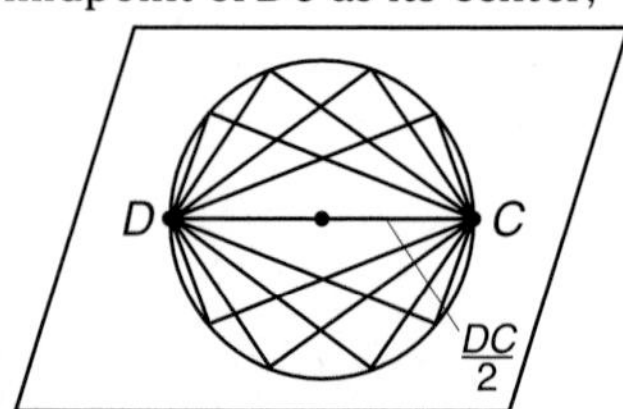

39. 0 **41.** trapezoid; Sample answer:

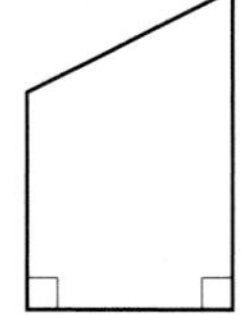

43. Sample answer:

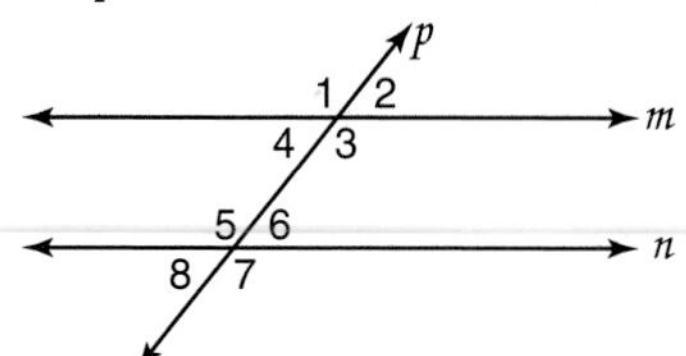

all odd-numbered angles have measure 145 **45.** 6

Pages 712–714 Lesson 13-3

5. none, a point, two points, or a circle
7. point on the angle bisector 7 in. from the vertex;

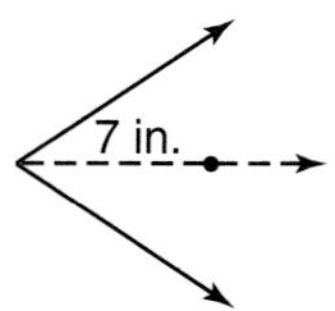

9. a plane that is perpendicular to the *xy*-plane intersecting the *xy*-plane at $x = 5$ **11.** Construct the perpendicular bisectors of each side of the triangle formed by these three cities. This intersection area is the area that should be considered; Columbus, Ohio, area. **13.** none or a point **15.** none, a point, two points, or a line segment

17. four points whose coordinates are $(\pm 3\sqrt{2}, \pm 3\sqrt{2})$

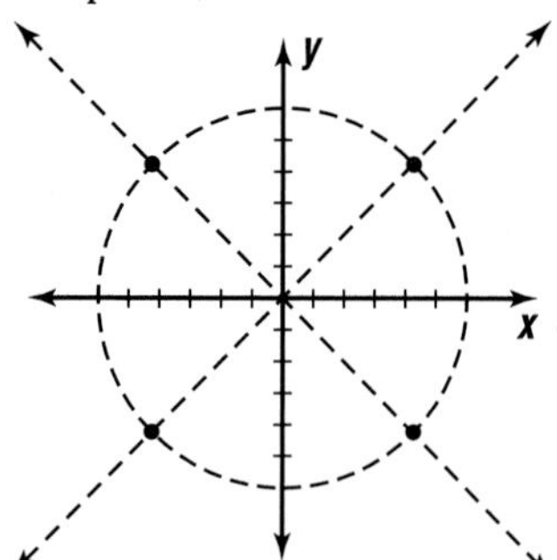

19. the axis of the cylinder

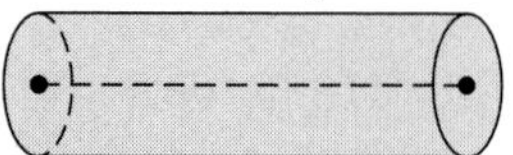

21. two circles with a radius of 5 cm that each lie in the parallel planes and whose centers are the intersection of the perpendicular line with each plane

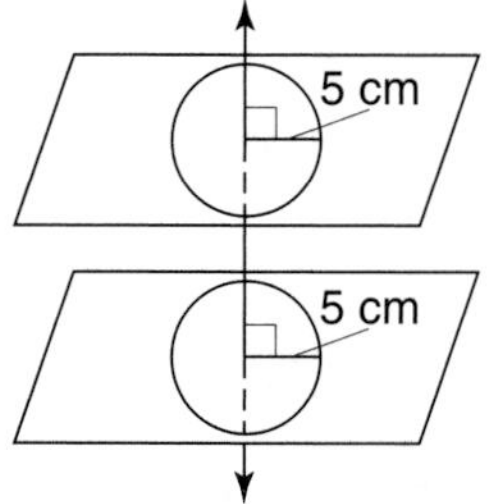

23. set of two intersecting planes;

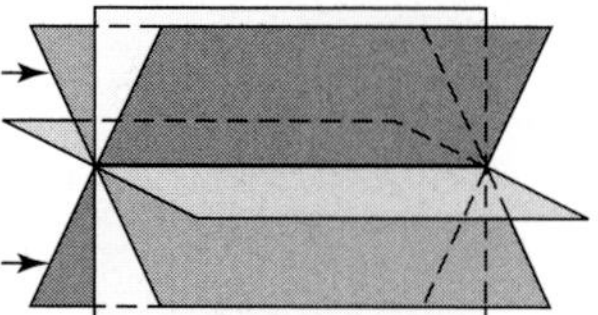

25. a parabola that opens up with vertex at (9, 0) **27.** circle with center (4, 1) and radius 5 **29.** a sphere with center (3, 4, 5) and radius 4 **31.** (−6, 5), (−3, −4) **33.** no points

35.

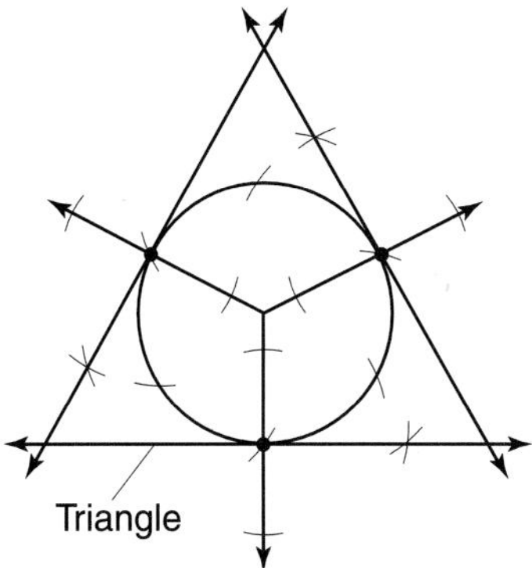

37. There would be many solutions. **39.** the corner of E St. and 3rd St. **41.** a semicircle with radius 80 ft and center at the center of the goal

43. ⊥ bisector of the segment connecting the two points of intersection;

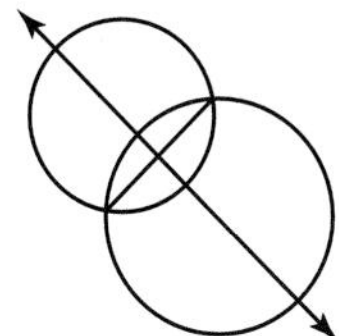

45. $T \approx 78.5\ m^2$; $V \approx 65.4\ m^3$ **47.** 10.8 miles **49.** $\angle A \cong \angle C$, $\angle B \cong \angle D$, $\angle C \cong \angle E$, $\overline{AB} \cong \overline{CD}$, $\overline{BC} \cong \overline{DE}$, $\overline{AC} \cong \overline{CE}$ **51.** −5.54, 0.54

Pages 718–721 Lesson 13–4

5. $\overline{CD}$ **7.** $\angle STU$ **9.** two reflections or a rotation **11.** Yes; all angles and sides are congruent. **13.** $\angle BAC$ **15.** $\overline{ED}$ **17.** enlargement and reflection **19.** enlargement, reflection **21.** Yes; all angles and sides are congruent. **23a.** rotation;

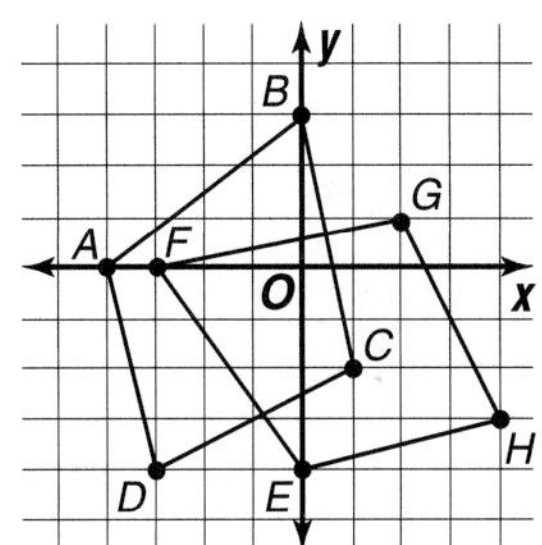

23b. yes **25.** quadrilateral *SRUT* **27.** △*GIH* **29.** quadrilateral *HIJK* **31.** *A*′(3, −3), *B*′(1, −2), *C*′(2, −4), *D*′(4, −5) **33.** two reflections or a rotation;

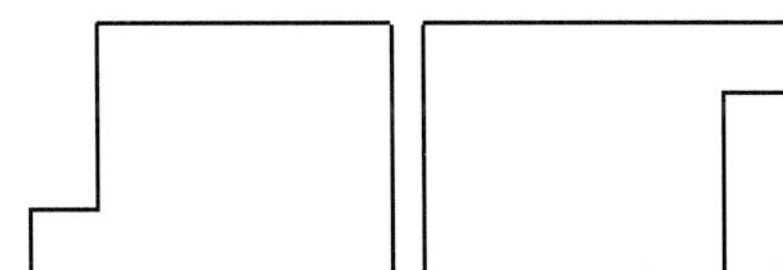

35. reflection of order and refection of L and J **37.** reflections **39.** (−1, 2) **41.** $y = 2x - 8$ **43.** 5, $5\sqrt{3}$ **45.** 35%

Page 721 Self Test

1. the circle that is the intersection of the cylindrical surface with axis ℓ and radius 8 cm, and the sphere with center *R* and radius 8 m **3.** (6, −1)
5. one point;(−3, 1);

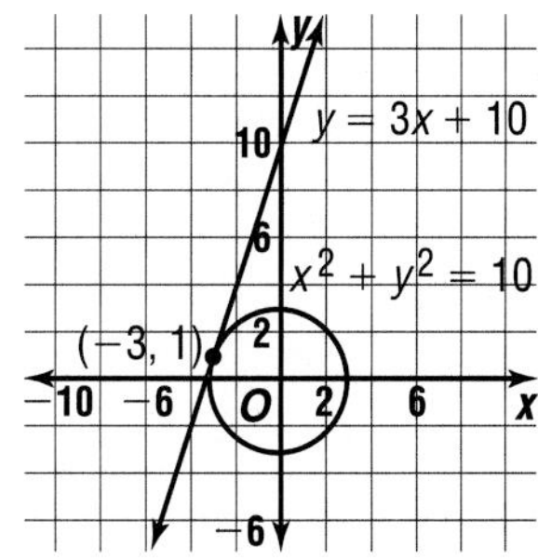

7. 0, 1, 2, 3, or 4 points **9.** $\overline{DB}$

Pages 726–729 Lesson 13–5

5. $\overline{PS}$ **7.** △*SQP*

9.

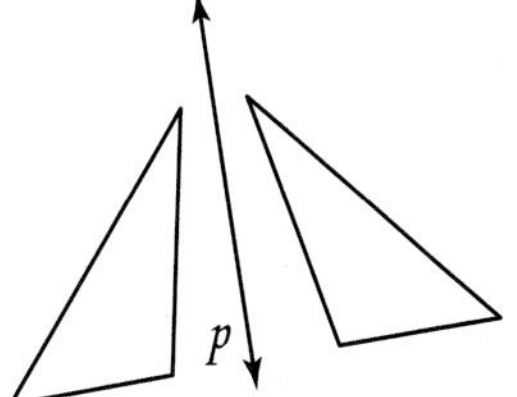

11. no lines of symmetry; no point symmetry **13.** four lines of symmetry; point symmetry **15.** 168 cm **17.** $\overline{HG}$ **19.** $\angle CBA$ **21.** △*BHG* **23.** quadrilateral *HGFB*

25. **27.**

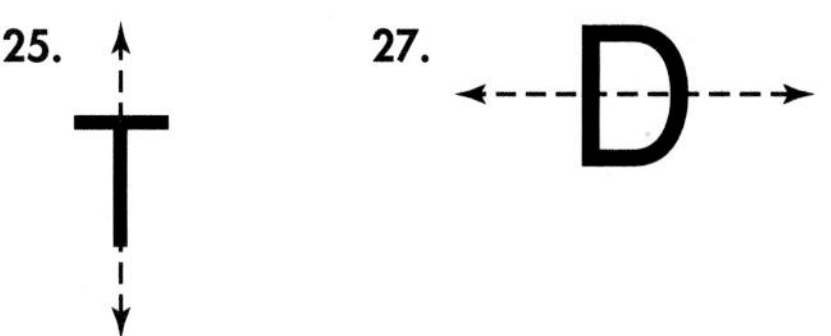

29.

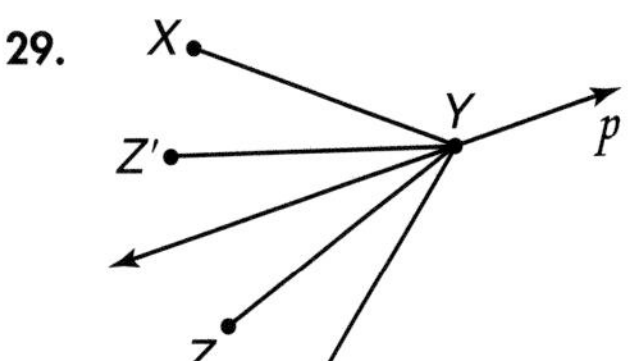

31.

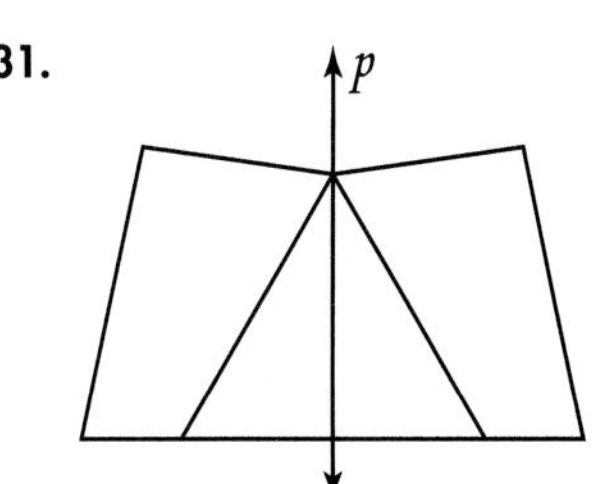

33.

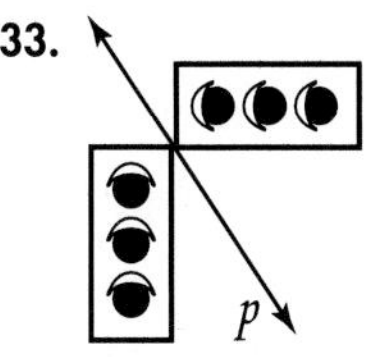

35. Yes; ℓ is the line of symmetry because it is possible to find, for every point A, another point B so that ℓ is the perpendicular bisector of $\overline{AB}$. **37.** Yes; ℓ is the line of symmetry because it is possible to find, for every point A, another point B so that ℓ is the perpendicular bisector of $\overline{AB}$. **39.** No; not all points are the same distance from ℓ.

41.

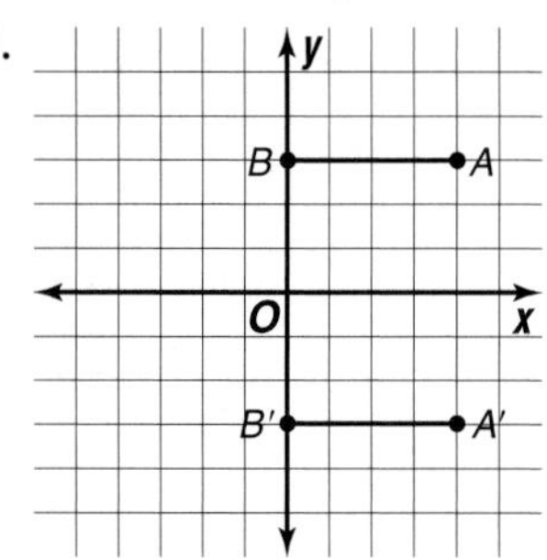

43.

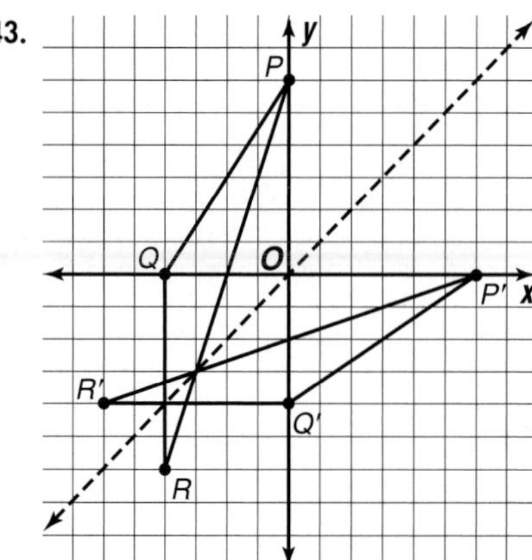

45. none **47.** none **49.** none **51.** both
53a. $(x, y) \rightarrow (-x, y)$ when reflected over the y-axis
53b. $(x, y) \rightarrow (x, -y)$ when reflected over the x-axis
53c. $(x, y) \rightarrow (y, x)$ when reflected over line $y = x$
55. Line m is a line of reflection, the vertex is a point of reflection. **57a.** The y-coordinates are negated and the x-coordinates are unchanged. It is reflected over the x-axis. **57b.** The x-coordinates are negated and the y-coordinates are unchanged. It is reflected over the y-axis. **59a.** $\angle S$ **59b.** $\overline{CD}$

61. (3, 3);

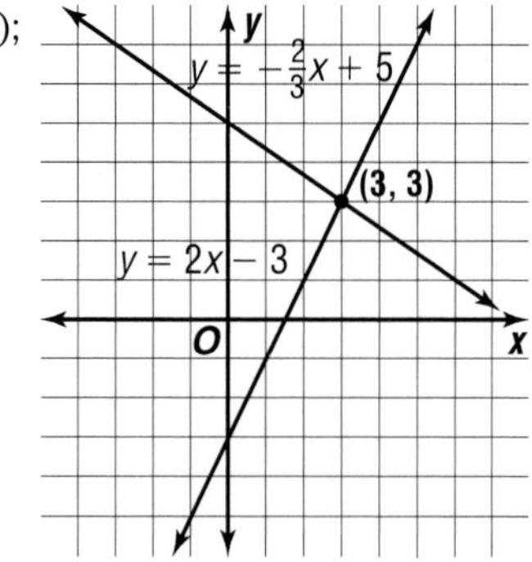

63. $y = -\frac{3}{4}x + \frac{7}{4}$ **65.** 62.89 units^2 **67.** diameters **69.** $\varnothing$

Page 730 Lesson 13–6A

1. The orientation remains unchanged. **3.** $AA'' = BB'' = CC'' = 2$(distance from ℓ to m) **5b.** The image is a translation of the preimage. **5c.** yes

Pages 734–737 Lesson 13–6

7. No; $\triangle HGK$ is not turned correctly.

9. $E'(4, 1)$, $F'(7, 3)$, $G'(7, 1)$;

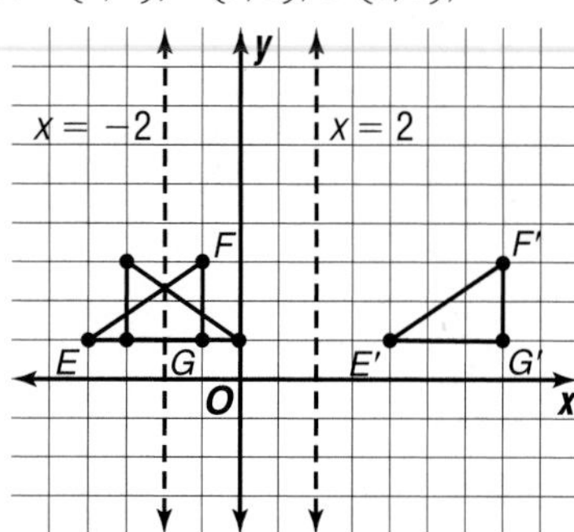

11. It is moved 15 ft from the wall and 21 ft to the left, then rotated. **13.** T **15.** U **17.** none **19.** Yes; it is one reflection after another with respect to two parallel lines. **21.** No; it is a translation and then a reflection with respect to a line.

23. $A'(-8, -4)$, $B'(-3, -2)$, $C'(-4, -5)$, $D'(-7, -6)$;

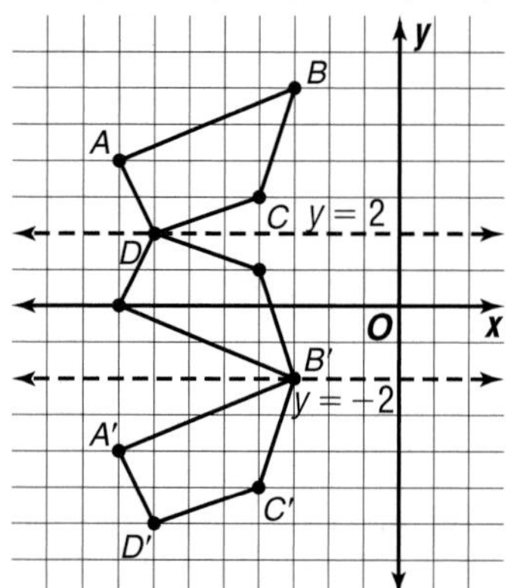

25. $N'(-4, 1)$, $I'(0, 0)$, $C'(0, -3)$, $K'(-8, -1)$;

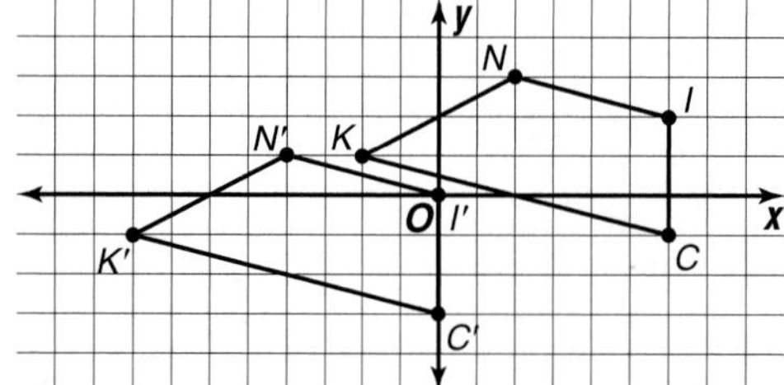

27.

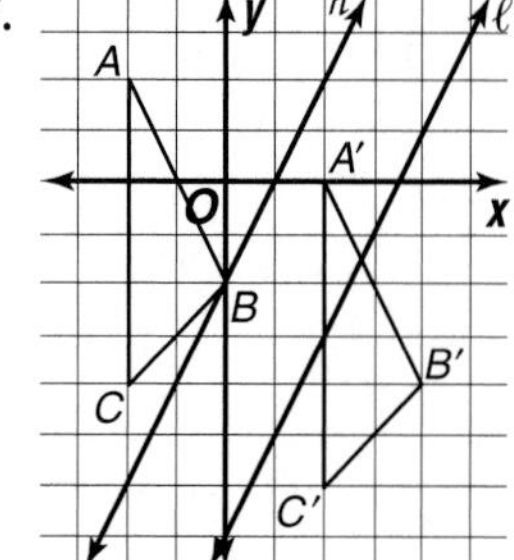

29.

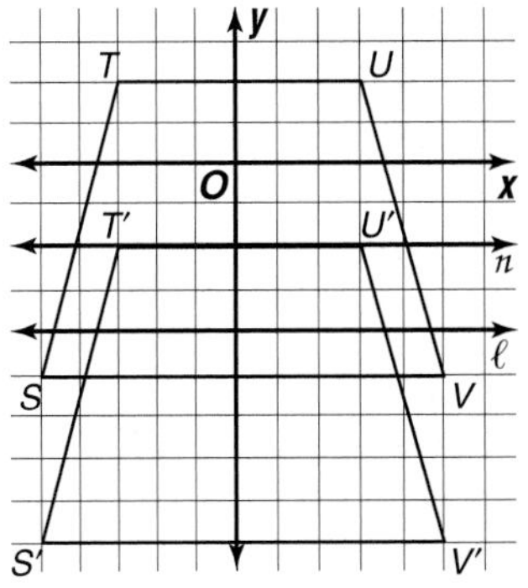

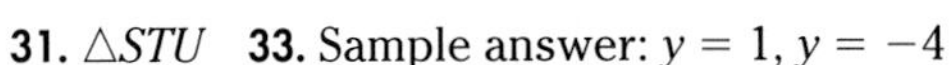

31. $\triangle STU$ 33. Sample answer: $y = 1, y = -4$

35.

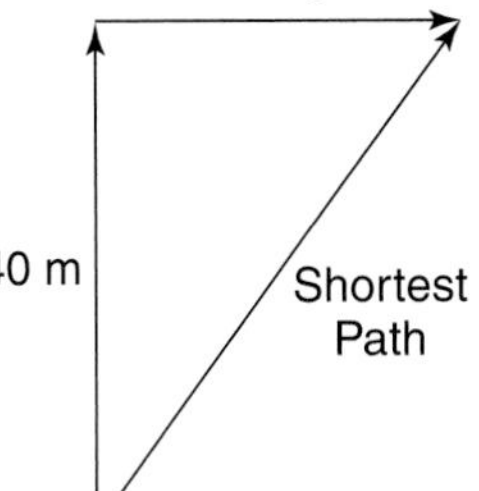

37. a circle 39a. $\overline{SQ}$ 39b. $\overrightarrow{QT}$ 39c. $\overline{TR}$ 41. yes; HL
43. $8(x + 3y)^2$

Pages 742–745 Lesson 13–7

5. 74° 7. pentagon *PKLMN* 9. $\angle CQN$ 11. $\overline{NK}$

13.

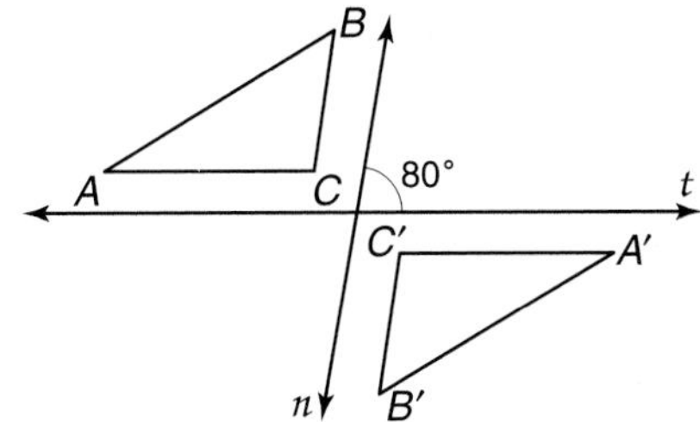

15. Yes; it is a proper successive reflection with respect to two intersecting lines. 17. 110 19. 148

21.

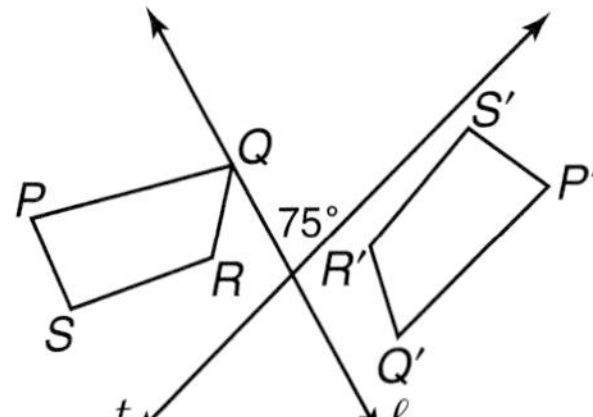

23.

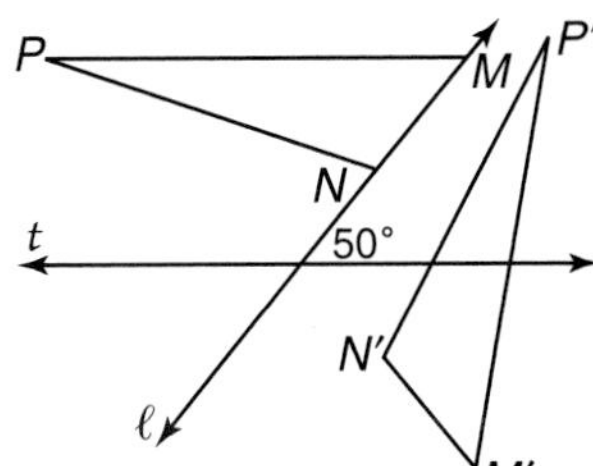

25.

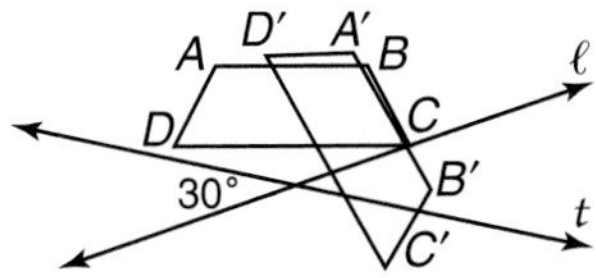

27.

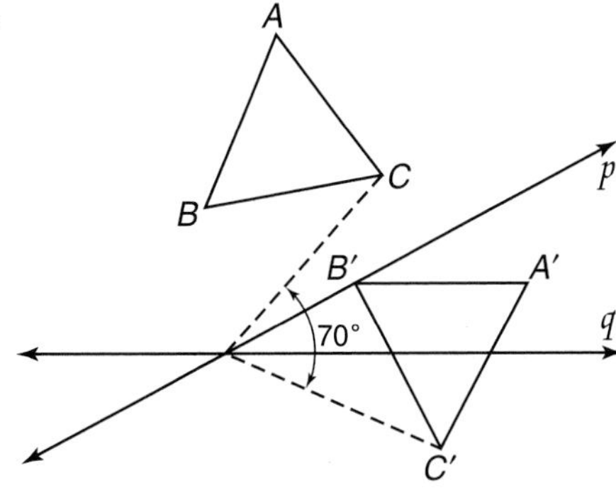

29. yes, yes, yes, yes, yes

33.

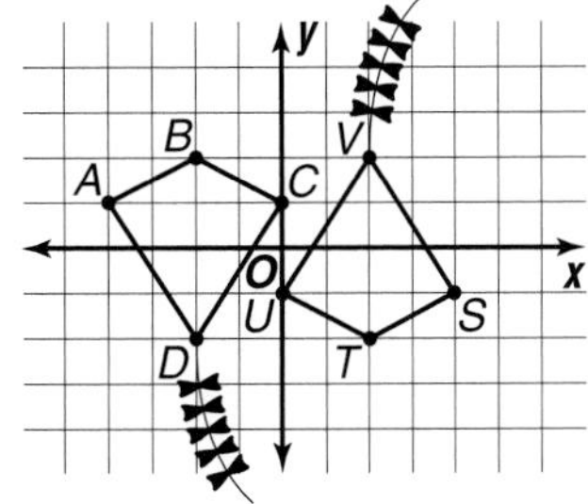

35. 120 37. reflection 39. translation 41. Angles of rotation with measures of 90 or 180 would be easier on a coordinate plane because of the grids used in graphing. 43a. 90 43b. seat 7 45. yes; the upper left, upper right, and center quadrilaterals 47. two circles concentric to the given circle, one with a radius of 2 in. and the other with a radius of 8 in. 49. $768\pi \approx 2412.7 \text{ ft}^3$ 51. 37.5 53. $\frac{1}{x + 9}$

Pages 749–753 Lesson 13–8

7. < 9. 62 11. congruence transformation
13. reduction
15. Scale factor of 2:

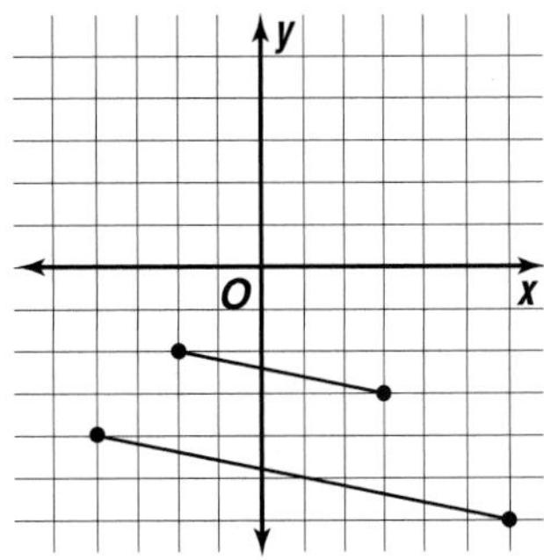

Scale factor of $\frac{1}{2}$:

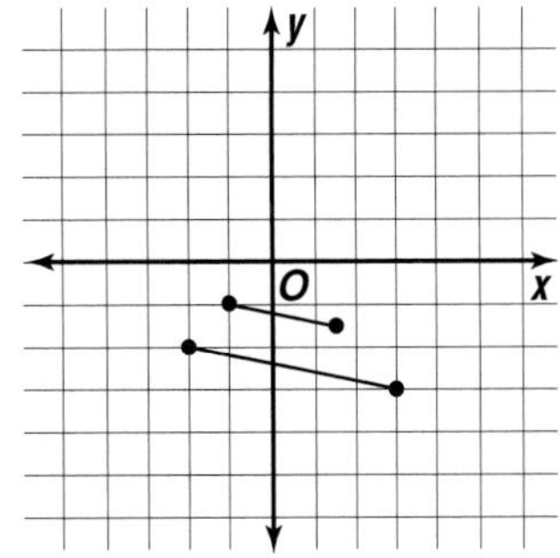

17. $\frac{1}{2}$ **19.** B **21.** S **23.** 30 **25.** 24 **27.** 3
29. 2, enlargement **31.** 1, congruence transformation
33. $\frac{1}{4}$, reduction **35.** $\frac{3}{4}$ **37.** 2

39. Scale factor of 2:

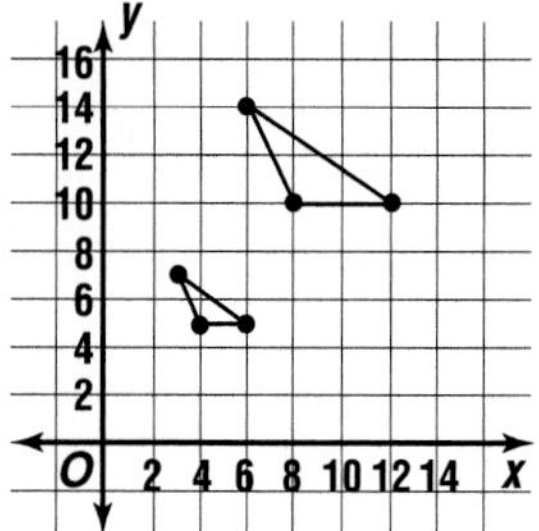

Scale factor of $\frac{1}{2}$:

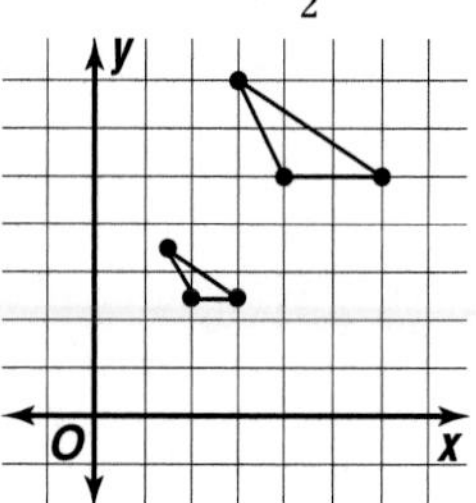

41. Scale factor of 2:

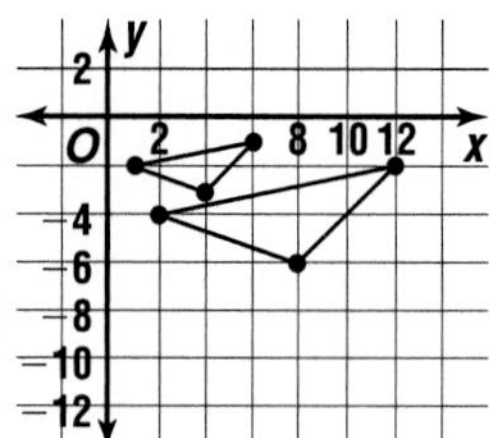

Scale factor of $\frac{1}{2}$:

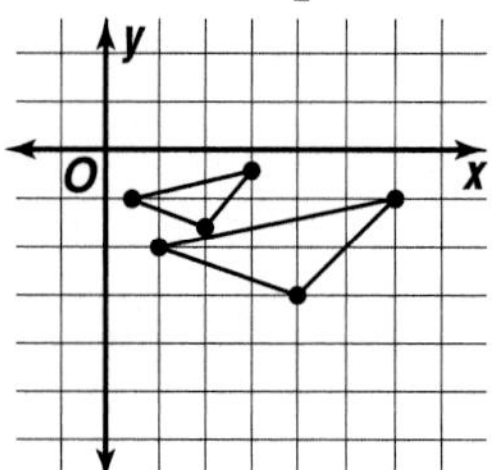

43. Scale factor of 2:

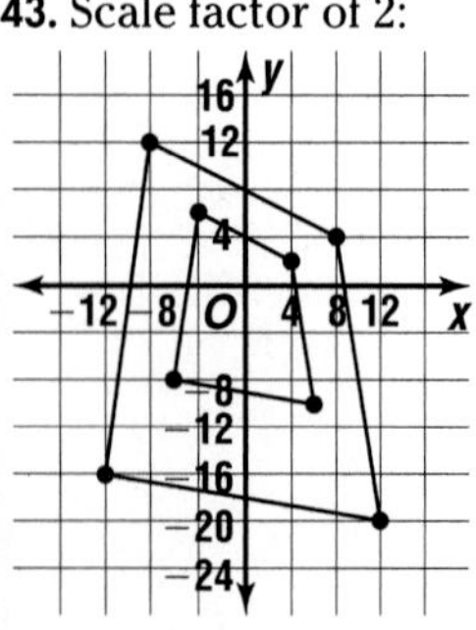

Scale factor of $\frac{1}{2}$:

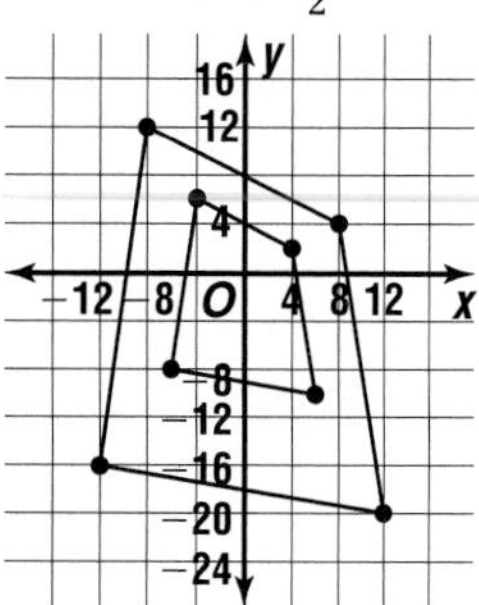

45a. The surface area of the image will be nine times the surface area of the preimage. **45b.** The volume of the image will be 27 times the volume of the preimage.
47. (0, 5); $\frac{5}{2}$ **49.** 337.5 miles

51.

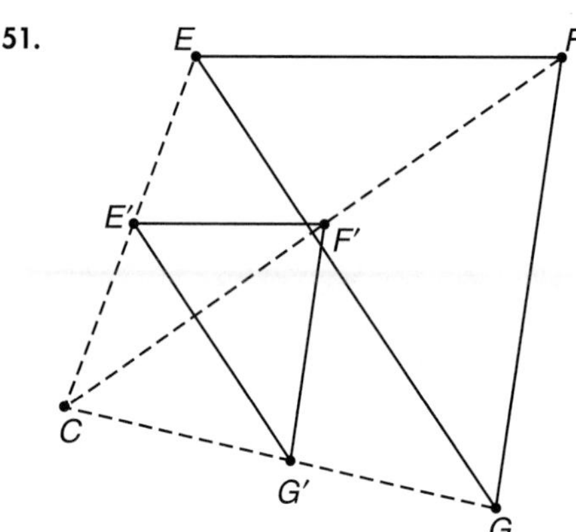

53.

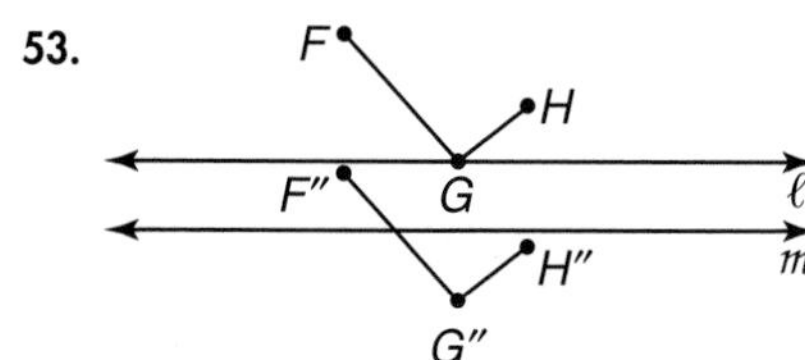

55. $\sqrt{179} \approx 13.4$ units **57.** 5100 mm^2 **59.** 7.5

61a.

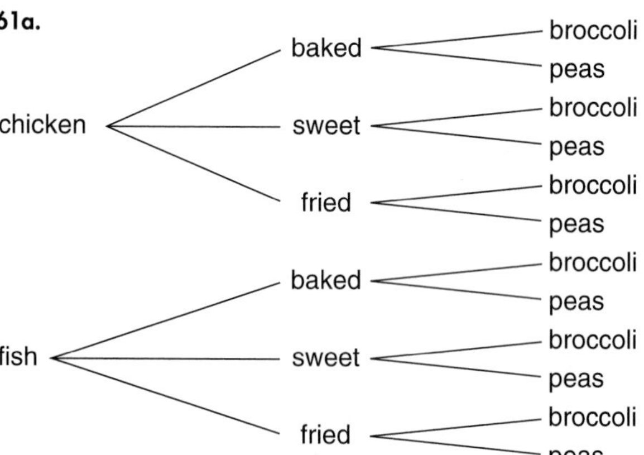

61b. $\frac{2}{12}$ or $\frac{1}{6}$

Page 755 Chapter 13 Highlights

1. true **3.** true **5.** false; image **7.** true **9.** false; preimage

Pages 756–758 Chapter 13 Study Guide and Assessment

11. the interior of a circle with center at the given point and radius 6 cm **13.** the circumcenter; the point where the three perpendicular bisectors of the segments joining A, B, and C intersect **15.** (1, 1) **17.** (1, 2)

19. the circle that is the intersection of spheres P and Q with radii 3 cm;

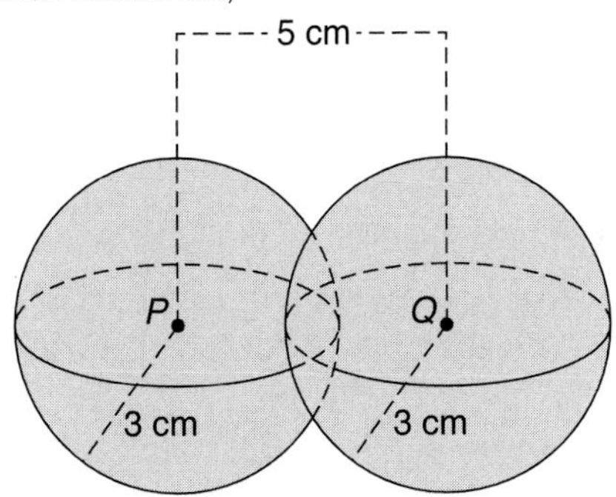

21. E **23.** $\overline{CD}$ **25.** reflection

27.

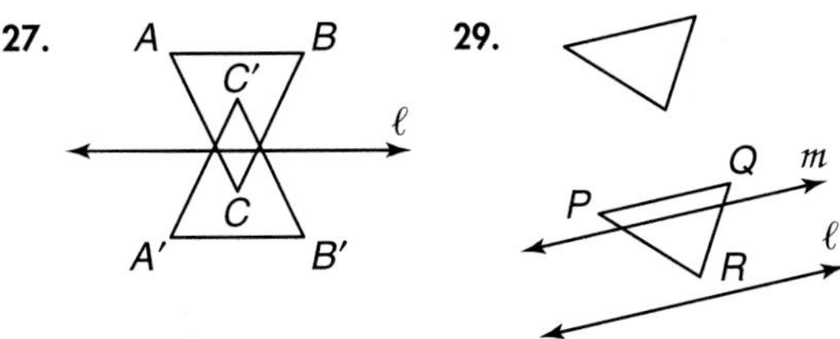

29.

31.

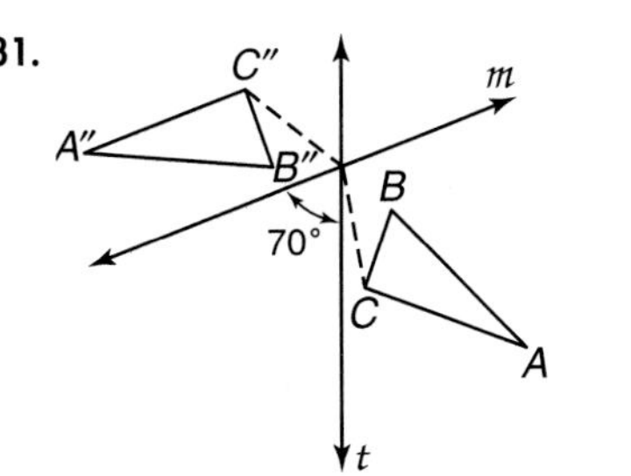

33. enlargement; $|3| > 1$ **35.** Imagine the reflection of the 5 ball with respect to the line formed by one side of the billiard table, and aim for the reflection. **37.** 144

PHOTO CREDITS

COVER (background) Paul Raftery/Arcaid, (center) Joanne Lotter/Tom Stack & Associates, (bottom) Mark Tomalty/Masterfile; **iii** (t)Aaron Haupt Photography, (b)SuperStock; **iv** (t)courtesy Cindy Boyd, (c)courtesy Gail Burrill, (b)courtesy Jerry Cummins; **v** (t)courtesy Tim Kanold, (c)courtesy Carol Malloy, (b)courtesy Mrs. Marie Yunker; **viii** (t)Doug Martin, (b)B. Graham/FPG; **ix** (t)Mark Burnett, (b)Mark Steinmetz; **x** (t)Mark Burnett, (b)North Wind Picture Archives; **xi** (tl)Sharon Kurgis, (r)John Evans Photography, (bl)Steve Lissau; **xii** (t)David Ball/The Stock Market, (b)Stock Montage/Newberry Library; **xiii** (tl)Scott Camazine/Photo Researchers, (tr)courtesy of Norma Merrick Sklarek, (b)SHOE © 2-22-89 Tribune Media Services, Inc. All rights reserved; **xiv** (t)David Madison/Tony Stone Images, (b)John Evans Photography; **xv** (t)courtesy Mark Clingan, (b)John Elk/Tony Stone Images; **xvi** (l)David Hamilton/The Image Bank, (r)David Noble/FPG; **xvii** (t)Flip Nicklin/Minden Pictures, (b)Aaron Haupt Photography; **4** (tl)Nissan Motor Corporation, USA, (tr)British Museum, Ref:1904-6-21-34, (bl)Corbis-Bettmann, (br)Mary Evans Picture Library; **5** (tl)(cr)Brian Payne/Boy's Life Magazine, April 1996, (cl)North Wind Picture Archives, (bl)Steve Dunwell/The Image Bank, (br)NASA; **6** (t)(br)Doug Martin, (bl)Giboux/Gamma-Liaison, International; **7** (l)Mark Steinmetz, (r)Doug Martin; **9** Kaku Kurita/Gamma-Liaison, International; **10** Arslanian-Liaison/Gamma-Liaison, International; **11** (l)Loviny Chrix/Gamma-Liaison, International, (r)John Veltri/Photo Researchers; **13** Doug Martin; **16** Mark Steinmetz; **18** Doug Martin; **19** (l)Mark Burnett, (r)Ric Ergenbright Photography; **20** (t)courtesy Johnson Publishing Company, (b)Mark Steinmetz; **22** Doug Martin; **24** (tl)Mark Burnett, (bl)Mark Steinmetz, (r)Steve Lissau; **28** (tr)Eric Sander/Gamma-Liaison, International, (c)Doug Martin, (bl)THE FAR SIDE, 5-25-93 © 1993 FarWorks, Inc./Dist. by Universal Press Syndicate; **30** Mark Steinmetz; **32** North Wind Picture Archives; **33** Martyn Goddard/Colorific; **34** PEANUTS ® reprinted by permission of United Feature Syndicate, Inc; **35** Jeremy Scott/International Stock; **38** Mark Burnett; **43** (t)Doug Martin, (b)Mark Burnett; **44, 46** Mark Burnett; **49** (t)B. Graham/FPG, (b)Dick Luria/Science Source/Photo Researchers; **51** The Kobal Collection; **53** (t)courtesy USGS Photographic Library, Denver, CO, (b)James Westwater; **58** Richard Martin/Agence Vandystadt/Duomo; **60** Doug Martin/Centex Homes, Columbus, OH; **64** Mark Steinmetz; **65** The Museum of Modern Art, New York. Acquired through the Lillie P. Bliss Bequest; **66** Aaron Haupt Photography; **67** (t)Rich Brommer, (bl)Cincinnati Art Museum, John J. Emery Fund, (br)file photo; **68** (tl)Bob Mullenix, (tr)(bl)Mary Evans Picture Library, (br)Ancient Art and Architecture Collection, Ltd; **69** (tl)photo by Youth Voice Collaborative, (tr)(bl)Mark Burnett, (c)The Schlesinger Library, Radcliffe College, (br)Marcel Thomas/FPG; **70** SHOE, 2-22-89. © 1989 Tribune Media Services, Inc. All rights reserved.; **71** Aaron Haupt Photography; **73** courtesy JPL; **74** Ancient Art & Architecture Collection, Ltd.; **75** (t)Aaron Haupt Photography, (b)Cyberware; **76** Stock Montage; **77** Mary Evans Picture Library; **79** PEOPLE Weekly © 1996/John Zich; **80** (t)John Madere/The Stock Market, (b)Stock Montage/Charles Walker Collection; **82** SuperStock; **83** Aaron Haupt Photography; **86** CBS Entertainment; **87** (t)Mike Hewitt/Allsport, (b)Ancient Art & Architecture Collection, Ltd; **88** (t)Larry Hamill, (b)LGI Photo Agency; **89** Mark Burnett; **90** (t)Ken Frick, (b)Aaron Haupt Photography; **92** George Anderson; **95** Bob Daemmrich Photo, Inc; **99** Aaron Haupt Photography; **100** David Frazier Photolibrary; **104, 106** (t)Aaron Haupt Photography, (b)Gerard Photography; **108** Boltin Picture Library; **113** Stock Montage; **114** Randy Schieber; **118** David Frazier Photolibrary; **119** Hess/The Image Bank; **122** (tl)DUOMO/William R. Sallaz, (tr)North Wind Picture Archives, (bl)Ancient Art and Architecture Collection, Ltd, (br)Bettmann Archive; **123** (tl)Doug Dukane, (tr)Aaron Haupt Photography, (c)DUOMO/William R. Sallaz, (bl)Bettmann Archive, (bc)Stock Montage, (br)Reuters/Corbis-Bettmann; **124** (tl)Grant Heilman/Grant Heilman Photography, (tr)David Ball/The Stock Market, (b)Aaron Haupt Photography; **125** Charles O'Rear; **126** (t)file photo, (b)Charles O'Rear; **127** (t)file photo, (b)Mark E. Gibson; **128** Guido Rossi/The Image Bank; **129** Bob Daemmrich Photo, Inc; **132** William D. Popejoy; **137** (l)Mark Steinmetz, (r)Mark Scott/FPG; **138** Salt Lake City Convention & Visitors Bureau; **140** Doug Martin; **142** Eliot Cohen; **144, 145** Aaron Haupt Photography; **146** (t)file photo, (b)Roger K. Burnard; **150** Doug Martin; **153, 157, 158, 160, 161** (t)Aaron Haupt Photography; (bl)Architectural Association Photo Library, London/John Edward Linden, (br)Architectural Association Photo Library, London/V. Bennett; **163** (t)Aaron Haupt Photography, (b)Doug Martin; **164** Larry Kunkel/FPG; **165** (t)SSEC/University of Wisconsin, Madison, (bl)TRIP/Eric Smith, (br)David Noble/FPG; **167** Aaron Haupt Photography; **168** (t)Mark E. Gibson, (ct)Travelpix/FPG, (cb)Alan Schein/The Stock Market, (b)Rafael Macia/Photo Researchers; **169** (l)Aaron Haupt Photography, (r)Mary Evans Picture Library/Explorer; **170** (l)Ken Van Dyne, (r)Phyllis Picardi/International Stock; **174** (l)Aaron Haupt Photography, (r)Aaron Haupt Photography; **175** David Frazier Photolibrary; **176** William J. Weber; **177** Gail Shumway/FPG; **178** (tl)Tim Courlas, (tr)SuperStock, (c)North Wind Picture Archives, (bl)G.Tortoli/Ancient Art & Architecture Collection, (br)Bettmann Archive; **179** (tl)Cheryl L. Opperman/Opperman Photographics, Inc, (tr)Morton and White, (c)Doug Martin, (bl)Laurie Platt Winfrey, (br)Leonardo Cendamo/Woodfin Camp & Associates; **180** (t)Sharon Kurgis, (c)Aaron Haupt Photography, (b)Yousuf Karsh/Woodfin Camp & Associates; **181** Mark E. Gibson; **186** SuperStock; **189** courtesy Carolyn Shoemaker; **195** Michael Kellar/The Stock Market; **196** courtesy Diane Leighton; **200** Museum of Modern Art; **202** Doug Martin; **204** Mark Steinmetz; **206** Doug Martin; **212** SuperStock; **213** (t)file photo, (b)David Park/Science Photo Library/Photo Researchers; **221, 222** John Evans; **227** (t)Mark Burnett, (b)SuperStock; **228** Mark Burnett; **232** Doug Martin; **233** Patty Jedick; **236** (tl)Robert Frerck/The Stock Market, (tr)M. C. Escher © 1996 Cordon Art, Baarn, Holland. All rights reserved, (bl)North Wind Picture Archives, (br)Corbis-Bettmann; **237** (t)Daniel Westergren/National Geographic Society, Image Collection, (c)(br)M. C. Escher © 1996 Cordon Art, Baarn, Holland. All rights reserved, (bc)Sipa/Woodfin Camp & Associates, Inc, **241** (l)Marcia Keegan/The Stock Market, (r)courtesy Jack Friedman/Park Board of Parkville, MO; **242** Aaron Haupt Photography; **244** Aaron Haupt Photography/puppy provided by Jack's Aquarium & Pets; **245** (t)David Brownell, (b)Steve Lissau; **246** Aaron Haupt Photography; **249** Robert Frerck/Odyssey; **252, 254** Aaron Haupt Photography; **255** Doug Martin; **256** Aaron Haupt Photography; **258** The Art Collection, Harvard Law School; **263** file photo; **265** Aaron Haupt Photography; **267** Doug Martin; **268** Jim Brown/The Stock Market; **271, 272** (t) Aaron Haupt Photography, (b)Lynn Stone; **273** Aaron Haupt Photography; **278** Lynn M. Stone; **280** (t)Tracy Borland, (b)Lynn Stone; **285** Aaron Haupt Photography; **286** David Noble/FPG;

287 Aaron Haupt Photography; **288** (tl)Doug Martin, (tr)Kent State University Museum, gift of the Silverman/ Rodgers Collection, 1983. 1.2158, (bl)Ancient Art & Architecture Collection for Kakokawa, (br)Stock Montage; **289** (tl)courtesy Quilter's Newsletter Magazine, Leman Publications, CO, (tr)Tim Courlas, (bl)Stock Montage, (br) © 1995 Universal City Studios & Amblin/The Kobal Collection; **291** (t)courtesy Dalia Berlin, ASID, (b)Mark Burnett; **293** Mark Burnett; **297** (tl)William Caram/ International Stock Photo, (bl)Mark E. Gibson, (br)Frank Muller-May/Woodfin Camp & Associates; **301** (l)Margaret Kois/The Stock Market, (r)Mark Burnett; **303** (l)Jerry Jacka Photography, (r)Mark Burnett; **304** (t)Mark Steinmetz, (b)Neil Troiano/Arcaid; **306** (t)Jerry Irwin/SuperStock, (b)Hans Namuth/Photo Researchers; **308** Doug Martin; **309** DUOMO/Mitchell Layton; **313** Mark Burnett; **315** High Museum of Art, © 1997 Dorothea Rockburne/Artists Rights Society (ARS), New York; **319** Aaron Haupt Photography; **320** SuperStock; **321** Telegraph Colour Library/FPG International; **325** Gil Ullom; **327** (l)courtesy Norma Merrick Sklarek, (r)Doug Martin; **332** (tl)Mark E. Gibson, (tr)Yale University Art Gallery, gift of Anni Albers and The Josef Albers Foundation, Inc., (b)Colin Molyneux/ Molyneux Associates/The Image Bank; **333** Doug Martin; **336** (tl)Runk/Schoenberger from Grant Heilman, (tr)Mark Steinmetz, (bl)Ronald Sheridan/Ancient Art & Architecture Collection, (br)North Wind Picture Archives; **337** (tl)courtesy of Tree Musketeers founders, Sabrina Alimahomed and Tara Church, (tr)Mark Steinmetz, (cl)Doug Martin, (cr)Mark Steinmetz, (bl)The Metropolitan Museum of Art, Rogers Fund, 1919(19.164), (br)Western Reserve Historical Society; **338** (t)Scala/Art Resource, NY, (b)The Kobal Collection; **339** John Evans; **340** SuperStock; **341** (t)Stock Montage/Newberry Library, (b)Hulton Getty Collection/Woodfin Camp & Associates; **342** NASA; **343, 344** Mark Burnett; **345** Smithsonian Institution; **346** The Metropolitan Museum of Art, Alfred Stieglitz Collection, 1949(49.59.1); **348** John Evans; **350, 352** Mark Burnett; **353** Elaine Shay; **356** David W. Hamilton/The Image Bank; **361** Mark Burnett; **362** The Kobal Collection; **368** North Wind Picture Archives; **369** Stock Montage, Charles Walker Collection; **372** Doug Martin; **376** Tony Demin/International Stock; **377** Mark Burnett; **378** (t)Peitgen, Jurgens, Saupe, FRACTALS FOR THE CLASSROOM, Springer-Verlag, NY, © Image by Manfred Kage, Institut fur Wissenschaftliche Fotografie; from FRACTALS FOR THE CLASSROOM, Springer-Verlag, NY, (b)Michael D. McGuire; **379** North Wind Picture Archives; **382** Scott Camazine/Photo Researchers; **383** (l)Edward Berko, oil on wood 48" × 36" ©1991, courtesy private collection, NYC, fig.1 © 92 Lawrence Hudetz, fig.2 Loren Carpenter, fig.3 Peitgen, Jurgens, Saupe, FRACTALS FOR THE CLASSROOM, Springer-Verlag, NY; fig.4 © 1987 Lawrence Hudetz; **386** Aaron Haupt Photography; **390, 391** Mark Burnett; **394** (tl)courtesy Mercedes-Benz of North America, (tr)Philip Wallick/International Stock, (bl)Ronald Sheridan/Ancient Art & Architecture Collection, (br)North Wind Picture Archive; **395** (tl)courtesy Kim Todd, (tr)Michael Brown/Gamma-Liaison Network, (bl)North Wind Picture Archive, (bc)Corbis-Bettmann, (br)The Bettmann Archive; **398** Doug Martin; **401** Ed Wheeler/The Stock Market; **403** Hulton Getty Collection/Woodfin Camp & Associates; **404** (t)Kunio Owaki/The Stock Market, (b)The Kobal Collection; **405** (t)Mark Steinmetz, (b)David Madison/Tony Stone Images; **406** David Madison/Tony Stone Images; **409** Grant V. Faint/The Image Bank; **410** from THE NATURE OF CITIES © 1955 Paul Theobald, Chicago; **411** Doug Martin; **412** (t)AP/Wide World Photos, (b)Doug Martin; **414** (t)Joe Towers/The Stock Market, (b)Bob Daemmrich/ Stock Boston; **416** Doug Martin; **418** (t)NASA, (b)file photo; **420** Doug Martin; **421** NASA; **423** (l)Peter Menzel/Stock Boston, (r)ITTC Productions/The Image Bank; **424** (t)G. R. Roberts, (c)Jake Baker/The Image Bank, (b)Robert Brons/BPS/Tony Stone Images; **425** (t)file photo, (bl)David Frazier Photolibrary, (br)KS Studios; **426** Thomas Braise/The Stock Market; **428** Mastrorillo/The Stock Market; **429** (t)Philip Jon Bailey/Stock Boston, (b)Jim Dwyer/Stock Boston; **430** Aaron Haupt/Photo Researchers, Inc.; **431** Paul J. Sutton/DUOMO; **434, 440** Doug Martin; **441** Tom McGuire; **444** (tl)Mike Yamashita/Westlight, (tr)Chuck Mason/International Stock, (bl)Corbis-Bettmann, (br)Ancient Art & Architecture Collection, Ltd; **445** (t)Jim Jager © 1995/Anchorage Daily News, (c)Rick Weber, (bl)Stock Montage, (br)North Wind Picture Archives; **446** Doug Martin; **447** John Evans Photography; **448** (tl)Fisk University Archive, (tr)Neal Lauron/Sipa Press/ Woodfin Camp & Associates, (bl)Bob Daemmrich/Stock Boston; **451** (t)AP Photo/Journal Star, Randy Hampton/ Wide World Photos, (b)Aaron Haupt Photography; **454** Bavaria/Viesti Associates; **458** Claire Ferguson; **459** Larry Hamill; **462** Rick Weber; **465** Mark Burnett; **466** (t)Mark Steinmetz, (b)Doug Martin; **472** Hans Reinhard/ Okapia/ Photo Researchers; **475** file photo; **482** William J. Weber; **483** Guy Marche/FPG; **487** NASA/Science Photo Library/Photo Researchers; **491** Doug Martin; **498** (l)UPI/ Bettmann, (r)Benjamin Rondel/The Stock Market; **503** Edi Ann Otto; **504** The Kobal Collection; **508** Rick Weber; **509** JPL; **511** (t)Bob Daemmrich/Stock Boston, (b)Chuck Mason/ International Stock; **512** (tl)J. Messerschmidt/ The Stock Market, (tr)Paul L. Ruben, (b)Mary Evans Picture Library; **513** (tl)courtesy Ryan Morgan, (tr)(c)Doug Martin, (bl)Ancient Art & Architecture Collection, (br)FPG International; **514** Photo Researchers, Inc.; **516** Mark Burnett; **519** Elaine Shay; **520** (l)Aaron Haupt Photography, (r)Tim Courlas; **523, 527** (t)Aaron Haupt Photography, (b)courtesy Mark Clingan; **529** Joseph DiChello, Jr.; **531** Aaron Haupt Photography; **532** Tim Barnwell/Stock Boston; **534** Doug Martin; **537** Aaron Haupt Photography; **539** Doug Martin; **542** Steve Grohe/The Stock Market; **543** (t)courtesy Knox Parks Foundation, (b)Jack McConnell; **547** Mark Burnett; **549** (l)Ruth Dixon, (r)American Photo Library, Tokyo; **551** (t)Lou Jones/The Image Bank, (top inset)Aaron Haupt Photography, (b)FPG International; **552** courtesy St. Joseph's Historic Foundation; **553** Doug Martin; **554** (t)Mark Burnett, (b)Mark Steinmetz; **555** Mark Burnett; **556** Jon Feingersh/Stock Boston; **557** (t)Mark Burnett, (b)file photo; **558** Philip Rosenberg/Pacific Stock; **560** Joseph DiChello, Jr.; **562, 568** Mark Steinmetz; **569** Rudy VonBriel; **572** (tl)AP Photo/Damian Dovarganes/ Wide World Photos, (tr)(br)North Wind Picture Archives, (bl)Mary Evans Picture Library; **573** (tl)courtesy Adalee Valesquez, (tr)Barbara Filet/Tony Stone Images, (c)Mark Burnett, (bl)Grazia Neri/Woodfin Camp & Associates, (br)Gaslight Advertising Archives, Inc.; **574** Mark Burnett; **575** Ancient Art & Architecture Collection; **576** Mark Burnett; **577** (t)David Parker/Science Photo Library/Photo Researchers, (b)Laurie Platt Winfrey; **578** Mark Burnett; **579** Tate Gallery, London/Art Resource, NY; **584** (t)Jeff Gnass/The Stock Market, (b)Woodfin Camp & Associates; **585** Mark Burnett; **587** Mercury Archives/The Image Bank; **589** Mark Burnett; **590** Aaron Haupt Photography; **591** (t)Jeff Gnass/The Stock Market, (l)(b)Mark Burnett; **593** Mark E. Gibson; **596** Mark Steinmetz; **597** (l)file photo, (r) © Tribune Media Services. All rights reserved. Reprinted with permission.; **598** Aaron Haupt Photography; **601** Grant V. Faint/The Image Bank; **605** SuperStock; **608** Bob Daemmrich/Stock Boston; **613** Doug Martin; **615** (tr)Library of Congress, (l)The Lowe Art Museum, The University of Miami/SuperStock, (br)John Elk/Tony Stone Images; **616** Timothy Hursley; **617** Mark Steinmetz; **620** Bill Keogh; **621** (t)Library of Congress, (b)Egyptian National Museum, Cairo, Egypt/SuperStock; **623** Mark Steinmetz; **627, 628** (t)Doug Martin, (b)Mark Steinmetz; **629** Matt Meadows; **631** AP Photo/Des Moines Register, Harry Baumert; **632** Nippondenso Company, Ltd; **635** Mark Burnett; **642, 643** The Kobal Collection;

644 (tl)Mark Burnett, (tr)John E. Evans Photography, (bl)North Wind Picture Archives, (bc)Ancient Art & Architecture Collection, (br)Bettmann Archive; **645** (t)courtesy Leander Morgan, (c)(br)Mark Burnett, (bl)Larry Hamill; **649** John E. Evans Photography; **650** Mark Steinmetz; **651** Les Losego/Crown Studio; **653** (t)Telegraph Colour Library/FPG, (b)Les Losego/Crown Studio; **655** Mark Burnett; **657** John E. Evans Photography; **658** Les Losego/Crown Studio; **659** The Living Earth, Inc; **663** (t)DUOMO/Ben Van Hook, (b)Fiona Pragoff/The Image Bank; **671** (t)Peter Gridley/FPG, (b)Les Losego/Crown Studio; **673** (t)Corbis-Bettmann, (b)UPI/Corbis/Bettmann; **679** John E. Evans Photography; **680** Flip Nicklin/Minden Pictures; **684** Kim Westerskov/ Tony Stone Images; **685** Aaron Haupt Photography; **691** file photo; **694** (tl)Art Trade, Bonhams, London/ Bridgeman Art Library, London/SuperStock, (tr)(br)Mark Burnett, (bl)Corbis-Bettmann; **695** (tl)Ian R. Howarth, (tr)Lynn M. Stone, (c)(bl)Roger K. Burnard, (br)courtesy Borden, Inc/COSI, Columbus, OH; **696** Tom McHugh/Science Source/Photo Researchers; **698, 699, 701, 703, 704** Aaron Haupt Photography; **706** (t)Brownie Harris/The Stock Market, (b)Mark Downey Photographer; **707** Mark Burnett; **708** Doug Martin; **710** Mark Burnett; **712** Rafael Macia/Photo Researchers; **714** courtesy Volkhart Caro; **715** Greg Probst/Tony Stone Images; **720** Dumbarton Oaks; **722, 729** (l)Aaron Haupt Photography, (r)Tim Courlas; **730, 731** (t)Mark Burnett, (b)courtesy Alcy Haden/Reynoldsburg, OH, Chamber of Commerce; **733** Mark Burnett; **734** Richard Gross Photo; **736** SuperStock; **739** A. & L. Sinibaldi/Tony Stone Images; **741** Mark Burnett; **745, 746, 749, 750** Aaron Haupt Photography; **752** M. C. Escher © 1996 Cordon Art, Baarn, Holland. All rights reserved.; **753** Aaron Haupt Photography; **754** The Kobal Collection; **758** (t)David Ball/The Stock Market, (b)Mark Burnett; **759** Brooks/Brown/Photo Researchers.

TEACHER'S WRAPAROUND EDITION
T1 (t)David Brownell, (c)A. & L. Sinibaldi/Tony Stone Images, (b)Matt Meadows; **T2** (l)Doug Martin, (r)David Ball/The Stock Market; **T3, T6** Doug Martin; **T7** (tl)Mark Steinmetz, (tr)Aaron Haupt Photography, (b)Doug Martin; **T12,T13** Doug Martin; **T14** Mark Burnett; **T15** (t)Photo by Youth Voice Collaborative, (b)Ian R. Howarth; **T16** George Haling/Photo Researchers.

APPLICATIONS & CONNECTIONS INDEX

A

B

C

D

E

F

G

H

I

K

L

M

N

O

INDEX

Red type denotes items only in the Teacher's Wraparound Edition.

A

B

C

Index

Index